Broadcast Engineer's Reference Book

Broadcast Engineer's Reference Book

Edited by
E.P.J. Tozer

With specialist contributors

AMSTERDAM • BOSTON • HEIDELBERG • LONDON
NEW YORK • OXFORD • PARIS • SAN DIEGO
SAN FRANCISCO • SINGAPORE • SYDNEY • TOKYO
Focal Press is an imprint of Elsevier

Focal Press is an imprint of Elsevier
200 Wheeler Road, Burlington, MA 01803, USA
Linacre House, Jordan Hill, Oxford OX2 8DP, UK

Library of Congress Cataloging-in-Publication Data
APPLICATION SUBMITTED

British Library Cataloguing-in-Publication Data
A catalogue record for this book is available from the British Library.

ISBN: 0-2405-1908-6

For information on all Focal Press publications
visit our website at www.focalpress.com

04 05 06 07 08 10 9 8 7 6 5 4 3 2 1

Printed and bound by CPI Antony Rowe, Eastbourne

This book is dedicated to the memory of Les Weaver.

Contents

Section 4 Studio and Production Systems

Section 5 Outside Broadcast Systems and Hardware

Section 6 Transmitter Systems and Hardware

Section 7 Test and Measurement

List of Contributors

I G Aizlewood
Managing Director
Continental Microwave Ltd

G Anderson
Technical Training
Tandberg Television Ltd

B Arbon
Just Technicalities Ltd

D Bancroft
Manager, Advanced Technology
Thomson Broadcast & Media Solutions

K Barritt
Professional Services
Sony Business Europe

J Barron BA, MA (Cantab)
University of Cambridge

P Bruce
Marketing Manager, Production Mixers
Thomson Grass Valley

P Cameron

C K P Clarke
Senior Engineer, BBC Research
Department

J Colpaert
Director of Technology
Scientific-Atlanta Europe

C Dale MIIE
Previously BBC Broadcast Networks
Engineer

P Davies
Broadcast Systems Manager
Dolby Laboratories, Inc.

K Davison
Manager, Communications
Thames Television

I Dennis BSc
Technical Director
Prism Sound group

J Doeven
Senior Technical Manager
Nozema (The Netherlands)

M A Dolan
Television Broadcast Technology, Inc.

P Dubery

R Duggan

J Emmett
Technical Director and Chief Executive
Broadcast Project Research Ltd

B Flowers MIEE, MRTS, MNYAS
Senior Engineer (retired)
European Broadcasting Union

D Hardy
Technical & Quality Manager
PAG Ltd London

P Harman AMIQA, Dip IDM, Dip BE
Dip PC (Business)
Marketing & Training Manager
Vinten Broadcast Ltd

A Hingley
Sony Broadcast & Professional
Europe

P Horne
Omneon Video Networks

R G Hunt D Sc, FRPS, FRSA, MRTS
Professor of Physiological Optics
The City University

P I' Anson BSc

Y Imahori
Chief Engineer, NHK

G A Johnson BSc, C Eng, MIEE
Deputy Head of Engineering services
ITV Association

A Jones BSc, CEng, MIEE
Engineering Learning Executive
BBC Training & Development

P Kafno

P Kemble C Eng, MIEE, B Sc
Principal Engineer, IBA

J Kerr
Educator
Discreet Logic

W H Klemmer
Broadcast Television Systems, GmbH

WJ Leathem
Consultant

J Levett BSc, CEng, MIEE
Formerly Senior Project Manager
Fixed Links, BBC, and Crown Castle International

A Major
Snell & Wilcox

R G Manton B Sc(Eng), PhD, C Eng, MIEE
Transmission Engineering Department, BBC

G Markarian, Prof.
CEO, Rinicom Ltd
Chair in Communication Systems
University of Leeds (UK)

J D Millward BSc, CEng, MIEE
Head of Research
Rank Cintel Ltd

R Pape
Advanced Broadcast Solutions

R S Roberts CEng, FIEE, SenMIEEE
Consultant Electronics Engineer

J T P Robinson

R Schiller

P Schmidt, Ph.D. (Physics)
Sony Broadcast & Professional Europe, UK

S Sidat MBA, DBA, BSc (Hons), CEng, MIEE
Customer Systems Manager, mmO2, Ltd

P Simpson MISM, AITT
Technical Training
Tandberg Television Ltd

B L Smith
Chief Technical Writer, Thomson-CSF

P Sproxton
Alpha Image Ltd

J Summers
Formerly Lighting Director, BBC

P Swinson BKSTS, SMPTE, RTS
Peter Swinson Associates Ltd (UK)

M Talbot-Smith BSc, Cphys, MInstP
Formerly BBC Engineering Training Department

C Thorpe
CTP Systems Ltd

A Todorovic
Chairman of the Board, Kompani

E P J Tozer
Zetrox Broadcast Communications

S Tranter

L W Turner FIEE
Consultant Engineer

J Wardle BSc (Hons)
John Wardle Associates

L E Weaver BSc, CEng, MIEE
Formerly Head of Measurements
Laboratory, BBC Designs Department

P Weitzel BSc(Eng), CEng, MIEE, MSMPTE
Manager, Broadcast Technology Developments
BBC Technology

J Whitaker
Technical Director
Advanced Television Systems Committee

J P Whiting MSc, CEng, FIEE
Head of Power Systems, IBA

R Wilson BSc
Continental Microwave

G W Wiskin BSc, CEng, MICE, MIStructE
Architectural and Civil Engineering Department, BBC

G Young BSc
Managing Director
Antares (Europe) Ltd

Preface

The *Broadcast Engineer's Reference Book* has been compiled at a time of great change in our industry, though as it was 2,500 years ago that Heraclitus observed 'Nothing endures but change,' this perhaps shouldn't come as too great a surprise. Just as digital systems encompassing audio and video have become the standard and staple components of broadcast systems, so technology has taken another step forward. Now compressed audio and video systems and IT are making their mark on the industry as part of the great Convergence. Traditionally, television and radio broadcast has been a fixed quality system whether this was 270 Mbits^{-1}, 10 bits, or 6.75 MHz bandwidth in the studio, 525 or 625 lines composite in a 6, 7 or 8 MHz broadcast transmission slot, or a fixed AM or FM channel bandwidth. Now, digital compression has introduced a whole new choice of variable quality to the engineer, and with it comes a new challenge in maintaining technical standards whilst working with marketing and commercial staff who demand more channels in the same transmission space, all for the same budget... and now Interactive services need to be fitted in too.

So, today, it is a greater task than ever for any one engineer to be an expert in all facets of the broadcast industry as the great scope of technologies and systems seemingly expand faster than ever. Yet with the reliability and low maintenance needed by modern hardware, the scope of every engineer's role does expand to encompass a broader and broader range of technologies, hardware and systems. This expansion of roles is particularly true in the convergence arena where new skills are needed as equipment becomes networked and programme transmission becomes data transfer.

The content of the *Broadcast Engineer's Reference Book* has been compiled to reflect these industry changes. As well as encompassing traditional audio and video technologies, coverage is given to the fast emerging data and telecommunications aspects of the industry. Above all, the *Broadcast Engineer's Reference Book* is designed as a practical reference work written by engineers working in the broadcast industry for engineers working in the broadcast industry. Each chapter author is an expert in their subject area and shares their wealth of knowledge and experience.

The *Broadcast Engineer's Reference Book* has been laid out as seven sections, brought together according to the major engineering groupings and completed by test and measurement, it moves forward from foundation technologies and information, through industry standards, hardware components, and systems. Each chapter of the book is designed both to stand alone and to integrate with others to provide a comprehensive whole, and will prove a valuable resource to the engineer, the student, and all those who have an interest in the technique and technology of modern broadcast.

Finally, I must express my very great thanks to all of the contributors to this book for finding time from their busy schedules to provide their own unique and valuable insight into the technology of broadcast.

E.P.Joe Tozer
Basingstoke
January 2004

Section 1 Supporting Technologies and Reference Material

L W Turner FIEE
Consultant Engineer

1.1 Quantities and Units

The International System of Units (SI) is the modern form of the metric system agreed at an international conference in 1960. It has been adopted by the International Standards Organisation (ISO) and the International Electrotechnical Commission (IEC) and its use is recommended wherever the metric system is applied. It is now being adopted throughout most of the world and is likely to remain the primary world system of units of measurement for a very long time. The indications are that SI units will supersede the units of existing metric systems and all systems based on Imperial units.

SI units and the rules for their application are contained in *ISO Resolution* R1000 (1969, updated 1973) and an informatory document *SI-Le Système International d' Unités*, published by the Bureau International des Poids et Mesures (BIPM). An abridged version of the former is given in British Standards Institution (BSI) publication PD 5686 *The Use of SI Units* (1969, updated 1973) and BS 3763 *International System (SI) Units*; BSI (1964) incorporates information from the BIPM document.

The adoption of SI presents less of a problem to the electronics engineer and the electrical engineer than to those concerned with other engineering disciplines as all the practical electrical units were long ago incorporated in the metre-kilogram-second (MKS) unit system and these remain unaffected in SI.

The SI was developed from the metric system as a fully coherent set of units for science, technology and engineering. A coherent system has the property that corresponding equations between quantities and between numerical values have exactly the same form, because the relations between units do not involve numerical conversion factors. In constructing a coherent unit system, the starting point is the selection and definition of a minimum set of independent 'base' units. From these, 'derived' units are obtained by forming products or quotients in various combinations, again without numerical factors. Thus the base units of length (metre), time (second) and mass (kilogram) yield the SI units of velocity (metre/second), force (kilogram-metre/second-squared) and so on. As a result there is, for any given physical quantity, only one SI unit with no alternatives and with no numerical conversion factors. A single SI unit (joule = kilogram metre-squared/second-squared) serves for energy of any kind, whether it be kinetic, potential, thermal, electrical, chemical..., thus unifying the usage in all branches of science and technology.

The SI has seven base units, and two supplementary units of angle. Certain important derived units have special names and can themselves be employed in combination to form alternative names for further derivations.

Each physical quantity has a quantity-symbol (e.g. *m* for mass) that represents it in equations, and a unit-symbol (e.g. kg for kilogram) to indicate its SI unit of measure.

1.1.1 Base Units

Definitions of the seven base units have been laid down in the following terms. The quantity-symbol is given in italics, the unit-symbol (and its abbreviation) in roman type.

Length: *l*; metre (m). The length equal to 1,650,763.73 wavelengths in vacuum of the radiation corresponding to the transition between the levels $2p_{10}$ and $5d_5$ of the krypton-86 atom.

Mass: *m*; kilogram (kg). The mass of the international prototype kilogram (a block of platinum preserved at the International Bureau of Weights and Measures at Sèvres).

Time: *t*; second (s). The duration of 9,192,631,770 periods of the radiation corresponding to the transition between the two hyperfine levels of the ground state of the caesium-133 atom.

Electric current: *i*; ampere (A). The current which, maintained in two straight parallel conductors of infinite length, of negligible circular cross-section and 1 m apart in vacuum, produces a force equal to 2×10^{-7} newton per metre of length.

Thermodynamic temperature: *T*; kelvin (K). The fraction 1/273.16 of the thermodynamic (absolute) temperature of the triple point of water.

Luminous intensity: *I*; candela (cd). The luminous intensity in the perpendicular direction of a surface of 1/600,000 m^2 of a black body at the temperature of freezing platinum under a pressure of 101,325 newtons per square metre.

Amount of substance: Q; mole (mol). The amount of substance of a system which contains as many elementary entities as there are atoms in 0.012 kg of carbon-12. The elementary entity must be specified and may be an atom, a molecule, an ion, an electron, etc., or a specified group of such entities.

1.1.2 Supplementary Units

Plane angle: $\alpha, \beta \ldots$; radian (rad). The plane angle between two radii of a circle which cut off on the circumference an arc of length equal to the radius.
Solid angle: Ω; steradian (sr). The solid angle which, having its vertex at the centre of a sphere, cuts off an area of the surface of the sphere equal to a square having sides equal to the radius.
Force: The base SI unit of electric current is in terms of force in newtons (N). A force of 1 N is that which endows unit mass (1 kg) with unit acceleration (1 m/s^2). The newton is thus not only a coherent unit; it is also devoid of any association with gravitational effects.

1.1.3 Temperature

The base SI unit of thermodynamic temperature is referred to as a point of 'absolute zero' at which bodies possess zero thermal energy. For practical convenience two points on the Kelvin temperature scale, namely 273.15 K and 373.15 K, are used to define the Celsius (or Centigrade) scale (0 °C and 100 °C). Thus in terms of temperature *intervals*, 1 K = 1°C; but in terms of temperature *levels*, a Celsius temperature 0 corresponds to a Kelvin temperature (0 + 273.15) K.

1.1.4 Derived Units

Nine of the more important SI derived units with their definitions are given below.
Newton That force which gives to a mass of 1 kilogram an acceleration of 1 metre per second squared.
Joule The work done when the point of application of 1 newton is displaced a distance of 1 metre in the direction of the force.
Watt The power which gives rise to the production of energy at the rate of 1 joule per second.
Coulomb The quantity of electricity transported in 1 second by a current of 1 ampere.
Volt The difference of electric potential between two points of a conducting wire carrying a constant current of 1 ampere, when the power dissipated between these points is equal to 1 watt.
Ohm The electric resistance between two points of a conductor when a constant difference of potential of 1 volt, applied between these two points, produces in this conductor a current of 1 ampere, this conductor not being the source of any electromotive force.
Farad The capacitance of a capacitor between the plates of which there appears a difference of potential of 1 volt when it is charged by a quantity of electricity equal to 1 coulomb.
Henry The inductance of a closed circuit in which an electromotive force of 1 volt is produced when the electric current in the circuit varies uniformly at a rate of 1 ampere per second.
Weber The magnet flux which, linking a circuit of one turn, produces in it an electromotive force of 1 volt as it is reduced to zero at a uniform rate in 1 second.

Some of the simpler derived units are expressed in terms of the seven basic and two supplementary units directly. Examples are listed in Table 1.1.1.

Table 1.1.1 Directly derived units

Quantity	*Unit name*	*Unit symbol*
Force	newton	N
Energy	joule	J
Power	watt	W
Electric charge	coulomb	C
Electrical potential difference and EMF	volt	V
Electric resistance	ohm	Ω
Electric capacitance	farad	F
Electric inductance	henry	H
Magnetic flux	weber	Wb
Area	square metre	m^2
Volume	cubic metre	m^3
Mass density	kilogram per cubic metre	kg/m^3
Linear velocity	metre per second	m/s
Linear acceleration	metre per second squared	m/s^2
Angular velocity	radian per second	rad/s
Angular acceleration	radian per second squared	rad/s^2
Force	kilogram metre per second squared	kg m/s^2
Magnetic field strength	ampere per metre	A/m
Concentration	mole per cubic metre	mol/m^3
Luminance	candela per square metre	cd/m^2

Units in common use, particularly those for which a statement in base units would be lengthy or complicated, have been given special shortened names (see Table 1.1.2). Those that are

Table 1.1.2 Named derived units

Quantity	*Unit name*	*Unit symbol*	*Derivation*
Force	newton	N	kg m/s^2
Pressure	pascal	Pa	N/m^2
Power	watt	W	J/s
Energy	joule	J	N m, W s
Electric charge	coulomb	C	A s
Electric flux	coulomb	C	A s
Magnetic flux	weber	Wb	V s
Magnetic flux density	tesla	T	Wb/m^2
Electric potential	volt	V	J/C, W/A
Resistance	ohm	Ω	V/A
Conductance	siemens	S	A/V
Capacitance	farad	F	A s/V, C/V
Inductance	henry	H	V s/A, Wb/A
Luminous flux	lumen	lm	cd sr
Illuminance	lux	lx	lm/m^2
Frequency	hertz	Hz	l/s

Table 1.1.3 Further derived units

Quantity	*Unit name*	*Unit symbol*
Torque	newton metre	N m
Dynamic viscosity	pascal second	Pa s
Surface tension	newton per metre	N/m
Power density	watt per square metre	W/m^2
Energy density	joule per cubic metre	J/m^3
Heat capacity	joule per kelvin	J/K
Specific heat capacity	joule per kilogram kelvin	J/(kg K)
Thermal conductivity	watt per metre kelvin	W/(m K)
Electric field strength	volt per metre	V/m
magnetic field strength	ampere per metre	A/m
Electric flux density	coulomb per square metre	C/m^2
Current density	ampere per square metre	A/m^2
Resistivity	ohm metre	Ω m
Permittivity	farad per metre	F/m
Permeability	henry per metre	H/m

named from scientists and engineers are abbreviated to an initial capital letter: all others are in lower-case letters.

The named derived units are used to form further derivations. Examples are given in Table 1.1.3.

Names of SI units and the corresponding EMU and ESU CGS units are given in Table 1.1.4.

1.1.5 Gravitational and Absolute Systems

There may be some difficulty in understanding the difference between SI and the Metric Technical System of units which has been used principally in Europe. The main difference is that while mass is expressed in kg in both systems, weight (representing a force) is expressed as kgf, a gravitational unit, in the MKSA system and as N in SI. An absolute unit of force differs from a gravitational unit of force because it induces unit acceleration in a unit mass whereas a gravitational unit imparts gravitational acceleration to a unit mass.

A comparison of the more commonly known systems and SI is shown in Table 1.1.5.

1.1.6 Expressing Magnitudes of SI Units

To express magnitudes of a unit, decimal multiples and submultiples are formed using the prefixes shown in Table 1.1.6. This method of expressing magnitudes ensures complete adherence to a decimal system.

1.1.7 Auxiliary Units

Certain auxiliary units may be adopted where they have application in special fields. Some are acceptable on a temporary basis, pending a more widespread adoption of the SI system. Table 1.1.7 lists some of these.

Table 1.1.4 Unit names

Quantity	*Symbol*	*SI*	*EMU and ESU*
Length	l	metre (m)	centimetre (cm)
Time	t	second (s)	second
Mass	m	kilogram (kg)	gram (g)
Force	F	newton (N)	dyne (dyn)
Frequency	f, v	hertz (Hz)	hertz
Energy	E, W	joule (J)	erg (erg)
Power	P	watt (W)	erg/second (erg/s)
Pressure	p	newton/metre^2 (N/m^2)	dyne/centimetre^2 ($dyne/cm^2$)
Electric charge	Q	coulomb (C)	coulomb
Electric potential	V	volt (V)	volt
Electric current	I	ampere (A)	ampere
Magnetic flux	ϕ	weber (Wb)	maxwell (Mx)
Magnetic induction	B	tesla (T)	gauss (G)
Magnetic field strength	H	ampere turn/metre (At/m)	oersted (Oe)
Magnetomotive force	Fm	ampere turn (At)	gilbert (Gb)
Resistance	R	ohm (Ω)	ohm
Inductance	L	henry (H)	henry
Conductance	G	mho (Ω^{-1}) (siemens)	mho
Capacitance	C	farad (F)	farad

Table 1.1.5 Commonly used units of measurement

	SI (absolute)	*FPS (gravitational)*	*FPS (absolute)*	*cgs (absolute)*	*Metric technical units (gravitational)*
Length	metre (m)	ft	ft	cm	metre
Force	newton (N)	lbf	poundal (pdl)	dyne	kgf
Mass	kg	lb or slug	lb	gram	kg
Time	s	s	s	s	s
Temperature	°C K	°F	°F °R	°C K	°C K
		ft lbf	ft pdl	dyne cm = erg	kgf m
Energy					
heat		hp	hp		metric hp
mech.	joule*	Btu	Btu	calorie	kcal
Power					
elec.					
mech.	watt	watt	watt	erg/s	watt
Electric current	amp	amp	amp	amp	amp
Pressure	N/m^2	lbf/ft^2	pdl/ft^2	$dyne/cm^2$	kgf/cm^2

*1 joule = 1 newton metre or 1 watt second.

Table 1.1.6 The internationally agreed multiples and submultiples

Factor by which the unit is multiplied		*Prefix*	*Symbol*	*Common everyday examples*
One million million (billion)	10^{12}	tera	T	
One thousand million	10^9	giga	G	gigahertz (GHz)
One million	10^6	mega	M	megawatt (MW)
One thousand	10^3	kilo	k	kilometre (km)
One hundred	10^2	hecto*	h	
Ten	10^1	deca*	da	decagram (dag)
UNITY	1			
One tenth	10^{-1}	deci*	d	decimetre (dm)
One hundredth	10^{-2}	centi*	c	centimetre (cm)
One thousandth	10^{-3}	milli	m	milligram (mg)
One millionth	10^{-6}	micro	μ	microsecond (μs)
One thousand millionth	10^{-9}	nano	n	nanosecond (ns)
One million millionth	10^{-12}	pico	p	picofarad (pF)
One thousand million millionth	10^{-15}	femto	f	
One million million millionth	10^{-18}	atto	a	

*To be avoided wherever possible.

Table 1.1.7 Auxiliary units

Quantity	*Unit symbol*	*SI equivalent*
Day	d	86,400 s
Hour	h	3600 s
Minute (time)	min	60 s
Degree (angle)	°	π/180 rad
Minute (angle)	′	π/10,800 rad
Second (angle)	″	π/648,000 rad
Acre	a	$1\ dam^2 = 10^2\ m^2$
Hectare	ha	$1\ hm^2 = 10^4\ m^2$
Barn	b	$100\ fm^2 = 10^{-28}\ m^2$
Standard atmosphere	atm	101,325 Pa
Bar	bar	$0.1\ MPa = 10^5\ Pa$
Litre	l	$1\ dm^3 = 10^{-3}\ m^3$
Tonne	t	$10^3\ kg = 1\ Mg$
Atomic mass unit	u	1.66053×10^{-27} kg
Angström	A	$0.1\ nm = 10^{-10}\ m$
Electron-volt	eV	1.60219×10^{-19} J
Curie	Ci	$3.7 \times 10^{10}\ s^{-1}$
Röntgen	R	2.58×10^{-4} C/kg

1.1.8 Universal Constants in SI Units

Table 1.1.8 Universal constants

The digits in parentheses following each quoted value represent the standard deviation error in the final digits of the quoted value as computed on the criterion of internal consistency. The unified scale of atomic weights is used throughout ($^{12}C = 12$). C = coulomb; G = gauss; Hz = hertz; J = joule; N = newton; T = tesla; u = unified nuclidic mass unit; W = watt; Wb = weber. For result multiply the numerical value by the SI unit.

Constant	*Symbol*	*Numerical value*	*SI unit*
Speed of light in vacuum	c	2.997925(1)	10^{8} m/s
Gravitational constant	G	6.670(5)*	10^{-11} N m^{2} kg^{2}
Elementary charge	e	1.60210(2)	10^{-19} C
Avogadro constant	N_A	6.02252(9)	10^{26} kmol^{-1}
Mass unit	u	1.66043(2)	10^{-27} kg
Electron rest mass	m_e	9.10908(13)	10^{-31} kg
		5.48597(3)	10^{-4} u
Proton rest mass	m_p	1.67252(3)	10^{-27} kg
		1.00727663(8)	u
Neutron rest mass	m_n	1.67482(3)	10^{-27} kg
		1.0086654(4)	u
Faraday constant	F	9.68470(5)	10^{4} C/mol
Planck constant	h	6.62559(16)	10^{-34} J s
	$h/2\pi$	1.054494(25)	10^{-34} J s
Fine-structure constant	$\propto$	7.29720(3)	10^{-3}
	$1/\alpha$	137.0388(6)	
Charge-to-mass ratio for electron	e/m_e	1.758796(6)	10^{11} C/kg
Quantum of magnetic flux	hc/e	4.13556(4)	10^{11} Wb
Rydberg constant	$R_\propto$	1.0973731(1)	10^{7} m^{-1}
Bohr radius	a_0	5.29167(2)	10^{-11} m
Compton wavelength of electron	$h/m_e c$	2.42621(2)	10^{-12} m
	$\lambda C/2\pi$	3.86144(3)	10^{-13} m
Electron radius	$e^2/m_e c^2 = r_e$	2.81777(4)	10^{-15} m
Thomson cross-section	$8\pi r_e^2/3$	6.6516(2)	10^{-29} m^{2}
Compton wavelength of proton	$\lambda c,p$	1.321398(13)	10^{-15} m
	$\lambda c,p/2\pi$	2.10307(2)	10^{-16} m
Gyromagnetic ratio of proton	γ	2.675192(7)	10^{8} rad/(s T)
	$\gamma/2\pi$	4.25770(1)	10^{7} Hz/T
(uncorrected for diamagnetism of H_2O)	γ'	2.675123(7)	10^{8} rad/(s T)
	$\gamma'/2\pi$	4.25759(1)	10^{7} Hz/T
Bohr magneton	μB	9.2732(2)	10^{-24} J/T
Nuclear magneton	μN	5.05050(13)	10^{-27} J/T
Proton magnetic moment	μp	1.41049(4)	10^{-26} J/T
	$\mu p/\mu N$	2.79276(2)	
(uncorrected for diamagnetism in H_2O sample)	$\mu' p/\mu N$	2.79268(2)	
Gas constant	R_0	8.31434(35)	J/K mol
Boltzmann constant	k	1.38054(6)	10^{-23} J/K
First radiation constant ($2\pi hc^2$)	c_1	3.74150(9)	10^{-16} W/m^{2}
Second radiation constant (hc/k)	c_2	1.43879(6)	10^{-2} m K
Stefan-Boltzmann constant	σ	5.6697(10)	10^{-8} W/m^{2} K^{4}

*The universal gravitational constant is not, and cannot in our present state of knowledge, be expressed in terms of other fundamental constants. The value given here is a direct determination by P.R. Heyland and P. Chrzanowski, *J. Res. Natl. Bur. Std.* (U.S.), **29**, 1 (1942).

The above values are extracts from *Review of Modern Physics*, **37**, No. 4 (October 1965) published by the American Institute of Physics.

1.1.9 Metric to Imperial Conversion Factors

Table 1.1.9 Conversion factors

SI units	*British units*
SPACE AND TIME	
Length:	
1 μm (micron)	$= 39.37 \times 10^{-6}$ in
1 mm	= 0.0393701 in
1 cm	= 0.393701 in
1 m	= 3.28084 ft
1 m	= 1.09361 yd
1 km	= 0.621371 mile
Area:	
1 mm^2	$= 1.550 \times 10^{-3}$ in^2
1 cm^2	= 0.1550 in^2
1 m^2	= 10.7639 ft^2
1 m^2	= 1.19599 yd^2
1 ha	= 2.47105 acre
Volume:	
1 mm^3	$= 61.0237 \times 10^{-6}$ in^3
1 cm^3	$= 61.0237 \times 10^{-3}$ in^3
1 m^3	= 35.3147 ft^3
1 m^3	= 1.30795 yd^3
Capacity:	
10^6 m^3	$= 219.969 \times 10^6$ gal
1 m^3	= 219.969 gal
1 litre (l)	= 0.219969 gal
	= 1.75980 pint
Capacity flow:	
10^3 m^3/s	$= 791.9 \times 10^6$ gal/h
1 m^3/s	$= 13.20 \times 10^3$ gal/min
1 litre/s	= 13.20 gal/min
1 m^3/kW h	= 219.969 gal/kW h
1 m^3/s	= 35.3147 ft^3/s (cusecs)
1 litre/s	$= 0.58858 \times 10^{-3}$ ft^3/min (cfm)
Velocity:	
1 m/s	= 3.28084 ft/s = 2.23694 mile/h
1 km/h	= 0.621371 mile/h
Acceleration:	
1 m/s^2	= 3.28084 ft/s^2
MECHANICS	
Mass:	
1 g	= 0.035274 oz
1 kg	= 2.20462 lb
1 t	= 0.984207 ton = 19.6841 cwt
Mass flow:	
1 kg/s	= 2.20462 lb/s = 7.93664 klb/h
Mass density:	
1 kg/m^3	= 0.062428 lb/ft^3
1 kg/litre	= 10.022119 lb/gal

Table 1.1.9 *continued*

SI units	*British units*
Mass per unit length:	
1 kg/m	= 0.671969 lb/ft = 2.01591 lb/yd
Mass per unit area:	
1 kg/m^2	= 0.204816 lb/ft^2
Specific volume:	
1 m^3/kg	= 16.0185 ft^3/lb
1 litre/tonne	= 0.223495 gal/ton
Momentum:	
1 kg m/s	= 7.23301 lb ft/s
Angular momentum:	
1 kg m^2/s	= 23.7304 lb ft^2/s
Moment of inertia:	
1 kg m^2	= 23.7304 lb ft^2
MECHANICS	
Force:	
1 N	= 0.224809 lbf
Weight (force) per unit length:	
1 N/m	= 0.068521 lbf/ft
	= 0.205566 lbf/yd
Moment of force (or torque):	
1 N m	= 0.737562 lbf ft
Weight (force) per unit area:	
1 N/m^2	= 0.020885 lbf/ft^2
Pressure:	
1 N/m^2	$= 1.45038 \times 10^{-4}$ lbf/in^2
1 bar	= 14.5038 lbf/in^2
1 bar	= 0.986923 atmosphere
1 mbar	= 0.401463 in H_2O
	= 0.02953 in Hg
Stress:	
1 N/mm^2	$= 6.47490 \times 10^{-2}$ $tonf/in^2$
1 MN/m^2	$= 6.47490 \times 10^{-2}$ $tonf/in^2$
1 hbar	= 0.647490 $tonf/in^2$
Second moment of area:	
1 cm^4	= 0.024025 in^4
Section modulus:	
1 m^3	= 61,023.7 in^3
1 cm^3	= 0.0610237 in^3
Kinematic viscosity:	
1 m^2/s	= 10.76275 ft^2/s = 10^6 cSt
1 cSt	= 0.03875 ft^2/h
Energy, work:	
1 J	= 0.737562 ft lbf
1 MJ	= 0.3725 hph
1 MJ	= 0.27778 kW h
Power:	
1 W	= 0.737562 ft lbf/s
1 kW	= 1.341 hp = 737.562 ft lbf/s

Table 1.1.9 *continued*

SI units	*British units*
Fluid mass:	
(Ordinary) 1 kg/s	= 2.20462 lb/s = 7936.64 lb/h
(Velocity) 1 kg/m^{2}s	= 0.204815 lb/ft^{2}s
HEAT	
Temperature:	
(Interval) 1 K	= 9/5 deg R (Rankine)
1 °C	= 9/5 deg F
(Coefficient) 1°R^{-1}	= 1 deg F^{-1} = 5/9 deg C
1 °C^{-1}	= 5/9 deg F^{-1}
Quantity of heat:	
1 J	= 9.47817 × 10^{-4} Btu
1 J	= 0.238846 cal
1 kJ	= 947.817 Btu
1 GJ	= 947.817 × 10^3 Btu
1 kJ	= 526.565 CHU
1 GJ	= 526.565 × 10^3 CHU
1 GJ	= 9.47817 therm
Heat flow rate:	
1 W (J/s)	= 3.41214 Btu/h
1 W/m^2	= 0.316998 Btu/ft^2 h
Thermal conductivity:	
1 W/m °C	= 6.93347 Btu in/ft^2 h °F
Coefficient and heat transfer:	
1 W/m^2 °C	= 0.176110 Btu/ft^2 h °F
Heat capacity:	
1 J/°C	= 0.52657 × 10^{-3} Btu/°R
Specific heat capacity:	
1 J/g °C	= 0.238846 Btu/lb °F
1 kJ/kg °C	= 0.238846 Btu/lb °F
Entropy:	
1 J/K	= 0.52657 × 10^{-3} Btu/°R
Specific entropy:	
1 J/kg °C	= 0.238846 × 10^{-3} Btu/lb °F
1 J/kg K	= 0.238846 × 10^{-3} Btu/lb °R
Specific energy/specific latent heat:	
1 J/g	= 0.429923 Btu/lb
1 J/kg	= 0.429923 × 10^{-3} Btu/lb
Calorific value:	
1 kJ/kg	= 0.429923 Btu/lb
1 kJ/kg	= 0.7738614 CHU/lb
1 J/m^3	= 0.0268392 × 10^{-3} Btu/ft^3
1 kJ/m^3	= 0.0268392 Btu/ft^3
1 kJ/litre	= 4.30886 Btu/gal
1 kJ/kg	= 0.0096302 therm/ton
ELECTRICITY	
Permeability:	
1 H/m	= 10^7/4$\pi\mu_0$
Magnetic flux density:	
1 tesla	= 10^4 gauss = 1 Wb/m^2

Table 1.1.9 *continued*

SI units	*British units*
Conductivity:	
1 mho	= 1 reciprocal ohm
1 siemens	= 1 reciprocal ohm
Electric stress:	
1 kV/mm	= 25.4 kV/in
1 kV/m	= 0.0254 kV/in

1.1.10 Symbols and Abbreviations

Table 1.1.10 Quantities and units of periodic and related phenomena (based on ISO Recommendation R31)

Symbol	*Quantity*
T	periodic time
τ, (T)	time constant of an exponentially varying quantity
f, ν	frequency
η	rotational frequency
ω	angular frequency
λ	wavelength
σ ($\bar{\nu}$)	wavenumber
K	circular wavenumber
$\log_e (A_1/A_2)$	natural logarithm of the ratio of two amplitudes
$10 \log_{10} (P_1/P_2)$	ten times the common logarithm of the ratio of two powers
δ	damping coefficient
Λ	logarithmic decrement
α	attenuation coefficient
β	phase coefficient
γ	propagation coefficient

Table 1.1.11 Symbols for quantities and units of electricity and magnetism (based on ISO Recommendation R31)

Symbol	*Quantity*
I	electric current
Q	electric charge, quantity of electricity
P	volume density of charge, charge density (Q/V)
Σ	surface density of charge (Q/A)
E, (K)	electric field strength
V, (φ)	electric potential
U, (V)	potential difference, tension
E	electromotive force
D	displacement (rationalised displacement)
D'	non-rationalised displacement

Table 1.1.11 *continued*

Symbol	Quantity
ψ	electric flux, flux of displacement (flux of rationalised displacement)
ψ'	flux of non-rationalised displacement
C	capacitance
E	permittivity
ε_0	permittivity of vacuum
ε'	non-rationalised permittivity
ε'_0	non-rationalised permittivity of vacuum
ε_r	relative permittivity
χ_e	electric susceptibility
χ'_e	non-rationalised electric susceptibility
P	electric polarisation
p, (P_e)	electric dipole moment
J, (S)	current density
A, (α)	linear current density
H	magnetic field strength
H'	non-rationalised magnetic field strength
U_m	magnetic potential difference
F, (F_m)	magnetomotive force
B	magentic flux density, magnetic induction
ϕ	magnetic flux
A	magnetic vector potential
L	self-inductance
M, (L)	mutual inductance
k, (x)	coupling coefficient
Σ	leakage coefficient
M	permeability
μ_o	permeability of vacuum
μ'	non-rationalised permeability
μ'_o	non-rationalised permeability of vacuum
μ_r	relative permeability
k, (χ_m)	magnetic susceptibility
k', (χ'_m)	non-rationalised magnetic susceptibility
M	electromagnetic moment (magnetic moment)
H_ν (M)	magnetisation
J, (B_i)	magnetic polarisation
J'	non-rationalised magnetic polarisation
W	electromagnetic energy density
S	Poynting vector
C	velocity of propagation of electromagnetic waves *in vacuo*
R	resistance (to direct current)

Table 1.1.11 *continued*

Symbol	Quantity
G	conductance (to direct current)
P	resistivity
y, σ	conductivity
R, R_m	reluctance
A, (P)	permeance
N	number of turns in winding
M	number of phases
P	number of pairs of poles
ϕ	phase displacement
Z	impedance (complex impedance)
$[Z]$	modulus of impedance (impedance)
X	reactance
R	resistance
Q	quality factor
Y	admittance (complex admittance)
$[Y]$	modulus of admittance (admittance)
B	susceptance
G	conductance
P	active power
S, (P_s)	apparent power
Q, (P_q)	reactive power

Table 1.1.12 Symbols for quantities and units of acoustics (based on ISO Recommendation R31)

Symbol	Quantity
T	period, periodic time
f, ν	frequency, frequency interval
ω	angular frequency, circular frequency
λ	wavelength
K	circular wavenumber
P	density (mass density)
P_s	static pressure
P	(instantaneous) sound pressure
ε, (x)	(instantaneous) sound particle displacement
U, ν	(instantaneous) sound particle velocity
A	(instantaneous) sound particle acceleration
q, U	(instantaneous) volume velocity
C	velocity of sound
E	sound energy density
P, (N, W)	sound energy flux, sound power
I, J	sound intensity
Z_s, (W)	specific acoustic impedance

Table 1.1.12 *continued*

Symbol	*Quantity*
Z_a, (Z)	acoustic impedance
Z_m, (w)	mechanical impedance
L_p, (L_N, L_W)	sound power level
L_p, (L)	sound pressure level
δ	damping coefficient
Λ	logarithmic decrement
α	attenuation coefficient
β	phase coefficient
γ	propagation coefficient
δ	dissipation coefficient
r, τ	reflection coefficient
γ	transmission coefficient
α, (α_a)	acoustic absorption coefficient
R	{ sound reduction index sound transmission loss
A	equivalent absorption area of a surface or object
T	reverberation time
L_N, (Λ)	loudness level
N	loudness

Table 1.1.13 Some technical abbreviations and symbols

Quantity	*Abbreviation*	*Symbol*
Alternating current	ac	
Ampere	A or amp	
Amplification factor		μ
Amplitude modulation	am	
Angular velocity		ω
Audio frequency	af	
Automatic frequency control	afc	
Automatic gain control	agc	
Bandwidth		Δf
Beat frequency oscillator	bfo	
British thermal unit	Btu	
Cathode-ray oscilloscope	cro	
Cathode-ray tube	crt	
Celsius	C	
Centi-	c	
Centimetre	cm	
Square centimetre	cm^2 or sq cm	
Cubic centimetre	cm^3 or cu cm or cc	
Centimetre-gram-second	cgs	
Continuous wave	cw	
Coulomb	C	
Deci-	d	
Decibel	dB	

Table 1.1.13 *continued*

Quantity	*Abbreviation*	*Symbol*
Direct current	dc	
Direction finding	df	
Double sideband	dsb	
Efficiency		η
Equivalent isotropic radiated power	eirp	
Electromagnetic unit	emu	
Electromotive force instantaneous value	emf	E or V, e or v
Electron-volt	eV	
Electrostatic unit	esu	
Fahrenheit	F	
Farad	F	
Frequency	freq.	f
Frequency modulation	fm	
Gauss	G	
Giga-	G	
Gram	g	
Henry	H	
Hertz	Hz	
High frequency	hf	
Independent sideband	isb	
Inductance-capacitance		L-C
Intermediate frequency	if	
Kelvin	K	
Kilo-	k	
Knot	Kn	
Length		l
Local oscillator	lo	
Logarithm, common		log or $\log_{10}$
Logarithm, natural		ln or $\log_e$
Low frequency	lf	
Low tension	lt	
Magnetomotive force	mmf	F or M
Mass		m
Medium frequency	mf	
Mega-	M	
Metre	m	
Metre-kilogram-second	mks	
Micro-	μ	
Micromicro-	p	
Micron		μ
Milli-	m	
Modulated continuous wave	mcw	
Nano-	n	
Neper	N	
Noise factor		N
Ohm		Ω
Peak to peak	p-p	

Table 1.1.13 *continued*

Quantity	*Abbreviation*	*Symbol*
Phase modulation	pm	
Pico	p	
Plan-position indication	PPI	
Potential difference	pd	V
Power factor	pf	
Pulse repetition frequency	prf	
Radian	rad	
Radio frequency	rf	
Radio telephony	R/T	
Root mean square	rms	
Short-wave	sw	
Single sideband	ssb	
Signal frequency	sf	
Standing wave ratio	swr	
Super-high frequency	shf	
Susceptance		*B*
Travelling-wave tube	twt	
Ultra-high frequency	uhf	
Very high frequency	vhf	
Very low frequency	vlf	
Volt	V	
Voltage standing wave ratio	vswr	
Watt	W	
Weber	Wb	
Wireless telegraphy	W/T	

Table 1.1.14 Greek alphabet and symbols

Name	*Symbol*		*Quantities used for*
Alpha	Α	α	angles, coefficients, area
Beta	Β	β	angles, coefficients
Gamma	Γ	γ	specific gravity
Delta	Δ	δ	density, increment, finite difference operator
Epsilon	Ε	ε	Nepierian logarithm, linear strain, permittivity, error, small quantity
Zeta	Ζ	ζ	coordinates, coefficients, impedance (capital)
Eta	Η	η	magnetic field strength, efficiency
Theta	Θ	θ	angular displacement, time
Iota	Ι	ι	inertia
Kappa	Κ	κ	bulk modulus, magnetic susceptibility
Lambda	Λ	λ	permeance, conductivity, wavelength
Mu	Μ	μ	bending moment, coefficient of friction, permeability
Nu	Ν	ν	kinematic viscosity, frequency, reluctivity
Xi	Ξ	ξ	output coefficient
Omicron	Ο	ο	
Pi	Π	π	circumference ÷ diameter
Rho	Ρ	ρ	specific resistance
Sigma	Σ	σ	summation (capital), radar cross-section
Tau	Τ	τ	time constant, pulse length
Upsilon	Υ	υ	
Phi	Φ	ϕ	flux, phase
Chi	Χ	χ	reactance (capital)
Psi	Ψ	ψ	angles
Omega	Ω	ω	angular velocity, ohms

References

1. Cohen, E.R. and Taylor, B.N. *Journal of Physical and Chemical Reference Data*, **2**, 663 (1973).
2. *Recommended Values of Physical Constants*. CODATA (1973).
3. McGlashan, M.L. *Physicochemical Quantities and Units*, The Royal Institute of Chemistry, London (1971).

J Barron BA, MA (Cantab)
University of Cambridge

1.2 Engineering Mathematics, Formulas and Calculations

1.2.1 Mathematical Signs and Symbols

Sign, symbol	*Quantity*
$=$	equal to
$\neq \neq$	not equal to
$\equiv$	identically equal to
Δ	corresponds to
$\approx$	approximately equal to
$\rightarrow$	approaches
$\sqcup$	asymptotically equal to
$\sim$	proportional to
∞	infinity
$<$	smaller than
$>$	larger than
$\leqslant \leqq$	smaller than or equal to
$\geqslant \geq \geqq$	larger than or equal to
	much smaller than
	much larger than
$+$	plus
$-$	minus
$. \times$	mutiplied by
$\frac{a}{b}$ a/b	*a* divided by *b*
$\|a\|$	magnitude of *a*
a^n	*a* raised to the power of *n*
$a^{1/2} \sqrt{a}$	square root of *a*
$a^{1/n} \sqrt[n]{a}$	*n*th root of *a*
$a\langle a \rangle$	mean value of *a*

Sign, symbol	*Quantity*
$p!$	factorial *p*, $1 \times 2 \times 3 \times \cdots \times p$
$\binom{n}{p}$	binomial coefficient, $\frac{n(n-1)\ldots(n-p+1)}{1 \times 2 \times 3 \times \cdots \times p}$
Σ	sum
$\prod$	product
$f(x)$	function *f* of the variable *x*
$[f(x)]_a^b$	$f(b) - f(a)$
$\lim\limits_{x \to a} f(x); \lim_{x \to a} f(x)$	the limit to which $f(x)$ tends as *x* approaches *a*
Δx	delta $x =$ finite increment of *x*
δx	delta $x =$ variation of *x*
$\frac{df}{dx}; df/dx; f'(x)$	differential coefficient of $f(x)$ with respect to *x*
$\frac{d^n f}{dx^n}; f^{(n)}(x)$	differential coefficient of order *n* of $f(x)$
$\frac{\delta f(x,y,\ldots)}{\delta x}; \left(\frac{\delta f}{\delta x}\right)_{y,\ldots}$	partial differential coefficient of $f(x,y,\ldots)$ with respect to *x*, when $y, \ldots$ are held constant
df	the total differential of *f*
$\int f(x)dx$	indefinite integral of $f(x)$ with respect to *x*
$\int_a^b f(x)dx$	definite integral of $f(x)$ from $x = a$ to $x = b$
e	base of natural logarithms
e^x, $\exp x$	e raised to the power *x*
$\log_a x$	logarithm to the base *a* of *x*
$\lg x; \log x; \log_{10} x$	common (Briggsian) logarithm of *x*

Sign, symbol	*Quantity*
lb x; $\log_2 x$	binary logarithm of x
$\sin x$	sine of x
$\cos x$	cosine of x
$\tan x$; tg x	tangent of x
$\cot x$; ctg x	cotangent of x
$\sec x$	secant of x
cosec x	cosecant of x
arcsin x, etc.	arc sine of x, etc.
sinh x, etc.	hyperbolic sine of x, etc.
arsinh x, etc.	inverse hyperbolic sine of x, etc.
i, j	imaginary unity, $i^2 = -1$
Re z	real part of z
Im z	imaginary part of z
$\lvert z\rvert$	modulus of z
arg z	argument of z
z^*	conjugate of z, complex conjugate of z
$\tilde{A}, A', A^t$	transpose of matrix A
A, **a**	vector
$\lvert\mathbf{A}\rvert$, A	magnitude of vector
$A \cdot B$	scalar product
$\mathbf{A} \times \mathbf{B}, \mathbf{A} \wedge \mathbf{B}$	vector product
∇	differential vector operator
$\nabla\varphi$, gradφ	gradient of φ
$\nabla^2\varphi, \Delta\varphi$	Laplacian of φ

1.2.2 Trigonometric Formulas

$$\sin^2 A + \cos^2 A = \sin A \,\mathrm{cosec} A = 1$$

$$\sin A = \frac{\cos A}{\cot A} = \frac{1}{\mathrm{cosec}\ A} = (1 - \cos^2 A)^{1/2}$$

$$\cos A = \frac{\sin A}{\tan A} = \frac{1}{\sec A} = (1 - \sin^2 A)^{1/2}$$

$$\tan A = \frac{\sin A}{\cos A} = \frac{1}{\cot A}$$

$$1 + \tan^2 A = \sec^2 A$$

$$1 + \cot^2 A = \mathrm{cosec}^2 A$$

$$1 - \sin A = \mathrm{coversin}\ A$$

$$1 - \cos A = \mathrm{versin}\ A$$

$$\tan \tfrac{1}{2}\theta = t;\ \sin\theta = 2t/(1 + t^2);\quad \cos\theta = (1 - t^2)/(1 + t^2)$$

$$\cot A = 1/\tan A$$

$$\sec A = 1/\cos A$$

$$\mathrm{cosec}\ A = 1/\sin A$$

$$\cos(A \pm B) = \cos A \cos B \mp \sin A \sin B$$

$$\sin(A \pm B) = \sin A \cos B \pm \cos A \sin B$$

$$\tan(A \pm B) = \frac{\tan A \pm \tan B}{1 \mp \tan A \tan B}$$

$$\cot(A \pm B) = \frac{\cot A \cot B \mp 1}{\cot B \pm \cot A}$$

$$\sin A \pm \sin B = 2 \sin \tfrac{1}{2}(A \pm B) \cos \tfrac{1}{2}(A \mp B)$$

$$\cos A + \cos B = 2 \cos \tfrac{1}{2}(A + B) \cos \tfrac{1}{2}(A - B)$$

$$\cos A - \cos B = 2 \sin \tfrac{1}{2}(A + B) \sin \tfrac{1}{2}(B - A)$$

$$\tan A \pm B = \frac{\sin(A \pm B)}{\cos A \cos B}$$

$$\cot A \pm \cot B = \frac{\sin(B \pm A)}{\sin A \sin B}$$

$$\sin 2A = 2 \sin A \cos A$$

$$\cos 2A = \cos^2 A - \sin^2 A = 2\cos^2 A - 1 = 1 - 2\sin^2 A$$

$$\cos^2 A - \sin^2 B = \cos(A + B)\cos(A - B)$$

$$\tan 2A = 2\tan A/(1 - \tan^2 A)$$

$$\sin \tfrac{1}{2}A = \left(\frac{1 - \cos A}{2}\right)^{1/2}$$

$$\cos \tfrac{1}{2}A = \pm\left(\frac{1 + \cos A}{2}\right)^{1/2}$$

$$\tan \tfrac{1}{2}A = \frac{\sin A}{1 + \cos A}$$

$$\sin^2 A = \tfrac{1}{2}(1 - \cos 2A)$$

$$\cos^2 A = \tfrac{1}{2}(1 + \cos 2A)$$

$$\tan^2 A = \frac{1 - \cos 2A}{1 + \cos 2A}$$

$$\tan \tfrac{1}{2}(A \pm B) = \frac{\sin A \pm \sin B}{\cos A + \cos B}$$

$$\cot \tfrac{1}{2}(A \pm B) = \frac{\sin A \pm \sin B}{\cos B - \cos A}$$

1.2.3 Trigonometric Values

Angle:	*0°*	*30°*	*45°*	*60°*	*90°*	*180°*	*270°*	*360°*
Radians	0	$\pi/6$	$\pi/4$	$\pi/3$	$\pi/2$	π	$3\pi/2$	2π
Sine	0	$\frac{1}{2}$	$\frac{1}{2}\sqrt{2}$	$\frac{1}{2}\sqrt{3}$	1	0	−1	0
Cosine	1	$\frac{1}{3}\sqrt{3}$	$\frac{1}{2}\sqrt{2}$	$\frac{1}{2}$	0	−1	0	1
Tangent	0	$\frac{1}{3}\sqrt{3}$	1	$\sqrt{3}$	∝	0	∝	0

1.2.4 Approximations for Small Angles

$\sin\theta = \theta^3/6;\quad \cos\theta = 1 - \theta^2/2;\quad \tan\theta = \theta + \theta^3/3;$

(θ in radians)

1.2.5 Solution of Triangles

$$\frac{\sin A}{a} = \frac{\sin B}{b} = \frac{\sin C}{c} \qquad \cos A = \frac{b^2 + c^2 - a^2}{2bc}$$

$$\cos B = \frac{c^2 + a^2 - b^2}{2ca} \qquad \cos C = \frac{a^2 + b^2 - c^2}{2ab}$$

where A, B, C and a, b, c are shown in Figure 1.2.1. If $s = \frac{1}{2}(a + b + c)$,

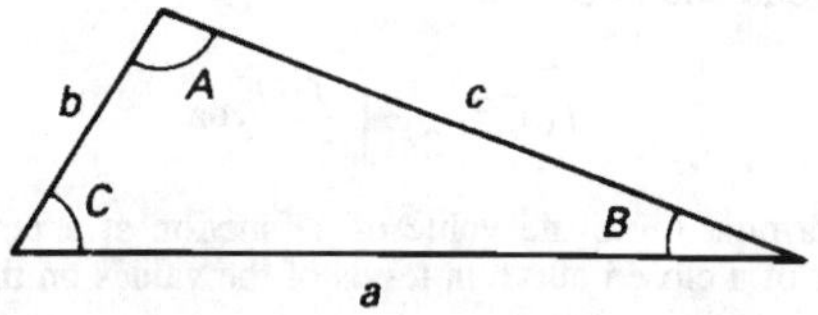

Figure 1.2.1 Triangle.

$$\sin\frac{A}{2} = \sqrt{\frac{(s-b)(s-c)}{bc}} \qquad \sin\frac{B}{2} = \sqrt{\frac{(s-c)(s-a)}{ca}}$$

$$\sin\frac{C}{2} = \sqrt{\frac{(s-a)(s-b)}{ab}}$$

$$\cos\frac{A}{2} = \sqrt{\frac{s(s-a)}{bc}} \qquad \cos\frac{B}{2} = \sqrt{\frac{s(s-b)}{ca}}$$

$$\cos\frac{C}{2} = \sqrt{\frac{s(s-c)}{ab}}$$

$$\tan\frac{A}{2} = \sqrt{\frac{(s-b)(s-c)}{s(s-a)}} \qquad \tan\frac{B}{2} = \sqrt{\frac{(s-c)(s-a)}{s(s-b)}}$$

$$\tan\frac{C}{2} = \sqrt{\frac{(s-a)(s-b)}{s(s-c)}}$$

1.2.6 Spherical Triangle

$$\frac{\sin A}{\sin a} = \frac{\sin B}{\sin b} = \frac{\sin C}{\sin c}$$

$$\cos a = \cos b \cos c + \sin b \sin c \cos A$$

$$\cos b = \cos c \cos a + \sin c \sin a \cos B$$

$$\cos c = \cos a \cos b + \sin a \sin b \cos C$$

where A, B, C and a, b, c are shown in Figure 1.2.2.

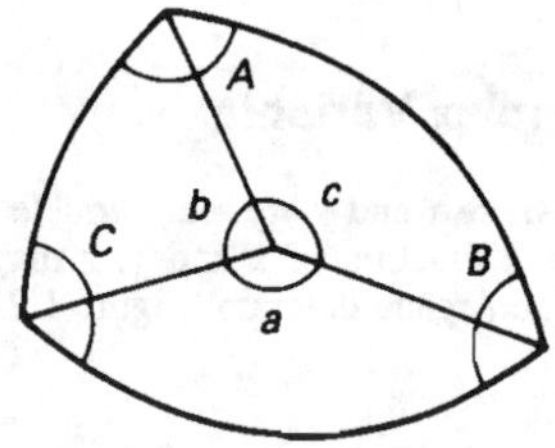

Figure 1.2.2 Spherical triangle.

1.2.7 Exponential Form

$$\sin\theta = \frac{e^{i\theta} - e^{-i\theta}}{2i} \qquad \cos\theta = \frac{e^{i\theta} + e^{-\theta}}{2}$$

$$e^{i\theta} = \cos\theta + i\sin\theta \quad e^{-i\theta} = \cos\theta - i\sin\theta$$

1.2.8 De Moivre's Theorem

$$(\cos A + i\sin A)(\cos B + i\sin B) = \cos(A + B) + i\sin(A + B)$$

1.2.9 Euler's Relation

$$(\cos\theta + i\sin\theta)^n = \cos n\theta + i\sin n\theta = e^{in\theta}$$

1.2.10 Hyperbolic Functions

$$\sinh x = (e^x - e^{-x})/2 \qquad \cosh x = (e^x + e^{-x})/2$$

$$\tanh x = \sinh x / \cosh x$$

Relations between hyperbolic functions can be obtained from the corresponding relations between trigonometric functions by reversing the sign of any term containing the product or implied product of two sines, e.g.:

$$\cosh^2 A - \sinh^2 A = 1$$

$$\cosh 2A = 2\cosh^2 A - 1 = 1 + 2\sinh^2 A$$
$$= \cosh^2 A + \sinh^2 A$$

$$\cosh(A \pm B) = \cosh A \cosh B \pm \sinh A \sinh B$$

$$\sinh(A \pm B) = \sinh A \cosh B \pm \cosh A \sinh B$$

$$e^x = \cosh x + \sinh x \quad e^{-x} = \cosh x - \sinh x$$

1.2.11 Complex Variable

If $z = x + iy$, where x and y are real variables, z is a complex variable and is a function of x and y, z may be represented graphically in an Argand diagram (Figure 1.2.3).

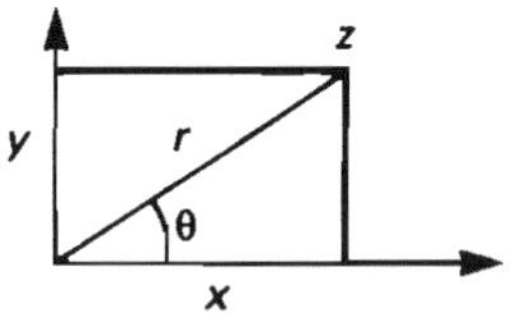

Figure 1.2.3 Argand diagram.

Polar form:

$$z = x + iy = |z|e^{i\theta} = |z|(\cos\theta + i\sin\theta)$$

$$(x = r\cos\theta \quad y = r\sin\theta)$$

where $r = |z|$.

Complex arithmetic:

$$z_1 = x_1 + iy_1 \quad z_2 = x_2 + iy_2$$

$$z_1 \pm z_2 = (x_1 \pm x_2) + i(y_1 \pm y_2)$$

$$z_1 . z_2 = (x_1x_2 - y_1y_2) + i(x_1y_2 + x_2y_1)$$

Conjugate:

$$z^* = x - iy \quad z.z^* = x^2 + y^2 = |z|^2$$

Function: another complex variable $w = u + iv$ may be related functionally to z by

$$w = u + iv = f(x + iy) = f(z)$$

which implies

$$u = u(x, y) \quad v = v(x, y)$$

e.g.,

$$\cosh z = \cosh(x + iy) = \cosh x \cosh iy + \sinh x \sinh iy$$
$$= \cosh x \cos y + i \sinh x \sin y$$

$$u = \cosh x \cos y \quad v = \sinh x \sin y$$

1.2.12 Cauchy–Riemann Equations

If $u(x, y)$ and $v(x, y)$ are continuously differentiable with respect to x and y,

$$\frac{\delta u}{\delta x} = \frac{\delta v}{\delta y} \quad \frac{\delta u}{\delta y} = -\frac{\delta v}{\delta x}$$

$w = f(z)$ is continuously differentiable with respect to z and its derivative is

$$f'(z) = \frac{\delta u}{\delta x} + i\frac{\delta v}{\delta x} = \frac{\delta v}{\delta y} - i\frac{\delta u}{\delta y} = \frac{1}{i}\left(\frac{\delta u}{\delta y} + i\frac{\delta v}{\delta y}\right)$$

It is also easy to show that $\nabla^2 u = \nabla^2 v = 0$. Since the transformation from z to w is conformal, the curves $u =$ constant and $v =$ constant intersect each other at right angles, so that one set may be used as equipotentials and the other as field lines in a vector field.

1.2.13 Cauchy's Theorem

If $f(z)$ is analytic everywhere inside a region bounded by C and a is a point within C

$$f(a) = \frac{1}{2\pi i}\int_c \frac{f(z)}{z - a}\,dz$$

This formula gives the value of a function at a point in the interior of a closed curve in terms of the values on that curve.

1.2.14 Zeros, Poles and Residues

If $f(z)$ vanishes at the point z_0 the Taylor series for z in the region of z_0 has its first two terms zero, perhaps others also: $f(z)$ may then be written

$$f(z) = (z - z_0)^n g(z)$$

where $g(z_0) \neq 0$. Then $f(z)$ has a *zero* of order n at z_0. The reciprocal

$$q(z) = 1/f(z) = h(z)/(z - z_0)^n$$

where $h(z) = 1/g(z) \neq 0$ at $z_0 \cdot q(z)$ becomes infinite at $z = z_0$ and is said to have a *pole* of order n at $z_0 \cdot q(z)$ may be expanded in the form.

$$q(z) = c_{-n}(z - z_0)^n + \cdots + c_{-1}(z - z_0)^{-1} + c_0 + \cdots$$

where c_{-1} is the *residue* of $q(z)$ at $z = z_0$. From Cauchy's theorem, it may be shown that if a function $f(z)$ is analytic throughout a region enclosed by a curve C except at a finite number of poles, the integral of the function around C has a value of $2\pi i$ times the sum of the residues of the function at its poles within C. This fact can be used to evaluate many definite integrals whose indefinite form cannot be found.

1.2.15 Some Standard Forms

$$\int_{0^*}^{2} e^{\cos\theta}\cos(n\theta - \sin\theta)\,d\theta = 2\pi/n!$$

$$\int_0^{\alpha} \frac{x^{a-1}}{1+x}\,dx = \pi\operatorname{cosec} a\pi$$

$$\int_0^{\alpha} \frac{\sin\theta}{\theta}\,d\theta = \frac{\pi}{2}$$

$$\int_0^{\alpha} x\exp(-h^2x^2)\,dx = \frac{1}{2h^2}$$

$$\int_0^{\alpha} \frac{x^{a-1}}{1-x}\,dx = \pi\cot a\pi$$

$$\int_0^{\alpha} \exp(-h^2x^2)\,dx = \frac{\sqrt{\pi}}{2h}$$

$$\int_0^{\alpha} x^2\exp(-h^2x^2)\,dx = \frac{\sqrt{\pi}}{4h^3}$$

1.2.16 Coordinate Systems

The basic system is the rectangular Cartesian system (x, y, z) to which all other systems are referred. Two other commonly used systems are as follows.

1.2.16.1 Cylindrical coordinates

Coordinates of point P are (x, y, z) or (r, θ, z) (see Figure 1.2.4), where

$$x = r\cos\theta \quad y = r\sin\theta \quad z = z$$

In these coordinates the volume element is $r\,dr\,d\theta\,dz$.

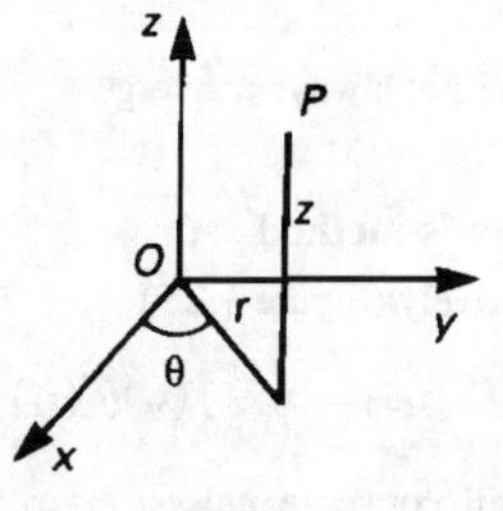

Figure 1.2.4 Cylindrical coordinates.

1.2.16.2 Spherical polar coordinates

Coordinates of point P are (x, y, z) or (r, θ, ϕ) (see Figure 1.2.5), where

$$x = r\sin\theta\cos\phi \quad y = r\sin\theta\sin\phi \quad z = r\cos\theta$$

In these coordinates the volume element is $r^2\sin\theta\,dr\,d\theta\,d\phi$.

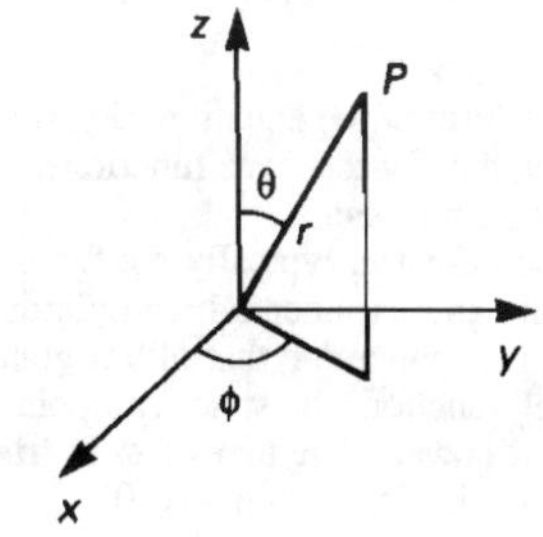

Figure 1.2.5 Spherical polar coordinates.

1.2.17 Transformation of Integrals

$$\iiint f(x,y,z)\,dx\,dy\,dz = \iiint \phi(u,v,w)|J|\,du\,dv\,dw$$

where

$$J = \begin{vmatrix} \frac{\delta x}{\delta u} & \frac{\delta y}{\delta u} & \frac{\delta z}{\delta u} \\ \frac{\delta x}{\delta v} & \frac{\delta y}{\delta v} & \frac{\delta z}{\delta v} \\ \frac{\delta x}{\delta w} & \frac{\delta y}{\delta w} & \frac{\delta z}{\delta w} \end{vmatrix} = \frac{\delta(x,y,z)}{\delta(u,v,w)}$$

is the Jacobian of the transformation of coordinates. For Cartesian to cylindrical coordinates, $J = r$, and for Cartesian to spherical polars, it is $r^2\sin\theta$.

1.2.18 Laplace's Equation

The equation satisfied by the scalar potential from which a vector field may be derived by taking the gradient is Laplace's equation, written as:

$$\Delta^2\phi = \frac{\delta^2\phi}{\delta x^2} + \frac{\delta^2\phi}{\delta y^2} + \frac{\delta^2\phi}{\delta z^2} = 0$$

In cylindrical coordinates:

$$\Delta^2\phi = \frac{1}{r}\frac{\delta}{\delta r}\left(r\frac{\delta\phi}{\delta r}\right) + \frac{1}{r^2}\frac{\delta^2\phi}{\delta\theta^2} + \frac{\delta^2\phi}{\delta z^2}$$

In spherical polars:

$$\nabla^2\phi = \frac{1}{r^2}\frac{\delta}{\delta r}\left(r^2\frac{\delta\phi}{\delta r}\right) + \frac{1}{r^2 \sin\theta}\frac{\delta\phi}{\delta\theta} + \frac{1}{r^2\sin^2\theta}\frac{\delta^2\phi}{\delta\phi^2}$$

The equation is solved by setting

$$\phi = U(u)V(v)W(w)$$

in the appropriate form of the equation, separating the variables and solving separately for the three functions, where (u, v, w) is the coordinate system in use.

In Cartesian coordinates, typically the functions are trigonometric, hyperbolic and exponential; in cylindrical coordinates the function of z is exponential, that of θ trigonometric and that of r is a Bessel function. In spherical polars, typically the function of r is a power of r, that of φ is trigonometric, and that of θ is a Legendre function of cos θ.

1.2.19 Solution of Equations

1.2.19.1 Quadratic equation

$$ax^2 + bx + c = 0$$

$$x = -\frac{b}{2a} \pm \frac{\sqrt{b^2 - 4ac}}{2a}$$

In practical calculations if $b^2 > 4ac$, so that the roots are real and unequal, calculate the root of larger modulus first, using the same sign for both terms in the formula, then use the fact that $x_1x_2 = c/a$ where x_1 and x_2 are the roots. This avoids the severe cancellation of significant digits which may otherwise occur in calculating the smaller root.

For polynomials other than quadratics, and for other functions, several methods of successive approximation are available.

1.2.19.2 Bisection method

By trial find x_0 and x_1 such that $f(x_0)$ and $f(x_1)$ have opposite signs (see Figure 1.2.6). Set $x_2 = (x_0 + x_1)/2$ and calculate $f(x_2)$. If $f(x_0)f(x_2)$ is positive, the root lies in the interval (x_1, x_2); if negative in the interval (x_0, x_2); and if zero, x_2 is the root. Continue if necessary using the new interval.

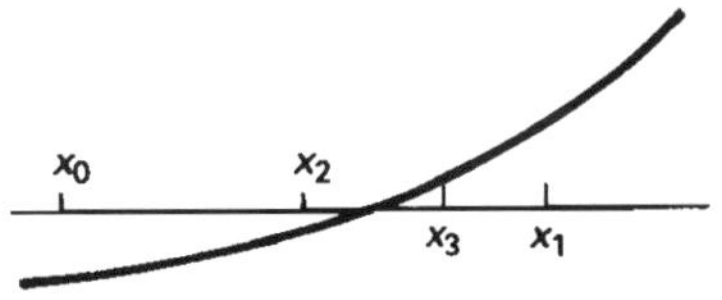

Figure 1.2.6 Bisection method.

1.2.19.3 Regula falsi

By trial, find x_0 and x_1 as for the bisection method; these two values define two points $(x_0, f(x_0))$ and $(x_1, f(x_1))$. The straight line joining these two points cuts the x-axis at the point (see Figure 1.2.7):

$$x_2 = \frac{x_0 f(x_1) - x_1 f(x_0)}{f(x_1) - f(x_0)}$$

Evaluate $f(x_2)$ and repeat the process for whichever of the intervals (x_0, x_2) or (x_1, x_2) contains the root. This method can be accelerated by halving at each step the function value at the retained end of the interval, as shown in Figure 1.2.8.

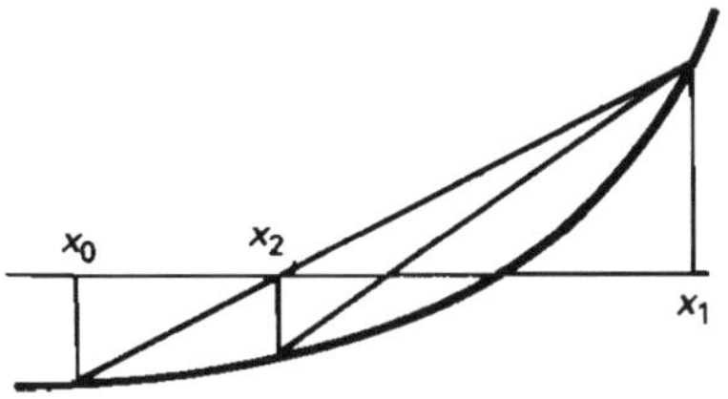

Figure 1.2.7 Regula falsi.

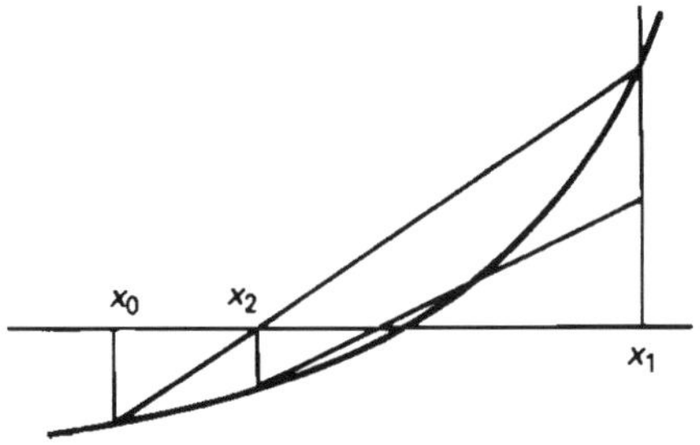

Figure 1.2.8 Accelerated method.

1.2.19.4 Fixed-point iteration

Arrange the equation in the form

$$x = f(x)$$

Choose an initial value of x by trial, and calculate repetitively

$$x_{k+1} = f(x_k)$$

This process will not always converge.

1.2.19.5 Newton's method

Calculate repetitively (Figure 1.2.9)

$$x_{k+1} = x_k - f(x_k)/f'(x_k)$$

This method will converge unless: (a) x_k is near a point of inflexion of the function; or (b) x_k is near a local minimum;

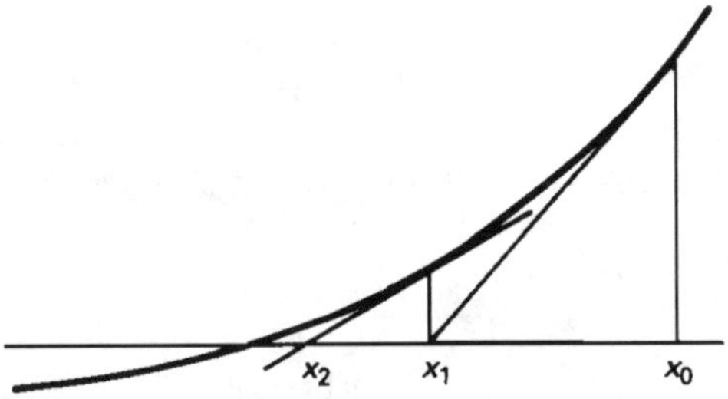

Figure 1.2.9 Newton's method.

or (c) the root is multiple. If one of these cases arises, most of the trouble can be overcome by checking at each stage that

$$f(x_{k+1}) > f(x_k)$$

and, if not, halving the preceding value of $|x_{k+1} - x_k|$.

1.2.20 Method of Least Squares

To obtain the best fit between a straight line $ax + by = 1$ and several points $(x_1,y_1),(x_2,y_2),\ldots,(x_n,y_n)$ found by observation, the coefficients a and b are to be chosen so that the sum of the squares of the errors

$$e_i = ax_i + by_i - 1$$

is a minimum. To do this, first write the set of inconsistent equations

$$ax_1 + by_1 - 1 = 0$$

$$ax_2 + by_2 - 1 = 0$$

$$\vdots$$

$$ax_n + by_n - 1 = 0$$

Multiply each equation by the value of x it contains, and add, obtaining

$$a\sum_{i=1}^{n} x_i^2 + b\sum_{i=1}^{n} x_i y_i - \sum_{i=1}^{n} x_i = 0$$

Similarly multiply by y and add, obtaining

$$a\sum_{i=1}^{n} x_i y_i + b\sum_{i=1}^{n} y_i^2 - \sum_{i=1}^{n} y_i = 0$$

Lastly, solve these two equations for a and b, which will be the required values giving the least squares fit.

1.2.21 Relation Between Decibels, Current and Voltage Ratio, and Power Ratio

$$\text{dB} = 10\log\frac{P_1}{P_2} = 20\log\frac{V_1}{V_2} = 20\log\frac{I_1}{I_2}$$

dB	I_1/I_2 or V_1/V_2	I_2/I_1 or V_2/V_1	P_1/P_2	P_2/P_1
0.1	1.012	0.989	1.023	0.977
0.2	1.023	0.977	1.047	0.955
0.3	1.035	0.966	1.072	0.933
0.4	1.047	0.955	1.096	0.912
0.5	1.059	0.944	1.122	0.891
0.6	1.072	0.933	1.148	0.871
0.7	1.084	0.923	1.175	0.851
0.8	1.096	0.912	1.202	0.832
0.9	1.109	0.902	1.230	0.813
1.0	1.122	0.891	1.259	0.794
1.1	1.135	0.881	1.288	0.776
1.2	1.148	0.871	1.318	0.759
1.3	1.162	0.861	1.349	0.741
1.4	1.175	0.851	1.380	0.724
1.5	1.188	0.841	1.413	0.708
1.6	1.202	0.832	1.445	0.692
1.7	1.216	0.822	1.479	0.676
1.8	1.230	0.813	1.514	0.661
1.9	1.245	0.804	1.549	0.645
2.0	1.259	0.794	1.585	0.631
2.5	1.334	0.750	1.778	0.562
3.0	1.413	0.708	1.995	0.501
3.5	1.496	0.668	2.24	0.447
4.0	1.585	0.631	2.51	0.398
4.5	1.679	0.596	2.82	0.355
5.0	1.778	0.562	3.16	0.316
5.5	1.884	0.531	3.55	0.282
6.0	1.995	0.501	3.98	0.251
6.5	2.11	0.473	4.47	0.224
7.0	2.24	0.447	5.01	0.200
7.5	2.37	0.422	5.62	0.178
8.0	2.51	0.398	6.31	0.158
8.5	2.66	0.376	7.08	0.141
9.0	2.82	0.355	7.94	0.126
9.5	2.98	0.335	8.91	0.112
10.0	3.16	0.316	10.0	0.100
10.5	3.35	0.298	11.2	0.0891
11.0	3.55	0.282	12.6	0.0794
15.0	5.62	0.178	31.6	0.0316
15.5	5.96	0.168	35.5	0.0282
16.0	6.31	0.158	39.8	0.0251
16.5	6.68	0.150	44.7	0.0224

dB	I_1/I_2 or V_1/V_2	I_2/I_1 or V_2/V_1	P_1/P_2	P_2/P_1
17.0	7.08	0.141	50.1	0.0200
17.5	7.50	0.133	56.2	0.0178
18.0	7.94	0.126	63.1	0.0158
18.5	8.41	0.119	70.8	0.0141
19.0	8.91	0.112	79.4	0.0126
19.5	9.44	0.106	89.1	0.0112
20.0	10.00	0.1000	100	0.0100
20.5	10.59	0.0944	112	0.00891
21.0	11.22	0.0891	126	0.00794
21.5	11.88	0.0841	141	0.00708
22.0	12.59	0.0794	158	0.00631
22.5	13.34	0.0750	178	0.00562
23.0	14.13	0.0708	200	0.00501
23.5	14.96	0.0668	224	0.00447
24.0	15.85	0.0631	251	0.00398
24.5	16.79	0.0596	282	0.00355
25.0	17.78	0.0562	316	0.00316
25.5	18.84	0.0531	355	0.00282
26.0	19.95	0.0501	398	0.00251
26.5	21.1	0.0473	447	0.00224
27.0	22.4	0.0447	501	0.00200
27.5	23.7	0.0422	562	0.00178
28.0	25.1	0.0398	631	0.00158
28.5	26.6	0.0376	708	0.00141
29.0	28.2	0.0355	794	0.00126
29.5	29.8	0.0335	891	0.00112
30.0	31.6	0.0316	1000	0.00100
31.0	35.5	0.0282	1260	7.94×10^{-4}
32.0	39.8	0.0251	1580	6.31×10^{-4}
33.0	44.7	0.0224	2000	5.01×10^{-4}
34.0	50.1	0.0200	2510	3.98×10^{-4}
35.0	56.2	0.0178		3.16×10^{-4}
36.0	63.1	0.0158	3980	2.51×10^{-4}
37.0	70.8	0.0141	5010	2.00×10^{-4}

1.2.22 Calculus

1.2.22.1 Derivative

$$f'(x) = \lim_{\delta \to 0} \frac{f(x + \delta x) - f(x)}{\delta x}$$

If u and v are functions of x,

$$(uv)' = u'v + uv'$$

$$\left(\frac{u}{v}\right)' = \frac{u'v - uv'}{v^2}$$

$$(uv)^{(n)} = u^{(n)}v + nu^{(n-1)}v^{(1)} + \cdots + {}^nC_p u^{(n-p)}v^{(p)} + \cdots + uv^{(n)}$$

where

$${}^nC_p = \frac{n!}{p!(n-p)!}$$

If $z = f(x)$ and $y = g(z)$, then

$$\frac{\mathrm{d}y}{\mathrm{d}x} = \frac{\mathrm{d}y}{\mathrm{d}z} = \frac{\mathrm{d}z}{\mathrm{d}x}$$

1.2.22.2 Maxima and minima

$f(x)$ has a stationary point wherever $f'(x) = 0$: the point is a maximum, minimum or point of inflexion according as $f''(x) <, >$ or $= 0$.

$f(x, y)$ has a stationary point wherever

$$\frac{\delta f}{\delta x} = \frac{\delta f}{\delta y} = 0$$

Let (a, b) be such a point, and let

$$\frac{\delta^2 f}{\delta x^2} = A, \qquad \frac{\delta^2 f}{\delta x \delta y} = H, \qquad \frac{\delta^2 f}{\delta y^2} = B$$

all at that point, then:

If $H^2 - AB > 0$, $f(x, y)$ has a saddle point at (a, b).
If $H^2 - AB < 0$ and if $A < 0$, $f(x, y)$ has a maximum at (a, b), but if $A > 0$, $f(x, y)$ has a minimum at (a, b).
If $H^2 = AB$, higher derivatives need to be considered.

1.2.22.3 Integral

$$\int_a^b f(x)\mathrm{d}x = \lim_{N \to \infty} \sum_{n=0}^{n-1} f\left(a + \frac{n(b-a)}{N}\right)\left(\frac{b-a}{N}\right)$$

$$= \lim_{N \to \infty} \sum_{\text{if } n=1}^{N} f(a + (n-1)\delta x)\delta x$$

where $\delta x = (b - a)/N$.

If u and v are functions of x, then

$$\int uv' = uv - \int u'v\mathrm{d}x \text{ (integration by parts)}$$

1.2.22.4 Derivatives and integrals

y	dy/dx	$\int y dx$
x^n	nx^{n-1}	$x^{n+1}/(n+1)$
$1/x$	$-1/x^2$	$\ln(x)$
e^{ax}	ae^{ax}	e^{ax}/a
$\ln(x)$	$1/x$	$x[\ln(x) - 1]$
$\log_a x$	$\frac{1}{x}\log_a e$	$x \log_a(\frac{x}{e})$
$\sin ax$	$a \cos ax$	$-\frac{1}{a}\cos ax$
$\cos ax$	$-a \sin ax$	$\frac{1}{a}\sin ax$
$\tan ax$	$a \sec^2 ax$	$-\frac{1}{a}\ln(\cos ax)$
$\cot ax$	$-a\ \mathrm{cosec}^2 ax$	$\frac{1}{a}\ln(\sin ax)$
$\sec ax$	$a \tan ax \sec ax$	$\frac{1}{a}\ln(\sec ax + \tan ax)$
$\mathrm{cosec}\ ax$	$-a \cot ax\ \mathrm{cosec}\ ax$	$\frac{1}{a}\ln(\mathrm{cosec}\ ax - \cot ax)$
$\arcsin(x/a)$	$1/(a^2 - x^2)^{1/2}$	$x \arcsin(x/a) + (a^2 - x^2)^{1/2}$
$\arccos(x/a)$	$-1/(a^2 - x^2)^{1/2}$	$x \arccos(x/a) - (a^2 - x^2)^{1/2}$
$\arctan(x/a)$	$a/(a^2 + x^2)$	$x \arctan(x/a) - \frac{1}{2} a \ln(a^2 + x^2)$
$\mathrm{arccot}(x/a)$	$-a/(a^2 + x^2)$	$x\ \mathrm{arccot}(x/a) + \frac{1}{2} a \ln(a^2 + x^2)$
$\mathrm{arcsec}(x/a)$	$a(x^2 - a^2)^{-1/2}/x$	$x\ \mathrm{arcsec}(x/a) = a \ln[x + (x^2 - a^2)^{1/2}]$
$\mathrm{arccosec}(x/a)$	$-a(x^2 - a^2)^{-1/2}/x$	$x\ \mathrm{arccosec}(x/a) + a \ln[x + (x^2 - a^2)^{1/2}]$
$\sinh ax$	$a \cosh ax$	$\frac{1}{a}\cosh ax$
$\coth ax$	$a \sinh ax$	$\frac{1}{a}\sinh ax$
$\tanh ax$	$a\ \mathrm{sech}^2 ax$	$\frac{1}{a}\ln(\cosh ax)$
$\coth ax$	$-a\ \mathrm{cosech}^2 ax$	$\frac{1}{a}\ln(\sinh ax)$
$\mathrm{sech}\ ax$	$-a \tanh ax\ \mathrm{sech}\ ax$	$\frac{2}{a}\arctan(e^{ax})$
$\mathrm{cosech}\ ax$	$-a \coth ax\ \mathrm{cosech}\ ax$	$\frac{1}{a}\ln(\tanh \frac{ax}{2})$
$\mathrm{arsinh}(x/a)$	$(x^2 + a^2)^{-1/2}$	$x\ \mathrm{arsinh}(x/a) - (x^2 + a^2)^{1/2}$
$\mathrm{arcosh}(x/a)$	$(x^2 - a^2)^{-1/2}$	$x\ \mathrm{arcosh}(x/a) - (x^2 - a^2)^{1/2}$
$\mathrm{artanh}(x/a)$	$a(a^2 - x^2)^{-1}$	$x\ \mathrm{artanh}(x/a) + \frac{1}{2} a \ln(a^2 - x^2)$
$\mathrm{arcoth}(x/a)$	$-a(x^2 - a^2)^{-1}$	$x\ \mathrm{arcoth}(x/a) + \frac{1}{2} a \ln(x^2 - a^2)$
$\mathrm{arsech}(x/a)$	$-a(a^2 - x^2)^{-1/2}/x$	$x\ \mathrm{arsech}(x/a) + a \arcsin(x/a)$
$\mathrm{arcosech}(x/a)$	$-a(x^2 + a^2)^{-1/2}/x$	$x\ \mathrm{arcosech}(x/a) + a\ \mathrm{arsinh}(x/a)$
$(x^2 \pm a^2)^{1/2}$		$\begin{cases} \frac{1}{2} x(x^2 \pm a^2)^{1/2} \pm \frac{1}{2} a^2 \mathrm{arsinh}(x/a) \\ \frac{1}{2} x(a^2 - x^2)^{1/2} + \frac{1}{2} a^2 \arcsin(x/a) \end{cases}$
$(a^2 - x^2)^{1/2}$		$\begin{cases} \frac{1}{2}(x^2 \pm a^2)^{p+1}/(p+1) & (p \neq -1) \\ \frac{1}{2}\ln(x^2 \pm a^2) & (p = -1) \end{cases}$
$(x^2 \pm a^2)^p x$		$\begin{cases} -\frac{1}{2}(a^2 - x^2)^{p+1}/(p+1) & (p \neq -1) \\ -\frac{1}{2}\ln(a^2 - x^2) & (p = -1) \end{cases}$
$(a^2 - x^2)px$		$(ax^2 + b)^{p+1}/2a(p+1) \quad (p \neq -1)$
$x(ax^2 + b)^p$		$[\ln(ax^2 + b)]/2a \quad (p = -1)$
$(2ax - x^2)^{-1/2}$		$\arccos(\frac{a-x}{a})$
$(a^2 \sin^2 x + b^2 \cos^2 x)^{-1}$		$\frac{1}{ab}\arctan(\frac{a}{b}\tan x)$
$(a^2 \sin^2 x - b^2 \cos^2 x)^{-1}$		$-\frac{1}{ab}\mathrm{artanh}(\frac{a}{b}\tan x)$

y	dy/dx	$\int y dx$	
$e^{ax}\sin bx$		$e^{ax}\dfrac{a\sin bx - b\cos bx}{a^2+b^2}$	
$e^{ax}\cos bx$		$e^{ax}\dfrac{(a\cos bx + b\sin bx)}{a^2+b^2}$	
$\sin mx\ \sin nx$		$\dfrac{1}{2}\dfrac{\sin(m-n)x}{m-n} - \dfrac{1}{2}\dfrac{\sin(m+n)x}{m+n}$	$(m \neq n)$
		$\dfrac{1}{2}\left(x - \dfrac{\sin 2mx}{2m}\right)$	$(m = n)$
$\sin mx\ \cos nx$		$-\dfrac{1}{2}\dfrac{\cos(m+n)x}{m+n} - \dfrac{1}{2}\dfrac{\cos(m-n)x}{m-n}$	$(m \neq n)$
		$-\dfrac{1}{2}\dfrac{\cos 2mx}{2m}$	$(m = n)$
$\cos mx\ \cos nx$		$\dfrac{1}{2}\dfrac{\sin(m+n)x}{m+n} + \dfrac{1}{2}\dfrac{\sin(m-n)x}{m-n}$	$(m \neq n)$
		$\dfrac{1}{2}\left(x + \dfrac{\sin 2mx}{2m}\right)$	$(m = n)$

1.2.22.5 Standard substitutions

Integral a function of	*Substitute*
$a^2 - x^2$	$x = a\sin\theta$ or $x = a\cos\theta$
$a^2 + x^2$	$x = a\tan\theta$ or $x = a\sinh\theta$
$x^2 - a^2$	$x = a\sec\theta$ or $x = a\cosh\theta$

1.2.22.6 Reduction formulas

$$\int \sin^m x\,dx = -\frac{1}{m}\sin^{m-1}x\cos x + \frac{m-1}{m}\int \sin^{m-2}x\,dx$$

$$\int \cos^m x\,dx = \frac{1}{m}\cos^{m-1}x\sin x + \frac{m-1}{m}\int \cos^{m-2}x\,dx$$

$$\int \sin^m x\cos^n x\,dx = \frac{\sin^{m+1}x\cos n^{-1}x}{m+n} + \frac{n-1}{m+n}\int \sin^m x\cos^{n-2}x\,dx$$

If the integrand is a rational function of sin x and/or cos x, substitute $t = \tan\frac{1}{2}x$, then

$$\sin x = \frac{1}{1+t^2}, \quad \cos x = \frac{1-t^2}{1+t^2}, \quad dx = \frac{2dt}{1+t^2}$$

1.2.22.7 Numerical integration

1.2.22.7.1 Trapezoidal rule (Figure 1.2.10)

$$\int_{x1}^{x2} y\,dx = \frac{1}{2}h(y_1 + y_2) + O(h^3)$$

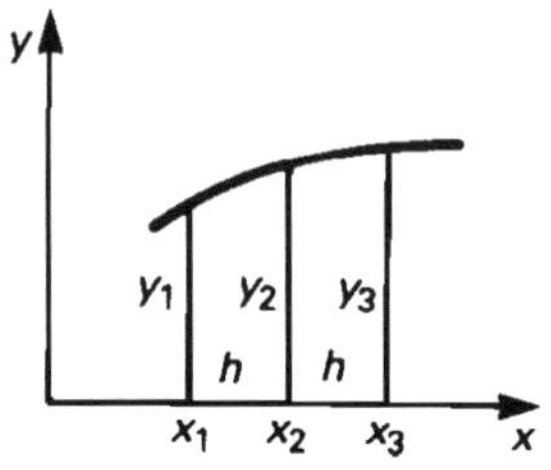

Figure 1.2.10 Numerical integration.

1.2.22.7.2 Simpson's rule (Figure 1.2.10)

$$\int_{x1}^{x2} y\,dx = 2h(y_1 + 4y_2 + y_3)/6 + O(h^5)$$

1.2.22.7.3 Change of variable in double integral

$$\iint f(x,y)\,dx\,dy = \iint F(u,v)|J|\,du\,dv$$

where

$$J = \frac{\delta(x,y)}{\delta(u,v)} = \begin{vmatrix} \dfrac{dx}{\delta u} & \dfrac{dx}{\delta v} \\ \dfrac{\delta y}{\delta u} & \dfrac{\delta y}{\delta v} \end{vmatrix} = \begin{vmatrix} \dfrac{dx}{\delta u} & \dfrac{dy}{\delta u} \\ \dfrac{\delta x}{\delta v} & \dfrac{\delta y}{\delta v} \end{vmatrix}$$

is the Jacobian of the transformation.

1.2.22.7.4 Differential mean value theorem

$$\frac{f(x+h)-f(x)}{h} = f'(x+\theta h) \quad 0<\theta<1$$

1.2.22.7.5 Integral mean value theorem

$$\int_a^b f(x)g(x)dx = g(a+\theta h) \int_a^b f(x)dx$$

$h = b - a, \quad 0<\theta<1$

1.2.22.8 Vector calculus

Let $s(x, y, z)$ be a scalar function of position and let

$$\mathbf{v}(x,y,z) = \mathbf{i}v_x(x,y,z) + \mathbf{j}v_y(x,y,z) + \mathbf{k}v_z(x,y,z)$$

be a vector function of position. Define

$$\nabla = \mathbf{i}\frac{\delta}{\delta x} + \mathbf{j}\frac{\delta}{\delta y} + \mathbf{k}\frac{\delta}{\delta z}$$

so that

$$\nabla.\nabla = \nabla^2 = \frac{\delta^2}{\delta x^2} + \frac{\delta^2}{\delta y^2} + \frac{\delta^2}{\delta z^2}$$

then

$$\text{grad } s = \nabla s = \mathbf{i}\frac{\delta s}{\delta x} = \mathbf{j}\frac{\delta s}{\delta y} + \mathbf{k}\frac{\delta s}{\delta z}$$

$$\text{div } \mathbf{v} = \nabla.\mathbf{v} = \frac{\delta v_x}{\delta x} + \frac{\delta v_y}{\delta y} + \frac{\delta v_z}{\delta z}$$

$$\text{curl } \mathbf{v} = \nabla \times \mathbf{v} = \mathbf{i}\left(\frac{\delta v_z}{\delta y} - \frac{\delta v_y}{\delta z}\right) + \mathbf{j}\left(\frac{\delta v_x}{\delta z} - \frac{\delta v_z}{\delta x}\right) + \mathbf{k}\left(\frac{\delta v_y}{\delta x} - \frac{\delta v_x}{\delta y}\right)$$

The following identities are then true:

$$\text{div}(s\mathbf{v}) = s \text{ div } \mathbf{v} + (\text{grad } s).\mathbf{v}$$

$$\text{curl}(s\mathbf{v}) = s \text{ curl } \mathbf{v} + (\text{grad } s) \times \mathbf{v}$$

$$\text{div}(\mathbf{u} \times \mathbf{v}) = \mathbf{v}.\text{curl } \mathbf{u} - \mathbf{u}.\text{curl } \mathbf{v}$$

$$\text{curl}(\mathbf{u} \times \mathbf{v}) = \mathbf{u} \text{ div } \mathbf{v} - \mathbf{v} \text{ div } \mathbf{u} + (\mathbf{v}.\nabla)\mathbf{u} - (\mathbf{u}.\nabla)\mathbf{v}$$

$$\text{div grad } s = \nabla^2 s$$

$$\text{div curl } \mathbf{v} = 0$$

$$\text{curl grad } s = 0$$

$$\text{curl curl } \mathbf{v} = \text{grad}(\text{div } \mathbf{v}) - \nabla^2\mathbf{v}$$

where ∇^2 operates on each component of $\mathbf{v}$.

$$\mathbf{v}x \text{ curl } \mathbf{v} + (\mathbf{v}.\nabla)\mathbf{v} = \text{grad } \tfrac{1}{2}\mathbf{v}^2$$

Potentials:

If curl $\mathbf{v} = 0, \mathbf{v} = \text{grad } \varphi$ where φ is a scalar potential.
If div $\mathbf{v} = 0, \mathbf{v} = \text{curl } \mathbf{A}$ where $\mathbf{A}$ is a vector potential.

P Sproxton
Alpha Image Ltd

1.3 Analogue and Digital Circuit Theory

1.3.1 Analogue Circuit Theory

1.3.1.1 Resistors

Resistors are the simplest linear components encountered when applying circuit theory techniques. The symbol for a resistor is shown in Figure 1.3.1(a). If an emf V (measured in volts) is applied across the terminals of the resistor (resistance R, measured in ohms), it can be shown, as in (b), that the current flowing through the resistor is linearly proportional to the current I (measured in amperes).

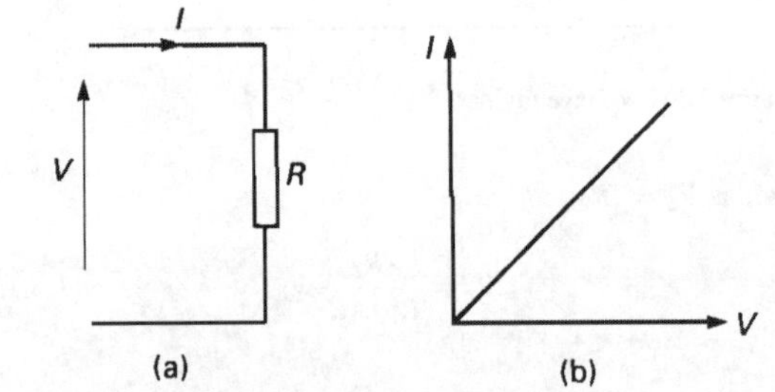

Figure 1.3.1 Resistor and voltage/current characteristic.

This relationship is better known as Ohm's law, i.e.:

$$I = \frac{V}{R} \tag{1.3.1}$$

Another important relationship here is the power P (in watts) which is dissipated in the resistor. This is given by:

$$\begin{aligned} P &= VI \\ &= I^2R \end{aligned} \tag{1.3.2}$$

1.3.1.2 Series resistance circuits

A series configuration for resistors is illustrated in Figure 1.3.2. In this circuit, the current flowing through all three resistors is the same. Therefore according to equation (1.3.1) the potential difference across each resistance is given by:

$$V_1 = R_1I; \; V_2 = R_2I; \; V_3 = R_3I$$

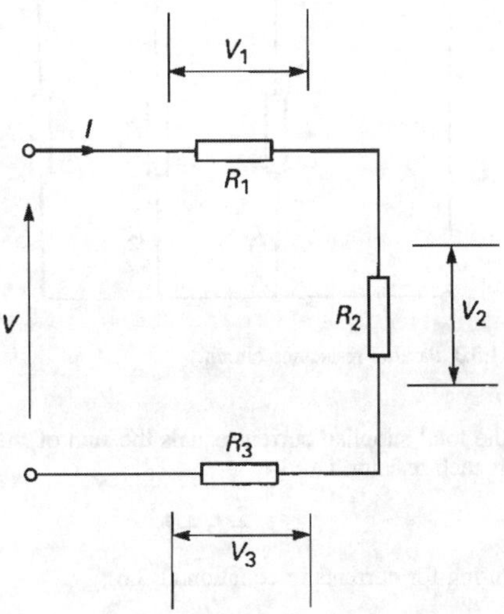

Figure 1.3.2 Series resistance circuit.

The sum of these three potential differences is equal to the applied voltage V, i.e.:

$$V = V_1 + V_2 + V_3 \tag{1.3.3}$$

Equation (1.3.3) indicates that the algebraic sum of the potential differences around any complete circuit is equal to zero.

Using equation (1.3.1) we can rearrange equation (1.3.3) to give:

$$V = IR_1 + IR_2 + IR_3$$
$$= I(R_1 + R_2 + R_3)$$

Therefore

$$I = \frac{V}{R_1 + R_2 + R_3} = \frac{V}{R_e} \quad (1.3.4)$$

The equivalent resistance of the circuit in Figure 1.3.2 is therefore R_e, which is given by

$$R_e = R_1 + R_2 + R_3 \quad (1.3.5)$$

Equation (1.3.5) can be stated more generally: If any number of resistors are connected in series, then the equivalent resistance is the sum of the individual values.

1.3.1.3 Parallel resistance circuits

A second way of configuring resistors is shown in Figure 1.3.3. In this circuit, the voltage across each resistor is the same, but the current in each is given by:

$$I_1 = \frac{V}{R_1}; \quad I_2 = \frac{V}{R_2}; \quad I_3 = \frac{V}{R_3} \quad (1.3.6)$$

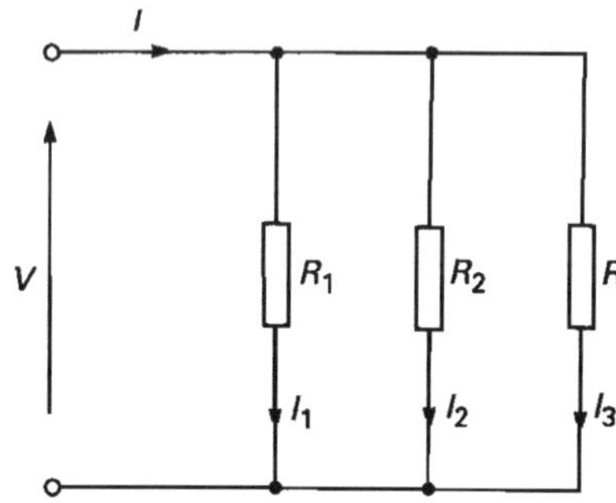

Figure 1.3.3 Parallel resistance circuit.

Here, the total supplied current equals the sum of the currents through each resistor. i.e.:

$$I = I_1 + I_2 + I_3 \quad (1.3.7)$$

Substituting for currents in equation (1.3.6):

$$I = \frac{V}{R_1} + \frac{V}{R_2} + \frac{V}{R_3} = V\left(\frac{1}{R_1} + \frac{1}{R_2} + \frac{1}{R_3}\right) \quad (1.3.8)$$

From Ohm's law (equation (1.3.1)) we have

$$I = \frac{V}{R_e} = V\left(\frac{1}{R_1} + \frac{1}{R_2} + \frac{1}{R_3}\right) \quad (1.3.9)$$

Therefore

$$\frac{1}{R_e} = \frac{1}{R_1} + \frac{1}{R_2} + \frac{1}{R_3} \quad (1.3.10)$$

This can be more generally stated by:

$$\frac{1}{R_e} = \frac{1}{R_1} + \frac{1}{R_2} + \frac{1}{R_3} + \cdots + \frac{1}{R_n} \quad (1.3.11)$$

i.e. the reciprocal of equivalent resistance is equal to the sum of the reciprocals of the component resistances.

1.3.1.4 Voltage dividers

One every common implementation of resistor networks is the voltage divider (see Figure 1.3.4). For this circuit, the output is always smaller than the input, in a ration which is determined by the value of R_1 and R_2. The current through R_1 and R_2 is given by:

$$I = \frac{V_i}{R_1 + R_2} \quad (1.3.12)$$

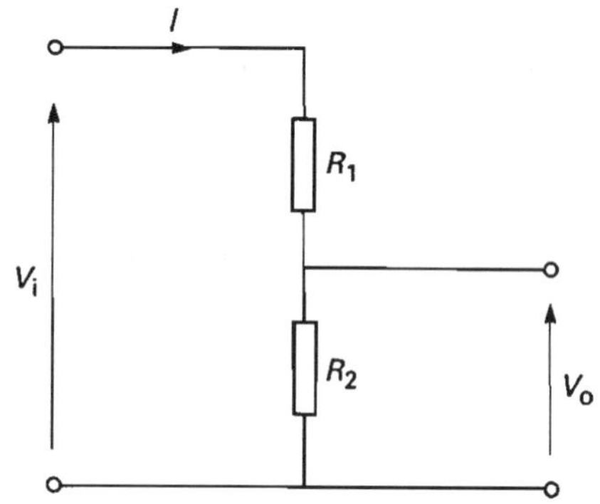

Figure 1.3.4 Voltage divider.

and, as $V_o = IR_2$:

$$V_o = \frac{R_2}{R_1 + R_2} V_i \quad (1.3.13)$$

1.3.1.5 Kirchoff's laws

The circuit in Figure 1.3.5(a) contains a parallel set of resistances with two *nodes* indicated. A node is a point where three or more conductors are joined. Kirchoff's first law states: The total current flowing towards a node is equal to the total current flowing away from the node, i.e. the algebraic sum of currents flowing towards a node is zero. Thus in Figure 1.3.5(b) at node B:

$$I_1 + I_2 + I_3 = I$$

or

$$I_1 + I_2 + I_3 - I = 0$$

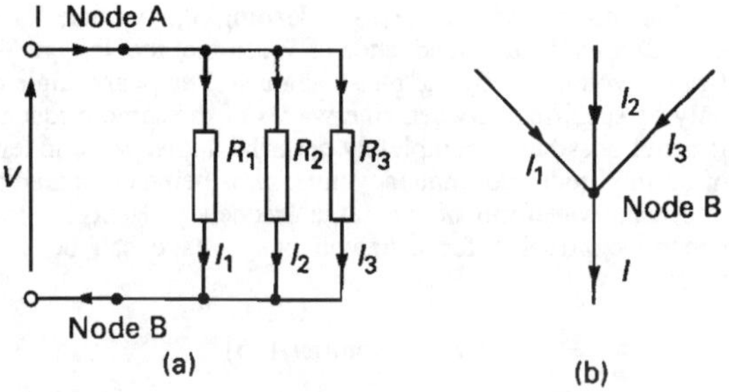

Figure 1.3.5 Kirchoff's current law.

In general, for any node,

$$\Sigma I = 0 \tag{1.3.14}$$

The circuit in Figure 1.3.6 illustrates Kirchoff's second law, which states: The algebraic sum of the potential differences is zero, i.e.:

$$V = V_1 + V_2 + V_3$$

or

$$V_1 + V_2 + V_3 - V = 0$$

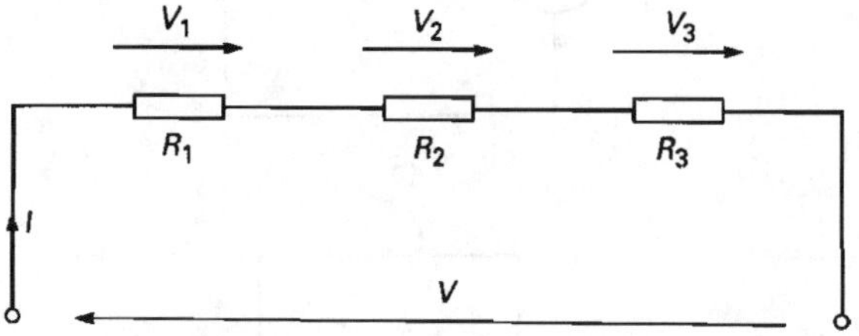

Figure 1.3.6 Kirchoff's voltage law.

Generally, for any node,

$$\Sigma V = 0 \tag{1.3.15}$$

1.3.1.6 Equivalent circuits

To simplify circuit analysis, it is often necessary to reduce portions of a circuit to a simpler equivalent form. This is in order to clarify areas of the circuit that are of particular interest. The following two sections describe two theorems that enable some networks to be simplified.

1.3.1.6.1 Thevenin's theorem

Consider the network and load resistance in Figure 1.3.7(a). Thevenin's theorem can be stated as follows: An active network, having two terminals, A and B, can be replaced by a constant voltage source having an emf V_o and an internal resistance r. The value of V_o is equal to the open circuit potential between A and B, and r is the resistance of the network measured between A and B, with the load disconnected and the sources of emf replaced by their internal resistances. On this basis, the network in Figure 1.3.7(a) can be redrawn as in (b).

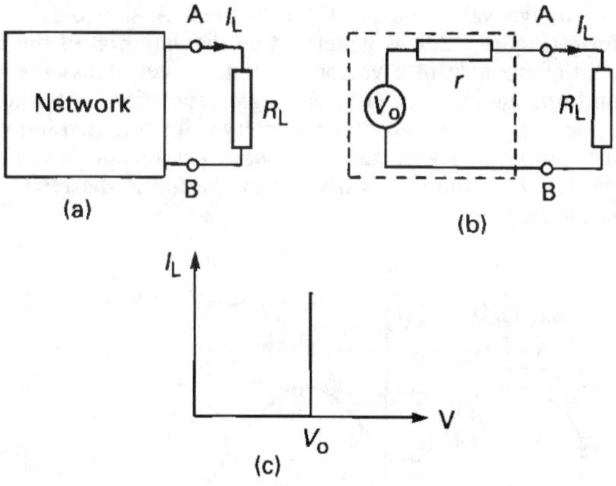

Figure 1.3.7 Thevenin's equivalent circuit.

1.3.1.6.2 Norton's theorem

This is another theorem for an equivalent circuit. Norton's theorem states that: Any two-terminal network consisting of dc voltage sources and resistors can be replaced by a parallel combination of a current source I_e and a resistance R_e (see Figure 1.3.8). The current source is the short-circuit current at the output terminals and R_e is the same as for Thevenin's theorem. The Norton equivalent is illustrated in Figure 1.3.8, with the characteristic of the current generator in (c).

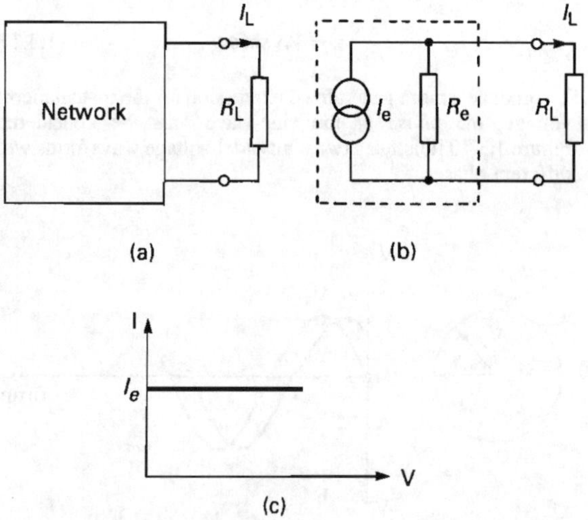

Figure 1.3.8 Norton's equivalent circuit.

1.3.2 Alternating Current Circuits

In practice, the currents and voltages most often found in electronic circuits vary with time. The simplest time-varying waveform is one where the current or voltage changes direction periodically; this is termed an *alternating current* (ac).

The simplest ac waveform is the sine-wave, where current or voltage vary sinusoidally with time. A sinusoidal wave form (Figure 1.3.9) is generated by the variation of the vertical component of a vector rotating counter-clockwise with uniform angular velocity. A single revolution is termed a *cycle*, where the elapsed time interval for one revolution is the period T. The number of cycles per second is the *frequency*, f, of the sine-wave and the *period* is the reciprocal of the frequency.

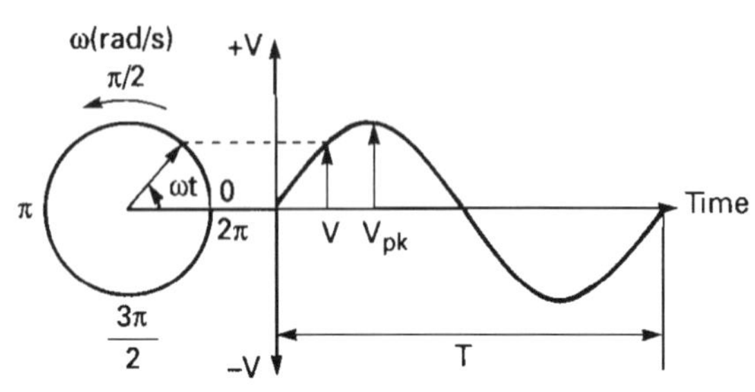

Figure 1.3.9 Rotating vector representation of a sinusoidal quantity.

For a complete cycle, the vector will turn through 2π radians, therefore the angular frequency (in radians per second) is given by:

$$\omega = 2\pi f \quad (1.3.16)$$

If the magnitude of the rotating vector is V_{pk}, the instantaneous value of v, at any time t, is given by:

$$v = V_{pk} \sin \omega t \quad (1.3.17)$$

To produce a more generalized expression for the instantaneous voltage, the *phase* of the sine-wave must be considered. Figure 1.3.10 illustrates two sinusoidal voltage waveforms with different phases.

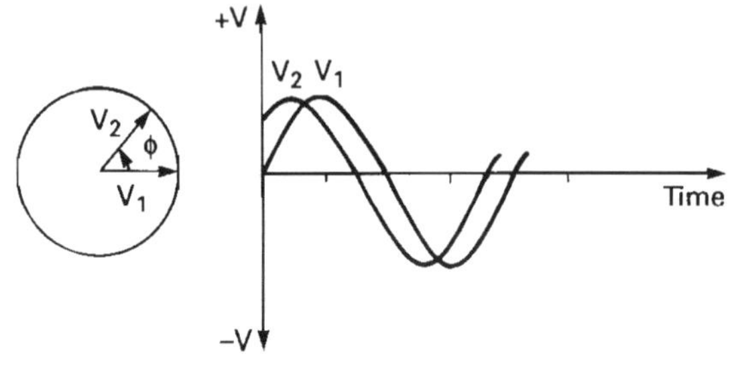

Figure 1.3.10 Vector representation of phase sinusoidal quantity.

This shows that voltage V_2 is leading V_1, because V_2 has passed through zero in advance of V_1. In fact the voltage V_2 is *leading* voltage V_1 by the phase angle ϕ. The phase angle can only be specified between sine-waves of the same frequency. It is not possible to completely describe a sine-wave in terms of its amplitude and frequency unless it is being compared to a reference waveform of the same frequency. Hence, a more general expression for instantaneous voltage can be stated thus:

$$v = V_{pk} \sin(\omega t + \phi) \quad (1.3.18)$$

In working with ac circuits, there are two circuit elements that become particularly significant. These are capacitors and inductors.

1.3.2.1 Capacitors

The circuit in Figure 1.3.11(a) contains a capacitor C consisting of two parallel plates separated by an insulator (e.g. air) with a dc voltage applied across the plates.

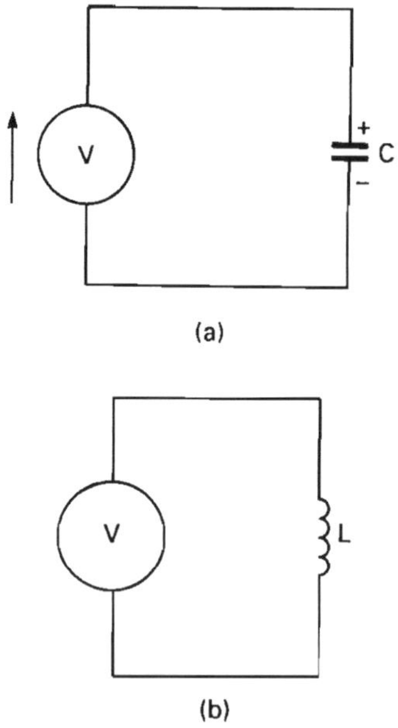

Figure 1.3.11 Capacitor and inductor circuits.

A potential difference exists between the plates of the capacitor, and hence a positive charge will develop on the plate connected to the positive battery terminal and a negative charge on the plate connected to the negative terminal. The charge, q, on the plates is proportional to the voltage across them, i.e.:

$$q = CV \quad (1.3.19)$$

where C is a constant, called the capacitance, which depends upon the size, shape and separation of the plates, and the type of insulator between the plates.

In an ac circuit, the effect of the voltage changing with time will give rise to a time-varying charge on the capacitor plates. This is equivalent to a current through the circuit:

$$i = \frac{dq}{dt} \tag{1.3.20}$$

Substituting for q from equation (1.3.19),

$$i = C\frac{dv}{dt} \tag{1.3.21}$$

For a sinusoidal exciting voltage v, equations (1.3.19) and (1.3.20) give the current i in the circuit as:

$$\begin{aligned} i &= C\frac{d}{dt}(V_{pk} \sin \omega t) \\ &= \omega C V_{pk} \cos \omega t \\ &= \omega C V_{pk} \sin(\omega t + \pi/2) \end{aligned} \tag{1.3.22}$$

Equation (1.3.22) shows the current to be sinusoidal and leading the voltage by a phase $\pi/2$.

1.3.2.2 Inductors

An inductor consists of a coil of wire around a magnetic circuit, a circuit which may, for example, consist of iron or air. The symbol for the inductor is shown connected to a voltage source V in Figure 1.3.11(b). The current in an inductor produces a magnetic flux around that inductor. The magnetic field will vary as the current changes, which will, in turn, induce an emf in the circuit. This emf is given by:

$$v = L\frac{di}{dt} \tag{1.3.23}$$

Equation (1.3.23) indicates that the voltage v across the inductance is proportional to the rate of change of current through it. The proportional constant, L, is determined by the size, shape and magnetic properties of the inductor. Substituting a sinusoidal exciting current for the current in equation (1.3.23):

$$\begin{aligned} V &= L\frac{d}{dt}(I_{pk} \sin \omega t) \\ &= \omega L I_{pk} \cos \omega t \\ &= \omega L I_{pk} \sin(\omega t + \pi/2) \end{aligned} \tag{1.3.24}$$

Equation (1.3.24) shows that the current through the inductor *lags* the voltage by a phase angle of $\pi/2$.

1.3.2.2.1 Complex representation of sinusoidal quantities

We have seen in Section 1.3.2 that sinusoidally varying voltages or currents can be represented by a rotating vector (Figure 1.3.9). By representing this vector on an Argand diagram, it can be described by a complex number.

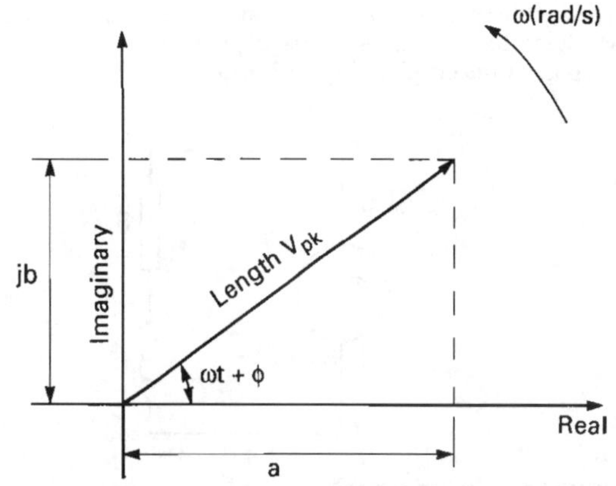

Figure 1.3.12 Complex representation of a sinusoidal quantity.

For a complex number of the form $a + jb$, Figure 1.3.12 shows that

$$a = V_{pk} \cos(\omega t + \phi)$$
$$b = V_{pk} \sin(\omega t + \phi)$$

Therefore

$$v = V_{pk} \cos(\omega t + \phi) + jV_{pk} \sin(\omega t + \phi) \tag{1.3.25}$$

This result can also be represented exponentially:

$$v = V_{pk}\, e^{j(\omega t + \phi)} \tag{1.3.26}$$

By this means, the inductor and capacitor can be represented in terms of their complex impedance. Consider first the inductor:

$$\begin{aligned} V_{pk}\, e^{j(\omega t+\phi)} &= L\frac{d}{dt}(I_{pk}\, e^{j\omega t}) \\ &= j\omega L I_{pk}\, e^{j\omega t} \\ \therefore V &= j\omega L I \end{aligned} \tag{1.3.27}$$

By comparing equation (1.3.27) with Ohm's law (equation (1.3.1)), it can be reduced to

$$V = ZI \tag{1.3.28}$$

This is the ac form of Ohm's law, where Z is the complex impedance. Similarly, for the capacitor:

$$\begin{aligned} I_{pk}\, e^{(j\omega+j)} &= C\frac{d}{dt}(V_{pk}\, e^{j\omega t}) \\ &= j\omega C V_{pk}\, e^{j\omega t} \\ \therefore I &= j\omega C V \end{aligned} \tag{1.3.29}$$

Here the complex impedance for the capacitor is $Z = 1/j\omega C$. From equation (1.3.29) it can be seen that the impedance for the capacitor is $1/j\omega C$, and from equation (1.3.27) the impedance for the inductor is $j\omega L$. This indicates that they are both imaginary quantities. This means that the voltage and current are always 90° out of phase, i.e. these circuits are purely *reactive*. In practice circuits always have some resistance, which results in a real part to the impedance.

Consider the circuit in Figure 1.3.13.

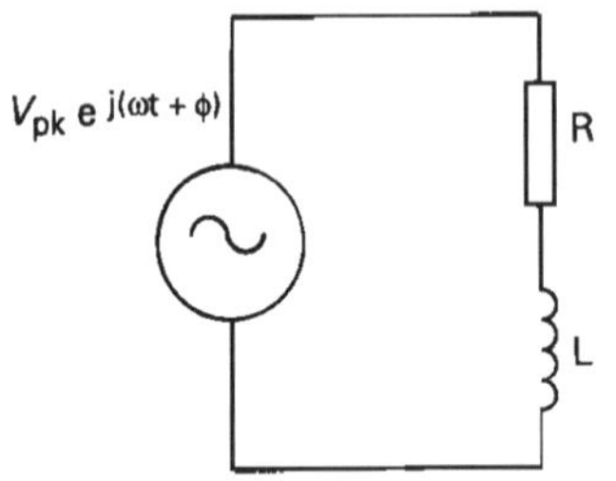

Figure 1.3.13 RL series circuit.

The differential equation for this circuit is:

$$v = R_i + \frac{di}{dt} \quad (1.3.30)$$

Allowing for a phase angle between the current and voltage, equation (1.3.30) can be expressed in its complex form, i.e.:

$$V_{pk}\ e^{j(\omega t + j)} = RI_{pk}\ e^{j\omega t} + j\omega L I_{pk}\ e^{j\omega t}$$
$$= (R + j\omega L) I_{pk}\ e^{j\omega t}$$
$$\therefore V = (R + j\omega L) I \quad (1.3.31)$$

Equation (1.3.31) shows that the impedance for the RL series circuit is given by

$$Z = R + j\omega L \quad (1.3.32)$$

which has a real part R and an imaginary part $j\omega L$. The impedance phase angle can be found for equation (1.3.32), as generally, for a complex number $a + jb$, the angle is arctan a/b, or for the equation (1.3.32):

$$\phi = \arctan \frac{\omega L}{R}$$

or more generally:

$$\phi = \arctan \frac{X}{R} \quad (1.3.33)$$

where X is the pure reactance and R is the pure resistance.

To summarize, sinusoidally excited circuits, consisting of inductance, capacitance and resistance can be analysed using the component complex impedance:

pure resistive impedance	$Z_r = R$
pure capacitive impedance	$Z_c = 1/j\omega C$
pure inductive impedance	$Z_i = j\omega L$

For complex impedance, Ohm's law can be rewritten:

$$V = IZ$$

where Z is the equivalent impedance of the circuit. The complex impedance obeys the same rules as for parallel and series resistance circuits, i.e.:

$$\text{series}: \quad Z = Z_1 + Z_2 + Z_3$$
$$\text{parallel}: Z = 1(1/Z_1 + 1/Z_2 + Z_3)$$

1.3.2.2.2 RLC series circuits

Consider the circuit in Figure 1.3.14. The complex impedance method of determining the total equivalent impedance in the circuit gives:

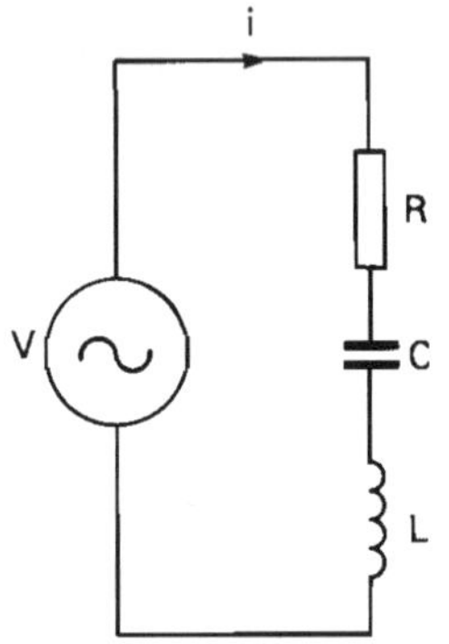

Figure 1.3.14 RLC series circuit.

$$Z = R + j\omega L + \frac{1}{j\omega C} \quad (1.3.34)$$

Separating into real and imaginary parts:

$$Z = R + j\left(\omega L - \frac{1}{\omega C}\right) \quad (1.3.35)$$

Therefore the current in the circuit is:

$$I = \frac{V}{Z} = \frac{V}{\{R + j[\omega L - (1/\omega C)]\}} \quad (1.3.36)$$

and the impedance angle is given by:

$$\phi = \arctan \frac{[\omega L - (1/\omega C)]}{R} \quad (1.3.37)$$

Observation of equation (1.3.36) indicates that, as $\omega \to 0$, the current will be very small due to a high capacitive reactance. Also, as $\omega \to$ infinity, the inductive reactance becomes very high and hence the current very small. However, at a value of ω where the capacitive reactance is equal to the inductive reactance, the current is only dependent on the pure resistance R, i.e.:

$$\text{for } \omega L = \frac{1}{\omega C}, \ I = \frac{V}{R}$$

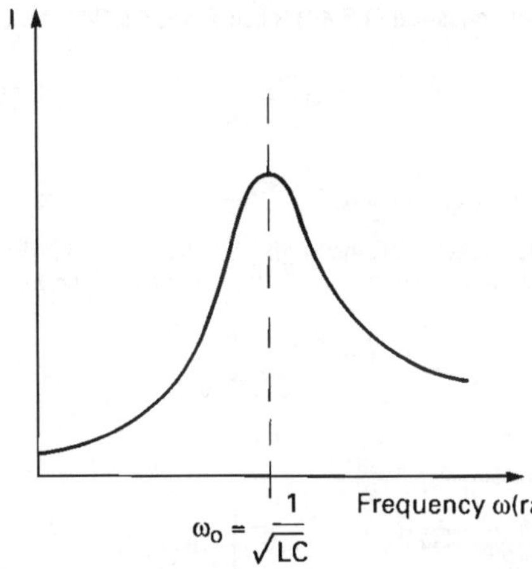

Figure 1.3.15 Resonance condition for the series circuit.

The value of ω for this condition is the *resonant frequency*, ω_o; i.e.:

$$\omega_o = \frac{1}{\sqrt{LC}} \tag{1.3.38}$$

In Figure 1.3.15 the RLC series current is plotted against frequency, showing the current peaking at resonant frequency. Two points to notice about the series resonance condition are that the circuit appears as a pure resistance, and its current is in phase with the applied voltage. Also, in calculating the voltage drops across C and L, Kirchoff's laws would appear to be breached. This is explained, however, by indicating that the voltages are in antiphase, and hence cancel each other.

1.3.2.2.3 Parallel RLC circuits

In the circuit shown in Figure 1.3.16(a), the impedance for the parallel LC combinations is given by:

$$Z = j\frac{\omega L}{1 - \omega^2 LC} \tag{1.3.39}$$

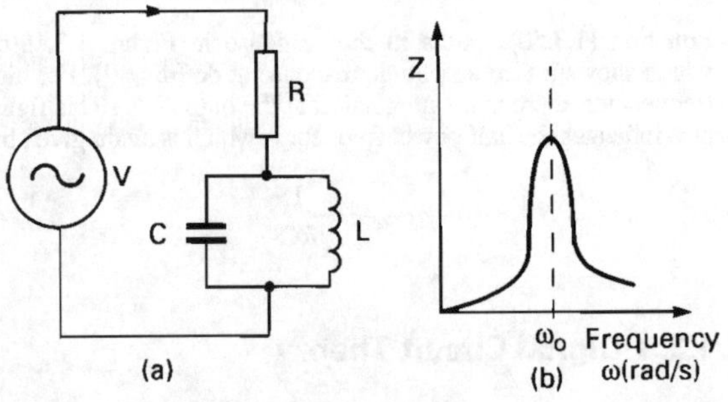

Figure 1.3.16 RLC parallel circuit.

Equation (1.3.39) shows that the impedance is infinite when

$$\omega_o^2 LC = 1 \tag{1.3.40}$$

Here, ω_o is the resonant frequency. Furthermore, equation (1.3.40) can be rewritten:

$$\omega_o = \frac{1}{\sqrt{LC}} \tag{1.3.41}$$

Equation (1.3.41) is identical to equation (1.3.38) for the series LC combination. The difference between series and parallel resonance, however, is that the impedance is a ***minimum*** for series circuits at resonance, whereas it is a ***maximum*** for parallel circuits at resonance. The impedance characteristic for the parallel circuit is shown in Figure 1.3.16(b).

1.3.2.2.4 Q-factor

About the resonant frequency for a series circuit, there is a voltage magnification across the inductor (or capacitor), which is given by (voltage across inductor)/(supply voltage).

Thus

$$Q = \frac{\omega_o L I}{R I} = \frac{\omega_o L}{R} \tag{1.3.42}$$

Equation (1.3.42) indicates that the voltage magnification factor (the *Q*-factor) is equal to the ratio of the inductive reactance to the resistance. This means that near resonance, the resistance inherent in all inductors (due to multiple turns of wire) becomes significant in determining the sharpness or selectivity of the resonant circuit (Figure 1.3.17).

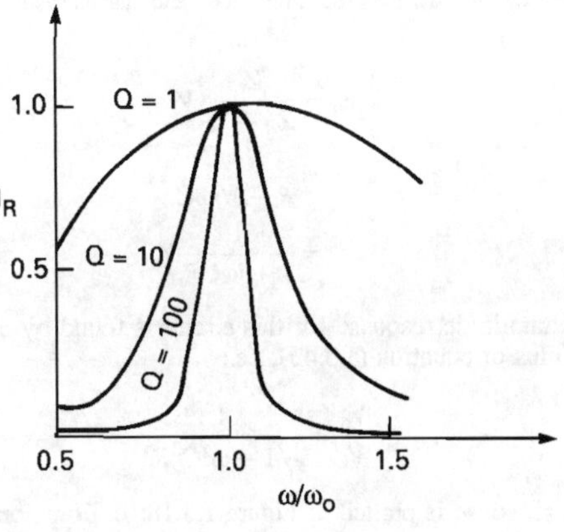

Figure 1.3.17 *Q*-factor.

For circuits that require very high frequency selectivity, crystals are usually used instead of inductors and capacitors. This is because the *Q*-factor for LC circuits is typically in the range 10–100, whereas the *Q* for crystals can be as high as several thousand.

1.3.2.3 RC circuits as filters

RC circuits can be arranged to discriminate selected frequency bands by using low-pass and high-pass configurations.

1.3.2.3.1 Low-pass RC filters

Consider the circuit in Figure 1.3.18(a). The output, V_o, of this circuit can be defined as a proportion of the input V_i:

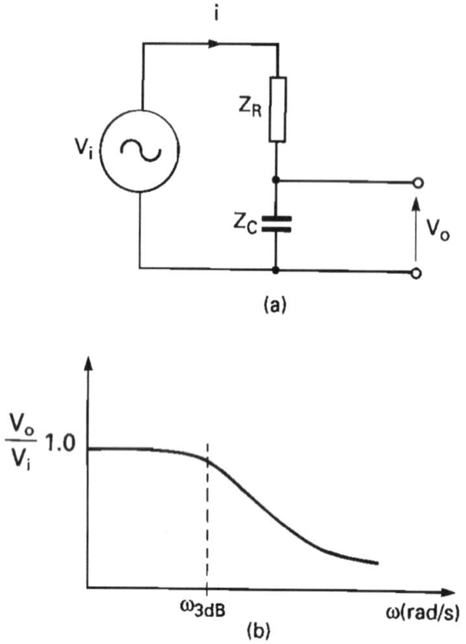

Figure 1.3.18 Low-pass RC filter circuit and characteristic.

$$V_o = \frac{Z_C}{Z_R + Z_C} V_i = \frac{1/j\omega C}{R + 1/j\omega C} V_i = \frac{1}{1 + j\omega CR} V_i \tag{1.3.43}$$

The amplitude response for this circuit is found by taking the modulus of equation (1.3.43), i.e.:

$$\frac{V_o}{V_i} = \frac{1}{\sqrt{1 + \omega^2 C^2 R^2}} \tag{1.3.44}$$

This response is plotted in Figure 1.3.18(b). Equation (1.3.44) shows that as $\omega \to 0, V_i = V_o$, and that as $\omega \to \infty, V_o = 0$. Thus, the larger the frequency the lower the magnitude of the output V_o. For this reason, this configuration is called the *low-pass filter*. Figure 1.3.18(b) indicates the *half power* frequency, where:

$$V_o = \frac{1}{\sqrt{2}} V_i \tag{1.3.45}$$

which, from equation (1.3.44), occurs when $\omega CR = 1$, or

$$\omega_{3\,\mathrm{dB}} = \frac{1}{RC} \tag{1.3.46}$$

1.3.2.3.2 High-pass RC filters

The configuration for the high-pass RC filter is shown in Figure 1.3.19(a). The current in this circuit is given by:

$$I = \frac{V_i}{Z} = \frac{V_i}{R + \frac{1}{j\omega C}} \tag{1.3.47}$$

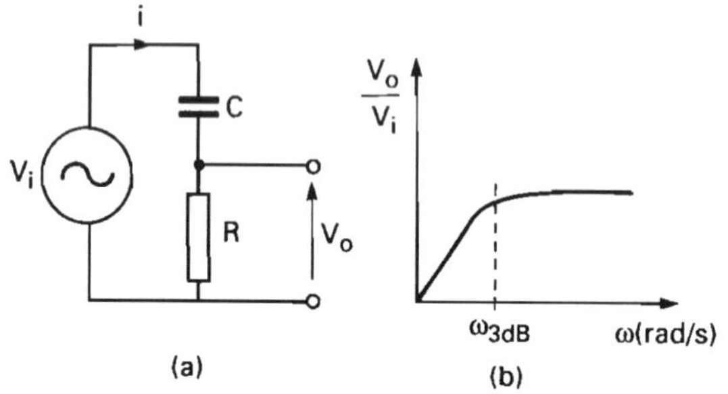

Figure 1.3.19 High-pass RC filter circuit and characteristic.

Therefore, multiplying numerator and denominator by the complex conjugate of the denominator

$$I = V_i \left\{ \frac{j\omega C}{1 + j\omega CR} \cdot \frac{1 - j\omega CR}{1 - j\omega CR} \right\} = V_i \frac{[R + j/\omega C]}{R^2 + (1/\omega^2 C^2)} \tag{1.3.48}$$

The voltage across R in Figure 1.3.19(a) is given by:

$$V_o = IR = V_i \frac{[R + (j/\omega C)]R}{\sqrt{R^2 + (1/\omega^2 C^2)}} \tag{1.3.49}$$

Therefore, the magnitude response is given by

$$\frac{V_o}{V_i} = \frac{R}{\sqrt{R^2 + (1/\omega^2 C^2)}} \tag{1.3.50}$$

Equation (1.3.50) results in the response in Figure 1.3.19(b), which shows a zero amplitude response at dc ($\omega = 0$). For high frequencies, there is no attenuation at the output, V_o. This figure also indicates the half power frequency, which is again given by:

$$\omega_{3\,\mathrm{dB}} = \frac{1}{RC}$$

1.3.3 Digital Circuit Theory

Electronic circuits used for digital systems are designed to generate only two recognized output voltage levels, and probably the most common definition for these voltage levels is 5 V for the high level and 0 V for the lower level. In practice, a certain range is allowed for each of the two levels, again the most common being 0–0.8 V for 0 V, and 2.7–5 V for 5 V. Any voltage levels that exist outside these two ranges are invalid, and if present in a digital system will give rise to error conditions.

These voltage ranges are as defined for the ttl (transistor-transistor logic) family of digital circuits; other digital circuit families can use different voltage ranges, e.g. ecl (emitter coupled logic) uses levels −2.1 to −1.7 V and −1.3 to −0.9 V.

Ensuring digital circuits operate on two distinct voltage ranges, these two ranges can be equated to the two binary conditions 1 and 0 which are used in logic circuits. For ttl type circuits, binary 1 can be equated to the range 2.7–5 V and the binary 0 condition can can be equated to the range 0–0.8 V. This is known as the *positive logic convention*, where the negative logic definition would equate 1 to the range 0–0.8 V and 0 to the range 2.7–5 V. Most digital systems use positive or mixed logic conventions.

1.3.3.1 Logic gates

The construction of the logic gate from the transistor or fet type devices is beyond the scope of this section, as the aim is to describe the use of the logic gates, not their construction. Any logic gate manufacturer provides this information in his data sheets.

All digital systems consist, at the most fundamental level, of individual logic gates. Although there are a number of different gate characteristics, there are three types from which all other logic functions can be synthesized: OR, AND and NOT gates.

Logic gates are circuits with one output and one or more inputs. The gate in Figure 1.3.20 illustrates a two input logic gate. The truth table in (b) lists the gate output for all possible combinations of binary inputs.

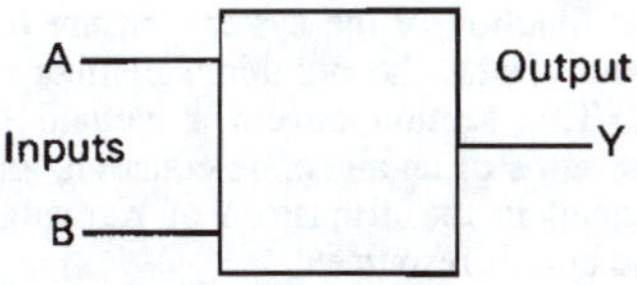

(a)

A	B	Y
0	0	0
0	1	0
1	0	0
1	1	1

(b)

Figure 1.3.20 Logic gate and truth table.

From this truth table we can see that the output of the gate will be binary 0 unless both input A and input B are binary 1. Manufacturers will normally quote their truth tables in terms of positive logic.

The truth table is used in the following descriptions of the basic logic gates.

1.3.3.1.1 OR gates

The symbol for the OR gate along with its truth table is shown in Figure 1.3.21. Considering the truth table for the two input OR gate, it can be seen that the output Y will be a binary 1 if input A is 1, *or* input B is 1, *or* both are 1, hence its name.

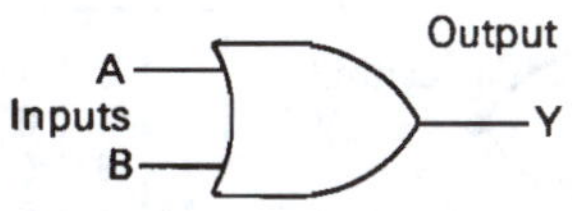

(a)

A	B	Y
0	0	0
0	1	1
1	0	1
1	1	1

(b)

Figure 1.3.21 OR gate and truth table.

1.3.3.1.2 AND gates

The symbol for the AND gate along with its truth table is shown in Figure 1.3.22. Observing the truth table for this device, we see that the output Y will only be binary 1 when input A is 1 *and* input B is 1.

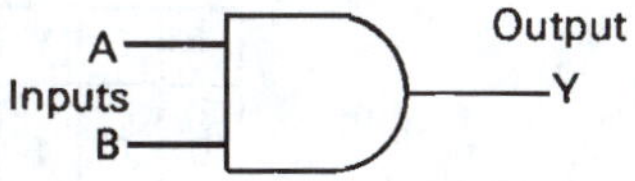

(a)

A	B	Y
0	0	0
0	1	0
1	0	0
1	1	1

(b)

Figure 1.3.22 AND gate and truth table.

1.3.3.1.3 The inverter

The symbol for the NOT gate along with its truth table is shown in Figure 1.3.23. Observation of the truth table indicates the output Y to be the logical inverse of whichever state is present on the input A.

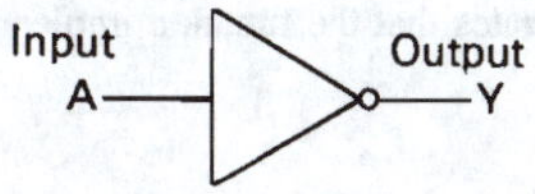

(a)

A	Y
0	1
1	0

(b)

Figure 1.3.23 Inverter and truth table.

1.3.3.1.4 NOR gates

The symbol for the NOR gate along with its truth table is shown in Figure 1.3.24. The NOR gate is simply an OR gate followed by an inverter; this can be seen from the truth table in Figure 1.3.24(b) where the output Y is the inverse of the output Y for the OR gate in Figure 1.3.21(b).

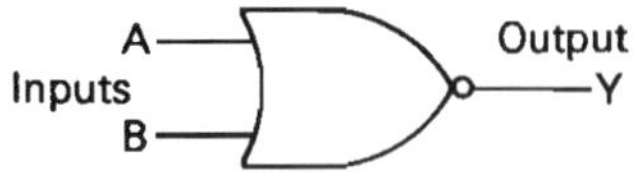

(a)

A	B	Y
0	0	1
0	1	0
1	0	0
1	1	0

(b)

Figure 1.3.24 NOR gate and truth table.

1.3.3.1.5 NAND gates

The symbol for the NAND gate, and its truth table, is shown in Figure 1.3.25. As with the NOR function, the NAND gate is an AND gate followed by an inverter as can be seen from the two truth tables which show the output Y of the NAND gate to be the inverse of the output of the AND gate.

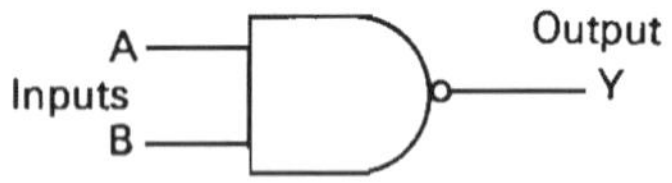

(a)

A	B	Y
0	0	1
0	1	1
1	0	1
1	1	0

(b)

Figure 1.3.25 NAND gate and truth table.

1.3.3.2 Implementing AND/OR functions from NAND/NOR gates

Both the NAND and the NOR gates can be used to implement inverters (see Figure 1.3.26). In Figure 1.3.27 a NAND gate has been followed by a second NAND configured as an inverter. The truth table demonstrates that the function implemented is

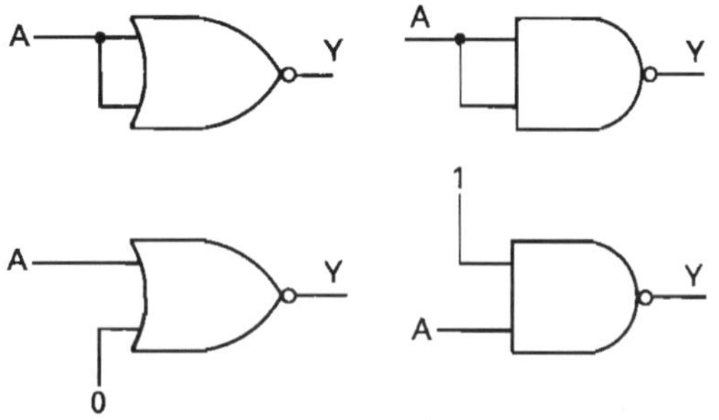

Figure 1.3.26 Implementing inverters with NAND and NOR gates.

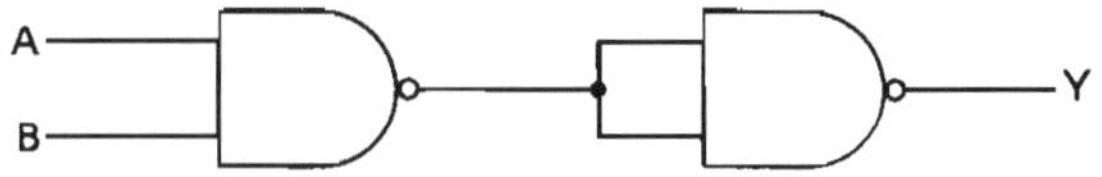

A	B	Y
0	0	0
0	1	0
1	0	0
1	1	1

Figure 1.3.27 Synthesizing an AND gate with NAND gates.

that of an AND gate. This means that NAND elements can be used to implement AND functions, and that NOR functions can be used to implement OR functions, hence the NAND and NOR gates are functionally complete. This becomes particularly important when considering that most manufacturers provide multiple gates within a single integrated circuit.

1.3.4 Boolean Algebra

1.3.4.1 Combinational and sequential circuits

All logic circuits can be subdivided into two types: combinational logic and sequential logic. A *combinational* circuit can be described by stating that its output will be true for only certain combinations of input variables; all other input combinations will cause the output to be false. The output(s) for a *sequential* circuit depend upon current input variables, time and past input variables. Sequential circuits use combinational circuits as building blocks, so it is essential to understand these elements. Examples of the implementation of sequential logic circuits are covered in some later chapters.

For effective design of digital systems, the designer must be able to specify clearly the function of the system, ensure the design will be reliable, and ensure the minimum number of logic elements are used. This section covers a method of defining a logic system in terms of an algebraic equation, and this method will be extended in the discussion of Karnaugh maps for minimizing logic circuit resources.

Consider the logic circuit in Figure 1.3.28. In this example, we have a four bit parallel data transmission link, where there is a requirement to recognize the presence of particular binary codes. The detection circuit is designed such that its output will be logic 1 for the presence of any of the decimal codes 16, 11, 9, 7, 4, 3 or 6 and logic 0 for any other combination of codes on the link.

Each of the logic system input variables are designated an alphabetic character (in this case A, B, C, D). Each of these variables implies its complement, i.e. $\bar{A}, \bar{B}, \bar{C}, \bar{D}$. For a positive logic convention, A = 1 (true) and $\bar{A} = 0$ (false). Thus, for the binary code 0100 on the transmission link in Figure 1.3.28, it is required to recognize $\bar{A}B\bar{C}\bar{D}$.

To recognize the code 0100, in this example, we can use a four input AND gate to detect the presence of $\bar{A}B\bar{C}\bar{D}$. The other required codes are similarly detected.

Instead of using unwieldy grammar to describe this example, Boolean algebra can be used. The basic rules for Boolean algebra are covered in the following sections.

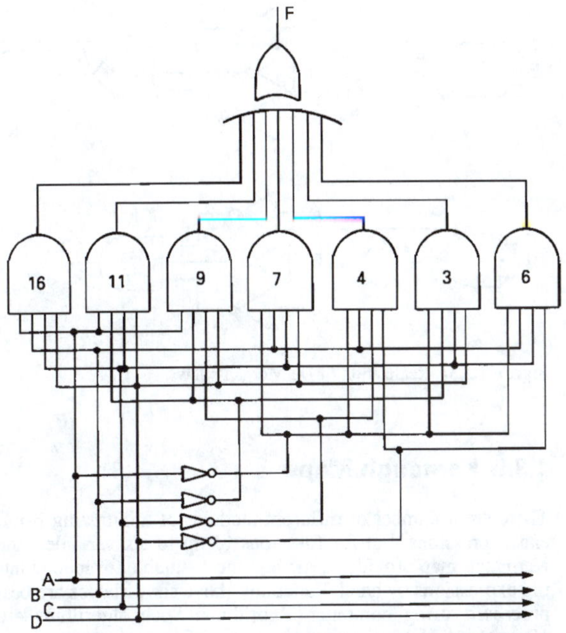

Figure 1.3.28 Logic system example.

1.3.4.2 Boolean OR/AND identities

The truth table for the logical OR relation is shown in Figure 1.3.21(b). It can be represented in terms of a Boolean expression:

$$A + B = X \qquad (1.3.51)$$

where the + symbol indicates the Boolean OR operation, A and B are the input variables, and X is the output. There are a number of important Boolean identities associated with the OR function, which can all be verified using the OR truth table.

$$A + 0 = A \qquad (1.3.52)$$

$$A + 1 = 1 \qquad (1.3.53)$$

$$A + A = A \qquad (1.3.54)$$

$$A + B + C = (A + B) + C = A + (B + C) \qquad (1.3.55)$$

$$A + B = B + A \qquad (1.3.56)$$

All these relations can be directly realized using OR, as illustrated in Figure 1.3.29.

The truth table for AND gate is shown in Figure 1.3.22(b). It can be written in the Boolean expression

$$A \cdot B = X \qquad (1.3.57)$$

As with the OR gate, there are a number of important AND identities, these are:

$$A0 = 0 \qquad (1.3.58)$$

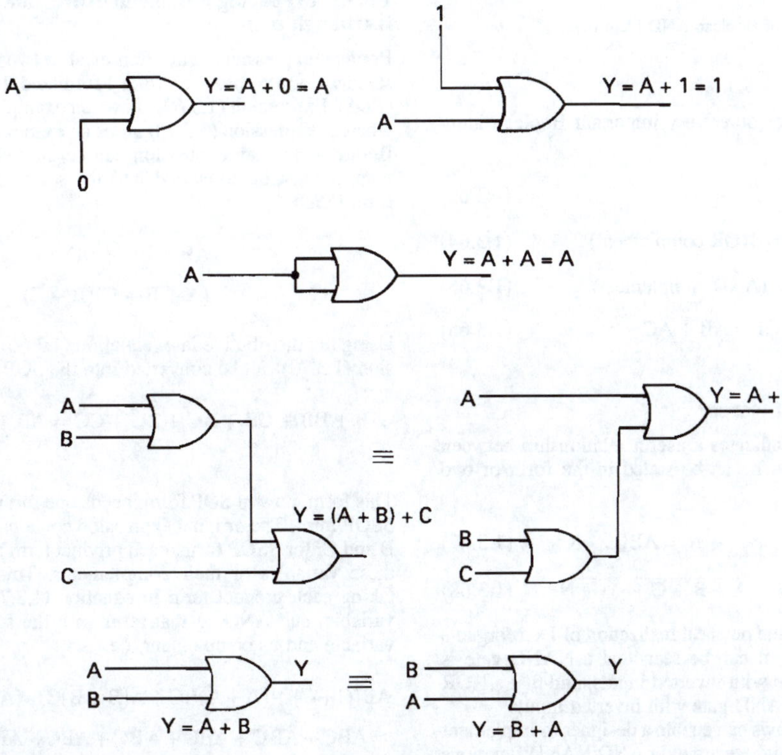

Figure 1.3.29 Realization of Boolean OR identities.

$$A1 = A \quad (1.3.59)$$

$$AA = A \quad (1.3.60)$$

$$ABC = (AB)C = A(BC) \quad (1.3.61)$$

$$AB = BA \quad (1.3.62)$$

All these relations can be directly realized using AND, as illustrated in Figure 1.3.30.

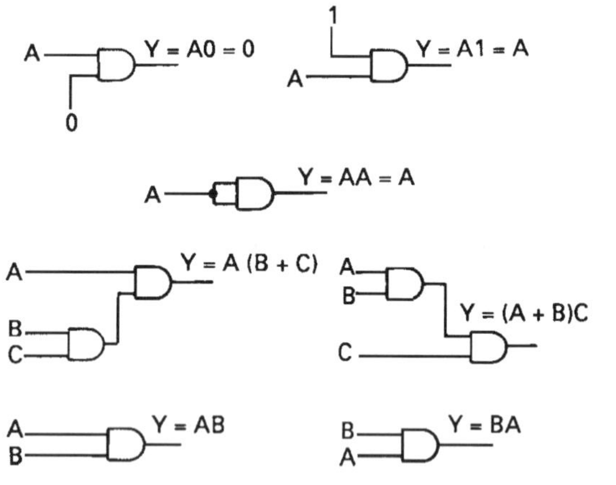

Figure 1.3.30 Realization of Boolean AND identities.

There are a number of other very important Boolean identities:

$$\overline{A} = A \quad (1.3.63)$$

$$A + \overline{A} = 1 \text{(OR complement)} \quad (1.3.64)$$

$$A\overline{A} = 0 \text{ (AND complement)} \quad (1.3.65)$$

$$A(B + C) = AB + AC \quad (1.3.66)$$

1.3.4.3 De Morgan's theorem

De Morgan's theorem indicates a useful relationship between AND and OR functions. It can be stated in the form of two laws, i.e.:

$$\overline{A} + \overline{B} + \overline{C} + \cdots + \overline{N} = \overline{ABC \ldots N} \quad (1.3.67)$$

$$\overline{A}\,\overline{B}\,\overline{C} \ldots \overline{N} = \overline{A + B + C + \cdots + N} \quad (1.3.68)$$

Figure 1.3.31 illustrates the physical realization of De Morgan's laws. In terms of gates, it can be seen that a NAND gate is equivalent to an OR gate with inverted inputs, and that a NOR gate is equivalent to an AND gate with inverted inputs.

Application of these laws can enable a designer to implement OR/NOR functions when there are only AND/NAND functions available, and vice versa.

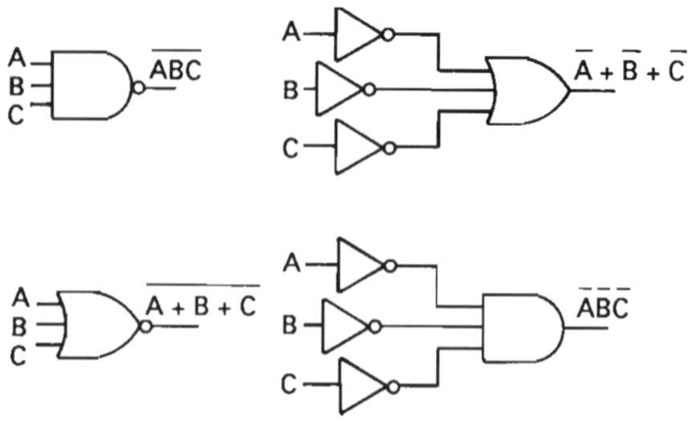

Figure 1.3.31 Realization of De Morgan's laws.

1.3.5 Karnaugh Maps

There are a number of different methods of minimizing Boolean expressions, but for functions of up to six variables, the Karnaugh map provides a method most suitable for manipulating expressions by hand. Functions above six variables are best processed using computer algorithms. Such algorithms are often supplied by the manufacturers of programmable logic devices.

1.3.5.1 Preparing a Boolean expression for plotting on a Karnaugh map

Boolean expressions will often exist in two distinct forms: the standard sum of products (*SOP*) form and the product of sums (*POS*). Expression (1.3.69) shows an example of the SOP form, whereas expression (1.3.70) gives an example of the POS form. Before a Boolean expression can be plotted on the Karnaugh map, it must be converted into the standard sum of products form (*SSOP*).

$$AB + BC + \overline{B}D \quad (1.3.69)$$

$$(A + B + C)(\overline{B} + \overline{C}) \quad (1.3.70)$$

Using the distributive law (equation (1.3.66)), the POS expression (1.3.70) can be converted into the SOP form:

$$A\overline{B} + B\overline{B} + C\overline{B} + A\overline{C} + B\overline{C} + C\overline{C} = A\overline{B} + C\overline{B} + A\overline{C} + B\overline{C} \quad (1.3.71)$$

This form, now in SOP form, needs one further conversion into SSOP form. The original expression contains three variables A, B and C; for SSOP form, each product term must include all of these variables or their complements. This can be done by taking each product term in equation (1.3.71) with a missing variable, and ANDing that term with the sum of the missing variable and its complement, i.e.:

$$A\overline{B}(C + \overline{C}) + (A + \overline{A})\overline{B}C + A(B + \overline{B})\overline{C} + (A + \overline{A})B\overline{C}$$
$$= A\overline{B}C + A\overline{B}\,\overline{C} + A\overline{B}C + \overline{A}\,\overline{B}C + AB\overline{C} + A\overline{B}\,\overline{C} + AB\overline{C} + \overline{A}B\overline{C} \quad (1.3.72)$$

This operation does not affect the expression, as the sum of a variable and its own complement is 1 (see equation (1.3.64)). Duplicate terms can now be removed from equation (1.3.72), giving:

$$A\overline{B}C + A\overline{B}\,\overline{C} + \overline{A}\,\overline{B}C + AB\overline{C} + \overline{A}B\overline{C} \qquad (1.3.73)$$

The SSOP terms in an expression are called *minterms*. Minterms are numbered according to the decimal code they represent, e.g.:

$$F = \overline{A}\,\overline{B}C + \overline{A}BC + A\overline{B}\,\overline{C} \qquad (1.3.74)$$

contains the minterms m1, m3 and m4. This shows that $F = 1$ when minterms m1, m2 or m4 are present and that $F = 0$ for all other possible minterms. (For equation (1.3.74), F will be 1 when, for example, the third term, minterm 4 is 1, i.e. $A = 1, B = 0$ and $C = 0$.)

1.3.5.2 Entering an expression on the Karnaugh map

A Karnaugh map is plotted by entering each term of the SSOP expression in one of the map locations. The map for a two variable function is shown in Figure 1.3.32. It contains four locations, as there are four possible combinations of the two input variables, A and B. The minterm numbers are also shown in Figure 1.3.32, although they are not normally drawn in.

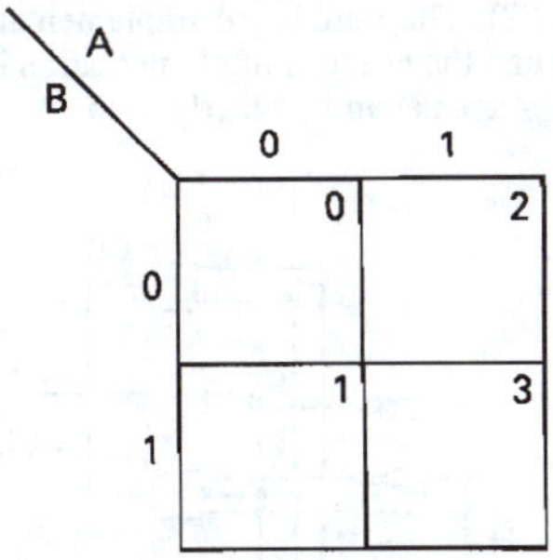

Figure 1.3.32 Two-variable Karnaugh map.

Consider the Boolean function

$$F = AB + A\overline{B} \qquad (1.3.75)$$

which indicates that F will be 1 when A is 1 and B is 1 and when A is 1 and B is 0. F will be 0 for the other two combinations of variables A and B. This can be plotted on the Karnaugh map as in Figure 1.3.33. Note the minterms present and their number and position on the Karnaugh map.

1.3.5.3 Reducing an expression using a Karnaugh map

Once an expression has been plotted, it can be reduced by forming the entries into logically adjacent groups. Consider Figure 1.3.33, where the function $F = AB + A\overline{B}$ has been plotted. In this example, the bottom left and right entries are logically adjacent, forming a *couple*. This couple indicates that the function will be a 1 regardless of the state of A, because A

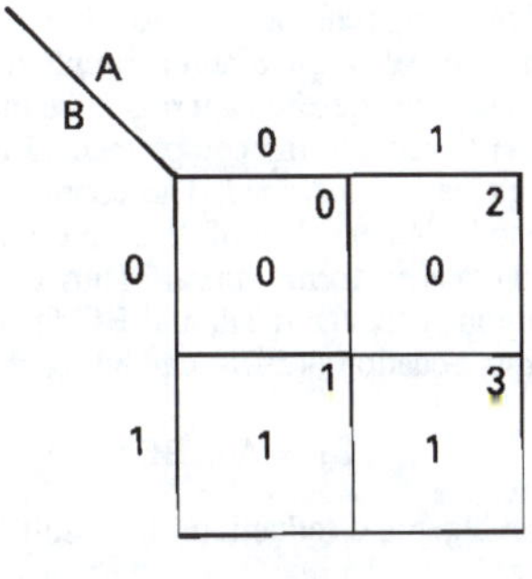

Figure 1.3.33 Karnaugh map of the function $F = AB + A\overline{B}$.

is present in its complemented and uncomplemented form. Therefore

$$F_R = B \qquad (1.3.76)$$

Diagonal coupling is not valid, as it would mean more than one variable changing state at one time, e.g., minterm 2 is not logically adjacent to minterm 1. The result in equation (1.3.76) can be verified using the Boolean identities, i.e.:

$$F = BA + B\overline{A} = B(A + \overline{A}) = B \qquad (1.3.77)$$

where $(A + \overline{A})$ is the OR complement.

A three variable Karnaugh map is shown in Figure 1.3.34. It contains eight locations to represent each of the three variable combinations. The locations on the map are arranged so that all entries are logically adjacent. This means that only one variable will change state between any two adjacent map locations. Logical adjacency also exists between the extreme opposite squares for a row or column; for the example in Figure 1.3.34, location $A\overline{B}\,\overline{C}$ (minterm 4) is adjacent to the location $\overline{A}\,\overline{B}\,\overline{C}$ (minterm 0). Note that due to the logical adjacency, the minterm numbers do not follow in order.

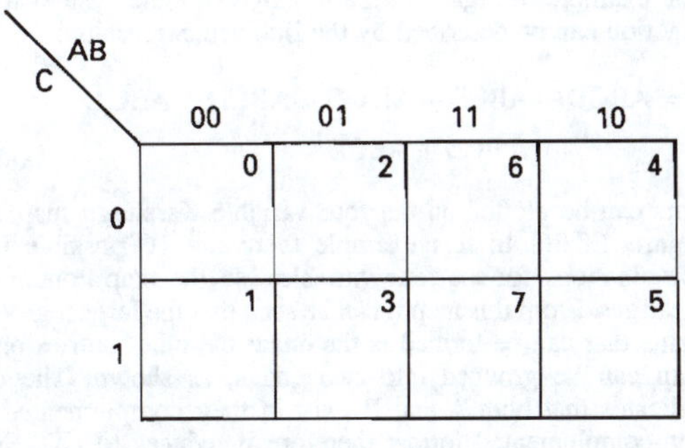

Figure 1.3.34 Three-variable Karnaugh map.

An example of a three variable expression reduction is shown in Figure 1.3.35, which maps the following expression:

$$F = \overline{A}\,\overline{B}C + A\overline{B}\,\overline{C} + A\overline{B}C + AB\overline{C} + ABC \qquad (1.3.78)$$

Grouping logically adjacent locations to form the largest possible groups, the map reveals a *quad* and a *couple* of adjacent terms. The quad indicates the constituent terms to be independent of the variables B and C because the complemented and non-complemented form exists for this quad. The couple indicates its constituent terms to be independent of A, because its complemented and non-complemented forms appear. Thus the function F will depend only upon A for the quad, and $\bar{B}C$ for the couple. Thus the expression in equation (1.3.78) can be reduced to:

$$F_R = A + \bar{B}C \qquad (1.3.79)$$

Using Boolean algebraic reduction, the result for the quad can be verified, i.e.:

$$\begin{aligned} &A\bar{B}\bar{C} + A\bar{B}C + AB\bar{C} + ABC \\ &= A(\bar{B}\bar{C} + \bar{B}C + B\bar{C} + BC) \\ &= A[\bar{B}(\bar{C} + C) + B(\bar{C} + C)] \\ &= A(\bar{B} + B) \\ &= A \end{aligned} \qquad (1.3.80)$$

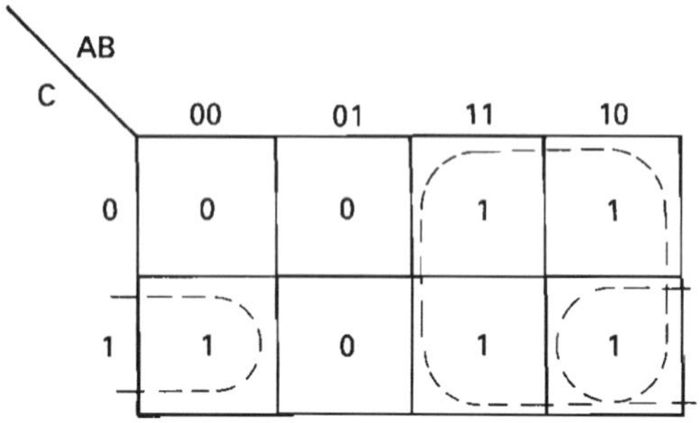

Figure 1.3.35 Karnaugh map of the function $F = \bar{A}\bar{B}C + A\bar{B}\bar{C} + A\bar{B}C + AB\bar{C} + ABC$.

The example in Figure 1.3.28 describes a logic system whose function can be described by the Boolean expression:

$$\begin{aligned} F = {} & ABCD + A\bar{B}CD + A\bar{B}\bar{C}D + \bar{A}BCD + \bar{A}B\bar{C}\bar{D} \\ & + \bar{A}\bar{B}CD + \bar{A}BC\bar{D} \end{aligned} \qquad (1.3.81)$$

This can be plotted on the four variable Karnaugh map as in Figure 1.3.36. In this example there are 16 possible input combinations for the four variables, so the map contains 16 locations. From this map it can be seen that the largest group of terms that can be formed is the quad; the other entries on the map can be grouped into two quads, as shown. The quad indicates that both A and B exist in their complemented and non-complemented forms; therefore it reduces to CD. Reducing the two couples, the resultant reduced expression becomes:

$$F_R = CD + \bar{A}B\bar{C} + A\bar{B}D \qquad (1.3.82)$$

The power of this technique can be illustrated by considering the implementation of the unreduced expression in equation (1.3.81) and comparing it with the reduced expression in

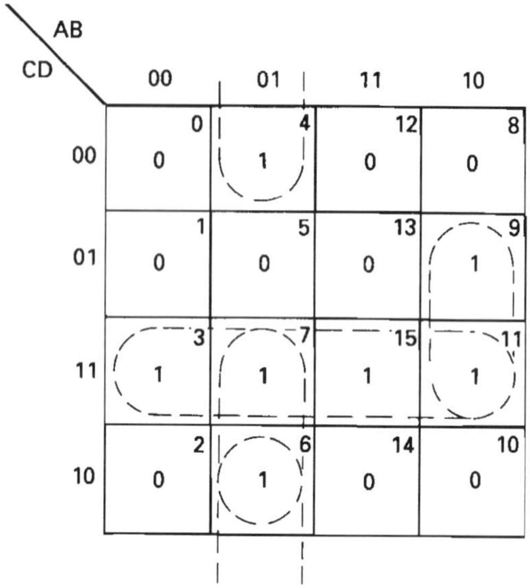

Figure 1.3.36 Karnaugh map of the function $F = ABCD + A\bar{B}CD + A\bar{B}\bar{C}D + \bar{A}BCD + \bar{A}B\bar{C}\bar{D} + \bar{A}\bar{B}CD + \bar{A}BC\bar{D}$.

equation (1.3.82). The unreduced implementation is shown in Figure 1.3.28 and the reduced implementation in Figure 1.3.37; the reduction in gates can be clearly seen.

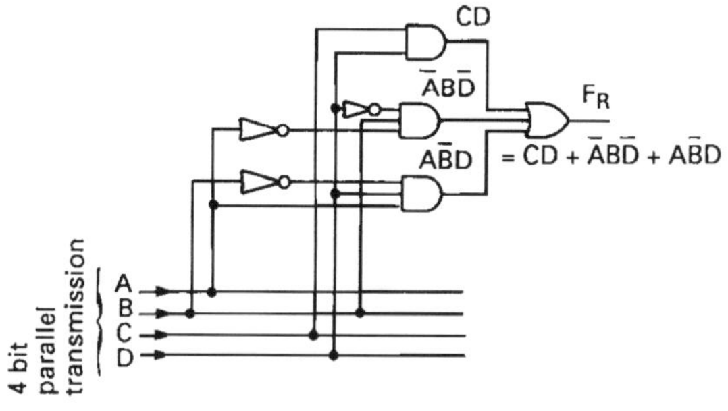

Figure 1.3.37 Implementation of function $F_R = CD + \bar{A}B\bar{C} + A\bar{B}D$.

The operation of using a Karnaugh map for reducing an expression can be summarized as follows:

- Form standard sum of products of the unreduced expression.
- Plot the minterms on the map.
- Form the largest and least number of groups of logically adjacent entries (these groups will always contain a number which is a power of 2, e.g. 2, 4, 8).

Generally, a couple allows two original terms to be reduced to one smaller term, a quad allows four original terms to be reduced to one smaller term, etc. A map entry surrounded by 0s cannot be reduced, and will appear in the final expression without any reduction.

1.3.5.4 Prime implicants

Given the reduction method described in Section 1.3.5.3, it is still possible to produce a non-minimal result. This can be illustrated by considering the *prime implicants*.

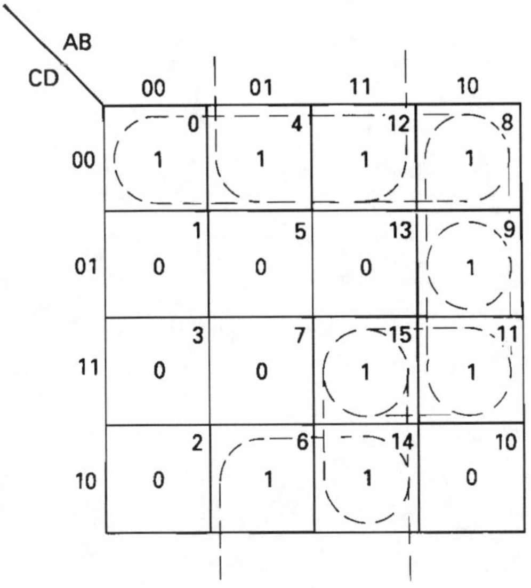

Figure 1.3.38 Function with non-essential prime implicants.

Once the Karnaugh map has been plotted, logically adjacent entries are grouped together; these groups are the prime implicants. The four variable example in Figure 1.3.38 shows the following prime implicants are formed:

$$\overline{C}\,\overline{D} \quad A\overline{B}D$$
$$B\overline{D} \quad ACD$$
$$A\overline{B}\,\overline{C} \quad ABC$$

The figure shows that soon entries on the map are covered by more than one prime implicant, e.g. minterm 15. If any of the overlapping prime implicants results in all entries of another prime implicant being covered, then that prime implicant is non-essential. For example, the couple covering minterms 15 and 11 is covered by couple 9/11 and couple 14/15, hence it is not an essential implicant. For a minimal expression all of these non-essential implicants must be removed. Figure 1.3.38 indicates three possible results, depending upon which essential implicants are chosen. These are:

$$F_R = \overline{C}\,\overline{D} + B\overline{D} + A\overline{B}D + ABC$$

$$F_R = \overline{C}\,\overline{D} + B\overline{D} + A\overline{B}\,\overline{C} + ACD$$

$$F_R = \overline{C}\,\overline{D} + B\overline{D} + A\overline{B}D + ACD$$

The results of these three equations indicate that there can be more than one result for a minimized expression; each one is valid.

Garik Markarian, Prof.
CEO, Rinicom Ltd.,
Chair in Communication Systems,
University of Leeds, UK

1.4 Information Theory and Error Correction

1.4.1 Elements of Information Theory

In TV broadcasting (both digital and analogue), the information source is a video whose output is a moving image. These information sources produce signals, which contain redundancy and without appropriate measures will require significant bandwidth for a broadcasting with the required quality. In order to make a broadcasting system practical and economically viable, the output of the analogue information source is transformed to a format that can be successfully transmitted with the reduced (or in many cases completely removed) redundancy. This process is known as *source coding* and its rules are governed by the principles of information theory.

Shannon developed the main elements of information theory in a series of classical articles.[1,2] In this section we briefly describe basic elements of information theory and show how they are applicable to the broadcasting.

We start with the definition of the term *information* and its differentiation from the *data*, as there is a misconception that these terms are interchangeable.

Definition:[3] Data is the raw quantity, whereas information is the same data plus some meaning, which is drawn from the context of the data.

Table 1.4.1 demonstrates the differences between the raw data and information contained within it.

Table 1.4.1 Examples of data and information

No.	*Data*	*Information*
1	07763706399	Mobile telephone number
2	SO16 7NP	UK post code
3	XJT36-B8T7W-9C3FV-9C9Y8-MJ226	Serial number of a CD with the licensed software

It follows from this table that pure data has little meaning if its content is not known. The concept of information content is related to predictability and information is measured by its unexpectedness. That is, the more predictable or probable a particular message, the less information is conveyed by transmitting that message. In other words, information (I) is inversely proportional to some function (f) of the probability of an event.

Shannon suggested the properties of an information measure listed in Table 1.4.2 for messages (m) and their probabilities ($P(m)$).

Table 1.4.2 Information measures[4]

Message	*Probability of message*	*Information content*
m_1	$P(m_1)$	I_1
m_2	$P(m_2)$	I_2
$m_1 + m_2$	$P(m_1)P(m_2)$	$I_1 + I_2$

Shannon defined the information content I_m of the message (data) m as:

$$I_m = \log \frac{1}{P(m)} = -\log P(m) \quad (1.4.1)$$

where the base 2 of the logarithm is usually chosen and the resulting quantity of information is measured in *bits*.

Definition: The rate at which information is transmitted is given by:

$$R = \frac{I}{T} \quad (1.4.2)$$

where T represents time taken to transmit I and R is measured in *bit/s*.

Definition: Entropy (*H*) is defined as the average amount of information per symbol:

$$H = \sum_{m=1}^{n} P(m) \log \frac{1}{P(m)} \tag{1.4.3}$$

where *n* defines a number of symbols in the alphabet and *H* is measured in *bit/symbol.*

Definition: Channel capacity (*C*) is defined as the maximum amount of information that could be transmitted error-free over a noisy channel:

$$C = B \log[1 + S/N] \tag{1.4.4}$$

where *S/N* represents Signal-to-Noise Ratio (SNR) in the Additive White Gaussian Noise (AWGN) channel, *B* represents channel bandwidth and *C* is measured in *bit/s.*

Channel capacity is a fundamental parameter for any communications or broadcasting system as it defines the maximum rate at which information can be transmitted over a given channel:

$$R_{max} \leq C \tag{1.4.5}$$

1.4.2 Block Forward Error Correction Codes in Digital Broadcasting

1.4.2.1 Basic Elements of Error Control Coding

To protect data transmitted over a noisy channel, two different forms of Forward Error Correction (FEC) are proposed. The first, which is not used in digital broadcasting, is known as Automatic Repeat Request (ARQ).[5] This scheme requires a return channel that will be used to acknowledge correct receipt of data and request a resend for incorrectly received data. The second technique is known as Forward Error Correction (FEC) and allows correction/detection of errors in the received data by utilising redundancy added in a special way at the transmitter side.[5,6] In block forward error correction codes, a stream of *k* information bits is encoded into a stream on *n* coded bits ($n > k$) according to a certain given algorithm.

The forward error correction techniques are widely used in digital broadcast applications, where a broadcaster omnidirectionally transmits the data, and these will be outlined in this section.

1.4.2.2 Linear Block Codes and Their Major Parameters

These are the first class of codes that have been introduced after the inception of information theory. In block forward error correction codes, a stream of *k* information bits is encoded into a stream on *n* coded bits ($n > k$) according to a certain given algorithm. The simplest class of these codes is known as *repetition codes*, where every information bit a_i is repeated *n* times at the transmission side and at the decoder the decision regarding the transmitted bit is made based on a majority logic. The simplest example of a repetition code is shown in Table 1.4.3, where $k = 1$ and $n = 5$.

Table 1.4.3 Encoding table for $k = 1$, $n = 5$ repetition code

No.	*Information bits*	*Coded bits*
1	0	00000
2	1	11111

The linear block codes could be binary, non-binary, cyclic, systematic, etc. Each of these types will be described in this section and each of these is defined by a set of parameters specific to this particular type. However, there is a generic set of parameters that is applicable to all linear block codes. These are:

Code rate, *R*
Minimum distance of the code, d_{min}
Generator matrix, *G*
Parity check matrix, *H*
Number of corrected errors, *t*.

In this section we briefly describe these parameters and illustrate them by the way of simple examples.

The ***code rate, R***, of a block code specifies the redundancy of the code and is defined as the ratio between the number of information bits and the number of encoded bits:

$$R = \frac{k}{n} \tag{1.4.6}$$

Repetition code described in Table 1.4.3 has a code rate of $R = 1/3$.

The ***Hamming distance of two code words***, $x = (x_1, x_2, \ldots, x_n)$ and $y = (y_1, y_2, \ldots, y_n)$ is defined as the number of positions where vectors *x* and *y* differ and is denoted as $d(x, y)$.

The ***minimum distance of the code***, also known as *minimum Hamming distance of the code C*, is defined as the minimum Hamming distance between any two admissible code words of the code *C*:

$$d_{min}(C) = \min\{d(x, y) : x, y \in C, x \neq y\} \tag{1.4.7}$$

It is easy to see that the minimum Hamming distance of the code given in Table 1.4.3 is equal to $d_{min} = 5$.

Minimum Hamming distance of the code characterises its error correction capabilities, as the number of errors that a code *C* can correct is given as:

$$t = \left\lfloor \frac{d_{min} - 1}{2} \right\rfloor \tag{1.4.8}$$

From equation (1.4.8) it follows that the repetition code described in Table 1.4.3 is capable of correcting any two errors in the received code word.

It was mentioned above that in block forward error correction codes a stream of *k* information bits is encoded into a stream on *n* coded bits according to a certain given algorithm. This is usually achieved by multiplying the input vector *u* of information bits by a parameter, known as the *generator matrix, G*:

$$c = uG \tag{1.4.9}$$

where *c* is the encoded code word.

The generator matrix, G, of the linear block code is defined as a $k \times n$ matrix with linearly independent rows.

If a code is defined by the generator matrix $G = [I|A]$, where $[I]$ is the identity matrix of size $k \times k$, the code is known as systematic, since the encoded code word c contains the exact copy of the original information vector, u.

The parity check matrix H of a linear block code C is defined as an $(n-k) \times n$ matrix that satisfies the following condition:

$$GH^T = 0 \tag{1.4.10}$$

where the T denotes the transpose of H.

Example. The following matrices:

$$C = \begin{bmatrix} 1 & 0 & 0 & 1 & 1 & 1 \\ 0 & 1 & 0 & 0 & 1 & 1 \\ 0 & 0 & 1 & 1 & 0 & 1 \end{bmatrix} \quad H = \begin{bmatrix} 1 & 0 & 1 & 1 & 0 & 0 \\ 1 & 1 & 0 & 0 & 1 & 0 \\ 1 & 1 & 1 & 0 & 0 & 1 \end{bmatrix}$$

define a (6,3) linear systematic block code with minimum Hamming distance $d_{\min} = 3$.

1.4.2.3 Hamming Codes

Hamming codes represent a special class of perfect codes with the following parameters:

$$n = 2^r - 1;\ k = 2^r - 1 - r;\ d_{\min} = 3 \tag{1.4.11}$$

where r is any integer number ≥ 2. Since $d_{\min} = 3$, they represent a class of single error correcting codes. Some examples of Hamming codes are: (7,4), (15,11), (21,26), etc. Hamming codes were among the first codes introduced for forward error correction; however, their use in broadcasting is limited to TELETEX applications of analogue TV broadcasting and is gradually diminishing.

1.4.2.4 Cyclic Codes

Definition: Cyclic shift to the right of the *n-tuple* vector $c = (c_0, c_1, c_2, \ldots, c_{n-2}, c_{n-1})$ is defined as $c^R = (c_{n-1}, c_0, c_1, c_2, \ldots, c_{n-2})$.

Note: *The left shift is defined in a similar way. In this chapter we consider only right shifts.*

Definition: Consider an (n, k, d) linear block code, C. If every cyclic shift of a code word is also a code word then the code C is known as a *cyclic code*.

Note: *Cyclic codes do not possess the all-zero code word in the code table.*

It is convenient to express every code word of a cyclic (n, k, d) code C as a polynomial of degree n. For example, the above code word c and its right shift c^R can be expressed as:

$$c = c(x) = c_0 + c_1x + c_2x^2 + \cdots + c_{n-1}x^{n-1}$$

$$c^R = c^R(x) = c_{n-1} + c_0x + c_1x^2 + \cdots + c_{n-2}x^{n-1}$$

Similar to the encoding procedure of block codes, described above, encoding of cyclic codes can be represented in a polynomial form as:

$$c(x) = u(x)g(x)$$

where

$$u(x) = u_0 + u_1x + u_2x^2 + \cdots + u_{k-1}x^{k-1}$$

is a polynomial representation of the input *k-tuple* information vector and

$$g(x) = g_0 + g_1x + g_2x^2 + \cdots + g_{n-k-1}x^{n-k-1}$$

is a generator polynomial of the cyclic code C. In addition, since cyclic code C is also a block code, its generator matrix G can be represented via a generator polynomial as follows:

$$G = \begin{bmatrix} g_0 & g_1 & g_2 & \cdots & g_{n-k-1} & 0 & 0 & \cdots & 0 & 0 \\ 0 & g_0 & g_1 & \cdots & g_{n-k-2} & g_{n-k-1} & 0 & \cdots & 0 & 0 \\ \cdot & \cdot & \cdot & & \cdot & \cdot & \cdot & & \cdot & \cdot \\ 0 & 0 & 0 & \cdots & g_0 & g_1 & g_2 & \cdots & g_{n-k-1} & 0 \\ 0 & 0 & 0 & \cdots & 0 & g_0 & g_1 & \cdots & g_{n-k-2} & g_{n-k} \end{bmatrix} = \begin{bmatrix} G_0 \\ G_1 \\ \cdots \\ G_{k-2} \\ G_{k-1} \end{bmatrix} \tag{1.4.12}$$

Cyclic code structure allows very low complexity implementation based on shift registers (this was particularly important during the early days of FEC theory, when hardware complexity implementation was defining the feasibility of codes implementation). In digital TV broadcasting systems cyclic codes are widely used in the form of Reed–Solomon codes.[7,8]

1.4.2.5 Reed–Solomon Codes

Reed–Solomon (RS) codes are widely used in digital TV broadcasting systems, such as DVB-S, DVB-T, DVB-DSNG, DVB-RCS and DVB-RCT, to name just a few. The RS codes represent a class of non-binary cyclic codes that can also be considered as non-binary: Bose–Chaudhuri–Hocquenghem (BCH) codes.[5,9–12]

In general, an RS code word consists of n symbols, where each symbol is composed of m bits. In theory, m can be any integer number; however, the most widely used RS codes are based on $m = 8$ symbols. RS codes have the following parameters:

Code alphabet:	$q = 2^m$
Block length:	$n = q - 1$
Minimum distance:	$d_{\min} = n - k + 1$
Number of corrected errors:	$t = (d_{\min} - 1)/2$

Since $d_{\min} = 2t + 1$, RS codes are also known as maxim distance separable codes.[7,8] Given $r = n - k$ parity bytes, a Reed–Solomon code can correct up to r byte errors in known positions (erasures), or detect and correct up to $r/2$ byte errors in unknown positions.

Since RS code is a cyclic code, the entire encoding procedure can be defined by a generator polynomial:

$$g(x) = (x - \alpha)(x - \alpha^2)\ldots(x - \alpha^{2t}) \quad (1.4.13)$$

where α is a primitive element in the field $F(2^m)$.[6,7] The code word is generated such that:

$$c(x) = g(x)i(x) \quad (1.4.14)$$

where $g(x)$ is the generator polynomial, $i(x)$ is the information block and $c(x)$ is a valid code word. The basic principle of encoding is to find the remainder of the message divided by a generator polynomial $g(x)$. The encoder works by simulating a Linear Feedback Shift Register with degree equal to $g(x)$, and feedback taps with the coefficients of the generating polynomial of the code.

At the receiver side the received code word is a valid code word, $c(x)$, corrupted by an error polynomial $e(x)$. The Reed–Solomon decoder will identify the position and magnitude of up to t errors and correct them.

The symbol error location is found by solving a simultaneous equation with t unknowns. Algorithms that do this take advantage of the matrix structure of Reed–Solomon codes. Two steps are required to find the error location. First an error locator polynomial is found. The two algorithms that find this special polynomial are the Berlekamp–Massey algorithm and Euclid's algorithm. By analysing the errors as the elements of a finite field, the Berlekamp–Massey algorithm finds the shortest linear recurrence that will produce those elements. This algorithm usually leads to more efficient software and hardware, but Euclid's algorithm is most often used because it is easier to implement. The roots of the error location polynomial are found using the Chien search algorithm. Once the errors are located, the proper symbol is found by solving the equation with t unknowns using the Forney algorithm.

As mentioned above, RS codes are widely used in digital TV broadcasting systems. In particular, the (204,188) shortened RS code is used in DVB-S, DSB-DSNG, DVB-C and DVB-T[13–16] standards, to name just a few.

The (204,188) shortened code is derived from the original (255,239) code, with the code generator polynomial given as:

$$g(x) = (x + \alpha^0)(x + \alpha^1)\ldots(x + \alpha^{15})$$

and the field generator polynomial given as:

$$p(x) = x^8 + x^4 + x^3 + x^2 + 1$$

The code may be implemented by adding 51 all-zero bytes to each randomised packet of 188 bytes and is also applied to the packet synchronisation byte. The use of RS code in the DVB standards provides a 'Quasi Error Free' (QEF) quality target, which means less than one uncorrected error effect per transmission hour, corresponding to Bit Error Ratio (BER) $= 10^{-11}$ at the output of the MPEG-2 demultiplexer.

1.4.2.6 Cross Interleaved Reed–Solomon Codes

Cross Interleaved Reed–Solomon Code (CIRC) is both the error detection and correction technique used in digital VTRs and all CD and DVD players. CIRCs were developed by Sony and Philips in 1980 and published in the Red Book Standard, which describes how information on a disc would be stored. The standard was printed in a red binder, hence its name.

The CIRC is designed as a concatenation of two shortened RS codes (*C1* and *C2*), such that the *C1* code corrects most of the random errors and *C2* code corrects most of the burst errors. The interleaving between the *C1* and *C2* codes is performed to make it easier for the *C2* Reed–Solomon code to correct burst errors. This means that the data is distributed over a relatively large physical area of the disc. If the data were recorded sequentially, a small defect could easily wipe out an entire word (byte). With CIRC, the bits are interleaved before recording and de-interleaved during playback. One data block (frame) of 24 data bytes is distributed over 109 adjacent blocks. To destroy 1 byte, one needs to destroy these other bytes. With scratches, dust, fingerprints, and even holes in the disc, there is usually enough data left to reconstruct any data bytes that have been damaged or caused the disc to become unreadable.

The CIRC encoding begins by taking the 24 symbols of 8-bit words and encoding them in an RS(28,24) code by adding four parity symbols (this code is capable of correcting $t = (n - k)/2 = 2$ symbols). At the next stage the encoded 28 symbols are interleaved and the information bits from the original frame are distributed over 109 frames. This is demonstrated in the following simplistic example.

Three symbols of 3-bit words must be transmitted, [010][001][011]. In order to prevent this sequence from the burst errors, the information bits are interleaved as shown in Figure 1.4.1: if the first word is incorrectly received as [110], when the information is de-interleaved, the received words would be [110][101][111]. Through error correction, the words would be decoded as [010][001][011]. If the data were not interleaved, it would have been corrupted and unreadable.

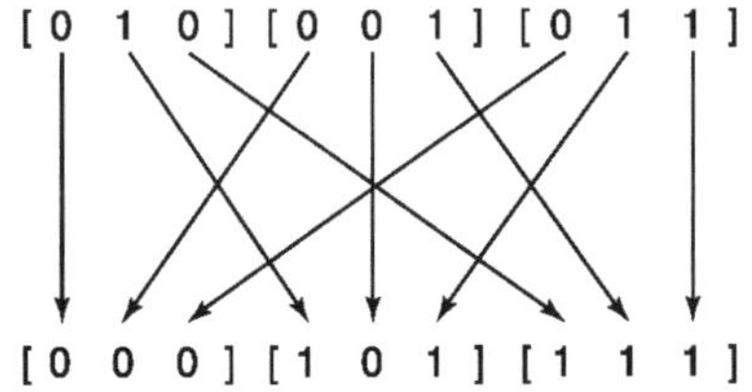

Figure 1.4.1 An example of interleaving for CIRC.

After interleaving, the 28-bit encoded code word is further encoded in an RS(32,28) code (since $n - k = 4$, this is also a $t = 2$ error correcting code), and the data is once again interleaved with a new pattern.

The first code, RS(28,24), is called the *C2* level of encoding. It corrects errors due to the physical condition of the CD and the way that the CD was recorded. The RS(32,28) code is the *C1* level and it corrects errors due to fingerprints and scratches. This completes the FEC process of the audio information, and control information is added during the next stage.

The CIRC code described above is a very powerful code, which is based on simple component codes. Interleaving works well for CIRC because one frame is spread over 109 frames. Since each frame is 32 bytes long and corrects up to four symbol errors, 109 frames can correct 436 errors. With interleaving, 13.625 frames can be corrected, whereas it would be impossible to do this without the interleaving.

1.4.3 Convolutional Codes

Convolutional codes represent another class of linear codes which can be classified as codes with the memory, such that the output at the particular instant is dependent not only on the current input, but also some finite memory.

A convolutional encoder can be thought of as a finite state machine, which maps a k-dimensional space input sequence to an n-dimensional output sequence, where the output is determined by the input and the state of the machine.

Definition:[6] An (n, k, m) convolutional code is defined as a k-input, n-output, time-invariant, causal finite state machine of encoder memory order m.

Definition: The *constraint length* ν of the convolutional encoder is defined as the number of shifts over which a single information word can influence the system.

Similar to block codes, convolutional code can be represented by a generator matrix. However, unlike block codes, this generator matrix will be infinite as the code operates on the sequences that are infinite. Such a representation of convolutional codes is not very convenient and other means of representation are often used. For example, Figure 1.4.2 shows shift register representation of the (2,1,2) convolutional code.

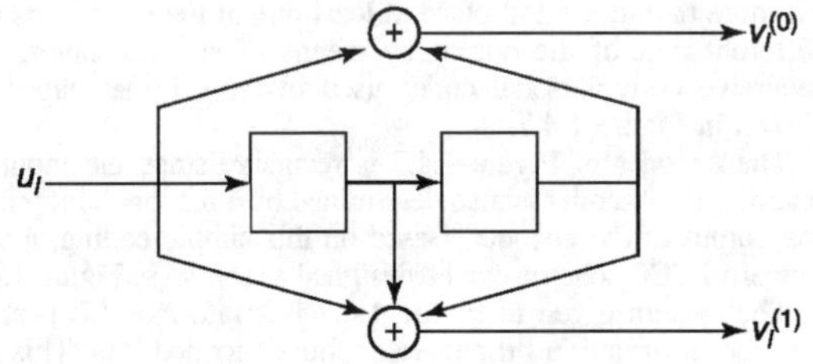

Figure 1.4.2 A (2,1,2) convolutional encoder.

The same (2,1,2) convolutional code can be represented by its *tree diagram*, presented in Figure 1.4.3.

In this diagram we assume that initially the encoder memory (the state) is set to zero and the branches of the tree represent both input bits (a '0' corresponds to branch up and a '1' to branch down) and the output pair of bits (which are labelled on the branches). A path on the tree from the initial zero state to any state on the tree will provide both the input and the corresponding encoded sequence. However, the tree grows exponentially and is not convenient for long input sequences.

This can be simplified further by representing the state of the encoder as the content of its shift register elements. Such a representation is known as the *state diagram* of the convolutional code. For example, the state diagram of the (2,1,2) convolutional code of Figure 1.4.2 is shown in Figure 1.4.4.

If a state diagram takes into account the time domain of the encoding process, it can be expanded into a *trellis diagram*, which shows how the content of encoder memory and its output changes with every time shift (clock). Figure 1.4.5 illustrates the trellis diagram of the same (2,1,2) convolutional encoder from Figure 1.4.2.

The trellis representation of the convolutional codes allows us to simplify not only the description of the encoding procedure, but provides means for very efficient decoding algorithm, known as

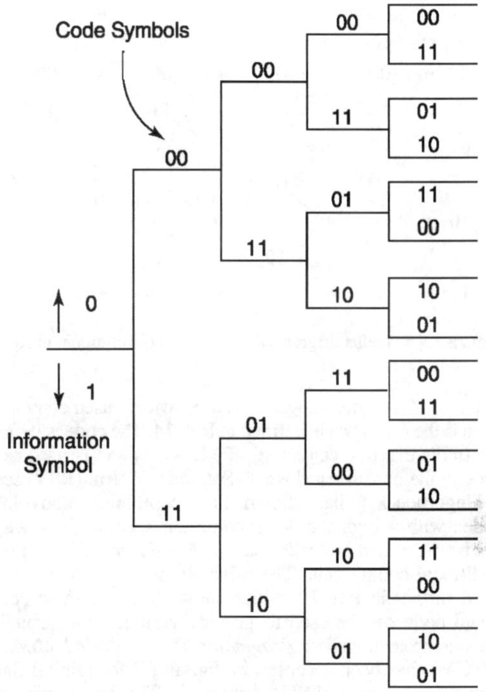

Figure 1.4.3 Tree diagram of the (2,1,2) code.

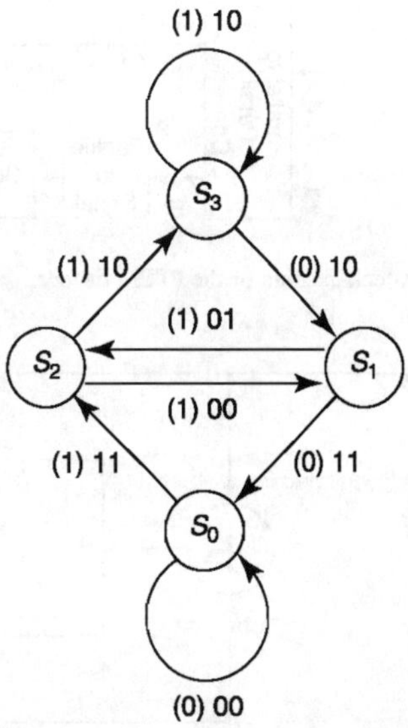

Figure 1.4.4 State diagram of the (2,1,2) convolutional encoder.

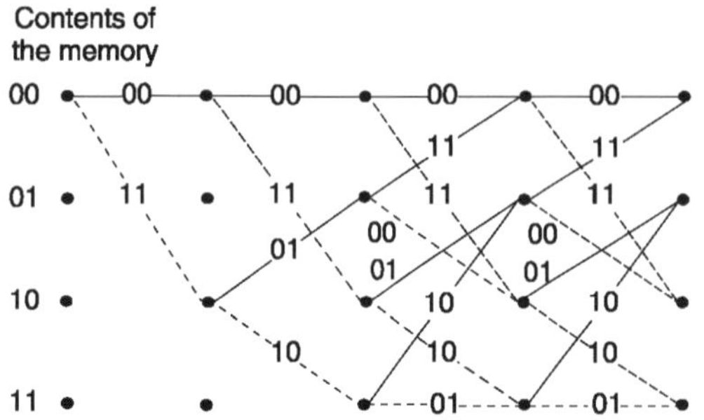

Figure 1.4.5 Trellis diagram of the (2,1,2) convolutional encoder.

trellis or *Viterbi* decoding.[17] The most widely used convolutional code is the (2,1,7) code defined in Ref. 14. The code is defined by the trellis diagram consisting of 128 states and is used as inner code in the concatenated Reed–Solomon–Viterbi (RSV) scheme.

Ungerboeck[18] has shown that combining convolutional codes with a higher order modulation scheme, in a way that has been defined as *Trellis Coded Modulation* (TCM), provides additional coding gain. The principle of TCM has been developed further in Ref. 19, which shows how the same convolutional code can be used to provide various data throughputs. This approach, called *Pragmatic Trellis Coded Modulation* (PTCM), has been accepted in digital TV for Digital Satellite News Gathering (DSNG) systems.[15] The basic principle of PTCM can be illustrated with the help of Figure 1.4.6.

The input stream of the PTCM encoder is split into two branches, called Encoded (E) and Non-Encoded (NE) streams. The NE signals are passed directly to the input of the modulator, thus avoiding the convolutional encoder. They are protected only by the large Euclidean distance in the signal space due to a special mapping of bits into a modulated signal. The E signal is encoded by the conventional convolutional code and is protected by the error correction properties (free distance) of the code.

1.4.4 Turbo Codes

In 1993, Claude Berrou demonstrated that a parallel concatenation of convolutional codes together with an iterative decoding allows the achievement of near-Shannon limit performance.[20] The developed codes, called *Turbo Convolutional Codes* (TCCs), have been widely investigated over the past decade and significant performance improvements have been achieved. For a more detailed explanation of TCCs, the reader is addressed to a well-written book by Heegard and Wicker.[21]

The TCCs have found applications in various fields of communications, including digital TV broadcasting. In particular, TCCs have been recommended by the DVB for interactive services over satellite (DVB-RCS) and terrestrial (DVB-RCT) links. In this section we provide a brief overview of the TCCs; for a more detailed explanation of turbo convolutional codes for digital TV broadcasting, the reader is addressed to Refs. 6 and 22.

In the original article and patent,[23] Berrou and others proposed a coding scheme based on the parallel concatenation of two simple recursive systematic convolutional codes. The parallel concatenation would allow the encoding of the incoming bit more than once and place at least one of the coded bits in a different part of the output by means of an interleaver. The recursive convolutional codes used in the original paper are shown in Figure 1.4.7.

The encoder of Figure 1.4.7 is recursive since the input bit entering the encoder is also determined by a feedback loop from the output of the encoder. Based on this simple coding, a very powerful TCC scheme can be designed as shown in Figure 1.4.8.

The system given in Figure 1.4.7 is a rate $R = 1/3$ system, i.e. one information bit generates three encoded bits. This can

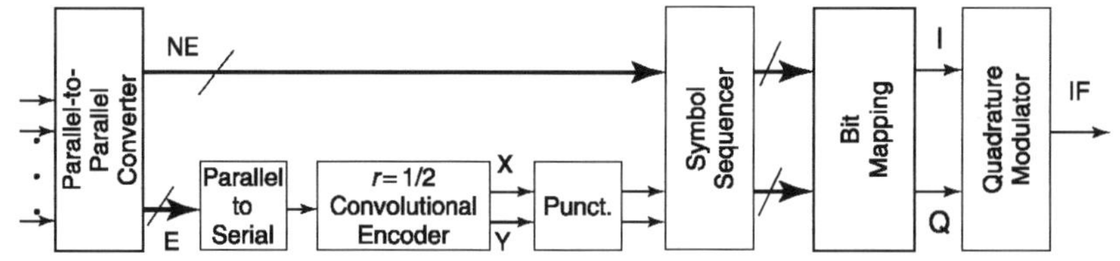

Figure 1.4.6 Block diagram of the PTCM encoder.

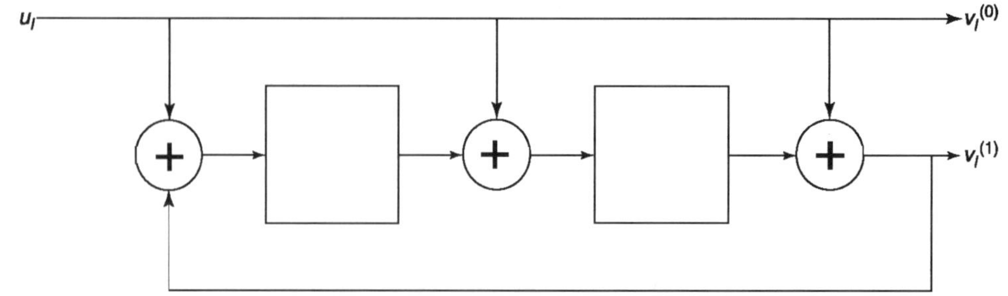

Figure 1.4.7 Recursive systematic convolutional encoder.

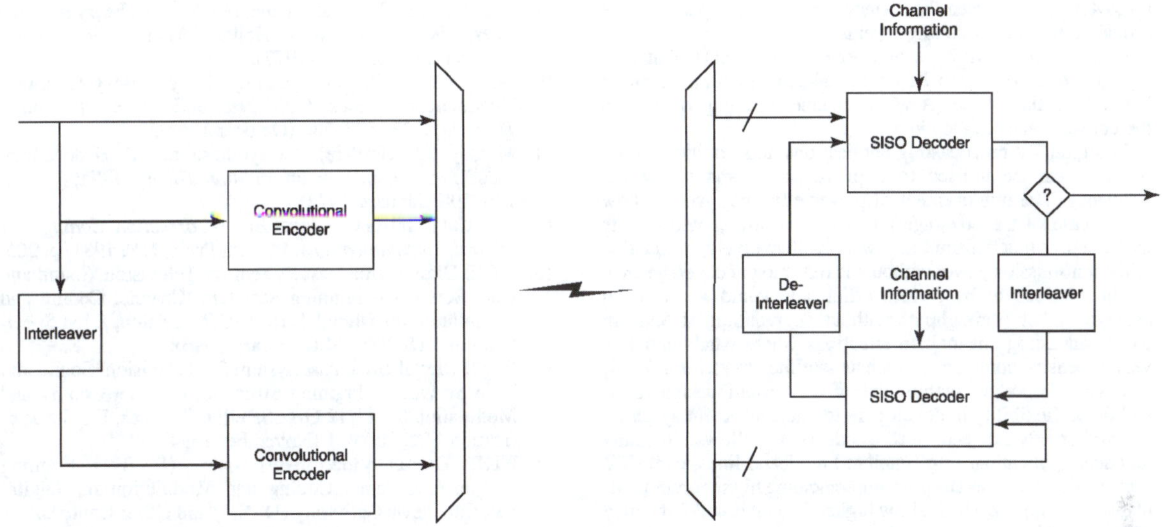

Figure 1.4.8 Turbo convolutional encoder and decoder.

easily be converted to any higher rate by simply puncturing the coded bits. The choice of interleaver is essential for achieving good error performance, as the size and the type of the interleaver specify the randomness of the code and define the overall performance.

At the receiver (decoder) side, two convolutional decoders with the *Soft In Soft Out* (SISO) feature are used in iterative fashion. The SISO decoder of convolutional codes is an extension of the conventional Viterbi decoding to provide additional reliability estimates of the decoded symbols. A number of SISO algorithms are known,[24,25] and the choice of the algorithm should be determined by the system requirements and allowed complexity of implementation.

1.4.5 Practical Benefits of Using Turbo Codes

Currently, the most widely used forward error correction technique is based on concatenated convolutional and Reed–Solomon codes, as described in a number of DVB standards. Figure 1.4.9 illustrates the efficiency of this solution and its comparison with the Shannon bound. As follows from this figure, there is

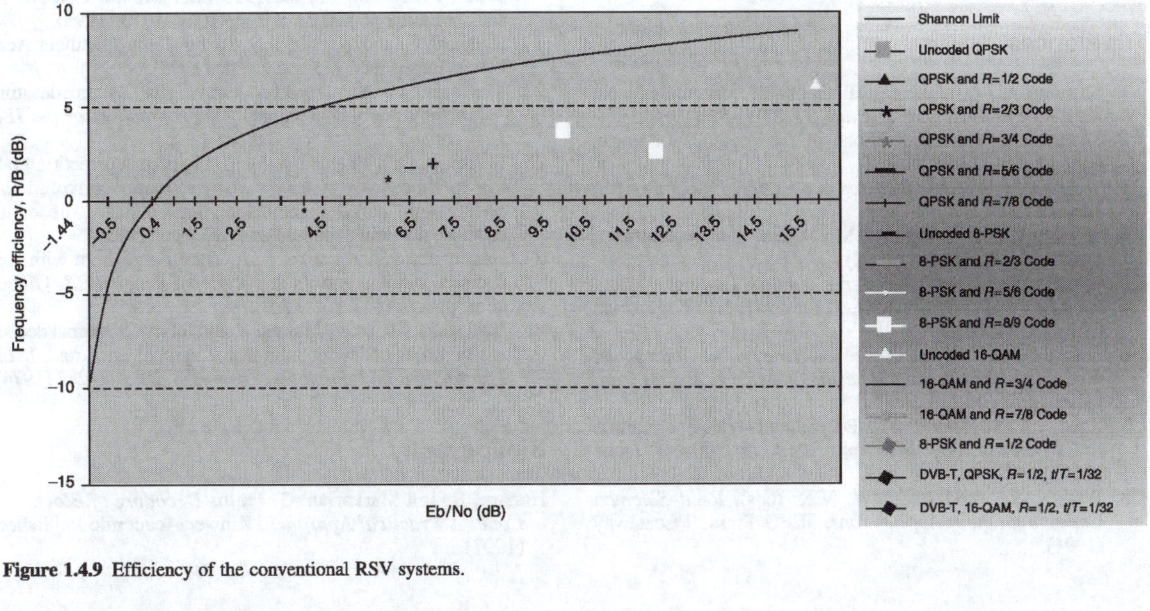

Figure 1.4.9 Efficiency of the conventional RSV systems.

up to 4 dB gap between the Shannon limit and the performance provided by the RSV coding scheme.

It has been shown by a number of researchers[21] that it is possible to achieve up to 2.5 additional coding gain in Additive White Gaussian Noise (AWGN) channel in comparison with the conventional RSV coding.

In digital TV broadcasting applications, this additional coding gain can be utilised to improve power and bandwidth efficiency. The improvement of power efficiency would allow an increase of the coverage area of broadcasting without the increase of output transmitted power or, alternatively, a reduction of the transmission power without the reduction of coverage area.

The increase of bandwidth efficiency would allow us to achieve higher (throughput) without degradation of SNR in the broadcasting channel. In situations where bandwidth is a very expensive commodity (such as satellite broadcasting), this is a very important feature that gives system designers an additional flexibility in developing the most efficient system.

Another advantage of turbo code is as follows: in many practical applications (e.g. satellite broadband links, cable TV systems), system operators are implementing higher order modulation techniques, which allow higher data rates and frequency efficiency. However, increase in modulation order is always associated with higher implementation margins and higher systems costs due to additional equalisers and more sophisticated clock and carrier recovery systems (e.g. cable TV systems moving from 128-QAM to 256-QAM). The situation is worse in the case of satellite links, which are based on 16-QAM modulation formats and conventional DVB-type RSV FEC. While providing higher bandwidth efficiency, these systems demand 6 dB power back-off to compensate effects of nonlinearity. Therefore, a 20–30% increase in bandwidth efficiency is associated with up to 6 dB of power losses and eventually in increased systems cost. It has proved that by applying turbo codes to 8-PSK modulation it is possible to achieve bandwidth efficiency similar to 16-QAM DVB systems without increase of required Eb/No and the need for additional 6 dB additional power back-off.

References

1. Shannon, C. A mathematical theory of communications, *Bell Systems Technical Journal*, **27**, 379–432 (1948).
2. Shannon, C. Communications in the presence of noise, *IRE*, **37**, 10–21 (1949).
3. Sloane, A. *Multimedia Communication*, McGraw-Hill, London (1996).
4. Glover, I.A. and Grant, P.M. *Digital Communications*, Prentice-Hall, London (1998).
5. Lin, S. and Costello, D. *Error Control Coding: Fundamentals and Applications*, Prentice-Hall, Englewood Cliffs, NJ (1983).
6. Drury, G., Markarian, G. and Pickavance, K. *Coding and Modulation for Digital Television*, Kluwer Academic Publishers, Boston (2001).
7. Reed, I.S. and Solomon, G. Polynomial codes over certain finite fields, *J. Soc. Ind. Appl. Math.*, **8**, 300–304 (June 1960).
8. Wicker, S.B. and Bhargava, V.K. (eds). *Reed–Solomon Codes and their Applications*, IEEE Press, Piscataway (1994).
9. MacWilliams, F.J. and Sloane, N.J.A. *The Theory of Error Correcting Codes*, North Holland Mathematic Library, Elsevier Science B.V. (1977).
10. Berlekamp, E.R. On decoding binary Bose–Chaudhuri–Hocquenghem codes, *IEEE Transactions on Information Theory*, **IT-11**, 577–580 (October 1965).
11. Massey, J.L. Shift register synthesis and BCH decoding, *IEEE Transactions on Information Theory*, **IT-15**, No. 1, 196–198 (January 1972).
12. J.B. Cain and G.C. Clark. *Error-Correction Coding For Digital Communications*, Plenum Press, NY, 1981, p. 205.
13. ETSI. Digital Broadcast System for Television, Sound and Data Services – Framing Structure, Channel Coding and Modulation for Digital Terrestrial Television, ETSI Specification ETS 300 744 (February 1996).
14. ETSI. Digital Broadcast System for Television, Sound and Data Services – Framing Structure, Channel Coding and Modulation for 11/12 GHz Satellite Services, ETSI Specification ETS 300 421 (November 1994).
15. ETSI. Digital Video Broadcasting (DVB) – Framing Structure, Channel Coding and Modulation for Digital Satellite News Gathering (DSNG) and Other Contribution Applications by Satellite, ETSI Specification EN 301 210 (December 1998).
16. ETSI. Digital Video Broadcasting (DVB) – Framing Structure, Channel Coding and Modulation for Cable Systems, ETSI Specification EN 300 429 (April 1998).
17. Viterbi A.J. Error bounds for convolutional codes and an asymptotically optimum decoding algorithm, *IEEE Transactions on Information Theory*, **IT-25**, No. 1, 97–100 (January 1979).
18. Ungerboeck, G. Channel coding with multilevel/phase signals, *IEEE Transactions on Information Theory*, **IT-28**, 555–567 (1982).
19. Viterbi, A.J., Wolf, J.K., Zehavi, E. and Padovani, R. A pragmatic approach to trellis-coded modulation, *IEEE Communications Magazine*, **27**, No. 7 (1989).
20. Berrou, C., Glavieux, A. and Thitimajshima, P. Near Shannon limit error-correcting coding and decoding: turbo codes, *Proceedings of the 1993 International Conference on Communications*, Zurich, pp.1064–1070 (1993).
21. Heegard, C. and Wicker, S.B. *Turbo Coding*, Kluwer Academic Publishers, Massachusetts (1999).
22. Markarian, G. (ed.). Turbo codes in digital TV broadcasting – can they double the capacity?, *Proceedings of the IEE Conference*, London (November 1998).
23. Berrou, C. US Patent No. 5446747, Error-correction coding method with at least two systematic convolutional codings in parallel, corresponding iterative decoding method, decoding module and decoder (1995).
24. Hagenauer, J. and Papke, L. A Viterbi algorithm with soft decision outputs and its applications, *Proc. IEEE*, Globecom, pp. 47.1–47.1.7 (1989).
25. Bahl, L.R., Cocke, J., Jeinnek, F. and Raviv J. Optimal decoding of linear codes for minimising symbol error rate, *IEEE Transactions on Information Theory*, **IT-20**, 248–287 (1974).

Bibliography

Honary, B. and Markarian, G. *Trellis Decoding of Block Code: A Practical Approach*, Kluwer Academic Publishers (1997).

Internet Resources

Algebraic Decoding, http://math.berkeley.edu/~berlek/alg.html

Coding and Modulation Schemes for Digital Audio, http://www.stanford.edu/courses/192b/lectures/5/5.html

Compact Discs, http://www.cs.tut.fi/~ypsilon/80545/CD.html

CD Data Coding, http://www.disctronics.co.uk/technology/cdbasics/cd_frames.htm

CD Physical Specification, http://www.disctronics.co.uk/technology/cdbasics/cd_specs.htm

Reed–Solomon Codes, http://www.4i2i.com/reed_solomon_codes.htm

R S Roberts C Eng, FIEE, Sen MIEEE
Consultant Electronics Engineer

1.5 Coaxial Cable and Optical Fibres

1.5.1 Cable Transmission

The use of cable for distribution of broadcast signals is not new. Provision of a central reception point for radio sound programmes and a distribution network dates from 1924 in the UK. Modest networks distributed television programmes before 1939. More than 13 per cent of television viewers in the UK receive their programmes by means of cable, with much higher percentages in other parts of Europe, Canada and the USA. Existing UK networks must conform to various technical requirements, and must be licensed. Various BSI Specifications and other documents exist (see Bibliography).

Cable distribution systems must conform to the broad principles governing the transmission of energy from a source to a load. A cable distribution system involves three main parts:

- the *head end* assembles and combines the signals to be distributed, and feeds them to
- the *network*, from which they are tapped off to final cables to supply
- the *subscriber outlets* in individual premises.

Network requirements are that:

- losses must be as low as possible;
- no radiation should take place that could cause interference to other services remote from the network;
- the entire system must be screened from possible interference by the fields radiated by other services.

1.5.1.1 The transmission line

The pair of conductors shown in Figure 1.5.1(a) constitute a transmission line, connected to an rf source *e*, the energy progressing along the line as shown towards the termination at the end of the line.

There is a capacitance *C* between the two conductors and, if the line is terminated by a short-circuit, an inductive loop *L* is formed. If the line length is doubled, the capacitance is doubled and the inductance of the loop is doubled, but the ratio *L/C* remains constant. The line may be considered to consist of a succession of unit lengths, each with capacitance and inductance as shown in Figure 1.5.1(b). The line is *balanced*, i.e. neither input terminal is at earth potential.

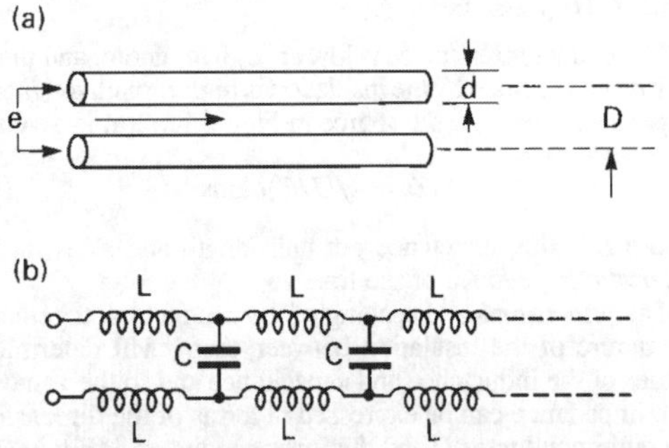

Figure 1.5.1 Balanced transmission line (a) and its equivalent circuit (b). Its characteristic impedance is double that of the coaxial line in Figure 1.5.2.

The currents flowing in the two conductors will be in opposite directions, and their magnetic fields will therefore tend to cancel. Radiation of any energy from the line would be a loss of energy from the system, and the field-cancellation effect helps to minimize losses from the line.

The most widely used form of the transmission line is the *coaxial* construction shown in Figure 1.5.2. The two-conductor line now takes the form of an outer conductor enclosing an inner, axial conductor.

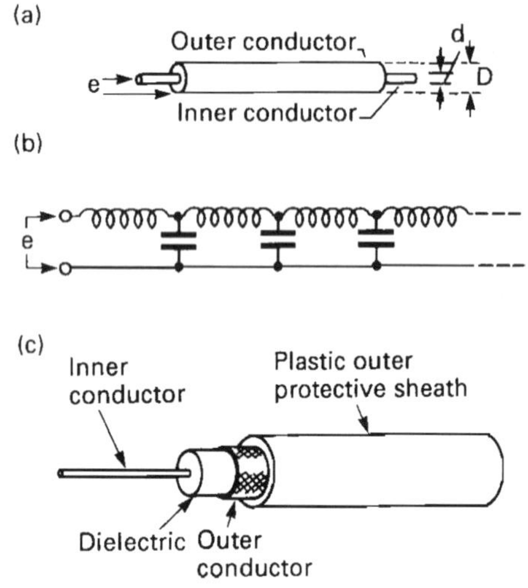

Figure 1.5.2 The two-conductor transmission line in coaxial form (a), its electrical equivalent (b) and its physical construction (c).

Balanced lines are not used in cable distribution networks, and we will be concerned only with the unbalanced coaxial line of Figure 1.5.2.

The *unbalanced* line has an immense practical advantage over the balanced line in that the outer conductor can be at earth potential, thus fully screening the inner conductor. This feature prevents radiation from the line, and prevents interfering signals from being received by the line inner conductor.

1.5.1.2 Impedance

If the conductor resistance is low enough to ignore, and the shunt resistance provided by the insulation is high enough to ignore, the impedance 'seen' by the source in Figure 1.5.2(a) is given by:

$$Z_o = \sqrt{(L/C)}\,\text{ohms} \tag{1.5.1}$$

where L is the inductance per unit length and Z_o is the *characteristic impedance* of the line.

The dimensions and spacing of the conductors, together with the nature of the insulation between them, will determine the values of the inductance and capacitance and so the value of Z_o. The impedance can be expressed in terms of the dimensions of the outer conductor D and the inner conductor d (Figure 1.5.2). The permeability μ and permittivity ε of the dielectric must be taken into account, and equation (1.5.1) becomes:

$$Z_o = 138\log_{10}\frac{D}{d}\sqrt{\left(\frac{\mu}{\varepsilon}\right)}\,\text{ohms} \tag{1.5.2}$$

For completeness, the impedance for the balanced line of Figure 1.5.1(a) is given by:

$$Z_o = 276\log_{10}\frac{D}{d}\sqrt{\frac{\mu}{i}}\,\text{ohms} \tag{1.5.3}$$

For air spacing, μ and ε are approximately equal to 1, and this part of equation (1.5.2) can be ignored. For other insulating materials, such as are used in practical cables, μ will probably be near 1, but ε will have higher values.

Equations (1.5.1) and (1.5.2) do not include frequency; a transmission line will transmit all frequencies from infinitely high to direct current. The signal source will 'see' an impedance Z_o, but as this impedance consists of reactive elements, the impedance will absorb no power, provided that the resistance and dielectric losses are zero.

The characteristic impedance can have any value, determined by the ratio D/d, but in practice, manufacturing limitations exist. The clearance between the inner and outer conductors cannot be too small, and the inner conductor diameter cannot be reduced without increasing resistance losses. These considerations limit the available range of impedance for coaxial lines to about 20–200 ohms. (For a balanced line the range is about 150–800 ohms.)

The value of cable impedance used for distribution networks has been standardized in the UK at 75 ohms, but 50 ohms has been used elsewhere in Europe. With air spacing, a ratio D/d of 3.6 gives an impedance of 75 ohms, but the practical cable using an insulator for spacing will have a different ratio to provide a Z_o of 75 ohms, as defined by equation (1.5.2).

1.5.1.3 Cable losses

Any cable will have some conductor resistance and insulator loss. This loss would be a minimum if the dielectric between the conductors were air and the conducting surfaces had a large area. Unfortunately, some insulation is required between the inner and outer conductors to ensure that the inner is accurately centred in the outer, and a large surface area for a given impedance requires a large overall diameter for the cable.

Power loss from a balanced cable has been mentioned, where energy radiated represents a 'radiation resistance' loss that would add to the total losses by the cable. Some coaxial cables use a woven braid for the outer conductor, and some energy leakage can take place through the meshes of the weave. This may not be serious if the power levels are low but, where the power levels on the cable are large, some regulations require a double-woven cable or a solid sheath to be used.

Cable losses are usually expressed in decibels per unit length, and they increase as frequency is raised in a complicated manner, mainly due to the increase in *skin resistance* of the conductors. The increase in decibel loss is approximately proportional to $\sqrt{f}$. Thus, for example, a length of cable that has a loss of, say, 2 dB at 50 MHz, will have a loss of the order of 4 dB at 200 MHz and 8 dB at 800 MHz.

1.5.1.4 Matching and termination

For maximum transfer of power from a source e into a load R_L consider the example shown in Figure 1.5.3. Assuming a value of 75 ohms for R_s and 10 V for e, the resulting circuit voltages and currents are shown in Figure 1.5.4.

It is seen that, for maximum transfer of power from the source to the load, the load impedance must equal in value the internal impedance of the source. This is termed a *matched* condition and, in a simple case where a source feeds a line and the far end of the line feeds the input terminals of a receiver, is a two-fold process: the source feeds its power into the line, and the line feeds the receiver at its input. The line must match the source, and the receiver input impedance must match the line. For the usual UK network, the source output impedance will be

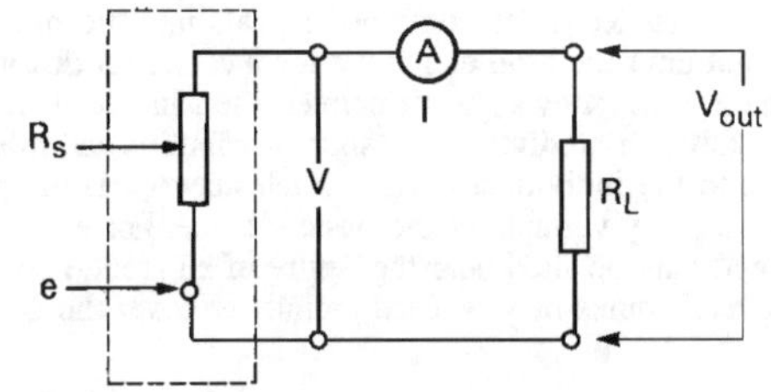

Figure 1.5.3 A power source connected to a load.

75 ohms, the network impedance will be 75 ohms and the receiver input impedance will be 75 ohms.

Figure 1.5.4 shows that the matched condition is not sharply defined, and it extends over a broad band. In fact, if the load is half or twice the correct value required for matching, the power in the load changes by only 0.5 dB, but mismatching produces other effects, and must be avoided as far as possible.

A line with a source of power *e* at one end, and a terminating load Z at the other, is depicted in Figure 1.5.5. Consider a transmission of rf energy advancing along the line from *e* to Z. If the line is terminated with an open circuit (an infinitely high impedance) the energy cannot be dissipated when it reaches the end of the line, and it is returned (reflected) back along the line towards the source. No current can flow in the termination (b), and the voltage can be a maximum (c). The energy advancing along the line towards the termination will now experience interference with that reflected from the termination, this effect producing positions along the line where additive and subtractive values of current or voltage give rise to 'stationary' patterns or *standing waves.*

In Figure 1.5.5, (d) and (e) show similar standing wave effects for a termination consisting of a short-circuit. In this case, no voltage can exist at the termination, but current can have a maximum value. All four standing wave patterns show that the line now has a distribution of voltage and current along its length that is not uniform.

A voltmeter connected across the line between the two conductors will indicate maximum, minimum or intermediate values, depending on its position along the line. The currents and voltages in adjacent half wavelength sections of the standing wave will be in phase opposition.

If we now examine the open circuit line for amplitude only (e.g. by sliding a voltmeter along it), we obtain the variations shown in (f) and (g). The connection of a load to the line would have to be at a point where the voltage and current have suitable values, and the system would then function on one frequency only. This is not very practical, but the example serves to show that standing waves must be eliminated as far as possible.

This can be done in two ways. One is a theoretical concept only, and simply recognizes that a line of infinite length would absorb all the energy that is sent along the line, and would not reflect any back to the source. The second, practical method is to note that a termination with a finite value of impedance somewhere between zero and infinity can simulate a line of infinite length and absorb all the energy it receives without any reflection back to the source. Such a termination would have the same impedance value as the characteristic impedance Z_o of the cable, and would equal $\sqrt{(L/C)}$ ohms.

The resulting current and voltage along the line would then be as shown in (h) and (i), except that in the practical case, where line losses will exist, there would be a downward slope of the current and voltage values as the source energy moves away along the line.

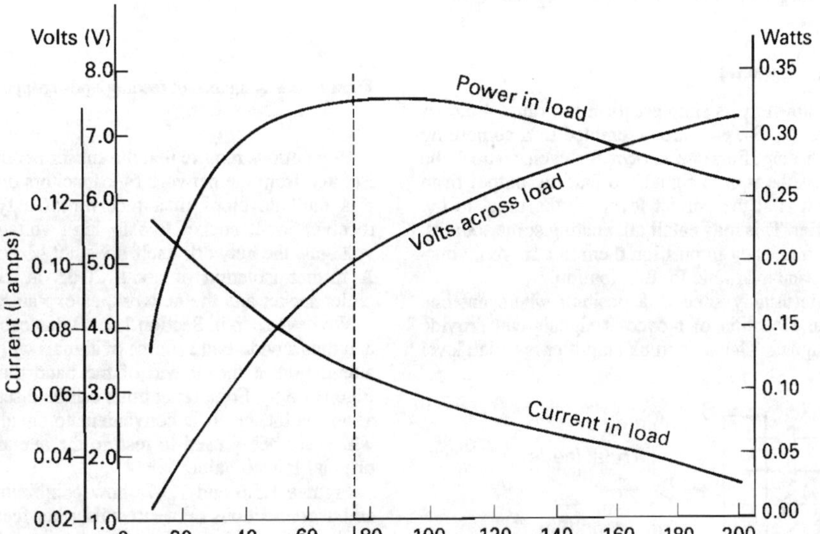

Figure 1.5.4 The current, voltage and power relationships in Figure 1.5.3, as the load is varied.

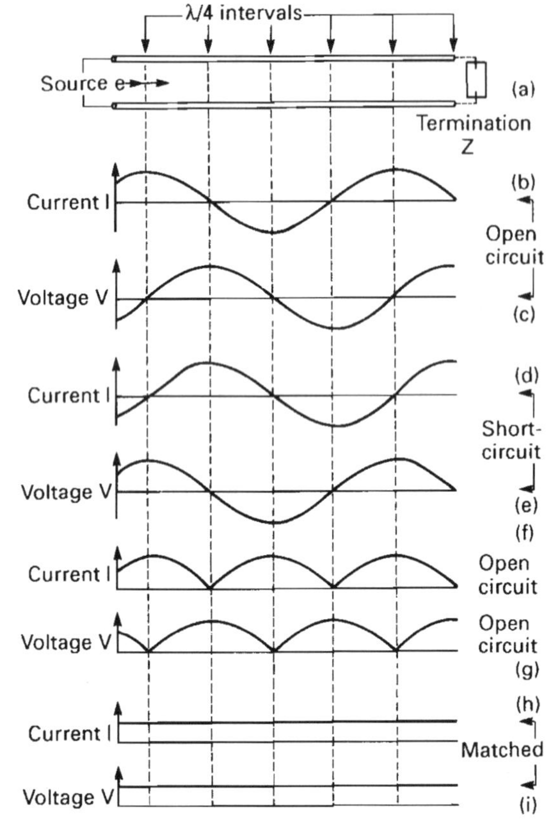

Figure 1.5.5 A long timeline fed from an rf source showing current and voltage waveforms along it for various termination conditions (see text).

1.5.1.5 The cable network

The head end is shown in its simplest form in Figure 1.5.6. A number of sources, S_1, S_2, etc., are assembled in a combining unit and, although amplifiers are shown for each source, the object is to process the source signals so that the output from the combining unit is at the correct level for the input to the main trunk amplifier. This may entail attenuating some sources, or changing their frequency to position them in a more favourable place in the band available for distribution.

The head end is usually sited in a position where antenna systems for off-air reception of broadcast signals can provide the best possible signals. The main trunk amplifier is a high level amplifier that feeds the combined signals into the main trunk cable that links the head end to the main centre for distribution.

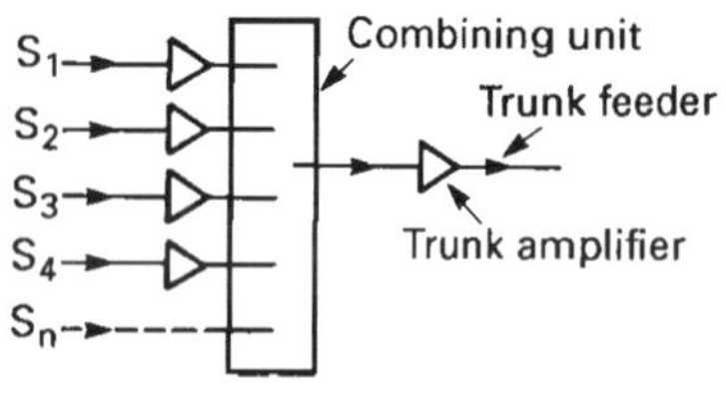

Figure 1.5.6 Head and essentials.

Figure 1.5.7 shows the principles used in the distribution network, which subdivides into spur, distribution and subscriber feeders to the individual outlets. Each network is unique and there are many variants of the basic scheme. For example, the head end may be sited near the centre of an area to be served, and several trunks may be used radially to cover the area.

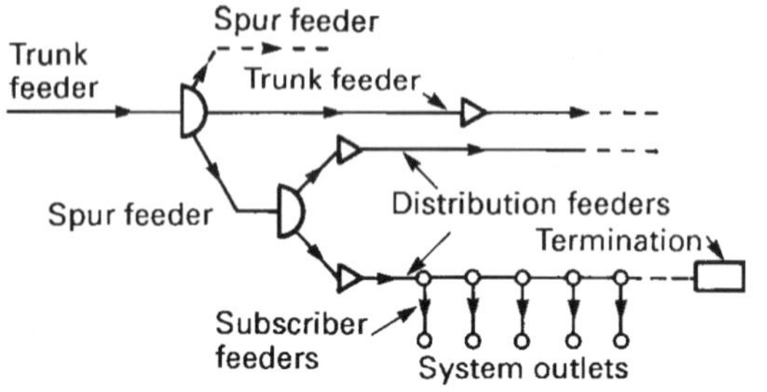

Figure 1.5.7 The network supplied by a high level, main trunk feeder.

Another feature concerns the many amplifiers on the network and their power supply. Locally derived mains power may be difficult to provide at some remote amplifiers and, in these cases, it is possible to feed power along the network cable as shown in Figure 1.5.8.

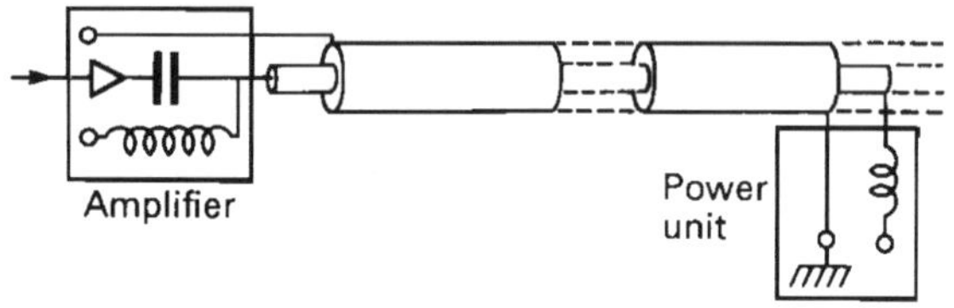

Figure 1.5.8 A method of feeding a power supply along a signal feeder.

Regulations require that the subscriber outlet sockets must be isolated from the network by capacitors or transformers so that if a fault develops in a receiver of a type that renders the receiver input socket live to high voltages or mains supply voltages, the network itself does not become dangerously live. A further isolation of about 20 dB is required between any outlet socket and the network, as explained in Section 1.5.1.6.

We have seen in Section 1.5.1.3 that cable losses vary as $\sqrt{f}$, and thus a wide-band source of signals will experience a greater attenuation at the hf end of the band than at the lf end (see Figure 1.5.9). Equalizing units can be inserted into the network at any point, but it is convenient to equalize at the amplifiers which are being used to restore the attenuated signals to their original launch value.

Figures 1.5.6 and 1.5.7 show combining and splitting units and T connections between subscriber feeders and distribution feeders. All of these must have input and output impedances that match the system impedance, and must not introduce any undue loss. A principle used for splitting from one cable into two is shown in Figure 1.5.10 where the system's 75 ohm impedance is maintained on each line.

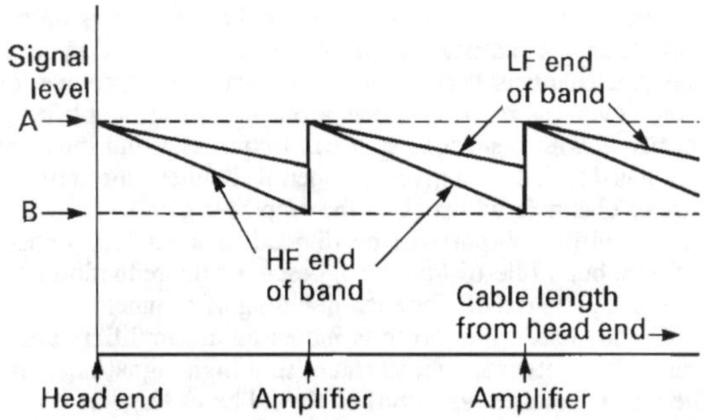

Figure 1.5.9 Cable attenuation and network amplifiers.

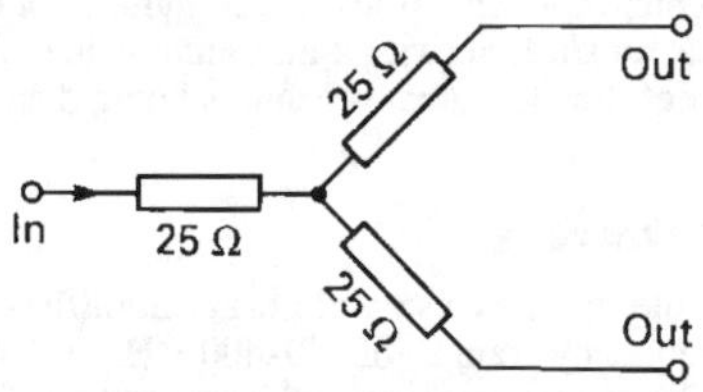

Figure 1.5.10 Basic principle for dividing a cable signal source to feed two cables while maintaining matching.

A splitter can operate with its function reversed as a combiner. However, such a system that uses resistors would absorb some power and, in practice, a transformer would be used to effect the same result. A directional coupler can also be used for combining or splitting.

1.5.1.6 Cable amplifiers and signal levels

The most significant aspect of cable operation is the amplifiers that are used on the network. Figure 1.5.9 shows how signals are launched at a level A. As they progress along the cable, some energy is lost due to cable attenuation, and a minimum level B is reached. At this point an amplifier is used to raise the level back to A. The associated circuitry includes that necessary to rebalance the levels of high and low frequency signals in the band.

Four basic factors that determine the values of levels A and B are:

- Any outlet socket used for the connection of a standard television receiver to the system must supply any of the distributed television channels at a minimum level of 1–2 mV, if the signal/noise ratio of the receiver is to be acceptably high. The receiver input voltage must not exceed 5–10 mV if receiver overload effects are to be avoided. Any broadcast fm sound channels being distributed must provide signals at a minimum of 0.5 mV and a maximum of 5 mV.
- Cable standards require that isolation of about 20 dB must be provided between any outlet socket and the network. This ensures that, if a fault develops in a receiver of a type that might feed into the network and affect other receivers (e.g. instability or oscillation), a minimum isolation of about 40 dB exists between any two receivers.
- Many cables use a woven conducting braid for the outer conductor, and so the screening is not complete. A double-woven braid is often used to increase the screening efficiency, but some radiation takes place from any braided cable. It is important that signal levels on the network are not so high that radiation constitutes an interfering signal to other services. A cable with a solid tubular outer conductor is the most satisfactory and is widely used for trunk feeders where the signal levels are at their maximum. The screening of the cable from interfering fields from other services is particularly important because, if an interfering signal finds its way into one of the amplifiers on the system, it may not be eliminated by any subsequent filtering.
- A large network requires some form of *automatic gain control* (agc). Cable attenuation is temperature sensitive. Copper has a positive temperature coefficient of resistance, and the network attenuation due to resistance will increase during hot weather, becoming lower in a colder ambient temperature. The agc control signal is usually derived from one or more pilot signals, distributed from the head end along the signal channels. The control is applied to a few selected amplifiers on the network.

If a required broadcast signal level of 2 mV is required at the remote outlet on a distribution feeder, a level of 20 mV is required on the distribution feeder at this point. If the feeder loss is, say, 6 dB, the level at the distribution amplifier output must be 40 mV.

The input to the distribution amplifier will be attenuated by splitting. The input power to the splitter will divide between the two output paths, each of which will be −3 dB on the input level. The output from the trunk splitter will need to be 80 mV and the level on the trunk feeder in the region of 100 mV. The head end launch level needs to be higher than 100 mV to allow for the trunk cable attenuation and to provide an ample design margin.

The maximum launch level A in Figure 1.5.9 is determined by:

- signal/noise ratio;
- signal level;
- amplifier distortion effects.

The s/n ratio will be determined by the network noise which will be amplified along with the signals, and the noise generated by the amplifiers.

Any resistor generates noise which is given by:

$$E = \sqrt{(4kTBR)}$$

where k is Boltzmann's constant, T the temperature (kelvin), B the bandwidth and R the resistance.

The network resistance is 75 ohms and, for a bandwidth of 5.5 MHz, E becomes about 2.5 μV which, matched into 75 ohms, becomes a 1.25 μV noise level at an amplifier input.

Figure 1.5.11 shows the two sources of noise, and their relationship to an amplifier. The contribution by the amplifier gives it a *noise factor* (nf) given by:

$$\text{nf} = \frac{G(N_n + N_a)}{GN_n} = 1 + \frac{N_a}{N_N} \qquad (1.5.4)$$

With no signals input to the system, the output at any outlet socket will consist of noise. With a signal input to the system,

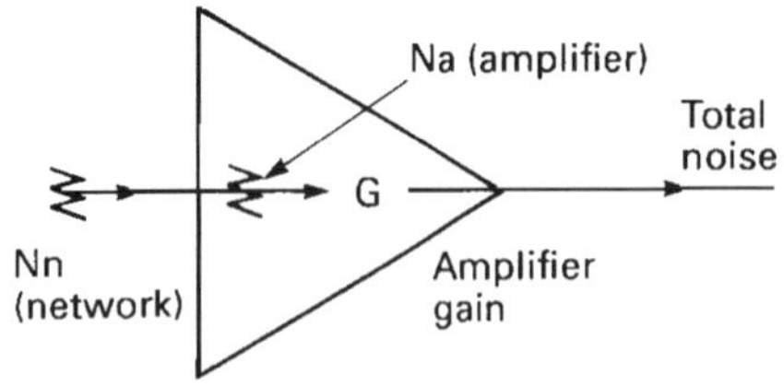

Figure 1.5.11 Network and amplifier noise.

the signal level at the output sockets needs to be high enough to provide an acceptable s/n ratio. A minimum value of this ratio is usually taken as about 45 dB.

There are several constraints that prevent the s/n ratio being as high as we would choose to make it. One has been mentioned: signal levels must not be so high that possible radiation from the system becomes a source of serious interference to other services. Interference can be set up at nearby receivers that are not connected to the network, but receive their programmes off-air, using their own antenna system. The most serious design consideration for the network is the degree of non-linearity in the operation of the network amplifiers.

All amplifiers are non-linear in their operation to some extent. A single input to an amplifier will, if the amplifier has a linear input/output transfer characteristic, be reproduced at the output as an amplified and faithful copy of the input waveform. Any non-linearity will result in waveform distortion and consequent harmonic generation.

If two or more signals are supplied to the input of a broadband linear amplifier, the signals will be amplified in a distortionless fashion with no mutual interference effects. However, if non-linear operation takes place in the amplifier, mutual interference effects arise between the signals, and a number of spurious signals are generated.

Consider an input consisting of two signals:

$$E = A\cos\omega_1 t + B\cos\omega_2 t \qquad (1.5.5)$$

applied to an amplifier having a gain G. The output due to non-linearity may be expressed as:

$$E_{\text{out}} = G_1 E_{\text{in}} + G_2 E_{\text{in}}^2 + G_3 E_{\text{in}}^3 \ldots \qquad (1.5.6)$$

G_1, G_2, G_3 etc. do not have the same meaning as the gain G, and they will have different values from G, the extent of the difference being determined by the levels A and B, and the non-linear law of the amplifying device when it is being driven under these particular conditions.

The output will consist of:

First order — $G_{1\text{in}} = G_{1A}\cos\omega_{1t} + G_{1B}\cos\omega_{2t}$

Second order — These will include a dc component, second harmonics of ω_1 and ω_2 and other even harmonics, and difference frequencies with intermodulation between the two signals

Third order — These will include third and odd harmonics, original frequencies with cross modulation, and difference frequencies with intermodulation.

If the input signal consists of three frequencies (as in a single television channel, with three carrier frequencies), a further set of output signals is generated. To handle two television channels (i.e. six carrier frequencies), the interaction effects between channels become more serious and, of course, even more severe as more channels are added. If an amplifier performance with respect to spurious frequencies and intermodulation is decided for a given number of channels, the addition of further channels will require the amplifier gain to be reduced. The amplifier output will be divided in a random statistical fashion, but a rule-of-thumb suggests that the reduction in gain should be about 3 dB for each doubling of channels.

We have seen that noise is increased as amplifiers are cascaded in a network and, to establish a high signal/noise ratio, the signal input to an amplifier should be as large as possible. To keep spurious frequency levels and effects as low as possible, the signal inputs to an amplifier must be kept as low as possible. These two conflicting requirements for amplifier performance require that a compromise must be accepted for amplifier gain, input signal levels and number of channels that will provide an s/n ratio with a minimum value of about 45 dB at each outlet, for each channel that is being distributed.

1.5.1.7 Talkback

Modern cable systems use wide-band amplifiers that operate over a frequency range of 40–400 MHz or 40–850 MHz, although the recent removal of broadcasting from Band 1 (40–68 MHz) has tended to make the lowest frequency for a distribution system around 75 MHz. In either case, a wide range of frequencies, from dc up to the lowest distributed frequency, is available for a subscriber to 'talk back' to the head end.

The network attenuation is reduced, and a few low-power amplifiers would be required, e.g. a cable attenuation figure at 40 MHz is halved at 10 MHz and reduced to a quarter at 2.5 MHz. Relatively simple filters will ensure that the forward and backward signal information will not interfere with each other.

1.5.2 Optical Fibre Transmission

Development of transmission systems using light waves within optically transparent fibres may be considered to have commenced in the 1960s, when silica glass fibres were produced with an attenuation in the region of 1000 dB/km. A decade later, attenuation had been reduced to about 20 dB/km. By 1989, fibre cables were available off the shelf with attenuation of about 0.5 dB/km, with several manufacturers producing cable with much lower values of attenuation.

Light waves and radio waves differ only in their wavelength. Radio waves may be propagated along confining tubes in waveguides, and be guided in the ionosphere (as in 3–30 MHz band radiation) through refraction and reflection.

Visible light spans a frequency range from about 385–790 THz bordered by invisible bands in the infrared and ultraviolet. It is now possible, using lasers, to generate a light beam of a single, discrete wavelength.

Such a light source can be a *carrier wave* which can be modulated by any of the usual methods. The carrier frequency is so high that the bandwidth resulting from carrier modulation can be very wide and still be only a small fraction of the carrier frequency. As a consequence, the effects shown in Figure 1.5.9, although present to some degree, will show negligible differences between the low and higher frequencies resulting from signal modulation.

Modulation can be by analogue or digital means, but most fibre systems use digital modulation, with all the familiar advantages of digital transmission, such as operation at a constant amplitude and the many methods that can be used for modulation.

1.5.2.1 The fibre transmission line

A transmission fibre can be of any material, solid, liquid or gas, that conducts light with minimum attenuation. Of all the possible transmission media, glass offers the most advantages. Glass can be drawn into fibres relatively easily, and has a mechanical robustness that is not readily available with many other substances, such as plastics.

Figure 1.5.12 shows a glass fibre core surrounded by a glass sheath or *cladding*. The refractive index of the core is higher than that of the cladding, and the transmission paths are confined to the core.

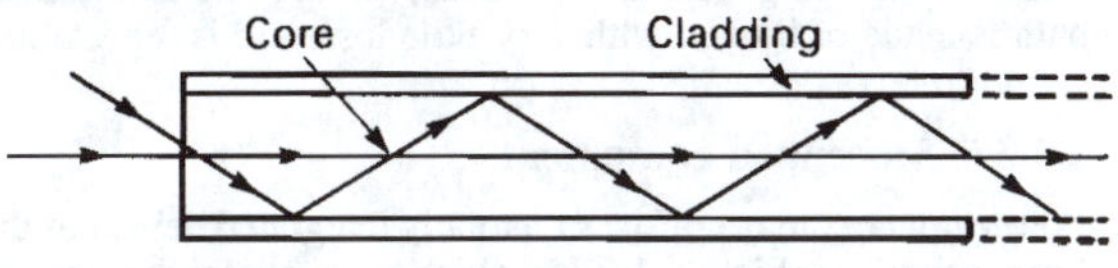

Figure 1.5.12 Transmission along a multi-mode fibre cable.

A limitation is that light rays entering the core at an angle will take a longer time for transmission along the fibre than those parallel to the axis. Each bit of a digital pulse signal contains many component frequencies, and these must be retained if the bit is to have short rise and fall transmission times. The late arrival of some of the component frequencies will have the effect of 'broadening' the signal bit and thus reducing the possible bit rate (an effect known as *dispersion*).

This transmission is termed *multi-mode*, and the pulse broadening effect sets a limit to the transmission bit rate that can be used for a given length of cable. An associated effect is *attenuation*, the combined effect being termed *modal dispersion*.

There are two possible solutions to the problem of dispersion. One is to fabricate the fibre so that its refractive index varies from a high value at the centre of the core, where the velocity of propagation is lower, to a low value at the outer diameter where the velocity is comparatively high. The changes in index values are obtained either in steps or in a continuous manner.

The second method is to reduce the core diameter to a value comparable with the wavelength of the light source, typically 3–10 μm. Such a core can support only one mode of propagation, and is termed *monomode*. Two practical difficulties exist with the use of monomode cable. The light source must be a laser, as a beam is needed small enough in diameter to couple into the fibre core with reasonable efficiency. With such a concentration of the light source, the laser power must be kept low to avoid excessive temperature rise of the core at the point of beam injection.

1.5.2.2 Attenuation and dispersion

Any medium used for light transmission will introduce some attenuation, i.e. the *transparency* can vary. Figure 1.5.13 shows how a typical silica fibre may attenuate light sources of different wavelengths. The broken line shows the *Rayleigh scatter* or *Rayleigh dispersion*, but the solid curve is representative of a cable that might be used today. The solid curve shows the combined effects of Rayleigh scatter and other effects.

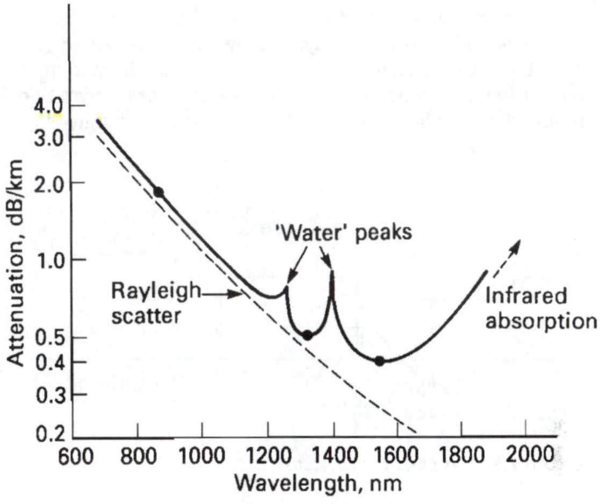

Figure 1.5.13 Example of the attenuation of a silicon fibre cable.

The Rayleigh curve is the result of imperfections in the glass such as cracks, bubbles and, surprisingly, water ions that result in scatter and absorption with attenuation peaks at 1.25 and 1.39 μm. The minute imperfections have dimensions comparable with the wavelength of the light being transmitted, and the attenuation decreases in proportion to $1/\lambda^4$.

Optical materials also experience *chromatic dispersion*, caused by the fact that the refractive index varies for different wavelengths. A light pulse includes various wavelengths, and no source is perfectly monochromatic, so that the various wavelengths in the light pulse travel along the fibre at different speeds.

At any particular wavelength, dispersion is measured as the amount of delay in picoseconds per kilometre, per nanometre difference in wavelength. At 850 μm, for example, chromatic dispersion in silica fibre is typically about 90 ps/km/nm. This figure falls with increasing wavelength to a minimum at about 1.3 μm.

Towards the long wave end of the curve, the falling attenuation is met by a rising infrared absorption loss due to molecular resonance with silicon oxide, and similar causes. The combined falling and rising attenuations as wavelength is increased result in the minimum around 1.55 μm.

1.5.2.3 Communication systems

A communication system requires a fibre cable with low attenuation at clearly defined wavelengths. It then requires a light source generator that will operate at maximum efficiency at wavelengths of low attenuation. At the end of the cable, a photo detector is required that functions at the same wavelength as the light source, with peak efficiency.

Historically, the early systems had difficulty in equating the need to have light sources and detectors each operating at the same wavelength with maximum efficiency at a wavelength suitable for transmission with minimum loss. 850 μm was the general choice, as this provided the necessary compromise, using available light sources and detectors. More recently, the lower cable attenuations at 1.31 μm and 1.55 μm have become useable through the development of appropriate semiconductor devices. 850 μm is often referred to as *short wave*, with 1.31 and 1.55 μm being termed *long wave*.

The essentials of a basic system are shown in Figure 1.5.14. The light source can be an led, where the light output is proportional to diode current, and where linear operation is required, as might be the case for some analogue signals.

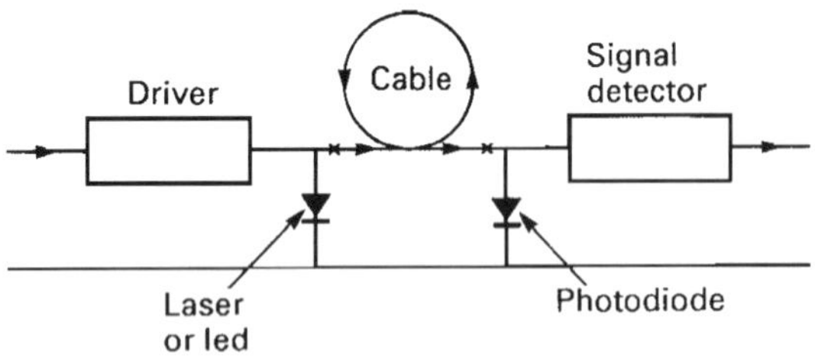

Figure 1.5.14 A basic fibre system.

A laser can provide more light output, and so provide a better signal/noise ratio, but its operation is non-linear and is a good choice for a digital system. The non-linearity results from its behaving as a diode up to a threshold where laser action starts, with a massive increase in output.

While coupling into a monomode fibre is carried out more efficiently with the small diameter light beam of a laser, the concentration of power that results may mean that the available laser output may need to be reduced to limit the temperature rise of the glass at the point of power, injection. In practice, a trade-off has to be made between the low coupling loss due to the narrow beam, and the power loss that might have to be made to limit the temperature. As an example of laser operation, one particular make, operating at 850 μm, has its threshold current at about 50 mA, and can launch more than 1 mW into the fibre.

Detectors were usually reverse-biassed p-n photodiodes, which rely on the production of electron holes by photons. The *quantum efficiency* is the ratio of the number of electron holes to the number of photons. For example, 100 photons creating 95 holes would give the device a quantum efficiency of 95 per cent. The quantum efficiencies of silicon and germanium vary with frequency; as a result, silicon detectors are more efficient at 850 μm, while germanium would be used in long wave systems.

1.5.2.4 Splicing and connecting

Fibre cables are mechanically quite rugged, but they are sensitive to crushing, longitudinal strain and internal stresses caused by sharp bends. Temperature changes can change their characteristics. For cable laying, the fibre cables are often housed loosely in a semi-flexible tube. This outer tube can take all the tension stresses involved in drawing through ducts, and the tube is tough enough to prevent any attempt to bend it round a small radius.

Splicing posed many early problems but there are now many designs for use in the field. The procedure is to diamond-cut, cleave and polish the cut ends. These are held in a jig that butts the ends, which are then welded with an oxy-hydrogen micro torch. With care, such a joint can achieve a loss lower than 0.1 dB. Clearly, splicing losses must be as low as possible because they add to the general system loss.

There are four possible causes for splicing loss:

- *separation* of the cable ends produces negligible attenuation for separation up to one-tenth of the core diameter, but 1 dB for a separation of half a diameter;
- an *angle* at the butt of 2° can produce 0.2 dB loss;
- a *shift* of the butt ends up to one-tenth of the diameter introduces 0.3 dB;
- if the *diameters* of the butt ends differ by 10 per cent, the loss can be 0.5 dB.

Plug-and-socket or other forms of *connectors* pose similar problems to those of splicing, but in a more acute form due to the need for mechanical accuracy that will ensure that the losses are low but, above all, are repeatable. One form uses lenses at the 'plug' and at the 'socket' so that the actual light path is made or broken with very little loss, and is repeatable.

1.5.2.5 Associated equipment

Power dividers may consist of a totally integrated device of the form shown in Figure 1.5.15 which is a photolitho optical waveguide that splits a light beam into two paths.

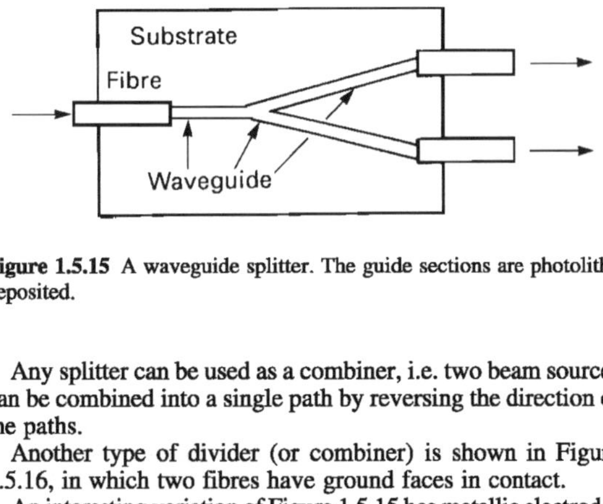

Figure 1.5.15 A waveguide splitter. The guide sections are photolitho deposited.

Any splitter can be used as a combiner, i.e. two beam sources can be combined into a single path by reversing the direction of the paths.

Another type of divider (or combiner) is shown in Figure 1.5.16, in which two fibres have ground faces in contact.

An interesting variation of Figure 1.5.15 has metallic electrodes adjacent to each of the optical waveguides. The refractive index in each guide can be changed by the application of an electric field across the guides. The device can thus be used as a switch whereby a choice of either of the two output paths may be made.

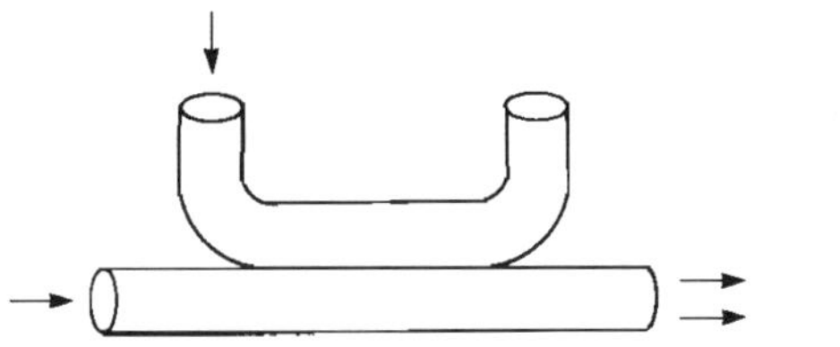

Figure 1.5.16 A method for combining two signals into a common output, similar to the coupler often used with coaxial lines.

A further application of this device is to increase the signalling speed. Digital modulation of a laser diode in *on/off* fashion requires a relatively long time and, at about 500 Mbit and above, the laser can no longer respond accurately. The device used as a switch can permit the diode to remain *on* continuously, the bit signals switching the light beam from the transmission path to a sink, with almost no constraints on switching speed.

A more recent development has seen the appearance of light-frequency generators which have a high order of frequency stability, suitable for use as the local oscillator of a super-heterodyne receiver. The receiver changes the frequency of the detected signal to an intermediate frequency of about 1–2 GHz for subsequent processing. The usual laser is not suitable for superhet purposes, the bandwidth being too wide and the frequency stability quite inadequate. Viable oscillators have now been produced, and receivers can now provide a very useful gain of 20 dB or so over the use of a simple diode detector.

It has been seen that basic path loss can be very low. It is now possible, with that extra 20 dB of gain, to obtain very long paths (e.g. 100–200 km) without the use of repeater stages, and it is probable that all future fibre systems will use superhet receivers.

1.5.3 Future Developments

It is probable that coaxial copper systems will not receive any spectacular changes in design or technique, but these systems will always find a place in 'short-haul' networks for economic reasons. Repeaters are amplifiers of relatively simple design, compared with those for a fibre optic system where a repeater entails demodulating the signal to baseband, and remodulating a new carrier — a relatively expensive operation. Fortunately, fibre systems will use very few, if any, repeaters.

Fibre systems are capable of very low system loss, but many lines of research are proceeding, each of which is aimed at reducing attenuation to the point where no repeaters will be required. Figure 1.5.13 shows the general trend of reducing attenuation as wavelength is increased, with infrared attenuation predominating over Rayleigh scatter as the main source of loss for wavelengths beyond about 1.5 μm. Other materials than silica are being investigated, some of which can have very low losses when operated at wavelengths in excess of 2 μm. The new glasses are metal oxides and halides, and calcogenide glasses, all with very optimistic theoretical figures for attenuation, such as 3×10^{-4} dB/km at wavelengths in the range 2–12 μm.

Improvements in s/n ratios are the object of investigations into the use of higher powered sources, and reductions in the noise generated by lasers.

Wavelength multiplexing, using two or more sources, is being developed, using dichroic filters or gratings to extract a required source from a wide-band carrier of several channels.

Bibliography

BS 415, Safety Requirements for Mains-connected Apparatus.
BS 5425, Coaxial Cable for Wideband Distribution Systems (1986).
BS 6330, Code of Practice for Reception of Sound and Television Broadcasting (1983).
BS 6513, Wideband Cabled Distribution Systems (1984).
BS 6558, Optical Fibres and Cables (1985).
IEC Pubn 96-1, Radio-frequency Cables (1986).
IEC Pubn 728, Cabled Distribution Systems (1982).
IEC Pubn 794, Optical Fibre Cables (1984).

E. P. J. Tozer
Zetrox Broadcast Communications

1.6 TCP/IP Networking

1.6.1 Introduction

The merging of the broadcast, telco and computer industries has brought a lot of new technologies into broadcast systems. This chapter looks at the key aspects of the Internet Protocol (IP) and Ethernet that are used extensively in broadcast control systems and applications.

The intention is to provide a practical understanding of the key concepts of working with and networking IP hosts, but does not explore the concepts needed to build an associated LAN or WAN infrastructure.

This chapter concentrates on IP over Ethernet as the networking method, as today this is the de facto standard for interconnecting hosts.

1.6.2 OSI Seven-Layer Model

1.6.2.1 Open Systems Interconnection (OSI) Model

The OSI seven-layer model is a theoretical view of a communication connection which breaks up the process into seven subprocesses. Whilst in practice real world network systems don't map directly to the OSI model, it's a useful starting point for understanding TCP/IP.

The reason for splitting the communications channel into seven layers is to avoid the need for each application to have to communicate directly with the application on the remote host. Instead, each layer communicates with the layers above and below through defined connections.

The OSI layers with the associated four layers that normally appear in a TCP/IP system are shown in Figure 1.6.1.

1.6.2.1.1 Layer 1 – Physical

Layer 1 defines the connection to the physical medium, for example in Ethernet 10 base T, 10 base 2, etc. These would define the connectors and cables used, e.g. RJ-45 connectors, using unshielded twisted pair, wired 1-1, 2-2, . . . , 6-6, using pairs 1-2 and 3-6. Other non-Ethernet connections include FDDI and Token Ring.

OSI layer	IP function
7 Application	Application
6 Presentation	Application
5 Session	Application
4 Transport	Transport or service
3 Network	Network or routing
2 Data Link	Network Access or Physical
1 Physical	Network Access or Physical

Figure 1.6.1 OSI seven-layer model and IP function.

1.6.2.1.2 Layer 2 – Data Link Layer

The Data Link Layer defines the protocol for data transmission across the Physical Layer.

In the case of Ethernet a unique MAC (Media Access Control) address is provided for each device in order that data packets on the physical connection can be processed by the target device.

CSMA/CD (Carrier Sense Multiple Access/Collision Detect) provides packet management at the physical level. CSMA is a process for preventing a device accessing the network whilst another device is using the network. CD provides a process for managing the transmission when CSMA fails(!).

1.6.2.1.3 Layer 3 – Network

The Network Layer provides a method for independent systems to interact through network addressing schemes and routing between networks. This interconnection is possible even if the underlying Data and Physical Layers are different, e.g. Ethernet-based systems can communicate with Token Ring-based systems. Network Protocols include IP, NETBUEI, IPX and Banyan Vines. In a TCP/IP network devices are identified by IP addresses.

1.6.2.1.4 Layer 4 – Transport

The Transport Layer provides a Quality of Service (QoS) as requested by the Session Layer; if required it can add error correction schemes to the connection. The Transport Layer can also multiplex multiple sessions over the same physical connection. Two of the IP protocols are TCP (Transmission Control Protocol) and UDP (Universal Datagram Protocol).

1.6.2.1.5 Layer 5 – Session

The Session Layer sets up the communication between the two ends of the connection, e.g. creating a duplex (two-way) or simplex (one-way connection).

1.6.2.1.6 Layer 6 – Presentation

The Presentation Layer interfaces between differing formats which may be in use at each end of the connection, e.g. converting between character representations such as ASCII (American Standard Code for Information Interchange) and EBCDIC (Extended Binary Coded Decimal Interchange Code).

1.6.2.1.7 Layer 7 – Application

The Application Layer provides the services required by the system user, such as email (SMTP) and file transfer (FTP).

1.6.3 Introduction to TCP/IP over Ethernet

The concept of applications on hosts interacting over a network appears fairly straightforward (Figure 1.6.2). At first view, however, the applications need to overcome a number of hurdles to interact.

Firstly they will packetise their information at transport level. This process adds headers indicating the port or socket that the application needs to communicate with on the remote host, a particular port being associated with a particular application, e.g. port 80d is used for requesting a web page via HTTP (Hyper Text Transfer Protocol) or port 25d for sending email via SMTP (Simple Mail Transfer Protocol). At transport and service levels the type of transport is also selected, for example UDP (Universal Datagram Protocol) or TCP (Transmission Control Protocol). UDP is used, for example, for data multicast and TCP for HTTP (see Figure 1.6.3).

The next layer of packetisation is for network and routing. Here the source and destination IP addresses are added to the packet so that it can be routed over the network to its destination, and so that the remote host knows where to reply. Note that the destination IP address may be discovered from the destination domain name using DNS (Domain Name System) protocols. As an example the IP address 80.94.193.236 could be found using DNS with the address www.zetrox.com.

The final layer is for physical network access. This is the Ethernet layer where hosts communicate over a physical connection, which may be electrical, optical or RF. At this level physical source and destination addresses are added along with synchronising and check sum data.

1.6.4 Ethernet

Today the most common method for connecting networked devices is Ethernet, though many other protocols such as Token Ring and FDDI are used. Ethernet is essentially a bus communication method where all devices are connected to the same communication channel (though many modern implementations using hubs may blur this definition slightly) (see Figure 1.6.4).

The MAC address is a 6-byte, 48-bit number that is unique to every Ethernet adapter manufactured; if a host wishes to send a data packet to another host it is addressed using the MAC address. The first three bytes of the MAC address are the 'vendor ID' and unique to the manufacturer (e.g. 04008c is NTL) and the last three bytes are the device ID, unique to that device made by a particular manufacturer.

1.6.4.1 Ethernet Physical Implementations

1.6.4.1.1 XXX Base T physical connection

There are a number of physical implementations of Ethernet. Today the most common is 10 or 100 base T – 'ten/hundred base T' (IEEE 802.3I) – which is most commonly cabled using 110 Ω 'UTP' (Unshielded Twisted Pair); 10/100 base T uses eight-pin RJ45 connectors using pairs 1-2 and 3-6 only. N.B. If wiring connectors, great care needs to be taken to ensure that pins 3 and 6 use a twisted pair within the actual cable. One pair of the cable is used for transmit and one pair of the cable is used for receive; 10 base T communicates at a bit rate of 10 Mbit/s and 100 base T at 100 Mbit/s (Figures 1.6.5–1.6.7).

When using 10 or 100 base T hosts connect each connect to a hub to interconnect and communicate (Figure 1.6.8).

1.6.4.1.2 XXX Base T Crossover cable

It is possible to interconnect a pair of hosts without using a hub. In this case a 'crossover' cable is used in which at one end the

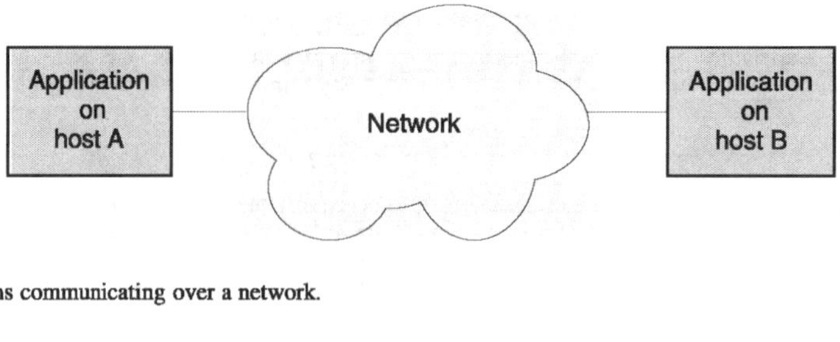

Figure 1.6.2 Two applications communicating over a network.

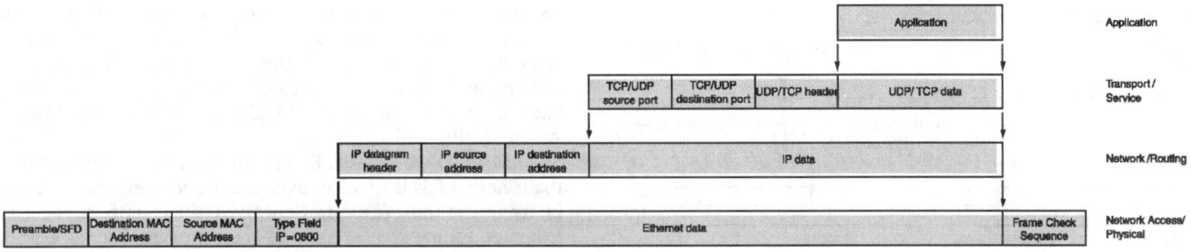

Figure 1.6.3 Data layering/packetisation in TCP/IP/Ethernet.

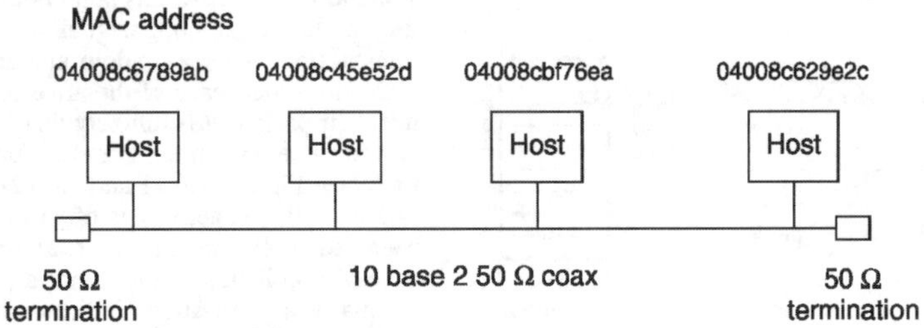

Figure 1.6.4 MAC addresses on 10 base 2 Ethernet bus.

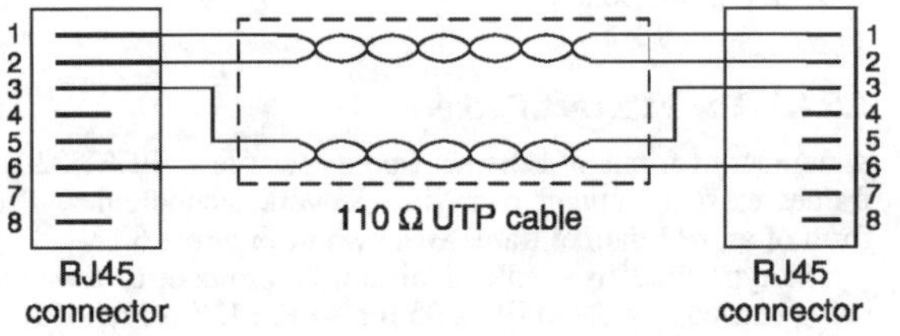

Figure 1.6.5 Cable connection details for 10 base T UTP Ethernet.

Figure 1.6.6 RJ45 connector.

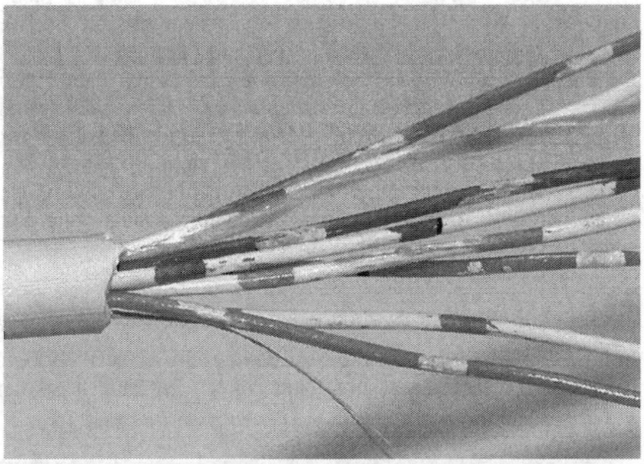

Figure 1.6.7 Wire colours for standard RJ45 cable. N.B. The nylon cord in the cable can be used as an aid to stripping back the cable sleeving.

pairs 1-2 and 3-6 are swapped over (Figure 1.6.9). A crossover cable can also be used to interconnect hubs, though more normally a special 'daisy chain' output is provided or they will auto sense the appropriate pins to use for transmit and receive. Most commonly the maximum cable length for UTP is 100 m, though this may vary according to cable type.

1.6.4.1.3 Other Ethernet physical implementations

Historically 10 base 2 'thin Ethernet' (IEEE 802.3b) and 10 base 5 'thick Ethernet' (IEEE 802.3) 50 Ω co-axial cable Ethernet systems have been common. In thin Ethernet the hosts are

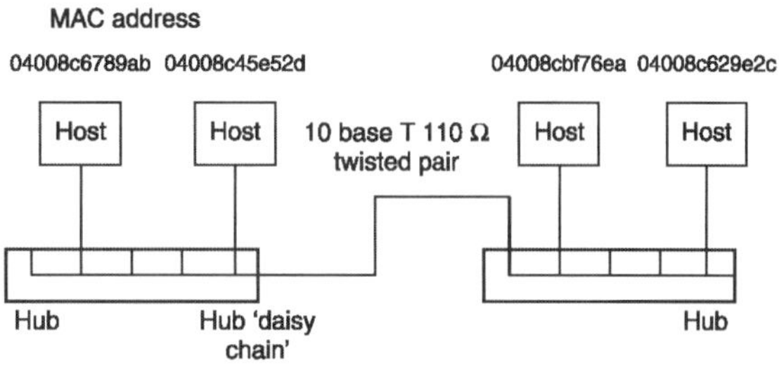

Figure 1.6.8 MAC addresses on 10 base T UTP/hub Ethernet bus.

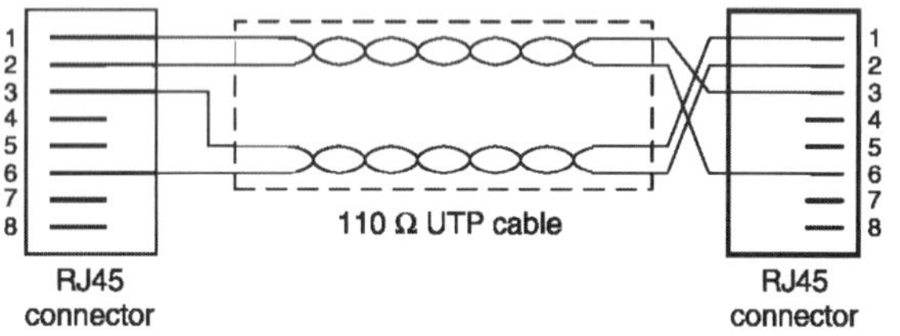

Figure 1.6.9 Ethernet UTP 'crossover' cable.

simply 'Teed' into the co-ax cable (Figure 1.6.4) using 'T-piece' connectors. With the thick Ethernet implementation a 'bee-sting' or 'vampire tap' connector is used to connect the host to the cable via an AUI (Attachment Unit Interface) (see Figure 1.6.10). The maximum cable lengths are 185 m (200 yards, 10 base 2) and 462.5 m (500 yards, 10 base 5); 10 base 2 and 10 base 5 both operate at a maximum bit rate of 10 Mbit/s.

There are numerous other variations in Ethernet cabling systems, including 10 base F (IEEE 802.j) using a fibre-optic connection. These will often be connected via appropriate AUIs.

1.6.4.2 Ethernet in Action

In an Ethernet system all the host devices are effectively connected to the same cable bus. None of the Ethernet hosts has any knowledge of when other hosts may wish to talk to the bus, and clearly if two hosts attempt to put data on to the cable segment at the same time there will be conflict and data corruption. To overcome the cable contention problem Ethernet uses a system called CSMA/CD (Carrier Sense Multiple Access/Collision Detection).

CSMA/CD works as follows: when a host wishes to talk to the Ethernet bus it first listens to the bus to check no other host is using the bus (Carrier Sense); it then starts to send its Ethernet frame. If a second host on the bus listens simultaneously and sends a packet to the bus at the same time the two packets 'collide' on the bus. Two hosts simultaneously transmitting to the bus will cause an abnormal signal level (Collision Detect), which indicates to the two affected hosts that a collision has occurred. At this point both hosts 'back off', transmitting a 32-bit jamming signal onto the bus, and then stop transmitting. After a random amount of time, between 0 and 1024 slot times, each of the affected hosts attempts to retransmit their packet. It is unlikely that the two random pauses will be the same and so the hosts are unlikely to collide a second time (see Figures 1.6.11 and 1.6.12).

This collision activity is often indicated on a hub, typically by a red 'collision' indicator, and so long as there is not an excess of collisions there is not a problem with transmission. Normally a 'utilisation' indicator is provided on a hub and generally with an Ethernet segment that is simultaneously accessed by many hosts the maximum utilisation achievable is around 40–70%, though in a point-to-point implementation of Ethernet with a single host pair access levels approaching 100% are achievable.

1.6.4.3 The Ethernet Packet

A number of forms of Ethernet (more accurately IEEE 802.X) frames exist to support overlying network technologies. The form of an IP Ethernet frame is shown in Figure 1.6.13.

The type field[1] is used to indicate the protocol of the data being carried, e.g. 0800 IP, 0806 for ARP, 8137 Novell IPX.

1.6.4.4 Connecting Ethernet cable segments

For most modern networks the UTP limitation of 100 m maximum cable length is a severe limitation. To overcome the

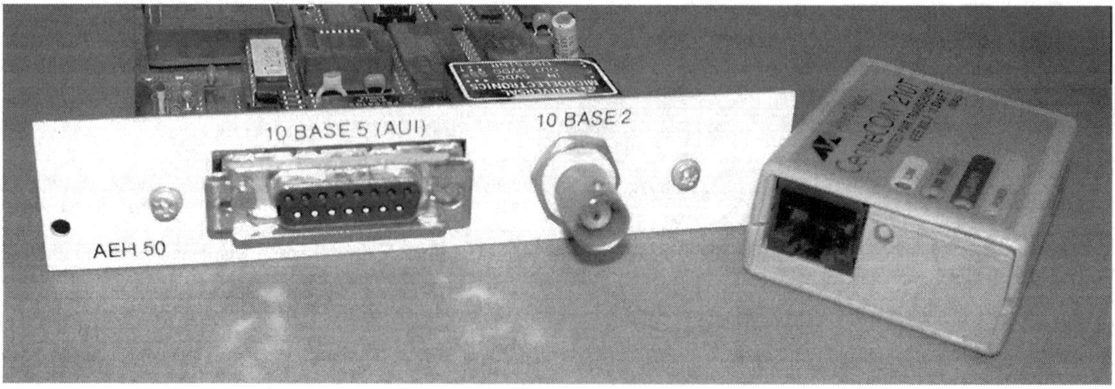

Figure 1.6.10 NIC with 10 base 5 and 10 base 2 connectors, with AUI for 10 base T.

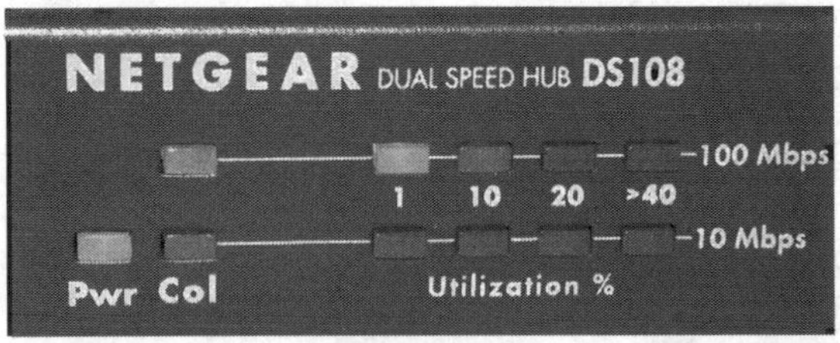

Figure 1.6.11 Ethernet packet collision occurring at low utilisation.

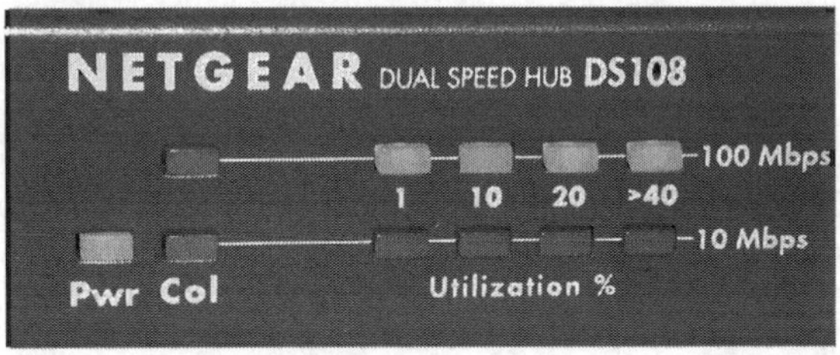

Figure 1.6.12 High network bit rate,with no collisions.

problem a number of solutions for interconnecting networks are possible.

1.6.4.4.1 Repeaters

Repeaters are essentially OSI Physical Layer devices that operate at the electrical (or optical) level and simply act to buffer and re-amplify the signal to be passed on to the next cable segment (Figure 1.6.14). Repeaters are unintelligent devices passing all packets between segments.

1.6.4.4.2 Bridges/Switches

Bridges/switches are essentially OSI Data Link Layer devices that operate at the MAC address level (Figure 1.6.15).

A bridge makes an 'intelligent' decision whether to pass an Ethernet packet between cable segments based on the destination MAC address of the packet, plus a knowledge of which physical connection has which MAC address present. Thus a packet sent from host A to host B would not be forwarded to hosts C and D.

Bridges and routers can be used to segment networks into VLANs (Virtual Local Area Networks). This can allow, say, hosts A, B, C and F to form an independent network from hosts D, E, G and H. The two VLANs are isolated from one another by the switch, which can bring benefits in terms of security and traffic separation. If a need does arise to communicate between hosts on the two VLANs this could be achieved through the use of a router (see Figure 1.6.16).

1.6.4.4.3 Repeater/Hub/Bridge/Switch

In modern Ethernet connecting hardware the functionality of devices is often blurred, a hub operating as a bridge and maybe containing low-level Ethernet isolating ability to prevent non-functioning or 'jabbering' sections of network taking out the network sections on other hub connectors. Switches can also contain intelligence that could, for example, disable an Ethernet connection that contained a duplicate address.

In a broadcast environment some care needs to be taken when selecting a network connection device; the more intelligence a device contains the longer it is likely to take to 'come up' after a power down. It may be acceptable in an office environment of networked PCs and servers for the network to be down for a couple of minutes while a switch reboots, but this downtime is unlikely to be considered acceptable in a live broadcast transmission environment.

1.6.5 Internet Protocol

Whilst MAC addressing at cable level is sufficient to move Ethernet packets around a single piece of network cable, a higher level addressing scheme is needed once different Ethernet cabling becomes interconnected. The de facto standard today is IP (Internet Protocol).

1.6.5.1 Internet Protocol RFCs (Requests for Comment)

The Internet and Internet Protocol standards are defined in documents called RFCs (Requests for Comment). The RFC process is managed by the IETF (Internet Engineering Task Force) and the RFC documents are freely available from them[2] via Internet download.

Length	Content
7	Preamble
1	Start frame delimiter
6	Destination address (MAC unicast/multicast/broadcast)
6	Source address (MAC)
2	Ethernet data type (length in the IEEE 802.3 protocol)
46–1500	IP (or other) data
4	Frame check sequence

Preamble/SFD	Destination MAC Address	Source MAC Address	Type Field IP=0800	Ethernet data	Frame Check Sequence

Figure 1.6.13 Ethernet packet content and structure.

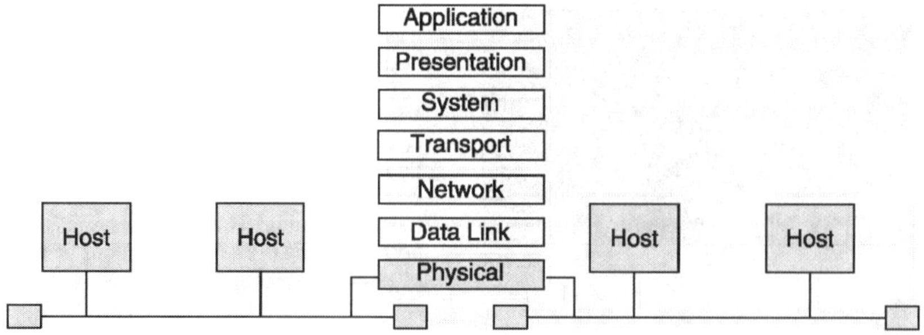

Figure 1.6.14 Repeater linking Ethernet cable segments.

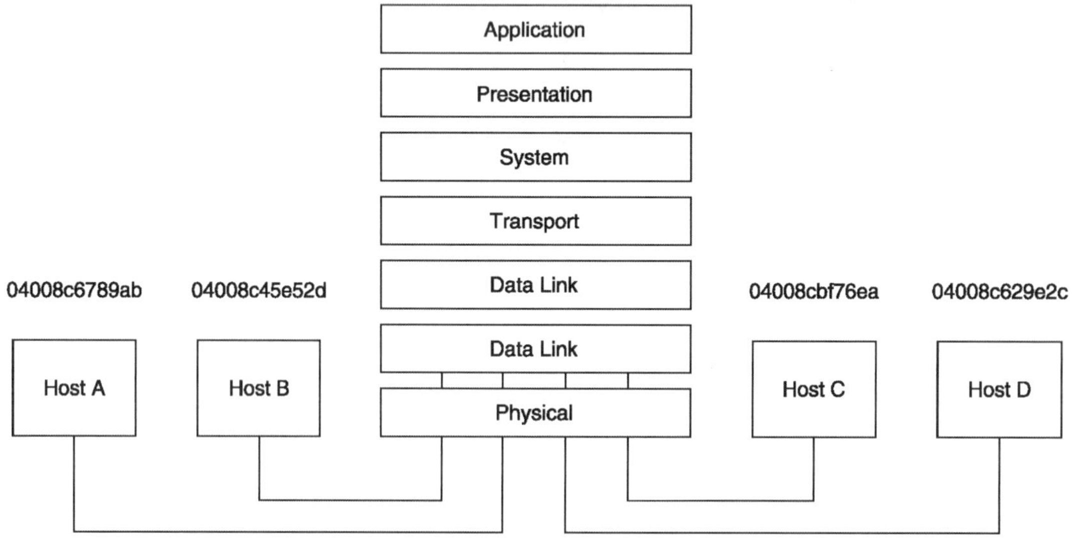

Figure 1.6.15 Bridge linking Ethernet cable segments.

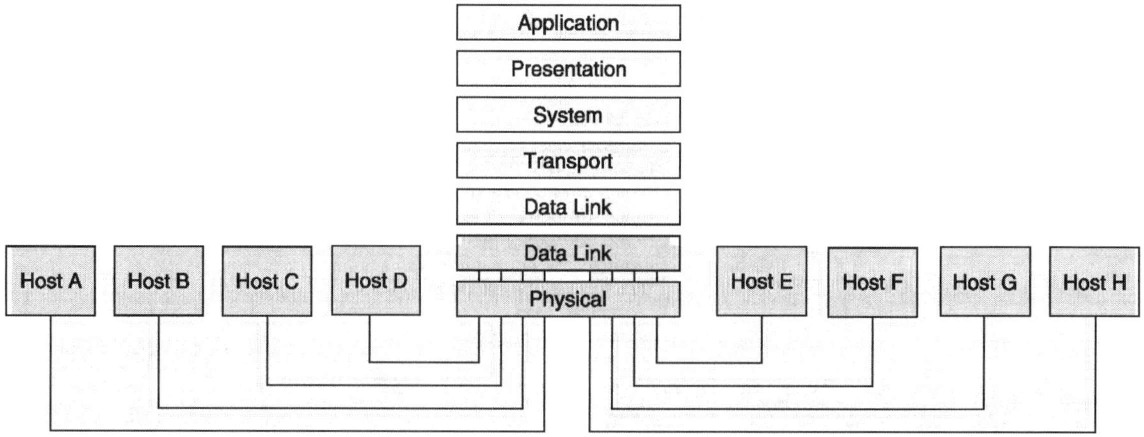

Figure 1.6.16 Network segmented into VLANs.

1.6.5.2 The IP Datagram

The IP datagram is the packet of IP information to be transmitted by Internet Protocol. It occupies the 46 to 1500 bytes of data in an Ethernet packet and depending upon its size may be spread across more than two or more Ethernet packets. The construction of an IP datagram is shown in Figure 1.6.17.

An explanation of the full detail and use of the IP header content is beyond the scope of this chapter. Further details can be found in Ref. 3. Some of the more interesting items are:

- TTL (Time To Live) is set when the packet is created and indicates how many seconds this packet may stay alive on the Internet; it should be decremented by at least one when the packet passes through a router. If the count reaches 0 the packet is discarded and an ICMP (Internet Control Message Protocol) message is returned to the sender.
- IP Protocol indicates the protocol type in use, e.g. 0x6 for TCP, 0x11 for UDP, 0x01 for ICMP.
- IP source and destination addresses, the 4-byte IP addresses of the communicating hosts.

1.6.5.3 IP Addressing

1.6.5.3.1 Introduction to IP Addressing

In an IP network each device has a unique 'IP address', often referred to as a 'dotted quad', for example 172.17.246.157. The addresses range from 0.0.0.0 to 255.255.255.255, though some addresses are reserved for special purposes.

The address is split into two parts. The first part is the network address (in the example above, 172.17) and the second part the device or host address (in the case above, 246.157). Generally only devices that have the same network address can communicate directly with each other. If a device needs to communicate with a device with a different network address (i.e. another network) the two networks need to be connected by routers (gateways).

1.6.5.3.2 IP Address classes

The full range of Ethernet addresses is split into a number of 'classes'. In the descriptions below, nn is the 'network ID' component of the address and hh the 'host ID' component of the address. Note that on every network the all 0s address (lowest address) is reserved for the network address and the all 1s address (highest address) for the host address, thus on any IP network two addresses are unusable by hosts.

Class A addresses range from 1.hh.hh.hh to 126.hh.hh.hh. There are 126 class A networks (network address 0 has no meaning, network 127 is reserved for loopback testing), each of which can have 16,777,214 ($256^3 - 2$) different hosts attached. In the early days of IP address allocation the class A addresses were largely allocated to government bodies and the large corporations, e.g. UK DHSS (Department of Health and Social Security) has Class A address 51 and German car manufacturer, Mercedes Benz, Class A address 53.

Class B addresses range from 128.nn.hh.hh to 191.nn.hh.hh. Thus a device with an IP address of 157.123.78.32 would have a network ID of 157.123 and a host ID of 78.32. Each of the networks can have 65,534 ($256^2 - 2$) hosts.

Class C addresses range from 192.nn.nn.hh to 233.nn.nn.hh. A device with an IP address of 196.123.78.32 would have a network ID of 196.123.78 and a host ID of 32. Each of the networks can have 254 hosts.

Class D addresses range from 224.xm.mm.mm to 239.xm.mm.mm. Class D addresses are reserved for multicasting purposes, the last 23 bits of the IP address being mapped to the MAC address of the host.

Class E addresses range from 240.xx.xx.xx to 255.xx.xx.xx. Class E addresses are reserved for experimental use.

1.6.5.4 IP communication over Ethernet

As described previously, at Physical and Data Link levels the communication between hosts takes place using the hosts' MAC address. If a device wishes to communicate at IP address

Length	Content
4 bits	IP version type
4 bits	Internet header type
1 byte	Type of service
2 bytes	Total datagram length
2 bytes	Identification
3 bits	Flags
13 bits	Fragment offset
1 byte	Time To Live (TTL)
1 byte	IP Protocol
2 bytes	Header checksum
4 bytes	Source IP address
4 bytes	Destination IP address
variable	Options
1–3 byte	Padding
variable	IP data

IP datagram header	IP source address	IP destination address	IP datagram header	IP data

Figure 1.6.17 IP datagram content and structure.

level then some translation is required between the IP (OSI Network Layer) and the MAC level (OSI Data Link Layer) addressing. This linking of IP to MAC addresses is done through ARP (Address Resolution Protocol).

1.6.5.4.1 Address Resolution Protocol (ARP) (RFC 826)

ARP is the first of the Internet Protocols required to use IP over Ethernet and allows a host's MAC address to be associated with a host's IP address (Figure 1.6.18).

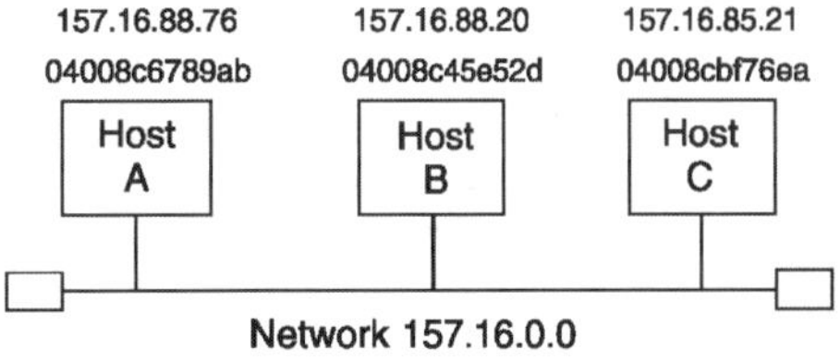

Figure 1.6.18 ARP associates an MAC address and an IP address.

When a host wishes to communicate with another host at IP level an ARP sequence is used. In the example above, host A wishes to communicate with the host with IP address 157.16.85.21. Host A issues an ARP packet onto the Ethernet bus. An ARP packet is an Ethernet broadcast packet, so all hosts on the network receive the packet and read the content. The ARP packet contains information indicating that it is an ARP packet, the MAC address (04008c6789ab) and IP address (157.16.88.76) of the sending host A, and also the IP address of the destination host. All hosts with the exception of the one with the correct destination IP address ignore the packet. The host with the matching IP address responds with a unicast Ethernet packet to the MAC/IP address of host A. This packet in turn contains the required MAC address of the host with the known IP address. All further communication between hosts A and C can now continue in unicast, so that other stations on the network can ignore the packets at Data Link (MAC) level.

Most devices contain an ARP cache (ARP table) of known IP/MAC address pairings so that undesirable broadcast messages can be kept to a minimum (Figure 1.6.19). A listing of the ARP table can be produced using, for example, the Windows command line command arp –a (see later section for further details on the arp command).

Potentially the caching of MAC/IP address pairs can cause problems should a faulty device be replaced on the network. Typically the replacement device will have the same IP address but a different MAC address, causing the ARP tables of other devices on the network to contain incorrect MAC address information and so to communicate with the wrong MAC address for the cached IP address.

To overcome this problem devices, typically, will periodically clear their ARP tables if no communication has occurred with a particular MAC address for some time. ARP tables can also often be 'flushed' manually using an ARP command.

1.6.5.5 IP Routing

IP address routing takes place at OSI Network Layer 3 (Figure 1.6.20).

If host A wishes to talk to host B it can send a message to IP address 157.16.88.20 as, through the ARP process, it will know the MAC address of the host. If, however, host A wants to communicate with host C it will not be able to do so directly as the two devices are on different physical systems and consequently will be unaware of each other's MAC addresses.

The function of a router is to be able to connect to two different physical networks. In Figure 1.6.21 the router is

```
Interface: 172.17.248.20 on Interface 2
  Internet Address      Physical Address      Type
  172.17.248.1          00-20-aa-04-00-93     dynamic
  172.17.248.2          00-20-aa-04-00-e2     dynamic
  172.17.248.13         00-20-aa-06-04-d2     dynamic
  172.17.248.14         00-20-aa-06-04-d2     dynamic
  172.17.248.100        00-80-d4-00-f7-33     dynamic
```

Figure 1.6.19 Example ARP table listing.

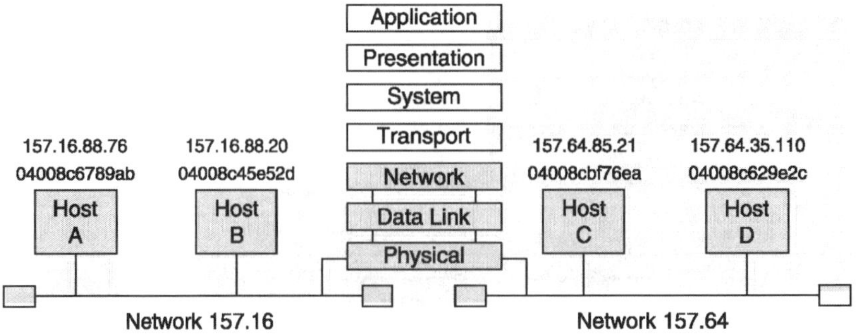

Figure 1.6.20 Routing at IP address level.

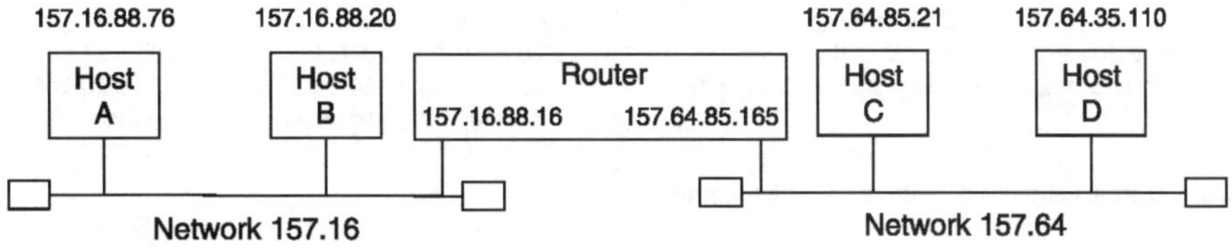

Figure 1.6.21 Router interconnecting networks.

connected to both network 157.16 and to network 157.64. All hosts on network 157.16 are configured with a 'default gateway' address of 157.16.88.16. The function of the default gateway is that any IP packet that is addressed to a device on a different network is sent to the default gateway. The same is true for network 157.64; all devices are configured with a default gateway address of 157.64.85.165, and all packets for other networks are sent to this address.

With a router connecting the two IP networks, if host A wishes to communicate with host C it will send an IP packet to its default gateway address, that of the router. The router will receive the packet and pass it onto its other connected network, where it will be received by host C. The reverse communication uses the same process to send packets form host C to host A.

1.6.5.5.1 Multiple hop routing

In the case where a device wishes to communicate with a host that is not on an adjacent network, multiple hop routing can be used. In the example in Figure 1.6.22, if host A wishes to communicate with host E then, as host E is not on the same network, it will send an IP packet to its default gateway, router A. Router A will see that the packet is not destined for network 157.64 and pass the packet on to its default gateway, router B. Router B will

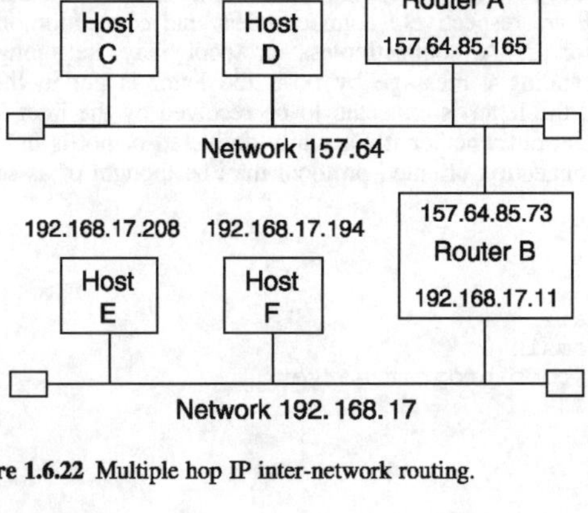

Figure 1.6.22 Multiple hop IP inter-network routing.

in turn see that the destination address for the packet is network 192.168.17 and forward the packet to host E. Thus packets can traverse interconnected networks until reaching their destination.

1.6.5.5.2 Subnets and subnet masks

In the case of large networks, e.g. class A with up to 16,777,214 ($2^{24} - 2$) connected hosts, it is unfeasible to connect them all to the same physical system, with interconnections by repeaters and bridges. To make handling of large networks easier, subnetworks can be used.

The concept of a subnetwork is to break down the address range of a large network into smaller parts and interconnect the parts using routers. Thus, for example, network 157.16.0.0 could be broken into 256 subnetworks 157.16.0.0, 157.16.1.0, 157.16.2.0 ... to ... 157.16.255.0. In order that a device understands that it is connected to a subnetwork rather than its main class X network, a 'subnet mask' is used on the device.

In the example of splitting network 157.16.0.0 into 256 subnetworks the subnet mask would be 255.255.255.0. In the subnet mask the 255s indicate the network addresses and the 0s indicate the host addresses. Thus the device with IP address 157.16.123.76 would be on network 157.16.123 with a host address of 76.

It is possible to split networks into smaller chunks than 256. To understand how the subnet mask works under these circumstances it is best to consider it as a binary number. Figure 1.6.23 shows, in binary, the example IP address of 157.16.123.76 with subnet mask 255.255.255.0.

To create smaller subnets, the relative number of 1s and 0s in the subnet mask is changed. Consider the example of class C network 192.168.17.0. This could be split into 16 subnetworks using the subnet mask 255.255.255.240. Figure 1.6.24 shows the IP address, subnet mask, network and host addresses for the IP address 192.168.17.76.

Thus using the subnet mask 255.255.255.240, the class C network can be subdivided into 16 smaller networks, 192.168.17.16, 192.168.17.32, 192.168.17.48, ..., to 192.168.17.240.

The number of hosts that each of these 16 networks can support is 16. However, the address with all bits set to 0 is reserved for the network identity and the address with all bits set to 1 (this would be 1111b, 15d in the example above) is used as the network broadcast address, thus only 14 hosts are actually available on the network.

An alternative description of a network and mask can be used. The network 192.168.17.16 with a subnet mask of 255.255.255.240 may also be referred to as the network 192.168.17.16/28, indicating that the network address is 192.168.17.16 and the subnet mask can be represented by the 32-bit binary number starting with 28 1s followed by four 0s (Figure 1.6.25).

```
                 Dotted quad        Binary notation
IP address  : 157. 16.123. 76 = 10011101.00010000.01111011.01001100
Subnet mask : 255.255.255.  0 = 11111111.11111111.11111111.00000000
                                |-----network address-----|-host ID
```

Figure 1.6.23 Binary representation of IP address and subnet mask.

```
                 Dotted quad        Binary notation
IP address  : 192.168. 17. 76 = 10011101.00010000.01111011.01001100
Subnet mask : 255.255.255.240 = 11111111.11111111.11111111.11110000
Network     : 192.168. 17. 64 = 10011101.00010000.01111011.01000000
Host        :              12 = 00000000.00000000.00000000.00001100
```

Figure 1.6.24 Splitting a class C network into subnets.

```
                 Dotted quad        Binary notation
Subnet mask : 255.255.255.240 = 11111111.11111111.11111111.11110000
                                ----------- 28 1s -----------
```

Figure 1.6.25 Derivation of alternative subnetwork annotation.

1.6.5.6 Host names and Aliases

Thirty-two-bit, 4-byte dotted quad IP addresses are used for IP network connections and are a good solution for computer systems; however, for humans it can be difficult remembering the IP address 57.1.16.99 when what we want to do is connect to Fred Smith's computer to copy a file. To help humans cope with IP addressing, computers can contain a 'hosts' file, which is a list of IP addresses and 'alias' names associated with those addresses (Figure 1.6.26). The hosts file will be /etc/hosts in UNIX systems and C:\WinNT\System 32\drivers\etc\hosts on Windows NT systems.

```
57.1.16.98        jenny        jennyjones  jjones
#172.17.248.101   termserver1  termserv1
57.1.16.100       termserver1  termserv1
57.1.16.1         encoder1     enc1        enc01
57.1.16.2         encoder2     enc2        enc02
57.1.16.99        fred         fredsmith   fsmith
57.1.16.4         encoder4     enc4        enc04
```

Figure 1.6.26 Sample hosts file.

The hosts file allows us humans to use one of the alias names instead of the IP address; thus to get a file from Fred Smith's machine one could equally use any of the commands: ftp 57.1.16.99, ftp fredsmith, ftp fred etc.

The # in the second line of the sample hosts file indicates that this entry should be ignored; it has been 'hashed out'.

1.6.5.7 Sockets, Ports and Services

It is very common in a networked system to need to establish more than one IP connection between two machines. To enable this IP supports services (also known as sockets and ports, the names are freely interchangeable).

Services allow multiple connections between two machines by identifying each with a unique number, the 'service number'. Some service numbers are established for use in particular circumstances, as defined in RFC 1060, e.g. Telnet is defined as using port 23, others are undefined. Port numbers have a maximum value of 65,535. The IANA (Internet Assigned Numbers Authority) website[4] is a useful source of information.

Thus a Telnet connection could be established to Fred Smith as: 57.1.16.99:21. At the same time a second connection could be established on port 65,432 as 57.1.16.99:65432 for some other, RFC 1060, undefined purpose (see Figure 1.6.27).

1.6.5.8 TCP and UDP

1.6.5.8.1 Introduction

UDP (Universal Datagram Protocol) and TCP (Transmission Control Protocol) are OSI layer 4, Transport functions which control how IP packets are sent over an IP network. UDP and TCP are respectively connectionless and connection-oriented protocols. A connectionless protocol may be thought of as sending a message by post; the letter is put in the post box, the letter is expected to be received by the intended recipient but whether it's actually delivered or not is unknown. A connection-oriented protocol may be thought of as sending

```
ftp          21/tcp
telnet       23/tcp
smtp         25/tcp    mail
nameserver   53/udp    domain
bootp        67/udp    # boot program server
tftp         69/udp
```

Figure 1.6.27 Sample section from services file.

a message by fax; here a connection is established between the sending and receiving machines, if the message send completes successfully the message is known to have been received by the receiving machine using a handshaking process.

1.6.5.8.2 Universal Datagram Protocol (UDP)

UDP is a connectionless protocol which does not guarantee delivery; however, its low overhead, potentially higher data throughput and the fact that it does not need a bidirectional connection make it the better choice for some applications.

One of the key uses for UDP is in data broadcast applications over a satellite or terrestrial transmission channel where, typically, the data is sent at high speed, over a reliable connection with no return handshaking path.

1.6.5.8.3 Transmission Control Protocol (TCP)

TCP is a 'connection-oriented' protocol, which means a connection and 'handshaking' process is established between the transmitting and receiving devices' TCP, guaranteeing IP packet delivery. The vast majority of IP applications use TCP.

The connection-oriented aspect of TCP provides the possibility of a flow control mechanism that can optimise the packet rate over the medium, but also adds overhead and slows down transmission. The connection-oriented aspect also means that TCP must use a two-way communication path.

1.6.5.9 IP Multicasting (RFC 1112)

Multicasting is an IP technique that allows a transmitted IP packet to be received by multiple hosts; this should not be confused with IP broadcast, where the transmitted packet is to be received by all hosts.

The concept of IP multicast is rather different to unicast and broadcast. In a multicast hosts choose to join (and leave) multicast groups. The multicast group is defined by an address in the same way that a unicast or broadcast IP address defines its recipients. Multicast IP addresses are group D addresses, so can range from 224.0.0.0 to 239.255.255.255. RFC 1700 defines a set of standard and reserved multicast addresses.

To create an Ethernet multicast frame the destination MAC address starts 01005E and has the last 23 bits of the IP address appended. Thus an IP multicast to IP address 225.16.1.57 (E1.10.01.39 hex) would be carried in an Ethernet frame with a destination MAC address of 01005E100139.

The multicast receiving host 'joins' the group of multicast receivers and will receive IP packets destined for 225.16.1.57 and receive Ethernet packets destined for MAC address 01005E100139.

For an IP network to function with multicast transmissions, all its routers must be multicast enabled (generally routers do not have multicast ability), as multicast routers must support specific multicast routing protocols, DVMRP (Distance Vector Multicast Routing Protocol) or MOSPF (Multicast Open Shortest Path First).

Figure 1.6.28 shows an IP network enabled for multicast. Hosts A and S are multicast transmitters, hosts M, J and R are multicast receivers. IGMP (Internet Group Management Protocol) is used by the hosts and edge routers to determine multicast group memberships. A router broadcasts an IGMP 'query message' and hosts respond with IGMP 'host membership reports', thus a router knows which multicast packets to forward to a particular network.

In Figure 1.6.28 host A is sending multicast packets to the 'vertical stripe' group, of which hosts J and M are receiving members. The multicast packets will be propagated from host

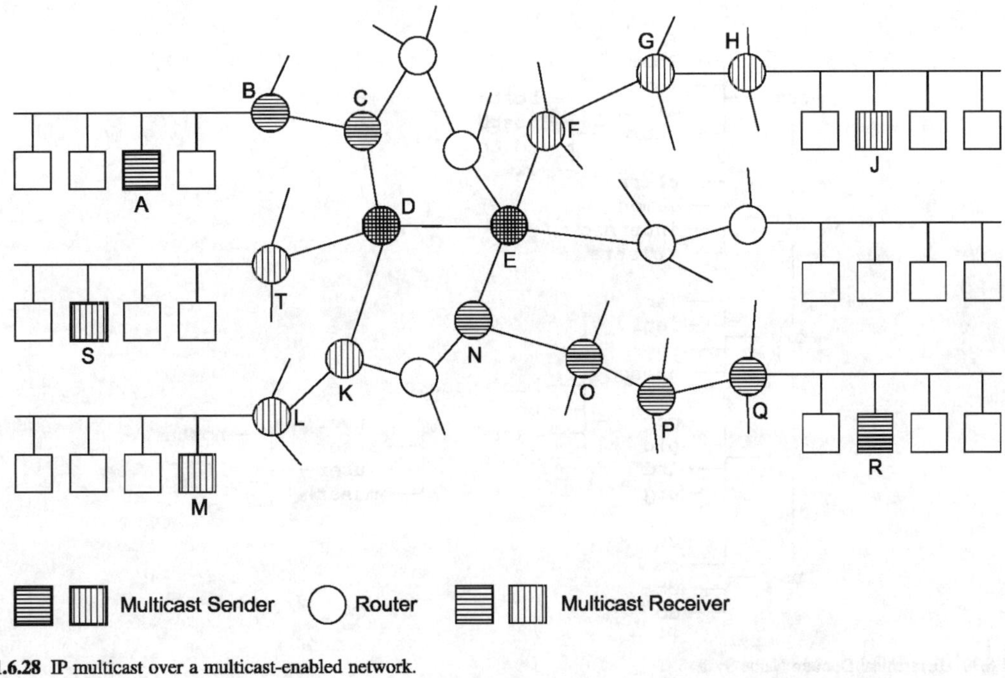

Figure 1.6.28 IP multicast over a multicast-enabled network.

A over the network via routers T, D, K and L to host M, and via T, D, E, F, G and H to host J. It should be noted that when the multicast packet reaches router D it is duplicated and forwarded to both the routers E and K, to continue its journey to the receiving hosts.

As IP multicasting sends each IP packet to multiple hosts, the handshaking TCP protocol cannot be used; therefore multicasting uses UDP datagrams.

Generally, in a broadcast environment where the IP multicast is being transmitted over, say, a DVB path, the multicast UDP data will be generated locally to the DVB multiplexer, avoiding the need to consider multicast IP routers and network traffic issues (Figure 1.6.29).

1.6.5.10 Domain Name System (DNS)

DNS is a method of associating user-friendly host names with IP addresses, the most familiar of which is the now ubiquitous www.mysite.com type name. The domain name system does not just apply to WWW (World Wide Web) addresses but to any host for any IP purpose, provided the system supports DNS. IP addresses and domain names may be used interchangeably.

DNS is a flexible hierarchical system that allows distributed management of the name space; there is a root level containing root domains such as com, net, org and also the two-letter national domains, e.g. uk and us. Each domain can be independently administered and is required to be supported by a primary and secondary name server (Figure 1.6.30).

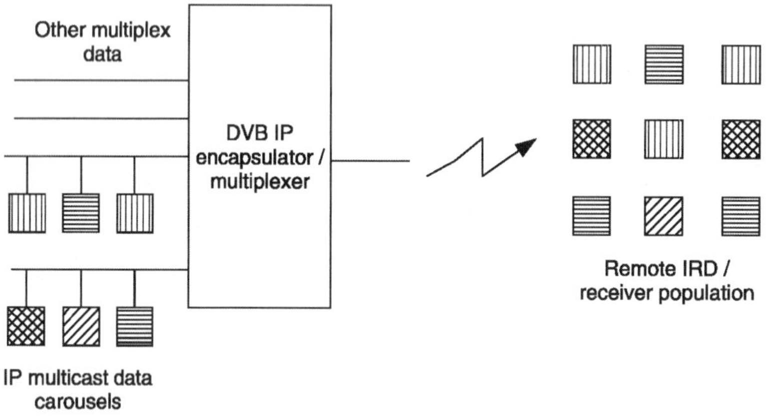

Figure 1.6.29 Typical multicast structure in DVB data broadcast environment.

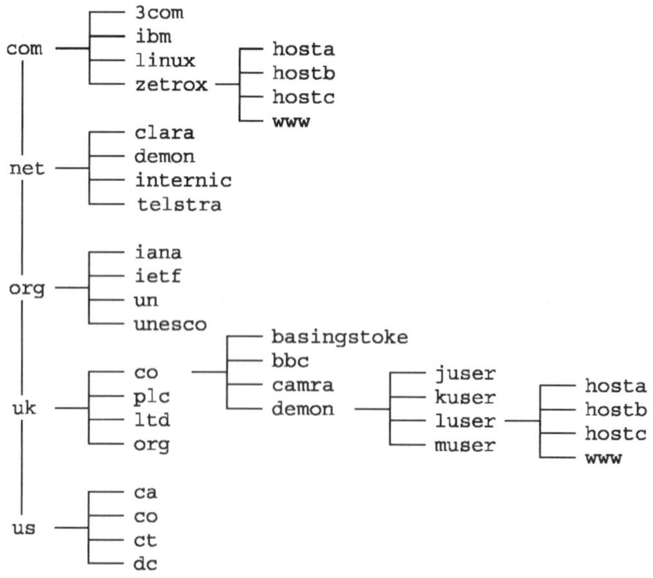

Figure 1.6.30 Hierarchical Domain Name System.

All DNS names are case insensitive and at any level a label (name) may appear only once, though the same label may appear in different branches of the structure; thus co appears as co.uk and as co.us; the domain administrators use co for unrelated purposes, COmpany in uk and COlorado in us. Each node in the tree structure is mapped to an IP address by the name server responsible for that node. The mappings to IP addresses do not need to be unique; both zetrox.com and www.zetrox.com are mapped to the IP address 80.94.193.236.

The process of converting a DNS name to an IP address is executed by querying the name with a Resolver. It is impractical for a local Resolver to know the mapping of every DNS name in the worldwide structure, so if a name is unknown to a Resolver the query is passed up and down the structure until the resolver containing the domain name is located.

The process of locating the record may be done 'iteratively', in which case the host requesting the IP address is passed up and down the DNS structure from name server to name server until the record is found, or 'recursively', where the name server manages the search and passes the discovered IP address directly to the querying host.

1.6.5.11 FTP and TFTP

1.6.5.11.1 FTP (File Transfer Protocol)

FTP (File Transfer Protocol) provides a method to transfer files between machines. FTP requires an FTP server machine that an FTP client can log into. The FTP server is normally provided by default on a UNIX machine; on Windows NT machines it can be installed as one of the Peer to Peer Web Services.

To start an FTP session use the command ftp hostaliasname or ftp ipaddress. The FTP server will request a userid and password to establish the session.

Once logged into an FTP server there are many commands to help transfer files; these can be listed by typing help at the FTP command prompt. Some of the more commonly used commands are:

help	lists ftp commands
put filename	sends file from client to server
get filename	gets file from server to client
lcd dirname	local change directory (on the client)
cd dirname	change directory (on the server)
dir	list files on the server
ls	list files on the server
mput filename	multiple file put, can be used with wild cards?*
mget filename	multiple file get, can be used with wild cards?*
binary/ascii	before transferring files using FTP the transfer type should be set to binary; ASCII type file transfer will convert between Windows and Unix text formats, changing some of the file data.

1.6.5.11.2 Anonymous FTP

Many FTP servers support 'anonymous FTP'. This is a method of allowing log-in and file transfer where the client doesn't have an account on the FTP server. To log-in using anonymous FTP the log-in userid is 'anonymous' and the log-in password is the client's email address, e.g. 'fred.smith@abc.com'.

1.6.5.11.3 TFTP (Trivial File Transfer Protocol)

FTP works using TCP. TFTP (Trivial File Transfer Protocol) is a simplified version of FTP that uses UDP, to allow fast file transfer over a unidirectional connection, without handshaking or error correction.

1.6.5.12 Telnet

If supported by the remote host, Telnet allows remote command line access to that host. Normally there is a log-in process requiring a userid and password to be submitted, e.g. telnet 157.1.16.11 or telnet hostaliasname.

1.6.6 Useful Network Commands

There are very many commands available to work with networked machines and help diagnose network problems; this section contains some of the more useful and frequently used commands for working with and troubleshooting IP network systems.

The commands and their responses illustrated below are all taken from a DOS/Windows machine; on a UNIX or Linux platform the detail of the command and response text may vary.

For more details on these commands, at the command line enter command – help.

1.6.6.1 arp

The ARP (Address Resolution Protocol) command allows the user to view and edit the ARP table of the machine. In order to stop a machine attempting to communicate with a non-existent host it can sometimes be useful to delete an ARP table entry when an IP host has been replaced with another on the same IP address but having a different MAC address.

1.6.6.2 ping

The ping (Packet InterNet Groper) command gets its name from its analogous use to a submarine 'pinging' with sonar to detect a reflection, for example from another submarine. Ping fires an ICMP packet to the destination host, which will usually return a packet to the sending host. Though note that, for reasons of security, the target host may be configured not to respond to pings. If a ping response is received then it is known that the network between the two hosts is functioning. The successful ping response will contain ping loop trip time and time to live information.

Below are a number of possible responses to ping commands:

- Reply from 209.17.248.1: bytes = 32 time = 508 ms TTL = 110 – indicates successful communication with the remote host.
- Destination host unreachable – indicates that the network on which the remote host resides could not be contacted;

this would indicate the remote host was on a different network/subnetwork and that there was no default gateway configured.

- Unknown host xxxxxx – indicates that the alias name used for the ping could not be found in the hosts file or on the configured Domain Name Server.
- Request timed out – there was no reply from the host pinged.

1.6.6.3 ipconfig

Ipconfig displays the configuration of all network adaptors on a host (Figure 1.6.31).

1.6.6.4 netstat

The netstat command provides a view of all existing IP connections on a machine (Figure 1.6.32).

1.6.6.5 tracert

Tracert uses an ICMP packet with a varying TTL to return router information from all routers between hosts (Figure 1.6.33). Initially tracert sends a packet with a TTL of 1; this packet is killed at the first router, which returns kill notification to the sending host, plus details about itself. Tracert then sends a packet with a TTL of 2, which is terminated by the second

```
Windows NT IP Configuration
Ethernet adapter El90x1:
      IP Address. . . . . . . . . : 172.17.248.20
      Subnet Mask . . . . . . . . : 255.255.0.0
      Default Gateway . . . . . . : 172.17.248.1
```

Figure 1.6.31 Sample ipconfig report.

```
Active Connections

  Proto  Local Address     Foreign Address                State
  TCP    ws10999:1031      berp-fm11.dial.aol.com:13784   ESTABLISHED
  TCP    ws10999:1037      205.188.45.197:5190            ESTABLISHED
  UDP    ws10999:1001           *:*
  UDP    ws10999:1900           *:*
```

Figure 1.6.32 Sample netstat report.

```
> tracert www.zetrox.com

Tracing route to www.zetrox.com [209.81.157.205]
over a maximum of 30 hops:
  1  2499 ms  2582 ms  2458 ms  rt-syd06.proxy.aol.com [202.67.64.185]
  2  2513 ms  2307 ms  2513 ms  accesss2-syd-FA5-0-0.proxy.aol.com 202.67.64.252]
  3   275 ms   384 ms   261 ms  pos1-1-0.bdr1.hay.connect.com.au [203.8.183.122]
  4   357 ms   261 ms   357 ms  g2-0-15.sybr2.netgate.net.nz [202.50.116.201]
  5   261 ms   260 ms   248 ms  p1-1.sybr3.global-gateway.net.nz [202.50.119.85]
  6   934 ms   646 ms   672 ms  p1-1.pabr2.global-gateway.net.nz [202.50.116.194]
  7   412 ms   618 ms   508 ms  g0-0-0.pabr3.netgate.net.nz [202.50.245.24]
  8   672 ms   714 ms   440 ms  ge-1-1-0.128.br2.PAO2.gblx.net [64.209.88.233]
  9   426 ms   425 ms   701 ms  so6-0-0-2488M.cr2.PAO2.gblx.net [207.136.163.125]
 10  1222 ms   975 ms   837 ms  pos0-0-2488m.cr1.CHI1.gblx.net [208.49.59.242]
 11   948 ms   480 ms   701 ms  so0-0-0-2488M.br2.CHI1.gblx.net [208.49.59.206]
 12   549 ms   536 ms   521 ms  atm0-0.225ohio.megsinet.net [204.246.198.18]
 13   467 ms   522 ms   522 ms  core1-ATM-FA1-0.grandrapids.corecomm.net
[216.214.85.16]
 14   494 ms   522 ms   480 ms  209.81.240.75
 15   480 ms   522 ms   508 ms  209.81.240.74
 16   508 ms   522 ms   535 ms  node8.fastdnsservers.com [209.81.157.205]

 Trace complete.
```

Figure 1.6.33 Sample report from tracert command.

router. This process repeats until the packet has sufficient TTL to reach its intended destination.

References

1. http://standards.ieee.org/regauth/ethertype/type-pub.html
2. Internet Engineering Task Force, http://www.ietf.org/
3. Miller, P. *TCP/IP Explained*, Digital Press, ISBN 1-55558-166-8, 1997.
4. Internet Assigned Numbers Authority, http://www.iana.org/assignments/port-numbers

Further Reading

Miller, P. and Cummins, M. *LAN Technologies Explained*, Digital Press, ISBN 1-55558-234-6, 2000.

Phil Horne
Omneon Video Networks

1.7 SAN and NAS Technologies

1.7.1 Introduction

As broadcast facilities transition from analogue to digital, new disc-based storage systems and integrated networking technologies are changing the landscape of these installations. In the past 5 years, many different types of storage and networking technologies have been developed to improve the efficiency of media storage and the speed in which that media can be distributed. Videotape is rapidly being augmented by random access devices such as laser disc recorders, still stores, digital disc recorders, standard PCs, and other disc-based storage devices. The advent of cost-effective video compression hardware and low-cost, high-capacity disc drives has made many of the above devices pervasive in today's broadcast environment.

The primary advantage of disc-based systems over tape-based systems is the speed and flexibility in which users can access the stored material. The significant advantage of a well-designed networking topology is low-latency distribution of the material. Video servers are valuable products because multiple users can access any material quickly and simultaneously. Now larger, more flexible disc-based systems are providing storage capacities greater than 1000 hours and shared access to this material through multiple I/O interfaces such as SDI, SDTI, Ethernet, Fibre Channel, ATM and IEEE 1394. Even though the new storage devices provide significant capacity increases and some shared access to the stored material, the primary mode of accessing content stored on these devices is through a traditional point-to-point video network, usually based on a large, multi-layered video and audio routing switcher. As the number of video compression formats increase and as the capacity of storage systems increase, the networks that provide access to this content become more important. By providing simple and efficient access to stored material, control and facility operations are simplified, costs are reduced, and higher-quality, reliable service is possible.

This chapter explores the characteristics and benefits of several existing network-based storage architectures, including traditional storage shared via a standard analogue/digital router, networked video file servers, Storage Area Networks (SANs) and enterprise data networks (utilising both Storage Area Networks and Network Attached Storage (NAS)). For each system architecture details of specific device I/O and control, redundancy strategies, and scalability will be discussed. It is this combination of SAN and NAS coupled with high-speed networking at data rates suitable for broadcast applications that truly unlocks the potential of disc-based storage systems.

1.7.2 Storage System Architectures

The proliferation of digital devices, better compression methods and the increase in capacity of digital media storage systems creates a need to examine the different storage architectures and access methods as they are implemented today. The chapter begins by examining the traditional broadcast storage and playback system, looking at the control, I/O, redundancy and scalability attributes. It then takes a look at storage architectures involving networked video file servers and how they employ Storage Area Networking. It will then move on to compare these network architectures to a traditional corporate enterprise computer network architecture employing Network Attached Storage. Each of these architectures will be evaluated on the basis of three critical system design attributes which are common to each approach: device I/O and control, redundancy and scalability.

1.7.3 Traditional Broadcast System

Many broadcast facilities utilise multiple VTRs and Digital Disc Recorders (DDRs) or video servers for commercial and programme playout. These devices are interconnected or, loosely speaking, 'networked' via standard coaxial cable to a central router that connects these storage devices to the master control switcher, cameras, Q/A monitors, etc. The two primary storage components of a typical facility are VTRs and video servers.

1.7.3.1 Device I/O and control

In most cases, both VTRs and DDRs have video inputs (SDI in a digital facility), video outputs (SDI), audio inputs (AES, two pairs), audio outputs (AES, two pairs), sync input and loop-through, LTC input, LTC output, RS-422 port for serial control, and a parallel port for GPI control. Digital disc recorders may also have an Ethernet interface. For a single video stream to play out of a VTR or a DDR, there could be as many as six cables connected to each device. As the number of VTRs and/or DDRs increase in a facility, the number of video, audio, timecode, sync and control cables (serial/parallel) become very large and hard to manage. This is certainly not news to anyone familiar with a typical broadcast facility.

1.7.3.2 Redundancy

The most common redundancy strategy in a traditional broadcast architecture is to have two of everything operating simultaneously (see Figure 1.7.1). In its simplest form, this translates to having a spare VTR standing by in case the primary VTR fails. In many cases, the redundant VTRs are actually 'rolled' simultaneously, giving the station a 'hot' backup. If the primary VTR fails, then the master control operator can just switch to the backup feed and continue operations as normal. The method of using 'hot' rolling backups has been used for many years.

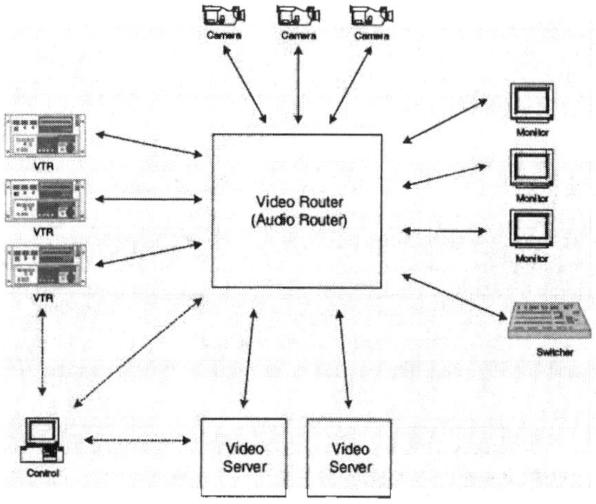

Figure 1.7.1 Broadcast facility architecture.

At the system component level most broadcasters require redundant parts, such as power supplies, fans and system controllers. It is also generally required that these parts be 'hot-swappable'.

1.7.3.3 Scalability

Traditional broadcast facilities exist in a wide range of sizes, so this architecture can scale to very large installations. Adding routing capacity typically involves upgrading to or purchasing a larger router, or placing a second router next to the first and cross-connecting some of the ports. Similarly, adding storage capacity is achieved by purchasing additional video servers and the associated router connections to integrate them into the network. These types of upgrades can be very expensive when forced by small incremental capacity requirements.

1.7.3.4 Summary

The traditional broadcast facility architecture is the starting point for all analogue facilities more than a few years old. Digital technology has emerged here in a manner that largely replaces the analogue components while keeping the basic architecture the same. Routers now route SDI or SDTI signals instead of analogue video, audio signals are now digital and disc-based servers are replacing VTRs and flexicarts. This approach to migration has assured a smooth transition and allowed analogue and digital components to coexist in the same facility, but it has failed to exploit the cost and technology trends in digital storage and networking that have been driven by the enterprise network computing market, something that will be explored in more detail later in the chapter.

1.7.4 Networked Video Servers

A video file server is much the same as a data file server in concept. It provides the capability of a remote device to connect to the server, request a video file, read and manipulate the video file, and store the video file. The primary difference is that the video file server provides the capability to receive real-time video streams, convert them to data files and store them to disc. Conversely it has the capability to read a file from disc and generate a real-time video stream as an output. In most cases the video server also incorporates hardware capable of compressing and decompressing the video stream. A typical video server contains the following functional components: a control CPU (usually running a Real-Time Operating System – RTOS), a disc controller (SCSI or Fibre Channel), interface cards (Codecs, Ethernet, SCSI, ATM, etc.) and a bus for exchanging data between these elements. In a networked video server architecture, the most common network interconnects used are Ethernet or Fibre Channel. Ethernet is the most widely deployed networking standard, operating at 100 Mbit/s (Fast Ethernet) or 1000 Mbit/s (Gigabit Ethernet). Fibre Channel operates at either 1 or 2 Gbit/s and is commonly used for connecting discs to computer systems. This storage network is sometimes exploited to provide a server-to-server network, although these networks are typically proprietary to a single vendor.

1.7.4.1 Device I/O and control

Video file servers provide a variety of interfaces to external devices. For each video input or output, there is an SDI (or analogue) video port and ports for at least two pairs of audio (two AES ports or four analogue audio ports), an RS-422 port for serial control and sometimes an LTC port for timecode. Each video server must also have an analogue black and burst input for synchronising the server to house reference. In most cases, video servers also have an Ethernet port for error monitoring, control, status, etc. For a video server that has 10 video I/O, this means that it must have a minimum of 10 SDI video connections, 20 AES audio connections, 10 RS-422 control connections, 10 LTC connections, a sync connection and an

Ethernet connection. This results in a total of 52 cables connected to this one box for control and access to information stored on the server (see Figure 1.7.2).

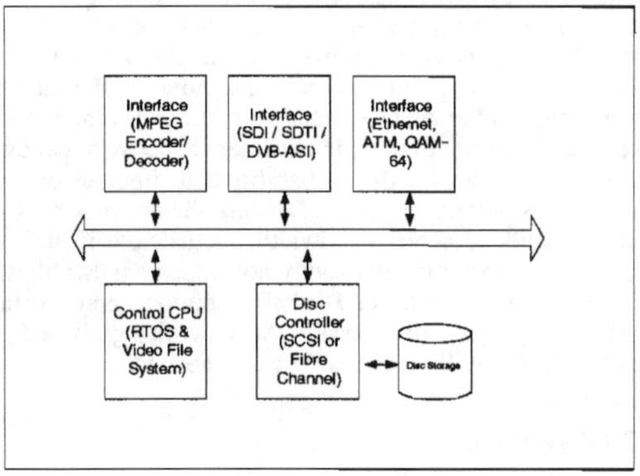

Figure 1.7.2 Video server architecture.

In order to get video into or out of the server network, an SDI video port must be available. As a video server network gets larger and larger, there are more ports available for use and there are more devices that must access those ports, making the management of the access to those ports very important. As the video server network gets larger and larger, the applications that control the individual video servers, the router and the devices connected to the router become very important. A control application must keep track of all of the content stored on each server, as well as the content required by each playout channel for each video server. If there is some content that is required by a playout channel that does not exist on that server, the control application must locate a copy of that content and transfer it to the server that needs it. This control application must also keep track of each RS-422 control port and which video/audio channel that it corresponds to, as well as the cable connection from each video/audio channel to the router.

As broadcasting migrates to an all-digital environment, the management of resources moves from managing physical resources to managing digital (or virtual) resources. As it becomes cost-effective to move more and more functions and content into digital video file servers, there are fewer and fewer VTRs, switchers, routers, etc. to manage. With only a few video file servers, control applications do not have to manage so many resources. As the number of video file servers grow, the control applications are required to manage bandwidth utilisation, file system coherency, file movement, redundancy, fault monitoring, automatic fail-over, automatic resource reallocation and several other aspects that lend more to computer network management than video facility management.

1.7.4.2 Redundancy

There are two ways to approach redundancy and fault tolerance of video systems. Component redundancy refers to redundancy inside the 'box' of a video server and system redundancy refers to outside of the 'box' redundancy. System redundancy requires multiple independent devices that can be utilised in a redundant manner, such as when two VTRs play two identical tapes simultaneously. System redundancy is utilised to protect against catastrophic failure of a device.

Networking together multiple video servers is a way to increase the system redundancy options for stand-alone video servers, as well as extend the storage capacity and I/O capacity. This approach allows broadcasters to configure storage systems that provide as little or as much redundancy as their operation requires. This approach also allows a system to be designed that can expand to add more video stream access and storage capacity. Figure 1.7.3

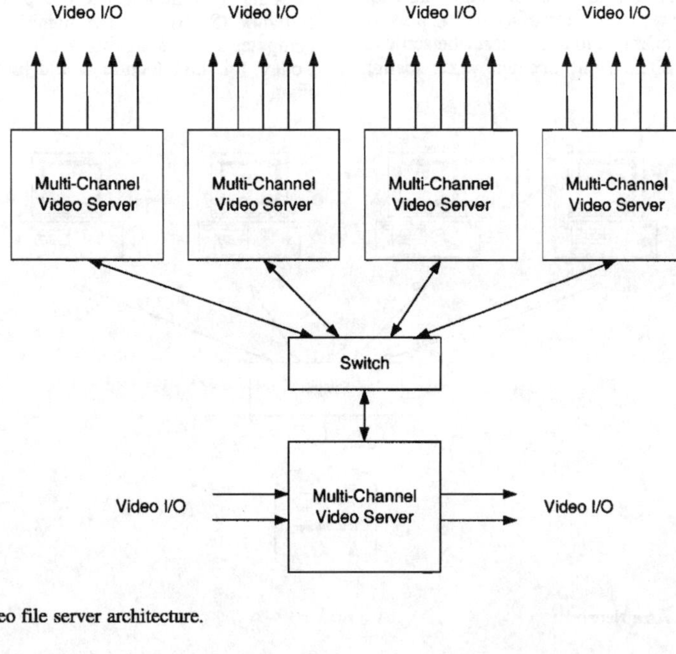

Figure 1.7.3 Networked video file server architecture.

shows an example of a cluster of multi-channel video servers networked to provide video stream access and storage capacity.

This diagram demonstrates how stand-alone video file servers can be networked to increase channel capacity, increase storage capacity and increase redundancy within the system operational model.

There are many network configurations that can be designed to guarantee storage and access. A single video file server can be networked to a second identical video file server for redundancy. Video can be recorded once and electronically copied to the 'backup' file server. A caching model can be applied such that a large video file server can be used as the primary record/storage server and a smaller server can be networked to the larger one for caching of material as required for air. In this cache configuration, multiple caches can be networked to a central storage server. This model is shown in the example above. The ingest server can store material for multiple caches or playout servers while simultaneously providing backup for any of the playout servers that may fail. For higher bandwidth applications such as programme playout applications, multiple content storage servers can serve content over the network to multiple playout servers while providing backup for any playout server. In this networked approach to redundancy, because each video file server maintains its own file system and local storage, there will always be at least two copies of all material available for playing to air at any given time. This guarantees that if there is a catastrophic failure of any single server, it will not affect the function of any other server. The backup server can then take over for the failed server and continue operation with minimal interruption of service. These server architectures can be utilised along with many other network configurations to develop extensive storage systems for any level of redundancy, storage capacity and I/O requirements.

1.7.4.3 Scalability

The networked file server architecture can be highly scalable given a thorough understanding of the operation constraints of the network, the file server and the facility. The reference architecture above is a good representation of a caching playout application and it is a good example to use to describe some of the constraints that must be taken into account when scaling this type of architecture.

There are many factors to consider when designing a server network. Each server has specific operational parameters such as the number of I/O channels each will support at a given bit rate. Most video servers manage internal bandwidth from disc such that video/audio I/O requirements are the highest priority and other I/O traffic such as file transfers are at a lower priority and utilise the internal bandwidth available, when it is available. The point-to-point file transfer bandwidth depends heavily on the number of I/O channels operating simultaneously. The other important factor is the internal network bandwidth. As the network grows, the probability that simultaneous multiple transfers will occur grows. Because the network bandwidth and the available server bandwidth fluctuates, it is difficult to specify with any certainty exactly how much bandwidth will be available at any one time. The most common practice in networks of any type is to overbuild the network significantly such that bandwidth will be available when needed.

1.7.4.4 Summary

The distributed video server architecture has the advantage of providing a modular solution that is scalable to a large degree and can be designed to avoid any single points of failure. The disadvantages of this approach lie in the need to spend video server bandwidth on copying assets between servers, as well as a redundancy strategy that dictates the need to duplicate entire servers rather than only the critical components.

1.7.5 Storage Area Networks

Storage Area Networks are a relatively new technology that were developed to allow data servers to take advantage of new high-speed disc I/O technology such as Ultra SCSI and Fibre Channel. Several initial applications such as video production, relational database access and corporate data centres required higher and higher capacity storage systems, as well as high throughput to the attached computer systems. A Storage Area Network (SAN) was the name coined to describe multiple computer systems sharing a common set of storage resources. Today, this architecture would look much like that shown in Figure 1.7.4.

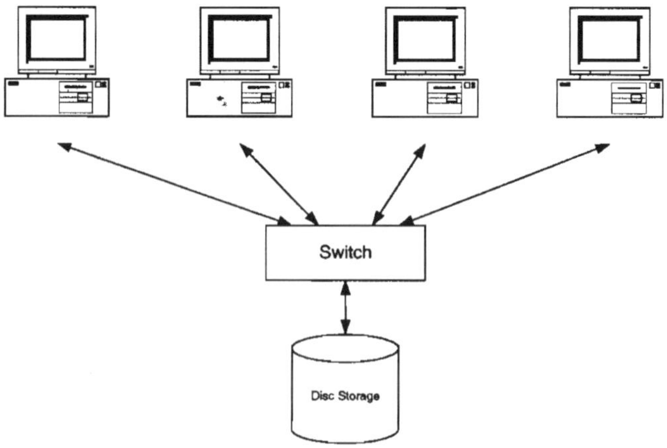

Figure 1.7.4 Four-PC Storage Area Network.

The intelligence for controlling how the discs are formatted, what type of file system is used to access the data and who can allocate storage for writing the data is controlled by the PCs attached to the storage. All PCs, servers or workstations typically have local storage and local file systems based on the type of operating system they run. For example, Windows NT computers use FAT file systems or NTFS file systems. The difficulty in implementing a Storage Area Network is in developing a method of sharing access to the actual shared storage resources. This is particularly important when writing data to the disc storage. This requires a resource locking mechanism and communication of the status of the lock to all computers connected. This can become very complicated when the number of computers connected to the SAN becomes large.

To date, there is no standard method for sharing access to common storage resources across vendors. Each implementation of software to allow storage resources to be shared is proprietary. This makes it difficult for connecting different types of computers to a single network to share resources. In some cases, an extra computer is used as a file system controller and access synchroniser such that each computer on the SAN must obtain permission from the control computer before accessing the storage. Because of the 'intelligence' required to access the shared storage resources, any device wishing to connect to the SAN must be capable of running an appropriate operating system and file system. This limits the SAN to connecting storage resources to workstations and computers. In video applications, this means there is no means for connecting a VTR, switcher, router or camera to an SAN directly.

To get video onto the storage medium, it must be encoded or acquired by one of the workstations on the Storage Area Network. The video must enter and exit the network from standard SDI (or analogue) video ports on the video server itself. In order to move the finished video segment out to a place where it can be distributed for viewing, the video segment must be decoded or played out of one of the workstations to a VTR or transmission source. The server can control these video devices by using a standard protocol such as Sony BVW over an RS-232 or RS-422 serial connection.

Fibre Channel storage is becoming the standard for interconnecting high-performance storage to these high-performance computers for editing of uncompressed and compressed video and audio. Fibre Channel provides a cost-effective method for accessing storage at bandwidths capable of meeting the requirements for editing and playback applications. In many cases, video production can be streamlined, allowing multiple editors to work on the same file simultaneously without maintaining multiple copies of the file.

Several companies have extended the basic video file server architecture to incorporate Fibre Channel storage (Figure 1.7.5). Others have extended the Fibre Channel storage to include characteristics of a Storage Area Network (Figure 1.7.6). The use of Fibre Channel as the primary interface to storage makes it possible to network multiple video file servers to a single storage pool through a Fibre Channel switch. Since a video file server contains a CPU, operating system and a local file system, it is possible to implement in software the ability to share a common set of storage resources in a Storage Area Network architecture. By extending the video file server architecture to include a Storage Area Network, multiple video file servers can now access the same clip simultaneously without having to transfer clips from server to server.

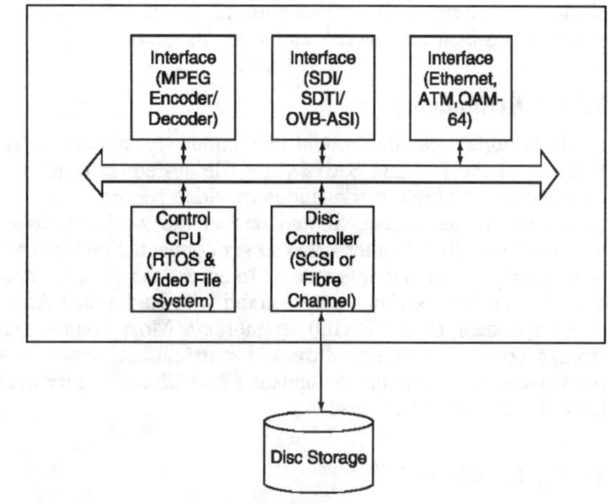

Figure 1.7.5 Video server with external storage.

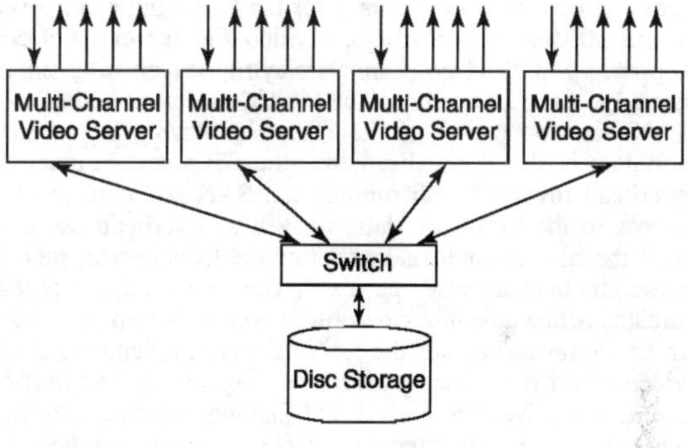

Figure 1.7.6 Video server SAN.

1.7.5.1 Device I/O and control

While a video server SAN is easily expanded by adding additional video servers, the number of video connections required to access the new resources increases proportionally. Audio connections and control connections increase proportionally as well. As the access bandwidth to the SAN increases by adding new servers, so does the requirements for the video, audio and control networks that allow resources to be accessed and controlled. The growth in server channels also requires a corresponding number of router inputs to be made available such that these new resources can be shared.

As I/O requirements increase, so does the control network. For each video/audio channel, another control port must be added to the control application. While the I/O control applications scale with video I/O in this architecture, media management resource control does not increase as the system stores all of the material in a large central storage system. For very large configurations, multiple video servers within SANs will be required such that material must be moved from one SAN to another. This is accomplished in much the same fashion as in the networked video server application. High-bandwidth networks connecting video servers or workstations between SANs allow for material to be passed through the video servers or workstations as if they were stand-alone devices.

1.7.5.2 Redundancy

SANs provide a certain amount of redundancy automatically. Because of the fact that SANs share file system information, they provide multiple workstations or video servers access to the same storage system. Should one of the workstations or video servers fail, the other video server connected on the SAN contains all appropriate information to access the existing storage. The storage itself can be protected with Redundant Array of Independent Discs (RAID) technology. Most commercial storage array chasses also provide for redundant power supplies, redundant fans and redundant Fibre Channel interfaces (or redundant RAID controllers).

1.7.5.3 Scalability

If it is assumed that Fibre Channel can sustain a throughput of approximately 600 Mbit/s then the number of video channels that can access this common storage pool for recording and playout can be calculated. The minimum and maximum compression rates that will be used for the following examples are 8 and 50 Mbit/s. At 8 Mbit/s, a video file server SAN could support up to 75 video channels playing or recording simultaneously. At a maximum of 50 Mbit/s, the same video server SAN can support up to 12 channels playing or recording simultaneously. The reality is that there is a certain amount of overhead involved with running the SAN software to obtain access to the video file data, as well as overhead associated with the file system management on the local server, such that these numbers are very optimistic. The actual number of video streams achievable in a video file server SAN depends heavily on the compression rate, the SAN software implementation, the video server file system performance capabilities and traffic on the network. Most recently Fibre Channel interfaces to drives and cards has moved from 1 to 2 Gbps, doubling the theoretical drive access bandwidth and thereby doubling the number of channels that can be supported by the server.

1.7.5.4 Summary

In order for video and audio data to be placed into, or transferred out of, the SAN, it must pass through one of the computers connected to the SAN. This also means that content management of any content on the SAN must be performed in or through one of the connected computers. As the number of computers connected to the SAN grows, the ability to manage access to the storage and to synchronise file system information grows in complexity. Also, as more computers are connected to the SAN, the bandwidth of a single Fibre Channel loop gets saturated and other more complex architectures involving control, access and synchronisation of multiple Fibre Channel loops are required. Storage Area Networks connecting six, eight or even 10 computers on a network offer an effective solution for collaborative video editing applications, but the lack of ability to connect non-computer devices to the network makes it an inflexible architecture.

The video server SAN does present some very tangible benefits over the networked file server approach. An SAN-based video server provides the capability of multiple video file servers to have one consistent file system. This allows multiple independent software applications to control independent video file servers and all applications can see all of the video files within the SAN. SANs are limited in their total bandwidth, so in order to increase the bandwidth of the network, multiple SANs must be networked. By reducing the number of file servers that must be networked together, the management of the network is far less complex and much easier to manage.

1.7.6 Enterprise Computer Networks

Enterprise computer networks actually display characteristics of both Storage Area Networks and networked file servers mentioned above. A typical computer network incorporates multiple PCs, workstations, printers, routers, switches and hubs, all connected to a central server or several servers. However, one important difference from the other storage architectures is that the Ethernet network carries all control and data over the same network. Because of this inherent ability for a single standard connection to carry all information, different storage architectures can be developed to provide access to storage based on the requirements of the applications. In the past few years, the increase in network bandwidth on Ethernet from 10 base T to 100 base T and now to 1000 base T have fostered different server and storage architectures. The traditional server model has dictated that a single high-performance computer would act as the centralised server and feed data over Ethernet to many PCs connected to the network (see Figure 1.7.7). As networks became larger, with interconnections to other networks around the country and around the world, the amount of data stored began to grow significantly, yet the need for timely access to this data has remained.

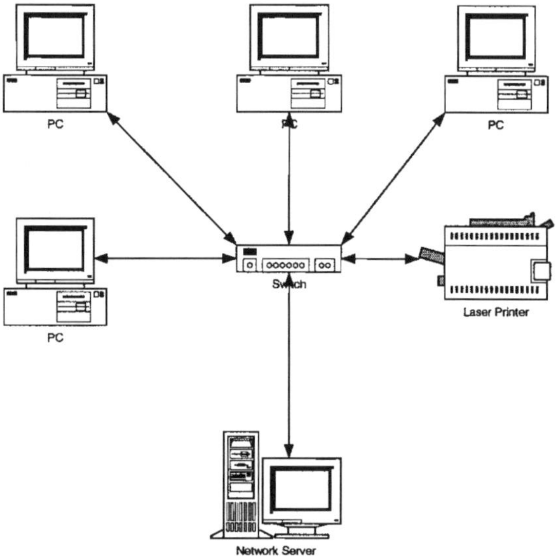

Figure 1.7.7 Simple Ethernet network.

1.7.6.1 Network Control

In order to connect users and devices to storage systems or servers, the IT industry has standardised on Ethernet as its

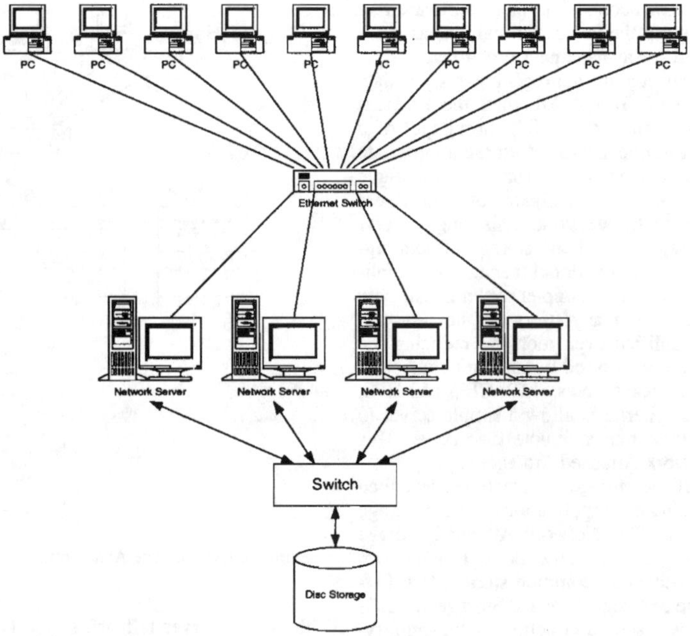

Figure 1.7.8 Ethernet network and data SAN.

networking technology. The vast majority of enterprise networks use 10 base T and 100 base T Ethernet, which are capable of 10 and 100 Mbit/s respectively. As mentioned above, one enviable quality of Ethernet networks is their ability to carry multiple protocols simultaneously, including control and data information. Ethernet networks can be used to move files from place to place; they can be used to access information stored on remote storage devices; and they can control devices on the network such as printers, PCs and servers. These networks are even beginning to be used for streaming video and audio, although at lower bit rates, such as those defined by MPEG-4 for example.

It is the ability to carry control and data simultaneously on the same transport that allows such flexibility in network architectures. It is this same ability that allows for new, more flexible storage architectures. As CPU speeds increased in line with the much quoted Moore's Law, the bottleneck for access to storage quickly became that of the network interconnections to the server. Several new server and storage architectures developed to address these access issues.

1.7.6.2 Data storage area networks

Large centralised data server architectures are now beginning to utilise Storage Area Networks to provide multiple data server access to a single storage entity for increasing the data throughput to the network. A simple example of this type of architecture could have four high-performance data servers all attached to a common storage array through a Fibre Channel hub or switch. Each server can distribute data to devices on the network by using standard 100 base T Ethernet. This approach to providing high-bandwidth access to devices on the network allows for a great deal of flexibility within the design of the system itself (see Figure 1.7.8).

1.7.6.3 Data SAN redundancy

As described above in the video server SAN example, the data SAN architecture can provide a high level of redundancy just by adding an additional server to the SAN. The additional server can be a hot standby server ready to take over for any other failed server on the SAN. The data architecture proves to be more reliable than the video server SAN because it can take advantage of the Ethernet I/O connections. Since the Ethernet network carries both control and data over the same connection, in the event of a server failure, the backup server can automatically reconfigure itself and 'rename' itself as the failed server, thus re-establishing the original connection to all devices accessing this failed server. This fail-over protection scenario can correct itself in milliseconds, thus appearing to the end-user as if nothing ever happened. This level of reconfiguration is not possible with a video server SAN as there are many control connections, video connections, audio connections, etc. that all must be reconfigured or re-routed.

1.7.6.4 Network attached storage

The SAN example above describes how multiple network servers can attach to a common pool of storage by utilising SAN architecture and SAN software running on the servers. This system is moderately complex in the fact that multiple network servers must be set up, configured, managed and maintained. The network servers must provide the capability to run server applications, provide data server capability and manage the SAN file system access. The more software that runs on the server, the lower the actual throughput, as more CPU cycles are being used for management functions.

Now that networking speeds are approaching gigabit data rates with the emergence of Gigabit Ethernet, the Ethernet connection on a server can support the full throughput of a single Fibre Channel storage connection. With the appropriate processor and memory architectures in a data server, a single server is now capable of providing a gigabit of data access bandwidth to any device on the network. This single server model simplifies the management of the storage system significantly, reducing the need for additional management applications to run on the server. These simpler systems can now provide gigabit data access to many users on the network. These systems can be simplified even more by reducing the amount of server functionality required by implementing standard network storage interfaces, such as SMB, CIFS and NFS. These standard file system interfaces allow a simple server to sit on the network and look like a remote disc drive. This architecture is called Network Attached Storage.

In the SAN and centralised storage architectures described above, the servers are running both applications and the storage management software itself. The Network Attached Storage approach allows each user on the network to run its own application locally while accessing common storage (see Figures 1.7.9 and 1.7.10). The Network Attached Storage server is optimised for providing file system transactions only, simplifying the server requirements even more and therefore reducing the cost of the overall network. By limiting the NAS server functionality to providing standard file system interfaces, there is no need to use a full commercial operating system like Windows NT® or UNIX®. As long as the NAS appears to network devices as a standard file system, the operating system can be anything.

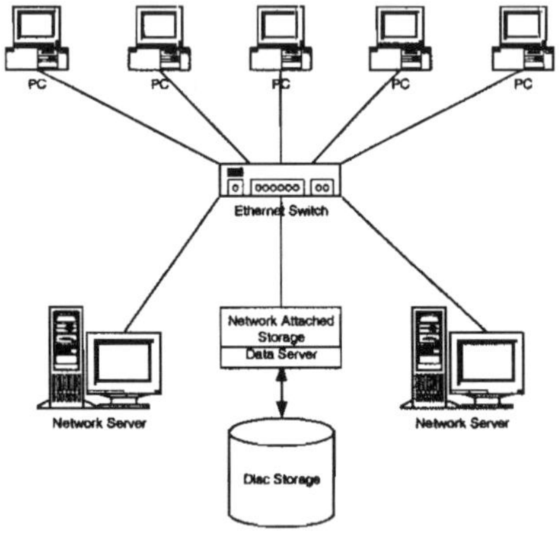

Figure 1.7.9 Network Attached Storage.

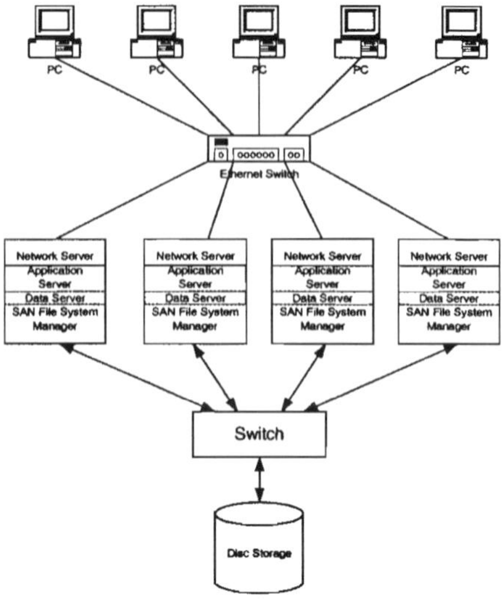

Figure 1.7.10 Storage Area Network.

1.7.6.5 NAS redundancy

With the storage architecture simplified, the redundancy requirements are also simplified. In the SAN redundancy scheme, if the server were running a specific application, the backup server would also have to be configured to run that application as well. For some applications, the backup server would also have to keep track of the state of the running applications on other servers such that it could pick up where the failed server left off. In the NAS model, the applications actually run on the remote PCs; the NAS server has only to keep track of the file system transactions occurring on the system. The fail-over scenario is similar to the SAN fail-over scenario in that there must be a second server attached to the storage system to take over for the failed server. This fail-over scenario is also much simpler than in the SAN scenario. The redundant NAS server must only keep track of the file system transactions and need not be concerned with file system synchronisation, since it does not actually share access to the storage with the other NAS server. Only one server reads and writes to storage at a time. NAS server state information can be mirrored to the backup server, and if the primary server fails, the backup server can reconfigure itself quickly to take over as the failed server.

1.7.6.6 Summary

For data networks no specialised hardware is necessary and no modifications to the network are required to scale an NAS server environment. Customers can quickly expand the capacity and functionality of their enterprise networks. Likewise storage capacity can be added to an NAS server with minimal administration effort. However, the distribution of video data is more challenging and cost sensitive. For each media stream expensive video and audio ports must be added to the server, and routers and switchers must be added to the networking infrastructure.

1.7.7 Using SAN and NAS Architectures

Combining SAN and NAS storage architectures in a server design brings significant benefits (see Figure 1.7.11), as discussed earlier. Storage Area Networking allows workstations to share a common pool of storage. If those workstations are standard off-the-shelf, IT-based workstations, then applications running on those workstations can see the shared storage but not, for example, a VTR or other dedicated broadcast device.

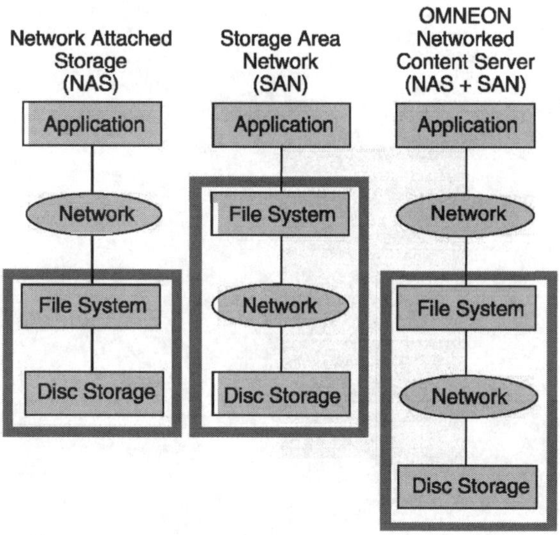

Figure 1.7.11 Flexibility of SAN and NAS.

In most broadcast server applications it's important to be able to connect standard broadcast devices such as VTRs to the shared storage, and this would be achieved by using dedicated 'broadcast'-specific I/O connections such as SDI and AES/EBU. Here broadcast-specific SAN 'devices' would be used in the server design.

Sharing a common pool of storage amongst SAN aware workstations is a very convenient way to modulate the size of a server and, depending on the granularity of the server, building blocks using a wide range of channel counts and storage capacities can be achieved. When demand for server ports exceeds the bandwidth available, another broadcast aware SAN unit is added to the existing SAN along with more disc drives if necessary.

If SAN brings us the flexibility to design servers with a wide range of sizes, then NAS gives us the capability of networking individual servers using standard IT protocols and technologies. Very few broadcast facilities will decide to use just one broadcast server. The attraction will be there from a media management point of view as a single server significantly minimises material movement; however, the question of security will come to the fore and typically programme backups and archives will be required, increasingly so into the future as videotape usage falls. To copy material between two VTRs, for example, a number of connections would be required: SDI for the video, AES/EBU for the audio and RS-422 for control. When using Ethernet and TCP/IP, the video, audio and copy control mechanism is conveyed by a single wire (or twisted pair in reality). Until recently a limit has been imposed by using 100 base T, good for many IT requirements but too slow for anything other than small broadcast applications. With the advent of 1000 base T, moving programme content between broadcast equipment has become a practical reality and a cost-effective one as 1000 base T Ethernet networking technology costs rapidly decrease.

In very broad terms SAN-centric server designs could be argued to be more suited to the needs of post-production, where a group of non-linear workstations might share a common pool of storage. Conversely NAS server 'modules' are potentially most suited to transmission applications, where individual playout caches often supported by a backup cache are required to provide a 100% reliable on-air service. Here a steady feed of transmission-ready programme content is fed to the NAS TX modules by 1000 base T Ethernet ready for inclusion in the on-air schedule.

Phil Simpson MISM AITT
Tandberg Technical Training
Tandberg Television Ltd

1.8 Telco Technologies

The word 'telecommunication' is derived from two Greek words, which can be translated as 'the passage of information at a distance'. Historically the only means of communication available were visual communications, e.g. beacons, smoke signals and semaphore signalling. Modern telecommunications networks enable different types of data to be passed: Voice (by telephony), Paper images (by facsimile), Computer systems (by using digital data networks), Plain text (by telex), Formatted Text (by electronic mail and related systems) and Video Systems (by Integrated Services Digital Network – ISDN). The different forms of telecommunication are distinguished in two ways: firstly by the type of device used (telephone, fax machine, telex machine, etc.) and secondly by the kind of network used to interconnect the devices (telephone network, telex network, data network, etc.) For example, telephones, fax and modems would use the Public Switched Telephone Network (PSTN) and telex machines would use a dedicated telex network. Digital data would have to use a digital network like X.25, Frame Relay, Switched Multimegabit Data Service (SMDS) or Asynchronous Transfer Mode (ATM), to name a few. All PTTs (Post, Telephone and Telegraph) and all other Licensed Operators in the UK operate a digital network. The bandwidth of PSTN is determined by various speech components. The speech input from a PSTN telephone consists of a sum of a number of different frequencies components, from about 50 Hz to about 7 kHz depending on who is speaking. The speech output from a PSTN telephone also consists of a number of frequency components, but this time only from 300 to 3400 Hz. Thus the telephone network cuts out all frequency components below 300 Hz and above 3400 Hz. Hence the bandwidth of PSTN is 3.1 kHz (3.4 kHz to 300 Hz). Voice being transmitted through the PSTN is called 3.1 k or 'commercial speech'. Music frequencies can be in the range of 30 kHz (Double Bass) to 20 kHz (Flute). The telephone service is a result of this compromise. Transferring all speech components would result in a high quality but expensive service; progressively reducing the bandwidth lowers both the quality and cost.

There are several different methods to achieve this. Two main methods used in telecommunications are Pulse Code Modulation (PCM; 64 kbit/s) and Adaptive Differential Pulse Code Modulation (ADPCM; 16–64 kbit/s). The basic coding technique used is PCM, but this coding technique is expensive on networks because each digitally encoded channel will require 64,000 bit/s of bandwidth. ADPCM is a predictive coding which converts the analogue signal to PCM first. Then this coding scheme will predict a new value with the reference to the previous sample point and it will look at the actual point and work out the difference between the two points: if the change is positive then the 3-bit code is prefixed with a 1. If the change is negative then the 3-bit code is prefixed with a 0. Thus the speed of ADPCM is calculated as follows: 8000 samples per second multiplied by 4 bits per sample = 32,000 bit/s. ADPCM will use 32 kbit/s for voice compression. Note: Recommendation G.721 ADPCM, which will only support 32 kbit/s, is now obsolete and is replaced by G.726, which supports 5-, 4-, 3- and 2-bit sampling for speeds of 40, 32, 24 and 16 kbit/s respectively.

1.8.1 Modems

Codecs encode analogue voice into a digital bit stream to be transmitted over a digital line. This is no good for transmitting a digital bit stream over an analogue line. To be able to transmit a digital bit stream over an analogue line we need a device called a modem. The word MODEM stands for MODulator and DEModulator. Modem speeds typically range from 300 bit/s to 56 kbit/s.[1] There are higher speeds than these but these are proprietary. One disadvantage of using a modem is the introduction of noise on the line. As a general rule the longer the local loop, the slower the speed of the modem. V.92 is the latest recommendation from the ITU, with downstream and upstream speeds of 56,000 bit/s. Table 1.8.1 shows some of the modem specifications.

1.8.2 Local Access

The access layer of the PTT consists of all the equipment a domestics or business subscriber requires in order to connect to

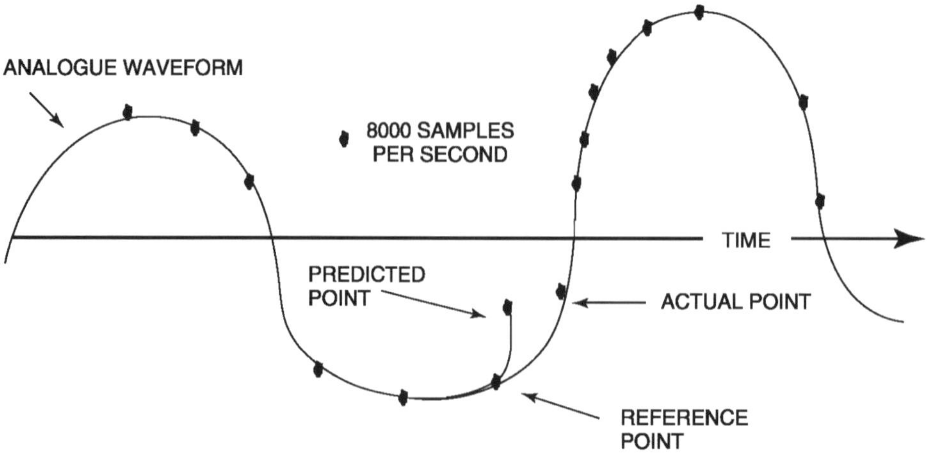

Figure 1.8.1 Voice Sampling

their Digital Local Exchange (DLE). The path that a subscriber takes to access these resources of the network is called the 'local loop'. The network of local loop communication is collectively called the 'local line'. Access can be split into three main categories: Domestic Access Telephony is the most common requirement a domestic user will need; however, because of the massive increase in Internet use, users now demand improved access. For many users access is improved by using a V.90 modem over their existing copper two-wire local loop. With the explosion of another technology, ADSL (Asymmetric Digital Subscriber Line), it is now possible to utilise high bandwidth on existing phone lines to homes and business.

Access is available from a single 64 Kbit/s data link to 30 × 64 Kbit/s channels, which together form a 2 Mbit/s data link (E1), depending on the needs of the subscriber. A Private Automatic Branch eXchange (PABX) on a business subscriber's premises accesses the network through an E1 trunk or sometimes using an ISDN Primary Rate Interface (PRI). A business subscriber's local loop will most likely be provided through a fibre optic connection, not copper wire.

Table 1.8.1 Modem specifications

ITU-T Standard	*Speed (bit/s)*	*Two-wire/four-wire*
V.21	300	Two
V.27	4800	Four
V.27 bis	4800	Two to four
V.27 ter	4800	Two
V.29	9600	Four
V.32	9600	Two
V.32	14,400	Two
V.33	14,400	Four
V.34	33,600	Two
V.90	56,000	Two

1.8.3 Digital Main Switching Units

The DLEs are responsible for collecting calls from customers and passing calls to the Digital Main Switching Units (DMSUs). DMSUs are the heart of the trunked network and within BT's network there are over 50 DMSUs supporting thousands of DLEs. The DMSUs are also where the network owner allows Other Licensed Operators (OLOs), such as Energis, to connect to the network. Nowadays national governments issue telecommunications licences to compete with the main network owners. Because it is uneconomical for these OLOs to install their own local loops, a typical scenario is that licensed operators would have BT install and maintain their local loop whilst they just rent it. The layout of a typical DMSU Interface is shown in Figure 1.8.2.

1.8.4 Public Switched Telephone Network (PSTN)

PSTN is the world's collection of interconnected voice-oriented public telephone networks, both commercial and government owned. It is commonly referred to as Plain Old Telephone Service (POTS).[2] POTS is a term sometimes used in discussion of new telephone technologies in which the question of whether and how existing voice transmission for ordinary phone communication can be accommodated. For example, ADSL and ISDN connections provide some part of their channels for 'plain old telephone service', while providing most of their bandwidth for digital data and multimedia transmissions. The PSTN is essentially a network supplied by a PTT company. A PTT can also be called a Public Telecommunications Operator (PTO). The largest cost to the infrastructure to install a PSTN network is the millions of two-wire copper connections to the domestic user. This will be done normally using the cheapest cable possible, but also must be able to get the maximum distance possible. Also, PTTs normally do not amplify the local loop because this would also increase the cost. In the UK the longest standard two-wire connection can go up to 8 km. The standard two-wire connections average a distance

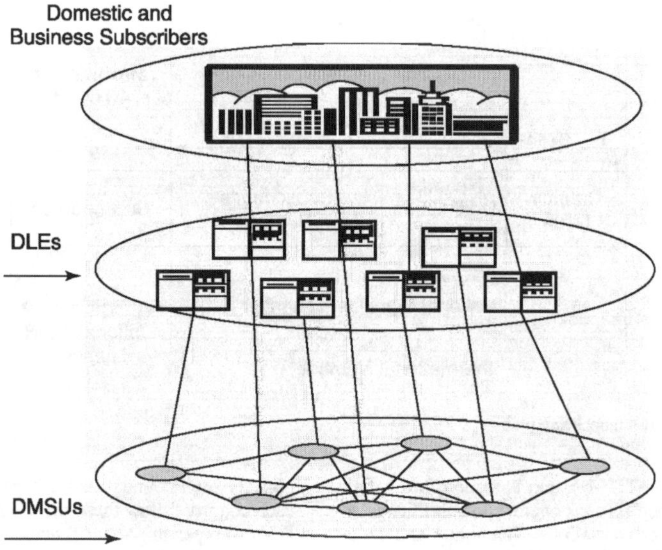

Figure 1.8.2 Layout of a typical DMSU Interface.

of around 3.5 km. Because of this limitation, about 4500 exchanges are needed throughout the UK.

1.8.5 Dealing with Echo

Although the local loop is always two-wire, transmission within the old analogue networks is mostly four-wire. The advantages of a four-wire system is that the transmit and receive wires are separate; this makes it much easier to amplify. Four-wire circuits are of higher quality, but much more expensive (twice as many wires) than two-wire. To convert from two- to four-wire there is a need for a special transformer called a 'hybrid'. Two-wire systems suffer from a kind of distortion called 'echo' due to the reflection of a hybrid. Essentially a telephone is a four-wire system, two wires for the mouthpiece and two wires for the earpiece. The hybrid is a passive device that makes two-wire telephony possible. Firstly, it directs all but a small fraction of the transmitted speech signal out to the line. Secondly, when the signal reaches the distant end the hybrid directs it to the earpiece. Hybrid systems produce 'feedback'. One type of feedback that is required is at the handset hybrid. When you speak in to the mouthpiece a certain amount of your voice is reflected back to the earpiece, which is called a 'comfort zone' or more commonly referred to as 'side-tone'. If this process was not carried out during quiet periods the earpiece would be dead, and hence one would keep stating: "Are you still there?" Echo is based on delay, so the longer the circuit the more chance of delay. In a small country the echo is so close to the original speech that this is not a problem. Echo can be due to long circuits with many amplifiers or a satellite connection; this can make the circuit unusable. There are two remedies for this echo suppression or the more modern approach, echo cancellation. Echo Suppression clips the speech in the reverse direction; however, speech has enough built-in redundancy for this not to be a problem. Echo Cancellation works by working out the round trip delay between devices, and then feeds a negative signal into the voice in the opposite direction, including the delay variation. To overcome this, the transmitting device will produce a tone (2100 Hz) on initial set-up, which temporarily disables the suppression technique used by the carrier. Figure 1.8.3 shows a basic path of these two techniques.

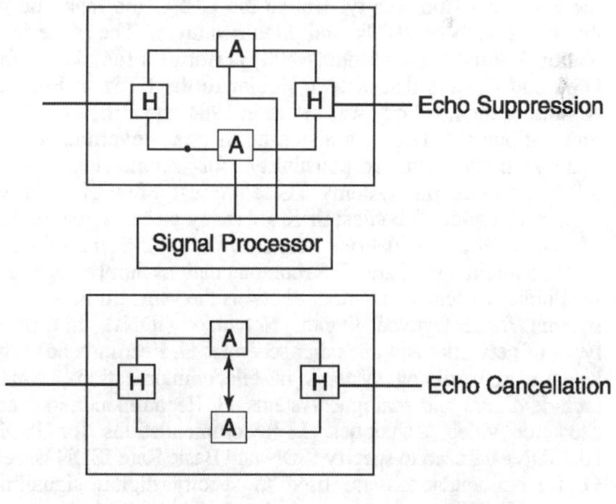

Figure 1.8.3 Echo Suppression and Echo Cancellation.

1.8.6 Standards and Numbering Plans

As in the Digital Broadcasting world with bodies looking after compression (MPEG), transmission (DVB), etc., there are several within the telecommunications industry.

So, as in all industries there are many standards bodies, which look after particular areas of the telecommunications vendors and operators. The International Telecommunications Union (ITU) is part of a worldwide body that looks at telecommunications

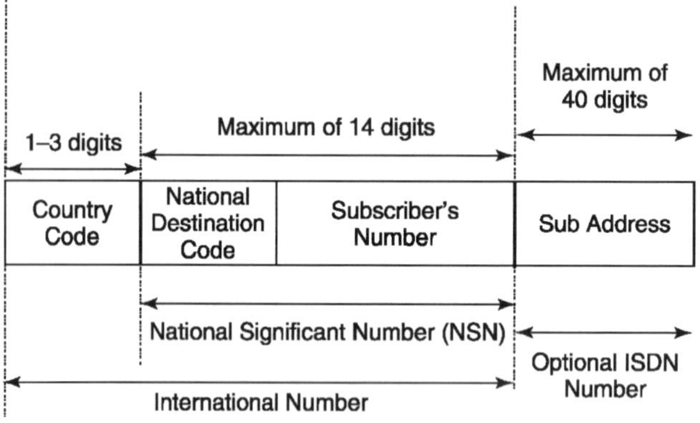

Figure 1.8.4 Layout of the E.164 numbering plan.

(ITU-T) and radio networks (ITU-R), and is responsible for all aspects within telecoms. The ITU took over responsibilities from the Consultative Committee International Telephone and Telegraph (CCITT) in March 1993. The ITU issues recommendations, which are not mandatory, but manufacturers treat them as *de jure* (legal) standards. The Conference of European Post and Telegraph (CEPT) is a European body set up in the late 1960s to enable Europe to discuss key issues of the time and speak with one voice. Their role in the 1990s has been reduced to consideration of European Regulatory Issues. CEPT recommendations are called NETs and follow closely the ITU recommendations. The European Telecommunications Standards Institute (ETSI) was formed in 1987 by the European Commission for standards within the European Community. Unlike the CEPT, the make up of the ETSI is both PTOs and Manufacturers. The American National Standards Institute (ANSI) performs a function in the USA and Canada that is very similar to the ETSI in Europe. Another standards body was set up in 1984, the Office of Communications (OFTEL), as a non-ministerial government regulatory body with responsibility for monitoring public telecommunications systems. These are just a few of the most important standards bodies; there are many more. A general list of ITU-T recommendations runs from A to Z. Some of these most commonly used are: E. Recommendations numbering plans for Public Switched Networks. E.164 is the standard numbering scheme for Integrated Digital Networks (IDNs), but other types of networks will use other schemes. G. Recommendations are for transmission systems in telecommunications, which include digital and analogue systems. H. Recommendations are used for video techniques. I. Recommendations for ISDN (B-ISDN) are used to specify ISDN and Basic Rate ISDN issues. Q. Recommendations are used to specify digital signalling systems, V. Recommendations deal with data transmissions over telephone circuits (analogue communication techniques) and X. Recommendations are used for Data Networks (X.1–X.96).

If a call over a telecommunications network is to be set up it must be possible for the originator to specify the identity of the other user. This is achieved in telephone networks by allocating each line a number and requiring the caller to signal the number of the party to his exchange by dialling the number wanted. Because every country's telephone network is connected to that of every other country (either directly or via another country's network), there must be some consistency in the way the numbers are allocated. This is achieved by defining a numbering scheme or plan for each country, and ensuring that the numbering plans of each country follow these rules. These rules are laid down by the ITU. Another important function of a numbering plan within a single country is to ensure that sufficient telephone numbers are available to serve the demand in each centre of population, while keeping the length of a telephone number within reasonable bounds. The objective of a national numbering plan is to ensure that the total number of digits dialled is approximately the same, while allowing local calls to use a lesser number of digits by requiring only the local field to be dialled. The ITU-T has published recommendations for telephony numbering plans as number E.164. This specifies that a complete telephone number should consist of the follows information. A Country Code (CC), which is used to select the destination country and can vary in length from one to three digits. When calling an international number, the CC will usually be prefixed by the digits 00 to indicate an international call. Most countries with modern digital exchanges will use 00, but older type PTT networks may use 0 or 010 (E.163). The National Significant Number (NSN) is used to connect to the destination subscriber. It is broken into two parts, The National Destination Code (NDC) and the Subscriber's Number (SN). The maximum number of digits for the NSN is 14. The NDC is responsible for routing the call through the correct network, i.e. BT, Mercury, Energis etc., to the correct area. The NDC is also known as the AFN (All Figure Number) and the NNG (National Number Group). The SN is the final connection from the PTT network to the subscriber's premises. There is also another technique to use Sub Addressing (SA), which provides a separate additional addressing capability outside the E.164 numbering plan, but constitutes an intrinsic part of the ISDN capabilities. The SA can be set up to 40 digits and is passed on call set-up transparently through a PTT network to the destination. The layout of the E.164 numbering plan is shown in Figure 1.8.4.

1.8.7 Integrated Services Digital Network (ISDN)

The first steps to digitise the PSTN into the IDN were started in the late 1960s and have been completed, so that BT's network only contains digital DLEs.

Digital communications means that a connection from an ISPBX through the PTT network and back to another ISPBX

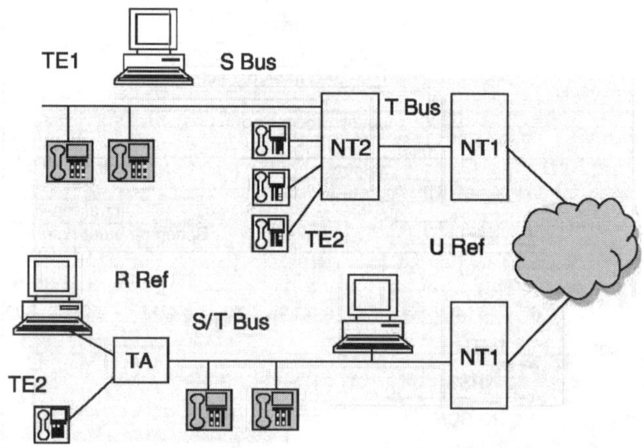

Figure 1.8.5 Topology of ISDN System and terminology.

must be digital. For most companies this type of connection has been available since the early 1980s. There are two types of channels in the ISDN service. The first is a Bearer channel or 'B'-channel, which is always at a speed of 64 kbit/s. This channel carries information or digitised voice signals. The other type of channel is the Delta or 'D'-channel. This channel is used for the signalling and can be 64 or 16 kbit/s. In the USA the Primary Rate consists of 23 B-channels and one D-channel of 64 kbit/s. ISDN offers two different types of service: Primary Rate Access (PRA) and Basic Rate Access (BRA). Basic Rate Access ISDN is primary aimed at the SOHO (Small Office–Home Office) and the domestic user. One of the major problems is how to transmit high-speed digital data over a two-wire connection. This can be achieved by using an encoding technique called 2B1Q (two Binary one Quaternary); 2B1Q allows 2 bits of data for every voltage change and there are four discrete voltages used. The Basic Rate ISDN will supply a 2B + D service, which enables two channels at 64 kbit/s for calls and a signalling channel of 16 kbit/s, which is designated as Q931.[3] Prior to 1997 this was called ISDN2. Since 1997, because it is fully ETSI compliant, it has been called ISDN2e. This means that if you were using the full 128 kbit/s (both 64 kbit/s) for a download over the Internet, and a telephone call came through, one of the 64 kbit/s would be released to allow the call through. Once the call has terminated your download would again increase to 128 kbit/s. The Basic Rate Interface (BRI) has a number of Reference Points, indicating what type of equipment can be connecting to BRA. The U reference point defines the interfaces on the line between the exchange and subscriber. The T reference point is only used for the connection between Network Termination 1 (NT1) and Terminal Equipment (TE). This is sometimes called the S/T reference point. The S reference point is used to connect up to eight different types of equipment to the NT2 or NT1 equipment. There is also an R reference point for non-ISDN equipment. Some of the basic terminologies used for ISDN equipment are Network Termination (NT1), which is used to connect the PTT service from the ISDN equipment. If an NT1 is connected to an S Bus, then no local connection between devices can be made. Network Termination 2 (NT2) equipment is used to connect from an NT1 to the S Bus, for local interconnections. Examples of NT2 equipment are PBXs, Local Area Networks and terminal controllers. Terminal Equipment type 1 (TE1) is designed to connect onto the S Bus or to an NT1 device. Terminal Equipment type 2 (TE2) is non-ISDN equipment and must be connected using a TA unit, which is a Terminal Adapter that itself is used to connect TE2 onto the S Bus at 64 kbit/s. An example of a TA is a MoDec (Modulator/Decoder), which is used to connect a PC to an analogue line through an ISDN network. A typical topology of the above terminology is shown in Figure 1.8.5.

Primary Rate ISDN is based on the G.704 framing structure and the basic configuration is 30B + D. Other variations are 6B + D, 8B + D and 15B + D. Within the framing structure Timeslot 0 is used for synchronisation of the equipment to the exchange and the signalling will be sent in timeslot 16. The international standard for connecting private networks to public networks is Q.931. PRI/ISDN will only use the T and U reference points. The ISPABX is generally classified as an NT2 unit. Some ISPABX can supply an S Bus to other users if required, but the ISPABX has to run Q.931 to achieve this. Historically networks had to build into their existing WAN some Spare Link Redundancy to take account of Link Failure. During peak times if a BRA was being used then 128 kbit/s can be used. In this way ISDN has become more flexible. To summarise ISDN it provides a basic circuit switched infrastructure of 64 kbit/s speed (bearer services), which transports information worldwide. And buildings on the bearer services are supplementary services, which allow more information, or services between callers, e.g. DDI, call waiting.

1.8.8 Broadband Integrated Services Digital Network (B-ISDN)

The planning for ISDN, or Narrowband ISDN (N-ISDN) as it is correctly termed, began as far back as 1976. Only in the 1980s did ISDN start to be implemented. Work on the relevant recommendations by the ITU-T is still going on, although the majority of the work has been completed. On its initialisation ISDN revolutionised telecommunications with respect to the services that are supported and features that can now be supplied to both the business world and domestic users. Since 1988

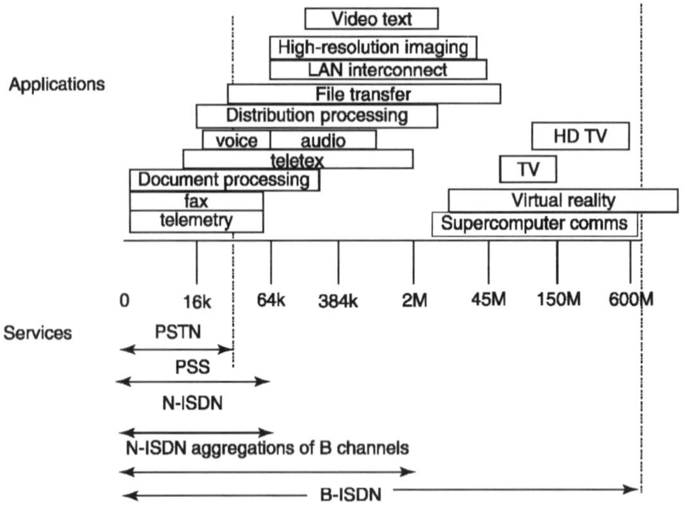

Figure 1.8.6 Bandwidth usage: applications and services.

much of the planning and design effort has become directed towards a network concept that would be far more revolutionary than N-ISDN itself. This concept is referred to as Broadband ISDN or B-ISDN. The ITU-T defines B-ISDN as "a service requiring transmission channels capable of supporting rates greater than the primary rate" (greater than 2.048 Mbit/s). B-ISDN offers three main types of transmission services: symmetrical 155.52 Mbit/s and 622.08 Mbit/s; asymmetrical user to network 155.52 Mbit/s; and network to user 622.08 Mbit/s. Signalling within B-ISDN is achieved through the same method as with N-ISDN in that it is out-band signalling. The signalling system used for B-ISDN is Q.2931. Figure 1.8.6 shows typical bandwidth usage by applications and what services are capable of carrying these applications.

1.8.9 Asynchronous Transfer Mode (ATM)

ATM is the technology that was chosen by the ITU-T to support the Broadband ISDN (B-ISDN). ATM allows a single network to handle all types of traffic, including voice, video and data, all with a common access point. Historically different technologies have been used for different types of traffic. ATM allows for the integration of networks, improving efficiency, manageability and, ultimately, cost. ATM can be run over fibre, twisted pair cable, coaxial cable etc., on such systems as SONET, SDH, DS3, E3, E1, T1 etc. ATM allows for interfaces operating at a wide range of speeds. Initially, it was defined to operate only at 155 and 622 Mbit/s. As time moved on, many other rates have been defined, as high as gigabits and as low as the two Primary rates of 1.544 and 2.048 Mbit/s. This range of operating speeds allows the development of new high-speed applications without needing to change the communications infrastructure. Furthermore it allows the easy migration from lower-speed systems used currently to the faster newer ones in the future. The ITU-T started development of standards for ATM in the late 1980s. The ATM Forum was formed in 1991 to speed up the development of standards so that the networks could be deployed earlier to satisfy demand.

ATM is a technology that has been developed to allow the integration of voice, video and data into one seamless network. All traffic types are carried in the same unit of transmission, called 'The ATM Cell'.[4] The network will carry the cells from the source to the destination based upon a connection that was established prior to traffic flow. Each connection will be associated with a Quality of Service (QoS), which ensures that the traffic being carried on that particular connection is treated in a way appropriate to the application. So really the asynchronous nature of ATM does not fit into the synchronous nature of a physical link. For the concept to work consider the following example: a PC is connected to an ATM hub via a 155.52 Mbit/s link. And when the PC has data to send, it puts this data into ATM cells and transmits it to the hub. However, when the PC has no data to send, it will insert dummy cells (known as idle cells) into the link in order to maintain the link's synchronous nature. The ATM cell has a fixed length of 53 bytes, which are divided into two sections, the Header and Payload. Why did the implementation vendors of ATM decide on the size of 53 cells? It could be that the Internet people liked long packets, say 64 bytes, and the telecom people like smaller packets, say 32 bytes (for faster switching and processing), and that they both agreed on somewhere in between. However, more realistically it may be that from the standard payload of 48 bytes, 1 byte is taken away for the ATM adaptation mechanism, leaving 47 bytes. This, multiplied by 4, gives a total of 188 bytes, a number which corresponds to a simple mapping of an MPEG Transport Stream into four ATM cells using two of the ATM Adaptation Layers, AAL 1 and AAL 5. The header is 5 bytes long and is basically used as an addressing mechanism for the cell on a link-by-link basis. The payload is 48 bytes long and carries whatever the data is (video/voice/data). Because of these fixed length cells it is easy for switches to manage and re-route through the use of dedicated hardware. Should variable length cells be used, which would be called, say, frames, then a lot more processing would be needed. ATM relies on low error rate lines because it is designed to operate over digital communication links that have a very low error rate. The Higher Level data needs to be cut up into the correct size for transportation. This function is carried out by the AAL, or ATM Adaptation Layer, so that this higher layer information can be mapped into ATM cells. From here the ATM Layer takes the 48-byte payloads and adds the connection identifier before handing over the cell to the Physical Layer. The cells are then transmitted through the ATM network via a path, which has either been configured to deal with a manually set connection known as a PVC (Permanent Virtual Circuit) or SVC (Switched Virtual Circuit). A Permanent Virtual Circuit is a circuit similar to a lease line, yet over some form of packet switched network. Once a PVC is set up, usually through a management system, which is supporting the network that it intends to use, it will exist until it is removed (again normally by the management system). A Switched Virtual Circuit is a circuit that only exists for the duration of a session, after which it is normally disconnected. ATM offers both PVC and SVC, and a service provider will inform the user as to the connection numbers to use to send traffic to a particular connection. Normally for the PVC, which is like a leased line, the transmit number and the receive number do not change, but with the SVC the number will almost certainly change on nearly every link or segment of the network, or at least part of the number will change because of routing, queuing and congestion.

There are two basic types of connection within an ATM network. Point-to-Point connections are simply common connections between two devices across the network, and these may be uni- or bidirectional in that the data may flow in one or both directions. Point-to-Multipoint is where the single source (known as the root node) sends the ATM cells to multiple destinations (known as leaf nodes). To do this the network uses a function called Cell Replication, where switches at a connection split into two or more 'branches' that copy the cell and send to multiple destinations. Because of this the connections are only unidirectional, allowing the root to transmit to the leaves, but not vice versa. An ATM network normally consists of a number of ATM switches connected by links. The four main layers, which make up the 'ATM Stack' in any adapter device, the Higher Layer, the ATM Adaptation Layer (which chops up the data into 48-byte chunks), the ATM Layer (which adds a header and is really implementing how ATM works) and the Physical Layer (which is the interface to whatever type of medium is being used). To understand the way ATM works, it may be easier to firstly deal with the ATM Layer as critical features are determined here and then look at the ATM Adaptation Layer and Physical Layer afterwards.

There are two types of ATM cell formats, the UNI and the NNI. The only difference in these cell structures is that in the UNI cell there are 4 bits used for Generic Flow Control and in the NNI cell these bits are used for VPIs. Figure 1.8.7 describes the layout and the order of bit transmission of a UNI cell.

Generic Flow Control (GFC) was originally thought to allow for multiple ATM devices to be multidropped onto a single UNI, but the use of GFC is another matter. ATM is a connection-orientated protocol and as such there is a connection identifier in every cell header, which associates a cell with a particular virtual connection on a physical link. This identifier comprises of two parts, the Virtual Path Identifier (VPI) and the Virtual Channel Identifier (VCI), sometimes called Indicators, that are used by the system for multiplexing, demultiplexing and switching a cell through the network. They are not addresses that identify locations of a particular device on the network; instead they identify a single connection within the network along which a cell travels. On reaching a switch they will then be issued with another VPI/VCI and continue to its next routed switch or device. Overall, the idea of switching is to

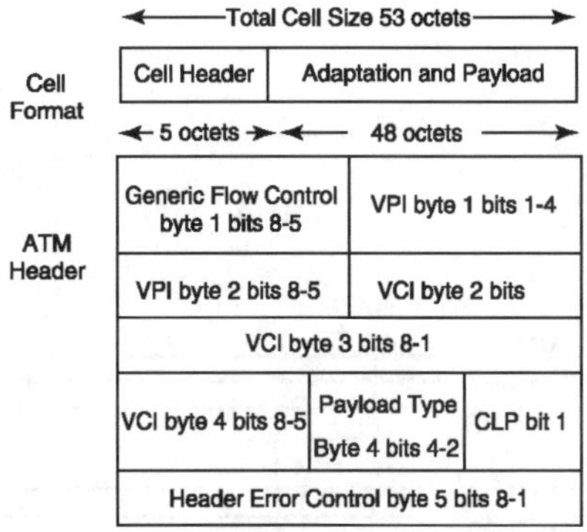

Figure 1.8.7 Layout and bit transmission of a UNI cell.

take in cells on one port, and after inspecting the VPI/VCIs and adding a new header, transmit them from another port to the next link. In practice there are two types of switching: Virtual Channel Switching takes in a cell on one port and inspects the VPI/VCI to determine where to route that particular connection. As shown in Figure 1.8.8, consider a physical link (A), which carries two VPIs, each with three VCIs. The switch port takes VPI 5/VCI 32, which the switch inspects for the correct information that matches and, in accordance with the translation table, knows that this virtual connection needs to be switched through to link C, VPI 7/VCI 86. Virtual Path Switching takes the cell in on one port and inspects only the VPI to determine which path to route that particular connection. In this way the same VPI coming in will always be switched to the same destination port. The Payload Type (PT) is used to differentiate between cells that contain user information from those that do not (i.e. Management Information). The first bit of the PT indicates whether it is a user or non-user cell. If it is a user cell, then bit 2 is used to indicate whether a network element is congested or not. When AAL 5 is used, bit 3 is used to indicate the last cell associated with the user packet. A switch sometimes uses this to perform EPD (Early Packet Discard) and PPD (Partial Packet Discard). The Cell Loss Priority within the ATM cell is very important and probably the most import single bit within the ATM Traffic Management. Simply it allows the network to discard a cell (providing CLP is set) in the event of there not being enough bandwidth available to pass all the other cells that are contending for bandwidth. The purpose of the CLP bit is to identify the cells that should be discarded before cells that do not have the bit set. It can be set by either an ATM terminal or by the network; for example, a user may be buying a low-cost service from a carrier, which requires that users traffic be dropped first in the event of a problem, or possibly, the user may wish to use it as a method prioritisation of traffic when they know they are exceeding an agreed traffic contract. If the bit is set to '0', then it is deemed higher priority than a cell carrying a PT of '1'. To produce the cells in the first place, a function known as ATM Adaptation is performed, taking the original information from the real user. Since this process converts or translates information that can be put into an ATM cell (and the other way round), it is only performed at the end-user's location. The ATM Adaptation Layer takes these packets, chops them up and puts them into the 48-octet payload size. The ATM Layer puts in the VPI/VCI information and hands the 'Package' to the Physical Layer.

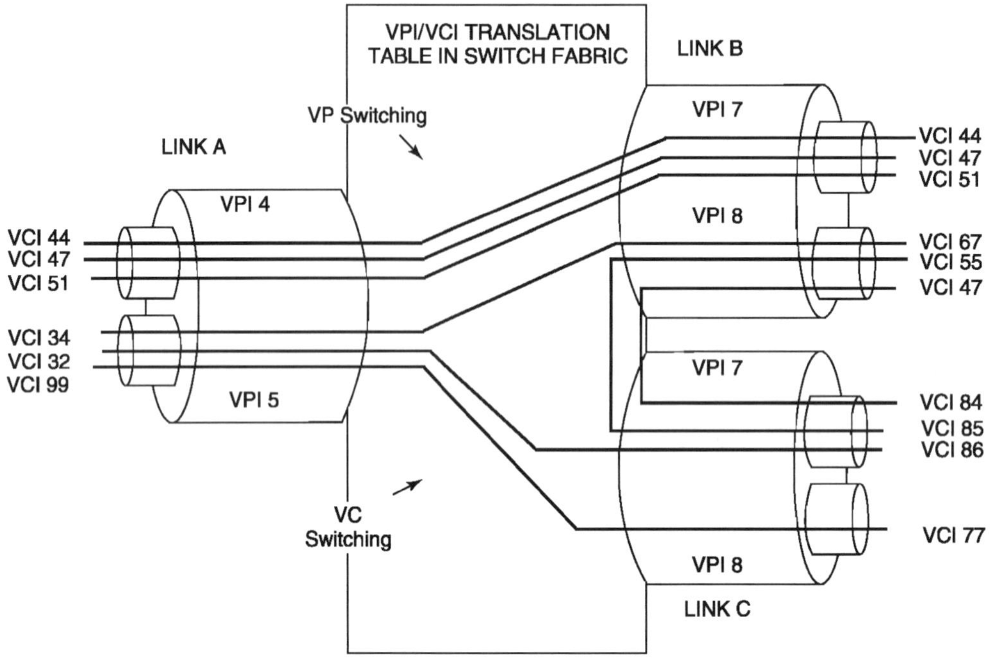

Figure 1.8.8 Virtual Path and Virtual Channel Switching.

Obviously ATM cells of 53 octets, where only 48 bytes represent the payload, are sufficient to cope with the wide range of types of data and applications that has to be supported in a communications network. As such, the AAL is used to translate between the user data and the data within the payload cells.

The only two devices that are concerned with AAL are the two end devices. The two types, which are used for digital broadcasting at high speeds over fibre or electrical media, are AAL 1 and AAL 5. AAL 1 is designed for Constant Bit Rate with end-to-end timing support. To format the data stream or video stream into cells, AAL 1 does the following things. It takes the 47-octet chunk and adds a header which comprises of two parts, the SN (Sequence Number Field) and the SNP (Sequence Number Protect Field). The SN Field is broken down into two parts, the CSI (Convergence Sublayer Indication) bit, which for unstructured data transfer is set to '0', and 3 bits for the sequence itself. These sequence number cycles identify missing or mis-inserted cells and are called Sequence Count Field. The SN Field is protected by the SNP Field, which comprises of a 3-bit CRC check and a 1-bit parity field. Figure 1.8.9 shows the 'segmentation' carried out by AAL 1 and in turn the 'reassembly' by the layer at the receive end. The sequence number can detect up to seven lost or mis-inserted cells.

Forward Error Correction (FEC) RS (199-204) coding and frame interleaving can also be used providing protection for four-cell losses in each 128-cell block. Effectively an MPEG-2 Transport Stream 188 packet will map into four ATM cells (4×47 byte payload $= 188$). AAL 5 started life being called 'SEAL' (Simple Efficient Adaptation Layer) and was developed by the data industry. It still does error protection, but unlike other AALs, which work on a cell-by-cell basis, it does it on a complete frame of data, as presented by the higher level. AAL 5 is ideal for sending video data over IPoA (IP over ATM). It maps into a frame size of eight 48-byte cells. An 8-byte trailer is inserted into the last cells, including a 4-byte CRC which enables cell losses and bit error to be detected, not corrected. Error corrections are based on a frame as opposed to a cell.

Therefore, using this mapping technique the ATM overheads in both AAL 1 and AAL 5 are the same. To summarise the AAL 5 mechanism, it requires very little processing and all payloads are data except for the last one. The 8-byte trailer on the end of the last cell contains a 4-byte field CRC; a 2-byte length field, which allows the receiver to calculate how many padding bytes were used, so they can be subtracted; a 1-byte CPI field (which is currently under further study); a 1-byte CPCS-UU (Common Part Convergence Sublayer User-to-User) field, which transfers information end-to-end; and finally a PAD field of up to 47 bytes to ensure the PDU (Packetised Data Unit) carrying the data is segmented into 48-byte ATM payloads. The Physical Layer is the point where the data actually leaves the edge device to enter the telcom network.

Cells are cells whether they are carried on fibre at high speeds of 155 Mbit/s or copper at lower speeds of 2.048 Mbit/s. But the way the information is carried from A to B can greatly differ. HEC (Header Error Control) generation/verification is a function in which the ATM layer passes the cells to the Physical Layer, which calculates the HEC and inserts it into the relevant position within them. Cell scrambling may be used.

Some Physical Medium-Dependent Sublayers are not deemed robust enough to handle excessive rows of 1s and 0s, and possibly synchronisation is not guaranteed. The HEC is

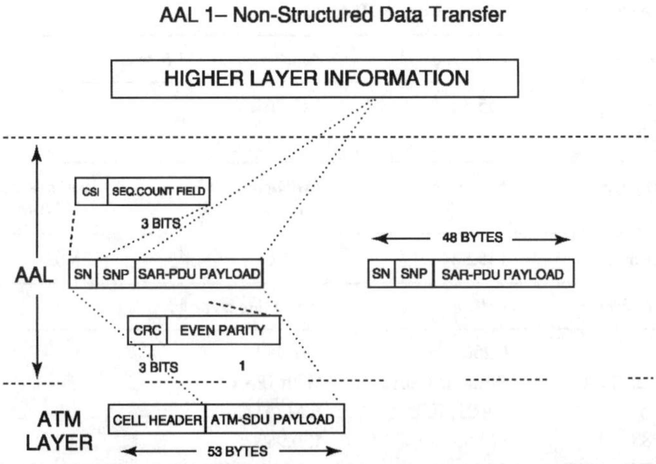

Figure 1.8.9 AAL 1: non-structured data transfer.

used at the start of cells. Most physical interfaces are synchronous and require that cells be sent continuously. Cell Rate Decoupling ensures cells are transmitted on a link even if there are no traffic cells being sent. It basically adds in unassigned cells to make up any shortfall. Figure 1.8.10 shows a list of some of the Physical Layer Interfaces currently in use. Traffic Control is one of the biggest challenges facing the operation of ATM networks. Since an ATM network works on an asynchronous basis and bandwidth is only used when traffic is required to be sent to and from a device, when the bandwidth is not being used by one user it can be used by someone else (Statistical Multiplexing). Once the network has been set up and connections have been established across the ATM network, in order to ensure the traffic does not exceed the traffic contract and thus affect others, a function known as UPC Usage Parameter Control is performed. UPC basically monitors the traffic sent to the network on a particular connection (VC or VP) and if it exceeds the traffic contract then cells may be discarded. Along with the ATM Forum Categories[5] and Traffic Parameters are ATM Service Parameters, which define a quality of connection between two end-points of a user connection. These fall into two subcategories, Negotiated and Not–Negotiated (Network Dependent). The Negotiated parameters are Peak-to-Peak Cell Delay Variation (CDV), which is the variation in delay between the transferred cells across an ATM network.

PHYSICAL LAYER INTERFACES

SPEED	FORMAT	TRANSMISSION MEDIA
622.08 Mbits	STS-12/STM-4	SMF, MMF
155.52 Mbits	STS-3C/STM-1	SMF, MMF, Coax, UTP-5, UTP-3
100 Mbits	FDDI	MMF
51.84 Mbits	STS-1	SMF, MMF, Coax, UTP-5, UTP-3
44.736 Mbits	DS-3	Coax
34.368 Mbits	E3	Coax
25.6 Mbits		UTP-5, UTP-3, STP
6.312 Mbits	T2	Coax
2.048 Mbits	E1	Coax, Twisted Pair
1.544 Mbits	T1	Twisted Pair

Figure 1.8.10 Physical Layer Interfaces.

CDV refers to the difference between the best and worst cases of CTD (Cell Transfer Delay). Maximum Cell Transfer Delay (CTD) is the time taken between a cell leaving, for example, the source UNI and arriving at the destination UNI for a particular connection. Cell Loss Ratio (CLR) is the ratio of errored cells to the number of transmitted cells on a particular connection. Not–Negotiated parameters are Cell Error Ratio (CER), essentially the ratio of errored cells to the total number of cells transmitted on a particular connection; Severity Errored Cell Block Ratio (SECBR), which is the ratio of severely errored cell blocks to the total number of transmitted cell blocks; and Cell Mis-insertion Rate (CMR), the number of mis-inserted cells (e.g. due to undetected error in the cell header) in a given time interval. Calculations for these ratios are displayed in Figure 1.8.11. Causes of CDV are various factors including Switching Fabric, which because of contention within a switch may result in cells from one connection slowing down cells from another. Buffering occurs where buffers at the input

The ITU-T Recommendation 1356 defines the parameters that are used to produce the results of a QoS test:

$$\text{Cell Loss Ratio (CLR)} = \frac{\text{Number of cells lost}}{\text{Total number of cells transmitted}}$$

$$\text{Cell Error Ratio (CER)} = \frac{\text{Number of errored cells}}{\text{Total number of cells transmitted (including errored cells)}}$$

$$\text{Cell Mis-insertion Rate (CMR)} = \frac{\text{Number of wrongly inserted cells}}{\text{Time interval}}$$

Cell Transfer Delay (CTD) is the time between t_1 and t_2 of a test cell
Where:
t_1 = time the cell enters the device under test
t_2 = time the cell leaves the device under test

Figure 1.8.11 Definitions of ITU-T parameters.

Table 1.8.2 Example of overheads (STM-1, 155 Mbit/s) using SDH framing

Framing	*Frame payload*	*Base rate*	*Available rate (53-byte call)*		*Note*
STM-1	Direct mapped 260/270	155.52	149.76		Simple mapping
ATM header	*Frame payload*	*Base rate*	*Available rate (53-byte cell)*	*Available rate (48-byte payload)*	*Note*
STM-1	Direct mapped	155.52	149.76	135.63	Simple mapping
Switched system	*ILMI (2.5%)*	*UNI*	*Remaining payload*		*Note*
STM-1	3.391	0.256	131.983		(Approx.)
AAL 1, FEC	Without IPoA	Without IPoA	With IPoA		With IPoA
AAL	AAL 5	AAL1, FEC	AAL 5		AAL1, FEC
STM-1	131.983	125.95	130.983		124.246

and output of the switch are at different fill states at different times depending upon loading of the switch and the output link. This will result in different delays for the cells. In practice, mechanisms will probably be put in place to prioritise cells from different connections depending upon the service category they are associated with (CBR, UBR etc.) and the QoS parameters requested from the network.

The ITU-T have defined a number of QoS classes, namely Classes 1 to 4, in which the CBR commonly used by digital broadcasters is Class 1. Traffic Management Specification V.4.0 changes the situation by defining some parameters which may be optionally negotiated with the network at call set-up and some which are network dependent, because ATM in most cases is connection orientated and therefore requires a connection to be established before information can flow. In the case of PVCs then the connections are established via the management system. However, to use SVCs (connections set up on demand) signalling functions are required. Both UNI V.3.1 and V.4.0 use a signalling protocol that is based on, and compatible with, ITU-T Q.2931. All UNI signalling is performed using a pre-defined VPI/VCI combination of VPI 0/VCI 5. This signalling is performed in-band and will incur a certain amount of overhead, which along with other overheads will be discussed at the end of this subsection. Indeed, there are several reserved VPI/VCI combinations. It is quite common for broadcasting equipment for adaptation to ATM cells to have default settings of above the reserved range, typically VPI 0/VCI 32. So very rarely will a combination of below this be used for an active service. Further control protocols are used to initialise address registration. Integrated Local Management Interface (ILMI) is based on SNMP version 1 and the ILMI MIB. ILMI uses a reserved combination of VPI 0/VCI 16. This will incur further overheads. On initial set-up an exchange of information between the user and the UNI (switch on edge of network) will result in the network allocating a 13-byte field that includes the AFI (Authority and Format Identifier), which identifies the address type – private, British Standards Institute, ANSI etc.; an IDI (Initial Domain Identifier), which identifies the country, region or enterprise etc.; a DSP (Domain Specific Part), which is really the subnet ID; and an HO-DSP (higher order), which identifies the destination switch port address. The user will then reply with a 7-byte field containing a 6-byte ESI (End-System Identifier), which is used by the switch to find the end-system interface, and a 1-byte field to identify the service/application within the end system called the SEL (Selector). A typical private address could look like this: 490000.0000000000003366BB04.002468000111.00, where 490000. is the AFI, 0000000000003366BB is the subnet, 04 is the switch, 002468000111 is the hardware or MAC address and 00 is the SEL, making a full ATM address of 20 bytes. Various overheads associated with framing, the ATM Header, ATM Adaptation Layers, Signalling and IPoA (ATM over IP) are shown in Table 1.8.2.

1.8.10 Pleisochronous Digital Hierarchy (PDH)

Plesiochronous means 'nearly synchronous', which in turn means that the internal clocks of the equipment are 'free running', not locked as in synchronous equipment. This causes problems today that were not envisaged when this particular hierarchy was introduced 20–30 years ago. Multiplexes in one hierarchy are demultiplexed then remultiplexed into another hierarchy. What service operators would like to do back in the mid-1990s was to run an optical fibre cable through an industrial area dropping of 140 Mbit/s here, 34 Mbit/s there and even the odd 2 Mbit/s to a PBX. But PDH is not suitable for this task; the odd 2 or 1.5 Mbit/s cannot be dropped off without demultiplexing the entire high-bit-rate stream. Every manhole in the estate would have to be crammed with expensive PDH multiplexers. PDH is now being gradually replaced with SDH (Synchronous Digital Hierarchy). Bit rates for European Standards for PDH systems are 2.048 Mbit/s (E1), 8.448 Mbit/s (E2), 34.368 Mbit/s (E3) and 139.264 Mbit/s (E4). The North American Standard bit rates are 1.544 Mbit/s (DS1), 6.312 Mbit/s (DS2), 44.736 Mbit/s (DS3) and 374.176 Mbit/s (DS4). As PDH connections were to be multiplexed up to the higher speeds of SDH networks there was a problem in that the E2 8.448 Mbit/s connection could not map into the STM-1 (Synchronous Transport Module) container. PDH multiplexes are normally found in urban or rural street cabinets patching over from SDH connections and small business concerns. It is quite common for PDH multiplexes that connect subscribers across an internal bus called the P Bus (Plesiochronous Bus)[6] to have an S Bus implemented in the same device and have an STM-1 optical output of 155.52 Mbit/s, which is the first common transport rate for an SDH ring. The North American Standard for the common transport rate is OC-1 (Optical Carrier 1), which has a capacity of 51.84 Mbit/s.

1.8.11 Synchronous Digital Hierarchy (SDH)

Synchronous Digital Hierarchy (SDH) and Synchronous Optical NETwork (SONET) refer to a group of fibre-optic transmission rates that can transport digital signals with different capacities. SDH has provided transmission networks with a vendor-independent and sophisticated signal structure that has a rich feature set. This has resulted in new network applications, the deployment of new equipment in new network topologies, and management by operations systems of much greater power than previously seen in transmission networks. As digital networks increased in complexity in the early 1980s, demand from network operators and their customers grew for features that could not be readily provided within the existing transmission standards. These features were based on high-order multiplexing through a hierarchy of increasing bit rates up to 140 or 565 Mbit/s in Europe and had been defined in the late 1960s and early 1970s along with the introduction of digital transmission over coaxial cables. Their features were constrained by the high costs of transmission bandwidth and digital devices. The development of optical fibre transmission and large-scale integrated circuits made more complex standards possible. There were demands for improved and increasingly sophisticated services that required large bandwidth, better performance monitoring facilities and greater network flexibility. They were to unify transmission networks worldwide. The new standard appeared first as SONET, drafted by Bellcore in the United States, and then went through revisions before it emerged in a new form compatible with the international SDH. Both SDH and SONET emerged between 1988 and 1992. SONET is an ANSI standard; it can carry as payloads the North American PDH hierarchy of bit rates: 1.5/6/45 Mbps, plus 2 Mbps (known in the United States as E-1). SDH embraces most of SONET and is an international standard,[7] but it is often regarded as a European standard because its suppliers – with one or two exceptions – carry only the ETSI-defined European PDH bit rates of 2/34/140 Mbps (8 Mbps is omitted from SDH). Both ETSI and ANSI have defined, detailed SDH/SONET feature options for use within their geographical spheres of influence. The original SDH standard defined the transport of 2/34/45/140 Mbps within a transmission rate of 155.52 Mbps and is being developed to carry other types of traffic, such as Asynchronous Transfer Mode (ATM) and Internet Protocol (IP), within rates that are integer multiples of 155.52 Mbps. The basic unit of transmission in SONET is at 51.84 Mbps, but in order to carry 140 Mbps, SDH is based on three times this (i.e. 155.52 Mbps (155 Mbps)). Through an appropriate choice of options, a subset of SDH is compatible with a subset of SONET; therefore, traffic interworking is possible. It is only possible in a few cases for some features between vendors of SDH and slightly more between vendors of SONET. Bit rates in long-haul systems are now using up to STM-16 2.5 Gbit/s, STM-64 10 Gbit/s and beyond. SDH defines traffic interfaces that are independent of vendors. At 155 Mbps they are defined for both optical and copper interfaces and at higher rates for optical ones only. These higher rates are defined as integer multiples of 155.52 Mbps in an $n \times 4$ sequence, giving, for example, 622.08 Mbps (622 Mbps) and 2488.32 Mbps (2.5 Gbps). To support network growth and the demand for broadband services, multiplexing to even higher rates such as 10 Gbps continues in the same way, with upper limits set by technology rather than by lack of standards as was the case with PDH. Each interface rate contains overheads to support a range of facilities and a payload capacity for traffic. Both the overhead and payload areas can be fully or partially filled. Rates below 155 Mbps can be supported by using a 155 Mbps interface with only a partially filled payload area. Terminal multiplexers provide access to the SDH network for various types of traffic using traditional interfaces such as 2 Mbps G.703 or in data-oriented forms such as Fibre Distributed Data Interface (FDDI). Via an appropriate bridge or router, ADM can offer the same facilities as terminal multiplexers, but they can also provide low-cost access to a portion of the traffic passing along a bearer. Most designs of ADM are suitable for incorporation in rings to provide increased service flexibility in both urban and rural areas (spans between ADMs are typically 60 km). An SDH cross-connect performs this function for SDH virtual containers (VCs); that is, when connecting a PDH signal, the SDH cross-connect also connects the associated SDH path overhead (POH)[8] for network management. In contrast, with telephony exchanges (central offices (COs) in North America), which respond primarily to individual customer demands, cross-connects are the major flexibility points for network management. The nature of SDH/SONET is like a very complex 'Russian Doll' technique, which fits smaller bit rates into virtual containers, which grow in size until they reach their maximum payload and overheads, and due to the mechanism, which is based on locating any one individual volume of connection at any time during transportation, this mechanism can drop-off any of these connections at any physical multiplexer. Table 1.8.3 shows the relevant bit rate, container and hierarchy for this technology.

Table 1.8.3 Bit rate, container and hierarchy for SDH/SONET

Bit rate (Mbit/s)	*PDH*	*Container*	*SDH*	*SONET*
9953.28			STM-64	OC-192
4976.64			STM-32	OC-96
2488.32			STM-16	OC-48
1866.24				OC-36
1244.16				OC-24
933.12				OC-18
622.08			STM-4	OC-12
466.56				OC-9
155.52			STM-1	OC-3
51.84				OC-1
139.264	E4	VC4		
44.736	T3	VC3		
34.368	E3	VC3		
8.448	E2			
6.312	T2	VC2		
2.048	E1	VC12		
1.544	T1	VC11		

1.8.12 Asymmetric Digital Subscriber Lines (ADSLs)

Asymmetric Digital Subscriber Lines (ADSLs) are used to deliver high-rate digital data over existing ordinary phone lines. A new modulation technology called Discrete MultiTone (DMT) allows the transmission of high-speed data. ADSLs facilitate the simultaneous use of normal telephone services, ISDN and high-speed data transmission, e.g. video. DMT-based ADSLs can be seen as the transition from existing copper lines to the future fibre cables. This makes ADSLs economically interesting for the local telephone companies. They can offer customers high-speed data services even before switching to fibre-optics. The ADSL is a newly standardised transmission technology facilitating simultaneous use of normal telephone services, data transmission of 6 Mbit/s in the downstream and Basic Rate Access (BRA). ADSLs can be seen as an FDM system in which the available bandwidth of a single copper loop is divided into three parts. The baseband occupied by POTS is split from the data channels by using a method that guarantees POTS services in the case of ADSL system failure (e.g. passive filters). Figure 1.8.12 shows the frequency spectrum of ADSLs. A possible ADSL system is illustrated in Figure 1.8.13.

A flexible way to connect various servers to a corresponding applications device is to use ATM-switches. A local ATM-switch is connected to an access module in a telephone central office. The access module is used to connect the ATM network to phone lines. In the access module the ATM data stream from the server is decomposed and routed to the corresponding

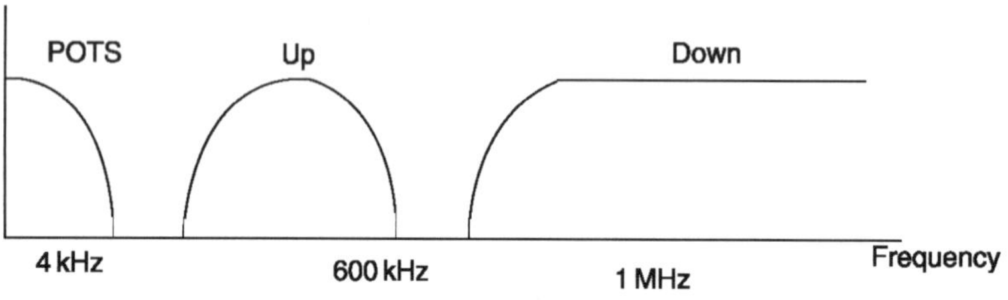

Figure 1.8.12 Frequency spectrum of ADSLs.

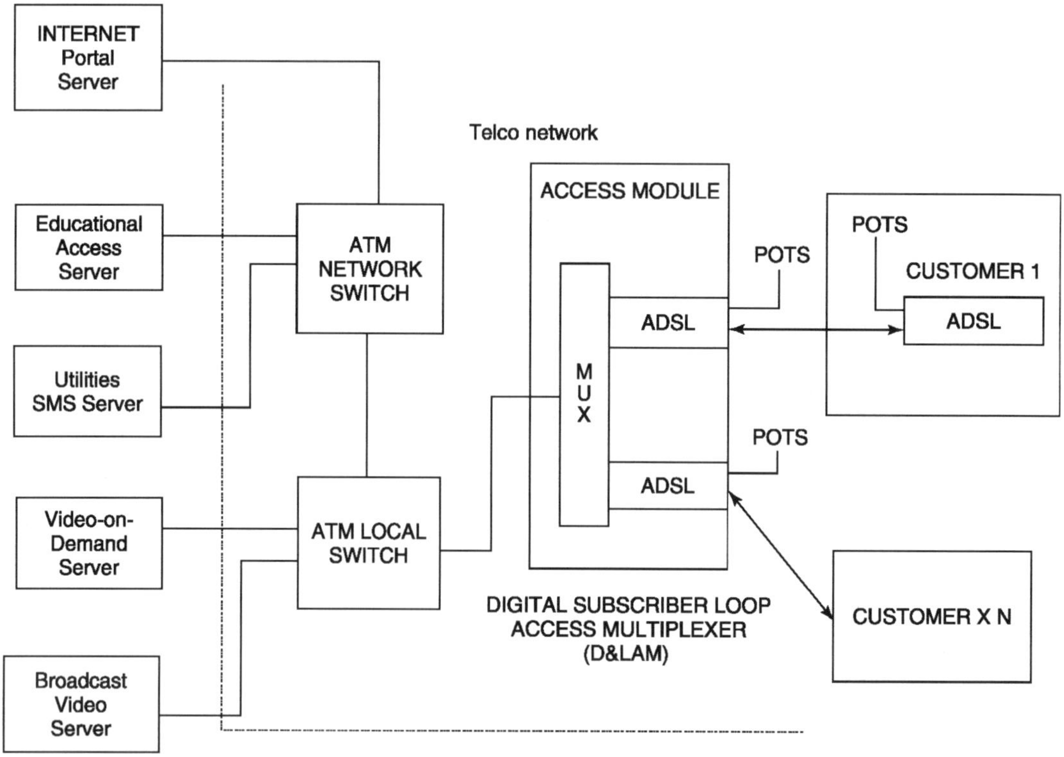

Figure 1.8.13 A possible ADSL system.

phone lines. An employee using a work-at-home server can take full advantage of the high-speed capabilities of an ADSL system in many ways, e.g. running licensed software, downloading CAD, documents etc. The Video-on-Demand (VoD) service is one of the most interesting aspects of ADSLs. By using MPEG-coded video it is possible to deliver video-quality movies over existing copper loops to customers. Video quality can be achieved by using only a 1.5 Mbps data rate. Together with pure VoD services, there might exist combined movie/information/advertiser services in which commercial and non-commercial information providers and advertisers can deliver their information. It is possible to achieve higher data rates of 52 and 155 Mbps, corresponding to ranges of one mile and a quarter mile, if the used transmission medium is fibre. ADSL upstream transport capacity is 0–640 kbit/s depending on transport class. The downstream and upstream data channels are synchronised to the 4 kHz ADSL DMT (Discrete MultiTone) symbol rate, and multiplexed into two separate data buffers (fast and interleaved). ADSLs[9] use the superframe structure (See Figure 1.8.12). Each superframe is composed of 68 ADSL data frames, which are encoded and modulated into DMT symbols. From the bit level and user data perspective, the DMT symbol rate is 4000 baud (period = 250 μs). Because of the sync symbol inserted at the end of each superframe, the transmitted DMT symbol rate is 69/68 × 4000 baud. Eight bits per ADSL superframe are reserved for the CRC and 24 indicator bits (ib0–ib23) are assigned for OAM functions. The 'fast' byte of the fast data buffer carries either CRC or synchronisation bits. Each user data stream is assigned to either the fast or the interleaved buffer during initialisation. Forward Error Correction (FEC) is used to assure optimal performance. It is based on Reed–Solomon coding and it must be implemented. The size of the Reed–Solomon code word is $N = K + R$, in which the number of check bytes R and code word size N vary depending on the number of bits assigned to either fast or interleaved buffer. The Reed–Solomon code words in the interleave buffer are convolutionally interleaved. The interleaving depth values are either 16, 32 or 64 (32 or 64 for 2.048 Mbit/s based systems). A DMT[10] time-domain signal has a high peak-to-average ratio (its amplitude distribution is almost Gaussian) and large values may be clipped by the D/A-converter. The error signal caused by clipping can be considered as an additive negative impulse for the time sample that was clipped. The clipping error power is almost equally distributed across all tones in the symbol in which clipping occurs. Clipping is therefore most likely to cause errors on those tones that have been assigned the largest number of bits (and therefore have the densest constellation).

The system performance can be improved by block processing of Wei's 16-state four-dimensional trellis code. It is possible to achieve 2–3 dB better coding gain and the overall improvement in coding gain by a well-designed ADSL system can be about 5.5 dB. The transmitter includes all analogue transmitter functions: the D/A-converter, the anti-aliasing filter, the hybrid circuitry and the MTS splitter. The channel attribute values determined by the initialisation procedure include the number of bits and relative power levels to be used on each DMT subcarrier, as well as any messages and final data rate information. Bit swapping enables an ADSL system to change the number of bits assigned to a subcarrier, or change the transmit energy of a subcarrier without interrupting data flow. The basic idea of DMT is to split the available bandwidth into a large number of subchannels. DMT is able to allocate data so that the throughput of every single subchannel is maximised. If some subchannel cannot carry any data, it can be turned off and the use of available bandwidth is optimised. First an equal number per tone is transmitted to measure the characteristics of the line. The processing of the signal takes place in ATU-R, and the optimised bit distribution information will be delivered for ATU-C by using the same phone line at a secure low speed. The first example describes a segment of 24-gauge twisted pair phone line. Low frequencies are eliminated by the transformer coupling. The attenuation at the higher frequencies depends on the length of the phone line. The second example includes the notch in spectrum that is illustrative of bridge taps and also the interference of an AM radio station. A third example shows that DMT is also an interesting possibility for other transmission channels, such as coaxial cable-TV networks, as well. In ADSL DMT systems the downstream channels are divided into 256 4-kHz-wide tones. The upstream channels are divided into 32 subchannels. Carrier 64 ($f = 276$ kHz) is reserved for a pilot. The data modulated onto the pilot subcarrier shall be constant 0,0. Use of this pilot allows resolution of sample timing in receiver modulo-8 samples.

1.8.13 Very-High-Bit-Rate Asymmetrical Digital Subscriber Line (VDSL)

The use of fast Internet connections has grown rapidly over the last few years. As more people buy home computers and create home networks the demand for broadband (high-speed) connections steadily increases. Two technologies, cable modems and ADSLs, currently dominate the industry. However, another DSL technology known as Very-high-bit-rate DSL (VDSL) is seen by many as the next step in providing a complete home communications/entertainment package. There are already some companies, such as US West (now part of Qwest), that offer a VDSL service in selected areas. VDSL provides an incredible amount of bandwidth, with speeds up to about 52 megabits per second (Mbps). Compare that with a maximum speed of 8–10 Mbps for ADSL or cable modem and it's clear that the move from current broadband technology to VDSL could be as significant as the migration from a 56K modem to broadband. Incredible speeds are achievable, as high as 52 Mbps downstream (to your home) and 16 Mbps upstream (from your home). That is much faster than ADSL, which provides up to 8 Mbps downstream and 800 Kbps (kilobits per second) upstream. But VDSL's amazing performance comes at a price: it can only operate over the copper line for a short distance, about 4000 feet (1200 m). The key to VDSL is that the telephone companies are replacing many of their main feeds with fibre-optic cable. In fact, many phone companies are planning Fibre To The Curb (FTTC), which means that they will replace all existing copper lines right up to the point where your phone line branches off at your house. At the least, most companies expect to implement Fibre To The Neighbourhood (FTTN). Instead of installing fibre-optic cable along each street, FTTN has fibre going to the main junction box for a particular neighbourhood.

By placing a VDSL transceiver in your home and a VDSL gateway in the junction box, the distance limitation is neatly overcome. The VDSL alliance, a partnership between Alcatel, Texas Instruments and others, supports VDSL using a carrier system called Discrete MultiTone (DMT). According to equipment manufacturers, most of the ADSL equipment installed today uses DMT. The other VDSL group is called the VDSL

Coalition. Led by Lucent and Broadcom, the Coalition proposes a carrier system that uses a pair of technologies called Quadrature Amplitude Modulation (QAM) and Carrierless Amplitude Phase (CAP). There is a possibility that VDSL will encompass both standards, with providers selecting which technology they will implement across their system.

1.8.14 Universal Mobile Telecommunications Service (UMTS)

UMTS (Universal Mobile Telecommunications Service) is a so-called 'third-generation (3G)', broadband packet-based transmission of text, digitised voice, video and multimedia at data rates up to 2 Mbps that will offer a service to mobile computer and phone users no matter where they are located in the world. Based on the Global System for Mobile Communication standard, UMTS, endorsed by major standards bodies and manufacturers, is the planned standard for mobile users around the world by 2002. Once UMTS is fully implemented, computer and phone users can be constantly attached to the Internet as they travel and have the same set of capabilities no matter where they travel to. Users will have access through a combination of terrestrial wireless and satellite transmissions. Until UMTS is fully implemented, users can have multi-mode devices that switch to the currently available technology (such as GSM 900 and 1800) where UMTS is not yet available. Today's cellular telephone systems are mainly circuit switched, with connections always dependent on circuit availability; packet switched connection using the Internet Protocol means that a virtual connection is always available to any other end-point in the network. It will also make it possible to provide new services, such as alternative billing methods (pay-per-bit, pay-per-session, flat rate, asymmetric bandwidth and others). The higher bandwidth of UMTS also promises new services, such as video conferencing. UMTS promises to realise the Virtual Home Environment, in which a roaming user can have the same services to which the user is accustomed when at home or in the office, through a combination of transparent terrestrial and satellite connections. The electromagnetic radiation spectrum for UMTS has been identified as frequency bands 1885–2025 MHz for future IMT-2000 systems, and 1980–2010 and 2170–2200 MHz for the satellite portion of UMTS systems.[11]

1.8.15 Dense Wave Division Multiplexing (DWDM)

Wave Division Multiplexing (WDM) increases the carrying capacity of the physical medium (fibre) using a completely different method from Time Division Multiplexing (TDM). WDM assigns incoming optical signals to specific frequencies of light (wavelengths, or lambdas) within a certain frequency band. This multiplexing closely resembles the way radio stations broadcast on different wavelengths without interfering with each other. Because each channel is transmitted at a different frequency, we can select from them using a tuner. Another way to think about WDM is that each channel is a different colour of light; several channels then make up a 'rainbow'. As shown in Figure 1.8.14, in a WDM system, each of the wavelengths is launched into the fibre and the signals are demultiplexed at the receiving end. Like TDM, the resulting capacity is an aggregate of the input signals, but WDM carries each input signal independently of the others. This means that each channel has its own dedicated bandwidth; all signals arrive at the same time, rather than being broken up and carried in timeslots. The difference between WDM and Dense Wavelength Division Multiplexing (DWDM) is fundamentally one of only degree. DWDM spaces the wavelengths more closely than does WDM, and therefore has a greater overall capacity. The limits of this spacing are not precisely known, and have probably not been reached, though systems are available in mid-year 2000 with a capacity of 128 lambdas on one fibre. WDM takes multiple optical signals, maps them to individual wavelengths and multiplexes the wavelengths over a single fibre. Another fundamental difference between the two technologies is that WDM can carry multiple protocols without a common signal format, while SONET cannot. There is also rapidly increasing demand on access networks, which function primarily to connect end-users over low-speed connections, such as dial-up lines, DSL, cable and wireless, to a local POP. These connections are typically aggregated and carried over a SONET ring, which at some point attaches to a local POP that serves as an Internet gateway for long hauls. As a result, it is increasingly likely that a customer now obtains many high-speed services directly from the POP, without ever using the core segment of the Internet. DWDM is the clear winner in the backbone. It was first deployed on long-haul routes in a time of fibre scarcity. Then the equipment savings made it the solution of choice for new long-haul routes, even

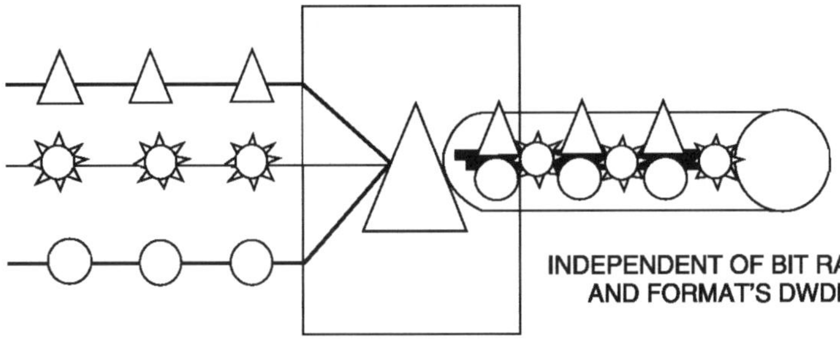

Figure 1.8.14 Dense Wave Division Multiplexing.

when ample fibre was available. While DWDM[12] can relieve fibre exhaust in the metropolitan area, its value in this market extends beyond this single advantage. Alternatives for capacity enhancement exist, such as pulling new cable and SONET overlays, but DWDM can do more. What delivers additional value in the metropolitan market is DWDM's fast and flexible provisioning of protocol- and bit-rate-transparent, data-centric, protected services, along with the need to provision services of varying types in a rapid and efficient manner in response to the changing demands of customers; this is a distinguishing characteristic of the metropolitan networks. With SONET, which is the foundation of the vast majority of existing MANs, service provisioning is a lengthy and complex process. Network planning and analysis, ADM provisioning, Digital Cross-connect System (DCS) reconfiguration, path and circuit verification, and service creation can take several weeks. By contrast, with DWDM equipment in place, provisioning a new service can be as simple as turning on another lightwave in an existing fibre pair.

Potential providers of DWDM-based services in metropolitan areas, where abundant fibre plant already exists, or is being built, include Incumbent Local Exchange Carrier (ILEC), Competitive Local Exchange Carrier (CLEC) and Inter-eXchange Carrier (IXC) network architectures. Carriers can create revenue today by providing protocol-transparent, high-speed LAN and SAN services to large organisations, as well as a mixture of lower-speed services (Token Ring, FDDI, Ethernet) to smaller organisations. In implementing an optical network, they are ensuring that they can play in the competitive field of the future. From both technical and economic perspectives, the ability to provide potentially unlimited transmission capacity is the most obvious advantage of DWDM technology. The current investment in fibre plant can not only be preserved, but optimised by a factor of at least 32. As demands change, more capacity can be added, either by simple equipment upgrades or by increasing the number of lambdas on the fibre, without expensive upgrades. Capacity can be obtained for the cost of the equipment, and existing fibre plant investment is retained. With DWDM, the transport network is theoretically unconstrained by the speed of available electronics. There is no need for Optical-Electrical-Optical (OEO) conversion when using optical amplifiers, rather than regenerators, on the physical link. Although not yet prevalent, direct optical interfaces to DWDM equipment can also eliminate the need for an OEO function. While optical amplifiers are a major factor in the ability to extend the effective range of DWDM, other factors also come into play. For example, DWDM is subject to dispersion and non-linear effects. Many components, such as the Optical Add/Drop Multiplexer (OADM), are passive and therefore continue to work, even if there is a power cut. In addition, these components tend to have a very high Mean Time Between Failures (MTBF). Protection schemes implemented on DWDM equipment and in the network designs are at least as robust as those built into SONET. All these factors contribute to better performance and lower maintenance in the optical network.

1.8.16 Mobile Communications

Whether by microwave or satellite, or by another means of transmission, there is always a telephone wire at some point. Eliminating this wire gives a wireless system. No longer restricted to a physical space, users are now able to 'roam' anywhere they wish and still communicate, whether by phone or laptop. Mobile phones are now well established, with an estimated two-thirds of the UK population owning one. The passing of data over a mobile system like text messaging is very commonplace. However, because of the low transfer rate and possibly poor marketing, more complex data transmission using these methods has led to this still being regarded as an evolving technology. These issues are now being addressed, as the growth in sales of smart phones and Personal Digital Assistants (PDAs) has recently increased. Research shows the GSM's Short Messsage Service (SMS) increased particularly among students and the younger generation. But sometimes with limitations of a maximum of typically 160 characters and speeds of 9.6 kbit/s, this is not good enough. Some providers may offer 14.4 kbit/s but this is a trade-off, as speed is achieved by decreasing the redundancy in the data streams. A service called GPRS (General Packet Radio Service) is ideal for the sending and receiving of email. Using packet-switching techniques, speeds of up to 384 kbit/s can be achieved by linking timeslots in series. Furthermore, GPRS can also travel over IP networks. In competition with GPRS is another service, HSCSD (High-Speed Circuit Switched Data). HSCSD is based on GSM (Global System for Mobile Communication). However, instead of dividing a channel into eight timeslots, with each user offered one for transmit and one for receive, HSCSD uses multiple timeslots ($N \times 9.6$ or 14.4 kbit/s), with the user given a choice of a two-slot (28.8 kbit/s) service or an asymmetric service of four timeslots (one upstream and three downstream), ideal for Internet/intranet access. Whether GPRS or HSCSD is the used technology, mobile communications providers will have to research and develop real 'data' solutions in order to meet this demand.

1.8.17 Cellular Communications

A cellular radio transmits radio signals to a base station, then onto a cellular switch that connects the wireless network with fixed PSTN or IDN. The PSTN or IDN is connected by fibre or coaxial cable to devices like telephones or other cellular phones, or to modems in host computers that may be mini or mainframes or other PCs. The base station can serve tens of wireless terminals, whilst the cellular switch may serve up to 100 base stations. The cell sizes range from 0.5 to 10 km and each adjacent cell must operate on different frequencies than the other adjacent cells, to avoid interference. The first generation of cellular systems was analogue and the UK used a system called ETACS (Enhanced Total Access Communication System).

Different countries around the world used different systems, so a pan-European system was not possible. With the advance to digital, cellular ETSI developed GSM which uses Time Division Multiple Access (TDMA) to divide each air channel into eight timeslots. Originally, 32 countries adopted GSM[13] and it is now one of the most popular implemented standards worldwide, fulfilling the need for a pan-European system. GSM is the de facto wireless telephone standard in Europe. GSM has over 120 million users worldwide and is available in 120 countries, according to the GSM MoU Association. Since many GSM network operators have roaming agreements with foreign operators, users can often continue to use their mobile phones when they travel to other countries. Figure 1.8.15 shows a schematic of a cell layout and the GSM frequency bands.

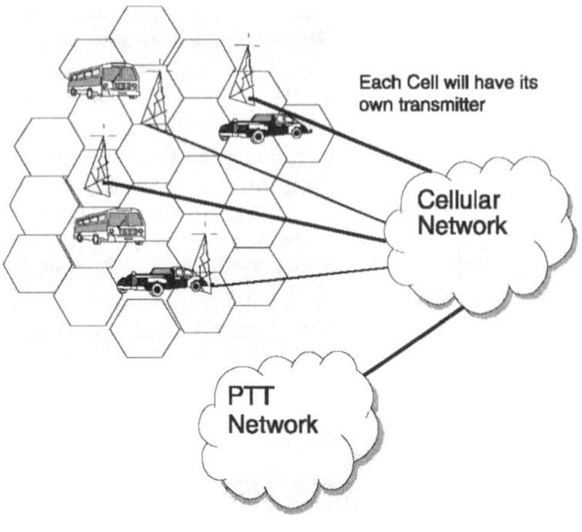

Figure 1.8.15 Schematic of a cell layout.

1.8.18 Bluetooth

Bluetooth is a computing and telecommunications industry specification that describes how mobile phones, computers and Personal Digital Assistants (PDAs) can send data using a short-range wireless connection. Using this technology, users of cellular phones, pagers and personal digital assistants such as the PalmPilot will be able to buy a three-in-one phone that can double as a portable phone at home or in the office, get quickly synchronised with information in a desktop or notebook computer, initiate the sending or receiving of a fax, initiate a printout and, in general, have all mobile and fixed computer devices totally coordinated. Bluetooth requires that a low-cost transceiver chip be included in each device. The transceiver transmits and receives in a previously unused frequency band of 2.45 GHz that is available globally (with some variation of bandwidth in different countries). In addition to data, up to three voice channels are available. Each device has a unique 48-bit address from the IEEE 802 standard. Connections can be point-to-point or multipoint. The maximum range is 10 metres. Data can be exchanged at a rate of 1 Mbps (up to 2 Mbps in the second generation of the technology). A frequency hop scheme allows devices to communicate even in areas with a great deal of electromagnetic interference. Built-in encryption and verification is provided.

1.8.19 Telco Networks

This section looks at some of the different types of digital networks that are common to the telecommunications industry and how they are accessed. Most Wide Area Networks (WANs) have been developed for use by the PTTs. A Public Data Network (PDN) is a network established and run by any PTT operator for the purpose of transmitting data. There are two types of PDN, Circuit Switched Public Data Network (CSPDN) and Packet Switched Public Data Network (PSPDN). Each type of network will have its own standards and connections. And in turn there are four basic methods of transferring data between two devices: Circuit Switching (CSPDN), which uses TDM (Time Division Multiplexing), Synchronous Digital Hierarchy (SDH) and Pleisochronous Digital Hierarchy (PDH); Packet Switching (PSPDN), which uses X.25, Frame Relay and Statistical Time Division Multiplexing (STDM); Cell Switching, primarily for ATM and SMDS; and Message Switching, which is used for Telex, electronic mail and X.400 etc. Selecting the correct method depends on the type of data to be carried and the speed of the data. Real Time Data is deemed when the data must be carried through the network with a fixed delay. Examples are voice, which has a constant transfer rate, and video conferencing, which can be bursty, but cannot tolerate delays, whereas computer generated data can tolerate delays. LAN traffic is also bursty, but can stand variable delays, as are file transfers, which are constant but can still stand variable delays. Circuit Switched Networks work by allowing multiple sessions to run between end-users by simply switching to the most direct or available path. Because a circuit switch dedicates bandwidth to each session, it is a less efficient way of working than packet switching, which provides bandwidth-on-demand. Although data transfer would normally take place via a circuit switch in block mode, i.e. blocks of octets in one direction and acknowledgements in the other, in fact any method can be used, even one octet at a time. Thus the advantage of a circuit switch is that it is transparent. And the path is always open, hence the delay through a CSPDN is constant, which is ideal for real time applications. One of the main disadvantages is that CSPDN networks are very wasteful of bandwidth, because if the device is not transmitting data, the bandwidth would be still available. Effectively, TDMs (Time Division Multiplexers) are circuit switched systems. An advantage of TDM is consistent network delays due to the fact that response times can be defined in the network and are independent of network loading. TDM would be used intensely over the 'Transport Plane', which is the backbone or core of the PTT network. Packet Switching differs from Circuit Switching in that all data is transmitted in frame/packets of data over a trunk, where each packet or frame can belong to a different session. In Packet Switching, all of the data being sent into the network is marked with a sequence number. This means that many PSPDNs can correct any data corrupted or lost within a network. Also, unlike CSPDNs, they only use the bandwidth when there is data to transmit. One of the main disadvantages of PSPDNs is due to buffering. Buffering happens when two or more packets of data are to be transmitted over the same trunk at the same time. This cannot happen, so one frame is transmitted and all the other frames have to wait and are stored in memory. Statistical Time Division Multiplexing (STDM) uses a data link protocol to guarantee the data over the STDM link. STDM systems are frequently used for asynchronous data transfers, where a high degree of compression can be achieved by removing the start, stop and parity bits (a 30% saving). Data integrity is maintained by the use of a High-Level Link Control (HDLC)-based protocol on the aggregate link for error detection and correction techniques. Some of the advantages of STDM are that it will only reserve bandwidth on the data link when a device is being used. If a device is idle, then the STDM will not reserve bandwidth, making it available for other users.

Another protocol common to networks like a simplified version of X.25 is Frame Relay, which is an end-to-end protocol and hence any errors that occur within the frame network result in the frame being discarded. A device that converts normal data to packetised data is called a Frame Relay Access Device (FRAD). Cell Switching was really developed to answer the following requirements. Because of the variety of networks that have evolved over the years many have their own specific requirements.

Telephony requires a constant bit stream normally associated with circuit switching, whereas bursty data is normally associated with packet systems. Also, certain applications like high-quality moving pictures require high bandwidth in a range of tens of megabits per second. The fastest cell switching speed developed in laboratories is 1,270,000,000,000 bit/s = 1.27 terabit/s (Tbit/s). Cells are brought into the switch, the header (Destination Address) is read and then the cell is transmitted out, as the payload is still being brought in. This is more commonly known as 'Cut Through' switching. Because of this, although essentially cell switching is still a packetised system, very low delays are achievable because of the cut through system.

A basic cell system is SMDS (Switched Mutlimegabit Data Service), which is a connectionless high-speed digital network service based on cell relay, aimed at the LAN/WAN markets.

1.8.20 Network Access

A PTT network is broken down into three planes, known as Access, Switching and Transport. Both analogue and digital communication systems are supported by PTTs, but now all communications within most PTT networks are carried out digitally.The access plane collects traffic from both residential and business customers and then passes it into the switching layer. Thus the access plane is in a constant state of change, reflecting individual customers changing demands and traffic patterns. Essentially the access plane has to be capable of passing different types of traffic on to the switching plane, which in turn uses the transport plane, i.e. data networks would be Frame Relay, X.25, SMDS or ATM. Telephony would be PSTN or ISDN and on top of this there would be leased lines as a dedicated point-to-point link. So it is really in the access network where the most effort and money has to be spent in order to give the customers the service they demand. There are many different Digital Datalinks that can be supplied by various PTTs. Throughout Europe E1 systems (2.048 Mbit/s) to E3 (34.368 Mbit/s) are commonplace; E1 systems come in two forms, unstructured and structured. An unstructured E1 is the normal type of connection for a 2.048 Mbit/s leased line for point-to-point communications. Structured E1 links supplied by POTs use a specific ITU-T framing format called G.704, to connect a PABX into a PTT exchange for normal call connections. All E1 Systems will use the G.703 interface specification to supply a digital data link. G.704 specifies how the 64 kbit/s timeslots (TS) are distributed on a frame, the numbering of bits in a TS and the structure of the frame alignment signal form synchronising the trunk. At 2.048 Mbit/s the frame is made up of 32 timeslots, each of 64 kbit/s. These are numbered from 0 to 31. TS16 can be optionally used for passing signalling information, but when connecting to a PABX it will be always carry either analogue or digital signalling information. This allows PBXs to transmit dialling or special features depending on the protocols in use. TS0 is used for synchronisation between the two end devices. Also, it can carry a simple alarm bit to indicate line problems or no line problems and is also used for carrying the frame alignment or not the frame alignment word. TS1–15 and 17–31 can carry data or voice, and have a total bandwidth of 1.92 Mbit/s. Digital Signalling can really be defined by two main methods, Common Channel Signalling (CCS) and Channel Associated Signalling (CAS). CCS is used where a single digital link is used to interconnect two PABXs or a PABX with an exchange. Used with specific PABX protocols for features such as Three-Way Party etc., CAS is normally used for analogue Ear and Mouth Signalling to digital communications, although you could use CAS for digital-to-digital communications. CAS cannot support any features that a modern PABX supports. Both the above communications systems have to be able to communicate between different manufacturers over a digital link, so a common method must be used. There are several ITU-T Recommendations covering this area. As well as the G.703 and G.704 for electrical and framing specifications, there is G.706 for optional CRC-4 cyclic redundancy checking, G.711 for PCM encoding, and G.732 for alarm and error detection. CAS uses 4 bits in TS16 to show the signalling information for the frames sent. The 4 bits are directly associated with the frames, hence the name. With only 4 bits called, the ABCD bits to indicate signalling, only limited features can be implemented between PABXs. The first frame transmitted, Frame 0, has the MFAS (Multi-Frame Alignment Signal) signal 0000 1A11 sent in it and it has an alarm bit to indicate loss of multi-frame alignment. TS16 is disassembled by the aggregate module and the individual signalling for each channel is passed across the link with its associated voice channel. CCS means that a common protocol is used to pass signalling information for all the channels over a single timeslot. This common protocol is sent in a framing structure similar to that of an HDLC frame. Timeslot 16 signalling is sent across the network as a single 64 kbit/s data channel.

1.8.21 Other Services

This subsection looks at some of the other services which can be provided over a telecommunications network. Centrex is a public service provider's answer to the private voice network. It enables all the commonly needed PABX facilities to be provided from the public exchange, and in addition allows users at different locations to share a common numbering scheme. It is this feature in particular that can make Centrex attractive to large multi-site telecommunications users. The principal hardware feature of a Centrex installation at the user's site is the Remote Line Concentrator Module (RLCM).[14] This terminates the exchange connection, which typically is a Mega stream link or similar, and concentrates the traffic from the extensions on the site into the 30 channels available on the Mega stream connection. It also contains provision for the connection of local operators' consoles and for the local switching of on-site calls. All other facilities are provided from the public exchange, including the mapping of local telephone numbers onto the global numbering scheme of the customer's Centrex service. Centrex and Direct Dialling In (DDI) both require the allocation of a block of numbers on the public network to on-site extensions: the principal difference here is that DDI allows a block of numbers on the local exchange, while Centrex allocates numbers with reference to the overall numbering scheme of the customer's Centrex Service. Featurenet provides enhanced telephone service 'Featurephones' that are required to access many features. Ranges of featurephones are available, from the expensive (presumably for senior staff or an operator's position), to the fairly basic. The difference is mainly ease of use. The top-of-the-range will have, in addition to the usual DTMF keypad, programmable keys, programmable status lights and a large LCD. Even the cheapest will be able to access most features, but by entering actual codes from the DTMF keypad, rather than by pressing pre-programmable keys. Virtual Private Networks (VPNs) do not use a leased line, they use a public-switched network for switching between offices and for the connection into the PSTN. VPNs can give a customer both interoffice and PSTN access, with only one link per PABX, which is a more efficent application.

References

1. Website dealing with modem functionalities from an article on modem specifications, http://www.56k.com/
2. Information from training course 'Understanding Telecommunications', describing functionality of PSTN, http://www.nortelnetworks.com/
3. Website article on Q391 ISDN Specification, http://www.whatis.com/
4. McDysan, D.E. and Spohn, D.L. *ATM Theory and Applications* (Signature Edition), McGraw-Hill Series on Computer Communications (1999).
5. The ATM Forum Website definition of the User Network Interface 53-byte cell.
6. Description of PDH, http://www.nortelnetworks.com
7. Website on Synchronous Digital Hierarchy, http://www.iec.org/online/tutorials/sdh/
8. Fast Guide to ADSL, *IEE Electronics and Communication Engineering Journal* (June 1994).
9. Describing ADSL, http://www.whatis.com/
10. Website article describing DMT, http://www.cs.tut./fi/tlt/stuff/adsl/node6.html
11. Article on UMTS, http://www.whatis.com/
12. Article by Cisco Systems on introduction on DWDM for metropolitan networks, http://www.bitpipe.com/
13. Nortel Networks Training Manual, *Understanding Telecommunications*. Description of GSM fundamentals.
14. Nortel Networks Training Manual, *Understanding Telecommunications*. Description of Centrex installations.

R G Hunt D Sc, FRPS, FRSA, MRTS
Professor of Physiological Optics
The City University

1.9 Colour Displays and Colorimetry

1.9.1 Types of Colour Display

The display device in a colour television system has to be capable of receiving the red, green and blue picture signals and using them to produce the appropriate amounts of red, green and blue light at each point of the picture. The most widely used display devices are of three main types: shadow-mask tubes, Trinitron type tubes, and triple projection devices.

1.9.1.1 Shadow-mask tubes

The principle of the shadow-mask tube is illustrated in Figure 1.9.1. The red, green and blue picture signals are applied to the electron guns marked R, G and B, respectively, and all three electron beams from the guns scan the phosphor screen together. However, the screen consists of triads of dots of three different phosphors and, between the guns and the screen, a metal plate with holes in it (the shadow-mask) which ensures that the electron beam from gun R lands only on phosphor dots that produce red light, that from gun G only on dots producing green light, and that from gun B only on dots producing blue light.

The rows of dots do not have to be aligned with the lines of the picture, but moiré patterns caused by beats between the line structure and the dot pattern arise at certain angles. As these are worst at ±30° and negligible at 0° it is arranged for the lines of the picture and the lines of the dots to be more or less parallel. For 525-line displays, the shadow-mask usually has about 357,000 holes, which provide about 520 lines of holes with about 690 holes in each line; hence the maximum definition of the tube amounts to about 345 picture-point pairs along a line and 260 picture-point pairs vertically.

If the three electron beams were small enough to irradiate, on the average, not more than one line of holes and its associated triad of phosphor dots, then the tube would not restrict the definition much in a 525-line system. However this would be a rather critical condition in which to operate, and each electron beam normally irradiates more than one line of these holes and their associated triads of dots. There is therefore some theoretical loss of definition, but other factors, such as interlacing, may make the loss unimportant in practice.[1]

For 625-line displays, the tubes usually have about 440,000 holes providing about 575 lines of holes, with about 770 holes in each line. In a 56 cm tube, the distance between adjacent dots is usually only about 0.4 mm, so that very great accuracy is required in constructing these tubes.[2]

Shadow-mask tubes are often used in colour video display units (VDUs) for viewing data generated by computers. In this case the viewing distance is usually only about 0.5–1 m, instead

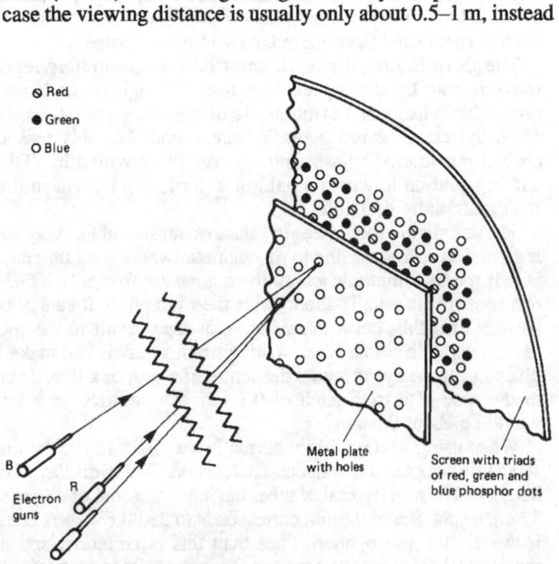

Figure 1.9.1 Principle of the shadow-mask tube.

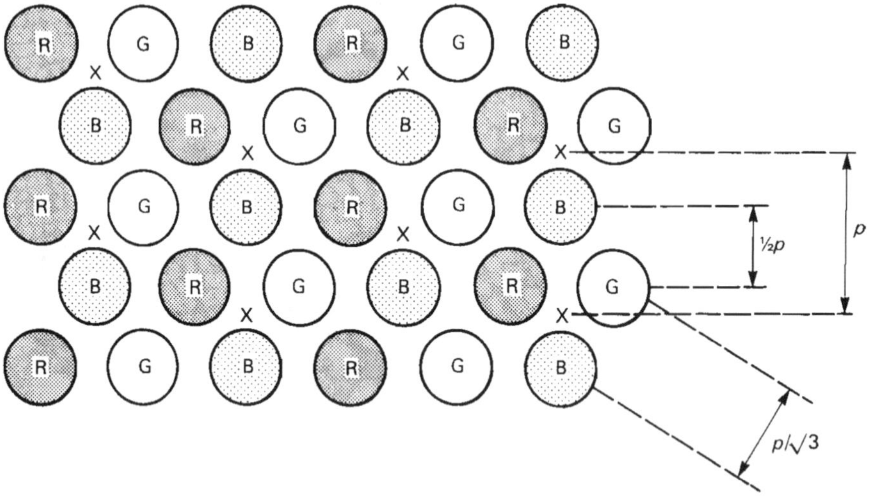

Figure 1.9.2 Arrangement of phosphor dots in shadow-mask tubes. Positions of shadow-mask holes are indicated by crosses. The triad pitch is p.

of about 2–3 m typical for viewing normal pictorial television. It is therefore necessary to use tubes having finer dot structures in VDUs. The size of the dot structure is usually quoted as the *triad pitch*; by this is meant the distance, p, between adjacent holes in the mask. Adjacent rows of phosphor dots are then separated by $p/2$ (see Figure 1.9.2), and the distance between adjacent phosphor dots is $p/\sqrt{3}$. Thus, in the case where (for typical pictorial television) adjacent phosphor dots are separated by 0.4 mm, the triad pitch is given by:

$$p = 0.4 \times \sqrt{3}$$

which is equal to about 0.7 mm. For VDUs, triad pitches of about 0.3 mm (or sometimes about 0.2 mm) are usually used. In each vertical triad pitch there are two lines of holes.

The three beams of a VDU must be very accurately registered over all the display area, because mis-registration is very noticeable when small symbols are displayed, especially, as is often the case, against a black background. For this reason, special registration adjustments are usually provided in VDUs and registration to within a half, or a third, of a television line width is usually desirable.

The nominal spot size in shadow-mask tubes may be regarded as corresponding to the diameter where the luminance is half the maximum, when all three guns are firing. In VDUs, this spot size is usually about twice the triad pitch. It cannot be smaller than this, because smaller spot sizes result in the spot having variable colour when writing small symbols and make it difficult for the eye to locate the centre of a spot or a line. Thus, in the case of a triad pitch of 0.3 mm, the nominal spot size would be about 0.6 mm.

When using VDUs it is not normally necessary to use the line standards adopted for broadcast television. For a display tube height of 280 mm (typical of tubes having a diagonal of 48 cm or 19 in), a spot size of 0.6 mm corresponds to 280/0.6 = 467 lines. However, the use of more lines than this is common, and as many as 1000 are sometimes used. The excess lines are useful in reducing the incidence of spurious patterns (aliasing) and of jagged edges to lines that should be smooth.

For pictorial television, spot sizes may be similar in diameter to the triad pitch, because small symbols are not often displayed. Thus, for a display height of 325 mm (typical of tubes having a diagonal of 56 cm or 22 in), a triad pitch and nominal spot size of 0.7 mm corresponds to 325/0.7 = 464 lines. The use of more lines than this (525 or 625) in practice, again reduces the incidence of aliasing and jagged edges.

1.9.1.2 Trinitron type tubes

A three-gun tube in which the phosphors are laid down in stripes, instead of in dots, is the Trinitron. In this tube, the three electron guns lie in the same horizontal plane, and a metal plate with vertical slots in it is positioned so that the electrons from one gun can reach only vertical stripes of a phosphor that produces red light, those from another gun only the stripes that produce green light, and those from the third gun only the stripes that produce blue light (see Figure 1.9.3).

This tube has certain advantages over the shadow-mask tube:

- Deflection of the three electron beams is easier because the gun construction enables the neck of the tube to be smaller.
- The displayed picture emits twice as much light per unit area because, for the same spot size, the beam current can be increased by a factor of 1.5 times, and the stripes of phosphor cover 1.33 times as much area of the tube faceplate.
- Vertical resolution is not affected by the screen structure so that there is moiré pattern or loss of vertical resolution by the screen.
- Adjusting the convergence to obtain registration of the three images is easier because the three beams are in a single plane.

The triads of phosphor stripes may be up to about half a millimetre wide, giving about 600 triads in a tube of 300 mm width. For equal horizontal and vertical definition the luminance signal should be able to resolve about 350 cycles per line (e.g. $525 \times \frac{1}{2} \times \frac{4}{3}$ black/white pairs in a system having 525 actual picture lines). The number of triads of vertical lines required is therefore ideally not less than about 700, but, as in

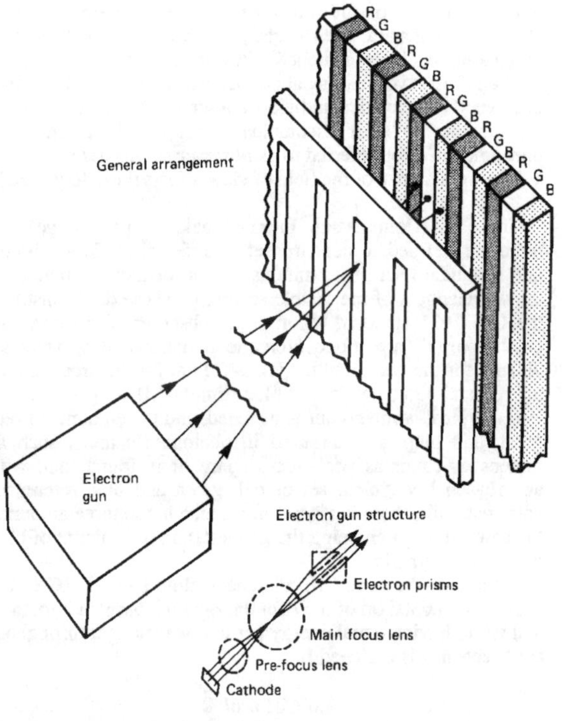

Figure 1.9.3 Principle of the Trinitron tube.

the shadow-mask tube, smaller numbers can be used without too much apparent loss of definition because the actual visual appearance is complicated by interlacing and various other factors.[1]

The Trinitron tends to be used for smaller displays than the shadow-mask tube.

1.9.1.3 Self-converging tubes

In Trinitron and conventional shadow-mask tubes, it is necessary to provide dynamic convergence correction. This is required because stronger magnetic fields are needed to bring the three electron beams into coincidence around the centre of the picture, than those required for the corners, which are farther away and therefore have longer electron paths. As the three electron beams scan the picture, the amount of convergence is therefore adjusted dynamically according to their position in the scan.

In the *precision in-line* tube,[3] the three electron guns are arranged parallel to one another in the same horizontal plane, as in the Trinitron tube. However, they do not have dynamic convergence correction, but a special deflection coil is accurately cemented to the neck of the tube. This coil is designed to converge the three electron beams on to the shadow-mask at all positions in the picture. Such a coil can be made to do this only for horizontal or for vertical fans of electron beams; in this case the horizontal fans are converged, and the vertical fans converge before the mask is reached and then separate out into short vertical lines. However, by making the shadow-mask with vertical slots, instead of holes, the efficiency with which it allows the electrons through is about 16 per cent, which is similar to that of a conventional shadow-mask tube (although less than the 20 per cent of a Trinitron tube).

After passing through the slots, the electrons hit the red, green and blue phosphors, which are laid down in stripes as in the Trinitron tube. By making the slots in the shadow-mask discontinuous, the mask is sufficiently rigid to enable it to be made with a spherical profile, as in the conventional shadow-mask tube, rather than cylindrical as in the Trinitron tube. The stripes of phosphor are 0.27 mm (0.0108 in) wide, so that each colour is repeated every 0.81 mm (0.0324 in). The geometry of the phosphor stripes, the slots and the electron guns, is arranged to result in the electrons from each gun landing on phosphor of only one colour. The electron guns are mounted 5.08 mm (0.200 in) apart from one another. The precision in-line tube is particularly suitable for small and medium picture sizes; it can also be made using dots rather than short vertical lines.

1.9.1.4 Triple projection devices

When projection television devices are being used, it is possible to use the triple projection principle by having three projection television tubes arranged so that they throw red, green and blue images onto a single reflecting screen.[4] The problems of registration of the three images have to be overcome, but when the final display is wanted in projected form, the method is usually the best to adopt. Triple projection is used for the display of colour television pictures to large audiences, and sometimes also for displaying the terrain in simulators used for training the crews of aircraft and ships.

1.9.1.5 The luminance of reproduced white

The introduction of rare-earth red phosphors enabled whites of luminance about 50 cd/m^2 to be attained. If the screen has the same luminance at all angles of viewing, this corresponds to the emission of about 50π, or 160, lm/m^2. For a screen area of 0.15 m^2, the total emission is therefore about 24 lm. At 3 lm/W this would require about 8 W, or an effective beam current of about 0.3 mA at 25 kV anode voltage.

Subsequent improvements to the phosphors, such as the use of europium activated yttrium oxysulphide for the red, and copper activated zinc cadmium sulphide for the green, made whites of about 85 cd/m^2 attainable; slight modification to the chromaticities used, and improved screening techniques, have further increased the luminance to about 120 cd/m^2 in modern tubes.[2] By filling the interstices between phosphor dots with a black absorbing material, it is possible to increase the transmission of the faceplate of the tube (which is normally grey to reduce the effects of ambient illumination), and this further increases the luminance.

High luminance in the display is desirable both because colourfulness increases with luminance, and because a given level of ambient illumination will be less harmful. Flicker caused by the field frequency, however, becomes more noticeable as the luminance rises.

1.9.2 Colorimetric Principles

1.9.2.1 Trichromatic matching

Colorimetry[5] is based on the experimental fact that observers can match colours with additive mixtures of three reference colour stimuli, normally a red, a green and a blue. This is possible because, in colour vision, the retina of the human eye transduces the incident radiant power of the light to electrical signals by means of only three spectrally different types of receptor, known as *cones*. There is a fourth spectrally different type of receptor, known as a *rod*, but the rods give only monochromatic vision with low levels of illumination. At levels high enough for colour perception to be operating effectively, it can be assumed, for the purpose of practical colorimetry, that the rods are inoperative. In Figure 1.9.4, spectral sensitivity curves typical of those believed to be characteristic of the cones are shown.

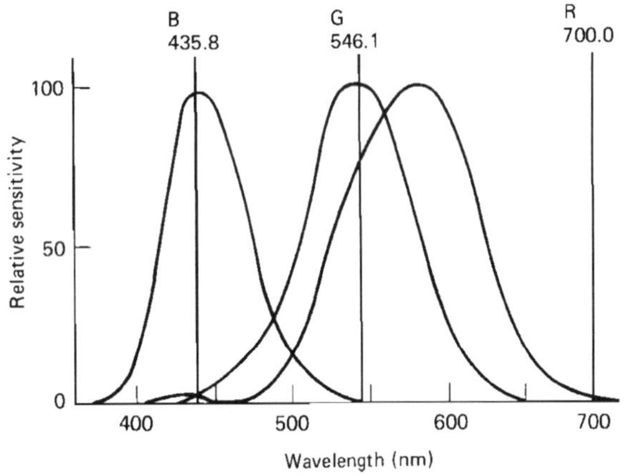

Figure 1.9.4 Probable spectral sensitivity curves of the cones of the eye, with the wavelengths of the CIE RGB primaries.

It is clear that one type of cone has a peak sensitivity at about 580 nm, another at about 540 nm, and the third at about 440 nm. Hence, red light stimulates mainly the first type of cone, green light mainly the second type and blue light mainly the third type. Therefore, if beams of red, green and blue light can be varied in their amounts, and additively mixed together, the combination can be made to produce a very wide range of excitations of the three different types of cone. By adjusting the amounts until the cone excitations are the same as those produced by another colour stimulus, a match can be made. The amounts of the red, green and blue needed to make the match can then serve as a measure of the colour of the other stimulus, and these amounts are known as *tristimulus* values.

If yet another colour stimulus, having a different spectral *radiant power distribution*, was also matched, different tristimulus values would indicate that the colour looked different, while identical tristimulus values would indicate that it looked the same. Colours having identical tristimulus values but different spectral radiant power distributions are called *metamers*, and the phenomenon *metamerism*. The greater the difference in spectral radiant power distribution between two matching colours, the greater is said to be the degree of metamerism.

1.9.2.2 The CIE 1931 standard colorimetric observer

For tristimulus values to provide a satisfactory basis for the measurements of colour, various elements of the system must be standardized.

- If the colours of the red, green and blue reference-colour stimuli are changed, even slightly, the tristimulus values for a given stimulus being matched will also change.
- Even if observers having abnormal colour vision (colour blind observers) are excluded, individual observers differ slightly from one another in their tristimulus values for a match — a phenomenon often referred to as *observer metamerism*.
- The angular size of the field of view affects the colour match.

The CIE (Commission Internationale de l'Eclairage) has therefore defined a standard set of reference-colour stimuli, and a standard set of tristimulus values for them to match all the wavelengths of the visible spectrum. These data constitute the CIE 1931 standard colorimetric observer. The reference-colour stimuli are monochromatic radiations of wavelength 700 nm for the red stimulus (R), 546.1 nm for the green stimulus (G) and 435.8 nm for the blue stimulus (B).

If a typical white colour is matched, and the amounts of red, green and blue are measured in photometric units, such as lumens or candelas per square metre, it is found that, with any reasonably typical set of red, green and blue reference-colour stimuli, there is a great imbalance in the three amounts, the amount of green being the greatest, and the amount of blue being much smaller.

Thus, with the three CIE reference-colour stimuli, RGB, it is found that 5.6508 lm of the ***equi-energy illuminant*** (a hypothetical white having equal energy per unit wavelength throughout the spectrum) is matched by:

1.0000 lm of R,
4.5907 lm of G,
0.0601 lm of B.

Because white is a colour that is not biased towards either red, green or blue, it is desirable in a colorimetric system for whites to be matched by roughly equal amounts of the three reference-colour stimuli. This is achieved by using units of different photometric magnitudes for each of the three reference colour stimuli. In the above case, 1.000 lm is still used for R, but 4.5907 lm is used for G, and 0.0601 lm is used for B. Then 5.6508 lm of the equi-energy illuminant would be matched by:

1.0000 lm of R,
1.0000 new unit of G,
1.0000 new unit of B.

Other whites, of slightly different colours, would then be matched by amounts of R, G and B that, although not exactly equal to one another, would be not very greatly different.

Using these new units for the CIE reference-colour stimuli, RGB, the amounts of them required to match a constant amount of power (per unit wavelength) of each wavelength of the visible spectrum are shown in Figure 1.9.5.

It is clear from this set of curves that, at some wavelengths, one of the three amounts is negative. Some colours cannot be matched by an additive mixture of the three reference-colour stimuli. This is most obviously the case for the blue-green part of the spectrum around 500 nm. The reason for this can be seen by referring again to Figure 1.9.4. It is clear that the R stimulus will excite only the cones whose peak sensitivity is at around 580 nm (the 580 cones), and the B stimulus will excite mainly

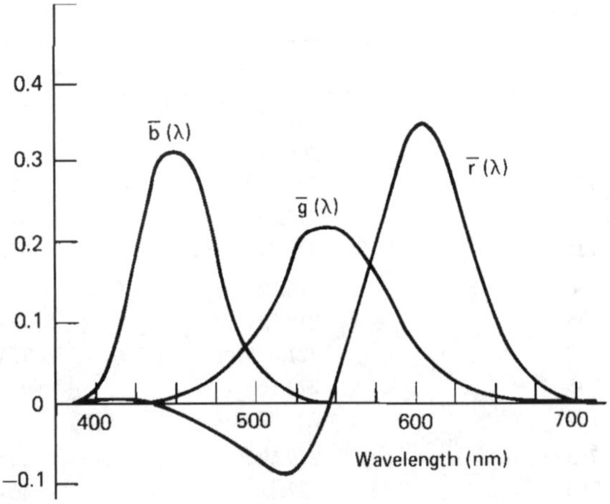

Figure 1.9.5 Colour-matching function for the CIE RGB primaries.

the 440 cones, but the G stimulus, although exciting the 540 cones most strongly, also excites the 580 cones to a considerable extent. At 500 nm, the cone excitations are approximately in the ratios of (1 of 580) : (2 of 540) : (1 of 440). But the G stimulus excites the cones in the ratio of about (1½ of 580): (2 of 540). This means that, when trying to match the blue-green of 500 nm, even without any R present, the G stimulus produces too high a ration of 580 cone to 540 cone stimulation.

The only way to make a match is therefore to add some R stimulus to the 500 nm colour, so that the combination then produces a higher ratio of 580 cone response to 540 cone response. By adding just the right amount of R stimulus to the 500 nm colour, it is found that a match can then be made by adjusting the amounts of the G and B stimuli appropriately. When this is done, the amount of R in the match is counted as negative. This problem of unmatchable colours occurs (although to different extents) with all sets of reference-colour stimuli, and is caused by the degree of overlap of the three curves of Figure 1.9.4.

The curves of Figure 1.9.5 are known as colour-matching functions and are of great importance in colorimetry. They are denoted by symbols of the type $\bar{r}(\lambda)$, $\bar{g}(\lambda)$, $\bar{b}(\lambda)$, and they enable tristimulus values to be calculated from spectral radiant power distributions (see Section 1.9.2.4).

1.9.2.3 CIE standard illuminants

The CIE defines a *source* as a physical emitter of light, such as a lamp or the sun and sky. The term *illuminant* refers to a specific spectral radiant power distribution, not necessarily provided directly by a source, and not necessarily realizable.

The following standard illuminants have been defined by the CIE for colorimetric purposes:

Illuminant A	representing light from a Planckian radiator at 2856 K
Illuminant C	representing average daylight with a correlated colour temperature of approximately 6774 K
Illuminant D_{65}	representing a phase of daylight with a correlated colour temperature of approximately 6504 K

A *Planckian radiator* is an illuminant whose spectral radiant power distribution is in accordance with Planck's radiation law, the nature of the radiation depending only on the temperature, usually expressed in kelvins.

An artificial light source of the incandescent type (e.g. a tungsten filament lamp) usually emits light than can be closely matched in colour by that from a Planckian radiator at a particular temperature, and this is said to be the *colour temperature* of the source. The spectral radiant power distributions of incandescent sources are also usually similar to those of Planckian radiators. However, the spectral radiant power distributions of discharge lamps (including fluorescent lamps) and of daylight are considerably different from those of Planckian radiators, and it is frequently impossible to achieve a close colour match with the light from a Planckian radiator at any temperature. In such cases, the light source can be given a *correlated colour temperature*, which is the temperature of the Planckian radiator yielding the nearest possible colour match.

Some correlated colour temperatures typical of sources often met with in practice are as follows:

North sky light	7500 K
Average daylight	6500 K
Xenon (arc or flash)	6000 K
Sunlight plus skylight	5500 K
Fluorescent lamps	3000–6500 K
Studio tungsten lamps	3200 K
Floodlights	3000 K
Domestic tungsten lamps	2800–2900 K
Sunlight at sunset	2000 K
Candle flame	1800 K

Spectral radiant power distributions of CIE standard illuminants A and D_{65} are tabulated at 5 nm intervals in Table 1.9.1.

Table 1.9.1 The spectral power distributions of CIE standard illuminants A and D_{65}

λ(nm)	*A*	D_{65}
380	9.80	49.98
385	10.90	52.31
390	12.09	54.65
395	13.35	68.70
400	14.71	82.75
405	16.15	87.12
410	17.68	91.49
415	19.29	92.46
420	20.99	93.43
425	22.79	90.06
430	24.67	86.68
435	26.64	95.77
440	28.70	104.86
445	30.85	110.94
450	33.09	117.01
455	35.41	117.41
460	37.81	117.81
465	40.30	116.34

Table 1.9.1 *continued*

λ(nm)	A	D_{65}
470	42.87	114.86
475	45.52	115.39
480	48.24	115.92
485	51.04	112.37
490	53.91	108.81
495	56.85	109.08
500	59.86	109.35
505	62.93	108.58
510	66.06	107.80
515	69.25	106.30
520	72.50	104.79
525	75.79	106.24
530	79.13	107.69
535	82.52	106.05
540	85.95	104.41
545	89.41	104.23
550	92.91	104.05
555	96.44	102.02
560	100.00	100.00
565	103.58	98.17
570	107.18	96.33
575	110.80	96.06
580	114.44	95.79
585	118.08	92.24
590	121.73	88.69
595	125.39	89.35
600	129.04	90.01
605	132.70	89.80
610	136.35	89.60
615	139.99	88.65
620	143.62	87.70
625	147.24	85.49
630	150.84	83.29
635	154.42	83.49
640	157.98	83.70
645	161.52	81.86
650	165.03	80.03
655	168.51	80.12
660	171.96	80.21
665	175.38	81.25
670	178.77	82.28
675	182.12	80.28
680	185.43	78.28
685	188.70	74.00
690	191.93	69.72
695	195.12	70.67
700	198.26	71.61

Table 1.9.1 *continued*

λ(nm)	A	D_{65}
705	201.36	72.98
710	204.41	74.35
715	207.41	67.98
720	210.36	61.60
725	213.27	65.74
730	216.12	69.89
735	218.92	72.49
740	221.67	75.09
745	224.36	69.34
750	227.00	63.59
755	229.59	55.01
760	232.12	46.42
765	234.59	56.61
770	237.01	66.81
775	239.37	65.09
780	241.68	63.38

1.9.2.4 The XYZ system of colour specification

The presence of negative tristimulus values in red, green and blue systems of colorimetry has led the CIE to adopt a system in which a new set of tristimulus values, XYZ, are obtained from RGB by using the following equations:

$$X = 0.49000\ R + 0.31000\ G + 0.20000\ B$$
$$Y = 0.17697\ R + 0.81240\ G + 0.01063\ B$$
$$Z = 0.00000\ R + 0.01000\ G + 0.99000\ B$$

This simple transformation was carefully designed so that all colour stimuli would have all positive values of X, Y and Z. It was also designed so that the coefficients in the equation for Y, that is 0.17697, 0.81240, and 0.01063, are in the same ratios as 1.0000, 4.5907, and 0.0601, the photometric values of the units used for expressing the amounts of R, G and B. This means that the tristimulus value, Y, is proportional to the luminance, and hence the ratio of the values of Y for any two colours, Y_1 and Y_2, is the same as the ratio of their luminances, L_1 and L_2. Hence:

$$\frac{Y_1}{Y_2} = \frac{L_1}{L_2}$$

The transformation was also designed so that, for the equi-energy illuminant, the values of X, Y and Z are equal to one another. The above equations can also be used to transform the colour-matching functions to the XYZ system, thus:

$$\bar{x}(\lambda) = 0.49000\,\bar{r}(\lambda) + 0.31000\,\bar{g}(\lambda) + 0.20000\,\bar{b}(\lambda)$$
$$\bar{y}(\lambda) = 0.17697\,\bar{r}(\lambda) + 0.81240\,\bar{g}(\lambda) + 0.01063\,\bar{b}(\lambda)$$
$$\bar{z}(\lambda) = 0.00000\,\bar{r}(\lambda) + 0.01000\,\bar{g}(\lambda) + 0.99000\,\bar{b}(\lambda)$$

These colour-matching functions, $\bar{x}(\lambda)$, $\bar{y}(\lambda)$ and $\bar{z}(\lambda)$, are shown in Figure 1.9.6; they are the most important spectral functions in colorimetry. They enable tristimulus values, XYZ, to be calculated directly from spectral radiant power data. If the radiant powers at wavelengths 1, 2, 3, ..., $P_1, P_2, P_3, \ldots$, and the values of these colour-matching functions are $\bar{x}_1, \bar{x}_2, \bar{x}_3, \ldots, \bar{y}_1, \bar{y}_2, \bar{y}_3, \ldots,$ and $\bar{z}_1, \bar{z}_2, \bar{z}_3, \ldots,$ at the same wavelengths, then the tristimulus values are given by:

$$X = k(P_1\bar{x}_1 + P_2\bar{x}_2 + P_3\bar{x}_3 + \ldots)$$
$$Y = k(P_1\bar{y}_1 + P_2\bar{y}_2 + P_3\bar{y}_3 + \ldots)$$
$$Y = k(P_1\bar{z}_1 + P_2\bar{z}_2 + P_3\bar{z}_3 + \ldots)$$

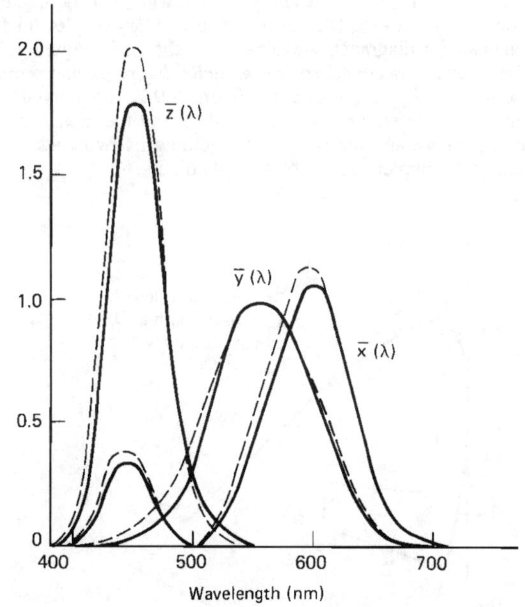

Figure 1.9.6 Colour-matching functions for the CIE XYZ primaries. Full lines for the 1931(2°) observer; broken lines for the 1964 (10°) observer.

For reflecting and transmitting samples, the constant k is usually chosen so that X, Y and Z are all equal to 100 for the perfect reflecting or transmitting diffuser (which reflects or transmits all the light at every wavelength) when similarly illuminated. The values of Y then usually give the luminance factor, reflectance factor, reflectance, or transmittance, in all cases as a percentage.

For self-luminous sources, k can be chosen so that $Y = 100$ when they are used to illuminate the perfect reflecting or transmitting diffuser. For self-luminous objects, such as typical television displays, k can be chosen so that $Y = 100$ for a suitably chosen reference white in the scene considered. In all these cases, the absolute photometric level can be indicated by quoting the luminous flux, luminous intensity, illuminance, luminance, luminous exitance or light exposure, as appropriate, in addition to the tristimulus values, X, Y and Z. However, if k is set equal to 683, and $P(\lambda)$ is the spectral radiometric quantity corresponding to the photometric measure required, then this will be given directly by the Y tristimulus value. The symbols X_a, Y_a and Z_a, can be used for such absolute tristimulus values to distinguish them from the usual relative tristimulus values X, Y and Z.

When calculating tristimulus values XYZ, or $X_aY_aZ_a$, the summations are usually carried out at 5 nm intervals throughout the visible spectrum, but intervals of 1, 10 or 20 nm may sometimes be used instead. A table of values of $\bar{x}(\lambda)$, $\bar{y}(\lambda)$ and $\bar{z}(\lambda)$ at 5 nm intervals is given in Table 1.9.2.

Table 1.9.2 The CIE colour-matching function $\bar{x}(\lambda)$, $\bar{y}(\lambda)$ and $\bar{z}(\lambda)$

λ*(nm)*	$\bar{x}(\lambda)$	$\bar{y}(\lambda)$	$\bar{z}(\lambda)$
380	0.0014	0.0000	0.0065
385	0.0022	0.0001	0.0105
390	0.0042	0.0001	0.0201
395	0.0076	0.0002	0.0362
400	0.0143	0.0004	0.0679
405	0.0232	0.0006	0.1102
410	0.0435	0.0012	0.2074
415	0.0776	0.0022	0.3713
420	0.1344	0.0040	0.6456
425	0.2148	0.0073	1.0391
430	0.2839	0.0116	1.3856
435	0.3285	0.0168	1.6230
440	0.3483	0.0230	1.7471
445	0.3481	0.0298	1.7826
450	0.3362	0.0380	1.7721
455	0.3187	0.0480	1.7441
460	0.2908	0.0600	1.6692
465	0.2511	0.0739	1.5281
470	0.1954	0.0910	1.2876
475	0.1421	0.1126	1.0419
480	0.0956	0.1390	0.8130
485	0.0580	0.1693	0.6162
490	0.0320	0.2080	0.4652
495	0.0147	0.2586	0.3533
500	0.0049	0.3230	0.2720
505	0.0024	0.4073	0.2123
510	0.0093	0.5030	0.1582
515	0.0291	0.6082	0.1117
520	0.0633	0.7100	0.0782
525	0.1096	0.7932	0.0573
530	0.1655	0.8620	0.0422
535	0.2257	0.9149	0.0298
540	0.2904	0.9540	0.0203
545	0.3597	0.9803	0.0134
550	0.4334	0.9950	0.0087
555	0.5121	1.0000	0.0057
560	0.5945	0.9950	0.0039
565	0.6784	0.9786	0.0027
570	0.7621	0.9520	0.0021
575	0.8425	0.9154	0.0018
580	0.9163	0.8700	0.0017
585	0.9786	0.8163	0.0014
590	1.0263	0.7570	0.0011
595	1.0567	0.6949	0.0010

Table 1.9.2 *continued*

$\lambda(nm)$	$\bar{x}(\lambda)$	$\bar{y}(\lambda)$	$\bar{z}(\lambda)$
600	1.0622	0.6310	0.0008
605	1.0456	0.5668	0.0006
610	1.0026	0.5030	0.0003
615	0.9384	0.4412	0.0002
620	0.8544	0.3810	0.0002
625	0.7514	0.3210	0.0001
630	0.6424	0.2650	0.0000
635	0.5419	0.2170	0.0000
640	0.4479	0.1750	0.0000
645	0.3608	0.1382	0.0000
650	0.2835	0.1070	0.0000
655	0.2187	0.0816	0.0000
660	0.1649	0.0610	0.0000
665	0.1212	0.0446	0.0000
670	0.0874	0.0320	0.0000
675	0.0636	0.0232	0.0000
680	0.0468	0.0170	0.0000
685	0.0329	0.0119	0.0000
690	0.0227	0.0082	0.0000
695	0.0158	0.0057	0.0000
700	0.0114	0.0041	0.0000
705	0.0081	0.0029	0.0000
710	0.0058	0.0021	0.0000
715	0.0041	0.0015	0.0000
720	0.0029	0.0010	0.0000
725	0.0020	0.0007	0.0000
730	0.0014	0.0005	0.0000
735	0.0010	0.0004	0.0000
740	0.0007	0.0002	0.0000
745	0.0005	0.0002	0.0000
750	0.0003	0.0001	0.0000
755	0.0002	0.0001	0.0000
760	0.0002	0.0001	0.0000
765	0.0001	0.0000	0.0000
770	0.0001	0.0000	0.0000
775	0.0001	0.0000	0.0000
780	0.0000	0.0000	0.0000

1.9.2.4.1 The CIE 1964 supplementary standard colorimetric observer

As mentioned earlier, colour matches are affected by the angular subtense of the field of observation, and CIE has a set of colour-matching functions, $\bar{x}_{10}(\lambda)$, $\bar{y}_{10}(\lambda)$ and $\bar{z}_{10}(\lambda)$ for fields of view in excess of about 4° (see Figure 1.9.6); they constitute the colour-matching properties of the CIE 1964 supplementary standard colorimetric observer. In television applications, the areas of colours of interest in displays usually have angular subtenses less than 4° and the CIE 1931 standard colorimetric observer is therefore the appropriate one to use.

1.9.2.4.2 Chromaticity coordinates

Important colour properties are related to the relative magnitudes of tristimulus values. It is therefore useful to calculate chromaticity coordinates, and this can be done as follows:

$$x = X/(X+Y+Z)$$
$$y = Y/(X+Y+Z)$$
$$z = Z/(X+Y+Z)$$

Since $x+y+z=1$, if x and y are known, z can be deduced from $z = 1 - x - y$. It is therefore, customary to plot, in two dimensional diagrams, y against x, as shown in Figure 1.9.7. These diagrams are called *chromaticity diagrams*, and provide useful 'maps' of colours. In Figure 1.9.7, the curved line represents the colours of the spectrum, and the area bounded by this curve and the straight line joining its two ends represents the complete gamut of all real colours.

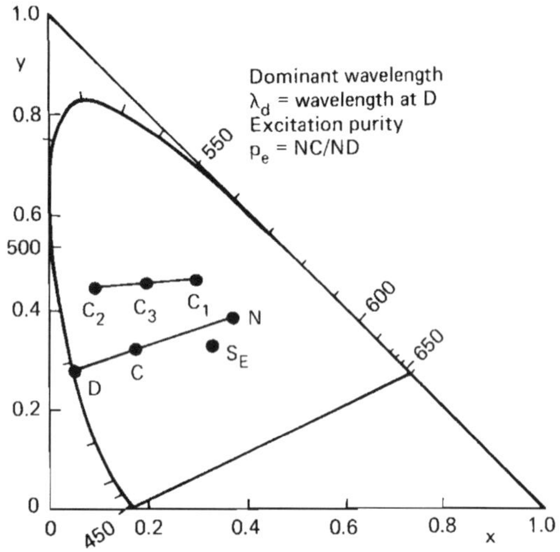

Figure 1.9.7 x,y chromaticity diagram.

If, in a chromaticity diagram, a colour C_1 plots at x_1y_1, and another colour C_2 plots at x_2y_2, then the position of C_3, the colour formed by the additive mixture of C_1 and C_2, is such that it lies on the straight line joining the points x_1y_1 and x_2y_2, as shown in Figure 1.9.7. The point representing C_3 divides the line joining the points representing C_1 and C_2 in the ratio such that:

$$\frac{C_1C_3}{C_2C_3} = \frac{L_2/y_2}{L_1/y_1}$$

where L_1 and L_2 are the luminances of C_1 and C_2 respectively.

1.9.2.4.3 Dominant wavelength and excitation purity

In Figure 1.9.7 are illustrated the derivations of two measures that correlate more closely with perceptual attributes of colours than tristimulus values or chromaticity coordinates. The point C represents the chromaticity of the colour considered; the point N represents that of a suitably chosen reference white or grey (usually the chromaticity of the illuminant, but this is normally different from the equi-energy illuminant, S_E); and the point D lies on the spectral locus intersected by the line NC produced. The wavelength corresponding to the point D is then termed the ***dominant wavelength***, λ_d (if the point D lies on the line joining the two ends of the spectrum, then it is produced in the other direction to give the complementary wavelength, λ_c).

Dominant wavelength provides a measure that correlates approximately with the hue of the colour. The ratio NC:ND is termed the excitation purity, p_e, and correlates approximately with saturation (colourfulness judged in proportion to brightness; see Table 1.9.3).

Table 1.9.3 Colour attributes and the correlates

Colour attributes	*Correlates*
Hue. Denotes whether the colour appears reddish, yellowish, greenish or bluish.	Dominant wavelength λ_d. †CIE 1976 hue-angle, h_{uv} or h_{ab}.
Brightness. Denotes the extent to which the colour appears to be emitting or reflecting more or less light.	Luminance, L.
‡*Colourfulness*. Denotes the extent to which the colour appears to exhibit a hue.	Not yet available.
‡*Saturation*. Denotes colourfulness judged in proportion to brightness.	Excitation purity, P_e. †CIE 1976 saturation, s_{uv}.
Lightness. Denotes brightness judged relative to the brightness of a similarly illuminated area that appears to be white.	Luminance factor, L/L_n. †CIE 1976 lightness. L^*.
‡*Chroma*. Denotes colourfulness judged as a proportion of the brightness of a similarly illuminated area that appears to be white.	†CIE 1976 chroma, C^*_{uv} or C^*_{ab}.

†These correlates are approximately uniform with the attribute.
‡Saturation and chroma are both relative colourfulness. In a series of colours of constant chromaticity but reducing luminance factor (a *shadow series*), the saturation remains constant (because the falling colourfulness is judged relative to the falling brightness of the samples), but the chroma reduces (because the falling colourfulness is judged relative to the constant brightness of the reference white).

For reflecting or transmitting samples, ***luminance factor*** can also be evaluated. Luminance factor is equal to L/L_n where L is the luminance of the sample and L_n that of the reference white. Luminance factor correlates with lightness. Hence, dominant wavelength, excitation purity and luminance factor, provide correlates of hue, saturation and lightness, respectively.

1.9.2.5 Approximately uniform colour systems

1.9.2.5.1 Uniform chromaticity coordinates u′,v′

Chromaticity diagrams are very useful in colorimetry, but, although the CIE x,y diagram has been widely used in the past, it does suffer from one important disadvantage: the colours in it are not uniformly distributed.

In Figure 1.9.8 is shown a chromaticity diagram in which are plotted:

$$u' = \frac{4X}{X + 15Y + 3Z} = \frac{4X}{-2x + 12y + 3}$$

$$v' = \frac{9Y}{X + 15Y + 3Z} = \frac{qY}{-2x + 12y + 3}$$

In this chromaticity diagram the colours are more nearly uniformly distributed. It is known as the CIE 1976 uniform-chromaticity scale diagram, or the CIE 1976 UCS diagram, often referred to as the u′,v′ diagram. (In 1960 the CIE introduced a similar diagram in which u and v were plotted, where u = u′ and v = 2/3v′; this u,v diagram has now been superseded by the u′,v′ diagram.) The u′,v′ diagram is very useful for representing the additive mixtures of the light emitted by phosphor primaries in television displays.

1.9.2.5.2 CIE 1976 hue-angle and CIE 1976 saturation

The u′,v′ diagram can provide better correlates of hue and saturation than dominant wavelength and excitation purity. In Figure 1.9.9, the point C represents the chromaticity of the

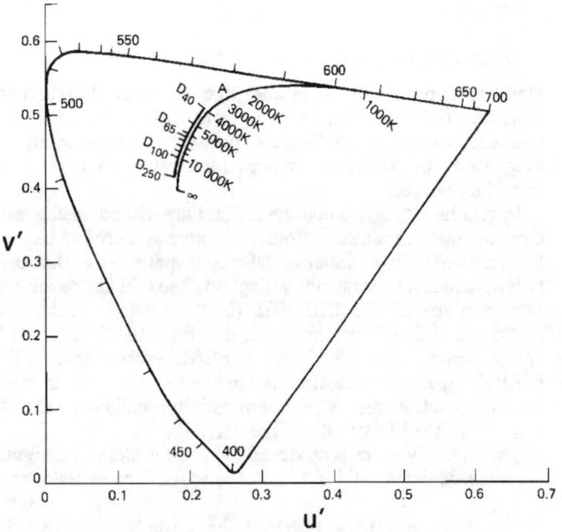

Figure 1.9.8 u′,v′ chromaticity diagram, with locus for planckian radiators having colour temperatures from 1000 K to infinity, and for daylight illuminants having correlated colour temperatures from 4000 K(D_{50}) to 25,000 K(D_{250}).

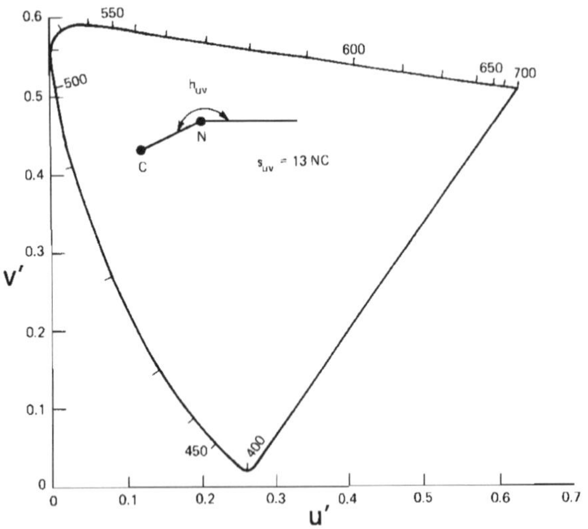

Figure 1.9.9 The derivation of hue-angle, h_{uv}, and saturation, s_{uv}, in the u′,v′ diagram.

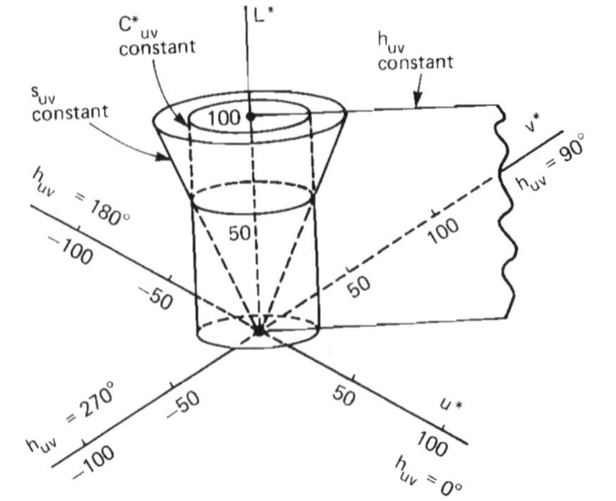

Figure 1.9.10 CIELUV colour space.

colour considered, and the point N that of a suitably chosen reference white or grey. The angle between a line from N horizontally to the right and the line NC is the CIE 1976 hue-angle, h_{uv}, and correlates with perceived hue better than dominant wavelength. The distance NC (when multiplied by 13) is the CIE 1976 saturation, s_{uv}, and correlates with perceived saturation better than excitation purity.

These better correlates arise in part from the better uniformity of colours in the u′,v′ diagram, as compared with the x,y diagram, and in part from the different type of formulation of h_{uv} and s_{uv}.

1.9.2.5.3 Uniform colour spaces

Uniform chromaticity diagrams, like all other chromaticity diagrams, only represent proportions of tristimulus values, not their actual values. They therefore only represent uniformly the magnitudes of colour differences for stimuli all having the same luminance.

In general, when two colours differ, they will not necessarily have the same luminance. Colour differences therefore have to be evaluated in three-dimensional colour space, rather than on a two-dimensional chromaticity diagram. The CIE has developed two such spaces: the CIE 1976 (L*u*v*) colour space, also called the CIELUV colour space, and the CIE 1976 (L*a*b*) colour space, also called the CIELAB colour space. The CIELUV space is more directly applicable to television, since it incorporates the u′,v′ chromaticity diagram already described. It is illustrated in Figure 1.9.10.

The CIELUV space is produced by plotting, along rectangular coordinates, the quantities, L*, u* and v*, defined as follows:

$$L^* = 116(Y/Y_n)^{1/3} - 16$$

$$u^* = 13L^*(u' - u'_n)$$

$$v^* = 13L^*(v' - v'_n)$$

where Y, u′ and v′, refer to the colour considered, and Y_n, u_n and v'_n, refer to a suitably chosen reference white. (If Y/Y_n is less than 0.008856, then L* is evaluated as 903.3 Y/Y_n, instead of by the formula for L* given above.)

The total difference between two colours whose differences in L*, u*, and v* are ΔL*, Δu* and Δv* respectively, is then evaluated as:

$$\Delta E^*_{uv} = \{(\Delta L)^{*2} + (\Delta u^*)^2 + (\Delta v^*)^2\}^{1/2}$$

In this L*u*v* system, approximate correlates of perceptually important colour attributes, as shown in Figure 1.9.10, may be calculated as follows:

CIE 1976 lightness:

$$L^* = 116(Y/Y_n)^{1/3} - 16$$

where Y/Y_n must not be less than 0.008856.

CIE 1976 u,v saturation:

$$s_{uv} = 13\{(u' - u'_n)^2 + (v' - v'_n)^2\}^{1/2}$$

CIE 1976 u,v chroma:

$$C^*_{uv} = (u^{*2} + v^{*2})^{1/2} = s_{uv}L^*$$

CIE 1976 u,v hue-angle:

$$h_{uv} = \arctan\{(v' - v'_n/(u' - u'_n)\}$$
$$= \arctan(v^*/u^*)$$

CIE 1976 u,v hue-difference:

$$\Delta H^*_{uv} = \{(\Delta E^*_{uv})^2 - (\Delta L^*)^2 - (\Delta C^*_{uv})^2\}^{1/2}$$

h_{uv} lies between 0° and 90° if v* and u* are both positive, between 90° and 180° if v* is positive and u* is negative, between 180° and 270° if v* and u* are both negative, and between 270° and 360° if v* is negative and u* is positive. CIE 1976 u,v hue-difference is introduced so that a colour difference ΔE^* can be broken up into components ΔL^*, ΔC^* and ΔH^*, whose squares add up to the square of ΔE^*. The hue-difference, ΔH^*_{uv}, is to be regarded as positive if indicating an increase in h_{uv} and negative if indicating a decrease in h_{uv}.

CIE 1976 u,v chroma, C^*_{uv}, has been designed to correlate with perceived chroma. This is the perceptual attribute defined as colourfulness judged as a proportion of the brightness of a similarly illuminated area that appears white or highly transmitting. It is equal to the product s_{uv} L^*, and the multiplication of the correlate of saturation, s_{uv}, by L^* allows for the fact that, for a given difference in the chromaticity of a colour from that of the reference white, its colourfulness decreases as the luminance factor is reduced. By being based on relative tristimulus values (X, Y, Z) and not on absolute tristimulus values (X_a, Y_a, Z_a), this measure C^*_{uv} does not change as the level of illumination if changed.

Thus, an orange and a brown may have the same chromaticities, and therefore the same values of s_{uv}, and the same saturation. But the lower value of L^* for the brown will result in it having a lower C^*_{uv}, and it appears of lower chroma. If the illuminance level is changed, the values of C^*_{uv} will not change, and this represents the fact that the perceived chromas of the orange and brown remain fairly constant over a wide range of illuminances. At lower illuminances both the orange and the brown will look less colourful than at higher illuminances; they will also look less bright at the lower illuminances.

Under specified viewing conditions, luminance can usually provide an approximate correlate with brightness, but does not provide a perceptually uniform scale. However, there is at present no agreed measure that provides a correlate for colourfulness (see Table 1.9.3).

The CIELAB system, which was designed to be similar to certain systems used widely in the colorant industries, is similar to the CIELUV system, but has no associated chromaticity diagram and no correlate of saturation. The CIELAB space is produced by plotting along rectangular coordinates the quantities, L^*, a^*, and b^*, defined as follows:

$$L^* = 116(Y/Y_n)^{1/3} - 16$$

where $Y/\ Y_n \geq 0.008856$.

$$a^* = 500\{(X/X_n)^{1/3} - (Y/Y_n)^{1/3}\}$$

where X/X_n, Y/Y_n and $Z/Z_n \geq 0.008856$.

$$b^* = 200\{(Y/Y_n)^{1/3} - (Z/Z_n)^{1/3}\}$$

where X/X_n, Y/Y_n and $Z/Z_n \geq 0.008\,856$. X, Y and Z refer to the colour considered, and X_n, Y_n and Z_n refer to a suitably chosen reference white. Colour differences in this system are evaluated as:

$$\Delta E^*_{ab} = \{(\Delta L^*)^2 + (\Delta a^*)^2 + (\Delta b^*)^2\}^{1/2}$$

Approximate correlates of lightness, chroma, and hue in this system are calculated as follows:

CIE 1976 lightness:

$$L^* = 116(Y/Y_n)^{1/3} - 16$$

where $Y/Y_n \geq 0.008856$.

CIE 1976 a,b chroma:

$$C^*_{ab} = (a^{*2} + b^{*2})^{1/2}$$

CIE 1976 a,b hue-angle:

$$h_{ab} = \arctan(b^*/a^*)$$

CIE 1976 a,b hue-difference:

$$\Delta H^* = \{(\Delta E^*_{ab})^2 - (\Delta L^*)^2 - (\Delta C^*_{ab})^2\}^{1/2}$$

These spaces are intended to apply to comparisons of differences between reflecting object colours of the same size and shape, viewed in identical white to middle-grey surroundings, by an observer photopically adapted to a field not too different from that of average daylight. They are not necessarily applicable to self-luminous displays such as are used in television, without appropriate modifications.

A summary of colour attributes and their correlates is given in Table 1.9.3.

1.9.3 Chromaticities of Display Phosphors

1.9.3.1 Introduction

The choice of the colours emitted by reproduction phosphors is important because it affects:

- the gamut of colours that can be reproduced;
- the spectral sensitivities that are optimum for the three colour channels of the camera;
- the maximum luminance attainable on the display.

The phosphors are usually chosen to strike the best compromise between these three factors.

1.9.3.2 NTSC phosphors

When the NTSC (National Television Systems Committee) system was originally set up in the USA in 1953, the best phosphors then available were such as to produce primary colours having the following chromaticities:

	x	y	u′	v′
Red	0.67	0.33	0.477	0.528
Green	0.21	0.71	0.076	0.576
Blue	0.14	0.08	0.152	0.195
Illuminant C	0.3101	0.3162	0.2009	0.4610

The system used illuminant C as the reference white.

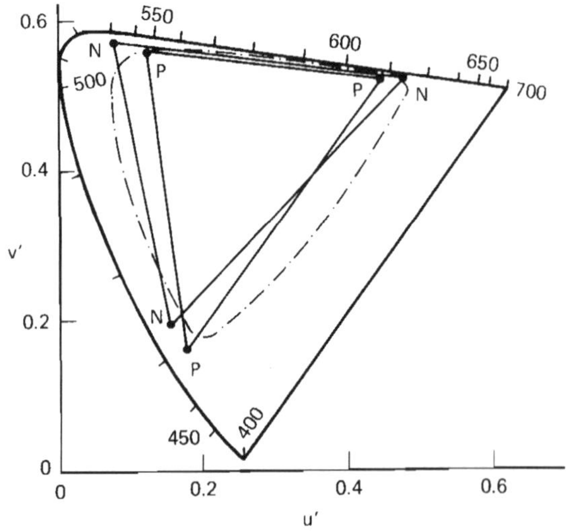

Figure 1.9.11 Chromaticity gamuts for NTSC phosphors (N), EBU phosphors (P) and real colours (broken line curve), in the u′,v′ diagram.

The positions of these primaries in the u′,v′ diagram are shown in Figure 1.9.11 by the points marked N. The triangle connecting the N points represents the gamut of chromaticities that can be reproduced. It is clear that there is a region of colours near the blue-green part of the spectrum, 470–530 nm, which lie outside the triangle and therefore cannot ever be displayed. There is also an even larger area between the edge of the triangle and the line joining the two ends of the spectrum. This area represents red, magenta, purple and violet colours that cannot be displayed. Although these two areas comprise quite a large proportion of the total gamut of real colours, the gamut of typical surface colours is considerably smaller, as shown by the broken line in Figure 1.9.11, and most of that area is covered by the triangle.

1.9.3.3 EBU phosphors

Since 1953, a much wider range of phosphors has become available, and the EBU (European Broadcast Union) has adopted the following set of chromaticities to represent typical phosphors that are used currently:[6]

	x	y	u′	v′
Red	0.64	0.33	0.451	0.523
Green	0.29	0.60	0.121	0.561
Blue	0.15	0.06	0.175	0.157
Illuminant D_{65}	0.3127	0.3290	0.1978	0.4683

This system uses illuminant D_{65} as the reference white.

The position of these primaries is also shown in the u′,v′ diagram in Figure 1.9.11 by the points marked P. The corresponding triangle shows that the displayed gamut is even more restricted for blue-green colours. However, these phosphors are capable of giving pictures of much higher luminance, and this increases the colourfulness of the displayed colours sufficiently to offset the loss of saturation of the blue-greens.

The tolerances for the EBU phosphors are such that their chromaticities should lie somewhere between the following four points specified for each colour:

Red		*Green*		*Blue*	
u′	v′	u′	v′	u′	v′
0.441	0.530	0.115	0.562	0.157	0.159
0.441	0.520	0.119	0.570	0.174	0.170
0.461	0.518	0.128	0.560	0.183	0.154
0.461	0.526	0.124	0.552	0.176	0.146

1.9.3.4 Camera spectral sensitivities

For any set of primaries, there will be a corresponding set of colour-matching functions showing the amounts of the primaries needed to match each wavelength of the spectrum. This set of colour-matching functions then shows what the spectral sensitivities of the three colour channels of the camera should be. The two sets of colour-matching functions for the NTSC and EBU primaries are shown in Figure 1.9.12. Both sets have negative portions, which are slightly more pronounced in the case of the EBU phosphors because, in the case of the green, the EBU phosphor is more yellow than the NTSC phosphor.

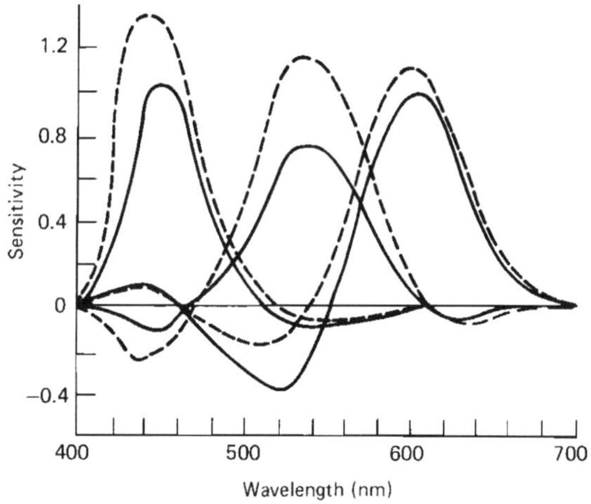

Figure 1.9.12 Colour-matching functions for the NTSC phosphors (broken lines) and the EBU phosphors (full lines).

1.9.3.5 Matrixing

It is possible to realize negative portions in camera sensitivity curves by the technique of *matrixing*. Modified red, green and blue signals $R_mG_mB_m$ are obtained from the camera tube signals RGB by a circuit whose algebraic equivalent is a matrix:

$$R_m = +1.14\,R - 0.18\,G + 0.04\,B$$

$$G_m = -0.06\,R + 1.23\,G - 0.17\,B$$

$$B_m = -0.03\,R + 0.02\,G + 1.01\,B$$

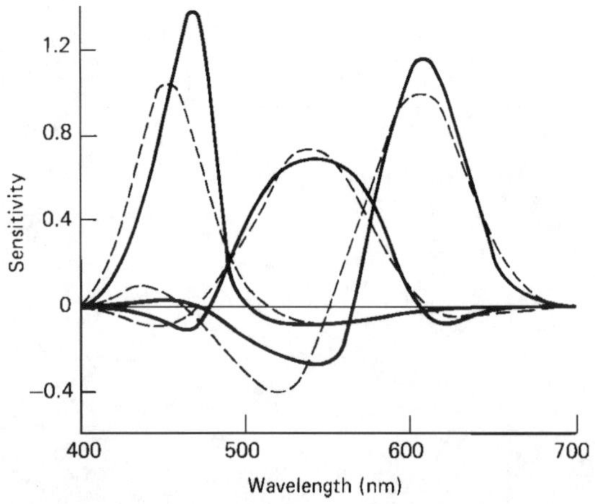

Figure 1.9.13 Typical matrixed camera sensitivities (full lines) and colour-matching functions for EBU phosphors (broken lines).

The numerical coefficients in the above set of equations are only given as an example of typical values that may be used.

In Figure 1.9.13 are shown the spectral sensitivities corresponding to the matrixed signals of a typical camera, together with the colour-matching functions for the EBU phosphors. The two sets of curves are only roughly similar, but the matrixing step usually produces very significant improvements in colour reproduction. What matrixing can do is to give correct colorimetric reproduction within the phosphor gamut; colours lying outside the gamut still cannot be displayed, and usually move to the edges of the triangle.

1.9.3.6 Effect of gamma correction on colour reproduction

If a system is designed to give correct colour reproduction within the phosphor gamut for an overall system gamma of 1, and is then used with a display device that results in an overall gamma of 1.27, then the distortions shown in Figure 1.9.14 are produced.

In practice, this can occur because the transmitted signals are down-gammaed by the power of 1/2.2, and typical receivers operate at gammas of about 2.8.

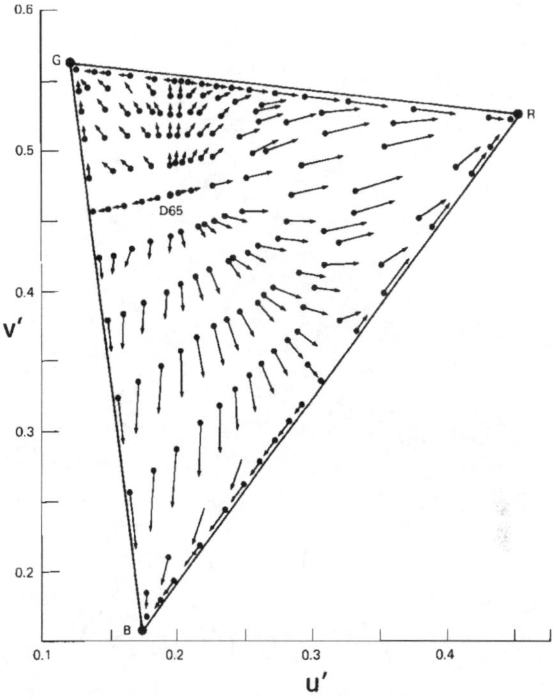

Figure 1.9.14 Arrowheads show how the chromaticities of the dots are distorted by an increase in system gamma of 1.27.

References

1. Jesty, L.C. *Proc. IEE*, **105B**, 425 (1958).
2. Wright, W.W. *J. RTS*, **13**, 221 (1971).
3. Neate, J. *Television*, **23**, 344 (1973).
4. Federman, F. and Pomicter, D. *J. RTS*, **16**, vii (May–June 1977).
5. *IBA Technical Review* **22**, Light and colour principles (1984).
6. BREMA. *Radio and Electronic Eng*., **38**, 201 (1969).

Bibliography

Hunt, R.W.G. *The Reproduction of Colour*, 4th Edn, Fountain Press, London (1987).

Section 2 Broadcast Technologies and Standards

E. P. J. Tozer
Zetrox Broadcast Communications

2.1 Linear Digital Audio

To understand the advantages of processing audio digitally, it is necessary to look at the fundamental differences between analogue and digital signals (Figure 2.1.1).

An *analogue* signal can, within the limits of peak level and signal bandwidth, exist at any level and at any time. This means that if, during processing, there is any level variation, caused by distortion or the addition of noise, or any time variation, caused by wow and flutter, the new signal is a valid one.

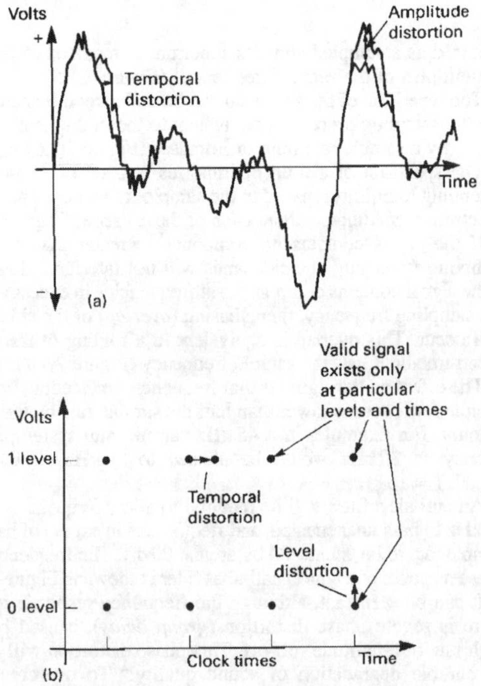

Figure 2.1.1 Signal distortion: (a) analogue; (b) digital.

A *digital* signal differs from an analogue one in that it is constrained to be valid only at particular levels, generally one and zero, and particular times, clock intervals. Digital signals thus have an inherent immunity to change. So long as the temporal or amplitude variation is small they may be brought to the nearest allowable value. This perfect regenerating ability of digital signals means that digital audio may be duplicated or transmitted with zero degradation.

Figure 2.1.2 shows the path of an audio signal through a typical digital audio recorder which contains virtually all the different types of processing applied to a digital audio signal.

The layout of a typical mixing console is shown in Figure 2.1.3. Here the particular aspects of processing are not explicit, most processes being performed as software execution.

2.1.1 Digital Audio Concepts

2.1.1.1 Analogue/digital interface

2.1.1.1.1 Linear pre-emphasis

Linear pre-emphasis is used with digital audio processing in order to improve the overall signal/noise ratio of signals containing high frequencies at only low levels. It is used in precisely the same manner as for analogue systems, but is generally less useful for digital recording. Digital meters read even the shortest transient, so causing the operator to reduce input level to the system, in turn reducing any benefit obtained from the pre-emphasis. The most commonly used form of pre-emphasis is EIAJ (Electronic Industries Association of Japan) pre-emphasis (Figure 2.1.4).

2.1.1.1.2 Anti-alias filtering

The function of the anti-alias filter is to remove any audio signal in excess of half the sample rate. This filtering is

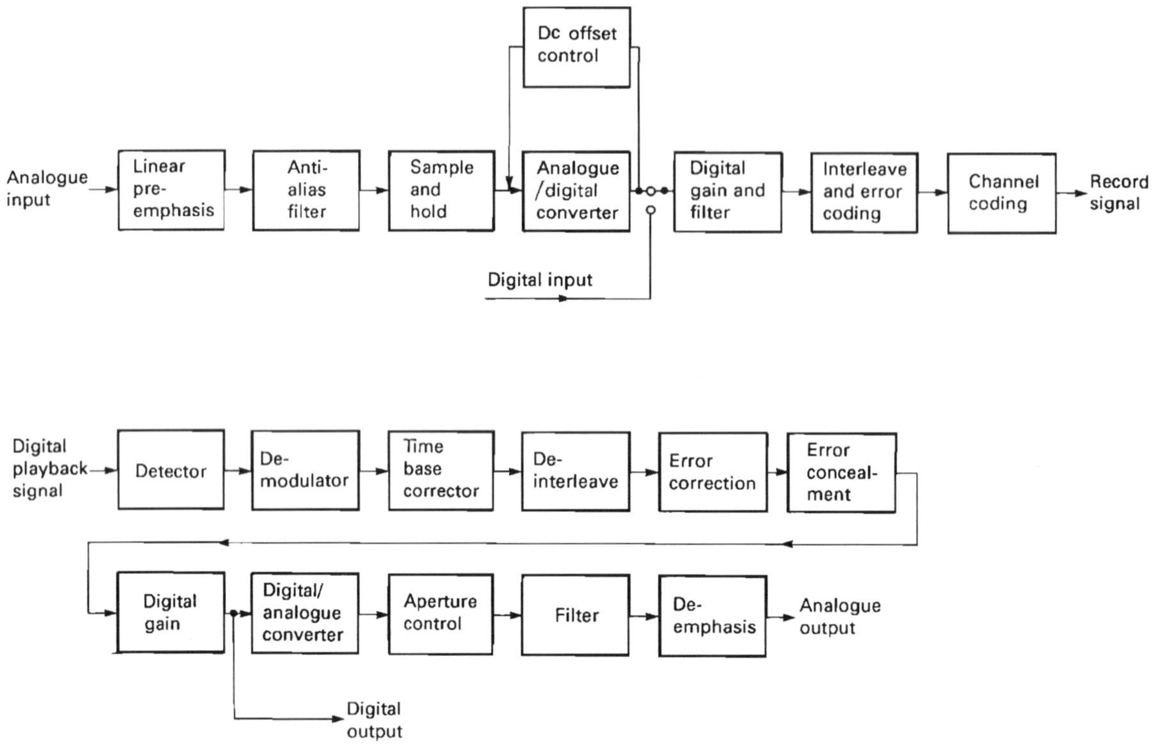

Figure 2.1.2 The path of an audio signal through a digital recorder.

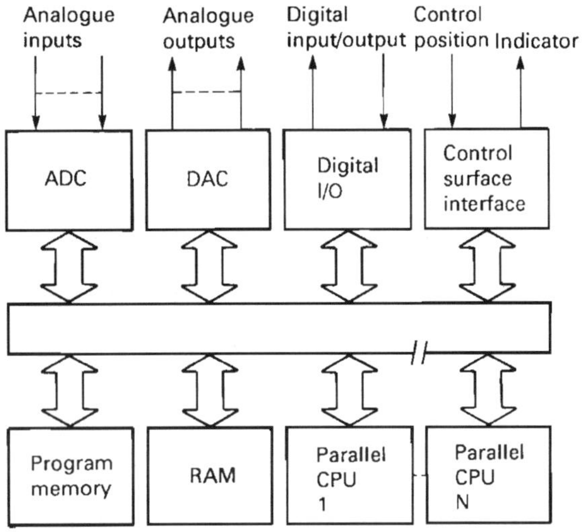

Figure 2.1.3 Electronic structure of a digital console.

required as a sampled signal's spectrum is repeated to infinity at multiples of the sample frequency[1] (Figure 2.1.5).

The repetition of the spectrum can be understood by considering the sampling process as equivalent to the multiplication of a signal by a sample rate train of impulses (Figure 2.1.6).

The spectrum of a train of impulses is a set of frequencies extending to infinity, spaced at the sample frequency. The signal spectrum is modulated about each of these carrier frequencies.

If the signal contains no components greater than half the sampling frequency, the sidebands will not interfere. However, if the signal contains components at frequencies in excess of half the sampling frequency, then aliasing (*overlap*) of the sidebands will occur. This overlap is equivalent to a folding of the audio spectrum about half the sample frequency (Figure 2.1.7).

The effect on the signal is that frequencies exceeding half the sample rate become lower than half the sample rate by the same amount. For example, in a 48 kHz sample rate system, a frequency of 30 kHz would be aliased to 24 kHz − (30 kHz− 24 kHz) = 18 kHz.

An anti-alias filter will be required to allow frequencies up to 20 kHz to pass unattenuated, and frequencies in excess of half the sample rate to be attenuated by around 90 dB. The frequency and phase response of a typical anti-alias filter is shown in Figure 2.1.8.

It can be seen that, although the frequency response is flat, there is severe phase distortion (*group delay*), caused by the high rate of amplitude roll-off. This phase distortion will cause an audible degradation of sound quality.[2] To overcome the problems caused by group delay in the anti-alias filter, it is possible instead to use an oversampling system.

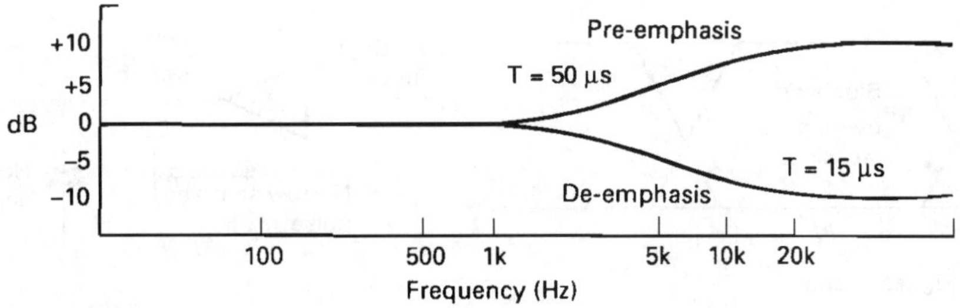

Figure 2.1.4 EIAJ pre-emphasis curve.

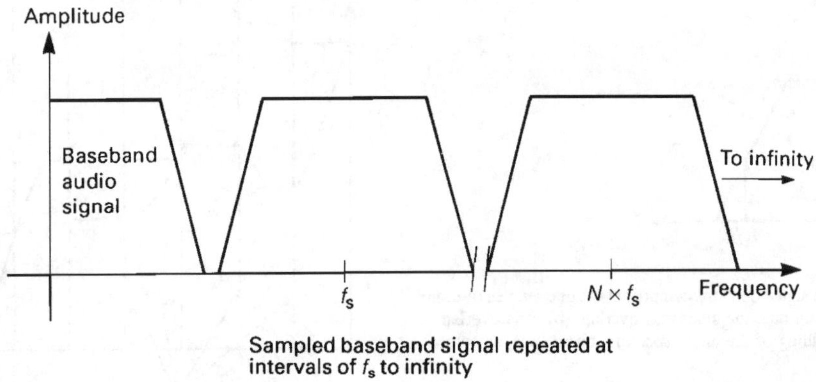

Figure 2.1.5 Sampled baseband signal repeated at intervals of f_s to infinity.

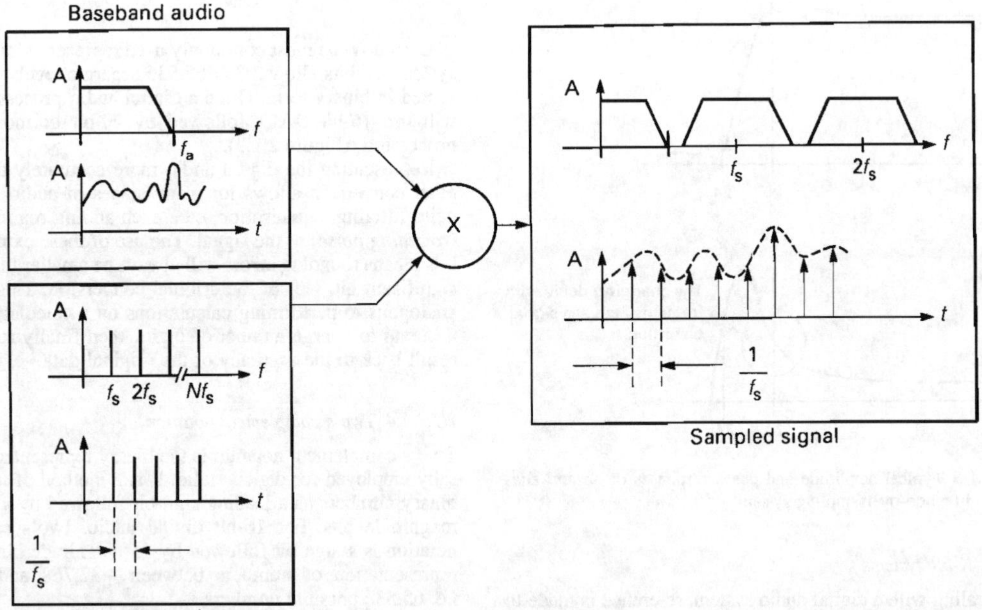

Figure 2.1.6 The sampling process may be considered equivalent to the multiplication of a signal by a sample rate train of impulses.

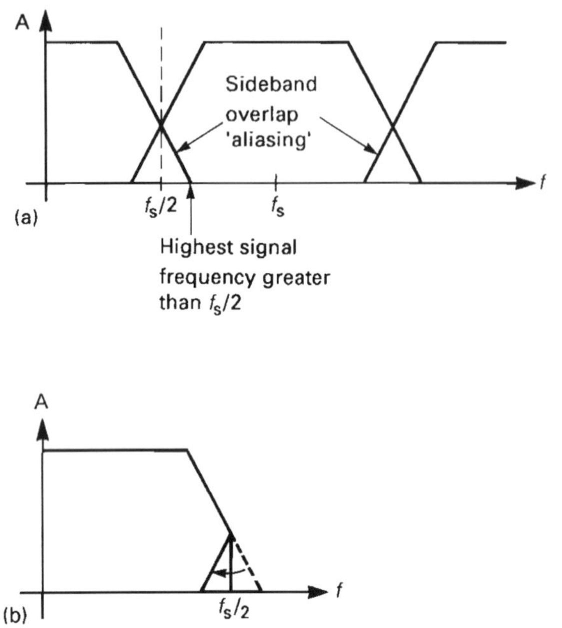

Figure 2.1.7 (a) If a signal contains components at frequencies of more than half the sampling rate, the sidebands overlap. (b) This overlap is equivalent to the folding of the audio spectrum about half the sample frequency.

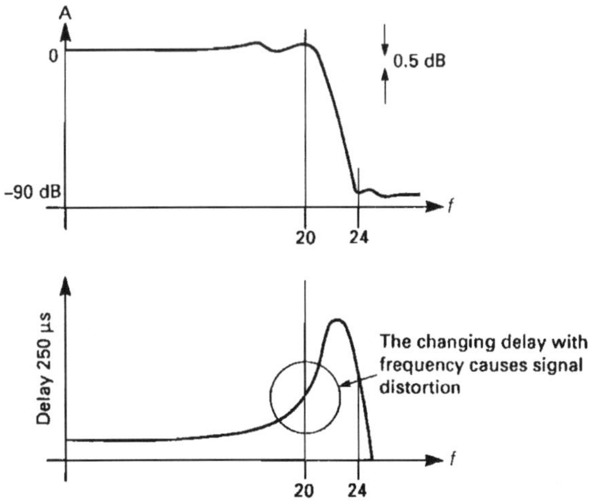

Figure 2.1.8 Typical amplitude and phase responses of an anti-alias filter used in a non-oversampling system.

2.1.1.1.3 *N-bit systems*

When dealing with a digital audio system, reference is made to an *N*-bit system. *N* refers to the number of binary digits used to represent the digitised signal. *N* bits will allow 2^N discrete levels of signal to be represented.

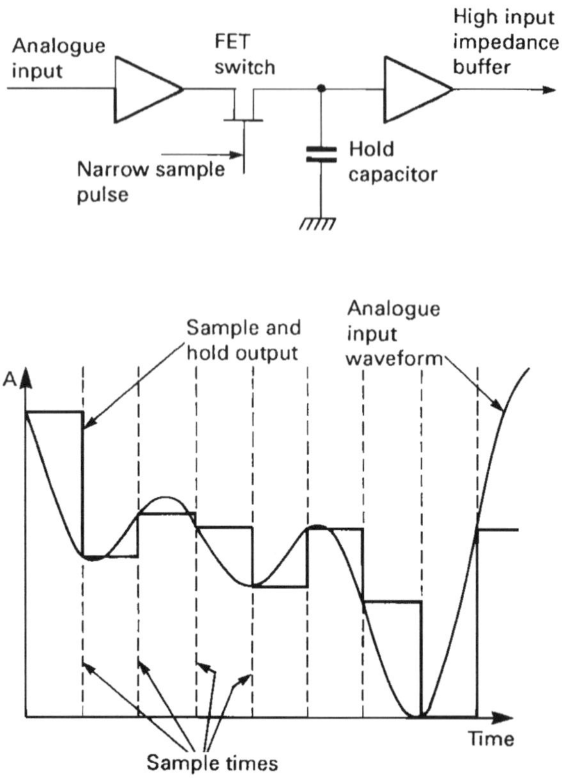

Figure 2.1.9 Sample and hold circuit and associated waveforms.

Currently the most commonly used representation is a 16-bit system; 16 bits allow $2^{16} = 65{,}536$ separate levels to be represented in binary form. Often a digital audio processing system will use 16-bit DACs followed by 20-bit or more 'internal processing' (Figure 2.1.9).

Representing the digital audio more accurately than its original conversion allows for many stages of audio processing, gain, filtering, equalisation, etc., each adding rounding errors[3] (*rounding noise*) to the signal. The use of these extra 'internal' bits means rounding errors will always be smaller than the least significant bit (lsb) of the original conversion. This process is analogous to performing calculations on a calculator which is accurate to a large number of digits, then finally rounding the result back to the accuracy of the original data.

2.1.1.1.4 *Two's complement notation*

Two's complement notation is the binary representation universally employed for digital audio. It is a method of representing binary numbers as a leading sign bit followed by a number of magnitude bits. For 16-bit digital audio, two's complement notation is a sign bit followed by 15 magnitude bits, allowing representation of numbers between −32,768 and +32,767, i.e. 65,536 possible numbers.

The convention for the sign bit is 0 for positive, 1 for negative. In order to make two's complement notation a mathematically useful system, the magnitude bits are inverted for

Table 2.1.1 Some 16-bit two's complement numbers

Decimal	*Two's complement binary*	
−32,786	1000000000000000	1 bit overload
+32,767	0111111111111111	+maximum
+32,766	0111111111111110	
:	:	
:	:	
+2	0000000000000010	
+1	0000000000000001	
0	0000000000000000	
−1	1111111111111111	
−2	1111111111111110	
−3	1111111111111101	
:	:	
:	:	
−32,767	1000000000000001	
−32,768	1000000000000000	−maximum
+32,767	0111111111111111	−1 bit overflow

negative numbers, and in order to eliminate two zero representations (+0 and −0), one is added to all negative numbers. Table 2.1.1 shows the 16-bit two's complement numbers around 0 and around plus or minus maximum.

As may be seen from Table 2.1.1, a count between plus maximum and minus maximum involves only a one bit change, and a change between −1 and 0 involves only a one bit overflow. It is therefore convenient to envisage two's complement numbers as a ring of numbers (Figure 2.1.10).

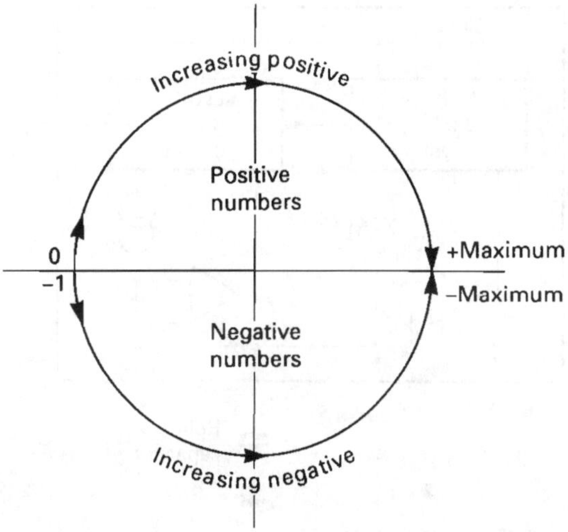

Figure 2.1.10 Two's complement ring.

2.1.1.1.5 Analogue/digital conversion

Analogue/digital conversion is the starting point of most current digital systems. The vast majority of digital audio signals start life as analogue and must therefore pass through an analogue/digital converter (ADC).

The ADC is the section of circuitry which causes the most problems for digitally processed audio in terms of noise and distortion. For digital audio, ADCs are often based around a comparator, counter and constant current source (Figure 2.1.11).

At the start of the conversion process, the counter is set to zero; the constant current source is turned on and starts discharging the hold capacitor of the preceding sample and hold gate; at the same time the counter is clocked at a high frequency. These processes continue until the hold capacitor is discharged, the comparator detects this and the counter stops. The count reached by the counter will be proportional to the time taken to discharge the hold capacitor, which in turn, as discharge is by a constant current source, will be proportional to the initial hold voltage.

The clock frequency required for an actual system of this type is not practicable (for a 48 kHz 16-bit system, the counter must be capable of counting to 65,536, 48,000 times a second; this gives a clock frequency of 48,000 × 65,536 Hz = 3.15 GHz.

To reduce the required clock rate, a dual slope converter is used which splits the counter into two parts, an upper and a lower, each with its respective current source and comparator.

A typical circuit arrangement is seen in Figure 2.1.12. Firstly the upper 8-bit counter current source discharges the hold capacitor to within 1 lsb (*least significant bit*) of 0, followed by the lower counter current source discharging to exactly 0. The clock frequency is now reduced to 2 × 256 × 48,000 Hz = 24.6 MHz. If the counter clock is not locked to sample rate, low-level audible beats may be heard, and the clock frequency should be trimmed for optimum audible effect.

The two major problems associated with ADCs are distortion and dc offset. *Distortion* can be caused by a mismatch between the two current sources in the converter. This mismatch causes

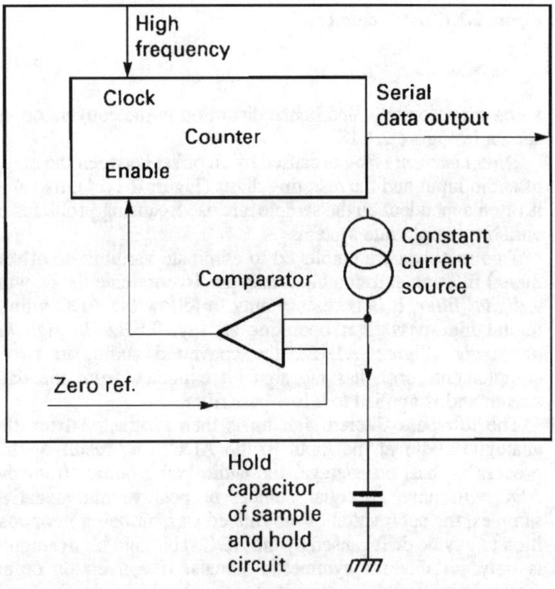

Figure 2.1.11 Simple analogue/digital converter.

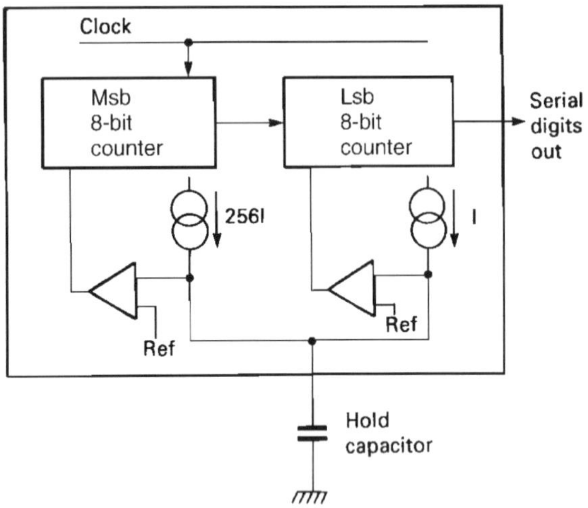

Figure 2.1.12 Dual slope analogue/digital converter.

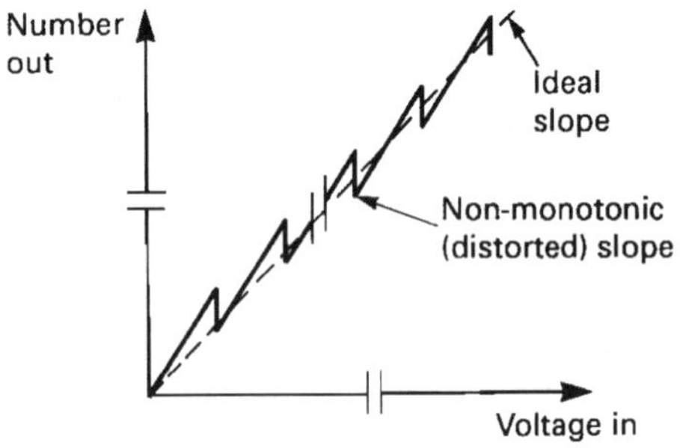

Figure 2.1.13 ADC distortion.

a non-monotonicity, and hence distortion in the conversion, as shown in Figure 2.1.13.

Direct current offset is caused by an offset between the range of audio input and the resulting digits (Figure 2.1.14). Its value is often dependent on the sample rate used, causing problems in variable sample rate systems.

Two methods are employed to eliminate variable dc offset: digital filtering and sign bit averaging. To eliminate the dc with a *digital filter*, it is necessary only to follow the ADC with a digital high-pass filter operating at, say, 0.5 Hz. In *sign bit averaging* (Figure 2.1.15), the converted audio, in two's complement form, has the sign bit extracted from the data stream and is applied to a low-pass filter.

The low-pass filtered sign bit is then subtracted from the analogue audio at the input to the ADC. The result of this process is that, on average, the audio being output from the ADC will have an equal number of positive and negative samples, the subtraction of the filtered sign bit being in opposition to any dc drift caused by the ADC. This sign bit averaging is only suitable for symmetric signals. If conversion of an asymmetric signal is attempted, the sign bit averaging will instead add a dc offset (Figure 2.1.16).

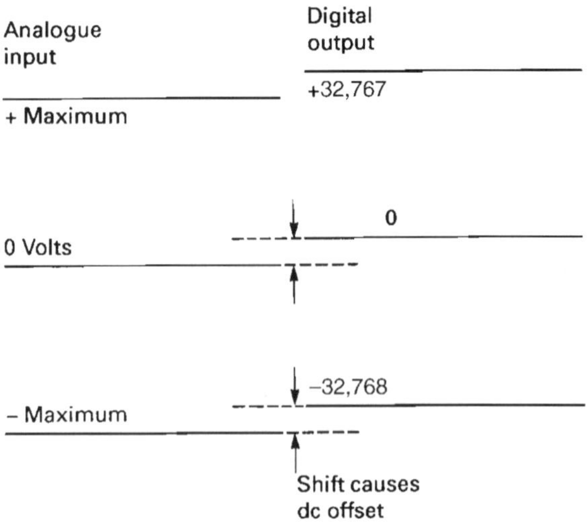

Figure 2.1.14 ADC dc offset.

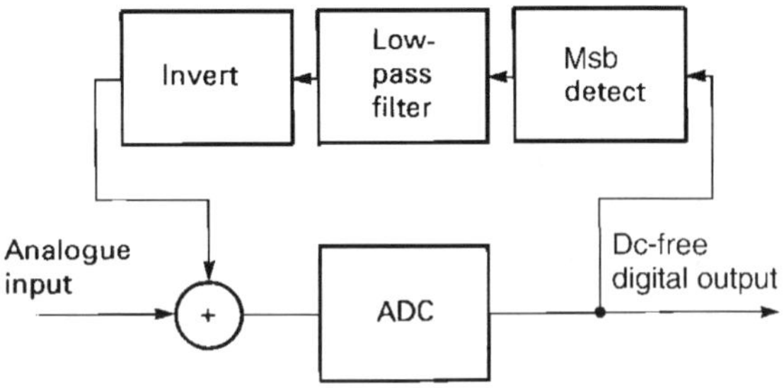

Figure 2.1.15 Most significant bit averaging to remove dc.

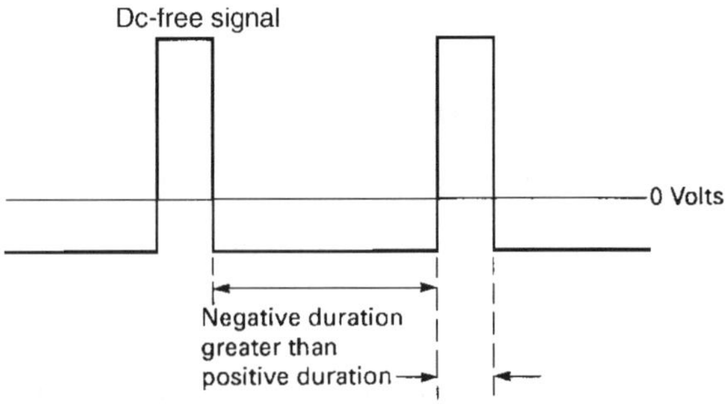

Figure 2.1.16 Asymmetric dc-free signal.

When converting from analogue to digital, a known dc offset is sometimes intentionally added to the analogue signal before conversion and subtracted in digital form after conversion. This dc offset prevents analogue noise causing switching between digital 0, two's complement 000...0, and digital −1, two's complement 1111...1. This switching between all 0s and all 1s could modulate the supply voltage of the converter, which in

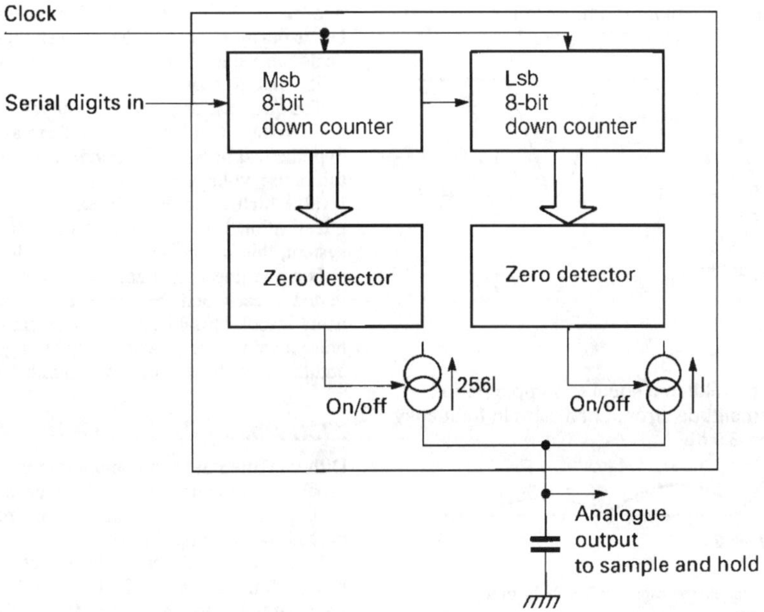

Figure 2.1.17 Dual slope digital/analogue converter (cf. Figure 2.1.12).

turn could cause modulation of the analogue signal, leading to increased noise levels.

2.1.1.1.6 Digital/analogue conversion

Digital/analogue converters (DACs) generally operate in a similar manner to ADCs, utilising a counter and constant current source (see Figure 2.1.17). Again, unlocked clock sources may cause beat frequencies.

2.1.1.1.7 Sample and hold, and aperture effect

Sample and hold circuits are required for both ADC and DAC conversions. A simple form of sample and hole circuit and the waveforms associated with it are shown in Figure 2.1.18. For use with an ADC, the sample and hold gate is required to hold the signal level constant while the ADC calculates the digital representation. When used with DACs, the sample and hold gate has a more complex function. The signal output from a DAC should be an infinitely narrow pulse, if the frequency response of the analogue signal is to be correct.[2] It is also necessary that the analogue signal be low-pass filtered to remove unwanted harmonics that might cause intermodulation distortion in later stages (Figure 2.1.19).

The signal from the DAC must therefore be held for some finite time before low-pass filtering so that there is a finite amount of energy to filter. The hold duration has an effect on the frequency response of the filtered signal, as shown in Figure 2.1.20. This frequency response error is known as the *aperture effect.*[3]

A hold for the full duration of the sample period will lead to a frequency response error of approximately −3.9 dB at half the sample frequency.

For most systems, the aperture effect causes no problems. It is of a known value and may be corrected with a simple RC

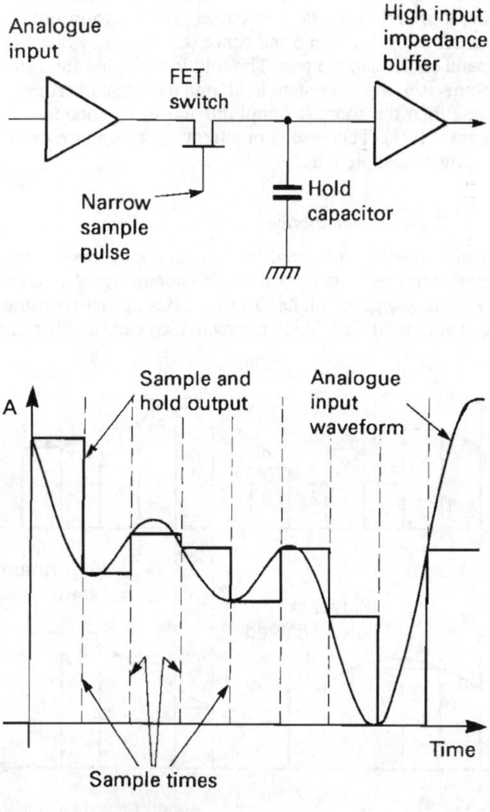

Figure 2.1.18 Simple sample and hold circuit and its associated waveforms.

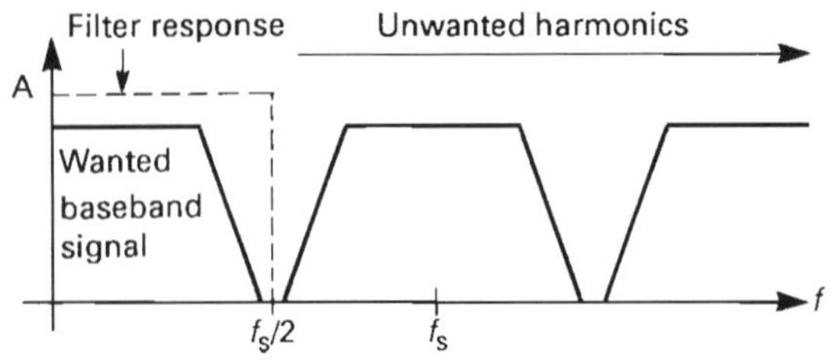

Figure 2.1.19 DAC output filtering.

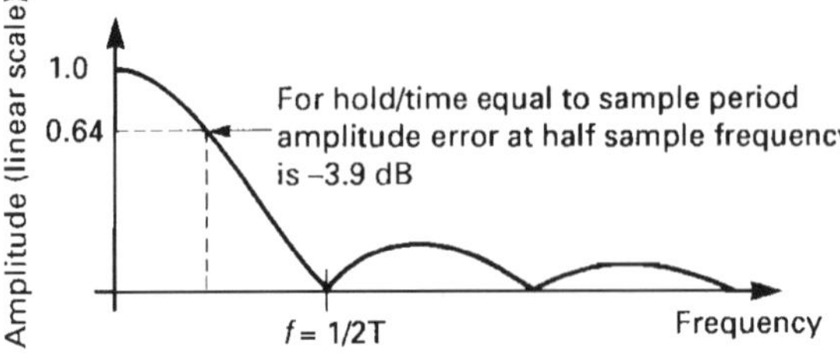

Figure 2.1.20 Aperture frequency response; T = hold time.

network. However, for systems capable of variable rate playback, e.g. vari-speed tape machines, full duration hold cannot be used, as the hold time and hence the frequency response will depend on the sample rate. The solution adopted for vari-speed systems is to use a constant hold time (constant aperture) which is less than the shortest sample period to be encountered (see Figure 2.1.21). This results in a frequency response error independent of sample rate.

2.1.1.1.8 Signal quantisation

When a signal is converted from an analogue representation to digital, the continuously varying analogue signal is converted to a finite range of digits. The relationship between the analogue voltage and its digital representation can be either linear or

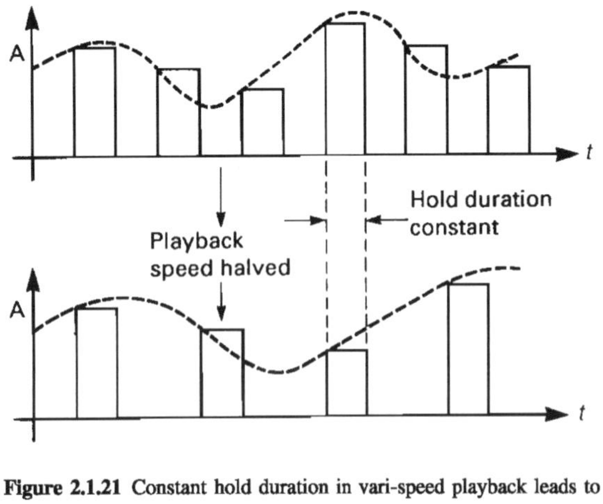

Figure 2.1.21 Constant hold duration in vari-speed playback leads to constant response errors.

non-linear. Whichever method is chosen, an approximation will be made by choosing the closest digital value to the actual analogue value (Figure 2.1.22). This approximation is known as the *quantisation error.*

The quantisation process is equivalent to adding to the signal a noise voltage that has the same form as the approximation. This hypothetical noise is the *quantisation noise*. The peak level of this noise voltage is plus or minus one-half the quantisation level, which in an N-bit linear quantisation system leads to a quantisation noise level of $6.02 \times N + 1.76\,\text{dB}$.[4] In a 16-bit system, this is a noise level of −98 dB.

For non-linear quantisation systems, the signal is first converted linearly and then converted to non-linear representation using a look-up table. In a non-linear quantisation system, the noise level will be dependent on the signal level (Figure 2.1.23), and the noise floor for a given number of bits will be lower.

2.1.1.1.9 Dither

Dither is often used in conjunction with an ADC, to overcome problems of distortion when converting low-level signals, and in digital signal processing to overcome similar problems caused by rounding errors.

Figure 2.1.24(a) shows the effect of applying ±1 lsb sinusoidal signal to an ADC. The low-level sine-wave has been approximated to the nearest quantisation levels, and the digital representation is now closer to a square-wave than a sine-wave.

Figure 2.1.24(b) shows the harmonic spectrum of the converted sine-wave, which contains the fundamental plus a high level of audibly objectionable odd harmonics. This digital representation of the signal is a distorted version of the input signal. A second feature of this distortion is that the level of harmonic distortion varies suddenly as the number of quantisation levels covered by the signal varies. This stepping of distortion is also audibly apparent.

To overcome the problem of quantisation (*granulation*), noise dither is added to the analogue signal before quantisation. Dither is most commonly a white noise signal of ±0.5 lsb amplitude. The probability density function (PDF) is shown in Figure 2.1.25.

The effect of adding dither before the ADC is shown in Figure 2.1.26(a). The quantisation of the signal is randomised, causing an effect similar to random pulse width modulation on the converted signal. The dramatic effect on the harmonic spectrum is shown in Figure 2.1.26(b), the harmonic distortion having completely disappeared, and been replaced by the audibly more pleasant white noise floor.

This type of dither has the advantage that, in an inactive digital filter, ±0.5 lsb of dither will not alter the sample value, thus enabling cloning of material. It has the disadvantage of level modulation by the analogue signal.

To overcome the problem of noise modulation by the programme material, *triangular dither*[5] is employed. Triangular dither is generated by adding together the outputs of two white noise generators, to produce a dither signal with a white spectrum and PDF as in Figure 2.1.27.

The effects of adding triangular dither to a low-level signal are shown in Figure 2.1.28. The effect is the same as for square dither, except that the noise level is 3 dB worse and the noise floor is no longer signal level dependent. However, ±1 lsb triangular dither has the disadvantage that in an inactive filter the sample values will be changed, making cloning impossible.

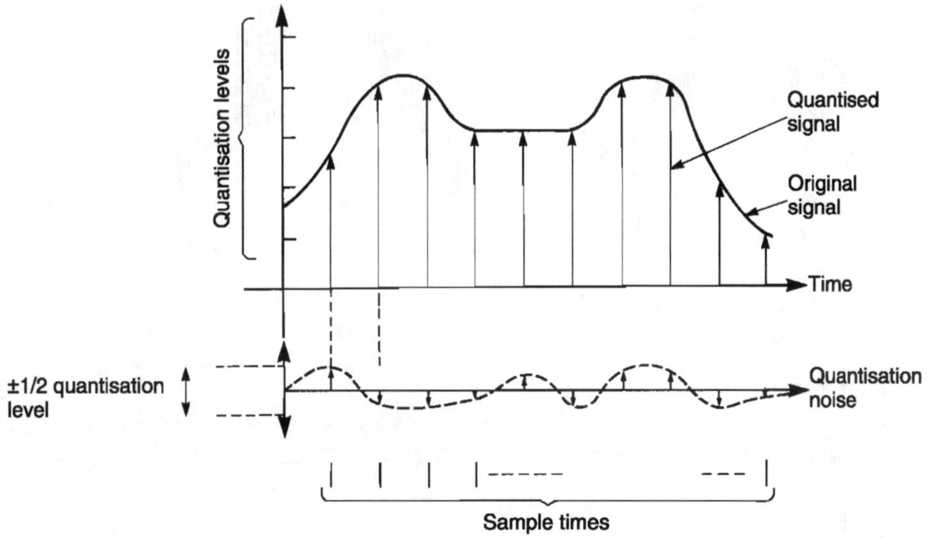

Figure 2.1.22 Signal quantisation and associated quantisation noise.

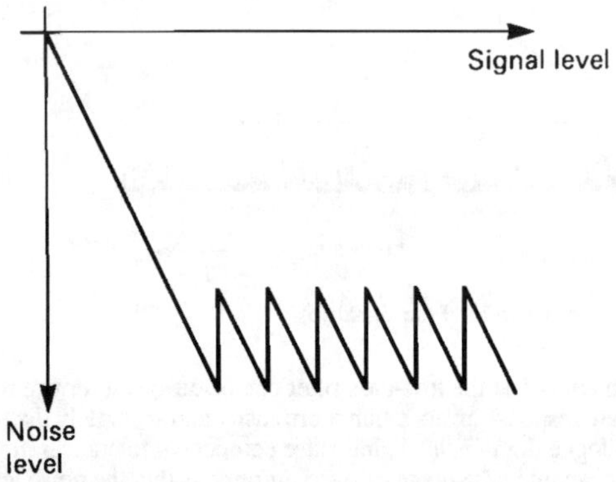

Figure 2.1.23 Noise level in a non-linearly quantised signal.

2.1.1.1.10 Oversampling conversion

Oversampling is used both to eliminate the group delay distortion caused by the anti-alias filter, and to improve the signal/noise ratio of the digitised signal. In an oversampling system, the ADC is operating at a multiple, normally a power of two, of the actual sample rate (Figure 2.1.29).

The first image frequency of the converted signal is now centred around the higher sample rate, which means that the rate of roll-off of the anti-alias filter can be drastically reduced, to virtually eliminate phase distortion.

The sample rate of the oversampled signal is now higher than necessary, and must be reduced before the digitised signal is further processed. To do this a digital low-pass filter is used (Figure 2.1.30).

In an oversampling system, the band limiting of the signal to half the sample rate is moved from the analogue domain to the digital. The advantage of performing the filtering in the digital

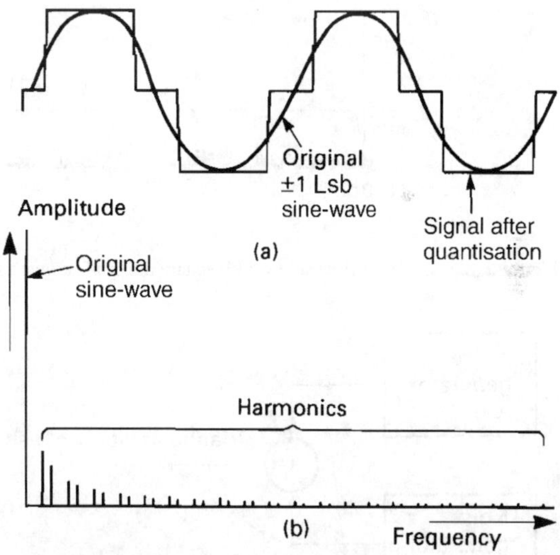

Figure 2.1.24 Quantisation of low-level sine-wave: (a) time domain; (b) frequency domain.

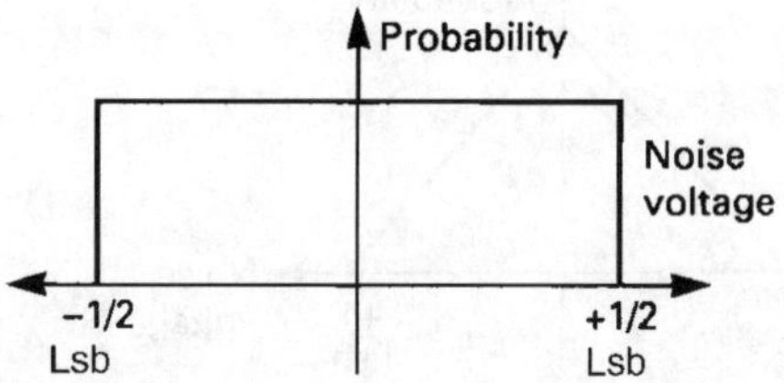

Figure 2.1.25 The probability density of square dither noise.

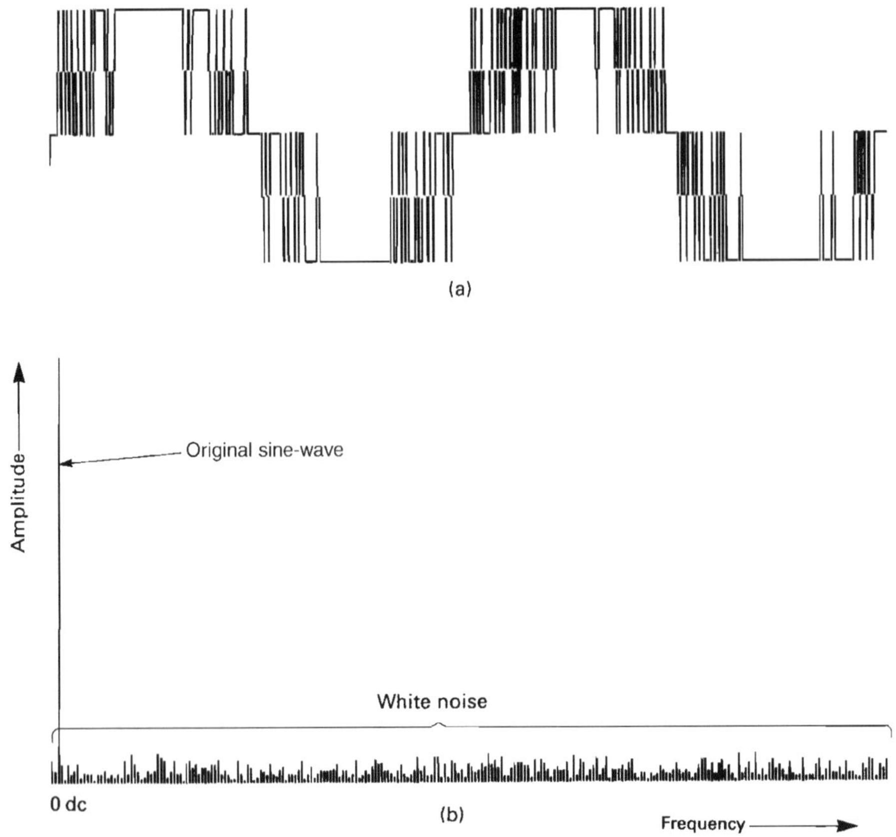

Figure 2.1.26 ±1 lsb sine-wave ±0.5 lsb square dither after quantisation (a) and the spectrum of the signal (b).

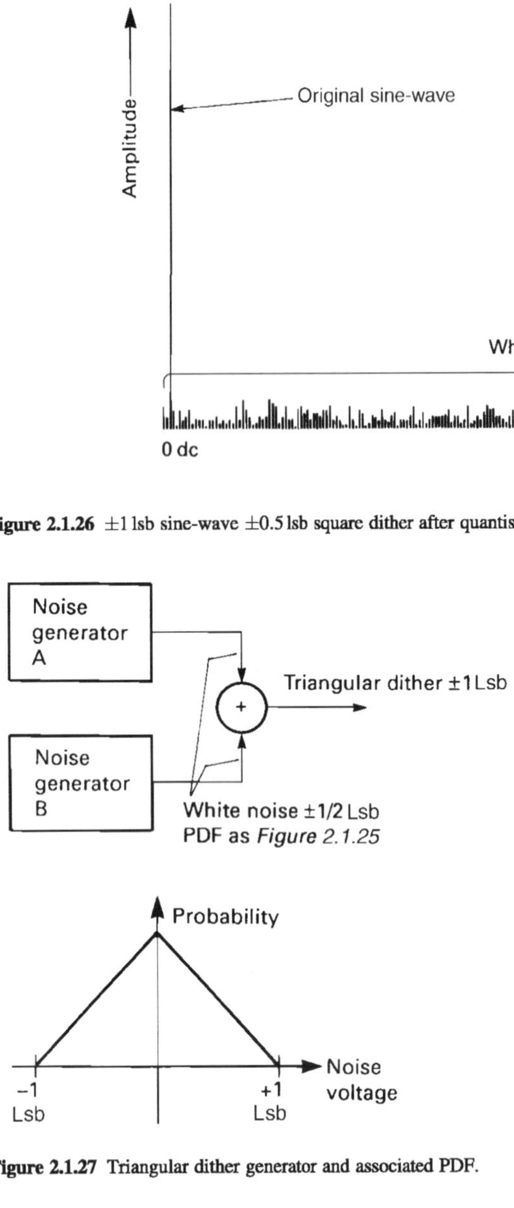

Figure 2.1.27 Triangular dither generator and associated PDF.

domain is that the low-pass filter can be designed for fast roll-off phase response errors much more easily and accurately than in the analogue domain, as digital filter component tolerances are zero.

A second advantage of oversampling is that the noise level of the converted signal is reduced. The noise caused by signal conversion in a 16-bit system is a white noise floor at approximately 96 dB below peak signal level. If the conversion is at, say, twice the sample rate, then this noise is spread over twice the bandwidth, and consequently only half the noise power is in the signal band. The digital filter following the oversampling ADC removes this out-of-band noise, consequently improving the s/n ratio and increasing the number of bits representing the signal (Figure 2.1.31).

2.1.1.2 Digital signal processing

2.1.1.2.1 Digital gain

Digital audio gain control is effected through the use of binary multipliers. Multiplying a signal by numbers greater than unity will achieve gain, whilst multiplying a signal by numbers less than unity will achieve attenuation.

Multiplying an x-bit signal by a y-bit coefficient will produce a result of $x+y$ bits. This means, for example, that when multiplying a 16-bit signal by a 16-bit gain coefficient, the true result will consist of 32 bits. If the result is to be returned to 16-bit

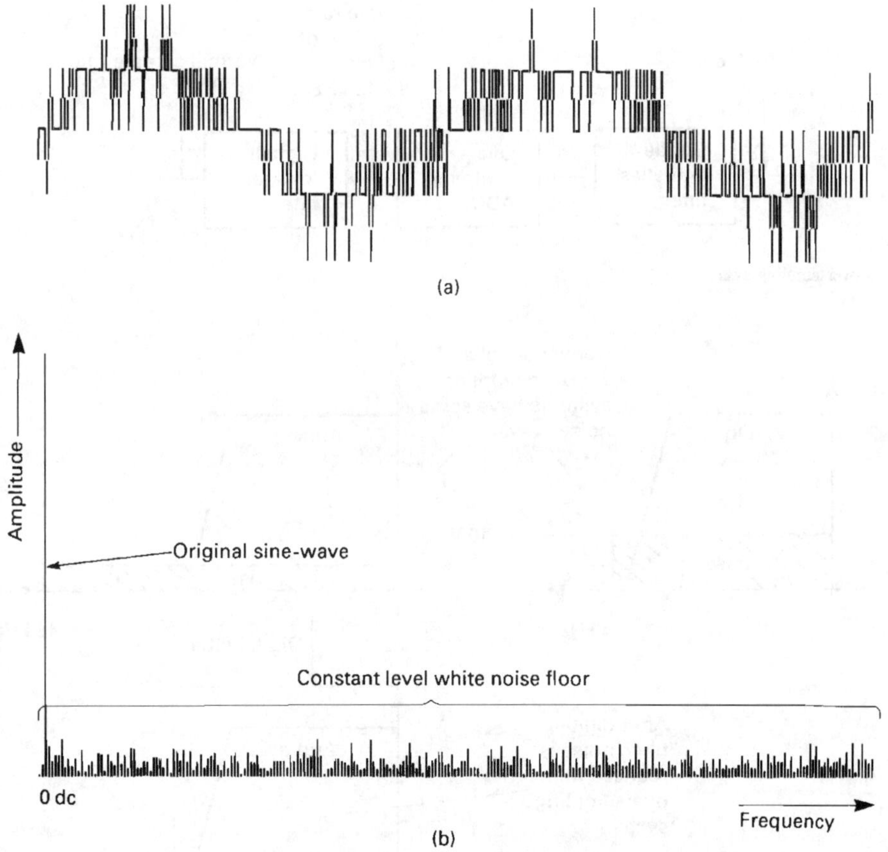

Figure 2.1.28 ±1 lsb sine-wave ±1 lsb triangular dither after quantisation (a) and the spectrum of the signal (b).

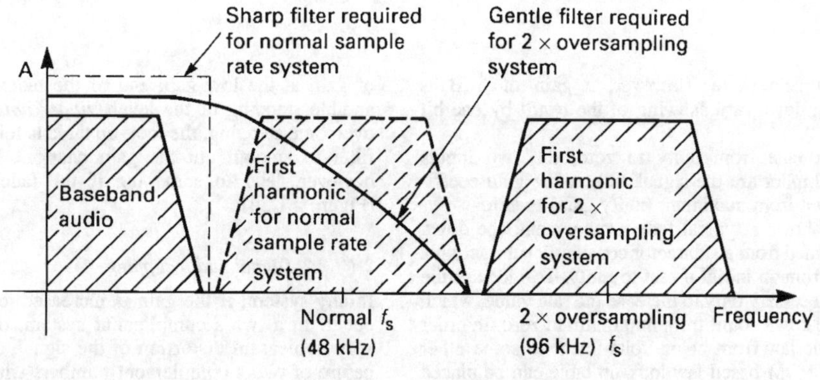

Figure 2.1.29 Comparison of filters required for a 2 × oversampling system and a normal sample rate system.

format, the result must either be *truncated*, the least significant bits being thrown away, or *rounded*, using the nearest 16-bit number. Both truncation and rounding will distort the signal. This approximation distortion may, however, be masked by the use of dither.

The result of a multiplication consists of more bits than the original signal. Choosing which bits of the result are subsequently used for the signal will decide whether gain or attenuation is achieved; by using only the most significant bits, unity is

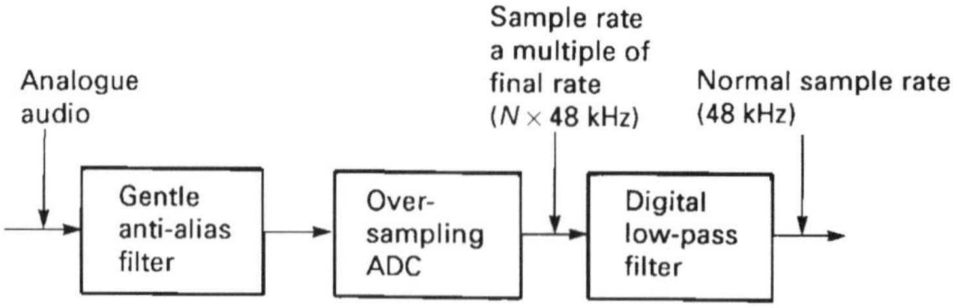

Figure 2.1.30 An oversampling system.

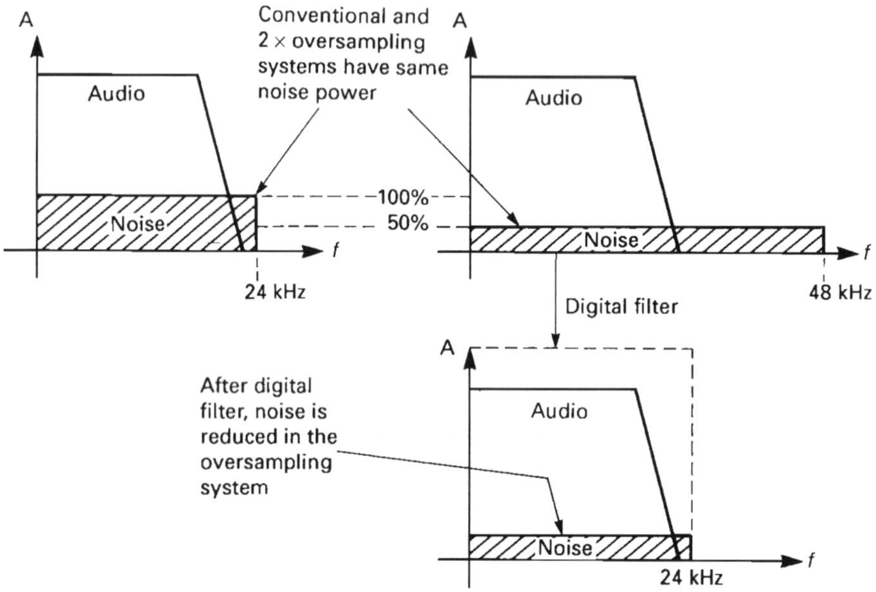

Figure 2.1.31 Comparison of noise levels in conventional and 2 × oversampling systems.

the highest gain achievable. However, a gain of 6 dB is achieved for each downward skewing of the result by one bit (see Figure 2.1.32).

To generate a fade from unity to zero, the two inputs required to a multiplier are the signal and, as the gain coefficient input, a count from maximum (unity gain) to zero.

Figure 2.1.33 shows a typical fade arrangement, the down counter being formed from a subtractor constantly subtracting a fixed rate value from an initial preset to unity. To increase the rate of fade it is necessary only to increase the rate value, which in turn speeds the down count from maximum to zero. In order to change the fade law from being volts-linear to some other characteristic, an ROM-based law look-up table can be placed between the counter and the multiplier.

For manual fades, there must be an input device which can either be a binary fader or a conventional fader followed by an ADC. The digital gain value is then applied to the fader law look-up table and hence to the multiplier as coefficient input.

If a binary fader is used, there may be only 256 steps to cover a range of +12 dB to −∞ dB, which will necessitate large steps of gain at the low gain end of the fader. This will result in audible stepping of the level (*zipper noise*) at low gains. To overcome stepping, the look-up table is followed by a low-pass filter to smooth out the gain changes. This approach will, however, lead to some lag if the fader is moved rapidly (Figure 2.1.34).

2.1.1.2.2 Digital audio clipping

In any system, if the gain is increased too far, overload will occur. In a two's complement system, overload generates a very unpleasant distortion of the signal, caused by the cyclic nature of two's complement numbers where a 1-bit overload of maximum positive value generates maximum negative value.

Figure 2.1.35 shows the effect of overload on a sine-wave in two's complement notation. To improve the subjective effect of overload in a two's complement system, it is normal to follow any process capable of supplying gain by a limiter, driven from an overload detector. When overload occurs, the

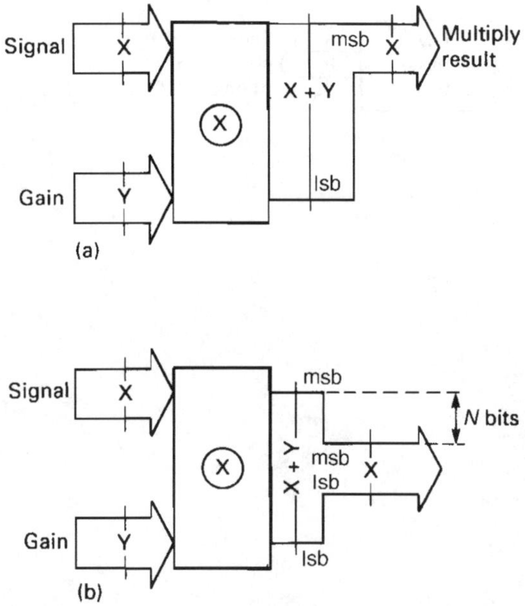

Figure 2.1.32 Multiplier systems having maximum gain of 0 dB (a) and 6*N* dB (b).

limiter will supply either maximum positive or maximum negative value in place of the multiplier output, depending on the polarity of the input signal, returning the overall effect to normal clipping.

2.1.1.2.3 Levels, metering and overlevel indication

Digital audio metering is performed in the digital domain using peak reading meters with zero attack time. A peak hold facility and overload indication are usually provided. The use of peak reading meters is essential for controlling levels in a digital audio environment as, unlike analogue tape machines, where distortion gradually becomes worse as tape saturation effects occur, distortion in a digital recorder or system is negligible up to peak level, and from that point on the signal is clipped as the system overloads (Figure 2.1.36).

Overload indication in a digital audio meter is performed by detecting consecutive samples at peak level. A single sample at peak level cannot be considered as an overload, but consecutive samples at peak level will normally have been caused by signal clipping.

In a digital audio overload indicator, it is normal to be able to set the number of consecutive samples which cause overload indication. When metering low-frequency peak level tone, as may be the case when playing a test tape or test disc, it is possible that overload indication occurs even though the signal is at maximum level and not over. This spurious overload indication is caused by full level low-frequency signals having several consecutive samples at peak level (see Figure 2.1.37).

Historically digital audio was metered using a number of different level indications; today metering is standardised to using dBFS (dB Full Scale). dBFS is a level indication relative to full digital audio level, i.e. to clip level, thus a level of 0 dBFS is the maximum digital audio level achievable in a system. The ADC/DAC relationship between dBFS and dBU/dBm may vary, though commonly three levels are used: EBU (European Broadcast Union) levels, where 0 dBFS = +18 dBU; 0 dBFS = +20 dBU is commonly used in the USA and Japan; and 0 dBFS = +24 dBU is common in the music industry. This relationship is illustrated in Figure 2.1.37(b).

2.1.1.2.4 Digital filtering

Digital filtering of signals is necessary for a number of reasons. Equalisation, oversampling converters and dc removal all require the use of digital filters. Digital filters divide into two classes, infinite impulse response (IIR) and finite impulse response (FIR). IIR filters contain feedback paths while FIR filters have none. All filters consist of different arrangements of only three basic elements: the multiplier, the adder and the delay.

A basic form of *infinite impulse response* low-pass filter is shown in Figure 2.1.38. The input signal is fed through an adder to the output, a portion of the output, determined by the multiplier coefficient, being fed back to the adder through the delay.

Figure 2.1.38(b) shows the circuit output for an impulsive input, a delay of one clock cycle and a gain of 0.8. It is clear that, if the multiplier coefficient is set to 1, then the output will continue at the same level for all time. This is possible due to the feedback path in the circuit, thus the name of *infinite* impulse response.

A basic *finite impulse response* filter structure is shown in Figure 2.1.39. The response of any filter is uniquely determined by its impulse response.[6] For an FIR filter, this impulse response will be identical to the multiplier coefficients. The filter coefficients may therefore be determined by transforming

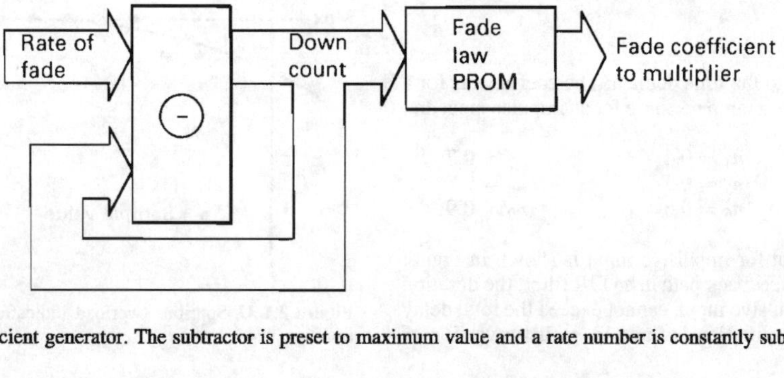

Figure 2.1.33 Fade coefficient generator. The subtractor is preset to maximum value and a rate number is constantly subtracted.

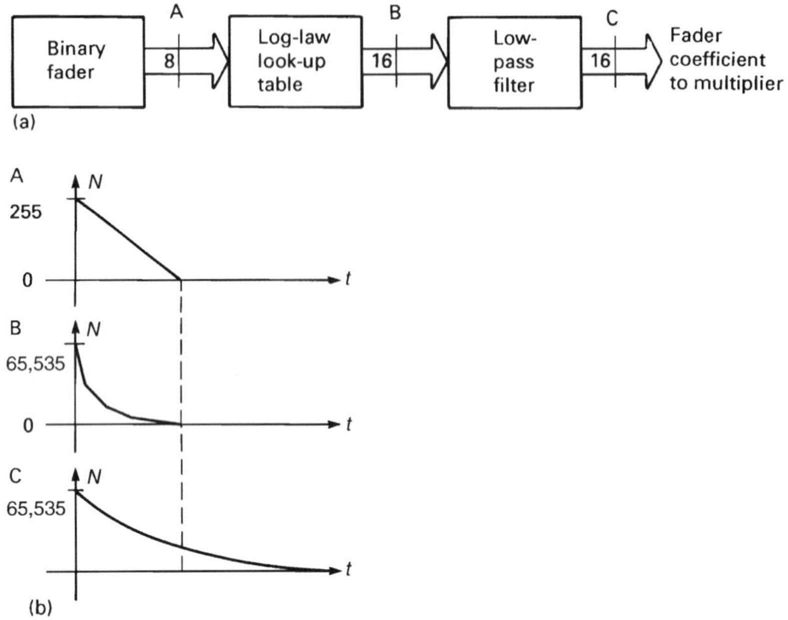

Figure 2.1.34 (a) Using a log-law look-up table and low-pass filter to generate a smooth logarithmic fader action. (b) The associated waveforms.

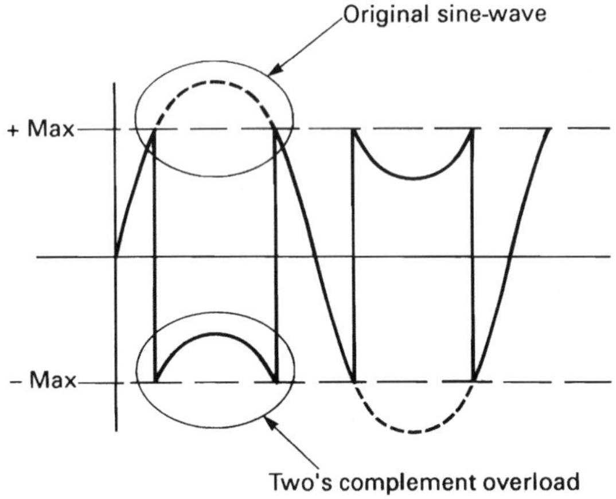

Figure 2.1.35 Overload effect with two's complement representation.

the desired frequency to the time domain. The coefficients for a simple filter to generate a ramp response for an impulse input are:

$n_1 = 0.1$ $n_4 = 0.4$ $n_7 = 0.7$
$n_2 = 0.2$ $n_5 = 0.5$ $n_8 = 0.8$
$n_3 = 0.3$ $n_6 = 0.6$ $n_9 = 0.9$

The resulting output for impulsive input is shown in Figure 2.1.40. As there is no feedback path in an FIR filter, the duration of any output, for impulsive input, cannot exceed the total delay time. Thus the output duration is finite. A realistic version of

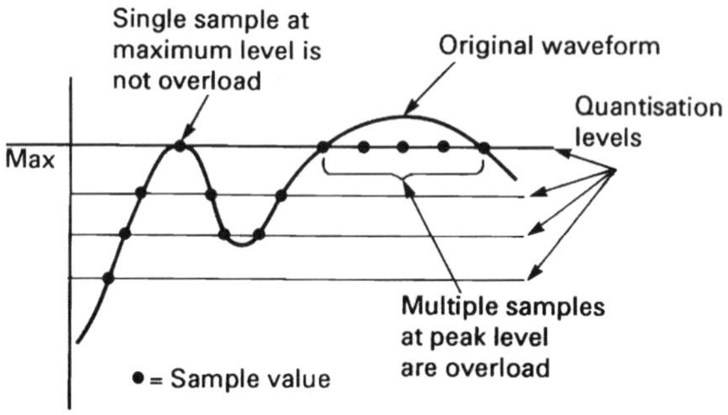

Figure 2.1.36 Overload detection in digital metering.

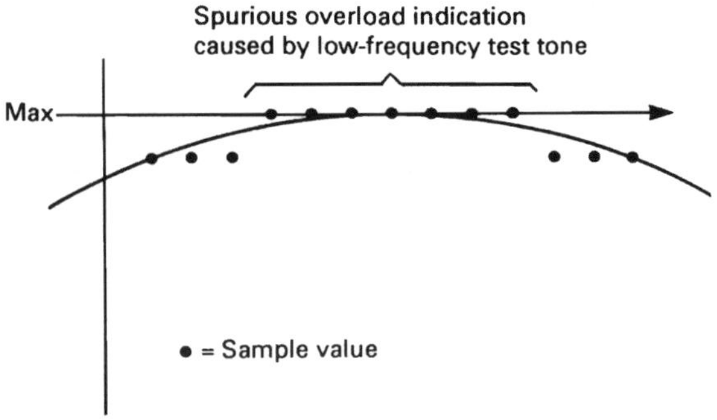

Figure 2.1.37 Spurious overload indication caused by low-frequency test signal.

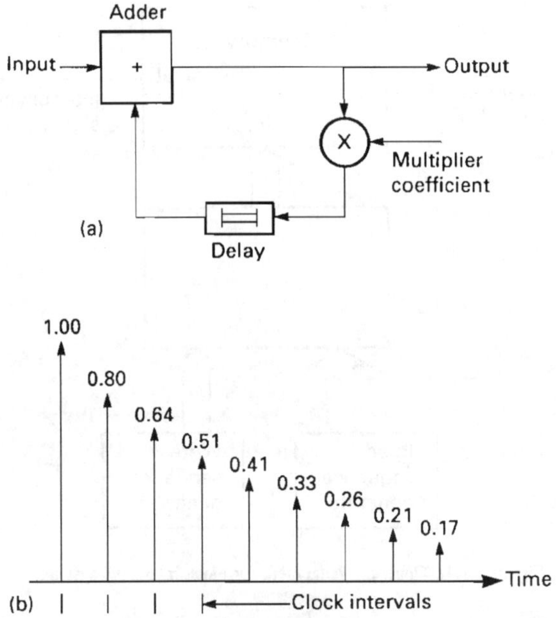

Figure 2.1.38 (a) Basic IIR. (b) The filter output for unity impulse input and coefficient of 0.8.

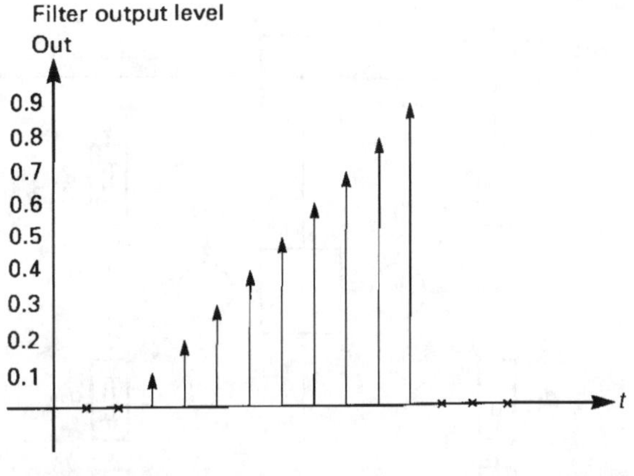

Figure 2.1.40 Nine-element FIR filter impulse response for multiplier coefficients n_1–n_9 = 0.1–0.9.

the filter would typically have 96 delay elements, the delay consisting of RAM. The filter coefficients would be stored as a look-up table in ROM, with the multiplier and adder time division multiplexed between filter sections.

A general structure for a complex digital filter is shown in Figure 2.1.41, the actual filter characteristics being determined by the multiplier coefficients and delay lengths.[7]

2.1.1.2.5 Interleave and error coding

Interleave is used in digital audio record systems to:

- disperse the error bursts caused by dropouts or contamination;
- allow splice (razor blade) editing of open reel format digital tapes.

Error burst dispersal is necessary for the error correction systems to work more effectively. Should the error correction system become overwhelmed, it is also easier to interpolate from the remaining correct samples, to construct replacement samples, if the errors are dispersed.

Either block-based interleave or convolutional interleave can be used.

Block-based interleave is used in the Pro-Digi, Sony PCM 1630 and DAT formats. In a block-based interleave, a fixed number of audio samples are read into a memory, reordered and then recorded on tape (Figure 2.1.42). Editing a block-based interleave system requires edit accuracy to be limited to block ends. Block-based interleaves are convenient for rotary head systems as the interleave length can be related to integral numbers of rotations of the drum. For example, in DAT one interleave period is equal to one drum rotation. In PCM 1630 format, 14 interleave periods occur in rotation. In PCM 1630 format, 14 interleave periods occur in one drum rotation (one frame of video).

Convolutional interleave is a continuous one, where different audio samples are delayed by varying times. The longer the interleave distance, the further the errors will be spread and the better the error correction may be performed (Figure 2.1.43).

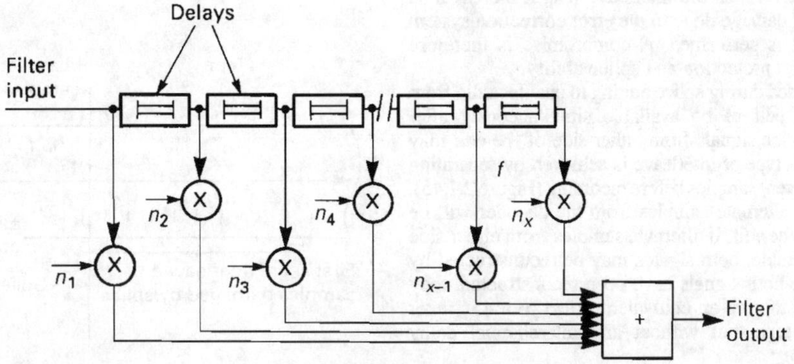

Figure 2.1.39 Basic FIR filter structure.

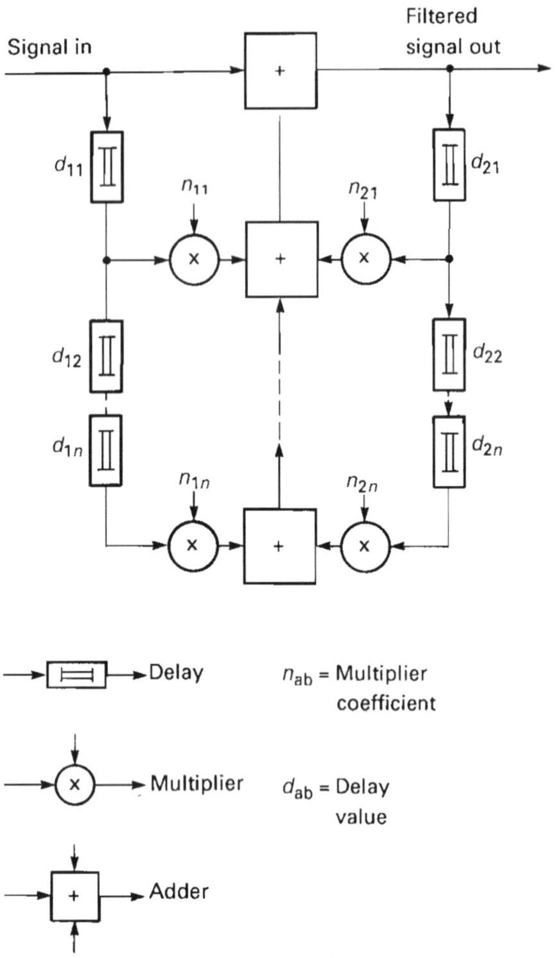

Figure 2.1.41 General form of digital filter.

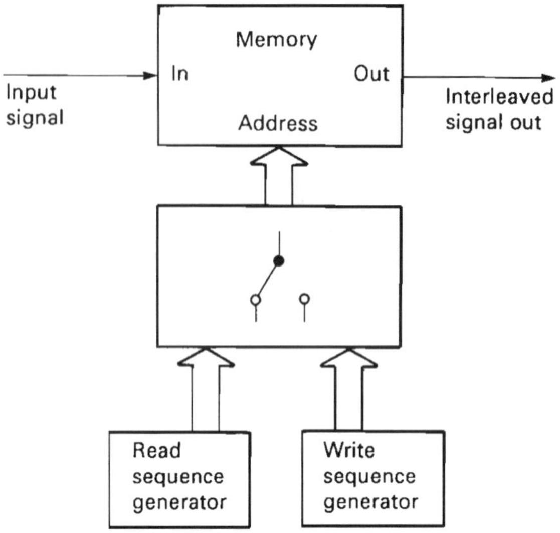

Figure 2.1.42 General arrangement for block interleave system.

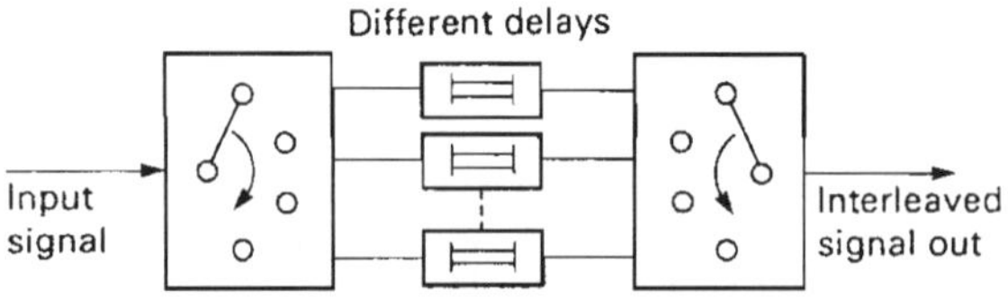

Figure 2.1.43 General form of convolutional interleave.

However, there is a conflicting requirement for the interleave system. This is its ability to deal with splice edits. As may be seen from Figure 2.1.44, as the interleave length increases so does the amount of damage done to the error correction system when a splice edit is performed. A compromise is therefore drawn between burst protection and splice ability.

Interleave is utilised during splice editing to enable audio from both sides of the edit to be available simultaneously after de-interleave. Thus the signals from either side of the edit may be cross-faded. This type of interleave is achieved by separating alternate (odd and even) samples before recording (Figure 2.1.45).

At a splice edit, alternate samples from the decoder will be from either side of the edit. If alternate samples from either side of the edit are available, both signals may be reconstructed by interpolation. Once both signals have been reconstructed, they can be cross-faded to give an equivalent effect to a 45° razor edit on analogue tape, but without any interchannel delay (Figure 2.1.46).

Interleave has an effect on electronic tape editing, due to the time delay required for decoding the playback audio and for encoding the record audio. To enable a cross-fade between playback and record audio at an electronic edit, it is normal to play back the audio from a head earlier in the tape path than the record head. This enables the playback audio to be decoded and re-encoded in the time it takes the tape to travel between the two heads.

The cross-faded audio is thus re-recorded on the tape in exactly the same physical position that it was played from. The record head downstream from the playback head is referred to as the *sync record* head (Figure 2.1.47).

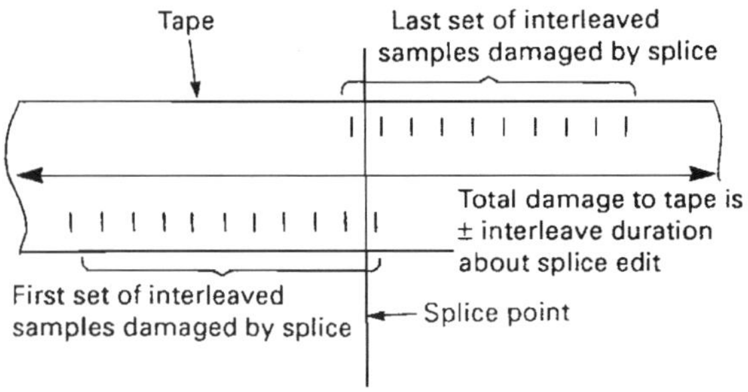

Figure 2.1.44 Disruption to interleave system caused by splice. Tape motion is from right to left.

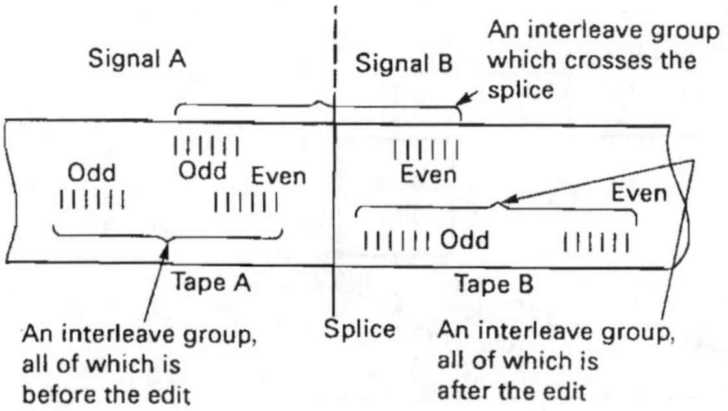

Figure 2.1.45 After de-interleave the group of data which crosses the splice will become alternate samples from either side of the splice.

2.1.1.2.6 *Error correction*

Error detection and correction systems are required in digital audio transmission and recording systems for detection and prevention of data corruption. All error correction systems use additional redundant information for error detection and correction, but there are many ways of implementing the redundancy.[8]

The three commonly used forms of redundancy are CRCC (cyclic redundancy check code), parity and Reed–Solomon codes.

A CRCC *code* is a number generated from and appended to the digital data stream which is used to detect playback errors.

Parity words are generated by an exclusive-or (XOR) of data words and are used to correct errors indicated by a CRCC check circuit.

Reed–Solomon codes use galois field[9] processing techniques. Reed–Solomon codes are basically a hybrid between CRCC and parity codes, and may be designed to both detect and correct errors. Overall error correction techniques result in vast improvements in error rates.[10] Random error rates of hundreds per second can be reduced to single errors per hundreds of minutes.

2.1.1.2.7 *Channel coding*

Channel coding[11,12] is used by record systems to change the binary, two's complement, signal to a form more suitable for recording on a particular medium.

A signal to be recorded has numerous requirements. Firstly, the recorded signal should have a strong clock content, so that it may be easily extracted on replay (on a multi-track tape, each channel will require its own clock signal as tape path instability will cause interchannel timing variations). Secondly, tape record/replay systems have poor low- and high-frequency responses, and consequently any signal recorded on tape should have limited lf and hf content. Thirdly, high-density magnetic recording systems suffer from peak shift[13] distortion – a widening of narrow asymmetric pulses.

Problems with binary signals are numerous. For example, digital silence, a data stream of all zeros, has a strong lf content, no clock content and very high asymmetry, rendering it unplayable. A channel coding system attempts to match the record signal to the medium.

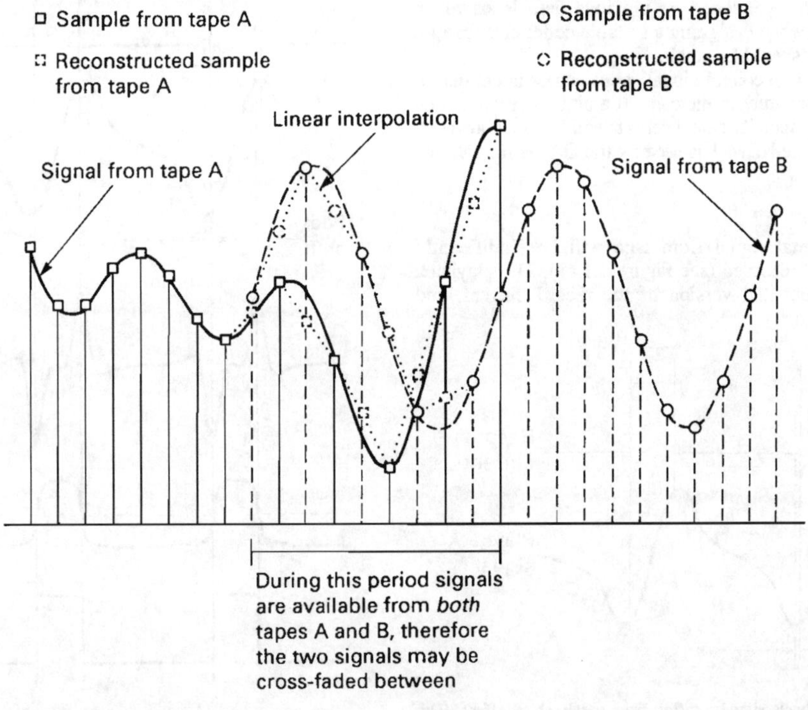

Figure 2.1.46 Signal reconstruction at a splice edit.

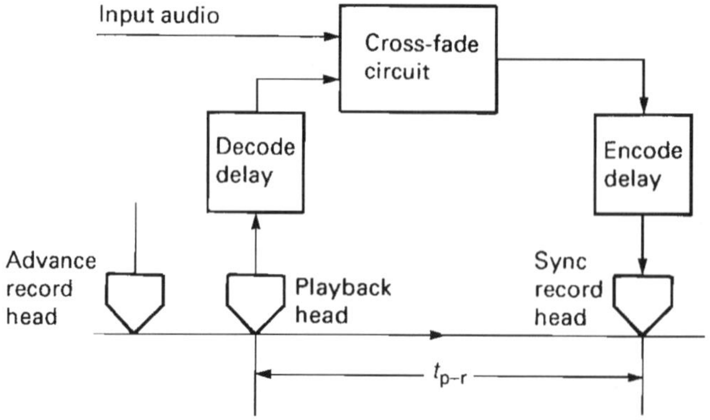

Figure 2.1.47 The use of a sync record head to enable electronic editing. The time taken for the tape to travel from the *play* head to the *record* head (t_{p-r}) must equal the sum of the encode and decode delays.

All channel codes work in a similar manner, by converting one group of data into another, more desirable set of record data. The codes break down into two basic types: bit or word substitution codes and convolutional codes.

In *substitution codes*, a group of bits, generally a byte, are converted into an alternative group of bits with more desirable characteristics. These are found by using a look-up table. Substitution codes are used by DAT (8–10 modulation), Pro-Digi (4/6M) and compact disc (EFM).

Convolutional codes differ from substitution codes in that the channel code for a particular group of data bits depends not only on the data bits themselves, but also on the data that have gone before. This dependence on previous data leads to an extra requirement when designing a suitable code: error propagation in the code should be small.

Error propagation in convolutional codes comes about due to the prior dependent nature of the code. If a bit is corrupted, then as later bits depend upon its state, later bits will also be corrupted. A convolutional code HDM-1 is used by the DASH format.

2.1.1.2.8 Signal detection

The playback digital signal from tape will be significantly different from that recorded (see Figure 2.1.48). The playback signal is a differentiated version of the record current, and suffers from peak shift caused by the spreading of asymmetric record signals. Before the playback signal can be converted into a true digital signal for playback processing, it must undergo equalisation to eliminate the peak shift, followed by integration to restore the wave shape, and finally slicing at a 1/0 decision level to recreate a true digital signal (Figure 2.1.49).

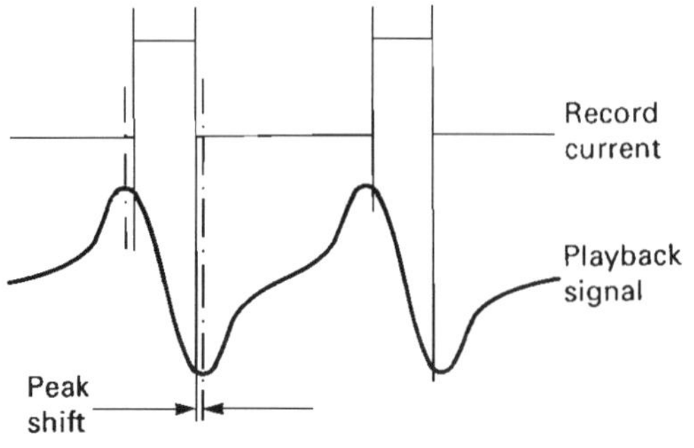

Figure 2.1.48 Playback signals suffer from peak shift effect. The playback signal is also a differentiated version of the record current.

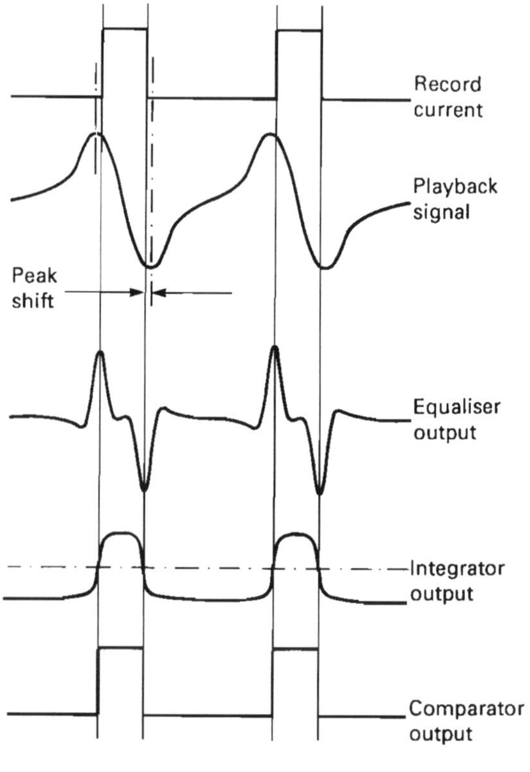

Figure 2.1.49 Processes required to reform the playback signal.

2.1.1.2.9 Timebase correction

Timebase correction of the playback signal in a recording system is necessary to eliminate playback timing instabilities. These instabilities are caused by a variety of problems, such as servo lock errors or tape weave. Timebase correction is performed by extracting the timing information from the playback signal, then writing the playback data into a memory using the extracted clock (see Figure 2.1.50).

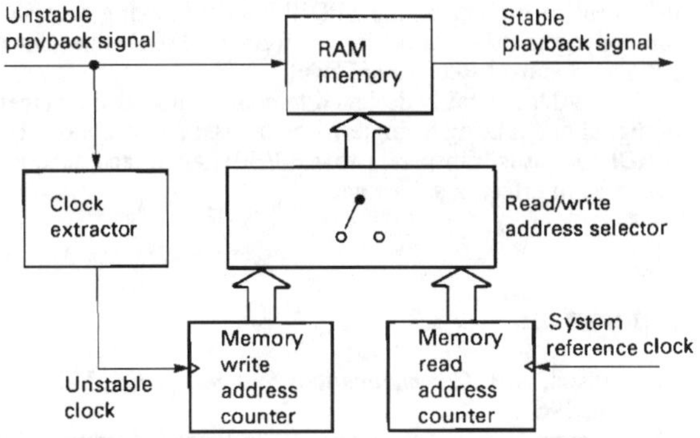

Figure 2.1.50 Timebase corrector.

The data may then be read from the memory at a constant, stable rate using the reference clock of the system. The size of memory must be greater than maximum instability, i.e. if the maximum playback jitter encountered is ±40 samples, then an 80-sample memory will be required.

2.1.1.2.10 Error concealment

If random errors occur in a digital system, the audio samples will be replaced by random numbers. The analogue equivalent of a continuous stream of random numbers is full level white noise. It is therefore necessary to employ an error concealment system in case the error correction systems are overwhelmed and unable to correct corrupted data.

Error concealment systems normally operate in a number of different manners according to the severity of the error. The first level of concealment occurs when only a single sample is in error. The usual strategy for concealment is to replace the sample by an average of the preceding and the following samples (Figure 2.1.51(a)). This is also referred to as *interpolation.*

If multiple errors occur, the next level of concealment is hold, followed by average (Figure 2.1.51(b)). Here the value of the last correct sample is repeated until the corruption is finished, the last corrupt sample being replaced by an average of the held value and the first uncorrupted sample.

For severe errors, the strategy of concealment by hold is inappropriate, as the many transitions between held data and good data will produce large numbers of transients which are audibly objectionable. For severe data corruption, the normal approach to concealment is to mute the output, replacing the corrupted data with a fade to and from zeros for a given duration. This has the advantage over hold that, if a single uncorrupted sample is output, then it will not produce a transient, audible as a click.

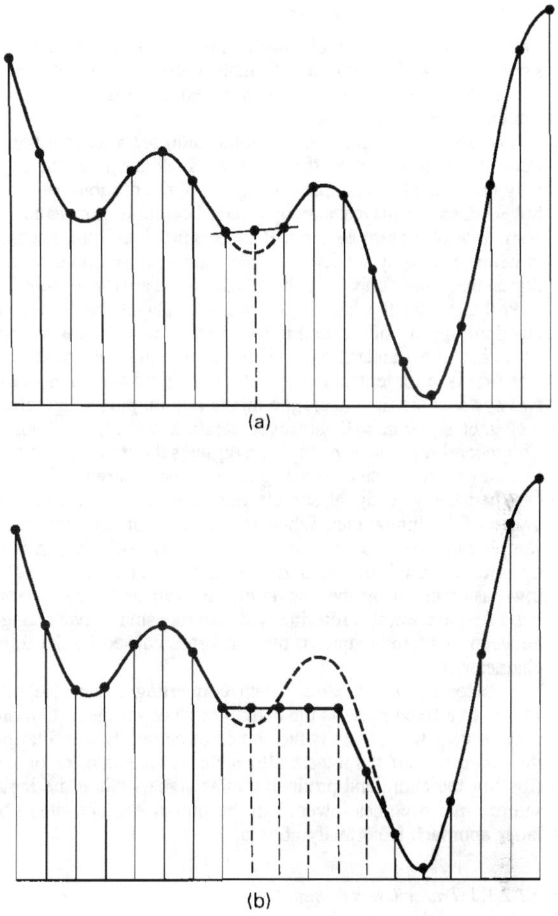

Figure 2.1.51 Concealment of errors: (a) concealment of single error by averaging (interpolation); (b) concealment of multiple errors by hold and average.

2.1.2 Digital Audio in Application

2.1.2.1 Standard formats and their conversion

2.1.2.1.1 Sample rates

Sample rates for digital recording have had a much debated history. In 1981, 60, 54, 52.5, 50.4, 50.35, 50, 48, 47.25, 47.203, 45, 44.1, 44.056, 32 and 30 kHz were all considered as possible contenders for a digital frequency standard.[14]

Now there are four rates in normal usage: 48 kHz is used by convention in sound recording studios and for digital sound on VTRs, this rate being a convenient multiple of both the 25 Hz and the 30 Hz television frame rates; 44.1 kHz is the standard for compact disc and its associated mastering equipment; 44.056 kHz evolved from 44.1 kHz and enables equipment designed for 44.1 kHz to be used with drop frame NTSC video (29.97 Hz frame rate); 32 kHz is used for PTT, TV sound and fm radio, where the audio bandwidth is limited to 15 kHz.

2.1.2.1.2 Sample rate and format conversion

The use of vari-speed playback, along with the variety of sample rates and transmission formats in use, means that digital audio standards converters are required to convert between standards and sample rates.

Vari-speed operation with a digital radio recorder is a rather more complex operation than with an analogue machine. Not only does the playback speed and pitch of the programme vary, but so does the playback sample rate, becoming non-standard. Variations of the sample rate will not be a problem if the machine is being used only via its analogue inputs and outputs, as the digital machine is effectively being used as an analogue recorder.

Problems occur when digital audio is played back at non-standard speed and is required as input for a digital system operating at a standard rate. There are two possible solutions. The first is to forget that the operation is digital and connect the two systems via their analogue inputs and outputs. This rather inelegant solution will, however, degrade the signal quality. The second solution is to employ a digital sample rate converter to change rates without entering the analogue domain.

When using a digital sample rate converter, one must be aware of its limitations. When changing from a known high sample rate, say 48 kHz, to a lower rate, say 44.1 kHz, then in order to avoid aliasing of high frequencies, the audio must be low-pass filtered by the converter. As well as losing out-of-band frequencies, this filtering will lead to a small level change between input and output of the converter, caused by the filter characteristics.

A second problem arises when converting a variable rate source to a fixed rate. As the actual range of sample rate input may be very wide, there is the choice of either always filtering the audio to half the lowest allowable sample rate, or of not filtering the audio and producing alias signals when the input sample rate becomes lower than the output sample rate. The latter approach is normally chosen.

2.1.2.1.3 Transmission formats

There are four standard formats used to interconnect digital audio equipment. The transmission standards are single-channel SDIF[15] (Sony Digital Interface Format), twice channel AES/EBU[16] format, its domestic variation SPDIF[17](Sony Philips Interface Format) and multi-channel MADI format (multi-channel digital audio interface).

SDIF is based on a transmission format of one audio channel per connection. It is not a self-clocking signal and will therefore also require connection of a synchronising signal (word clock, a square-wave sample rate) between the transmitter and receiver.

An SDIF connection is capable of handling a single channel of audio of up to 20 bits. Emphasis information is transmitted with the audio. The format also provides a facility for user data, although in practice this is unused. An SDIF connection uses 75-ohm coaxial cable with BNC connections.

The AES/EBU transmission format sends one or two audio channels of up to 24 bits per connection along with considerable quantities of subcode information. The subcode indicates amongst other things sample rate, two-channel mode and clock status. The AES/EBU format also transmits two additional bits of information with each audio word. The first is a *parity* bit, used to assess transmission link quality; the second is a *validity* bit intended to indicate whether the audio word is a genuine sample, or whether the sample has been generated by interpolating from other data.

The AES/EBU format is a balanced self-clocking format designed to use existing audio cabling. The standard connection is via 110-ohm cabling with XLR connections. As AES/EBU format is self-clocking, there is no requirement to send additional synchronising information with the data. However, it is good practice to synchronise equipment using a dedicated clock, in preference to the data stream.

SPDIF is an unbalanced variant of the AES/EBU format, intended for domestic use, utilising phono plugs for connection. The major difference between AES/EBU and SPDIF formats is in the subcode channel, the SPDIF standard having provision for copyright information in the form of ISRC data (International Standard Recording Code).

The MADI format is designed to connect up to 56 channels of digital audio using a single 75-ohm coaxial connection. The MADI format is transparent to AES/EBU audio and subcode, which is, in effect, a subformat.

References

1. Carlson, A.B. *Communications Systems*, p. 299, McGraw-Hill, 1968.
2. Hoshino, Y. and Takegahara, T. Influence of group delay distortion of low-pass filters on tone quality for digital audio systems, *Proc. AES 3rd Int. Conf. on Present and Future of Digital Audio*, p. 115 (1985).
3. Taub, H. and Schilling, D. *Principles of Communication Systems*, pp. 5.4–5.6, McGraw-Hill, 1971.
4. Taub, H. and Schilling, D. *Principles of Communication Systems*, p. 71, McGraw-Hill, 1971.
5. Carlson, A.B. *Communication Systems*, p. 2.5, McGraw-Hill, 1968.
6. Rabiner, L. and Rader, C. (eds). *Digital Signal Processing*, IEEE Press, publication year: unknown.
7. Lin, S. and Costello, D.J. Jr. *Error Control Coding: Fundamentals and Applications*, Prentice-Hall, publication year: not known.
8. Birkhoff, G. and Maclane, S. *A Survey of Modern Algebra*, p. 15, Macmillan, 1977.
9. Doi, T. Error correction for digital audio recordings, *AES Prem. Conf. Collected Papers* (1982).
10. Blesser, B., Locanthi, B. and Stockham, T.G. Jr (eds). Digital audio, *AES Prem. Conf. Collected Papers*, p. 5 (1982).
11. Doi, T.T. Channel coding for digital audio recordings, *AES 70th Convention* (1981).
12. Jorgensen, F. *The Complete Handbook of Magnetic Recording*, p. 262, Tab Books, publication year: unknown.
13. Gibson, J.J. A review of issues related to the choice of sample rates from digital audio, SMPTE Digital Television Group (1981).
14. Pohlman, K. *Principles of Digital Audio*, p. 156, Howard W. Sams, publication year: unknown.
15. AES recommended practice for digital audio engineering – serial transmission format of linearly represented digital audio data, *J. Audio Eng. Soc.*, **33**, 975–984 (1985).
16. *Draft Standard for a Digital Audio Interface*, IEC (February 1987).
17. Wilkinson, J., Easty, P., Ward, D.G. and Lidbetter, P. Proposal for a Serial Multichannel Digital Interface, *AES and EBU Paper*, publication year: unknown.

Paul Davies
Broadcast Systems Manager
Dolby Laboratories, Inc.

2.2 Non-linear Audio Systems

Non-linear Audio Systems are those systems which encode audio to reduce the data rate, allowing the signal to pass through a network or system at that lower data rate, whilst allowing as complete as possible a recovery of the audio on decoding. This usually involves taking a linear PCM signal and processing it to remove certain components of the signal. The components to be removed are determined by the coding scheme employed. There are typically two types of coding systems existing that are in everyday usage, namely lossy and lossless coding. As their descriptions indicate, the lossless coding systems use techniques to pack the data more efficiently, allowing a bit-for-bit recovery of the data. This is a variable data rate compression scheme and, depending on the entropy of the signal being coded, can either not compress the data rate at all, or compress it to near zero data rate. Meridian Lossless Packing (MLP) as used in DVD-Audio standard is a typical example of this type of coding scheme. In lossy coding schemes, bit-for-bit preservation through the encode/decode signal path is not a requirement. As such, a variety of techniques are employed to reduce the data rate, such as to remove those parts of the audio spectrum that the human ear is not capable of hearing, using psychoacoustic techniques. The goal of any lossy audio coding system is to achieve perceptually lossless audio coding, where sound quality is fully preserved through the signal chain. With these lossy schemes we are able to control the data rate, which makes them more easily handled through the broadcast chain. In these processes the signal ends up being quantised, which generates noise, and how the noise is hidden within the coding scheme is one of the discerning factors between the many schemes available. These lossy schemes can further be divided into two sections: those best suited for contribution and distribution through the broadcast chain, and those designed for emission to the consumer. In the former case it is necessary to use a less compressed data rate, as these signals are the ones that are likely to require encoding and decoding over a number of generations as required in broadcast post-production and point-to-point broadcast links. As the scheme is lossy it is important that after several generations the remaining signal is as near to the original as possible and suffers from the minimum acceptable degradation. Different schemes have varying abilities concerning generation loss. Also of importance to the broadcaster is the ability to edit and manipulate these signals in the coded domain, and to retain the synchronisation with the associated video signal. Systems usually employed in the distribution chain are MPEG-2 Layer 2, APT-X and Dolby E. In the latter case of emission to the consumer, a more aggressive data rate reduction can be employed because the signal will, in most cases, only be encoded and decoded a single time. Typically these will be MPEG-2 Layer 2, AC-3 and AAC. There are several comparison tests that have been performed in recent years to investigate the subjective quality of these various coding schemes, such as that reported by Soulodre et al.[1]

2.2.1 Perceptual Coding

Many of the techniques used to reduce the bit rate of the audio signal rely on perceptual coding techniques. These techniques are based on the ability of the basilar membrane within the inner ear, to distinguish between the amplitude and frequency of different signals.

Consider, in Figure 2.2.1, the ear being stimulated by a sound pressure level of a given signal at a particular frequency. This signal will excite the basilar membrane within a narrowband region of audio spectrum referred to as a critical band. The human basilar membrane system consists of 24 such critical bands from 0 to 20 kHz, where each higher critical band is wider in bandwidth than the preceding lower band. Consequently, a lower level signal in the same critical band and below the masking level will not be able to stimulate the basilar membrane and will be effectively masked or unheard. As can be seen in Figure 2.2.1, a masking curve is imparted by a masked signal onto nearby signals in such a way that signals that fall below the mask threshold can be

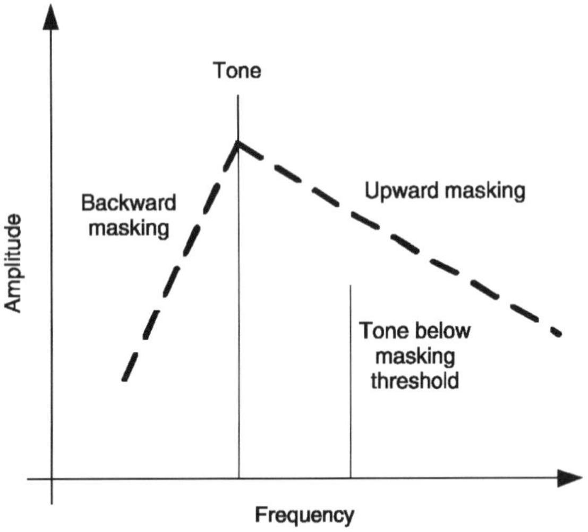

Figure 2.2.1 Example of masking threshold.

considered imperceptible. The upward masking curve is not as steep as the downward masking curve. These critical bands vary in width from a few hundred hertz with a centre frequency in the hundreds of hertz up to several kilohertz when the centre frequency is around 5 kHz or above. The critical bands are continuously variable in frequency and any given audible tone will create a mask threshold centred on it. An aggregate mask threshold curve can then be formed for the entire audible spectrum via linear superposition of individual masking curve contributions from all masking signals. Temporal masking occurs when signals arrive close in time. These can be pre-masked, with the masking occurring immediately before the main signal, or post-masked, where the masking occurs immediately after the main signal. These various masking criteria can be applied to remove those portions of the signal that the ear will not be able to discriminate, either due to time-domain or frequency-domain masking. In removing these parts of the signal we generate quantisation noise; however, as the quantisation of the signal is performed in the critical bands it is possible to hide this quantisation noise below the masking curve.

Coupling is a technique used by many of the coding systems described here. It is based on the psychoacoustic understanding that high-frequency signals are detected as an envelope rather than detailed signal waveform. When the coding scheme runs out of bits, the high-frequency content of individual channels can be combined and sent as the individual channel signal envelopes along with the combined coupling channel.

2.2.2 NICAM 728

Compressed or data rate reduced digital audio first appeared on the analogue terrestrial broadcast services of Europe in the later part of the 1980s. Based on an earlier similar system invented by the BBC, NICAM 728 (Near Instantaneous Companding and Audio Multiplexing) is capable of delivering near CD quality sound in stereo, mono or dual mono (for dual language transmission). NICAM is still found in PAL I, B, G and H systems mainly in Europe, and shares a commonality with the MAC packet family common in Europe on the DBS system.

There are three main stages of processing: analogue-to-digital (A-D) conversion, near instantaneous companding and multiplexing of the data stream. In the A-D process, pre-emphasis is added to the signal according to CCITT J.17. Then it is sampled to 14-bit resolution at 32 kHz sampling rate, the samples being coded in two's complement. The sampling frequency of 32 kHz was chosen for its commonality with other contribution circuits and MAC systems. To prevent aliasing, a 15 kHz filter is introduced before the A-D stage.

The signal then undergoes a near instantaneous companding process, which is achieved by assigning five companding coding ranges that can be signalled by just 3 bits. For details of the exact method, refer to the BBC R&D document RD 1990/6 by Bower.[2] The companding effectively reduces the data rate to 704 kbps, comprising 64 10-bit samples each with a parity bit. The coding information indicating the companding applied to all 32 samples in each channel in each 1-ms block is signalled to the receiver in a 3-bit scale factor code, which is applied using 'Signalling-in-Parity'.

The bit stream is assembled of the 704 sound/data bits which are bit-interleaved, 11 bits of additional data, 5 bits for control information and an 8-bit frame alignment word, totalling 728 kbps.

2.2.3 MPEG Audio

The MPEG (Motion Picture Experts Group) was set up by the IEC and ISO, and is responsible for setting the MPEG standards in use today. The first audio standard was MPEG-1 Layer –1, which was a simplified version of MUSICAM. MUSICAM (Masking-pattern Universal Sub-band Integrated Coding And Multiplexing) was one of the early perceptual coding algorithms and was in turn based on MASCAM (Masking-pattern Adapted Sub-band Coding And Multiplexing). MUSICAM divides the incoming audio into 32 sub-bands and uses perceptual coding and masking techniques to achieve data rate reduction. Originally formalised in ISO/IEC 11172, the audio portion of the standard (ISO/IEC 11172-3 and ISO/IEC 13818-3) describes all three of the current MPEG audio coding schemes. The standard describes the bit stream format and the reference decoder model, it does not describe the encoder model. The MPEG-2 standard owes much to the MPEG-1 coding scheme and consequently a reasonable amount of compatibility exists between the two, such that MPEG-1 equipment can partially decode an MPEG-2 signal, and MPEG-2 equipment can fully decode MPEG-1 signals. With each generation of the coding schemes, a lower data rate and higher quality is achieved, with Layer 3 giving the higher quality. Multi-channel audio capability was added to the MPEG-1 standard and was formalised into the MPEG-2 standard in 1995. The standard was revised to allow transmission of MPEG audio with many of the attributes found in Dolby AC-3 multi-channel audio coding system (Dialogue Normalisation, dynamic range control, etc.); however, the MPEG Layer 2 multi-channel system has not been a commercial success. MPEG-1/2 Layer 3 has thus far been greatly used in Internet applications but has not seen use in broadcast applications. More recently AAC (Advanced Audio Coding) has been standardised by the MPEG group as MPEG-2 AAC and MPEG-4 AAC, and MPEG-4 audio offers a much wider range of audio tools, which should make a significant difference to broadcast audio in the coming years. There are two different psychoacoustic models used for Layers 1–3 which are described in Annex D of the MPEG-1 standard (ISO/IEC 11172-3). Although either may be used for any layer, in general Model 1 is used for both Layers 1 and 2 and Model 2 for Layer 3.

2.2.3.1 MPEG-1 Layer 1

In MPEG-1 Layer 1 audio coding, the wideband signal is first split into 32 equal width sub-bands using a polyphase filter. As critical bands are not of a fixed width, this has to be compensated for in the way the encoder allocates bits to each sub-band. The adjacent sub-bands overlap and consequently a given signal can affect two sub-bands. The filter, a 512-sample FFT (Fast Fourier Transform), outputs 32 samples, one for each band of the 32 input samples. Twelve of these sub-band samples from each of the 32 sub-band samples are grouped in a frame which represents 384 wideband samples. Each 12 sub-band samples is given a bit allocation. Where the sub-band is considered to be inaudible the bit allocation is zero. Bit allocation is determined by the number of bits required to quantise the sample based on the calculated masking threshold. Floating point notation is used where the resolution is given by the mantissa and the dynamic range by the exponent. A fixed scale factor exponent is computed for each group of 12 sub-band samples that are non-zero, and the group of 12 samples is divided by this factor to optimise quantiser resolution.

Four possible channel configurations can be used in Layer 1: mono, a single channel; dual mono, two separate audio channels; stereo; and joint stereo. The joint stereo configuration exploits certain irrelevant elements of the stereo signal. Intensity stereo combines the left and right channel information above a certain frequency (or sub-band) and can gain up to a 30 kbps reduction in overall data rate. The technique used is to code only the envelope for both channels, but code separate scale factor values for each of the channels. Joint stereo coding is normally selectable at 3, 6, 9 and 12 kHz (assuming a 48 kHz sample rate). When using this method in transmission it is important to remember that a stereo signal already 4:2 matrix encoded as Dolby stereo (surround) can be corrupted by this technique as important phase information is required at up to 7 kHz, and normal stereo coding is to be preferred in such cases. Another technique of joint stereo coding is known as MS stereo, which has similarity with FM stereo transmissions, whereby the sum and difference signals are transmitted and de-matrixed in the decoder. MS stereo can be found in later versions of the MPEG codec.

2.2.3.2 MPEG-1 Layer 1 Bit stream

The bit stream consists of four main parts, the frame header, the CRC error check, audio data and ancillary data (Figure 2.2.3), and it is assembled as follows: each Layer 1 audio frame (8 ms long) begins with a frame header. The frame header consists of two parts: the sync word and the system word. The sync word is 12 bits long and all bits are set to 1s, and the system word is 20 bits long. Within the system word is informational and control data to describe the following attributes: ID (1), Layer (2), Error protection (1), Bit Rate Index (4), Sampling frequency (2), Padding bit (1), Private bit (1), Mode (2), Mode extension (2), Copyright (1), Original or copy (1) and Emphasis (2). If the protection bit in the frame header is set to zero, then the CRC error check is active and will contain a 16-bit error check word to enable error detection in the bit stream. The audio data section can be divided into three subsections containing the bit allocation, scale factors and sub-band samples. The length of the sub-band sample section will vary as 12 audio samples are allocated to each of the 32 sub-bands, which are represented by between 2 and 15 bits, determined by the calculated masking requirements. The ancillary bits are user definable and of a varying length, and it is in this part of the bit stream in Layer 2 the multi-channel extension is located.

MPEG-1 Layer 1 specifications are as follows:

- Data rates 32–448 kbps.
- 8 ms coded audio frames.
- Supports mono (single channel), dual mono (two independent channels), stereo and joint stereo.
- Designed for studio and recording applications.

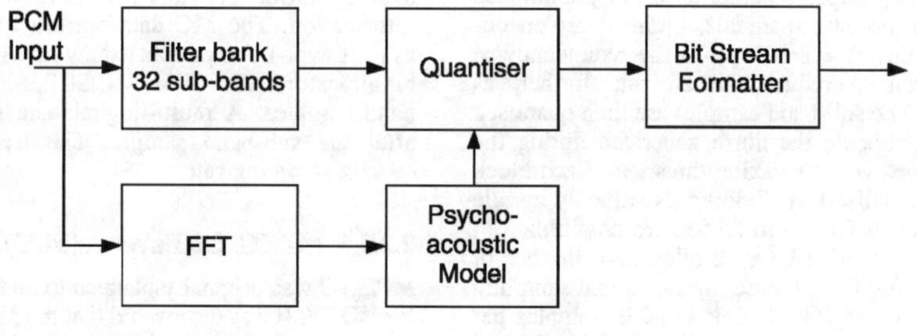

Figure 2.2.2 General MPEG audio encoding process.

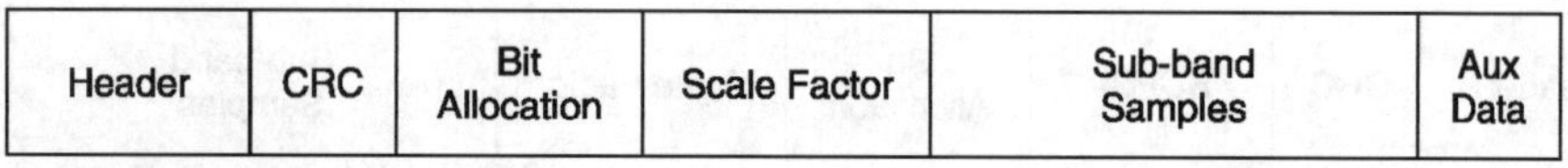

Figure 2.2.3 MPEG-1 Layer 1 bit stream.

- 32 sub-bands (equal width).
- 512-sample FFT.

2.2.3.3 MPEG-1/2 Layer 2

Originally, MPEG-1/2 Layer 2 was designed to give an extension to Layer 1, allowing an increase in the number of audio channels from 2 up to 5.1, and be backwards compatible with Layer 1. Additionally, lower sampling rates were introduced to improve the audio quality at bit rates less than 64 kbps, which is particularly useful for speech.

As with Layer 1, the input signal is mapped from the time domain into subsampled spectral components via the polyphase filter bank consisting of 32 equally spaced sub-bands. These sub-bands, due to the non-uniform nature of critical band spacing, provide more spectral discrimination in the lower frequency region and wider, less spectral discrimination in the higher frequency bands. This allows optimal adaptation of the audio blocks to assist masking in the temporal domain and helps reduce delay and complexity. Simultaneously, the same input signal is processed by the FFT to estimate the actual frequency-dependent masking threshold. The FFT analysis block size is increased in Layer 2 to 1024, which is used to better distinguish between the tonal (sinusoidal) and non-tonal (noise-like) components of the signal, and helps compensate for the limited accuracy of the polyphase filter bank. This assists in applying the correct masking threshold, as tonal and non-tonal components have differing influences on the masked threshold. The individual and absolute masking thresholds are calculated according to frequency, loudness level and tonal criteria. The polyphase filter bank giving 32 sub-bands has a frame length that is three times longer than that used in Layer 1, giving 36 sub-band samples instead of 12, coded as three groups of 12 samples for each of the 32 sub-bands. This corresponds to 1152 input PCM samples. Once the maximum signal level is calculated based on both the scale factors and power density of the FFT, the difference between this maximum signal level and the minimum masked threshold is calculated giving the signal-to-mask threshold (SMR), and this is input to the bit allocation routine. A single bit allocation is given to each group of 12 sub-band samples with up to three scale factors for each band, and these can be shared between groups when the difference is small. For greater coding efficiency, three successive samples for all 32 sub-bands are grouped together to form a granule and are quantised together. For stereo signals, a joint stereo encoding mode can be employed which exploits the redundancy of typical stereo material to reduce the bit rate further, as described in Layer 1. The sub-band samples are then quantised such as to optimally allocate the noise generated during the quantisation process below the masking threshold. Each block of 12 sub-bands is normalised by dividing its value by a scale factor. Quantisation levels from 3 to 65,535 are possible.

Sample rate enhancement in Layer 2 allows for the use of half sample rate encoding for all three layers, so that sampling can additionally be done at 16k, 22.05k and 24k samples per second. This reduces the upper frequency limit to 7.5, 10.5 and 11.5 kHz respectively. As a result, bit rates can be as low as 8 kbps for Layer 2 and Layer 3.

The multi-channel extension was introduced to give Layer 2 the possibility to transmit up to 5.1 channels of audio and still be backwards compatible with Layer 1 decoders. Information on this extension is described in ITU-R BS.1196-1. In order to make a multi-channel extension to the Layer 2 scheme and ensure its backwards compatibility with Layer 1 decoders, a backwards compatible (BC) version of the standard was created, namely MPEG-2 BC audio coding standard (ISO 13818-3). The backwards compatibility is created by generating a channel mix for left and right channels from the multi-channel audio, and encoding the left and right signals with an MPEG-1 encoder. This allows an MPEG-1 decoder to produce a 'down-mixed' version of the multi-channel audio. The mix would be $\text{left}' = \text{left} + (x \times \text{centre}) + (y \times \text{left surround})$, and similarly $\text{right}' = \text{right} + (x \times \text{centre}) + (y \times \text{right surround})$. Thus, the MPEG BC for Layer 1 consists of Left′ and Right′ mixes, and the Centre (C), Left surround (Ls) and Right surround (Rs) are transmitted in the MPEG-1 ancillary data field. All five original channels can, in principle, be derived from the R′ and L′ and a combination of the other channels depending on which will give the lower bit rate. The side-effect of this re-matrixing process is that some of the channel information will naturally be cancelled, but the quantisation noise will not. This leads to dematrixing artefacts, as the quantisation that takes place between matrixing and dematrixing gives quantisation noise (errors) from all the multi-channel extension.

2.2.3.4 MPEG-1/2 Layer 2 Bit stream

The Layer 2 bit stream naturally has many similarities with Layer 1 for backwards compatibility reasons (Figure 2.2.4). The basic structure can be seen to include one new element over Layer 1, the introduction of dynamic scale factor selection information (SCFSI). In Layer –1, the scale factors are 6-bit linear for every 12 sub-band samples, but Layer 2 has three times the amount, 36 sub-band samples. Each three successive scale factors of one frame are considered together to form a scale factor pattern which may consist of one, two or three scale factors. These are transmitted along with the additional scale factor selection information consisting of 2 bits. Within the ancillary data, consisting of data 1 and data 2, the data 1 portion contains the multi-channel (MC) audio data information. The MC data portion follows the same scheme as the Layer 1 compatible data, with MC header, MC CRC, MC bit allocator, MC SCFSI, MC SCF, MC predictor and MC sub-band samples. A multi-lingual data section is also included after the sub-band sample. The frame length is 24 ms at 48 kHz sampling rate.

2.2.3.5 MPEG-1/2 Layer 3 (MP3)

MPEG-3 was originally planned to be the successor to MPEG-2 for HDTV. It was discovered that the MPEG-2 video coding was capable of doing everything required for HDTV, so the audio

Header	CRC	SCFSI	Bit Allocation	Scale Factor	Sub-band Samples	Aux Data

Figure 2.2.4 MPEG-1/2 Layer 2 bit stream.

section was rolled into the MPEG-1/2 standard as MPEG-1/2 Layer 3. Although a capable coding scheme which found primary use in Internet use as mp3, Layer 3 was soon overtaken by the standardisation of AAC in MPEG-2 and MPEG-4. The Layer 3 algorithm would later benefit from the addition of SBR (Spectral Band Replication) technology, which is known as mp3PRO. mp3PRO has not been standardised by the MPEG and remains a proprietary extension to MPEG-1/2, Layer 3.

The Layer 3 coder at first sight has a number of similarities with its predecessors. The signal to be encoded is first divided into 32 sub-bands using a polyphase filter bank as used in Layer 1 and Layer 2. The following MDCT (Modified Discrete Cosine Transform) further divides each sub-band into 18 finer sub-bands to increase coding efficiency for tonal signals. The same input signal is applied to a 1024-point FFT and psychoacoustic model which controls the window switching on the MDCT. Quantisation and coding is done with a system of two nested loops. The inner iteration loop uses Huffman code tables to assign shorter code words to smaller quantised values. This reduces the overall coder rate and is sometimes referred to as the 'rate loop'. Should the resulting number of bits exceed the number of bits available, a correction to the global gain can be made to increase the quantisation step size, resulting in smaller quantised values. This can be repeated with different step sizes until the bit rate demand is small enough. The outer iteration loop (noise control loop) applies scale factors to each scale factor band. If the quantisation noise is found to exceed the masking threshold, the scale factor is adjusted and checked again. Both the inner and outer iteration loops repeat application of codes or scale factors until the best fit is achieved; however, several conditions can be applied to prevent this from becoming excessive, which is important in coder delay. Finally the bit stream is formatted and appropriate CRCs added.

As with other MPEG audio structures, the Layer 3 bit stream contains a mandatory header, consisting of a sync word, bit rate, sampling frequency, layer, coding mode and copy protection. Unlike other MPEG audio formats, it is possible for the sync word to occur within the audio data portion of the stream. The bit rate is given for the stream, not per channel, and can be switched on the fly, giving variable bit rate encoding. The Layer indicates whether the stream is Layer 1, Layer 2 or Layer 3 (all share the same header structure). Similarly, the coding indicates whether mono, dual mono, stereo or joint stereo coding mode is being used.

Table 2.2.1 Comparison of MPEG audio coders

	Layer -1	*Layer 2*	*Layer 3*
Bit rates (kbps)	32–448	8–160 (LSF)* 32–384 for stereo up to 1066	8–320
Sample rates (kHz)	32–48	16–48	8–48
Frame length (ms)	8	24	24

*LSF, low sampling frequency.

2.2.3.6 MPEG-4

The MPEG-4 standard contains a wide range of audio possibilities, from low bandwidth (4 kHz) and low data rates (2 kbps) up to high-quality multi-channel broadcast quality audio as found in MPEG-4 AAC. It is not anticipated in the near term that MPEG-4 will supersede MPEG-2 in mainstream broadcast applications, but the possibilities it offers are likely to make a significant contribution and impact when it matures. The ISO/IEC WG11 document details a wide variety of possible audio profiles that can be deployed within the standard. These fall broadly into four areas: Main profile, containing all the other profiles and tools for natural and synthetic audio; Scalable (6–24 kbps and 3.5–9 kHz); Synthesis (SAOL, wavetables and Text-To-Speech); and Speech (HVXC, CELP and Text-To-Speech).

Speech coding at data rates between 2 and 4 kbps is supported with the use of Harmonic Vector eXcitation Coding (HVXC), and between bit rates of 4 and 24 kbps with Code Excited Linear Predictive Coding (CELP), although other bit rates are also possible. Bit Slice Arithmetic Coding (BSAC), Twin VQ and Harmonic Individual Lines and Noise (HILN) are also included in the MPEG-4 audio toolbox, with MPEG-4 AAC the multi-channel audio format. The MPEG-4 General Audio (GA) Coder, as described by Grill,[3] gives an insight as to how AAC and Twin VQ systems can be used together to give a scalable solution.

2.2.3.7 MPEG-2 AAC

MPEG AAC was developed using technologies from AT&T, Fraunhofer, Sony and Dolby, and standardised by the MPEG group in 1997 (ISO 13818-7). Unlike the other MPEG audio systems, MPEG AAC is not backwards compatible with the other systems, which allows the coding system to use an even lower bit rate than MP3. Typically, an AAC bit stream at 128 kbps will perform as well as an MP3 bit stream at 160 kbps. The MPEG Audio subgroup reported in 1998 that: "Overall, all AAC profiles at 128 kbps gave a significantly better performance than MPEG Layer 2 at 192 kbps or Layer 3 at 128 kbps." Additionally, AAC can support sampling rates up to 96 kHz and up to 48 audio channels. MPEG AAC was the multi-channel audio coding system of choice for Japanese Digital Television, and is also used by the Digital Radio Mondial system. There are three specific application types of AAC, known as profiles. These are Main, Low Complexity (LC) and Scalable Sampling Rate (SSR). A further improvement in the AAC technology uses Spectral Band Replication and is known as CT-aacPlus. Generally the three AAC profiles are not compatible; however, the Main profile is capable of decoding a Low Complexity bit stream. AAC is perhaps best considered to be a suite of coding tools, the choice of which varies depending on the type of profile required. In general, the Main profile enables coding at the highest quality sound, the LC profile (as used in Japan) has less decoding complexity but has some slight signal degradation, and the SSR profile allows for decoding at several audio bandwidths.

The basic structure of the AAC encoder is not very different from its predecessors in the MPEG group of coders, in that it still uses signal masking properties to reduce the amount of data and employs a plain MDCT. Similarly the noise produced in the quantisation process is again distributed across the frequency bands in such a way as to be masked by the total signal content. However, a number of other tools are employed to further reduce the data rate and improve the quality of the decoded signal. The MDCT uses an increased window length of 2048 instead of the hierarchical sub-band, block transform filter bank used in Layer 3. Prediction is also employed, but is limited to the Main profile only, and uses the fact that some audio signals are easy to predict, and hence improves the coding efficiency. Temporal Noise Shaping (TNS) shapes the distribution of the quantisation noise in time by doing an open loop prediction in the frequency domain followed by coding the residual signal after application of the TNS linear predictor. This is especially good at improving the speech quality of low-bit-rate encoded signals. Pre-echo artefacts occur when there is an inappropriate temporal spread of quantisation noise, causing the noise to arrive ahead of the signal unmasked. This occurs when coding a transient signal, and the noise is spread out over the whole window length of the filter bank and is not masked by the signal. TNS reduces this noise by shaping it in time using prediction in the frequency domain. Reduction in the pre-echo artefacts is achieved by reducing the impulse response to 5.3 ms at 48 kHz as compared to Layer 3 (18.6 ms).

MPEG-2 AAC, while defining the audio format and transport syntax for synchronisation and coding parameters, allows a choice for the transport syntax dependent on application. The ADIF (Audio Data Interchange Format) contains all decoder control data in a single header before the audio stream. Although good for file exchange formats, it does not allow for start of decoding at any point. The ADTS (Audio Data Transport Stream) formats the header in a way similar to MPEG-1/2 formats. The audio data for a complete frame is contained in the audio data payload between two sync words, which allows for decoding in the middle of the audio stream. The ADTS format has become the more popular format for most applications.

The ADIF file format sequence consists of header, byte alignment and raw data stream. The ADIF header consists of ID (32), copyright_ID_present (1), copyright_ID (72), original_copy (1), home (1), bitstream_type (1), bitrate (23), num_program_config_elements (4) and adif_buffer_fullness (20). The copyright_id consists of a 64-bit copyright number.

The ADTS format supports only a single program stream raw_data_stream, which can contain up to seven audio channels with an independently switched coupling channel. However, unlike the ADIF file format it does not include the average bit rate. The stream consists of a sync word and adts frame. Within the adts frame is the fixed_header, variable_header, error_check and byte_alignment. The fixed header giving sync word (12), ID (1), Layer (2), Protection absent (1), profile (2), sampling frequency index (4), private bit (1), channel configuration (3), Original copy (1) and home (1). Profile is indicated by '0' for Main, '1' for LC and '2' for SSR.

The adts variable header consists of the copyright_identification_bit (1), copyright_identification_start (1), frame_length (13), adts_buffer_fullness (11) and number_of_raw_data_blocks_in_frame (2).

MPEG-2/4 AAC characteristics are as follows:

- Sampling frequency 8–96 kHz.
- Support up to 48 channels.
- Support for Low-Frequency Enhancement (LFE) channels.

2.2.4 AC-3 (Dolby Digital)

Dolby Digital (AC-3) was first introduced into the film world in 1991. This application, based on the AC-3 multi-channel digital audio coding scheme, found uses in Laser Disc, DVD-V, DVD-A, PC Games, Standard Definition Digital TV, HDTV and numerous other applications. Standardised in ATSC and DVB, AC-3 is broadcast via satellite, cable and terrestrial in many areas where multi-channel audio is required (except Japan, which uses AAC).

To appreciate the scheme fully, it is necessary to understand what the coding scheme actually does before looking at how this is achieved. AC-3 carries the full dynamic range of all channels to the decoding device. It is at the decoder that the information carried in the metadata section of the stream is used to modify the audio to produce a suitable audio output determined by the capabilities of the decoding equipment and user input. It is worth noting that although other schemes are equally capable of supplying the full dynamic range, to allow for the lowest common denominator in decoding product, broadcast practice is currently to pre-process the audio and apply a certain amount of compression prior to transmission. This means all decoding product receives a pre-compressed signal.

Metadata, which is best described as data about the audio, is an intrinsic part of Dolby Digital. Metadata is used to describe and control all areas of the compressed audio stream, and an understanding of the key features of this is essential to being able to apply the scheme properly. Some of these key features are: Dialogue Normalisation (DialNorm), Dynamic Range Control (DynRange) and DownMixing (mixing the 5.1 channels down to fewer channels). Although Dolby Digital is better known for its 5.1 channel configuration, it is worth noting that the system can code from mono to 5.1, and is regularly used in 2.0 and 5.1 modes. There are two conventions for indicating the number of channels present in a Dolby Digital stream, 5.1 and 3/2L. Both of these indicate the same programme, the 5.1 indicating five full bandwidth channels and one LFE channel (limited to 120Hz), and the 3/2L indicating the presence of three front channels, two surround channels and an LFE channel.

2.2.4.1 Dialogue Normalisation (DialNorm)

To understand why Dialogue Normalisation is needed in a coding system we need to look at current broadcast practice. Reference level has long been used to line up the broadcast chain and is usually given as 0 VU or 0 dBu (0.775 V). This allows checking of unity gain throughout the system. Peak level is given as 8 dBu (or national equivalent) and in analogue broadcasting is the level which must not be exceeded to ensure clipping does not occur. Peak level in digital systems can typically be +18 or +20 dBu and it is common practice where audio signals are fed to both analogue and digital broadcast systems for the receiving set-top box to apply suitable gain, in the region of +10 dB. As most program material is adjusted to reach just below peak level on transmission, the average level of the signal can vary considerably, resulting in the often complained about jump in sound level at the consumer's receiving equipment. This average level is a good approximation for the average speech level of the content, hence the name Dialogue Normalisation (DialNorm) Value. The value of the program material is measured (an Leq(A) measurement is a good approximation) and given a DialNorm value, which is encoded within the Bit Stream Information (BSI) section of the AC-3 stream without altering the actual level of the program at all. It is on decoding the AC-3 stream that the DialNorm value is used to reduce the average level of the program to −31 dBFS. The reason this value has been used can be seen in Figure 2.2.5, where the standard Hollywood action movie has a high dynamic range above its DialNorm value of 27 dB. To allow full dynamic reproduction of this, and allow some headroom, it

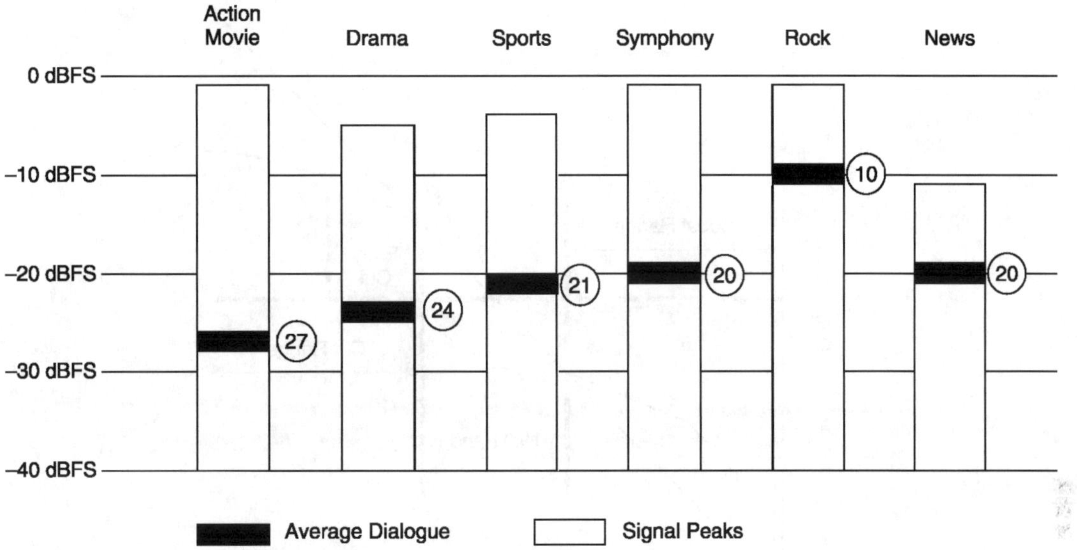

Figure 2.2.5 Difference in dialogue level for different program genres.

is necessary to reduce this level at the decoding stage to −31 dBFS (Figure 2.2.6). Reduction of all program material to this level ensures that all program reproduction should sound approximately the same.

2.2.4.2 Dynamic Range Compression

High-quality programming typically has a high dynamic range, but it is necessary for most broadcast systems to reduce this dynamic range in order to allow replay on equipment that is not able to reproduce this full dynamic range. A broadcaster will usually have to apply dynamic range compression to make the signal suitable for all environments and replay equipment. The AC-3 encoding system does not apply any Dynamic Range Compression (DRC) to the audio content of the signal, instead it generates boost and cut data that is carried in the metadata part of the bit stream. A dynamic range control value is supplied to each audio block in the AC-3 system (every 5 ms) in order to alter the level of each block on replay. This compression works in conjunction with a correctly assigned DialNorm

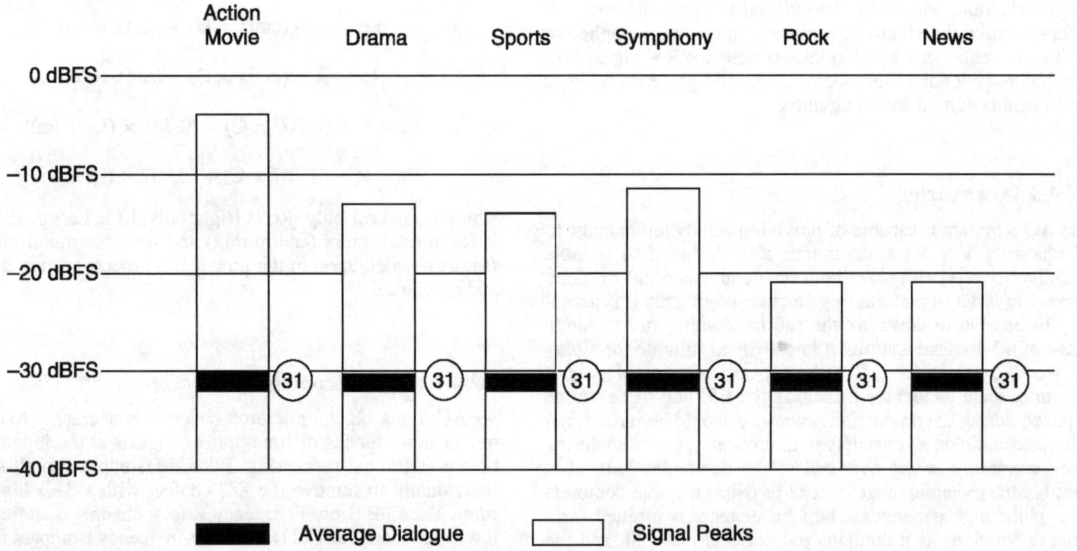

Figure 2.2.6 Normalised signal levels after decoding.

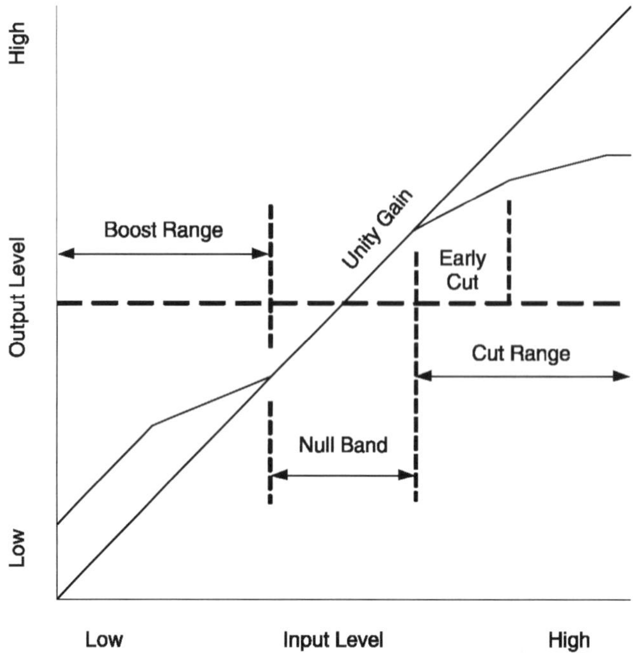

Figure 2.2.7 Typical compression profile and its gain/attenuation structure.

value, where the part of the signal not affected is the dialogue region. Above a certain threshold a gain reduction will be introduced, and below a certain level gain boost (Figure 2.2.7).

Various profiles are available for the DRC, and in the decoder two distinct compression modes will be used to configure the dynamics processing. These are named Line and RF modes. In line mode the consumer may control the degree of dynamic range reduction, which may be reduced to zero, allowing full dynamic range. In RF mode, compression is always applied to ensure the audio that is remodulated onto the TV signal will match conventional TV broadcast levels and prevent problems with antenna demodulation circuitry.

2.2.4.3 Downmixing

The AC-3 system is capable of providing full dynamic range to all channels of a 5.1 system. It is also designed to be able to deliver stereo (with or without surround encoding) or mono signals. In order to perform this function adequately it is necessary to be able to downmix the full bandwidth, full dynamic range of a 5.1 signal and mix it into a signal suitable for replay (at worst case) on a small mono speaker. A set of default downmix scale factors are present in the BSI and these can be adjusted during the production process. It would be rare to find full-scale audio on all channels simultaneously, but even during those conditions where overload of the downmixed signal is possible the dynamic range control provides suitable compression of the high dynamics. The LFE channel is omitted from most downmixes as it contains only effects material, and the other channels contain all the necessary bass or low-frequency information that is required. The LFE is not meant to carry low frequencies that would otherwise be carried in the full bandwidth channels. Instead, the LFE channel is used exclusively as a low-frequency effects channel. During the encode process it is possible to apply a phase shift to the surround channels, such that the decoder downmix is surround (Pro-Logic) decodable.

Examples of stereo (Lo, Ro) and surround encoded stereo (Lt, Rt) are given below:

$$\text{Lo} = \text{L} + (\text{cmix} \times \text{C}) + (\text{smix} \times \text{Ls})$$

$$\text{Ro} = \text{R} + (\text{cmix} \times \text{C}) + (\text{smix} \times \text{Rs})$$

$$\text{Lt} = \text{L} + (0.707 \times \text{C}) - (0.707 \times (\text{Ls} + \text{Rs}))$$

$$\text{Rt} = \text{R} + (0.707 \times \text{C}) + (0.707 \times (\text{Ls} + \text{Rs}))$$

where Lo is Left only, Ro is Right only, Lt is Left total and Rt is Right total; cmix (centre mix) and smix (surround mix) are factors decided upon in the production process to give the best audio balance.

2.2.4.4 AC-3 Encoding

As AC-3 is a block-structured coder, it is necessary to collect one or more blocks of time-domain signals at the input buffer before additional processing. Filtering is applied to all signals individually to remove the DC content with a 3 Hz high-pass filter. The LFE (Low-Frequency Effect) channel is additionally low-pass filtered at 120 Hz. A high-frequency bandpass filter is then employed to detect transients. This information is used to adjust the block size of the TDAC (Time Domain Alias

Cancellation) such that any quantisation noise is restricted to a small temporal region to prevent temporal unmasking. The TDAC is built around a 512-point MDCT using a 50% overlap which is employed in the decoding stage to provide a form of automatic crossfade which enables a smooth transition from one block to the next. The block length is 256 samples and has a frequency resolution of 93.75 Hz at 48 kHz. Should the signal be a transient one, a block switching technique is employed where two 256-point transforms are computed instead of the 512-point transform. The transform coefficients are then processed into exponent/mantissa pairs using a floating point conversion process. The mantissa is then quantised based on a parametric bit allocation model, giving a variable number of bits. This allocation is based on the psychoacoustic masking model which determines how many bits to provide for each mantissa within a given frequency band. However, with AC-3 this bit allocation is not passed on to the decoder, instead the encoder constructs a masking model based on the transform coefficient exponents and some key parameters, and it is these exponents and parameters that are used in reconstruction at the decoder. A global bit pool is then used to allocate bits where most needed. The bits are split between the channels and, where a channel is momentarily inactive, those bits will be distributed to the other channels. Coupling is then used to further reduce the bit rate for the high-frequency part of the spectrum. Again the limitations of the human ear are used. In this case the ear localises at high frequency based on the 'envelopes' of the critically filtered bands of the signal rather than the signal itself. The high-frequency sub-band signals are separated into envelope and carrier components, and these can be selectively combined or coupled across channels, and the localisation information preserved in the envelope information. In applying the masking curve and psychoacoustic model, the AC-3 system uses both forward and backward adaptive techniques, which allow a better allocation of bits and subsequently a more sophisticated psychoacoustic model is attained over the basic algorithm. Finally, the stream is assembled by the bit stream packer into the smallest independently decodable access units (see below). A further processing step can occur when the encoder is working in stereo mode, which enhances the interoperability with Dolby Surround 4:2:4 matrix encoded programs, called rematrixing. Here the signal spectrum is divided into four frequency bands, and within each band the left, right, sum and difference signals are checked for signal energy. If the dominant signal energy appears in the left or right channel the signal is encoded normally; however, if the sum or difference signals contain the dominant signal energy then these are encoded instead, the decision being made on a band-by-band basis.

2.2.4.5 AC-3 Bit stream

The primary method of conveying AC-3 bit streams is using standard IEC 60958 (SPDIF) or AES-3 interfaces which are capable of carrying bit streams of up to 2.3 Mbps. As the AC-3 stream will only carry a maximum data rate of 640 kbps stuffing packets are added to the stream, as described in IEC 61937. The AC-3 frame is 32 ms long (1536 audio samples) at 48 kHz (Figure 2.2.8). It consists of a sync word that allows decoders to recognise the frame boundaries, which is then followed by the first of two CRC words. The Bit Stream Information (BSI) word then conveys control data to indicate the number of coded channels, sample rate, data rate and several other parameters.

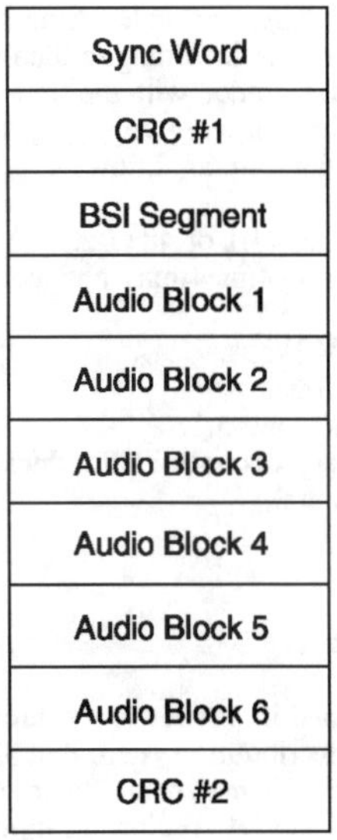

Figure 2.2.8 AC-3 frame structure.

Following the BSI are the six audio blocks, each representing 256 samples, followed by the second CRC word (Figure 2.2.9).

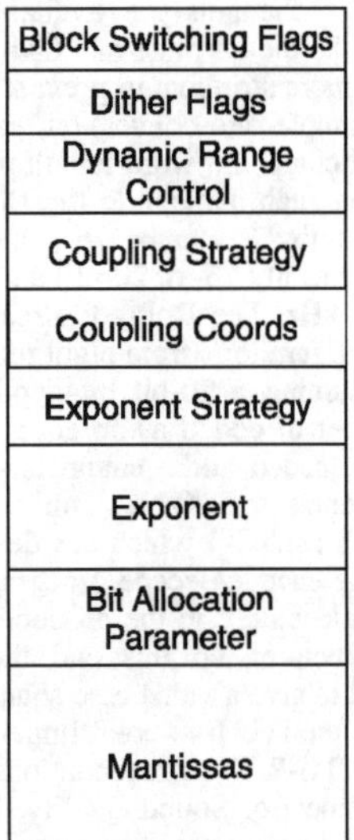

Figure 2.2.9 Construction of AC-3 audio blocks.

In consumer applications the Dolby Digital bit stream is packed into both left and right channels of the IEC 60958 transport, as specified in IEC 61937. In professional applications the bit stream may be packed into either or both channels of the AES-3 interface, as specified in SMPTE 340M. In professional applications SMTPE 12M timecode information can be included with the AC-3 packets. As the audio access packets are 32 ms long, and video time stamps will typically occur every 33 or 40 ms according to the frame rate being used, an indicator is also included with the time stamp to point to the audio sample in the following access packet to which it refers.

AC-3 specifications are as follows:

- Sample rate of 32, 44.1 or 48 kHz.
- Encode latency 187 ms (min) and can be adjusted up to 450 ms.
- Data rates 32–640 kbps.
- Frame length 32 ms.
- No. of channels mono-5.1.
- Includes metadata for control of loudness level, downmixing, dynamic range control.

2.2.5 Dolby E

Dolby E was created in answer to the problem of supplying a contribution and distribution system that allows the carriage of both coded audio and metadata. Since AC-3 was standardised by the ATSC and DVB allowing multi-channel audio TV broadcasts, it has been necessary to provide a system that will supply the AC-3 coders with the necessary multi-channel information and metadata through the existing TV broadcast infrastructure without incurring the costs of increased cabling. The important aspects of the AC-3 metadata are described in the previous paragraphs on AC-3, detailing dynamic range control, dialogue level, language identifiers and how to perform the best downmix if fewer channels are available on the consumer's decoder. Unlike the carriage of PCM on AES-3, Dolby E requirements are more stringent to prevent damage to the data. Events such as sample rate conversion, gain changing, cross-fading during switching and word-length truncation have to be avoided to prevent such damage to the Dolby E packets. The word length of Dolby E can be 16-, 20- or 24-bit (current implementations are only 16- or 20-bit) and works at a standard sample rate of 48 kHz. The Dolby E stream can contain multiple program configurations from eight mono channels up to a 7.1 program (assuming a 20-bit bit depth).The word length governs the number of coded audio channels that can be carried, typically six coded audio channels for 16-bit and eight coded audio channels for 20-bit. Unlike the final broadcast audio codecs such as AC-3 which are designed to give good results for a single encode/decode process, Dolby E has to be able to perform adequately in the production processes of the broadcast infrastructure. To this end the Dolby E system has been designed to give a worst case sound quality equal to or better than 4.5 on the ITU five-point impairment scale after 10 generations (see ITU-R Recommendations BS.562-3 (06/90), Subjective Assessment of Sound Quality, 1997). A worst case example might be the carriage of eight mono channels (each a separate program) with separate metadata for each channel in 20-bit format. Carriage of a single 5.1 program on a 20-bit stream would increase the number of generations as more bits are available to accurately code the audio.

The carriage of metadata with the coded audio must be performed in an exact time-aligned fashion, as the Dolby E metadata also includes professional metadata that will be used by other broadcast equipment. Such metadata includes metering information which can be extracted by third party products to display the exact content of the coded audio without the necessity of decoding it. SMPTE 12M timecode information can also be placed in the metadata stream to allow exact synchronisation with associated video content.

2.2.5.1 Dolby E A/V Synchronisation

To properly implement a contribution and distribution coding system it is necessary to ensure A/V synchronisation is maintained at all times (see ITU-R BT.1359-1, Relative Timing of Sound and Vision for Broadcasting, 1998). Table 2.2.2 shows characteristic timings for a number of common bit rate reduction codec systems at 48 kHz sample rates and Table 2.2.3 lists some common video frame rates/sample counts.

Table 2.2.2 Characteristic timings of common coders

Audio coding	*Frame size (samples)*	*Time resolution (48k)*
PCM	1	20.8 µs
AC-3	1536	32 ms
MPEG Layer 2	1152	24 ms
MPEG AAC	1024	21.33 ms

Table 2.2.3 Common video frame rates/sample counts

Video format	*48 kHz samples*	*Time resolution (ms)*
30 Hz DTV	1600	33.33
NTSC	1601.6	33.37
PAL	1920	40
Film	2000	41.67
23.98 Film	2002	41.71

It is clear from the above tables that there is no simple correlation between conventional audio frame sizes and video frame sizes. In order to ensure the coded audio exactly matches the video frame sizes, Dolby E uses an internal sample rate that is 1792 times that of the video frame rate. For the above examples this gives an internal sampling frequency of between 53.76 and 42.965 kHz. This allows the Dolby E coders to work at video frame rates of 23.98, 24, 25, 29.97 and 30 fps. The encoded output is sent out at 48 kHz and is memory buffered and synchronised to the video frame clock.

Editing of the encoded audio is necessary during routine production processes and switching is required on live feeds. As Dolby E is always locked to the video frame rate and the switching point for video is always known (line 6 for PAL and line 10 for NTSC), the Dolby E stream is packetised according to SMPTE 337M such that the switching point occurs during the Dolby E guard band. Dolby E decoding offers two particularly useful operational modes for post-production and live switching of PCM/Dolby E streams. In Program Play mode, the Dolby E decoder can accept off-speed input for decoding. Program Play is defined as using an input that varies from the 48 kHz input. An example of this might be replaying a 24 fps Dolby E encoded signal into a Dolby E decoder running at 25 fps. Switching Dolby E at off-speed data rates should be done with caution as the frame synchronisation would no longer apply.

The second useful operational mode is the ability to switch between Dolby E streams and PCM, and vice versa, assuming the switch point is correctly located with reference to the video frame position.

2.2.5.2 Dolby E Encoding

Each Dolby E channel is coded independently using a predetermined quantity of the available data rate. An MDCT provides a critically sampled filter bank utilising a time-domain alias cancellation technique. This provides the basis for the system's ability to be edited and switched in the encoded domain. The transform lengths vary from 256 samples for the short lengths to 2048 samples for the long lengths, while 512 samples are used for the bridging lengths; 2048 samples are used for the steady-state conditions and 256 samples for the transient transform window block length. There are four types of these transform window block lengths: the short for frame boundaries (stop and start); the mid-short, which is a conventional short transform window; the long length for frame boundaries (start and stop); and the bridging transform windows for interfacing between the short and long versions. The individual block sequences and their interrelationships are described by Fielder and Davidson.[4] The transform windows have five possible combinations depending on the signal characteristics. The frames are divided into two equal length audio segments to reduce encode/decode latency. Each segment is constructed from the three transform window lengths described above depending on the position of the transient part of the signal within the frame.

The transform coefficients are grouped together in frequency bands that approximate the ear's critical bands to perform bit allocations and quantisation. Each band has a single exponent value and a group of compressed transform coefficient values representing the maximum of whole bands' transform coefficient magnitudes. The mantissas are then generated and quantised, the accuracy of which is determined by the bit allocation process. The mantissa quantisation is performed using 'gain adaptive quantisation' to improve coding efficiency over linear quantisation.

2.2.5.3 Dolby E Bit stream

The Dolby E bit stream was designed to be able to be used in current broadcast infrastructures by using the AES-3 interface, thus enabling compatibility with most contribution and distribution equipment. Figure 2.2.10 shows how the Dolby E data is conveyed in the AES-3 transport.

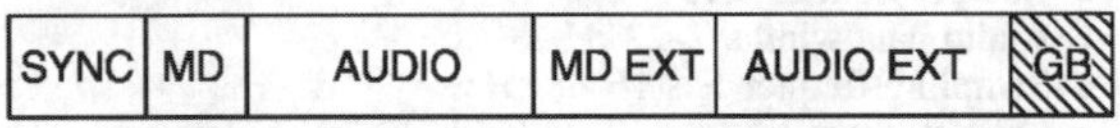

Figure 2.2.10 Dolby E frame structure.

The Dolby E frame is constructed of six segments: sync, metadata, audio, metadata extension, audio extension and meter information. Each segment begins on an AES-3 word boundary and has common bit depth. The overall audio frame is contained within about 95% of the available data rate, and this allows a guard band to be placed about the video switch point to ensure data is not corrupted during switching to either the previous or next frame. The sync segment allows decoders to maintain synchronisation. The sync segment has a unique sync word for each bit stream depth, which allows decoders to determine the bit stream depth dynamically. The metadata portion carries parameters that describe the coded audio: frame rate, number of programmes and timecode. Also included here are gain words that can be applied to the decoded output, and a frame counter to determine whether a splice has occurred in the bit stream. Metadata that will be used by the AC-3 encoder are also included. These parameters are not used by the Dolby E decoder but are passed through to the emission AC-3 encoder. The audio data is passed through in the audio and audio extension segments. By breaking the audio data into two segments the encoder and decoder can be designed to have exactly one frame of latency. In practice the decoder can work with less than one frame, so a signal arriving slightly late will be corrected and output in sync with the video frame. The metadata extension segment carries only metadata that applies to the audio extension segment. The metering segment carries metering information of signal levels in both peak and rms values. These metering values can be accessed without the need to decode the audio and cover a range of almost 100 dB on a logarithmic scale.

2.2.6 SBR

Spectral Band Replication (SBR), developed by Coding Technologies, is an addition to the available audio technologies for enhancing low-bit-rate coding in MPEG-4 audio. Combining this technology with MPEG-2 Layer 3 audio (mp3PRO) and MPEG-2/4 AAC (CT-aacPlus) can effectively reduce the data rates of these technologies by around 30%. SBR is not a codec by itself, but a useful enhancement technology for use with other schemes.

The principle of operation of this scheme is to code only the lower frequency part of the audio signal, and to use this to recreate the higher frequency portion on decoding. This is based on the principle that there are usually large dependencies between the lower and higher portions of the audio frequency spectrum. In order to perform this task adequately, the core coding scheme needs to only concern itself with the lower frequency part of the audio spectrum, and the SBR works with the higher frequency portion. SBR control information is carried in the coded bit stream to allow the decoder to correctly shape the high-frequency spectrum that is copied from the lower frequency spectrum. The data rate of the SBR side chain information is significantly less than coding the same bandwidth using standard AAC.

The SBR technology ideally works as part of a dual rate system, with the core coder using half the original sampling rate and the SBR coder using the original sampling frequency (Figure 2.2.11). The SBR coder estimates the spectral envelope of the SBR range to find a suitable time and frequency resolution given the current input signal characteristics. This is performed using a complex QMF (Quadrature Mirror Filter) analysis and energy calculation. Thus, the SBR side chain extracts all the necessary data that will need to be used to control the decoder post-process of reconstruction, then remove the envelope and package the bit stream with the original core coder. At each stage of the process, the SBR coder needs to make decisions on which frequency range it needs to cover at any given time. The control parameter extraction algorithm requires tuning to the core coder to overcome the different characteristics of the core algorithm when bit demand is high.

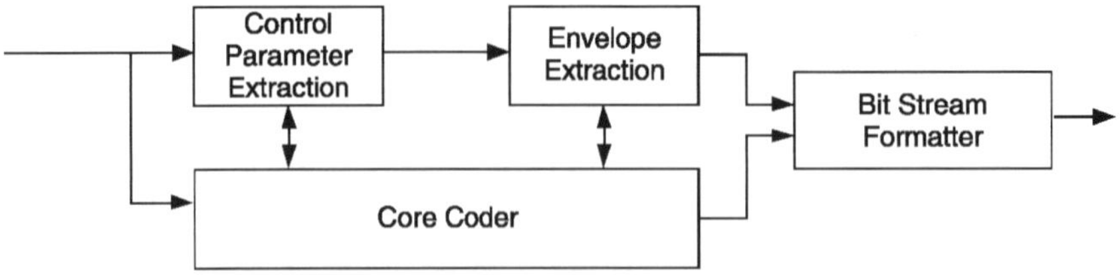

Figure 2.2.11 Block diagram of an SBR and core encoder.

Decoding of the SBR enhanced coder audio is performed first by unpacking the bit stream. The core coder low-frequency data is fed to the core decoder for decoding at the lower sample rate, and the control data for the SBR reconstruction is separately decoded. This control data extraction module first obtains the time–frequency matrix prior to reading the envelope data. The decoded audio frame from the core decoder is then passed into the High-Frequency Reconstruction (HFR) module. From here additional high-frequency components may be added and then the envelope is adjusted by using a QMF filter bank and performing adjustments on the sub-band samples. The reconstructed high-band signal is also adaptively filtered, at the sub-band level, based on the control data, to ensure spectral characteristics for a given time–frequency region are appropriate. When the low-band signal is added to the high-band reconstructed signal it is necessary to up-convert to the required sampling rate.

Comparisons of the use of SBR with other core encoders and against other similar low-bit-rate coding schemes can be found in Ref. 5. This demonstrates the quality of the AAC coding scheme against similar coders and the improvements that SBR adds to the coders.

2.2.7 APT-X and DTS

APT-X from Audio Processing Technology is a 4:1 data reduction technology based on an implementation of sub-band Adaptive Differential Pulse Code Modulation (Sub-band PCM). In this coding scheme the differences between successive PCM samples are coded, with the quantisation steps variable, which adapt to the energy of the audio signal.

Coding is achieved by taking four successive PCM samples and filtering them into four equal bandwidth frequency sub-bands. The 64-tap QMF is used to divide the signal into low, low-to-medium, high-to-medium and high frequency bands and is followed by a 32-tap filter further dividing each of these into four sub-bands. The sub-band bit allocation is 7, 4, 3 and 2 bits (LF–HF). By using a QMF to perform this task, the short-term coding resolution of each band can be altered to exploit spectral redundancies. Further noise generated within each of these sub-bands will be kept within these bands, effectively decoupling them from each other. These four 16-bit signals are simultaneously processed in four separate chains. Each chain has a backwards linear prediction loop to estimate the incoming signal. This prediction is based on the previous PCM samples, and is subtracted from the input signal to give a difference or error signal. This method is highly dependent on the periodicity of the incoming signals. Signals such as music are likely to give a very low error signal as they are easier to predict. Noise, by virtue of its nature, is very difficult to predict and likely to result is significantly larger error signals. Each signal is then re-quantised using a backwards adaptive Laplacian quantiser, which allows the step sizes to be adapted to the magnitude of the error signal, based on an analysis of the previous samples. This exploits the slow time-varying nature which audio exhibits, such that little level change in the sub-band samples will give a stable quantiser step size and help increase the efficiency of the coder. Likewise, when the sub-band samples change quickly in level, so also the quantiser tries to accommodate these changes, resulting in lower coder efficiency. Temporal masking is also exploited here, and the quantiser efficiency immediately after an impulse sound is reduced. The sub-band samples are then multiplexed together into the APT-X stream with additional sync and aux data bits. With the auto-sync function active, a 10-bit sync word is spread over the first 10 16-bit APT-X code words in a data frame of 128 samples. Additional aux data, up to 1/16 of the bit rate, can be inserted (in the left channel only).

An enhanced version of APT-X allows the use of 20- and 24-bit resolution. A different 32-tap filter used in the second stage of processing reduces the number of audio samples required from 122 to 90 and reduces the latency of coding to under 2 ms at 48 kHz, or under 7 ms if SRCs are engaged. Bit resolution is increased over the standard APT-X (7, 4, 3, 2) to 8-, 5-, 4- and 3-bit resolution for the 20-bit version and 9, 6, 5 and 4 bits for the 24-bit version. Subtractive dither is employed which is added before the quantisation stage and removed at the decoder. Sync bits are buried in each encoded sample, which allows for a much reduced lock-up time (from 50 ms for the standard version to around 3 ms for the enhanced version) and carries information about the format of the encoded sample.

APT-X characteristics are as follows:

- 56–384 kbps bit rates.
- Compression of 4:1.
- Audio bandwidths 7–22 kHz.
- Sampling frequencies 16–48 kHz.
- Latency from 1.8 up to 7 ms (with SRCs).

Early in 2003, DTS audio was standardised for optional use in DVB by the DVB committee. Although not in use for DTV at the time of writing, it is worth mentioning along with the APT-X system. The APT-X100 system is used in DTS theatrical audio, and the DTS system as used in DVD and DTV is similarly based on the same algorithm from AlgoRhythmic Technology, called Coherent Acoustics. Data rates for DTS vary from 384 kbps to 1.5 Mbps, and using the core plus extension for 96 kHz, 24-bit audio up to 4.5 Mbps.

Table 2.2.4 shows a general comparison of codes.

Table 2.2.4 General comparison of codecs

	MPEG-1	*MPEG-2*	*MPEG-3*	*AC-3*	*APT-X*	*Dolby E*	*AAC*
Data rates (kbps)	32–448	32–384	32–320	32–640	Fixed 4:1	Full AES-3	8–320
Filter bank	PQMF	PQMF	PQMF/MDCT	MDCT	PQFM	MDCT	MDCT
Frame length (ms)	8	24	24	32	2.54	One video frame	21.33
Processing delay (ms)	<50	100	150	187	2.7–7	One video frame	From 20

References

1. Soulodre et al. Subjective evaluation of state-of-the art two channel audio codecs from the Communications Research Centre, Ottawa, Canada, *AES Journal*, **46**, No. 3 (March 1998).
2. Bower, A.J. BBC Research Department Report, BBC RD 1990/6, NICAM 728 – Digital Two-Channel Sound for Terrestrial Television (1990).
3. Grill, B. The MPEG-4 General Audio Coder, *AES 17th International Conference* (1999).
4. Fielder, L.D. and Davidson, G.A. Audio coding tools for digital television distribution, *AES 108th Convention*, 5104 (F-5) (2000).
5. Dietz, M. and Meltzer, S. CT-aacPlus, *EBU Technical Review* (July 2002).

Bibliography

Brandenburg, K. MP3 and AAC explained, *AES 17th International Conference* (1999).

Brandenburg, K. and Bosi, M. Overview of MPEG-audio: current and future standards for low bit-rate audio coding, *AES 99th Convention* (1995).

Davis, M.F. The AC-3 multichannel coder, *AES 95th Convention* (1993).

Dietz, M. et al. Spectral Band Replication, a novel approach in audio coding, *AES 112th Convention*, 5553 (2002).

Fielder, L.D. and Todd, C. The design of a video friendly audio coding system for distribution applications, *AES 17th International Conference* (1999).

International Telecommunications Union, ITU-R BS.1196-1.

Kramer, L. DTS: Brief history and technical overview.

Vernon, S. Design and implementation of AC-3 coders, *IEEE Trans. Consumer Electronics*, **41**, No. 3 (August 1995).

Vernon, S. Dolby Digital: audio coding for digital television and storage applications, *AES 17th International Conference* (1999).

Vernon, S. and Spath, T. An integrated multichannel audio coding system for digital television distribution and emission, *AES 108th Convention*, 5154 (R-1) (2000).

R S Roberts C Eng, FIEE
Sen MIEEE Consultant Electronics Engineer

Revised by
Chris Dale MIIE
Previously BBC Broadcast Networks Engineer

2.3 Television Standards and Broadcasting Spectrum

Television systems can be divided into two groups for the purpose of describing how they function: analogue TV systems have been with us for over 70 years, and all depend on being able to transmit a continuously varying signal level over a distance, representing the picture information. Several digital TV systems have been developed in the last 20 years. All transmit the picture information by changing the continuously varying signal levels from the camera, or other source equipment, into a series of codes consisting of two or more discrete signal states. Some digital systems are suitable for local distribution in studios, and others, described here, are suitable for broadcast transmission.

2.3.1 Analogue Television Systems

Every analogue colour television channel consists of three modulated carriers:

- The *vision* information, derived from a camera or other signal source, is used to amplitude modulate a carrier with the electrical equivalents of the basic 'black and white' variations that are encountered during transmission of the scene.
- A subcarrier, situated within the bandwidth of the vision modulated carrier, is itself modulated with information related to the *colour* information in the scene.
- A separate adjacent carrier is modulated with the *sound* information contained in the scene.

The eye, as a visual communication system, 'sees' a large amount of detail simultaneously, by virtue of the fact that it has several million communication channels operating in parallel at any instant. The electrical signals that are generated by the millions of sensors in the eye are partly processed in the retina at the back of the eye, and further processed in the brain to provide the familiar human experience of normal vision. The mass of detail forming the visual scene consists of variations in light and shade, colour and, because we have two eyes, perspective.

Picture transmission, using electronic means to convey information of a scene, cannot be carried out as a simultaneous process embracing the total field of view. Any telecommunication system can process only a single item of information at a time, and hence the data relating to any visual scene must be analysed in such a way that the complete scene can be transmitted as separate items of electrical information. At the receiver, the individual bits of information are recovered and processed for display.

2.3.2 Scanning and Aspect Ratio

The visual scene is explored by examining the small areas of detail that are contained in it, a process known as *scanning*. When we read the page of a book, our eyes scan it line by line to extract the total visual information. Electronic scanning carries out a similar line-by-line scan process, the detail encountered being translated into voltage variations that can be used to modulate a radio transmitter. At the receiver, the received signals are demodulated and used to vary the beam current(s) of a display tube, the beam of which is sweeping in synchronism with the transmitter scanning beam.

A constraint of the electronic scanning system is the need to put a *frame* round the field of view to be transmitted. In the human seeing process, the eye is quite unrestricted in its movements, and it roams freely over a very wide angular range which, with head and body movements, provides an unlimited field of view. In the electronic process, a finite limitation must be imposed by means of a frame, within which the picture can be analysed line by line.

2.3.2.1 Aspect ratios

Early television engineers did not 're-invent the wheel', so very wisely adopted many of the principles and standards that have evolved in film presentation. One of these concerns the shape of the frame. In the film industry, a standard rectangular shape with an *aspect ratio* (ar) of 4 (horizontal):3 (vertical) was the norm. If a system has a standard aspect ratio at both the transmitter and the receiver, picture size is irrelevant. The relative dimensions of objects in the field will be correct.

Early experimental systems by Baird scanned the frame vertically with an aspect ratio of 1:2. The first television engineers concerned with the need to establish standards had no reason to depart from the 4:3 ar, particularly as it was realised that film would constitute a large proportion of programme material. These engineers were the team that created the standards for the world's first regular broadcast service of television programmes in 1936. This ratio has been adopted by all the systems that followed until the mid 1990s, when the 16:9 aspect ratio was added.

16:9 ar pictures provide a wider horizontal field of view which is closer to, but obviously still less than, that seen by the human eyes. This gives new opportunities for providing the viewer with more visual information, and for displaying widescreen cinema films without cropping the sides of the picture, or needing to 'pan and scan'.

2.3.2.2 Aspect Ratio Converters (ARCs)

Because many viewers own only 4:3 aspect ratio TV displays, there is a requirement for 16:9 ar programme material to be produced in such a way that at least some of the sides of the picture can be lost without losing important information. Also, aspect ratio converters have been designed to provide several methods for displaying 16:9 ar material on 4:3 ar TV displays. These converters will often be inserted into transmission networks at the input to analogue distribution networks designed for 4:3 ar display. The converter operating modes are: cropping the sides of the picture; 'letter-box' where a reduced size 16:9 ar picture is displayed with black bands above and below the picture; and 14:9 ar, which is a combination of side cropping and 'letter-box'. Without conversion, the picture will be distorted so that people on screen look taller and thinner than they should. Aspect ratio converters are also used to provide effects opposite to those described above when 4:3 ar programme material is to be transmitted on digital networks which support 16:9 ar displays. Also, for digital transmission networks, an Active Format Descriptor (AFD) switching signal can be transmitted with the pictures. This can be used in viewers' television displays to cause the raster width to be switched to display each programme in the correct aspect ratio.

2.3.3 Still and Moving Pictures

2.3.3.1 Still pictures: facsimile systems

One method of still picture transmission is the facsimile (fax) system. In a simplified example of a basic fax system, the document or picture to be transmitted is scanned horizontally line by line. The A4 page to be scanned is moved vertically past a horizontal row of, typically, 1728 photo detectors, giving 203 picture elements (dots) per inch. The electrical output of each detector is selected sequentially along the row, then the paper is moved down one line (a line feed) and the process repeated. This process of scanning and stepping the paper down is repeated until the whole page has been scanned. The paper is then ejected and the next page moved to the start position (a page feed). The selected output from the row of detectors is digitised, coded and fed to a modem (modulator–demodulator), where it is converted to a signal suitable for transmission over a telephone line. (Modern fax systems actually use data compression before modulation.) A synchronising signal is added to the signal fed to the modem so that the scanning process at the receiver can be synchronised both for line scanning and paper movement.

At the receiver, the demodulated and decoded signal is returned to its original form, which is a copy of the waveform from the photo detector selector output at the transmitter. The synchronising signal is also extracted from the digital signal at the output of the modem. The rest of the signal containing the decoded image data is switched sequentially to each of a row of tiny heat elements, equal in number to the number of photo detectors at the transmitter. Special heat-sensitive paper that turns black when heated is stepped past the row of heat elements line by line in synchronism with the movement of the scanned document at the transmitter. Thus, where a photo detector at the transmitter 'sees' black, the paper at the corresponding point on the paper at the receiver will be heated and so turn black. (This printing process is described because of its simplicity; however, most modern fax machines use inkjet or laser printing.)

The image on one page of A4 can typically be transmitted and reproduced in between 15 and 30 seconds using group 3 fax machines.

2.3.3.2 Moving pictures

The difference between scanning and transmission of a still or moving picture is one of time. Transmission of a still picture can take as long as is necessary for the required quality, but a moving field of view must be totally scanned in a time that is very short compared with the time being taken by any movements in the field of view. In other words, complete scanning of the moving picture must be so fast that we are concerned with what is, virtually, a still picture.

One of the properties exhibited by the human eye is *persistence of vision*. When the image of a still picture is impressed on the eye, removal of the visual stimulus does not result in an immediate cessation of the signals passed to the brain. An exponential lag takes place with a relatively long time for a total decay of the image. The cinema, and television, exploits this effect by presenting to the eye a succession of still pictures (or *frames*) one after the other, each frame differing from the previous one only by the change in position of any moving objects in the field of view. The presentation of one frame after

---

*Depending on whether shades of grey are required, or just a black and white image as for printed text, there will be either a number of grey shade values coded, or just two: one for black and one for white.

another must not allow time for the image decay to become obvious and, provided that the presentation is sufficiently rapid and not too bright, an illusion of continuous movement is maintained.

In the cinema, still pictures are projected in succession onto a screen. A frame is drawn into position with light cut off by means of a rotating shutter. As the shutter opens, the frame is stationary and the projected image illuminates the screen. The shutter cuts off the light, the next frame is drawn into position, the light is re-exposed through the frame, and so on in a continuous sequence. A great deal of early work with film showed that, for most people, a projection rate as low as 10–12 frames/second is adequate to present a complete illusion of movement.

However, at this projection rate another property of the eye becomes significant. The eye is extremely sensitive to the interruption of light at this rate, and the viewer would be very aware of *flicker*. As a consequence, the standard rate adopted for projection was 16 frames/second, well above that necessary for the presentation of continuous movement and, with the low-level illuminants of those days, adequate to minimise any awareness of flicker.

The eye sensitivity to flicker is a function of picture brightness and interruption rate, the rate needing to be higher if brightness is increased. By exposing each frame twice, for a picture projection rate of 16 frames/second, the interruption frequency is raised to 32 per second, and the visibility of flicker is very much reduced without doubling the length of film. As time passed, better light sources came into use and flicker reappeared. This problem was solved along with another concerned with the soundtrack that films now required. Film was not passing through the projector system fast enough for good sound quality. The standard was changed to that in use today. The frame rate was raised from 16 to 24 frames/second. This raised the interruption frequency to 48, and increased the physical film length by 50%.

2.3.4 Television Picture Frequency

The engineers who had the task of establishing standards for the first broadcast system (system A; see Table 2.3.1) adopted the aspect ratio of 4:3, but were concerned by the film projection rate of 24 frames/second. It was feared that, with a 50 Hz power supply frequency, any residual 50 Hz or 100 Hz power supply ripple in the receiver might modulate the beam current of the display tube with a subharmonic at 25 Hz which would produce a visible 'bar' across the picture. If film was being transmitted at the film standard rate of 24 frames/second, the difference frequency of 1 Hz would result in the bar sweeping down the picture once per second.

It was decided that the picture rate would be 25 pictures/second instead of the film rate of 24. It was considered that the effect on sound would not be serious. It was further considered that, in the event of interference from the power supply, a stationary bar across the picture would be less offensive than a bar sweeping down the picture at the 'beat' frequency of 1 per second.

Figure 2.3.1 shows a simple six-line picture consisting of a black bar on a white background and the voltage output from the scanning system during a one-line scan. The scanning spot has a diameter equal to the width of one line, and sweeps across the picture from right to left along line 1, returns to its starting point displaced vertically by one line, sweeps line 2, and so on down the field of view to line 6. It is then returned to the top of the picture for the second picture scan, and so on. The voltage output may be as shown with maximum voltage indicating peak white and minimum voltage corresponding to black, termed

Table 2.3.1 Standard television systems

Parameter	*System*										
	A	*B*	*C*	*D(K)*	*E*	*G*	*H*	*I*	*L*	*M*	*N*
Lines per picture	405	625	625	625	819	625	625	625	625	525	625
Field frequency (Hz)	50	50	50	50	50	50	50	50	50	60	50
Line frequency (kHz)	10.125	15.625	15.625	15.625	20.475	15.625	15.625	15.625	15.625	15.734	15.625
Video bandwidth (MHz)	3	5	5	6	10	5	5	5.5	6	4.2	4.2
Channel bandwidth (MHz)	5	7	7	8	14	8	8	8	8	6	6
Sound/vision carrier spacing (MHz)	3.5	5.5	5.5	6.5	11.15	5.5	5.5	6	6.5	4.5	4.5
Vestigial sideband width (MHz)	0.75	0.75	0.75	0.75	2	0.75	1.25	1.25	1.25	0.75	0.75
Vision modulation polarity	Positive	Positive	Positive	Negative	Positive	Negative	Negative	Negative	Positive	Negative	Negative
Sound modulation	am	fm	am	fm	am	fm	fm	fm	am	fm	fm
Deviation (kHz)		50		50		50	50	50		25	25
Pre-emphasis (µs)		50		50		50	50	50		75	75
Colour system(s) used (not part of TV Standards A–N)	–	PAL SECAM		SECAM PAL		PAL SECAM	PAL SECAM	PAL	SECAM	NTSC	NTSC

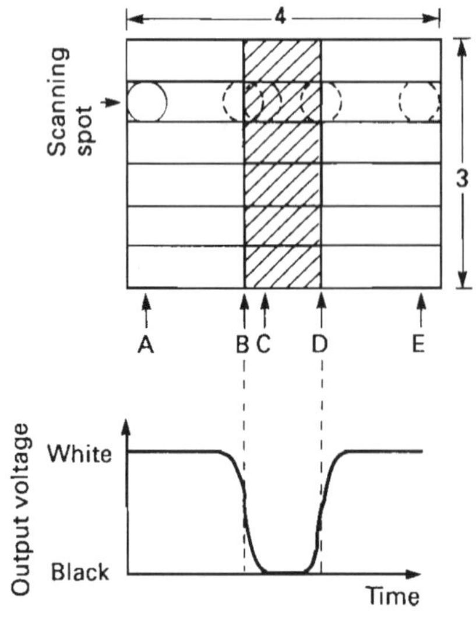

Figure 2.3.1 A six-line scan of a 4:3 field, showing aperture distortion.

positive modulation. In a system, the polarity may be inverted, so that minimum voltage equals white and maximum is black. This would be *negative modulation*, as exemplified in the current UK system I (see Table 2.3.1). This method of scanning is termed *sequential scanning*, and results in a very obvious flicker because the field rate and picture frequency are the same, i.e. 25 per second.

A scanning system was standardised which provided a similar effect to the double-shuttering used in film projection. Instead of scanning the lines in sequence, the picture field is scanned by one field using lines 1, 3 and 5, and a second scan fills the gaps by re-scanning the field using lines 2, 4 and 6. This is *interlaced scanning*, and constitutes two sweeps of each field for a complete picture field. This has the same effect as double-shuttering of film, and raises the flicker frequency to 50 per second. Some modern large television displays further reduce the flicker by displaying each field twice. This doubles the display rate to 100 Hz.

The interruption rate imposes a limitation on the brightness level at which the display tube can be operated before flicker becomes visible. The relationship between flicker and brightness is expressed by the Ferry–Porter law:

$$f_c = F + 12.6\ \log_{10} B$$

where f_c is the critical frequency below which flicker is observed, F is a constant related to viewing conditions and B is the luminance of picture highlights.

Tests on the viewing conditions of television pictures have suggested a value of about 37 for F and, with $f_c = 50$ (as in all European and some other systems), a picture highlight value of about 10 foot-lamberts is obtained.

Television standards in the USA adopted the same general principles. The picture rate was related to a power supply frequency of 60 Hz, resulting in a picture rate of 30 per second and a light-interruption rate of 60 per second. This increase of interruption frequency, compared with the UK rate, results in a permissible increase in highlight value by 6.8 times.

The six-line system illustrated in Figure 2.3.1 would have very poor picture quality. The picture has a sharp transition at the edges of the black bar, but the voltage output does not change instantly from the white value to the black value. At position A (Figure 2.3.1) the scanning spot 'sees' peak white. As the spot reaches the bar at B it 'sees' half white and half black, the output being half of the peak value, as shown. The output only reaches the value due to black at position C. At D, the half value is derived as shown, and the remainder of the scan gives a white output. The resulting effect is termed *aperture distortion*; it prevents any small detail in the picture being reproduced accurately.

In any practical television system, the picture quality will depend on the ability of the system to reproduce at the display tube all the sharp edges and fine detail. This requires the scanning spot size to be reduced, with a consequent increase in the number of lines necessary for a complete scan of the field of view. It is shown in Section 2.3.6 that the channel bandwidth is determined by the scanning spot size. The smaller the spot, the more lines that are required for a complete scan, and channel space is at a premium. This means that a 'standard' spot size has to be a compromise between the ability of the system to provide a picture quality that is acceptable and the minimum demand for channel space. Such standards are quoted in terms of the number of lines that are required in the vertical dimension for complete scanning of the entire field of view.

2.3.5 The Analogue Video Signal

The video signal derived by the camera or other scanning device for a practical black and white system will consist of random voltages generated by the scanning of black, white and grey images during the line scan. A possible scan output is shown in Figure 2.3.2.

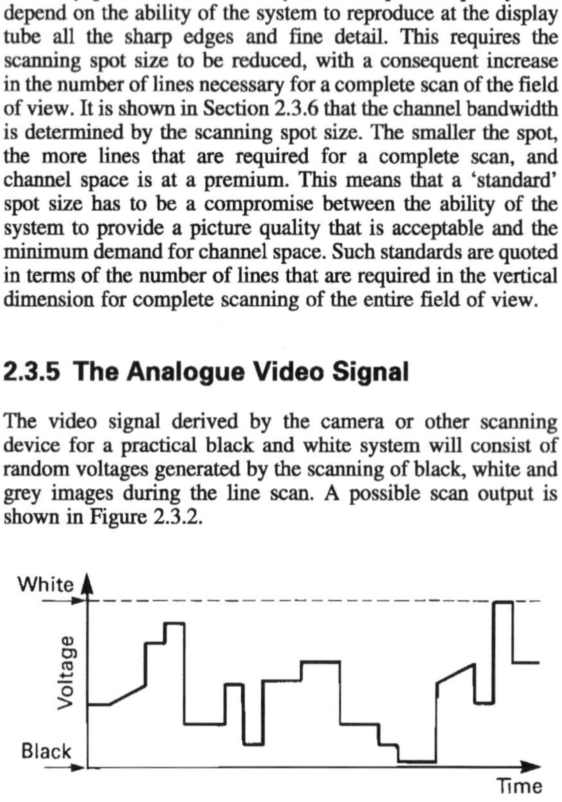

Figure 2.3.2 A possible video output from a camera during a one-line scan.

Two important conclusions arise from consideration of this type of voltage waveform:

- Voltage variations will generally consist of 'step' changes from one value to another. Smooth transitions from white to black, or from black to white, will be rare.
- Alternating current voltage variations are extremely unlikely, and their rare appearance might arise from a scan across regular bars, such as the black and white bars on a test card.

Figure 2.3.3 illustrates another important feature that arises from this type of waveform. The two line scans each show the same

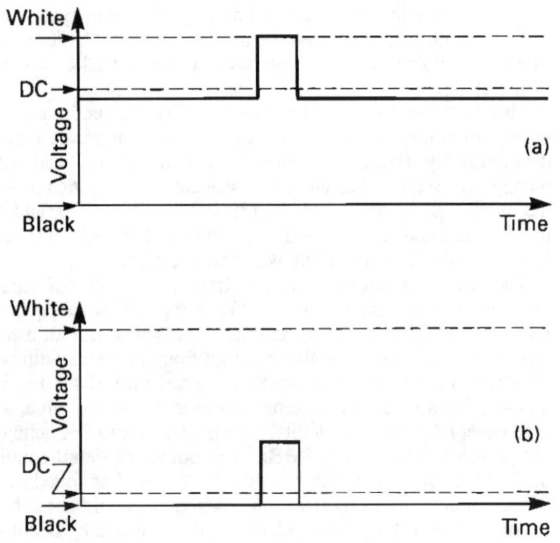

Figure 2.3.3 Similar video output signals are shown in (a) and (b), but with different dc levels.

signal voltage variation. In (a) a white bar is shown on a grey background, and in (b) the same output voltage variation shows a grey bar on a black background. The difference between the two identical signal variations is due to there being, in each, an average dc voltage component. The dc level determines picture brightness.

The video signals resulting from the scanning operation are processed and used to amplitude modulate the transmitter output. All standard analogue TV broadcast transmitters use am, but fm is employed for certain links and for satellite systems.

The principles of amplitude modulation are well known, but there are some important differences between sound and video as modulating signals. Figure 2.3.4(a) shows an alternating current variation that might be measured in the antenna system of a sound transmitter. Initially, the carrier is not modulated, then one cycle of an audio modulating tone is applied. The familiar features of this process are:

- The unmodulated carrier is radiated, and its mean level is constant with or without modulation being present.
- The carrier peak level is varied at the modulating frequency. The audio variation in carrier peak values during modulation is termed the *envelope*.
- An obvious limit exists in the modulating process whereby the carrier peak must not exceed twice the unmodulated level if distortion due to *clipping* is to be avoided.

Figure 2.3.4(b) shows a similar situation; it is amplitude modulation by a video signal similar to Figure 2.3.2. The envelope is now of a random character, and there is no mean carrier of

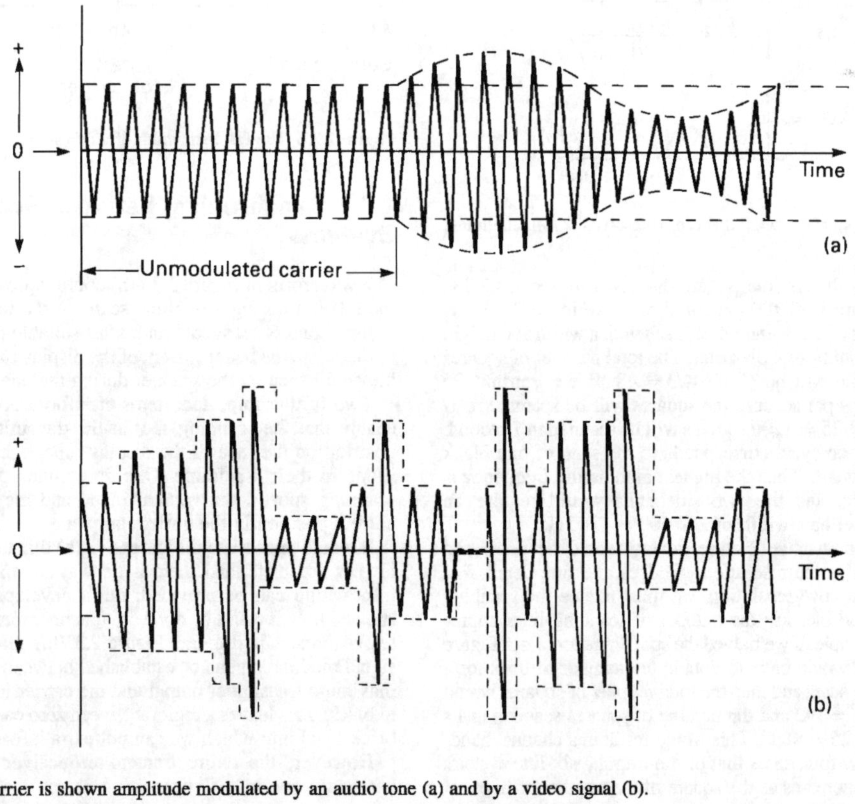

Figure 2.3.4 A carrier is shown amplitude modulated by an audio tone (a) and by a video signal (b).

constant level during modulation. When no modulating signal is present, no transmitter output is radiated.

2.3.6 Channel Bandwidth

If a carrier is amplitude modulated with, for example, a 1 kHz tone, three frequencies are produced: the carrier, a frequency lower than the carrier by 1 kHz (a *lower side frequency*), and a frequency higher than the carrier by 1 kHz. If the modulating signal is a band of frequencies such as voice, music or video, a band of frequencies is generated each side of the carrier, termed *sidebands*, extending on each side of the carrier frequency to a limit determined by the highest frequency in the range of modulating frequencies.

There are several ways in which the highest frequency component in a modulating video signal may be determined, and thereby the channel bandwidth. One is indicated in Figure 2.3.5, in which (a) is the top left-hand corner of a picture consisting of a regular pattern of alternate black and white squares. The sides of the squares are equal in length to the scanning spot diameter, and consequently the output signal generated will be a sine-wave, the frequency being the highest that will be generated at full amplitude. Any detail of smaller dimensions will not generate maximum output.

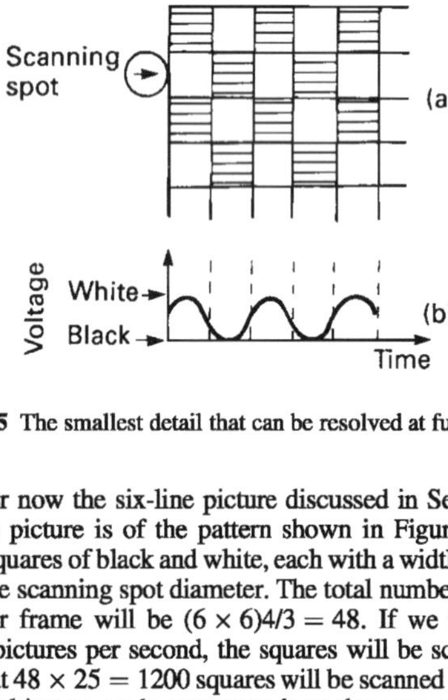

Figure 2.3.5 The smallest detail that can be resolved at full amplitude.

Consider now the six-line picture discussed in Section 2.3.4, where this picture is of the pattern shown in Figure 2.3.5, i.e. alternate squares of black and white, each with a width and height equal to the scanning spot diameter. The total number of squares in a 4:3 ar frame will be $(6 \times 6)4/3 = 48$. If we transmit 25 complete pictures per second, the squares will be scanned in 1/25 s, so that $48 \times 25 = 1200$ squares will be scanned in 1 second.

The resulting ac cycle corresponds to the scan of one black and one white square. Thus the highest modulating frequency is $1200/2 = 600$ Hz, and the two sidebands would require an overall rf channel bandwidth of $2 \times 600 = 1200$ Hz.

We have seen that the picture quality would be poor, and severely lacking in the resolution of edges and fine detail. We know that, to improve resolution, we must reduce the scanning spot diameter and increase the number of scanning lines appropriately. For example, if we halved the spot diameter of our Figure 2.3.1 model, we would have to double the number of lines for a complete picture scan, and then the total number of squares would be $12 \times 12 \times 4/3 = 192$ and the number of squares scanned in 1 s would be $192 \times 25 = 4800$. This would result in a channel bandwidth of 4.8 kHz, four times that of the original six-line system. The bandwidth increases as the square of the number of lines.

A standard adopted for a good quality television system has to be a compromise between the need for acceptable definition of edges and fine detail in a picture and the overall bandwidth of the channel.

There are two historical examples of 'line standards' that are worth consideration. The first UK television broadcast system developed by Baird used an aspect ratio of 1:2 and $12\frac{1}{2}$ pictures per second. Scanning was sequential and vertical. Thus, the number of squares would be $(30 \times 30) \times 2 = 1800$, and the number scanned in 1 second was $1800 \times 12.5 = 22{,}500$. The highest modulating frequency was thus 11,250 Hz.

The second example was the first ever 'high-definition' broadcasts that commenced in 1936 in the UK with a picture frequency of 25 per second, interlaced scanning and an aspect ratio of 4:3. The compromise on definition and bandwidth was decided on the basis of the scanning spot being of such a size that 405 lines would be required to cover the picture area, and give acceptable picture quality. The highest video frequency is $405 \times 405 \times 4/3 \times 25/2 = 27$ MHz. The output of detail smaller than 1/405 of picture height would be less than maximum. Detail generating 3.0 MHz, for example, would be about -3 dB. The total video channel rf bandwidth becomes 3.4 MHz and the lowest modulating frequency is zero or dc. The total radio spectrum space occupied by this first *system A* (as it became known) is shown in Figure 2.3.6.

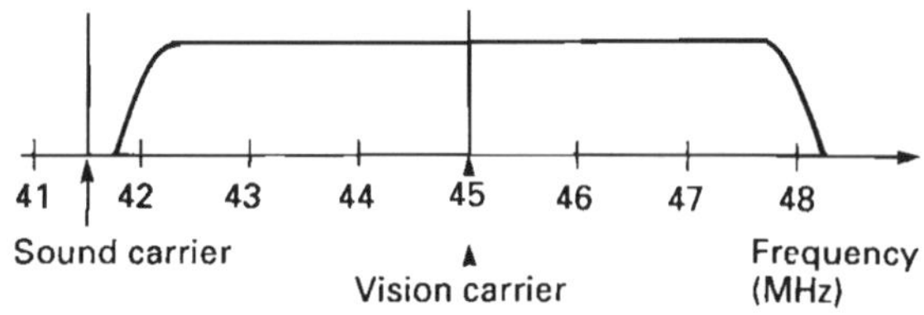

Figure 2.3.6 The full bandwidth of a dsb system A channel.

2.3.7 Synchronism between Scanning Systems

The waveform in Figure 2.3.4(b) could represent the transmitter modulation during a one-line scan. At the receiver, the modulating signal is recovered and, after suitable processing, is used to modulate the beam current of the display tube, thus recreating the detail seen by the scanner during the one-line scan.

Two further important items of information are necessary to ensure that the scanning spot at the transmitter, and the beam position on the face of the display tube, occupy identical positions in their 4:3 frame. One determines the position of the scanning spot in the vertical plane and the other ensures the correct position in the horizontal plane.

It would appear from Figure 2.3.4(b) that the video waveform is very complete, and there is no way in which any additional information can be provided, but a development of system A showed how it could be done, in a manner that forms part of any television standard today. Figure 2.3.7(a) shows how the video signal modulation can be established between the two limits: the maximum transmitter output and the carrier level corresponding to black. This leaves a region between zero carrier output and the black level into which we can put extra information.

However, the entire transmitter/receiver system can deal with only one bit of information at any instant, and we must

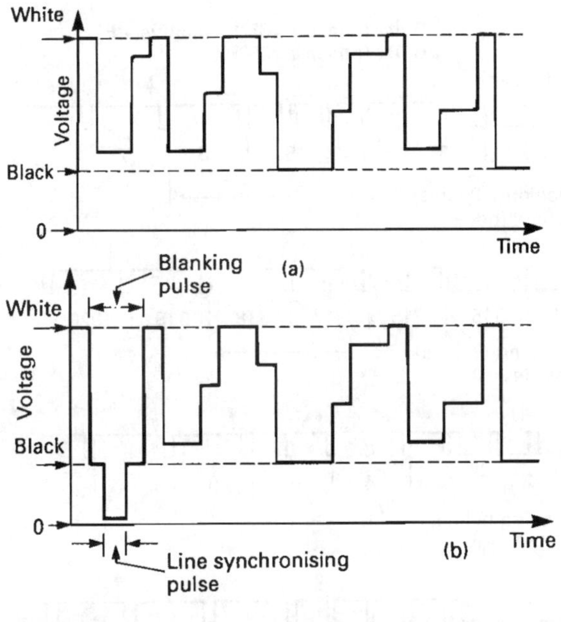

Figure 2.3.7 A typical camera output before (a) and after (b) the insertion of blanking and line sync pulses.

remove the video information while we provide any extra information. A *blanking pulse* blanks out all video information down to the black level, at the start of the line scan. A narrow *line synchronising pulse* is then inserted from the foot of the blanking pulse down to zero. The leading edge of this pulse is used to start the line scan on its traverse across the field from left to right. At the receiver, the leading edge of the demodulated narrow pulse is used to start the sweep of the display tube beam across the face of the tube. At the end of the line scan, the next blanking/synchronising pulse triggers the return of the spot to the left-hand side, and the scanning cycle starts again, this time slightly displaced vertically, to trace a new line path.

The waveform of a line of the UK system I is shown in Figure 2.3.8. The heavy black line is the waveform of a black and white system, and the shaded areas are concerned with the colour information. The timing of all the pulses and other features is with reference to the leading edge of the narrow sync pulse. All television standards use the same type of waveform, but their timing and pulse widths may differ.

To effect vertical displacement of the scanning spot in synchronism, it is necessary to use a different type of pulse. The line sync pulses are narrow and relatively infrequent, so we can suppress video information for several lines, and transmit either a single, long pulse or a train of relatively broad pulses to effect vertical synchronism. Figure 2.3.9 shows the four-field sequence of field sync pulses used in the UK system I. The 'burst' sequence is necessary for the PAL colour system.

At the receiver, the two types of pulse can be separated from the composite waveform by relatively simple forms of amplitude discrimination, and they can then be separated from each other by the passive circuitry shown in Figures 2.3.10 and 2.3.11.

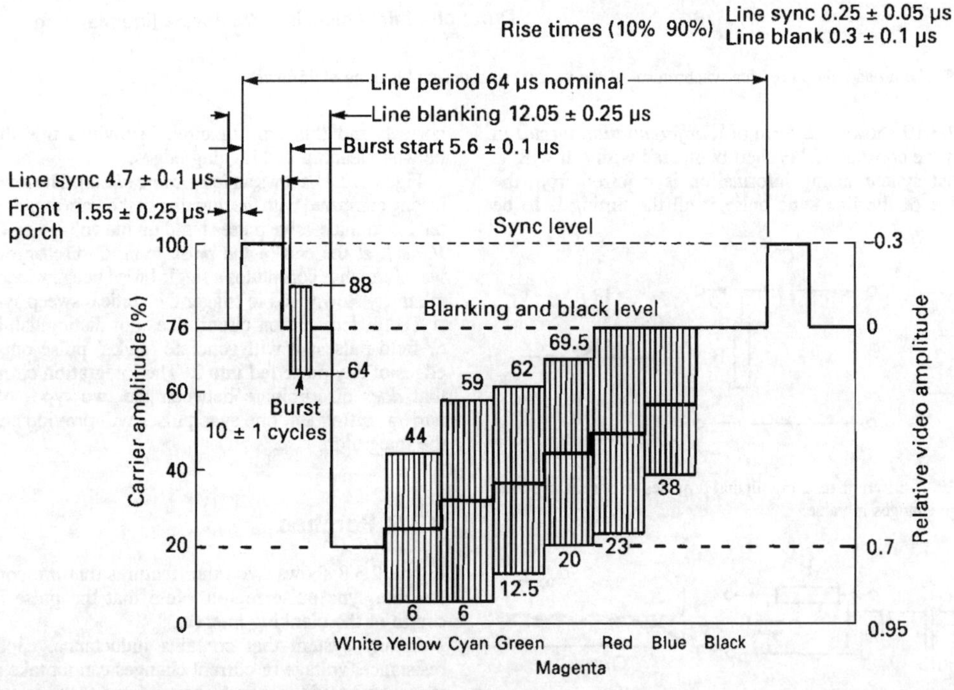

Figure 2.3.8 The thick line show a possible signal output resulting from a single line scan of a black and white system, the image being a grey 'staircase' from white to black. The shaded areas are concerned with the addition of colour (BBC).

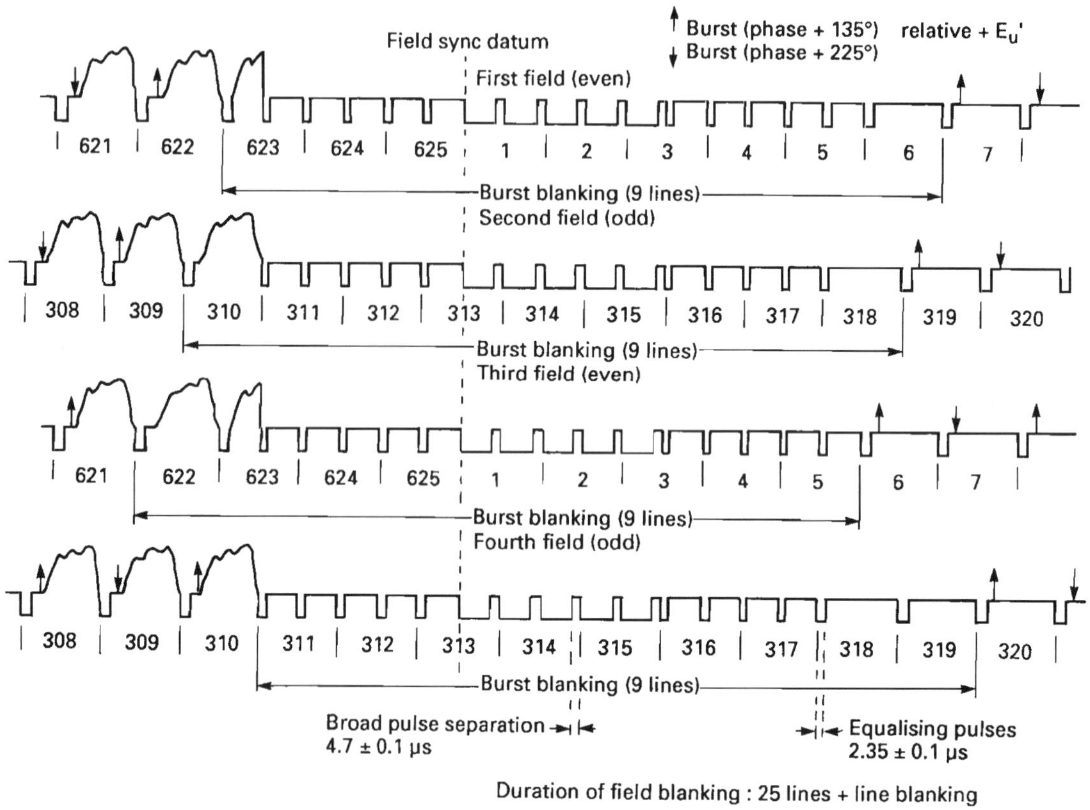

Figure 2.3.9 The broad pulses used for synchronism of the vertical scan during blanking of 25 lines.

Figure 2.3.10 shows one form of a *differentiation circuit* in which the time constant *CR* is short compared with *t*. It will be recalled that synchronising information is required from the leading edge of the line sync pulse if all the timing is to be correct, and this type of circuit provides this discrimination between leading and lagging edges.

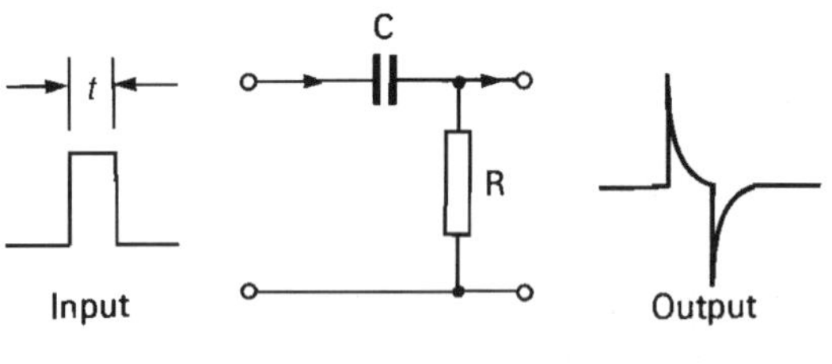

Figure 2.3.10 A differentiation circuit that provides an output when the input voltage changes in value.

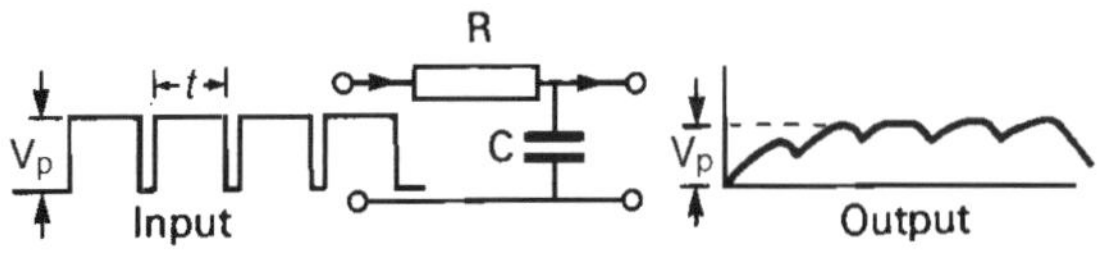

Figure 2.3.11 An integration system that enables a train of broad pulses to build up to a peak value.

Figure 2.3.11 shows an *integrating circuit*. The time constant *CR* is long compared with the duration of the train of pulses provided at the input. Successive pulses build up the voltage on C to the value V_p and, at the end of the pulse train, C discharges. The output waveform thus constitutes a single broad pulse, which operates at a relatively slow speed to trigger the vertical sweep system.

The differentiation circuit does not distinguish between line or field pulses; it will generate 'spike' pulse output from the edges of any pulse fed into it. The integration circuit is the one that *does* discriminate between the two types of pulse. The narrow, infrequent line sync pulses will provide no output from the integrator.

2.3.8 Porches

Figure 2.3.8 shows two other features that are concerned with the line sync pulse region. Note that the pulse is not in the centre of the blanking interval.

In any system that contains inductance, capacitance and resistance, voltage or current changes cannot take place instantaneously. A video signal change from black to peak white, or white to black, takes time for completion, requiring possibly a capacitor to charge or discharge. A black/white edge results in a voltage change, as shown in Figure 2.3.12 (not to be confused with the aperture distortion shown in Figure 2.3.1). Ahead of the pulse in Figure 2.3.8 is a narrow plateau, the *front porch.* Its purpose is to allow the video signal of the previous line (which might have been at peak white, for example) to reach black level before the sync pulse starts.

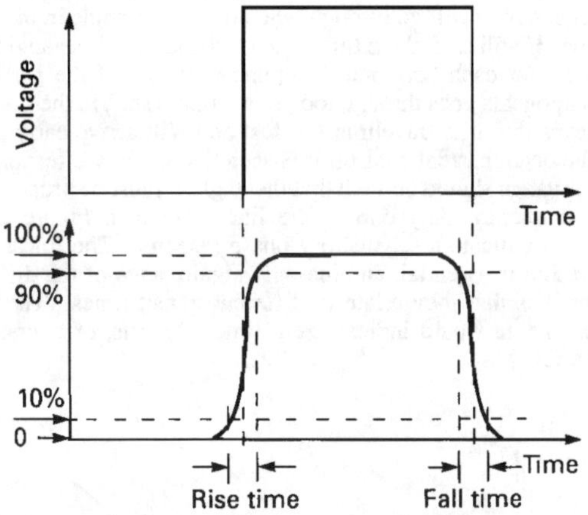

Figure 2.3.12 Voltage and current values cannot change instantly. Rise or fall times are determined by the values at 10% and 90% of peak values.

Behind the sync pulse is another plateau, the *back porch.* The original purpose of the back porch, when first used in system A, was to make sure that there was plenty of time for the receiver line scan circuitry to effect complete retrace of the line scan, and for the beam to be in its correct position for commencement of the next line scan. By comparison with current technology, receiver scan circuits in 1936 were crude, ponderous and extravagant of power, and they required a lot of time for retrace. Today, the back porch provides plenty of time for retrace, and it also provides the space necessary for colour information to be transmitted and extracted at the receiver.

2.3.9 Dsb, ssb, asb and vsb

The UK system A operated from 1936 to 1939, and then had to close down due to the outbreak of World War II. The USA standardised its television system and immediately found a serious problem. The double sideband (dsb) system was very extravagant of channel space and, as many channels were required, a new system of modulation was devised. It was known as a *vestigial sideband* (vsb) system, sometimes termed an *asymmetric sideband* (asb) system, and is now used throughout the world because it saves a considerable amount of channel space.

To appreciate the operation of vsb, it is necessary to digress into some important differences between sound and video amplitude modulation. Single sideband (ssb) am systems have been used for radio transmission of speech and music for at least 60 years. Why cannot ssb be used for video modulation and thus save half the dsb bandwidth?

There are two sources of distortion in a dsb speech or music system. An amplitude constraint exists whereby any attempt at over-modulation results in severe waveform distortion, with unacceptable audible quality. The second effect occurs in single-channel sound reproduction, where the quality of the reproduced sound is unaffected by changes of phase response. Many do not accept that our normal binaural experience distorts. Nevertheless, simply turning one's head, for example, results in a severe phase change of, say, a 5 kHz tone, where about 4 cm movement represents about 180°. We live with this effect and do not notice it, even when we use it for directional information.

Figure 2.3.13(a) shows the vectors relating to the carrier side frequencies during one cycle of a modulating tone of dsb am at maximum modulation; (b) shows the resultant vector addition of the vectors in (a), causing the rise and fall of carrier amplitude, and the phase of the resultant vector, which remains that of the carrier at all times during the modulating cycle.

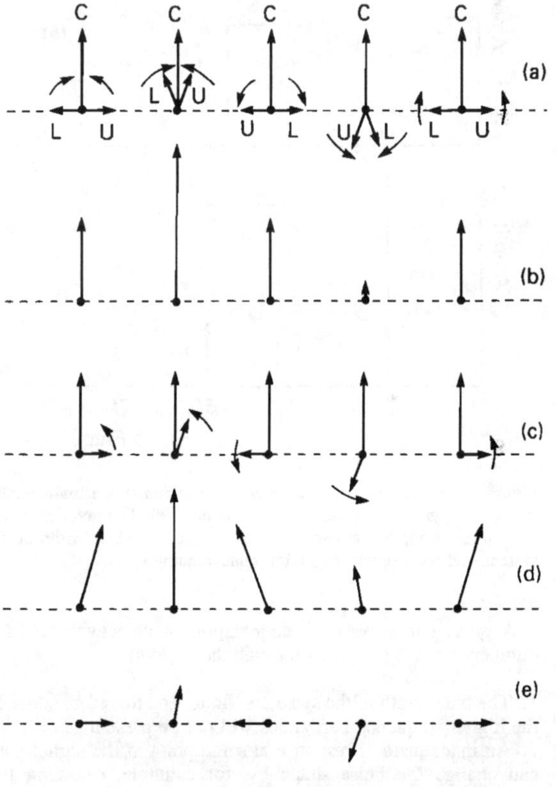

Figure 2.3.13 The vector relationships of dsb (a and b), ssb (c and d) and a single side frequency (e).

Figure 2.3.13(c) is as (a), but with the lower side frequency suppressed, and (d) is the new resultant. Two features can be seen:

- The modulation, considered in terms of carrier amplitude, is halved.
- The angle of the resultant vector is now swinging with respect to the carrier vector.

We have produced phase modulation.

To determine the requirements of a video signal used for amplitude modulation, consider the simple video modulating signal shown in Figure 2.3.14(a). Such a pulse can be analysed into a number of discrete harmonic components, which extend out to infinity with a descending order of amplitude. The square waveform of (a) could be synthesised by addition of the frequency components of (b) at the correct amplitudes and phases. In a television system, we require the pulse signal to progress through the system in such a manner that the shape of the waveform that finally modulates the display tube beam current is a faithful copy of the signal derived by the scanning process at the transmitter.

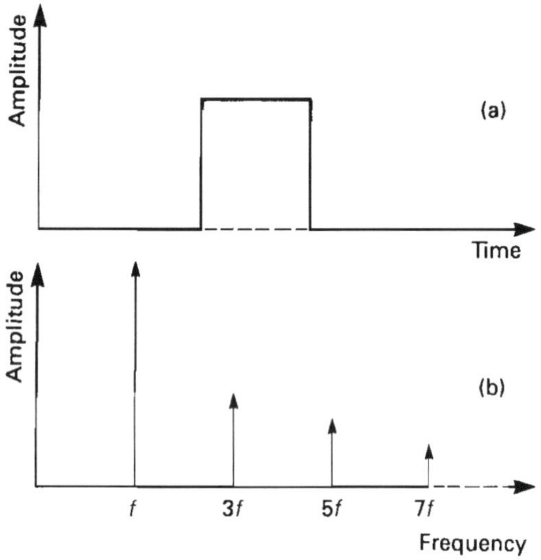

Figure 2.3.14 A theoretical rectangular pulse contains a fundamental frequency f, and all the odd harmonic frequencies. The practical pulse loses some of the higher harmonics due to the finite bandwidth of the system, and has sloping sides with rounded corners.

A pulse can experience degradation of its waveform in a number of ways as it goes through the system:

1. The bandwidth of the system is finite, and therefore some of the higher frequency harmonics will not be present.
2. An inadequate hf response at some point in the signal path can change the pulse shape by, for example, rounding the corners and sloping the sides of the pulse.
3. Inadequate lf response can produce a 'tilt' at the top of a pulse. The dc voltage may not hold up for the duration of the pulse.
4. The various discrete frequency components that constitute the signal waveform may experience differing transit times as they travel through the system. Some may go through faster or slower than others, distorting the wave shapes in varying degrees at the point where they are intended to arrive together, i.e. at the point where the display tube beam current is modulated.

Of the above, point 1 is determined by a definition compromise as discussed in Section 2.3.6, and points 2 and 3 are concerned with circuit behaviour and any deficiencies can be resolved. The timing effect of point 4 is much more important. The phase response of the system, clearly, determines the preservation of pulse shape. Each of the signal frequency components must travel through the system in the same time, although the actual time of transit is, within reason, of no importance.

If a fundamental frequency f passes through the system in time t, this time can be interpreted as a phase change ϕ. The frequency $2f$ will go through twice this phase angle in the same time; $3f$ will shift three times the fundamental phase angle, and so on for each harmonic component. If one of the harmonic components goes through too great a phase shift in the time t, it means that it is travelling too fast and will arrive early at the tube beam current control. It is seen that the phase response of the system should be such that the angle ϕ must be proportional to frequency. Any one of the lines shown in Figure 2.3.15 would indicate a satisfactory phase response. The linear relationship is essential, and the only significance of the differing slopes is that they relate to different transit times. (The horizontal line would indicate zero time which is, of course, not possible.)

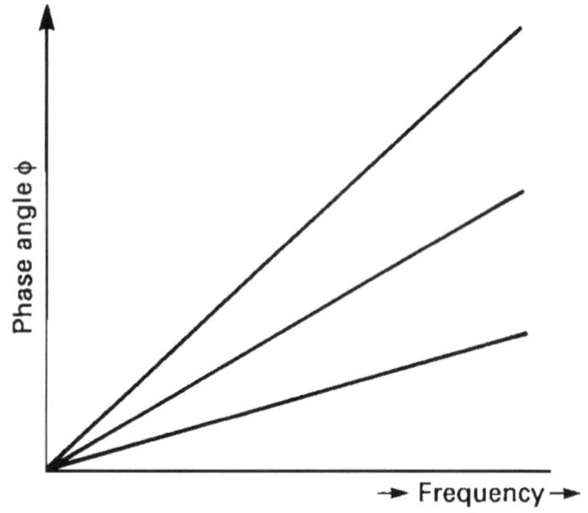

Figure 2.3.15 A video system phase response must be proportional to frequency.

The main differences between an audio channel and video signal processing are now more clearly seen. Audio signal channels can tolerate a high degree of phase distortion, but amplitude distortion must be kept to very low limits. Video signal processing must ensure that phase distortion is minimised, but overload effects are far less serious. The peak value of Figure 2.3.14 may represent peak white, for example, and any increase above this value would not be visually of much significance, although it may drive a transmitter into an overload condition. Alternatively, the peak value may correspond to a black signal, in which case 'blacker than black' is of no visual consequence.

The phase distortion produced by ssb (Figure 2.3.13(c, d)) may be acceptable for audio, but is quite unacceptable for video. However, it *is* possible to use partial suppression of one sideband (vsb) for a television system and save considerable channel space. Whereas phase distortion can be quite unacceptable where large picture areas are involved, phase errors become difficult to see on small detail in the picture. Thus, a practical system must have a very low phase distortion at low video frequencies, while the high frequencies generated by the fine detail in the picture can be severely distorted, but still acceptable.

Figure 2.3.16 shows, at (a), the full dsb channel bandwidth that would result from video amplitude modulation, together with the sound channel, for the UK system I. The transmitted signals are shown in (b), where the lower sideband has had about 4 MHz filtered off, leaving a 'vestigial' 1.25 MHz. The 12 MHz channel has been reduced to 8 MHz.

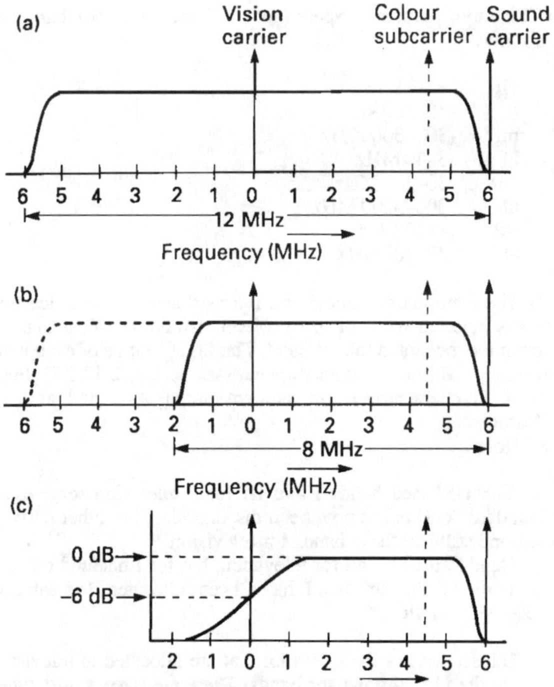

Figure 2.3.16 A full channel bandwidth is shown width, at (b), the radiated signal of vsb modulation, and at (c), the required receiver if response.

For vsb to function correctly, it is necessary for the received signals to be processed before being presented to the demodulator. The pass-band of the receiver IF channel must be shaped as shown in Figure 2.3.16(c). Consider, for example, a 50 Hz modulating frequency, and its position in Figure 2.3.16. Such a frequency would result from the scan of a large area, indeed a complete field, and the resulting side frequencies would be very close to the carrier. The signals presented to the demodulator by the receiver IF amplifier are shown as vectors in Figure 2.3.17(a). The carrier and side frequencies are cut by 6 dB, but no phase distortion would result because the signal is a full double sideband am signal at 100% modulation.

Consider next a modulating frequency resulting from the scan of a smaller detail, producing a frequency around, say, 500 kHz. Figure 2.3.17(b) shows the vector relationship between side frequencies and carrier. The overall length of the upper plus lower side vectors add to the same value as in (a), showing that the modulation factor has not changed. However, one vector is longer than the other, and some phase distortion is beginning to appear in the resultant addition of the three vectors. At frequencies of about 1.5 MHz and above, the vector situation is as shown in (c). The single side frequency resulting from the scan of a small detail has an amplitude equal to the attenuated carrier and, again, the level of modulation is unchanged. Phase distortion is present, but the detail is small and the errors become more difficult to see as the modulating frequency increases.

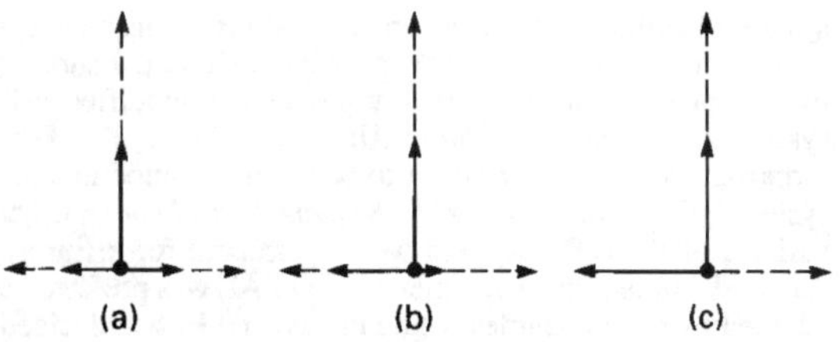

Figure 2.3.17 Demodulated output from the vsb input to a receiver. See description in text.

The US problem of minimising channel space was eased considerably by the use of vsb, and all standard analogue systems now use vsb. All UK system A stations that followed Channel I used vsb and, when the first station moved from its original site to Crystal Palace, the opportunity was taken to bring it into line and change it from dsb to vsb. The vestigial band was about 1 MHz, and the overall channel width was 5 MHz.

2.3.10 National Standards

Many of the world's analogue television standards use 625 lines, the exceptions being the 525-line system M, used by North and South America, Japan and a few other countries, and the French system E on 819 lines. System E was developed before World War II in an attempt to provide a better picture quality than the 405-line standard then prevailing. This virtual doubling of the line structure does, indeed, provide an excellent picture quality, although at the expense of channel space. Consideration has been given to the adoption of new line standards for High-Definition TV (HDTV) systems that would approximately double the number of lines. A few countries, such as Japan, are experimenting with HDTV transmissions using 1125 or 1250 lines per frame.

The almost universal adoption of a 625-line standard resulted from the first international conference on television standards after World War II. All countries planned to start a television service, and it was hoped that the 625-line standard would permit international links between European and other countries that used 625 lines. Unfortunately, widespread adoption of 625 lines did not result in a universal standard. There are many 625-line systems, but few are compatible. Some have positive modulation, others negative. Some have am sound, and others fm sound. There are further differences in the colour systems.

The US system M was the first practical broadcast colour system. A later European conference attempted to standardise a European colour system. It was considered that the NTSC system M could be improved, and several proposals were made for systems based on the NTSC principle that used a subcarrier for colour information. Germany proposed a modified NTSC system known as PAL (*Phase Alternation Line*). The French proposal was for the subcarrier to be frequency modulated, the system being termed SECAM (*Sequentiel á Mémoire*). Other variants of the NTSC system were considered for a European standard, but the final outcome was that PAL was preferred and adopted by most countries, while France and Russia decided to use SECAM. (They actually have different versions of SECAM.)

Table 2.3.1 gives details of most of the standards in use throughout the world. (System A has been included, although it is no longer in use in the UK.) Many countries now include a second sound carrier for stereo sound. For example, in the UK, France and several other countries, this carries NICAM digital sound. This is 6.552 MHz above the vision carrier in the UK. In the USA, MTS stereo sound is used. These are not the only differences between systems: the radio-frequencies on which they operate also differ between countries. The frequency allocations on which radio services operate are decided by the International Telecommunications Union (ITU). The world is divided by pole-to-pole boundaries into three regions:

- *Region 1* includes Europe, Africa and Russia. The eastern boundary includes the whole of Russia and Mongolia.
- *Region 2* includes North and South America, Greenland and Alaska.
- *Region 3* includes Australia, New Zealand, India and Pakistan, China and Japan.

This partition is intended to ensure that the various radio services should operate with minimum interference to each other and other services. However, this does not ensure that all television services will operate on the same frequency bands. Considerable differences exist, not only between regions, but also within regions.

For example, Table 2.3.2 shows the UK and European allocations for television services, but Africa and Scandinavia, also in Region 1, differ. South Africa has two bands, one on vhf 175–255 MHz and the other a uhf band on 471–632 MHz. The Danish bands are 55–68, 175–216 and 615–856 MHz. Both countries use PAL. South Africa uses system I, as in the UK, and Denmark uses PAL on systems B and G, which differ only in channel bandwidth.

Table 2.3.2 European bands and designations

Band	*European band*	*Frequency range*
vhf	I	41–68 MHz
vhf	II	88–108 MHz (fm sound)
vhf	III	174–216 MHz
uhf	IV	470–582 MHz
uhf	V	614–854 MHz
shf		11.7–12.5 GHz

Many similar examples can be quoted. A notable one is that Japan uses NTSC system M, but a receiver manufactured by the Japanese for home use could not be used in the USA because the two countries are in different regions and their radio-frequencies are different.

Further examination of Table 2.3.1 shows several very similar standards. Many have derived from basic systems, and differences are small. Systems D and K, for example, are the same for the listed parameters, but the D version uses PAL and the K version uses SECAM. The K system, in turn, has spawned the K′ system, which is similar to the K system but with a wider vestigial sideband.

2.3.11 Bands and Channels

The radio-frequency spectrum is divided up into bands as follows:

vlf	< 30 kHz
lf	30–300 kHz
mf	300–3000 kHz
hf	3–30 MHz
vhf	30–300 MHz
uhf	300–3000 MHz
shf	3–30 GHz
ehf	30–300 GHz

The band classifications are international, and broadcasting takes place in some of them. Medium-wave broadcasting, for example, occurs in the mf band. The '3–30' range of each band is not so arbitrary as it may appear (see Section 2.13.2.1). Each band is characterised by its own propagation and antenna features.

Notes:

1. The UK used bands I and III for a television service, on standard A. This has now been discontinued, but other parts of Europe still use these bands for television.
2. Band II is not used for television, but for fm sound only.
3. Parts of the shf band have been allocated for satellite television services.

The sections of the spectrum that are allocated to television are in the vhf, uhf and shf bands. There are three world zones with different frequency allocations for broadcasting, and Table 2.3.2 shows the European bands and their designations.

The UK television service functions in Bands IV and V only. Channels 21–34 are in Band IV, and channels 39–68 are in Band V. The various standards in use have different channel bandwidths so, for example, a UK channel 40 does not necessarily occupy the same frequency space as a European channel 40.

The actual frequencies in the shf band allocated to television broadcasting in the UK are shown in Table 2.3.3. There are 40 channels, each 20 MHz wide.

Table 2.3.3 UK shf frequencies

Shf channel	*Frequency (GHz)*	
4	11.78502	All left-hand circular polarisation
8	11.86174	
12	11.93846	
16	12.01518	
20	12.09190	

2.3.12 Adding Colour to a Monochrome System

It is shown in the Colour Encoding and Decoding Systems chapter that colour requires three types of information. It is necessary to determine *luminance, hue* and *saturation* of any colour at any instant. Fortunately, existing monochrome systems are luminance-only systems, and they can supply the luminance information. This leaves two extra items of information to be transmitted and received which, when suitably processed, can determine the hue and saturation encountered during scanning.

Apart from the obvious difficulties, some severe constraints exist on the addition of colour information:

- It must not require extra channel space.
- There must be no interference between the existing luminance information and the added information.
- It must be totally compatible, i.e. a monochrome receiver must display a colour transmission in black and white, and a colour receiver must be capable of displaying a monochrome transmission in black and white.

2.3.12.1 Colour bandwidth

There are some aspects of human colour perception that can be used to advantage. One of these concerns the bandwidth required for colour display. Our ability to perceive fine detail in colour is considerably inferior to our awareness of fine detail as a brightness or luminance variation, thus colour information does not require the same wide bandwidth that is needed by a luminance system.

The eye can resolve small variations of luminance detail over the very small angle of 0.5–1.0′ (the precise figure varies with individuals). Visual acuity for colour is far less sensitive than for luminance, and depends to some degree on the range and saturation of the colours involved. Over an orange/cyan range of colours the resolution angle becomes 1.5–3.0′, and over a green/magenta range, detail resolution requires about 4–8′. The bandwidth required for colour information can thus be significantly less than the bandwidth required for luminance. Whatever methods we are going to use to process colour, it is only necessary to use about 1 MHz for the relatively low-detail colour information.

2.3.12.2 Spectrum utilisation

Let us examine more closely the distribution of energy over the band of frequencies occupied by a carrier which is modulated by an analogue video waveform. Without any video modulation, the carrier is already modulated by a number of frequencies. The line scanning rate is at a constant frequency (15,625 lines/second in system I), and it establishes side frequencies on each side of the carrier, spaced at line frequency and from each other, as shown in Figure 2.3.18. Interlaced scanning modulates the carrier at 50 Hz, and the side frequency of 50 Hz and the harmonics are spaced on both sides of the carrier and the line scanning harmonics. The picture frequency is another modulating signal at 25 Hz, and its energy joins the clusters around each line frequency harmonic.

The line frequency harmonics with their side frequencies extend out to the bandwidth limits of the video channel, and they are of a descending order of energy level as they become

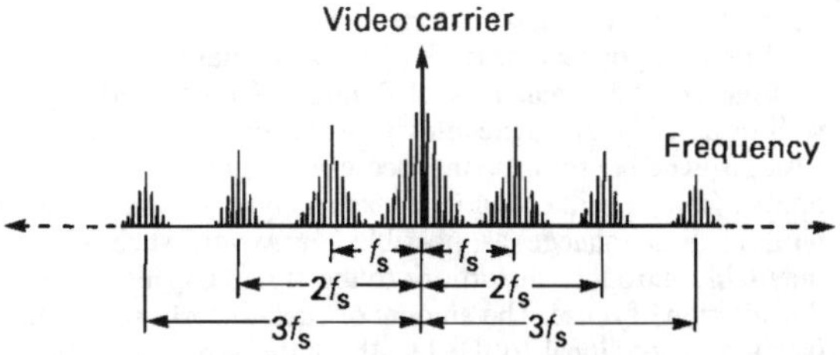

Figure 2.3.18 The signal components resulting from amplitude modulation; f_s is the line-scanning frequency.

more remote from the carrier. Over the channel bandwidth, the energy generated across the band, without video modulation, is contained in clusters, spaced at line-scanning frequency from the carrier and from each other. Large regions of the channel space contain little or no energy.

In 1934, long before any broadcast television system existed, two mathematicians, Merz and Gray, studied the situation when video signals were included in the modulation process and concluded that the video energy simply joined the existing clusters round each line-scan harmonic, leaving the overall pattern substantially unchanged. The only differences between one picture and another, or whether one is moving or stationary, are relatively small variations in the magnitude of the energy in the clusters. The low- or zero-energy gaps between the line frequency harmonics remain with all types of video modulation.

The original work of Merz and Gray was directed at reducing interference between two adjacent transmitters, operating on nominally the same frequency. If their carrier frequencies were 'offset' by half the line frequency, the line harmonics of each would drop into the low-energy gaps of the other, thus minimising mutual interference. However, in the years 1951–1953, the gaps were seen by the NTSC as the place for colour information.

A subcarrier can be inserted in the video frequency band, at a frequency about 1 MHz lower than the top of the video band, where the line-scan harmonics are low in amplitude. This subcarrier can then be modulated with colour information in such a manner that, when extracted and processed in the receiver, the colour display tube is driven with appropriate display information.

With the exception of Multiplexed Analogue Component (MAC) systems, all existing standard analogue colour systems are derived from the concept of the use of a subcarrier in the video band to supply colour information.

2.3.13 Digital Television Systems

What is digital television? All practical TV systems must be able to reproduce pictures relating to the 'real world'. These pictures are usually essentially analogue in nature. That is, the visual information consists of continuously varying light intensities and hues (colours). However, in modern parlance, 'digital' usually means binary codes consisting of 'ones' and 'zeros'. Thus, a digital TV system will need to change analogue picture information into digital codes suitable for transmission over digital channels, and then to decode the signal to recover an analogue signal representing the original picture and display it to the viewer. As with analogue TV, digital TV signals may be broadcast by means of terrestrial vhf or uhf transmitters, satellite transponders, or fibre-optic or coaxial copper cable.

What are the advantages of digital TV compared with the well-established analogue television systems? All transmission systems, whether they use the medium of radio, copper cable or optical fibre, are intended to transport signals from 'A' to 'B' with as little change as possible. However, such channels inevitably introduce distortions to the signal carried, including the addition of noise. The amount of such distortion introduced is often proportional to the length of the signal path in each link. Each link in a chain of transmission channels may not cause any perceptible degradation of analogue television picture signals, but the cumulative effect on such signals of several links, which may have a total route length of 1000 km for example, could well be noticeable to viewers. For terrestrial transmissions, routes will consist of one or more telecommunication links from the studio to the broadcast transmitter and one or more broadcast or re-broadcast links to the viewer's receiver.

For digital signal transmission, the receiver is only required to distinguish between a limited number of discrete signal states, i.e. a minimum of two and usually a maximum of 256, depending on the type of modulation used. Thus, digital receivers can be designed to be insensitive to signal distortions as long as they are not large enough to mask the correct signal state. The digital signals can also be regenerated at intermediate points to correct errors due to cumulative distortions in a chain of links, if required. Thus, for well-designed digital transmission systems, television (or any other) signals can be transmitted over greater distances than the equivalent analogue signals with little or no degradation. Many errors can be introduced into digital links by, for example, bursts of electrical noise. As will be seen later, some but not all of these can be corrected by the use of error correction techniques.

A further advantage of digital television is that several services can be more easily compressed and then carried in the same uhf terrestrial, shf satellite or cable TV bandwidth as one analogue service. This is achieved by time-division-multiplexing several MPEG compressed signals together. Also, for terrestrial transmissions, greater re-use of uhf channels is possible because a greater level of co-channel interference can be tolerated before picture or sound impairments become apparent.

One disadvantage of digital television systems is that more equipment is necessary to produce the signal for transmission than is required for analogue television, and that this equipment is much more complex and requires complex software to control it. Another disadvantage is that viewers require either replacement television sets or adapter boxes in order to receive the programmes.

2.3.13.1 Transmission of Digital Television Signals

The MPEG compression of digital television signals is described in another chapter. This section describes the additional signal processing and coding of the MPEG coded Digital Video Broadcasting (DVB) multiplex Transport Stream signal that is necessary for broadcast to viewers in the UK, and other countries where European DVB transmission standards are used. The North American ATSC transmission system is covered in a separate chapter. Three DVB transmission systems are considered here: Digital Terrestrial Television (DTT) or DVB-T; Digital Satellite Television (DSAT) or DVB-S; and Digital Cable Television (DCable) or DVB-C. The block dia-

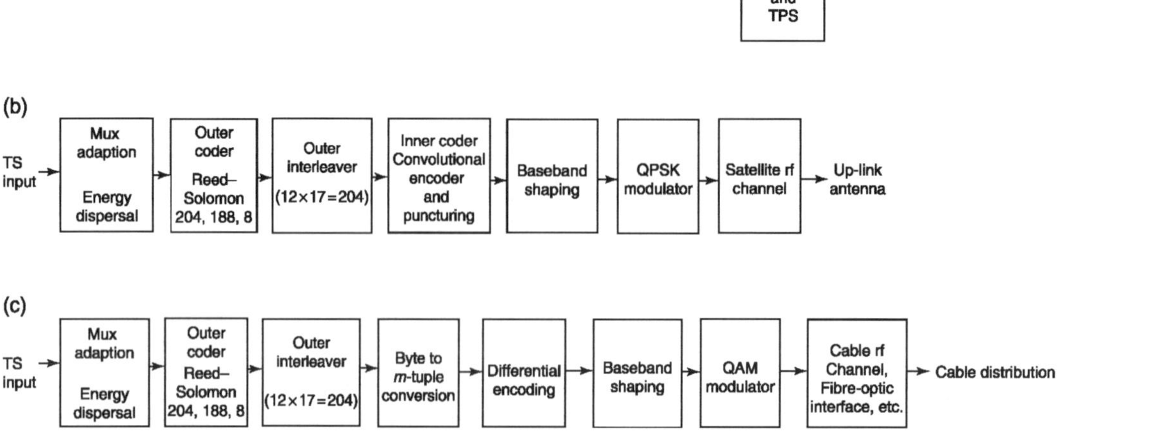

Figure 2.3.19 (a) Digital Terrestrial Television (DTT) transmission system. (b) Digital Terrestrial Television (DSAT) transmission system. (c) Digital Terrestrial Television (DCable) transmission system.

grams for these systems are shown in Figure 2.3.19(a, b and c), respectively.

Although digital signals can be inherently more robust than analogue signals, very complex processing of the MPEG multiplex signal is still necessary to reduce received bit errors to acceptable levels. If not corrected, such errors result in unwanted effects such as 'blocking', where the picture momentarily breaks up into squares.

DTT signals are particularly vulnerable to errors. These are introduced into the received broadcast signal by the addition of interfering signals such as random electrical noise, other co-channel transmissions and also by multipath reception, where the wanted signal is delayed by reflection from objects such as hills and high buildings.

As can be seen by comparing Figure 2.3.19(a, b and c), the first three processes in each system are common to all three systems and the first four are common to DTT and DSAT. All the following processes are designed to produce as rugged a signal as necessary; however, there have to be some compromises if the required data payload is to be accommodated within the allocated rf bandwidth. These processes will be described first, followed by separate descriptions for the remaining processes for each system.

It should be noted that all, or most, of the processes described for all three systems are usually carried out by Digital Signal Processors (DSPs) controlled by microprocessors running special software. Diagrams are for explanation only, and do not necessarily represent the actual hardware implementation method. In practice, all the signal processing described will usually be carried out in a DTT, DSAT or DCable modulator unit.

2.3.13.2 DVB Transmission: The First Three Processes

The input to the transmission system will usually be in the form of a DVB multiplex transport stream signal made up of several MPEG coded video and audio streams, encrypted or non-encrypted, and Service Information (SI) data such as: information about the programmes being transmitted; tables associating the correct audio data stream with the video; and timing information. This is covered in detail in the chapter on DVB Standards. The interface to the transmission system input is usually Asynchronous Serial Interface (ASI), where the Transport Stream (TS) data rate is typically 18–26 Mbit/s. It consists of 188- or 204-byte packets, each of which starts with a synchronising byte. 'Stuffing' bits are added to give the standard ASI rate of 270 Mbit/s. Note that unlike 270 Mbit/s Serial Digital Interface (SDI) signals used for uncompressed digital video, ASI signals are polarity sensitive and may not be inverted. The stuffing bits are only used for the interface and are removed prior to signal processing.

Forward Error Correction (FEC) is a technique used to correct a proportion of data errors introduced by unidirectional transmission paths, or by recording systems. Most telecommunication links are bidirectional and point-to-point, so although protocols for identifying errored data packets at the receiver are still required, an automatic request can then be sent back to the sending terminal equipment for such data packets to be repeated. Clearly, such protocol systems cannot be used for digital broadcast transmission, which provide unidirectional point-to-multipoint paths. Error correction systems, including FEC, require an increase in data rate in order to carry the extra information needed to identify errors. The DVB specification provides for two layers of FEC for UK DTT and DSAT, known as outer and inner coding, and one layer for DCable.

2.3.13.2.1 Energy dispersal

Although not an error correction system, energy dispersal is the first process. Scrambling of the bit stream according to a predetermined Pseudo-Random Bit Sequence (PRBS) reduces the chances of repetitive bit patterns causing peaks in the transmitted frequency spectrum and has the effect of creating a more noise-like signal with very little low-frequency energy. This scrambling is achieved by inverting pseudo-randomly selected bits. Also, every eighth synchronising byte is inverted so that the receiver can lock to the signal and re-invert the inverted bits. The remaining sync bytes are not inverted. Re-inversion at the receiver is acheived by adding an identical PRBS to the signal.

2.3.13.2.2 Outer coding

The second process is a form of Forward Error Correction (FEC) known as Reed–Solomon error protection RS(204,188,8). This means that an 'overhead' of 16 bytes is added to every 188-byte DVB Transport Stream packet to make 204-byte packets, and up to 8 random bytes per packet can be corrected. The value of the 16 extra bytes is calculated from the payload data according to a complex algorithm, and this information is processed at the receiver to identify and correct some errored bytes. This process can reduce an error rate of 2×10^{-4} to 1×10^{-11}.

2.3.13.2.3 Outer interleaving

Next, convolutional interleaving is applied. This process delays all bytes, except sync bytes and every 12th byte, by multiples of 17 bytes successively, as follows. For the purposes of explanation only, the input bytes are numbered 0 to 217 starting with a sync byte. Bytes 0, 12 and every 12th byte retain their original position in the bitstream. Then bytes 1, 13 and every 12th byte are delayed by 17 bytes. Next, bytes 2, 14 and every 12th byte are delayed by 2×17 bytes, and so on until bytes 11, 23, etc. are delayed by 11×17 bytes. This process results in the bytes being scattered and interleaved in a predictable way so that they are never close to their original neighbours. The process is synchronised with the packet structure so that sync bytes retain their original position. This process continues because 12×17 bytes = 204 bytes, which is the number of bytes in each packet after the addition of the 16 Reed–Solomon error protection bytes. Table 2.3.4 illustrates how 16 bytes at the end of one packet and the beginning of the next have been scattered. If impulsive interference causes several consecutive

Table 2.3.4 Convolutional Interleaving of bytes. Sync. bytes are shown in bold type.

Original position	200	201	202	203	**204**	205	206	207	208	209	210	211	212	213	214	215	**216**	217
Interleaved positon	136	57	170	91	**204**	125	46	159	80	193	114	35	148	69	182	103	**216**	137

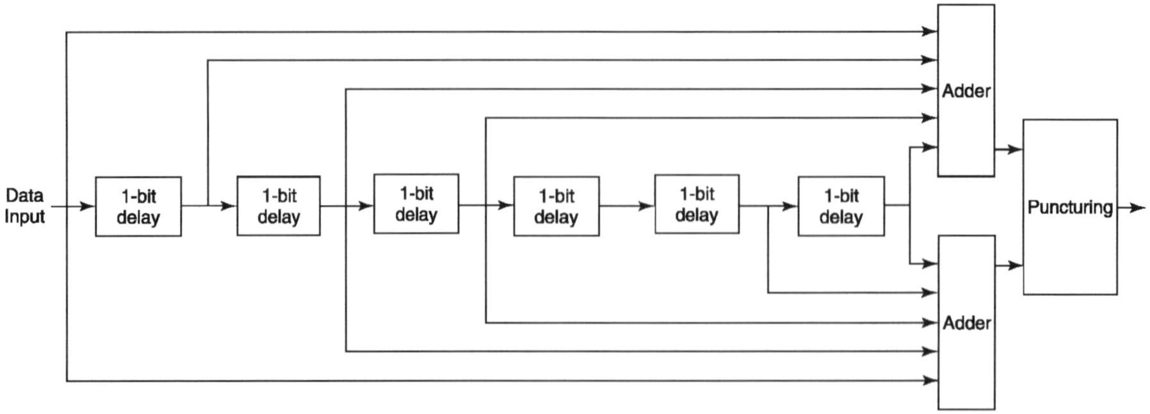

Figure 2.3.20 Convolutional encoder: conceptual block diagram.

bytes to be corrupted, then when the bytes are restored to their original positions at the receiver, the errored bytes will be widely distributed. This makes it more likely that the Reed–Solomon process will correct the shifted bytes than if they remained next to each other.

2.3.13.3 Digital Terrestrial Television (DTT) and Digital Satellite Television (DSAT) Transmission

2.3.13.3.1 Inner coding

The outer coding processes are insufficient to correct enough errors received on terrestrial and satellite television transmissions to ensure acceptable pictures and sound in all cases, so further inner coding error correction techniques are required (see Figure 2.3.19(a, b)).

The next process is known as convolutional encoding and is intended to mitigate the effects of a low signal-to-noise ratio of the received rf signal. This coding system treats the signal as a continuous bitstream rather than as bytes and packets. The theoretical block diagram of the convolutional coder is shown in Figure 2.3.20. The two adders will produce different outputs due to the different taps used. The decoder in the receiver can calculate the coder input bits from the combined X and Y coder outputs, and because there is redundancy,† a proportion of errors can be corrected. The X and Y outputs from the coder are combined into what would be a bitstream with twice the bit rate. Such a two-for-one system would produce a code rate of 1/2 (i.e. 2 output bits for each input bit). This system could provide a good error correction rate, but such a high bit rate could not be accommodated within the available rf bandwidth without reducing the decoding margin, so a proportion of the checking bits are omitted. This proportion can be set according to the requirements of the system, and is a trade-off between payload bit rate and signal ruggedness. For example, a code rate of 2/3 means that only 2 out of every 3 bits of a 1/2 code are sent. These are known as 'punctured' codes, and the pattern of omitted bits is the 'puncturing vector'. Information about the code rate used is added to the Transmission Parameter Signalling (TPS) (see Section 2.3.13.4.7). Other available code rates are 3/4, 5/6 and 7/8. In the receiver, the convolutional encoding is decoded and errors corrected by a 'Viterbi' decoder. The Viterbi decoder makes use of 'trellis diagram' theory to calculate error corrections.

2.3.13.4 Digital Terrestrial Television (DTT) Transmission

EBU standards document EN 300 744[1] provides the specifications for DVB framing structure, channel coding and modulation for digital terrestrial television.

A brief explanation of the modulation systems used is offered here to explain the purpose of the inner interleaving processes that follow the inner coding. As will be explained in more detail later, the modulating bit stream is demultiplexed into 1512 or 6048 low-bit-rate streams, each of which modulates a separate carrier, depending on the modulation mode to be used. The 1512 or 6048 data carriers, together with 176 pilot carriers and 17 signalling carriers, are evenly spaced across the 8 MHz channel in the uhf band. Each low-bit-rate stream feeds a quadrature modulator which produces a Quaternary Phase Shift Keyed (QPSK), 16- or 64-Quadrature Amplitude Modulation (16-QAM or 64-QAM) modulated output, depending on the mode to be used. For QPSK, 16-QAM and 64-QAM, the symbol rate‡ is a half, a quarter and a sixth of the bit rate respectively, and each symbol carries 2, 4 and 6 bits respectively.

2.3.13.4.1 Bit interleaver

Selective fading can cause errors in the bits carried by one or more adjacent carriers. One symbol error can result in any or all of the bits represented by that symbol being errored. So to avoid such selective fades causing adjacent

†Redundancy: the inclusion of information (data) which, in some way, conveys the same information as other data already present in the signal.

‡The symbol rate is the rate at which the modulated carrier changes state. The 'state' of the modulated carrier in this case is its amplitude and/or phase relative to the amplitude and phase of the unmodulated carrier input.

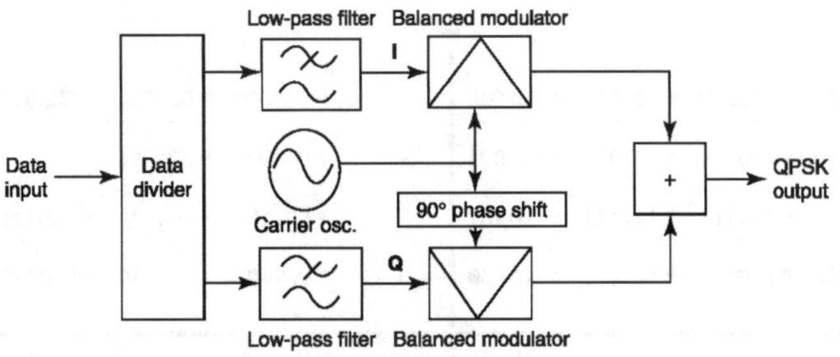

Figure 2.3.21 Quadrature modulator. The I and Q signals are each half the input rate.

bit errors, bit interleaving is used to spread adjacent bits to non-adjacent carriers. The signal from the convolutional encoder puncturing is demultiplexed into two, four or six parallel streams, depending on whether QPSK, 16-QAM or 64-QAM is being used, and each stream is processed separately in blocks of 126 bits so that adjacent bits are shifted by at least 21 bits away from their original positions. This results in the shifted bits modulating different carriers. After bit interleaving, the two, four or six streams are combined into 2-, 4- or 6-bit symbols so that each symbol contains 1 bit from each stream.

2.3.13.4.2 Symbol interleaver

The symbol interleaver maps the symbols from the bit interleaver on to the 1512 or 6048 carriers, depending on whether 2k or 8k carrier mode is to be used. The mapping is carried out in 12 (for 2k mode) or 48 (for 8k mode) groups of 126 bits, and the algorithm ensures that adjacent symbols are allocated to widely separated carriers.

2.3.13.4.3 Modulation systems and bandwidth

In all modulation systems, one or more parameters of a carrier signal are changed in sympathy with a lower frequency modulating signal which carries the information that is to be conveyed. The parameters of the carrier that can be changed are amplitude, frequency and phase. In the case of digital modulation systems, the modulating signal will have two or more discrete states rather than the continuously changing amplitude and/or frequency/phase of analogue systems. In both cases, the modulation process produces sidebands which are a band of frequencies either side of the carrier. For complex modulation waveforms with unrestricted bandwidth, these sidebands would extend to infinity in theory. However, the useful width of the sidebands are proportional to the rate of change of the amplitude, frequency or phase of the carrier, so it is necessary to limit these sidebands if efficient use is to be made of the available frequency spectrum. Filtering of the modulating or modulated signal can be used to reduce the bandwidth of the modulated signal, but the more the bandwidth is restricted, the more distortion is introduced into the demodulated signal, so there is a need for compromise and to make the best possible use of the bandwidth available.

2.3.13.4.4 The quadrature modulator

The modulating signal is separated into two streams of, for example, alternate bits, and each bit stream is used to modulate the same carrier in separate balanced modulators, but with the phase of the carrier input to one modulator delayed by 90°. The signal modulating the undelayed carrier is known as the 'I' (In-phase) signal, and that modulating the delayed carrier as the 'Q' (Quadrature) signal. Each modulator produces a signal at the carrier frequency, but with the phase switched 180° by the modulating signal. The signals from the two modulators are then combined to produce a QPSK signal (see Figure 2.3.21). This process is very similar to the quadrature modulation used to produce PAL and NTSC chrominance signals. The QPSK signal can have any one of four different phases relative to the carrier input signal (see Figure 2.3.22). The original modulating signal can be recovered by feeding the signal into two balanced demodulators. The carrier inputs to the demodulators are fed with carriers of the same frequency and phase as those fed to the balance modulators. In practice, these modulation and demodulation processes can be implemented in digital

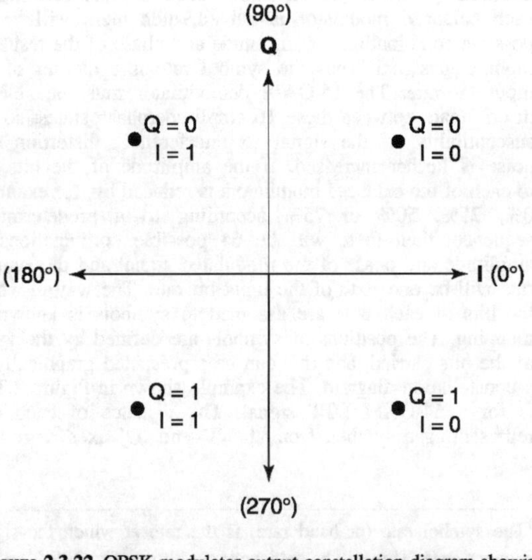

Figure 2.3.22 QPSK modulator output constellation diagram showing the phases of the four symbol states relative to the I and Q axes.

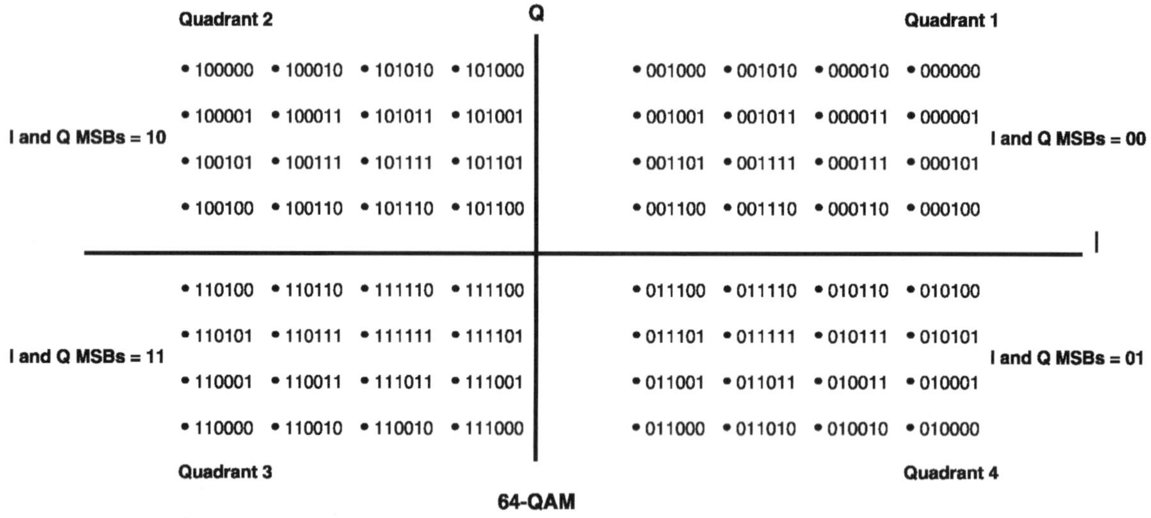

Figure 2.3.23 64-QAM DTT constellation diagram, where 6 bits are carried by each symbol.

signal processing chips. The advantage of this system over simple amplitude, frequency or phase modulation is that the symbol rate[§] is halved, thereby reducing the bandwidth of the modulated signal. A disadvantage is that the signal is more susceptible to the effects of interference, distortion and noise on the transmission path than a two-state signal.

2.3.13.4.5 16- and 64-Quadrature Amplitude Modulation (QAM)

The concept of quadrature modulation can be extended to further reduce the symbol rate by introducing amplitude modulation in addition to the phase modulation. If the same QPSK modulation process is applied, but the amplitude of alternate (or other predetermined sequence of) bits fed to each balanced modulator is halved, then there will be 16 possible combinations of amplitude and phase of the resulting modulated signal. Thus the symbol rate is a quarter of the input bit rate. The 16-QAM demodulator must be able to discriminate between these 16 amplitude/phase states, so the susceptibility of the signal to interference, distortion and noise is further increased. If the amplitude of the bits fed to each of the balanced modulators is reduced by, for example, 0%, 25%, 50% or 75% according to a predetermined sequence, then there will be 64 possible combinations of amplitude and phase of the modulated signal and the symbol rate will be one-sixth of the input bit rate. The way in which the bits of each byte are assigned to symbols is known as mapping. The positions of symbols are defined by the value of the bits carried, and this can be represented graphically in a constellation diagram. The example shown in Figure 2.3.23 is for a 64-QAM DTT signal. The distance of each dot, representing a symbol, from the 'I' and 'Q' axes represents amplitude, and rotation about the origin represents phase angle. Amplitude spacings can be either uniform or non-uniform. It can be seen in the example that only one bit changes between any two horizontally or vertically adjacent symbols, so a small demodulating error will only cause one bit error. This is known as Gray mapping. The closer together that the adjacent amplitudes and phases are in the modulated signal, the more likely it is that noise and distortion on the transmission path will cause the demodulator to produce an error, so there has to be a compromise between bit rate, noise and distortion immunity, and bandwidth. UK DTT transmission systems use both 16- and 64-QAM at the time of writing.

2.3.13.4.6 Coded Orthogonal Frequency Division Multiplex (COFDM)

Two important requirements for a Digital Terrestrial Television (DTT) transmission signal are that it needs to be robust and that it should cause minimal co-channel interference to other DTT or analogue TV transmissions. Both of these requirements are due to the need for the DTT signals to coexist in a radio-frequency environment already crowded with analogue TV transmissions. The DTT signal also needs to be robust because it not only needs to be decodable in the presence of distant co-channel transmissions from other TV services, but it also needs to be decodable in the presence of delayed versions of itself reflected off tall buildings and the terrain. A signal meeting this latter requirement will also be suitable for use in Single Frequency Networks (SFNs), where two or more synchronous transmitters carrying identical signals, and with overlapping service areas, are operating on the same uhf channel. This is because, in both cases, two or more identical signals of different amplitudes, and with different relative delays, arrive at receiving aerials.

A single 16- or 64-QAM signal would not meet the above requirements. The reception of such a signal would be liable to

[§]The symbol rate (or baud rate) is the rate at which the signal changes state. This will be half the input bit rate for QPSK, so that each symbol carries 2 bits of input data.

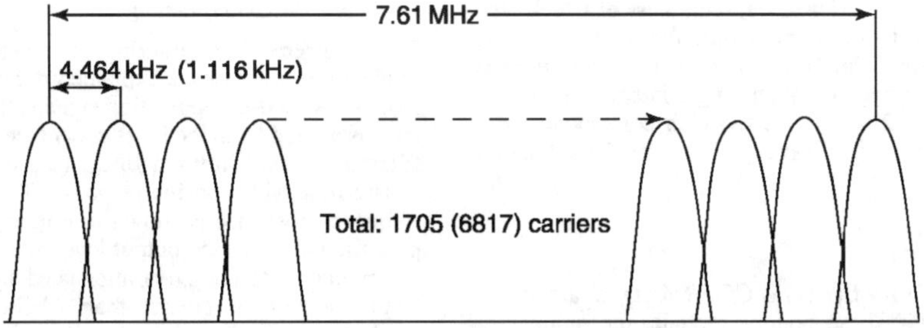

Figure 2.3.24 2k carrier COFDM signal spectrum carrier spacing and number of carriers (8k carrier COFDM signal carrier spacing and number of carriers are shown in parentheses).

suffer from intersymbol interference.¶ Similarly, frequency selective fading** could knock out parts of the signal bandwidth where a delayed version of the wanted signal was received due to reflections, or from adjacent transmitters in an SFN. A single QAM signal would also have strong sideband components at multiples of the symbol rate which could cause very noticeable interference to co-channel analogue TV services.

In the Coded Orthogonal Frequency Division Multiplex (COFDM) system, the high-bit-rate digital modulating signal is divided up into a large number of lower-bit-rate parallel bit streams. Each low-bit-rate stream is the input to a separate QPSK, 16-QAM or 64-QAM modulator. Each Quadrature modulator is fed with a different equally spaced carrier frequency, so that each band of frequencies occupies a different slot in the overall rf band occupied by the multiplex. All the modulator outputs are combined and added to the pilot carriers and TPS described below, to form the transmitted multiplex signal. In the UK, each multiplex occupies an 8 MHz uhf channel (actually a 7.61 MHz band) in the same band-plan as analogue TV channels. Each 8 MHz channel has 1705 carriers spaced at 4.464 kHz intervals. For convenience, the number of carriers is referred to as 2k carriers. The specification also provides for an alternative standard of 8k carriers (actually 6817 carriers with a spacing of 1.116 kHz). The spectrum of a COFDM signal is shown in Figure 2.3.24.

¶Intersymbol interference can occur when adjacent symbols arrive at the receiving antenna at the same time, one symbol being delayed by reflection from a tall building or the terrain. If the direct and delayed signals are of similar amplitudes, demodulation errors occur. Intersymbol interference can also be caused by group delay distortion, which distorts the shape of symbol waveforms so that the duration is increased, and causes them to start to merge with the next symbol.

**Frequency selective fading occurs when a direct signal, and the same signal delayed by reflection from any object, arrive at a receiving antenna with such a carrier phase relationship that partial cancellation occurs at particular frequency(s) within the signal bandwidth, causing a much reduced signal amplitude at these frequencies.

2.3.13.4.7 Pilot, and Transmission Parameter Signalling (TPS), carriers

In the 2k carrier system, only 1512 carriers are used for the DTT data. Of the remaining 193 carriers, 176 are used at pilot carriers and 17 for Transmission Parameter Signalling (TPS). The 176 pilot carriers are at a higher amplitude than the data carriers, and are used for fine frequency control, and for the recovery of timing information used in the receiver demodulators. Forty-five of the pilot carriers are always at the same positions (frequencies) in the multiplex, but the remaining 131 are inserted every 12th data carrier, and move up three carriers at the symbol rate. Some will therefore coincide with a fixed pilot carrier every 12th symbol and the pattern repeats every fourth symbol. The 17 TPS carriers give the demodulators information about whether 2k or 8k carriers are used, whether QPSK, 16-QAM or 64-QAM is used, and the length of the guard interval (see below). The TPS always uses Bi-polar Phase-Shift Keying (BPSK), which only has two states: 0° and 180° carrier phases. The same data is carried by all 17 TPS carriers to ensure that the information can be recovered in the presence of frequency selective fading.

2.3.13.4.8 The guard interval

As mentioned above, one of the qualities of the DTT signal is that it should be decodable in the presence of one or more attenuated and delayed versions of itself, as can occur if the transmitted signal is reflected off a tall building or the terrain (multipath), or in an SFN (see Section 2.3.13.4.6). As well as destructive and constructive interference between the wanted and unwanted uhf carriers, there would be a problem if the symbols in the wanted signal were to coincide in time with adjacent symbols in the unwanted delayed signal. If such a delayed signal were to be of sufficient amplitude, the demodulator could produce an errored 'bit'. However, the use of many low symbol rate carriers in the multiplex instead of one high symbol rate carrier results in the symbol duration being long enough so that the signal delay would need to be longer than would occur in the service area. In other words, the amplitude of a signal delayed by one or more symbol durations will be too small to cause an error in the demodulated signal. As long as the demodulator does not sample the symbol until a specified period after the start of the symbol, to allow for a small overlap with the previous symbol from the delayed signal, we can be sure that a bit error due to the delayed signal will not occur. This period, from the start of the symbol to the point where sampling for demodulation can commence, is known as the guard interval. In the 2k carrier mode, regardless of which type of modulation is used, the symbol duration will be 224 μs (896 μs for 8k carriers). The guard interval commonly used is 1/32 of this symbol period, although longer intervals up to 1/4 are available; 1/32 of 224 μs is 7 μs, which corresponds to a signal path length of 2.1 km, leaving 217 μs for demodulation of the signal.

2.3.13.4.9 DTT options

The options available for the DVB COFDM signal are specified in ETS 300 744. These options include: the number of carriers (2k or 8k); the guard interval (1/32, 1/16, 1/8 or 1/4); the modulation system (QPSK, 16-QAM or 64-QAM); and the puncturing vector of the inner coding (1/2, 2/3, 3/4, 5/6 or 7/8). 8k carriers are more suitable for SFNs because of the longer symbol duration (896 μs), but require more processing power in the receiver demodulators than for 2k carriers. Longer guard intervals give better protection against multipath, but reduce the period available for accurate demodulation of each symbol. QPSK provides a more robust signal than 16-QAM or 64-QAM, but the payload bit rate is less. The option information is carried by the 17 Transmission Parameter Signalling carriers.

2.3.13.4.10 DTT transmitters

The combined signal, consisting of all 1705 or 6817 carriers, is up-converted from a fixed intermediate frequency (IF) signal to the final broadcast uhf frequency in the transmitter drive, and then amplified to provide the transmitter output to the combiner and antenna system. The signal path through the transmitter needs to be as linear as possible to minimise distortion of the signal. Any amplitude/amplitude, amplitude/frequency or group delay distortion, or the addition of noise or extraneous signals within the transmitted bandwidth, will reduce the decoding margin of the signal. The decoding margin is a measure of the quality of the signal.

2.3.13.4.11 Co-channel interference

Because of the multiple carriers, and the various scrambling processes used for error protection, the transmitted signal will have its energy spread fairly evenly across its band, and is said to be noise-like. Thus this signal will be much less likely to cause visible interference to analogue TV pictures than would another analogue TV signal, which has strong energy components at multiples of the TV line frequency. However, because of the need in the UK to use many extra channels in the existing uhf band-plan to implement DTT alongside analogue transmissions, some extra co-channel interference was inevitable. Some visible interference in the form of patterning can still occur to analogue transmissions if multiples of the DTT symbol rate coincide with multiples of the analogue TV line frequency, so small uhf offsets can be applied, where necessary, to either the analogue or DTT transmissions to prevent this.

2.3.13.5 Digital Satellite Television (DSAT) Transmission

EBU standards document EN 300 421[2] provides the specifications for DVB framing structure, channel coding and modulation for 11/12 GHz satellite services.

2.3.13.5.1 Baseband shaping

The requirement for limiting the sidebands produced by the modulation processes is explained in Section 2.3.13.4.3. In the case of the DSAT modulating signals, the baseband 'I' and 'Q' signals are fed from the inner coder (see Section 2.3.13.3.1) and filtered by band limiting filters (see Figure 2.3.19(b)) with a square-root raised cosine response. The calculated amplitude/frequency response is shown in Figure 2.3.25. These are low-pass filters, where the output level at a frequency equal to half the symbol rate has been attenuated by 3 dB, and an out-of-band rejection of greater than 40 dB. Pass band amplitude ripple has to be less than 0.4 dB, and group delay ripple not more than ±0.07 times twice the symbol period, up to half the symbol rate.

2.3.13.5.2 The QPSK modulator

The modulation process used is as described in Sections 2.3.13.4.3 and 2.3.13.4.4.

2.3.13.5.3 Satellite up-link interface

The output of the QPSK modulator is centred on a fixed intermediate frequency (IF) and feeds the satellite rf up-link equipment, where it is up-converted to the up-link frequency, amplified, combined with other signals occupying other rf channels, and fed to the transmitting antenna.

2.3.13.6 Digital Cable Television (DCable) Transmission

EBU standards document EN 300 429[3] provides the specifications for DVB framing structure, channel coding and modulation for cable systems.

The signal sent to the cable is a Quadrature Amplitude Modulated (QAM) signal. The DVB standard specifies either 16-, 32-, 64-, 128- or 256-QAM. Receiver demodulators must support at least 16-, 32- and 64-QAM.

2.3.13.6.1 Byte to m-tuple conversion

The signal from the interleaver (see Section 2.3.13.2.3) is mapped onto the symbols. The number of bits in each symbol depends on which QAM system is to be used, as explained in Section 2.3.13.4.5. The group of bits carried by each symbol is known as an '*m*-tuple' where '*m*' is the number of bits per symbol. This is done sequentially starting with the most significant bit of a byte, which is mapped to the first bit of a symbol. Thus for 64-QAM, the 6 most significant bits of the first byte considered will be mapped in sequence, to the first symbol. The remaining 2 bits of the first byte and the 4 most significant bits of the second byte are then mapped to the second symbol in sequence, and so on. This is illustrated in Table 2.3.5.

2.3.13.6.2 Differential coding

Differential coding is applied to the 2 most significant bits only of each symbol before the final mapping of the symbols to the 'I' and 'Q' inputs to the quadrature amplitude modulator. This results in the quadrant of the constellation diagram where any symbol is positioned, being determined by the value of the 2 most significant bits. This is illustrated in Table 2.3.6.

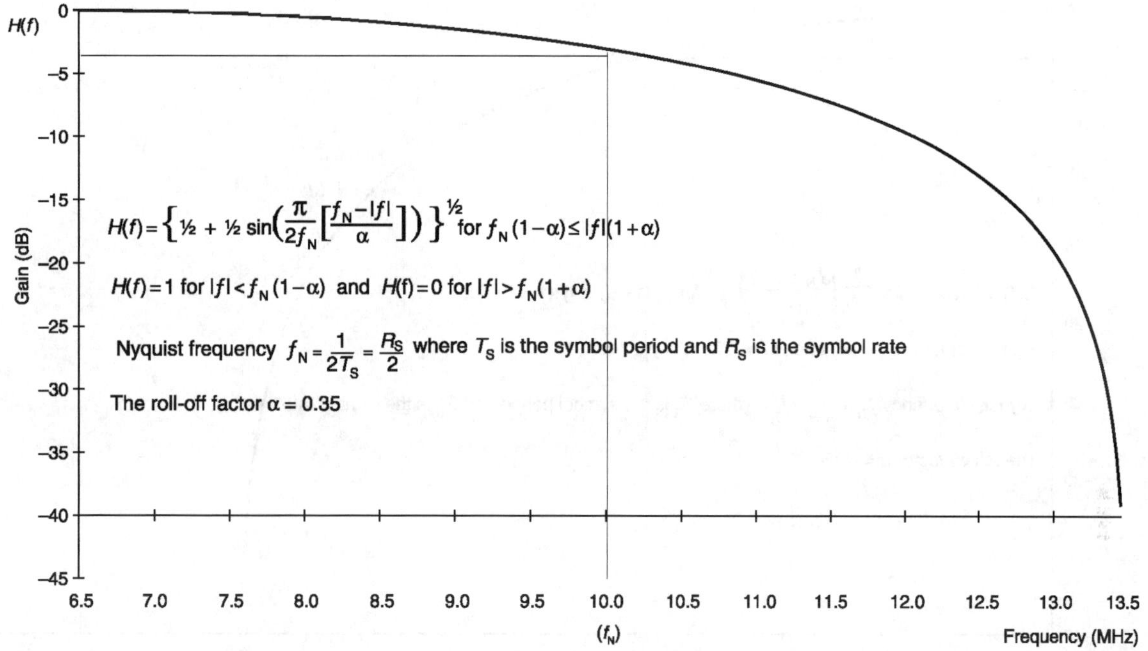

Figure 2.3.25 DSAT baseband low-pass filter: calculated amplitude/frequency response for $f_N = 10$ MHz.

Table 2.3.5 Mapping of bytes to symbols. Byte boundaries shown as broken lines. The bits carried by each symbol is known as an m-tuple.

	Byte 0	Byte 1	Byte 2	3		
Bytes (8 bits)	7 6 5 4 3 2 1 0	7 6 5 4 3 2 1 0	7 6 5 4 3 2 1 0	7		
32QAM	Symbol 0	Symbol 1	Symbol 2	Symbol 3	Symbol 4	
Symbols (5 bits)	4 3 2 1 0	4 3 2 1 0	4 3 2 1 0	4 3 2 1 0	4 3 2 1	0
64QAM	Symbol 0	Symbol 1	Symbol 2	Symbol 3		
Symbols (6 bits)	5 4 3 2 1 0	5 4 3 2 1 0	5 4 3 2 1 0	5 4 3 2 1 0		

Table 2.3.6 Differential encoding: Rotation of constellation points from quadrant 1 to quadrants 2, 3, and 4. The two MSBs have the values shown. All other bits are rotated as shown.

Quadrant	*2 MSBs*	*LSBs Rotation*
1	00	–
2	10	+90°
3	11	+180°
4	01	+270°

2.3.13.6.3 Baseband shaping

The requirement for limiting the sidebands produced by the modulation processes is explained in Section 2.3.13.4.3. In the case of the DCable modulating signals, the baseband 'I' and 'Q' signals are filtered by band limiting filters with a square-root raised cosine response. The calculated amplitude/frequency response is shown in Figure 2.3.26. These are low-pass filters, where the output level at a frequency equal to half the symbol rate has been attenuated by 3 dB, and an out-of-band rejection of greater than 40 dB. Pass band amplitude ripple has to be less than 0.4 dB, and group delay ripple not more than $^1/_{10}$ of twice the symbol period, up to half the symbol rate.

2.3.13.6.4 16-, 32-, 64-, 128- and 256-Quadrature Amplitude Modulation (QAM)

The concept of quadrature modulation described in Sections 2.3.13.4.3 and 2.3.13.4.4 can be extended to further reduce the symbol rate by introducing amplitude modulation in addition to the phase modulation; 16- and 64-QAM are described in Section 2.3.13.4.5. However, for DCable systems, 32-, 128- and 256-, as well as 16- and 64-QAM, are specified. The bit mapping for 64-QAM DCable is shown in Figure 2.3.27. The upper practical limit for QAM is 256 states, where each symbol carries 8 bits and the symbol rate is $^1/_8$ of the bit rate.

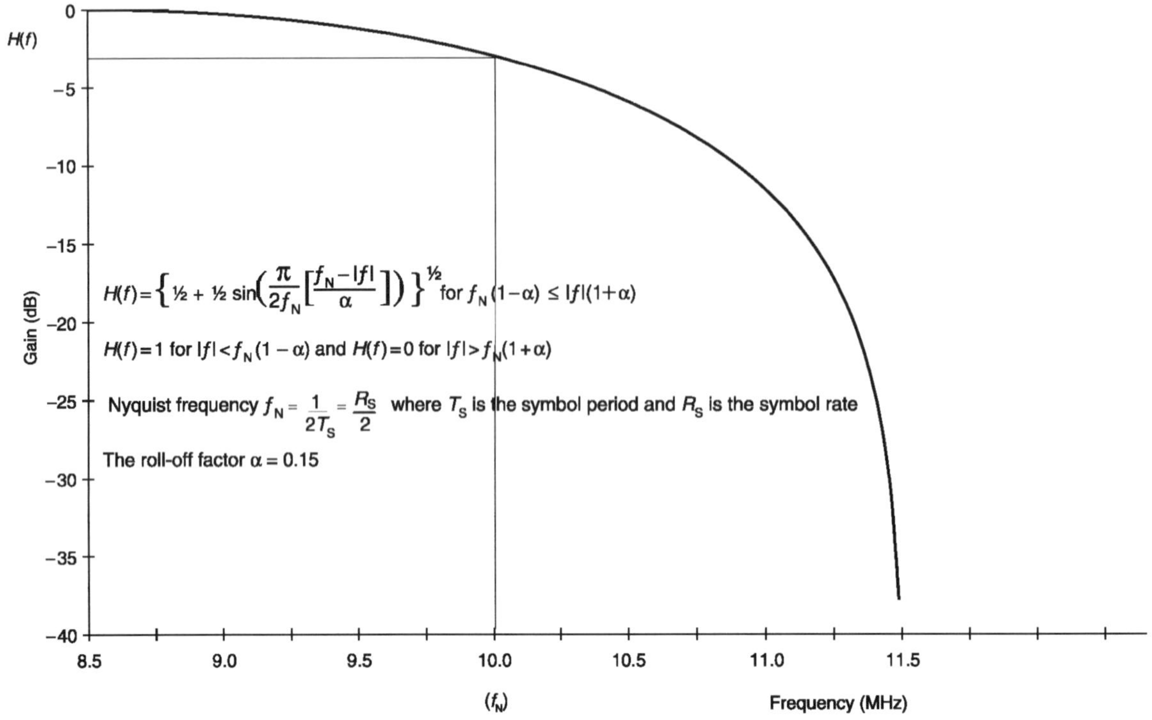

Figure 2.3.26 DCable baseband low-pass filter: calculated amplitude/frequency response for $f_N = 10$ MHz.

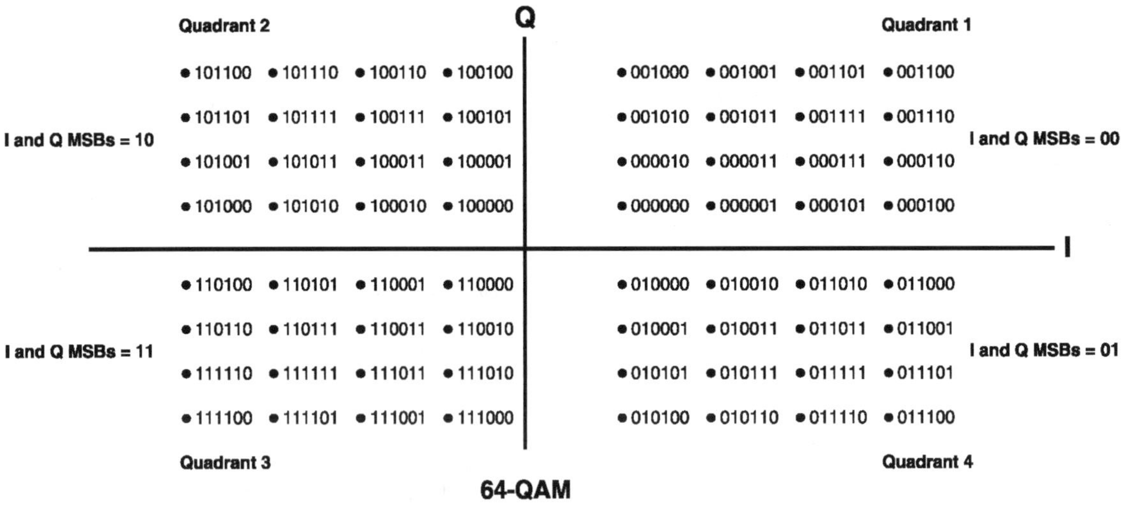

Figure 2.3.27 64-QAM DCable constellation diagram showing the 6-tuples of bits carried by each symbol.

2.3.13.6.5 Distribution cable interface

The signal from the QAM modulator is centred on a fixed intermediate frequency (IF). This is connected to the cable head-end rf channel equipment or fibre-optic interfaces, and then to the distribution cable system.

References

1. EN 300 744 V1.4.1, Digital Video Broadcasting (DVB); Framing Structure, Channel Coding and Modulation for Digital Television (2001-01). © European Broadcasting Union 2001 and European Telecommunications Standards Institute 2001.
2. EN 300 421 V1.1.2, Digital Video Broadcasting (DVB); Framing Structure, Channel Coding and Modulation for 11/12 GHz Satellite Services (1997-08). © European Broadcasting Union 1997 and European Telecommunications Standards Institute 1997.
3. EN 300 429 V1.2.1, Digital Video Broadcasting (DVB); Framing Structure, Channel Coding and Modulation for Cable Systems (1998–04). © European Broadcasting Union 1998 and European Telecommunications Standards Institute 1998.

Acknowledgements

The original chapter by R.S. Roberts, covering analogue television standards and the broadcast spectrum, derives from a series of lectures given by the author during courses on Television Engineering, organised for the Royal Television Society (RTS). The RTS course lectures were arranged into a book, *Television Engineering*, and published by Pentech Press. We acknowledge permission given by Pentech Press for the use of some diagrams and text from the RTS book.

Digital Television Transmission: Kit Norfolk, BBC Centre for Broadcasting Skills Training. © BBC 1997.

C K P Clarke
Senior Engineer, BBC Research Department

Revised by
Chris Dale MIIE
Previously BBC Broadcast Networks Engineer

2.4 Colour Encoding and Decoding Systems

2.4.1 Introduction

The signals produced by colour television cameras are in the form of red, green and blue (RGB) colour signals. Cathode ray tube colour displays also operate with RGB. Although RGB signals can be of high quality, the three channels required represent an inefficient use of bandwidth and circuitry. Furthermore, impairments can arise if the three channels are not accurately matched. More efficient and rugged methods of colour encoding are therefore required at other points in the signal chain, such as for studio processing, recording, distribution and broadcast emission.

Several additional methods of colour encoding have been developed. Some produce composite signals, such as PAL, NTSC and SECAM, which retain compatibility with monochrome receivers. Such systems have been in use for many years. Now, however, component coding methods such as 4:2:2 digital components have been introduced. These methods sacrifice direct compatibility for advantages such as improved signal processing and picture quality.

Compatibility between systems is increasingly a problem in television broadcasting, especially for the new digital media: satellites, cable and new terrestrial services. This is particularly so in the case of the coding systems now in use. Here, all the main systems are presented in common terms, making the similarities and differences more apparent.

2.4.2 Colour Signal Relationships

All the colour systems described here are based on encoding of a luminance signal and two colour-difference signals, instead of the red, green and blue colour separation signals.

2.4.2.1 Gamma

The methods use signals pre-corrected for the assumed gamma of the crt display. Thus, the gamma-corrected colour separation signals are denoted R′, G′ and B′, with the prime (′) signifying that pre-correction has been applied. PAL and SECAM systems are generally matched to a display gamma of 2.8, while a gamma of 2.2 is assumed for the NTSC system. The 4:2:2 digital components system uses signals appropriate for the associated composite colour standard, hence generally 2.2 for the 525-line, 60 fields/second scanning standard and 2.8 for the 625-line, 50 fields/second standard.

2.4.2.2 Luminance and colour-difference equations

The use of luminance Y and the two colour-difference signals B − Y and R − Y provides improved compatibility with monochrome systems by extracting the common luminance content of the colour separation signals. Also, the bandwidth of the colour-difference signals can be reduced, resulting in improved coding efficiency. In all the encoding systems described here, the luminance signal is defined as:

$$Y' = 0.299R' + 0.587G' + 0.114B' \qquad (2.4.1)$$

based on the NTSC signal primary colour chromaticities.[1] The colour-difference signal relationships derived from this are:

$$B' - Y' = -0.299R' - 0.589G' + 0.886B' \qquad (2.4.2)$$

and

$$R' - Y' = 0.701R' - 0.587G' - 0.114B' \qquad (2.4.3)$$

At the display, the colour separation signals can be regained by applying the inverse relationships:

$$R' = Y' + (R' - Y') \quad (2.4.4)$$

$$G' = Y' - 0.194(B' - Y') - 0.509(R' - Y') \quad (2.4.5)$$

$$B' = Y' + (B' - Y') \quad (2.4.6)$$

It should be noted that the analysis characteristics used in cameras for $^{625}/_{50}$ colour signals are based on different primary colour chromaticities from those chosen for the NTSC system (on which equation (2.4.1) is based). These match the chromaticities of present-day display phosphors more accurately. Retention of the coefficients in equation (2.4.1) results only in slight grey-scale inaccuracies in the compatible monochrome picture.[2] In the NTSC system, it is assumed that correction is applied in the display circuitry of the receiver.

2.4.2.3 Constant luminance coding

In principle, colour coding using a luminance signal and two colour-difference signals produces a system in which distortion or perturbation of the colour-difference signals leaves the displayed luminance unaffected. Systems that maintain this principle are termed *constant luminance*.

The luminance signal of equation (2.4.1) is synthesised by matrixing the gamma-corrected colour separation signals, and consequently the luminance produced is not equivalent to that of a monochrome system. As a result, the compatibility of the colour signal reproduced on monochrome receivers is adversely affected, so making areas containing highly saturated colours darker than they should be. In modulated systems, however, a further factor influencing monochrome compatibility is the presence of the colour subcarrier. In this case, the effect of the tube brightness non-linearity on the subcarrier signals is to produce an additional average brightness component in highly coloured areas. To a first approximation, this offsets the losses resulting from matrixing non-linear signals.

Although the luminance component of a colour signal is not identical to a monochrome signal, the correct luminance can be reproduced by the colour display. This can be explained by considering that a correction for the inaccurate luminance signal is carried as part of the colour-difference signals. However, when the colour-difference signals are limited to a narrower bandwidth than the luminance, the high-frequency content of the correcting component is lost. This results in a failure of constant luminance on colour transitions. Even so, the presence on colour transitions of subcarrier signals not removed by the notch filter tends to offset this effect, as explained above for the case of the compatible monochrome signal.

A closer equivalent to the monochrome signal could be obtained by matrixing linear colour separation signals to produce luminance. This would avoid the failure of constant luminance on colour transitions and would improve the compatibility of the monochrome signal. Such techniques are considered for high-definition television systems. Nevertheless, because linear signals are much more susceptible to noise, gamma correction or some form of non-linear pre-emphasis is still necessary to obtain satisfactory noise performance over the transmission path.

2.4.3 Composite Colour Systems

Broadcast composite colour signals are based on three systems of colour encoding: PAL, NTSC and SECAM. The three systems have the similarity that each consists of a broad-band luminance signal with the higher frequencies sharing the upper part of the band with chrominance components modulated onto subcarriers. Figure 2.4.1 shows a typical distribution of luminance and chrominance frequency components for a PAL signal.

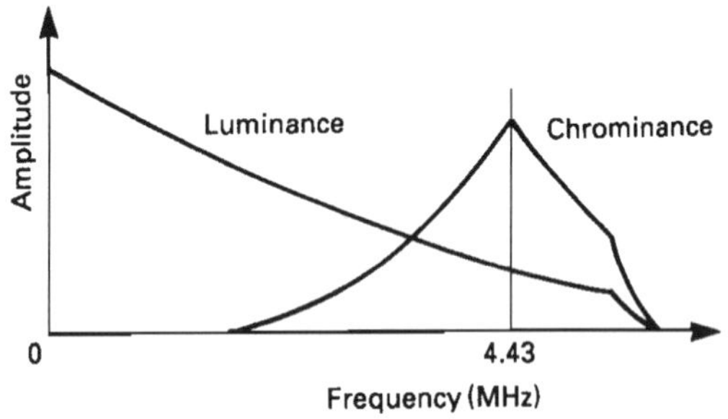

Figure 2.4.1 Positions of the main frequency components in a composite PAL signal.

The composite approach provides a form of signal which is directly compatible with monochrome receivers. Many features of the colour signal have been chosen to optimise the compatibility of the systems with monochrome operation, particularly the visibility of the subcarriers on the monochrome picture.

The following sections describe the principal features of the three systems, highlighting their individual differences. While PAL and NTSC use amplitude modulation for the colour-difference signals, PAL with an additional offset, SECAM uses frequency modulation. Each of the systems is capable of providing high-quality pictures and each has its own strengths and weaknesses. Although a development of the earlier NTSC system, the PAL system is simpler in some respects and is therefore described first.

2.4.3.1 PAL

2.4.3.1.1 Development

The distinguishing feature of the PAL colour system developed by Bruch[3] is that it overcomes the inherent sensitivity of suppressed carrier amplitude modulation to differential phase distortion. This is achieved by encoding the two colour-difference signals, U and V, with the phase of the V subcarrier reversed on alternate television lines, thus leading to the name *Phase Alternation Line*. The PAL system was developed primarily for the 625-line, 50 fields/second scanning standard, first used in Europe.

2.4.3.1.2 Colour subcarrier frequency

The colour subcarrier frequency f_{SC} used for system I PAL signals is 4.43361875 MHz ± 1 Hz. Systems B, D, G and H use the same frequency, but with a wider tolerance of ± 5 Hz.

The subcarrier frequency and the line frequency (f_H) are linked by the relationship:

$$f_{SC} = \left(283\tfrac{3}{4} + \frac{1}{625}\right) f_H = \frac{709{,}379}{2500} f_H \qquad (2.4.7)$$

Therefore the subcarrier phase at a point in the picture is subject to a cycle that repeats every eight field periods (2500 lines). PAL signals in which the specified relationship is not maintained are termed *non-mathematical*. The relationship of equation (2.4.7) was chosen to minimise the subcarrier visibility by providing the maximum phase offset from line to line and from picture to picture, subject to the constraint that the alternate-line phase inversion of the V signal has already introduced a half-line frequency offset between the two subcarriers. Thus, in the high-frequency region of the PAL signal, the main components of the Y, U and V signals are interleaved, as shown in Figure 2.4.2.

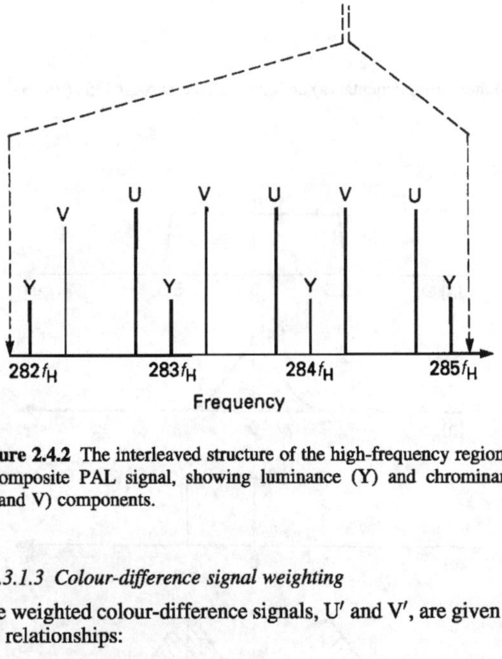

Figure 2.4.2 The interleaved structure of the high-frequency region of a composite PAL signal, showing luminance (Y) and chrominance (U and V) components.

2.4.3.1.3 Colour-difference signal weighting

The weighted colour-difference signals, U′ and V′, are given by the relationships:

$$U' = 0.493(B' - Y') \qquad (2.4.8)$$

and

$$V' = 0.877(R' - Y') \qquad (2.4.9)$$

The weighting factors have been chosen to limit the amplitude of the colour-difference signal excursions outside the black-to-white range to one-third of that range at each end. The inverse relationships required in a decoder are:

$$B' - Y' = 2.028U' \qquad (2.4.10)$$

and

$$R' - Y' = 1.140V' \qquad (2.4.11)$$

2.4.3.1.4 Colour-difference signal filters

The PAL colour-difference signals are filtered with a low-pass characteristic approximating to the Gaussian template shown in Figure 2.4.3 to provide optimum compatibility with monochrome receivers. This wide-band, slow roll-off characteristic is needed to minimise the disturbance visible at the edges of coloured objects because a monochrome receiver includes no further filtering. A narrower, sharper-cut characteristic would emphasise the subcarrier signal at these edges, widening the transitions and introducing ringing.

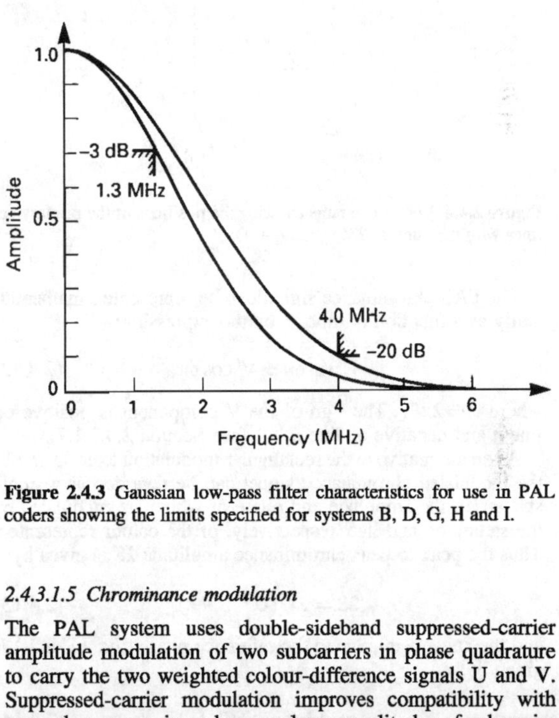

Figure 2.4.3 Gaussian low-pass filter characteristics for use in PAL coders showing the limits specified for systems B, D, G, H and I.

2.4.3.1.5 Chrominance modulation

The PAL system uses double-sideband suppressed-carrier amplitude modulation of two subcarriers in phase quadrature to carry the two weighted colour-difference signals U and V. Suppressed-carrier modulation improves compatibility with monochrome receivers because large amplitudes of subcarrier occur only in areas of highly saturated colour. Because the subcarriers are orthogonal, the U and V signals can be separated perfectly from each other. However, if the modulated chrominance signal is distorted, either through asymmetrical attenuation of the sidebands or by differential phase distortion, the orthogonality is degraded, resulting in crosstalk between the U and V signals.

The alternate-line switching of PAL protects against crosstalk by providing a frequency offset between the U and V subcarriers in addition to the phase offset. Thus, when decoded, any crosstalk components appear modulated onto the alternate-line carrier frequency, in plain coloured areas producing the moving pattern known as *Hanover bars*. This pattern can be suppressed at the decoder by a comb filter averaging equal contributions from switched and unswitched lines.

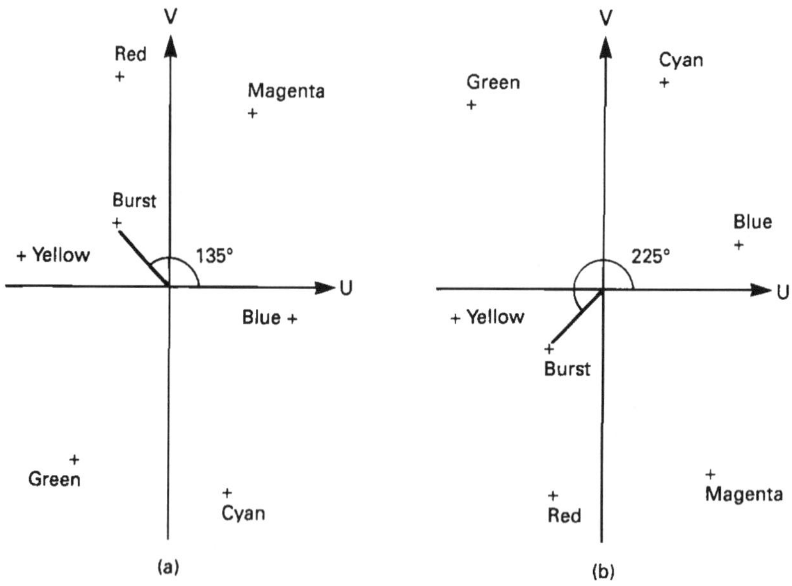

Figure 2.4.4 Vector diagrams showing the positions of the primary colours and their complements: (a) on lines with the burst at 135° (line n); (b) on lines with the burst at 225° (line $n+1$).

The PAL chrominance signal can be represented mathematically as a function of time, t, by the expression:

$$U' \sin \omega t \pm V' \cos \omega t \qquad (2.4.12)$$

where $\omega = 2\pi f_{sc}$. The sign of the V component is positive on line n and negative on line $n+1$ (see Section 2.4.3.1.7).

As an alternative to the rectangular modulation axes, U and V, the modulated chrominance signal can be regarded as a single subcarrier, the amplitude and phase of which are modulated by the saturation and hue, respectively, of the colour represented. Thus the peak-to-peak chrominance amplitude 2S is given by:

$$2S = 2\sqrt{(U'^2 + V'^2)} \qquad (2.4.13)$$

and the angle α relative to the reference phase (the +U axis) is given by:

$$\alpha = \pm \tan^{-1} (V'/U') \qquad (2.4.14)$$

on alternate lines. Vector representations of the primary colours and their complements are shown in Figure 2.4.4 for the two senses of the V-axis switch.

The modulation process is shown in spectral terms in Figure 2.4.5. In this figure, (a) represents the baseband spectrum of a full bandwidth colour-difference signal, the high-frequency components of which are attenuated by Gaussian low-pass filtering to produce the spectrum of (b). When convolved with the line spectrum of the subcarrier signal, (c), this produces the modulated chrominance signal spectrum shown in (d).

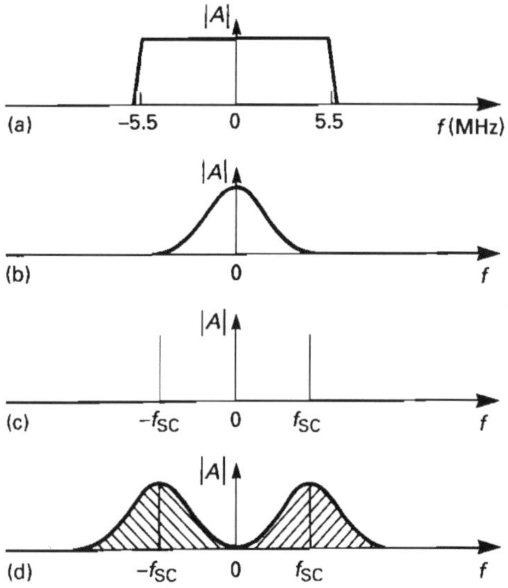

Figure 2.4.5 Frequency spectra in PAL chrominance modulation: (a) the baseband colour-difference signal; (b) the Gaussian filtered colour-difference signal; (c) the subcarrier sine-wave; (d) the modulated chrominance spectrum produced by convolving (b) and (c).

2.4.3.1.6 PAL encoding

The main processes of a PAL coder are shown in Figure 2.4.6. Gamma-corrected colour separation signals are converted to YUV form by a matrix combining the relationships of equations (2.4.1)–(2.4.3) with those of equations (2.4.8) and (2.4.9). The chrominance signal is formed by a pair of product modulators, which multiply the orthogonally phased subcarrier waveforms by the low-pass-filtered baseband U and V signals. The modulated

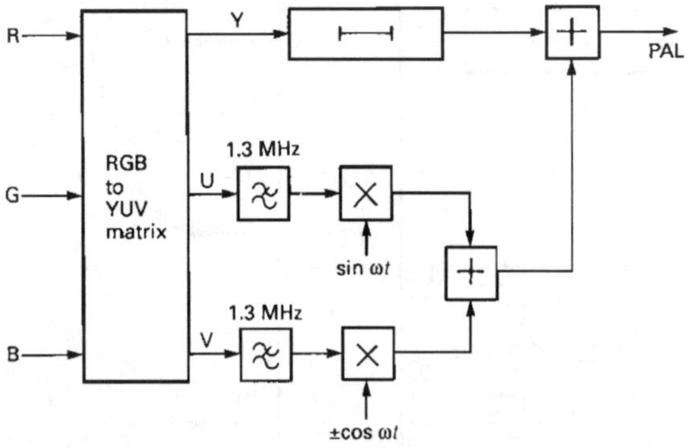

Figure 2.4.6 The main processes of a PAL encoder.

subcarrier signals are added to a delayed version of the luminance signal Y, timed to compensate for delays in the chrominance circuitry, producing a composite PAL signal at the output.

In practice, PAL coders contain no band-limiting low-pass filter at their outputs. Because of this, the upper chrominance sidebands, which extend well beyond normal luminance frequencies, may sometimes be retained up to the broadcast transmitter, at which point the portion of the upper sideband above the nominal luminance bandwidth (5.5 MHz for system I) is removed. The partial loss of the upper sideband causes ringing and desaturation at sharp chrominance transitions. The resulting impairments to the picture are usually only noticeable in systems B, G and H.

2.4.3.1.7 Colour synchronisation

Information about the subcarrier reference phase and the V-axis switch sense is transmitted with the signal by means of a colour synchronising burst situated on the back porch of the video waveform, following the line synchronising pulse. The signal timings and amplitudes of the colour reference burst are shown in Figure 2.4.7.

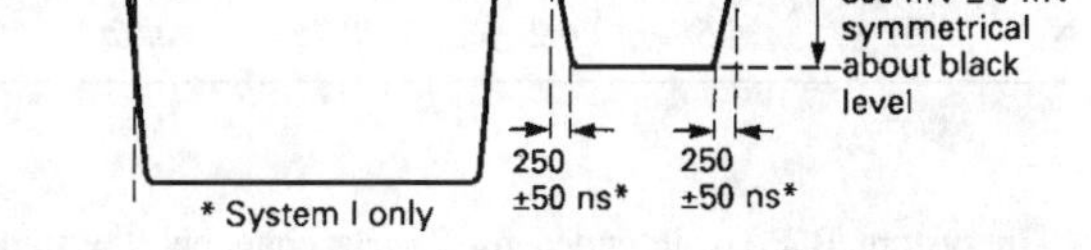

Figure 2.4.7 Waveform amplitudes and timings for the PAL colour burst in systems B, D, G, H and I. Items marked with an asterisk apply only to system I.

The burst phase is 135° on lines where the V-signal is in the true sense (known as line n), and 225° on lines where the V-signal is inverted (line $n+1$). As a complete picture contains an odd number of lines, the V-switch sense changes from one picture to the next, being positive on line 1 of the first field and negative on line 1 of the third field. In the eight-field cycle of the PAL signal, the first four fields are identified by having a subcarrier reference phase ϕ in the range $-90° \leqslant \phi < 90°$ at the beginning of the first field. To facilitate editing, the EBU[4] has defined the value of ϕ at the beginning of the first field to be $0° \pm 20°$.

Colour bursts are omitted from nine consecutive lines during each field blanking interval, in a sequence that repeats every four fields. The lines with no burst are detailed in Table 2.4.1.

Table 2.4.1 Lines with no burst in the PAL signal

	Field timing reference
Field 3>	Field 4>
311 312 313	314 315 316 317 318 319
Field 4>	Field 1>
623 624 625	1 2 3 4 5 6
Field 1>	Field 2>
310 311 312 313	314 315 316 317 318
Field 2>	Field 3>
622 623 624 625	1 2 3 4 5

2.4.3.1.8 PAL decoding

The main processes of a PAL decoder, shown in Figure 2.4.8, consist of separating the luminance and chrominance signals and demodulating the chrominance to retrieve the colour-difference signals U and V. Then, for display, the Y, U and V signals are converted to RGB by a matrix circuit embodying the relationships of equations (2.4.10) and (2.4.11), combined with those of equations (2.4.4)–(2.4.6).

In a conventional PAL decoder, as normally used in a domestic receiver, the modulated chrominance signals are first separated from the low-frequency luminance by a Gaussian bandpass filter centred on the subcarrier frequency. The subcarrier signals are then comb filtered by averaging across a delay of duration 283½ or 284 cycles of subcarrier, as shown in Figure 2.4.8. With a 284-cycle delay, the V subcarrier signals are in anti-phase across the delay and so cancel at the output of the adder, while the U subcarrier signals are co-phased and are added. Similarly, the subtractor cancels the U components to leave the V signal. In each case, the averaging process suppresses any U–V crosstalk components (*Hanover bars*) caused by distortion of the quadrature subcarrier signals. Product demodulators fed with appropriately phased subcarrier signals are used to regain the baseband U and V signals, with subsequent low-pass filters to suppress the twice subcarrier components. The Gaussian bandpass filter and the post-demodulation low-pass filters combine to provide somewhat less chrominance bandwidth than the PAL signal contains. Although this sacrifices some chrominance resolution, it results in less cross-colour, caused by luminance signals being demodulated as chrominance.

The luminance signal is obtained by suppressing the main subcarrier frequency components with a simple notch filter, usually having a maximum attenuation at about 20 dB. The

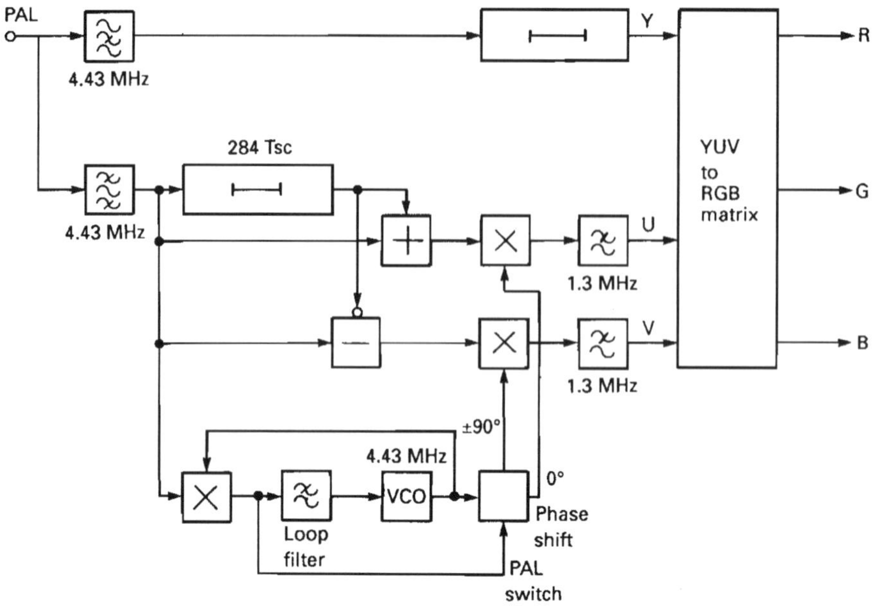

Figure 2.4.8 The main processes of a conventional delay-line PAL decoder.

luminance path includes a delay to compensate for the chrominance processing circuits, so that the signals remain aligned horizontally. However, no attempt is made to compensate for the vertical delay of the chrominance comb filter. This introduces a mean delay of half a line relative to the luminance, so that coloured objects are displaced down the picture by this amount.

Quadrature subcarrier signals for the demodulators are generated by a voltage-controlled crystal oscillator locked to the reference phase of the incoming colour bursts. A method frequently used is to allow the alternating bursts at 135° and 225° to pull the oscillator on alternate lines. With a relatively long loop time constant (usually between 30 and 100 line periods), the oscillator takes up the mean phase of 180° relative to the +U reference axis. The error signal of the phase detector then provides the alternating V-switch square-wave signal directly.

In difficult reception conditions, the received signal may have significant variations of gain across the video band, which could lead to incorrect colour saturation in the displayed picture. To reduce this effect, receivers incorporate automatic colour control circuitry to adjust the gain of the colour-difference amplifiers in response to the received amplitude of the colour burst.

2.4.3.1.9 System performance

The phase alternation property of the PAL signal substantially overcomes the susceptibility of suppressed carrier amplitude modulation to differential phase distortion. With severe distortion, however, PAL averaging causes a desaturation of the resulting picture. Also, the PAL switching halves the vertical chrominance resolution obtainable from the system. Another slight penalty is the subcarrier visibility that results from the subcarrier offsets of $^1/_4$ and $^3/_4$ line frequency.

2.4.3.1.10 PAL system variations

The baseband PAL coding parameters of the B, D, G, H and I standards are substantially the same. Some differences do arise, however, as a result of the constraints of the radiated signal parameters shown in Table 2.4.2.

Table 2.4.2 Radiated signal parameters of PAL systems

PAL system	*Channel spacing (MHz)*	*Sound spacing (MHz)*	*Luminance bandwidth (MHz)*	*Vestigial sideband (MHz)*	*Chrominance sideband (MHz)*
B	7	5.5	5	0.75	0.57
D	8	6.5	6	0.75	1.57
G	8	5.5	5	0.75	0.57
H	8	5.5	5	1.25	0.57
I	8	5.9996	5.5	1.25	1.07
M	6	4.5	4.2	0.75	0.62
N	6	4.5	4.2	0.75	0.62

In system B/PAL, intended for 7 MHz channels, the upper chrominance sideband is constrained to 570 kHz by the 5 MHz nominal signal bandwidth. System G/PAL, although intended for 8 MHz channels, maintains the same frequency allocations within the channel as B/PAL for compatibility. Systems B/ and G/PAL are widely used in Western Europe, Scandinavia, Australia, New Zealand and in parts of Africa. System H/PAL, used in Belgium, retains the 5 MHz bandwidth in the main sideband, but uses some of the extra bandwidth of the 8 MHz channel to extend the width of the vestigial sideband. On the other hand, system D/PAL, used in China, extends the signal bandwidth to 6 MHz, which provides capacity for the full 1.3 MHz of the upper chrominance sideband. System I/PAL, used principally in the British Isles, Hong Kong and South Africa, combines the two features, both extending the vestigial sideband and providing 5.5 MHz in the main sideband. Thus the upper chrominance sideband extends to 1.07 MHz. The full specification is given in ITU-R Recommendation BT.470-6.

Variants of PAL for 6 MHz channels incorporate greater differences. System N/PAL, used in Argentina, has a subcarrier frequency of 3.58205625 MHz to match the 4.2 MHz bandwidth. This conforms to the relationship:

$$f_{SC} = \left(229\tfrac{1}{4} + \frac{1}{625}\right) f_H \qquad (2.4.15)$$

System M/PAL, used in Brazil, is a 525-line, 60 fields per second version of PAL for a 4.2 MHz signal bandwidth. M/PAL has a subcarrier frequency of 3.57561149 MHz given by the relationship:

$$f_{SC} = 227\tfrac{1}{4} f_H = \frac{909}{4} f_U \qquad (2.4.16)$$

This system ignores the picture rate component of the subcarrier frequency, making the subcarrier more obtrusive on monochrome receivers.

2.4.3.2 NTSC

2.4.3.2.1 Development

The NTSC system, adopted for use in the USA in 1953, takes its name from the National Television System Committee[5] convened to recommend its parameters. It is based on the system M monochrome television standard, having 525 lines per picture and 60 fields/second, with a 4.2 MHz bandwidth and designed to occupy channels at a 6 MHz spacing. The relatively low horizontal resolution of this system placed severe constraints on the development of a compatible chrominance signal.

Colour information is transmitted using suppressed carrier amplitude modulation of two orthogonally phased subcarriers to carry two weighted colour-difference signals, known as I and Q (*in-phase* and *quadrature*).

2.4.3.2.2 Colour subcarrier frequency

The NTSC subcarrier frequency of 3.579545 MHz ±10 Hz was chosen to fall in the upper part of the 4.2 MHz luminance band to reduce its visibility on monochrome receivers. It was also chosen to have a half-line frequency offset so that peaks and troughs of adjacent lines would provide a degree of visual cancellation. Accordingly, the subcarrier frequency is related to the line frequency by the following relationship:

$$f_{SC} = 229\tfrac{1}{4} f_H = \frac{455}{2} f_H \qquad (2.4.17)$$

This causes the main luminance and chrominance components in the high-frequency part of the video signal spectrum to be interleaved, as shown in Figure 2.4.9. The subcarrier phase at a point in the picture repeats every four fields.

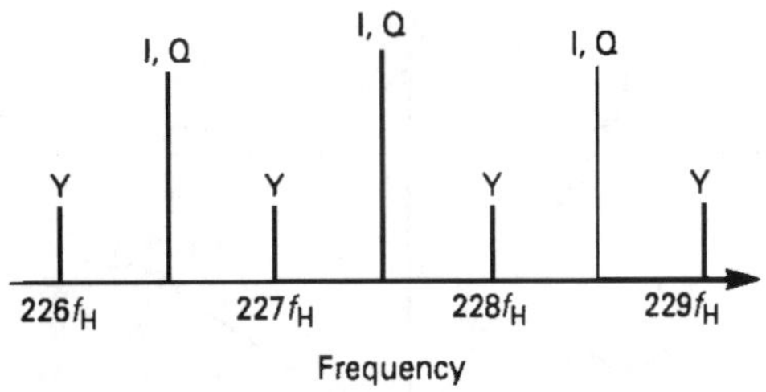

Figure 2.4.9 Positions of the main luminance (Y) and chrominance (I, Q) components in the high-frequency part of an NTSC signal.

A further factor in the choice of colour subcarrier frequency was that any beat between the system M sound carrier (mean frequency 4.5 MHz) and the subcarrier would result in a frequency also having a half-line offset (approximately 920 kHz). This choice of subcarrier produced small changes in the nominal line and field frequencies to 15,734.264 Hz and 59.94 Hz, respectively, for colour operation, although these values remain within the original tolerances of system M.

2.4.3.2.3 Colour-difference signals

The NTSC system is based on the principle that significantly lower resolution is acceptable for some colours, notably for blue/magenta and yellow/green. So, instead of using the B − Y and R − Y signals directly, the signals are first weighted as in PAL, to limit their maximum excursions, and then phase rotated by 33°. Thus, the I and Q colour-difference signals, used to modulate the two orthogonal subcarriers, are given by:

$$I' = -U' \sin 33^\circ + V' \cos 33^\circ \qquad (2.4.18)$$

and

$$Q' = U' \cos 33^\circ + V' \sin 33^\circ \qquad (2.4.19)$$

with U and V given by equations (2.4.8) and (2.4.9). As can be seen from Figure 2.4.10, this rotation aligns the Q-axis with the regions of colour least affected by a loss of resolution, so allowing the signal to be encoded with a low bandwidth.

In terms of direct relationships to B − Y and R − Y, the I and Q signals are given by:

$$I' = -0.269(B' - Y') + 0.736(R' - Y') \qquad (2.4.20)$$

and

$$Q' = 0.413(B' - Y') + 0.478(R' - Y') \qquad (2.4.21)$$

The inverse relationships are:

$$B' - Y' = 1.105I' + 1.701Q' \qquad (2.4.22)$$

$$R' - Y' = 0.956I' + 0.621Q' \qquad (2.4.23)$$

2.4.3.2.4 I and Q filters

The wide-band I signal uses a slow roll-off characteristic (Figure 2.4.11(a)) very similar to that used for the U and V

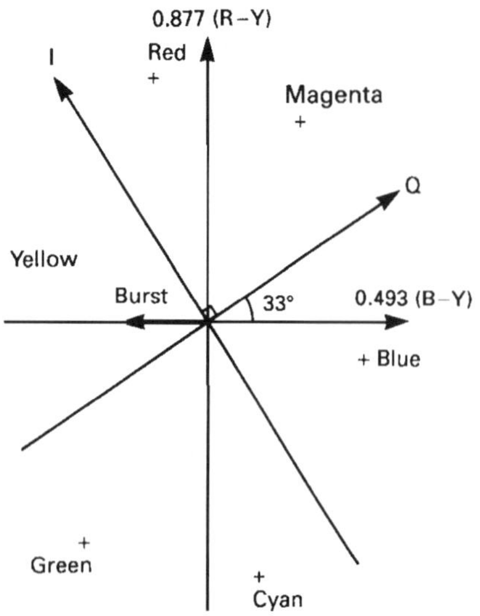

Figure 2.4.10 Positions of the I and Q modulation axes of the NTSC system relative to the weighted colour-difference axes (U and V). The positions of the colour primaries and their complements and the NTSC colour burst are shown.

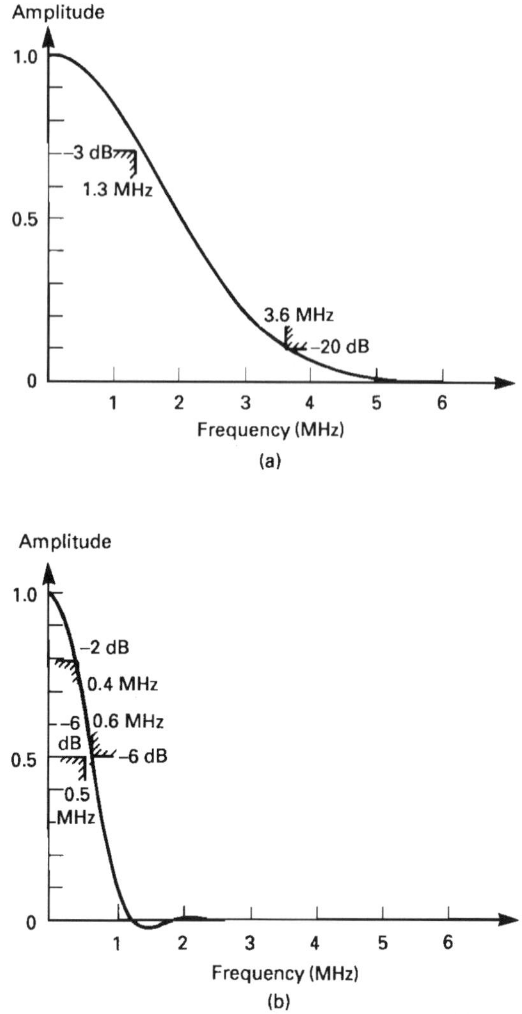

Figure 2.4.11 Limits for colour-difference low-pass filters for NTSC coders, showing some appropriate characteristics: (a) for the I channel; (b) for the Q channel.

signals in PAL. The Q signal, however, is limited to about 0.5 MHz bandwidth by the template of Figure 2.4.11(b).

2.4.3.2.5 Chrominance modulation

The NTSC modulated chrominance signal can be represented mathematically by the expression:

$$Q' \sin \omega t + I' \cos \omega t \qquad (2.4.24)$$

where $\omega = 2\pi f_{sc}$.

The peak-to-peak chrominance amplitude 2S is given by:

$$2S = 2\sqrt{(I'^2 + Q'^2)} \qquad (2.4.25)$$

As the signal bandwidth of the NTSC system is only 4.2 MHz, the spectrum space available for the modulated chrominance components is relatively limited. In particular, making the subcarrier a fine, high-frequency pattern to reduce its visibility results in a significant loss of the upper chrominance sidebands. If both colour-difference components were wide-band, the asymmetry would cause crosstalk between the two signals. However, as the Q signal is low bandwidth, it remains a double-sideband signal and causes no crosstalk. Part of the upper sideband of the wide-band I signal, on the other hand, is lost. When demodulated, some I components do cross into the Q signal, but at higher frequencies than the true Q components. Therefore, the I–Q crosstalk components can be removed by low-pass filtering the demodulated Q signal.

2.4.3.2.6 NTSC encoding

The main processes of an NTSC encoder are shown in Figure 2.4.12. RGB signals are matrixed to YIQ form using the relationships of equations (2.4.1)–(2.4.3), combined with those of equations (2.4.20) and (2.4.21). The I and Q signals are filtered according to the templates of Figure 2.4.11 and modulated onto orthogonal subcarriers. The modulated chrominance signals are combined with the luminance signal, appropriately delayed to compensate for the chrominance processing.

2.4.3.2.7 Colour synchronisation

Subcarrier reference phase information is carried in the composite NTSC signal by a colour burst occupying a position on the back porch, following the line synchronising pulse, as shown in Figure 2.4.13. The phase of the burst is 180° relative to the B – Y reference axis (see Figure 2.4.10). Colour bursts are omitted from those lines in the field interval starting with an equalising pulse or a broad pulse.

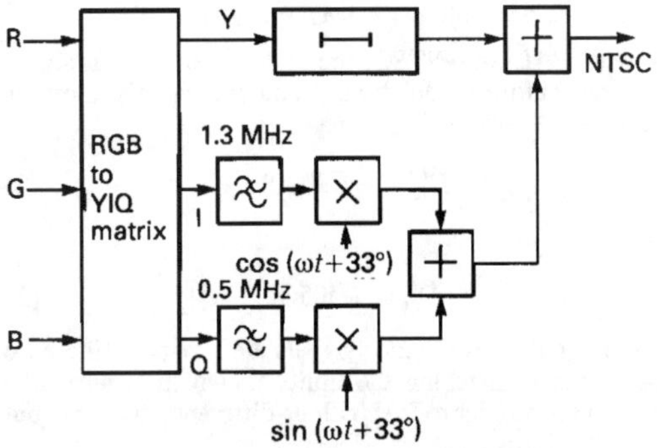

Figure 2.4.12 The main processes of an NTSC encoder.

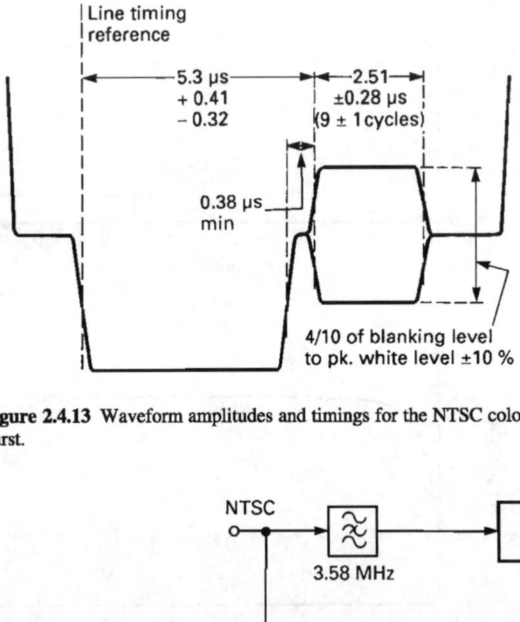

Figure 2.4.13 Waveform amplitudes and timings for the NTSC colour burst.

The NTSC reference subcarrier and the line timing reference point (the half-amplitude point of the falling edge of the line synchronising pulse) are not only related in frequency by equation (2.4.17), but are also related in phase. The line timing reference point is defined as being coincident with zero crossings of the reference subcarrier.

The first field in the four-field sequence of an NTSC signal is defined by two factors. The first equalising pulse of the field interval must:

- occur at a line pulse position;
- coincide with a negative-going zero crossing of the reference subcarrier.

2.4.3.2.8 NTSC decoding

The main processes of a simple NTSC decoder are shown in Figure 2.4.14. The colour-difference signals are separated by product demodulators, with the low-bandwidth Q filter used to remove crosstalk resulting from the asymmetrical sideband I signals. A 3.58 MHz notch filter is used to remove the main subcarrier components from the luminance signal. Y, I and Q signals are matrixed to RGB according to the relationships of equations (2.4.22), (2.4.23) and (2.4.4)–(2.4.6).

The locally generated subcarrier reference, sin ωt, is synchronised to the incoming colour burst waveform, sin (ωt + 180°). Phase-shifted versions of the reference phase are fed to the product demodulators.

With NTSC signals, comb filters based on line delays are frequently used in decoders to improve luminance resolution and to reduce cross-colour impairments.[6]

2.4.3.2.9 System performance

The NTSC signal is particularly susceptible to differential phase distortion. This causes high frequencies at one average level to be delayed more than those at another level, so altering the phase relationship between the burst and the chrominance signals of the active-line video. This results in hue errors in the final picture. Also, sideband asymmetry causes hue errors on colour transitions. This makes the NTSC signal much less robust than PAL or SECAM.

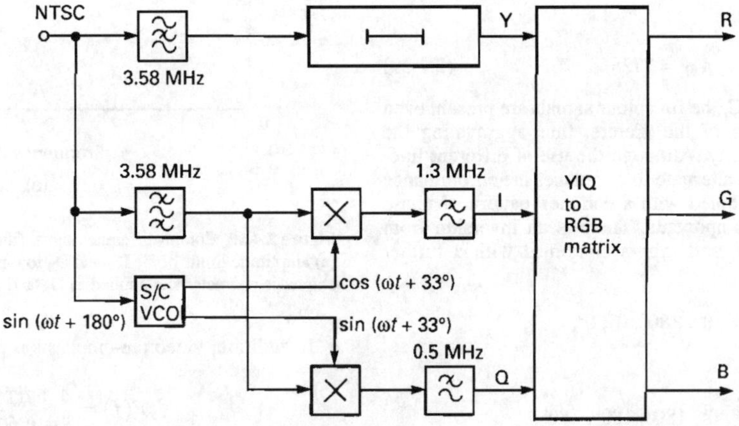

Figure 2.4.14 The main processes of an NTSC decoder.

2.4.3.2.10 Countries using system M/NTSC

The NTSC system is used in North America, most of Central America and the Caribbean, and much of South America with the notable exceptions of Argentina and Brazil. The system is also used in Japan, South Korea and Burma.

The radiated signal is substantially similar in all cases, using a 0.75 MHz vestigial sideband and a 4.2 MHz main sideband within a 6 MHz channel.

2.4.3.3 SECAM

2.4.3.3.1 Development

The SECAM system, invented by De France,[7] was developed in Europe for 625/50 scanning, like PAL, as a means of overcoming the susceptibility of the NTSC system to differential phase distortion. Instead of relying on two subcarriers remaining orthogonal, the colour-difference signals are transmitted separately on alternate lines, so that the signal from the previous line has to be held over in the decoder by a line memory. This led to the name *Sequential Colour with Memory.*

Early versions of SECAM used amplitude modulation in common with PAL and NTSC, but this proved unsatisfactory. This was because the substantial differences in the signals on alternate lines resulted in poor compatibility for monochrome receivers. The use of frequency modulation gave some improvement, but considerable optimisation of the signal parameters was needed to achieve acceptable compatibility.

2.4.3.3.2 Colour subcarriers

The subcarriers on alternate lines use different undeviated frequencies:

$$f_{OR} = 4.406250\,\text{MHz} \pm 2\,\text{kHz} \qquad (2.4.26)$$

and

$$f_{OB} = 4.250000\,\text{MHz} \pm 2\,\text{kHz} \qquad (2.4.27)$$

These frequencies have the following relationships to the 625/50 line frequency:

$$f_{OR} = 282 f_H \qquad (2.4.28)$$

and

$$f_{OB} = 272 f_H \qquad (2.4.29)$$

Unlike PAL and NTSC, the fm colour signals are present even in monochrome areas of the picture, thus aggravating the problem of compatibility. Although the use of different line-locked frequencies on alternate lines reduces the chrominance visibility, this is combined with a complex pattern of phase reversals with two components. There is an inversion from one field to the next, and this is combined with a further pattern, either:

0°, 0°, 180°, 0°, 0°,

or

0°, 0°, 0°, 180°, 180°, 180°,

thus resulting in a 12-field sequence[8] for the SECAM signal.

2.4.3.3.3 Colour-difference signal processing

As with PAL and NTSC, the R − Y and B − Y signals are weighted before modulation, producing signals known as D_R and D_B, given by:

$$D'_R = -1.902(R' - Y') \qquad (2.4.30)$$

and

$$D'_B = 1.505(B' - Y') \qquad (2.4.31)$$

Also, the colour-difference signals are low-pass filtered with a characteristic matching the limits shown in Figure 2.4.15(a). This is very similar to PAL colour-difference filtering, but with a slightly sharper cut-off.

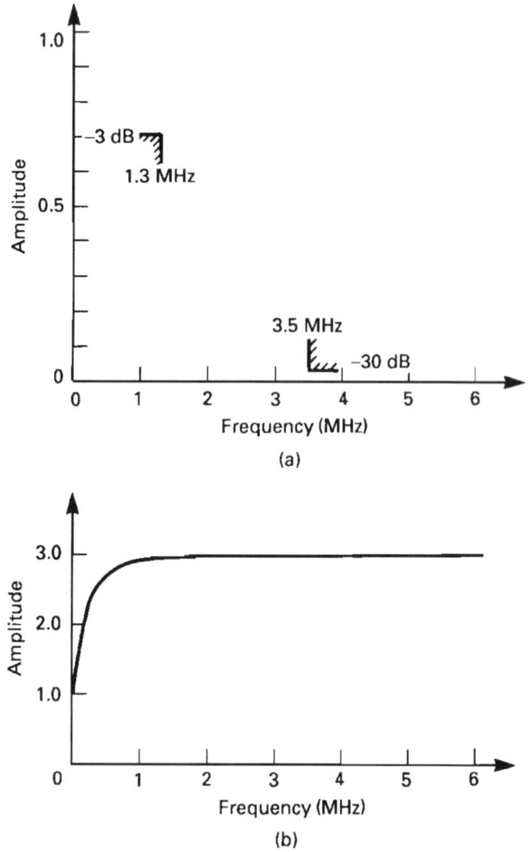

Figure 2.4.15 Colour-difference signal filtering in the SECAM system: (a) amplitude limits of the D_R and D_B low-pass filters; (b) the video pre-emphasis characteristic applied to D_R and D_B.

In addition, video pre-emphasis is applied according to:

$$F(f) = \frac{1 + j(f/f_1)}{1 + j(f/3f_1)} \qquad (2.4.32)$$

where $f_1 = 85\,\text{kHz}$.

This produces a lift of about 9.5 dB above 750 kHz, as shown in Figure 2.4.15(b). The pre-emphasis allows a relatively low amplitude of subcarrier to be used in plain coloured areas, thus further assisting monochrome compatibility.

2.4.3.3.4 Frequency modulation

The weighted filtered and pre-emphasised colour-difference signals D_R^* and D_B^* are used to frequency modulate the subcarriers of Section 2.4.3.3.2, producing nominal deviations of 280 ± 9 kHz for D_R^* and 230 ± 7 kHz for D_B^*, corresponding to the saturations of 75% colour bars.[9] Limiters prevent deviations greater than +350 or −506 kHz for D_R^*, and +506 or −350 kHz for D_B^*. This restricts the rise time of fast, large-amplitude colour transitions. The asymmetry, combined with the offset between the two subcarriers, limits the modulated chrominance signals to the range 3.900–4.756 MHz. The modulated chrominance signal is then filtered with the rf pre-emphasis characteristic of Figure 2.4.16, which further reduces the visibility of the subcarrier near the rest frequencies. This is given by the relationship:

$$G = M_0 \frac{1 + j16F}{1 + j1.26F} \qquad (2.4.33)$$

where $F = f/f_0 - f_0/f$ and $f_0 = 4.286$ MHz.

M_0 sets the minimum subcarrier amplitude at f_0 to be 161 mV_{p-p} in a standard level signal. Thus the chrominance components are represented mathematically by the expressions:

$$G \cos 2\pi (f_{OR} + \Delta f_{OR} \int_0^t D_R^* dt) \qquad (2.4.34)$$

or

$$G \cos 2\pi (f_{OR} + \Delta f_{OR} \int_0^t D_B^* dt) \qquad (2.4.35)$$

on alternate lines.

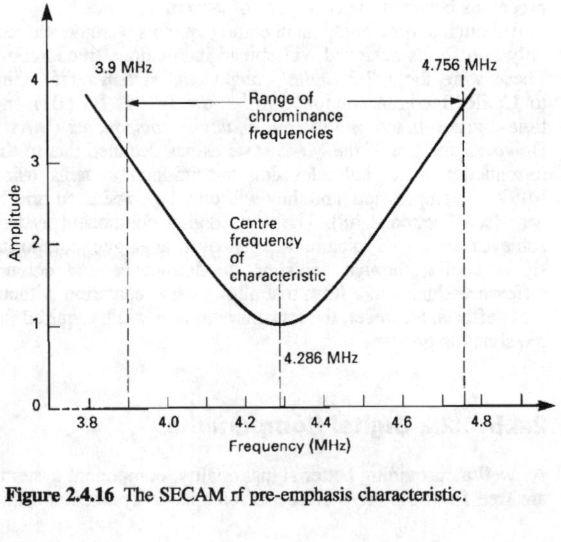

Figure 2.4.16 The SECAM rf pre-emphasis characteristic.

2.4.3.3.5 SECAM encoding

The main features of a SECAM encoder are shown in Figure 2.4.17. After matrixing to Y, D_R, D_B form, the colour-difference signals are filtered with a combined low-pass and pre-emphasis characteristic. Limiters then restrict the amplitude range of the resulting signals, before they are applied to voltage-controlled oscillators for the two subcarrier frequencies. The initial oscillator phase is selected by a sequence generator according to one of the relationships in Section 2.4.3.3.2. Each modulated colour-difference signal is then selected on alternate lines and weighted by the rf pre-emphasis characteristic before being added to an appropriately delayed luminance signal.

Because of the fm capture effect, cross-colour in the SECAM system becomes a problem only if there is a significant amplitude of luminance. For this reason, SECAM coders sometimes incorporate level-sensitive filters to retain some high-frequency luminance, while limiting the larger amplitude components that might produce cross-colour.

2.4.3.3.6 Colour synchronisation

The SECAM signal provides for two methods of synchronising the bistable switching sequence in the coder and decoder. First, the subcarrier signals on each line begin before the active-line period, as shown in Figure 2.4.18. From this, the differing frequencies of the two carriers can be sensed on a line-by-line basis.

Alternatively, the field intervals contain a sequence of nine lines of identification signals, occupying lines 7–15 on the first and third fields, and lines 320–328 on the second and fourth fields. Line 7 contains D_R^* signals on the first field and D_B^* signals on the third, while line 320 contains D_B^* signals on the second field and D_R^* signals on the fourth.

The large deviation and long duration of the field identification signals provide a very rugged means of regenerating the colour sequence in the decoder, whereas the line reference signals are short and have a much smaller frequency difference. However, because the field blanking lines represent a valuable resource for other uses, line reference synchronisation is being encouraged in preparation for suppression of the field identification signals.

2.4.3.3.7 SECAM decoding

Figure 2.4.19 shows the main units of a SECAM decoder. The chrominance is filtered with a bell-shaped bandpass characteristic to remove the rf pre-emphasis and then applied to a line delay. The delayed and undelayed signals are switched into the D_R and D_B discriminators on alternate lines, so that each receives lines of the appropriate signal continuously. After demodulation, the low-pass filter provides video de-emphasis.

The luminance channel includes a filter with two notches at the rest subcarrier positions, thus providing good suppression of the subcarrier in monochrome areas. The luminance signal is delayed to match the processing delays of the chrominance circuits before being matrixed to RGB using the relationships in equations (2.4.4)–(2.4.6), combined with the inverse colour difference weighting factors for D_R and D_B signals, given by:

$$R' - Y' = -0.526 D_R \qquad (2.4.36)$$

$$B' - Y' = 0.664 D_B \qquad (2.4.37)$$

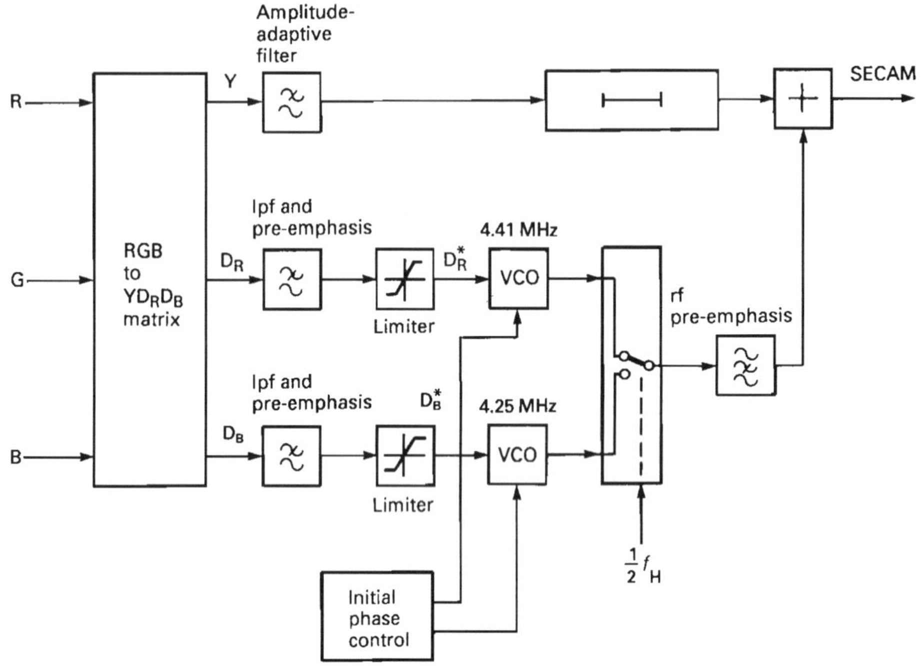

Figure 2.4.17 The main processes of a SECAM encoder.

2.4.3.3.8 System performance

The resistance of the SECAM system to differential gain and differential phase distortions makes it particularly advantageous for poor quality transmission links and magnetic recording. Also, the modulation parameters have been optimised to balance chrominance noise performance against subcarrier visibility to obtain reasonably satisfactory monochrome compatibility. However, the main disadvantage of the SECAM system is the non-linearity of the fm signal. This prevents mixing SECAM signals without separating and demodulating the chrominance. Also, the combined effect of luminance filtering in the coder and the decoder results in rather less high-frequency resolution than for PAL. The fm chrominance signals prevent the use of comb filtering in the decoder to improve luminance resolution.

2.4.3.3.9 System variations

SECAM encoding is used in virtually the same form in several broadcast signal standards. System B/G SECAM, used mostly in North Africa, the Mediterranean area and the Middle East, has only 5 MHz bandwidth. System D/K SECAM, used mostly in Eastern Europe, is similar, but provides 6 MHz bandwidth. System K1 also has 6 MHz bandwidth, but with the vestigial sideband extended to 1.25 MHz. This standard is used in much of Africa. SECAM L, used mainly in France, is similar to system K1, although it differs from all the other standards through its use of positive vision modulation and am sound.

2.4.4 Component Colour Systems

While the composite colour signals described in the previous section were developed for compatible transmission through existing monochrome channels, the development of component systems is less constrained. With component systems, the extent of compatibility is limited to the need to coexist with current studio and broadcast standards. In practical terms, this depends on maintaining reasonably straightforward conversion processes between the two types of system.

Although several component coding systems were developed, only two forms achieved widespread international recognition. These were the *4:2:2 digital components system* conforming to CCIR Recommendation 601[10] (now ITU-R BT.601) and time-division-based *multiplexed analogue components* (MAC). However, the use of the MAC systems has declined due to the introduction of digital television transmission systems using MPEG-2 compression, and they will only be considered briefly here (see Section 2.4.8). The 4:2:2 digital component system achieves improved picture quality over analogue composite signal coding through encoding the luminance and colour-difference signals in a form that allows easy separation without cross-effects. However, the transmission bandwidth required for the signal is greater.

2.4.5 4:2:2 Digital Components

As well as providing better signal quality, component systems are free from the cross-effects of composite systems and this

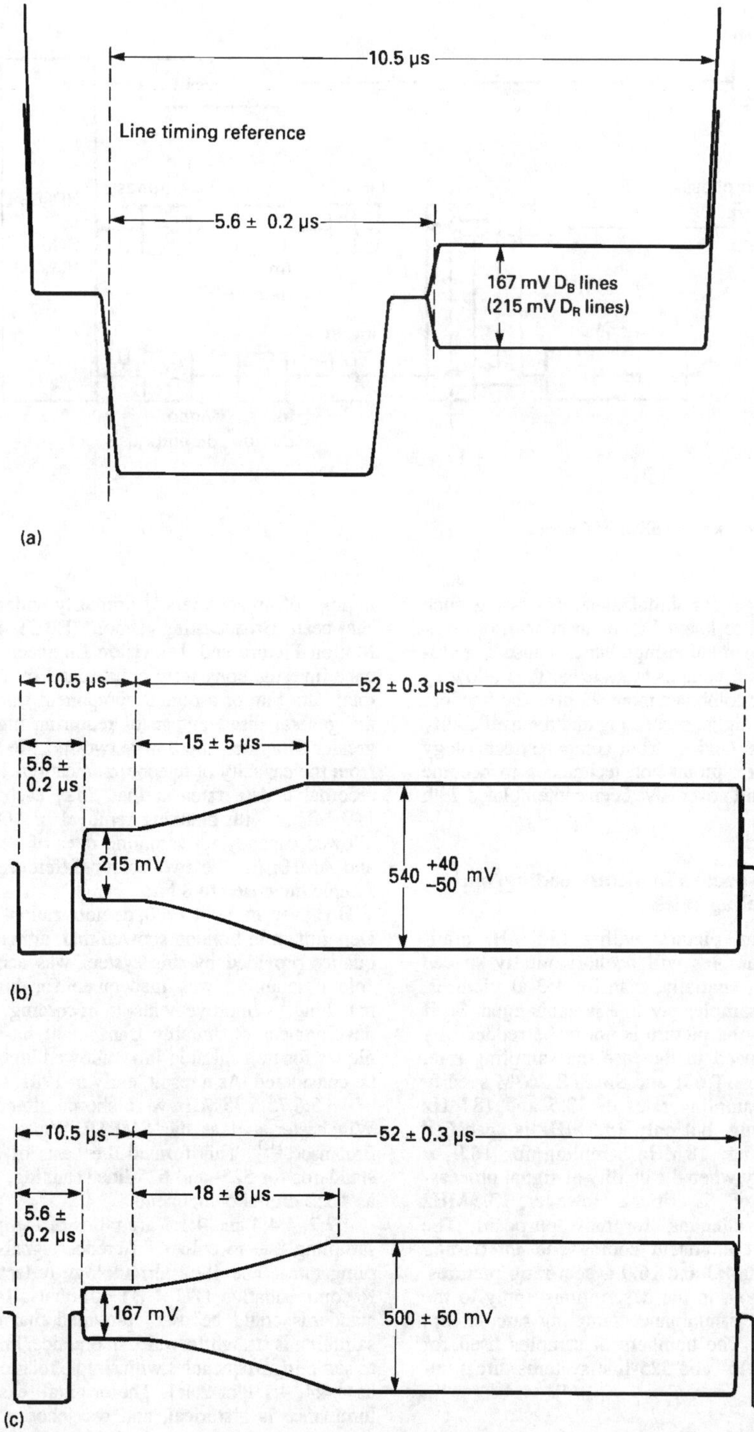

Figure 2.4.18 Colour synchronising waveforms in the SECAM signal: (a) line synchronising signal (D_B line); (b) D_R field synchronising signal; (c) D_B field synchronising signal.

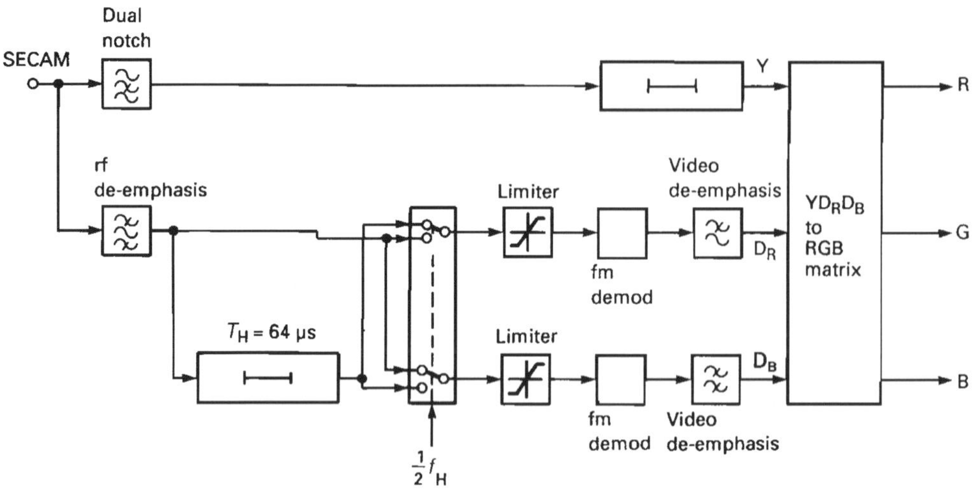

Figure 2.4.19 The main processes of a SECAM decoder.

has considerable advantages for digital signal processing, such as for 'special effects'. The lack of a colour subcarrier eliminates the complex phase relationships which cause a multi-field sequence in analogue composite systems, thus allowing the signals to be stored or combined more simply. The freedom from distortion of digital signal processing and the availability of arithmetic and storage devices from computer technology has allowed very complex production techniques to become commonplace, which could never have been contemplated with analogue circuitry.

2.4.5.1 4:3 and 16:9 aspect ratio picture coding, and 13.5 and 18 MHz sampling rates

For 16:9 aspect ratio (ar) pictures with a 13.5 MHz luminance sampling rate, samples will be horizontally spaced 1.33 times farther apart, spatially, than for 4:3 ar pictures, because the number of samples per line is unchanged. So if the spatial resolution of the picture is not to be reduced by about 25%, we would need to increase the sampling rate. ITU-R Recommendation BT.601 and SMPTE 267M specify alternative luminance sampling rates of 13.5 and 18 MHz for 16:9 ar picture coding, but only 13.5 MHz is specified for 4:3 ar picture coding; 18 MHz sampling for 16:9 ar pictures is only necessary when the quality of signal processing, such as 'chroma key', is critical. However, 13.5 MHz luminance sampling is adequate for most purposes. The descriptions of digital component coding and interfacing that follow apply to both 4:3 and 16:9 aspect ratio pictures. However, the values given in the descriptions apply to the commonly used 13.5 MHz luminance sampling rate, except where otherwise stated. The numbers of samples used for both sample rates, for 625- and 525-line systems, are summarised in Table 2.4.8.

2.4.5.2 Background

The parameters of the 4:2:2 components standard were determined as the result of lengthy discussions between a large number of broadcasters,[11] primarily under the auspices of the European Broadcasting Union (EBU) and the Society of Motion Picture and Television Engineers (SMPTE). Preliminary investigations were constrained by the need to limit the total data rate of a digital component standard. In addition to the general disadvantage of requiring higher speed logic for greater data rates, there were two specific limitations, resulting from the capacity of telecommunications links and the limit of recordable data rates at that time, each of which provided 140 Mbit/s with blanking removal or 160 Mbit/s gross. This allowed capacity for sampling rates of 12 MHz for luminance and 4 MHz for the two colour-difference signals, with each sample quantised to 8 bits.

However, in April 1980, demonstrations at the BBC Designs Department in London showed that, although the basic picture quality provided by this system was acceptable, the 4 MHz colour sampling was inadequate for high-quality colour – matching.[12] Improvements in recording techniques and the development of suitably transparent bit-rate reduction techniques for transmission links allowed higher sampling rates to be considered. As a result, early in 1981, sampling frequencies of 13.5:6.75:6.75 MHz were chosen after further tests at IBA, Winchester and at the SMPTE Winter Conference in San Francisco.[11,13] This formed the basis of the compatible dual standards for 525- and 625-line scanning, subsequently known as 4:2:2 digital components.

4:2:2, 4:4:4 and 4:1:1 are ratios of the luminance signal (Y) sampling rate to colour difference signals (C_B and C_R) sampling rates. The 4:2:2 terminology reflects the assumption in Recommendation ITU-R BT.601 that a family of compatible standards could be used, provided that conversion to 4:2:2 sampling is straightforward. In practice, this restricts the choice to sampling frequencies with simple relationships to 4:2:2, such as 4:4:4, 4:1:1 or 2:1:1. The original reason for using '4' for luminance is historical, and was chosen when consideration was being given to using the NTSC colour subcarrier frequency as the luminance sampling frequency. (This rate was not used.) However, it is necessary to retain the '4' because of the 4:1:1 system.

2.4.5.2.1 Sampling structure

The digital coding standards of Recommendation ITU-R BT.601 are based on line-locked sampling. This produces a static orthogonal sampling grid in which samples on the current line fall directly beneath those on previous lines and fields, and exactly overlay samples on the previous picture. This orthogonal sampling structure has many advantages for signal processing, including the simplification of filters and repetitive control waveforms.

The sampling is not only locked in frequency, but also in phase, so that one sample is coincident with the line timing reference point (the half-amplitude point of the falling edge of the line synchronising pulse), as shown in Figure 2.4.20. This feature is also applied to the lower rate colour-difference signal samples, known as C_B and C_R. This ensures that different sources produce samples nominally at the same positions in the picture.

The fundamental sampling frequency of 13.5 MHz gives a measure of commonality between 625/50 and 525/60 systems because this frequency has the property of being an exact harmonic of the line rate on both scanning standards. This produces 864 or 858 luminance samples and 432 or 429 colour-difference samples per line, respectively. When the 13.5 MHz sampling frequency was chosen, the problem of the ninth and 18th harmonics causing interference to the international distress frequencies of 121.5 and 243 MHz was not considered. It is believed, however, that interference can be avoided by careful equipment design.[14]

The different formats of the digital line for the two standards are shown in Figure 2.4.21. Each digital line consists of a blanking period followed by an active-line period. The analogue line timing reference falls in the blanking period, a few samples after the beginning of the digital line. The active-line period of 720 samples for luminance and 360 samples for each colour-difference signal is common to both standards. This is sufficiently long to accommodate the full range of analogue blanking tolerances on the two standards. As a result, analogue blanking need only be applied once, preferably in picture monitors or at the conversion to composite signals for broadcast transmission. Samples at the beginning and end of the blanking period are set aside for synchronising codes, details of which are given in Section 2.4.5.7.

2.4.5.3 Conversion methods

Two approaches can be used for the conversions between the analogue RGB signals produced by cameras and accepted by displays, and the digital YC_BC_R signals used for processing. These are shown in Figure 2.4.22. In the figure, (a) shows digital conversion of YC_BC_R signals with the colour-difference signals sampled directly at 6.75 MHz. While providing a

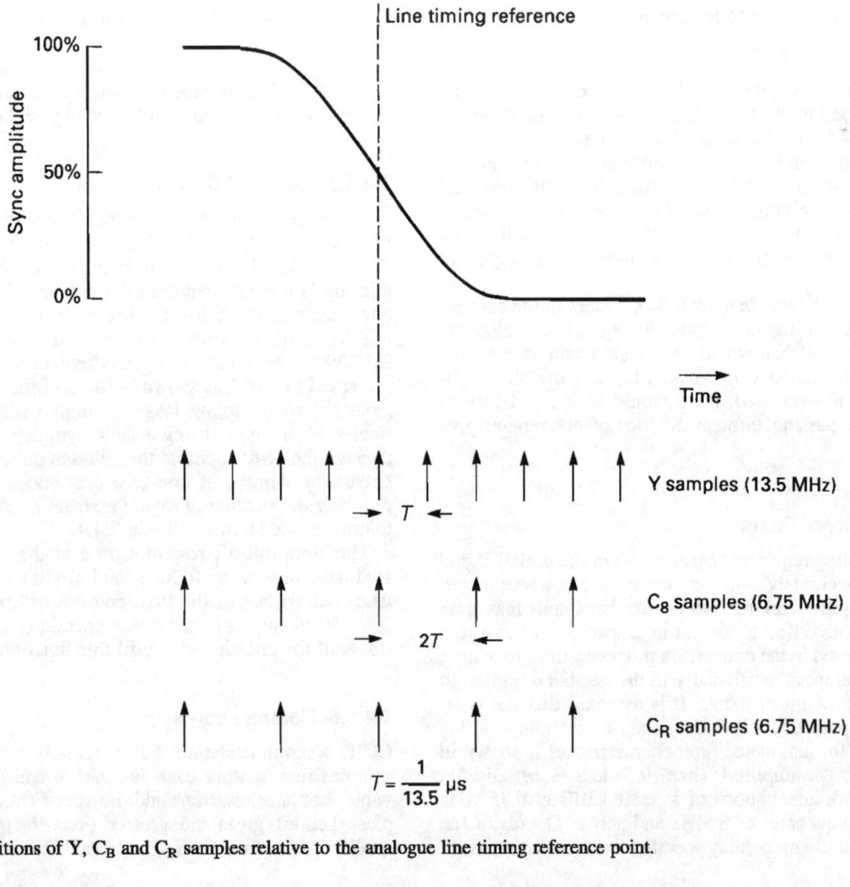

Figure 2.4.20 Positions of Y, C_B and C_R samples relative to the analogue line timing reference point.

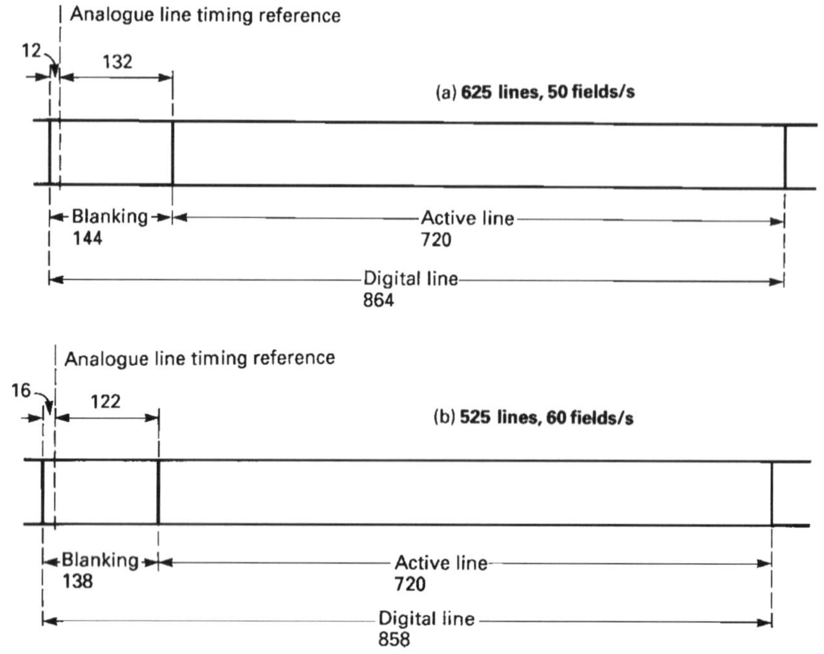

Figure 2.4.21 Number of 13.5 MHz samples in each section of the digital line for 625/50 and 525/60 scanning standards.

relatively simple implementation, this arrangement depends on the accuracy of setting the individual conversion circuits for its colour balance stability. Also, the widely differing delays in the analogue luminance and colour-difference signal filters need to be accurately compensated. In contrast, converting the RGB signals as shown in Figure 2.4.22(b) produces matched delays in the analogue circuitry and includes the gain factors digitally, although the digital matrixing results in a greater degree of complication.

Both methods can include slow roll-off filters in the conversions for display or for composite coding. This avoids the ringing impairments that would otherwise occur due to the sharp-cut colour-difference low-pass filters. If the slow roll-off filters were always used, there would be a rapid loss of resolution when passing through a series of conversion processes.

2.4.5.4 Luminance filters

The use of sampling requires the bandwidth of the digital signal to be specified accurately. Also, in circumstances where mixed analogue and digital working may occur, the signals may pass through many conversion processes in cascade. For these reasons, the filters used in the conversion processes have to adhere to very tight tolerances, particularly in the passband region, to avoid a build-up of impairments. It is assumed that the post-conversion low-pass filters include sin x/x correction.

The template for luminance filter characteristics is shown in Figure 2.4.23(a). The amplitude characteristic is essentially flat to 5.75 MHz, with attenuations of at least 12 dB at 6.75 MHz and 40 dB for frequencies of 8 MHz and above. Details of the passband ripple and group delay specifications are summarised in Table 2.4.3. The same template applies for RGB filters used in the conversion arrangements of Figure 2.4.22(b).

2.4.5.5 Colour-difference filters

The template for colour-difference filters is similar to a scaled version of the luminance filter, with the response essentially flat to 2.75 MHz, as shown in Figure 2.4.23(b). However, as aliasing is less noticeable in the colour-difference signals, the attenuation at the half-sampling frequency is relaxed to 6 dB. For digital filters as would be used in the arrangement of Figure 2.4.22(b), the attenuation requirement in the stopband is increased to 55 dB as shown because of the lack of attenuation from sin x/x sampling loss. For digital filters there is a particular advantage in using a skew-symmetric response passing through the –6 dB point at the half-sampling frequency, as this halves the number of non-zero coefficients in the filter, thus reducing the number of taps. Passband ripple and group delay tolerances are shown in Table 2.4.4.

The slow roll-off response used at the final conversion to analogue for viewing follows the form of a Gaussian characteristic such as that of the PAL coder response shown in Figure 2.4.3. With outputs to composite encoders, the normal low-pass filters of the encoder will fulfil this function.

2.4.5.6 Coding ranges

CCIR Recommendation 601 originally only specified 8-bit samples representing each luminance and chrominance level value, but later revisions add the specification for 10-bit samples. Ten-bit gives the greater precision required for good quality processing such as chroma key and special effects.

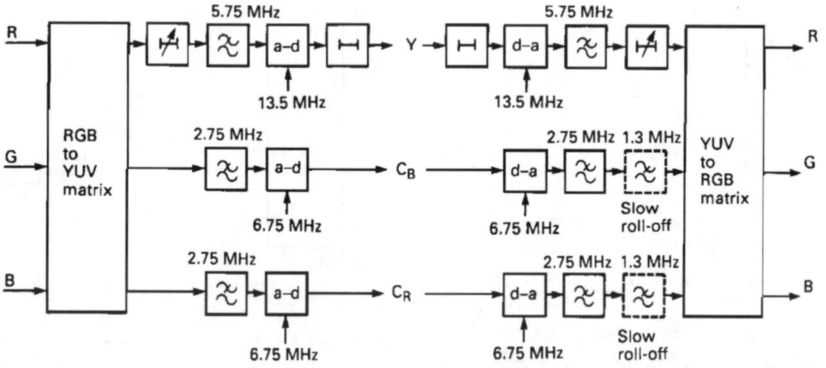

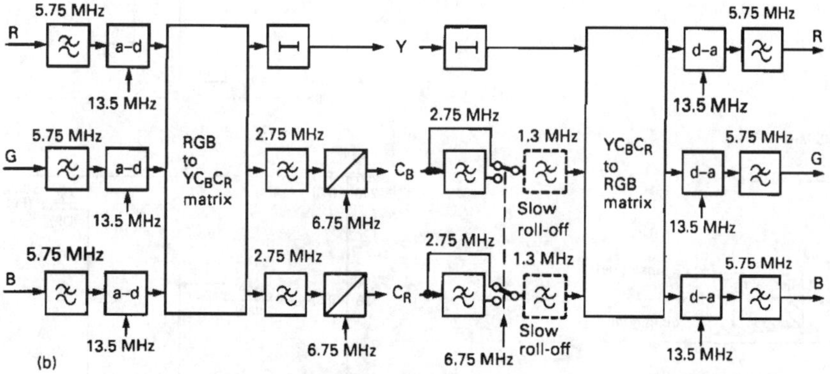

Figure 2.4.22 Alternative methods of conversion between analogue RGB and digital YC_BC_R signals: (a) with the C_B and C_R signals sampled directly at 6.75 MHz; (b) with digital encoding of the RGB signals.

Table 2.4.3 Luminance filter ripple and group delay tolerances

Frequency range	*Design limits*	*Practical limits*
Ripple:		
1 kHz–1 MHz	±0.005 dB	Increasing from ±0.01 to ±0.025 dB
1–5.5 MHz	±0.005 dB	±0.025 dB
5.5–5.75 MHz	±0.005 dB	±0.025 dB
Group delay:		
1 kHz–5.75 MHz	0	±1 ns
	Increasing to ±2 ns	Increasing to ±3 ns

Provision is made for interworking between 8- and 10-bit systems. The two least significant bits from 10-bit systems are ignored by 8-bit systems, and 10-bit systems read the missing two LSBs from 8-bit systems as zeros.

The 8-bit coding of Recommendation ITU-R BT.601 corresponds to a range of 256 codes, referred to as 0–255. Levels 0 and 255 are reserved for synchronising information, leaving levels 1–254 for signal values. For luminance, a standard level signal extends from black at 16 to white at 235. Thus a small margin for overshoots is retained at each end of the coding range. The colour-difference signals occupy a symmetrical range about a zero chrominance level of 128, thus extending down to level 16 and up to level 240. The same coding levels are specified in ITU-R BT.601 for the 10-bit system, but the increments are 0.25 of each integer level, e.g. codes (decimal notation) 16.00; 16.25; 16.50; 16.75; 17.00, etc. The 8-bit coding levels corresponding to a 100% colour bars waveform[9] are shown in Figure 2.4.24 for the Y, C_B and C_R signals.

Additional gain factors have to be used in the conversion to digital form to normalise the range of the B – Y and R – Y signals, given by:

$$C'_B = 0.546(B' - Y') \qquad (2.4.38)$$

and

$$C'_R = 0.713(R' - Y') \qquad (2.4.39)$$

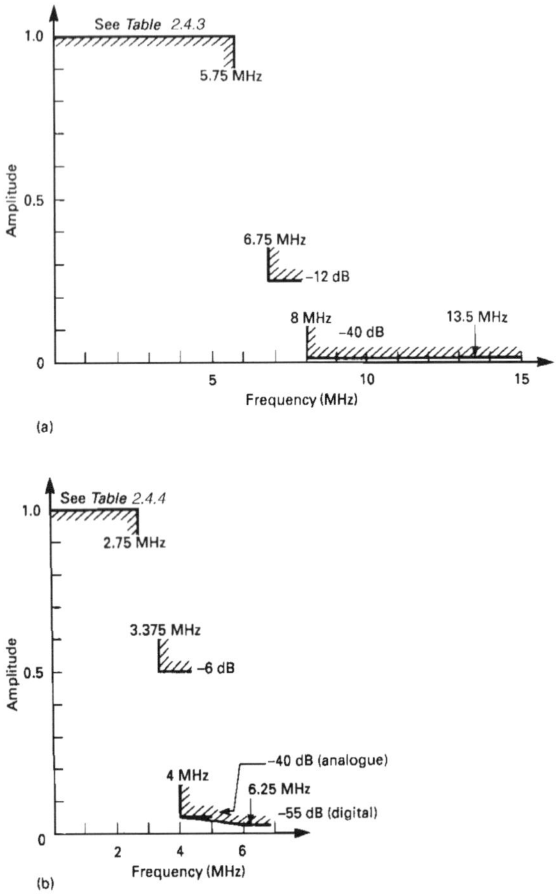

Figure 2.4.23 Amplitude limits for luminance (a) and for colour-difference filters (b) for 4:2:2 component signals. When used for analogue post-sampling conversions, it is assumed that sin x/x correction is included to give, overall, the passband response shown. In (b), a greater stopband attenuation is specified for digital filters because no sin x/x attenuation occurs.

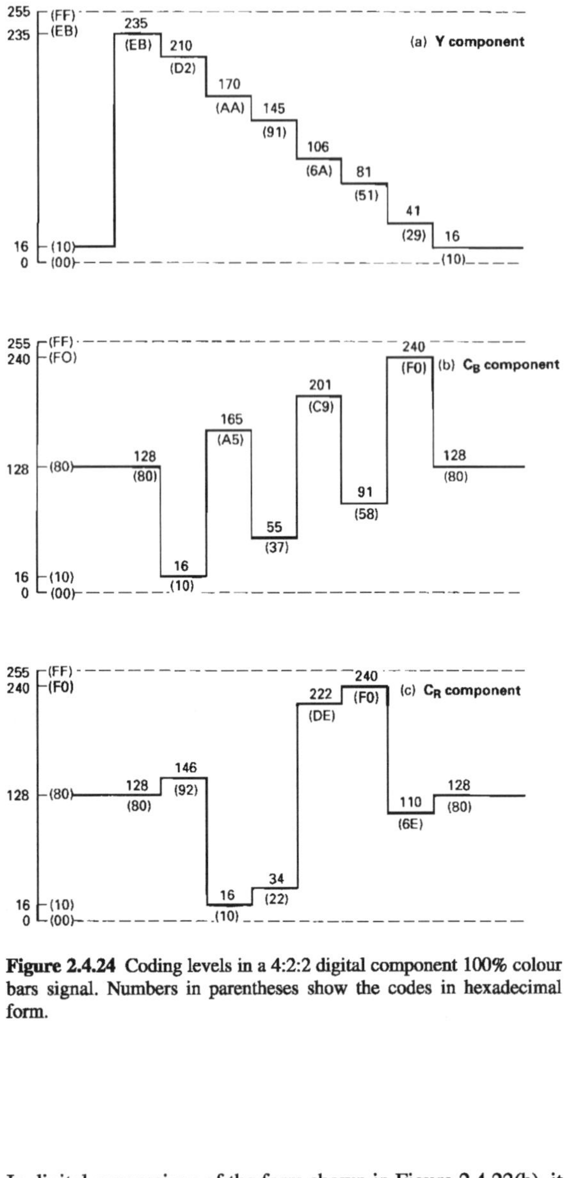

Figure 2.4.24 Coding levels in a 4:2:2 digital component 100% colour bars signal. Numbers in parentheses show the codes in hexadecimal form.

Table 2.4.4 Colour-difference filter ripple and group delay tolerances

Frequency range	*Design limits*	*Practical limits*
Ripple:		
1 kHz–1 MHz	±0.01 dB	Increasing from ±0.01 to ±0.05 dB
1–2.75 MHz	±0.01 dB	±0.05 dB
Group delay:		
1 kHz–2.75 MHz	Increasing from 0 to ±4 ns	Increasing from ±2 to ±6 ns
2.75 MHz–$f_{-3\text{dB}}$		±12 ns

The use of transversal digital filters ensures zero group delay distortion.

In digital conversions of the form shown in Figure 2.4.22(b), it is also necessary to take account of the factor 224/219 to allow for the different quantising ranges of the luminance and colour-difference signals.

2.4.5.7 Synchronising codes

The positions of the active-line samples are marked by *End of Active Video* (EAV) and *Start of Active Video* (SAV) codes, otherwise known as the *Timing Reference Signal* (TRS), which occupy the first and last sample positions respectively in the digital blanking period. Originally defined as a four-word sequence in a 27 MHz multiplexed signal, the format of the codes in the corresponding separate component Y, C_B and C_R signals is shown in Figure 2.4.25.

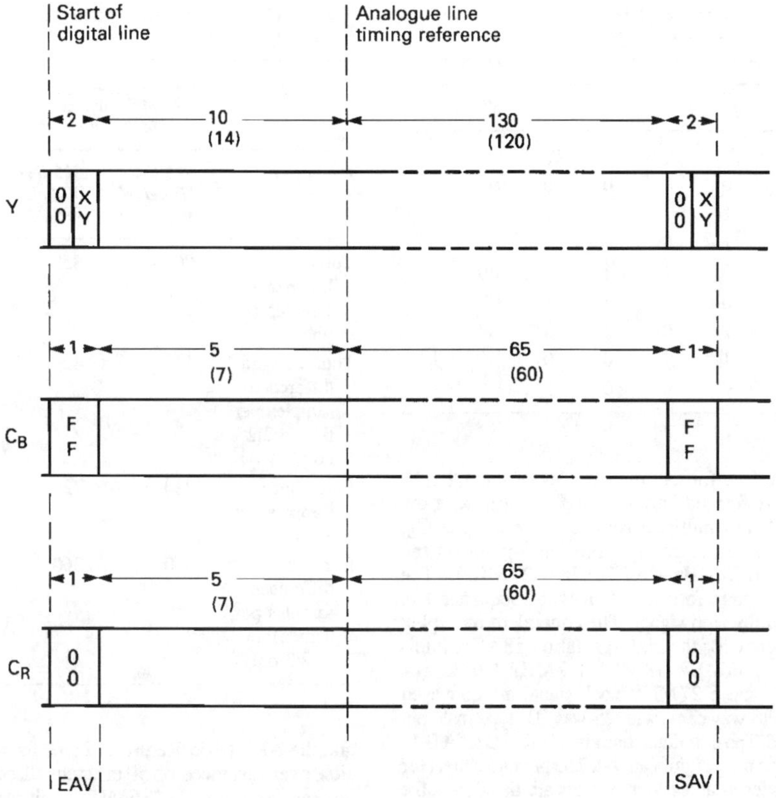

Figure 2.4.25 Positions of the EAV and SAV code words in Y, C_B and C_R digital signals for 625/50 scanning. Values in parentheses apply for 525/60 scanning.

The preamble codes, often referred to by their hexadecimal values FF, 00 and 00, mark the position of the line label word known as XY. In binary form, X has values 1FVH, with F giving odd/even field information, V denoting vertical blanking and H differentiating between EAV and SAV codes. The values of F, V and H are defined in Table 2.4.5. The Y portion of the XY word provides a 4-bit parity check word $P_3P_2P_1P_0$ for the X data, as defined in Table 2.4.6.

The beginning of the EAV code marks the beginning of the digital line period, which starts slightly before the corresponding analogue line. The start of the digital field occurs at the start of the digital line, which includes the start of the corresponding analogue field. The half lines of active picture in the analogue signal are accommodated as full lines in the digital signal. Unused samples in the digital blanking intervals are set to blanking level, i.e. 16 (10 hex) for Y, and 128 (80 hex) for C_B and C_R.

2.4.5.8 Component parallel interface

It should be noted that the parallel interface is not commonly used in modern equipment, but is described for completeness, and also because the more widely used serial interface is derived from it.

Table 2.4.5 Timing reference identification in codes

Line numbers	*F*	*V*	*H(EAV)*	*H(SAV)*
625/50				
1–22	0	1	1	0
23–310	0	0	1	0
311–312	0	1	1	0
313–335	1	1	1	0
336–623	1	0	1	0
624–625	1	1	1	0
*525/60**				
1–3	1	1	1	0
4–9	0	1	1	0
10–263	0	0	1	0
264–265	0	1	1	0
266–272	1	1	1	0
273–525	1	0	1	0

*Line numbers are in accordance with current engineering practice for 525/60 in which numbering starts from the first line of equalising pulses, instead of the first line of broad pulses as had previously been used.

Table 2.4.6 Parity values for timing reference codes

Binary								Hexadecimal
X	Y							XY
1	F	V	H	P_3	P_2	P_1	P_0	
1	0	0	0	0	0	0	0	80
1	0	0	1	1	1	0	1	9D
1	0	1	0	1	0	1	1	AB
1	0	1	1	0	1	1	0	B6
1	1	0	0	0	1	1	1	C7
1	1	0	1	1	0	1	0	DA
1	1	1	0	1	1	0	0	EC
1	1	1	1	0	0	0	1	F1

In the parallel interface format, the samples of the 4:2:2 Y, C_B and C_R signals are formed into a 27 MHz multiplex in the order C_B, Y, C_R, Y. In the multiplex sequence, the samples C_B, Y, C_R are co-sited (all correspond to the same point on the picture: 13.5 MHz + 6.75 MHz + 6.75 MHz = 27 MHz). The EAV and SAV codes thus form the four-word sequence FF, 00, 00, XY in the multiplexed signal. The equivalent multiplex rate and clock frequency for the high-resolution 18 MHz luminance sampling rate is 36 MHz (18 MHz + 9 MHz + 9 MHz).

The 10-bit samples and a 27 MHz clock signal are conveyed down 11 pairs in a multi-way cable with 25-way 'D' type male pin connectors at each end. The individual bits are labelled DATA 0–9, with DATA 9 being the most significant bit. The pin allocations for the individual signals for 8- and 10-bit systems are listed in Table 2.4.7. Equipment inputs and outputs both use female sockets.

Signals on the parallel interface use logic levels compatible with ecl (emitter-coupled logic) balanced drivers and receivers,

Table 2.4.7 Pin connections in the 25-way parallel interface (component and composite)

Pin	Eight-bit	Ten-bit	Pin	Eight-bit	Ten-bit
1	Clock A	Clock A	14	Clock B	Clock B
2	System ground	System ground	15	System ground	System ground
3	Data 7A	Data 9A	16	Data 7B	Data 9B
4	Data 6A	Data 8A	17	Data 6B	Data 8B
5	Data 5A	Data 7A	18	Data 5B	Data 7B
6	Data 4A	Data 6A	19	Data 4B	Data 6B
7	Data 3A	Data 5A	20	Data 3B	Data 5B
8	Data 2A	Data 4A	21	Data 2B	Data 4B
9	Data 1A	Data 3A	22	Data 1B	Data 3B
10	Data 0A	Data 2A	23	Data 0B	Data 2B
11	Spare A-A	Data 1A	24	Spare A-B	Data 1B
12	Spare B-A	Data 0A	25	Spare B-B	Data 0B
13	Cable shield	Cable shield			

The notation A and B is used to denote the two terminals of a balanced pair. For a logic 0, A is negative with respect to B, and for a logic 1, A is positive with respect to B.

Table 2.4.8 Sampling rates, and numbers of samples per line for 625 and 525 line systems

	Sampling rate: 13.5 MHz lum. and 6.75 MHz colour diff.		Sampling rate: 18 MHz lum. and 9 MHz colour diff.	
Parameter	625 line, 50 fields/sec.	525 line, 60 fields/sec.	625 line, 50 fields/sec.	525 line, 60 fields/sec.
Total luminance samples per line	864	858	1152	1144
Total chroma difference samples per line (4:2:2 only)	432	429	576	572
Luminance samples per active line	720	720	960	960
Chroma difference samples per active line (4:2:2 only)	360	360	480	480

and the NRZ (Non-Return to Zero) format allows for transmission over distances of 50 m unequalised or 200 m with equalisation (less for the 36 MHz multiplex signal). The rising edge of the essentially square, 27 MHz clock waveform is the active edge and is timed to occur in the middle of the bit-cell of the data waveforms at the sending end.

The parallel component digital interface standards for 525- and 625-line signals for the 13.5 MHz sampling frequency are specified in Recommendation ITU-R BT.656. It is also specified for 625-line signals in EBU Tech. 3267, and for 525-line signals in SMPTE 244M. The equivalent interface for the 18 MHz luminance sampling rate is specified in ITU-R Recommendation BT.1302.

2.4.5.9 Component Serial Digital Interface (SDI)

Conversion to the serial interface format allows the signals of the 27 MHz multiplex to be conveyed down a single bearer, either coaxial cable or optical fibre. The 8- or 10-bit signal is first converted to serial form, least significant bit first, using a 270 MHz clock derived from the 27 MHz clock. For the 18 MHz luminance sampling frequency, the serial clock frequency is 360 MHz, which can be derived from the 36 MHz clock. For 8-bit signals, two least significant bit zeros are added to make up the 10-bit signal. Auxiliary data such as AES-EBU digital audio can be added at this point if required. It should be noted that there are several different, and incompatible, systems for embedding AES-EBU digital audio. The main difference is that some systems insert the audio data in both the line and field blanking intervals and some insert it only during line blanking intervals. Details can be found in the SMPTE 272M document. The resulting multiplexed Non-Return to Zero (NRZ) signal is then scrambled to provide the Non-Return to Zero Inverted

(NRZI) 270 Mbit/s (or 360 MHz) output signal. The scrambling process prevents long sequences of 'ones' and 'zeros' from occurring in the output, which would otherwise make accurate clock recovery at the receiver difficult and increase the bandwidth of the SDI signal. The NRZI signal also has the property of being polarity insensitive because a '1' is represented by a transition in either direction, and a '0' by no transition.

The serial digital interface standards for component coding for the 13.5 MHz sampling frequency are specified in EBU Tech. 3267 for 625-line and SMPTE 259M for 525-line signals. The parallel and serial digital interfaces for 625- and 525-line component coding for the 13.5 MHz sampling frequency for ITU-R BT.601 coded signals are also specified in ITU-R recommendation BT.656. The equivalent interface for the 18 MHz luminance sampling rate is specified in ITU-R Recommendation BT.1302.

The SDI electrical interface specifies a coaxial cable impedance of 75 ohms, with a nominal signal amplitude of 800 mV peak-to-peak. The signal can be transmitted over lengths of suitable serial digital video cable up to about 300 m (less for the 360 MHz multiplex signal) when equalisers are used to compensate for the $1/\sqrt{f}$ loss of the cable. A fibre-optic interface is also specified. This uses an LED emitter with a wavelength of 1280–1380 nm.

2.4.5.10 Error Detection and Handling (EDH)

Errors in the transmission of SDI data can be detected if a Cyclic Redundancy Check (CRC) checksum is calculated from all the Y, C_B, and C_R samples in the active picture area of each field, and added to the signal at the sending end of a transmission path such as a length of cable. A second EDH CRC packet is calculated and added to the signal for all the data in each field period, including embedded data such as audio. The CRC packets are inserted in the line blanking interval immediately prior to the SAV TRS for lines 5 and 318. At the receiving equipment, the two CRCs are again calculated from the incoming video samples, and compared with the previously inserted CRC packets. Any difference between the two checksums indicates that one or more data errors have occurred on the transmission path. Where two or more transmission paths are cascaded, flags can be set by CRC checksum error detectors, and added to new CRC packets which are inserted onto the ongoing signal path, to indicate whether upstream errors were present. Monitoring equipment connected to a CRC checker at the final receiving equipment can thus be used to indicate the source of errors, and whether the errors will affect just the active picture area or the whole signal. The quantity of errors are measured in errored seconds.

2.4.6 Digital Encoding of Composite Analogue Signals

ITU-R BT.601 and the equivalent SMPTE standards for digital component coding and component SDI are the common standards for modern TV studios. However, there will be instances where vision signal sources remain as 625-line PAL or 525-line NTSC format signals. There are several options when it is required to convert these signals to digital formats: they can be converted directly to component SDI (see Section 2.4.5.9); they can be converted to digital compressed formats such as MPEG-2 (see MPEG-2 chapter); or they can be converted to one of the composite digital formats. One proprietary composite format uses 10-bit, 13.5 MHz sampling, which is the same as that used for ITU-R BT.601 luminance sampling. The resulting 135 Mbit/s serial signal is suitable for transmission over 140 or 155 Mbit/s telecommunications links. The surplus capacity can be used for ancillary data such as multi-channel audio. Further alternatives are the digital composite encoding and interface standards described in detail below (see Section 2.4.6.1). It should be noted that such converted signals will not be of as high a precision as component originated signals because of the analogue chrominance band-limiting filtering, and also because of the coarser quantising level steps, as explained below. The use of digital composite encoding will become less common as analogue equipment is phased out.

2.4.6.1 Digital Composite Encoding

Unlike component coding, for composite coding the whole line and field is sampled so that timing information and the chroma burst phase can be recovered. The quantising level steps will be larger (i.e. coarser) than for 10-bit component coding because of the need to code levels above peak white and below black level where there is high saturation colour information, as well as to include the sync pulse tips. The maximum coded level is 903.3 mV for 625-line PAL and 937.7 mV for 525-line NTSC signals, above black level, and the minimum is 300 mV for 625-line PAL and 285.7 mV for 525-line NTSC signals, below black level. The sampling levels are shown in Figure 2.4.26 and in Table 2.4.9.

After clamping, the analogue composite signal is sampled at four times the colour subcarrier frequency. This is 17.7 MHz for PAL, giving 1135 samples per line, 948 of which are in the active line, and 14.3 MHz for NTSC, giving 910 samples per line, 786 of which are in the active line, both with 8- or 10-bit resolution. If 8-bit sampling is used, the two least significant bits of the required 10-bit output signal are set to zero. The sampling points are locked to the colour subcarrier, with a timing relationship for PAL signals such that sampling never occurs when the subcarrier in the active picture period is at or near a positive peak. This technique allows high colour saturation picture information, which has chroma subcarrier signals with peaks that exceed the maximum coded value of 903.3 mV, to be correctly sampled.

2.4.6.2 Composite Parallel Digital Interface

The parallel composite interface is similar to the 10-bit parallel component interface in that it uses 25-way 'D' connectors with the same pin number allocations (see Table 2.4.7). The data rate is equal to the sampling frequency (17.7 MHz for 625-line PAL and 14.3 MHz for 525-line NTSC). It will be appreciated from the above description of composite coding that component and composite signals are not compatible with each other. The parallel composite digital interface standard is specified in EBU Tech. 3280 for 625-line PAL and SMPTE 244M for 525-line NTSC coded signals.

2.4.6.3 Composite Serial Digital Interface (Composite SDI)

As with the serial component interface, the serial composite digital interface signal is NRZI, polarity insensitive, with the

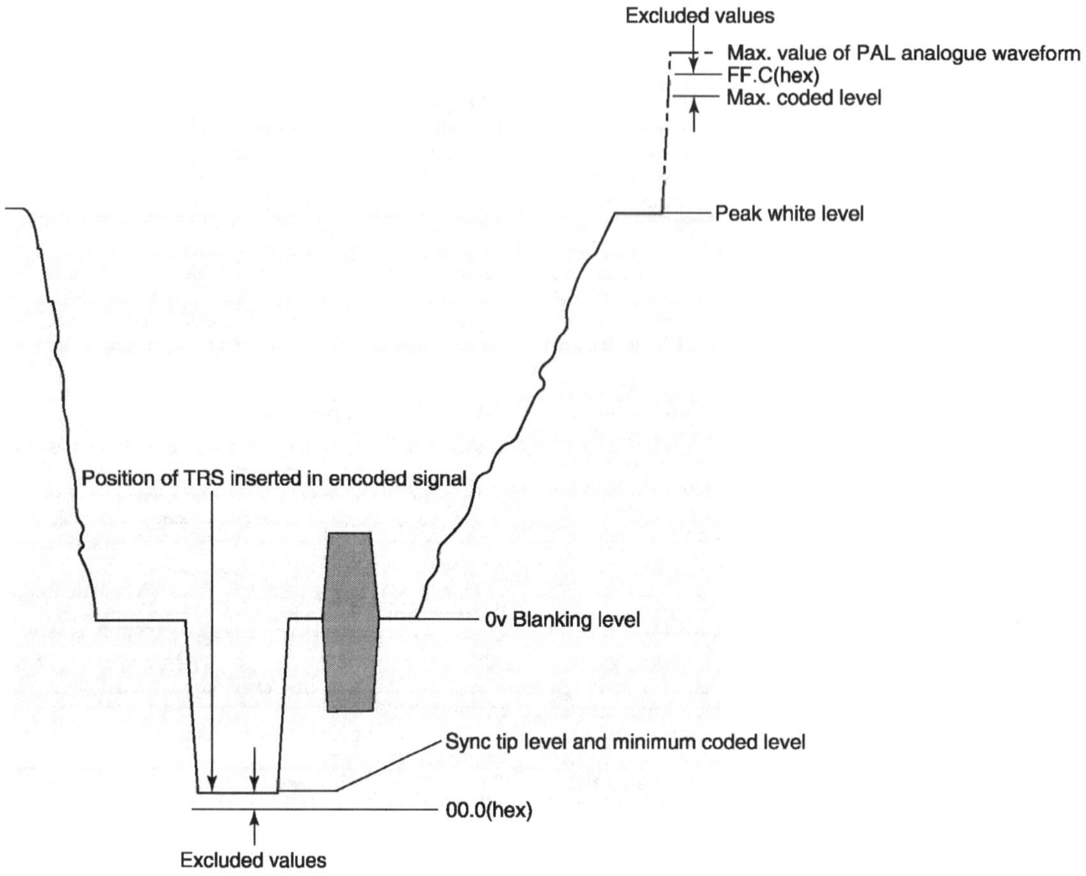

Figure 2.4.26 Coding levels for sampling of PAL and NTSC signals.

Table 2.4.9 Composite PAL and NTSC signal levels and corresponding coded values

Signal/codes	*Level (mV)*	*PAL Hex Code*	*Binary*	*Level (mV)*	*NTSC Hex Code*	*Binary*
Excluded codes	931.1	FF.C	1111 1111 11	998.7	FF.C	1111 1111 11
		FF.8	1111 1111 10		FF.8	1111 1111 10
		FF.4	1111 1111 01		FF.4	1111 1111 01
	909.5	FF.0	1111 1111 00		FF.0	1111 1111 00
Maximum coded value	903.3	FE.C	1111 1110 11	937.7	F3.C	1111 0011 11
Peak white	700	D3.0	1101 0011 00	714.3	C8.0	1100 1000 00
Blanking level	0.0	40.0	0100 0000 00	0.0	3C.0	0011 1100 00
Minimum coded value (Sync tips)	−300.0	01.0	0000 0001 00	−285.7	04.0	0000 0100 00
Excluded codes	−304.8	00.C	0000 0000 11		00.C	0000 0000 11
		00.8	0000 0000 10		00.8	0000 0000 10
		00.4	0000 0000 01		00.4	0000 0000 01
	−308.3	00.0	0000 0000 00	−306.1	00.0	0000 0000 00

Note: The first two digits of the hex code, and the first eight digits of the binary code, are the integer value and the last digit of the hex code, and the last two digits of the binary code are the fractional value same voltage levels. However, the serial bit rate is 177.34 Mbit/s for 625-line PAL and 143 Mbit/s for 525-line NTSC (10 times the parallel composite data rate). Due to the sampling of the whole composite line, there are no timeslots into which auxiliary data such as audio can be inserted. There is no Start of Active Video (SAV) or End of Active Video (EAV) signal as in component coding, but a similar Timing Reference Signal (TRS) is inserted for synchronising purposes by replacing sample numbers 967–970 for PAL, or 790–794 for NTSC, during the line sync-pulse bottoms (see Figure 2.4.26). This TRS consists of a four-word group using the excluded codes as follows: FF.C, 00.0, 00.0 and 00.0. (The first two digits of these hex codes are the integer part and the last digits the fractional part, using the same notation as for the video sample codes.) This allows decoders, and other equipment, to lock on to the signal and so identify the boundaries between 10-bit words, in what is essentially a synchronous signal (i.e. sample words do not include start and stop bits). A line ID word is added after the TRS to identify the current field in the eight-field PAL or four-field NTSC sequence, and the line numbers in field blanking, for video synchronisation purposes. The line ID word codes for PAL signals are shown in Table 2.4.10.

The component and composite serial digital signals are, of course, not compatible with each other. The electrical interface has the same 800 mV peak-to-peak nominal send level, and uses 75-ohm coaxial cable as for the component SDI. The fibre-optic interface also has the same physical characteristics as specified for component. The serial digital interface standards for composite coding are specified in EBU Tech. 3280 for 625-line PAL and SMPTE 259M for 525-line NTSC signals.

Table 2.4.10 Line ID word codes for composite serial digital PAL signals

Field	*Lines*	*Bit 0 (LSB)*	*Bit 1*	*Bit 2*
1	1–313	0	0	0
2	314–625	0	0	1
3	1–313	0	1	0
4	314–625	0	1	1
5	1–313	1	0	0
6	314–625	1	0	1
7	1–313	1	1	0
8	314–625	1	1	1

Line ID word: Field number codes.

Lines	*Bit 3*	*Bit 4*	*Bit 5*	*Bit 6*	*Bit 7*	*Bit 8*	*Bit 9 (MSB)*
(not used)	0	0	0	0	0	*	*
1 & 314	1	0	0	0	0	*	*
2 & 315	0	1	0	0	0	*	*
3 & 316	1	1	0	0	0	*	*
(4–28 and 317–342: binary count sequence continues)						*	*
29 & 343	1	0	1	1	1	*	*
30 & 344	0	1	1	1	1	*	*
All other lines:	1	1	1	1	1	*	*

* Bit 8 is even parity for bits 0 to 7, and bit 9 is the complement of bit 8
Line ID word: Line number codes during and following field blanking.

2.4.7 Other Video Encoding Formats

There are several other video colour encoding formats not covered elsewhere that should be mentioned:

- 34 Mbit/s Video. This can be used where a small amount of compression is acceptable, and there is a requirement for programme material to be routed to distant studios over 34 Mbit/s telecommunications links. Inputs and outputs to the encoders and decoders are usually analogue video and audio, or component SDI and AES-EBU digital audio.
- DV, DVCAM, DVCPRO and DVCPRO50. These are compressed formats commonly used to connect portable equipment such as camcorders to computers for editing purposes. The interface for these signals is often IEEE 1394, also known as 'Firewire'.
- M-JPEG. This format uses the same JPEG compression system that is often used to compress still images for storage as computer files. Unlike MPEG coding, the data for each M-JPEG frame is a complete picture, and is independent of data contained in adjacent frames. This makes it particularly suitable for off-line editing using computer software. M-JPEG signals are usually routed, at various bit rates, on high-capacity ethernet LANs. M-JPEG is also a computer file type. The inputs and outputs to the encoders and decoders are usually analogue video and audio, or component SDI and AES-EBU digital audio.

2.4.8 Multiplexed Analogue Component (MAC)

Although its use is now in decline, one version of the MAC system was used for Direct Broadcasting by Satellite (DBS) in the UK for a few years in the late 1980s and early 1990s. The MAC system avoids the unwanted cross-colour effects of the PAL system by time-compressing the luminance and chrominance signals for each line separately, and sending them sequentially, together with digital sound, by time-division multiplexing within the standard PAL/SECAM 64 µs line period. A further advantage was that wide-screen 16:9 aspect ratio pictures can be accommodated due to the time-compression technique. The D-MAC signal used for DBS occupied a wider baseband width (over 8 MHz) than PAL 'I' signals (5.5 MHz), so could not have been transmitted in the available 8 MHz uhf UK terrestrial TV channels. Also, the signal was not compatible with most existing telecommunications links. The development of MPEG-2 digital compression techniques, which allowed multiple TV channels to be broadcast on one carrier, was one of the factors that resulted in the decline of MAC systems for broadcast TV transmission.

References

1. Bingley, F.J. Colorimetry in color television – parts 1, 2 and 3, *Proc. IRE*, Part 1, **41**, 838–851 (1953), Parts 2 and 3, **42**, 48–58 (1954).
2. CCIR Report 476-1, Colorimetric Standards in Colour Television, *XVIth Plenary Assembly*, Dubrovnik, XI-1, pp. 42–43 (1986).
3. Bruch, W. The PAL colour TV system – basic principles of modulation and demodulation, *NTZ Communications Journal*, **3**, 255–268 (1964).
4. EBU Technical Statement No. D 23-1984 (E), Timing Relationship Between the Subcarrier Reference and the Line Synchronizing Pulses for PAL Recordings, Technical Centre, Brussels (1984).
5. Federal Communications Commission Document No. 53-1663, Compatible Color Television (1953).
6. Kaiser, A. Comb filter improvement with spurious chroma deletion, *SMPTE Journal*, **86**, 1–5 (1977).
7. De France, H. Le systéme de télévision en couleurs sequentiel simultané, *L'Ondé Electr.*, **38**, 479–483 (1958).
8. Sabatier, J. and Chatel, J. Qualité des signaux de télévision en bande de base, *Revue Radiodiffussion-télévision*, **70**, 12–21 (1981).
9. CCIR Recommendation 471-1, Nomenclature and Description of Colour Bar Signals, *XVIth Plenary Assembly*, Dubrovnik, XI-1, pp. 39–41 (1986).
10. CCIR Recommendation 601-1, Encoding Parameters of Digital Television for Studios, *XVIth Plenary Assembly*, Dubrovnik, XI-1, pp. 319–328 (1986).
11. Guinet, Y. Evolution of the EBU's position in respect of the digital coding of television, *EBU Rev. Tech.*, **187**, 111–117 (1981).
12. Jones, A.H. Digital television standards, IBC80, *IEE Conf. Publ.*, No. 191, 79–82 (1980).
13. A report of digital video demonstrations using component coding, Special Issue, *SMPTE Journal*, **90**, 922–971 (1981).
14. ITU-R Recommendation BT656-4, Interfaces for Digital Component Video Signals in 525-line and 625-line Television Systems, p. 13 (1998).

Bibliography

Composite systems – general

CCIR Report 624-3, Characteristics of Television Systems, *XVIth Plenary Assembly*, Dubrovnik, XI-1, pp. 1–33 (1986).

Clarke, C.K.P. BBC Research Dept Report RD 1986/2, Colour Encoding and Decoding Techniques for Line-locked Sampled PAL and NTSC Television Signals (1986).

Sims, H.V. *Principles of PAL Colour Television and Related Systems*, Iliffe, London (1969).

Colorimetry

Hunt, R.W.G. *The Reproduction of Colour in Photography, Printing and Television*, 4th Edn, Fountain Press, Tolworth (1987).

PAL system

Bruch, W. PAL, a variant of the NTSC colour TV system – selected papers I and II, *Telefunken-Zeitung* (1965 and 1966).

ITU-R Recommendation BT.470–6(1998). Conventional Television Systems.

Specification of Television Standards for 625-line System I Transmissions in the United Kingdom (second impression), Radio Regulatory Division, Dept. of Trade and Industry, London (1985).

NTSC System

Fink, D.G. (ed.). *Colour Television Standards – Selected Papers and Records of the NTSC*, McGraw-Hill, New York (1955).

Pritchard, D.H. U.S. colour television fundamentals: a review, *SMPTE Journal*, **86**, 819–828 (1977).

SECAM system

Weaver, L.E. *The SECAM Color Television System*, Tektronix, USA (1982).

4:2:2 Digital components and Digital composite

ITU-R Recommendations BT601-5, Encoding Parameters of Digital Television for Studios (1995).

ITU-R Recommendation BT656-4, Interfaces for Digital Component Video Signals in 525-line and 625-line Television Systems (1998).

CCIR Report 629-3, Digital Coding of Colour Television Signals, *XVIth Plenary Assembly*, Dubrovnik, XI-1, pp. 329–337 (1986).

Devereux, V.G. Performance of cascaded video PCM codecs, *EBU Rev. Tech.*, **199**, 114–130 (1983).

EBU Tech. 3267-E, Interfaces for 625-line Digital Video Signals at the 4:2:2 level of CCIR Recommendation 601, 2nd Edn (January 1992).

EBU Tech. 3280-E, Specification of Interfaces for 625-line Digital PAL Signals (April 1995).

Fibush, K.D. *A Guide to Digital Television Systems and Measurements*, Tektronix (1994).

Harris, P.J. Untitled handout, ref. TIC 3929, BBC Centre for Broadcasting Skills Training (1998).

Recommendation ITU-R BT.601-5, Studio Encoding Parameters of Digital Television for Standard 4:3 and Widescreen 16:9 Aspect Ratios.

Recommendation ITU-R BT.656-4, Interfaces for Digital Component Video Signals in 525-line and 625-line Television Systems Operating at the 4:2:2 level of Recommendation ITU-R BT.601 (Part A).

Recommendation ITU-R BT.1302, Interfaces for Digital Component Video Signals in 525-line and 625-line Television Systems Operating at the 4:2:2 level of Recommendation ITU-R BT.601 (Part B).

Stickler, M.J., Nasse, D. and Bradshaw, D.J. The EBU bit-serial interface for 625-line digital video signals, *EBU Rev. Tech.*, **212**, 181–187 (1984).

Acknowledgements

The original author wishes to thank his colleague Andrew Oliphant for reading the manuscript and the Director of Engineering of the British Broadcasting Corporation for permission to contribute this section.

E. P. J. Tozer
Zetrox Broadcast Communications

2.5 Timecode

2.5.1 Introduction

Timecode is an essential component of the professional broadcast system, providing a reference and synchronising signal for audio and video programme material.

2.5.2 EBU and SMPTE, DF and NDF Timecodes

Timecode indicates time to video frame accuracy with a 24-hour range as HH:MM:SS:FF, typically displayed as either +/− 12 hours or 0–24 hours, i.e. 00:00:00:00 to 23:59:59:24(29) or −12:00:00:00 to +11:59:58:24(29).

Timecode which counts 0 to 24 frames is variously referred to as EBU (European Broadcast Union), 25 frame or PAL timecode, and is used with 25 FPS (frames per second) video. Timecode which counts 0 to 29 FPS comes in two varieties, 'Drop Frame' (DF) and 'Non-Drop Frame' (NDF) timecode. Drop frame timecode is in most common use today with the NTSC 29.97 frame rate video. NDF timecode would normally be used with (uncommon) 30 FPS 'non-drop frame' video. Both DF and NDF timecode are referred to as SMPTE (Society of Motion Picture and Television Engineers) timecode.

NDF timecode counts 0 to 29 frames. If this count is used with 29.97 Hz (30 × 1000/1001 Hz) video there will be approximately a 0.1% count error of timecode time with respect to real time; this is approximately equivalent to timecode running 3.6 seconds per hour too slow. To correct this discrepancy DF timecode misses the count of the first two frames of the first second of each minute, so 01:57:59:29 is followed by DF time value 01:58:00:02; unfortunately if this is done every minute the error will be over-corrected, so every tenth minute frames are not dropped, thus 01:49:59:29 is followed by DF time 01:50:00:00. This correction will match DF time to real time to within approximately 2.6 frames per day; to eliminate the residual error the timecode generator can be reset each midnight.

2.5.3 Timecode Frames

To represent the timecode data a timecode 'frame' is created consisting of 80 bits of information (see Figures 2.5.1 and 2.5.2).

The information in the timecode frame not only contains the timecode itself but also the following.

2.5.3.1 Timecode bits

The individual HH:MM:SS:FF digits of the timecode data are encoded as 4-bit binary BCD (Binary Coded Decimal) values; tens of hours can only take the value 0, 1 or 2 and so only require 2 bits of the timecode frame, whereas hours units can be any value from 0 to 9 and thus use 4 bits of the timecode frame (see Figure 2.5.3).

2.5.3.2 Thirty-two User Bits

The purpose of these is not defined but typically they would be set during the initial programme recording and might be set to contain the date of the recording or the tape number.

2.5.3.3 Colour Frame Flag

Colour frame flag is used to indicate the start of the recorded video's PAL eight-field or NTSC four-field sequences, as an aid to later editing of the tape.

2.5.3.4 Drop Frame

This indicates whether the timecode format is Drop Frame or Non-Drop Frame (SMPTE timecode only).

2.5.3.5 Phase correction Bit

This may be set to a 1 or a 0 depending on the value of all the other bits in the timecode frame and is used to ensure that after

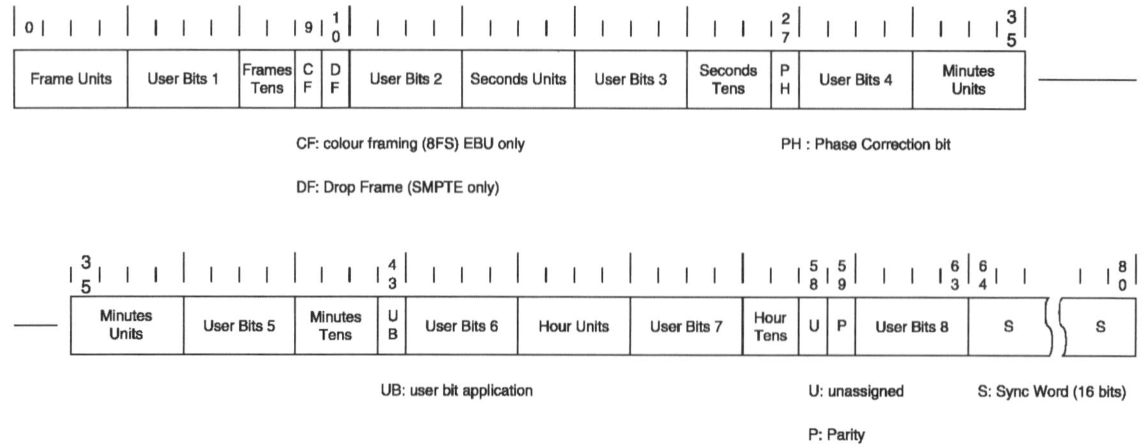

Figure 2.5.1 Linear timecode frame.

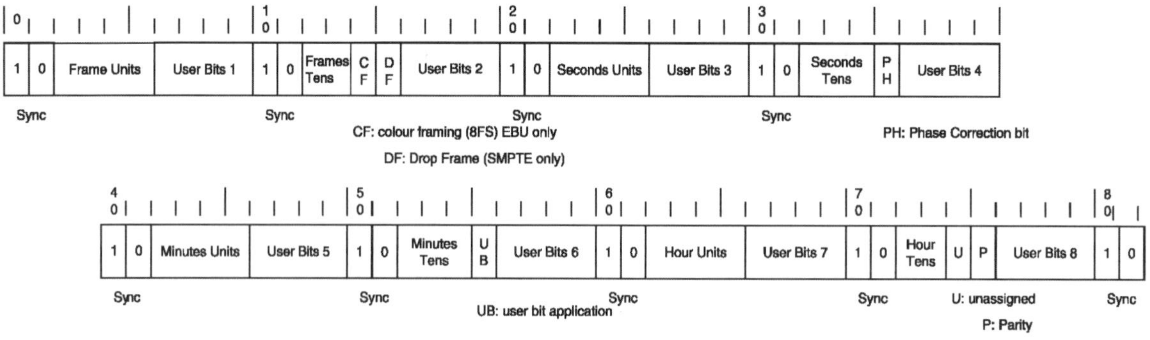

Figure 2.5.2 Vertical interval timecode frame.

Binary value	Decimal value
0000	0
0001	1
0010	2
0011	3
0100	4
0101	5
0110	6
0111	7
1000	8
1001	9
(1010)	Illegal BCD value
(1011)	Illegal BCD value
(1100)	Illegal BCD value
(1101)	Illegal BCD value
(1110)	Illegal BCD value
(1111)	Illegal BCD value

Figure 2.5.3 Binary coded decimal values and the decimal equivalents.

bi-phase mark modulation the first bit of the timecode starts with a positive edge.

2.5.3.6 Unassigned

This is unassigned.

2.5.3.7 Parity

The parity bit can be used as a simple error check on the timecode data.

2.5.3.8 Sync Word

The sync word pattern is 0011111111111101. The 12 1s in the centre of the pattern can be used by a timecode decoder to lock to the timecode frequency when the timecode is being played back fast or slow (e.g. in tape shuttle). The start 00 and end 01 can be used to detect the playback direction, i.e. forward or reverse.

2.5.4 Recording Timecode

Timecode is recorded in two different ways, LTC (Linear TimeCode) and VITC (Vertical Interval TimeCode).

2.5.4.1 LTC

LTC is recorded by modulating the timecode 1s and 0s using a 'bi-phase mark' system (Figure 2.5.4).

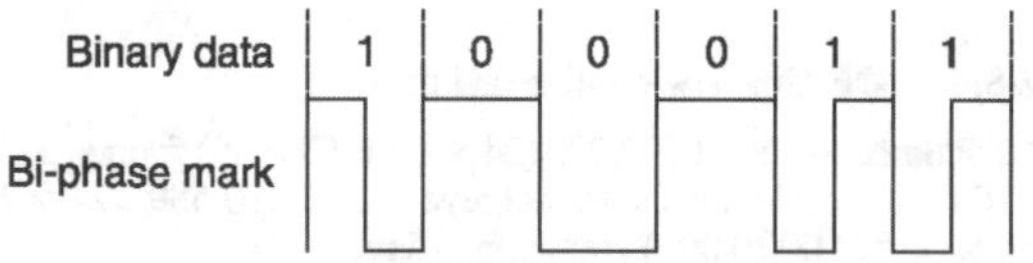

Figure 2.5.4 LTC bi-phase mark modulation.

Bi-phase mark is essentially a frequency modulation system where 1s are recorded at 1 kHz (1.2 kHz) and 0s at 2 kHz (2.4 kHz). LTC is an audio-like signal and can be recorded on an audio track or, more usually, on a dedicated timecode track on a VTR or as a narrow centre track, between audio tracks on a quarter-inch audio recorder.

2.5.4.2 Burnt-in Timecode

Timecode can be 'burnt' into the pictures it references to provide a user-readable display, where it may be necessary to manually note the timecode value of a particular scene or transition (Figure 2.5.5).

Figure 2.5.5 Timecode burnt into video.

The burn-in facility is normally provided as a VTR playback function, and is often on a supplementary video output, meaning that both clean and burnt-in versions of the video are available.

2.5.4.3 VITC

VITC is recorded by placing the timecode frame on one or more lines of the video vertical interval; typically this would be lines 19 and 21, though the actual video lines used varies from broadcaster to broadcaster (Figure 2.5.6).

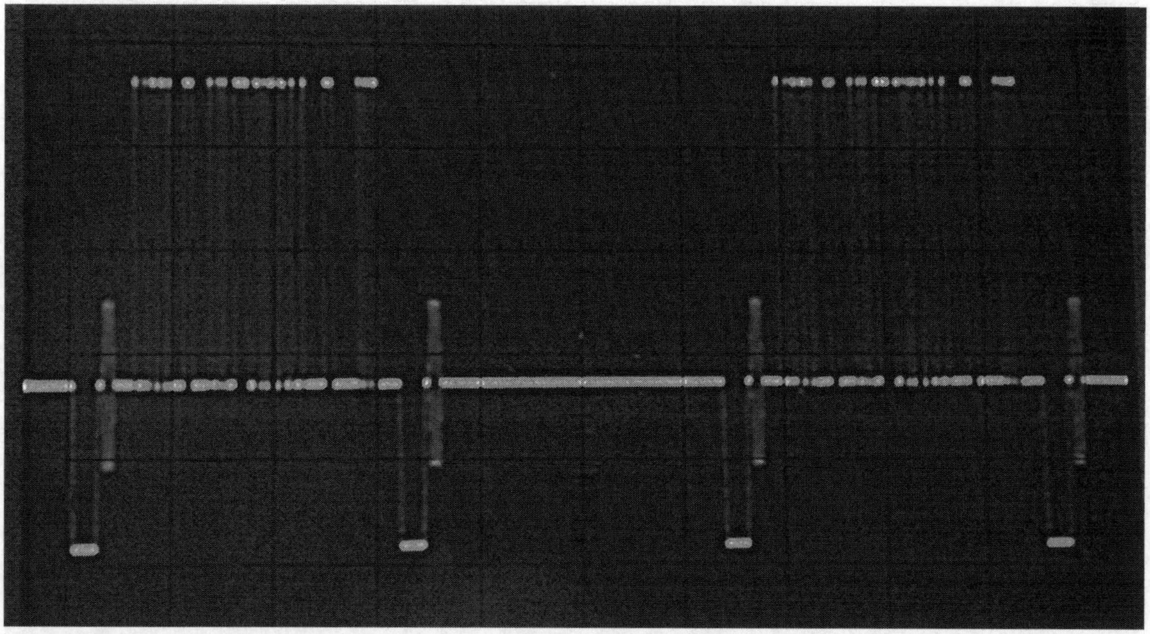

Figure 2.5.6 VITC on lines 19 and 21.

2.5.4.4 VITC and LTC

Normally in a videotape recording LTC is laid down as a linear track along the tape edge and VITC recorded in the video. When a VTR is locating a particular timecode value, maybe an edit point, the LTC can be read during the fast wind of the tape, and the VITC can be read as the VTR moves slowly to fine-adjust the tape position to the correct timecode location. Clearly for this timecode locate process to work, the VITC and LTC must have been recorded with the same values for the same video frames. Surprising as it may seem, it is not an uncommon problem for LTC and VITC to contain different timecodes, leading to problems with timecode location.

2.5.4.5 Timecode Master and Slave

In a production using multiple recorders it is normal to synchronise the timecode of all recorders in the system. To do this on timecode generator, the master and other timecode generators will be 'slaved' to the master timecode generator, meaning that their time will be derived from the master timecode source. This process is also known as timecode regeneration and jam synching timecode.

2.5.4.6 Timecode Free Run and Record Run

Timecode generators incorporated into recorders normally have two operating modes, 'free run' and 'record run'. In free run mode the timecode generator starts counting and continues irrespective of whether the recorder is recording or not; a normal use of this would be to record real time or time of day timecode on the tape as the recording stopped and started. In record run mode the timecode generator will run only when the recorder is running, producing a set of continuous timecode values from the beginning to the end of the tape recording.

2.5.4.7 DF timecode calculation

NDF timecode has 2,592,000 ($24 \times 60 \times 60 \times 30$) frames per day.

DF timecode needs to achieve 2,589,410.589 ($24 \times 60 \times 60 \times 30 \times 1000/1001$) frames per day.

Starting with 30 frames per second, 2,592,000 frames per day:

2880 ($2 \times 60 \times 24$) frames are dropped by the first second of every minute rule; this over-corrects to 2,589,120 frames per day.

288 ($2 \times 6 \times 24$) frames are restored by the first second of every 10th minute rule, under-correcting to 2,589,408 frames per day.

Gordon Anderson
Technical Training
Tandberg Television Ltd

2.6 Sound in Syncs

'Sound in Syncs' (SiS) is the name for the process of placing digitally encoded sound into the line sync portion of a composite TV signal. Mono SiS (then just called SiS) was developed in the late 1960s by the BBC and was the first hybrid multiplex of analogue and digital signals in common use in broadcasting.

SiS confers all the benefits of digitising a waveform (improved signal-to-noise, low distortion, no degradation with distance) on the audio signal. In addition, SiS also 'marries' the audio and video signals together, so that they can travel through a complex contribution or distribution system without becoming inadvertently separated and without 'lip-sync' problems occurring.

The challenge for the designers of these systems was to introduce a digital signal into a complex analogue waveform without introducing crosstalk into the video ('sound on vision').

2.6.1 Mono Sound in Syncs

Mono SiS was used by broadcasters in the United Kingdom for over 15 years (from 1970 onwards) before being gradually replaced by a 'Dual-Channel' (DC) system. The digital audio was converted back to analogue before being transmitted. The use of SiS improved sound quality everywhere in the UK but the major beneficiaries were those who lived furthest from the London studios.

The EBU Eurovision network used Mono SiS from 1974 until it was superseded by an MPEG-2 digital system in 1997.

Mono SiS was also used by Bell Canada on its terrestrial circuits.

2.6.1.1 Sampling

The audio signal is passed through an analogue low-pass filter with a cut-off frequency of 14 kHz to prevent aliasing. It is then sampled at 31.25 kHz and converted to 10-bit binary words. Applying the well-known formula $(6.02 \times N + 1.76)$ dB reveals that, without additional processing, the Signal-to-Noise Ratio (SNR) will be approximately 62 dB (Full Scale Deflection, (FSD)/quantising noise).

The sampling frequency is twice the video line frequency, and is frequency- and phase-locked to the incoming video waveform into which the digital audio is to be placed. Two audio samples are placed in each line sync pulse.

2.6.1.2 Pre-Emphasis

To improve the performance of the 10-bit Pulse Code Modulation (PCM)–encoding, the signal is processed (before analogue-to-digital conversion) in pre-emphasis and compression sections.

The pre-emphasis curve is similar in shape to the standard CCITT J-17 curve[1] but is level-shifted. Low frequencies pass through the pre-emphasis unaffected and high frequencies are boosted by up to 18 dB. (In the standard curves low frequencies are attenuated by up to 9 dB and high frequencies are boosted by up to 9 dB.)

2.6.1.3 Companding

The audio signal is filtered to remove any signal at 15.625 kHz. A 15.625 kHz pilot tone is then added to the audio signal. This pilot tone is synchronised to the sampling waveform so that its minimum and maximum values coincide with the sampling times.

The compressor reduces the total signal amplitude by up to 18 dB depending on the amount of high-frequency content. In a receiver the pilot tone is used to expand the audio to its original amplitude and a matching de-emphasis curve restores the original frequency response.

The use of companding improves the signal-to-noise ratio by 13 dB, giving an overall performance (75 dB) slightly better than a 12-bit PCM system (74 dB).

2.6.1.4 Sample Reversal, Bit Inversion and Interleaving

PCM-encoded audio has a time-varying average level. If the samples are merely placed in the sync pulse of a video waveform the result would be substantial crosstalk ('sound on vision').

Most of the time the Least Significant Bits (LSBs) of a succession of PCM-encoded audio samples are changing far more rapidly than the Most Significant Bits (MSBs). By reversing the order of the binary word the rapidly changing bits of the signal will be near the beginning of the sync pulse and the slowly changing bits will be near the back of the sync pulse. This is useful since the clamping waveform in an SiS system operates on the back porch, as the sync floor is no longer available (due to the presence of the audio digits).

Unlike video waveforms, most audio signals do not have abrupt changes in level. It follows that two adjacent audio samples will tend to have similar values and will usually have similar-looking binary representations (except across binary boundaries such as 128, 256, 512 etc.). By simply inverting the bits of one of the 10-bit binary words the average DC content of a pair of samples will remain fairly constant.

The final process is to interleave the two binary words bit-by-bit.

The three processes have taken two samples and produced a 20-bit word, which is now 'crosstalk-friendly'. The rapidly modulating parts of the word are at the beginning – as far away from the video-clamping signal as possible – and the mean voltage level of the signal will be approximately constant.

2.6.1.5 Inserting the Bits into the Video Waveform

A marker pulse is added to the beginning of the 20-bit word – this prevents timing ambiguity if the first few bits are all at zero. The 21 bits are shaped into sine-squared pulses with a Half-Amplitude Duration (HAD) of 182 ns and a spacing of 182 ns. The total pulse group has a half-amplitude duration of 3.82 µs (±10 ns) and occupies about 4 µs of the sync pulse base. The sine-squared pulses have negligible energy content above 5.5 MHz and should not be affected or distorted by standard video circuits (See Figure 2.6.1).

During field blanking the pulse groups are still inserted into the video waveform every 64 µs. The equalising pulses are widened when they need to accommodate the pulse groups; the other equalising pulses are unaffected. For display on a monitor the SiS pulses can merely be blanked – but, usually, all sync pulses are restored to their original shape and duration (see Figure 2.6.2).

In the original specification[2] the SiS digits had an amplitude of 1 V peak-to-peak, as this was the largest signal guaranteed to pass through a video processing chain unmolested. The digits were later reduced to 700 mV peak-to-peak to reduce audio crosstalk on vision.

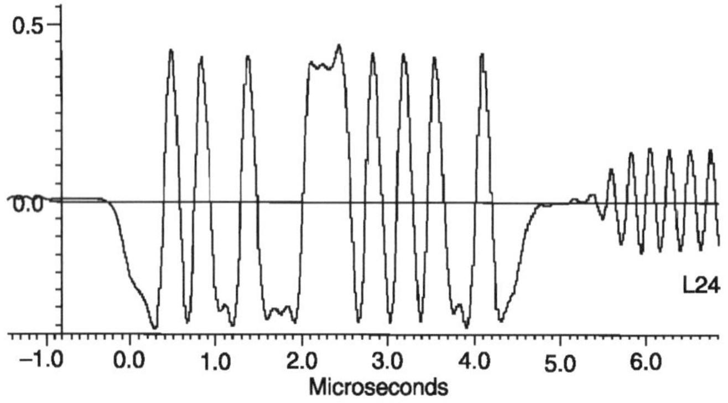

Figure 2.6.1 Mono SiS signal with slightly oversized digits. The two binary words correspond to samples of 500 and 494.

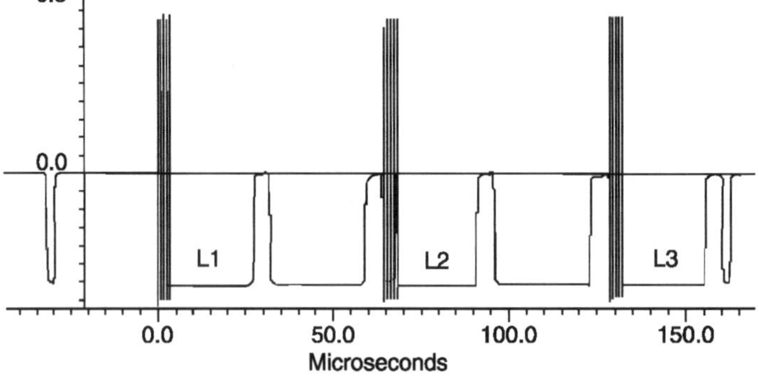

Figure 2.6.2 Mono SiS signal in lines 1 to 3 of the vertical blanking.

2.6.1.6 Recovery of Audio

Recovery of audio is essentially a reversal of the previously described processes. Twenty data bits are extracted from the video sync pulses (and sync pulses are restored to their original shape). The two words are de-interleaved and the inverted word is re-inverted back to its original value. The samples are used as inputs to a digital-to-analogue converter which feeds its output to an expander. The expander uses the pilot tone to set the correct output level and a de-emphasis circuit restores a flat frequency response to the signal.

The SiS decoder masks transmission faults by either 'holding' the last good sample or, in the case of severe impairment, by muting its output. Some early caption generators could cause problems for SiS decoders by generating waveforms with switching spikes below black level, but Mono SiS was virtually immune to most distortions introduced by vision circuits.

2.6.1.7 'Sound on Vision'

Due to the extreme care taken in designing the SiS waveform, in normal operation 'sound on vision' is minimal. It can be seen from Figure 2.6.1 that, in this case, the leading and trailing edge of line sync have been slightly modified by the SiS encoder. IBA engineers discovered that, even when no edge distortion is visible, it is possible to extract a 'low-fidelity' audio signal by phase-demodulating the jitter on the leading edge of line sync.

This illustrates the difficulty of combining digital and analogue signals without some impairment to the analogue one (however slight).

2.6.1.8 Satellite Distribution

SiS can be, and was, used successfully over satellite links. (Intelsat did not permit SiS waveforms on its satellites for a number of years.)

2.6.1.9 NTSC variant

There is an NTSC variant of Mono SiS with the sampling frequencies adjusted slightly for a 30 Hz waveform. It was not widely used in the USA, partly because the standard video tariffs included one or two audio circuits.

2.6.1.10 Audio Fidelity

Mono SiS was used for many years to carry all types of audio, including classical music, without any adverse comments on its quality from BBC production staff. When BBC Sports producers adopted the US idea of 'close-mikeing' tee-shots at golf tournaments for added impact, they noticed a complete lack of 'attack' on the transmitted audio signal. This was traced to the compander action of Mono SiS and some BBC SiS encoders were modified to allow better reproduction of sharp transients.

2.6.2 NICAM Compression Systems

In the late 1970s and early 1980s the BBC distributed audio signals to its domestic radio transmitters using a proprietary 13-channel PCM system. The system used unmodified PCM and employed weighting to avoid quantisation noise. The aggregate bit rate was around 6 Mbit/s and was carried over a standard FM microwave link designed for analogue video. Due to the BBC practice of having 'split networks' (the mono AM and stereo FM variants of a service carry different programme material at certain times of day) and regional expansion, it became necessary to increase the capacity of the distribution network beyond 13 channels. There were five main design criteria for the NICAM (Near Instantaneously Companded Audio Multiplex[3]) system, which was to be used to replace the 13-channel PCM distribution system. Firstly, it could be used in conjunction with the 13-channel PCM system to increase the number of channels. It could subsequently supplant the PCM equipment entirely, substantially increasing capacity to 24 channels without modification of the FM microwave links. It could be carried on standard E1 and E2 bearers (running at 2.048 and 8.448 Mbit/s) to increase its operational flexibility. Codec delay was to be minimal as the output of the system might be used for off-air cues, etc. Finally, it had to deliver the best quality audio technically achievable within the bounds of the other four constraints.

Two types of NICAM signal are used with Dual-Channel SiS – NICAM 676 (the basic two-channel mux of the original NICAM 3 audio system) and NICAM 728 (a system chosen for its compatibility with some MAC transmission formats).

2.6.2.1 Sampling

The audio is sampled at 32 kHz to produce a usable bandwidth of 15 kHz. The samples are converted to 14-bit digital words. Fourteen bits will produce a signal with a (FSD/quantising noise) ratio of 86 dB, considered satisfactory for a broadcast contribution and distribution system. Using these parameters, each audio channel requires 448 kbit/s of bandwidth, giving four channels of audio per E1 multiplex or 16 channels of audio when four E1 tributaries are combined into one E2. Sixteen channels were not a sufficient improvement over the 13-channel PCM system, so some form of bit rate reduction ('compression') was required. Analogue compression could not be used as tracking errors would cause the stereo image to move around when NICAM was being used for stereo channels.

2.6.2.2 Compression

Six channels of 32 kHz-sampled audio in a 2048 kbit/s multiplex gives an average of 10.667 bits per sample per channel. The samples must be compressed from 14 to 10 bits, with the 'fractional bits' used for multiplex overheads (start codes, signalling data, range codes, etc.).

2.6.2.3 Range Coding

The signal is linearly quantised and converted to two's complement format. The sign bit is always transmitted and the other 9 bits are selected depending on signal amplitude. If the sample is very small (less than 1/16 of FSD) the nine least significant bits are transmitted. If the sample is large (greater than 50% of FSD) then nine most significant bits are transmitted. Three other ranges are used for signals of intermediate amplitude. When using range 0 (9 LSBs) the full 14-bit accuracy is maintained. Signals in range 4 (9 MSBs) only have 10-bit resolution.

The NICAM receiver cannot guess which range is in use, so this information needs to be transmitted in some way. Five range codes would require a 3-bit range code, taking the word length up to 13 bits. To overcome this problem blocks of 32 samples are given the range code of the largest sample in the block. All samples in the block transmit bits from the same range. A block corresponds to 1 millisecond of time, hence the expression 'near instantaneous'.

The original NICAM 3 system combines three 1-ms blocks and sends a 7-bit range code with a 4-bit Hamming code to protect the range bits.

NICAM 728 uses one 1-ms block and three range coding bits; the range coding bits are carried in parity bits using a novel technique invented by the BBC.

2.6.2.4 Pre-Emphasis

All companding schemes produce modulation of the noise by the required audio signal. The NICAM system uses CCITT J17 pre-emphasis to reduce the audible effects of this 'noise pumping'.

2.6.2.5 Parity/Error Detection

In NICAM 728 a parity bit is added to each 10-bit sample to protect the six most significant bits. Because NICAM 728 uses bit interleaving bursts of errors are unlikely. The parity bit can detect, but not correct, one bit in error; the decoder can then substitute an interpolated value in place of the errored value.

In an error-free channel there is a known relationship between the parity bits and the data bits. NICAM 728 uses 'signalling in parity' to convey the range code information without increasing the bit overhead (at the expense of a slight weakening of the parity mechanism). Three groups of nine parity bits use odd or even parity depending on the value of the three range coding bits.[4] In the receiver the groups of parity bits are examined to determine whether they are using odd or even parity. If the parity bits do not agree a majority vote determines which type of parity is in use (which is why an odd number of parity bits is required). The three range coding bits can then be recovered and any parity bits indicating an error are used to indicate which samples need to use error concealment. Signalling in parity is a very robust technique, which is important when conveying information like range codes, where errors can produce unpleasant distortions.

2.6.2.6 Quiet Signal Protection

Since NICAM 728 uses a 3-bit range code to indicate five ranges, there are three unused combinations. Two of these are used to give additional protection to low-level signals. The two extra codes are used to signal that the first 5 or 6 bits of the sample (ignoring the sign bit) are the same. A decoder can use this information to detect (and correct) multiple errors which have overwhelmed the parity mechanism.

2.6.2.7 Stereo/Dual Language/Data Signalling

The NICAM 728 signal can carry a stereo pair (where frequency and phase response must be matched), dual languages (where separation between the channels is of utmost importance), mono audio and one data channel at 352 kbit/s or a single 704 kbit/s data channel.

A NICAM 728 system sends a 728-bit frame of data every millisecond. The first byte is the Frame Alignment Word (0x4E). The next 5 bits are the Application Control Bits, three of which are used to indicate whether the service is stereo, dual mono, mono and data or data only. Eleven additional data bits follow, which may be used as a private data channel by the broadcaster or for some other application. The rest of the frame consists of 64 11-bit sound and parity words (if the signal is not being used to convey data).

2.6.3 Dual-Channel Sound in Syncs

Although used exclusively in the UK for carrying stereo sound (if test transmissions are ignored), Dual-Channel (DC) Sound in Syncs can also be used to convey two unrelated audio signals. Once DC SiS arrived the original Sound in Syncs was renamed Mono SiS to avoid confusion.

A major difference between DC SiS and Mono SiS is that the NICAM 728 signal is actually radiated on the vision signal so that viewers can experience 'digital quality' audio in their own homes.[5]

2.6.3.1 Quaternary Bits

A Mono SiS system has to fit one marker pulse and 20 data bits into a 4.7 μs sync pulse. The Mono SiS pulses have an HAD of 182 ns, which corresponds to approximately 5.5 MHz. A NICAM encoded pair of audio signals would require more than 43 data bits to be inserted into each sync pulse. It is clearly impossible to reduce the duration of the pulses by a factor of more than two, as they would not be able to pass through an analogue transmission system. The solution was to use four-level signals rather than pure binary. A four-level signal can carry 2 bits of information, giving twice the bit rate of a binary signal for the same Baud rate. These four-level signals were called 'quaternary bits', a name which was normally shortened to 'quits' (see Figure 2.6.3).

2.6.3.2 Marker Pulses

The marker pulse serves two functions in DC SiS systems. It serves as a timing reference for the 'quits' and, by its position relative to the leading edge of sync, it indicates how many 'quits' are present in the sync pulse (see Figure 2.6.4). (The NICAM sampling clocks are not related to the video line rate, so the DC SiS systems have to carry a variable number of quits per sync pulse.)

2.6.3.3 676 DC SiS

The original DC SiS system was to carry a NICAM 676 signal to the transmitters. At the transmitters the NICAM 676 signal would be transcoded into a NICAM 728 signal before being added to the analogue video signal.

A DC SiS system needs to carry 43.264 data bits (21.632 quits) per sync pulse. It is clearly impossible to have a fractional 'quit'. The original DC SiS system carried 1 marker bit plus 22 quits per sync pulse. Every fifth or sixth line would carry 1 marker bit plus 20 quits. The exact position of the marker bit indicates how many quits are to follow. The marker pulse and quits have an HAD of 182 ns, which is the same as Mono SiS.

2.6.3.4 728 DC SiS

Transcoding from NICAM 676 to NICAM 728 at each transmitter would have been technically difficult and, hence,

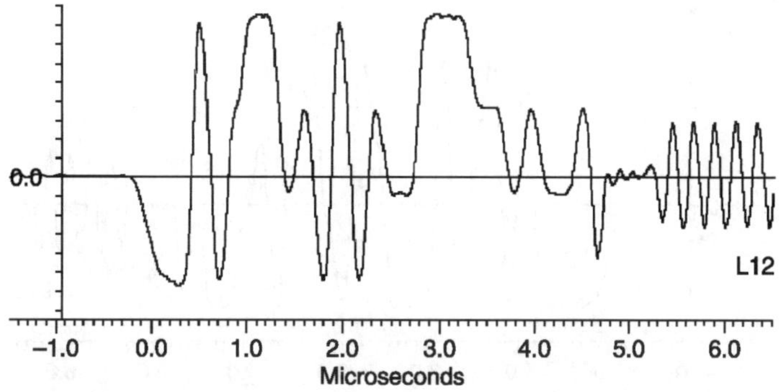

Figure 2.6.3 Dual-Channel Sound in Sync waveform with one marker pulse and 24 'quits'. Notice that the final 'quits' encroach on the trailing edge of sync, which has to be regenerated in a DC SiS decoder.

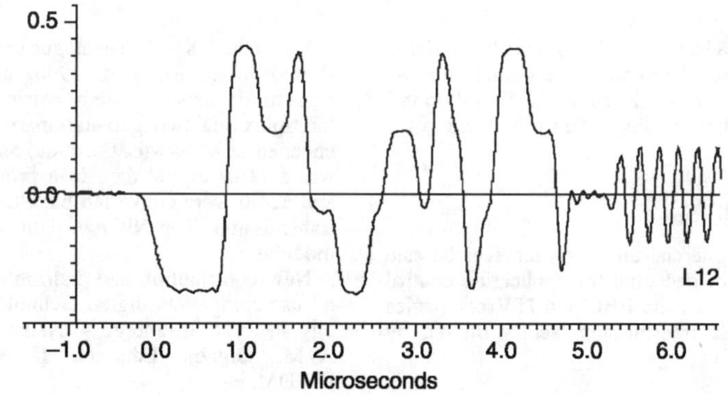

Figure 2.6.4 Dual-Channel Sound in Sync waveform with one marker pulse and 22 'quits'. The position of the marker pulse is delayed to indicate that fewer 'quits' are present (cf. Figure 2.6.3). Also notice that the raised cosine shape of the pulses is more accurate than those in Mono SiS (cf. Figure 2.6.1).

expensive with the technology available at the time. Because of the different block sizes it would also have increased audio delay relative to video.

It was decided to send a NICAM 728 signal direct to the transmitters inside the DC SiS waveform. This required more quits per sync pulse. Forty-eight data bits (24 quits) were sent in some sync pulses, 44 were sent on other lines. The average number of data bits per sync pulse was 46.592. The 22 data quits of the NICAM 676 signal were already very close to the sync trailing edge, so it was not possible to add two more quits with the same HAD. For NICAM 728 transmission the HAD of the quits was reduced to 172 ns – corresponding to a frequency of 5.814 MHz.

The two systems were not interoperable. NICAM 676 was mainly used for (TV and radio) contribution circuits and occasional circuits. NICAM 728 was originally only used for TV distribution to transmitters, as its quits were significantly less robust than those of the 676 system.

2.6.3.5 IBA variant

The IBA also decided to send a NICAM 728 signal directly to their transmitters without any need to transcode at the transmitter sites. The IBA variant of NICAM 728 DC SiS had a different solution to the variable number of quits problem. One marker pulse and 24 quits were inserted into each sync pulse until the NICAM data buffer emptied. The last eight or nine TV lines in a field would contain no NICAM data.

At the transmitter the radiated signal would have an identical format to the BBC signal to guarantee interoperability with domestic television receivers and VCRs.

The IBA variant of DC SiS was not interoperable with either of the BBC variants.

2.6.3.6 Eye height

The DC SiS signal is not as robust as the Mono SiS signal. The shorter duration (and hence higher frequency) pulses and the four levels of the 'quits' produced a waveform at the mercy of imperfections in the phase and frequency response of the transmission channel in the 5–6 MHz range (See Figure 2.6.5).

The BBC's permanent contribution and distribution circuits (the local ends of which were equalised by BBC engineers) were quite capable of carrying DC SiS traffic, usually with only minor modifications.

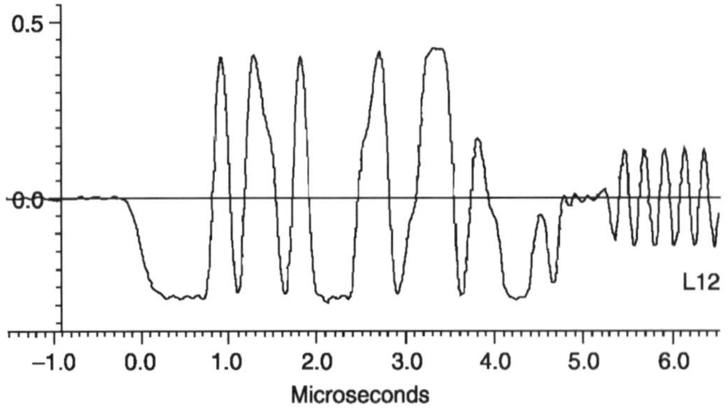

Figure 2.6.5 Dual-Channel Sound in Sync waveform with one marker pulse and 22 'quits'. It can easily be seen from the shape of the quits how minor phase distortions would cause decoding errors.

Temporary Outside Broadcast radio links (and the variable equalisers used to 'linearise' them) were not suitable for DC SiS. The BBC designed and produced a DC SiS 'top-up' equaliser to bring occasional circuits within specification for dual-channel use.

2.6.3.7 Modified Specification

At least one brand of commercial television receiver on sale in the UK was incapable of decoding the application control bits correctly. Because of this, the BBC and ITV companies agreed not to use the data transmission facility offered by NICAM 728.

2.6.3.8 All-Digital Systems

The BBC's analogue contribution and distribution network was replaced with an all-digital backbone using Asynchronous Transfer Mode (ATM) on fibre-optic cables wrapped around the overhead wires of the National Grid. Some of the spurs to the transmitters remained analogue. Once an all-digital network was in place, the use of hybrid signals such as DC SiS might be expected to have diminished, but this has not happened.

NICAM 728 DC SiS is still in daily use at the BBC. Distribution to the analogue transmitter network uses 728 DC SiS, which is PCM-encoded before travelling over 139 Mbit/s channels on the digital backbone. Contributions from Outside Broadcasts use 728 DC SiS for the main sound and vision, with clean effects sound being carried in a VIMCAS signal in the Vertical Blanking Interval (VBI) of the DC SiS waveform. VIMCAS (Vertical Interval Multi-Channel Audio System) is a proprietary method of carrying extra audio in the VBI of a video signal. Two audio signals are sampled and placed in a buffer. Twenty milliseconds' worth of audio are replayed at approximately 100 times normal speed ('time compression') and the resulting audio signals (with a 5.5 MHz bandwidth!) are placed on six mid-grey pedestals in the VBI. The combined 728 DC SiS with VIMCAS signal will travel from the OB to the studio centre on SHF radio links, analogue circuits or via the digital backbone (or some combination of these methods).

NICAM 676 DC SiS is no longer in regular use.

When the UK's fifth analogue terrestrial channel (Channel 5) started broadcasting in 1996, distribution signals to the transmitter network were carried on an MPEG-2 digital multiplex via two geo-stationary satellites. The video was encoded as MP@ML, the audio as Musicam and the NICAM was sent as digital data in a proprietary format. The video and audio were converted back to PAL and analogue before transmission. The NICAM data were fed to a NICAM 728 modulator.

New contribution and distribution networks will probably all use completely digital technologies. A dominant technology has yet to emerge – systems architects are proposing ATM, Gigabit Ethernet, IP-over-ATM (IPoA), native DWDM, etc.

2.6.3.9 NICAM Countries

DC SiS was never adopted by the EBU for its Eurovision network. The presence of three incompatible versions of the standard and the stringent phase and frequency performance of the video circuits required by DC SiS signals made them unsuitable for a news exchange network where the average item might only last for a few minutes. The EBU did produce a modified version of the (radiated) NICAM 728 standard for use in countries which did not use system I.[6]

NICAM transmission to the home was adopted by more than a dozen countries; many of these would have originally used DC SiS for the distribution network. Countries which use NICAM include Belgium, China, Denmark, Finland, France, Germany, Hungary, Ireland, Italy, New Zealand, Norway, Portugal, Spain, Sweden and the UK.

References

1. *CCITT Red Book*, Volume III, Fascicle III.4, Transmission of Sound Programme and Television Signals, Recommendation J.17: Pre-emphasis Used on Sound-programme Circuits.
2. BBC Technical Instruction P.15, Sound in Syncs (January 1972). Republished as BBC Wood Norton Information Sheet 49H (1989).

3. Kendall, R.L. BBC Wood Norton Information Sheet 27G, An Introduction to NICAM 3 (24 February 1983).
4. BBC Wood Norton Information Sheet 26R, Dual Channel Sound for Television.
5. NICAM 728, Specification for Two Additional Digital Sound Channels with System I Television, BBC/IBA/BREMA (August 1988).
6. EBU Specification for Transmission of Two-channel Digital Sound with Terrestrial Television Systems B, G, H and I, SPB 424, 3rd revised version, EBU Technical Centre, Geneva (1989).

Internet Resources

Hosgood, S. All you ever wanted to know about NICAM but were afraid to ask, http://tallyho.bc.nu/~steve/nicam.html (5 November 1997).

Gordon Anderson
Technical Training
Tandberg Television Ltd

2.7 VBI Data Carriage

The original purpose of the Vertical Blanking Interval (VBI) was to allow early television receivers to have very simple sync recovery circuits. Modern receivers are not affected if extra waveforms are placed in the VBI, so broadcast engineers can now use this precious resource for test, control and data signals (Figure 2.7.1).

In Europe (and other 25 Hz regions) a popular text and data distribution system called teletext has used spare capacity in the VBI (Figure 2.7.2). Variants of teletext have been used for broadcasting Internet, private data and control signals. Several advanced forms of teletext have been designed but have not been widely used for a number of reasons, including incompatibility with legacy decoders, bandwidth scarcity and the high cost of developing the newer decoders. The imminent arrival of newer, interactive, digital technologies means that the more advanced forms of teletext may never materialise. Teletext is in use in over 30 countries, including Australia, Austria, Belgium, China, Denmark, Finland, France, Germany, Hungary, India, Ireland, Italy, Malaysia, the Netherlands, Norway, New Zealand, Poland, Portugal, Singapore, Spain, Sweden, the UK and the United Arab Emirates.

Closed Captions also utilise space in the VBI and are in widespread use in the USA and Canada. Despite lacking some of the more advanced features of teletext, Closed Captions have the great advantage that they can be recorded onto domestic VHS tapes and played back perfectly. Tapes for sale and rent in Europe often come with a Closed Caption signal which a separate adapter can turn into captions which are cut in to the video signal. (Televisions in Europe do not generally have integral Closed Caption decoders.)

Other waveforms in the VBI containing data include Wide Screen Signalling (WSS), Video Programme System (VPS), Automated Measurement Of Line-ups (AMOL) and Vertical Interval TimeCode (VITC).

2.7.1 Teletext

Teletext[1] was jointly developed in the UK by the BBC and the IBA in the early 1970s. Described in modern terms, teletext is a carousel data, text and graphics service (see Figure 2.7.3).

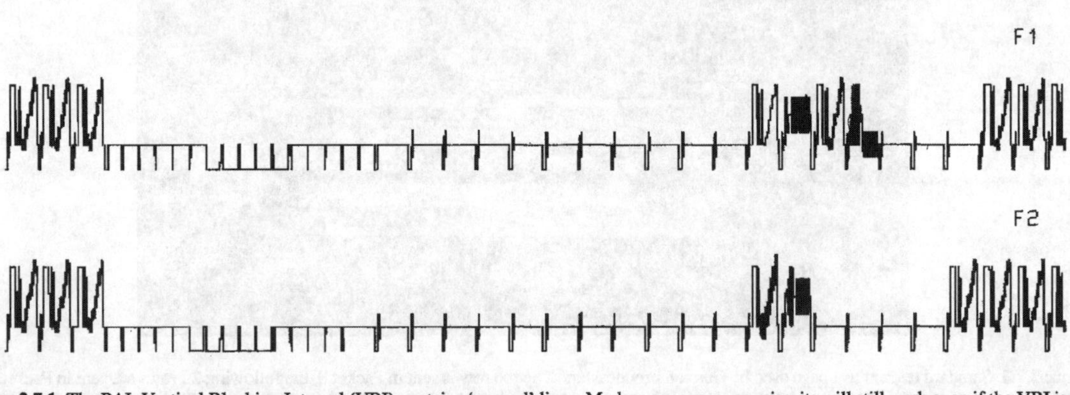

Figure 2.7.1 The PAL Vertical Blanking Interval (VBI) contains 'unused' lines. Modern sync recovery circuits will still work even if the VBI is filled with test waveforms and data.

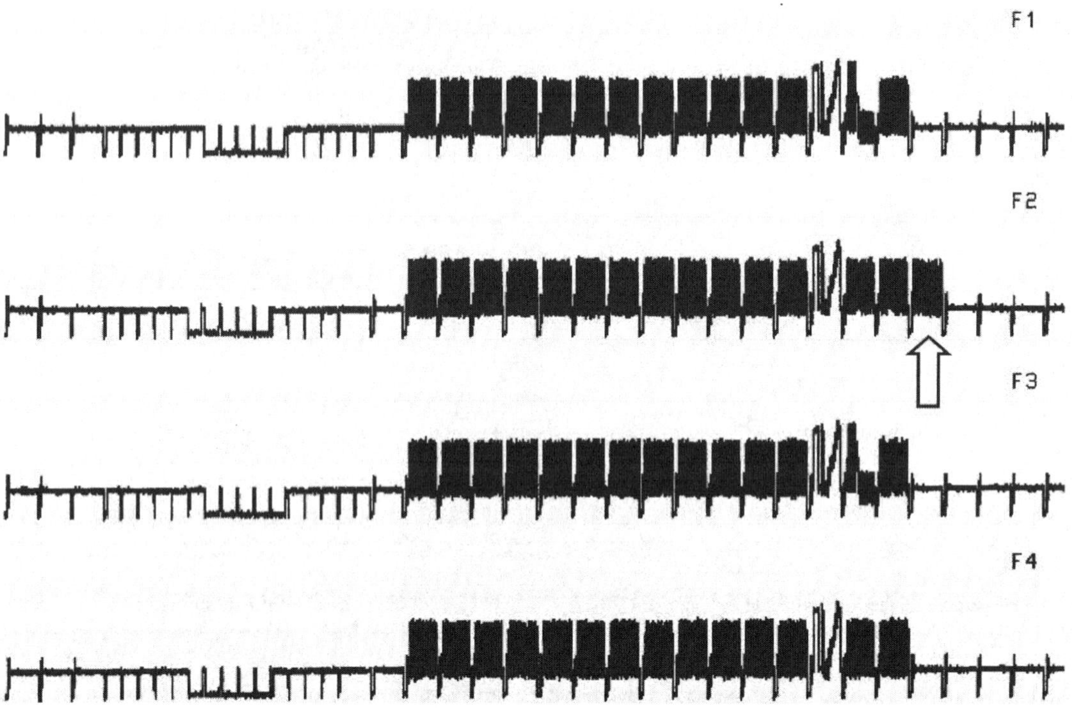

Figure 2.7.2 In some European countries the VBI is now virtually full of teletext and other data. The diagram shows VBI of ITV in the UK. Line 335 (arrowed) is used for subtitling. Note that it does not carry subtitles every time.

Figure 2.7.3 Standard teletext test page used by German broadcasters. The top row is sent in Packet 0, the following 23 rows are sent in Packets 1–23. Test page features complete German alphabet, chunky teletext graphics and double-height (Doppelte Höhe) text. Lines 02 and 09 feature 'clock-cracker' signals which produce the repetitive sequence /01111111/11111110/. (The ÷ sign replaces the ess-zett (β) in English-language test pages.)

Figure 2.7.4 The high data rate of the teletext waveform prevents it being reliably recorded on domestic VCRs.

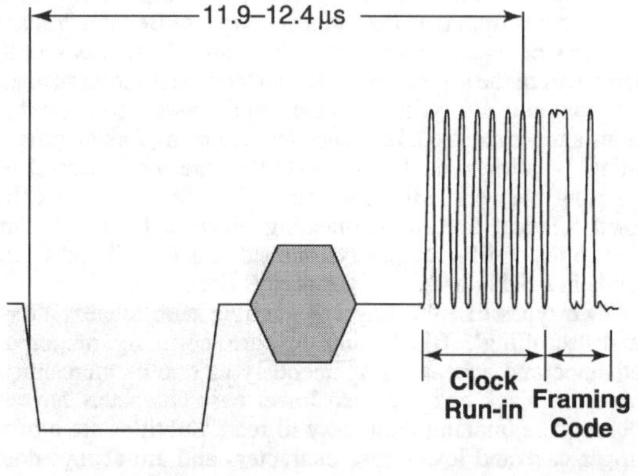

Figure 2.7.5 Clock Run-in and Frame Coding bytes, showing timing relative to sync front edge.

It consists of up to eight 'magazines', each of which can contain up to 100 pages. The pages can each contain up to 3200 sub-pages (or versions). Although computer monitors of the time could display up to 80 columns of text, televisions were not capable of displaying such 'high-resolution' images. Additionally, television viewers are not as close to the screen as computer users, so teletext was limited to 40 characters across the screen by 24 rows of characters. The characters (or graphics) could be any of the colour bar saturated colours (yellow, cyan, green, magenta, red, blue) or white and the background could be any of those colours or black. The characters could be upper or lower case and the lower case characters had true descenders (making them easy to read). The characters could be displayed 'double height' and could be made to flash on and off. Graphics elements were displayed on the same 40 × 24 grid as the letters but each character was divided into a 2 × 3 graphical element, giving a graphics resolution of 80 × 72 (rather primitive by current computer standards).

Teletext can support a 'newsflash' mode where breaking news is inset into the video inside a black box.

2.7.1.1 Packets

There are 32 types of teletext packets. In the original UK specification Packets 24–31 were not used.

Packet 0 contains the Page Header Row, which only contains 32 bytes of display information. The Page Header Row contains the magazine and page number (known as MPAG – Magazine and Page Address Group), a number between 100 and 899. This is the number which a viewer taps into a remote control to select a teletext 'page'. Information displayed on the top line usually includes magazine and page number, name of broadcaster or teletext source, day, date and time. When displaying a teletext page a decoder usually scans all the other Packet 0s so that the clock can be updated (pages are only updated every 10–30 seconds, depending on the number of pages transmitted, but the displayed teletext clock updates every second).

The next 23 rows (rows 01–23) are contained in Packets 1–23 and contain up to 40 bytes of displayable data. Rows which do not contain information do not have to be sent. Each row starts with single-height white alphanumerics on a black background. If a control code is present at the start of the line it will change the rest of the line until the next control code. Control codes can switch between the alphanumeric and graphics characters, change background and foreground colours, change characters to double height, make the row flash, etc. Rows from a page may be sent in any order and may be mixed with rows from other magazines. Packets 1–23 only contain Magazine and Row (MRAG) information – they do not contain the Page address. It is not possible to interleave rows from different pages on the same magazine as the decoder cannot work out which row belongs to which page. A page is sent in its entirety before a new page from the same magazine can be transmitted. There must be an active field of video between the start and end of a page – this is known as the Page Erasure Interval, the original purpose of which was to allow a receiver to erase a previous page from its memory. The result of all of these constraints is that teletext pages are difficult to multiplex efficiently without wasting some teletext bandwidth. Empty packets (usually with a page number of 0xFF) occur regularly and can be seen on an oscilloscope as they have a distinctive 'comb' appearance (see Figures 2.7.6 and 2.7.7). Because of the difficulties of multiplexing teletext, time-sensitive pages (such as subtitles) are often given their own line in the vertical blanking interval (Figure 2.7.2). Modern implementations tend to devote a VBI line to a particular magazine, but other bandwidth optimisation techniques are available. The original teletext specification only allowed teletext to occur on lines 7–22 and 320–335, but lines 6, 318 and 319 may now also be used. Originally lines 17, 18, 330 and 331 were used for UK teletext transmissions, giving a page rate of 4 pages per second.

2.7.1.2 Signal Coding

Teletext uses three different error detection schemes. Displayable (7-bit) characters are protected with a single odd parity bit – this can detect, but not correct, one, three, five or seven errors in a byte. A decoder will normally show a blank space instead of an incorrect character – an even number of errors will usually result in the wrong character being displayed.

Important data like the Page and Magazine numbers are protected with Hamming[2] codes. Hamming 8/4 produces four extra protection bits for every 4 bits of data. This code can correct one errored bit and detect (but not correct) 2, 4 or 6 errored bits in a byte. This code is used for the first two information bytes of the display Packets 1–23 containing the Magazine and Row (MRAG) information. It is also used for 10 bytes in Packet 0.

Some packets in the 24–31 range use a Hamming 24/18 code.

The transmitted physical teletext signal shares some features with an ethernet frame. A Clock Run-in waveform is transmitted to synchronise receiver clocks, followed by a Framing Code to achieve byte alignment. Bytes of data are sent least significant bit first. An odd parity code is used to prevent clock slippage by ensuring that there are never more than 14 ones or zeros in a row. The Clock Run-in sequence is only 2 bytes long in teletext, yet a receiver should still acquire a line of data even if one or two of the leading ones are absent (Figures 2.7.4–5).

2.7.1.3 Original Teletext Character Set

The original teletext characters were based on (but not identical to) the ASCII[3] character set (see Table 2.7.1). Characters 0x00 to 0x1F are non-printing control codes used for changing colours, changing from text to graphics, etc. Characters 0x23, 0x24, 0x40, 0x5B, 0x5C, 0x5D, 0x5E, 0x5F, 0x60, 0x7B, 0x7C, 0x7D and 0x7E are modified to produce the other Latin-based character sets.

Teletext also has two types of graphics characters which can be used to draw low-resolution (80 × 72) pictures. Alphanumerics and graphics can exist on the same page or line but alphanumerics cannot be overlaid on graphics (Figure 2.7.8).

2.7.1.4 Subtitles

Open captioning is the name given to captions or subtitles which are 'burnt in' to the video (i.e. they are always present and cannot be deselected by the viewer). Closed captioning (with two small 'c's) is the name given to captions which can be selected by the viewer. The expression 'closed captions' is used in Australia to refer to captions supplied by teletext; in North America there is a system called 'Closed Captions' (two big Cs) which supplies captions using a different type of VBI data mechanism. The words subtitling and captioning are often used interchangeably in Europe but they do have subtly different meanings in the USA. Subtitling refers to text placed on a video picture, which is often a translation of the original language on the soundtrack. Captioning contains sound effects (such as 'footsteps', 'knock on the door', 'telephone ringing', etc.), as well as the speech of the actors or presenters. It would theoretically be possible to use translated teletext subtitles for all foreign-language programming in Europe, but technical and cultural considerations prevent this from happening. Unlike North America, where the majority of television households now have one or more Closed Caption decoders, in Europe many televisions do not have teletext decoders. If teletext is used for translated subtitles, those viewers without suitable televisions cannot access the subtitles. Most imported European programming comes from the USA, so a country like Belgium has to use two lots of 'open captions' (Flemish and French) obscuring a large part of the screen. In the Nordic countries foreign-language programming always has subtitles on the video, so teletext subtitles are not required. In France and Germany foreign programming is always dubbed as subtitles are not acceptable to (hearing) viewers in those countries, but they can obviously be used for deaf and hard-of-hearing viewers. In the UK some categories of film are always dubbed, but French and German films invariably have subtitles on the video.

Two types of subtitling are possible with teletext, 'block' and 'scrolling'. Block subtitles are normally prepared in advance and appear instantaneously as one or more lines of text in a black box. Teletext lower case characters have true descenders, making them easy to read. Subtitles are a mix of upper case and lower case characters and are always double height (but not double width – 'normal' teletext characters look slightly squat and double-height characters look stretched – see Figure 2.7.3). By convention they normally appear near the bottom of the screen but they can be placed anywhere (this might be done for sound effects or to avoid covering something important at the bottom of the screen – like a 'burnt-in' caption identifying a speaker). Different speakers are indicated with different colour text, but voice-overs are normally in white. A hash (#) symbol is usually used to indicate singing or music.

The first high-profile use of live subtitles by the BBC was for the Outside Broadcast of the Royal Wedding in 1981. In live subtitling stenocaptioners listen to the broadcast sound and use a phonetic keyboard to input data to a computer. The computer looks at the phonetic characters and tries to work out which word or sound is required – it then generates the teletext characters for transmission. The computer contains a model of sentence structures and a large dictionary to convert phonetic

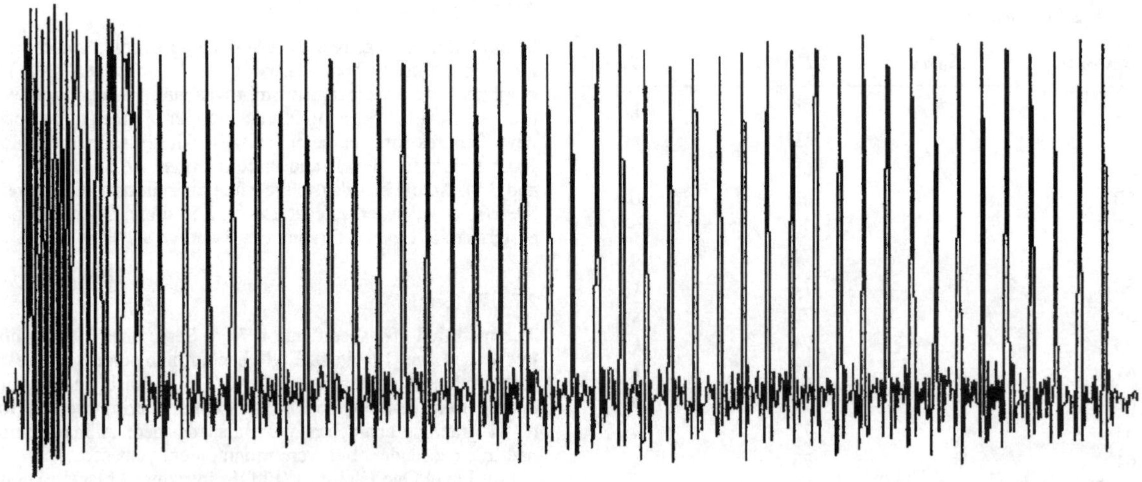

Figure 2.7.6 Empty teletext packet showing parity bits.

'words' to English text. The system now works quite well, but on the first broadcast it had a lot of trouble with proper nouns (London landmarks, Royal titles and foreign dignitaries) and uncommon words which were obviously not in its original dictionary. (The system rendered 'procession' as 'prosejun' or something similar.) Live subtitles may be presented as block or scrolling. Live block subtitles look very similar to the pre-recorded version except that the captioners may not be able to indicate who is talking by using different coloured text. Scrolling subtitles are written to the screen a word at a time. At the end of the line the entire line is moved up and new words appear from the left of the new bottom line.

Figures 2.7.9–2.7.13 show a teletext subtitling packet, which reads "why I'm here," and which will appear on (teletext) rows 22 and 23 of the television screen. The 'Start Box' and 'End Box' control codes are sent twice as a simple form of error correction to prevent 'spurious boxing'. Teletext row 23 does not have to be sent as it will contain identical data (other than the line number). Subtitles are time-sensitive and broadcasters sometimes allocate a whole line of the VBI to subtitles to ensure that they are not stuck in a queue of other, time-insensitive, teletext packets. This is an especially important consideration

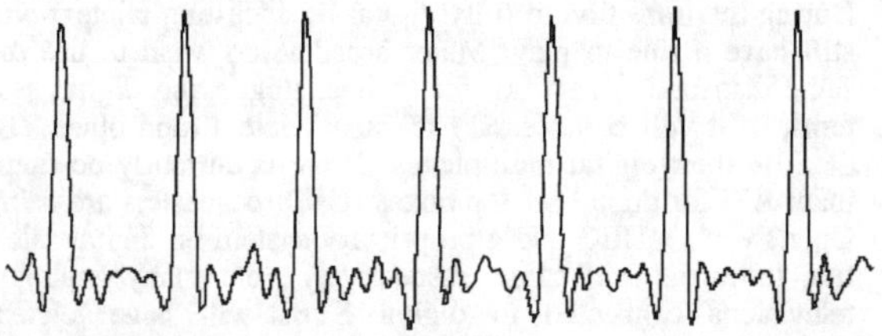

Figure 2.7.7 Expanded view of parity bits. Due to the coding and parity in use there can never be more than 14 bits between binary transitions. (This may not be true for some data transmission applications.)

Table 2.7.1 Original teletext character set

Hex code	*Symbol*	*Hex code*	*Symbol*
20	space	3C	<
21	!	3D	=
22	"	3E	>
23	£	3F	?
24	$	40	@
25	%	41	A
26	&	42	B
27	'	43	C
28	(	44	D
29	)	45	E
2A	*	46	F
2B	+	47	G
2C	,	48	H
2D	-	49	I
2E	.	4A	J
2F	/	4B	K
30	0	4C	L
31	1	4D	M
32	2	4E	N
33	3	4F	O
34	4	50	P
35	5	51	Q
36	6	52	R
37	7	53	S
38	8	54	T
39	9	55	U
3A	:	56	V
3B	;	57	W

Table 2.7.1 *continued*

Hex code	*Symbol*	*Hex code*	*Symbol*
58	X	6C	l
59	Y	6D	m
5A	Z	6E	n
5B	←	6F	o
5C	½	70	p
5D	→	71	q
5E	↑	72	r
5F	#	73	s
60	—	74	t
61	a	75	u
62	b	76	v
63	c	77	w
64	d	78	x
65	e	79	y
66	f	7A	z
67	g	7B	¼
68	h	7C	‖
69	i	7D	¾
6A	j	7E	÷
6B	k	7F	

All of these codes can also be used as graphics characters if they are preceded by a graphics control character; 0x7F is a graphics character.

with live subtitling which is being compiled after the words have been spoken.

In the UK line 335 is used by the ITV companies for subtitling. In Australia and New Zealand lines 21 and 334 are used. Live teletext subtitling is used in Australia, France and the UK.

Teletext can support multiple, multi-lingual subtitles.

Figure 2.7.8 Four of the 64 contiguous and separated teletext graphics primitives. Each shape occupies the same space as a teletext letter.

2.7.1.5 Teletext Levels

Current UK teletext is only a slight improvement on the original system used when the service was launched in 1974. (The graphics have been slightly improved and Fastext is now in use.) Teletext has been modified a number of times to introduce new features or character sets.[4–8] These refinements are grouped into five levels and three intermediate levels (1.5, 2.5 and 3.5). Actual broadcasts are often at an intermediate level – the UK broadcasters are at Level 1.5, some German broadcasters are at Level 2.5 (sometimes known as 'HighText').

2.7.1.5.1 Level 1

The modified UK specification from September 1976 is now known as Level 1 teletext. Eight background colours and seven text/graphics colours were allowed. Newsflash and Subtitle boxing were defined; 96 English alphanumerics and two lots of 64 graphics characters were also defined. Alphanumerics had true descenders but were monotonically spaced.

Full Level One Facilities (FLOF) introduced black text on a coloured background and Fastext (teletext's equivalent of hypertext links).

2.7.1.5.2 Level 2

Multi-language text defined. Proportional spacing of characters for easier legibility. More colours allowed plus an extended mosaic pictorial set.

2.7.1.5.3 Level 3

Dynamically Redefined Character Sets (DRCS) introduced. These were based on a 12 × 10 pixel matrix more suitable for non-European character sets. Teletext now theoretically capable of representing any written or pictorial character set.

2.7.1.5.4 Level 4

Vector graphics capability introduced – greatly reducing the amount of data needed to send complex images, the teletext decoder effectively becoming an electronic plotter. Very large colour palette introduced (>250,000).

2.7.1.5.5 Level 5

Full-definition, full-colour still pictures.

2.7.1.5.6 Future of teletext

During the transition to fully digital broadcasting teletext will still have a rôle to play. Many broadcasters wish to use the same distribution network for their analogue and digital systems, so it will be necessary to carry teletext (and other VBI data) in their digital multiplexes. There is currently no dominant API for digital set-top boxes (UK broadcasters are using OpenTV™, MHEG and a proprietary system on digital satellite, terrestrial and cable, respectively), but many analogue televisions connected to digital STBs will have teletext decoders already built in. Since DVB has defined a method for carrying VBI data inside MPEG-2 PES packets[9,10] (see Chapter 2.12), teletext can still be used on these televisions and may be necessary during the early phases of digital broadcasting when an EPG may not have been developed. A fairly sophisticated

Figure 2.7.9 Subtitling packet from reserved line 335. The 11 spaces before and after the subtitle are clearly visible.

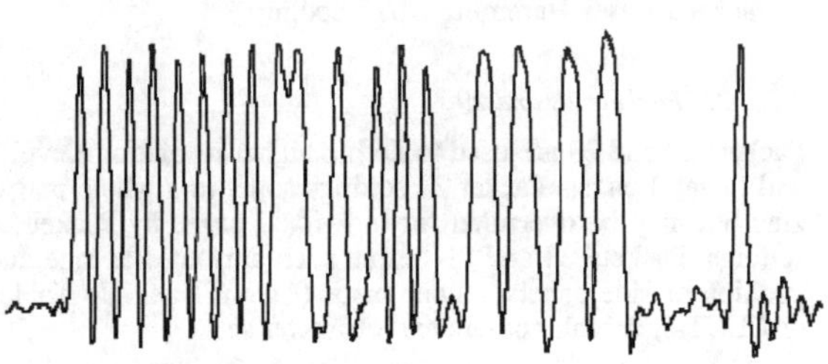

Figure 2.7.10 First 7 bytes of subtitling packet in Figure 2.7.9. Waveform can be translated as 2 bytes of Clock Run-in, Framing Code, magazine 8-row 22, 'Double height characters', SPACE.

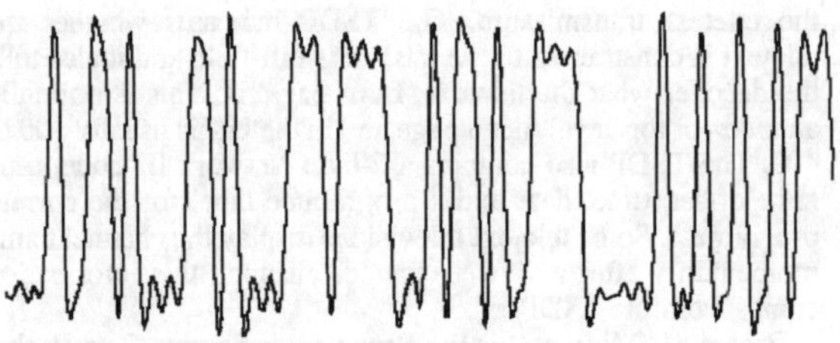

Figure 2.7.11 Central portion of Figure 2.7.9 expanded to show subtitling data. 'Start Box', 'Start Box', w, h, y, SPACE, I.

teletext-driven EPG[11] already exists and is in use in Germany under the name of 'NexTView'. This EPG offers features similar to a DVB-SI driven EPG, including the ability to highlight a programme from a TV listing and obtain further information about it (similar to the DVB-SI 'synopsis').

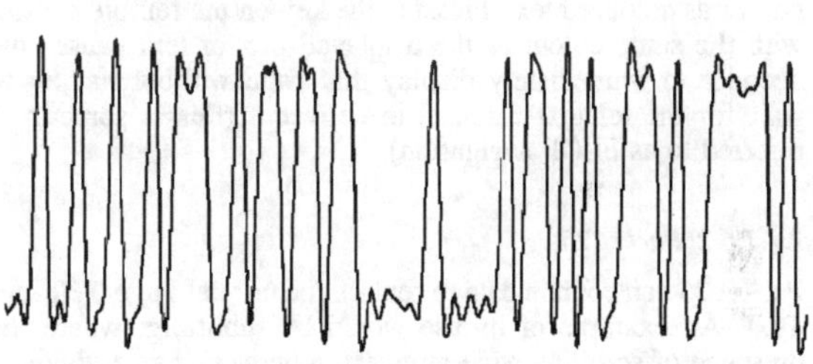

Figure 2.7.12 Central portion of Figure 2.7.9 expanded to show subtitling data. SPACE, I, ', m, SPACE, h, e, r.

Countries with non-Latin fonts (Middle East, India, China, most of Asia) could use Level 3 teletext, but it is more likely that their local development efforts will be targeted at the emerging digital technologies such as MHP. The sophisticated graphics capabilities of teletext were demonstrated by CBC in the 1980s in Canada, but it is unlikely that Levels 4 and 5 will ever be used by broadcasters.

2.7.1.6 Special Packets

Originally only Packets 0–23 were defined – decoders were to disregard packets in the range 24–31. To prevent any problems with a large installed base of legacy decoders, all of the newer functions of teletext are placed in the previously unused packets. It was common to refer to row 'X' packets (where X was a number between 0 and 23 inclusive); since the packets in the range 24–31 cannot usually be associated with a row, all packets are now usually described as packet followed by the number.

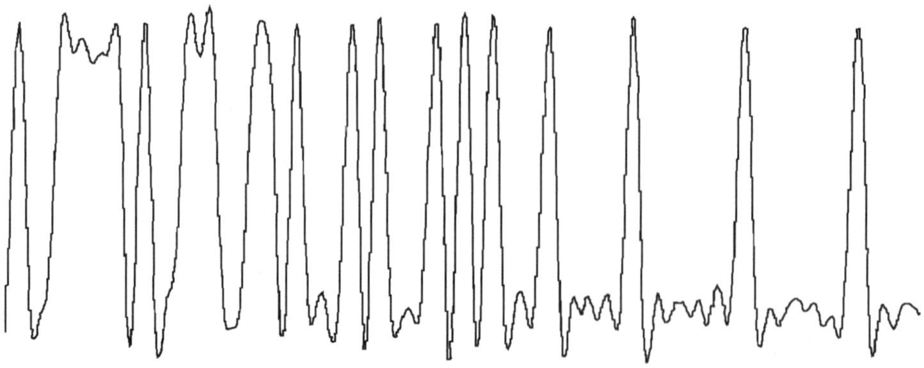

Figure 2.7.13 Central portion of Figure 2.7.9 expanded to show subtitling data. r, e, „ 'End Box', 'End Box'.

2.7.1.6.1 Packet 24

In 1987 a 'Fastext' service was launched. With the falling cost of memory it became feasible to store more than one page of teletext in a decoder. 'Fastext' provides a method of cacheing related pages locally in the television to reduce page access times for the viewer. Packet 24 contains up to six labels, four of which are displayed at the bottom of the viewer's screen. The four labels are usually displayed as some text on a colour-coded box or as coloured text. Pressing the key on the remote control with the same colour as the displayed box or text causes the decoder to immediately display that page without having to wait for the teletext carousel to deliver it (this is sometimes referred to as FLOF navigation).

2.7.1.6.2 Packet 25

Packet 25 can contain data to replace the normal Page 0 Header Text. An example of its use would be subtitling, where the presence of scrolling page numbers, a network name, day and date, and a digital clock would prove distracting and might just be replaced with 'XYZ subtitles' or a page number (888 in the UK, 801 in Australia).

2.7.1.6.3 Packet 26

Packet 26 contains 13 data groups of 3 bytes. The data groups associate some data with a position on the displayed screen and can be used for a variety of tasks. A PDC recorder could be programmed from the TV listing on a teletext page if the broadcaster was sending the correct information in Packet 26. The viewer could highlight a programme on the listings page and then press another button to tell a PDC-capable VHS recorder to record that programme. Another use of Packet 26 is to point the decoder to redefined characters that can be used to overwrite parts of a page with a different character set. Spanish teletext uses this mechanism for accented vowels, and 'ñ'. (This mechanism was required since the three 'spare' signalling bits in Packet 0 could not be used to designate more than eight languages. World Standard teletext uses a different – but compatible – mechanism for selecting character sets.)

Packet 26 can also address character positions in the 'side panels' possible with some of the enhanced forms of teletext (side panels are similar to frames on a web page). Packet 26 uses Hamming 24/18 coding.

2.7.1.6.4 Packet 27

Packet 27 works in conjunction with Packet 24. The labels in Packet 24 tell the decoder which bank of memory to display when a special remote control key is pressed (one of four colours, 'Next' or 'Index'). Packet 27 tells the decoder which teletext pages to put in the pages of cached memory. To use another Internet analogy – Packet 24 contains the displayable hyperlinks and Packet 27 contains the linked URLs.

Packet 27 uses Hamming 24/18 coding.

2.7.1.6.5 Packets 28 and 29

Packets 28 and 29 are used to define higher resolution Level 2 and Level 3 pages. Packet 29 settings apply to a whole magazine but may be overruled for individual pages by Packet 28 settings. Packets 28 and 29 define such attributes as size and position of side panels, colour maps (Colour Look-Up Tables or CLUTs), default colours and character sets.

2.7.1.6.6 Packet 30

Packet 30 is not logically associated with any page or magazine but it has to have a magazine address, so it is usually called Packet 8/30. Packet 8/30 format 1 is the Television Service Data Packet (TSDP), which describes some general features of the teletext transmission. The TSDP indicates whether the teletext is constrained to the VBI or is full-field and it also tells the decoder what the teletext 'front page' is. This is normally an index or top-level menu page and in the UK is usually 100 or 400. The TSDP also contains a 2-byte Network ID code, time zone offset, time, date and a programme label for the current programme. Some teletext televisions display the channel name momentarily after a viewer changes channel; this information comes from the TSDP.

Packet 8/30 format 2 is the Broadcaster Service Data Packet (BSDP), which is used for Programme Delivery Control (see PDC section).

Packet 30 uses Hamming 24/18 coding. Both types of Packet 8/30 are sent once a second.

2.7.1.6.7 Packet 31

Packet 31 is used for data broadcasting,[12] usually to private networks of users (this is sometimes referred to as Independent

Data Lines or Services); 28–36 bytes of user data can be sent per line of VBI, giving a data rate of 5600–7200 bit/s (if two fields are used this rate is doubled). Full-field transmission could theoretically deliver 3.4048–4.3776 Mbit/s (assuming lines 6, 318 and 319 are not used).

The reason for the variable number of bytes in the payload of Packet 31 is that the header is of variable length depending on the application (Figure 2.7.14). If the data are to be broadcast to all recipients (i.e. no addressing necessary) then the higher data rates can be achieved. Packet 31 can be associated with magazines 1, 2, 3 or 8, but the magazine number has no significance.

2.7.1.6.7.1 Packet structure

Bytes 1–3 are the Clock Run-in and Framing Code bytes as usual.

Byte 4 indicates which of four Data Channel Groups the message belongs to. This allows receivers to filter out unwanted messages without having to do too much decoding.

Byte 5 contains private Message Bits, the contents of which are not defined.

Byte 6 is the Format Type byte containing 4 bits signalling presence or absence of packet repetition, continuity indicator and data length indicator.

Byte 7 is the Interpretation and Address Length byte, which indicates how many of the following bytes carry further addressing information. The range is 0–6 bytes.

Optional Address bytes may be absent or occupy some, or all, of bytes 8–13. Since they are Hamming 8/4 protected, the Address bytes can contain up to 24 bits of addressing information; 24 bits can address 16,777,216 different recipients. Combining this number with the four data channel groups will provide over 67 million addressable receivers.

A packet can be sent more than once as a simple form of error rate reduction. If packet repetition is in use, a Repeat Indicator byte will increment for each transmission of the packet (packets can be sent up to 15 times).

If an explicit Continuity Indicator (CI) is present it will increment (from 0 to 255) for each transmission of a new packet for a given service. If an explicit CI is not present, the information is incorporated into the CRC at the end of the packet.

A Data Length byte points to the end of valid data. This is used for incompletely filled packets caused by 'bursty' data. A DL byte can contain numbers in the range 0 to 35. An optional stuffing byte (0xA5 = 10100101) can be used to prevent long runs of ones or zeros at the end of the packet, which could cause problems for some legacy receivers.

Data can be encrypted before being placed in the teletext packets for extra security.

Bytes 4, 6 and 7, the optional Address bytes and Repeat Indicator byte, are Hamming 8/4 encoded. The payload and explicit Continuity Indicator (if present) are protected with a 16-bit CRC.

2.7.1.7 Programme Delivery Control (PDC)

PDC[13] enables a viewer to record a programme even if the transmission schedule is changed (the DVB equivalent would be the Run Status Table). There are several methods that can be used to instruct the recorder. A programme can be selected from a teletext listings page (provided the broadcaster is transmitting the PDC information in Packet 26). A programme can be selected from a teletext listings page which has the PDC data 'hidden' on the page (the hidden data can be displayed by pressing the 'Reveal' button on the remote control). The PDC code might be inserted using a bar-code reader scanned across a printed TV listing from a newspaper. The code is called a Programme Ident Label (PIL) and contains the date, channel and start time.

The broadcaster uses Packet 8/30 format 2 to control the video recorder. Thirty seconds before the programme starts the broadcaster sets a Prepare to Record Flag (PRF) inside Packet 8/30. If the PIL matches the one pre-programmed into the VCR it starts recording. At the end of the programme the Record Inhibit/Terminate (RI/T) code in Packet 8/30 tells the VCR to stop recording. Packet 8/30 also includes a Programme TYpe indicator (PTY), which is a number describing the genre of programme material. This could be used to set a VCR to record (for instance) all science programmes, or all news programmes. Broadcasters can define their own PTYs so that a video recorder can be programmed to record all episodes of a particular series.

PDC is in use in Belgium, the Czech Republic, Denmark, Finland, France, the Netherlands and the United Kingdom.

2.7.1.8 Video Programme System (VPS)

PDC was largely based on VPS, which was a proprietary system developed in Germany. VPS still tends to be used in German-speaking countries in preference to PDC.

VPS data are encoded on line 16 in 13 bytes (104 bits). VPS does not use the teletext waveform to carry its data, it has a different frequency of data pulse and uses bi-phase encoding rather than the NRZ used by teletext.

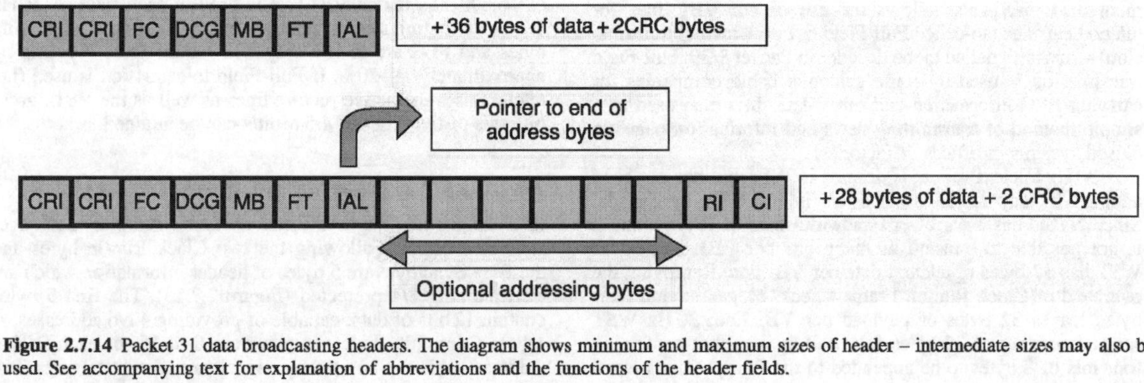

Figure 2.7.14 Packet 31 data broadcasting headers. The diagram shows minimum and maximum sizes of header – intermediate sizes may also be used. See accompanying text for explanation of abbreviations and the functions of the header fields.

VPS is in use in Austria, the Czech Republic, Germany and Switzerland. Some analogue satellite transmissions are using VPS.

2.7.1.9 Antiope

The French version of teletext is called Antiope.[14] In the UK teletext became very popular at the expense of a competing Viewdata service (Prestel) which was offered by British Telecom but was abandoned after a few years. In France the situation was reversed. The astonishing, and unique, success of the Minitel service has prevented Antiope being quite as popular as teletext is elsewhere.

Antiope uses the variable data format, whereas the UK and other countries use the fixed data format. The advantage of the variable format is that almost any text-based data can be transmitted on Antiope without any complex reformatting or pagination being necessary. The disadvantage is that the broadcaster has very little control over the final page display. The binary ones are 700 mV rather than the 462 mV used in the UK system.

Antiope decoders are called Didon (Diffusion de Données).

2.7.1.10 World Standard Teletext (WST)

WST[15] includes support for Latin, Cyrillic and Middle Eastern fonts. Latin-based character sets supported are Czech, English, Estonian, Finnish, French, German, Hungarian, Italian, Lettish, Lithuanian, Polish, Portuguese, Romanian, Slovak, Slovenian, Spanish, Swedish and Turkish. Cyrillic character sets supported are Bulgarian, Croatian, Russian, Serbian and Ukrainian. Arabic, Greek and Hebrew characters are also supported. To maintain compatibility with older decoders 13 of the characters from the original character set have been redefined for the Latin-based character sets. (For instance, the Pound Sterling character (£) can be a hash pound/octothorn (#), a lower case e acute (é), a lower case c cedilla (ç) or the symbol for Turkish Lira, depending on the font in use.) Level 1 decoders will only display the correct characters for the country they were originally manufactured for.

Four 96-entry character sets are defined: two (designated G0 and G2) are for alphanumerics, the other two (G1 and G3) are for graphical characters. A combination of the Packet 0 control bits (C12, C13 and C14) and information in Packets 28 and 29 is used to select the default fonts (this preserves compatibility with older decoders).

Older decoders may not support teletext on lines 6, 318 and 319. WST suggests that these lines are only used for enhancement data. WST also allows the use of non-VBI lines for teletext carriage (so-called Full-Field or Full-Channel transmission) – this is signalled to the decoder in Packet 8/30. Full-Field transmission is used by some analogue cable companies for carrying EPG information and other data. It is also used as a simple method of transmitting news and information pages in closed systems in offices or factories.

A 30 Hz teletext format is defined in WST but has not been adopted by any major broadcasters in the USA or Canada. Since NTSC has a narrower bandwidth than 25 Hz systems it is not possible to transmit as many bits per VBI line; 30 Hz WST has 37 bytes of teletext data per VBI line. Removing the overhead of Clock Run-in, Frame Check, Magazine and Row bytes leaves 32 bytes of payload per VBI line; 30 Hz WST sends four rows of 32 bytes followed by a packet containing four lots of 8 bytes to be appended to the four previous rows. An extra 'Tab' bit has been slotted in to the Magazine and Row byte to indicate which type of packet is being transmitted (front of row or four ends of four rows).

2.7.1.11 Chinese Character Set Teletext (CCST)

CCST[16] supports modern (or modified) Chinese characters as used in mainland China. A variant exists which supports traditional characters[17] – as still used in Hong Kong, Taiwan and Singapore. The use of an odd parity bit for letters suffices for European languages, whose written forms contain a lot of redundancy. If the parity is wrong the decoder does not display the character. If an even number of errors occur in a byte the wrong character will be displayed, but the next time the row is transmitted the correct character will be shown. *lph*b*t-b *s*d t*xt c*n st**l b* *nd*rst**d w*th * c*ns*d*r*bl* n*mb*r *f e*r*rs. Each Chinese character (representing a complete word) is represented by just 2 bytes. If one of the bytes is in error the word (and possibly the entire sentence) will become incomprehensible. In the two Chinese teletext systems Packet 25 contains vertical parity data which can be used by the decoder to check for (and correct) errors.

Fastext can also be used with CCST.

China and Malaysia have both experimented with services delivering Internet pages via teletext.

2.7.1.12 NABTS

Although there is a 30 Hz variant of World Standard Teletext, it has apparently never been used for transmitting teletext pages in North America (except for experimental purposes). North American Basic Teletext Standard[18] (NABTS) was developed from Antiope and a French Canadian 30 Hz teletext system called Telidon. NABTS is widely used for data broadcasting rather than sending teletext pages – it transmits 36 bytes per VBI line and uses the variable data format.

2.7.2 Datacasting

Many proprietary systems are in use for transmitting data inside teletext packets. In WST teletext systems the data will usually be carried in Packet 31. Some form of Forward Error Correction (FEC) must be used to protect the data. This may vary from vendor to vendor, so interoperability between transmitters and receivers from different manufacturers is not guaranteed.

A 25 Hz WST datacasting system can deliver a gross bit rate of 16,800 bit/s if one VBI line is used in each field. A 30 Hz WST datacasting system can deliver 16,320 bit/s. If addressing bytes and FEC are in use they will reduce these bit rates by approximately 10–20%. If Full-Field transmission is used (i.e. teletext uses the active picture lines as well as the VBI), gross bit rates of the order of 4.5 Mbit/s can be attained.

2.7.2.1 RFC 2728

RFC 2728[19] defines a method of transmitting IP data over NABTS teletext. Following the two Clock Run-in bytes and the Byte Sync byte are 5 bytes of header information which are all Hamming 8/4 protected (Figure 2.7.15). The first 3 bytes contain 12 bits of data, capable of providing 4096 addresses. A 1-byte Continuity Index counts from 0 to 15 for each Packet Address – this can be used by a receiver to detect erasures (lost

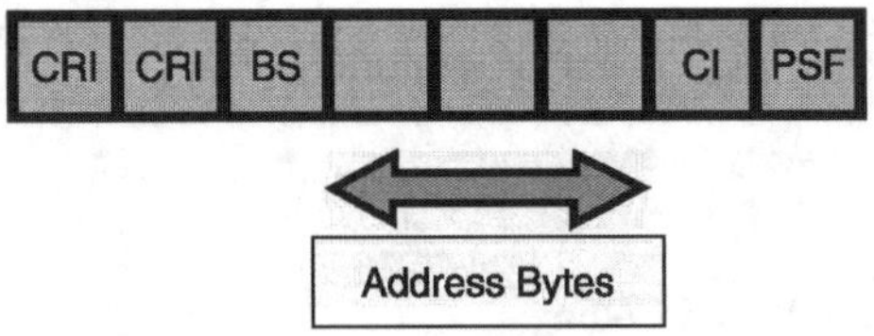

Figure 2.7.15 Data broadcasting header for NABTS. See text for explanation of header fields. NABTS and RFC 2728 use the expression Byte Sync in place of the more usual Framing Code.

packets), which may be recoverable with the FEC bytes which are transmitted in other packets. The packet structure field contains 4 bits which describe the type of payload that follows. The payload is either 26 bytes followed by 2 bytes of FEC data or 28 bytes of FEC data.

Since IP packets might not be a multiple of 26 bytes a stuffing mechanism is required: 0×15 is used to indicate the start of stuffing – all bytes to the end of the 26 bytes of payload data will be filled with 0xEA.

The first stage of the FEC process is to split incoming data into 364-byte bundles. The bundle of data is sent in 14 teletext packets with 2 bytes of FEC data at the end of each packet. ($14 \times 26 = 364$). The following two teletext packets contain 28 bytes of FEC data only. The first packet from the bundle has a Continuity Index of 0 and the FEC-only packets have Continuity Indices of 14 and 15. The FEC used can correct for all random single-bit errors, most double-bit errors, all single-byte errors and one or two lost packets from the 16-packet block.

The Maximum Transmission Unit (MTU) size is 1500 bytes and the SLIP[20] encapsulation protocol is used. The IP header may be compressed;[21] one scheme is proposed in the RFC but others are allowed.

A 4-byte CRC (identical to the ISO/IEC 13818 (MPEG-2) CRC but different from the standard CRC used on ethernet) is appended to the end of the SLIP encapsulated packet.

Assuming one line in each field is used to send IP data in this way, a bit rate of 10,920 bit/s is theoretically achievable in a 30 Hz system. Increasing the number of lines dedicated to datacasting will increase the bit rate attainable. Constraining the size of the input IP packets so that they exactly fill the teletext packets with no wasteful stuffing is a sensible practice.

2.7.3 PCM Audio

There are proprietary schemes for transmitting digitised audio on VBI lines. These use similar clock rates to teletext but can be distinguished from it in two ways. The Framing Code should be different (i.e. not 0x11100100) and the 'comb' pattern should never be seen (Figure 2.7.6).

2.7.4 Telecommand Systems

Many broadcasters use the VBI to send messages around their contribution and distribution networks and telecommands to their studios and transmitters. These signals are sometimes sent as inverted teletext, where all of the ones are transmitted as zeros and vice versa. This kind of signal will be ignored by a standard teletext decoder.

RAI use a system built on teletext signalling principles but with a different clock rate. Line 335 contains 19 bytes (152 bits) of data. The signalling elements are 333.333 ns wide, corresponding to a frequency of 3 MHz ($192 \times$ line frequency). NRZ signalling is used with zero represented by 0 V and one by 490 mV. The first bit is 11 μs after the leading edge of line sync. The Clock Run-in and Framing Code bytes are the same as standard teletext. The remaining 16 bytes are all Hamming 8/4 encoded. (In a telecommand system data integrity is more important than throughput.)

Other broadcasters use similar systems. VPS was originally a telecommand system before it was developed into a method for controlling home video recorders.

2.7.5 Closed Captions

The Television Decoder Circuitry Act of 1990 required that after 1 July 1993 all televisions sold in the USA with a screen size of 13 inches or greater would have to be fitted with an integral Closed Caption decoder. Most televisions on sale in Canada after that date would also have a Closed Caption decoder, since manufacturers would sell the same models in both markets. The Telecommunications Act of 1996 introduced an 8-year transition period (ending on 1 January 2006), after which 95% of new, non-exempt, English language programming must be captioned. (Different percentages apply to Spanish language and archive programming. Exemptions were intended to minimise the financial burden on smaller stations, but also included categories of programming such as music, adverts and public service announcements.) Closed Captioning was developed by the PBS stations in the USA in the 1970s but no formal technical specification was written by them. The Television Data Systems Subcommittee (TDSS) subsequently produced the Closed Captioning standards.[22,23] (The TDSS has also produced a Closed Captioning standard for ATSC transmissions[24] – this will not be dealt with here.) The system is capable of supporting four data channels on field 1 line 21, known as CC1, CC2, T1 and T2 (Figure 2.7.16). Field 2 line 21 can carry four more data channels known as CC3, CC4, T3 and T4. A fifth channel, Extended Data Services, can utilise excess capacity in field 2. CC1 and CC3 support synchronous captions which can be placed anywhere on the screen. CC2 and CC4 can carry non-synchronous captions and the text (T) channels can carry a half screen or full screen of scrolling text. In practice CC1 and CC3 seem to be the only channels in common use. They carry English and Spanish subtitles in the USA and English and French subtitles in Canada. An English speaker can produce about 10 words per second. The average

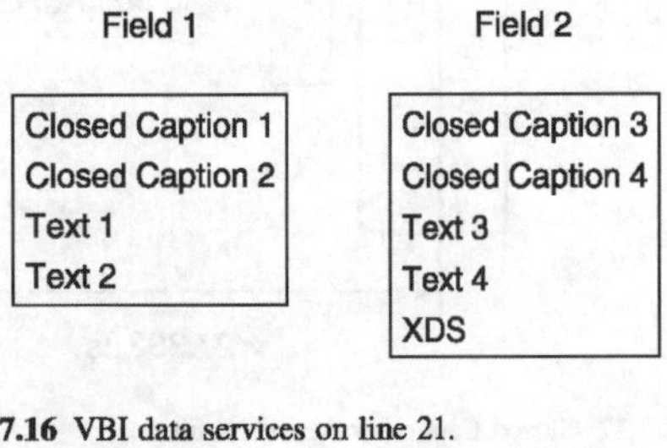

Figure 2.7.16 VBI data services on line 21.

English word is about 5.5 letters (plus a character for the space!), so a subtitling system has to be able to support 65 bytes per second to keep up with fast speakers. The Closed Caption system can support 60 bytes per second per field. Add in the extra control data for placing captions, removing captions, changing text colour, etc. and it can be seen that, in reality, CC2, CC4, T1, T2, T3 and T4 will rarely be used if a programme is captioned in two languages. (Many programmes will still only be captioned in one language, however.)

Three types of captioning modes are supported, 'Roll Up', 'Paint On' and 'Pop Up'. Roll Up mode is intended for live captioning. Captions are wiped on from the left, a word at a time, and then rolled up the screen with a new line appearing at the bottom. In Paint On mode a single line of text is wiped onto the screen, a letter at a time; they are mostly used for adverts and special effects. Pop Up mode is used for pre-prepared captions which appear and disappear when required. All three modes allow the captions to be placed in different parts of the screen.

The Closed Caption character set supports the accented characters needed for French, Spanish and Portuguese (the major languages needed for North and South America – other than English). Some French Canadian characters are not supported. Although coloured text is possible, many early decoders do not support it so it cannot be used to distinguish between speakers; captioners tend to use ≫ instead to indicate that someone new is speaking. Closed Captions can also be placed on different parts of the screen to indicate where a sound (speech or effect) is originating from. Since the early Closed Caption decoders did not have (easily readable) true descenders on their lower case letters, captions in the USA tend to be upper case only, with upper case italics used for emphasis. Captions normally appear as white letters inside a black box, but newer decoders offer white characters with a black edge and other enhancements. A number of additional 2-byte codes were added to the original captioning codes for letters and symbols which might be useful to captioners (such as ¢).

Closed Captions are sometimes referred to as 'LITO' (from LIne Twenty One).

2.7.5.1 Signal Coding

The Closed Caption signal consists of seven cycles of Clock Run-in (at 0.5035 MHz), a start or marker pulse and then 2 bytes consisting of 7 data bits with odd parity protection (Figure 2.7.17). The Clock Run-in peak and the data pulses are between 48 and 52 IRE high and the data pulses use NRZ encoding. This is a very robust signal which will survive most transmission impairments and still be recordable on a domestic VHS recorder for later playback.

There should be no set up (pedestal) on line 21 when it is carrying a Closed Caption waveform. Many SDI to NTSC converters add 7.5 IRE to line 21. One workaround is to have the SDI Closed Caption encoder sit the line 21 waveform down by 7.5 IRE; this will, however, produce an illegal waveform in the SDI domain.

2.7.5.2 Closed Caption Character Set

The Closed Caption character set and 2-byte symbols are listed in Tables 2.7.2 and 2.7.3.

Control codes have hex values from 0x00 to 0x1F. Control codes are sent twice as a simple form of error protection. The first byte of the 2-byte codes will vary depending on which captioning channel it is used in. A few of the characters have been changed since the system was introduced, causing older decoders to sometimes display incorrect characters. The ®, ™, ç, Ñ and ° symbols were originally ¼, ¾, {, } and /, respectively.

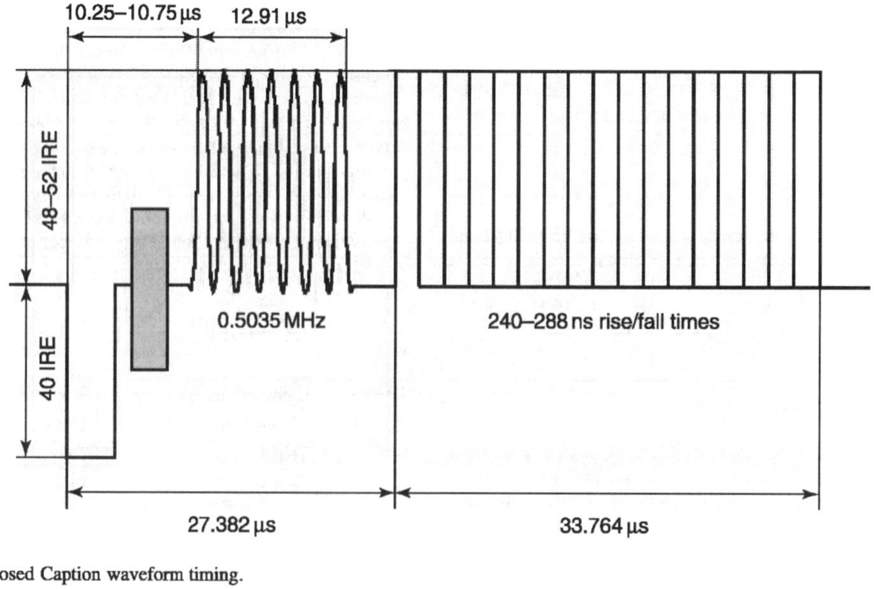

Figure 2.7.17 Closed Caption waveform timing.

Table 2.7.2 Closed Caption character set, based on the ASCII Latin-1 character set with minor modifications

Hex code	*Symbol*	*Hex code*	*Symbol*
20		50	P
21	!	51	Q
22	“	52	R
23	#	53	S
24	$	54	T
25	%	55	U
26	&	56	V
27	‘	57	W
28	(	58	X
29	)	59	Y
2A	á	5A	Z
2B	+	5B	[
2C	,	5C	é
2D	–	5D	]
2E	.	5E	í
2F	/	5F	ó
30	0	60	ú
31	1	61	a
32	2	62	b
33	3	63	c
34	4	64	d
35	5	65	e
36	6	66	f
37	7	67	g
38	8	68	h
39	9	69	i
3A	:	6A	j
3B	;	6B	k
3C	<	6C	l
3D	=	6D	m
3E	>	6E	n
3F	?	6F	o
40	@	70	p
41	A	71	q
42	B	72	r
43	C	73	s
44	D	74	t
45	E	75	u
46	F	76	v
47	G	77	w
48	H	78	x
49	I	79	y
4A	J	7A	z
4B	K	7B	ç
4C	L	7C	÷
4D	M	7D	Ñ
4E	N	7E	ñ
4F	O	7F	█

Table 2.7.3 Two-byte Closed Caption symbols

Hex code	*Symbol*
11,30	®
11,31	°
11,32	½
11,33	¿
11,34	™
11,35	¢
11,36	£
11,37	♪
11,38	à
11,39	
11,3A	è
11,3B	â
11,3C	ê
11,3D	î
11,3E	ô
11,3F	û

2.7.5.3 Korean Closed Captions

All new Korean televisions contain Closed Caption decoders but these are only capable of decoding US (English and Spanish) Closed Captions. Korean viewers use an external box to decode Korean Closed Captions.

2.7.5.4 Japanese Closed Captions

American programmes shown on Japanese TV usually have Japanese subtitles on the video. To facilitate the learning of English, a company in Japan produces an external decoder which reads the US Closed Captions (if these are still present) and overlays the English captions on top of the Japanese subtitles.

2.7.5.5 Profanity Removal

At least two companies in North America are offering devices which read the Closed Captions and remove objectionable words from the captions or the soundtrack, or both. Depending on the system, profanities in the captions are replaced with ‘X’s or a milder epithet. The systems will also optionally mute the soundtrack at the correct place (this will not work with live captions, only pre-recorded ones).

2.7.5.6 Non-broadcast Closed Captions

Laser discs were the analogue precursor to DVDs, similar in size to long-playing records (the analogue precursor to CDs). Closed Captions are often found on laser discs.

Low-bandwidth signals such as Closed Captions can be placed on pre-recorded VHS tapes. Since about 1995 Closed Captions have been available on line 22 of PAL VHS tapes in the UK and elsewhere in Europe (Figure 2.7.18). Closed Captions on 30 Hz material can deliver 60 bytes of data per second, but on 25 Hz systems they can only deliver 50 bytes per second.

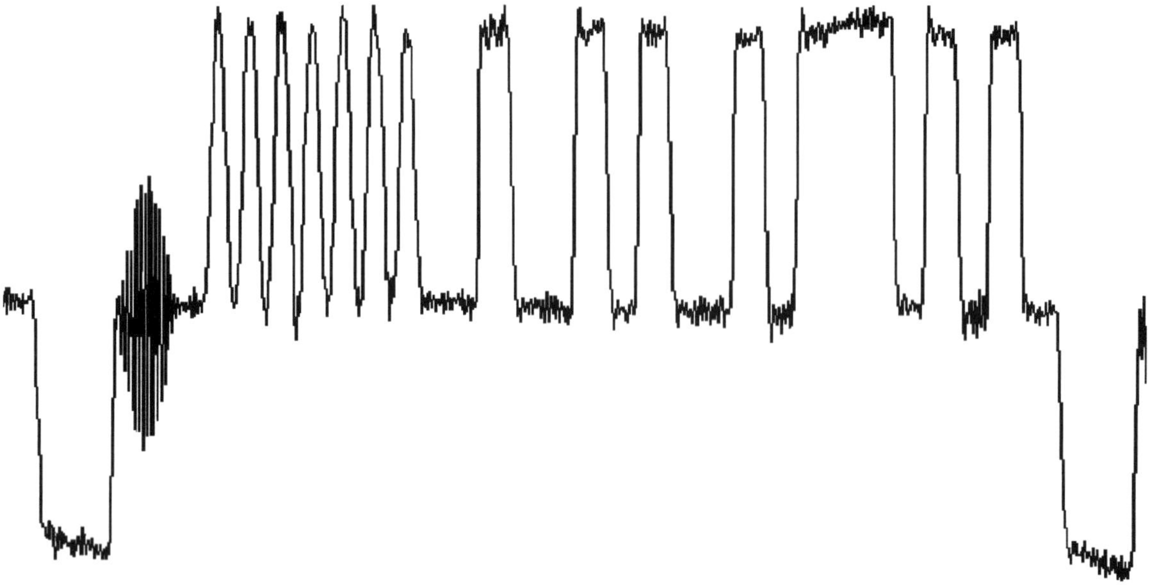

Figure 2.7.18 Closed Captions from a PAL VHS tape.

The US captions cannot be transferred directly but will normally require editing and/or translating. External Closed Caption decoders for pre-recorded tapes are used in Denmark, Finland, Germany, Norway, Sweden and the UK.

2.7.6 Extended Data Services (XDS)

XDS (previously EDS) is carried on field 2 line 21. XDS information is carried in packets and contains details such as time of day, station ID, network, current programme name, etc. This could be used for PDC type functions with a suitably equipped VHS recorder. XDS also carries V-chip information.

2.7.7 V-chip

Under FCC rules all televisions with a screen 13 inches or larger sold in the USA since 1 January 2000 must be equipped with V-chip technology. The violence control chip, or V-chip,[25] allows parental control of material that might be viewed by minors. The V-chip data is part of the XDS signal and is carried on field 2 line 21. The television decodes the V-chip data and can block the display of the programme. Access to programming above the parentally defined age threshold requires the use of a password. Three different classification schemes are in use in North America, one in the USA and two in Canada.

Table 2.7.4 shows the three ratings systems in use in North America. The Canadian systems have an Exempt/Exemptées category which does not exist in the USA. Exempt programming includes news, sports and documentaries. Other than this the three systems differ only in detail. The TV-Y (All Children), C (Children) and G (Général) ratings denote programming which can be watched by all ages. The TV-MA (Mature Audience Only), 18+ (Adult programming) and 18 ans+ (Cette émission est réservées aux adultes) indicate programming intended for adult audiences. Further details can be obtained from the three websites listed at the end of this chapter.

Table 2.7.4 V-chip ratings in use in North America

US ratings	*English Canadian ratings*	*French Canadian ratings*
	E	E
TV-Y	C	G
TV-Y7		
	C8+	8 ans+
TV-G	G	
TV-PG	PG	
		13 ans+
TV-14	14+	
		16 ans+
TV-MA	18+	18 ans+

Near equivalent categories are shown on the same horizontal row.

2.7.8 Non-Standard ITS

As the vertical blanking interval has become full of teletext and other waveforms over the years, every effort has been made to make the best use of test signal lines. The BBC has been using the VBI for internal data transmission since 1972. A system known as ICE (Internal Communication Equipment) was present on lines 21 and 334 and provided up to 12 signalling channels for such purposes as network identification, opt-out

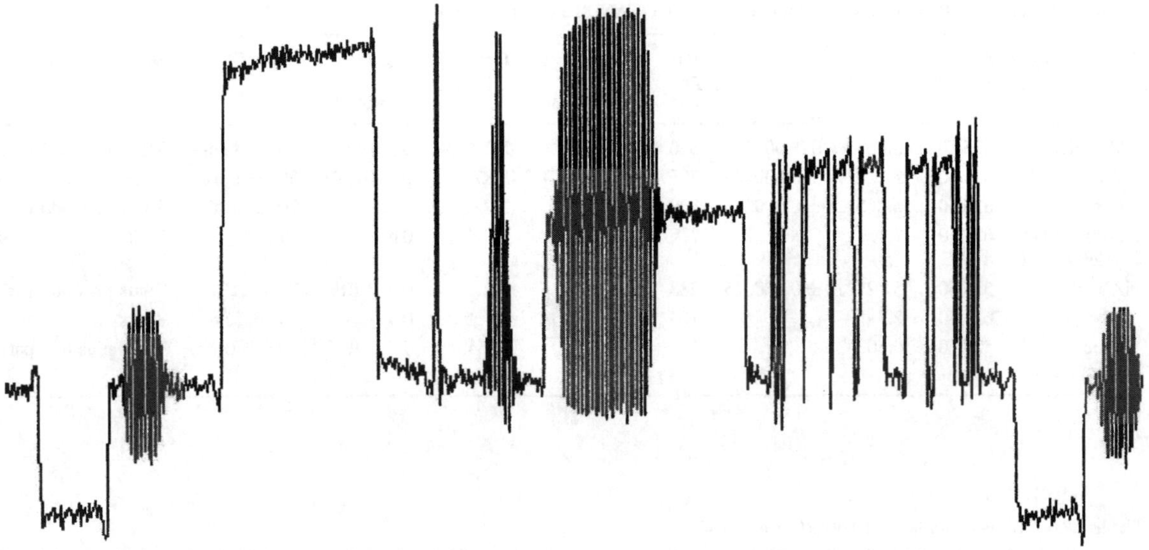

Figure 2.7.19 BBC2 line 334 showing modified VITS signal.

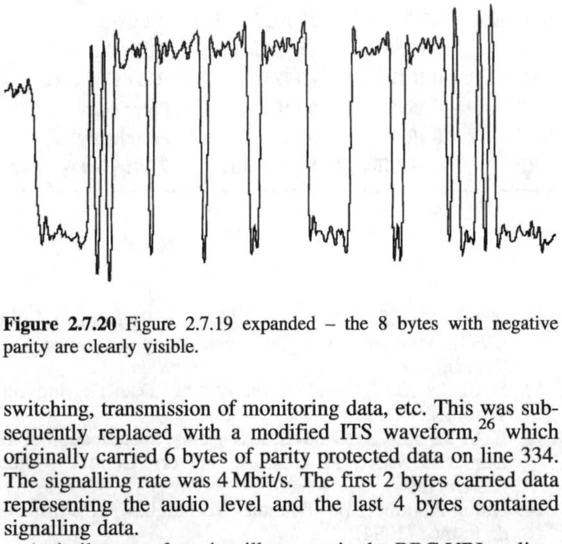

Figure 2.7.20 Figure 2.7.19 expanded – the 8 bytes with negative parity are clearly visible.

switching, transmission of monitoring data, etc. This was subsequently replaced with a modified ITS waveform,[26] which originally carried 6 bytes of parity protected data on line 334. The signalling rate was 4 Mbit/s. The first 2 bytes carried data representing the audio level and the last 4 bytes contained signalling data.

A similar waveform is still present in the BBC VBI on lines 334 and 335 (Figures 2.7.19 and 2.7.20). It now carries 8 bytes of data but none of these are modulated by the transmitted audio level. The signal is probably used for opt-out switching.

Other European broadcasters also use non-standard ITS signals, but North American broadcasters tend to use standard waveforms.

2.7.9 VITC

Vertical Interval TimeCode (VITC) is covered in depth in Chapter 2.5 but is mentioned here for completeness. VITC puts an hours, minutes, seconds and frames count into the VBI. In 30 Hz systems it is usually on line 14. In 25 Hz systems it is usually on line 16 or 22. It can, however, appear on almost any line (and can sometimes be seen 'in vision' during the active picture).

2.7.10 AMOL (Automated Measurement of Line-Ups)

AMOL-I and AMOL-II are two systems that Nielsen use for assessing audience numbers and other purposes. A set-top box reads the Nielsen data and can then record which programme or advert is being tuned to at any given time.

AMOL-I[27] has an instantaneous data rate of 1 Mbit/s and is present on field 1 line 20 and/or field 2 line 22 (line 285).

AMOL-II[28] can work in a backwards compatible mode with AMOL-I. It also has a 2 Mbit/s signal which can be present on lines 20 or 22 in either field (lines 20, 22, 283 or 285)

2.7.11 Wide Screen Signalling (WSS)

WSS[29] is present on the first half of line 23 in PAL systems. The 14 bits of data describe the aspect ratio of the video, whether material is field or frame derived and the presence or absence of subtitles in teletext. Widescreen televisions can use this information to automatically switch between different display modes. Line doubling televisions can use the information to decide whether to interpolate between physically adjacent lines or lines from one field.

2.7.12 Summary of Technical Data

A summary of technical data is given in Tables 2.7.5 and 2.7.6.

Table 2.7.5 Teletext, VPS and Closed Caption technical data (TANDBERG Television)

Data	*Line standard*	*Bit rate (Mbit/s)*	*Bits per line*	*Bit rate w.r.t. line frequency*	*Data*	*Logic 0*	*Logic 1*	*Data format*
WST teletext	625/50	6.9375 ± 0.000025	360	444	NRZ	0 ± 14 mV	462 ± 42 mV	7 bits plus odd parity
WST teletext	525/60	5.727272 ± 0.000025	296	364	NRZ	0 ± 14 mV	500 ± 1 mV	7 bits plus odd parity
Antiope	625/50	6.203125 ± 0.000025	320	397	NRZ	0 mV	700 ± 7 mV	7 bits plus odd parity
Australian/NZ teletext	625/50	?	?	?	NRZ	0 mV	490 mV	7 bits plus odd parity
NABTS	525/60	5.727272 ± 0.000025	288	364	NRZ	0 ± 2 IRE	70 ± 2 IRE	7 bits plus odd parity
VPS	625/50	2.5	104	160	Bi-phase	0 V	500 ± 25 mV	8 bits
Closed Caption	525/60	0.50352	17	32	NRZ	0 ± 2 IRE	50 ± 2 IRE	7 bits plus odd parity

Table 2.7.6 VBI data signals (TANDBERG Television)

Data	*Line standard*	*Bit rate (Mbit/s)*	*Bits per line*	*Bit rate w.r.t. line frequency*	*Data*	*Logic 0*	*Logic 1*	*Data format*
WSS	625/50	0.833	14	53.3	Bi-phase	0 V	500 ± 25 mV	Bit field
VITC (EBU)	625/50	1.8125	90	116	NRZ	0 V	550 ± 50 mV	BCD plus CRC
VITC (SMPTE)	525/60	1.7915625	90	113.75	NRZ	0 IRE	80 IRE	BCD plus CRC
AMOL-I	525/60	1.0	48	N/A	NRZ	0 IRE	50 IRE	Proprietary
AMOL-II	525/60	2.0	96	N/A	NRZ	0 IRE	50 IRE	Proprietary
XDS/V-chip	525/60	0.50352	17	32	NRZ	0 ± 2 IRE	50 ± 2 IRE	7 bits plus odd parity

References

1. BBC, IBA and BREMA. *Broadcast Teletext Specification*, ISBN 0 563 17261 4 (October 1976).
2. Hamming, R.W. Error detecting and error correcting codes, *Bell System Tech J.*, **29**, 2 (1950).
3. ISO/IEC 8859, Information Technology – 8-bit Single-byte Coded Graphic Character Sets.
4. EBU SPB 492, Teletext Specification (625-line Television Systems) (1992).
5. EACEM Technical Report No. 8, Enhanced Teletext Specification, Draft 3, EACEM (November 1994).
6. ITU-R Recommendation 653, System B, 625/50 Television Systems.
7. NZS 6606, Broadcast Teletext (New Zealand) (1988).
8. International Telecommunications Union Recommendation ITU-R BT.653–3, System C (February 1998).
9. ETSI EN 300 472 V1.2.2, Digital Video Broadcasting (DVB); Specification for Conveying ITU-R System B Teletext in DVB Bitstreams (August 1997).
10. ETSI EN 301 775 V1.1.1, Digital Video Broadcasting (DVB); Specification for the Carriage of Vertical Blanking Information (VBI) Data in DVB Bitstreams (November 2000).
11. ETSI EN 300 707 V1.2.1, Electronic Programme Guide (EPG); Protocol for a TV Guide using Electronic Data Transmission.
12. ETSI EN 300 708, Television Systems; Data Transmission Within Teletext.
13. ETSI TS 300 231, Television Systems; Specification of the Domestic Video Programme Delivery Control System (PDC) (1996).
14. Didon-Antiope Specifications Techniques, Télédiffusion de France (1984).
15. ETSI TS 300 706 V1.2.1, Enhanced Teletext Specification (December 2002).
16. Chinese Character System Teletext, SARFT (March 1993).
17. HKTA 1106, Technical Standard for Public Teletext Services in Hong Kong (March 1996).
18. EIA-516, Joint EIA/CVCC Recommended Practice for Teletext: North American Basic Teletext Specification (NABTS), Electronic Industries Association, Washington (1988).
19. Panabaker, P., Wegerif, S. and Zigmond, D. RFC 2728, The Transmission of IP Over the Vertical Blanking Interval of a Television Signal (November 1999).
20. Romkey, J. RFC 1055, A Nonstandard for Transmission of IP Datagrams Over Serial Lines: SLIP, STD 47 (June 1988).

21. Jacobson, V. RFC 1144, Compressing TCP/IP Headers for Low-Speed Serial Links (February 1990).
22. EIA-608-B, Line 21 Data Service.
23. EIA-608 R-4.3, Recommended Practice for Line 21 Data Services (September 1992).
24. EIA-708-B, Digital Television (DTV) Closed Captioning (1995).
25. EIA-744, Content Advisories (V-chip Data) using XDS.
26. Holder, J.E. and Spicer, C.R. BBC internal document, A New Insertion Test Signal with Audio Monitoring Facilities.
27. Nielsen Engineering and Technology. AMOL Signal Specification, Document Number ACN 403-1122-000, Revision Level 1.4 (16 January 1995).
28. Nielsen Engineering and Technology. AMOL-II Signal Specification, Document Number ACN 403-193-024, Revision Level 3.2 (19 May 1995).
29. ETSI EN 300 294, 625-Line Television – Wide Screen Signalling (WSS) (1996).

Bibliography

Mothersole, P.L. and White, N.W. *Broadcast Data Systems: Teletext and RDS*, Butterworths, London (1990).

Internet Resources

MRG Systems, Teletext Tutorial, http://www.mrgsystems.co.uk
625 Andrew Wiseman's Television Room, PDC (Programme Delivery Control) explained, http://625.uk.com/pdc
http://pdc.ro.nu/teletext.html
http://www.fcc.gov/vchip
http://www.puceantiviolence.ca
http://www.vchipcanada.ca
http://www.robson.org/capfaq

Eur Ing Andrew W. Jones BSc CEng MIEE
Engineering Learning Executive
BBC Training & Development

2.8 Digital Interfaces for Broadcast Signals

2.8.1 Layered Approach to Broadcast Media Interfaces

It is common to talk of a 'stack' when referring to interfaces and protocols for transfer of IT data across a computer network. This is a reference to layers of processing within software and perhaps associated hardware, necessary to provide a reliable information link between computer applications. Some communications engineers refer to an idealised 'OSI seven-layer model'.

Transmission of data on the broadcast industry standard interfaces described here is in many ways simpler than transfer of data over IT style *packet* networks. These broadcast interfaces carry digital video, audio and related data in one direction, usually over a dedicated point-to-point or point-to-multi-point circuit. Such data interconnections offer very high reliability, freedom from errors, stable reference clocks and consistently low latency. The use of dedicated physical media data pathways is suited to operational practices that make it straightforward to distribute, route and monitor video and audio. The physical routes are robust and reliable, but if they do fail it is usually quick to isolate a fault. Standards for these broadcast interfaces are well known, stable and widely supported by equipment manufacturers.

Having sung the praises of specialist broadcast interfaces over dedicated links there is an increasing trend towards use of IT networks to carry broadcast data. The attraction of less specialist, and therefore lower cost, fast Ethernet network infrastructure carrying broadcast media essence (sound and vision) as well as metadata (data about that sound and vision, e.g. programme scripts, broadcast rights, etc.) is obvious. Broadcast media is increasingly stored on fileservers, alongside the file servers hosting automation and playout systems and more traditional business 'back-office' packages. It will rapidly become the norm to interconnect systems outside the studio or operational island with whatever high-speed networking technology the IT industry has to offer at the time. The layered approach that I am using to take the reader through this chapter is a very common way of understanding and structuring IT or Telecomm style networks (see Figure 2.8.1). The Internet Protocol stack reigns supreme! Asynchronous Transfer Mode with its well-established and inbuilt control Quality of Service offers advantages, albeit at a cost, in the Wide Area Network, with proven gateways allowing IP traffic to be carried on ATM on a teleco (telecommunications service provider) long haul Synchronous Digital Hierarchy bearer.

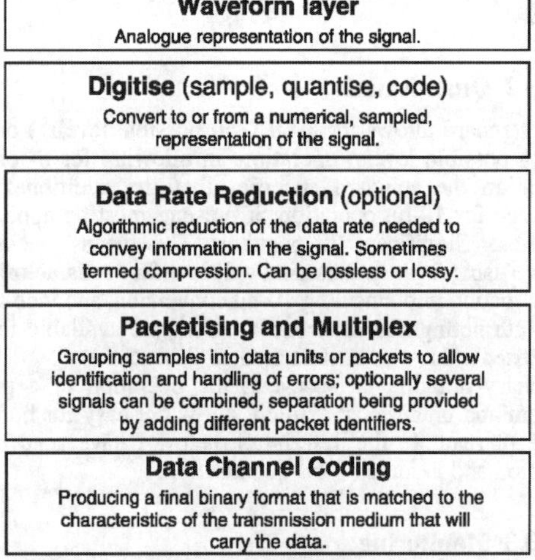
Waveform layer
Analogue representation of the signal.
Digitise (sample, quantise, code)
Convert to or from a numerical, sampled, representation of the signal.
Data Rate Reduction (optional)
Algorithmic reduction of the data rate needed to convey information in the signal. Sometimes termed compression. Can be lossless or lossy.
Packetising and Multiplex
Grouping samples into data units or packets to allow identification and handling of errors; optionally several signals can be combined, separation being provided by adding different packet identifiers.
Data Channel Coding
Producing a final binary format that is matched to the characteristics of the transmission medium that will carry the data.
Physical layer
Handles the physical launching of the waveform representing the data onto the transmission medium. Includes amplitude, characteristic impedance and line matching. Optical parameters such as wavelength (for fibre transmission) and connector types, etc.

Figure 2.8.1 Broadcast media interface layers.

The broadcast layer is applicable to IT style LAN/WAN networks too. Sampling standards for broadcast digital audio and video described in the following pages are essential to the transmission of undistorted and high-quality pictures and sound.

Data that is related to this 'essence' such as timecode and source aspect ratio information or audio channel status may also be required to pass across an IT style LAN.

2.8.2 Waveform Layer

Chapter 2.4.4 gives details of component video waveforms while Chapter 2.1 describes audio waveforms and levels.

2.8.3 Digitising Layer

The process of converting an analogue waveform to a digital representation is sometimes called *digitising*, or analogue-to-digital conversion. The process is one of representing a continuous waveform as a discrete series of numbers.

Chapter 2.4.5 gives details of sampling, quantising and coding video, while Chapter 2.1.2 gives equivalent details for digital audio sampling standards.

2.8.3.1 Component Colour Systems

Chapter 000 provides details of component colour systems. The most fundamental colour components of a TV system are signals conveying quantities of the **R**ed, **G**reen and **B**lue colour television primaries, though most digital video interfaces work with Y, Cb and Cr (luminance and colour difference) component signals.

Figures 2.8.2 and 2.8.3 give a useful summary of digital video sampling and quantising for standard-definition television.

2.8.3.2 Quantisation

The standard allows for 8-bit (256 possible levels*) or 10-bit (1024 possible levels) operation. In allowing for 8- or 10-bit operation the standard specifies that the additional 2 bits required for 10-bit operation, if present, must be appended in the **L**east **S**ignificant **B**it positions. This means that a signal source using 8-bit quantisation will interface satisfactorily with a destination implementing 10-bit conversion, and vice versa. It is worth noting that different notations are available for these quantised values, as summarised in Table 2.8.1.

Binary notation is closest to the operation of a practical system and is the most explicit, but is not easy for humans to write or read! In the description below I have used decimal notation and assumed 8-bit quantisation.

2.8.3.3 Monitoring

Viewing the signal in numeric form is not the best way to monitor its operational parameters in day-to-day usage. Digital waveform monitors will show the signal on a graphical display and this is usually the best way to check line-up levels, signal timing, etc. Most waveform monitors will also indicate colour gamut errors, check the validity of the Timing Reference Signals and check for bit error rate or count 'error seconds' if an Error Detection and Handling system is in use.

*In practice, a few values are reserved for use within the Timing Reference Signal.

2.8.3.4 Handling Widescreen

Increasing the aspect ratio from 4:3 to 16:9 has proved popular in some countries, including the UK, where it is the *aspect ratio* (ratio of picture width to picture height) of choice for many of the newer digital TV services. Increasing the width of the television picture while keeping the video bandwidth constant reduces the resolution of the picture seen by the viewer. Each line is 'stretched' across a greater distance on the screen. Some manufacturers (notably Grass Valley Group – now part of Thomson) lobbied for the higher sample rate of 18 MHz for such systems and this has been included in the standards. Few have produced, or bought, equipment that operates at 18 MHz however, perhaps because the increased cost of replacing studio and post-production equipment could not be justified. Experience in the UK and elsewhere has shown that viewers seem content to watch widescreen pictures with a bandwidth of 5.5 MHz. Many domestic widescreen receivers have problems resolving fine detail in any event due to compromises made in the larger deflection angles required in the CRT display.

In practice, widescreen production is handled by digital equipment using exactly the same 13.5 MHz 4:2:2 sampling structure as 4:3.

An important part of handling widescreen pictures in production, post-production and transmission is signalling the shape and active portion of the picture to equipment in the production chain. Several systems are available to do this (Table 2.8.2).

2.8.3.5 Pixel Aspect Ratio

In common usage a *pixel* is a sample point in the picture. For standard-definition television digital active line has 720 luminance pixels, hence '720 pixels per line'. In this context a pixel corresponds the instantaneous signal amplitude at sample time. A 625/50 sampled signal is said to have 720 × 576 pixels in each picture.

Computer graphics adapter resolutions are commonly referred to by their resolution in pixels. For instance, SVGA resolution is 800 × 600 and XVGA is 1024 × 768; both may be considered as a grid of picture sampling points. The sampling structure is said to be *orthogonal* as the equivalent samples on each row of pixels form columns that are at right angles to those rows. Orthogonal sampling results from arranging a whole number of samples across each display line, and a whole number of samples in each field and frame.

Computer displays usually have an aspect ratio of 4:3, the number of pixels horizontally and vertically also possessing this 4:3 ratio, 800:600 = 4:3. This means that the horizontal and vertical spacings of the sample grid are equal, and the system is said to have 'square pixels' or a pixel aspect ratio of 1:1.

Rec. 601 sampling does not use the 'square pixels' of the computer display.[1] The number of active lines was fixed by analogue television standards, while the number of samples in each line was chosen to obey the Nyquist criteria for system bandwidth and to provide a convenient sample frequency for both 625- and 525-line systems. The 13.5 MHz luminance

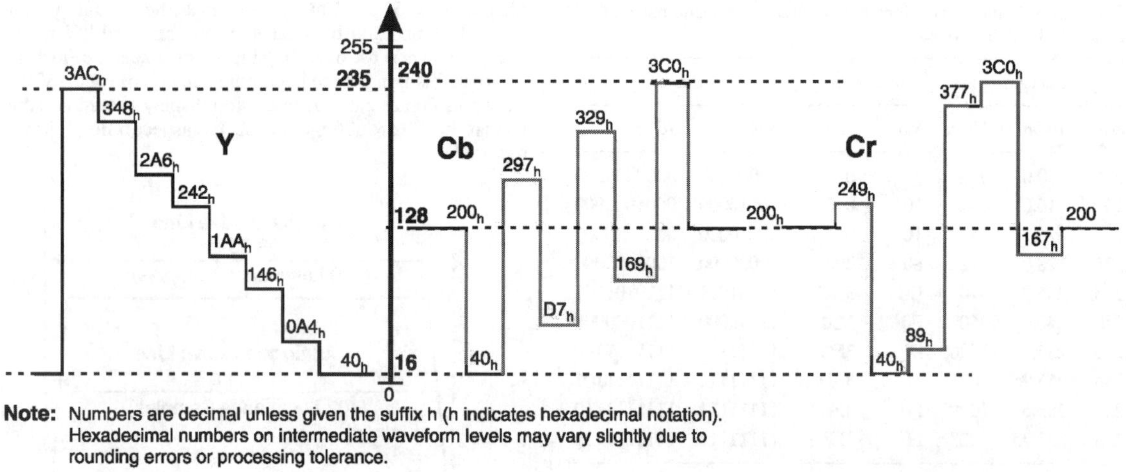

Figure 2.8.2 Ranges of Y, Cr and Cb colour components for ITU-R BR.601[1] sampled waveforms.

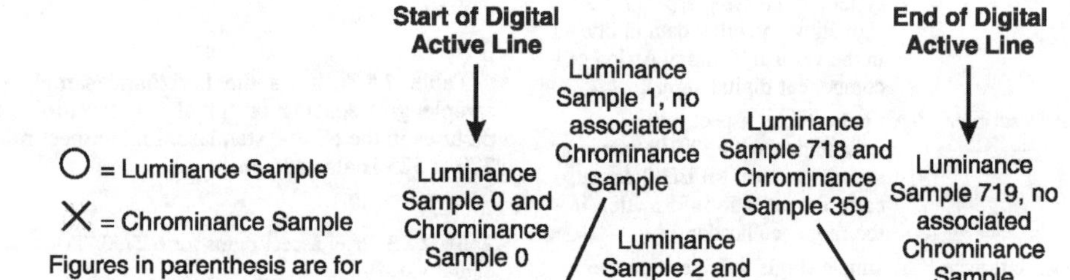

Figure 2.8.3 Sampling structure of 4:2:2 digital video.

Table 2.8.1 Common notations used to describe numeric range of 8- and 10-bit digital video

Decimal			Hexadecimal		Binary	
8-bit	*10-bit*	*10-bit*	*8-bit*	*10-bit*	*8-bit*	*10-bit*
0	0.0	0	0	0	00000000	0000000000
16	16.0	64	10	040	00010000	0001000000
16	16.25	65	10	041	00010000	0001000001
128	128.0	512	80	200	10000000	1000000000
235	235.0	940	EB	3AC	11101011	1110101100
240	240.0	960	F0	3C0	11110000	1111000000
255	255.0	1020	FF	3FC	11111111	1111111100
255	255.25	1021	FF	3FD	11111111	1111111101
255	255.5	1022	FF	3FE	11111111	1111111110
255	255.75	1023	FF	3FF	11111111	1111111111

Table 2.8.2 Methods used to signal aspect ratio in television systems

Name	*Summary*
Widescreen Signalling	An analogue waveform carried on the first half of line 23.
Video Index	A system for carrying aspect ratio signalling and other data in line 11 in the vertical blanking period of a component digital signal.
Active Format Descriptor	An indicator of aspect ratio information within DVB transport stream packets. MPEG-2 can also carry aspect ratio information in the sequence header.
General Purpose Interface	A simple single wire per function interface that can indicate widescreen mode, or any other control function.

sampling frequency chosen is an integer multiple of both 625/50 and 525/60 line frequency and frame rate, hence the sample grid is orthogonal for these major world TV scanning standards.

Converting to and from the 'square pixels' of the computer world to the pixel aspect ratio of TV sampling standards requires *interpolation* – digital filtering and resampling – of the data. If this is not done the picture will change shape in the conversion process. Some computer drawing packages will perform this conversion automatically, others require more manual intervention. In all cases careful consideration needs to be given to the choice of pixel resolution in the 'square pixel' world if unacceptable losses or picture shape distortion are to be avoided.

If the number of horizontal and vertical pixels and the shape of the picture are known, calculation of the pixel aspect ratio is straightforward. Care needs to be taken to remember that, while Rec. 601 sampled grids are 720 × 576 (720 × 488 for NTSC), when correctly blanked in the analogue domain the width of the picture is reduced slightly; 720 pixels per line at a 13.5 MHz sampling clock results in a line duration of 53.33 μs. Analogue active line time, measured to the half-amplitude points of blanking, is 52 μs. This means that the visible part of a correctly timed and blanked Rec. 601 sampled TV picture is 702 pixels wide for 625/50. Nine 74 ns pixels are 'lost' in the analogue blanking at the beginning and end of each TV line, as shown in Figure 2.8.4. A television display should be adjusted so that this 702 × 576 picture has the aspect ratio (4:3 or 16:9) required.

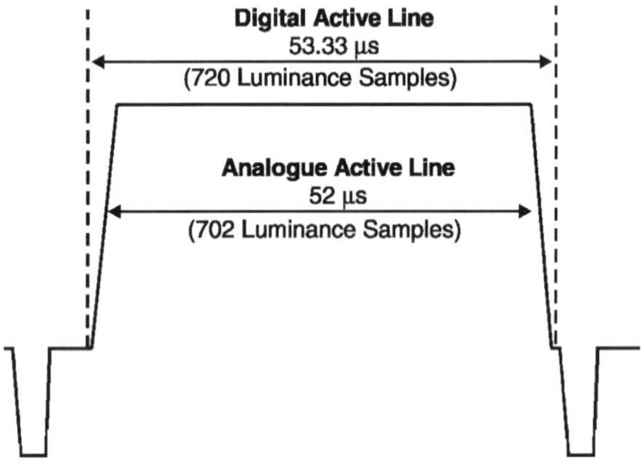

Figure 2.8.4 Analogue blanking and digital active line for 625/50 television.

Table 2.8.3 shows the horizontal sample grid to vertical sample grid spacing or 'pixel aspect ratio' for 4:3 and 16:9 pictures in the 625/50 standard. Pixel aspect ratio of unblanked 720 × 525 is also shown.

Table 2.8.3 Pixel aspect ratios for 625/50 ITU-R BT.601 sampled digital video

Viewport aspect ratio	*Rec. 601 sampling pixel grid*	*Pixel aspect ratio*	*'Square pixels' pixel grid*
4:3	702:576	1.094:1	768:576
4.1026:3	720:576	1.094:1	787.69:576
16:9	702:576	1.459:1	1024:576
16.4103:9	720:576	1.459:1	1050.25:576

Table 2.8.3 refers to a viewport aspect ratio. The only aspect ratios applicable to 'real' television displays are 4:3 and 16:9 'widescreen'. The *viewport* concept is that of the aspect ratio necessary before blanking if a correctly shaped display is to be produced after the picture is blanked, cropping the line width from 720 pixels to 702 pixels in the case of the 625/50 standard.

Many computer packages recommend working on a 'square pixel' grid of 768:576 if the pictures are intended for transfer to an SD TV display. This is correct if the end result is mapped to the central 702 Rec. 601 pixels and intended for 4:3 display. If the computer graphic is mapped to the entire 720:576 pixels of Rec. 601 the 'square pixel' artwork is best produced in 788:576 resolution if a 4:3 display is anticipated or 1050:576 if working in widescreen. Many of the most current high-definition standards use square pixels, in part to ease the issues of conversion to and from computer display resolutions.

2.8.4 Multiplex Layer – Component Video

The modern studio uses uncompressed video component signals for real-time signal handling (compression may be used in recording or long-haul transmission systems). Luminance sampling of 13.5 MHz and 4:2:2 sample structure is the industry standard. ITU-R BT.656 describes[2] the multiplexing and synchronising requirements for transmitting this over parallel and serial links, serial data over coaxial cable being the most commonly used. A separate standard describes 4:4:4 operation.[3]

The multiplexing of 4:2:2 component video data into a single communication channel is quite straightforward and summarised in Figure 2.8.5.

Data for each active line is transmitted as 1440 sample values (8- or 10-bit operation is allowed for) with luminance and chrominance samples interleaved. The first sample value sent is Cb, followed by Y then Cr. These three values are taken from a single point in the picture so they are said to be co-sited. The next sample sent is Y, though this has no accompanying chrominance

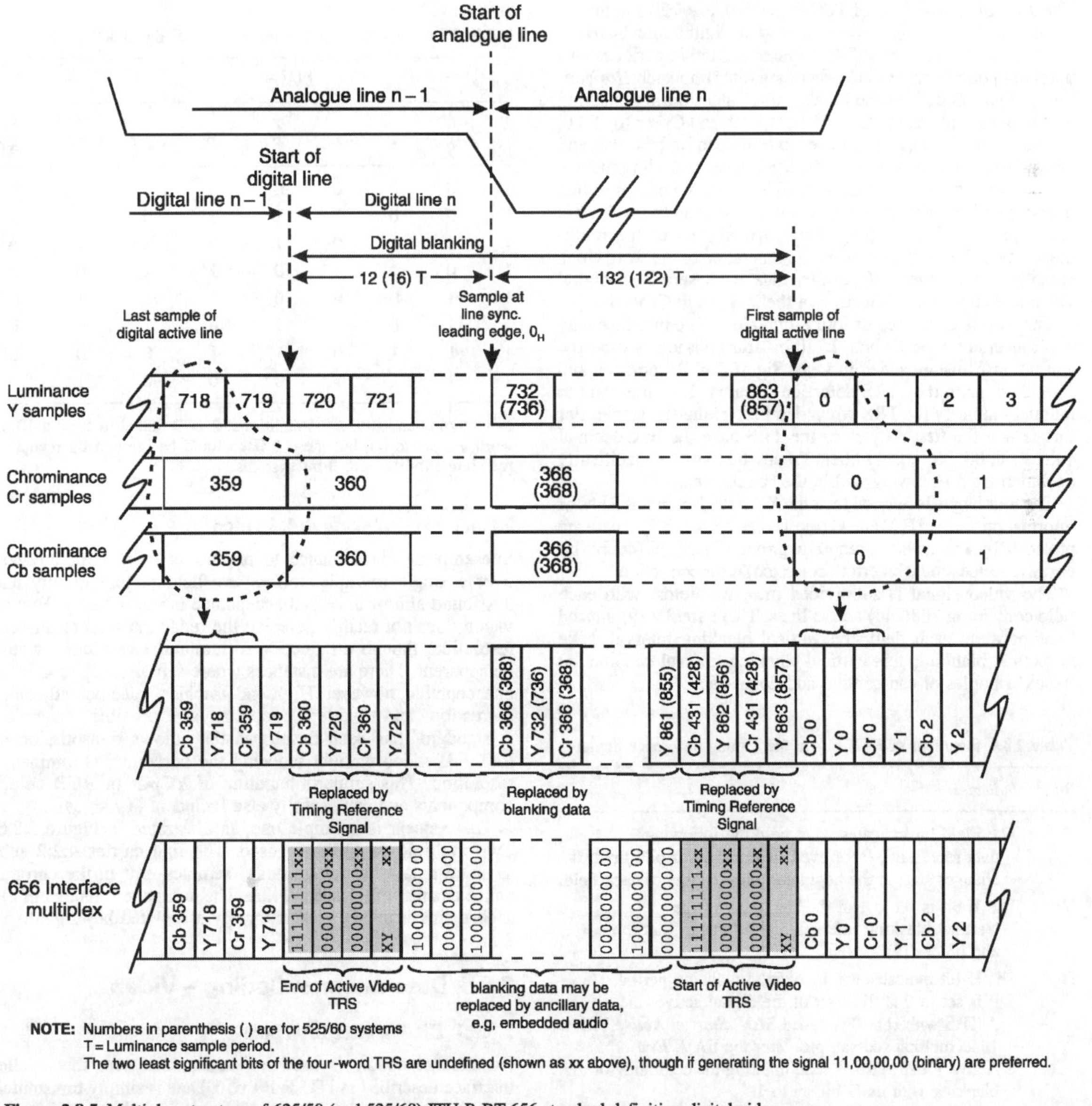

Figure 2.8.5 Multiplex structure of 625/50 (and 525/60) ITU-R BT.656 standard-definition digital video.

information. Assuming 13.5 MHz sampling, the two colour components Cb and Cr are sampled at 6.75 MHz and the overall clock rate of the multiplex is 27 MHz. Transmission of this multiplex requires sending 27 million 10-bit (or 8-bit) values each second, corresponding to a raw data rate of 270 Mbit/second. This headline bit rate includes blanking periods. For the European 625/50 scanning standard the bit rate used by 10-bit picture information alone is 207.36 Mbit/second or 165.888 Mbit/second if 8-bit quantising were used. The blanking intervals may contain values corresponding to 'digital black' though they could be used to carry ancillary data, as described on p 247.

2.8.4.1 Timing Reference Sequence

The *Digital Active Line* of ITU-R BT.601 has 720 luminance picture samples. This is preceded by a digital line blanking interval of 144 (138 for 525-line systems) sample periods. This blanking period can contain other data, but if no such *Horizontal Ancillary Data* is carried the sample values during blanking will represent black, i.e. Y = 16.0 with Cb and Cr set to 128.0.

Many pieces of equipment need to be able to find the start and end of each active line of 1440 luminance and chrominance samples so that they can access the video data, or perhaps so that they can embed ancillary data from within the multiplex. This is achieved by adding **T**iming **R**eference **S**ignals to the multiplexed data. The TRS also allows regeneration of the word clock essential for recovery of parallel data from serialised digital video and correct identification of the Y, Cb and Cr values.

The TRS is composed of four 8-bit values. The interface may work at either 8 or 10 bits. If 10-bit transmission is used the additional 2 bits are added in Least Significant Bit position and should be ignored by TRS detection circuitry. It is important to uniquely identify the TRS when demultiplexing the signal. For this reason the first 3 bytes of the TRS have the hexadecimal values FF, 00, 00. The numbers FF and 00 must be specifically excluded from occurring within the sample data.

The fourth byte is referred to as the XY byte. It contains 3 bits of information, termed F, V and H (see Tables 2.8.4–2.8.6). These are protected by 4 parity bits arranged to provide DEDSEC (double bit errors detected single bit errors corrected) error correction.

The video signal is subdivided into two fields, with each field containing 288(244) active lines. There are 25(19) unused lines between each field, the vertical blanking interval. Like horizontal blanking, this vertical blanking interval can contain 'black' samples or can contain additional data.

Table 2.8.4 Functions of FVH bits in the Timing Reference Signal

Bit	*Indicates*
F	The **F**ield bit indicates what field the following data represents, 0 for even field and 1 for odd field. It changes state at the beginning the first line of each field.
V	The **V** bit is set to 1 at the beginning of the **V**ertical blanking interval, and returns to 0 at the end of 22 (19).
H	The **H** bit indicates the **H**orizontal blanking period; it is set to 0 at the start of the digital active line. A TRS with H = 0 is called SAV, *Start of Active Video*. 1440 multiplexed samples later the EAV, *End of Active Video*, TRS is sent, marking the beginning of line blanking with its H bit set to 1.

Table 2.8.5 FVH bits for each line in 625/50 and 525/60

Line numbers					
625/50	*525/60*	*F*	*V*	*H(EAV)*	*H(SAV)*
1–22	4–19	0	1	1	0
23–310	20–264	0	0	1	0
311–312	264–265	0	1	1	0
313–335	266–282	1	1	1	0
336–623	283–525	1	0	1	0
624–625	1–3	1	1	1	0

Table 2.8.6 Details of the Timing Reference Signal XY word

Binary								
X				*Y*				*Hex*
1	*F*	*V*	*H*	*P3*	*P2*	*P1*	*P0*	*XY*
1	0	0	0	0	0	0	0	80
1	0	0	1	1	1	0	1	9D
1	0	1	0	1	0	1	1	AB
1	0	1	1	0	1	1	0	B6
1	1	0	0	0	1	1	1	C7
1	1	0	1	1	0	1	0	DA
1	1	1	0	1	1	0	0	EC
1	1	1	1	0	0	0	1	F1

It is recommended that the additional 2 LSB added to form a 10-bit word are set to 00, but these LSBs should be ignored on reception for compatibility with 8-bit systems.

2.8.4.2 Multiplexing 4:4:4 video

Subsampling chrominance to produce a 4:2:2 structure sent over a single multiplex offers excellent picture quality and has found almost universal acceptance in the industry. Human vision does not readily perceive the reduction in chrominance resolution, though a reduction in luminance resolution would be apparent. There are instances where subsampling to 4:2:2 is unacceptable, however. High-end graphics, telecine and colour correction and overlay operation are possible examples. A standard[3] has been developed that allows transmission of Rec. 601 video without reducing the horizontal chrominance resolution. This supports handling of YCbCr or RGB colour components and can usefully also include a key signal.

The concept is a simple one, as illustrated in Figure 2.8.6. Two multiplexed links are used. The first carries 4:2:2 subsampled data with the 'missing' samples sent in the chrominance channel of the second link. The luminance channel of the second link is available to carry any associated key signal.

2.8.5 Data Channel Coding – Video

2.8.5.1 Bit-Parallel Interface

No channel coding is used for data passed over the parallel interface described in ITU-R BT.656. Data is simply transmitted as 10 (or 8)-bit words with each bit being carried on a differential

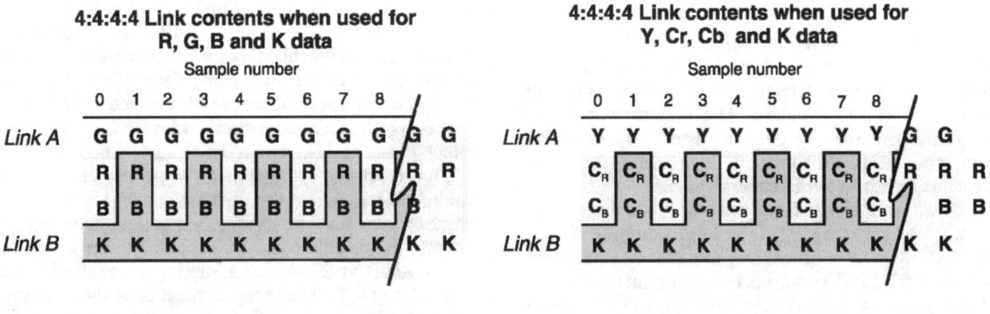

Figure 2.8.6 Multiplexing 4:4:4:4 video data using two multiplexed video links.

transmission line. The 27 MHz word clock is carried on a further balanced pair. The 25-pin D-type connectors used are cumbersome and the Serial Digital Interface that uses coaxial cable or optical fibre has largely superseded the parallel interface.

2.8.5.2 Serial Digital Interface

Transmission of digital video as serial data has proved more popular than bit-parallel implementations. Serial transmission offers the advantage of smaller, simpler connectors and the freedom from data skew problems means that transmission over longer cable lengths (or increased bit rate) is possible. The multiplex structure shown in Figure 2.8.5 is serialised transmitting the LSB of each 10-bit word first. Assuming 13.5 MHz luminance sampling, the resulting 27 MHz multiplex is serialised to a bit rate of 270 Mbit/s.

A very early EBU serial standard[4] based on 8B9B was rapidly superseded by the SDI system now standardised by ITU-R[2] and SMPTE.[5] Channel coding is required in this serial transmission to provide spectral shaping and reliable clock recovery. There is no separate channel to convey bit and word clocks. The bit clock must be recovered from the data channel. It helps to have plenty of transitions in the data.

The technique used is one of pseudo-random scrambling using a ninth order polynomial and NRZI (Non-Return to Zero Inverted) conditioning to provide a polarity insensitive signal (see Figure 2.8.7).

SDI video can be carried over medium distance (<280 metres) via 75 Ω coaxial cable and over longer distances via optical fibre. The <280 metre quoted for coaxial cable assumes that the commonplace Belden 8281 cable is being used; operation over greater distances is possible if using a cable designed for serial digital operation. Typical cable equalisers can correct for about 30 dB of loss at 135 MHz.

The broadcast industry has given many names to this commonplace signal to the ITU-R BR.601/656 and SMPTE 259M standard:

"SDI" (Serial Digital Interface)	"DSC" (Digital Serial) Component
"270 megabit video"	"4:2:2 video"
"SIF" (Serial Interface)	"Rec. 601 serial" (or parallel)
"Serial Component"	"ITU R BT.601/656"

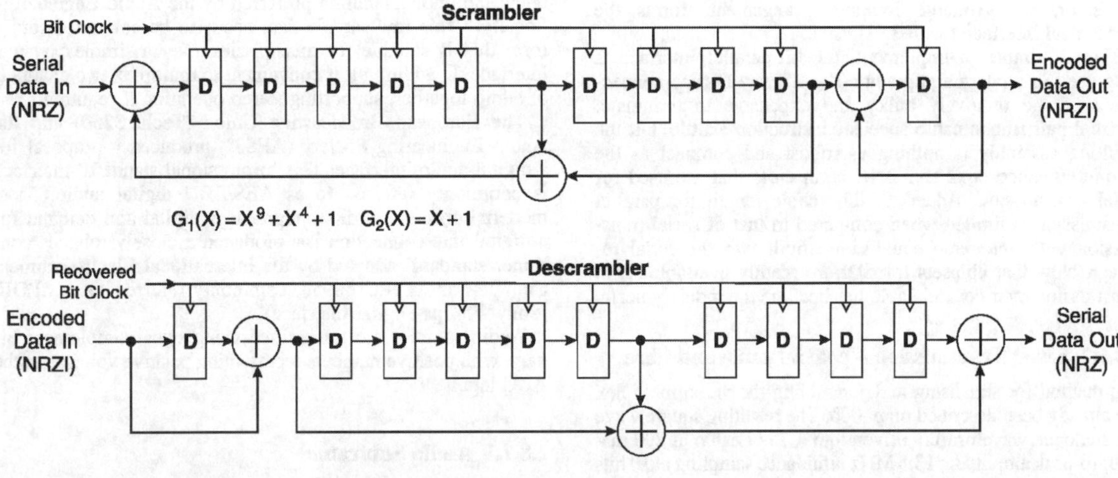

Figure 2.8.7 Data channel coding used by SDI video.

2.8.5.3 Error Handling

With the exception of the parity bits built into the TRS that can detect and correct errors in the three (FVH) signalling bits, there is absolutely no error detection or correction built into this video interconnection standard. The system works well because it is designed for use over an essentially error-free transmission medium (well screened, high-quality cable of sufficiently short length to avoid excessive loss).

An error detection and handling scheme has subsequently been standardised by the SMPTE[6] and is becoming widely supported by manufacturers of test and monitoring equipment. More importantly the EDH system is becoming integrated into equipment designed to source, process and transmit SDI video. EDH works by embedding CRC (Cyclic Redundancy Check) checkwords as ancillary data within the multiplex. Two checkwords are included, one allowing detection of errors within the active video, the other checking the full field, including any data carried in the blanking intervals. Though it is not possible to correct any errors the equipment can set flag bits in the EDH data and generate alarms to indicate a problem that needs to be investigated.

2.8.6 Physical Layer – Video

The final layer involved in connecting the digitised signal to the rest of the world involves converting it into an analogue voltage waveform! This analogue signal takes the form of a square wave as only two states, '0' and '1', need to be conveyed, with the signal having a controlled slew rate between these two states.

2.8.6.1 Bit Parallel

As has already been mentioned this interface will generally be found only on older items of equipment and perhaps within items of equipment as an internal bus for digital video data.

The interface is based on ECL (emitter coupled logic) voltage levels and uses balanced transmission, each bit being conveyed as a differential signal on a twisted pair transmission line with a nominal impedance of 110 Ω (see Figure 2.8.8). The word clock is sent in a similar manner. A 25-way D-type connector, with suitable locking arrangement, forms the mechanical interface for these signals.

Though simple to implement, the bit-parallel interface is little used in modern equipment. The 25-way D-type connectors used are relatively bulky and expensive to terminate. Twisted pair ribbon cable speeds construction a little, but the resulting assembly is nothing as robust and compact as the ubiquitous video coax and BNC connectors that are used for serial transmission. Added to this, cable length for parallel transmission is limited when compared to that of serial transmission with automatic equalisation built into the serial receiver. Now that chipsets for SDI are readily available serial transmission over coaxial cable has become the industry norm.

2.8.6.2 Serial Transmission – coaxial cable and fibre

The method for serialising and scrambling the video multiplex has already been described on p. 000. The resulting square wave (an analogue waveform) is driven into a 75 Ω cable at 800 mV peak-to-peak amplitude; 13.5 MHz luminance sampling at 10 bits gives rise to a data rate of 270 Mbit/s on the cable. Care must be taken to correctly terminate and drive the cable at such high frequencies; a return loss (measure of reflected power at the termination) must be at least 15 dB over a frequency range of 5–270 MHz. In practice this means taking care over PCB layout around the serialiser, driver and deserialiser chips and close to the connector itself. The coaxial connector specified is a BNC socket which can mate correctly with BNC plugs constructed for either 75 Ω or 50 Ω, though the characteristic impedance is 75 Ω.

The receiving arrangement must correctly terminate the cable and correct for the high-frequency loss of the cable used to carry the data. The expectation is that the receiver will have an automatic cable equaliser capable of correcting for a loss of up to 40 dB at 270 MHz, assuming a cable with a loss characteristic of $\sqrt{f}$. The receiver must then slice the waveform to produce a binary signal and convert this to NRZ, descramble as shown in Figure 2.8.7 and probably convert from serial to parallel. A key requirement of this process is to correctly recover the original bit clock by locking a Phase Locked Loop to the data transitions in the incoming signal. All of this can be accomplished in a single i.c. Single-chip solutions for SDI sources and destinations are available from a number of manufacturers, including Sony, Gennum National and Cyprus.

High-quality video coax cable can provide for essentially error-free transmission over 280 metres of cable, perhaps a little more. Using specialist cables can provide for transmission over longer distances. Another solution is to regenerate the signal with a distribution amplifier at 280 m intervals; this may be inconvenient or expensive. Fibre-optic transmission over multimode fibre is an economic solution and provides for distances of 7–70 km depending on the configuration, with single-mode fibre providing even greater distances. Both involve modulating a semiconductor laser or LED source with the same binary data that is used for transmission on coax. A laser source is essential if single-mode fibre is employed, as the smaller core requires an intense light source. ITU-R BT.1367 describes the fibre interface parameters for both single-mode and multi-mode applications.

2.8.7 Multiplex Layer – Professional Audio

Table 2.8.7 lists the common sampling standards, with 48 kHz sampling quantised to 20- or exceptionally 24-bit resolution being the implementation preferred by the Audio Engineering Society.[7] This raw sample data needs to be converted into a form that is suitable for transmission over a simple physical interface. In addition it is convenient to multiplex two channels of audio together, supporting stereo operation if required.

The European Broadcasting Union (Tech. 3250) and the Audio Engineering Society (AES3[8]) produced a proposal for a digital audio interface. This 'professional standard' interface is commonly referred to as AES/EBU digital audio. Most modern consumer audio equipment is digital and demand for a digital interconnection has produced a closely related 'consumer standard' adopted by the International Electrotechnical Commission as IEC 60958, commonly referred to as SPDIF (Sony–Philips Digital Interface).

Audio samples are transmitted using twos complement binary, with positive numbers representing positive voltages at the ADC input.

2.8.7.1 Audio Subframe

The first step in producing the required multiplex, or 'frame', is to wrap each audio sample into a packet or 'subframe' (Figure 2.8.9).

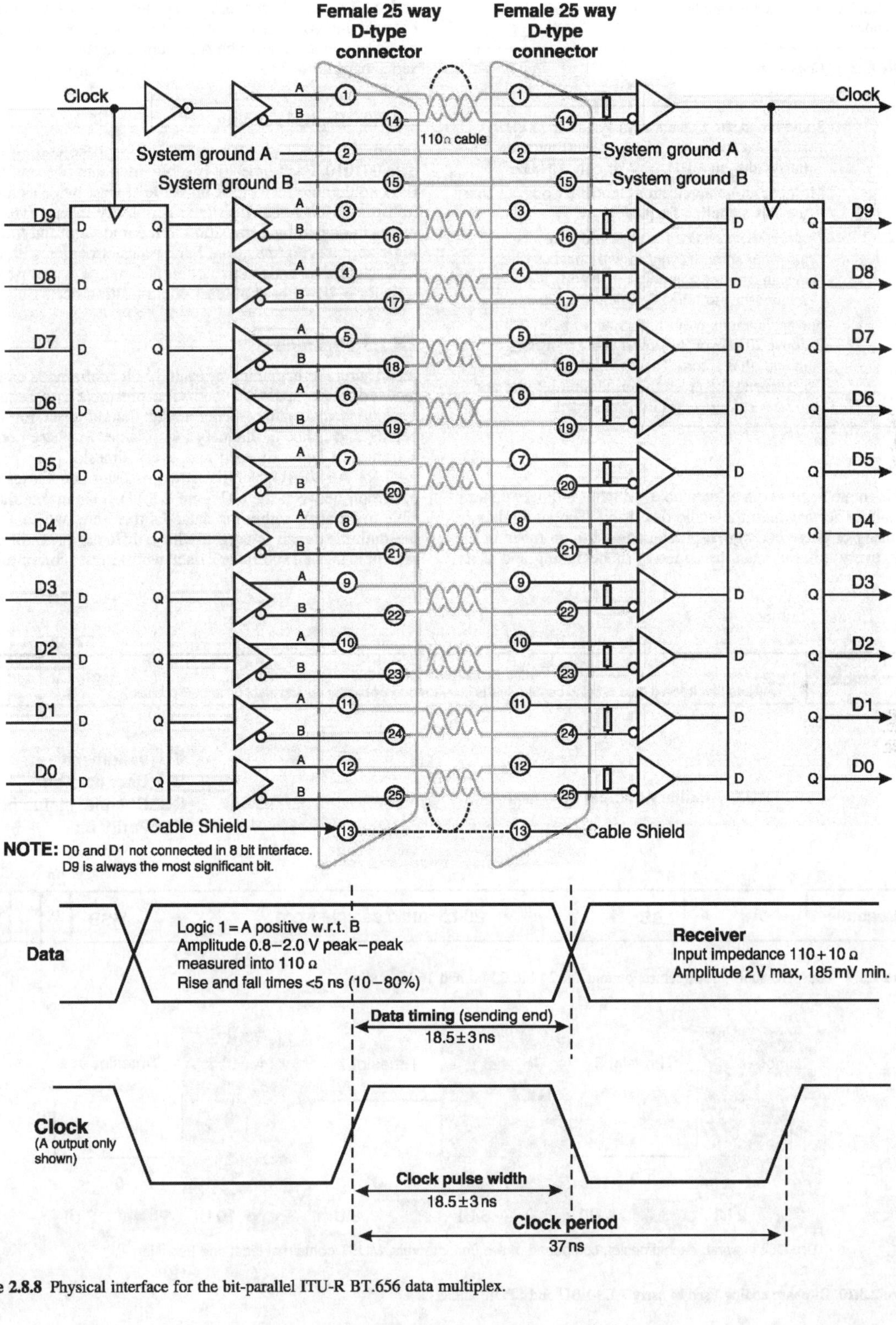

Figure 2.8.8 Physical interface for the bit-parallel ITU-R BT.656 data multiplex.

Table 2.8.7 Common audio sampling standards for professional applications

Sample rate (kHz)	*Comment*
32	Some broadcast transmission systems; 32 kHz is sufficient for most radio broadcasting, with a bandwidth on VHF 'FM' of only 15 kHz.
44.1	Most consumer applications, notably Compact Disc, use this sampling frequency.
48	*The preferred choice* for most broadcasters, allowing some leeway in anti-aliasing filter design and, most importantly having a convenient relationship to television field frequency.
96	For applications requiring an audio bandwidth above 20 kHz or to allow a wider transition region in anti-alias filters. The downside of this standard is its need for greater transmission bit rate and greater storage capacity if recorded.

This frame begins with a 'preamble' which is a binary pattern that cannot occur within the audio data itself. The audio data is sent using bi-phase coding. Bi-phase coding has an inherent bit clock, every timeslot has a transition at its beginning and at its end. This makes it very easy to regenerate a clock from the audio data stream. Another advantage of bi-phase coding is that it has no DC content, so it can be AC coupled through a transformer or series capacitor. Finally, it is polarity insensitive.

2.8.7.2 Bi-phase coding

Figure 2.8.10 shows an example of bi-phase coding carrying data 001010. Each timeslot is subdivided into two unit intervals (UI), so the interface clock must run at least twice as fast as the transmitted bit rate. The signal is polarity insensitive. A logic one is indicated by a transition in the middle of the time slot, a logic zero has no transition. Each timeslot begins with a transition. Note that under this coding scheme it is not possible to wait for 3 UI without at least one transition occurring.

2.8.7.3 Preambles

Identifying the beginning of each 32-bit subframe is essential to recovering the audio payload. The preamble is a sequence of transitions that cannot occur under the rules set for bi-phase coding and hence is uniquely identifiable. Bi-phase coding has a transition between each and every timeslot. The preambles used for AES/EBU each have two timeslot boundaries with no transition between them (Figure 2.8.11). This makes them possible to identify within the data. As there are two subframes in the multiplex each is tagged with a different preamble. The Y preamble tags all subframe 2 data, whilst most subframe 1 data is

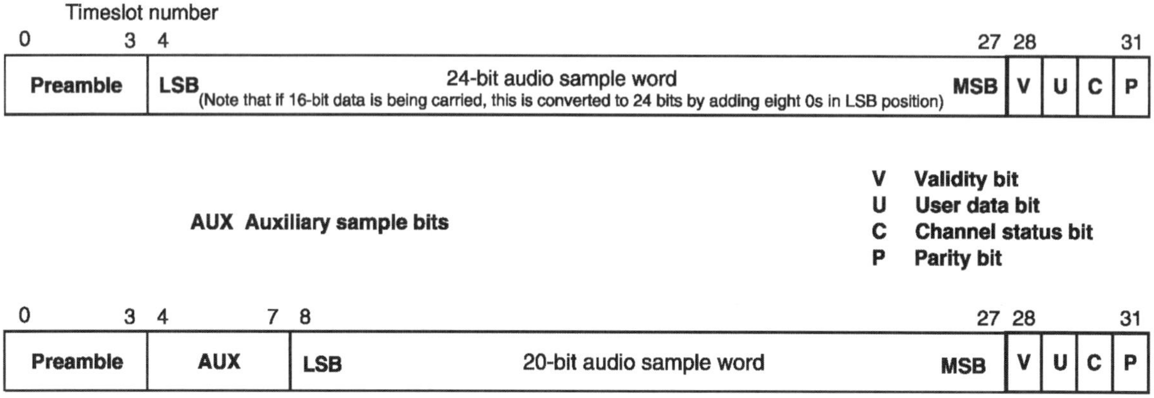

Figure 2.8.9 AES/EBU subframe structure, options for 24-bit, 20-bit and 16-bit audio.

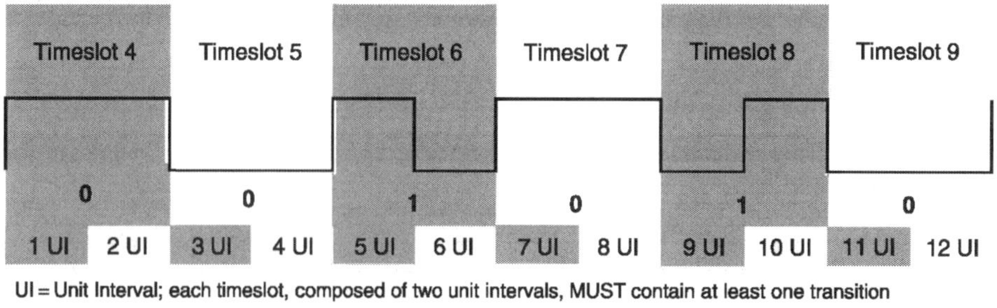

Figure 2.8.10 Bi-phase coding used to carry AES-EBU and SPDIF audio data.

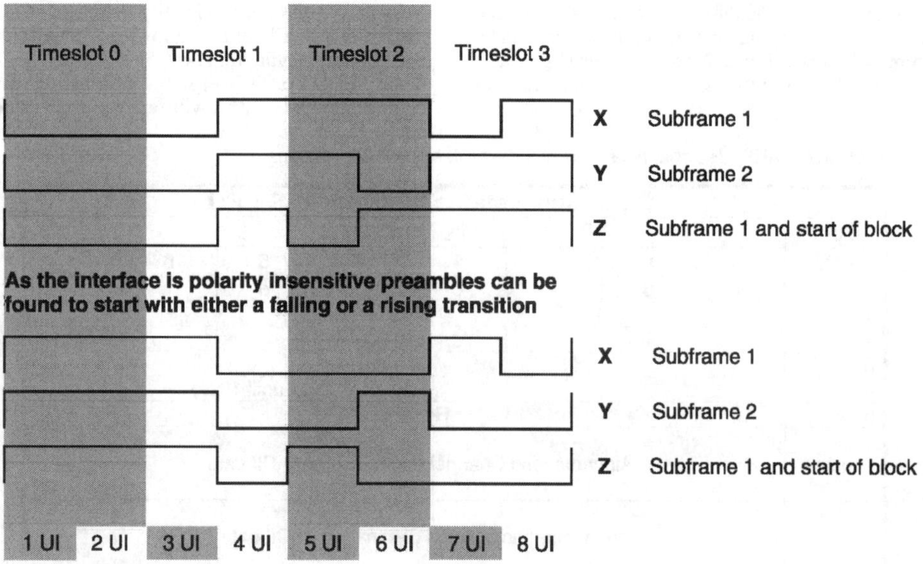

Figure 2.8.11 AES/EBU preambles; the same preambles are used in SPDIF.

identified by beginning with preamble X. Figure 2.8.12 shows the overall format of the AES/EBU multiplex; 192 frames are grouped together into a block of data. The beginning of this 192-frame block is identified by a Z preamble. This Z preamble indicates the first frame of each block, frame 0 in Figure 2.8.12.

2.8.7.4 AES/EBU Frame and Block Format

We shall see in later paragraphs that some of the single bits carried in the subframe structure of Figure 2.8.9 are concatenated into larger structures, blocks of 192 bits of data. In order to achieve this the subframes themselves are grouped into blocks of 192 subframes each. Frame 0 begins with a Z preamble, all other frames begin with preamble X, as shown in Figure 2.8.12.

Figure 2.8.9 shows the subframe structure for AES/EBU digital audio. Four single bits are associated with each subframe.

2.8.7.4.1 Validity bit

According to AES3, if this bit is set the accompanying data is identified as being unsuitable for conversion to analogue audio. However, some equipment, notably some CD players, set the validity bit if an error has been found and concealed. This means that it may be difficult to decide how a digital-to-analogue converter should behave when a subframe is marked as invalid. Perhaps because of this the V bit is ignored by many DACs, though some may mute if they find the V bit set within a subframe.

Signalling V bit status on an operational device such as a digital mixing desk could be useful as it would indicate error concealment occurring in a signal source (if this source behaves as some CD players do) and give the operator some warning that all is not as it should be.

The validity bit should certainly be set if the interface is carrying a payload that is not linear PCM. If data-rate-reduced audio is carried on the interface the validity bit should be set, requiring that digital-to-analogue converters designed for linear PCM mute instantly if they are configured to use the V bit. The non-audio flag within the channel status block should also be set in these circumstances.

2.8.7.4.2 User bit

This can carry user-specific information. The professional and consumer standards define its usage differently.

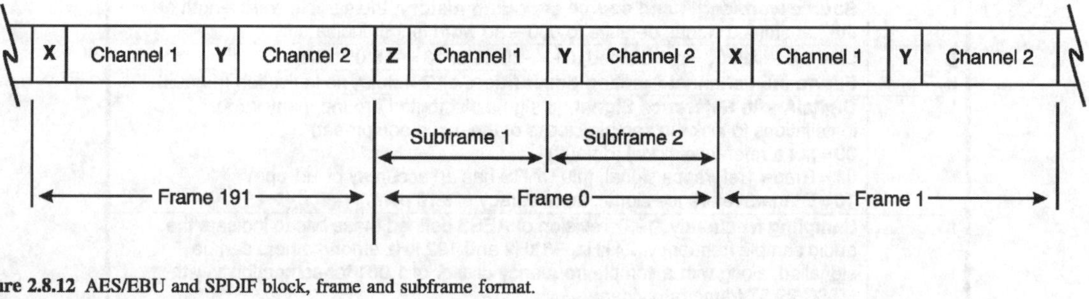

Figure 2.8.12 AES/EBU and SPDIF block, frame and subframe format.

Consumer standard IEC 60958-5[9] describes an optional packet-based format carrying program-related metadata. User bits from subframes 1 and 2 are combined, giving a user data rate of 88,200 bit/s assuming a frame rate of 44.1 kHz.

Some systems use this facility to transmit CD subcode information over the interface, allowing a remote display of track number and playing time, etc.

AES/EBU has channel status information carrying data that defines how the user bit will function (see Table 2.8.8).

Table 2.8.8 AES/EBU channel status block – professional applications

Channel status block forAES-EBU								
Byte No.	**Bit Number**							
	0	1	2	3	4	5	6	7
0	a	b	c			d	e	
1	f				g			
2	h			i			j	
3	k							
4	L		m (was reserved until 1997)					
5	Reserved							
6 to 9	Alphanumeric Channel origin (source) ACII data							
10 to 13	Alphanumeric channel destination ASCII data							
14 to 17	Local sample address code							
18 to 21	Time-of-day sample address code (32 bit binary)							
Reliability flags, show which bits in this table contain reliable information, set if unreliable								
22	0	0	0	0	Bytes 0–5	Bytes 6–13	Bytes 14–17	Bytes 18–21
23	Channel status cyclic redundancy check character, used to detect errors							

Label	Comment
a	**Use of channel status channel** set to indicate professional use (this table summarises professional use) clear to indicate consumer use of channel status block
b	**Audio/non-audio use** Set to indicate that the audio sample word contains linear PCM samples Clear to indicate other use and mute simple digital-to-analogue converters
c	**Audio Signal emphasis,** 000 – no emphasis signalled, manual control enabled
d	Locking of source sample frequency
e	**Sampling Frequency**, indicates 48 kHz, 44.1 kHz or 32 kHz sampling
f	Channel Mode, see Table 2.8.7
g	**User bit management**, indicates how user bits in the subframe are to be interpreted. All bits cleared if no user information is indicated.
h	**Use ofAuxiliary sample bits,** 000 indicates AUX bits undefined and 20 bit audio, 001 indicates no AUX bits and 24 bit audio, 010 indicates AUX bits in this channel carry a single coordination signal.
i	**Source wordlength and source encoding history**, the sample word length of the transmitted signal, defaults to 000 – no word length indicated.
j	**Line-up level,** 00 = not defined, 01 = –20 dBfs, 10 = –18.05 dBfs
k	**Future multichannel function description**, bit 7 = clear, no indication (default)
L	**DigitalAudio Reference Signal**, (a signal distributed around many larger installations to lock the sample clocks of the audio equipment) 00 = not a reference signal (default) 01 = Grade 1reference signal, this DARS has an accuracy of ±1 ppm 10 = Grade 2 reference signal, an accuracy of ±10 ppm
m	**Sampling frequency**, 1997 revision of AES3 defined these bits to indicate the audio sample frequency. 24 kHz, 96 kHz and 192 kHz among others can be signalled, along with a sample frequency divisor of 1.001 for compatibility with NTSC 29.97 frame rate video.

2.8.7.4.3 Channel status bit

The channel status bits for each subframe are concatenated into two blocks of 192 bits each. These blocks usually carry identical data, so many receivers will only examine the data from one of the subframes. Some channel status information affects how the audio payload is treated. Flags for non-audio data and emphasis information will be particularly important, particularly to those DACs that do not mute if the V bit is set.

The channel data information is defined very differently in the commercial and professional formats. Only the first 2 bits have a common interpretation. If the first bit of the channel status block is set to one 'professional' channel status format is indicated. If the first bit is cleared to zero the channel status block is formatted to the commercial standard.

Interpretation of the channel status block for AES/EBU is summarised in Table 2.8.8.

2.8.7.4.4 Parity bit

The parity bit is used to ensure even parity for each subframe in the multiplex. This provides a means of error detection. In fact, the receiver does not really need to check for parity in the usual fashion. The fact that even parity is maintained means that there are an even number of mid timeslot transitions (logic 1), as there are an even number of timeslots in each frame and in each block there are an even number of all other transitions. This means that each subframe begins with a transition in the same direction every time. Checking parity is simply a matter of confirming that the logic state of the second half of the parity bit is the same as that of the previous subframe.

Checking for violations of the bi-phase coding scheme is also an important way of detecting errors in the received data.

2.8.7.4.5 Auxiliary bits

The 4 auxiliary bits, 'AUX' in Figure 2.8.9, can be used to carry another signal known as auxiliary audio data. Use of subframes with this structure must be signalled in the channel status information for that subframe (byte 2, labelled h in Table 2.8.8).

Use of these AUX bits is rare in practice. The AES has defined a usage as a 'coordination channel', for instance carrying talkback. Of course, one reason for using talkback is for operators to talk so that they can rapidly identify a problem. If the fault or misconfiguration includes loss of the AES/EBU link, talkback travelling over this link will not be of much use! Another problem with use of these bits is that they require a receiver to mask off their contents before conversion of the main audio channel. Many receivers do not support this.

2.8.7.4.6 Mode

Several modes of operation are defined for AES/EBU. *Stereophonic mode* presumes that the two channels have been simultaneously sampled. The Left (or 'A') audio is carried in subframe 1 while the Right (or 'B') channel is within subframe 2. *Two-channel mode* can handle two completely independent audio channels: channel 1 in subframe 1 and channel 2 in subframe 2. The mode is indicated in byte 1 of the channel status block. *Single-channel double sampling frequency mode* has channels 1 and 2 carrying successive samples of the same signal. This means that the sampling frequency is double the frame rate, for instance allowing a '48 kHz' AES/EBU link to carry monophonic 96 kHz sampled audio data.

2.8.8 Channel Status Block – Consumer Format

Table 2.8.9 provides a summary of the bit fields within the channel status block that is intended for consumer applications. Refer to the appropriate standard[9] for more details, particularly for details of the Serial Copy Management System (SCMS), which provides control over copying operations. The set of rules defined by SCMS can govern operation of consumer digital recording equipment such as DAT and MiniDisc. The rules are complex, partly to allow reverse compatibility with the system initially specified for CD. In general, the L bit is set to indicate the signal is from pre-recorded material and cleared to indicate a 'home copy' if the L bit is clear *and* the copyright bit is zero; no further copying is allowed by equipment that implements SCMS.

2.8.9 Physical Layer – Professional Audio

2.8.9.1 Twisted pair cable, balanced transmission

Figure 2.8.13 illustrates typical circuit arrangements that fulfil the requirements of the AES/EBU specification. Balanced transmission over high-quality twisted pair cables is used, the connectors specified being three-pin XLR, also in common use for professional analogue audio. Operation over several hundred metres of cable is possible. While AES3 does not demand the use of a transformer, these do provide better balance (common-mode impedance) and are mandatory in the EBU specification. This helps minimise crosstalk, a particular concern for broadcast installations with large numbers of twisted pair cables running parallel to each other. The capacitor shown in the line receiver circuit blocks the DC path between the twisted pair.

Rise times specified for the digital output are between 5 and 30 ns, measured working into a 110 Ω resistor but without the capacitive loading of an interconnecting cable, operation towards the faster 5 ns end improving the eye-height of the transmitted data and therefore tending to make the link more robust. Some practical implementations of Figure 2.8.13 have a capacitor across the line, with some additional series resistors. These may be included to control (reduce) the rise time of the signal to meet the specification. Practical circuits may also have two identical high-value (approx. 1 MΩ) resistors, joining a common signal earth point to pins 2 and 3 respectively. These reduce the common-mode termination impedance of the line and hence reduce any common-mode voltage that may be present.

2.8.9.2 Coaxial cable, unbalanced transmission

The original AES3 specification only describes balanced transmission over twisted pair cable. Carrying the signal unbalanced over coaxial cable has advantages in some circumstances, and AES3id[8] describes and standardises this electrical interface.

AES/EBU data carried on coaxial cable is identical to that on the twisted pair interface of Figure 2.8.13, the only differences being electrical and physical parameters. Transmission over coaxial cable offers increased maximum cable length (operation up to approximately 1 km is possible) and usage of existing tielines if the audio is being handled as part of a television installation (75 Ω cable is used). The BNC connector specified

Table 2.8.9 Consumer format channel status block – summary

Consumer format channel status block								
Byte No.	**Bit Number**							
	0	1	2	3	4	5	6	7
0	a	b	c	d			e	
1	**Category code** Depends on equipment type, most codes contain the "L bit" used to by Serial Copy Management System SCMS to indicate the generation of copyright protected material. CD Player = 1000 0000 DAT player = 1100 000L DCC Player = 1100 001L Mini Disc = 1001 001L							
2	**Source number**				**Channel number**			
3	**Sampling Frequency**				f			
4	**Word Length**							
5–23	**Reserved**							

Label	Comment
a	**Use of channel status channel** Set to indicate professional use Clear to indicate consumer use of channel status block (this table summarises professional use)
b	**Audio/non-audio use** Set to indicate that the audio sample word contains linear PCM samples Clear to indicate other use and mute simple digital to analogue converters
c	**Copyright bit, set if no copyright.** If set to zero copyright is asserted and a set of rules defined in the Serial Copy Management System (SCMS) is applied. SCMS uses an additional bit, the "L bit" in the category code in such a way as to allow limited copying of material.
d	**Audio Signal emphasis,** 000 – no emphasis signalled, 100 CD type
e	**Channel status mode** = 00
f	**Clock accuracy,** 10 = Level 1 ±50ppm 00 = Level 2 ±1000ppm 01 = Level 3 Variable pitch shifted

is smaller, cheaper and quicker to fit to a cable than the three-pin XLR standardised in AES3. BNC connectors take up less space than XLRs on equipment rear panels. A disadvantage of coaxial transmission is that the cable itself is more bulky and costly than twisted pair.

Table 2.8.10 summarises the electrical characteristics used when carrying AES3 data on coax. These can be readily achieved by altering the line driving and line receive circuitry for unbalanced operation. It is possible to buy passive adaptors that convert to and from coaxial line transmission, allowing equipment built for balanced operation to be readily adapted. These are based on small wideband transformers and are little bigger than an XLR connector themselves.

2.8.10 Physical Layer – Consumer Audio

2.8.10.1 Coaxial cable

The consumer version of the standard 'SPDIF' (IEC 60958-3) defines an unbalanced interface over coaxial cable. An RCA 'phono' connector is used, already ubiquitous in analogue consumer audio interconnects. Driver stages will be very similar to those for the professional unbalanced electrical interface, but the signal driven onto the cable is 0.5 V peak to peak.

This standard anticipates interconnection over just a few metres, and at such short cable lengths the low-performance thin coaxial cabling common for analogue consumer applications will perform adequately. More careful choice of cable should provide for cable lengths of many tens of metres, perhaps more, though this is not anticipated by the consumer application.

2.8.10.2 Optical cable

A more recent addition to the IEC standard has included optical interconnect using low-cost optical light guide (usually using a plastic rather than a glass fibre-optic cable) and connector called TOSlink (Figure 2.8.14).

This low-cost optical interconnect is easy to implement as the equipment part of the connector contains optical source and drive electronics within a single PCB mount component. All that is required is a 5 V supply and a TTL or CMOS drive signal. The interconnect is short distance only, operating reliably only below 10 m.

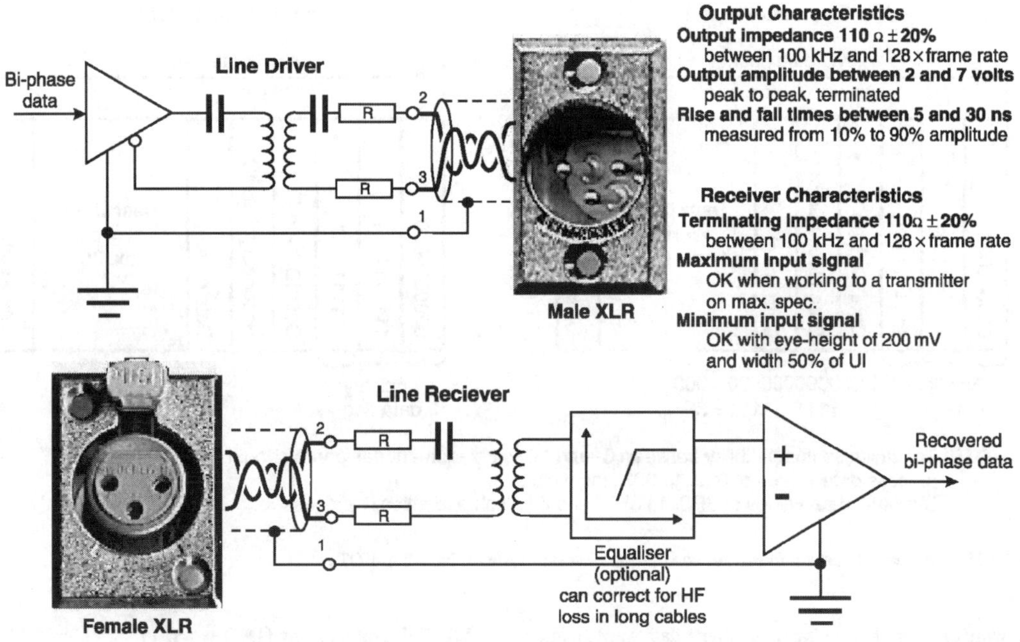

Figure 2.8.13 AES/EBU line driver and receiver generalised circuits.

Table 2.8.10 Electrical characteristics for coaxial transmission – AES/EBU only

Parameter	*Minimum*	*Typical*	*Maximum*	*Unit*
Output voltage (pk-pk)	800	1000	1200	mV
DC offset			<50	mV
Rise and fall time	30	37	44	ns
Impedance		75		ohms
Return loss	15 dB (measured from 100 kHz to 6 MHz)			

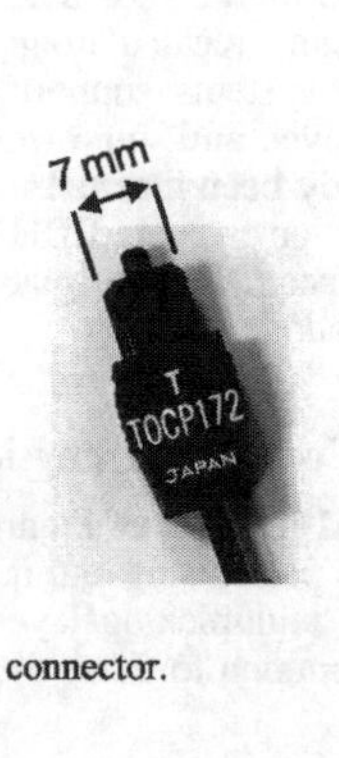

Figure 2.8.14 TOSlink connector.

2.8.11 Embedding Data in SDI Video

The horizontal and vertical blanking periods of an SDI video signal represent a significant 'spare' data capacity. It is possible to use these parts of the signal, which would otherwise carry only *digital black* (Y = 16, Cb and Cr = 0), to carry other data. This *embedded* data could be completely independent of the video, though as it is travelling over the same path it usually has some sort of association. Common uses of embedded data include:

- Audio channels – up to eight AES/EBU audio multiplexes (16 channels) can be included.
- Error detection and handling – allows detection of errors in the associated SDI data.
- Timecode – a standard exists for carrying timecode data embedded in the SDI multiplex.

The basis format of embedded data, the Ancillary (ANC) Data packet structure, has been further extended to allow the SDI multiplex to carry other types of data in the active video region. This is called SDTI.

2.8.12 Ancillary Data (ANC)

The system described here has been standardised by the SMPE[10] and ITU-R. Data is carried as packets, usually in the horizontal blanking interval of the video multiplex. When ancillary data is carried in the horizontal blanking period it is often referred to as HANC data. Two types of ANC packet formats are described by the SMPTE standard (Figure 2.8.15).

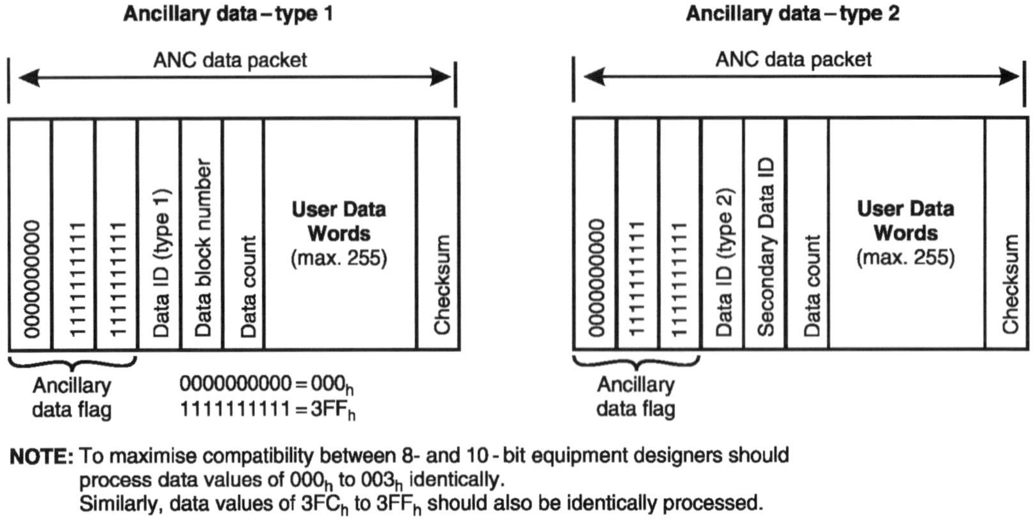

Figure 2.8.15 Ancillary data packet format for component television systems, including HDTV.

It is worth noting that an early ancillary data specification adopted by the ITU for Rec. 656 had a format that differed slightly, allowing a 2-byte word count and 8-bit operation. Data of this format is seldom found; if it does occur, modern test equipment may flag an ANC formatting error.

Type 1 ANC data requires 10-bit interconnection. Any item of equipment that truncates the word length to 8 bits could destroy the ancillary data. Type 2 packets may operate over 8- or 10-bit word interconnects (see Table 2.8.11). The data ID (DID) word indicates what type of ancillary data is being carried (bit 7 is set to indicate a type 1 packet, clear for type 2). Type 2 packets have a secondary data ID word (SDID), providing for a large number of types of data to be carried as ANC data packets.

2.8.12.1 Eight-bit or 10-bit working

A total of 189 data identification (DID) values are reserved for 8-bit applications while approximately 29,000 are available for 10-bit operation.

2.8.12.2 Excluded values

Values $3FC_h$ to $3FF_h$ and 000_h to 003_h must be specifically excluded from the ANC data packets, to prevent false detection of ancillary data flags. This must be achieved by the application or process that is generating the *payload*, the user data words (UDW) held inside each packet.

2.8.12.3 Data block number (DBN)

The DBN word is present in type 1 data packets. Its least significant 8 bits increment from 1 to 255, with bit 9 indicating even parity of the least significant 8 bits and bit 9 being the inverse of bit 8. This prevents any occurrences of excluded values for the system. Block numbering is not in use if the least significant 8 bits of the DBN are zero.

2.8.12.4 Data count (DC) word

The DC word indicates the number of user data words contained in the packet. It can indicate values from 0 to 255 and uses the same strategy as the DBN to avoid occurrences of excluded values.

2.8.12.5 ANC packet space formatting

The first ANC packet should follow immediately after the EAV if in the horizontal blanking interval or alternatively directly after SAV (unless the data is being carried in an HD-SDI stream; see Figure 2.8.21). Packets should be contiguous and left justified within this space. ANC packets must be wholly contained within the ancillary space in which they are inserted. The SMPTE recommends[11] avoiding using lines 6 and 319 (10 and 273 in 525/60 systems) and the horizontal blanking period of the line immediately following these to carry ANC data. This is intended to avoid corruption of ANC packets during vertical interval switching in a routing matrix.

It is optional to insert a packet with DID = 84_h to mark the last packet of a contiguous ancillary space. If new ancillary data is inserted these packets will be overwritten by new data. ANC packets may be marked for deletion by replacing their DID with value 80_h and recalculating the checksum for that packet. Few current systems support these *ancillary space formatting* options as yet, and some will simply overwrite any HANC that has already been inserted.

ANC packets used for carrying EDH data and for the Serial Data Transport Interface (SDTI) format do not have to comply with the restrictions above.

2.8.12.6 Channel Coding and Physical layers

In terms of the layered structure of Figure 2.8.1 ancillary packet coding sits within the packetising and multiplex layer, immediately above the SDI multiplexing layer. Channel coding and physical layers are common to those used for SDI as the ANC

Table 2.8.11 Data identification word assignment for type 1 and type 2 SMPTE ANC packets

Data type	*Data value*	*Data assignment*
	00_h	Undefined format
	01_h	Reserved
	03_h	
	04_h	Reserved for 8-bit applications
	$0F_h$	
Type 2 (two word ID)	10_h	Reserved
	$3F_h$	
	40_h	Internationally registered
	$4F_h$	
	50_h	User application
	$5F_h$	
	60_h	
	$7F_h$	
	80_h	Ancillary packet marked for deletion
	81_h	Reserved (8-bit equipment will also consider these as marked for deletion)
	83_h	
	84_h	Optional ancillary packet data end marker
Type 1 (one word ID)	85_h	Reserved (8-bit equipment will see these as data end markers)
	87_h	
	88_h	Optional ancillary data start marker
	89_h	Reserved (8-bit equipment will see these as data start markers)
	$9F_h$	
	$A0_h$	Internationally registered
	BF_h	
	$C0_h$	User application
	CF_h	
	$D0_h$	Internationally registered
	DF_h	
	$E0_h$	Internationally registered
	EF_h	

Few of the secondary data ID (SDID) words have been assigned for use at the time of writing.

data is passing over the same link, though care needs to be taken to use the correct ANC type if 8-bit SDI operation is selected.

2.8.12.7 Embedded Audio Application

The additional data carrying capacity provided by ancillary data packets carried in the horizontal blanking interval of an SDI signal, HANC, has many uses. Among the most obvious and common application is the use of HANC to carry the audio associated with the video carried on the link.[12]

2.8.12.7.1 Number and type of channels

The standard[12] provides for carrying between two and 16 channels of audio in the HANC space of a component video signal. It is also possible to carry embedded audio over a composite digital link, though only four channels can be embedded as there is less ancillary space in this format. A sample rate of 48 kHz locked to the video clock is preferred, though the standard provides for operation between 32 and 48 kHz, and for asynchronous operation if necessary.

Several modes of operation are defined, indicated by a letter suffix. The modes are arranged to be backwards compatible. Audio sample words of 20 bits in length with synchronous 48 kHz sampling are the default, though carriage of 24-bit audio or AES auxiliary data is provided by some of the more advanced modes. Many items of current equipment do not support the higher modes of operation, however.

2.8.12.7.2 Identifying audio groups and embedded audio ANC packets

Audio channels are transmitted in pairs and combined where necessary into groups of four. Each group of four channels is carried by an ancillary data packet that can be identified by a unique data ID defined in the standard (see Table 2.8.12). These are called ANC audio data packets and typically carry three or four sample values for each of the audio channels in each group.

Table 2.8.12 Ancillary packet data ID defined for embedded audio applications

Group	Audio packet data ID	Extended packet data ID	Audio control packet data ID
1	$2FF_h$	$1FE_h$	$1EF_h$
2	$1FD_h$	$2FC_h$	$2EE_h$
3	$1FB_h$	$2FA_h$	$2ED_h$
4	$2F9_h$	$1F8_h$	$1EC_h$

The default mode, mode A, does not require extended packets or control packets.

If 24-bit audio is being carried, further ANC data packets, called extended data packets, will be required to carry the additional 4 bits for each audio sample. Presence of extended data packets is optional, and the default mode for embedded audio assumes 20-bit audio samples. Any extended ANC data packets must follow immediately after the ANC audio data packets that they are associated with (see Figure 2.8.16).

Control packets are defined and used to indicate various parameters for more advanced modes of operation (see Table 2.8.13).

2.8.12.7.3 Mapping audio data into 10-bit ANC words

The audio data from a subframe, less **AUX** and **P**arity bits are mapped into three contiguous ANC data words as shown in Table 2.8.14. Word 1, bits 1 and 2 identify the audio channel within a group. So binary 01 would indicate audio channel 2 or 6 or 10 or 14 depending on the embedded data group. The group is identified by the data ID used for the ANC audio data packet that contains the samples. The Z bit is derived from the AES Z preamble and indicates the start of a channel status block. Both Z bits of a channel pair are set to 1 at the same sample, coincident with the beginning of a new AES channel status block. The Z bit is cleared for all other samples. Level J operation allows operation with AES sources whose Z preambles are not coincident. The Z bits for each channel must operate independently under level J operation.

2.8.12.7.4 AES sample frequency versus TV line and field rate

Figure 2.8.16 notes the number of 48 kHz samples per frame for various standards. In most cases there are a whole number of samples for each picture if the 48 kHz audio sample clock is synchronous with the video. For the 625/50 system, if every horizontal blanking period were used to carry audio HANC packets it would be necessary to carry 1920/625 samples per line, 3.072 samples for each channel within the HANC for that group. This results in three samples per channel being sent in most HANC spaces, representing audio samples that were converted during the previous line. A buffer in the embedder accumulates these samples until the right time to release them into HANC data space. Every 15 lines or so there will be an additional sample in the buffer, as there are fractionally more than three audio samples per TV line. So every 15 lines or so there will be an HANC packet containing four samples. A buffer in the de-embedder holds the samples recovered from each HANC until the correct time to release it, measured against the audio 48 kHz clock. Some systems that predate the SMPTE standard[12] insert audio packets into every HANC blanking space, but normal practice is to avoid embedding audio in HANC spaces that may be affected by field interval switching in a routing matrix[11] – that is, lines 7 and 330 for a 625/50 system. The HANC space used by EDH packets must also be avoided. The standard stipulates that audio samples should be spread as evenly as possible amongst the HANC spaces that are available in order to minimise the size of the buffer required at the receiver. Receivers must look for and recover all valid HANC audio packets, whichever line they are carried on.

The default level of operation, level A or 'SMPTE 272M-A', will only work with 48 kHz synchronous sampling. Receivers with a buffer size of 48 samples per channel should be capable of operating with a level A data stream, though 64 samples per channel is the minimum buffer size recommended. Higher levels allow operation at other sample frequencies with some providing for asynchronous audio sampling.

Levels A, B and C can only operate correctly with audio sampling that is synchronous to the video frame rate. Passing asynchronous audio over such links is likely to lead to clicks or mutes of the sound as samples are lost during buffer overflow or underflow conditions. Audio sampling is said to be synchronous when the number of audio samples occurring within an integer number of frames is itself a constant integer number, as shown in Table 2.8.15. A simple test for synchronism of 625/50 operation is to view the HANC (line blanking) period on a picture monitor, after digital-to-analogue conversion, ensuring that the D-to-A does not blank the HANC audio packets. A repeating pattern of short (three-sample) and long (four-sample) HANC packets will be seen. If this pattern is static, that is it does not move slowly up or down the picture, the audio sampling is synchronous. The multi-frame repetition sequence required for 29.95 fps operation makes this simple test invalid.

2.8.12.7.5 Audio control packets

Presence of audio control packets is optional for level A but mandatory for all other levels. When present, these packets should be transmitted in the second HANC space after the video switching point[11] (lines 8 and 321 for 625/50) and must precede any audio packets in this HANC space. The ANC data ID used for the audio control packets is given in Table 2.8.12; note that each audio group has an associated control packet. If audio control packets are not transmitted a default operating mode of 48 kHz synchronous audio is assumed.

The audio control packets themselves contain information necessary for operation at some of the higher levels. For instance, 3 bits of information are sent to convey the audio sampling rate. An *audio frame number* is contained in the control packet to provide a sequential frame numbering system for the progression of non-integer samples per frame inherent for systems with a 29.95 frame rate, as shown in Table 2.8.15.

Level I is an advanced level of operation that is very useful in modern television systems, where the presence of video timing synchronisers or equipment that uses framestores for effects etc. can lead to a condition where the video is delayed with respect to the audio. While a delay in sound compared to vision can occur naturally if the viewer is some distance from the source of the sound, a delay in vision with respect to audio is an unnatural occurrence. Human perception finds 'audio leading video' quite a disturbing effect and many will find one frame of delay very noticeable. It is possible to recover the audio samples, delay them appropriately and re-embed them at every processing stage that causes video delay, though this is expensive. Level I operation of the embedded audio standard offers an alternative: any delay required for each audio channel is transmitted down the chain of video equipment within the audio control packet as a 26-bit twos complement number, indicating the amount of audio delay

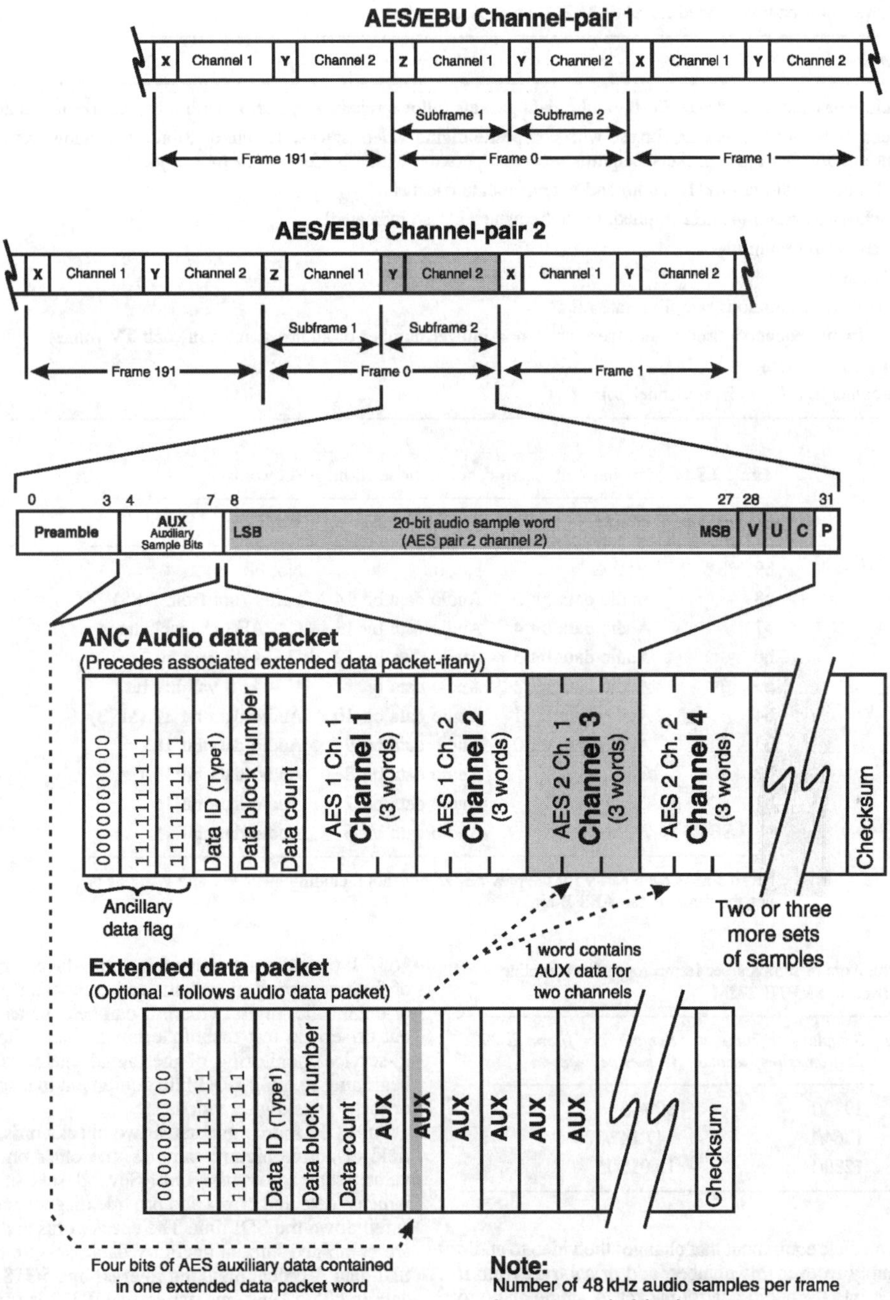

Figure 2.8.16 Mapping AES data into ANC audio data packets.

Table 2.8.13 Levels of operation defined for SMPTE 272M embedded audio

Level	*Summary*
A	Synchronous audio at 48 kHz, 20-bit audio data packets (allows receiver operation with a buffer size less than the 64 samples)
B	Synchronous audio at 48 kHz, for use with composite digital video signals, sample distribution to allow extended data packets, but not utilising those packets (requires receiver operation with a buffer size of 64 samples)
C	Synchronous audio at 48 kHz, audio and extended data packets
D	Asynchronous audio (48 kHz implied, other frequencies if so indicated)
E	44.1 kHz audio sampling
F	32 kHz audio
G	32 to 48 kHz continuous sampling rate range
H	Audio frame sequence (used where there are a non-integer number of audio samples in each TV frame)
I	Time delay tracking
J	Non-coincident Z bits in a channel pair

Table 2.8.14 Mapping AES audio data into three 10-bit ANC words

Bit location	*Word 1*	*Word 2*	*Word 3*
b9 **MSB**	Not b8	Not b8	Not b8
b8	Audio data bit 5	Audio data bit 14	Parity (not from AES)
b7	Audio data bit 4	Audio data bit 13	C – AES channel status
b6	Audio data bit 3	Audio data bit 12	U – AES user bit
b5	Audio data bit 2	Audio data bit 11	V – AES validity bit
b4	Audio data bit 1	Audio data bit 10	Audio data bit 19 (MSB)
b3	Audio data bit 0	Audio data bit 9	Audio data bit 18
b2	Ch. 1	Audio data bit 8	Audio data bit 17
b1	Ch. 0	Audio data bit 7	Audio data bit 16
b0 **LSB**	Z	Audio data bit 6	Audio data bit 15

Word 3 b8 is even parity for the previous 26 bits, not including b9 of W1 and W2; this is not the same as the AES P-bit.

Table 2.8.15 Numbers of samples per frame for audio sampling frequencies allowed in SMPTE 272M

Audio sampling rate (kHz)	*Samples per frame at 25 frames per second*	*Samples per frame at 29.95 frames per second*
48.0	1920/1	8008/5
44.1	1764/1	147,147/100
32.0	1280/1	16,016/15

required. Intermediate equipment that changes the video to audio timing can simply recover this number, add or subtract from it and re-embed it into the audio control packet. A single audio to video timing or 'lip-sync' corrector at the end of the processing chain can act on this information in the audio control packet correcting any A/V delay caused by equipment in the chain.

2.8.13 Error Detection and Handling (EDH)

The EDH system was developed by Tektronix, but has been subsequently recognised in SMPTE Recommended Practice RP 165.[6] It provides a means of detecting bit errors and other types of fault occurring in systems and equipment using component, or composite, digital video. It can be used as a stand-alone or out-of-service test technique but is also suited to continuous in-service monitoring of the signal chain. Manufacturers are beginning to build EDH into equipment that sources, processes or distributes digital video.

An EDH source generates two checkwords, one based on a field of active picture samples, the other on the full field of values sent across the link. Several sets of single-bit flags indicate detected errors and provide diagnostic information that is fed down the SDI link. The checkwords and diagnostic flags are combined into an ancillary data packet embedded into the SDI data. Vertical blanking interval line 5/318 carries the EDH data in 625/50 systems, while line 9/272 is used for 525/59.94 operation. The data ID (DID) of the ANC packet used by EDH is $1F4_h$, with block number 200_h and data count 110_h.

At the receiver the full field and active picture checkwords are recalculated and compared with those contained in the error detection packet to determine if a transmission error has occurred. Reception errors can be indicated by triggering a local alarm on the equipment or by passing information over a separate command and control system, though the detection of an error must also result in setting the 'error detected here' (edh) flag. If the input SDI already had this edh flag set, the equipment will recalculate and replace the appropriate checkword in the output SDI, clear the edh flag but set the eda 'error detected already' flag bit. This error detected already flag indicates to all downstream equipment that an error has occurred somewhere in the signal chain. Any equipment that monitors and reports on its own status can use the 'internal error detected here' and 'internal error detected already' to indicate a fault condition. A final flag, 'unknown error status', indicates that the incoming signal does not include EDH information.

An addition to the EDH concept, called 'Active Picture Zero CRC' (APØ), modifies the last five words in the active picture of each field in such a way as to force the active picture checkword to zero. Many current vision mixers re-blank signals passing through them, destroying EDH and any other ancillary data packets. Such systems can still be checked by APØ as the active picture checksum can be recalculated; if it is not zero an error has occurred or the active picture data altered in some way. The five active pixels corrupted by the APØ process will be removed by analogue blanking, so are unlikely to cause any visible artefacts.

Popular digital waveform monitors can read and check EDH status, indicating an alarm if an incoming error occurs. One way of monitoring these errors is to count the number of 'errored seconds' of video. The count will be incremented for each second of incoming video that contains one or more errors. Errors are usually indicated by EDH monitoring or by detectable format errors in the TRS sequence. Errored seconds are not the same as bit error rate (BER) but the two are related.

The large variation between worst case BER and best case BER in Table 2.8.16 is due to the fact that a single full-field errored second could be caused by a bit error in just 1 bit of the 270 Mbits that will have passed through the system in that second; on the other hand, it could have been caused by a failure in all 270 million bits! The table is included mainly for those communication engineers who are familiar with BER and BERT (bit error rate testing). The errored second is probably a better measure of likely disruption to the signal and the ability of EDH to make this measurement possible on links that are in service makes the system more useful to system operators.

Table 2.8.16 Relationship between full-field errored seconds and bit error rate

Number of errored seconds	*Best case bit error rate*	*Worst case bit error rate*
1 every 10 seconds	3.7×10^{-10}	1×10^{-1}
1 each minute	6.2×10^{-11}	1.6×10^{-2}
1 each hour	1×10^{-12}	2.7×10^{-4}
1 each day	4.3×10^{-14}	1.2×10^{-5}
1 each week	6×10^{-15}	1.7×10^{-6}
1 each year	1.2×10^{-16}	3×10^{-8}

2.8.14 Video Index and D-VITC

Video index data carries various classes of programme-related source data and is coded for transport within a component digital signal (see Table 2.8.17). The well-established Sony Digital Betacam tape format carries video index data in the vertical blanking interval. A 'Digi Beta' camcorder will flag its aspect ratio setting over the data on this line.

Table 2.8.17 Classification of video index data

Classification of data	*Description*
Class 1.1	This is the only class that *must* be carried by the video index data and contains information required to display the signal, not including pan and scan. It indicates the aspect ratio used.
Class 1.2	First three octets of pan and scan data.
Class 1.2	Last three octets of pan and scan data.
Class 2.1	Field rate technical heritage information useful for further signal processing.
Class 2.2	Slow rate technical heritage information useful for further signal processing.
Class 2.3	Reserved for other technical heritage information.
Class 3.x	Undefined data, could include program identification number, tape length, purchase date and serial number, data of production, studio colour temperature, etc.

Carriage of Class 1.1 data in the video index structure is mandatory, all other classes are optional.

Class 1.1 data must be carried in the video index structure. Perhaps its most important function is to signal the aspect ratio and scanning standard of the picture it accompanies. It can also indicate the sampling structure of the picture data, i.e. 4:2:2 or 4:4:4, etc. It can carry data on which composite standard, if any, the component video data originated from and what colour space (i.e. YCbCr or RGB) is used for the component data.

2.8.14.1 Carrying Video Index data

Video index data is carried by modifying the two least significant bits of the chrominance signal of lines 11 and 324 (14 and 277 for 525/60). The first chrominance sample following SAV represents the least significant bit of video index word 0, the next chrominance sample (usually a Cr value) carries bit 1 of video index word 0 and so on. A chrominance value of 200h indicates binary 0, while a value of 204h indicates binary 1. A total of 90 8-bit words can be carried in this way and as only chrominance least significant bits are affected, the signal can be converted to analogue without the need to blank the video index information.

2.8.14.2 Digital Vertical Interval TimeCode (D-VITC)

Timecode is essential to most post-production processes. It carries time as hours: minutes: seconds: frames and is used to provide a reference number for every frame in a video sequence. Such a reference is essential to most editing processes. Timecode can be carried on tape on a longitudinal track like that used for the audio in analogue tape formats; this is called Longitudinal TimeCode (L-TC). Timecode can also be carried on lines in the vertical blanking interval of analogue video, as Vertical Interval TimeCode (VITC). Digital Vertical Interval TimeCode (D-VITC) is carried by modifying the luminance samples of a lines in the vertical blanking interval, usually lines

11 and 324 (14 and 277 for 525/60). These same lines may carry Video Index information coded into the chrominance least significant bits. The luminance samples are synthesised in such a way as to produce an acceptable analogue VITC signal if they are passed through an analogue-to-digital converter.

2.8.15 Serial Data Transport Interface (SDTI)

The Serial Data Transport Interface (SDTI) was developed to allow high-bit-rate data to be sent over an SDI link using the active line period (Figure 2.8.17). The resulting multiplex has the same data rate and the same synchronising structure as that used for 4:2:2 component video, so the SDTI packets can be passed through systems intended to handle and route SDI video. SDTI can provide a high-bit-rate communications channel, allowing many different types of data to be carried over a broadcast facility's SDI infrastructure, including high-definition television pictures that have been data-rate-reduced by compression.

SDTI is a data stream protocol intended to transport packetised data over a unidirectional 10-bit SDI link. The data to be transported is contained in the part of the SDI multiplex that was intended for video active line. SDTI sits above SDI in the packetising and multiplex section of the media interface layers of Figure 2.8.1. It uses channel coding and physical layers identical to those of SDI, though 10-bit operation is mandatory.

It is important to note that the SDTI protocol itself is just a *container* or *wrapper* for this payload data. Associated standards define how the payload itself is to be handled. Many different types of SDTI data are allowed for, compatibility between source and destination being indicated by a suffix to the SDTI label, e.g. SDTI-CP. The first applications of SDTI have been to provide for output of compressed video and audio from a VTR using that compression. This is useful when copying tapes, or transferring tapes to a server using an identical compression standard, as it avoids the losses associated with decoding the video and audio back to a 4:2:2 (and AES3) data simply for the transfer process, requiring that these signals are data-rate-reduced by compression at the destination machine. Cascading data-rate-reduction techniques can result in unacceptable performance. Transferring compressed data also has the advantage that the data rate is lower, allowing the possibility that operations such as transfer from tape to server can run faster than real time. If the compressed data is actually a high-definition television picture SDTI provides a means of distributing and routing HDTV over a standard-definition television infrastructure.

Figure 2.8.18 shows the format of a single line carrying SDTI data. The signal is identical to SDI in respect of data rate, TRS timings, etc., so will pass across systems intended to distribute SDI. A type 2 HANC packets is used to provide a header, while the space between SAV and EAV that usually carries picture information is used for the SDTI payload.

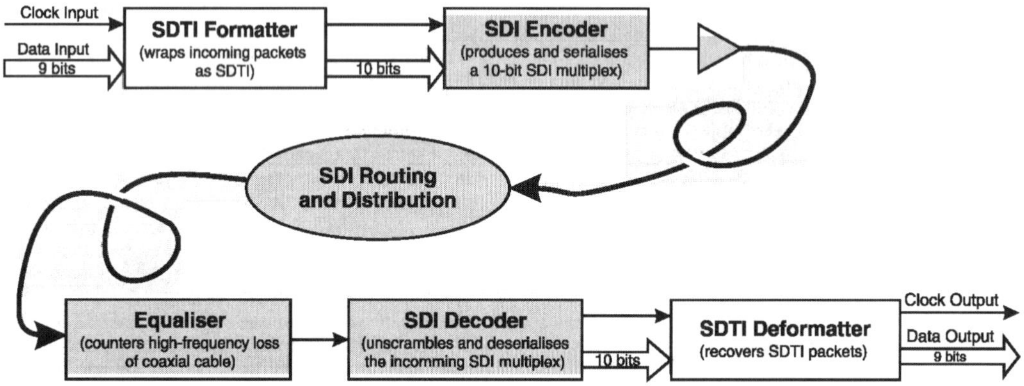

Figure 2.8.17 System block diagram of SDTI.

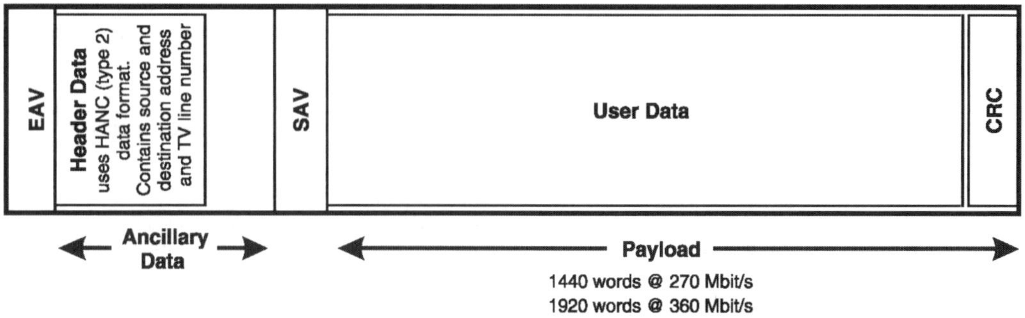

Figure 2.8.18 SDTI signal format (one television line).

2.8.15.1 SDTI Header

This is a type 2 ANC data packet with data ID of 40_h and a secondary data ID of 01_h. It contains a source address and a destination address for the data, opening the possibility of automated routing of SDTI packets via specially adapted SDI routing matricies. A word in the SDTI indicates how the payload data is to be segmented. The payload data may be a fixed length block with or without error correction or the payload may be segmented into blocks of length defined by the SDTI header. This makes for flexible use of the payload data space; how this space is used is not defined by the SDTI specification itself, but determined by the application. This is why SDTI is called a 'wrapper' for a wide range of applications that need to transport data over an SDI interface. The TV line number is also encoded in the SDTI header. It is not mandatory to include an SDTI header on every line.

2.8.15.2 SDTI Block Structure

Figure 2.8.19 shows the simple block structure of SDTI. These blocks are carried within the payload of each SDTI line (Figure 2.8.18); one data block may be placed immediately after the next, or on the following line. A variable length data block may be longer than the payload area of a single TV line. The type code indicates the application whose data is being carried in the block. Type codes are defined for use by various digital videotape formats, including DVCPRO, Digital S, DV Cam, HD Cam and HD-D5. Data types for carriage of MPEG stream data are also included. SDTI Content Package (SDTI-CP) is an important application of SDTI that allows for transport of pictures, sound and *metadata* (information associated with the sound and pictures).

2.8.15.3 SDTI-Content Package format

Known as SDTI-CP, this application provides for transport of picture and audio information, along with auxiliary data such as teletext or ANC data packets, within the payload (active line) area of an SDI signal in the form of SDTI data blocks. The format allows metadata (e.g. time, date and timecode information, etc.) to accompany the video and audio data. Control information can also be passed across the SDTI-CP format.

A content package consists of up to four items:

- A *picture item* contains data, specifically an assembly of up to 255 picture stream elements. These elements contain data-rate-reduced video in the form of an MPEG-2 video elementary stream.
- An *audio stream item* is composed of elements carrying full data rate (uncompressed) audio in a format very like AES3, but supporting up to eight audio channels.
- An *auxiliary data item* is an assembly of up to 255 auxiliary elements. These elements can carry representations of specific vertical blanking interval lines, allowing simple transport of teletext waveforms, for instance. They can also carry ANC data packets of the form described in Figure 2.8.15. Transport of still images as TIFF or SPIFF files (TIFF and SPIFF are still image compression formats) or EBU format WAV audio files is provided for. A final auxiliary element format provides for the carriage of 'free form' data – anything that needs to be transmitted via this data channel.
- Finally, the *system item* is assembled from elements that can carry metadata. This is the first item to be transmitted in the field containing the content package and the format allows this metadata to be tagged as a descriptor to any of the other element types. An obvious use of this metadata is for timecode information. SMPTE Unique Media Identifier (UMID) can be carried as metadata elements. Information on editing processes that have been carried out on the picture and audio can also be conveyed as metadata elements.

An SDTI-CP decoder is a unit that has the ability to receive and decode an element, or a set of elements and associated metadata, carried within an SDTI-CP stream (Figure 2.8.20).

The SDTI-CP layer lies partly within the 'Data Rate Reduction' layer of Figure 2.8.1 and partly in the Packetising and Multiplex layer. It lies immediately above the Multiplexing, Channel Coding and Physical layers of SDI video as it shares all of these layers with the ubiquitous SDI format.

The SMPTE has standardised SDTI and SDTI-CP in standards 305.2M, 326M and 331M, all released in 2000.

2.8.16 Interface Formats for High-Definition Television

There is no single format for High-Definition Television (HDTV) and this section can only give an overview of the common interface standards.

Many scanning standards and sampling standards exist (Table 2.8.18).

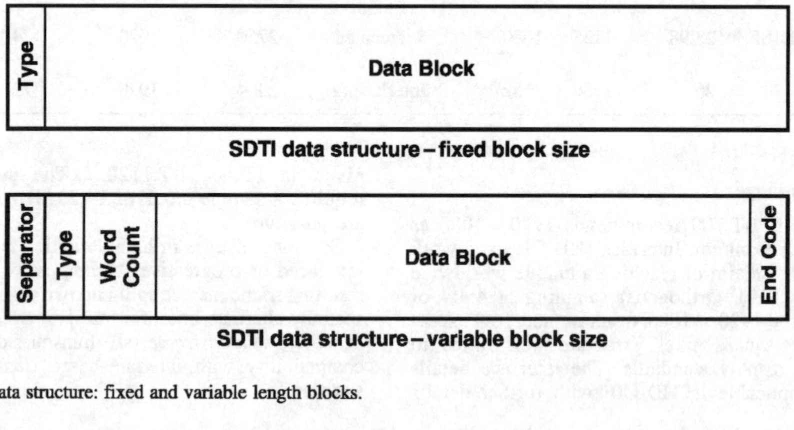

Figure 2.8.19 SDTI data structure: fixed and variable length blocks.

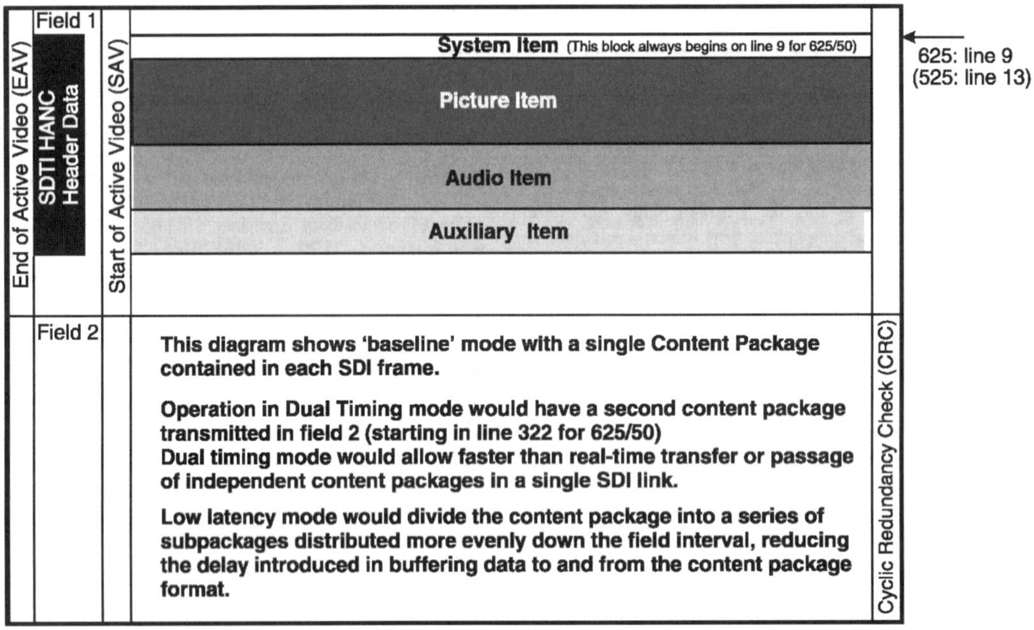

Figure 2.8.20 Arrangement of system, picture audio and auxiliary items in the content package.

Table 2.8.18 Some high-definition scanning and sampling standards

Line standard	*Frame rate*	*Lines frame*	*Active lines*	*Scanning*	*Y samples per line*	*Y samples per active line*	*Y sample freq. (MHz)*	*Aspect ratio*
625/50/2:1	25	625	576 (575 analogue)	Interlace (2:1)	864	720	13.5	4:3 or 16:9
525/59.94/2:1	29.97	525	483	Interlace (2:1)	858	720	13.5	4:3
1920 × 1080/50/2:1	25	1125	1080	Interlace (2:1)	2640	1920	74.25	16:9
1920 × 1080/60/2:1	30	1125	1035	Interlace (2:1)	2200	1920	74.25	16:9
1920 × 1080/59.94/2:1	29.97	1125	1035	Interlace (2:1)	2200	1920	74.176	16:9
1920 × 1080/25/1:1	25	1125	1080	Progressive	2640	1920	74.25	16:9
1920 × 1080/25/1:1SF	25	1125	1080	Segmented frame	2640	1920	74.25	16:9
1920 × 1080/24/1:1SF	24	1125	1080	Segmented frame	2750	1920	74.25	16:9
1920 × 1080/23.98/1:1SF	23.98	1125	1080	Segmented frame	2750	1920	74.176	16:9
1920 × 1152/50/2:1	25	1250	1152	Interlace (2:1)	2304	1920	72	16:9

2.8.16.1 Digitising layer

The standard ITU-R BT.709 recommends 1920 × 1080 as the High-Definition Common Interface (HD-CIF) and work is progressing with the aim of reaching a unique worldwide standard (Table 2.8.19). Orthogonal sampling at 4:4:4 or 4:2:2 is used and the 1920 × 1080 structure and 16:9 aspect ratio results in a 'square pixel' structure well suited to modern computer display standards. The interface details given here are applicable to HD-CIF with further details given in ITU-R BT.1120.[13] The system outlined here requires a sample clock of 74.25 MHz but other clock rates are possible.

Scanning of lines in high-definition television can be either interlaced or progressive. If progressive scanning is used the resulting frame may be split into two segments, each containing spatially alternate lines, for transport over the interface. This is called *segmented frame* (sF) transmission, offering increased compatibility with interlace-based transmission and display systems.

Table 2.8.19 HD-CIF systems recommended in ITU-R BT.709

System	*Capture (Hz)*	*Transport*
60/P	60 progressive	Progressive
30/P	30 progressive	Progressive
30/PsF	30 progressive	Segmented frame
60/I	30 interlace	Interlace
50/P	50 progressive	Progressive
25/P	25 progressive	Progressive
25/PsF	25 progressive	Segmented frame
50/I	25 interlace	Interlace
24/P	24 progressive	Progressive
24/PsF	24 progressive	Segmented frame

2.8.16.2 Multiplexing layer

HDTV interfaces operate at a higher data rate than their standard-definition counterparts. The Y, Cb and Cr signals may be quantised with 8-bit or 10-bit resolution, though 10-bit operation is assumed here. The Cb and Cr components are time-multiplexed with luminance to form 20-bit words as indicated by the brackets in the sequence below:

$$(C_{b1}Y_1)(C_{r1}Y_2)(C_{b3}Y_3)(C_{r3}Y_4), \text{ etc.}$$

The valid ranges for quantised video are the same as those of Rec. 601 with the same exclusion of ranges 0.00 to 0.75 (0_h to 3_h) and 255 to 255.75 ($3FC_h$ to $3FF_h$) to provide reserved words for the TRS.

The timing reference signal (TRS) itself is very similar to that used by standard-definition component television (see Table 2.8.4) though obviously the F and V bits change states to suit the high-definition line structure. The fourth (last) word of the TRS is termed the XYZ word in Figure 2.8.21. Byte XY has the same function as while Z represents 2 bits recommended to be transmitted as 00 but to be ignored when detecting the TRS.

2.8.16.2.1 Information in the horizontal blanking interval

The horizontal blanking interval is available to carry HANC packets of similar structure to that of SDTV (see Figure 2.8.15); however, the four words immediately following the EAV at the beginning of the digital line must be avoided as they carry important information. Words LN0 and LN1 carry line number data while words YCR0 and YCR1 are error detection words for the luminance channel. CCR0 and CCR1 are error detection words for the chrominance multiplex.

2.8.16.3 Bit-Parallel Interface for HDTV

Data representing Y, Cb and Cr in the format shown above is transmitted over 20 pairs, using the same (ECL) voltage levels and impedance as used for the Rec. 656 SDTV parallel interface. A separate pair carries the 74.25 MHz clock. If RGB data is carried over the interface then 30 pairs plus clock are required. A 93 contact equipment mounted socket is specified for the equipment connector, though 1250/50/2:1 may use a 50-way D-type socket. The parallel interface is no longer recommended for new designs.

2.8.16.4 Serial Digital Interface for HDTV (HD-SDI)

Serial interconnection is recommended for new designs, dubbed HD-SDI. The multiplexed parallel data stream of Figure 2.8.21 is serialised for transmission over a 75 Ω coaxial cable or an optical fibre. The interface uses the same principles of parallel-to-serial conversion followed by channel coding (data 'scrambling') and NRZI transmission, much as shown in Figure 2.8.7. As the clock rate of the 10-bit multiplexed parallel data stream is 148.5 MHz, the resulting serial data has a bit clock of 1.485 GHz. Increased cable losses at this frequency limit transmission to 50–80 m of coaxial cable, so longer runs will be best provided by optical fibre connection.

Packetised data can be carried over the active picture area of an HD-SDI link if required, a practice recently standardised as High-Data-Rate Serial Transport Interface (HD-SDTI) SMPTE 348M-2000. The format is similar to that of SDTI (Figure 2.8.18).

2.8.17 Carrying Digital Audio and Video over IT Data Networks

2.8.17.1 The problems of routing – circuit switching

Audio signals can be carried from a single source to a single destination across a simple AES/EBU link; similarly, SDI can be used for video. Providing for a more complex operation, such as a broadcast centre with many AV sources that need to be connected to a range of destinations to suit minute-by-minute operational requirements, has been traditionally achieved by *circuit switching*. At its simplest, circuit switching requires bringing all the audio pairs, and video coax, to a central area where an operator could plug a source to a destination with a patch lead. This antiquated technique, used for early 20th century telephone exchanges, is impractical in anything but the smallest installation. More modern practice would be to run all the audio pairs back to an audio router, and video coax to an SDI router. A router is a matrix of switches or 'cross points' that allows any source to be routed to any of one or more destinations. The router, or matrix, can be controlled from a number of remote locations, e.g. a studio could have a control panel that allowed it to select audio for its outside source lines. Each source has a dedicated wire circuit through the switch to its destination. This solution works very well, but can require a very large number of cables to be installed around a building to connect every single possible source or destination to the central audio router. This makes it expensive to install or move a studio installation, or to rig a small temporary facility, though any one of the circuits installed may get little use. Costs and inconvenience escalate if the central router is 'full', necessitating some sort of upgrade or satellite switching arrangement to be installed. Added to this, as the numbers of signals to be routed extends into the hundreds, constructing the routing switch itself is not easy. A partial solution to a few of the problems of scale in routing audio signals can be found by multiplexing up to 56 digital audio channels together using a standard called MADI (Multi-channel Audio Digital Interface), developed for digital multi-track tape machines and standardised by the audio engineering society as AES10. MADI allows these 56 channels, sampled at up to 48 kHz with 20- or 24-bit resolution, to be carried over a single 75 Ω coax cable or an optical fibre.

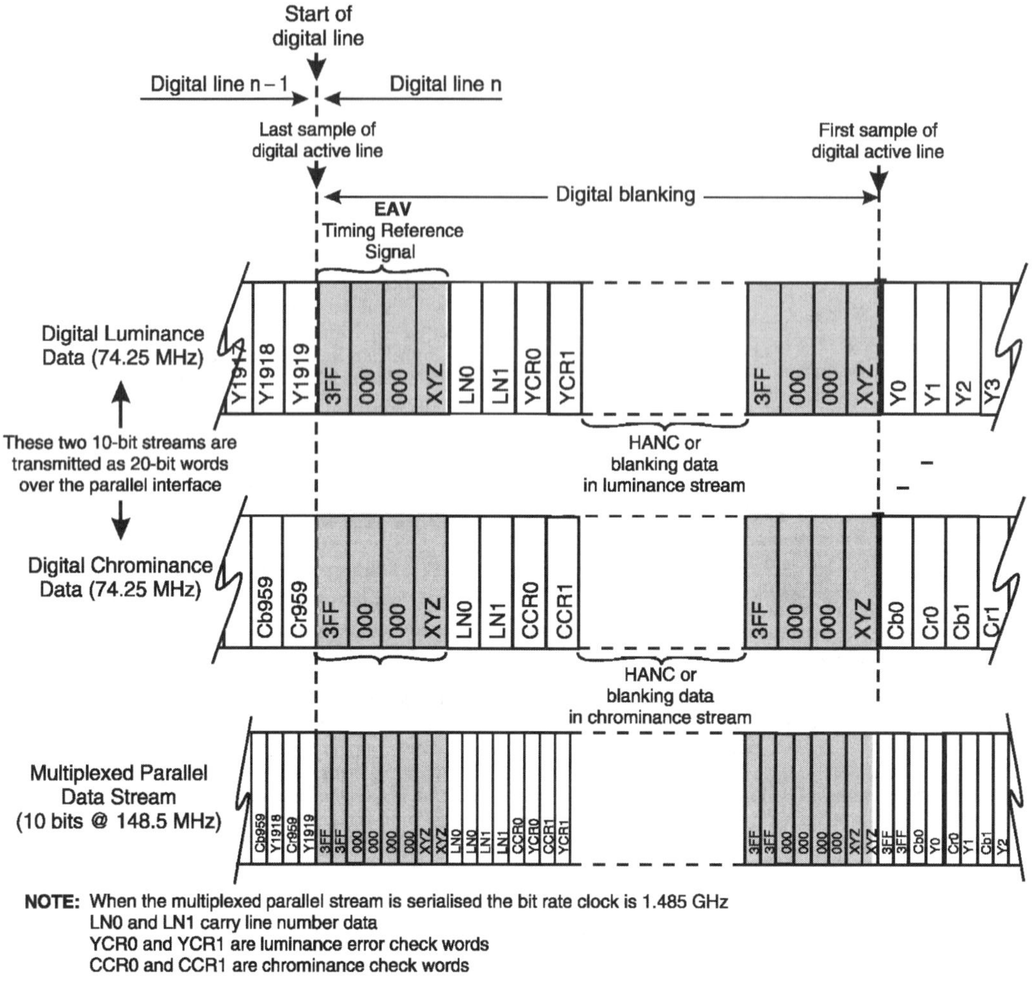

Figure 2.8.21 High-definition data stream (ITU-R BT.1120-3).

2.8.17.2 IT networks and packet switching

Chapter 000 deals with the alternative to circuit switching. Packet switched systems were deployed in long-haul telecom systems. IT Local Area Networks are packet based and usually use Internet Protocol (IP), sometimes combined with Transmission Control Protocol (TCP), and carried over one of the Ethernet standards. Asynchronous Transfer Mode is a network technology developed for both LAN and Wide Area Network (WAN) operation, though it should be noted that ATM and IP are not really in competition. IP traffic passes across ATM and SDH/SONET for many worldwide Internet connections.

2.8.17.3 Audio Distribution and routing over ATM

The Audio Engineering Society has standardised a method, referred to as AES47, for carrying digital audio across an ATM network. The audio data is sent in uncompressed form and the standard provides for operation at any of the bit rates defined for AES3.

Systems like that in Figure 2.8.22 are already showing that they can handle live broadcast operation. They are robust, as the ATM network can be engineered with redundant paths that will maintain operation if a single link or switch fails. Talkback channels and control information can also pass across the ATM network, as can the TCP-IP traffic that runs the broadcaster's desktop PCs. Ethernet and IP cannot readily combine the demands of IT traffic with those of carrying broadcast audio. ATM can handle both, thanks to its ability to set well-defined QoS parameters for each service.

The audio interfaces can include features such as the ability to mix together several AES47 input channels if required or to cross-fade between one signal source and another. The distributed nature of the ATM network and its compatibility with long-haul telecommunications standards such as SDH would allow national broadcasters to build a distributed routing system, making it simple to assemble elements of audio that have

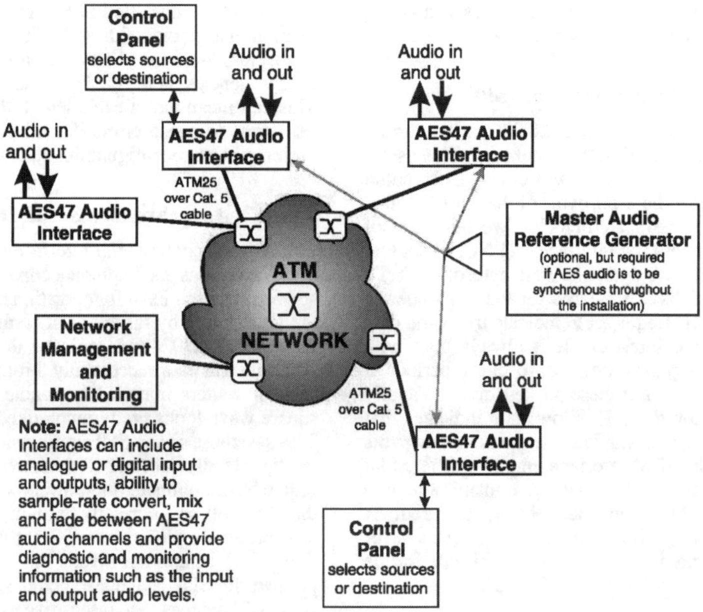

Figure 2.8.22 Conceptual diagram of an AES47 audio distribution and routing system.

originated from sources that are separated geographically. The high cost of ATM network hardware may be reduced if future systems include the consumer 'firewire' (IEEE 1394) standard as an alternative interconnect technology for edge of network operation, connections to individual audio interfaces, though this is not included in the current standard.

2.8.18 Physical Interfaces for DVB and MPEG Transport Streams

I shall not attempt to detail or explain the entire operation of MPEG-2 or its DVB (Digital Video Broadcasting) application in this chapter, as they are dealt with elsewhere. This section will confine itself to dealing with the handling and interfacing of the DVB transport stream that contains MPEG-2 programme and service information. Such transport streams can carry audio, video and data for one or more television or radio programmes.

Transport streams are composed of fixed length packets, as shown in Figure 2.8.23. The packets can be 188 or 204 bytes in length. This apparent ambiguity lies in the fact that DVB systems need to provide for the addition of forward error correction. An MPEG coder, or compressor, provides MPEG data, along with associated headers or PIDs (Packet IDs). These have 188 bytes of useful information. These transport streams may pass through further multiplexing equipment, for instance to add additional programme material, and be routed around a building or transmitted across a telecommunication link, for instance to get to a satellite ground station for up-link. At some point the transport stream will reach a DVB modulator. The DVB system requires the modulator to add forward error correction capability in the form of 16 Reed–Solomon (RS) bytes appended to the 188-byte transport packets. It can be more straightforward to allow for this by making the packets 204

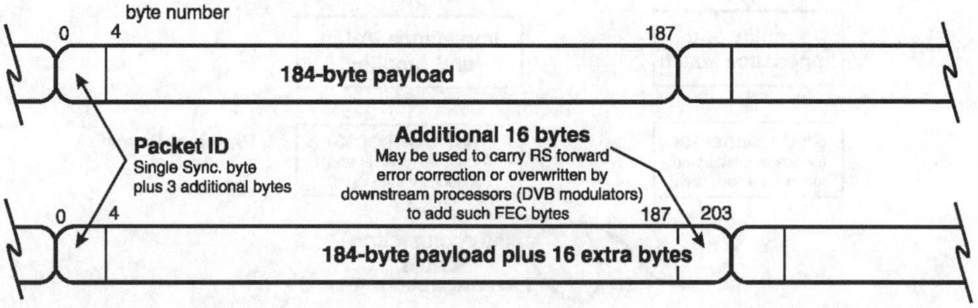

Figure 2.8.23 DVB transport stream, showing options for both 188- and 204-byte packets.

bytes long to begin with – with the last 16 bytes containing dummy information destined to be overwritten by the modulator.

2.8.18.1 Synchronous Parallel Interface (SPI)

The transport stream packets of Figure 2.8.23 can be carried across a parallel interface defined in the standard.[14] This uses a common electrical interface called Low Voltage Differential Signalling (LVDS) to carry the 8-bit bytes of the MPEG transport stream along with its associated clock. Two other bits of information are carried across the interface PSYNC, a flag that is set when the *synchronising byte*, the first byte of each TS packet, is transmitted. PSYNC is cleared for the remainder of the TS bytes. The DVALID flag is set to indicate that valid data is being passed across the interface. It is always high with 188-byte TS packets, but pulses low for 16 clock periods if 204-byte packets are being sent if these packets do not actually contain RS error correction data. This low state indicates that no attempt should be made to validate the packets using this FEC data. The LVDS electrical interface employs differential data being carried across 110 Ω balanced transmission line cabling, with 11 such balanced pairs being required. A 25-way D-type socket is specified for the equipment side of the interface. The clock rate is one-eighth of the bit rate for the transport stream, with an upper limit of 13.5 MHz.

2.8.18.2 Synchronous Serial Interface (SSI)

This is an extension of the parallel interface, involving parallel-to-serial conversion of the 8-bit TS data and bi-phase mark coding similar to that illustrated in Figure 2.8.10, though this particular diagram was drawn with AES3 in mind. Coaxial cable and BNC connectors are used. As synchronous transmission is employed the frequencies on the cable increase as the TS bit rate increases. This can mean that a cable length that is satisfactory at low bit rate starts to cause errors if the bit rate of the TS is increased during system reconfiguration, due to increased cable losses.

2.8.18.3 Asynchronous Serial Interface (ASI)

The Asynchronous Serial Interface has become one of the most popular options for interconnecting DVB or MPEG-2 transport streams. Options exist for operation over fibre or 75 Ω coax, with the latter being by far the most common choice. The bit rate on the cable is 270 Mbit/s, offering the significant advantage that ASI data can pass successfully through many distribution amplifiers or routers intended to handle SDI video. One 270 MHz square wave looks pretty much like another, though ASI has no TRS structure, so any SDI equipment that performs processing on the signal will reject an ASI data stream. ASI is polarity sensitive, unlike SDI, confusing the unwary when they discover, perhaps, that ASI will pass correctly through outputs 1, 3 and 5 of an SDI distribution amplifier, but will not pass via outputs 2, 4 and 6. Some SDI equipment inverts the signal passing through it, no problem for SDI data, but something that destroys ASI. That said, some ASI receivers are insensitive to such inversion, as they have the ability to automatically correct for this in their input stage.

Figure 2.8.24 shows the processing steps necessary to convert parallel packet-synchronous transport stream data to the ASI format; 8-bit to 10-bit conversion is done using a mapping

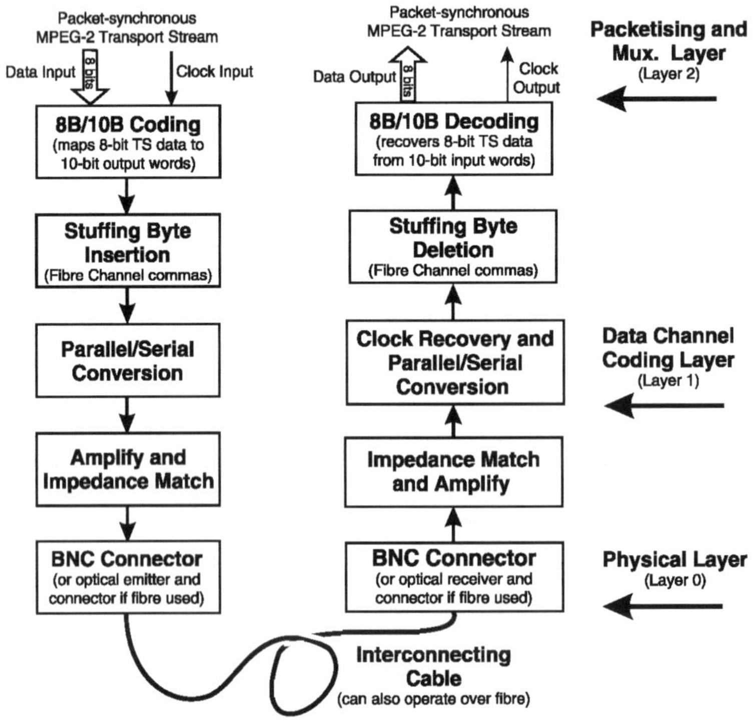

Figure 2.8.24 Processing and layers for Asynchronous Serial Interface transport stream data.

that is also used by the Fibre Channel IT interconnect format. Details of this mapping can be found in the ASI standard document.[14] The 8B/10B code has the property that there are never more than four consecutive 0s or 1s in the binary stream, improving clock recovery at the receiver. Minimising the running disparity of the data, a property that is useful in detecting ASI data errors, also provides for minimum DC component. *Running disparity*, the difference in the number of '1' bits to '0' bits, is never greater than ± 1 for the ASI stream.

The physical channel for the data (Layer 0) can be provided by 75 Ω coaxial cable or fibre. When cable is used, the peak-to-peak data amplitude is 800 mV, transformer coupled.

Transport stream data rate can vary dramatically depending on the configuration, so 'stuffing words' are added to achieve a constant 270 Mbit/s output data rate. The stuffing words are a 10-bit number that does not occur in the 8B/10B mapping (Fibre Channel comma character K28.5 in the coding table), so they can be readily detected and removed by the ASI receiver. These stuffing words also provide for byte alignment at the receiver, detection of the transport packet sync word (47_h) providing packet synchronisation. Every transport packet begins with this synchronisation byte.

Figure 2.8.25 shows that the interface can operate in two modes. *Packet mode* transmits each byte in a transport stream packet contiguously, adding sufficient stuffing words between each packet to increase the data rate to the 270 Mbit required. *Burst mode* transmits single TS bytes as soon as they are presented to Layer 2 of the interface, adding stuffing characters as necessary between each byte.

2.8.19 Broadcast Interfaces between IT Systems

There is more to building a broadcast system than simply getting video and audio from one place to another. It is necessary to provide something that will handle the other information in the broadcaster's workflow. Early systems did this with bespoke software that communicated non-standard messages across private data channels and perhaps held information in a central database in a way only understood by the application programmers. Though useful, such systems almost inevitably end up as islands within the overall workflow. It was possible to get the finished audio and video programme information out, but little else. *Metadata* (information about the sound and pictures) was lost, rushes and clips were difficult to identify or categorise after they were used, and the archiving of material or the production of a similar, but not identical, version of a programme involved labour-intensive repetition of some tasks that had already been performed once.

Much of this can be solved by finding ways for the increasing number of IT applications used along the broadcast chain to 'talk' to each other and to exchange *essence* (video and audio), metadata and control information in standard ways. It is this standardisation that allows systems from one manufacturer to communicate with those of another; it is these standard ways of exchanging information that will interconnect the islands into an efficient workflow for the broadcasters. What is needed is a 'software communication' equivalent of SDI and AES3 that have been hugely successful as interfaces between one bit of hardware equipment and another.

2.8.20 Standards from the IT World

An operating system is itself a 'standard' for communicating information between applications and between any application and the hardware that it runs on. The reader will be aware of the dominance of various Windows OS, along with the presence of more open standards such as Linux and Unix. The list is too long to continue. Application-to-application communication across a network needs to be OS independent, however. It is important that niche applications such as broadcast do not re-invent the wheel in seeing a need to allow exchange of information and control messages between software-based systems. Many well-supported solutions to this problem are emerging from the IT industry, much based on Object-Orientated software techniques. CORBA (Common Object Request Broker Architecture – see www.omg.org) allows software objects on one computer platform to interact with each other, probably across a computer network, exchanging control information and perhaps data. The objects code can be written in whatever way and whatever language is appropriate to that particular application and platform; CORBA defines only the interface. A CORBA-compliant object advertises this interface for use by

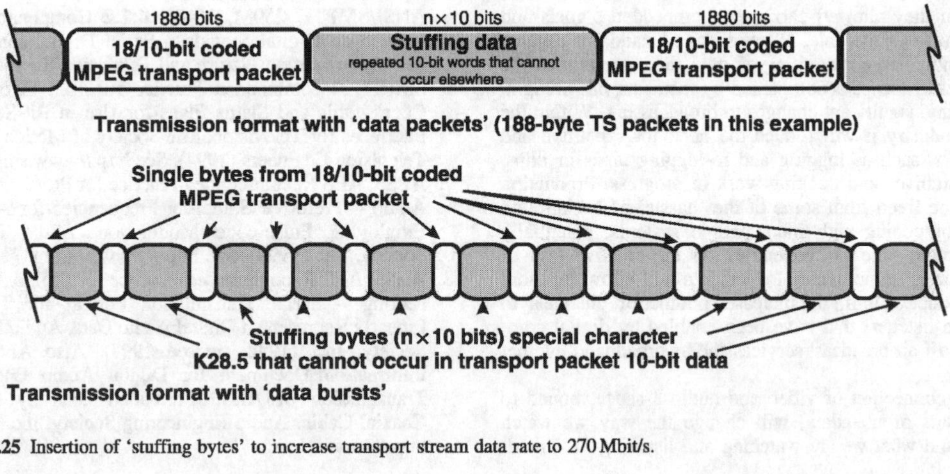

Figure 2.8.25 Insertion of 'stuffing bytes' to increase transport stream data rate to 270 Mbit/s.

other applications; the code and data contained within the object are encapsulated behind a boundary that no CORBA client application can cross. CORBA is about control and interaction of objects within applications that may be built by different companies and running on different hardware under different operating systems. If CORBA is for control, SOAP (Simple Object Access Protocol) is about the exchange of information. SOAP is based on XML (Extensible Mark-up Language). It does not define any application semantics so it is highly customisable, for instance for broadcast use. Further details can be found from www.w3.org. A final well-established standard, SNMP (Simple Network Management Protocol), provides for configuration, diagnosis and control of infrastructure hardware such as hubs routers and switches and the servers and workstations that sit on such an infrastructure and are the IT system. Many manufacturers of specialist broadcast hardware have also adopted SNMP.

There are many more such IT standards. Chapter 000 describes the file formats[15] developed for broadcast applications.

2.8.21 Digital Interfaces, the Railroads of a New Media

Arthur C. Clarke maintained that we tend to overestimate the impact of technology in the short term and underestimate it in the long term; I think that he was right. Digital interfaces between equipment were simply replacements for their analogue counterparts at their large-scale inception in the 1980s. They offered slightly better quality, especially in the case of DVE (Digital Video Effects) boxes that worked digital component internally. Component digital interfaces avoided the need to decode and recode PAL or NTSC, with attendant quality loss. I have made little mention of composite digital standards, though these have their place, in particular when accompanying digital composite video recorders or in long-haul transmission of PAL or NTSC video over fibre telecommunications networks. Relatively small increases in quality, and little requirement for equalisation adjustment, made life easier for engineers, though parallel interconnects for video were cumbersome and offered limited cable length.

Digital interfaces for both audio and video became more useful when they allowed carriage of embedded signals and metadata. At the same time, IT systems and standards became able to handle large quantities of data, and sometimes the essence itself. Essence accompanied by metadata has brought, and will bring, significant changes to broadcasting. Within the broadcast industry it will reduce the need for repetitive and tedious tasks, such as logging and re-logging tapes or clips, searching archives and tracking work in progress. Production teams will be freed from some of the 'hassle' of background tasks and of coping with incompatible systems, potentially providing more time to concentrate on the creative task of producing programme material. Metadata will allow for more essence production to fill the increased number of 'channels' to viewers and listeners that have been enabled by digital compression of off-air broadcast services and by broadband Internet streaming.

The interconnection of video and audio essence, bound to large amounts of metadata, will change the way we watch and listen and what we are watching and listening to. Broadcasters currently build playout automation systems that simply churn out linear streams of essence, stripping out much of the metadata before transmission. Video, audio and data will increasingly be carried by digital interfaces direct to in-home 'playout servers'. These servers are dumb disk recorders at the moment, but with plentiful and cheap supplies of metadata they will become media centres that will formulate individualised news 'programmes', sell us entertainment customised to our taste and supply us with the information we choose, or others pay service providers to get us to watch. Marshall McLuhan's book *The Medium is the Massage* tells us that: "Each medium, independent of the content it mediates, has its own intrinsic effects which are its unique message." Engineering and technology is the creation of something new, an art-form of its own and a media with a message that can change the world we live in. American and European societies have already been shaped by 'media' such as the printing press, electrical power and railway interconnections. This chapter is about the nuts and bolts, the binary formats, which are essential to fixing the 'railroad' of a new media together. It's impossible to know where these new lines of information will take us, but the track has been laid and the journey begun.

References

1. Recommendation ITU-R BT.601-5, Studio Encoding Parameters of Digital Television for Standard 4:3 and Wide-Screen 16:9 Aspect Ratios, International Telecommunications Institute Radiocommunication Assembly.
2. Recommendation ITU-R BT.656-4, Interfaces for Digital Component Video Signals in 525-Line and 625-Line Television Systems Operating at the 4:2:2 Level of Recommendation ITU-R BT.601 (Part A), International Telecommunications Institute Radiocommunication Assembly.
3. Recommendation ITU-R BT.799-3, Interfaces for Digital Component Video Signals in 525-Line and 625-Line Television Systems Operating at the 4:4:4 Level of Recommendation ITU-R BT.601 (Part A), International Tele-communications Institute Radiocommunication Assembly.
4. Sandbank, C.P. (ed.). *Digital Television*, John Wiley & Sons, ISBN 0 471 923 60 5.
5. ANSI/SMPTE 259M, 10-Bit 4:2:2 Component and 4fsc Composite Digital Signals – Serial Digital Interface, The Society of Motion Picture and Television Engineers (1997).
6. SMPTE Recommended Practice RP 165, Error Detection Checkwords and Status Flags for Use in Bit-Serial Digital Interfaces for Television, The Society of Motion Picture and Television Engineers (1994). See http://www.smpte.com
7. AES5, AES Recommended Practice for Professional Digital Audio – Preferred Sampling Frequencies for Applications Employing Pulse-code Modulation, Audio Engineering Society, Inc. (1998). See http://www.aes.org
8. AES3, AES Recommended Practice for Digital Audio Engineering – Serial Transmission Format for Two-channel Linearly Represented Digital Audio Data, Audio Engineering Society, Inc. (1992, revised 1997). Also AES-3id, AES Information Document for Digital Audio Engineering – Transmission of AES3 Formatted Data by Unbalanced Coaxial Cable, Audio Engineering Society, Inc. (2001). See http://www.aes.org

9. IEC 60958-3, Digital Audio Interface – Part 3: Consumer Applications, The International Electrotechnical Commission. See http://www.iec.ch/
10. SMPTE Standard 291M, Ancillary Data Packet and Space Formatting, The Society of Motion Picture and Television Engineers (1998). See http://www.smpte.com
11. SMPTE Recommended Practice RP 168, Definition of Vertical Interval Switching Point for Synchronous Video Switching, The Society of Motion Picture and Television Engineers (1993). See http://www.smpte.com
12. SMPTE Standard 272M, Formatting AES/EBU Audio and Auxiliary Data into Digital Video Ancillary Data Space, The Society of Motion Picture and Television Engineers (1994). See http://www.smpte.com
13. Recommendation ITU-R BT.1120-3, Digital Interfaces for HDTV Studio Signals, International Telecommunications Institute Radiocommunication Assembly.
14. European CENELEC Standard EN 500083-9, Part 9: Interfaces for CATV/SMATV Headends and Similar Equipment for DVB/MPEG-2 Transport Streams, The European Committee for Electrotechnical Standardisation.
15. Useful websites for broadcast control and file formats: www.mosprotocol.com, www.ebu.ch, www.pro-mpeg.org, www.pro-mpeg.org, www.aafassociation.org

Acknowledgements

Thanks to many colleagues in BBC People Development Engineering Training, particularly Andy Woodhouse and Michael Hunter. This chapter is dedicated to my good friend John MacKay.

Bibliography

Nyquist, H. Certain topics in telegraph transmission theory, *IEEE Transactions* (April 1928).
Reeves, A.H. British Patent 535 860 (1939).
Shannon, C.E. The mathematical theory of communication, *Bell Systems Technical Journal* (Oct. 1948).

Peter Schmidt Ph.D. (Physics)
Sony Broadcast & Professional Europe, UK

2.9 File Formats for Storage

2.9.1 Introduction

The advances in digital technologies have transformed the broadcasting industry dramatically. Traditional analogue techniques are gradually being replaced with digital systems in almost every aspect of a broadcasting environment. Only a few decades ago all audiovisual (AV) content was analogue and Edit Decision Lists (EDLs) were nearly the only digital content produced.[1,2] AV raw material has been stored on analogue tapes, and information pertaining to its content kept separately, often in a non-digital form.

Today, digital media have found their way into editing, production and archiving systems. With improved compression algorithms and increased bandwidth digital content is increasingly used for transmission and will indeed replace traditional broadcasting in the near future.

Storage and material exchange is one of the many areas where the use of digital content is of advantage. For instance, a variety of physical devices can be utilised for storing digital data, such as optical discs, digital tapes or standard computer hard discs. Once in the computer domain, data may be easily transferred using standard network protocols such as TCP/IP,[3,4] or may be even used for Internet broadcasting using the real-time protocol RTP/RTSP.[5,6]

Computer systems store data in binary form, called a file. Typically, the actual raw data within a file are stored in a structured way. Additional information, in many cases located at the top of a file (hence called a 'header'), is provided to allow data access and processing. This type of information is called metadata. A file format specification contains a complete definition of both data structure and content.

Many different multimedia file formats are already available.[7] Not all of them are of use in broadcasting, where high image and sound quality is required. As a consequence, formats such as MPEG and DV have been created to meet those AV quality requirements. However, optimising storage by designing appropriate compression techniques is only part of a digital content and workflow solution. To take full advantage of digital technologies, content has to be efficiently accessible and manageable. To facilitate this, additional descriptions and unique identification is needed. This type of content-related metadata is known as 'descriptive' metadata (as opposed to 'structural' metadata describing the digital material). 'Rich' metadata contains both structural and descriptive metadata. Considerable efforts have been undertaken to define satisfactory descriptive metadata models such as P/Meta[8] and MPEG-7.[9] Some of these efforts have gone into the formats described below. The provision and inclusion of rich metadata is an essential requirement for any new data format in broadcasting.

Over the years, manufacturers have been designing their own digital formats to accommodate broadcasters' needs. This has been an acceptable solution for single self-contained systems. However, most broadcasters wish to choose their equipment and systems from different manufacturers. In addition, digital content is no longer produced, edited and transmitted by one organisation alone. Rather, the process may involve external departments and studios. Therefore, exchanging digital content is an essential part of any broadcasting workflow today. To exchange material from different origins and in different formats requires, often costly, transcoding processes and equipment.

To resolve this issue the broadcasting industry has been working on defining and standardising storage and data exchange file formats. As a consequence, new standards have emerged. To give a complete account of all existing proprietary standards is beyond the scope of this chapter. Instead, an introduction to the most recent and relevant formats is given. The selection includes:

- The General Exchange Format (GXF).
- MPEG-4 File Format.
- Advanced Authoring Format (AAF), together with the Material Exchange Format (MXF).

Of the formats listed above, only MPEG-4 specifications include the AV compression technique. The other formats define methods to encapsulate digital media and metadata content. This approach allows the use of different existing digital AV coding formats and future extensions for new optimised compression technologies.

Metadata are included in the file as binary data. However, at various points in a broadcasting workflow, metadata will be reused separately from the AV content. It has been recognised that a standardised format is needed to allow exchange of metadata between different applications. For that reason, the eXtensible Mark-up Language (XML) has been identified as the suitable solution.[10] A detailed discussion of the various XML implementations is beyond the scope of this chapter. Suffice to say that XML metadata representations have been developed for most discussed standards in this chapter.

2.9.2 The General Exchange Format

The General Exchange Format (GXF) has been developed by Grass Valley Group (GVG) and then adopted as an SMPTE standard (SMPTE 360M). Originally, GXF was developed to transfer digital AV and metadata content across Fibre Channel using standard File Transfer Protocol (FTP),[11] although any reliable network protocol can be utilised for transfer. GXF is based on data encapsulation, i.e. the actual raw AV content remains unchanged during the coding process.

2.9.2.1 GXF Applications

The format is designed for interchanging material and is not intended for persistent storage on servers. In a typical application a video server (e.g. Grass Valley profile video server) generates a GXF stream during transfer. This process is called serialisation. The received GXF stream is then being de-serialised by the receiving video server. Transcoding essence while transferring can be a demanding task. To optimise this process, GXF files are constructed based on a simplified audio/video interleaved stream using data packets (see below). A GXF stream may also be used for playout during transfer.

Several compression types are supported by GXF:

- A stream of JPEG fields.
- DV-based frames.
- MPEG elementary streams, including temporal compressed image sequences (long Group Of Pictures – GOP) or only spatially compressed image sequences (intraframe only).

2.9.2.2 GXF Data Packets

For streaming purposes, all metadata and AV content in GXF are encoded as data packets. The structure of a GXF data packet is as follows:

A packet is preceded by a 16-byte packet header (see Table 2.9.1). This is followed by the packet content, i.e. the payload.

The packet header specifies the packet identity and the total packet length. Five different packet types are defined in GXF:

- MAP packet – containing information to decode and interpret media data.
- Media packet – for digital video, audio data and timecodes.
- Field Locator Table (FLT) packet – provides up to 1000-byte offsets into the stream.
- Unified Material Format (UMF) packet – this contains detailed material description and used defined information.
- End Of Stream (EOS) packet.

Table 2.9.1 Structure of a 16-byte GXF packet header (SMPTE 360M)

Byte position	*Value*	*Usage*
0	0x00	Packet leader, used for synchronisation
1		
2		
3		
4	0x01	
5	Variable	Packet type 0xbc = map 0xbf = media 0xfa = reserved 0xfb = end of stream 0xfc = file locator table 0xfd = UMF file 0xfe = reserved 0xff = reserved
6	Variable	Packet length (MSB)
7		Packet length
8		Packet length
9		Packet length (LSB)
10	0x00	Reserved
11	0x00	Reserved
12	0x00	Reserved
13	0x00	Reserved
14	0xe1	Trailer
15	0xe2	Trailer

The values of each byte position in the header are given in hexadecimal values.

The MAP, FLT and UMF packets contain details necessary to decode and present the material. All three packet types have an internal structure described below:

- The MAP packet has two main structures: a material data section and a track description section (see Figure 2.9.1). Both are preceded by a preamble, which immediately follows the MAP packet header. The material section contains a list of first and last fields or frames in the stream. The digital format of the AV material, the media file name and some auxiliary information (useful, for example, for decoding MPEG-2 data) is listed in the track description section.
- The FLT packet contains up to 1000-byte offsets into the GXF stream and may be used as a media locator.
- The UMF (see Figure 2.9.1) packet consists of a preamble of 5 bytes and four different sections. The sections are:
  - Payload – a fixed size packet of 48 bytes. It defines the size and location of the various UMF sections in the stream.

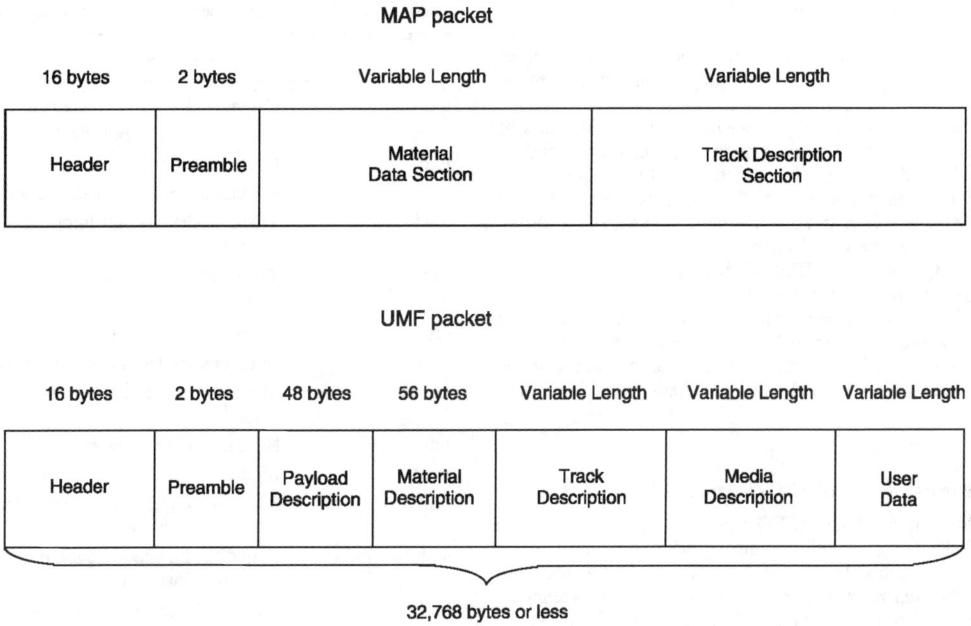

Figure 2.9.1 The structures of the MAP and UMF packets in GXF.

- Material description – a fixed size packet of 56 bytes. This contains information about the AV content (e.g. sampling rate).
- Track description – a packet of variable length. It describes the track (using a track ID) and the number of media files referenced by each track.
- Media description – a packet of variable length. A more detailed decoder/encoder of specific information on the AV content.
- User data – packets of variable length.

2.9.2.3 GXF Stream Composition

A GXF data stream consists of a defined order of GXF data packets (see Figure 2.9.2). Every stream starts with an MAP packet and ends with an EOS packet.

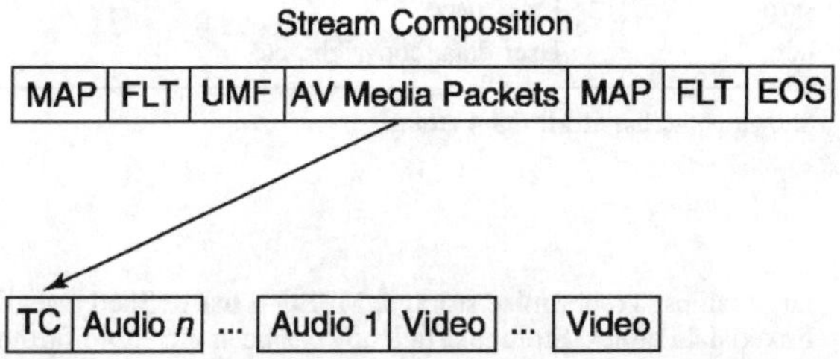

Figure 2.9.2 An example of a GXF stream composition. In media packets, timecode packets precede any audio packets. Audio packets are followed by video packets.

At least one UMF packet must be included in the stream. FLT packets are placed following the first MAP packet, in cases where the location of media data in the stream is known. Alternatively, the FLT follows the last MAP packet in the stream. MAP packets are repeated in the stream after 100 media packets. This enables dynamic update of structural metadata while streaming.

Video, audio or timecode data form the content of media packets. Depending on the source material (e.g. audio only), a GXF stream may not contain all media packet types. In the case of all three types being present, timecode packets must precede any audio and video packets. Audio packets have to be streamed before video packets (see Figure 2.9.2). This ensures that all the digital content is available at the time of presenting a given video frame or field.

A GXF stream may contain material from different sources. The media tracks from each source are comprised into media segments. A simple clip in GXF is defined as a number of timecode, audio and video originating from one source. A compound clip consists of several media segments, where each segment originates from a single source.

MAP packets are used to indicate a change in media segments.

Transitional video effects are not allowed in GXF other than those already embedded in the elementary data stream.

2.9.3 The MPEG-4 File Format

Building on the success of, for example, the MPEG-2 compression format, the Moving Picture Expert Group (MPEG) has developed MPEG-4. The format defines a new range of compression technologies for both high and low bit rates. An object-oriented approach has been taken to encode information in media samples, allowing for new technologies such as picture overlay or the encoding of individual image objects.

However, MPEG-4 is not just an improvement in terms of compression efficiency and an extension to existing encoding practices. In contrast to its predecessors, MPEG-4 also defines its own file format for storage.[12] This step has been taken to allow better data access and management. The MPEG-4 format has been based on the QuickTime format.[13]

As with QuickTime, MPEG-4 files contain metadata to access, decode and present the AV material. In order to assist media search and management and to provide extensive content description, another MPEG standard, MPEG-7, has been developed.[9] MPEG-7 rich metadata associated with an MPEG-4 scene may be included in a dedicated elementary stream.

2.9.3.1 MPEG-4 and MPEG-7 Applications

Compared with previous MPEG specifications, MPEG-4 has increased the range of compression algorithms towards lower and higher bit rates. In particular, low-bit-rate transfers are increasingly used for a wide range of mobile digital technologies, such as mobile phones, Personal Digital Assistants (PDAs) and others. Another important application is broadcasting of live streams to a network of computers. These are typically connected using standard TCP/IP protocols. AV material is broadcasted using the Real-Time Protocol (RTP). This technology is already successfully used for QuickTime file streams.

The extension of compression efficiency to allow for better image and sound quality (MPEG-4 Studio profile) and bit streams of 50 Mbit/s and more make MPEG-4 suitable for technologies such as High-Definition TV.

Including rich metadata as defined with MPEG-7 widens the range of applications considerably. The range of applications using metadata content is practically limitless, but a few spring to mind, e.g. archives, educational purposes, legal institutions, copyright protection.

2.9.3.2 MPEG-4 Atoms

An MPEG-4 file consists of elements called 'atoms'. Atoms are used to encapsulate both metadata and elementary streams. Each atom is preceded by an 8-byte 'header' followed by the atom content, the payload. The header specifies the atom type and the total atom length.

Atoms may be nested within other atoms. A comprehensive list of MPEG-4 atom types is given in Table 2.9.2.

Within a nesting level, atoms may be stored in any order. The 'free' and 'skip' atoms can be used to align data to byte boundaries or even to include non-MPEG-4 data.

Table 2.9.2 List of MPEG-4 atoms and their nesting order

	Atom description
moov	Container for all the metadata
mvhd	Movie header, overall declarations
iods	Object descriptor
trak	Container for an individual track or stream
tkhd	Track header, overall information about the track
tref	Track reference container
edts	Edit list container
elst	An edit list
mdia	Container for the media information in a track
mdhd	Media header, overall information about the media
hdlr	Handler, at this level, the media (handler) type
minf	Media information container
vmhd	Video media header, overall information (video track only)
smhd	Sound media header, overall information (sound track only)
hmhd	Hint media header, overall information (hint track only)
<mpeg>	mpeg stream headers
dinf	Data information atom, container
dref	Data reference atom, declares source(s) of media in track
stbl	Sample table atom, container for the time/space map
stts	(Decoding) time-to-sample
ctts	Composition time-to-sample table
stss	Sync (key, I frame) sample map
stsd	Sample descriptions (codec types, initialisation, etc.)
stsz	Sample sizes (framing)
stsc	Sample-to-chunk, partial data-offset information
stco	Chunk offset, partial data-offset information
stsh	Shadow sync
stdp	Degradation priority
mdat	Media data container
free	Free space
skip	Free space
udta	User data, copyright, etc.

Shown is the list of MPEG 4 atoms.

2.9.3.3 Data Access and Data Descriptors

Video, audio and other metadata are described in separate tracks. Access to the raw AV elementary stream is facilitated through a series of data access and synchronisation tables. They are stored within the 'stbl' atom of each track. Storing access information of every video or audio sample may result in very large tables. To optimise storage, MPEG-4 has defined a set of linked data tables. Storage efficiency has been increased further by condensing information and avoiding redundancies.

For this, AV data have been defined as a series of 'samples' and 'chunks'. A sample could be one video frame or 1 byte of 8-bit mono audio.

The number of all samples and their sizes are listed in the 'stsz' atom. Samples are comprised into 'chunks'. The 'stsc' atom contains a sample-to-chunk table, which maps the number of samples to each chunk. Finally, the 'stco' atom contains a table listing the absolute byte positions of each chunk in the file.

As mentioned above, data redundancy is avoided. For instance, in the 'stsz' table only samples of different sizes are listed. Another example is given in Table 2.9.3. In this sample-to-chunk table ('stsc' atom), seven chunks are listed. However, since the first six chunks contain the same number of samples, only two entries are needed.

Table 2.9.3 An example of a sample-to-chunk table stored within an 'stsc' atom

Chunk index	*Samples per chunk*
1	10
7	6

The first six chunks contain 10 samples, the last only six. The table contains two entries; the index in the first row indicates a change in chunk sizes.

Together with the sample size listed in the 'stsz' and the absolute byte position of each chunk, all media data can be easily accessed. Further synchronisation and timing information is provided in other tables (e.g. in the 'stts' atom).

The 'stds' atom contains essential information needed for decoding the elementary stream. This is the location where the elementary stream descriptors ('esds') are stored, containing decoder-specific details.

All digital video and audio material is stored in the 'mdat' atom. This atom has no further nesting structure. The order of video and audio samples within this container is not specified in the MPEG-4 format.

2.9.3.4 MPEG-7 Metadata

The 'Multimedia Content Description Interface' (MPEG-7) provides a rich set of standardised tools to describe AV content. At the centre of these tools is the Description Definition Language (DDL). This is based on the eXtensible Mark-up Language (XML). XML documents consist of structured text. The structure is facilitated by XML elements and their attributes. To standardise the use of elements and attributes in XML documents, XML provides another tool called XML-Schema. The XML-Schema document (itself written in the XML language) may be regarded as a dictionary and grammar reference for the use of XML in a specific application.

XML is an internationally standardised tool designed for data transfer. It is already used in a wide range of computer applications (many of which provide an XML interface per default, e.g. databases). Being a structured text document, XML has the additional benefit of being largely hardware and systems independent.

MPEG-7 provides several XML-Schemas, called description schemes. Description schemes exist for visual and audio content (structural metadata), as well as for content-related descriptions (descriptive metadata).[9] A set of generic entity and Multimedia Description Schemes (MDS) is also provided. The use of XML-Schema technology makes MPEG-7 very flexible and suitable for further extensions and amendments.

Using textual data for transfer may not be of advantage for all cases. Transfer of binary data allows compression, which can be considerable for text data. For this reason, MPEG-7 has introduced the 'Binary Format for MPEG-7 Description Streams' (BiM). The standard defines a set of tools to transcode between BiM and XML representations of metadata.

2.9.4 Advanced Authoring Format and Material Exchange Format

Authoring and editing audiovisual content is a complex process. This is made more difficult by the large number of different AV formats currently available. As interoperability between different applications and systems has become essential, the need for a platform-independent authoring format has arisen. As a consequence, the Advanced Authoring Format Association has been established to develop the Advanced Authoring Format (AAF) interchange format and promote its use. Together with the Material eXchange Format (MXF) described below, it offers a comprehensive and standardised solution for editing, authoring (AAF) and the exchange (MXF) of digital AV content and related metadata.

2.9.4.1 AAF and MXF Applications

Both AAF, and in particular MXF, are relatively new formats in broadcasting. However, they have been broadly accepted in the broadcasting industry. This has largely been due to the close cooperation between the AAF association, the Pro-MPEG forum, which has been promoting MXF and the Society of Motion Picture and Television Engineers (SMPTE). Several companies such as Avid, Grass Valley Group, Leitch, Panasonic, Sony and many others were involved in the development process. As a result of these efforts, technical AAF and MXF solutions have already begun to emerge.

It is important to realise that both formats address different aspects in broadcasting. AAF is aimed at editing and authoring solutions, e.g. non-linear editors. Because MXF metadata are based on the AAF data model, the result from the editing and authoring process can be exported to an MXF file. In many applications MXF is not just used for exchange of data, but also for persistent storage on digital media.

As with MPEG-7, AAF/MXF metadata are extensible. This may be of interest to those organisations that need to persist metadata not covered by the current standard or don't intend to share certain data (e.g. contract details, etc.) with other parties.

2.9.4.2 AAF

The specification of AAF[14] has three major components:

- AAF Object Specification, which is described in more detail below.
- AAF Low-Level Container Specification, describing the data storage of an AAF file.
- AAF Software Development Kit (SDK) Reference Implementation. This open source and cross-platform tool, written in C++, allows engineers to develop their own AAF applications. File storage is based on the 'Structured Storage' technology, developed by Microsoft Corporation.

An AAF file contains a complete set of metadata and related AV material, called 'essence'. The object-oriented data model for metadata is based on a set of defined metadata classes and their relationships to each other. In each class a set of properties, or metadata items, is defined. Class inheritance optimises the reuse of properties for similar types of data sets. Hierarchical data structures are built using class aggregation. Ownership, also called strong reference, is achieved by the use of unique instance identifiers (see MXF).

Instances of classes are called objects. The hierarchical structure of metadata sets and essence objects forms the basis of an AAF file. The root of the object hierarchy within an AAF file is shown in Figure 2.9.3.

The AAF data model is very complex and flexible. Only a brief overview can be given. At the root of the AAF object hierarchy is an instance of the Header class. Only one Header instance is permitted. The Header class provides vital information of, for example, byte order within the file. The header contains an important metadata set, called Content Storage (see Figure 2.9.3). This is the container for the main bulk of defined metadata.

AAF metadata fall into different categories depending on whether they relate to physical storage, synchronisation between different sources or creative edit decisions. Each of these categories is comprised in a metadata 'package'. The following packages exist in AAF:

- Physical Source package, where information on raw material storage, e.g. videotapes, is stored.
- File or Source package. This describes the digital essence, e.g. colour schemes, sound sampling rate, compression algorithms, etc. It also contains location and access details of the AV content.
- Composite packages contain metadata relating to creative decisions, e.g. edit cuts and effects used for transitions.
- Material packages synchronise the metadata from file and source packages and the creative decisions stored in the composite packages. Material packages may be regarded as describing the output timeline.

The separation of metadata into different packages enables us to describe the physical media, the content or any transitional effects to a high degree of granularity.

Packages may be identified using a 32-byte identifier, the package ID. This binary key is otherwise known as the Unique Material Identifier (UMID). The procedure to create a UMID is internationally standardised (SMPTE 298M). The UMID has been recognised as a valuable tool to retrieve and archive digital media.

Each metadata package is highly structured. They contain a sequence of tracks, sometimes called 'slots'. Video, audio data and the timeline describing the entire content are described separately. These tracks are called 'timeline slots'. Other slots are also defined, e.g. to mark the beginning of a transitional effect (event slot) or to insert time-independent metadata (static slots).

To allow further complex audiovisual compositions or inclusion of media from different sources, each slot may be segmented. An example of a slot/segment structure within a package is given in Figure 2.9.4.

Tracks and their subsequent structures are used for synchronisation of source material, composition and the final product. They describe the timeline of each individual component in the AV content. However, to decode and present digital AV material further details are needed. This is provided by a set of 'Descriptor' classes. A range of descriptors exists to help describing the AV digital audiovisual content, e.g. compression parameters, aspect ratios, etc.

AV material is called 'essence'. It may be located within the file or externally. If the AV source material is stored separately (e.g. on a digital tape), 'Locator' classes provide the necessary details to access the source.

2.9.4.2.1 Metadata Item Definitions And Metadata Dictionary

AAF metadata items are based on standards published by the SMPTE. The SMPTE provides a common registry for its metadata, defined in the SMPTE RP210 document. Each metadata item has an associated data type (e.g. string, integer or more complex types like vectors) and unique key of 16 bytes, known as the Universal Label (UL). The set of ULs is maintained by the SMPTE.

Each AAF file is self-contained, i.e. dictionaries for all metadata in the file are included (see Figure 2.9.3).

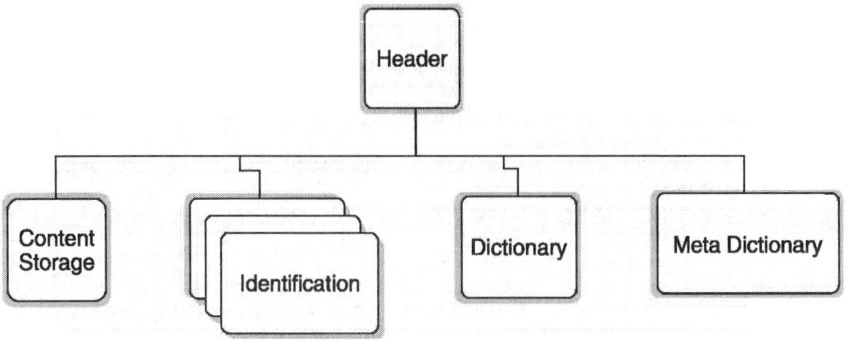

Figure 2.9.3 The metadata object aggregation in an AAF file. The header object is at the root of the file. It owns the objects shown below. The Content Storage contains the metadata and references to the AV essence.

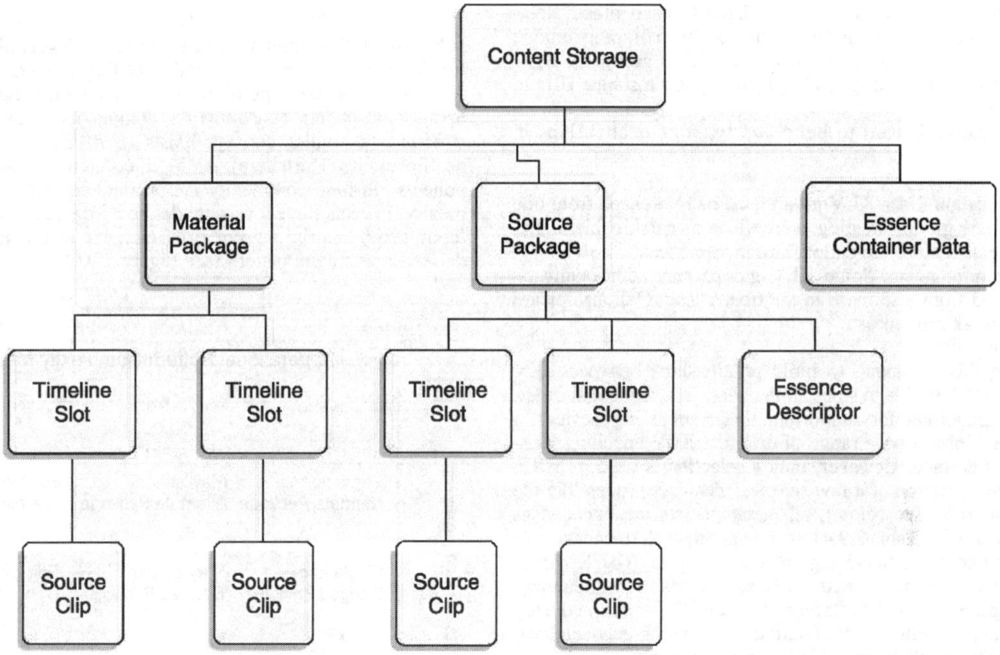

Figure 2.9.4 An example of the structure of AAF Content Storage. Timeline slots may refer to timeline, video or audio material.

2.9.4.3 MXF

MXF has emerged as the new exchange file format in broadcasting. It has been developed and promoted by various companies and broadcasting institutions gathered in the Pro-MPEG forum and is currently being standardised by the SMPTE. MXF is a file format that is optimised for streaming and the exchange of AV material and metadata.

In addition, MXF defines sets of rich and extensible metadata. MXF metadata are based on the AAF data model (see above) and are sometimes regarded as a subset of AAF.

The MXF standard consists of several normative parts:

- The format specification. This is the central document covering all encoding and storage aspects of MXF (SMPTE 377M).
- Operational Pattern (see below) specification documents.
- Generic Container (see below) specification and all subsequent AV format-dependent Generic Container instance specifications (known as Generic Container mapping documents) (SMPTE 379M).
- A specification document for descriptive metadata (SMPTE 380M).

2.9.4.4 Basic Encoding Techniques

All MXF metadata and AV content are defined and encoded as Key Length Value (KLV) entities (SMPTE 336M). Each key, or Universal Label (UL), is a binary identifier for a specific metadata item. As with AAF metadata all keys are defined in a global SMPTE registry (SMPTE RP210). The length is the number of bytes necessary to represent the value. The value is the actual content of a given metadata item or the original digital AV bit stream.

MXF organises the large number of metadata items by defining data groups of KLV items (SMPTE 336M). These data groups are also encoded in KLV (See Table 2.9.4). The value of the group is a set of individual KLV items (ordered or unordered, depending on the type of KLV group). The main structure of a KLV group is illustrated in Figure 2.9.5. KLV groups in MXF are the equivalent of AAF metadata classes.

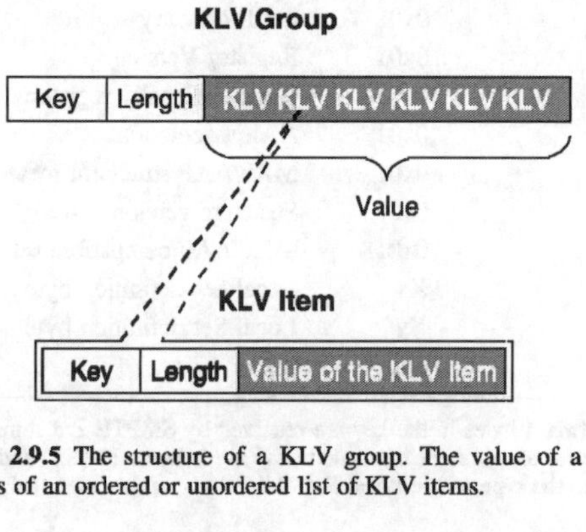

Figure 2.9.5 The structure of a KLV group. The value of a group consists of an ordered or unordered list of KLV items.

As in AAF, KLV groups are related to each other. Relationships between groups are achieved by assigning unique identifiers (an Instance UID) to each KLV group. A KLV item refers to another group by including the Instance UID in its value.

This feature is used to build two types of relationships in MXF:

- Aggregation. Each KLV group must be referenced from one other KLV group, creating a tree-like ownership hierarchy. These relationships are called 'strong references'.
- Cross-referencing. Some KLV groups may additionally be referenced from elsewhere in the tree. These relationships are called 'weak references'.

Using this technique to build relationships between KLV groups, MXF is able to store all metadata in a sequential order. This is important for supporting linear streaming devices.In SMPTE 336M, a wide range of different KLV encoding techniques is defined. However, only a selection is used in MXF. With the exception of a few selected metadata groups like the partition pack (see below), all metadata sets are encoded as 'local sets' (See Table 2.9.4). In a local set, each metadata item is identified by a 'local tag' of only 2 bytes. The length is expressed as an integer with 2 bytes. This has been implemented to optimise metadata storage. A special KLV pack at the top of the file, the primer pack, facilitates a mapping between local tags and their full 16-byte SMPTE representation.

The length of KLV groups is expressed in Abstract Syntax Notation number 1 (ASN.1) Basic Encoding Rules (BER) notation.[15] MXF recommends the use of 4 bytes for the length (although up to 9 bytes are permissible).

All data within an MXF file are in big Endian notation.

Table 2.9.4 An example of an SMPTE key for a local metadata set in MXF

Byte position	*Value*	*Description*
0	0x06	Object Identifier
1	0x0e	Label Size
2	0x2b	Designator (ISO/ORG)
3	0x34	Designator (SMPTE)
4	0x02	KLV group (set or pack)
5	0x53	Local set: 2-byte tag, 2-byte length
6	0x01	Set Dictionary
7	0x01	Registry Version
8	0x0d	Organisationally registered
9	0x01	AAF association
10	0x01	MXF/AAF structural metadata set
11	0x01	Structure version
12	0x01	MXF/AAF compatible set
13	Xx	Local Set definition byte
14	Yy	Local Set definition byte
15	00	

The first 4 bytes in the key are reserved by SMPTE and remain the same for all different labels and keys used. Bytes 13 (Xx) and 14 (Yy) define the type of the Local Set. All values are hexadecimal.

2.9.4.4.1 MXF Operational Patterns

MXF allows file structures of very different levels of complexity. To optimise the coding and decoding process, MXF has introduced the concept of operational patterns (OPs). Each operational pattern constrains the complexity of an MXF file structure (a similar concept exists in MPEG – the MPEG profile and level structure). An MXF complexity has two components: an item complexity and a timeline or package complexity. In each, three levels are defined. Together they define a 'complexity matrix', where each element in the matrix is a defined operational pattern (see Figure 2.9.6).

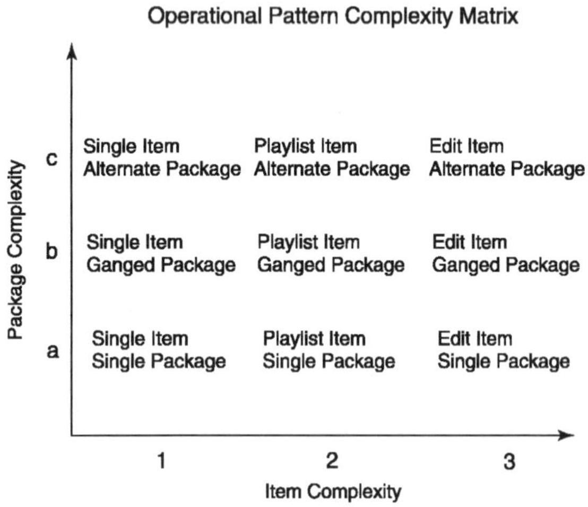

Figure 2.9.6 The complexity matrix for MXF operational patterns. The simplest structure is defined by operational pattern 1a, the most complex is operational pattern 3c.

Item complexity:

- Only one source material is contained in the file (single item).
- Several different source materials may be stored in the file (playlist items).
- There are several edited source materials in the file (edit items).

Timeline complexity:

- The output timeline is based on referencing a single source (single package).
- Several source timelines may contribute to the output timeline (ganged package).
- There may be several output timelines each referencing several timelines from different sources (alternate package).

Each operational pattern is defined in its own standard document. For instance, the simplest operational pattern is OP 1a. An MXF file based on OP1a contains a single playable source. The content storage contains one source package and one material package only.

2.9.4.5 MXF Partitions and File Structure

Each MXF file is divided into partitions. There are three different types of partitions:

1. Header partition. This is a mandatory partition. Each MXF file or stream must start with a header partition.
2. Body partition. This is an optional partition type and may be included in the file as required.
3. Footer partition. All standard MXF files must end with a footer partition.

Special operational patterns may be designed to define different structures, e.g. a run-in sequence or a file without a footer partition. However, these are not covered by the MXF standard.

Header and body partitions may contain both metadata and AV material. Header and body partitions may include one AV stream only. On the other hand, a specified AV material may extend over several partitions. The footer partition can contain only metadata.

An MXF partition is headed by a special KLV group, the partition pack. This KLV pack defines several important values, e.g. the AV content format, the descriptive metadata scheme used (if any), byte offsets into the file stream for the beginning of the header metadata and beginning of AV content. It has no AAF equivalent.

An example of a file structure in OP 1a is shown in Figure 2.9.7. MXF metadata and AV content can be aligned to specified binary boundaries. This may assist coding and decoding processors. The alignment is called the KLV alignment grid. A special KLV item, 'KLV filler', is used for this purpose.

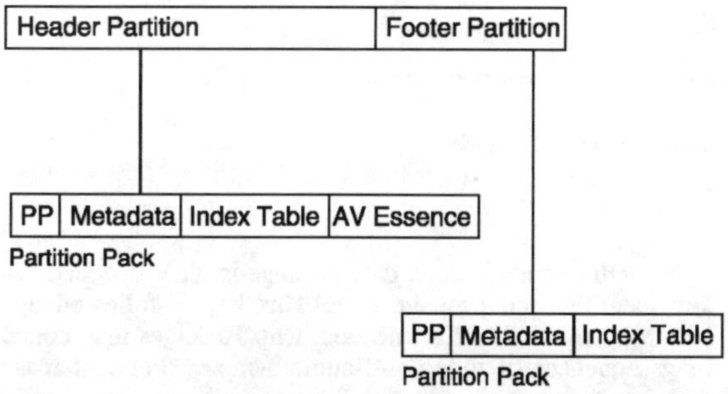

Figure 2.9.7 An example of an MXF file structure. Each standard MXF file must contain at least two partitions, the header partition and a footer partition. Each partition is preceded by a partition pack KLV group. Metadata and index table can be included in both partitions. The footer partition cannot contain any AV essence.

2.9.4.5.1 MXF Metadata

MXF has defined a rich set of metadata based on the AAF model (see previous section). In addition to structural metadata, MXF has also provided a data model for descriptive metadata. Although MXF metadata provides many interesting features, some users may wish to include different metadata. In order to ease inclusion of metadata from different origins a modular approach has been taken. Metadata based on a single defined model are comprised in a metadata 'scheme'. MXF structural metadata form the core scheme required by all valid MXF files. Other schemes, such as descriptive metadata, may be optionally added. Sometimes organisations wish to include data not meant for public use. MXF has allowed inclusion of such 'dark' metadata.

The basic structure and concept of MXF/AAF metadata has already been described (see Figures 2.9.3 and 2.9.4). At the root of the MXF metadata hierarchy is the 'Preface' (called Header in AAF) KLV set. Like AAF, MXF distinguishes between metadata describing the source material, 'Source Package', and that describing the output synchronisation from various sources, 'Material Package'.

A list of 'Source Packages' may be included in case there is material from different sources (this depends on the operational pattern – see above). Each track in the source packages is associated with a 4-byte identifier. This and the duration of AV media clips or segments help to trace the content through the potentially complex package structure.

Structural metadata help to decode and present the digital content. Further details, however, are required to allow content search and archive retrieval of stored material. MXF has designed a complex descriptive metadata model, called 'descriptive metadata scheme 1' (SMPTE 380M). It is a self-contained data model, extending the existing AAF model described above. It offers a variety of metadata to include details on, for example, production title, awards, scene locations, administrative, contract and financial details.

To include Descriptive Metadata (DM) schemes in an MXF file a special track, a DM track, has been defined. The technique for including descriptive metadata schemes is shown in Figure 2.9.8.

A DM track is the link between structural MXF metadata and a descriptive metadata scheme. It also allows inclusion of other metadata schemes not defined by MXF.

All publicly accessible DM schemes are identified using a unique label (UL) of 16 bytes. This label is registered with the SMPTE. The label must be included both in the Partition Pack and the Preface Set (i.e. the root of the metadata). An MXF file may contain metadata from different sources or schemes.

Generally, a DM scheme consists of different 'frameworks'. Each framework describes a particular aspect of the content-related metadata. In DMS 1 three different frameworks are defined:

- Production Frameworks. Metadata in these frameworks relate to the entire production of the material.
- Clip Frameworks. These are source material based. They allow detailed description of individual AV clips.
- Scene Frameworks. These are creative notes on the production.

Many MXF developers have recognised the usefulness of XML in representing MXF metadata. At the time of writing, an XML dictionary is being developed to standardise the use of XML in MXF applications.

2.9.4.5.2 The Generic Container For Audiovisual Data In MXF

The Generic Container specification in MXF defines a technique to encapsulate AV essence applicable to all MXF supported AV formats. The details for encapsulating each of the MXF supported digital AV formats are defined in extensions to

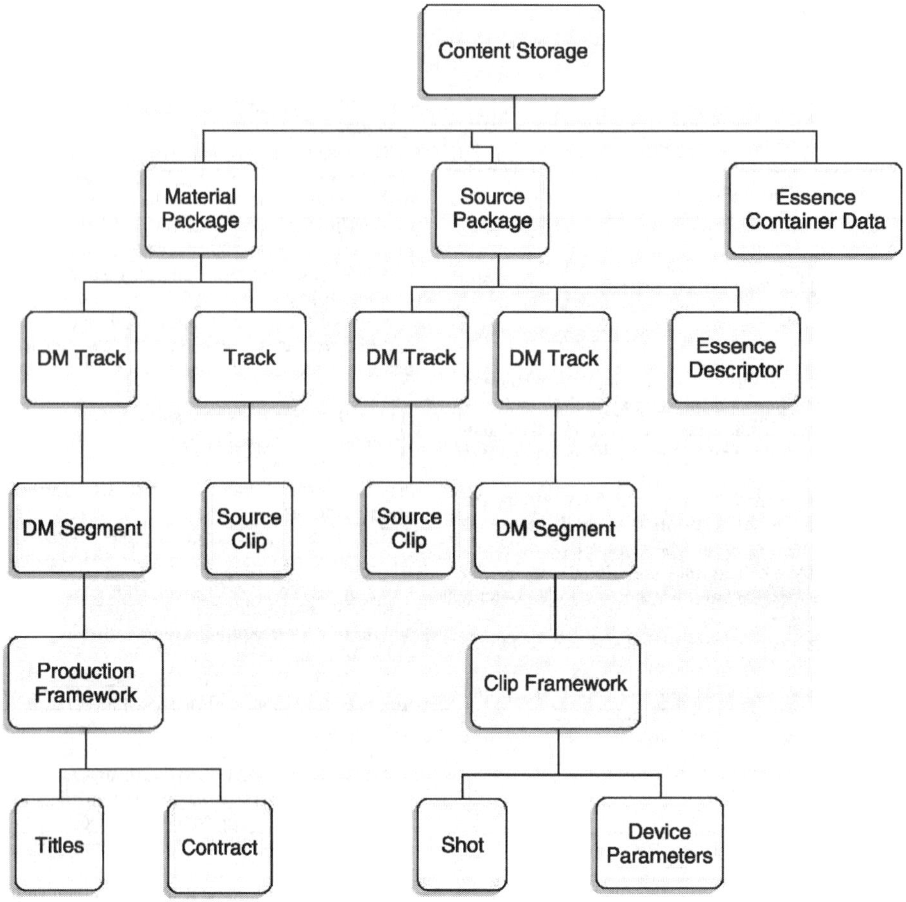

Figure 2.9.8 An example of DMS 1 metadata in the Material and Source Packages of an MXF header.

the Generic Container document ('mapping' documents). Defining a Generic Container technique for encapsulating AV content allows future extensions to the standard.

The type of essence (D10, DV, D11, etc.) is indicated by an SMPTE registered UL. This UL must be included as a metadata item in the Partition Pack and in the Preface Set of the header metadata.

The AV encapsulation model is based on the Serial Device Transport Interface, SDTI (SMPTE 305M). Digital essence is encapsulated into packets called 'content packages' (SMPTE 326M). The following content package types are defined:

- System Package.
- Video Package.
- Audio Package.
- Data Package.
- Clip Package.

The Clip Package has been introduced in MXF to improve encoding of formats such as DV or uncompressed audio.

As with metadata, each data package is KLV encoded. The key identifies the package type. This key is followed by 4 length bytes ASN1.BER encoded. Clip Packages may contain long sequences of video and audio samples. For that reason their length is expressed using 8 bytes.

The available packages in MXF allow for two different ways of essence encoding:

- Frame or field based.
- Clip based.

Frame- or field-based encoding is typically used for D10, for example, whereas clip-based wrapping may be useful for DV encapsulation. An example of frame-based encapsulation is given in Figure 2.9.9.

At the time of writing, MXF has supported the following formats: D10 (MPEG-2 I frame interleaved with uncompressed audio), D11 (High Definition), DV and uncompressed audio (broadcast WAV). However, further extensions (e.g. MPEG-2 long GOP) are being developed and will be published in the near future.

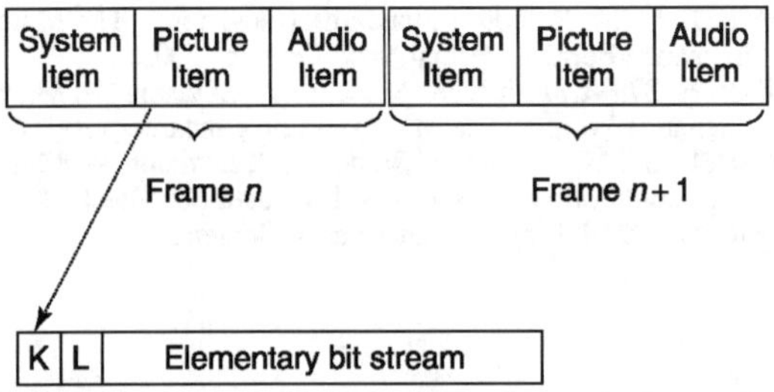

Figure 2.9.9 An example of frame-based MXF essence coding. Each frame contains a defined sequence of content packages, called items. Each item is KLV encoded, where the value consists of the picture or audio elementary stream. The system item contains additional metadata (e.g. timecode).

2.9.4.5.3 MXF Data Indexing and Data Access

In order to optimise data access MXF has defined a special KLV set, called Index Table Segment. This set may be included in the MXF file following the header metadata in a partition. Index tables provide the means to compute byte positions of each individual content package in the stream. Depending on the composition of the AV material, the basic index unit is called an 'edit unit'. This can be, for example, a single frame comprising a system, picture and audio item (see Figure 2.9.9). If the content packages are of variable length, as is the case in MPEG-2 video compression, index tables may grow substantially in size. In many applications, however, a more practical approach is taken and the size of each edit unit is fixed. This can be accomplished by using KLV fillers to align the essence to well-defined boundaries. Index tables, like headers, may be repeated in the file to allow data recovery while streaming. Although the use of index tables is optional, their use is strongly recommended for most supported formats.

2.9.5 Summary

In this chapter different file formats have been discussed. Each solution offers a number of benefits to a broadcasting environment. However, digital technologies are advancing fast and with them technical requirements for data storage. To meet those challenges all discussed formats have the potential to adapting to new paradigms. This is considerably aided by the fact that they are engineered, monitored and maintained by internationally recognised agencies. This has required close collaboration between, often competing, manufacturers and broadcasters. Organisations, such as IEEE, SMPTE, the EBU and many others have done invaluable work. Their efforts have helped to create sound and robust storage solutions.

References

1. Anderson, G.H. *Video Editing*, Focal Press, Boston (1997).
2. Rubin, M. *Nonlinear – A Guide To Digital Film And Video Editing*, Triad Publishing Company, Gainesville (1995).
3. RFC 791, Internet Protocol, The Internet Engineering Task Force (1981), http://www.ietf.org/rfc/rfc0791.txt
4. RFC 793, Transmission Control Protocol, The Internet Engineering Task Force (1981), http://www.ietf.org/rfc/rfc0793.txt
5. Schulzrinne, H. et al. RFC 1889, RTP: A Transport Protocol for Real-Time Applications, The Internet Engineering Task Force (1996), http://www.ietf.org/rfc/rfc1889.txt
6. Schulzrinne, H. et al. RFC 2326, Real-Time Streaming Protocol (RTSP), The Internet Engineering Task Force (1998).
7. Born, G. *The Files Formats Handbook*, International Thomson Computer Press, London (1995).
8. P/Meta, PMC Project, Metadata Exchange Standards, European Broadcasting Union (EBU), http://www.ebu.ch/departments/technical/pmc/pmc_meta.html
9. Manjunath, B.S., Salembier, P. and Sikora, Th. *Introduction to MPEG-7: Multimedia Content Description Language*, John Wiley and Sons (2002), http://mpeg.telecomitalialab.com/standards/mpeg-7/mpeg-7.htm
10. XML, the extensible mark-up language, http://www.w3c.org/XML
11. RFC 959, File Transfer Protocol (FTP), The Internet Engineering Task Force (1985), http://www.ietf.org/rfc/rfc0959.txt
12. Pereira, F. and Egrahimi, T. *The MPEG-4 Book*, IMSC Press Multimedia Series, Prentice-Hall (2002).
13. QuickTime, The File Format, Apple Computer, Inc. (2000), http://developer.apple.com/techpubs/quicktime/quicktime.html
14. AAF, The Advanced Authoring Format Specification Version 1.0.1 (2000), http://www.aafassociation.org
15. Dubuisson, O. *ASN.1 – Communication Between Heterogeneous Systems*, Morgan Kaufmann Publishers (2000), http://www.oss.com/asn1/dubuisson.html

Bibliography

ISO/IEC 13818-2, MPEG-2 Video Compression.

ISO/IEC 14496-1, Chapter 13, MPEG-4 File Format Specification.

ISO/IEC 14496-2, MPEG-4 Video Compression.

ISO/IEC 61834-2, DV/DIF Format Specification.

SMPTE 12M, For Television, Audio and Film – Time and Control Code (1995).

SMPTE 298M, For Television – Universal Labels for Unique Identification of Digital Data (1997).

SMPTE 305M, Television – Serial Data Transport Interface (SDTI) (1998).

SMPTE 314M, Data Structure for DV-based Audio, Data and Compressed Video, 25 Mb/s and 50 Mb/s (1999).

SMPTE 326M, Television – SDTI Content Package Format (SDTI-CP) (2000).

SMPTE 331M, Television – Element and Metadata Definitions for the SDTI-CP (2000).

SMPTE 336M, For Television – Data Encoding Protocol Using KLV (2001).

SMPTE 356M, Television – Type D10 Stream Specifications – MPEG2 4:2:2 P@ML for 525/60 and 625/50 (2000).

SMPTE 360M, Television – The General Exchange Format (2001).

SMPTE 367M, Television: Type D-11 Picture Compression and Data Stream Format (2002).
SMPTE 369M, Television: Type D-11 Data Stream and AES3 Data Mapping over SDTI (2002).
SMPTE 370M, Data Structure for DV Based Audio, Data and Compressed Video at 100 Mb/s 1080/60i, 1080/50i, 720/60p.
SMPTE 377M, Proposed Standard, Television – The Material Exchange Format, Format Specification (2003).
SMPTE 379M, Proposed Standard, Television – Material Exchange Format, The MXF Generic Container (2003).
SMPTE 380M, Proposed Standard, Television – Material Exchange Format, Descriptive Metadata Scheme 1 (2003).
SMPTE RP210, Metadata Dictionary Contents.

Dave Bancroft
Manager, Advanced Technology
Thomson Broadcast & Media Solutions

2.10 HDTV Standards

HDTV (High Definition Television) as a technology for representing moving images with much more convincing fidelity than SDTV (Standard Definition Television) has been more than three decades in the making. This period of time has seen information technology (IT) arrive and displace much of the unique and isolated engineering used to create the first electronic television systems in the first half of the twentieth century.

The effect has been to make HDTV technology a complex blend of legacy and IT constituents, as it has sought to achieve a comfortable introduction through compatibility with SDTV on key parameters such as frame rate, yet reap the advantages of the newer digital and IT-based approaches not available to the earlier SDTV pioneers. The standards that underpin HDTV technology reflect this blend, combining as they do both "computer-friendly" and apparently "computer-hostile" parameter values in their specifications. Many of these numeric values make no sense without some degree of archeological exploration. For example, some values will be found to derive from compromises made to achieve political agreement on international exchange between inherently incompatible television distribution systems. This chapter aims to assist and speed the digging process.

2.10.1 Conventions used in HDTV Standards Documents and this Chapter

2.10.1.1 60 Hz Temporal Frequency

Reference to "60 Hz" in the context of video temporal frequencies implies the 59.94 Hz frequency arising from application of the NTSC-derived 1/1.001 factor, instead of or as well as an exact 60.00 Hz. Exceptionally, 59.94 Hz will be mentioned where a distinction vis à vis 60.00 Hz is important. Other pairs of temporal frequencies occur: 23.98 Hz is paired with 24.00 Hz and 29.97 Hz is paired with 30.00 Hz. Where they occur, the non-integer values are shown rounded to two decimal places (using the fraction 1/1.001 gives the exact value).

2.10.1.2 74.25 MHz Sampling or Interface Clock Frequency

In the same way, reference to "74.25 MHz" also refers to 74.25/1.001 MHz (74.18 MHz).

2.10.1.3 Temporal Frequency versus Picture, Frame or Field Frequency

Where appropriate, temporal frequency is used as a more generic term than picture, frame or field frequency. Picture frequency is synonymous with frame frequency; the former term is preferred by IT professionals to avoid confusion with frames in data communication, HTML and other contexts; the latter is traditionally preferred by television engineers.

2.10.1.4 Hertz versus Frames/Second

Where there is a strong connection between an HDTV format and motion picture film as an image source, the latter may be referenced by "frames/second" rather than Hz, to emphasize its non-electrical nature.

2.10.1.5 Scanning Format Notation

Formats in the various standards are named in this chapter according to the most common usage that has evolved for them. This inevitably results in some inconsistency. Early HDTV formats were referenced, like SDTV formats, simply by the total number of scan lines per picture and the temporal frequency, e.g. 1125/60.

The notation of formats developed later has become less consistent; sometimes the number of active lines per picture and temporal frequency is quoted; sometimes the form (number of active pixels per scan line) × (number of active scan lines per picture) is used, often without reference to temporal frequency. At one time, the distinction between progressive and interlaced scanning was indicated by the use of "1:1" (progressive) versus "2:1" (interlaced), but recently this has become simplified to the use of a single letter "i" or "p" inserted in the notation, but without consistency as to placement or use of upper/lower case. *Poynton* has proposed a consistent notation that takes into account a number of other parameters, but it is likely that the habits of many years will prove difficult to displace.

Examples:

- "1080*p*24" or "1080/24P" means 1080 active scan lines per picture, and 24 progressively scanned pictures per second;
- "1280 × 720" or 720*p*60 means 1280 active pixels per scan line and 720 active scan lines per picture, without reference to temporal rate or scanning order;
- "1920 × 1080/60.00/2:1" means 1920 active pixels per scan line and 1080 active scan lines per picture, with a temporal rate of 60.00 Hz. The "2:1" means that the 1080 scan lines are interlaced 2:1 between two fields in each picture. Some opinions hold that the "60.00" in such a notation is incorrect and should be replaced by "30.00" because in this particular format *complete* pictures comprising *both* fields are occurring only 30 times per second, not 60.

2.10.1.6 The Term "Scanning"

The term "scanning" is used in this text because of its familiarity. However, it should be remembered that most image capture devices today do not really "scan" the image; they perform something closer to a "flash" exposure to it. The rate or order of exposure of each photo-site in an image sensor may then be different from that specified at the interface; many television cameras, for example, expose all pixels in the frame simultaneously and subject them to sequential ordering only at the output of the sensor chip or even later in the image processing stages. Compliance with the "scanning" standard therefore does not occur until the external interface.

2.10.1.7 "Luma" versus "Luminance"

This chapter follows the convention proposed by Poynton and gradually being adopted by the society of Motion Picture and Television Engineers (SMPTE) and the International Telecommunications Union – Radiocommunications Sector (ITU-R) of distinguishing carefully between true *luminance*, denoted *Y*, which is a linear light measure, and the symbol Y′, which in the past was incorrectly referred to as luminance and is frequently written without the necessary prime symbol. Y′ does not accurately represent luminance, because of its derivation by matrixing from non-linear (gamma-corrected) rather than linear (tristimulus) red, green and blue components. The preferred term for Y′ is *luma* and that is what is used here.

2.10.1.8 Abbreviation of Standards Designations

Where specific standards are referenced in this chapter, the nomenclature of their designations may be abbreviated. For example, SMPTE 274M may be referenced simply as 274M. In such cases, the context of the reference should indicate the full designation.

2.10.2 Scope of Chapter

This chapter catalogues existing published HDTV standards, but also explores the composition of a typical standard (SMPTE 274M) and breaks it down into its constituent parts, distinguishing between attributes such as sampling structure, image representation, timing issues and interface issues. The reason for this is that while these quite different areas are clearly separated and explicitly described in comparable documents in the IT world, for various reasons television standards including HDTV have traditionally bundled them together into monolithic documents. This analysis of the functional parts of an HDTV standard will be helpful for understanding and making comparisons as increasing convergence occurs between the IT and television disciplines.

But to begin, a short history of the development of today's HDTV standards will reveal why certain things are done in certain ways; sometimes there were good reasons that need some explanation; sometimes there were not such good reasons that deserve comments too.

2.10.3 Historical Background

2.10.3.1 First Era

When discussions on HDTV standardisation first started in the 1980's, they were characterized by the same assumption that had guided existing television standards (525-line and 625-line), namely, that the same scanning format should persist from the camera or telecine all the way through the production, distribution and transmission chain to the home receiver display. Since HDTV was defined as having a simple relationship to standard-definition television (twice the horizontal and twice the vertical resolution were commonly-quoted factors), HDTV standards in those days were largely based on taking an SDTV standard and multiplying up fundamental parameters such as the number of scan lines by the appropriate scaling factor.

A further characteristic of existing television standards was perpetuated: the various regions of the world differed in the scanning and colour encoding formats they had chosen. Expressed simply, there were 50 Hz regions and 60 Hz regions in SDTV, and this schism was repeated in HDTV despite some early efforts to seize the opportunity to replace the divided world of SDTV with a new world of HDTV based on a unified standard. Thus, formats such as 1050/60 and 1250/50 appeared in the U.S. and Europe, respectively.

In fact, the very first proposed HDTV format, from NHK in Japan, was neither of these; it was 1125/60. Although this appeared to be unrelated to either 525 or 625 scanning, it was deliberately chosen to have a common factor of 25 in the total line count. It was thought this would facilitate standards conversion between the new format and both 525 and 625 versions of existing SDTV and thus unify the television world, but the much more difficult issue of temporal conversion between 60 Hz and 50 Hz was not helped by this concept. However, the 1125-line form factor stuck and has remained the basis for one of the two major HDTV scanning families to this day.

All of these early HD formats inherited another characteristic from their SDTV predecessors: none of them was designed to have what are referred to today as "square sampling" (*or "square pixels," more accurately described as: "a spatial image sampling lattice with equal sample spacing in horizontal and vertical axes"*). "Square pixels" became important in the second era (see below), when the amenability of television signals to processing by computers started to become much more important. In the 1980's and into the early 1990's, however, most of the effort was concentrated on dealing with the problem of the transmission bandwidth required for HDTV, since digital compression was still in development. Bandwidth was an analogue quantity: the calculations were performed in the analogue domain, and the horizontal and vertical pixel and line counts that emerged from the format design process were thus arbitrary in the resulting "pixel squareness." For example, the European 1250/50 format (with 1152 lines active out of the total of 1250) had a pixel aspect ratio of 1152/1080, or 1.07:1, meaning that the pixels were slightly broader than tall. Conversely, the NHK 1125/60 format had a pixel aspect ratio of 1035/1080, or 0.96:1, meaning the pixels were slightly taller than broad. *(This discussion of pixel aspect ratio should not be confused with the issue of image aspect ratio, i.e. the ratio of image width to height; all HDTV standards today have a common aspect ratio of 16:9, giving a "widescreen" effect that contrasts distinctively with the 4:3 aspect ratio of SDTV. The NHK format mentioned above had an aspect ratio of 5:3; the efforts of the SMPTE's Working Group on High-Definition Electronic Production then led to the internationally agreed shift to 16:9 which was a better fit to motion picture film aspect ratios in common use. A given HDTV image format can be quickly examined to see if it has "square pixels:" if the ratio of its horizontal active pixels to its active scan lines is the same as its image aspect ratio, i.e. 16:9, then it has "square pixels.")*

Towards the end of this era, HDTV started to acquire the necessity for digital representation in studio standards, mirroring the developments that had been taking place in SDTV. SMPTE, after standardizing the 1125/60 format as 240M, prepared a digital version 260M. But it was a development in transmission technology that was to give digital studio standards a big boost.

2.10.3.2 Second Era

Accounts may differ as to what marked the onset of the "second Era" of HDTV, but two events in the United States were significant. The first was the introduction of the General Instrument "Digicypher" system in 1990, which while not the first proposal for compressed digital transmission in television, was certainly the first for HDTV. Analogue schemes were quickly displaced by the new method. The second major event was the coming together of the competing proponents for the U.S. HDTV standard, in the form of the "Grand Alliance" announcement in 1993. Not only did this embody the radical concept of compressed digital coding; it also introduced square sampling in at least one of the allowed video transmission formats, the 1920 × 1080 format.

In fact, the Advanced Television Systems Committee (ATSC), the body responsible for the formats and systems adopted for HDTV terrestrial transmission in the U.S., defined only the input of finished material to the compression encoder prior to transmission. The task of specifying the corresponding formats in studios, so that programme material could be captured and edited to feed the new ATSC transmission standards, was given to the SMPTE.

SMPTE had already standardized the original NHK 1125/60 analogue format as SMPTE 240M (with associated digital representation in SMPTE 260M). Since SMPTE 240M faithfully represented the associated transmission format, the studio image was in the format of 1920 × 1035. Since this was at odds with the new square sampling lattice requirement, a new studio standard had to be devised. The result was SMPTE 274M, which was first published in 1994 (274M will be covered in more detail later).

2.10.3.2.1 Role of ATSC T4, and CCIR/ITU-R

SMPTE's work at this time was focused on the requirements of North American HDTV production. At the same time, other bodies were liaising to try to preserve as much as possible of the original dream of HDTV: a single worldwide standard for this "new" television to replace the parochial formats of SDTV. The International Radio Consultative Committee (CCIR, later renamed ITU-R) had been very successful in achieving common international agreement on the digital signal representation standard for SDTV (the famous Recommendation 601), and many of the same experts were engaged to follow a similar path into HDTV standards for international programme exchange. Just as Rec. 601 had bridged the 50 Hz and 60 Hz regions with a common horizontal active pixel count of 720, and a common sampling frequency of 13.5 MHz, there was a desire to do something similar for HDTV, but possibly also to take the opportunity to unify more parameters than had been achieved in SDTV. This concept became known as the "Common Image Format" and became an important feature of the international Recommendation 709. This format established a common sample clock and interface clock frequency of 74.25 MHz, and a common spatial structure in horizontal and vertical sampling of 1920 × 1080, as well as a defined system colorimetry. Of course, at the global level, two versions of the so-called common image format persisted: a 50 Hz version and a 60 Hz version. Considered on a still-frame basis, however, it was indeed a common format in the spatial axes, and as such represented a significant step forward for international programme exchange. The clock frequency decision also represented the consensus that it was better to have a common frequency than to have the variations that had preceded it, namely 72 MHz in the European system and 54 MHz in some other proposals as well as the more widely-used 74.25 MHz. This had advantages when it came to serializing the digital signal for studio distribution (see later).

The role of the ATSC should be pointed out here. Beyond its core work of finalizing standards for North American HDTV, the ATSC was a very active partner to CCIR/ITU-R in the work on the Common Image Format, through its T4 committee.

2.10.3.2.2 The 720p/60 Alternative

The HDTV standards that were designed in this era took as one of their starting points the available bandwidth for terrestrial transmission. In North America, this was approximately 19 Mbits/second, based on the use of heavy compression. The corresponding uncompressed studio standards in the 1920 × 1080 family that provided the input to the compressor required about a gigabit per second (excluding interface overhead). Interfaces and recorders were designed for this level of bit rate, which reflected a particular combination of compromises in the various video parameters used in the design of the studio standards. For example, the use of interlace allowed a 1920 × 1080 spatial sampling format with a temporal rate of 60 fields/sec to fit into the studio bit rate budget, whereas 60 whole progressive frames per second at the same spatial resolution would have exceeded the budget, preventing use of the same interfaces and recorders. The use of progressive scanning at slower frame rates such as 24 Hertz would also solve the bit rate budget problem (see "Third Era," below), but another approach arose.

The new approach was to say that, while it was valid that a high temporal rate was needed to represent motion adequately in fast-moving action such as sports, it was unfortunate that the spatial resolution obtained with interlaced scanning on stationary objects was effectively halved when they were moving, this being the classic disadvantage of interlaced scanning. The alternative approach therefore was to declare that a somewhat lower static spatial resolution should be used, but one that would remain the same rather than halving when objects moved, by the use of progressive instead of interlaced scanning. The lower spatial resolution would then permit a full temporal rate of 60 frames/second while fitting the bit stream within the same gigabit or so of studio bit rate budget. The result was the format known as 720*p*60, which in full means 1280 × 720 pixels at 60 progressive frames per second. This was a descendent of the 787-line format (see **Legacy Formats**, below). A simple derivation of the numbers is that 1280 is two-thirds of 1920 and 720 is two-thirds of 1080. Two-thirds squared is 4/9, i.e. just under a half, which compensates for the bit rate doubling factor of progressive versus interlaced.

The 1280 × 720 image fits into a 750-line raster and uses the same sample and interface clock rate as the 1920 × 1080 family, so is compatible with the same interfaces. The standard is in use today at a major U.S. network.

2.10.3.3 Third Era

The third era can be characterized by the concept of "source-adaptive encoding." The Grand Alliance proposal of 1993 allowed eighteen video scanning formats. Significant freedom was available in the temporal rate parameter, with both non-traditional television frame rates such as 24 frames/second being permitted as well as the familiar 30 frames/second (*usually interlaced into 60 fields*). Much of the content for television originates on film, which is normally shot at 24 frames/second, not 30 frames/second. The normal method of accommodating this discrepancy in 60 Hz television broadcasting is the established "2/3 pulldown" method utilizing repeated fields. It would become wasteful of compression efficiency to deliver such a signal to a digital encoder, so instead the repeated fields are removed before compression to allow the encoder to run at the frame rate of the original source, hence the expression "source-adaptive encoding." For comfortable viewing, the repeat fields are then inserted by the receiver's decoder instead.

It then soon became apparent that it was also wasteful to carry the repeated fields through the studio production infrastructure if they were only to be discarded at the compressor. Thus, there was a need for temporal rate variants in the studio standard, not just in the transmission standard, to accommodate the different native source temporal rates.

This led to a request to SMPTE to expand the scope of the 274 M studio standard from its original "60 Hz only" basis to add 24 Hz and other temporal rates. This expansion was started in 1995 and completed in 1998 with the first multi-frame-rate version of SMPTE 274M. The original four temporal rate options had to be increased finally to a total of eleven, to accommodate 1/1.001 variants, representation of other film frame rates (25 fps and 30 fps), and a 50 Hz broadcast temporal rate to make the standard truly international in application.

This created a dilemma: the common image format (1920 × 1080) concept had become firmly established as a desirable attribute, but what could be done about the sample/interface clock rate with the big frequency range of temporal rates now to be accommodated? One opinion held that not only was it important to maintain the common image format in the active pixels, it was also important to preserve the *total* number of sample points in the format constant, for the sake of interface hardware. This could be done by allowing the clock rate to rise or fall according to temporal frequency, thereby keeping the total samples per line constant at 2200. Thus, the clock rate would range from the maximum of 74.25 MHz at 30 Hz temporal frequencies down to as low as 59.40 MHz at 24 Hz.

This scheme, however, could have caused problems for the HDTV serial interface, where the receiver phase lock loop would have had to swing over a proportionate frequency range when performing clock recovery instead of locking only to a fixed (or almost fixed) frequency of 1.485 GHz. An alternative proposal therefore was to keep the clock frequency constant by allowing a variable number of vertical blanking lines, the number increasing as the frame period lengthened (temporal rate falling). This idea was discarded on the grounds of excessive complexity.

The scheme that was finally adopted was to maintain the clock rate constant for all temporal frequencies by padding out the longer scan lines of the slower temporal rate variants with extra inactive samples following the 1920 active samples (Figure 2.10.1). In analogue argot, this would be called "extending the front porch*." For example, at 60 Hz interlaced operation, the duration of one total scan line is 29.63 μS. At a clock frequency of 74.25 MHz, this holds 2200 sampling periods, i.e. in addition to the 1920 active image samples specified in the common image format there are 280 inactive samples per line. Compare this with 24 Hz progressive scan operation: the scan line duration is now 37.04 μS. At the same 74.25 MHz clock frequency, the number of sample periods in this time is 2750, i.e. 1920 active image samples plus 830 inactive samples. 25 Hz progressive scanning gives a total of 2640 samples per line (1920 active and 720 inactive).

This might seem to be a wasteful scheme when operated at the slower temporal frequencies, but the advantages of the common clock frequency outweigh the disadvantages in the studio and production infrastructure. Even where a bandwidth saving is important, e.g. in recording operations, this can be achieved by simply discarding some or all of the inactive samples; this works just as well as using a slower clock frequency.

2.10.3.3.1 Extension to 720p

After completing the work on 274M, SMPTE then applied the same frame rate extensions to 296M, the 720*p* standard. Despite an even greater proportion of total line time being allocated to inactive samples at the slower frame rates, the

*"Front porch" and "back porch" are analogue terms describing certain portions of the waveform of one horizontal line of video. The "front porch" is the part of the horizontal blanking between the end of active video and the start of the synchronizing pulse, and the "back porch" is the part between the end of the synchronizing pulse and the start of active video. The terms derive from the appearance of the waveforms on a waveform monitor.

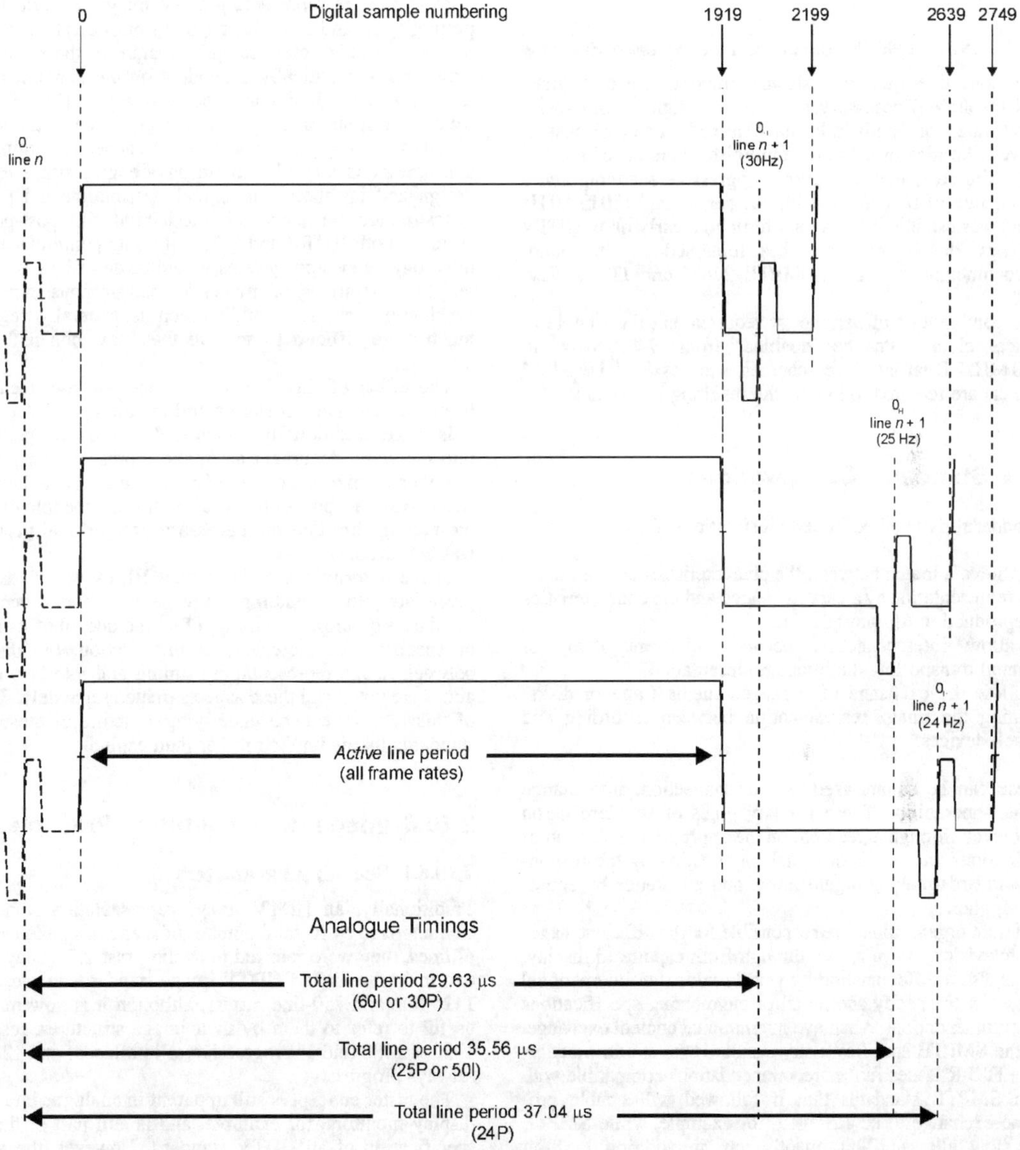

Figure 2.10.1 Variation in total number of sample periods per line with temporal frequency (constant clock rate principle).

NOTE 1: 0_H is the horizontal timing point for aligning digital sampling to the analogue waveform.
NOTE 2: Analogue timings are measured from 0_H, but the numbering of digital samples in each line begins with the first sample representing the active portion of the line.
NOTE 3: The sampling points indicated by vertical arrows are the time instants representing the start of each sample period. For example, the active line period of 1920 sample periods begins with the start of sample number 0 aligned with the mid-point of the *leading* edge of the analogue active line period and ends with the start of sample number 1919 aligned with the mid-point of the *trailing* edge of the active line period. This means the digital sampling of this analogue period of time does not end until one sample period *after* the trailing edge. However, the timing instant representing the start of sample 2199, 2639 or 2749, i.e. the last sample of the whole line, is located one sample period *before* the end of the line, so that the *end* of this last sample aligns with the end of the line.

constant clock principle was maintained here, too. More details are given in the 296M standard description later.

2.10.3.3.2 1920 × 1080 60 Frames/Second Progressive Scanning

The compromises (interlace, slower temporal rate or reduced spatial resolution) necessary to fit HDTV signals into specified bit rates or bandwidths have already been discussed. However, the idea of a "no-compromise" version of HDTV, with maximum spatial resolution, progressive scanning and a temporal rate of 60 complete frames per second (50 in 50 Hz regions) was established as a long-term goal early on in HDTV technology development and has remained in the standards documentation, e.g. in SMPTE 274M and ITU-R Rec. BT.709.

This combination of parameters requires that the sample or interface clock rate be doubled from 74.25 MHz to 148.50 MHz. Dual interface schemes, such as SMPTE 372M (see later) are required to handle the resulting bit streams.

2.10.4 Standards Organisations

A "standard" is required in certain situations, for example:

- to achieve a match between the characteristics applied to an image representation by a capture device and the characteristics of a reproduction (display) device;
- to allow interconnection between different items of equipment transporting the image representation;
- to allow the exchange of packaged media (tape or disks) containing the image representation between recording and playback devices.

These can be summarized as interconnection, interchange and inter-operability. These primary goals of standardisation are achieved through agreement on the appropriate selection of specifications and provisions, their publication by the responsible standards-making organisation, and adherence by equipment suppliers.

Different organisations are responsible for the different stages that a television image passes through from capture to display. There is also a differentiation by world region, but international bodies exist to specify additional or augmenting specifications and recommendations to achieve international content exchange. Thus the SMPTE specifies many standards for studio practice, but the ITU-R issues further recommendations compatible with certain SMPTE standards that if followed will enable programme exchange to be achieved (for example, while SMPTE 274M now allows 12-bit quantisation, in addition to 8-bit and 10-bit, 12-bit operation is not included in ITU-R Rec. 709). Conversely, SMPTE image standards utilise international image sampling and colorimetry specifications developed by the ITU-R.

Once an item of HDTV content has been created (e.g. to SMPTE standards) and exchanged (ITU-R), it is transmitted according to standards that may be more specific to the locality. In the U.S. and Canada, the specification for the input to the terrestrial transmission encoder/compressor has been established by the ATSC; in Japan, the equivalent body is ARIB. In the first 50 Hz region to transmit HDTV — Australia — transmission specifications derived from the SDTV equivalents published by DVB/ETSI are in use.

2.10.5 General Principles

As a method of representing scene images electrically, transporting them and reproducing them on a display, television's reliance on the discrete sampling nature of the scanning principle makes it critically dependent on the maintenance of an exact timebase: individual spatial samples ("pixels") taken from the original image must be reproduced at the display in exactly the same relative positions. In general, "timing" issues consume a considerable proportion of engineering efforts in the design and installation of signal distribution and processing infrastructure for television production and post-production systems, both HDTV and SDTV. In data communications terminology, television systems would be described as "handling real time streaming of images via isochronous transport in a synchronous environment," which is generally regarded as much more difficult to achieve than, say, non real time file transfers.

The effect of this dependency on timebase accuracy has been that television scanning and image representation standards have traditionally "bundled" interface specifications into the same document as image representation parameters (number of pixels, colour primaries, etc.). This has ensured successful transport of the images through the television plant by making sure that the necessary time-critical performance was achieved.

In the descriptions of individual HDTV standards that are given later, this "bundling" is respected where it occurs; each standard's description will therefore include all of its key areas of specification. However, in this section, the distinctions between image representation, timing and interface considerations are noted and these aspects treated separately. The intent of this is to reveal the underlying structure of these bundled standards that is implicit rather than explicit.

2.10.6 Image Representation Principles

2.10.6.1 Scanning Parameters

Traditionally, an HDTV image representation standard was characterized by its total number of scan lines; taken as a group of lines, they were referred to as the "raster." Today there are two main families of HDTV image standards in use, based on 1125-line and 750-line rasters, although it is now much more useful to refer to them by their image structures, respectively 1920 × 1080 (either progressive or interlaced) and 1280 × 720 (always progressive).

The raster concept is still apparent in analogue interfaces for display monitors, for example, and is still part of the general specification of an HDTV standard. However, the advent of almost-total digitization of the television production chain has made it more important to stress the number of *active* lines within the raster and this is reflected in the way standards are now written. For example, the SMPTE 274M HDTV scanning standard refers to 1080 active lines rather than 1125 total lines in most of its specification, and likewise the SMPTE 296M standard refers to 720 active lines rather than 750 total. Another shift in the nomenclature has been caused by the change from tube to solid state sensing (e.g. CCDs) in cameras. In tube cameras, it was considered that only the vertical scanning was discrete, the horizontal scanning being considered continuous; but in solid-state sensor cameras, the horizontal scanning is also discrete. Thus, standards now call out the number of active pixels in the horizontal axis as a characterizing parameter, as well as the number of active scanning lines.

Each standard then has to specify how the remaining space in the raster is used, i.e. define horizontal and vertical blanking intervals. Interlaced formats must also declare which active lines fall into the first field of the image frame and which into the second field. This may seem strange; in the analogue world, most interlaced rasters were designed with an odd number of total lines (active plus inactive) in the frame, achieving interlace automatically via the half-line which then occurs prior to the vertical retrace of one of the two fields. The digital world, however, cannot handle half-lines, so an arbitrary but consistent definition must be made as to which of the two interlaced fields contains one more line than the other. A timing relationship between the analogue and digital representations of the same image must also be defined. This is done by referencing the two forms to a common timing point "0_H," shown in Figure 2.10.1.

2.10.6.2 Temporal Rate

The raster structure and the image samples within it create the electronic representation of a single frame. As in other image reproduction systems, motion is then conveyed by exploiting the characteristics of human vision through the repeated application of this structure in time, i.e. a series of single frames is captured and displayed in rapid succession.

HDTV standards in use today cover frame rates from 24 frames/second to 60 frames/second, often with a choice of several offered within the same document.

While there is no disagreement on the temporal rate of a progressive scan system, there are different opinions concerning interlaced scan systems. For example, when HDTV images are transmitted with the 1080 lines/60 fields variant of SMPTE 274M, is the temporal rate 60 Hz because updates of large areas of the image are occurring 60 times a second, or is it only 30 Hz because it takes two fields (the whole frame) for *all* the image information, including small areas, to be updated? There is disagreement on this issue. However, all HDTV standards documents make it clear what the frame rate and field rate are in each case, and the reader can then choose which nomenclature is preferred.

2.10.6.3 Image Coding and System Colorimetry

The previous section dealt with the establishment of an ordered array of image sampling points, repeated at a specified temporal rate. This section now deals with the coding of image values at each of these sampling points ("coding" embraces both analogue representation and digital coding).

Just as it is not possible to specify an infinite number of spatial sampling points because of bandwidth limitations, it is also necessary to limit the extremes of image representation values, such as luminance. All HDTV standards assume colour reproduction via three primary components. For any fixed set of primaries, there is a limit to the range of colours (the colour gamut) as well as luminance values, that can be represented by the chosen number of coding bits per sample. At the same time, all television systems, including HDTV, apply dynamic range companding by imposing a non-linear transfer characteristic between linear light in and signal value encoded. This is commonly referred to as "gamma correction," although it should be understood that the mechanism does far more than that, being fortuitously well optimised to perception (*Poynton*).

Taken together, the expression of colour primaries, the colour chosen for reference white (the "white point") and the coefficients used to form the luma (Y′) signal are referred to as "system colorimetry."

2.10.6.3.1 Colour Encoding

Fortunately HDTV standards are simplified by the absence of any subcarrier-based chroma/luma interleaving schemes equivalent to PAL, NTSC or SECAM; the three colour primary values or their derived luma/colour difference components remain separate signals in HDTV.

HDTV standards specify the chromaticities (see Figure 2.10.2 in 274M section) of the red, green and blue primaries used for colour representation at the display and the chromaticity of the white colour obtained with these primaries when they are set to be equal and of unity value. Aiming for these *interchange primaries* (formerly called transmission or display primaries) allows equipment designers to optimise spectral analysis curves for camera sensors and coefficients for associated encoding matrices such that images from a variety of capture products will achieve the intended colour reproduction. (*The primaries are called interchange primaries rather than display primaries today, because instead of implementing the interchange primaries directly, a particular display may incorporate its own signal processing that maps the interchange signals to a different set of signals optimised to drive the particular physical display primaries of the device. This is becoming important as display technologies based on a wide range of physical methods are rapidly eroding the former dominance of the colour cathode ray tube; in contrast to SDTV, HDTV standards were designed in anticipation of this trend.*)

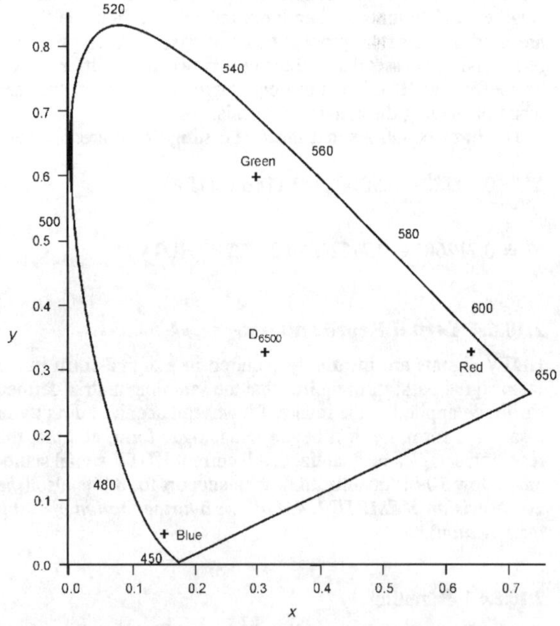

Figure 2.10.2 Rec. 709 Primaries and White Point on CIE 1931 Chromaticity Diagram.

At this stage, only trichromatic stimulus values have been computed; it is necessary to specify their transformation (via the non-linear transfer function mentioned earlier) into real R′, G′, B′ signals. This transfer function applies a fractional power law to the linear signals that can be thought of as a "black-stretching" function. However, such a law, if applied to the whole of the signal amplitude range, would imply infinite gain at black level, with associated impairment of signal-to-noise ratio. HDTV standards, like other video standards, therefore specify a *linear segment* close to black level before the power law takes over at a specified *break point*. The power law portion, the linear portion, and the break point are specified within the body of the standard. The correct nomenclature for the resulting non-linear signals resulting is R′, G′ and B′, the prime mark denoting non-linearity (see Figure 2.10.2).

It is necessary in all HDTV imaging standards to define R′, G′ and B′ components in this way; optionally they can then be utilised directly, operating in so-called "RGB" or "4:4:4" mode, or transformed further, as described next.

2.10.6.3.2 Luma and Colour Difference Components

The image representation portions of HDTV standards permit direct R′G′B′ operation, but it is more common to perform the additional step of converting these signals into *luma* and *colour difference* signals first, since this allows the opportunity to reduce the bandwidth/bit rate requirement at studio interfaces by 33%. Matrix equations specify how the luma (Y′) and colour difference signals (C'_R/C'_B or P'_R/P'_B) are derived from the non-linear R′, G′ and B′ signals. In all current HDTV systems, these equations use different coefficients from those used in SDTV standards. This is of concern when downconversions and upconversions are made between HDTV and SDTV signals, either in the studio or in consumer products. Although colour errors in HDTV and SDTV conversions using R′, G′ and B′ signals are small enough to ignore, when luma and colour difference signals are used a non-trivial process of de-matrixing, resampling and re-matrixing is essential to transform between the different sets of coefficients. If this is not done, large colour errors are seen when displaying the converted signals.

The luma equations that must be distinguished are:

$$Y' = 0.299R' + 0.587G' + 0.114B'(\text{SDTV})$$

$$Y' = 0.2126R' + 0.7152G' + 0.0722B'(\text{HDTV})$$

2.10.6.4 Digital Representation

HDTV signals are invariably required in a digital form in the main signal path; this requires that the sampling matrix defined earlier be applied to the image if it was not acquired directly in a sampled form, i.e. if it began in analogue form, and that the resulting samples be quantized. All current HDTV signal standards allow 10-bit quantization, with support for 8 bits also (*the 2003 revision of SMPTE 274M allows a further option of 12-bit quantization*).

2.10.6.4.1 Sampling

The application of the sampling lattice is via a sampling clock, usually at a frequency of 74.25 MHz. Appropriate pre-filtering must be applied to prevent aliasing becoming apparent during image reconstruction. Advice on this is usually provided in the standard.

2.10.6.4.2 Quantization

In computer graphics, the whole of the available digital coding range (typically codes 0 through 255 in an 8-bit system) is used to represent the reference black to white range of each colour component. Television does not use all of the available coding range to span the reference range of each of the component signals. Instead, small reserves are left near the extreme bottom ("footroom") and top ("headroom") of the range to prevent clipping of overshoots caused by filtering or blanking application. The very bottom and top 1 or 4 (*for 8-bit and 10-bit quantization, respectively*) digital codes are further set aside for synchronisation and other purposes (see below) so are also out of bounds to image data. See Figure 2.10.7 for the allocation of the total coding range to these various functions.

2.10.6.5 Synchronisation and Timing

2.10.6.5.1 Analog

Just like their SDTV counterparts, HDTV signals in analogue form employ horizontal and vertical sync pulses to establish timing points for displays and other devices. However, HDTV uses tri-level rather than bi-level sync waveforms. (Figures 2.10.8 and 2.10.9).

2.10.6.5.2 Digital

"Digital sync pulses" along with "digital blanking," take the form of special digital codewords that are prohibited in image information. These sequences are generically referred to as "timing reference sequences" (TRS). Specific sequences designate "Start of Active Video" (SAV) and "End of Active Video" (EAV), which could be considered the equivalent of horizontal blanking. Combinations of special bits indicate the extent of vertical blanking and identify each of the two fields in an interlaced system. See Table 2.10.2 and Figure 2.10.5 for the details of these sequences in an example standard.

2.10.7 Transport and Interfaces

Image samples always used to be transported in the order in which they were captured. It was shown earlier that in recent years this is not necessarily the case where solid state image sensors are involved. However, there is another type of re-ordering which can occur and this has come about because of the addition in recent years of progressive scan formats to an industry that hitherto was based exclusively on interlaced scanning.

When new formats such as progressive scan are added to existing infrastructures, the changes can be inconvenient in areas such as interfaces and recording. For example, a 25 frames/second progressive scan video signal would have a frame period twice as long as the field period of a 50 fields/second interlaced scan signal, all other parameters being kept the same. This would theoretically require major changes to VTR formats and interface design.

However, a small modification solves the problem very easily. All that is required is for the scan lines of the progressive scan signal to be reordered into two parts or segments per frame, with the odd numbered lines being assigned to the first segment and the even-numbered lines to the second segment. Perhaps this sounds remarkably like the recipe for an interlaced signal and that is the whole point: VTRs and interfaces are able to treat this re-ordered progressive scan signal exactly as if it were indeed an interlaced signal, so do not require any modification from their orignal form. All that is necessary is for the system to "know" that the signal has origins different from those of a natively-interlaced signal—it began life as a progressive scan signal (* *and therefore has the same motion phase in the two segments, not different motion phases as in the interlaced signal*)—and is most likely to be restored to that form after leaving the VTR or at some point downstream from the interface, probably for display purposes. (It is of course essential to avoid any vertical or spatial processing while the signal is in this form, any such operations requiring a temporary restoration to the native progressive form). This option of re-ordering the scan lines of a progressive signal is known, not surprisingly, as segmented scanning, or, more formally, "progressive segmented frame" (PsF), and is documented in the 2003 version of the SMPTE 274M standard and in ITU-R Recommendation 709.

2.10.7.1 Analogue Interfaces

Although the main signal flow path in HDTV systems is now invariably digital, analogue interfaces are still required for other connections, mostly to picture monitors. These interfaces are component, therefore triple connections are required.

For R′G′B′ interfaces, signal levels will already be nominally equal (normally 1 volt peak-to-peak, including sync), so the interface is quite straightforward. Sync pulses are specified to be normally added to all three signals – this is standardised in HDTV to try to avoid repeating the chaos that occurred with SDTV component analogue interfaces. For luma and colour difference signals, however, differential weighting of the components is required to produce nominally-equal voltage excursions on the interface. In the SMPTE RP160 standards, the B-Y signal is weighted to produce an interface component P'_B:

$$P'_B = \frac{0.5}{1 - 0.0722}(B' - Y')$$

and the R-Y signal is weighted to produce component P'_R:

$$P'_R = \frac{0.5}{1 - 0.2/26}(R' - Y')$$

Note that the scaling factors necessary to achieve identical (unity) excursion for these two components are different. The result is to scale them to an analogue range of ±350 mV.

The Y′ signal is:

$$Y' = 0.2126\,R' + 0.7152\,G' + 0.0722\,B'$$

(no weighting required)

Synchronisation is provided by addition of the appropriate sync waveforms; unlike SDTV practice, sync pulses are added to all components.

2.10.7.2 Parallel Digital Interface

Having three signals to interconnect is awkward in analogue and produces further issues in digital. A small economy is possible in that the digital domain allows the two half-sample-clock-frequency colour difference signals to be multiplexed together making two components in total rather than three to send to the interface. However, that still leaves the 8 or 10 codeword bits of each component as separate signals to be sent over a parallel interface. Parallel interfaces were defined as part of the documentation of several HDTV scanning standards, although recent practice has been to move them to annexes in the documents to emphasise their separate function from that of the scanning and signal definition part of the standard.

2.10.7.3 Serial Digital Interface

At HDTV clock frequencies the distance over which interconnection over a parallel digital interface remains reliable is extremely limited, primarily because differential bit skew soon becomes a problem. However, the fact that the signals are digital provides the opportunity to multiplex them into a more convenient form; this is the serial interface, requiring only a single coaxial or fibre cable connection, and it has taken over almost entirely from the parallel interface. The primary standard document for the HDTV serial interface is SMPTE 292M.

Because of the strict timing requirements of video signals mentioned earlier, the serial interface has to specify which input signal formats are allowed.

2.10.7.4 Multiple Serial Interfaces

A single serial interface as typified by SMPTE 292M has a bit rate capacity sufficient only for the $Y'C'_BC'_R$ (4:2:2) signal format; R′G′B′ operation, for example, would exceed this bit rate.

Some installations therefore employ a pair of SMPTE 292M interfaces to double the bit rate and thereby accommodate serialized R′G′B′ signals. This was initially an ad hoc arrangement but a standardised method ("dual link") has now been documented by SMPTE in standard 372M (see later description).

Some applications use the additional bit rate for other purposes. For example, when transferring film-derived images which convey the full-height camera aspect ratio of 4:3 rather than 16:9 or other widescreen aspect ratio, the extra height requires extra scan lines and therefore extra bits in the interface. However, it should be emphasised that no standards body has yet formally documented the packaging for any of these schemes.

2.10.8 Specific Standards – Legacy

2.10.8.1 787 lines/60P Scanning

This was an early entrant from Zenith in the run-up to the Grand Alliance announcement. Its distinctive feature was that it offered 60 full progressive frames per second and as such was unbeaten for motion rendition of fast-moving action such as sports events. The reduction in spatial resolution from that of the 1080 line systems was a necessary trade-off to get the signal to fit into the same transmission bandwidth.

The benefits of the format were realised and eventually Zenith agreed to a modification suggested by the ABC Television Network, which was to change the total line count to 750 and to fit 720 active lines within that. This created the

"720*p*60" standard – see below – making the 787-line format obsolete.

2.10.8.2 SMPTE 240M and 260M

SMPTE 240M and 260M are the studio analogue scanning document and digital representation respectively for the original 1125 lines (total) system that came from NHK in Japan. Consequently they reflect that system in having 1035 active lines (non-square sampling) per frame, not 1080. SMPTE 240M also has slightly different colorimetry from the standard that superseded it, SMPTE 274M.

These standards are however somewhat more copious in their descriptive text than later documents so are a valuable source of information to students of HDTV technology, when combined with the updates provided by later standards such as 274M.

The parameter values of 240M/260M remain documented in part 1 of ITU-R Recommendation BT.709 (see below) on a legacy support basis, but with the recommendation that new programme production should use the later Common Image Format values shown in part 2 of the same document.

2.10.8.3 1050/60i

This was based on the simple principle of "doubling-up" 525/60 television's scan lines to make a 1050-line raster. It had non-square sampling and a non-standard clock rate. It is now obsolete.

2.10.8.4 EU 95 1250/50i

This standard was also based on the "doubling-up" principle, in this case based on the 625/50 raster. Thus, it had 1152 active lines (twice the 576 active lines of 625/50) and hence non-square sampling. 1250/50i was intended to be the studio format feeding Eureka HD-MAC (High Definition Multiplexed Analogue Components), the transmission scheme chosen for the unsuccessful attempt to establish European HDTV broadcasting in the early 1990s.

In its digital representation, 1250/50i had a sample/interface clock frequency of 72 MHz, creating an incompatibility with the 74.25 MHz clock standardised by SMPTE and ITU-R. 1250/50i also used luma coefficients matched to SDTV practice rather than the newer values of Rec. 709. This did however make it more colour-compatible than Rec. 709 in downconversions to 625/50 and 525/60 video standards, requiring no luma/chroma component re-matrixing as part of the process.

Like the 1125/60 (1035 active lines) format, this format remains documented on a legacy basis in ITU-R Rec. BT.709 part 1, but otherwise should be considered an obsolete format.

2.10.8.5 SMPTE 295M 1920×1080 50 Hz – Scanning and Interfaces

The notion of creating an HDTV standard that had a simple relationship to a local SDTV standard persisted for a time. SMPTE 295M was perhaps the last fling of this idea. Instead of the 1125-line raster that was growing in popularity, it retained the 1250-line raster of the earlier EU 95 format (see above), but fitted within it the 1080 active line Common Image Format in place of 1152 active lines, thus achieving square sampling.

Confusion with the later 274M standard (see below) sometimes occurs because the title of this document does not make it clear that the total number of lines is 1250, not 1125.

There is no evidence of this format ever being implemented; the 1125-line raster was adopted even in 625/50 countries such as Australia. SMPTE 295M can therefore be considered obsolete.

2.10.9 Specific Standards – Current

2.10.9.1 Current Image Standards (some with bundled interface definitions)

2.10.9.1.1 SMPTE 274M 1920 × 1080 Image Sample Structure, Digital Representation and Digital Timing Reference Sequences for Multiple Picture Rates

(from "Image Structure" onwards, the subheadings of this section correspond in name to those of the SMPTE 274M document)

SMPTE 274M can be thought of as the "square sampling version" of 240M/260M, because the number of active lines was changed from 1035 to 1080, while continuing to fit them into the same 1125-line raster.

As described earlier, SMPTE 274M was also adapted to the multiple temporal rate requirement of representing content from a variety of sources (notably scanned film) that has different frame rates from the normal transmission frame rate. SMPTE 274M therefore allows a total of eleven temporal rate options, referred to in the document as "Systems," as shown in Table 2.10.1.

Note that with the exception of Systems 1 – 3, these different temporal rates are accommodated while keeping the nominal sampling frequency constant; while the number of **active** luma samples per line remains constant (*luma is cited, because in most cases, colour difference signals are each conveyed at half the luma rate on the interface*), the number of **total** samples per line shows a big variation with temporal rate.

The total number of temporal rate options becomes large in 274M because of the variants that have to be covered:

- Systems 1, 2 and 3 are the "double clock rate" versions, allowing 50 or 60 full progressive frames per second instead of 50 or 60 interlaced fields per second.
- Systems 2, 5, 8 and 11 are the same as Systems 1, 4, 7, and 10, except that the temporal rate has been reduced by a factor of 1/1.001 (see earlier note), with a proportional down shift in the interface sampling frequency.

Some of these systems may appear confusingly similar to one another; the reader should take care to note that, as an example, while Systems 4 and 7 are both shown as having a "Frame Rate" of 30 Hz, System 4 interlaces each frame into two fields, while System 7 keeps it as a single progressive frame. This is reflected in the nomenclature column. Other pairs of systems will be found in the table with this kind of relationship.

2.10.9.1.1.1 Image Structure (called "Scanning" in pre-2003 versions of 274M)

This section defines how the active lines of image samples (pixels) are mapped into the total 1125 line raster in relation

Table 2.10.1 Temporal rate options (reproduced from SMPTE 274M)

System no.	*System Nomenclature*	*Luma or R′G′B′ Samples per active line (S/AL)*	*Active lines per frame (AL/F)*	*Frame rate (Hz)*	*Interface sampling frequency f_s (MHz)*	*Luma Sample periods per total line (S/TL)*	*Total lines per frame*
1	1920 × 1080/60/P	1920	1080	60	148.5	2200	1125
2	1920 × 1080/59.94/P	1920	1080	$\frac{60}{1.001}$	$\frac{148.5}{1.001}$	2200	1125
3	1920 × 1080/50/P	1920	1080	50	148.5	2640	1125
4	1920 × 1080/60/I	1920	1080	30	74.25	2200	1125
5	1920 × 1080/59.94/I	1920	1080	$\frac{30}{1.001}$	$\frac{74.25}{1.001}$	2200	1125
6	1920 × 1080/50/I	1920	1080	25	74.25	2640	1125
7	1920 × 1080/30/P	1920	1080	30	74.25	2200	1125
8	1920 × 1080/29.97/P	1920	1080	$\frac{30}{1.001}$	$\frac{74.25}{1.001}$	2200	1125
9	1920 × 1080/25/P	1920	1080	25	74.25	2640	1125
10	1920 × 1080/24/P	1920	1080	24	74.25	2750	1125
11	1920 × 1080/23.98/P	1920	1080	$\frac{24}{1.001}$	$\frac{74.25}{1.001}$	2750	1125

to vertical blanking. 274M allows both progressive scanning and interlaced scanning structures. The latest revision (2003) also allows segmented frame operation. This has sometimes been referred to incorrectly as segmented scanning, when it is actually an interface mode. For this reason, its definition in 274M is in an Annex, rather than in this section.

2.10.9.1.1.2 System Colorimetry

SMPTE 274M uses ITU-R Rec. BT.709 colorimetry to define the relationship between red, green and blue linear tristimulus values and the chromaticities they produce in the CIE x,y chromaticity diagram, i.e. these are the *interchange primaries*. The *white point* is also defined by linking equal values in the three primaries to the production of a white illuminant of specified colour temperature (CIE D_{65}).

Rec. 709 defines these values as follows:

	x	*y*
Red	0.640	0.330
Green	0.300	0.600
Blue	0.150	0.060
White Point	0.3127	0.3290

The next specification is how the non-linear signals R′, G′ and B′ are produced from the linear tristimulus values, using the Rec. 709 equation:

$$V' = \begin{Bmatrix} 4.5L, & 0 \le L < 0.018 \\ 1.099L^{0.45} - 0.099, & 0.018 \le L \le 1 \end{Bmatrix}$$

This equation means:

- a linear tristimulus value L is the input, ranging from 0 to 1 in value and representing linear R, G, or B
- the non-linear value V′ is the output primary signal, ranging from 0 to 1 in value and representing non-linear R′, G′ or B′
- when the input is 0 or more but less than 0.018, the output is 4.5 times the input, i.e. this is the linear portion with a slope of 4.5
- when the input is 0.018 or greater, a power law of 0.45 is applied, with scaling and offset applied so as to achieve tangent continuity at the breakpoint.

This is illustrated in Figure 2.10.3:

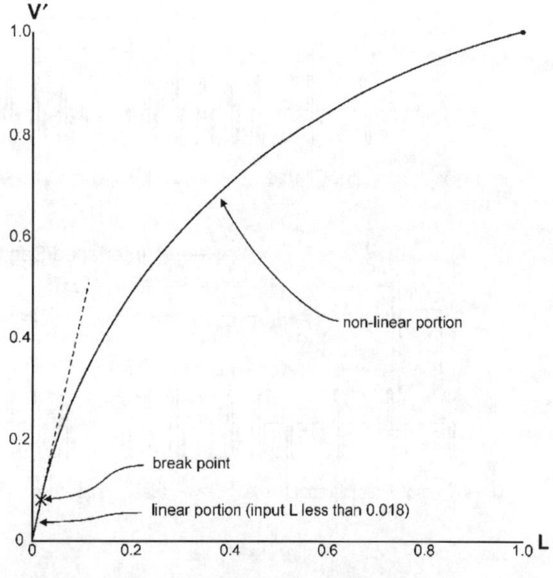

Figure 2.10.3 Rec. 709 Non-Linear Transfer Characteristic.

Next, this section specifies how to make the luma signal Y′ from the non-linear R′G′B′ components. SMPTE Recommended Practice RP177 is cited and an equation based on its principles is used:

$$Y' = 0.2126R' + 0.7152G' + 0.0722B'$$

It is very important to note that the proportions of R′, G′ or B′ used in this equation in 274M and in all other current HDTV image standards are quite different from those—perhaps more familiar—values used in SDTV (i.e. 0.299 R′ + 0.587 G′ + 0.114 B′). Note that the sum of the coefficients is unity. At the analogue interface, Y′ ranges from 0 to 700 mV.

The Y′ signal is one of the three components used when representing the image in "4:2:2" mode, the others being the colour difference signals B-Y and R-Y. The derivation of these is not explicitly shown in 274M, but they are easily calculated by substituting the equation for Y′:

$$B'\text{-}Y' = B' - (0.2126R' + 0.7152G' + 0.0722B') = 0.9278B' - 0.2126R' - 0.7152G'$$

$$R'\text{-}Y' = R' - (0.2126R' + 0.7152G' + 0.0722B') = 0.7874R' - 0.7152G' - 0.0722B'$$

These signals will be different from one another in reference excursion and both will be larger than the Y′ signal, so they are scaled to make colour difference signals P'_B and P'_R:

$$P'_B = \frac{0.5}{1 - 0.0722}(B' - Y')$$

$$P'_R = \frac{0.5}{1 - 0.2126}(R' - Y')$$

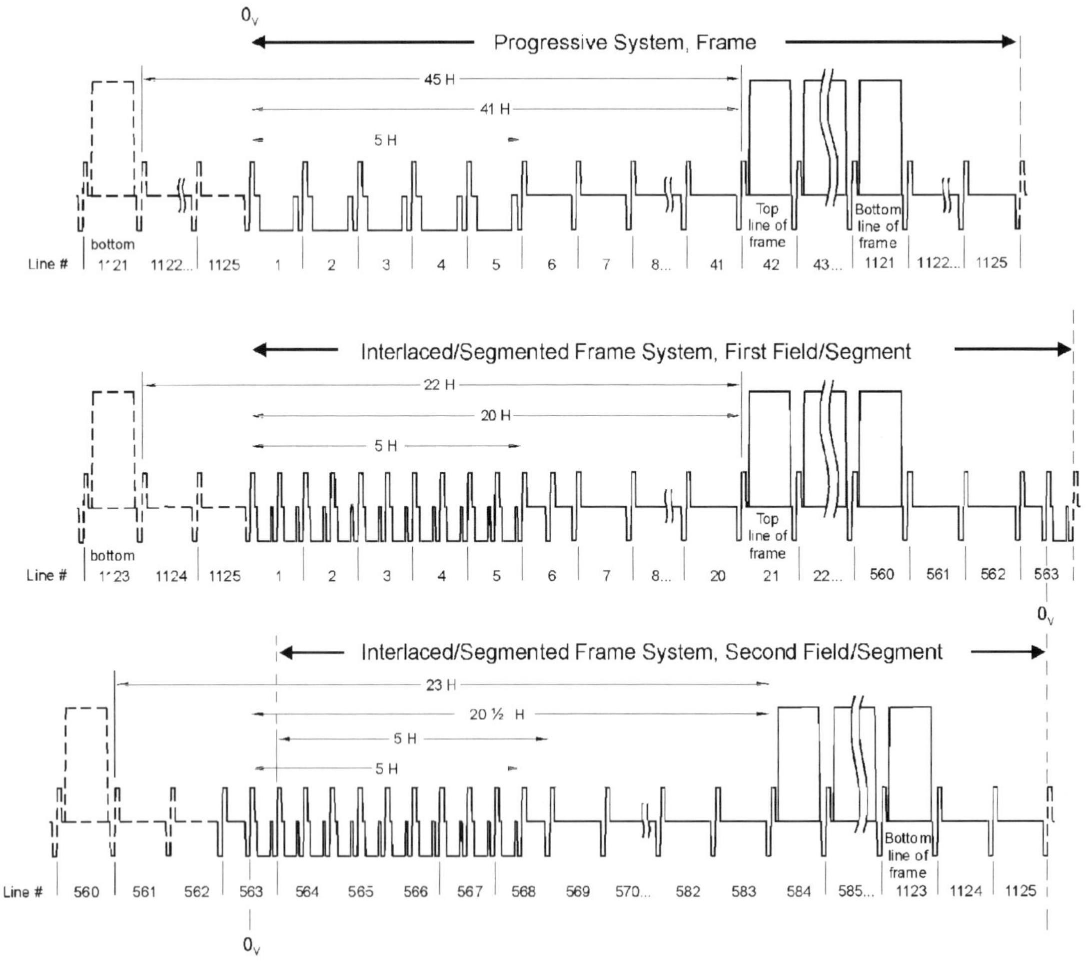

Figure 2.10.4 Organisation of Raster Lines (from SMPTE 274M, Figure 2.10.1).

At the analogue interface, P'_B and P'_R have a range of ±350 mV. P'_B and P'_R along with the Y' signal are then available for use directly as analogue interface outputs.

2.10.9.1.1.3 Raster Structure

The Raster Structure section of 274M should be read in conjunction with the Image Structure section described earlier. It describes the total raster, part of which is used to contain the image structure. It defines the assignment of lines within the raster between active image lines and vertical blanking lines and shows where the H, V and F bit patterns in the Timing Reference Sequences (see below) are placed that assert the changeover points (see figures). In the case of interlaced scanning, it also defines the unequal distribution of lines (see explanation earlier) between the two fields of a frame. It further defines those parts of the raster used for other, non-active-image purposes (ancillary signals and ancillary data), and other housekeeping details.

2.10.9.1.1.4 Digital Picture Representation

This section specifies:

- what the analogue inputs to this stage are (choice of $R'G'B'$ or $Y'\ C'_BC'_R$)
- what the bandwidth limits of these inputs are (to prevent aliasing)
- how the analogue signals are sampled in time (varies with temporal rate) and how co-siting is organized between the different components
- how the samples are quantized into digital code values
- how code values from 8-bit, 10-bit and 12-bit quantization inter-relate
- which code values are prohibited and which are allocated to undershoots and overshoots in the video waveforms that are being quantized.

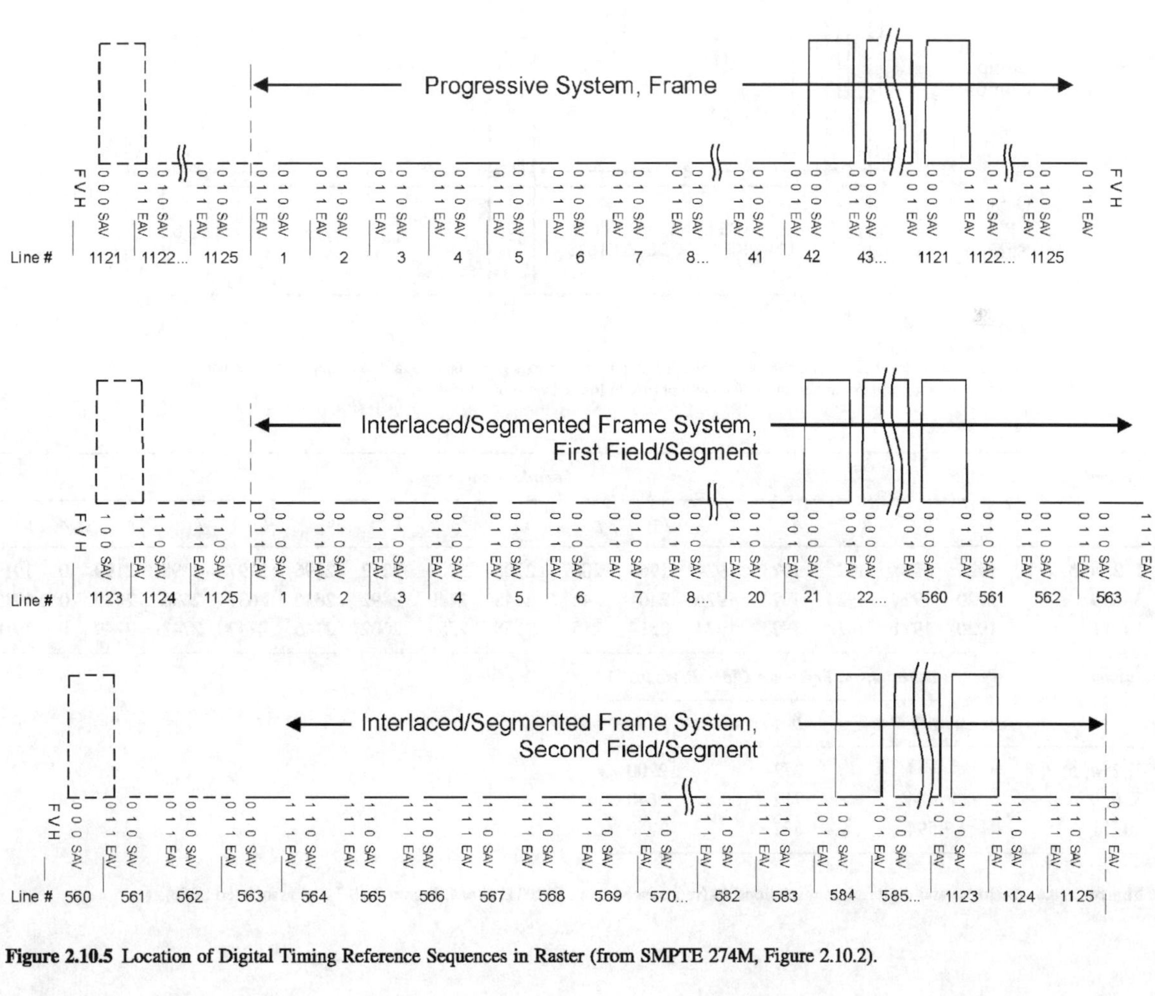

Figure 2.10.5 Location of Digital Timing Reference Sequences in Raster (from SMPTE 274M, Figure 2.10.2).

2.10.9.1.1.4.1 Sample Timing

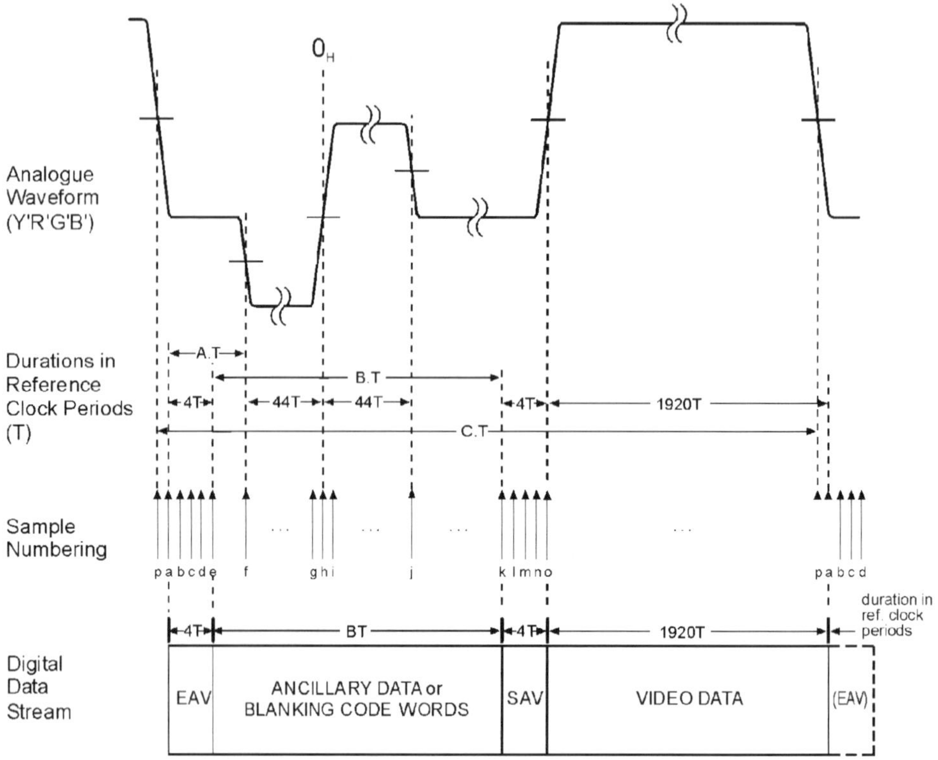

NOTES
1 Horizontal axis not to scale.
2 O_H is the analogue horizontal timing reference point, and in the analogue domain is regarded as the start of the line.
3 A line of digital video extends from the first word of EAV to the last word of video data.

System	*Sample Numbering*															
	a	*b*	*c*	*d*	*e*	*f*	*g*	*h*	*i*	*j*	*k*	*l*	*m*	*n*	*o*	*p*
1, 2, 4, 5, 7, 8	1920	1921	1922	1923	1924	1964	2007	2008	2009	2052	2196	2197	2198	2199	0	1919
3, 6, 9	1920	1921	1922	1923	1924	2404	2447	2448	2449	2492	2636	2637	2638	2639	0	1919
10, 11	1920	1921	1922	1923	1924	2514	2557	2558	2559	2602	2746	2747	2748	2749	0	1919

System	*Durations in Reference Clock Periods (T)*		
	A	*B*	*C*
1, 2, 4, 5, 7, 8	44	272	2200
3, 6, 9	484	712	2640
10, 11	594	822	2750

Figure 2.10.6 Analogue and digital timing relationship (reproduced from SMPTE 274M, Figure 2.10.3 and associated table).

2.10.9.1.1.4.2 Quantisation

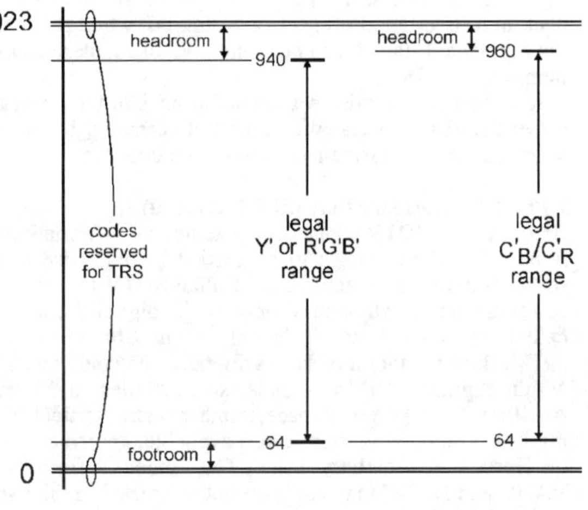

Figure 2.10.7 Allocation of Digital Coding Range (10-bit example).

Figure 2.10.7 represents an example 10-bit application of the 274M quantization equations:

$$L'_d = \lfloor 219DL' + 16D + 0.5 \rfloor; D = 2^{n-8}$$

$$C'_d = \lfloor 224DC' + 128D + 0.5 \rfloor; D = 2^{n-8}$$

- L' represents the input analogue value Y' or R', G' or B',
- C' is C'_B or C'_R,
- n is 8, 10 or 12 according to the quantizing bit depth
- L'_d or C'_d is the resulting output digital code value.

Note:

- two equations are required because the legal code range for Y' or R', G' or B' is different than for C'_B or C'_R
- consequently the headroom is different
- headroom is different from footroom (for all components)
- the equations adapt to 8, 10, 12 or any other bit depth in use

2.10.9.1.1.5 Digital Timing Reference Sequences (TRS)
These are periodic combinations of unique digital code values that define the start and finish of timing events in the raster and

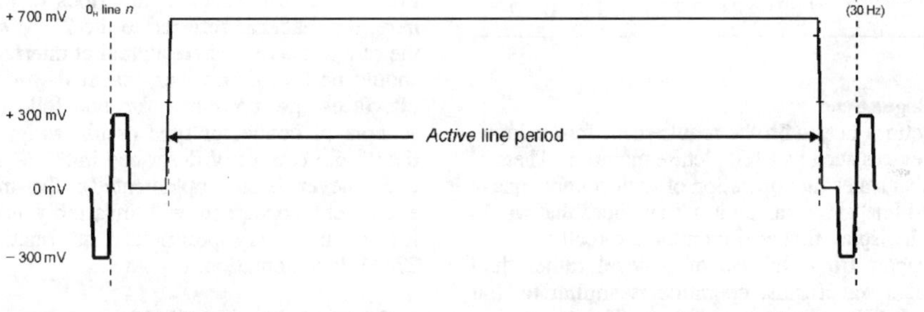

Figure 2.10.8 Horizontal sync (30 Hz version shown).

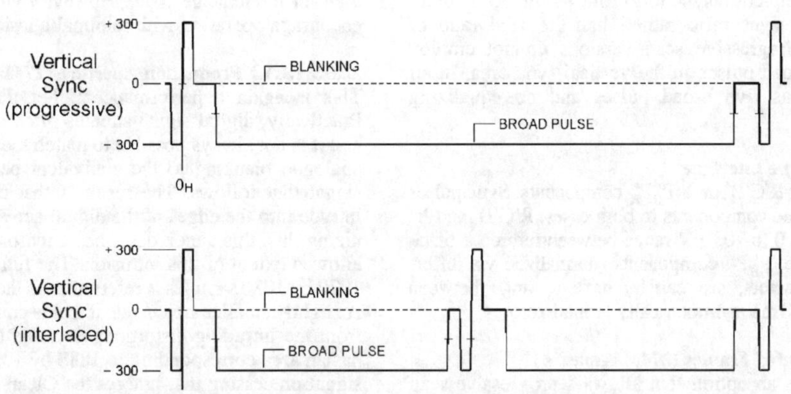

Figure 2.10.9 Vertical sync (part of Figure 2.10.5 from SMPTE 274M).

are essential to the correct functioning of the digital interface, especially a serial interface. These events are:

- start and end of active image samples within a line
- start and end of each interlaced field within a frame
- start and end of the extent of active lines within the total raster

One code pattern is common to all these and occurs at the beginning of each TRS and that is the sequence: code word of all ones, a code word of all zeros and another code word of all zeros. This is then used in combination with a fourth code word containing three special bits, H, V and F to distinguish each of the six events. Because of the extreme importance of the TRS, parity bits are also included in the fourth code word to make the final sequence in each instance. The make up of the TRS code words is shown in Table 2.10.2 (10-bit example shown).

Table 2.10.2 Composition of TRS code words

Bit number		*9*	*8*	*7*	*6*	*5*	*4*	*3*	*2*	*1*	*0*
Word	*Value*	*(MSB)*									*(LSB)*
0	1023	1	1	1	1	1	1	1	1	1	1
1	0	0	0	0	0	0	0	0	0	0	0
2	0	0	0	0	0	0	0	0	0	0	0
3		1	F	V	H	P3	P2	P1	P0	0	0

2.10.9.1.1.6 Analogue Sync

The next two sections deal with the requirement for analogue connections to devices such as CRT picture monitors. There is a note in 274M that states that operation of such connections is not recommended for "slow frame rates," i.e. those that would produce too much display flicker if monitored directly.

In principle, apart from the use of tri-level rather than bi-level sync pulse waveforms, operation is similar to that of syncs on an SDTV analogue interface. The 0_H timing reference point is symmetrically in the middle of the tri-level pulse and therefore at blanking level. Note that even in the 30 Hz versions, the picture/sync ratio follows the 50 Hz practice of 7:3 picture:sync ratio rather than the 10:4 ratio of 525-line SDTV. Progressive scan versions do not employ serration of the broad pulses in the vertical sync area. In all frame rate versions five broad pulses and no equalizing pulses are used.

2.10.9.1.1.7 Analogue Interface

This may use either R′G′B′ or $Y'P'_BP'_R$ components. Sync pulses are added to all three components in both cases. R′G′B′ and Y′ components have a 0 to 700 mV range between reference black and reference white; $P'_BP'_R$ components (normalized via differential weighting factors, see earlier) have a range between −300 mV and +300 mV symmetrically around zero.

2.10.9.1.1.8 Segmented Frames (274M Annex A)

This annex defines an option that allows a progressive scan signal's active scan lines to be re-ordered so that an interface or recording device, for example, can handle the signal exactly as if it were interlaced. Because of its different origin, the two constituent parts of each frame are called segments rather than fields, but the ordering of the lines is identical to interlace. It is assumed that after passing through the interfaces and recorders, the signal will eventually be re-ordered back into its original progressive form, but while it is still in the segmented state, it can be monitored on an interlaced scan monitor.

It is best to consider segmentation an interface operation rather than as a process that is part of scanning, because the scanning (in progressive mode) has to be done first.

2.10.9.1.1.9 Ancillary Data (274M Annex B)

As in SDTV, HDTV signals allow additional information that is not part of the image to be carried in otherwise unused space in the main image signal. Unlike SDTV however, this additional information may only be in digital form and is called Ancillary Data. It is defined in SMPTE standard 291M. Because ancillary data is digital, it may only be carried within digitized HDTV signals, so it resides in "Vertical Ancillary Data (VANC) space," rather than "vertical blanking." A horizontal version is also available, known as HANC or Horizontal Ancillary Data. This annex defines where VANC and HANC may and may not be carried in the overall digital HDTV signal.

2.10.9.1.1.10 Parallel Digital Interface (274M Annex C)

For the reasons stated earlier, parallel digital interfaces have been superseded almost entirely by serial interfaces, but it is necessary to maintain documentation for "legacy" installations, so 274M Annex C describes it. Rather than repeat it here, the reader is referred to the 274M standard directly for the rare occasions where a parallel interface is encountered. It should be noted, however, that real-world instances of these interfaces are notorious for not following any consistent scheme of connector type or pin assignment (for example, the 93-pin connector described in 274M Annex C has apparently never been implemented); the manufacturer of the equipment concerned will invariably have to be consulted for the necessary proprietary information to augment the 274M documentation.

2.10.9.1.1.11 Filtering (274M Annex D)

This is in an annex rather than the main normative part of the standard, because correct filtering is a performance issue rather than an interchange issue. However, guidance is provided to encourage operation with minimal aliasing.

2.10.9.1.1.12 Production Aperture (274M Annex E)

This is again a performance rather than interchange issue. Practically, digital signal streams often have analogue sources and it is not always possible to match the position and extent of analogue blanking to the equivalent periods in the digitized signal that follows. The result is that certain distortions can intrude into the edges of the digital active picture area. Recognizing this, this annex describes a reasonable tolerance for the allowed extent of this intrusion. The full active picture area of 1920 by 1080 samples is referred to as the Production Aperture. A slightly smaller area inside it that is considered to remain free from the intruding distortions is called the Clean Aperture; it has an area corresponding to 1888 by 1062 samples. However, signal processing that ignores the Clean Aperture and uses the full extent of the Production Aperture would still be maintaining a valid 274M signal, because this part of the document is only an informative annex.

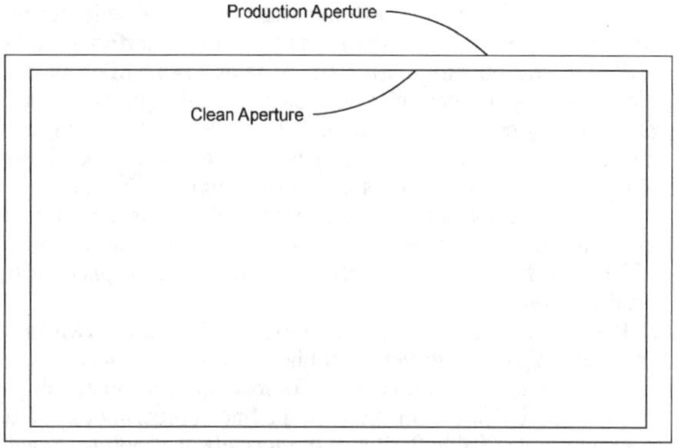

Figure 2.10.10 Production and Clean Apertures.

2.10.9.1.2 ITU-R Rec. BT.709 Parameter Values for the HDTV Standards for Production and International Programme Exchange

Recommendation 709 has two parts. The first part specifies the parameters of two formats which are now obsolete but which are documented to support legacy equipment and inventories of recorded material. These formats are the 1125/60 format (with 1035 active lines) and the 1250/50 format (1152 active lines). The document makes it clear that these formats should not be used for new productions.

The second part of Rec. 709 describes the preferred Common Image Format of 1920 pixels by 1080 active lines and this will be summarized in this section.

2.10.9.1.2.1 Purpose and Relationship to SMPTE 274M

Rec. 709 is the source of system colorimetry, sampling and quantizing specifications for SMPTE 274M. Conversely, subsequent 274M enhancements such as multiple-frame rate options were reflected in later revisions of Rec. 709. However, segmented frame options are listed in the main table of "picture rates and transports" in the introduction to Part 1, instead of being described in an Annex. Rec. 709 also does not include any description of physical interfaces, these appearing in a separate ITU-R Recommendation (Rec. BT.1120). It also excludes provision for the 12-bit quantisation option that is allowed in 274M, because the whole purpose of 709, as with its SDTV counterparts, is to apply the constraints in parameter values necessary to ensure interchange of content on the broadest possible international basis.

2.10.9.1.3 SMPTE 296M 1280 × 720 Progressive Image Sample Structure – Analogue and Digital Representation and Analogue Interface

296M is the 720P equivalent to 274M, in that it is the studio standard for 720*p* transmission and respects the same principle of "source-adaptive encoding" expressed in its same support for multiple temporal rates. However, the number of temporal rate options is simplified compared to 274M, since all options use progressive scan only. Notably, however, it supports 50P and 25P operation (see section on ITU-R Rec. 1543 below).

Importantly, 296M uses the same interface sampling frequency as 274M (not including 274M's 148.5 MHz option). This allows it to use the same serial interface.

The following sections of 296M will be found to be identical or equivalent in principle to those in 274M (some sections will be found in the main document body, rather than in an Annex):

- System Colorimetry
- Digital Representation
- Digital Timing Reference Sequences
- Parallel Digital Interface (by reference to 274M)
- Analogue Sync
- Analogue Interface
- Production Aperture
- Filtering

The section called "Timing" in 296M corresponds to "Image Structure" or "Scanning" in 274M. "Segmented Frames" are absent because they are not required.

"Raster Structure" and the space allocation within it for Ancillary Data are naturally different in 296M because of the different number of scan lines and the simplification due to the absence of interlace. Timing values in "Digital Representation," Digital Timing Reference Sequences and "Analogue Interface" are also different, but are based on the same principles for derivation.

2.10.9.1.4 ITU-R Rec. BT.1543 1280 × 720, 16 × 9 Progressively-Captured Image Format for Production and International Exchange in the 60 Hz Environment

This recommendation is the ITU-R equivalent to SMPTE 296M but with an important difference; 296M allows a choice of temporal frequencies similar to 274M and ITU-R Rec. 709, but BT. 1543 does not, restricting operation to 60 frames/second (and 59.94 frames/second) only. ITU-R considers the 720P format to be a studio origination format, but limited to 60 Hz regions as an international exchange format, since it does not comply with the fixed parameters of 1920 by 1080 in the Common Image Format. Limiting its frame rate to 60 Hz regions but excluding 50 Hz regions then effectively prevents its use for global distribution. (The technical basis for this decision is not readily apparent.)

Otherwise, with the exception of there being no section on legacy formats, the document layout and scope is similar to that of Rec. 709.

2.10.9.2 Current Interface Standards

2.10.9.2.1 SMPTE 292M Bit-Serial Interface for High Definition Television Systems

Often known as "HD-SDI," this is the most important studio interface in HDTV as it has almost entirely replaced analogue interfaces and the parallel digital interface. It operates at a serial data rate of 1.485 Gigabits/second (and 1.485/1.001 Gb/s) and conveys most HDTV video signals compliant with SMPTE 260M, 274M, 295M and 296M, and ITU-R Recommendations 709 and 1543. Its randomized NRZI channel code was inherited from its SDTV counterpart 259M and it can operate over either coaxial cable for short distances or single-mode fibre for up to 2 km.

2.10.9.2.1.1 Limitations

A single 292M serial interface will carry only the digitised $Y'C'_BC'_R$ component set with 8 or 10 bits quantization because of its capacity limitation. Transporting the $R'G'B'$ or $Y'C'_BC'_{RA}$ component sets or using 12-bit quantization requires more than one 292M interface (see SMPTE 372M below). 292M will also not transport the 148.5 MHz sample clock variants in 274M or Rec.709, either as a single interface or as multiple interfaces.

More generally, 292M has to specify known source formats as its input, because it cannot create a decodable serial output unless critical timing information (TRS) is in tightly defined locations in the input signal. These source formats are listed as "Reference SMPTE Standards" in 292M's Table 2.10.3 (reproduced below) and notated "A" through "M."

Serialization onto a single path is effected by interleaving the 8 or 10 bits of each signal component and also the signal components themselves. Ancillary data inserted according to the provisions of the allowed source formats will be carried transparently through the serial bitstream.

2.10.9.2.2 ITU-R Rec. 1120 Digital Interfaces for HDTV Studio Signals

This recommendation's scope is different from that of SMPTE 292M. Like Rec. 709 it has a Part 1 for legacy systems (1035i and 1250/50) as well as a Part 2 for current common image format systems. Each of these parts also specifies the signal representation for interface purposes and a parallel interface as well as a serial interface. The serial interface in Part 1 includes provision for a serial data rate of 1.400 Gb/s (for 1250/50) as well as the 1.485 Gb/s rate that has superseded it. The serial interface specification in Part 2 however, although covering the same temporal rates as 292M for 1920 × 1080 signals, makes no mention of 1280 × 720 signals, even though these have the same clock rate and would be compatible generally with the interface. Like SMPTE 292M, the limitations on source formats exceeding the bitstream capacity apply.

2.10.9.2.3 SMPTE 372M Dual-Link 292M Interface for 1920 × 1080 Picture Raster

SMPTE 372M specifies how two 292M serial interfaces may be used in combination to transport some of the formats whose combinations of signal parameters exceed the capacity of a single interface, as described above.

2.10.9.2.3.1 Allowable Source Formats

372M has its own limitations. In the current (2002) version, only sources based on 274M formats are covered (for example, 296M 720P sources are excluded). The allowed source formats are:

Table 2.10.3 372M source formats

Signal format sampling structure/pixel depth	*Frame/Field Rates*
4:2:2 ($Y'C'_BC'_R$)/10-bit	60, 60/1.001 and 50 Progressive
4:4:4 ($R'G'B'$), 4:4:4:4: ($R'G'B'$ +A$^{\text{see note 1}}$)/10-bit	30, 30/1.001, 25, 24, and 24/1.001 Progressive, PsF
4:4:4 ($R'G'B'$)/12-bit	
4:4:4 ($Y'C'_BC'_R$), 4:4:4 ($Y'C'_BC'_R$ +A$^{\text{see note 1}}$)/10-bit	60, 60/1.001, and 50 Fields Interlaced
4:4:4 ($Y'C'_BC'_R$) 12-bit	
4:2:2 ($Y'C'_BC'_R$) 12-bit	

Note 1 Definition of the "A" channel is application-dependent. In the cases when the A channel is used for non-picture data, the payload is constrained to 8-bit words maximum.

Some "worst case" combinations remain absent because they would exceed even the capacity of two interfaces. At 60, 60/1.001 and 50 progressive frame rates (148.5 MHz parallel clock), $R'G'B'$ operation is not possible at all, nor 12-bit operation in any component format. At the slower temporal rates, 12-bit quantization becomes possible for both $R'G'B'$ and $Y'C'_BC'_R$, but prevents transport of the Auxiliary signal A.

Early in the standardisation process, some consideration was given to covering these extreme cases by allowing up to four 292M interfaces to be specified, but this was dropped in the final version.

Because the multiplexing operations between the two interfaces exhibit a considerable number of variations according to source format (multiplexing by component, scan line, sample, or individual bit), the format exploits line numbering and also a mapping table (Table 2.10.3 in the format). that defines exactly what source format is being split across the two interfaces so that its multiplex structure can be determined. Table 2.10.3's values are then carried in a Payload Identifier, defined in SMPTE 352M.

2.10.10 Videotape Recording Standards for HDTV

The development of HDTV technology took place over a period when the parallel world of SDTV was making a significant transition away from tape towards disk-based non-linear systems for post-production. Consequently, the number of videotape formats for HDTV has not exhibited the explosion seen in the earlier SDTV era and it seems likely that in the long run, disk-based storage will take over in HDTV post-production too. However, tape remains an important capture and exchange medium, because of the easy removability of the medium and its infinitely extendable capacity just by using more blank tapes. Some HDTV videotape formats that have emerged for these functions are therefore described below.

2.10.10.1 Capture

The capture formats described here are used in so-called "camcorders" where the recorder is integral with an HDTV camera. The consequent size, weight and power consumption constraints have dictated the use of modified SDTV recorder transports which then need image pre-filtering and compression to fit the HDTV image into the limited bit rate capacity of the recorder. Bit depth may be limited effectively to 8-bits by the compression schemes used, even when the non-compressed interfaces may be 10-bit. Studio versions are available from the same manufacturers, offering the usual extended search and slo-mo modes, but they are still subject to the same dependency on bit rate reduction since they have to be compatible with the recordings captured on the camcorder versions. This tends to limit the number of copy generations these formats can be used for; often they will be used only for the single generation involved in capture and transfer into post-production.

2.10.10.1.1 HDCAM™† (SMPTE D-11 Format)

The HDCAM™ tape format accepts a scanning format of 1920 × 1080, at various temporal rates (Table 2.10.4). SMPTE

†Trademark of Sony (SMPTE D-11 Format).

Table 2.10.4 Allowable D-11 temporal rates (from SMPTE 367M Table 1)

Picture rate	*Base data rate (Mb/s)*
24 ÷ 1.001/PsF	111.863
24/PsF	111.975
25/PsF	116.640
30 ÷ 1.001/PsF	139.828
50/I	116.640
60 ÷ 1.001/I	139.828

has standardized it with a family of separate documents that together constitute the D-11 format.

SMPTE 367M specifies the compression format. This takes a 292M 1.485 Gbits/sec bit stream and reduces it to approximately 112–140 Mbits/second by means of pre-filtering in luma and chrominance and a lossy compression scheme accessed via a DCT transform.

SMPTE 368M then takes the 367M bit stream and creates the channel coding and functions necessary to record and play it. 368M also specifies the track locations and dimensions and the 12.65 mm (0.5 inch) cassette used.

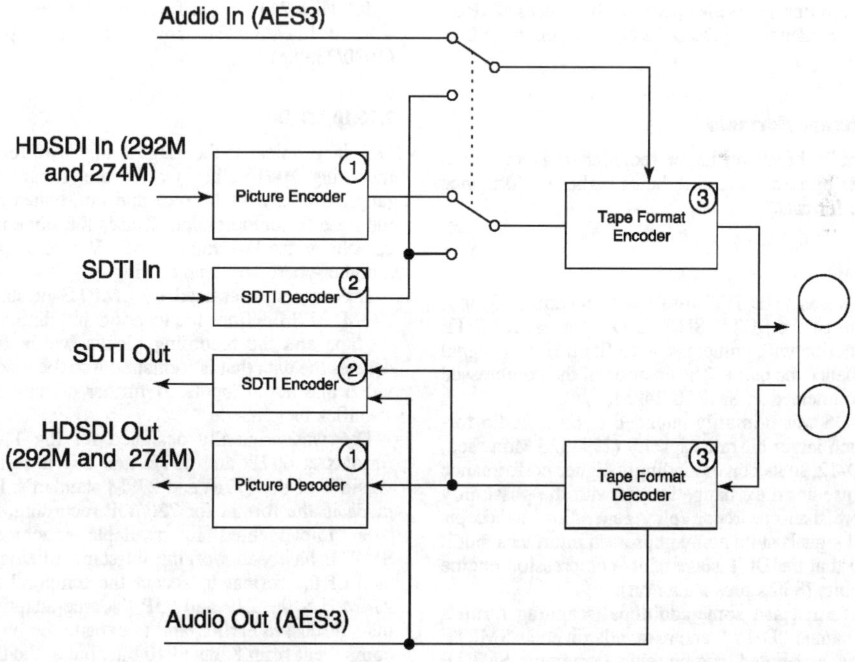

Figure 2.10.11 (Blocks (1) represent SMPTE 367M, (2) represent 369M and (3) represent 368M documents) D-11 Interconnection Options and Document Relationships (from SMPTE 367M, Figure E.1).

Direct tape-to-tape dubbing and other data stream connections are also possible, bypassing 367M, since the images will already be compressed. In this mode SMPTE 368M interfaces with a mapping format, SMPTE 369M, instead of 367M. 369M in turn uses the SMPTE 305M SDTI method of carrying data bytes in place of image bytes on a SMPTE 259M serial interface (note that this is an SDTV interface, since compression allows the HDTV images to occupy a sufficiently small bit rate).

The various connection options for recording and playing back D-11 material are summarized in Figure 2.10.11 reproduced from SMPTE 367M.

2.10.10.1.2 DVCPRO-HD™‡ (SMPTE standards 370M and 371M, D-12 format)

Like HD-CAM™, DVCPRO-HD™ is an HD camcorder format subject to similar physical constraints. It is documented by a similar family of SMPTE standards, including a compression format, grouped as the D-12 format. The D-12 tape format can record either a 1920 × 1080 format (SMPTE 274M) or a 1280 × 720 60P scanning format, (SMPTE 296M), but the DVCPRO-HD™ camcorders currently available use only the 1280 × 720 format. The output rate for the compression stage is 100 Mbits/sec, so again image pre-filtering and compression is required.

Pre-filtering and compression is documented in SMPTE 370M and tape formatting in SMPTE 371M (the 6.35 mm tape cassette is documented separately in SMPTE 307M).

‡ Trademark of Panasonic.

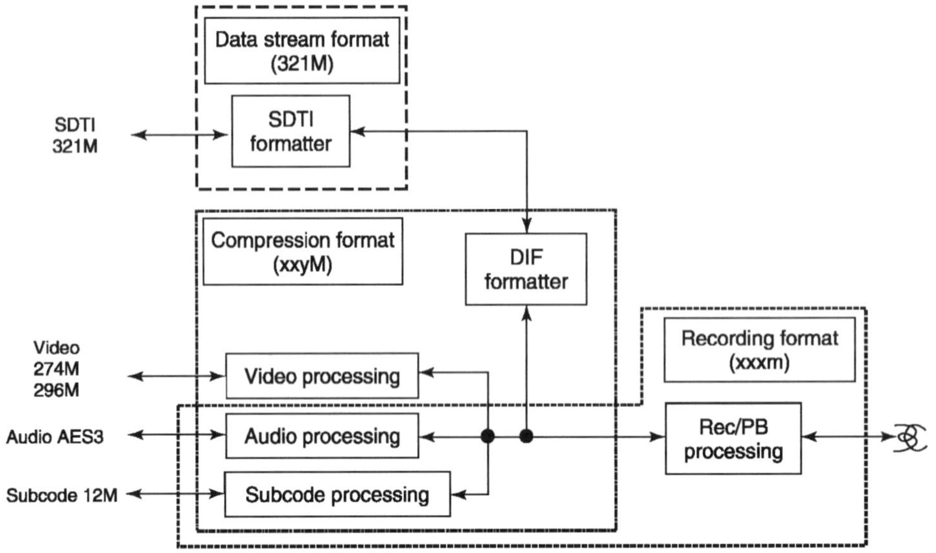

Figure 2.10.12 D-12 Interconnection Options and Document Relationships (from SMPTE 370M Figure A.1).

Again, bypassing the compression stage with direct bit-stream-to-bitstream transfer is also possible. This uses SMPTE 321M to define the mapping of the D-12 information into SDTI (Figure 2.10.12).

2.10.10.2 Exchange Formats

These are generally based on larger recorder transports with inherently higher bit rate, so tend to be of higher performance than the capture formats.

2.10.10.2.1 HD-D5

The format known as "HD-D5" uses the 12.65 mm (0.5 inch) tape and cassette of the SDTV SMPTE D-5 format (SMPTE 279M) in conjunction with compression to fit an HDTV signal onto D-5, tape, hence the name. The creation of the compressed bit stream is documented in SMPTE 342M.

Because HD-D5 was primarily intended to be a studio format, it has a much larger bit rate capacity (259–323 Mbits/sec) than D-11 and D-12, so its correspondingly higher performance has enabled its use as an exchange format with the possibility of enduring more than one record/play generation. It accepts 10-bit-quantized signals at its non-compressed interfaces, but it should be noted that the DCT stage of its compression engine operates uses 9 bits (8 bits plus a sign bit).

In 2002, the format had some additional scanning formats added so was renamed "D-15," and re-standardized as SMPTE 399M-2003. The associated compression document SMPTE 342M was updated at the same time.

The scanning formats accepted are (quoting from SMPTE document):

- 1080 line/59.94 Hz field frequency interlace system (1080/59.94i)
- 720 line/59.94 Hz frame frequency progressive system (720/59.94p)
- 1080 line/50 Hz field frequency interlace system (1080/50i)
- 1080 line/24 Hz frame frequency progressive system (1080/24p)
- 1080 line/23.98 Hz frame frequency progressive system (1080/23.98p)

2.10.10.2.2 D-6

D-6 is physically the largest physical format available for recording HDTV, has by far the greatest bit rate capacity (approximately 1 Gbit/sec) and consequently records without the need for compression. It uses the same form-factor 19 mm cassette as the D-1 and D-2 SDTV recording formats, but with a higher-coercivity tape coating.

D-6 was standardized by SMPTE in standards 277M and 278M. 277M defines the location and dimensions of the tracks on tape and the recording blocks within them, while 278M defines the data that is formatted into the recording blocks from video and audio inputs. A further document, SMPTE 226M, specifies the cassette.

D-6 was originally designed for the 1125 and 1250 line formats at 60 Hz and 50 Hz and this is reflected in the 1996 versions of the 277M and 278M standards. Provision was also made in the format for 720/60P recording, but this has never been implemented in available machines. Subsequently, SMPTE has been working on standardizing an extended version of the format to accept the temporal rate extensions of 274M, i.e. the 24P and 25P "source-adaptive encoding" formats discussed earlier, and to expand the bit depth in the luma component from 8 bits to 10 bits, but at the time of writing this work had not been completed.

Acknowledgments

Thanks are extended to the Director of Engineering, SMPTE, for permission to reproduce portions of the SMPTE standards described in this chapter (for further study, the reader is urged to obtain copies of the complete standards via the excellent CD-ROM and hard copy service available from SMPTE).

Particular thanks are due also to the extremely diligent efforts of Charles Poynton, Johann Safar and Merrill Weiss, in pointing out errors and suggesting improvements in my text.

Reference – Standards Documents

ITU-R Recommendations

BT.709 Parameter Values for the HDTV Standards for Production and International Programme Exchange
BT.1120 Digital Interfaces for HDTV Studio Signals
BT.1361 Worldwide Unified Colorimetry and Related Characteristics of Future Television and Imaging Systems
BT.1543 1280 × 720, 16 × 9 Progressively-Captured Image Format for Production and International Programme Exchange in the 60 Hz Environment
(ITU-R recommendations are available by download from http://www.itu.int)

SMPTE Standards

240M-1999 – 1125-Line High-Definition Production Systems - Signal Parameters
260M-1999 – 1125/60 High-Definition Production System - Digital Representation and Bit-Parallel Interface
274M-2003 1920 × 1080 Image Sample Structure, Digital Representation and Digital Timing Reference Sequences for Multiple Picture Rates
277M-1996 19-mm Type D-6 – Helical Data, Longitudinal Index, Cue and Control Records
278M-1996 19-mm Type D-6 – Content of Helical Data and Time and Control Code Records
279M-2001 1/2-in Type D-5 Standard-Definition Component Video and Type HD-D5 High-Definition Video Compressed Data
292M-1998 Bit-Serial Digital Interface for High-Definition Television Systems
296M-2001 1280 × 720 Progressive Image Sample Structure – Analog and Digital Representation and Analog Interface
305.2M-2000 Serial Data Transport Interface (SDTI)
321M-2002 Data Stream Format for the Exchange of DV-Based Audio, Data and Compressed Video over a Serial Data Transport Interface
342M-2000 HD-D5 Compressed Video 1080i and 720p Systems — Encoding Process and Data Format
367M-2002 Type D-11 Picture Compression and Data Stream Format
368M-2002 12.65-mm Type D-11 Format
369M-2002 Type D-11 Data Stream and AES3 Data Mapping over SDTI
370M-2002 Data Structure for DV-Based Audio, Data and Compressed Video at 100 Mb/s 1080/60i, 1080/50i, 720/60p
371M-2002 6.35-mm Type D-12 Component Format – Digital Recording at 100 Mb/s 1080/60i, 1080/50i, 720/60p
372M-2002 Dual Link 292M Interface for 1920 × 1080 Picture Raster
399M-2003 1/2 in Type D-15 High-Definition Compressed Video Data Format

SMPTE Recommended Practices

RP 160-1997, Three-Channel Parallel Analog Component High-Definition Video Interface
RP 177-1993 Derivation of Basic Television Color Equations
(SMPTE Standards, Recommended Practices and Engineering Guidelines are available from SMPTE in hard copy or CD-ROM form. Details can be found at http://www.smpte.org).

Bibliography

Hunt, R.W.G., "The Reproduction of Colour in Photography, Printing & Television," 6th Edition, Fountain Press, 2003
Lang, H., "Colour and Its Reproduction, Part 1: Colorimetry," Muster-Schmidt, 2002
Poynton, C., Digital Video and HDTV Algorithms and Interfaces, Morgan Kaufmann, 2003
Watkinson, J., "Convergence in Broadcast and Communications Media," Focal Press, 2001

E. P. J. Tozer
Zetrox Broadcast Communications

2.11 MPEG-2

2.11.1 Introduction

Whilst many other methods exist, MPEG-2 is today the de facto standard method for carrying compressed audio and video signals. The standard provides a highly flexible set of techniques that enable a wide variety of bit rates and programme structures to be carried under a single standard technique.

At its simplest an MPEG-2 signal may carry just a single audio or video component; at its most complex it may carry several complete television services, each comprising audio, video, data and text components.

One of the key features of MPEG-2 is that it is a standard, defined by ISO 13818, meaning that one manufacturer's equipment complying with the standard is compatible with another manufacturer's equipment complying with the standard, at least in theory, and in most instances today in practice too.

2.11.2 MPEG Video Coding

MPEG-2 video coding is a highly complex and flexible process that allows studio quality, 270 Mbit/s video to be compressed to transmission bit rates of the order 1–6 Mbit/s, with compression ratios of 300:1 to 50:1 that transmit at between 0.3% and 2% of the original studio bit rate, yet are still able to maintain viewable picture quality.

2.11.2.1 Levels and Profiles

MPEG-2 is a vastly scoped standard that allows for video resolutions of up to 16,384 × 16,384 samples, video coding in 4:2:0, 4:2:2 and 4:4:4 modes, with various aspect ratios, panning modes, 3:2 pull-down and a plethora of other features supported. Clearly it is not feasible for all equipment to support all the possibilities of MPEG-2 so, in order to make compatibility between vendors a more manageable task, MPEG-2 is broken down into a number of subsets, called levels and profiles, which place different constraints on the scope of the standard.

Levels define the resolution of the signal, in terms of a video frame's spatial resolution, i.e. the number of pixels that define it horizontally and vertically, in terms of its temporal resolution, i.e. the number of frames per second, and in terms of the bit rate of the compressed video stream.

Profiles define the toolbox that may be used in the compression of the video, for example the chroma structure 4:2:0, 4:2:2 or 4:4:4, prediction frame types, i.e. I, B and P frames (see later in the chapter), and various other detailed compression tools.

Whilst an MPEG signal of 16,384 × 16,384 sampled 4:4:4 at 60 FPS (frames per second) may be within the scope of the MPEG-2 standard, in practice this mode remains unused.

Levels and profiles are derived from an agreed application.

Some of the common defined levels and profiles are shown in Figure 2.11.1. For general transmission use, Main Profile at Main Level is used, highlighted; this is written as MP@ML. The profile allows the use of IP and B frames and the level allows up to 720 samples per picture line, up to 576 lines per frame, up to 30 FPS and up to 15 Mbit/s in the compressed video bit stream. In practice a typical broadcast stream in a DTH (Direct To Home) application might be 704H by 576V at 25 FPS and 4Mbit/s for PAL and 704H × 480V at 29.97 FPS and 4 Mbit/s for NTSC.

For contribution and distribution applications, 422P@ML would typically be used with 4:2:2 chroma sampling, 720H × 608V pixels at 25 FPS and 20 Mbit/s.[1]

For the full details of levels and profiles, the reader should see Ref. 2.

High-definition DTH transmission might typically use MP@HL with typically used resolutions of 1080V × 1920H and 720V × 1280H, at frame rates of 24–60 FPS; 480V × 720H × 60 FPS is also used to provide high definition, in the temporal domain only, for sports transmissions and similar.

		High Level	High 1440 Level	Main Level	Low Level
Simple Profile	I P			720H	
	4:2:0			576V	
				30 FPS	
				15 Mbit/s	
Main Profile	I P B	1920H	1440H	720H	352H
	4:2:0	1152V	1152V	576V	288V
		60 FPS	60 FPS	30 FPS	30 FPS
		80 Mbit/s	60 Mbit/s	15 Mbit/s	4 Mbit/s
SNR Profile	I P B			720H	352H
	4:2:0			576V	288V
				30 FPS	30 FPS
				15 Mbit/s	4 Mbit/s
Spatial Profile	I P B		1440H		
	4:2:0		1152V		
			60 FPS		
			60 Mbit/s		
High Profile	I B P	1920H	1440H	720H	
	4:2:0 / 4:2:2	1152V	1152V	576V	
		60 FPS	60 FPS	30 FPS	
		100 Mbit/s	80 Mbit/s	20 Mbit/s	
422 Profile	I P B			720H	
	4:2:0 / 4:2:2			576V	
				30 FPS	
				50 Mbit/s	

Figure 2.11.1 Some MPEG-2 levels and profiles.

2.11.2.2 Sampling in space and time

MPEG-2 supports the use of 4:4:4, 4:2:2 and 4:2:0 chroma sampling structures, though in practice 4:4:4 is unused, and 4:2:2 only used in contribution and distribution applications, 4:2:0 being the de facto standard for DTH transmission.

Figure 2.11.2 shows the sample points from a number of lines within a frame used by MPEG-2; the sample points for 4:2:2 are the same as CCIR 601/656.[3]

Note that the sample points defined for 4:2:0 fall between lines of fields 1 and 2 and represent an average of the line above and the line below.

MPEG-2 video is purely a component system. Figure 2.11.3 shows the three sampling modes in terms of the three component planes, Y, Cr and Cb. Note that in 4:2:2 mode the Cr and Cb pixels are the same height and twice as wide as the luminance pixels. In 4:2:0 mode the Cr and Cb pixels are both twice as wide and twice as high as the Y pixels.

MPEG-2 supports both Interlace and Progressive (non-interlaced) scan modes.

2.11.3 MPEG Encoding

2.11.3.1 Overview

MPEG-2 encoding is both a highly effective and highly complex process, using techniques and processes optimised for broadcast type moving image sequences. The MPEG-2 encoding process can be broken down into two key processes: Inter-Frame coding and IntraFrame coding (Figure 2.11.4). Both these processes attempt to remove unnecessary (redundant)

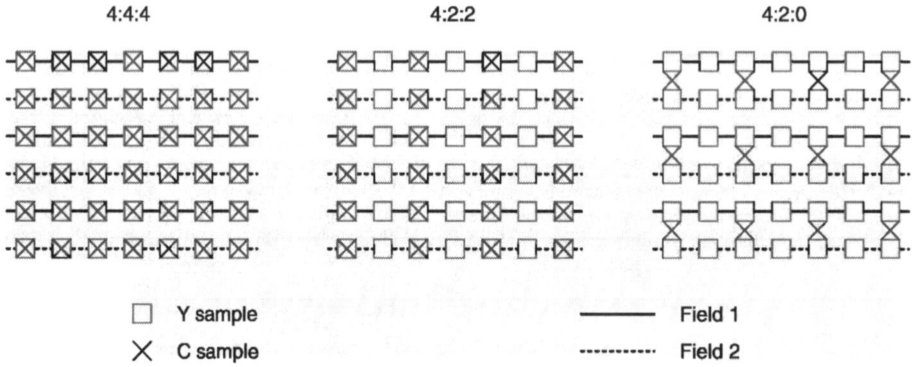

Figure 2.11.2 MPEG-2 sample points on picture frame.

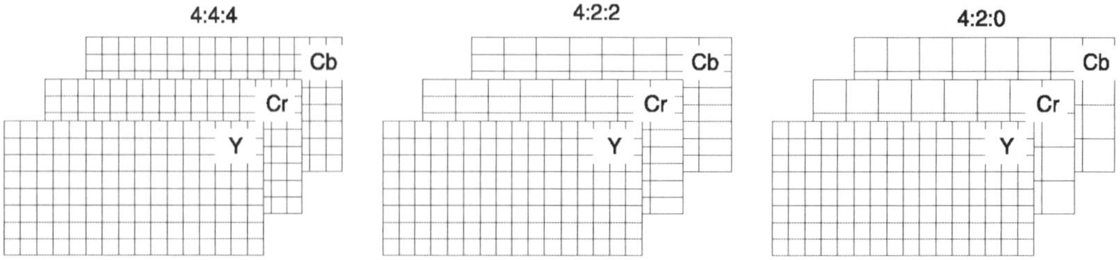

Figure 2.11.3 MPEG-2 component field viewed as Y, Cr and Cb planes for three sample modes.

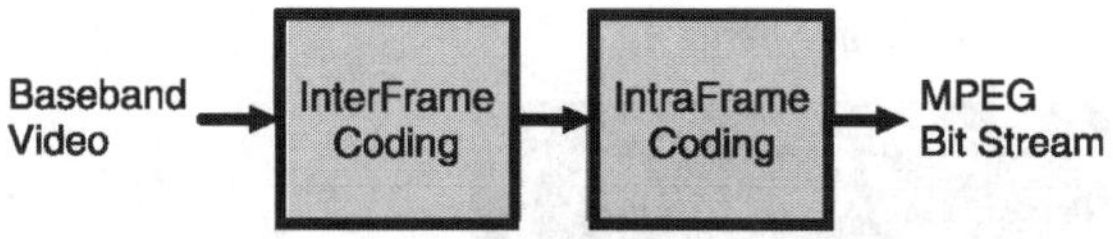

Figure 2.11.4 MPEG-2 InterFrame and IntraFrame processes.

information from the picture in order to reduce the data needed to represent the moving image sequence, i.e. to increase the entropy, or information carried by each bit of the signal that represents the moving image.

InterFrame coding processes the image sequence by removing information that is common between frames, e.g. subsequent images in a colour bars sequence contain no new information, and could be represented by reusing the first frame of the sequence.

IntraFrame coding processes the image sequence by removing information that is common within an image frame, e.g. within a colour bars frame large areas of the image are rendered in the same colour and could be represented by a single item of colour and luminance information.

2.11.3.2 Video InterFrame coding, and B and P frame creation

InterFrame coding in MPEG-2 consists of a number of processes, not all of which may be employed by an encoder:

- I, B and P frame generation.
- GOP creation.
- Motion estimation.

2.11.3.2.1 I, P and B frames

The creation of I, P and B frames is the first process in an encoder. These are:

- I – Intra frame.
- P – Predicted frame.
- B – Bidirectionally predicted frame.

An I frame is essentially an unmodified video frame.

A P frame is the difference between the current frame and a preceding frame. In simplistic terms, P frames are generated by subtracting the sample values of the preceding frame from the current frame (Figure 2.11.5).

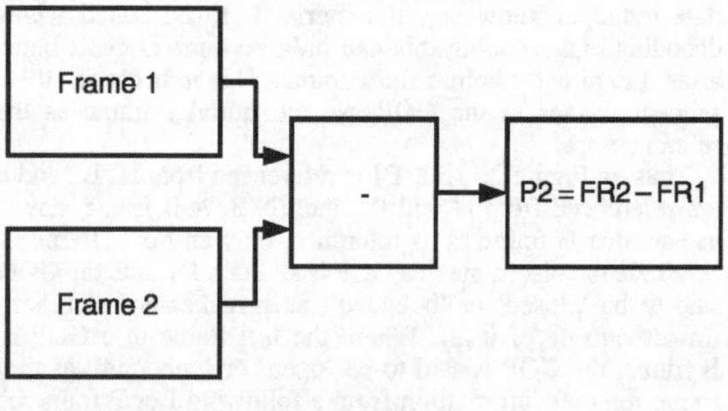

Figure 2.11.5 MPEG P frame generation.

Figure 2.11.6 shows I and P frames in terms of picture content; in the example the orange has moved between frames 1 and 2, so a positive and negative orange are included in the P frame. To recreate frame 2 the decoder adds the P frame data to the decoded frame 1, the negative orange subtracts from the orange of frame 1 and the positive orange is added to it.

B frames can be created as shown in Figure 2.11.7, where the average of the preceding and following frames are subtracted from the current frame; as an alternative B frames may use information that is only from the preceding or following frames.

In practice MPEG P and B frame generation is done at macroblock level; a macroblock is a 16 × 16 pixel area of the source image. An I frame may only consist of I macroblocks, a P frame may consist of both I and P macroblocks, and a B frame may consist of I and B macroblocks, predicted from prior (which is equivalent to P prediction), following or averaged macroblocks. In the case of B macroblocks it is allowable not to use information from the preceding frame, and just use the following frame as a reference, i.e. a P frame in reverse; this type of frame is sometimes referred to as an R frame.

2.11.3.2.2 Motion estimation

In practice the use of B frames will provide a good data reduction on relatively static images; however, if there is a lot of movement in the image, or the scene is a panning shot, then the difference frame will contain a great deal of information.

Figure 2.11.8 shows the P frame that results when the camera is panning a scene; the displacement between frames causes a large amount of information to be present in the P or B frame. As panning and object movement are common material for broadcast video, MPEG-2 provides a special tool, 'motion vectors', to reduce the data content.

Figure 2.11.9 shows the resulting P macroblock when an object has moved within an image; here the P macroblock contains a large amount of information.

The process of motion estimation allows the MPEG-2 encoder to search the reference frame for a matching object (Figure 2.11.10), and uses a different area to generate the difference P macroblock; the location of the matching object is indicated to the decoder by including motion vector (displacement) information along with the P macroblock in the MPEG-2 data stream. Motion vector information can be indicated to a resolution of 0.5 pixels.

The ability to perform good motion estimation is key to achieving good compression ratios; however, it is an extremely processor-intensive process.

As an indication of the potential level of complexity in the case of a typical 704 × 576 pixel frame of video there are 1584 (44 × 36) macroblocks, each of which could be matched to approximately 704 × 576 × 4 half-pixel positions in the reference frame. In order to calculate the match of each macroblock 16 × 16 comparisons must be made, all done for each of the 25 FPS and each of the Y, Cr and Cb channels. This gives a process rate around 32 × 1012 (1584 × 704 × 576 × 4 × 16 ×16 × 25 × 2) comparisons per second. In order to reduce the processor burden encoder will typically employ a number of methods to more readily find object matches. Typically an encoder can limit its search to a small range, maybe 20%, of the image area, as it is unlikely a pan will be greater than this in viewable picture material. In the case of a pan, once an encoder has found a positional match it can use this motion vector

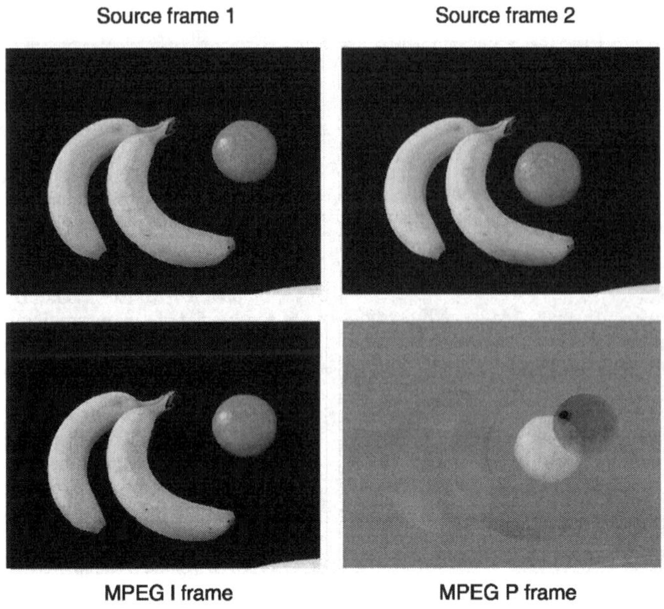

Figure 2.11.6 Creation of I and P frames from video source frames.

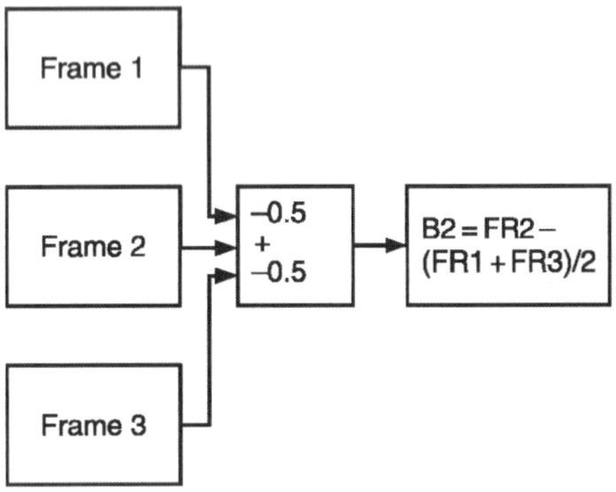

Figure 2.11.7 MPEG B frame generation.

information for all subsequent macroblocks. Instead of using an exhaustive search of all areas of interest, an encoder can employ a successive approximation method where the initial search step is large then, once an approximate match has been found, the step size can be progressively reduced until the best match is found. Additionally, motion estimation will normally be colour blind, using only the luminance channel to calculate motion vectors.

Whilst motion estimation can dramatically reduce the information content of P and B frames, there are limitations. In Figure 2.11.11 the shape has both moved and rotated; in this case motion estimation would not be able to match the object.

2.11.3.2.3 GOP structures

I, P and B frames can be used in various combinations to produce GOPs (Groups Of Pictures). Some typical GOP structures are:

Single-frame GOP
II ...IIIIIIIIIIII...
Two-frame GOP
IB ...IBIBIBIB...
12-frame GOP
IP ...PPP**IPPPPPPPPPPP**IPPP...
IBP ...BPB**IBPBPBPBPBPB**IBPB...
IBBP ...PBB**IBBPBBPBBPBB**IBBP...
IBBBP ...BBB**IBBBPBBBPBBB**IBBB...
15-frame GOP
IBBP ...PBB**IBBPBBPBBPBBPBB**IBBP.

The choice of GOP structure to use is affected by a number of factors. In general the more B and P frames the better the data reduction; however, if material is to be edited without decoding and recoding this can only be done at GOP boundaries, i.e. directly before the I frame. This is because all subsequent frames in the GOP use the initial I frame as their reference.

Thus, in Figure 2.11.12, P4 is referenced from I1, B2 and B3 are referenced from I1 and P4, etc. (N.B. A B frame may not use another B frame as its reference, only an I or P frame.)

Where the last frame in a GOP is an I or a P frame the GOP is said to be 'closed' or 'bounded', as it makes no reference to frames outside of itself. Where the last frame in a GOP is a B frame, the GOP is said to be 'open' or 'unbound', as the B frame requires information from a following I or P frame outside of its GOP to be decoded.

Figure 2.11.8 Left: image from a panning scene. Right: resultant P frame.

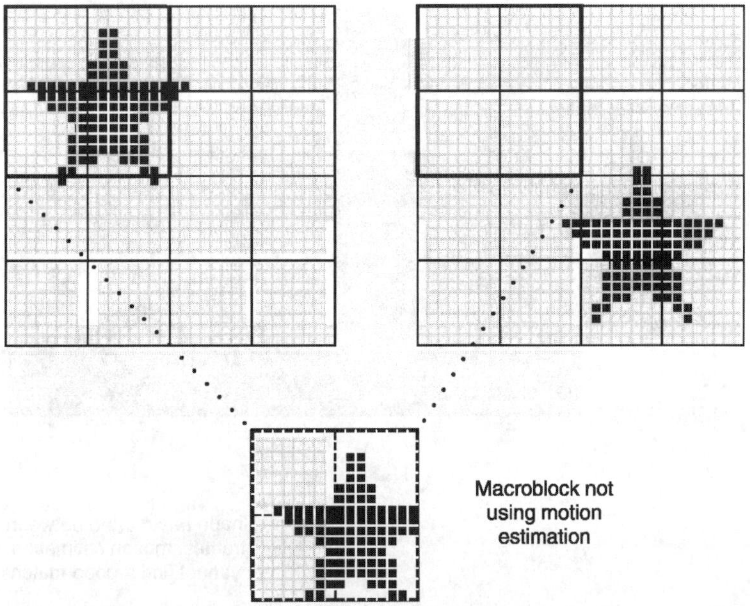

Figure 2.11.9 Macroblock generated when an object has moved between frames.

Typically for emission use a long GOP is used, 12 frames in 25 FPS applications and 15 frames in 30 FPS applications, which gives a maximum (MPEG) latency of 0.5 seconds when switching between MPEG encoded channels (these GOP lengths are generally doubled for 50 and 60 FPS systems). N.B. In practice channel switching latency may be affected by a number of other factors, including MPEG PSI (Program Specific Information) acquisition time and CA (Conditional Access) decryption time.

For emission the choice of IP structure would typically be determined by the receiver capability, SP@ML not supporting the use of B frames, and some, largely historical, set-top boxes exhibiting 'B frame judder', i.e. the inability to decode B frames when presented with a full resolution (576 × 704 pixel) video stream, due to insufficient decode memory (16 Mbit).

The choice between IBP, IBBP and IBBBP structures is determined by the programme content, IBBBP being most efficient for material with little movement between scenes, e.g. landscape channel, and IBP being most efficient for material with a large movement content, e.g. sport.

For studio use either I frames only or two-frame IB (Sony Betacam SX) would be used to give edit resolutions of 1 and 2 frames respectively.

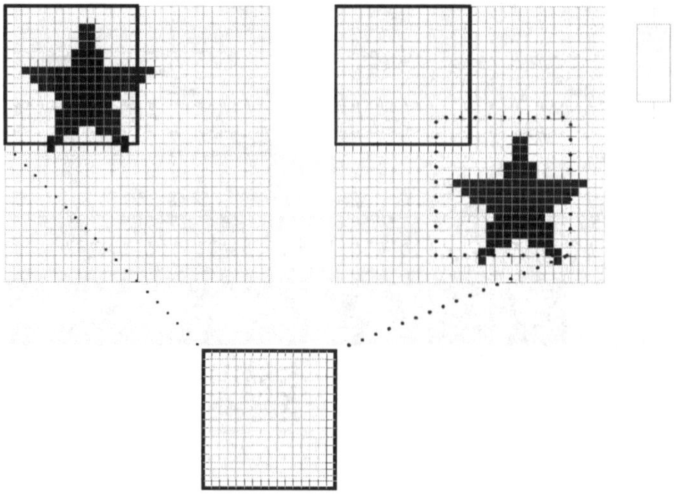

Figure 2.11.10 Macroblock generated when an object has moved between frames and motion estimation locates the moved object.

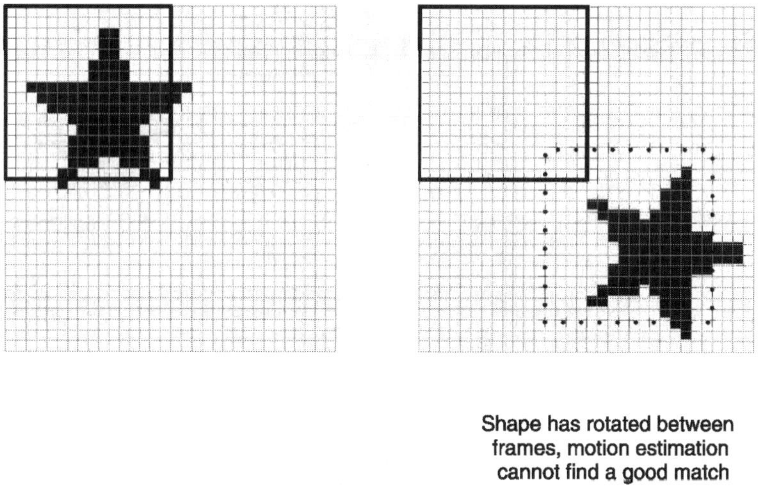

Figure 2.11.11 When an object rotates, motion estimation is unable to find a matching shape.

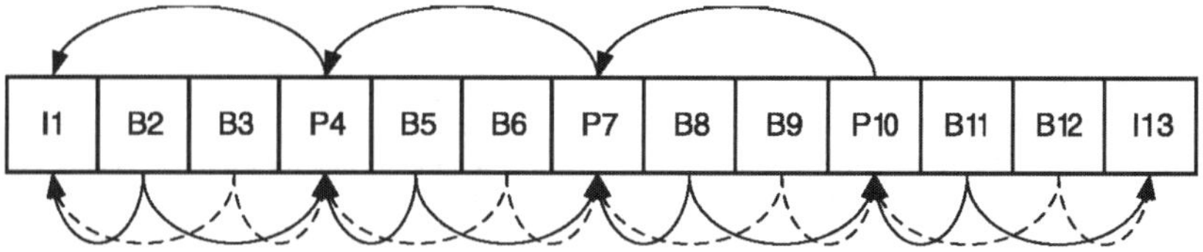

Figure 2.11.12 Twelve-frame IBBP GOP.

2.11.3.2.4 I, P and B frame transmission sequence

Where B frames are used in the Interframe coding process, the decoder requires later I or P frames from the video sequence before it can decode the B frame(s). In Figure 2.11.12 frames B2 and B3 cannot be decoded until frame P4 has been decoded, as both B2 and B3 reference I1 and P4. Similarly frames B11 and B12 cannot be decoded until frame I13 has been decoded.

In order that the decoder does not need to use valuable buffer space storing yet to be decoded B frames, where B frames are included in the encoded video frame data is sent out of its natural sequence (Figure 2.11.13).

2.11.3.3 Encoder structure (and noise build-up)

If an encoder simply generates P frames from the source video, this causes noise build-up in the decoded picture (Figure 2.11.14).

As each P frame is dependent on the preceding frame, the decoded frames move further away from the reference I frame so more and more noise is added to the decoded frame. If an encoder functioned in this manner there would be highly visible noise build-up through the GOP. To overcome this effect, an encoder incorporates a decoder (Figure 2.11.15).

Where the MPEG encoder incorporates a decoder to provide the reference frame for P frame generation, the decoded encode

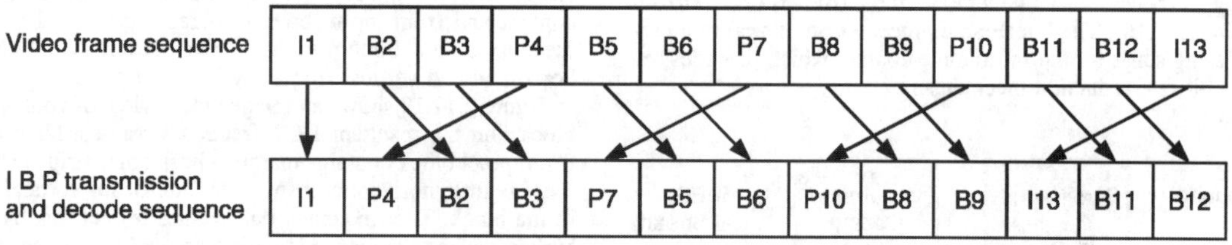

Figure 2.11.13 Reordered frame sequence where an encoding includes B frames.

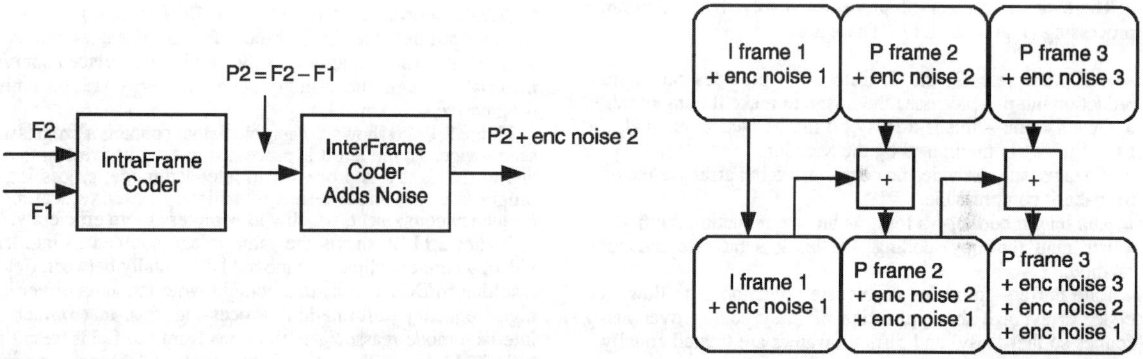

Figure 2.11.14 Noise build-up through GOP with simple encode method.

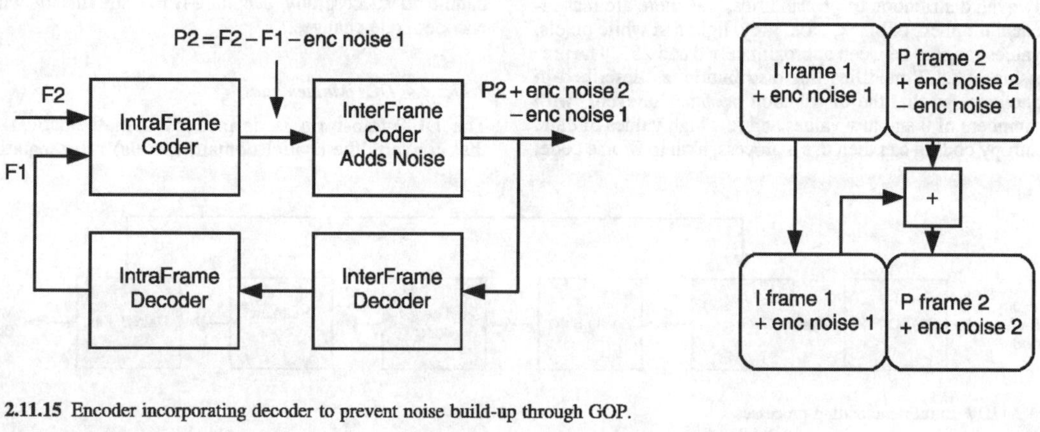

Figure 2.11.15 Encoder incorporating decoder to prevent noise build-up through GOP.

noise is subtracted from the P frame, thus, when the decoder recreates the video frame from the I and P frames, the I frame noise cancels, eliminating the build-up of noise through the GOP.

If there is any difference between the decode process within the encoder and the decode process in the actual decoder, this can lead to a noticeable GOP noise build-up, visible as a pulsing of the noise level in the viewed pictures.

2.11.3.4 Video IntraFrame coding

2.11.3.4.1 Overview

The MPEG IntraFrame coding process follows the InterFrame coding process that generates the IBP frame stream (Figure 2.11.16). The InterFrame processes are in reality a pre-processing stage for the IntraFrame coding, which provides the actual bit rate reduction mechanisms.

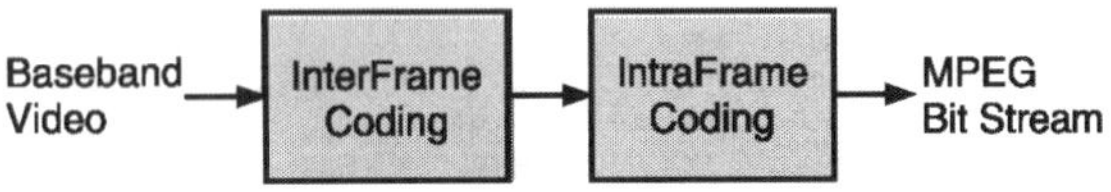

Figure 2.11.16 MPEG Intra- and InterFrame coders.

There are a number of processes involved in IntraFrame processing (Figure 2.11.17). These are:

- DCT (Discrete Cosine Transform) – provides no bit rate reduction but pre-processes the image to make it compressible.
- Quantisation – this is a lossy bit rate reduction method, the use of which is minimised by the encoder.
- Zigzag scan – provides no compression but arranges the data to make it compressible.
- Run length coding – a lossless bit rate reduction method.
- Huffman (entropy) coding – a lossless bit rate reduction method.
- Rate buffer – provides no bit rate reduction but allows the coder to average the application of lossy coding over many frames so that 'easy' and difficult frames are treated equally.

In order for the lossless entropy (Huffman) coding to work, the data to be coded must have an uneven distribution of probabilities. The data values of normal broadcast pictures have a fairly even distribution of probabilities,[4] i.e. there are reasonably equal numbers of black, dark, grey, light and white pixels, with values ranging between approximately 0 and 255. The pre-processing of DCT modifies this distribution, as described in the following, so that the distribution becomes uneven,[4] with large numbers of 0 and low values and few high values of data. The entropy coding can then use a process, akin to Morse code, to allocate short code sequences to the more common, low values and longer sequences to the rarer, high values, overall resulting in fewer bits being used to describe the scene.

2.11.3.4.2 DCT (Discrete Cosine Transform)

DCT is a spatial to frequency transform. In MPEG the DCT works on block of 8 × 8 pixels, DCT blocks converting the picture's spatial representation into a frequency-based representation.

Figure 2.11.18 shows the set of 64 8 × 8 pixel frequency patterns that DCT converts the spatial image to. In the top left is the DC (average picture brightness) component; going from left to right are patterns with increasing horizontal frequency content and from top to bottom increasing vertical frequency content. In the bottom right is the pattern with both high horizontal and vertical frequency information.

Figure 2.11.19 shows an image with a view of an 8 × 8 DCT block and the resultant DCT frequency values. Note that the 8 × 8 pixel block contains mainly a horizontal stripe pattern, so there is little horizontal change but significant vertical change in the block. The frequency data values reflect this by having higher values to the left, indicating vertical frequency information.

2.11.3.4.3 Progressive and non-progressive DCT

MPEG-2 allows for two different DCT processing modes which optimise the DCT block for film/progressive scan sourced material and alternatively for video/interlaced sourced material (though increasingly video cameras are becoming progressive scan capable).

Figure 2.11.20 shows a macroblock that contains a progressive scan image. To the left it is processed in progressive scan mode and to the right it is processed in interlace mode; processing in progressive scan mode results in a slightly lower vertical high-frequency component, which will compress more efficiently.

Figure 2.11.21 shows the same image sourced in interlace video, where the object has moved horizontally between fields, resulting in DCT blocks that contain large amounts of vertical high-frequency information. Processing this macroblock in interlace mode reassociates the lines from the fields, resulting in DCT blocks with a much lower vertical frequency component that will compress more efficiently.

MPEG-2 encoders are free to choose whether to use interlace or progressive scan DCT modes; however, many decoders are unable to successfully decode MPEG streams in which the encode mode changes.

2.11.3.4.4 DCT Mathematics

The DCT transform is defined by a mathematical process that converts the spatial domain (pixels) representation to a

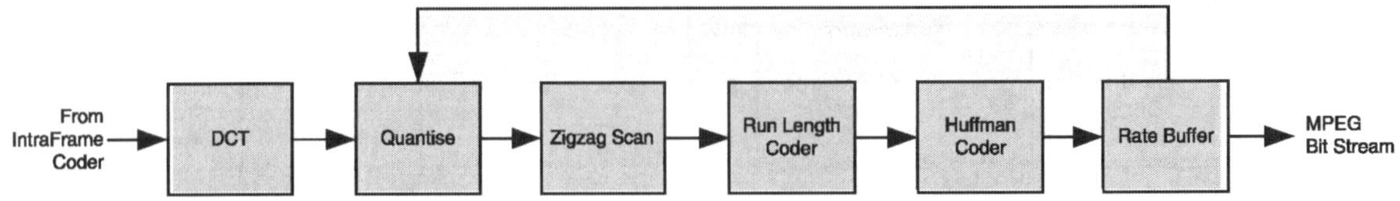

Figure 2.11.17 IntraFrame coding processes.

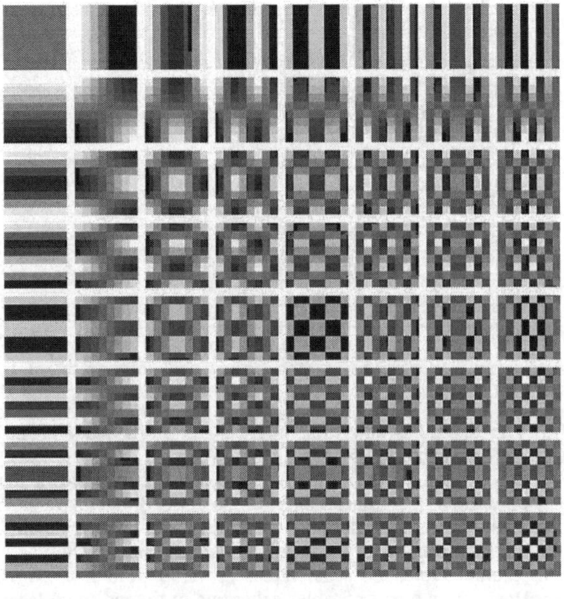

Figure 2.11.18 DCT pattern set; from left to right increasing horizontal frequency, from top to bottom increasing vertical frequency.

frequency domain representation, a process which is somewhat analogous to viewing a signal either on an oscilloscope (time domain) or on a spectrum analyser (frequency domain).

The DCT process is:

$$F(u,v) = 2C(u)C(v)/N \sum_{x=0}^{N-1}\sum_{y=0}^{N-1} f(x,y)\cos((2x+1)u\pi/2N)$$
$$\cos((2y+1)v\pi/2N)$$

for $u, v, x, y = 0, 1, 2, \ldots, N-1$ (for MPEG-2, N is 8)

x,y are the spatial (pixel) sample points, i.e. 0–7 for an 8×8 macroblock;

u,v are the coordinates in the frequency domain, again 0–7 for MPEG-2;

$F(u,v)$ is the amplitude of the frequency component;

$f(x,y)$ is the amplitude (brightness) of the luminance/chrominance of the pixel;

$C(u)C(v) = 1/\sqrt{2}$ for $u,v = 0$, else $= 1$.

The IDCT process is:

$$f(u,v) = 2/N \sum_{n=0}^{N-1}\sum_{v=0}^{N-1} C(u)C(v)F(u,v)\cos((2x+1)u\pi/2N)$$
$$\cos((2y+1)v\pi/2N)$$

2.11.3.4.5 *Quantisation*

Once the DCT has been performed the data can be quantised; the degree of quantisation necessary depends on how much

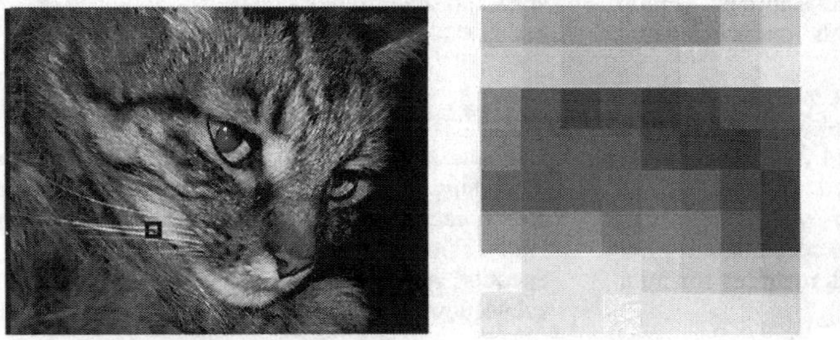

646	66	9	17	14	10	3	0
-87	-41	70	-14	22	6	-3	-6
314	-74	18	-25	-12	-9	-12	1
7	14	-52	13	-22	-10	1	14
-40	-14	-8	-19	12	-3	4	2
-38	34	34	-22	18	5	0	-2
-129	29	21	15	-1	14	-7	-1
-35	-22	15	7	7	8	-9	-6

Figure 2.11.19 Left: full image with DCT block highlighted. Centre: 8×8 pixel area of DCT block. Right: DCT frequency space values.

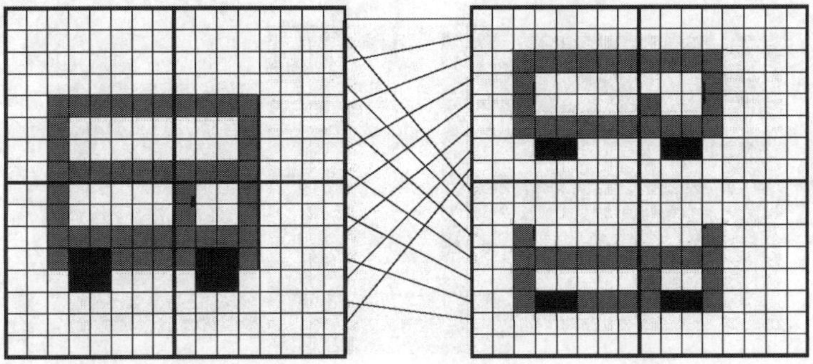

Figure 2.11.20 Progressive scan (film) sourced macroblock processed in DCT field mode.

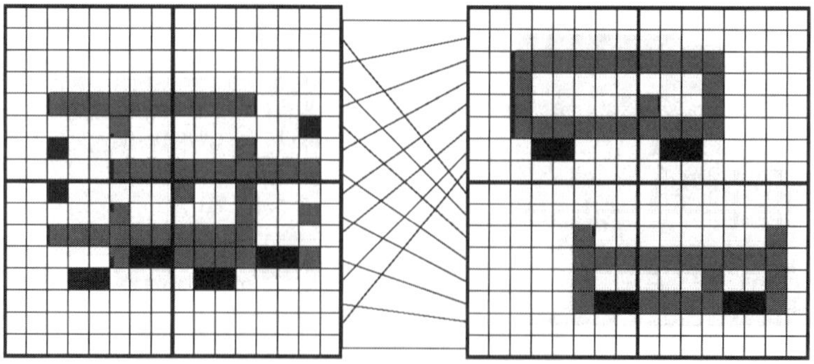

Figure 2.11.21 Interlaced (video) sourced macroblock processed in DCT field mode.

compression the lossless bit rate reduction process, run length and Huffman coding have achieved, and the final required video bit rate.

Figure 2.11.22 shows the effect on the picture and data of quantising the DCT block of Figure 2.11.19 by applying a quantising value of 32 to all DCT values (note that, in Figure 2.11.22, the values have been rescaled, multiplied by 32, to indicate the grey scale values in the image); the result is a significant reduction in the amount and range of values, which allows the following run length and Huffman coders to dramatically reduce the quantity of data needed to represent the image. As can be seen by comparing the images in Figures 2.11.19 and 2.11.22, the quantisation causes a change in the picture content; this is the lossy aspect of the coding.

In general terms the quantisation artefacts are most visible on hard edges, e.g. text or captions as 'mosquitoes' (Figure 2.11.23), or on areas of the images with low-level detail, e.g. grass or skin (Figure 2.11.24).

There are a number of different default quantisation tables defined by MPEG-2; for 4:2:0 encoded data quantisation matrices are defined for Intra and non-Intra frames (i.e. P and B frames). Different matrices are defined as there will be considerable difference in the structure of data contained in the two types of frames. For 4:2:2 and 4:4:4 coded data MPEG-2 additionally defines separate quantisation matrices for luminance and chrominance.

The default quantisation matrices for Intra and non-Intra frames are shown in Figure 2.11.25.

In addition to the default matrices, MPEG-2 allows that user-defined quantisation matrices be defined and also matrix extensions can be defined to modify an existing matrix. Further flexibility to quantisation is given through the use of a 'scale factor'; the scale factor (a value from 1 to 31) is selected from a pair of tables according to the indicated 'scale code value'. The quantiser scale value is then used to multiply the quantisation values, increasing the degree of quantisation.[5] As an example, if the applicable quantisation matrix value were 8 and the scale value 6, then the actual quantisation value used would be 48.

A new quantisation matrix can be defined in the bit stream during a sequence header and the matrix may be rescaled at slice and macroblock levels.

2.11.3.4.6 *Multi-generation*

Concatenation of encode decode phases leads to rapidly increasing levels of artefact in the image. The degree of degradation can be greatly reduced if the downstream encoders are GOP frame locked to the upstream encoders, i.e. what was encoded as an I frame originally is encoded as an I frame subsequently, rather than as a P or B frame.

640	64	0	0	0	0	0	0
-64	-32	64	0	0	0	0	0
288	-64	0	0	0	0	0	0
0	0	-32	0	0	0	0	0
-32	0	0	0	0	0	0	0
-32	32	32	0	0	0	0	0
-128	0	0	0	0	0	0	0
-32	0	0	0	0	0	0	0

Figure 2.11.22 Left: 8 × 8 pixel block after quantisation. Right: rescaled DCT values after quantisation.

Figure 2.11.23 Original image (left) and after compression (right), showing artefacts on text, detail and hard edges.

Figure 2.11.24 Increasing compression artefacts on skin.

8	16	19	22	26	27	29	34
16	16	22	24	27	29	34	37
19	22	26	27	29	34	34	38
22	22	26	27	29	34	37	40
22	26	27	29	32	35	40	48
26	27	29	32	35	40	48	58
26	27	29	34	38	46	56	69
27	29	35	38	46	56	69	83

16	16	16	16	16	16	16	16
16	16	16	16	16	16	16	16
16	16	16	16	16	16	16	16
16	16	16	16	16	16	16	16
16	16	16	16	16	16	16	16
16	16	16	16	16	16	16	16
16	16	16	16	16	16	16	16
16	16	16	16	16	16	16	16

Figure 2.11.25 Default Intra and non-Intra quantisation matrices.

2.11.3.4.7 Zigzag Scan

After the DCT and quantisation have been performed, the data is reordered using 'zigzag' scan. The purpose of the reordering is to arrange the data in a sequence that allows more efficient bit rate reduction by the following run length and Huffman processes. Two zigzag scan modes are available to the MPEG-2 encoder (Figure 2.11.26): the first is generally optimum for progressive scan sourced images, the second for interlace scanned images. The encoder is free to use either of the scan patterns.

The zigzag scan process reads the data starting in the top left position and reads the data following the path shown to the bottom right (see Figure 2.11.18 for the patterns corresponding to the scan positions).

2.11.3.4.8 Run Length Encoding and Entropy Coding

The final bit rate reduction process is the lossless coding; this consists of two methods, run length coding and Huffman coding.

Figure 2.11.27 shows the run length coding applied to data sequences generated by zigzag scan. The left shows a run of 10 zero words terminated by a +1 value; this is coded using a look-up table of values to the 8-bit sequence 001001110, where the first 8 bits are the run length code and the terminating 0 indicates a positive value (1 for negative values). The right of the diagram shows the special EOB (End Of Block) code, which codes all remaining zero words up to the end of the data sequence into the 2-bit pattern 01.

Overall there are around 100 run length codes defined in two alternate look-up tables (see Figure 2.11.28); as this is insufficient to define all possible DCT value sequences, an 'escape' code, 000001, is provided for sequences not included in the table. If the DCT value sequence is not contained in the look-up table the escape sequence is included, followed by an 18-bit pattern (Figure 2.11.29) to indicate the DCT value.

Thus the DCT value word sequence 0 0 0 0 +157 would be represented by the bit sequence 000001 000100 000010011101.

2.11.3.4.9 Rate Buffer

The rate buffer is an essential component of the compression process which allows the encoder to allocate more bits to some coded frames than others as, typically, I frames will require substantially more bits to encode for a given quality than P and B frames. During the encoding of a GOP, the buffer will be relatively full after encoding an I frame, then gradually empty as the P and B frames are encoded and the bit stream output.

If a sequence of video is input to the encoder that is difficult to encode, the buffer will start to fill, and the encoder will then increase the degree of quantisation to reduce the number of bits required to encode a frame; conversely, when the pictures become easier to encode, the buffer starts to empty and the encoder can reduce the degree of quantisation.

Figure 2.11.30 shows how the encoder buffer fullness varies over a number of frames and GOPs. Once a frame has been

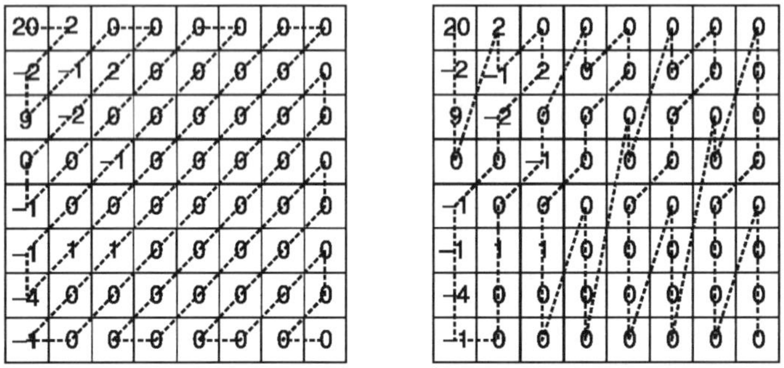

Figure 2.11.26 Zigzag scan patterns 0 and 1.

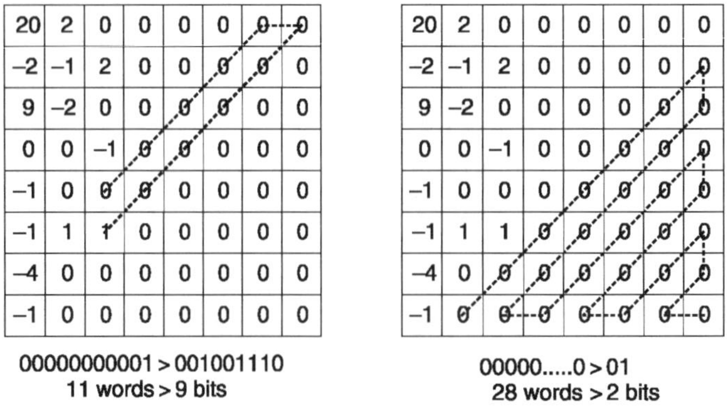

Figure 2.11.27 Example run length code and End of Block code.

0s run length	Final DCT value	Code word
End of Block		10
0	1	1
0	1	11
2	2	0000100
8	1	0000111
9	1	0000101
Escape		000001
0	5	00100110
3	2	00100100
10	1	00100111

Figure 2.11.28 Some example values from the DCT values look-up table (zero). An additional sign bit is added to the end of each sequence, 0 for +, 1 for −.

encoded its data is transferred to the buffer, causing the fullness level to jump then, while the next frame is being coded, the buffer empties at a constant rate, then jumps again as the coded data is moved to the buffer. When I frames are encoded the buffer fills to a greater degree, I frames being less efficient at bit rate reduction; conversely, as P and B frames are coded the buffer empties, as they require less bits for a given picture quality.

During GOP 1 the buffer is relatively empty so the encoder can use a relatively low degree of quantisation; during GOP 3 the buffer is relatively full so the encoder increases the degree of quantisation.

To decode the bit stream successfully, the receiving decoder requires an equivalent decode buffer. Details of standard buffer sizes are specified by the MPEG-2 standard.

2.11.3.5 Audio Coding ISO13818-3

MPEG audio coding is covered in the chapter covering non-linear audio systems.

2.11.3.6 Elementary Stream

The MPEG-2 elementary stream is the raw coded bit stream from the encoder containing a single element, e.g. audio or video.

Figure 2.11.31 shows the MPEG-2 bit stream structure from its highest level sequence down to DCT block level; each of the different headers starts with an identifying start code followed by information appropriate to and about that level.

Figure 2.11.32 shows some of the typical information carried at each level; there is a wealth of other mandatory and optional information carried by the various headers. For full details see Ref. 2.

2.11.4 MPEG System Layer ISO 13818-1

2.11.4.1 PES (Packetised Elementary Stream)

Once the various elementary streams, audio, video etc., have been coded they are converted to an intermediate level, the Packetised Elementary Stream (PES). The PES structure takes the raw bit stream and breaks it into smaller chunks (packets). Each of these packets is up to a maximum of 65,535 bytes in length, with the exception of video, which is unconstrained in length.

6-bit escape, 000001	6-bit run length, 0–63	12-bit value, −2047 to +2047

Figure 2.11.29 Representation of DCT values not included in look-up table.

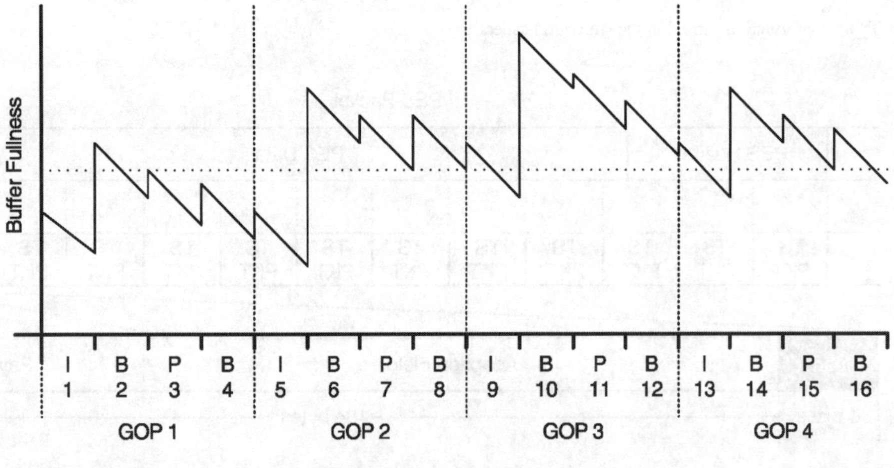

Figure 2.11.30 Encoder buffer fullness over four GOP/16 frame time periods.

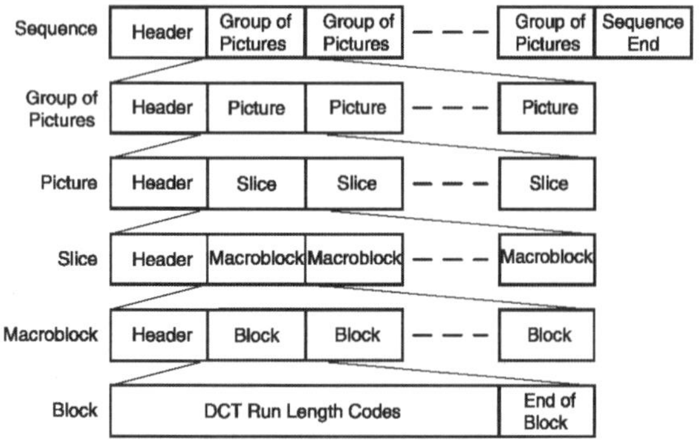

Figure 2.11.31 MPEG structures.

The PES header starts with the sequence 0x000001 and contains a variety of reference information, including:

- Stream ID, identifies the content as audio, video (0x01) etc., encryption data etc.
- Packet length, size in bytes, or 0 for unbounded.
- PES level scrambling control.
- Copyright information.
- PTS and DTS (Program and Decode Time Stamps; see Section 2.11.4.3.4).
- ESCR (Elementary Stream Clock Reference).
- Buffer fullness information (prevents buffer overflow if changing streams).

A wealth of other information is included (see Ref. 2).

The PES stream is only an intermediate MPEG-2 coding layer; in practice signals encountered outside equipment will be in the form of Transport Streams or, possibly, Program Streams.

2.11.4.2 Program Stream

A structure of Program Stream is defined by MPEG-2, originally intended for use within studios or other 'robust' environments; in practice, today, in a broadcast environment, the MPEG-2 Transport Stream is used in its place. Program Stream is commonly used for Internet MPEG files and for DVDs. For a description of the MPEG-2 Program Stream, see Ref. 6.

2.11.4.3 Transport Stream

Transport Stream (TS) was originally defined as the 'strong' method for carrying MPEG-2 data over transmission and emission systems. Today Transport Stream is the de facto standard used for interconnecting MPEG-2-based equipment, carried on a variety of physical interfaces (see Broadcast Interconnects chapter), the most common of which is ASI (Asynchronous Serial Interface).

The intention of the Transport Stream is to carry multiplexed PES streams, though it is legal just to carry a single PES stream.

2.11.4.3.1 TS Packet

Figure 2.11.33 shows how PES data is broken down into 184-byte chunks for inclusion in 188-byte TS packets. The

Sequence	Horizontal size, vertical size, aspect ratio, frames per second, bits per second, profile, level
Group of Pictures	Timecode information, closed GOP, broken link (GOP)
Picture	Temporal reference (frame in GOP), picture structure (field/frame)
Slice	Quantiser scaling
Macroblock	Quantiser scaling, motion vectors, DCT type
Block	Run length codes, EOB

Figure 2.11.32 Typical information carried in the structure headers.

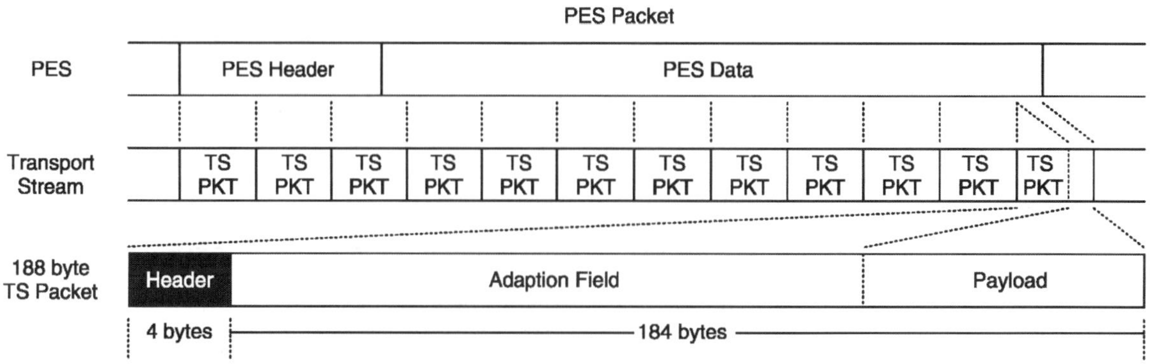

Figure 2.11.33 Transport Stream packet.

TS packet header is 4 bytes and contains the following information:

8-bit sync	0x47/01000111
1-bit TEI	Transport Error Indicator to flag corrupt packets.
1-bit PUSI	Payload Unit Start Indicator; 0 indicates PES header content starts at beginning of TS packet, otherwise a pointer to the start of the PES header information.
1-bit TP	Transport Priority, can flag high priority packets.
13-bit PID	Packet ID, used to associate packets with like content; see later.
2-bit TSC	Transport Scramble control; 00 indicates an unencrypted packet; 01/10 is used to flag encrypted Odd/Even packets (see Conditional Access, Simulcrypt and Encryption Systems chapter for explanation of operation).
2-bit AFC	Adaptation Field Control, indicates the presence of an Adaptation field, (which would often be the PCR).[7]
4-bit CC	Continuity Counter; counts 0 to 15 sequentially for packets of the same PID value, acting as a simple error check mechanism.

2.11.4.3.2 TS Multiplex

In practice a TS multiplexer will combine packets from many different sources to create usable TS.

Figure 2.11.34 shows a multiplexer creating an MPEG-2 Transport Stream by combining TS packets from various sources; packets incoming to the multiplexer will typically be from independent, unsynchronised sources, so the multiplexer buffers the incoming packets before creating the TS. As the TS bit rate is precisely defined, if there is an excess of packets incoming to the multiplexer some must be dropped; if there are insufficient packets coming into the multiplexer 'null' packets are inserted into the TS.

In order that equipment receiving the TS is able to associate packets from the same PES, each source of packets is stamped with an identifying PID value in the header, thus video A packets might be 128 and video B packets 512; with the exception of a few reserved packet values the PID values are freely assignable (Figure 2.11.35).

The ATSC also recommends the use of particular PID values (see ATSC chapter for details).

2.11.4.3.3 MPEG PSI

In addition to null packets the multiplexer also inserts PSI (Program Specific Information) packets. PSI contains the linking information a TS decoder requires to rebuild complete services.

Figure 2.11.36 shows an example TS containing three MPEG-2 programs. To decode the TS and recreate the original programs, a decoder first reads the PAT, which is defined to be on PID 0. The PAT contains the PID values of the PMT packets. The PMT packets in turn contain the PID values of the components needed to recreate the program. Thus if the decoder has been instructed to display program 2 it first decodes the PAT, known to be on PID 0, reads the PID value for PMT 2, 193, decodes packets with PID 193 to read the PMT, which in turn contains the PID values of the components needed to display program 2, Video from PID 1025 packets, Audio from PID 1090 packets, Teletext from PID 1152 packets and PCR (Program Clock Reference; see later) timing information from PID 8190.

Note that programs may have multiple audio channels, typically languages, associated with them, ISO 639 language descriptors being carried to enable users to choose their preferred language, program 1 in Figure 2.11.36.

Note also that programs can share components; in Figure 2.11.36 video on PID 1025 is used by both programs 2 and 3 to create English and Finnish language versions of the same programme.

An NIT (Network Information Table) is defined by MPEG, which may carry tuning information such as frequency satellite position station name etc. If the NIT is carried it is referenced in the PAT as program 0. MPEG-2 defines the NIT as being optional, but is made mandatory by DVB (see the DVB chapter for further details).

MPEG-2 defines the existence of an optional CAT (Conditional Access Table) on PID 1, used to link encryption data and program components (see Encryption chapter for further details).

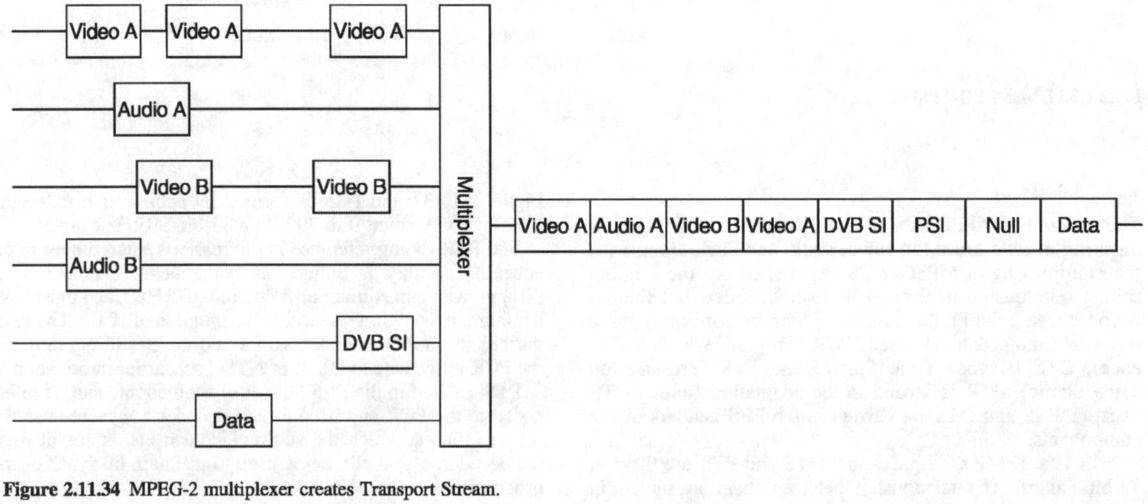

Figure 2.11.34 MPEG-2 multiplexer creates Transport Stream.

PID range	Use
0x0000	PAT (Programme Association Table)
0x0001	CAT (Conditional Access Table)
0x0002	TSDT (Transport Stream Descriptor Table)
0x0003–0x000F	reserved for MPEG
0x0010–0x001F	reserved for DVB
0x0020–0x1FFE	
0x1FFF	Null packets

Figure 2.11.35 PID value allocation.

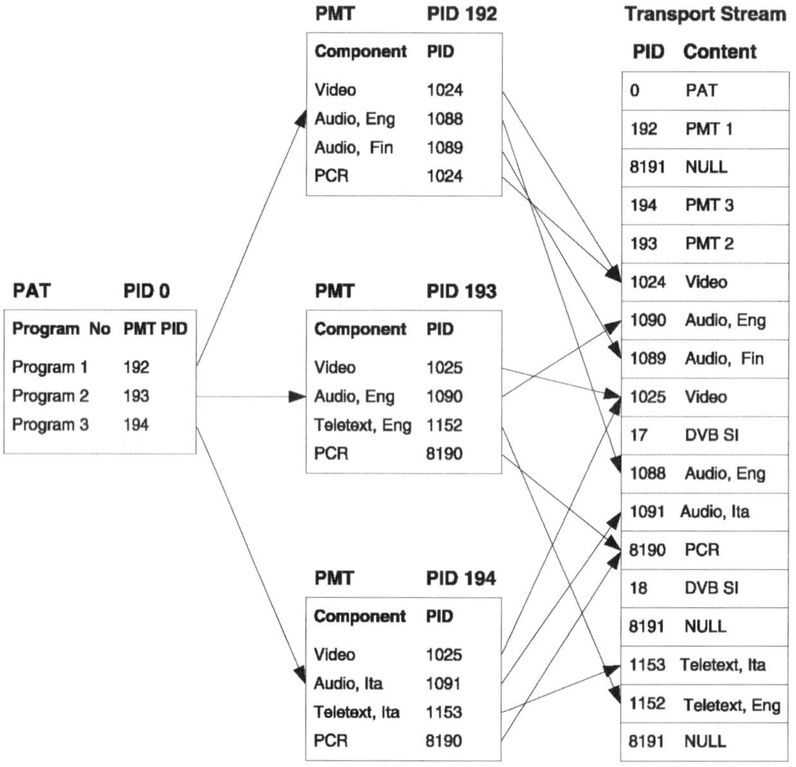

Figure 2.11.36 MPEG-2 PSI structures.

2.11.4.3.4 Timing

The creation of MPEG PES streams, particularly with greatly differing encode times for video, audio and data, along with the multiplexing of MPEG-2 TS packets, causes the original timing relationship of data to be lost. In order that the TS decoder can rebuild the original program correctly, three types of timing data are used, PCR (Program Clock Reference), DTS (Decode Time Stamp) and PTS (Presentation Time Stamp). PCR is carried in the adaptation fields of TS packets, PTS and DTS are carried in the PES headers of the components.

PCR is a 27 MHz, 42-bit, value; DTS and PTS are 90 kHz, 33-bit values. The relationship between them is shown in Figure 2.11.37; the PCR is formed by adding nine extension bits, that only count 0 to 299, to a 33-bit, 90 kHz count.

The PCR timing reference is the master system timing reference; its stability is defined as being better than ±810 Hz, 30 ppm, with a maximum drift rate of 0.075 Hz/s, and must have a minimum repetition rate within the program of 10 Hz. The strict definitions on PCR are necessary as a decoder will typically use the PCR to recreate the PAL or NTSC subcarrier information.

PTS, carried in the PES header of each component, is referenced to the PCR and used by the decoder to synchronise the relative time at which the audio/video/text etc. is output from the decoder, essentially being used to maintain lip synch of the programme.

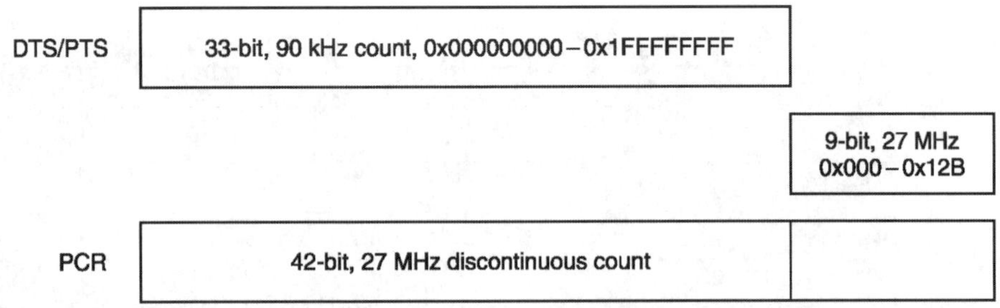

Figure 2.11.37 Relationship of PCR, PTS and DTS.

DTS is necessary for the decoder as MPEG video frames may be transmitted out of sequence; before a decoder can rebuild a B frame it requires the following I or P frame, thus it is more efficient in terms of buffer usage for an encoder to transmit the I or P frame out of sequence, before the B frame referencing it. DTS allows the decoder to know in which sequence to use the received I, P and B frames.

2.11.4.3.5 PCR carriage methods

The PCR is defined as being carried in the adaptation field of a TS packet; a number of different methods exist for carrying PCRs in a transport stream.

The TS can carry a single PCR stream. In a dedicated PCR PID, the TS can carry multiple PCR streams, typically one per program; the PCR may be embedded in the adaptation field of a program component carrying component data.

In a TS carrying many programs, using separate PCR streams, where the PCR is not embedded in a packet carrying data, the bandwidth used can be significant. For example, 16 programs each with a separate PCR stream would use 240.64 kbit/s of bandwidth (16 programs × 10 PCRs/second ×188 bytes/packet × 8 bits/byte).

2.11.4.4 Statistical Multiplexing

Statistical multiplexing is a technique commonly used in MPEG-2 Transport Streams carrying multiple video components to improve the average picture quality. A normal video programme will contains scenes of varying difficulty to encode – for example, a news programme might contain easy-to-code head and shoulders shots interspersed with sport or action footage. If this programme material is encoded at a constant bit rate the allocated bit rate might be excessively generous for the head and shoulders scenes but insufficient for the sports or action scenes. Statistical multiplexing takes a number of video sources and allocates an overall bit rate to the group; if diverse video sources are grouped it is likely that when one source requires a high bit rate others will not. An alternative to allocating a fixed group bit rate is to encode the video sources at constant quality, i.e. the encoder only uses the bit rate it needs to achieve the desired encode quality. Figure 2.11.38 shows the three encode methods. In the case of constant quality encoding there will be times when all the available TS bandwidth goes unused; this spare bandwidth could be used to carry opportunistic data services or other data that does not need to be transmitted as real time information.

Figure 2.11.39 shows the variation in encoded picture quality vs. used bit rate for a broadcast news programme. Note how the bit rate varies from 1.5 to 8 Mbit/s as the encoder attempts to achieve constant picture quality.

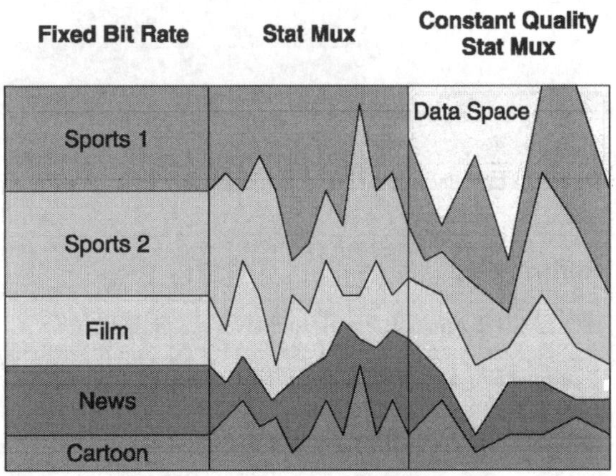

Figure 2.11.38 Constant bit rate, fixed group bit rate and constant quality encoding.

Note also that at some times the picture quality is relatively low even though the bit rate is relatively high; this indicates scenes that are particularly difficult to encode. At other times both the quality and bit rate are relatively low; this indicates other encoders sharing the transport stream are also encoding difficult material.

2.11.4.5 Remultiplexing

It is often necessary to remultiplex MPEG-2 transport streams – for example, complete programmes or components may be extracted from a broadcast source and be rebroadcast mixed with local content. The process of remultiplexing requires some fairly complex processing and rebuilding of the transport stream.

Packets for audio, video, etc. components are inserted or deleted; as a consequence of this PMTs need to be added, modified or deleted, which in turn requires the PAT to be modified.

Packets to be remultiplexed may be incoming on the same PID values, which requires PID values to be rewritten.

Due to differences in transport stream bit rates remultiplexed PCRs are not able to be inserted into the new transport stream at exactly the correct points in time, meaning they need to have their time values rewritten (to be de-jittered).

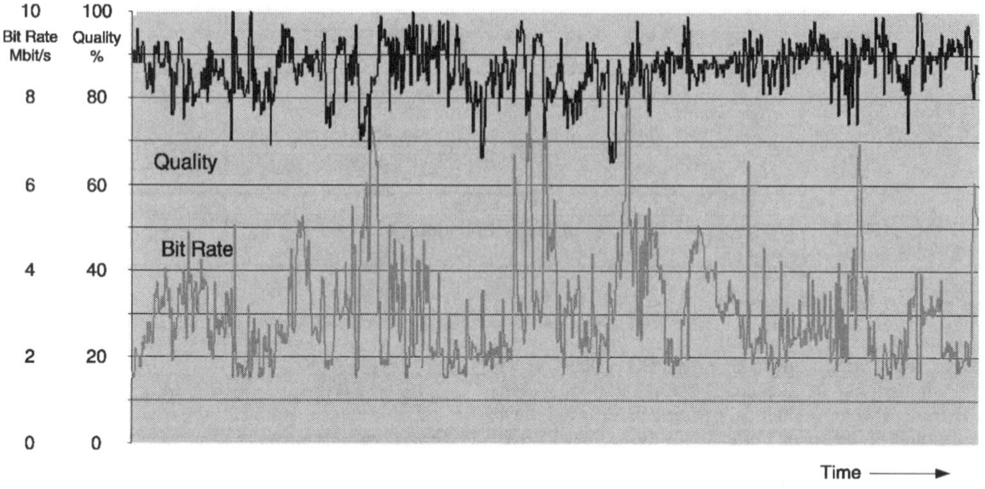

Figure 2.11.39 Encoded bit rate (lower trace) and achieved picture quality (upper trace) for a statistically multiplexed video source.

References

ISO 13818-2 Annex E Profile and level restrictions.

ITU-R Recommendation BT.601-5 (10/95) *Studio encoding parameters of digital television for standard 4:3 and wide-screen 16:9 aspect ratios.*

ISO 13818-2 section 7.4.2.2 Quantiser scale factor

ISO 13818-1 section 2.5 Program bitstream requirements

ISO 13818-1 section 2.4.3.4 Adaptation Field

Gordon Anderson
Technical Training
Tandberg Television Ltd

2.12 DVB Standards

The European Launching Group (ELG) was formed in 1991 to set a digital TV broadcasting standard following the growing realisation that analogue HDTV would not be a commercially successful proposition. ELG, in turn, created the Digital Video Broadcasting (DVB) Project in September 1993.

DVB specifications originally only applied to European broadcasters, but have now become de facto standards in many other parts of the world. Interoperability between equipment from different vendors was uncertain in the early days of digital compression (1995–1997), but a DVB-S modulator was virtually guaranteed to work with a DVB-S receiver, especially after the Intelsat Satellite Operators Group (ISOG) sponsored a series of interoperability tests. DVB also defined baseband interfaces so that Conditional Access systems, encoders, multiplexers and modulators from different vendors could be connected together in complex systems.

The DVB Project now comprises over 260 organisations from more than 30 countries (including non-European countries). Members include public and private broadcasters, telcos, cable TV operators, consumer electronics and computer companies. Learning from the experience of the fragmentation of the European MAC system[1] into several incompatible variants, DVB standards are produced in a timely, market-led manner. The DVB Project attempts to draft future-proof standards which can be implemented (at least partially) within months of their publication date.

The General Assembly meets every year to present progress reports on all the modules, budgets and other matters. Every second year it holds the elections of members of the Steering Board.

The Steering Board controls the activities of the four modules beneath it.

The Commercial Module is the usual starting point for the standards process. If enough members are in favour of starting a new standard, a subgroup is formed to create draft Commercial Requirements. When these have been approved by the Commercial Module they are passed to the Technical Module.

The Technical Module forms a subgroup of specialists to create Technical Specifications (and, sometimes, Implementation Guidelines) which are then passed back to the Commercial Module for approval.

The Intellectual Property Rights Module deals with issues such as patents and copyright. DVB members are required to licence essential IPR under Fair, Reasonable and Non-Discriminatory (FRND) terms.

The Promotions and Communications Module deals with promotion, advertising, attendance at trade shows, etc. (see Figure 2.12.1).

DVB standards are agreed by consensus rather than a majority vote in the 'pre-competitive' phase of development, before any

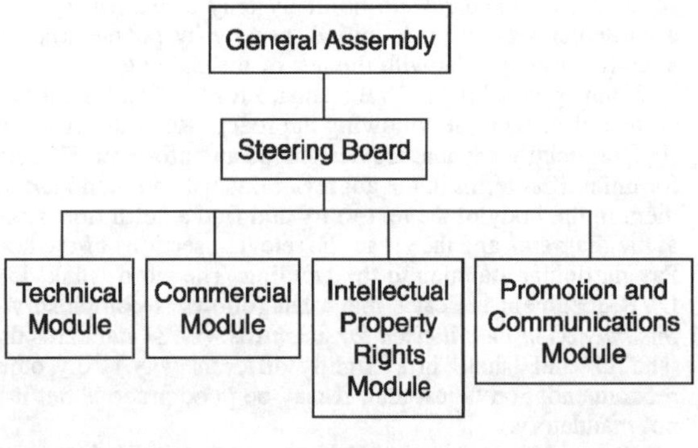

Figure 2.12.1 Structure of DVB Project.

manufacturer has been able to dominate the market with a proprietary product or system. Despite the requirement for unanimity, standards are produced fairly rapidly without protracted wrangling. Both MPEG-2 and DVB are examples of a new breed of 'fast-track' standardisation body which should be able to forestall a recurrence of the costly 'format wars' of the past – although the current situation with three recordable DVD formats shows what can still happen when manufacturers rush to market.

Although the DVB Project drafts technical standards, it is not an official standards body. DVB proposals only become official standards when published by a body such as the European Telecommunications Standards Institute (ETSI) or the International Telecommunication Union (ITU-T). CENELEC (Comité Européen de Normalisation ELECtrotechnique) produces official standards for consumer items such as Set-Top Boxes (STBs) and Personal Video Recorders (PVRs).

DVB documents usually have a document number with an Axxx format, where xxx is a number. When published by the ETSI they will have an ETR or ETS prefix. ETR indicates an ETSI Technical Report, ETS indicates an ETSI Technical Standard. ETSI and CENELEC publish EN standards (European Norms).

The DVB Project has generated over 60 standards. Rather than listing them numerically or chronologically, they are grouped according to function in this chapter. Originally each standard was known by its official title, but gradually colloquial 'nicknames' replaced the proper names (e.g. ETS 300 421 became known as DVB-S). DVB documents now tend to use these shorter forms, and they will usually be used here as headings. After each heading the most recent DVB document title is printed in full. This will be followed by the ETSI, CENELEC or ITU-T standard name in brackets. If more than one DVB proposal applies to a particular topic they will all be listed. Obsolete DVB documents will not be listed.

The use of *italicised_underscored_phrases* indicates a variable name taken directly from a DVB or MPEG-2 specification. MPEG-2 and DVB documents are generally available in electronic formats. Using the 'Find' function it should be easy to find these variables in the relevant documents.

2.12.1 Reading a Technical Standard

Standards documents are intended to be an unambiguous definition of an interface, process, signal, etc., but they are often neither easy to read nor understand the first time they are encountered. Standards are normally only consulted to check a particular section or definition, usually by people who are already fairly familiar with the rest of the contents.

When approaching a DVB standard for the first time it may be useful to take the following approach. Read the Annexes first, especially any that are designated as 'informative'. Look for unfamiliar terms in the abbreviations list and then look for them in the body of the text to try and find a definition. Look at the diagrams and then read the relevant sections of the text. Pay particular attention to the wording. The word 'shall' in a DVB document indicates that what follows is compulsory – *this may often be italicised for emphasis.* (ETSI standards use 'should' and 'shall' in a slightly different way.) Any other recommendation is just that, it may be good practice but it is not mandatory.

UK and US readers should be aware that DVB documents normally use the European convention of representing a decimal point with a 'comma'.

2.12.2 DVB Cookbook

DVB Document A020 Rev. 1: *Digital Video Broadcasting (DVB); A Guideline for the Use of DVB Specifications and Standards*, May 2000
(**Draft ETSI TR 101 200 V1.2.1:** *Digital Video Broadcasting (DVB); A Guideline for the Use of DVB Specifications and Standards*, September 1997)

This document serves as an overview of the DVB standards but does not explain any of them in detail. Professor Ulrich Reimers is the technical guiding hand of the DVB and in his introduction he provides a brief overview of the three phases of DVB activity. Phase 1 was the production of standards covering modulation, service advertisement/acquisition and encryption. Phase 2 concentrated on interactivity and phase 3 is moving into the multi-media arena.

The text has sections covering baseband processing, transmission, Conditional Access, interactive services and 'miscellaneous'. Each section gives a brief background to the reason for having particular standards and lists all of the standards for that subject by document number.

The final six pages contain a comprehensive listing of published specifications and other DVB documents. The official document number is followed by the DVB nickname (e.g. TS 101 191 is DVB-SFN) and the full document title.

The ETSI Technical Report adds an abbreviations list and a short bibliography.

2.12.3 DVB Baseband

DVB Baseband refers to all DVB processes carried out before or in a multiplexer. In this chapter, Service Information and Conditional Access have separate sections of their own although, strictly speaking, they are baseband processes.

2.12.3.1 Direct To Home (DTH)

DVB Document A001 Rev. 6: *Digital Video Broadcasting (DVB); Implementation Guidelines for the Use of MPEG-2 Systems, Video and Audio in Satellite, Cable and Terrestrial Broadcasting Applications*, May 2000
(**ETSI TR 101 154 V1.4.1:** *Digital Video Broadcasting (DVB); Implementation Guidelines for the use of MPEG-2 Systems, Video and Audio in Satellite, Cable and Terrestrial Broadcasting Applications*, July 2000)

This document specifies permitted syntax in DVB compliant Transport Streams and minimum performance of domestic Set-Top Boxes (STBs). Most of the standard refers to the MPEG-2[2–5] and MPEG-1[6,7] specifications.

2.12.3.1.1 General

Transport Streams will be used rather than Program Streams. DVB Service Information (SI) is to use the same table structure as MPEG-2 Program Specific Information (PSI). IRDs should ignore valid descriptors or data structures which they do not understand.

2.12.3.1.2 PCRs

It is recommended that the inter-PCR interval is no greater than 40 ms. Clock tolerance is recommended to be within 5 parts per million.

2.12.3.1.3 PSI Repetition Rates

It is recommended that PATs and PMTs are sent 10 or more times per second. (For a single packet table this gives a transmitted rate of approximately 15 kbit/s per table.)

It is recommended that a Transport Stream Descriptor Table is sent at least every 10 seconds. (The TSDT contains the ASCII code for 'DVB' to tell a receiver what type of Transport Stream it is in.)

2.12.3.1.4 ISO 639 Language Descriptor

The ISO 639[8] descriptor is usually placed in the PMT to describe audio, teletext or subtitling components. Its use is mandatory if different language components are present in a program. The descriptor contains three ASCII bytes detailing the language of the component (eng for English, fre for French and so on). ISO 639 has abbreviations for most major European languages, but does not describe non-European languages as comprehensively. Broadcasters can (and do) make up their own descriptors – but it is essential that the receiving STB understands novel descriptors or they may not work properly.

2.12.3.1.5 Video

Main Profile at Main Level is to be used for SDTV. Main Profile at High Level is to be used for HDTV.

Thirty hertz IRDs are to support drop-frame rates as well as 30 Hz and 24 Hz. IRDs must support a number of video resolutions, aspect ratios and the use of pan vectors. They may support the use of the Active Format Descriptor as well.

It is recommended that a video sequence header and an I frame are sent at least twice a second (i.e. short Group Of Pictures (GOP) with an I frame at the start of each GOP).

2.12.3.1.6 Audio

MPEG-1 or MPEG-2 stereo Layer I and Layer II are to be supported by all IRDs. Layer II coding is recommended. J.17 pre-emphasis is not to be used. Half sampling rates are not mandatory. IRDs may also support Dolby AC-3 decoding.[9] IRDs shall support mono, dual mono, stereo and joint stereo decoding. (There has been a suggestion to remove dual mono from the DVB standard, but this would cause problems for some legacy systems, so it will probably be kept.)

Originally DVB only specified MUSICAM for audio compression (as used in the MPEG-1 and MPEG-2 standards). When Australia started simulcasting SDTV and HDTV programmes they used both MUSICAM and Dolby AC-3 to simplify STB design. DVB subsequently incorporated Dolby AC-3 into the baseband specification and some European broadcasters are experimenting with Dolby AC-3 transmissions. Dolby AC-3 has the same equivalent quality as MUSICAM Layer II at half the transmitted bit rate and is an attractive proposition for multi-language broadcasters. (Advanced Audio Coding[10] has a similar efficiency to Dolby AC-3 but is not specified by DVB.)

The recommended level for reference tones is 18 dB below clipping.[11]

2.12.3.1.7 Annexes

Annex A is a table containing all of the allowable video resolutions and frame rates.

Annex B contains a detailed description of the Active Format Descriptor (AFD) and its relationship to Wide Screen Signalling (WSS).[12] It also contains a table showing what sort of image should be displayed on 4:3 and 16:9 displays for each AFD value.

Annex C gives details of AC-3 audio in DVB Transport Streams.

Annex D gives details of ancillary data for MPEG audio.

Annex E describes the use of private data bytes in adaptation fields.

2.12.3.2 Contribution and Distribution

DVB Document A058: *Digital Video Broadcasting (DVB); Implementation Guidelines for the use of MPEG-2 Systems, Video and Audio in Contribution and Primary Distribution Applications*, March 2001
(ETSI TR 102 154 V1.1.1: *Digital Video Broadcasting (DVB); Implementation Guidelines for the use of MPEG-2 Systems, Video and Audio in Contribution and Primary Distribution Applications*, April 2001)

This standard is the professional equivalent of the previous standard and covers the types of Transport Streams likely to be encountered by a professional receiver rather than a domestic one. Many of the requirements or recommendations are identical to those in ETR 101 154.

2.12.3.2.1 General

The Basic Interoperable Scrambling System[13] (BISS) may be used.

2.12.3.2.2 PSI

The TSDT rate recommendation is increased to 10 times per second.

2.12.3.2.3 Video

SDTV encoding shall be 4:2:2 Profile at Main Level or Main Profile at Main Level (or simpler). HDTV encoding shall be 4:2:2 Profile at High Level or Main Profile at High Level.

HDTV decoders should be able to decode all SDTV pictures. (This is not the case with many domestic HDTV decoders or early professional HDTV decoders.)

2.12.3.2.4 Audio

Contribution IRDs shall support uncompressed audio[14] or SMPTE/AES data[15] via an AES3 interface.[16] IRDs shall be able to support SMPTE 302M[17] mapping of AES3 signals into Transport Stream packets. A 302M elementary stream has a *stream_id* of 0xBD (indicating a private stream) and a *stream_type* of 0x06 (private data). (The former indicates the service type and the latter indicates the component type.) The PMT will include a *format_identifier* of 0x42535344 (ASCII for 'BSSD') to indicate the presence of AES PCM or SMPTE/AES data.

Lip-sync is not tightly specified in 'standard' MPEG-2. SMPTE 302M data should be reproduced by an IRD with 'lip-sync' better than ±1 ms. (This is half the ITU-R Recommendation BT.1359-1[18] tolerance.)

2.12.3.2.5 Annexes

The Annexes A–E are similar to those in ETSI TR 101 154.

Annex F gives details of the BISS system.

2.12.3.3 Teletext (DVB-TXT)

DVB Document A041: *Digital Video Broadcasting (DVB); Specification for Conveying ITU-R System B Teletext in DVB Bitstreams*, June 1999
(ETSI EN 300 472 V1.3.1: *Digital Video Broadcasting (DVB); Specification for Conveying ITU-R System B Teletext in DVB Bitstreams*, January 2003)

Teletext[19–21] is a method of carrying digital data in the Vertical Blanking Interval (VBI) of a 25 Hz video waveform. (Experimental systems were tried in the USA and Canada but it was never widely adopted in 30 Hz countries.) The data normally contain TV listings but may also contain subtitles, news, weather or other information services. A suitably equipped receiver will allow a viewer to select a 'page' of teletext data which is then converted to simple, eight-colour text or graphics for display on the TV screen (for further details, see Chapter 2.7).

Each VBI line containing teletext carries 360 bits of data (45 bytes). The first 3 bytes are 0xAAAAE4, the rest are variable and contain address and control information as well as the actual data. The first 2 bytes are the clock run-in sequence – these are used to synchronise the clock in a teletext receiver and do not need to be transmitted in a synchronous system. If the original teletext waveform is to be regenerated in an STB the clock run-in sequence will be recreated. If the teletext is to be decoded in the STB this is not necessary. The remaining 43 bytes are placed in a PES packet which is then placed in a Transport Packet. PES packets containing teletext subtitles may have a PTS in the PES header.

DVB teletext can also indicate on which VBI line the teletext data were present in the original analogue waveform so that a decoder can re-insert them on the correct line. Lines 6, 318 and 319 cannot be signalled in this way. (The original (UK) teletext specification did not allow the use of these lines but the newer World Standard Teletext does.)

A Transport Packet containing a PES header can carry one, two or three lines of teletext data. A Transport Packet without a PES header can carry one, two, three or four lines of teletext data. For efficient use of transmission bandwidth PES packets should contain a multiple of four (minus one) teletext lines (3, 7, 11, 15, 19 etc.). If an analogue waveform had three lines of teletext on each field of video they would occupy 75.2 kbit/s of Transport Stream bandwidth (given by 188 × 8 × 50).

2.12.3.3.1 Example Packet

Figure 2.12.2 shows a Transport Packet containing DVB teletext. The first four (highlighted) bytes are the Transport Packet header (0x47 40 82 16). From these bytes it can be seen that the PID of this packet is 130 (0x082) and that there is a payload but no adaptation field. The Payload Unit Start Indicator reveals that a PES packet starts in this Transport Packet (see Chapter 2.11 for details of Transport Packet structure).

```
47 40 82 16 00 00 01 bd 05 02 84 80 24 21 78 d1
2f ed ff ff ff ff ff ff ff ff ff ff ff ff ff ff
ff ff ff ff ff ff ff ff ff ff ff ff ff ff ff ff
ff 10 02 2c e7 e4 f4 f4 04 04 0d ec f4 9d ec 04
83 75 75 75 75 75 75 75 75 75 04 a2 c1 04 ec 4c
0d 75 ec cd 40 04 04 d5 2c a4 e0 cd 0d f4 0d 2c
02 2c e8 e4 57 f4 04 04 8c 0d f4 9d ec 04 83 75
75 75 75 75 75 75 75 75 04 a2 c1 04 ec 4c 1c 75
8c ad 40 04 04 d5 2c a4 e0 cd 0d f4 0d 2c 02 2c
e9 e4 f4 0b 04 04 0d 8c f4 9d 1c 04 83 75 75 75
75 75 75 75 75 75 04 a2 c1 04 ec 0d 9d 75 6d ec
40 04 04 d5 ad a4 e0 cd 0d f4 0d 2c
```

Figure 2.12.2 Transport Packet containing PES header and three teletext packs.

The next 45 bytes are the PES header (Figure 2.12.3). The first 3 bytes (0x000001) are the PES start code prefix. The next byte (0xBD) indicates that this is a *private_stream_1*. The following 2 bytes (0x0502) indicate the length of the rest of the PES packet (1282 in decimal). From this it can be calculated that this PES packet contains 27 teletext packets and occupies seven Transport Packets in total. The next 2 bytes (0x8480) contain a number of flags and indicators – one of which indicates the presence of a PTS. The next byte (0x24 = 36 decimal) indicates the length of the remainder of the PES header. The next 5 bytes are the PTS (when the three stuffing bits are removed from 0x1 78 D1 2F ED what remains is the PTS, i.e. 0x1E 34 17 F6 or 506,730,486 in decimal). The following 31 bytes are stuffing bytes (0xFF) (see Chapter 2.11 for details of PES Packet structure).

```
            00 00 01 bd 05 02 84 80 24 21 78 d1
2f ed ff ff ff ff ff ff ff ff ff ff ff ff ff ff
ff ff ff ff ff ff ff ff ff ff ff ff ff ff ff ff
ff 10
```

Figure 2.12.3 PES header and *data_identifier*.

The next byte (0x10) is a *data_identifier*. EBU teletext data always has values in the range 0x10 to 0x1F.

The rest of the packet consists of three 46-byte sections. Each section contains a teletext packet (a packet is a complete line of teletext data from the VBI). The *data_unit_id* is 0x02, indicating that this is EBU teletext with non-subtitle data. The *data_unit_length* is 0x2C (44 decimal), indicating that there are 44 bytes to follow. The bytes 0xE7, 0xE8 and 0xE9 indicate that these three teletext packets are from field 1 lines 7, 8 and 9. The *framing_code* 0xE4 is the same as the framing code at the start of a line of teletext after the two clock run-in bytes. The next 2 bytes are Hamming-protected[22] magazine and line numbers and the remaining 40 bytes are parity-protected display bytes. (The teletext packets are all from magazine 7 and correspond to teletext rows 14, 15 and 16 (see Figure 2.12.4). See Chapter 2.7 for details of teletext.)

```
   02 2c e7 e4 f4 f4 04 04 0d ec f4 9d ec 04
83 75 75 75 75 75 75 75 75 75 04 a2 c1 04 ec 4c
0d 75 ec cd 40 04 04 d5 2c a4 e0 cd 0d f4 0d 2c

02 2c e8 e4 57 f4 04 04 8c 0d f4 9d ec 04 83 75
75 75 75 75 75 75 75 75 04 a2 c1 04 ec 4c 1c 75
8c ad 40 04 04 d5 2c a4 e0 cd 0d f4 0d 2c

                                          02 2c
e9 e4 f4 0b 04 04 0d 8c f4 9d 1c 04 83 75 75 75
75 75 75 75 75 75 04 a2 c1 04 ec 0d 9d 75 6d ec
40 04 04 d5 ad a4 e0 cd 0d f4 0d 2c
```

Figure 2.12.4 Three teletext packets (magazine 7, rows 14–16).

The 40 remaining bytes do not 'look' like ASCII text. There are two reasons for this. These packets do not contain much text – they appear to be share price listings or some form of table. The data are sent in what MPEG and DVB specifications refer to as *bslbf* (bit string, left bit first) format and they have an odd parity bit (*bslbf* is sometimes referred to as 'little-endian').

2.12.3.4 VBI carriage (DVB-VBI)

DVB Document A056: *Digital Video Broadcasting (DVB); Standard for Conveying VBI Data in DVB Bitstreams*, May 2000
(ETSI EN 301 775 V1.1.1: *Digital Video Broadcasting (DVB); Specification for the Carriage of Vertical Blanking Information (VBI) Data in DVB Bitstreams*, November 2000)

This standard is an enhanced version of EN 300 472 with support for further VBI signals. All VBI PES packets carry a PTS. In EN 300 472 teletext packets the PTS was not supposed to be optional but ambiguous wording in the original version of the specification (subsequently rectified) caused some manufacturers to not include a PTS – so the two standards are not necessarily interoperable. EN 301 775 should supplant EN 300 472 eventually. The PES packet contains a *data_identifier* indicating the type(s) of data following and one or more *data_unit_ids* indicating the data contained in each loop.

2.12.3.4.1 Teletext

Three distinct types of teletext are described in EN 301 775; these are EBU teletext, EBU teletext used for subtitling and inverted teletext.[23] They are indicated with *data_unit_ids* of 0x02, 0x03 and 0xC0 respectively. The *data_unit_id* is followed by a byte detailing the field and line number, a byte containing the framing code and 42 bytes (336 bits) of teletext data. (Inverted teletext has an inverted framing code, i.e. 0x1B instead of 0xE4.)

2.12.3.4.2 Video Programme System (VPS)

VPS[24] has a *data_unit_id* of 0xC3. The *data_unit_id* is followed by a byte detailing the field and line number and 13 bytes (104 bits) of VPS data. The run-in and start code for VPS are not sent in the PES packet. VPS is only present on line 16 of field 1.

2.12.3.4.3 Wide Screen Signalling (WSS)

WSS[12] has a *data_unit_id* of 0xC4. The *data_unit_id* is followed by a byte detailing the field and line number and 2 bytes containing 14 bits of WSS data and two reserved bits. WSS is only ever present on the first half of line 23.

2.12.3.4.4 Closed Captioning

Closed Captioning[25] has a *data_unit_id* of 0xC5. The *data_unit_id* is followed by a byte detailing the field and line number, and 2 bytes of Closed Captioning data. Closed Captioning subtitles are only present on line 21 of field 1. 'V-chip' data are present on line 21 of field 2.

2.12.3.4.5 Luminance Samples

If the VBI contains a non-standard waveform it can be sampled and sent as (up to) 720 8-bit luminance samples. Monochrome 4:2:2 samples have a *data_unit_id* of 0xC6. The *data_unit_id* is followed by a byte detailing the field and line number and 2 bytes to indicate the position of the first pixel (a number between 0 and 719). This is followed by up to 720 8-bit luminance samples. A full line of luminance samples will occupy more than one Transport Packet, so the PES packet will be fragmented in the usual way. If 32 lines were encoded in this manner (in a 25 Hz system) the total TS bandwidth required could be more than 4.6 Mbit/s. Even one line would require just over 144 kbit/s, so this technique should only be used when no others are available.

2.12.3.4.6 Combining VBI data with EN 300 472 teletext data

Some older (DVB teletext) receivers may not be able to cope with DVB VBI data. Where a mixed population of receivers is in use it is recommended that two PIDs are used, one for EN 300 472 teletext data and the other for DVB VBI data. The teletext data must be simulcast on both PIDs.

2.12.3.5 DVB Subtitles (DVB-SUB)

DVB Document A009 Rev. 1: *Digital Video Broadcasting (DVB); Subtitling Systems*, February 2002
(Draft ETSI EN 300 743 V1.2.1: *Digital Video Broadcasting (DVB); Subtitling Systems*)

DVB Subtitling is a method of transmitting graphical objects to an STB for display on regions of the TV screen. The regions may have varying sizes, positions and durations – these are conveyed to the STB in a syntactic element called a *page_composition_segment*. The objects may contain graphical or textual information and may be used for subtitling, on-screen logos, maps, etc. The regions are rectangular and two or more regions may be showing objects at the same time. No two regions can share scan lines, so regions may not be located at the same height on the screen nor can they partially overlap. Run-length encoding can be used to reduce the amount of transmitted data. A PTS is always present in the PES packet to tell the STB when to present the graphical object.

2.12.3.5.1 CLUTs

In each region a Colour Look-Up Table (CLUT) translates logical colours into display colours. Three forms of CLUT exist to cope with STBs with differing graphical processing capabilities. Pixel depths of 2, 4 and 8 bits are supported, giving four, 16 or 256 different colours for each region. Mappings between logical and display colours allow dynamic colour changes on-screen without the need to redraw objects. Smaller parts of regions may be coded with fewer colours than the total required for the whole region. Reduction schemes are defined so that, for instance, a 2-bit IRD can decode 8-bit CLUTs.

2.12.3.5.2 Subtitles

Different language subtitles may be distinguished by having the languages on separate PIDs or by using a *page_id* mechanism inside one PID. As well as graphical objects, DVB Subtitling supports the use of character codes (e.g. ASCII, Unicode, etc.) to convey subtitling strings. Character codes initially seem very attractive as they consume far less bandwidth than graphical subtitles (approximately 500 bit/s compared with 70–80 kbit/s) but, in practice, they are not that popular. An STB which can cope with graphical subtitles can display any subtitle in any language or font. A character code-driven STB must have one or more fonts resident as well as sophisticated rendering software. If a broadcaster decides to use a new subtitling language new fonts must be downloaded to the STBs. Some languages are written from left to right, others are written from right to left. Starting to use new subtitles with a different convention would require downloading new rendering and scrolling software into the STB.

Widescreen switching can present problems for both types of subtitles.

2.12.3.6 Data Broadcasting (DVB-DATA)

DVB Document A027 Rev. 1: *Digital Video Broadcasting (DVB); DVB Specification for Data Broadcasting*, June 1999
(ETSI EN 301 192 V1.2.1: *Digital Video Broadcasting (DVB); DVB Specification for Data Broadcasting*, June 1999)
DVB Document A047: *Digital Video Broadcasting (DVB); Implementation Guidelines for Data Broadcasting*, June 1999
(ETSI TR 101 202 V1.1.1: *Digital Video Broadcasting (DVB); Implementation Guidelines for Data Broadcasting*, February 1999)

Data broadcasting is the placing of non-MPEG-2 data inside MPEG-2 Transport Stream packets. (This is not strictly true since the data being transported might be MPEG-2 files.) This standard defines five main types of data broadcasting: data piping, data streaming, multi-protocol encapsulation, data carousels and object carousels.

2.12.3.6.1 Data Piping

In data piping the data to be transmitted are fragmented into 184-byte pieces and placed into the payloads of Transport Packets. There is no PES header, timing mechanism (e.g. PTS) or defined reassembly mechanism. In piping systems the receiver has to do a lot of work to reassemble the original data stream and/or packets.

A data pipe is indicated in SI by the presence of a *data_broadcast_descriptor*. The descriptor is associated with a *component_tag* which is also present in the PMT (see Chapter 2.11 for a description of how Program Map Tables point to components of a service). The *data_broadcast_descriptor* contains a *data_broadcast_id* set to 0x0001 to indicate a DVB data pipe.

2.12.3.6.2 Data Streaming

Three types of data streaming are defined – asynchronous, synchronous and synchronised.

2.12.3.6.2.1 Asynchronous Streaming

In asynchronous streaming the data are placed in PES packets which are then placed in Transport Stream packets. The PES header does not contain a PTS, just a start code, a *stream_id* of 0xBF and a non-zero *pes_packet_length*.

Asynchronous streaming is indicated in SI by the presence of a *data_broadcast_descriptor*. The descriptor is associated with a *component_tag* which is also present in the PMT. The *data_broadcast_descriptor* contains a *data_broadcast_id* set to 0x0002 to indicate an async data stream.

Several mutually incompatible proprietary (non-DVB compliant) asynchronous protocols exist.

2.12.3.6.2.2 Synchronous Streaming

In synchronous streaming the data are placed in PES packets which are then placed in Transport Stream packets. The PES header contains a start code, a *stream_id* of 0xBD, a non-zero *pes_packet_length*, a *data_identifier* of 0x21, a PTS and an *output_data_rate* indicator (a 28-bit number which can be used by the receiver to set its rate clock).

Synchronous streaming is indicated in SI by the presence of a *data_broadcast_descriptor*. The descriptor is associated with a *component_tag* which is also present in the PMT. The *data_broadcast_descriptor* contains a *data_broadcast_id* set to 0x0003 to indicate a synchronous data stream. The *stream_type* in the PMT will be 0x06.

2.12.3.6.2.3 Synchronised Streaming

In synchronised streaming the data have to be co-timed with another MPEG-2 PES stream.

The data are placed in PES packets which are then placed in Transport Stream packets. The PES header contains a start code, a *stream_id* of 0xBD, a non-zero *pes_packet_length*, a *data_identifier* of 0x22, a PTS and an *output_data_rate* indicator (a 28-bit number which can be used by the receiver to set its rate clock).

Synchronous streaming is indicated in SI by the presence of a *data_broadcast_descriptor*. The descriptor is associated with a *component_tag* which is also present in the PMT. The *data_broadcast_descriptor* contains a *data_broadcast_id* set to 0x0004 to indicate a synchronised data stream. The *stream_type* in the PMT will be 0x06.

2.12.3.6.3 Multi-protocol Encapsulation

Multi-protocol encapsulation is used where the data may have to be routed or addressed downstream from the receiving device. The destination MAC address and IP[26] address are both included in the Transport Packet payload. The IP and MAC addresses may be unicast, broadcast or multicast[27] addresses. The format is compliant with the DSM-CC_section format for private data.[28] Both the payload and the MAC address may be scrambled. Scrambling MAC addresses can be used to disguise the recipients of a data broadcast, which may be necessary for commercial reasons. There is no value in scrambling the MAC address by itself as the IP address will reveal the intended recipient(s). Users may wish to scramble the payload without scrambling the MAC address, however. DVB does not insist that the standard multicast mapping of MAC and IP addresses is used, but it is allowed.

Protocols other than IP can be encapsulated by using LLC/SNAP encapsulation.

MPE is used by data broadcast operators offering high-speed Internet access over satellite links. For efficient utilisation of satellite bandwidth there are optimum sizes of IP packets which require no wasteful stuffing when encapsulated. (Bandwidth efficiency varies between approximately 22.5% and 89%.) Alternatively, multiple MPE packets may be placed in a Transport Packet to increase bandwidth efficiency – but not all receivers support this (see Chapter 2.13 for more detailed information on MPE).

2.12.3.6.4 Data Carousels

Data carousels repetitively play out information which can be selected by a user. This type of system can be used for software or application downloads to Set-Top Boxes. Within the carousel data is resident in Modules. These Modules can be divided up to form the payloads of 'download data messages'.

2.12.3.6.5 Object Carousels

An object carousel is a much more complex mechanism than a data carousel. Structured groups of objects can be transmitted to a receiver population using directory objects, file objects and stream objects. DSM-CC events can be broadcast to trigger DSM-CC applications which have been previously downloaded. Although there is no return path, an object carousel 'spoofs' interactivity rather like a very sophisticated form of teletext (but with applications and content – rather than just text pages and simple graphics).

2.12.4 DVB Modulation Schemes

Having chosen MPEG-2 as its first compression standard ('source coding'), one of the early challenges for DVB was to specify modulation and error correction schemes for digital broadcasting ('channel coding'). Compression removes redundant information from a signal, but this information is regenerated in the Set-Top Box before presentation to the viewer. Compressed signals are very sensitive to noise and distortion introduced on the transmission path.

A single incorrect bit in an uncompressed digital video bit stream in SDI format[29,30] will probably only affect one pixel for one frame (40 ms in a 25 Hz system). The most that a single bit error could affect would be one line of video for one frame (this could happen if the bit error corrupted a Start of Active Video (SAV) signal).

A single bit error in a compressed signal, in contrast, will almost certainly affect more than one pixel with a good possibility of the effect lasting for more than one frame. Usually errored bits will affect blocks (8×8 pixels), macroblocks (16×16 pixels) or macroblock stripes (16-line-high strips of the picture). A system on the verge of noise-induced failure will usually exhibit 'blockiness', misplaced strips of the picture and partial or complete freezing of the picture. In theory an errored bit could affect an entire Group Of Pictures (GOP) causing a freeze frame for half a second or more in a DTH system (where the GOPs are usually 12–15 frames long).

The challenge when specifying channel coding for a digital broadcasting scheme is to reduce the Bit Error Ratio (BER) at the consumer device to an acceptable level. DVB modulation schemes are designed to achieve Quasi Error Free (QEF) transmission – defined by the DVB as a BER of 10^{-10} to 10^{-11}. This is often referred to as an error rate of about one per hour in a compressed video stream – but, of course, this will depend on the bit rate of the compressed video.

2.12.4.1 V4MOD

All early DVB transmission schemes (DVB-S, DVB-C and DVB-T) specify V4MOD[31] for the first part of the channel coding (Figure 2.12.5).The V4MOD unit comprises three functional blocks – Energy Dispersal, Reed–Solomon FEC and Byte Interleaving.

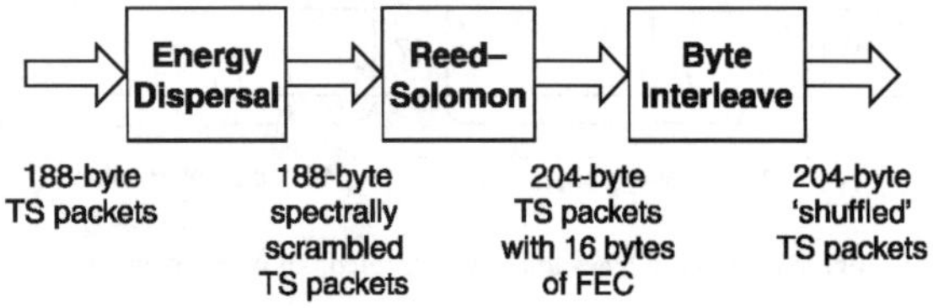

Figure 2.12.5 V4MOD processes.

2.12.4.1.1 Energy Dispersal

The purpose of the Energy Dispersal block is to 'spectrally scramble' the bit stream, removing long repetitive runs of bits. In a simple modulation scheme (such as non-differential QPSK) a long run of ones would cause the modulator to become 'stuck' on one symbol. The output of the modulator would then be a large-amplitude sine-wave (unmodulated carrier) rather than the required low-amplitude, 'noise-like' signal expected. The Energy Dispersal block is the digital equivalent of the triangular waveform used for energy dispersal in analogue TV satellite systems.

The Transport Stream is applied to an exclusive-OR gate which has a Pseudo-Random Binary Signal (PRBS) on its other input. An exclusive-OR gate can be thought of as a programmable inverter. When the PRBS is at 'zero' the corresponding TS bit travels through the logic gate unaffected. When the PRBS is at 'one' the TS bit is inverted at the output of the gate. The PRBS is used to 'scramble' eight TS packets before it is reset. The Transport Packet sync bytes (0x47) are not scrambled with the PRBS – only the 187 bytes which follow them. Recovery of the original TS data can be achieved by applying the spectrally scrambled data to another exclusive-OR gate with an identical PRBS applied to the other input (Figure 2.12.6).

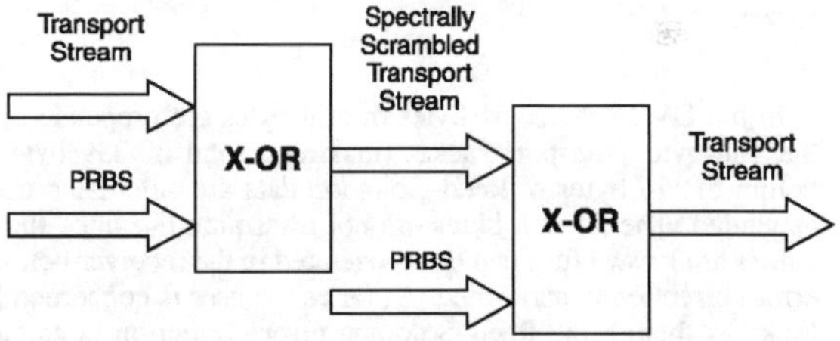

Figure 2.12.6 Energy Dispersal is reversible if the PRBS streams are synchronised.

The descrambling PRBS has to be synchronised to the original PRBS to recover the original TS. Synchronisation is achieved by inverting the first sync byte of the first Transport Packet in the group of eight (0x47 becomes 0xB8). This inverted byte tells the receiver that the next byte is scrambled with the first byte of the PRBS.

The bit stream from an Energy Dispersal gate would have the form 0xB8, 187 scrambled bytes, 0x47, 187 scrambled bytes, 0x47, 187 scrambled bytes, 0x47, etc. (Figure 2.12.7).

2.12.4.1.2 Reed–Solomon Forward Error Correction

The second block in a V4MOD section is the Reed–Solomon Forward Error Correction. (In a concatenated system this is

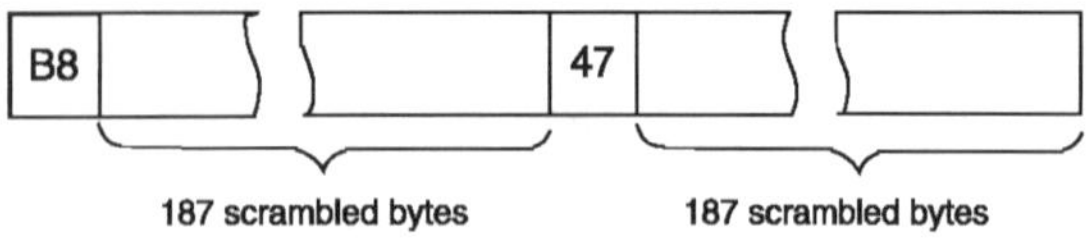

Figure 2.12.7 Sync bytes are not scrambled, all other bytes are.

often referred to as the 'Outer Code'.) Forward Error Correction must be used as negative acknowledgements followed by retransmission (as used in point-to-point data transfer) are impractical in a broadcast system. (The details of modern error correction techniques are covered in depth in Chapter 1.4, so we will merely confine our description to the main impact of the techniques used on system performance.)

Reed–Solomon is a block-based error correcting code, which means that a block of data to be protected has another block of error correcting data added to it. These extra data bits occupy channel bandwidth and reduce the information capacity of the channel. In classical Reed–Solomon[32] a block of 239 bytes of data has 16 bytes of extra data added, making the total block size 255 bytes – an approximately 6% overhead (Figure 2.12.8).

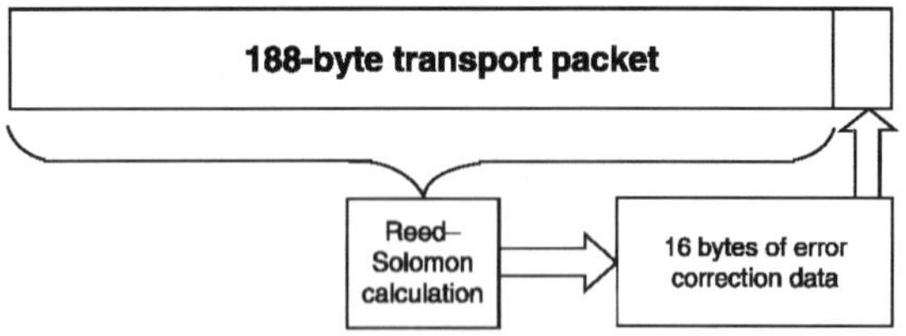

Figure 2.12.8 DVB Reed–Solomon process applies to whole Transport Packet.

In the DVB scheme 51 bytes of null bytes are prepended to the 188-byte Transport Packet (making a total of 239 bytes) before the 16 bytes of Reed–Solomon data are calculated and appended. The 51 null bytes are not transmitted – since their values are known they can be regenerated in the receiver before error correction is performed. As far as the user is concerned it looks as though the Reed–Solomon error correction is calculated on the 188-byte transport packet and 16 bytes of error detection are added to the end of the packet. This has the effect of expanding the packet size by a factor of 204/188 (8.5%) and reducing channel capacity by 7.84%.

With random bit or byte errors this form of Reed–Solomon can correct up to eight errored bytes per TS packet. (Since it is byte based, this form of Reed–Solomon does not distinguish between bytes with one or more errored bits and those which are totally inverted, all are regarded as errored bytes.) The FEC can detect up to 16 errored bytes per packet but cannot correct this many errors. If the FEC is overwhelmed in this way it will set the Transport Error Indicator (TEI) flag in the TS packet header, but will not attempt to correct the errors. Downstream devices should not discard packets marked in this way, as the TEI flag may be useful to an end device for error concealment. (Even if the packet has a PID which is not required downstream it cannot be safely discarded as the PID itself may be errored.)

2.12.4.1.3 Byte Interleaving

In a practical transmission system errors often occur in groups rather than sporadically. This is particularly true if the block decoder is being fed from the output of a trellis decoder (such as a Viterbi decoder[33]), which is often the case in DVB systems. When the error rate overwhelms a trellis decoder it can produce long streams of corrupted data on its output until it can reacquire synchronisation in some way. By shuffling the contents of the packets together prior to transmission the effect of bursts of errors can be mitigated.[34]

The V4MOD Byte Interleaver takes 17 bytes from 12 TS packets to create one new 204-byte packet ($17 \times 12 = 204$). The transmitted packet will have bytes 1, 13, 25, 37 etc. from one packet, bytes 2, 14, 26, 38 etc. from the previous packet and so on. Should a large transmission error occur which corrupts up to 96 bytes in a row, after de-interleaving the receiver will have 12 consecutive packets, each of which has eight corrupted bytes. The Reed–Solomon code should be able to correct all of these errors completely. The 12-packet interleave can be seen to have 'multiplied' the effectiveness of the Reed–Solomon code by a factor of 12. (This is only true if there are no other errors within the interleaving 'window' before and after the burst of errors.)

Some authorities state that DVB uses a two-packet interleave, which would have little practical benefit. This interpretation results from a misunderstanding of the interleaving diagrams in the DVB specifications.

2.12.4.1.4 Performance of V4MOD

A V4MOD receiver consists of a de-interleaver, a Reed–Solomon decoder and an energy dispersal descrambler – the three blocks being in reverse order to their corresponding sections in a V4MOD encoder (Figure 2.12.9).

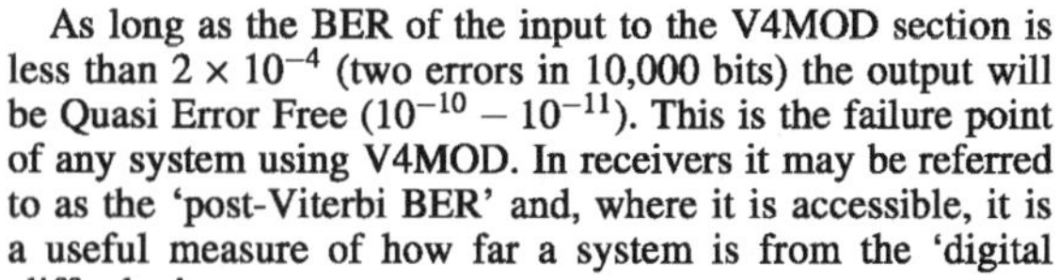

Figure 2.12.9 Inverse V4MOD on the verge of failure.

As long as the BER of the input to the V4MOD section is less than 2×10^{-4} (two errors in 10,000 bits) the output will be Quasi Error Free ($10^{-10} - 10^{-11}$). This is the failure point of any system using V4MOD. In receivers it may be referred to as the 'post-Viterbi BER' and, where it is accessible, it is a useful measure of how far a system is from the 'digital cliff-edge'.

2.12.4.2 DVB-S

DVB Document A040: *Digital Video Broadcasting (DVB); Framing Structure, Channel Coding and Modulation for 11/12 GHz Satellite Services*, June 1999
(ETSI EN 300 421 V1.1.2: *Digital Video Broadcasting (DVB); Framing Structure, Channel Coding and Modulation for 11/12 GHz Satellite Services*, August 1997)

DVB Document A003 rev. 1: *User Requirements for Cable and Satellite Delivery of DVB Services Including Comparison with Technical Specifications*, May 1995

DVB-S is one of the DVB Project's greatest successes. All of the digital satellite DTH systems launched between 1995 and the time of writing (2003) use the DVB-S modulation system (two US systems launched before the DVB-S standard was finalised used a proprietary standard). Nearly all satellite contribution and distribution systems also use some form of DVB-S or DVB-SNG (DVB-S is used for C-band transmissions as well as the Ku-band it was intended for).

DVB committees do not generally 're-invent the wheel' if an existing technology is available. DVB-S was based largely on channel coding schemes pioneered by NASA in the 1970s.[35–37]. The major difference is that – whilst NASA needed mainframe computers to perform their error correction calculations – DVB needed a chip that would fit in a reasonably priced consumer receiver (20 years of the application of Moore's law ensured that the silicon was fast enough, small enough and – most importantly – cheap enough). NASA now uses turbo-codes[38] rather than concatenated block and convolutional codes to approach the Shannon limit[39] more closely.

In digital broadcasting early success can lead to unwanted legacy issues later. It is not clear how the existing, large installed base of domestic receivers could be migrated to a new modulation standard. Most existing receivers use demodulation and error correction units which cannot be modified by software upgrade. The second generation of DVB specifications (DVB2.0) will include new source coding (probably MPEG-4 part 10) and channel coding schemes. The new satellite modulation standard will be called DVB-S2 but it is currently at the draft stage. (The description which follows is taken from publicly available documents/announcements from members of the Technical Module rather than the draft specification itself, which is far from complete and subject to change.) Seven proposals were evaluated by performance simulation on the powerful computers of RAI's Centro Ricerche Innovazione Tecnologica in Turin over a 10-month period. There will be five modulation schemes (BPSK/QPSK/8-PSK/16-APSK/32-APSK) and up to eight FEC code rates. Surprisingly, the main code used will not be a turbo code but a Low-Density Parity Check code. (LDPC codes were discovered by Gallager 30 years before turbo coding but have recently been 'rediscovered'. They can outperform turbo codes, have a lower decoding complexity and – importantly – are patent-free.) DVB-S2 will deliver bit rates up to 2.6 times those of DVB-S. DVB-S systems have constant robustness defined by worst-case propagation characteristics, so most of the time they are wasting channel capacity. DVB-S2 will be able to dynamically change coding and modulation parameters to 'ride' the prevailing weather and individual receiver characteristics to enable optimum use of channel capacity for unicast and multicast applications. (The return channel could be DVB-PSTN, DVB-RCC, DVB-RCS, etc.). DVB-S receivers may be able to decode, or partially decode, some DVB-S2 modulation schemes, giving a possible upgrade path for legacy systems. DVB-S2 will also be able to carry non-TS data and multiple TS sources.

A003 compares the features of DVB-S and DVB-C with user requirements. This has not been done for other DVB specifications.

2.12.4.2.1 Overall View of DVB-S processes

The DVB-S processes consist of V4MOD, trellis encoding and Gray-coded QPSK modulation (Figure 2.12.10). V4MOD has been described previously, it has the effect of expanding the bit rate of the original 188-byte Transport Stream by a factor of 204/188. The trellis encoder is a convolutional encoder which acts on a stream of data rather than blocks. The combination of a block-based encoder and a trellis encoder with an interleaver between them is particularly powerful and, until the advent of turbo encoding, was the best practically realisable solution for a system containing many low-cost domestic receivers.

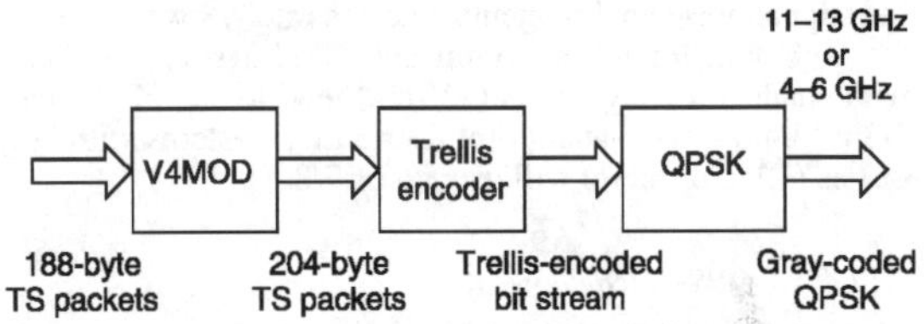

Figure 2.12.10 DVB-S modulation scheme. Up-conversion and high-power amplification are not shown. In satellite systems the down-link frequency is different from the up-link frequency. DVB-S modulators usually produce IF rather than RF outputs. The IF frequency is often, but not always, 70 MHz.

2.12.4.2.2 Trellis Encoding

A trellis encoder can be thought of as a device which spreads the energy contained in one information bit over a number of transmitted bits. (The trellis encoder produces the 'Inner Code' of the DVB-S system – Reed–Solomon being the 'Outer Code'.) The trellis decoder recovers the data bits and has effectively averaged the unwanted transmission noise power over several received bits. This has the effect of increasing the signal-to-noise performance of a trellis-encoded channel by a factor of 3 dB over an unencoded channel at the expense of reduced channel capacity.

A soft decision decoder takes account of the 'channel state' to decide which of the received bits are more or less reliable. By using this 'reliability' information in the decoding process a further 2 dB coding gain can be achieved. With simple modulation schemes such as BPSK and QPSK channel estimation can be easily calculated by measuring the amplitude of the received signal and determining its distance from the slicing point.

A 5 dB coding gain means that, for a given required BER, a coded channel can operate at a signal/noise ratio 5 dB below that of an uncoded channel.

A coded channel uses some of its channel capacity for error correction and is more sensitive to signal/noise variations than an uncoded channel.

A trellis encoder is implemented by applying the bit stream to be transmitted to the input of a shift register. Taps are taken off at various points along the shift register and these are exclusive-ORed to give two bit streams which are then multiplexed together. The outgoing bit stream has twice the bit rate of the incident bit stream – so the trellis encoder is called a rate 1/2 encoder. It is possible to design encoders with differing rates and, hence, differing levels of error correction. There are five different trellis-encoding rates in DVB-S but these have not been achieved by having five different types of encoder (which could have delivered optimum performance but would have been more expensive). DVB-S uses the same shift register and taps for all five rates but uses a technique called 'puncturing' for four of the rates. If 7 bits enter a rate 1/2 trellis encoder 14 bits will be output. A rate 7/8 encoder can be implemented by only transmitting 8 of the 14 output bits (a receiver has to know which 6 bits have not been sent). A rate 7/8 code does not have the error correcting capacity of a rate 1/2 code but it has the advantage of allowing a greater data rate through a channel of a given bandwidth (i.e. only 12.5% of channel capacity has been used for error correction rather than 50%).

The code rate chosen in practice will depend on the link budget of the transmission channel. Rate 1/2 would be used for low signal/noise environments whereas rate 7/8 would be used for high signal/noise environments. DTH systems in Europe with small receiving dishes (~50 cm) tend to use FEC rates of 2/3 or 3/4. Point-to-multipoint distribution systems with larger dishes (~2 m or more) will use 5/6 or 7/8.

2.12.4.2.3 QPSK Modulation

In DVB-S the V4MOD and trellis encoding sections are followed by QPSK modulation. Differential QPSK (DQPSK) is not used as it requires a greater signal-to-noise ratio (~2 dB) than absolute QPSK. Gray coding is used to improve the bit error rate for a given signal to noise by about 3 dB (Figure 2.12.11) – equivalent to the performance of a BPSK signal of the same amplitude (at low SNR this improvement disappears).

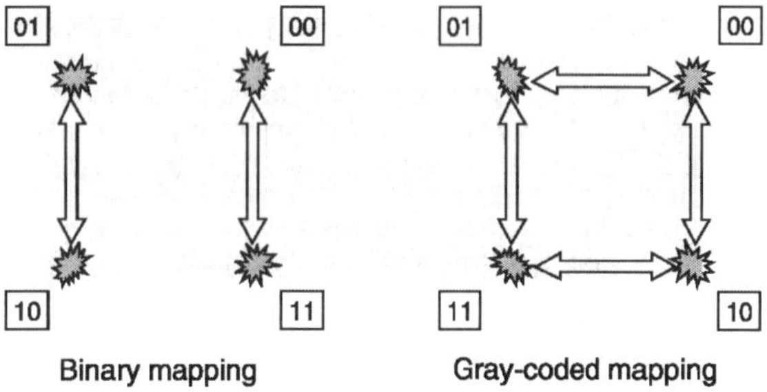

Figure 2.12.11 Gray code mapping increases the noise immunity of QPSK. The two symbol errors shown on the left would each produce two bit errors. The four symbol errors on the right will only produce one bit error each. Symbol errors between opposite corners of the right-hand diagram will produce two bit errors, so at low signal-to-noise ratios QPSK will be more susceptible to noise than BPSK.

2.12.4.2.4 Overall Performance of DVB-S

In theory the channel coding scheme used by DVB-S allows us to get within 4.5 dB of the Shannon limit. More modern turbo coding schemes should be able to approach within 1 dB of the Shannon limit. It should be remembered, however, that Ku band transmissions suffer from rain fades of up to 12 dB (and indications from satellite operators are that rain fades are 'deepening') so a 12 dB margin is needed above the theoretical failure point. Since no modulator, transponder or receiver is perfect, a small implementation margin must also be allowed for. With an FEC of 1/2 a DVB-S system will theoretically work with an Eb/No of just 4.5 dB. (Eb/No is a 'normalised' form of S/N that enables mathematicians to compare different modulation and channel coding schemes. Eb/No is 'energy per bit'/ 'noise per hertz', so for QPSK Eb/No = 0.5 × C/N. See Chapter 1.4 for further details of Eb/No.) Most modern Ku-band DTH systems can use very small receiving dishes (~50 cm diameter) with an FEC of 2/3 or 3/4. Professional distribution networks with large (~2 m diameter) receive dishes usually use an FEC of 7/8 – this requires an Eb/No of 6.4 dB (see Figure 2.12.12).

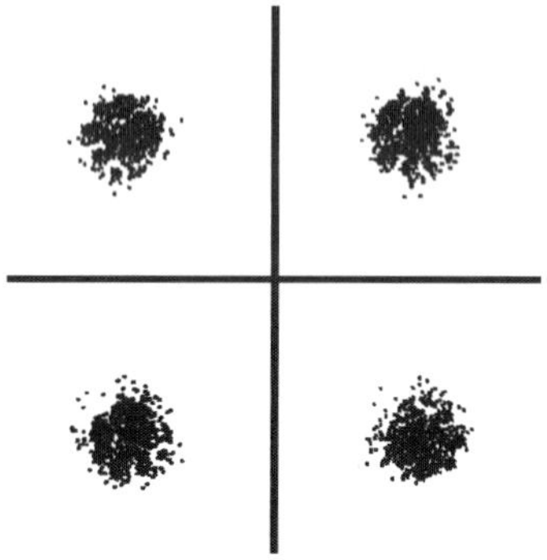

Figure 2.12.12 DVB-S signal 13.4 dB above failure point.

DVB-S is also used for C-band transmissions in the tropics. Although C-band suffers from more attenuation than Ku-band it is not so susceptible to rain fade during the monsoon season. Rain fades at C-band are of the order of 2 dB only, but larger receive dishes are often required than with Ku-band systems (1.3 m or greater).

2.12.4.2.5 Converting Symbol Rates to Bit Rates (and Vice Versa)

The formulae for converting symbol rates into TS bit rates and vice versa are very simple. Tables 2.12.1 and 2.12.2 summarise the conversion factors.

Table 2.12.1 Factors for converting symbol rates to bit rates

FEC rate	*Conversion factor for 188-byte rate*	*Conversion factor for 204-byte rate*
1/2	0.921569	1.0
2/3	1.228758	1.333333
3/4	1.382353	1.50
5/6	1.535948	1.666667
7/8	1.612745	1.750

Table 2.12.2 Factors for converting bit rates to symbol rates

FEC rate	*Conversion factor for 188-byte streams*	*Conversion factor for 204-byte streams*
1/2	1.085106	1.0
2/3	0.813830	0.75
3/4	0.723404	0.666667
5/6	0.651064	0.60
7/8	0.620061	0.571429

Table 2.12.1 shows the factors for converting symbol rates to bit rates. (As an example – if the symbol rate is 30 Msymbols/s and the FEC is 3/4, the 204-byte TS rate is 1.5×30, i.e. 45 Mbit/s.)

Table 2.12.2 shows the factors for converting bit rates to symbol rates. A (204-byte) 41.25 Mbit/s TS with an FEC of 3/4 will have a symbol rate of 27.5 Msymbols/s (given by: symbol rate $= 41.25 \times 0.666667$). A symbol rate of 27.5 Msymbols/s will correspond to a modulated signal with a 3 dB bandwidth of 27.5 MHz (i.e. 3 dB bandwidth equals numerical value of symbol rate). In a practical system the occupied bandwidth will be between 15% and 30% greater than 3 dB bandwidth.

2.12.4.3 BPSK Variant of DVB-S

DVB Document A036: *Implementation of BPSK in Digital Video Broadcasting Satellite Transmission Systems*, March 1998
ETSI TR 101 198 V1.1.1: *Implementation of Binary Phase Shift Keying (BPSK) Modulation in DVB Satellite Transmission Systems*, September 1997

Normally BPSK cannot compete with Gray-coded DVB-S in terms of bandwidth efficiency for a given Eb/No. Under some exceptional circumstances (such as strong narrowband co-channel interference into a QPSK signal occupying only part of a transponder) BPSK may require less Eb/No and cause less interference.

For these unusual situations a modified form of DVB-S using BPSK was defined (Figure 2.12.13). The trellis encoding, puncturing and RF filtering are identical to those in 'classical' DVB-S. The modulation scheme used is non-differential BPSK.

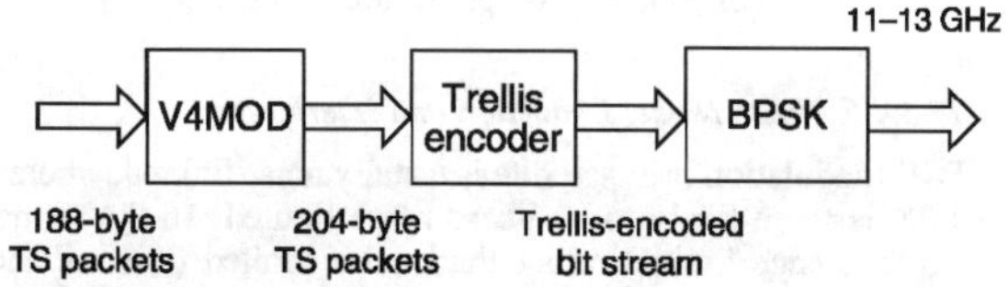

Figure 2.12.13 BPSK variant of DVB-S.

2.12.4.4 DVB-DSNG

DVB Document A033: *DSNG Commercial Users' Requirements*, March 1998 (User's requirements documents are written by the Commercial Module for the use of the Technical Module of DVB – they are not ratified/published by the ETSI)
DVB Document A049: *Digital Video Broadcasting (DVB); Framing Structure, Channel Coding and Modulation for Digital Satellite News Gathering (DSNG) and Other Contribution Applications by Satellite*, June 1999
(EN 301 210 V1.1.1: *Framing Structure, Channel Coding and Modulation for DSNG and Other Contribution Applications by Satellite*, March 1999)
DVB Document A051: *Digital Video Broadcasting (DVB); User Guidelines for Digital Satellite News Gathering (DSNG) and Other Contribution Applications by Satellite*, June 1999
(ETSI TR 101 221 V1.1.1: *User Guideline for Digital Satellite News Gathering (DSNG) and Other Contribution Applications by Satellite*, March 1999)
DVB Document A050: *Digital Video Broadcasting (DVB); Co-ordination Channels Associated with Digital Satellite News Gathering (DSNG)*, June 1999
(EN 301 222 V1.1.1: *Digital Video Broadcasting (DVB); Co-ordination Channels Associated with Digital Satellite News Gathering (DSNG)*, July 1999)

DVB DSNG uses standard DVB-S QPSK modulation and (optionally) pragmatic Trellis Coded Modulation (TCM).[40–42] 8-PSK and 16-QAM are the optional modulation schemes, and these require modification of the use of the trellis encoder from the DVB-S norm. The standards suggest that QPSK and 8-PSK can be used at the transponder saturation point, but that 16-QAM must be operated in a linear portion of the transponder characteristic by using 'back-off'. It is, in fact, possible to operate 16-QAM with the outer constellation points at saturation by using dynamic pre-correction, but this may be difficult to implement in a DSNG environment as an iterative set-up procedure may be required.

The suggested standard interface between modulators and encoders or multiplexers is 188-byte ASI (see Section 2.12.6).

Use of the Transport Stream Descriptor Table (TSDT) is mandatory.

2.12.4.4.1 DSNG Users' Requirements

Compared with analogue SNG equipment, DSNG is smaller and lighter and requires less transponder bandwidth. Equipment should be fast to set up and should require no more than two people to operate it.

Flyaway DSNG units will probably use MP@ML at about 8 Mbit/s; transportable stations should support 422P@ML at 8–34 Mbit/s.

Two or more 64 kbit/s full-duplex speech circuits should be provided on the same transponder as the sound and vision multiplex.

Higher Order Modulation (HOM) such as 8-PSK and 16-QAM should be investigated as possible alternatives to QPSK in order to reduce transponder rental costs.

2.12.4.4.2 QPSK

This mode of operation is identical to DVB-S (Figure 2.12.14).

2.12.4.4.3 8-PSK Variant

8-PSK uses standard V4MOD followed by a standard DVB-S trellis encoder. Unlike DVB-S, however, 1 or 2 bits (depending on the FEC) in the 8-PSK symbol bypass the trellis coder completely. The mathematics behind pragmatic trellis coding are quite complex, but the basic idea is that if the coding is done properly, then transitions between adjacent symbols should only happen rarely, thus increasing the noise immunity. The 8-PSK mapping is absolute rather than differential.

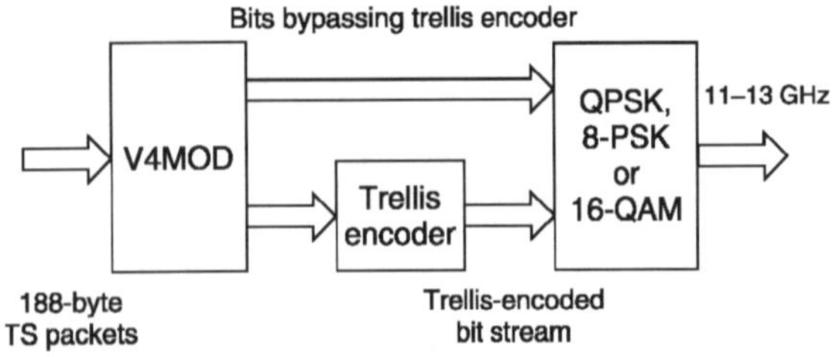

Figure 2.12.14 DVB-DSNG modulation. The QPSK version is identical to DVB-S with the entire bit stream passing through the trellis encoder. In the 8-PSK and 16-QAM variants some bits bypass the trellis encoder depending on the FEC rate employed.

Table 2.12.3 shows the factors for converting symbol rates to bit rates. (As an example – if the symbol rate is 30 Msymbols/s and the FEC is 8/9 the 204-byte TS rate is 2.666667 × 30, i.e. 80 Mbit/s.)

Table 2.12.3 Factors for converting symbol rates to bit rates

FEC rate	*Conversion factor for 188-byte rate*	*Conversion factor for 204-byte rate*
2/3	1.843137	2.0
5/6	2.303922	2.50
8/9	2.457516	2.666667

Table 2.12.4 shows the factors for converting bit rates to symbol rates. A (188-byte) 25 Mbit/s TS with an FEC of 5/6 will have a symbol rate of 12.76595 Msymbols/s (given by: symbol rate = 25 × 0.510638). A symbol rate of 12.76595 Msymbols/s will correspond to a modulated signal with a 3 dB bandwidth of 12.76595 MHz (i.e. 3 dB bandwidth equals numerical value of symbol rate). In a practical system the occupied bandwidth will be between 25% and 35% greater than 3 dB bandwidth (depending on value of roll-off factor α selected – 0.25 or 0.35).

Table 2.12.4 Factors for converting bit rates to symbol rates

FEC rate	*Conversion factor for 188-byte streams*	*Conversion factor for 204-byte streams*
2/3	0.542553	0.50
5/6	0.510638	0.40
8/9	0.406915	0.375

For FEC rates of 2/3, 5/6 and 8/9 the required Eb/No figures are 6.9, 8.9 and 9.4 dB respectively. The modem implementation margin for 8-PSK is slightly higher than that for QPSK (1–1.6 dB versus 0.8 dB).

2.12.4.4.4 16-QAM Variant

16-QAM also uses standard V4MOD followed by a trellis encoder which is partially bypassed to implement pragmatic TCM.

Table 2.12.5 shows the factors for converting symbol rates to bit rates. (As an example – if the symbol rate is 30 Msymbols/s and the FEC is 3/4 the 204-byte TS rate is 3 × 30, i.e. 90 Mbit/s.)

Table 2.12.5 Factors for converting symbol rates to bit rates

FEC rate	*Conversion factor for 188-byte rate*	*Conversion factor for 204-byte rate*
3/4	2.764706	3.0
7/8	3.225490	3.50

Table 2.12.6 shows the factors for converting bit rates to symbol rates. A (204-byte) 45 Mbit/s TS with an FEC of 3/4 will have a symbol rate of 15 Msymbols/s (given by: symbol rate = 45 × 0.333333). A symbol rate of 15 Msymbols/s will correspond to a modulated signal with a 3 dB bandwidth of 15 MHz (i.e. 3 dB bandwidth equals numerical value of symbol rate). In a practical system the occupied bandwidth will be between 25% and 35% greater than 3 dB bandwidth (depending on roll-off factor selected).

Table 2.12.6 Factors for converting bit rates to symbol rates

FEC rate	*Conversion factor for 188-byte streams*	*Conversion factor for 204-byte streams*
3/4	0.361702	0.333333
7/8	0.310030	0.285714

For FEC rates of 3/4 and 7/8 the required Eb/No figures are 9.0 and 10.7 dB respectively. The modem implementation margin for 16-QAM is 1.5–2.1 dB but the major factors affecting performance will probably be phase noise and linearity.

2.12.4.4.5 Phase Noise, Distortion and Interference

QPSK modulation over satellite is usually noise limited, whereas 8-PSK is more likely to be phase noise limited. 16-QAM may be interference limited rather than noise limited (this will not be known for certain until more 16-QAM systems are in operation).

Modulation schemes like QPSK and 8-PSK are usually relatively unaffected by non-linearities in the transmission chain (although this may cause them to produce unwanted out-of-band components – so-called 'spectral regrowth'). 16-QAM is not a constant-amplitude scheme, so non-linear effects will distort the shape of the constellation (Figure 2.12.15). Amplitude distortion will usually reduce the size of the larger symbol points relative to the inner points. Phase distortion will cause the inner and outer points to rotate relative to each other. Group delay will 'defocus' the constellation points.

2.12.4.4.6 Pre-Correction

One solution to the problem of non-linear distortion is to back off the power of the transmitted signal until it is operating in a linear part of the satellite transponder characteristic. This is

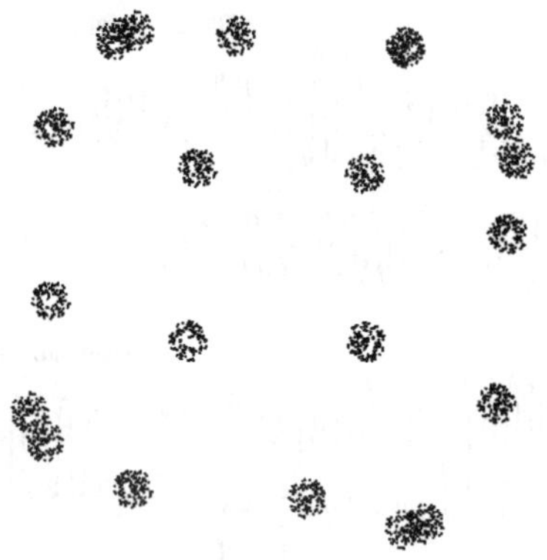

Figure 2.12.15 Effect on 16-QAM constellation of non-linear transmission channel. Vectorscope displays usually show the central portion rotated relative to the corners, but this is because they 'assume' the corners are correct.

not usually practical as the 16-QAM signal would be sunk too far into the noise to be receivable with a reasonable sized dish. A large receive dish may be able to overcome this problem.

Another solution is to pre-correct the transmitted signal. Pre-correction is really pre-distortion of the transmitted signal. If done properly, pre-correction will exactly cancel out the distortions introduced by the transmission chain, producing a 'square', unrotated constellation at the input of a receiver. Link improvements of 6–8 dB are claimed for 16-QAM modulation as well as increased capacity if group delay is compensated for.

Pre-correction also has some benefit in 8-PSK and QPSK systems (1–2 dB in the case of 8-PSK). (It should be noted that measuring the Modulation Error Ratio (MER) of a pre-corrected signal will produce very poor results on the up-link signal. The down-link signal, however, should produce much better results. MER is discussed more fully in Section 2.12.9).

2.12.4.4.7 Transport Stream Descriptor Table

TS packets with a PID of 0x2 carry the TSDT, which contains ASCII strings describing the content of the TS. The TSDT will be sent at least once in 10 seconds.

In DSNG transmissions the TSDT will contain a *transport_stream_descriptor* identifying the TS as a CONtribution Application. The descriptor is 0x67 04 43 4F 4E 41, which is ASCII for 'CONA'.

A *DSNG_descriptor* will also be present. The *DSNG_descriptor* has a descriptor tag of 0x68 and contains the station code, the SNG HQ and the SNG provider – all coded as ASCII and separated by ASCII commas.

Optionally, if compatibility with consumer IRDs is required (and all SI tables are present) the TSDT will contain three descriptors. The first is 0x67 03 44 56 42 – signifying that this is a DVB-compliant TS. The second is 0x67 04 43 4F 4E 54 – indicating that this is a CONTribution feed and the third is the *DSNG_descriptor*, as previously described.

2.12.4.4.8 Annexes

EN 301 210 contains five Annexes:

- Annex A contains a spectral template for the RF signal.
- Annex B proposes some interoperability and emergency set-up parameters for DSNG units.
- Annex C defines the meaning of optional in the context of optional modulation modes (in essence 8-PSK and 16-QAM are optional but if they are offered they must conform to the specification).
- Annex D details SI requirements (actually PSI requirements)
- Annex E has tables listing usable bit rates versus various transponder (or slot) bandwidths for all combinations of FEC and modulation scheme.

2.12.4.4.9 Coordination Channels

Use of DVB coordination channels is optional since other communication mechanisms (e.g. cell phones) may be preferred for operational reasons. Voice traffic is first filtered (ITU-T Recommendation G.712[43]) then encoded at 8 kbit/s using Algebraic Coded Excited Linear Prediction (ACELP).[44] Normal uncompressed PCM-encoded speech at 64 kbit/s[45] can be input if it is first converted to 16-bit PCM. Up to four voice channels coded at 8 kbit/s may be multiplexed together. The mux frames data into 204-byte packets. The first 4 bytes are a 2-byte Sync Word (0x47B8), a mux configuration signalling byte and a reserved byte. The remaining 200 bytes are input data to the mux. The mux output is fed to an energy dispersal section (identical to the one used in DVB-S). No Reed–Solomon FEC or interleaving is required as ACELP encoded speech is more tolerant of errors than compressed video. The 'spectrally scrambled' bit stream is fed to a rate 1/2 trellis encoder (again, identical to that used in DVB-S but without the 'punctured' codes option). Direct-Sequence Spread-Spectrum (DS-SS)[46,47] coding is then applied before the signal passes to a QPSK modulator section. Use of DS-SS means that the coordination channels may all occupy the same portion of spectrum without mutual interference. The coordination channels may even occupy the same spectrum as the DSNG signal without significant degradation to either waveform.

An Eb/No of 3.6 dB is required for a BER of 10^{-3}, considered to be adequate for voice services. Optionally RS-232[48] data may be input to the mux at 9.6, 19.2 or 38.4 kbit/s. The encoder strips off start, parity and stop bits before transmission.

2.12.4.4.10 User Guidelines

ETR 101 221 reproduces some of the diagrams from EN301 210 and EN301 222 with the addition of explanatory text. It also contains a number of calculations, tables and graphs showing usable bit rates, Eb/No requirements, link budgets and clear sky margins.

2.12.4.5 DVB-T

DVB Document A004: *User Requirements for Terrestrial Digital Broadcasting Services*, December 1994 (User requirements documents are written by the Commercial Module for the use of the Technical Module of DVB – they are not ratified/published by the ETSI)

DVB Document A012 Rev. 2: *Digital Video Broadcasting (DVB); Framing Structure, Channel Coding and Modulation for Digital Terrestrial Television*, March 2001
(ETSI EN 300 744 V1.4.1: *Digital Video Broadcasting (DVB); Framing Structure, Channel Coding and Modulation for Digital Terrestrial Television*, January 2001)

Currently (2003) the terrestrial modulation standard (DVB-T) is in use in Australia, Finland, Germany, the Netherlands, Spain, Singapore, Sweden, Taiwan and the UK. DVB-T test transmissions are taking place or are planned in most European countries, and reception of DVB-T broadcast signals in buses is being demonstrated in Shanghai and Singapore. Various manufacturers and broadcasters have demonstrated reception of experimental broadcasts in trams, sports cars and buses, and have also received pictures from aeroplanes, helicopters, buses and vans.

DVB-T was only intended for fixed and so-called 'fixed-mobile' (i.e. stationary but portable) receivers. The fact that it can also be used to and from moving vehicles indicates that the DVB-T committee slightly exceeded their original design brief, but this has been a useful feature for promoting the standard.

2.12.4.5.1 *Taboo Channels*

Analogue terrestrial broadcasting uses spectrum in the VHF and UHF bands. In countries with a 30 Hz electrical supply analogue TV channels usually occupy 6 MHz of spectrum, in other countries the channel spacing is 7 or 8 MHz.

It is difficult to produce two high-power signals in adjacent channels so the channels either side of one containing a TV signal are not used. The superheterodyne front-ends in television receivers produce interfering signals which are radiated from the viewers' aerials. These signals would interfere with analogue transmissions and so these channels cannot be used. Each channel of television creates up to six 'taboo' channels. When a terrestrial frequency plan is produced channels are chosen so that the 'taboo' channels overlap.

The DVB-T scheme is unaffected by the interfering signals in 'taboo' channels, allowing broadcasters to use spectrum which is unavailable for analogue transmissions.

2.12.4.5.2 *Inter-Symbol Interference*

Since terrestrial transmissions are subject to multi-path ('ghosting') a Guard Interval (GI) is defined which allows the delayed versions of the previous symbol to be ignored by a receiver. If a single carrier were to be used either the symbol rate would have to be too slow to be useful or the guard interval would be too short. The solution is to use multiple, low-symbol rate carriers (see Figure 2.12.16).

2.12.4.5.3 *Multiple Carriers*

The DVB-T system uses 2048 or 8192 mutually orthogonal carriers. The two schemes are usually referred to as 2000 carrier (2K) or 8000 carrier (8K). Both schemes have exactly the same payload for a given set of transmission parameters (i.e. 8K does not have four times the data rate of 2K). The 8K scheme will tolerate echoes of up to 224 µs – four times the duration of those the 2K scheme can cope with.

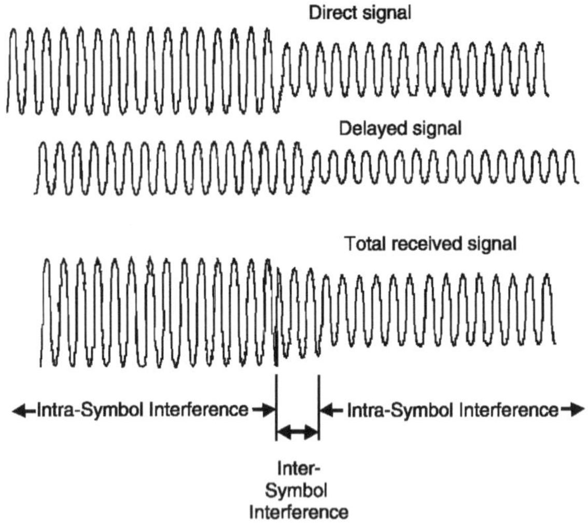

Figure 2.12.16 As long as the duration of the Inter-Symbol Interference is less than that of the guard interval a DVB-T receiver can ignore it. (Intra-Symbol Interference is compensated for by using pilot tones for channel estimation. The resultant data are fed into the Viterbi decoder to assist the 'soft decision' process.)

2.12.4.5.4 *Pilot Tones*

'Intra-Symbol Interference' (an echo interfering with its own symbol rather than a subsequent one) is measured using static pilot tones scattered throughout the channel. The pilot tones are boosted in amplitude relative to the average power of the data carriers and are modulated with a known PRBS. These pilot tones can also be used to estimate Doppler shift.

2.12.4.5.5 *Scattered Pilots*

In addition to the continuous pilots there are a number of scattered pilots which 'march through' the waveform every symbol (Figure 2.12.17). These carriers are also modulated with a PRBS and are at a boosted amplitude. After four symbols have been received the receiver will have information about the amplitude and phase of every third carrier, which can be used to correct for Intra-Symbol Interference. (The data are not used to drive an equaliser but are fed into the trellis decoder as 'confidence' information.)

2.12.4.5.6 *Transmission Parameter Signalling*

Additional pilots (called TPS pilots) carry information about the modulation scheme in use. This allows a receiver to tune in rapidly without having to try all of the possible modulation schemes allowed by DVB-T.

2.12.4.5.7 *Modulation Schemes*

There are three modulation schemes used by DVB-T for carrying Transport Stream data. These are QPSK, 16-QAM and 64-QAM. QPSK carries the fewest bits per second but is the most robust and will mostly be used for mobile transmission and reception. 64-QAM produces the greatest transmitted bandwidth, but requires a good signal-to-noise ratio.

BPSK is used to carry pilot tones and TPS.

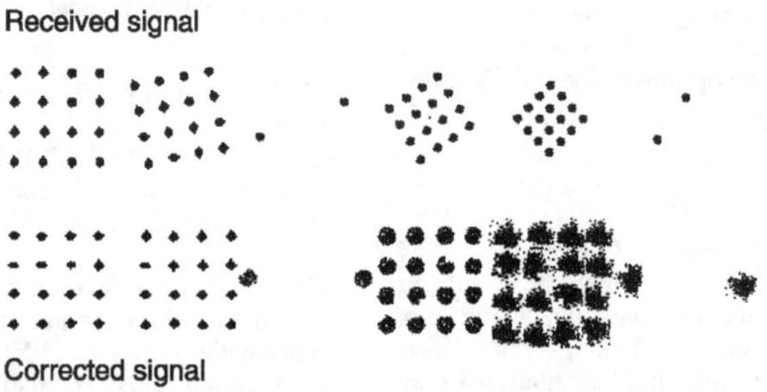

Figure 2.12.17 Pilot tones are used to dynamically measure the transmission channel phase and frequency response. COFDM carriers may be amplified or diminished by the presence of multi-path signals ('echoes'). The diagram is a simplification – scattered pilots do not occupy every third carrier as this would be too wasteful of bandwidth. After four symbol periods the scattered and continuous pilots will have measured the amplitude and phase of every third carrier.

2.12.4.5.8 64-QAM with 68 'Spots'

On a constellation monitor a 64-QAM signal will have 68 'spots'. The extra spots are the pilot tones and TPS signals (Figure 2.12.18). The pilot tones are boosted in amplitude relative to the average power. The TPS signals are the same amplitude as the constellation average power.

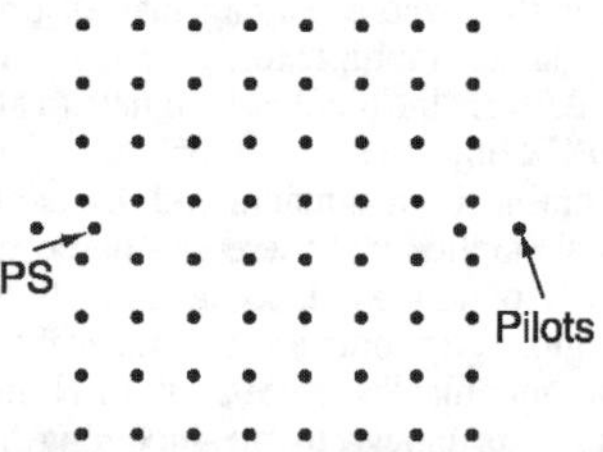

Figure 2.12.18 Undistorted COFDM signal showing extra 'spots'.

Similarly 16-QAM will have 20 'spots' and QPSK will have eight 'spots'.

2.12.4.5.9 DVB-T Bit Rates

DVB-T supports a wide range of transmitted bit rates, from 4.976471 to 31.668449 Mbit/s in an 8 MHz channel. (One feature of all digital terrestrial modulation schemes is that the fixed channel spacing results in very accurately defined TS data rates.) Generally, the higher the bit rate the less robust the signal is to noise and interference. Although the system uses the same FEC as DVB-S (plus additional bit interleaving) it is unlikely that FECs of 5/6 and 7/8 will be robust enough for DTH systems, making some of the higher data rates unrealisable in practice (5/6 and 7/8 might be used for point-to-point SHF COFDM links, however). European broadcasters are experimenting with modulation parameters which yield a 14.5–26.5 Mbit/s payload (approximate figures).

2.12.4.5.10 DVB-T Reception

DVB-T is being radiated at 15–20 dB below the power level of the analogue services. Coverage is similar to that of the analogue transmissions and DVB-T signals can be received on set-top aerials ('rabbit's ears') as DVB-T is fairly immune to dynamic multi-path interference. Some susceptibility to wideband RF noise has been noticed. This usually emanates from unsuppressed electric motors (drills) and household appliances switching on and off (freezers, central heating). A properly earthed, double-screened aerial downlead and filtering of mains-borne interference should cure these problems.

2.12.4.5.11 Other Digital Terrestrial Standards

The Advanced Television Systems Committee (ATSC) is the American equivalent of DVB. The ATSC has defined modulation schemes designed to carry HDTV signals in the terrestrial and cable environments. The terrestrial scheme is known as 8-VSB and uses a single carrier with eight digital levels modulated by 3 bits of data. Forward Error Correction is similar, but not identical, to the DVB scheme.

Japan's equivalent to DVB-T is the Integrated Services Digital Broadcasting-Terrestrial (ISDB-T). ISDB-T is very similar to the DVB-T system, but is optimised for mobile reception. It uses time interleaving, whereas DVB-T uses frequency interleaving, to reduce the effects of Intra-Symbol Interference. (Investigation into the possibility of incorporating time interleaving into the DVB-T standard was carried out – but this would not be compatible with existing systems and it may have been abandoned since it would increase delay (latency) and the cost of receivers for no appreciable benefit. It is possible that a similar proposal might be made for a DVB-T-DENG standard if one appears). ISDB-T also uses 64-QAM and D-QPSK at the same time to carry the different 'segments' of the transmitted data. (Differential QPSK is preferred for systems with mobile reception, which was a major design criterion for the Japanese system.)

ISDB-T has not yet been formally adopted outside Japan, although it has been tested side by side with DVB-T and 8-VSB in several countries, including China and Brazil.

China is believed to be designing its own transmission scheme, which may incorporate some of the best features of the existing three proposals and be optimised for HDTV transmission.

2.12.4.5.12 1705 Is Not the Same as 2048

A Fast Fourier Transform (FFT) using the Cooley and Tukey algorithm[49] must use a number of carriers that are an integer power of 2 (other FFT algorithms are available, but are not widely used); 2048 is 2 to the power of 11 (2^{11}) – but 1705 is not on the 'binary ladder'. An early implementation of an OFDM modem was susceptible to Co-Channel Interference (CCI), so it was decided to choose a carrier spacing so that OFDM carriers could be placed 'on top of' harmonics of line frequency from a CCI interferer. The solution chosen does not allow the OFDM carriers to mask line frequency components and the vision carrier at the same time, but, in any case, CCI is not considered to be a problem with modern chipset receivers.

2.12.4.5.13 Hierarchical Modulation

Hierarchical modulation in the DVB-T scheme can be thought of as a QPSK COFDM signal with a different QPSK (or 16-QAM) COFDM signal modulated on top of the four QPSK constellation points (Figure 2.12.19). The two COFDM signals occupy the same RF channel and must have the same number of carriers and same guard intervals but can otherwise be regarded as totally unconnected signals. (Their symbol rates, whilst they may be different, have to be locked to the same frequency reference or one must be rate adapted to the other.)

Australian broadcasters have proposed carrying an SDTV (MP@ML) service on the High Priority (QPSK) signal and an HDTV (MP@HL) service on the Low Priority (16-QAM) signal. The HP SDTV service would be accessible to mobile receivers and the LP HDTV service would be accessible to receivers connected to fixed aerials. All DVB-T receivers can receive the HP signal but LP signals are only accessible to receivers which are 'hierarchy-capable'.

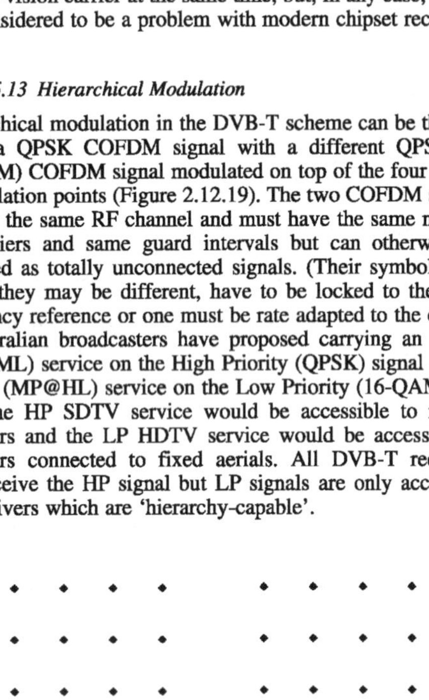

Figure 2.12.19 16-QAM signal hierarchically modulated onto a QPSK signal. This signal has an $\alpha = 2$. DVB-T also allows an $\alpha = 4$ (constellation clouds further from centre) and an $\alpha = 1$ (uniform spacing).

2.12.4.6 DVB-T-SFN

DVB Document A024 Rev. 1: *Digital Video Broadcasting (DVB)*, February 2001
(ETSI TS 101 191 V1.3.1: *Digital Video Broadcasting (DVB); DVB Mega-Frame for Single Frequency Network (SFN) Synchronization*, January 2001)

The DVB-T modulation scheme can use Single Frequency Networks (SFNs) which may be metropolitan, regional or nationwide in extent. SFNs are in operation in Finland, Spain and Sweden. SFNs are also planned for Germany, Taiwan and Australia.

Broadcasters have some experience of SFN systems from Digital Audio Broadcast[50] (DAB) networks. DAB systems are in use in Belgium, Canada, Denmark, Finland, France, Germany, Greece, Norway, Portugal, Spain, Sweden, Switzerland, Taiwan and the UK. Many other countries have experimental systems in operation.

SFNs have also been used by military communications engineers for many years but this experience is not publicly available.

2.12.4.6.1 The Three Engineering Constraints of SFNs

These can be expressed very simply as 'same bits at the same time on the same frequency' everywhere inside the SFN.

2.12.4.6.2 Same Bits Everywhere

If identical bits must be radiated from all of the transmitters in an SFN this is the same as saying that all transmissions must originate in a single multiplexer. Any telco links to the transmitters must deliver 'bit-identical' signals to all sites, so no rate adaption or PCR adjustment is possible.

This constraint is misunderstood by some broadcasters. 'Identical bits' implies that there can be no regional variation within an SFN of any kind. Local adverts, OOV announcements and regional opt-outs are all impossible within an SFN.

All of the modulators within an SFN must either slave their transport rate clocks to the incoming Transport Stream or be using the same clock source as the multiplexer. In practice SFNs may synchronise transport rate clocks to a GPS 10 MHz reference, although this is not required by the DVB-T specification.

2.12.4.6.3 Synchronising the Transmitters

The transmitters all need a synchronisation mechanism which will work anywhere within the SFN. The basic principle is that all transmitters store incoming bits in a buffer until the time for radiation is reached when all transmitters will radiate the same bits at the same instant.

2.12.4.6.4 The Mega-frame Initialisation Packet

The mechanism for synchronising the transmitters is the Mega-frame Initialisation Packet (MIP). A Mega-frame has been defined to guarantee that all transmitters map the same Transport Packets into the eight-packet Energy Dispersal sequence and byte and bit interleavers. If this had not been done the signals radiated from the transmitters would not have been identical. The MIP contains five types of information. These are a pointer to the start of the next Mega-frame, the time of generation of the MIP, the time delay before the Mega-frame is to be transmitted, the TPS signals and private data which can be addressed to an individual transmitter by means of a unique numerical address.

The time is specified as counts of the GPS 10 MHz clock since the last GPS second boundary. The delay is specified as a number of 'ticks' of the GPS 10 MHz clock.

The MIP can sit anywhere inside a Mega-frame, which is why a pointer is required. A Mega-frame is approximately half a second's worth of TS – the exact duration is dependent on the Guard Interval and the channel spacing. The timing information refers to the next Mega-frame, not the one which the MIP is in. The TPS information (which can be used to change the modulation parameters of an entire SFN) applies to the Mega-frame after the one which the timing information applies to.

2.12.4.6.5 Synchronising the output frequencies

Within a 2000 carrier SFN the output frequencies of the transmitters must be within 6 or 7 Hz of each other. Obviously the tolerance for an individual transmitter must be tighter than this.

2.12.4.6.6 Transmitter Siting

SFN transmitters are usually sited at existing main transmitter locations. Because of the 'SFN gain' effect it may not be necessary to have as many repeaters as in a normal MFN.

2.12.4.6.7 On-Frequency Repeaters

On-frequency repeaters are difficult to use in a traditional MFN. Either the service area must be completely shadowed from the main transmitter (as might be the case in a valley) or the on-frequency repeater must use the opposite polarity to prevent aerials picking up both signals. On-frequency retransmission poses no problems for DVB-T SFNs – this can ease network planning considerably.

2.12.4.6.8 Cold Spots and Dead Spots

If two transmitters are radiating an identical signal there is the unwelcome possibility of total cancellation where the two signals are of equal power but opposite polarity. This is unlikely to affect a large part of the service area – typically a narrow (~100 m) corridor approximately midway between two equal-power transmitters.

In practice this is unlikely to affect static receivers – even if they are inside the cold spot a simple remedy is to use a directional aerial. Cold spots or dead spots are more likely to be a problem for mobile and hand-held receivers. If a viewer is watching a portable receiver near to a hard reflecting surface, such as a wall, a reflection may interfere destructively with the direct signal. Although the interference is frequency selective it can be regarded as a 'flat fade' as the COFDM signal effectively sinks into the noise. Multiple aerials for diverse reception with an intelligent front-end to select the best signal is one promising solution being investigated.

2.12.4.6.9 Doppler Shift

A moving receiver will receive a Doppler shifted version of the transmitted signal. It may also receive time-varying echoes with different amounts of Doppler shift from the main signal. The faster a receiver is moving the larger the Doppler shift will be (this principle is used by RADAR systems to estimate speeds of moving objects). Doppler shift is also directly related to the transmitted frequency. Doppler has less effect on VHF than UHF and less effect on low-end UHF (~500 MHz) than high-end UHF (~1 GHz). The inter-carrier spacing between 2000 carrier systems is four times as great as in 8000 carrier systems so 2000 carrier systems are more robust in the presence of multiple Doppler echoes.

2.12.4.6.10 Constellation Patterns

DVB-T constellation patterns can be used to diagnose some simple SFN faults (Figures 2.12.20–2.12.22).

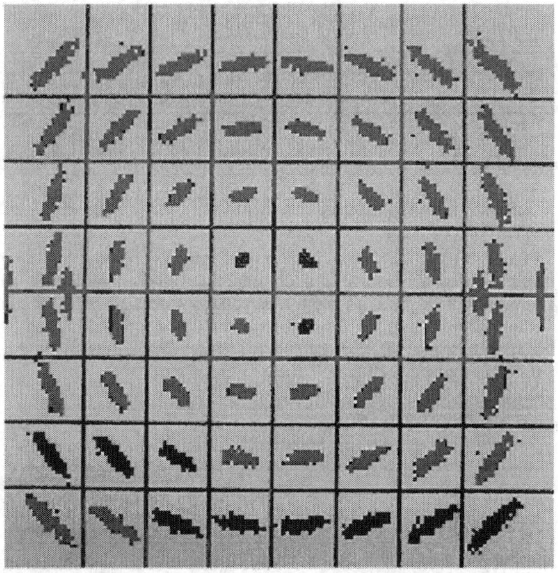

Figure 2.12.20 SFN with phase noise. This diagram shows phase noise on the local signal, so the phase noise arcs are centred on the centre of the constellation. If the phase noise were to be present on the signal from a remote transmitter the arcs would be centred on the individual symbols (TANDBERG Television).

Since the DVB-T signal is effectively an 8 MHz sweep it can be used as an 'in-service' test to diagnose some simple RF faults, such as VSWR problems at a transmitter.

2.12.4.6.11 Driving the 'Sweet Spot'

If the guard interval is insufficiently long to provide ISI immunity between two transmitters it is possible to delay the transmission time from one of the transmitters to bring the 'sweet spot' (or, more properly, corridor) closer to the delayed transmitter.

2.12.4.6.12 Satellite Distribution to SFNs

Satellite distribution to DVB-T SFNs is possible, but it must be remembered that so-called geostationary satellites actually move around slightly relative to the earth. Geostationary satellites are kept inside a cube with 75 km sides by 'station-keeping' signals sent from a ground station which actuate

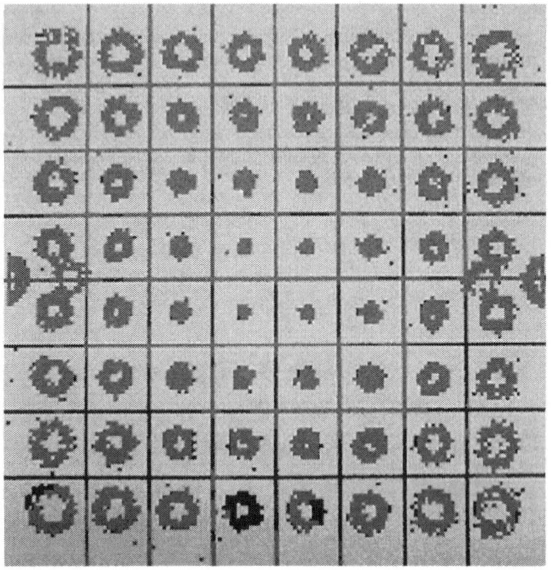

Figure 2.12.21 SFN with unlocked RF signals. Notice that small-amplitude symbols have a small-amplitude interfering symbol and large-amplitude symbols have a large-amplitude interfering symbol ('dots in the centre, circles at the corners'). This shows that the two transmitters have different output frequencies ('same bits, same time, different frequencies') (TANDBERG Television).

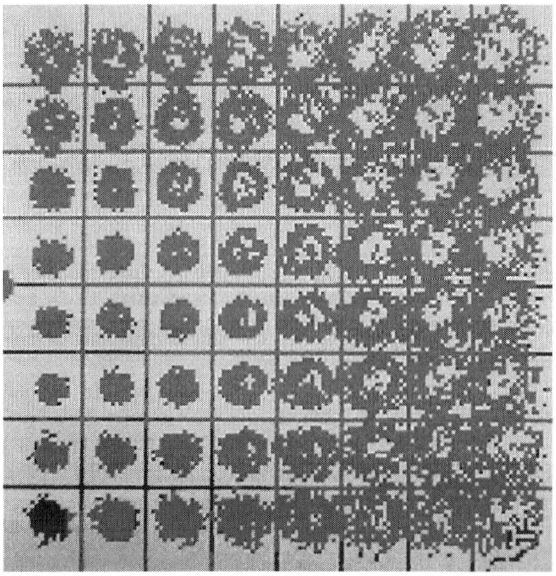

Figure 2.12.22 SFN with CCI from vision carrier. Multi-carrier systems are immune to the interference from single tone CCI as it only affects a few of the OFDM carriers (TANDBERG Television).

propulsion systems on the satellite. In theory a satellite moving between opposite corners of this box would correspond to a change in delay of up to 780 μs along the diagonal. An earth-based observer could not experience this magnitude of delay variation as the diagonal does not point at the earth, but diurnal, sinusoidal variations of about 200 μs peak-to-peak have been observed, and larger variations are possible. The DVB-T transmitters must be able to track both maximum variations in distribution delay (maximum guard interval for 8K systems is only 224 μs) and the maximum rate of delay change. This effect has also been observed in satellite-fed DAB networks, where – because of the smaller buffers in use – interruptions to service have occurred. When a satellite runs out of station-keeping fuel it is sometimes allowed to degenerate into an inclined orbit before being finally 'retired'; this would increase the maximum delay variation further.

2.12.4.7 DVB-T Implementation Guidelines

DVB Document A037: *Implementation Guideline for DVB-T Transmission Aspects*, March 1998
(ETSI TR 101 190 V1.1.1: *Digital Video Broadcasting (DVB); Implementation Guidelines for DVB Terrestrial Services; Transmission Aspects*, December 1997)

This document starts with a short tutorial on the DVB-T standard and progresses to network planning, gap-fillers, RF issues, DVB-T distribution, SFNs and protection ratios. ETR 101 190 is essential reading for any transmission group contemplating a transition to DVB-T, particularly if they plan to have any SFNs.

2.12.4.8 DVB-MT

DVB Document A052: *Digital Video Broadcasting (DVB); OFDM Modulation for Microwave Digital Terrestrial Television*, June 1999
(ETSI EN 301 701 V1.1.1: *Digital Video Broadcasting (DVB); OFDM Modulation for Microwave Digital Terrestrial Television*, August 2000)

Existing analogue broadcast TV distribution networks often use point-to-point microwave links to distribute the baseband video and audio signals to the transmitter sites. Outside broadcasts often use point-to-point SHF links as well, although there is increasing use of satellite feeds where these are cheaper. (In the UK 2, 5, 7 and 12 GHz have been used by broadcasters for contribution and distribution, but there is increasing pressure to release some of these bands.)

This standard describes a method of distributing DVB-T signals to remote transmitters using a microwave transport layer. The DVB-T signal can then be down-converted to the correct UHF frequency before radiation. This architecture removes the need to have a costly COFDM modulator at each transmitter site.

The standard can also be applied to MMDS and MVDS systems, which may benefit from the extra robustness of the COFDM signal – but EN 301 701 is not intended to supplant DVB-MC (EN 300 749) or DVB-MS (EN 300 748), it is merely an alternative modulation scheme. (DVB-MT has been shown to be more robust than DVB-MC but it does carry less bits/s/Hz.)

The Annex describes the phase noise and frequency stability requirements of DVB-T at microwave frequencies (these are more onerous for designers than the equivalent requirements for DVB-T on UHF).

2.12.4.9 DVB-C

DVB Document A035: *Digital Broadcasting Systems for Television Sound and Data Services; Framing Structure, Channel Coding and Modulation for Cable Systems*, March 1998
(**ETSI EN 300 429 V1.2.1:** *Digital Video Broadcasting (DVB); Framing Structure, Channel Coding and Modulation for Cable Systems*, April 1998)
(**ITU-T Recommendation J.83:** *Digital Multi-Programme Systems for Television, Sound and Data Services for Cable Distribution Annexe A*, April 1997)

The cable environment is much less hostile to RF signals than satellite or terrestrial – being essentially linear and virtually noise free. The engineering challenges in cable distribution are frequency response of the cables and static echoes caused by mismatches. A DVB-C modulator consists of a V4MOD section and a QAM modulator. Trellis encoding is not required since cable is inherently 'quiet' and high-order QAM constellations can be used for the same reason; 16-, 32-, 64-, 128- and 256-QAM are supported but STBs only have to work with 16-, 32- and 64-QAM to be compliant (Figure 2.12.23).

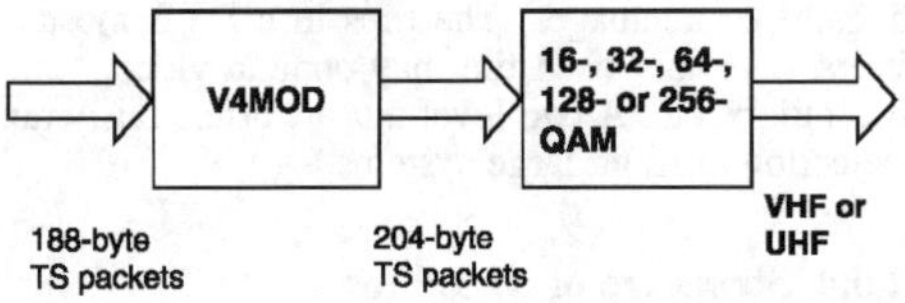

Figure 2.12.23 DVB-C modulation. Trellis encoding is not required for cable systems as they generally have a good signal-to-noise ratio on all carriers.

The two MSBs of each symbol are differentially encoded to produce a rotation-invariant constellation. This produces a phase reference for the demodulator to lock on to.

Cable systems in Europe generally use an 8 MHz channel spacing (some use 6 MHz and at least one uses a mixture of 6 MHz and 8 MHz). With a roll-off factor of 0.15 this limits the symbol rate to 6.96 Mbaud, giving a maximum TS rate of 51.31 Mbit/s. In practice cable networks often acquire their programming from a master head-end via ATM, PDH or satellite.

Annex A has a spectral mask for the baseband filter characteristics.

Annex B contains a table showing how a TS can be seamlessly transferred to a cable system without the need for service-dropping or bit rate adjustment. Because of the integrated nature of DVB specifications the transition of a TS from satellite to cable can be achieved by merely changing 13 bytes in the Network Information Table (NIT).

2.12.4.10 MMDS (DVB-MC)

DVB Document A015: *Framing Structure, Channel Coding and Modulation for MMDS Systems below 10 GHz*, February 1997
(**ETSI EN 300 749 V1.1.2:** *Digital Video Broadcasting (DVB); Microwave Multipoint Distribution Systems (MMDS) below 10 GHz*, August 1997)

Microwave Multipoint Distribution Systems (also known as Multi-channel Multi-point Distribution Systems or 'wireless cable') were originally used for rural areas that were difficult to serve with 'traditional' cable technology but they are now also being used in urban environments. In rural areas a low-power transmitter operating in the 2.5–2.7 GHz band can provide coverage up to 40 km away in an analogue MMDS system – digital systems may require less power for the same resulting coverage. Modern MMDS systems usually offer some form of two-way interactive services, such as Internet access and bidirectional high-speed data transmission (for business use). DVB-MC is virtually identical to DVB-C, the major difference being that DVB-MC is radiated at microwave frequencies and DVB-C uses VHF or UHF constrained within a coaxial distribution system. DVB-MC uses 16-, 32- and 64-QAM modulation.

2.12.4.11 MVDS (DVB-MS)

DVB Document A013: *Specification for Framing Structure, Channel Coding and Modulation for Multipoint Video Distribution Systems at 10 GHz and above*, May 1996
(**ETSI EN 300 748 V1.1.2:** *Digital Video Broadcasting (DVB); Multipoint Video Distribution Systems (MVDS) at 10 GHz and above*, August 1997)

Multipoint Video Distribution Systems (or Microwave Video Distribution Systems) are similar to MMDS except that they use a quite different transmission frequency with different propagation properties. A low-power (~1 W) signal is radiated for local reception (up to 6 km in rural areas for analogue systems – further for digital systems). DVB-MS is virtually identical to DVB-S, the major difference being that DVB-MS is radiated at millimetric frequencies (40.5–42.5 GHz) and DVB-S uses 11–13 GHz. DVB-MS may be used for other frequency bands above 10 GHz.

Annex C contains a table showing bit rates versus MVDS bandwidth for all FEC combinations.

2.12.4.12 SMATV (DVB-CS)

DVB Document A042: *Digital Video Broadcasting (DVB); Satellite Master Antenna Television (SMATV) Distribution Systems*, June 1999
(**ETSI EN 300 473 V1.1.2:** *Digital Video Broadcasting (DVB); Satellite Master Antenna Television (SMATV) Distribution Systems*, August 1997)
DVB Document A065: *Digital Video Broadcasting (DVB); Baseline Specification for a Control Channel for SMATV/MATV Distribution Systems*, April 2001

(**ETSI TS 101 964 V1.1.1:** *Digital Video Broadcasting (DVB); Control Channel for SMATV/MATV Distribution Systems; Baseline Specification*, August 2001)

Satellite Master Antenna Television (SMATV) systems consist of communal satellite and/or terrestrial antennas with a local cable distribution network. They are sometimes known as 'community antenna installations' or 'domestic TV cable networks'.

System A consists of 'transparent transmodulation' between incoming DVB-S and outgoing DVB-C signals where the incoming signal is decoded to Transport Stream level and then remodulated. System B has two variants – SMATV-IF and SMATV-S. In SMATV-IF the incoming signal is demodulated down to L-band and then distributed as a DVB-S signal at 950 MHz or above. In SMATV-S the incoming signal is transmodulated to VHF or UHF (230–470 MHz).

EN 300 473 specifies a control channel which allows a consumer to request a particular multiplex using DiSEqC[51] signalling.

2.12.5 Service Discovery and Acquisition

MPEG-2 defines Program Specific Information (PSI), which are tables a receiver can use to find the components of an MPEG-2 Program (video, audio, CA information, subtitles, etc.). PSI tables are sufficient for professional applications such as contribution and distribution systems, but for DTH broadcasting something more user-friendly is required. PSI tables also only refer to the Transport Stream which they inhabit, but modern digital systems may consist of many Transport Streams. A viewer is not usually interested in the technicalities of satellite transponder frequencies or which multiplex contains which Programs. DVB Service Information (SI) is a group of tables which are used to drive the Electronic Programme Guides in a viewer's domestic Set-Top Box (STB) or integrated receiver. A viewer selects a Service or Event from the EPG and the STB automatically finds the correct multiplex and Service without the viewer having to know anything about Packet IDentifiers (PIDs) or frequencies.

2.12.5.1 New Digital Vocabulary

MPEG-2 uses the word 'Program' to refer to what used to be known as a 'channel' or 'network' (e.g. BBC2, RAI Due, ABC, CCTV, etc.). In DVB terminology a Program is always referred to as a 'Service'. DVB terminology is summarised in Table 2.12.7.

Table 2.12.7 New digital vocabulary

	Analogue	*MPEG-2 PSI*	*DVB SI*
Europasat	Network/Channel	Network	Network
Europasat News	Network/Channel	Program	Service
Midday News	Programme/ Program	Not defined	Event
Video, audio, subtitles, etc.	No name in common use	Elements	Components

Uncompressed video and audio are sometimes referred to as VandA or VANDA.

2.12.5.2 Linking PSI to SI

An MPEG-2 Program is uniquely referenced by a 16-bit *program_number*. A DVB Service is uniquely referenced by a 16-bit *service_id* (Figure 2.12.24). The *service_id* in an SI table links the DVB Service to the *program_number* in MPEG-2 PSI, i.e. it is the same number in both sets of tables. Use of this number by SI tables means that the SI generator does not have to know anything about PIDs in Transport Streams.

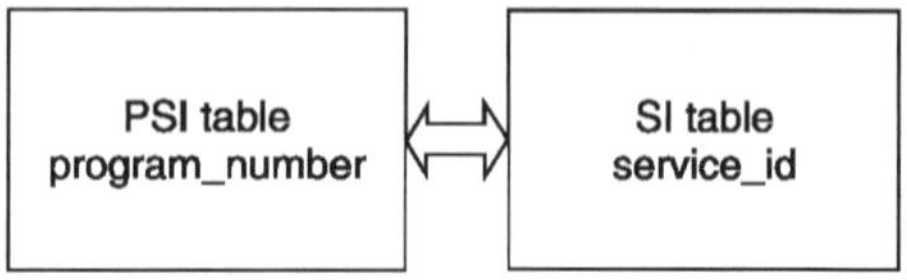

Figure 2.12.24 *program_number* is identical to *service_id*.

2.12.5.3 DVB-SI

DVB Document A038 Rev. 1: *Digital Video Broadcasting (DVB); Specification for Service Information (SI) in DVB Systems*, May 2000
(**ETSI EN 300 468 V1.4.1:** *Digital Video Broadcasting (DVB); Specification for Service Information (SI) in DVB Systems*, November 2000)

DVB SI tables support text and mosaic-based EPGs (Figure 2.12.25). Text-based EPGs cannot be used in countries with low literacy levels and may be difficult to implement in regions with multiple languages. The tiles in a DVB mosaic may be different sizes and each tile may contain video, text, channel logos, trailers, etc. A two-level mosaic offers fairly rapid channel selection even on large systems.

2.12.5.4 Structure of SI Tables

SI tables have the same structure as PSI tables. Each table is composed of 1–256 sections. Each section is a maximum of 1024 or 4096 bytes (depending on the table). A *table_id* indicates what type of table is in the TS packet.

2.12.5.4.1 NIT

The Network Information Table (NIT) is used by the STB for automatic tuning to a new Transport Stream. The NIT contains the name of the network and will normally contain the tuning information for all of the multiplexes forming the broadcast network. It may also contain tuning information for other broadcast networks. Initially, some broadcasters were concerned about a 'lobster pot' effect where an STB retuned to another network which did not point back to the referring network in its own NIT, but this fear seems to have subsided.

As well as tuning information an NIT can also contain a *service_list_descriptor*, which is a list of all the services present in a particular multiplex. An STB can scan the *service_list_descriptor* for each multiplex until it finds the Service a viewer wants to watch. It can then use the associated tuning information to find the correct multiplex, followed by the PAT/PMT mechanism to find the components of the required Service.

Figure 2.12.25 DVB SI tables support navigation via 'mosaics'. A descriptor in the SDT informs the STB that the service is a mosaic. The mosaic descriptor tells the STB which service corresponds to each section of the video image. When the viewer highlights a section of the mosaic and presses 'select' the STB will automatically tune to the required channel.

In the UK DVB-T platform the NIT also contains the *logical_channel_number* descriptor. This descriptor links a Service ID to a Logical Channel Number (LCN). In different parts of the UK the Service 'BBC2' has different Service IDs. On an STB EPG it always appears as service number 2, because that is what its LCN is.

The tuning information in an NIT is always contained in a 13-byte descriptor. If a cable broadcaster wishes to receive a DVB-S satellite transponder and rebroadcast it digitally it is merely necessary to remove the 13 bytes with the satellite tuning parameters and replace them with 13 bytes describing the cable tuning parameters. Of course, if any Services are being dropped or added further modifications to the other SI tables will be necessary.

Since the satellite descriptor includes information about the satellite orbital position it is possible to automatically move a steerable dish to point at a different satellite using information from the NIT. A viewer would not necessarily be aware that this was happening, except that changing satellites would take longer than retuning to another transponder on the same satellite. At least one Scandinavian broadcaster is believed to be using this mechanism. Satellite operators such as Astra and Eutelsat have multiple satellites in an orbital slot – this should provide plenty of capacity for any broadcaster so that moving a dish becomes unnecessary for a given network. (Astra pioneered the idea of multiple satellites in one slot; it will probably be copied by other satellite operators as slots on the Clarke belt become harder to obtain.)

2.12.5.4.2 SDT

The Service Descriptor Table (SDT) can be used by an STB to display a list of available Services to a viewer (Figure 2.12.26). It is also used in this way by most professional receivers in contribution applications.

The SDT comes in two varieties – SDT-Actual and SDT-Other (often shortened to SDT-A and SDT-O). SDT-Actual lists the Services on the multiplex it is present in. SDT-Other lists Services in other Transport Streams. SDT-Actual is mandatory in DVB systems, SDT-Other is not.

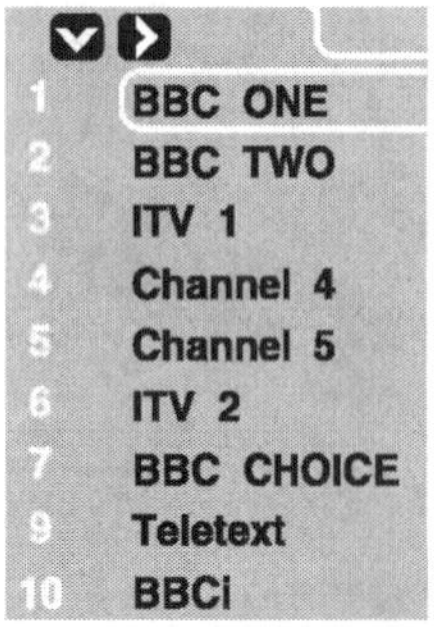

Figure 2.12.26 Section of EPG from early UK DVB-T Set-Top Box. Service names (BBC1, ITV, etc.) are from SDT. Channel numbers (1, 2, 3, 4, etc.) are from *logical_channel_number* descriptor in NIT.

The SDT can be thought of as a table containing three columns. The columns list Service IDs, Service Names and the Service Providers. A domestic STB will only normally display the Service Names. On a domestic STB the Service IDs may be mapped to a Logical Channel Number, usually by information contained in the NIT or BAT. Many professional receivers will be able to display the contents of all three 'columns'.

If a Service is a 'mosaic service' there will be further descriptors linking it to other mosaics or services.

2.12.5.4.3 TDT and TOT

The Time and Date Table (TDT) is the only SI table that everyone can remember the function of! It contains the time in Universal Coordinated Time (UTC) format and the date as a Modified Julian Date (MJD).

UTC is the modern equivalent of Greenwich Mean Time, with leap seconds added in or subtracted when deemed necessary by a collection of atomic clocks.

Julian Date[52] is the number of days elapsed since *noon* on 1 January 4713 BC. Julian Date was mainly used by archaeologists who did not want the inconvenience of expressing dates in BC format, where dates count backwards as you move forward in time. Modified Julian Date[53] is the number of days elapsed since *midnight* on 17 November 1858 AD. Both Julian Date and Modified Julian Date systems are used by astronomers. DVB systems only need to refer back to dates after 1995 so MJD was adopted.

To be useful in a consumer environment the date must be converted by the STB into the familiar day, date, month, year format before display on an EPG (30/06/1997 would be 50629 in MJD format – which is not instantly comprehensible).

The Time Offset Table gives the time offset from UTC. Time zones East of the UK have a positive number for the offset, time zones to the West have a negative number. The TOT can also be used for summer time or 'daylight savings time' adjustments and for multiple time zones in one broadcast region (e.g. Australia).

The STB uses the TDT and TOT to calculate the local time for display on the EPG.

The TDT is mandatory in DVB multiplexes.

2.12.5.4.4 EIT(p/f)

The Event Information Table (present/following) can be used by the STB for very simple EPG functions (Figure 2.12.27). 'Event' is the DVB term for what used to be called a 'programme'. EIT(p/f) contains the name of the current 'Event' together with its start time (in UTC) and duration. It may also contain a *short_descriptor* (usually referred to as 'the synopsis'), which contains up to 255 bytes of information about an Event – such as a precis of the plot and the names of some of the actors for a soap opera. Short descriptors may be linked together to form longer synopses.

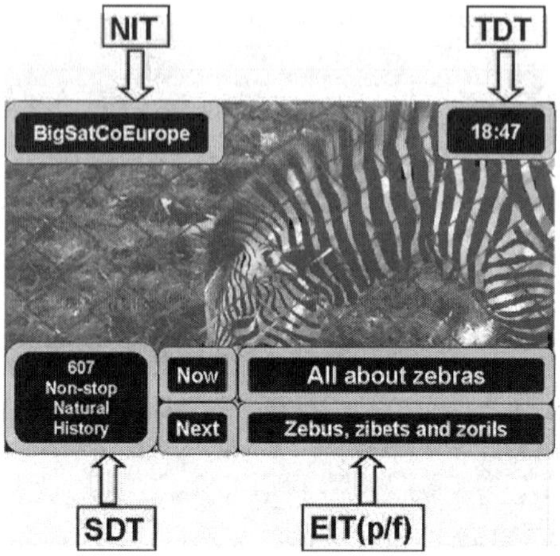

Figure 2.12.27 Zebras Now, Zorils Next. The diagram shows which SI tables are responsible for generating different parts of an EPG. Early terrestrial EPGs may be as simple as this (fictional) example due to bandwidth constraints.

The EIT(p/f) also contains the same type of information for the next Event.

In an ideal broadcasting system, the contents of the EIT(p/f) should change at every Event boundary, with the old following Event becoming the new present Event and a new following Event appearing in the table. DVB does not have any rules about how accurately EIT(p/f) should track Event boundaries. Even if EIT(p/f) is several hours adrift this does not appear to be in breach of any rules, although this is undesirable.

EIT(p/f) has two variants, EIT(p/f)-Actual and EIT(p/f)-Other. EIT(p/f)-Actual refers to Events on the 'actual' multiplex, i.e. that mux which contains the EIT(p/f)-Actual. EIT-(p/f)-Other refers to Events on 'other' multiplexes.

EIT(p/f)-Actual is mandatory in DVB multiplexes.

2.12.5.4.5 EIT(schedule)

EIT(schedule) is used to produce the 'programme grid' on an EPG (Figure 2.12.28). EIT(schedule) may occupy a large amount of bandwidth (more than 1 Mbit/s on some satellite systems which use it) and is often not used by terrestrial broadcasters as they do not have as much bandwidth to devote to EPG information.

EIT(schedule) is not mandatory.

2.12.5.4.6 BAT

The Bouquet Association Table (BAT) can be used to group Services into thematic groups (such as 'football' or 'soap

---

BigSatCoEurope
EPG Plus!
18:43
Channel
18:30
19:00
19:30
601
The Dog Detectives
Man's Best Friend
602
Shadow People
Plan 9 from Outer Space
603
Politics Today
PM's Question time
604
The Mystery of Hill Farm
Murder at
605
Fields of Ice
Volcanoes of Europe
606
Return to the Moon
Killer Asteroids
607
All about Zebras
Zebus, Zibets & Zorils
608
Vernacular Furniture of England
Going, going,
609
The Gentle Giant
The Best Man

Figure 2.12.28 EPG grid generated from contents of EIT (schedule). A real EPG grid will normally show Service Names (from SDT).

operas') on the EPG. A viewer could change from one news Service to another and be totally unaware of whether the transmission mechanism was terrestrial, cable, satellite or something else. It would require unprecedented coordination and cooperation between broadcasters for a BAT to be used in this way.

In practice BATs have been used by Conditional Access (CA) companies to associate entitlements with smart cards in domestic STBs (at least two CA companies have used BATs like this). When used in this way it would be possible to only display Services and Events which a viewer is entitled to watch on an EPG. It is also possible to remap Service IDs to Logical Channel Numbers – depending on which 'bouquet' a smart card belongs to. BATs used for CA purposes are usually 'bandwidth-hungry' and may need 1–2 Mbit/s (per multiplex) on very complex systems.

The expression 'bouquet' is used by many broadcasters to refer to a multiplex (usually on a satellite) containing Services from only one broadcaster. This usage comes from DVB Document A003, where a bouquet is described as: "A collection of services marketed as a single entity." In the case of a single multiplex or transponder a BAT is not required, since the SDT can adequately describe such a Transport Stream.

The BAT is not mandatory.

2.12.5.4.7 RST

The Running Status Table (RST) can be used to update portions of an EIT(schedule) if an Event over- or under-runs (or is cancelled) without having to re-send the whole EIT(schedule). If an EIT(schedule) contains several days worth of programming this might be a useful way to save bandwidth.

The RST could also be used to duplicate Programme Delivery Control[24] (PDC) or Video Programme System (VPS) functionality in a digital system. (PDC and VPS are signalling systems, transmitted in the VBI, which can be used by VCRs to record a 'programme' even if it starts late or early, or is extended. VPS is used in German-speaking countries, PDC is used elsewhere. PDC is carried in a teletext signal, VPS is carried in a signal similar, but not identical, to teletext.)

The author has never seen an RST table.

Transmission of the RST is not mandatory.

2.12.5.4.8 ST

The Stuffing Table (ST) can be used by the broadcaster to invalidate sections of SI tables when these are no longer correct. An example of where this might be necessary would be a broadcaster re-multiplexing a Transport Stream and dropping some of the Services or changing the modulation parameters (or both). The contents of the SDT, EIT(p/f) and NIT would no longer be valid.

Stuffing Tables are certainly not in common use and, indeed, may never have been used. Manufacturers of re-multiplexing equipment usually provide fairly sophisticated SI-processing functions, making Stuffing Tables unnecessary. If a remux device is dropping Services it is a relatively trivial matter to remove references to those Services from the PSI and SI tables which contain them.

Use of ST is not mandatory.

2.12.5.4.9 SIT

The Selection Information Table (SIT) is never broadcast. It resides on a server with a partial Transport Stream and contains a summary of the information required to produce compliant SI in a broadcast stream. For example, an SIT may contain a clip length which could be used to generate EIT(p/f) information.

2.12.5.4.10 DIT

The Discontinuity Information Table (DIT) is also never broadcast and will normally be found on servers. It is inserted at points where the SI is discontinuous (e.g. at Event boundaries).

2.12.5.5 Non-European Character Sets

DVB-SI supports Latin, Cyrillic, Arabic, Greek, Hebrew, Thai, Celtic, Korean, and Simplified and Traditional Chinese.[54–58] It may also support Indian and other alphabets.

2.12.5.6 Descriptors

Descriptors use a tag/length/variable structure. DVB descriptors have tags of 0x40 or greater. The data in a descriptor may be binary bit fields, ASCII strings, UNICODE, Binary Coded Decimal or some combination. Descriptors are used by STBs to tune into a new transponder, display a service name, choose an audio channel with a preferred language, etc.

2.12.5.7 Bandwidth occupied by SI Tables

The BSkyB satellite system broadcasting to the UK has approximately 200 TV channels (plus a large number of radio channels). The BAT – which is mostly used for CA purposes – has a bandwidth of about 200 kbit/s. The EIT(p/f) bandwidth is approximately 50 kbit/s. Other SI table bandwidths are negligible.

2.12.5.8 Alternatives to SI

The BSkyB system generates enough SI tables to be DVB-compliant, but the main EPG is driven by private tables called eXtended Service Information (XSI). Text compression can be used in XSI to reduce the bandwidth required to transmit a complex, large EPG. BSkyB transponders contain about 700 kbit/s of XSI, but for more detailed EPG information the STB tunes to a special transponder. This mechanism increases the total amount of bandwidth available for video and audio services.

2.12.5.9 Guidelines

DVB Document A005 Rev. 2: *Digital Video Broadcasting (DVB); Guidelines on Implementation and Usage of Service Information (SI)*, May 2000
(ETSI TR 101 211 V1.4.1: *Digital Video Broadcasting (DVB); Guidelines on Implementation and Usage of Service Information (SI)*, July 2000)

This document contains a comprehensive list of the descriptors used in DVB systems. This information has been collated from a number of other DVB documents. It also suggests mechanisms for auto-tuning of mobile DVB-T receivers passing from one transmitter region to another. Annex A contains advice on PID, table ID and descriptor ranges which should be avoided to ensure usability of DVB TS in ATSC systems. (The ATSC has made similar provisions; if the Program Paradigm is in use ATSC TS should not contain Program 1 as this would clash with DVB SI allocated PIDs.)

2.12.5.10 Allocation of Codes

(Draft ETSI TR 101 162 V1.2.1: *Digital Video Broadcasting (DVB); Allocation of Service Information (SI) and Data Broadcasting Codes for Digital Video Broadcasting (DVB) Systems)*

ETR 101 162 lists all of the Network IDs and Original Network IDs allocated to broadcasters. Special provision is made for reuse of Network IDs by terrestrial broadcasters. There is no DVB document associated with this ETR.

2.12.6 DVB Interfaces

DVB has defined interfaces for professional and domestic equipment. It is now possible to connect encoders, decoders, multiplexers and modulators (from different manufacturers) to form complex broadcast systems.

2.12.6.1 ASI, SPI and SSI (DVB-PI)

DVB Document A010 Rev. 1: *Interfaces for CATV/SMATV Headends and Similar Professional Equipment*, 26 May 1997
(CENELEC EN 50083-9 V2: *Cable Networks for Television Signals, Sound Signals and Interactive Services Part 9: Interfaces for CATV/SMATV Headends and Similar Professional Equipment for DVB/MPEG-2 Transport Streams*, June 1998)
DVB Document A055: *Digital Video Broadcasting (DVB); Implementation Guidelines for the Asynchronous Serial Interface*, May 2000
(ETSI TR 101 891 V1.1.1: *Digital Video Broadcasting (DVB); Professional Interfaces: Guidelines for the Implementation and Usage of the DVB Asynchronous Serial Interface (ASI)*, February 2001)

DVB defines three types of Transport Stream interface on two different media (optical and copper) for interconnecting such devices as encoders, multiplexers, modulators, etc. The guidelines were issued to overcome some early interoperability problems encountered with the ASI interface. The three copper interfaces use Low-Voltage Differential Signalling (LVDS).[59]

2.12.6.1.1 ASI (Asynchronous Serial Interface)

ASI is described in Annexes B, C and E of EN500083. ASI has a constant clock rate of 270 Mbit/s and has copper and optical variants. Data bytes are converted to 10-bit words before transmission (using the X3T11[60] 8-bit to 10-bit conversion). The transmitter has a choice of two 10-bit words for every 8-bit data word; it chooses the one which maintains DC balance to eliminate deterministic jitter. The interface operates in two modes, Byte and Packet burst (different aliases will be encountered for these modes). If operating in Byte mode there will be two or more 'byte alignment synchronisation patterns' (also known as K28.5 or 'comma') between bytes. There will be two or more 'comma' characters between packets in 'Packet burst' mode. From these rules, maximum transmission rates can be calculated. The results are shown in Table 2.12.8. ASI is the most commonly encountered DVB interface.

Table 2.12.8 ASI maximum bit rates

Format	*188-Byte bit rate*	*204-Byte bit rate*
188-Byte (Byte mode)	72	N/A
204-Byte (Byte mode)	66.352941	72
188-Byte (Packet mode)	213.726315	N/A
204-Byte (Packet mode)	197.126213	213.902912

2.12.6.1.2 SSI (Synchronous Serial Interface)

SSI is described in Annexes A and D of EN500083. There are copper and optical variants. The signals are Biphase Mark Coded with an inverted Sync Byte (0xB8) indicating the presence of valid Reed–Solomon FEC Bytes. SSI interfaces are rarely encountered in DVB systems, although a similar interface (SMPTE 310M) is used in ATSC systems.

2.12.6.1.3 SPI (Synchronous Parallel Interface)

SPI is used for short to medium distances and consists of three signalling wires and eight data wires. There is a (Byte) CLOCK, a PSYNC (to indicate start of a Transport packet) and a DVALID line (which is low to indicate 16 stuffing bytes – otherwise high). A 25-way connector is used. Connectors on equipment are always female. Maximum bit rate is 108 Mbit/s.

2.12.6.2 STB Interfaces (DVB-IRDI)

DVB Document A016 Rev. 2: *Digital Video Broadcasting (DVB); Interfaces for DVB Integrated Receiver Decoder (DVB-IRD)*, 7 June 1999
(ETSI TS 102 201 V1.1.1: *Digital Video Broadcasting (DVB); Interfaces for DVB Integrated Receiver Decoder (DVB-IRD)*, March 1999)

(**CENELEC EN 50201:** *Interfaces for Digital Video Broadcast Integrated Receiver Decoder (DVB-IRD)*)

TS 102 201 specifies the electrical interface and types of connectors to be used on STBs. Annex A contains a number of diagrams showing how an STB can be connected to other devices and Annex B lists the characteristics of RF channels encountered in Europe.

The input RF connector will be an F-type[61] (for satellite) or a Belling–Lee type (for VHF/UHF reception). The modem interface will be a nine-way D-type or an RJ11. The STB will support V.21, V.22, V.22 bis and V.23. Support for V.32, V.32 bis, V.34, V.25 and V.42 is optional. Video can be output as RGB, Y/C (S-VHS), CVBS (composite) or modulated onto an RF carrier. Video connectors can be Peritelevision[62] (aka Peritel or SCART), phono, IEC 933–5[63] or Belling–Lee. Audio can be output on Peritel or phono (or modulated onto an RF carrier). Linear PCM encoded audio is allowed. Low-bit-rate data will use a nine-way, female D-type connector, high-speed data will use a 25-way, female D-type connector. IEEE 1284 interface may be used up to 10 Mbit/s.

2.12.7 Conditional Access

Conditional Access (CA) is the process of scrambling broadcast material so that it can only be descrambled when certain conditions are met (the 'access criteria') (Figure 2.12.29). DVB has defined a Common Scrambling Algorithm, a Common Interface and a method for using two or more Conditional Access systems to descramble one encrypted stream. (Further details of modern CA systems can be found in Chapter 2.16).

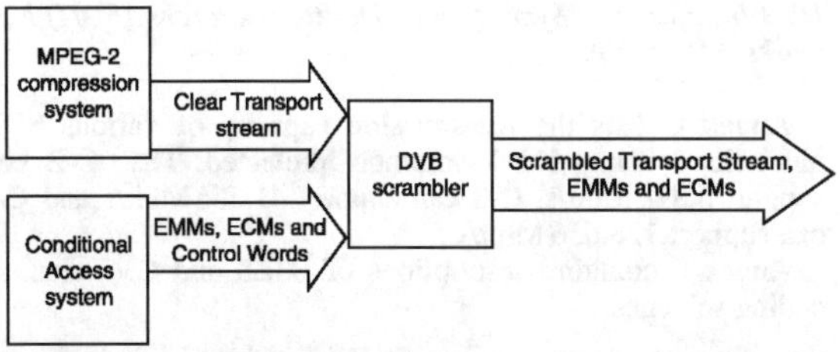

Figure 2.12.29 CA system sending EMMs, ECMs and Control Words to scrambling device. The scrambling device is often inside a Transport level multiplexer. Note that Control Words are never sent explicitly in the broadcast TS.

2.12.7.1 Common Scrambling Algorithm (DVB-CSA)

DVB Document A007: *Support for Use of Scrambling and Conditional Access within Digital Broadcasting Systems*, 28 February 1997
(**ETSI TR 289 V.1:** *Digital Video Broadcasting (DVB); Support for Use of Scrambling and Conditional Access (CA) within Digital Broadcasting Systems*, October 1996)

DVB supports Transport level scrambling (where the Transport packet payload is scrambled but the header and any adaptation field are left unscrambled) and a limited form of PES level scrambling (where 184-byte chunks of the PES payload are scrambled). The DVB Common Scrambling Algorithm is not publicly available but can be obtained from ETSI by bona fide CA companies. The Control Word ('key') used is 64 bits long (or, less commonly, 48 bits) and some details of the algorithm have been revealed by Benoit (see Bibliography). Typically a Control Word is used for 10–30 seconds before being replaced with a new Control Word.

In DVB CA synchronisation between the ECM and Control Word used for scrambling is achieved by using two different *table-ids* of 0x80 and 0x81 for the ECMs. These correspond to even and odd keys, signalled by the *transport_scrambling_control* bits being set to 0x2 and 0x3 in the Transport packet header (see Chapters 2.11 and 2.16).

2.12.7.2 Simulcrypt (DVB-SIM)

DVB Document A045 Rev. 1: *Digital Video Broadcasting (DVB); Head-end Implementation of DVB Simulcrypt*, February 2002
(**ETSI TS 103 197 V1.2.1:** *Digital Video Broadcasting (DVB); Head-end Implementation of DVB SimulCrypt*, January 2002)
DVB Document A028 Rev. 1: *Digital Video Broadcasting (DVB); DVB SimulCrypt; Head-end Architecture and Synchronization*, February 2002
(**ETSI TS 101 197 V1.2.1:** *Digital Video Broadcasting (DVB); DVB SimulCrypt; Head-end Architecture and Synchronization*, February 2002)
DVB Document A070: *Implementation Guidelines of the DVB Simulcrypt Standard*, February 2002
(**Draft ETSI TR 102 035 V1.1.1:** *Digital Video Broadcasting (DVB); Implementation Guidelines of the DVB Simulcrypt Standard*)

DVB Simulcrypt allows two or more CA systems to describe the Control Word(s) used in a Transport Stream (Figure 2.12.30). This allows two completely different populations of STBs with different CA to decrypt the same encrypted TS. Simulcrypt has three main uses. Since the Simulcrypt interfaces have been defined it is now easier to integrate a CA system with a compression system (even if only one CA system is in use). Simulcrypt is useful if two broadcasters merge (or one

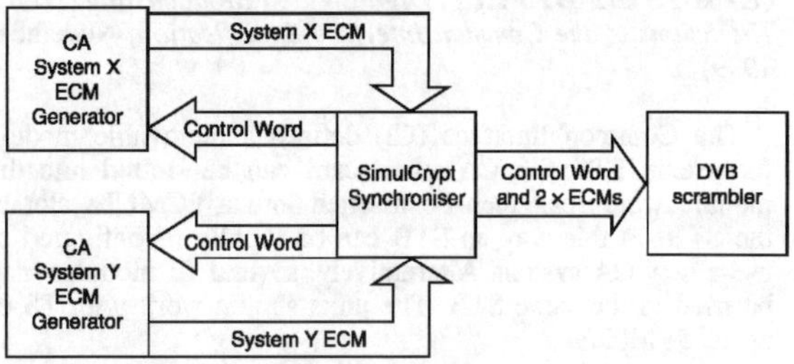

Figure 2.12.30 In a Simulcrypt system the SimulCrypt Synchroniser (SCS) sends the Control Word to one or more CA systems. Each CA system sends the ECM for the Control Word back to the SCS. The SCS then sends the Control Word and its associated ECMs to the DVB scrambler.

acquires the other) and the resulting network is composed of two sets of STBs with incompatible CA systems. Lastly, Simulcrypt is sometimes required by European regulatory authorities.

A SimulCrypt Synchroniser (SCS) makes a TCP/IP connection to one or more ECM Generators (ECMGs). A Channel is set up to each ECMG and within the Channel there will be a Stream for each Control Word (CW). The SCS sends Control Words to the ECMGs and the ECMGs send the corresponding ECMs back to the SCS. Each ECMG may take a different amount of time to generate an ECM from a Control Word and each STB may take a different amount of time to regenerate a Control Word from an ECM. A major function of the SCS is to synchronise transmission of Control Words to the ECMGs and the scrambler so that the STBs receive the ECMs in time to produce the next required Control Word.

ETS 101 197 is Part 1 (or V1) of the Simulcrypt standard. ETS 103 197 is the 'Final' version (V2) of the Simulcrypt standard. ETR 102 035 gives details of the incompatibilities between the two versions.

2.12.7.3 Multicrypt

Multicrypt is the practice of having two or more services in a TS encrypted by different CA systems. Each service can only be decrypted by a STB with the correct CA installed. This was used on the SkyPlex system operated by Eutelsat but is not in common use.

2.12.7.4 Common Interface (DVB-CI)

DVB Document A017: *Common Interface Specification for Conditional Access and Other Digital Video Broadcasting Decoder Applications*, 31 May 1996
(CENELEC EN 50221: *Common Interface Specification for Conditional Access and Other Digital Video Broadcasting Decoder Applications*, February 1997)
DVB Document A025: *Guideline for Implementation and Use of the Common Interface for DVB Decoder Applications*, May 1997
(CENELEC R206-001 V1: *Digital Video Broadcasting (DVB); Guidelines for Implementation and Use of the Common Interface for DVB Decoder Applications*, March 1997)
DVB Document DVB TM2088r3, CIT 037r11: *Extensions to the Common Interface Specification*, May 1999
(ETSI TS 101 699 V1.1.1: *Digital Video Broadcasting (DVB); Extensions to the Common Interface Specification*, November 1999)

The Common Interface (CI) defines a removable module for use in STBs. A CA smart card can be slotted into the module which can then be inserted into a PCMCIA slot in the STB. In this way an STB can be rapidly reconfigured to use a new CA system. Alternatively, several CI modules may be used in the same STB. The units should work with TS of up to 58 Mbit/s.

2.12.8 DVB Telco Interfaces

DVB recommends the use of ATM Adaptation Layer AAL-1[64] (DAVIC recommends AAL-5). The long interleave version of AAL-1 with FEC[65] is also supported. An FEC overhead of just over 3% (128/124) will correct up to four cell losses in a group of 128 cells or up to two octets in a block of 128 octets. VPI values of 17–24 (decimal) are suggested along with a VCI value of 32. (A Transport packet of 188 bytes will fit exactly into four AAL-1 ATM cells.)

2.12.8.1 DVB PDH

DVB Document A018: *DVB Interfaces to PDH Networks*, February 1997
(ETSI TS 300 813 V1: *Digital Video Broadcasting (DVB); DVB Interfaces to Plesiochronous Digital Hierarchy (PDH) Networks*, December 1997)

Annex C lists the transmission capacity of various PDH hierarchies using Reed–Solomon protected TSs. E3 can support 29.14 Mbit/s, DS-3 can support 37.980 Mbit/s and E4 can support 118.759 Mbit/s. (There are a number of proprietary (non DVB-compliant) adaptation schemes in use. A common approach is to apply V4MOD to the TS before placing it directly onto an unframed PDH link. This approach means that E3 circuits can transmit 31.672 Mbit/s and DS-3 circuits can transmit 41.227 Mbit/s.)

Annex D contains descriptions of AMI, HDB3, B3ZS, B6ZS, B8ZS and CMI channel coding schemes.

2.12.8.2 DVB SDH

DVB Document A019: *DVB Interfaces to SDH Networks*, February 1997
(ETSI TR 300 814 V1: *Digital Video Broadcasting (DVB); DVB Interfaces to Synchronous Digital Hierarchy (SDH) Networks*, March 1998)

Annex D lists the transmission capacity of various SDH hierarchies using Reed–Solomon protected TSs. C-2 can support 5.828 Mbit/s, C-3 can support 41.566 Mbit/s and C-4 can support 128.656 Mbit/s.

Annex F contains descriptions of B3ZS and CMI channel coding schemes.

2.12.8.3 DVB ATM

DVB Document A044: *Digital Video Broadcasting (DVB); Guidelines for the Handling of Asynchronous Transfer Mode (ATM) Signals in DVB Systems*, June 1999
TR 100 815 V1.1.1: *Digital Video Broadcasting (DVB); Guidelines for the Handling of Asynchronous Transfer Mode (ATM) Signals in DVB Systems*, February 1999

Although MPEG-2 Transport packets were designed to fit exactly into four ATM cells, ATM cells do not fit neatly into Transport packet payloads. (Four ATM cells occupy 212 bytes and three ATM cells occupy 159 bytes.) Rate adaptation is performed by using ATM idle cells or MPEG-2 adaptation fields (or a combination of both methods). ATM cell payload scrambling is applied to guard against false ATM cell delineation or spurious Transport packet sync byte emulation. Conditional Access scrambling may be applied at the TS level or the ATM level.

2.12.9 DVB Measurements (DVB-M)

2.12.9.1 ETR 290

DVB Document A014 Rev. 2: *Digital Video Broadcasting (DVB); Measurement Guidelines for DVB Systems*, 26 February 2001
(**ETSI TR 101 290 V1.2.1:** *Digital Video Broadcasting (DVB); Measurement Guidelines for DVB Systems*, May 2001)
DVB Document A072: *Digital Video Broadcasting (DVB); SNMP MIB for Test and Measurement Applications in DVB Systems*, February 2002
(**Draft ETSI TR 102 032 V1.1.1:** *Digital Video Broadcasting (DVB); SNMP MIB for Test and Measurement Applications in DVB Systems*)

ETR 101 290 describes a large suite of tests which can be performed on DVB Transport Streams both at baseband and in the RF domain. The baseband tests are almost universally referred to as 'ETR 290' from the number of the original document.

The baseband tests are divided into three levels, with level 1 faults being the most serious. All Transport Stream Analysers offer ETR 290 analysis.

The standard describes five profiles (MGB1–MGB5) for bit rate measurement, which is always problematic with packet-based, bursty data sources.

The standard defines Quality Of Service (QOS) parameters such as System Availability and Link Availability which can be used to characterise the performance of contribution and distribution networks.

The RF part of the standard defines a Modulation Error Ratio (MER) parameter. MER is a measure of total signal degradation and is expressed in deciBels. A number of specific errors which contribute to the total MER can be measured separately. These include phase jitter, Equivalent Noise Degradation (END) and digital modulation-specific errors such as System Target Error (STE – a measure of the displacement of the centres of the symbol clouds from their ideal position).

A number of measurements are devoted to checking consistency of DVB-T SFN Transport Streams.

A072 defines a standard MIB (Management Information Base) which can be incorporated into test equipment to facilitate interoperability between test and monitoring solutions from different manufacturers. (An MIB is a database structure which can be browsed with SNMP.)

2.12.9.2 Test PIDs

DVB Document A046: *Digital Video Broadcasting (DVB); Usage of the DVB Test and Measurement Signalling Channel (PID 0x001D) Embedded in an MPEG-2 Transport Stream (TS)*, 4 June 1999
(**ETSI TR 101 291 V1.1.1:** *Digital Video Broadcasting (DVB); Usage of the DVB Test and Measurement Signalling Channel (PID 0x001D) Embedded in an MPEG-2 Transport Stream (TS)*, June 1998)

TR 101 291 defines types of test and measurement information which can be carried in a packet with a PID of 0x1D. Test signals include a PRBS ($2^{23} - 1$) for measuring Bit Error Ratios on transmission links and three 'pathological' signals. GPS signals can be sent as a time and frequency reference (the GPS source's position is also sent to counteract the 'dither' signal sometimes used with GPS). A Network Status Table is defined which can contain details of missing Service components or SI tables. A Reception Status Table can contain details of carrier level, lock status, frequency drift and other reception-related measurements. PID 0x1D can also contain measurements of video and audio quality, etc. Provision is made for source identification using an ISO 639[8] language descriptor followed by an ASCII string detailing the source and types of measurement, etc.

Despite its useful features, test packet 0x1D does not seem to be in common use.

2.12.10 DVB Return Channels

DVB Document A008: *Commercial Requirements for Asymmetric Interactive Services Supporting Broadcast to the Home with Narrowband Return Channels*, October 1995 (This document has no ETSI equivalent as it is not a standard.)
DVB Document A021: *Network Indepenent Protocols for Interactive Services*, February 1997
(**ETSI TS 300 802 V1:** *Digital Video Broadcasting (DVB); Network-independent Protocols for DVB Interactive Services*, November 1997)
DVB Document A026: *Guidelines for Use of DVB-SIS Specification – Network Independent Protocols for Interactive Services*, May 1997
(**ETSI TR 101 194 V1.1.1:** *Digital Video Broadcasting (DVB); Guidelines for Implementation and Usage of the Specification of Network Independent Protocols for DVB Interactive Services*, June 1997)

The DVB interactive services use the same protocols, regardless of the underlying physical channels. One or more broadcast channels (of up to 50 Mbit/s each) carry video, audio and possibly forward interactive data. A bidirectional interaction channel (150 kbit/s) is used for the return channel from the user and, optionally, for a narrowband forward interaction path (Figure 2.12.31). The interaction channels can use RTP,[66] UDP,[67] TCP/IP[68] and DSM-CC[28] protocols. Many of the following specifications duplicate ETR 101 194, adding a few pages dealing with the specifics of the interaction channel. These specifications will not be described in detail.

2.12.10.1 DVB-RCS

DVB Document A054 Rev. 1: *Digital Video Broadcasting (DVB); Interaction Channel for Satellite Distribution Systems*, May 2000
(**ETSI EN 301 790 V1.2.2:** *Digital Video Broadcasting (DVB); Interaction Channel for Satellite Distribution Systems*, December 2000)
DVB Document A063: *Digital Video Broadcasting (DVB); Interaction Channel for Satellite Distribution Systems; Guidelines for the Use of EN301 790*, April 2001

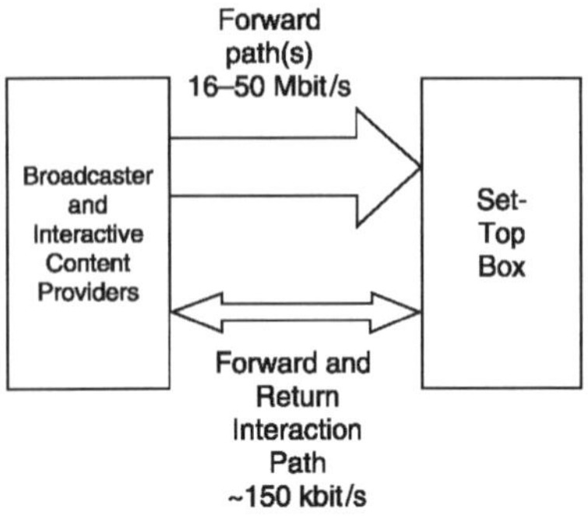

Figure 2.12.31 Greatly simplified diagram of DVB return channel architecture. The broadcasters and interactive service providers may be separate entities.

(**ETSI TR 101 790 V1.1.1:** *Digital Video Broadcasting (DVB); Interaction Channel for Satellite Distribution Systems; Guidelines for the Use of EN 301 790*, September 2001)

DVB-RCS uses VSAT technology for the return channel. A Network Control Centre (NCC) provides monitoring and control for the entire system. (An NCC may control more than one network.) A 'Feeder' transmits the forward link signal, which is a standard DVB-S signal with some additional tables and timing signals. Receivers are called Return Channel Satellite Terminals (RCSTs). RCSTs are allocated timeslots and frequencies on the reverse channel (by the NCC) for sending messages/requests to the Gateway Stations, which pass these signals on to external service providers. A Network Clock Reference (NCR) is defined – this has an identical format to a PCR, except that it is reset to zero at UTC midnight and is carried in the Adaptation Field of an otherwise empty Transport Stream packet. RCSTs can use the time and frequency reference provided by the NCR to adjust their transmit frequencies, transmission burst and symbol clock synchronisation.

RCSTs send four types of traffic burst back to the Gateway. These are traffic (TRF), acquisition (ACQ), synchronisation (SYN) and common signalling channel (CSC). CSC is used by an RCST during log-on to identify itself and its processing/modulation capabilities.

Four cyclic tables are defined. The Satellite Position Table (SPT) contains ephemeris data for the satellite which can be used to 'fine-tune' network synchronisation. The Superframe Composition Table (SCT) provides timing data for superframes and frames to enable timeslot synchronisation. The Frame Composition Table (FCT) partitions frames into timeslots, which may be of variable length. The Timeslot Composition Table (TCT) defines transmission parameters for each timeslot. Parameters defined include symbol rate, code rate, code type (concatenated or turbo) and payload type (ATM or MPEG-2).

Three dynamic tables are defined. The Terminal Information Message (TIM) is either unicast from the NCC to one particular RCST or broadcast to all RCSTs. The Terminal Burst Time Plan (TBTP) is used to allocate timeslots to RCSTs. The Correction Message Table (CMT) provides correction values for burst frequency, timing and amplitude of RCST bursts. The NCC calculates these correction figures by examining ACQ and SYN bursts from individual RCSTs.

The net result of this complexity is a synchronous VSAT system with efficient use of reverse channel capacity.

2.12.10.2 DVB-RCT

DVB Document A064: *Digital Video Broadcasting (DVB); Interaction Channel for Digital Terrestrial Television (RCT) Incorporating Multiple Access OFDM*, April 2001
(**ETSI EN 301 958 V1.1.1:** *Digital Video Broadcasting (DVB); Interaction Channel for Digital Terrestrial Television (RCT) Incorporating Multiple Access OFDM*, March 2003)

In DVB-RCT receivers are called Return Channel Terrestrial Terminals (RCTTs). RCTTs derive their frequency reference from the DVB-T (forward channel) system clock and their time sync from MPEG-2 (Upstream Sync Field) packets with a PID of 0x1C, which contain start time and slot index information. The return channel modulation scheme is similar to DVB-T. Return channels contain 1712 or 842 orthogonal carriers using QPSK, 16-QAM or 64-QAM modulation. Randomisation, turbo (or concatenated) coding and interleaving are applied before modulation. Each RCTT is allocated one, four or 29 carriers and sends a data burst of an integer number of ATM cells on 144 modulated symbols.

2.12.10.3 DVB-RCC

DVB Document A023 Rev. 2: *Digital Video Broadcasting (DVB); DVB Interaction Channel for Cable TV Distribution Systems (CATV)*, February 2002
(**ETSI ES 200 800 V1.3.1:** *Digital Video Broadcasting (DVB); DVB Interaction Channel for Cable TV Distribution Systems (CATV)*, November 2001)
DVB Document A031: *Guidelines for the Use of the Specification – DVB Interaction Channel for Cable TV Distribution Systems (CATV)*, March 1998
(**ETSI TR 101 196 V1.1.1:** *Digital Video Broadcasting (DVB); Interaction Channel for Cable TV Distribution Systems (CATV); Guidelines for the Use of ETS 300 800*, December 1997)

The forward interactive path uses a frequency band of 70–130 or 300–862 MHz, with 5–65 MHz being used for the return path. A mixture of QPSK and 16-QAM modulation schemes is used. Three access modes are used to allocate return bandwidth efficiently – but they cannot all be used at the same time. Three cryptographic schemes (Diffie–Hellman, HMAC-SHA1 and DES) are proposed for (optional) security.

DVB-RCC has not been widely adopted by digital cable systems in Europe, where many have adopted a rival DOCSIS scheme.

2.12.10.4 DVB-RCP

DVB Document A022: *DVB Interaction Channel through the Public Switched Telecommunications System (PSTN)/Integrated Services Digital Network (ISDN)*, February 1997

(**ETSI TS 300 801:** *Digital Video Broadcasting (DVB); Interaction Channel through the Public Switched Telecommunications System (PSTN)/Integrated Services Digital Network (ISDN)*, August 1997)

Standard specifies PSTN interfaces[69] and protocols to be used as return channel.

2.12.10.5 DVB-RCD

DVB Document A030: *Interaction Channel through the Digital Enhanced Cordless Telecommunications (DECT)*, March 1998
(**ETSI EN 301 193 V1.1.1:** *Digital Video Broadcasting (DVB); Interaction Channel through the Digital Enhanced Cordless Telecommunications (DECT)*, July 1998)

Standard specifies DECT interfaces[70,71] and protocols to be used as return channel. STB may use an external Wireless Relay Station[72] (WRS). (DECT uses a TDMA mechanism to allocate bandwidth to users.)

2.12.10.6 DVB-RCL

DVB Document A032 Rev. 1: *DVB Interaction Channel for LMDS Distribution Systems (CATV)*, June 1999
(**ETSI EN 301 199 V1.2.1:** *Digital Video Broadcasting (DVB); DVB Interaction Channel for Local Multi-point Distribution Systems (LMDS)*, June 1999)
DVB Document A059: *Digital Video Broadcasting (DVB); LMDS Base Station & User Terminal Implementation Guidelines for EN 301 199*, March 2001
(**ETSI TR 101 205 V1.1.2:** *Digital Video Broadcasting (DVB); LMDS Base Station and User Terminal Implementation Guidelines for ETSI EN 301 199*, June 1999)

Bandwidth is allocated to users with a mixture of TDMA and FDMA. Downstream signalling may be In Band (IB) or Out Of Band (OOB). OOB downstream and upstream signalling utilises ATM cells with RS FEC and D-QPSK modulation inside a 2 or 4 MHz channel. Each downstream channel can synchronise up to eight upstream channels.

The implementation guidelines discuss modulation, capacity, spectral planning, propagation, cell clusters and radiation safety. Since most programme material will be sourced from digital satellite and/or digital terrestrial transmissions much of the document is given over to discussing optimum use of downstream capacity within the confines of 40.5–43.5 GHz. (The network designers may have no knowledge of exactly which multiplex is associated with a particular broadcast interactive channel, so all multiplexes on a network must be delivered to every subscriber.)

2.12.10.7 DVB-RCG

DVB Document A043: *Digital Video Broadcasting (DVB); Interaction Channel through the Global System for Mobile Communications (GSM)*, June 1999
(**ETSI EN 301 195 V1.1.1:** *Digital Video Broadcasting (DVB); Interaction Channel through the Global System for Mobile Communications (GSM)*, February 1999)

Standard specifies GSM interfaces and protocols to be used as return channel. STB may use an external Mobile Station (MS). (GSM uses a TDMA mechanism to allocate 9.6 kbit/s of bandwidth to users.)

2.12.10.8 DVB-RCCS

DVB Document A034: *DVB Interaction Channel for Satellite Master Antenna Television (SMATV) Systems; Guidelines for Versions Based on Satellite and Coaxial Sections*, March 1998
(**ETSI TR 101 201 V1.1.1:** *Digital Video Broadcasting (DVB); Interaction Channel for Satellite Master Antenna TV (SMATV) Distribution Systems; Guidelines for Versions Based on Satellite and Coaxial Sections*, October 1997)

Three types of forward channel are envisaged. SMATV-DTM transmodulates the incoming DVB-S signal into a DVB-C signal without decoding to baseband. SMATV-IF uses the down-converted L-band output (950–2150 MHz) of the Low Noise Block (LNB). SMATV-S converts the incoming DVB-S signal to extended S-band (230–470 MHz).

The coaxial part of the SMATV system uses a simplified form of DVB-RCC using 3.088 Mbit/s transmission in both directions. TDMA and CDMA with Slotted Aloha are recommended for allocating bandwidth on the satellite return path. Annex A contains link budget examples for both access protocols.

2.12.11 Home Networks (DVB-IHDN)

DVB Document A029: *Digital Video Broadcasting (DVB); User and Market Requirements for In-Home Digital Networks*, May 1997 (This document has no ETSI equivalent as it is not a standard.)

The In-Home Digital Network (IHDN) consists of two parts defined in the following two specifications. The Home Access Network (HAN) is the unit which connects the Home Network to the telco (or other) network. The Home Local Network (HLN) is the network of gateways, cabling, clusters and end devices within the home.

2.12.11.1 DVB-HAN

DVB Document A039: *Digital Video Broadcasting (DVB); Home Access Network (HAN) with an Active Telco Network Termination*, August 1998
(**ETSI TS 101 224 V1.1.1:** *Digital Video Broadcasting (DVB); Home Access Network (HAN) with an Active Network Termination (NT)*, July 1998)

The HAN may be used (in conjunction with a 'traditional' DVB broadcast delivery network) to provide two-way, interactive multi-media delivery (such as VOD) over xDSL, CATV/HFC, ISDN, MMDS, etc. The HAN is based on an ATM[73–75] interface operating at 25.6 or 51.2 Mbit/s running over up to 50 m of Category 5 Unshielded Twisted Pair (UTP) with RJ-45 connectors.[76] The HAN will provide an 8 kHz clock to the connected devices. MPEG Transport Streams are to be carried in AAL-5 PDUs. IP can be tunnelled via MPEG-2 TS packets or use LLC/SNAP format in AAL-5.[77]

2.12.11.2 DVB-HLN

DVB Document A060: *Digital Video Broadcasting (DVB); Home Local Network Specification Based on IEEE 1394*, March 2001
(ETSI TS 101 225 V1.1.1: *Digital Video Broadcasting (DVB); Home Local Network Specification Based on IEEE 1394*, May 2001)

The HLN will be based on IEEE 1394 technology[78–83] and the Home Audio/Video Interoperability (HAVi) Architecture.[84,85] The specification describes various protocol stacks[86–90] and the processes required to translate MPEG-2 and DVB tables and descriptors into HAVi equivalents. (DVB uses ASCII strings; these must be converted into Unicode.[55] DVB Service names can be 254 bytes – the HAVi equivalent is only 32 characters, etc.)

2.12.12 Data Downloads (DVB-SSU)

DVB Document A067: *Digital Video Broadcasting (DVB); DVB Data Download Specification; Part 1: Simple Profile*, February 2002
(ETSI TS 102 006 V1.1.1: *Digital Video Broadcasting (DVB); DVB Data Download Specification; Part 1: Simple Profile*, December 2001)

TS 102 006 specifies mechanisms that enable an STB to locate a *system software update service*. The PMT carries a *data_broadcast_id* descriptor (0x000A) which indicates where the data download service is located. A *system_software_update* linkage descriptor (0x09) is present in the NIT or BAT. Three types of download are allowed – proprietary, a two-layer DVB data carousel and a sophisticated system using an Update Notification Table (UNT), which can contain update scheduling information, selection and targeting information, action notification and filtering descriptors. It is possible to use all of these systems with multiple STB models from multiple manufacturers.

The specification also defines a standard format for STB manufacturers to send update code to the broadcasters.

Part 2 of the specification ('Extended Profile') has not appeared.

2.12.13 MHP or DVB-MHP

DVB Document A057 Rev. 2: *Multimedia Home Platform, 1.0.2, DVB BlueBook A057 Rev. 2*, 26 February 2002
(ETSI TS 101 812 V1.1.3: *Digital Video Broadcasting (DVB); Multimedia Home Platform (MHP)*)
DVB Document A062: *Digital Video Broadcasting (DVB); DVB Commercial Module – Multimedia Home Platform User and Market Requirements, Enhanced and Interactive Digital Broadcasting in the Local Cluster*, April 2001
DVB Document A066: *The DVB Project, Multimedia Home Platform (MHP), MHP Implementation Arrangements and Associated Agreements*, October 2001
DVB Document A068: *Multimedia Home Platform, 1.1, Dvb Blue Book A068*, November 2001
(ETSI TS 102 812 V1.1.1: *Digital Video Broadcasting (DVB); Multimedia Home Platform (MHP) in HTML Extensions*, November 2002)

The Multimedia Home Platform (MHP) is DVB's Application Programming Interface (API). An API (also known as 'middleware') is an abstraction layer between Set-Top Box hardware and programs intended to run on the STB. A programmer does not have to know the details of STB graphics drivers or machine code, merely how to write for the API. (APIs are similar to 'C', Java and other programming languages in this respect.) MHP is a group of technical standards of over 1500 pages. Much of MHP is based on existing standards from the worlds of graphics (JPEG,[91] PNG,[92] GIF[93]), video compression (MPEG-2 video,[3] MPEG-1 audio[7]) and the Internet (HTTP,[94] XHTML,[95] Cascading Style Sheets, DOM, ECMAScript, Java,[96,97] RSA,[98] ISO 639.2,[8] Unicode,[55] etc.). MHP will allow a wide variety of applications to run on a viewer's STB and TV. These will include EPGs, information services, applications related to TV content (playing TV quizzes at home, sports statistics, etc.), educational services, e-commerce, etc.

Three profiles are defined: 'Enhanced Broadcast' for one-way services, 'Interactive' profile for services incorporating a return channel and 'Internet' profile for browser-type services.

Despite the presence of so much 'Internet technology' in the standard it will not be possible to 'surf the web' with a television. Standard (non-HDTV) televisions have a low refresh rate (25 or 30 Hz), interlaced scanning and low resolution compared with a PC monitor. Typical web content consists of small fonts, horizontal lines in frames, lines and boxes and fast-changing graphics intended for frame-based displays running at 72–90 frames per second. Much of the material on the web is unwatchable on a normal television due to unreadable letters, flicker, aliasing ('jaggies') and jerky animations. (Experiments with Internet access by cable TV networks in the UK have been a disappointment.) MHP uses the same RNIB/DTG 'Tiresias' font specified in the UK for subtitles on DVB-T STBs. Tiresias was designed to be easy to read on a television even for people with impaired vision. It will have large letters about the same size as those chosen for teletext, for similar reasons of normal viewing distance and display resolution. (DVB teletext is also used in MHP.)

Web pages will have to be authored especially for display on domestic televisions with large letters and no flicker-inducing horizontal lines. Some broadband implementations may use web portals for service selection and VOD applications as the normal DVB SI mechanisms are too bandwidth hungry for a narrowband (!) environment.

The presence of MHP components is signalled in the PMT as private data sections with a *stream_type* of 0x0B. An Application Information Table (AIT) is present for every DVB Service that is associated with an MHP application. AITs have separate PIDs and a *stream_type* of 0x05.

MHP is currently in use in Finland, Germany and Spain. There are a number of competing APIs – most of which are proprietary – but DVB expects MHP to eventually replace them all as STB hardware speed and memory size improves and older STBs are replaced.

The MHP Umbrella Group (MUG) has recently developed Globally Executable MHP (GEM), which allows MHP to be used in markets like the USA and Japan, where minor technical differences – such as vertical resolution – have prevented it from being implemented. This was submitted to ETSI in January 2003 and will be published as ETSI TS 102 819 v1.1.1.

2.12.14 DVB Broadband (DVB-IPI)

DVB Document A071: *Digital Video Broadcasting (DVB); Architectural Framework for the Delivery of DVB-Services over IP-based Networks*, February 2002

A071 contains a layer model describing the distribution of services over IP from a content provider to the home. Annex A describes different types of service and discusses their requirements with reference to this model. Annex B lists seven provisional documents which will form the basis of the eventual DVB-IPI standard.

IPI2001-012 Architectural framework.
IPI2001-016 Encapsulation and transport.
IPI2001-059 Service discovery and selection.
IPI2001-071 IP address allocation and management.
IPI2001-079 IEEE1394 Home Network.
IPI2001-072 Ethernet Home Network.
IPI2001-073 Security.

These documents are not finalised, but a brief description of IPI2001-016 can be given. Multiple- or Single-Program Transport Streams can be encapsulated using RTP,[66,99,100] UDP[67] and IP[26] packets. One to seven transport packets can be carried inside each IP packet, which can be unicast or multicast.[27] Unicast services will be initiated and controlled using a subset of RTSP.[101] (Optional) FEC is based on standard DVB FEC with RS(204,188) and a large byte interleaver. Five FEC modes are defined, which will cope with (up to) two missing IP packets. Quality of Service (QoS) will be implemented with the Differentiated Services[102,103] model.

2.12.15 DVB Policy Documents

2.12.15.1 Technical Regulations

DVB Document A002: *A Fundamental Review of the Policy of European Technical Regulations Applied to the Emerging Digital Broadcasting Technology*, November 1994 (This document has no ETSI equivalent as it is not a standard.)

A002 compares the relative success and failure of the GSM (digital cellular telephony) and MAC (analogue satellite and cable TV) standards and makes recommendations on the types of standardisation activity that will ensure a healthy digital broadcasting environment in Europe. The document also addresses issues such as erosion of cultural identity and the possibility of having a common scrambling system for Europe.

2.12.15.2 Antipiracy Legislation

DVB Document A006 Rev.1: *Recommendations of the European Project – Digital Video Broadcasting Antipiracy Legislation for Digital Video Broadcasting*, October 1995

(This document has no ETSI equivalent as it is not a standard.)

A006 contains a number of recommendations to tighten up existing European legislation on audiovisual piracy, to take account of the differences between analogue and digital systems. Standardised DVB receivers across Europe with a Common Scrambling Algorithm inside become an attractive target for cross-border piracy. A006 recommends that European laws should be harmonised to criminalise the manufacture, importation, promotion, distribution and possession of pirate decoders.

2.12.15.3 Convergence

DVB Document A048: *DVB Response to the Commission Green Paper on the Convergence of the Telecommunications, Media and Information Technology Sectors and the Implication for Regulation*, April 1998

(This document has no ETSI equivalent as it is not a standard.)

A048 attempts to predict the nature of the convergent world 4–5 years from the date of publication (i.e. 2002–2003). The document asserts that convergence is not an objective but a consequence of the evolution of various technologies, which consumers will only adopt if there is sufficient compelling content. A048 suggests that there are three main regulatory areas – the market, content provision and infrastructure/terminals. Broadcast and telecommunications infrastructures in Europe are moving from state-regulated monopolies to competitive markets. It is important that regulators understand the new technologies and the problems of applying competition law to new, convergent markets. A048 also suggests that provision of education and training via digital TV, the Internet and PCs will be a key factor in Europe's future competitiveness.

2.12.16 Australasian Standards

Australia and New Zealand produced their own versions of some of the DVB standards. The original DVB-T standard did not include explicit calculations for bit rates, guard interval durations, etc. for 7 MHz channels, but the Australian variant does. (Annex E of EN 300 744 now contains tables for 6 MHz and 7 MHz channels.)

Australia had a requirement to simulcast HDTV and SDTV signals, but wished to modify existing STBs rather than design their own from scratch. US standard STBs used Dolby AC-3 audio rather than MUSICAM, but Dolby AC-3 was not mentioned in any of the DVB specifications. Dolby AC-3 descriptors had to be specified for the Australian market as STBs might have to switch between services using MUSICAM and services using Dolby AC-3. The standards are listed below but are not described.

AS/NZS 4540:1999: *Digital Broadcasting Systems for Television, Sound and Data Services – Framing Structure, Channel Coding and Modulation for Cable Systems*
AS/NZS 4541:1999: *Digital Broadcasting Systems for Television, Sound and Data Services – Framing Structure, Channel Coding and Modulation for 11/12 GHz Satellite Services*
AS 4599-1999: *Digital Television – Terrestrial Broadcasting – Characteristics of Digital Terrestrial Television Transmissions*
AS 4599-1999/Amdt 1-2001: *Digital Television – Terrestrial Broadcasting – Characteristics of Digital Terrestrial Television Transmissions*

DR 00224: *Supplement 1 to AS 4599-1999, Digital Television – Terrestrial Broadcasting – Characteristics of Digital Terrestrial Television Transmissions*
DR 00325: *Supplement 2 to AS 4599-1999, Digital Television – Terrestrial Broadcasting – Characteristics of Digital Terrestrial Television Transmissions*

References

1. EBU Tech 3258-E, Specification of the Systems of the MAC/Packet Family, Technical Centre, Brussels (1986).
2. ISO/IEC 13818-1, Information Technology – Generic Coding of Moving Pictures and Associated Audio – Part 1: Systems (1996).
3. ISO/IEC 13818-2, Information Technology – Generic Coding of Moving Pictures and Associated Audio – Part 2: Video (1996).
4. ISO/IEC 13818-3, Information Technology – Generic Coding of Moving Pictures and Associated Audio – Part 3: Audio (1998).
5. ISO/IEC 13818-9, Information Technology – Generic Coding of Moving Pictures and Associated Audio – Part 9: Extension for Real Time Interface for Systems Decoders (1996).
6. ISO/IEC 11172-1, Information Technology – Coding of Moving Pictures and Associated Audio for Digital Storage Media up to about 1,5Mbit/s – Part 1: Systems.
7. ISO/IEC 11172-3, Information Technology – Coding of Moving Pictures and Associated Audio for Digital Storage Media up to about 1,5Mbit/s – Part 3: Audio (1993)
8. ISO 639-2, Codes for the Representation of Names of Languages – Part 2: Alpha-3 Code.
9. ITU-R Recommendation BS.1196-E, Annex 2: Digital Audio Compression (AC-3) Standard (ATSC Standard) (1995).
10. ISO/IEC 13818-7 Information Technology – Generic Coding of Moving Pictures and Associated Audio – Part 7: Advanced Audio Coding, AAC (1997).
11. EBU Recommendation R.68, Alignment Level in Digital Audio Production Equipment and in Digital Audio Recorders.
12. ETSI EN 300 294, 625-Line Television – Wide Screen Signalling (WSS) (1996).
13. EBU Tech 3290, Basic Interoperable Scrambling System (BISS), Technical Centre, Brussels (March 2000).
14. SMPTE 276M, Television – Transmission of AES/EBU Digital Audio Signals Over Coaxial Cable.
15. SMPTE 337M, Television – Format for Non-PCM Audio and Data in an AES3 Serial Digital Interface.
16. ANSI S4.40-1992, Digital Audio Engineering: Serial Transmission Format for Two-Channel Linearly Represented Digital Audio Data (AES3).
17. SMPTE 302M, Television – Mapping of AES3 Data into MPEG-2 Transport Stream.
18. ITU-R Recommendation BT.1359-1, Relative Timing of Sound and Vision for Broadcasting.
19. BBC, IBA and BREMA. *Broadcast Teletext Specification*, ISBN 0 563 17261 4 (October 1976).
20. ITU-R Recommendation 653, System B, 625/50 Television Systems.
21. EBU SPB 492, Teletext Specification (625-line Television Systems) (1992).
22. Hamming, R.W. Error detecting and error correcting codes, *Bell Syst. Tech. J.*, **29**, 2 (1950).
23. EACEM Technical Report No. 8, Enhanced Teletext Specification, Draft 3, EACEM (November 1994).
24. ETSI TS 300 231, Television Systems; Specification of the Domestic Video Programme Delivery Control System (PDC) (1996).
25. EIA-608 R-4.3, Recommended Practice for Line 21 Data Services (September 1992).
26. Postel, J. RFC 791 (IP), Internet Protocol (1981).
27. Deering, S.E. RFC 1112, Host Extensions for IP Multicasting (1989).
28. ISO/IEC 13818-6, Information Technology – Generic Coding of Moving Pictures and Associated Audio – Part 6: Extensions for DSM-CC.
29. ITU-R Recommendation BT 601-3, Encoding Parameters of Digital Television for Studios (1992).
30. ITU-R Recommendation BT 656-4, Interfaces for Digital Component Video Signals in 525-line and 625-line Television Systems Operating at the 4:2:2 Level of Recommendation ITU-R BT601.
31. DTVB 1110/GT V4/MOD 252/ DTVC 18, 7th revised version: Baseline Modulation/Channel Coding System for Digital Multi-programme Television by Satellite (January 1994).
32. Reed, I.S. and Solomon, G. Polynomial codes over certain finite fields, *J. Soc. Ind. Appl. Math.*, **8**, 300–304 (1960).
33. Forney, G.D., Jr. The Viterbi algorithm. *Proc. IEEE*, **61**, 268–278 (1973).
34. Forney, G.D. Burst correcting codes for the classic bursty channel, *IEEE Trans. Comm. Tech.*, **COM-19**, 772–781 (October 1971).
35. McEliece, R.J. and Swanson, L. Reed–Solomon codes and the exploration of the Solar System, in *Reed–Solomon Codes and their Applications*, IEE Press, September 1999 (Wicker, S.B. and Bhargava, V.K., eds), Chapter 3.
36. Cheung, K.-M., Deutsch, L.J., Dolinar, S.J., McEliece, R.J., Pollara, F., Shahshahani, M. and Swanson, L. Recent advances in coding theory for near error-free communications, in *Technology 2000*, NASA Conference Publication 3109, Vol. 2, pp. 229–241, Scientific and Technical Information Division (1991).
37. Consultative Committee for Space Data Systems. Recommendations for space data standard: Telemetry channel coding, *Blue Book*, Issue 2, CCSDS 101.0-B2 (Jan. 1987).
38. Berrou, C., Glavieux, A. and Thitimajshima, P. Near Shannon limit error-correcting coding and decoding: Turbo codes, *Proc. 1993 IEEE Int. Communications Conf.*, Geneva, Switzerland, pp. 1064–1070 (May 1993).
39. Shannon, C.E. A mathematical theory of communication, *Bell Syst. Tech. J.*, **27**, 379 (1948).
40. Morello, A. and Visintin, M. Transmission of TC-8PSK digital television signals over Eurovision satellite links, *EBU Technical Review*, No. 269 (Autumn 1996).
41. Viterbi, A. et al. A pragmatic approach to trellis-coded modulation, *IEEE Communications Magazine* (July 1989).
42. Delaruelle. A pragmatic coding scheme for transmission of 155 Mbits/s and 140 Mbits/s PDH over 72 MHz transponders, *Proc. ICDSC-10 Conf.*, Brighton (May 1995).
43. ITU-T Recommendation G.712, Transmission Performance Characteristics of Pulse Code Modulation Channels.
44. ITU-T Recommendation G.729, Coding of Speech at 8 kbits/s using Algebraic-Coded-Excited-Linear-Prediction (CS-ACELP).
45. ITU-T Recommendation G.711, Pulse Code Modulation (PCM) of Voice Frequencies.
46. Holmes, J.K. *Coherent Spread Spectrum Systems*, Krieger Publishing Company, 1990, Malabar, FL.

47. Gaudenzi, R., De, Elia, C. and Viola, R. Bandlimited quasi-synchronous CDMA: a novel satellite access technique for mobile and personal communications, *IEEE J. Selected Areas Comm.*, **10**, No. 2, 328–343 (February 1992).
48. TIA/EIA RS-232, Interface Between Data Terminal Equipment and Data-circuit Terminating Equipment Employing Serial Binary Data Interchange.
49. Cooley, J.W. and Tukey, J.W. An algorithm for the machine calculation of complex Fourier series, *Math. Comput.*, **19**, 297–301 (1965).
50. ETS 300 401, Digital Audio Broadcasting.
51. Eutelsat, Digital Satellite Equipment Control (DiSEqC) Bus Functional Specification, version 4.2 (February 1998).
52. Seidelmann, P.K. (ed.). Julian Date, in *Explanatory Supplement to the Astronomical Almanac*, Section 2.26, pp. 55–56, University Science Books, Mill Valley, CA (1992).
53. Duffett-Smith, P. *Practical Astronomy with Your Calculator*, 3rd Edn, pp. 6–9, Cambridge University Press, Cambridge, UK (1992).
54. ISO 8859.
55. ISO/IEC 10646-1, Information Technology – Universal Multiple Octet Coded Character Set (UCS) – Part 1: Architecture and Basic Multilingual Plane (1993).
56. KSC5601, Code for Information Interchange (Hangul and Hanja), Korea Industrial Standards Association (1987).
57. Simplified Chinese Character Set, GB-2312–1980.
58. Big5 subset of ISO/IEC 10646-1.
59. EIA/TIA SP 3357, Low Voltage Differential Signalling.
60. ANSI X3T11, Fibre Channel Physical Level Working Draft Proposed American National Standard for Information Systems, Rev. 4.3, Levels FC-0 and FC-1 (1 June 1994).
61. IEC 60169-24, Radio-frequency Connectors, Part 24: Radio-frequency Coaxial Connectors with Screw Coupling, Typically for Use in 75 Ω Cable Distribution Systems (Type F).
62. IEC 60807-9, Rectangular Connectors for Frequencies Below 3 MHz, Part 9: Detail Specification for a Range of Peritelevision Connectors.
63. IEC 60933-5, Audio, Video and Audiovisual Systems – Interconnections and Matching Values – Part 5: Y/C Connector for Video Systems – Electrical Matching Values and Description of the Connector.
64. ITU-T Recommendation I.363.1, B-ISDN ATM Adaptation Layer (AAL) Specification.
65. ITU-T Recommendation J.82, Transport of MPEG-2 Constant Bit Rate Television Signals in B-ISDN.
66. RFC 1889, RTP: A Transport Protocol for Real-Time Applications (January 1996).
67. Postel, J. RFC 768 (UDP), User Datagram Protocol (August 1980).
68. Postel, J. RFC 793 (TCP), Transmission Control Protocol (September 1981).
69. ETS 300 001, Attachments to the Public Switched Telephone Network (PSTN); General Technical Requirements for Equipment Connected to an Analogue Subscriber Interface in the PSTN.
70. EN 300 765-2, Digital Enhanced Cordless Telecommunications (DECT); Radio in the Local Loop (RLL) Access Profile (RAP); Part 2: Advanced Telephony Services.
71. EN 301 240, Digital Enhanced Cordless Telecommunications (DECT); Data Services Profile (DSP); Point-to-Point Protocol (PPP) Interworking for Internet Access and General Multi-protocol Datagram Transport.
72. ETS 300 700, Digital Enhanced Cordless Telecommunications (DECT); Wireless Relay Station (WRS).
73. ATM Forum, Residential Broadband Physical Interfaces Specification.
74. ATM Forum, Physical Interface Specification for 25,6 Mbit/s over Twisted Pair Cable, af-phy-0040.000.
75. ITU-T Recommendation I.610, B-ISDN Operations & Maintenance Principles & Functions.
76. IEC 60603-7, Connectors for Frequencies Below 3 MHz for Use with Printed Boards, Part 7: Detail Specification for Connectors, 8-way, including Fixed and Free Connectors with Common Mating Features.
77. DAVIC 1.3 Specification Part 7: High and Mid Layer Protocols (Rev. 6.3 – 29 October 1997).
78. IEEE 1394, IEEE Standard for a High Performance Serial Bus (1995).
79. P1394a (V4.0), Draft Standard for a High Performance Serial Bus (Supplement) (15 September 1999).
80. P1394b (V0.9), Draft Standard for a High Performance Serial Bus (Supplement) (6 October 1999).
81. 1394 Trade Association (Revision 0.96), Power Specification Part 1: Cable Power Distribution (9 December 1998).
82. 1394 Trade Association (Revision 0.90), Power Specification Part 2: Suspend/Resume Implementation Guidelines (4 January 1999).
83. 1394 Trade Association (Revision 0.72), Power Specification Part 3: Power State Management (21 October 1998).
84. IEC 61883-2, Consumer Audio/Video Equipment – Digital Interface – Parts 1–5 (1998–02).
85. Specification of the Home Audio/Video Interoperability (HAVi) Architecture (Version 1.0) (January 2000).
86. RFC 2734, Ipv4 Over IEEE 1394.
87. RFC 1918, Address Allocation for Private Internets.
88. RFC 1631, The IP Network Address Translator (NAT).
89. RFC 2131, Dynamic Host Configuration Protocol.
90. IETF draft, DHCP for IEEE 1394 (February 2000).
91. ISO/IEC 10918-1, Digital Compression of Continuous-tone Still Images (JPEG).
92. Portable Network Graphics V1, 1 October 1996, available at http://www.w3.org/TR/REC-png.html
93. GIF 89a, Graphics Interchange Format (sm), Version 89a, © 1987, 1988, 1989, 1990. Copyright CompuServe Incorporated Columbus, OH.
94. RFC 2616, IETF Hypertext Transfer Protocol – HTTP/1.1 (June 1999).
95. XHTML™ 1.0, The Extensible HyperText Markup Language, available at http://www.w3.org/TR/xhtml1
96. Gosling, J., Joy, W. and Steele, G. *The Java Language Specification*, ISBN 0-201-63451-1, 2000.
97. Lindholm, T. and Yellin, F. *The Java Virtual Machine Specification*, Java VM, Addison-Wesley, ISBN 0-201-63452-X, 1996.
98. RFC 2313, PKCS #1: RSA Encryption Version 1.5 (March 1998).
99. RFC 1890, RTP Profile for Audio and Video Conferences with Minimal Control (January 1996).
100. RFC 2250, RTP Payload Format for MPEG1/MPEG2 Video (January 1998).
101. RFC 2326, Real Time Streaming Protocol (RTSP) (April 1998).
102. RFC 2474, Definition of the Differentiated Services Field (DS Field) in the IPv4 and IPv6 Headers (December 1998).
103. RFC 2475, An Architecture for Differentiated Services (December 1998).

Bibliography

Benoit, H. *La Télévision Numérique; MPEG-1, MPEG-2 et les Principes du Système Européen DVB*, Dunod, Paris, 1996.

Benoit, H. *Digital Television; MPEG-1, MPEG-2 and Principles of the DVB System*, Arnold, ISBN 0-340-69190-5, 1997.

Reimers, U. (ed.). *Digitale Fernsehtechnik – Datenkompression und Übertragung für DVB*, Springer, Berlin (1997).

Viterbi, A.J. and Omura, J.K. *Principles of Digital Communication and Coding*, McGraw-Hill (1979).

Wood, D. Satellites, science and success – The DVB story, *EBU Technical Review*, No. 266, pp. 4–11 (Winter 1995).

Internet Resources

General Interest

www.bbc.co.uk/rd
www.bbc.co.uk/validate
www.davic.org
www.digitag.org
www.dtg.org.uk
www.dvb.org
www.ebu.ch/technical.html
www.etsi.org
www.eutelsat.org
www.intelsat.int
www.itu.int
www.ses-astra.com
www.unicode.org

IHDN-Specific URLs

ftp://ftp.t10.org/1394/P1394a/Drafts/P1394a40.pdf
http://www.zayante.com/p1394b/index.shtml
http://www.iec.ch
http://www.havi.org/techinfo/index.html
http://www.ietf.org/rfc
ftp://ftp.ietf.org/internet-drafts/draft-ietf-ip1394-dhcp-03.txt

MHP-Specific URLs

http://www.w3.org/Graphics/GIF/spec-gif89a.txt
http://www.w3.org/Graphics/JPEG/itu-t81.pdf
ftp://ftp.javasoft.com/docs/specs/langspec-1.0.pdf
http://java.sun.com/docs/books/jls/clarify.html
http://www.w3.org/Graphics/JPEG/jfif3.pdf
http://www.w3.org/TR/xhtml1

Michael A Dolan
Television Broadcast Technology, Inc.

2.13 Data Broadcast

Data broadcast has generally referred to any information carried along with the video or audio signal that is not the video or audio coding itself. The classic example of 'data' associated with television is teletext.[1] However, there is other information that falls into this category, including subtitling and captioning. 'Data' is often also generally used to refer to literally anything encoded in the Vertical Blanking Interval (VBI) of the analogue television signal, including what we today recognise more formally as metadata.

More commonly today, there is a distinction drawn between metadata (information about the video and audio) and data 'essence'.[2] The latter refers to data items that convey their own information to the viewer and not just information about the video or audio. For example, a digital programme identifier encoded into the VBI relates to the video and audio, has no intrinsic value on its own, and would rarely be displayed to the viewer. It is thus considered metadata. In contrast, teletext is considered data essence since it has value on its own. Electronic Programme Guide (EPG) data is technically metadata (information about the video and audio programming), but is less clearly so due to the sheer volume of information and its intrinsic value to the viewer. This chapter, data broadcast, is concerned with data essence and not metadata, and the use of the term data here means data essence.

The introduction of digital television systems enables a robust array of data broadcasting possibilities. This is primarily due both to the fact that the broadcast is inherently digital, and that there is so much more bandwidth available. Being inherently digital makes the insertion and extraction of data easier, and provides a more robust (less error prone) delivery. In contrast, analogue VBI has very limited bandwidth and is quite error prone, especially when the signal was not strong at the receiving equipment. Reliable delivery in the analogue environment often required various FEC techniques, which used up even more bandwidth. Digital television systems have defined a set of building blocks, or infrastructure, in which to carry some basic data models. While every digital transport has defined their own unique details, the basic data models that have been universally defined are: Internet Protocol (IP) data packets, files (analogous to a computer file system), streams of bytes, and 'triggers'. All of these can be thought of in aggregate as a data broadcast infrastructure, which can then be used for the design and delivery of specific *data services* to the viewer. *Data services* are an application of data broadcasting that results in some experience for the viewer. Also, using this data infrastructure, interactive television (ITV) systems such as Digital Video Broadcast Multimedia Home Platform (DVB MHP)[3] and the Advanced Television Systems Committee DTV Application Software Environment (ATSC DASE)[4] can add another layer of 'tools' for the ultimate data service provider.

2.13.1 Traditional Analogue System Data Essence

In order to understand current developments in data broadcasting, it is important to review analogue data broadcasting, as it ultimately forms the basic requirements of the digital systems. There are three main analogue data services in use today: teletext,[1,5] subtitling and captioning.[6] They are all text-based systems – that is, they deliver electronically encoded text, but for different specific purposes. The bit-level character encodings are all different, but that is mostly historical in their evolution. The main difference is in their application and viewer experience.

Teletext is a stand-alone, asynchronous data service. That is, it provides a set of text and limited character-cell graphics symbols for viewing on the television display, usually unrelated to any video and audio programming. The information is organised into discrete pages, or blocks, and is not fundamentally a stream. The viewer may then navigate manually and asynchronously (relative to any video and audio) through these pages of information. Typical services that teletext provides are airplane and train schedules, weather, and sports scores. While teletext is common in Europe and other continents, it was never widely deployed in the USA.

Teletext pages are normally *carouselled* in the broadcast. A carousel is when the page is repeated at some time interval in order to facilitate its capture by the receiving equipment. This is needed for three main reasons. The first is when a viewer first tunes to the service that is sending the pages, the receiving equipment must be given a chance to acquire all the relevant pages. Since tuning can occur at any time, the pages are repeated. The second reason is to permit inexpensive receivers with virtually no memory to acquire the pages on demand from the viewer, thus not having to cache anything. Finally, a carousel provides some amount of error recovery. If a page or portion of a page is received with an error, it can be acquired again the next time it is sent. Carouselling algorithms often vary the frequency and order of the pages to try and improve the viewer experience. For example, the 'initial' pages the viewer will see are sent faster than pages 'deeper' in the organisation of the pages. The general notion of carouselling remains important in the digital domain.

Subtitling and captioning are similar to each other and both associated with the video and audio. They are also *synchronised* data services. That is, they provide a set of text and graphics that can be viewed simultaneously with its related video and audio. They are *synchronised* services since the information is viewed along with specific video or audio segments, or even frames. Captioning is also *streaming* since it defines a stream of bytes not organised into pages that can be interpreted with very little context or structure. The difference between subtitling and captioning is a bit subtle and comes from their intended use, as well as encoding. Subtitling is typically encoded using graphic overlays at a specific location on the screen defined by the author (not the viewer). Its primary use is for providing alternative language text from the language found in the primary audio. Captioning is encoded as text, can be turned on and off by the viewer, is normally *in the same language* as the primary audio, and is intended primarily as an aid to the hearing impaired. Subtitling is common in Europe and other continents, but rarely found in the USA, while captioning is the exact opposite.

This review of analogue data broadcasting is important in providing the context for the development of the digital data models. From the above three systems we learn that we need both stand-alone and associated data services, both synchronised and asynchronous services, both streams of data and organised blocks of data, and the ability to carousel the blocks (pages) of information. This is summarised in Table 2.13.1.

Table 2.13.1 Summary of features of traditional data services

	Stand-alone	*Synchronised*	*Stream*	*Block/ carousel*
Teletext	X			X
Subtitling and captioning		X	X	

2.13.2 Digital System Data Models

The introduction of digital television systems has offered the ability to define more formal building blocks for the carriage of data services. These data models are: IP packets, files, streams and triggers. Each of these provides an important component for building data services, and in aggregate form a data broadcasting infrastructure.

Sadly, the primary analogue data services (teletext, subtitling and captioning), in their transition to the digital transports, have not been built upon these formal data models, but were designed as individual special cases. This is primarily due to their integration early in the digital system design process before the data models and infrastructure were designed.

The ISO has defined many of the basic building block mechanisms used by data broadcasting. Most come from the 13818 series, part 6 and its amendments on Digital Storage Media – Command and Control (DSM-CC).[7] The whole of the DSM-CC standard is complex and designed as an architecture and control mechanism for Video on Demand (VOD) systems. However, it includes some special MPEG-2 private sections (as defined in 13818-1) to carry a set of message formats that support the carriage of files and network packets. More on the specific uses of the DSM-CC messages is covered below. First, the data models that apply generically to all of the transports are discussed, along with more details of their carriage using the DSM-CC mechanisms. Then, there is a brief introduction to the ITV systems that use the data models; and then the specific details for each of the digital facility, the DVB and the ATSC digital transports is discussed. It is assumed the reader is generally familiar with MPEG-2 and computer system concepts.

2.13.2.1 Internet Protocol (IP) Packets

IP packets are defined by a variety of Internet Engineering Task Force (IETF) standards and are beyond the scope of this section. But due to the extensive amount of interoperable equipment available that can encode, decode and route IP packets, their migration into broadcast television should not be surprising. The television signal is effectively used as a single hop, unidirectional subnetwork (in the Internet use of the term) route. When used in the forward channel of broadcast systems, it is common to use IP multicast addressing, as it is generally not efficient to carry point-to-point (unicast) addressed packets in a broadcast medium. When used in MPEG-2 transports, IP packets are usually constructed according to the ISO MPEG-2, Part 6 standard commonly known as DSM-CC[7] (amendment #1). The encapsulation is a private table in MPEG-2 known as a DSM-CC addressable section. The encapsulation includes fields for physical layer addressing (MAC address), as well as the IP packet payload itself.

IP packets are typically asynchronous. A synchronised version of the encapsulation is defined, but quite uncommon in practice. IP packets form the basis for many of the early new data services today found in several start-up 'data broadcasting' service companies' designs. In fact, IP packets are often used today to carry proprietary file and stream formats, even though standard file and stream mechanisms are defined.

2.13.2.2 Files

Files are blocks of data just like files on a computer system. When used in MPEG-2 transports, they are packaged and carried according to two kinds of mechanisms also defined in DSM-CC:[7] the data carousel and the object carousel. The term 'data carousel' is used interchangeably here with 'download' to loosely describe all variant forms of the DSM-CC download protocols, which include the data carousel. The data carousel provides a basic packaging of a single file into a module, along with some signalling to identify and locate all the modules of the carousel. Each module is broken into one or more blocks. The organisation of the modules is, at most, a two-layer organisation and is thus flat (no useful hierarchy) and, for small files, carrying them one per module can be relatively inefficient. The module blocks are carried in a DSM-CC message called a Download Data Block (DDB). The DDBs are sequentially numbered and, when combined, form a complete module. Metadata for the modules, as well as additional signalling that assists in their acquisition, is communicated in a message called a Download Information Indication (DII). This message contains a list of modules, their size and optionally a name. Some implementations support a two-layer hierarchy with the addition of the message, Download Server Initiate (DSI). This message references one or more DII messages. The organisation of a simple two-layer data carousel including the DSI message, the DII messages, modules and DDB messages is shown in Figure 2.13.1.

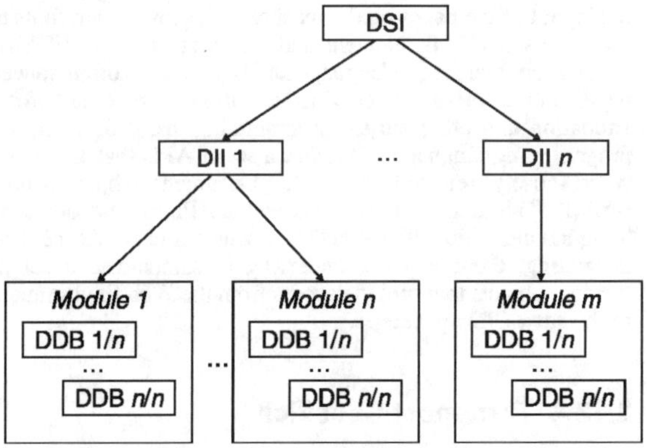

Figure 2.13.1 DSM-CC data carousel organisation.

The object carousel builds on the data carousel structure defined above and provides a more robust, although more complex, system. Relative to the data carousel, it adds a full hierarchical organisation through the addition of directories (analogous to those found in computer file systems). This is done through the definition of another layer of structure generically called objects. The object carousel is an open-ended mechanism in which to deliver objects of any kind. Two important intrinsic objects defined by DSM-CC are File and Directory, which contain the obvious contents. Additionally there is a support object called the ServiceGateway, which basically provides the entry point to the object carousel Directory hierarchy. The organisation of these basic objects is shown in Figure 2.13.2.

The objects are mapped onto the elements of the data carousel as follows. The ServiceGateway object is signalled in the DSI. The ServiceGateway is also a Directory and provides the 'root' directory of the file system. All other objects (including Directory and File) are carried in data carousel modules. Objects may be packed more than one per module, thus making it more efficient for broadcasting many small (File) objects. This packing strategy provides another layer of opportunity to increase the acquisition efficiency. For example, one could

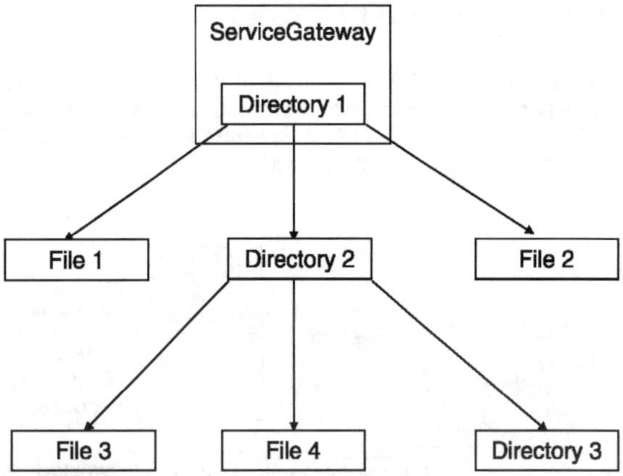

Figure 2.13.2 DSM-CC object carousel object organisation.

aggregate all the Directory objects into a single module so that a receiver could more efficiently acquire Files more randomly.

Other DSM-CC intrinsic objects that are common to add to the object carousel in ITV systems are the Stream and StreamEvent. See the DSM-CC standard[7] for more information on these. Some of the new ITV designs have chosen to support only the object carousel rather than both that and the data carousel.

2.13.2.3 Streams

Streams of data bytes can be carried just like video and audio elementary streams. In the case of MPEG-2-based transports, they are most often Packetised Elementary Stream (PES)-encapsulated. However, they can also be transmitted using a blocked format with the DSM-CC Download protocol. This form of download is not a carousel, but a single emission of data blocked into modules. This blocking offers some advantages over PES. First there is error detection through CRC and block numbering. The latter allows for detection of entire missing blocks. Data cannot arrive out of order due to the block sequencing. Finally, there is a more well-formed buffer model for the 'chunks' of the stream.

Streams can also be carried in what is known as 'data piping'. Data piping is a formal name for carrying data in the raw transport packet without any further standard structure (i.e. not PES or Download chunks). Streams are the least well-formalised carriage of all the models.

2.13.2.4 Triggers

Triggers are a special type of data model that are intended to link data to the video and audio elements, deliver an event to a receiver data application, and/or provide a timed sequence of data essence stand-alone. Triggers are inherently synchronised, although in practice they can be sent as a 'best effort' and not linked to specific frames of video or audio. The trigger can also be used to get around some MPEG-2 PCR and decoder model buffering issues when the data chunks are large and transmitted over a long period of time. Triggers have the most widely varying implementation across the transports, including URL strings, special DSM-CC Download payloads and Object Carousel StreamEvent objects.

2.13.2.5 Summary of Data Models on MPEG-2 Transports

The general MPEG-2 encapsulations for all the above data models can be summarised in a stack of transport protocols defined by 13818-1 (MPEG-2 Systems) and 13818-6 (DSM-CC) and are shown in Figure 2.13.3.

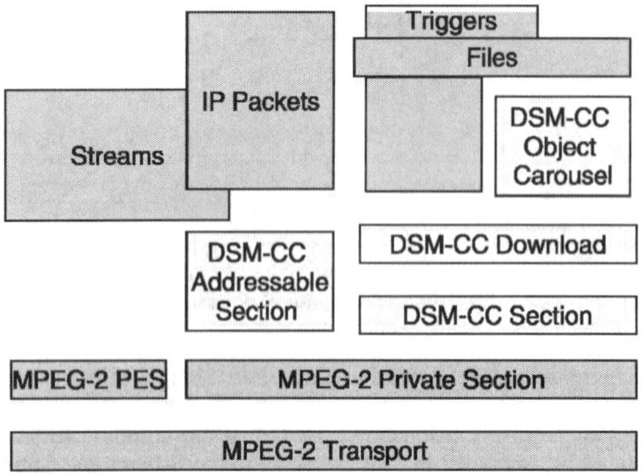

Figure 2.13.3 Summary of MPEG-2 data broadcast protocols.

2.13.3 ITV Environments

Interactive television, or 'ITV', means many different things today and generally applies to the collection of digital system viewer experiences ranging from EPG and VOD to the technology of various middleware systems. One use of the term considered here is the standard ITV environments defined by regional digital television standards organisations: ATSC, DVB and US Cable. The ATSC ITV environment is known as the DTV Application Software Environment (DASE); the DVB ITV environment is the Multimedia Home Platform (MHP); and the US Cable ITV environment is the OpenCable Application Platform (OCAP). This section will provide a high-level overview of the technology generally common to these systems, that can be grouped into two main environments – the Java™ Application Environment and the Web Application Environment. There are also some special file formats supported that include: JPEG, PNG and often the Bitstream™ font format.

2.13.3.1 Java Environment Technology

The Java Environment includes a Java Virtual Machine (JVM)[8] and a set of standard *Java packages*. The JVM is a programming language (Java) interpreter that executes the Java *byte codes* in *class files* on the receiver's processor. The byte codes are the compiled version of the text Java programming language. The class files contain the compiled byte codes for the packages. The advantage to an interpretive language like Java is that it provides a hardware-independent environment for the ITV authors.

The Java packages (i.e. application programming interfaces) consist of several main components: Personal Java,[9] Java Media Framework (JMF)[10] and Java TV.[11] Personal Java is a profile of the more widely known Java Development Kit (JDK), and includes access for file data and Internet packets. JMF is a package used for stream data access and control. The Java TV package is an abstraction of the MPEG-2 environment of tables and descriptors defined by DVB and ATSC for signalling and announcing the programmes and services in the MPEG-2 transport. These packages have been aggregated recently by the DVB and given the name Globally Executable MHP (GEM).[12]

2.13.3.2 Web Environment Technology

The Web Environment usually consists of a set of basic web browser technology to process what amounts to a set of web pages not unlike today's desktop web browser. The standards for this environment are derivations of those published by the World Wide Web Consortium (W3C). The web technology is usually HTML, Cascading Style Sheets (CSS), and often a script interpreter called ECMAScript[13] (aka Javascript, but not to be confused with Java). The HTML derivation is more lately a profile of XHTML,[14] although there are some existing systems that use HTML 3.2 and 4.0. A profile of CSS has been defined by the W3C, called CSS-TV.[15] CSS is a way to control the layout of the pages on the receiver's display – what fonts to use for each HTML element and that sort of thing. ECMAScript is an interpretive language and is used quite often on web pages on the Internet to provide animation, receiver-side form validation and other simple programming tasks. Since it is a programming language, it too has a set of APIs that are somewhat loosely referred to as the Document Object Model (DOM),[16] although strictly speaking the API goes beyond anything having to do with the HTML document itself. At the time of writing, there is work underway to standardise a set of 'bridge' objects that provide access from the Web Environment to the Java GEM packages.

2.13.4 Transport Specifics

Each of the data models and ITV environments described above has variations depending on the transport. This section discusses the variations in each of the facility, the DVB transport and the ATSC transport.

2.13.4.1 Facility

For the purposes of this section, it is assumed that the facility uses serial digital 'base-band' links defined by SMPTE 259 and 292. Unlike the MPEG-2-based digital systems, the facility can carry data in multiple transports. Today's facilities generally route IP packets, not on 259 or 292, but by using traditional IP networks, with CAT-5 cable or similar. When data is carried on the serial digital systems, it is packaged into the ancillary data space (ANC), generally according to SMPTE 291M.[17]

Captioning (US) is carried as defined by SMPTE 334M.[18] Note that there is work underway at the time of writing to extend 334M to address the carriage of teletext and subtitling. However, there is currently no standard way to carry any of the new data models described above (IP packets, files, streams and triggers) in the serial digital interfaces. Nor is there any standard way to manage ITV systems, whether they are built on these layers or not. The assumption by the industry today is that there is a special device (a 'data server') that formats these data models into the appropriate MPEG-2 transport format based on the data arriving at the server from unspecified means. This situation is less than desirable as it prevents any standard tight association of data on peer with its related video and audio, and it encourages proprietary solutions surrounding the 'data server'. More standardisation work can be expected in this area.

2.13.4.2 DVB

DVB has addressed the digital data broadcast needs through several standards ultimately published by the European Telecommunications Standards Institute (ETSI). The foundation data broadcasting specification for DVB is ETSI 301-192.[19] It defines the basic packaging for IP packets, files and streams. Files may be carried in either the data carousel or the object carousel. DVB MHP[3] builds on and extends the basic object carousel as the exclusive means to carry files. MHP also defines a trigger data model through the use of the DSM-CC Stream Event object. MHP defines both a Java Environment as well as Web Environment for the receivers, although the Java Environment is required and the Web Environment is optional.

2.13.4.3 ATSC

The ATSC publishes its own standards. The foundational work for data broadcasting is A/90[20] that provides the basic encapsulations, including the IP packets and files in the DSM-CC data carousel. All of the data models described here are more formally defined in A/94.[21] IP packet carriage for multicast sessions is defined in more detail in A/92.[22] The ATSC object carousel that can be used for files (in addition to the data carousel) is defined in A/95.[23] Triggers are defined in A/93.[24] A more thorough discussion of the ATSC data broadcast specification can be found in the Chernock, Crinon and Dolan text.[25] ATSC's DASE ITV standard builds on this infrastructure and, like DVB, includes both a Java Environment and a Web Environment. At the level we have touched on the ITV systems here, they are virtually identical to the DVB MHP system, except both Environments are optional. There are many detailed differences, however, thus making common authoring nearly impossible.

2.13.4.4 US Cable

US digital cable standards are developed jointly between the Society of Cable and Telecommunications Engineers (SCTE) and a private organisation, Cablelabs. US cable standards are, in practice, a mixture of ATSC and DVB standards, and the carriage of data is no exception. Cable has defined the carriage of IP packets over MPEG-2 following the DVB encapsulation in SCTE 42.[26] While SCTE has a defined standard for IP packets in the MPEG-2 transport, due to the existence of bidirectional communication inherent in cable, IP packets will most likely be carried in the out-of-band channel and not in the forward MPEG-2 transport. It appears that data standards on US cable will be driven by the deployment of a modified version of the ITV standards work, DVB MHP, known as OCAP.[27]

At the time of writing, there is an effort underway to harmonise the two ITV specifications between US cable (OCAP) and ATSC (DASE), which also include some of the data model elements (but not IP packets or streams). The end result may form the basis of a core set of data models in the USA, and, given the close technical relationship between OCAP and MHP, perhaps worldwide.

2.13.4.5 Other ITV Standards

There are three other international ITV standards that are worth mentioning: MHEG,[28] Declarative Data Essence (DDE)[29] and Teleweb.[30] MHEG was standardised some time ago by ISO and is a simple, small footprint system not based on many other standards. It is currently deployed in the United Kingdom. DDE was standardised by the SMPTE and is a formalisation of the earlier work from the Advanced Television Enhancement Forum (ATVEF).[31] It is based on HTML 4.0, CSS-1 and ECMAScript, and is currently deployed in the USA and France. Teleweb, as the name implies, is an extension of teletext concepts, only with the use of more modern web technologies described above. Like DDE, Teleweb also started as an industry consortium,[32] but its work was moved to the IEC, where its publication is still pending at the time of writing. It consists of several profiles, the first being HTML 3.2. Additional profiles add CSS and ECMAScript, and thus are very similar to DDE. There is some work underway in the IEC to try and merge the advanced Teleweb profile(s) with a version of DDE.

MHEG is interesting due to its extremely low receiver implementation cost. DDE and Teleweb are interesting due to their being based on teletext concepts extended to modern web technology. While DDE and Teleweb are more difficult to implement than MHEG, they are significantly simpler than MHP, DASE and OCAP discussed here.

2.13.4.6 Summary of Digital MPEG-2 Transport Standards

A summary of the standards that cover the transport encapsulations is shown in Table 2.13.2. IP encapsulation varies only slightly between DVB (including SCTE) and ATSC. ATSC's object carousel is based on the MHP object carousel, so they are compatible. Triggers are entirely different, mainly since the ATSC trigger is not dependent on the object carousel as is the DVB trigger.

Table 2.13.2 Summary of transport standards

	DVB (ETSI)	*ATSC*	*SCTE*
IP Packets	301-192	A/90, A/92	42
Files	301-192, 102-812	A/90, A/95	90-1
Streams	301-192	A/90	–
Triggers	102-812	A/93	90-1
Java	102-812	A/100	90-1
HTML	102-812	A/100	–

References

1. ITU-R BT.653-3, Teletext Systems.
2. SMPTE/EBU Joint Task Force for Harmonized Standards for the Exchange of Program Material as Bit Streams, Final Report: Analyses and Results (July 1998).

3. ETSI 102-812, Digital Video Broadcasting (DVB); Multimedia Home Platform (MHP) Specification 1.1.
4. ATSC A/100, DTV Application Software Environment.
5. ETSI 300-706, Digital Video Broadcasting (DVB); Enhanced Teletext Specification.
6. CEA/EAI 608B, Line 21 Data Services.
7. ISO/IEC 13818-6, Information Technology – Generic Coding of Moving Pictures and Associated Audio Information – Part 6: Extensions for DSM-CC.
8. Lindholm, T. and Yellin, F. *The Java Virtual Machine Specification*, Addison-Wesley, ISBN 0-201-63452-X.
9. *The OEM Personal Java Application Environment, Version 1.2a Specification*, ISBN 1-892488-25-6.
10. *Java Media Framework API, Version 1.0 Specification*, ISBN 1-892488-25-6.
11. *Java TV API, Version 1.0 Specification*, ISBN 1-892488-25-6.
12. DVB, DVB MUG Work in Process, Globally Executable MHP.
13. ISO Standard 16262, Information Technology – Ecmascript Language Specification.
14. W3C, *XHTML™ 1.0: The Extensible HyperText Markup Language*, http://www.w3.org/TR/xhtml1
15. W3C, Candidate Recommendation, CSS TV Profile 1.0, 14 May 2003, http://www.w3.org/TR/2003/CR-css-tv-20030514
16. W3C, Document Object Model (DOM) Level 2 Core Specification, http://www.w3.org/TR/DOM-Level-2-Core
17. SMPTE 291M, Ancillary Data Packet and Space Formatting.
18. SMPTE 334M, Vertical Ancillary Data Mapping for Bit-Serial Interface.
19. ETSI 301-192, Digital Video Broadcasting (DVB); DVB Specification for Data Broadcasting.
20. ATSC A/90, Data Broadcast Standard.
21. ATSC A/94, Data Application Reference Model.
22. ATSC A/92, Delivery of IP Multicast Sessions over Data Broadcast Standard.
23. ATSC A/95, Transport Stream File System Standard.
24. ATSC A/93, Synchronized/Asynchronous Trigger Standard.
25. Chernock, R., Crinon, R., Dolan, M. et al. *Data Broadcasting: Understanding the ATSC Data Broadcast Standard*. McGraw-Hill (2001).
26. SCTE 42, IP Multicast for Digital MPEG Networks.
27. SCTE 90-1, SCTE Application Platform Standard, Part 1: OCAP 1.0 Profile.
28. ISO Standard 13522-5, Information Technology – Coding of Multimedia and Hypermedia Information – Part 5: Support for Base-level Interactive Applications.
29. SMPTE 363M, Standard for Television: Declarative Data Essence, Content Level 1.
30. IEC Committee Drafts and Projects 62297 and 62298 on Teleweb.
31. Teleweb Project, http://www.superteletext.tv
32. Advanced Television Enhancement Forum, http://www.atvef.com

Jerry C Whitaker
Technical Director
Advanced Television Systems Committee

2.14 ATSC Video, Audio and PSIP Transmission*

2.14.1 Overview of the ATSC Digital Television System

The Digital Television (DTV) standard has ushered in a new era in television broadcasting. The impact of DTV is more significant than simply moving from an analogue system to a digital system. Rather, DTV permits a level of flexibility wholly unattainable with analogue broadcasting. An important element of this flexibility is the ability to expand system functions by building upon the technical foundations specified in ATSC standards such as the ATSC Digital Television Standard (A/53)[1] and the Digital Audio Compression (AC-3) Standard.[2]

With NTSC, and its PAL and SECAM counterparts, the video, audio and some limited data information are conveyed by modulating an RF carrier in such a way that a receiver of relatively simple design can decode and reassemble the various elements of the signal to produce a program consisting of video and audio, and perhaps related data (e.g. closed captioning). As such, a complete program is transmitted by the broadcaster that is essentially in finished form. In the DTV system, however, additional levels of processing are required after the receiver demodulates the RF signal. The receiver processes the digital bit stream extracted from the received signal to yield a collection of program elements (video, audio and/or data) that match the service(s) that the consumer has selected. This selection is made using system and service information that is also transmitted. Audio and video are delivered in digitally compressed form and must be decoded for presentation. Audio may be monophonic, stereo or multi-channel. Data may supplement the main video/audio program (e.g. closed captioning, descriptive text or commentary) or it may be a stand-alone service (e.g. a stock or news ticker).

The nature of the DTV system is such that it is possible to provide new features that build upon the infrastructure within the broadcast plant and the receiver. One of the major enabling developments of digital television, in fact, is the integration of significant processing power in the receiving device itself. Historically, in the design of any broadcast system – be it radio or television – the goal has always been to concentrate technical sophistication (when needed) at the transmission end and thereby facilitate simpler receivers. Because there are far more receivers than transmitters, this approach has obvious business advantages. While this trend continues to be true, the complexity of the transmitted bit stream and compression of the audio and video components require a significant amount of processing power in the receiver, which is practical because of the enormous advancements made in computing technology. Once a receiver reaches a certain level of sophistication (and market success) additional processing power is essentially 'free'.

2.14.1.1 DTV System Overview

The DTV standard describes a system designed to transmit high-quality video and audio and ancillary data over a single 6 MHz channel. The system can deliver about 19 Mbps in a 6 MHz terrestrial broadcasting channel and about 38 Mbps in a 6 MHz cable television channel. This means that encoding High-Definition (HD) video sources at 885† Mbps (highest rate

*This chapter is based on: ATSC, *Guide to the Digital Television Standard*, Doc. A/54A, Advanced Television Systems Committee, Washington, DC (2003).

†$720 \times 1280 \times 60 \times 8 \times 2 = 884.736$ Mbps (the 2 represents the 4:2:2 or colour subsampling; an RGB with full bandwidth would be a 3).

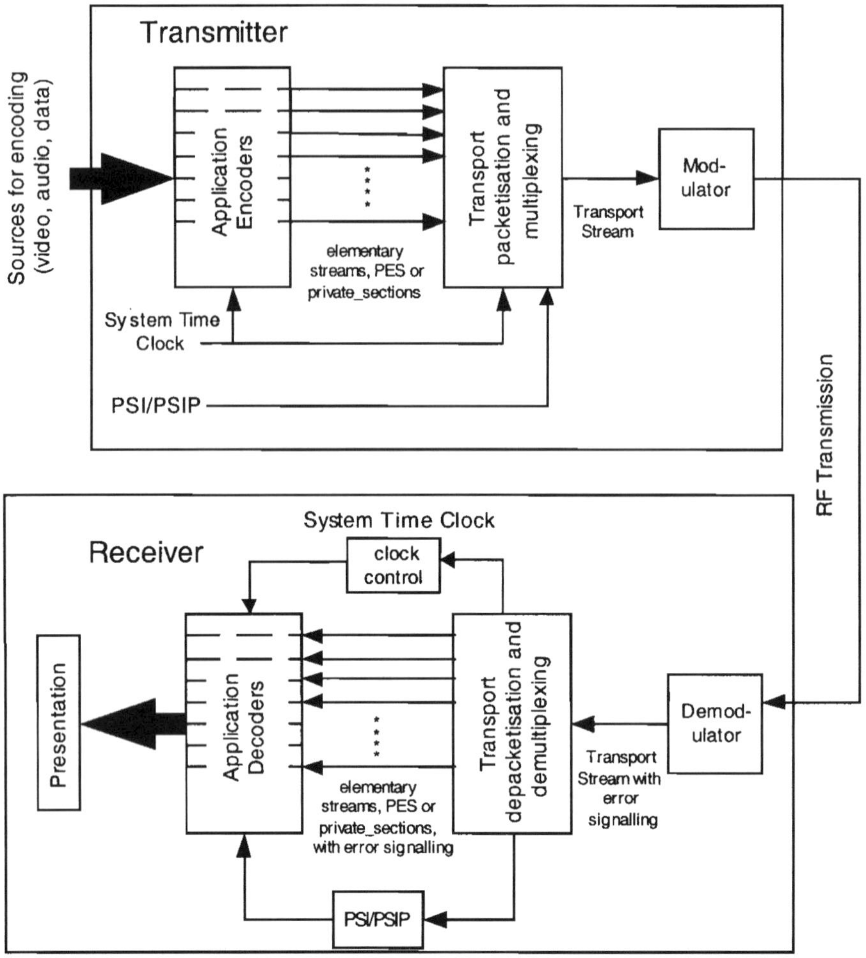

Figure 2.14.1 Block diagram of functionality in a transmitter/receiver pair. Note: PSIP refers to ATSC Standard A/65, which was developed subsequent to ATSC Standards A/52, A/53 and this chapter. For details regarding PSIP, refer to ATSC Standard A/65.

progressive input) or 994[‡] Mbps (highest rate interlaced picture input) requires a bit rate reduction by about a factor of 50 (when the overhead numbers are added, the rates become closer). To achieve this bit rate reduction, the system is designed to be efficient in utilising available channel capacity by exploiting complex video and audio compression technology.

The compression scheme used for DTV service optimises the throughput of the transmission channel by representing the video, audio and data sources with as few bits as possible while preserving the level of quality required for the given application.

The RF/transmission subsystems described in the DTV standard are designed specifically for terrestrial and cable applications. The structure is such that the video, audio and service multiplex/transport subsystems are useful in other applications as well.

2.14.1.2 System Block Diagram

A basic block diagram representation of the DTV system is shown in Figure 2.14.1. According to this model, the digital television system consists of four major elements, three within the broadcast plant plus the receiver.

2.14.1.2.1 Application Encoders/Decoders

The 'application encoders/decoders', as shown in Figure 2.14.1, refer to the bit rate reduction methods (also known as data compression) appropriate for application to the video, audio and ancillary digital data streams. The purpose of compression is to minimise the number of bits needed to represent the audio and video information. The DTV system employs the

‡ $1080 \times 1920 \times 29.97 \times 8 \times 2 = 994.333$ Mbps (the 2 represents the 4:2:2 or colour subsampling; an RGB with full bandwidth would be a 3).

MPEG-2 video stream syntax for the coding of video and the ATSC Standard 'Digital Audio Compression (AC-3)' for the coding of audio.[2]

2.14.1.2.2 Transport (de)Packetisation and (de)Multiplexing

'Transport (de)packetisation and (de)multiplexing' refers to the means of dividing each bit stream into 'packets' of information, the means of uniquely identifying each packet or packet type, and the appropriate methods of interleaving or multiplexing video bit stream packets, audio bit stream packets and data bit stream packets into a single transport mechanism. The structure and relationships of these essence bit streams is carried in service information bit streams, also multiplexed in the single transport mechanism. In developing the transport mechanism, interoperability among digital media – such as terrestrial broadcasting, cable distribution, satellite distribution, recording media and computer interfaces – was a prime consideration. The DTV system employs the MPEG-2 transport stream syntax for the packetisation and multiplexing of video, audio and data signals for digital broadcasting systems. The MPEG-2 transport stream syntax was developed for applications where channel bandwidth or recording media capacity is limited and the requirement for an efficient transport mechanism is paramount.

In the DTV standard, the carriage of program and system information is specified in ATSC Standard A/65, 'Program and System Information Protocol'.[3]

2.14.1.2.3 RF Transmission

'RF transmission' refers to channel coding and modulation. The channel coder takes the digital bit stream and adds additional information that can be used by the receiver to reconstruct the data from the received signal that, due to transmission impairments, may not accurately represent the transmitted signal. The modulation (or physical layer) uses the digital bit stream information to modulate a carrier for the transmitted signal. The modulation subsystem offers two modes: an 8-VSB mode and a 16-VSB mode.

2.14.1.2.4 Receiver

The ATSC receiver must recover the bits representing the original video, audio and other data from the modulated signal. In particular, the receiver must:

- Tune the selected 6 MHz channel.
- Reject adjacent channels and other sources of interference.
- Demodulate (equalise as necessary) the received signal, applying error correction to produce a transport bit stream.
- Identify the elements of the bit stream using a transport layer processor.
- Select each desired element and send it to its appropriate processor.
- Decode and synchronise each element.
- Present the programming.

2.14.2 Video System

The MPEG-2 specification is organised into a system of profiles and levels, so that applications can ensure interoperability by using equipment and processing that adhere to a common set of coding tools and parameters.[4] The DTV standard is based on the MPEG-2 Main Profile. The Main Profile includes three types of frames (I frames, P frames and B frames), and an organisation of luminance and chrominance samples (designated 4:2:0) within the frame. The Main Profile does not include a scalable algorithm, where scalability implies that a subset of the compressed data can be decoded without decoding the entire data stream. The High Level includes formats with up to 1152 active lines and up to 1920 samples per active line, and for the Main Profile is limited to a compressed data rate of no more than 80 Mbps. The parameters specified by the DTV standard represent specific choices within these constraints.

2.14.2.1 Compatibility with MPEG-2

The DTV video compression system does not include algorithmic elements that fall outside the specifications for MPEG-2 Main Profile. Thus, video decoders that conform to the MPEG-2 Main Profile at High Level (MP@HL) can be expected to decode bit streams produced in accordance with the DTV standard. Note that it is not necessarily the case that all video decoders which are based on the DTV standard will be able to properly decode all video bit streams that comply to MPEG-2 MP@HL.

2.14.2.2 Overview of Video Compression

The DTV video compression system takes in an analogue video source signal and outputs a compressed digital signal that contains information that can be decoded to produce an approximate version of the original image sequence. The goal is for the reconstructed approximation to be imperceptibly different from the original for most viewers, for most images, for most of the time. In order to approach such fidelity, the algorithms are flexible, allowing for frequent adaptive changes in the algorithm depending on scene content, history of the processing, estimates of image complexity and perceptibility of distortions introduced by the compression.

Figure 2.14.2 shows the overall flow of signals in the ATSC DTV system. Note that analogue signals presented to the system are digitised and sent to the encoder for compression, and the compressed data then are transmitted over a communications channel. On being received, the compressed signal is decompressed in the decoder and reconstructed for display (following error correction, as necessary).

2.14.2.3 Video Pre-processing

Video pre-processing converts the analogue input signals to digital samples in the form needed for subsequent compression. The analogue input signals are typically composite for standard-definition signals or components consisting of luminance (Y) and chrominance (Cb and Cr) signals.

2.14.2.3.1 Video Compression Formats

Table 2.14.1 lists the compression formats allowed in the DTV standard.

In Table 2.14.1, 'vertical lines' refers to the number of active lines in the picture. 'Pixels' refers to the number of pixels during the active line. 'Aspect ratio' refers to the picture aspect ratio. 'Picture rate' refers to the number of frames or fields per second. In the values for picture rate, 'P' refers to progressive scanning and 'I' refers to interlaced scanning. Note that both

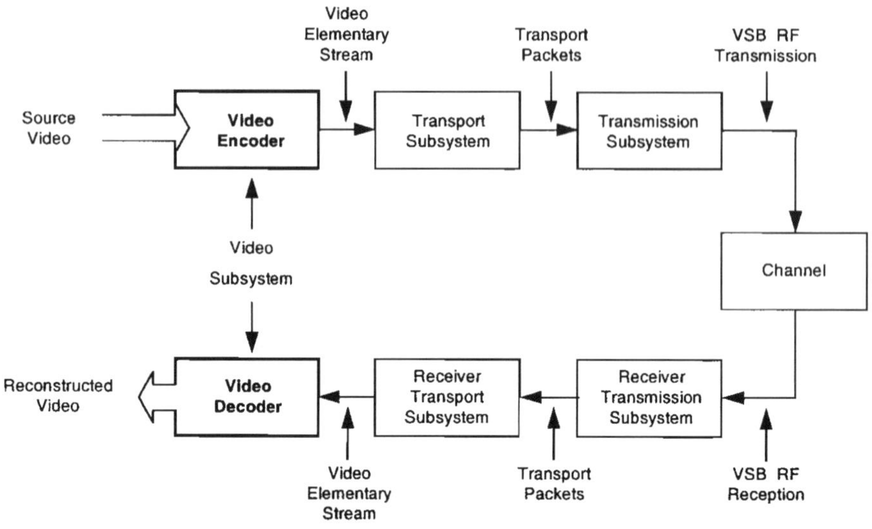

Figure 2.14.2 Video coding in relation to the ATSC DTV system.

Table 2.14.1 DTV compression formats

Vertical lines	*Pixels*	*Aspect ratio*	*Picture rate*
1080	1920	16:9	60I, 30P, 24P
720	1280	16:9	60P, 30P, 24P
480	704	16:9 and 4:3	60P, 60I, 30P, 24P
480	640	4:3	60P, 60I, 30P, 24P

60.00 Hz and 59.94 ($60 \times 1000/1001$) Hz picture rates are allowed. Dual rates are also allowed at the picture rates of 30 and 24 Hz.

Designers should be aware that a larger range of video formats is allowed under SCTE 43,[5] and that consumers may expect receivers to decode and display these as well. One format likely to be frequently encountered is 720 pixels by 480 lines, matching the format of ITU-R BT.601-5.[6]

2.14.2.3.2 *Possible Video Inputs*

While not required by the DTV standard, there are certain digital television production standards, shown in Table 2.14.2, that define video formats which relate to compression formats specified by the standard.

Table 2.14.2 Standardised video input formats

Video standard	*Active lines*	*Active samples/line*	*Picture rate*
SMPTE 274M–(1998)	1080	1920	24P, 30P, 60I
SMPTE 296M–(2001)	720	1280	24P, 30P, 60P
SMPTE 293M–(2003)	483	720	60P
ITU-R BT.601-5	483	720	60I

The compression formats may be derived from one or more appropriate video input formats. It may be anticipated that additional video production standards will be developed in the future that extend the number of possible input formats.

2.14.2.3.3 *Sampling Rates*

For the 1080-line format, with 1125 total lines per frame and 2200 total samples per line, the sampling frequency will be 74.25 MHz for the 30.00 frames per second (fps) frame rate. For the 720-line format, with 750 total lines per frame and 1650 total samples per line, the sampling frequency will be 74.25 MHz for the 60.00 fps frame rate. For the 480-line format using 704 pixels, with 525 total lines per frame and 858 total samples per line, the sampling frequency will be 13.5 MHz for the 59.94 Hz field rate. Note that both 59.94 fps and 60.00 fps are acceptable as frame or field rates for the system.

For both the 1080- and 720-line formats, other frame rates, specifically 23.976, 24.00, 29.97 and 30.00 fps rates, are acceptable as input to the system. The sample frequency will be either 74.25 MHz (for 24.00 and 30.00 fps) or 74.25/1.001 MHz for the other rates. The number of total samples per line is the same for either of the paired picture rates (see SMPTE 274M[7] and SMPTE 296M[8]).

The six frame rates noted here are the only allowed frame rates for the DTV standard. In this discussion, any references to 24 fps include both 23.976 and 24.00 fps, references to 30 fps include both 29.97 and 30.00 fps, and references to 60 fps include both 59.94 and 60.00 fps.

For the 480-line format, there may be 704 or 640 pixels in the active line. The interlaced formats are based on ITU-R BT.601-5;[6] the progressive formats are based on SMPTE 294M.[9]

If the input is based on ITU-R BT.601-5 or SMPTE 294M, it will have 483 active lines with 720 pixels in the active line.

Only 480 of the 483 active lines are used for encoding. Only 704 of the 720 pixels are used for encoding; the first eight and the last eight are dropped. The 480-line, 640-pixel picture format is not related to any current video production format. It does, however, correspond to the IBM VGA graphics format and may be used with ITU-R BT.601-5 sources by using appropriate resampling techniques.

2.14.2.3.4 Colorimetry

For the purposes of the DTV standard, 'colorimetry' means the combination of colour primaries, transfer characteristics and matrix coefficients. Video inputs conforming to SMPTE 274M and SMPTE 296M have the same colorimetry. Video inputs corresponding to ITU-R BT.601-5 should have SMPTE 170M colorimetry.[10]

2.14.2.4 Active Format Description (AFD)

The DTV standard makes provisions for conveying *Active Format Description* (AFD) data.[1] The term 'active format' in this context refers to that portion of the coded video frame containing 'useful information'. For example, when 16:9 aspect ratio material is coded in a 4:3 format (such as 480i), letter-boxing may be used to avoid cropping the left and right edges of the widescreen image. The black horizontal bars at the top and bottom of the screen contain no useful information, and in this case the AFD data would indicate 16:9 video carried inside the 4:3 rectangle. The AFD data solves a troublesome problem in the transition from conventional 4:3 display devices to wide-screen 16:9 displays, and also addresses the variety of aspect ratios that have been used over the years by the motion picture industry to produce feature films.

There are, of course, a number of different types of video displays in common usage – ranging from 4:3 CRTs to wide-screen projection devices and flat-panel displays of various design. Each of these devices may have varying abilities to process incoming video. In terms of input interfaces, these displays may likewise support a range of input signal formats – from composite analogue video to IEEE 1394.

Possible video source devices include cable, satellite or terrestrial broadcast set-top (or integrated receiver-decoder) boxes, media players (such as DVDs), analogue or digital VHS tape players and personal video recorders.

Although choice is good, this wide range of consumer options presented two problems to be solved:

1. No standard method had been agreed upon to communicate to the display device the 'active area' of the video signal. Such a method would be able, for example, to signal that the 4:3 signal contains within it a letter-boxed 16:9 video image.
2. No standard method had been agreed upon to communicate to the display device, for all interface types, that a given image is intended for 16:9 display.

The AFD solves these problems and, in the process, provides the following benefits:

- Active area signalling allows the display device to process the incoming signal to make the highest-resolution and most accurate picture possible. Furthermore, the display can take advantage of the knowledge that certain areas of video are currently unused and can implement algorithms that reduce the effects of uneven screen ageing.
- Aspect ratio signalling allows the display device to produce the best image possible. In some scenarios, lack of a signalling method translates to restrictions in the ability of the source device to deliver certain otherwise desirable output formats.

2.14.2.4.1 Active Area Signalling

A consumer device such as a cable or satellite set-top box cannot reliably determine the active area of video on its own. Even though certain lines at the top and bottom of the screen may be black for periods of time, the situation could change without warning. The only sure way to know active area is for the service provider to include this data at the time of video compression and to embed it into the video stream.

Figure 2.14.3 shows 4:3- and 16:9-coded images with various possible active areas. The group on the left is either coded explicitly in the MPEG-2 video syntax as 4:3, or the uncompressed signal provided in NTSC timing and aspect ratio information (if present) indicates 4:3. The group on the right is coded explicitly in the MPEG-2 video syntax as 16:9, provided with NTSC timing and an aspect ratio signal indicating 16:9, or provided uncompressed with 16:9 timing across the interface.

As can be seen in the figure, a pillar-boxed display results when a 4:3 active area is displayed within a 16:9 area, and a letter-boxed display results when a 16:9 active area is displayed within a 4:3 area. It is also apparent that double-boxing can occur, for example when 4:3 material is delivered within a 16:9 letter-box to a 4:3 display, or when 16:9 material is delivered within a 4:3 pillar-box to a 16:9 display.

For the straight letter- or pillar-box cases, if the display is aware of the active area it may take steps to mitigate the effects of uneven screen ageing. Such steps could, for example, involve using grey instead of black bars. Some amount of linear or non-linear stretching and/or zooming may be done as well using the knowledge that video outside the active area can safely be discarded.

2.14.3 Audio System

As illustrated in Figure 2.14.4, the audio subsystem comprises the audio encoding/decoding function and resides between the audio inputs/outputs and the transport subsystem. The audio encoder(s) is (are) responsible for generating the audio elementary stream(s), which are encoded representations of the baseband audio input signals. The flexibility of the transport system allows multiple audio elementary streams to be delivered to the receiver. At the receiver, the transport subsystem is responsible for selecting which audio streams(s) to deliver to the audio subsystem. The audio subsystem is responsible for decoding the audio elementary stream(s) back into baseband audio.

An audio program source is encoded by a digital television audio encoder. The output of the audio encoder is a string of bits that represent the audio source, and is referred to as an *audio elementary stream*. The transport subsystem packetises the audio data into PES packets, which are then further packetised into transport packets. The transmission subsystem converts the transport packets into a modulated RF signal for transmission to the receiver. At the receiver, the signal is demodulated by the receiver transmission subsystem. The receiver transport subsystem converts the received audio

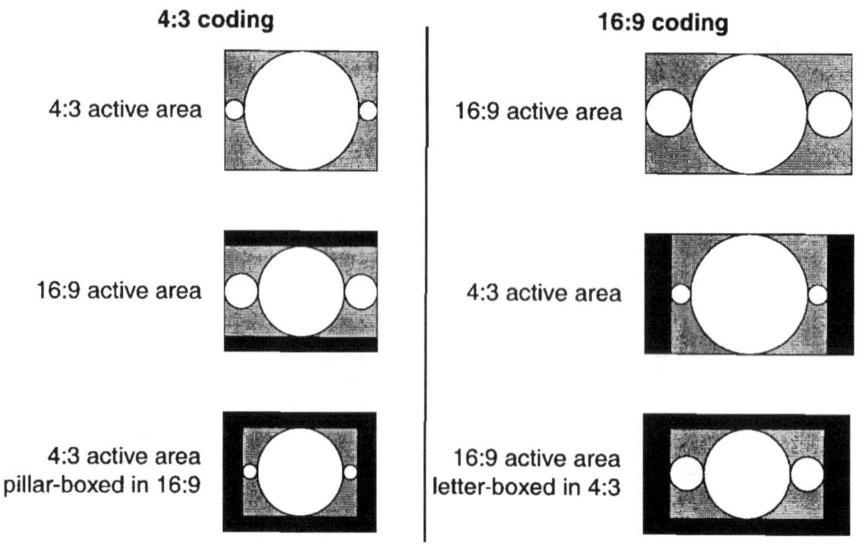

Figure 2.14.3 Coding and active area.

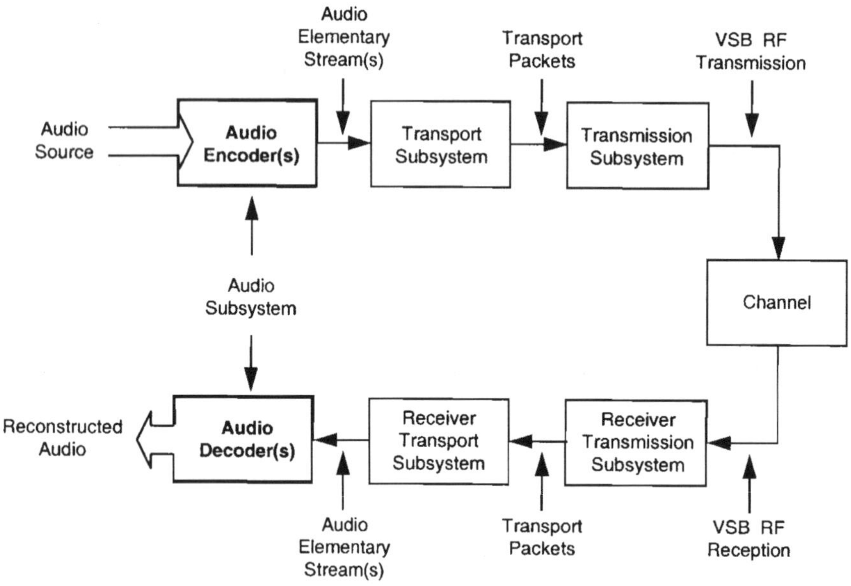

Figure 2.14.4 Audio subsystem within the DTV system.

packets back into an audio elementary stream, which is decoded by the digital television audio decoder. The partitioning shown is conceptual, and practical implementations may differ. For example, the transport processing may be broken into two blocks: one to perform PES packetisation and the second to perform transport packetisation. Alternatively, some of the transport functionality may be included in either the audio coder or the transmission subsystem.

2.14.3.1 Audio Encoder Interface

The audio system accepts baseband audio inputs with up to six audio channels per audio program bit stream. The channelisation is consistent with ITU-R Recommendation BS-775, 'Multi-Channel Stereophonic Sound System With and Without Accompanying Picture'. The six audio channels are: Left, Centre, Right, Left Surround, Right Surround and

Low-Frequency Enhancement (LFE). Multiple audio elementary bit streams may be conveyed by the transport system.

The bandwidth of the LFE channel is limited to 120 Hz. The bandwidth of the other (main) channels is limited to 20 kHz. Low-frequency response may extend to dc, but is more typically limited to approximately 3 Hz (−3 dB) by a dc blocking high-pass filter. Audio coding efficiency (and thus audio quality) is improved by removing dc offset from audio signals before they are encoded.

2.14.3.2 Input Source Signal Specification

Audio signals that are input to the audio system may be in analogue or digital form. Audio signals should have any dc offset removed before being encoded. If the audio encoder does not include a dc blocking high-pass filter, the audio signals should be high-pass filtered before being applied to the audio encoder. For analogue input signals, the input connector and signal level are not specified. Conventional broadcast practice may be followed. One commonly used input connector is the three-pin XLR female (the incoming audio cable uses the male connector) with pin 1 ground, pin 2 hot or positive and pin 3 neutral or negative.

For digital input signals, the input connector and signal format are not specified. Commonly used formats such as the AES3 two-channel interface may be used.[11] When multiple two-channel inputs are used, the preferred channel assignment is:

Pair 1: Left, Right
Pair 2: Centre, LFE
Pair 3: Left Surround, Right Surround

2.14.3.3 Sampling Frequency

The system conveys digital audio sampled at a frequency of 48 kHz, locked to the 27 MHz system clock. If analogue signal inputs are employed, the A/D converters should sample at 48 kHz. If digital inputs are employed, the input sampling rate should be 48 kHz, or the audio encoder should contain sampling rate converters that convert the sampling rate to 48 kHz. The sampling rate at the input to the audio encoder must be locked to the video clock for proper operation of the audio subsystem.

In general, input signals should be quantised to at least 16-bit resolution. The audio compression system can convey audio signals with up to 24-bit resolution.

2.14.3.4 Summary of Service Types

The following service types are defined in the Digital Audio Compression (AC-3) Standard[2] and in the ATSC Digital Television Standard:[1]

- **Complete Main Audio Service (CM).** This is the normal mode of operation. All elements of a complete audio program are present. The audio program may be any number of channels from 1 to 5.1.[§]
- **Main Audio Service, Music and Effects (ME).** All elements of an audio program are present except for dialogue. This audio program may contain from 1 to 5.1 channels. Dialogue may be provided by a D associated service (that may be simultaneously decoded and added to form a complete program).
- **Associated Service: Visually Impaired (VI).** This is typically a single-channel service, intended to convey a narrative description of the picture content for use by the visually impaired, and intended to be decoded along with the main audio service. The VI service also may be provided as a complete mix of all program elements, in which case it may use any number of channels (up to 5.1).
- **Associated Service: Hearing Impaired (HI).** This is typically a single-channel service, intended to convey dialogue that has been processed for increased intelligibility for the hearing impaired, and intended to be decoded along with the main audio service. The HI service also may be provided as a complete mix of all program elements, in which case it may use any number of channels (up to 5.1).
- **Associated Service: Dialogue (D).** This service conveys dialogue intended to be mixed into a main audio service (ME) that does not contain dialogue.
- **Associated Service: Commentary (C).** This service typically conveys a single channel of commentary intended to be optionally decoded along with the main audio service. This commentary channel differs from a dialogue service in that it contains optional instead of necessary program content. The C service also may be provided as a complete mix of all program elements, in which case it may use any number of channels (up to 5.1).
- **Associated Service: Emergency Message (E).** This is a single-channel service, which is given priority in reproduction. If this service type appears in the transport multiplex, it is routed to the audio decoder. If the audio decoder receives this service type, it will decode and reproduce the E channel while muting the main service.
- **Associated Service: Voice-Over (VO).** This is a single-channel service intended to be decoded and added into the centre loudspeaker channel.

2.14.3.5 Multi-Lingual Services

Each audio bit stream may be in any language. In order to provide audio services in multiple languages a number of main audio services may be provided, each in a different language. This is the (artistically) preferred method, because it allows unrestricted placement of dialogue along with the dialogue reverberation. The disadvantage of this method is that as much as 384 kbps is needed to provide a full 5.1-channel service for each language. One way to reduce the required bit rate is to reduce the number of audio channels provided for languages with a limited audience. For instance, alternative language versions could be provided in two-channel stereo with a bit rate of 128 kbps or a mono version could be supplied at a bit rate of approximately 64–96 kbps.

Another way to offer service in multiple languages is to provide a main multi-channel audio service (ME) that does not contain dialogue. Multiple single-channel dialogue associated services (D) can then be provided, each at a bit rate of approximately 64–96 kbps. Formation of a complete audio program requires that the appropriate language D service be simultaneously decoded and mixed into the ME service. This method allows a large number of languages to be efficiently

[§]5.1 channel sound refers to a service providing the following signals: right, centre, left, right surround, left surround, and low-frequency effects (LFE).

provided, but at the expense of artistic limitations. The single channel of dialogue would be mixed into the centre reproduction channel, and could not be panned. Also, reverberation would be confined to the centre channel, which is not optimum. Nevertheless, for some types of programming (sports, etc.) this method is attractive due to the savings in bit rate it offers. Some receivers may not have the capability to simultaneously decode an ME and a D service.

Stereo (two-channel) service without artistic limitation can be provided in multiple languages with added efficiency by transmitting a stereo ME main service along with stereo D services. The D and appropriate language ME services are simply combined in the receiver into a complete stereo program. Dialogue may be panned and reverberation may be included in both channels. A stereo ME service can be sent with high quality at 192 kbps, while the stereo D services (voice only) can make use of lower bit rates, such as 128 or 96 kbps per language. Some receivers may not have the capability to simultaneously decode an ME and a D service.

Note that, during those times when dialogue is not present, the D services can be momentarily removed, and their data capacity used for other purposes.

2.14.3.6 Audio Bit Rates

The information in Table 2.14.3 provides a general guideline as to the audio bit rates that are expected to be most useful. For main services, the use of the LFE channel is optional and will not affect the indicated data rates.

Table 2.14.3 Typical audio bit rates

Type of service	*Number of channels*	*Typical bit rates (kbps)*
CM, ME or associated audio service containing all necessary program elements	5	384–448
CM, ME or associated audio service containing all necessary program elements	4	320–384
CM, ME or associated audio service containing all necessary program elements	3	192–320
CM, ME or associated audio service containing all necessary program elements	2	128–256
VI, narrative only	1	64–128
HI, narrative only	1	64–96
D	1	64–128
D	2	96–192
C, commentary only	1	64–128
E	1	64–128
VO	1	64–128

The audio decoder input buffer size (and thus part of the decoder cost) is determined by the maximum bit rate that must be decoded. The syntax of the AC-3 standard supports bit rates ranging from a minimum of 32 kbps up to a maximum of 640 kbps per individual elementary bit stream. The bit rate utilised in the DTV system is restricted in order to reduce the size of the input buffer in the audio decoder, and thus the receiver cost. Receivers can be expected to support the decoding of a main audio service, or an associated audio service that is a complete service (containing all necessary program elements), at a bit rate up to and including 448 kbps. Transmissions may contain main audio services, or associated audio services that are complete services (containing all necessary program elements), encoded at a bit rate up to and including 448 kbps. Transmissions may contain single-channel associated audio services intended to be simultaneously decoded along with a main service encoded at a bit rate up to and including 128 kbps. Transmissions may contain dual-channel dialogue associated services intended to be simultaneously decoded along with a main service encoded at a bit rate up to and including 192 kbps. Transmissions have a further limitation that the combined bit rate of a main and an associated service that are intended to be simultaneously reproduced is less than or equal to 576 kbps.

2.14.3.7 DTV Transport

The transport subsystem employs the fixed-length transport stream packetisation approach defined in ISO/IEC 13818-1,[12] which is usually referred to as the MPEG-2 Systems Standard. This approach is well suited to the needs of terrestrial broadcast and cable television transmission of digital television. The use of relatively short, fixed-length packets matches well with the needs and techniques for error protection in both terrestrial broadcast and cable television distribution environments.

The ATSC DTV transport may carry a number of television programs. The MPEG-2 term 'program' corresponds to an individual digital TV channel or data service, where each program is composed of a number of MPEG-2 program elements (i.e. related video, audio and data streams). The MPEG-2 Systems Standard support for multiple channels or services within a single, multiplexed bit stream enables the deployment of practical, bandwidth-efficient digital broadcasting systems. It also provides great flexibility to accommodate the initial needs of the service to multiplex video, audio and data while providing a well-defined path to add additional services in the future in a fully backward-compatible manner. By basing the transport subsystem on MPEG-2, maximum interoperability with other media and standards is maintained.

Referring to Figure 2.14.1, the transport subsystem resides between the application (e.g. audio or video), encoding and decoding functions and the transmission subsystem. At its lowest layer, the encoder transport subsystem is responsible for formatting the encoded bits and multiplexing the different components of the program for transmission. At the receiver, it is responsible for recovering the bit streams for the individual application decoders and for the corresponding error signalling. The transport subsystem also incorporates other higher-level functionality related to identification of applications and synchronisation of the receiver.

2.14.4 PSIP

The 'Program and System Information Protocol' (PSIP) is data that is transmitted along with a station's DTV signal that tells DTV receivers important information about the station and what is being broadcast. Described in ATSC Standard A/65,[3] the most important function of PSIP is to provide a method for DTV receivers to identify a DTV station and to determine how a receiver can tune to it. PSIP identifies both the DTV channel and the associated NTSC (analogue) channel. It helps maintain the current channel branding because DTV receivers will electronically associate the two channels, making it easy for viewers to tune to the DTV station even if they do not know the RF channel number.

In addition to identifying the channel number, PSIP tells the receiver whether multiple program channels are being broadcast and, if so, how to find them. It identifies, for example, whether the programs are closed-captioned, conveys V-chip information, and if data is associated with the program.

2.14.4.1 PSIP Structure

PSIP is a collection of tables, each of which describes elements of typical digital television services.[13] Figure 2.14.5 shows the primary components and the notation used to describe them. The packets of the base tables are all labelled with a base *Packet Identifier* (PID). The base tables are:

- System Time Table (STT).
- Rating Region Table (RRT).
- Master Guide Table (MGT).
- Virtual Channel Table (VCT).

The Event Information Tables (EITs) are a second set of tables, whose packet identifiers are defined in the MGT. The Extended Text Tables (ETTs) are a third set of tables, and similarly, their PIDs are defined in the MGT.

The System Time Table is a small data structure that fits in one Transport Stream packet and serves as a reference for time-of-day functions. Receivers can use this table to manage various operations and scheduled events, as well as display time-of-day.

The Rating Region Table has been designed to transmit the rating system in use for each country using the ratings. In the United States, this is incorrectly but frequently referred to as the 'V-chip' system; the proper title is Television Parental Guidelines (TVPG). Provisions have been made for multi-country systems.

The Master Guide Table provides indexing information for the other tables that comprise the PSIP standard. It also defines table sizes necessary for memory allocation during decoding, defines version numbers to identify those tables that need to be updated, and generates the packet identifiers that label the tables.

The Virtual Channel Table, also referred to as the Terrestrial VCT (TVCT), contains a list of all the channels that are or will be on-line, plus their attributes. Among the attributes given are the channel name and channel number.

There are up to 128 Event Information Tables, EIT-0 to EIT-127, each of which describes the events or television

Figure 2.14.5 Overall structure of the PSIP tables.

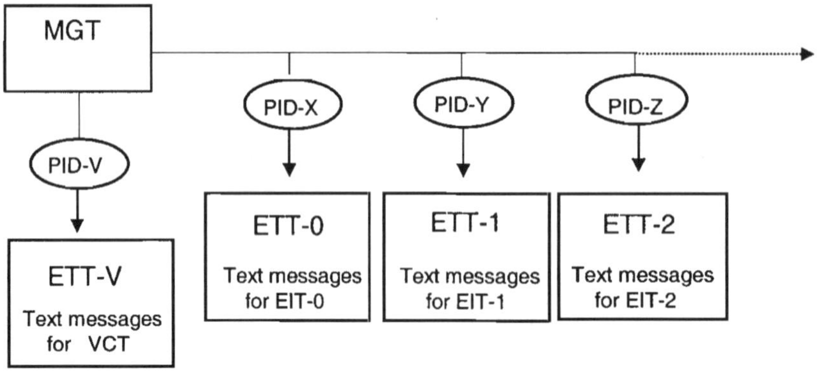

Figure 2.14.6 Extended text tables in the PSIP hierarchy.

programmes for a time interval of 3 hours. Because the maximum number of EITs is 128, up to 16 days of programming may be advertised in advance. At minimum, the first four EITs must always be present in every transport stream, and 24 are recommended. Each EIT-k may have multiple instances, one for each virtual channel in the VCT.

As illustrated in Figure 2.14.6, there may be multiple Extended Text Tables, one or more channel ETT sections describing the virtual channels in the VCT, and an ETT-k for each EIT-k, describing the events in the EIT-k. These are all listed in the MGT. An ETT-k contains a table instance for each event in the associated EIT-k. As the name implies, the purpose of the ETT is to carry text messages. For example, for channels in the VCT, the messages can describe channel information, cost, coming attractions and other related data. Similarly, for an event such as a movie listed in the EIT, the typical message would be a short paragraph that describes the movie itself. Extended Text Tables are optional in the ATSC system.

2.14.4.2 PSIP Requirements for Broadcasters

The three main tables (VCT, EIT, STT) contain information to facilitate suitably equipped receivers to find the components needed to present a program (event). Although receivers are expected to use stored information to speed channel acquisition, sometimes parameters must change and the VCT and EIT-0 are the tables that must be accurate each instant as they provide the actual connection path and critical information that can affect the display of the events. If nothing has changed since an EIT was sent for an event, then the anticipatory use of the data is expected to proceed, and when there is a change the new parts would be used. Additional tables provide TV parental advisory information and extended text messages about certain events. These relationships – and the tables that carry them – are designed to be kept with the DTV signal when it is carried by a cable system.

2.14.4.2.1 The Basics

There are certain 'must have' items and 'must do' rules of operation. If the PSIP elements are missing or wrong, there may be severe consequences, which vary depending on the design of receiver. The following are key elements that must be set and/or checked by each station:

- **Transport Stream Identification (TSID).** The pre-assigned TSIDs must be set correctly in all three locations (PAT, VCT common information and virtual channel-specific information).
- **System Time Table (SST).** The STT time should be checked daily and locked to house time. Ideally, the STT should be inserted into the TS within a frame before each seconds-count increment of the house time with the to-be-valid value.
- **Short Channel Name.** This is a seven-character name that can be set to any desired name indicating the virtual channel name. For example, a station's call letters followed by SD1, SD2, SD3 and SD4 to indicate various SDTV virtual channels or anything else to represent the station's identity (e.g. WNABSD1, KNABSD2, WNAB-HD, KIDS, etc.).
- **Major Channel.** The previously assigned, paired NTSC channel is the major channel number.
- **Service Type.** The service type selects DTV, NTSC, audio only, data, etc., and must be set as operating modes require.
- **Modulation Mode.** A code for the RF modulation of the virtual channel.
- **Source ID.** The Source ID is a number that associates virtual channels to events on those channels. It typically is automatically updated by PSIP equipment or updated from an outside vendor. Proper operation of this feature should be confirmed.
- **Service Location Descriptor (SLD).** Contains the MPEG references to the contents of each component of the programs plus a language code for audio. The PID values for the components identified here and in the PMT must be the same for the elements of an event/program. Some deployed systems require separate manual set-up, but PID values assigned to a VC should seldom change.

The maximum cycle time/repetition rate of the tables should be set or confirmed to conform with the suggested guidelines given in Table 2.14.4 for mandatory PSIP tables and Table 2.14.5 for optional PSIP tables.

It is recommended that broadcasters send populated EITs covering at least 3 days. The primary cycle time guidelines are illustrated in Figure 2.14.7.

Table 2.14.4 Mandatory PSIP table suggested repetition rates

PSIP table	*Transmission cycle*
MGT	Once every 150 ms
TVCT	Once every 400 ms
EIT-0	Once every 0.5 seconds
EIT-1	Once every 3 seconds
EIT-2 and EIT-3	Once every minute
STT	Once every second
RRT (not required in some areas*)	Once every minute

*The US is one such area.

Table 2.14.5 Suggested repetition rates for optional PSIP tables

PSIP table	*Transmission cycle*
A/65A specifies the following repetition rates for DCC per specified conditions.	
DCC request in progress	150 ms
DCCT	
2 seconds prior to DCC request	400 ms
No DCC	n/a
DCCSCT	Once per hour
ETT	Once every minute
EIT-4 and higher	Once every minute
DET	A later version of this Recommended Practice will address data services

The recommended table cycle times given in this section result in a minimal demand on overall system bandwidth. Considering the importance of the information that these PSIP tables provide to the receiver, the bandwidth penalty is trivial.

2.14.4.2.2 Program Guide

Support for an Electronic Program Guide (EPG) or Interactive Program Guide (IPG) is another important function enabled by the PSIP standard. The concept is to provide a way for viewers to find out 'what's on' directly from their television sets, similar to the *guide channels* that are typically available for cable and satellite broadcast services. In a terrestrial broadcast environment, there is no single authority that determines what programs are on all the channels, so each broadcaster needs to include this type of information within the broadcast stream. Viewers' receivers will be able to scan all the available channels and create a program guide channel from the consolidated information. The value of this type of guide information is high, and it will continue to increase in the DTV environment. Viewers will have the ability to choose not only what channel to watch, but also be able to select from multiple options within a broadcast. Examples include selecting from a set of alternative audio tracks in different languages, or choosing one of several SDTV programs shown at the same time on different virtual channels.

The more viewers rely on these guides, the more critical the accuracy and detail of the information they contain will become. Figure 2.14.8 shows what a typical electronic programme guide might look like.

2.14.4.2.3 Building the On-Screen Display or EPG

The EIT has the dual functionality of announcing future programs and providing critical information about the current program. It contains program names and planned broadcast times as well as other information about an event. The data can be combined to build a receiver on-screen display such as that shown in Figure 2.14.9.

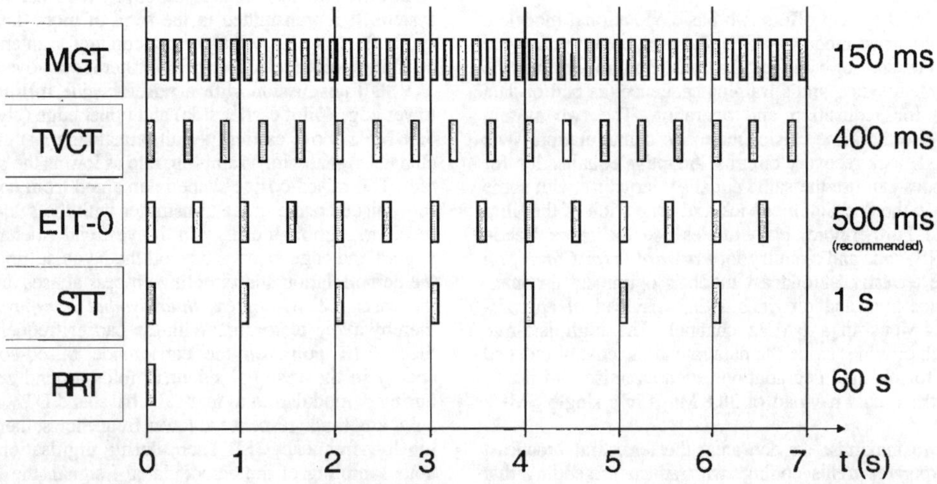

Figure 2.14.7 Recommended PSIP table cycle times.

Chan	Name	6:00 PM	6:30 PM	7:00 PM	7:30 PM	8:00 PM	8:30 PM
6-0	XYZ	City Scene		Travel Log		Movie: Speed II	
6-1	XYZ	City Scene		Travel Log		Movie: Speed II (HD)	
6-2	XYZ	Movie: Star Trek—The Voyage Home				Tune 6-1 for Movie: Speed II (HD)	
6-3	LNC	Local News		Airport Info		HD Program on 6-1	

Figure 2.14.8 Example electronic program guide.

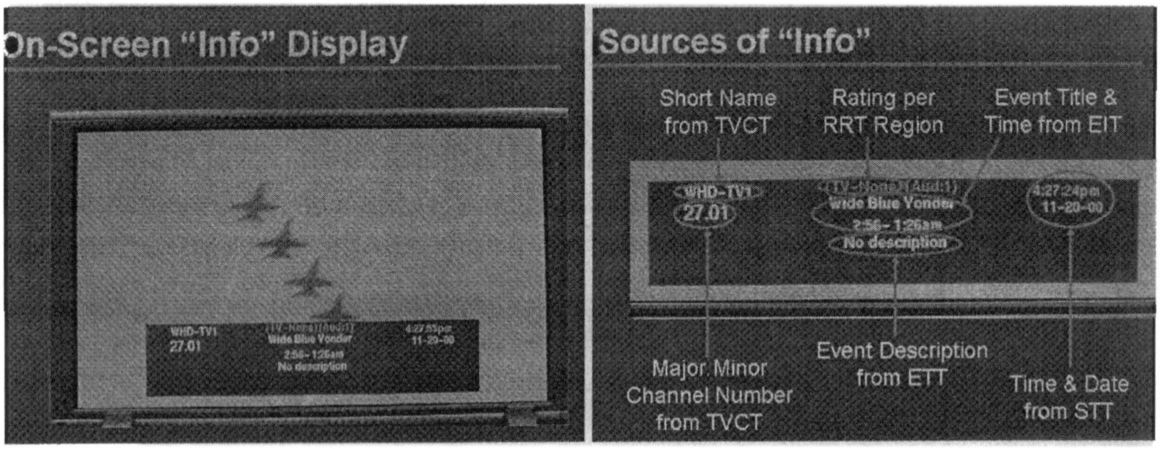

Figure 2.14.9 Illustration of how the various PSIP tables could combine to produce the on-screen display at the receiver.

2.14.5 RF Transmission

The ATSC VSB system offers two basic operational modes: a terrestrial broadcast mode (8-VSB) and a high-data-rate mode (16-VSB) intended for cable applications. Both modes provide a pilot, segment syncs, and a training sequence (as part of data field sync) for acquisition and operation. The two system modes can use the same carrier recovery, demodulation, sync recovery and clock recovery circuits. Adaptive equalisation for the two modes can use the same equaliser structure with some differences in the decision feedback and adaptation of the filter coefficients. Furthermore, both modes use the same Reed–Solomon (RS) code and circuitry for *Forward Error Correction* (FEC). The terrestrial broadcast mode is optimised for maximum service area and provides a data payload of approximately 19.4 Mbps in a 6 MHz channel. The high-data-rate mode, which provides twice the data rate at the cost of reduced robustness for channel degradations such as noise and multipath, provides a data payload of 38.8 Mbps in a single 6 MHz channel.

In order to maximise service area, the terrestrial broadcast mode incorporates trellis coding, with added pre-coding that allows the data to be decoded after passing through a receiver comb filter, used selectively to suppress analogue co-channel interference. The high-data-rate mode is designed to work in a cable environment, which is less severe than that of the terrestrial system. It is transmitted in the form of more data levels (bits/symbol). No trellis coding or pre-coding for an analogue broadcast interference rejection (comb) filter is employed in this mode.

VSB transmission with a raised-cosine roll-off at both the lower edge (pilot carrier side) and upper edge (Nyquist slope at 5.38 MHz above carrier) permits equalising just the in-phase (I) channel signal with a sampling rate as low as the symbol (baud) rate. The raised-cosine shape is obtained from concatenating a root-raised cosine in the transmitter with the same shape in the receiver. Although energy in the vestigial sideband and in the upper band edge extends beyond the Nyquist limit frequencies, the demodulation and sampling process aliases this energy into the baseband to suppress *Inter-Symbol Interference* (ISI) and thereby avoid distortion. With the carrier frequency located at the −6 dB point on the carrier-side raised-cosine roll-off, energy in the vestigial sideband folds around zero frequency during demodulation to make the baseband DTV signal exhibit a flat amplitude response at lower frequencies, thereby suppressing low-frequency ISI. Then, during digitisation by synchronous sampling of the demodulated I signal, the Nyquist slope through 5.38 MHz suppresses the remnant higher-frequency ISI. With ISI due to aliasing thus eliminated, equalisation of linear distortions can be done using a single A/D converter sampling at the symbol rate of 10.76 Msamples/s and a real-only (not complex) equaliser also operating at the symbol rate. In this simple case, equalisation of the signal beyond the −6 dB points in the raised-cosine roll-offs at channel edges is dependent on the in-band equalisation and cannot be set independently. A complex equaliser does not have this limitation, nor does a fractional equaliser sampling at a rate sufficiently above the symbol rate.

The 8-VSB signal is designed to minimise interference and RF channel allocation problems. The VSB signal is designed to minimise peak-energy-to-average-energy ratio, thereby minimising interference into other signals, especially adjacent and 'taboo' channels. To counter the man-made noise that often accompanies over-the-air broadcast signals, the VSB system includes an interleaver that allows correction of an isolated single burst of noise up to 190 microseconds in length by the (207,187) RS FEC circuitry, which locates as well as corrects up to 10-byte errors per data segment. This was done to allow VHF channels, which are often substantially affected by man-made noise, to be used for DTV broadcasting. If soft-decision techniques are used in the trellis decoder preceding the RS circuitry, the location of errors can be flagged, and twice as many byte errors per data segment can be corrected, allowing correction of an isolated burst of up to 380 microseconds in length.

The parameters for the two VSB transmission modes are shown in Table 2.14.6.

Table 2.14.6 Parameters for VSB transmission modes

Parameter	*Terrestrial mode*	*High-data-rate mode*
Channel bandwidth (MHz)	6	6
Excess bandwidth (%)	11.5	11.5
Symbol rate (Msymbols/s)	10.76	10.76
Bits per symbol	3	4
Trellis FEC	2/3 rate	None
Reed–Solomon FEC	$T = 10$ (207,187)	$T = 10$ (207,187)
Segment length (symbols)	832	832
Segment sync	4 symbols per segment	4 symbols per segment
Frame sync	1 per 313 segments	1 per 313 segments
Payload data rate (Mbps)	19.39	38.78
Analogue co-channel rejection	Analogue rejection filter in receiver	N/A
Pilot power contribution (dB)	0.3	0.3
C/N threshold (dB)	14.9	28.3

2.14.5.1 Bit Rate Delivered to a Transport Decoder by the Transmission Subsystem

As outlined previously, all data in the ATSC DTV system is transported in MPEG-2 transport packets. The useful data rate is the amount of MPEG-2 transport data carried end-to-end, including MPEG-2 packet headers and sync bytes. The exact symbol rate of the transmission subsystem is given by:

$$\frac{4.5}{286} \times 684 = 10.7\ldots \text{ million symbols/second (megabaud)} \qquad (2.14.1)$$

The symbol rate must be locked in frequency to the transport rate.

The numbers in the formula for the ATSC symbol rate in 6 MHz systems are related to NTSC scanning and colour frequencies. Because of this relationship, the symbol clock can be used as a basis for generating an NTSC colour subcarrier for analogue output from a set-top box. The repetition rates of data segments and data frames are deliberately chosen not to have an integer relationship to NTSC or PAL scanning rates, to ensure that there will be no discernible pattern in co-channel interference.

The particular numbers used are:

- 4.5 MHz = the centre frequency of the audio carrier offset in NTSC. This number was traditionally used in NTSC literature to derive the colour subcarrier frequency and scanning rates. In modern equipment, this may start with a precision 10 MHz reference, which is then multiplied by 9/20.
- 4.5 MHz/286 = the horizontal scan rate of NTSC, 15,734.2657 + ... Hz (note that the colour subcarrier is 455/2 times this, or 3,579,545 + 5/11 Hz).
- 684. This multiplier gives a symbol rate for an efficient use of bandwidth in 6 MHz. It requires a filter with Nyquist roll-off that is a fairly sharp cut-off (11% excess bandwidth), which is still realisable with a reasonable surface acoustic wave (SAW) filter or digital filter.

In the terrestrial broadcast mode, channel symbols carry 3 bits/symbol of trellis-coded data. The trellis code rate is 2/3, providing 2 bits/symbol of gross payload. Therefore the gross payload is:

$$10.76\ldots \times 2 = 21.52\ldots \text{ Mbps (megabits/second)} \qquad (2.14.2)$$

To find the net payload delivered to a decoder it is necessary to adjust equation (2.14.2) for the overhead of the *data segment sync, data field sync* and Reed–Solomon FEC.

To get the net bit rate for an MPEG-2 stream carried by the system (and supplied to an MPEG transport decoder), it is first noted that the MPEG sync bytes are removed from the data stream input to the 8-VSB transmitter and replaced with segment sync, and later reconstituted at the receiver. For throughput of MPEG packets (the only allowed transport mechanism), segment sync is simply equivalent to transmitting the MPEG sync byte, and does not reduce the net data rate. The net bit rate of an MPEG-2 stream carried by the system and delivered to the transport decoder is accordingly reduced by the data field sync (one segment of every 313) and the Reed–Solomon coding (20 bytes of every 208):

$$21.52\ldots \text{ Mbps} \times \frac{312}{313} \times \frac{188}{208} = 19.39\ldots \text{ Mbps} \qquad (2.14.3)$$

The net bit rate supplied to the transport decoder for the high data rate mode is:

$$19.39\ldots \text{ Mbps} \times 2 = 38.78\ldots \text{ Mbps} \qquad (2.14.4)$$

2.14.5.2 Performance Characteristics of Terrestrial Broadcast Mode

The terrestrial 8-VSB system can operate in a signal-to-additive-white-Gaussian-noise (S/N) environment of 14.9 dB. The 8-VSB segment error probability curve, including four-state trellis decoding and (207,187) Reed–Solomon decoding, in Figure 2.14.10 shows a segment error probability of 1.93×10^{-4}. This is equivalent to 2.5 segment errors/second, which was established by measurement as the *Threshold Of Visibility* (TOV) of errors in the prototype equipment.[14] Particular product designs may achieve somewhat better performance for subjective TOV by means of error masking.

The *Cumulative Distribution Function* (CDF) of the peak-to-average power ratio, as measured on a low-power transmitted signal with no non-linearities, is plotted in Figure 2.14.11.

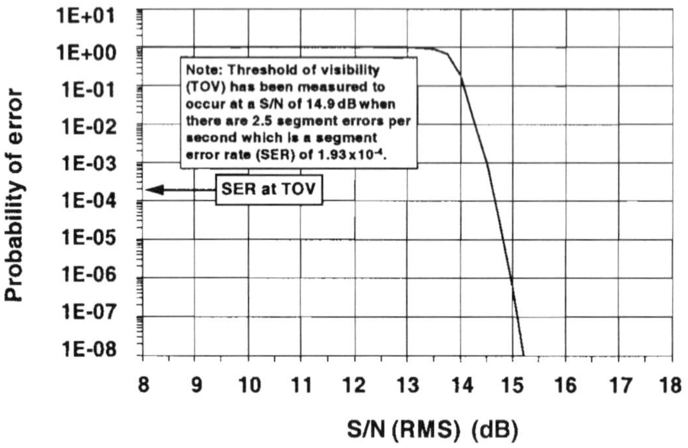

Figure 2.14.10 Segment error probability, 8-VSB with four-state trellis decoding, RS (207,187).

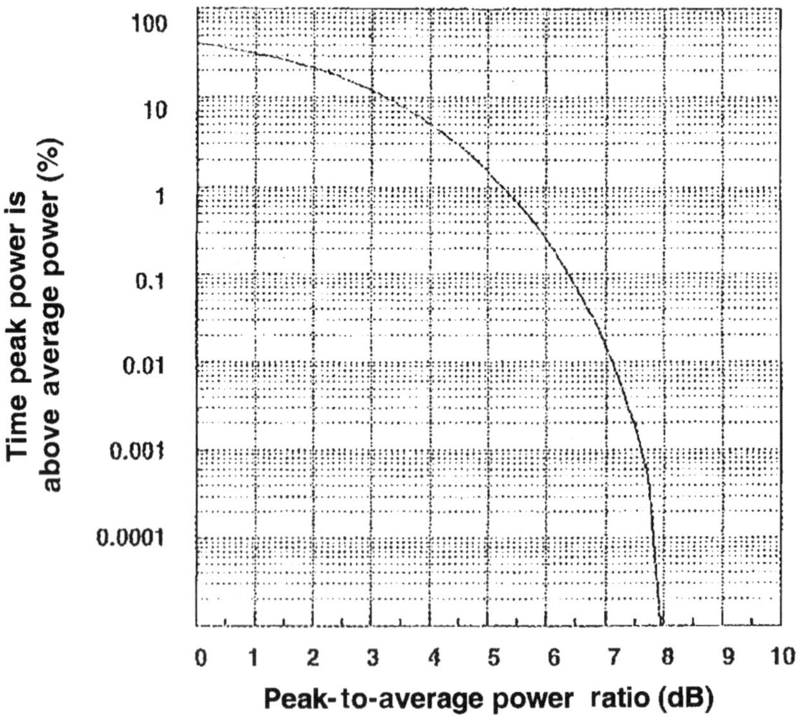

Figure 2.14.11 Cumulative distribution function of 8-VSB peak-to-average power ratio (in ideal linear system).

The plot shows that 99.9% of the time the transient peak power is within 6.3 dB of the average power.

2.14.5.3 Transmitter Signal Processing

A pre-equaliser filter is recommended for use in over-the-air broadcasts where the high-power transmitter may have significant in-band ripple or significant roll-off at band edges. Pre-equalisation is typically required in order to compensate the high-order filter used to meet a stringent out-of-band emission mask, such as the US FCC required mask.[15] This linear distortion can be measured by an equaliser in a reference demodulator ('ideal' receiver) employed at the transmitter site. A directional coupler, which is recommended to be located at the sending end of the antenna feed transmission line, supplies the reference demodulator a small sample of the antenna signal feed. The equaliser tap weights of the reference demodulator are transferred to the transmitter pre-equaliser for pre-correction of transmitter linear distortion. This is a one-time procedure of measurement and transmitter pre-equaliser adjustment. Alternatively, the transmitter pre-equaliser can be made continuously adaptive. In this arrangement, the reference demodulator is provided with a fixed-coefficient equaliser compensating for its own deficiencies in ideal response.

A pre-equaliser suitable for many applications is an 80-tap, feed–forward transversal filter. The taps are symbol-spaced (93 ns) with the main tap being approximately at the centre, giving approximately ±3.7 microsecond correction range. The pre-equaliser operates on the I channel data signal (there is no Q channel data signal in the transmitter), and shapes the frequency spectrum of the IF signal so that there is a flat in-band spectrum at the output of the high-power transmitter that feeds the antenna for transmission. There is no effect on the out-of-band spectrum of the transmitted signal. If desired, complex equalisers or fractional equalisers (with closer-spaced taps) can provide independent control of the outer portions of the spectrum (beyond the Nyquist slopes).

The transmitter vestigial sideband filtering is sometimes implemented by sideband cancellation, using the phasing method. In this method, the baseband data signal is supplied to digital filtering that generates in-phase and quadrature-phase digital modulation signals for application to respective D/A converters. This filtering process provides the root-raised cosine Nyquist filtering and provides compensation for the (sin x)/x frequency responses of the D/A converters as well. The baseband signals are converted to analogue form. The in-phase signal modulates the amplitude of the IF carrier at zero degrees phase, while the quadrature signal modulates a 90-degree shifted version of the carrier. The amplitude-modulated quadrature IF carriers are added to create the vestigial sideband IF signal, cancelling the unwanted sideband and increasing the desired sideband by 6 dB.

2.14.5.4 Up-converter and RF Carrier Frequency Offsets

Modern analogue TV transmitters use a two-step modulation process. The first step usually is modulation of the data onto an IF carrier, which is the same frequency for all channels, followed by translation to the desired RF channel. The digital 8-VSB transmitter applies this same two-step modulation process. The RF up-converter translates the filtered flat IF data signal spectrum to the desired RF channel. For the same approximate coverage as an analogue transmitter (at the same frequency), the average power of the DTV signal is of the order of 12 dB less than the analogue peak sync power (when operating on the same frequency).

The nominal frequency of the RF up-converter oscillator in DTV terrestrial broadcasts will typically be the same as that used for analogue transmitters (except for offsets required in particular situations).

Note that all examples in this section relate to a 6 MHz DTV system. Values may be modified easily for other channel widths.

2.14.5.5 Nominal DTV Pilot Carrier Frequency

The nominal DTV pilot carrier frequency is determined by fitting the DTV spectrum symmetrically into the RF channel. This is obtained by taking the bandwidth of the DTV signal – 5381.1189 kHz (the Nyquist frequency difference or one-half the symbol clock frequency of 10,762.2378 kHz) – and centering it in the 6 MHz TV channel. Subtracting 5381.1189 kHz from 6000 kHz leaves 618.881119 kHz. Half of that is 309.440559 kHz, precisely the standard pilot offset above the lower channel edge. For example, on channel 45 (656–662 MHz), the nominal pilot frequency is 656.309440559 MHz.

2.14.5.6 Requirements for Offsets

There are two categories of requirements for pilot frequency offsets:

1. Offsets to protect lower adjacent channel analogue broadcasts, mandated by FCC rules in the United States, and which override other offset considerations.
2. Recommended offsets for other considerations such as co-channel interference between DTV stations or between DTV and analogue stations.

2.14.5.6.1 Upper DTV Channel into Lower Analogue Channel

This is the overriding case mandated by the FCC rules in the United States – precision offset with a lower adjacent analogue station, full service or Low-Power Television (LPTV).

The FCC rules, Section 73.622(g)(1), state that:

> "DTV stations operating on a channel allotment designated with a 'c' in paragraph (b) of this section must maintain the pilot carrier frequency of the DTV signal 5.082138 MHz above the visual carrier frequency of any analog TV broadcast station that operates on the lower adjacent channel and is located within 88 kilometers. This frequency difference must be maintained within a tolerance of ±3 Hz."

This precise offset is necessary to reduce the colour beat and high-frequency luminance beat created by the DTV pilot carrier in some receivers tuned to the lower adjacent analogue channel. The tight tolerance assures that the beat will be visually cancelled, since it will be out of phase on successive video frames.

Note that the frequency is expressed with respect to the lower adjacent analogue video carrier, rather than the nominal channel edge. This is because the beat frequency depends on this relationship, and therefore the DTV pilot frequency must track any offsets in the analogue video carrier frequency. The offset in the FCC rules is related to the particular horizontal scanning rate of NTSC, and can easily be modified for PAL. The offset O_f was obtained from:

$$O_f = 455 \times \left(\frac{F_h}{2}\right) + 191 \times \left(\frac{F_h}{2}\right) - 29.97 = 5{,}082{,}138\,\text{Hz} \tag{2.14.5}$$

where F_h = NTSC horizontal scanning frequency = 15,734.264 Hz.

The equation indicates that the offset with respect to the lower adjacent chroma is an odd multiple (191) of one-half the line rate to eliminate the colour beat. However, this choice leaves the possibility of a luma beat. The offset is additionally adjusted by one-half the analogue field rate to eliminate the luma beat. While satisfying the exact adjacent channel criteria, this offset is also as close as possible to optimal comb filtering of the analogue co-channel in the digital receiver. Note additionally that offsets are to higher frequencies rather than lower, to avoid any possibility of encroaching on the lower adjacent sound. (It also reduces the likelihood of the Automatic Fine Tuning (AFT) in the analogue receiver experiencing lock-out because the signal energy including the pilot is moved further from the analogue receiver bandpass.)

2.14.5.6.2 *Other Offset Cases*

The FCC rules do not consider other interference cases where offsets help. The offset for protecting lower-adjacent analogue signals takes precedence. If that offset is not required, other offsets can minimise interference to co-channel analogue or DTV signals.

In co-channel cases, DTV interference into analogue TV appears noise-like. The pilot carrier is low on the Nyquist slope of the IF filter in the analogue receiver, so no discernible beat is generated. In this case, offsets to protect the analogue channel are not required. Offsets are useful, however, to reduce co-channel interference from analogue TV into DTV. The performance of the analogue rejection filter and clock recovery in the DTV receiver will be improved if the DTV carrier is 911.944 kHz below the NTSC visual carrier. In other words, in the case of a 6 MHz NTSC system, if the analogue TV station is not offset, the DTV pilot carrier frequency will be 338.0556 kHz above the lower channel edge instead of the nominal 309.44056 kHz. As before, if the NTSC station is operating with a ±10 kHz offset, the DTV frequency will have to be adjusted in the same direction. The formula for calculating this offset is:

$$F_{pilot} = F_{vis(n)} - 70.5 \times F_{seg} = 338.0556\,\text{Hz} \quad \text{(for no NTSC analogue offset)} \tag{2.14.6}$$

where:

F_{pilot} = DTV pilot frequency above lower channel edge
$F_{vis(n)}$ = NTSC visual carrier frequency above lower channel edge
= 1250 kHz for no NTSC offset (as shown)
= 1240 kHz for minus offset
= 1260 kHz for plus offset
F_{seg} = ATSC data segment rate = symbol clock frequency/832 = 12,935.381971 Hz.

The factor of 70.5 is chosen to provide the best overall comb filtering of analogue colour TV co-channel interference. The use of a value equal to an integer +0.5 results in co-channel analogue TV interference being out of phase on successive data segment syncs.

Note that in this case the frequency tolerance is ±1 kHz. More precision is not required. Also note that a different data segment rate would be used for calculating offsets for 7 or 8 MHz systems.

2.14.5.6.3 *Co-channel DTV into DTV*

If two DTV stations share the same channel, interference between the two stations can be reduced if the pilot is offset by one and a half times the data segment rate. This ensures that the frame and segment syncs of the interfering signal will each alternate polarity and be averaged out in the receiver tuned to the desired signal.

The formula for this offset is:

$$F_{offset} = 1.5 \times F_{seg} = 19.4031\,\text{kHz} \tag{2.14.7}$$

where:

F_{offset} = offset to be added to one of the two DTV carriers
F_{seg} = 12,935.381971 Hz (as defined previously).

This results in a pilot carrier 328.84363 kHz above the lower band edge, provided neither DTV station has any other offset.

Use of the factor 1.5 results in the best co-channel rejection, as determined experimentally with the prototype equipment. The use of an integer +0.5 results in co-channel interference alternating phase on successive segment syncs.

2.14.5.7 Summary: DTV Frequency

Table 2.14.7 summarises the various pilot carrier offsets for different interference situations in a 6 MHz system (NTSC environment). Note that if more than two stations are involved the number of potential frequencies will increase. For example, if one DTV station operates at an offset because of a lower-adjacent-channel NTSC station, a co-channel DTV station may have to adjust its frequency to maintain a 19.403 kHz pilot offset. If the NTSC analogue station operates at an offset of ±10 kHz, both DTV stations should compensate for that. Cooperation between stations will be essential in order to reduce interference.

2.14.5.8 Frequency Tolerances

The tightest specification is for a DTV station with a lower adjacent NTSC analogue station. If both NTSC and DTV stations are at the same location, they may simply be locked to the same reference. The co-located DTV station carrier should be 5.082138 MHz above the NTSC visual carrier (22.697 kHz above the normal pilot frequency). The co-channel DTV station should set its carrier 19.403 kHz above the co-located DTV carrier.

If there is interference with a co-channel DTV station, the analogue station is expected to be stable within 10 Hz of its assigned frequency.

While it is possible to lock the frequency of the DTV station to the relevant NTSC station, this may not be the best option if the two stations are not at the same location. It will likely be easier to maintain the frequency of each station within the

Table 2.14.7 DTV pilot carrier frequencies for two stations (normal offset above lower channel edge: 309.440559 kHz)

Channel relationship	*DTV pilot carrier frequency above lower channel edge*			
	NTSC station zero offset	*NTSC station +10 kHz offset*	*NTSC station −10 kHz offset*	*DTV station no offset*
DTV with lower adjacent NTSC	332.138 kHz ± 3 Hz	342.138 kHz ± 3 Hz	322.138 kHz ± 3 Hz	
DTV co-channel with NTSC	338.056 kHz ± 1 kHz	348.056 kHz ± 1 kHz	328.056 kHz ± 1 kHz	
DTV co-channel with DTV	+19.403 kHz above DTV	+19.403 kHz above DTV	+19.403 kHz above DTV	328.8436 kHz ± 10 Hz

necessary tolerances. Where co-channel interference is a problem, that will be the only option.

In cases where no type of interference is expected, a pilot carrier frequency tolerance of ±1 kHz is acceptable, but in all cases, good practice is to use a tighter tolerance if practicable.

References

1. ATSC Standard A/53B, ATSC Digital Television Standard, Advanced Television Systems Committee, Washington, DC (7 August 2001).
2. ATSC Standard A/52A, Digital Audio Compression (AC-3), Advanced Television Systems Committee, Washington, DC (20 August 2001).
3. ATSC Standard A/65B, Program and System Information Protocol, Advanced Television Systems Committee, Washington, DC (2003).
4. ISO/IEC IS 13818-2, International Standard, MPEG-2 Video (1996).
5. ANSI/SCTE 43, Digital Video Systems Characteristics Standard for Cable Television, Society of Cable Telecommunications Engineers, Exton, PA.
6. ITU-R BT.601-5, Encoding Parameters of Digital Television for Studios.
7. SMPTE 274M, Standard for Television – 1920 × 1080 Scanning and Analog and Parallel Digital Interfaces for Multiple Picture Rates, Society of Motion Picture and Television Engineers, White Plains, NY (1998).
8. SMPTE 296M, Standard for Television – 1280 × 720 Progressive Image Sample Structure, Analog and Digital Representation and Analog Interface, Society of Motion Picture and Television Engineers, White Plains, NY (2001).
9. SMPTE 294M, 720 × 483 Active Line at 59.94-Hz Progressive Scan Production – Bit-Serial Interfaces, Society of Motion Picture and Television Engineers, White Plains, NY (2001).
10. SMPTE 170M, Standard for Television – Composite Analog Video Signal, NTSC for Studio Applications, Society of Motion Picture and Television Engineers, White Plains, NY (1999).
11. AES3, Serial Transmission Format for Two-Channel Linearly Represented Digital Audio Data, Audio Engineering Society, New York, NY (1997).
12. ISO/IEC IS 13818-1 (E), International Standard, Information Technology – Generic Coding of Moving Pictures and Associated Audio Information: Systems (2000).
13. ATSC Recommended Practice A/69, Program and System Information Protocol Implementation Guidelines for Broadcasters, Advanced Television Systems Committee, Washington, DC (25 June 2002).
14. ATSC Recommended Practice A/54A, Guide to the Use of the Digital Television Standard, Advanced Television Systems Committee, Washington, DC (2003).
15. FCC, Memorandum Opinion and Order on Reconsideration of the Sixth Report and Order, Federal Communications Commission, Washington, DC (17 February 1998).

Peter Weitzel BSc (Eng) CEng MIEE MSMPTE
Manager, Broadcast Technology Developments
BBC Technology

2.15 Interactive TV

2.15.1 Introduction

'Interactive TV' can mean a variety of things to each broadcaster, and for the engineers covers a wide range of technologies. There are two main categories of service when interacting with the TV:

- Enhanced TV, where streams of information are ancillary to the programme and the selection is made in the TV or Set-Top Box (STB).
- Interactive TV, where additional information is provided and can be selected in the TV or STB but may engage on a one-to-one basis with that viewer. This implies a back channel from the TV or STB to a central server.

Both forms of interactivity use text as well as the moving pictures and sound that the television normally provides.

Although the roots of interactivity lie in the technology, it is the benefits to the viewer that force the business case for its adoption. Taking a viewer-centric view is key to an understanding of what interactivity can do to enhance a programme, provide more information or increase revenues. Thus, the way in which a service is navigated is of key importance to maximise its friendliness and therefore use.

2.15.2 Interactive TV

Interactive TV is the selection by the viewer of additional information in a compelling way which keeps them engaged and gives them a personal service either from broadcast or on demand material. Interactivity in TV usually drags the viewer away from the linear programming, by providing different information which is not connected with the programme.

2.15.2.1 'Unidirectional/Wireless'

The bulk of systems use the large broadcast 'pipe to the home' to provide additional content, usually in some form of carousel of pages or screens of information which may be cached in the STB/TV set to provide a near instantaneous access.

2.15.2.1.1 Teletext

Teletext is at the foundation of interactive TV, and being launched in 1972 predates the terminology. A collaboration between the BBC (Ceefax), IBA/ITV (Oracle) and the British Post office (Viewdata – with a telephone connection), the simple 24 rows of text with 40 non-proportionally spaced characters and only eight colours has been a great success both in Public Service broadcasting, and in the UK and a few other countries for advertising. About one-third of the holidays booked in the UK are selected on a teletext service.

Teletext is a transport mechanism using 'spare' capacity at the top of an analogue TV picture (lines 6/3190 to 22/225). Editorially its structure of up to eight magazines of up to 256 page slots (each of which could have 4k subpages individually addressed) give a large canvas on which to carry news, sport, TV listings, weather and other information. With limited display capabilities, it also found favour with tabular information such as stock exchange prices and airport landing information via automated feeds. In addition, it is used to carry subtitles and other ancillary services.

Although textual in nature, teletext (in Level 1.5) has basic graphical shapes based on a 3 × 2 grid within each character space. These mosaic shapes can be either contiguous or each 'blob' can be separated from each other. This can give some dramatic if rather crude graphics and also weather maps, etc. It has a Latin font, with 96 alphabetic characters, of which 13 can be selected on a page-by-page basis to be from one of 11 national character sets covering the languages of Eastern and

Western Europe. In addition, it has Arabic, Hebrew and Cyrillic fonts, which are used in the Middle Eastern countries.

Navigation splits into two main methods – the Germanic countries use Table Of Pages (TOP), which uses a 'filing cabinet' type of model, while the rest of the world uses Fastext (FLOF), where the coloured buttons at the bottom of the page move the viewer onto any page in the service. Various conventions have been established – the red button being 'Next'.

There are two types of transmission. The first is serial, where all the pages from any magazine are transmitted one after the other on all the TV lines allocated for transmission (usually with TOP navigation). The other is parallel, where pages from one or more magazine are transmitted on one or more TV lines – this gives better control of page cycle times.

At a data rate of about 2 page slots per second per TV line pair, and a typical cycle time of about 15–20 seconds – there are probably about 500 page slots in transmission on most services, sometimes supporting up to 20 subpages which are transmitted in sequence in the slot; thus there is a total of perhaps 2000–4000 pages in the service.

Teletext, although an old technology, received a rewording of its specification and the expansion to more flexible and colourful display in the standard ETS 300-706 (available from www.etsi.org). This standard is also a very good primer on how teletext works.

As well as the text service, teletext can be the transport mechanism for data services (ETS 300-708), which includes the control of Home VCRs by Programme Delivery Control (PDC) (ETS 300-231).

Thus, there is ample opportunity for services that may be used in the 'digital world' to be tried out – both editorially and in terms of viewer expectation—in the analogue world. This prototyping is an important prerequisite of any move into more advanced forms of interactive TV.

Finally, at an engineering level, teletext is the only signal which has its form in the analogue, Serial Digital Video (BT.601-5) and MPEG worlds, and thus may have a place as a data service for broadcasters intimately linked to the programme video for more years than public consumption of a text service.

2.15.2.1.2 Digital Text

Digital text services are a natural evolution of analogue teletext. Each platform's implementations (DTT, DCable, DSAT, xDSL) have their individual strengths and weaknesses.

In comparison to analogue teletext, all platforms provide significantly improved text quality. In most cases the addition of images, animation or even video and VOD are possible. Another advantage is the ability for the user to view only the page they want. Although digital text services do provide multiple pages for individual items, the user now controls when the next pages are accessed. This is a significant step forward in functionality.

While the new functionality may offer advantages from a presentational point of view, it does create issues. The main problem centres on bandwidth. The bandwidth limitations of DSAT and DTT create the need for a balance between textual content and other features. In many cases images are sacrificed in order to improve the speed of a service or increase the number of text pages available. Theoretically DCable and xDSL do not suffer the problem of limited page numbers, since this type of service is based on a client-server model. However, most wired platforms do allocate capped bandwidth to digital text services, which creates the same issues of speed over functionality and size of file. Popular services may also suffer from contention issues.

Another issue to consider relates to end-user hardware. The infrastructure may have enough bandwidth to handle distribution, but can the users' STBs render the on-screen display at an acceptable speed?

2.15.2.2 Back channel

For a digital TV platform to be considered fully interactive it requires some form of return path that the viewers can use. Voting, gambling, shopping, quizzes, etc. have all proved to be popular uses of the return path and offer significant revenue streams for broadcasters.

Cable and xDSL platforms have these as standard, but platforms using wireless transmission (satellite and terrestrial) usually use the phone network. The return path significantly increases the type of interactivity available.

Although the return path bandwidth available via the dial-up phone network is small in comparison to that of cable or xDSL, it does have the advantage of automatic billing. Any call made can be charged at standard or premium rates. This has led to landline-based and mobile phone (SMS) return path voting becoming one of the largest revenue generators within interactive TV. Cable and xDSL operators are now instigating micro-payment systems to replicate this process.

Note: In the UK the original digital terrestrial platform OnDigital had no return path. This is generally seen as one of the reasons for the platform's failure.

2.15.2.3 Phone Vote

One of the oldest forms of interactivity is phone voting. Viewers simply call a number and either talk to an operator or in modern implementations contact an automated counting system. It is possible to generate very high volumes of votes via this technique. As a result, significant revenues can also be derived from premium phone lines.

Due to the potential high call volumes generated by popular voting services (e.g. *Big Brother*), a suitable infrastructure partner is of utmost importance.

2.15.2.4 SMS

SMS has become one of the most popular and effective forms of interactivity available. Simple voting by sending an SMS to a specified number or group of numbers has been used to great effect on several TV events. One of the most interesting aspects of SMS voting is that it offers multiple return paths per household. With around 70% penetration of mobile phones it is feasible for every member of a household to be able to send their own votes.

SMS votes can also be charged at premium rates for enhanced revenues.

2.15.2.5 STB dial back?

Using the phone network via a modem in the STB has proved to be an effective way to extend interactivity and generate revenues on wireless platforms. In the UK only DSAT employs this technique, but it has been a major success.

A direct one-to-one connection between the STB and the services provides an interactive infrastructure allowing services including email, voting, shopping and gambling. Due to the potential high call volumes generated by popular voting services (e.g. *Big Brother*) a suitable infrastructure partner is of utmost importance.

A significant advantage of this approach over xDSL and DCable is the well-defined use of premium phone lines. To achieve the same revenue-generating benefits, an additional micro-payment layer needs to be added to the infrastructure.

2.15.3 Enhanced TV

Enhanced TV is where the information is ancillary to the main programme stream and enhances the viewer's experience of the programme. These ancillary services include additional language audio tracks and the access services of subtitling and audio description, which are usually considered part of most programmes and therefore not covered in this chapter.

2.15.3.1 Text only

2.15.3.1.1 Teletext

Teletext provides a way of signalling and inserting (short) text into the picture, rather like subtitles. Thus, news flashes or scores in associated sports games can be viewed whilst watching the TV programme. In addition, the main text service can hold additional information to do with the programme – for instance, the recipes for a cooking programme or more details about the holidays in a travel show.

Some broadcasters have used this to provide extra information about a programme timed to events within the programme.

2.15.3.1.2 Digital Text

Using similar techniques to those described in Section 2.15.2.1.2 it is possible to enhance traditional broadcast programmes with additional text and images. This usually takes the form of overlays that are either controlled by the viewer or by triggers associated to programme events.

The ability to do this type of service is entirely dependent on available bandwidth and the ability of the target STB to display text and images over video.

2.15.3.2 Alternative Vision and sound

One of the most simple but effective uses of enhanced TV is in the provision of additional audio and video. Viewers are presented with a simple interface that allows access to alternative video/audio streams which complement the traditional broadcast channel. Some of the most effective uses of this technique have been in sport. Coverage by the BBC and BSkyB of soccer, tennis, athletics, rugby, etc. have significantly enhanced the viewer experience.

Normal coverage of the Wimbledon Tennis Tournament would offer a limited view of the multiple matches being played at any one time. Viewers are only able to see the main match as selected by the programmes director. This is despite all matches being recorded. The BBC's interactive service allows viewers to pick which match they watch. This has been found to increase significantly the amount of time they stay with the service.

This style of services has been found to work best with live events.

An alternative example is the BBC's Walking With Beasts (WWB) service. The interactive service attached to the documentary *Walking With Beasts* allowed viewers to see additional content relating to the main programmes. This included evidence behind the show, additional facts, a "making of" and alternative audio (see Section 3.3). These were all presented in sync with the main broadcast. After the initial broadcast the whole service was made available on a continuous loop for the next 7 days. This allowed viewers to access any of the additional streams as required.

2.15.3.3 Alternative Audio and captioning

Alternative audio streams are another effective use of interactive TV. Although DVB MPEG-2 allows alternative audio streams to be provided, the user interface has never been easy and as a result broadcasters have rarely used the facility. Interactive TV applications allow the creation of simple user interfaces. Applications with alternative language streams and commentary are becoming increasingly common. Coverage of sport and documentaries has used this to great effect.

Another factor to take into account is cost. Additional audio streams use a fraction of the bandwidth required for video and require considerably less production costs. This approach can offer a cost-effective way to enhance programming for broadcasters with limited budget and bandwidth.

2.15.3.4 Loops

On wireless platforms true VOD is not easily possible. The closest that can be achieved uses either multiple channels broadcasting the same content at slight time offsets for large content such as movies, or alternatively short loops for short items such as news stories. By offering several loops a viewer only has to wait a short amount of time for the content they are interested in to start.

Many applications using the loop technique also use alternative audio streams.

2.15.3.5 Quarter-screen

A variation of the mosaic technique is to broadcast only one quarter-screen with a text-based application. This approach uses relatively small amounts of bandwidth and can be an effective alternative to a full broadcast channel.

2.15.4 Bidirectionally Wired DSL/Cable

Having a wired back channel connection enables a personal selection to be made and then the information requested delivered to just that viewer. In addition, the STB is permanently connected and thus can be addressed and have material downloaded, in a trickle, during the day. DSL with a dedicated one-to-one connection has some benefits against cable, which immediately has contention between multiple subscribers on the same cable segment, but as most applications require little data to be sent, this is not the same issue as affects a Broadband Internet connection.

All of the TV interactivity is within a walled garden environment.

With a permanently on connection, viewer behaviour can be monitored, and subject to legislation, records kept of how the viewer is using the services.

2.15.4.1 Basic VOD

Basic VOD consists of entire TV programmes, movies, sporting events, etc. being made available via a simple user interface. Most VOD services operate by some form of Pay Per View (PPV) with costing models including single pay per play, fixed time period rental (i.e. view a movie multiple times within a 24-hour period) and full purchase. The full purchase option does not supply the viewer with any physical entity but does allow them infinitely to view a VOD file for a single payment.

When directly compared to Near Video On Demand, true VOD has been found to generate significantly higher buy rates. Some examples have seen a ratio of four to one.

2.15.4.2 Interactive VOD

Interactive VOD takes concepts from the Internet, DVD and Basic VOD to create unique service propositions. The textual content of a page can be linked to a quality replay of a video clip – for instance, to have a new report which is clicked through from the news headlines text pages.

The BBC successfully provided a back library of national and local news stories in Hull.

2.15.4.3 Clip Services

As mentioned in Section 2.15.4.1 VOD is usually targeted at whole programmes. However, access to short clips within specific genres can be an effective use of the format. This would typically focus on news and education.

The BBC in October 2001 launched an Interactive VOD TV service on the Kingston Communication broadband platform in Hull. Local and national news stories were made available through a simple user interface. Content was broken down into categories that allowed the viewer to access only the content they wanted.

2.15.4.4 DVD Style Extras

Most DVD releases consist of a main feature (i.e. a movie) with additional content such as "making of" documentaries, deleted scenes, etc. DVD is a fixed format with no ability to expand beyond its initial content. This can result in multiple re-issues with new content. It is not unusual for fans of specific movies to have multiple DVDs with similar content.

Interactive VOD allows DVD features to be replicated but also allows for additional content to be continually added. This approach offers the double advantage of less overall financial outlay for the viewer but potentially increased revenue for studios through the reduction in distribution costs (DVD pressing, physical distribution, no over-production).

2.15.4.5 Enhanced VOD

VOD has two features which potentially allow creation of unique applications. Since VOD platforms are usually based around a client server architecture, a viewer does not have to watch a VOD file sequentially.

2.15.4.6 Network PVR

Using the concept of VOD it is possible to create a network PVR. This works by automatically storing all programmes broadcast within a cabled network and then making them available via a simple interface. Theoretically only a single instance of each programme needs to be stored for an entire network to be able to view it at a future point. This allows multiple household profiles to be created for little additional cost to the network operator.

Another advantage of this approach is the ability fully to monitor viewer usage. Full digital rights management (DRM) is possible, although many programme makers and broadcasters are still cautious about the implementation of such systems. It is DRM rather than technology that is the major sticking point for Network PVR implementation.

2.15.5 EPGS and TV Anytime/Where

In a multi-channel world, the need for a broadcaster to point the viewer at programmes and for the viewer to be able to find the programmes that they wish to use has never been higher. There is much work both technical and regulatory in this area and the few points below give a good introduction to the process.

2.15.5.1 EPG Layout

After the technology, the key to an effective EPG is its presentation and navigation. EPG layout is defined by three factors, outlined below.

2.15.5.1.1 Number of Channels

The two most successful types of EPG have either focused on grids listings channels and broadcast schedules or alternatively by displaying thumbnail video of available channels.

Most EPGS provide some type of grid. This is a simple technique which is bandwidth efficient, easy for viewers to understand and navigate, and allows an almost unlimited number of channels to be displayed.

The use of video thumbnails is most effective when only used with a small number of channels. A single channel is broadcast with small video windows showing what is currently being broadcast on the various channels. Viewers access the channel by selecting the appropriate window.

Intelligent structuring of genres, channel groupings and channel types (i.e. free to air, subscription, PPV) should always be considered.

2.15.5.1.2 Available Bandwidth

As with all digital services, EPG functionality is limited by available bandwidth. Some platforms only provide a simple Now/Next feature due to this issue. The number of channels displayed on both a grid or thumbnail mosaic, amount of synopsis information and the number of days available are all determined by this factor.

2.15.5.1.3 Set Top Box Capabilities

In many cases, due to financial factors, the interactive capabilities of STB always take second place to the ability to decode video and audio. In some cases bandwidth is not an issue but the available STBs are not capable of any advanced functionality.

2.15.5.2 TV anytime

Just as the EPG allows the viewer to select the programme they want to watch now or in the future, the TV anytime forum (www.tv-anytime.org) is working to provide a universal standardised system for recording the programmes and their metadata. It is seen that in the future the viewer may well not view from the broadcasts directly but via a hard disk appliance in the home. This will act as the portal, selecting what the viewer has desired to view, and perhaps recording and then suggesting what the viewer may enjoy. This is likely to be a great growth area and one where standardisation is most needed.

2.15.6 Technologies

2.15.6.1 Teletext

Teletext uses the vertical banking interval of an analogue TV waveform (and its digital equivalent) in 625/50 TV standards. Each TV line contains 5 bytes of Clock run-in and addressing, and a payload of 40 characters as a 40-byte string. One row of displayed text is derived from one transmitted TV line. It is because of the length of the line that teletext was not easy to implement in 525/60-line countries, where they saw no commercial gains.

For page-based transmission there are 25 displayed rows of text – which use the packet address 0 to 24. Packets up to 29 are used to give enhancements. Packet 30 and 31 are used for data broadcasting.

Each character is 7 bits plus odd parity, and colours and graphics switching is done by control attributes, which take up one character space and are rendered as a blank space on the displayed page.

A page consist of the header Packet 0, which contains the page number and details such as the language set used and the status (subtitle, news flash, etc.) of the page. There then must be a field interval before the other packets of a page can be transmitted (unless they are the more esoteric enhancement packets).

The remaining packets of the page only have the magazine and row address, inheriting the page number from the previous header of that magazine.

There is a lot of wisdom about how teletext can be used which is contained in the specification ETS 300-7906 and its codes of practice, ETS 287 and ETR 288 from the ETSI (www.etsi.org).

Teletext is the easiest way of embedding data in a TV signal and there could be much use made of it for broadcasters' own data signalling.

2.15.6.2 OpenTV

OpenTV are an American company with offices in or close to all significant iTV markets. Their product suite has made them one of the biggest players in the DTV market with over 35 million STB installations worldwide.

The OpenTV middleware is similar in appearance to C/C++ and allows extensive use of STB capabilities. Development resources with the appropriate skills to use OpenTV can be relatively expensive due to the complexity of the programming language (see Section 2.15.7).

For further information, see www.opentv.com.

In the UK, OpenTV middleware is used on the BSkyB platform.

2.15.6.3 Liberate

Liberate Technologies is an American provider of digital infrastructure software and services for cable networks. Its middleware is based on HTML and Javascript with specific extensions for television. Due to its use of web-based standards it is relatively simple to create applications.

For further information, see www.liberate.com.

In the UK, Liberate is used by both NTL and Telewest.

2.15.6.4 MHEG

MHEG itself is a simple mark-up language similar to HTML with limited capabilities, but optimised for broadcast transmission and the TV display space. It is an international standard and was used in the UK on the Freeview DTT Platform.

For further information, see www.dtg.org.uk.

2.15.6.5 MHP

The Multimedia Home Platform (MHP) defines a generic interface between interactive digital applications and the terminals on which those applications execute. This interface decouples different providers' applications from the specific hardware and software details of different MHP terminal implementations. It enables digital content providers to address all types of terminals, ranging from low-end to high-end set-top boxes, integrated digital TV sets and multi-media PCs. The MHP extends the existing, successful DVB open standards for broadcast and interactive services in all transmission networks, including satellite, cable, terrestrial and microwave systems.

MHP is an attempt to create a standardised format that will allow reuse of applications and resources on as many platforms and in as many different countries as possible. The standard is based around HTML and Java, with the intention that an application is written once and will work anywhere with little or no alteration. This is a developing standard which requires a powerful STB to run.

Implementations of MHP have taken place in Germany and South Korea and will become the open standard in Europe. MHP forms the basis of the Cable labs standards for the API on cable and terrestrial TV in the USA.

2.15.7 Authoring Tools

Creation of interactive TV content has followed the same route as the Internet. Initially, skilled developers manually writing code were the only option available to create applications. This is both time-consuming and expensive. Toolkits that allow applications to be created based around pre-defined templates and functionality are becoming increasingly available. Many toolkits provide the ability to design an application once and then have it run across several platforms (OpenTV, Liberate, etc). This has the advantage of reducing costs and development time. However, it is unlikely that any toolkit will be able to take 100% advantage of a platform's capabilities in the way a skilled coder can. Toolkits are useful for volume production of applications across multiple platforms but will never create truly cutting-edge applications.

Note: The market is now being flooded with these types of application, each with differing capabilities. No one toolkit is the solution to all development requirements.

2.15.8 Servers and Transmission Systems

All broadcast text systems use a form of carouselling of data – sending the same content every so often, usually in sequence. In teletext these are usually referred to as inserters or transmission systems, but in the digital world the phrase broadcast server or streamer is more often applied.

The standard for interactive TV systems is the DSM-CC extension to MPEG specifications. DSM-CC stands for Digital Storage Media – Command and Control and includes two different methods of broadcasting data files to a receiver. The complete DSM-CC specification is very complex and as a result over 1000 pages long. The idealised way would be to define and download a file structure for the content in the set-top box. This is what the Object Carousel does and it is built onto the flat structure of a data carousel.

The data carousel can be used on its own for getting file data to a receiver and is used by some digital TV systems, but is more limited and is less efficient in this case.

The object carousel has a tree structure which contains modules – payload elements of 64 Kbytes which can contain smaller feeds from any part of the tree structure. The 64 Kbytes is the largest file that the system can carry – larger files cannot be spanned across more than one module.

DSM-CC is a complex system, but there are many streamer servers available and the model used is very well understood by interactive specialists.

2.15.9 STB/STB+

2.15.9.1 Dumb

The most basic type of STB has the ability to decode video and tune to different channels. While this may not appear to offer any form of interactivity it is possible to use this type of box on wired networks. By putting all interactivity at a network's head-end and simply sending a video stream to the STB, the impression of interactivity can be created. This has the advantages of STBs being cheap and any head-end upgrade instantly affecting all boxes. The disadvantages centre on head-end and infrastructure costs. Since each iTV application is a one-to-one video stream in its own right, the required infrastructure needs to be considerably more robust.

2.15.9.2 Average, e.g. Freeview/Sky

The most common type of STB has a degree of processing power and middleware inside it. This allows applications to be broadcast directly to the box with no additional stress on the head-end. Boxes can be expensive (see MHP sections) and legacy issues can arise over time. The performance of iTV applications on early adopter boxes can be significantly slower than on more recent boxes.

2.15.9.3 Hard disk, e.g. Sky+PVR

An increasingly common feature on STBs is the inclusion of a hard disk which is used to record TV programmes. For interactive applications the hard disk can be used to pre-store video, audio, text images, etc. to create a more advanced service than is possible on 'average' boxes (see Section 2.15.9.2).

2.15.10 Some Examples

The following are all BBC-based examples.

2.15.10.1 Voting

Voting applications have proven to be extremely popular across all types of content, but significant differences exist between each platform.

2.15.10.1.1 DSAT

The BBC's voting application has been used on several occasions. It allows viewers to take part in a vote displayed as a banner running along the bottom of the TV picture (see Figure 2.15.1). Viewers can be presented with several categories on which to vote. In each category they can choose to make a selection from the options offered or to skip the category. They submit their selection(s) via the Set-Top Box modem, and the results can then be compiled. Results can be displayed as part of the application if required.

DSAT requires the use of the viewer's telephone line. Calls can be charged at a premium rate.

2.15.10.1.2 DCable

In general, Digital Cable's way of enhancing TV differs markedly from the formats available on DSAT and DTT.

Since DCable is based on Internet technologies it is relatively easy to create voting applications. The BBC has created a single template that is reused across all services and can be used for both voting and quizzes (see Figures 2.15.2 and 2.15.3).

On-screen/in-vision voting has not yet been attempted. This is technically possible but work with the cable operators has proved to be difficult.

2.15.10.1.3 DTT

DTT currently has no return path, so no true voting applications are possible. Results of votes from other sources can be displayed on screen as part of an application (i.e. phone vote, SMS, etc.).

2.15.10.2 Quizzes

2.15.10.2.1 DSAT

The BBC has used several quiz applications. The two most well known are *Antiques Roadshow* (Figure 2.15.4) and *Test The Nation* (Figure 2.15.5). The applications are both capable of using the return path register results.

2.15.10.2.2 DCable

There is little difference between voting and quiz applications.

2.15.10.2.3 DTT

Similar to DSAT application but with no fixed return path, but telephone or SMS are used.

Figure 2.15.1 Example of a DSAT voting application.

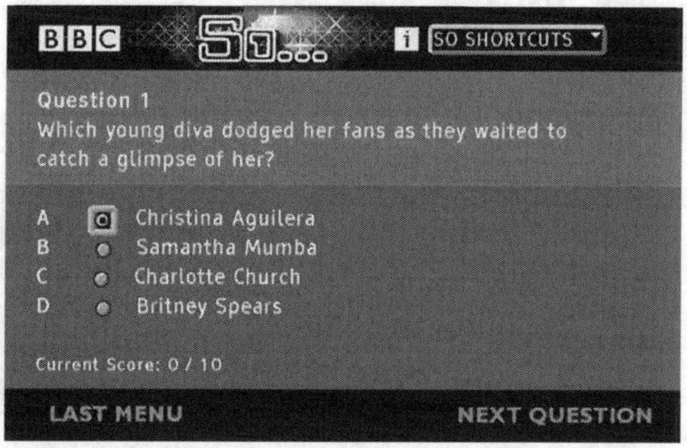

Figure 2.15.2 A DCable voting application.

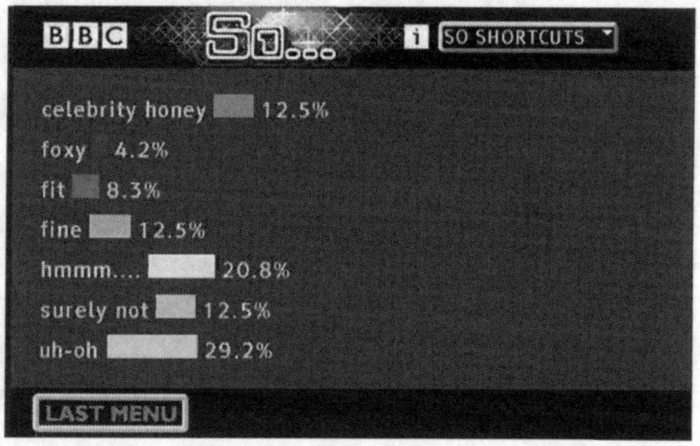

Figure 2.15.3 Results of a DCable vote.

2.15.10.3 Multi-stream

Multi-stream applications use multiple video channels and audio streams which can only be accessed via an application to provide an enhanced experience. This type of application is dependent on bandwidth being available.

2.15.10.3.1 DSAT

Several different multi-stream applications exist on DSAT. At the most simple level is Walking With Beasts (WWB; see Figure 2.15.6).

Viewers are given the option to access alternative video streams via the coloured buttons on the remote control or alternatively alter the style of narration by using the left and right buttons.

The majority of on-screen text seen is not generated by the STB and is instead burnt into the video. This offers a simple application that can be modified quickly for a variety of uses.

The most common use for multi-stream so far is sport (see Figure 2.15.7). The BBC has covered most sports using a similar application with a simple re-skin of the presentation.

Similar navigation to WWB is used; however, on-screen graphics are generated by the STB and can be updated on the fly.

Many DSAT multi-stream applications use a mosaic as a navigation tool. This requires an additional video stream.

Figure 2.15.4 *Antiques Roadshow* quiz application.

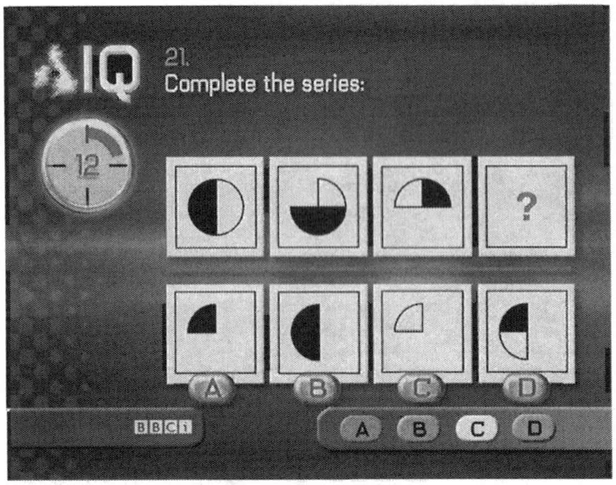

Figure 2.15.5 *Test The Nation* quiz application.

Figure 2.15.6 Walking With Beasts multi-stream application.

2.15.10.3.2 DCable

Services have now been written to appear almost identical to the DSAT service, featuring switchable audio, video 'Highlights' switchable between a Live stream and match highlights, the 'Extra' video stream featuring the tactical view, live match and moderated SMS/Email contributions, which are pre-composited into the video stream. Full integration with data feeds is also possible.

2.15.10.3.3 DTT

DTT multi-stream is now available (Wimbledon 2002), but due to recent events in the industry a maximum of two additional video streams are available to the BBC.

The BBC's current applications have full integration with data feeds.

2.15.10.4 Mosaic

Mosaic applications use a single video stream and multiple audio streams to give an experience similar to multi-stream applications. The single video stream that is created uses input from four others. This creates a 'mosaic' effect. The viewer never sees this channel. Each application uses the ability of the STB to 'cut' a section of the screen and move or enlarge it. The application will also automatically select the appropriate audio stream.

BBC, BBCi, Antiques Roadshow and Walking with Beasts are trademarks of the British Broadcasting Corporation and are used under licence.

Figure 2.15.7 Multi-stream application for sport.

It should be noted that any increase in video size via the abilities of the STB would result in loss of quality.

2.15.10.4.1 DSAT

Mainly used by news at this point and is integrated into the BBC text service (see Figure 2.15.8).

More recent applications have taken the presentation of this type of service forward.

2.15.10.4.2 DCable

No applications exist at this time. The DCable boxes do not have the ability to resize or move video.

2.15.10.4.3 DTT

No applications exist at this time. The DTT boxes do not have the ability to cut video.

2.15.10.5 Data

Data applications allow viewers to access pages of information, including text and graphics, while the TV channel they are watching remains in quarter-screen.

2.15.10.5.1 DSAT

This type of functionality is usually integrated into a hybrid application (see Figure 2.15.9). Stand-alone applications can be created though.

2.15.10.5.2 DCable

If Liberate 1.2 is available then similar applications (with live video) can be created on DCable boxes. Otherwise, only text and graphics are possible.

2.15.10.5.3 DTT

This has been the predominant type of text service on DTT and is extremely well developed (see Figure 2.15.10).

Figure 2.15.8 DSAT mosaic application.

Figure 2.15.9 Example of a DSAT data service.

Figure 2.15.10 Example of a DTT data service.

Steve Tranter

2.16 Conditional Access, Simulcrypt and Encryption Systems

2.16.1 Introduction

Conditional Access (CA), as the term implies, is a mechanism by which television services can only be *accessed* on the provision that certain predetermined *conditions* have been met. CA is required wherever there is an open transmission of video or data content to a target set of receivers over an insecure broadcast channel. Using an insecure broadcast channel means that other receivers, in addition to the intended audience, could gain access to the transmission. CA systems are usually associated with securing traditional, broadcast pay television services, such as those provided by satellite, cable and terrestrial transmission systems, and are used to ensure that viewers can only watch the services for which they have paid. In fact, CA plays a broader role in the broadcast environment, by not only securing the revenues generated by a broadcaster, but also providing:[1]

- Protection of digital content from unlawful copying, which is required if a broadcaster wishes to carry premium content within their channel line-up.
- Additional revenue generating tools that can be used by a broadcaster to sell other services in addition to the traditional subscription packages. This includes being able to assign multiple conditions on which services may be accessed, in order to offer services as part of different bundled packages as well as changing viewing conditions over time, synchronised to the programme schedule to enable part day subscription channels and Pay Per View (PPV) services. These advanced methods of purchasing services make use of a return channel to the broadcaster provided by a modem in the receiver.
- A trusted device in the viewers' homes that can authenticate communications with the broadcast head-end when a return path is available (such as reporting back PPV transactions made using the remote control).

There are four stages to implementing CA on a digital broadcast, as shown in Table 2.16.1. The first two stages are associated with the broadcast process and take place in the broadcast head-end, while the second two stages are associated with the reception of the content and take place at the receiving device, known as the Set-Top Box (STB).

Table 2.16.1 Four stages of implementing CA on a digital broadcast

Stage	*CA process*	*Location*	*Description*
1	Scrambling	Broadcast head-end	Scrambling (encrypting) the individual content streams (usually video and audio) using an encryption algorithm to prevent any meaningful interpretation of the data.
2	Key delivery/ management	Broadcast head-end	Securely delivering to the STB the method by which each stream of content was scrambled along with the access criteria that must be fulfilled by an STB in order to be able to view that content.
3	Authorisation	STB	Checking whether a viewer is authorised to view a selected stream of content, based on entitlements associated with that viewer, delivered and stored locally in the STB.
4	Descrambling	STB	If the viewer satisfies the access criteria associated with the scrambled content, the method by which the content was scrambled is made available within the STB to enable it to descramble the content and make the data meaningful to the STB.

2.16.2 Scrambling

2.16.2.1 Scrambling Process

For CA to be effective, digital content must be protected throughout the entire digital broadcast chain, from the point the multiplexed transport stream of services is created up to the point a selected service's digital stream is decoded back into analogue format at the STB. This is referred to as content protection and ensures that digital content is not accessible in unscrambled, digital format at any point in the broadcast transmission. This is a particular requirement of premium content providers to prevent high-quality, digital copies (equivalent to DVD quality) being made of their content.

Scrambling of MPEG-2 digital content is performed using a scrambling algorithm. The purpose of the scrambling algorithm is to provide enough protection of the content such that it would not be economical for a hacker to compromise the security of the system. Therefore the sophistication of the scrambling scheme used only has to be such that it would be cheaper for somebody to buy the service legitimately rather than purchase expensive computing equipment with which to overcome the scrambling.

Broadcast systems must also avoid overprotection of broadcast services by using scrambling methods that are more sophisticated than practical for their purpose. For example, a highly sophisticated, military-style, scrambling system would secure broadcast services beyond the reach of any potential hacker. However, the system would require high-cost components to support the descrambling process in the STB, which would increase the cost of the STB and the service, thus deterring new subscribers and incurring expensive running costs for the broadcaster.

2.16.2.2 Scrambling Algorithm Theory

In order to provide a secure scrambling process for the protection of broadcast services, without enduring high system costs necessary for military-style protection, scrambling algorithms used for television broadcasts tend to comprise an encryption cipher with two layers, each of which conceal the weakness in the other layer:

- a *block layer*, using between 7- and 16-byte blocks – reverse cipher block chaining mode;
- a *stream layer* – pseudo-random byte generator.

The block layer part of the cipher determines the mechanism by which the digital signal is scrambled. The stream layer part of the cipher introduces a dynamic element into the scrambling process that can be changed over time to strengthen the security of the scrambling.[2]

A simple example of a block cipher would be to scramble the message 'THIS TEXT IS TO BE SCRAMBLED USING A BLOCK CIPHER'. The message is first arranged in the form of a block (removing all spaces first). For this example an 8 × 5 array is used, although any size array would be valid:[3]

THIST
EXTIS
TOBES
CRAMB
LEDUS
INGAB
LOCKC
IPHER

Note the division of the transmission stream into sections and formation of each section into *blocks* requires processing and temporary storage at both the scrambling and descrambling phases. A block is scrambled (or encrypted) by reading out each column from the block, while maintaining the block length (5 in this example):

TETCL
ILIHX
ORENO
PITBA
DGCHS
IEMUA
KETSS
BSBCR

As a variation, the information could be read from diagonal lines, starting at one corner:

TEHTX
ICOTS
LRBIT
IEAES
LNDMS
IOGUB
PCASH
KBECR

The stream layer part of the cipher works with the block layer to specify the sequence in which the columns are read from the block. The stream layer uses a *control word* or *key*, the length of which is the same size as the number of columns in the block and the value of which determines the output delivery sequence. The longer the control word, the more columns there are in the block and the more combinations in which the data can be reordered. Therefore, the longer the control word, the more secure the scrambling algorithm, but the greater the processing required to scramble/descramble the content.

A block cipher can be made further secure by combining the output of one message block with the next message block. This ensures that if any message block is altered, it will affect all the blocks that follow, providing a message authentication mechanism that can be used to prevent the transmission being altered.

An advantage of utilising a control word for scrambling is that, while the core block cipher remains unchanged, the control word can be changed as often as required, dynamically altering the scrambling process. The more often the control word is changed, the more secure the scrambling process. This is a useful way of ensuring the long life of a CA system by preventing long-term analysis of scrambled broadcasts, which could eventually result in the compromise of the scrambling algorithm. Dynamically changing the scrambling process restricts any analysis of scrambled broadcasts to only the length of time each control word is in effect. The frequency with which the control word is changed can vary per system and depends on the capabilities of the scrambling device and the CA system, as well as the level of security required by the broadcaster. For example, premium content providers might demand the scrambling control word change at a higher frequency than for non-premium content. Typically the control word is changed every fraction of a second up to every 20 seconds, depending on the system. The time between each control word change is referred to as a *crypto period*.

Control words are defined as either *odd* or *even*, the definition of which is alternated for each crypto period (see

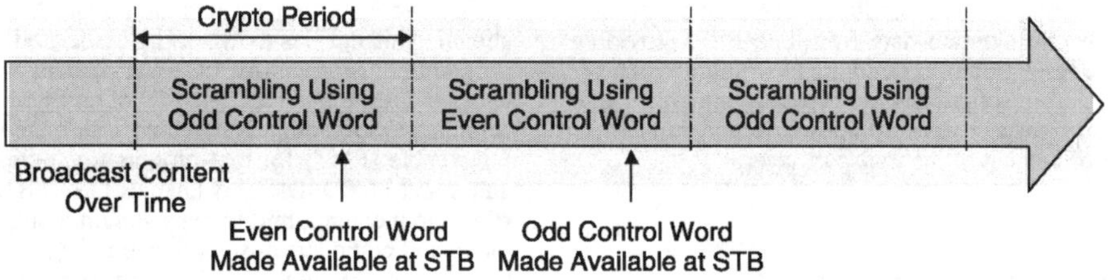

Figure 2.16.1 Use of odd and even control words.

Figure 2.16.1). Being able to distinguish between control words for adjacent crypto periods enables the control word for the next crypto period to be made available in the STB ahead of time (while still using the current control word to descramble content). Thus, the control word for the next crypto period can be loaded into the STB register, ready to descramble the content at the next crypto period boundary. This method ensures the next control word is always available and prevents any glitches in the content descrambling process over crypto period boundaries.

There are a number of standard encryption algorithms currently being used today. Some of the more common algorithms used to scramble broadcast television content are:

- DVB Common Scrambling Algorithm (CSA) – 64-bit (8-byte) control word.[4]
- Advanced Encryption Standard (AES) – 128-bit (16-byte) control word.[5]
- Data Encryption Standard (DES) – 56-bit (7-byte) control word.

In addition to these algorithms, there are also proprietary algorithms that are used only by vendors who invented them or who have been licensed to use them.

Only the general principles of the above algorithms are available in the public domain, due to the necessity to maintain security and keep the detailed information away from potential hackers of systems that have implemented these algorithms. The details of these algorithms are only available to broadcasters and equipment manufacturers who have signed a non-disclosure agreement with the standards body or company controlling access to the algorithm and paid any associated licence fees.

Making the choice of which algorithm to use for protecting pay TV, broadcast services is based on:

- The level of security required (against a brute force attack on an algorithm).
- The availability of scrambling devices for the head-end that support each scrambling algorithm.
- The number of STBs that have integrated each algorithm into their descrambling chips. The greater the choice of STBs that have implemented the same scrambling algorithm, the greater the competition between STB manufacturers, which drives down prices and increases functionality for the broadcaster or service operator.

2.16.2.3 Scrambling Layer

When scrambling content, it is important that the scrambling process takes place at a sufficient layer to ensure that when descrambled, the digital, in-the-clear content is not available in the STB for copying and/or distribution to other devices. The DVB standard, widely used around the world for the broadcast of digital transmissions, defines two layers at which digital content can be encrypted:

- PES layer;
- transport layer.

When encrypting at the PES layer, the encoded, elementary stream is encrypted prior to being multiplexed into a multi-programme transport stream for transmission. The 2-bit PES_scrambling_ control field in the PES packet header is used to indicate whether the content is scrambled or not (see Table 2.16.2).[6]

Table 2.16.2 PES_scrambling_control field options

Value	*Description*
00	No scrambling (in the clear)
01	No scrambling (in the clear)
10	Scrambling with an even control word
11	Scrambling with an odd control word

The PES header cannot be scrambled. When descrambling the stream, the starting point for the scrambled data is determined by information in the PES_header length field and the end-point is determined by the packet_length field.

Scrambling is applied to 184-byte portions and only the last transport packet may include an adaptation field. The PES packet header cannot exceed 184 bytes so that it will fit into one transport packet.

When scrambling at the transport layer, the encoded elementary streams are scrambled after they have been multiplexed into a multi-programme transport stream for transmission. The 2-bit transport_scrambling_flags field in the transport packet header is used to indicate whether the content is scrambled or not (see Table 2.16.3).

Table 2.16.3 Transport_scrambling_flags field options

Value	*Description*
00	No scrambling (in the clear)
01	Scrambling with the default control word (free access) – not generally used
10	Scrambling with an even control word
11	Scrambling with an odd control word

The transport packets in the multiple-programme transport stream can each only contain data from one PES, providing exactly the same resolution of scrambling as with the PES layer scrambling. Therefore scramblers are typically integrated into multiplexers and transport stream transcoders and implement the scrambling of content at the transport layer.

2.16.3 Key Delivery/Management

The process of scrambling content only prevents access to that content. Information related to the scrambling process must also be transmitted along with the content. This additional CA descrambling information enables the STB to determine whether it is authorised to access the content and also provides the *secrets* that the STB must use to recover the control words, which are required to descramble the content.

The descrambling information required by the STB is generated by the CA system and delivered within the multiplexed transport stream along with the scrambled content. The MPEG defines a message format and transmission parameters by which CA systems can transmit this information, known as *Entitlement Control Messages* (ECMs).[6] An independent stream of ECMs is associated with each scrambled programme within a multiplexed transport stream. The rate at which ECMs are transmitted depends on the maximum wait a broadcaster is prepared to endure for the content to be descrambled when a channel is first selected (given that the descrambling process cannot be started until an ECM is acquired). Typically ECMs are transmitted every 0.1–0.5 seconds.

ECM streams are allocated their own PID within a transport stream. This enables ECMs to be associated with a scrambled programme using the MPEG PSI mechanism (as described in Chapter 2.12, MPEG-2). The PID of an ECM associated with a scrambled programme is signalled within the Programme Mapping Table (PMT) for that programme, along with the video, audio and PCR PIDs. This enables the STB to filter the ECM from the incoming multiplexed stream along with the selected video and audio streams, as shown in Figure 2.16.2.

Although the use of ECMs has been standardised by the MPEG as a mechanism by which to deliver CA descrambling information, the payload of the ECMs is proprietary to the system providing the CA. Typically the ECM is used to carry two types of information securely from the broadcast head-end to the STB:

- access criteria that must be satisfied in order for an STB to be able to descramble the service;
- secrets that enable the receiving device to recover the control words required to descramble the service.

The way CA systems generate and use ECMs to deliver the above information differs and is related to each CA system's confidential security practices. In general, there are two techniques for delivering descrambling information within ECMs, referred to as key-based and algorithm-based CA systems.

Key-based CA systems insert an encrypted form of the control word and Access Criteria (AC) into the ECM payload (see Figure 2.16.3). The encryption of the control word and AC is performed using a *service key*. The service key is periodically changed, usually once a month. In order for the STB to access the control word within an ECM, it must first acquire the current

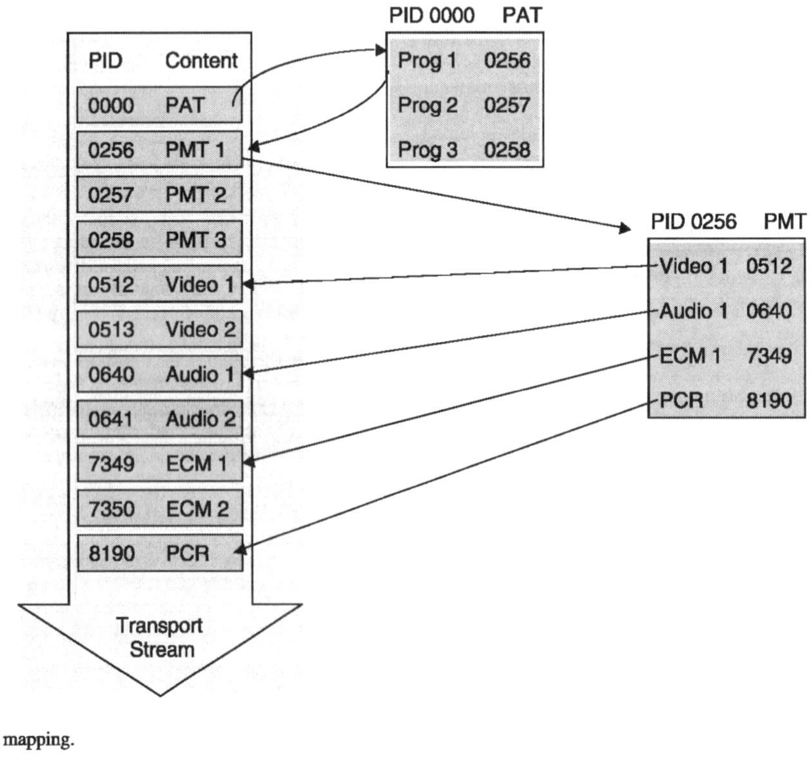

Figure 2.16.2 ECM mapping.

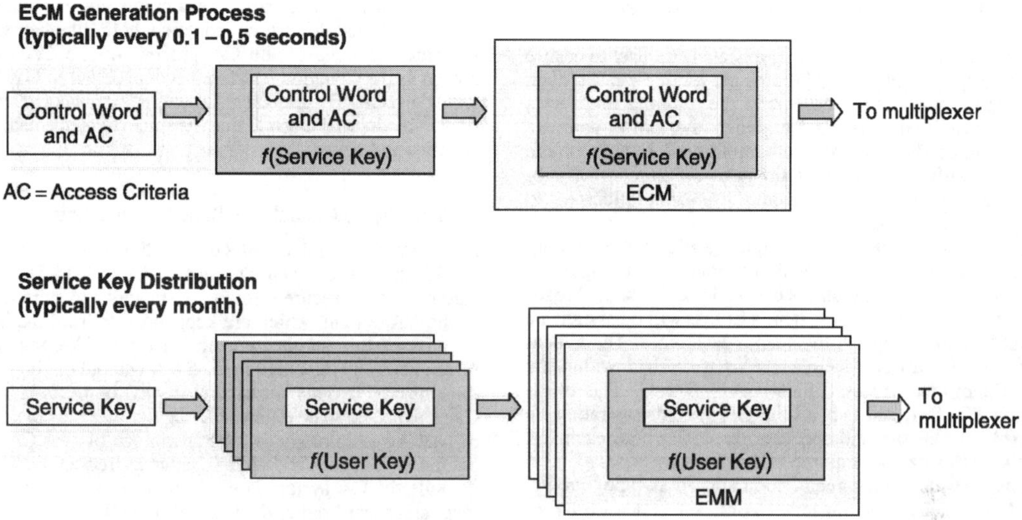

Figure 2.16.3 Key-based CA ECM generation.

service key. An updated service key is delivered on a regular basis (usually once a month) to all STBs as part of an authorisation message, known as an *Entitlement Management Message* (EMM) (see Section 2.16.4). The service key is itself encrypted using a *user key*, for which there is a unique user key for every viewer, stored securely within the CA module of the STB (often on a replaceable smart card). As each viewer has their own user key, the service key must be encrypted independently for every viewer and delivered separately to every STB. The user key is used to access the updated service key whenever it is transmitted.

Algorithm-based CA systems do not insert the control word into the ECM payload. Instead, the ECM payload is signed with a digital signature, generated by passing the ECM through an encryption algorithm combined with the control word (see Figure 2.16.4). The ECM now acts as a seed for the algorithm. The same algorithm is stored in a secure CA module within every STB (often on a replaceable smart card). At the STB, the ECM is passed through the algorithm, which regenerates the control word.

2.16.4 Authorisation

Whether or not a viewer can access a particular broadcast service depends on what access authorisations they have been granted. The authorisations relate to the product offerings that are defined and packaged by the broadcaster and are available to viewers for subscription or purchase, such as a movie package or tier, containing a number of bundled channels, or an individual PPV event. Authorisations are allocated identity values (IDs) within the CA system that map to each of the services or packages offered by a broadcaster.

The allocation of authorisations is controlled by the subscriber management or billing system, which generates messages that are translated by the CA system into authorisation messages that can be transmitted and targeted to the STBs of the viewers to which they relate. In the same way that the MPEG defined a mechanism for delivering access criteria along with scrambled content (ECMs), there is also a message format defined that CA systems can use to deliver authorisation information within the broadcast transport stream, known as *Entitlement Management Messages* (EMMs).

For wireless broadcasts, such as satellite and terrestrial transmissions, EMMs are duplicated and delivered on all transport streams. This is to ensure that a viewer will receive an EMM regardless of what transport stream they are tuned to at any given time. In wired broadcasts, such as cable, there is an option to utilise the out-of-band channel (to which the STBs are always connected) to deliver EMMs, thus avoiding

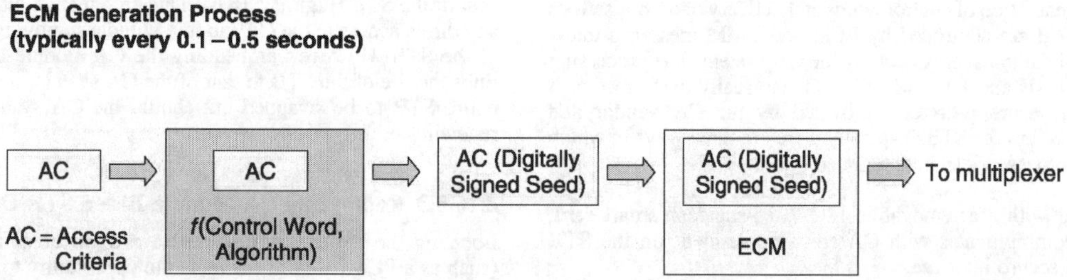

Figure 2.16.4 Algorithm-based CA ECM generation.

the need for taking up bandwidth on broadcast transport streams.

EMMs are usually delivered in a repeated carousel to ensure that an STB will acquire an EMM targeted for that viewer, even if the receiver was not powered on at the time the EMM was initially transmitted. The rate at which the EMM carousel repeats is dependent on the total number of EMMs in the carousel (usually directly proportional to the number of subscribers) and how much bandwidth has been allocated to EMMs on each transport stream.

Similar to ECMs, EMMs are assigned a PID, although only one PID is assigned for all EMMs generated by a single CA system. If a second CA system is operating on the same broadcast network, the second CA system's EMMs will be assigned a different PID to distinguish them from each other. The allocation of PIDs for a CA system's EMMs is defined within the MPEG Conditional Access Table (CAT). The CAT is transported on PID 0x1, and lists all the CA systems operating on the broadcast network and specifies the PID where each CA system's EMMs can be acquired.

EMMs provide a continuous communication pipe through which CA systems can transmit CA-related data directly to STBs. EMMs can be targeted to individual STBs or groups of STBs, depending on targeting criteria defined by the CA system. The payload of the EMM is normally used to carry authorisations aimed at specific viewers, such as granting or removing access to subscription services or PPV events. EMMs can also be used to transmit any other information that needs to be received and possibly stored on the STB that is specific to a CA system.

STBs continually filter the EMM PID stream related to the CA system resident in the STB (defined within the CAT) on whatever transport stream they are currently receiving. An STB will only act upon an EMM if it contains target information in its header that relates to that STB. When an STB acquires a relevant EMM, the EMM is passed to the CA module for processing. Any new authorisations are stored (or removed) from the secure CA module within the receiver (usually a smart card).

2.16.5 Descrambling

The descrambling of content requires the STB to reverse the scrambling process performed at the head-end. To do this, the STB passes the scrambled content through a reverse function of the scrambling algorithm along with the control word applicable for the scrambled content at that time. As described in Section 2.16.3, different CA systems use different methods by which to deliver the control word to the STB, based on proprietary, secure mechanisms.

The generation of control words and delivery of authorisations to the STB are controlled by ECM and EMM messages transmitted in the transport stream by the CA system. The processing of the ECMs and EMMs in the STB is usually performed in a separate, secure processor, provided by the CA vendor and integrated into the STB. Typically there are three ways in which the CA processor can be integrated into the STB:[7]

- Stored within a removable, ISO 7816-standard smart card, which communicates with CA software resident on the STB across a secure interface.
- Embedded CA hardware and software, integrated within the STB.
- CA hardware and software completely contained on a removable module, which contains both the hardware and software elements of the CA processor. The CA module is known as the Common Interface Module (CIM) in DVB systems and CableCARD, formally known as Point of Deployment (POD) module in Open Cable systems (implemented in North America and Korea).

2.16.5.1 Smart Card CA-Based STB Design

An STB/smart card architecture is the most widely adopted architecture by CA vendors today (Figure 2.16.5). The smart card provides a secure form factor on which to store the secrets of the CA system, which are kept separate from the STB. The benefits of this architecture are that the CA component can be replaced and upgraded in the STB, without having to replace the entire STB. This is essential in environments in which the CA is expected to be renewed at regular intervals to keep ahead of hacking technology or take advantage of new CA features, but the STB is not expected to change as frequently. The cost to upgrade the CA system is minimised to the cost of a replacement smart card rather than an entire STB.

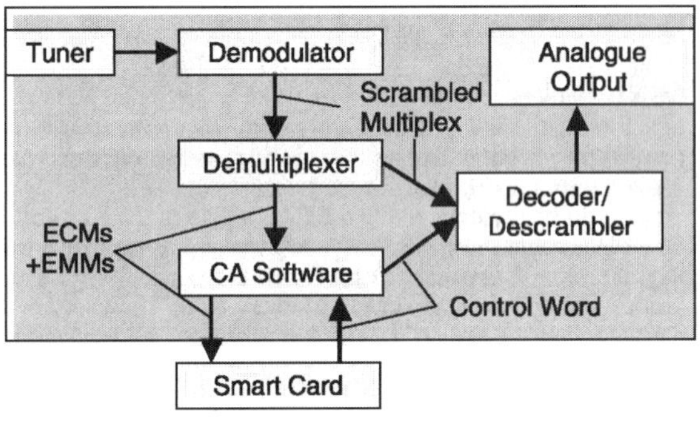

Figure 2.16.5 Smart card STB design.

Although the smart card communicates with software in the STB across a standard interface, which can be analysed, the messages themselves are encrypted to prevent any secrets from being learned.

2.16.5.2 Embedded CA-Based STB Design

The advantage of embedding the CA processor and software into the STB is to remove the cost of a smart card in the overall cost of the STB (Figure 2.16.6). This architecture also prevents any direct access to the CA module without removing the cover of the STB. However, embedding the CA module in the STB links the life of the STB to that of the CA system, requiring the entire STB to be swapped out should the CA system require renewing.

2.16.5.3 Removable CA Module-Based STB Design

Locating the entire CA module on a removable form factor (such as a PCMIA card) prevents the opportunity to access the CA system via an open interface (Figure 2.16.7). This makes for a very flexible architecture, making it easy to entirely

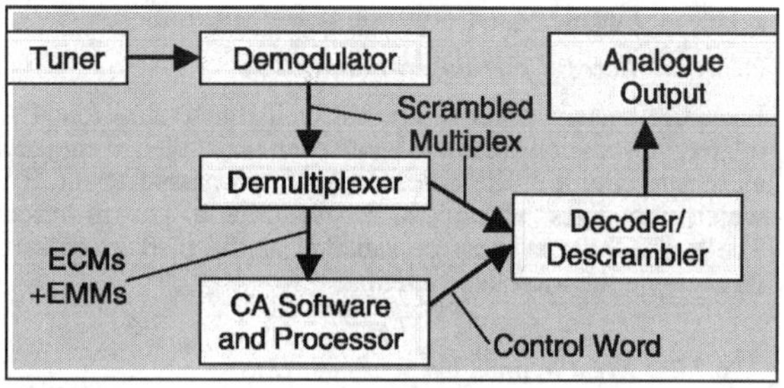

Figure 2.16.6 Embedded STB design.

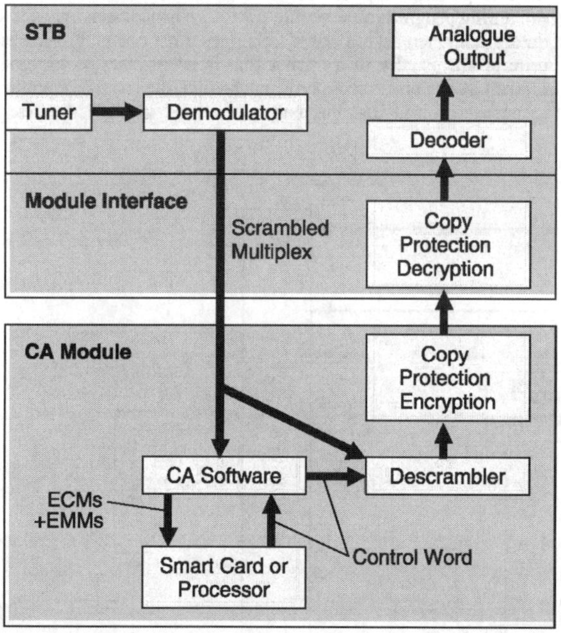

Figure 2.16.7 Removable CA module STB design.

change or remove the CA vendor present in the STB, and also extends the implementation of CA modules into other consumer electronic devices, such as televisions.

However, the disadvantage of this implementation is that the entire multiplex transport must be passed through the CA module for descrambling. This requires that the descrambled digital content must be re-encrypted before being passed back across the module interface to the STB to prevent access to in-the-clear digital content, which could be redistributed. This therefore requires the use of a separate scrambling/descrambling mechanism between the module and the STB. Another drawback of the architecture is the limited processing capabilities of the CA module form factor, which prevents the simultaneous descrambling of multiple streams that may be required for Personal Video Recorders (PVRs) and access across home networks.

2.16.6 CA Systems

Table 2.16.4 lists some of the major CA system vendors that have deployments around the world.

Table 2.16.4 Major CA system vendors

CA system vendor	*Product name*	*Website*
Canal+ Technologies	MediaGuard	www.canalplus-technologies.com
Irdeto	Irdeto Access	www.irdetoaccess.com
Motorola	Digicipher	www.broadband.motorola.com
Nagravision	Nagravision	www.nagravision.com
NDS	VideoGuard	www.nds.com
Philips	CryptoWorks	www.digitalnetworks.philips.com
Scientific Atlanta	PowerKey	www.sciatl.com
ViaAccess (a France Telecom company)	ViaAccess	www.viaccess.fr

2.16.7 DVB Simulcrypt Standard

2.16.7.1 Overview[8]

The main purpose of a CA system is to provide security of broadcast services through the use of proprietary and confidential mechanisms, specific to the CA system. As such, CA systems require integration with other systems in the broadcast infrastructure, such as the following:

- the encoding and multiplexing (compression) system in the head-end at the point that the scrambling takes place and CA messages are inserted into the broadcast stream;
- the programme scheduling and management system, which can be used to determine changes to access criteria for specific channels and programmes;
- the STB, where the descrambling takes place, based on information and secrets transmitted along with the services by the CA system.

Due to the extensive integration of the CA system in both the head-end and STB, it can be seen that, in the past, once a broadcaster chose a CA system, they were also typically locked into the compression system and STBs that were integrated with that CA system. Thus, once deployed, a broadcaster would be restricted from deploying other CA systems or switching out components on the same network. For that reason, the Digital Video Broadcasting (DVB) standards group developed the *Simulcrypt* standard to standardise these interfaces, without compromising the security of the CA systems. The Simulcrypt standard defines the interfaces by which CA systems can interface to other head-end systems and each other, and is widely available in CA and compression systems available today.

A DVB Simulcrypt system is based on the concept of a shared scrambling and descrambling method. To do this, DVB Simulcrypt defines both the algorithm to be used for the scrambling and the method by which the control word is passed between systems in the head-end.

Figure 2.16.8 shows the logical components that comprise a DVB Simulcrypt head-end system and how they interface with one another.

The components specified within the DVB Simulcrypt architecture are categorised into three areas:

- **Host head-end components**. The existing head-end system prior to CA being introduced, predominantly comprising compression equipment.
- **Simulcrypt CA components**. The CA-specific components introduced into the head-end by each CA vendor. CA components are duplicated for each CA system present in the head-end.
- **Simulcrypt Integrated Management Framework (SIMF)**. Optional network management function, used to control the components of multiple CA systems in the head-end.

There is no assumption that the components in each of these areas must physically reside on the same platform or be supplied by the same manufacturer.

2.16.7.2 Simulcrypt Component Definitions[9]

2.16.7.2.1 Access Criteria Generator (ACG)

Each CA system supplies one ACG for generating the CA-related information associated with each scrambled event (such as access criteria). The access criteria are passed to the EIS, which schedules when the access criteria are in effect. The access criteria must be satisfied at the STB in order to descramble the associated event.

2.16.7.2.2 Event Information Scheduler (EIS)

This component is responsible for storing all schedule information, configurations and CA-specific information for the complete system. Amongst its roles, the EIS provides access criteria at pre-scheduled times to the ECMGs (via the SCS), which are used to generate ECMs. The EIS is usually loaded with the upcoming programme schedule for future broadcasts in order to determine when access criteria changes must occur. The access criteria are usually in a format that is proprietary to its associated CA system. Access criteria can take the form of a pointer to access criteria that has been pre-stored in the CA system

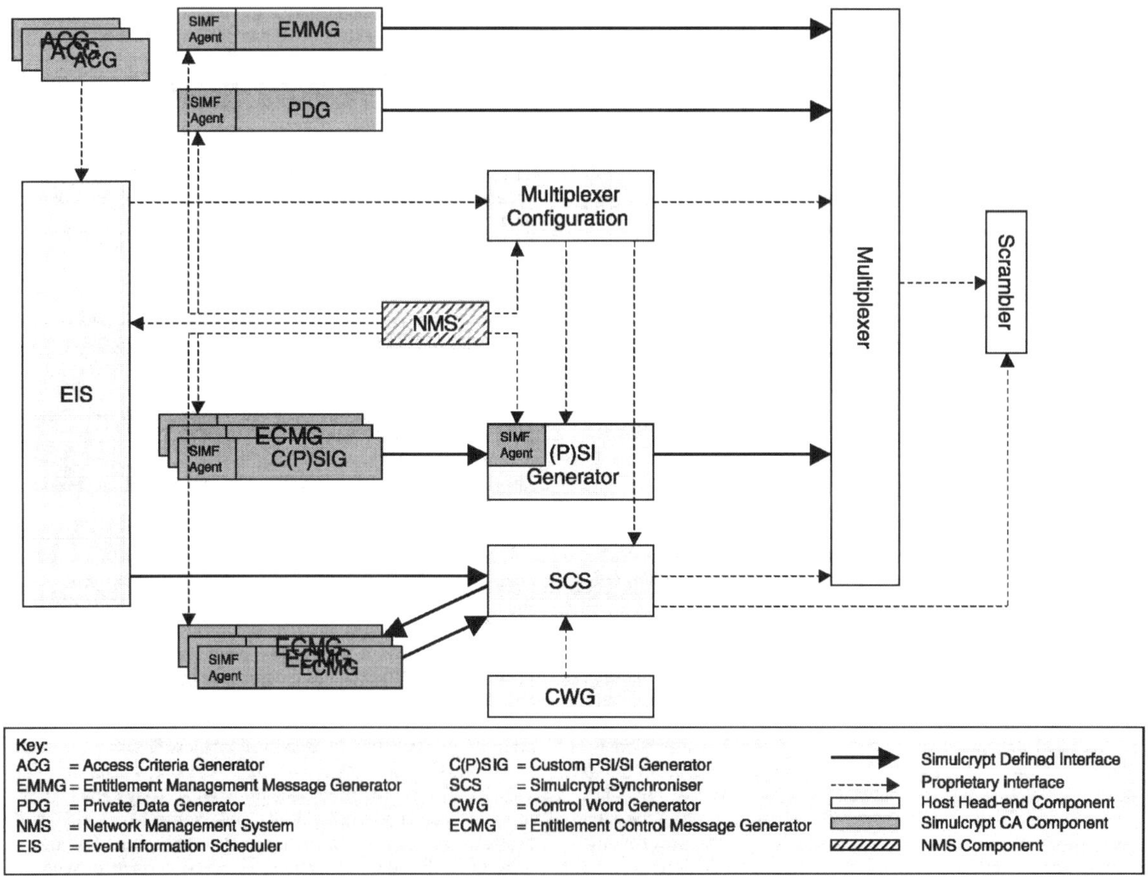

Figure 2.16.8 DVB Simulcrypt system architecture.

(such as a channel, start date and time for a given programme). This method of allocating access criteria is particularly useful when defining access criteria for multiple CA systems.

The EIS function may be distributed over several physical units, storage locations and/or input terminals, and may interface with any other functional unit in the architecture diagram shown in Figure 2.16.5. The EIS functionality is often supplied in the form of a scheduler by either the CA system or incorporated into broadcast traffic and scheduling systems.

2.16.7.2.3 Control Word Generator (CWG)

This component is responsible for randomly generating control words for injection into the scrambler as part of the scrambling process. The CWG will generate new control words at a frequency as specified by the crypto period. The control words are passed to the SCS to make them available to the CA systems' ECMGs, before the control word is injected into and implemented by the scrambler.

The CWG is often integrated, along with the scrambler, in the multiplexing device.

2.16.7.2.4 Simulcrypt Synchroniser (SCS)

The SCS is located at the heart of the Simulcrypt architecture and manages the synchronisation of the scrambling of multiple programme streams by multiple scramblers for multiple CA systems. There can be more than one SCS in a Simulcrypt architecture, each assigned to manage a specific set of transport streams. The SCS is responsible for:

- Establishing network connections with the ECMGs from each CA system and an EIS from which it receives access criteria, associated with each scrambled service.
- Receiving access criteria updates from the EIS.
- Receiving the control word at a configurable time prior to it being used to determine the scrambling for the next crypto period.
- Distributing the control word and access criteria to each connected ECMG. This is done regularly, whenever a new control word is received (every crypto period) and also whenever new access criteria are received. Whenever new access criteria are received, the SCS will shorten or lengthen the current crypto period to ensure the change is implemented in the stream as soon as possible.
- Acquiring an ECM from each ECMG.
- Associating each ECM with its control word and crypto period. Note: ECMs could be generated in advance and stored to be used for future control word/crypto periods.
- Inserting the ECMs into the multiplexer for inclusion into the transport stream, specifying the repetition rate and odd or even scrambling flag value.
- Injecting the control word into the scrambler for the specified crypto period. This is only done after a configurable delay, during which time the associated ECMs have been present in the transport stream long enough for the STBs to have acquired the ECM and generate the control word in readiness for the next crypto period. The delay is usually of the order of 0.5 seconds, but depends on the CA system.

2.16.7.2.5 ECM Generator (ECMG)

Each ECMG receives a control word and access criteria for each scrambled programme stream from an SCS. The ECMG will generate an ECM based on the access criteria and control word and return the ECM to the SCS. This is only performed once per crypto period. One ECMG will support ECM generation for multiple programme streams.

Multiple ECMGs from the same CA system can be connected to an SCS. This can be to divide processing of ECM generation between more than one server and/or to provide redundancy. The SCS can be configured to pass control words and access criteria intended for one ECMG to a backup ECMG, should communication with the first be lost.

2.16.7.2.6 Multiplexer (Mux)

The Multiplexer performs time multiplexing of incoming elementary stream packets (from encoded video and audio sources) and other data sources, including ECMs from an SCS. The Multiplexer also receives:

- programme-specific and service information from a (P)SIG, which it uses to build the PAT and PMT for the scrambled services (including a pointer to the associated ECM for each scrambled service);
- EMMs from each CA system's EMMG.

2.16.7.2.7 Scrambler (SCR)

The Scrambler normally resides within the Multiplexer and is responsible for scrambling the data in the MPEG-2 transport stream packets. The Scrambler receives control words from the SCS, which are used to vary the scrambling over time.

2.16.7.2.8 Multiplexer Configuration

This component is part of the compression system and is responsible for configuring the Multiplexer and synchronising the generation of PSI by the (P)SIG.

2.16.7.2.9 EMM Generator (EMMG)

An EMMG is provided by each CA system and is used to generate messages containing authorisations and other CA-related instructions specific to each CA system. The EMMs are broadcast to the STB population via the multiplexed stream, by way of an interface between each EMMG and the Multiplexer. EMMs can also be delivered using an alternative out-of-band channel when available on certain networks (such as cable). Simulcrypt only defines the broadcast of EMMs through the interface to the Multiplexer.

2.16.7.2.10 Private Data Generator (PDG)

A PDG can be used by a CA system to generate private data, in addition to EMMs, which can be inserted into the multiplexed streams through the same interface into the Multiplexer, used by the EMMG.

2.16.7.2.11 (P)SI Generator ((P)SIG)

The (P)SIG is used to generate MPEG-2 PSI and/or DVB Service Information (SI) using data primarily from the EIS, supplemented with data from the C(P)SIG (neither interface is defined by Simulcrypt).

(P)SIG refers to a process in the head-end that serves as a PSIG, SIG or both.

Table 2.16.5 Simulcrypt-defined interfaces

Interface	*Description*
ECMG <> SCS	Enables a CA system to generate ECMs for scrambled services under the control of an SCS. The SCS is responsible for establishing and maintaining communication with multiple ECMGs, which may include managing redundancy policies where there are multiple ECMGs from the same CA system.
EMMG <> Mux and PDG <> Mux	Defines TCP and UDP protocols, either of which can be used by an EMMG and PDG to connect to a multiplexer in order to pass CA-specific messages for insertion into the multiplex stream. The UDP protocol is generally used where network performance is paramount, whereas the TCP protocol is used where network reliability is more important.
EIS <> SCS	Defines the message protocol used by an EIS to specify which elementary streams are to be grouped into a single programme for scrambling using a common control word, known as a Scrambling Control Group (SCG). The EIS also specifies the access criteria associated with each SCG at any given time.
(P)SIG <> Mux	Defines the interface through which either separate PSI and SI generators or combined PSI/SI generators pass PSI and SI tables into the multiplexer for inclusion in the transport stream. One PSI and/or SI generator can supply tables to more than one multiplexer. However, a multiplexer can receive tables from only one (P)SIG.
C(P)SIG <> (P)SIG	Defines the interface used by CA systems to insert private data into standard MPEG-2 PSI and DVB SI tables.

2.16.7.2.12 Custom PSI/SI Generator (C(P)SIG)

The C(P)SIG is responsible for generating private PSI descriptors and/or SI descriptors that are required by each CA system.

2.16.7.2.13 Network Management System (NMS)

The NMS is responsible for monitoring and controlling the SIMF agents. The function of the NMS varies, and is dependent on the role of the host component in which the agent is located (ECMG, EMMG, PDG, etc.) and the type of management function required of the NMS (fault detection, configuration, etc.).

2.16.7.2.14 SIMF Agent

The SIMF supports network management protocol transactions on the Simulcrypt Management Information Base (SMIB), which it implements. The SIMF agent provides the SMIB with monitoring and control functionality for the host component (ECMG, EMMG, PDG, etc.).

2.16.7.3 Simulcrypt-defined Interfaces

Table 2.16.5 describes each of the interfaces defined by the Simulcrypt standard. All other interfaces shown in the Simulcrypt system architecture (Figure 2.16.5) are proprietary and are proprietary to the vendors of the respective components.

References

1. Bar-Haim, P. *The NDS Guide to Conditional Access*, NDS (2001).
2. Stinson, D. *Cryptography: Theory and Practice*, CRC Press (February 2002).
3. Cherowi, W. *Lecture Notes on Classical Cryptology*, www-math.cudenver.edu
4. DVB. Support for use of scrambling and Conditional Access (CA) within digital broadcasting systems, www.DVB.org
5. Daemen, J. and Rijmen, V. AES Specification: Rijndael, www.csrc.nist.gov/encryption/aes/rijndael/
6. Benoit, H. *Digital Television, MPEG-1, MPEG-2 and Principles of the DVB System*, Focal Press (2002).
7. Stein, M.J. *Implementing Conditional Access*, CED (February 2003).
8. ETSI 103 2.177, rev. 1.3.1, DVB: Head-end Implementation of SimulCrypt.
9. ETSI TR 102 035, rev. 1.1.1, DVB: Implementation Guidelines of the DVB Simulcrypt Standard.

Section 3 Broadcast Components

M Talbot-Smith BSc, C Phys, M Inst P
Formerly BBC Engineering Training Department

3.1 Sound Origination Equipment

Equipment for sound origination can conveniently be divided into two categories: *primary* sources and *secondary* sources. Primary sources are those which are, as it were, right at the start of the chain and convert acoustic signals into electrical signals. Microphones are, of course, primary devices. Secondary sources are essentially devices which store the outputs of the primary sources, i.e. recording and reproducing equipment.

3.1.1 Primary Sources

There are a number of basic features that should be present in any professional microphone. While compromises may be necessary in practice, the following items form a basic check-list:

- The *frequency response* should be as flat as possible, although it is desirable where a microphone is to be mounted in a boom or on a hand-held pole that there should be some bass cut below about 150 Hz to reduce the effects of rumble. On some microphones this bass cut is switchable.
- There should be good *transient response*, i.e. response to the important short-lived frequencies present in the first few milliseconds of a sound. This response can only be judged by ear.
- *Sensitivity* may be very significant. An approximate but useful guide is to see what the output is for normal speech at a distance of about 0.5 m. In such conditions a sensitive *electrostatic* microphone will produce a peak output of some 55–60 dB below a reference level of 0.775 V. This is of the order of 1 mV. A typical *moving coil* (dynamic) microphone gives around −70 dB, and a *ribbon* microphone or a low sensitivity moving coil may produce about −80 dB, i.e. about one tenth of the voltage of an electrostatic microphone.

(0.775 V is a standard audio reference voltage. It originated from the fact that it is the voltage across a 600 ohm resistance when 1 mW is dissipated in it.)

- The stated *polar diagram* should be well-maintained over the majority of the audio range.
- The microphone should not be unduly affected by *environmental influences* such as humidity, temperature, stray magnetic fields (which may cause 'hum' in the output), radio frequency pick-up, rumble and vibration.
- The electrical load into which the microphone will be connected should be compatible with existing equipment. It is the usual practice among manufacturers of professional microphones to give them an impedance in the region of 150–200 ohms, but specify that the electrical load should not be less than five or six times this value (around 1000 ohms or more). The microphone inputs on professional mixing consoles generally have an impedance of 1000–1500 ohms, for this reason.
- Electrostatic microphones need a *power supply*. This may be provided by batteries (in which case it is desirable that the batteries are readily available) or by a 'phantom power' system (see Section 3.1.1.2.3.1) and this also should be compatible with existing equipment.
- Susceptibility to wind noise, or *popping*, is very important if the microphone is to be used out of doors, or if it is to be used close to the mouth as with vocalists' hand microphones. Some microphones have a built-in windshielding. Others may need a separate foam or gauze shield. This should not affect the quality of the microphone's output, but at the same time should do its job satisfactorily.
- Microphones used in professional work need to be as *robust* as possible. However carefully microphones are handled, there is always a risk of accidental damage. Therefore a good repair service by the maker is most desirable. This is most likely to be the case if the makers are well-known as long-standing suppliers to the professional studios.
- The availability of suitable *holders* such as clips (for stands) boom cradles, etc., must be looked into. This is not likely to be a problem with good manufacturers.
- There are sometimes advantages in having built-in *attenuators* or frequency correction circuits, selected by switches on the

microphone body (see the first item). Some microphones have on/off switches. The latter are often more trouble than help as it can be quite easy to discover, after the microphone has been rigged in a place difficult of access, that the switch is in the off position!

- *Cost* is usually a factor, but not one for which much general advice can be given. It is, however, worth noting that good professional microphones are not cheap to buy but low cost microphones may prove more expensive in the long run.

3.1.1.1 Microphone sensitivities

Different manufacturers tend to use different units to describe the sensitivity of their products. Table 3.1.1 shows with sufficient accuracy the relationship between some of the commonly used set of units.

Table 3.1.1 Relation between scales of units of sensitivity

dB rel. 1 V/0.1 N/m²	*dB rel. 1 V/N/m²*	*mV/µbar*	*mV/10 µbar*
−40	−20	9.5	95
−45	−25	5.5	55
−50	−30	3.0	30
−55	−35	1.8	18
−60	−40	1.0	10
−65	−45	0.55	5.5
−70	−50	0.3	3.0
−75	−55	0.18	1.8
−80	−60	0.10	1.0
−85	−65	0.055	0.55

The first column, showing decibels relative to 1 volt/newton/10(metre)^2, is included because average loudness speech at a distance of half a metre produces very roughly a sound pressure of 0.1 newton/m^2. Thus a microphone with a sensitivity of −60 dB rel.1 V/0.1N/m^2 will give a peak output of approximately 60 dB below 1 V.

3.1.1.2 Microphone transducers

The transducer converts diaphragm movements into an electrical voltage. In principle there are many mechanisms that could be used. In practice professional microphones use no more than three: moving coil, ribbon and electrostatic.

3.1.1.2.1 Moving coil transducers

These are sometimes known as *dynamic*. Figure 3.1.1 shows the simplified construction. The coil, typically some 20 turns of wire (often aluminium for the sake of lightness), moves in the magnetic field and an emf is generated. A typical impedance is of the order of 30 ohms, although, as stated in Section 3.1.1, the intended electrical load may be as much as 1000–1500 ohms.

Two main advantages of the moving coil transducer are:

- It is generally robust.
- No external power source is needed.

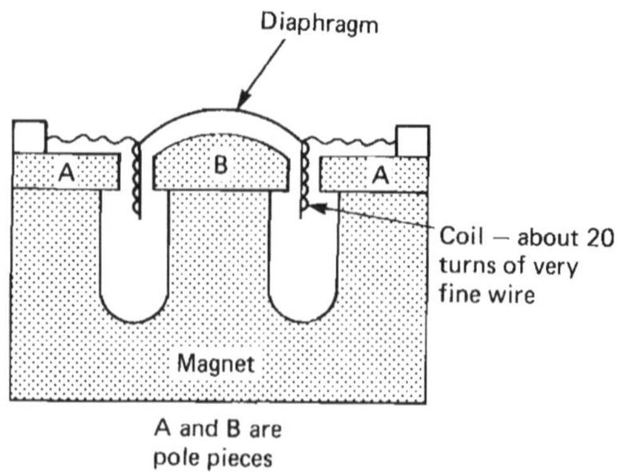

Figure 3.1.1 Moving coil transducer.

Three disadvantages are:

- While the quality of reproduction may be very satisfactory it is rarely as good as can be obtained with other systems. This is because light through the diaphragm and its attached coil are nevertheless not as light as those in electrostatic and ribbon microphones.
- The sensitivity is generally some 10–15 dB (or more) below that of an electrostatic microphone.
- The delicate work involved in assembling moving coil microphones makes them relatively expensive.

3.1.1.2.2 Ribbon transducers

As with the moving coil device, the emf results from the movement of a conductor in a magnetic field, but this time the conductor is linear, being formed of corrugated aluminium. The length of the ribbon is of the order of 2 cm, and it may be from 0.5 cm wide (Figure 3.1.2).

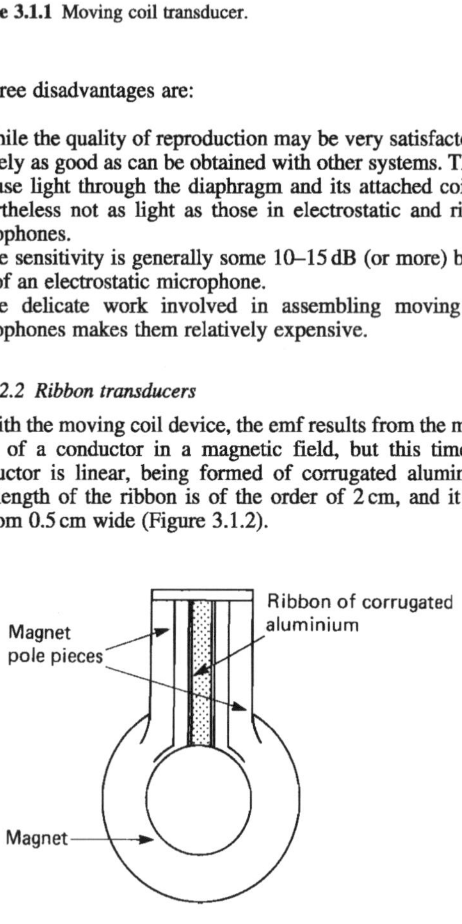

Figure 3.1.2 Ribbon transducer.

The impedance of the ribbon is very low, usually less than 1 ohm, but an integral transformer is used to produce an impedance at the output of something much higher than this. The emf is also raised, of course. A typical load impedance is, as with other transducers, 1000–1500 ohms.

Relatively few ribbon microphones are made now, and what are available tend to be expensive. Their main advantages are:

- It is easy to produce a very good figure-of-eight polar diagram (see Section 3.1.1.3.2).
- The lightness of the ribbon results in a very good response to sound transients.

The disadvantages of a ribbon transducer are:

- It has a relatively high cost.
- The sensitivity is low, being some 20–25 dB below that of an electrostatic microphone.
- The tendency is for it to be large and heavy.
- It is fragile.

Despite these drawbacks many professional organizations still find applications for ribbon microphones.

3.1.1.2.3 Electrostatic transducers

Essentially an electrostatic transducer consists of a very light circular diaphragm close to a metal plate. The two are typically 0.02 mm apart, insulated from each other and forming a capacitor whose value is generally of the order of 10 pF. The combination is known as the *capsule* (Figure 3.1.3).

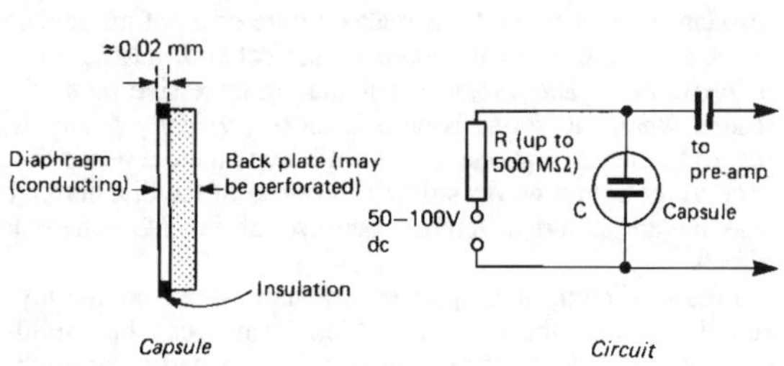

Figure 3.1.3 Electrostatic transducer.

Older types of electrostatic microphones (sometimes called *condenser* microphones) had a dc potential of around 50 V applied between the conductive diaphragm and the back plate. More modern microphones use *electret* materials, i.e. a permanent electrostatic charge is carried by either plate or diaphragm. In the former system the direct currency is supplied in series with a very high resistance (500 megohm is typical) so that the time-constant of the CR combination is sufficiently long for the charge on the capacitor to be effectively constant. With an electret microphone the charge is constant anyway.

From

$$Q = CV$$

where Q is constant and C, the capacitance, varies as a result of sound wave pressures affecting the diaphragm, then V, the voltage across the capacitor, also varies. Because of the very high impedance involved it is necessary for a pre-amplifier, usually an fet device, to be mounted as close as possible to the capsule.

Advantages of an electrostatic transducer are:

- It generally has very good frequency response and good transient response, because of the lightness of the diaphragm.
- The sensitivity is normally high — around −35 dB relative to 1 V/N/m^2.
- It can be made very small.
- It is fairly easy for electrostatic microphones to be designed to have interchangeable capsules, etc.
- Despite the apparent complexity, manufacture is often not as difficult as is the case with, say, moving coil microphones; hence costs are competitive.

The main disadvantage of the electrostatic transducer is:

- Because of the very high insulations needed, proneness to humidity is sometimes a problem.

Condensation on the capsule and its associated components can occur if a cold microphone is brought into a warm studio. This usually manifests itself as a marked 'frying' on the output, which will normally disappear after several minutes if the microphone is put in a warm place.

3.1.1.2.3.1 Powering arrangements for electrostatic microphones

The pre-amplifier for the capsule needs a little power for its operation. Some electrostatic microphones have a battery in a suitable case. More commonly a *phantom power* system is used (Figure 3.1.4). The standard system uses three-core microphone cable to carry an earth (ground) and two programme wires for the audio signal, while the programme wires both take +48 V to the microphone, the earth wire being the return for the power.

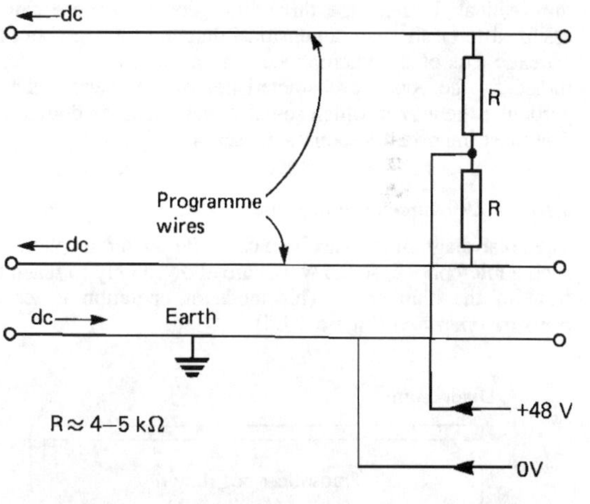

Figure 3.1.4 Standard 45 V phantom power system.

In studio installations it is usual for one power supply to feed all the microphone sockets. Note that non-powered microphones can be plugged into these sockets without ill effects, although it is better to ensure that the appropriate faders are out if loud clicks or bangs in the loudspeaker(s) are to be avoided.

An alternative system, usually for feeding only one microphone at a time, is termed *modulation-lead* powering, or, more commonly, *A-B powering*. Here a low voltage, usually less than 12 V and derived from batteries, is carried to the microphone on the two programme wires. A drawback is that the microphone fails to work if there is an accidental phase reversal in the microphone cable.

3.1.1.2.4 RF/electrostatic microphones

The capsule is basically similar in design to that in a conventional electrostatic microphone described above. However it is not charged. Instead it forms part of a tuned circuit controlling a frequency modulation discriminator. The latter is fed with rf at several megahertz from a stable oscillator. Changes in capsule capacitance cause variations in the tuning control of the discriminator and consequently produce an audio output.

Advantages are:

- high sensitivity;
- very good freedom from humidity problems.

RF electrostatic microphones are much used for outdoor news gathering, exterior filming, and so on.

A disadvantage can be:

- relatively high cost.

3.1.1.3 Acoustic characteristics of microphones

The polar response, i.e. the sensitivity of the microphone to sounds arriving at different angles, is usually of great importance to a sound engineer. This response is best represented by a *polar diagram* in which the microphone is considered to be at the centre and the distance from there to the graph is a measure of the microphone's sensitivity at each angle. Such diagrams are, of course, shown as two-dimensional plots, but it should be remembered that the polar response of a microphone is three-dimensional. Usually the three-dimensional response can be assumed to be the two-dimensional diagram rotated about the acoustic axis of the microphone. The polar response is determined by the acoustic characteristics of the microphone, in particular the way in which sound waves reach the diaphragm. The most important responses are set out below.

3.1.1.3.1 Omnidirectional response

The polar diagram is basically a circle (in reality, a sphere). In such a microphone, sound waves are allowed only to reach the front of the diaphragm. This mode of operation is termed *pressure operation* (Figure 3.1.5).

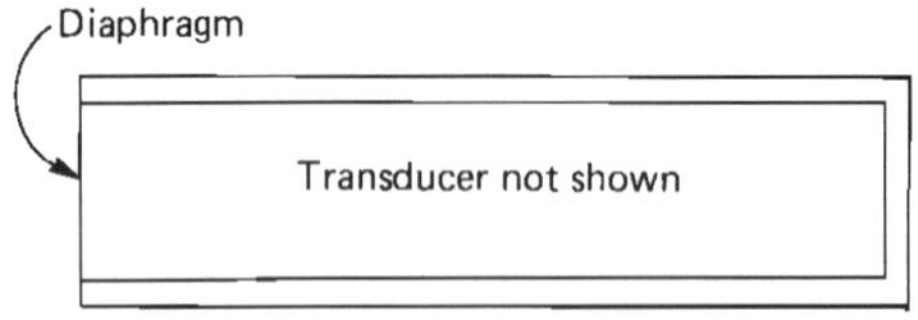

Figure 3.1.5 A pressure operated microphone (simplified).

It can be assumed that at low frequencies, when the wavelengths are much greater than the dimensions of the microphone, sound waves will diffract round onto the diaphragm no matter what their angle of incidence is. The microphone will thus be truly omnidirectional.

However at higher frequencies, when the sound wavelengths are similar to the diameter of the microphone, this diffraction process does not occur so effectively, and the response of the microphone to sounds from the side and rear falls off. For a microphone 2 cm in diameter, the departure from truly omnidirectional behaviour starts to occur at around 3 kHz. For a 1 cm diameter microphone it will be about 6 kHz. It is a useful characteristic of omnidirectional microphones that they are freer from rumble and vibration effects than most other types.

Omnidirectional microphones have limited applications in studios because they tend to pick up unwanted sounds, excessive reverberation, and so on. However, they are perfectly satisfactory for small 'personal' microphones clipped to clothing, and they are also well-suited as hand-held microphones for interviews, when their lack of directionality and relative freedom from rumble can both be advantageous.

3.1.1.3.2 Figure-of-eight-response

The shape of the polar diagram of a figure-of-eight (or *bidirectional* microphone) is self-explanatory. The effect is achieved by allowing sound to reach both sides of a diaphragm. The force on the diaphragm results from the difference in acoustic pressures on the two sides. This difference, in turn, is a consequence of the path difference travelled by sound waves reaching front and back of the diaphragm. The process is generally referred to as *pressure gradient operation.* The obvious type of transducer to permit this is the ribbon type, and indeed most ribbon microphones are figure-of-eight. Electrostatic transducers can, however, be used (see Section 3.1.1.3.4). In order to have a flat frequency response it is necessary for the diaphragm to have a low mechanical resonant frequency, and this makes figure-of-eight microphones prone to rumble. They also exhibit an effect known as *bass tip up* or *proximity effect*, which means that an excessive bass output results when the sound source is close (typically nearer than 0.5 m) to the microphone, when not only is there a phase difference in the sound waves striking the sides of the diaphragm but also an amplitude difference because of inverse square law effects.

Figure-of-eight microphones are used most commonly in sound studios, where the two 'dead' sides can be useful in reducing the effects of unwanted noises. A particular application is in some so-called *noise-cancelling* microphones. Used close to the mouth the otherwise excessive bass-rise due to proximity effects is removed by a built-in bass-cut circuit. Distant noise, which will not have a bass-rise, is also reduced at the low frequency end by the equalizer.

3.1.1.3.3 Cardioid response

The heart-shaped polar diagram of a cardioid microphone is produced by allowing some sound to enter the microphone and reach the back of the diaphragm. Such microphones are usually recognizable by slots or other apertures to the rear of the diaphragm. To achieve the cardioid pattern an acoustic labyrinth introduces a delay, or phase shift, into sounds that have entered the aperture. Cardioid microphones are partly pressure-gradient operated (see Section 3.1.1.3.2) and are thus apt to exhibit a degree of proximity effect and also a tendency to be rumble-prone. It is virtually impossible to design cardioid microphones that maintain their polar diagram well over the whole audio range. The front/back ratio, which expresses in decibels the difference between front and rear sensitivities, is rarely better than 25–30 dB, and may be as little as 5–10 dB at some frequencies (Figure 3.1.6).

Cardioid microphones are probably the most widely used type in professional work because of their insensitivity to sounds arriving from behind the microphone.

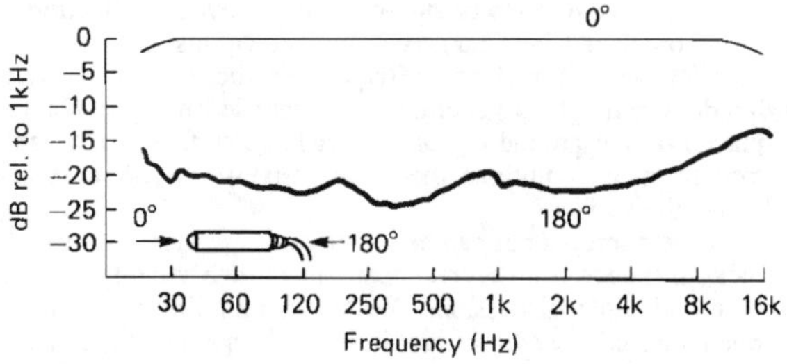

Figure 3.1.6 Typical front/back ratios for a high-grade cardioid microphone.

Television booms, vocalists' microphones, musical instrument pick-up, and so on, are frequently cardioids.

Their use out-of-doors can be limited unless well windshielded because the apertures behind the diaphragm can make them vulnerable to wind effects.

3.1.1.3.4 Variable response

With microphones with a variable polar diagram, a range of responses is available, typically omnidirectional, cardioid, figure-of-eight and maybe *hypercardioid* (a pattern between cardioid and figure-of-eight). Selection of pattern is usually by a switch on the body of the microphone, but remote selection is sometimes available.

The operation depends on two electrostatic cardioid capsules mounted back-to-back. The overall polar pattern depends on how the outputs of the capsules are combined. Switching out one cardioid results in the response of the remaining cardioid; if the two cardioid outputs are added, an omnidirectional response is obtained; subtraction gives a figure-of-eight, while partial addition of one cardioid to the other gives a hypercardioid (Figure 3.1.7).

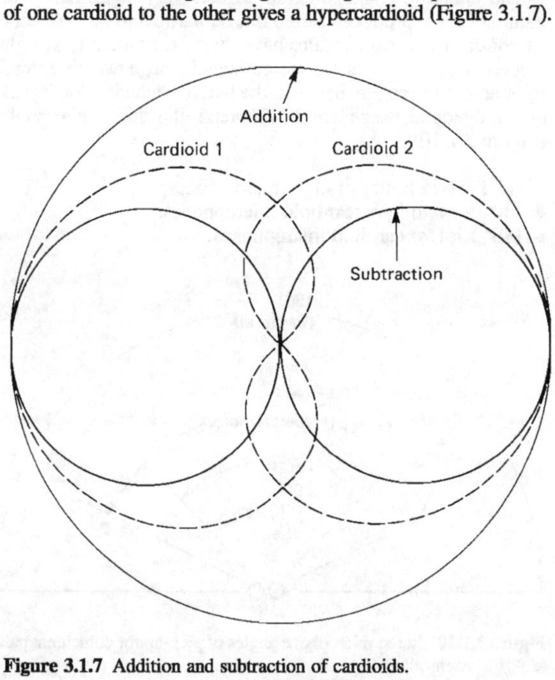

Figure 3.1.7 Addition and subtraction of cardioids.

These microphones are useful for studio work where the ability to vary the polar response is an advantage. They tend to be expensive.

3.1.1.3.5 'Gun' or 'rifle' microphones

These consist of a slotted or perforated tube, usually about 0.5 m in length, behind which is a normal capsule, often cardioid with either an electrostatic or moving coil transducer. The slotted tube in front of the diaphragm has little effect on sound waves arriving from along its axis. Sounds arriving at an angle greater than about 20° tend to undergo phase cancellations inside the tube because their path length to the diaphragm depends on where they enter the tube. The result is that the microphone is very directional, provided that the sound wavelengths are comparable with, or less than, the length of the tube. At the higher audio frequencies, a good gun microphone will have a flat frequency response over a total front angle of about 30°. At low frequencies, gun microphones tend to be omnidirectional, but the use of a cardioid capsule can help to give some directional effect even then.

Gun microphones are used mostly for exterior work such as location shooting and news gathering. For such purposes an rf electrostatic capsule (see Section 3.1.1.2.4) is preferred for its sensitivity and freedom from humidity effects. In a large studio, gun microphones are used for audience contributions, but it should be noted that directionality is generally poor in small rooms. Shorter tubes are often favoured for location shooting as they are lighter and less unwieldy for the operator. Their polar response is wider and approximates to a hypercardioid (Figure 3.1.8).

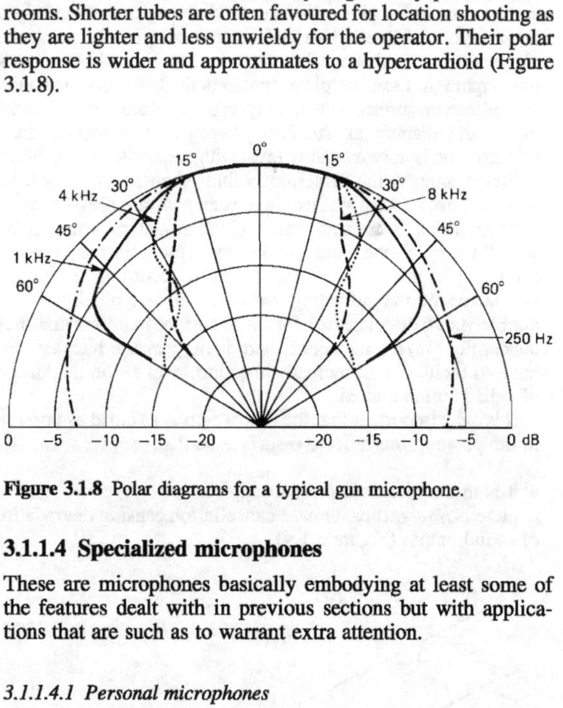

Figure 3.1.8 Polar diagrams for a typical gun microphone.

3.1.1.4 Specialized microphones

These are microphones basically embodying at least some of the features dealt with in previous sections but with applications that are such as to warrant extra attention.

3.1.1.4.1 Personal microphones

Personal microphones are very small microphones which can be clipped inconspicuously to clothing, or concealed inside it. The majority have electrostatic (electret) transducers and are pressure-operated (i.e. omnidirectional). Pressure operation is generally employed because such microphones can be made smaller more easily than can cardioids. Also, the relatively rumble-free character of the omnidirectional microphone is an advantage when the microphone is to be worn. Some personal microphones have a battery pack which also needs concealing in clothing. Care has to be taken in fitting personal microphones to artists as the nature of adjacent clothing can have a marked effect on the performance of the microphone.

If circumstances allow the microphone to be visible it is usually sufficient to position it so that it, and possibly its cable, does not rub against clothing. There can be problems, however, when it must be invisible. Man-made fibre materials can generate static electric charges which may discharge through the microphone, resulting in audible effects. Also some materials, especially heavy or closely woven ones, can attenuate the higher audio frequencies. It is therefore very important to carry out tests with the microphone in place at the rehearsal. A measure of equalization is often needed with these microphones, even if outside the clothing, as the microphone is off the high-frequency axis of the voice. It is interesting to note that it is sometimes possible for a conversation between two people, such as an interview, to be picked up with only one of the persons wearing a microphone of this type. This technique cannot, of course, be used in stereo and it is unlikely to be satisfactory for television purposes.

3.1.1.4.2 Pressure zone microphones

These are devices in which there is fairly conventional transducer, usually electrostatic and pressure-operated, but which differ from conventional microphones in that the diaphragm is effectively in the plane of a hard reflective surface. In some types the transducer element is mounted above, but very close to, a metal plate so that the diaphragm faces the plate. In others the transducer is behind the reflective surface which may take the form of a wooden block; the diaphragm, suitably protected by a screen, faces upwards and is approximately level with the surface of the block.

Stated simply, the principle is that when sound waves are reflected from a hard surface there is an acoustic pressure at the surface which is greater than the maximum pressure in the sound wave in free field conditions. This pressure increase, occurring in the 'pressure zone', varies according to the size and shape of the reflecting surface, but is typically around double (i.e. 6 dB increase) for large surfaces and normal incidence. For grazing incidence there is no pressure increase. For random incidence an increase in sound level at the diaphragm of 3 dB is often quoted.

It is very important that the diaphragm is as close as possible to the plane of the reflective surface so that:

- it is in the increased pressure region;
- there is no significant phase cancellation causing degradation of sound output (Figure 3.1.9).

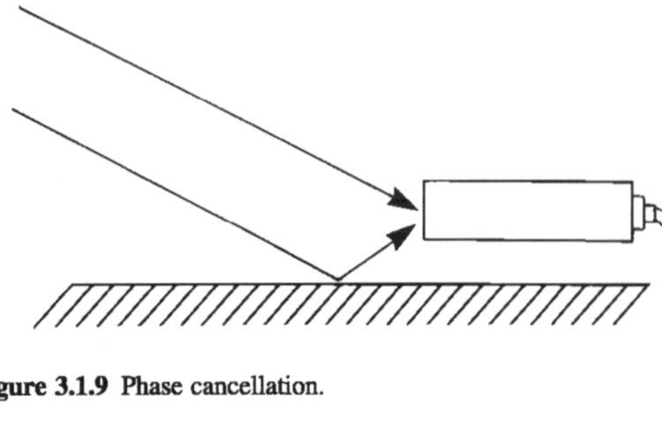

Figure 3.1.9 Phase cancellation.

The reflective plate or surface supplied by manufacturers is of the order of 15–25 cm across; on its own this is too small to be effective except at audio frequencies above at least several hundred hertz. It is generally recommended that the plate is placed on the ground or some other large surface, e.g. a table top. In such conditions the polar response is a hemisphere above the surface.

These microphones can be effective where there is a need to pick up speech from several positions in their vicinity, such as in round table discussions. Within certain limits pick-up is relatively independent of proximity of the source to the microphone.

A reasonably effective pressure zone microphone can be created by taping a small personal microphone to the top of a table.

3.1.1.4.3 Stereo microphones

These and their use are a large subject and can only be dealt with here in outline. A *coincident pair* has a number of applications in sound recording and broadcasting. Essentially it consists of a pair of microphones mounted as close together as possible. This may be achieved in one of two ways:

- A specifically stereo microphone having two separate capsules, generally mounted one above the other. Their polar responses can be individually adjusted and one capsule can be rotated relative to the other.
- Two conventional mono microphones fitted onto a suitable mounting such as a 'stereo bar'. This is a metal bar some 20 cm long and fitted with clips or other means of attachment. The bar itself can be screwed onto a microphone stand. There may be adjustments so that the two microphones can have their spacing and the angle between them varied.

The choice of polar diagram affects very significantly the nature of the reproduced stereo image, particularly in respect of its width. If the two capsules have their axes mutually at right-angles, then to produce a stereo sound image which extends right across the region between the two reproducing loudspeakers, the sound stage should subtend the following angles (Figure 3.1.10):

- 90° for two figure-of-eight microphones;
- 130° for two hypercardioid microphones;
- 180° for two cardioid microphones.

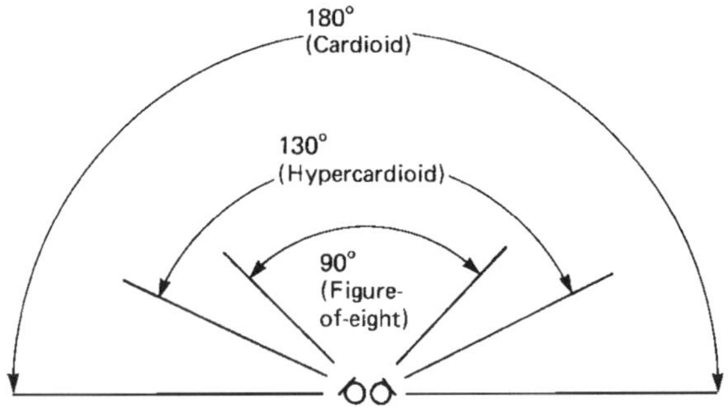

Figure 3.1.10 Stereo microphone angles of pick-up for coincident pairs at 90° to each other.

3.1.1.4.4 Direct injection boxes

These are included as alternatives to microphones. Very simply, a direct injection (DI) box is inserted somewhere between an electrical or electronic musical instrument and its loudspeaker(s). A socket on the box can then be used to feed the instrument's signal to the sound mixing console instead of placing a microphone near the loudspeaker.

DI boxes are not normally expensive but it is advisable to check on their internal electrical insulation properties. A power fault in a mains-powered instrument may in itself be serious enough, but there could be further unfortunate consequences if the output of the DI box were not suitably isolated.

The nature of the sound signal obtained with a DI box may be preferred to that picked up by a microphone in front of the instrument's loudspeaker; there is no risk of the sound of other instruments being picked up and consequently 'separation' is greatly improved. This can be a big advantage where musicians and their instruments are in a cramped space.

3.1.1.4.5 Soundfield microphones

A soundfield microphone is very complex. The microphone capsule assembly contains four capsules arranged in a regular tetrahedron and is inside a unit that is only slightly larger than many conventional microphones. The output goes to a control box. This may be connected directly to monitoring loudspeakers and line, or a four-track tape machine can be connected to it. The complete unit can be thought of as a coincident pair, the characteristics of which — angle, polar diagram, tilt, etc. — can be controlled on the box. If a four-track recording is made, it can be processed 'off-line', so that the parameters may be adjusted subsequently. The soundfield microphone can be used for ambisonic recordings, although this is of limited interest at present.

3.1.1.5 Radio microphones

There are two broad categories of radio microphones: the miniature type intended to be concealed in the artist's clothing or incorporated in a hand microphone, and the larger pattern which is strapped to a belt or worn as a backpack.

3.1.1.5.1 Miniature radio microphones

Usually the transmitter and battery compartment are built into the same housing. The aerial is typically a dipole formed from a thin wire plugged into the casing and the screen of the microphone cable, although the radio hand microphone usually has only an aerial protruding from the base. Where the microphone is separate from the transmitter it is normal to have a small personal microphone, although in principle other types can be used. There may be difficulties, however, if the separate microphone needs 48 V phantom power. An audio limiter is generally built into the system to avoid over-modulation. The receiver may be battery or mains powered. Some types are fitted with signal strength indicators, and there is often a headphone-listening jack as well as a microphone-level output intended to be connected directly into the microphone inputs of a mixing console.

Because of the small size of this type of radio microphone, the batteries which power them also have to be small and this means the radiated power is not great — typically of the order of a few tens of milliwatts. In turn the working range is limited. Under very good conditions, such as line-of-sight in the open air, a range of 300–400 m is possible, but may not be reliable. Inside a studio the range can be very much less, not just because of the dimensions of the studio but because the presence of reflecting metalwork can result in regions where there is interference between direct and reflected signals. A range of as little as 5–10 m is not unknown.

The problem of this interference can be greatly reduced by the use of *diversity reception*. More than one receiving aerial is used and these are placed a few metres apart. The diversity reception unit automatically switches to the output of whichever aerial system is receiving the best rf signal. Most of the miniature radio microphones use a frequency modulated carrier in the vhf bands. The audio performance of the microphone to receiver output link is normally good; the quality of the microphone capsule is often the limiting factor.

The most obvious advantage bestowed by a miniature radio microphone is freedom of movement for the artist, and for many television applications this is very valuable. It must, nevertheless, be recognized that a radio microphone is less reliable than a conventional microphone on the end of a cable. Also if more than one radio microphone is used in a studio, care has to be taken that each is on a different channel. Further, in some environments, there can be serious interference from nearby radio sources.

Careful checks of a radio microphone's performance in all the places where it is going to be used should be carried out before recording or transmission whenever possible. It should not be assumed that anywhere in a studio will be satisfactory. If difficulties are encountered it may help to change the position of the receiver aerial.

3.1.1.5.2 High power radio microphones

These are more in the nature of small radio link systems. A conventional microphone may be plugged into the transmitter, or, in some cases, a 'head-and-breast set' consisting of a small microphone on a rod (or *boom*) attached to the headphones is used. In the latter case the transmitter/receiver system is obviously a two-way arrangement. Batteries for these units are frequently in a separate pack and are then large enough to provide relatively high power. The rf output may be several watts, giving a range of up to 2–3 km. Radio frequencies used may be in Band 1 or they may be very much higher.

3.1.2 Secondary Sources

3.1.2.1 Tape recording and reproducing systems (analogue)

The general principles of magnetic tape recording are dealt with in other books (see Bibliography); particular points and a typical specification for a professional recorder will be given here.

3.1.2.1.1 Tape

The standard tape width for mono or stereo recordings is 6.25 mm, frequently referred to as *quarter-inch tape*. The base is polyester, commonly some 30 μm thick with a coating of suitable oxide of perhaps half this thickness. For stereo recording, the two tracks are usually 2.75 mm wide, with a guard track of about 0.75 mm between them. Twin-track recordings, where there may be totally different types of programmes on the two tracks, have a wider guard track of about 2 mm, leaving the active tracks 2.1 mm each. There is then better separation between the tracks but noise figures are less good. For multi-track recording work, the standard tape width is *two inch* (50.8 mm).

Standard tape speeds are:

- 38.1 cm/s (15 ips) for general recording;
- 19.05 cm/s (7.5 ips) where tape economy is important and the lower performance can be accepted; this speed is generally quite adequate for speech;
- 76.2 cm/s (30 ips) as a preferred alternative to the use of noise reduction systems at 38.1 cm/s; however, standard NAB spools of tape carry only about 16 minutes of recording at this speed.

3.1.2.1.2 Bias

Satisfactory analogue recordings cannot be made unless a high frequency signal is mixed with the programme signal before it is applied to the record head. This bias frequency is not critical, but is usually in the range 150–200 kHz. The bias level, however, is critical. For optimum performance, the bias current should normally be slightly higher than is needed for maximum recording sensitivity (see Figure 3.1.11).

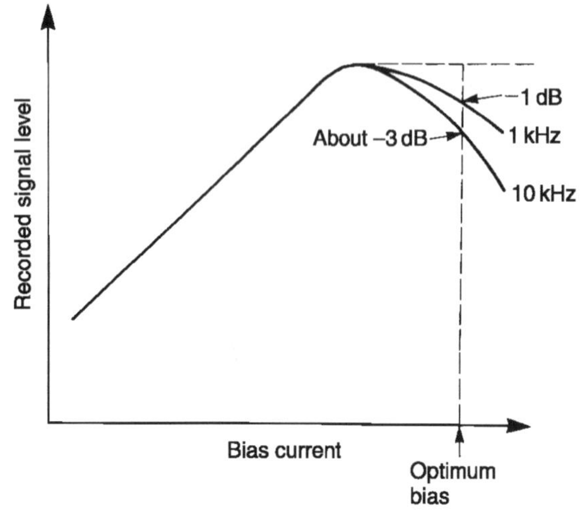

Figure 3.1.11 Optimum basis.

3.1.2.1.3 Equalization

For further optimization of recording, equalization is needed. The exact characteristics vary; the IEC, for example, sets standards that are common in Europe. In North America, NAB standards are adopted. The IEC recording characteristic is flat up to a frequency which depends on tape speed, being 4.5 kHz for 38.1 cm/s, and then falls off at 6 dB/octave. The corresponding time constant is 35 µs.

Replay equalization, which compensates for the fact that the emf induced in the replay head is proportional to rate of change of flux, and hence to frequency, is shown by a curve falling at 6 dB/octave (Figure 3.1.12).

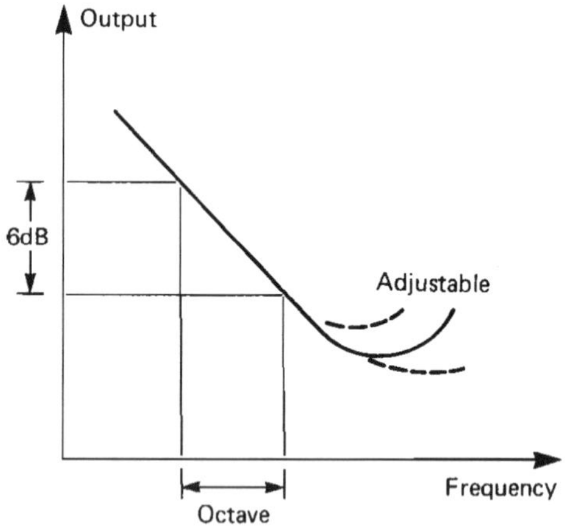

Figure 3.1.12 Replay equalization curve.

3.1.2.1.4 Compact cassettes

Until relatively recently these cassettes and their machines had no place in the professional world. However, with modern recording and replay machines, marked improvements in tape materials and the use of noise reduction systems, cassettes can be useful, especially in broadcasting where the small size is useful for news gathering, for example. A major problem is not so much the poorer quality compared with full-size machines using 38 cm/s tape, but the virtual impossibility of editing cassettes. The width of cassette tape is 3.81 mm (0.15 in) and the track layout is shown in Figure 3.1.13.

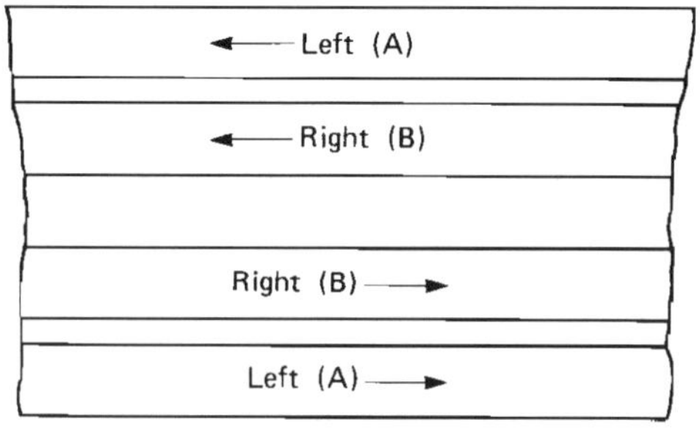

Figure 3.1.13 Cassette track layout.

3.1.2.1.5 Typical specifications for a professional tape machine

A professional machine may have the following specifications:

Tape speed deviation	±0.2%
Tape slip	0.1%
Wow and flutter, weighted, at 38 cm/s	0.04%
Effective start time	0.5 s
Rewind time for full NAB spool (approx.)	90 s
Input impedance	8–10 k
Input level	+22 dBm max

Output impedance	30 Ω
Load impedance	200 Ω min
Output level	+24 dBm max
Frequency response at 38 cm/s	
30 Hz–18 kHz	±2 dB
60 Hz–15 kHz	±1 dB
Signal/noise ratio, CCIR weighting at 38 cm/s, full track	56 dB
Stereo crosstalk rejection at 1 kHz	45 dB
Erase efficiency at 1 kHz	>75 dB
Erase and bias frequency	150 kHz

3.1.2.2 Digital tape recording

Brief generalizations cannot be made here as there are several different systems. Almost the only common factors are the sampling rates (44.1 kHz or 48 kHz)! Basically the systems can be divided into two broad categories: stationary head machines and rotary head machines.

Stationary head machines usually use 6.25 mm ('quarter-inch') tape running at 38 cm/s. A typical machine has several data tracks, including SMPTE time code, plus eight digital tracks carrying two 'music' channels. Each channel is thus spread over four digital tracks using very complex coding to reduce the effects of tape 'drop-outs' and also to record the very high bit rate (approaching 1 Mb/s).

Rotary head machines achieve high bit rate recording by using the principle of video tape machines. One popular low-cost system in fact uses a domestic video recorder in conjunction with a unit that converts the input stereo audio signals into digital form and then arranges the digital data into a format that simulates video signals and is thus acceptable to the recorder.

Advantages of digital tape recording are:

- very high signal/noise ratios—some 30 dB better than analogue recordings;
- multiple copying—up to 100–200 times compared with less than half a dozen for analogue;
- print-through and crosstalk between tracks virtually eliminated;
- tape speed constancy less of a problem with a digital system as small cyclic fluctuations can be corrected by storing the digital data and 'clocking' it out at the correct rate;
- bias and equalization unnecessary;
- rotary head machines relatively cheap.

Disadvantages are:

- with rotary head machines, only dub-editing possible;
- stationary head machines expensive.

Cut editing can, however, be carried out with a razor blade on stationary head machines because the digital data are 'scattered' around the tracks. Complex circuitry can be used to interpolate missing data in the region of the cut.

3.1.2.3 Disc systems

Although analogue discs are threatened by the advent of the compact disc (cd) they are likely to be used or produced by professional studios for some time yet.

Typical performances of digital and analogue systems are given in Table 3.1.2. This listing appears to suggest that the vinyl (analogue) disc is a very inferior alternative to the compact disc. In terms of quality of reproduction this may be true, although a new vinyl disc can produce excellent quality. However, broadcast studios often rely on discs being accurately cued-up for sound effects purposes in radio or television dramas as well as for musical programmes, and many operators feel that in some circumstances they have better control over vinyl discs, which they can see, than over compact discs. The relative invulnerability of compact discs over vinyl discs remains, however, a major point in their favour.

Table 3.1.2 Comparison of digital and analogue systems

	Digital	*Analogue*
Frequency range	20 Hz–20 kHz	30 Hz–18 kHz
	0.5 dB	1 dB
Signal/noise ratio	±90 dB	±60 dB
Separation between channels	90 dB	better than 25 dB at 1 kHz
Total harmonic distortion	0.005%	0.2%
Life expectancy of recordings	Disc: probably infinite*	Disc: about 100 playings
	Laser: several thousand playings	

*The long-term effects of age on compact discs is not yet known. Shrinkage of the materials may occur, for example, although there seems to be no evidence of this at present.

Bibliography

Alkin, G. *Sound with Vision*, Newnes-Butterworths (1973).

Alkin, G. *Sound Recording and Reproduction*, Butterworths (1981).

Amos, S.W. (ed.). *Radio, TV and Audio Technical Reference Book*, Newnes-Butterworths (1977).

Nisbett, A. *The Use of Microphones*, 2nd Edn, Focal Press (1983).

Robertson, A.E. *Microphones*, 2nd Edn, Iliffe (1963). Although published so long ago this is still one of the most exhaustive works on the subject.

John D. Wardle BSc (Hons)
John Wardle Associates

3.2 Lens Systems and Optics

3.2.1 General Principles of Lenses

Most people who will at some time have used a lens to focus an image of the sun or some bright object know the principles of a simple lens. At the same time they will have observed that as the distance from the object to the lens is changed, the image focuses at a different distance and the size of the image changes. This principle of magnification is used in the production of a zoom lens. In a zoom lens the focal length can be changed continuously to alter the magnification whilst the image still remains in focus.

3.2.1.1 Magnification in a lens

If a combination of two lenses is used, and they are moved independently of each other in the correct ratio, it is possible to change the magnification whilst retaining the same focal point (Figure 3.2.1). For this, a combination of a convergent and a divergent lens is used. In real lenses each of these will in themselves be made up of groups of convergent and divergent lenses. This very simple type of zoom lens is used for both still and film photography but would not be suitable for a TV broadcast camera lens.

A TV lens is much more complex and usually has four sections: two moving groups, one called a variator, which changes the image size, and a compensator, which maintains the focus during the zooming, a front focusing group and a rear back focus or relay group, which is used to maintain the image the correct distance from the back of the lens and onto the pick-up device. In total there may be over 20 separate glass elements used (Figure 3.2.2).

The trick or skill in designing a zoom lens is to achieve the correct movement of the compensator relative to the variator in accordance with the laws of geometric optics. This used to be achieved by a barrel cam mechanism. In this, an inner barrel has a linear guide groove (with a pin following a linear slot), and an outer barrel with two curved guide slots. As the outer barrel is turned each of the lens groups follows a predetermined path determined by its own curved slot, which is designed to keep the two groups in the correct relationship. The tolerances for the positions of the lenses must be kept within micrometres so the guides must be very accurate. In many modern lenses motors drive the two lens groups, instead of cams, and the positions are set by a microprocessor and memory.

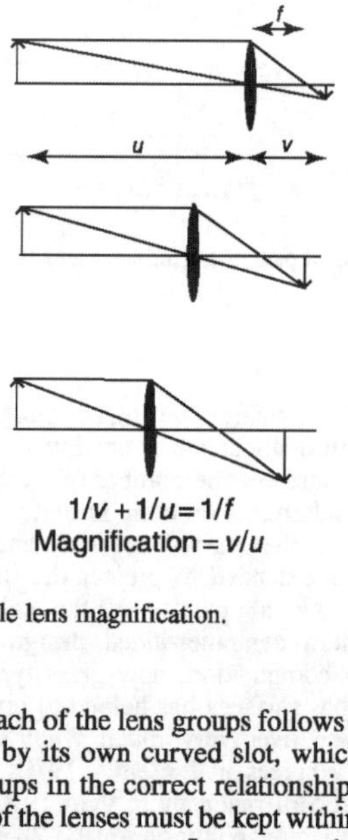

Figure 3.2.1 Simple lens magnification.

There are various options for the movement of the lens groups. It is possible to have both variator and compensator divergent when the compensator will have to move both forwards and backwards through the zoom range. This method is used in compact lightweight lenses. If the variator is divergent and a convergent compensator is used, the compensator will always move in one direction. This is used in studio/OB lenses.

Lenses suffer from various aberrations and these must be corrected both statically and dynamically as the light path

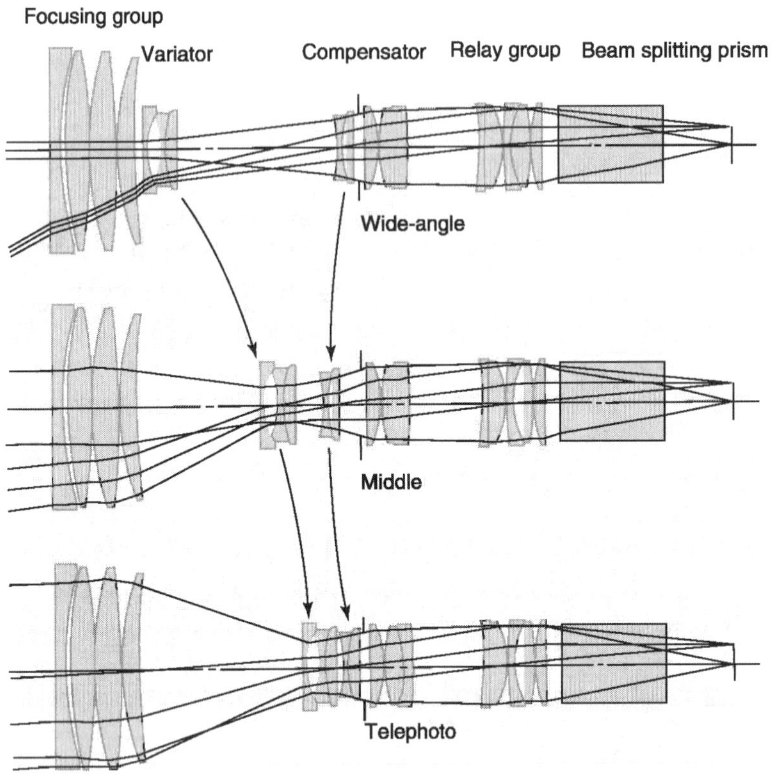

Figure 3.2.2 Optical path of a hand-held lens (Canon).

undergoes complex changes during zooming, so that the image remains sharp throughout the zoom range. It is this need for correction that increases the number of elements used in a TV zoom lens. An additional problem is that a TV camera has an optical splitter to the rear of the lens and the optical path length has to be extended to project the image through this. This increases the complexity of the calculations and the power of modern computer-aided design is much needed. The use of the computation, new glass types and the use of non-spherical lens surfaces has helped to give much improved lens performance over very much wider zoom ranges (see Figure 3.2.3). Whereas in the early 1970s it was unusual to find a zoom ratio of much more than 10× in the studio and 20× on OBs, it is now common to find 20× in studios and up to 101× on OBs.

3.2.1.2 Angle of view

As the focal length of a zoom lens is changed the angle of view will also change. It is this change in angle and magnification that makes the lens so useful in a wide variety of conditions.

At the wide-angle end of the zoom the angle of view will be at a maximum and the image size will be smallest. This can be used to view distant panoramic scenes, or in a studio where there is little room for movement away from the subject matter. Often when choosing a lens for studio use it is the wide-angle capability that determines the choice.

A telephoto zoom gives a narrow angle of view and will pick out a distant subject and make it look as if it had been viewed close to. Telephoto lenses have long focal lengths and are used for sports coverage. One effect of using a lens at the telephoto end is to compress the depth of the image and merge the background and foreground so that they seem to be at similar distances. The angle of view of a lens depends both on its focal length and the size of the image format that is being used. For example, a focal length of 8 mm on the 18 mm format represents a narrower angle of view on the 13 mm format.

3.2.1.3 Zoom ratio

The ratio between the maximum and minimum focal lengths of a zoom lens is known as the zoom ratio and may range from about five times for a lightweight lens to 100 times for a box-type OB lens at the present time. There seem to be ever-increasing zoom ratios with very little increase in lens size or weight. Lenses are usually specified by both their zoom ratio and their minimum focal length. Examples are 13 × 4.5, 20 × 7 and 101 × 8.9. These represent a very-wide-angle lightweight lens (maximum angle of view of about 93°), a typical studio lens with a fairly wide angle and an OB lens with extremely narrow angle of view (0.3°).

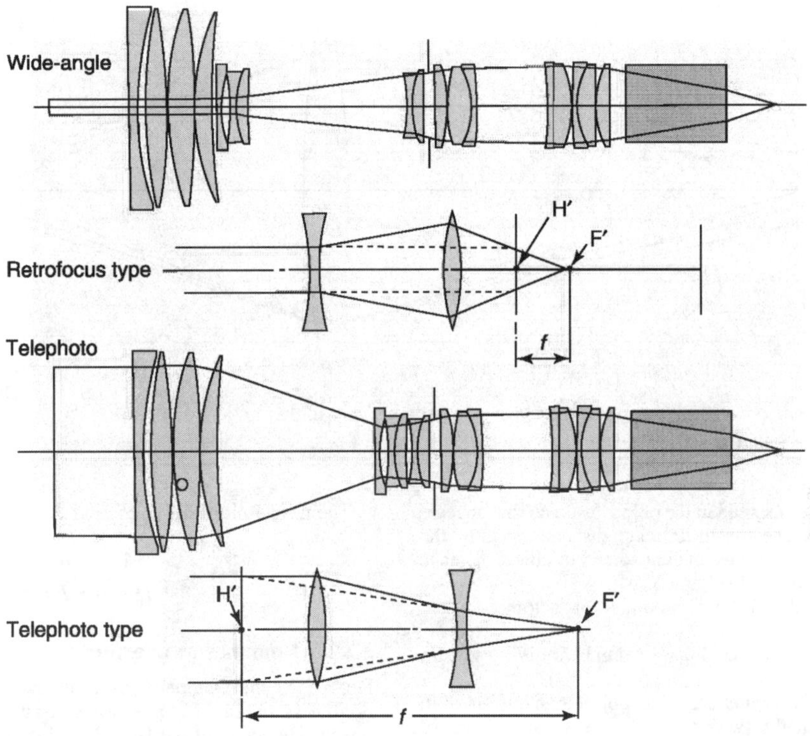

Figure 3.2.3 Lens positions at wide-angle and telephoto ends of zoom (Canon).

3.2.2 Fundamental Optics

3.2.2.1 Minimum object distance (MOD)

The minimum object distance or MOD of a lens is the closest distance at which the image can be focused. It is measured from the frontmost surface of the lens to the object. This distance depends on a number of factors. A wide-angle lens will usually have a shorter MOD, say 0.4 m, than a telephoto OB lens, which may have an MOD of 2.5 m. This difference is because of the action of the front focusing group in focusing the image. As the object moves nearer to the lens, the front focusing group must extend further out to maintain focus. At the wide-angle end of the lens this would require the size of the front elements to be very large to avoid cutting off the outer rays of light. A very large front element would be very expensive and difficult to make but would also be very heavy because of the increased volume of the glass. To avoid this the MOD is kept longer. In a studio wide-angle lens the focusing group may be split into a fixed convergent group and a moving divergent group to give a shorter MOD. This method of construction also helps reduce the lens aberrations caused by moving the front group during focusing. It also reduces the interaction between focusing and angle of view where, after zooming in to fill the frame, the action of focusing then changes the image size. A fixed front group also helps when mounting effects filters on the front of a lens as the front unit no longer rotates or moves.

To enable a lens to focus closer than the MOD, some lenses have a macro facility. When focusing a close object the image distance increases. Macro enables the lens to focus such objects without increasing the overall length. A different group of lenses is moved to bring the image back into focus. This is usually the rear focal or relay group, which normally sets the back focal distance. Another alternative where there is no macro facility would be to change the back focus or flange-back adjustment where this is adjustable. When a lens is in macro mode it is the action of zooming that is used to carry out focusing.

3.2.2.2 Depth of field

The depth of field defines the range within which a subject will appear to be in focus to the camera. All real subjects have depth and so each part of the subject will be focused at a different image distance. The spread of image distance for which the image appears to remain in focus depends on the sampling at the image plane and the smallest change that can be resolved. With film this dimension is called the 'circle of confusion' and is determined by the silver grain size. In video the smallest distance that can be resolved is approximately the distance between two scanning lines. This will also be slightly affected by the monitor quality and the viewing conditions. As each image format size has to accommodate the full number of scanning lines the 'circle of confusion' will change with the format. For instance, with the 25 mm format the 'circle of confusion' is 0.03 mm and for the 18 mm format it is 0.021 mm. The spread of image distance that appears to remain

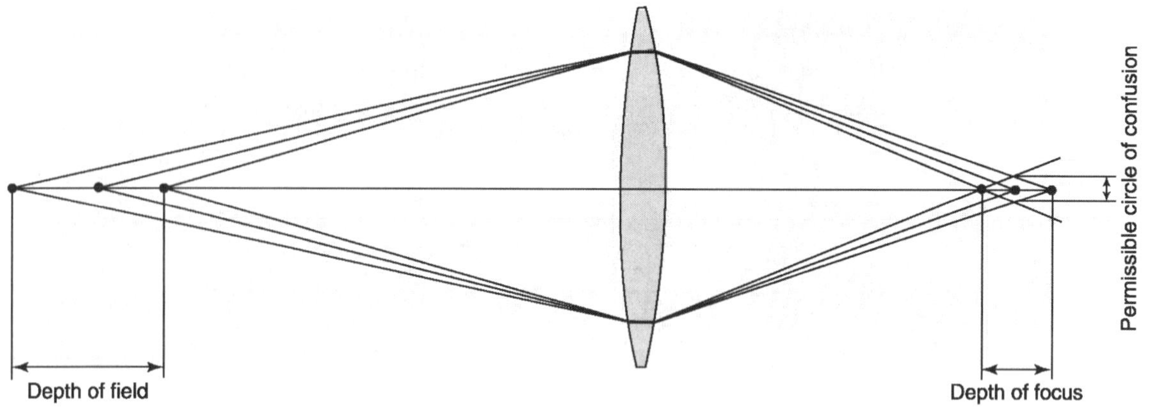

Figure 3.2.4 Depth of field (Canon).

in focus translates to a spread in the object distance that appears to remain in focus. The spread in image distance is called the depth of focus and the corresponding spread in object distance is called the depth of field (Figure 3.2.4).

The depth of field varies with a number of factors:

- lens aperture or f-number, higher f-numbers giving greater depth of field;
- the depth of field is greater behind an object than in front of it;
- shorter focal lengths (wider angle) give greater depth of field.

This variation in depth of field can be put to good use in programme production. The ability to throw the background out of focus can emphasise a subject and is easier to achieve at the telephoto end of the lens range. The ability to change the emphasis on the subject, who is speaking on a two shot, by rocking focus between the two persons is often used on drama. However, this can only be achieved by working at a wide aperture or the depth of field will be too great. This presents problems with modern CCD cameras, which are very much more sensitive, as with normal studio lighting levels the lenses are well stopped down. To achieve wide aperture it is necessary to use neutral density filters to reduce the camera sensitivity. Some broadcasters refer camera sensitivity to a standard depth of field. This standard depth of field was that attained whilst working with 4½-inch Image Orthicon cameras, where the standard aperture was f5.6. Referring this back to 25 mm cameras gives an aperture of f2.8 and f2 for 18 mm, in which case full camera output might be achieved with an illumination of less than 100 lux.

3.2.2.2.1 Calculation of depth of field

Let object distance be U, the far limit of depth of field d_1, the near limit of depth of field d_2, the aperture of f-number be Fno, the focal length f and the acceptable circle of confusion δ, then:

The far limit of the depth of field

$$d_1 = \frac{\delta \times Fno \times U^2}{f^2 - \delta \times Fno \times U}$$

The near limit of depth of field

$$d_2 = \frac{\delta \times Fno \times U^2}{f^2 + \delta \times Fno \times U}$$

3.2.2.3 f-number or aperture

For a lens the f-number gives an indication of the relative brightness of an image. A lower f-number gives a brighter image. The value of the f-number also influences the depth of field. In a simple lens the f-number is given by a simple ratio of the focal length of the lens (f) and its effective diameter (D):

$$Fno = \frac{f}{D}$$

If the focal length is fixed, the larger the aperture the smaller its f-number.

If the diameter of the lens opening is doubled its f-number is halved. As doubling the diameter quadruples the area of the opening, four times as much light flux will pass through the lens. This means that the brightness of the image is inversely proportional to the square of the f-number. The aperture ring of a lens is marked by a series of numbers with a ratio of root 2: 1.4, 2, 2.8, 4, 5.6, 8, 11, 16 and 22. Each successive number represents a doubling or halving of the brightness of the image.

However, the effective diameter of a real lens is not the same as its physical diameter but is the diameter of the image of the diaphragm as seen through the front of the lens. It is called the 'entrance pupil' (Figure 3.2.5). The position and the diameter of the entrance pupil also change with the amount of zoom or magnification. The diameter increases in proportion to the focal length and is thus larger at the telephoto end. This leads to what is known as F Drop or ramping.

The f-number is a relative measurement of the light through a lens and assumes a lens has 100% transmission. Real lenses all have transmission loss but of differing amounts depending on construction and coatings, so two lenses with the same f-number may not produce the same camera sensitivity. In the film world T-number is used and this takes account of the transmission loss. Two lenses of the same T-number will have equal sensitivity.

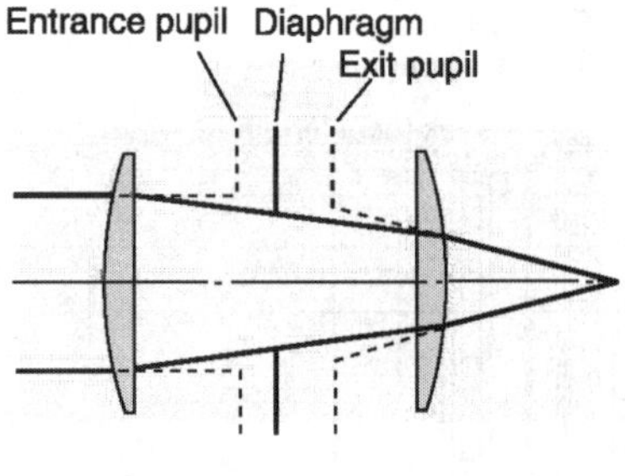

Figure 3.2.5 Entrance pupil (Canon).

$$\text{T-number} = \frac{\text{f-number}}{\sqrt{\text{Transmittance}}} \times 10$$

3.2.2.4 F Drop or ramping

As a zoom lens is zoomed towards the telephoto end, whilst maintaining the aperture setting the same, at some point the video level will be seen to fall. This effect is called F Drop or ramping and usually occurs at a zoom ratio of over 15× (Figure 3.2.6).

As the lens zooms to the telephoto end the entrance pupil enlarges. When it becomes equal to the diameter of the front focusing group it cannot get any larger, so the f-number actually falls. The only way to eliminate this would be to increase the size of the front elements of the focusing group. This would mean a large increase in weight and cost of the lens.

To keep lenses within an acceptable size it is normal to allow a certain amount of F Drop, especially on OB lenses with a high zoom range; however, in failing light conditions the effect may limit the maximum usable focal length and zoom range. Some lens designs automatically open up the iris during the zoom, to compensate for the loss until the limit is reached. Other lenses have an optional switch that can be set to limit the maximum focal length when the lighting conditions are such as to cause ramping.

3.2.2.5 Back focal length or flange-back

The flange-back is the distance from the flange at the rear of the lens mount to the image plane (Figure 3.2.7).

This distance must be identical to the distance from the camera lens mount flange to the surface of the CCD or pick-up tube. A camera has specified flange-back and a corresponding lens must be used. Typically the figure will be 48 mm in air or 58 mm in air, for the 18 mm format. Quoting the figure in air means it is independent of the optical glass used in the effects or neutral density filters, splitter block and, in the case of a pick-up tube, the faceplate. Placing a glass block behind the lens has the effect of lengthening the path.

If a block of thickness d and refractive index n is placed behind the lens the flange-back (FB) relationship is:

$$\text{FB (in air)} = \text{FB (actual)} - \left(1 - \frac{1}{n}\right) \times d$$

Thus, if $n = 1.5$ the path is lengthened by one-third of the length of the block.

A camera manufacturer will specify which flange-back is required. Most lenses will have an adjustment ring at the rear which can adjust the rear focal point by ±0.5 mm to take up any tolerances and wear. With a tubed camera this adjustment can be set to the datum and the individual camera tubes adjusted for optimum focus. However, with a CCD camera the only possible adjustment is the lens ring.

The type of glass affects the lens aberrations and compensation for this is designed into the lens. The fact that a lens is compensated, and for which type of glass, is denoted by the last two digits of the lens type number, e.g. 14 × 8B3 or 14 × 8B4.

The back focal length is the distance from the rearmost lens surface to the image plane and again is quoted as the air equivalent value.

3.2.2.6 Chromatic aberration

The refractive index of glass varies with the wavelength of the light. This is called 'dispersion' and causes 'chromatic aberration'

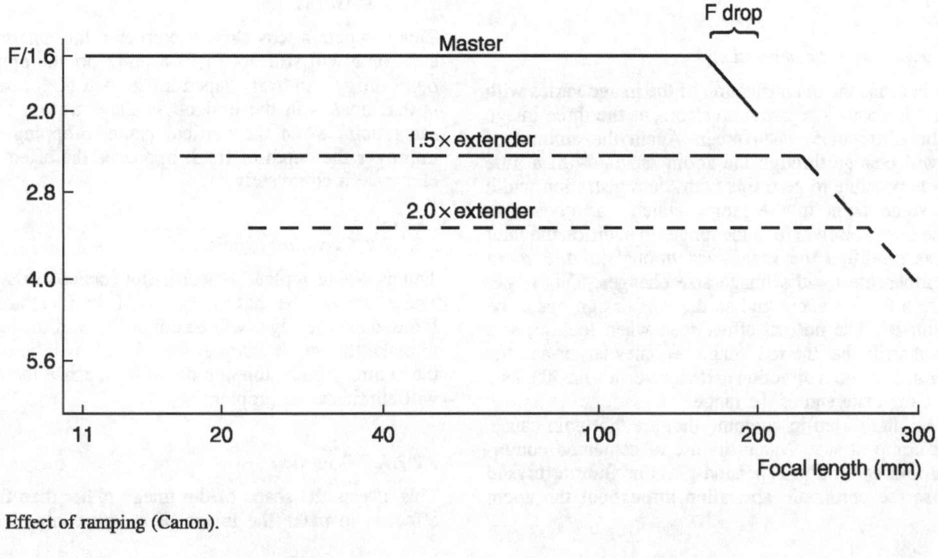

Figure 3.2.6 Effect of ramping (Canon).

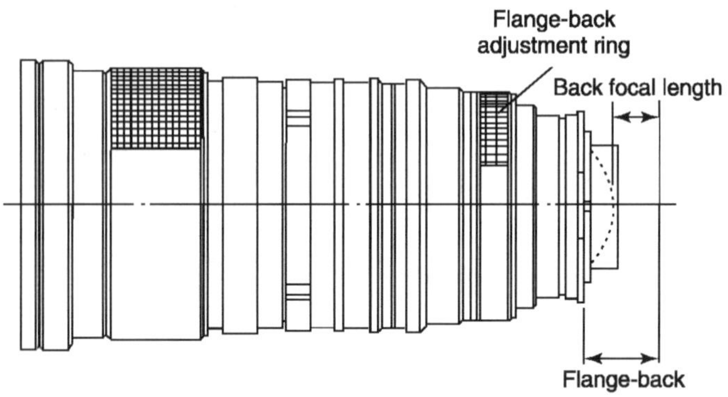

Figure 3.2.7 Back focal length and flange-back (Canon).

where light rays of a different colour are focused at different points.

There are two types of chromatic aberration: longitudinal aberration, which occurs along the axis and gives rise to tracking error, and lateral aberration, where the image size changes and causes registration errors.

3.2.2.6.1 Longitudinal chromatic aberration

This causes different wavelengths to be focused at different points along the axis. With a zoom lens this aberration changes with zoom angle and is largest at the telephoto end. If the red and blue focus at different points to green, this will result in loss of resolution and image blurring. Normal lens design corrects the chromatic aberration for two wavelengths but leaves secondary aberrations at wavelengths in between. This is inadequate for TV zoom lenses with three-colour analysis and further correction must take place. It is the focus group which is the main cause of the residual aberration, and the use of extraordinary dispersion glass or fluorite crystal in cemented combinations can be used to improve this.

3.2.2.6.2 Lateral chromatic aberration

This occurs because the magnification of the image varies with wavelength. This causes registration errors, as the three image sizes will be different to each other. Again the amount of aberration will change through the zoom range. With a tube camera it was possible to zero this with the registration width controls at some point in the range, which was chosen to minimise the errors elsewhere in the range. In addition the later tube cameras modified the scan sizes throughout the zoom range to compensate for the image size changes. This is not possible with a CCD camera and so the lens design has to be further optimised. The normal effect seen when looking at a registration chart is that the red image becomes larger and the blue smaller in one zoom direction and vice versa when the lens zooms to the opposite end of the range.

It is the variators used in zooming that are the main cause, due to their compact size. Again the use of cemented combinations with extraordinary dispersion glass or fluorite crystal can minimise the chromatic aberration throughout the zoom range.

3.2.2.7 Lens distortions

Seidel classified the five basic aberrations of lenses that are present in addition to chromatic aberration. These are outlined below.

3.2.2.7.1 Spherical aberration

This is where the rays of light which leave the outer edge of the lens are focused at a different point to the rays which are near the axis. Spherical aberration is improved by stopping down the lens and is minimised at about two to three stops below full aperture. Further stopping down causes a reduction in sharpness due to diffraction effects.

3.2.2.7.2 Coma

This is where rays of light not parallel to the axis are focused at a point not on the image plane. These give rise to a comet-like image with a tail. Coma is also improved by stopping down the lens.

3.2.2.7.3 Astigmatism

This is where a lens already corrected for spherical aberration and coma will still not focus a point off the optical axis to a point image. An oval-shaped image will be formed. The effect is that images in the horizontal plane are not focused at the same point as in the vertical plane. Stopping down, which improves the depth of field, improves the effect but does not eliminate it completely.

3.2.2.7.4 Curvature of field

This is where a plane object is not focused as a plane image. The effect will be that the centre of the image can be sharply focused but the edges will be out of focus. If the lens is adjusted to make the image sharp at the edges it will be out of focus at the centre. Again stopping down to increase the depth of field will eliminate the problem.

3.2.2.7.5 Distortion

This affects the shape of the image rather than the focus. The effect is to make the image of a square object appear barrel

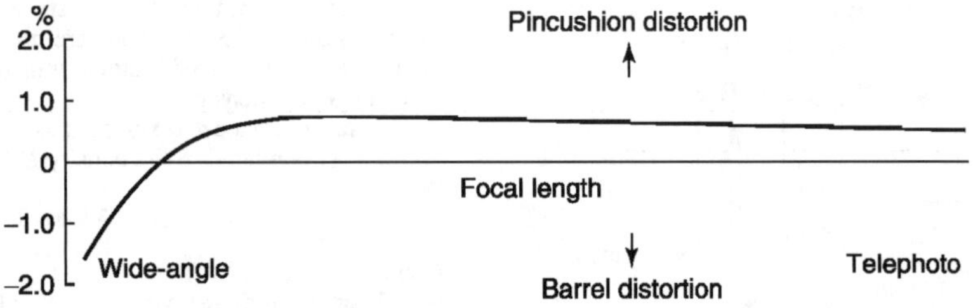

Figure 3.2.8 Distortion in a TV lens (Canon).

shaped or pincushion shaped. The distortion is usually expressed as a percentage and is a measure of the error in height at the edge of the picture compared to the centre line. For a zoom lens the distortion is usually barrel (negative) at the wide-angle end and pincushion (positive) at the telephoto end (Figure 3.2.8).

$$\text{Distortion}(\%) = \frac{\Delta h}{h} \times 100$$

3.2.2.8 Resolving power and MTF

The resolving power of a lens is used to express its performance and the sharpness of image it will reproduce. The simplest way of measuring this is by using a chart with patterns of vertical or horizontal lines of differing widths. The different widths represent different frequencies in television terms. Camera manufacturers often quote camera resolution as the number of lines and as with lenses this tells us very little about the performance. A lens may have a high resolving power in the number of line pairs it will resolve but it should be noted that this is the limiting resolution and tells us nothing about the shape of the response characteristic. At worst, there could be a hole in the middle of the response, where it matters a great deal.

The television system has a defined bandwidth and we need to maintain a good flat response within that band. So a lens with high performance outside this band may not necessarily give the best results. What matters is the performance within the TV bandwidth and the lens should be optimised for this.

The bandwidth for the 625-line system is nominally 5 MHz and the spatial frequency equivalent to this is given by the formula:

$$\text{Spatial frequency} = 65.873 \times \frac{\text{video frequency (MHz)}}{\text{image diagonal (mm)}}$$

Thus, for the 18 mm format, 5 MHz is produced by a spatial frequency of 29.9 line pairs/mm.

As resolving power gives no indication of the overall performance of a lens, a measure called the Modulation Transfer Function (MTF) is used instead. To generate this characteristic the lens is swept through its frequency range using a sine-wave chart and its output measured by a sensor at each frequency. The response curve can then be plotted in the same way as the frequency response of a camera or amplifier (Figure 3.2.9).

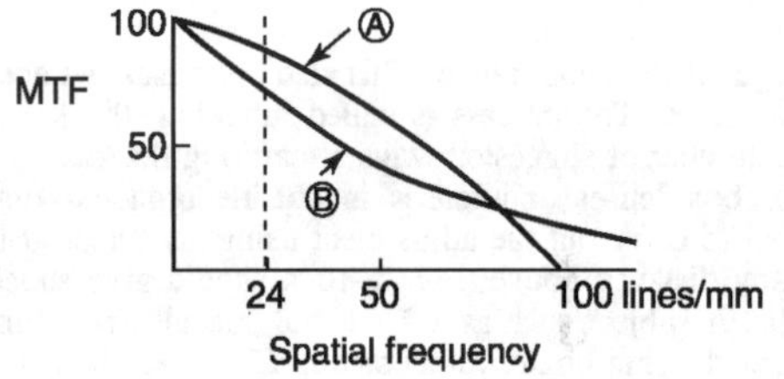

Figure 3.2.9 Typical MTF plot of two lens showing lenses A and B, A being better (Canon).

3.2.2.9 Transmission and coatings

As has been stated earlier, a TV lens has many elements and light is lost at each surface by reflection. The loss is from 4% to 10% unless other steps are taken to reduce this. Also the reflected light is scattered and causes flares and ghost images that reduce the contrast and resolving power.

To reduce the reflection and loss at each surface, special coatings are used which use interference effects to reduce the reflection. Multi-layer coatings are deposited onto the lens surfaces in a vacuum chamber, each layer having a different characteristic.

To reduce the ghost images and flares a combination of anti-reflection paint, anti-reflection grooves and hair flocking is used at different points within the lens barrel.

3.2.3 Standardisation of Lens Back Focus for CCD Cameras

As discussed earlier, it was possible to take up the tolerances in the back focus of the three colours with tubed cameras by adjusting the tube positions. CCDs are cemented into a fixed position and cannot be adjusted, so it was necessary to tightly define the image positions of CCD lenses when these came into use. It would be expected that setting their foci at the same distance from the rear of the lens would optimise the three images. However, this is not the case and a better compromise for the aberrations can be achieved with the red, green and blue all being focused at different points. The optimum distances were standardised for the 18 mm image size and 48 mm back focus in the *SMPTE Journal* in September 1989 (Figure 3.2.10).

An EBU Standard has also been set for HDTV lenses for the 25 mm format.

3.2.4 Setting of the Back Focus or Tracking (flange-back adjustment)

To get the best performance out of a lens it is necessary to set the back focus so that the image remains at optimum focus

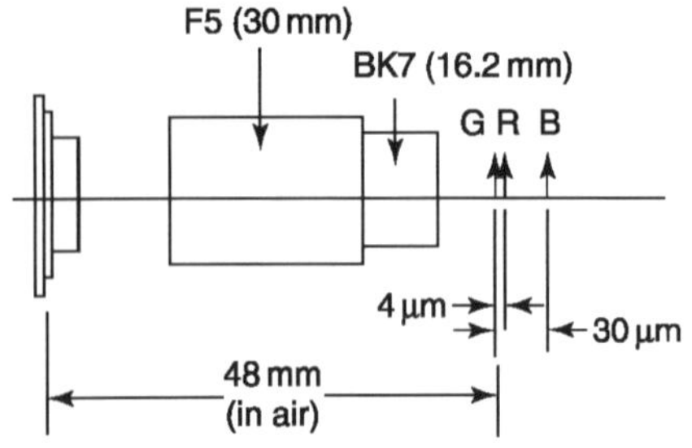

Figure 3.2.10 Standardisation of CCD mounting positions.

throughout the zoom range. This also optimises the aberration performance. The process is called 'tracking' the lens. There are a number of suggested ways of carrying this out.

For box lenses and lenses in outside broadcast use it is normal to carry out the adjustment using an object at infinity but any distance above, say, 5–10 m should give satisfactory results. A subject such as a TV aerial, distant tree, aluminium camera carrying box or Siemens Star Chart may be chosen. To get the most accurate results the lens aperture should be fully open. In bright conditions this may mean the use of the neutral density filters or the use of the electronic shutter to reduce the camera sensitivity.

The lens is first zoomed to its most narrow angle (telephoto end) and the lens front focus adjusted for best image focus using the focus demand control. The lens is then set to the wide-angle end and the lens back focus ring adjusted for optimum focus. Then zoom to narrow again, readjust the front focus for optimum image quality, zoom to wide and readjust back focus if necessary. The process is iterative. Partly tighten the rear focus adjustment then zoom to both ends again and check the image before fully locking the ring. Do not over-tighten!

For lightweight lenses used for News, PSC, etc. it is said that a better result is obtained by tracking the lens at a distance of about 3 metres as it will be used mainly at these distances. The author has never been able to verify the difference. The process is otherwise identical except that the focus ring of the lens may be adjusted manually instead of by a focus demand control.

3.2.5 Optical Accessories

3.2.5.1 Filters

Most cameras have positions for one or two filter wheels for using neutral density, colour correction or effects filters.

Any filter placed in the optical path must be optically flat if it is not to affect the overall resolution of the lens/camera. It must also have a defined thickness, as a different path length will change the back focus position. Switching, say, between the clear filter and a two-stop ND filter should not cause the lens tracking to fail. So a thickness tolerance has to be set which is of the order of 12 μm.

Neutral density filters are used to control exposure in bright lighting situations, as the range of lighting outside will be much too great for the usual seven-stop range of the iris (a camera may have to handle a range of 10,000 to 1). Also lenses should ideally not be used in the fully stopped down condition because of diffraction effects. A typical neutral density filter selection may be a two-stop, four-stop and six-stop or 1/4, 1/16 and 1/64 transmission. Various methods are used to denote the strength of the filter: number of F stops, transmission as a fraction or percentage and density.

The transmittance T (expressed as a decimal fraction, where $100\% = 1$) is related to the density D by the formula:

$$D = -\log_{10} T$$

Neutral density filters must have a flat transmission to all wavelengths within the camera's optical response range, so that a change of filter does not affect the colour balance of the camera. This is difficult to achieve for the densest filters.

Colour correction filters are used to correct the cameras response for different lighting conditions. Camera processing is nominally balanced for tungsten lighting conditions. Tungsten lighting has a colour temperature of about 3.200 K. Colour temperature expresses the balance of colours and refers to the colour of a perfect black body when heated to that temperature (in degrees Kelvin). A low colour temperature is reddish and a high colour temperature is blue. If a camera balanced for 3.200 K observes a white chart lit by daylight it will appear very blue, whereas the eye has a complex auto white balancing function which makes the chart still look white in daylight. An amber filter in front of the camera will correct the response and its output will now look white.

The colour temperature the camera may be required to work under may range from 1500 to 20,000 K and will be controlled by the use of the camera gain balance and colour correction filters.

A number of colour correction filters may be fitted in the filter wheel and are often combined with the neutral density filter, so for instance you may find 5600 + 1/4 ND and 5600 + 1/16 ND on an ENG camera wheel. This gives correction for nominal daylight.

The colour conversion capability of a colour correction filter is also expressed in terms of mireds or microreciprocal degrees, which is 1,000,000 divided by the Kelvin temperature, or decamired (10 mired) units. This makes it easy to calculate the shift by adding the mired values together.

A filter which converts a colour temperature T_1 to a colour temperature T_2 has a mired value of:

$$\frac{1,000,000}{T_1} - \frac{1,000,000}{T_2}$$

Effects filters provide specific effects that a production may wish to use; these may typically be soft-focus, fog and cross or star filters.

Soft-focus filters either soften the whole image or soften the edges whilst leaving the centre sharp. Fog filters soften the whole image and create a misty dramatic effect; they come in varying grades. Star filters are very popular and cause areas of highlights (such as lamps) to form a star-like image with either a four-, six- or eight-pointed effect. They also slightly degrade the whole image quality.

3.2.5.2 Aspect ratio and minifiers

When the widescreen 16:9 format was introduced it was necessary to modify the lenses to make them suitable for displaying the wider image, as a larger image width has to be covered. The first consideration was to modify the shape of any masks which had been used to minimise flares and out-of-frame information in the 4:3 format – for instance, the mask on the front of the focus group.

The other disadvantage for a switchable format camera with 16:9 sensors is when it is used in the 4:3 format and only the centre 4:3 cut out is used; the angle of view is reduced for the same focal length, removing the advantage offered by wide-angle lenses. This may be overcome by the use of a 'minifier' lens within the rear group of the lens. The minifier has the opposite effect to a range extender and is a lens with 0.8 times magnification. The result of switching in the minifier is to concentrate the image onto the smaller area. This also results in a small increase in the camera sensitivity. The minifier is also called a 'crossover' by Canon and 'WidePower' by Fujinon.

In lightweight lenses the minifier may sometimes be provided in addition to the ×2 extender, or alternatively may replace it.

3.2.6 HDTV Lenses

Lenses for HDTV have to be of a higher specification, as the 'circle of confusion' is smaller and the camera bandwidth greater, the depth of field being halved. As HD techniques are used for film-making, the depth of field is very important and wide aperture lenses are used to minimise this. To get the same depth of field as in standard TV the f-number must be doubled, which also reduces the sensitivity by a factor of four. With tubed cameras this was a problem but modern CCDs are very sensitive.

As the pixel size is smaller the various aberrations are much more important and special techniques are used to minimise these, including the increased use of low dispersion glasses and fluorite.

The EBU produced a specification for a common interface for 25 mm format HDTV lenses in 1991.

3.2.7 Lens Control and Mechanics

All lenses, except some fixed focal length lenses, have motor-driven control. Motor-driven zoom and optional remote control for the zoom and focus drives are available for lightweight lenses, whereas box lenses almost always have full servo control.

The control originally took a simple form in lightweight lenses but larger lenses have very accurate servo drives for smooth and responsive control. The systems were analogue but more recently digital drive systems have come into use, even on lightweight lenses. Digital control using microprocessors gives more memory options and the ability to vary the drive characteristics and control laws.

A basic lens servo package will consist of a drive motor, a tacho-generator for measurement of the speed and a positional sensor. The lens controller will be set to achieve a certain iris or focus position, a difference amplifier will then measure the difference between the current position and the required position and the output used to drive the motor. As the difference becomes smaller the driving voltage is reduced until it becomes zero and the motor stops. The overall gain of the loop must be set as a compromise between speed of control and overshoots. The tacho-generator is also fed into the feedback loop to modify the response characteristic and damping.

The iris and focus drives work this way. For the zoom control a shot box will select a preset zoom angle this way, but the zoom demand control operates in a slightly different way. The zoom demand is on a rocker where only the direction and speed of control are determined. There is no positional information and the lens only ceases to be driven when the control returns to the centre null point, or reaches either end of its focal length range. The potentiometer on the demand unit usually has a square law so that zoom speed is not linearly related to its displacement from the centre null point, but increases rapidly at greater displacement. The potentiometers used in analogue servo controls were very expensive to replace.

With the introduction of microprocessors into lenses for the control of the variators during zooming, the next logical step was to use digital control for the motor drives. All positional, directional and speed information is fed into the 3.2-bit processor for optimisation of the control. The potentiometers in the zoom demand and other controllers are replaced by digipots, which are much cheaper and more reliable. The digipots are linear but the use of a microprocessor, within the controller, allows the zoom demand law to be modified by a number of characteristics stored in a PROM. The focus controller also has a processor. Many new features are also introduced, such as connection to a PC for diagnostic purposes. Canon have also introduced 'shuttle shot' (full speed movement between two focal lengths which can be selected), 'speed preset' (choice of memorising a particular zoom speed) and 'framing preset' (for memorising a particular angle of view without using a shot box). All of these are for lightweight lenses and are controlled by three extra buttons on the lens adjacent to the zoom rocker. They are also repeated on a new zoom demand unit.

3.2.8 Image Stabilisers

It is very difficult to prevent image shake at long focal lengths even when the camera is mounted on a pedestal. As the maximum focal lengths available become ever greater the problem will not go away. Some years ago, Canon introduced an image stabiliser on a lightweight lens to overcome this problem. There is also a version to mount on the front of a standard lens. The 100× Ultra Zoom Ratio box lens also has a built-in image stabiliser but of a different design. Fuji also have an image stabiliser lens and adaptor.

There are two techniques used for image stabilisers. The lightweight lenses (and the add-on unit) use a 'Vari-Angle Prism' image stabiliser and the box lens uses a 'Shift' image stabiliser (Figures 3.2.11 and 3.2.12).

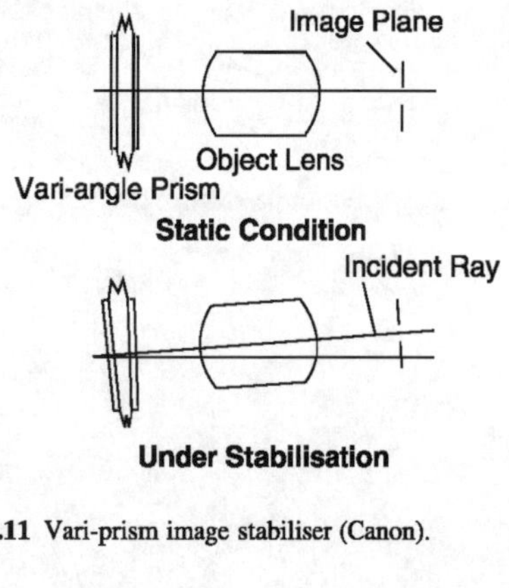

Figure 3.2.11 Vari-prism image stabiliser (Canon).

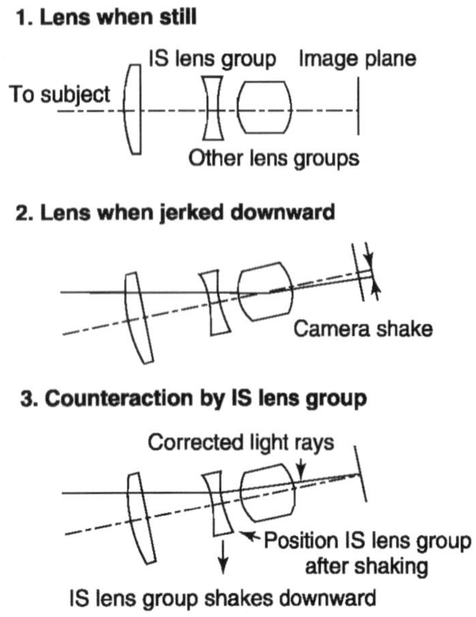

Figure 3.2.12 Optical shift image stabilisers (Canon).

A Vari-Angle IS is mounted in front of the lens and covers a higher bandwidth of movement, including the high frequencies that are found in vehicles and helicopters. Shake detecting sensors detect the speed and angle of movement, and feed the information to a microprocessor, which then drives actuators that are used to modify the shape of a variable angle prism in the IS.

The Optical shift IS is mounted internally and allows for a compact design and lighter weight. This system is more effective for lower frequency movements caused by handshake, platform vibration or wind effect.

Here two shake detecting sensors, one for yaw and one for pitch, send information concerning angle and speed to the 3.2-bit processor. This then drives the IS lens group to minimise the shake. When a lens moves, the light rays from a subject are bent relative to the optical axis, causing a blurred image because the light rays are deflected. By shifting the IS lens group on a plane perpendicular to the optical axis to counter the degree of image shake, the light rays reaching the image can be steadied. When the lens is deflected downwards by shake, the centre of the image is also moved downwards in the image plane. If the IS lens shifts in the vertical plane the light rays may be refracted back to their correct position so that the image centre returns to the centre of the image plane. The IS lens can move in both the horizontal and vertical planes.

Bibliography

Canon. *TV Optics 2. The Canon Guide Book of Optics for Television Systems* (1992).

Paul Cameron

3.3 Optical Sensors

3.3.1 Early Image Sensors

3.3.1.1 Selenium detectors

Selenium was the first photoelectric material to be found, in 1873. It was used in the first mechanical experiments in television, like the Nipkow disk system. Selenium is classed as a photoconductive material because its resistance changes when exposed to light.

3.3.1.2 The Ionoscope

The Ionoscope was the first image sensor of any commercial importance. It consisted of an evacuated glass enclosure with a tube fixed to it enclosing an electron gun (Figure 3.3.1). The main enclosure had a screen made from a sandwich of photosensitive particles, called a mosaic, a thin mica insulation layer and a conductive sheet backing. The plate acted like a capacitor.

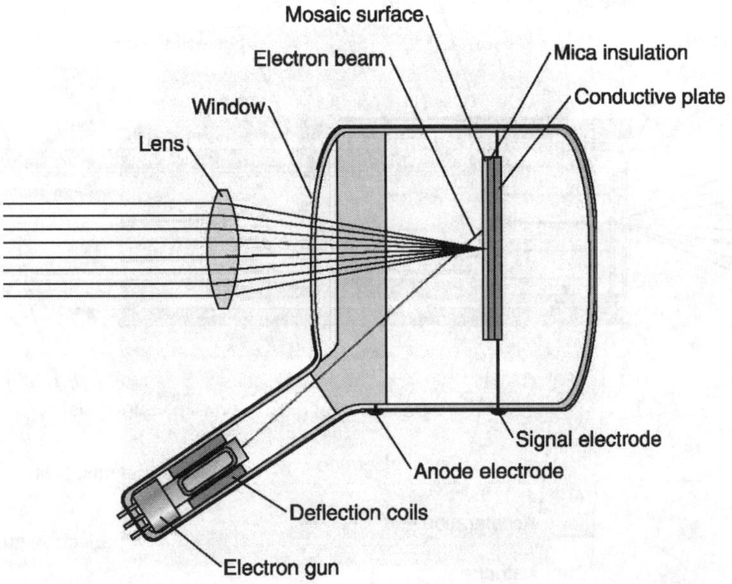

Figure 3.3.1 Schematic of the Ionoscope.

Light from the lens could enter the enclosure through a window and land on the mosaic, releasing electrons which were attracted away towards the anode. Thus, a positive charge image built up on the surface of the mosaic. The charge was proportional to the intensity of light. The electron gun fired electrons in a raster scan at the mosaic. Any positive charge was cancelled by absorption of electrons from the beam. This absorption was detected by the conductive plate and output as a signal at the signal electrode. The rest of the electrons bounced off the mosaic to be picked up by the anode and were drawn away to the anode electrode.

The Ionoscope used a high voltage between the electron gun and the anode. This resulted in a high-velocity electron beam, which tended to 'splash' off the mosaic surface, causing what is termed secondary emission. This tended to reduce the tube's ability to catch all the electrons released purely by photoemission.

The Ionoscope's geometry was not ideal. It was difficult to focus the electron beam because it was striking the mosaic at an angle. The angled electron gun also made it difficult to produce an accurate raster scan.

The Orthicon tube design helped eliminate these deficiencies.

3.3.1.3 The Orthicon tube

The Orthicon tube was invented by Iams and Rose at RCA in 1939 (see Figure 3.3.2).

The Orthicon tube had a much reduced anode voltage. This tended to saturate the mosaic with electrons when no light was present. Any more electrons would not strike the surface. Thus, no signal appears when there is no light. This produces better black recognition. Any electrons not striking the surface of the mosaic return back down the same path as the electron beam to soak away in a collector next to the electron gun.

The beam focusing and deflection was better than for the Ionoscope. A long focus coil was used, and either electrostatic or electromagnetic deflection. The beam retained its helical nature. This aided focusing. The beam was also deflected such that it always struck the target at a perpendicular angle.

The low anode voltage and the resulting low velocity of the electron beam meant that the beam was subject to interference by stray electric fields near the mosaic. This resulted in a loss in resolution compared to the Ionoscope.

3.3.1.4 The Image Orthicon tube

The Image Orthicon was an improvement over the Orthicon. In this design light was focused onto a photocathode plate. Light striking the plate released electrons which were attracted by an accelerating grid back into the tube towards a two-sided glass target plate. Thus, an image in electrons was formed on the target.

A thin mesh was placed in front of the target. This removed secondary emission. The electrons from the photocathode penetrated straight through the mesh and onto the target. However, any secondary electron emission was soaked up by the mesh.

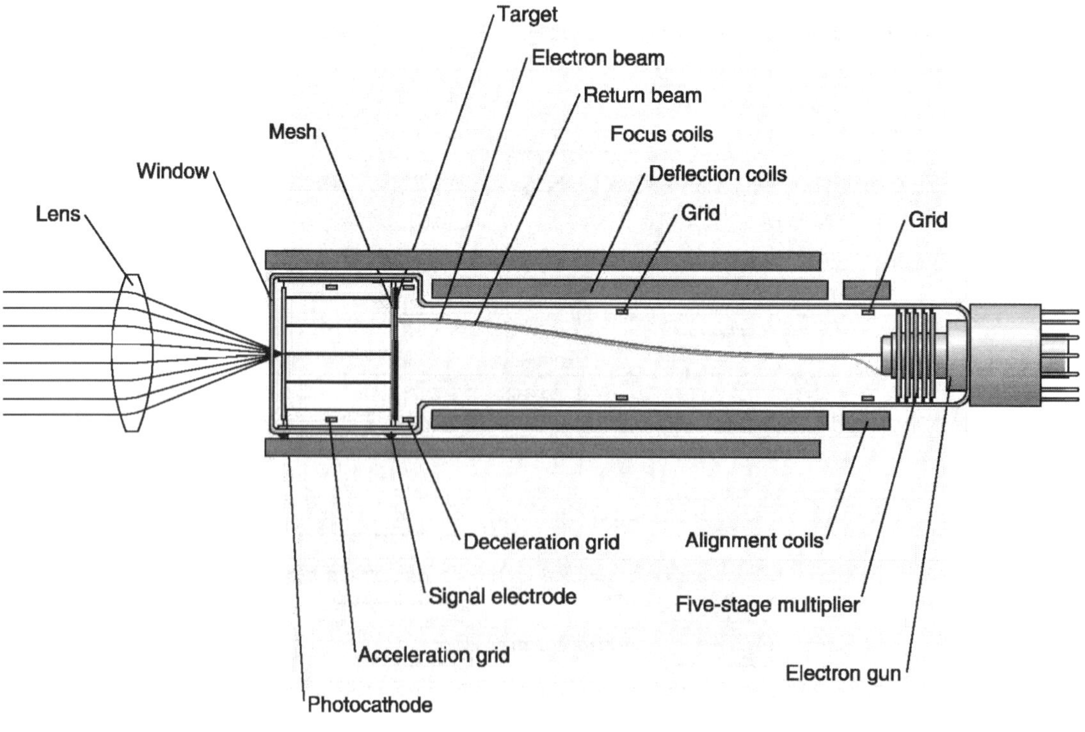

Figure 3.3.2 Schematic of the Orthicon tube.

The electron beam was a similar low-velocity perpendicular design as the Orthicon. It scanned a raster image on the back of the target. Any electrons not being soaked up by the target were returned back down the beam to be collected by the anode, next to the electron gun. Thus, the return beam was a raster scan of the charge, and thus of the image.

3.3.1.5 The Vidicon tube

The Vidicon tube was introduced by RCA in 1950 (Figure 3.3.3). It used an antimony trisulphide photoconductive target. The resistance across the target changes when it is exposed to light. Within certain limits the change of resistance is proportional to the intensity of the light.

The back of the target is scanned by an electron gun with the same basic design features as the low-velocity perpendicular design used in the Orthicon tube. The target is biased to the anode voltage. As the beam strikes the back of the target current flow to the front is inversely proportional to the resistance, which is inversely proportional to the light intensity. Thus, the anode bias voltage will alter as a raster scan of the image.

3.3.1.6 Variations on the Vidicon design

There were various improvements on the basic Vidicon tube design. The Plumbicon was introduced by Philips in 1962. It used lead oxide as a target material. The Saticon was another design with a target made from arsenic, selenium and tellurium. The Diode Gun Plumbicon used a photo diode instead of an electron gun. These later designs offered better resolution, greater contrast and better colour balance than the basic Vidicon.

3.3.1.7 Impending revolution

The discoveries and advances in semiconductor technology during the 1950s and 1960s suggested that semiconductors could be used as a sensor in television cameras. Early semiconductor sensor designs did not achieve the quality required by television. Semiconductor sensors attracted a reputation for being an inferior alternative to tube sensors. This reputation stuck even though semiconductor sensors continued to improve.

Eventually, during the 1980s the broadcast community eventually realised that semiconductor camera sensors had some very real advantages over tube sensors. Within a few years the whole broadcast industry stopped using tube sensors and over 100 years of development was dropped.

3.3.2 Dichroic Blocks

3.3.2.1 The purpose of a dichroic block

The purpose of a dichroic block is to split an incoming colour image into its three primary colours. Most of the block is coated in black paint to stop light getting in, except for a window to let the incoming colour image in and three windows to let the outgoing primary images out. They are fitted just behind the lens of a colour video camera. A sensor is placed on each outgoing window, one for each primary. Each sensor measures the brightness of each primary and sends out a video signal for each primary.

3.3.2.2 Mirrors and filters

Various designs have been created over the years. Most designs are now beginning to look very similar. Two basic design patterns are now in use. The first is generally simply called a prism block or dichroic block. The other is called a cross block or X block.

3.3.2.2.1 Conventional dichroic blocks

The conventional prism block consists of at least three prisms, glued together with a transparent epoxy cement (Figure 3.3.4). Light enters the incoming window at the front and passes through the first prism. The back surface is angled, and is coated with a red dichroic mirror. The choice of material, and

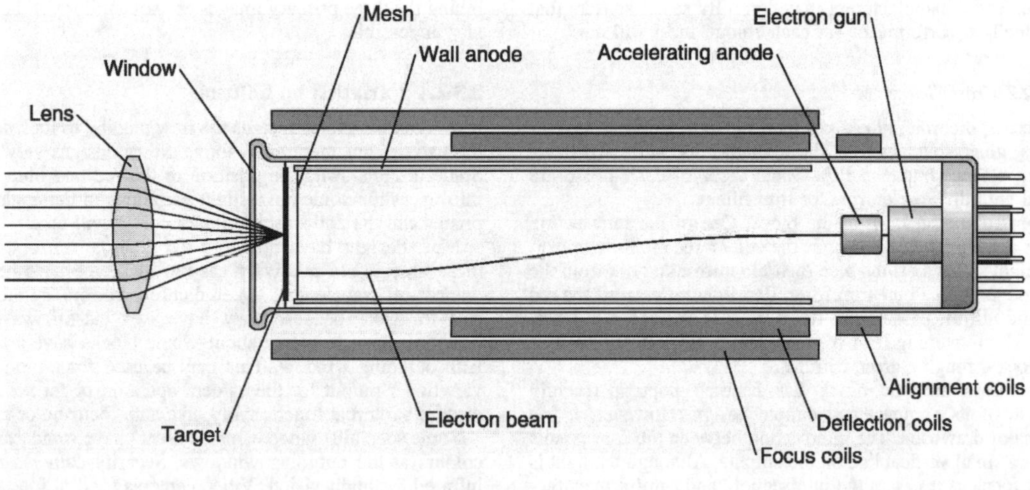

Figure 3.3.3 Schematic of the Vidicon tube.

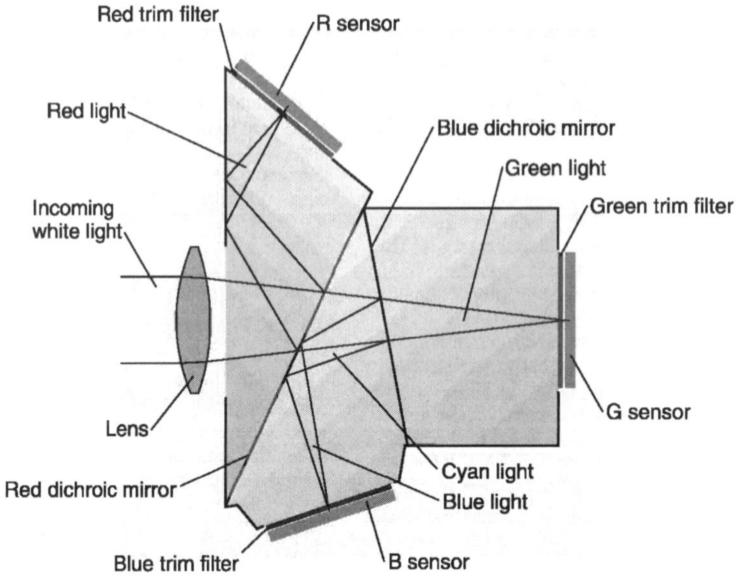

Figure 3.3.4 Conventional dichroic block.

its thickness, define the colour that is reflected. Manufacturers can 'tune' the dichroic mirror by altering the coating thickness. The red light is reflected once more off the front of the first prism and out through the red outgoing window. There is a red filter to trim the light before it strikes the red sensor.

The cyan light passes through the second prism. The back of this prism is coated with a blue dichroic mirror that reflects blue light, letting everything else through (green). The blue light passes out through the blue outgoing window, through a blue trim filter and onto the blue sensor. Likewise, the remaining green light passes out through the green outgoing window, through a green trim filter and onto the green sensor.

It is worth noting that the sensors are basically the same device. Some manufacturers may carefully select sensors that have the best performance for each colour, most will not.

3.3.2.2.2 Cross dichroic block

The cross dichroic block consists of four small triangular prisms, glued together to make a small cube with two intersecting planes (Figure 3.3.5). Some faces of each prism are coated with dichroic mirrors, or trim filters.

Light enters the front of the block. One of the intersecting planes is a blue dichroic mirror, the other a red dichroic mirror. Blue light reflects off the blue dichroic mirror and out from the left side through a blue trim filter. Red light reflects off the red dichroic mirror, passing out the right side through a red trim filter. The remaining light is green, and passes out the back of the block through a green trim filter.

The cross dichroic block has become popular recently because of its compact and simple design. However, it has one major drawback. The intersection between the four prisms causes a small vertical line on the outputs. Although the light is out of focus as it passes the intersection, and careful manufacture can make this intersection as tight as possible, this is the main reason the cross block is not used on professional and broadcast video cameras.

3.3.2.3 Optical requirements of a dichroic block

Every optical path, from the incoming window to each outgoing window, is identical. This is essential, because the lens will focus through the dichroic block and onto the surface of the sensors behind. If one of the optical paths is different, that particular primary colour will be out of focus.

The position of the sensors is critical. All three sensors must be mounted in exactly the same place relative to its own window. If there is any error, that particular primary colour image will be in a different position to the other two. Recombining the three primary images on the monitor will be practically impossible.

3.3.2.4 Variation on a theme

Most dichroic block designs are now tending to look similar to the two designs mentioned above. Some designs vary slightly. Some designs swap the position of the red and blue dichroic mirrors, while some have slight variations in the angles of the prisms and the paths each primary colour will take.

Most designs have blue and red dichroic mirrors. Both of these mirrors are relatively easy to make, because both have one cut-off wavelength. Green dichroic mirrors are more difficult to make because they have two cut-off wavelengths designers have to worry about. Some blocks have a fourth or fifth outgoing window. This may be used for a monochrome viewfinder output for the camera operator, or for some of the camera's internal functionality, like auto focusing or metering.

Some specialist video cameras do not have standard primary colours at the outgoing windows. Security cameras may use infrared for night vision. Video cameras used in food processing and monitoring also use infrared to check the quality of

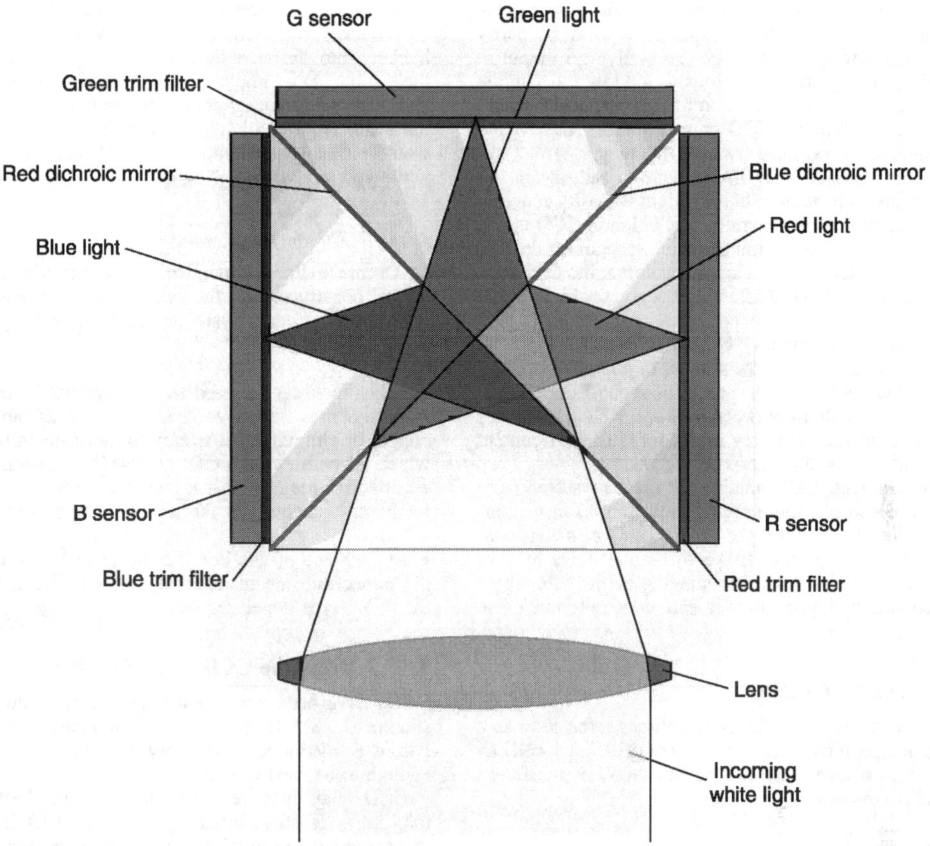

Figure 3.3.5 Cross dichroic block.

food. These cameras may have one window in the dichroic block dedicated to infrared.

3.3.2.5 Using dichroic blocks in projectors

The increased popularity of low-cost video projectors has led to an explosion in the need for cheap compact dichroic blocks. Quality is not so much of an issue with projectors, and the cross dichroic block is therefore very popular.

Dichroic blocks are used the opposite way round from video cameras. Simple filters split light from the lamp into three primary beams. These three beams are passed into the dichroic blocks through light valves, where the sensors would be in a video camera. The light valves build up an image for each primary by shutting light on or off for each pixel. The dichroic block then combines the three primary images into one colour image that is projected out to the screen.

3.3.3 Charge-Coupled Device (CCD) Sensors

3.3.3.1 Advantages of CCD image sensors

When looking at the advantages of CCD image sensors, one has to realise what alternatives there are and what was used before these devices became available. Before CCD image sensors became popular, video and television cameras used some form of tube sensor. The Vidicon tube and its variations were very popular before CCD sensors became popular. Bearing these devices in mind, let us consider the advantages of CCD image sensors.

- *Compact design*. The first and most obvious advantage of CCD image sensors is that they are considerably smaller than tube sensors. They allow very compact cameras to be made which can be used in discreet surveillance and remote investigation in dangerous or confined places.
- *Light design*. CCD image sensors are considerably lighter than tube sensors. They can weigh only a few ounces. This allows them to be designed into portable cameras without contributing to the overall weight of the camera by any undue amount.
- *High shock resistance*. CCD image sensors have no moving parts. They also have a very light-duty mechanical construction that is highly resistant to acceleration and deceleration damage.
- *Low power consumption*. CCD image sensors use a lot less power than older tube sensors. This makes them suitable for any battery-powered device.
- *Good linearity*. Linearity is important in measuring light levels accurately. Linearity means that the output signal is

proportional to the number of photons of light entering the device. Film and tube sensors are highly non-linear, partly because of their low dynamic range. They give no output at all if the light level (number of photons) is too low and saturate if the light level is too high, giving no further output if the light intensity increases further. CCD image sensors have good dynamic range, and good linearity over this range.

- *Good dynamic range*. CCD image sensors saturate in the same way as any light sensor, but the light intensity required to saturate these devices is generally much higher. CCD image sensors have no effective minimum. Some specialised devices can measure near total dark. Typically, photographic film has a dynamic range of about 100. CCD image sensors achieve about 10,000.
- *High QE (quantum efficiency)*. QE is the ratio of the number of photons of light detected to the number of photons that enter the device. Photographic film has a QE of about 5–20%. CCD image sensors have a QE of between 50% and 90%. This makes them very efficient and thus very useful for dark environment monitoring and studies of deep space.
- *Low noise*. In fact, CCD image sensors can suffer from thermal noise. However, this noise if predictable can be controlled or reduced. Cooling the image sensor using conventional cooling and a fan can keep thermal noise to very low levels. For specialist scientific imaging Peltier effect heat pump and cooling by liquid nitrogen can reduce thermal noise to virtually zero.

3.3.3.2 The basics of a CCD

A charge-coupled device (CCD) is sometimes referred to as a bucket brigade line. It consists of a series of cells. Each cell can store an electric charge. The charge can then be transferred from one cell to the next.

3.3.3.2.1 The line of buckets

A very good way of thinking of a CCD is to imagine a line of buckets (Figure 3.3.6). At one end is a set of digital scales where you can measure the amount of water you pour into the first bucket.

As soon as you pour the water from the digital scales into the first bucket you will transfer the water from all the other buckets into the next bucket down the line. The water from the last bucket will be poured into the digital scales at the other end, and measured. Of course, it would be impossible to move the water from one bucket to the next at the same time. You would probably need another set of buckets to store the water while you were transferring it.

3.3.3.2.2 The electronic reality

CCDs use a line of metal oxide semiconductor (MOS) elements, constructed on the same chip. Each element contains two, three or four polysilicon regions sitting on top of a thin layer of silicon oxide.

- Polysilicon can be used as a charge holder or a conductor. Although it is not as good a conductor as other metals like copper or aluminium, it is easy to fabricate and is transparent, which is useful when CCDs are used in cameras.
- Silicon oxide (glass) is a good insulator.
- These elements are fabricated on a p-type doped silicon substrate.
- At each end of this line is a region of n-type doped silicon.
- Connections are made to all the polysilicon regions and to the two n-type doped regions.

3.3.3.3 Using the CCD as a delay line

CCDs have been very popular as a semiconductor delay line (Figure 3.3.8). They were used in many electronic designs before semiconductor memory became cheap and complex enough to be used instead.

CCD delay lines are essentially analogue. That is, the charge they carry is an analogue quantity. If a CCD delay line is to be used in a digital environment there must be a digital-to-analogue converter fitted to the input and an analogue-to-digital converter fitted to the output. The transfer of charge is however digital, and the CCD will have a clock input which is used to transfer the charge from one MOS element to the next in the line.

Figure 3.3.6 The CCD as two line of buckets.

Figure 3.3.7

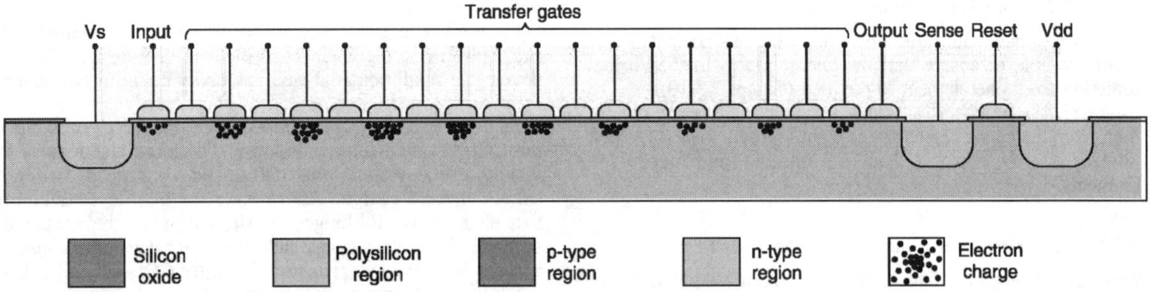

Figure 3.3.8 Using the CCD as a semiconductor delay line.

3.3.3.3.1 How does the CCD delay line work?

The input signal is fed into the first polysilicon region. Using field effect principles electrons are pulled from the n-type doped region and collect under the insulation layer. The potential on the first region creates a potential 'well' that the electrons effectively fall into. Although there is a maximum charge that can be held in this potential well, the amount of charge is proportional to the amount of time and the potential applied to the first region. The potential between the first and second regions is switched. This effectively moves the potential well from just underneath the first region to just underneath the second region. The first region becomes a potential barrier. The charge is attracted to the second region. The potential between the second and third regions is switched. The charge is now attracted to the third region.

By switching the potential from one region to the next the charge can be transferred from one region to the next, sitting just underneath the insulation layer. This leaves the first region clear. And the next charge packet can be input to the line. When the charge reaches the last region it transfers to the n-type doped region at the other end of the line and appears as an output signal.

3.3.3.3.2 Two-region elements

CCD delay lines with two polysilicon regions per MOS element use the second region in each element in the same way you might use the spare buckets in the line of buckets (Figure 3.3.9).

The charge is transferred to the second region before being transferred to the first region of the next element. The disadvantage of the two-region element is that the charge could flow the wrong way. Two-region elements employ special gates in the element and use a stepping transfer voltage to ensure the charge flows correctly. This all adds to the complexity and cost of this type of CCD. However, two-region elements offer higher density than three- or four-region elements.

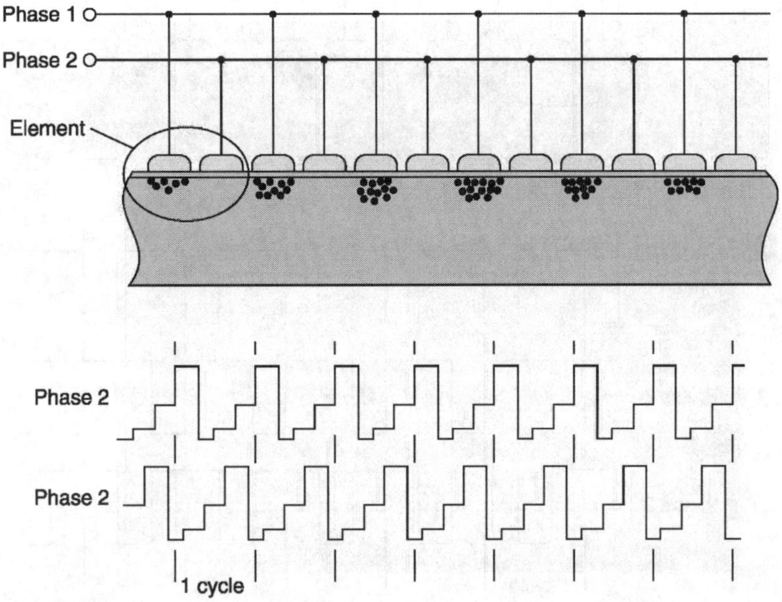

Figure 3.3.9 CCD delay line with two polysilicon regions.

3.3.3.3.3 Three-region elements

CCD delay lines with three polysilicon regions per MOS element are able to ensure that the charge flows in the correct direction from one element to the next (Figure 3.3.10).

The charge is pulled from the left region to the centre region, then from the centre region to the right region. However, clock phasing is more complex than both the two- and four-region elements.

3.3.3.3.4 Four-region elements

Four-region elements have simpler clocking signal arrangements than three-region designs and have better charge transfer capabilities, but it is more difficult to achieve high-density devices (Figure 3.3.11).

3.3.3.4 Using CCDs as image sensors

3.3.3.4.1 The basic principles

Metal oxide semiconductors are sensitive to light. If light enters the substrate of an MOS device, under certain conditions it excites electrons in the silicon into the conduction band. Put simply, electrons are shaken loose by light. The elements used in image sensors are similar to those used on CCD delay lines. They can be two-, three- or four-region elements.

3.3.3.4.1.1 Sensing light

When used as an image sensor a positive voltage is applied to the first polysilicon region in each element (Figure 3.3.12). This develops a small potential well just under the insulation layer.

- *Step A – Exposure*. As light penetrates the p type substrate of the CCD it shakes electrons loose. The loose electrons in the vicinity of the potential well fall in and are trapped, forming a small collected charge. The stronger the light level falling on that element, or the longer the time allowed, the greater the number of loose electrons, and the greater the stored charge.
- *Steps B, C and D – Transfer*. When the CCD sensor has been exposed to the image for the required time the charges stored under each element have to be transferred to the end of the row, where they can be sensed and output. In Step B the potential of region 2 is raised. Now the potential well extends over two regions and the charge spreads to fill the space. In Step C the potential of region 1 is lowered and region 3 is raised. The potential well now occupies regions 2 and 3. The charge is pulled across so that it sits under regions 2 and 3. In Step D the potential of region 2 is lowered and the potential of region 1 is raised. The potential well now occupies regions 3 and 1 of the next element. The charge is pulled across so that it sits under regions 3 and 1. Steps B, C and D are repeated until the whole row has been transferred, element by element, to the output gate at the end of the row. When this has been done and the whole row is empty of charge, exposure can begin again.

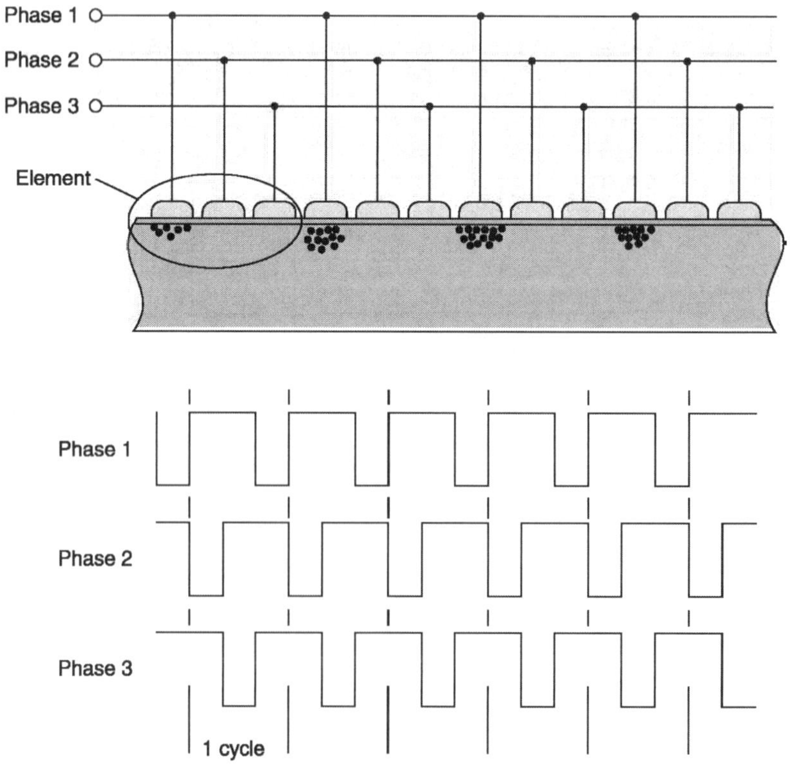

Figure 3.3.10 CCD delay line with three polysilicon regions.

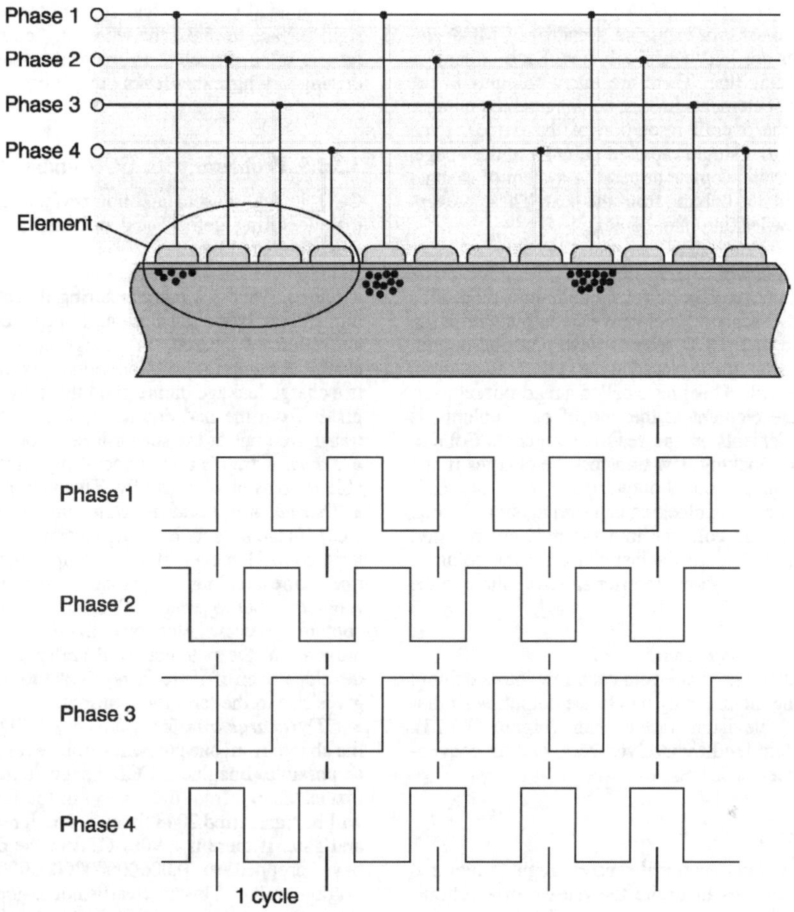

Figure 3.3.11 CCD delay line with four polysilicon regions.

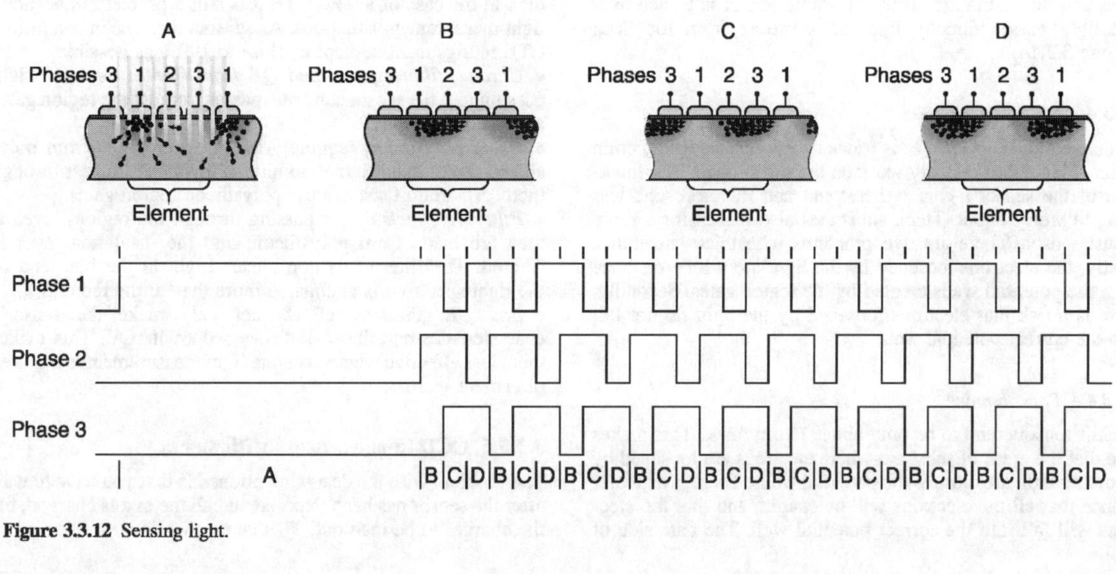

Figure 3.3.12 Sensing light.

3.3.3.4.1.2 The arrangement of MOS elements

CCDs used as image sensors comprise a matrix of MOS elements. The elements are laid out in columns. Each column is similar to a CCD delay line. There are many columns in the matrix. The number of elements in each column and the number of columns define the overall resolution of the device. Each element corresponds to a single captured point from the image, otherwise called a picture element or *pixel*. A system of channel stops is used to guard one column from the next. These prevent charges from one row leaking into the next.

3.3.3.4.1.3 Reading columns

Steps B, C and D above explain how each column is read. This would imply that CCD sensors would have an output gate at the end of each row. In fact, CCD sensors have just one output. Therefore, another CCD line is placed at the end of the column, perpendicular to them all. This line is called a read-out register. The charge from the element at the end of each column is transferred to the elements in the read-out register. Column clocking now stops. Clocking now transfers the charges in the read-out register to the sense and output gate. When the read-out register is empty, column clocking can start again and clock the next charge from the columns into the read-out register. This procedure carries on until the last charge in the columns has been clocked into the read-out register and from there to the output.

3.3.3.4.1.4 Similarity to raster scans

This method of reading one pixel from each row into a column line, then transferring them one by one to the output, is similar to a conventional television raster scan (Figure 3.3.13). CCD sensors therefore lend themselves very well as conventional television camera sensors.

3.3.3.4.2 Backlit sensors

Backlighting improved conventional sensors. Light is required in the substrate, where it can excite the release of electrons. Conventionally light has to pass though the MOS gate structure before reaching the substrate. Rather than allowing light to enter the front of the sensor, passing through the region gates and the insulation layer, the whole sensor is turned over and light passes directly into the substrate from the back (Figure 3.3.14).

3.3.3.4.3 Substrate thickness

A conventional substrate is thick. This makes production easier. Manufacturers only work on the top surface. The thickness of the sensor's chip is irrelevant and therefore one less thing to worry about. Thick substrates also makes for a more robust sensor. There are two problems with thick substrates. Firstly, the electrons loosened by the light are a long distance from the potential wells created by the region gates. Secondly, there is a risk that electrons loosened by the light do not fall into the correct potential well.

3.3.3.4.4 Back thinning

Backlit sensors tend to be only about 15 µm thick. This makes sure that the areas of substrate where electrons are loosened by incoming light are close to the potential wells. There is a greater chance that all the electrons will be caught, and that the electrons will fall into the correct potential well. The gate side of back-thinned CCD optical sensors tends to be mounted on a rigid surface to make the whole device more robust. This surface is often reflective to make the sensor more efficient by driving any light that leaks out the back into the substrate.

3.3.3.5 Problems with CCD image sensors

CCD image sensors are not perfect. They can suffer from manufacturing defects and operational anomalies. A few of these are listed here.

- *Shorts*. This is a manufacturing defect. Shorts can occur in the silicon oxide insulation. Shorts result in the improper collection of charge, or charge loss. If the collection of charge is compromised, individual pixels may be lost. If there is a charge leakage in one pixel this may result in loss of all the pixels from the bad one to the end of the line. As charge is transferred out of the sensor it leaks out of the bad pixel.
- *Traps*. A trap is a manufacturing defect where charge is not able to transfer successfully. This may be due to a gate.
- *Thermal noise and dark current*. As previously mentioned CCD image sensors have very noisy characteristics if they are kept cool. However, if their temperature rises thermal noise rises correspondingly. This can give rise to a number of other problems, but overall will affect the quality of the image capture process. Electrons freed by thermal activity are attracted to the potential well under each pixel. Thus, charge develops even if there is no light falling on the sensor. This gives rise to the term dark current.
- *CTE (charge transfer efficiency)*. CCD sensors must transfer the charge from one element to the next in the line as efficiently as possible. Imagine a CCD image sensor with 1024 by 1024 pixels. Charge from the far end of the furthest line of a device will be transferred 2048 times before it reaches the output sense and gate. If there is a 90% CTE in the device the charge will have dropped to 0.0000000000000000000000008 of its original value! This is clearly not a good thing. CCD image sensors generally have CTEs better than 99.999%. With a 1024 by 1024 sensor this still means that the charge in the furthest pixel has dropped by 0.02%. While this is significantly better than in the case of a 90% CTE it is still a problem in accurate light measurement situations. As sensors increase in resolution, CTE ratings must be kept as close to 100% as possible.
- *Chroma filtering and bad QE from frontlit devices*. Light passing into the sensor substrate passes through the region gates and insulation layer.
- *Using polysilicon regions*. Making the regions from polysilicon rather than from aluminium allows light to pass through them. All frontlit sensors use polysilicon region gates.
- *Filtering effects*. Light passing through the regions, even if they are made from polysilicon, and the insulation layer is filtered. The filtering is non-linear. Light at the blue end of the light spectrum is attenuated more than at the red end.
- *Bad QE*. Filtering effects not only make the sensor's characteristics non-linear, but they reduce its QE. This makes them less effective where accurate light measurement is required of camera sensors.

3.3.3.6 CCD image sensors with stores

The problem with the design mentioned is that you have to wait after the sensor has been exposed and all the pixels charged, for the charges to be read out. This takes a while.

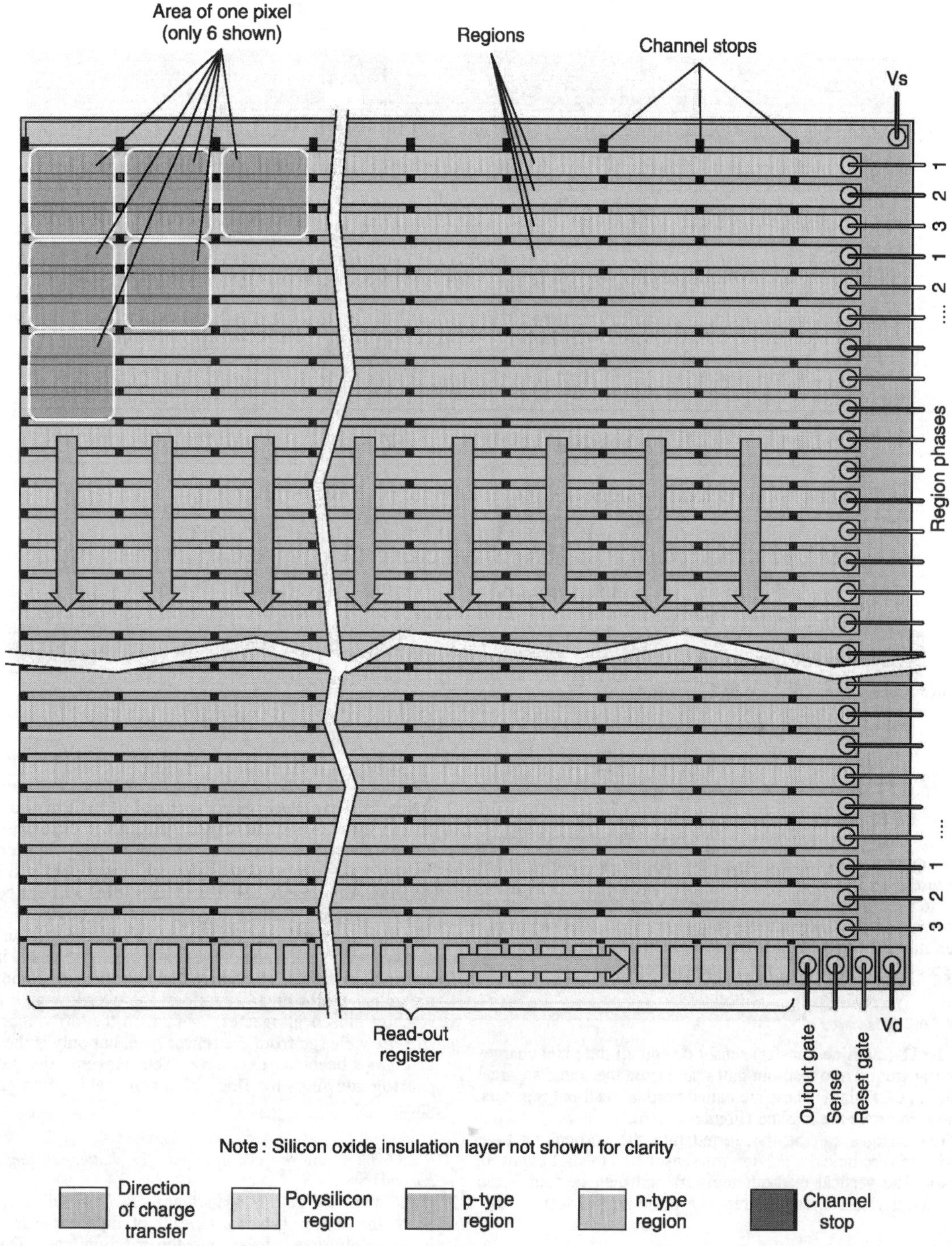

Figure 3.3.13 Reading columns.

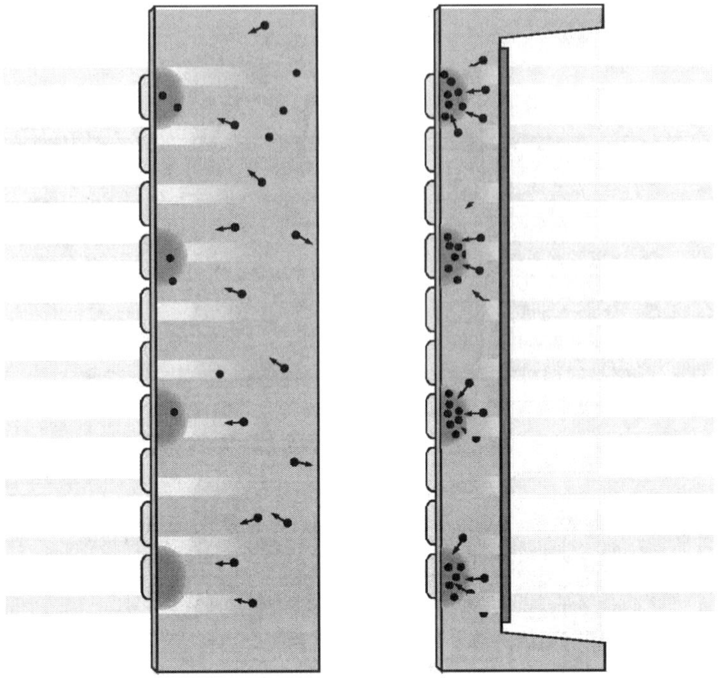

Figure 3.3.14 Conventional and backlit sensors.

3.3.3.6.1 FT sensors

In the FT (frame transfer) sensor design each column is twice as long. Half of the columns are exposed. The other half acts as a temporary store, and are covered by an aluminium mask (Figure 3.3.15).

After the image has been exposed to the sensor the charges are transferred quickly into the temporary store. The sensor can then start exposing the next frame while the frame that was just exposed is output through the read-out register.

3.3.3.6.2 IT sensors

In the IT (interline transfer) sensor design all the pixel charges are transferred into read-out gates, and from there into separate column CCD lines. These are called vertical read-out registers. These registers are masked (Figure 3.3.16).

The charges can be transferred into the vertical read-out registers very quickly, leaving the sensor free to start exposure again. The vertical read-out registers can then be transferred into the horizontal read-out register in the normal way.

3.3.3.6.3 Overflow gate technology and shuttering

With the introduction of IT sensors came the introduction of an overflow gate. This gate is placed on the opposite side of each sensor gate from the vertical read-out register. It can be used in a number of ways.

3.3.3.6.3.1 Using the overflow gate to eliminate flare and burnout

When the overflow gate is closed it will not draw any charge away from the sensor gate. After exposure all this charge can be drawn away by the vertical register. However, if light levels get too high the sensor gate will become flooded. Any charge above a certain level will not be drawn away from the vertical register and the device will peak, causing 'burnout' in the image. Furthermore, the excess charge will leak out of the affected gate and into the surrounding gates, spreading the perceived brightness from the actual bright area. Therefore, the overflow gate is never actually closed altogether. In fact, in its 'off' mode it will still draw charge from the sensor gate, but only if the amount of charge becomes excessive. This prevents the gate from peaking and stops the flood of charge leaking into any other gates.

3.3.3.6.3.2 Using the overflow gate for shutter opening and iris control

If the overflow gate is opened any charge building up underneath the sensor gate as a result of light exposure will be immediately drawn away into the overflow gate. This effectively switches the device off in the same way as a mechanical shutter would do. This can make the sensor behave a little like movie cameras with variable shutter wheels. It is also used in applications like CCTV as an electronic iris, assisting the mechanical iris in the lens itself.

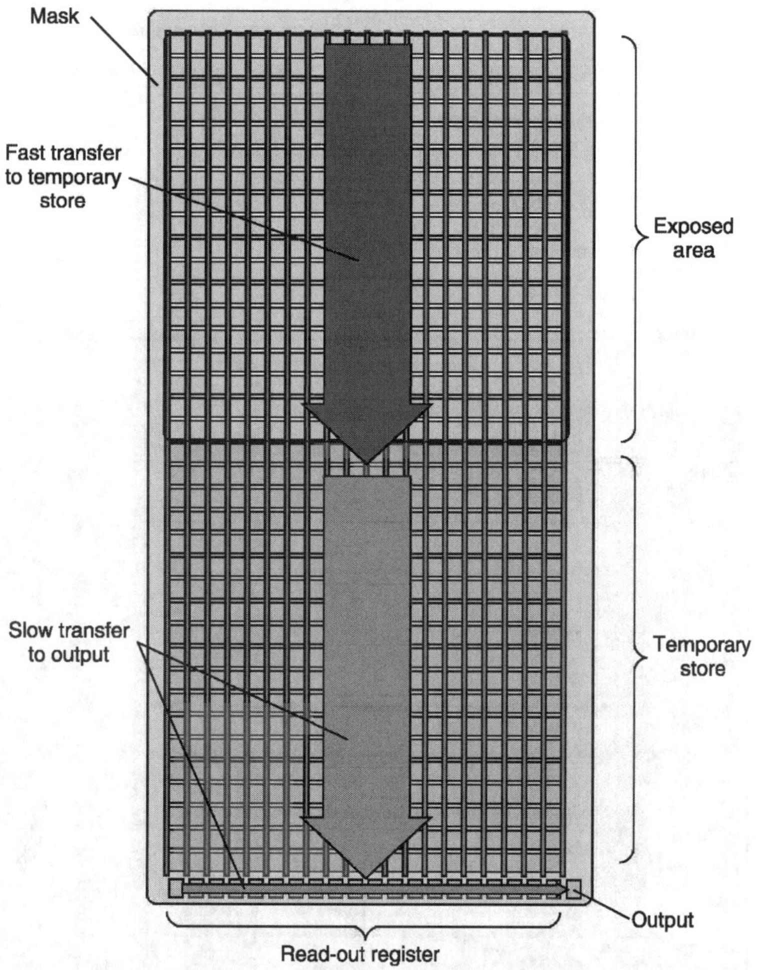

Figure 3.3.15 The FT sensor design.

3.3.3.6.3.3 Using the vertical register and overflow gate for shutter closing

To simulate the shutter closing the accumulated charge under the sensor gate can be drawn into the vertical register. At the same time the overflow gate is opened, so that any further charge is drawn away from the sensor gate.

3.3.3.6.4 FIT sensors

One problem with IT designs is that the vertical read-out registers are very close to the exposed regions of the device. If light levels are very strong it is possible for charge to leak from the exposed region of the sensor into the vertical register regions.

In the FIT (frame interline transfer) sensor design both FT and IT design philosophies are used. The charges built up in the exposed areas are transferred quickly to the vertical read-out registers, and then into an FT-type store. This takes them away from the vertical read-out store so that they cannot be corrupted if light levels are very high.

3.3.3.6.5 HAD technology

Sony introduced a new technology for image sensors in 1984. This technology was a real departure from the conventional designs up to that date. Rather than using the photo-excitation technology of older designs, this new design uses an embedded photo diode for each pixel. The photo diode has a heavily doped p-type region called a p++ region. p++ doped regions have a high number of accumulated holes. Hence the name Hole Accumulated Diode or simply HAD (Figure 3.3.17).

The HAD increases the number of electrons that are released as light enters the device. These electrons flow down into the device's substrate and collect underneath the HAD. HAD sensors also have the advantage that light does not have to pass through polysilicon regions to reach the diode. This makes the sensor more efficient and linear.

3.3.3.6.5.1 HAD sensor operation

HAD sensors comprise an array of HADs. An insulation layer of silicon dioxide is laid on top of the HADs and a

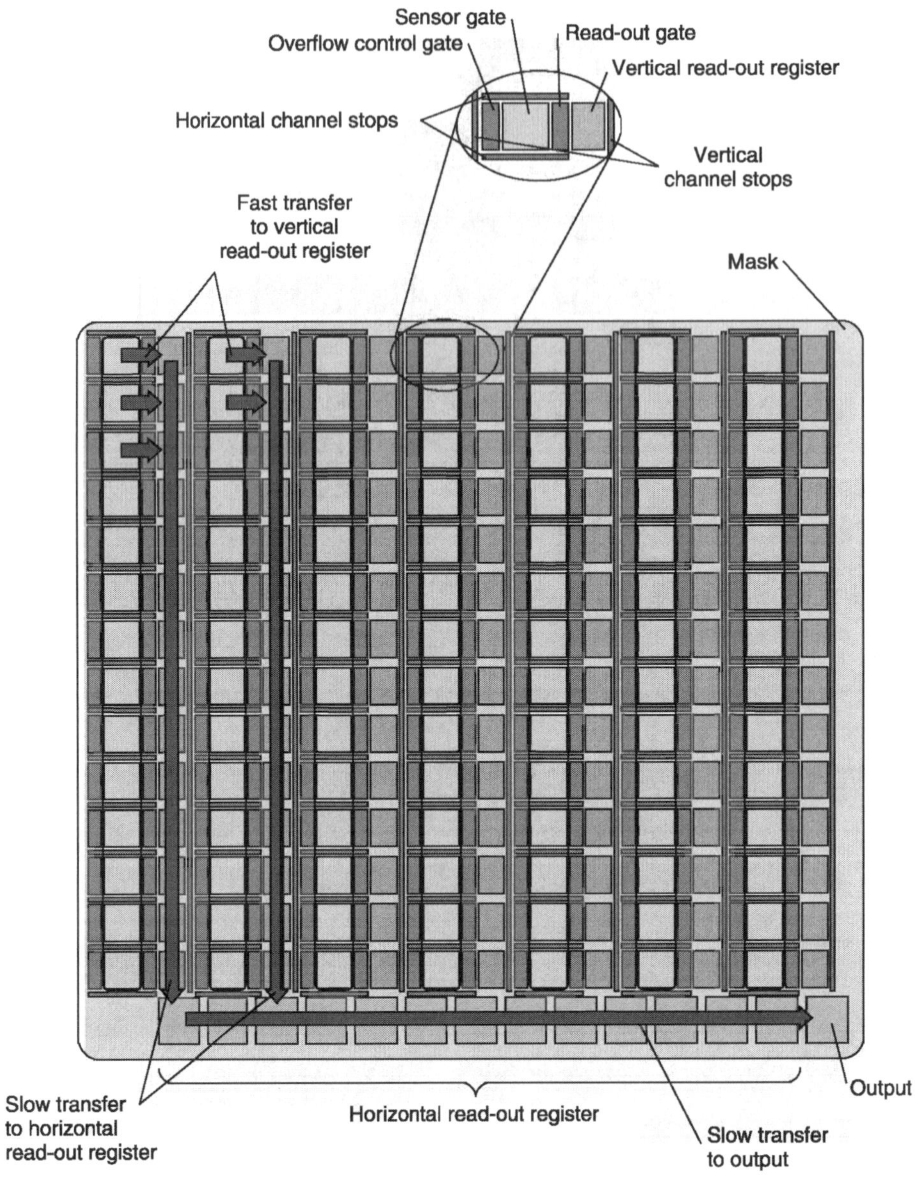

Figure 3.3.16 The IT sensor design.

thin aluminium photo mask is printed on top of the insulation. The photo mask prevents light from getting into the sensor except where there is an HAD. Light passes into the HAD and excites electron flow down into the n-type substrate. At a certain time, when the image is sampled, the voltage on the polysilicon gate, to the right of the HAD, switches, creating a potential well that attracts the accumulated charge away from underneath the HAD. The polysilicon region is part of a chain of polysilicon regions that form a vertical register to transfer the accumulated charges out of the device. Charge is therefore transferred out of the device in the same way as any other IT or FIT device. A region of p++ material is fabricated deeper into the device just to the left of each HAD. This acts as a channel stop, preventing charge from leaking from underneath each HAD into the vertical register to the left.

3.3.3.6.5.2 Problems with HAD devices

The first problem with HAD sensors is the increased manufacturing complexity. However, in commercial terms this increased complexity and its resulting higher cost is more than offset by the increase in performance. Manufacturing

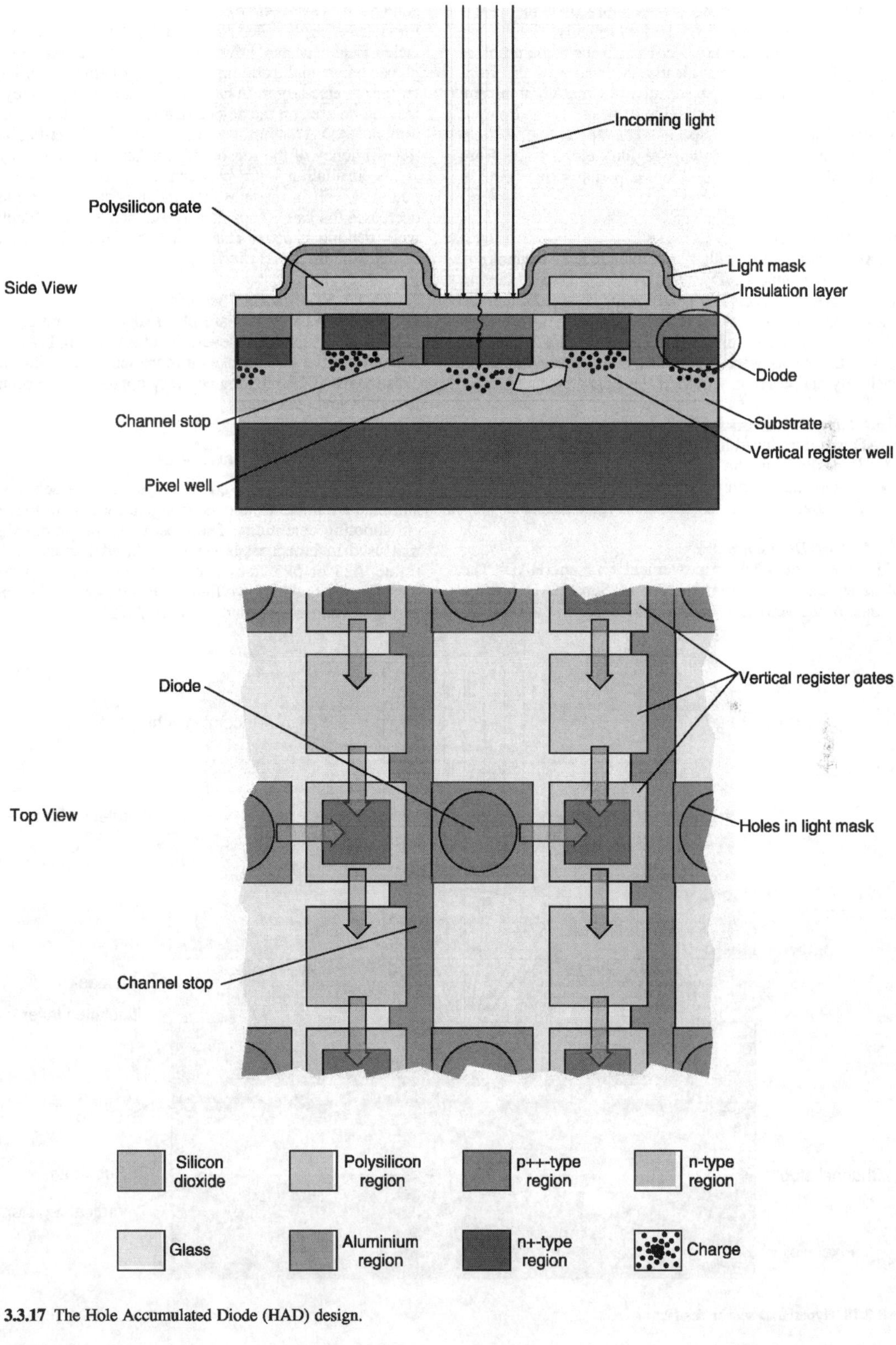

Figure 3.3.17 The Hole Accumulated Diode (HAD) design.

techniques have also improved considerably over the year's making it easier to produce HAD devices reliably.

The second problem with HAD devices is the same problem facing any IT or FIT device. Ideally the whole of the front of the device should be light sensitive so that light hitting anywhere on the surface of the device is picked up and output as a signal. The amount of space taken up by the vertical registers, channel stops, polysilicon regions, etc. detracts from this light-sensitive area. This problem is partially overcome in later designs.

3.3.3.6.5.3 HyperHAD

HyperHAD, sometimes called microlenticular technology, improves on the simple HAD device by fitting a small lens in front of each HAD. This channels light from the area around the actual gate that would otherwise be lost, increasing the effective area of each HAD outside of the HAD itself (Figure 3.3.18).

HyperHAD sensors were introduced in 1989 and increased the sensitivity and efficiency of HAD sensors.

3.3.3.6.5.4 SuperHAD sensors

SuperHAD was introduced in 1997. It is basically similar to the HyperHAD design, but the actual lenses are larger, and are therefore able to capture more light, making SuperHAD sensors more sensitive than HyperHAD sensors (Figure 3.3.19).

3.3.3.6.5.5 PowerHAD sensors

PowerHAD is a marginal improvement on SuperHAD. The microlens structure is similar to that of SuperHAD but the capacitance of the vertical registers is reduced.

3.3.3.6.5.6 PowerHAD EX (Eagle) sensors

Previously simply called New Structure CCD and now sometimes called Eagle sensors, PowerHAD EX sensors have another lens placed between the on-chip microlens and the HAD. The microlenses are also larger. In fact, they are so large that they overlap, leaving no area on the device that is not somehow concentrated into an HAD. This further concentrates light capture, increasing the efficiency of the sensor still further (Figure 3.3.20).

The insulation layer between the polysilicon gate and the potential well in the substrate underneath is also thinner. This decreases the gate's capacitance and increases the 'depth' of the well, making it better able to collect the HAD's accumulated charge (Figure 3.3.21).

3.3.3.6.5.7 EX View HAD sensors

EX View HAD sensors are physically the same as any other HAD-based sensor. However, the exact doping levels and construction of the HAD makes it more sensitive to infrared light. This makes EX View devices very appropriate to security and low light level cameras.

3.3.3.7 Reading out variations

CCD sensors can be used in professional video cameras to simulate a number of television signal formats or to suit different shooting conditions. The most common output signal format used in broadcast is field interlaced frames. A complete frame, 625 or 525 lines depending on where you are in the world, is divided into two fields. The odd lines are scanned first and the even lines second (Figure 3.3.22).

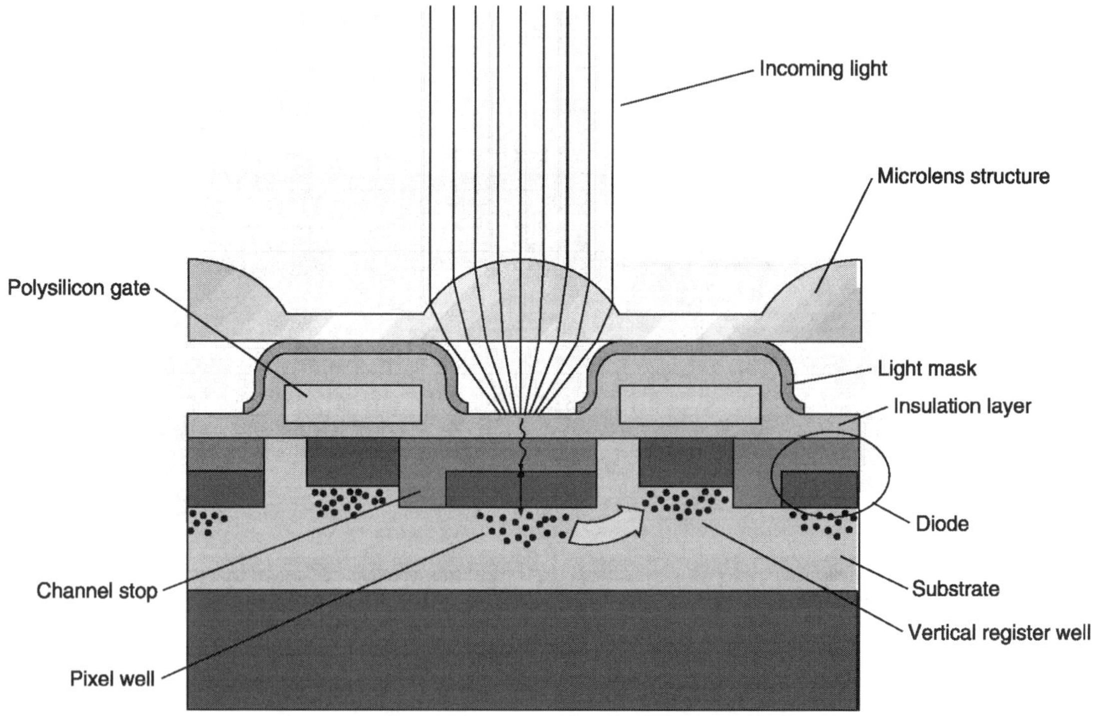

Figure 3.3.18 HyperHAD sensor design.

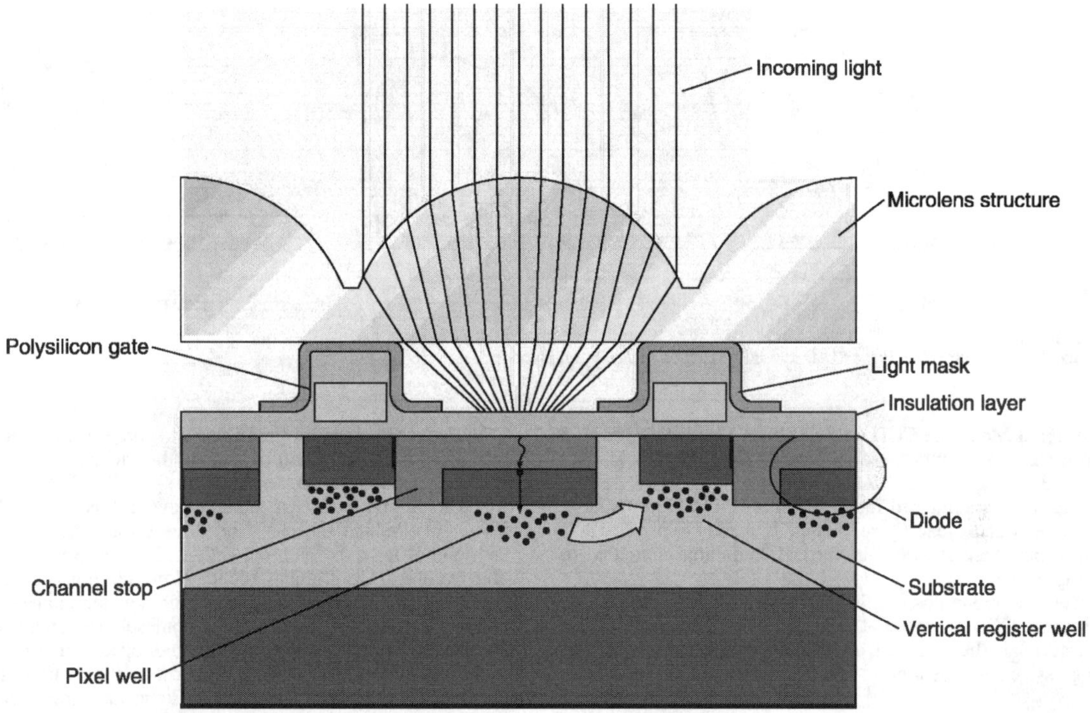

Figure 3.3.19 SuperHAD sensor design.

It is important to remember that the two fields occur one after another. They are temporally displaced. If the CCD is to simulate this it must split field 1 and field 2 in time (Figure 3.3.23).

This is easy to accomplish at the design stage. CCD sensors have a number of different output modes. In Frame Mode the CCD will output the odd lines first and the even lines next, therefore temporally displacing the two fields (Figure 3.3.24).

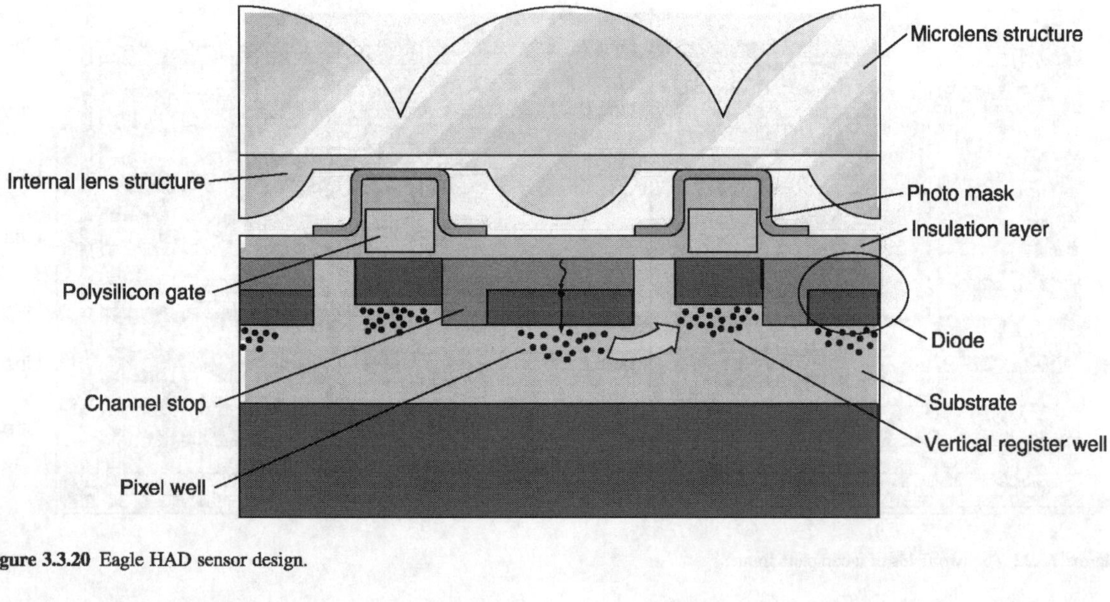

Figure 3.3.20 Eagle HAD sensor design.

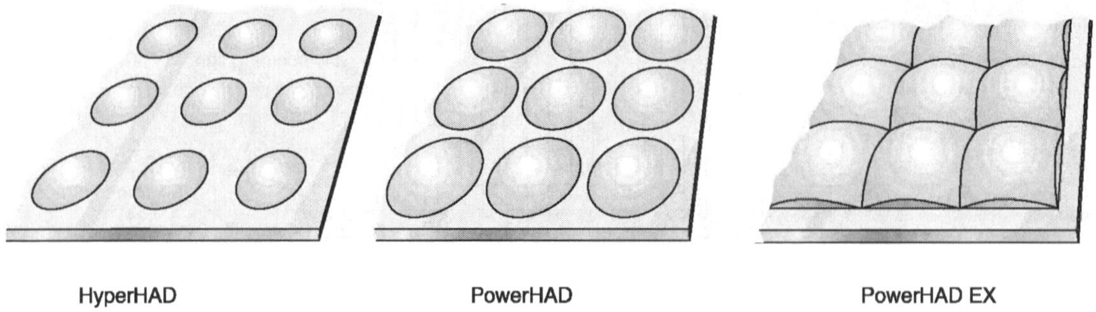

Figure 3.3.21 HyperHAD, PowerHAD and PowerHAD EX insulation layers.

In Field Mode the CCD will add lines 1 and 2, 3 and 4, 5 and 6 etc., and output these as field 1. Then it will add lines 2 and 3, 4 and 5, 6 and 7 etc., and output these as field 2. The result is a simulated interlaced scan. Field Mode is more sensitive because each pixel output is made from two CCD pixels, but there is a loss in vertical resolution compared to Frame Mode.

The final mode possible is Progressive Mode. In this mode the whole CCD is sensed at once, and all the lines are output one after another, instead of splitting them into two fields. Progressive Mode is used to create a filmic look.

3.3.3.8 Single-chip CCD designs

Professional and broadcast camera systems normally process the image they are looking at as three primary colours. This is important for good colour matching, and is essential if the camera's outputs are to comply with broadcast signal standards.

The split from the original image to three images, one for each primary colour, could be done in the camera's electronics. However, it is better to do the split optically. Therefore, all professional and broadcast cameras have a three-way dichroic splitter just behind the lens, and three CCD sensors, one on each output from the dichroic splitter. Each sensor is responsible for one of the primary colours. However, this is either too expensive or simply not possible for smaller cameras and cameras intended for industrial and domestic use. Small security cameras simply do not have enough space for a dichroic block and three CCD sensors. The cost of the dichroic block and three CCD sensors would make domestic cameras simply too expensive. In any case the increase in quality would almost certainly not be appreciated.

Therefore, all these types of cameras have one CCD sensor. The split from the original image to its primaries is still required and is still best done optically. Consequently, single-chip CCD cameras have a filter screen fitted over the sensor (Figure 3.3.25).

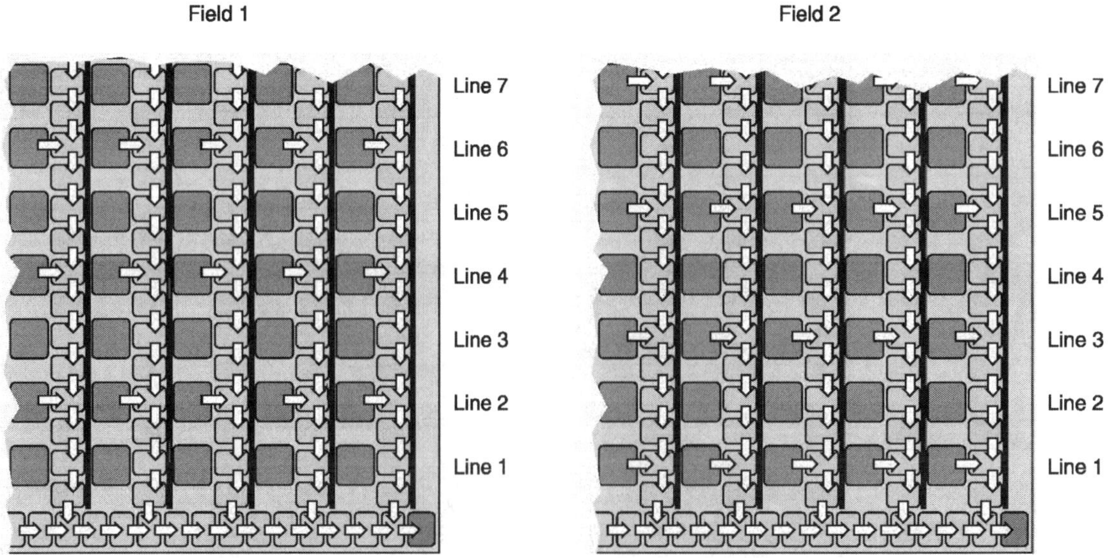

Figure 3.3.22 The two fields of a complete frame.

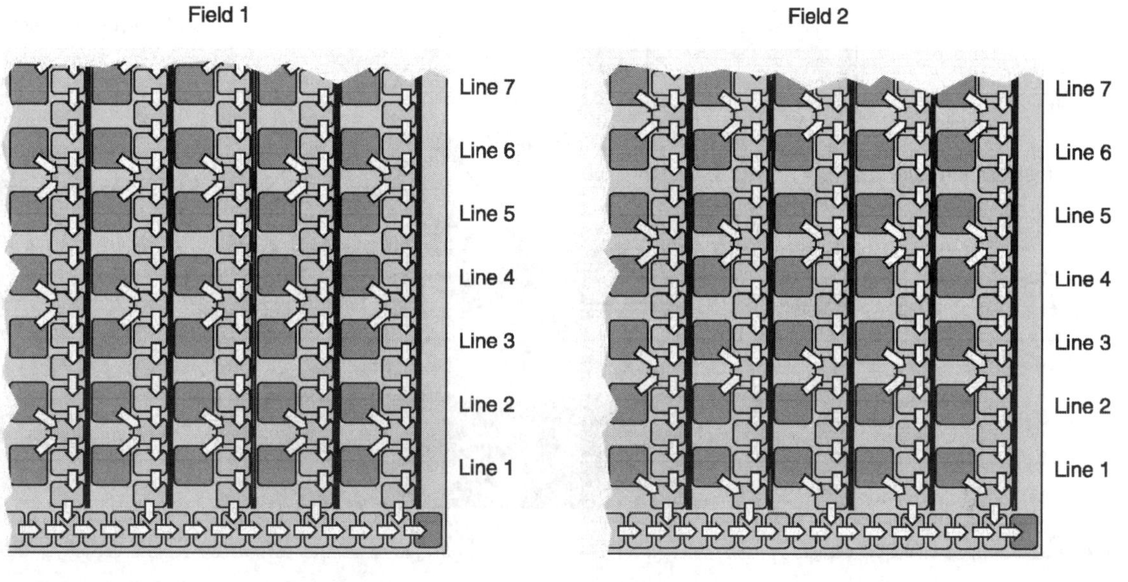

Figure 3.3.23 Splitting of fields 1 and 2 in time.

3.3.3.8.1 CCD filter screens

CCD filter screens consist of an array of small coloured squares. The resolution of the CCD sensor and the filter squares is the same. Thus, each pixel in the sensor has one filter square.

3.3.3.8.1.1 Random filter screen

The human eye is very good at recognising patterns. Thus, it may seem a good idea to design a filter screen with a random design of squares of the three primary colours (Figure 3.3.26). However, there is a chance that there will be discernible areas of one colour.

3.3.3.8.1.2 The Bayer filter screen

A popular screen design is the Bayer screen (Figure 3.3.27). This screen has a greater number of green squares, because of the human eye's relative high sensitivity to green areas of the colour spectrum.

The Bayer screen is very popular in single CCD cameras.

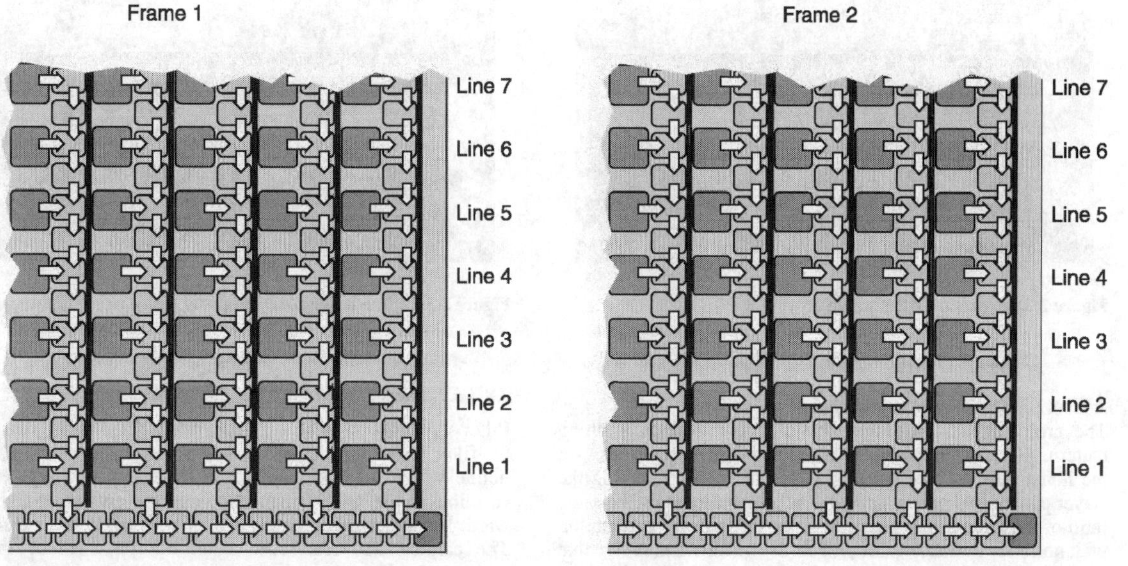

Figure 3.3.24 Temporal displacement of the two fields in Frame Mode.

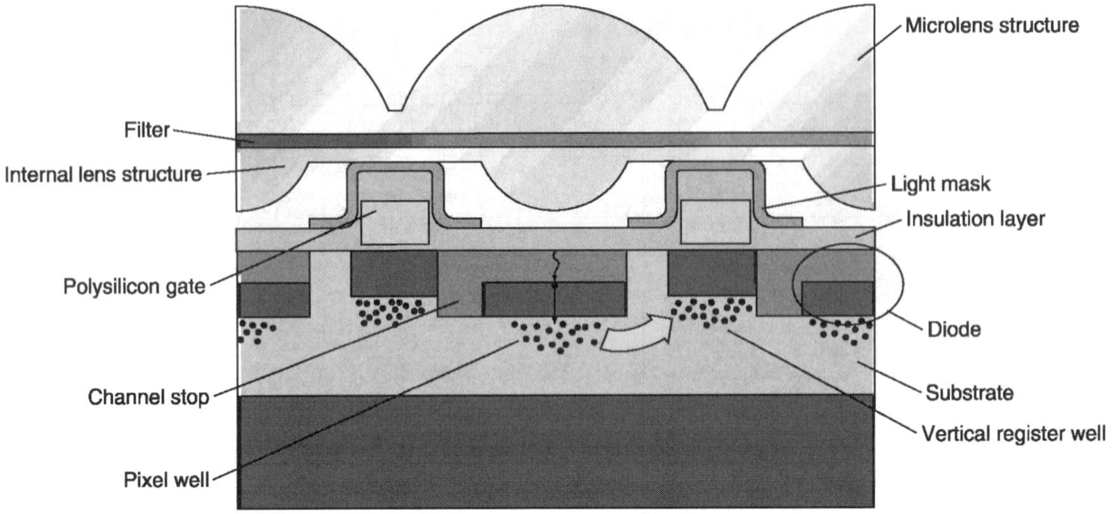

Figure 3.3.25 Single-chip CCD design.

Figure 3.3.26 Random filter screen.

Figure 3.3.27 The Bayer filter screen.

3.3.3.8.1.3 The pseudo-random Bayer screen

The problem with the Bayer screen is that there is a strong pattern. It may be possible, at certain low resolutions, for the human eye to pick out this pattern. Thus, by jumbling the Bayer pattern in a particular way, it is possible to retain the same ratio of the three primary colours as the basic Bayer pattern, but with no easily definable pattern. This design also makes sure that there are no large areas of one colour by ensuring that there are no squares of the same colour next to each other (Figure 3.3.28).

3.3.3.8.2 Reading out from a single CCD camera

Filter screens need to be very accurately fitted to the sensor. If the filter position is defined and fitted exactly, it is possible to define which pixel is responsible for which primary colour. Reading single CCDs then becomes reasonably straightforward, if a little more complex, than with triple CCD designs. The camera will read each pixel out in sequence in the same way. However, the read-out circuitry knows which pixel is responsible for each primary colour, and sequentially

Figure 3.3.28 The pseudo-random Bayer screen.

demultiplexes all the pixels for each primary to a different part of the electronics for further processing.

3.3.3.8.3 Pixel interpolation and sharpening

With a Bayer filter screen half the pixels are responsible for the green primary, a quarter for red and a quarter for blue. Single CCD cameras rebuild a full image of pixels for each primary by interpolating the pixels it has to make up the missing pixels. Putting these three separate images back together gives a much more pleasing result. A newer approach is to follow the interpolation process by a small amount of sharpening to improve the perceived quality of the image.

3.3.3.8.4 Noise reduction

Noise is a problem in any image sensor. The first area where noise is introduced is in the pixel itself, where thermal noise is an enduring problem. The only way of eliminating this kind of noise is to freeze the sensor to absolute zero to eliminate thermal electron movement. This is not practical and the sensor would fail to work at all anyway! It is really down to good image sensor design to accept that thermal noise will exist and reduce its effect.

Another area where noise can be introduced is during the charge transfer period, where the charge collected under each pixel is transferred to the output.

The last area where noise can be a problem is in the output gate itself, where the charge is placed into a capacitor and measured as a voltage. This capacitor needs to be carefully and quickly drained of any charge from a previous pixel, or from anywhere else, before the pixel charge is put into it.

Updated by **John D. Wardle** BSc (Hons)
John Wardle Associates

Original by **W H Klemmer**
Broadcast Television Systems GmbH

3.4 Studio Cameras and Camcorders

The electronic colour camera is one of two important signal sources for the creation of television signals. The other is the telecine (see Chapter 4.8). The process of replacing the classic 35 mm film camera by a High-Definition (HDTV) electronic camera still seems very distant in spite of the increasing use of HD cameras. The range of professional electronic cameras can be split into the following categories:

- Studio.
- Electronic Cinematography (EC).
- Electronic Field Production (EFP).
- Electronic News Gathering (ENG).
- Industrial and other uses.

This list shows the cameras in descending order, with respect to their picture quality and their design features. There are actually no real EFP cameras, so this application is covered by suitable adaptation of studio cameras or by high-quality ENG cameras (Figure 3.4.1).

At the top of the list are EC and studio cameras, both claimed to have the highest possible picture quality within a chosen scanning mode. In addition, studio cameras are fully fitted out from an operational point of view, and are, as far as possible, automated. Manufacturers of modern studio camera systems offer not only large full-facility cameras, but also portable camera heads. These can be integrated into the system as portable heads or mounted in cradles with large lenses where they become full-facility cameras with full compatibility, with only small reductions in quality and features. Modern studio cameras almost exclusively use the three channels red, green and blue. One camera has a four-channel system where two green channels are used to give increased dynamic range and resolution.

Whether PAL, SECAM, NTSC or SDI is used, the difference does not affect the general system configuration. It only affects the encoder at the output of the camera system, and this decreases in importance with the use of component television. The system configuration described in Section 3.4.1 also applies to HDTV cameras, even though they do not have any modulating encoders, as only component outputs are used.

The majority of cameras in use and all those in current production have CCD sensors which now match or better camera tubes in all technical quality. A few older studio cameras that are still in use today use camera tubes with image sizes of ⅔ inch, 1 inch and 1¼ inches, and with a plumbicon or saticon layer. However, there are two disadvantages with CCD in comparison to tube cameras which must be mentioned: less flexibility with respect to multi-standard camera operations as their line standard and picture shape are defined at manufacture, and the impossibility of correcting lens-dependent colour registration errors.

All modern cameras make extensive use of microprocessor systems for their line-up and control, both within the camera and for communication with their remote control panels. The remote control systems vary from the very simple to the very complex for multi-camera operations (Figure 3.4.2).

3.4.1 System Structures

Every studio camera system includes a camera head, with zoom lens and viewfinder, which is connected through a camera cable to the camera control unit (CCU). The operational control panel (OCP) and, in the case of multi-camera operations also a master control panel (MCP), is connected to this CCU as well. Figure 3.4.3 shows a typical block diagram of a studio camera,

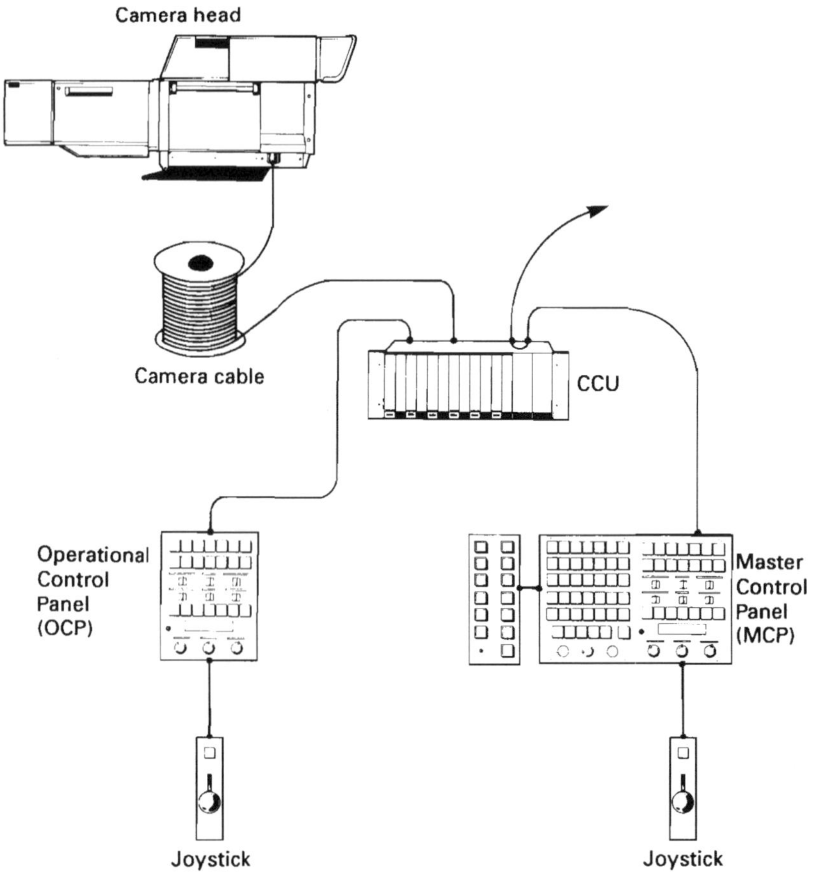

Figure 3.4.1 Components of a studio camera chain (Broadcast Television Systems GmbH).

illustrating one of the many possible arrangements of the cable interface.

3.4.1.1 Camera head

After optical colour separation in the prism, the signals for the three colour channels R, G and B are created by the CCD read-out or by line scanning for camera tubes. After pre-amplification, this acquired signal is then linearly and non-linearly processed in the head signal processor.

Where camera tubes are still used a microprocessor-controlled correction system may create analogue correction voltages, which are mixed into the scanning beam deflection unit for correction of colour registration errors. The scanning beam intensity in the camera tubes is guided by the *dynamic beam control* (DBC) or *automatic beam control* (ABC) so that the improved scanning of high lights is possible, leading to the reduction of comet tails.

All types of camera then introduce additive and multiplicative correction waveforms into the analogue video signal to correct for shading errors. After this point the sequence of the signal processing may vary with different manufacturers.

3.4.1.2 Camera cable

There are various ways of transmitting the fundamental signals:

- Parallel transmission in baseband via a multi-wire cable – baseband multi-core mode.
- Partly modulated transmission via a multi-wire cable – RF multi-core mode.
- Time division multiplex transmission, i.e. data and impulses mixed in the blanking intervals.
- Fully modulated bidirectional transmission via a triax cable – triax mode.
- Transmission in analogue or digital form via one or more optical fibres – fibre mode.

3.4.1.2.1 Multi-core mode

When baseband signals are used the cable length is restricted to about 300 m maximum and switched equalisers are required to maintain a flat frequency response. Generally this mode is only used in lower cost systems. However, if the three signals are modulated onto carriers (usually amplitude modulation at about 30 MHz) the system will perform over longer cable lengths such

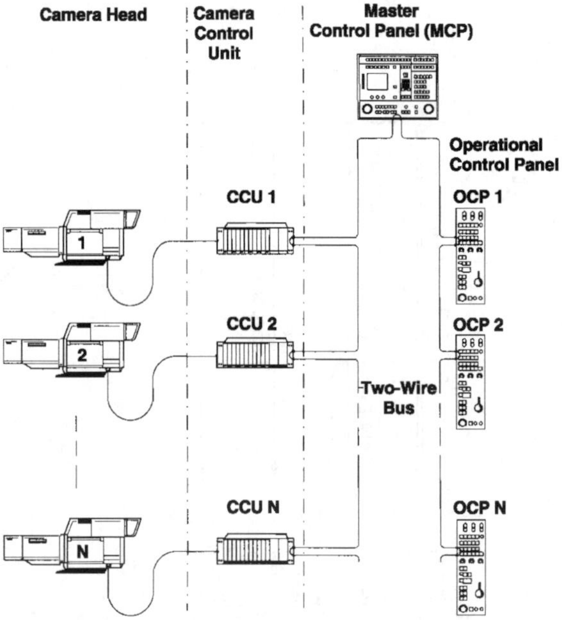

Figure 3.4.2 Multi-camera system (Broadcast Television Systems GmbH).

as 600–1000 m and may be self-equalising. Multi-core cable is expensive and difficult to repair. There are also several different lays of the internal cables, which can cause crosstalk problems if a type is chosen which does not match the camera system.

3.4.1.2.2 Triax mode

Triax cable has the advantage that it has a much lower cost and is easy to repair even in the field. The outer screen serves only as a safety earth, all power and signals being carried by the inner core and inner screen. The video is modulated onto carriers that can be up to 100 MHz. There will usually be up to five video carriers, the three camera output channels plus teleprompt and external video in the other direction. The data control, communications and audio will also be modulated onto other lower frequency carriers. Cable lengths of about 1500 m of 11 mm triax are possible, longer with larger diameter cable.

The video signals from the camera and on the triax may be full bandwidth RGB; Y, R – Y, B – Y or in some earlier cases encoded with B – Y for Chroma Key. One manufacturer has a lower cost system that transmits SDI down a limited length cable. Where R – Y, B – Y are used these may be narrowband or wideband and may also be quadrature modulated onto the same carrier. There has been a move to wideband with increased component, 16:9 and SDI working. Chroma key is also improved where the wider bandwidth is used. In another case the channels are G and R/B quadrature modulated. The important design factors are: linearity, noise, gain stability, frequency response and, particularly where quadrature modulation of R – Y/B – Y is used, crosstalk.

With good practice any camera should work on any CCU/Base Station without significant change in gain and frequency response. The frequency response is not cable length dependent and so no switched equaliser is needed, though there may be a cable length detector, which switches in an attenuator at short cable lengths to prevent overload of the demodulators. The power systems should prevent the presence of voltages above 30–50 V when the cable is open, and should have camera type recognition to prevent damage to the system, or the other product, if it is accidentally connected to the wrong type of CCU during rigging, especially on large OBs.

3.4.1.2.3 Fibre systems

Ideally, to maintain the optimum performance of a digital camera a digital transmission system should be used. The camera cable transmission system has long been the most unstable and limiting part of the chain. However, digital transmission needs a high data rate and the triax length would be severely restricted for OB use. A number of manufacturers now have a fibre option available. Following the use of this for the Winter Olympics, a cable and connector was standardised (SMPTE 3K.93C). This cable has two fibres, power conductors and a number of additional wires which may be used for security and communications.

3.4.1.3 Camera control unit

In theory, the units for the colour correction matrix, the frequency response correction and the gamma correction (see Section 3.4.2.4) should be arranged in the camera head to utilise fully the gamma stage for companding before the camera cable. At one time this was not possible due to the limitations of space and power dissipation. However, with modern circuitry the optimum method is used and in most cases the full processing takes place in the camera head, except for any variable saturation, luminance black stretch and any coding prior to the output distribution amplifiers.

If the signal transmission is via a baseband multi-core cable, the length-dependent frequency response of the camera cable must then be adjusted. RF-based systems, whether multi-core or triax, do not require this correction.

After colorimetric correction and two-dimensional frequency response correction, the RGB signal is non-linearly pre-compensated in the gamma correction stage in order to compensate for the subsequent inverse non-linear transmission characteristic of the receiver. This pre-compensation stage has the advantageous and vital effect of transmission companding.

Following it, the signal output exists either as the component triplet RGB or YC_RC_B, or, after appropriate coding, as a composite video signal. The component triplets RGB or YC_RC_B enable transmission which is theoretically correct, at least if they are output in the full bandwidth and as long as appropriate measures have been undertaken in the camera and receiver.

Connected with the signal coded in a frequency division multiplex are various defects dependent on this principle. Modern studio cameras tend to use component or serial digital outputs. They do so completely for HDTV applications.

3.4.1.4 Other units

Both the camera head and the CCU use the units: data transceiver, synch pulse generator, audio intercom and power supply.

The *data transceivers* are not normally stand-alone units, but rather microprocessor interfaces in the head and CCU requiring a large amount of software. The data path from CCU to head carries the control and switching signals, which originated at the control panel, to the head. The data path from head to CCU,

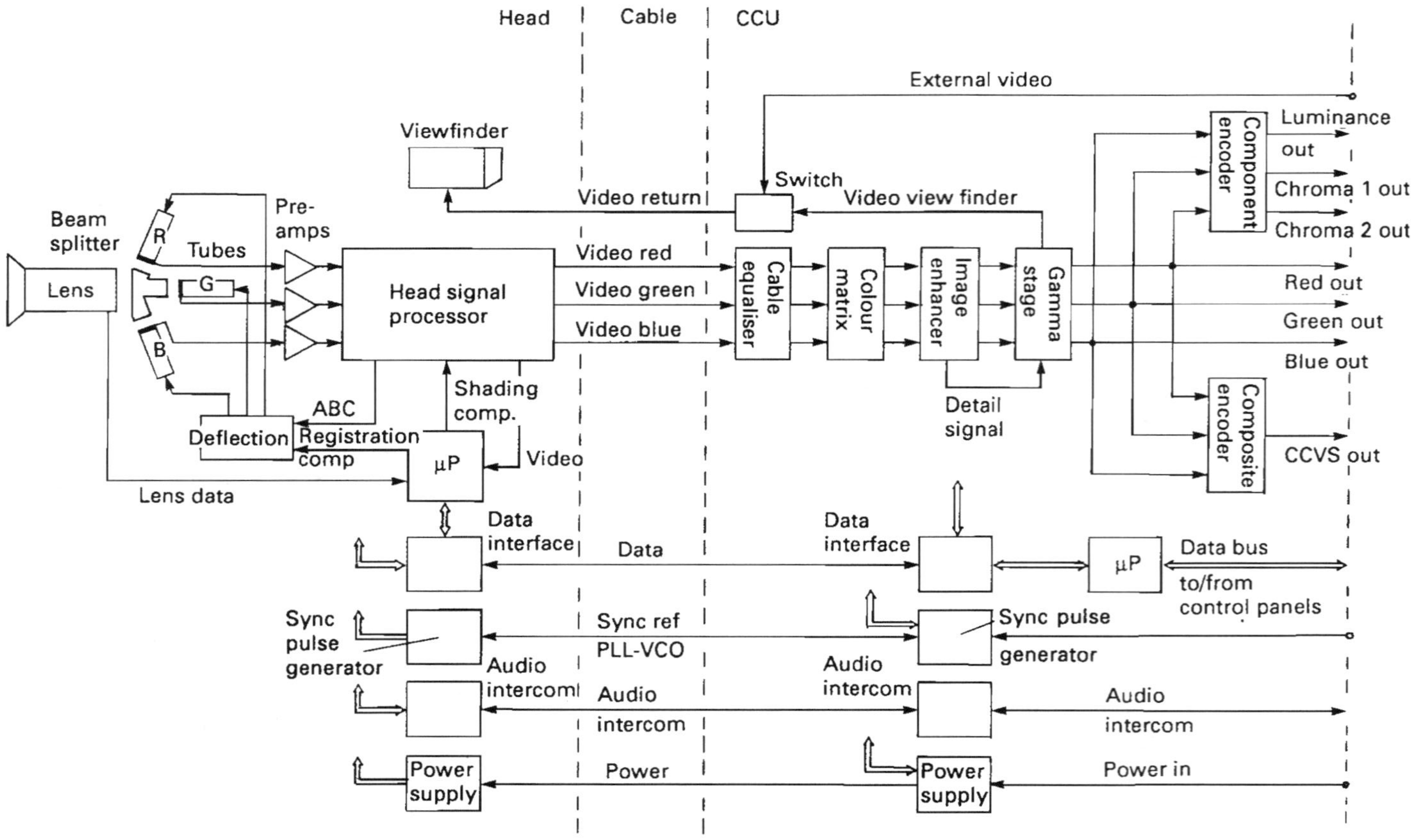

Figure 3.4.3 Block diagram of a studio/OB camera, including head, cable and CCU.

on the other hand, carries information on the camera head status and any error diagnostic data.

The *pulse generators* in the head and CCU are connected via a phase-locked loop (pll). The CCU clock generator is also coupled to an external clock via a further pll. Most studio/OB camera systems provide the facility for the transmission of two high-quality microphone channels down the camera cable, which are output at the CCU. These will usually have a remote gain control facility.

For easy communication between operators, bidirectional intercom paths are provided; ideally, these should have independent circuits for production talkback, engineering talkback and programme sound to the camera head, and allow reverse camera talkback to the CCU.

The power supply used in the camera head is derived either from a high direct voltage (200–400 V) or from an alternating voltage with a high frequency. These supply voltages are then converted via dc/dc or ac/dc converters into the required dc voltages for the camera head.

3.4.2 System Components

3.4.2.1 Optical block

The optical block is the unit for optical/electrical signal conversion, i.e. lenses, filter wheels, beam splitters and camera sensors. A detailed description of camera tubes and solid-state CCD imagers can be found in Chapter 3.3 Optical Sensors.

The fundamental arrangement of the optical block is shown in Figure 3.4.4. The lens reproduces the scene to be observed on the sensors (tubes or CCDs) via the beam splitters. Zoom lenses are almost exclusively used, with focal length variations of up to 1:100 and f-numbers of 1:2.1 (image size 1¼ inch), 1:1.6 (image size 1 inch) and usually 1:1.4 (image size ⅔ inch as 1:1:2 is very difficult and expensive to make). The three adjustable parameters of focal length, iris and focusing distance are usually altered via servo motors, which are built into the lens (see Chapter 3.2).

The operator at the camera head controls focal length and distance, whereas the aperture opening is controlled from the control panel or automatically, so that a constant image signal level is produced. The requirement for optimal, constant picture quality over the whole focal length range necessitates, for example, having the simultaneous complex shifting of several lens groups within the lens (see Chapter 3.2).

CCD cameras do not usually use diascopes as there are no registration controls to set up, and other methods are used for any automatic line-up.

For the last generation of tubed cameras, an important requirement for automatic camera alignment is that diascopes are built into the lens, which are allowed to swing into the beam path. With the help of such a projected test slide, it is possible to adjust the parameters of image geometry, colour registration and white shading.

It can be recognised from the arrangement of the projector in the lens that dynamic (i.e. setting-dependent) chromatic errors of the main lens cannot be corrected at that point. For this reason, the lens gives actual values of the setting parameters or even correction data to the camera head, so that an electronic correction can take place there. Even so, the test projector simulates lens aberrations which occur at a medium operating focal length or distance.

Section 3.2 deals with the electronic correction of lens errors and with sensitivity. Figure 3.4.5 shows the spectral transmittance, denoted by $\tau_0(\lambda)$, of a studio lens in the wavelength range 380–780 nm, relevant to colour transmission.

Modern studio cameras contain two *filter wheels*, each with up to five filter plates. One filter wheel has neutral density filters to reduce light (manipulation of the depth of field). The second wheel is for colour conversion, i.e. adaption to various types of illumination. The depolarising plate, also shown in Figure 3.4.4, is present in order to prevent colour falsifications

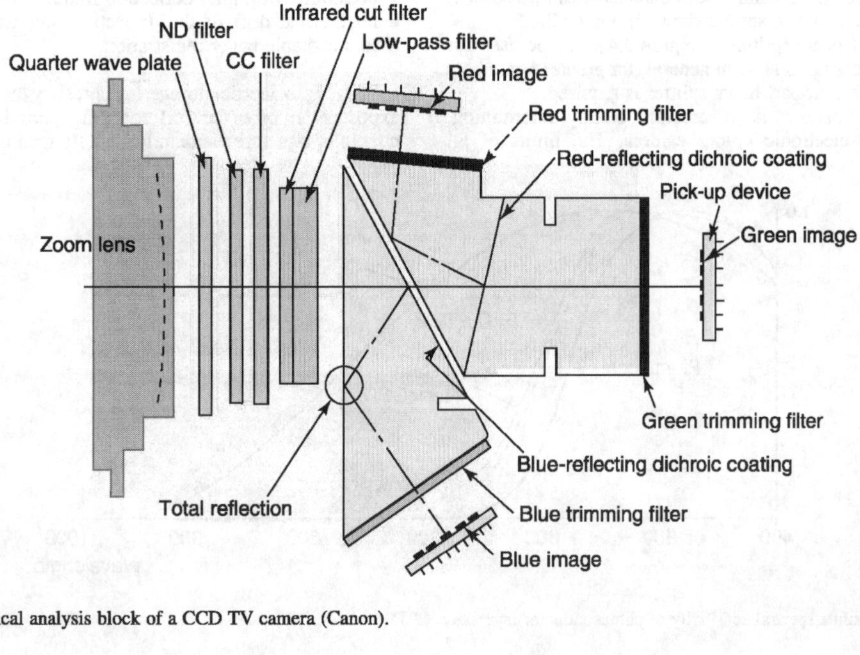

Figure 3.4.4 Optical analysis block of a CCD TV camera (Canon).

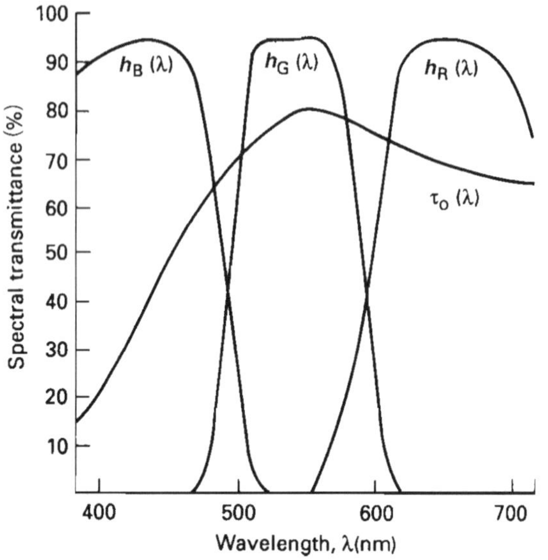

Figure 3.4.5 Spectral transmittance of a lens $\lambda_0(\lambda)$ and spectral filtering function $h(\lambda)$ of a beam splitter.

on the dichroic layers in the beam splitter, due to eventual incoming linearly polarised light.

Three glass prisms, with dichroic layers at the connecting surfaces, form the *beam splitter*. The selective reflection at the dichroic layers is utilised to separate spatially the light currents (Figure 3.4.4). Correction filters are necessary at the three output boundary surfaces of the splitter. This is due to the transmission of partly unwanted wavelength ranges as a result of the use of only the wavelength selection for the dichroic layers. This leads to the typical spectral transmittance characteristics $h_B(\lambda)$,$h_G(\lambda)$ and $h_R(\lambda)$ of the complete splitter shown in Figure 3.4.5.

The three-part beam splitter of Figure 3.4.4 can be used for relative f-numbers up to f1.4. In general, for greater f-numbers of around f1.2, a four-part beam splitter is required.

The *camera sensor* is a crucial, quality-determining, element in the electronic colour camera. The limits of all the quality-determining parameters are set by the camera sensors, i.e.:

- static resolution in horizontal and vertical directions;
- dynamic resolution, i.e. temporal resolution and lag;
- signal/noise ratio;
- sensitivity (absolute and spectral sensitivity);
- signal dynamics;
- stability of the scanning raster (colour registration) – not applicable to CCDs.

CCD types may be interline transfer (IT), frame interline transfer (FIT) or frame transfer (FT). Each of these has variants and may or may not be fitted with microlenses. Each type has its advantages and disadvantages (see Chapter 3.4). Where tubes are still used in studio cameras they may have plumbicon, saticon II or saticon III layers.

Figure 3.4.6 shows the relative spectral sensitivity distributions of a plumbicon tube compared to CCD sensors. Note that the plumbicon tube has a higher blue sensitivity than a CCD and a CCD has a much higher infrared sensitivity.

The image format sizes to scale, which have been produced for the image sizes $1^1/_4$ inch, 1 inch and $^2/_3$ inch, as well as for aspect ratios 4:3 (standard television) and 16:9, are shown in Figure 3.4.7. The numerical values can be obtained from Table 3.4.1.

In the past, camera tubes have had two coils for horizontal and vertical deflection, one coil for beam focusing, and possibly coils for beam alignment. With the development of electrostatic deflection, particularly in HDTV cameras, the deflection coils are not required. Instead, there are tube envelopes containing vapour-deposited and laser-cut deflection electrodes.

The concept of electrostatic deflection has fundamental advantages:

- uniform resolution distribution over the whole picture;
- less geometrical distortion;
- possibility of higher deflection frequencies (HDTV);
- no thermal drift of the deflection unit with respect to the tube, i.e. stable raster registration.

The photoelectric storage layer resists temporal changes in exposure. It takes several scanning periods (fields) for the charge of a picture element to be fully recombined. This effect

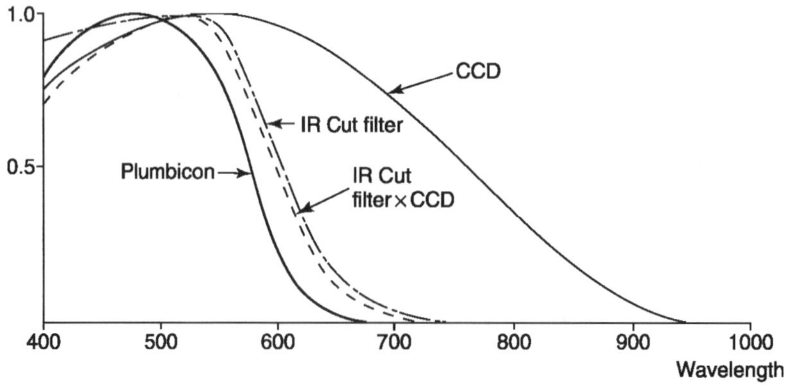

Figure 3.4.6 Absolute spectral sensitivity of photoconductor layers and CCD.

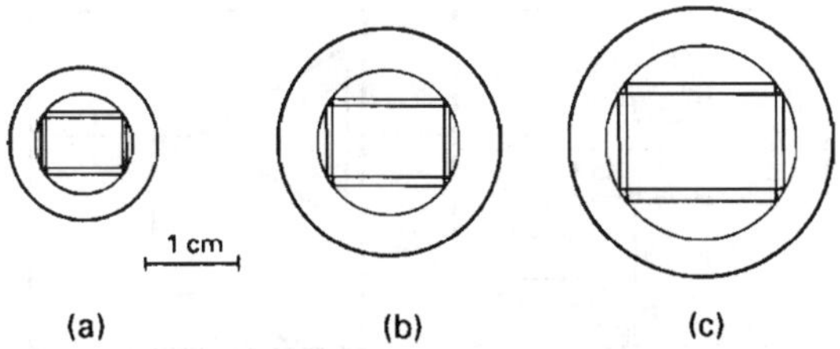

Figure 3.4.7 Relative scanning raster for different image sizes for 18 mm (a), 25 mm (b) and 30 mm (c). Ignore outer ring for CCDs.

Table 3.4.1 Scanning raster or image format dimensions

Nominal diameter (inch)	(mm)	Image diagonal (mm)	4:3 Image size (mm × mm)	Image area (mm^2)	16:9 Image size (mm × mm)	Image area (mm^2)
2/3	18	11	8.8 × 6.6	58.1	9.6 × 5.4	51.8
1	25.4	16	12.8 × 9.6	122.9	14 × 7.8	109
1 1/4	30	21.4	17.1 × 12.8	218.9	18.7 × 10.5	196.4

is characterised by lag and depends on the structure and capacitance of the storage layer, and on the non-linear landing behaviour of the scanning electron beam, as well as on the signal level. Plumbicon layers are better here than saticon I and II layers.

To improve the behaviour of the lag, a bias light of approximately 3% of the nominal light level is made to strike the camera tubes (Figure 3.4.8).

This bias light can either be injected from the front via the beam splitter, or it can be transported from behind via the tube

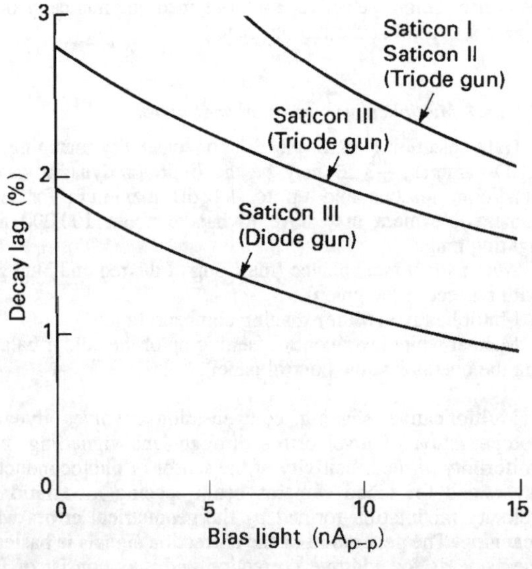

Figure 3.4.8 Decay lag characteristics for saticon tubes with several gun types, as a function of the bias light level (Hitachi).

envelope or special light conductor. It is necessary to be able to adjust separately the bias light in the three colour channels, in order to be able to set up uncoloured residual lags.

CCDs do not suffer from lag (except for some very early types) and so do not use bias light. Their read-out is determined by complex pulse trains and is highly stable in terms of the geometry, which is determined solely by the masks used in the manufacture of the sensor.

3.4.2.2 Pixel correction

The appearance of white spot blemishes on the picture output is a problem on both tubes and CCDs.

White spots on CCDs (which are supposed to have a very long life) may appear randomly at any time, even in the first months of use. On some occasions the blemish will appear and then disappear at the next switch off/on cycle, others continuously vary in amplitude and in either case they increase with ambient temperature (doubling for each 5 °C rise in temperature). The blemishes may be called 'pixel defects', 'dead pixels' or 'residual point noise'. In fact they are not dead but are still light sensitive, with the output sat on a dc offset. Broadcasters require a very high specification of the spot amplitude as it is made much worse by the gamma stretch and their presence may be very perceptible in low key scenes as coloured or white spots. A typical specification for blemish amplitude is less than 2% of peak white measured in the luminance with gamma on, with camera capped and sat up. So a number of different methods of compensation or correction have evolved to minimise their effect on the picture.

The simplest method of compensation for defects that are present at manufacture is to store a correction waveform and subtract it from the CCD output. If two conditions are stored for different temperatures the correction may be weighted according to the ambient. However, as defects constantly appear this takes no account of ageing. Some cameras have a system where the output is observed on a monitor and a cursor is moved over the defect. This position is stored and used to modify the CCD read-out and effectively repeat the previous good cell instead of the faulty one. Other cameras carry out an automatic check for defects when the auto black balance is operated. At its simplest this may again repeat the last pixel output but additional complexity on the DSP cameras is used to take the mean of surrounding pixels in both horizontal and vertical directions (Sony AAPR, Adaptive Automatic Pixel Restoration). In all cases the correction is added to all colours to eliminate edge effects. One DSP chipset (Thomson / Philips/Grass Valley LPX or LPC) continually monitors the live output and makes a correction where there is a change in the picture content from one pixel to its surrounding pixels at a rate unlikely in nature. The noise floor sets the minimum blemish level that can be corrected.

3.4.2.3 Video pre-processor

3.4.2.3.1 Pre-amplifier

Typical signal currents for tubed cameras, at a studio illumination of 3200 K, are about 240 nA (R), 300 nA (G) and 160 nA (B) for picture white. One of the tasks of the video processor that follows is to create, for a picture white to be defined, the same signal level in the three channels via selective amplification.

CCDs have a voltage output as they have an on-chip amplifier. However, the signal representing the image has to be extracted from all the various pulses and reset noise. Correlated double sampling (CDS) or delay line processing (DDS) is often used to extract the signal.

CDS samples the signal at both the part of the output waveform representing light and the blanked area, both of which have reset noise present. The two measurements of signal level are held on capacitors and the two voltages are then fed into a differential amplifier; the reset noise which is on both the signals is cancelled, the final output of the amplifier being the wanted signal. Account must be taken of the timing difference in the green channel due to the spatial offset and the sampling pulses for green must be offset by half a cycle. The RGB timings are then corrected by delays in red and blue to make the images coincident. Some CCDs have the CDS circuit integrated on-chip. Some further filtering is used to remove any remaining clock frequency. This filter should be phase corrected to retain pulse symmetry (Figure 3.4.9).

DDS may be used as an alternative to reduce out-of-band noise (above the Nyquist limit), which is folded back in-band by the sample and hold process (Figure 3.4.10).

The principal structure of the camera tube pre-amplifier is shown in Figure 3.4.11. The amplifier has the task of converting the above small signal currents into signal voltages of the order of magnitude of 0.3–0.7 V. For practical reasons, the transimpedance concept illustrated in Figure 3.4.11 is exclusively used. The demand for an optimal signal/noise ratio results in the separation of the pre-amplifier into two subassemblies. The front part, consisting of a field effect transistor and a feedback resistor, is fixed directly to the top of the camera tube. It is possible to couple inductively the pre-amplifier via the so-called *Percival coil*, beside the direct coupling. The structure of the pre-amplifier determines the obtainable camera signal/noise ratio.

The functions of the processor usually residing in the camera head can be further split up, as shown in Sections 3.4.2.3.2–3.4.2.3.5.

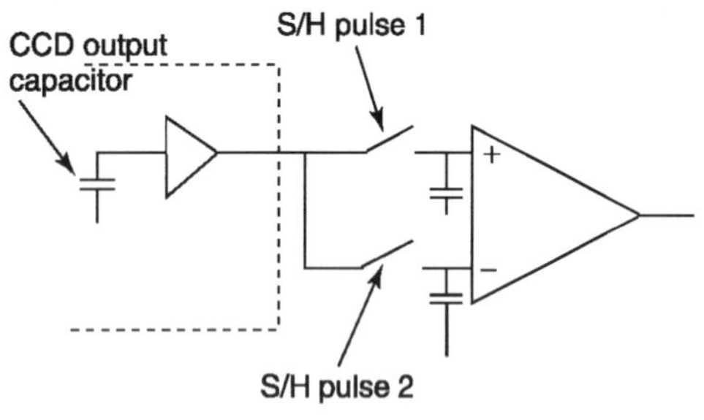

Figure 3.4.9 Block diagram of correlated double sampling (CDS).

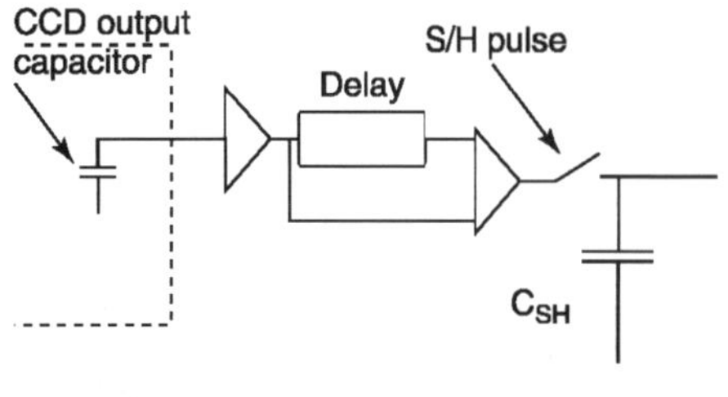

Figure 3.4.10 Block diagram of delay line sampling (DDS).

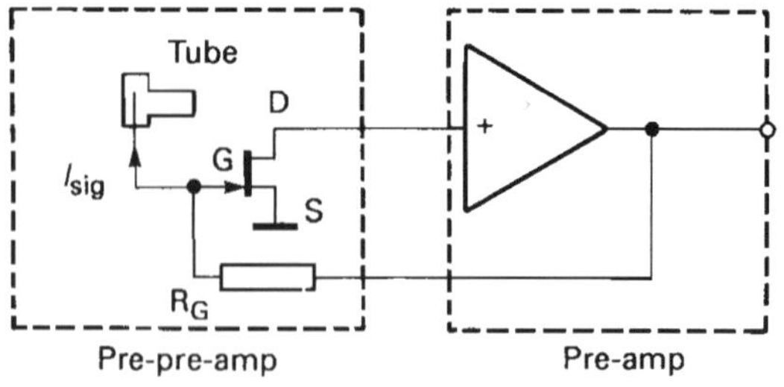

Figure 3.4.11 Schematic diagram of a tube camera pre-amplifier or head amplifier.

3.4.2.3.2 Frequency compensation

- Compensation of the frequency response of the pre-amplifier.
- Horizontal aperture correction (out-band correction, i.e. part compensation of the camera or sensor frequency response due to sampling), occasionally used at this stage.

3.4.2.3.3 Additive correction and control

- Flare compensation (in the optical components, an increase in the black level is produced, approximately proportional to the mean signal value). The compensation is based on the average picture level and corrects for overall but not localised flare.
- Black shading compensation is still used on CCD cameras (on tubed cameras dynamic compensation of the non-uniform image illumination caused by the bias light was very necessary). This may be achieved by adding H,V sawtooth and parabolas or by area-based digitally generated correction signals from the camera microprocessor.
- Black level control (master black or master lift, i.e. colour neutral or colour selective, each obtained automatically or by operator control at the remote panel).

3.4.2.3.4 Multiplicative correction and control

- Gain selection (−6 to +24 dB, for sensitivity matching, for CCD cameras; −3 dB may be the limit for dynamic range/headroom reasons and up to +42 dB maximum for ENG cameras); camera may have to handle about 100,000 to 1 lighting range.
- Automatic white balance (matching of the red and blue gain with respect to the green).
- Multiplicative (white) shading compensation.
- Manual white level control (setting up of the colour balance via the operator at the control panel).

Multiplicative shading compensation involves dynamic compensation of level errors through lens vignetting, non-uniformity of the sensitivity of the sensor or photoconductive layer, and for tubed cameras beam spacing modulation or velocity modulation formed by the geometrical errors when scanning. The generation of the correction signals is basically the same as for additive correction and may consist of H,V sawtooth and parabola waveforms or again by an area-based system.

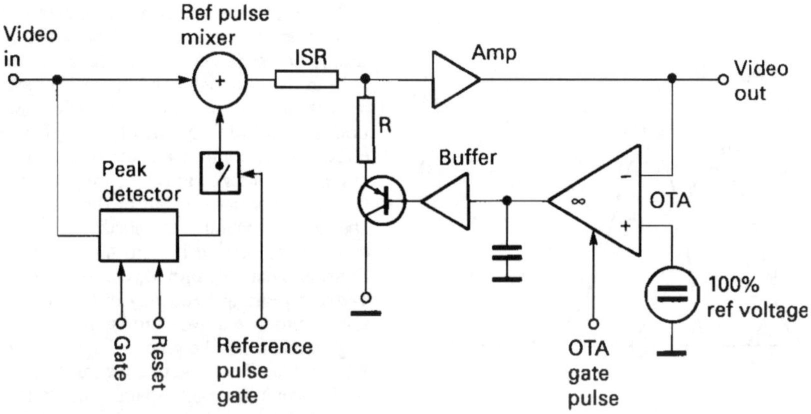

Figure 3.4.12 Schematic diagram of a dynamic knee processor.

3.4.2.3.5 Non-linear control

- Knee processor.

The knee processor possesses a compressing transfer characteristic, through which the available contrast ratio of the camera can be increased by a factor of 4–6 with respect to conventional systems (Figures 3.4.12 and 3.4.13).

As Figure 3.4.13 shows, the input signal values, which exceed 100%, are compressed into the output range of approximately 80–100%. In the pivoting knee shown, this is dynamically produced (auto-knee), i.e. controlled from the peak value of the input signal, so that the maximum input value (assuming that it is greater than 100%) is always transferred to 100% at the output. An auto-knee may either vary the slope with the knee point fixed, vary the point with the slope fixed or vary both point and slope. A manual knee may also be used which may be more predictable with multi-camera operation. The knee point is set to about 85% and one and a half stops of overload compressed between 85% and 100%.

There are various different methods of carrying this out, some of which give better results than others. The best methods retain the correct hue in the overload area of the picture.

3.4.2.4 Main video processor

As can be deduced from Figure 3.4.3, the stages for the colour correction matrix, two-dimensional frequency response correction and gamma correction form the core of the main video processor.

3.4.2.4.1 Colour correction matrix

For a colorimetrically correct picture reproduction onto a screen, the camera's primary signals have to be matched to the screen phosphor. This occurs via a linear transformation, i.e. a weighted addition and subtraction of the input primary signals into an electronic matrix. Figure 3.4.14 shows that the spectral sensitivities of the camera channels can be matched to a good approximation to the negative components of the spectral curves of the screen phosphor.

Although calculations give a theoretical set of coefficients for the optimum response (e.g. EBU colorimetry), different broadcasters have their own preferred colorimetry. Most modern cameras will have a selection of settings programmed into memory (e.g. EBU, BBC, RAI, ARD), as well as variable controls for special characteristics or matching to other camera types.

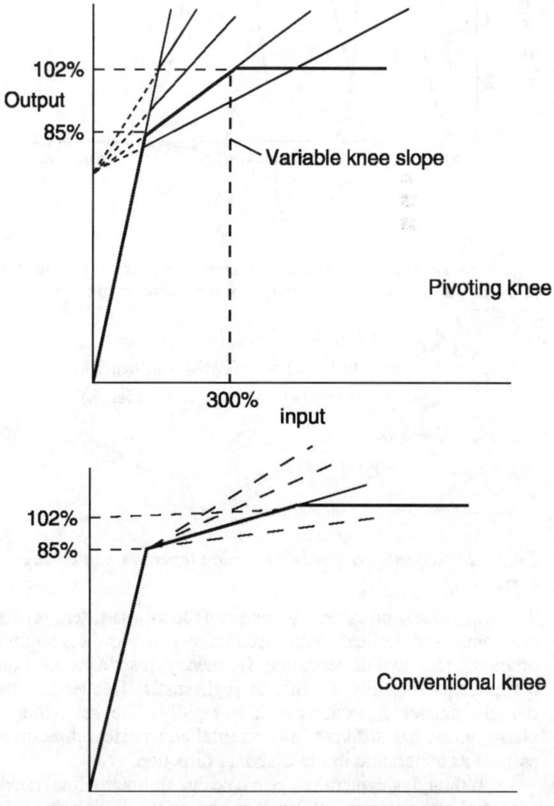

Figure 3.4.13 Transfer characteristic of knee circuit using either a pivoting knee or conventional knee.

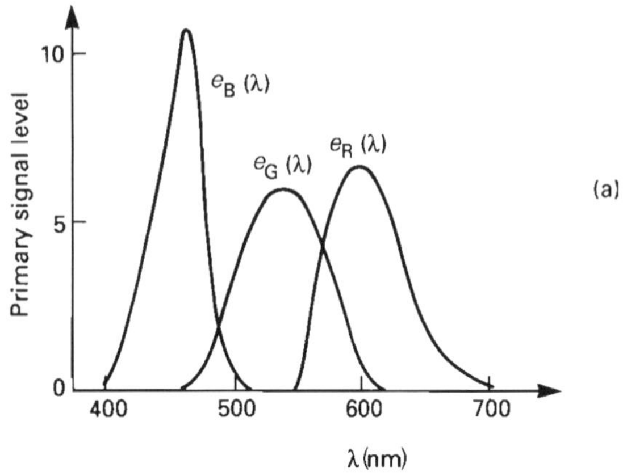

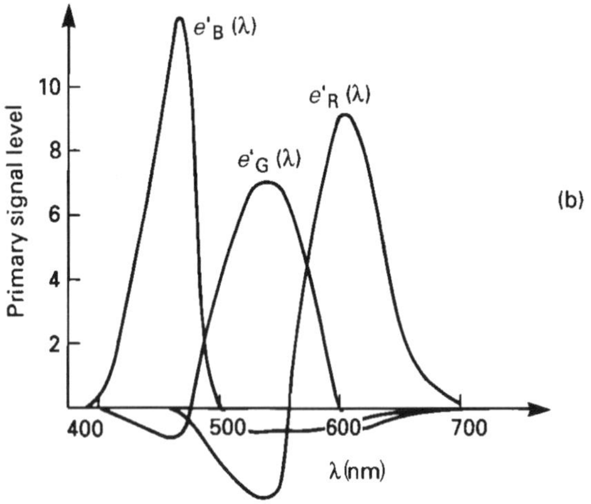

Figure 3.4.14 Spectral sensitivities of an RGB camera: (a) at the input of the colour matrix; (b) at the output of the colour matrix.

$$e'_R(\lambda) = 1.45e_R(\lambda) - 0.45e_G(\lambda)$$
$$e'_G(\lambda) = -0.10e_R(\lambda) + 1.20e_G(\lambda) - 0.10e_B(\lambda)$$
$$e'_B(gl) = -0.05e_R(\lambda) - 0.10e_G(\lambda) + 1.15e_B(\lambda)$$

(Lang[1]).

3.4.2.4.2 Frequency response correction (aperture and contour correction)

The lens, CCDs or camera tubes act as low-pass filters for the horizontal and vertical spatial frequencies due to the sampling process. The two-dimensional frequency responses of both components multiply, so that at high spatial frequencies the overall transfer characteristic falls rapidly. The behaviour is phase linear, but different in horizontal and vertical directions, as well as astigmatic in the diagonal direction.

Even though a camera (a system scanning line by line) needs a spatial pre-filtering, at least in the vertical direction, the frequency response must be raised in the horizontal and vertical directions to give a satisfactory image.

The terms aperture correction, detail correction and contour correction are loosely used. Theoretically, aperture correction should be used to correct for the aperture loss due to sampling in both the horizontal and vertical directions. This flattens the total frequency response, after which detail or contour correction is used to lift the middle or higher spatial frequencies for effect, or to compensate for losses in the final reproduction system (monitor). However, many circuits used by manufacturers do not carry it out in this way and they often provide aperture correction of the horizontals only, followed by contour or detail correction in both the horizontal and vertical directions, or even only provide contour correction (H and V) with a variable peaking frequency in the horizontal path. It is necessary to produce a two-dimensional phase linear high-pass filter to generate a detail correction signal, which is then added to the main signal. The principle for the creation of this detail signal is the same for both spatial coordinates, i.e. displaced and weighted parts of the signal are subtracted in the x and y directions from this signal (Figure 3.4.15). A spatial discrete filtering then takes place, which is, as required, phase linear.

The creation of the circuit obviously takes place in the time domain, i.e. in the horizontal direction, via delays with analogue or digital delay time elements, and in the vertical direction via line delays. The latter are conventionally achieved by ultrasonic delay lines (with the video modulated onto a high-frequency carrier), digitally by line stores, or in the case of the latest cameras, in the digital signal processor. The vertical displacement can naturally only occur in increments of one line, which implies a displacement of two spatial lines in interlace scanning.

Usually the horizontal filter can be connected in series with the vertical filter, as in Figure 3.4.16.

In the past the correction signal was made from the green signal and was added equally to all three of the colour channels. This reduced problems due to registration errors. It is now more usual (because of the accurate registration possible with CCDs) to generate a luminance signal from a weighted sum of the three colours and to generate the correction signal from this, and then add it to all channels. Another alternative would be to carry out full contour correction in each of the three colours, but the cost of glass delay lines for the vertical correction previously made this prohibitive in all but the highest cost cameras. The use of digital signal processors has now made this possible without significant extra cost.

The correction should be generated and added back before gamma correction, though for noise reasons early CCD cameras, in particular, both generated and added the correction after gamma. Some DSP cameras generate the signal before gamma but add it back after gamma, whilst others provide a control that allows correction to be added both before and after gamma. The sequence of detail and gamma affects the resolution in the dark areas of the picture. One technique in use, low-pass, filters the RGB channels and processes all the highs in a side chain before adding this back to all channels; this becomes obvious as, when the detail and aperture correction are switched off, the resolution falls to about 10% at 5 MHz (Figure 3.4.17). Another camera type processes the luminance and contour correction at a higher data rate for superior performance.

The raising of the frequency response naturally worsens the signal/noise ratio at high frequencies. To minimise this problem, non-linear processing of the detail signal is carried out. Two methods are used: level dependence and coring (crispening). These both reduce the noise in the blacks, where it is more obvious. Level dependence removes the lower part of

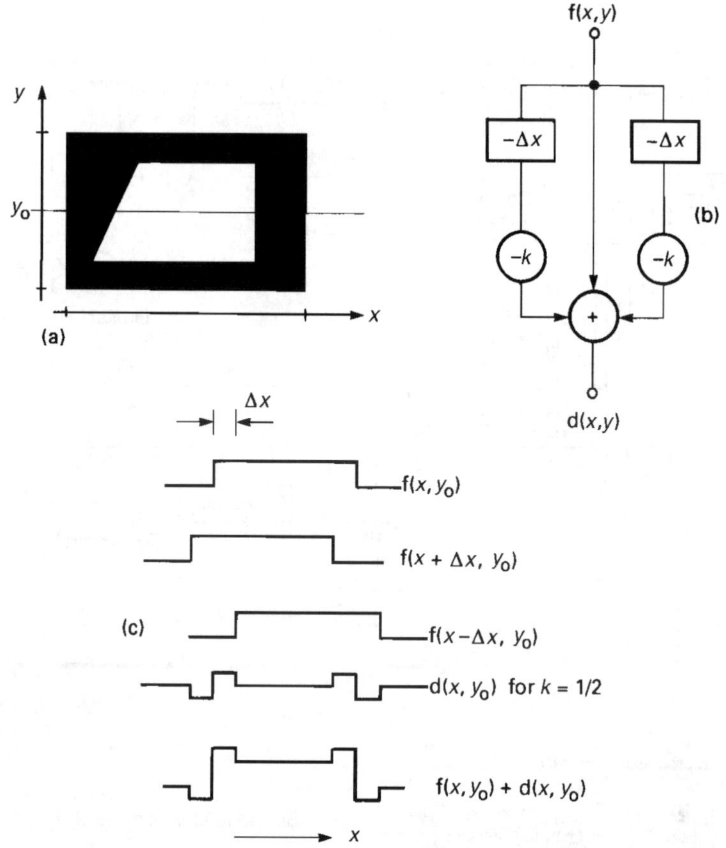

Figure 3.4.15 Generation of a horizontal details signal for contour or aperture correction: (a) image at the input f(*x*,*y*) of detail processor; (b) principle of a horizontal detail processor; (c) horizontal waveforms at the constant vertical position y_0.

the input signal and some of the noise before summing the signals from the delays to generate the correction signal. It is usually used on the horizontal signal path. Coring is performed on the correction waveform (H and V) and takes a slice out of the middle of the waveform, removing all small signals. Note that the small signals removed are not only noise, but also small corrections from anywhere on the wanted signal. With modern cameras having a high signal/noise ratio it should not normally be necessary to use either level dependence or coring.

With the advent of digital signal processing many other controls have been added to the contour corrector circuit. Examples are: detail black and white compress (clip or limit) to reduce excessive overshoots and black edges on high contrast scenes; corner detail to improve the corner losses associated with lenses; skin detail to reduce the enhancement of skin defects; slim detail and negative detail for softening the image.

3.4.2.4.3 Gamma correction

The transfer characteristic of colour display tubes follows approximately a power law (from 2.2 to 2.8). Therefore, for a correct reproduction of colour, an inverse alignment has to take place at one point in the transmission system, i.e. the function $y = x^{\gamma}(\gamma \approx 0.4)$ must be realised. The use of such a power law suggests infinite gain at the bottom of the curve, which would lead to very noisy pictures. A compromise must be used and the initial slope is usually limited to between 3:1 and 5:1 and the power law is often 0.45. The lower values of slope tend to make the pictures look over-saturated.

Traditionally and for companding reasons (the low-amplitude areas with noise are boosted before any noise is added in the transmission system and then compressed at the receiver), gamma correction is carried out in the source equipment, i.e. the camera. The gamma law used has a far greater effect on the colorimetric response than the colour correction matrix. The frequency response of the gamma corrector is often modified to allow a fall-off in HF response at low signal levels to reduce the noise. However, this can also lead to pictures that lack detail in the dark areas. Some typical methods used to generate characteristic are shown in Figure 3.4.18. Different laws and variable control range are shown in Figure 3.4.19.

The corrections currently carried out for colour reproduction and gamma pre-compensation in the image source equipment are with respect to a particular standard reproduction

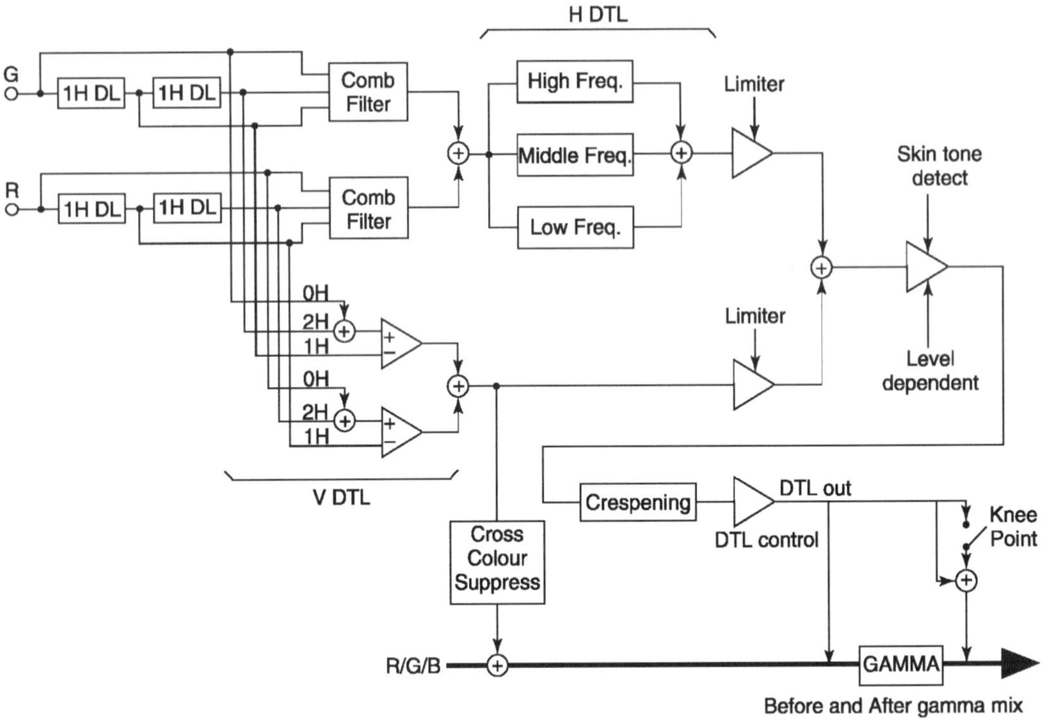

Figure 3.4.16 Typical detail processor chain used in a DSP camera (Sony).

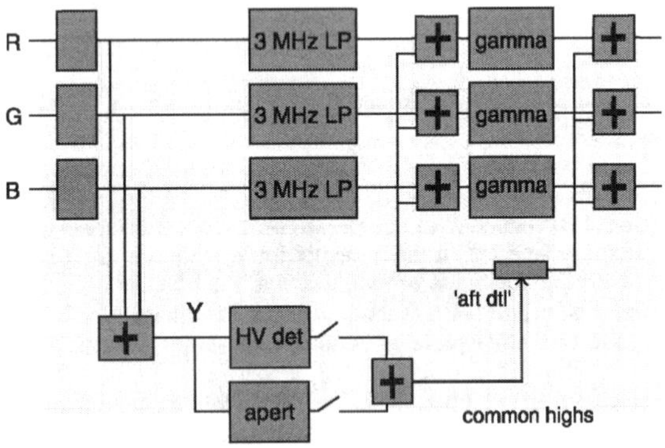

Figure 3.4.17 Detail process used in a lower cost camcorder.

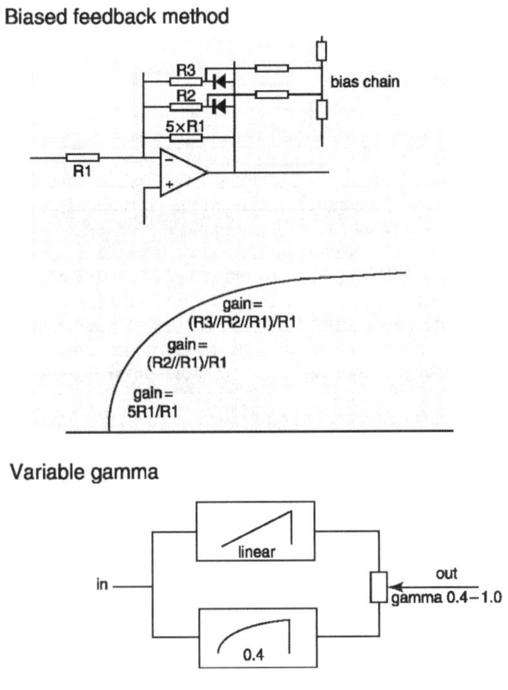

Figure 3.4.18 Typical methods of generating the gamma characteristic.

device (a monitor) with given phosphors and transfer characteristics. In the process of defining and introducing new television systems, it is desirable to have a standard output interface for the source equipment and a standard input interface for the image reproduction equipment. In Europe this standard has been defined by the EBU and monitors with EBU phosphors are used to determine colour matching. In the USA a different set of phosphors defines the colour reproduction of the system. Only then can a random coupling of the wide variety of source and display devices lead to optimal colour reproduction.

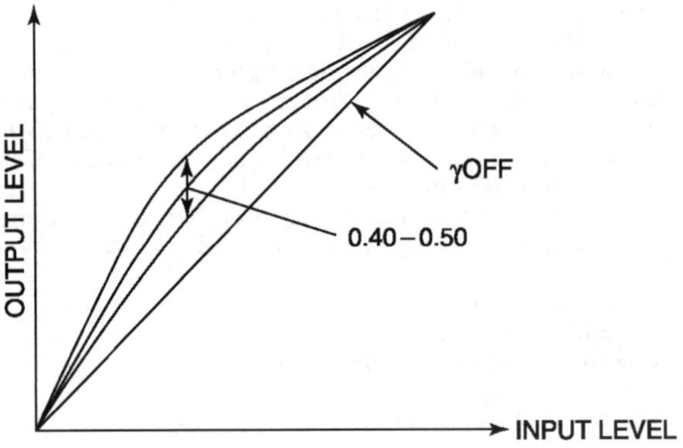

Figure 3.4.19 Variable gamma control (Sony).

3.4.3 Operational Characteristics

The characteristics of an electronic camera which determine its quality are:

- signal/noise (s/n) ratio;
- sensitivity;
- static and dynamic resolution;
- colour reproduction;
- colour registration of the three images.

These characteristics interact and cannot be studied in isolation. For example, static and dynamic resolution is electronically improved at the cost of the s/n ratio, or a bad colour registration worsens the luminance resolution. An overall analysis is especially required for sensitivity and the s/n ratio. In selecting the nominal signal current for picture white with tubes, the limits for the s/n ratio and sensitivity are set at the same time.

3.4.3.1 Signal/noise ratio

For a camera using CCD sensors a range of factors is responsible for the signal/noise performance:

- Noise is linearly related to the quantum efficiency of the CCD; increasing the light level also improves the signal/noise but the maximum light level is limited by the dynamic range of the sensor.
- The aperture ratio of the CCD, i.e. the light-sensitive area compared to the masked transfer area (use of microlenses improves this).
- Absorption of photons in the surface layers.
- Substrate type and construction.
- Thermal, reset, shot, residual point, 1/f and output amplifier noise in the CCD.

Most of these factors come from the design of the CCD itself and the only area affected by the circuitry is the sampling circuit used (CDS or delay-line processing).

The signal/noise ratio of an electronic camera with tubes is determined by:

- the nominal signal current;
- internal noise of the first fet in the pre-amplifier;
- effective capacitance at the amplifier input (i.e. the sum of the target source capacitance, the fet input capacitance and unavoidable stray capacitances);
- noise due to the effective ohmic resistances at the amplifier input.

The internal noise of the camera tube is negligible.

The signal/noise ratio of modern cameras is about 61 dB measured in the luminance channel (PAL System I, 5 MHz bandwidth, without subcarrier notch). In 1984 a typical signal/noise ratio for the same condition was about 50 dB. At present the limiting factor in the signal/noise at the output of the CCU is the signal /noise of the RF triax transmission system rather than that of the camera head.

3.4.3.2 Sensitivity

The sensitivity of an electronic camera is defined by the choice of lens, CCD or tube type. The sensitivity of a charge storing camera tube is, for example, defined as the quotient of the signal current, I_S, and the illumination on the tube target, E_{TG} (measured in lux). The tube is used in a TV system, with a total picture duration T_{tot} and an active picture duration T_{act}. The camera tubes have an integral layer sensitivity S_i' (measured in amps per lumen) and a scanned target area A_{TG}. The signal current is calculated as follows:

$$I_s = S_i' \cdot T_{tot}/T_{act} \cdot A_{TG} \cdot E_{TG} = S_i \cdot A_{TG} \cdot E_{TG}$$

The sensitivity of the tubes does *not* depend on the number of lines, on the frequency of image changes, or on the type of scanning (interlace or progressive). During scanning, whenever a stabilised state has been reached between the charging and discharging of the photoconductive layer, the charge which has flowed in via the photocurrent will also be completely removed again by the signal current. In other words, with a decrease in the integration time (less charge stored), a similar decrease in the read-out time occurs (the charge is read out quicker). Hence the quotient, $I = \Delta Q/\Delta T$, remains constant.

The different television standards are only reflected by the signal current through the relationship between the total and active picture periods, and through the scanned target area. Of course, an increase in the number of lines or the transition from interlace to progressive scanning leads to a decrease in the amplitude deviation of the target surface potential. Hence, with the same layer capacitance, an unchanged signal current is produced. The relationship between illumination, E_{SC}, and the aperture number, F, is shown in Figure 3.4.20 for the aspect ratios 4:3 and 16:9, as well as for the image sizes $^2/_3$ inch, 1 inch and $1^1/_4$ inch. The aspect ratio 16:9 has, for the same target diagonals (due to the smaller area), 89% of the sensitivity of 4:3 scanning.

At first sight, according to Figure 3.4.20, the sensitivity of a camera appears to increase with an increase in the tube diagonals. In fact, the $1^1/_4$ inch camera, for the same aperture number, F, delivers a signal current which is a factor 1.8 times that of a 1 inch camera. The false deduction arises from the completely different reproduction geometry of the $1^1/_4$ inch camera. If two cameras with picture diagonals d_1 and d_2 are operated with the same layer sensitivity, aspect ratio, lens transmission factor, etc., then the ratio of their signal currents is:

$$I_{S2}/I_{S1} = (d_2/d_1)^2 \cdot (F_1/F_2)^2$$

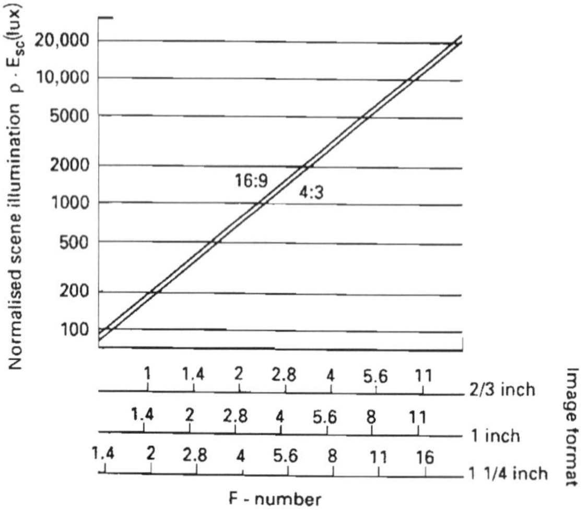

Figure 3.4.20 Sensitivity of a tube TV camera for different f-numbers, image formats and aspect ratios. $I_s = 300$ nA, $S_i = 150$ μA/lm, $\tau_{lens} = 0.8$.

If the same visual angle and depth of field are required, the aperture numbers can be calculated as follows:

$$F_1/F_2 = d_1/d_2$$

If this is then substituted into the previous equation, the same signal currents are produced, i.e.

$$I_{S2}/I_{S1} = 1$$

Hence the sensitivity of the electronic camera is, for the same picture composition, independent of image size!

CCD cameras also have their sensitivity affected by the integration time; hence, when an electronic shutter is in use, which shortens the integration time, the sensitivity is reduced in direct proportion. Since CCDs were first produced in 1984 their sensitivity has increased by some eight times (three F stops).

3.4.4 Automation Functions

The automation functions so far realised for modern studio cameras can be divided into basic set-up, pre-operational set-up and continuous automatic functions. The *basic set-up* is carried out the first time the system is put into operation and, with tubed cameras, after a tube change, as well as periodically over large cycles. The *pre-operational set-up* includes algorithms, which optimise the camera state to the actual requirements of a scene. The *continuous automatic functions* are active during the actual picture operation. It is possible to realise the following individual functions.

3.4.4.1 Automatic basic set-up

With CCD cameras there is very little to align regularly as the sensor performance is determined by the clocking and sample waveforms. Camera menus will have some of the CCD timing and voltage functions adjustable in the deeper engineering menus. The main parameter (in the writer's experience), which may be adjusted during acceptance, is the CCD substrate voltage. This may be adjusted to optimise the dynamic range of the sensor to that claimed by the manufacturer and ensure that all colours are linear in the highlight areas. It must be set carefully as it will also affect any overload breakdown streaking. Also the new setting must be checked for any deterioration of the response to small speculars.

For tubed cameras the following operations may be carried out automatically by use of a diascope and test pattern:

- mechanical back focus;
- electrical focusing;
- beam landing or beam alignment;
- beam current;
- beam current control (ABC).

Optimisation of the scanning raster:

- mechanical rotation;
- coarse registration;
- geometry;
- fine registration.

Both CCD and tube cameras may have automatic alignment of the following using either an internal test waveform inserted early in the processing, external charts or possibly a diascope:

- channel gains (internal test signal).
- shape of the gamma correction (one or two point adjustment rather than total curve shape using internal test signal).
- flare compensation (external chart).
- black shading.
- white shading (full shading with area-based correction using a very evenly lit chart or basic tilt and parabola on a grey scale-type chart).

3.4.4.2 Automatic pre-operational set-up

- White balance.
- Black balance.
- Selection of conversion filters.

3.4.4.3 Continuous automatic functions

- Iris control if desired.
- Continuous auto white balance (usually with ENG/PSC camcorders).
- Dynamic centring of the red and blue rasters for tubed cameras only.
- Dynamic lens error corrections for tubed cameras.

3.4.4.4 Error corrections for tube cameras

From the functions above, two important processes will be described in more detail as examples – namely, fine registration corrections of the scanning raster and dynamic lens error corrections. The same principle is used for aligning auto white shading and, though the principle described here is for tube cameras, the same method is used for CCD camera shading.

3.4.4.4.1 Fine registration correction

The introduction of microcomputers into studio camera technology has led to conventional registration correction processes being superseded by computer-supported digital algorithms. In conventional processes, the horizontal and vertical signals of a given form (e.g. sawtooth or parabola) were mixed with the deflection currents. However, firstly an extensive balancing of numerous potentiometers was necessary, and secondly the obtainable quality, especially in the border areas of the picture, was limited due to a finite supply of periodic correction signals.

On the other hand, modern computer-supported processes start by segmenting the image field, from which, by definition, an almost freely programmable correction signal can be created from the matrix of correction points containing $N_H \times N_V$ values. The flexibility of this signal is limited by the number of correction points and by the necessary horizontal and vertical interpolation processes between the correction points (Figure 3.4.21). A high number of correction points, e.g. 16, in the horizontal direction leads to a bandwidth of several hundred kilohertz for the correction signal.

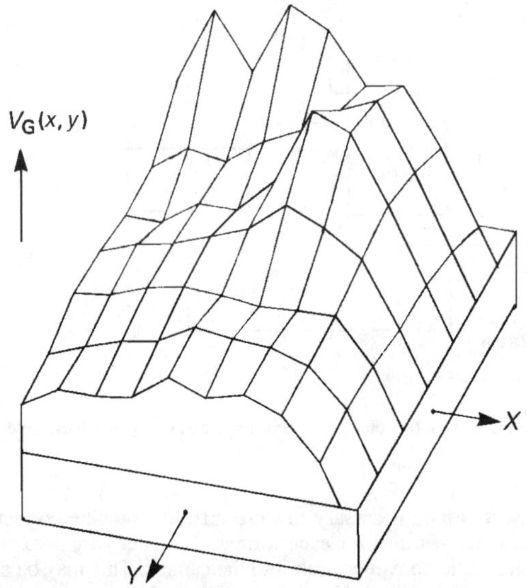

Figure 3.4.21 Typical waveform of a geometric correction signal as a function of spatial coordinates *x* and *y*.

Even though the horizontal correction signal is dependent on the vertical coordinates, and the vertical correction signal is dependent on the horizontal coordinates, two similar and (in the bandwidth) equivalent signals are obtained. The vertical deflection must also be able to transmit this additively mixed broadband correction signal.

The correction points can now be found, either by menu-driven manual tuning or by a fully automatic process. Figure 3.4.22 shows the principal structure of the tuning system.

This, with the help of a *diascope* (test slide in the lens), measures and minimises the registration errors. The test slide shown in Figure 3.4.23, as well as measuring two-dimensional registration errors (vertical depositing positions are also converted into temporal errors, due to the diagonal lines), also determines white shading errors above the wide white lines. The actual correction is via a control process, which determines the preliminary depositing position errors between, for example, red and green, and then iteratively changes the correction value until the remaining error is within the tolerance limits. The correction values found from correction point to correction point by this method are deposited in memory.

The signals interpolated from these values are always available during the actual camera operation. As already indicated, such a process can also sensibly be used for the correction of image geometry, and black and white shading.

3.4.4.4.2 Dynamic lens error corrections

Using a registration correction process as described above, no errors can be corrected which result from the three operational parameters focal length (f), setting distance (a) and aperture (F) of the lens. Of course, to a very good approximation, the error here is purely an error in the magnitude of the red and blue rasters with respect to the green one. Hence the following functions should be stored in the correction computer:

$$\text{H magnitude (R)} = f_1(f, a, F)$$
$$\text{V magnitude (R)} = f_2(f, a, F)$$
$$\text{H magnitude (B)} = f_3(f, a, F)$$
$$\text{V magnitude (B)} = f_4(f, a, F)$$

The computer must also be given here the current actual values of f, a and F. The obtained correction values are then added to the deflection currents or voltage. Figure 3.4.24 shows the typical errors dependent on the lens before and after correction.

3.4.5 Digital Signal Processing (DSP)

As has been mentioned, modern cameras exclusively use digital processing where the signals from the pre-amplifier are converted to digital signals by an A-to-D converter. Initially 10-bit A-to-Ds were used in conjunction with a pre-knee to limit the dynamic range of the input. At present most full specification cameras use a 12-bit A-to-D, one chipset being ready for a 14-bit input. The A-to-D may be separate or part of the DSP chipset, which consists of one or two integrated circuits.

Generally the master gain, individual gain, flare and shading circuits remain in the analogue domain and the rest of the processing is digital. However, the shading (and flare) is often measured and the waveforms generated in digital and fed back via a D-to-A to the analogue modulators and clamps.

Digital processing is completely stable and so any controls may be set entirely as numbers which are repeatable. This allows the use of set-up memories and transfer of standard settings between cameras. This gives good matching between cameras of gamma law and matrix coefficients. However, the accuracy of this matching will be dependent on the analogue stages at the front end being accurately set up and matched.

3.4.6 Aspect Ratio Conversion

With the advent of widescreen there has been a need for 16:9 cameras. At the present time we still live in a mixed world of 4:3 and 16:9, so camera sensors need to have switchable format.

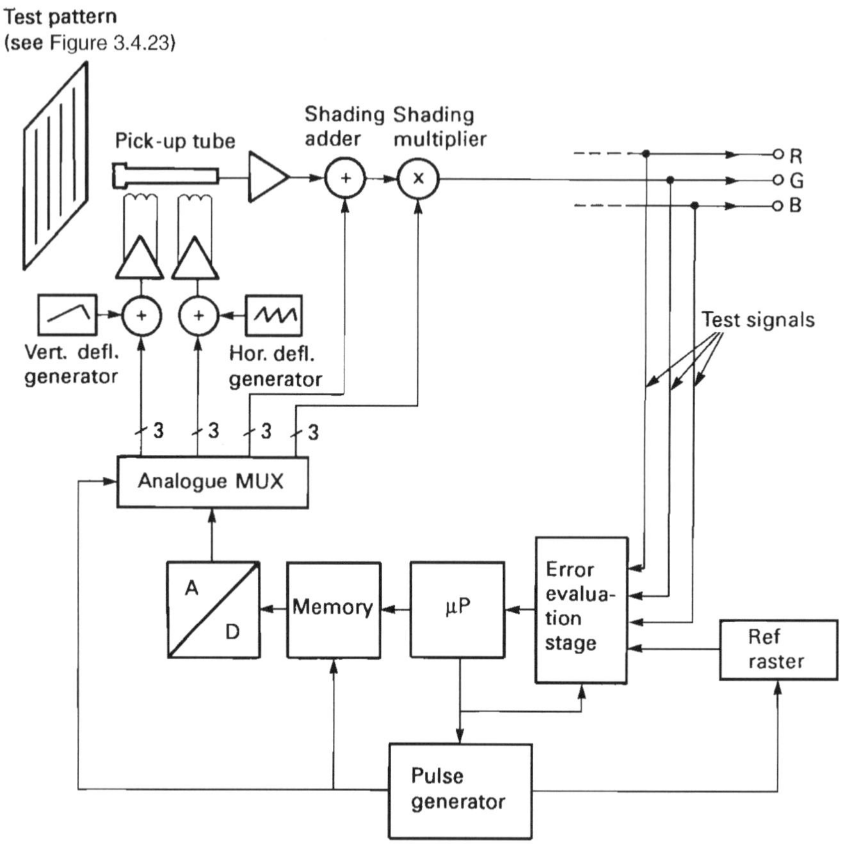

Figure 3.4.22 Block diagram of automatic registration and shading compensation, using a test chart shown in Figure 3.4.23 (Broadcast Television Systems GmbH).

Figure 3.4.23 Test slide used in lens diascopes for automatic registration and shading set-up (Broadcast Television Systems GmbH).

There are two ways of switching between the formats with a 16:9 sensor. The original method was to take a 4:3 centre cut-out horizontally when working in 4:3. Increasing the clock speed for the first and last part of the horizontal read-out allows this. It is then necessary to store different shading corrections and detail settings for each format. The later way involves an aspect ratio converter built into the camera. This may be a part of the DSP chipset. In this case the need for separate line-up memories is minimised. The other method is to start with a 4:3 sensor (Philips) and take a 16:9 section out of this vertically. The construction of this type of Frame Transfer sensor allows this without any loss of vertical resolution.

As the format is changed the camera viewfinder must also have its scans switched to present the correct shape of picture for the camera operator. The viewfinder will have cursors available to show the area of the other format to aid composition of the picture.

3.4.7 Portable Cameras

Most of the above relates to full-facility studio/OB cameras. Most manufacturers make compatible lightweight portable cameras that will have exactly the same processing chain and performance and are fully interchangeable with the large heads. The only concessions will usually be in the operational facilities, such as the number of filter wheels or the communications chain. They will of course have a smaller viewfinder and all the

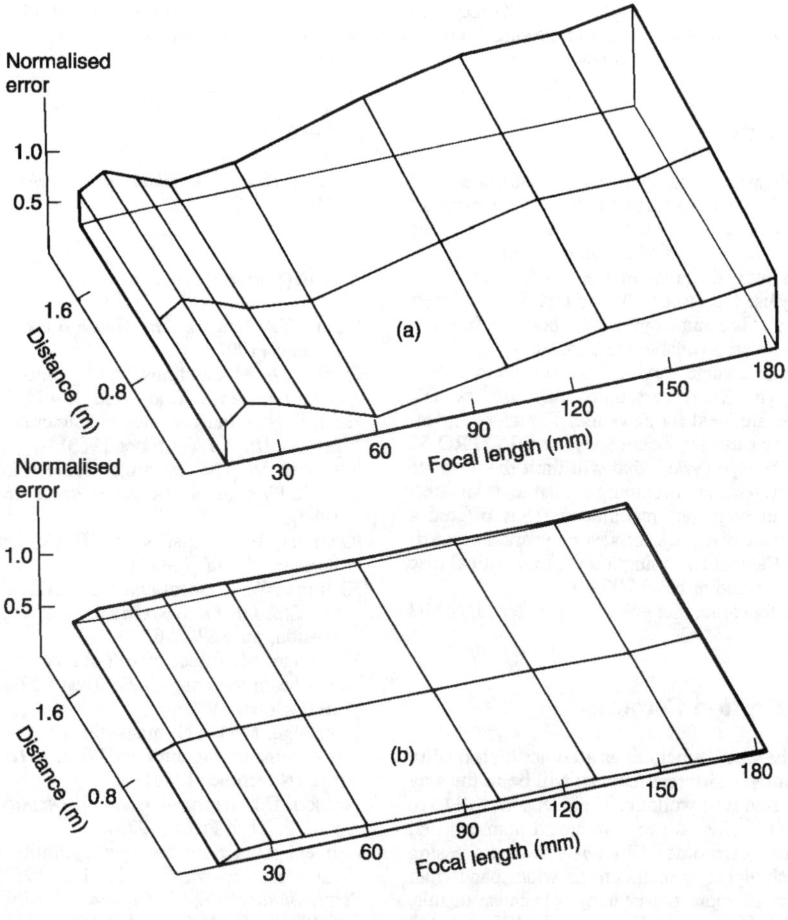

Figure 3.4.24 Typical lateral chromatic errors of a studio TV lens: (a) without dynamic lens error correction; (b) with dynamic lens error correction (Broadcast Television Systems GmbH).

operational controls on the camera will be smaller and closer together.

These cameras may either operate on triax or cable or be powered locally by battery (11–17 volts) and used with a separate or on-board VCR adaptor.

There is a general production move to small camera use either using lightweight or large lenses, as they are easier to move about and transport.

3.4.8 Cradle Systems

With the increasing use of portable lightweight cameras, particularly on OBs and with some broadcasters in studios, there was a need to improve the operational facilities provided. The first option was to allow the use of a large lens. A mounting adaptor is used to attach the lens and hang the camera behind. It will often also have a power supply to provide any increased power needed.

To develop this further provision was made for mounting a large (17 cm) viewfinder, provision of larger controls on a rear control panel, adding a teleprompt channel, power break-out and tracker, cue facilities. A number of manufacturers have offered this 'sportcam' type of system. The BBC London Studios are entirely equipped with these as they offer an easy choice for the producer, with the minimum down time changing from portable to full-facility mode.

3.4.9 Radio Mode Operation

There is also the option of radio mode, where the cable is replaced by a radio link. The camera may be at a very large distance from the vehicle, in a helicopter or car mounted. The camera will usually retain remote control from the OB vehicle via a data link and will be synchronised either via the data system or by a synchroniser in the vehicle. Early radio mode used analogue links that were very susceptible to multi-path effects, which caused either break-up, loss of picture or, if a synchroniser was in use, a frame freeze. Rotating/tracking aerial systems gave some improvement but new systems are in use using the principles of Digital Terrestrial Television (DTT) and

MPEG-2. These new systems use multiple carrier COFDM and are very rugged. The main problem is delay in the signal processing, which needs to be minimised for lip-sync reasons.

3.4.10 Camcorders

The stability, size, low power and lightweight possibilities with the introduction of CCDs have allowed the development of many camcorders, cameras with recorder integrated into one package. These invariably use digital signal processing and the quality of output is little different to the full-facility studio camera (in fact, they use the same CCDs and the DSP chipsets are probably identical). The same options for both engineering and operational line-up are available via a menu.

Camcorders come in a range varying from simple to expensive and complex. The lower cost units may use the DV recording system and are ideal for news use. The more expensive versions will either use Digibeta (Sony) or DVC PRO 50 (Panasonic) as it is the tape system that will limit the ultimate quality. Disc-based systems are becoming available as an alternative to tape, but although one manufacturer has offered a camera using a magnetic disc pack for several years the transition to disc is slow. Camcorders using a writeable optical disc are expected to be launched at NAB 2003.

The batteries used for camcorder power may be NiCd, NiMH or Lithium.

3.4.11 High-Definition Cameras

HD cameras share the same principles as all other high-quality cameras and most of their different features will be in the way they are adapted for film type working. Their sensors will have about 2 million pixels for a tripod-mounted camera and 1.2 million pixels for a camcorder. Obviously their processing speed has to be much higher to maintain the wider bandwidth needed and they may be more power hungry because of this. The standards were 1250 lines for Europe and 1125 in Japan and the USA, but new standards have been defined by the ATSC and there is a range of standards (including 1080 and 720) depending also on whether interlace or progressive scanning is used. Most of the new cameras will switch between a number of these standards.

Reference

1. Lang, H. *Farbmetrik u. Farben sehen*, R. Oldenburg Verlag, Munich (1978).

Bibliography

Canon. *TV Optics 2. The Guide Book of Optics for Television Systems* (1992).

Franken, A. Modulationsubertragungs-funktion einer Kamerarohre, *Fernseh u. Kino Tecknik*, 4 (1978).

Kato, S. et al. Performance characteristics of improved pickup tube, *SMPTE J.* (October 1983).

Klemmer, W. The two dimensional resolution of pickup tubes in HDTV camera systems, *Image Technology*, 328–332 (July 1987).

Klemmer, W. Multistandard HDTV camera, *Electronics and Wireless World*, 708–711.

Klemmer, W. Concept and realization of a HDTV studio camera, *14th Int. TV Symposium*, Montreux, Symp. Record, Joint Session, pp. 382–396.

Kurashige, M. Effect of self-sharpening in low-velocity electron beam scanning, *IEEE Trans. Electron. Devices*, **ED-25**, 10 (October 1982).

Kurashige, M. 1 inch magnetic focus and electrostatic deflection compact saticon for HDTV, *NHK Laboratory Notes No. 322* (November 1985).

Pearson, D.E. *Transmission and Display of Pictorial Information*, Pentech Press (1975).

Reimers, U. Origin and perceptibility of noise in a HDTV camera, *Frequenz*, **37**, 316–323 (1983).

Sony. *Studio/OB/EFP Camera Family BVP-500/550P/500P/550P Product Information Manual.*

Theuwissen, A.J.P. *Solid-State Imaging with Charge-Coupled Devices*, Kluwer Academic Publishers (1995).

Paul Cameron

3.5 VTR Technology

3.5.1 The Videotape Recorder: A Short History

There are many different videotape formats. Some have been more successful than others. Some have been technically more superior than others. There is often very little correlation between commercial success and quality or technical excellence.

3.5.1.1 Beginnings

3.5.1.1.1 AEG Magnetophons, Bing Crosby and the beginnings of Ampex

The videotape recorder has been in existence for about 50 years. The only way to record video before videotape recorders was to use film. Film did not lend itself well to television. It required a telecine to convert the film to a television signal.

During the Second World War, John (Jack) Mullin was stationed in England as part of the Signal Corps. He became intrigued by the fact that the Germans were able to transmit propaganda and music in the middle of the night. The music was particularly interesting, as the quality was very good, but the music was orchestral. It seemed unlikely that this quality came from 78 rpm record discs, and it seems even more unlikely that entire orchestras were being employed to play every night.

As the Allied forces pushed forward into Germany, Jack found the machine that was able to play out with such quality. It was an AEG Magnetophon. He grabbed two and proceeded to ship the mechanical parts back to the States using the war souvenir parcel service. He reassembled a machine, rebuilding the electronics and making a few improvements on the original AEG design. He demonstrated the improved Magnetophon at a number of venues.

Bing Crosby saw one of the demonstrations and became interested. He desperately wanted to avoid doing live radio shows every day, and knew that quality of the shellac-based recording methods available at the time were so poor that audiences at home could tell the difference between a live show and a recorded one.

Bing Crosby invested $50,000 in Ampex to have John Mullin's machines produced on a more commercial basis. At the time, Ampex was a very small company, set up by Alexander Poniatoff in 1944 in Redwood City, California. It had been making electric motors for aircraft during the war and was looking for new projects to get involved in. Bing Crosby agreed to the investments on the basis that his company Crosby Enterprises would become the sole marketing channel for Ampex machines.

Ampex audio recorders were very successful and allowed radio programmes to be pre-recorded and edited before going live, but still maintaining such quality that radio listeners could not tell the difference.

In the meantime 3M had made significant advances in the formulation of magnetic tape, and Ampex had set up a division specifically to make tape to supply the rapid increase in demand.

3.5.1.1.2 The first commercial video recorders

At the end of the 1940s television was enjoying a jump in popularity. Television producers had to use modified film technology to record television shows. The quality was poor. A new method of recording video was desperately required, so that producers could edit shows and delay transmission for different time zones.

In 1949 John Mullin approached Bing Crosby and proposed that the same plan that was used to design a commercial a audio tape machine be used for video. He demonstrated a prototype machine in the early 1950s. A team, led by Alexander Poniatoff, was put together and the first working machine was demonstrated in 1956.

3.5.1.1.3 Other early machines

In England Axton began research work in 1952 on a video recorder they called Vera (Vision Electronic Recording

Apparatus). By the mid 1950s other companies like RCA had started projects to develop a videotape recorder.

3.5.1.2 The 1970s

During the early 1970s several companies, including Sony, Teac and JVC, introduced a semi-professional format based on a $^3/_4$-inch tape, called U-matic. The 2-inch machine was superseded by machines using a 1-inch tape during the mid 1970s, with companies like Sony and Bosch entering the fray. This format was truly helical, with the tape now wrapped around the drum, which spun almost in line with the tape, rather than at right angles to it. The Bosch machines were ratified as the B format, while the SMPTE ratified the 1-inch tape standard as the more successful C format.

3.5.1.3 The 1980s

By the early 1980s, $^1/_2$-inch tape-based broadcast machines began to appear. The most successful of these were the Sony Betacam and Betamax formats. Betacam, the broadcast format, was later improved with the introduction of Betacam SP (Superior Performance). The domestic $^1/_2$-inch format, Betamax, eventually lost the commercial battle with the VHS (Video Home System) format, although it was technically superior.

In the late 1980s, digital videotape recorders had begun to appear with the D1 and D2 tape formats. A digital version of Betacam SP was introduced. Called Digital Betacam, this format became a widely accepted standard for high-quality digital television recording.

3.5.1.4 The 1990s

Half-inch tape formats based on the original Betacam format were introduced during the second half of the 1990s. They introduced MPEG compression to mainstream broadcast tape recording and the idea of high-quality, low-bit-rate recordings, metadata, and offering a bridge between streaming technology (tape) and file-based technology (computers).

The DV format was introduced during the mid 1990s. Originally designed as a digital replacement for VHS and Hi8, it has been more popular as a domestic camcorder format rather than for recording television programmes at home. In the professional and broadcast arenas, manufacturers have squeezed extra quality and performance out of the DV format to produce the DVCAM (Sony) and DVCPRO (Panasonic) formats suitable for more professional and broadcast use. DV, and its high-quality derivatives, are helical scan systems; indeed, there is actually little fundamental mechanical difference from the very first true helical scan video recorders. They are just a lot smaller.

3.5.2 The Present Day

3.5.2.1 The domestic arena

In the domestic environment VHS is still king, although its days are almost certainly numbered. The general quality of television output (from an image point of view) has been steadily increasing over the last few years and people are starting to realise just how bad VHS is. Even from a convenience point of view VHS is starting to look cumbersome and fragile. DV was designed as a possible replacement to VHS. However, manufacturers have never produced DV equivalents of the ubiquitous domestic VHS recorder, with remote control and timer functions. It looks likely that VHS will be superseded by recordable optical disc rather than tape.

3.5.2.2 The professional and broadcast arenas

3.5.2.2.1 The transition to hard disk

Broadcast television has seen an increased use of hard disk technology. Indeed, people have predicted that tape will be replaced by disk for many years, and yet new tape formats keep appearing, and broadcasters are still buying tape-based technology. It is true that hard disk technology is being used a lot more than it used to be, and it is slowly replacing areas of the broadcast chain previously occupied by tape technology. Hard disk is now used heavily in post-production and editing.

However, tape is still cheaper, and more robust than hard disk. It is still a popular choice in acquisition and archive storage. All popular camcorders in use today use tape. It is removable, can be treated with a fair amount of disrespect and is readily available. Archive and long-term storage systems use tape, although video and audio material is now generally stored as digital data, and is often compressed to further save space on the tape. Tape robotics machines allow large amounts of tapes to be stored safely, with the advantage of automatic scheduling and database support, so that video and audio archives can be searched. It is unlikely that hard disk will entirely replace tape. It is more likely that optical disc, using blue laser technology, will replace tape.

3.5.2.2.2 The cost/quality balance

Cost is now more of an issue than it ever was. The newer tape formats are a careful balance between cost and quality. 'Cost' means total cost, not just the price of the equipment, but also maintenance costs and running costs, commonly referred to as 'total cost of ownership'. 'Quality' means the image quality, as well as manufacturing quality and quality of after-sales service and support.

Digital tape technology satisfies this careful balance much better than analogue tape technology. All major television companies now use digital tape technology, and almost all of these companies record new material in digital format. However, with large analogue tape archives, analogue tape players are still in popular use.

For very high quality DI is still used. It is expensive but offers the kind of quality not attainable by any other broadcast format. Many companies use Digital Betacam, and a few the M2 format. While these formats are slightly compressed they still offer superb image quality at a much more realistic price.

Betacam SX, DVCAM and DVCPRO are popular for news gathering where convenience and price are more important. Indeed, domestic DV is being used in many professional areas.

3.5.2.2.3 The stream/file bridge

The major thrust in digital tape technology is in bridging the very difficult gap between streams and files. Video and audio are basically streams. They have no beginning and no end, and do not contain any kind of header, label or other information. Video and audio are also strongly related to time. They are continuous and must be played at the correct speed without breaks. Files, on the other hand, are contained chunks of data. They have a header, information about the contents of the file, etc. Files are also not related to time. When copying files from one location to another it really does not matter how long it takes, how the data is actually transferred, or if the beginning of the file gets to its destination before the end.

With increasing use of computer technology in broadcast, television companies require a way of bridging this gap between these two basic methods of storing media. Manufacturers are starting to produce tape recorders that can place extra information into the video or audio stream, much like a computer file can. This so-called 'metadata' is the focus of a lot of research work. Manufacturers are also starting to introduce tape formats that can output video and audio in packets, or in file structures, so that they can be saved as files on hard disk, and treated as files within the television station.

The MPEG IMX format is a good example of this. Although this format is essentially a stream recording system, just like the original Ampex VR-1000 of the mid 1950s, it is able to output a stream as a series of chunks, or packets of data.

The E-VTR takes this one step further and allows bits of video or audio to be marked, and played out as a file. Although the material has been recorded from a video or audio connection, as a stream, locked to time, it is now being played out to a computer network, as a file with associated metadata, and at a speed governed by the network, in bursts, faster or slower than real time.

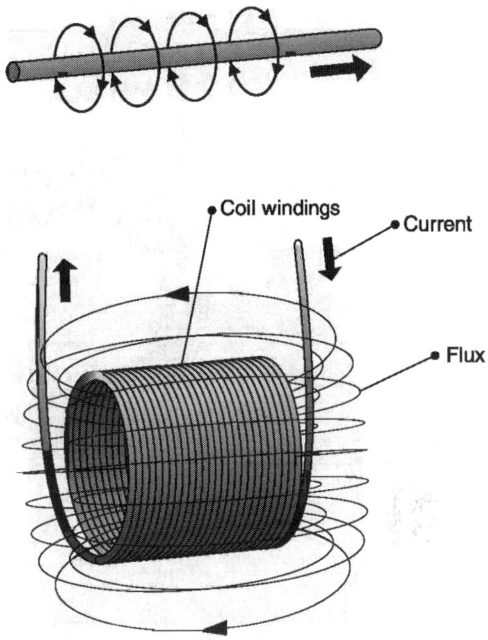

Figure 3.5.1 Magnetic flux in a straight wire and coil.

3.5.3 Magnetic Recording Principles

Although magnetic recording heads may have got a lot smaller, the materials used are better, and manufacturing tolerances a lot higher, all videotape recorders depend on the same basic principles of recoding a signal to magnetic tape.

3.5.3.1 Principle of a magnetic field

Man has known of the existence of magnets for several thousands of years. The Chinese used them to invent the compass about 1000 years ago. However, the fact that an electric current develops a magnetic field was not discovered until much later. In 1820 Hans Christian Oersted discovered that a straight wire, carrying an electric current, developed a magnetic field, which circulates around the wire. Andre-Marie Ampere discovered that the magnetic field could be concentrated and magnified by winding the wire into a coil. William Sturgeon later discovered that placing an iron core inside the coil greatly increased the strength of the magnetic field, and bending the coil and iron into a 'U' shape further concentrated the field at the two ends of the 'U' shape. A little later Joseph Henry insulated the wire, thus enabling larger and tighter coils to be wound (see Figure 3.5.1).

It is perhaps a pity that Oersted, Ampere and Henry have all had their names immortalised as units of magnetic flux density, current and inductance, but Sturgeon's name has sunk into relative obscurity. The magnetic field is known as flux, and its strength the magnetic flux density (see Figure 3.5.2). Flux finds less resistance through some materials than through others. Many materials are magnetic. This means that they become magnetised if subjected to a magnetic field. The ability to become magnetised is called the remenance.

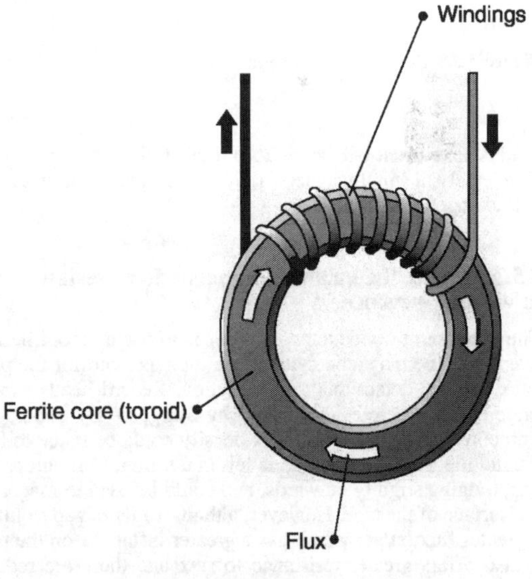

Figure 3.5.2 Magnetic flux in a toroid.

3.5.3.2 Principle of electromagnetic induction

The opposite of the principle of a magnetic field is that of electromagnetic induction (see Figure 3.5.3). When a magnetic field is applied to a wire it induces a current to flow in the wire. To be more exact, when a magnetic field changes, it induces a current. Indeed, it does not matter how strong the magnetic

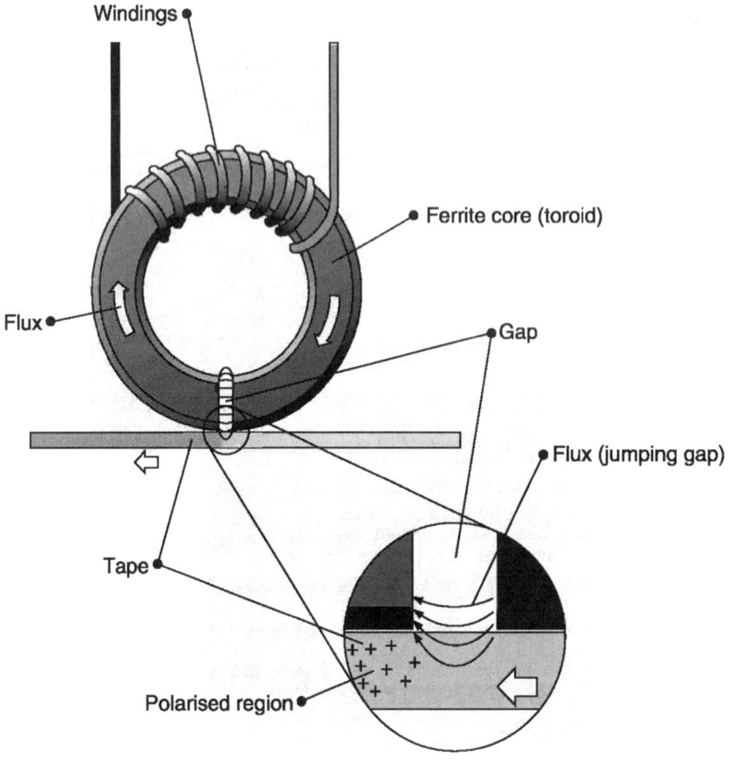

Figure 3.5.3 Recording to magnetic tape.

field is, no current will be induced if this field remains constant. Conversely, a small magnetic field could induce a high current if it changes rapidly.

3.5.3.3 Using the qualities of magnetic materials in videotape recorders

The purpose of a videotape recorder is to use a record head to magnetise the magnetic material on the tape, and for the playback head to detect this magnetisation. Record heads use the same basic principles discovered by Sturgeon. By bending an iron core into a ring a high flux density could be made to flow around the ring. A small gap is left in the ring. Flux jumps this gap, bulging slightly outwards, and could be used to magnetise the surface of the tape. However, although a large gap results in a greater flux bulge, which has a greater influence on the tape, it also offers greater resistance to flux and therefore reduces flux density. Thus, the design of the record head meets its first compromise.

Playback heads use the same design. When the magnetised tape passes across the head gap, it induces a small magnetic flux in the head core. As the flux changes, so a small current is induced in the coils. Thus, record and playback head cores need to be made from magnetic material with low flux resistance and low remenance. Conversely, the magnetic material used in tape needed high remenance so that the maximum amount of signal could be recorded.

3.5.4 The Essentials of Helical Scan

3.5.4.1 The bandwidth problem

Humans can hear audio from about 20 Hz to about 20 kHz. The bandwidth of audio is therefore about 20 kHz. If we consider modulation or sampling, the Nyquist criteria doubles this bandwidth to about 40 kHz. No matter how we look at it, the frequencies involved are well within the capability of magnetic tape and recording head technology, using stationary record and playback heads.

Video is very different. Broadcast channels have a total bandwidth of about 6 MHz. In component form we would expect to retain as much of the quality as possible, and give the luminance signal as near to 6 MHz as we can. Each colour difference signal may have a bandwidth of about 3 MHz. Add these bandwidths, and take into account the Nyquist criteria, and any recording system will need at least 24 MHz of bandwidth! If nothing else, these somewhat crude calculations show us that we cannot record video on magnetic tape in the same way we do with audio. Either the recording system has to be radically different, or the bandwidth must be reduced.

3.5.4.1.1 Head-to-tape speed

Key to the problem of bandwidth is the relative head-to-tape speed. In analogue audio recorders this is achieved by pulling the tape across a static head. If the bandwidth of the video signal is higher one could simply increase the tape speed. This idea was tried in the first video recorders.

Ampex were at the forefront of videotape recorder development and demonstrated video recorders with high-speed tape at the beginning of the 1950s. Other groups were working on the design of a video recorder, like the BBC Vera (Vision Electronic Recording Apparatus), which ran through tape at 21 metres per second. With reels 21 inches in diameter just 15 minutes could be recorded. In the States RCA built a prototype that ran through tape at 9 metres per second. An improvement, maybe, but it only gave 9 minutes of recording time.

It became clear to all those groups working on video recorder designs that the kind of tape speed required for the kind of bandwidth normally found in video made the machine difficult to control and used up vast amounts of tape. Designers eventually decided that, to achieve the relative head-to-tape speeds required, the recording head itself could not stay still.

3.5.4.2 Early scanning techniques

Many of the earliest video recorders used a moving head to increase the relative head-to-tape speed. From the very first prototypes this was achieved by mounting the heads on a rotating drum.

3.5.4.2.1 Ampex Mark 1 arcuate recorder

An early notable attempt to increase the head-to-tape speed was the Ampex Mark 1 arcuate recorder (Figure 3.5.4). Built in 1952, this machine wrote the video information onto tape as arcs using three heads on the drum. It proved unreliable and difficult to regulate. The geometry also wasted more tape than was necessary. Arctuate tape machines were not successful. The Ampex Mark 1 did give rise to the transversal scanning technique used by their quadruplex machines.

3.5.4.2.2 The Ampex VR-1000 Quad recorder

The original Ampex VR-1000 machine used four heads fitted 90 degrees apart on a spinning drum. (Hence the name 'Quadruplex' or simply 'Quad'.) The drum spins in line with the tape. The tape itself is 2 inches wide and is curved by a vacuum chamber, to fit around the drum (see Figure 3.5.5). Each head records a stripe of video across the tape, called a track. As soon as one head breaks contact with the tape the next one is ready to carry on. The tape moves a little more than the width of one of these tracks before the next head comes along. This keeps tape speed slow. The drum spins quickly. This makes the head-to-tape speed high and allows a high-bandwidth signal to be recorded.

The drum spins at 14,400 rpm (USA machines), recording 960 tracks per second. Sixteen tracks make up each video field. Each head recorded or played back just 16 lines of video. Quad recorders manage to use up 15 inches of tape per second. With a 4800-foot reel of tape it was possible to record just over 1 hour of video.

The Quad machine is generally not considered a true helical scan machine, because the tracks' angle is almost perpendicular to the tape's direction. The scanning method is generally referred to as transversal. A longitudinal control track is also recorded along the edge of the tape so that the machine can lock to it, with the playback heads exactly following the tracks as they were recorded. Audio can also be recorded with a static head and a longitudinal track, and there is also provision for a lesser quality audio track called cue (see Figure 3.5.6).

Quad remains the most long-lasting videotape format of all time. This is partly because there was no alternative video recording format for many years, and partly because no video recording format since has been able to last the length of time Quad was used for, before it has been superseded by something else.

However, Quad has its quirks. Tape operators would have to regularly ensure that the machine was aligned correctly. The drum and vacuum chamber were particularly sensitive.

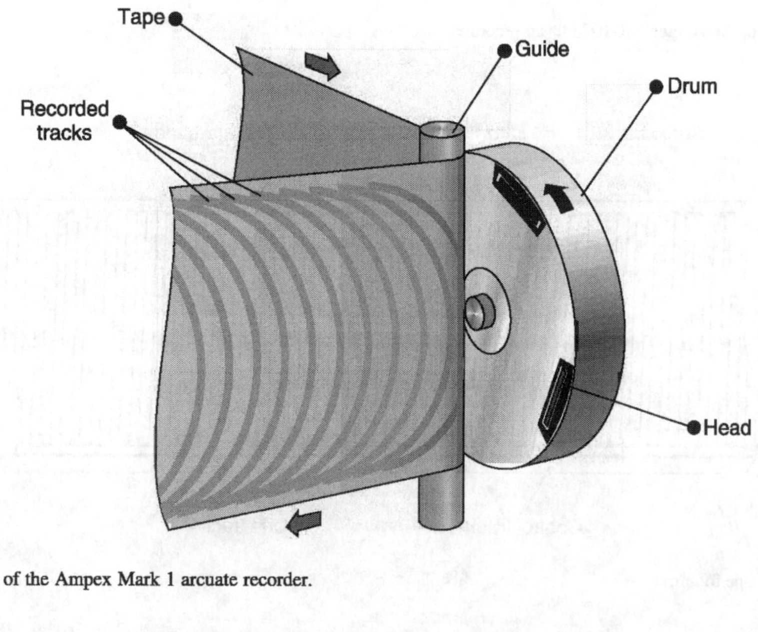

Figure 3.5.4 Schematic of the Ampex Mark 1 arcuate recorder.

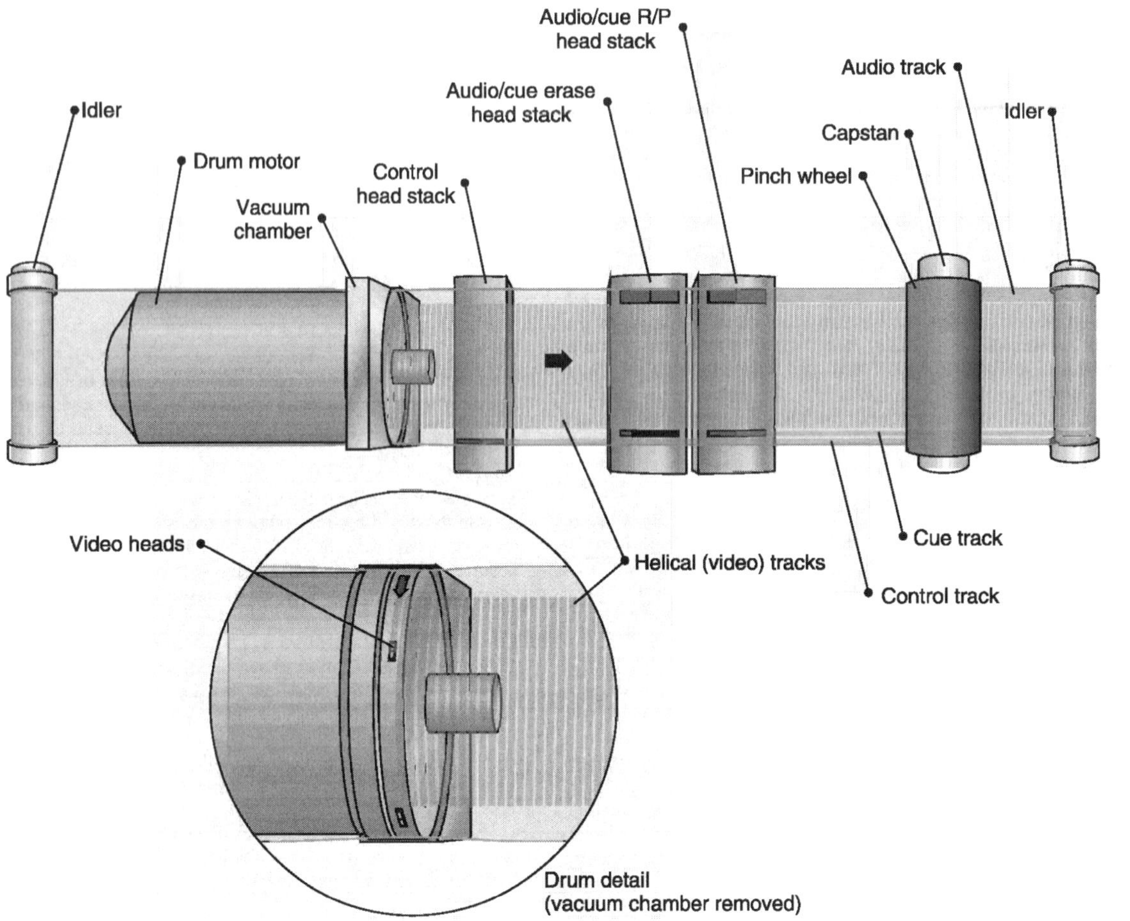

Figure 3.5.5 Schematic of the Ampex VR-1000 Quad recorder.

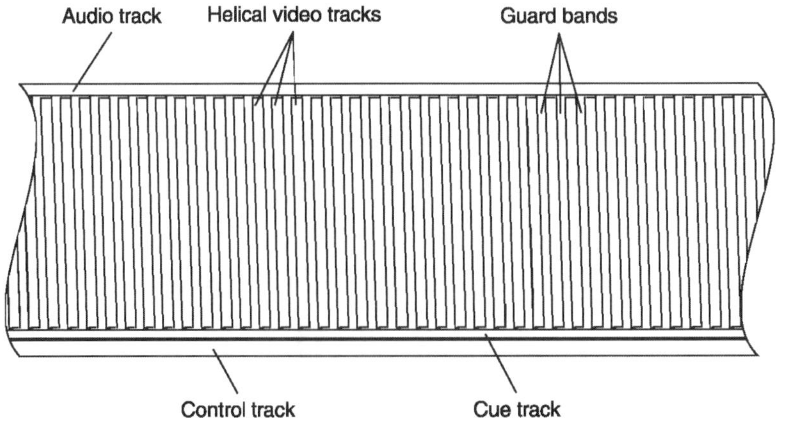

Figure 3.5.6 The Quad tape footprint.

The forces applied literally stretched the tape. Applying exactly the same level of brutal treatment to the tape every time it was played was not easy.

3.5.4.3 True helical scanning

Helical scanning uses the same basic geometry as transversal scanning. However, instead of the drum spinning vertically in transversal scanning, producing tracks on tape that are almost vertical, helical scanning uses a drum that is nearly horizontal. Another difference between transversal and helical scanning is the tape wrap. In transversal scan the vacuum chamber achieves a slight wrap around the drum, stretching the tape as it does so. However, in helical scan designs the wrap is huge by comparison. In some formats wraps of almost 360 degrees have been used, although a little over 180 degrees is more common. Even though the wrap is large none of the stresses common in Quad machines is placed on the tape in helical scan machines. During recording and playback the tape moves slowly through the machine and round the drum. The point where the tape meets the drum is called the entrance side. The point where the tape leaves the drum is called the exit side (see Figure 3.5.7).

The drum assembly itself sits in the machine at a slight angle, and in most cases consists of two halves. The top half spins and the bottom half is static. The record and playback heads are fitted to the bottom edge of the top half. The bottom half has a rebate cut into it, called a rabbet. The rabbet is cut at an angle, and in most machines is very close to the top of the lower drum near to the entrance side, and much lower at the exit side. Taking into account the angle of the drum assembly as a whole, the rabbet is effectively horizontal. The bottom edge of the tape rests on the rabbet as it passes around the drum assembly (see Figure 3.5.8).

The angled drum assembly and the way that the rabbet is cut mean that the heads prescribe a helical scan across the tape as the drum spins. In some formats the track angle goes upwards, in others it goes downwards, depending on the angle of the rabbet and the rotational direction of the drum. In most modern tape formats the drum spins anticlockwise. The tracks recorded on tape are about 5 degrees from the line of the tape, and very long compared to those in transversal scanning machines.

3.5.5 Modern Video Recorder Mechadeck Design

The mechadeck is the mechanical part of a video recorder, consisting of the tape reels, any servo mechanism, including the pinch wheel and capstan mechanism, tension regulation, the drum and all the record and playback heads, tape cleaners, head cleaners and cassette handling mechanism.

Most modern videotape recorders now have similar mechadeck designs. Tape, normally enclosed in a cassette, travels from the left (supply) reel to the right (take-up) reel during normal recording or playback. The route taken by the tape from the supply reel to the take-up reel is called the tape path. The tape path from the supply reel to the capstan/pinch wheel is called the supply side. From the capstan/pinch wheel to the take-up reel is called the take-up side. The supply side of the tape path is by far the most important part. It contains all the record and playback heads. Good supply side tension regulation is important. There is often no take-up side tension regulation at all.

Many machines have a tape cleaner placed in the tape path as the tape leaves the supply reel. There will also be a tension

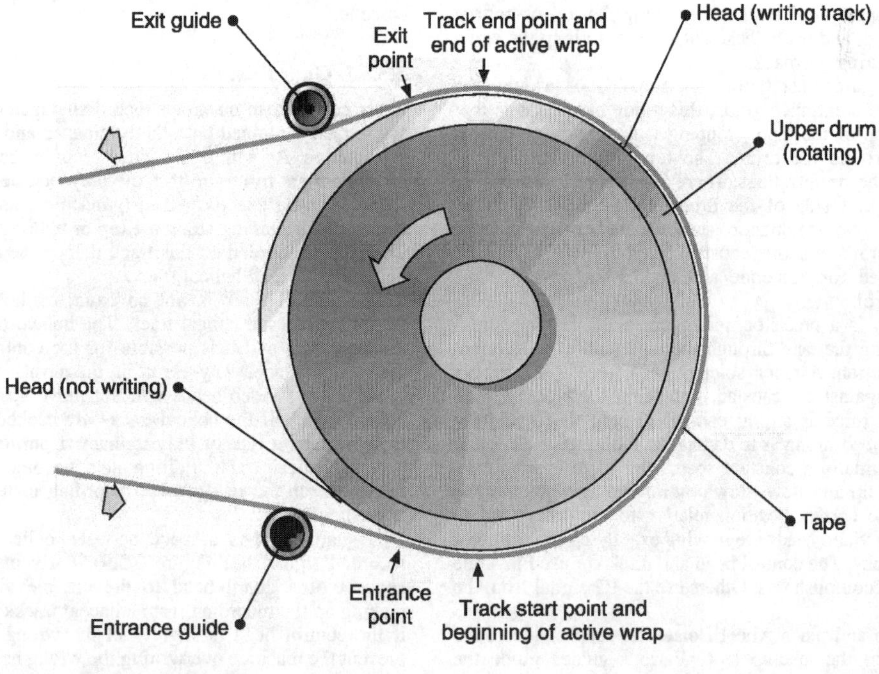

Figure 3.5.7 Top view of a simple helical wrap.

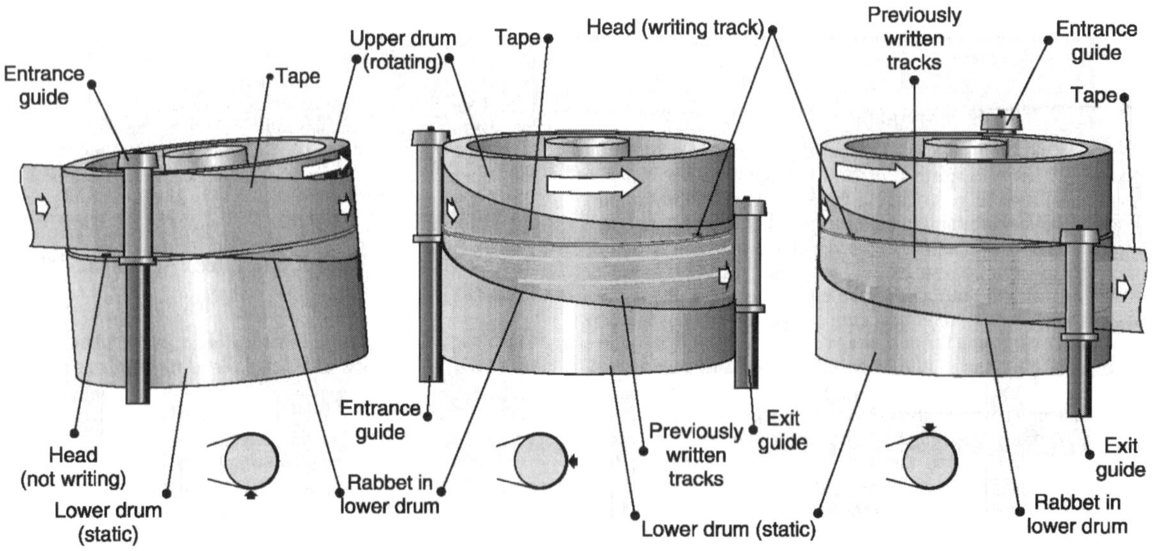

Figure 3.5.8 Side view of a simple helical wrap.

regulator in the tape path between the supply reel and the drum, to ensure that the tension around the drum is correct.

Many videotape machines have static heads before the drum. A full erase head will be fitted to all recorders to erase everything on the tape before any new recording is made. Some machines also include a control head. This head records special pulses on a longitudinal track either along the top or bottom edge of the tape, and plays these pulses back to help the servo system lock during playback.

There is a guide just before the tape wraps around the drum. Called the entrance guide, this guide has a flange that touches the top of the tape, stopping it from riding up as it is wrapped around the drum. The tape is prevented from dropping by the drum rabbet. There is another top-touching guide on the exit side of the drum, called the exit guide. There may also be one or more static heads between the exit side of the drum and the capstan/pinch wheel. These are commonly used for timecode and audio, but may also be used for control.

The capstan is a precision servo-controlled motor responsible for pulling the tape through the tape path at the correct speed and position. A pinch solenoid will force a soft rubber pinch wheel against the capstan, squeezing the tape between the two. This force is strong enough to stop the tape from slipping but not so strong as to damage it. Although the capstan rotates at essentially a constant speed, the servo system constantly speeds up and slows down by minute amounts to keep the tape in the correct position relative to the drum, and to ensure that the video heads are moving exactly up the centre of the helical tracks. The control head and track are used in some machines to accomplish this. Others use the RF signal from the helical tracks.

The capstan and pinch wheel isolate the supply side of the tape path from the take-up side. Various guides guide the tape back into the cassette and onto the take-up reel. Some machines include a take-up tension regulator to ensure that there is a small amount of slack on the take-up side to allow for sudden changes in direction during normal operation, but not so much as to start throwing tape loops. The drum assembly is at an angle and some machines include an angled guide to ensure that the tape is straight as it re-enters the cassette.

3.5.5.1 Guard bands

Older helical scan machines record analogue composite video. Each track contained both the luminance and the colour video information. As with every video recorder before and since it was important to ensure that the playback heads followed the recorded tracks exactly. All early machines used a longitudinal control track running along the top or bottom edge of the tape. The pulses recorded on this track helped the machine find the beginning of each helical track.

The control track is not an exact method of finding the beginning of each helical track. The helical tracks themselves are very thin, and it is possible for the control head to be in the wrong place. Any error in the position of the control head, and the video heads will not move up the centre of the helical tracks. If the helical tracks are packed together tightly there will be a risk of playing back a portion of video from another helical track. Editing also becomes problematic as recorders run the risk of over-recording material on tape that they should not.

A guard band is a space between helical tracks with no recorded signal (see Figure 3.5.9). Early machines used the concept of a guard band to prevent the video heads from picking up the recording from adjacent tracks during playback, if the control head is slightly in the wrong position, and to prevent the machine overwriting the wrong helical track during edits.

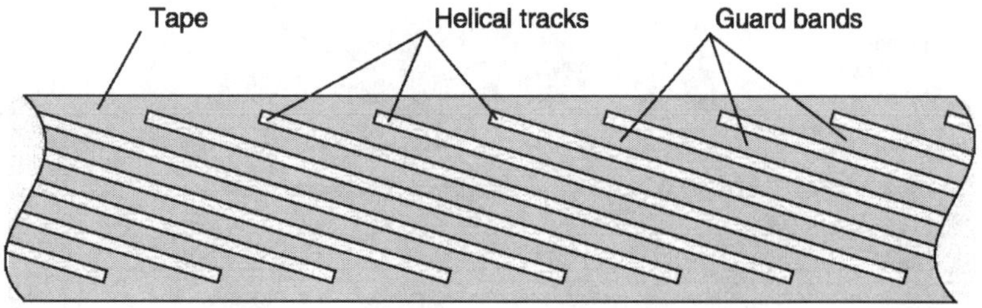

Figure 3.5.9 Guard bands.

However, guard bands use up tape. Later machines abandoned guard bands in favour of track azimuth, and thus saved tape.

3.5.5.2 Helical tracks with track azimuth

Later video machines recorded component video, with separate circuitry, record heads, playback heads, and tracks for luminance and colour. Making a tape machine that is able to record and play back entirely in component increases the quality of recording over older composite machines, and removes the problems associated with editing composite material. However, component video recorders are more expensive than composite ones as they are effectively two video recorders in one.

Designers needed a way for these machines to differentiate the helical tracks responsible for luminance from those for colour. The method adopted was track azimuth. Track azimuth involves tilting the head gaps over at an angle. The luminance head gaps are tilted over positively and the colour head gaps negatively (see Figure 3.5.10).

During recording the luminance tracks are recorded with a positive azimuth and the colour tracks with a negative azimuth. If the machine is badly aligned and a luminance playback head is trying to play back a colour track, the angle of the recording will be incorrect. In fact, it will be incorrect by twice the azimuth angle. This will severely reduce the signal. Azimuth angles of about 15 degrees are popular. This gives a total error, if each head is on the wrong track, of about 30 degrees.

Azimuth replaces the need for a guard band. The colour tracks are effectively guard bands for the luminance heads and vice versa. Helical tracks can be packed next to each other, saving a lot of tape and increasing the tape's recording capacity.

3.5.5.3 Video head design

The principles used by videotape recorders to record a signal onto magnetic tape have not changed since the very first tape recorders. They still rely on a doughnut-shaped head made from ferrite, or some similar material, with a slot cut in its front face and a coil wrapped around its back. The dimensions used in modern video recorders may be a lot smaller, but you can still find a doughnut idea somewhere in every video record and playback head.

The video heads, and the tracks they record, are thin. Older formats used heads close to 100 μm thick. Modern formats are using heads less than 10 μm thick. Video heads are no longer the classical round doughnut shape, they are square. The surface in contact with the tape is a long rectangle. This reduces tape bounce at the head gap and reduces wear. The coils are wound on the sides of the head. Only a few turns are required on each side for the head to be effective.

3.5.5.3.1 Channelling flux

One of the challenges facing head designers is to channel as much flux to the front of the record head gap as possible, where the head is in contact with the tape, and where recording and playback will take place. Likewise, the front of the playback head gap needs to be as sensitive as possible to achieve maximum signal off the pre-recorded tape.

The first modification is to cut away at the back of the head, where the head gap is. This forces flux lines forwards to the front of the head. The second modification is to introduce a different material with low reluctance to the front face of the head, just where the head gap is. Flux prefers to jump the gap at this material, rather than the ferrite behind. Several materials are used, often with exotic names to hide their true composition. Materials like Softmax and Sendust are used. However, these materials tend to be softer than ferrite and therefore tend to wear away quicker. Only a thin slither is placed on the head, and only close to the head gap, rather than across the whole front face.

3.5.5.4 Automatic tracking

Another important technology that has been a vital part of modern professional and broadcast videotape machines is the automatic tracking playback head. While servo systems using a control track were able to bring the video heads, in particular the playback heads, close to the centre of the helical tracks, there was a certain degree of error due to badly adjusted servo electronics, or an imperfectly adjusted control head.

Another problem is even more annoying. The geometry of all helical scan video recorders will play back correctly at play speed, because that was how the tape was recorded, at play speed. If the machine is speeded up slightly, slowed down or paused altogether, the geometry changes. Now the playback heads will not travel exactly up the centre of the helical tracks, and will wander off track and possibly cut across adjacent helical tracks. This is annoying for editors who regularly want to play back at other than play speed, or pause the video machine altogether and look at one frame or field on its own.

Automatic tracking video playback heads eliminate these problems. Introduced by Ampex in 1977 as the Automatic Scan Tracking (AST) system and by Sony in 1984 as the Dynamic Tracking (DT) system, both systems relied on moving the playback heads to keep them in the centre of the helical tracks.

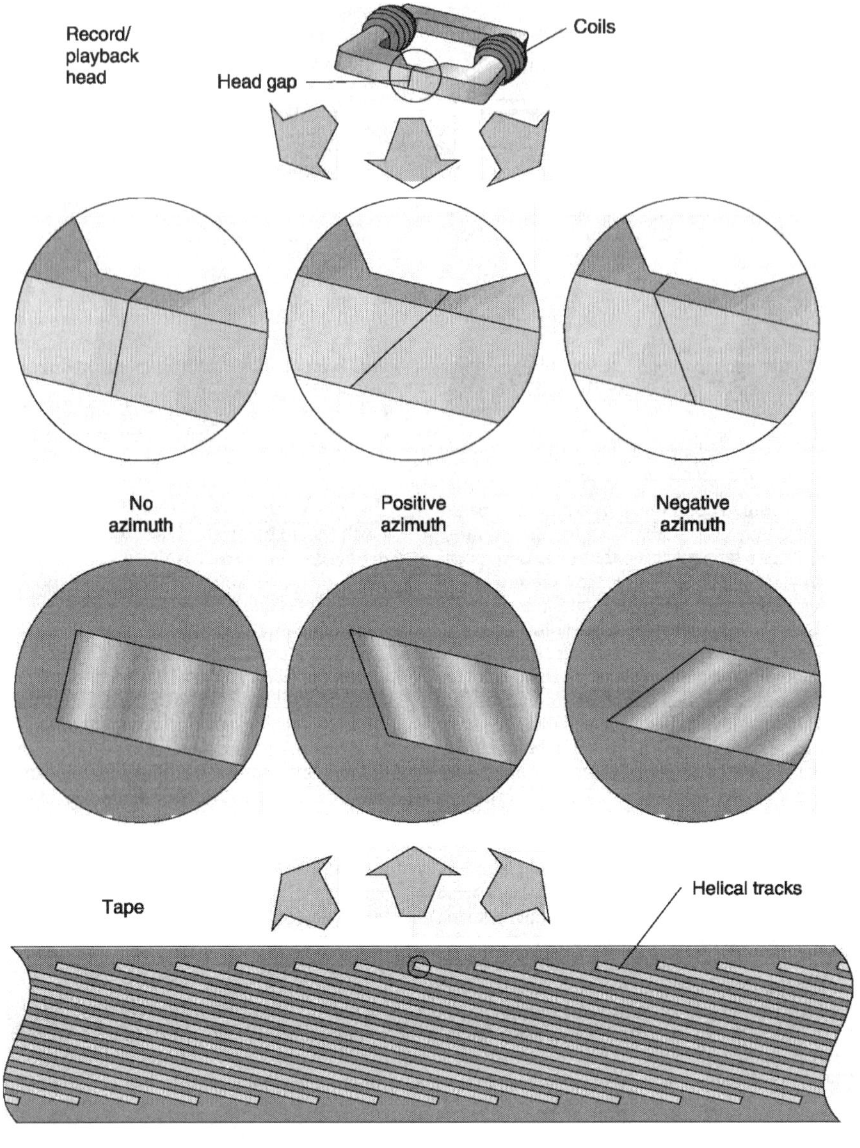

Figure 3.5.10 Track azimuth.

3.5.5.4.1 Automatic tracking playback heads

Automatic tracking video playback heads use piezoelectric crystal bimorphs. The bimorph consisted of two piezoelectric crystals, bonded together. When a voltage is applied across the bimorph one crystal expands while the other contracts. This causes the bimorph to bend. Reversing the voltage reverses the bend. One end of the bimorph is fixed to the drum. The playback head is placed on the other end. In early designs, including the Ampex AST designs, one bimorph was used. Two bimorphs are used in later designs, because this keeps the head itself perpendicular with the tape surface.

One disadvantage with this kind of tracking system is that the bimorphs require a high voltage to bend sufficiently. Any machine with automatic tracking heads needs brushes and slip rings to transmit these high voltages to the drum. Furthermore, the brushes and slip rings must maintain good contact and the drum contain smoothing circuitry. Any intermittence in the supply to the bimorphs could generate electromagnetic radiation that could be disastrous to the delicate record and playback process.

3.5.5.4.2 Automatic tracking in operation

A small alternating voltage of about 450 kHz is applied to the bimorphs, causing the heads to continually wobble. The wobble continually takes the head slightly off track, causing a slight drop in the RF signal. The servo system continually checks the level of the RF signal from the heads, keeping the drops in RF as small as possible, by adding a DC voltage to the wobble voltage. Automatic tracking playback heads allow operators to change the playback speed of a helical scan tape machine and still maintain a steady picture. They have become an essential part of professional and broadcast tape machines.

3.5.5.5 Tension regulation

It is vital that the tape tension around the drum is correct and maintained within a small range. If the tape is too tight the video heads and tape will wear out rapidly. If the tape is too loose the video heads will not be able to maintain good contact with the tape surface and there is a risk that the machine will throw tape loops or stick around the drum.

There are two types of tension regulator, the purely mechanical type and the electromechanical type. Most domestic machines, cheaper and smaller professional machines, use purely mechanical tension regulators. They are simple, light and cheap.

All high-end professional and broadcast tape machines, especially those intended for studio use, use electromechanical tension regulators. Although they are generally heavier, more complex and more expensive, they offer much finer tension regulation, and a change for the servo system to monitor the tension regulation process. This in turn allows machines to have different tension regulation response times for different modes of operation and fault detection in case the tape sticks or breaks.

3.5.5.5.1 The principles behind good tension regulation

All tension regulators operate in the same basic way. During recording or playback the capstan and pinch wheel pull the tape out of the supply reel and around the drum. The take-up reel motor will apply a constant pull on the tape. This pull is very light but it ensures that any tape that has come through the capstan and pinch wheel is drawn into the take-up reel in a tidy fashion. The supply reel motor will be trying to resist tape from being drawn out of the supply reel. This is what produces the tension. The higher the resistance, the higher the tension.

3.5.5.5.2 Mechanical tension regulators

Mechanical tension regulators have a sensing arm with a roller on the end of it, around which tape moves. The arm is connected to a spring and to a friction belt which is wrapped around the supply reel table. If the tape tension drops the spring will pull the arm further out, tightening the friction belt around the supply reel table, increasing its resistance. The capstan will continue to pull more tape out and the tension will increase. Likewise, if the tape tension increases, the arm will be pulled in against the spring, loosening the friction belt around the supply reel table and decreasing its resistance to allow tape out.

Mechanical tension regulators cannot handle loose tape by pulling it back into the supply reel. This is because the tension regulator can only stop the supply side reel motor, it cannot make it turn backwards.

3.5.5.5.3 Electromechanical tension regulators

Electromechanical tension regulators use a sensing arm with a roller on the end of it, just like mechanical tension regulators. Likewise, the arm is connected to a spring; however, the spring tends to be better quality, and in some cases may be more than one spring to give a more accurate response. The arm will also have a strong magnet attached to it. One or more hall effect detectors are fixed to a circuit board either on the mechadeck or on the tension regulator assembly. The hall effect detector will output a signal corresponding to the position of the tension regulator arm. With a properly aligned spring the position of the arm will also provide a measure of the tension in the tape.

The signal from the hall effect detector is sent to the machine's servo system, which controls the supply side reel motor. Reel motors in this kind of machine are more complex than those in machines with mechanical tension regulators. The servo system is able to control the direction, speed and amount or torque very precisely. Rather than using friction, the supply reel is effectively trying to turn backwards. This ability to control the backward rotation of the supply reel motor also allows electromechanical tension regulators to draw loose tape quickly back into the supply reel.

3.5.6 Variation in Tape Path Designs

Before the universal acceptance of cassettes for videotape recorders, manufacturers designed several exotic tape paths that often required the tape operator to spend a while lacing up before the machine could be used. Indeed, we often take it for granted as we slam another cassette into the machine that it was not always that easy.

Tape path designs have now settled to a just a few variations since the introduction of cassettes, because the machine must be able to automatically draw the tape out of the cassette before recording and playing can take place, and put it neatly back into the cassette before it is ejected. Any complicated lacing cannot be performed.

3.5.6.1 Terms of confusion

Various terms have been given to various tape wrap patterns, and there appears to be a fair degree of confusion as to which one is which. The term 'omega wrap' has been associated with many wrap patterns that are not that similar.

3.5.6.1.1 Alpha wrap

The alpha wrap takes its name from the Greek letter α. The tape passes around the drum for a full 360 degrees. The wrap is sideways, with the entrance and exit guides on the left or right. Alpha wrap would be very difficult to achieve with cassettes because the tape passes over itself. The tape must be manually laced. It is only used in machines with spools. As an example, alpha wrap is performed by the old Philips EL3402 1-inch machine.

3.5.6.1.2 Omega wrap

Omega wrap takes its name from the Greek letter Ω. The tape passes around the drum for almost 360 degrees. The active wrap is about 270 degrees. Although the term omega is used with many cassette machines it is not actually possible to perform a true omega wrap with a cassette. The tape must be laced. As an example, omega wrap is employed by 1-inch C format machines.

3.5.6.1.3 C wrap

The wrap pattern is actually in the shape of a backward 'C'. C wrap is possible and popular with cassette machines. The tape is drawn from the cassette at one point and taken between 200 and 300 degrees round the drum in an anticlockwise direction, giving an active wrap of anything between 180 and 270 degrees. As an example, C wrap is popular with broadcast studio machines using the Betacam SP and Digital Betacam formats and other similar tape formats.

3.5.6.1.4 M wrap

This is the most popular wrap pattern, and is used in cassette-based machines. Tape is drawn from the cassette at two points. It is drawn round the left side of the drum, and round the right side of the drum, to give a total wrap of between 250 and 300 degrees, and an active wrap of anything between 180 and 270 degrees. As an example, M wrap is popular with domestic VHS machines and some broadcast machines like the Sony D1 and D2 machines and the PVW range of Betacam SP machines.

3.5.6.2 Definition of a good tape path

A perfect tape path will contain a perfectly circular supply and take-up reel. The tape would move from the supply reel to the take-up reel in a straight line without touching anything. Video and audio would be recorded and played back without any heads touching the tape. Clearly this is an impossibility! Compromises have to be made. The record and playback heads must touch the tape. Furthermore, helical scanning requires that the tape be wrapped around the drum. Thus, the tape must change direction dramatically. Helical scanning also requires accurate tape tension control.

The speed of the tape must be governed and regulated. Reel motors are simply not good enough to accomplish this. A capstan is required. Any item like a guide, drum or static head, cleaner of capstan changes the tapes direction and adds friction. Spinning guides, and the drum itself, are never absolutely central and always add a slight wobble to the tape's motion. There are therefore opportunities for the tape to stick, be forced into the wrong position or the timing to be altered. The important part of any videotape recorder tape path is the distance between the supply side reel and the capstan. This is where all the heads are and this is where the tape must be at the correct tension and in the correct position. This length of tape should be as short as possible, and should pass across as few items as possible. Static heads, the cleaner, the drum, entrance and exit guides, the supply side tension regulator and the capstan/pinch wheel are vital and therefore always present. Designers will ensure that any other guides will only be added to the design if they are absolutely necessary.

A perfect tape path does not need top-touching or bottom-touching guides, or a rabbet on the lower drum. The tape would pass around the various items on the mechadeck in exactly the correct position. Designers calculate the angle of guides, drum, static heads, etc. so that the tape runs smoothly through the tape path. Although it is impractical to expect a perfect level of mechanical accuracy, the rabbet and any guides touching the top or bottom of the tape should do so very lightly.

It is not very critical what happens to the tape after the pinch wheel and capstan. The pinch wheel and capstan act as a wall, isolating the drum and static heads from any minor wobbles in the tape afterwards. Therefore, the amount of tape, number of guides and other hardware is not important.

3.5.7 The Servo System

Modern video machines consist of a number of servo loops. Normally one item is the master, and servo loops slave off the master. When a video machine is playing back the master is the drum. It obeys the incoming reference, taking no regard of anything else, and spins at a constant rate, somehow related to the reference itself. The drums in early analogue machines spin at frame rate, 25 Hz for PAL (625-line)-based machines and 29.97 Hz for NTSC (525-line)-based machines. Later digital machine drums spin at a multiple of frame rate.

The rest of the servo system slaves off the drum. The first servo loop to consider uses signals from the drum and the control head and uses the capstan as a control. Although the machine's servo system will control the capstan to pull the tape through at almost a constant rate, signals from the drum and the control head inform the servo system of the relative position of the tape and the spinning drum. By slightly altering the speed of the capstan, the servo system will ensure that the timing between the pulses from the drum and the control head is correct, thus ensuring that each playback head finds the beginning of each helical track.

Another servo loop slaves off the capstan servo loop. This loop uses a signal from the tension regulator to control the supply reel, trying to maintain the tension around the drum at a constant predefined level. In simpler machines this is done mechanically. In more complex machines this is done electronically. In some machines there is also a servo loop between a take-up tension regulator and the take-up reel, to maintain the take-up tension.

3.5.8 Analogue Videotape Recorder Signal Processing

Videotape recorder signal processing can be divided into a number of distinct areas. The first division is between record and playback processing.

3.5.8.1 The problems of recording to tape

The earliest pioneers of tape recording technology discovered that it was impossible to record an audio or video signal directly to tape and expect a reasonable playback signal. Two characteristics ensured that the recording process was not going to be that straightforward, the basic behaviour characteristics of inductors and magnetic hysteresis.

When a record head records a signal the current applied to the head generates a flux which magnetises the tape. The strength and direction of magnetisation is directly related to the current. When the playback head plays the tape back, current is generated at the output of the head that is proportional to the rate of change of magnetisation in the tape.

This term 'rate of change' is crucial. If a high DC signal is recorded to tape, it will magnetise the tape a lot, but nothing will be played back, because the rate of change is zero. Conversely, if a smaller high-frequency signal is recorded to tape a large high-frequency signal will be played back because the rate of change will be high.

This characteristic is evident when looking at the control track of most professional VTRs. The control head records a 25 Hz or 29.95 Hz square wave signal on tape. The resulting control track consists of positively and negatively magnetised regions. The resulting playback signal is a series of large negative and positive spikes for each negative and positive transition. This 'rate of change' characteristic makes the record/playback process non-linear, it integrates the record signal, and introduces a phase shift between the record and playback sine waves.

The second characteristic is hysteresis. This defines the 'memory' magnetic materials have. Apply a magnetic flux to a magnetic material and it will remember this by becoming magnetised. The answer to these problems is modulation. Modulation is the process of combining a low- and high-frequency signal together into one signal. There are two types of modulation, amplitude modulation (AM) and frequency modulation (FM). AM involves changing the amplitude of the high-frequency signal with the low-frequency signal. FM involves changing the frequency of the high-frequency signal with the low-frequency signal.

AM is the easier modulation system to design, and was used in the first attempts at modulating the video signal before recording it to tape. However, FM is more resilient and was chosen as the modulation of choice for videotape recorders.

3.5.8.2 Input processing

3.5.8.2.1 Reference input selection

Another important part of the input circuitry is the reference input. The machine should be able to play a tape back on its own, maintaining good and consistent timing. It should also be able to lock to an incoming reference when playing back, locking the entire playback process to the incoming reference. The machine should also be able to lock to an incoming reference while recording, or lock to the incoming video signal it is recording.

Therefore, every tape machine will include a precision oscillator and sync pulse generator (SPG). This module can either free run to provide a good reference for the machine, or it can be genlocked to either an incoming reference signal or video input. Part of the input processing will include a switch to select which input will be directed into the oscillator and SPG.

3.5.8.2.2 Video input processing

All videotape recorders have input circuitry. This is required to convert the input video into a common form appropriate for recording on tape. For instance, component video recorders require any video input be in component form before any final encoding or modulation can occur prior to recording to tape. Therefore, input circuits will include a composite decoder, or S-video decoder, and a selection switch to allow the operator to select which type of input to record.

3.5.8.2.3 Input audio processing

As with video inputs, all videotape recorders will include switches, equalisers, noise reduction encoders (Dolby for instance) and even provide microphone power, to convert and process the incoming audio.

3.5.8.2.4 Tape encoding

The video signal needs processing prior to recording to tape. This will include FM. It may also include pre-emphasis to improve the recorder's ability to capture sharp transitions and detail in the image. It may also include a small amount of AM after the FM to reduce the possibility of over-modulation problems that sometimes manifest themselves as bearding on the playback image. The audio signal needs little further processing other than standard bias modulation, before being recorded to longitudinal tracks to tape.

3.5.8.2.5 Signal transfer to the drum

Once a decision had been made to use a rotating drum to increase the relative head-to-tape speed, a way was needed to transfer the video signals onto the spinning drum and to the record heads. Wire connection could hardly be used. They would very quickly tie themselves in knots and wrench themselves free. Slip rings and brushes also presented problems. It was impossible to maintain good enough connection.

The answer lies in the rotary transformer. This operates in a similar way to a standard transformer, with two windings (coils) sitting close to one another. Current in the input coil produces a magnetic flux. As the current changes the flux changes. The rate of change of this flux excites a current in the output coil. If an AC signal is input to the input coil an AC output will appear in the output coil, albeit phase shifted.

Rotary transformers have one coil built into the static lower drum and the other built into the spinning upper drum. An RF signal can be transferred from the lower to upper drum during recording, and from the upper to lower drum during playback. Modern videotape recorders have many rotary transformers for transferring more than one signal onto and off the upper drum. This is essential in component video machines, where there is a separate path for the luminance and colour signals. A separate transformer is often also used to transfer switching information to the upper drum, so that the drum itself can switch record or playback signals between different heads either as the drum rotates, or for multi-format machines where different playback heads are used to play back tapes from different formats.

3.5.8.3 Output processing

One of the challenges facing designers of early tape recorders was how to play the tape back with smooth consistent timing. The timing requirements of a standard broadcast video signal are very accurate. Videotape recorders are essentially mechanical. No matter how well the tape machine is built, and no matter how good the servo system is, there will still be a slight amount of mechanical wobble that will introduce huge timing fluctuations compared to the timing accuracy required of broadcast video signals. The answer lies in a clever piece of circuitry called a timebase corrector, which is explained in a separate section below.

3.5.8.3.1 Signal transfer from the drum

RF signals from the playback heads are transferred off the drum through the rotary heads described above. Head switching is required for those machines with a drum wrap of less than 360 degrees. Many machines use an active wrap of 180 degrees. Two sets of playback heads are used, 180 degrees apart. The machine will switch between heads at exactly the correct point, and thus maintain a continuous video signal from the tape.

This switching can be performed on or off the drum. Switching on the drum means that fewer rotary transformers are required to transfer the video signals from the upper to lower drums. However, an extra rotary transformer is required to transfer switching information to the upper drum. Switching off the drum removes the need to transfer switching information to the upper drum, but increases the number of rotary transformers required to transfer the video signals off the upper drum. Once the signal is off the drum it is buffered and equalised. There may also be an automatic gain control (AGC) to automatically correct small irregularities in the amplitude of the playback RF signal.

3.5.8.3.2 Output video and audio processing

The final piece of circuitry before the outputs processes the video and audio signals to provide whatever outputs the machine is designed to provide. Component analogue machines may include a composite encoder for a composite output, or analogue-to-digital converters for either digital audio or digital video outputs.

3.5.8.4 The timebase corrector (TBC)

Sitting between the playback equalisers and the final output processing is the TBC. A TBC evens out the irregularities in timing of the signal coming from the helical tracks of a videotape recorder. All TBCs do this by storing a certain amount of the signal and then playing it out at a constant rate.

3.5.8.4.1 Clock generation

An important part of any TBC is the ability to generate accurate clocks. Clocks are used to write video into the store and out of it. The read clock needs to accurately follow the timing irregularities in the signal coming from tape. The write clock needs to be locked to the machine's SPG, keeping constant smooth timing.

Of the two clocks the read clock presents the greatest challenge. A horizontal syncs detector sends the horizontal sync pulses off tape to timed monostables, which output a voltage depending on the rate of the sync pulses. The detector may also include a window discriminator, which will ignore any false horizontal syncs and half line pulses during the vertical interval. The signal from the timed monostables is fed into a voltage-controlled oscillator (VCO). The VCO is designed to run at the same rate as the read clock when there is no signal from the timed monostables. Timing irregularities in the signal off tape will increase or decrease the horizontal sync rate. This will cause the control voltage from the timed monostables to increase or decrease, shifting the VCO frequency up or down.

3.5.8.4.2 Charge-coupled device (CCD) delay line TBC

The first TBCs used CCD delay lines. These half analogue, half digital devices consist of a long line of cells. Each cell could contain an analogue charge. A clock input transfers all the charges one cell towards the end of the line. The input is connected to the first cell and the output to the last cell.

CCD delay lines cannot input and output at the same time. Therefore, two delay lines are used, one for writing and the other for reading. They are designed to store enough for one line of video. One line later and the delay lines are switched. The one that was writing is now reading, and vice versa. The clocks are also switched. The delay line that is writing uses the write clock and the one that is reading uses the read clock.

3.5.8.4.3 Semiconductor TBC

All new TBC designs use semiconductor memory devices instead of CCD delay lines. Semiconductor memory devices are totally digital. They therefore require a digital input and give out a digital output. All semiconductor memory TBCs used in analogue video recorders use analogue-to-digital converters at the TBC input and digital-to-analogue converters at the output.

3.5.8.4.4 Dual TBC designs

All modern broadcast analogue videotape recorders record component video, keeping the luminance and colour parts of the video signal separate throughout the whole record/playback process, even on tape. Thus, the luminance and colour playback signals experience their own timing inconsistencies. Each signal must be timebase corrected independently if quality is to be maintained.

A dual TBC has a separate horizontal sync detector, timed monostable, and VCO for luminance and colour. It also has two stores. The read clock is the same for both luminance and colour.

3.5.9 Popular Analogue Video Recording Formats

This is by no means an exhaustive list. There are many analogue videotape formats not mentioned here that were only moderately successful and others that were more of a failure.

3.5.9.1 Quadruplex (1956)

Ampex introduced the Quadruplex tape format, commonly known as Quad. Quad is a professional four-head transversal composite format. It uses spools of 2-inch tape. Helical tracks were originally 10 mils wide, 33 minutes from vertical. The drum is just over 2 inches in diameter, spinning at 1400 rpm for the original NTSC machines.

The most long-lasting of videotape formats, the Ampex VR-1000 machine was the first commercial videotape machine. There are still archives of Quad tape, and it is still in use in a few places, although it has been superseded for new recordings by other formats.

3.5.9.2 U-matic (1970)

Developed by JVC, Matsushita and Sony in 1971, and sometimes called Type E, U-matic is a professional two-head helical scan composite format. It uses spools of $^3/_4$-inch (19 mm) tape. Helical tracks are 84 µm wide and 4.95 degrees from horizontal. The drum is 110 mm in diameter, spinning at 1500 rpm (1800 rpm for NTSC). U-matic will record two longitudinal audio tracks. LTC was not designed in as a separate track to start with but was given a dedicated track under the helical tracks. This meant that LTC had to be recorded first and could not be re-recorded without overwriting part of the helical tracks. Later provision for VITC was available.

U-matic was a very successful format because of its wide user base, from high-end broadcast to professional and industrial use. Although not as long lasting as Quad, it was probably more popular. Eventually available in higher quality SP form, in lo-band and hi-band modes.

3.5.9.3 Betamax (1975)

Developed by Sony, Betamax is a domestic two-head composite helical scan format using the colour under system. It uses cassettes containing $^1/_2$-inch tape. Helical tracks are just over 30 µm wide and 5.85 degrees from horizontal. The drum is 74.487 mm in diameter, spinning at 1500 rpm (1800 rpm for NTSC). Betamax will record one longitudinal audio track and no timecode.

Betamax was head to head with VHS in the late 1970s, but eventually lost. Many reasons have been given for this: the reluctance of Sony to license the format, the reluctance of video rental firms to accept pre-recorded Betamax tapes, the lesser record times and features of Betamax machines.

3.5.9.4 1-inch type C (1976)

Developed by Sony and Ampex, C is a professional helical scan composite format. It uses spools of 1-inch tape. Helical tracks are 5.1 mils wide, and almost flat at 2.5 degrees from horizontal. The format uses a large drum of 132 mm diameter spinning at 3000 rpm (3600 rpm for NTSC). C will record three longitudinal audio tracks (four in Europe) and LTC is normally recorded on the last audio track.

3.5.9.5 VHS (Video Home System) (1976)

Developed by JVC, and adopted by many other manufacturers, VHS is a domestic two-head composite helical scan format using the colour under system. It uses cassettes containing $^1/_2$-inch tape. Helical tracks are 2.3 mils wide in standard play mode, 1.15 mils wide in long play mode and 5.96 degrees from horizontal. The drum is 60.5 mm in diameter, spinning at 1500 rpm (1800 rpm for NTSC). VHS will record two longitudinal audio tracks and no timecode.

VHS was head to head with Betamax in the late 1970s, but eventually won. VHS went on to become the most popular domestic and industrial format.

3.5.9.6 Video 2000 (1979)

Developed by Philips and Grundig, Video 2000 is a domestic two-head composite helical scan format using the colour under system. It uses cassettes containing $^1/_2$-inch tape. Helical tracks are 22.6 µm wide and 15 degrees from horizontal. The drum is 65 mm in diameter, spinning at 1500 rpm (1800 rpm for NTSC). Video 2000 will record two longitudinal audio tracks and no timecode.

In Europe Video 2000 was the 'other domestic format' while VHS and Betamax battled for supremacy. It boasted automatic tracking, using bimorphs similar to those used in professional machines, and dual-sided cassettes. However, Video 2000 was never going to win against either VHS or Betamax. While video rental firms were pushed to provide two versions of each movie in their shops, VHS and Betamax, it was inconceivable that they would supply three.

3.5.9.7 Betacam (1982)

Developed by Sony, and sometimes called Type L, Betacam is a professional four-head component helical scan format. It uses cassettes containing $^1/_2$-inch tape. Helical tracks are 86 µm wide, and +15.25 degree azimuth, for luminance and 72 µm, and −15.25 degree azimuth, for colour tracks, and 4.679 degrees from horizontal. The drum is 74.487 mm in diameter, spinning at 1500 rpm (1800 rpm for NTSC). Betacam will record two longitudinal audio tracks, LTC and VITC.

Betacam became a popular professional format using oxide tape similar to that used by domestic Betamax. However, Betacam was really just a 'rehearsal' for the improved version, Betacam SP, which became a workhorse professional and broadcast videotape format.

3.5.9.8 8 mm (1983) and Hi8 (1989)

Developed by a Japanese consortium, 8 mm is a domestic two-head composite helical scan format using the colour under system. It uses cassettes containing 8 mm tape. Helical tracks are 20.6 µm wide and 4.88 degrees from horizontal. The drum is 1.6 inches in diameter, spinning at 1500 rpm (1800 rpm for NTSC). 8 mm will record two PCM audio channels and two AFM audio channels and no timecode.

Hi8 is an enhancement of 8 mm, developed by Sony, using metal oxide tape.

Both 8 mm and Hi8 have gained reasonable success as a domestic camcorder tape format.

3.5.9.9 Betacam SP (1986)

Developed by Sony, and sometimes called Type L. An improvement over the Betacam format, Betacam SP, with the same format dimensions, uses higher FM frequencies and metal instead of oxide tape. Betacam SP introduced two AFM audio tracks inserted into the colour helical track signal, providing the format with four audio tracks altogether.

The BVW-75 and BVW-75P machines became workhorse machines within the broadcast industry, selling thousands of machines and millions of tapes worldwide.

3.5.9.10 M2 (1986)

Developed by Panasonic, M2 is a professional four-head component helical scan format. It uses cassettes containing $^1/_2$-inch tape. Helical tracks are 44 µm wide for luminance and 36 µm for colour tracks, with a 15 degree azimuth, and 4.29 degrees from horizontal. The drum is 76 mm in diameter, spinning at 1500 rpm (1800 rpm for NTSC). M2 will record two longitudinal audio tracks and two AFM audio tracks, LTC and VITC.

M2 was introduced as a competitor to Betacam SP and some broadcasters adopted it as a standard. Although technically very similar, the machines gained a reputation for unreliability, probably due more to spare parts availability and service rather than the machine's reliability. M2 did not gain the universal acceptance that Betacam SP gained.

3.5.9.11 S-VHS (1987)

Developed by JVC, and adopted by every other manufacturer, VHS is an enhancement of the VHS format with improved luminance bandwidth. It gained popularity because of its compatibility with VHS.

3.5.10 Digital Videotape Recorders

Practical broadcast digital video recorders began to appear at the beginning of the 1980s with the publication of CCIR 601 in 1982 and CCIR 656 in 1986. These two documents proposed a method of digitising component video signals and conveying them in digital form over a multi-core cable. Sony designed the D1 video recorder specifically to record CCIR 601 signals without any loss.

The original CCIR 601 document specified 8-bit samples. However, the CCIR 656 document also specified two spare data bits which were for 'future development'. The industry grabbed these two spare bits, using them at $^1/_2$ and $^1/_4$ resolution, increasing the samples sizes to 10 bits. About the same time as the transition from 8 bits to 10 bits, there was transition from the original multi-core cable, parallel method of conveying CCIR 601 data to a serial version using standard 75 ohm coaxial cable and BNC connectors.

Sony and other manufacturers, notably Panasonic and Ampex, followed over the years by producing broadcast-quality digital video recorders to record either 8- or 10-bit CCIR 601 samples, either entirely transparently or with compression.

Although digital video recorders have gained almost universal acceptance in broadcast, the domestic and industrial markets continue to use analogue tape formats, due to the overwhelming use of VHS, and the introduction of analogue formats like Hi-8, which have sufficient quality for most peoples' needs.

DV has gained wide acceptance as a camcorder standard for domestic use. The lack of any domestic DV television recorders has helped to keep VHS as the only practical home television recording format.

It is unlikely that there will ever be a de facto standard digital tape recording format for recording television in the home. The increased popularity of DVD recorders and the imminent release of Blu-ray will certainly now kill any chance for a manufacturer to introduce one.

3.5.10.1 The advantages of digital videotape recorders

Digital video recorders have a number of distinct advantages over analogue machines. The first is the record transparency. A digital video signal can retain all its quality through the record playback process. In theory, exactly the same digital data that is recorded to tape can play back. Although this is not exactly true, it is certainly true that professional digital video recorders allow video to be recorded, played back and re-recorded many more times than is possible with analogue recorders. This is important for editing and post-production.

The second is robustness. Digital data can be protected with error correction data far more easily than an analogue signal. Furthermore, digital data can be shuffled and scrambled before recording to tape. If there is a large error on tape, either during recording or during playback, due to, for instance, dust, the highly concentrated group of errors can be diluted over a large amount of data, as a widely spread group of small errors that can easily be corrected one by one.

Other advantages have become apparent over the years, as digital videotape formats have developed, and with the introduction of computers into the broadcast production chain. Later digital video recording formats now offer long recording time and small tape sizes. They also offer the possibility of transferring digital data directly to the IT world, without loss, where computer-based non-linear editors and effects processors can perform a wide range of previously unavailable creative possibilities.

3.5.10.2 Digital video recorder or digital data recorder

It is important to remember that all digital video recorders are not recording digital video, or audio. The data is always processed, scrambled, shuffled, sometimes compressed, and has extra error correction data added. What is actually recorded to tape is just digital data and bears very little resemblance to the original video and audio it came from.

Manufacturers are now using stripping the video and audio input and output processing out of their digital video recorders to produce very competent data recorders for the IT backup and archive markets.

3.5.10.3 Digital video recorder mechadeck design

There is no difference in the requirements of a digital video recorder mechadeck from that of an analogue one. Digital video recorder mechadecks differ more as a result of general developments in mechadeck design rather than any special requirements. Most digital video recorder mechadecks use a spinning upper drum and static lower drum. They employ either M wrap or C wrap, and they all incorporate sophisticated supply side tension regulation, and capstans on the exit side of the drum.

A notable difference employed by the Sony Digital Betacam, D1 and D2 machines was the rotating mid drum assembly. The lower drum is static, as normal, but these machines also have an upper drum fixed to the lower drum, leaving a narrow slot between the two. A mid drum assembly spins inside and between the upper and lower drums, with the record, playback and flying erase heads fixed to its circumference and protruding through the slot to touch the tape. This technique is more expensive but produces equal strain on the tape at every point round the drum, resulting in very straight helical tracks on tape.

From the start, broadcast digital video recorders have recorded audio as digital data somewhere on the helical tracks. Although some digital formats still retain a low-quality longitudinal cue track, this development has resulted in very high quality audio recording and the removal of much of the mechadeck hardware required for analogue longitudinal audio recording.

Some digital formats have even removed the need for a conventional longitudinal control track transferring all the servo lockup to the helical tracks. These formats only have one longitudinal heads on the mechadeck, the full erase head.

3.5.10.4 Digital video recorder channel coding

Analogue video recorders use FM as a method of coding the video signal before recording it to tape, and decoding it on playback, to overcome the problems of recoding to magnetic tape. In general terms this technique of coding and decoding is called channel coding. The tape is the channel.

Digital video recorders cannot use FM. It is both inappropriate and impossible considering the available bandwidth on tape and the required recording bandwidth. Digital video recorders use a combination of Partial Response type 4 (PR4 or PRIV) and Viterbi as a channel coding scheme.

3.5.11 Popular Digital Videotape Formats

3.5.11.1 D1 (1987)

Developed by Sony, D1 is a professional digital four-head component helical scan format. It records 8-bit CCIR 601 video data with no compression. It uses cassettes containing 19 mm tape. Helical tracks are 40 μm wide and 5.4 degrees from horizontal. The drum is 75 mm in diameter, spinning at 9000 rpm (10800 rpm for NTSC). D1 will record four audio channels on the helical tracks and one on a longitudinal track. It will also record LTC and VITC.

D1 is expensive, both for machines and tape cassettes, but is used in post-production, where quality is the prime concern.

3.5.11.2 D2 (1989)

Developed by Sony, D2 is a professional digital four-head composite helical scan format. It records a digitised PAL or NTSC (depending on the machine version) composite video signal with no compression. It uses cassettes containing 19 mm tape. Helical tracks are 35.2 μm wide and 6.1326 degrees from horizontal. The drum is 96.4 mm in diameter, spinning at 6000 rpm (7200 rpm for NTSC). D1 will record four audio tracks on the helical tracks and one on a longitudinal track. It will also record LTC and VITC.

D1 is expensive, both for machines and tape cassettes, but is used in post-production, where quality is the prime concern.

3.5.11.3 Digital Betacam (1993)

Developed by Sony, Digital Betacam is a professional digital four-head component helical scan format. It records 10-bit CCIR 601 video data with DCT-based compression at just over 2:1. It uses cassettes containing ½-inch tape. Helical tracks are 24 μm wide and 5 degrees from horizontal. The drum is 81 mm in diameter, spinning at 4500 rpm (5400 rpm for NTSC). Digital Betacam will record four audio tracks on the helical tracks and a low-quality cue channel on a longitudinal track. It will also record LTC and VITC.

Certain machines are capable of playing back Betacam and Betacam SP tapes. Digital Betacam is cheaper than D1 but offers indistinguishable image quality and 10-bit sample recording. It is widely used in post-production, where quality is the prime concern. However, the compression scheme is closed and proprietary. Present broadcasters are now looking to output the digital stream directly from the tape machine.

3.5.11.4 DV and MiniDV (1995)

A consortium of 10 companies agreed and created DV, sometimes called MiniDV. DV is a domestic digital four-head component helical scan format. It records intraframe 4:1:1 or 4:2:0 compressed video data with 5:1 compression ratio. It uses cassettes containing ¼-inch tape. Helical tracks are 10 μm wide and 9.18 degrees from horizontal. The drum is 21.7 mm in diameter, spinning at 9000 rpm. DV will record two audio tracks on the helical tracks. Timecode is recorded on the helical track as data (not VITC).

DV has become the most popular domestic digital videotape format, and is available from a wide range of manufacturers. Camcorders and decks offer direct compressed data outputs via the IEEE 1394 interface, otherwise known as FireWire (Apple) and iLink (Sony). Software companies also offer good support for DV, with drivers and plug-ins for DV data input to graphics, editing and rendering software.

3.5.11.5 DVCPRO (1995)

Developed by Panasonic and based on the DV format, DVCPRO is a professional digital four-head component helical scan format. It records DV data but with a wider track on metal evaporated tape to increase robustness and quality. It uses cassettes containing ¼-inch tape. Helical tracks are 18 μm wide and 9.18 degrees from horizontal, with a +20.03 – 19.97 degree azimuth. The drum is 21.7 mm in diameter, spinning at 9000 rpm. DVCPRO will record two audio tracks on the helical tracks and one longitudinal cue track. It will also record LTC and VITC.

DVCPRO is the Panasonic professional DV format, and initially gained wide acceptance due to its low price and compact design.

3.5.11.6 DVCAM (1996)

Developed by Sony and based on the DV format, DVCAM is a professional digital four-head component helical scan format. It records DV data. It uses cassettes containing ¼-inch tape. Helical tracks are 15 μm wide and 9.18 degrees from horizontal with a +20.03 – 19.97 degree azimuth. The drum is 21.7 mm in diameter, spinning at 9000 rpm. DVCAM will record two audio tracks on the helical tracks and one longitudinal cue track. It will also record LTC and VITC.

DVCAM is the Sony professional DV format. Introduced after DVCPRO, it lagged behind in popularity but is now beginning to gain widespread support as an industrial format and for low-budget television work. Machines like the DSR-PD-150 have almost gained 'classical' status.

3.5.11.7 Betacam SX (1996)

Developed by Sony using the Digital Betacam mechadeck, Betacam SX is a professional digital four-head component helical scan format. It records 8-bit CCIR 601 video data with MPEG 4:2:2P@ML-based compression. Betacam SX uses IB frame compression to maintain broadcast quality at 18 Mbps and 10:1 compression ratio. It uses cassettes containing ½-inch tape. Helical tracks are 32 μm wide and 5 degrees from horizontal with a 15.25 degree azimuth. The drum is 81 mm in diameter, spinning at 4500 rpm (5400 rpm for NTSC). Betacam SX will record four audio tracks on the helical tracks. It will also record LTC and VITC.

Certain machines are capable of playing back Betacam and Betacam SP tapes. Sony introduced a hybrid Betacam SX machine combining a conventional tape mechadeck with hard disks. Compressed video and audio material could be transferred to and from the tape and disks. This allowed for linear and non-linear editing in one unit. However, the hybrid machine proved too complex for many and was not widely adopted.

Betacam SX was introduced as a replacement to Betacam SP and has comparable digital quality. It is widely used as a news gathering format. However, although the 18 Mbps compressed digital stream is available at the output for direct high-speed transfer, the compression scheme was not ratified, with the standards authorities preferring 50 Mbps data instead. Sony responded with IMX.

3.5.11.8 Digital S (1996)

Developed by JVC and otherwise known as D9, Digital S is a professional digital four-head component helical scan format. It uses 4:2:2 sampling, like MPEG, making it technically better than DV and the same as MicroMV. It uses cassettes containing ½-inch tape. Helical tracks are 20 μm wide. The drum is 62 mm in diameter. Digital S will record four audio tracks on the helical tracks and two longitudinal tracks. It will also record LTC and VITC.

3.5.11.9 HDCAM (1997)

Developed by Sony and based on the Digital Betacam mechadeck, HDCAM is a professional digital four-head component helical scan format. It records high-definition video data with mild 4.4:1 compression. It uses cassettes containing ½-inch tape. Helical tracks are 22 μm wide and 5 degrees from horizontal with a 15.25 degree azimuth. The drum is 80 mm in diameter. HDCAM will record four audio tracks on the helical tracks and one longitudinal cue track. It will also record LTC and VITC.

HDCAM was introduced as an alternative to film, and thus records progressive 24 fps (24P), but can be switched to a number of television-based recording methods. HDCAM is expensive and exclusive, but offers very high quality recording.

3.5.11.10 DVCPRO 50 (1998)

Developed by Panasonic as an enhancement of the original DVCPRO format, with 50 Mbps recorded data to comply with the requirements of standards authorities. Machines can now be equipped with an IEEE 1394 interface, allowing high-speed transfer of 50 Mbps DV data.

3.5.11.11 IMX (2000)

Developed by Sony using a new design of mechadeck loosely based on the Digital Betacam mechadeck, IMX is a professional digital four-head component helical scan format. It records 8-bit CCIR 601 video data with MPEG 4:2:2P@ML I-frame-only based compression at 50 Mbps. It uses cassettes containing ½-inch tape. Helical tracks are 22 μm wide and 5 degrees from horizontal with a 15.25 degree azimuth. The drum is 80 mm in diameter, spinning at 4500 rpm (5400 rpm for NTSC). IMX will record eight audio tracks on the helical tracks. It will also record LTC and VITC.

Certain machines are capable of playing back Betacam, Betacam SP and Digital Betacam tapes. IMX was introduced to comply with industry requests, and standards authorities requirements for a 50 Mbps I-frame-only MPEG video recorder. Although recording 8-bit samples (a requirement of MPEG) IMX quality is indistinguishable from Digital Betacam. However, unlike Digital Betacam, the compressed stream is available at the output for direct transfer to other machines, computer hard disk or video servers.

A later modification, the E-VTR, allows video and audio material from tape to be packaged and sent directly out on a computer network cable.

3.5.11.12 MicroMV (2001)

Developed by Sony, MicroMV is a new format intended for the domestic market. However, it records true MPEG data on a tiny cassette, giving it comparable, if not better, quality than DV. Although technically superior than DV, MicroMV has a lot of work to do to gain any ground on DV and DV-based formats like DVCPRO and DVCAM. Software manufacturers have still to offer the kind of support for MicroMV that DV enjoys.

Andy Major
Snell & Wilcox

3.6 Television Standards Conversion

3.6.1 Introduction

Standards conversion is the process of taking one television system and converting to another television system. The classical standards converter came into being for international programme exchange where conversion was performed between NTSC and PAL/SECAM and the reverse. Today the need for standards conversion is even greater with the uptake of high-definition (HD) formats for both capture and broadcast. The proliferation of standards has led to a requirement for converters that can interface between different HD standards and standard-definition (SD) standards.

The most challenging standards conversion process is where there is a field rate change, for example changing from NTSC to PAL, where NTSC is at 60 Hz and PAL is at 50 Hz. The field rate conversion will be the main focus of this chapter. Colour encoding systems, such as NTSC and SECAM, will not be covered; it is assumed that all signals are in digital component form.

The goal with standards conversion is to produce a transparent representation of the input. Today's high-end motion-compensated converters achieve this, but this has not always been the case.

3.6.2 Historical Background

Early standards converters were based around optical systems – that is, a camera pointing at a monitor; although these were refined over the years, the resolution and motion portrayal were of poor quality by today's standards. Although experimental analogue systems were produced, the first significant digital converter was the DICE (Digital Intercontinental Conversion Equipment) designed by the IBA; only a few systems were made but it set the way for digital converters.

The next significant development was the ACE developed by the BBC. The ACE provides the blueprint for most modern converters; its filtering system involved taking contributions from four fields and four fields, known as a 16-point aperture. Not exactly a portable product, it occupied two 19-inch racks. The ACE was a linear converter, where linear in this context means that it will apply the same type of filtering irrespective of the input video. The main downside of this system was its ability to portray motion accurately. The next significant development was the AVS – ADAC. This quickly revolutionised the world of conversion; it was relatively compact at only 6 RU and was motion adaptive – that is, it was capable of charging its filtering operation, depending on the amount of motion in the source. During the early 1990s the race was on to produce a motion-compensated converter – that is, a converter that produces not only high resolution / low artefacts on static scenes, but accurate portrayal of motion.

The Thomson 7830 and Vistek Vector were the first units on the market; both these systems demonstrated what was possible with motion compensation. However, they were prone to falling over in dramatic fashion and could look a lot worst than the traditional linear approach to standards conversion. Snell & Wilcox Alchemist Ph.C came out shortly after the Thomson and Vistek and quickly became the industry standard, offering levels of performance and robustness previously unobtainable.

3.6.3 Motion Portrayal in Video Systems

The most challenging aspect of standards conversion is performance following the temporal rate change. This is changing the number of fields, e.g from 50 Hz to 60 Hz. To understand why it's necessary to resort to motion-compensated conversion first requires an understanding of temporal sampling and the human eye's response to motion.

3.6.3.1 Temporal sampling and display

Television signals run at 50 or 60 Hz. In theory, neither of these frequencies is sufficiently high to provide smooth or accurate motion portrayal; however, in practice, problems caused by the low temporal rates are rarely experienced. Problems do occur, however, when one tries to convert from one frame rate to another.

So, the question is why are the problems of low sampling rate not apparent? The explanation is related to the temporal response of the eye. Figure 3.6.1 shows the temporal response of the eye; it can be seen that the response falls off quite rapidly with temporal frequency.

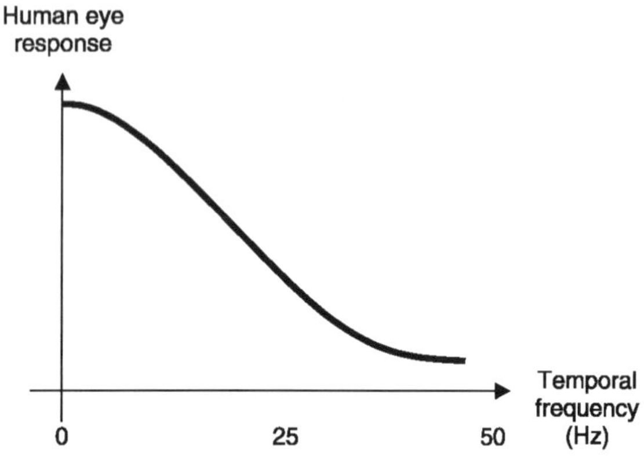

Figure 3.6.1 Temporal response of the eye.

Clearly this graph does not give the full picture; if this were the case, it would not be possible to see fast-moving objects. Many years ago this was used as an argument that man would never be able to fly at the speeds needed for flight.

The solution to seeing high-speed objects is for the eye to track them. Figure 3.6.2 shows the effects of a fixed eye viewing a moving object. The temporal frequencies experienced by the eye are very high and hence the eye would not be able to see the detail on the moving object, there will be motion blur.

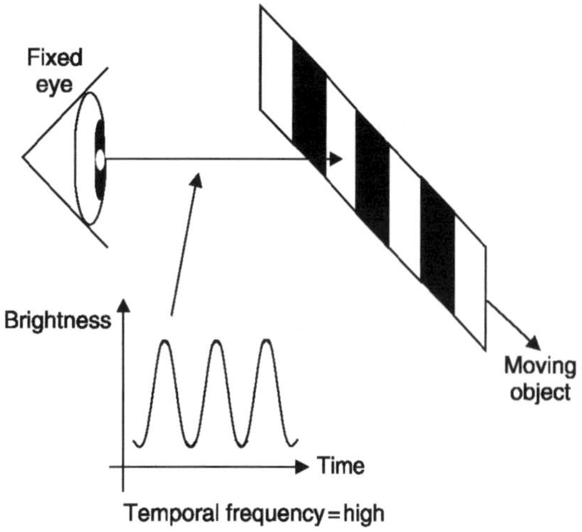

Figure 3.6.2 Fixed eye viewing a moving object.

Figure 3.6.3 shows the difference that tracking motion makes to the eye's ability to see moving objects. The eye is following the moving object and as a result the temporal frequency at the eye is zero; the full resolution is then available because the image is stationary with respect to the eye. In real life it's possible to see moving objects in some detail unless they move faster than the eye can follow.

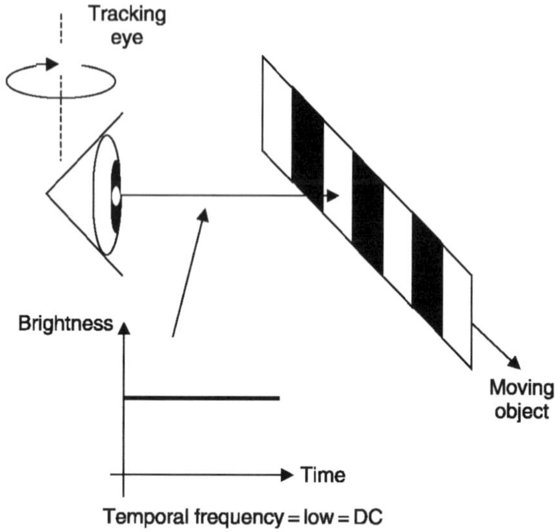

Figure 3.6.3 Tracking eye viewing a moving object.

How does the poor temporal response of the eye relate to motion portrayal in video systems? Television systems are not continuous, but are sampled at 50 or 60 Hz. The sampling rates of 50 or 60 Hz in theory are far too low to capture motion and distortion of the motion profile should be present, but in practice distortions are rarely noticeable.

The trick is that the eye tracks moving objects. An object moves past a camera, and is tracked on a monitor by the eye. The temporal frequencies should cause aliasing in the TV signals, but the tracking eye does not perceive these as tracking the motion reduces the temporal frequency to zero.

Figure 3.6.4 shows what happens when the eye follows correctly. The original scene and the retina are stationary with respect to one another, but the camera and display are both moving through the field of view. As a result, the temporal frequency is brought to zero and accurate motion portrayal is achieved.

The two points to remember from this section are:

1. The human eye has a poor temporal frequency response. However, it's able to see high-speed objects by tracking their motion path.
2. Television signals are under-sampled and contain temporal aliases, but viewers are oblivious to these problems, because:

- Moving objects are tracked by the eye.
- Television signals are displayed at the same rate as the capture rate.

This last point is key to why TV systems look okay. Problems start when the display rate is changed, e.g. from 50 to 60 Hz,

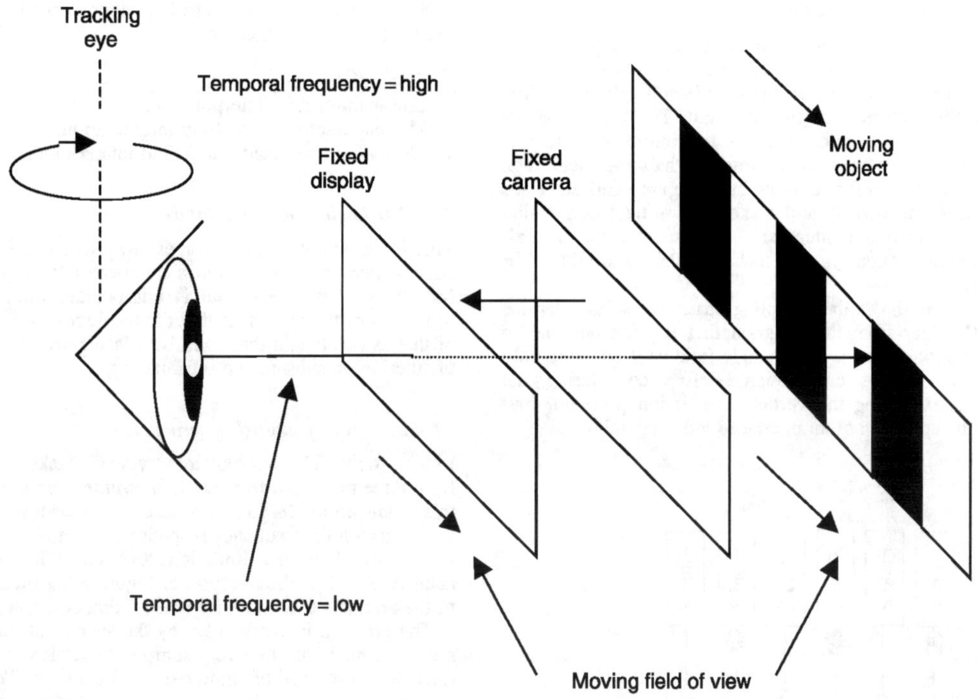

Figure 3.6.4 Motion capture and display.

then all of the processes that are relied upon for accurate motion portrayal break down. Hence temporal conversion of TV pictures is extremely challenging.

3.6.4 Standards Conversion Process

There are three aspects to standards conversion (see Figure 3.6.5):

1. Horizontal – changing the number of pixels, e.g. 704 to 710.
2. Vertical – changing the number of lines, e.g. 625 to 525.
3. Temporal – change of frame rate, e.g. 50 Hz to 60 Hz.

In many standards converters the processes of vertical and temporal conversion are combined, for practical and economic reasons.

3.6.4.1 Horizontal conversion

There is always a need to resample the image horizontally when performing standards conversion. When up-converting – that is, converting from a standard-definition (SD) to a high-definition (HD) television standard – more samples are required. When down-converting (converting from an HD to SD TV standard), then the number of samples per line is decreased. The difference in line length between NTSC (52.86 μs) and PAL (52 μs), although small, should be accounted for. Because the digital component sampling rate for both standards is 27 MHz, this is equivalent to re-sampling between 714 and 702 active picture samples.

To perform the change in line length, interpolation filters are required. The theory and implementation of interpolation are covered in Section 3.6.5.

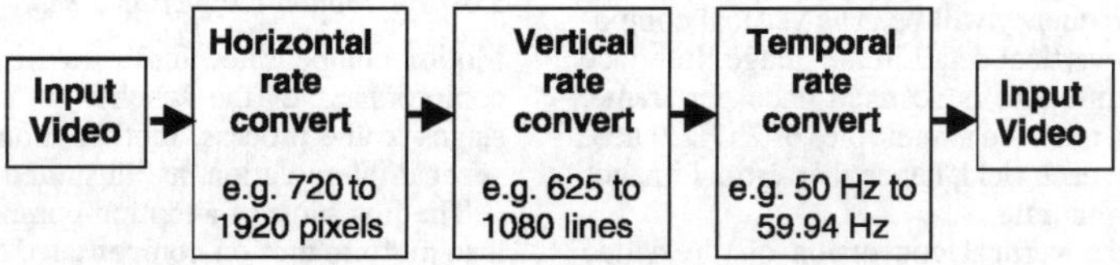

Figure 3.6.5 The standards conversion processes.

3.6.4.2 Vertical conversion

The vertical conversion process is made more complex due to the interlace format that is used with the majority of TV standards. Interlace is used to conserve bandwidth by sending only half the picture lines in each field. The flicker rate is perceived to be the field rate, but the information rate is determined by the frame rate, which is half the field rate. Although the reasons for adopting interlace are valid, interlace has numerous drawbacks and makes the vertical conversion process more difficult. Interlace also presents a lot of challenges in compression systems such as MPEG and with DVE processing.

Figure 3.6.6 shows the sampling structure of an interlace signal. It's clear from this diagram that it's not possible to perform interpolation on just a single field as it contains only half of the vertical information. Before considering the approaches to solving the vertical conversion problem, first consider the spectrum of an interlaced video signal.

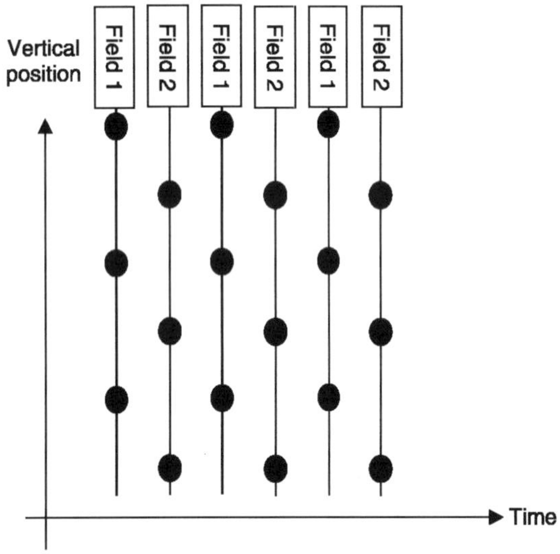

Figure 3.6.6 Interlace video format.

Figure 3.6.7 shows the vertical temporal spectrum of a 625/50 Hz signal. The black area in the corner is the wanted signal and the grey areas of energy are aliases that are unwanted components. How does this spectrum arise? The horizontal component of the star-shaped spectra is due to image movement, where the higher the speed and the more detail present, the higher the temporal frequency will be. The vertical component of the stars is due to vertical detail in the image. Interlace means that the same picture line is scanned once per frame, hence the spectra image is repeated at multiples of 25 Hz. There are 312.5 lines scanned in each field, hence the vertical images repeating at multiples of that rate.

In order to perform the vertical conversion on the video signal, there is a need to filter with a two-dimensional filter with a triangular response, shown in Figure 3.6.8.

Broadly speaking, there are four approaches to solving the interlace problem, which are:

1. Single-field interpolation.
2. Linear multi-field interpolation.
3. Motion-adaptive multi-field interpolation.
4. Motion-compensated multi-field interpolation.

3.6.4.2.1 Single-field interpolation

This is the simplest and cheapest way to solve the problem; contributions are taken from a number of lines (greater than four lines), then the output is interpolated from these. The limitation of the process is that the resolution will only be half of that available and there will be aliasing artefacts as a result of filtering on subsampled information.

3.6.4.2.2 Linear multi-field interpolation

Linear multi-field interpolators typically take contributions from three fields and from six lines within each of these fields; this is known an 18-point aperture. The response of the filter should match the frequency response of Figure 3.6.7; however, in practice there are limitations that make it impossible to achieve the ideal filter response. Figure 3.6.9 illustrates some of the artefacts that can arise when a practical filter is used.

The problem is confounded by the spectra of today's video signals, which are often very sharp both vertically and temporally due to the use of shuttered CCD cameras. The resulting spectra do not resemble the nicely separated quincunx pattern (resembling the five of dice), but in practice extend over a much wider area, almost merging with the alias spectra. Hence the demands on the filter are even more stringent.

A practical multi-field interpolating filter makes some compromises; a slight loss in vertical resolution is typically tolerated to avoid any motion artefacts. The results from multi-field interpolating filters are significantly better than the single-field filters.

3.6.4.2.3 Motion-adaptive multi-field interpolation

A small gain in performance is achieved if motion adaption is used in conjunction with the multi-field interpolation filters described in Section 3.6.5.2.2.

Motion-adaptive vertical interpolators use different filtering modes depending on how much motion there is in the scene (Figure 3.6.10). If the scene is static, the filter used will maximise the vertical resolution. In the event of motion, vertical resolution is compromised to minimise motion blur. In practice, motion interpolation offers little advantage over conventional linear approaches. The motion detection is rarely perfect and failures to accurately detect motion can result in artefacts such as resolution pumping.

3.6.4.2.4 Motion-compensated multi-field interpolation

Motion-compensated multi-field interpolation removes all the compromises of the scheme so far outlined. There are two stages to the process, motion-compensated de-interlacing and vertical interpolation, as illustrated in Figure 3.6.11.

The first stage of a motion-compensated vertical interpolator is to perform motion-compensated de-interlacing. In the case of a 625 input, the output of the process will produce 625 output lines for every field of 312 input lines. Figure 3.6.12 shows

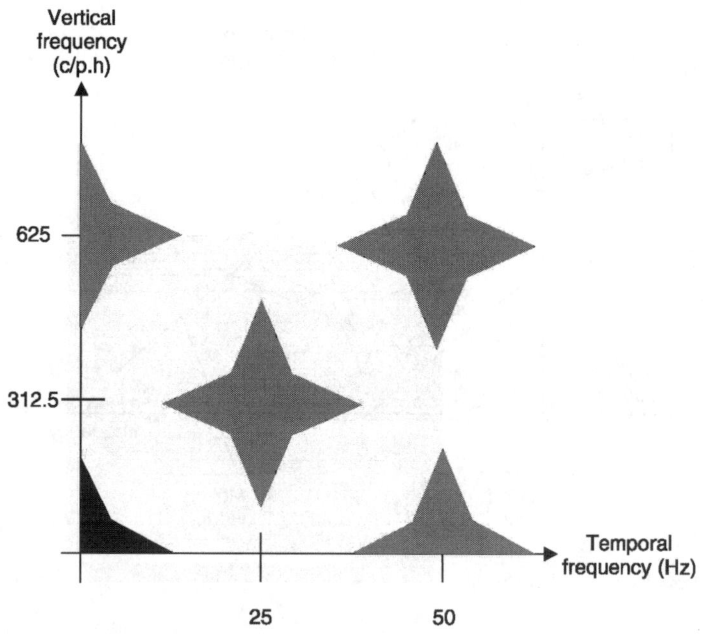

Figure 3.6.7 The vertical temporal spectrum of a 625 interlaced video signal.

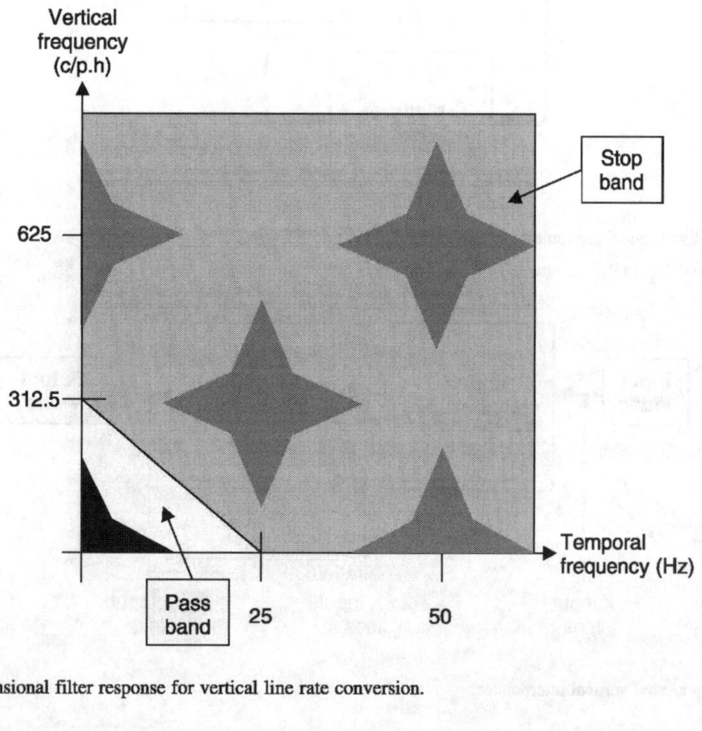

Figure 3.6.8 The two-dimensional filter response for vertical line rate conversion.

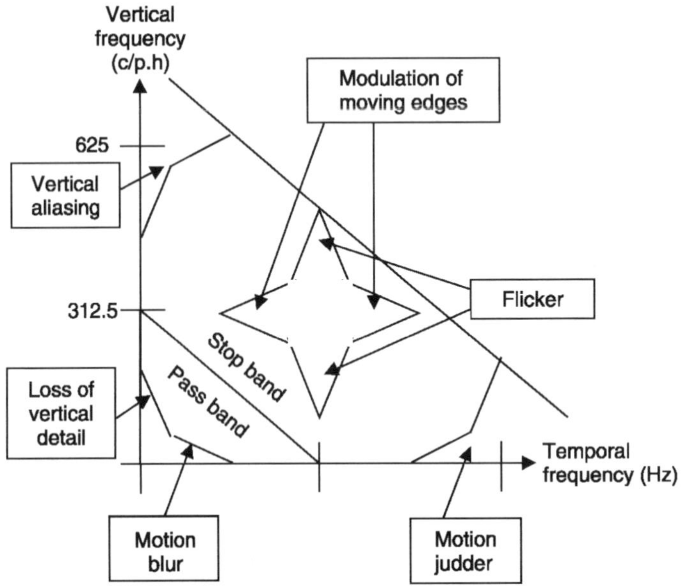

Figure 3.6.9 Potential problems due to non-ideal filtering.

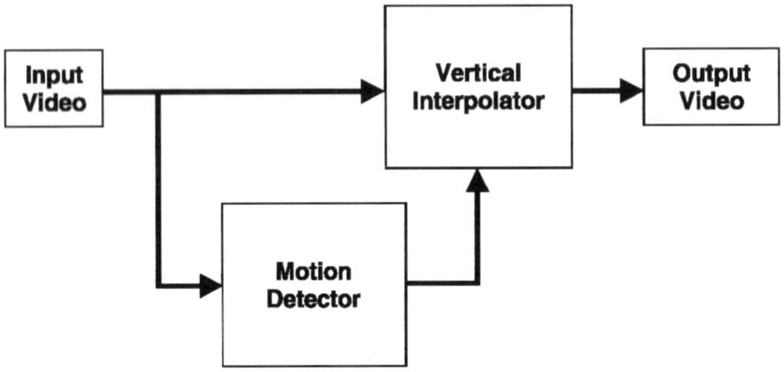

Figure 3.6.10 Motion-adaptive vertical interpolator.

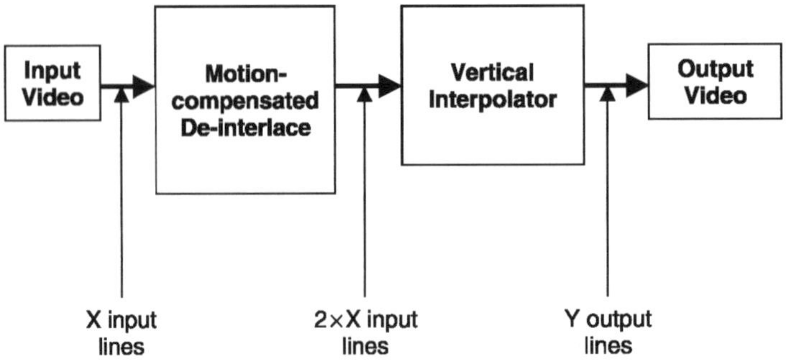

Figure 3.6.11 Motion-compensated vertical interpolator.

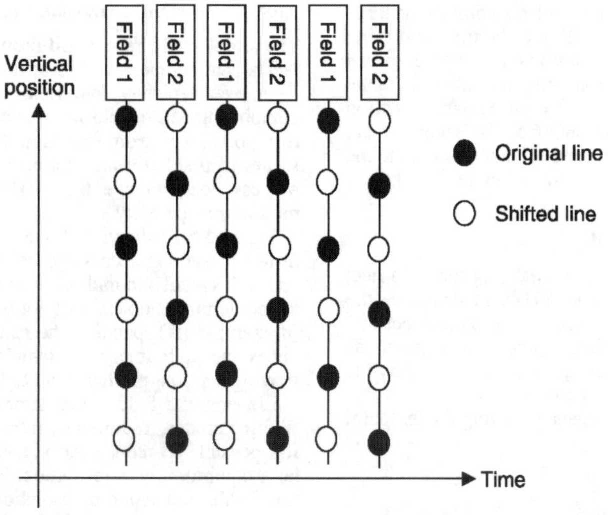

Figure 3.6.12 The de-interlaced sampling grid.

where the additional lines will be positioned relative to the original sampling grid.

To generate the additional samples, information is required from three input fields. Figure 3.6.13 shows a practical example of the de-interlacing process. Each field has motion vectors for every pixel within the image, where a motion vector represents the speed and direction of that pixel. With motion vectors it's possible to move the picture content onto the missing samples. The lower half of Figure 3.6.13 shows a cross-section through the scene and the arrows indicate where the additional

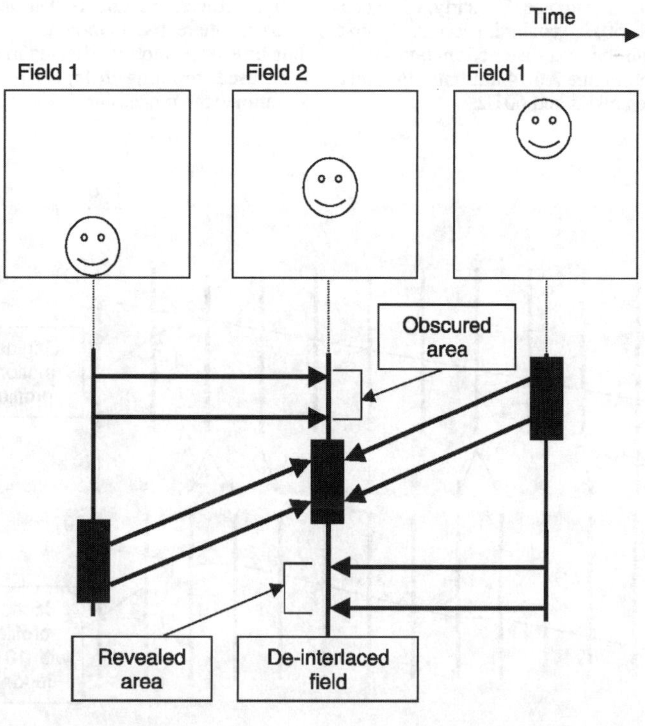

Figure 3.6.13 Motion-compensated de-interlacing, line doubling.

information will come from. The reason for having to use three fields is that areas of the picture will get obscured and other areas will be revealed. When there are revealed and obscured areas of the picture, information from only one side is suitable.

The process of deriving the motion vectors, known as motion estimation, will be covered in Section 3.6.5. The second stage, the interpolation process, that changes the number of lines to the desired output line rate, will also be covered in Section 3.6.5.

3.6.4.3 Temporal rate conversion

Temporal conversion is the process of changing the field rate, for example changing from 50 Hz to 60 Hz, as would be the case when converting from PAL to NTSC. Temporal conversion is the most demanding of all the processes in standards conversion due to the inherent aliasing problems caused by the low sampling rate used in TV standards.

There are in essence four approaches to solving the temporal conversion problem:

1. Field hold and repeats.
2. Linear multi-field interpolation.
3. Motion-adaptive multi-field interpolation.
4. Motion-compensated interpolation.

3.6.4.3.1 Temporal conversion by field holds and repeats

This process is very crude and is unsuitable for broadcast applications; the process is effectively one of synchronisation. Take the example of converting between 50 Hz and 60 Hz standards. Every five input fields will result in one extra repeated output field; the result will be a distorted motion profile with a 10 Hz beat component, an unacceptable outcome. Similarly, when converting between a 60 Hz and 50 Hz standard, every sixth input field is discarded, again resulting in a distorted motion profile with a 10 Hz beat component. Figure 3.6.14 illustrates the situation when converting between 50 Hz and 60 Hz.

3.6.4.3.2 Temporal conversion – linear multi-field interpolation

The simple field repeat and drop technique suffered from the 10 Hz beat component. Linear multi-field standards converters filter over typically four fields, to filter out the 10 Hz beat component. Multi-field interpolators interpolate through time, using pixel data from four input fields in order to compute the values of pixels for an intermediate output field. The filtering process does remove the 10 Hz effects, but cannot handle motion transparently.

Figure 3.6.15 shows that if an object is moving, it will be in a different place in successive fields. Interpolation between several fields results in multiple images of the object. The position of the dominant object will not move smoothly, it will judder. In example (A), panning the camera over a static scene produces multiple images. Example (B), a static camera with moving caption, produces multiple images on the caption.

Linear multi-field interpolators typically have a number of filtering modes, traditionally referred to as 'sports' and 'studio'. It's possible to reduce the judder effect; however, there will be a compromise in resolution. Hence the sports modes have low judder but reduced resolution, whereas the studio modes have more resolution but higher levels of judder.

3.6.4.3.3 Temporal conversion – motion-adaptive multi-field interpolation

Linear converters typically have a number of modes optimised for different applications. Typically the modes are referred to as 'sports' and 'studio'. The principle behind the motion-adaptive converter is its ability to automatically switch between modes depending on the amount of motion detected in the scene. If significant motion is detected a 'sports' aperture will be used, where the motion judder will be minimised. If there is little or no motion detected in a scene a 'studio' aperture will be used, resulting in higher resolution images. Figure 3.6.16 outlines the processing.

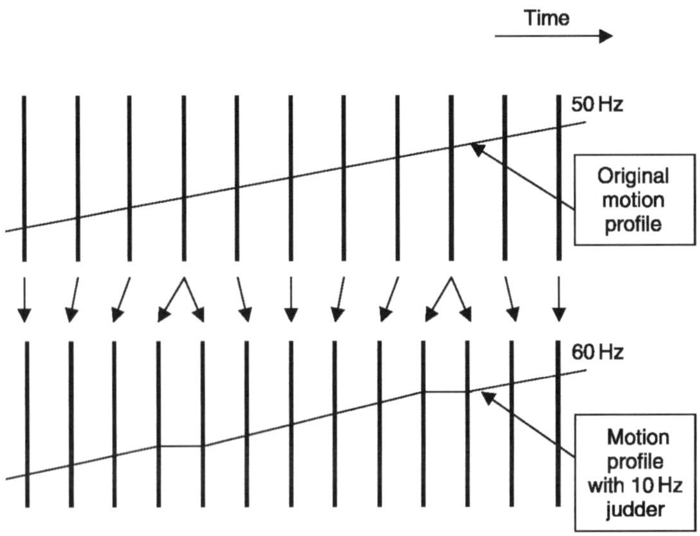

Figure 3.6.14 Conversion by field repeating, illustrating the unacceptable jerky motion.

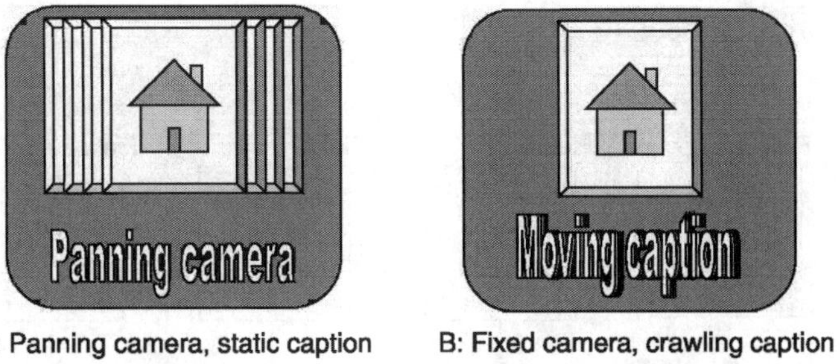

Figure 3.6.15 Limitations of the four-field interpolation process.

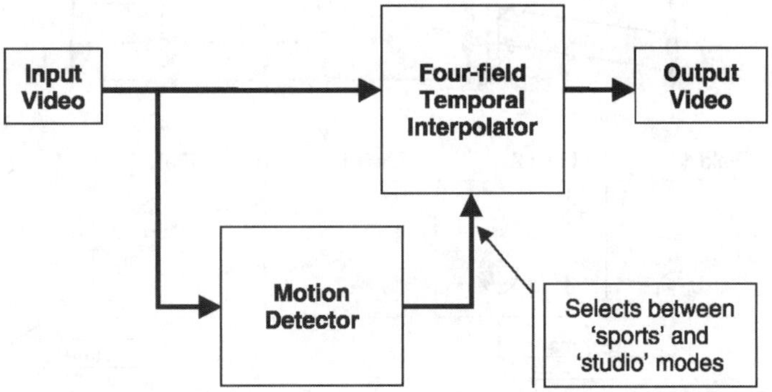

Figure 3.6.16 Motion-adaptive four-field interpolator.

Motion-adaptive standards converters do produce marginally improved results over the linear filters; however, they still suffer from some judder on moving objects and compromise the resolution of the moving images.

3.6.4.3.4 Temporal conversion by motion-compensated interpolation

All the schemes outlined so far suffer from poor motion portrayal; however, they all produce high-quality results if there is no motion. The goal of the motion-compensated converter is to compensate for motion such that the images can be filtered as if static. The result is high-resolution, judder-free images. To achieve this goal requires very sophisticated motion estimation.

Figure 3.6.17 outlines the process of converting between 50 Hz and 60 Hz standards. The top line shows a sequence of images with a moving object. The goal is to convert this to a 60 Hz grid, but maintain the correct motion profile. The second line is a cross-section through the sequence of images, the bold lines corresponding to the 50 Hz samples and the dashed lines to the required 60 Hz sampling. The arrows projecting towards the 60 Hz samples represent the motion vectors. The process of motion-compensated standards conversion projects picture elements by the appropriately scaled motion vector. Picture elements are projected from both 50 Hz fields either side of the new 60 Hz field. It is necessary to use information from either side as elements of the picture will be 'revealed' as objects move and similarly elements of the picture will be 'obscured'. Figure 3.6.18 outlines a system-level diagram of a motion-compensated temporal interpolator. The system consists of two channels, backwards and forwards; this enables the output field to be made up of contributions from both the input field before and after the required output field. Once the forward and backward input fields have been shifted, it's then the job of the forward backward selector to determine which channel to use to make the output field. The decision process is driven by control signals generated in the motion estimator.

3.6.5 Signal Processing

3.6.5.1 Interpolation

In Section 3.6.4 the requirement to change the number of lines and to change the number of pixels when performing standards conversion was introduced. The concepts behind interpolation will be described by two examples, first by considering sample doubling, then deriving a fully variable interpolator.

3.6.5.1.1 Sample doubling

A simple form of interpolator is one that exactly doubles the number of samples. For example, such an interpolator may form the basis of a line doubling display. In line doubling half the samples are identical to the input samples and only the intermediate values need to be computed. Figure 3.6.19 illustrates the problem.

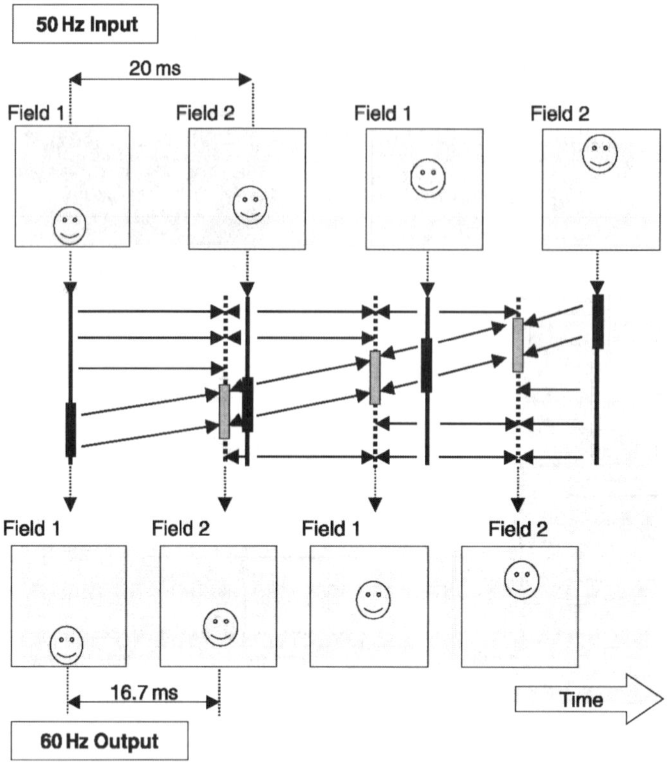

Figure 3.6.17 Motion-compensated temporal interpolation between 50 Hz and 60 Hz.

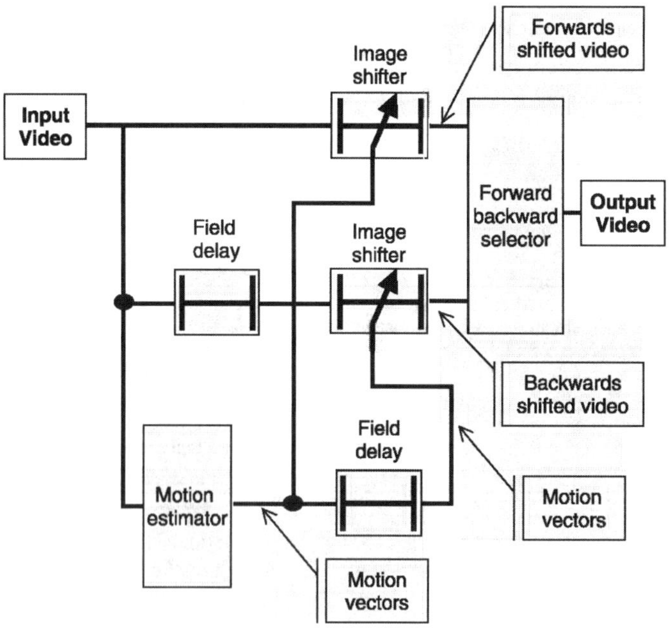

Figure 3.6.18 Motion-compensated temporal interpolator functional block diagram.

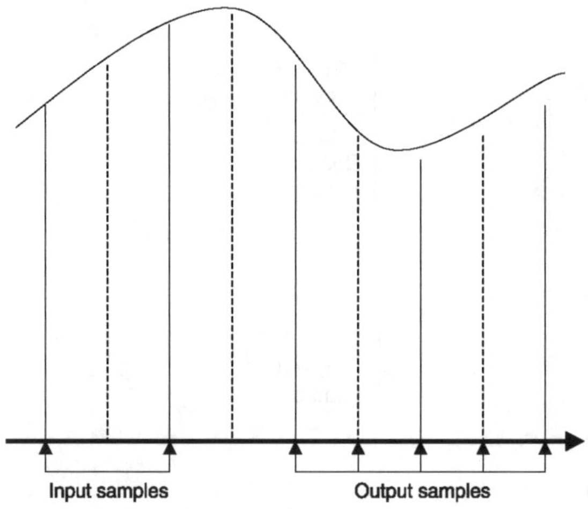

Figure 3.6.19 Sample doubling.

An intuitive approach to solving this problem would be to make each of the new output samples from 0.5 of the input sample before and 0.5 of the input sample after. This is illustrated in Figure 3.6.20. This simple approach performs interpolation; however, the drawback is the frequency response of such a filter. Also included in Figure 3.6.20 is the frequency response of a two-tap filter alongside the ideal response. In practice a filter with this response will result in softening of the image, which would be unacceptable for broadcast applications. Also included in Figure 3.6.20 are the filter coefficients plotted against time, a format that will be used in the next example.

A filter that has a response close to the ideal response requires contributions from many input samples. For explanatory purposes consider a filter with contributions from seven input samples. The design of such a filter is based on frequency domain techniques using Fourier transform, the details of which are beyond the scope of this chapter. Figure 3.6.21 is a block diagram of a seven-tap filter and its frequency response plotted against the ideal frequency response. It's clear that this is much closer to an ideal filter response than was achieved with the two-tap filter. Also included in Figure 3.6.21 is the impulse response of the filter; this provides the fixed values (coefficients) that are multiplied by the input data.

The seven-point filter shown in Figure 3.6.21 now has close to the correct frequency response and will not unduly soften images, but it does not provide interpolation to the 0.5 sample point. The next exercise is to derive a set of filter coefficients that have the same frequency response as the original seven-tap filter but now provide interpolation to the 0.5 sample position. Figure 3.6.22 shows how to achieve this. The original plot of the filter coefficients against time is known as the filter's impulse response. The impulse response is then over-sampled by four times to provide additional coefficient positions. The over-sampling process is achieved in the frequency domain using Fourier transform. Figure 3.6.22 shows two example sets of coefficients, one producing a shift of 0.5 pixels, the other a shift of 0.25.

To summarise the need for multi-tap filters, Figure 3.6.23 has been included. This compares the effect of using a simple two-tap filter to using a seven-tap filter when doubling the size of an image. The resolution loss is clearly seen with the two-tap filter.

The next example, which is close to the practical application of changing the number of lines or number of pixels in a TV system, describes variable interpolation. For example when changing from 525 lines to 625 lines, for every input line there will be 1.19 output lines. To achieve this will require continuously changing the shift value in the interpolator. Figure 3.6.24 illustrates an example where one input sample is equivalent to 1.33 output samples. The process works by measuring the phase difference between the input and output samples; this phase value is then used to look up the appropriate set of coefficients for the correct shift. Once the correct coefficients have been selected these are then used to filter the data.

3.6.5.2 Motion estimation

Motion estimation is essential for the temporal conversion process and beneficial to the de-interlacing process. Temporal conversion requires very high quality motion estimation, unlike other application such as MPEG compression systems. The reason temporal conversion is so challenging is that the system is open loop, there is no error signal. In comparison with MPEG there is an error signal, so in the event of the motion vectors being inaccurate, the error will help minimise any artefacts.

There are broadly three types of motion estimation:

1. Block matching.
2. Gradient method.
3. Phase correlation.

Each of these techniques will be described in the following sections.

3.6.5.2.1 Block matching

Figure 3.6.25 shows a block of pixels in an image. The block is compared with the same block on the next image. If there is no motion between fields, there will be high correlation between the pixel values. However, in the case of motion, the same or similar block of the image will be elsewhere and it will be necessary to search for them by moving a search block to all possible locations in the search area. The location that gives the best correlation is assumed to be the new location of a moving object.

Whilst simple in concept, to perform a full search over the image for each pixel is extremely intensive, so for practical systems a hierarchical system is usually adopted, where large blocks are used to find the large motion component, then smaller blocks are used to refine these values. The problem with hierarchical schemes is that they are prone to missing small fast-moving objects, as small objects may be smaller than the initial block size. Examples of fast small objects would be balls in sports events and small captions.

Block matching systems have not proved to be up to the challenges of high-end standards conversion, although they are adequate for MPEG compression systems.

3.6.5.2.2 Gradient method

Within an image, the function of brightness with respect to distance across the screen will have a particular slope, known as a spatial luminance gradient. It is possible to estimate the motion between images by locating a similar gradient in the

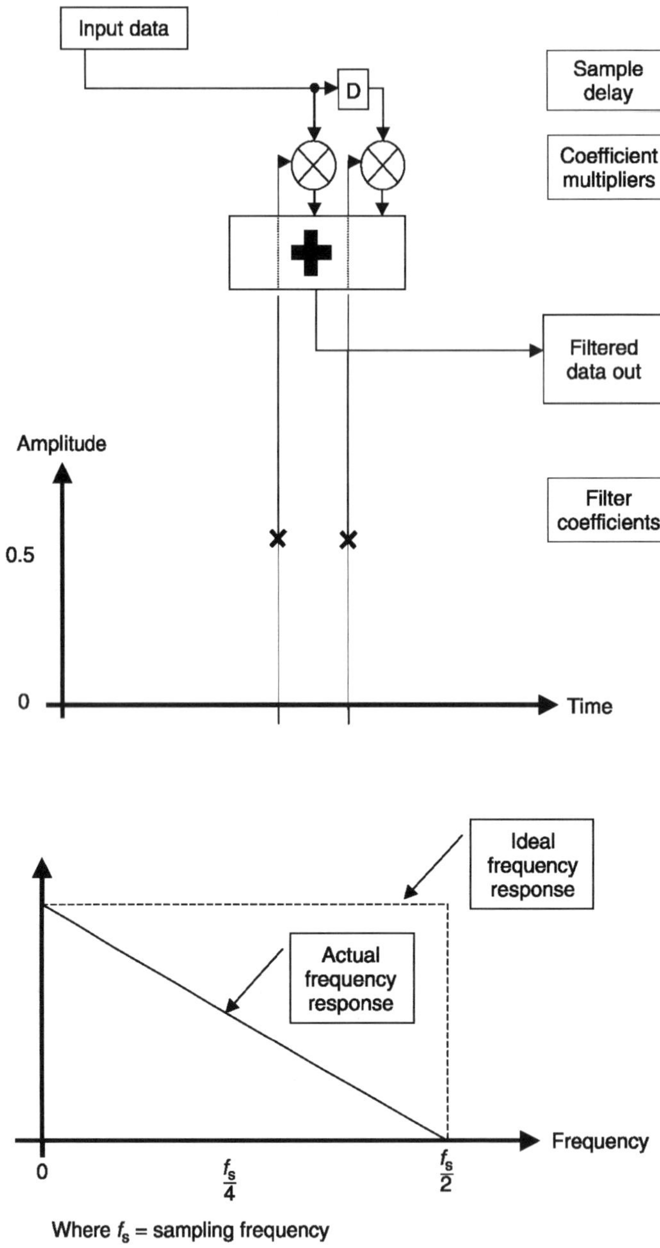

Figure 3.6.20 Simple two-point interpolation filter and its frequency response.

next image. Alternatively, the gradient may be compared with the temporal gradient, as shown in Figure 3.6.26. If the associated picture area is moving, the slope will traverse a fixed point on the scene and the result will be that the brightness now changes with respect to time. For a given spatial gradient, the temporal gradient becomes steeper as the motion speed increases. Thus, motion speed can be estimated from the ratio of the spatial and temporal gradients.

In practice there are a number of limitations. A similar spatial gradient may be found in the next image, which is due to a different object and false match results. When an object moves in a direction parallel to its side, there is no temporal gradient. Variations in illumination, such as when an object moves into the shade, also cause difficulties. The greatest problem of the temporal gradient method occurs in the presence of temporal aliases. If temporal aliasing due to moving objects

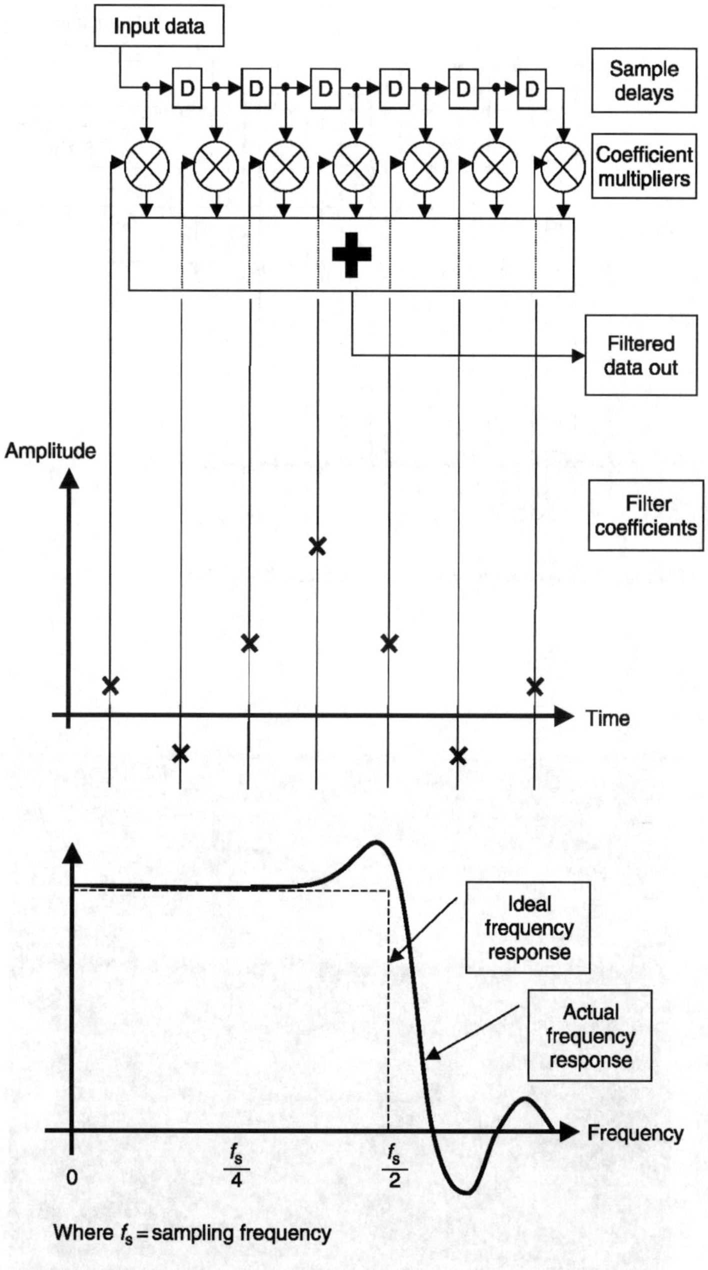

Figure 3.6.21 Seven-point interpolation filter and its frequency response.

were absent, gradient measurement would be easy, but motion compensation would not be necessary. Stated differently, the reason why motion compensation is necessary is the same reason that temporal gradient measurement is inaccurate. The accuracy of both gradient methods can be improved by using recursion. In this case the original estimate of the velocity is used as a basis for successive calculations. Recursion is essential with highly detailed images; the motion range that can be handled between fields becomes very small and it becomes essential to use prediction from previous fields. Cuts will cause problems for recursive systems where it can take several fields for the vectors to recover any level of accuracy.

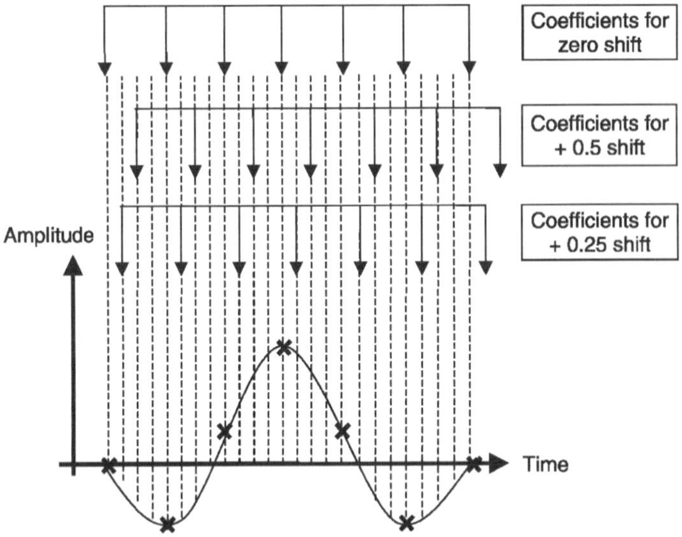

Figure 3.6.22 Over-sampled filter impulse response, providing coefficients for different shift values.

(a)

Figure 3.6.23 Comparison of the resolutions achieved using two- and seven-tap filters.

(b)

(c)

Figure 3.6.23 *continued*

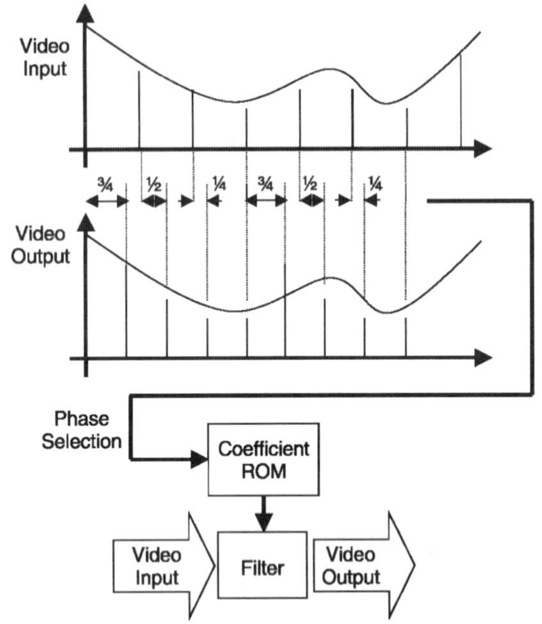

Figure 3.6.24 Sample rate conversion by variable interpolation.

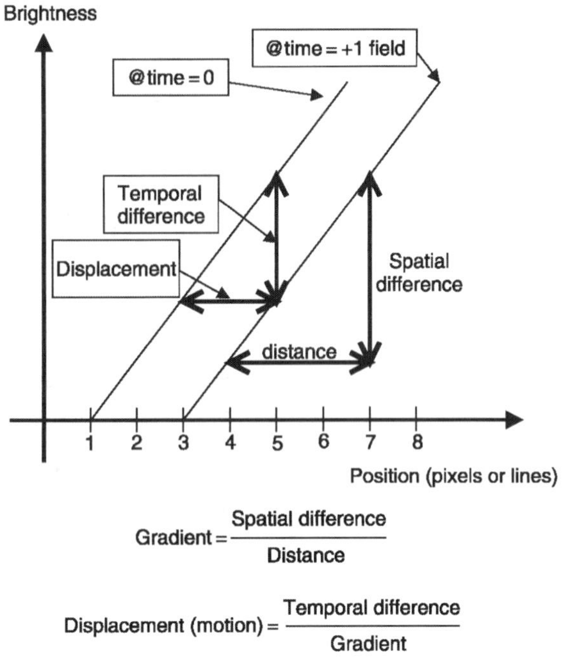

Figure 3.6.26 Gradient method.

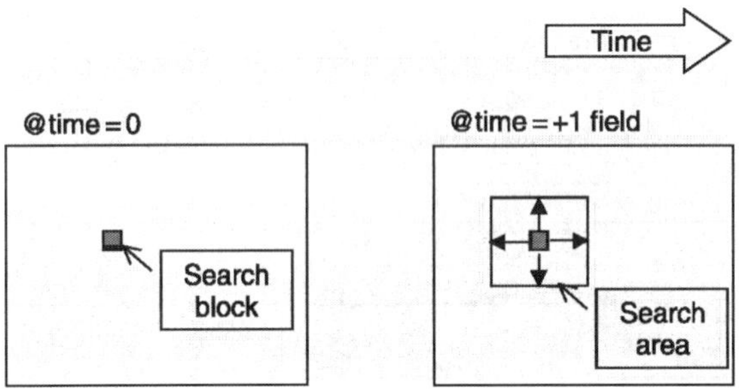

Figure 3.6.25 Block matching motion estimation.

3.6.5.2.3 Phase correlation

The shift theory is the basis behind phase correlation. The shift theory states that a displacement in the time domain equates to a phase shift in the frequency domain. Phase correlation uses phase difference information to determine motion speed. In the application of motion estimation, displacement = motion. The technique to analyse signals in the frequency domain is via the Fourier transform.

Figure 3.6.27 shows two sine waves, one displaced from the other. The displacement or motion can be described as 20 pixels or 45 degrees.

Television signals are somewhat more complex than a single sine wave, but by applying Fourier transforms to an image it's

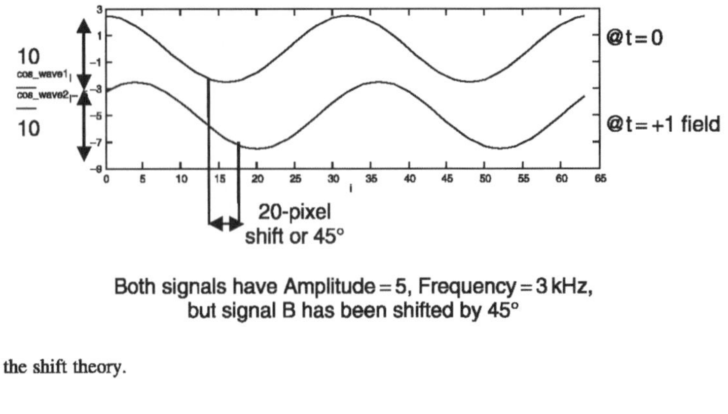

Figure 3.6.27 Example of the shift theory.

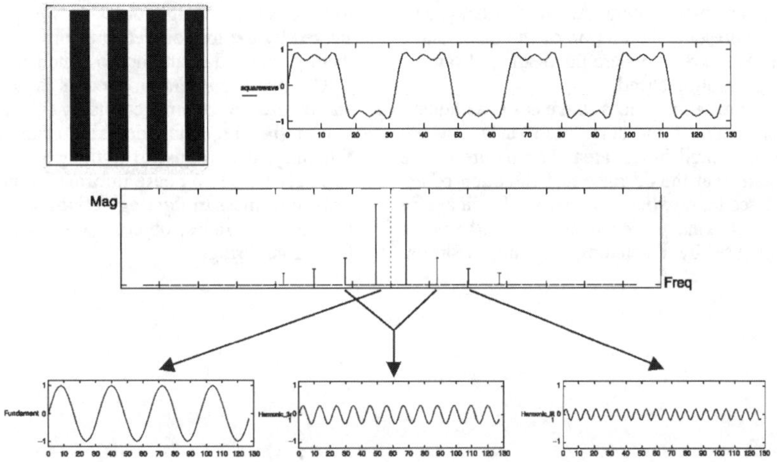

Figure 3.6.28 The Fourier transform of a square wave pattern.

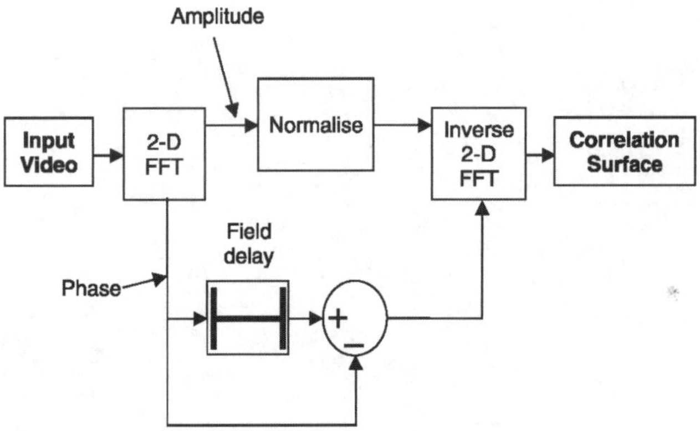

Figure 3.6.29 Phase correlator block diagram.

possible to break the image down into a series of sine waves. Figure 3.6.28 shows how a square wave pattern can be broken down into a series of sine waves via a Fourier transform. The phase of each sine wave component can be determined, then phase correlation can be applied.

3.6.5.2.3.1 The phase correlation system

The phase correlation system is illustrated in Figure 3.6.29. Phase correlation works by performing a spectral analysis on two successive fields, then subtracting all of the individual phases of the spectral components. The phase differences are then subject to a reverse transform, which directly reveals peaks whose positions correspond to motion between the fields. In Figure 3.6.29 the transform blocks have been labelled FFT, which stands for fast Fourier transform. The FFT is a computation-efficient method of implementing the Fourier transform. Figure 3.6.30 shows an example of the phase correlation process, including the resulting output correlation surface.

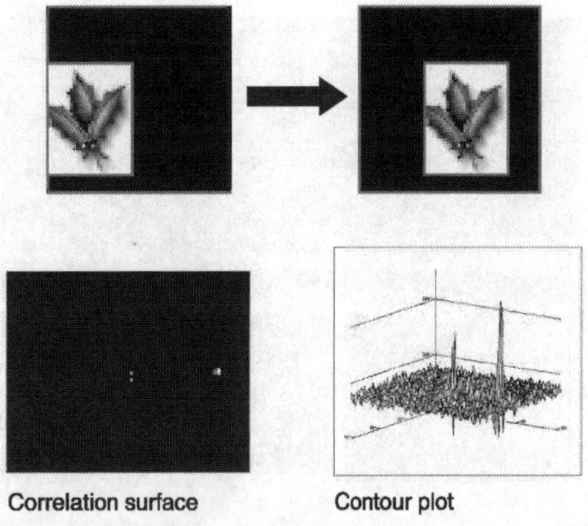

Figure 3.6.30 Example of phase correlation.

On the final correlation surface there are two distinct peaks, one at the centre that represents the motion of the black static background, the other to the right, where the distance from the centre to the peak is the motion speed.

Following the phase correlation process there is a peak hunter; this finds the highest points on the correlation surface which represent the motion measured in the area. The nature of the transform domain means that the distance and direction of the motion are measured accurately, but the area of the image in which it took place is not. Thus, in practical systems the phase correlation stage is followed by a matching stage not dissimilar to the block matching process. However, the matching process is steered by the motion vectors from the phase correlator and so there is no need to attempt to match every possible motion.

The phase correlation process is very accurate and robust, and unaffected by image noise, as the processing is all carried out in the frequency domain using the phase components. Similarly, it is robust to lighting variations as the processing is purely based on phase information not the brightness levels. This robustness to lighting variations ensures high-quality performance on fades, objects moving in and out of shade, and flash-guns firing.

Section 4 Studio and Production Systems

Peter Weitzel BSc (Eng) CEng MIEE MSMPTE
Manager, Broadcast Technology Developments
BBC Technology

4.1 Television Studio Centres

Over the last 20 years or so, many things have changed in television – more channels giving more output; digital technology providing lighter and cheaper cameras and VT; and more IT-based production, meaning that a smaller proportion of TV is made in studios. This has changed the landscape of the studio centre beyond recognition and provides a range of new ways of making TV programmes.

This chapter solely looks to provide signposts as to how this can be done – digital always seems to provide more questions than answers. Today's programme making is far more 'joined up' so a holistic approach is required, to look beyond the boundaries of technology and to consider the wider business issues. TV technology today can address many business requirements at an affordable price.

This chapter is split into three parts – scope setting; a more detailed look at the requirements for each part of the production chain; and finally some common information, applicable to any facility.

4.1.1 Scope

This chapter covers the various types of studios and looks at the special needs of news and sport. Following on, it touches on post-production, infrastructure/connectivity issues, playout and finally archiving. What is doesn't cover is the complex topic of coding and multiplexing digital services.

To close the chapter some of the current and future issues – such as High Definition (HD) – are addressed, ending with some thoughts on 'being digital'.

The business of broadcasting is handled by so many different organisations that there is not a single view. Successful broadcasters are those who have got the most fit-for-purpose solution with the flexibility to grow or change as the need develops. Thus, implementations are not directly transferable, as they used to be in the days of analogue television.

Production is much more joined up, and needs metadata to function and give the extra richness that is now being demanded on television and other platforms. It is a good idea to have someone with 'end-to-end' knowledge to look at the whole process from idea to the viewer – or at least for a very large part of this.

Best practice starts with the collection of requirements and their incorporation within an overall vision, providing some validation of both the project requirements and the overall vision. Above all, this will ensure the integration of all projects across the business, encompassing programme making, resources, operational procedures and technologies. From this, the long-term architecture and the principles of operation can be defined; more specific requirements and proposed implementations can also be validated.

4.1.2 Studios

The studio today is not the only place where television is made; more lightweight cameras and cheaper post-production mean there are alternative ways to make programmes. With this the greatest change is the demise of a general purpose studio and the growth of specialist studios – not just for news (which always tended to need its own facilities) but also sport, soap operas, game shows, etc.

So what are the drivers for studios?

- Need for an available and controllable space.
- Connectivity.
- Proximity to programme-making facilities.
- Programmes made live or 'as live' – audience shows.

Traditionally the TV studio has had five architectural spaces:

- The studio itself.
- A control suite containing:
  - Sound.

- Production.
- Lighting/vision control rooms.
- An apparatus area.

Although in a particular studio implementation the rooms may be merged, it is helpful to consider each one in turn.

4.1.2.1 Studio

This area needs to have enough space – horizontally for access and camera shots, and vertically to have the lighting in a suitable place. It should be acoustically separated from the outside world – an appropriate curve – e.g. NC20, and have acoustic treatment to reduce the reverberation. Technically it should have wall (and grid) boxes where equipment can be plugged up and connection for other services, such as mains power and computer networks, can be made. The location of these should take into account any cyclorama and typical scenery placing, rather than architectural elegance.

It should have a lighting grid which should be strong enough to take tie-off points to secure scenery and slung equipment. Whether the lighting luminaries are permanently assigned to the studio or brought in for the production depends on use/commercial considerations, but it is helpful to have some permanent luminaries – a mix of soft and spot (or dual source).

4.1.2.2 Control suite

The three control rooms have different uses and although these brief comments are based on a traditional layout, the basic functionality remains the same.

4.1.2.2.1 Sound control room

Sound is often overlooked in television and so it does not always get the standard of quality accommodation and equipment it deserves. The Sound Control Room (SCR) needs to be an island of quiet, with minimal through traffic. It needs to be acoustically treated to stop noise getting in or out and should be acoustically neutral. The sound desk should be positioned putting the operator at the optimum acoustic position with respect to its loudspeakers. As well as the sound desk there will be outboard equipment and recorders (what used to be known as 'tapes and grams'), which may need a second operator. The sound team also tend to look after communications, so any audio patching facilities should be included here.

4.1.2.2.2 Production control room

This is the heart of the operation and it also tends to have the largest group of 'hangers on' who need to be accommodated. One wall of the Production Control Room (PCR) will be a monitor stack – with most (but not all) of the sources displayed. It would be normal for all these monitors to be colour and to display the correct aspect ratio pictures.

There are likely to be four roles conducted at the main desk:

- the production assistant, who assists the director, counts and call shots;
- the director, who is responsible for what is seen shot by shot;
- the vision mixer, who is directed to select sources and to mix/wipe/cut between them;
- the senior technical person, who looks after all technical issues affecting the production.

All need good eye contact with each other and all the control surfaces should be conveniently to hand.

Sitting behind (or elsewhere a bit out of the way) are any editorial staff that may need to see and communicate into the production process – they may need a good view and some previewing and talkback systems. The 'hangers on' who love to see the programme being made should be located well out of the way – it may be helpful to give them a studio out monitor so that they are drawn to this rather than interfering with the production team.

The architecture of this room needs to be sufficiently muted to enable the director to choose the right shots and be aware of the technical quality, but it also needs to be vibrant to encourage the creative process.

There needs to be good access to the production control room both to the studio and also to the outside world.

4.1.2.2.3 Lighting and vision control room

This should be a haven of technical calm and have controlled lighting so that the technical quality of the individual sources and particularly the cameras can be assessed. For this reason the Lighting and Vision Control Room (LVCR) monitor stack should contain monochrome monitors for each of the cameras and good quality colour monitors to check the overall rendition.

Some waveform monitors and a close viewing colour monitor switchable to each of the camera outputs may be necessary to check and set up the standard exposure or to ease camera matching. The lighting operator and the vision operator (controlling the cameras) should sit with a good common view of the monitor stack.There should be good access to the studio floor and also to the apparatus area.

4.1.2.3 The apparatus area

This contains all the equipment required to operate the studio. Much of the equipment will be fan blown and thus noisy, and hot air will need to be removed efficiently. It is always better to provide heat extraction or cooling close to where the heat is being generated. Equipment from different manufacturers often blows in opposing directions, which can generate local hot spots if it is not properly managed. At a technical level local monitoring of all the signals should be provided so that fault-finding can continue whilst the studio is in use.

Circuits from outside the studio will be terminated in this room. As recording equipment has become smaller and cheaper, it is now normal to have these within the apparatus room, thus removing any need to have a studio connected to anywhere – unless it is working live to somewhere else.

4.1.3 Types of Studio

As the style of system has changed it becomes more difficult to generalise about the ways it is used and thus the facilities required.

4.1.3.1 Traditional General Purpose Studios

These used to be the core of any studio centre, and covered a range of sizes which could cope with more or less any

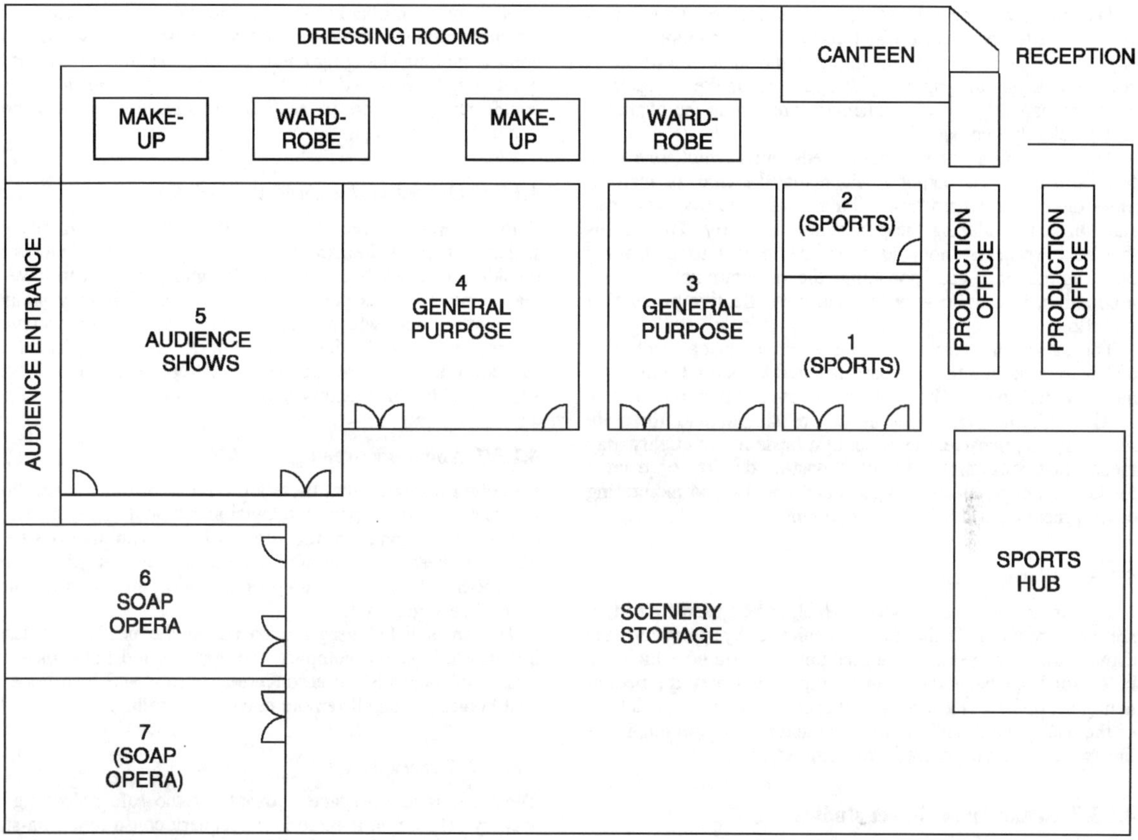

Figure 4.1.1 A Typical Large Broadcaster's Studio Area.

production requirements. They now cover a range of light entertainment, often with an audience, or magazine programmes. They are generally the most traditional in layout and facilities.

These are normally purpose-built structures, duly solid for sound isolation and specifically set up with a great deal of flexibility in how and where scenery, lights, cameras and sound equipment can be used.

The studio volume would have ample scenery access, permanent (foldaway) audience seating if this type of show was required, and be of spacious dimensions. Acoustic treatment would be one every wall and perhaps even the roof.

The control suite would be of the traditional form, with the three control rooms usually adjacent along one wall. They could be at studio floor level, but could also be at first floor level, allowing ancillary areas such as wardrobe, make-up and dressing rooms or audience access to the studio to be at ground floor level. If this is the case there will need to be a way into the studio and stairs down to the studio floor.

4.1.3.1.1 Typical Floor areas

These are categorised as: large – in excess of 900 m^2 with audience size of 300; medium and the most general purpose, about 750 m^2 with audience size of 220; smaller, about 250 m^2 and audience size of about 150. See Figure 4.1.1.

4.1.3.2 Four wallers

These are very similar to a film stage, usually with fixed lighting, which can be used by an outside broadcast as a 'drive in'. They come in the same range of sizes as traditional general purpose studios. Because there are some wired facilities, such as power, lighting systems and technical cabling, these can be very attractive for some types of production, and can give a broadcast centre a studio environment without the expense of a studio sound and vision system.

It is very useful to have some technical spaces available so that some partial de-rig could be done or a simpler system set up from 'Flyway' equipment rather than OB trucks. Ancillary areas for the performers may be needed, but again much of this, for instance catering, could be driven in.

4.1.3.3 Soap operas (continuing drama)

A full studio is likely to be overequipped for long running, fairly intensive productions like soap operas, which require a permanent set, but use single (or two)-camera production.

The studio volume itself needs to be of solid construction to cut down extraneous noise which could make for more retakes. A good design of the facilities in the studio such as wall boxes communications and lighting facilities will make it easier to work intensively. Likewise, ancillary areas must be available and efficiently laid out.

The production control room needs only switch one of a few cameras to the output – which may be tape, or increasingly directly to a hard disk editing system, so that work can start on post-producing the programme instantly. This means that each scene is shot and compiled in real time, leaving only assembling and polishing the programme in post-production, leading to very fast and cost-effective production workflow.

Thus, a vision mixer may not be required – just a 'cut box' – and it may be helpful to record isolated cameras (with alternative audio) to cover any gaps discovered in post-production.

Although the equipment needs are simple and therefore relatively inexpensive, the effect of a lapse in the quality may mean time-consuming additional post-production or even a retake on a different day, so good audio and video monitoring environments and staffing are important.

4.1.3.4 Virtual reality

It is often useful to have a space which can be used for virtual reality production. This can be achieved by providing the camera tracking systems in a medium-sized studio which can be set for VR when required or having a smaller studio permanently set up. VR is a very skilled operation and the modelling of the backgrounds and the way that actors or presenters interface with the environment is very important.

4.1.3.5 Remote news inject studios

Many organisations have small television studios where an interviewee can sit for a 'piece down the line'. Banks and companies often make their staff available in this way, as well as broadcasters needing a point of presence in a significant locality. Most of these are in standard office buildings and care must be taken to reduce sound transfer into the studio. If the rooms are small, action is needed to reduce the boominess of the echo – the bathroom effect. Thus, walls should have good acoustic treatment, and scenery or backings need to be of soft acoustically absorbent or transparent materials. If the system is set up in an open office such as a bank dealing floor, then separation and screening to reduce the noise spill will be needed. Any hard surfaces such as glass should be positioned so as not to focus the sound – which just gives the bathroom effect all over again.

There will be typically a single camera with some form of remote control, so either the far end can frame the shot or more likely the interviewee will be asked to adjust the shot framing by looking in a preview monitor.

Sound equipment comprises one or two tie-clip microphones and a simple mixer – with some form of automatic level taker to set the microphone gains and a changeover mechanism. There will be an ear piece or loudspeaker for cue programme and talkback.

Lighting tends to be likewise simple, and can be cool fluorescent or more conventional spotlights. These would normally only be used during the actual interview, as most room air-conditioning will not be able to cope with the continual load of equipment, person and lights for long periods. Thus, for short uses there is probably no need to take special care over air-conditioning, but for a more heavily used facility additional cooling and air changing could be very useful. It need not be quiet if it is switched off during takes – i.e. lights on, air-conditioning off. See the section on external circuits for how this can be connected.

4.1.3.6 Offices and around the building

Some programmes feel that their office makes a good space in which to present their programme. This could be done as an outside broadcast, but it is often far more efficient to have at least some facilities permanently installed. This may include the lighting sets which are a problem to rig, and technical cabling and power hidden in the floor or ceiling void for safety. In addition, some strategic acoustic treatment can make the sound quality considerably better.

4.1.3.7 Ancillary areas

It is often forgotten that there are a range of other spaces that are required to support a television studio beside the usual office accommodation for the staff that run the business. Most of these areas require technical facilities which may be associated with the studio or provided on a more central basis. See Figure 4.1.1.

The areas listed here are based on the needs of a large broadcaster's studio complex, but apply to most studios – but 'make-up' may just be a convenient mirror and 'wardrobe' a coat hook for a small remote news inject studio.

4.1.3.7.1 Scenery storage

There needs to be space allowed for the safe unloading of scenery off lorries, if there is no scenery construction on-site. This area can also be used for large props – like cars and buses. The scenery will need to be stored undercover either to await transport to the studio or to do minor repairs or adaptations. It is usual to have a scenery runway as a partially enclosed area outside the studio doors. This area may also have technical facilities for an OB unit to connect to, or for noisy equipment such as air compressors.

4.1.3.7.2 Make-up

An area very close to the studio floor with suitable lighting and general fitting out may be needed. The make-up artist may wish to be in touch with the production and thus a TV monitor of good quality for studio output should be provided. Good communications to the production control room would enable the director to find the actors or presenters. The make-up artist, may need to view in quality conditions, and they will often use the lighting control room.

4.1.3.7.3 Wardrobe

Although very rapid changes of apparel are not now usually required, like make-up, a wardrobe area is usually required close to the studio floor and the dressing rooms. As this may also be used to store costumes there should be suitable handing and good access for wheeled clothes rails. The wardrobe supervisor, like the make-up artist, will need to be in touch with the studio production activity, so a TV monitor with studio out and communication to the studio will be required.

4.1.3.7.4 *Dressing rooms*

These can range in style from a suite of rooms for a pop star to a large sparse room with just clothes rails, benches and mirrors. They need to be secure, with easy access to the studio itself and the ancillary areas, make-up, wardrobe and other key facilities like showers and toilets.

4.1.3.7.5 *Production office*

If the production team do not have an office on the site (or even if they do), they need some place to meet, leave their belongings and keep in contact with others – so a large comfortable space with some desks and lots of telephones is almost essential. This could double as dressing room space or even a conference room.

4.1.3.7.6 *Reception areas*

Artists and contributors will need welcoming; this could be done by the main receptionists, but a separate entrance or a main coordination point for the backstage may be needed. This – much like a stage door in the theatre – can become a focal point for all working in and around the studios, thus it needs to have details of everything that is going on in the studio areas.

If there are audience-based shows, then the audience will need a security protected entrance, then welcoming, cloakroom and perhaps access to a shop or food bar. It is a good opportunity to convey information to the audience, so TV sets showing suitable material will help them enjoy the programme.

The whole site reception is the first contact that visitors have of the organisation. Thus, it is a showcase of what goes on and needs a very capable reception team. All reception areas can benefit from having large screens with moving pictures on them – which could be studio output, promotional activity or just off-air TV.

4.1.4 News Centre

News is at the heart of many a broadcaster's output. Because of the different editorial and technical processes of news, it often has its own areas, resources and culture. There have been many changes in the way in which television news is compiled.

At the heart of the news operation is the editorial system, from which all the journalists create their scripts. News is an instantaneous medium and so there is a great need for all processes to be linked up. The overall workflow needs to be defined and reviewed, as different styles of news and changes in technology can affect the detail of the production process.

Books can be written on how newsrooms should work – but a few brief comments about the technical system that supports the editorial (newsroom system) process may give a sufficient outline.

4.1.4.1 Intake live feeds (streams)

The live feeds are recorded and if required passed to a studio for live transmission. If metadata is entered now, then the recording and its content will be available to anyone in the future. Large organisations with ample infrastructures may have more complex ways of handling live feeds, but communication with the remote journalist or contributor is very important. The use of field editing has meant that edited packages can be streamed in, but this is starting to decline as file-based delivery increases.

4.1.4.2 Intake tape/file

As the tape or file is ingested into the system (which could be tape based) the metadata should be captured and entered – usually into the main newsroom system. Increasingly, material is being edited in the field and sent as files over the Internet or internal IP networks. Thus, some of the ingest may be done almost at acquisition; here the metadata should be entered in, leaving the ingest operator to check it and add the system details as to where it is stored.

4.1.4.3 Editing

If there is a large newsroom system it is often the practice for journalists to do rough or simple editing using a proxy copy of the material, which is then conformed with the real material on the server. More difficult items will need craft editing, with a specialist editor being directed by the journalist in suites just off the main newsroom floor. Also, it is normal for the newsroom system to compile the captions and other graphics that will be used in the studio and in the editing process.

4.1.4.4 Studio/transmission

The studio volume is likely to be permanently set, but may contain a number of sets with a common branding to give a different look to the breakfast, lunchtime, early evening and late night weekday or weekend news programmes. Often, back projection screens are used to give a Window on the World (WOW), which can be an external shot or a camera looking at the newsroom or even a computer-generated image. Although many news operations use robotic camera systems, either with just pan or tilt heads or with fully moving pedestals, this may not be money well spent if only simple shots are required or that there are staff available.

Often, for a new operation there is more than one studio and more than one control suite. So the system design should allow any control suite to work with any studio with both flexibility and resilience.

The news output is a mix between live studio and a playout operation, and should be well integrated into the newsroom editorial systems. Thus, for every item in the running order, which is stored on the newsroom systems, the assets such as tape/file, captions and script are all available to the director. As it is a live operation, the triggering of an event will be done by a person, but the stacking up and management of individual items should be done by the one system so that everyone can get a common view.

The control room may be a single room to give the eye contact that is needed, but it is important to try to give the sound area some isolation. The news bulletin is an intensive time and thus even more care should be given to the ergonomics of the control room layout. It is also crucial to analyse the workflow and the roles that the various operators and journalists can take.

4.1.4.5 Continuous news

Continuous news is broadcast from what is frequently a corner of the newsroom; this gives an immediacy of look but can give problems with lighting – now partly overcome by 'cool' fluorescent lights, ambient noise which can affect sound quality, and robotic cameras which give safety issues. From a production point of view there is little or no time for rehearsals. The control rooms will be inhabited 24/7 and thus should be perhaps more spacious but fenced off from the rest of the newsroom so that there is not a continuous flow of staff visitors, etc.

Staffing levels are likely to be lower than for a news bulletin, and editorial decisions may be made on the fly in the studio control rooms. From a technical point of view, it is often very useful to have a means of bypassing most of the system – so that rehearsals can be done, or maintenance carried out, while a long form programme is being played out.

Continuous news has an insatiable demand for content and thus there will a large number of live feeds coming in to the broadcast centre, all needing communication and data feeds.

4.1.4.6 Archiving and reuse

News will store all (or most of) its material and thus some form of archiving policy needs to be installed. It is obvious that most of the collected information needs to be stored on a running story and there would need to be a lot of information about people in the public eye, politicians, heads of state, etc. Many organisations review and move to deeper archive material every month or so and have a strategy for acquiring material for the archive from future events.

News has a number of outlets; beside radio and television there is also the World Wide Web – news sites are a key form of information. The website can include both still pictures grabbed from the material held in the archive and also edited packages of new reports. With some cable VOD services, both news bulletins and stories can be accessed by the viewer. Third generation phone companies are starting to provide video news on demand to the phone.

4.1.5 Sports Information Hub

Sport programming attracts mass audiences and can be one of the most competitive areas in the programme schedule. TV stations will pay a lot of money for the sports rights and thus expect that the action itself is well packaged. Sport programming is fundamentally an event which happens away from the studio centre. Great swaths of the schedule will be sport, and this varies from major all-day events like cricket or all-afternoon events like football cup finals to a mix of outside broadcast like racing, mixed with other events or insert programme items like a football preview. The studio will also do interviews, comment and discussion, and present the 'results sequence'.

In order to support multi-channel sport output, a concept of a 'sports information hub' has been developed by a number of broadcasters. This links video recording, editing, graphics and the studio in a tight-knit system with communication to all the OBs and each other, with access to a wide number of information sources. It has similarities with the news operation but is usually seen to be (and is) very different. To highlight two of them: sport produces a large amount of text/tabular information – which is used to create captions – and the video editing can be used to provide quick half time summary packages, as well as longer form edited matches and highlights.

The sports information hub may just be brought together for the sports events, or may be used for other programmes when not in use for sport. But if the output goes up, both in hours per week and how many channels are being handled at once, permanently allocated systems will be more efficient.

Although sport is almost entirely an outside broadcast activity, the studio is a visible part of the output, anchoring the programme. There may be a number of studio spaces and control suites accessing the total technical and editorial resources of the sports information hub.

The studio may only have three cameras and may be small – yet with scenery and perhaps with the use of virtual reality techniques it can appear bigger. Because of the resources of the sports hub, the control rooms can be compact and often there is only a single control room containing production, (simple) lighting, etc. and (isolated) sound. Like continuous news there should be a way of bypassing much of the studio equipment so that rehearsals or the recording of inserts can take place whilst the main outside broadcast is being transmitted directly from the OB.

Like news, the outlets for sports material are very large and although rights may prevent the broadcaster from maximising the use of the video and audio of the event, material created by the broadcaster can be exploited. This can extend into the use of alternative features/views of the event on interactive television.

As the broadcaster has access to the video and audio it may be that they are in a very good position to aggregate and produce packages for the other rights holders – such as highlights packages for the World Wide Web and for third generation mobile phones. The mobile phone companies offer text alerts of goals being scored and then even video clips of the goal being scored.

4.1.6 Post-Production

Post-production has migrated from a tape-based to an almost entirely server-based, tapeless environment. This does not mean that tape has died; it is still a very effective medium for the recording at acquisition and transport for programme delivery, and probably the most efficient storage medium for (deep) archiving. Thus, the editing system has to work with tape input (Ingest) and perhaps output (Outgest).

Tape editing has almost died out, as computer-based editing – either Apple Mac or PC based – has come down in price. The three components of a tapeless editing environment are the edit tool, the (bulk) storage and the archive system. All should meld together to give the editor a seamless view of the material that is available to the edit, and conversely no access to anything else. This media management is commercially the most important feature of a shared system.

The edit tools perhaps lie in three broad bands: simple, craft and specialist.

The simple edit system will enable assembly editing with mixes and wipes. Simple captioning will be available as a plug-in. Sound will be handled in about four stereo channels. This is equivalent to a two (or three)-machine VT edit suite.

The craft edit system extends the range of mixes and wipes and will render most of these in real time. Picture manipulation, crop, DVE and ARC will be provided. Captioning and other plug-ins will be available. Audio is handled within the video edit application or for more complex work with an audio plug-in or tool. These systems can be 5–10 times the price of the simple system. Many areas have equipment well beyond both the needs of the client and often the creativity of the editor. It is not necessary to have a full craft editing tool if the entire requirement is to cut a 30-minute version of a 90-minute football match.

The specialist systems cover a wide range of systems which are the ultimate 'high-end' systems. These would include animation, frame-by-frame colour grading, multi-layer compositing, etc. This category can encompass single purpose devices like a 'weather computer' or animation (CGI) system.

Each system will have different capabilities and can cost about 5–10 times the price of the craft edit set. Specialist post-production is a very different business to the routine of cutting programmes. Each area is very specialised and it is likely specialist and independent advice would be required before moving into any of these areas.

Each system will have its own storage but major efficiencies come from having a shared storage system. The needs for this can vary, from wanting secure storage of the part finished edit until the next day, the ability to use a different editing sets/tool at different times or enabling multiple edits to be done on the same material. Obviously the storage required depends on the bit rate at which the material has been stored. Increasingly this is at the 'full SDI' rate of 270 Mbits per second for SD material, but typically 50 Mbit/s I frame is becoming the norm. If material has been acquired on tape – where DV compression at 25 Mbit/s is the usual standard, then to edit natively within a DV environment is probably the best. Many systems offer the option of working on material stored in a variety of formats.

With any storage system there should be adequate backup and repair if material gets corrupted. In past times good editing staff always kept a safety copy of the most valuable clips or partial work, usually because things did go wrong. With more reliable systems, although it is less likely to go wrong, if a drive is lost, it is still very important to be able to recover material.

This is why most drive arrays are RAID protected. The level of RAID is really mainly a commercial decision. Disk storage should be treated as a commodity and with improvements in disk drive technology, the difference between a good business system's drive and a broadcast-quality drive is narrowing. So often a high-end business system RAID array from a leading computer supplier will be as good as and considerably cheaper than the equivalent product from a broadcast editing supplier.

With vast volumes of storage it is key that there is a management system not only to index the material to enable it to be searched and found, but also to allow only those who should have access to do so. The metadata entered in any system is key, but in the post-production environment, where there are usually a few moments to spare in a pleasant environment, there should be no excuse not to enter the required information.

4.1.6.1 Building space

As many post-production operations can now be done on the office PC or a laptop, the needs for specialist rooms to do all the editing are changing. However, the editing environment does require some care, if only to make the workstation and its surroundings ergonomic and suitably screened from other activities so that sound spilling does not annoy people. Craft editing using the skills of the editor will always require a special room. If the client is involved in the edit, they appreciate being cosseted, and thus good soft furnishings and stylish look will make them more attracted to work with you. Post-production is a very competitive business so customer care is essential.

There is always the need to ingest the material; the idea that the tape has been ingested prior to the editing session starting probably forces a central ingest function within the facility. This can also be equipped with 'one of each' of the common tape formats so that any material can be handled. Also, by putting one area in charge of inputting material, the metadata requirement and business processes can be followed without impinging on the work of the editor.

4.1.7 Infrastructure and Central Technical Areas

The technical infrastructure is the more or less unseen connectivity which holds the whole of the broadcasting process together. As television is an instant medium, most connectivity is for streams, although file transfer is increasing.

Broadcasting relies on interoperability, which is gained by the use of standards by all the equipment in the chain. Many broadcasters have shunned anything other than open standards where the use of the IPR is available on fair and reasonable terms. The key standards are laid down by the SMPTE for video and the AES for audio, plus a number of international standards by non-broadcasting organisations like the ETSI and ITU. Increasingly the broadcaster is using standards which are used and set by others, for instance telcos or IT manufacturers.

The starting point standards are:

- Timecode – SMPTE 12M.
- Video – ITU-R BT.601-5 ('Rec. 601') or equivalent SMPTE 259M for 625/50 or 525/60 systems; SMPTE 292M for high-definition systems.
- Audio – AES3 (1995).
- DVB – ASI EN50083-9.

There is a full list of standards bodies later in this chapter; however, most specifications do not make easy reading.

Large manufacturers have in the past laid down their own proprietary standards, but unceasingly they have seen the benefits of being able to interwork and have cooperated in the standards-making process.

Infrastructure can be divided into internal – on the site – and external – usually provided by telephone companies/telcos.

Internally, the infrastructure is a mix of connectivity, distribution, switching, control, monitoring and provision of reference signals and synchronisation/timing. Although traditionally this meant that everything was routed through a central apparatus room, increasingly more distributed architectures are being used. This would connect between geographically separate nodes such as the satellite up/down link, the news area, the production studios, post-production and the central apparatus room, which is where telco connectivity exists, with each node then splitting its own connectivity, if required, within its area.

The planning of these nodes and the amount of connectivity of each type – video, audio, communications (non-programme sound audio), control/monitoring, IT traffic, file transfer – needs to be very carefully worked out.

When considering connectivity both resilience – such as dual routing – and basic architecture should be considered. Resilience usually increases connectivity; an architecture which passes the signal from one node to the next usually reduces it. However, having two ways to get to any node from any other can be very helpful, as it often means that there is a way from

A to D, albeit not direct but via B and C. Despite the emphasis on fibre-optics, copper pairs or coax have an equally important role and tend not to need complex and therefore expensive terminal equipment. However, the earth loop issues and grounding need to be considered.

It is always a good idea to run in more fibre and cable than will be used initially, with a mix of mono- and multi-mode fibre and coax and multi-pair cable. It is always easier to add more equipment than to dig up floors, ceilings and roadways to add an extra circuit. Thus, extra duct space should be allowed for. There should always be a way of disconnecting and over-plugging any circuits; in fibre-optic and video it tends to be the connector on the equipment or some similar connection point, for audio on multi-pair cables some form of interface like a jack field may be required, but a well laid out intercept frame/terminal block may suffice. IT circuits always have CAT 5/6 patching systems.

At each node there should be enough equipment permanently installed to enable quick checks to be done without having to bring any more equipment – this may be just a TV set and a loudspeaker, but a source of test signals' intercom/telephone and probably IT network connectivity would also make things easier. Simple means of looping signals back can be invaluable, enabling one person to test both ends of a link.

Moving from the basic connectivity, signals need to be distributed and/or switched. Again, there is a move in the overall architecture to distribute if the recipient needs the signal most of the time rather than to switch it. So, for instance, the output of the receiver at the down-link can be fed to the news node and the central apparatus room, rather than being fed from the central apparatus room, which then switches it to the news node. Within the news node it may be that it is distributed onto each vision/audio mixer rather than being switched.

Video switching systems (matrices) have grown in connections and reduced in size and price, so being able to switch a signal is probably no more expensive than not – and gives an easy way to monitor the signal. In the planning stages a rough allocation of sources and destinations on units of about eight ports will usually enable the size of the matrix to be determined. As matrices have fairly large granularity, it is easy to go to the next size up, which will give enough expansion space – if the planned size is just over the size needed for 80 sources, for instance, when matrices come in 64 or 128 inputs only. Many systems can be economic with a larger frame sub-equipped. Then if the system grows only the cards required at each stage need adding – but check if the whole frame has to be depowered before any changes have to be made. Audio is often carried as embedded audio in the SDV signal (SMPTE 272M). This has the advantage that it is difficult to lose the audio and cuts down the cost of a separate matrix, but can cause some operational problems, so it is normal for there to be some audio-only routing alongside the video routing.

Most equipment can be remote controlled, and thus can be installed at an optimum geographic site to meet the system needs. However, having a separate control surface – even if it is only a web page – for each piece of equipment soon gets confusing and thus is inefficient and prone to operator errors. Often, every variable can be changed when all that is required is to change one parameter. Thus, a federated control system can be used which provides a uniform interface to equipment from different suppliers, the ability for the equipment to be controlled from more than one point, and an integrated and annotated system view of the equipment.

Federated control is a key to efficient and reliable operation. As each screen can be designed to meet a particular operation – for instance, setting up the receiver, setting a standards converter and routing – can all be done at one time on one screen. Other screens could show the allocation of signals to the standards converter, etc. Useful metadata can be added – such as the number or times of the booking, the contact phone numbers, etc.

A federated control system also means that, during quiet periods, one operational position can handle all the workload, thus reducing the number of staff just sitting around. Conversely, for major events a large number of operators can be used at any one time, reflecting the different ways of operation which may be required – for instance, someone dealing with all the satellite operations, yet keeping everyone informed about the status of all parts of the system.

Note that the federated system does not mean that everyone can control everything, often it just means that control functionality can be given or shared with someone else.

A communications system is another area where the federated approach can pay great benefits – the key point is to ensure that everyone who needs to speak or listen to someone else can do so. This may be a director at an outside broadcast being able to speak to the director in the studio and the network editor in playout. Here, because the signal is simple four-wire audio plus a bit of control and signalling, there may be a number of different communication systems interworking over a number of different links. The communication circuit requirements are often forgotten when the infrastructure is designed. But often the only communications system required is the telephone.

Between them, federated communications and control systems enable operator and production staff to work very efficiently and remain talking.

It is likely that there will be one room where much of the central equipment and the staff that operate it will be located. This area, which can have a number of names – Master Control Room (MCR) or Central Apparatus Room (CAR) or Central Technical Area (CTA) – has two functions: the location for lots of equipment and an operational area. With more equipment being remote there may be less equipment in the CAR, but it is all controlled and monitored in one place. In a large broadcasting facility, there may be central areas for TV, radio and news.

4.1.7.1 Equipment area

The main broadcasting plant is likely to be the line termination, distribution synchronisation and conversion equipment, as well as the main routing systems. There will also be the system reference signal generators – typically some form of sync pulse generator and its distribution. A logical layout must be adhered to as it is very easy, particularly with the more equipment, just to put it in the next hole, rather than where it is best put to give operational benefits. There will also be many ancillary systems like off-air receivers, RF ring main and perhaps business IT.

4.1.7.2 The operational area

The operational area is likely to have operational desk positions with computer screens, some simple measurement equipment and monitor stacks, so that the sources can be seen. Increasingly, presence and reasonableness but not detailed quality assessments are made by using a large screen system, plasma or projector with a multi-viewer allowing, say, 16 sources to be displayed on one screen. Often the display can also take PC screen displays, so that the bookings or alarm status can be seen in the same glance as whether the sources are present or not.

As the workloads can change the operational area should be set out to work with minimal as well as typical/maximum staffing. This means that all sources can be seen from one point if required and that the control surfaces close to each person allow them to do their tasks. This sounds very trite, but so often something which looks attractive is unworkable because of the basic layout of the desk and stack. One feature that can make the area more flexible is to feed each monitor (or virtual monitor) from a matrix and use automatic under-monitor displays to announce what the source is in terms of source number – e.g. I/C2 (incoming 2) – but also in terms of the booked – Scot Foot M (Scottish football Main). Clarity and good flow of information is key.

There should be one position where all key measurements are made – this one measurement position means that there is no doubt that a measurement made has been done on the same equipment as last time and so one of the variables is removed. This also means that operational desk position may have simpler but still fit-for-purpose monitoring. Likewise, it is always good practice to be able to look at or listen to a single source in an isolated room – which has another use – e.g. an office under quieter and more controlled conditions.

One final point that needs close attention is what happens if the CAR is made unworkable, by mains loss damage to the building, etc. Can the facilities work from somewhere else? There are many ways of making the system more resilient, but the disaster recovery plan is a very important operational document.

4.1.7.3 External connectivity

Circuits to and from the outside world including regional offices and other organisations tend to be used either permanently – like the output to the transmitter – or temporarily – for instance, from an outside broadcast or from other broadcasters. There is also the business communications needs like telephone and Internet connections, which is outside the scope of this article.

The first question should be 'is connectivity actually required?' For instance, most television is transferred at the moment by tape – but this may be as files on the Internet in the future. Thus, the need to connect to the outside world may not be that great.

Assuming that some real-time/streamed connectivity is required there are a number of options, outlined below.

4.1.7.3.1 A telco connection

This is usually a permanently rented circuit to the telco switching centre, for which you will be charged installation and rental. Any connectivity and connection charge will be on a per booking basis, paid for by the user of the circuit. It is worth noting that telco circuits are often bidirectional, so that you can receive as well as send. The precise deal can be arranged with the telco, which could be a specialised company rather than the national telephone company. The more specialised provider may be able to offer you a more tailored service and a different charging regime. This is often the best means of connecting a remote news inject studio.

4.1.7.3.2 A satellite connection

While a telco will be able to offer at a reasonable cost a full 270 Mbit/s circuit with no great problem over a short distance, the satellite bit rate is limited to about 12 Mbit/s; hence, compression will have to be applied to the signal.

The receiving dish, which may be, say, 3 m in diameter or more, will require space with a good view of the satellite and usually planning permission or building control permission. Some planners find dishes unattractive, others think that they add to the 'hi-tech' look of an area. For reception all you would need in most countries is the dish and the receiver, although some permits may be required. To operate a transmitter requires licensing and much more expensive equipment – so it may be worthwhile to hire the transmitting service using an SNG vehicle. The service provider would take care of all the satellite booking – usually charged per minute with some time for line-up and a minimum booking time.

The one beauty about (single-hop) satellite transmission is that, unlike telco, you are not charged by the distance. Some satellite up-link operators offer a range of services, including recording a transmission.

Incoming circuits will need synchronisation to the 'station syncs'; thus, it is highly desirable to lock any MPEG decoders to the station reference. SDV 270 Mbit/s circuits will also require synchronisation and it is essential that the sound is both synchronised (rate changed) and delayed to remain in synchronism to the video.

4.1.8 Playout

With an ever-increasing number of channels available, the business of playing out programmes has increased greatly. The presentation and broadcasting of TV programmes covers a wide range of activities from managing schedules, commercial sales, scheduling, channel branding, creative services, compliance, EPG and a host of related activities in addition to just playing the programmes out.

As in any live system the minimum of work should be done at the point of transmission. So work needs to be done to ensure that all components of a programme – such as the video, audio, subtitles, audio (video) description, PDC code and interactivity – are all assembled before the programme enters the transmission suite. If the programme is being streamed, the components must exist in the stream; if it is coming from tape or server, the components must also be as integrated in one stream as possible before transmission. Thus, the live transmission activity is reduced to switching the one stream – the programme – to air.

The playout suites can work from tape, but ingesting to a server can give flexibility and the certainty that the programme is actually the area and the right one (as it has been ingested). The Ingest is also the key point at which all the components can be brought together and checked. The output of the server could be put to air directly, and many simple services can do this very efficiently, but often there is a need to add logos and other screen furniture, so some form of integrated logo generator linked to the schedule held in the server may be needed.

When more complex interstitials and complexity in presentation are required, then playout automation will be needed to link the servers, the logo generator/mixer, and other devices, such as an audio store for announcements. These can range in complexity greatly and the more complex should be integrated into the whole media management systems that are looking from programmes being delivered to the playout area to them actually being transmitted.

All that has been described so far is using pre-recorded programmes, to a greater or lesser extent being transmitted automatically with someone keeping an eye out for breakdowns and other issues. Often, one person can be looking after 4–6 channel outputs.

With live material the channel will need to be reactive to the changes that will occur and thus will need people to intervene when a programme does not run to time or does not arrive at all. This will still require automation, but with a lot of ways in which the network director can change the schedule to ensure that the channel runs to time. These systems are far more complex and require much workflow and process planning to get the right solution for a particular distinctive look and feel of the channel.

After transmission the 'as run' log will need to be provided, to be processed to bill the advertisers, to provide a legal record of what was transmitted and to provide insight into what went wrong. Material will need to be demounted/purged from servers and tape returned to the archive. The signal will pass out of the playout to the central technical area, where other signals such as teletext are added, and then if it is for digital emission, to the coding and multiplex equipment, where all the additional and ancillary services can be added before it leaves the building.

4.1.9 Archive

An archive should be the place where everything is kept in a way that makes it accessible to everyone. It has often been said that programmes should be made for the archive and then playout can just pick them up from there. This is a very healthy way of looking at the archive, as it is an integral part of the programme-making process, not a dusty area full of tapes.

Material that you don't know about is just wasting space, so a good index and strict input (and output) logging system needs to be in use. A good librarian can help a great deal in the setting up of what needs to be recorded, so that the metadata can be searched to find the item required and to know how it can be played. This will include the rights in any of the material – for instance, agreements on what can be used 'topically in news' are very different from use later in a 'long form' programme. A programme or clip without rights is useless.

Many broadcasting organisations do not capture the material they have acquired and thus cannot get the maximum benefit from it. One centrally indexed archive is essential for efficient use, but sports and news both require specialist indexing and searching, and so may have their own systems with good federated search middleware, making the various archives appear as one to a producer trying to find material.

Every programme transmitted should be archived – whether at broadcast quality or at lower 'VHS' quality is a commercial decision. Many regulators require copies of transmitted programmes to be kept for a fixed period. Many legal actions for defamation have resulted years after the programme has been transmitted. If there is a restriction on the use of material it should be clearly mentioned in the database and perhaps even on the physical media.

Like playout, broadcast archiving can be a useful source of income – probably at the margin of other activity, but a good archive system and a keen team of librarians and researchers could make the archive become a source of income.

Now and in the future there are so many ways of getting content to the viewer that the more content that there is available, the more likely it is that it can be used. The archive is a living store of what is being worked on now and in the past, and is one of a broadcaster's greatest assets.

Physically, a tape archive requires controlled conditions and a good racking system for the tapes. Not only does it need to be secure, it should also have systems and layout which mean that it is impossible for someone to withdraw or add an asset without being logged – and some of the logging of metadata can be done by others than the archive staff. It is worth noting that tapes (or any archive material) are heavy and thus a very solid floor may be needed.

4.1.10 Current and Future Trends

As the broadcasting world expands and changes there are many topics that will come and go; in 2003 the most high-profile topics have been widescreen/HD, streams and files, collaborative production and 'going digital'.

4.1.10.1 Widescreen Production

The use of the 16×9 aspect ratio over the traditional 4×3 (12×9) has been widely adopted in the UK and part of Europe for standard definition 625/50, and worldwide for high definition. Most cameras, and downstream processing equipment like vision mixers and character generators, can be switched to work with either format. Very strong process has to be used to ensure that the correct aspect ratio is used throughout the production process.

There are a few key matters that can help with this: from issuing camera staff with balls to line up on – so that the image will always be circular, to the use of the same aspect ratio VT Clock as the material, to the clear labelling of material and the correct use of Aspect Ratio Converters (ARCs).

The BBC publishes a widescreen guide which is a practical guide to how the problems can be overcome (http://www.bbc.co.uk/guidelines/delivering_quality/delivery_widescreen.shtml). Many 16×9 programmes are played out at 14×9, so the use of VI signalling (SMPTE RP192) prevents many errors.

4.1.10.2 High Definition

High Definition (HD) is used for emission to the home in three main areas of the world – Japan (analogue), and the USA and Australia in digital. In Europe the drive has tended to be for more channels and thus, whilst MPEG-2 compression is used, then HD is unlikely to take a major hold. However, in the production process HD camcorders are now at an affordable price and offer a picture quality which can be thought of as being future proof. Certainly HD acquisition is of a quality that film practitioners feel happy with and is not more expensive.

Although there are a vast range of standards, a few seem to be settling down and becoming the favoured systems – 1080 24p or 25p/50i frames. Thus, 'film' is moving to HD but other than where the production specifies HD there has not been much studio HD production. As different compression systems are used – particularly MPEG-4 AVC – there will be changes in the use of HD in production and perhaps in emission and on blue laser DVD. In the feature film industry digital cinematography has taken off, with all the major studios operating very-high-definition systems.

4.1.10.3 Video networks

In the 'analogue world' the format of the vision signal was the same everywhere – as it came out of the camera then through the studio onto videotape, which just sort of stored it in linear form, through editing to the final tape which got played out and to the transmitters and to the home. The only difference was the standard 625/50 PAL European and the 525/60 (59.97) NTSC American.

In the digital world it is all different! Starting with the signal coming out of the camera to the mixer and then sending it short distances uses the same signal format. If we want it on tape or going long distances, we usually compress the signal by either DV, which comes in two flavours, or MPEG, which comes in a lot more!

If we send a stream anywhere it has to get there in the right order and at a more or less constant rate and get there more or less instantly. Streams are not of the same sort at any stage in the production process as they were in analogue times.

But digitised compressed video can also be stored as a file. If it is a file we can send it slower or faster than real time, then on reception check that it is all there and in the right order, and request retransmission of any missing portions and not really care too much when it arrives. There has been much work done to standardise how files can be constructed to carry all the components and metadata of a programme, and the format known as MXF will begin to gain widespread acceptance.

As soon as the material is a file it becomes logical to think of using IP to transport it. Thus, files can be delivered over the Internet. Within a broadcaster's facility it is usual to create a good quality LAN by partitioning off the media network from the business IT network, either physically or by the use of layer 3 switches, which are to all intents an internal firewall.

And it is a very little jump to see that packet technology can be used for transmitting compressed video. Packet networks come in two types – IP or ATM (both of these can take files or streams with some shortcomings) or DVB, which is good for streams. With ATM and DVB it is very easy to allocate the bit rate required for each service, a requirement if error-free constant latency streams are being handled, but less easy in the IP world. The world of packet networks for broadcasting is developing, and with more or less infinite bit rate being available, it is not unusual to see a telco link which is ATM carrying DVB services within an IP encapsulation – this is the backbone for a home DSL system.

4.1.10.4 Collaborative working

With the advent of more powerful personal computers and the network infrastructure to support them, the concept of collaborative working for production teams, not only in the news/sport area, has been developed. The use of collaborative media handling, merging not only the media but also its rights and contractual elements, has come into being.

This is a new workflow for production teams – who can now ingest and view all material gathered – so replacing VHS tapes and scraps of paper. Full access can also be given to archive systems so that the production team can work on one integrated system and be able to share information and ideas though their streamlined workflow.

It is important to remember that much of the richness in a programme comes from collaboration, and systems which offer integrated collaboration are a great aid to achieving better programmes quicker. Good, well-planned IT facilities can mean that material can be shared, viewed and accessed from multiple sites.

Building on this – there is a great area of content aggregation, whether for playout or for other operation, where the handling of multi-media material and most importantly its metadata in a communal manner will speed up work flows and generate much better derived content. Other assets can be linked together – such as subtitling and audio (video) description or other sound or interactive video tracks.

All collaborative working systems can support simple VT style editing, with embedded craft editing available if it is needed. The production team can also off-line edit with a good quality proxy system – often at more than 10 Mbit/s data rates, which are easily supported on well managed IT networks.

Collaborative working is the incoming way in which programmes can be made better, richer and cheaper and thus is being developed and adopted by leading broadcasters.

4.1.11 Sources of Information

Broadcasting is a very varied industry and encompasses a very great range of companies. The following are some websites from which information can be found and more importantly links can be found to the people who can provide you with the answers.

4.1.11.1 Exhibitions

There are two main exhibitions in the world each year – The National Association of Broadcasters (NAB) in April in Las Vegas and the International Broadcasting Convention (IBC), held in September in Amsterdam. Both of these events are very large and mix exhibitions with lectures and more formal sessions (particularly at IBC). Both exhibitions have very full websites (http://www.nab.org/conventions/ and www.ibc.org).

4.1.11.2 Key organisations and their websites

The following are key websites to know about and bookmarks for some of them.

4.1.11.2.1 Standards bodies

International Telecommunications Union – the top standards body – www.itu.int

ETSI – the European Standards body – you can download individual specifications free – www.etsi.org

4.1.11.2.2 Major associations

European Broadcasting Union – www.ebu.ch

Society of Motion Picture and Television engineers (SMPTE) – www.smpte.org

Royal Television Society – www.rts.org.uk

National Association of Broadcasters – www.nab.org

ProMPEG – www.pro-mpeg.org

4.1.11.2.3 Other useful links

The UK Digital Television group – www.dtg.org.uk

The DVB Project – www.dvb.org

The Digital Project – www.digitag.org

The International Association of Broadcast Manufacturers – www.theiabm.org

Major broadcasters have websites – but few broadcasters have the range of services that the BBC offers:

www.bbc.co.uk
www.bbcresources.com
www.bbcbroadcast.com
www.bbctechnology.com

4.1.12 Digitisation Philosophy

The following is taken from advice to a major world broadcaster, but it has a generality of impact on all TV broadcasting.

Analogue television is declining.

TV equipment which works on analogue signals is becoming increasingly unobtainable. Although simple devices such as distribution amplifiers and some specialist equipment may still be available in an analogue format, most of the other items – such as vision mixers – can no longer be purchased in analogue form. Other key equipment like VTs and telecom links do not handle an analogue signal internally but often present their interfaces as analogue as well as a digital SDV, usually with embedded audio.

Radio/audio equipment which is at the start of the chain – typically mixing desks – works well in either analogue or digital, and at this point whichever is cheapest is probably a good pointer, remembering the costs of conversion. This may mean that a complex sound desk is analogue whilst a simpler desk is digital. All audio infrastructure can be digital and thus performance fit-for-purpose in the whole system is the clear criterion for selection.

Thus, there is a compelling force that when new equipment is purchased, the only economic or even available equipment is digital.

Analogue PAL equipment if properly maintained has a long life and there is usually no need to replace it so long as spares and maintenance expertise are still available and that it is fit for purpose.

Digital is different – there are more choices.

Digital equipment often has very different functionality from analogue, and it is often the fact that the business or media system flows require some element to use digital technology that provokes the start of the migration process. One key feature is that digital technology always introduces a choice – different and more ways of solving the problem.

Just 'going digital' rarely solves a problem.

When attention is drawn to 'replacing a piece of equipment' it is very important to understand the business case and the media flows so that a system solution which makes the new digital facility efficient is taken. This often speeds the adoption of digital technology in an area – by considering what is and can be used with digital interconnection.

This could be termed 'getting a critical mass' in a particular area.

For instance, new incoming lines have the ability to be digital; the VT format chosen has digital inputs and outputs – if one studio is digital, then any new equipment will/could also be digital. As the next move the MCR can become digital. The critical mass can cause a 'snowball effect'. This also meets an aim for a more (technically) reliable output, and may mean that more output can be gained from the same equipment.

It should be noted that it is rare for digitalisation to be done for solely technical reasons.

The following methodology has been applied to each project to consider if, why, how and when systems should be migrated from analogue to digital.

- Something – which might just be an idea – brings digital into an area. Looking to see what other systems in the media or business flow that could benefit from being changed to digital is the normal immediate response.
- Changes to the operational process may also satisfy the current and near future requirements.
- Are the changes required just fit for purpose – and not (grossly) over-engineered?
- If there is equipment upstream or downstream which is digital then it is likely that to convert the current system to digital is wise. However, it is also important to look for other systems which have a need to be changed by adapting digital technology – as it is different.
- At a technical level it is very important to ensure that there are minimal conversions to and from analogue as the signal passes through the system.
- This tends to make it more important to convert central equipment and systems to digital as the signal tends to have to pass through a number of times – so MCR and VT/editing are likely, yet playout and studios less so.
- Studios are often converted for commercial rather than technical reasons.
- In the overall plan, it is important to pay attention to the place at which analogue-to-digital and digital-to-analogue conversion will be taken to minimise unnecessary conversions. Ideally only one decode–recode cycle should take place, but whilst there is still much analogue equipment – two cycles may be acceptable.
- A similar consideration affects audio, and in both audio and video consideration has to be given to synchronising digital clocks between sources, usually by a synchroniser/rate changer. Clever design of the architecture may mitigate the issues in this area.

Peter Harman, AMIQA, Dip IDM, Dip BE,
Dip PC (Business)
Marketing & Training Manager
Vinten Broadcast Ltd

4.2 Studio Cameras and Mountings – Mounts

4.2.1 Positioning Equipment

This is equipment that enables the operator to place the camera in the right place at the right height. It varies from simple tripods to pedestals and simple cranes. Section 4.2.2 deals with pan and tilt heads, and includes a look at different designs that have particular features or benefits.

4.2.1.1 Tripods

The simplest portable mounting for any camera is a tripod. There are many built to professional standards, but few achieve all the desirable characteristics, which are:

- lightness;
- torsional rigidity;
- a wide range of height adjustment;
- simple and easy to operate leg locking mechanisms;
- a stable floor or mid-level spreader or triangle;
- usable on both smooth and uneven surfaces, and on soft as well as hard ground.

Most tripods have spikes that can be used to dig into soft, uneven ground to obtain a stable foothold or with rubber feet attached for hard level surfaces. In both cases, it would be usual to use a mid-level spreader to ensure overall stability (see Figure 4.2.1). Alternatively, for a hard level surface, a floor spreader could be used as it provides maximum stability. All tripods are a compromise, but modern materials have enabled designers to make extremely rigid structures that are not too heavy to carry, and the newer of such tripods give a height range of 42–157 cm (16.6–61.7 inches), and collapse down to a package only 71 cm (28 inches) long.

When buying a tripod and spreader, check that:

- When fully extended with its spreader, the top platform resists horizontal rotation; such rotation (induced primarily by the drag system and wind buffeting) is called *wind up*. Clearances in the design will also produce a 'dead spot' known as *backlash*, and is noticeable when the direction of the camera pan movement is changed. If one or both exits in any degree, it makes predictable framing very difficult at the start and finish of a move.
- The points or feet on the tripod fit very snugly into the ends of the floor spreader, rubber feet or points of the triangle.
- Vertical downward forces do not deflect the legs in any way.
- All fittings and clamps are designed for ease of use with gloved hands.

A triangle is more rigid than a Y-shaped spreader, but is very difficult to fold up quickly, and much more difficult to use on uneven ground or a staircase, which requires using either a flexible Y-shaped spreader or, preferably, a mid-level spreader.

Most tripods in Europe use either a 75 mm, 100 mm or a 150 mm bowl interface for pan and tilt heads. However, in the USA, and the film industry in particular, Mitchell and even Junior Mitchell fittings are used with clamps levelling the tripod by adjusting the legs.

4.2.1.2 Heavy-duty tripods

Heavy-duty tripods are a much more substantial construction and use four bolt fittings with the same Mitchell alternatives in the USA (see Figure 4.2.2).

In order to make elevation changes, heavy-duty tripods can be fitted with a simple crank-operated elevation unit, giving additional height adjustment of approximately 46 cm (18 inches).

4.2.1.2.1 Rolling base

By the addition of a rolling base (a *skid* or a *dolly*), a tripod becomes a moving camera platform that can be adjusted for height throughout the range of the tripod. If the skid has

Figure 4.2.1 ENG Fibertec tripod and mid-level spreader (Vinten).

castoring wheels, some simple tracking moves are possible, but elevation changes are difficult in a continuous shooting studio environment, and certainly not possible on-shot.

4.2.1.3 Portable pedestals

Small portable pedestals form the next category of mounting. They differ significantly from simple tripods and skids through the addition of a balanced elevation unit. A pneumatic telescopic column supports the weight of the camera, and pan and tilt head in poise. Portable pedestals vary in sophistication from devices with simple castors to steerable systems, as shown in Figure 4.2.3.

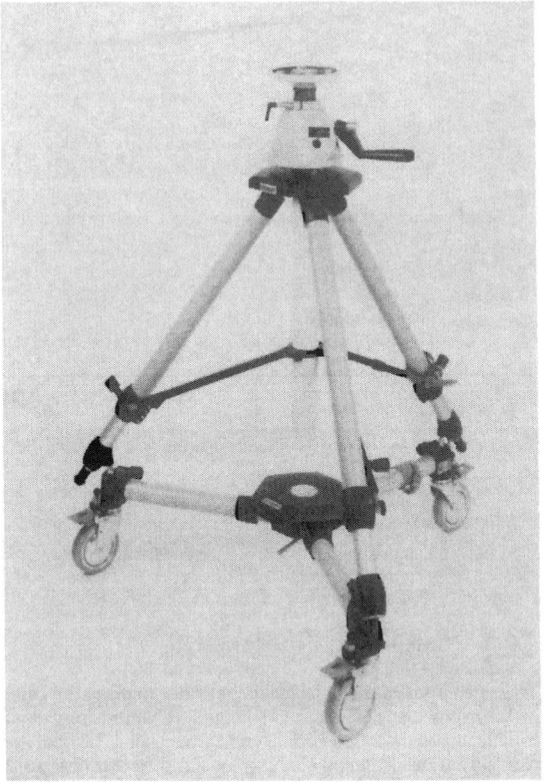

Figure 4.2.2 Heavy-duty tripod, elevation unit and skid (Vinten).

The most sophisticated of these can make steer and crab movements similar to big studio pedestals, but few provide 'on-shot' quality movement, and most have quite severe restrictions on the range of elevation possible.

Regardless of whether the balance system is hand cranked, hydraulic, pneumatic or spring, older, low-cost units are usually single stage, and the adjustable height range is limited to about 420 mm (16 inches).

In recent years, Vinten has developed the Osprey range of portable pedestals. Three versions offer single-stage, two-stage with one stage 'on-shot' and two-stage fully 'on-shot' elevation respectively. Osprey features low-pressure pneumatic counterbalance, crab and steer capability, and a wide range of operational heights from 65 cm (25.6 inches) to 147 cm (57.9 inches).

4.2.1.3.1 Studio pedestals

A studio television camera together with prompters, pan bar controls and viewfinders can weigh nearly 100 kg (220.5 lb). The most satisfactory way of mounting it is to use a pedestal with crab and steer facilities.

These are traditionally six-wheeled devices on a triangular base of approximately 1 m side and are designed to carry loads of between 80 kg (176.4 lb) and 120 kg (264.4 lb). Both lightweight and heavy-duty studio pedestals can be single-, two- or four-stage versions (see Figure 4.2.4), the former having a 71–122 cm (28–48 inches) height range, the two-stage

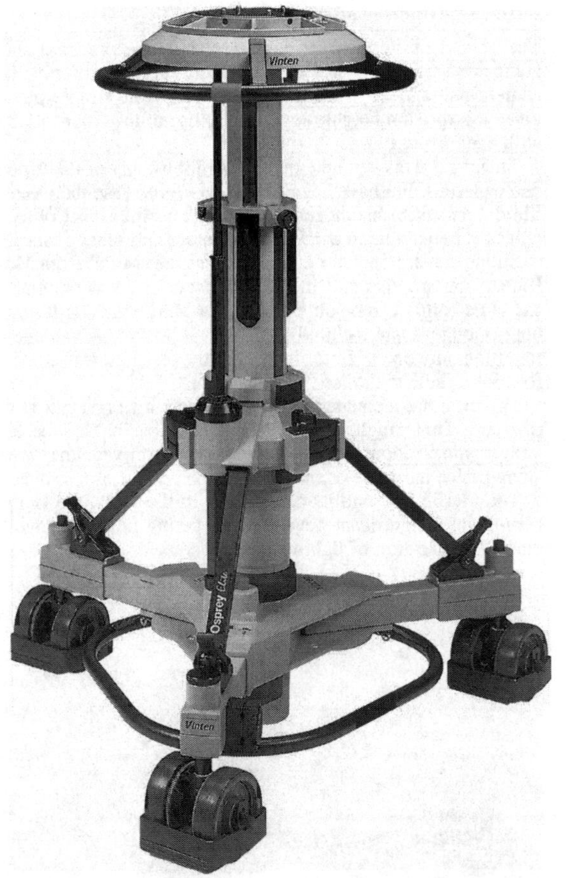

Figure 4.2.3 Osprey Elite pedestal. This gives a height from a minimum of 66 cm (26 inches) to a maximum of 1.43 m (56 inches) (Vinten).

Figure 4.2.4 Quattro four-stage studio pedestal (Vinten).

having a 66–143 cm (26–56 inches) height range and the four-stage pedestals having a height range of 45.2–148 cm (17.8–58.2 inches).

On all studio pedestals, steering and elevation is by means of a steering ring immediately below the pan and tilt head. When correctly balanced, the cameraman should be able to raise and lower the camera on-shot, and skilled operators will be able to manoeuvre the pedestal almost as an extension of themselves. Tracking in and out, traversing left to right or any combination makes it practical to follow a performer walking round a room.

The secret lies in the crab/steer mechanism. When all three wheels are locked together in parallel, the pedestal is said to be *in crab*. Crabbing does not affect the angular orientation of the pedestal, and therefore the pan action of the head is not affected by pedestal movement.

The pedestal travels in the direction of the steer spot or pip on the steering ring. It may be surprising that this has proved to be the most versatile way of using pedestals. The steering position where two wheels are locked together and only one rotates is used only for negotiating awkward obstacles or turning the pedestal round on its base in order to orientate the base to suit the operator.

There are four types of pedestal:

- weight balanced;
- spring balanced;
- hydraulic operating against compressed gas;
- pneumatic compressed gas only.

Most older type, high-pressure pedestals use a welded construction pressure vessel, usually steel. Nitrogen, which is inert and moisture free, is the perfect choice when it is available in a compressed form. However, clean, dry compressed air can be used.

4.2.1.4 Cranes

The Hollywood film industry created the ride-on camera crane, and we are all familiar with the picture of one of these huge devices and the director with his megaphone shouting instructions to the camera operator and his focus puller.

Some of these cranes were used in the early days of TV, and indeed some are still in use for the older TV studio cameras, although the advent of lightweight TV cameras is rendering this type of crane obsolescent. It gives a much higher point of view than is possible with a pedestal, whose maximum height must be limited to the height of the cameraman's fully extended arms. The Nike and Tulip are examples of big cranes still in use.

A smaller two-man crane such as the Kestrel (still in use, but no longer manufactured) can be used for high and low shots, but the boom cannot be moved (*swung*) from side to side. The reduction in crew is the main advantage when using the older heavier cameras.

The Kestrel crane shown in Figure 4.2.5 was supplied with either a manual or powered elevation. The tracker steers and pushes the dolly, and in the case of the manual version elevates the platform. Jib elevation controls are extended to the cameraman in the case of the powered version.

While it might seem hard for the tracker to have to elevate his cameraman, the job is made much easier by the jib arm balancing system, which is a pneumatic balancing ram adjusted to take the total load of camera, operator, etc. This makes both the motor drive and the effort required to manually raise and lower the jib very practicable.

4.2.1.4.1 Short crane arms

The advent of lightweight cameras has made practical short crane arms that are operated by standing cameraman only. The camera pan and tilt head is balanced, and this simple device gives a surprising height range from close to the floor to 1.8 m with even a short arm.

The crane arms can be mounted on almost any of the tripods and pedestals that have been described, to increase their versatility. Apart from height range, the most useful aspect of short cranes is being able to *track over* furniture and other obstacles.

Single camera operator cranes such as the 'Mole' from Mole Richardson are also still in use, but becoming less popular as the same height can be obtained with a Merlin arm (still in use but no longer manufactured). The Merlin has the advantage of requiring only one operator instead of the usual crew of three or four for a mole crane (see Figure 4.2.6).

The mole has a counterbalanced arm with a weight box at the rear end. This weight box is the control position for the arm swayer who positions the camera in space. The driver either walks along beside the crane or sits on a seat behind the arm swayer.

The Merlin arm and its counterpart in the USA, the Barber boom, was a significant development, taking full advantage of the new generation of lightweight cameras.

Figure 4.2.5 Kestrel camera crane (Vinten).

Figure 4.2.6 Merlin crane (Vinten).

The camera is slung in a cradle at one end of a boom arm and the operator stands at the other shorter end of the boom with pan, tilt, elevation, zoom and focus controls extended to him as well as the viewfinder.

4.2.2 Pan and Tilt Heads

A pan and tilt head is a device that enables the cameraman to point his camera at a subject, compose the picture well and, if necessary, follow the subject when it moves. It is occasionally necessary to simply move left or right to see a greater field of view, but usually this movement is associated with following something.

The human eye is unable to pan smoothly unless it is following an object that is moving, and therefore a panning camera movement without something to follow seems unnatural. It is necessary to make camera movements as imperceptible to the viewer as possible.

Jerky movements, inaccurate movements and unnecessary movement are all extremely disturbing to the viewer, and professional equipment is designed to help the camera operator overcome these undesirable effects.

A tripod or pedestal, described in Section 4.2.1, carries the weight of the camera, but one of the prime functions of the pan and tilt head is to render the camera effectively weightless.

Different manufacturers use various methods to try to achieve this weightless balance or counterbalance. Since pan tilt is controlled by the single or double pan bars, the 'feel' of the movement in both axes has to be as near identical as possible. It is important to understand that balance and counterbalance do not impact 'feel' or 'drag' to the pan and tilt head. The solution to this problem is described in Section 4.2.2.5.

Figure 4.2.7 Swan Mk II post head (Vinten).

4.2.2.1 Post head

There are now five basic balance methods used in pan and tilt heads, and the simplest of these is the post head. It consists of two identical swivels: one is under, or near, the centre of the horizontal camera; the other is mounted on a post attached to the previous swivel (see Figure 4.2.7).

A correctly set up camera and lens can be balanced about its own centre of gravity, and therefore will remain where it is pointed until such time as the camera operator moves it to another position. As the two swivels are identical, the freedom of movement in both axes is the same.

Although the apparatus appears somewhat ungainly, the simplicity of the method of balancing has a lot to commend it, and post heads have earned high reputations and have dedicated users.

There are two disadvantages with post heads. The first is that, in order to carry heavy loads, they need to be massive and are only therefore practical for lightweight cameras. The second disadvantage is that access to the internal circuitry, or mechanism of the camera, is restricted on one side by the post itself.

Some camera operators would say that, unless the post head is designed with the horizontal immediately beneath the centre of the lens axis, panning is an unnatural movement. Although this would appear to be so, tests have shown that, for most normal movement, being off axis by up to 15 cm is virtually undetectable.

All these disadvantages can be overcome by the second type of pan and tilt head, which is often referred to as a *geared* head and is one of the most popular mountings in the film industry.

4.2.2.2 Geared heads

A gear head consists of a quadrant of a circle. The camera is arranged to have its centre of gravity placed at the centre of that circle (see Figure 4.2.8).

This method achieves perfect balance, and movement is restricted only by the friction of the mechanism and the practical size of the quadrant. This type of head is particularly suited when the camera centre of balance alters significantly, as can be the case with film moving in a camera magazine, and when manipulating very heavy 35 mm or 70 mm film cameras.

The control of the camera movement using the gear wheels is a technique that needs practice, but the wide adoption for film cameras is an endorsement for the method. There are very few applications for this type of mounting in television operations, although there is some suggestion that the move to high-definition video production may see the geared head finding some application in television production. However, it should be noted that geared heads are generally large, quite heavy and have a restricted tilt range, potentially limiting the scope of the shoot and the speed of video production.

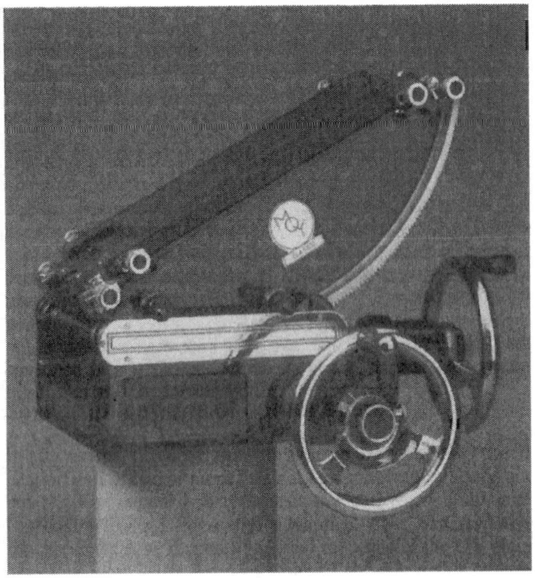

Figure 4.2.8 Moy classic geared (Ernest Moy Ltd).

4.2.2.3 Cam heads and the Vinten Vector system

Cam heads keep the centre of gravity of the camera in the same position throughout the tilt range. Perfect balance can also be achieved by the use of a developed cam, which allows greater tilt by letting the centre of gravity of the load move forwards and backwards without moving up and down.

The camera is, in effect, pivoting about its centre of gravity, and although its mass has to be shifted fore and aft a few centimetres, this requires very little effort. As with the two previous heads, the camera will remain at any tilt angle, and will only move to another one by the action of the camera operator.

The disadvantage of this method of camera mounting is that, if different loads and different cameras are used, new cams have to be fitted for a normal studio operation using standard cameras and lenses. This takes time, as changing cams to accommodate prompters, etc., necessitates a strip down of the head.

The fourth and most recent counterbalance system comes in the form of the Vector system (See Figure 4.2.9), which relies on the same principle of the cam system, i.e. if the centre of gravity of a camera can be forced to scribe a horizontal line (the centre of gravity does not fall when tilted and therefore there is no change of 'potential energy'), then the camera will remain in balance. The patented Vinten Vector balance mechanism provides the lift necessary to maintain the camera's centre of

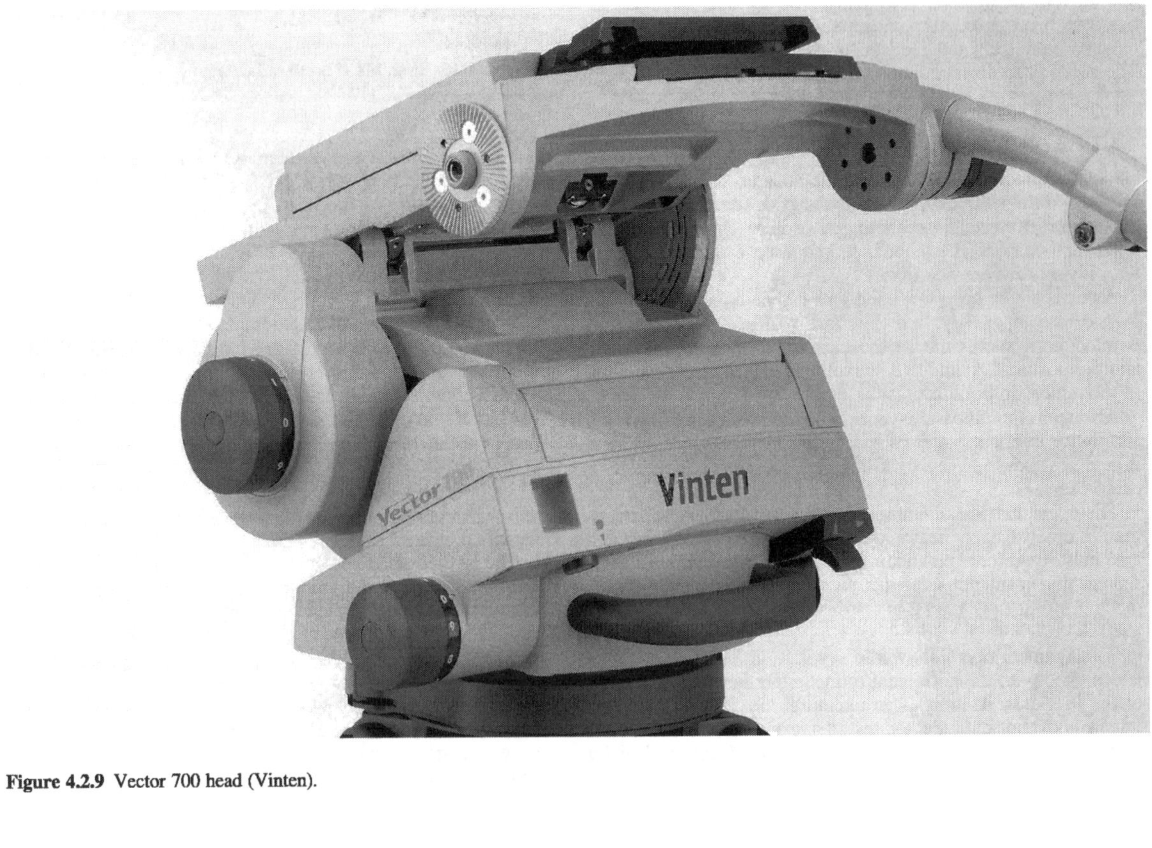

Figure 4.2.9 Vector 700 head (Vinten).

gravity in a horizontal plane, as the head tilts from horizontal. The rate at which the mechanism lifts is infinitely adjustable to match any camera that falls within the head's centre of gravity range and weight capacity.

Therefore, as the system is infinitely adjustable, the operator can trim or make major adjustments whenever necessary without having to strip the head down, saving time and improving camera control.

All preceding pan and tilt types have been developed to handle a wide range of full-facility film and TV cameras with large lens configurations.

4.2.2.4 Spring-balanced pan and tilt heads

The post head became more practical when lightweight (hand-held) cameras came along, but it also became possible to use spring-balanced heads (see Figure 4.2.10). Improvements in the use of springs have made it possible to design this type of head for up to 40 kg of load.

The key benefit of the spring-balanced head is the ease of adjusting the balance when the camera load is changed. This requirement becomes more common in the field when interchange of, say, eyepieces and viewfinders may occur, or when batteries or recorders are added.

In this type of pan and tilt head, some form of spring mechanism counterbalances the weight of the camera, which may be fixed or fully variable. The amount of spring counterbalance generated (torque) depends on both the overall weight of the camera and the height of its centre of gravity. Hence significant effect is felt when adding, for example, a viewfinder on top of the camera.

Spring-balanced heads frequently have the additional advantage that their mechanism is totally encased. This is necessary to provide adequate protection for the operating mechanism, but also means that they are particularly suited for operating in adverse climatic conditions on location.

They are also often compact and light, making them very easy to transport to and from location.

Figure 4.2.10 Vision 100 spring-balanced head (Vinten).

4.2.2.5 Drag mechanisms

As was mentioned in Section 4.2.2, the balance achieved by the pan and tilt head does not introduce drag to provide the right 'feel' for the camera, and indeed should not. However, particularly as cameras get lighter, the inertia of the camera gets very small, potentially leading to shake and jerky movement as the cameraman starts or stops a movement. Similarly, as lenses get longer, controlling the smallest of movements becomes even more critical.

To ensure complete control, it is usual to introduce artificial drag or inertia, which makes the job of the cameraman much easier when making smooth movements with the camera.

The lowest cost way of achieving this is to use a simple friction plate and a brake that can be applied or released depending on the speed of movement needed. The problem with this type of mechanism, particularly with lighter cameras, is that sticking can occur before the friction plate starts to move against the brake. This can lead to grabbing and hence jerking of the movement at start and stop.

A more reliable method is to use lubricated friction wherein a fluid is forced between friction plates, the pressure between which can be varied to increase or decrease the amount of drag. Precision construction is required for this type of mechanism, and hence it is more costly than non-lubricated friction, but the performance achieved is considerably better.

This type of fluid damping has benefits over simple hydraulic systems in rapid moves create hydrodynamic lubrication, enabling, for example, a whip pan to be performed.

Newer developments in drag technology rely on the principles of achieving resistance to movement through fluid shear using a sandwich of fixed and moving plates, separated by a thin film of grease. As this type of sealed system can accommodate virtually any type of fluid, extreme temperature ranges can be accommodated. However, one characteristic of this type of system is undesirable: resistance increases proportionally to speed. Where this is used, whip pan becomes impossible. Another problem is that infinitely variable adjustment is extremely difficult to achieve. Many manufacturers approach this by making their systems modular, with levels of drag selected by switching in different combinations of modules. The Vinten TF system, however, is infinitely variable and has design features that modify the output/speed characteristics so that whip pan is still achievable.

4.2.3 System Stability

In summary, the key requirement of the camera mounting is that it can provide a smooth movement of the camera as might be seen through the eye of an observer. This quality of movement relies on a very stable platform, be it a tripod, a pedestal or a crane, and a balanced and controlled action through the mounting and the pan and tilt head. While at the end of the day the mounting frequently costs far less than the camera on it, no cameraman should forget that his ability to shoot quality pictures relies more often on the quality of his mounting than anything else.

J Summers
Formerly Lighting Director, BBC

4.3 Studio Lighting

4.3.1 The Purpose of Lighting

Artificial lighting is needed for four main purposes:

- To ensure the presence of sufficient light to obtain an intelligible and technically satisfactory reproduction of the original scene. The level required will depend on the sensitivity of the system used, and will be affected by the stop number of the lens, the presence of colour filters or diffusers and the use of additional equipment such as an artist prompting device.
- To keep the scene contrast within the limits acceptable to the TV system. Colour cameras currently in use handle a contrast range of 30:1, but a sunlit exterior scene may have an inherent contrast range of 500:1. Additional light is needed in the shadow areas to reduce this to an acceptable level.
- To give shape to objects and depth to a picture. A television image is two-dimensional. By using light and shade the illusion of the missing third dimension can be created.
- To interpret the emotional content of a scene. A lighting director has the opportunity to use creative skill to enhance the scene with a chosen style of lighting. It is possible to heighten the pictorial interest, or increase the emotional response of the viewer, by the subtle use of light and shadow.

4.3.2 Lighting Sources

Light sources used to produce a photographic image fall into one of two categories: *hard* or *soft*. Hard sources produce a hard shadow; soft sources a soft one. The sun is an example of a hard source, and skylight on an overcast day an illustration of a soft. Soft sources are large in area relative to the distance they are from the scene. On a dull day the whole hemisphere of the sky is the source of light; as there is no one direction of light, no object can cast a shadow. A hard source is small in area relative to its distance from the scene. If an object is placed in the path of unobscured sunlight, it casts a sharply defined shadow.

4.3.2.1 Soft sources

A commercially produced soft source should have as large an area as is practicable and transmit an even light over its whole surface. Those that are smaller will cast soft, discernible shadows and are designed to be used in situations where space is limited. The lamps in soft sources are usually small, and therefore have dimpled reflectors situated behind them to scatter the light. In addition they are usually masked in some way to prevent any direct shadow-forming light leaving the luminaire.

Many matte, white reflecting materials can be used to provide an efficient soft source. If they are illuminated by one or more hard sources the reflected light will be soft. Cyclorama cloths, out of vision walls of a set, polystyrene sheets and card are a few examples of devices used in practice to obtain soft light. Large area soft sources become virtually shadowless and cannot be easily contained within limits. The principal use of soft light is to control the depth of shadow created by a hard light, whilst minimizing additional, conflicting shadows from itself.

4.3.2.2 Hard sources

4.3.2.2.1 Soft edged fresnel spotlight

This is a light source fitted with a lens, designed to focus light into an adjustable beam angle. When the lamp and reflector are moved away from the lens (*spotting*), the intensity of the beam is increased but coverage is reduced. If the luminaire is set for maximum beam angle it is said to be *fully flood*. It is fitted with *barn doors* (shadow casting blades) whose purpose is to control the shape of the light beam. The ability of the barn doors to control the beam reduces as the lamp is spotted. Colour filters, diffusers or *jellies* (light attenuators) may be fitted, as well as a *flag*. A flag is a blade similar to a barn door, but clamped to the lamp housing by a ball and socket arm. It is used as an adjustable shield to prevent stray light entering a camera lens and causing flare.

Spotlights are used to shape and texture objects by creating shade and shadow.

4.3.2.2.2 Profile spot

In optical design, a profile spot is similar to a photographic projector, but it is intended to project sharp images over shorter distances. These images can be produced from specially designed *gobos* in sheet steel, or from hand-made ones in aluminium foil.

Effects projectors give moving patterns, obtained from a continuous photographic image deposited on a plate glass disc, rotated by an electric motor. Another type rotates a circular metal gobo. If profile spots are suspended, care has to be taken with studio ventilation so that the effect does not sway.

4.3.2.2.3 Follow spot

A follow spot is a narrow angle spotlight that can be used over long throws, and panned to follow moving artists. It can be focused to give very sharp shadows, and cut off. Provision is made for beam shaping shutters, iris diaphragms with blackout discs, and colour frames.

4.3.2.3 Other sources

Some studios use dual source luminaires, which can be transformed fairly quickly from hard sources to soft ones, or vice versa. They can be dual wattage, which enables the light output to be altered whilst maintaining a constant colour temperature. They are most usefully employed as a saturated rig, where suspended luminaires are permanently installed over the whole studio area at a high density. Although they have disadvantages, such as heavy weight and compromised softlight size, they also have the advantages of speedy rigging, and flexibility in use.

A number of luminaires have been specially designed to light cycloramas, and they can be divided into two groups:

- *Top cyclorama units*. These can be floodlights used to light the top portion of the cyclorama as evenly as possible. An alternative design, employing a multi-curve reflector in each of four compartments, is capable of giving even illumination from top to bottom of the cyclorama. Each compartment is separately controlled so that, by fitting a different colour filter in each, many hues can be synthesized.
- *Ground row*. These floor units are frequently placed close to the bottom of the cyclorama, deliberately using the fall off effect of the light to create an artificial horizon. A unit may consist of four compartments in line, individually wired to facilitate colour mixing. One design of ground row unit overcomes the problem of the curve of the cyclorama by hingeing the rear of the compartments so that the unit can be adjusted to fit the bend.

There is a group of luminaires available that can be generally described as *disco lighting*. The range is large, and an exhaustive list could not be given as they are in an ever-changing market that responds to fashion. A common factor is that they are designed to be used in conjunction with smoke effect. The light sources themselves have a very low thermal inertia, so that they respond almost instantaneously to changes of input voltage.

Narrow angle or parallel beam sources are used to illuminate the smoke, creating beams of coloured light that can be made to pulse to music, or sequence in any predetermined manner. Some units are available with remotely controlled colour changing, and remote pan and tilt mechanisms. A large group of kinetic effects employs parallel beam sources mounted on a structure that can be rotated in either horizontal or vertical plane, or in both simultaneously, using variable speed motors.

There are units designed to fire a number of photographic flashguns in sequence, while others will give an effect similar to twinkling stars, by randomly firing a string of small discharge tubes.

Automated lighting systems are available that can remotely control all the operational functions of a luminaire. There is a spotlight unit using a metal halide arc, and a wash version with a tungsten source. Having its own microprocessor, each unit can store up to 1000 cues, controlling light hue, saturation, intensity, beam angle, beam edge and gobo patterns, together with pan and tilt movement. They can be linked to a control console to permit individual manual operation or preprogrammed cues, combining up to 1000 units.

Automated lighting has a clear application in disco and rock concerts, but because they are so versatile, they are a useful tool in any situation where remote control is needed.

4.3.2.4 Lighting console

Television is a photographic medium that gives instant pictures. To utilize this fact fully it is important to be able to make instant adjustments to the brightness of any luminaire, and electronically memorize the levels for future recall. This is the function of a *lighting console*. The combination of picture monitor and lighting console considerably extends the artistic choice of the lighting director, as any combination of luminaires may be selected at various dimmer levels, and the effect instantly seen.

Changes to occur in vision may be set up, as either an instant or punch change or a fade across from one state to another. More subtle lighting changes, meant to go unnoticed by the viewer, may be made by using a group fader or by using manual adjustment.

4.3.3 Static Portraiture

There are no rules in television lighting, only objectives. How they are achieved is up to each individual, and although custom and fashion may shape what is generally accepted, there is no reason for these to become a straightjacket worn by the lighting director. The following sections should be regarded as a guide rather than a rule book to the way that certain re-occurring situations can be dealt with.

4.3.3.1 Straight to camera

An arrangement of lights for a single sitting subject is shown in Figure 4.3.1. In this arrangement, **A** is the modelling, or keylight, and would normally be a fresnel spotlight. The light beam can be controlled by the barn doors to keep it off the backing. It can be placed either side of the camera. The shadow area of the face becomes greater as the keylight is raised in height, or the horizontal angle increased. The nose shadow should not be distractingly long, nor the eye sockets too dark. The position of the keylight is set to give roundness and depth to the face, probably within 0–30° horizontal angle to the camera.

B is the fill light and will be a soft source. It is placed on the opposite side of the camera to the key, as close to it as possible,

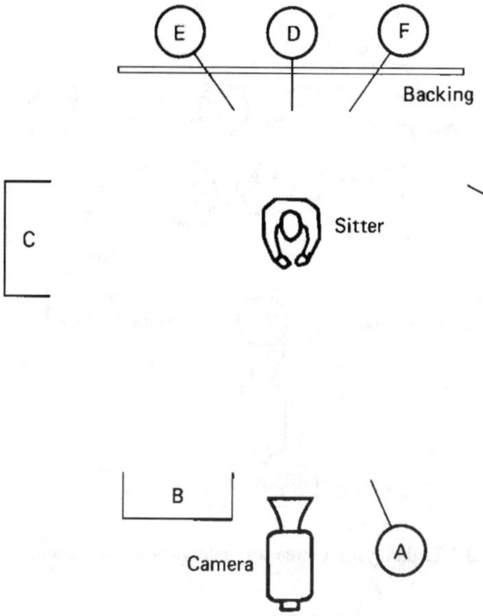

Figure 4.3.1 Lighting for a single static subject.

and set at the same height as the camera lens. It has a number of functions. It controls the depth of shadow created by the key-light on the front of the face, neck and clothes. It acts as an eye light, ensuring that if the sitter looks down, light will get into the eyes. If there are unflattering lines or wrinkles, this light will minimize them.

C is a soft source and is another fill light to control the shadow from the key on the side of the face. It is placed at eye height, and at 90° to the camera position, so that it does not cast a nose shadow.

D is a single backlight, a fresnel spot suspended over the top of the backing, directly behind the sitter. This gives specular highlights in the hair from the camera position, and adds depth by rimming the shoulders. The barn doors, or a flag, should be adjusted to keep any light off the camera lens. Sometimes studio space is limited, but the subject should not be placed too close to the backing, as this increases the angle of the backlight, and causes the background to be excessively sharp, and possibly distracting. A backlight of 30–40° is ideal.

An alternative to single backlight is double backlight. Two spotlights (**E** and **F**) light the back of the sitters head at a horizontal angle of approximately 40° to the axis. Interesting specular highlights are now obtained on the sides of the face as well as the hair, defining the cheeks and jawline with more clarity. A highlight instead of as shadow is used to suggest the third dimension. Because backlight **E** lights the part of the face unlit by the keylight, the effect is stronger on this side of the face. Fill light **C** may not now be needed. Double backlight is often used on light entertainment productions, where the fuller effect is more suitable.

G is shown symbolically to light the backing. How the backing is actually lit will depend on its colour, luminance and texture. It was suggested earlier that the keylight is kept off the backing by the barn doors so that the backing can be lit separately. In this way, control of the backing may be obtained independently of the key. For maximum control of a lighting set-up, it is desirable to have only one function for each light source.

When all the luminairies have been set, they should be balanced, i.e. the final brightness of each should be determined by operating the lighting console. First the key is adjusted to obtain full exposure on the camera. Fill lights **B** and **C** must be used with discretion, as too high a level will destroy the modelling created by the keylight. The backlight and background levels chosen will be determined largely by the type of programme being made, and should be adjusted to suit its style. The picture monitor should be examined to see if there are signs of an unwanted colour shift due to excessive dimming of a light source. This is most likely to occur with fill lights, as very small quantities are needed. If necessary a smaller wattage luminaire should be used, or the light attenuated by means of a neutral density filter.

Consideration has been given to static portraiture involving one camera. Floor mounted lights can be used wherever practicable for this, for ease of adjustment. When more cameras are used, floor lights are no longer viable, as they impede camera movement, and there is more risk of lens flares. Suspended luminaires should be used, and because distances are increased to approximately four or five metres, higher wattage units must be employed.

4.3.3.2 Two-way interview

Lighting for a two-way interview is illustrated in Figure 4.3.2. Cameras 1 and 3 are positioned to take close-ups of participants while camera 2 takes a two shot. The two spotlights **A** and **B** each have two functions: to backlight the nearest head, and key the other. The barn doors are set to keep the light off the backing. Soft sources **C**, **D** and **E** are fill, and the background is lit with spotlights **F**, **G**, **H** and **J**.

Although quick to rig, this lighting set-up has two disadvantages. The balanced level of **A** and **B** is adjusted to give keying levels on faces for correct exposure, and this may result in too much backlight on one or both participants. This problem can be reduced by using material called a 'jelly' in the bottom half of each of these spotlights, to attenuate the brightness of the backlight only. The other problem is that the positions of **A** and **B** are chosen for optimum backlight, and this may not give

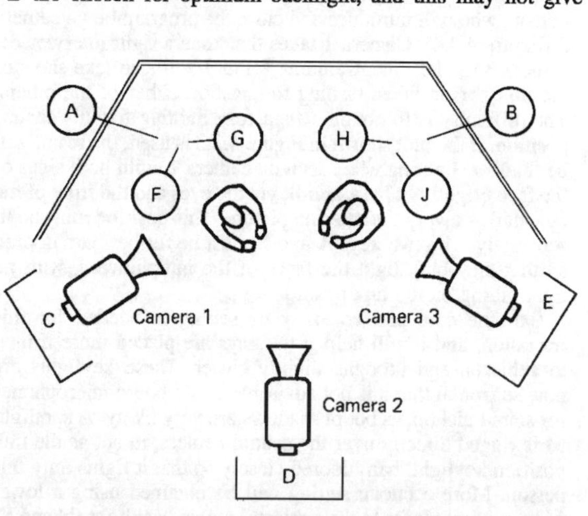

Figure 4.3.2 Basic lighting for a two-way interview.

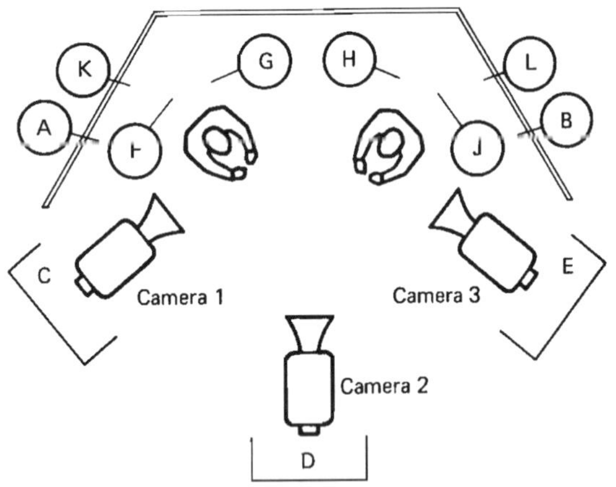

Figure 4.3.3 Improved lighting for a two-way interview with separate backlight.

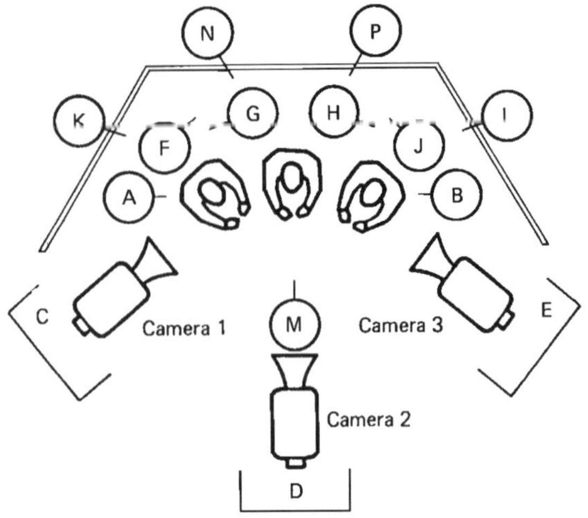

Figure 4.3.4 Lighting for a three-way interview or discussion.

a satisfactory key angle. These difficulties are overcome by rigging separate backlights, the spotlights **K** and **L** shown in Figure 4.3.3.

These backlights are placed in the ideal position, directly in line with each cross camera lens, barn doored to light only the back of the participants. The keylights **A** and **B** may be moved to a more frontal position to give better modelling and smaller 'nose shadow'. Their barn doors are set so that their light falls only on faces, and not on the backs of heads or backing. The set-up can be adapted for as many participants as necessary, so long as they are in two groups facing each other. The keylights are widened to cover each side and backlights added as necessary.

4.3.3.3 Three-way interview

The set-up for a three-way interview includes a central link person, who will introduce and close the programme on camera 2 (Figure 4.3.4). Camera 1 takes the camera right interviewee, camera 3 the left one. Cameras 1 and 3 will also take shots of the link person when turning to question either of the others. The difficulty is to obtain satisfactory lighting for this central position. If the plot shown in Figure 4.3.3 is used, the result will be 'badger' lighting when seen on camera 2, with both sides of the face brightly lit by **A** and **B**, yet the eyes and the front of the face dark—a very unflattering picture. This position must be lit separately. The two keys **A** and **B** must be further barn doored so that they only light the faces of the interviewees, with no light falling on the link position.

Extreme care and accuracy of setting is needed for this operation, and it will help if the keys are placed more frontal to subjects, and brought slightly closer. These keylights are now so frontal that it is not advisable to use boom microphones for sound pickup, as boom shadows are very likely. A spotlight **M** is rigged directly over the central camera, to act as the link position keylight, barn doored closely so that it lights only this person. More accurate setting will be obtained using a lower powered unit closer to the subject. Double backlight (**N** and **P**) that is effective from all camera positions is useful.

Sufficient separation between the link position and the interviewees must be allowed in order to light this set-up satisfactorily.

Some quiz games take the same form as a three-way interview, but are on a large scale. Instead of two interviewees, there are two teams, and a quiz master replaces the link person. The principles suggested can be adapted to lighting this situation.

4.3.4 Moving Portraiture

In circumstances where there is movement of both artists and cameras, as in a drama or a situation comedy production, the lighting requirements become more complex.

Many different treatments may be employed, but suggestions made here will continue to be based on applying portrait techniques. For simplicity the following diagrams do not show doors or windows. The effect of windows as light sources is dealt with later. It is assumed that a boom microphone will be used for sound pick-up, and therefore precautions must be taken to avoid boom shadows. All acting areas, known as *sets*, can be categorized as single-sided, two-sided or three-sided.

4.3.4.1 Single-sided sets

The single-sided set is the simplest and most limited in production values. It will be used when economy of space or cost is desirable. The backing shape is shown diagramatically in Figure 4.3.5. It may be more complex, but essentially it consists of one wall of a building. It is apparent that if either the camera or the artist is given much movement, there is insufficient backing to prevent 'shoot off' as shown from camera position 'b'. If another artist is brought in, shots can only be taken in profile from the front, on a central axis, for the same reason.

Because the camera is restricted to frontal shooting, a keylight can be placed at a suitable angle to it to portray to the desired dramatic mood. If it is positioned very frontally to the camera, it will give a high-key picture, and as it is moved

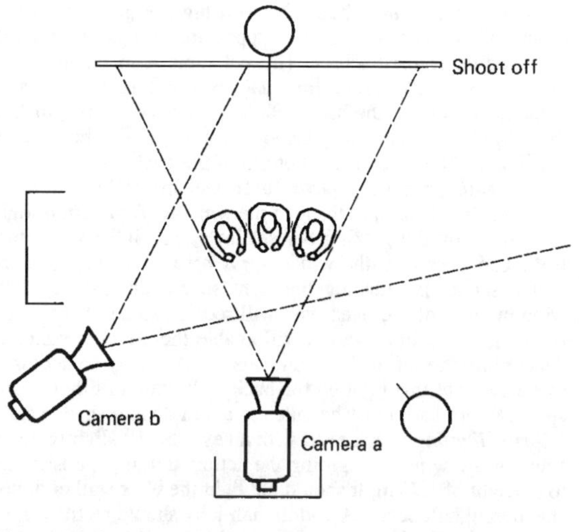

Figure 4.3.5 A single-sided set with one moving camera.

to the side, more shadow will appear, and the effect will be low-key. Softlight is placed in the same relative positions as shown in Figure 4.3.1 to control the depth of keylight shadow. Backlight is optional. If there is no window or other apparent light source behind the artist, the result may appear odd if backlight is used. Separate background lights should not be used unless very localized, as the artist may move close to the backing and distracting double shadows appear. It is better to use the keylight for this purpose, and plan in advance with the scenic designer a suitable tone for the backing.

4.3.4.2 Two-sided or L-shaped sets

These are sets consisting of two adjacent walls forming a corner. They give greater production flexibility, as now there is backing to enable cameras 1 and 2 (Figure 4.3.6) to take frontal shots of two or more artists facing each other. This is called *cross shooting*. A third camera takes a two shot.

Movement of artists is severely restricted. Action at the sides of the set will result in either shots with bad eye lines, or shoot off problems. If the action moves upstage, deeper into the corner, the cameras become cramped by the walls.

The lighting plot is based on that used for a static two-way (shown in Figure 4.3.2) but separate background lights have not been used. Lights **A** and **B**, fitted with bottom half jellies, each have three functions; keylights, backlights and background lights. It was advised earlier to have only one function to each luminaire, but here lack of space forces otherwise. If there is no artist movement these functions can be separated by using additional luminaires. Soft fill lights **C**, **D** and **E** are placed behind the cameras, and rigged as low as practicable. Keylights have to be steep in these small sets; keeping the frontal softlight low in elevation helps to get light into the eyes. Bounce clothes, large white reflectors approximately 3.5 m long and 2.5 m wide, may be used as an alternative soft fill, but care is needed to position them so that they do not impede camera or boom movement.

There is another position that can be used for action, and that is downstage, shot on camera 3 in position 'b'. In effect this can be regarded as a single-sided set, and lit as shown in Figure 4.3.5,

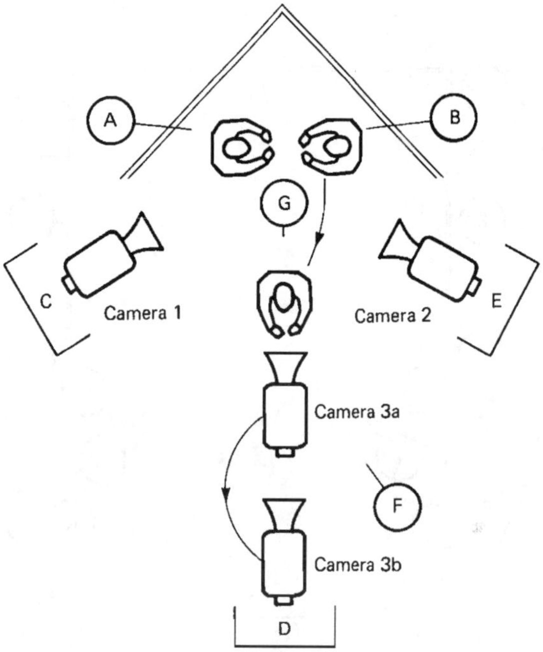

Figure 4.3.6 A two-sided or L-shaped set.

using luminaires **C**, **D**, **F** and **G**. The keylight **F** must be barn doored so that it only lights this downstage area, or boom shadows may occur in the cross shooting part. Alternatively **F** can be controlled by using the lighting console, fading it up as the action approaches this position, and taking it out afterwards if the action returns upstage.

4.3.4.3 Three-sided set

Three-walled sets offer maximum production flexibility, allowing the possibility of considerable artist and camera movement, and shots with good eyelines. The lighting plot must be equally flexible to meet these production requirements. The principle suggested is to rig keylights for all areas appropriate for the planned action, then at any moment use only those immediately necessary. By using the lighting and console in this way, good modelling should result and boom shadows avoided. A high degree of skill is needed for this way of working, but the effort is rewarded by the quality of the final pictures.

The set-up is shown in Figure 4.3.7. **A**, **B** and **C** function as keylights, backlights and background lights, each covering the same area but from different angles. **D**, **E** and **F** perform the same functions from the other side, and all six luminaires have bottom half jellies fitted. At no time is more than one from each side in use, but they can be used in any combination. Choice depends on the mood to be created and the camera positions at the time. These six should offer suitable alternative keylights for the central area of the set.

When action approaches the side walls, these spotlights become unacceptably steep, and keylights **G** and **H** cover the upstage part, while **J** and **K** light the downstage area. These spotlights must light only the side portions, or doubling with

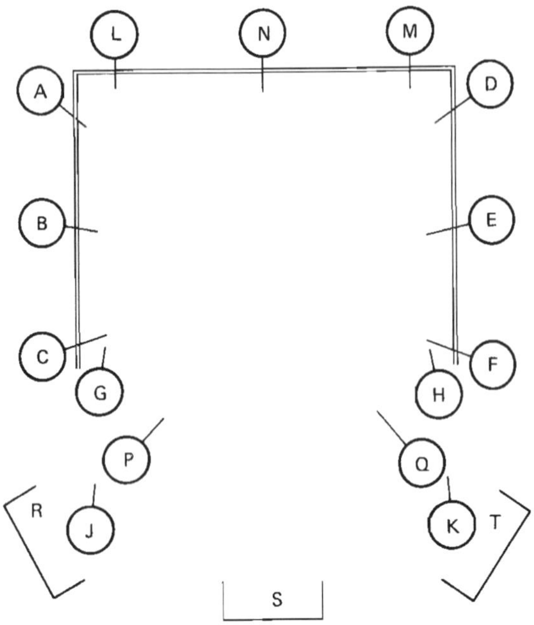

Figure 4.3.7 Lighting for a three-sided set.

the cross lights will occur. These side areas have **L** and **M** as backlights, and they should be fitted with bottom half jellies. **P** and **Q** are alternative frontal keylights, one of which should be used when the action in the middle is played out to a central camera. Backlight **N** covers this position. Softlights **R**, **S** and **T** are rigged at the front of the set, keeping their elevation low.

This plot will work well for productions employing continuous recording, and using several cameras. Care must be constantly exercised to ensure that the modelling created by a keylight is not destroyed by another, left up inadvertently. The aim should be to have only one keylight on the front of an artist at any time. The cross keys present a problem as they are also background lights, and it may become necessary on occasion to split these two functions, in order to fade out the key element. Where this is so, extra set lights are rigged.

4.3.5 Creative Lighting

So far we have considered lighting only to achieve portraiture; no attempt has been made to create a feeling of realism in the pictures. Lighting may be modified to take this into account. Some productions will call for more emphasis on glamorous portrayal than on reality, while the converse may hold for others. Additional luminaires should be added to the plot to simulate the effect from light sources, such as windows, or artificial lights that appear in the set.

4.3.5.1 Simulated daylight

Sets with windows, seen in daylight, require three groups of lighting: backcloth lighting, set lighting and artist lighting.

Backcloth lighting should be to a high brightness level in order to simulate an exterior appearance. It is most easily achieved using hard sources (fresnel spots **A** in Figure 4.3.8), but care must be taken to illuminate evenly. Space between the back of the set and the backcloth is often very limited, making these lights steep, and so ground-row units, **B**, can be added to supplement the level at the bottom of the backing.

Set lighting requires a powerful fresnel spot (**C**) positioned to simulate direct sunlight through the window. A pattern of light, modulated by the glazing bar shadows, will fall on furniture, walls and floor near the window, giving a realistic, sunny feel to the scene. By placing the light source downstage of the window, part of the back wall will receive some of the effect. Keeping it low in elevation will enable the effect to penetrate deeper into the set. If the window is small, it may be necessary to supplement this light on the back wall with a further fresnel spot (**D**), but care must be taken to avoid double shadows.

Artist lighting is achieved with a keylight (**E**) slightly downstage of the window, lighting the action and the opposite wall to a height of 1.75 m. It should not light the back wall or double shadowing will occur. A bottom half jelly should be fitted if the set is small. The key from the opposite direction (counter-key) should be well upstage (**F**), and if necessary a bottom half jelly fitted. It is barn doored off the back wall, and lights the action, but as little as possible should fall on the window wall. Lens flare can appear from camera 1 position, caused by spotlight **F**, so it may be necessary to use a flag on this luminaire. By counter keying from an upstage position, artists with their back to the window seen from camera 3 position will appear darker than those facing the window, seen in camera 1.

If more realism is required, soft sources **C** may be used instead of keylight **F**. Soft frontal fill **H** and **J** complete the plot. More soft light is rigged on the side farthest from the window (**H**), to give greater control of the shadow area created by upstage key **F**.

4.3.5.2 Artificial light sources

If a set has no specific light source visible, but is assumed to be lit conventionally by overhead and wall lighting, suitable pictures will be obtained by selecting keylights appropriate for camera and artist positions. An interior evening feeling will probably best be obtained by increasing the soft frontal light level, and so achieving a softer contrast in the pictures. Window walls can be brighter than shown in daylight scenes. If suitable to the design of the set, lit wall lights will reinforce the evening mood.

Sometimes a script will call for specific light sources in a scene, such as candle light, or firelight. Where this is so, the problems of keeping the light sources giving the effect out of camera shot must be solved at the planning stage. The effect of candle light can be simulated effectively by using a 150 W projector bulb, suitably mounted and protected, close to the candle, but hidden from camera view. A realistic firelight flicker can be obtained by feeding suitably placed floor spotlights to a sound-to-light modulator, a unit more often used on pop music programmes. Excellent pictures can result from using these effects, but all shots must be planned in detail for successful results.

Television lighting should, by its chosen style, always complement the production and create imperceptibly the appropriate mood for the viewer. The production style should be discussed and fully understood by all members of the production team, at the initial planning stage. Words are often

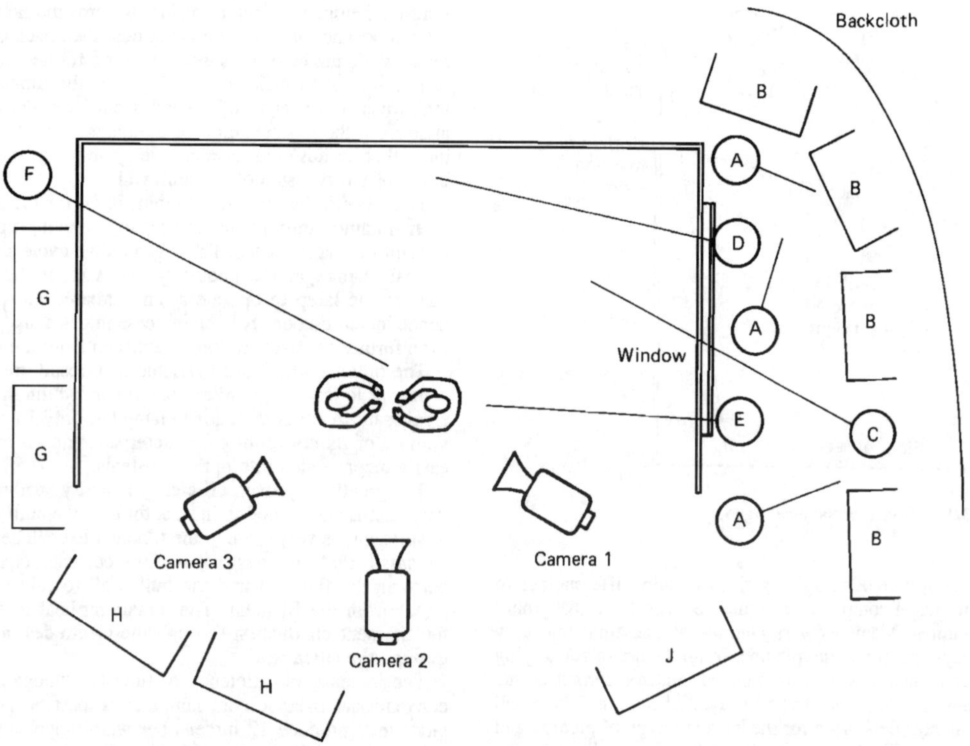

Figure 4.3.8 Lighting for a set with window light.

inadequate to describe visual concepts, and books of paintings or photographs can often assist the exchange of ideas.

The suggested treatments might range through documentary realism, theatrical (or 'heightened') realism, a delicate high key ethereal approach, or any shade or combination of these. When a coordinated style has been agreed, the lighting director is able to design a lighting plot, and balance lighting during rehearsals, constantly guided by the mental images conceived at the planning stage.

4.3.5.3 Recent techniques

By definition, creative lighting cannot be taught, but it is possible by considering the latest techniques to develop a new idea or concept to solve a particular lighting problem. Some practitioners have abandoned the use of hard key portrait lighting, believing it to be too unreal. Except in areas close to windows admitting direct sunlight, rooms are softly lit by light reflected from walls. Some lighting directors have developed ways of reproducing this effect, obtaining modelling in faces solely by the use of soft light. In practice, soft light can be obtained in a number of ways:

- Soft sources approximately 1.25 m square, mounted on floor stands or rigged overhead, can be used either singly or in a group, to produce a high intensity soft source. Floor stands are best used on large sets, when the studio is working in the rehears/record mode (i.e. breaking the production down to a series of short recordings by rehearsing part of a scene then immediately recording it). Because the production is broken down into a series of short takes, the floor stands should be less of an obstruction to camera movement.
- Sheets of fireproofed polystyrene 1.25 m by 2.5 m can be fixed horizontally to the top of set flattage. When these are lit with powerful hard sources, they reflect the light into the set, as shown in Figure 4.3.9. The light may be modified by fixing a reflecting surface to the polystyrene. If this has a metallic finish that is dimpled or perforated, a higher intensity of soft light is obtained. Some form of masking is necessary at the bottom of the reflectors, to prevent light streaking down the wall of the set below.
- Set walls that are light toned can be used as reflectors while they are out of vision. Pieces of white card can be hidden in sets, and used as reflectors when lit from above. Smaller pieces of polystyrene, 1.25 m square, fixed to lightweight stands and lit from above can be moved around more quickly and quietly than a soft luminaire.
- On some occasions it is necessary to reproduce the effect of a very powerful soft source over a large area. This can be achieved by using a large number of hard sources from the same direction. There are pitfalls in this technique, as multiple shadows will be visible if the floor or backing is plain and unbroken. It is more likely to be successful when portraying a woodland exterior, where the studio floor is covered with turf and leaves, and the background is trees and bushes. By covering the whole set with hardlights at high density from the same direction, the effect of diffused sunshine can be obtained.

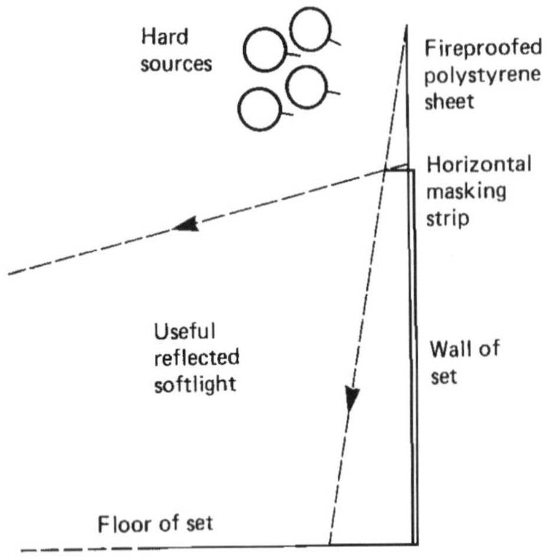

Figure 4.3.9 Reflected or 'bounce' lighting.

The soft key method of lighting gives a totally different look to the picture, when compared with that obtained from using hard key techniques. Many viewers claimed at one time that they could always spot the studio pictures. With the use of soft keying far greater realism has been introduced, and for certain drama productions this has proved to be an essential element. There will however always be a place for the hard key type of picture, and both styles are to be found in modern television studio lighting.

4.3.6 Lamps

4.3.6.1 Studio lamps

4.3.6.1.1 Incandescent filament lamps

These produce visible light as well as infrared energy as a result of the heating effect of the electric current flowing through the filament wire. Tungsten is particularly suitable for a filament material because of its high melting point and low evaporation rate at high temperatures. The filament wire is coiled to shorten the overall length and to reduce the thermal loss.

The proportion of the radiation from a filament which gives visible light increases sharply with temperature, so it is advantageous to operate the filament at the highest possible temperature. In a vacuum, this is limited by the evaporation of tungsten. The evaporation can be greatly reduced by filling the bulb with an inert gas, so the temperature can be raised without the life becoming too short, and the radiation efficiency is increased.

4.3.6.1.2 Tungsten halogen lamps

The conventional incandescent gas-filled lamp loses filament material by evaporation, much of which is deposited on the bulb wall. When halogen is added to the filling gas (and certain temperature and design conditions established), a reversible chemical reaction occurs between the tungsten and the halogen. Tungsten is evaporated from the incandescent filament, and some diffuses towards the bulb wall. Within a specified zone between the filament and the bulb wall, where the temperature conditions are favourable, the tungsten combines with the halogen. The tungsten halogen molecules diffuse towards the filament, where they dissociate, the tungsten being deposited back onto the filament, while the halogen is available for a further reaction cycle.

The improved efficiency and life of the tungsten halogen lamp over a conventional incandescent lamp does not in fact arise from the re-deposition of the tungsten onto the filament, but rather because the regenerative cycle prevents the accumulation of the tungsten of the bulb wall.

This allows the luminous output figure for the new lamp to be maintained throughout its life, so achieving up to 100 per cent lumen maintenance. The regenerative cycle also permits a radical change in the geometry and size of the lamp. This enables the lamp to operate at an increased gas pressure and hence increased density. The increase in gas density suppresses even further the evaporation of tungsten from the filament.

The options which are available as a result are higher efficiency or longer life. Furthermore, the reduction in size is quite significant. A tungsten halogen lamp has only 1 per cent of the volume of its conventional counterpart, and so may be more easily incorporated into optical systems.

The greatly improved efficiency is partly attributable to this very significant reduction in size, for now the bulb wall is made of silica and is very close to the filament itself. The clearance is so small that the gas convection currents cannot operate between the filament and the bulb wall to cause further heat losses from the filament. The remaining heat losses are from the filament conduction through the electrodes, and radiation through the silica wall.

The increase in efficiency of tungsten halogen lamps over conventional incandescent lamps is at least 50 per cent. The latter may produce 12 lumens per watt, whereas the tungsten halogen lamp will produce 15–35 lm/W.

4.3.6.2 Gas discharge lamps

4.3.6.2.1 Metal halide lamps

The few lines of visible radiation from high pressure mercury discharge lamps (MB) are shown in Figure 4.3.10(a). The absence of energy in other areas of the visible spectrum results in only moderately efficient luminosities of around 50 lm/W and extremely poor colour rendition indices of about 16–50 (out of the possible maximum of 100). It is this colour rendition aspect which is in most need of improvement when considering gas discharge lamps for use in television and film studios.

The method of improving the colour rendition of a mercury lamp, type MB, is to include more than one metal within the discharge tube, so that the emission lines occur over a wide range of the visible spectrum. However, the use of other metals as well as mercury, in the arc tube of a discharge lamp introduces problems, namely:

- The vapour pressure must be sufficiently high for these metals at the temperature of the arc tube wall to be excited into the discharge.
- The metals themselves must not react with the arc tube material or its electrodes at these temperatures.

The arc core temperature may reach 5700 K, so a silica body is used for the tube itself. The problems listed have been overcome by using halides of the metals rather than the metals themselves. The vapour pressure of the metal halide is generally higher than that of the metal itself; the reactivity in the case of alkali metals is less.

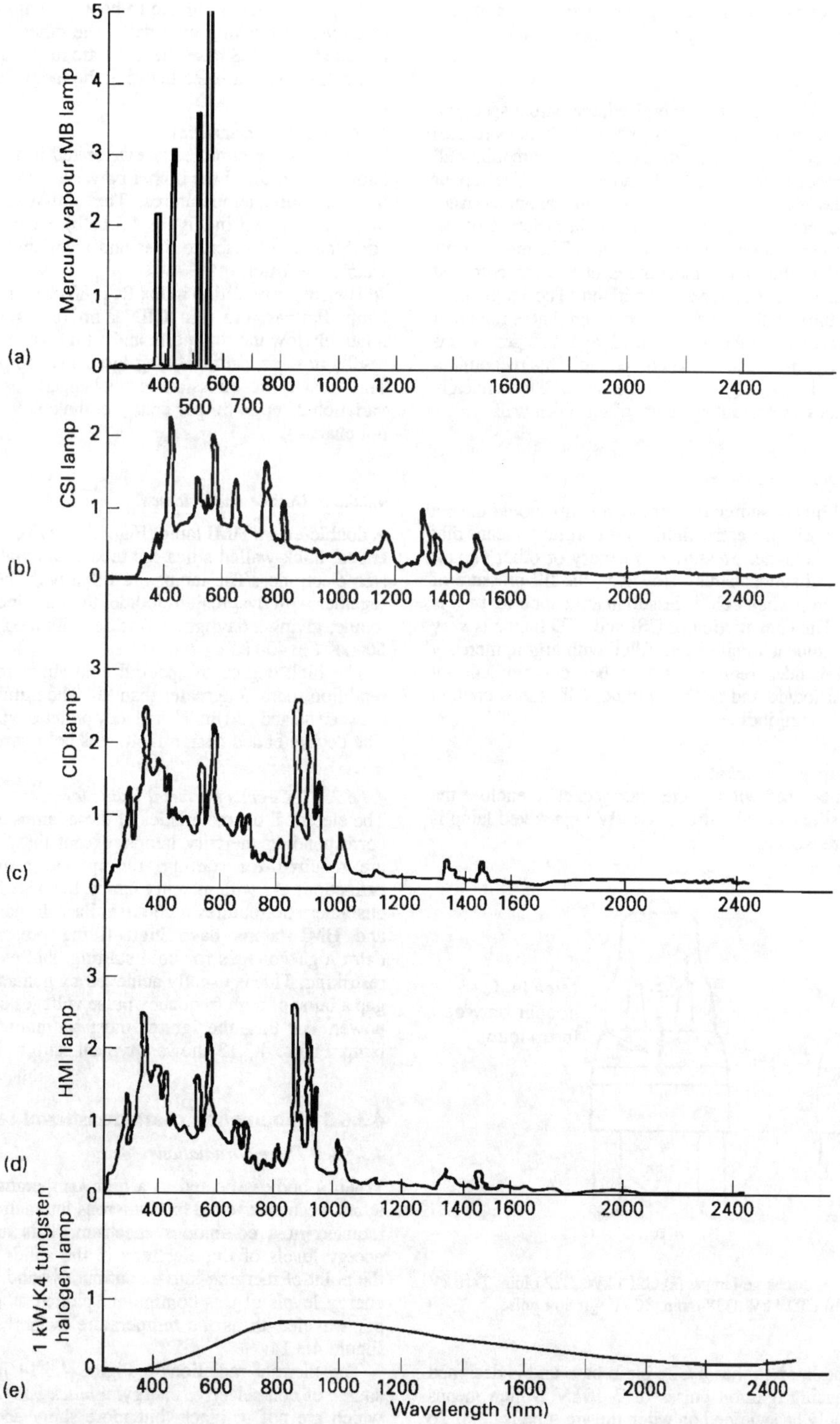

Figure 4.3.10 Lamp characteristics.

The iodide compound is used in practice. The compact source lamps, CID and CSI, may also use the chloride.

4.3.6.2.1.1 Halide cycle

When a metal halide lamp is first energized, the output spectrum is initially that due to mercury vapour, since the halides remain solidified on the relatively cool arc tube wall. As the arc tube wall temperature increases, the halides melt and vaporize. The vapour is carried into the hot region of the arc by diffusion and convection, and the temperature of the arc causes dissociation of the halide compound into halogen and metal atoms. The metal atoms are then excited in the high temperatures of the arc core and produce their characteristic spectral emission. The atoms continue to diffuse through the arc tube volume and, in the region of the relatively cool arc tube wall, metal and halogen atoms recombine in the form of the halide compound. This recombination process is particularly significant, in the case of the chemically active alkali metals, in preventing attack of the silica wall.

4.3.6.2.2 Compact source lamps

To achieve still higher source brightness, a lamp should have a short arc gap and a high electric field. This normally means that lamps operate at a higher pressure of mercury or other vapour (usually of several atmospheres). However, the fill pressure of a metal halide lamp when cold is less than atmospheric, so it is safe to handle. The construction of CSI and CID lamps is very similar. The arc tube is of silica and filled with argon, mercury and the metal halides required. CSI tubes contain gallium iodide, thallium iodide and sodium iodide. CID tubes contain tin iodide and indium iodide.

4.3.6.2.2.1 Arc tube and jackets

The lamps can be used without the glass jacket to enclose the extremely hot silica arc tube itself. A 1 kW unjacketed lamp is shown in Figure 4.3.11.

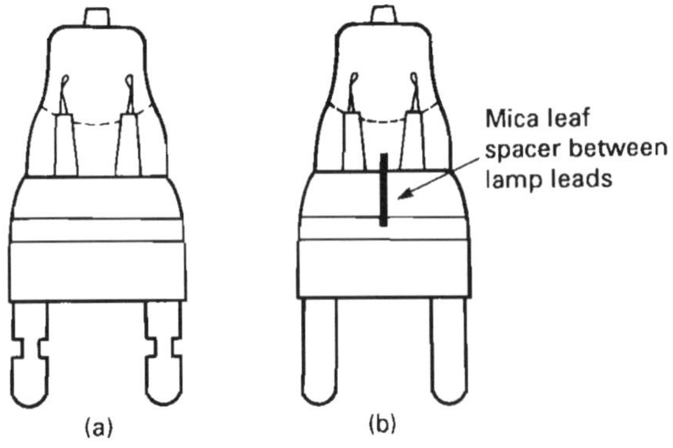

Figure 4.3.11 Gas discharge lamps: (a) CSI 1 kW, G22 bipin, 7–10 kV ignition pulse; (b) CID 1 kW G38 bipin, 30 kV ignition pulse.

The left-hand lamp (a) has a G22 bipin base and is designed for switch-on using ignition pulses of 7–10 kV, which means the lamp can only be switched on when the arc tube is cold. To achieve re-strike of the lamp while it is still hot requires an ignitor pulse of 30 kV.

To permit such a voltage to be used, without the possibility of arcing or tracking externally, the other lamp shown (b) is mounted on a G38 base. Here the arc tube base pinch is slotted and a mica leaf inserted between the lamps leads.

4.3.6.2.2.2 Characteristics

The halides are completely evaporated in the compact source lamps as a result of the higher power loading and consequently high arc tube temperatures. The emission spectra for both lamps are shown in Figure 4.3.10(b) and (c). They exhibit a combination of discrete lines and a continuum throughout the visible spectrum.

The colour rendition index R_a is 80 or more for both types of lamp. Furthermore, the CID lamp (using tin halide) has a relatively low melting point and a high vapour pressure, which results in the spectrum being insensitive to power dissipation. This means the emission is unchanged for a mains voltage variation of up to 50 per cent (i.e. the colour temperature does not change).

4.3.6.2.3 Double-ended lamps

A double-ended HMI lamp (Figure 4.3.12) consists of a double-ended, thick-walled silica arc tube with axially mounted tungsten electrodes. The lamps are filled with mercury and argon, together with dysprosium iodide, thallium iodide and holmium iodide, giving a daylight spectrum, with a color temperature of 5600 K ($\pm$400 K).

The high degree of spectral continuum results in a colour rendition index R_a greater than 90. The luminous efficiency is between 80 and 120 lm/W, and they produce typically 1100 cd/m^2. The double-ended design facilitates hot restrike operation.

4.3.6.2.3.1 Circuits for metal halide lamps

The electrical characteristics of these lamps are very similar to corresponding mercury lamps, except that increased voltages are required for starting, run-up and restriking (following extinction). As with mercury lamps, the negative slope volt/amp characteristic requires a choke ballast. In particular, CSI, CID and HMI lamps have high filling pressures and require extra high voltages for cold starting, and even higher for hot restriking. This is usually achieved by generating from a spark gap a burst of high frequency pulse voltages every half-cycle of power. For this, the ignitor must be mounted very near the lamp. Figure 4.3.13 shows a typical starter circuit.

4.3.6.3 Illumination characteristics of television lamps

4.3.6.3.1 Thermal radiation

When a body is heated to a high temperature, its constituent atoms become excited by numerous interactions, and energy is radiated in a continuous spectrum. This arises because the energy levels of the electrons in the solids are broadened to the point of merging into a continuous band. The extent of the energy levels of this continuum, plotted as packets of energy per wavelength as the temperature is increased, is shown in Figure 4.3.14.

The thermal radiation in Figure 4.3.14 is from the purest source of non-selective energy: a *black body*. For other bodies, which are not so black, but more shiny and more reflective, such a curve would not be so continuous or calculable; they cannot therefore be used as bases for reference.

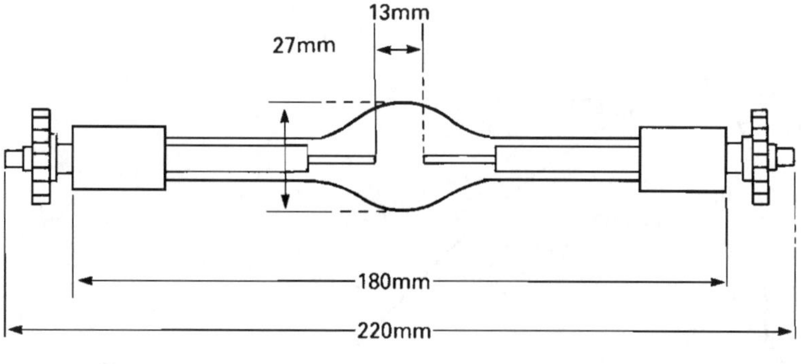

Figure 4.3.12 HMI 1200 W gas discharge lamp.

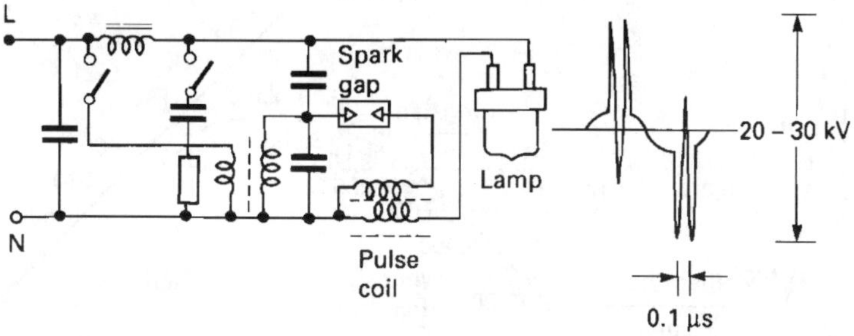

Figure 4.3.13 CSI ignition circuit for hot restart.

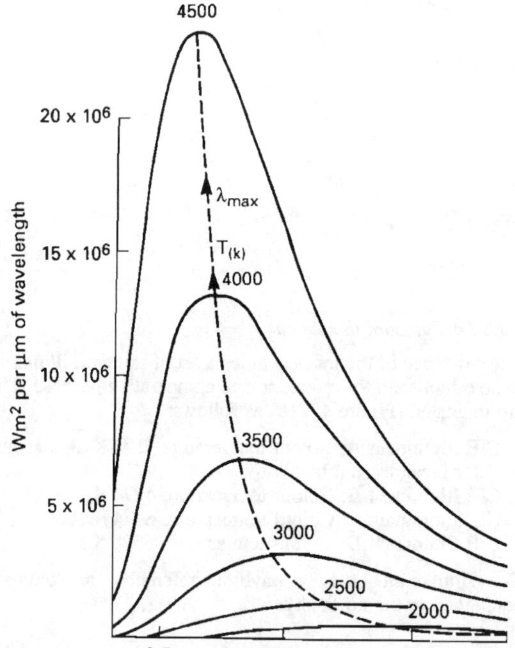

Figure 4.3.14 Black body radiation. The spectral power distribution curve for a black body radiator at various temperatures of incandescence.

4.3.6.3.2 Thermal radiation peak

It will be seen from Figure 4.3.14 that the packets of energy increase in value rapidly with increasing temperature. In fact, the wavelength of they maximum energy packet is inversely proportional to the temperature, T.

4.3.6.3.3 Colour temperature

As the temperature increases, the peak of the power radiation shifts from the red end of the spectrum, through the yellow, to the blue, and this agrees with the visual colour observation.

It is possible to calibrate this colour shift by colorimetric means, and to illustrate it on a chromaticity diagram. Such a diagram is the CIE 1970 version, which uses U and V coordinates. These chromaticity coordinates are obtained from the special power distribution curve of the black body radiator, and plotted for the various temperatures as a smooth curve called the ***black body temperature locus*** (see Figure 4.3.15).

The chromaticity measurement of a light source places it on a locus that identifies its colour temperature.

A light source that is not on this locus can nevertheless be quoted as a figure of ***correlated colour temperature***. This is the temperature of the black body radiator whose perceived colour most closely resembles that of the source.

The recommended method of determining the correlated colour temperature of a source, is to draw a line (the *ISO temperature line*) from the chromaticity of the source to cut the locus at right-angles (on the CIE 1970, UV chromaticity diagram). The ISO temperature line is shown in Figure 4.3.16.

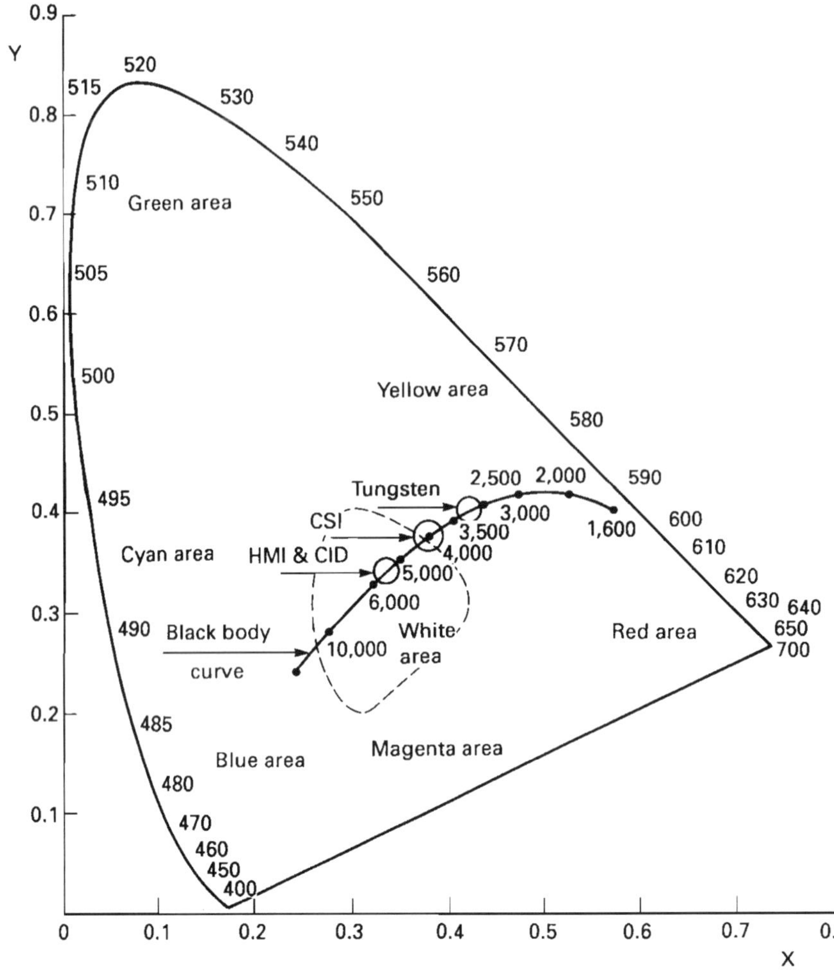

Figure 4.3.15 Colour temperature.

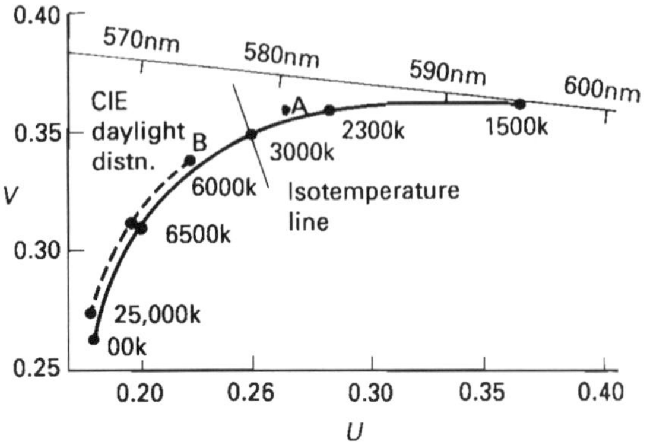

Figure 4.3.16 CIE (U, V) chromaticity diagram with black body locus and ISO temperature line; also, locus of CIE daylight illuminants.

4.3.6.3.4 Standard illuminants

The existence of the locus enables a set of standard illuminants to be established for reference and comparative purpose. These are indicated (Figure 4.3.16) as follows:

CIE Illuminant A: Colour temperature 2856 K. A standard for incandescent lamps.
CIE Illuminant B: Colour temperature 4874 K.
CIE Illuminant C: Colour temperature 6774 K.
CIE Illuminant D: Colour temperature 6500 K.

The Illuminant D is a daylight reference, approximately equivalent to the north sky.

4.3.6.3.5 Colour rendition of lamps

This is an important consideration in television operations where lamps, other than incandescent lamps, are used, i.e. gas discharge lamps. The latter do not have a continuum of radiated energy output, but large peaks of energy (or *lines*) between which little radiant energy is visible. Although they can be quoted as sources with chromaticity coordinates, e.g. with a correlated colour temperature, this information is insufficient without further details about the missing radiant energies. It results in a false representation of the colours from these lamps. However, an attempt has been made to introduce a *figure of merit*. This has had some acceptance in general lighting circles, and is called the *colour rendition index*, R_a.

The colour rendition index is derived from the colorimetric measurements made on selected Munsell chips, using a specified test lamp. The same measurements are made on the unknown lamp. The unknown lamp readings are then subtracted from the test lamp readings to establish the difference between the two sets. These colour differences are rms averaged and then labelled as ΔE. This is a single figure describing the overall error around the complete hue circle for the lamp in question for the colour samples used. In this form, however, it has not received a wide acceptance in television where a more usual unit is the *just noticeable difference* (JND). The JND approximately equals 4.6 times ΔE. Accordingly, the colour rendition index, R_a, which is a percentage scale, is introduced by subtracting this augmented colour difference error from 100, i.e. $R_a = 100$ 4.6 AE.

The colour rendition indices for studio lamps are:

tungsten halogen:	$R_a = 100$
high pressure mercury vapour:	$R_a = 16-48$
metal halide discharge:	$R_a = 70-90$

The metal halide lamps, such as CSI, CID and HMI, are considered to be satisfactory for outside broadcast television use.

Chris Thorpe
CTP Systems Ltd

4.4 Talkback and Communications Systems

Talkback systems derive their roots from the public exchange telephone system. In many cases the systems are not dissimilar today with the concept of a central switching unit, or exchange, connecting and routing many incoming and outgoing circuits.

In the television production environment different technical and production areas have their own priorities and so must work in acoustic isolation. The studio floor or recording area must be quiet so all operators require headsets for communications with other areas. All areas must be able to communicate efficiently if the production process is to operate smoothly. Talkback systems have evolved dramatically over the years. Systems which were set up and operated by the sound supervisor or sound engineer have now increased in complexity to the point of requiring at least one communications engineer, and often more, to operate and service the system.

A talkback system in a studio or outside broadcast environment is normally based around a mainframe audio switching matrix which may operate in the analogue or digital domain, where inputs and outputs, sources and destinations, may be selected to provide communication paths between areas. The system is analogous to a telephone exchange with its switched lines between telephone subscribers.

Talkback panels, cameras, four-wire boxes, radio talkback units, Integrated Service Digital Network (ISDN) lines and belt packs are all connected together via the talkback matrix.

Talkback systems have evolved in two different formats, two-wire and four-wire. American systems have in the past been based around two-wire talkback with users wearing headsets, while European productions have generally preferred four-wire talkback systems using loudspeakers and stalk microphones, although there is now a large amount of overlap between the two.

4.4.1 Four-Wire Circuits

The four-wire circuit is so called because it requires four wires to complete a two-way communications circuit. Figure 4.4.1 shows the principles of the basic four-wire circuit. Each direction of the circuit requires two wires and is a balanced line, although an earth screen is often included to help prevent interference pick-up, particularly on long cable runs. Signal levels are normally the order of 2 volts peak to peak, which also helps to keep interference levels low.

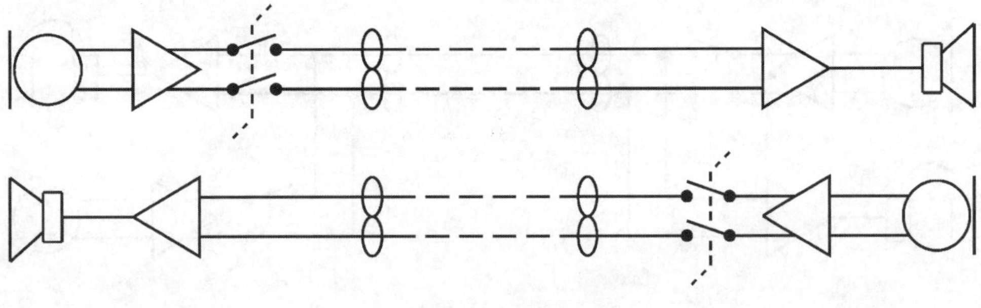

Figure 4.4.1 Basic four-wire circuit.

Each party has a microphone with microphone amplifier and a talk key to connect the microphone amplifier to the circuit. When the key is not pressed, the circuit is not made and silence is heard at the remote end. Press the key to make the circuit and the microphone output will be heard at the remote end through the loudspeaker.

This circuit works perfectly on the condition that only one party is keying to the other at a given time. Should both parties key at once then there will be audio 'howl-round' when the output of one party's loudspeaker will feed into their microphone and be sent to the other party, where the same will occur, making a continuous audio loop. For this reason, when keying to talk, the local loudspeaker will either be 'dimmed' (dropped in level) or cut for the duration of the key press.

Dimming of the loudspeaker is preferable so the user can still hear reverse conversation from the remote end. A good quality, well-positioned, directional talkback microphone can help by partially rejecting the loudspeaker output from the rear and enabling a higher loudspeaker level to be achieved while keying.

Four-wire circuits require more intensive cabling than a two-wire system as each circuit is individually connected to its destination. For this reason it is also known as P-P or point-to-point. Circuits are normally connected together using three-pin XLRs, where pin 1 of the XLR will be a screen to help reject interference.

Figure 4.4.2 shows a typical small four-wire box. This unit will handle two four-wire circuits and has a headphone output for studio floor use. Refinements include 'audio present' LEDs which illuminate when a remote user is talking on the line.

4.4.2 Two-Wire Circuits

The two-wire circuit, shown in Figure 4.4.3, has many operational similarities to a standard telephone line and was designed primarily for headset use. Rather than employing separate circuits for each direction as with the four-wire system, the two-wire connection modulates many voice channels onto the same line. The two-wire interconnection only requires one live connection and one reference (or earth) to support a number of users communicating; each circuit used is commonly called a 'loop'. One advantage of two-wire operation is that interconnection between users is easy. With the four-wire system each user must be individually connected to the circuit destination, while with a two-wire system only one pair need be run out

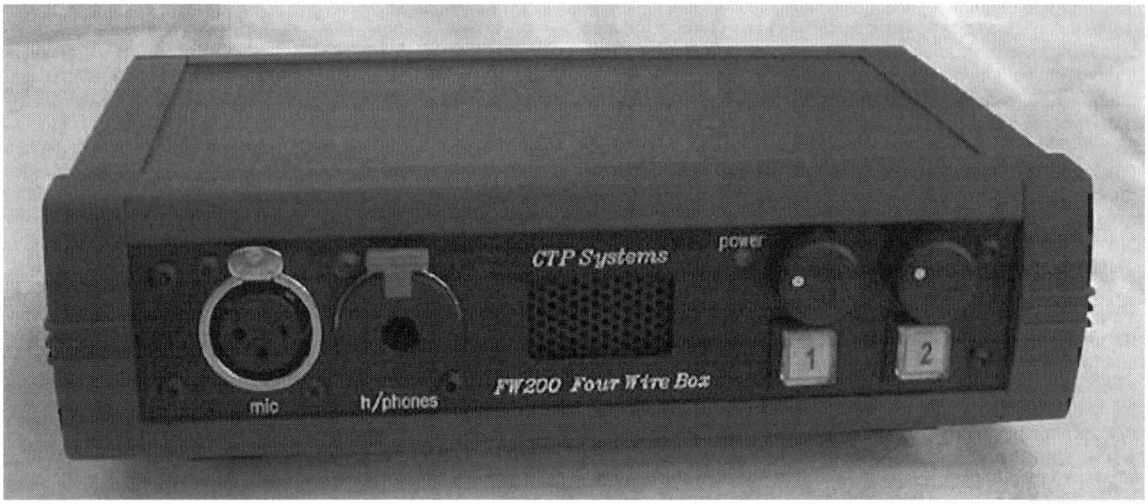

Figure 4.4.2 A typical four-wire box.

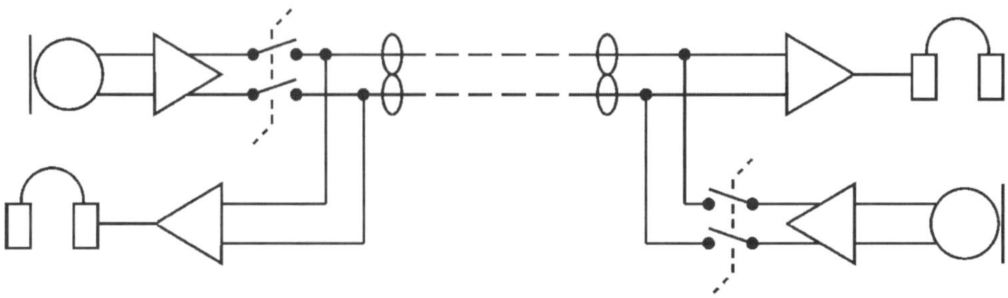

Figure 4.4.3 Basic two-wire circuit.

from the system power supply unit and individual users may be connected with cabling looped from one user to another. Audio signals are unbalanced and circuits are at a relatively low impedance of 200 ohms to help keep interference levels low. Although generally known as a two-wire system, practical units normally use three connections, invariably using the standard three-pin XLR connector. Clearcom systems use two wires for the communication circuit with the third wire supplying power. RTS systems use a design enabling two wires to carry both power and one audio circuit. The third wire is then available to carry a second audio only circuit. This is made possible by using 'impedance generator electronics',[1] making the power supply line appear as a 200-ohm resistor at audio frequencies. Pin 2 of the XLR will carry audio and power and pin 3 audio only, with pin 1 as common earth. Powered channels are known as 'wet' channels and audio only as 'dry' channels. Power supply voltage is nominally 24–32 volts dc and operating range will vary according to the gauge of the wire used and the number of units on the line. A range in excess of 1 km can be expected with two belt packs on the line.

If all users are wearing headsets the two-wire system works perfectly. Disadvantages of the two-wire system become more apparent when many individual talkback channels are required or if users are on microphone/loudspeaker panels rather than headsets. Microphone and loudspeaker combinations, when used with two-wire circuits, must include electronic circuitry to prevent the user's voice returning through their own loudspeaker and creating an audio 'howl-round'. This is analogous to using a speakerphone on a telephone line, and requires level reduction on the loudspeaker whenever the user talks on the line, or 'ducking'. Although it works successfully with one or two microphone/speaker units on the line, it becomes increasing difficult to balance the line as more units are connected and the system can become acoustically unstable. For this reason a complete talkback system often comprises microphone/loudspeaker panels working into a four-wire mainframe with two-wire belt pack units connected into the system using two- to four-wire converters.

Two-wire systems from different manufacturers are rarely compatible and will require an interface unit for interconnection. For this reason it is preferred to use two-wire products from one source in an installation.

4.4.3 IFB Circuits

The IFB or 'interruptible foldback' circuit (Figure 4.4.4) is a variation on the four-wire circuit and is often used for presenter talkback. A one-way balanced circuit is used, with an additional input circuit, or 'standing feed', which is normally present on the circuit output line, unless the talk key is pressed when the IFB through circuit is cut or 'interrupted' and replaced with the operator's voice. This arrangement is frequently used to feed programme sound or another audio source to a presenter with the facility for a producer or director to 'over key' or talk over the programme audio to the presenter to issue instructions or directions.

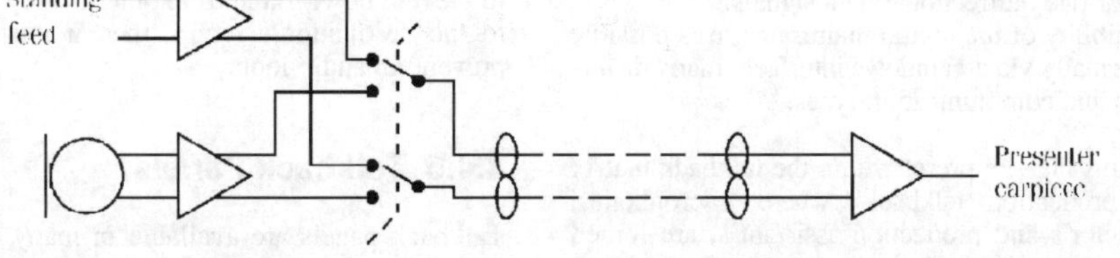

Figure 4.4.4 IFB circuit.

4.4.4 The Talkback Matrix

The four-wire mainframe switching matrix is the hub of a four-wire talkback system with all talkback panels, incoming and outgoing four wires, radio talkback, ISDN lines and cameras connected to this unit.

The mainframe may be a hard-wired unit where pressing a key on a talkback panel will make a circuit dedicated to a given destination, or may be a PC programmable system where the destination of each key on a talkback panel may be programmed.

Studio and outside broadcast requirements of a talkback matrix can vary dramatically. Most studio environments are reasonably static, with a fixed number of cameras and other facilities, the only additional requirement normally being external four wires for communication with remote locations. Once set up, a studio system will rarely require alteration. By contrast, outside broadcast use requires enormous flexibility in a system where requirements will vary on a day-to-day basis and the system will constantly be changing. Four-wire mainframe systems are now invariably PC programmable.

The programmable mainframe consists of a point-to-point matrix of switched cross points. These cross points may be analogue field effect transistor (FET) switches, electronically connecting input and output lines together in the analogue domain, or digital, where signal paths are routed and manipulated in the digital domain. The principle remains the same. Figure 4.4.5 shows the basic principle of cross point operation.

For smaller systems, digitally controlled analogue frames can be more cost-effective, while for larger systems (48 inputs/outputs and above) analogue systems would require far too many analogue switches, with a 64 square system requiring a total of 4096 FET switches or 128 square system requiring over 16,000 switches. The solution here is to go digital and perform analogue-to-digital conversion on all audio inputs, pass all digits through a digital signal processor (DSP) connecting all required cross points in software, and passing all routed outputs thorough digital-to-analogue converters. Once the audio signals are in the digital domain it becomes possible to control many aspects of the audio path, including cross point levels, all with near CD quality.

Typically a digital matrix mainframe consists of an array of circuit boards, each processing eight input and output ports,

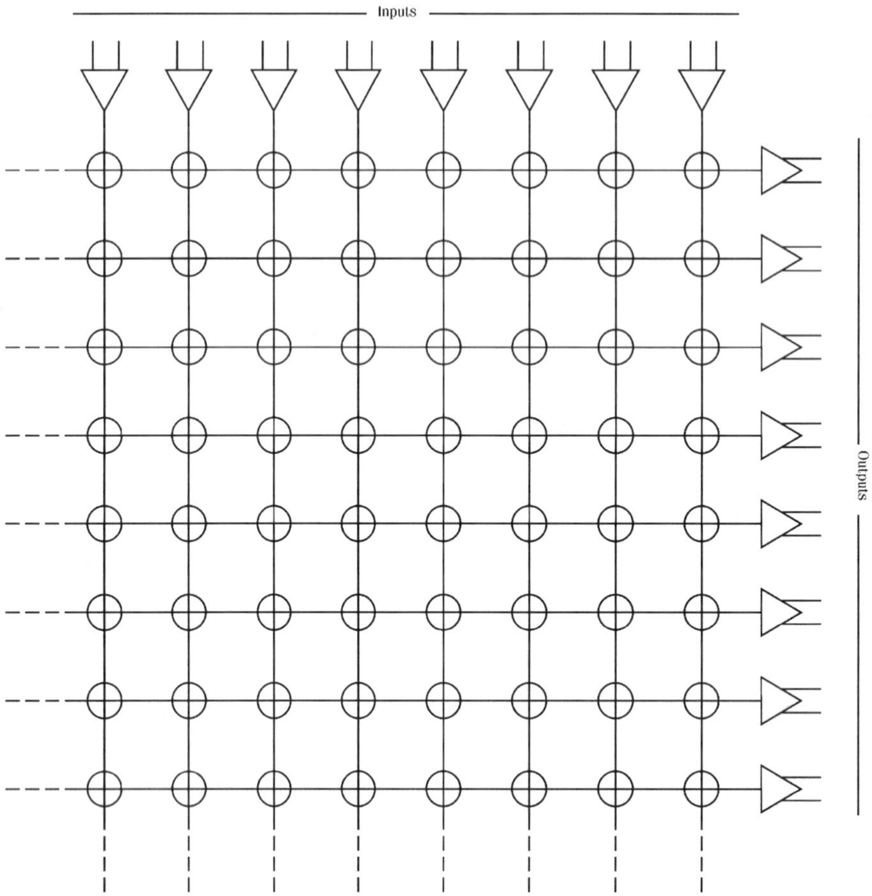

Figure 4.4.5 Four-wire talkback matrix.

allowing in excess of 136 users or input/output ports in a seven-rack unit mainframe size. Interconnection of mainframes may be achieved using just two coaxial cables for larger matrix sizes up to 1000 ports. Because so much of a production relies on efficient communications, the unit will include dual redundant power supply units with automatic switchover in case of failure, this being considered the most vulnerable part of the electronic system. Audio inputs and outputs to the frame are made via ports, which are either nine-pin 'd' type connectors, RJ11 or RJ45 connectors. For radio talkback, ISDN, camera and similar four-wire circuits these ports will carry just audio signals. For the talkback panels these ports will also carry also either two single or one bidirectional data signals.

Such is the flexibility of the digital mainframe, it is possible to programme, normally via a Windows interface, many different configurations and communication types:

- Fixed cross points may be preset within the talkback matrix to form 'open production talkback' where microphones, normally the director's and production assistant's, are forced on at all times to selected destinations so operators can hear directions without any users having to press keys.
- ISOs, or isolated communications, are easily provided for where two operators may maintain private communications between themselves. A typical example is where the audio operator may require a private conversation with the vision engineer without disturbing production talkback or other communications.
- Grouped communications, where one operator may talk to many listeners using only one key.
- Party line, where many people may talk and listen on the same line. This is similar in use to a two-wire party line.
- Group dimming can be important in areas fitted with more than one panel. When a destination is keyed, then the return from that destination will be dimmed or cut on the keying panel to prevent howl-round. If another nearby panel is also listening to the destination's return, then it must also be dimmed to prevent an audio loop.

4.4.5 Talkback Panels

Talkback panels are available in many formats and sizes for desktop and 19-inch rack mounting in one, two and three rack units. Facilities include stalk microphone and loudspeaker or headset use. Push buttons or keys are provided to talk to other stations or external devices and the user can control incoming volume levels and select which inputs to listen to.

PC programmability allows flexibility of panel layout as all keys may be set to talk to any destination and so may be set up to the user's preference. On more advanced systems, panels may include LED or LCD displays with port names, where the destination port number of the key is replaced with a 'label' where the name or title of the destination is displayed. Some allow the user to reprogramme keys and select listen sources. If the system is fitted with ISDN dial-up facilities it may be possible to initiate the circuit from the talkback panel.

Connections from the mainframe to the talkback panels may either be fully digital, with all audio and control information passed between panel and mainframe in a serial digital stream, or, more commonly, a mixture of analogue and digital, where the audio to and from the talkback panel is carried over a pair of balanced lines with the data carried on either one bidirectional RS485 balanced line or over a pair of single direction balanced lines. For outside broadcast use remote panel connections are often broken out to individual XLRs for multi-way connection.

4.4.6 Cameras

Camera talkback systems vary wildly from model to model and manufacturer to manufacturer. Some models may be switched to operate with either two- or four-wire talkback systems, some offer just one four-wire circuit (one talk and one listen) and some include separate four-wire circuits for talkback to production and to engineering. Cameras with two separate four-wire circuits are simple to interface to a talkback system where the destination of each circuit is either to the director or engineering. Less simple is dealing with a camera with a single four-wire circuit where the reverse talkback may be destined either for production or for engineering talk. The solution to this problem is a talkback divert system.

Production talkback is normally sent to the camera. When the engineer keys to the camera his voice will also be heard at the camera. Whenever the engineer's key is pressed, any reverse talkback from the camera will be diverted away from production and heard by the vision engineer. On releasing the talkback key the diversion will be disabled and the cameraman will have normal talkback with the director.

This sequence may be programmed into the more advanced talkback systems. Dedicated units to deal with camera talkback tend to be more cost-effective than using relatively expensive talkback ports. These units only require one port carrying production talkback, with all vision engineer feeds and reverses being derived from the camera talkback switcher.

Two-wire interfacing of cameras can be more difficult and less stable. Often the camera talkback systems have been built as an afterthought and the line impedance of some camera systems is 600 ohms as against the more normal two-wire system impedance of 200 ohms.

The most effective interface to a two-wire-based camera is by use of a two- to four-wire converter, where the system can be manually balanced to help prevent system instability.

4.4.7 Belt Packs

The two-wire belt pack system is immensely popular because of its inherent simplicity. Belt packs may allow either one or two channels of PL (party line) conversation over one XLR cable. They may be 'daisy chain' connected and operational distances in excess of 1 km are possible.

Units have one or two volume controls and talk keys, often also with a call light, which illuminates when audio is on the loop. The talk keys are pressed to talk or tapped to latch on; the latch-on feature can be disabled via internal dipswitches. Belt packs are designed to operate with headsets only and some units have an additional balanced input for a programme sound feed.

A complete two-channel talkback system may be made up using just belt packs and a power supply unit. This may be enough for very simple television productions and is used extensively in the theatre world. More commonly the units are used in conjunction with two- to four-wire converters and interfaced to a matrix talkback system.

Rack mount versions of the belt packs are available, often with four channels for the more complicated two-wire set-ups.

4.4.8 Source Assignment Switcher

In some circumstances many two-wire loops may be operated at once and more flexibility is required than simply hard-wiring systems together. The source assignment switcher or 'SAP' unit is a passive switchbox enabling the user to switch any available loop to any output. Typically this unit allows two of 16 loops to be switched to any belt pack unit supporting up to 26 dual-channel belt packs. Care must be taken that channel 1 to each belt pack is a powered channel. The SAP unit can effectively make the simple two-wire circuit an assignable intercom system.

4.4.9 Two-Wire IFB Systems

Dedicated presenter IFB units are manufactured and work in much the same manner as a normal two-wire circuit but are one way; the presenter cannot talk back on the line. These are particularly popular with American production teams. The IFB standing feed is normally programme sound and the panel operator can over-key the programme in the normal IFB manner. Where the system differs is that many talk key panels can be connected, each having switches to select that panel's priority. If a producer tries to talk to the presenter at the same time as the director, if the director's panel has a higher priority than the producer's, then the director will be heard by the presenter, not the producer. When used with stereo headphones it is normal for the system to be set up not to cut the programme sound, but for the programme to be in one earphone and the panel user in the other.

4.4.10 Two- to Four-Wire Converters

As two-wire systems carry speech for both directions on a single pair and four-wire systems use two discrete pairs they cannot be simply connected together and so signal conversion is required.

This is achieved by using a two- to four-wire converter unit, which compares the four-wire audio being sent to the two-wire circuit with the signal being returned from the two-wire circuit and rejects or nulls any common components of the audio signals. This system is somewhat imperfect as changing line conditions, which can be caused by changing the number of units on the two-wire line or operators altering volume controls, can throw the system off balance. The converters have a 'null' control to alter the signal comparison characteristics to allow for these changes with varying degrees of success. When the system is not properly balanced a familiar 'ringing' will be heard on the circuits or when badly unbalanced the system will howl. Careful set-up will result in a usable system. Best results are produced by the newer generation digital converter units, which use a digital signal processor algorithm to constantly analyse the line and adjust the audio signals dynamically, eliminating the need to readjust the system as more users are added.

4.4.11 Telephone Balance Units

Telephone balance units (TBUs) are used for interfacing four-wire systems to the public telephone network. In essence, TBU operation is similar to the two- to four-wire converter, but is designed to operate more efficiently with telephone line characteristics. A simple but effective TBU may be made using just one or two transformers;[2] the circuit is shown in Figure 4.4.6. The transformer makes use of the out-of-phase windings to reject any common signal between the four-wire send and the telephone line, and the four-wire return and the telephone line.

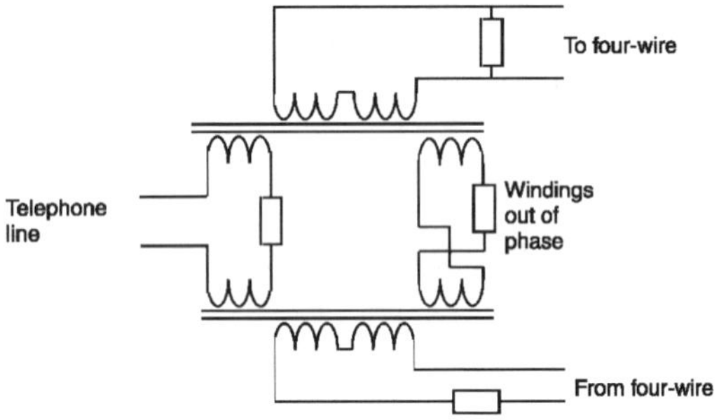

Figure 4.4.6 Telephone hybrid.

A rejection ratio of 40–50 dB is possible with this circuit. Most professional analogue TBUs are based around this kind of circuit, but with some active circuitry added to improve performance. As with the two- to four-wire converter, much improvement on this circuit can be achieved by digitising the signals and performing signal rejection using digital signal processing algorithms before converting back to analogue.

A telephone instrument is connected to the TBU to establish a call and often the TBU may be fitted with a ring detector to automatically answer incoming calls and switch them to the four-wire interface.

4.4.12 Radio Talkback

Several different methods of utilising radio talkback exist. Its simplest form is the walkie-talkie or half-duplex operation where two users can talk to each other, but only one at a time and working on one single frequency. Useful for setting up but not for on-air use.

More common is the use of a base station working in full-duplex mode with half-duplex portables. This allows all users to hear continuously transmitted talkback, otherwise known as constant carrier talkback. Two frequencies are required for this form of operation, one for base station transmit, another for hand portable transmit. The disadvantage of this system is the handset operator cannot hear the base station transmissions while transmitting themselves. It is preferable for some users, particularly floor managers, to have two units, one handset for transmit and another for receive. As the floor manager will still hear production while transmitting, this is almost, but not quite, a full-duplex system. If two handsets transmit at the same time intermodulation will occur[3] and neither handset will be heard, just an unpleasant noise.

A true full-duplex system requires a separate base receiver and different frequency transmitter for each handset user. Although ideal, due to the equipment and licensing costs, this system is rarely used.

Talk-through is commonly used on base station units, allowing all handset users to hear all transmissions to the base station as well as the normal talkback feed. This is achieved by mixing the base station receiver audio output back into the transmitter audio input, along with transmitter talkback feed. Level controls can be added to allow adjustment of the mix.

Operating power and radio frequencies for radio talkback use are restricted and care must be taken to work within guidelines.[4] A licence is required for each frequency used. For outside broadcast use it is important that the radio talkback equipment is frequency agile. Often, events may have many units on site and licensed frequencies may be in short supply. Maximum radiated power from a constant carrier base station unit is 5 watts and the maximum handset power is 1 watt. Operating frequencies range from 455 to 468 MHz. If a good operating range is to be realised it is important to site aerials carefully in free air, and particularly away from metal objects. Base station receiver aerials need particular attention due to the relatively low-power output of hand portable units.

References

1. Professional Intercom. RTS PS31 Power Supply Technical Manual/TM4437, *The Reference Book* II, RTS Systems.
2. www.lundahi.se/typelist.html
3. Turkington, T. *A User's Guide to Wireless Intercom.*
4. www.jfmg.co.uk

Peter Bruce
Marketing Manager, Production Mixers
Thomson Grass Valley

4.5 Mixers and Switchers

Production switchers have benefited greatly since the early 1990s from digitisation. Analogue vision mixers suffered heavily with the requirement of many adjustments through the whole signal process. Analogue drifting was inevitable through the production mixer. If invisible effects were required critical adjustments were essential for luminance timing, gain, dc levels, horizontal and sc phase of all inputs, Mix Effect banks, keys, internal colour mattes, etc. Additionally, since the advent of digital production mixers tedious alignment of wipe generators is now a thing of the past. Present-day digital mixers have by nature no signal alignment and perfectly transparent source paths for invisible mixes or effects. Parallel digital input mixers have been replaced by serial input mixers since the early 1990s. However, to reduce processing speeds most production switchers use parallel processing internally after de-serialisation. The digitisation of production mixers has also enabled an easy switch from 4:3 or 16:9 and 625 to 525 formats and in high definition can easily process in many variation formats with a quick change of set-up or automatic detection of reference signals. The advent of the digital production mixer has also meant that the processing can be made in Component Serial Digital with full bandwidth 10-bit 4:2:2 for standard and high definition processing with a compact electronics unit. New technology such as FPGA (Field Programmable Gate Array) technology has meant that the physical hardware is rapidly reducing in size and increasing in features and functionality. However, the principles of production mixers have remained the same.

4.5.1 Panel Description

The production mixer has its own terminology and it is important to understand the functionality of a production mixer and the terminology that goes with it. Production mixers can be seen to have two main functions, the live and the post-production mixers. The two functions are not mutually exclusive. However, there are several abilities a live production mixer is capable of that a post-production mixer is not, and vice versa. For a live production mixer the main importance is the ability to get to main operational functions directly and quickly without the use of submenus. For the post-production mixer this is not such an issue. However, remote editor control and layering is of greater importance.

When we describe the panel there are several things that are common and typical to all production mixers. Referring to Figure 4.5.1 you will see two distinct areas called M/E and PP. The M/E or Mix Effect banks (section A) are the area in a live production mixer in which effects can be set up or programmed off-line. The PP stage refers to the Program/Preset (section B) stage. The 'Program Bus' (usually the upper bus selection buttons) is always live to air. The Preset selection Bus is where the video sources are selected prior to going to 'live on air'. Above the main video source selection buttons, the top row of buttons are usually the key source selection buttons. Above each bus name, source displays indicate the source associated to each button. Sometimes inlays are set into the button. However, electronic source name displays allow dynamic set-up of the mixer. As the operational conditions change the mnemonic will follow the source on the panel.

The Transition Module (section C) is where the operator preselects the type of transition required. This selection would be classified as a Cut or Take in which an instantaneous switch of the Program/Preset Bus would happen. A Mix is a transition in which one video signal dissolves or ramps to another. A Wipe is a transition in which a preselected pattern forms the transition between the two sources. The Auto or Take button will trigger the Mix or Wipe transition at a predefined rate, usually set in frames. The type of wipe can sometimes be selected directly on the panel (section D). The Transition Module also preselects any type of transition combination of direct video sources or keys. The fader allows manual transition of the selected effects.

Section E shows the E-Mem recall. This area stores and recalls operational set-ups of the mixer used during the live production. Section C is the key selection. This is the area in

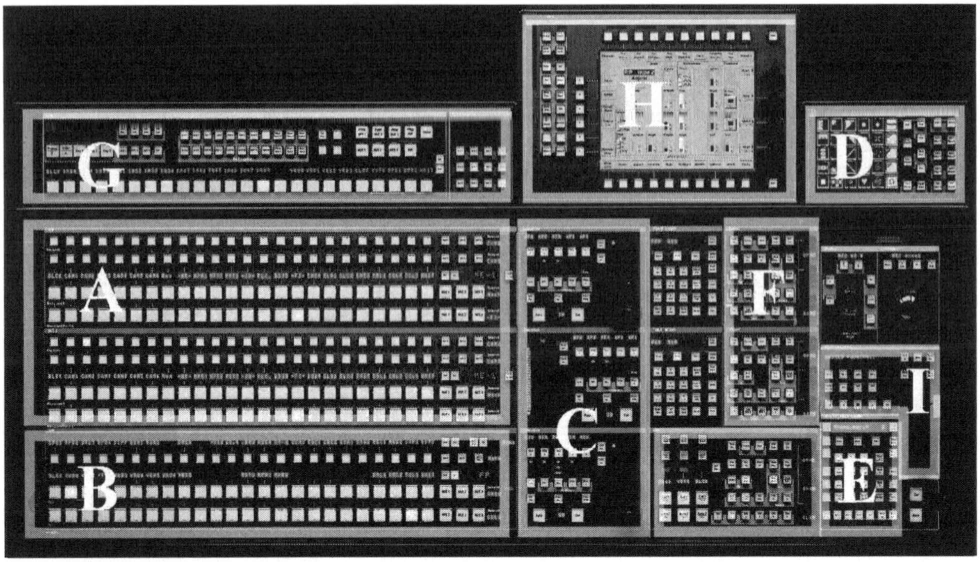

Figure 4.5.1 Control panel of a production mixer.

which all key set-up adjustments are made. The top and often raised panel (Section G) is sometimes called the 'splash back' area and is typically used to control Auxiliary outputs, Pre-router cross-points, Preview selection, internal and external DVE input selections. The screen area or Graphical User Interface (Section H) is used for installation/configuration of the mixer as well as critical set-up and running of operational effects. Most production mixers incorporate a Joystick or Track ball for re-positioning wipes and creating internal DVE moves. As the vision mixer is now able to control external digital disk recorders and VTRs, the bottom right area enables the use of control with timecode. The bottom right also illustrates 'FTB' or Fade To Black button which, when selected, will force black only on the program outputs. This is used for emergencies or when the production is finished.

4.5.2 M/E Structure

The 'M/E Bus' or Mix Effect Cross-point Bus Row is a row of push buttons that allows immediate selection of video or routing of sources. These could be the M/Es, Program, Preset, key, mask, Aux external or internal routing.

The Program Bus or 'PGM' is usually the upper row of sources in the downstream bus row, called Program because it is always 'a live output'. Whatever is selected here is seen in the output of the program of the mixer. The Program Bus 'flips' selected sources with the Preset Bus once the transition has occurred. The Preset Bus or 'PST' is the lower row of sources in the 'downstream' bus row. It is called the Preset Bus as it allows the operator to make preselection and preview of the video sources before transitions are made to air. Pushing the Cut or Auto Transition buttons activates the transitions. The 'Cut' or 'Take' button allows an instantaneous switch of the source between the two selected sources. The Auto Trans button initiates a preselected transition over a given time. This could be any selection of Mix, which is a cross-fade of levels between the Program and Preset sources; a wipe, which is a transition made from an internally generated pattern; or a DVE transition, which uses an internal or external Digital Video Effects generator for producing the effects.

4.5.3 Engineering

Timing of sources has changed dramatically since the days of analogue, compared with the more recent digital production mixers. Analogue has meant that the input video must be timed to a very close tolerance. Timing inputs first meant adjusting the horizontal phase of each input compared to reference. If the horizontal phase is not exactly set then a horizontal picture jump is visible on the program output when making a mix or wipe between different sources. Then the subcarrier phase would have to be adjusted per source until no chrominance phase errors were visible when making transitions. To allow stable recordings of the mixer output most analogue mixers would reinsert synchronising and burst pulses on to the program output.

Digital timing is different. All digital mixers have an auto-phasing range in which the serial digital video inputs should fall within. There is always a trade-off. The larger the auto-phase range window the larger the delay through the mixer. Several production mixers incorporate a frame synchroniser on each input. This eliminates the need for any timing of the input source. The disadvantage of this is that the incoming video is now a frame late and therefore produces audio/video lip sync problems and causes problems if the production requires read before right editing. For mixers that do not have an input frame synchroniser there is still a timing to be made for each input. However, if this timing is incorrect the input source will drop lines. Unlike analogue, the digital timing window will be many microseconds and more tolerant.

Throughout any mixer the propagation delays produce timing planes at the various stages through the mixer. Due to the electronic processing the timing of the Mix Effect bank will inevitably be earlier then the program output. Any studio designer should take this into account. However, some digital production mixers incorporate auto-phasers on the Auxiliary and main outputs. This then allows any direct input (before the delay) to be switched with the Mix Effect bank output or program output without a change of phase. The timing of the phased outputs are normally as close to the main output as possible (see Figure 4.5.2).

4.5.4 Production Mixer Workflow

With production mixers we often use the terms upstream and downstream. Upstream refers to the switcher functions nearest the mixer's inputs such as the M/Es, while downstream refers to the switcher functions nearest the switcher output towards the Program/Preset or PGM/PST Buses, which are nearest to the switcher output (downstream from the other buses). Within the system each M/E has its own key layers on its own bus (see Figure 4.5.3). After the PGM/PST stage the keyers are referred to as DSKs or DownStream Keys. This is because they are, after all, the other video processing and therefore their key layers are always on top of everything previous.

As described there are several levels of internal buses. Typically each M/E has a primary source for background A and background B: a key selection bus and a video mask selection. Modern digital mixers use the term for another type of internal bus selection called the Utility Bus. This has two main purposes. The first is a separate bus with the Mix Effect Bus that is used for routing of internal video mattes, video in wipe boarders or video used for creating special wipes. The other use is for

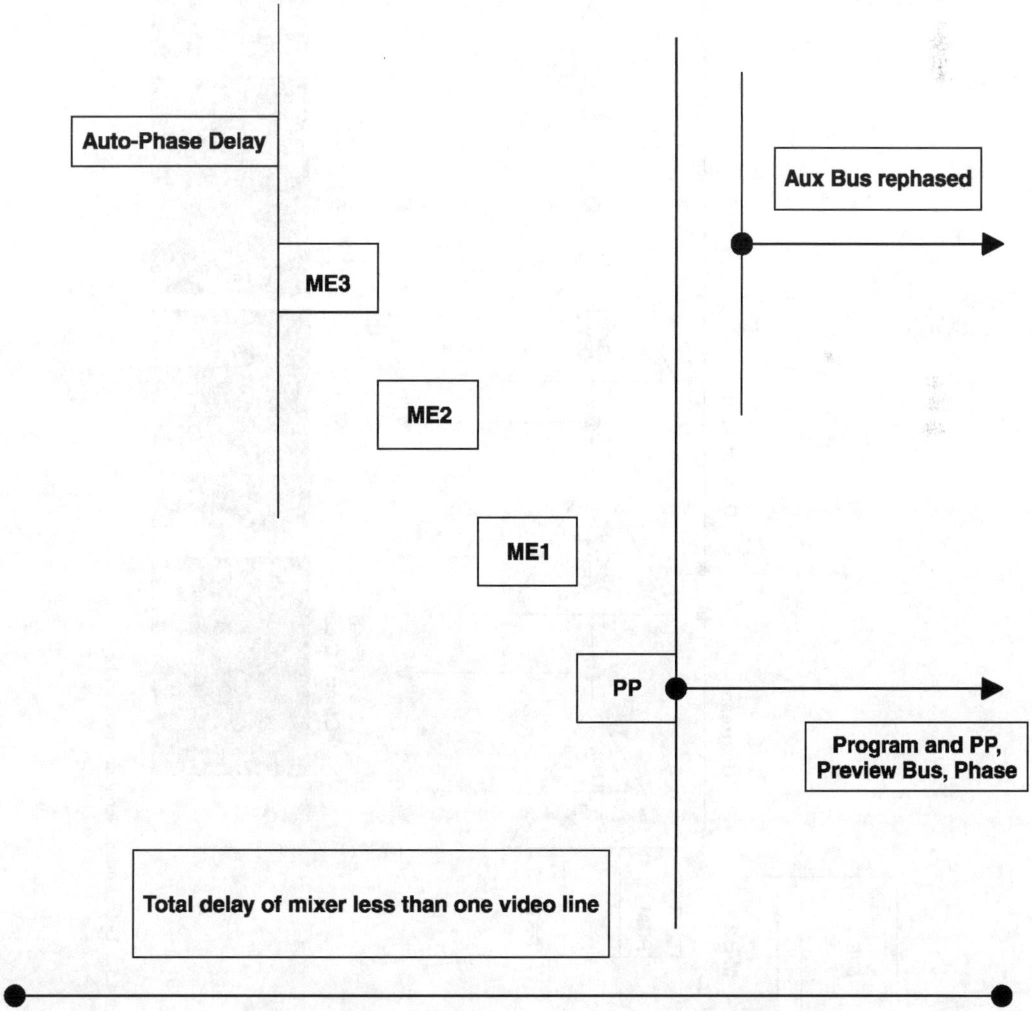

Figure 4.5.2 A typical digital vision mixer's timing plane.

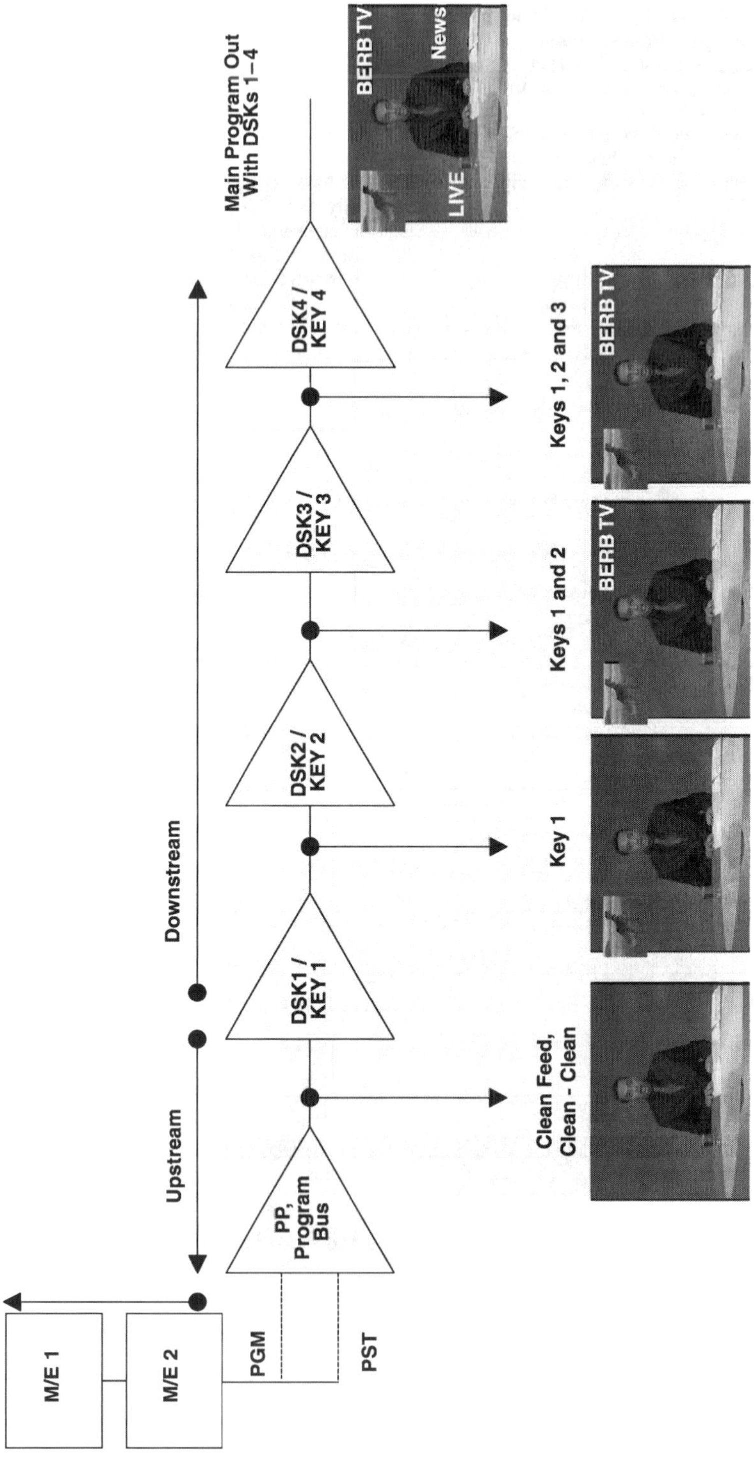

Figure 4.5.3 Upstream and serial downstream keys.

a separate bus on the output to the mix effect for assignable clean feeds, DVE sends, etc.

Clean feeds are separate outputs that are 'clean' of all keys. On early mixers the clean feed would be an output before the DownStream Keys. As demands of producers have grown mixers may give several clean feeds out at different points of the mixer. For example, a production may require several clean feeds of the program out. One clean feed recording of the production without any keys. This can sometimes be called 'clean clean'. A second clean feed may be needed with Key 1 showing the logo of the network but not Key 2 showing a logo that the output is live, so that the recording may be replayed without the 'Live' logo at a later time.

As more modern production mixers allow four or more downstream keys, new requirements have meant that a cross-point bus for the clean feeds have been incorporated such that any combinations can be output on to several clean feed outputs. For example, Keys 1 and 3 are required on one clean feed output and Keys 1 and 4 on the other. Meanwhile the program would take all keys. As the number of cross-point buses increased, developments have now allowed each M/E to be split with two program outputs, one with Background A and B with Key 1 and Key 2, the other output with Utility Buses 1 and 2 using Keys 3 and 4. Splitting the M/Es in this way then doubles the power of each M/E in certain applications (see Figure 4.5.4).

4.5.5 Transition Module

The Transition Module is critical to the operational control of the mixer. This section of the panel interface controls the running of effects. The Cut button allows immediate switching of one signal source to another. The Dissolve or Mix button

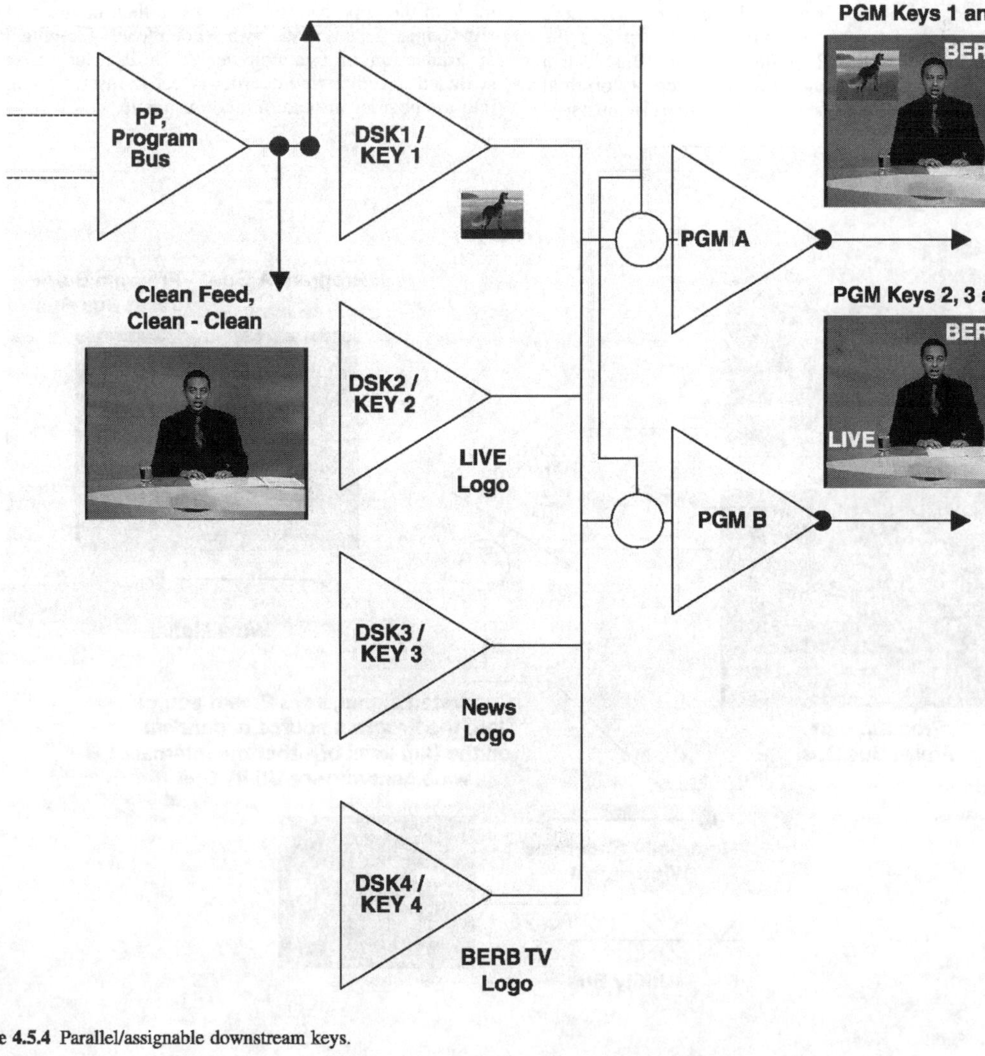

Figure 4.5.4 Parallel/assignable downstream keys.

allows what is a more common term of an additive mix in which gradual transitions in equal amounts of luminance are added from both video sources. A Non-Additive Mix could also be selected which is a mix where the source with more luminance is selected to predominate.

The Auto Transition activates any predefined effect, e.g. Cut, Wipe, DVE or Mix. Transitions will have associated transition duration in frames set by the operator. A manual transition is made by moving the fader arm. If the operator does not require the wipe or dissolve to be fully completed the Preset pattern Limit button allows the transition to only continue to a predefined point. Moving the fader to the required position and then pressing the Limit Set button defines this point.

4.5.6 Mix, Wipe Patterns and Effects

A Mix is a more common term for an additive mix, in which gradual transitions in equal amounts of luminance are added from both video sources. A Non-Additive Mix (NAM) could also be selected, which is a mix in which the source with more luminance is selected to predominate. An NAM passes the higher levels of either the Program or Preset Bus. When starting the NAM, the Program Bus will keep its original brightness while the preset source mixes to full intensity. The higher luminance of the Preset Bus replaces the lower luminance of the Program Bus. At the crossover point the preset picture mixes out the darker areas of the Program Bus until the original program source fades away. The bus then switches and the preset is now the program source. Wipes are effects traditionally only available in the Mix Effects banks. Modern mixers allow wipe effects on the Program Preset bank. The wipe is an internally created pattern using an internal key to transition between the Program A and Program B Bus. For example, the operator requires a vertical wipe transition between the on-air source and the preset. The mixer would produce an internally generated ramp. As the auto transition is made the mixer internally keys the Preset Bus further up the key ramp level, thus producing a vertical transition across the screen (see Figure 4.5.5). As the demands for effects has increased most mixers have increased the complexity of processing. This has become easier as effects are now digitally generated and therefore precise and stable compared to their analogue past. If a soft boarder is required between the Program and Preset an internal graduation is made to the wipe pattern. Colour can then be inserted on the wipe edges. This is made by a second level of keying a colour or graduation of two colours. A Utility Bus can also be switched into the wipe boarder effect to insert moving video into the boarder instead of a colour matte.

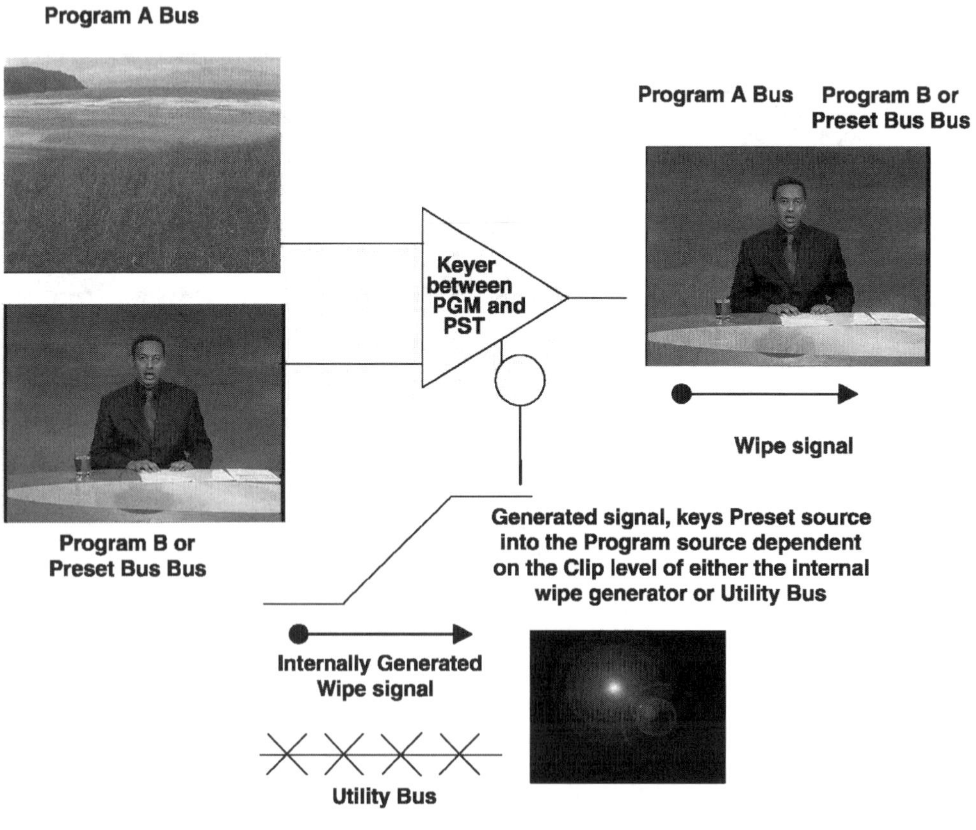

Figure 4.5.5

4.5.7 Utility Wipe

Another way of branding or creating effects is by use of what is known as a utility wipe. Instead of using an internal created pattern, a graphic or moving source is used to increasingly transition between the Program and Preset Buses of that Mix Effect bank. This is done by keying the background from the source on the Utility Bus. As the transition starts the background is keyed on to the foreground, depending on the clip level of keying from the Utility Bus' luminance level. As the fader arm or Auto Transition moves from 0% the keying starts to key on black. As the transition moves in percentage from 0 to 100% so the clip level moves transitioning to key a greater amount of the Preset Bus. The video in the Utility Bus is keyed until the transition is 100%, thus keying the preset source fully onto the background until the bus switches (see Figure 4.5.5).

A flying key transition is another form of transition that has changed the way in which production mixers create effects. Since vision mixers have increased external machine control they have been used in collaboration with hard disk recorders. As the hard disk recorders allow almost instantaneous cue and play for video and key signals, this has allowed more creative ways of transitioning. Graphics departments create 'flying videos with associated keys' loaded onto the hard disk recorder. These graphics are made up such that they start off with no visible video. As the mixer cues and then plays the effect from the Digital Disk Recorder (DDR) the graphic increases in size to either full frame or transition across the picture frame. An E-Mem is then recalled to Cue up the DDR – recalling a background user wipe. Then the key incorporating the graphics is switched on. As the DDR is triggered to play the background user wipe or cut effect is activated. The background source is discretely transitioned behind the keyed graphics.

4.5.8 Previewing

For the live on-air production mixer the ability to preview effects before going to air is essential. There are three main modes of previewing: these are Look Ahead Preview, Transition Preview and Key Preview.

For Look Ahead Preview each M/E of a production mixer normally has a 'look ahead preview' output. This output shows you what will happen if you were to move the lever arm and is based on what is currently on air and which next transition buttons you have selected. For the PGM-PST bank, that's what's next to air if you do a transition on that bank.

Some production mixers have what is referred to as the switched preview or Aux Preview Bus. This is a selector row that allows you to rapidly get to the look ahead previews or perhaps any source or row or cross-point bus within the mixer. In many production mixers all M/E outputs are routed back to the cross-point matrix, so any Aux Bus has access to the M/E preview signals. This allows any Aux Bus to be used for additional preview outputs if desired. The switched preview or Aux preview output can be programmed to follow certain operations automatically (push to preview and show key signal; Chroma key auto set-up; mask, etc.).

Some production mixers' M/E preview outputs have an 'Auto' mode that shows the look ahead when the M/E was on air and the program output when it was not. The assumption is that if the M/E is on air, you see that signal on the switcher's main program output *and* you may be planning to do a transition on that M/E. If the M/E is not on air, the next 'logical' operation is to put it on air, so we show you the program output.

4.5.9 Keying

A key is a video that overlays or cuts a hole in the background. A key is a way of layering video. Every key consists of a hole cutter or key signal and a key fill. Traditionally each Mix Effect bank will have two keys. Modern digital mixers have enabled up to four keys. The key priority allows the layers to change. The selection of keying can be Luminance, Linear, Chrominance, Pattern and Mask keys. Once the hole is cut from the choice keying types, the decision by the operator is to choose what video source will fill the hole. The fill can be a Self key, internally generated matte, or a video bus.

A Luminance key is simply a key that is dependent on the luminance or brightness of the key source. There are two ways of adjusting a Luminance key. These are: 'Clip and Gain' and 'Clean Up' and 'Density'. With Clip and Gain the key is generated by adjusting the Clip or level of luminance at which a key cuts its hole. As the clip level is adjusted higher a greater amount of video fill will happen. The Gain adjustment electronically varies the graduation angle between the key on and key off. As the Gain is adjusted higher the harsher the edge will appear on the output of the key (see Figure 4.5.6). When a Luminance key is adjusted using 'Clean Up' and 'Density' the two adjustments work such that the 'Clean Up' adjusts the key from the highest part of the luminance going down, while the Density adjusts from the darkest part of the luminance up. This mode is implemented when keying on highlights such as sky or low dark levels. An Inverted key is a key in which the whole is based on a low luminance. On screen the background and key fill video are interchanged. A Linear key is a Luminance key with specific Clip values of 50% and Gain of 100%. The purpose is that the key will perfectly reflect the incoming key signal. What you enter as a key value will be reflected linearly as the output. Therefore, if the input key is 50% the video fill will be 50% transparent. Typically Linear keys are used for graphics when the key setting is predefined. Older analogue mixers use a separate linear key input known as ISO key or 'Isolated key'. In more modern digital production mixers the graphics key signal is entered with the primary video sources and then assigned as a key.

Consideration of the keying video source is important. If an external video coming, for example, from a graphics station is not shaped, a linear non-equalised or unshaped key is required. This is a multiplicative key. If the incoming video is already shaped then an additive key is required which is unshaped. If the wrong key is selected then either a light halo is created around the key or a dark surround, as illustrated in Figure 4.5.7.

A Pattern key is a key in which the key is generated using the production mixer's internally generated wipes to create the key layer.

A Chroma key is a key that uses colour in the key source to differentiate between the foreground and background. Typically the colour in a studio for keying would be blue or green. Blue is most common. However, green is often used as this colour is almost 100% opposite to skin tone in the colour circle. When making a Chroma key the mixer will replace all differentiated colour with the background and retain all opposite parts. As a requirement the chrominance level of the background must be as highly saturated as possible. The Chroma

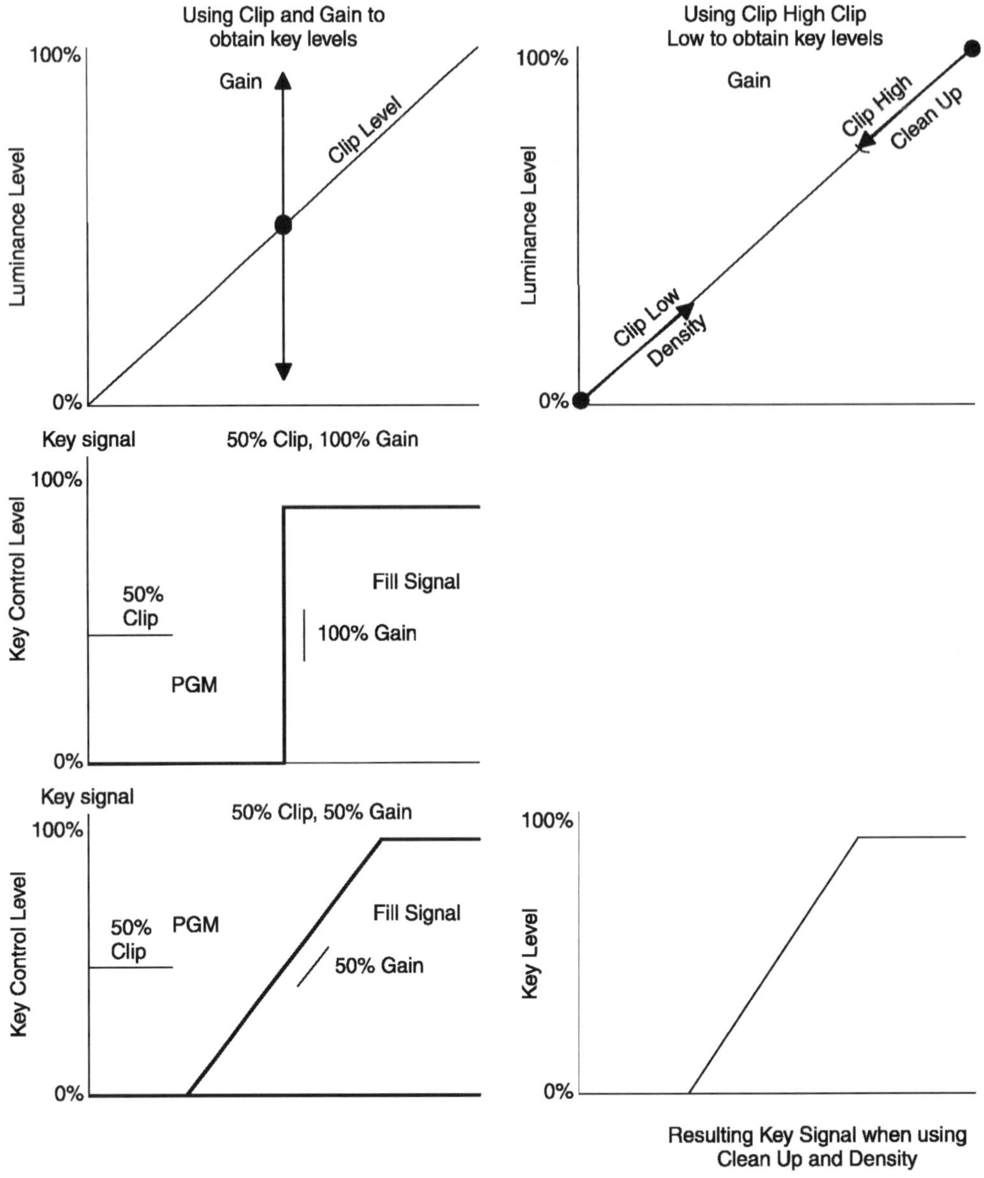

Figure 4.5.6 Differences of luminance keying using clip and gain (LEFT) and clean up and density (RIGHT).

key process works by suppressing the background colour in the original scene and then creating a key hole in which to layer the original key signal onto. A high-quality Chroma key requires several stages for a seamless key. This is because it is harder to differentiate a Chroma key source with varying levels and the key edge will inevitably contain a proportion of the original background colour. A high-quality Chroma key will typically contain a primary and secondary colour suppression. The primary colour suppression replaces the selected colours before replacing them with black, then replacing it with the new background video. The primary colour suppression has low selectivity, suppressing a wide range of colours. The secondary suppression is normally of a higher selectivity so as to recreate the detail of the edges (see Figure 4.5.8).

Key memory is a system with the control in which the key set-up is stored per source. When this is enabled all settings are remembered for each incoming source.

A DSK or downstream key is the last layer in a sequence of keys on the switcher. Older analogue mixers had specific inputs for these DSKs which were assigned for Titleing. Digital production mixers have the facility to key from any primary input. If a video and key are being sent from a caption generator a

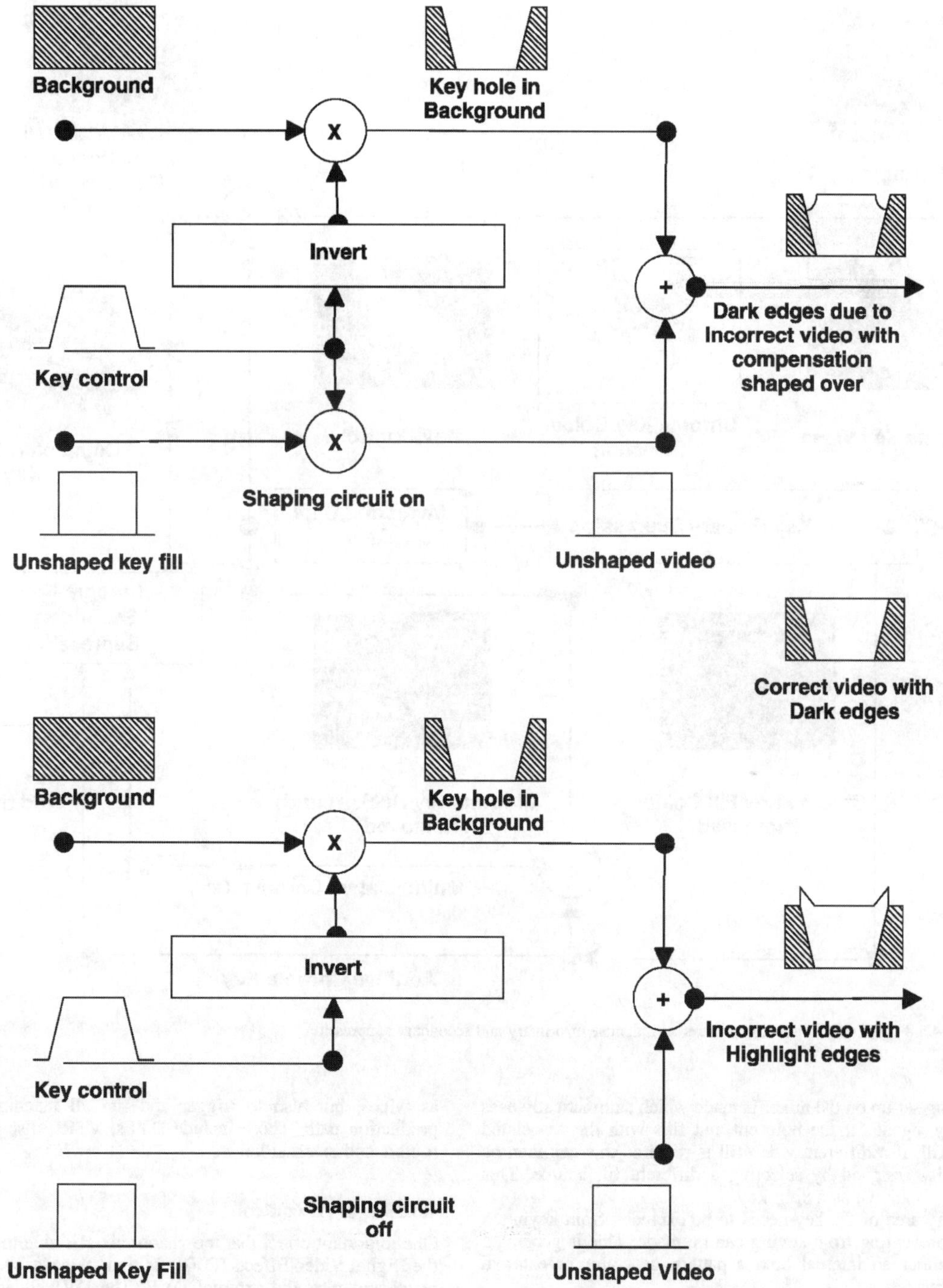

Figure 4.5.7 The result of incorrect shaped and unshaped keying.

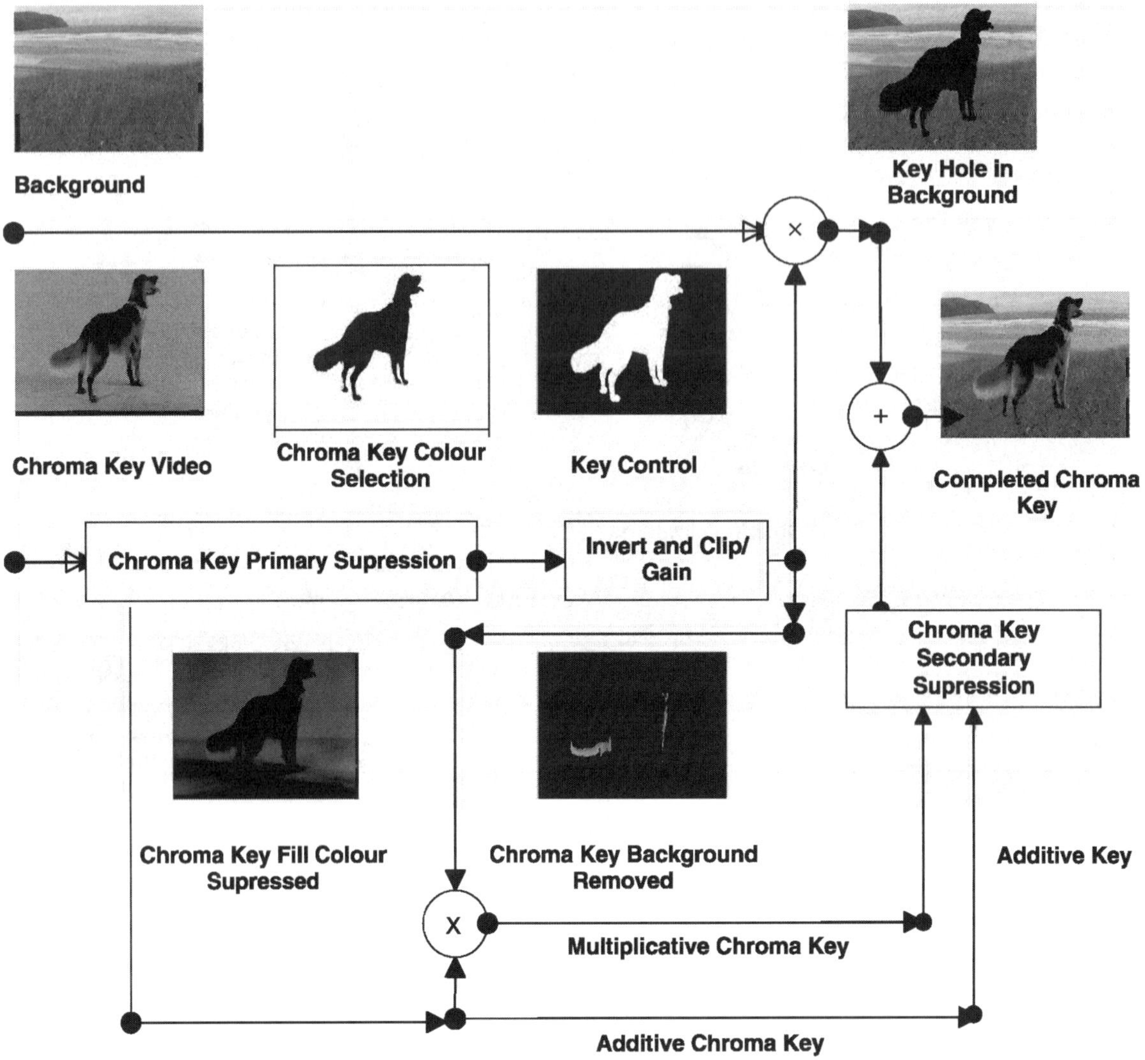

Figure 4.5.8 The full Chroma key process with the use of primary and secondary suppression.

coupling set-up on the mixer is made which automatically uses the key signal for the hole cut and fills with the associated video fill. If a different video fill is required this can then be manually assigned by selecting a different fill source. This process is known as a Split key.

If any area of the key needs to be excluded from keying, a mask preventing from keying can be made. This is generated from either an internal box, a pattern generator or external video bus.

4.5.10 Machine Control from the Mixer

The production switcher has taken on a central role in the gallery, not only to control the mixer's internal effects such as wipes, but also to trigger and run all machines in the production path. These include DVEs, VTRs, disk recorders, routers and external keyers.

4.5.10.1 DVE control

One important effect that the vision mixer is able to control is the Digital Video Effects (DVE). Initially this was the ability to recall and play the external DVE. The DVE video and key would then be inserted on to the M/E using normal keying techniques on one of the main keys. Closer integration has been available as the recalling of DVE effects is selectable from the mixer panel. Even more tighter integration has been available since the mixer has been able to create an effect send of video from either its background or internally generated key.

The returned video and fill from the DVE is then automatically rekeyed either back on to the main Program B Bus or associated key. The DVE effect is then rescaled in time to run for the length of the auto trans time of the production mixer or manually from the fader arm, thus running the whole effect as a DVE loop from anywhere in the mixer and back into the same place from the mixer. When a background DVE transition is required instead of a wipe effect the selected Program source is routed out to effect send. The condition of a seamless transition effect is that the DVE effect starts with a full frame. As the effect runs the returned video and key is reinserted on to the background or preset source. The second condition for a seamless DVE transition is that the DVE effect should end out of frame. Once the DVE effect is run the Preset and Program cross-points are exchanged and the DVE's video and key are switched out of the loop. One problem can be that the DVE by nature has a frame delay through the system, therefore a frame jump might be perceived when switching in and out of the DVE loop. To counteract this most systems allow for that bank to permanently loop through the M/E DVE.

4.5.10.2 Router Control

High-end production mixers also have the ability to control the external pre-router. This can be particularly useful when there is a limited amount of inputs to the mixer or the operator would need to reconfigure the studio rapidly. From the mixer panel it is possible to reassign the output cross-points from the router into the mixer.

4.5.11 External Devices Controlling the Mixer

External control of the production mixer may take many forms. The most basic is a GPI or General Purpose Interface. Most production mixers can accept a GPI in the form of a contact closure or TTL driven pulse. The pulse can then be associated with an action on the vision mixer. This could be an auto transition, cut, run, an E-Mem or initiate a key. For control from an external editor most productions comply with the GVG 100 or GVG 200 protocol. This is an RS422 control and allows control in the form of switching Program and Preset Mix effect sources, recalling SMPTE wipe patterns, key selections and E-Mem effect recall. Although control is important another critical feature is for the editor to upload and download E-Mems to and from the editor. The importance is that when the editor makes an auto assembly the production mixer will recall the exact state when the original effect was made. Although the GVG 100 and 200 protocols are about the only real open protocols, many manufacturers allow control from their own native commands to more closely control their feature set.

4.5.11.1 Audio Follow Video

As control becomes more extensive in production mixers, Audio Follow Video has been developed in some production mixers. The idea is not to replace sound production crew but an aid to enhance audio production. Some productions use this feature to raise the audio attached to the video sources. For example, imagine microphones attached to a camera during a tennis match. When the tennis player hits the ball the microphone connected to the camera will be switched on and the level is raised. When the camera on the wide shot is taken to air then the microphone on the audience is then switched on. In areas in which studios need a more powerful transmission mixer, Audio Follow Video has enabled some studios to allow the remote audio desk to follow the video sources. The production mixer sends a string of commands using standard protocols when each cross-point changes. If an auto transition is made on the assigned Mix Effect bank of the production mixer a string will be sent to include the transition rate, forcing the audio mixer to do a cross-fade.

4.5.12 VTR and Disk Recorder Control

From around 1990 external machine control has been incorporated in production mixers. Initially the control was basic, sending simple play, stop, rewind, fast forward, etc. using a standard VTR protocol. Since around 1998 more complete control has been implemented within high-level production mixers. This has enabled the operator to read specific timecode from the external VTRs, and cue and trigger videotape machines within the program. The recent revolution has come when the RS422 and LAN control ports have been connected to hard disk recorders. The advancement to production has come due to the almost instantaneous cueing of the hard disk recorder and the reading of a clip list. The ability to make many cues and recall these cues from one push of a button using E-Mems or macro has been particularly prevalent in sports. The production mixer can now trigger the hard disk recorder to cue a video and key graphics. The cue points may be a graphic 'Sting' that indicates the replay.

4.5.13 Application Backup and System Set-Up Storage

The set-up storage of a production should not only store the engineering set-ups of the production mixer, but also the operational set-up of all effects. Almost all production mixers have some form of storage, so that after shutdown the mixer can then be reloaded to a known state. Usually this basic backup is in the form of Non-Volatile RAM. Ideally the RAM is updated every time the mixer condition changes, so that when the mixer is unexpectedly shut down, it can then be re-powered to the exact state before power down. As the effects during productions have increased in complexity and the need to transfer effects and set-ups has increased, removable and reloadable set-ups have been required. These set-ups can now be transferred using standard $3^1/_4$-inch floppy disks. The production mixer's set-up allows the total reconfiguration of engineering parameters such as input source names, assignment to buttons, etc. and operational states such as E-Mems, user wipes, macros, internal DVE effects and key memory. Several mixers have integrated hard disk drives into the panel for storage. The main advantage of the HDD is that the storage is large. This then allows many configurations which could reflect the operator's preferences, set-ups and effects or applications per show, or the individual engineering set-ups for each situation.

4.5.14 Networking

Traditionally a production mixer's mainframe communicates with the panel using serial communication such as RS422. For this application manufacturers have found capacity in the communication chain to slave a smaller panel off a main panel communicating to the mainframe. This has allowed a second operator to split complicated operations or work independently to have a separate output to feed a video screen or record a separate edit output. As LAN communication technology has advanced, production mixers have utilised 10 base T and then 100 base T communications. The use of computer-based TCP/IP communications has had several advantages. Usually the panel will have two control systems, the first a panel control driving the panel electronics and secondly an internal PC Graphical User Interface (GUI) displaying the set-ups (see Figure 4.5.9). The first advantage is that PCs have been included into the mixer panel. The PC normally drives the Graphical User Interface, reflecting the status of the electronics box. Importantly the PC should not be the single source of control due to reliability of PC-based products. However, the advantages are that it is easier to store mixer set-ups, transfer graphic images and utilise internal hard drives. The cost of the common PC means a very inexpensive redundancy can be used for back and failure. Also, several PCs can be used on the network for set-up and the main advantage has been the ability for many control surfaces to communicate on the LAN control at the same time. Therefore, many panels can connect to the network and attach to any mainframe available. True networking allows several panels to work and mimic the main operations or segregate the mixers into several discrete independent sections. A Four Mix Effect electronics box can then be separated as two virtual mixers of two M/Es (see Figure 4.5.10).

LAN connectivity has led the way for simulcast productions in which two mixer frames follow each other. This is particularly useful for HD and SD simulcast productions in which the operator can control both systems from one control panel. The system works with one master mainframe, with the slave following the master mainframe's commands. As the two systems may not have the same number of sources or sources connected in the same input cross-point location, a re-mapping between the master and slave can be made to resolve the differences. The production would then have two separate output streams, one in SD and the other in HD (see Figure 4.5.11).

Other controls of the mixer are remote Aux Bus panels for routing sources for camera control monitoring, VTR sources, etc. These panels will be connected via normal RS422 control onto the main panel or electronics frame.

4.5.15 Tally

Tally outputs are either serial or parallel triggers to indicate to the operators of the sources such as camera or VTRs that they are live on air. Serial tally is a string of commands indicating which sources are live on air, while parallel tally is a contact relay closure on the output of the production mixer. When any source is on air the contact closure will trigger. This is known as a Red tally. More recently, cameras accept a second tally level or Yellow tally, which indicates that the source is selected on the Preset Bus and is ready to go on air. A third level of tally, sometimes known as the Green tally, indicates to ISO or Isolated outputs that they are on air. An example of this would be a camera being recorded separately from the main production. Although not on air, the tally gives the camera operator notice that they are being recorded.

4.5.16 Effect Recall or E-Mems

Effect recall or Effect Memories are now standard on many production mixers. Initially the effect storage and recall on

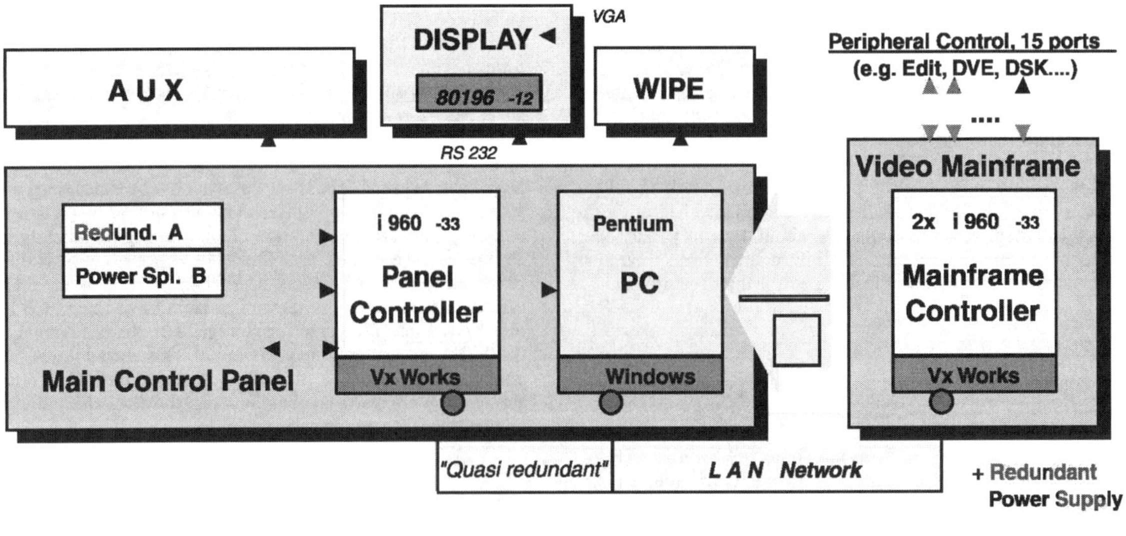

Figure 4.5.9 Panel Redundancy showing that the panel has dual control. Primary control with a more stable platform such as Vx works and PC control to operate the GUI and injection of graphics etc.

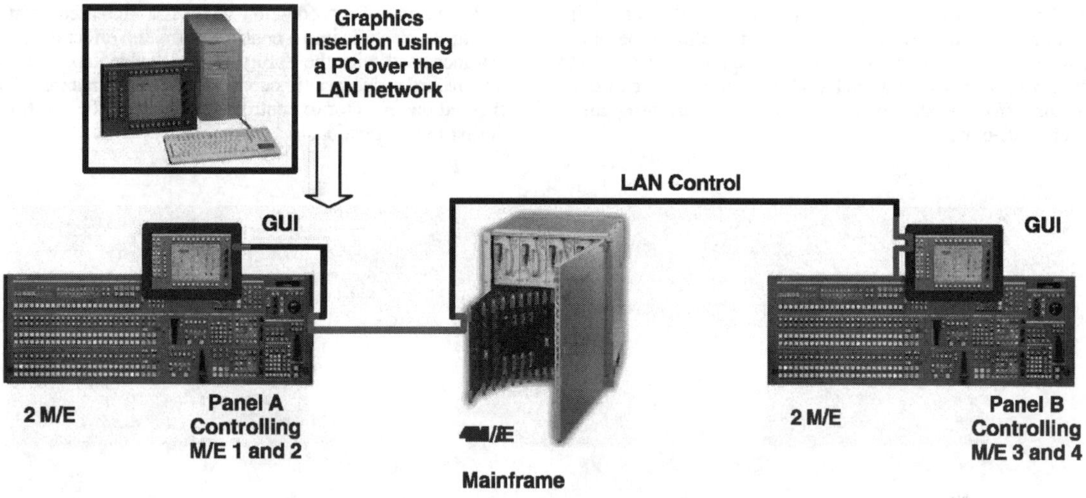

Figure 4.5.10 Networking with two panels and one mainframe. A separate PC GUI is used to inject graphics into the video stream.

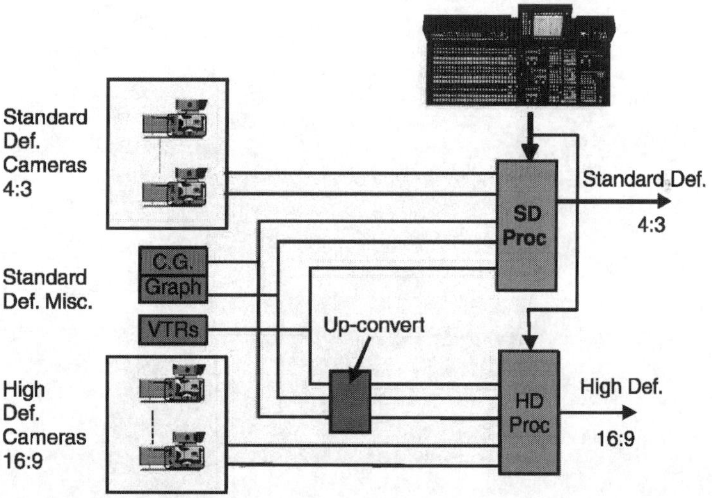

Figure 4.5.11 SD and HD output streams/Symalcast Production.

mixers could only work with a single mixer set-up. Often these stores were very limiting. The modern production mixer allows timelines that can perform the insertion of many stored effect set-ups separated by time. With modern production mixers the use of E-Mems or Snapshots or Timelines is an essential part of the operation. As productions have become more complicated the E-Mem has allowed the vision mixer or technical director to recall whole set-ups with one touch of a button, particularly useful when the operator can recall one Mix Effect bank while not on air.

When building effects a selection or definition of stored parameters can be made. For example, the operator can select whether only Mix Effects bank 1 set-up is stored. This could then be defined as the transition, cross-points and Key 1 only. When recalling the effect only these parameters would then be reloaded. This feature would be enabled with the Auto recall switched on. With a moving timeline, Auto Run disabled will allow the effect to be loaded at the first key frame. When the operator is ready the effect could then be triggered manually. With Auto Run on, the timelined effect would be loaded and run at the same frame. While recalling effects the define or E-Mem enable section allows the operator to recall the sections of the mixer required. On each Mix Effect bank a cross-point enable allows the operator to override the effect recall, not to change the on-air program or preset cross-points.

As the role of a production mixer has increased from being an effects machine into a control centre, the effect recall power has increased with the ability to insert videotape machine or disk recorder cue per timecode and play commands. This has enabled the operator to control complicated effects effortlessly during the production.

Jeremy Kerr
Educator, Discreet Logic

4.6 Visual Effects Systems

The final stages of post-production typically involve various amounts of special effects generation. Whether it is basic colour correction (post-film transfer) compositing, or more complicated effects such as particle generation, morphing or 3D camera tracking, many projects make use of visual effects.

Visual effects systems are utilised in the post-production phase of the project. These systems run on a variety of platforms such as PC, Mac, SGI or even dedicated hardware and offer a wide range of toolsets. Determining the visual effects system to use for a given project is dependent on many factors, such as budget, format of footage, time and the complexity of work required.

Complex special effects are typically storyboarded and pre-visualised during earlier stages of production, but the actual creation and finishing of the effects occurs during the on-line process. A typical process of special effects generation for broadcast is illustrated in Figure 4.6.1. A typical process of special effects generation for film is illustrated in Figure 4.6.2.

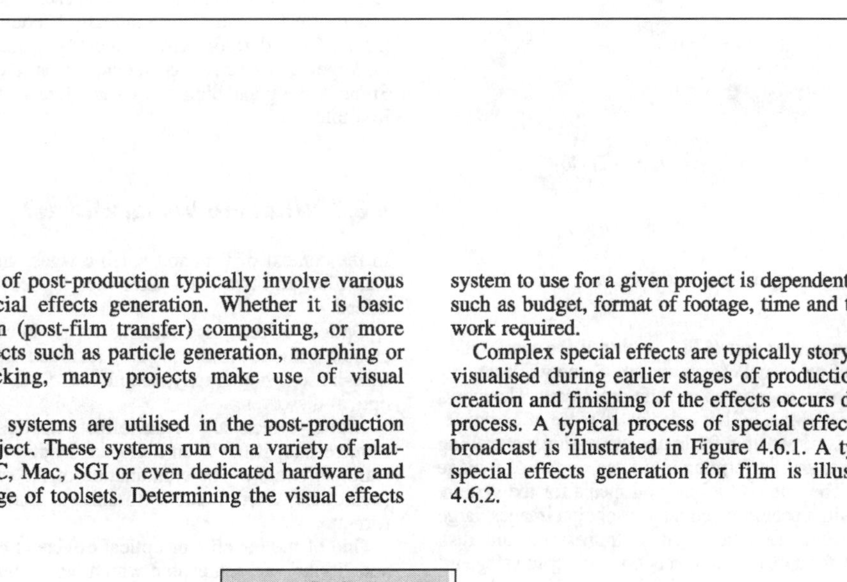

Figure 4.6.1 Typical process of special effects generation for broadcast.

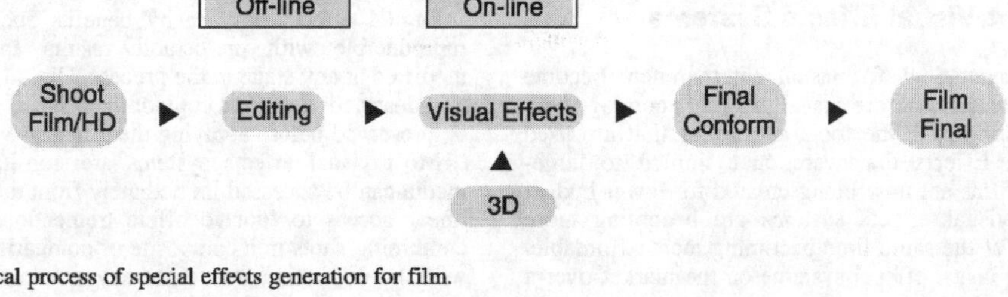

Figure 4.6.2 Typical process of special effects generation for film.

4.6.1 Components of Visual Effects Systems

Visual effects systems operate on many different computer platforms or even dedicated hardware; however, many components are common among various systems. Figure 4.6.3 illustrates components found in typical visual editing systems.

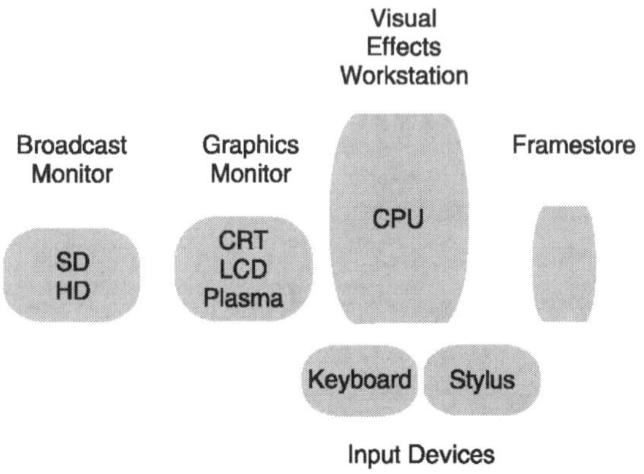

Figure 4.6.3 Typical components of visual effects systems.

The central processing unit (CPU) is part of the workstation and is the brain of the visual effects system. Some workstations, such as an SGI Onyx, support up to eight processors. The workstation may be a PC, Mac, SGI or a dedicated piece of hardware. The workstation is responsible for image processing, controlling graphics hardware, and performing input and output of footage into the system. The framestore holds all media for the system. When working with uncompressed high-resolution images, large capacity framestores are crucial. Most framestores are disk arrays with some form of RAID protection in order to safeguard against lost data. The workstation also has many input devices such as a keyboard and mouse. Many visual effects systems also utilise a stylus and tablet for more precise control of image manipulation, such as using a paint program, or modifying points on a garbage mask – tasks which are easier to perform with a stylus than with a mouse. The workstation monitor is typically a CRT; however, it may also be an LCD or plasma display. The workstation monitor can usually support multiple resolutions, which is essential when working with high-resolution film images and various scan rates. The broadcast monitor is essential to verify the final output if the destination for the spot is video. If the final is film, a separate screening room is required to preview the results.

4.6.2 Current Visual Effects Systems

As audience expectations for visual entertainment become more sophisticated, so too are visual effects becoming more sophisticated, along with the tools and artists that are used to create them. Effects that were once limited to large-budget feature films are now being created for lower-budget projects since visual effects systems are becoming more advanced while at the same time becoming more affordable. There are many visual effects systems on the market over a wide range of price points, offering producers, directors, artists and post-production facilities tremendous flexibility when deciding how to incorporate visual effects into their projects.

Visual effects systems can be separated into two main categories: desktop systems and high-end systems. Desktop systems operate on standard computer platforms such as Intel-based and Macintosh systems. High-end systems operate on high-performance workstations such as SGI supercomputers, or require dedicated hardware. High-end systems are designed for highly interactive and superior quality visual effects generation.

As tight budgets become more of a concern for visual effects production, and processing power and graphic capabilities of cheaper systems continue to increase, desktop solutions for visual effects creation are gradually becoming more commonplace. The desktop environment includes many software packages that operate on Macintosh or Intel-based platforms. Some desktop and high-end visual effects systems are included in Table 4.6.1.

4.6.3 What are Visual Effects?

In the context of film and video images, visual effects can be classified into two main categories: optical effects and digital effects.

Optical effects, also referred to as 'in-camera' effects, are created while the image is being recorded or filmed and include time-lapse photography, depth of field manipulation and optical dissolves.

Digital effects are visual effects applied to film or video footage after it has been digitised. Digital effects utilise hardware, software or the combination of the two to apply digital (numerical) modifications to images that exist in a digital format.

One of the benefits of optical effects is that specific results and 'looks' are achievable which would otherwise be impossible using only digital techniques. For example, creating a super slow motion effect optically involves overcranking the camera. When played back at the standard frame rate, the result is a smooth slow motion sequence. Creating this slow motion effect digitally could result in a stuttering of the image since less temporal information is available. Digital techniques can try to simulate the optical slow motion by interpolating image data, but the look of an optical and digital slow motion will differ.

One drawback of optical effects is that they are not easily reproducible. There are many dynamic variables when shooting footage which make it inefficient to create in-camera effects. Another drawback of optical effects is that they can be considered destructive in that they are applied directly to the recorded image and cannot be undone.

Digital effects offer many benefits such as being easily reproducible with predictable results and can be easily modified at any stage in the process. Digital effects are usually non-destructive, since a copy of the original source footage can be preserved before applying the effect.

Most visual effects systems are non-linear, meaning all media can be accessed immediately from the framestore. Non-linear access to footage offers tremendous flexibility when combining shots in a composite or comparing an affected shot with the original source, since any shot can be brought to the

Table 4.6.1 Some desktop and high-end visual effects systems

Visual effect system	*Description*
After Effects (Adobe)	2D/3D compositor for film, broadcast and multi-media, running on PC and Macintosh platforms. With animatable parameters and plug-in support, After Effects is often used for creating graphics and effects such as network branding or animatics.
Combustion (Discreet)	Paint, animation and 3D compositor for film, HDTV, SD and multi-media running on PC and Macintosh platforms. Vector-based paint, keyer, colour correction and particle generation, as well as close integration with Discreet High-End systems, allow Combustion to be utilised in any post-production environment.
Digital Fusion (Eyeon)	2D/3D compositor for film and broadcast. Tracking, keying, colour correction, particles and warping tools can generate complex visual effects and utilise network rendering to speed up workflow.
Elastic Reality (Avid)	Warping and morphing software running on PC, Macintosh and SGI platforms. 3D distortion tools allow for complex manipulation of objects and images.
eQ (Quantel)	SD/HD non-linear editor. Quantel has incorporated the effects and compositing tools from Henry and Paintbox into eQ to create a combined editor and visual effects system for video.
fire/smoke (Discreet)	SD/HD/film on-line editors running on SGI platforms. These systems are primarily editing systems, but because they integrate many of the rich feature sets from flint, flame and inferno, they can also be used to create a significant amount of visual effects during on-line sessions.
flint/flame/inferno (Discreet)	2D/3D compositors for film, HDTV and SD running on SGI platforms. Rich feature sets for colour correction, particle generation, advanced compositing, 2D/3D tracking, paint and morphing are integrated into a node-based workflow.
Nuke (Digital Domain)	2D/3D compositor for film and broadcast running on PC, Linux and SGI platforms. Once a proprietary system, Digital Domain began selling Nuke in late 2002.
Photoshop/Illustrator (Adobe)	Image manipulation and vector drawing programs for PC and Macintosh platforms. Since both programs are designed to work with still images, full animations are not created with these products; however, they are used to design highly detailed graphic elements which can be composited in another software, such as flame or inferno from Discreet.
Shake (Apple)	2D/3D compositor for film and broadcast, running on Macintosh and UNIX platforms. Keying, colour correction, film tools and distributed rendering allow Shake to be used effectively and efficiently in feature-film post-production environments.
Symphony/Media Composer/Film Composer (Avid)	SD/HD/film conformers with some visual effects capabilities such as colour correction, paint and compositing tools.
Proprietary systems (various)	Many facilities develop and use their own proprietary systems for visual effects creation. These systems are usually not for resale and details about their feature sets are typically subject to high levels of confidentiality.

workspace at any time. Working with media in a digital format means that unlimited numbers of copies can be made with no generation loss. This enhances the creative process since many versions of an effect can be tried before the desired result is achieved.

Some examples of various types of digital visual effects include the following.

4.6.3.1 Colour Correction

Colour correction is the process of adjusting chroma and luma properties of an image. Colour correction can be used to create a special effect such as altering luma and saturation levels of a bright daylight shot to make it look as if it was shot at night (called day-for-night), or colour correction can be used for colour matching between scenes to give a commercial, film or documentary a consistent look.

Colour correction can also be used to correct white balance problems in an image. The example in Figures 4.6.4 and 4.6.5 shows the colour corrector in fire being used to fix a shot in which a blue filter was accidentally left on the camera during shooting. Using the internal vectorscope, the Blue Offset values are modified to restore the original chroma attributes of the image.

4.6.3.2 Keying

Keying is the process of building a matte (alpha) based on the chroma or luma properties of an image for the purpose of compositing. Keyers use algorithms to isolate pixels of an image and assign those pixels to black, white or a gradient value for the alpha based on user-defined settings. Advanced keyers have multiple algorithms for analysing pixels in several different modes, such as RGB, YCrCb, HLS and Luma. Some keyers also have colour correction tools for suppressing or hue shifting colours. Colour suppression and hue shifting is an effective way to remove unwanted colour spill on the foreground image. The keyer in fire is integrated directly into the DVE module, meaning multiple layers can be keyed and composited simultaneously. In the example shown in Figures 4.6.6–4.6.8, a model plane was prepared on bluescreen; fire's keyer was used to build the alpha for this image and replace the background with the sky image as the new background.

Figure 4.6.4 Before colour correction. In this shot, the white balance needs to be corrected because the vectors are shifted towards blue and the entire image has a blue tint.

4.6.3.3 Compositing

Compositing is the process of integrating multiple elements over a common background. These elements can be 3D models, green/blue screen footage or full raster images. Simple compositing tools only operate in 2D, meaning the only axis manipulations that can be made are on the X and Y axes. Most compositing for the purposes of visual effects require the ability to composite in true 3D space. Advanced compositing tools are 3D environments that provide an additional Z axis for layer manipulations. Compositing in a three-dimensional space means an effect can have authentic perspective changes that can add to the realism of the final shot.

4.6.3.4 Graphic Design

Graphic design is the process of creating graphic elements to integrate with existing footage. For example, in order to brand a network station, a specific package which includes logos, overlays and other design elements is created for a news broadcast. Graphics departments use visual effects systems to create these elements and then provide them to the post-production department.

4.6.3.5 Particle Generation

Particle generation is the process of generating and manipulating 3D geometries in a controlled manner. These particles can be used to simulate natural phenomena such as rain, fog or even tornadoes. Particles can be controlled with mathematical operands called manipulators, which can simulate gravity, wind or even collision properties that cause particles to interact with other objects in the scene.

4.6.3.6 Stabilising/Tracking/3D Camera Tracking

Stabilising is the process of analysing motion in a shot and then applying the inverse of that motion to the axis of the image in order to remove unwanted camera motion. Stabilising is often required if a steady cam was not used during filming or if the camera was not otherwise locked off.

Tracking is the process of calculating the motion of an object in the shot, typically used to composite new elements into the

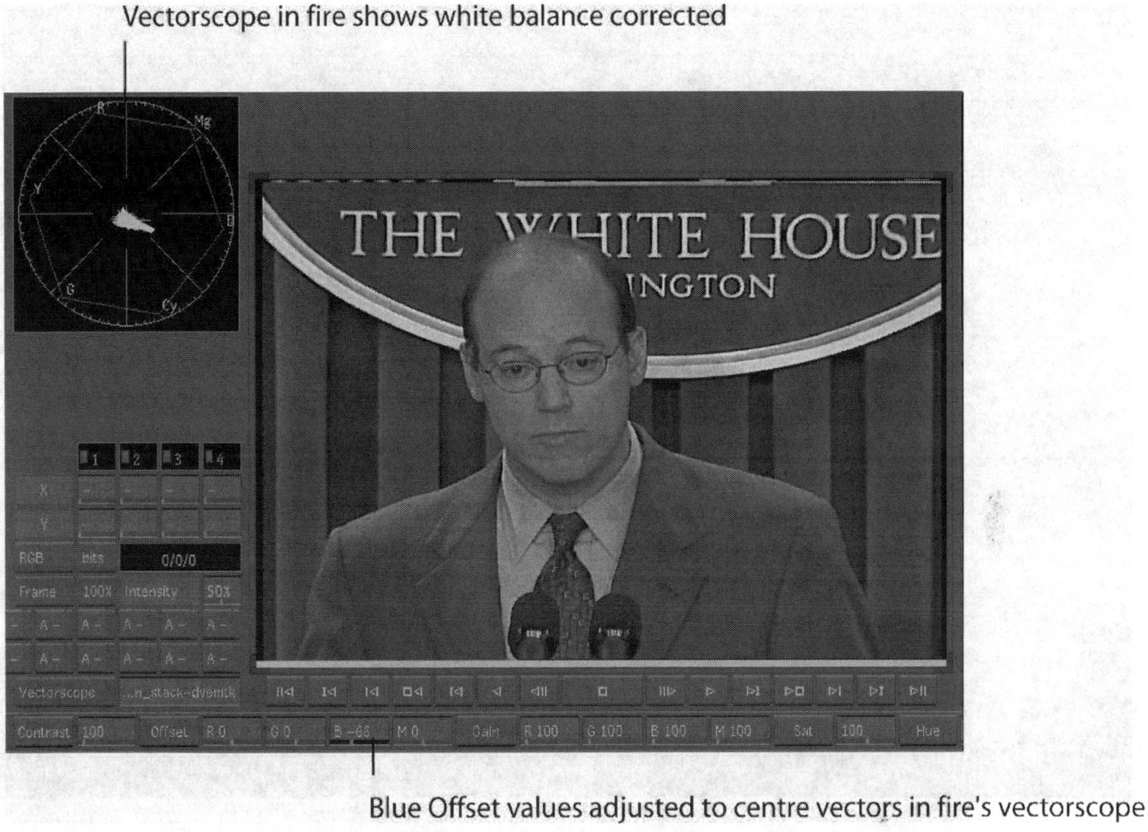

Figure 4.6.5 After colour correction. Adjusting the Blue Offset values centres the vectors and fixes the white balance of the shot.

scene using the same motion as the tracked object. For example, if a car is moving across the frame, a tracker can be used to determine the exact path of the object, and then a separate layer such as a company logo can use the same motion on its axis so that the logo follows the motion of the car.

3D camera tracking is the process of obtaining and reproducing camera motion of a digitised scene in a simulated 3D environment. First, the live-motion scene is shot with a camera capable of recording its own motion (called e-motion), such as a MILO. The camera data is then imported into a 3D modelling system where 3D elements are placed within the scene. The live-motion and 3D elements are then composited together in the visual effects system.

4.6.3.7 Timewarping

Timewarping a shot is the process of affecting the speed of a shot so it plays faster or slower than the original source. A timewarp can be either a constant speed timewarp or a variable speed timewarp (speed ramp). In a variable speed timewarp, the playback speed is animated over time so the shot either slows down or speeds up as it is played back. Some visual effects systems also include options to add trail to the speed effect. Trail is a method of simulating camera blur in which surrounding frames are blended into the current frames. The example in Figure 4.6.9 shows a timewarp with a trail effect created in fire.

4.6.3.8 Morphing

Morphing is the process of transitioning from one image to another image by matching the morphology of an object in the outgoing shot with that of another object in the incoming shot. The transition is similar to a dissolve; however, a morph also uses a mesh to isolate pixels in both images and has additional interpolation options to control the blending of pixels at any given frame. An example of morphing is turning a human face into the face of an animal. By precisely drawing a mesh around each object, precise control of the transition is possible using morphing tools.

4.6.3.9 Film Look

Film look is the process of affecting video footage to simulate characteristics of film, such as grain, contrast and film strobe. Several processes are required to achieve film look, such as colour correction, timewarps and regrain tools.

The following example details the workflow for giving film look to a video image in fire.

Figure 4.6.6 Model plane on blue background.

First, the colour corrector in fire is used to adjust black and white levels in order to simulate film look. The source footage in this example is from video, and does not have the same contrast or black level as would be found in a film-exposed image. The colour corrector is used to lift the black and white levels to increase the contrast of the image (Figure 4.6.10).

The histogram shows the luma distribution of pixels in the image. In this example, the footage has no black values at absolute zero, and the shadow areas in the image are too bright. The colour corrector is able to remap the black input levels to darker values in order to darken shadows in the shot. White values are also remapped to increase the overall contrast of the image.

If the final result will be broadcast in an NTSC environment, a film strobe can be added to simulate the 2:3 sequence introduced by the telecine process (Figure 4.6.11). The timewarp tool in fire includes a film strobe option that adjusts the timing curve of a source clip to replicate the field order when 24 fps film is transferred to 30 fps.

Finally, to provide a video image with the texture of film footage, grain can be added; fire also includes regrain tools which can apply grain to simulate Kodak 35 mm film stocks or any other custom-defined grain (Figure 4.6.12). Adding grain to video footage allows it to be composited more seamlessly with actual film footage.

4.6.3.10 Photorealism

Photorealism is the process of using advanced rendering and compositing techniques to make digital elements seamlessly fit into a realistic environment. These techniques include adding lights, shading and reflections to a composite. The following example illustrates enhancing the photorealistic properties of a texture in fire (see Figures 4.6.13 and 4.6.14). The compositing tool in fire, called DVE, provides the ability to add lights to the scene, modify the reflective properties of the surface and displace the texture. Displacement is the extrusion of a surface in three-dimensional space based on the chroma or luma of a displacement source. In Figure 4.6.14, a light is added to the scene, the shine of the surface is increased and the texture is displaced slightly on the Z axis. The displacement causes the light to shade and highlight areas of the surface based on their position relative to the light source. The resulting clip has more reflective properties than the original texture (Figure 4.6.13).

4.6.3.11 Previsualisation

While not a visual effect in and of itself, previsualisation (previz) is often an essential component of visual effects creation, especially for effects that involve large amounts of 3D elements, or which integrate virtual camera motion with live-action footage. Previz is the process of building a 3D

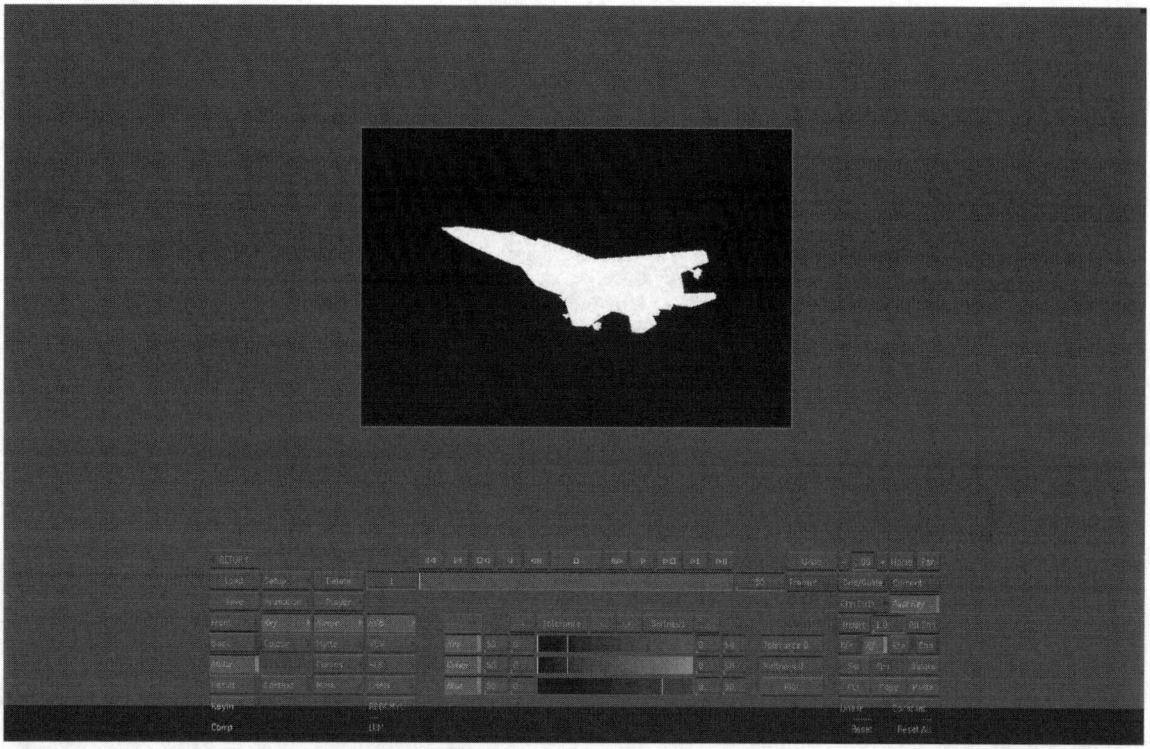

Figure 4.6.7 Fire keyer pulls a chroma key in RGB mode. This is the resulting matte.

scene and testing many variations of a shot before actually filming it on set. This can save time and money by allowing opportunities to troubleshoot potential problems in a virtual environment and resolve technical issues before arriving on set.

One example of using previz to assist the post-production process is using a 3D modelling software, such as 3ds max, to prepare 3D elements for a motion-control shot.

In order to prepare a complex shoot for a motion control project, 3D animation programs can be utilised to recreate a virtual copy of the soundstage to be used for the shoot. Testing the shoot in a virtual environment allows the production and post-production crews opportunities to test variations of a shot without spending money for studio time or film stock. Pre-visualisation in 3D software also has the benefit of allowing the art director additional flexibility to test a large number of camera angles and enhances the creative process, since a much wider range of possibilities can be tested in a virtual environment than would be achievable with a real camera. Moving a physical camera to test a new angle is time consuming and involves several people, whereas moving a virtual camera can be done almost instantaneously within the 3D program.

The first step of previz in 3D software is recreating the set precisely by applying exact dimensions of the soundstage, furniture and props. 3D models are used to simulate all objects in the scene.

If people will be used in the shot, a character modeller such as Character Studio can be used to design scale models of the actors and integrate them with animation (kinematics) into the virtual set.

Once the virtual set and characters have been designed, a virtual camera can be added to the set. A motion-control shoot might utilise a robotic camera such as a MILO. Since owning or renting a MILO is costly, it is more cost efficient to simulate the camera in 3ds max. By importing an accurate model of the MILO into 3ds max and using a feature called Bone Structure, the virtual camera model can be assigned the same ranges of motion and physical limitations as the real camera. Even accurate lens attributes can be simulated in the 3D environment.

Camera moves can then be tested in the virtual studio in order to determine the optimal camera angles and motion. The final virtual camera motion can be exported directly to the MILO for the final shoot. Since the final camera data also exists in the 3D software, it is easy to integrate 3D models into the final composite which exactly match the camera moves of the real-world camera.

4.6.4 Visual Effects Industries

There are three primary streams for visual effects creation: broadcast, commercial/episodic and film. The approach each industry takes for creating visual effects is significantly different, and each industry operates in unique production and post-production environments. Although some of the techniques used to create effects and even the final results of some effects

Figure 4.6.8 Final result after compositing model over sky background with generated matte.

might be similar, these industries have very different structures and are subject to their own dependencies and constraints.

4.6.4.1 Broadcast Visual Effects

In the broadcast industry, the majority of special effects generation is done in-house, since most networks have dedicated graphics and post-production departments and own their own visual effects and editing systems. In this context, budget is not as much a factor for broadcast visual effects production as it would be for film work. Instead, time is a critical factor for broadcast visual effects, since topicals, promos and network ads often need to be turned around in days if not hours.

Complex visual effects that require elaborate production techniques or time-consuming compositing are typically not developed in a broadcast environment. Instead, visual effects systems are used for graphic design such as network branding or promotional spots. Most graphic elements, such as fills, mattes, logos and overlays, are created in the visual effects system and then sent to the editorial department, where they are integrated into the final spot.

4.6.4.1.1 Common Visual Effects Workflows for Broadcast

A typical process for using a visual effects system to design graphics for broadcast involves using text generation, paint and compositing tools to create still and animated graphics. Mattes, or alpha channels, are also created for these elements. All elements are then sent to the editing system to build the final spot. The following example details the creation of a topical for a news network (see Figure 4.6.15).

Topicals are short promotional teasers that introduce stories for upcoming newscasts. While the stories and footage change for almost every topical, the graphic elements such as animated transitions and overlays usually remain constant. The graphic elements are created in the visual effects system and are used over and over again in the editing system, where the stories and footage are updated within the same package designed by the graphics department.

The station ID, or bug, was created in the Text module of flame. In this instance, the corporate logo exists in a custom TrueType font (TTF) and is easily integrated with a text generator (Figure 4.6.16). Another method for creating a corporate logo is with a vector-based illustration program. The vector graphic can then be imported into the visual effect system for compositing.

The alpha for the bug in this example is filled with light grey instead of white so that the bug is partially transparent when composited with the background video. Adding transparency to an element such as this station ID allows other elements to be visible under the graphic.

Figure 4.6.9 Timewarped shot with trail effect.

The second element for the topical is an animated overlay, which serves as the primary graphic for the spot. This overlay is a keyable element that has a matte that matches the design of the front animation. The main graphic is known as the fill, and the combination of fill and matte is composited with the video for the topical to create the final result.

In the example shown in Figure 4.6.17, the geometric elements for the animation were created in the Paint module of flame and composited in flame's Action module. For this overlay, the matte is static and all animation occurs in the fill.

Because each geometric element was created separately, they can be combined in Action using transfer modes. Transfer modes are ways of combining different layers in a compositor by using mathematical operations to merge luma and chroma values of pixels in each image.

The thin linear elements in this animation use the Screen transfer mode, which inverts and multiplies the difference between the two layers. When the lines intersect, they create a subtle flare effect because the intersecting areas become lighter. Visual effects systems make it easy to create these types of effects because instead of manually painting an effect frame by frame, a compositor can be used to animate graphic elements and combine them with transfer modes to automatically produce complex-looking results.

The third element for the topical is a 15-frame animated transition. Since a topical may contain several stories for the upcoming newscast, transitions are required to provide a visual separation between each story. This transition was created in the same way as the overlay by building graphic elements in flame's Paint module and compositing them in flame's Action module; however, for this element the matte is also animated. The fill spins on the X axis and its matte animates from full raster white to full raster black. An animated matte allows the incoming video to be revealed as the outgoing video exits. This transition has an associated sound effect, and since flame has full audio support, the timing of the transition can easily be aligned with its audio.

The final elements for this topical are all of the text graphics required for each version. Since the topical may introduce newscasts at different days or times, several variations of text elements need to be created. For example, if the topical is used for three evening newscasts, separate 'Tonight at 5', 'Tonight at 6' and 'Tonight at 11' text elements are required. In addition, if the topical also relates to following day broadcasts, 'Tomorrow at 5' (Figure 4.6.18), 'Tomorrow at 6' (Figure 4.6.19) and 'Tomorrow at 11' graphics are also designed.

A fill and matte for each graphic is created in the character generator module of the visual effects system. These elements can be transferred through an internal network to the editing system, or exported as an image file. Many image file formats, such as TIFF and SGI (a pure RGB format), allow the matte to be embedded with the fill in a single 16-bit file. Embedding fills and mattes in the same file simplifies the process of

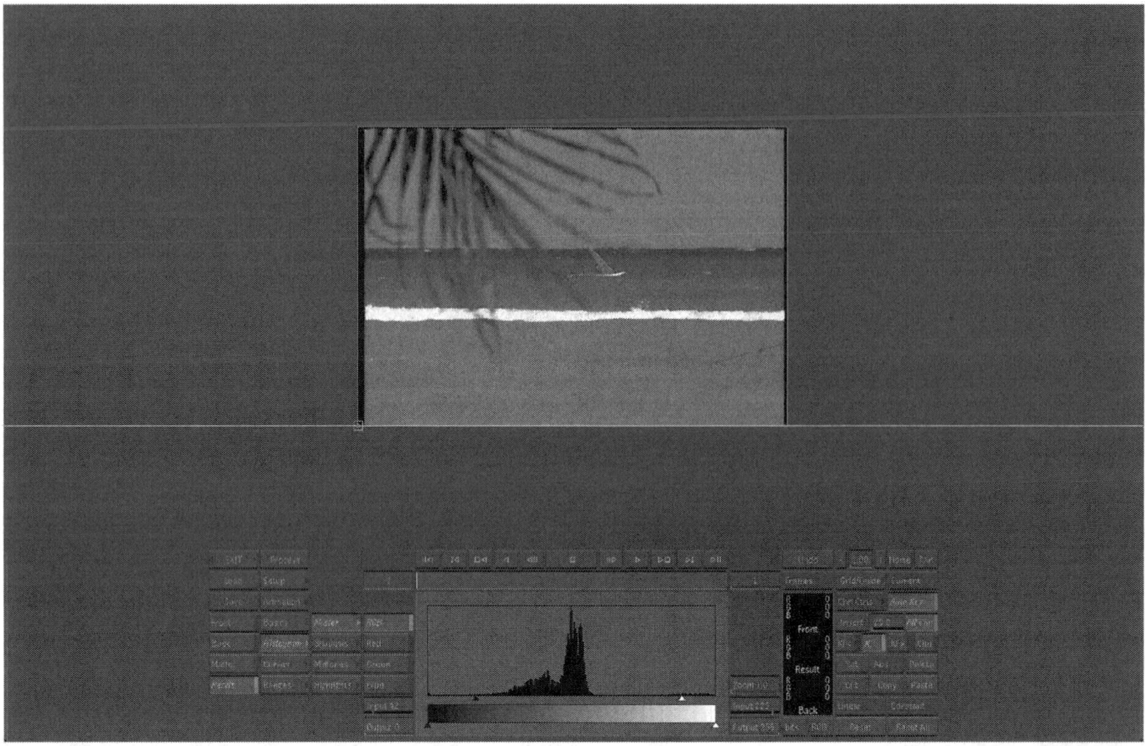

Figure 4.6.10 Adjusting black and white levels to increase contrast.

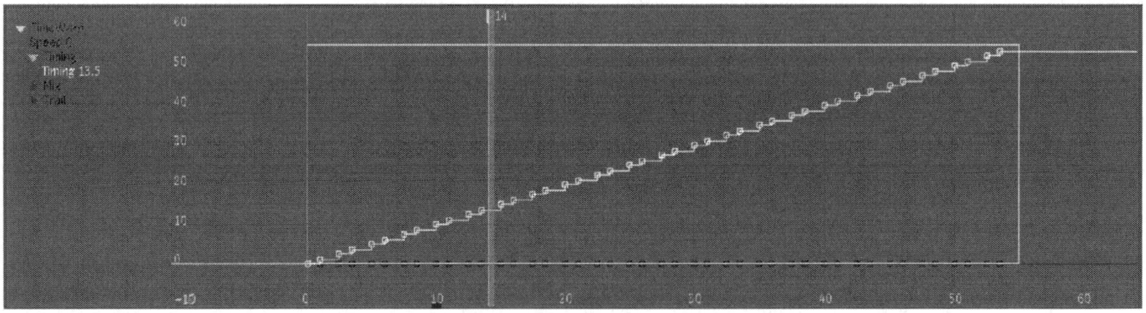

Figure 4.6.11 Animation channel of a film strobe simulating a 2:3 sequence.

transferring these elements between visual effects systems and editing systems.

The use of visual effects systems to create individual elements for the station ID, overlay, transition and text simplifies the versioning process because multiple composites can quickly be created from the generic elements. When new video is shot for a topical, it can be composited below the overlay and then the text elements are added for each version. If the topical contains several stories, the animated transitions are also composited into the final spot.

4.6.4.2 Commercial/Episodic Visual Effects

The industry of visual effects generation for commercial and episodic projects is significantly different from the broadcast industry and more closely aligned with the film industry. In both the commercial/episodic and film industries, visual effects systems are typically owned and operated by large post-production facilities or smaller boutiques which design the visual effects on a project-by-project basis. Unlike a network, which creates visual effects for itself, post-production facilities

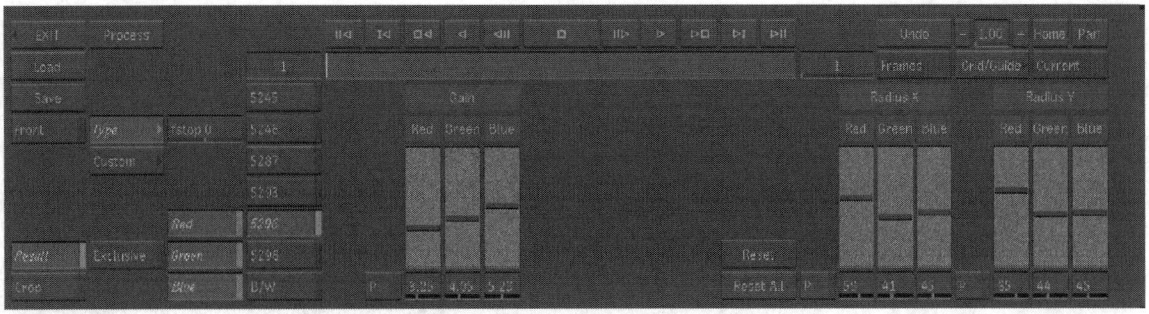

Figure 4.6.12 Adding grain to a shot with fire's regrain tool.

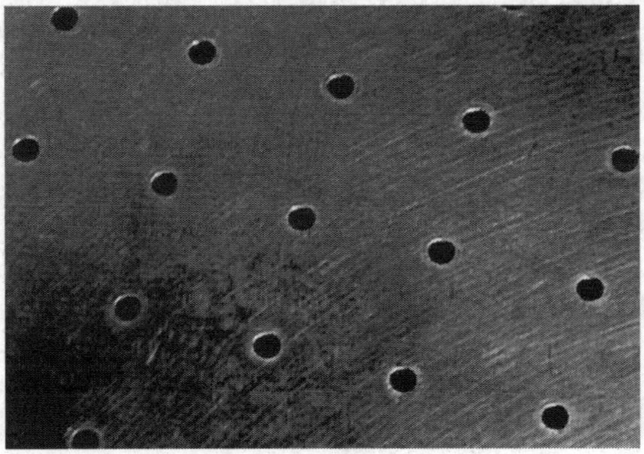

Figure 4.6.13 Original texture.

Figure 4.6.14 Texture with light, shine and displacement in DVE.

and boutiques create visual effects as a service for ad agencies, producers and other organisations developing commercials, television episodes or even documentary work.

Budget becomes a primary concern for commercial and episodic visual effects creation, since post-production facilities and boutiques charge by the hour, day or project. For high-end visual effects systems, this rate may be quite high and post-production costs for lengthy or complex projects could become substantial.

Commercial visual effects typically involve large amounts of compositing and colour correction, and require visual effects systems with versatile toolsets. In today's market, some commercials are as elaborate and wide in scope in their visual effects design as some feature films and utilise high-end visual effects systems to achieve complex results.

A simple example of compositing for commercial visual effects is sky replacement. Often, all environmental variables during a shoot may not be exactly as required for the final look of the shot. For example, a location might be chosen to shoot a car commercial because of certain geographical features of that location; however, the sky may not be as bright or the clouds might not be as full as the director of photography desires. A common use for visual effects systems in this scenario is to composite a new sky shot under the right conditions into the original scene shot on location. Sky replacements, and even other element replacements or additions, such as compositing new buildings or people into the original scene, are common techniques in order to present a product in the best possible environment.

More complex commercials may incorporate a large amount of green screen shots, 3D elements or sophisticated compositing to create for the viewer an elaborate visual experience.

Episodic and documentary work (longform) may not rely as heavily as commercials on intensive visual effects; however, their duration and number of shots bring unique requirements to the post-production process. Even longform pieces which only require colour correction place high demands on visual effects systems in terms of rendering capabilities and speed of interaction. High-end visual effects systems are usually required for this type of work. The ability to manage large amounts of image data and effects information is crucial in order to optimise the cost-effectiveness of longform work. Inferno and fire from Discreet have many advanced tools and capabilities that make them well suited for longform projects; fire is a non-linear editor which also includes many visual effects tools.

Colour correction for multiple shots is simplified by utilising soft effects in fire's timeline. A soft effect is a visual effect that can be applied to an element in a timeline as metadata. This means it can be easily modified, and also easily copied from one shot to another. For episodic work that contains many shots requiring the same colour correction – for example, exterior

Figure 4.6.15 Example of a topical for a news station.

Figure 4.6.16 Network logo created with a TrueType font in flame's Text tool.

Figure 4.6.17 Some elements used for the overlay.

shots that need the same white balance correction – copying the effect from one shot to another greatly increases workflow. In fire, the white balance can be corrected for one shot and then the soft effect colour correction can be copied to all other shots in a single step. In the fire timeline, the colour correction can be dragged from the first shot to all other selected shots in the timeline. All highlighted shots immediately acquire the same white balance correction. Without this capability, manually colour correcting a large number of shots could be a time-consuming process.

Figure 4.6.18 Elements created for versioning topical.

Figure 4.6.19 Elements created for versioning topical.

In inferno, colour corrections can be linked between shots with the use of expressions. By linking parameters of a colour correction between shots, one shot can be assigned the master element and all linked shots will dynamically update as changes are made to the master element. Using expressions to link parameters is a flexible way to share effects information since specific parameters can be isolated for linking. For example, it is possible to link only the Blue Offset channel between shots to correct white balance problems caused by a camera filter. In inferno all common shots have colour correction parameters linked together. One shot acts as a master element. As the master shot is colour corrected, all linked shots are dynamically updated.

4.6.4.3 Film Visual Effects

The film industry, like the commercial/episodic industry, also relies heavily on post-production facilities and boutiques for the majority of visual effects creation. Some large film companies do own and operate their own visual effects systems which are utilised for some feature film work; however, most special effects are outsourced to independent post houses. If a feature requires a large number of visual effects, many facilities may work in parallel in order to speed up the post-production process.

While budgets are still an important concern for film-based visual effects, monetary resources allocated for special effects are usually high, especially for action or sci-fi features that are compositing-intensive. The critical components of visual effects for film are speed and interactivity with the software and quality of the final result. Since the resolution of film images is very high, visual effects systems require large amounts of processing power to allow good interaction for the operator. Also, precise rendering features are essential to deal with aliasing and other artefacts that could be easily noticed in a theatre viewing environment.

4.6.4.3.1 Common Visual Effects Workflows for Film

A typical process for building a visual effect for film involves shooting all the elements such as backplates and foreground elements, creating masks for foreground elements, creating 3D elements such as particles, and compositing the final shot. The following example (see Figures 4.6.20–4.6.23) details the creation of the *Prague* scene involving the elements listed in the captions.

Because the backplate and foreground shots were filmed at different locations and under different lighting conditions, the shots were colour corrected for a better scene match. A garbage mask was created for the foreground shot so that its background could be replaced with the backplate. Finally, snow was created using flame's particle system and integrated into the scene. All elements were combined into the final composite and sent to the final conform.

4.6.4.3.2 Colour Correction

Colour correction is the process of adjusting chroma and luma properties of an image, and is an essential component of almost every visual effect. Shots may be colour corrected to modify the contrast of an image, shift the hue of specific regions of the shot or even simulate other film looks.

In the *Prague* scene, the backplate and foreground shots were filmed at different locations and required colour correction before they could be composited together. Figure 4.6.24 illustrates using the colour corrector in flame to adjust the gain of RGB channels to match hues between shots.

Both shots appear in the image window using a split bar to preview the foreground and background images simultaneously. Plotting tools are used to compare chroma and luma differences between the images. In this example, the Gain for all RGB channels in the background image is adjusted to make the lighting between the two shots match.

4.6.4.3.3 Mask Creation

There are many ways to build a matte (alpha channel) for the purposes of compositing. If the foreground is shot on blue or green screen, a chroma keyer can be used to build the matte. However, if the foreground shot is live action with no blue or green screen, garbage masks are required (Figure 4.6.25). Creating a garbage mask, also called rotoscoping, is the process of manually drawing vector-based geometries around the regions of the foreground image to be included in the composite. Rotoscoping is a time-consuming process since points on the mask may need to be adjusted at every single frame in the shot if the foreground image is moving or the shot contains

Figure 4.6.20 Backplate.

Figure 4.6.21 Wall foreground.

Figure 4.6.22 Wall matte.

Figure 4.6.23 Particle matte.

Figure 4.6.24 Colour matching foreground with backplate.

Figure 4.6.25 Foreground with garbage masks.

camera motion. Most visual effects systems have mask creation tools, although many differ in their complexity.

In the *Prague* scene, a garbage mask isolates the wall from the background of the shot. Additional garbage masks appear in the windows to allow the new background to appear. The matte is created by assigning colour values (white or black) to the garbage masks (Figure 4.6.26). Black keys the background image and white keys the foreground image. A gradient of white to black allows the foreground to mix with the background and therefore the edges of all masks have a slight gradient in order to blend the two images together and enhance the realism of the final composite.

The foreground image, or fill, is now ready to be composited over the new backplate by using the matte created from the garbage masks.

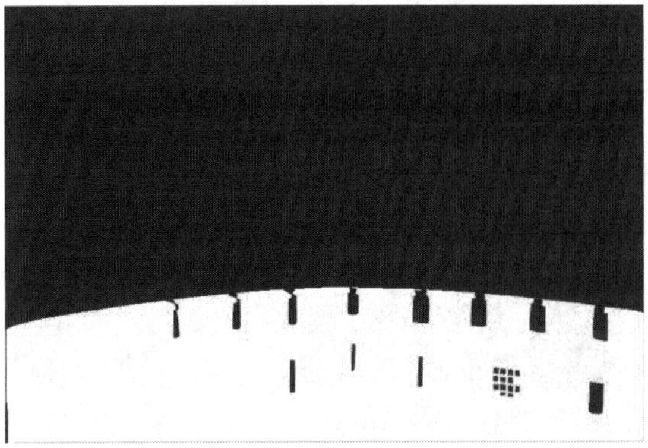

Figure 4.6.26 Matte created by garbage masks.

4.6.4.3.4 Particle Generation

One common feature of many visual effects systems is a particle generation system. Particle generation is the automated creation of 3D objects (particles) which can be controlled and manipulated to exhibit various types of behaviours (Figure 4.6.27). For example, particles can simulate rain, snow or fog, and can be manipulated within the system to behave as if affected by gravity, wind or even other particles. When particles are integrated into a live motion scene, they can add to, or enhance, the realism of the shot.

In the *Prague* scene, snow needed to be added to the shot. Since it was not snowing on location during the shoot, the snow needed to be computer-generated. The particle system in flame was used to achieve this.

Two light sources generate the 3D particles. The size, speed and lifetime of the particles are random, so that each snow flake appears to be unique. A particle manipulator in

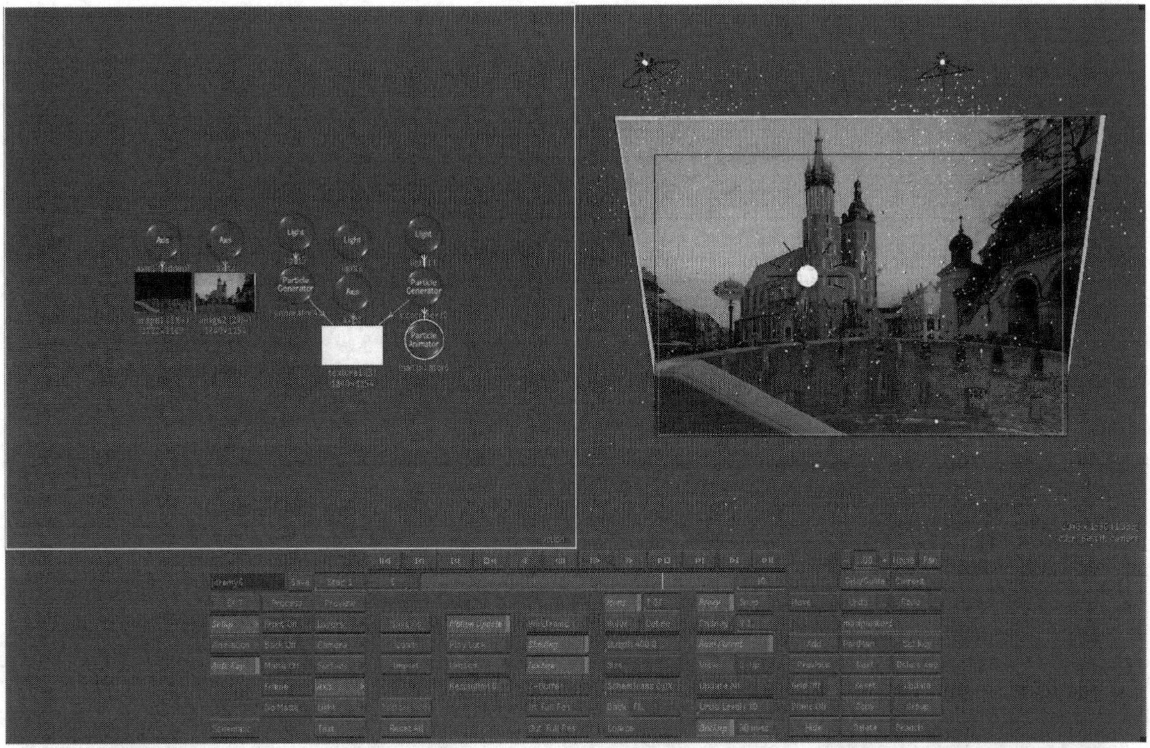

Figure 4.6.27 Particle generation in inferno's Action module.

the scene operates as a vortex to simulate the behaviour of wind blowing the snow flakes in a swirling pattern in the scene.

4.6.4.3.5 Compositing

Compositing is the process of combining multiple images or elements into a single image. This is accomplished by using mattes to key images and elements into the same scene. A compositing tool is an essential component of most visual effects systems, since almost every visual effect uses more than one image or element. In flame, the compositing tool is called Action and is a true 3D environment that allows an unlimited number of layers and elements to be composited into the scene. Action is one of the most advanced compositing tools of any visual effect system on the market and has a very rich feature set for manipulating images and elements within the scene.

In the *Prague* scene, several elements needed to be composited into the shot. Garbage masks were created for the foreground image so the wall could be composited over the new background plate. The particle system was used to generate a matte for the snow, which was also composited into the scene using Action. Figure 4.6.28 shows the composited shot, prior to its final rendering.

When played in real time, the particle snow exhibits the attributes and physics of real snowfall, and the colour-corrected composite allowed this scene to be used as an establishing shot for a later scene shot inside the church.

4.6.5 Integrating Visual Effects into the Final Conform

Integration of multiple systems in a facility can greatly improve efficiency for projects. Systems can be integrated with a network to enable sharing of media between framestores. Several advantages of network connectivity between systems include avoiding generation loss from RGB-to-YCrCb conversion that might be introduced during output to a tape medium such as Digibeta, as well as faster-than-real-time transfers of media if using a fast network protocol such as Gigabit Ethernet or High-Performance Parallel Interface (HIPPI).

In many workflows, visual effects are generated on systems that are independent from the system used for the final conform. Once the final cut list is determined during the off-line edit, shots are sent to the visual effects system for special effects generation. The final stage of the visual effects process is integrating the affected shots back into the final edit (see Figure 4.6.29).

Finalised effects sequences are sent to the final conform either by recording the effects back to film, digitising to HD or 601, or transferring them directly across an internal network connected to the on-line editing system.

Multiple Discreet visual effects and editing systems can easily be integrated together using the wire network. The wire network allows all framestores and their media to be available to any other system. This allows all media to be transferred in their native RGB format without being subject to any generation loss. With a HIPPI network, transfer speeds are many

Figure 4.6.28 Final composite of the *Prague* scene.

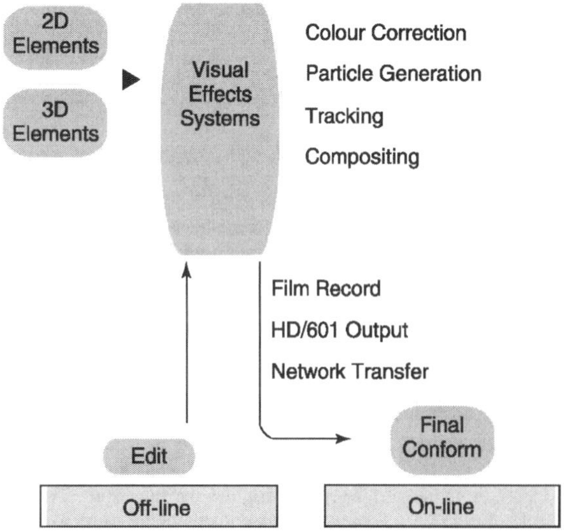

Figure 4.6.29 Process for integration of visual effects into final conform.

times faster than real time and allow artists to share materials and footage quickly from different locations. When visual effects are completed, the wire network can be used to transfer all finished shots to a fire system for the final conform.

4.6.6 The Future of Visual Effects Systems

The post-production industry is constantly changing and technological advances in CPU processing power and graphics capabilities will continue to provide directors, producers and artists with increasing amounts of tools to enhance the process of visual effects generation. Even within the past few years, the introduction of powerful visual effects systems at accessible price-points, such as Combustion from Discreet, has enabled projects with smaller budgets to design and produce stunning special effects that were once left to the domain of high-budget feature film work. It is difficult to predict the future landscape for visual effects systems, but without question, feature sets will continue to get richer and turnkey solutions will continue to get cheaper, meaning the possibility for sophisticated effects will grow in the years to come.

Although the trend towards desktop solutions is gaining momentum, advances in HD and e-cinema will still place enormous demands on bandwidth, storage and graphics capabilities of visual effects systems. High-end visual effects systems will be the cornerstone of complex visual effects generation for some time to come.

Desktop solutions will become more predominant in the post-production industry and play an increasing role in broadcast, commercial and even film work. As high-capacity storage devices become more compact, desktop solutions may even become more portable, meaning more visual effects systems might find their way on set to assist with previsualisation and even test some visual effects before the final footage is brought to the final post-production phase.

Advances in networking technology will allow larger amounts of data to be transferred at much faster speeds, which will optimise workflows for high-resolution images such as 1080i HD or 2 K or 4 K film. Long-distance networking might even advance to the point where projects can be worked on in different cities or countries while still maintaining fast and reliable transfer speeds.

Change is the only predictable aspect of an environment as dynamic and fast-paced as the post-production industry. Facilities, artists and developers of visual effects systems all must adapt quickly to emerging demands in this realm, and those who do not will quickly fall behind.

Kit Barritt
Professional Services
Sony Business Europe

4.7 Editing Systems

4.7.1 Introduction

4.7.1.1 Film Editing

All television programmes are edited. Even the earliest live transmissions used more than one camera and switching from one camera to another is a form of editing. However, the term edit is more commonly used to mean a process of compiling previously recorded scenes into a finished programme.

The earliest movies simply recorded an entire event from a single viewpoint as if one were a viewer at that location. It became apparent that it was possible to add impact and make the movie more visually interesting by switching from one viewpoint to another. It is also possible to create effects, such as cutting from an arrow being fired to someone with an arrow sticking out of them, generating the illusion that the actor had been shot.

With film, the cuts are achieved by simply splicing (cutting and cementing) one piece of film to another. This process is still performed today, even on the biggest budget Hollywood movies, although the process is split into two parts. A rough edit is done using a print copy of the original negative, changing the edits until the director was happy. Then a final cut is performed on the original negative.

Often this final edit is performed using an optical printer. This would have alternate shots loaded onto an A-roll and a B-roll. The printer would copy from the A-roll then the B-roll. This technique is also used in videotape editing and will be described later.

4.7.1.2 The Early Days of Videotape Editing

The earliest videotape format that was successful and widely used was the Ampex Quadraplex system. Introduced in 1956, it recorded onto 2-inch-wide tape using a rotating cylinder fitted with four recording heads. This recorded tracks transversally across the tape, perpendicular to the direction of tape movement. Each track recorded 16 lines of video. A full field of video was therefore 'segmented' and composed of many tracks.

To edit a Quad recording, a cut was made across the tape using a razor blade and the two ends cemented together. It was important to know where to make the cut so that the TV signal was not interrupted. A developing fluid, called 'Ediview', was applied to the tape and allowed to dry. The fluid contained magnetically sensitive iron particles that formed visible patterns when dry. These patterns revealed the underlying video and control tracks, and identified the point to make the cut.

This process was slow, difficult and unreliable. The resultant tape was also liable to come apart at the edit points and could not be reused. It was a destructive process at a time when videotape was very expensive. The alternative to physically cutting the tape is to selectively copy from source tape to edit master in a process analogous to the optical printing technique used with film. Control electronics cause the play machine and the record machine to run in synchronism such that, at the point the cut is to be made on the recorder, the player starts to play the first frame of the new shot. At that point the recorder switches from play to record. This procedure forms the basis of all tape-based editing systems.

4.7.2 Traditional Linear Edit Systems

4.7.2.1 Timecode

For any tape-based edit system to work it is necessary to be able to identify each video frame uniquely. This identifier is then used to 'mark' the points of interest on the tape, such as the point a new recording is to start. The identifier used is timecode. This code is recorded in one of two ways: longitudinally down the tape (LTC) on a separate track, or in the vertical interval of the video signal (VITC), typically on lines 19 and 21.

Timecode is recorded as 4 bytes of binary coded decimal digits, representing hours, minutes, seconds and frames. Also

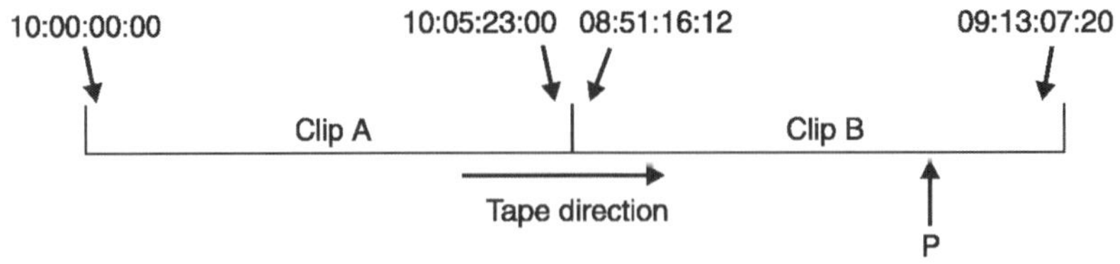

Figure 4.7.1 Discontinuous timecode.

recorded in the timecode data-space are various flags and user-bit data. The timecode runs from 00:00:00:00 to 23:59:59:24.

Timecode is normally recorded concurrently with the video recording. The edit control system has to assume that the timecode is consistent and consecutive. That is, the same timecode must not occur at more than one place on the tape and the time value must get greater as the tape is played. If, for example, the timecode jumped from 10:05:23:00 to 08:51:16:12, as shown in Figure 4.7.1, and the tape was in still at point P, the control system would be unable to locate to Clip A as it would assume it was after Clip B.

4.7.2.2 Recording Modes

A videotape recorder can make a recording in one of three modes: Crash record, Assemble edit or Insert edit. To be able to play back correctly, a control track must be recorded in addition to the video. In Crash record, new recordings are made of video, audio, timecode and control (CTL). In Assemble mode new recordings are made as in Crash record, but the VTR first synchronises itself to the existing recording on the tape.

Crash record and Assemble edit modes both erase everything previously on the tape, using the full, or wide, erase head, before making the new recording. In Insert edit mode the existing CTL recording is played, with video or audio tracks being recorded over the previously recorded tracks.

Unfortunately the term 'insert' has opposite meanings in linear and non-linear editing. In linear editing, an Insert edit replaces part of the material with something different. The edit process might, for example, record the audio tracks then 'insert' the video. This process is called 'overlay' in non-linear editing. In non-linear editing, however, an 'insert' edit is where the new material extends for the duration of the programme. This is analogous to 'insert' mode in a word processor, as opposed to 'overtype' mode.

4.7.2.3 Tape Preparation

For the recorder to be able to synchronise during the edit there must be something recorded on the tape for the recorder to lock on to. It is common practice to record a test signal such as colour bars for 2 minutes on the start of the tape to allow playback alignment. A black signal is also often recorded after the bars and this can be used to lock to during the edit.

More typically, black is recorded for the duration of the tape in a process called 'striping'. This ensures that there is good timecode all the way through the tape and that there is something for the servo system to lock to at any point on the tape. A tape that has been recorded in this way is called 'striped', or 'blacked'.

The value of the timecode that is recorded at the start of the tape is arbitrary, but a common practice is to start at 09:57:00:00. With 2 minutes of colour bars and a minute of black the programme will start at 10:00:00:00. This is a convenient round number that allows the programme duration to be determined very easily.

4.7.2.4 The Two-Machine Edit

The control electronics of videotape recorders as early as 1-inch C-format machines were able to control another machine via a control cable. The simplest editing system is two VTRs connected as shown in Figure 4.7.2.

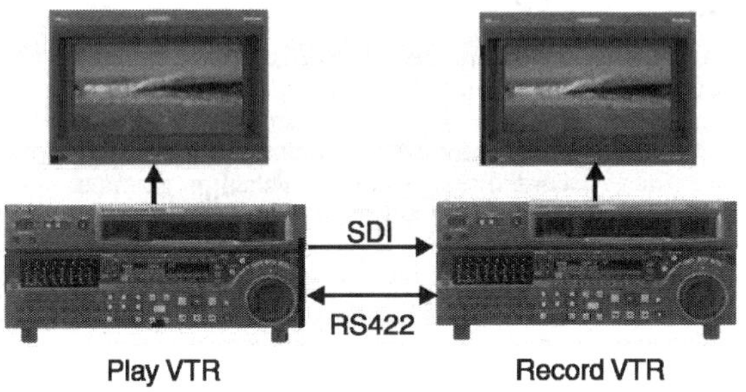

Figure 4.7.2 Two-machine edit.

The machine on the right (the recorder) takes control of the player by issuing commands over the RS422 control cable. The minimum editing procedure is as follows:

- Select Insert or Assemble edit mode.
- Mark an in-point on the recorder to indicate the point the new recording should start.
- Mark an in-point on the player to indicate the first frame of the new recording.
- Execute the edit by pressing the 'Auto Edit' button.
- Press stop at the end of the required new shot.

In addition, the editor may choose to:

- Select just video or audio insert.
- Mark an out-point on the recorder or player.
- Select an in-point or out-point for the audio different from the video point.
- Alter the audio playback or record levels during the edit.
- Preview the edit prior to committing it.

Although a two-machine edit can only perform cut edits it is common to have an external audio mixer and to use it to perform audio cross-fades and mixes. The vast majority of video edits in a typical TV programme are cuts. However, most programmes require mixes or cross-fades on the audio. Even when editing news and sports it is the audio that is the tricky thing to get right.

4.7.2.5 Editing Problems

Most problems that occur in linear tape editing are due to timecode. Firstly, timecode value 00:00:00:00 is avoided because the servo system is unable to cue to 5 seconds before as this crosses the start/end boundary. Similarly, any timecode discontinuity will cause problems so a point cannot be marked for 5–10 seconds after such a disturbance. Secondly, it is virtually impossible to use or repair a tape that has gaps in the control track. Selecting Assemble edit mode at a point other than at the end of the edit will result in a control track hole at the end of the edit. This is normally only recoverable by dubbing the entire tape onto a newly prepared striped tape.

4.7.2.6 A/B-Roll Editing

The two-machine edit is only able to produce video cut edits because there is only one source of video. To produce a mix, wipe or other effect, multiple sources of video must be available. These will be connected to a vision mixer. The VTRs and mixer will be connected to an edit controller (or editor). This is a computer system either based on proprietary hardware, such as the Sony BVE-9100, or based on a PC platform, such as the Ampex ACE-10.

A typical system may be composed of three source VTRs, a record VTR, a vision mixer, an audio mixer and an edit controller, as shown in Figure 4.7.3. The editor selects in- and out-points and the required effects. The edit controller then cues all the machines, including the mixer, and makes everything synchronise up to the Recorder in-point. The process is called A/B-roll editing by analogy to the printing process in film.

A mix edit is typically achieved using the following process, illustrated in Figure 4.7.4:

- The first shot is copied from source tape (VTR A) to recorder.
- The point of the mix is marked as the in-point on the recorder.
- The corresponding point is found and marked on the VTR A. This will produce a so-called 'invisible' edit at the in-point, an edit to exactly the same shot.
- The selected frame to mix to is marked on the VTR B.
- An effect and duration are selected.
- An out-point may be marked on any one of the VTRs.
- The edit is executed.

4.7.2.6.1 The EDL

The edit system records all of the information about the edits, such as the timecode and effects, in a file called an Edit Decision List or EDL. A number of different formats for this file exist, but the most common are CMX and Sony. By

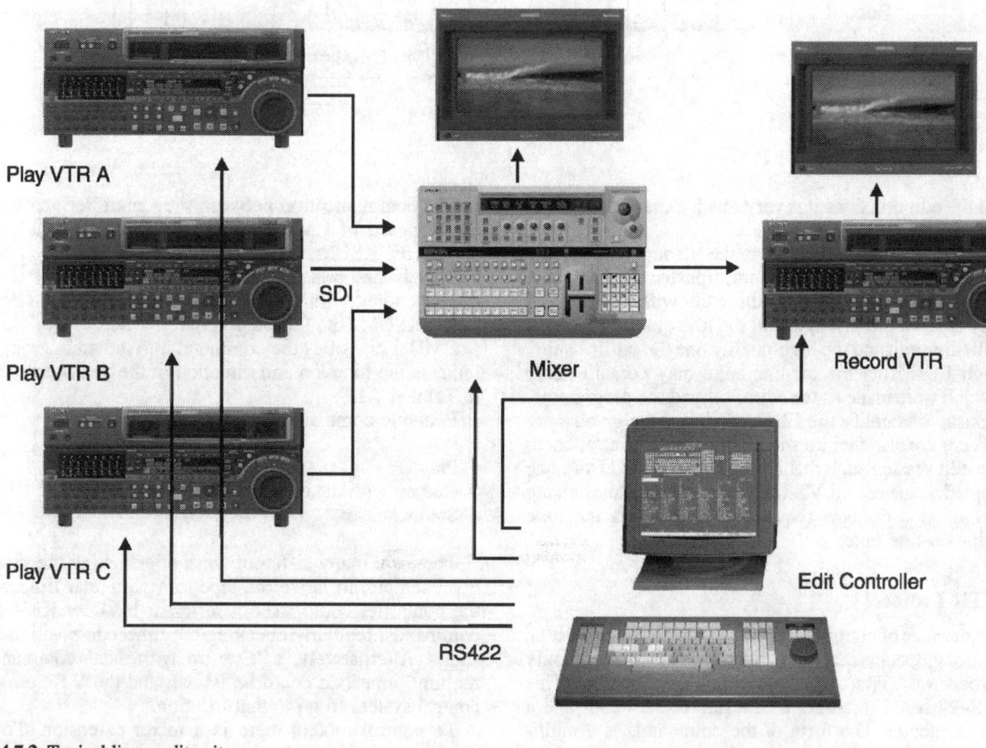

Figure 4.7.3 Typical linear edit suite.

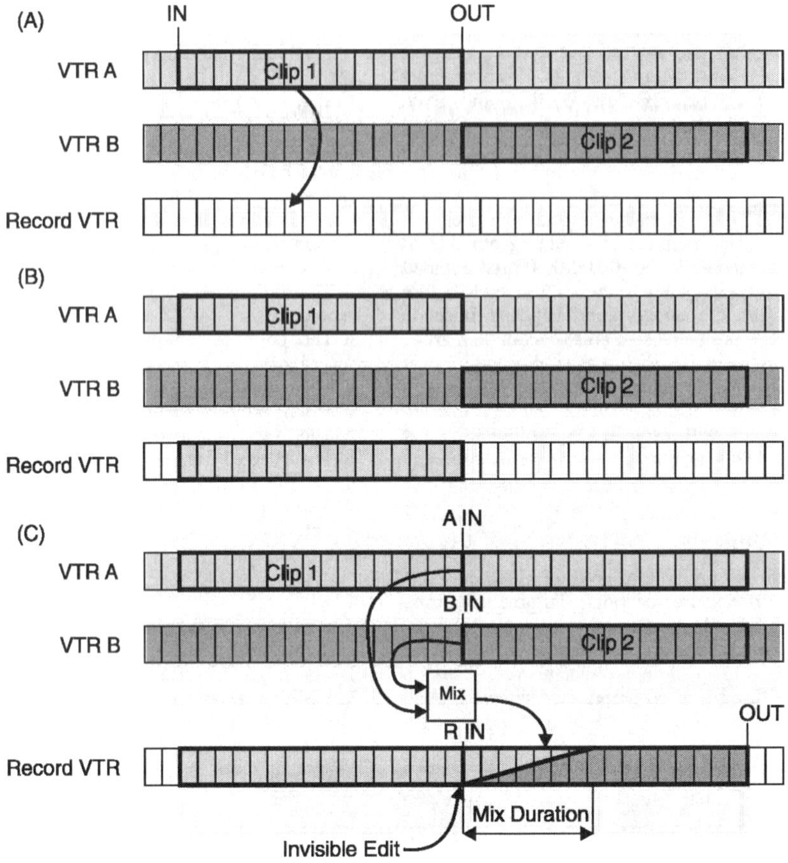

Figure 4.7.4 A/B editing.

tracking all the edit decisions it is very much easier to amend or redo an edit as the programme is made.

A common editing process is to make a rough edit of the programme using a low-quality system, transfer the EDL to a high-quality system and then redo the edits with the original high-quality tape. In this arrangement the low-quality system is called the off-line edit and the high-quality one the on-line suite.

The rough EDL from the off-line suite may contain many changes, which were made as the editor refined the programme. Software exists to simplify the EDL into just the clips that are required. There are also facilities in high-end edit controllers to organise an edit session such that all the shots needed from one tape are copied to the record VTR in one pass, thus minimising the number of tape changes required. This reduces the time needed in the on-line suite.

4.7.2.7 VTR Protocol

There are a number of communication languages, or protocols, used to control videotape machines, but the one most commonly used by broadcast VTRs is the Sony Nine-pin Protocol. This uses the RS422 serial standard, a data rate of 38.4 kbps and a DB9 serial connector. The form of the commands is a multi-byte message which may be acknowledged or expect a reply.

The communication between the Controller and the Device is composed of CMD-1 + DATA COUNT, CMD-2 + DATA and CHECKSUM. When the DATA COUNT is zero, the DATA is not transmitted. When it is not zero, the DATA corresponding with the value is inserted between CMD-2 and CHECKSUM (see Figure 4.7.5).

CMD-1 classifies the command into the main groups, which indicate the function and direction of the data words, as shown in Table 4.7.1.

Example commands are:

Play:	20 01 h
Cue up with data:	24 31 h
Status Sense:	61 20 h

There are many different ways of achieving the same result. For example, to move the tape to a particular timecode point the controller could issue a series of FWD or REV and JOG commands, regularly checking the timecode position from the status. Alternatively, a 'Cue up with data' command to the required timecode could be issued, and the VTR uses its own control system to go to that location.

To control mixers there is a mixer extension. To achieve proper control of all the facilities of the mixer the edit

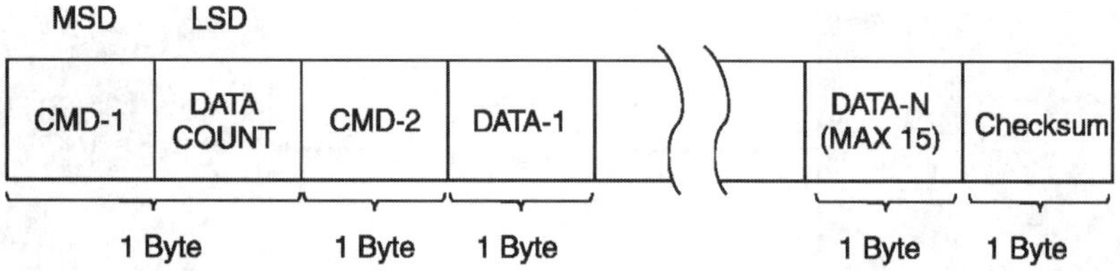

Figure 4.7.5 Protocol structure.

CMD-1	*Function*	*Direction* (*Controller* / *Device*)
0	System Control	→
1	System Control Request	←
2	Transport Control	→
4	Preset and Select Control	→
6	Sense Request	→
7	Sense Return	←
DATA COUNT	Indicates the number of data words that exist following the CMD-2 (0 to F Hex)	
CMD-2	The designated COMMAND to the Device or the COMMAND return from the DEVICE	
DATA	The number of data words and their contents are defined by the specific CMD-2	
CHECKSUM	The CHECKSUM is the sum of the DATA (D0 to D7) contained in each data word, from CMD-1/DATA COUNT to last data word before CHECKSUM	

Table 4.7.1 CMD-1 classification

controller needs to know what facilities the mixer offers. Alternatively, the mixer may be able to emulate a popular simple mixer such as the GVG-100.

There are extended versions of the protocol for controlling disk devices (Disk Protocol), allowing for file selection, for example. This is used to control servers such as the Sony MAV-555 or the Textronix Profile.

4.7.2.8 Multi-VTR, Multi-Layer Systems

Each recording can only be composed of a combination of the video sources, and it is not normally possible to combine new images with what is on the recorder. If a complex image is made up of more layers than video sources then the process must have multiple stages. Suppose we wish to have a background image with 10 independent keys (such as captions) laid over it. With only one character generator, only one key can be added at a time. The first stage is to record the background and the first key, as shown in Figure 4.7.6.

The tape is then removed from the recorder and placed in player A. A new striped tape is placed in the recorder and an in-point marked. The background to the edit is the original recording and a new key is added to it, as shown in Figure 4.7.7.

Once recorded, the tape in the recorder is once again placed in VTR A and a new tape is placed in the recorder. Alternatively,

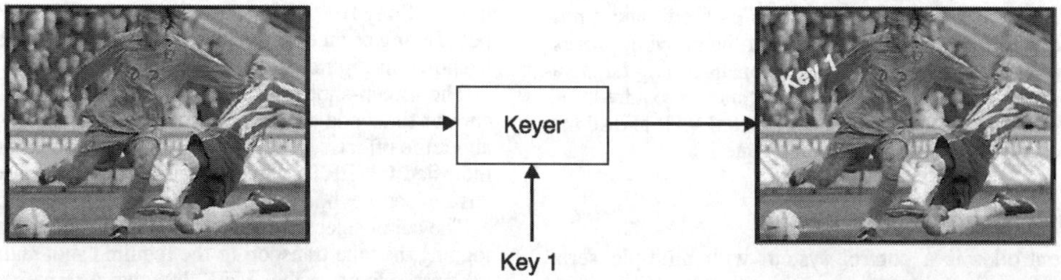

Figure 4.7.6 First layer.

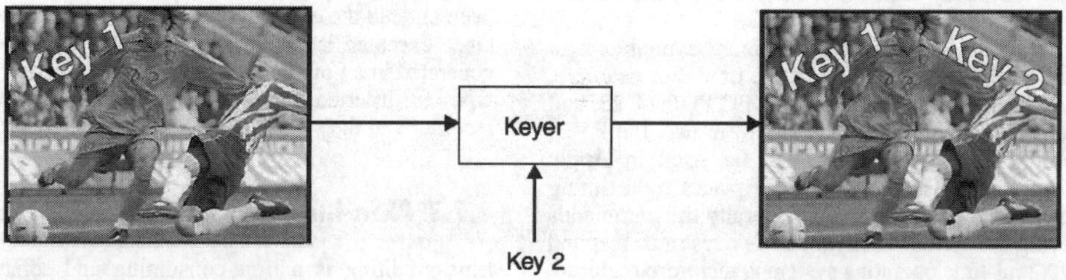

Figure 4.7.7 Second layer.

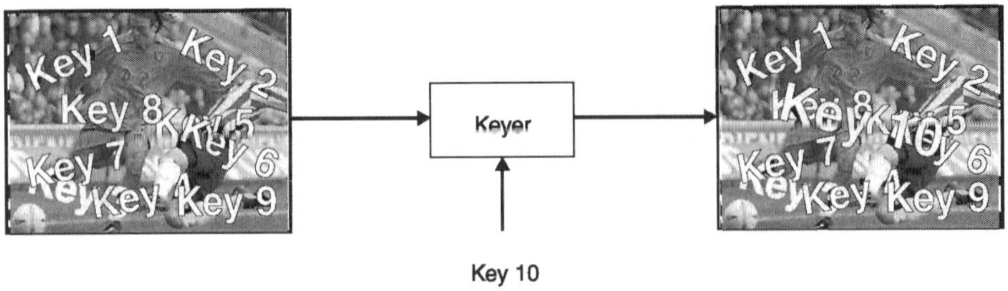

Figure 4.7.8 Last layer.

the tapes in the recorder and VTR A may be swapped or the system reconfigured so that VTR A becomes the recorder. This process is repeated until all the required layers have been added, as in Figure 4.7.8.

It is important to organise the edit from the back to the front, so that each new image will be laid over the previous images. Careful examination of Figure 4.7.8 will illustrate that higher numbered keys are in front of lower numbered ones. It is very important to get the edit correct the first time, as it is virtually impossible to correct an error on a lower layer.

Through this process the original background image will have been copied 10 times. If the system uses analogue tape recorders it is likely that this will have caused significant deterioration of the image quality. It is therefore preferable to try to add as many layers as possible at each stage. This requires a very large and complex edit suite.

4.7.2.9 Pre-Read and Multi-Layering

Some digital VTRs such as the Sony DVW-500 Digital Betacam VTR have pre-read playback heads. These allow the image on the record machine to be played back as a new image is recorded over it. This significantly simplifies the process of multi-layer editing. The recorder is the background to each edit and a new layer is added to the recorder each pass. If the copying process does not produce a generation loss, as happens in Digital Betacam, this process can continue for as many times as required. The in- and out-points remain the same on the record VTR and all that happens each pass is a change of source material.

4.7.2.10 Edit Controllers

An edit controller is a control system with multiple serial connections to each of the devices in the edit suite. The earliest edit controllers were based on proprietary hardware, as there were no computers that were powerful enough for the job. They were therefore extremely expensive: for example, the Ampex EDM-1 introduced in 1971 cost $95,000.

Later, as computer hardware became more available, edit controllers used the microprocessors of the time. For example, the Ampex ACE-2000 was based on the DEC PDP 11-23, and a popular Sony edit controller which is still in use, the BVE-9100, is based on the Motorola 68030 (as used in Apple Macintosh computers). The VTR protocol places tight timing requirements on the communication. Generally the commands are issued during the vertical interval and devices must respond within 9 ms. A real-time operating system is therefore required, or a computer quick enough to transmit the commands at the exact moment they are needed.

For many years now commodity PCs have been powerful enough to do the job, although they generally need expansion cards to provide the additional RS422 ports and video timing reference signals interfaces. The earliest edit controllers used either custom or QWERTY keyboards. The output was either a numerical display or a text display on a monitor. Even the latest controllers continue to use this human interface. The advent of GUIs and mouse input is of little benefit to experienced editors, who can drive an edit controller like a concert pianist plays the piano.

A major manufacturer in the early days, CMX, ceased trading in 1998. The EDL file format used by CMX machines, however, is still in use today as a mechanism for exchanging information between non-linear off-line suites and linear on-line systems.

Modern low-end edit controllers such as the Panasonic AG-A850 and the Sony PVE-500 have returned to proprietary hardware and simple LED-based displays. They are, however, hundreds of times cheaper than their predecessors.

4.7.2.11 Linear Edit Operations

The principles of linear edit operations using different controllers are similar, the main variation being the use of either LED indicators or a CRT type screen. The more powerful devices like the Sony BE-9100 use a screen, whereas the PVE-500 does not. Taking as an example the BVE-9100, a typical screen-shot is shown in Figure 4.7.9.

The screen-shot shows the data important to the editor: the current timecode positions, in- and out-points, and information about the effects. Data is entered into this screen using either a modified QWERTY keyboard as shown in Figure 4.7.3 or the customised keyboard shown in Figure 4.7.10.

The editor selects a point on the tape by shuttling, playing and jogging the tape transport to the required shot and then entering the timecode using the 'Mark' buttons. Alternatively, the points could be entered directly using timecode values, if known. The effects are similarly entered, with, for example, wipe patterns being entered by number. When the required information has been entered the edit can be rehearsed (previewed) or executed. Once executed it is added to the EDL. To execute an edit list generated in an off-line suite the EDL is loaded, the appropriate tapes are inserted into the players, a striped tape placed in the recorder and the system told to perform all the edits.

4.7.3 Non-Linear Editing

Linear editing is a time-consuming and sometimes difficult process. For simple tasks such as editing news, in the hands of experienced editors it can be quick. For other programmes

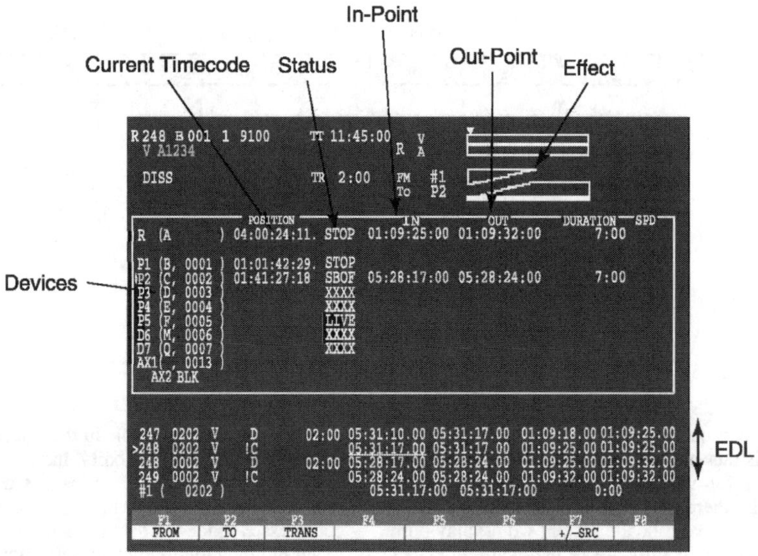

Figure 4.7.9 BVE-9100 screen.

Figure 4.7.10 BVE-9100 dedicated keyboard.

such as drama and documentaries, there are time constraints for the programme and aesthetic considerations. The editing process for these programmes often requires many changes to achieve the desired result.

Each edit requires the material to be copied from source VTR to record VTR. To insert a 5-second clip into the middle of a 20-minute programme requires the copying of a significant portion of the programme. Editing on film is a much quicker process, as sections can just be inserted or removed as required. Many editors prefer this as they can concentrate on the end result rather than the mechanism. Non-linear systems were developed to address this need.

Before true non-linear edit systems there were devices that recorded onto hard disks, but which behaved like tape machines. Abekas and Quantel were early manufacturers of hard disk recorders, with machines such as the Abekas A64. These systems allowed rapid cueing to any point and very short pre-roll times, and could be controlled using conventional tape edit controllers. The early machines, however, used customised hard disks and could only record a few minutes of uncompressed video. As hard disks got cheaper and compression techniques were developed, it became possible to record a whole programme to the disk and edit it in a more user-friendly way: non-linearly.

Editing on tape is called linear because a tape machine can only play linearly from start to finish. By contrast, non-linear systems can access the material in any order with random access to the clips.

A hard disk system has a buffer (the cache), which is filled as required by the playback system. To edit the material it is only necessary therefore to play the clips in the required order. No copying of material is necessary. This process is illustrated in Figure 4.7.11.

Five clips have been copied to the disk. The editor decides to trim the length of each of these clips and edit them in the order D, A, B. The non-linear edit system stores the required ‘in’ and ‘out’ marks as pointers to the material (numbered 1 to 6) and plays back the edit using those pointers.

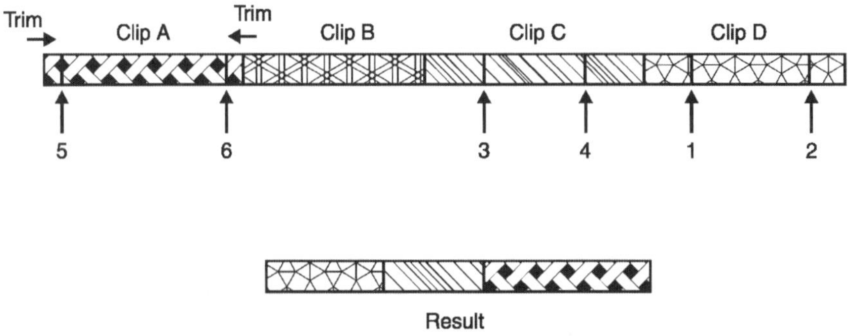

Figure 4.7.11 Non-linear editing.

The edit decisions can be changed as quickly as the editor can make those decisions and the computer can change the pointers. The pointers for video and audio are independent, and in the case of audio there are likely to be several tracks. To achieve an audio mix or cross-fade the system would play both tracks of audio and, either in software or hardware, alter the gain and add the tracks together to produce the desired result.

Video effects are created in a similar way. If the system is sufficiently powerful it will be able to play back both sources of video and process them to generate the required effect in real time. In many systems, however, the effects are rendered more slowly than real time and stored on the disk as a transitional clip. This is shown in Figure 4.7.12.

On playback the system plays to point A, cuts to the start of the rendered effect clip (point B), plays that clip to point C, and then cuts to point D and plays clip 2.

4.7.3.1 Hardware Requirements

4.7.3.1.1 Broadcast Systems

The data rate requirements for broadcast-quality video are immense. Uncompressed 8-bit video has a data rate of 180 Mbit/s. To be able to record at this rate onto hard disk, the early systems used customised drives to give parallel recording to multiple platters, 1 bit per platter.

One way to achieve the required data transfer rate and also provide some redundancy is to use a disk array system such as RAID (Redundant Array of Independent Disks). These systems distribute the data over multiple hard disks, reducing the data rate to any individual drive.

It has only been possible in the last few years to transfer data at this rate onto commodity hard drives. For example, the Seagate Barracuda 7200 series has a sustained data transfer rate of greater than 58 Mbyte/s. If a system is working with compressed video the data transfer requirements are, of course, lower but the input/output circuitry must be able to perform the compression in real time.

Two popular broadcast-quality compression systems are DV and MPEG IMX. These run at 25 and 50 Mbps respectively. At these rates the storage requirements are 5 minutes per Gbyte and 21/2 minutes per Gbyte respectively. Many current broadcast systems offer multiple concurrent channels to the same hard disk array. This increases the transfer requirements so disk arrays are normally used. A disk array also has greater storage time than possible on a single disk.

Many RAID systems have a degree of protection that allows for failure of one of the drives. The process of transferring the shot video material from tape to disk is often called digitisation. This name comes from time when the video was analogue on tape and digital on the hard disk. A more common modern term is capture.

With many current videotape systems this transfer is not only digital but the data is transferred in compressed from. This has the advantage of avoiding the decompression–recompression process and the attendant quality loss that this involves. Some formats and VTRs also allow the material to be transferred faster than real time, and in some cases, such as the Sony DSR-85 and ES-3, up to four times faster than real time. Thus, a 20-minute clip can be transferred in 5 minutes.

In addition to the disk capacity and speed, the processing power of the computer system is also significant. The earliest

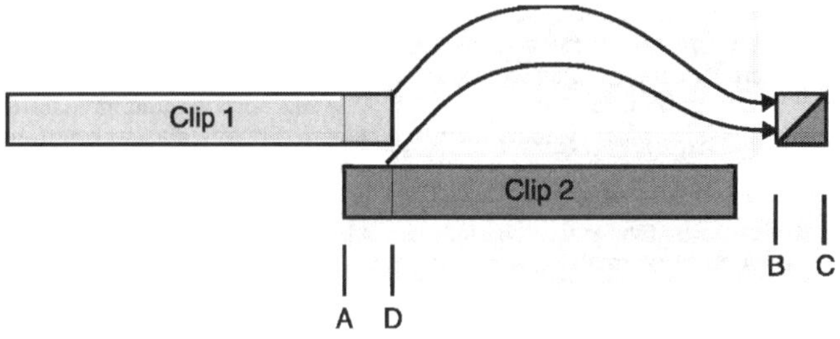

Figure 4.7.12 Rendering an effect.

systems were unable to generate or preview effects in real time using software and either had to output the signal to external effects hardware or the user had to wait for the system to render the effects. PC hardware is now able to render simple effects in real time but high-performance computers and dedicated hardware are needed for high-definition productions.

4.7.3.1.2 Non-Broadcast Systems

As early as 1995 the Quadra 840av from Apple Computers was able to capture video to disk and edit it into a programme. Contemporary hard disks had capacities of 50–200 Mbytes. The QuickTime compression system subsampled the video and reduced the data rate to around 4 Mbps. Even at this low quality, the system could only store tens of minutes of video.

Many modern PCs and Macs have built-in DV interfaces (iLink or Firewire), and IEEE-1394 expansion cards are cheap and simple to fit. Hard disks of most current PCs are able to store many hours of high-quality video. This has become a selling feature of consumer computers, and both Apple and Microsoft have video editing facilities built into their software.

Consumer cameras are readily available with DV interfaces and it is a simple job to connect these to the PC, control the camera and capture the video. The principal limitation of consumer PC systems is the inability to perform effects in real time. Often even the simplest of effects such as a mix or wipe can take many seconds to generate.

4.7.3.1.3 Compression Issues

Compression of the video signal is widely used at all stages of TV production, from Digital Betacam acquisition through post-production to digital satellite transmission. Compression is used because it is a data-bandwidth-efficient tool. It reduces transmission bandwidth and storage requirements.

It is, however, virtually impossible to manipulate a compressed image without first decompressing it. MJPEG, I-frame only MPEG and DV are three popular compression schemes used in post-production systems. Each of these systems permits cuts at arbitrary points in the video without problem. The vast majority of editing can therefore be performed without additional processing.

As described earlier, effects such as cuts and wipes are generated in the non-linear systems either in hardware or in software, and may be real time or slower than real time. As it is not possible to perform this processing on compressed signals they must be decoded prior to processing and recoded after processing.

This is exactly what happens in, for example, the Sony MAV-555, which is a multi-channel audio/video recorder with a built-in non-linear edit system. To perform a mix, two MPEG decoder channels are used to play the clips to a hardware effects generator. The output then goes to an MPEG encoder that recodes the clip, as shown in Figure 4.7.13. This rendered clip is inserted into the finished edit at the appropriate point and in so doing the result can be played using only one decoder. This rendered fragment will have used up a little of the hard disk space but this is not usually significant.

The decoding–recoding process has to happen even for the simplest of effects, such as a mix or addition of a caption. This will, of course, introduce a generation loss and may affect the quality of the end result.

Non-linear editing is a very good and easy way to produce the highly complex multi-layer clips used in title sequences and pop videos. For this use it is far better to work on the material in the uncompressed domain to avoid problems of quality loss and this is what happens on the highest quality systems.

4.7.3.2 Software Systems

There are many software packages varying from the simple, such as Windows Movie Maker as built into Windows XP, through to packages such as Avid Media Composer operating on standard PCs and on to high-end systems such as Fire from Discreet operating on high-performance workstations or specialist hardware. All of these packages offer a similar way of working, with the differences being the facilities, functionality and speed. The principle of operation is similar and Adobe Premier is a representative example to use. The user is faced with a GUI, with a project or 'clips' window, a timeline window, a monitor window and other ancillary windows. Items are loaded into the project, either by importing them as files or capturing them from tape. To speed this process, the clips are often captured in batches.

For simple cut editing, clips are trimmed as required and assembled into the required order either on the timeline or on the storyboard. The timeline gives a 'linear' representation of the final result as if it were videotape. The storyboard shows each a still from each clip (the thumbnail) arranged in the required order. Audio tracks can be similarly assembled. These windows are shown in Figure 4.7.14.

For more complex edits the video is arranged to overlap and an effect is applied to that transition. Keys such as captions are added by dragging them to a key track and setting the transparency mode to achieve the desired effect. It is relatively simple

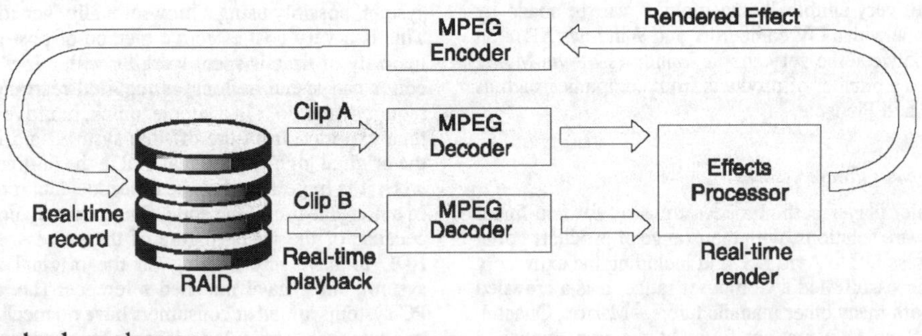

Figure 4.7.13 The decode–recode process.

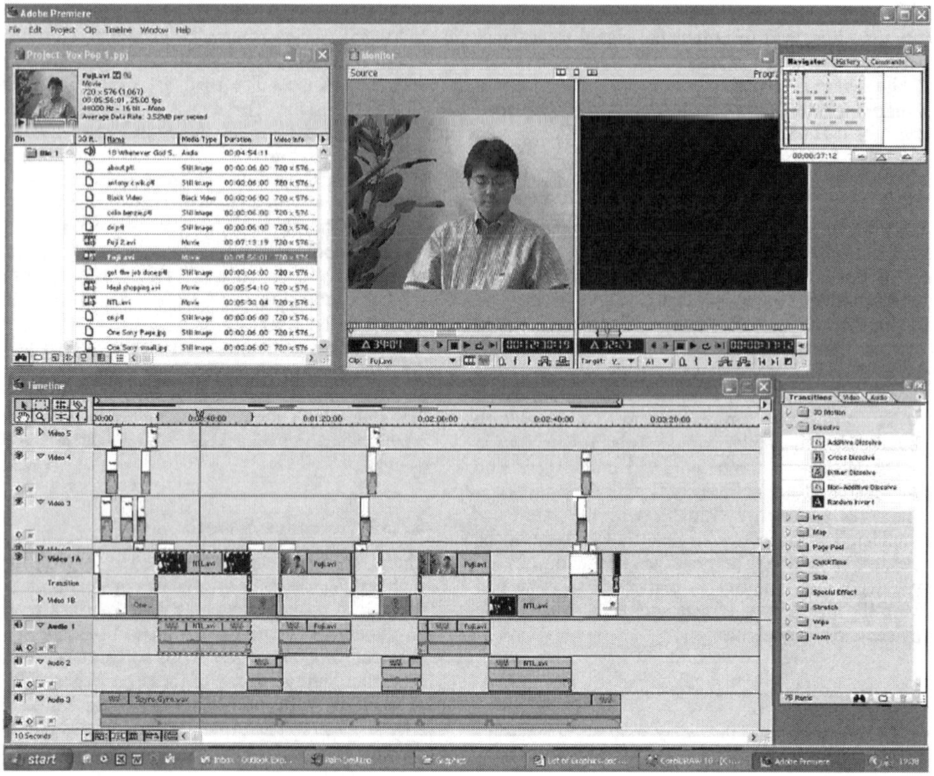

Figure 4.7.14 Premier GUIs.

to incorporate graphics produced by other software packages provided they are the correct resolution and aspect ratio.

Audio editing is performed by arranging the audio on parallel tracks in a visual representation of a multi-track recorder. The levels of the tracks are then set to achieve the desired mix. Levels can be faded up and down to produce a cross-fade.

4.7.3.2.1 Low-end software systems

Adobe Premier used to be the first name in high-end domestic video production on a Mac/PC platform. Now as PC systems have improved in their facilities and performance, Premier is being used not just for off-line editing but even for low-budget broadcast TV programmes. With the power of modern computer hardware, very simple 'home movies' can be made by just about anyone with a DV camcorder and Windows XP or an iMac. Sitting between the very simple Windows Movie Maker and Premier are a number of products from companies such as Ulead, Dazzle and Pinnacle.

4.7.3.2.2 High-end software systems

Avid is the major player in the broadcast market for non-linear hardware/software solutions, with their range of products going from Avid Xpress DV to Avid DS, and including the extremely popular and successful Media Composer range. It is a crowded marketplace, with many other manufacturers – Matrox, Quantel, Pinnacle and Sony to name but four. At the very high end there are systems aimed at film post-production for compositing the computer graphics and real-life images of a movie like *Star Wars*. Companies such as Discreet Logic produce such a range of products, most of which run on high-performance workstations like those made by Silicon Graphics.

4.7.4 Combining Linear and Non-Linear

Post-production is time-consuming and costly. A significant part of that cost is the equipment cost. One technique used to reduce these costs is to split the process between a low-quality off-line edit and a high-quality on-line edit.

The off-line system is increasingly a computer-based edit system, possibly using a browse-quality version of the material. This is a very cost-effective method of post-production as the majority of time is spent working with a low-cost system. The editor can spend as long as required rearranging the shots as required, and do it in a simple, quick, intuitive way. The EDL is then exported from the off-line system, and is then used with the original high-quality material in the on-line suite. For this to work it is important that the timecode data remains unchanged. In a non-linear off-line suite, timecode may not be needed at all because of the visual nature of the process. However, for the EDL to have any bearing on the original material the two systems must have matched references (i.e. timecode). Many PC systems aimed at consumers have no mechanism for recording timecode, particularly if recording from an analogue source such as a VHS deck. It is possible, however, to set the starting timecode of the captured clip by reading from the source deck and entering that data into the timecode properties of that clip.

4.7.5 Server-Based Systems

It is becoming increasingly common to store the audio/video material on a central server rather than the hard disk of the workstation. Working in this way allows many editors access to the same material at the same time. This makes it particularly popular in a news or sports environment, where a very quick turnaround is required and the material may need to be re-purposed.

With this method of working, the workstation that controls the transfer of material to the server is often called the ingest station. In a news or sports environment this could be working on either live or tape-based material. Typically, the filing systems of the server allow for not only multiple workstations to access the material at the same time, but also to allow that access during the recording. The ingest system may also be simultaneously providing outputs in multiple resolutions, qualities and formats of the material at the same time. For example, a football match may need to be provided in the following forms:

- A browse-quality feed to enable a journalist to write the story.
- A low-resolution, low-data-rate feed to web service stream over the Internet.
- A high-quality feed I-frame only for a half-time highlights package.
- A high-quality, long GOP version for archiving purposes.

The relentless move toward computer and file-based editing is likely to continue, particularly as file exchange standards such as MXF become adopted. The use of non-linear techniques permits greater efficiency in the post-production process, obviating the need for time-consuming tape copying. However, there are still savings to be made, facilitated by the use of metadata. A log sheet is the simplest form of metadata. Transferring the log sheet to the computer system alongside the audio/video material allows quick access to the required clips, as the editor does not have to determine which one is which. This process becomes more powerful as the metadata becomes more rich and meaningful. This is particularly true for archive material. It is possible to imagine, for example, a nature programme being made by:

- Searching the metadata of the library.
- Browsing low-quality copies.
- Copying the high-quality copies to a local edit system.
- Adding a narrative track and filing the result to the transmission server.

Not a tape in sight!

Tape will still have a home as an archive medium, but linear editing will rapidly have no benefit and no place in broadcast.

Original document by **J D Millward**
BSc, C Eng, MIEE
Head of Research, Rank Cintel Ltd
Major revision by **P R Swinson**
BKSTS, SMPTE, RTS
Peter Swinson Associates Ltd, UK

4.8 Telecines

4.8.1 Telecine Types

Telecines may also be described as motion picture film scanners. The term *telecine* refers to a film scanner designed to provide television signals from the film image. The term *film scanner* is presently used to describe devices that can also provide images in a digital data form more closely related to computer image files than television signals.

Telecines can be grouped into three distinct types: photoconductive, flying spot and CCD.

The term *photoconductive* applies to those telecines that incorporate an electronic tube camera as the pick-up device; a variety of camera tubes can be used, such as plumbicon, vidicon, saticon, etc. The light source is usually a tungsten halogen lamp and the projector will have an intermittent film transport so that the image on the film frame can be transferred to the camera tube target while the film is stationary, a necessary requirement since the camera tubes exhibit storage characteristics. The colour photoconductive camera has three separate tubes to produce the red, green and blue signals, and the scans on these tubes have to be accurately matched to obtain colour registration. Photoconductive telecines have not been manufactured since the 1970s and are therefore of historic interest only.

Flying spot telecines can be grouped into two distinct types: Cathode Ray Tube (CRT) and Laser-based telecines.

The CRT-based *flying spot* telecine uses a single, white light, flying spot cathode ray tube with an unmodulated raster as a light source. An image of the raster is projected onto the film by an object lens. Light passing through the film is modulated by the film image, and the light is then collected and directed through colour splitting optics to photomultipliers or in more recent telecines to Large Area Avalanche PhotoDiodes (LAAPDs) which produce the electrical signal. Such detectors are generally referred to as photoelectric cells (PECs).

Since the colour splitting process occurs after the imaging process, there is no requirement for colour registration when using a white light CRT-based scan (see Figure 4.8.1).

Laser telecines have been built experimentally; such devices rely on three unmodulated RGB lasers whose *flying spot* beams are combined to project onto the film image. As with the CRT-based telecine, light passing through the film is modulated by the film image and is directed through colour splitting optics to PECs which produce the electrical signal. At the time of writing the author knows of no laser-based telecines that are commercially available.

CRT-based *flying spot* telecines at present being manufactured use continuous motion film transports, although there may be a small number still in use which have a 16 mm transport with intermittent motion and a film frame pull-down time of 1.2 ms.

CCD telecines can be grouped into two distinct types: Area or Field Array and Linear or Line Array.

Area Array CCD telecines use similar principles to the photoconductive telecine with a tungsten halogen, xenon, HMI or triple RGB LED light source, but with the three photoconductive tubes replaced by three CCD area arrays. An Area Array CCD (*charge-coupled device*) consists, as the name implies, of an area matrix of CCD photosites. These arrays are usually adaptations of arrays used in modern CCD video cameras with the most significant difference being the quite different colour separation responses (see Chapter 3.3; also see CCD cameras for a detailed description of CCD Area Array devices).

Such telecines have intermittent operation transports where the film is held stationary while the CCD arrays are exposed;

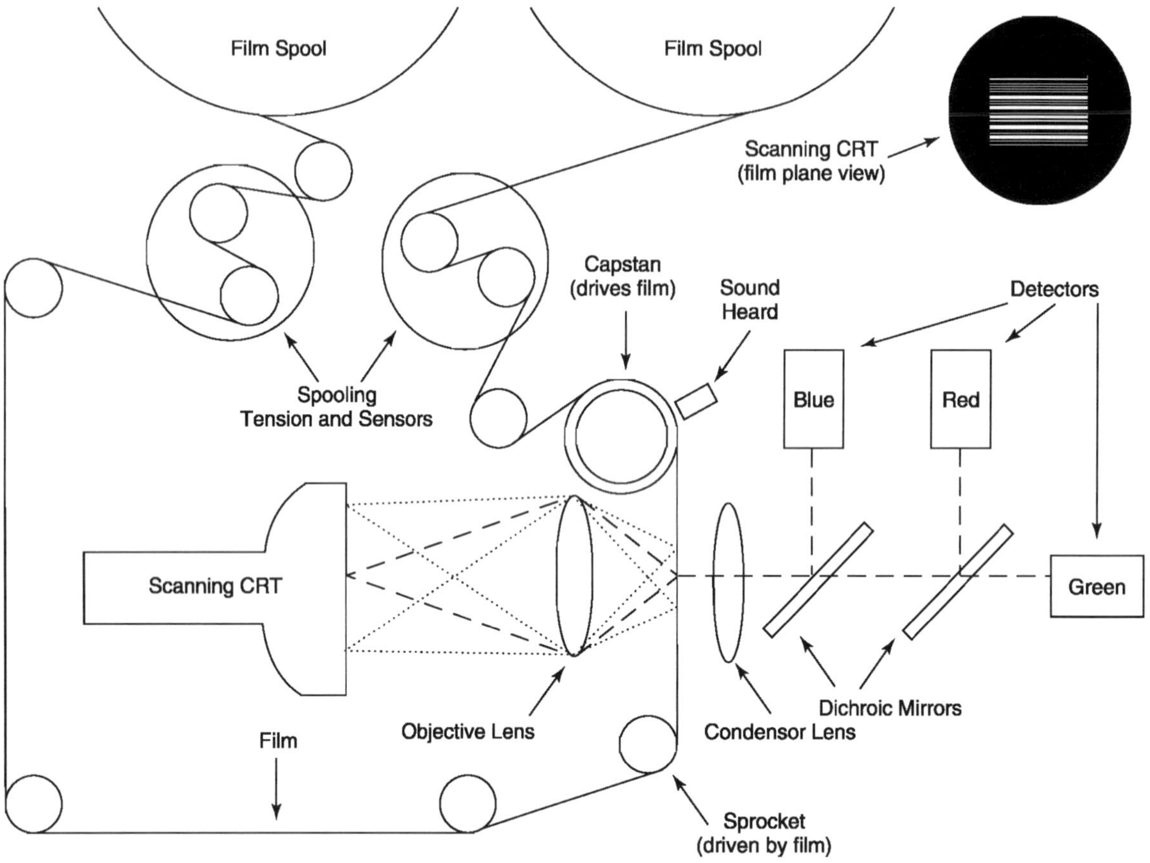

Figure 4.8.1 CRT flying spot transport and scanner head.

the film is then advanced for each subsequent frame. The film transport for such telecines is necessarily somewhat complex due to the requirement to stop and start at rates up to 30 frames per second, while maintaining very high positional precision.

Such precision is accomplished in real-time telecines by means of intermittent sprocket drive or in older systems by using intermittent pull-down claws. Additional image stability methods used in current machines are optically based, where the film's image is electro-optically repositioned prior to capture by the imaging head. This is achieved in real time, based on capacitive detection of the film perforation position in the film gate.

Non-real-time film scanners that require very high positional precision tend to utilise pin registration gates; while these cannot normally operate at real-time speeds they provide positional accuracy that often exceeds that of the original film camera (see Section 4.8.8.3).

The entire film image is illuminated by a white light source and the resultant image is focused by the object lens and RGB beam splitter onto RGB area arrays. These arrays directly generate the electrical image signal (see Figure 4.8.2).

It is worth noting that the output of a CCD, Area or Linear Array, is analogue, and is subsequently converted to a digital signal by analogue-to-digital converters (see Section 4.8.7.2).

Also, since this method of scanning the film produces a sequential or progressive scan, a sequential-to-interlaced converter is required before the signal can be recorded or transmitted as an interlaced format (see Section 4.8.7.13).

Today many mastering transfers do not require conversion to interlace as HD mastering standards include progressive recording formats.

Line Array CCD telecines typically are comprised of three sensors (RGB), each being a single strip of photosites arranged to image across the width of the film. A small horizontal portion of the film, at the same position as the CCD, is illuminated with a tungsten halogen, xenon or HMI light source; this allows the light to be concentrated at the CCD imaging area. An object lens then focuses the entire image width over the small slit height through a beam splitter to the Line Array CCDs. The Line Array CCDs comprise a single line of photosites adjacent to a charge-coupled shift register, into which the charge is transferred from the photosites at frequent intervals. Then, while new charges are being generated in the photosites, the transferred charges are clocked out serially from the shift register to produce the electrical signal.

Since the CCD used in Line Array telecine has only one line of photosites, it normally scans the film only in the horizontal

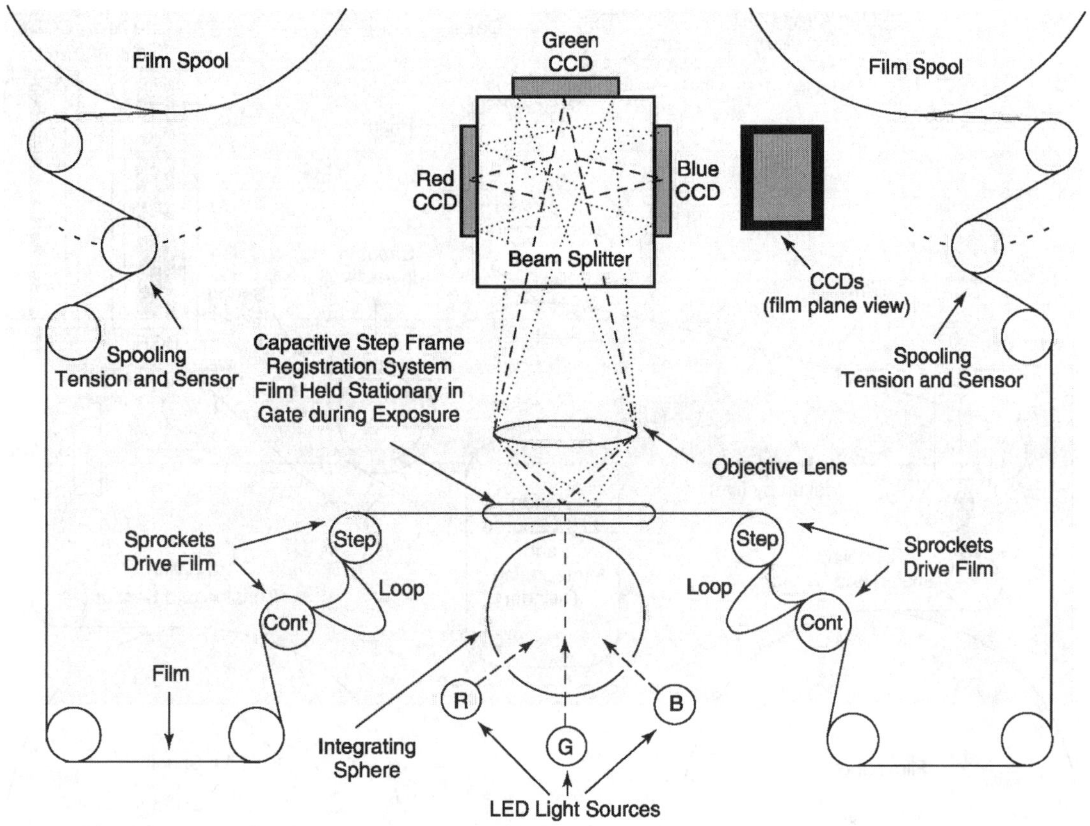

Figure 4.8.2 Area Array CCD transport and scanner head (Sony Vialta).

direction, and film movement provides the vertical scan. Therefore, as with flying spot telecines, continuous motion film transports are used, but there is a requirement for colour registration of the three CCDs, as is the case for the photoconductive and Area Array CCD telecines.

Several variations of Line Array CCD 'scan heads' exist. To improve light gathering performance some telecines/film scanners utilise a fourth array that has a high resolution and detects the very sharp luminance detail of the film while using RGB arrays with larger photosites to gather more light in the separate colours. This offers advantages in signal-to-noise quality and maintains visual sharpness by matrixing the luminance detail into the final RGB signals. An additional structural benefit of this scan head is that the RGB colour line arrays are positioned in a single vertical plane with the luminance detail sensor mounted on the opposite side of a simple broadband beam splitter prism. As such telecines use continuous film transports with very accurate speed controls, it is a relatively simple matter to compensate for the delay between sensors when reconstructing the image (see Figure 4.8.3).

4.8.2 Film formats

Although there are various film gauges, the vast majority of telecines and film scanners handle S35/35 mm and S16/16 mm cine film. Other gauges that are catered for on some devices are 8 mm, S8 mm and 65/70 mm.

There are two standard aspect ratios (a.r.) of the transmitted television signal, 4:3 for Terrestrial standard-definition TV and 16:9 for High-Definition TV and most digital deliveries of standard definition. Many films produced for the theatre have a wider a.r. varying from 1.66:1 to 2.35:1. Also, although the displayed a.r. may be 2.35:1, the film frame a.r. may be 1.17:1; this is known as *anamorphic film*, the image having been squeezed 2:1 in the horizontal plane only. The image must be unsqueezed during telecine transfer to obtain original image geometry.

For photoconductive telecines, special anamorphic lenses were required to unsqueeze and transmit this film. To avoid this, the film was usually printed in the displayed aspect ratio. Modern flying spot and CCD telecines can transmit anamorphic film without the need for special lenses, the correction being performed electrically during the scanning process (see Figure 4.8.4).

Having obtained the correct display aspect ratio for the particular film, it still has to be decided whether to transmit a widescreen format exactly as intended or as a 4:3 or 16:9 a.r. to match the home receiver.

In the former case, part of the CRT is not utilised and the picture is smaller than it could be; in the latter case, although all the screen is utilised, some film information is lost.

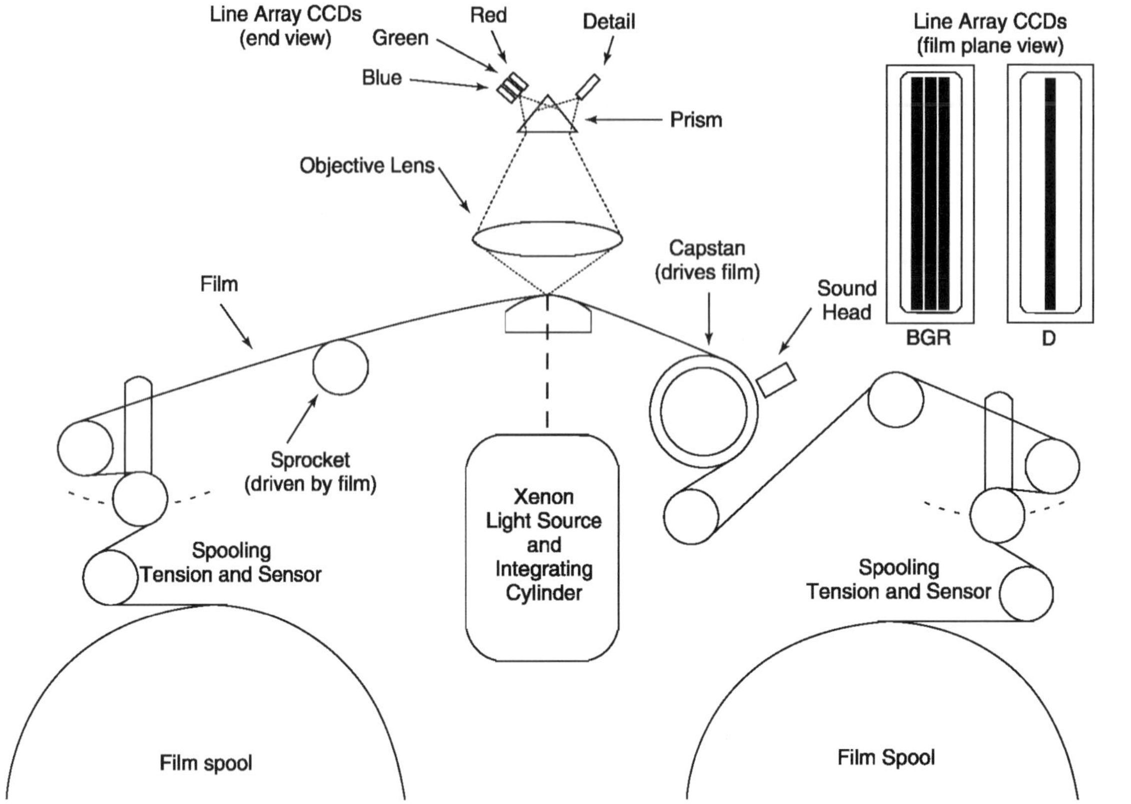

Figure 4.8.3 Line Array CCD transport and scanner head (Thomson Spirit).

To insure that important information is not lost, the part of the frame to be transmitted is selected frame by frame prior to transmission.

The former method of transmission is called *letter-box* and the latter *pan scan*. Obviously the letter-box mode does not require editing, but the pan scan mode does, and this editing is performed prior to transmission by pre-programming the scan position of each scene with the aid of a telecine remote controller (see Section 4.8.7.12). Two forms of pan scan can be selected during pre-programming, either a *pan cut*, i.e. an instantaneous shift during field blanking, or a *dynamic pan* with variable rates.

One big advantage that telecines have over optical projection is that negative film can be used directly by means of electronic inversion. This avoids several forms of degradation arising from the printing process and so provides better picture quality.

Two easily understood improvements are detail resolution, and vertical and horizontal stability, especially if the original camera negative is available.

A third important improvement, which is not immediately obvious, is grey scale resolution or linearity. Negative film is manufactured with a very low gamma so that it is very tolerant of a wide range of exposure variations without loss of information. When the negative is printed, the exposure can be adjusted to take account of the density of the negative, but since the print is normally intended for optical projection it must have a gamma of at least unity to reproduce the original scene. Therefore, scenes which have a contrast range in excess of 100:1 require a print density of at least 2. Under these circumstances, with existing film dye technology, the highlights and shadows become compressed even when the print is correctly exposed. The negative in effect compresses all the information equally from highlight to shadow, so that when the telecine electronically decompresses the information there is no loss of linearity from highlight to shadow.

4.8.3 Film Transports

The film speed in the majority of telecines must be related to the television system in which they are operating for various reasons connected with the type of telecine. The two standard-definition television picture rates are 60 fields/30 pictures per second and 50 fields/25 pictures per second.

Prior to the advent of television, the standard film rate was 24 frames/second and this speed is still the most common. Obviously film produced at 24 frames/second could not be transmitted at 30 frames/second for that particular system because the increase in sound pitch and picture speed would have been unacceptable. Fortunately, there is a direct if not ideal relationship between 24 frames/second and 60 fields/30 pictures per second, i.e. two and then three television fields per film

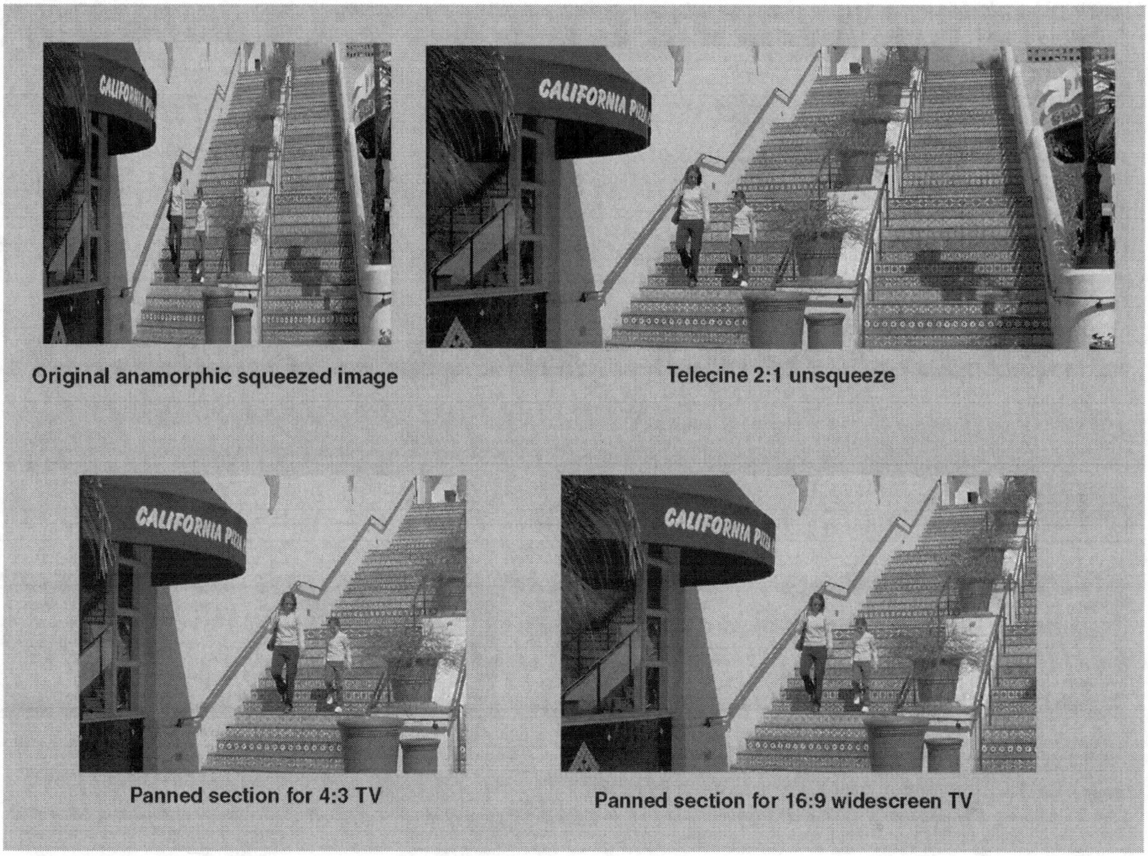

Figure 4.8.4 Anamorphic unsqueeze and pan scan.

frame alternately. This gives synchronisation between the telecine and the television system, but moving objects in the scene or camera panning produce just noticeable non-uniform movement under this arrangement. The alternating field ratios of this system are known as the 3:2 sequence and the sequence repeats every four film frames (see Figure 4.8.5). Therefore, although most film shot for the 30-frame television standard is still 24 frames/second, some film has been produced at 30 frames/second.

The other television standard of 50 fields/25 pictures per second can manage by operating the film at 25 frames/second, which gives a more direct relationship between telecine and television system and an increase in sound pitch and picture speed of 4%, which is quite acceptable to most observers. Nevertheless, film shot for this particular television standard has been produced at 25 frames/second, which again is acceptable on the 30-frame system or for optical projection at 24 frames/second.

The introduction of frame stores and digital sound compression/expansion equipment which correct the sound pitch has enabled the use of a much wider range of film speed whilst still retaining a whole number of television fields per whole number of film frames.

In recent years High-Definition Television standards that originally adopted the 60 fields/30 pictures per second standard have evolved. At the time of writing telecine transfers to High Definition are often mastered to a HD format known as 24PsF. The format recognises that most films' native frame rates are 24 and that a film frame is not in itself interlaced, as is the case of television cameras. PsF refers to 'Progressive, segmented Frame'. The telecine scans the film frame in HD at 24 per second, each frame is scanned progressively, the scanned result is held in a frame store memory and the store is read out in a similar fashion to an interlace video signal, all odd lines followed by all even lines; in other words the frame is read out segmented.

Film scanners as distinct from telecines deliver scanned images as computer image files and, as these files are initially not required at real time or indeed synchronously, the scan rates can vary from seconds per frame to 30 frames per second or more. Such freedom of scan speed offers much flexibility in scan size and quality.

4.8.3.1 Intermittent motion

Telecines designed on Area Array CCD principles can be regarded as very-high-quality electronic cameras photographing each film frame; exposure times are therefore measured in milliseconds. To avoid image blur, the film image must be stationary during the exposure, and hence an intermittent film motion system is required for Area Array CCD telecines.

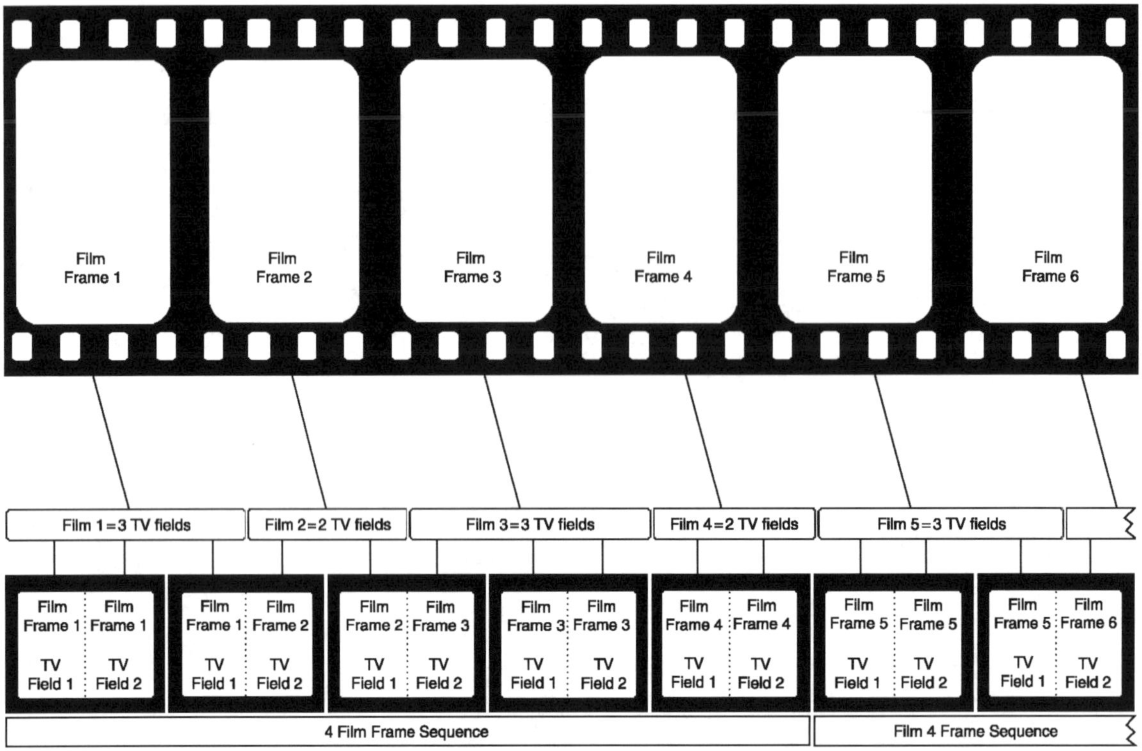

Figure 4.8.5 The 3:2 sequence (60-field video from 24 fps film scan).

Historically the most common type of intermittent mechanism was the Maltese cross or Geneva movement, as described by Wheeler.[1] This provided intermittent rotational motion to a film sprocket that advanced the film.

A modern Area Array telecine retains the use of one or more sprockets to advance the film intermittently; however, with modern electronics the sprockets are intermittently advanced using digital servos or stepping motors. Such technical advances offer gentler film handling and more control than mechanical systems (see Figure 4.8.2).

Film scanners dedicated for image data output only are almost exclusively based on intermittent transport principles. Often, such transports use 'pull-down' mechanisms similar to those found in film cameras. The film is then held in place within a gate aperture and the accuracy of position is assured by lowering the film perforations onto exact fitting pins. Such a device then scans the film either as described above using an Area Array CCD or by means of a Line Array CCD scanning head, which is either moved across each film frame or the head is fixed and the 'pinned' film frame is moved complete with the film gate in front of the Line Array CCD head.

4.8.3.2 Continuous motion

All modern Line Array CCD and CRT flying spot telecines rely on continuous motion capstan-driven film transports.

The film is partially wrapped around a capstan which is driven at a uniform velocity. To ensure precise synchronous speeds are attained the film is also wrapped around a freewheeling film sprocket. The sprocket generates speed information that is compared to the desired speed, the capstan velocity, and phase relative to the sprocket is then servo controlled. Such 'servoing' accommodates the variable amount of film shrinkage or other physical film size defects that may occur in archive material and most importantly provides a constant film speed.

To ensure a good wrap at the capstan the entire film path is held under tension by applying torque to the film feed and take-up rolls. These torques are 'servoed' to preset film tension parameters such that almost instant transport speed changes and very fast shuttle speeds can be achieved.

To accommodate sound tracks recorded on the film, optical and in some instances magnetic sound heads are mounted at the capstan location. Historically additional lacing rollers were used to accommodate the correct positional delay between picture and sound location on the film. Modern digital audio delay techniques have made such rollers redundant (see Figures 4.8.1 and 4.8.3).

4.8.3.3 Image forming with continuous motion

Vertical film image stability in capstan-driven telecines is a function of two factors:

- uniformity of the film sprocket hole pitch, which determines the spacing of the film frames;
- uniformity of the film velocity in the telecine.

Since timebase correctors cannot be applied to telecine to correct vertical stability, we depend on accurate control of these factors. The first is controlled by the film manufacturer and the second by the telecine manufacturer.

In the telecine, the capstan is the main controlling element, but because of the gate spacing between sprocket and capstan and the need for constant velocity at all points, any intervening film guiding surfaces must also be near perfect. The capstan is coated with synthetic rubber and in one telecine example has a diameter of about 50 mm, giving good traction with moderate surface pressure and film tension while eliminating the need for a pinch roller (see Figures 4.8.1 and 4.8.3).

Besides the capstan motor, there are two spooling motors, not coupled to the capstan or each other electrically or mechanically except by the film. The capstan is driven at a constant velocity to synchronise the film frame rate.

To maintain good film stability the capstan servo has two feedback loops: velocity and phase. The velocity loop is responsible for stability and the phase loop determines the film frame rate. The reference for the velocity loop is an optical tachometer attached to the capstan shaft, which supplies several thousand pulses per revolution.

The reference for the phase loop is a film sprocket hole detector. This usually takes the form of a sprocket driven by the film. It is mounted in free bearings and so does not impart velocity modulation.

A mechanical sprocket detector is not absolutely necessary but has several practical advantages over a purely optical device. When 35 mm film is correctly laced on such a detector, it can be arranged to give an index pulse per film frame correctly phased, whereas an optical detector has difficulty in determining which of the four sprocket holes provides the correct phase relationship. Also, if a number of sprocket holes are damaged, the mechanical sprocket will continue to provide correctly phased pulses.

Each film spool motor is controlled by its own servo, to maintain constant film tension in all modes of operation. The tension rollers provide the tension reference for the servos.

Figure 4.8.3 shows the sound heads situated very close to the capstan to give the best possible wow and flutter performance. The capstan is able to handle 35 mm, 16 mm and 8 mm film formats with a common optical centre line, so that all film formats are supported over their full width. At least one telecine type also has extended this handling range to include 65/70 mm. Typically the optical sound head detectors are located in narrow slots below the capstan surface which is machined from one piece.

On older telecines the film path for 35 mm film includes one extra roller between the vision gate and the capstan to obtain correct sound synchronisation. All rollers are multi-gauge, so that when changing from one film gauge to another it is necessary only to change the film spools or back plate and change the vision gate.

4.8.4 Sound Reproduction

Since the 1980s most telecines have been equipped for Stereo Dolby Combined Optical tracks in 35 mm and mono for 16 mm. Several telecines also include 16 mm magnetic combined magnetic pick-ups.

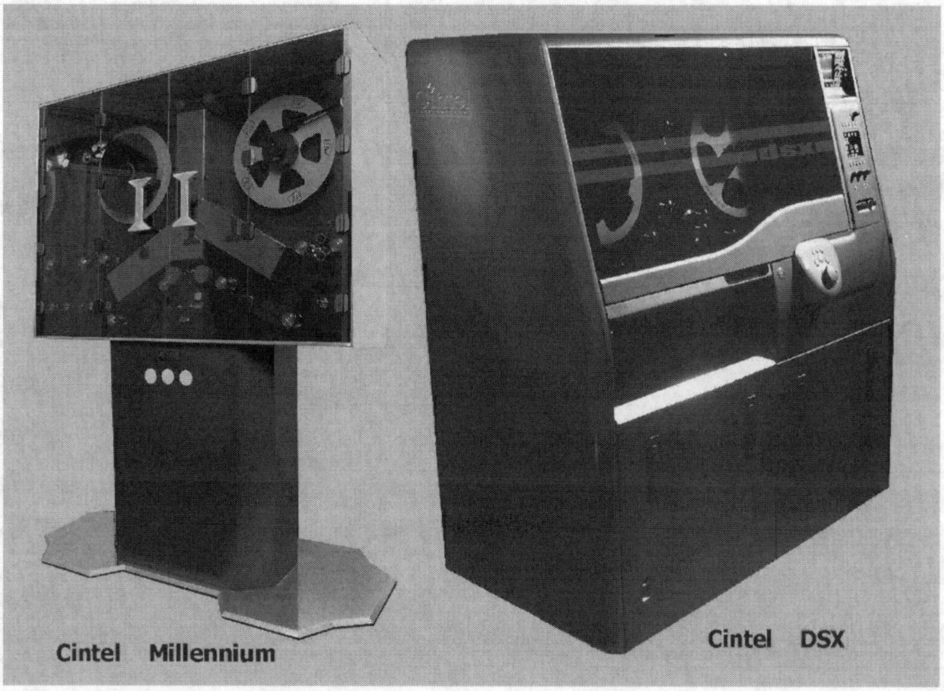

Figure 4.8.6 CRT flying spot telecine/scanners.

The soundtracks can be married to the film image or on separate sound-only films.

Married soundtracks are most commonly 35 mm optical, 16 mm optical or 16 mm magnetic; heads for these tracks are shown in Figures 4.8.1 and 4.8.3. The 35 mm optical soundtrack is 21 film frames in advance of the picture, the 16 mm optical sound track is 26 frames in advance and the 16 mm magnetic sound is 28 frames in advance. Before the days of Digital Audio delay systems the film lacing path needed to take these distances into account; modern audio delay techniques allow common film paths with electronic delays to compensate.

Optical sound tracks can be variable in density or variable in area, the latter being by far the most common. The optical head consists of an illuminated slit, imaged by an objective lens onto the soundtrack. Light passing through and modulated by the film falls on a photocell to produce the electrical signal.

Soundtracks on separate film are reproduced on *sound followers*, i.e. film transports very similar to the telecine without the picture components. The sound follower is synchronised to the telecine by comparing the sprocket hole pulses from the two machines and adjusting the speed of the sound follower accordingly. Both machines need to be synchronised manually at the beginning of the programme.

Modern sound techniques have largely overtaken these 35/16 mm sound followers. The audio element in film origination is now likely to be recorded digitally and stored on a digital computer medium. In such instances the telecine needs to be run synchronously with the digital audio playback devices.

4.8.5 Optical and Scanning Systems

4.8.5.1 Flying Spot 'JumpScan' (Field scanning)

All CRT-based flying spot telecines built since the 1970s utilise continuous motion transports. Early wholly analogue versions used a principle known as 'JumpScan'.

JumpScan refers to the requirement, as the film is continuously moving, to scan the two video fields from the film at different positions. This is accomplished by offsetting the two field scans on the CRT by the distance the film has moved during each field. Due to non-linearities in 1970s scanners, it was necessary to apply analogue geometry correction and shading correction to each field scan, so as to ensure that each raster scanned the correct part of the film with the same illumination. Without geometry correction alternate fields could be misplaced, causing the picture to 'twitter' with differential movement where the error existed. Likewise, differential shading could cause brightness flicker.

Figure 4.8.7 CCD Line Array telecine/scanners.

For 625-line, 50-field TV with two fields per film frame the correction was relatively simple, requiring just two field geometry and shading correctors. However, for 525-line, 60-field systems a total of five scan patches were needed to accommodate the film's continuously moving position during 3:2 field sequences when the film was running at 24 fps. With geometry and shading correction required for each of the five scan positions, alignment became difficult.

A further complication with field scanning a continuously moving film is the shrinkage of the material. As the film shrinks its linear motion per frame reduces; therefore, at a constant frame rate the linear speed reduces. In such cases the position of each field scan needs to be reduced by the amount of shrinkage. This further complicated JumpScan telecines.

4.8.5.2 Introduction of Frame Stores (for more details see Section 4.8.7.13)

The problem of JumpScan systems was totally eliminated when the Digital Frame Store was added to telecines. The frame store allows sequential or progressive film scanning, without recourse to scanning individual TV fields. The film is scanned progressively and the image is written to the store as a complete film frame, including all TV frame lines. The store is then read out in an interlaced fashion, odd lines being read followed by even lines. The store is usually arranged in groups of at least two frames capacity as one frame needs to be written into the store at the same time a previous frame is being read out of the store. However, modern stores utilise dual port memory chips that can be written to and read at the same time. This has greatly simplified frame store designs in recent years. For 3:2 sequences a store needs to hold two frames to generate the additional third field every other film frame (see Figure 4.8.5).

The introduction of frame stores offered progressive principles to be used for all film scanning, with other image detector devices including Line Array CCDs. While in theory Area Array CCDs could be made to operate without frame stores, as far as the author knows, none ever were. In practice Area Array CCD telecines utilise the advantages of such stores.

4.8.5.3 Progressive scan techniques

All modern telecines and film scanners can be regarded as using progressive scan techniques, also known as sequential scan. This involves scanning the entire film frame in one pass, unlike a conventional video camera that acquires the video frame image as two interlaced fields captured one after the other.

4.8.5.4 Flying spot progressive scan (a.k.a. sequential scan)

This involves scanning the film with all the horizontal lines in one vertical sweep rather than two sweeps as in a 2:1 interlaced scan or five sweeps in 3:2 field sequence systems.

Sequential scanning was introduced to overcome the problems and reduce the alignment procedure associated with the JumpScan system, while retaining its simple optical arrangement.

As each film frame is scanned only once, there is no possibility of field-to-field registration errors or field-to-field shading, and no requirement for shrinkage correction.

The CRT beam provides an instantaneous scan illuminating the film at each 'pixel' location for a duration that can be calculated approximately as the total frame time divided by the total number of pixels required per film frame (see Figure 4.8.1). The signal is then digitised and, as with most other telecines and film scanners, the scanned frame is stored in a frame store just after the digitisation process, often known as the 'input' store.

It is necessary obviously to convert the sequential scan to interlaced for transmission purposes, and for this an 'output' frame store is used (see Section 4.8.7.13).

4.8.5.5 Area Array CCD

For Area Array CCD scanning the film is held still in the film gate while the Area Array CCD is exposed to the image for many milliseconds. The scanning can be regarded as a parallel capture of all the image elements over the period of exposure. Therefore the actual capture is instantaneous over the whole image rather than progressive. However, the exposure level could be regarded as progressive from an illumination level point of view; a better term here is that the exposure is accumulative (see Figure 4.8.2). Once the exposure is complete the CCD either acts as an electronic shutter and ceases acquiring the image, or a physical shutter prevents further exposure. The image now residing as a charge within the CCD is read out progressively to the scanner's electronics. Invariably, these days the signal is digitised and fed to a frame store that resides at the 'front end' of the telecine or scanner.

4.8.5.6 Line Array CCD

The optical arrangements of the Line Array CCD telecine are very similar to those of the Area Array telecine except that, since the film moves continuously rather than intermittently, the vision gate is curved instead of flat. Also, as the CCD head consists of only one horizontal line of photosites at the centre of the optical axis, the radius of curvature of the film in the vertical direction does not affect resolution in the vertical axis. Wide aperture lenses can therefore be used, with the advantage of more light throughput. Modern top-of-the-range Line Array CCD telecines and scanners use high-power Xenon or HMI light sources, providing as much light as possible at the 'slit' aperture, where the film scan takes place. A focusing lens then images the film onto the CCD's photosites. As the film is in continuous motion the exposure for each line of the image is very small; it can be calculated as the number of milliseconds the entire frame needs to transit the slit divided by the number of lines required for the scanned video standard (see Figure 4.8.3). The efficiency of bright light sources, wide aperture optics and modern CCDs makes such scanning successful. The image signal is clocked out of the CCDs for each line scanned, the signal is digitised and built up to complete frames in an 'input' front end frame store.

4.8.6 Colour Response

Telecines, like electronic cameras, usually have three colour channels: red, green and blue. The main difference between them is the spectral width of each channel.

Electronic cameras need to be sensitive to all wavelengths in the visible spectrum and so require overlapping spectral characteristics. In the case of telecine, the film is sensitive to all

Figure 4.8.8 CCD Area Array telecine/scanner.

wavelengths and, during its exposure and processing, the film image itself is converted to three dye colours. Therefore the telecine needs only to determine the density of each colour dye. For this purpose a single narrow-line response for each dye would suffice. In practice, because of signal/noise considerations, the responses are usually made as wide as possible without overlapping.

Light can be split into three components by several methods: dichroic mirrors, dichroic prisms or, in the case of modern Area Array CCD scanners, by utilising individually shuttered RGB LED sources. Mirrors are the simplest arrangement but can only be used outside the imaging optics because of secondary reflections at the second glass/air interface of the mirror. Therefore dichroic mirrors tend only to be used in flying spot systems.

On the glass surface of dichroic mirrors and prisms there are several subwavelength glass layers of various refractive indices to give the required characteristic for a defined angle of incidence.

The dichroic glass/air interface reflects all light above or below a designated wavelength and transmits the rest with virtually no absorption. The change from total reflection to total transmission occurs over approximately 20 nm of wavelength. The wavelength at which half the light is reflected and half transmitted (known as the *50% point*) is the design wavelength for the dichroic surface.

The 50% point changes with angle of incidence, and therefore off-axis light rays from the edges of the picture suffer a shift of the spectral characteristic, which is manifested as a change in amplitude resulting in colour shading. The rate of change of the 50% point increases with increase of angle of incidence, so angles of incidence less than 45° are normally selected – typically between 25° and 35°.

Using just two dichroic surfaces will split the visible light into three components, as shown in Figure 4.8.1, but the spectral characteristics for the three channels will be overlapping since there is very little absorption. To give some separation between the three channels, additional filters are added. It is also necessary to add an infrared filter in the red channel because, even though there is normally very little sensitivity at these wavelengths, it is enough to cause red shadows with some films.

Figure 4.8.9 shows a typical telecine/scanner colour response. Each manufacturer chooses their own precise

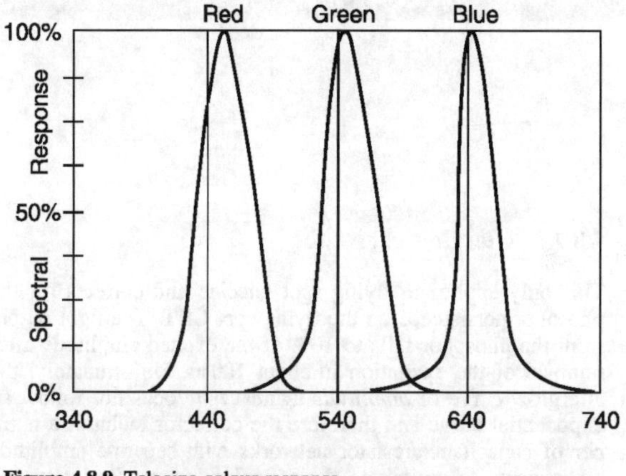

Figure 4.8.9 Telecine colour response.

response; however, in all instances the response is as close to that of the film dyes so as to reproduce the film image as faithfully as possible. All of today's scanners have good sensitivity to the required spectrum; some achieve it by means of powerful light sources, others by selection of very sensitive detectors, or a combination of both.

4.8.7 Signal Processing

4.8.7.1 Noise sources and signal gain control

Figure 4.8.10 shows the various image detector devices used in modern telecines and scanners.

4.8.7.1.1 Flying Spot Detectors: PhotoMultipliers (PMT) and Large Area Avalanche PhotoDiodes (LAAPD)

The signal in a flying spot telecine is generated in PMTs or LAAPDs and is greater than 50 μA, or some 300 times larger than photoconductive signals. The head amplifier noise is therefore negligible, the noise being shot noise, which is a function of the number of photons striking the photosensitive regions and the number of electrons released in the detector. The noise is therefore a function of incident light and is proportional to the square root of the signal current, falling to zero at black level.

Gain control is obtained by varying the supplies to the PMT or LAAPD to maintain a constant signal output as the light output decreases. The gain of the detector tends to increase more rapidly with increasing supply voltage. This non-linearity does not affect the linearity of the light input/signal current output characteristic, which is linear at all normal operating voltages. It affects the colour balance with gain change since the three detectors are operating at different mean voltages due to the variation in the light output from the CRT, the colour splitting optics and the film's base colour. To maintain colour balance as the master gain control is adjusted, it is preferable to preset the detector gain of the red and blue detectors to match the green signal.

4.8.7.1.2 Area and Line Array CCDs

These produce a nominal voltage output of about 1 V rather than current. Here again, most noise is generated in the CCD, and as this is a semiconductor, lf noise predominates, measuring approximately −60 dB. As the noise is so low and the signal high, it is possible to use electronic means of gain, which is advantageous for rapid change from film frame to film frame. However, recent CCD scanners, utilising efficient illumination sources, take considerable advantage of varying the light level and balance to optimise the CCD exposure; this further minimises introduction of noise by reducing the requirement to add electronic gain to the image signal. Scanners utilising LED light sources can apply frame-by-frame illumination level and colour balance changes.

4.8.7.2 Analogue-to-Digital conversion, 14- to 16-bit

All modern telecines and film scanners utilise the advantages of digital signal processing. The analogue signals from the scanning

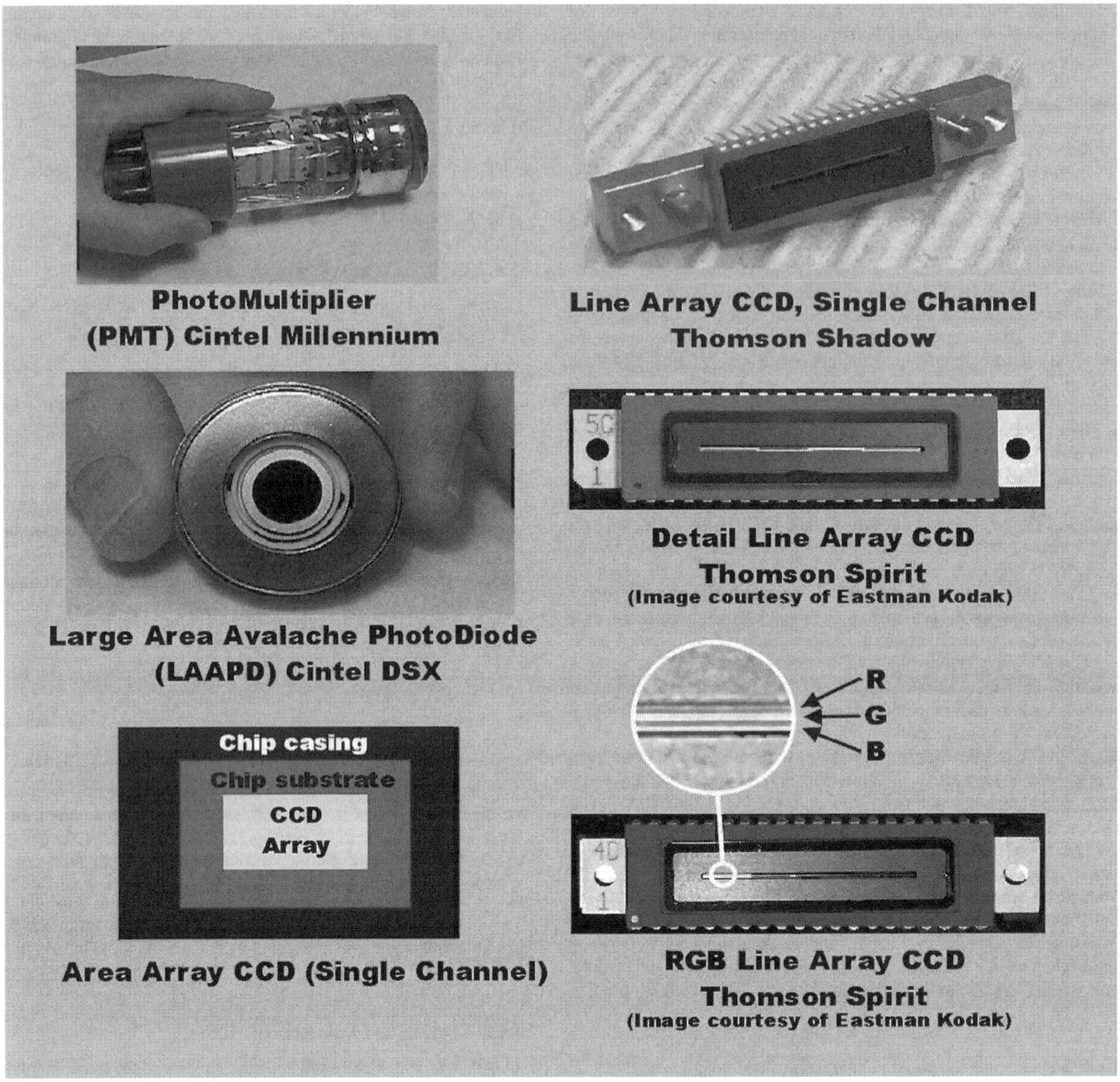

Figure 4.8.10 Telecine image detectors.

head must be converted to accurate digital representations of the film's density range. Film density represents a very large range and very subtle shading of the original image. To represent the film's quality the A-to-D converts the signal to 12, 14 or even 16 bits for each colour channel. Most telecine digital colour correction manipulates a large range of colorimetry, which at even 12 bits can represent a total of 68 billion colour shades within the image. For TV purposes the bit depth is reduced to 10 bits at the output and for other film scanning purposes logarithmic conversion from 16-bit to 10-bit is applied (see Section 4.8.8.2).

4.8.7.3 Afterglow correction

This only applies to flying spot telecine and corrects for the phosphor persistence on the flying spot CRT. The light output from the phosphor falls to 10% of the excited amplitude after removal of the excitation in about 100 ns. Unfortunately, the afterglow curve of *amplitude* against *time* does not follow an exponential curve and therefore the corrector includes a number of capacitance/resistor networks with separate amplitude adjustments to assimilate the afterglow curve. Basically this corrector attempts to null the afterglow, thereby effectively increasing the horizontal resolution.

4.8.7.4 Burn correction

Burn correction is needed only with flying spot telecine, where the raster on the CRT will reduce the efficiency of the phosphor and faceplate with time. If the raster were always in a fixed position and of fixed amplitude, there would be less need for burn correction, but facilities such as zoom and variable film frame rate alter the size of the raster. Thus, some parts of the phosphor screen are used more than others and so give less light output after a time. The burn corrector consists of detectors which measure the light output from the CRT, and a reciprocal process which drives shading correctors in each colour channel. The result is that as the CRT 'wears' the reduction in brightness is compensated for, ensuring consistent signal level throughout the CRT's life.

4.8.7.5 Shading correction

Every type of telecine is affected by variation in brightness across the picture. Whether or not correction is necessary, and what amount, depends on the type of shading error.

If the shading is only in luminance, then the human eye can tolerate very large variations before it is evident that something may be wrong. The eye is much more sensitive to variations in colour shading. Additionally, for older JumpScan telecines the eye is extremely sensitive to field-to-field shading, which can cause flicker at 25 Hz and lower frequencies.

Shading correction is therefore essential for JumpScan telecines, and highly desirable in colour telecines. In practice all telecines include shading correctors of some kind.

Shading errors arise from the light source, objective lens, and colour splitting dichroic mirrors and prisms. The errors are usually constant and can be automatically sensed, digitised, stored in a separate memory and applied reciprocally to the scanned image.

For older JumpScan telecines the shading can be corrected by developing waveforms from scan currents in the deflection system. These waveforms are adjusted individually in amplitude and shape, and used to modulate the individual red, green and blue signal channels.

4.8.7.6 Fixed Pattern Correction

This particular process is only relevant to the CCD telecines and film scanners. The term refers to vertical stripes that are seen if the errors in the Line Array CCDs remain uncorrected and have different brightness in pixels in Area Array telecines.

A CCD consists of a number of separate photosites, and the variation in sensitivity between adjacent sites can be as much as 5% and is quite random. In a Line Array CCD telecine, as the film moves through the telecine, each photosite scans a vertical line of the film frame and therefore if all photosites are not equally sensitive, a vertical line pattern will appear in the picture. The same effect occurs in Area Array telecines, except that here the result is a random variation in brightness over the whole image.

To overcome these CCD sensitivity differences, during the initial line-up procedure an electronic measuring feedback arrangement can correct these and other shading errors, storing the corrections in a memory. This has the advantage that somewhat variable shading errors due to the light source can also be corrected day by day.

4.8.7.7 Gamma Correction

Often confused with logarithmic-to-linear film density conversion, sometimes known as film gamma correction, which is specific to the telecine process, gamma correction is an additional function.

Like electronic cameras, telecines incorporate a gamma corrector which is a power law transfer characteristic to compensate for the power law transfer characteristic of the receiving CRT. There is one corrector in each colour channel.

In telecine, the gamma corrector fulfils another function, i.e. the correction of colour film errors from scene to scene or film to film, errors which electronic cameras do not normally encounter.

Film colour balance errors are not normally objectionable when projected optically in a dark theatre, because the human eye adapts its own colour balance to the predominant wavelength in the incident light.[2] When viewing a television receiver, the observer usually has a reference colour near the receiver and only a small proportion of the light incident on the observer's eye comes from the receiver. Therefore, colour balance errors in film shown on television need correction, and differential gamma variation between the colour channels is part of this correction (see Section 4.8.7.9).

Continuously variable gamma is obtained by using logarithmic and antilog manipulations. If the manipulation of gain between the log and antilog stages is unity, the channel will be linear, or in mathematical terms:

$$y = x^{\gamma}$$

By varying the gain between log and antilog stages, a range of gamma can be obtained.

4.8.7.8 Log masking correction

Film dye characteristics are never perfect, and while the telecine spectral responses are designed to closely match the film dye characteristics, they can never be exactly placed with respect to the film dyes since light sources and pick-up devices limit complete flexibility.

Consequently, when combining two imperfect media, the film and the telecine, there will inevitably be some saturation and hue errors. Correcting these errors involves multiplicative cross-modulation between the channels, a function which can be realised with logarithmic masking,[3] i.e. addition or subtraction of logarithmic signals. The function is specific to telecines as film density is logarithmic, as opposed to linear masking that applies to video camera sources.

The logarithmic matrix can be calculated,[4] or alternatively a colour bar film can be placed in the telecine and the matrix equations adjusted to simulate an electronic colour bar pattern.[5]

4.8.7.9 Primary colour correction

Telecines are now provided, by popular demand, with a whole array of controls for colour correction.

Historically such correction comprised of what has become referred to as primary colour correction. This includes lift, gamma and gain, with overall controls and colour differential controls. Basically the image can be made lighter or darker, have its contrast adjusted and its highlight and lowlight points set, both overall and in colour balance. Primary correction has an effect on the entire range of image hues and is generally used to establish the correct desired colour balance of the whole picture.

The adjustment means for these controls have evolved to be, in the main, by three trackerballs, lift, gamma and gain for colour differential adjustment and three linear controls for overall adjustment.

4.8.7.10 Secondary colour correction

While primary colour correction has existed since the inception of colour telecines, it soon became apparent that a further need for a secondary function to control individual hues and saturations existed. Modern telecines include secondary colour correction. Here individual hue, saturation and luminance ranges can be adjusted without affecting unselected ranges of the image. An example would be the desire to make an apple held in a hand more green and have a greater colour saturation, while not affecting the skin colour of the hand. To achieve this result, primary correction would firstly be used to establish the correct skin tone. The secondary selection would then be made by choosing the range of hue, saturation and luminance of the apple, as it appeared in the primary corrected image. This process could be referred to as isolating the area to be corrected. Having established the area it is then a matter of adjusting for the desired result by changing the 'Effect Hue, Saturation and Luminance'. Because this secondary colour correction is selective the 'effect' controls will not cause the skin tones to change. This powerful colour manipulation/correction feature is usually available with six or more individual sets of isolation and effect. Therefore, if the hand in the example above was holding many different coloured objects, they could all be treated individually without affecting each other or the hand that held them.

4.8.7.11 Aperture correction

Aperture correction is compensation for detail or resolution losses due to film camera lenses, printing processes, telecine lenses, scanning spot or sensor size.

Aperture correction is not a device designed to falsely sharpen or edge enhance the image, it is for the purpose of compensating for a finite aperture causing a lowering of resolution. Therefore the correction should be used with care and adjusted just sufficiently to compensate for the limit of aperture. Excessive levels of correction produce false-looking images with haloes around edges.

Aperture correction enhances transient edges in the image. Where an edge exists within the image the edge sharpness will be determined by the system's aperture. The larger the aperture the softer the edge will be. Applying aperture correction boosts the edge contrast and this appears to sharpen the edge.

Normally any correction should be applied symmetrically in both the horizontal and vertical planes. However, modern telecines often allow individual adjustment to compensate for differences in both planes. Additionally, the amount of aperture correction will depend very much on whether the transfer is for SDTV or HDTV.

4.8.7.12 Telecine Controller

Modern telecines have a plethora of controlled functionality (Figure 4.8.11). Primary and secondary colour correction, pan scan and other positional decisions, film type selection, aperture correction and many other parameters. Obviously, adjustment of

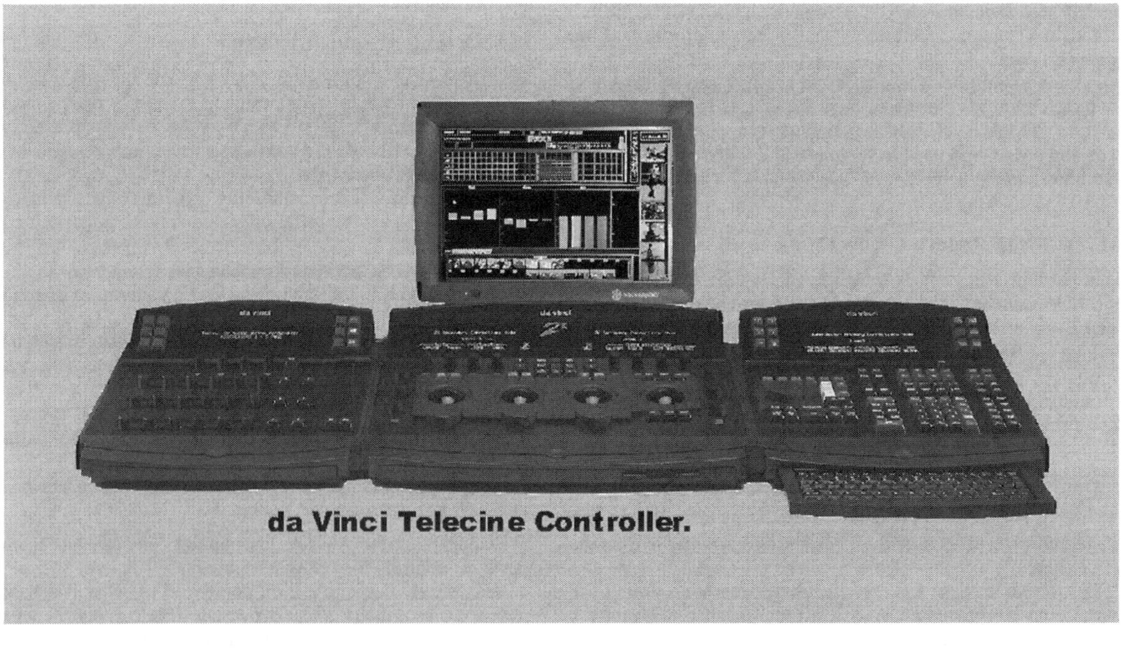

Figure 4.8.11 Typical telecine controller.

this number of controls 'on air' or during a telecine transfer would be impossible. The film is therefore previewed, and at each scene where correction is necessary, the film is held stationary in the film gate while the necessary adjustments are made at leisure. When a satisfactory balance has been obtained, all the control settings are stored in a computer together with film frame numbers at the beginning and end of each section for which the control settings apply. Each scene is calibrated in this way, and when the film is transferred, the control settings are extracted automatically, the change of control values occurring during frame blanking as instructed by the film frame counter. Gradual changes over time can also be accomplished in this way by interpolating between two sets of parameters as the film runs in the gate.

For storage purposes, the control settings for any particular film can be stored in the controller's archive or off-line. For further replay, the control values are simply loaded back into the telecine controller and resynchronised by resetting the controller frame counters.

4.8.7.13 Frame stores

All modern telecines rely on digital techniques. This includes frame stores, which were first introduced in telecine to convert sequential scan to interlaced for reasons described in Section 4.8.5.6.

Most modern telecines include both 'input' and 'output' frame stores. The input frame store holds a scanned film image at the 'front end' of the telecine, before the colour corrector. This store is used by Line Array telecines to regenerate the image while the telecine is stopped and is used in some CRT-based telecines to segment the image for parallel processing across three colour channels.

All modern telecines include output frame stores. These are used to convert sequential scans to interlaced output, to provide 3:2 scan sequences for 24 fps scanning to 60-field video (see Figure 4.8.5), to provide variable speed operation, and in some telecines, to provide conversion from HD to SD.

Present TV standards provide for full digital bandwidth Red, Green and Blue signals, commonly referred to as 4:4:4, and half colour bandwidth signals, known as 4:2:2, where the signal is in a Y,Cr,Cb format. These conversions are carried out within the telecine frame store.

Finally the output store generally includes a digital parallel-to-serial conversion function. (At the time of writing most every digital TV i/o in the industry is serial.)

4.8.8 Recent Developments in Film Scanning

4.8.8.1 Reduction of film surface defects

For all of film's high-quality reputation, it has always suffered from a sensitivity to physical damage.

Today's telecines include methods to minimise or eliminate the effects of surface damage on the film. Bearing in mind that the surface damage rarely affects the actual image buried in the film, it is only necessary to avoid the effects of surface damage (see Figure 4.8.12).

CCD telecines achieve this by means of diffuse illumination of the film image.

CRT-based telecines, which are unable to create diffuse illumination, overcome the surface effects either by utilising very wide aperture detectors which gather the light diffused by the dust and scratches or they use additional means to detect the diffused light and add it back to the main image signal.

4.8.8.2 Data scanning

The telecine industry is in a transient stage. With SDTV and HDTV plus film digital effects and the demand for digital mastering at film quality, the worlds of telecine and non-TV film scanning are merging.

The film scanning data market differs from the TV market in several basic ways. The images are normally scanned as computer data files. Here each film frame is treated as an individual file and a set of files in a folder can be regarded as a film scene.

Resolutions are usually higher than TV, 2 K (2048 × 1556) and 4 K (4096 × 3112) being examples. Images are invariably coded Red, Green and Blue, and bit depths of more than 10 are expected. To achieve the higher bit depth a standard has arisen that offers effectively 14- to 16-bit linear, while scanner outputs are held at 10-bit. This is achieved by converting the 14- to 16-bit linear image density to a logarithmic form before converting to 10-bit. At the digital image's destination the log image density is converted back to 14- to 16-bit linear.

These more stringent requirements are tempered by the fact that real-time transfers are not essential. There is no 'sync' heartbeat in scanner transfers, they just take as long as they take. In general the speed of transfer is inversely proportional to the required resolution. As technology progresses scanning speeds increase. With each film frame scan a 2 K image occupies 12 Mbytes while a 4 K scan occupies 48 Mbytes. Such file sizes require large amounts of data storage. A typical feature film scanned at 2 K requires around 2 terabytes of storage and a 4 K scan of the same material would require 8 terabytes, a terabyte being 1000 Gbytes. Such storage is becoming economical and even petabyte (1000 terabyte) storage is now being considered in the film scanning industry.

4.8.8.3 Data scanners (excluding telecine capabilities)

In many respects telecine quality has converged on the higher resolution requirements of film quality scanners. Today, most all telecine manufacturers offer 2 K scanning capabilities and at least two manufacturers offer 4 K scanning. While most telecines have data scanning capabilities there also exist film scanners dedicated to the high resolution that do not include telecine video facilities.

The principles of operation are almost identical to CCD telecines, either Line or Area Array. They differ from telecines in the fact that they offer very high resolutions, increased bit depth and make use of ultra-stable film transports. Keep in mind that at 4 K scan resolution image positional stability needs to be held to better than 0.2 of a pixel and this is about 1 micrometre. Such scanners often use pin registration to hold the film image accurately during the scan process.

In general scanners of this type are less expensive than a telecine/scanner combination; however, they are normally somewhat slower in operation, especially above 2 K scanning.

Today, there is great debate regarding what processing should be involved at the film scanning stage. Should colour correction be performed during scans or, as the entire density gamut of the film is transferred, should the colour correction be performed in the digital domain at a later stage?

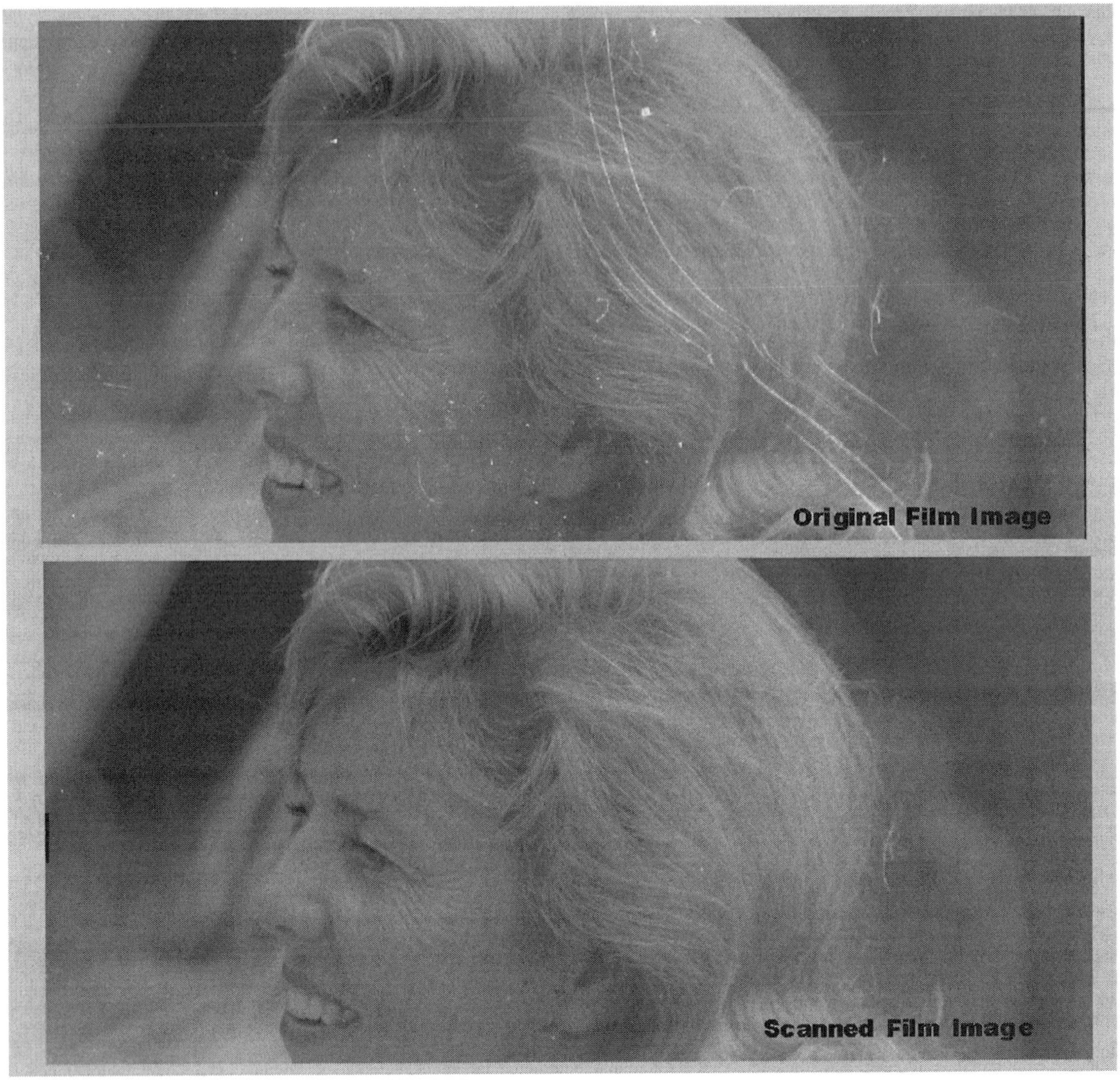

Figure 4.8.12 Minimising dirt and scratches on film.

Debate also ranges around issues such as scan resolution and archive media; while 2 K is good enough for today's Cinema and HDTV, future digital D Cinema may well provide 4 K quality and today's digital transfers should take this into account.

References

1. Wheeler, L.J. *Principles of Cinematography*, 3rd Edn, p. 224, Fountain Press, London (1953).
2. Hunt, R.W.G. *The Reproduction of Colour*, 2nd Edn, p. 68, Fountain Press, London (1957).
3. Burr, R.P. The use of electronic masking in colour television, *Proc. IRE*, **42**, 192 (1954).
4. Griffiths, F.A. The calculation of electronic masking for use in telecine, BBC Research Dept Report 1972/24 (1972).
5. Hunt, R.W.G. *The Reproduction of Colour*, 2nd Edn, p. 426, Fountain Press, London (1957).

Bibliography

Cinematography

Wheeler, L.J. *Principles of Cinematography*, 3rd Edn, Fountain Press, London (1953).

Colour

Hunt, R.W.G. *The Reproduction of Colour*, 2nd Edn, Fountain Press, London (1957).

Film standards

BS 677, Parts 1 and 2 (1958).
Davies, H. *Colloquium on Sound on Film*, IEE, London (1966).
SMPTE Standards, New York.

Telecine Principles and Operations

Blake Jones, S., Kallenberger, R.H. and Cvjetnicanin, G.D. *Film into Video*, 2nd Edn, Focal Press (2000), ISBN 0-240-80411-2.

Flying spot telecine for 525/60

Millward, J.D. Flying spot scanner on 525-line NTSC standards, *J. SMPTE* (Sept 1981).

Frame stores

Jesson, G.S. A variable speed frame store for flying spot telecine, *Professional Video* (Sept 1982).

Logarithmic masking

Burr, R.P. The use of electronic masking in colour television, *Proc. IRE*, **42**, 192 (1954).
Griffiths, F.A. The calculation of electronic masking for use in telecine, BBC Research Dept Report 1972/24 (1972).
Jones, A.H. A theoretical study of the application of electronic masking to television, BBC Research Dept Report PH-16 (1968).

Twin lens telecine

Nuttall, T.C. The development of a high quality 35 mm film scanner, *Proc. IEE*, **99**, Part IIA, No. 17 (1952).
Nuttall, T.C., Boston, D.W., Askew, G.H. and Lowry, P. The Rank Cintel twin claw twin lens flying spot 16 mm film scanner, *BKSTS J.*, **48**, 1, 2 (1966).

John Emmett
Technical Director and
Chief Executive
Broadcast Project Research Ltd

4.9 Sound Recording

4.9.1 Sound Recording as a Process

Recording delays an audio signal. For a radio interview the delay might involve only a few seconds, just enough for a 'top and tail' edit. On the other hand, a transcription recording of a live concert could well be recorded and then lay dormant for a century or more.

Nowadays, economic pressures in the broadcast industry demand a recording process that is far more complex than these two examples represent. For instance, if a recording can be made available to multiple operators soon after the start of that recording, the editing processes for different distribution paths and different programmes could take place in parallel.

Therefore, we now need to think of broadcast sound recording as a part of general 'workflow', and in this context the audio will in all probability accompany other 'essence' such as text and pictures. These essence items are all linked together by the 'metadata', which takes the place of the information once carried on the label and package of a disc or tape recording, although the metadata by itself possesses a much greater potential power than a label.

For the overall workflow process to be a success, there are several presumptions that are made about the sound recording. Firstly, the technical quality of the recording must be adequate to survive the numerous signal processing and editing procedures that may be applied at any stage along the workflow route. Another way of putting this requirement is to demand the 'transparency' of the recording stage, meaning that the other items in the signal chain produce greater level of impairments than the recording. This does not mean that current sound recording practices are perfect, it simply means that the recording should be better than the worst other items in *current use* in the audio signal chain. This requirement leaves implications about any bit rate reduction of the recording. The highest source quality should be preserved as far along the workflow chain as possible. This then preserves the possibilities for later processing, or for future (and possibly lucrative) applications of the recorded material. The days of needing bit-rate-reduced recordings for the convenience or economic use of Information Technology (IT) applications are long over, although low-bit-rate contribution of interviews, and similar recordings over dial-up telephone circuits, is likely to remain a prime use of bit rate reduction for many years to come.

This should not give the impression that sterility has crept into sound recording. In reality, the creative opportunities available in the most basic computer recording equipment well exceed those existing in the most complex analogue studio facilities of 10 years ago. What has gone are the subtle perceptual sound effects that were inherent in analogue recording or noise reduction processes. However much of a cult built up around some of those effects over the years, there can be no doubt about the technical efficiency of a digital recording process. Digital recording has enabled big and helpful changes in audio production, mainly as a result of the 'transparency' factor. For instance, the recording of two-channel stereo in the form of Mixed and Side (M and S) channels was not considered using analogue equipment, because any recording artefacts such as noise on the M channel would appear as a coherent centre image in the reproduced sound field. 'A and B' (Left and Right channel) analogue recording resulted in a much more diffusely reproduced field of any recording noise; therefore, this method became the standard for two-channel stereo recording. Now M and S recording can prove very useful, and there are other examples of such potential useful recording techniques being discovered (or rediscovered) as the freedoms of digital recording become apparent.

Perhaps this is a good point to step back from the actual details of the recording technology, and ask if we have actually experienced some kind of Digital Audio Revolution.

4.9.2 A Digital Audio Revolution?

Revolutions take place suddenly, and afterwards the remnants of the old guard are usually removed entirely. A quick look at any broadcast sound area (Figure 4.9.1) shows no evidence of

Figure 4.9.1 The sound control room in a large television studio gallery. The mixer is digitally controlled, allowing for a smaller physical desk within the restricted monitoring sound field, yet maintaining many instantly available controls for live operation. There are a selection of play-in devices to cope with a wide range and age of programme inserts and effects.

this; note, for example, the ¼-inch analogue tape deck on the right hand side of the main mixer desk. What has taken place in the field of sound recording has therefore been *evolutional* not revolutional, and what is more it has taken place in two distinct steps. Firstly, digital encoding of the recorded signal circumvented the imperfections inherent in analogue recording systems. This led to the development of dedicated digital audio recording formats, mostly using magnetic tape as the storage medium.

The second stage of the revolution came when the ready encoded digital audio signals from stage 1 became available for computer-based editing. This led to an obvious use of generic IT storage for sound recordings. However, many of the stage 1 dedicated recording formats, such as the multi-track DASH and DA88 systems covered later in this chapter, are still in widespread use where low cost and conveniently exchangeable signal storage are needed. The main disadvantage of any tape-based system lies in the inevitable slow access to any given part on the tape, although the archive life of tapes may not be exceptional. On the other hand, IT-based storage media are often not so easy to exchange, sometimes for physical reasons and sometimes for format incompatibilities. We will return to the battle between economics of generic storage against the desired longevity of audio material later in this chapter.

Returning to the 'audio' part of the digital audio evolution, the key stage was the initial encoding of the audio in a digital form. The later digital recording and manipulation technologies were both largely developed by the generic IT industry. Audio coding in the broadcast industry in the form of Pulse Code Modulation (PCM) was first put into practical use during the 1970s for the accurate long-distance transmission for the two channels of a stereo baseband signal. Prior to using this system, there was a limit of 100 kilometres or so for the distance over which you could maintain a sufficiently good match of the two landline paths that were necessary for stereo operation. At this time, a similar form of weakness was beginning to show up in the recording studio, as the number of generations or layers of recording that could be employed before the signal quality was compromised were limiting multi-track recording techniques. When accurate digital audio coding arrived (and accuracy in audio conversion did take some years to develop), it quickly bypassed the need for the expensive mechanical precision in professional analogue recorders. This need for precision had passed to the electronics, and that development was paid for not by the small audio market but by the millions of computers, mobile phones and countless other electronic items in the mass consumer market.

The new notion that there was no such thing as a 'professional' digit rapidly created an economic jump which reduced the cost of processing, mixing and editing of the audio. These processes have now become integrated in the form of what has become known as an audio 'workstation'.

Meanwhile, the advances made in distributing the audio material, in finished or unfinished form, have been crucial factors in creating even further jumps in recording economics. These advances are nowhere more visible than in the often 'invisible' contributions from radio reporters in the field. These contributions are sent via email on telephone lines or satellites, and have led to fast audio file transfer between workstations anywhere in the world. In this way the audio rushes from Hollywood can be sent during the evening to London, for editing during the day, and the finished material can then be sent back ready for the following morning on the West Coast.

In parallel with these operational advances, a previously mentioned but often overlooked (and slow) technical advance has taken place in the quality of the basic audio digital coding and decoding. If anyone doubts this, compare any early CD player with the much lower cost equivalent of today, where what you actually hear does at last approach the theoretical coding quality that was promised in 1982.

Revolution or not, one solid piece of advice when facing any new regime is to plan how to get out of it before you dive headlong into it. All too often, the economic advantages of one technical advance have been reversed when the next advance came along. Planning any sound recording process for broadcasting requires much thought, especially as the sound is now so closely linked with other workflow patterns such as video production, and these other patterns are still in the process of changing inside their own digital revolutions. A good start, however, in our own particular part of this planning process is to look at the sound recording toolkit currently at our disposal.

4.9.3 Streaming and File Formats, and How to Use Them

A microphone channel produces 'streaming' audio, and a real-time output data *stream* will be required in many stages along the signal route to feed loudspeakers or headphones. Any streaming format can be thought of as one with the capability for audio data delivery in real time with a low and controlled latency. Streaming audio cannot be slowed down or speeded up, and it cannot usually be interrupted without undesirable consequences. It can start up immediately the signal is available, and it can go on streaming indefinitely. As it only exists in real time, it effectively carries its own timing information with it. By now, you may have realised that we could be talking about any analogue audio signal, although when adapted to digital forms there arises a sensible need to declare the original sampling frequency, at the very least, in the associated metadata.

A file, on the other hand, never existed in the analogue world, although a finished physical recording was not a dissimilar concept. A file has to wait until a definable portion of the streaming programme is available to be packaged. The size of any file is limited, and needs to be declared, as do all the conditions necessary to rebuild a streaming output (i.e. replay it). Once this metadata information has been gathered and stored in a header 'chunk', tightly coupled with the audio 'essence' (bare data samples), the file is fully formed and only then can it be handled in the same way as any data file.

In a typical recording production chain it might at first seem easy to see where both streaming and files sit. This isn't necessarily true, as for instance microphone signals might stream into a recorder via a mixer of some kind. The recorder, however, may then record the signal as a series of tiny files, and even on a digital broadcasting system the signal will be formed of 'packets' which may be viewed as files of a tiny size. These packets must be delivered at the output as a streaming format. There is therefore some crossover between streaming and files, and the fundamental penalty for any use of intermediate files is signal latency (delay) between the input and the output streams.

File formats, on the other hand, are fairly well-defined and pre-agreed arrangements for storing and exchanging data of any kind at unspecified speeds. For audio, any file format must at the very minimum contain some form of header containing the information necessary to accurately rebuild the audio *stream* from which the files were initially built. Summing up, file formats and streaming formats are therefore inextricably linked, and in some cases, such as in those data formats used for packet transmission, the division between streaming and file may be blurred.

4.9.4 Practical Streaming Audio Formats

The AES/EBU format, defined in both the AES-3 standard[1] and the EBU Tech. document 3250,[2] as well as in the international Standards IEC 60958 and ITU-R,* is the most important streaming format at the heart of broadcast audio systems. It was originally designed around the existing cable and routing infrastructure in the analogue studios of the early 1980s, so that the connectors first specified were balanced XLR, allowing analogue patch cables to be used. Unbalanced signals on 75-ohm video coax using BNC connectors can also be found, and the consumer version of this interface (SPDIF) in the IEC[3] Standard was similarly (but not identically) electrically based on unbalanced signals on RCA phono connectors. Both in the PCM IEC60958 form and the packet-carrying version 60937, which is used for Dolby Digital and DTS carriage, an optical 'TOSLINK' is often used on consumer equipment today. Between these professional and consumer PCM interfaces, the audio 'essence' is identical, and the electrical interface differences are in reality no more problematic than those found in analogue practice. There are, however, different metadata formats (called Channel Status) in these two Standards.

4.9.5 Practical Audio File Formats

There are many types of possible audio file format that could be used for file exchange, and it is a relatively trivial task to convert between formats. However, there are always some penalties in conversion, and the EBU decided in 1996 that the demands of the professional broadcaster would be best met with an extensible generic format. They adopted an interesting approach, in the form of the Broadcast Wave File (BWF).

Now, no computer is the slightest bit interested in what your file is for, or what it actually does. The computer can be asked (by you) to look at the 'dot' suffix when given a file, and if a suitable application is available – either inside or connected to the computer – it will offer that file to the application. WAV-suffix audio sample files can be recognised by many different types of audio applications, and these applications will look at the 'chunks' of data in order to see if they recognise any data that is relevant to that application. It will leave all other chunks alone, a fact that enables us to insert broadcast-specific information into generic Wave files without disabling any low-cost existing applications such as simple players. If the mandatory file chunk known as 'fmt-ck' contains parameters such as sample rate and data information which suit that application, then the application will be satisfied that it can play the data. It will then send the data chunk itself (and no other content) to a buffer, from which it will play the file through a pre-arranged port. The beauty of the Wave procedure is that not only can we expand files by adding other chunks of broadcast-specific information, but different audio formats can be specified for the data chunk itself. Some of the first practical applications of the BWF used MPEG layer II coding, for example.

The full BWF file format is defined in the EBU Tech. document 3285, and this basic recommendation is further extended into the complexities of a 'native' file format for use in computer editing systems, within the AES Standard AES-31. The beauty of the whole procedure is that it lends itself to being extended even further, allowing industry users to incorporate their own enhancements to the basic arrangement. For example, when the BWF file goes to form part of a multi-media package, the BWF file can easily become a component of an exchange format such as MXF, or be incorporated in an assembly of files and metadata such as in the AAF structure.

4.9.5.1 Practical Piggy-back Audio Networks

'Piggy-back' Audio Networks are relatively new Standards, which have arisen out of the simple economics of using generic IT- or telecom-based distribution methods for carrying multiple streams of digital audio. All the current proposals such as AES-47 use multiple AES/EBU streaming formats as a signal source, although the MADI interface might be seen as an earlier stripped-down version of a piggy-back network.

4.9.6 Practical Sound Recording Formats

Because of the long history of sound recording, any broadcast sound operator is likely to be presented for many years to come with source recordings in many and varied forms, unlike video, where the value of the material on obsolete formats has led to at least a partial transfer programme. In order to recognise the majority of sound recording formats for what they are, the EBU has worked on the International Broadcast Tape Number (IBTN) scheme, and an associated bar-code label specification given in the EBU document Tech. 3279.

The IBTN scheme can be applied to any broadcast tape and related items, and enables these to be uniquely identified, from the earliest stages of the production process. The bar-code representation of the IBTN allows broadcast tapes to be scanned as they move from production facilities to broadcasting outlets, and during transfers between broadcasters.

For convenience, recording formats can be divided into those carrying analogue or digital signals, and then subdivided by the type of mechanical carrier. Table 4.9.1 provides an extract of the most common IBTN sound recording format codes, from which these subdivisions can be made.

4.9.7 Analogue Recording Formats

4.9.7.1 Quarter-Inch tape formats

Table 4.9.2 and Figure 4.9.2 provide an analysis to help visually identify typical types of ¼-inch broadcast analogue tape material. This list of tape, cassette and cartridge formats currently form the bulk of broadcast industry sound archives, although there are two possibly confusing digital formats

*The ITU (www.itu.int) is the legal international Standards body for both telecoms (ITU-T) and radiocommunications (ITU-R).

Table 4.9.1 The most common International Broadcast Tape Number (IBTN) scheme sound recording format codes

Code	*Material*
16T	16 mm SEPMAG analogue audio film
17T	17.5 mm SEPMAG analogue audio film
33L	33 rpm LP phonogram analogue audio disc
35T	35 mm SEPMAG analogue audio film
45D	45 rpm phonogram analogue audio disc
78D	78 rpm phonogram analogue audio disc
A01	6.3 mm (1/4″) analogue audio tape, full track
A02	6.3 mm (1/4″) analogue audio tape, two-channel
A04	6.3 mm (1/4″) track half-width analogue audio tape, stereo
A08	12.5 mm (1/4″) analogue audio tape, eight-channel
A16	25.4 mm (1″) analogue audio tape, 16-channel
A32	25.4 mm (1″) analogue audio tape, 32-channel
AI1	AIT (Advanced Intelligent Tape) digital data tape, 25 GB capacity
AI2	AIT (Advanced Intelligent Tape) digital data tape, 50 GB capacity
AI3	AIT (Advanced Intelligent Tape) digital data tape, 36 GB capacity
AIX	AIT (Advanced Intelligent Tape) digital data tape, extended length, 35 GB capacity
AS2	6.3 mm (1/4″) analogue audio tape, two-channel stereo
AT2	6.3 mm (1/4″) analogue audio tape, two-channel stereo and TC
CCA	Compact Cassette format analogue audio tape, cassette
CDA	Compact Disc Audio digital audio disc
CDD	CD-ROM digital data disc
CDR	Recordable CD digital data disc
D24	25.4 mm (1″) DASH format digital audio tape, 24-track
D32	25.4 mm (1″) PD format digital audio tape, 32-channel
D48	25.4 mm (1″) DASH format digital audio tape, 48-track
DA2	DAT format digital audio tape, two-channel
DAT	DAT format digital audio tape, stereo
DCC	DCC format digital audio tape
DD2	6.3 mm (1/4″) DASH format digital audio tape, two-channel
DP2	6.3 mm (1/4″) PD format digital audio tape, two-channel
H8A	Hi-8 format eight-channel digital audio tape, cassette
LAQ	Lacquer phonograph analogue audio disc
MDA	MD (MiniDisc) digital audio disc
NAB	NAB cartridge analogue audio tape
SVA	A-DAT eight-channel digital audio tape
WAX	Wax cylinder phonogram analogue audio disc

which also use open-reel ¼-inch tapes. These digital formats will only be found on tapes dating from the late 1980s and 1990s, and hopefully will be better documented than earlier recordings.

Unfortunately, even if an analogue recording can be established as having been made at a certain speed and track, layout, the equalisation characteristics applied during recording are not so easy to establish for European material. Table 4.9.3 lists the usual recording equalisation and noise reduction system characteristics for ¼-inch tapes. From this table, the characteristics of the correction equalisation can be worked out if a correct replay machine cannot be found. The correction equalisation can be applied post replay, maybe on a batch file basis inside an audio workstation. Indeed, if the correct noise reduction system cannot be found, it is not difficult to reproduce the characteristics using the dynamics available on quite modest audio workstations or computer editing programs. The only other difficulty that may be experienced with some ¼-inch analogue tapes is that it may be difficult to find a replay machine to cope with centre timecode or sync tracks.

Table 4.9.2 Typical quarter-inch analogue recording tape spool sizes and uses

Tape speed (cm/s) and uses	*Spool size and centre*	*Run time for LP tape*
38 (normal studio)		
19 (normal studio inc. carts)		
9.5 (reporter + home)		
4.75 (logging + home)		
4.75 (Fe-cassette)		
4.75 (Cr-cassette)		

4.9.7.2 SEPMAG film recording formats

Multi-track analogue audio tape, especially in the 2-inch-wide 24-track form, still forms a major interchange format within recording studios for remixing archive music sessions, but in broadcast use these tapes are rare. The same is not true of sprocketed analogue magnetic film, known as separate magnetic (SEPMAG) recordings, which for many years were associated with the huge quantities of 16 mm broadcast film material used for television recordings. Fortunately, as Table 4.9.1 shows, there are few variations in playback standards, although any SEPMAG transfer will require a specialised playback deck.

4.9.8 Dedicated Digital Audio Recording Formats

4.9.8.1 Fixed Head tape formats

4.9.8.1.1 DASH (two-channel stereo and multi-track formats)

During the 1980s several Digital Audio Stationary Head (DASH) recording formats were developed, often based on the platform of existing analogue tape decks.

The attraction at that time was the illusory economic advantages of razor blade editing, combined with digital audio quality. The most enduring of these DASH formats, especially for storage and exchange in the music recording industry, is probably the Sony multi-track version using 2-inch-wide tape, and capable of recording up to 24 audio channels. In broadcasting circles, multi-track machines such as this were only used for serious music recording backup, although even today, the

Figure 4.9.2 The external differences between the NAB cartridge (left) and the consumer eight-track cartridge (right). The NAB-standardised endless loop 'cartridge' was the staple contribution format for radio stations' commercials and inserts from the 1960s until the 1990s. They used ¼-inch tape with a lubricated backing, and the centre cue track carried several automation tones, which could be used to cue and trigger cart-players in order to automate commercial breaks. The tapes were usually recorded on open-reel recorders, or duplicators, and then physically transferred to the cartridge for distribution. They can be extracted for replay on a 38 cm/s open-reel recorder, and this is to be preferred for transfer, as the tape path control in cart-players was necessarily poor. For this reason, the advent of two-channel stereo commercials sidelined the NAB format in favour of digital storage, although a few specifically stereo analogue arrangements were briefly employed. In the background for scale is the Compact Cassette-sized 'DCC' cassette, whilst the consumer eight-track cartridge on the right was an endless loop format which used a mechanically stepping two-track head. As a format it was popular for in-car entertainment and background music from the mid 1960s through to the late 1970s.

obvious successor in terms of bulk multi-track storage has yet to be found.

4.9.8.1.2 DCC cassettes

The advent of the Compact Disc in 1982 exposed the inadequacies of the Compact Cassette in home recording applications, and as a result there was a three-sided development of digital recording formats for the consumer market. Ultimately, two of these (the R-DAT and the MiniDisc) were to find a ready acceptance in professional radio production applications. The third format started out life as the S-DAT (Stationary head Digital Audio Tape) format, using a compact cassette-sized tape, but with uncompressed 16-bit PCM stereo recording. It became the Digital Compact Cassette (DCC) when MPEG layer I bit rate reduction was applied. A few examples of these cassettes may be found in libraries, and Figure 4.9.2 shows what to look for. However, the recorders were discontinued around 1996.

4.9.8.2 Rotary head tape formats

4.9.8.2.1 Two-channel formats

4.9.8.2.1.1 '1610' type videotape recordings

The vital need in the mid 1970s, for Compact Disc source recordings in an editable digital form, created one of the most bizarre audio recording formats ever, and it has left us the curious legacy of a sample rate of 44.1k samples per second. As an idea, it started with the then emerging helical scan video

Table 4.9.3 The IEC and NAB equalisation characteristics for quarter-inch tape, and the characteristics of common noise reduction systems

Tape speed (cm/s) and uses	*CCIR/IEC*		*NAB*	
	Bass boost	*High roll-off*	*Bass boost*	*High roll-off*
38 (normal studio)	None	35 μs	3180 μs	50 μs
19 (normal studio inc. carts)	None	70 μs	3180 μs	50 μs
9.5 (reporter + home)	3180 μs	90 μs		
4.75 (logging + home)	3180 μs	120 μs		
4.75 (Fe-cassette)	3180 μs	120 μs		
4.75 (Cr-cassette)	3180 μs	70 μs		

The break frequencies of the filters are described by the time constants of simple RC low-pass networks. The basic difference between the IEC/CCIR and the NAB standards is therefore the NAB bass boost. This was intended to improve the hum performance of the recorder during playback.

recorders, which were capable of recording (via a Sony adapter type 1610 or 1630) three 16-bit (stereo) samples of digital audio on each television line. The most practical format in those days was the U-Matic cassette with (¾-inch tape. These tapes could be assemble edited using multiple machines and an edit controller. Because analogue recorders have no storage, the time when the rotary heads switched over at the tape edges needed to be avoided for recording, and the system therefore only used 588 lines of the 625-line, 25-frame system for recording the audio; 588 lines times three samples times 25 per second gave 44,100 samples per second. The equivalent US system used 490 lines out of 525, but the slightly lower frame rate of 29.94 gave 44,056 samples per second. The 0.1% difference in these sample rates was soon forgotten, and 44.1 kHz lives on.

4.9.8.2.1.2 R-DAT

Manufacturing experience of helical scan video recorders had reached such a stage by the mid 1980s that a dedicated audio recorder could be produced using tape only 13 micrometres thick and 3.81 mm wide. This could record at least 2 hours of two-channel 16-bit PCM audio at sample rates of 32, 44.1 or 48 kHz. The system incorporated many ingenious features, and Figure 4.9.3 shows the essentials of the track format. Each track is about 23.5 mm long and is at an angle of just

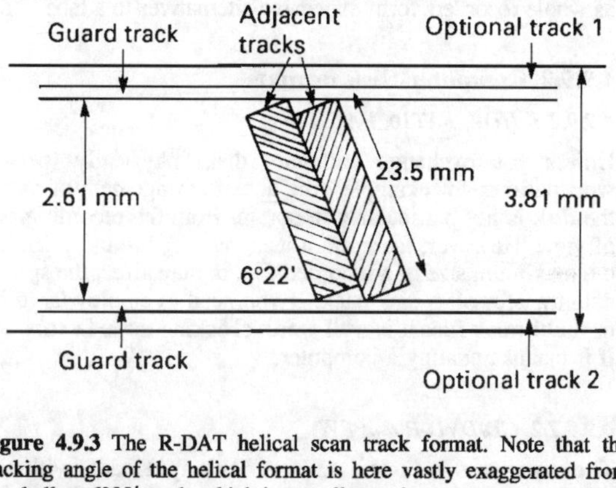

Figure 4.9.3 The R-DAT helical scan track format. Note that the tracking angle of the helical format is here vastly exaggerated from the shallow 6°22′ angle which is actually used.

over 6° to the direction of travel of the tape. The two linear tracks at the edge also serve as a buffer to protect the main tracks against edge damage. Because of the low tape speed they are of little real value as audio tracks, the highest recordable frequency being in the region of 3 kHz, but video time-code could be recorded on one of these edge tracks and used for a fast search. The integral sample count was more accurate, however.

The use of azimuth recording, in which each track is recorded with a 20° 'slant', alternate tracks having alternate 'slopes', reduces markedly the risk of crosstalk between tracks. The audio data block is shown in simplified form in Figure 4.9.5. In the 'preamble' to the 256 audio data bits, which include error correction, there are 4 bytes, the first of which is for synchronisation. The second, the ID code, specifies, for example, the sampling frequency and the number of channels. The third byte states whether the block consists of digital audio signals or whether it is a subcode, while the fourth provides parity checks. There then follow the 256 data bits (32 bytes), making so far a total of 36 bytes.

A complete track is shown in Figure 4.9.4. It is made up of 196 blocks performing the following functions:

1. Margins at beginning and end (11 blocks each). These are effectively guard bands.
2. Subcodes 1 and 2 (11 blocks each). These carry additional information (running time, contents, etc.) rather in the manner of a compact disc.
3. ATF 1 and 2 (Automatic Track Following) (five blocks each). These blocks provide information for the servo systems which cause the heads to follow the tracks accurately.
4. Four spacing sets of three blocks each – the Inter Guard Bands.
5. Blocks of audio data (130 in total), each of which is as in Figure 4.9.5.

4.9.8.2.2 Multi-track formats

4.9.8.2.2.1 ADAT and DA88

Once the R-DAT format was established and adopted by professional users, it was a small step to produce multi-track versions using S-VHS cassettes (ADAT) or 8 mm video cassettes (DA88 type). Professional users tended towards the generic DA88 series, which had track bounce and integral

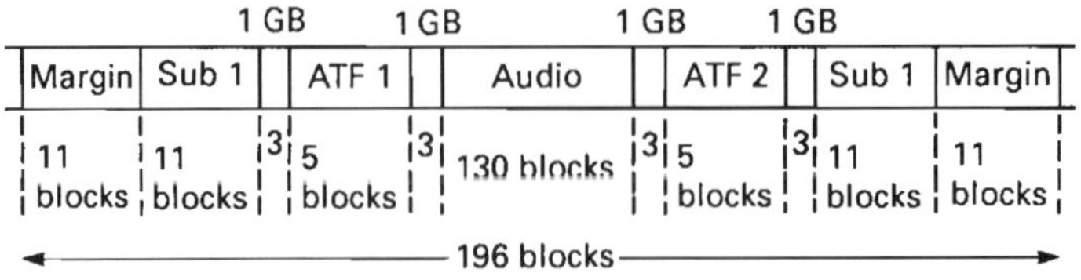

Figure 4.9.4 One audio track in the R-DAT format.

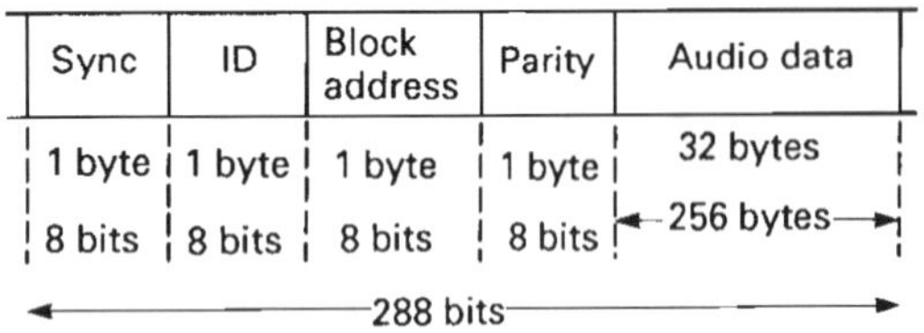

Figure 4.9.5 The format of the R-DAT audio data block.

timecode features. Originally carrying only eight 16-bit audio tracks (and integral timecode), a newer series of these machines will record 20-bit PCM tracks. Although these newer machines will play back 16-bit recordings, the earlier machines will not play back 20-bit recordings.

Interconnection of these multi-track recorders in the digital domain is via proprietary interfaces, a parallel electrical TDIF interface in the case of the DA88s and a serial optical 'lightpipe' for the ADAT.

4.9.8.3 Optical disc formats

4.9.8.3.1 CD and DVD Audio

CD-A (the commercial record format), whether in CD-R, RW or glass mastered form, is not a particularly good format to use for broadcast recording, although it may have attractions for easily played samplers, etc. It possesses limited error correction, a Table of Contents (TOC) must be written at the start of the disc, and the encoding is limited to 16-bit PCM. As a physical *carrier*, however, the optical medium of the CD (or equally the similar DVD) has a lot to commend it, not least the fact that the IT industry has reduced the recorder and blank media prices.

4.9.8.3.2 MiniDisc

The Sony MiniDisc format was developed from the 8 cm CD, by the use of the proprietary ATRAC audio data rate reduction system, based on block frequency transforms. The system was also based on the option of using rewritable media in affordable 'walkman' style recorders, a factor which endeared the format to reporters and radio playback applications, and MiniDisc formed a direct replacement for NAB tape cartridges in many cases. The system uses a TOC, very much like that on a CD-A, and when recording this TOC is written automatically *after* pressing the stop button. Physical vibration can sometimes impair recordings (including that important TOC), but ever larger buffer memories help to minimise this, and the low bit rate helps make playback access to any track particularly fast.

4.9.9 Generic Digital Audio Recording Formats

4.9.9.1 Hard Disk and Solid-state recorders

4.9.9.1.1 NAGRA and similar portable recorders

The NAGRA (Polish for recorded) series of ¼-inch analogue recorders made by the Swiss Kudelski company are still highly considered for location recording, and especially for feature film applications. Several digital versions of this old favourite have appeared using tape, disk packs or solid-state memory. In particular, the NAGRA-D offers four 24-bit PCM channels on tape, a highly attractive top-quality capture format. The only thing to consider is the need for real time to dump these recordings to computer storage, but also bear in mind that removable disk or solid-state packs can make for an expensive alternative to a tape library.

4.9.9.1.2 Digital Dubbers

The name is a film industry misnomer for solid-state or disk-based temporary stores for (typically) eight tracks of up to 24-bit PCM audio, used during audio editing and dubbing, hence the name. These recorders are rarely used for the actual interchange of recordings, and can be viewed as a sort of local server.

4.9.9.1.3 Reporters' solid-state recorders

An obvious solution to possible mechanical vulnerability of portable digital recorders is to record directly to a solid-state memory. In order to keep the power consumed down and the recording time up, bit rate reduction may be used. Here again, consider how quickly and conveniently you can dump the recording to a computer. This is vital so that you can reuse the recorder, as whole recorders form expensive alternatives to a tape library.

4.9.9.2 Computer Disk formats

4.9.9.2.1 UDF, FAT16, FAT32

Unless you exchange computer disks physically (between workstations, for example), the actual arrangement of data on the disk is not particularly important from the recording point of view. However, there are issues that, for instance, can limit the maximum size of a recorded file, or may affect the speed of data transfer off a disk pack. If you need to employ large files or multi-track files it is well worth checking these factors, even if it means updating a computer.

4.9.9.2.2 CD/DVD-R and RW

The computer industry, and particularly Kodak with their Photo-CD application, are entitled to our thanks for pushing forward the CD-R (and related RW) formats into affordable computer components, as these now form such economical carriers for fast audio file exchange. If you are using CD-R media you can 'close' the disc to ISO 9660 format, which is the basic universal CD-ROM format. However, CD-RW discs cannot be closed to the ISO 9660 format, and more recent CD-ROM formats such as UDF that are basic to the recordable DVD enable other benefits such as longer names to be entered.

4.9.10 Making a Practical Sound Recording

A broadcast sound recording team starts with the Studio Manager in a radio studio, or the equivalent Sound Supervisor in television. He, or she, will usually have at least one assistant present in the control room with them. This assistant or assistants are still known as 'gram ops' in the UK, named after the source of sound inserts, from gramophone records with the cue point marked with a chinagraph pencil.

On the studio floor, Sound Assistants manage the microphones, including radio microphones or microphone booms, whilst a studio audience will also have a PA operator who will be in charge of the local sound feeds for the audience. He or she often sits with the audience, and the audience sound feed will include several non-programme sound sources, such as music for the audience to enter and leave by, as well as feeds for the warm-up performers and playback of sound for any fill-in video sequences. If a music recording or DVD release of the programme is envisaged, a multi-channel recording may be made in parallel with the normal two-channel stereo mix programme output. In the case of television recordings, the Digibeta videotape format normally only allows for two AES/EBU bit streams to be recorded, the second bit stream pair often carrying the Music and Effects only, ready for International programme post-production. Any recording should include some kind of header with line-up tone, and frequently a voice ident with the take number, etc. This is called the 'slate' in film production, and mixers often include a path for the talkback microphone to be used to slate the recording. It is a good idea to run the sound recording a little while before the action and a little afterwards; the resultant 'atmosphere' can be very useful to patch over edits later.

4.9.11 Programme Recording Levels and Level Metering

The header on any recording needs to follow a fixed structure, so that exchange is easily possible between broadcasters who may not speak the same language or follow the same production procedures. In the case of a file format such as BWF, this header will be contained in the header data 'chunks' which serve the same purpose. In Europe EBU recommendations are normally followed for these headers, whilst the SMPTE[4] is the usual reference agency for US countries. For direct delivery to a broadcaster, there are a number of technical and programme delivery requirements that will need to be followed, and these can vary from broadcaster to broadcaster, as well as between world regions.

Listed amongst these recommendations are working practices, such as the requirement for broadcasting not to exceed the ITU-R 647 defined 'permitted maximum level',[*] how these levels should be metered and any line-up tone requirements,[†] as well as the preferred track allocations on multi-track formats.

Sound recordings that accompany pictures, be it for television or film, require a timecode to be recorded at the point of capture and the audio sample rate locked to the video. The timecode used refers to the *picture* format not the audio, and where there has been picture format conversion in post-production, this timecode may need to be changed, even if the audio essence is unaffected. Fortunately, much of the audio recording equipment and software used for post-production[‡] works happily with a number of timecode types, translating between them and using the locked audio sample rate as the timing reference.

Once the recording is delivered to the broadcaster or post-production stage, it may pass through a Quality Control check, and at this stage, some kind of an audit trail of the recording history[§] can be of enormous benefit in solving any problems.

4.9.12 Repair and Restoration of Sound Recordings

Lost or damaged audio files in IT-based equipment can be handled in much the same manner as any other 'lost' data, and maintaining simple (and often automatic) backup procedures will instil a degree of security within these systems. In the case of damaged physical media, however, it is well worth trying different playback machines. Especially within the many CD/DVD and digital tape formats, huge differences exist in the data recovery performance between different players.

Removing unwanted artefacts from recordings is a more controversial subject, and the availability of simple software 'clean-up' processes does not mean that they are necessarily beneficial to any recording, in any way. The only way to judge a correction process is to listen against the original several times, and even then make sure that the original material is still available in the original form afterwards. Having said that, frequency domain clean-up procedures can be used for previously impossible distortion and overload repairs. Here again though, it is too easy to persuade yourself of the value of such processing, rather than leaving things alone. Ultimately, the choice is up to you, your listeners and all your ears.

References

1. See www.aes.org
2. See www.ebu.ch
3. See www.iec.ch
4. www.smpte.org

[*]Typically, not to exceed a level 9 dB below digital FSD level.

[†]EBU line-up tone at a level 18 dB below digital FSD level can be audibly recognised from the left channel being interrupted every 3 seconds or so. SMPTE line-up tone for US markets will be at 20 dB below digital FSD.

[‡]Such as the DA88 family of recorders.

[§]Such as may be carried in the 'coding history' chunk in the BWF.

Andrew Hingley
Sony Broadcast and Professional Europe

4.10 Sound Mixing and Control

The sound mixer is often the centre-piece of the audio studio both in terms of its physical size and functionality. The purpose of the sound mixer is to process and combine multiple audio sources for live transmission, recording or presentation. Human hearing allows many simultaneous sounds to be 'meaningfully understood' and the largest audio projects may require hundreds of audio sources to be combined together by the sound mixer.

The first impression of a large-scale analogue mixer can be daunting; however, there are many shared features across virtually every type, shape and size of mixer, from the smallest to largest music/post-production mixers. For example, virtually every mixer has the same basic processing structure of input channels, output buses, output channels and monitoring/metering section. An understanding of this basic mixer structure allows an experienced operator to quickly assess the functions and control surface logic of a new model.

A good way to gain an understanding of this basic mixer structure is to study the control surface layout of an analogue mixer, characterised by the 'one knob per function' design. The control layout reflects the underlying signal flow as each knob and switch is mechanically linked to an element in the signal path. Typically, analogue mixers are constructed from a chassis frame with various plug-in module strips for different types of channels (input, output and monitor). The typical mixer processing has the following orientation:

- Signals are processed along the length of the module strip, starting with input selection at the top, moving through equalisation and routing and on to the ubiquitous fader at the bottom of the module.
- Output mix buses run across the mixer.
- Output and monitor signals are processed along the length of additional module strips.

By using this orientation, it should be possible to quickly identify a mixer's processing modules, the configuration of input and output channels, bussing structure as well as the individual processing sections within each type of module strip.

The majority of audio mixing consoles used in broadcast applications are analogue, although digital mixers have gained a wide acceptance. The control surfaces of digital sound mixers are often rationalised with a reduction in the number of physical controls, using a single set of controls and assignment modes. However, as underlying audio processing of a digital mixer is normally identical to an analogue mixer, it is normally a simple task to navigate the digital mixer's control surface – first by learning the 'assignability rules' and then relating functionality back to the 'one knob per function' structure of analogue designs. The basic principles of assignable digital control surfaces are covered in Section 4.10.4.1.

4.10.1 Audio Process of Sound Mixers

Figure 4.10.1 shows a highly stylised layout of an analogue mixer, highlighting seven processing functions of a sound desk:

1. Input signal selection and processing – converting a range of incoming signal types to the mixer's internal processing standard.
2. Input channel processing – typically involving equalisation (EQ) and dynamic gain control.
3. Input channel gain and routing control – allowing the operator to control the contribution of each input signal to various output buses, including monitor outputs, main outputs and 'effects sends'.
4. Output buses.
5. Output processing – allowing overall gain control, EQ and dynamic gain control to be adjusted on the output signals.
6. Output signal processing – converting the mixer's internal processing standard to the format required by external devices.
7. Monitoring functions – providing audio monitor control and metering.

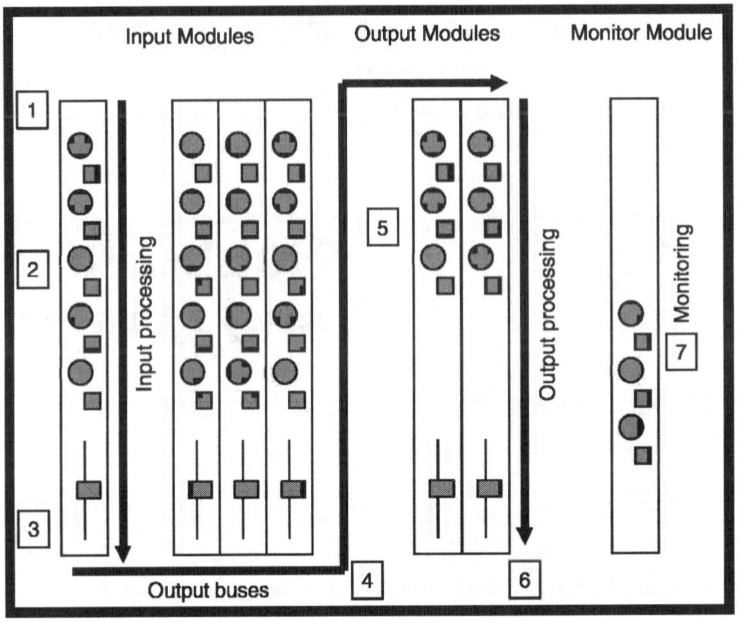

Figure 4.10.1 Stylised layout of an analogue mixing console control surface.

4.10.1.1 Input signal selection and processing in detail

Figure 4.10.2 shows a stylised diagram of a mixer's input module. The signal path starts with input selection and format conversion. The mixer may have to handle various analogue signals (microphone level, unbalanced and balanced line level) and digital formats (including standard and proprietary signals), and these have to be converted to the mixer's internal processing standard.

The input section commonly features gain controls (preamplifier to amplify microphones signals and a Pad switch to attenuate high level line signals) and phase reversal. Input impedance is normally switched between 1 k ohms for microphones and 10 k ohms for line level sources. Digital mixers are likely to feature ADCs (analogue-to-digital converters), digital sample rate and format converters. Often, external patch bays and format converters are built into the mixer chassis to provide additional signal management before interfacing the input module. In the case of some digital sound mixers, an integrated input router provides 'digital patching' and format conversion.

4.10.1.2 Input channel processing

The **Equaliser** (EQ) section imposes a frequency response on the signal, both to reduce unwanted spectral components (for example, low-frequency 'hum' or high-frequency 'hiss') and to enhance the sound quality (for example, boosting mid frequency energy in a voice signal). The overall functionality of the mixer's EQ section is defined by:

- the number of individual bands (typically there will be three or more EQ bands covering different frequency ranges);
- types of bands (filter, shelf or bell shape);
- number of variable EQ controls (frequency sweep, frequency steps, fully parametric, etc).

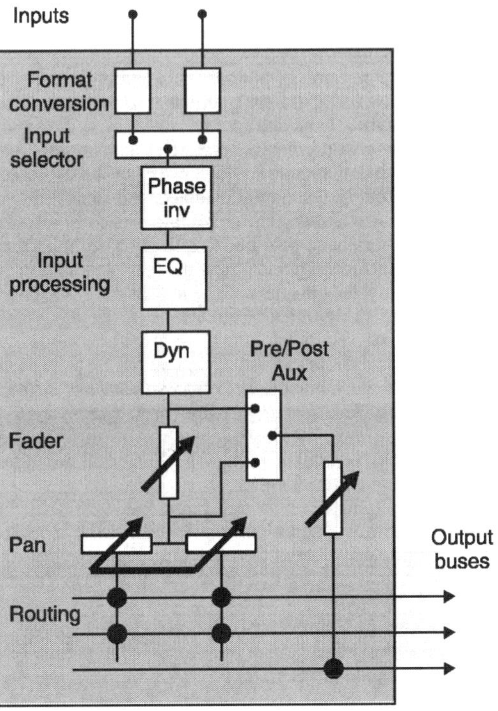

Figure 4.10.2 Stylised diagram of a sound mixer's input module.

The simplest EQ provides only basic Bass and Treble lift/cut, whereas the more powerful parametric type provides lift/cut, sweepable frequency and variable Q (Q refers to tightness of a bell-shaped frequency response – Quality Factor). Many mixers also have HCFs (High-frequency Cut Filters) and LCFs (Low-frequency Cut Filters) with steep (12 dB/octave) filters.

Many digital mixers display a graphical representation of the EQ's frequency response and these provide a useful tool for exploring the audible effect of different EQ curves.

Dynamics functions, such as limiters and compressors, are available on many sound mixers and provide automatic gain management of audio signals. Dynamics are used both 'artistically' to enhance sound quality and to help manage signals with unpredictable jumps in amplitude. There are four basic Dynamics processing types:

- Compression – controls the signal's maximum amplitude by compressing high level signals.
- Limiter – a special vision of a compressor that clamps maximum signal amplitude at a specified threshold level.
- Expansion – attenuates low level signal, often to reduce unwanted background noise.
- Gate – mutes the signal when its amplitude falls below a specified threshold.

Each Dynamics processing section is likely to have Attack and Release timing controls, providing tools to adjust the speed of response of the automatic gain management.

Many digital mixers provide a graphical representation of the Dynamics sections gain reduction profile (see Figure 4.10.3), often shown as an X/Y Graph. These types of display provide a useful tool for exploring the audible effect on signals as the dynamics is adjusted.

Insert points provide a break in the signal path for the insertion of 'outboard' devices such as graphic equalisers, limiters or any one of a wide range of processing units. Commonly there will be patch bay sockets for these signals; 'Insert Send' refers to the outgoing signal and 'Insert Return' to the incoming signal.

4.10.1.3 Input channel gain control and routing

Channel fader and Mute switch control the overall gain of the input module's signal. The vast majority of faders are linear, making use of durable conductive plastic tracks. However, on small mixers used for location recording the faders may be rotary. Most mixers also have a channel Mute switch with a status indicator. Whether the indicator is active for a muted or unmuted status depends on the traditions of different applications and manufacturer's preferences. The Mute switch may have one of several common reference labels, including: Mute, Cut or On.

The **Pan-pot** positions the audio signal in a stereo (or surround) image by simultaneously controlling the signal contribution to several output buses, effectively increasing the gain to one bus while decreasing it to others. The 'pan law' describes the interaction between these output contributions as the control is moved from a fully-left to a fully-right position. The Pan-pot is virtually always a rotary control with its physical position, indicated by controller's pointer, closely relating to

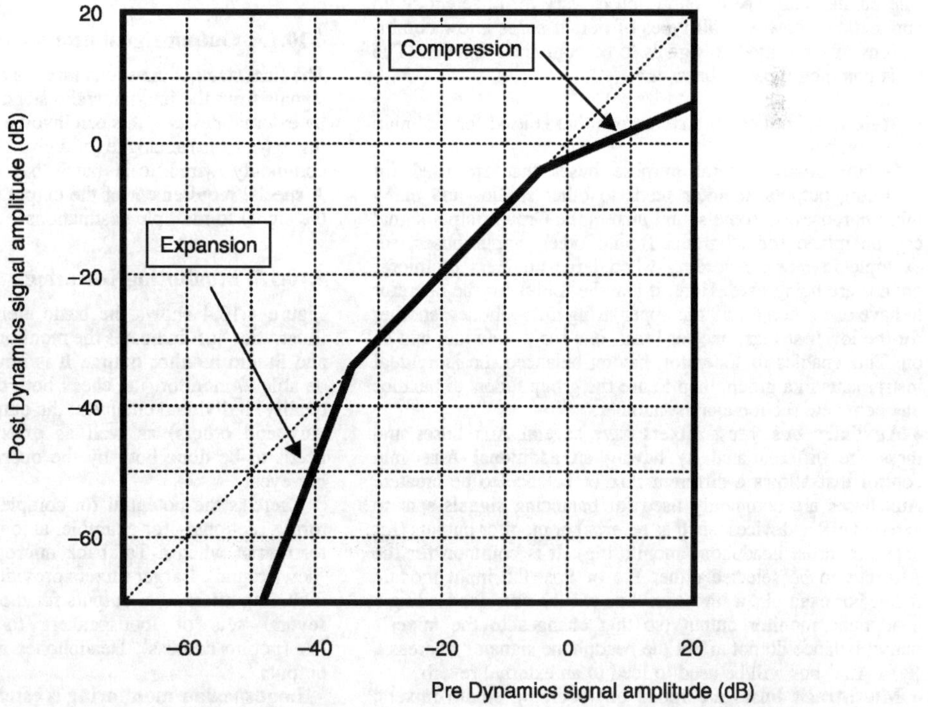

Figure 4.10.3 Dynamics section's gain reduction profile.

the position of the signal in the 'stereo image'. Many sound mixers will also have a Pan On/Off switch to bypass the pan.

Channel routing is used to send the module's signal to the selected output buses, routing being managed by sets of switches. Routing on some sound mixers will feature additional gain and pan controls.

Listen (PFL, AFL, Solo) switches allow the operator to audition signals from individual modules, either to identify 'problem noises' or to isolate audio from a single channel when adjusting its channel processing. Many sound mixers have a dedicated 'monitor' bus that allows the operator to audition signals while not disrupting the mixer's other output signals.

There are three standard Listen modes:

- **PFL** (Pre Fade Listen) allows an input module's signal to be monitored even when the input module is muted or faded down. On some broadcast mixers the PFL function is activated by 'over-pressing' the fader against a sprung microswitch in the faded-out position, allowing a quick method for checking the audio signal before fading it up.
- **AFL** (After Fade Listen) monitors the post fader signal. On some mixers the AFL signal is picked up after the module's Pan-pot and therefore monitored in stereo. Commonly this is known as 'Stereo In-place' Listen.
- **Solo** allows individual input modules to be monitored by automatically muting all other input channels. Solo is also known as 'Destructive listen' as this mode also affects the mixer's output. Therefore, the Solo mode should be avoided in broadcast applications.

4.10.1.4 Output buses

The number and type of output buses vary for different audio applications; however, all types of output buses allow combinations of input module signals to be combined together. The most common types of buses are:

- **Main bus**, commonly a stereo bus that is used for the main studio output.
- **Group buses**, general purpose buses that are used for recording outputs, monitor feeds to other studios and many other purposes. In some sound mixers the Group output signal can be mixed (or subgrouped) into other output buses, for example in music recording when large numbers of microphones are being used. Here, it may be useful for the operator to have one group for all the upper string microphones, another for the lower strings, one for brass, one for woodwind, and so on. This enables the operator, having balanced the individual instruments in a group, then to use the group faders to balance one complete section against another.
- **Auxiliary bus**. Most mixers have several Aux buses and these are differentiated by having an additional Aux gain control that allows a different mix or balance to be created. Aux buses are commonly used for balancing signals sent to external effect devices (such as reverbs) or monitor outputs (for example, artist headphone monitoring). It is common for the Aux bus to be selected either Pre or Post the input module fader. For example, a 'Pre Aux' bus will be used for feeding a headphone monitor output (so that changes to the mixer's output balance do not affect the headphone signals) whereas a 'Post Aux' bus will be used to feed to an external reverb.
- **Multi-track buses** are found on recording studio mixers; these buses are used to create eight (or more) individual outputs to feed signals to each input of the multi-track recorder. On large-scale music mixers, the Multi-track routing switches are often positioned at the top of the modules.
- **Clean Feed buses** are special buses used to create monitor return signals for 'contribution sources', such as telephone and remote studios. The associated monitor output signal required to send back to the 'contribution sources' must not contain the contribution input signal, otherwise howl-round occurs. These special monitor signals are also referred to as Mix-Minus and N-1.

4.10.1.5 Output processing

The available processing functions on output modules vary widely, ranging from a simple fader to replicating the same functions as the mixer's input module. In general the output audio processing will be simpler than inputs; however, it is often useful to have EQ for Aux outputs or the ability to compress a Group or Main output signal.

Main output will often have no additional audio processing apart from a fader. However, an Insert point, selectable Pre or Post the fader, is often provided to allow a high-quality external processor to be inserted into the Main output signal.

The output fader is a critical control in the signal path as it is this fader that will normally be used to fade in or fade out the studio, and it is also the point at which the operator will obtain a correct signal level, avoiding excessively high levels that may cause distortion in subsequent stages, and at the same time avoiding under-modulation.

Because the main fader is in a critical position in the signal path and any failure would be catastrophic, there may be safety measures such as dual tracks and a bypass switch to select the second track if the first becomes faulty.

4.10.1.6 Output signal processing

The final stage of the sound mixer's processing is to convert the signals from the internal audio standard to the format required by external devices; this can involve gain changing or converting to (or from) analogue (or digital) formats. Mixer outputs are commonly wired to a patch bay and external converters. A special requirement of the output processing is to distribute the signal to multiple destinations.

4.10.1.7 Monitoring/Metering

Figure 4.10.4 shows the basic elements of monitoring; the illustration just indicates the monitoring for the Control Room and Studio monitor output. It is important for an operator to be able to monitor (i.e. check both the signal level and sound quality) individual channels, the output signals (Main, Group, Aux and others) as well as external sources. Monitoring needs to be done both by the operator's ears and by his or her eyes.

There is the potential for complex switching in the monitoring sections – for example, automatic muting of the monitor output when a Talkback microphone is active, to avoid 'howl-round'. Larger mixers provide several monitor outputs, including alternative outputs for the Control Room (allowing several sets of loudspeakers to be interfaced), Studio PA (public address), Headphones and other communication outputs.

Loudspeaker monitoring is essential for checking balance and quality generally, including detecting any distortion. In some circumstances headphones have to be used – for

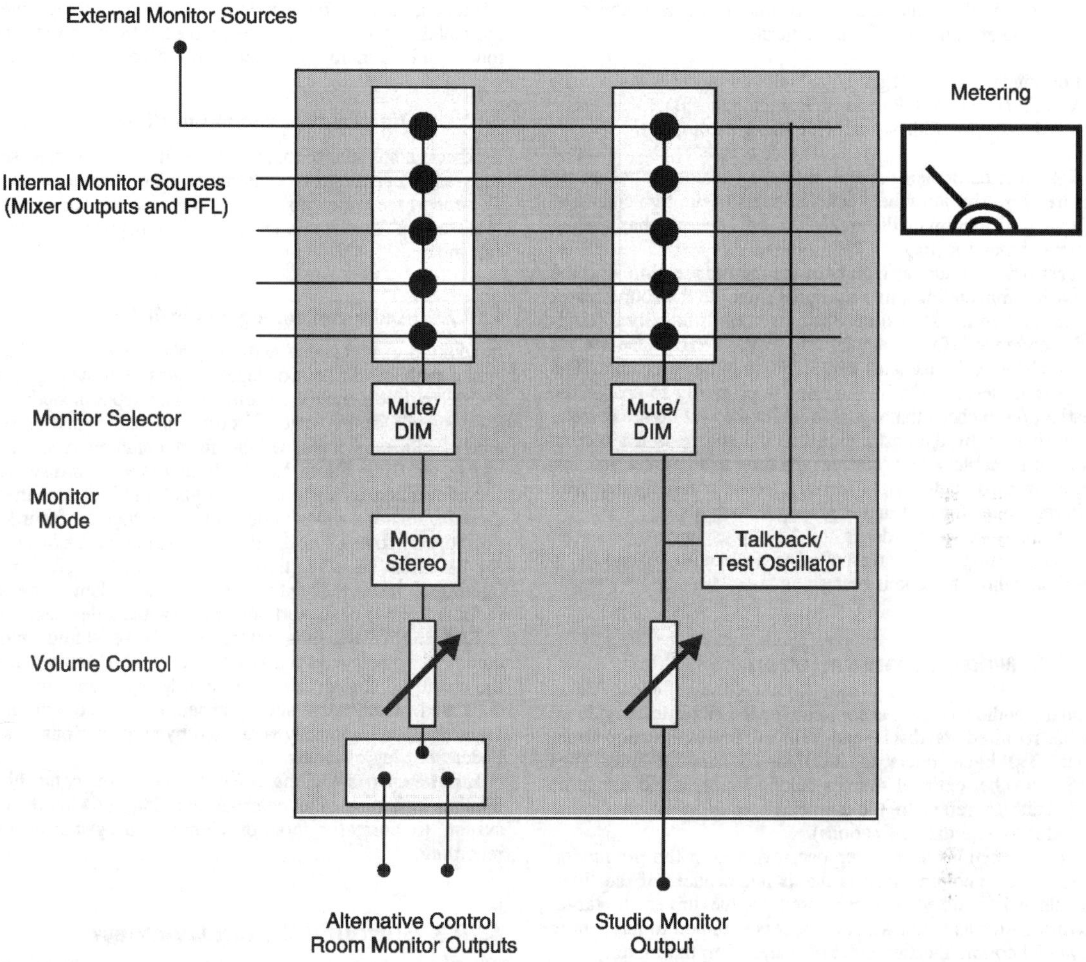

Figure 4.10.4 Stylised diagram of a sound mixer's monitor and metering.

example, when the mixing is done in the same space as the recording. It is probably fair to say that headphone monitoring is second best to the use of loudspeakers. This does not necessarily imply that the quality of reproduction of headphones is inferior; the reason is partly psychological, arising from the isolation of the operator. Stereo imaging with headphones is apt to be very different from that perceived using loudspeakers.

Important features of a good stereo monitoring system include volume control (stereo ganged controls), balance, to ensure equal sensitivity for both speakers, and phase reverse, operation of which is sometimes the only way of confirming the correct phase of material.

Mono on both speakers gives a central image which is necessary for setting the balance.

Mono on left simulates conditions for the mono listener better than Mono on both. In principle, it clearly does not matter on which loudspeaker the mono signal is reproduced, but in the UK the Left (A) speaker is generally preferred.

The Monitor Cut button provides an instant method of muting the loudspeaker without changing the monitor gain setting. The Dim button provides a similar function but attentuates the monitor output signal (typically 20 dB) and is useful, for example, when a studio telephone needs answering.

Metering is the only way of checking that signal levels are within correct upper and lower limits. Many sessions will start by 'lining up' signal paths using a test tone oscillator and adjusting gain controls to achieve the correct signal level on the output meters.

Popular types of metering standards include VU (Volume Unit) and PPM (Peak Programme Meter), with either moving coil or bargraph displays. There are many PPM metering standards based on national broadcasting standards and manufacturers' proprietary specifications. Meters, however, can give no indication of the quality of the sound nor of the balance between different sources.

For stereo metering there are many standards – for example, UK broadcasters commonly use a pair of double pointer PPM meters with the pointers mounted on coaxial shafts, rather like the hour and minute hands of a conventional clock. In this case the two meters side by side can indicate:

- Left signal (A) and Right signal (B).
- Middle signal (A + B) and Side signal (A − B).
- S + 20 (i.e. the Side signal with 20 dB extra gain).

S + 20 is particularly useful in line-up operations where, to ensure close similarity between the levels of the two channels, adjusting for minimum difference is more accurate than aiming for equal meter settings.

Test tone is a further feature of the monitor section with the provision for sending a line-up signal (tone) to the sound mixer output and on to external destinations. This is usually a standard reference (for example, 0.775 V, 1 kHz sine-wave), although other frequencies and levels may be selectable. The purpose of sending the line-up tone is primarily to enable the destination to check that signal levels in the system are correct. Identification by the destination that the source is the correct one can be achieved by 'cutting' the tone at an agreed instant; there are also standards to identify stereo compatibility with different tones for Left and Right signal outputs.

On many broadcast desks the line-up is injecting it after the monitoring, allowing monitoring for studio rehearsals to continue while the tone is being sent.

4.10.2 Mixer Communication

In many audio studios it is common for the mixer to provide all of the required 'talkback' and 'talk to' communication functions. **Talkback** refers to incoming communications lines (artist or other external sources talking to the sound operator) and **Talk to** refers to the outgoing communication (sound operator to external destinations).

These functions allow everyone involved in the production to have voice communication that is independent of the 'live' signals being mixed and processed by the mixer. In larger facilities such as broadcast studios, it is common to have more powerful communication systems, external to the mixer.

4.10.2.1 Studio communications

Normally there is two-way communication between the studio (recording area) and the control room (mixing area). The sound operator is likely to have control over studio communications but it is also possible for the artists to remotely activate 'reverse' talkback.

4.10.2.2 Talk to studio

This is normally via microphones available to the producer and sound operator, the microphones being operated by a switch or foot pedal. In rehearsal conditions, the talkback in the studio will usually be through a loudspeaker. In transmission conditions, i.e. when recording or on-air, the loudspeakers are muted and headphones are used instead. Talkback is frequently extended to other areas, such as recording suites or remote contributing studios.

4.10.2.3 Reverse talkback

During rehearsals, communication from studio to control room can often be via one of the studio microphones. This may be inappropriate on occasions, perhaps when part of an orchestra is rehearsing; in these cases a dedicated microphone attached to a musical director's desk and linked to a loudspeaker in the control room can be a more convenient form of communication.

4.10.2.4 Other remote communications

Talkback communication is also required for remote sources, such as telephone contributions or other broadcast facilities. To support the wide range of possibilities, it is common to find 'Talk to Aux', 'Talk to External' type of functions on many broadcast mixers.

4.10.2.5 Studio control logic interfacing

In addition to processing audio signals, the mixer is often integrated with studio control logic circuits, automating functions such as transport controls. There are many types of studio control logic signals, many standardised by national broadcasters, and other facilities. Several of the most common types of logic signals are Fader Starts, Red Light and Input/Repro switching.

Fader Start refers to a logic signal that indicates the open/closed status of a fader, 'open' meaning that the fader knob is positioned above a threshold 'off' position. Fader Start is used for functions such as 'live mic tally display' (for example, lighting an indicator next to a microphone when its associated fader is 'open') or to start and stop playback devices.

Red Light switching refers to a 'Live studio' indicator commonly displayed on a large, red 'On-Air' lamp that warns the studio is 'On-Air' or recording. Many sound mixers have dedicated switches that simply generate a logic output; in other cases the Red Light tally is created by combinations of several Fader Start logic signals.

Input/Repro switching refers to the mixer controlling the monitor switching of an external recorder, allowing the overall system to correctly monitor signals during rehearsals and recording.

4.10.3 Surround Sound Overview

Surround sound mixers have been used in film production for many years and, with the advent of DVD and digital broadcasting, surround sound production is now common in TV and music studios.

The most popular digital audio surround sound format (known as 5.1) requires a mixer to process and monitor a 6-wide output, in comparison to the 1 (or 2)-wide output for mono and stereo productions (5.1 refers to five 'full bandwidth channels' – Left, Centre, Right, Surround Left, Surround Right and a single 'band limited' sub-bass channel). In addition to the three-fold increase in number of output signals, the complexity of monitoring functions is greatly increased. There are also other surround sound format 'widths', ranging from 3- to 8-wide, that add further requirements for mixing and monitoring surround sound.

4.10.3.1 Input channel mixing

Surround sound production requires the mixer's input channel to route and pan signal contribution across the 'wide' output bus. However, in many situations surround sound is being mixed on stereo mixers and it is common to use multiple Group output buses to create the 'wide' output. Surround sound panning can only be achieved by using two-dimensional controllers, such as a set of Pan-pots (one handling left to right panning and a second handling front to back panning) or a joystick style controller. On a traditional stereo sound mixer the only possibility to Pan in surround is to double-up signals into several modules and route each module to different sections of the surround sound bus.

4.10.3.2 Surround sound monitoring

Surround sound monitoring has the potential to become complex; in addition to the 'wide' output bus, other functions are commonly required, including 'fold down', bass channel management, Stem monitoring and PEC/Direct switching.

Fold down refers to the ability to monitor the surround sound outputs in smaller 'widths' – for example, the operator may want to listen to a 5.1 output on stereo (or mono) loudspeaker(s). This is required to check compatibility with typical consumer replay systems – for example, checking how a DVD will sound on a television with only stereo speakers.

Bass management is also required to check compatibility with consumer replay systems. The monitor section may allow the sub-bass channel to be routed into other outputs (typically the Centre and/or Left and Right outputs), as this routing technique is used on some home systems.

Stem monitoring refers to the film soundtrack mixing technique which separates audio into different Music, Effects and Dialogue groups, often mixed by several sound operators. These different types of groups are known as 'Stems' and the monitoring section of film mixers allows for monitoring, muting and soloing of each individual Stem.

PEC/Direct switching is the same function as Input/Repro or EE/Repro monitor switching on audio and video recorders, where the mixer's monitor section can select to monitor either the external recording device's input (i.e. PEC, Input or EE) or output (Direct or Repro).

4.10.4 Digital Mixers

The audio processing capabilities of digital sound mixers are often identical to analogue models, sharing the same basic structure of input channels, output buses, output channels and monitoring. However, there are also differences – for example, control surface design, input/output management, additional audio processing functions and automation capabilities.

4.10.4.1 Control surface – assignability

Almost every digital mixer uses some level of control surface assignability, where a set of knobs and switches are assigned to control multiple audio processing functions. Assignability breaks the tradition of 'one knob per function' control layouts, and can cause concerns for sound operators who are used to instantly accessing any processing control. However, an assignable mixer can provide significant benefits to operators working even in the most critical, real-time environments.

Figure 4.10.5 shows two of the most common assignability functions:

1. **Fader paging** allows a set of faders to be allocated to different banks of channels. For example:

- Page 1 – assigns the set of faders to control input channels 1–24.
- Page 2 – assigns the set of faders to control input channels 25–48, etc.

An extension of fader paging is when the set of faders can be paged to other channel gain controls (for example, using the faders to quickly set up a headphone balance on an Aux bus). Fader paging often requires a 'motorised fader' system, where selecting a new Page automatically moves faders to the correct position.

2. **'Master' control section and Channel Access switches** – this arrangement is typically used on the functions that are repeated on each channel, such as the EQ. Rather than having multiple control sections for each fader, a 'master section' is provided and this is assigned to any channels by pressing an Access (or Attention) switch, typically positioned above the fader.

Assignability means that the status of switches and rotary controls on the 'master section' has to electronically displayed and updated by the mixer's CPU. There are several typical methods to show status information, including LED 'fuel gauges' around rotary controls, numerical parameter display or a separate computer graphic display.

To navigate an assignable sound mixer, it should be possible to identify the 'rules of assignability' by locating the Fader Paging and Channel Access switches, and then relating the overall control panel functionality to the basic mixer structure.

4.10.4.2 Input/Output routers

Many digital mixers have integrated input/output (I/O) signal routers that allow the signals physically connected to the system to be flexibly routed into any input channels, insert points, outputs and monitor I/O. The potential of these routers is to replace an external patch bay, so simplifying the installation and opening up the possibility of greater efficiency by including patching cross-points in mixer snapshot memories. In addition, many digital mixers offer a range of I/O cards supporting different audio formats, and these expand the router's functionality to include signal format conversion.

However, care must be taken with integrated input/output routers to ensure that signals are patched correctly, as the graphical displays normally used to control the router are more difficult to interrogate than the 'what you see is what you have' status of a patch bay. It is useful to fully check the functionality of input/output routing capabilities to avoid problems of missing signals or incorrect patching.

4.10.4.3 On-board FX processors

Some digital mixers offer on-board FX processes, such as reverb. In many cases internal FX buses are used to route signals to the processing, with processing outputs being routed to back into other input channels. On-board FX processing allows greater efficiency as less external signal patch is required and processor settings are stored in project snapshots.

4.10.4.4 Audio processing – copy functions and libraries

Most digital mixers provide the operator with tools for copying audio processing data between channels, or linking channels to handle 'wide' sources – for example, stereo inputs. Libraries allow commonly used process set-ups (for example, an EQ

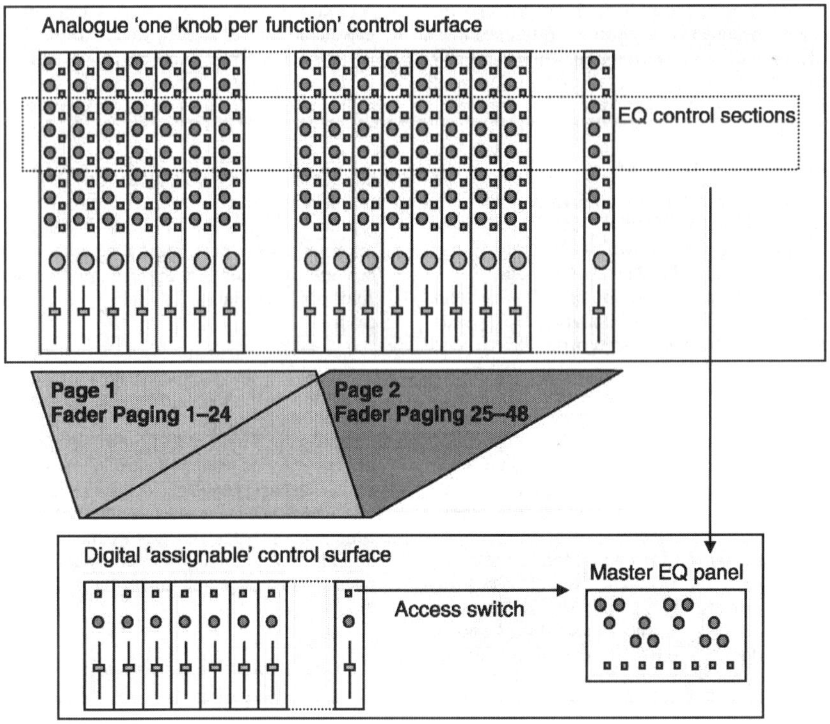

Figure 4.10.5 Typical control surface assignability functions.

curve for a particular voice-over artist) to be stored and recalled, so saving time and the risk of errors.

4.10.4.5 Snapshot and dynamic automation overview

Many digital mixers have extensive automation functionality. The following provides an overview of the most common elements of mixer automation.

Snapshot automation allows the operator to save and recall complete settings of the mixer processing. It is common for the digital sound mixer to provide management of the snapshots – for example, naming and editing. Many mixers will provide a mask to enable/disable the snapshot recall on selected channels and functions in the mixer.

Snapshots can be used to improve efficiency by decreasing the 'turnaround time' between projects or by automating the mixer processing during a live performance. Some digital sound mixers designed for broadcast extend snapshot management further by 'password' protecting individual operators' set-ups.

Dynamic automation refers to the ability to record, edit and replay audio processing 'moves' (for example, a fader movement) that are recorded in reference to a project's timecode. Dynamic automation is used in mixing projects where multiple channels of audio processing need to be simultaneously adjusted. Assignable mixers rely on using dynamic automation for complex mixing projects, as the operator cannot adjust parameters on 'hidden' controls.

Rodney Duggan

4.11 Surround Sound

4.11.1 The Origins and Development of Surround Sound

Multi-channel surround sound originated in the cinema industry, with the development in Hollywood of the 'Fantasound' system in the late 1930s for the film release of Walt Disney's *Fantasia* in 1941. A separate, synchronised sound film carried three channels of audio with a tone track that enabled the automated panning of sound around the auditorium to between three and eight loudspeakers depending on which system version was used.

A dedicated array of equipment was needed to be installed before each theatre could show the film, and development of the equipment showed considerable technical changes between each of the many versions. The MkX set-up, for example, used in a Los Angeles cinema, had left centre and right screen speakers with two speakers at each rear corner of the cinema, in a similar fashion to the current 5.1 channel digital cinema and home theatre systems. The signals to the rear speakers in this particular version were switched by notches on the edge of the film which operated relays.

After this great step forward in surround sound, no further commercial multi-channel development was made until the introduction of magnetic stripes on the film stock, which enabled some films in the 1950s and 1960s to carry up to four- and six-channel audio tracks on 35 mm and 70 mm stock respectively. This process was expensive to manufacture because of the magnetic coating process and also the extra recording stage. It was prone to wear (of both the magnetic stripe and the expensive playback heads) and other problems such as alignment and noise, so it was not until Dolby applied their A-type noise reduction to the existing dual mono optical stripes on 35 mm film in 1974 (*Callan*) and then encoded three- and, later, four-channel sound onto them in 1975 (*Lisztomania*) that surround sound emerged as a viable commercial success.

In the late 1960s, Ambisonics were developed in Britain by Michael Gerzon, Professor Peter Fellgett and others. Using a special phase coherent Soundfield microphone, a flexible playback system was produced where the number and placing of loudspeakers could be chosen to suit the requirements of the listener, and then this layout could be correspondingly selected in the playback decoder. Many music vinyl albums and CDs have been released with this format, and it is still in use today.

In the early 1970s, the recording industry was starting to release quadraphonic encoded vinyl albums and tape cartridges. These quadraphonic surround media could carry the four signals (Left, Right, Left rear, Right rear) in one of several competing encoding formats. The CBS SQ system carried the extra information on a supersonic carrier signal, whereas the Sansui QS system employed a matrix encoding technique which would later be adapted as the basis for encoding analogue cinema surround tracks.

Home enthusiasts in the 1970s would connect two speakers across the left and right positive output leads of a two-channel stereo system to give a difference signal, an invention attributed to David Hafler, and this was used to produce the rear or ambience signal, the first domestic instance of matrix decoding.

By employing the matrix encoding process (using the relative phase and amplitude differences), Dolby discovered that four audio tracks (Left, Centre, Right, Surround) could be reduced to two, Left total and Right total (Lt and Rt) signals, which when decoded would then reconstitute the four original signals fairly accurately. The surround signal was band limited (100 Hz–7 kHz), and represented the original use of the surround (or 'effects') channel for occasional ambient effects. The cinema industry traditionally strives to maintain the dominant part of the soundfield at the front of the auditorium in order not to distract the audience away from the screen action. The first surround channels were only occasionally used for subtle effects and the early cinema surround loudspeakers were quite small compared to the screen loudspeakers, hence they could not reproduce realistic levels of low frequencies, and so the LF roll-off was introduced at 100 Hz.

To enable domestic listening of taped releases of surround sound films, Dolby surround was introduced in 1982 for 'home theatre' systems, and superseded in 1987 by a more refined version called Pro-Logic, which itself was superseded more recently in 2000 by Dolby Pro-Logic II and in 2001 by DTS Neo 6.

The BBC experimented with stereo radio broadcasting as early as 1958, and introduced the first UK stereo television broadcasts using their Nicam digital audio coding scheme, with experimental broadcasts from 1988, developing into a full programme service from August 1991.

The cinema industry turned to digital audio technology for its next development of surround sound, with the launch of Dolby SRD in 1992 with *Batman Returns* (5.1 channel), followed soon after by DTS in 1993 with *Jurassic Park* (5.1 channel) and Sony's SDDS with *Last Action Hero* (7.1 channel) in 1994.

In the mid 1990s, the Digital Versatile Disc, or 'DVD', was launched which could carry 5.1 discrete channels of digital surround sound encoded in Dolby AC-3 (Audio Coding 3) or MPEG-2 compression formats in order to carry the digital surround soundtracks from film releases. In 1998, DTS also launched its own encoding format for DVD soundtracks, called Coherent Acoustics.

With the cinematic release of *Star Wars – The Phantom Menace* in 1999, and after a collaboration between THX and Dolby, a centre surround channel was added, giving a 6.1 channel format, and this format is known as either Dolby Digital EX or THX Surround EX (licenced separately by the two co-inventors). DTS also introduced a similar system known as DTS-ES.

To maintain compatibility with existing 5.1 systems, the extra channel is matrix encoded into the two existing surround channels, which then gives a phantom image of the centre channel. It can also be subsequently decoded in the cinema with a special EX (or ES) decoder to give a discrete centre channel signal. This feature has now been carried through to the DVD medium with the release of such films on DVD, and DTS has gone a stage further and developed 'ES Discrete', which carries the centre surround channel as discrete information in the DVD bit stream, which when subtracted from the matrix-encoded surround channels, is fed to the Centre rear speaker.

Currently, 5.1 channel digital surround sound is transmitted with digital television broadcasts from television stations around the world, for a wide variety of programmes. Future developments for surround sound may head towards the development of a 10.2 channel format.

4.11.2 Track Layout and Nomenclature

It is common to describe a surround system using the number of front channels, followed by the number of surround channels, separated by a forward slash, and then the LFE channels separated by a dot, e.g. 2/0 represents a regular two-channel stereo system, 3/1 represents a matrix-coded system with L, C, R, S channels, and 3/2.1 represents conventional 5.1 surround sytems with L, C, R, Ls, Rs, LFE channels.

The SMPTE and EBU organisations have recommended a track layout for delivery of multi-channel audio material which is shown in Figure 4.11.1.

4.11.3 Approaches to Multi-Channel Production

Multi-channel mixing for dramatic productions has been established in the film industry for many years, and the associated techniques and practices have now been developed to a mature stage. Generally speaking, dialogue is fixed in the centre channel, music is added as two-channel stereo on the left and right channels, foley and effects are panned to any channel as the action dictates, and low-frequency effects are sent to the LFE channel to add dramatic effect to crashes, explosions, etc.

Multi-channel mixing for music, however, is still in its early stages of development, with many different approaches currently being developed and refined. Music mixes for surround can vary in approach from giving the listener a seat in between the musicians or even on the stage with them, to more distant effects creating spatial ambient effects with the surround channels.

Whereas the film industry needs the centre channel in order for the majority of the audience to hear the dialogue from the centre of the cinema screen, many music mixers have found that putting the central vocalist in a music mix through a centre speaker gives it a prominence which is distracting compared to the traditional phantom centre channel given by two-channel stereo, and many music mixes, both classical and pop, currently omit the centre channel.

The film industry uses the Low-Frequency Effects channel to boost low frequencies when needed – for example, during dramatic crashes and explosions. No such sudden effects usually occur in music, however, and engineers go to great lengths to get the overall frequency range carefully balanced.

There appears to be no need for any extra low-frequency information on music mixes, and although experiments have been made on various mixes, no regular use of the LFE channel has been developed. Although a music mix might not use the LFE channel, the listener might well utilise bass management options to extend the low-frequency range of a domestic listening set-up by using the loudspeaker connected to the LFE output, whilst not actually using any signal from the LFE channel.

1	2	3	4	5	6	7	8
LEFT	RIGHT	CENTRE	LFE	Ls	Rs	(Lt)	(Rt)

Figure 4.11.1 SMPTE and EBU recommended track layout.

4.11.4 Bass Management

Consumer surround systems rarely have five full-range loudspeakers and a subwoofer capable of cinematic (or better) low-frequency reproduction. Some systems have small surround speakers, large front speakers and no subwoofer, and a popular type of loudspeaker system currently available from many manufacturers has five extremely compact satellite speakers and a small woofer unit with the amplifiers and decoder built into it.

In order for each loudspeaker system to reproduce as much of the original channel frequencies as possible, playback equipment manufacturers have applied a method of filtering and combining, called 'bass management', available as a series of options within the decoder/amplifier set-up menu.

As can be seen from Figure 4.11.2, the LFE signal can be redirected to the front speakers (assuming they are full-range units) if there is no dedicated loudspeaker connected for that channel, or the low-frequency content of each channel can be redirected to the LFE channel and hence the subwoofer. These frequencies are non-directional, and appear to extend the apparent frequency range of small loudspeakers. Commercial stand-alone bass management units which operate at line level are available for professional installations which may not use the consumer equipment with this function.

For 6.1 monitoring, a centre rear channel is added. However, as the hearing process can often confuse rear centre images with front centre images, a recommendation of two mono rear centre speakers offset from the centre line can give a more stable and predictable image at the rear, and is shown in dotted outlines in Figure 4.11.3.

Some playback equipment is specified as providing 7.1 channels, and in these instances seven full-range amplifiers are incorporated in the equipment, using four for the rear channels, which are switched automatically whenever two or three rear channels are selected for the 5.1 or 6.1 signal requirements, as shown in Figure 4.11.5.

4.11.5.1 Monitor alignment

Cinematic film mixing employs a system whereby a reference sound level from the loudspeakers on the mixing stage corresponds to a given reference recording level, and cinemas are then calibrated so that this recorded reference level then reproduces the same sound pressure level, which ensures that the film sound is heard at the same level as when it was mixed.

In recent years, however, this practice has been altered greatly by the subjectively excessive levels of digital audio soundtracks, and as a consequence projectionists around the

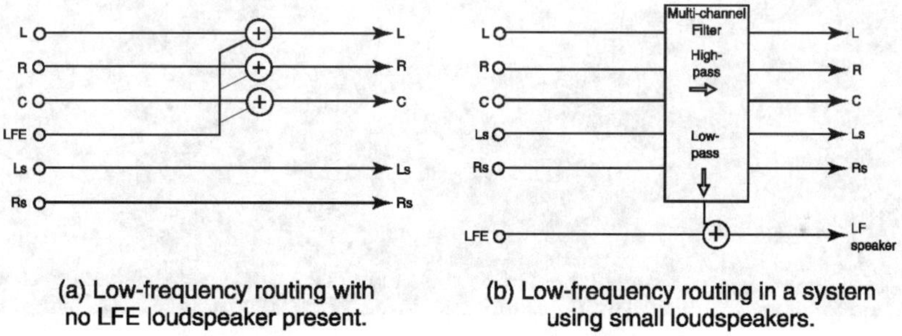

Figure 4.11.2 Two examples of bass management.

4.11.5 Monitoring

The ITU-recommended monitor loudspeaker layout for 5.1 channels is as shown in Figure 4.11.3.

The surrounds may be located between the 100 and 120 degree bearings, with the front loudspeakers located 30 degrees either side of the centre loudspeaker. The five loudspeakers should be equidistant from the ideal listening spot. The LFE loudspeaker should be situated in the best location for a smooth room response and may be sited close to a corner or room boundary to improve efficiency by limiting the solid angle that it operates into.

Full-range surrounds that match the front speakers are considered important, as music and full-range effects are nowadays being routed to the surround channels. Just as in two-channel stereo, good phase characteristics between all of the full-range speakers will ensure that stable phantom images are heard between each speaker and will ensure that effective surround sound is heard rather than isolated sounds coming from the individual loudspeakers. A five-channel surround loudspeaker broadcast monitoring system is shown in Figure 4.11.4.

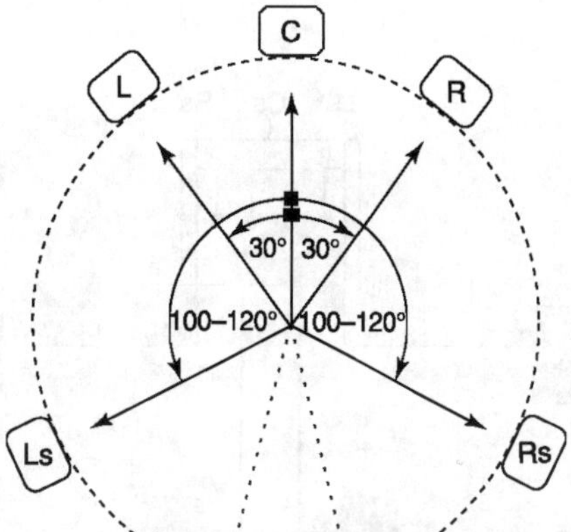

Figure 4.11.3 Monitor layout to ITU-R BS.775.1 recommendation.

Figure 4.11.4 A surround monitoring system installed at Yorin Television, The Netherlands (photo courtesy of PMC Ltd).

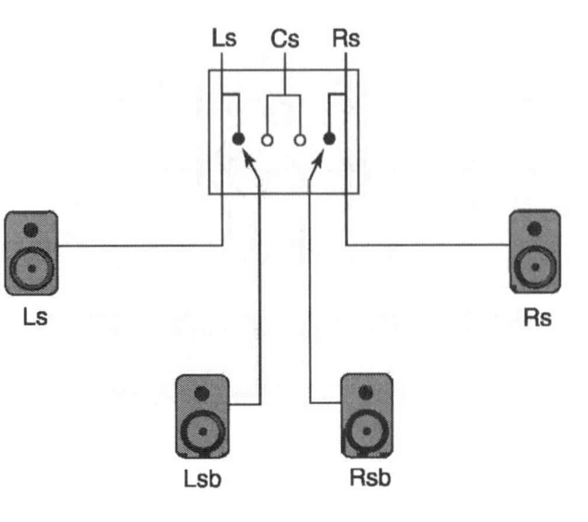

Figure 4.11.5 Rear monitor switching for 7.1 channel equipment.

world currently play back films at levels (as much as -14 dB) significantly below that intended.

The optimum sound level for dialogue reproduction in cinemas was decided, many years ago, to be 85 dBC spl, and it is this level that should normally correspond to the reference recording level on the master tape, DVD or CD. This sound level, however, is generally found to be too high for domestic listening, and figures of 79 and 80 dBC have been arrived at by several manufacturers to be a good compromise. This particular level is not as important with domestic playback or broadcast reception, as the listener has a volume control to adjust levels to their own needs.

Pink noise is used to align the levels of individual channels; accurate reference-encoded alignment CDs are available from all the various format suppliers and should be stocked and used both for acoustic calibration and to check the various internal alignment generators incorporated into equipment. Domestic/consumer equipment may have poorly aligned signal generators.

Reference recording level of -20 dBFS pink noise should give 79 dBC from each of the five full-bandwidth speaker channels for correct alignment. Band-limited pink noise can also be used which avoids inaccuracies in readings due to non-flat loudspeaker/room response.

N.B. Cinema surrounds are aligned to give 3 dB lower sound level than the front speakers as a legacy from the mono surround analogue system, and so cinematic master tapes will exhibit 3 dB higher surround track levels. It is often found that a 3 dB cut in level is needed when transferring cinema soundtracks for DVD or broadcast production, and encoder manufacturers fit just such a capability into their equipment. There are, however, some DVD audio tracks which have missed this point and have surround tracks which are 3 dB higher than intended. It may be prudent to watch out for this aspect when processing multi-channel material.

The LFE channel can only be aligned accurately by using a spectrum analyser, as a broadband SPL meter will not read the correct sound level on such bandwidth-limited noise (20–120 Hz). LFE 'pink' noise recorded at reference level should read 10 dB higher than each of the full-range channels, and this can be seen quite easily on a spectrum analyser display that has a dB amplitude scale. The reason for this 10 dB difference is so that the LFE channel has 10 dB more headroom for effects than the other channels using the same recording medium. Some reference setting tracks have the LFE level reduced to 10 dB below reference, so that the analyser display should then show equal sound pressure level (spl) readings with the main channels when correct alignment is made. This recorded level should be clearly marked on the reference information booklet.

Although the spl meter is not recommended for setting the LFE channel, the spl level can be subsequently read on a correctly aligned system and this empirical level used for day-to-day alignment purposes. A common figure is often around 4 dB higher than the main channels, so that when the centre channel gives 79 dBC spl, the subwoofer then gives 83 dBC spl for the same reference level pink noise. This empirical figure should be determined for each installation as it is unpredictable.

4.11.6 Analogue Surround Sound

4.11.6.1 Encoding

Film soundtracks are encoded with a matrix encoder. Four channels (L, C, R, S) are combined into two (Lt, Rt) channels for film prints and also broadcast and domestic release formats, as shown in the simplified block diagram of Figure 4.11.6.

The L and R channels are encoded directly into the Lt and Rt channels. The C channel is summed into the LT and RT channels with a 3 dB attenuation to maintain constant power. The S channel is summed into the Lt and RT channels also, but is first processed as follows:

- Frequency is band limited (100 Hz–7 kHz).
- Dolby B-type noise reduction is applied (modified to 5 dB instead of the regular 10 dB reduction).
- A delay is added to the surround channel.
- Opposite 90 degree phase shifts are applied to each output signal before summing into the Lt and Rt channels.

The HF roll-off and the noise reduction minimise crosstalk from the centre channel to the surround channel due to azimuth misalignment, which causes the problem to worsen with increasing frequency. Any audible dialogue (centre) leakage from the surround channel would be considered distracting by the listener.

The delay added to the surround channel ensures that if any crosstalk allows front signals through to the surround channel, then the listener will perceive the sound as coming from the front because of the Haas precedence effect. If a signal arrives at the listener from one direction before the same signal from a different direction, the listener perceives it to be only from the first direction.

Left and right separation is maintained because there is no interaction in the encoder, and centre to surround separation is maintained by the phase shift introduced into the surround signals, so that when Lt and Rt are summed to give the centre channel, the surround signals cancel each other out, being 180 degrees apart, i.e. out of phase. Similarly, when the surround channel is decoded by subtracting the Lt and Rt signals, the centre information is cancelled because of the two identical in-phase C signals contained within Lt and Rt.

The original cinema matrix encoding was designed to allow compatibility of soundtracks with mono and stereo cinema systems, and the same compatibility also maintains an acceptable sound balance when replayed with mono and stereo television/radio receivers. These matrix-encoded tracks are often broadcast on both analogue and digital stereo TV channels and can pass through the broadcast chain to be successfully decoded using matrix decoders, into the original L, C, R, S channels.

4.11.6.2 Decoding

The first simple surround decoder consisted of connecting loudspeakers across the hot feeds to the left and right front loudspeakers. This produces a difference signal, being the difference between left and right channel information, emphasising the out-of-phase content. However, a drawback is that

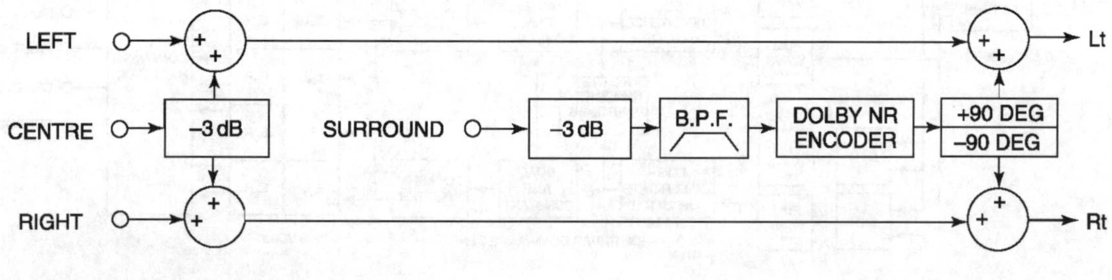

Figure 4.11.6 Surround sound matrix encoder.

the surround channel then reproduces any signal that is not correlated in both channels, e.g. left only or right only signals are decoded into the surround channel.

The next development was a simple passive decoder, shown in Figure 4.11.7, where the Lt and Rt signals pass directly through to the left and right loudspeakers, and are also processed to give the surround channel by subtraction circuitry.

Although an improvement, this still suffers from the characteristic that any difference signal will appear in the surround channel. The centre channel is produced as a phantom image from the left and right loudspeakers.

A significant improvement over this is the active matrix decoder shown in Figure 4.11.8, involving a considerable deal of signal processing to derive the decoding control signals. To produce these control signals, the two input signals are first filtered to remove frequencies which do not give directional cues (e.g. strong low frequencies and high frequencies which may contain phase errors). Then the two orthogonal signals (L, R and C, S) have their amplitudes derived and rectified into logarithmic-scaled voltages.

At this stage, these two bipolar voltages represent signal dominance along each of the two directions (L, R and C, S), having a value of zero for central images (e.g. positive volts for left and negative volts for right dominance). After conversion to four individual positive voltages corresponding to the four directions (L, C, R, S), these control signals are then applied to a matrix of VCAs, which determine the amplitude and combination of the Lt and Rt input signals that appear at the various decoded outputs.

The dominant sounds detected by the control signal processing are then reproduced with the correct position and sound level anywhere in the 360 degree soundfield, while the other non-dominant sounds are redistributed equally around the other channels, losing their own directionality but maintaining the original loudness and power levels. The dominant sound audibly masks the non-directionality of the others. If the dominance is found to be of a low level or relatively weak, then a slow control signal speed is automatically selected to avoid sudden changes in the soundfield which may be audibly distracting, and if the dominance signal is strong, a faster control signal speed is selected to locate the strong directional signal quickly.

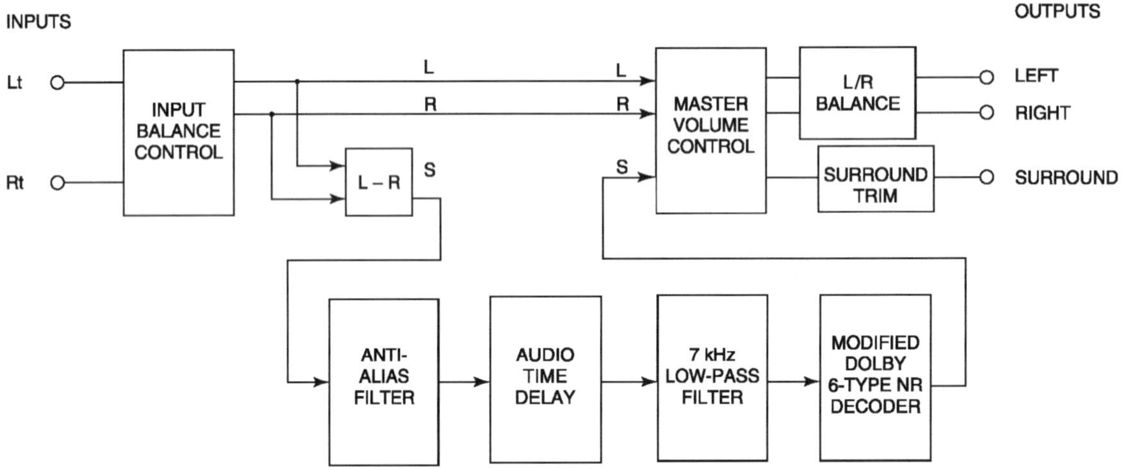

Figure 4.11.7 A simple surround sound matrix decoder.

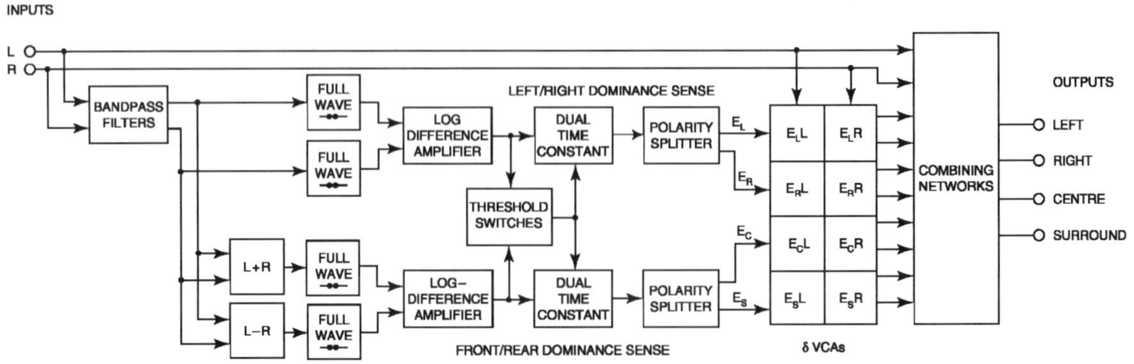

Figure 4.11.8 Pro-Logic matrix decoder.

The latest versions of these active matrix decoders utilise feedback servo methods to accurately match signal levels and hence improve any required signal cancellations; an example is shown in Figure 4.11.9. Up to seven channels of audio can be decoded from a two-channel input, with various listener-selected modes available to steer the soundfield imaging to suit the listening environment (e.g. for automotive applications). As well as decoding film soundtracks, this new generation of active matrix decoders[1–3] can derive acceptable multi-channel signals from two-channel non-encoded music tracks. For this mode, the delay in the rear channels is disabled, and the limited frequency bandwidth is increased.

4.11.6.3 A summary of current analogue surround formats

1. Dolby Pro-Logic – A four-channel matrix encoding scheme that decodes L, C, R, S signals from a two-channel motion picture matrix-encoded signal by detecting phase/amplitude differences.
2. Dolby Pro-Logic II – A refinement of Pro-Logic, which can produce up to five channels from stereo or matrix-encoded material, by detecting phase/amplitude differences.
3. DTS Neo 6 – A similar system to Pro-Logic II, producing up to six channels from stereo or matrix-encoded material.
4. Ambisonics – An encoding/decoding system producing a three-dimensional soundfield of up to 12 channels. Up to four inter-related tracks are recorded using either the special Soundfield microphone or an Ambisonics encoder, front minus back, left minus right, up minus down and a mono signal. The four-channel B-format can be encoded into UHJ format to give downwards compatibility with two-channel stereo decoded/undecoded playback systems.
5. Lexicon L7 – A synthesised coding system producing up to seven channels of matrix-encoded/decoded audio via a two-channel medium. This is aimed specifically at the automotive industry and in-car entertainment.

4.11.7 Digital Surround Sound

4.11.7.1 Encoding

The various digital encoding schemes available to broadcasters can be classified into two groups, which are contribution and delivery schemes.

4.11.7.1.1 Contribution schemes

These digital audio compression formats have been developed to fulfil specific needs within the broadcast industry. In order to convey multi-channel digital audio of up to six channels together with two stereo channels on existing digital audio equipment designed to convey only two or four AES/EBU channels of audio, some form of data compression is needed. Several characteristics are necessary:

- The format must compress the audio data to enable the use of existing equipment and installations.
- The format must be able to cope with several encode/decode operations in the production chain without causing any discernible audio degradation.
- It must be able to carry metadata.

A compression ratio of 4:1 has been achieved in both schemes, enabling eight channels to be carried on one AES/EBU data stream; a typical example of an encoding unit is shown in Figure 4.11.10.

These formats are intended for use throughout the production chain, with subsequent transcoding into a suitable delivery compression scheme at the end, ready for delivery to the consumer whether by broadcast or any other media.

4.11.7.1.2 Delivery schemes

Three of the many audio compression formats are of particular relevance to broadcasters in that they are currently suitable for

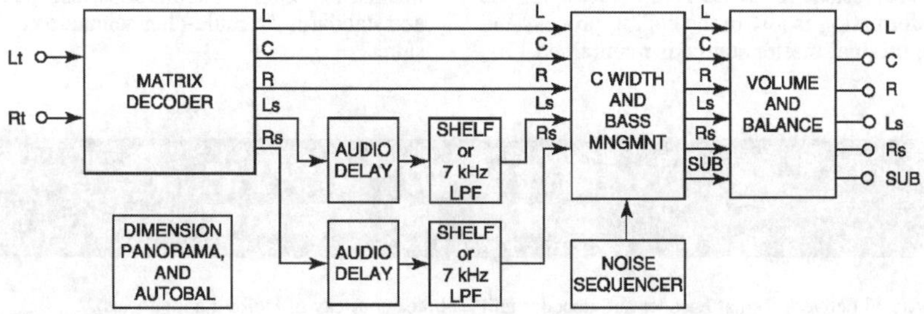

Figure 4.11.9 Dolby Pro-Logic II matrix decoder.

Figure 4.11.10 A contribution format audio encoder (photo courtesy of Leitch Ltd).

transmission and home decoding (Dolby AC-3, DTS, MPEG-2). They achieve similar results, but operate slightly differently.

Each coding scheme uses psychoacoustic analysis of the audio signals, after passing the signals through banks of filters in order to separate the signal into individual frequency bands. This analysis exploits perceptually irrelevant components of the signals which through research have been shown to be inaudible – for instance, the masking of similar sounds by a stronger sound, or the insensitivity of hearing at low levels and particular frequencies – and so these encoders ignore information which it is believed cannot be heard, and thus reduce the amount of data needed to represent the audio signal.

In signals where the directional information cannot be heard (usually in the 10–20 kHz region), joint coding or coupling is employed where the same signal is then shared by all channels, which will reduce the total necessary information.

The full details of each coding scheme and its implementation are beyond the scope of this chapter, but further in-depth reading may be found on the relevant websites.[4–6]

These coding schemes are known as lossy, because information in the audio signals is lost in the encoding stage which cannot be subsequently recovered. The opposite of this is known as lossless encoding, where a reduction in data is managed by the encoding scheme in a similar fashion to computer file storage (e.g. Winzip, PKZip, etc.) and all the original audio data can be recovered when decoded. Because of its nature, lossless encoding is usually associated with data reduction ratios of up to 2:1 depending on the signal content.

4.11.7.2 A summary of digital audio compression schemes

1. Dolby Digital – 5.1 channels of audio encoded with Dolby AC-3 compression method for film, DVD and TV/radio broadcast applications. Dolby Digital EX has a centre rear channel matrixed into the two rear surrounds for decoding as a sixth channel (also known as 6.1).
2. THX Surround EX – The consumer decoding scheme used for Dolby EX encoded 6.1 channel material.
3. MLP – Meridian Lossless Packing, a lossless coding scheme used as the mandatory format for DVD-Audio discs. As the name implies, no information is lost in the coding process and so the quality of the original master source is maintained.
4. MPEG-2 BC and AAC – BC stands for Backwards Compatible (with MPEG-1); this can carry 5.1 channels and includes matrixed channels to give left right stereo with MPEG-1 decoders. AAC stands for Advanced Audio Coding and can also carry 5.1 channels; whilst not being compatible with MPEG-1 decoders, it is of higher quality than the BC version.
5. DTS – 5.1 channels of audio encoded with DTS Coherent Acoustics method for DVD and TV/radio broadcast applications. This is not the same as the coding scheme used for cinematic release. DTS-ES has a centre rear channel matrixed into the two surround channels for decoding as a sixth channel, and DTS-ES discrete also has the rear centre channel included as a discrete channel in the digital bit stream (also known as 6.1 discrete). DTS 96/24 can include the extra information contained in a 96 kHz 24-bit 5.1 channel recording, and requires a DTS 96/24 decoder.
6. SACD – A multi-channel lossless encoding scheme used by Sony and Philips to encode surround sound onto special CDs.
7. SDDS – A Sony cinematic compression scheme using an adapted form of the Minidisc ATRAC compression scheme to encode up to eight digital audio channels on to 35 mm film stock.

A brief comparison of the various audio coding schemes is shown in Figure 4.11.12.

4.11.8 Metadata

Multi-channel coding schemes can incorporate extra data into their bit streams in order to carry information to subsequent equipment on how to process the signal. This is known as metadata.

4.11.9 Multi-Channel Production Standards

THX is a standards and certification company. Certification is available for equipment, mastering suites and broadcasters to ensure that high standards in multi-channel audio production are maintained. Dolby Laboratories also publish recommendations and standards for multi-channel audio equipment and production suites.

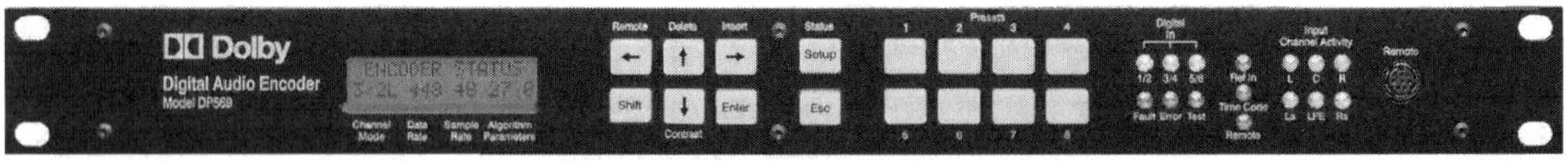

Figure 4.11.11 Professional delivery format encoder and decoder units (photos courtesy of Dolby Laboratories).

	Dolby E	Leitch Dmd	AC-3	MPEG-2 AAC	DTS	SDDS	MLP
Fs max (kHz)	48	48	48	96	96	96	192
Bit depth	24	24	24	24	24	24	24
Comp. ratio	4 : 1	4 : 1	12–15 : 1	10–12 : 1	4–8 : 1	5 : 1	1–2 : 1

Figure 4.11.12 A comparison of coding schemes.

References

1. Dressler, R. Dolby surround Pro-Logic II decoder principles of operation, http://www.dolby.co.uk/tech/1.wh.0007.PLIIops.html
2. http://www.dtsonline.com/neo6.pdf
3. http://www.lexicon.com
4. Todd, Davidson, Davis, Fielder, Link and Vernon, AC-3: Flexible perceptual coding for audio transmission and storage (February 1994), http://www.dolby.co.uk/tech/ac3flex.html
5. Thom, Purnhagen, Pfeiffer et al. MPEG audio FAQ, http://www.tnt.uni-hannover.de/project/mpeg/audio/faq/
6. http://www.dtsonline.com/whitepaper.pdf

Bibliography

BBC Internal DTS demo disk, *The Planets*.

Denon AVR-3803 AV receiver.

Elen, R. *Surround Professional* magazine (October 2002); also http://www.ambisonic.net/bassmgt1.html

Garity, Wm. E. and Hawkins, J.N.A. *Journal of the SMPE* (August 1941).

ITU-R Recommendation 775 (Rev. 1) Doc. 10/63, Multichannel Stereophonic Sound System With and Without Accompanying Picture (November 1993).

Nakahara, M. Sona Corporation/Yamaha, *Multichannel Monitoring Tutorial Booklet* (June 2002).

Sony HTK25 system.

The Eagles, *Hell Freezes Over*, DTS CD1006.

Internet Resources

Allen, I. Are movies too loud? (March 1997), http://www.dolby.co.uk/tech/toolouds.pdf

Dressler, R. Dolby surround Pro-Logic decoder principles of operation, http://www.dolby.co.uk/tech/whtppr.html

Elen, R. Ambisonics for the new millenium, http://www.ambisonic.net/gformat.html

Elliott, R. Simple surround sound decoder, http://www.sound.westhost.com/project18.htm Fritz, J. Outlaw audio bass management processor, http://www.soundstage.com/revequip/outlawaudio_icbm.htm.

Genelec. *Installation and Operation Manual HTS3 & HTS4*, http://www.genelec.com/ht/pdf/OMhts34.pdf

Holman, T. TMH bass manager, http://www.tmhlabs.com/products/products.html

Holman, T. http://www.tmhlabs.com/products/10_2.html

Hull, J. Surround sound past present and future, http://www.dolby.co.uk/ht/4340.l.br.9904.surhist.pdf

University of York, http://www.york.ac.uk/inst/mustech/3d_audio/ambison.htm

Concepts and development of multi-channel sound, http://www.mtsu.edu/~ dsmitche/rim456/Quad/Quad_Formats.html

Dolby installation guidelines for broadcast and professional products, http://www.dolby.com/pro/multiaud/wb.td.0103.install.pdf

Evolution of digital film sound, http://www.dolby.co.uk/tech/mp.br.0102.EvolutionOfSound.html

How THX audio standards optimise 5.1 digital sound, http://www.thx.com/mod/techLib/how.html

The EBU's multi-channel audio activities, http://www.ebu.ch/trev_292-multichannel_audio.pdf

http://www.dolby.com/ht/co_br_0110_ListenersGuideEx.pdf

http://www.dolby.co.uk/tech/m.br.9903.epaper.html

http://www.dolby.co.uk/tech/m.br.9904.metadata.html

http://www.dtsonline.com/dts-es2.pdf

http://www.dtsonline.com/history8.pdf

http://www.dtsonline.com/9624.pdf

http://www.interprod5.imgusa.com/son-637/technology.asp

http://www.leitch.com

http://www.meridian-audio.com

http://www.mmproductions.co.uk/faqgen3.html

http://www.sdds.com

http:/www.thx.com

http://www.vintagebroadcasting.org.uk

Paul Kafno

4.12 Working with HDTV Systems

This chapter outlines a producer's view of working with HD. It aims to give engineers an insight into the production process and help producers and directors working for the first time in the high-definition medium. The author shares his experience of producing different genres of high-definition programmes – musical events, sports, drama and documentary – and identifies some of the most common pitfalls.

The old definition of a producer's responsibility was to create programmes that educate, entertain and inform. Few professionals would deny that is still the aim, though giving excitement, delight and securing a good audience share are also requirements in a competitive world. The starting point in whatever genre – drama, light entertainment, documentary, music, news – will always be a narrative of some kind, whether factual of fictitious. Production decisions, artistic, commercial and technical, will be made to serve the needs of that story. Those needs will condition the choice of writers, actors, musicians, designers, camera, sound and post-production teams. They will certainly be factors in determining the level of budget that has to be raised, and the channel to broadcast it. Last, but not least, they will determine the acquisition format. Today, a major choice is increasingly whether to record in standard or high definition.

A producer may be attracted towards using high definition for aesthetic reasons, but the real motive will almost certainly be because the subject of the programme is especially suited to HD treatment. That insight will need to be recognised by a broadcaster who has decided to commission – and pay for – the programme in that format. Having made the decision to take the high-definition route, it is the producer's job to pull together the many strands, creative, technical, legal and commercial, and make sure that 'everyone is making the same programme'.

4.12.1 Where is the Market?

At the time of writing, there are television broadcasters commissioning high-definition programmes in Japan, the USA and Australia, as well as advanced plans for the UK. Alternatively, the producer may be intending the work for exploitation in different media – as a big screen presentation at a commercial show, for digital cinema or electronic theatre. Perhaps the programme is of such special archive value that it is worth making an investment to 'future-proof' the work.

Japan is the country with the longest commitment to high-definition broadcasting. Engineering work progressing since the 1960s resulted in the world's first analogue high-definition service (by satellite). Today's service is fully digital. Japanese broadcasters have a major commitment to HD and have provided an important market for producers from all over the world. US broadcasters are required by the FCC to produce some hours of digital broadcasts each year. Some activity is now taking place in Australia. These broadcasters all require HD programming and are possible commissioners.

The initiative will, however, almost certainly begin with the producer's decision that the programme's subject deserves a quality approach. It is worth remembering at this stage that producers are rarely engineers, though they may have acquired detailed technical knowledge. Even if they possess expertise, their concentration has to be on content, since their job is primarily to deliver a story that works well in its own right. Too much immersion in technical details can be a diversion. Some producers feign complete technical ignorance, since they believe that gives them the strongest position from which to look objectively at the production process. The technical means of recording, editing and distributing are the tools to reach the audience, and are best operated and supervised by those with detailed knowledge. The concert pianist's task is to communicate emotion to the audience. Tuning the piano is left to the expert.

4.12.2 So, Why HD?

The producer will be first drawn towards high definition because it serves the story. There are several aspects to this.

Maybe the programme has particularly exciting visual content: a journey through an exotic and beautiful landscape, or glorious period settings. Perhaps recording in high definition will give the production a cachet that will attract the best talent, or give the production an enhanced image that will help get the green light. Whatever the reason, however, the producer of a high-definition show will be working with technology which is in many ways familiar but with new, and sometimes unexpected, requirements. Across the whole range of planning, acquisition, post-production, re-versioning and marketing there are new lessons to be learned.

However, there is one aspect of the decision that will give concern. Even in those countries with the capacity for high-definition broadcasting, there are at present relatively few viewers. So it is usual for an HD programme to be down-converted and also shown in standard definition to reach a wider audience and 'earn' its budget. Many of the more expensive programmes likely to be made in HD are actually co-funded, and it is likely that a number of markets will only take the programme in standard definition. At present it is likely that the majority of viewers will actually be watching in standard definition. There is an unfortunate conflict here. High definition is a widescreen format with an aspect ratio of 16:9, while most standard definition (outside the UK) is 4:3. So the producer needs to make difficult choices. How are the shots to be framed? In widescreen, which optimises the medium for HD? Or in the compromise central portion that will convert easily to standard definition with normal aspect ratio?

Again, if most of the audience will be watching in standard definition, how does that affect editing decision about those shots that look so wonderful in high definition, but may be rather more ordinary in standard definition? And what about colour? And graphics? The list of questions is, actually, rather daunting.

It is at this point that the role of the engineer becomes so critical. Intelligent advice, based on an understanding of what the producer needs, is essential to keep problems in perspective and find intelligent solutions. Riding two horses is never very comfortable, and in time, high definition will be the predominant format in all television broadcasting. But that time is not yet, and each production needs to be seen as an individual case with different solutions.

4.12.2.1 Why High Definition?

When television began it was rather rudely referred to as 'radio with pictures'. In those days the actual viewing area was very small, and striking though its impact was, those 405 monochrome lines were not visually overwhelming. The story tended to remain in the words. As screens became larger, colour introduced and most recently widescreen, television could claim more and more to be a visual medium. High definition, however, has provided a great leap forward.

Producers generally respond very positively when experiencing high definition for the first time, believing that the images not only have improved clarity and colour, but also an almost three-dimensional quality. There are sound engineering reasons behind this. Improved resolution and colorimetry have been the preserve of 35 mm film, but those images suffered from the softening effects of multiple printing processes, and the degradation compounded by motion and judder in the transfer gate, together with unavoidable bits of dust and damage. It is not really surprising that the naked clarity of high definition seems to offer a new and desirable experience.There is also another effect. Producers start to have a new awareness of visual elements as a dynamic part of the story and this can lead to a different emphasis in storytelling, or even suggest new subject matter. It is necessary to keep a clear head, remember the necessity for compatibility with standard definition and concentrate on the hard-headed reasons to choose high definition:

- The first, already referred to, is money. If a programme idea has characteristics that make it particularly suitable for high definition, then it is likely to be commissioned. It then has the potential to generate revenues in both the HD and standard definition markets.
- Down-converted high-definition images have better quality than those directly acquired in standard definition. So, if the programme is likely to be released in VHS or DVD format, the resulting images will be competitive with the best feature films.
- If the programme is a drama, documentary or sports programme intended to be seen on a big screen, the images will be of high quality and competitive with those generated in 35 mm.

4.12.3 Choosing a System

High definition has never been easy in system terms. In the 1980s the Japanese proposed a global 1125-line solution which was hotly contested by the Europeans, who produced a 1250-line system. Enmity between the two systems persisted until the Lillehammer Winter Olympics, when for the first time both systems were used collaboratively side by side. The FCC's decision to implement digital television in the United States concentrated minds and a number of systems emerged based around common factors. The reference to line numbers was replaced by a description of the number of pixels and a quasi standard emerged of 1920×1080 pixels presented in interlaced form. Whether a producer chooses 1080i, 720p or 480p or whatever will probably be conditioned by the house choices of the broadcaster who commissions the programme, and the American majors all have differing opinions. However, one simple fact remains true. The higher the standard at which a programme can be acquired the more options are possible later. All the systems today employ some degree of compression. The general truth is – the less compression at source the better.

Part of the problem is that high definition itself is a very flexible term. As increasing numbers of cinema directors choose electronic means of acquiring images, the quality goal-posts keep moving. George Lucas acquires images for his films at a much higher resolution than directors working with systems designed for television, because his films are designed for display on very large theatric screens. It is also worth bearing in mind that cinema has different traditions of motion portrayal, and that needs for film may be different from those for television. So a progressive scan system will generally be the choice for cinema while interlace may be the favourite for television, where audiences (particularly in the USA) are sensitive to flicker.

The subject of interlace is indeed a troubled one. Apart from the intrinsic four-field problem with which engineers have grappled from the earliest days of television, there is an almost religious belief among some cinema-orientated producers that 24 progressive frames stimulate the imagination in a way which 50 or 60 interlaced frames cannot. The belief is that the lack of motion detail makes a particular demand on the brain that engages the imagination. Most of this is scientifically unprovable, but deeply held beliefs like this cannot be easily changed, and need to be respected. An artist chooses the paint that works best. Deep artistic convictions generally have valid, if subconscious, reasoning behind them.

Producers generally need guidance on the subject of compression, since the various algorithms and transforms in MPEG-2 are often seen as arithmetic black arts. Technical supervisors have the responsibility of pointing out the consequences of unwise choices. A producer aiming to make a piece of *cinéma vérité*-inspired drama might love low-light levels and wobbling hand-held shots, but be appalled when he sees what transmission compression does to the images. Equally, a sports producer taking a panning shot of a footballer running against a background of several thousand cheering spectators might be unpleasantly surprised by the consequences. On the other hand, static landscapes and art objects can bear considerable compression with equanimity.

4.12.3.1 Lighting

The subject of lighting is equally important. In the early years of television, it was necessary to light the subject in order to make a transmittable image, and the tradition of lighting continued even with current affairs interviews until the introduction of more sensitive digital recording devices. The consequence of controlling illumination levels through artificial lighting was the creation of noise-free images that could be acquired without turning up gain, maintaining high quality through the transmission chain. The introduction of digital camcorders encouraged producers to record using 'available' light, saving time and cost. Very soon it became accepted by broadcasters that in many cases lighting was an unjustifiable luxury. The effect of this has been a double negative. Images acquired with the gain control turned up are noisy. Coding systems like MPEG-2 hit on the noise, treat it as part of the picture and amplify it. But equally detrimental has been the artistic loss. Lighting is a way of 'shaping' the composition of images, putting light in a controlled way into shadow areas, bringing life to the eyes of presenters and interviewees.

This question of the value of lighting is reawakened by high-definition television. Because what is HD if not high quality? With that comes all kinds of cinematic expectations, not only that the technical quality of images and sound is higher, but that the artistic quality of the pictures is better. So, the producer wanting to put a higher content value into images should consider lighting them. This is not an issue of preciosity, but a sensible way of ensuring that the visual content value of each shot is worthy of its higher definition status.

4.12.3.2 Compatibility with Standard Definition

Basic economics determine that expensive high-definition shows seen by relatively few people will at some stage need to be down-converted to standard definition for the mass market. The issues need to be made clear right at the beginning of a project, because they are fundamental to the framing of all shots, with consequences right down the line into editing and graphics. Unless a philosophy for confronting the problem is developed early on, the effect can be wrenching conflict within the production team. The producer has to be responsible for confronting the issue, and will need guidance.

The first problem is the widescreen aspect ratio of high-definition television. In geometric terms, a 16:9 frame shows about one-third more picture sideways than a 4:3 frame. But the problem for the director and cameraman is not in numbers. The widescreen frame demands a different approach to composing shots. A single medium close-up in standard-definition television will probably be framed slightly left or right of centre. In high definition, the director will want to put the artiste or presenter much further towards the side of the screen, partly for compositional reasons and partly to achieve good cutting to the next shot, where the interlocutor will be in a 'mirror' position on the other side.

The consequence of the problem arrives when the high-definition image is aspect ratio-converted to standard definition. The aspect ratio converter (ARC) is set to a default which simply takes the central 4:3 portion of the image, which, in most cases, will neatly cut the human figure in two. There are two solutions. The first is to compose the high-definition shot within what makes a good 4:3 frame. But that will look uncomfortable, because in effect the human figure will be awkwardly just to one side of centre or the other. The second is to do the conversion shot by shot, re-framing for 4:3. But that is a slow (and therefore expensive) option with unpredictable effects, especially on foreground objects, which may jump from side to side on cuts.

Unfortunately there is no absolute solution to the problem. The worst problems can be avoided by being aware of the need for compatibility. The degree of artistic compromise will depend on other factors. A broadcaster may be so concerned to have a high value programme that shows the benefits of HD that a deliberate choice is made to 'go for it' in high definition and not worry about other exploitation. Part of the solution is budgetary. Enough money for proper aspect ratio conversion (and remembering to be careful with foreground objects) can ease the problem. However, if things are very tight, and both markets have to be fully satisfied, the only option is for the director to frame his shots to tell the story within the 4:3 frame. Some producers ask their directors of photography to put a marked 4:3 graticule inside the viewfinder, and keep all essential action within that.

Another issue with compatibility concerns the extra detail that can be seen in high definition, especially when shown on big screens. Small, unintended movements in the background of a shot can be highly distracting. The difficulty frequently results from location situation, where a director might have to work from low-resolution images. In mobile units, monitors are small to save space, and may actually be under-scanned 4:3 units running a down-converted image. In post-production, the images are compressed to save storage. The truth may not finally emerge until the final moment when the HD master is conformed. The answer, of course, is perpetual vigilance.

4.12.3.3 Lip Synch

A related problem here concerns lip synch. Keeping words and pictures in synchronisation is an essential requirement in all television, but a task that has actually become more challenging with digital technology. During acquisition, sound is locked alongside the picture on a tape. But in post-production the elements are logged as separate files, summoned into synchronicity by the timecode. As editing manipulates the files, so synch can drift. Unfortunately, watching and listening to the output of a non-linear editing device does not necessarily guarantee that everything is where it should be, since pictures and sound can be routed in different ways. Everything becomes focused on the final conform, where the full-definition images are assembled to make the programme. It is essential that this final tape is quality checked to ensure not only that there are no black holes, but that there are no infelicities of synchronisation. If things go wrong, the penalties are particularly acute in high definition because everything is seen so much more clearly.

For example, producers have noticed inconsistencies in lip synchronisation when watching projected versions of high-definition dramas. Generally sound is preceding picture, and in varying degrees. This kind of problem may result from an electronic projector's anti-key-stoning operation. Under pressure from heavy levels of data, the projector's processor can 'slow down' then catch up with itself. The effect on screen can be very disagreeable.

With experience, producers learn about these problems, but it is easier if they are warned about them early on. With all digital television there are problems resulting from the fact that sound generally passes easily through the system, while pictures may be subject to a great deal of processing. This is true in aspect ratio conversion, where the position of the sound always needs to be retarded to once again achieve true synch.

So there are a number of general lessons to learn in implementing high definition, but these become particularly focused during particular kinds of shooting.

4.12.4 Live Music and Theatre Events

One of the most popular uses for high definition during its early years has been for recording music and theatre events (see Figure 4.12.1). The reasons give an insight into the particular value of using higher quality acquisition systems:

- The events generally feature well-known stars who are 'brand names' with wide international appeal.
- Stars (and their agents) may believe that being recorded on a higher quality medium enhances their status.
- Music events have a broad international appeal that crosses language and cultural divides.
- The events (and stars) are expensive and production costs need to be covered through sales in different territories over a period of time.

Figure 4.12.1 'Master of the House' from the 10th anniversary concert of *Les Miserables* at the Royal Albert Hall, London. Still taken from the high-definition recording. Produced by the author.

- The events are generally not 'time-sensitive' and are seen as having long shelf-lives.
- The possibility of big-screen projection through digital cinema or electronic theatre further broadens market prospects.

Events like these tend to be recorded on location in large concert or theatre venues using multi-camera facilities. In many ways the production approach is identical to recording in standard definition. Careful surveys need to be made of the location to establish the best positions for cameras and cable runs, as well as access, parking and power supplies, together with the availability of refreshment and lavatory facilities for a large technical and creative team. However, high definition does bring some particular problems of its own.

4.12.4.1 Lighting

Anyone who has experience of recording in a theatre environment will be familiar with the problem of providing adequate light. Theatre lighting designers work closely with designer and director to create a sensitive light ambiance that reflects the atmosphere of a particular piece, and are often horrified by the need to increase lighting to levels that will make acceptable video pictures. In addition, there are three further difficulties. The first is that the lighting for a theatre event tends to throw light directly into the proscenium box. Though this works perfectly for an audience seated in the auditorium, it leaves enormous problems for the cross-shots, which will probably be most often used in the electronic coverage. The second is the lack of backlighting to separate performer from backing. The third is that spotlights, which are extensively used in theatre productions, will almost certainly be too bright. Adjust the racking to that and everything outside the spot-beam will be invisible.

So, the television lighting director faced with recording an existing musical or theatrical work will have to confront:

- Too little light.
- Lighting too frontal.
- Little, or no, backlight.
- Spotlights too bright.

These difficulties are accentuated when working in high definition. Without sufficient light, technicians have to increase gain, making for noisy and unsatisfactory pictures. Coded by a system like MPEG-2, the noise increases exponentially, to a point where the pictures can look less good than standard definition.

There are always solutions. Light levels can be turned up. A couple of cross-lights can do wonders to bring illumination into the cross-shots. Backlights can be rigged to separate performer from set. Spotlights can be 'knocked back' by a couple of points. However, it is essential that the television lighting director is given time to rig the lights, balance them for all camera angles, and view them with artistes present.

If the lighting is not adjusted to give a good all-round illumination, picture and colour quality will change as the vision mixer makes transitions from shot to shot. These difficulties can only be partly eased by the matching work of the 'racks' engineers. Unfortunately, a number of high-definition recordings of excellent theatrical performances have been marred by these kinds of problems.

The solution lies with the producer understanding the problems and being absolutely clear with theatre technicians (and producers) about the lighting needs of the production. Coming in to record an existing production at a late stage is never easy, especially when the theatre team feel they have achieved excellence and are suspicious of what the television 'outsiders' will do. However, lighting requirements do not go away. Engineers need to stress the penalties of driving high definition beyond its limits. In most cases, vastly more people will experience a show through television than live in the theatre, so it is in the interest of theatre, as well as television staff, to collaborate in achieving the best quality recording of a work.

4.12.4.2 Widescreen

The 16:9 aspect ratio of high-definition television is particularly well suited to theatrical events. The widescreen ratio was selected by system designers because it is very close to the main area of perception of the human eye, and this, of course, is what theatre designers and directors have worked to (without knowing it) throughout history. The consequence is that the full stage wide shot works extremely well in high definition. This may be of great significance for certain types of programme.

Ballet and theatre-originated dance have always been problematic in television. Theatre choreographers create their work for people sitting in an auditorium – effectively a wide shot – and find certain kinds of cross-shot objectionable, especially when they foreshorten body extensions (legs and arms extended sideways). High definition offers an apparent solution, in that the wide shot which shows the work from the 'correct' central angle is much more satisfying. However, it is not possible to sit on a single wide shot for the duration of a ballet. Audiences need changes of viewpoint. 16:9 format cross-shots have tremendous power, but foreshortening is inevitable. One possible solution is to take tighter shots from the centre, essentially placing the closer detail cameras adjacent to the central wide shot camera. Another is, additionally, to use tracking cameras moving laterally across the scene, though this is often impossible with a live audience because it impedes their view.

The combination of widescreen aspect ratio with enhanced detail (and apparent tactile qualities) does in fact change many of the characteristics of shots commonly used in multi-camera coverage. This is an exciting area, with much scope for creative camera work and direction. However, there are some points that need attention. Wobble can be more objectionable in widescreen pictures, so care needs to be taken with tracking, crane and hand-held shots. Focus decisions are critical. It might be possible to fudge a decision when taking, say, a deep two-shot in standard definition. In HD there is no place to hide. The focus has to be on the main point of interest, and the subsequent shots need to be focused so that the transition works effectively in terms of storytelling.

The approach to framing shots in widescreen is necessarily different to 4:3, where the dynamics of composition occur in a different shaped frame. Fortunately, since 16:9 is very close to the aspect ratio of the human eye, directors and camera operators find it easy and natural to move to this kind of shooting. Generally the difficulties arise when they have to move back to 4:3.

4.12.4.3 Sound

Digital stereo is now common for most recording and is of very good quality. But a natural consequence of recording pictures at higher quality is to look for a comparable advance in the audio area. Surround sound is an alluring option for those who can afford it, especially if a theatric showing is a possibility, but the inherent logistic problems need to be looked at carefully. Apart from the additional microphone positions needed for acquisition, achieving compatibility with stereo will almost certainly require additional mixing, either during recording or subsequently in audio dubbing. Producers need to think very carefully about the consequences of how they choose to record sound, and map out the routes and processes needed with an experienced sound engineer.

A popular show, starring well-known performers, will almost certainly need to earn money in several markets: high definition, standard-definition widescreen, standard-definition 4:3, video, DVD (in both wide and standard aspect ratios) and CD. To properly satisfy these markets, audio will almost certainly need to be remixed. Broadcast television compresses sound to a degree which is unacceptable to consumers who wish to experience VHS, DVD and CD at home. Certain segments of the market have invested heavily in high-quality audio systems, and expect top quality. Ironically, these 'segments' are almost certainly the early adopters who are attracted to high definition, though experience proves that if they really like a show on television they are also often prepared to buy it for the extra audio value on DVD.

Decisions about audio begin at the acquisition stage. For concert and theatre work, the producer will almost certainly take feeds directly from the theatre mixer desk, supplemented with additional microphones to pick up footfalls on stage and responses from the audience. With modern musicals, it is common for performers to work with miniature microphones hidden in their hair or clothing (or sometimes visibly on a 'headset' stalk from the earpiece) using small infrared transmitters that communicate the signal to reception points around the stage. Adding these sources to those from the orchestra and auditorium may total well in excess of 100. To simplify the management of these sources, the audio engineer will probably group and combine them into a smaller number. The high-definition producer will need to arrange facilities so that those sources are fed to a multi-track recording facility, so that all sources are recorded directly, as well as being combined in a working mix done 'live', which will be recorded onto the master recording tapes.

Post-production audio mixing can greatly enhance certain aspects of a show, but there are inbuilt perils. Because the emphasis is on artistic sound issues, vision is inevitably relegated to a minor reference role. Much dubbing is done with only a crude video reference copy, often derived from the compressed 'working' images used in non-linear editing, with reliance placed on timecode as the absolute point of reference. Producers really should insist on having the best quality video version possible for the audio mix, since judgement needs to be made on the basis of picture combined with sound. That can only happen adequately if the impact of the pictures can make itself properly felt.

It is quite common for the musical director of the theatre show to be involved in the mix. Useful though this is, the stage MD comes to the production with the memory of the theatre show, and sometimes may not be able to readjust to what is needed in a different medium. The producer needs to be able to concentrate hard, and have great reserves of tact – and firmness – to ensure that the best result is achieved.

However, the absolutely critical issue – artistic questions aside – is that synchronisation is retained. Track files can slip in the remixing process, but the absolutely key moment is when the finalised track is reunited with picture. It is essential that this stage is supervised, right through the process, and also checked to ensure that the routing of sound and picture in the suite is actually putting picture and sound together accurately. It scarcely needs to be said that in a high-definition medium, precise synchronisation is even more important than in standard definition. This requirement is heightened in musical coverage where, because of the acuity of the human ear, slight variations are instantly noticeable.

There are two other processes where vigilance is required. The first is during aspect ratio conversion, where a widescreen picture is reduced to 4:3. Because of the high level of processing, picture is generally delayed while sound 'goes straight through'. So an adjustment needs to be made, and the producer's eyes and ears have to be the final arbiter. Conversely, during a live transmission from an HD source, sound will need to be delayed appropriately to achieve correct synchronisation.

The second danger is during electronic projection. Most projectors are not directly aligned to the screen, and have built-in processing to 'squeeze' the key-stoning effect out of the picture. This processing can be variable, depending on the amount of data that needs to be handled, with resulting delays in picture. The effect of early sound is particularly objectionable on a big screen. The best advice is always to insist on having a dry run before an important projection.

4.12.4.4 Titles and Credits

High definition offers a great deal of creative scope in using graphics. The widescreen gives much more space for positioning credits to one side or the other. Smaller lettering is readable, and enables more lines of print to be put on screen, just 'like a movie'. However, there are a number of dangers here for the unwary, especially with regard to compatibility with standard definition. The elegant small lettering placed to one side of a high-definition end sequence will almost certainly be unreadable in standard definition. In addition, the aspect ratio converter will cut some off, unless the conversion is done shot by shot, adjusting the framing.

Titles and credits really have to be judged backwards from what will be the worst form of display. If a production is only for digital cinema or electronic theatre, then the producer can do whatever seems right, but if it also has to survive in the lower depths, then lettering style, size, number of lines and positioning need to obey the old rules. Unfortunately, the character sizes that look right for the 40-inch high-definition screen, the 32-inch standard-definition monitor and the projected 'alternative content' version in digital cinema are quite different. Television-style titles on a big screen look absurdly big and over bold. Nothing so well reveals the origin of a programme as its titles.

In order that programme titles and credits look artistically good and are readable in all versions, it is really necesssary to create them afresh for each viewing situation. The basic text input, once accomplished, can be expressed in appropriate styles, integrated with background shots and edited in as a complete sequence to a particular version.

4.12.5 Producing Sports in HD

There is no question that sporting events look utterly compelling in high definition. The wider aspect ratio and greater detail create a wonderful sense of engagement in the drama of competition, communicating an almost tactile sense of being part of the game. The viewer is able to read the terrain of a football pitch or golf course, feel an enhanced closeness to the competitors, almost have a tactile impression of contact with the ball. When HD Thames took the final of the 2002 World Cup in a live cable relay to London, there was an audible gasp from the audience as the wide shot of the stadium appeared on screen. Projections onto big screens are particularly compelling and clearly have a role in the future of digital cinema.

Sport has in fact been a staple of much high-definition work throughout the world, focused mainly on international competitions like the Olympic Games, the World Cup, World Ice Skating and Wimbledon. Some outstanding work has been produced, often with a relatively small number of cameras (six for the HD 2002 World Cup Final as opposed to 17 for host broadcaster standard definition). The experience is that the enhanced power of HD, applied in well-considered camera positions, actually compensates for multiple viewpoints.

The style of HD sports coverage has evolved differently to standard definition. The high wide shot, giving a view of the entire pitch, offers a unique overall impression of the game, revealing the strategies as they are applied by both sides. Because of its appeal, this shot tends to be used much more frequently, and for longer than comparable wide shots in standard definition. Audience feedback suggests a much better overall sense of the development of the game, and a sense of being able to anticipate moves. There also appears to be much more sense of drama from single or tight group shots of players. It would be foolish to suggest that there is a single 'correct' way for HD treatment of sports, but a common factor emerging does seem to be the dynamic richness of the wide shot seen in 16:9.

For the producer, much of the work of sports coverage is in preparation – the clearing of rights, obtaining camera positions, laying cables and links, as well as providing parking and creature comforts for the production teams. For most international events like the Olympics, the host broadcaster provides basic coverage supplemented by the national interests of the broadcasters of other nations. The International Broadcast Centre (IBC) is the point at which all this coverage converges for onward relay by terrestrial, satellite and cable. High-definition facilities are now beginning to become part of this picture. However, there are still hurdles to be overcome.

Sport is essentially a live entertainment, and this means that the HD sports producer must contrive to get pictures to his audience instantly. This may mean a combination of fibre, satellite and possibly a terrestrial final hop. Although equipment to do this exists, it is in short supply, and there is not a huge amount of experience in practical implementation, especially when the path is a hybrid of cable and satellite. As high definition becomes more commonly used, the situation will change, but for the moment, the producer of HD sports needs to work closely with engineering colleagues to implement a workable solution.

4.12.5.1 System to Choose

In the early days, the HD world was split between the Japanese 1125 system working at 60 Hz and the European 1250 system working at 50. Today, high definition has converged into a more user-friendly system defined by pixel array rather than line numbers, but there are several options, offering varying approaches to picture quality and motion portrayal. The picture has to some extent been complicated by the particular needs of motion picture makers who wish to acquire images in electronic form.

The choice of system used by a sports producer will almost certainly depend on the house choice of the major broadcaster involved. However, there is one basic principle that applies, which is always to acquire images at the best possible resolution. It is always possible to move down. It is generally problematic to move up. For sport it is also necessary to remember that motion is an essential element in the story, and therefore the choice of frame rate is important.

4.12.5.2 Ancillary Equipment

The television producer covering a major international sporting event like the Olympics may well supplement venue coverage with a number of small, single-camera units providing stories which can be sent in via local radio links. For events like athletics or speed skating, cameras will be mounted on a rapid tracking devices pulled by motorised cables or linear motors. Horse racing will probably use remote cameras mounted on a mobile tracking device to get dramatic motion pictures. The producer of American football may use a wire-mounted overhead tracking device. A cricket match may be improved by a view from the centre of the wickets. Some sports use helicopters sending in pictures via radio-horn.

We have all become used to using pieces of equipment which enhance the drama of sport and bring the viewer into the heart of the action. Some depend on miniaturised cameras, others on ingenious tracking devices. However, all have to get their pictures back to the control centre. Here high definition may currently appear to be at a disadvantage because the necessary pieces of equipment and links capable of handling the additional data are not 'off-the-shelf' items. This situation will change as HD is more widely implemented. In the meantime, however, the high-definition producer can rest comfortably on the knowledge that the intrinsic qualities of the system offers unique benefits.

4.12.5.3 Archive

Much sports coverage depends on live access to archives of previous matches. For the high-definition producer this demands certain decisions. The archive will almost certainly consist mostly of 4:3 standard-definition material (and in the UK some 16:9 standard definition). How can these be satisfactorily integrated, live, into high-definition coverage?

In fact, it is relatively easy to use standard-definition material in an HD context. Standard-definition recordings can be inserted into a quarter of the picture frame without loss of quality, or be up-converted using a device such as the excellent converter made by Snell and Wilcox (allowance must be made for conversion time, which will delay picture marginally). The 4:3 aspect ratio can be accommodated very satisfactorily by using the additional section of screen for graphics, logo or clock. What is not satisfactory is to distort 4:3 source material sideways to fill the widescreen.

4.12.5.4 Visual Approach

Although there has been a quantity of practical experience throughout the world, high-definition sports are essentially a new area with all kinds of exciting possibilities. The strength of sport lies in its ability to bring together the best qualities of several genres – drama, spectacle, news. Intentionally or not, all games have developed to conform with the aspect ratio of the human eye, and for that reason are natural for HD coverage. What the producer has to do is accept that established modes of coverage are based on getting the best from the 4:3 aspect ratio. With widescreen, the rules need to be changed, and new approaches tried. The high-definition medium offers a powerful new set of visual tools to capture the moving drama of sport and the possibility to rethink its coverage as pure spectacle.

4.12.6 Producing Drama in HD

In the United States, a great deal of television drama (including situation comedies and soaps) has traditionally been shot on 35 mm film. This made sense for all kinds of reasons. Image quality is high and survives translation into various electronic formats for other markets. Hollywood remains the world's principal production centre, and provides a highly skilled resource hub for 35 mm shooting for both television and feature films. As system after system of electronic acquisition rose and fell, Hollywood felt increasingly secure that it was master of the world's single base format. The issue became increasingly important as film and television surpassed aeronautics to become the USA's major export.

In Europe traditions were different. In the UK, a great deal of drama has always been shot on 16 mm film. Though the introduction of the portable camcorder in the 1980s showed there were other ways of working, film has survived as a popular acquisition medium for high-quality work; 16 mm film was used because stock, processing and equipment were cheaper than 35 mm, and the cameras more portable. For many years, British television did not particularly seek to export its productions, so there was no particular reason to aspire to a standard that could be re-versioned to foreign standards. The great strength of UK television drama meant that several generations of writers, directors and technicians grew to work with and love 16 mm film. When widescreen made its appearance, it was natural that ingenuity could find a little extra space on the celluloid and gave birth to Super Sixteen.

As the portable electronic camera continued to advance in quality and flexibility, so did 16 mm film strike back. Image resolution improved. Wet-gates helped minimise image dirt and damage during telecine transfer. Techniques were developed to digitise negative to get the maximum detail from the tiny frame, and post-production carried out on non-linear workstations. So strong is the feeling for film in the UK that television drama producers acquiring images with electronic cameras increasingly demand post-production techniques to give their work the 'film look'.

4.12.6.1 So Why Move to HD?

From the UK perspective, there are several aspects to the answer. The first is that there is a new emphasis on exporting programmes, and there is both a hunger for drama and, in some markets, a premium to be paid for it in HD. Second is the development of digital projection and its incipient use in digital cinema and electronic theatre. The quality of digital projection has come as a shock to many who regarded film projection as state of the art. The fact that some major creative talents have espoused electronic acquisition as well as projection has led to a reassessment of the potential of electronic ways of working. Although this will be a matter of years, it is clear that there is an exhibition market for what is referred to as 'alternative content', and electronically acquired material is the most obvious candidate to fill it.

For a drama producer, working in high definition is an opportunity to provide high-quality images that will survive being shown on large screens, and will look excellent on television. And because major creative talents have also now worked in HD, the technical industry has become much more sympathetic to providing the bells and whistles drama producers need. Meanwhile, broadcasters are eyeing the potential of the alternative content market for digital cinema.

4.12.6.2 Just Like Film

When high definition first arrived in Hollywood, the manufacturers thought producers would lap it up. They were wrong. Indeed, the film establishment became resentful of what they saw as a sullying of their art by new arrivals from the lowly world of television. After various pioneering forays, HD did finally gain a toe-hold, but mainly as an intermediate master for mass duping of VHS copies. Gradually, however, the equipment manufacturers began to listen to what they were told, and the message was that HD would have to become a lot more like film if it were ever to gain a real foothold.

There were many objections to HD as an acquisition medium. There was a limited range of lenses. Colours were different (less satisfactory) than on film. The straight gamma meant that, in low light levels, images rapidly plunged into invisibility. Finally, however, the animus against HD crystallised into two specific complaints:

- Its black levels could not match film.
- Its interlaced motion portrayal looked too much like television.

Black level is a tough nut to crack, but an answer has been found to interlace – 24P.

In fact, high-definition systems designers took Hollywood's criticisms very seriously indeed. 24P is an ingenious system which contrives to portray motion through a sequence of complete sequential images in a similar manner to film, and thus appeals, allegedly, to that part of the brain where imagination lies. It is specifically designed to appeal to the makers of motion pictures. But the work went further. HD cameras were designed to look like film cameras (often portrayed with wheel-focused lenses and 'French flags' attached) and engineered so that they could take a wide variety of specialist lenses.

So, the drama producer now has a choice of high-quality television-style interlaced systems like 1080i, or the more filmic 720P or 480P. Both offer extremely good quality, stable images, and both are supported through non-linear production devices. The decision will depend on the project, as well as the personal feelings of the creative teams involved.

4.12.6.3 Lenses

One of the fundamental differences between film and video cameras is that, whereas in a film camera light passes straight through the lens onto the film stock, in the video camera it has to be bent round corners to get to the three colour receptors. This difference in the length light travels has an influence on the use of lenses.

Most camcorders are designed for use with an electronically operated zoom lens, whose optics are calculated to give the best depth of field over the entire range. Film lenses tend to be fixed focal length objectives, and the great number available enables directors of photography to apply a personal look to their work. These lenses have the virtue that, unlike most zooms, they are easy to focus on specific points of interest, putting the rest of the image into softer focus.

These differences actually put the spotlight on major differences that have emerged between electronic and film ways of working. Television has generally tried to keep images brightly lit and clear, while film has often sought to create atmosphere by accentuating the main point of interest in the image, often softening (and darkening) the rest.

Manufacturers of high-definition equipment soon identified these differences, and set about engineering their cameras so that they would take standard film lenses. Today, the producer is able to fit a wide range of objectives to a high-definition camera, though it needs to be remembered that because of the inevitably longer light path, the focal length (and therefore resulting image size) will be slightly different.

4.12.6.4 Location Work

The great strength of an electronic acquisition system is that, provided monitoring systems are well set-up, the production team can be confident of the final result. With film, everything depends on the skills and experience of lighting director and operator, and the consequences of their work cannot be seen until after processing. With an electronic image the results are clear to see on the monitor, and many more of the production team can view that image than have access to the film viewfinder. The uncertainties – some might say mysteries – of image creation are removed. In the case of high definition, what has been achieved is visible in much greater clarity and detail, and probably encourages more artistic image creation.

The art of creating great images lies almost entirely in the manipulation of light. Outstanding cinematographers like Vittorio Storaro have conclusively proved that it possible to achieve wonderful pictures using high definition. He has made identical sequences on film and HD, and challenged professional audiences to identify which is which.

That said, high definition does have a particular quality of its own. This is partly to do with the lack of gate movement (discernible in almost all film images, however stabilised) and partly because the image acquired is the final image (unless subsequently processed). HD's ability to convey surface texture results from the lack of the intermediate stages film undergoes from negative to release print. For that reason it is always best to have a high-definition monitor on location, so that the real impact of the picture can be judged. Of course, modern cameras output a compatible standard-definition picture that can be seen adequately on a smaller, and much lighter, standard-definition monitor. However, to see the true richness of the image, it is necessary to view in the best quality.

4.12.6.5 Studio Work

Working with an electronic acquisition medium always seems easier in a studio. There is convenient power everywhere, lighting is totally controllable, the monitor can be set up where no light falls on it, and the rain cannot get in. However, it is not without problems. Great attention needs to be paid to sets to ensure that the finish is sufficiently good to survive high-definition scrutiny. Tiny chips in paintwork, unfilled screw-holes and creases in cyclorama cloths have a distressing way of making themselves known.

4.12.6.6 Post-production

Modern non-linear editing systems are able to handle high-definition images very adequately. But there are two dangers of which the producer needs to be aware.

The first results from working with compressed images. Most pictures have to be compressed in order to get the necessary footage onto the disk, and though they provide a perfectly acceptable guide for editing they can fail to show details that become apparent later, especially if the images are shown on big screens. The only protection against unpleasant surprises is to be vigilant in acquisition (having a proper HD monitor on location helps) and to keep shot notes. A production assistant keeping a detailed shot list with notes is actually an economy, paid back in saved time and frustration. Today, much assembly editing takes place as soon as the 'rushes' get back to the edit room, and certainly without the director being present until later in the process. Confident shot descriptions, with detail about timings, lenses and identification of bad takes, are enormously useful.

The second danger is the reverse. Images that have a particular weight and conviction in HD may lose some of their quality in standard definition. When forced to satisfy a double market there really is no substitute for constantly checking how things are working for both.

4.12.7 Documentary

High definition is a wonderful medium for showing the real world. The beauties of landscape, sea, snow and sky come alive. The animal kingdom emerges in breathtaking clarity. Processes involving natural materials – wood, metal, stone, paint – become compulsively absorbing.

The early high-definition cameras were weighty beasts, linked to their recorders with heavy cables. Today's generation of lightweight descendants are very different, the smallest working with DV-type cassettes, and eminently suited to the rigours of documentary. In the UK, there is an active movement to show documentaries on big screens, and these could be a very important kind of alternative content for the future.

One of the longest running HD documentary series made by NHK set out to record human crafts. Its mission was to become a kind of modern Encylopedia to show how cabinets are made and glass blown. In fact, the programmes demonstrated a strange transformative power, turning the images of these everyday manual crafts into a dance of objects and processes manipulated by human hands. Even 35 mm film could not have achieved the sense of presence with these materials.

4.12.8 In Conclusion

High definition is a fascinating extension to the programme-maker's tool-kit. It has unique qualities and has evolved, unlike most electronic television systems, with a real awareness of the qualities of film. Whether used in multi- or single-camera applications it has the power to enhance an audience's response by bringing the programme experience closer. We live in a world which in many ways is already high definition – digital photography, computer games and DVDs all present themselves in higher resolutions than standard-definition television programmes. Anyone who has tried to read an email on a television set will know the visual limitations of 626 or 525 lines. For the producer or engineer working with HD for the first time there are some new lessons to learn, but in the author's experience these are more than matched by the delight of new discoveries. High definition should come with a warning – it is addictive.

Bryan Arbon
Just Technicalities Ltd

4.13 Routers and Matrices

4.13.1 Routing: an Overview

A router, or routing switcher, can be thought of in the simplest terms as a grid comprising a number of inputs and outputs, or sources and destinations, as shown in Figure 4.13.1. Points within the grid where the input and output lines intersect are known as crosspoints and provide the switch enabling the required router input, when controlled correctly, to be connected to the selected output.

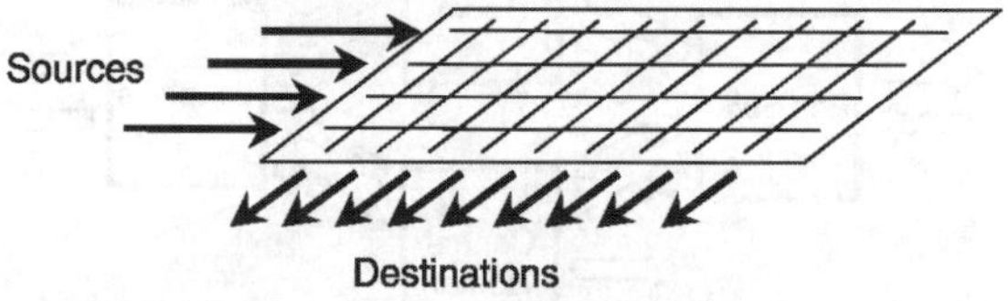

Figure 4.13.1 Representation of a single-level router.

Routers, also sometimes referred to as matrices, are generally made up of a number of modules. These are Input, Crosspoint, Output and Control; however, some matrices utilise a single card architecture combining all of these functions. While the distributive nature of routing systems provides the ability for a single input to be routed to one or more outputs, the switch architecture, in most router types, is designed to prevent more than one input being routed to a single output. There are, however, some exceptions to this – for example, intercom or talkback systems, where it is normal to connect a number of inputs to single or multiple outputs, and machine control routers which typically only permit one-to-one connections to be made.

Early routing switchers, while physically large, housed a small number of crosspoints per rack unit. A typical system provided between eight and 20 inputs and outputs, while assignment routers with between 50 and 100 inputs and outputs were considered to be large systems but still only provided approximately 200 crosspoints per rack unit. By comparison, today's routing switchers offer lower cost, compact solutions with much higher packing densities, benefiting from the use of multi-pin integrated circuits, surface mount technology and modern manufacturing techniques. These factors contribute to provide routers with a much higher ratio of crosspoints per rack unit, typically in excess of 1500, and are now almost at the point where frame size is virtually determined by the size and number of connectors required for the router inputs and outputs. While smaller routers are still in demand, they are considered to be utility switchers and are used mainly for purposes such as monitoring and inter area routing. Today, the average router size is between 128 and 256 inputs and outputs, while larger system routers can reach sizes of up to 1000×1000 and, in some cases, beyond.

The routers described so far have been considered as single-format, single-level devices, but today's routing requirements have become far more complex than that. Routers must be able to interface with, and more importantly pass transparently, a diverse range of signal types provided by the equipment they are being used to interconnect. By grouping a number of individual routers together using a common control system, multi-level routing switchers may be constructed (Figure 4.13.2).

These multi-level routers permit signals going to and from particular pieces of equipment to be grouped or associated with each other, and switched together, across all levels. This type of operation ensures, for example, that the video signal from a

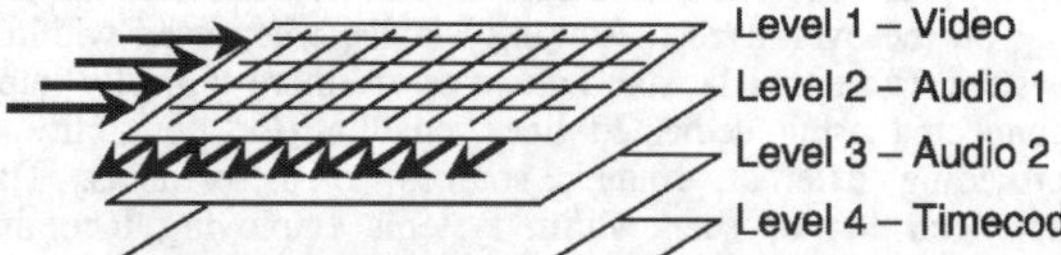

Figure 4.13.2 A four-level router.

VTR arrives at its destination with the correctly associated audio and timecode signal, even if they do not appear on the same numbered input on each level within the matrix. A further benefit of this type of grouping is that common signals, such as silence on an audio level, may be associated with, and used by, multiple inputs on the video level. Each level can therefore be specified to provide the correct number of sources and destinations required, rather than just using a router level the same size as the primary level.

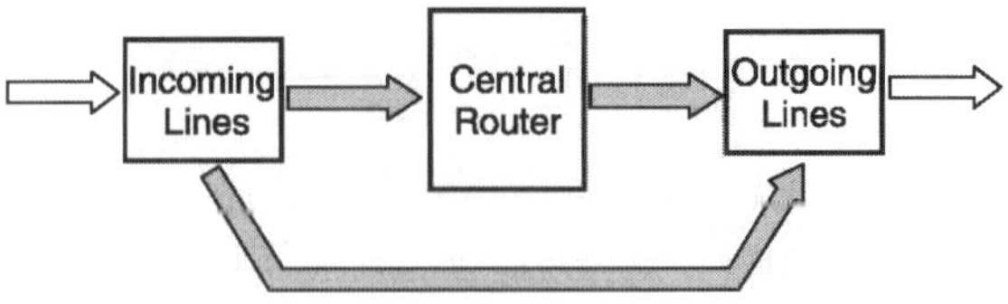

Figure 4.13.3 A typical multi-stage router with tie lines.

4.13.2 Central or Distributed Routing

There have been many discussions regarding the benefits of central routing systems over distributed routing architectures and vice versa. Looking at the trends seen during the past decade, installations have regularly moved between use of the distributed and central routing architecture models, or in some cases a combination of both. It is, however, apparent that the routing architecture adopted within any system depends entirely upon the operation required, the routing systems available for use with the required signal types and of course personal preference, where what might be right for one system may not necessarily be good for another.

Some systems require a guarantee that every source will be available for selection to any destination at any time. This type of operation should, of course, opt for the central router architecture. System environments where groups of sources and destinations can be treated as independent routing islands – for example, edit suites in a post-production centre – may benefit using a distributed routing architecture.

The central router architecture, while providing the ability to connect any source machine to any destination device, employs a single, multi-level routing switcher with enough inputs and outputs to accommodate each and every routable source or destination within the system. The increased packing density and reduced cost per crosspoint, coupled with the higher reliability offered by today's routers, make the central router more of an attractive proposition. When planning such a system, thought should be given at the outset to the possibility of future system expansion, where the addition of new equipment would require further sources and destinations to be added to the router. This expansion can be catered for by either part equipping a larger router frame and reducing the initial number of equipped inputs and outputs, or by using a router architecture which permits expansion through the addition of further source and/or destination frames. It is, however, generally accepted that, within a centralised router, there will be a number of sources and destinations that will never be routed to each other, possibly resulting in a large number of crosspoints that, in normal operation, may never be used.

A distributed architecture, however, maximises the router input/output count, while reducing the physical number of crosspoints required within each matrix and therefore impacting on the overall cost. Routing islands, when used within a system, can either be stand-alone or are more typically interconnected using dedicated lines, enabling the possibility of providing external, routed, sources to these areas. This approach can be seen within systems employing Incoming and Outgoing Lines matrices that feed, bypass or take the outputs from a Central Router. The system in Figure 4.13.3 allows signals from the Incoming Lines frame to be routed to either the Central Router or the Outgoing Lines matrix.

Tie lines used to interconnect island, or satellite, routers have typically been assigned manually using a router control panel, requiring sources and destinations to be selected for each route within the signal chain. In these cases, the signal may need to pass through three or more matrices, and require at least three routes to be made in order to get a signal through the system. Advances in control system operation have permitted tie lines to be defined within the structure of the routing system, indicating to the controller how the matrices are interconnected, permitting automatic tie line assignment across multiple router frames. Once defined within the database, the control system only requires a starting source and final destination to be specified in order for it to use the interconnection details to determine the best path between the selected points. Tie lines can be defined between any of the router frames within a system, as shown in Figure 4.13.4. Where there are no direct connections between two routers, for example, a source from the Edit Suite matrix that is to be routed to a destination on the Outgoing Lines matrix will be routed via the Central Matrix.

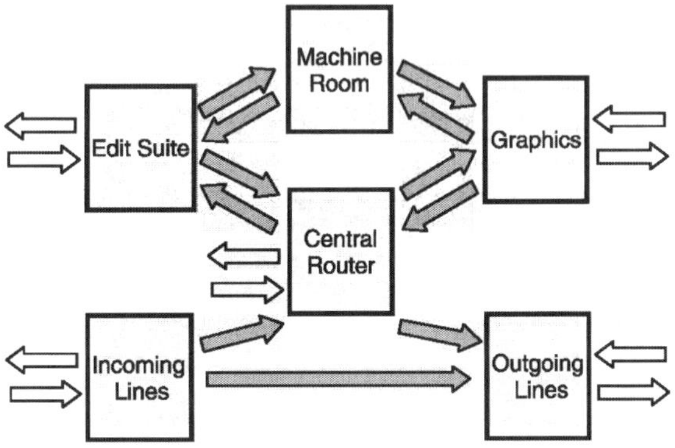

Figure 4.13.4 A multi-matrix with inter-area tie lines.

4.13.3 Hybrid Routing Environments

The proliferation of signal types, and in particular the arrival of digital signals, created an almost instant requirement for hybrid routing environments, where both analogue and digital equipment could coexist. Integrating new, digital equipment into analogue environments provided a new set of challenges to manufacturers and systems integrators. Initially these hybrid systems had much smaller pools of digital equipment available to them which, by exploiting the distributed routing architecture and tie line management systems, meant they could be integrated within an analogue infrastructure easily and cost-effectively.

Tie lines within these systems were established between the analogue and digital matrices. These lines were not direct interconnections, but went via banks of converters, either digital to analogue (DACs) or analogue to digital (ADCs), allowing the signals to be converted to the correct format before being presented at the router input (Figure 4.13.5). This approach is still utilised to provide a cost-effective method of integrating equipment within mixed operational environments, without the burden of providing dedicated converters or signal processing equipment for each router input and output.

Advances in routing technology have provided the ability to incorporate the analogue-to-digital and digital-to-analogue conversion processes within the input and output stages of routing switchers. Until very recently this has only been available within audio routers and was due mainly to the associated size and cost constraints of the circuitry required. Today's advanced circuit manufacturing techniques and lower component costs are now beginning to make the inclusion of analogue-to-digital converters within SDI routing systems viable. By using these converters composite, analogue, video signals can be presented at the input of a digital router, where they are converted into SDI signals and freely routed within the digital environment to digital or analogue destinations.

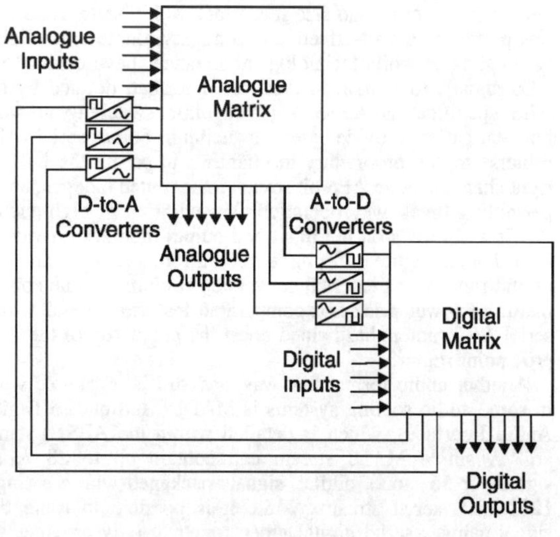

Figure 4.13.5 A hybrid system incorporating A-to-D and D-to-A converters.

4.13.4 Router Types

Early routing switchers provided the ability to switch analogue video and audio signals while offering a distributive signal infrastructure. Recent advances in technology have brought routing for standard- and high-definition signals into the digital video domain and, for audio, AES/EBU digital audio. It was thought by many that the introduction of these digital signal formats meant that the demise of analogue systems was in sight. In reality this has not happened as broadcasters, who over the years have made considerable investments in analogue equipment, were not willing or able to just replace their system infrastructure 'overnight'.

4.13.5 Analogue Routing

Analogue video and audio routers have traditionally been offered within fixed frame sizes. In the case of video routing a single video level could satisfy the analogue routing requirements, operating with composite signals requiring an 8 MHz bandwidth, and in some cases wideband signals at up to 30 MHz and above. Where it is necessary to route signals to and from equipment with component inputs and outputs, this simply means that where previously a single video router had been used for composite video, three are now required for YUV operation and four for RGBS. The architecture of many routers enables them to be 'soft' partitioned to provide three or four smaller matrices, specifically for use with component signals.

Audio moved from mono signals requiring one level, to stereo using two married router levels, and provided a bandwidth of up to 20 MHz. More recently, the stereo requirement has been addressed with routing systems providing stereo routing capability within a single router frame as a combined level, either as stereo routing modules or a pair of mono levels. The introduction of satellite and multi-region broadcasting brought with it a requirement for multi-language programming, requiring multiple mono or stereo levels to route the additional audio channels.

Timecode has traditionally been treated as an audio signal and, while it can be routed with an analogue audio router, the 20 kHz bandwidth provided by it will not pass a spooling timecode signal which can reach frequencies of up to 250 kHz; therefore, in timecode routers the output stage is fitted with a wideband output amplifier able to cope with the increased bandwidth requirement. More recently, the routing of longitudinal timecode has been achieved with a digital router core, as the signal is essentially binary data.

Due to the very nature of analogue routers, the signals they carried suffered from degradation after multiple passes through a router caused by noise introduced within the circuitry and crosstalk from adjacent channels. By designing circuits with signal-to-noise ratios and crosstalk performance typically in the range of −60 dB for video and −90 dB for audio, up to four or five passes were possible through the router.

4.13.6 Digital Routing

The advent of digital video and audio signal routing offered more robust signal formats capable of surviving many more passes through routing switchers than their analogue counterparts with little, or no, effect on signal quality, providing of course that they were correctly terminated and reclocked where necessary.

The first digital video routers operating with parallel CCIR 601 signals, with either 8 or 10 bits transmitted at 27 MHz, provided a small number of inputs and outputs, which, as the technology was in its infancy, were housed in large frames. With the introduction of serial digital routers, video inputs and outputs were presented on BNCs, permitting the return to a system architecture where video could once again be transported via a single coax. The serial digital interface, defined in SMPTE specification 259M-ABCD, for both 525- and 625-line Composite and Component digital video signals, can operate at frequencies up to 270 Mbit/s in the case of component digital signals. These signals can also carry up to eight channels of embedded audio.

The implementation of the serial digital interface, by different manufacturers, has generated a wide variety of matrices offering a mixture of reclocking and non-reclocking architectures. Routers providing input and/or output reclocking generally have the ability to configure inputs and outputs, either independently or in groups, to operate with different clock frequencies depending on the signal presented to it. This function may be selectable using jumper links, configuration software or may be auto-sensing. Matrices providing a non-reclocking architecture require no set-up for the input and output stages, and by employing a wideband crosspoint are capable of switching the digital signals presented to them within the specified bandwidth of the crosspoint itself and the input and output circuitry.

Use of these routers is not just limited to serial digital video signals; they can also be used to switch multiplexed signals such as ASI and SDTI. Care must be taken when using standard digital video routers with these signal formats as they do not necessarily fall into the 143, 177 or 270 Mbit/s categories, and reclocking circuitry employed within some routers may cause problems to downstream equipment. Because SDI signals use NRZI coding, some routers providing dual outputs at each destination provide the second output as an inverted signal, straight from the output driver circuit. While this practice is acceptable within SDI equipment, ASI equipment cannot operate with inverted signals.

With the increased availability of digital equipment there was also a greater demand for digital routing systems, and not only did the router input and output count gradually increase, but there was a new requirement emerging. Uncertainty within the United States over broadcasting standards left many US-based broadcasters and post-production facilities wondering just what their infrastructure would need to be capable of in the future. It became apparent, very quickly, that any new routing system being installed would have to be future-proof. These systems would need to operate with standard-definition signals, while providing an architecture which could be upgraded or was compatible with high-definition signals.

These HDTV systems would need to be capable of operating at bandwidths of up to 1.485 Gbit/s, with signals conforming to SMPTE specification 292M, the serial digital interface for high-definition television systems. However, due to the prohibitive cost of HDTV production equipment and availability of crosspoints, the first HDTV routers provided only a small number of inputs and outputs, typically in the range of 16 × 1 to 16 × 16, and while non-modular in construction, they were used to provide monitoring or routing within small HDTV production areas. Like the transition from analogue to digital video, the requirement for hybrid routing environments was, once again, in demand but this time between standard- and high-definition digital signals.

Advances within the telecommunications industry provided larger, more efficient and lower cost wideband crosspoint arrays, allowing larger HDTV routers to be designed. These routers, when used with signal-specific input and output modules, designed for operation with either high- or standard-definition signals, enabled routers to be supplied initially for SDV operation with the option to migrate all or part of the router to HDTV operation, should it be necessary, by replacing the input and output modules.

Early AES routers were offered with either a serial or parallel internal architecture. While serial AES routers provided low-cost routing solutions for digital audio, they had the disadvantage that they could only offer a crash switch between router sources. With the implementation of a parallel time division multiplex (TDM) architecture, input signals were converted from serial to parallel data before being clocked onto a parallel bus. As each input was assigned a dedicated timeslot on the bus, sources could be picked from the parallel bus and passed to the selected output. This approach also ensured that signals could be switched at the correct word boundaries, providing silent, or click-free, switch points. The addition of analogue input and output cards to TDM routers meant they could also be integrated within hybrid audio systems. Further development of this internal signal distribution method permitted the left and right channels for each signal to be identified and subsequently split, or broken away, allowing them to be routed via independent outputs as discrete left and right signals. One of the biggest benefits with the TDM architecture was the ability to increase the size of a router purely by adding input and output frames, as there was no central core of crosspoints. These systems were, however, large and only provided a cost-effective solution in systems requiring in excess of 300 sources and destinations.

Significant price and size reductions within serial AES routers permitted a cost-effective routing environment for digital audio signals, while technological advances have enabled the AES signals to remain in their serial format, defined by the AES specification AES3 – 1992, while switching at word boundaries to provide silent switching. Some serial AES routers employ proprietary mechanisms to permit the left and right channels of an AES bit stream to be routed independently, permitting breakaway operation. The addition of high-quality ADCs and DACs on the input and output modules provides a hybrid architecture offering a very affordable alternative to digital routers and external conversion equipment. The combination of lower price and comparable features offered within serial AES routers has helped boost the popularity of the AES crosspoint router.

Another audio format that was, and still is, exploited within some audio routing systems is MADI (Multiplexed Digital Audio Interface), which is detailed within the AES10 standard. A single MADI stream can contain up to 28 AES3 signals or 56 mono, digital, signals packaged within a single 125 Mbit/s serial stream. While it is possible to route this signal using a serial digital video router, it only provides the ability to switch a complete MADI stream between two points and does not provide synchronous switching. MADI routers, however, can accept a number of MADI inputs, enabling each one to be broken down into its discrete mono digital channels within the crosspoint module. Once separated, the sources are distributed throughout the frame before being selected for routing to particular destinations, where they are re-coded to the relevant MADI output. These output streams may contain signals from one or more of the MADI input signals. Outside the main router, input (encoder) and output (decoder) frames provide the interface between external equipment and the router, while some equipment such as sound desks, fitted with MADI inputs and outputs, can be connected directly to the router.

4.13.7 Timing Matters

Analogue video routing switchers can usually operate with signals of both 525/60 and 625/50 standards; however, in order for these signals to be switched cleanly, the switch must occur within the vertical interval, where no video or ancillary data is present. This is achieved by supplying the router with a reference signal, typically analogue colour black, in order for it to determine when to switch.

It is essential within any analogue routing environment where synchronous switching is required that each source should be locked to, and synchronous with, the system reference, ensuring it appears with all other router inputs as a co-timed signal.

From the reference signal the internal control module derives a timing pulse, locked to the reference, which is distributed to all of the crosspoint modules within the router. Upon receiving a switch command from the system controller, or control panel, the selected crosspoint will be forced to switch by the timing pulse at the point in time coinciding with the correct line for the video standard in use. Some routers are fitted with dual reference inputs, one for 525 and one for 625. By defining the line standard of each source, the router will use the correct reference to provide clean switching between signals of the same standard. This facility enables the router to be used in mixed 525/625 environments.

A common misconception is that digital systems do not need to be timed; however, digital routers behave no differently to their analogue counterparts as, unlike digital switching desks and some signal processing equipment, the HD and SD router inputs are not fitted with either line or frame synchroniser circuitry. Signal timing within the digital environment is still as crucial as it is within analogue video routing switchers.

Because SMPTE specifications 259 and 292 define a switch point within the video line on line 6 for 625 signals and line 10 for 525, signals arriving at the router input that are not synchronous with the reference may be switched during active video, causing bit stream disruption and lines containing more or less data than expected by the receiving equipment. Downstream effects caused by asynchronous switching of digital video signals range from signal bounce/rolling on monitors caused by the input circuitry re-locking to the new signal, to equipment such as VTRs recording or transmitting digital noise and flashes, when the input loses lock and the receiving input circuitry takes a number of frames to recover. It is also possible to see frozen images displayed where input frame synchronisers, fitted within receiving equipment, retain and display the last valid video frame received until the switched signal stabilises so that it can be decoded by the receiver.

Pure analogue audio signals can be switched at any point, with no effect on the output signal, and therefore do not require to be synchronised to a video reference signal to ensure clean switching. AES audio routing, however, is essentially no different to a digital video signal, as the signals being routed must be synchronous with an AES reference. Like the vertical interval in video signals, AES bit streams provide frame boundaries separating digital audio blocks. It is within these frame boundaries that AES signals should be switched in order to maintain audio word structures, their relationships while providing clean switches without clicks and pops. Switching AES signals asynchronously, in addition to providing clicks and pops, may also cause the outputs of downstream receiving equipment to mute until the signal is coherent and can be decoded.

4.13.8 Control Routing

Wide use is made, within broadcast and post-production environments, of RS422 control signals. While they are used within some routing architectures, to provide interconnections to control systems from control panels and router frames, they can also be used to provide control of equipment such as VTRs and video/audio servers from edit controllers and automation systems. Equipment under this type of remote control is known either as a machine or device, while the equipment providing the control is known as a controller. RS422 control is carried over a bidirectional, four-wire connection utilising dedicated transmit and receive pairs to enable controlled machines to receive commands and acknowledge receipt of them back to the controlling device. This type of communication is essential within any operational situation as it enables each controlled device to confirm the successful receipt of commands, while providing machine status and reporting failures or errors to the controlling equipment.

There are two types of RS422 router available, one providing conventional inputs and outputs and the second providing a port-based solution. Conventional RS422 routers provide dedicated connections offering fixed source and destination operation for machines and controllers respectively. This architecture requires devices that will be used as both a machine and a controller to be connected as both a source and a destination. Port-based systems, however, fitted with dynamically configurable ports, permit machines and controllers to be connected to any port on the matrix. Each port will, once selected, configure the transmit and receive pairs for either controller or machine operation. This approach means that machines, which are to be used as both a device and controller, only require to be connected to the router once.

RS422 routers are unlike conventional routing switchers as they typically operate in one-to-one mode, allowing a direct port-to-port connection between one controller and a single machine. This mode of operation ensures that both transmit and receive lines are correctly connected between both pieces of equipment, providing a bidirectional communication path. In some systems, it may be possible to connect one controller to multiple machines for ganged start and record operation. In this case, one machine, designated as the master, is connected to the controller via a four-wire link to provide status information to the controller. All remaining machines will be connected to the transmit pair from the controller, allowing them to only receive control commands.

4.13.9 Control Systems

Increasingly, routers are being looked upon as commodity items; however, with the exception of the simplest routing systems, it is the control system that provides the key to many of the complex routing operations that need to be undertaken within today's broadcast environments.

Router control systems are available as either single modules housed within the very frames they are controlling or as standalone, self-contained, system controllers. Regardless of type or manufacturer, most control systems can be supplied in either a single or dual configuration, providing a main and backup architecture with manual and automatic changeover and fault alarms, removing the possibility of a single point of failure.

An editable database contained within the controller holds configuration details for the complete routing system including, but not limited to, router sizes, source and destination names

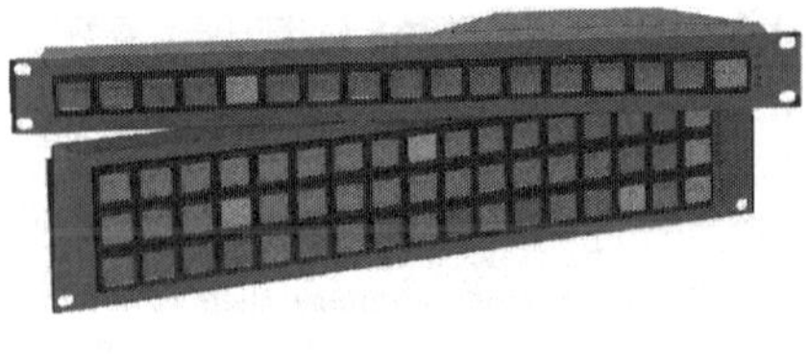

Figure 4.13.6 Router control panels fitted with LCD buttons.

and associations, control panel configurations including 'keypad dial-up' and tie line connection details.

Control panels can be divided into three types: button per crosspoint, multi-bus and X-Y. Each of these types can use either standard push-buttons with lamps, buttons with LCD displays (Figure 4.13.6) or keypad-style buttons for source or destination selection, and may also be fitted with alphanumeric displays to provide status feedback. Button per crosspoint panels provide control of a single destination and are fitted with a group of source buttons. Each button accesses a dedicated input, enabling it to be selected to the controlled output.

Multi-bus panels are configured to control groups of destinations, and can be fitted with either a dial-up keypad or a group of source buttons, enabling sources to be pre-selected before being switched to a selected destination, while an X-Y panel has access to all sources and destinations, typically using dial-up keypads, to provide access to every source and destination within the controlled matrices.

Panels are connected to the system controller, depending on the manufacturer, using RS422/485, Ethernet or other proprietary interconnection methods. Where routing switchers are used to provide router buses for Program and Preset selection, or even supplement crosspoints contained within production and master control switchers, their control surfaces can in some cases be connected to the control system either via the system remote control ports or as control panels to be used as an extension of the system control panels.

Remote control ports available on the system controllers and router frames enable external systems to control the routers remotely. These ports tend to be used by Automation systems and external, PC-based, control systems. These PC-based control platforms, when used with system controllers, supplement the hardware control panels while providing simple to understand and easy to operate graphical control panels, which can be configured to meet the operational demands of each and every system user.

Today's routing systems are, by far, technologically superior to the original systems developed 20 or 30 years ago (see Figure 4.13.7). While the wide availability of crosspoint chips used within other industries, such as telecoms, has enabled many companies to produce routing switchers, it is the innovations built into these routers and their associated control systems that enable some to stand out above the rest. As routing switchers continue to evolve, their shape, size and operational methods may change quite significantly, even down to the way signals are processed prior to being switched, but they will always be at the heart of the system.

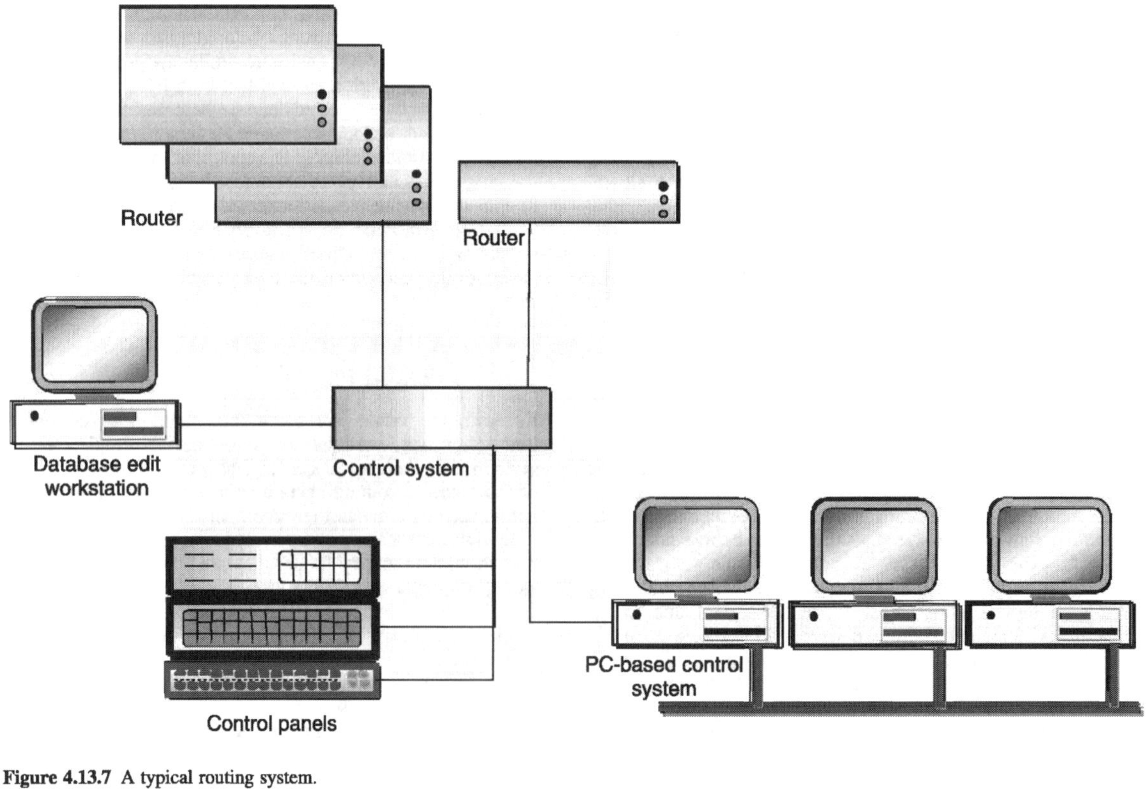

Figure 4.13.7 A typical routing system.

Richard Schiller

4.14 Transmission Systems

4.14.1 What is Transmission?

Transmission is the production line of broadcast television. Raw materials in the form of pre-recorded packages such as films, programmes and commercials are delivered to the operational centre and then sequenced together in a pre-defined order and delivered off the end of the production line to the consumer. It is the original streaming process. As much as new programmes can be freely created and scheduled by the broadcast organisation, it is only through the transmission process itself that ultimate revenue is secured. The process appears to the outside observer a pretty simple one and in its principles it certainly is. The additional complications for the unwary are defined in three aspects of the broadcast transmission operation. While not unique in industrial operations, they are each unusual and together make for a pretty challenging mix.

These three aspects of broadcast transmission are:

- The one-chance nature of the operation.
- The relatively high cost of failure.
- The need to track raw materials individually and not by class.

4.14.1.1 Just one chance

Unlike most of the processes up to the point of transmission, and indeed most industrial processes, there is no chance to go back and try again. For the general production line a failure results purely in the loss of time, which broadly means the loss of capacity. The immediate link to the consumer in broadcast transmission places it in a particular operational context, the one-chance service delivery operation. If the transmission sequence fails, then whatever has been lost cannot be reclaimed. This defines the requirement that transmission operates at very high levels of reliability. In fact, many broadcasters for many years would not try to define their reliability because the figures they achieved were in truth unacceptable. Now, real measurement and contracts securing the expected performance in writing are becoming commonplace and these discuss levels set at a small fraction of a percent of allowable failure. For example, a level of failure set at less than 0.3% could still allow a station to be off-air for an entire 24-hour day each year.

4.14.1.2 The cost of failure

Now, the quantity that is lost in the case of failure could simply be time, but in the commercial reality of broadcasting the consumer's service will be interrupted immediately. The consequences might just then be the consumer's view of the organisation or their messages. For commercial channels though, those messages are worth money and it would probably be income that is lost from the transmission of a valuable advertisement.

Commercial channels will value the advertising of course and perhaps the sponsorship of programming as well, but also now an increasing number of ancillary ways of deriving income such as real-time coordinated web events. The way non-commercial channels derive their value will be more dependent on their function and will vary between organisations. Subscription channels will place the premium on the programmes. State-sponsored channels or special interest broadcasting such as religious channels, however, might not make monetary judgements of value but instead be driven by pride and connection with the viewership. The one common link is that all stations will value some, and quite probably all, of their schedule.

For the broadcaster there is a second side to the equation though. Not only can the loss happen instantly, it can continue at a rate that reaches the unbelievable. It is this speed with which immense value can be lost that also plays a part in the design of transmission systems. This is of most direct value in commercial channels of course, but the principles apply across all channel types. In commercial terms, absolute value varies enormously between channels, by material type and by times of day. At its extreme though, millions of dollars can be lost per minute of failure for the most popular services in the highest value positions. It is important to reflect that broadcasting is driven by two monopolies: the monopoly on air time and the monopoly of rights to broadcast material. This keeps general values high in general proportion to the cost of purchasing or operating the production line equipment. Taking the highest values in the world then, a brief commercial that pays millions of dollars can pay for the entire transmission centre in minutes or even seconds. Very few industrial process chains amortise themselves at anything like this speed. This places particular stresses on those who design and operate broadcast transmission, and defines the nature of the products and systems that are used in the process.

4.14.1.3 Management of the raw material

As a production line there is one further aspect of the transmission operation that is key. The elements of raw material fed into that line have individual identities. It is not acceptable to provide the right type of material at the right point in the schedule, it is exactly the right material element that is required. Inventory management, media management in the broadcast context, is by individual piece of media not by class. Hence this last aspect of transmission is the need to track the media by these individual elements. This might seem an obvious statement to make but it makes a vast difference to the organisation of the processes.

Identification of material must be rigorous. The systems almost always run in a state where they cannot be allowed to fail. Broadcasting episodes out of sequence or the wrong commercial or placing adult material on a children's channel are examples of unacceptable failures. It is only in recent years that 'asset management' as a concept has begun to be foremost in the minds of broadcasters. However, it has always been an essential element of the process, it has just been conducted in a less formal way. Constant rechecking of the content at many different points in the chain has been one solution. Awareness is often heightened by the fact that, for many, the transmission of the wrong material was seen as the highest level of failure.

4.14.2 The Four Contracts

What are the requirements of the transmission operation? The broadcaster operates within the boundaries of four types of contract (Figure 4.14.1). These are:

- A licence to broadcast.
- A right to broadcast the material.
- Commercial agreements covering the transmission of advertisements or other promotional material.
- A contract or relationship with the viewer established through published programme guides.

4.14.2.1 A licence to broadcast

Most television stations operate under some kind of licence. The terms can be extremely involved and can be financial in nature. Whatever the exact plan, the broadcaster will have rules that they need to obey in order to continue their business. One of the objectives of the transmission process will therefore be to ensure that these rules are not violated. The type of material being broadcast, the balance between commercial and non-commercial content, and the technical quality are three different examples of conditions that may be imposed on the broadcaster.

4.14.2.2 A right to broadcast the material

Those who produce material make their money through their ownership of the intellectual property and the implementation of rights laws. Use of the material may be by gift, agreement or as a commercial contract. Whichever it is, it will generally come with conditions attached. These will often include payment terms for purchased material but can also include the

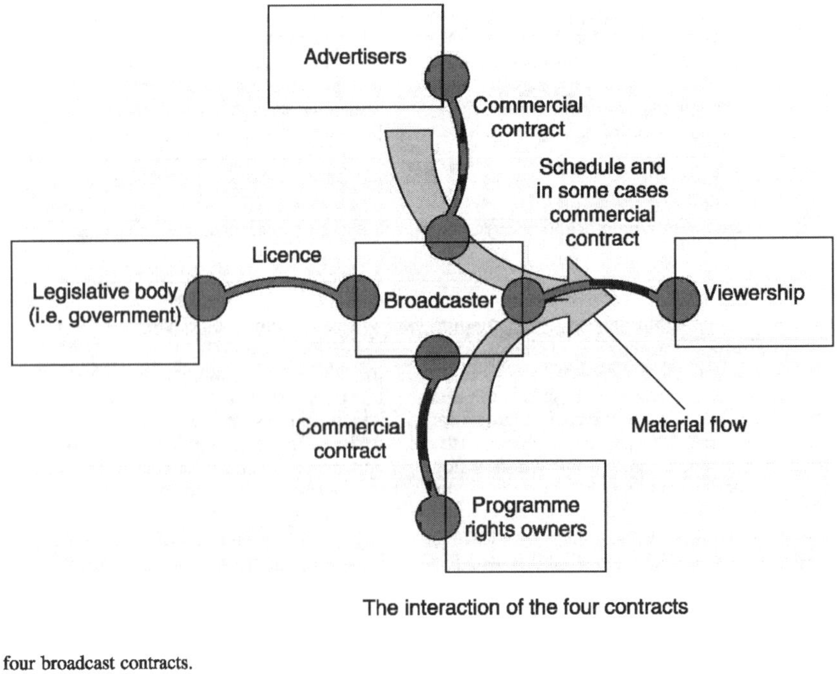

Figure 4.14.1 The four broadcast contracts.

number of repeats allowed, the times of day when the material can be aired and the nature of surrounding material.

4.14.2.3 Transmitting advertisements

For many broadcasters this is what pays for the whole operation. There are several ways that paid-for messages can be delivered, of which the most traditional is the commercial. Advertisers pre-purchase a space in the schedule for their commercials, generally known as a spot. The common basic advertising pattern operates by placing a series of commercial spots, perhaps four to eight, together in a commercial break. The programme is segmented. Before each segment is broadcast a commercial break is transmitted. Local rules, dictated by the licence to broadcast, might impose conditions such as the length of the break and how the break should be composed. As an illustration of the type of rules imposed, take the simple matter of beginning and ending a commercial break. Some broadcasters can go straight into the break and straight out at the end, then back into the next programme or programme segment. However, some licences impose that the break is announced or that a pause is placed between the programme and the commercials. A slide or animated announcement is common in mainland Europe, for example.

Advertising can take many forms though. Sponsorship might lead to particular products being used or displayed within the programme. Another sponsorship form is an announcement at the beginning or end of programme segments explaining that the programme was brought to the viewer courtesy of the sponsor. Text crawling across the screen is increasingly being used as a chargeable promotional tool. In fact, as the years go by, more and more ways of sponsoring the channel are being discovered.

4.14.2.4 The contract with the viewer

Most, but not all, channels announce their programme schedules in advance. Viewers will tune in to watch a specific programme and a contract is formed between the broadcaster and the end consumer. Like all businesses this supplier/consumer relationship will determine the return that the broadcaster receives. Consumers let down by not being provided with what was promised when it was promised will walk away and find their entertainment and information elsewhere. For a commercial operation this will result in loss of revenue. For special interest channels this will reduce the reach of their message. Good customer relations start with fulfilling these unwritten contracts and transmitting the programmes fully, without flaw and at the advertised time.

4.14.3 The Main Types of Transmission Channel

The transmission process consists of fulfilling the four contracts described above. With the exception of the contract with the consumer, most of the other contractual issues can be addressed in some way before transmission commences. The extent to which the actual transmission process is affected is determined by two factors. Firstly, just because issues can be addressed earlier does not mean that they necessarily are. Secondly, transmission is plainly the last place to correct any errors that have slipped through.

In any dispassionate analysis the observer would generally side with the view that as many issues as possible are cleared prior to the transmission process itself, simply because of its one-chance-only nature. In short, the fact that the transmission function is the highest-pressure part of the process makes it the best function to keep issue free. Surprisingly, many broadcasters still rely on this most pressured and difficult stage of the process for the solution to many of their contractual challenges. What can be achieved in terms of keeping issues away from transmission itself is also set by the type of transmission channel; more can be expected of, say, a film channel than a sports channel.

4.14.3.1 The packaged model

Considering first just packaged – pre-recorded – material. Because the transmission process is always live, broadcasters gradually realised that anything that can be done to pull the fulfilling of rules back from the transmission environment would reduce costs and increase reliability. At earlier stages in the system the organisation has the opportunity to take their own time and as many attempts as required to get it right. By the time the material is on-air, then there is only one chance to get it right and this always has to be done in real time. This has led to a gradual shift to an operational concept where everything is thoroughly prepared prior to transmission where possible. Transmission itself is then executed in as simple and as automatic a way as possible. This could be described as the full preparation model. Where the channel just features packaged material then this model is achievable in full. A total hands-off transmission can be implemented. Channels that can run this way include film channels, cartoon channels and general programming channels that exclude live content (see Figure 4.14.2).

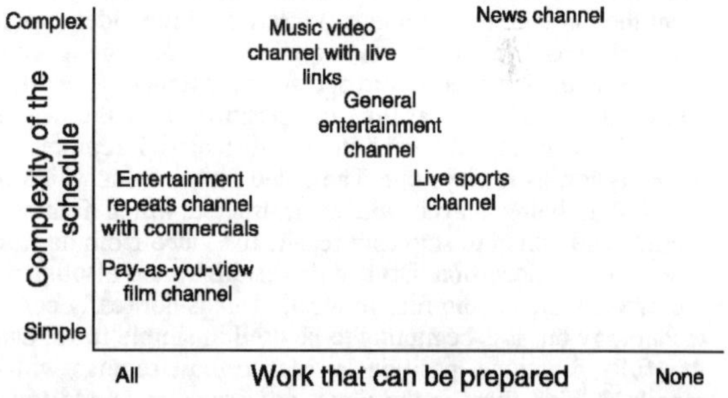

Figure 4.14.2 The complexity and preparation of different channel types.

4.14.3.2 The live and complex model

The exception to the packaged model is live broadcasting. Where some of the station output is live then not everything can be prepared prior to transmission commencing. More of the task of fulfilling the four contracts is down to the one chance in real time. For this reason, channels that operate live or operate with at least some live content are a very different business proposition from channels that only operate with packaged material. First of all, they cannot be fully automated and they will always carry more risk than fully packaged channels. In a practical sense this means that more people will be involved in the transmission function and more contingency to cope with failure is required. For these reasons the costs of running this kind of channel for the transmission function will be much greater than an efficiently organised packaged channel.

4.14.3.3 Evolving solution to the packaged model

As described above, channels broadcasting packaged material were evolving toward preparing everything prior to the transmission part of the process. As equipment started to become more reliable a new paradigm evolved. In this scheme, preparation is again carried out before transmission because of the lesser time constraints and the consequent lower operational pressure. However, this preparation extends to the planning processes and not the execution, at least not in its entirety. In this model the aim is to store the original material ready for transmission and not to store material in a pre-prepared form. Processing to make the material suitable is then executed live as it is played out. Examples include material stored in original 4:3 aspect ratio and then converted at playout to 16:9. Another would be an HD transmission carried out from an SD source via an up-conversion process. In the past these types of process would have been carried out prior to transmission, requiring the creation of a dedicated transmission copy of the converted material. If equipment is assumed unreliable then this was a sensible precaution but it is not now generally relevant. The saving from not pre-processing the material can be vast; storage and equipment savings are obvious but whole departments can disappear.

At the moment solutions encompass video processing. The evolution trend though is toward a different way of storing the material. In this system original versions of material are kept in random access storage, typically video servers. The transmission copies are generally edited versions of these originals. These will generally be edited for total duration, language and other culturally sensitive material, and then finally to segment the material for commercial insertion. Like video processing this has previously been carried out off-line prior to transmission. However, future operations will include real-time conforming. That is, real-time composition from the original files of the exact edit version of the material required for transmission as it plays out. The video server from which the material is being played will be instructed which frames to include and which to skip as it recalls the video from its store. The EDL (edit decision list) that drives this would also instruct the server which audio files to recall. This is not really current technology but it is beginning to be used in simple form. Once this fully functional position is reached, cost savings will be manifold. First there is the direct process savings. On top of this, however, are the savings made through only storing the original master material and not the multiple transmission versions. Space is saved of course, but with a massively reduced storage requirement material can be placed closer to where it is played out and kept there for longer, possibly forever. Whole areas of logistical work just disappear in this model with a consequent saving in administration, coordination and areas of potential failure.

4.14.4 The Transmission Process

Transmission proceeds according to a defined list known as a schedule or transmission log. Nowadays, this document is universally electronic, although a paper copy is still sometimes printed out for operational staff. Surprisingly for the casual observer though, through lack of integration the electronic version may not contain the entire detail required to execute the plan.

Schedules generally start life many months before transmission as a rough plan. For some this consists of a skeleton of the programme material, for others it begins with the first commercial sales. It can even start as both in separate operational areas of the organisation. However it starts though, the schedule will then gain definition with time as transmission approaches.

Apart from programmes and commercials the schedule will contain some additional content. For a commercial channel, at some point in the commercial break, broadcasters will often place their own self-promotional items such as announcements of forthcoming programmes. These additional but incidental pieces of material are generically known as interstitials. Non-commercial channels often still contain breaks but in this case entirely populated by self-promotional and informative material. It can be surprising to find that channels that have no need to break programmes into segments for commercial reasons still do so, simply because it is conventional.

Passing through perhaps several departments, the schedule gains definition with time. Major players within the station include those departments responsible for programming, airtime sales and interstitials. For commercial TV channels the sale of commercial air time is normally the most important pre-transmission stage of the process because it is where the revenue is generated. In the past, departments handling the sale of commercial slots were sometimes known as 'commercial traffic', and so now both the process and the software to assist the process are frequently known as traffic. Traffic systems schedule and sell the commercial air time. They usually also provide facilities for creating the programme schedule and to bring together those programme schedules with the commercials. The output of a traffic system takes the form of a list, the schedule or transmission log (see Figure 4.14.3).

The input to the traffic software consists of data describing programmes and commercial spots. A crucial element of any traffic application is its media database. In fact, traffic was the first real application of media asset management in the broadcast arena. Precise definition of the material within this database is key to creating a trouble-free schedule. A central database of commercials allows traffic systems to construct the commercial part of the schedules. Depending on their scope, the databases can also cover programmes and the miscellaneous interstitial material such as programme promotions mentioned above. Data stored in the database combines the necessary technical detail such as duration data with operational data such as the material's title and business information, e.g. ownership for programmes and billing data for commercials.

In operational terms one factor matters more than all others during preparation and that is the duration of the material. In order to run an accurate defined schedule the durations of the pieces of material must eventually be known with absolute accuracy. The accuracy and the precision of the definition of the material prior to transmission are rarely consistent. Commercials are sometimes the best defined, largely because in many channels the slot sizes are prescribed, meaning that commercial slots can only be bought in particular lengths. However, programmes are often less well defined, especially if they are to be broadcast live. During the composition of the schedule, rules pertinent to the broadcaster's licence must be

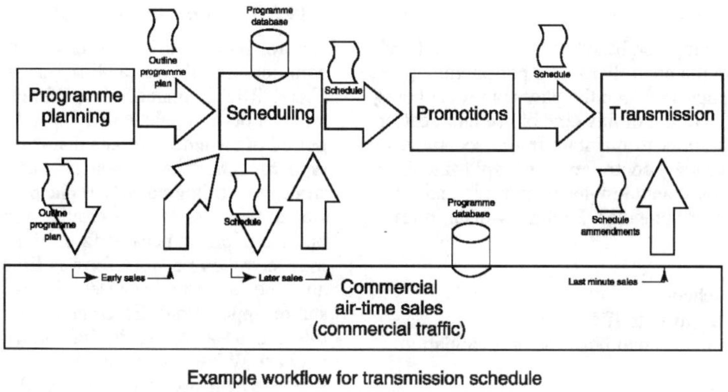

Figure 4.14.3 Example workflow for a transmission schedule.

enforced. If the traffic software can handle these, then there is a substantial benefit to the broadcaster. There are many solutions to how slots are sold, but most broadcasters will pre-sell some commercial space while allowing others to be sold later. For example, these could range from a year ahead through to as late as a few minutes before they are broadcast.

From the traffic process the schedule arrives at transmission. Conversion to a form suitable for the broadcast automation system allows traffic schedules to feed through without any need for human intervention and is generally standard.

4.14.4.1 Manual intervention

One visible irony of many automated TV channels is the amount of manual operation that is required. 'Why,' you might ask, 'does a channel that requires intervention also need automation?' The manual intervention allows for ultimate control, particularly for live programming. Even if the channel is not fully live all of the day, it will often run in multiple modes within a 24-hour period. Control can take many forms, but often what is required in terms of manual control is a little adjustment of timing. Holding onto a feed or package a few frames longer or alternatively cutting out slightly earlier is where human judgement is required. The same human intervention means that larger adjustments can be made by adding or removing a package from the schedule. This could consist of dropping an interstitial or adding a programme, for example.

In the beginning, automation systems were often designed only for fully automatic operation. Manual adjustment, such as it was, was often clumsy. In many cases it consisted of turning off the automatic control altogether and running manually until reverting to fully automatic operation again. This is a very inefficient way to operate and the cause of many on-air failures. In fact, a good relationship between operator and automation system mimics a good management mentoring strategy. The user should be able to leave the automation to run the schedule until the schedule becomes too much for it to handle. Then they should be able to dip in to take control. Ideally, they should be able to simply take control only of those elements that require their assistance, typically timing issues as described above. The operator should then be in a position to drive those aspects of operation for as long as required and then gracefully pull out to leave the automation system fully in charge again.

4.14.5 The Transmission Function

4.14.5.1 Equipment

The equipment that broadcasts the programme stream consists of:

- Sources such as VTRs, video servers, character generators and lines-in chains for live external feeds.
- Video switchers or routers to sequence together the sources.
- Keying and DVE (digital video effects) functions to provide logos, station idents, emergency announcements and secondary advertising.
- Video processing to enhance and manipulate the pictures to make them suitable for use.
- An automated control system to sequence the material and control ancillary actions and to provide the transmission as-run log.

In addition, it is usual for some or all of the above to be duplicated to provide redundancy in the case of equipment failure.

Most material is now sourced from video servers, which are generally more reliable and less expensive to run than VTRs. For the simplest channels showing packaged material, no subsequent switching is required. In other cases either a simple router or quite often a presentation switcher is used to sequence the material. The switcher allows different transition types for a more varied look to the channel or a stronger branding image. Switchers with no control surface are used where fully automated channels mean that no user intervention will be required. Presentation switchers resemble the units used in production except that they are simpler, have fewer inputs and buses, and handle sound as well as vision in one unit. They also have simpler control surfaces. Both transmission and production units need well laid out controls with large easy-to-use buttons of a certain quality.

Additions to the image being transmitted include text and logos from still stores and character generators keyed into the stream. Keying can be built into the presentation switcher or provided by external dedicated devices. The use of DVE moves to replace keying is becoming more common, with a squeeze back revealing the text or alternate pictures underneath the main feed. For this reason the DVE capability is increasingly found inside the switcher. Video processing consists of functions provided in either a box or modular form, up- or downstream from the switcher.

4.14.5.2 The control system

Automation control is computer based. At the lowest level, commands are sent from the controller to the peripheral equipment to activate it. Examples include the play command being sent to a server or VTR. Timing of this kind of communication is absolutely essential in order to maintain frame accuracy. In order to ensure frame accuracy controllers often split functions into higher level functions and low-level control, and then execute each in a different computer. Examples of high-level functions include:

- Creating and editing schedules.
- Downloading schedules from traffic systems.
- Asset management functions and business rule validation.

Low-level functions consist of clock watching, sending commands and in some systems responding to external events. Low-level controllers generally benefit from dedicated hardware that could be designed for real-time operation. They are also saved from being swamped by non-real-time functions that are executed in the high-level controllers. These low-level controllers might be invisibly mounted in the same box as the higher level computer so that the whole unit looks like one single computer. In other installations, they are mounted in their own separate enclosures. They might even be distributed closer to the units they control, so that the lowest level control connections are kept fairly short. High- to low-level connection within the controller architecture can take almost any form of course, but traditional serial connections are gradually giving way to networked topologies.

As desktop workstation hardware such as PCs became fast enough, the possibility of inexpensive architectures that were not tiered became realistic. Some automation controllers now consist just of a PC with the necessary interfaces.

Whether the control system consists of multiple control layers or just a workstation, the same constraints apply. With so much depending on their correct operation, most users will look for extremely reliable operation. Dual power supplies are one option but duplicated controllers are often used for even greater robustness. In case of ultimate failure, fast start-up times, boot times, are important. These reliability issues are difficult for workstation-based control systems. However, there is one area that challenges all automation controllers for reliability. Most legacy equipment operates with RS422 serial control and provides only one control port. It is extremely difficult to provide two control inputs to a device because, in normal operation, which one should the equipment listen to? It is easy to devise some scheme in any dual redundant system where one is used until it fails, at which point the other takes over. The trouble is that when things fail, they do not behave. With power supplies, failure generally results in a loss of power. By adding to power feeds together, failure can usually be accommodated. Control does not work like that. What if the failure consists of one controller sending the equipment wrong commands instead of no commands? Newer equipment might now have dual Ethernet connections for control, but the issue is essentially the same and has not gone away. That issue is, how do you switch over the control? Hence, even in dual redundant systems some form of switch-over mechanism will often be required. For RS422 that could be a mechanical switch or relay, for network connections it could be more subtle. Subtle systems often hide their own failure modes, however. Mechanical switches are nice and obvious.

4.14.5.3 Connections

Control interfaces largely take one of three forms. The most common was the serial link, usually RS422 running at 38.4 kbaud. RS232 connections are far less common in the broadcast arena. For unsophisticated links GPIs were used. These are pulsed or switched outputs that form simple on/off or start/stop type triggers. The newest connection technology is the network, which is gradually replacing serial control. Networking was at first less able to provide reliable real-time interfacing, but as equipment became faster this is now less of an issue. The most common network form is Ethernet, which is understandably the favourite because it is pretty well universal in workstation applications. Ethernet is not deterministic, however, and this can be a serious limitation when it is used as a control network. What this means is that there is no guarantee that a message will arrive in time to allow frame accuracy or any other precision of operation. However, deterministic network alternatives are available and may be preferable for some crucial connection roles.

4.14.5.4 Frame accuracy

The importance of frame accuracy is almost universally overestimated in television. Most of the time, quite sloppy practices are acceptable but few believe this. The odd extra frame here or there, a few clipped frames, starting a frame or two too late all rarely matter. When it does matter though, it is very obvious, vital and really unpleasant to watch when it fails. This is probably why so many overestimate it. The basic challenge is to bring together three accurately defined values in one event. These values are the end frame of one piece, the first frame of the next and the exact time of day at which the boundary between these occurs (see Figure 4.14.4).

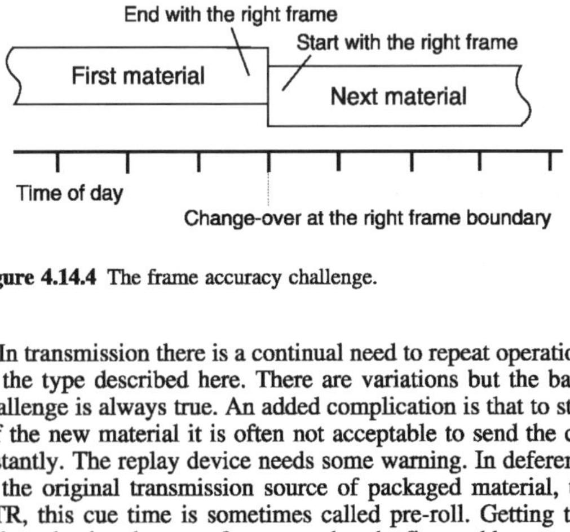

Figure 4.14.4 The frame accuracy challenge.

In transmission there is a continual need to repeat operations of the type described here. There are variations but the basic challenge is always true. An added complication is that to start off the new material it is often not acceptable to send the cue instantly. The replay device needs some warning. In deference to the original transmission source of packaged material, the VTR, this cue time is sometimes called pre-roll. Getting this right, selecting the exact frames used as the first and last, would intuitively seem to be vital. Most packaged material is prepared with some tolerance at both the start and end of the piece, however, and so exact timing is generally not as much of a challenge as it would seem. This way of building tolerance into the material is so traditional in many countries that it is almost built into the psyche, but it is not universal and cannot be assumed. As stated though, there will be times where frame accuracy is required, particularly where the material has not been prepared in a tolerant way as described above. One common cause of the need to select exactly the right frame is through audio starting 'hard' at the first frame or ending at the last. Another is through re-purposing the material. Here the broadcast output is derived live from the original recording rather than from specially prepared transmission versions, so there is no opportunity to include well-formed breaks where the transmission can cut in and cut out with a small error. In the future this practice will increase and make frame accuracy more vital.

Outside of material issues, there are other reasons to employ frame-accurate capabilities. These include the need to synchronise equipment along the length of the broadcast chain. So, for example, if there is an aspect ratio converter downstream of the source switch, then the transition from one source to another will probably need to be co-timed with changes in aspect ratio. This is the sort of frame accuracy that is often assumed by users but not always delivered. Ensuring that all parts of a transmission system can operate in harmony to give a frame-accurate result is an important part of specifying a transmission system.

4.14.5.4.1 Frame-accurate arithmetic in transmission

There are a set of rules for frame-accurate arithmetic. These are hardly ever stated and little understood but they are essential in order to design or test a frame-accurate operation. These relate to the definitions of the in-point, out-point and duration of a piece, and how these relate together (Figure 4.14.5). The first rule is that the in-point is the timecode that represents the frame address of the first frame of the piece that the viewer should see. The second rule is that the duration is entered and reported as a timecode but it is not used as a timecode. It is turned from a timecode into a count of the number of frames to be shown; we will return to this. The third rule is that this duration frame count is one more frame than the number actually shown. The fourth rule is that the out-point is the timecode address of the frame after the last one shown from the piece. In other words, the viewer never sees this frame. This convention is known as 'exclusive' since the out-frame is excluded.

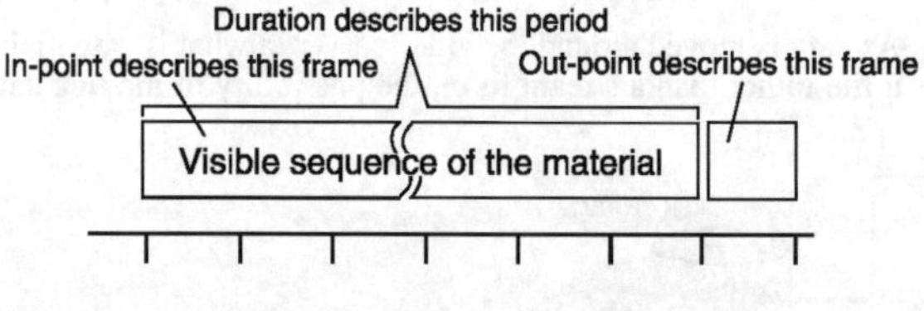

Figure 4.14.5 The meaning of in-point, out-point and duration.

So, an example would be a piece starting at an in-point defined by the timecode 10:00:00:00 and running for a duration of 00:00:00:10. Its out-point would be 10:00:0010. What the viewer will actually see is 10 video frames. The first frame they see will be the one with the address 10:00:00:00 and the last they see will be the one with the address 10:00:00:09.

This might all seem pretty simple and generally it is. What complicates the applications of these rules in practice is the translation of the duration which is expressed as a pseudo timecode address into a frame count. Even this would not be so challenging were it not for the NTSC frame rate of 29.97002997003 Hz and the added complication of drop-frame timecode. This conversion is covered in detail elsewhere in the book.

4.14.5.5 Reliability

The number one requirement from an automated transmission system, or from any transmission system, is reliability. Actually the transmission process is simple in principle and its challenges are almost all in the execution. Not losing revenue is of course the primary technical challenge of the whole operation. Reliability is a much misunderstood concept, however, largely because of people taking too narrow a view. Reliable systems are reliable because they provide a reliable structure and at the same time promote a reliable way of working. A reliable system will, for example, be constructed from reliable parts, but it requires more than just this. In the past keeping the equipment running was a substantial undertaking. Mechanical-based equipment such as VTRs were working at the limits of technology. Low standards of reliability were expected from the equipment. For many years the standard of performance across almost all functions has risen to make the question of equipment failure almost irrelevant. Unfortunately, that has not led to failure-free operation though. For a start, all parts of the system have to connect in an infallible manner. Next, no systems work entirely without human intervention and the presence of people will probably be the greatest source of loss of revenue. It is for this reason that the area requiring more attention now than any other is the ergonomic design of the system. In short, reliable equipment means that the focus should be at the macro level on reliable systems.

Reliability does not just stretch to the transmission function alone. The business process has to wrap around the complete operational system. In order to bill the clients for transmitted commercials, feedback is required indicating the successful broadcast. There is little difference operationally between not broadcasting a commercial and not billing for it. Feedback can be taken in the form of the as-run log from the transmission system. Feeding this data back to the traffic system completes the loop.

4.14.5.6 Ergonomics

The most impressive change in the broadcast transmission operation is possibly in the ergonomics of the systems being used. As has already been stated, the system as a whole needs to operate efficiently to ensure overall reliability.

A lot of this development has centred on monitoring multiple channels from a single operator's position. At-a-glance viewing of the complete output status is becoming essential. Fewer operators are looking after more and more channels. Timeline displays showing the current on-air programmes as well as the upcoming events represent the multiple schedules. Glass-cockpits that can display multiple feeds on one single display are becoming almost ubiquitous. These replace the conventional monitor stacks. Used well, they can display more channels more efficiently than separate monitors. They have an added advantage of being re-configurable so, for example, the size of screen dedicated to each feed can be varied according to immediate operational need. However, both of these features are insignificant compared to the main feature of glass-cockpits. Glass-cockpits provide for automatic detection and, most importantly, display of signal status. Currently this information is limited to basic signal condition at the output of the channel. In future though, combining this display capability with infrastructure monitoring systems such as RollCall will provide far greater depth to the displayed information.

With one individual assigned to look after four, 16 or even many tens of channels then some assistance is required. An operator tasked to look after one channel will be working against human nature. Concentrating for 8–12 hours straight is a challenge. Automatic detection of the simpler signal condition parameters at least helps. Still video, loss of signal and low-level audio are examples of useful parameters to detect. Deeper information would include subtleties such as whether the signals are being efficiently compressed for the transmission and if any are stealing too much bandwidth from a multiplex. The displays typically indicate a potential fault or warning by ringing the offending feed in red. Through this mechanism the operator can be quickly guided to the channel with the problem. The main weakness of the glass-cockpit solution is that few or even a single display device now substitutes for what once was many monitors in a traditional stack. This provides a single point of failure.

4.14.6 Video Servers

Video servers use hard disk storage to record video material and associated audio tracks (see Figure 4.14.6). They can store full bandwidth signals for post-production but for transmission operations are likely to use compression techniques to save space and increase performance. Early servers were either based upon adapted PC (personal computer) parts or a combination of dedicated hardware and PC parts.

A video server consists of four active elements plus another two supporting elements. These are:

Active

- Storage.
- A bus.
- Stream encoding.
- Stream decoding.

Support

- Control.
- Power supply.

4.14.6.1 Storage

The storage is the key to the server's existence of course and very much the heart of the device. Its most important qualities are its reliability, its access speed and bandwidth, and its size. We shall see that access speed and bandwidth can, to a degree, be compensated for in the way the server is constructed and operates.

At the moment storage is almost exclusively by magnetic disk, but essentially any near-random access media with sufficient bandwidth and short access times can be employed. The storage can be one or many disk drives working in parallel. Multiple drives allow for more storage but importantly also increases bandwidth and can reduce access time by providing multiple parallel I/O.

4.14.6.2 Bus

4.14.6.2.1 Internal buses

In fact, this should generally read buses rather than the singular bus. Data has to be martialled on and off the disks. First data has to be managed in the vicinity of the disk via a fibre channel, SCSI or other high-speed interface. Then the data is moved within the local area, perhaps using SCSI again or a non-standard proprietary bus, or perhaps a network connection. Then the data could be moved over a wider area. It is very dependent on the specific design of the server, but many cascaded buses might end up being used. The careful art of server design is largely related to optimising these buses and the management strategy for shepherding data on and off the disks at the maximum possible rate. The way data is written across multiple disk drives can aid the speed at which the data can be read back. This is related to the structure of the buses through which the data is written. For example, data words can be written in parallel to many drives. Or, blocks of data can be sent to each drive so that the block corresponds to the optimum size for quick access off that drive. Whichever strategy is chosen the buses have to accommodate breaking the data up and recombining it during the write and read operations.

4.14.6.2.2 External connections

As data is moved around the video server in what is essentially a file rather than a stream form, the possibility of moving data

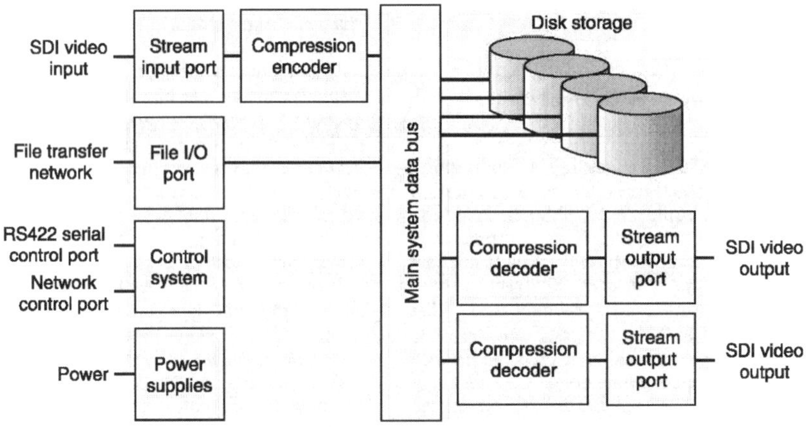

Figure 4.14.6 An example video server structure.

between servers in file form became a possibility. This has two main advantages. First, and most obvious, the data can be moved in its compressed state and therefore take less bandwidth to move. Second, the data is not being streamed and therefore does not have to be moved in real time. It can move slower than real time and take advantage of lower bandwidths than were useful for streamed video. It can also move at faster than real time with obvious benefits in being fully copied faster. To operate at anything like a useful speed the faster network connection systems are used. Fiberchannel or fast Ethernet are two examples.

The problem with first generation file-based connections was one of compatibility (see Figure 4.14.7). The fast network was a transport layer that might be compatible between devices but the files themselves were not. One manufacturer's files could not be decoded by another's server. The introduction of the Media Exchange Format or MXF standard is bringing a partial revolution in this area. If two devices can both handle MXF then data interchange is eased (see Figure 4.14.8). In essence this allows the sending device to wrap the data in a way the receiver can understand. This is not yet a complete solution, however. If one server is compressing the files as Motion-JPEG files and the other only has an MPEG decoder, the link is still essentially useless. So, to be completely compatible another layer of interoperability is required. This is compatibility at the compression system level. One of the first examples of this was video servers that could utilise DV files shot with camcorders according to this well-recognised standard. This layer is not the last, however. MPEG is a broad standard with many options, as users are discovering. The two servers need to be able to deal with an exactly similar version (or dialect) of the MPEG standard in order to work together. However, once this is achieved then not only servers can interoperate, but other devices can as well. Gradually, the movement of media migrates away from the streaming world to a file-based environment.

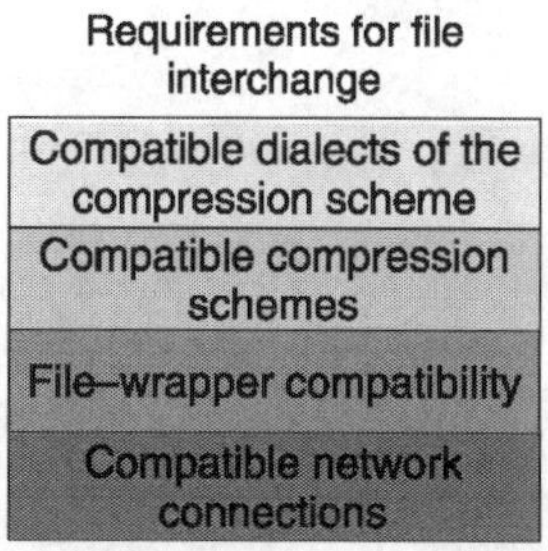

Figure 4.14.7 Levels of file interchange compatibility.

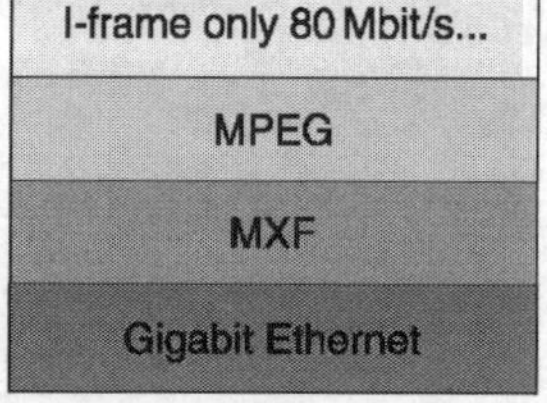

Figure 4.14.8 Example of file interchange compatibility.

4.14.6.3 Encoding and decoding

Information initially enters and primarily leaves the server in stream form. This usually takes the form of an SDI video stream. A conversion has to take place between the external streamed signal and the internal bussed data form. It is a specific form of stream-to-file conversion. In most cases the video is also compressed in order to reduce the data rate within the server. This has several advantages. Most obvious is the improved density of storage, which goes up pretty well proportionately with the compression ratio. This alone is a powerful argument for using compression but was not the primary reason for using it, however. More important in the early development of video servers is the fact that a reduced data rate improves the effective signal (media) bandwidth, as mentioned above. Achieving the bandwidth into and out of the server as well as on and off of the disks is quite a technical challenge. Reducing the data at the point of entry through compression simplifies greatly the technical challenge. It has a powerful operational benefit as well. It means that the number of feeds on and off of the storage within the server can be increased. For example, if 2:1 compression is used then the number of signal streams that can be used roughly doubles over an uncompressed storage subsystem.

4.14.6.3.1 Cue time

One key performance issue set by the decoder is cue time. When a new clip is required at the output of the server it cannot just be output. The clip has to be cued up. This consists of priming the control circuitry, output buffers and compression decoder. Cueing can take a few frames through to a few seconds. Once cued though, the signal can usually be started instantly and will flow continuously. This sets one of the inherent limitations of server technology, that clips cannot be instantly summoned. Cue times of many frames are common. They will set the shortest duration clip limit that will often be operationally a key parameter for the server. For example, if the server has a cue time of 20 frames and the user wishes to broadcast clips of no more than half a second long, then they will need to use either two server ports to achieve this, or another make of server.

4.14.6.4 Control

Somewhere in the server there is a control system that sorts out the functions mentioned above. The main requirements of the control system, apart from general reliability, are that it can work in real time and that it starts up quickly. Servers built from PC technology can inherit their poor start-up times, especially if they use desktop PC operating systems.

Essential, but well hidden from the user, is that the control system has real-time capability. This is very difficult to prove or disprove and often shows up as a lack of reliability rather than anything that can be tied down more precisely.

4.14.6.5 Fault tolerance

A major concern of many users of servers is fault tolerance. Disks can be configured in one of the standard IT redundant configurations. These systems, known as RAID, are well covered elsewhere and will not be detailed here. These provide for disk failure but not really for anything much more. However, because of their (the disks) mechanical nature, RAID does cover the most likely reason to lose output in a video server. There is no doubt that mechanical disk failure is the most common form. It should be remembered that the use of the disks in this application exceeds their typical use in almost any other, including as part of an IT server. Because of this, redundancy in disks is the number one feature on the specification that most will look for as a demonstrator of a reliable configuration. Dual power supplies are also key, however, and just as in other broadcast systems the need to hot-swap is also important.

All this kind of fault tolerance can seem hollow though if other potential failures have not been catered for. The control system is vulnerable but often not provided with redundancy. Indeed, ironically some server designs provide for redundancy in disks for storing the video but not for the disk used in the control system.

4.14.6.6 Storage Area Network (SAN) systems

Video servers are essentially built from separate components around the central data bus or structured set of buses, as described above. The addition of the capability to import and export data in file format changed their operation at quite a basic level. If you can move data as files between servers then there is now no real functional distinction between the internal data buses and the external network interfaces. Both move file-type information between storage devices. Storage Area Networks rely on an explosion of the internal structure of the video server and a merging of the internal and external data pathways.

Storage Area Networks need material to be acquired from the stream just as video servers do. In this case it could be a server acting as the acquisition element or a dedicated device essentially replicating the stream-to-file interface part of the server, complete with compression encoding. Disks with their own high-speed network connections are tied directly to a network backbone. Unlike with the video server, this backbone is not internal within a product box but an external connection, which can easily be added to. As already mentioned, devices to turn streams into files will form part of the system and connect across the same data backbone. So will devices that can turn the files back into streams again. The system still needs an overall controller that can coordinate the ingest and movement of files through the storage systems and across the network, keeping track of them at all times.

The advantages of SAN-style operation are difficult to express because they are so simple. They revolve around the fact that the components are all distributed and not collected within a box to create a product. This means that the user can optimise each component for its function. Then, as the system ages, particular components can be exchanged for newer, more capable or more reliable ones with ease. This is made all the more possible by choosing standard IT parts and technologies. Disks, for example, are known to date with great speed. They will be the first part of the system to need replacing but will have been surpassed by several generations by the time they need to be exchanged. SAN configurations were not easily possible when the combination of the internal buses and the way data was striped across the disks was so critical. The fact that these systems are now achievable shows that off-the-shelf IT products now have sufficient performance for broadcast storage applications. They are now operating at performance levels that mean they can be used with ease in an unmodified form.

Robert Pape
Advanced Broadcast Solutions

4.15 Media Asset Management Systems

Media Asset Management systems are an essential part of the infrastructure of a modern broadcast facility, as an MAM system provides the broadcaster with the ability to manage their asset in an efficient and consistent way across the 'enterprise'. Unlike many broadcast technologies, Media Asset Management systems are driven by working practices and the commercial requirements of a broadcaster; technological advances are the catalyst in the process, not the '*raison d'être*'.

The term Media Asset Management as it is used here encompasses the management of all Rich Media Assets from acquisition through to publishing. The principle of an MAM system is the integration of Information Technology with traditional broadcast technology throughout the broadcast 'enterprise' to streamline working practices in order to ensure video, audio and data (documentation, scripting, location information, etc.) are uniquely identifiable, accessible and usable. Metadata, the descriptive data associated with the essence (see later), plays a vital role in the identification of an asset, giving the ability to ingest, browse, utilise and publish the 'asset' seamlessly.

In order to fully understand the Media Asset Management system, it is imperative to understand the nature of an asset, the life cycle of that asset and its commercial value to the broadcast organisation. The workflow associated with the asset – pre-production, acquisition, ingestion, indexing, cataloguing and archiving – forms the basis of the operation of the system. Ownership and copyright of the asset must also be protected in order to protect the revenue of the broadcaster.

Recent trends towards IT-based production and advances in IT infrastructure – for example, Storage Area Networks, Fiber Channel and Gigabit Ethernet – have enabled faster-than-real-time transfer of an asset. Database technology has improved, storage capacity has increased and cost per byte has decreased significantly, all of which are factors that are bringing the enterprise-wide MAM system closer.

Using an 'enterprise-wide' Media Asset Management system to manage the rapidly increasing quantity of Rich Media Assets, to rationalise operations and create revenue is the panacea for all broadcasters. However, at the current state of maturity a more pragmatic, developmental approach has to be taken.

4.15.1 Definition of an Asset

In order to understand 'Media Asset Management' it is imperative to clearly define what an asset is. A joint task force was established by the SMPTE and EBU in 1998, one objective of which was to clearly define the nature of an asset. Their conclusion is that: 'an asset is defined by the SMPTE/EBU as being Contents + Rights, where content is Essence + Metadata' (Figure 4.15.1).

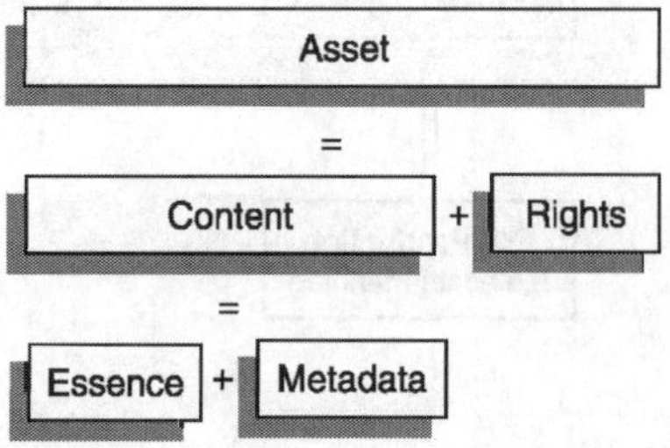

Figure 4.15.1 A representation of an asset as defined by the SMPTE.

Several important factors emerge from this definition; they are that the complete asset is a commercial entity – that is, it has rights and by definition a commercial value, that the metadata plays a major role in the construction of an asset and that the audio/visual material, the essence, alone cannot be defined as the asset. Metadata has an important role providing unique information about the essence that will form the basis of tracking, storage and searching of that asset.

4.15.2 Metadata for Media Asset Management Systems

The importance of metadata cannot be underestimated when considering Media Asset Management systems. Metadata provides *the* key information about the essence of the asset and will be used to identify and locate the asset throughout its life. Metadata can be handled independently or included in a file format such as UMID, MXF, AAF or DXF; please refer to Chapter 2.9 for a complete description of metadata and file formats.

4.15.2.1 Metadata Life Cycle

Metadata that is captured at source will accompany and be associated with the asset throughout its life, being modified, enhanced and supplemented with each downstream process. Metadata must be considered at each point of the production chain and be seen to have a life cycle (Figure 4.15.2). As an asset is modified, in post-production for example, further metadata and potentially a new asset will be generated.

4.15.2.2 Metadata Capture at Acquisition

The life of an asset starts at acquisition, but as can be seen in the SMPTE definition of an asset, it does not become an asset until metadata and rights are added; therefore, at this stage the material is best described as essence. Traditionally the information contained in metadata was a long-hand textual description of the essence and production, title, video and audio content, time and place, all being described during the production process. This method is simple but excessively time-consuming and impractical in an IT-based production environment. Ongoing developments in recording formats are enabling metadata to be added 'at source', thus including the generation of initial metadata in the production process at the same time as the essence material (see Figure 4.15.3). Adding metadata at source provides the MAM system with essential information about the essence at the point of entry into the system.

In the case of acquisition using a camcorder, typically news, metadata can be added to the essence within the camcorder, as data space is allocated within the data stream recorded on tape, not only technical settings of the camcorder and timecode, but information such as location and production information starting the metadata life cycle, reducing the need for subsequent manual intervention.

4.15.2.3 Metadata Enhancement

Metadata that is captured during acquisition will be enhanced during the ingestion and cataloguing processes. Both manual and automated processes will be applied during the ingestion process to achieve the optimum information within the metadata for subsequent retrieval of the asset. Figure 4.15.4 gives a basic list of 13 items that may be included in metadata. It is not unusual to have several hundred fields of data; most asset management systems have an infinite capacity for metadata, as it will grow rapidly with each entry enhancement.

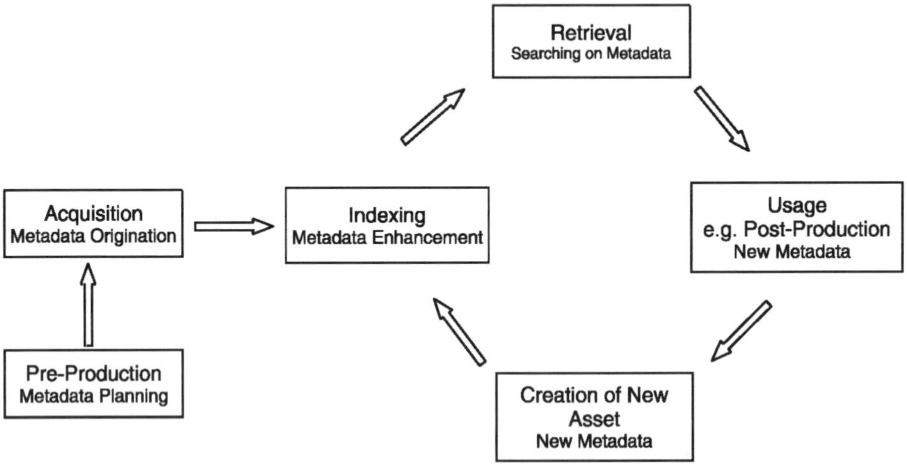

Figure 4.15.2 Metadata life cycle.

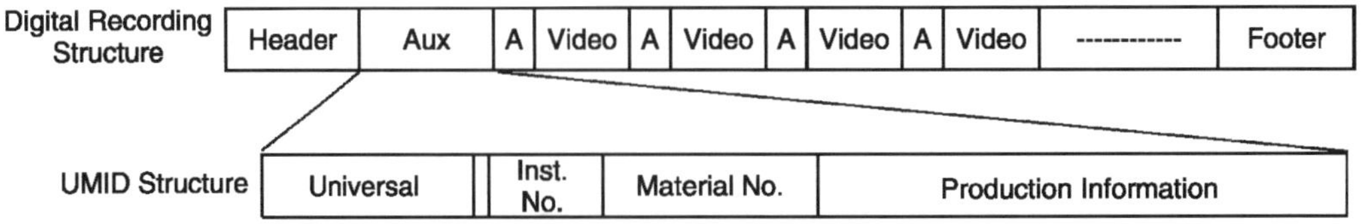

Figure 4.15.3 Digital recording structure with UMID recorded in Aux sector.

1	Title	Of the asset
2	Originator	Person and/or organisation
3	Subject	Keywords to describe topic
4	Description	Abstract, content description
5	Publisher	Company, department
6	Contributor	Person and/or organisation
7	Date	Dates, acquisition, post
8	Type	Category of the asset
9	Format	Recording format of the asset
10	Identifiers	URL, database ID
11	Language	Of the content
12	Relation	Relationship to other resources
13	Rights	Ownership and copyright info.

Figure 4.15.4 Metadata table.

Automatic processes, such as speech recognition, face recognition, closed captioning, Optical Character Recognition and Teletext capture, provide sources of metadata that can be produced automatically during the ingest and cataloguing processes. Many of these items are associated with keyframes that are also generated during ingestion. Keyframes are a single video frame generated on a time or event basis, each frame being a snapshot of the video being processed and stored. The keyframe itself may also be stored in Full resolution and 'Thumbnail' form, as a type of pictorial metadata, for use during the search and retrieval process.

Every broadcast enterprise has its own 'in-house' requirements for metadata; therefore, each Media Asset Management system must be sufficiently flexible to be customised to the end-user requirements. This flexibility, however, creates problems of incompatibility between systems. A significant amount of research by bodies such as the EBU is being undertaken to standardise metadata. MPEG-7 and UMID are examples of formats that are currently being used to standardise metadata.

4.15.3 Workflow

Workflow is the widely used term for the flow of decisions and processes that take place within the 'broadcast enterprise' from the point that an asset enters the system through to publishing and storage. Understanding how an organisation functions, commercially and technically, be it a national broadcaster or a post-production house, is essential to the scale and design of an appropriate Media Asset Management system. Accurately defining the workflow processes in a logical manner enables a suitable hardware and software solution to be developed to optimise each function. Each workflow process can be viewed individually but must be considered holistically, as workflow processes interact with one another.

Workflow in a typical broadcast environment can be represented by a seven-tier model covering the broadcast business requirements:

1. Business management.
2. Rights management.
3. Workflow management.
4. Content management.
5. Media management.
6. Automation.
7. Devices and storage.

As can be seen from this model, tiers 1, 2 and 3 address business- and management-related issues, again stressing the importance of understanding, and defining, the commercial requirements of the MAM system. The lower levels, 5–7, address technological issues, how the technology will enable business and production practices.

Current workflow practices within the broadcast enterprise may not, and most probably will not, fit exactly to the MAM workflow requirements, as they have evolved over time as technology has allowed change; therefore, when implementing an enterprise-wide or department-wide MAM system, it is essential to take a complete 'ground up' developmental approach. Legacy systems and existing equipment must be considered but not allowed to reduce the impact of the change and therefore workflow benefit of the new MAM system.

Having understood the higher-level requirements of business, the workflow model can then be translated into a technical model.

Each interface in Figure 4.15.5 indicates a process that is required by a typical broadcaster; at each interface there are workflow issues to be considered/resolved. The technological aspects of each process will be covered later in this chapter.

Current workflow is typically a hardware-based scheme, that has organically developed over time as technology has changed. The integration of IT and broadcast technology now allows workflow to be streamlined, hence improving the efficiency of the enterprise.

4.15.4 Media Asset Applications

A Media Asset Management system is fundamentally a suite of broadcast and IT hardware and software products that seamlessly integrate to manage assets as they flow through the broadcast production chain and are stored. The MAM system must have detailed knowledge of the asset, the hardware infrastructure and the workflow within that infrastructure. A Media Asset Management system is database-centric; that is, the database is the nucleus of the entity, maintaining key information about the asset. It must also interface directly with third-party applications that provide specialised and often legacy functionality.

4.15.4.1 The MAM application

Numerous MAM applications exist today, each with their own range of functionality; some stand alone as an IT management package, some fully integrate with the broadcast infrastructure and others are integrated within existing systems – for example, automation. A key feature that they have in common is that

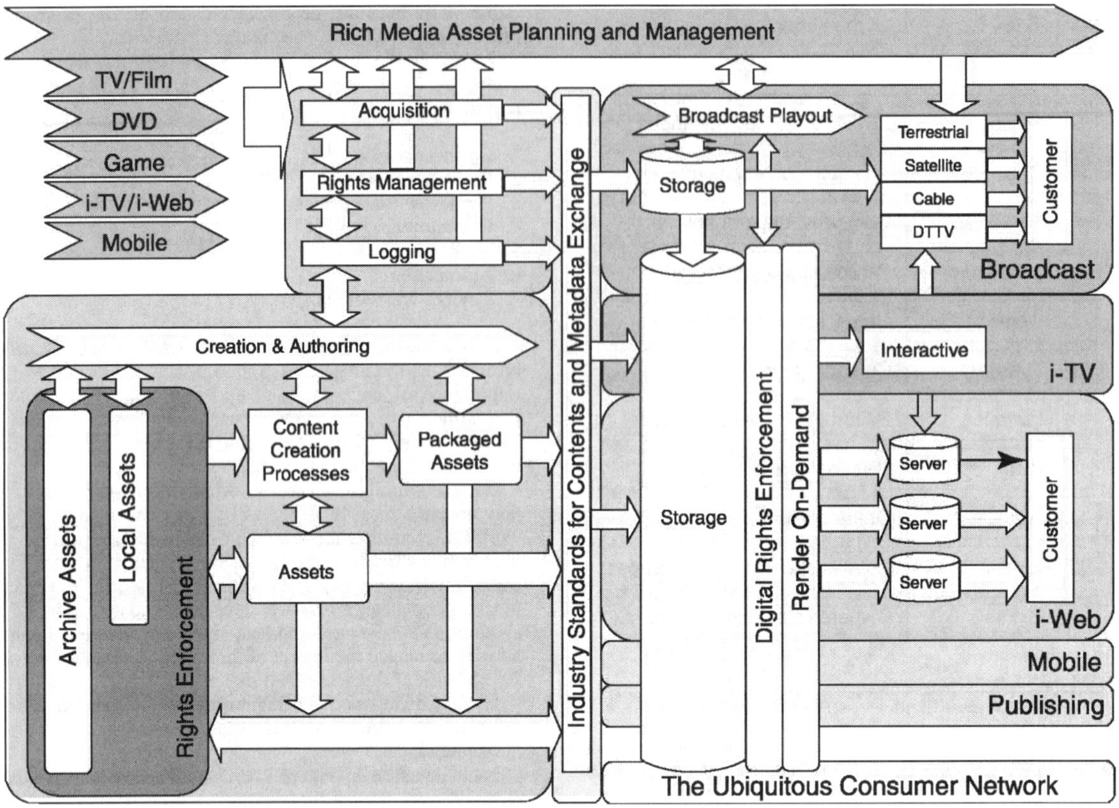

Figure 4.15.5 Overview of an integrated production environment – courtesy of Sony BPE.

the applications perform a management function, central to the workflow of the station.

The software architecture of a typical MAM system is a complex integration of applications, providing user-level operations such as ingestion through to device-level functionality such as data storage. It is the purpose of the MAM application to bring together this wide range of functions under the control and management of the application. The task itself involves a great deal of communication between individual applications to ensure consistency of data throughout. XML and HTML are typical of mark-up languages that are used to facilitate communication between modules.

An MAM system has a modular structure in order to fulfil the range of functions required of it. The software architecture is typically based on a layered or plane system whereby the individual applications perform a specific function and collectively make up the MAM system. Figure 4.15.6 illustrates several functions within the overall DAM system, each function being grouped logically by function. Using such a model gives a clear understanding of the range of functions and their relationship. The processes of ingestion and browsing are explained later. It is also possible to see that wrapped around the core functions of the MAM system are administration functions that monitor and manage the system.

At the lowest level, the system plane provides management of hardware associated with the functions shown. The service plane provides core functionaliy such as analysis engines. At the application level, key user functionality is provided.

In common with all 'windows'-based applications, the DAM application utilises a Graphics User Interface (GUI) to allow the user to access the levels of functionality within the MAM system. The functionality that is available to the user will depend upon which user community the operator is working within. Typically, user communities are based around the functionality provided by the applications within the application plane.

4.15.4.2 The Database

The core of the MAM application is its database. The system will have database management controlling traffic to and from the database. To ensure that the database is interoperable with other databases within the enterprise that the MAM system may interact with, for example an automation system, the database must be ANSI compliant. Microsoft SQL Server and Oracle 9i are typical examples of large databases that are used for MAM systems for their robustness and reliability. The application database may actually be split across several databases each with differing complexity and functionality. Different applications that the MAM system interfaces with may need their own database. The MAM system will handle different types of data, audio, video and documents; it is possible for all these types to

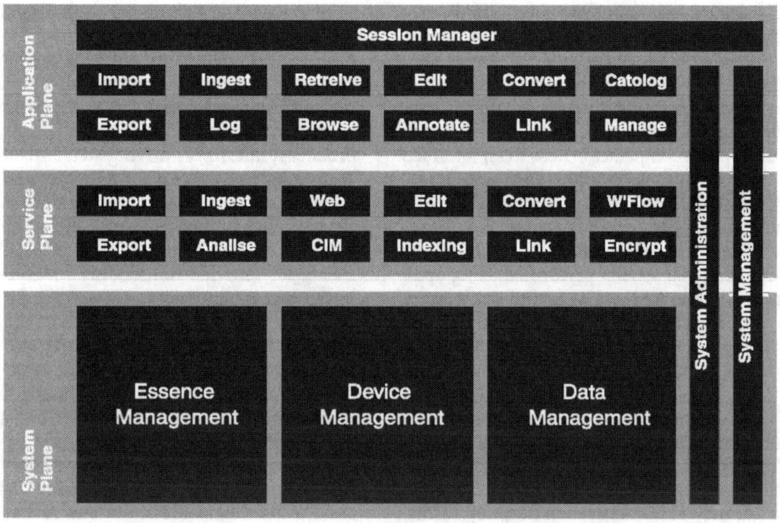

Figure 4.15.6 Software architecture.

be located in different dedicated databases under the management of the MAM application.

Each database will utilise a dedicated schema, a number of tables that describe the structure of the database, its constraints and the scalability of the database. The schema must be customised to the particular installation in which it is being used, to give a unique database profile only known by that system, to ensure the efficiency and security of the database.

The platform on which the database manager, and other critical devices, run must be highly reliable. Sun workstations are typical.

4.15.4.3 Processing

The central application of the MAM system will integrate and communicate with peripheral modules such as ingestion, browsers and editing. There are two main methods of providing this functionality to the peripheral modules: they are to perform processing centrally (server-side processing), or for each individual client to perform processing (client-side processing). There are benefits in each mode of operation; however, it is usual for an MAM system to employ both methods for different groups of users.

If processing is performed centrally at the server and a 'lite client' is used, there are benefits for remote users; they can browse over the WWW using a simple proprietary browser, the benefit being that the user is able to access the MAM system wherever a web connection is available, subject to access control. The main disadvantage is that the browse quality will be lower, and that editing functionality will be limited to the functions that can be supported over the narrow bandwidth available. An additional disadvantage is the reliance upon the performance of the central servers, serving many clients simultaneously. The availability of Broadband services is making this method more usable due to the increased bandwidth of the transport medium.

If client-side processing is used, processing, including editing of video and audio, takes place in the client's workstation. The performance of the workstation must be high, but there are many advantages of using this method if the client is on an LAN. Firstly, the time taken to browse and retrieve clips is much faster and at a higher quality compared to a web browser. The material can be edited using a fully featured editor and therefore a full EDL created for the programme at that workstation. Limitations of this method are the processing power of the workstation; each workstation must have a dedicated application and it must be connected to an LAN with sufficient bandwidth available to each client to provide high-quality browsable material.

As mentioned previously, an MAM system will be used by a multitude of users, some office based, some remote; therefore, an MAM system should employ a degree of both server- and client-side processing. Both systems rely upon high-speed transfer of data using either IP, JSP or an equivalent means of file transfer. The method is dependent upon the transport medium.

4.15.4.3.1 User communities

By definition the MAM system, be it enterprise wide or department wide, will be used by a wide range of personnel, each with different interests in the data held within the MAM system. This user community will be located throughout the enterprise and external to it. Each user must be able to access the information that is relevant to them without affecting other functions or data. Users will be identified as groups, e.g. editors, accountants, archivists, and within their group as individuals.

Each user will have specific rights related to their function – for example, an editor will have rights to modify content, whereas an accountant will not. Clearly identifying these users and their rights prevent inadvertent misuse of the data held within the MAM system. The system must track and amend assets that are being modified by a number of users simultaneously in order to maintain the integrity of the data held within it.

4.15.4.3.2 Integration with third parties

In order to perform the wider MAM function, synchronisation with an automation system for example, the core application must interface with software and hardware from third-party vendors that provide applications to perform specialised functions. There must be a flow of information between the applications to ensure that the MAM system and the third-party system are informed of any status elements that may affect them. Obviously manufacturers do not disclose the code for their applications but do allow third-party companies access to an interface layer. This interface layer, known as an SDK (Software Development Kit), allows the third party to access information from the MAM system without being able to access the application itself.

Having created interfaces with third parties, the files can be transferred between applications, usually performed using open file exchange protocols such as XML. The use of proprietary interchange formats is decreasing as the need for interchange and openness is increasing within the IT and broadcast worlds.

4.15.4.4 Ingestion

Entry of an asset into the MAM system is known as ingestion (see Figure 4.15.7). This process is the most crucial to the entire system as it is at this point that the MAM system is educated about the asset. An asset may enter the system in many forms, with or without metadata, and will be encoded or transcoded to comply with the 'house' format.

At the time of ingestion the following major processes occur:

- Encoding of broadcast-quality A/V material.
- Encoding of low-resolution 'proxy' copies.
- Entry/enhancement of metadata, including rights.
- Cataloguing/indexing entry into database.
- Storage of asset and proxy.

Essence material entering the system may be in many forms; the material may be live, from a studio or external source, or pre-recorded material in digital or analogue form. Whatever the format of the incoming material, the A/V infrastructure must present the essence material to the MAM system in the chosen format, typically MPEG-2 or DV (Digital Video) formats. The broadcast industry is driving standards such as AAF (Advanced Authoring Format) and MXF (Material Exchange Format) to increase standardisation and therefore interoperability between formats and platforms. Until such standards are ratified and adopted by the industry, products such as Telestream FlipFactory™ provide transcoding functionality that can be used in the ingestion process.

At the time of ingestion an encoding process takes place, creating proxy copies of the original material at a bit rate suitable for subsequent browsing. In many cases these are two or more proxy copies made of decreasing quality suitable for browsing on an LAN or, in the case of the lower bit rate, over the WWW.

In parallel with the encoding of the proxy, metadata entry and enhancement takes place. Many digital formats are able to handle basic metadata at present. A major development, taking place at present, is the ability of recording formats to include metadata such as UMID at time of acquisition. Metadata is the lifeblood of an MAM system; it provides usable information and for search and retrieval throughout the life of an asset, analysis of the incoming essence and enhancement of existing data during the ingestion process provides the source of metadata, including keyframes, text and voice recognition.

At the ingestion point, metadata could be as simple as the timecode recorded on the master tape, manually adding production

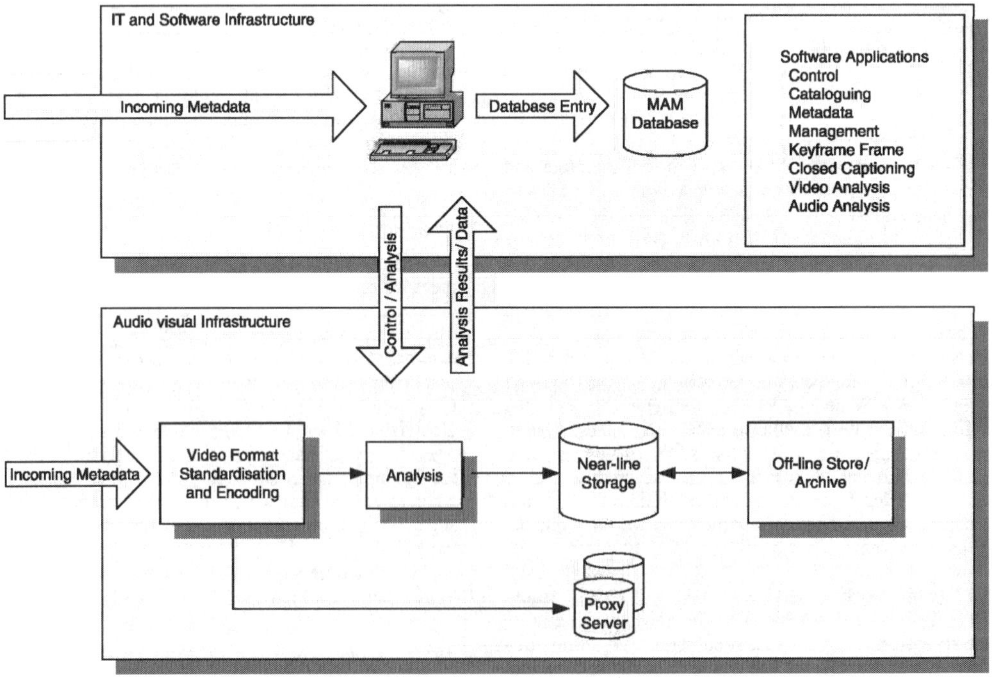

Figure 4.15.7 Ingestion and cataloguing processes.

information, title, location, date, actors, and ancillary data immediately enhance the value of that metadata. This data can be input manually by the operator of the ingestion station. Further enhancement of the metadata can be performed using software tools that automatically analyse the essence material as it is being entered into the system in real time.

Once the material is ingested, it will be catalogued and all information about the asset stored in the MAM system database. Essence material will be stored in suitable repositories; the high-resolution broadcast-quality material will be stored in a near line store and subsequently transferred to a tape library. The lower quality browse-quality proxy copies will be stored on dedicated servers, from where they can be viewed during a browse or searching process.

4.15.4.5 Search and Retrieval

Having ingested and stored assets within the Media Asset Management system, a multitude of users must be able to efficiently retrieve the asset, and information about the asset, when required. The MAM system will hold thousands of hours of material, many with similar content and titles – it is at this stage that the quality of the metadata that has been entered earlier becomes apparent. Researchers, Journalists and Producers are a small selection of the types of people that will be searching for material locally over an LAN or remotely over the WWW (Figure 4.15.8).

In order to retrieve the desired asset, it must first be located using a query. A hierarchical approach is taken to searching as the asset may or may not be uniquely identifiable. A simple scenario in the case of news production is that a piece that has been produced only a short time ago could be located immediately. In the case of material stored in an archive there may be several, tens, even hundreds, of items matching the search criteria. In this case it is necessary to step through the hierarchy to narrow the search and locate the material required. Techniques such as contextual searching and Boolean criteria are used within the search engine to enable the query to yield the desired result.

Having performed a search using the criteria entered by the user, the results will be displayed in the form of a hit list, usually most accurate down to least accurate. The GUI will display textual information about the clips and a number of thumbnails to give the user sufficient information to make their choice of clip. If the user is working on an LAN the quality of the thumbnails may be higher than over a web browser due to the available bandwidth. Once selected, a low-resolution copy of the asset can be browsed using a media player, Real Media for example.

Having selected a clip, the material can then be requested in full resolution or an off-line edit can be performed for later compositing. If an SAN exists within the enterprise with suitable storage capacity, the clip can be delivered in real time; in other cases it is usual to revert to a tape-based delivery system.

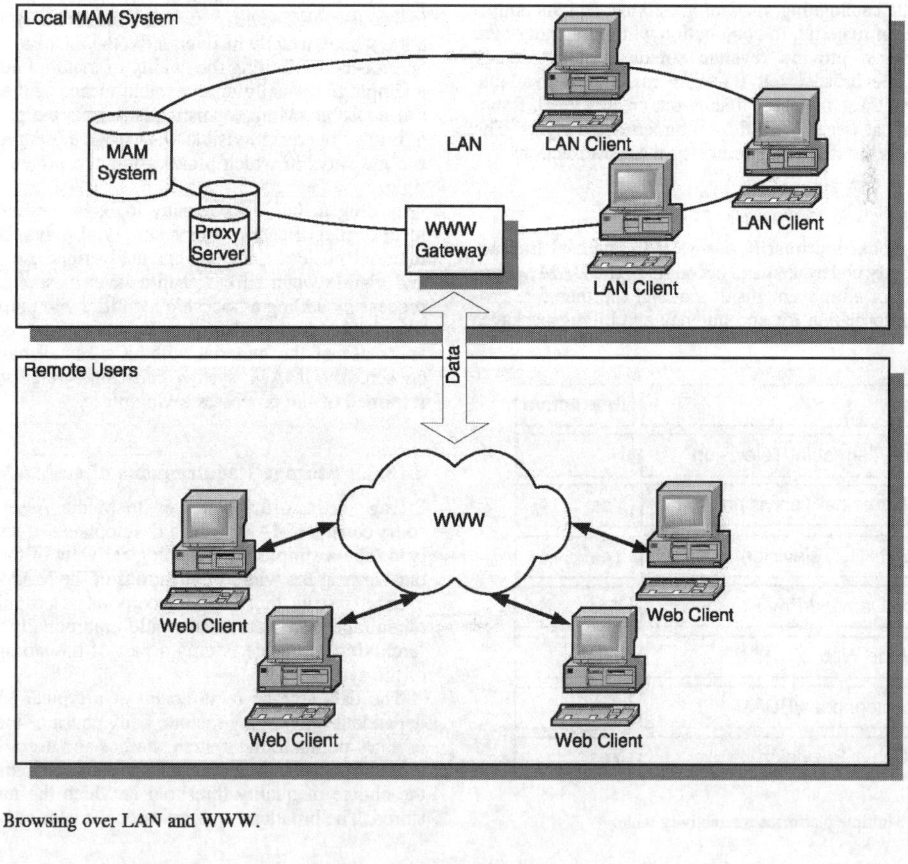

Figure 4.15.8 Browsing over LAN and WWW.

Once the clips have been edited, a new asset is produced. This is performed within the MAM system; therefore, minimal intervention is required to add metadata, versioning and other management data.

4.15.4.6 Publishing

Broadcasters are no longer dependent upon a single transmission platform; the broadcast business now 'publishes' content via many 'Multi-Media' (see Figure 4.15.9). The advent of the Internet and other bidirectional media, such as Interactive Television (iTV), have given many new opportunities to the viewer, and therefore many more challenges to the broadcaster. The broadcast model now includes platforms as diverse as analogue free-to-air TV and interactive viewing on PDAs (personal digital assistants). Each platform has to be catered for appropriately; therefore, programme material has to be repurposed to meet the need of the specific medium.

Programme versions have to be adapted for each platform. Duration and quality/bandwidth are specific to each medium; therefore, the MAM system must cater for these 'programme' types, storing each variant independently as a version, with the appropriate rights. The degree of interactivity also impacts upon the MAM system.

4.15.4.7 Commerce and Rights

Media Asset Management systems enable the broadcaster to fully utilise the commercial potential of their Rich Media Asset. Accurate cataloguing, meticulous storage and uncomplicated retrieval of material, in conjunction with publishing over several platforms, provide revenue streams not previously accessible to the broadcaster. Ensuring material is available for use enables these revenue streams but creates legal, financial and technical issues related to commerce and rights. The MAM system is the definitive source of these transactions.

4.15.4.7.1 Commercial Modules

Commercial modules primarily ensure that financial transactions are correctly undertaken and accounted. The DAM system has knowledge of all movements of material and therefore provides source information for accountancy and billing packages that are used by the enterprise. Commercial transactions not only form a financial, but also a legal, contract between vendor and purchaser; therefore, this highly complex area is usually handled by third-party specialists. The DAM system acts as a source of information only. For example, the clip sales division of a broadcaster with allow their clients to browse available material with a view to purchasing it. Once the material is chosen and 'sold', the DAM system will trigger an event informing the billing package that the clip has been sold; from this point the billing software within the company's sales/accountancy infrastructure completes all activities. The DAM system will have no further involvement until the process starts again. At present the level of integration that is required to complete these transactions is typically not in place; therefore, the majority of transactions take place manually.

Platform	Interactive
Analogue Terrestrial Television	No
Digital Terrestrial Television	Yes
Digital Satellite Television	Yes
Video on Demand/Pay Per View	Yes
World Wide Web	Yes
Mobile Telephones/PDAs	Yes
Digital Cable Television	Yes

Figure 4.15.9 Multiple platforms interactivity table.

4.15.4.7.2 Rights Management

As the provision of material becomes more automated using insecure transmission media, the potential for rights being abused increases significantly; therefore, the need to police and maintain copyright is essential in order for the publisher to maintain control of their asset, prevent pirating, and so maintaining the revenue from the asset. In order to maintain control over the material being made available for sale through the MAM system and maintain its value, the broadcaster must maintain copyright over the material.

Many systems either exist or are being developed to ensure that copyright is maintained. The process of allowing clients to browse high-quality material immediately creates a situation where, usually inadvertently, the material copyright could be breached. Limiting the quality of material being browsed is a simple but usually unacceptable means of making the material unusable. More sophisticated means are the norm; adding a 'bug', fingerprint, visible watermark or encrypting the material are ways in which broadcasters are protecting their copyright.

Adding a 'bug', a company logo, to the material is a first level of identification, which simply displays the source of the material but does not prevent inadvertent use. Fingerprinting and visible watermarking add a level of sophistication to the process by adding a traceable identifier and partially obscuring the visible picture. The most secure method, encryption, prevents use of the material without a key. Whatever system is chosen, the MAM system must integrate with it and be informed of the processes taking place.

4.15.4.8 Storage Requirements of an MAM system

Storage is inextricably linked to Media Asset Management; some consider MAM to be a development of archiving, which is to take a simplistic, technological view of the activity without viewing the wider implications of the MAM system. However, it is true that MAM incorporates a significant amount of storage hardware that would traditionally be considered 'archiving', but this is only a part of the storage requirement of the system.

The data storage requirement of a typical MAM system is dependent upon many factors: daily churn of material, number of entry points to the system, studios and users, to mention but a few; however, the greatest impact on storage requirement is the choice of quality threshold at which the material is to be stored. The broadcast-quality essence material will typically be stored at DVCPRO50, 50 Mbit/s, whereas the lower quality proxy copy may be stored using MPEG-1 at 1 Mbit/s for LAN applications and less that 200 Kbit/s for web-based browser applications.

4.15.4.8.1 Servers

Servers are required not only for the content, but also for the management elements of the system. Server-based storage is used in many areas, including the Application Database, Proxy Server, Ingest Server, Near-line Store, Transmission Server and Data Mover, usually in an SAN configuration (see Figure 4.15.10). Functions such as the application database are less storage hungry than the content elements mentioned previously; however, the database will grow rapidly as entries are added. Therefore, detailed consideration must be given to the current storage requirement and the annual growth of the database and content.

Content storage requirement is huge; a typical installation using DVCPRO50, 50 Mbit/s digital video recording with four channels of digital audio and VBI data requires approximately 30GB per hour compared to an MPEG-1 proxy recording, which is less than 0.5 GB per hour. In addition to simple capacity, the access time and writing speed are major factors that influence the type of storage device used. In order to provide the storage capacity required, individual servers typically have a total capacity in excess of 1 terabyte.

With data streams of 50 Mbit/s, it is essential to ensure that content can flow freely throughout the SAN. At each node sufficient storage must be provided and the interconnection between nodes must have sufficient bandwidth. The MAM system maintains communication with all storage devices in order to ensure that the location of each asset stored on the device is accurately tracked.

4.15.4.8.2 Tape and Disk Libraries

Tape libraries and disk libraries are used to provide the 'deep archive' functionality for the MAM system. They are used to store content in both long form and short form, and in many cases to provide a backup system for other storage elements in the system. The criteria that are used to determine which material will be stored on a deep archive are fully explained in Chapter 4.16.

4.15.4.8.3 Redundancy Strategies

The SAN configuration shown earlier contains hard disk servers and tape/DVD libraries, many of which are in a mission critical environment. It is of the utmost importance that a redundancy strategy is developed and adhered to. In an environment such as transmission, the expectation is that the system will be 'up' and providing programming 100% of the time! In this environment, a dual redundant system may be used where every item to be played out may be stored on a separate server in case of server failure. It is, however, more practical and more cost-effective to adopt a RAID strategy that will give a near 100% backup, for example RAID3.

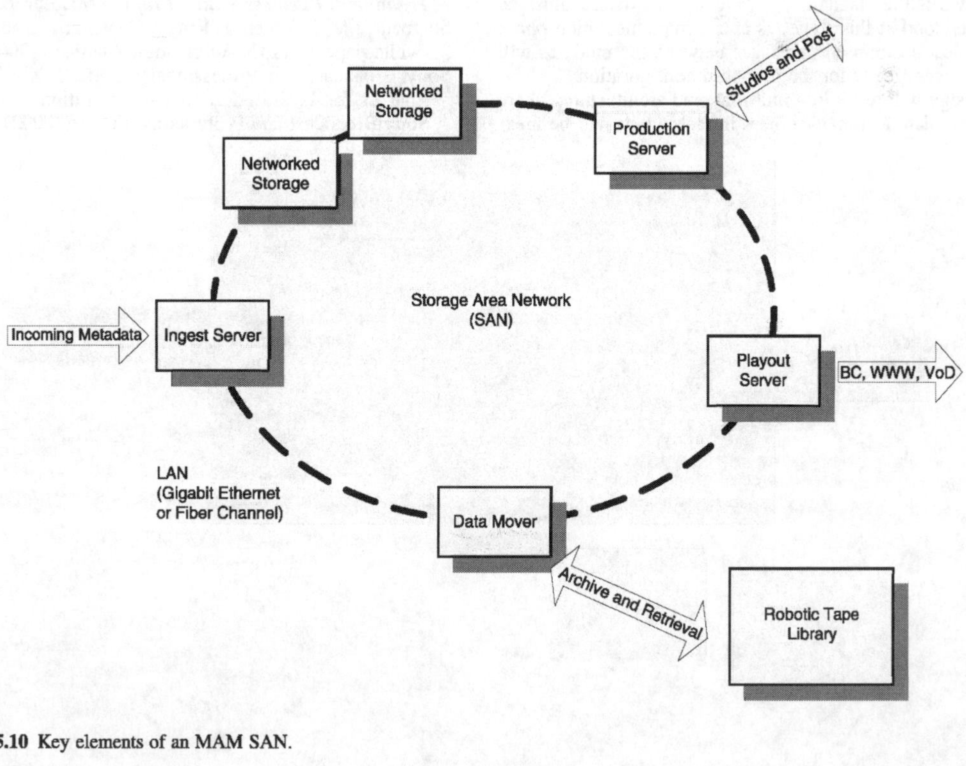

Figure 4.15.10 Key elements of an MAM SAN.

4.15.5 Implementation of an MAM System

In the final analysis, Media Asset Management is the physical integration of a wide variety of software and hardware from the broadcast and IT worlds that together make the MAM system. Decisions will have been made, based on current workflow practices, future requirements, audio and video formats, and ultimately cost, on how the system will fit together and how it will integrate with enterprise.

The process of designing and implementing a Media Asset Management system involves precise planning from the ground up, a clear understanding of the business processes involved and the function of the MAM system. Within the broadcast enterprise, the integration of 'traditional' hardware is well known, its integration with IT infrastructure less so. The challenge therefore is to ensure that the different technologies work correctly together. Involving the broadcast and IT engineering disciplines and the financial and administration communities from the commencement of the project is essential to achieve this first goal, a clearly defined specification for the system.

Stage 1 of all projects is the accurate capture of system requirements, traditionally the assessment of technical requirements, but in the case of MAM systems, workflow assessment must be understood prior to entering into technical design. Detailed workflow analysis must take place to ensure that each workflow function is accurately encapsulated, since if correctly implemented the MAM system will touch upon all functions and workflows within the broadcast enterprise. It is not until the workflow analysis is complete that the design of the system can commence.

In a project of the magnitude of an enterprise-wide MAM, the resultant system will be a 'one-off', constructed using standard 'components' in a unique configuration. The interfaces between the standard software and hardware must be fully understood at this stage, as at the implementation phase of the project customised interfaces between the modules will be written specifically for the proposed configuration.

The design will result in a multi-layered architecture, where the essence, data and management infrastructure will be inextricably linked but function independently. The result will be a management network, essence network and data network, each with clearly defined tasks. It is undesirable to implement networks with mixed functionality – that is, production (essence) networks must be separated from back-office and management networks.

The essence network, in the form of a high-bandwidth SAN, is of most interest to the Broadcast Engineer, as it forms the programme-making arm of the infrastructure; however, the MAM functionality is provided by the IT networks. Implementation brings together the design, tests that design and correct any errors prior to the commissioning, and the system going live.

Once the system is live a programme of support will be required, initially to capture any bugs that were not found during testing and finally to maintain the system in good order. The system will also grow over time as more material is generated. A programme to manage this scalability must be included early in the life of the system.

Bibliography

Austerberry, D. *Digital Asset Management*, Focal Press (2003).

Bancroft, J. Open All Hours Digital Video Archive Brings Metadata up to the Mark (white paper) (2003).

Dalet Digital Media Systems. Next Generation Digital Asset Management (white paper), Dalet Digital Media Systems (2001).

Edge, B. Attributes of File Formats for the Broadcast Operation and Archives (white paper), Grass Valley Group (2001).

Holst, S. *Digital Asset Management, XML, Rich Media and a Traditional Business Value: Profit*, XML Europe (2001).

Seigneur, J.M. "Content Is King", Content Management is Key (white paper), Harris Automation Solutions (2002).

Sony Broadcast and Professional Europe. MXF: Technology Enabler for IT Based Broadcast Operations (white paper), Sony Broadcast and Professional Europe (2002).

W J Leathem
Consultant

4.16 Electronic Newsroom Systems

4.16.1 The Newsroom Revolution

During the late 1990s, television newsrooms took a radical change in not only the output they provided, but more importantly in the workflow used to produce news stories. Many of the changes had already been being brought about by the adoption of lightweight electronic camcorder packages, a compact electronic alternative to the 16 mm film package which had been the workhorse of television news acquisition for over a half a century. However, although the change to Electronic News Gathering (ENG) was very rapid, the material to be edited still required the use of cassette-based VTRs and also the journalist still relied on systems that did not marry the technological and journalistic technologies. The skill sets of the journalist/reporter, the technician and the editor still remained separate.

In the mid 1990s, a revolution in television news started to take place. Based upon an electronic system, it boasted interconnectivity and interoperability, the system starting to weave together the skill sets of the journalist and the technician.

The revolution in the field began with the move from acquisition in analogue formats to digital ones. Camcorders had previously been based on the ¾-inch Sony BVU format based on the ubiquitous Sony U-Format, which by the nature of the tape format and cassette size made the camera/VTR package bulky and cumbersome. Even so, it was an improvement on the 16 mm film package as it gave instant replay and access to the material, therefore almost instantaneous access for editing! With the arrival of Betacam from Sony (the early formats being Betacam and Betacam SP, and the later formats being the digital Sony Betacam SX format), the material could now be acquired with a camcorder using ½-inch tapes. This change in tape size had an enormous impact in the use of the camcorder, removing the previously bulky package and making the complete package lightweight for field applications.

With every introduction of a ENG format, there always comes a format war, with Sony, Panasonic and others introducing alternative and competing tape formats. Clearly, in the introduction of the early tape formats, Sony was a clear winner with the U-Matic and Betacam formats. Panasonic tried very hard with M-Format with a few successes. In recent years the M-Format has given way to DVCPRO, which is offered in a number of different bit rates. Avid, in conjunction with Ikegami, launched a camcorder which was the first to offer recording onto a hard disk. However, this also failed to be taken up by users, preferring to continue on with the traditional means of recording using tape.

Acquisitions by some organisations are now even using ¼-inch formats based on the Panasonic DV format and some of these formats have gained popularity, especially for covert applications.

One of the other trends was to consider the recorder/player that news editors used in pairs to record and edit material in the field. With Betacam SX came the hybrid recorder, a device which provided not only a tape transport but also the addition of a hard disk included in the recorder. The advantage here was that within a single machine it was now possible to play back recorded material as well as copy it onto the hard disk all in one machine. The added advantage of this was that it was now also possible to play-off the machine into a satellite system at four times normal speed using SDDI, previously known as SDTI, thereby reducing the time taken to file a story back to base.

In America, the Remote SNG vehicle became the mainstay to the news operation. These vans were in themselves small news production vehicles, which with a large dish on the roof, enabled news to be reported from the scene as it happened. It also meant that live inserts of news material was also capable of being used as part of the news bulletins.

What this all meant for television news was that it was easier to acquire and produce.

Field editing units allowed the stories to be edited in the field and then, using satellite transmission, allowed it to be played in real time back to the studio. With the introduction of Betacam SX from Sony, this provided the ability to replay back to the studio using four times real-time playback, using SDTI transmission.

4.16.1.1 SDTI (SDDI) Offered Four Times Replay

SDTI (Serial Digital Transmission Interface) is a ratified SMPTE recommendation, identified as *SMPTE 305 2M 2000*. Originally Sony started off with SDDI (Serial Digital Data Interface), but we will refer to it in its SMPTE name – SDDI.

Based on SDI, it is a 270 Mbit/s stream, exactly the same as SDI (Serial Digital Interface), and uses a standard BNC connector and quality coaxial cable. One of the principles behind it was to ensure compatibility for those existing users of SDI and those who additionally wished to use SDDI compatibility with existing SDI routing/distribution equipment. So existing interface equipment such as distribution amplifiers, routing switchers, line and field synchronising devices, as examples, would pass both SDI and SDDI. Equipment which caused rearrangement of the bits, such as analogue-to-digital and digital-to-analogue conversion equipment, vision mixers and DVEs, would not be able to pass SDDI and therefore had to be avoided.

SDDI, as already mentioned, is based on 270 Mbit/s. The stream comprises a 220 Mbit payload and the balance set aside to the header. One of the other advantages, which was never really exploited, was that, like IP-based signals and ATM, source and destination addressing could be encapsulated in the header, so the signal could have been able to find its destination by each receiver equipment being able to pick out the destination addressing to direct the signal to a final address.

As previously mentioned, SMPTE took the original Sony SDTI into standardisation and produced SDDI. In practice, very few broadcasters warmed to the technology and it was never really a success. A few manufacturers did adopt this technology, but the overwhelming outcome was to wait for an IT-based news solution. One of the important factors to consider for using this technique with satellite is the amount of transponder space required. The more transponders used means the more money to be paid to get a story back to the station. Some systems put a lot of strain on the use of transponders, using more space, thus these systems did not gain sufficient popularity. Transponders have a bandwidth of 9 MHz and with some systems, at four times speed at 18 MHz, delivered a required bandwidth of 36 MHz! This would mean that you would have to stretch across four transponders!

This meant that material arrived in the station by a number of alternative methods, these being satellite, incoming lines, by bike and on foot, each designed as a method of speed, reliability and economy.

4.16.1.2 Properties and Advantages of SDI and SDDI

At this point, it is probably worth reiterating the values of an SDI/SDDI system over an IT-based system, especially when these systems were first starting to gain popularity.

SDI and SDDI have three main unique selling features (USPs) over an IT solution:

- SDI and SDDI have guaranteed bandwidth.
- SDI and SDDI have a guaranteed connection.
- SDI and SDDI have guaranteed timing.

What this means is that if you have a signal to connect to two points you can guarantee bandwidth, connection and timing. This really has become an important issue and even as silicon devices in IT-based systems become faster and faster, there remains little doubt that these three items will prolong our use of 270 Mbit/s technology. The fact that you could guarantee all of these properties is a major wining point for SDI/SDDI.

With some IT-based systems it is just impossible to provide any or all of these guarantees. Imagine a situation where you have to wait for the connection and then you lose all the material because you could not make the connection or indeed did not have bandwidth to transfer the material. However, that said, the new generation of newsroom systems which interconnect between devices using IT-based connections and rely on interfacing to outside of the system to incoming signals as well as transmission outputs are becoming more and more of a reality, and users in the future will have security in their investments by being able to reduce costs with this approach.

4.16.2 Early Newsroom Integration

The news system operation and journalists system in the station has also been changing. Prior to this revolution the management of news was transmitted in a similar way to programme output. News stories were edited as packages; the tape was then placed into (or dubbed to) a Sony Betacart. The changes in the newsroom had already changed the transmissions of news, as a play list or rundown of the programme was downloaded from the newsroom. Some of the familiar studio equipment and facilities had been replaced, such as robotics; news stories do not use much studio floor, especially if it is regional stories where it is only just a presenter to camera. Prompting systems were now loaded with the story from the newsroom. The balance of the production was (and still is) commanded from a gallery with Director, Producer, Editor, Vision Mixer and so forth.

One of the major changes that drove the news revolution was the use of server technology as a repository for material. Non-linear editors were already being used in the production of programme material and naturally found their way into news.

4.16.2.1 Central Material Storage on a Server

Material stored on a central server system, particularly for news, has to offer the following:

- A scaled approach to bandwidth to ensure a sufficient number of inputs to record from outputs to play back from.
- Be able to play back material as it is being recorded, thus allowing NLEs to start the process of logging shots for the edit.
- Resilience by using redundant facilities or the server system being capable of being mirrored and have a full database of the material being stored on the server.
- Be capable of generating a low-resolution (proxy) copy of the full-resolution file. This is used for newsroom editing.
- As part of the server or as an additional server, provide a means of playout of the edited programme material under the control of a play list.

A newsroom system, which does not wholly rely on IT connection, but uses SDI and SDDI, will normally be encapsulated by a routing switch. There are a number of purposes for incorporating such a switch.

Primarily, it provides the ability for sources and destinations in and out of the server system to be dynamically assigned based upon the demands placed upon it by the system management and, of course, the users. What this means is that from the initial scaling of the server system and through the addition of day-to-day connections, the management of the routing switcher and the server system will dynamically assign inputs and outputs to match user demands. This is better explained by the following example.

Consider a server system which has six inputs and three outputs. Without the routing switcher the inputs and outputs are to be hard-wired to the incoming and outgoing signal sources and destinations. Due to the fact of it being hard-wired, this system is unusable, the reason being that as the equipment is connected in this way there is no flexibility in how the server system connects to the inputs and outputs. Basically the server system management cannot make any choice of connections and to which equipment that is connected on these ports, and therefore can only connect to the server in this way.

One of the other expressions of bandwidth is how many connections you can safely demand without the system slowing down or the requests being rejected by the system's management system. Imagine a situation where you have a number of edit suites all making demands on the server. If the balance between editing suites and the number of server connections is not correct, then the result will be that when requests are made to the server to connect to the server, the ports will just not be available. Operating the server will no doubt eventually ensure a system resultant screen message of 'No ports available!' Therefore, it is extremely important to match the facilities with the likely demand on the server to ensure that resources will be available when needed.

Using a routing switcher around the server system really makes a significant difference as to how the system looks for its sources and destinations, the reason being that when a request is made for a connection, the Server Management System (SMS) will be able to work out from its map of connections what input to the server needs to be used.

What this leads onto is the amount of bandwidth that is available on the server and what this reflects is the number of I/O ports available. If the server system is to be used with a large number of inputs, then the structure must be scaled to accommodate this. This of course means a large number of connections, but reducing the number will compromise the operation and ability for the server system to sustain large amounts of use.

An important part of the scaling is therefore to balance the inputs and outputs of equipment to the number of sources and destinations. It would therefore be expected in a system that has a separate server for storage and for playback that the number of inputs into the server system will be high, with a smaller number of destinations to play back into the system in general as well as connect to the playback server. The playback server will only need a smaller number of outputs for feeding the signals out to the outside world and the transmission outputs.

Another important factor of the server system is that whilst material is being recorded on the server, being able to start to work on the material within a short period of time after the initial recording provides time saved, especially when a story arrives late or close to transmission time. Therefore, the advantage of this is clear. Non-linear editors connected to the server can therefore start to review material relatively quickly as it is loaded onto the server and can make the choice of shots. It goes without saying that news is all about immediacy, so being able to edit and package the material, especially when it is close to the transmission time, as quickly as possible is a clear benefit of going to a server-based news server. The older method of waiting until the material arrived at the station and then starting to log and then edit it in a long form is extremely antiquated in comparison.

4.16.2.2 Material Protection Whilst on the Server

Protection of the material whilst on the server is mandatory. The material is the most important asset in whatever form it is maintained in the system, be it live on the server or in archive.

Upon arrival at the station, each item going onto the server is logged into a database in the Ingest area. The Ingest area is normally populated with a number of tape decks, sometimes of different formats. Taped material may not always originate from the ENG crew, but may also come from the stringer or even the domestic camcorder! In some situations this area used to be referred to as the Lines area, as the material may come 'down the line' from a pool feed or even satellite. The Ingest area is therefore the point where the input of material can be made from any one of the sources already described onto the server system. In the Ingest process, the material information is logged onto database systems. The database is relational and is generally backed-up from time to time to ensure asset protection. The database records various items from the material, such as title, where it was shot, who shot it and when it was shot, as just some examples.

Hardware protection is an important part of the server solution and can be exploited by using redundant power supply units and, where the system permits, duplication of some of the hardware within the server.

One method that can be used for protection is to mirror the server that is the provision of a second physical server system that reflects the configuration, storage and protection characteristics of the main system. This mirroring can be right down to the database, and by using database replication between the two systems, they can be kept in sync as regards database content. In this way the system is protected against a catastrophic failure. However, the price to pay for this protection can be excessive. The purchase is for two server systems, additional power and air-conditioning facilities and of course increased rack space to hold all of this hardware. Of course, a smaller level of protection can be obtained by backing up the power supply with UPSU (Uninterrupted Power Supply units), provision of redundant power supply units, ensuring the server system has some form of RAID protection for the hard drives so that not all the data is stored on one set of drives, and also that in the case of failure the drive can be identified and replaced with the minimum of disruption and downtime.

One point that needs to be discussed is the compression schemes that the server system can use to store the material. Material coming into the station may already be in one of these compression schemes and therefore can be stored quickly and easily on the server. Signals entering the station may enter either as analogue or digital and therefore will require digitisation or conversion respectively into the chosen compression scheme. Some server systems will only adopt one particular compression scheme (nominally MPEG-2 4:2:2 MP@ML), whilst other server systems will offer a wide range of schemes.

4.16.3 Compression Schemes

In the initial offerings of server systems, these were based on MPEG-2 4:2:2 MP@ML. In addition to all of this, some material may now be available as AVI, or QuickTime™ or some form of graphics file which may need to be used, e.g. JPEG or TIFF. These may all need to be stored on the server system. Details of this and other compression systems can be found elsewhere in this book.

However, suffice to say, in line with the development of tape formats, servers themselves have also had to become more and more flexible to the different compression schemes available and today these can include DV 25 Mbits/DVCAM/DVCXPRO (50 Mbit/s), DVCPRO 50, MPEG-2 4:2:2P@ML IBP@15 Mbit/s, MPEG-2 4:2:2P@ML I-Frame 50 Mbit/s (D10) and MPEG-2 I-Frame at 1.5 Mbit/s exclusively used for proxy servers. For audio, the norm is generally four channels of uncompressed audio at 24 Kbit/s with 48 kHz sampling. For the proxy audio this again is normally four channels at MPEG-1 Layer 1 at 128 Kbit/s.

4.16.4 File Format Transfer

The first place that material arrives, as explained earlier, by whatever means it arrives, is the Ingest station. The Ingest station organises the copying of the material and that information accompanying the material, such as where it was recorded, by whom, etc., is added to the database entry. With the introduction of MXF this addition is now far easier, as the information found on the user area in the signal will be transferred to the material database.

The work around MXF and its introduction is set to revolutionise how material information is dealt with, referring to the information accompanying the recording placed away in the vertical interval. Of course, the Ingest area can also deal with sources that are known to be arriving in advance and therefore resources can be reserved to cater for this reasonably automatic recording. MXF is described in more detail elsewhere in this book.

Once material has been stored on the server of course, you will want to do something with it to produce it into a package or one package with a number of variations thereof. This is achieved by the editing of the material into one or a number of packages.

4.16.5 Editing Material

Editing of the material on the server, using Non-Linear Editing (NLE) systems, means that editing off the server has to be non-destructive. This means that the material is never physically cut into the sequence to make up the new package. Instead, the in- and out-points of the material can be selected, although as far as the NLE is concerned the material has been cut. In real terms, the material is never cut. All that is selected are the in- and out-points, with this edit list information uploaded to the server. When it comes to playing out the material, the server uses the in- and out-points to perform a pseudo edit, the material for each edit being quickly accessed and the material replayed between the two points. This of course has the effect of making it look, as far as the output of the server is concerned, like the material is being played out sequentially as one single item, whereas this is of course not the case. A ready comparison of this is the feeding of an on-line editing system with an Edit Decision List (EDL) which has been prepared in the off-line suite. Most important of all of this is that the core material on the server is still in its non-cut form. No destruction of the material has occurred.

Non-linear editing suites can vary between a simple cuts-only suite to moderately complex A/B suites. Some of these NLE suites also include some form of local storage. In these cases, material is selected from the server, usually preselected by a journalist in a low-resolution system.

At this point one will see separate editing taking place, between the simple cut-editing being made by the journalist, whilst the craft editor takes on the tightening up of the edit in the NLE suite.

The NLE suite allows craft editing to be performed: adding a new soundtrack, added blur or mosaic as effects in news to protect individual identities, maybe even to add a caption, although in some organisations this is added in the transmission area. Rolling news channels have caused stories edited in the NLE suite to become generic across the organisation, and individual channel branding therefore causes the material to be uniquely captioned in each news output in order to define the individual branding.

In some NLE suites, local storage and a VTR may well be located as part of the facility. Local storage is available for two reasons. Firstly, it can be used to download material from the server, ideally using high-speed techniques described earlier. Secondly, the local storage can be used to provide the means to have an A/B roll with involving the main server. This may involve taking a copy of the material to the local storage to allow the A/B roll. The VTR may be present to allow playback of material from the ENG crew as well as be part of the editing process itself.

4.16.6 Transmission of the Material

Transmission of the news bulletin or programme is made against a rundown or play list. In order to play back this may be from either a server specially configured for the purpose or else from a part of the main server system. In the case where the server is separate to the main server, a smaller number of inputs can be made to the transmission server, allowing for normal as well as high-speed interconnection.

4.16.6.1 Low-resolution Streams for the Journalist

With the development of these new news systems, journalists can now review a low-resolution version of the story where they are able to make the decision quite early on in the news process to how their story will in effect be finally cut. These stories can be viewed at the journalist's desktop, so providing them with a full-resolution version facility for viewing material is not required. This is maybe just as well, as newsroom systems tend to connect using standard network components and topology. In order to derive this low-resolution version of the material, two copies are made at the time when the material is initially stored. For the main server of course the material is at the full resolution, reflecting the requirement to have the highest possible quality; the lower resolution copy, sometimes referred to as the proxy copy, is used by the newsroom system.

When material has been edited in the non-linear suites, due to it being non-destructive editing, the original material is never compromised. So when it comes to placing the edited story on the transmission server, the edit list just sends the material from the main server to the transmission server, such that it is an auto-assembly from one server to the other, and as high speed is available for the transfer, it is possible to make this packaging at high speed. Using the EDL as an auto-conforming list means the sections of material selected by the journalist and also in the NLE suite can be passed to the transmission server, so that on playback the material takes the form of the complete item.

Once all the programme resources are on the transmission server, then it just becomes a straightforward process to run the news programme. All the timings, device control, package playback, etc. are all within and under control of the play list or rundown.

News systems of course continue to encompass the newsroom and the journalist and their role in the news making process. Significant development of newsroom systems has made great advances in the last few years to ensure that journalists can, by themselves, view and possibly edit material as well as produce their story items and, in some cases, produce the rundown or play list. The simultaneous production of news to the Internet is now also coming into its own, and newsroom systems need in some cases to be able to produce an output for both the television medium as well as the worldwide web.

Previously, the journalist would be able to write and edit only his script at his desktop and then move to the editing suite, with script in hand, to begin the edit in an editing suite with a craft editor. Once the item had been edited it would then be transferred to a Betacart for transmission. During the editing process, the journalist may have performed a voice-over to the material. In other cases, he would just write up his story and when the play list was constructed his piece would be included as an item.

Server-based newsroom systems changed this. Development of newsroom systems by many manufacturers such as Associated Press with ENPS (a joint development with the BBC), iNews from Avid and others have allowed the journalist to become involved with the workflow of the news material far earlier than ever before. Now, from his desktop, the journalist can still have his stories in front of him, access to wires, search database and archives.

In the mid 1990s, Associated Press undertook a joint development with the BBC to develop a newsroom product which became ENPS – Electronic News Production System. This has become a successful system, being rolled out in many newsroom operations worldwide. There are also a number of other newsroom suppliers who make high-quality systems which are in daily use in broadcast operations.

4.16.6.2 MOS Protocol

One of the aspects of ENPS was the introduction, adoption and use of the MOS Protocol, which is now used not only by newsroom manufacturers but also by manufacturers of other associated requirements, such as automation systems, server manufacturers, graphics and asset management.

MOS stands for Media Object Server and is a powerful communications protocol used to link and communicate, with equipment as far as we are concerned here, within the newsroom.

Basically, this protocol ensures a strict communication between devices such as Newsroom Computer Systems, Servers, Stills Servers and Character Generators to name just a few of the devices.

One way of looking at MOS is as a play list manipulator, manipulating the play list of media and media objects.

The journalist now has the ability to review material that has been added to the server and its database, being able to play the video material, write stories and scripts to attach to the material, and even make a simple cuts edit of the material. Where a simple cuts edit has been performed, this can then be picked up by the craft editor, where the cuts can be tightened, voice-over and captions added, and the complete package prepared for transmission. It should be appreciated that the Journalist System does interconnect using a standard IT network, so the use of material at low resolution is paramount to ensure that the system does not become overburdened.

4.16.7 Archiving Material

In any newsroom system there are a number of different connections for services that supply background information or function to the news. One of these is archive.

Considering that news has been stored both on film and also electronically for a number of years, then one can imagine the significant amount of archive material currently available. There really are masses of this material and somehow the task has to be undertaken to store and identify it for possible later retrieval.

Material, in whatever format it is currently recorded, needs to be re-recorded for archive storage and simultaneously needs to be indexed and catalogued so that the material can be searched by anyone who is keen to use this archive material in the production of a news story. Of course, it is not just news that the archive is restricted to, as any item of programme material can make use of suitable archive material.

Storage of archive material can normally be in near-line storage, where the material is not in deep storage and can therefore be recovered quite quickly. There is also deep archive where the material is held in a form and format that might take time to recover. Deep archive normally resides on the shelf in the archive vault, waiting for that time when it might be used again, if only just the once. This was really how archive systems dealt with material over the last few years until a decision was taken on how and where to store the material. However, the use of electronic and electro-robotic solutions has made it just become one archive.

Electronic archive can take a number of forms in the methods used to store material. However, most storage is on data tape. Tape still represents the cheapest form of storage compared to hard disk or other removal media such as DVD or CD-ROM. Usually, data tape storage is within a large robotics device such as a StorageWorks Silo or Sony Petasite. In these systems the tapes are stored within reach of the robotics and the robot itself is sent to seek where the tape resides. The system knows where the material is by using the coordinate data held on the database, and once at those coordinates can retrieve the tape by then reading the bar-coding on the side of the tape, which confirms to the robotic system the contents that are held on the database. Think of this as being in a normal library; you search the catalogue cards for the book you wish to read and once you have found the title, you then use the additional location information on the card to locate the place where the book is stored.

The problem for any archive system is that they continue to grow as each day passes and more and more material is placed into them. So a system like a Sony Petasite does have real advantages as it can grow into the room where the archive is installed. The replicating database has to become a fine-tuned engine and be able to work within the many different search criteria that the user may enter to enquire where the material exists. One of the growing developments for these systems is that they are now expanding to not just contain information on the archive database but also to develop and maintain information on all media, thus becoming the media management system, taking input from material being initially recorded through to programme material that has recently been transmitted and therefore requires archive.

Placing the material into archive relies on a number of different items. Initially, in order to make the material searchable and also recognisable, two tasks have to be performed simultaneously: firstly logging the material and secondly taking thumbnail snapshots of the material to be used for later retrieval. To this, searchable keywords can be added to be able to be used by the archivist or journalist to search the material on a particular subject. Material is then added to the archive with the database added to by the process described above. Finally, once this process is complete, the tape can then be bar-coded and then placed in the robotic archive.

The process that the archivist performs is to view the tape. At each shot change or indeed item of interest, a thumbnail is recorded to record that item of interest and descriptions and keywords are entered about each scene for later use in retrieval. In some systems, scene change or shot detection is also incorporated, so all that the archivist has to do is enter the description and keywords.

For a journalist or archivist, the search for the material now becomes significantly easier by being able to search by keywords and phrases to identify the part of the material being searched. This can be performed at the desktop of the journalist.

For some news stories the use of archive material may not be essential. However, in other cases it may require the journalist to be able to sit down and make searches on the archive database to locate the material. Once the material is located, then the journalist can recover the material to the main server for later use in the NLE suite. It can now be seen that having a fine-tuned database, precise keywords and description, robotic access to the material and also a secure place to store the material means that the recovery of archive material is now very straightforward.

Newsroom systems have undergone considerable changes in the last decade. The alliance of new technology, more defined workflows and new and exciting technology will continue and enhance the immediacy of news.

Section 5
Outside Broadcast Systems and Hardware

Original document by **J T P Robinson**
Major revision by **W J Leathem**

5.1 Outside Broadcast Vehicles and Mobile Control Rooms

5.1.1 Evolution of OB Vehicles and Mobile Control Rooms

Bringing outdoor events into the living room of the viewer is becoming more and more common. In order to record these events, be they sport, a concert or an important national event or ceremony, there is more and more reliance on the use of the outside broadcast or OB vehicle. Where these vehicles were once small control rooms on wheels, the move has been to build 'Super Vehicles' which are equipped for multi-camera/multi-recording situations. The design and construction of these vehicles requires a considerable concentration of effort from a group of designers, each a master of his or her own expertise in a particular aspect of the vehicle, which as a team come together to build this mammoth mobile television control room of the road.

The ability to televise any outside event, be it sport or a national event, is found in the original developments of television, when Sir Isaac Schoenberg led a team of EMI engineers in the 1930s.

Detailed plans were compiled, with a number of vehicles to be specifically constructed for this purpose. One vehicle housed the Emitron cameras and their associated control equipment, together with simple vision and sound mixers with the synchronising pulse generators. A second vehicle contained a transmitter for handling the video and audio signals from the 'scanner' vehicle and transmitting them on a carrier sufficiently far removed from the Alexandra Palace main transmitter frequency of 45 MHz (vision) to avoid mutual interference when both transmitters would be on the air simultaneously.

A Dennis fire brigade rescue ladder vehicle modified to take the transmitter aerial was provided so that the required height could be achieved for reaching the reception point in North London. A fourth vehicle was equipped with a diesel generator to provide the not inconsiderable power required for the scanner and transmitter vehicles when on site.

With this mobile fleet of equipment the complex technicalities of live outside broadcasts were mastered some 50 years ago. Their first notable occasion to witness was the coronation of King George VI and Queen Elizabeth in May 1937.

Since then, developments in camera pick-up technology, notably the CCD or Charge-Coupled Device, have resulted in highly sensitive and compact colour cameras. The accelerating move towards High Definition, an ideal match to a sport event or concert, has enabled technology to be exploited to record situations undreamt of before.

In addition to the changes in camera technology, recording technology has also made similar advances in quality, format and indeed size. Compared to the first 2-inch machines, 1-inch C-Format machines and in recent times the various cassette-based systems, the OB vehicle has taken from these advances the ability to fit more and more VTRs in the bays. In recent time, hard disk recorders have now become more and more prevalent, offering superb slow motion as well as rapid access to material.

In the early vehicles, recording was based around a quad head machine and here a dedicated vehicle would have been used to house two machines as an editing pair together with their noisy but necessary air-conditioners. This has now given way to vehicles with 10–12 VTR devices or equivalent hard disk systems.

Digital technology has in the last decade significantly advanced and this has had an impact on vehicle design, features and facilities. Apart from the high picture quality, the reduction in size of systems, married with lower power demands and less heat, have made the OB vehicle design become more relaxed in terms of weight and power consumption, allowing the designers to pay more attention to the amount of equipment which can be fitted, as well as special finishing such as expanded sides and increasing the comfort of the operational areas.

The ubiquitous microprocessor has become more and more common in all equipment within the vehicle, such as cameras, vision mixer, DVEs, VTRs and so on. However, it has also appeared in other areas of the vehicles for power control and air-conditioning, even into the advanced levelling and jack systems employed in some vehicles today. When the first vehicles moved off the drawing board and into construction, valves were commonplace in equipment. However, with the advances in semiconductor technology and the common usage of microprocessors, the operational and technical staff can now spend more time on production issues, in the safe mind that the equipment will report a problem through its self-monitoring system in the unlikely event a problem should arise.

It is with this security in technology, lower power consumption and also better operational areas that the staff on board these control rooms of the road can spend many hours making that vital recording of that important event. It really does not matter where the vehicle is located, be it Manchester, Moscow or Melbourne – the fact of reliability and comfort is now intrinsic in the design and operation.

5.1.2 Vehicle Design

Designing an OB vehicle is dependent on four factors. These are:

- vehicle purpose;
- choice of chassis and overall dimensions;
- country of destination;
- extent of facilities.

Let us look at each of these factors in turn.

5.1.2.1 Vehicle purpose

It may sound slightly absurd to start this section by questioning the vehicle purpose; with the wide range of chassis types, from panel van through to tractor and trailer unit and also including bus chassis, there are indeed a wide range of chassis types on which to build the OB vehicle upon.

The decision to build a vehicle is not only an issue of what technical inventory it should carry but also what work it should be used for and also what road conditions it may encounter. In some of the cities in Europe, the streets are so narrow and afford little turning area for taking a trailer-based vehicle that the choice may well have to be for a smaller fixed chassis to be able to negotiate these streets.

Likewise, if the final destination is going to be a desert location, a trailer vehicle is out of the question, but more relying on a four-wheel chassis and a panel van chassis may be the ideal selection, accepting the lack in most cases of four-wheel drive, but this type of chassis may well be ideal for the conditions. A stand-alone generator may also be required in some situations as the availability of electricity may be reduced or not reliable enough both in terms of availability but also in quality.

The use of bus chassis has provided some interesting approaches to design and construction. Without deviating from the basic chassis offered and taking into account the lower Gross Vehicle Weight (GVW), some very excellent designs and implementations have been achieved using these chassis. There is no golden rule of thumb for the selection of the chassis other than the choice should be made to reflect not only the end destination, but also the implied use that the vehicle will be put to.

5.1.2.2 Choice of chassis and overall dimensions

In North America, the regulations are different to those in Europe. The overall dimensions for large vehicles are clearly regulated.

Within Europe, the overall dimensions of a vehicle to travel legally on the roads are:

overall width (maximum)	2.50 m
overall height (maximum)	4.00 m
overall length (maximum)	
for a rigid chassis	11.00 m
for a combined tractor/trailer unit	15.50 m
for a draw bar trailer combination	18.00 m

Restrictions and relationships exist between wheelbase, rear overhang and axle weights, so the design must be observant of the wheelbase measurements.

A chassis of whatever size, for use on the roads anywhere in the world, must legally comply with the legal requirements of that country. In order to ensure compliance the recruitment of a local consultant, versed in the laws and regulations of that country, is an incredibly good investment. It is the consultant's role to ensure that the chassis is acceptable in that country and meet the traffic laws of that country. An additional responsibility to ask of the consultant is to ensure that the after-sales services exist for the chassis and tractor unit in the destination country. Should the vehicle need service or spare parts, being able to supply them locally is mandatory. The longer the vehicle remains off the road it is not making any revenue, so availability of service and parts is crucial.

There are considerable regulations concerning the height of lamps, lens colours, positions of reflectors and many more items. Throughout the world, these regulations do vary quite considerably, mainly determined by whether the vehicle will be driven on the left or right hand side of the road.

In some countries the axle weight is subject to legislation concerning the maximum weight limit on the rear axle. It is important that the weight on the axle does not exceed the maximum loading as defined by the manufacturer.

In order to ensure compliance, the weight distribution within the vehicles must be observed. It is important that, during the initial design phase, the weight loadings are calculated for the vehicle. These calculations should take into account all the equipment, cables, camera accessories, VTRs, etc. with respect to where the equipment will finally be located with the vehicle. The centre of gravity of the vehicle should also be kept as low as possible by restricting overhead weight to essential items only and making full use of skirt lockers for housing compressor motors and similar heavy units. Calculations should also take into account air-conditioning units, isolation transformers, batteries and other items.

Last, but by no means least, the weight of the coachwork making up the body of the vehicle together with all its fittings and materials must be calculated and added to the technical equipment weight, to give the payload. It is this payload that will figure prominently in deciding upon which chassis the body is to be built.

The result should be a balanced chassis; nose not dipping down indicating excess front axle load and equally not tail heavy. In addition, the vehicle should not be listing from either side. Upon completion, the vehicle should be weighed. However, at this point this is too late to make any corrections in the weight as the weight calculations should have already indicated the total weight with the vehicle weighting being no more than a confirmation of the design calculations.

The Gross Vehicle Weight (GVW) is how a manufacturer will rate the chassis. This is the total weight of the vehicle to be in contact and bearing on the road surface through the tyres. This combination includes the entire technical and non-technical payload and includes the weight of the chassis. The manufacturer's design of the chassis will also take into account the types, springs and other items that are represented by GVW and incorporating an appropriate margin of safety.

In normal commercial use the all-up weight can be allowed to approach the GVW limit since the payload will vary from day to day or week to week depending upon the vagaries of the type of work involved. This variable duty will allow the suspension a degree of recovery, since heavy abuse of the springs on one journey will be compensated by a lighter load or perhaps no load at all on the next journey.

A TV OB vehicle has a permanent load which, by the very nature of its role, will never or hardly ever vary. Therefore, care should be taken to make sure that the kerbside weight of the fully loaded vehicle is not much more that 80% of the GVW. This will ensure a good handling characteristic to the driver and will allow some degree of liveliness from a medium-sized engine in the range of chassis under consideration. It should also allow the suspension to still exercise its prime duty of shock absorption over bad road surfaces without bottoming.

As an example, consider a rigid (a vehicle with the cab and body on the same chassis) TV OB vehicle that is required to house six colour cameras and five cassette-based VTRs, and it is to be equipped to a high standard in terms of peripheral equipment. It is to be used in a Far East climate with seven operators, and is to be around 9 m in overall length. No on-board generator is required. The calculation is as follows:

Total weight of all electronic equipment, found from manufacturers' brochures, say:	1.5 tonnes (1500 kg)
Total weight of all interconnecting cables and connectors, say:	0.3 tonnes
Total weight of any cable and cable reels for external use, say:	0.5 tonnes
Total weight of air-conditioning equipment, say:	0.5 tonnes
Total weight of body structure to house all the above, say:	5.0 tonnes

Adding up these five items gives the total payload as 7.8 tonnes, which may be rounded up to 8 tonnes.

The manufacturer's data stipulates that for a payload of 8 tonnes a chassis is required which is at least 13 tonnes GVW. The kerbside weight of this chassis without payload is 4.5 tonnes, which for our example would give a margin of only 0.5 tonnes, arrived at by subtracting the all-up kerbside weight of the finished vehicle (12.5 tonnes) from the GVW (13 tonnes).

Unfortunately, the resultant 96% ratio of kerbside weight to GVW is too low a safety margin for a TV mobile, and therefore a higher GVW chassis rating will be necessary to meet the criteria outlined earlier. For another illustration, consider a 16-tonne GVW chassis which will have a chassis weight of 5 tonnes, but its payload capacity is nearly 11 tonnes. Therefore, if the example of body payload capacity of 8 tonnes given earlier is now added to the 5 tonnes of the basic chassis, a 13-tonne all-up kerbside weight will result. This gives a kerbside weight/GVW ratio of 81% which, coupled with a medium-sized engine in this particular model range, would fulfil the necessary requirements.

Panel vans (as mentioned earlier) have gained popularity as a variant to the coach-built specialist body. Several major chassis manufacturers market a range of small panel vans and these have proved popular with many broadcasters where the larger purpose-built vehicle is not required. Added to this, the panel van offers greater manoeuvrability and reduced parking demands, and therefore a practical, flexible and manoeuvrable alternative for TV work in a city.

These panel vans are generally 6 m in overall length, can be offered with a high roof to give an internal height of 2 m and have a maximum GVW of 6 tonnes. The kerbside weight of the empty van is usually about 3 tonnes, leaving a maximum of approximately 2 tonnes of payload if a figure of 80% is used as the ratio of kerbside weight/GVW.

Whilst 2 tonnes of payload may not seem a great deal, the advantage is that very little weight is added in the coach building/conversion stage, since the external shell already exists and is not included as part of the payload. Coupled with today's compact and increasingly lighter equipment in terms of cameras, VTRs, etc., a reasonably versatile OB unit can be constructed for a relatively modest cost, proved by the number of such vehicles giving excellent service in many parts of the world.

Demountable bodies are now a new approach to traditional coach building. This system allows the coach-built bodywork with all its technical equipment to be disengaged and raised above the chassis by means of inbuilt hydraulic or mechanical jacks placed at each corner of the body, allowing the chassis to be driven away from underneath. The chassis-less body is then lowered to near ground level to become a static studio control room.

This concept is of interest for the following reasons:

1. The unit can be taken to a site, demounted and left as a fully operational unit for days or weeks on end where continuous long-term coverage of events is required. This system has even been used in a naval capacity, where the demounted body has been slung aboard ship and taken to sea.
2. If two or more units are operated, then only one chassis with engine and driver is needed to deploy them.
3. For detailed chassis maintenance and inspection purposes, removal of the chassis away from the body creates greater access.
4. Only the bodywork need be constructed at the factory. The chassis may be locally supplied by the end-user and united with the body upon its arrival. This could have important tax and import duty advantages to the end-user.
5. Where chassis delivery is protracted, the bodywork may be constructed independently, thus cutting down on the overall delivery time.

5.1.2.3 Country of destination

Air-conditioning requirements within the vehicle will be wholly determined by the climate in which it is to be operated. A hotter climate means that more on-board air-conditioning equipment will be necessary for a given number of operators and for technical heat dissipation as well as any heat gains. Physical construction of the vehicle will affect the heat retention or loss.

Increasing the number of air-conditioning units that are carried on a vehicle increases the payload penalty. Since one air-conditioning unit having an output of 6000 kcal/h (24,000 Btu/h) will weigh about 100 kg and with the typical number of such units for a larger vehicle being at least four, then a weight penalty of between 400 and 500 kg can be expected.

To assess the necessary size of the air-conditioning system, the following information is required to perform the heat calculations:

1. The structural heat gain of the bodywork. The structural materials and construction of the walls, roof and floor must be identified and allocated a U value.
2. The internal dimensions and areas of the vehicle where air-conditioning is to be applied.
3. The setting of an acceptable interior temperature and humidity based upon the country of destination.
4. The heat dissipation of all equipment within the vehicle. This will be based mainly upon the technical equipment power consumption.
5. The occupancy level of the vehicle as an operational average.
6. The interior lighting heat dissipation.
7. The inclusion or otherwise of a roof platform.

The roof platform (item 7) in the above list is an example where occasionally two factors come together to assist each other rather than oppose. A roof platform may be called for in the specification to provide a high-level vantage point for cameras, commentators or even for a microwave link. By spacing the platform off the actual roof of the vehicle by about 50 mm and allowing a free flow of air to circulate beneath it, a very real assistance is given to the overall efficiency of the air-conditioning system, since the platform then acts as a solar shield.

It can be seen that an overall sensible heat gain figure has to be matched to an air-conditioning unit, or units, which are capable of rejecting that figure. It is worth mentioning that taking the cubic interior capacity of the vehicle and dividing this by 60 will return the number of clean air changes that will occur within the interior of the vehicle.

In practice, the rejection figure will normally exceed the sensible heat gain figure by around 20% in order to take care of latent heat gains due to deliberate external fresh air intake and heat gain from the opening of exterior doors, etc. during operation, which is an additional figure to heat gains through the bodywork.

By selecting a suitable air-conditioner to handle the interior heat gain, note should be taken of both its start-up current and normal running current for future calculations on required power intake to the vehicle. A useful guide is to note that generally the start current, lasting about 1 second, is about 3–4 times the normal run current. Due to these high peak currents upon switch-on of the compressor motor, the practice is now to allow the motor to continue to run for as long as the main power is switched on. If this were not so, then continued heavy short-duration pulses of intake current as the motors started up and stopped each time in response to the environmental demands would reflect back upon the vehicle technical power circuits, causing equipment supply voltage variations at the least.

It is also worth remembering that on initial switch-on the units will demand the same start current characteristics. If the vehicle has a number of compressors and evaporators, then switching all the units on at the same time can cause monumental problems in practical operation. To overcome this issue, the air-conditioning plant is normally started up in timed turn-ons by staggering the time between switch-on and is normally implemented by placing timers to control the time period on each air-conditioning plant.

In order to allow the compressor motor to run continuously an alternative method has been developed which varies the cooling demand. This is the *hot gas bypass system*. In this system a solenoid valve is placed across the vehicle interior evaporator coil and external compressor and is opened by the interior thermostat when the required vehicle interior temperature has been achieved. This allows the hot gas from the compressor to be routed direct to the evaporator rather than going through the cooling process of the external condenser coil and thus allows the evaporator temperature and hence the interior temperature to rise, in turn creating a demand for cooling. The solenoid valve is then closed by the thermostat and hot gas is routed to the condenser coil, cooled, liquefied and passed through the evaporator coil in the normal way, where it absorbs heat from and therefore cools the vehicle interior.

This cycle continues with the compressor motor running continuously. In the hot gas bypass mode it is normal for the total cooling capacity of the system to be reduced by more than 60% which, if the system has been sized correctly, will allow the interior temperature to build up and cause the cooling demand requirement to re-occur. It is therefore most important that the equipment selected for this cooling process is not excessively oversized. If there is over-specification on the air-conditioning, this cycle of allowing the interior heat build-up will not occur and the complete system will eventually ice up and malfunction.

Common use is made of the split system of configuration for deployment of air-conditioners. Here the evaporator coil with its fan is physically separated from the condenser coil and compressor, the units being connected only by rigid pipe work or, more usually for vehicles, a flexible pressure hose having a low effusion rate or loss for the refrigerant to be used (normally Freon, R22).

The advantage of a split system for vehicles is that the evaporator unit can be sited within areas inside the vehicle, where it is able to perform efficiently in terms of providing conditioned air flow. The compressor/condenser unit, on the other hand, requires access to an external air flow for cooling purposes and in any case is a source of noise and vibration. It is therefore normally located external to the operational shell, usually in a side skirt locker, and suitably treated to reduce both of these factors.

To ensure efficient operation of the air-conditioning system, all filters and radiator matrixes must be kept clean and free from a build-up of dust. Once a system has been charged and commissioned under operational conditions there is little else requiring attention, but as with all things mechanical, regular inspection of the system should be scheduled at specified intervals.

It is important to consider the way that the conditioned air is delivered into the operational areas. A direct flow of cold air from the evaporator unit via front, side or overhead outlets is not to be recommended for operational areas as the occupants will raise objections to the resultant draughts at ear, eye and top of the head levels!

A more sophisticated approach is to deliver the conditioned air overhead via a perforated ceiling. This ensures a very much more indirect delivery and satisfies the environmental criterion of moving a large mass of cold air relatively slowly.

Just as important as air delivery is air return. Any air delivered into an area must eventually find its way back to the evaporator, otherwise the system will be starved of air circulation. It is also a prerequisite that a portion of fresh air is mixed with the circulating air to prevent stale air building up inside the vehicle.

Unless there are items of equipment on board which are specifically temperature conscious, the normal practice for interior air circulation is to first deliver the air to the operational areas and then take the return air via the electronic equipment, usually housed in 19 in (482 mm) equipment racks. This ensures that the slightly warmer air, or more accurately the less cold air, than that delivered directly from the evaporator unit is passed through the equipment last and helps to prevent undesirable condensation forming on the equipment. Thus, the priority in terms of environment is that of treating the operational areas first and the technical equipment second. Not all that long ago the reverse would have been the case; this reflects the way that component manufacturing techniques have eliminated the heat problems of the 1950s and 1960s.

Hot daytime temperatures are often accompanied by low night temperatures and it is equally important to ensure that when the outside temperature falls below an acceptable minimum there is sufficient heat capability on board the vehicle to provide human comfort. This can be achieved by incorporating heater elements within the air-conditioning units on the premise that heating and cooling will not be required simultaneously. However, this would probably result in heated air being discharged at roof level, perhaps by means of a perforated ceiling. This is contrary to natural thermal circulation and not conducive to a healthy environment.

A more acceptable alternative is to place electric fan heaters at low level within the vehicle or to use a natural fuelled system of warm air heating. Such units will run on a variety of fuels, such as diesel, gas-oil, propane, etc., and will discharge the resultant warm air through purpose-built ducting within the vehicle to low-level outlets placed in the operational areas.

Since they can be operated from low-voltage dc batteries, they are independent of mains electrical power and can therefore operate when such power is not available on site. If the main vehicle engine is a diesel, then quite often the warm air heater is chosen to be diesel operated and will take its fuel from the vehicle tank. By siting the take-off feed near the half capacity point, the heater will shut off when that point is reached, leaving enough fuel in the tank for the vehicle to reach its fuelling point safely.

For very cold climates, recourse is sometimes made to a centrally heated hot water system of distribution within large vehicles, the heated water circulating to domestic radiators placed in the operational areas.

One final point is the use of air-conditioning units in environments where the mains power is inclined to 'dip' or run below the normal operating voltages. This can happen and the effect for the air-conditioning is that the compressors will be placed under such a strain that they will burn out. This can be very expensive as it is not just a matter of replacing the motors in each compressor, but also replacing the gas as well as re-balancing the system. This therefore can be extremely expensive. Under/over-voltage relays can be fitted to each compressor to remove the problems, as explained earlier.

5.1.2.4 Extent of facilities

The provision of the technical facilities placed on board the vehicle will dictate the overall size of that vehicle. Large deployments of cameras and VTRs, a larger sound desk, more operators and engineering staff in turn increase the demands for a greater air-conditioning capacity and an increased GVW of the vehicle.

Initial requirements and design objectives should not lose sight of the original requirements during the constructional phase. Additional items of equipment can be added to the system and even to alter the mode of the vehicle during the building, such that the margins hopefully allowed for during the planning stage are steadily eroded away. In extreme cases, it is not unknown for an extra load-bearing axle to have to be added at a late stage in construction with consequent disruption of the delivery time-scale and endless arguments as to how the situation arose. Some overweight can be compensated for by adding additional springs or indeed stiffening the springs. However, this can only really be done on a rigid vehicle.

A useful way of monitoring the weight of the vehicle as construction proceeds is to place a load-bearing pad under each wheel, which gives a readout in kilograms. This will not only give an overall picture of axle loading as a percentage of the maximum, but also indicate the lateral or side-to-side weight loading.

One of the first things to be done is to install an electrical infrastructure with both a non-technical and technical supply wired into the van. The non-technical supply is, as its name implies, for those items which are not used for technical equipment. Therefore, this supply can be used for items such as air-conditioning, battery chargers, lighting and power outlets in the interior of the van.

The technical supply is normally fed via an AVR or Automatic Voltage Regulator, designed to maintain the technical voltage at the normal operating value. From the output of this unit, the technical equipment is supplied and this unit ensures the voltage is always held constant on the output regardless of any voltage fluctuations on the input. In some countries, regulations may require the installation of an Isolation Transformer.

Supply to the vehicle can either be single-phase or three-phase. Indeed, in some installations, inputs for both single- and three-phase are often fitted.

Electrical safety and protection in an OB vehicle is very important. The use of RCCBs (Residual Current Circuit Breakers) on the input of the vehicle is important, as in essence this is an electrical installation which is operating outside the normal safeguards of a building. An additional device that should also be considered on the input of the vehicle, prior to the mains input breaker, is a Phase Indication of the incoming mains. When the vehicle is initially connected on-site, there is always the risk of the connection being made unintentionally incorrect. This device will quickly alert the engineers to the error, allowing the correction to be made without harm to the vehicle.

Inside the vehicle, non-technical and technical outlets can be easily distinguished from each other by using differently coloured outlet plates, aluminium for one and brass for the other.

The main switching for the equipment is controlled from a central resource, a mains distribution unit, which incorporates meters for mains voltage, current and frequency, with a switch to switch between input and output. The supplies to each rack or racks of equipment are controlled from circuit breakers, rated to the current in each rack with an overhead so that the breakers are not operating on the edge of the supply.

In some cases, mains electricity may need to be run to other equipment in remote locations from the OB vehicle – for example, for a commentator's box and monitor. These supplies need to use waterproof connections and are protected with an RCCB on each connector.

It does not need overstating that the dangers of electricity and in particular the risk can be greater without adherence to good planning and proper installation practices. Adherence to the latest IEE Electrical Installation requirements is mandatory.

5.1.3 Outline of Constructional Techniques for TV Mobiles

Whereas mobile OB units built in the 1950s used a considerable amount of hardwood for their framework, following the trend of most commercial vehicle box bodies of that period, the 1960s saw the move to aluminium for both framing and skinning of specialist vehicles, bringing with it a much needed reduction in body weight.

This trend has continued to the present day, with the only other real departure of note being the occasional use of a glass fibre or GRP sandwich construction for large expanses of side wall, supplemented by conventional construction for underskirt lockers, main entry doors and other similar areas.

The construction of an OB mobile starts off on the drawing board in the proposal stage, where consideration must be given to operational ergonomics, air-conditioning requirements and technical equipment housing based upon a known chassis. If this proposal is accepted then more detailed dimensional drawings of vehicle layout plan and elevations are made. From these, the various extrusion types and lengths, aluminium panelling and indeed all the many hundreds of items which go to make up the finished vehicle are listed. These comprise the material schedules which enable orders to be placed with suppliers. Since the timescale for the construction of a large TV mobile from placement of order to finished vehicle may be between 4 and 5 months, it is necessary to keep track of the various stages of construction for the benefit of the customer as well as the builder. To this end, use is made of the familiar bar chart, where the x-axis represents the time element, usually calibrated in week numbers, and the y-axis the 25 or so benchmarks of the construction from chassis modifications through to painting and sign writing.

The skeletal framework of aluminium is built upon aluminium transverse cross bearers, these in turn being bolted to the aluminium bearers running the length of the chassis and coincident with the vehicle chassis main steel members.

To provide a degree of 'give' between the steelwork of the chassis and the coach-built body, a semi-resilient packer is sandwiched between the two. This 'give' factor is very important for specialist vehicles, since it allows the various sections of the body to maintain their positions relative to each other as the vehicle settles down over its life.

It is important to choose construction materials that are not going to be problematical in tropical climates. Any hardwood used must be treated against attack by insects and a humid atmosphere, whilst fabrics should be wholly man-made and not derived from natural sources. Corrosion of metal fittings and screws used on external surfaces is best prevented by the use of stainless steel wherever possible. Otherwise, heavy chrome on brass will resist attack provided that the quality of plating is good.

When the detailed drawing is complete, the framework is in effect built from inside to out.

Thin aluminium sheet or exterior grade 6 mm plywood is fixed via a thermal break to the inside of the aluminium framework with a 50 mm build-up of very light but high thermal insulation material placed so that every cavity between the inner and the as yet absent outer aluminium skins will be completely filled. An alternative technique is to use a polyurethane foam sprayed into the cavity. Whichever method is used it is important to ensure that the result gives as high a U factor as possible to the body shell.

Similar treatment must be applied for the same reasons to the roof and the floor, but within these areas provision must be made for cable ducts in the floor and for delivery of conditioned air within the ceiling. If the preferred method of air delivery is via a perforated ceiling, then a suitably sized duct must be formed immediately below the insulated roof, faced on its lower level by perforated sheet. Such sheet will be of aluminium and should be treated with some form of anti-condensate finish. To avoid air turbulence and hence noise within the ducts – there may of course be more than one such duct within the ceiling void, depending upon the air-conditioning specification – they must be lined with non-toxic polyurethane sheet foam.

Prior to the external skinning of the vehicle it will be necessary to provide electrical conduits within the walls for all mains services for general and operational lighting switches, power outlet points and also for the various dc services such as skirt locker lighting, emergency lighting, compressors for any pneumatically operated telescopic masts, fire, smoke and other warning systems, and so on.

By now a roof platform, if called for, will have been constructed off-line and then fixed to the roof to give a 50 mm air gap, as described earlier. Equipment racks, partitions and console will have been constructed, and the air-conditioning equipment will have been installed and internal finishes will have started. To give an acceptable aesthetic finish and at the same time provide a reasonably warm acoustic feel to the interior, a high-quality man-made carpet is applied to the interior wall. The desk surfaces will have been treated with one of the many laminate finishes now available and apertures cut to accept the various sizes and shapes of the equipment to be positioned within them, edged with either a hardwood or an aluminium trim. Hardwood or soft cushioned edging will be fixed to the fronts of the various desks, and decisions on the type and positioning of fire extinguishers and first aid boxes will signal the end stages of coach building.

The interior should be vacuumed out and the floor covering protected for the next stage of the construction. This will be the technical installation of all necessary cables and equipment in accordance with the schedules, which will have been prepared in parallel with the coachwork construction.

Hopefully, the end of the constructional phase will coincide with the start of the installation phase, ensuring a gradual change of emphasis in the work content from mechanical construction to technical installation.

After this phase and preferably before commissioning and acceptance trials, the final stage in the coach-built activity is painting and sign writing. Some time during the building phase, and forming one of the many bars on the bar chart, will be the need to obtain the customer's requirement for the paint finish in terms of colour and coach line positions, as well as for sign writing of the station identification lettering and logos. Also, such things as tyre pressures, fuel identification, paint specification, vehicle length, height and width must all be determined and inscribed on or in the vehicle as appropriate.

If coach building only is the contractual requirement, then (as mentioned earlier) it is necessary to drive the vehicle to a public weighbridge and take front and rear axle loadings plus overall loading. The weighbridge tickets should then be handed over with the vehicle to the customer for his own processing of the necessary documentation to register the vehicle for road use.

If the contract is a turnkey, whereby both coachbuilding and the technical installation are carried out by the one contractor, then weighing should be made of the finished coach-built vehicle before the technical installation and again at the point prior to delivery. This will provide a useful record for future use, but as pointed out earlier, if continual weight monitoring can be integrated into the constructional phase by the use of load pads then this is by far the best way to avoid embarrassing last minute weight problems.

I G Aizlewood
Managing Director, Continental Microwave Ltd

5.2 Microwave Links for OB and ENG

5.2.1 System Concepts

5.2.1.1 Transmitters

Siting of the transmitter is largely dictated by location of the event to be televised and is therefore substantially outside the operator's choice. Consequently, the transmit terminal generally makes greater demands of the designer. Let us examine the main options available to the development engineer.

Two broad approaches are common: direct modulation and heterodyne.

5.2.1.1.1 Direct modulation

The generation of a UHF signal and its modulation and multiplication to the required output frequency is a technique generally termed *direct modulation*. Direct modulation has advantages of simplicity, small physical size and low cost.

The modulatable synthesiser eliminates the major difficulty in maintaining effective frequency stability while applying video modulation to the master oscillator, but there still remain significant disadvantages to this method of power generation, notably in multi-hop operation, which is now often demanded of portable point-to-point equipment.

FM modulation is applied to a relatively low-frequency oscillator and its modulated output subsequently multiplied to the desired final frequency. Deviation is therefore effectively multiplied by the same factor as the oscillator frequency. CCIR standards set the final deviation (usually 8 MHz peak-to-peak at the baseband crossover frequency), and it follows that deviation at the modulation frequency must be reduced by a factor F/N, where N is the multiplier and F is the final frequency.

Unfortunately the dominant system noise emanates from, or is prior to, modulation and is therefore multiplied with the deviation. Ultimate noise performance is thereby limited, and the technique becomes less practical as the operating frequency band increases.

Direct modulation has a further disadvantage. When line of sight is not available for the required path, a repeater is necessary. Transferring the signal from a receiver to a following transmitter without demodulation at a repeater is very attractive as non-linear distortions and noise associated with the demodulation and remodulation process are eliminated.

A receiver will generally have available a suitable IF output to feed the ongoing circuit (usually 70 MHz), but compatability with direct modulation transmitters is not possible as no matching frequency is available to access for injection. The modulated master oscillator frequency and subsequent multipliers are determined by the designated transmission frequency and hence are different in every case.

Recovery of at least composite video is therefore necessary at each intermediate station, followed by a repeat modulation process – giving rise to the term *remodulating system.*

5.2.1.1.2 Heterodyne

A more 'purist' solution is offered by the up-conversion or heterodyne philosophy. Here a separate video modulator, usually running at 70 MHz, is mixed with a suitable SHF pump frequency so that one sideband provides the required SHF channel frequency. FM deviation present at the modulator will be directly translated to the output frequency.

FM noise is inherently easier to control with this system concept as the pump generator largely responsible for generation of the microwave frequency is unmodulated and can have a narrow loop bandwidth. In addition, deviation is unaffected by the mixing process, and modulator noise contribution is hence not magnified (as with subsequent multiplication in the direct modulation system).

Perhaps the greatest advantage is yielded by the constancy of a 70 MHz modulator signal irrespective of the required SHF output frequency, which can be independently adjusted by changing the pump frequency. Availability of a 70 MHz interface in the transmitter permits the local modulator to be replaced by a compatible signal derived from the IF of a preceding receiver when required, so true non-demodulating repetition is now possible.

On the surface then, this transmission concept eliminates all the shortcomings of the direct modulation method for a modest increase in cost and physical size acceptable in point-to point portable applications. However, a new disadvantage now arises which greatly limits operational versatility of the single conversion concept.

In realising the SHF frequency, the pump frequency and unwanted sideband will also be present at the mixer output, only displaced by 70 and 140 MHz respectively from the wanted signal.

To remove the unwanted components, there exists the need for an output filter with very high rejection only 70 MHz from the wanted sideband, but with a passband of at least 20 MHz to pass the FM modulation. While this is realisable, it restricts transmitter operation to a single SHF channel without filter change, a major deficiency in congested operating environments with multiple co-located systems and high RFI.

The ultimate solution is achieved by double up-conversion, as shown in Figure 5.2.1.

The 70 MHz first IF is retained to provide IF inject facilities at a repeater, but the first pump oscillator translates this to a second intermediate frequency rather than to the final SHF channel. (Frequencies between 300 MHz and 1.5 GHz are popular.) The requirement remains for a filter at the second IF frequency capable of rejecting the UHF pump frequency while not distorting the modulation band, but as the ratio of bandwidth to absolute frequency is now much greater, and since the second IF is a fixed frequency, this filter is easy to realise and its presence not restrictive to final system frequency agility.

A second up-conversion then provides a modulated sideband at the required SHF channel. Rejection of the pump frequency is again required, but this time filter parameters are less stringent as pump offset will be significantly greater because of the higher IF. Indeed, it is quite possible to tune the second local

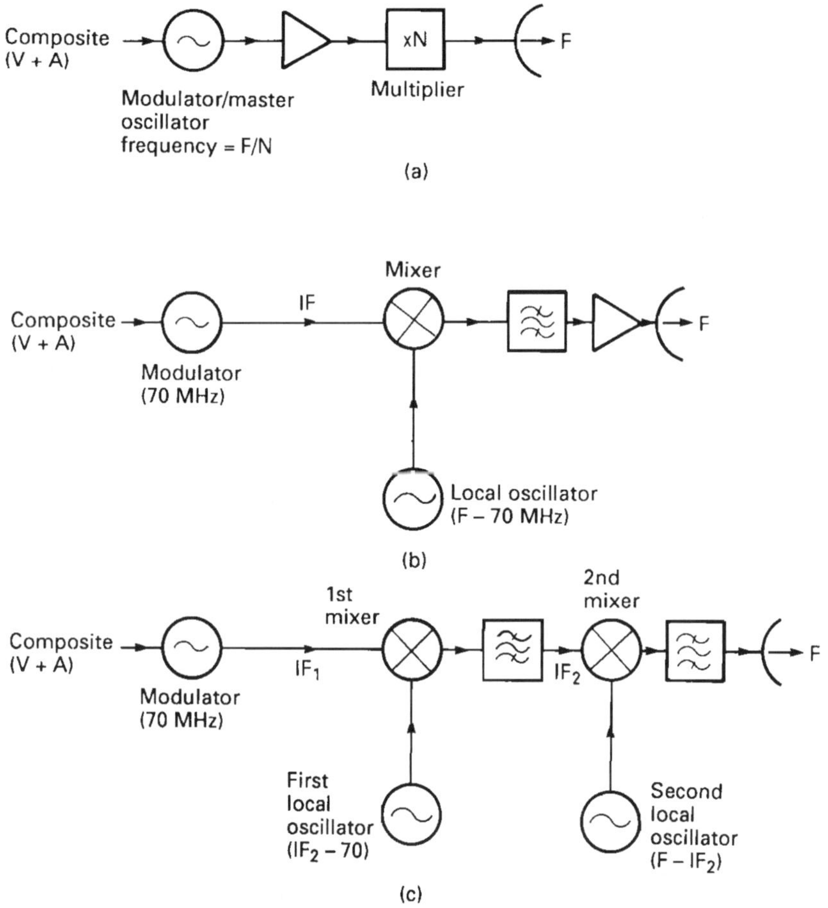

Figure 5.2.1 Simplified schematics of the transmission concepts: (a) remodulating; (b) single conversion heterodyne; (c) dual conversion heterodyne.

oscillator over bandwidths up to 80% of the UHF local oscillator frequency without the pump signal encroaching as a spurious output.

So now we have an elegant solution offering frequency agility, true IF repetition and excellent noise performance.

5.2.1.2 Receivers

Single and double conversion solutions are common (Figure 5.2.2). Down-conversion direct to 70 MHz offers economy of design but, with an image frequency only 140 MHz removed from the wanted signal, precludes operation without a relatively narrow band input filter. It renders frequency agility minimal (a similar situation to single up-conversion transmitters).

Use of a higher IF would extend frequency agility of the receiver but at the expense of losing a 70 MHz output for IF repetition.

5.2.2 Multiplexing

Operation of two or more independent links from a common antenna is now prevalent. Dual channels may be required for standby or for two separate vision channels, whilst a reverse channel for editing is also popular.

Several methods of *multiplexing* (*duplexing* or *diplexing*) multiple RF signals onto one antenna have been explored by manufacturers. The most common are: filter/circulator, hybrid and bipolar.

5.2.2.1 Filter/circulator multiplex

Filter/circulator multiplexing is traditional (derived from fixed system philosophy). It has one valuable advantage of positively protecting receivers from RF interference and a more dubious one of assumed low transmission loss. However, this method also drastically restricts frequency agility, thus destroying the most popular feature of wide-band tuning now available on most modern links.

In basic form, filter/circulator multiplexing is shown in the block schematic of Figure 5.2.3. The filter is chosen to have a pass-band for T_1.

T_1 output passes through the filter influenced only by its insertion loss and any minor distortions dictated by limitations of filter bandwidth. Generally, a group delay equalised filter with bandwidth 28 MHz to the −1 dB points will not significantly affect a (video + 4 audio) modulation to CCIR Rec. 405 (B). It is perfectly acceptable to increase the bandwidth of this filter to accommodate some variation in T_1 frequency, but at the expense of minimum channel spacing between T_1 and T_2.

The reason for this becomes clear if the T_2 transmission path is examined. T_2 enters the circulator and passes clockwise to the next available port, which coincides with T_1 entry via F_1. T_2 signal tries to exit from this port, but is reflected by the F_1 filter (which will appear as a mismatch at F_2) and re-enters the circulator to emerge again at the matched antenna port. Undistorted T_1 signal flow relies on F_1 appearing as a true short-circuit to F_2 frequency (including its full modulation bandwidth). However, in practice T_1 and T_2 frequencies may not be displaced adequately to guarantee uniform reflection for F_2 modulated bandwidth from the filter F_1. In this case F_2 will

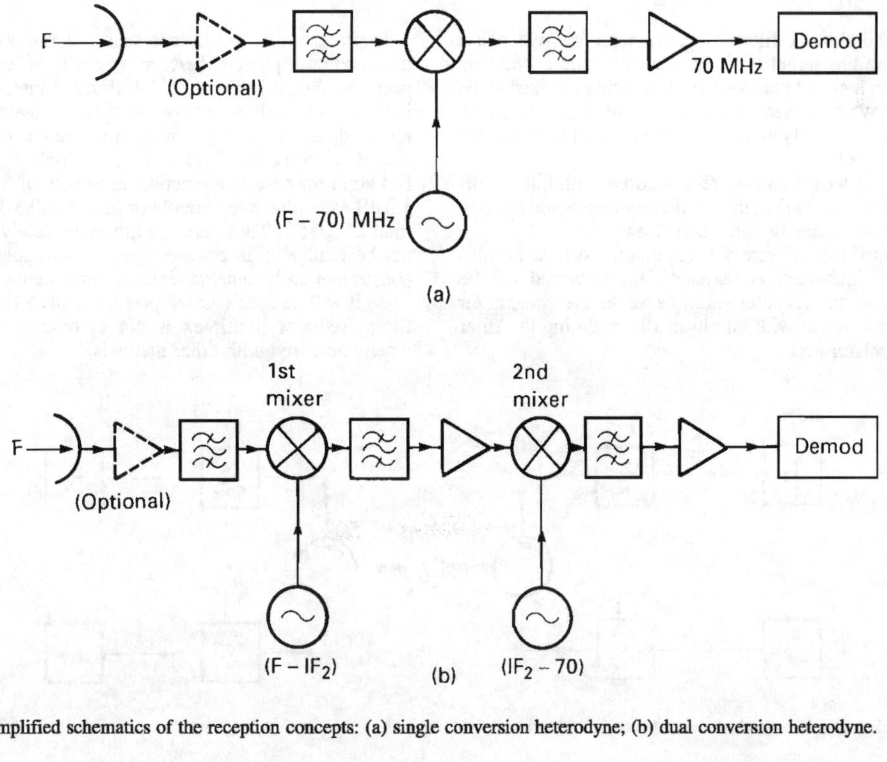

Figure 5.2.2 Simplified schematics of the reception concepts: (a) single conversion heterodyne; (b) dual conversion heterodyne.

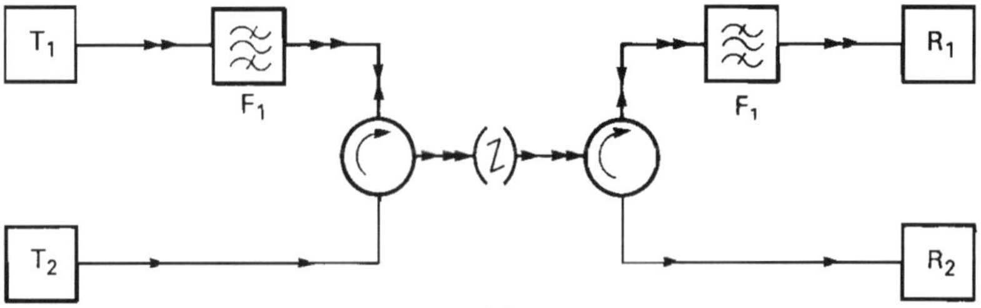

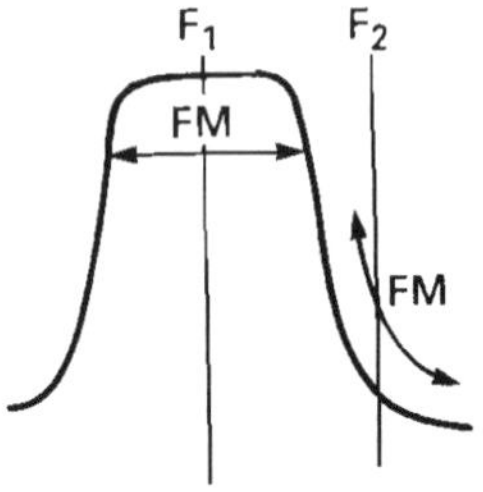

Figure 5.2.3 Filter/circulator multiplex schematic (a), showing the effect as the FM signal traverses the slope of the filter (b).

reflect from the skirt of filter F_1 causing progressive phase distortion as the FM signal traverses the slope and encounters a varying return loss and phase effect. Increasing the skirt slope of the F_1 filter by adding sections will improve distortion on the bounced channel, but only at the expense of insertion loss to the T_1 transmission path.

A typical filter would employ five sections with 0.8–1.2 dB insertion loss. Increasing the filter to six sections would steepen the skirt slope but typically add 0.5 dB loss.

It will be noted from Figure 5.2.3(a) that the simplest multiplex circuit as illustrated is 'handed', i.e. it would not be possible to swap the receiver multiplexer to the transmitter terminal (or vice versa) without physically changing the filter to another circulator port.

In practice this would represent a severe operational limitation, and multiplexers therefore typically include filters in both ports, as shown in Figure 5.2.4. Positive interference rejection is given to both receivers by this arrangement and, when required, the same equipment complement can of course be operated bidirectionally. However, multiplex insertion loss on a full hop now rises to a practical minimum of two filters at, say, 1.0 dB each plus three circulator passes at 0.3 dB each, giving a total of 2.9 dB. These are the minimum likely losses and may not be realisable in practice due, for example, to mechanical constraints on layout, waveguide components, etc.

So it will be seen that the presumed advantage of low loss in filter/circulator multiplex is not in practice very significant when compared with other methods.

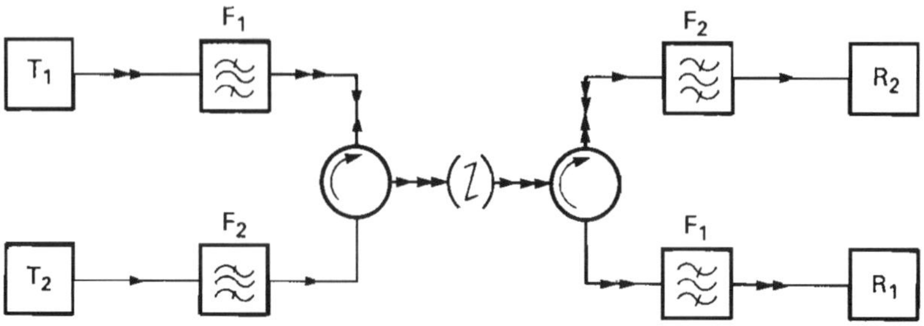

Figure 5.2.4 Schematic of multiplexes with filters in both ports.

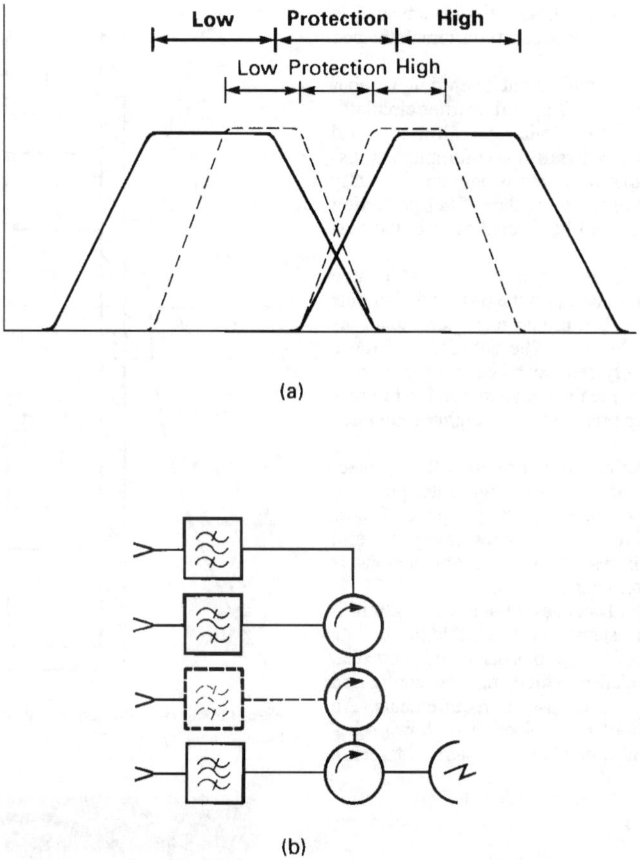

Figure 5.2.5 Variations of filter/circulator multiplex: (a) bandwidths increased to encompass a wide sub-band; (b) multi-channel combining arrangements.

In an attempt to regain some limited frequency agility, two 'variations' of filter/circulator multiplex are possible (see Figure 5.2.5):

- Using the same multiplex arrangement, filter bandwidths are increased from one channel of, say, 30 MHz to encompass a wide sub-band. As filter bandwidth increases, skirts become proportionally more shallow for any given number of filter elements, so it is not possible to divide the available band directly into two usable sections without a 'protection' bandwidth of typically 15% mid-band.
- Alternatively, multi-channel combining arrangements are quite practical (subject to size constraints) and can be constructed with a mixture of channel and sub-band elements tailored to offer the best user versatility for a particular environment and available operating frequencies. Such multiplexers are reversible between transmit and receive, so any combination of go/return traffic can be accommodated.

5.2.2.2 Hybrid multiplex

Provided the receivers in use are of double conversion superhet design and may therefore be operated in reasonable RF interference environments without external channel filters (or, alternatively, if the receivers are fitted with integral channel filters), hybrid combination and separation provides a cheap, wideband and physically small solution.

There will of course be a minimum theoretical loss over a full path comprising two such multiplexers of 6 dB with this method (typically 7 dB). However, when judged against the practical performance achieved by the filter circulator multiplex solution, the additional loss is unlikely to jeopardise significantly the overall system noise except at extreme range.

Hybrid multiplexing may not offer inter-port isolation and would therefore be unsuitable for duplex (bidirectional) operation unless receivers are separately protected by filters to prevent front-end damage from the high incident power of the adjacent transmitter.

5.2.2.3 Bipolar multiplex

Strictly, bipolar systems are not multiplexed at all. They operate from dual feeds which happen to share a common reflector and therefore behave as two single links. Bipolar advantages and disadvantages fall some way between filter/circulator and hybrid options.

The bipolar approach offers wide-band operation without any of the losses inherent in the 3 dB hybrid or filter/circulator multiplex solutions. Price is not dissimilar overall to a hybrid multiplexer, and without filter constraints no multiplexing distortions arise. However, bipolar operation is only conditionally able to support bidirectional operation without filter protection as antenna inter-port isolation will be limited even in the best designs to around 50 dB.

In practice, as system frequency increases, achievable transmission output power reduces, and the physical elements of the feed become smaller so that higher inter-port isolations are more realisable within the feed. The net effect of these parameters is to progressively reduce bleed energy from a transmitter into its adjacent receiver and to render bidirectional bipolar operation more practical in the higher frequency bands.

Bidirectional bipolar operation without receiver filter protection always has an element of risk, as antenna inter-port isolations can be significantly modified by the presence of local reflecting surfaces such as safety barriers or tower legs. Careful positioning of systems with respect to local obstructions is important in optimising system performance.

Some advantages and disadvantages of the various multiplexing options discussed are summarised in Table 5.2.1.

In practice, combinations of the options discussed above can provide highly versatile packages which may be configured quickly on-site for multiple transmission requirements. An example of a hybrid/filter/bipolar combination is shown schematically in Figure 5.2.6 and a practical realisation of such a system in Figure 5.2.7.

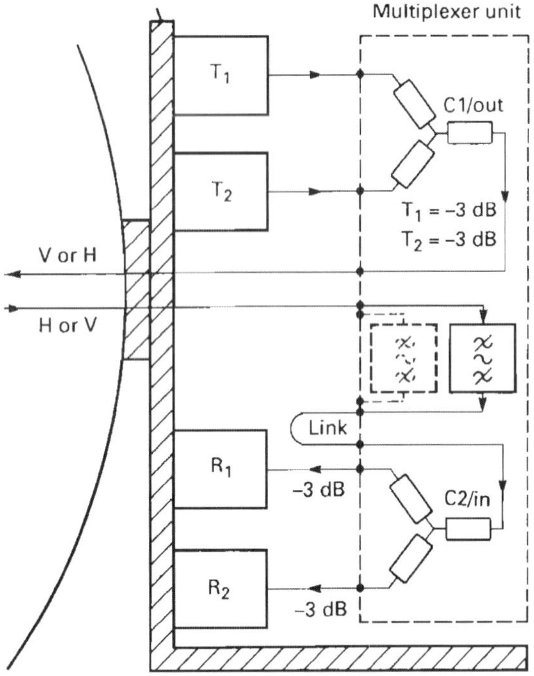

Figure 5.2.6 Hybrid/filter/bipolar multiplexer (arranged for 2 × bidirectional channels).

Figure 5.2.7 Three-channel mobile with remote control, hybrid multiplexer and bipolar feed, and 1.1 m antenna.

Table 5.2.1 Advantages and disadvantages of multiplexing methods

Type	*Advantages*	*Disadvantages*
Hybrid	Wide-band Compact Cheap Close adjacent channel use without distortion	High loss Diplex only unless used in association with filters or bipolar antenna
Bipolar	Wide-band No multiplexing loss No additional space required (integral to feed) Relatively cheap Close adjacent channel use without distortion	Conditional diplex operation unless receivers are filter protected
Filter/ circulator	Can be designed for any combination of multiple or two-way traffic with minimal interference risk	Restricted bandwidth operation Relatively high cost Less compact Distortion when used with very close adjacent channels

5.2.3 Antennas

5.2.3.1 Point-to-point

This category of link is required to operate over very long ranges (typically in excess of 50 km). It may well be co-sited with several other systems at starter or repeater sites when a major event is to be televised and is therefore potentially subject to high RF interference. In general, therefore, point-to-point operation of mobile links demands high antenna efficiency and a good polar diagram (see Figure 5.2.8) at the expense of some portability.

Derivations of parabolic antennas are generally chosen to provide high gain with narrow beam width, low side-lobe radiation and best front-to-back ratio.

Several manufacturers prefer offset paraboloids, which have the advantage of better clearance over local obstructions such as safety barriers, and lower aperture blocking if the feed is offset. However, it is doubtful if these advantages outweigh the relatively lower gain, increased mechanical complexity, storage difficulties and manufacturing cost over a simple centre-fed solution.

Since most mobile links operate in reasonable line-of-sight conditions, linear polarisation is traditional and still generally favoured, although circular polarisation is now sometimes used at lower frequencies.

Table 5.2.2 lists typical antenna size, frequency band and gain for use in the specimen path calculations in Section 5.2.5.

5.2.3.2 Horn

For short-range point-to-point operation, simple horn antennas providing vertical or horizontal polarisation can be a cheap and effective solution. These items are rarely seen in the commercial market, but can often be fabricated easily in the broadcaster's model shop. Table 5.2.3 provides dimensions and gain for some useful frequency bands.

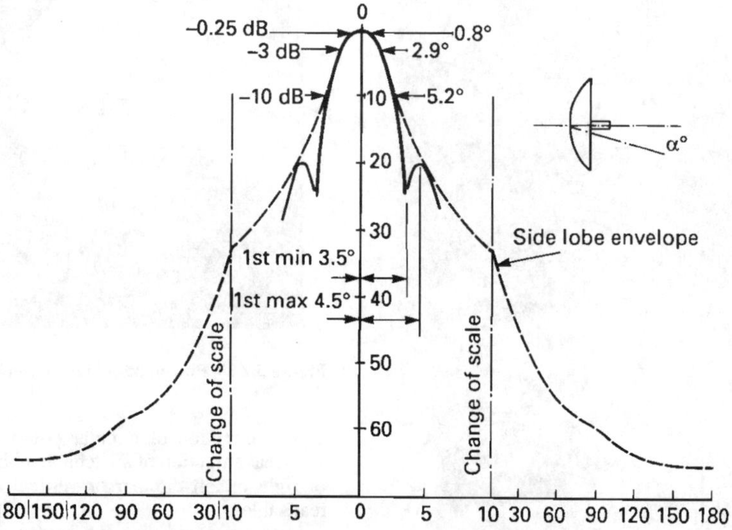

Figure 5.2.8 Polar diagram for a 7 GHz 1.1 m antenna.

Table 5.2.2 Parabolic antenna sizes versus gain for various frequency bands

Antenna size		*Gain (dB) at frequencies 2–23 (GHz)*									
ft	*m*	*2*	*2.5*	*3.5*	*5.5*	*7*	*8.5*	*10*	*13*	*14.5*	*23*
2	0.6	18.7	20.6	23.6	27.6	29.6	31.3	32.8	35.0	35.9	39.8
4	1.2	24.7	26.6	29.6	33.6	35.6	37.3	38.8	41.0	41.9	45.9
6	1.8	28.2	30.2	33.2	37.0	39.1	40.8	42.2	44.4	45.4	–
(8)	(2.4)	30.7	32.6	35.6	39.6	41.6	43.3	44.8	47.0	47.9	–
(10)	(3.0)	32.8	34.6	37.6	41.6	43.6	45.2	46.7	49.0	49.9	–
(12)	(3.7)	34.2	36.2	39.2	43.0	45.1	46.8	48.2	50.4	–	–

The frequencies chosen represent commonly used OB bands. For portable equipment an illumination efficiency of 45% has been assumed. All performance is 'typical' and will vary slightly with antenna bandwidth, method of illumination, mechanical configuration, etc. Sizes in parentheses are given for information only and are not generally appropriate to portable operation.

Table 5.2.3 Horn antenna parameters

Useful frequency band (GHz)	*WG Flange*	*10 dB gain Size (mm)**	*15 dB gain Size (mm)**	*20 dB gain Size (mm)**
2.6–4.0	10	120 × 90 × 225	220 × 150 × 420	–
4.0–6.0	12	75 × 55 × 150	220 × 150 × 420	230 × 170 × 420
5.9–8.2	70 × 40 × 110	90 × 95 × 280	160 × 120 × 340	
8.2–12.4	16	40 × 30 × 70	70 × 50 × 150	130 × 95 × 270
12.4–18.0	18	25 × 20 × 55	40 × 30 × 95	75 × 55 × 160
18.0–26.5	20	20 × 16 × 40	30 × 22 × 70	50 × 37 × 110

*Sizes are given in the order: flair width × flair height × front-to-back length.
Approximate beam widths to −3 dB points are: 10 dB gain = 50°; 15 dB gain = 30°; 20 dB gain = 20°.

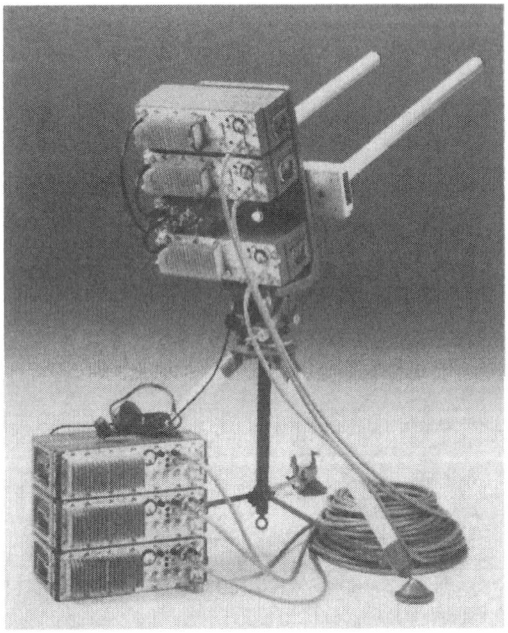

Figure 5.2.9 Triple mobile with remote control. Hybrid multiplex and disc-rod antenna (2.5 GHz/21 dB gain).

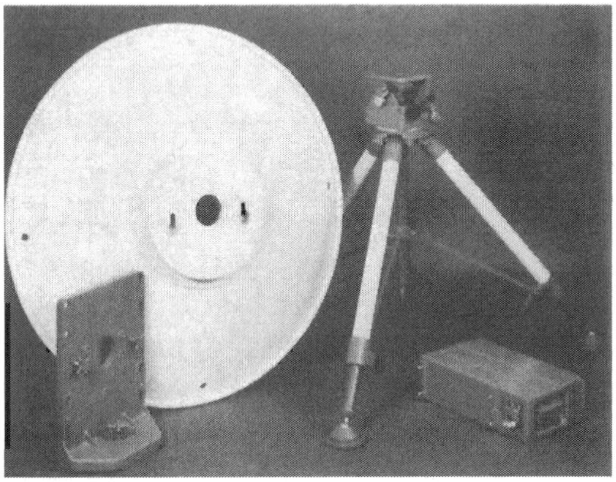

Figure 5.2.10 Broken down mobile mechanics set.

5.2.3.3 ENG systems

ENG systems are generally used in rapid deployment situations where the path is dictated by the origin of an event and a clear line of sight may not be available. In order to provide real-time television from such adverse environments, antennas need lightweight portability and the capability to utilise deliberate 'bounce' opportunities where line of sight is not achievable. Low installed wind resistance is also a requirement, so that such antennas may be operated on pump-up towers, lightweight tripods, etc.

Although small parabolic reflectors (0.3–0.6 m) are practical, most users seem to prefer helix or disc-rod solutions in single or combined form. Helix antennas have some nominal advantage of polar diagram over disc-rods, particularly in side-lobe performance. They are, however, restricted to a single pre-chosen direction of circular polarisation. Disc-rods have the very real attraction of switchable polarisation, any combination of right or left circular or vertical or horizontal linear being realisable.

Both helix and disc-rod antennas are relatively expensive due to their specialised, low volume nature, especially when supplied in multiple formats with integral combiners. Helices and disc-rods are practical to at least 8.5 GHz when required.

5.2.3.3.1 Transmission on the move

Effective antennas are a key component of any microwave transmission path, but nowhere is performance more critical than in the achievement of high-quality colour transmission from a moving source.

Two factors dominate the received signal quality:

- signal strength;
- distortions due to multiple received signals (multi-path).

Signal strength is enormously reduced when the source is masked from the receiver by a reflecting or absorbing surface. Even one tree close to and in line with the transmitting antenna is sufficient to impose unmanageable path fade at frequencies of 2 GHz and above. Worse, reflecting surfaces

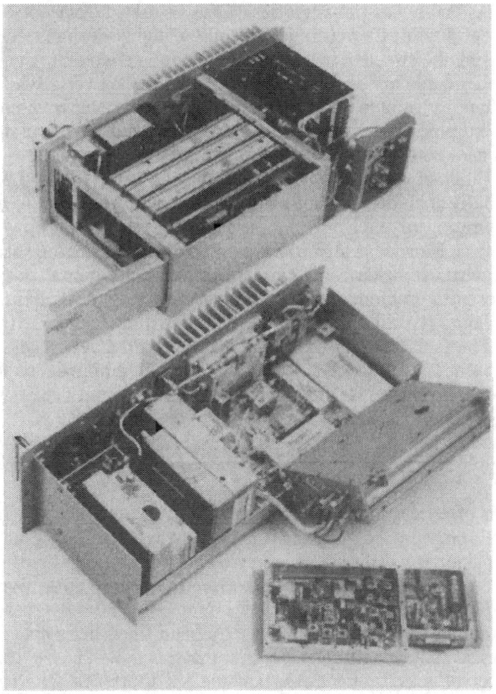

Figure 5.2.11 Typical interior layout of head and control electronics.

within the transmit antenna bandwidth can often present the receiver with two or more signal sources. Since these are derived from a common transmitter, the receiver will readily accept them all.

However, the direct and secondary (reflected) signals will travel different distances and hence arrive at the receiver in different phases. Because a wavelength at 2 GHz is only about 15 cm, even minor relative movement between transmitter and receiver will transcribe multiple full-cycle phase changes as seen by the receiver. When the multiple signals are in phase, enhancement will occur and the receiver will see a signal significantly in excess of that expected. Conversely at 180° phase conflict, signals will tend to cancel. It is quite possible in practice for microwave power to be reflected highly efficiently, and therefore for total signal cancellation to occur at the receiver.

Possibly the most objectionable effects occur when the receiver demodulator is presented with phase conflicts which translate into group delay distortion of the demodulated signals, with disturbance of colour parameters in particular.

Several methods are in common use to alleviate distortions, notably:

- if AGC characteristics with fast response times are able to react to flutter speeds of several kilohertz;
- powerful limiting for AM suppression in excess of 50 dB;
- video clamps;
- chroma AGC;
- component transmission.

The latter is a technically superior option. However effective these measures are, none can ever equal tackling the problem at source.

It follows that the best way to cope with multi-path distortions is by prevention, or at least reduction of multi-path itself, and this is where specialised antennas with parameters optimised to operational requirements play a vital part.

Linear antennas give rise to reflections of the same polarisation, and a receiving antenna is therefore equally receptive to the primary and any secondary signals. However, circular polarisation (following the laws of light) reverses its direction of rotation on reflection, and the receiving antenna can therefore discriminate against a first reflection signal, typically by a factor in excess of 20 dB.

Circular polarisation is not a total solution since many practical paths permit double reflections, where the direction of signal polarisation of course reverts to the original. Generally, however, signal strength of a double reflection is significantly less and does not present a major influence except in enclosed urban or sports stadium environments. For such situations, the use of semi-directional antennas to reduce the number of possible reflection opportunities, together with circular polarisation, generally yields acceptable results.

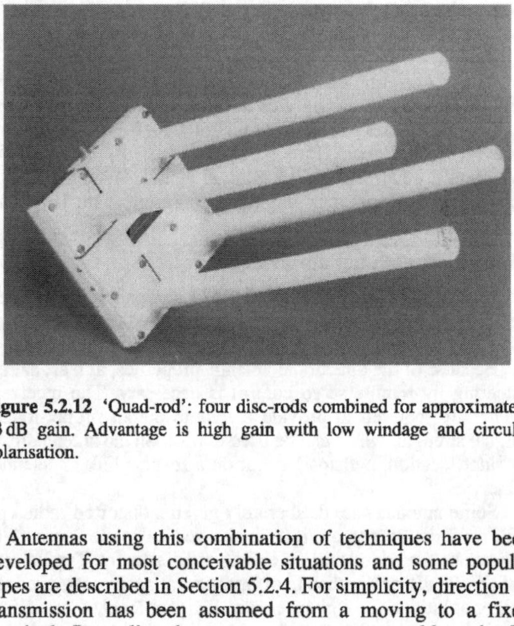

Figure 5.2.12 'Quad-rod': four disc-rods combined for approximately 23 dB gain. Advantage is high gain with low windage and circular polarisation.

Antennas using this combination of techniques have been developed for most conceivable situations and some popular types are described in Section 5.2.4. For simplicity, direction of transmission has been assumed from a moving to a fixed terminal. In reality the same commentary would apply for reversed transmission.

5.2.4 Central ENG

It is convenient to divide central ENG station considerations into *antenna* and *receiver* elements, since maintenance factors usually dictate sitting the receiver remote from its antenna in a more hospitable operational environment.

There are three fundamental approaches to remote mounted (masthead) central receiver antennas with 360° coverage: omni antennas, sector antennas operated with combiners, or switches and directional antennas with servo-driven remote steering systems. In general, the options are described in order of ascending cost!

5.2.4.1 Omni antennas

Omni antennas provide a very effective solution where the possibility for mounting at the topmost point of the tower exists to give full azimuth coverage without shadow from the tower structure.

Omnis can be produced with any polarisation mode but do not subsequently offer easy change of polarisation. The choice of circular polarisation generally offers superior multi-path performance (see Section 5.2.3.3.1), and co-linear combining can yield useful gain. In fact, in such fixed applications, omni-direction gains of 11–14 dB in azimuth are quite achievable without antenna size becoming excessively cumbersome.

However, this solution is not ideal for long-range operation or where local topography suggests a multi-path risk.

5.2.4.2 Sector antennas

Sector antennas such as *quad horns* are useful where tower-top mounting is not feasible. They are typically mounted on tower legs with each horn scanning a 90° sector. Choice of the relevant sector for any transmission is made by a remote switch.

Quad configuration generally permits selection of multiple polarisations: circular right or left or linear vertical or horizontal, giving a remote selection function of 1 from 16 to optimise the received signal. Horn gains of 14–16 dB are realisable, yielding a small but useful gain advantage over the simplest co-linear omni solutions and significant multi-path benefits.

5.2.4.3 Directional antennas

The most sophisticated central ENG antenna solution uses a narrow beam paraboloid (or offset paraboloid or cosec^2 variant) for optimum multi-path rejection and maximum gain (range).

Because of the directional antenna properties, at least azimuth steering by remote servo control is necessary. The receiver is generally required to provide a remote output proportional to signal strength, which can be used to position the antenna from a distant location, perhaps fed out on a reverse link or telephone line.

Some antennas are deliberately given a distorted radius profile to increase beam width in the vertical plate (cosec^2) and eliminate the complexity of elevation adjustment. Typical gains of 27–29 dB are realised, offering a range almost quadruple other solutions for a given received signal level.

Several remotely steered paraboloid solutions are commercially available. However, the superior performance available from this antenna type must be weighted against initial capital cost and increased maintenance of the moving parts.

5.2.4.4 Low-noise masthead pre-amplifiers

Because ENG transmission typically operates with small transmission antennas and less than perfect propagation conditions, received signal level is often close to threshold limits and any boost in signal/noise performance is welcome.

A low-noise pre-amplifier (LNA) situated adjacent to (or integral with) the receiving antenna is highly desirable in optimising the effective receiver noise factor by overcoming antenna feeder cable losses. This item, whilst almost universally of high reliability gallium arsenide fet design, is nevertheless vulnerable to extremes of temperature, water ingress and lightning strikes, so dual configuration or bypass facilities are preferable.

Channel or at least sub-band/filtering prior to the LNA is usually included for out-band interference protection to avoid saturation or inter-modulation occurring within this low-level device. Most central ENG receivers now are of double superhet concept, so wide-band noise from the LNA output does not present a problem and LNA output filtering is not required.

Some typical configurations are shown in Figure 5.2.13.

The isolator prior to the LNA buffers filter return loss, and subsequent to the LNA offers a good source impedance to the feeder to prevent reflections which could significantly affect colour performance.

5.2.4.4.1 Effective noise figure of masthead LNA

It is often assumed that a masthead LNA with 2 dB noise factor will provide 2 dB performance in a system. The reality is considerably different, as every element in the chain will introduce some modification of the effective system noise figure.

Figure 5.2.14 shows a typical example. It assumes a masthead remoting cable such as LDF5-50A foam filled helix and 100 m typical cable length. The loss over that length will be 6.6 dB. The receiver noise figure in this example is 9 dB, typical for a central ENG receiver without an integral LNA.

So, moving towards the antenna, the noise figure measured at the antenna end of the cable, including the isolator, will become $9.0 + 6.6 + 0.3 = 15.9$ dB. This will appear as the effective noise figure of the receiver as viewed from the LNA. The noise figure as seen from the LNA input will be 2 dB 'diluted' by the effective receiver noise figure, 15.9 dB (because of LNA transparency), and the dilution will be inversely proportional to the LNA gain, i.e. with very low LNA gain, the noise figure as viewed from the LNA input will not yield the expected improvement.

Choice of LNA gain is therefore largely based on the need to overcome the 'effective' receiver noise figure viewed down the antenna cable.

The actual effective noise figure can be deduced from:

$$N_{\text{effective}} = N_{\text{LNA}} + \frac{N_{\text{post LNA}} - 1}{\text{LNA gain}}$$

In our example, let us calculate effective LNA noise figures for LNA gains of 8 and 16 dB (being those typically realisable from one- and two-stage pre-amplifiers).
With LNA gain of 8 dB:

$$N_{\text{effective}} = 2 + \frac{15.9 - 1}{8} = 3.9\,\text{dB}$$

With LNA gain of 16 dB:

$$N_{\text{effective}} = 2 + \frac{15.9 - 1}{16} = 2.9\,\text{dB}$$

As can be seen, the difference is a useful 1 dB improvement in noise figure and may be well worth the additional LNA stage. However, in neither case was the 2 dB LNA noise factor realised.

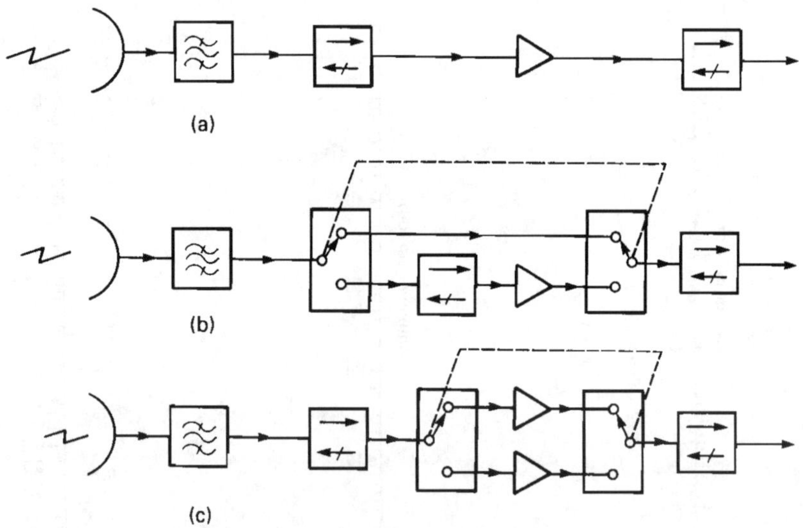

Figure 5.2.13 Some LNA configurations: (a) unduplicated; (b) unduplicated with bypass; (c) fully duplicated.

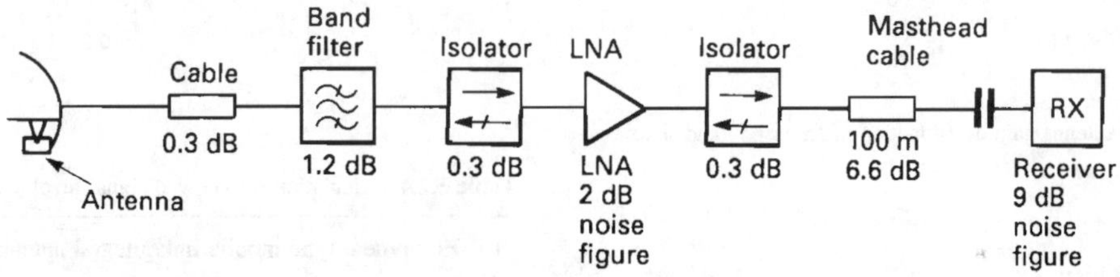

Figure 5.2.14 Typical calculation of effective noise figure. See text.

For both options, note that prior to the LNA, the effective noise figure viewed from the antenna will include the cable, band filter and isolator losses, yielding a final system noise figure for our example of $2.9 + 0.3 + 1.2 + 0.3 = 4.7$ dB.

5.2.5 System Calculations

5.2.5.1 By computation

The example assumed is:

1. A frequency of 7.5 GHz.
2. A path length of 40 km (25 miles).
3. An output power of 1 W (0 dBW/ + 30 dBm).
4. An antenna size of 0.6 m at the transmitter.
5. An antenna size of 1.2 m at the receiver.
6. A receiver with LNA (which therefore approximates to the 4 dB noise figure assumed for the full calculation).
7. A receiver threshold of −112 dB.

5.2.5.2 Short method using nomographs

The nomographs in Figures 5.2.15 and 5.2.16, and the graph in Figure 5.2.17, provide a quick everyday method of calculating expected system performance to a first approximation.

It is of course necessary to know some basic system parameters and, for the sake of useful comparison, we shall repeat the full calculation of Section 5.2.4.4.1 and assume again items 1–7 listed above.

To a first order approximation, items 6 and 7 can be assumed similar for most current OB links.

First, to find antenna gain use the graph in Figure 5.2.15. Place a straight edge between the frequency (assumed here at 7.5 GHz) and antenna size columns. Read off the gain from the centre column.

In the case being considered, the antenna gains are:

for the transmitter antenna of 0.6 m	31 dB
for the receiver antenna of 1.2 m	37 dB

Now to establish path loss, use the graph in Figure 5.2.16. With a straight edge from 7.5 GHz to 40 km, path loss is available from the centre column as 142 dB.

So received signal level equates to:

Transmitter output power +:	0 dBW
Transmitter antenna gain +:	31 dB
Receiver antenna gain −:	37 dB

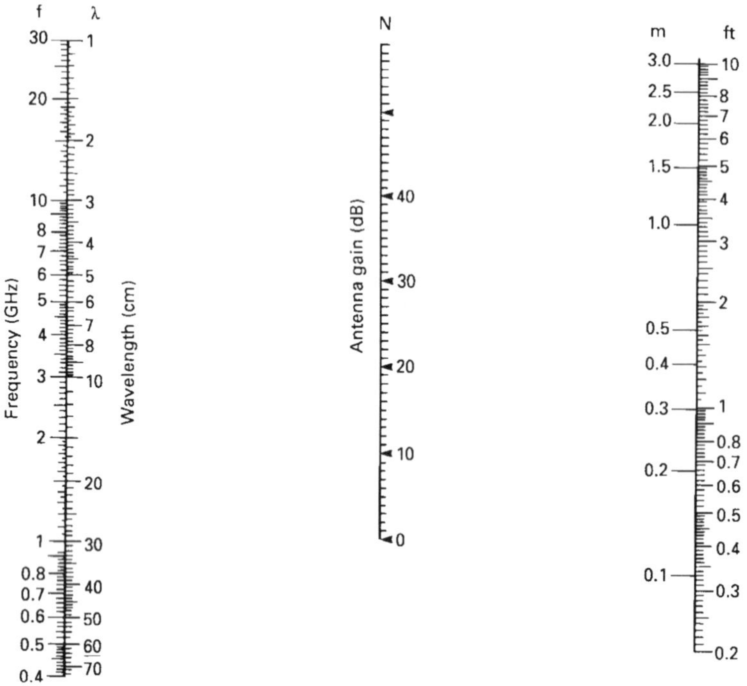

Figure 5.2.15 Antenna gain as a function of frequency and size.

Path loss:	−142 dB
Signal level to receiver =	−74 dBW

Since we know, for our example, that miscellaneous 'per hop' losses equate to 3.1 dB, we can now include these to improve correlation of our follow-on calculation. So:

Signal level to receiver = −74.0 − 3.1 dB = −77.1 dBW

Fade margin = received signal − threshold dBW

= −77.1 − (−)112 dBW = 34.9 dB

Finally, for signal/noise ratio, refer directly to Figure 5.2.17, where a −77.1 dBW received signal equates to 64 dB weighted s/n. This result correlates closely with our earlier full calculation.

So, for all practical purposes, a graphical computation is quite adequate for mobile performance prediction.

Note that the curves in Figure 5.2.17 also provide unweighted luminance and chrominance s/n figures with and without received LNA.

If first approximation audio performance is also required, refer to Figure 5.2.18.

Table 5.2.4 Calculation of received signal level

1.	Equipment type mobile link/integral antenna	
2.	Frequency	7.5 GHz
3.	Path length	25 miles
		40 km
4.	Path loss (free space)	142.0 dB
5.	Transmitter output power	1000 mW
		0 dBW
6.	Transmitter losses: multiplex	1.1 dB
	antenna cable loss	0.3 dB
	TOTAL	1.4 dB
7.	Transmitter antenna size	1 ft
		0.6 m
	gain	31.0 dB
8.	Effective radiated power (Item 5 − 6 + 7)	29.6 dBW
9.	Power to receiver antenna (Item 8 − 4)	−112.4 dBW
10.	Receiver antenna size	4 ft
		1.2 m
	gain	37.0 dB*
11.	Receiver losses: Antenna cable loss	0.3 dB
	Multiplex (filter +2	
	circulator passes)	1.4 dB
	TOTAL	1.7dB
12.	Signal level to receiver (Item 9 + 10 − 11)	77.1 dBW

*Transmitter and receiver antenna gain obtained from tables or Figure 5.2.15.

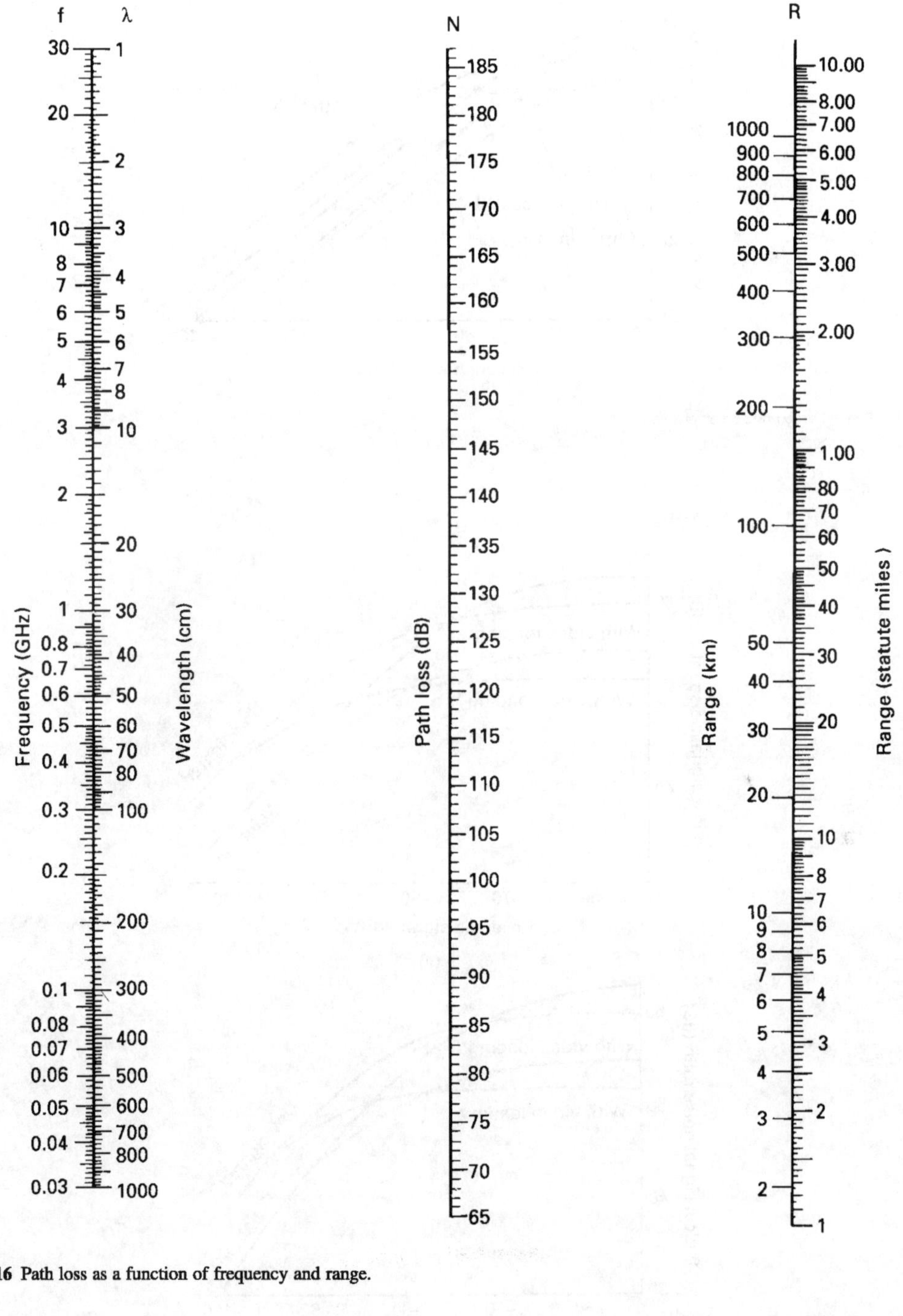

Figure 5.2.16 Path loss as a function of frequency and range.

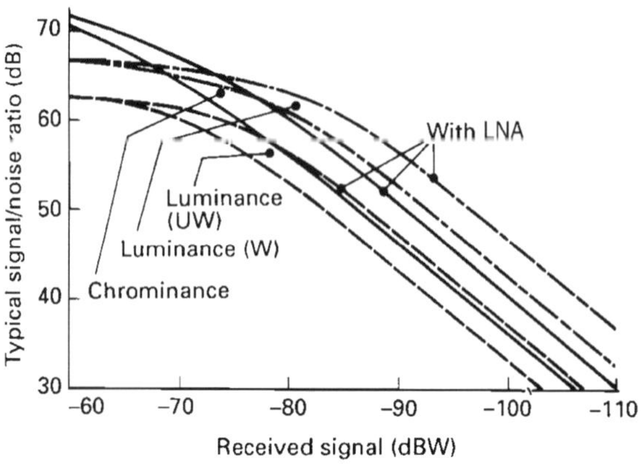

Figure 5.2.17 Typical video noise performance.

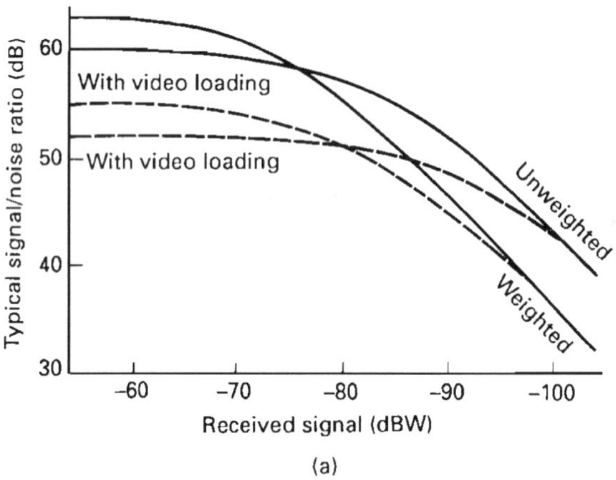

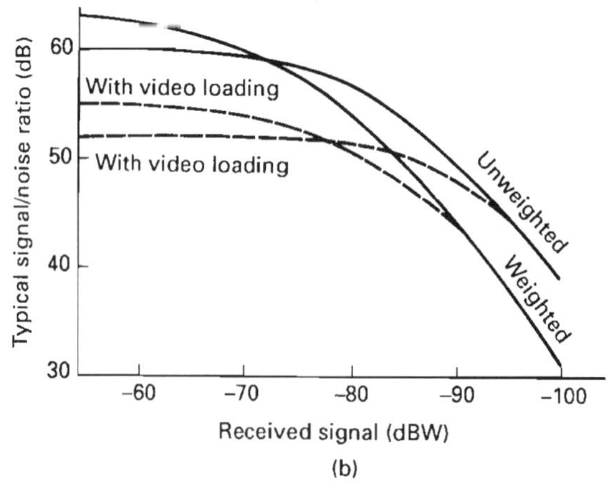

Figure 5.2.18 Typical audio noise performance: (a) with LNA; (b) without LNA. All measurements are with respect to 0 dBm audio loading, by ppm to CCIR 468-2; video loading is multi-burst.

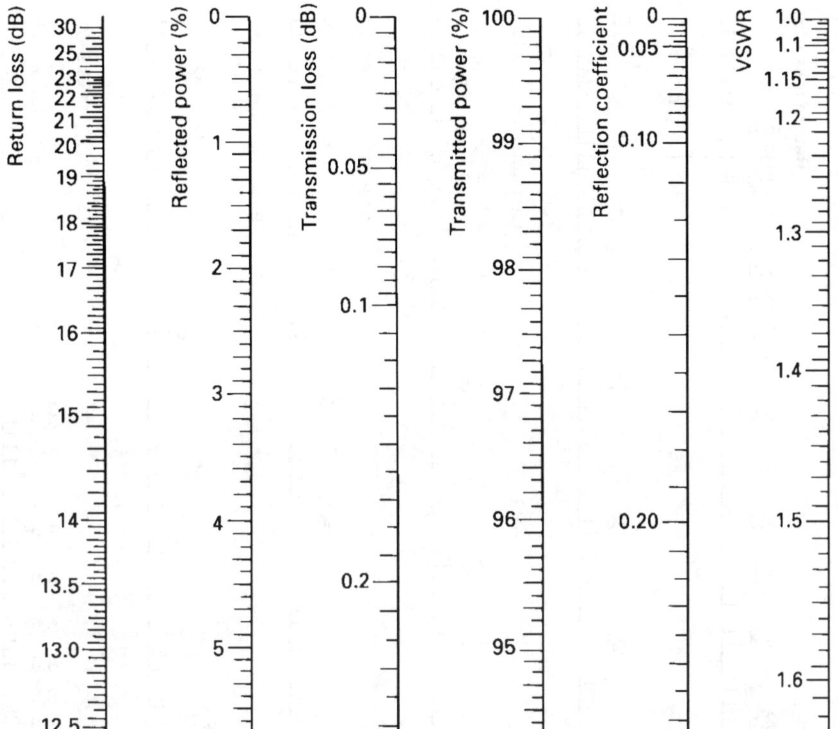

Figure 5.2.19 Return loss nomogram.

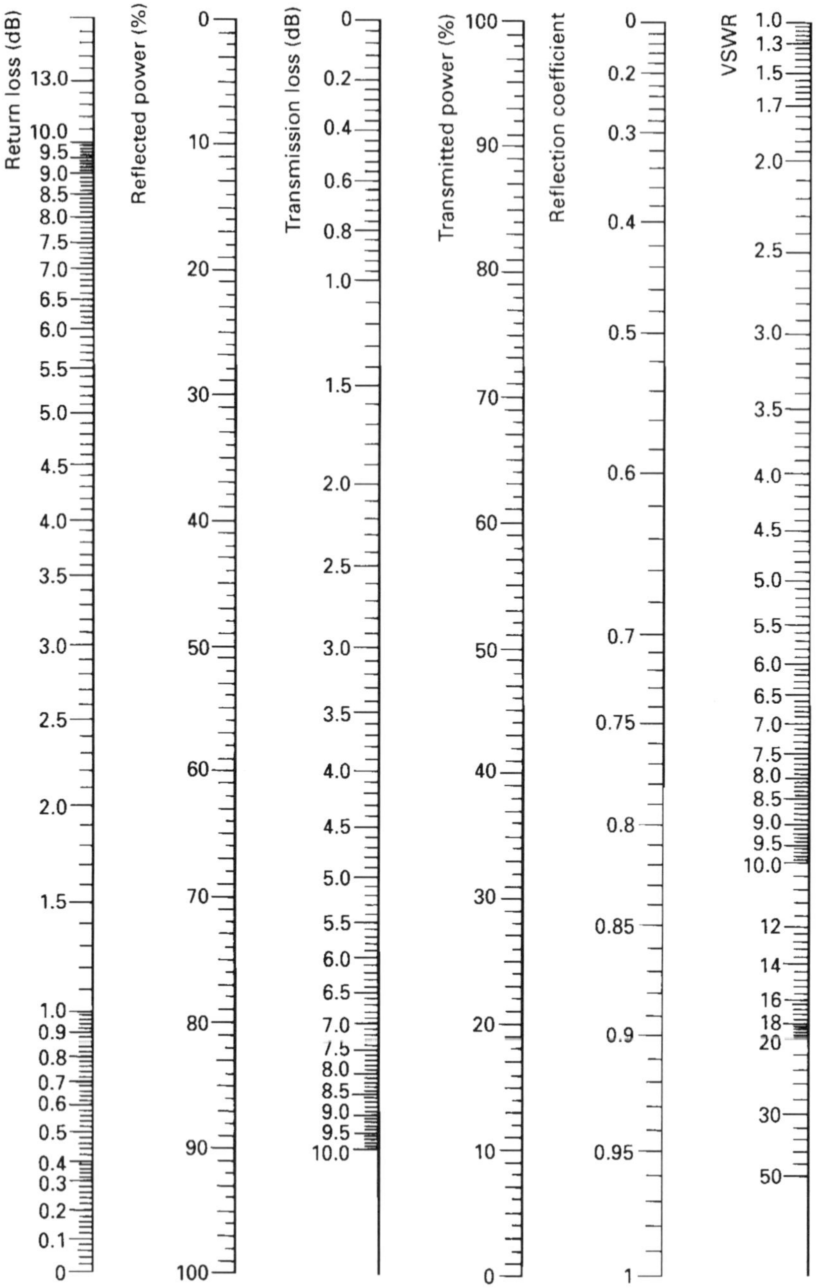

Figure 5.2.20 Return loss nomogram.

Table 5.2.5 Calculation of carrier/noise ratio and fade margin

13.	Receiver noise figure N	4.0
14.	Receiver bandwidth (between −3 dB points)	28 MHz
15.	Receiver noise power = K.T. + B + N	−125.0
	where K.T. = Boltzmann's constant (K) × absolute temp. (300 K)	
	= 203.5 dBW/Hz	−203.5
	B = receiver bandwidth($10\log_{10}$ Hz)	74.5
	N = receiver noise factor (dB)	4.0 dBW
16.	Carrier/noise ratio (Item 12–15)	47.9 dB
17.	Receiver threshold	−112.0 dBW
18.	Fade margin (free space, Item 12–17)	34.9 dB

Table 5.2.6 Calculation of video unweighted and weighted signal/noise ratio

19.	Total FM improvement	7.3 dB
	$= 10\log\frac{B}{2f} + 20\log\frac{SF_3}{f} = 4.47 + 2.83$	
	where B = receiver bandwidth	(28 MHz)
	f = highest modulation frequency	(5 MHz) (for noise measurement)
	6F = peak deviation	(4 MHz)
20.	Pre-emphasis degradation = 0.5	−0.5 dB
21.	Video signal/noise ratio (Item 16 + 19 + 20)	54.7 dB
22.	Picture signal/rms noise ratio due to phase thermal noise only	60.6 dB
	Video s/n ratio + picture signal/rms signal ratio	
	(Item 21 + 20 log 22 × 0.7)	(5.9)
23.	Picture signal/rms noise ratio due to equipment contribution	62.0 dB (dependent on equipment)
24.	Effective overall picture signal to rms noise ratio (Item 22 + 23)	58.2 58.4 dB
25.	Weighting improvement for noise (CCIR Rec. 267) = 16.3	9.8 dB
26.	Picture signal/weighted rms noise ratio due to phase thermal noise only (Item 22 + 25)	70.4 dB
27.	Picture signal/weighted rms noise ratio due to equipment contribution	65.0 dB (dependent on equipment)
28.	Effective overall picture signal/ weighted rms noise ratio (Item 26 + 27)	63.9 64.0 dB

Aleksandar Todorovic
Chairman of the Board, Kompani

5.3 Electronic News Gathering and Electronic Field Production

5.3.1 Electronic News Gathering

ENG – or Electronic News Gathering – is, as its name suggests, the gathering or collection of news stories intended for broadcast during different television newscasts. This technique is based on the use of portable units – camcorders, which combine electronic colour cameras and recorders in one single unit. In most cases the recorder part is based on analogue or digital magnetic tape recording, but in the 1990s other recording supports have appeared on the market, such as retractable computer hard disks.

A quarter of a century ago ENG was a great novelty. Like every novelty, it was greeted by some as a big step forward, a very important development in the field of television production, but also sneered at by some others ("... perhaps nothing is more innovative in ENG than its name and acronym").[1] However, this new way of television newsmaking swept the industry in no time and became very quickly a standard procedure (see Figure 5.3.1).

Today, some 25 or so years later, ENG is a 'mature', well-established and fully developed technology, to the point that it is now practically the only news production technique. At the same time, over these past years, ENG was permanently at the centre of developments. The high competitivity of news programmes lead to important investments in the area of news gathering and presentation. That effort resulted in a considerable advance of the ENG technology and of its operational practices. In addition, the dawn of the digital era and the resulting merging of technologies brought IT techniques into the field of news operations, thus opening new vistas of operational versatility.

5.3.1.1 Background

Preparing and transmitting news reports is certainly one of the most challenging tasks of broadcast organisations. Radio news reporting has a distinguished and brilliant history: in some of the crucial moments of the twentieth century, radio played an essential role. Still today, the ubiquity of radio receivers keeps radio very high on the list of important news disseminators.

The introduction of television was the next step in the development of the medium. While for radio news the voice of the reporter was, more or less, sufficient, in the case of television it was certainly not: the mandatory requirement for any television newscast is the picture of the event, or at least of the location where it happened and the visual presence of the reporter on that spot. Securing not only news as information, but also as a meaningful picture, meant searching desperately for technical means which could offer such a facility. Photographic techniques were already available, as well as the reversal film as a viable compromise between speed and quality. On the other hand, it was theoretically possible to use OB vans. However, the time required to bring the film back to base, to develop it (even with a pushed process), transfer the sound and edit all that together precluded the inclusion of pictures if the events had occurred less than 2 hours before the start of the newscast. An OB van requires even more time to be rigged and set to work, and then its presence on the spot may sometimes be more important than the event it is supposed to cover. Obviously electronic cameras and magnetic videotape recording were the right answer to that need. Unfortunately, cameras and recorders before the late 1970s were both bulky and heavy, in short a far cry from any thought of portability.

Rightly assessing the importance of immediacy in news operations and the lack of adequate equipment on the market to achieve it, one of the three US networks – CBS – instructed its R&D department to begin, in collaboration with world manufacturers, to develop the electronic equipment for news gathering and thus spearheaded the ENG revolution. A portable camera proved to be feasible, although its portability had not much in common with the present-day understanding of that term. The recorder, however, represented a much more delicate problem. The existing broadcast-quality format, based on 2-inch tapes, was practically inadaptable to a really portable version. The emerging 1-inch format B and C recorders were still at an early stage of development, so the only possible solution was to use an existing format, which was developed for industrial and educational applications – the

Figure 5.3.1 ENG at EFP: An ENG crew interviewing on the set the director of an EFP production (courtesy Lj. Kožul, Belgrade).

U-matic. For the relatively low subcarrier of NTSC, that 'colour under' format gave results which were considered as acceptable for short news stories. ENG was born.[2] Users all over the world realised quickly all the benefits and advantages of this new approach to newsmaking, but they also discovered the shortcomings of the first generation ENG equipment. The two most objectionable factors were the limited signal quality and the separate camera–recorder configuration, which was assessed as a very serious handicap. Consequently their first request addressed to manufacturers was to improve the quality of the recording process, and develop units which would combine in one piece the camera and the recorder.[1,3] The response to these requests came sooner than expected, with the advent of the half-inch analogue component recording principle, offering a clearly superior recording quality, while the use of narrower tapes enabled the construction of viable camcorders.

5.3.1.2 Operating Practices

Although the analogue component recording format is still the most widely used one in ENG operations, digital recording is quickly taking over the area of ENG. Since the advent of digital recording, manufacturers have developed a number of different formats. The first series of recording formats handled uncompressed digital component and composite video signals, but these formats were immediately assessed as 'too good' and especially too expensive for ENG. In addition, the use of $^3/_4$-inch tapes practically precluded the construction of viable camcorders, which definitively ruled them out as possible tools for ENG application. These formats were followed by several others based on $^1/_2$- and $^1/_4$-inch tapes and different compression schemes, resulting in different video bit rates. These formats can be classified roughly into four categories:

- format using a proprietary mild compression scheme (about 2.5:1);
- DV- and MPEG-2-based formats, offering a video bit rate of 50 Mbit/s;
- DV-based formats with a video bit rate of 25 Mbit/s;
- format using a proprietary high compression scheme (10:1).

5.3.1.2.1 Defining the Optimum Quality Level

The selection of the most appropriate compression scheme for a given application should be based on the assessment of the following key parameters:

- overall picture and sound quality;
- necessary multi-generation capacity; and
- required post-production margin.

Formal evaluations have shown that, in first generation recordings, it is difficult to detect a noticeable difference between all the contenders. The main differences reside in the multi-generation capability and the post-production margin of the formats in question. All of the existing formats permit frame-by-frame editing, as they are all based on intraframe compression schemes. While some types of programmes can be finalised just by performing a number of simple add-on or insert edits, others will require complex transitions, layered effects, etc. Such effects require access to individual pixels, therefore demanding the decompression of the video signal and, once the effect is performed, its recompression. That type of post-production leads to a larger number of recording generations and to more cascaded decompression–compression processes, and consequently imposes the adoption of milder compression ratios.

The choice of the most appropriate ENG acquisition equipment is always based on a trade-off between the performances on one side, and cost-effectiveness, robustness and ease of handling on the other. At the same time, it is necessary to take into account the relatively short duration of individual news items and modest post-production requirements in the news-making environment. Taking all these criteria into consideration and selecting as high reference point the quality offered by D 1 recordings (full 601 uncompressed signals), with Betacam SP analogue component recording as the low reference point, all internationally conducted formal evaluations showed extremely coherent results. These results, outlined below, were later confirmed by extensive field experience of numerous users around the world.

The Digital Betacam (based on a proprietary compression scheme with a 2.5:1 compression rate) has excellent performances from all points of view and is generally considered as the best compressed recording format appropriate for most complex television productions; however, there is also a general consensus that it is too expensive for ENG applications. On the other side of the compression rate scale, formats based upon 10:1 compression rates are slowly fading out of the picture, since successfully replaced by the formats that offer superior performances and remain in the same bracket of price and complexity.

The DV and MPEG-2 compression methods offering 50 Mbit/s and based on 4:2:2 sampling schemes ensure an adequate quality for all mainstream television productions. Specifically, after a series of evaluations, the EBU recommended that "all networked television production should focus on compression families based on DV and MPEG-2 4:2:2P@ML which had been identified as appropriate for television production applications". In addition, the EBU issued a statement discouraging the future use of MJPEG.[4]

All tests, the consequent field experience, and the market prices of camcorders and post-processing equipment based on DV compression schemes producing a 25 Mbit/s bit rate, lead to a general agreement that such equipment represents the most appropriate compromise for ENG applications. However, two different contenders have emerged in that domain – the DVCPRO and the DVCAM (alongside the original DV format, which is also used in some instances in news operations).

DV is an international standard created initially by a consortium of 10 companies as a consumer recording format. It uses ¼-inch (6.35 mm) metal evaporate tapes as the recording support. In DV-based camcorders video data are generated in the RGB space, and then have first to be converted, through a lossy process, to the YUV space. In the following step data are down-sampled according to the 4:1:1 or the 4:2:0 sampling raster. The sampled video is then compressed using a Discrete Cosine Transform (DCT). DV-based formats use intraframe with adaptive interfield compression, resulting in a nominal video data rate of 25 Mbit/s. Consequently it is possible to perform simple add-on and insert edits on the compressed signal, thus avoiding lossy decompression and recompression processes. The overall data rate recorded on tape (including audio, subcode and other information) is about 36 Mbit/s.

It should be stressed that the picture quality achievable with 4:1:1 and 4:2:0 sampling schemes is inferior to the one defined by the 601 standard, and that the post-production headroom is rather limited (mainly due to the effects of reduced chroma bandwidth and the progressive accumulation of compression artefacts). The main artefacts are the so-called 'mosquito noise' around fine diagonal details, the spatial 'quilting', also noticeable on long diagonals, and some degree of 'motion blocking' or loss of fine details in the immediate vicinity of moving objects.

In spite of the aforementioned shortcomings in overall picture quality, there is a unanimous agreement among broadcasting experts that DV-based 25 Mbit/s formats fully satisfy the requirements of news operations. In addition, the net data transfer rate and the storage capacity required for a 90-minute programme are well within the data transfer and volume storage capabilities of modern standard tape- and disk-based equipment, which makes easy the integration of 25 Mbit/s DV-based recordings into fully integrated networked production systems. It is also considered that, if kept in long-term archives, recordings made at that compression level will be reusable in the foreseeable future.

5.3.1.2.2 Acquisition

In the domain of digital ENG acquisition, it is possible to distinguish three main types of camcorders. The first one uses magnetic tapes. The second category is represented by camcorders which are based on other recording media, while the third group encompasses camcorders that record simultaneously on a video cassette and on a retractable hard drive.

The DV-based camcorders, which use magnetic tapes as the recording medium, offer a very good price/quality trade-off and represent certainly the most widely spread type of digital ENG acquisition equipment. There is, on the market, a wide selection of different DV-based camcorders, produced by several manufacturers. Their essential characteristics are given in Table 5.3.1.

As noted earlier, all these formats are founded on the same DV compression scheme, and the slight difference which exists between the DV, DVCAM and DVCPRO formats resides mainly in different codec implementations. In addition, all three formats use the same mechanical design of video cassettes. Owing to that, all manufacturers who produce DVCPRO and DVCAM compliant recorders (see Figure 5.3.2) can offer multi-format players capable of reproducing all three formats, i.e. DV (usually Standard Play only), DVCPRO and DVCAM.

By definition, the magnetic videotape is a linear recording medium and consequently it precludes the implementation of certain operational features, such as instant random access to any recorded frame or editing on a single machine. The desire to overcome these limitations prompted the development of a new category of camcorders conceived to use retractable hard disks as their recording support. First models offered to the market suffered from a number of drawbacks: weight, power consumption, recording capacity and price of retractable hard disks. Later developments considerably improved all of these factors, but nevertheless these camcorders did not become mainstream acquisition tools. However, they certainly initiated an intensive R&D work aimed at eliminating magnetic tapes as the main recording supports for acquisition. The result of such research efforts was the appearance of the first dockable DVD-RAM and DVD-R recorders offering 40–60 minutes recording time (depending on the required quality of the recording) and featuring special solutions for protection against vibration and shock-produced errors. It therefore seems realistic to expect that in the foreseeable future non-tape-based recording media could take centre stage in the ENG acquisition domain.

Finally, a combination of tape and hard drive recording is also available on the market. Such a camcorder has a dockable

Table 5.3.1 Comparison between DV, DVCAM and DVCPRO

	DV	*DVCAM*	*DVCPRO*
Suppliers	Consortium of 60 manufacturers	Sony	Panasonic
Tape used	Metal evaporate	Metal evaporate	Metal particle
Track pitch (μm)	10	15	18
Compression	5:1 intraframe video data rate: 25 Mbit/s	5:1 intraframe video data rate: 25 Mbit/s	5:1 intraframe video data rate: 25 Mbit/s
Resolution	720 × 480,4:1:1(525)	720 × 480,4:1:1(525)	720 × 480,4:1:1(525)
Sampling	720 × 576,4:2:0(625)	720 × 576,4:2:0(625)	720 × 576,4:1:1(625)
Audio	2 Ch. at 48 kHz, 16-bit 4 Ch. at 32 kHz, 12-bit	2 Ch. at 48 kHz, 16-bit 4 Ch. at 32 kHz, 12-bit	2 Ch. at 48 kHz, 16-bit reads 44.1 and 32 kHz
Digital interfaces	IEEE 1394	IEEE 1394 SDI	IEEE 1394 SDI
Analogue interfaces	Y/C Composite	Y/C, Component Composite	Y/C, Component Composite

hard disk drive and records simultaneously on tape and on disk, using either of these two supports for post-production. The recording on tape can be utilised in any linear editing system, or be transferred from a player into an NLE unit. The disk, for its part, can be used either for a fast data transfer rate on NLE units, or simply as an external disk storage of 3 hours capacity connected and directly controllable via an IEEE 1394 interface. In such a way all essence and metadata recorded in the field are immediately accessible for post-production, the data transfer process is eliminated and consequently the whole operation speeded up. Clearly such a solution widens the operational flexibility of ENG operations and could constitute a good intermediate step in the process of abandoning the tape in favour of some other recording supports.

Some developments of camera equipment were particularly significant for ENG operations. The list of such improvements is very long, from increased sensitivity of solid-state sensors (important for ENG operations, where, frequently, crews have to cover their stories under the existing, mostly unfavourable, lighting conditions) to electronic or optical image stabilisers, making it possible to eliminate the tripod (always heavy and cumbersome to carry around) and still to have good pictures (for more details see Chapter 3.2).

Figure 5.3.2 DVCAM camcorders: (a) shoulder mount model; (b) palm model (courtesy Sony).

Figure 5.3.2 *continued*

The key word in any sort of news operations is exclusivity, closely followed by speed, which explains the attractiveness (and sometimes misuse) of live connections between the studio and the location where the event is taking place. ENG camcorders are ideal tools for such live reports, since an extremely reduced crew can easily and relatively quickly establish the necessary connection. In order to set up such a connection, ENG crews have at their disposal a wide choice of different analogue or digital microwave links, as well as satellite communication equipment (SNG: Satellite News Gathering). More details on microwave and SNG communications can be found in Chapter 5.2.

Finally, a word of warning should be said about the choice and handling of batteries. It is certainly superfluous to stress the critical importance of battery power supply in ENG acquisition. In spite of all recent advancements of battery technology, that element still represents one of the most critical links in the ENG production chain. For that reason it is highly recommended to all those who want to have a full overview of ENG techniques to read carefully Chapter 5.5.

5.3.1.2.3 Post-production

Generally speaking, post-production in newsmaking is certainly less complex and demanding than in other fields of television production. It is most frequently limited to simple cut editing with the addition of a couple of subtitles and voice-over. Consequently, post-production of news can simply follow and utilise the general development of editing hardware and operational methods.

However, the specific requirements of ENG and newsmaking operations, like the portability of equipment and speed of operations, spearheaded the development of special news-oriented post-production solutions. In a television centre post-production is performed either on conventional linear editing systems or on non-linear units (NLE). These non-linear editing units are either products specially developed for newsmaking operations, or, lately, scaled-down versions of standard television production ones (see Figure 5.3.3). Namely, they are adapted for the news environment, where the speed of operations has to be traded off for sophistication of operational capabilities.

A typical NLE unit used in a newsmaking environment is adapted for fast-cutting editing. It can directly access native broadcast-quality (hi-res) media on central shared storage or store locally the camera material transferred either from a tape player or from the central storage. It can control several VTRs and capture direct to timeline as well as across timecode breaks. The NLE unit can add in real time some transitions, effects and titles, thus facilitating the creation of a fully completed news item. In an integrated newsroom environment, such NLE units can also, if needed, edit low-res proxy files and then switch directly to the high-resolution ones. In addition, they usually can handle more than one compression format and deliver the finished item in any required form (SDI, SDTI, etc.) as well as encode the signal in MPEG-2 format for direct digital playout.[5]

The need for a fast and efficient post-production process prompted the development of a new feature particularly well suited for news operations. During acquisition the in-point and out-point timecode information of each shot is recorded on a special cassette memory. At the same time a still frame of each in-point is recorded on tape. When such a tape is inserted in a player interfaced with a non-linear editing system, that log information is immediately downloaded, offering a quick overview of the raw material and speeding up the process of selecting shots which have to be ingested.

ENG field operations require lightweight portable editing units (see Figure 5.3.4). Such portable units ensure a full

Figure 5.3.3 Screen appearance of an NLE unit (courtesy INCITE).

Figure 5.3.4 Field laptop editing unit. Thanks to its dimensions (211 × 149 × 443 mm) and its weight (5.8 kg) it is easy to carry on assignment in the field (courtesy Sony).

independence of the reporting crew, but can also reduce transmission costs by shortening the length of the footage transferred over international lines. These field editing units consist of one or two tape transports, flat-screen monitors and an operational panel, all packed in a case which looks very much like a smaller attaché case. They can be powered either by batteries or by AC mains supply, are reasonably lightweight and offer more than adequate editing capabilities.

The increase of hard disk capacity has led to the adoption of notebooks as a base for field non-linear editing. Such units can be used either for off-line (proxy) editing and EDL creation, but also for on-line editing, where the only limitation is the storage capacity of the hard disk. An external storage can be added to the notebook, but at the price of extra weight which has to be carried around.

Taking full advantage of information technologies, special video store-and-forward units are now available for travelling crews. Weighing only several kilograms, such units are equipped with news-type non-linear editing software, and are designed to use camcorders for the tape-to-disk transfer. In addition, the software package is capable of generating MPEG-1 and MPEG-2 compliant streams that can be transmitted by store-and-forward techniques over satellite, ISDN lines or the Internet. Obviously, such units can handle email and Internet connections, which makes them complete communication centres for a travelling reporter.

5.3.1.2.4 Integrated Newsrooms

Traditionally all newsrooms were 'tape based'. In other words they were equipped with a number of individual stand-alone devices, such as VTRs, linear or non-linear editing units, Character Generators, etc. In addition, they were most frequently equipped with a newsroom computer system which featured a series of networked terminals for managing wires (textual news from news agencies), writing stories, creating rundowns, archiving and retrieving text material, messaging, etc. Such solutions were well established, the operations staff felt comfortable with them, but when analysed they showed a high degree of human and hardware inefficiency, excessive running costs and dubious reliability.[6]

Integrated, digital or server-based newsroom systems are designed to bring to news operations all the advantages of the merging of information and video/audio technologies. Basically they represent a seamless integration of the Newsroom Computer System, of the playout centre, of media devices, such as video or file servers, NLEs, proxy editors, graphic boxes, etc. and of a number of software solutions, such as the Media Asset Management (MAM) system, or Ingest Control, Management and Scheduling. The aim of that integration is to create a unified user-friendly system out of this plethora of different devices and subsystems. The conceptual framework of such solutions practically mirrors the usual workflow of news operations. However, the usual workflow does not mean a standard workflow, because in newsmaking, as for any other television production operation for that matter, it is not possible to speak about standard solutions. Each television organisation has its own practices and consequently server-based news production systems are custom made for each particular broadcaster. Nevertheless, as shown in Figure 5.3.5, it is possible to identify a number of standard processes which have to be part of any particular workflow (ingest, logging, storing, retrieving, editing, archiving, etc.). The precise details of each of them, their interrelations and hierarchical relationships will differ from station to station, thus leading to somehow different workflows and consequently to different newsroom systems, each optimised for a particular user.[7]

As shown in Figure 5.3.6, which represents one possible solution, the process starts with content acquisition. All content generated by ENG crews, studio recordings or originated by some external entities has to be ingested into the system in an appropriate format and logged. Essentially there are two types of ingesting procedures – the manual one when a cassette brought from the field is inserted in a player and the content ingested in the system (either through an ingest server or through any of the NLE stations), and the 'scheduled' operation, which concerns content coming through terrestrial and satellite feeds (news agencies or news exchanges), following some previously announced timetable. The ingest operator controls the ingest procedure and should detect any problem immediately. At the moment of ingest the material is catalogued and indexed, which is crucial for later easy retrieval. In order to allow sensible browsing and material research and recovery, it is essential to develop a comprehensive system of keywords, annotations and thumbnails, which will permit not only the retrieval of the needed material through the use of different search criteria, but will also display on the search result screen all relevant information concerning the retrieved essence (from content-related information to rights). Once the material is ingested it is transferred to a shared storage. The shared storage stores the material in two quality levels:

- the so-called 'high-resolution' ('hi-res'), in fact the selected broadcast-quality level (usually 25 Mbit/s), which will be later used for post-production and airing;
- at the same time, the ingested material is transcoded to a 'low-resolution' ('low-res') or 'browsing' level (usually MPEG-1) – since transcoded to a low bit rate these browse-proxy video clips require a relatively modest storage capacity.

The low-resolution material is accessible by journalists from their desktop workstations. They can browse, select and even perform rough-cut editing, thus generating edit decision lists (EDLs). Journalists would also be able to add their voice-over to the roughly assembled material, as well as add all information relevant to graphics, transitions, etc. It is customary to keep in the shared storage a certain quantity of well-selected 'current' archive material which the journalist can browse and, if needed, select pieces to be used as additional or background information (see Figure 5.3.7). In more elaborate and comprehensive installations, an on-line archive is connected to the system, offering to the journalist a large selection of archive materials. A further extension of the archive system could be achieved by a near-on-line archive. That additional archive is in fact a robotic streamer tape player, which can store at a reasonable price a huge quantity of programme material. The access time to the content stored on streamer tapes is not immediate as for the material stored in an archive server, but is

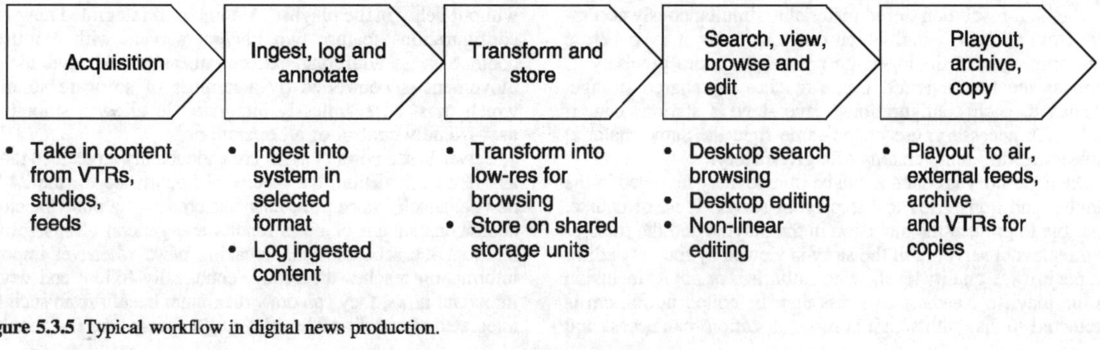
Acquisition
Ingest, log and annotate
Transform and store
Search, view, browse and edit
Playout, archive, copy
• Take in content from VTRs, studios, feeds
• Ingest into system in selected format
• Log ingested content
• Transform into low-res for browsing
• Store on shared storage units
• Desktop search browsing
• Desktop editing
• Non-linear editing
• Playout to air, external feeds, archive and VTRs for copies

Figure 5.3.5 Typical workflow in digital news production.

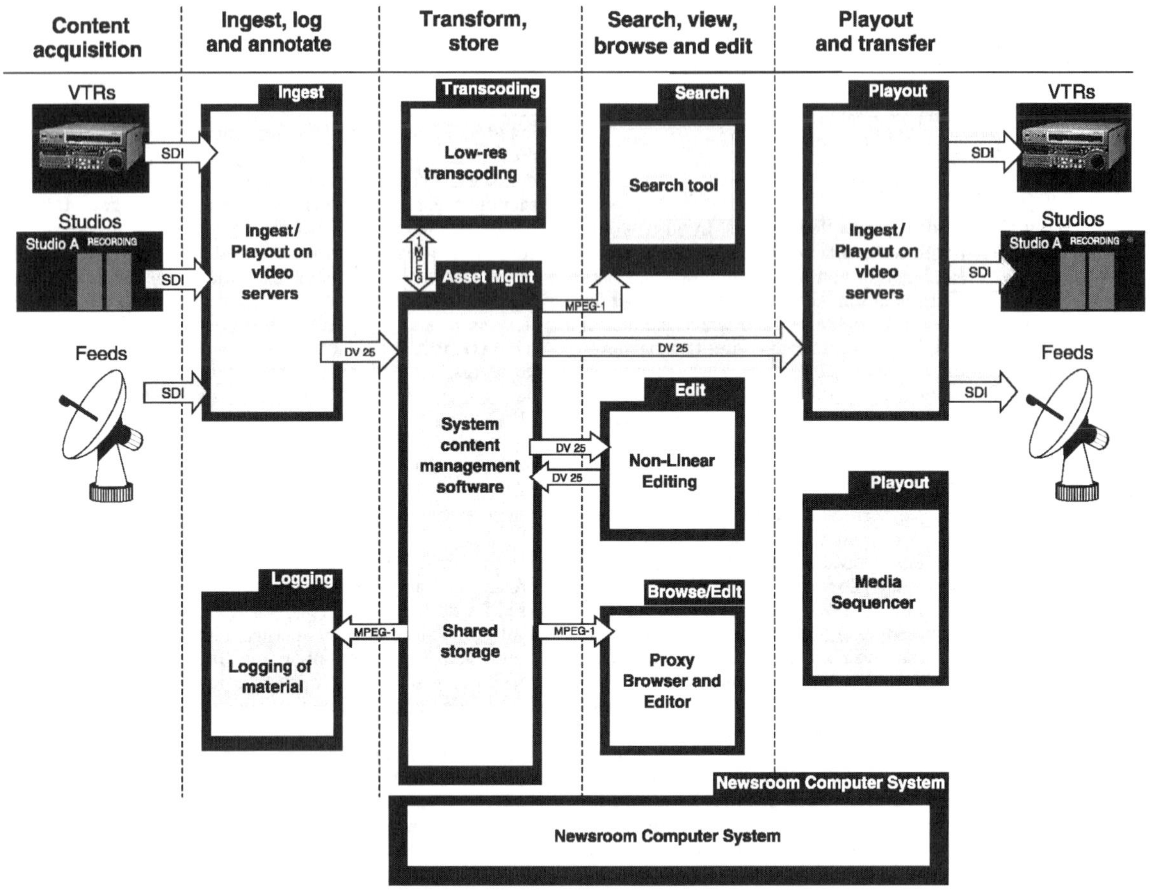

Figure 5.3.6 Overview of the processes in an integrated newsroom system (courtesy IBM).

sufficiently short not to seriously impede newsmaking operations. Once the EDL is generated, and possibly a live commentary added, that material is sent to an NLE station where, using information from the EDL and grabbing from the shared storage high-resolution material, the editor assembles the final version of the story, adding if required, graphics and subtitles, and sends the completed story back to the shared storage, where it is again stored both in high and low resolution.

The high-resolution stored material is simultaneously accessible from all NLE units, which can start using it even before the completion of the ingesting process, or more precisely as soon as the first ingested frame reaches the shared storage. Thanks to such configurations, several NLE stations can, if and when necessary, use at the same time the same material to produce different versions of a given story.

Once the story is edited it can be immediately included in the playlist and transferred to the playout server. It is, of course, possible to prevent the inclusion in the playlist and the transfer to the playout server until the story is viewed by the duty editor (at the browse-quality level), who authorises or not its inclusion in the playlist. Needless to stress that the edited news item is protected so that, although a number of stations can access and view it, only the authorised one can change its content or its position on the playlist. The playlist can control and air the material directly from the shared storage, but most frequently additional playout servers are used for the storage of listed material and its playout during the newscast. As breaking news can occur at any time, it is usual to implement such solutions that a story could be aired as soon as its first frame reaches the playout server, so that last-minute events can be integrated without delay in the playlist. A number of integrated newsroom configurations feature two playout servers with a mirrored content as an additional security measure. All that material movement is controlled by a number of software solutions, which must be seamlessly integrated to allow a smooth and user-friendly control of all operations.

Server-based news systems are undoubtedly a precious tool for any news production, but of crucial importance for the 24-hour news channels. Since these channels broadcast around the clock a continuous stream of news reports and general affairs features, interrupt the schedule with breaking news whenever important information reaches the centre, continually follow and develop important news, they can draw maximum benefit from such fully integrated IT and digital broadcasting technologies. The possibility

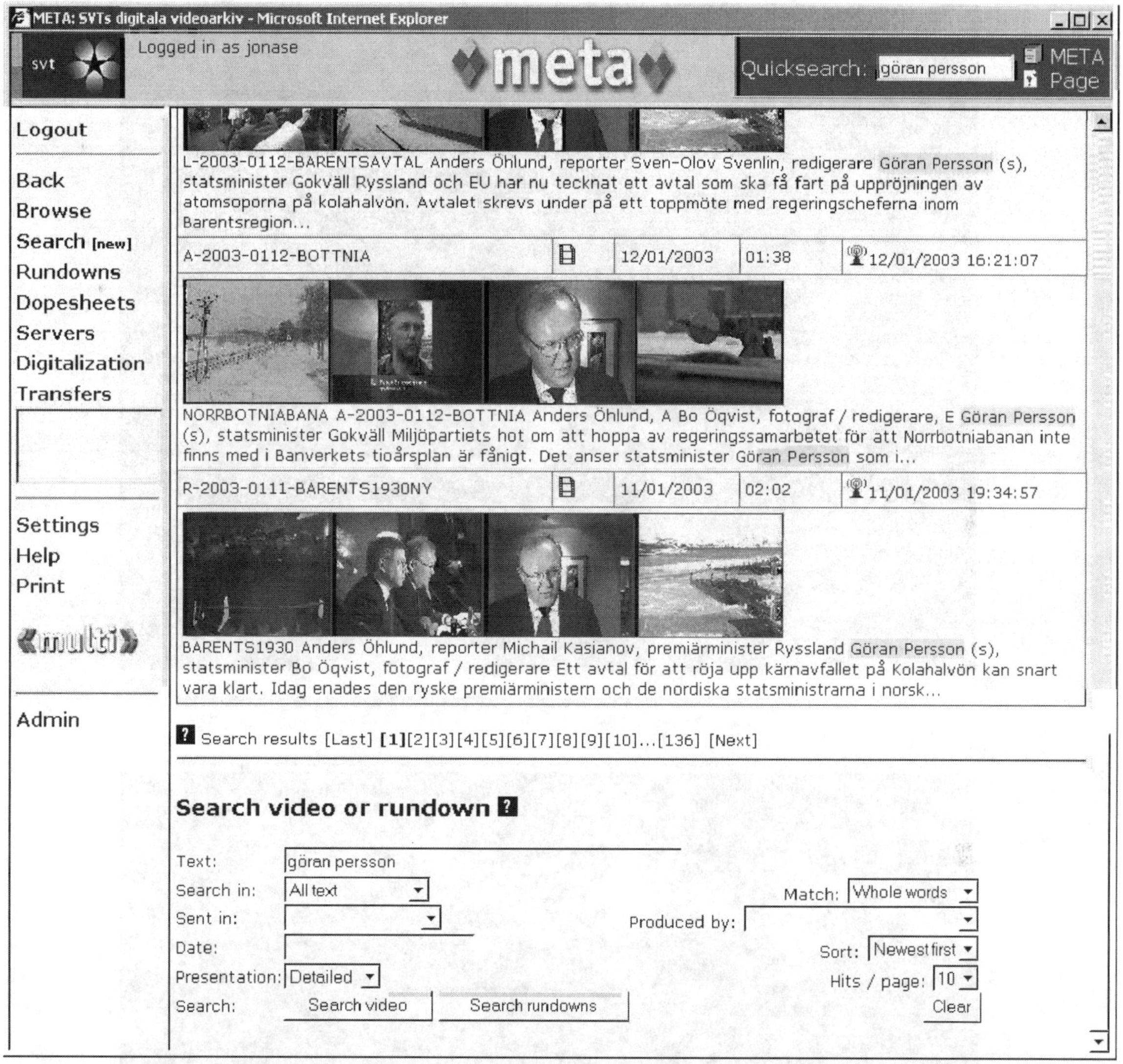

Figure 5.3.7 Appearance on screen of search results using the browse capabilities of an integrated news system (courtesy Ardendo).

to re-edit simply and quickly any news item at any time, to offer to journalists fingertip access to the whole breadth of materials from the latest incoming footage to archived background materials, to integrate easily feeds from numerous sources, all these features position the integrated newsroom system as an essential production tool for news channels (see Figure 5.3.8).

One important aspect of server-based systems is their ability to handle two-way communications with numerous bureaus and travelling crews. As 24-hour news channels have, by definition, a large number of bureaus scattered in key places around the world and at the same time a fleet of roaming crews covering 'hot spots' on the planet, communicating with those outside units and integrating all their feeds in a continuous newscast is best handled by server-based systems. Reports from around the world will arrive as satellite or terrestrial feeds, as digital video cassettes, as Internet downloads, even as video telephone reports. Once at the base they are all immediately stored in a central system, which then represents a single source of all available material for the journalists and editors in the newsroom. On the other hand, the reporter in the field can access via the Internet the central archive and download pieces of archive material in low-resolution form and insert them in the edited report. Once the editing is complete, the full report is sent back to the base, where an editor can easily replace the browse-quality material with high-resolution footage. The high-resolution footage is easily retrieved since both the browse-proxy and the high-resolution versions are unequivocally defined with the same set of timecode and metadata information.

Figure 5.3.8 News broadcast on Swedish television: (a) studio floor; (b) playout control (courtesy Ardendo).

5.3.1.2.5 Video Journalist

The increased handling simplicity of digital camcorders, their reliability and stability, the automation of camera settings and the enhancement of solid-state sensors' sensitivity prompted cost-conscious broadcasters to equip some of their correspondents as one-person crews, thus creating a new 'profession' in news operations – the video journalist.[8]

Video journalists are supposed to cover the story, to shoot on location and sometimes even to edit their own reports. It is obvious that all these factors prompted journalists working as stringers for television stations to purchase digital camcorders and join the new 'profession'. It is true that the video journalist profession requires a special sort of training. They have not only to be well-educated journalists, but must also pass through a serious and formal training in camerawork and editing. Although economically attractive, such a practice is not generally accepted, as it is felt that it might lead to an underachievement in all the concerned fields – reporting, shooting and editing – and that considerably better results are achieved by using specialists for each of these three posts.

Nevertheless, it seems that the present-day developments in newscasting and in the broadcasting industry in general will favour more and more cost-efficient solutions, even at the price of a certain compromise on the quality side. It is, therefore, interesting to see what video journalists would need for their solitary endeavour, what could be the minimum panoply which would support both acquisition and post-production.

On the acquisition side it would be realistic to equip the video journalist with a low-end professional or high-end industrial camcorder in one of the DV formats that fulfils some basic requirements:

- to be a shoulder mount configuration, as it ensures better shots and smoother camera movements than the palm-type one;
- to be fitted with high-sensitivity sensors;
- to be equipped with a 10×1 or 15×1 zoom lens, preferably with optical image stabilisation (electronic or digital image stabilisation is inadequate for professional applications);
- to have an on-camera microphone for ambient sound recording;
- to be equipped with XLR connectors for external microphones;
- to have at least an IEEE 1394 interface and analogue composite and component I/O.

In addition, the video journalist has to be equipped with accessories like:

- Batteries and battery chargers.
- AC adapter.
- Connecting cables.
- Cleaning cassette.
- One professional type lightweight tripod.
- Rain cover.
- Three microphones (one highly directional, a general purpose hand-held dynamic and one miniature lap microphone).
- Earphones.
- Battery-powered light (some camcorders offer a connection for such a light, thus eliminating the need for a separate battery; however, using such a connection will considerably reduce the recording autonomy). On the other hand, video journalists could be equipped with a complete set of reporter's lights; however, such a set can be used only for some previously planned long interviews, and even then its installation and utilisation would represent a serious strain on one single person. It is, therefore, more convenient for a video journalist to hire for such occasions a light operator and the necessary lighting equipment.
- Tools, cleaning material, gaffer tapes, etc.

Video journalists could also be equipped with post-production tools, allowing them to produce and deliver completed broadcast-quality news reports. Since such a post-production set has again to offer a best value for money ratio, non-linear editing should be the preferred solution.

Non-linear editing systems offer greater operational flexibility than the linear systems; they encompass in one software package a number of different functions, such as titling, effects, voice-over, sound processing. At the same time they can be configured in a number of different levels of price and complexity. It is possible to choose a non-brand name PC with a modestly priced editing software and inexpensive video and sound cards, which are halfway between professional and consumer type products, or a high-end hardware and correspondingly expensive and performant software. In addition, a video journalist who plans to roam the country can acquire a powerful notebook with a large memory and a good editing software, which will allow post-production in the field. In order to minimise the investment costs, and regardless of the choice of editing hardware and software, the camcorder should be used as the playback unit for material transfer.

5.3.2 Electronic Field Production

EFP – Electronic Field Production, which started as an offspring of ENG as soon as the picture and sound quality offered by portable equipment reached the level judged adequate for other programme genres – has also become today a standard production technique which confined OB operations to events where multi-camera coverage is a must. It is interesting to note that the full development of EFP coincided with the advent of analogue component recording, which was developed for ENG, but which became and remained for years the main recording standard of the broadcasting industry. In addition, EFP techniques, after many unsuccessful attempts, seem to have finally entered the select realm of movie production, and even, in a way, started a new phenomenon known as D-cinema.

5.3.2.1 What is EFP?

EFP is a method of production of mainstream television programmes based on the use of lightweight electronic acquisition equipment. That production method could be considered a result of cross-fertilisation between film and television production techniques. In fact, the appearance of high-quality portable acquisition equipment prompted the television community to turn towards 'single camera production' methods. The idea was to take benefit from all the artistic advantages of film production techniques as well as from all the economical advantages of electronic production tools. The results of the first EFP attempts were rather unconvincing due to a modest overall output quality of portable cameras and recorders, and to the severe limitations imposed by analogue post-production. However, with the advent of analogue component recording and especially with the development of digital production techniques, EFP very soon became the main on-location production technique.

5.3.2.2 EFP Operational Practices

As in ENG operations, analogue component recording is still used for EFP in a number of organisations around the world. However, digital techniques are quickly taking over, imposing a number of choices, the first one being, naturally, the selection of the most appropriate digital recording format. That choice is based on the evaluation of the already discussed parameters:

- the type of programme to be produced;
- the overall sound and picture quality required;
- the multi-generation capacity necessary; and
- the post-production margin needed.

Considering that EFP methods are used for a large gamut of programme genres, from documentary to drama, it is obvious that the overall picture and sound quality requirements are more stringent than in the case of news operations. The post-production of mainstream television programmes is also more demanding, as it requires a considerably larger number of quasi-transparent generations and creates more occasions when access to individual pixels is necessary, which will lead to more cascaded decompression–compression processes.

At the time of the introduction of digital production techniques, it seemed that only uncompressed video signals offered an adequate quality for mainstream television productions. However, it is now generally accepted (and even recommended by some organisations such as the EBU) that mainstream standard-definition television production can be successfully performed with compressed signals with a minimum bit rate of 50 Mbit/s, the result of either DV- or MPEG-2-based compression methods. The EFP acquisition is, therefore, generally confined to the use of Digital Betacam (compression ratio 2.5:1) or to the DVCPRO 50 and IMX Betacam (MPEG-2) formats, both offering a 50 Mbit/s video data rate.

The basic definition of EFP refers to the use of 'lightweight electronic equipment' for acquisition. However, EFP borrows a number of film production practices and, while the camcorder is indeed a 'lightweight' piece, the additional equipment, which typically can be found on any film shooting location, can hardly be defined in the same manner. EFP crews will use a large palette of lighting equipment, camera cranes, tracking rails, video monitors, a selection of optical attachments, etc. In short, it could be said that the only noticeable difference between a film and an EFP crew is the replacement of a film camera by a digital camcorder.

In the post-production domain it is difficult to discern EFP-specific requirements and operating practices, since the same linear and non-linear editing systems and post-production practices are used both for EFP and multi-camera productions. It could be said that what distinguishes an EFP post-production is the number of edits, and a more delicate and time-consuming colour matching of different shots.

5.3.2.3 D-cinema

Developed mainly during the last decades of the 20th century, High-Definition Television (HDTV) was, throughout that period of time, considered by a number of experts as a potential tool for movie cinema production.[9] However, its original analogue version (1125/60 or 1250/50) did not offer a quality which filmmakers considered acceptable. The advent of digital, the adoption of the common image format (1080 × 1920), and the development of powerful and highly performant digital effects made HDTV production tools 'interesting' for the cinema community. Finally, the appearance of a new production standard based on 24 progressively scanned frames (24p) finally tipped the balance in favour of the electronic production means.

Digital effects were the first large-scale application of digital techniques in the realm of motion pictures. Their operational flexibility, the enormous gamut of breathtaking effects which could be produced and the speed of production were main factors which attracted filmmakers. The fact that a number of blockbusters relied heavily on digital special effects, to the point that a good percentage of the whole movie was produced electronically, prompted some filmmakers to make the next step and adopt 24p EFP technology as their exclusive production tool.

However, while Hollywood and big European producers started assessing the appropriateness of high-end electronic technology for their mega projects, another, almost subversive current was gaining more and more importance. Its protagonists, talented and penniless moviemakers, adopted Digital Betacam, and even DV camcorders, for their shoestring financed productions. These movies were shot and post-produced in the digital domain and then transferred to 35 mm film for theatrical distribution. Thanks to the talent of those people, a number of such projects became surprising box-office hits, and consequently generated a stream of similar independent productions. Therefore, in the domain of D-cinema we can distinguish two main categories:

- the low-budget approach, relying on digital standard-definition acquisition and post-production equipment; and
- the medium- to high-budget productions using the most expensive 24p high-definition gear (See Figure 5.3.9).

However, the electronic system still has a number of adversaries. Their arguments are centred on several factors. They claim in general that the overall picture resolution of 35 mm film is still superior to the one offered by any of the proposed HDTV standards. Beside the questions of intrinsic picture quality, there is another argument permanently put forward by the movie-making community – the 'film look'. Although it is rather difficult to describe precisely what the 'film look' means, it seems that it is a combination of several objective characteristics and a number of indescribable sensations which film aficionados claim to feel when they watch a 35 mm product projected on a big screen.

It is clear that we can discuss objective characteristics, and among them the most frequently mentioned are the depth of field, the contrast ratio and the overall picture balance. The latest advances in lens technology, and special features developed by some manufacturers, combined with an adequate lighting of the stage, produce practically the same quality and appearance of the depth of field with electronic pick-up as with the film one. On the other hand, owing to the latest camera circuitry, the camera can be balanced in such a way as to imitate the look of a selected film stock. However, the contrast ratio of electronic pick-up remains more limited than the one offered by 35 mm film.

5.3.2.3.1 Distribution and Display

As indicated earlier, digital products have to be transferred to 35 mm film for theatrical distribution. However, a number of experts advocate the change of the whole cinema chain, from production to display, claiming that in all these stages significant advantages can be gained by using digital signals instead of 35 mm film.

These new projects assume that in movie theatres conventional mechanical film projectors will be replaced with video beamers

Figure 5.3.9 Cine Alta camcorder specially developed and equipped for D-cinema production. Note the typical cinematography accessories attached to the camcorder (courtesy Sony).

connected to a kind of server-based storage, and that the distribution should be done via broadband communication means. Such an approach has two basic advantages: it is more economical than the production of a large number of release copies and, if distributed via broadband networks, serious savings could be achieved on shipping and transportation costs, especially for worldwide distribution and in large countries like the USA, China or Russia. On the other hand, in this way the unsolvable problem of the deterioration of the film copy could be overcome. After a number of projections the film copy becomes scratched, it sometimes breaks and is subsequently simply shortened by a number of frames and then spliced together. This splice quite frequently provokes a projection irregularity. In addition, a number of currently used projectors demonstrate a certain amount of objectionable gate judder, which reduces the perceived resolution.

The electronic playback is considerably more robust to degradation. Highly efficient error correction and concealment circuits allow incomparably higher number of 'projections' or, to be more precise, playbacks before showing any visible degradation.

Projectors for electronic theatrical projection of D-cinema are in full development. There are a number of competing techniques, among which Light Valve Image Amplifiers and Digital Micromirror Devices offer at this moment the most appropriate resolution and light output for high-quality theatrical movie projection.

There is another important aspect of the move towards electronic projection – the question of piracy and copyright protection. One of the serious problems film producers have to face is illegal copying of the latest releases and their appearance on the video black market. Keeping the content in its digital form from acquisition to projection could facilitate the introduction of a number of data protection means, making unauthorised access much more difficult.

References

1. *EBU Report on Electronic News Gathering*, 2nd Edn, Technical Centre of the EBU, Brussels (1981).
2. Battista, T.A. and Flaherty, J.A. The all electronic news gathering station, *J. SMPTE*, **84**, No. 12 (1975).
3. Morrison, W.A. and Brogden, D. *ENG*, Australian Film and Television School, North Ryde, New South Wales (1981).
4. EBU–SMPTE Task Force for Harmonized Standards for the Exchange of Programme Materials as Bitstreams, Final Report, Analysis and Results, *EBU Technical Review*, Special Supplement (1998).
5. Ohanian, T.A. *Digital Non Linear Editing*, Focal Press (1998).
6. Schultz, F. Automation and live television news: enhanced support for complex workflow, *SMPTE Motion Imaging Journal*, **112**, No. 1 (2003).
7. Francis, T.J. Managing change in the digital newsroom, *ABU Technical Review*, No. 203 (2002).
8. Boyd, A. *Broadcast Journalism*, Focal Press (2000).
9. Stow, R. and Wyland, G. Electronic cinematography, progress towards high definition, *EBU Review*, No. 209 (1985).

Bibliography

Anderson, G.H. *Video Editing and Post Production*, Focal Press (1998).

Medigovic, M. *Digitalni Film*, FDU, Belgrade (1999).

Medoff, N.J. and Tanquary, T. *Portable Video, ENG and EFP*, Focal Press (2001).

Wheeler, P. *Digital Cinematography*, Focal Press (2001).

Graham Young BSc
Managing Director – Antares (Europe) Limited

5.4 Power Generators and Electrical Systems for Outside Broadcast

5.4.1 Introduction

Outside Broadcast vehicles contain electrical equipment to carry out their function; this can be anything from basic lighting, stabilisation jacks, heating, ventilation, air-conditioning through to the specialist broadcast equipment itself.

All this requires power to be generated, stored, converted and switched by a vehicle subsystem called the auxiliary electrical power system. This area is often overlooked when specifying the vehicle but has the most profound consequences if it fails to perform – this will invariably happen during a major cup final or a critical news broadcast!

The purpose of this chapter is to explore some of the key points in understanding and specifying such a system.

A major industry trend in the last 10 years has been a move to more flexible broadcast platforms. The large rigid-bodied 24 V truck broadcast unit was the staple vehicle to house the broadcast systems, though now vans, cars and even fly-away packs are commonplace – all require the same attention to the power system design.

5.4.1.1 Operational requirement

Before defining the system, it is very worthwhile bringing all the parties together to consider how the vehicle will be used, in particular what equipment is required to be operated and when.

5.4.1.2 Power balance calculation

The first stage is to list all the AC and DC loads on the vehicle. Each load needs to be defined in the following way:

- Description of load.
- Type of supply, DC or AC.
- Voltage.
- Frequency (if applicable).
- Input range.
- Input voltage range restrictions.
- Typical operating power/current and peak power/current requirements.
- Duty cycle (continuous or intermittent).

The loads are calculated for each of the following stages of operation: *travelling*, *setting up*, *long-term storage*, *operation*. Consideration should also be given for power consumption in the case when the system is running in *emergency standby* from backup batteries.

From this a power balance table is created. A simple example is shown in Table 5.4.1.

5.4.1.3 Typical layout

Figure 5.4.1 shows an illustration of a 24 V OB vehicle electrical system. To demonstrate the purpose of each of the techniques, the system is fully featured. Not every application will require this functionality.

The power system can be viewed in three parts: the on-board power generator system, the DC system and the AC system. Each is now discussed in more detail.

5.4.2 On-Board Power Generation

If the specification includes a requirement for generation of power, several factors must be considered.

The first is that of power capacity. If the generator is to supply the complete operational power requirement and the air-conditioning then this must be assessed. This gives the net power requirement and represents the minimum generator capacity. This is expressed as kW. It is sensible to allow for an additional 15–25% power to allow for starting difficult loads and also for future power demands.

Generator manufacturers usually express the kW power rating at sea level and 20°C; it can also be expressed as kVA; kVA relates to kW via the power factor. It is usual to use

Table 5.4.1 Power balance example

Description	*No.*	*Type*	*Power (W) running*	*Power (W) peak*	*Travelling*	*Setting up*	*Storage*	*Operation*	*Emergency*	*Comments*
Aircon	1	230 VAC	1200	2400				*		
Emergency lights	6	24 VDC	30	30		*	*	*	*	
Lights (main)	10	230 VAC	16	16				*		
Hydraulic mast	1	24 VDC	100	100		*		*	*	
Amplifiers	2	230 VAC	3000	3000				*	*	
Space heater	1	24 VDC	100	200			*	*		Sensitive to voltage
Communications	1	12 VDC	48	150	*	*		*	*	
Spare sockets	2	230 VAC	1000	1000				*	*	Backed up

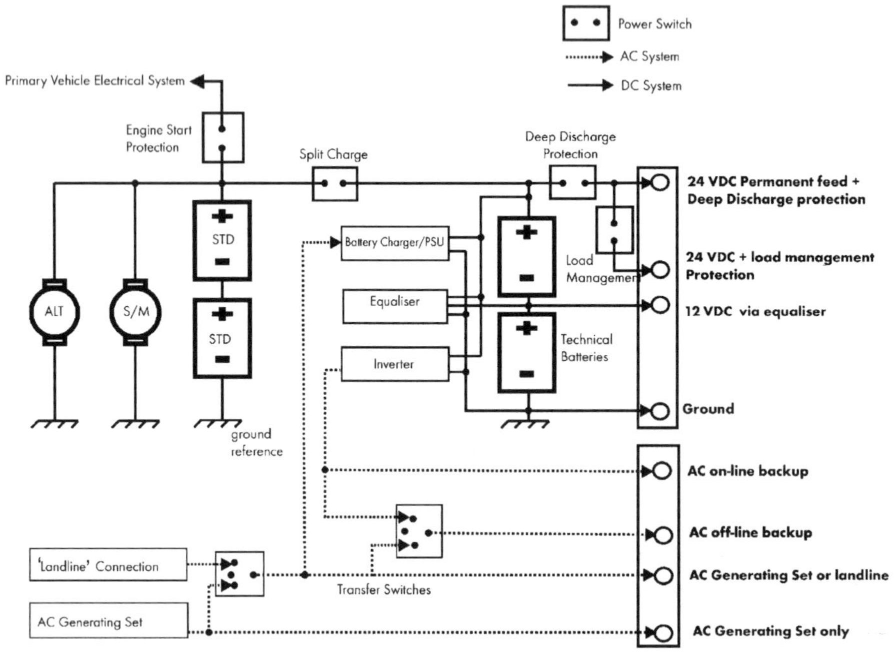

Figure 5.4.1 Typical Outside Broadcast vehicle layout. © Antares (Europe) Limited.

0.8 power factor as a general rule; therefore, 1 kW = 1.25 kVA. Power factor theory is beyond the scope of this chapter. Care should be taken to ensure that these are not confused when specifying the generating set.

Generator power is dependent on the density of the intake air. Where the set is to be operated at temperatures differing from the test conditions, a derating must by applied for high altitude and a derating factor for temperature. Note that cold temperatures increase the available power. Figure 5.4.2 shows typical rating adjustments for normally aspirated and turbocharged engines at various altitudes and temperatures.

The graph in Figure 5.4.2 is only a guide. Generating set manufacturers publish a wealth of information on the use of their products and it is wise to consult them if there is any doubt concerning the application.

Most modern generator prime movers used in OB are diesel-fuelled units which can safely rotate at 3000 rpm when directly coupled to a one-pole alternator to give an output frequency of 50 Hz. Earlier engines use 1500 rpm; however, the higher speeds allow the generator to be substantially smaller.

In 60 Hz applications the required engine speed is usually restricted to 1800 rpm, as diesel speeds of 3600 rpm are considered too high.

Rotational stability using a conventional mechanical governor is sometimes insufficient to avoid 'hum' or 'frame' bars. This problem is overcome by an electronic frequency comparator maintaining closer control of the generator speed.

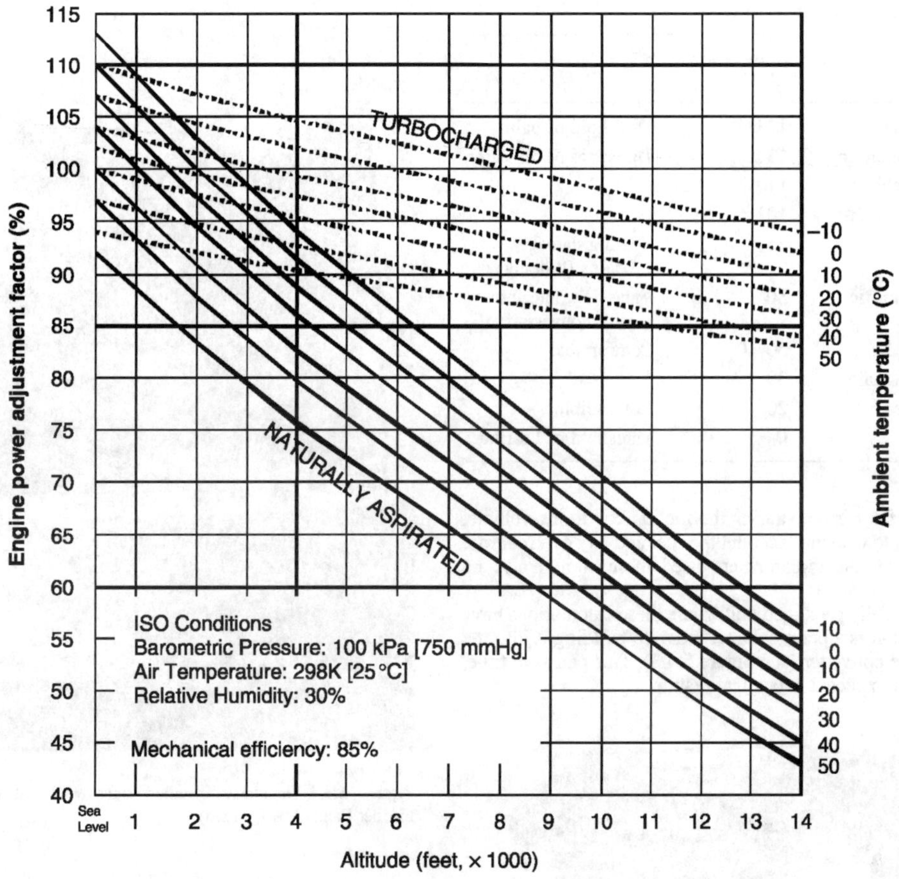

Figure 5.4.2 Altitude and ambient temperature rating adjustment curves relative to ISO 3046-1 reference conditions.

It is uncommon to find petrol-engined generators with power rating above 5 kW. Though they are free from the low-frequency diesel 'thump' and therefore more amenable to acoustic treatment, they suffer from factors such as economy in running, ruggedness and long life, and require the use of dual fuel systems on diesel vehicles. All of these factors weigh heavily in favour of the diesel, which will operate trouble free at a constant speed for very long periods and drive alternators having power ratings into the hundreds of kilowatts.

5.4.2.1 Noise attenuation

Once the power output required and the fuel type have been decided, the installation and noise attenuation must be specified.

Diesel engines are cooled by either air or water, but the higher output units will usually be water cooled. If an air-cooled unit is to be installed into a vehicle without any noise penalty inside and outside the vehicle, particular care must be taken to meet the manufacturer's airflow figures across the engine and alternator without creating undue airflow noise. This volume of airflow, however, must not be at the expense of the acoustic treatment and a careful balance must be struck.

Recent advances in generator set design allow for the alternator and the exhaust, in addition to the engine, to be water cooled, thereby allowing the unit to be completely acoustically sealed. This creates a very compact unit as it avoids the necessity to noise attenuate the alternator airflow using bulky and weighty materials, and allows the cooling radiator to be mounted remotely from the set.

Noise values are expressed as acoustic decibels (dBA) and are taken within the operational area external to the vehicle with the engine running at full load, typically at distances of 1, 3 and 7 m around the vehicle. Values of 70, 65 and 55 dBA respectively would be considered as good. Control room noise should be down to below 55 dBA, which is below quiet speech (see Table 5.4.2).

Finally, the generating set must be mounted on tuned anti-vibration mounts. This stops the engine vibration from being transmitted into the structure, causing attached panels to resonate. This is a specialised task and is integral to the design of the overall noise management.

To achieve the necessary noise levels on an outside broadcast vehicle the acoustic treatment requires steel, perforated zinc sheet, hard wood and dense rock wool, adding considerably to the weight. As a useful rule of thumb consider doubling the weight of the basic generator.

Table 5.4.2 Comparison of sound types

Sound	*Noise level (dBA)*	*Effect*
Near jet engines	125	Threshold of pain
Close-by thunderclap	120	Threshold of sensation
Symphony orchestra	110	
Jet flying over, 1000 ft	103	
Lawnmower	85	Hearing damage begins (8 hours)
Average city traffic	80	Annoying, interferes with conversation
Quiet office	50–60	Comfortable
Broadcasting studio	30	
Rustling leaves	20	Just audible
Silence	0	Threshold of hearing

If the generator is not sharing the fuel tank with the vehicle, you will need to take into account the fuel and any external fuel tank, and if remote radiators are used these should also be included.

Selecting a fully packaged unit where these calculations have been carried out is a practical alternative to building it into the vehicle by the converter (see Figure 5.4.3). This can save time, weight and unexpected acoustical results!

Figure 5.4.3 Fischer Panda 7 kW packaged engine (courtesy of Fischer Panda UK Limited).

5.4.2.2 Installation

The generator is commonly installed transversely at the rear of the vehicle (Figure 5.4.4). However, the maximum width of 2.5 m imposed upon road vehicles puts limits on the size of a generator that can be mounted this way. Faster compact sets with powers up to 80 kW can be mounted transversely. On smaller 12 V vehicles this is a more severe limitation. It is possible to locate the generating set under the floor, freeing up space. This is only really practical on the 24 V chassis due to necessary road clearance.

Figure 5.4.4 Generator installed transversely at rear (courtesy of Fischer Panda UK Limited).

The location should allow for access to all points on the engine requiring routine servicing and maintenance. This should be added to the overall dimensions of the generator. Where access is impossible the unit can be fixed on slides with flexible hoses; this is only really practical with the smaller sets. For major servicing it must be possible to easily disconnect the generator from fuel, electrical, water and exhaust connections, and lift the unit clear of the vehicle.

Where the application demands high power and the vehicle axle weights do not allow fitting on the vehicle, you should consider a trailer generator which is towed behind the vehicle. This arrangement has the advantages of removing the weight penalty on the vehicle and providing an independent power source for one or more vehicles. The trailer can be detached and parked remotely from the vehicle, away from microphones and the working area.

A trailer generating set can never match the ease of use of an integrated unit, in particular in Electronic News Gathering (ENG) applications.

A growing trend in the smaller power applications is the DC generating set. This system will automatically charge the technical battery bank when the voltage falls below a certain threshold. The engine is usually set to run for a minimum time to recharge the batteries. This reduces the engine run time considerably, thereby reducing costs, and avoids the engine running when light loads are required.

5.4.3 AC Distribution System

The size of the power requirement will determine the nature of the AC system. In Europe the systems are typically 230 V/50 Hz single phase for powers up to 30 kW and thereafter 400 V three phase. The other main voltage/frequency encountered is 110 V/60 Hz.

5.4.3.1 Landline

This is the input connection to the mains grid network and allows the vehicle to be plugged in whenever possible, reducing the need to run the generator. It can also be used for a remote generator. Thought should be exercised over what connections can be expected in the field of operation. A typical industrial single-phase connection is rated at 16 A, which will give 3.6 kW. Where this is insufficient, 32 A and 63 A connectors are available. The line should be rated for the current and the working temperature, and where appropriate protected from external damage from vehicles driving over it. Care should be taken to ensure that the input to any trailing cable has an earth leakage trip to ensure that damage to the cable will not cause risk of injury.

5.4.3.2 Inverters

Inverters are used to recreate an AC supply from the stored energy in a battery or from a DC source such as the vehicle alternator or a DC auxiliary generator. If the inverter is required for a short duration as a backup to a generator, it is commonly referred to as an Uninterrupted Power Supply (UPS). This time is used to safely 'power down' the equipment without loss of data. If longer duration is required it is a usually referred to as a standby power system. It can be configured in one of two ways, on-line or off-line.

The on-line approach connects the load permanently to the inverter. In turn, the inverter derives its power from a battery and battery charger. This arrangement is suitable for situations where the load is susceptible to a break in supply or the generator or landline supply is erratic. Since all the current flows through the inverter and battery charger, heat is generated, so it is generally preferred for smaller systems. In large systems the heat can be significant and is an extra burden on the air-conditioning. In this case the off-line system using a transfer switch should be used, as described below.

Choosing an inverter requires a knowledge of the loads you wish to operate, in particular what waveform the load requires, running and peak current.

When discussing the output of an inverter people refer to sine, quasi-sine and square waveforms. It is misleading to claim that any one type is better than another; it is all a question of compromise of cost, reliability and efficiency. Sine-wave output is closest to replicating the AC mains supply, but it is also likely to be the most costly and least efficient. Conversely, square wave is cheapest, probably the most efficient, but also the most likely to cause operational problems due to its waveform. The quasi-sine wave is an intermediate approach, replicating the key characteristics of the AC mains supply if properly controlled. Given proper control, and not all quasi-sine-wave inverters have this, the vast majority of loads can be successfully operated without the higher cost and power loss of full sine wave.

For this discussion we shall refer to the European system, which is 50 Hz and 230 VAC, but the same principles apply to other voltages and frequencies.

The mains has four important characteristics:

- **Peak voltage** – this is the peak voltage at the crest of a true sine wave. It reaches 325 volts on a 230-volt supply. This peak voltage needs to be maintained in order to successfully operate electronic equipment, principally to ensure that power supply input capacitors are fully charged
- **Average voltage** – this is the maximum value reached by the integral against time of the voltage waveform, or in other words, the average voltage. A '230-volt' sine-wave supply averages 207 volts. This characteristic is important for the successful operation of magnetic devices such as transformers and motors.
- **RMS voltage (230 V)** – this is a mathematically derived measure of the heating effect of a waveform when applied to a resistive load, and has traditionally been used to specify the magnitude of an AC voltage. This characteristic, which for a 230-volt mains supply is 230 volts, is important to ensure that heating and tungsten lighting equipment works to specification. If this is inaccurate or fluctuates then lighting may vary in intensity.
- **Stable frequency (50 Hz)** – this is necessary to ensure that timing devices using mains frequency operate accurately and that AC motors run at the correct speed.

It is recommended to carry out tests on the chosen inverter with the intended loads, thereby identifying potential problems.

5.4.3.3 Transfer switches

Switching between two sources can be achieved simply by moving the plug from one source to the other. Alternatively, use a manual switch but ensure that there is a centre off position. This is particularly important when one of the sources is an inverter whose electronics can be damaged by unsynchronised arcing from the other AC source. Transfer can be further improved by automating the transfer process using electronic control of contactors. Here the circuitry detects the failure of supply from the principal source and automatically switches over. This could be to a generator where it also auto-starts, or perhaps to an inverter which provides backup when all else fails.

In some cases the loads, such as computers, cannot tolerate seeing a break in supply. In this case the transfer must take place within one half-cycle or 10 ms. To do this the backup inverter supply runs parallel with the generating set and is synchronised to its frequency. When failure takes place the source is switched using a combination of electronic power switches backed up by contactors to ensure the requisite rating and safety. When power is restored, the transfer switch matches the inverter frequency and when synchronised changes over. The instantly loaded generator must settle to its correct frequency and voltage before the switch starts monitoring again. If not, the switch will oscillate between sources.

5.4.3.4 Earthing considerations

A key area of concern is the requirement to use an earthing spike. This is a spike driven into the ground to connect the vehicle chassis to the potential of earth. In some cases this is impractical, i.e. when parked over concrete, or impossible in the case of a vehicle requiring AC power when being driven.

Earthing is used for two different reasons:

- Electrical safety.
- Operation of radio communications equipment and information security.

Considering electrical safety, it is not always necessary to provide an earth spike, and may in some cases be detrimental. The need for an earth spike cannot be considered in isolation, but is part of the electrical system safety design. In general, however, an earth spike is not a reliable system and residual current circuit breakers should be used in preference, to provide protection.

Problems can also occur where vehicles are coupled, such as a trailer generating set or external portable generating set, where a break in the ground connection creates a shock hazard.

This is a specialist area and you should consult with an experienced practitioner.

5.4.4 DC Distribution System

5.4.4.1 Split Charging

This technique is used where the system requires power from the primary vehicle. Its main purpose is to allow the engine alternator to charge the technical system without the technical system drawing down the start battery. The voltage-controlled priority split charge is currently the most common approach; it uses a switch to link the systems together when the voltage on the start battery rises above a threshold. When this threshold is reached the switch deduces that the source must be an alternator or a battery charger.

Alternative methods use diodes, giving very unsatisfactory results. Firstly, they cause a voltage drop of around 0.6–1.0 volts, severely impairing the charging of the technical battery. They can also create gassing where the primary and technical load currents vary substantially. Newer systems use low-voltage-drop diodes; charging is still impaired, although to a lesser degree.

Using the alternator D+ to provide the control signal is not recommended, as modern alternators produce the D+ signal electronically and cannot sustain the relay coil current. Furthermore, the systems are linked irrespective of whether the alternator has capacity to power the loads and can lead to flat batteries.

The split charge system rating, including the cables and protection fuses, should match or exceed the maximum rating of the alternator output.

5.4.4.2 Mixed voltage systems

The auxiliary system voltage is usually the chassis voltage of the vehicle. Sometimes a 12 V supply is required to power loads. There are three methods used: the first and most common is direct DC conversion. This is suitable where the peak loads are below 20 A. Higher peak or intermittent loads, such as a powered 12 V motor, can be catered for either by direct connection to the centre 12 V tap, which is recharged via a DC equaliser, or using a separate 12 V battery, which is charged from a DC charger.

When 24 V is required on a 12 V vehicle, a DC up-converter or DC up-charger and battery is used. This is suitable for light loads or short intermittent loads. Where heavy DC power is required you will need to approach the problem another way, such as an AC battery charger or a separate 24 V alternator.

5.4.4.3 Technical Batteries

The battery bank should be sized to operate the DC and AC inverter loads for the standby period. As a rule the battery depth of discharge should be no more than a maximum 80%. If the batteries are to be used regularly, such as daily, you will need to specify a cyclic duty battery, the most common being the Valve Regulated Lead Acid (VRLA) GEL. If the battery is only required for an emergency standby and remains on float charge most of its life, then the Absorbed Glass Mat (AGM) technology can be used. Avoid the use of automotive batteries as they are not suited to cyclic use. Deep discharge protection should be used where it is anticipated that the batteries will be fully depleted in service. This will avoid sulphation damage.

Other technologies, such as NiMH, Li-ion and Ni–CAD, should only be considered for specialised applications where the weight, size and cyclic duty capabilities are necessary. Their application in vehicles is problematic and needs specialised skills.

Battery recharge currents are very sensitive to the applied voltage.

Where possible, all DC connections should be as short as possible and use cable thickness to avoid voltage drop. Care should be taken in the general layout of components. The sealed VRLA can be used in vehicle interiors without direct external venting.

The AGM or GEL valve regulated technology is sealed for life and so care must be taken with the charging regime applied to it. You must comply with the manufacturer's recommendations to avoid severely shortening the life.

Most applications are covered by GEL and AGM batteries, available in 12 V monobloc form up to 260 Ah. These can be combined in banks no larger than five in parallel to avoid imbalance and in parallel to form 24 V. Use traction cells where larger banks are required.

5.4.4.4 Battery Charging and Power Supplies

Technical batteries usually use a boost float temperature compensated regime (Figure 5.4.5). Typical charging regimes operate in two basic modes, creating five distinct phases. These charging characteristics are optimised to rapidly recharge the batteries over a wide temperature range.

- **Phase 1 – Bulk charging**. When the charger is initially connected to a discharged battery, current will flow up to the maximum current rating of the charging unit. This phase continues until the voltage reaches the float voltage.
- **Phase 2 – Boost charging**. The charger continues to allow the voltage to rise, thereby allowing the battery to accept charge at maximum current until the battery terminal voltage is just below the 'gassing' point. At this point the voltage is immediately reduced to float voltage. This corresponds to about 85% restored capacity.
- **Phase 3 – Taper charging**. During this phase the recharging current gradually declines to the float current, restoring the battery to 100% capacity.
- **Phase 4 – Float charging**. This is the reduced voltage phase, in which the charger provides enough current to overcome natural discharge and supports any connected loads.
- **Phase 5 – Reset**. This final phase is triggered when the charger is switched off, in which case it will revert to phase 1.

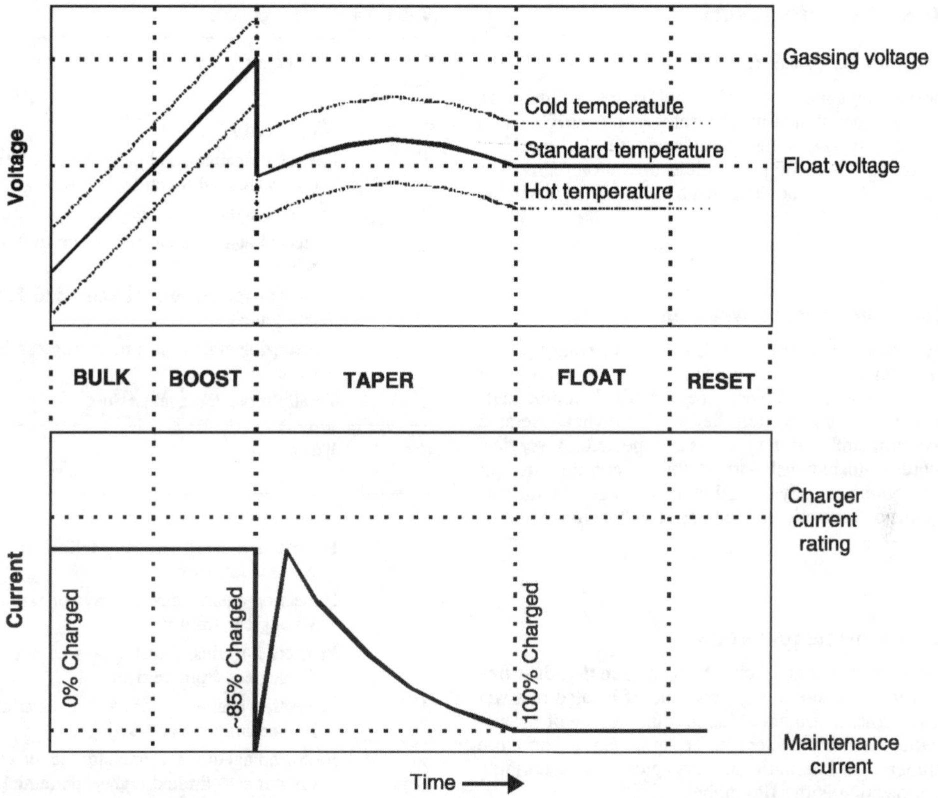

Figure 5.4.5 Boost/float/temperature compensated charging regime.

Alternatively, if the connected load exceeds the current rating of the charger it will revert to phase 2.

The charging regime is very accurately controlled, and 'temperature matched' to the battery. This results in fast recharge with minimal gassing and electrolyte loss.

These voltages are adjusted to account for the ambient temperature, raised in cold weather and reduced in hot. This regime can be used to leave the vehicle on charge for extended periods and can also be used as an 'opportunity' charger, being plugged in when AC power is available.

Some chargers have a timed period at the gassing point to speed charging. This can cause gassing problems in applications where there are connected loads during charging and should be avoided. Avoid the use of chargers with high-voltage ripple or constant-current chargers, as these will damage the batteries.

The charger size should be between 10% and 20% of the capacity of the battery plus the load current of any connected appliances. For example, a 200 Ah battery will require a 20 A charger for float duty and a 40 A charger for an accelerated boost system. If the connected load, such as a space heater, requires 10 A, the charger will be between 30 and 50 A. Note that larger chargers will not necessarily produce a quicker charging time, as the battery plate voltage will rise in response to the high charge rate and revert to the taper charge prematurely.

5.4.4.5 Task management and System Monitoring

The individual loads on a complex OB vehicle need to be controlled in specific ways. Most of this control can be automated, reducing driver and operator input. Task management is the drawing together of all these diverse requirements and viewing them as a single entity.

The key requirements of a system are:

- Manual switching of loads.
- Automatic switching of loads.
- Receiving signals from chassis.
- Receiving monitoring signals from auxiliary system.
- Interlocking load operation.
- Battery protection from deep discharge.
- Timing and delays.
- Providing dashboard information.
- Controlling and integrating third-party equipment.

Still the most common methods employed are relays and switches. These are appropriate for simple one-off systems. Where the task is more complex, programmable logic controllers (PLCs) or multi-node multiplexing can be used. These tend to be suitable for higher vehicle volumes or where the application requires logic functions requiring software.

5.4.5 Other Considerations

5.4.5.1 Statutory Regulation

As time passes, legislation in this field is becoming more complex, as are the standards for safety and EMC. Time should be taken to assess the compliance of the specified equipment for the scope of the vehicle operation. This is a specialist area and you should consult with an experienced practitioner.

5.4.5.2 Cooling and heat management

The electrical equipment in the truck and the auxiliary power system can develop considerable heat. Therefore, cooling air management is essential to keep the electrical components and batteries in the racks/lockers below 50°C when operated in the maximum ambient temperature expected in service. One technique is to exhaust part of the air-conditioning air through the equipment bays. Failing this, create an airflow via a fan. Do not hermetically seal electrical components in lockers.

5.4.5.3 Environmental protection

Where possible, components should be located in dry, dust-free areas. However, this is not always possible. If located in a wet or dusty environment, you should assess the degree of protection required and specify enclosures or equipment specified to this environment. An internationally accepted way of defining the type of protection is the IP number.

These ratings refer to specific tests. The IP number is made up of two components as follows: IP44. The first number refers to the protection against solid objects and the second against liquids. The higher the number, the better the protection (see Table 5.4.3 for a summary).

Note that there is a third number, commonly omitted, which refers to protection against mechanical impacts.

Table 5.4.3 IP ratings table

First number	*Solid objects*
0	No protection
1	Protected against solid objects up to 50 mm, e.g. accidental touch by hands
2	Protected against solid objects up to 12 mm, e.g. fingers
3	Protected against solid objects up to 2.5 mm (tools and wires)
4	Protected against solid objects up to 1 mm (small tools and wires)
5	Protected against dust, limited ingress (no harmful deposit)
6	Totally protected against dust
Second number	*Water*
0	No protection
1	Protection against vertically falling drops of water, e.g. condensation
2	Protection against direct sprays of water up to 15 degrees from vertical
3	Protection against direct sprays of water up to 60 degrees from vertical
4	Protection against water sprayed from all directions – limited ingress permitted
5	Protected against low-pressure jets of water from all directions – limited ingress permitted
6	Protected against low-pressure jets of water (use on shipdeck) – limited ingress permitted
7	Protected against the effect of immersion between 15 cm and 1 m
8	Protected against long periods of immersion under pressure

D Hardy
Technical & Quality Manager
PAG Ltd., London

5.5 Battery Systems

Portable broadcast equipment, including camcorders, field editors, monitors and associated equipment, is generally powered from rechargeable batteries. Digital broadcast technology demands more power from batteries than ever before, while the equipment itself becomes ever smaller and lighter, and the broadcast engineer is further challenged by the availability of several battery chemistries, each of which has specific advantages and disadvantages. This chapter explains the fundamentals of rechargeable battery performance, discusses the different chemistries and the trade-offs between them, and provides information concerning battery maintenance and disposal.

5.5.1 The Sealed Rechargeable Cell

Rechargeable cells used in professional broadcast batteries are invariably of cylindrical, sealed construction. Figure 5.5.1 shows the internal construction of a typical nickel–cadmium cell, and the other chemistries are broadly similar. The electrodes are very thin plates wound compactly in a roll, and insulated from each other by means of a porous separator. This roll fills almost the entire space within the cell casing, and a conductive electrolyte is absorbed into the separator and the surfaces of the plates. Because of the compact nature of the construction, there is almost no free electrolyte within the cell. The cell contains active materials that can be electrically oxidised and reduced repeatedly. The negative electrode is oxidised at the same time as the positive electrode is reduced, and this generates electric energy. Both of the electrode reactions are reversible; the application of a current from an external source drives the discharge reaction backwards to 'recharge' the electrodes.

The cell is designed to absorb and chemically re-combine any gas evolved during overcharge, but a safety vent is incorporated which opens automatically to release excessive pressure and then reseals again. Venting should only occur during extreme conditions such as severe overcharging, and eventually results in loss of capacity. Sealed cells can be used in any orientation without leakage, making them suitable for air transport.

5.5.2 Battery Chemistry

In the past, portable equipment was powered almost entirely by nickel–cadmium batteries, but nickel–metal hydride and lithium-ion alternatives are now available, each with its particular strengths and weaknesses. Lead–acid and silver–zinc batteries are little used within the broadcast industry, and are not discussed further.

5.5.2.1 Nickel–Cadmium

The sealed rechargeable nickel–cadmium cell was first developed in 1947 and continuous development has brought substantial capacity improvements in recent years. The environmental lobby has campaigned for the phasing out of nickel–cadmium because of the heavy metal content, but the battery industry has set up comprehensive reclamation and recycling programmes which substantially address these concerns.

The nickel–cadmium charging reaction is endothermic, with little heating during the main portion of the charge. The battery will begin to warm as it approaches full charge, with pronounced heating if serious overcharging occurs. Nickel–cadmium will withstand considerable abuse while offering arguably the best life of the three major chemistries. The discharge reaction liberates heat, but low internal resistance enables large currents to be drawn even from small batteries. The alternative chemistries do not possess the combination of high current capability, good cycle life and low resistance that gives the premium nickel–cadmium cell its durability, and it is still the mainstay of the portable power industry.

Several formulations of cell are produced, specialised for various applications, but all use an alkaline electrolyte of dilute potassium hydroxide. A single cell has a nominal voltage of 1.2 V, irrespective of the capacity. The most popular voltages

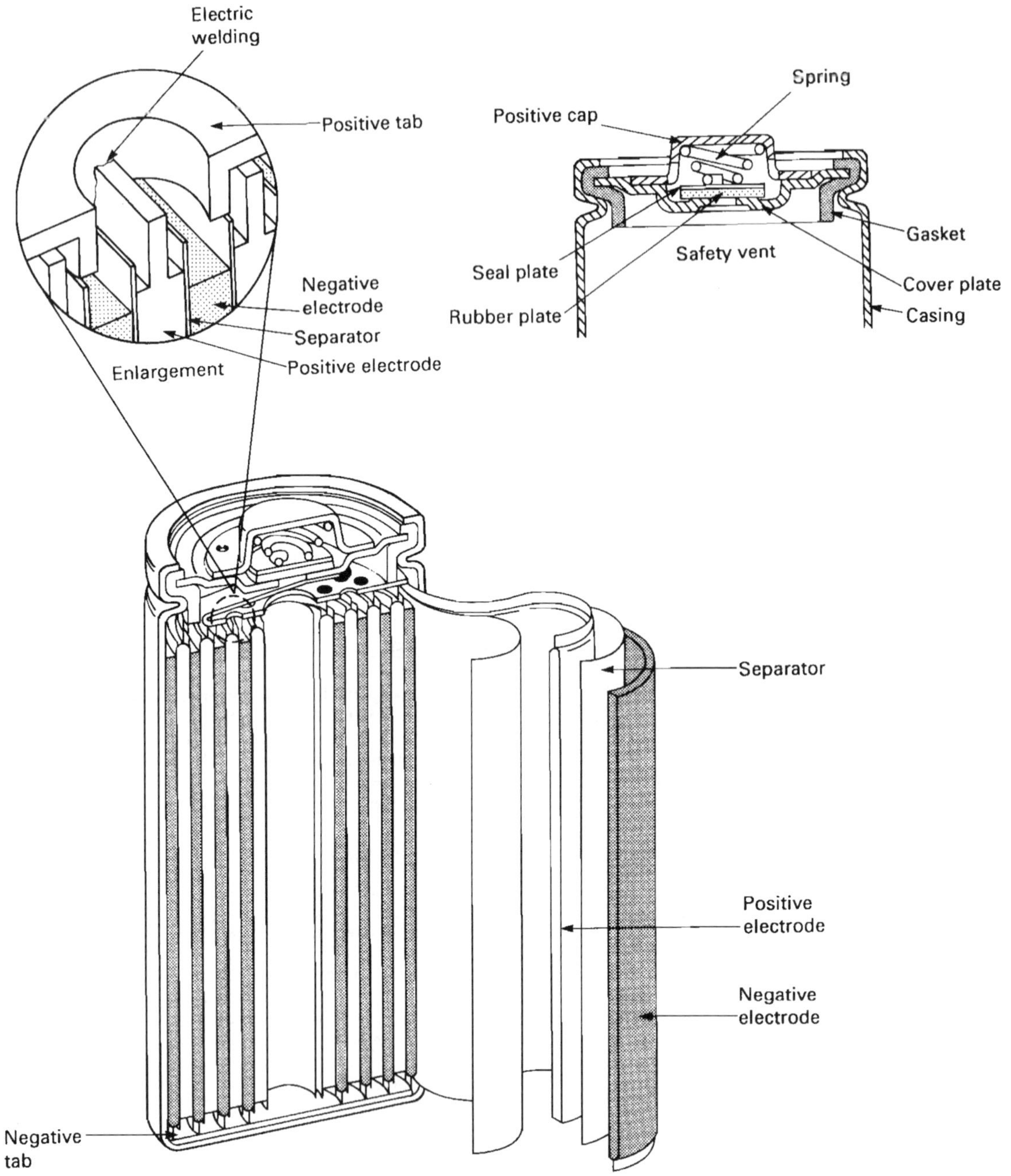

Figure 5.5.1 Cross-section of a typical sealed nickel–cadmium cell.

for broadcast cameras and associated equipment are 12, 13.2 and 14.4 V, so that batteries employ 10, 11 and 12 cells respectively, connected in series; 24 and 30 V batteries for lighting applications contain 20 or 25 cells respectively. Charging is with a constant-current regime, and constant-voltage chargers must never be used.

5.5.2.2 Nickel–Metal Hydride

The nickel–metal hydride cell is a hybrid of the positive electrode chemistry of the nickel–cadmium cell and a special hydrogen-absorbing negative electrode, which results in cells that demonstrate enhanced capacity while also having many of the characteristics of the proven nickel–cadmium cell. Environmental concerns over toxicity do not apply, since cadmium is not used in its construction.

Cells are similar in appearance and construction to the nickel–cadmium equivalents, with an electrolyte of dilute potassium hydroxide and a chemical voltage of 1.2 V per cell, but the greater energy density enables the production of batteries of increased capacity for a given size, or smaller batteries of the same capacity. However, a nickel–metal hydride battery is likely to yield a shorter life than a comparable nickel–cadmium battery, especially under harsh operating conditions. For nickel–metal hydride, ageing is partly a consequence of the number of charges undergone, so life is not necessarily extended by a regime of shallow discharges. Constant-current charging is required, using only a charger designed for nickel–metal hydride chemistry. A charger designed solely for nickel–cadmium batteries may not detect the fully charged condition of the nickel–metal hydride battery, and because the charging reaction is exothermic, the temperature will rise dramatically in overcharge. Severe overcharge will damage a battery permanently, and can be dangerous if the cells vent hydrogen gas. As a safety precaution, the batteries incorporate additional protection measures such as PTC over-current devices, thermostats and thermal fuses.

5.5.2.3 Lithium-Ion

Lithium-ion batteries have the highest gravimetric energy density of the popular battery systems (see Figure 5.5.2).

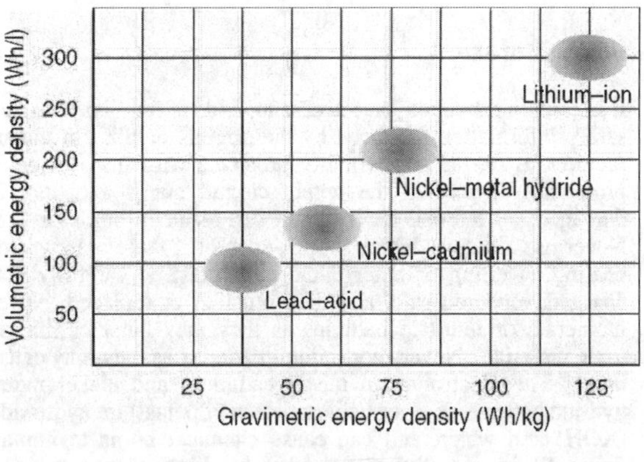

Figure 5.5.2 Comparative energy densities.

The earliest rechargeable lithium batteries employed metallic lithium as the anode material, but safety problems arose because of the high chemical reactivity of the metal. The present designs have completely overcome this disadvantage by using a method whereby lithium is not recombined as a metal, but is diffused into the plate structure in ionic form. The resulting device is a lithium-ion battery, and not a rechargeable lithium battery.

Cells are produced from spirally wound electrodes with an electrolyte consisting of a flammable organic solvent. Special measures are taken within the battery to guard against leakage of conductive electrolyte because of the close proximity of live circuits. Voltage is around 3.7 V per cell, with only four cells necessary to produce a 14.8 V battery. The cells are very closely matched, and multiples are sometimes connected in parallel to give the desired capacity. A special constant-voltage charge regime is required, and the battery incorporates sophisticated safety circuits and redundant protection devices to ensure there is no possibility of overcharge, over-discharge and over-current events that could be caused by failure of the charger or other systems.

The cells used by the majority of after-market manufacturers are based upon graphite-type chemistry as opposed to the coke-type chemistry used by a major Japanese manufacturer of batteries and broadcast equipment. The discharge curve of the graphite battery is significantly more flat than the coke type (see Figure 5.5.3). It can be seen that the discharge 'knee' for the graphite battery occurs at approximately 13.5 V for a 14.8 V (nominal) battery. The graphite battery is completely discharged at around 12 V, but the coke type may still retain approximately 20% charge at 12 V, and is fully discharged at 11 V. The higher average discharge voltage of the graphite battery results in a higher overall energy content. The steeper discharge curve of the coke-type battery enables the camera warning voltage to be set at about 12 V, whereas the graphite-type requires the warning point to be set higher, at around 13–13.5 V. Figure 5.5.3 shows how the lithium-ion discharge profiles compare with that of a typical 13.2 V nickel–cadmium battery.

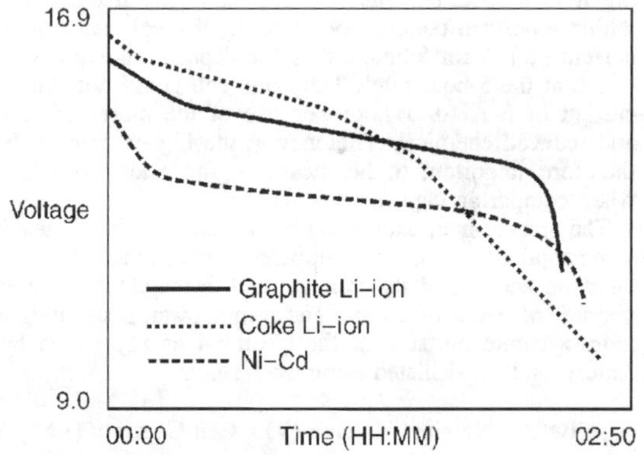

Figure 5.5.3 Coke and graphite lithium-ion batteries compared with nickel–cadmium.

5.5.3 Electrical Ratings

5.5.3.1 Voltage

A battery has a nominal voltage rating, and an operating voltage range. Nominal voltage is largely determined by the chemistry, and it is the mean voltage that is maintained during most of the discharge curve. A battery of 10 nickel–cadmium or nickel–metal hydride cells has a nominal voltage of 12 V. However, it can reach 14.5 V when fully charged, and it is fully discharged at 10 V on-load, and this represents the operating voltage range. A 14.8 V lithium-ion battery has an operating voltage range from 16.8 V down to 12 V.

On load, the initially high voltage of a fully charged battery falls steeply until the mean running voltage is reached. In a healthy battery, the mean running voltage constitutes the majority of the discharge curve and at normal discharge rates it remains very nearly flat. The average voltage of this portion of the discharge curve will approximately correspond to battery nominal voltage. Voltage may be depressed at high currents, reducing the capacity for that cycle, and conversely a very low discharge rate may result in a higher voltage. Other factors affecting voltage include the cell type and manufacturer, the age of the cell, the history of previous use and the temperature. Towards the end of discharge, the voltage curve will reach the 'knee' and fall away very rapidly. Below the fully discharged voltage there is virtually no energy remaining in the battery and continued discharge can result in permanent damage.

A battery must have the correct voltage range for the intended application. As previously noted, the voltage will vary considerably between the fully charged and the fully discharged conditions. When fully charged, it must not exceed the maximum rated input voltage of the load. Equally important, the end of discharge voltage should be higher than the minimum input voltage of the load, otherwise it will not be possible to fully discharge the battery and obtain the full run time.

5.5.3.2 Capacity

The traditional unit of measurement for batteries is 'ampere-hours'. Capacity is calculated by taking a single cell, charging it in a specified manner, and then fully discharging it within a defined period. For example, if a cell can supply a current of 1 A for 5 hours, then the capacity is expressed as '5 Ah at the 5-hour rate'. The same cell could not supply a current of 5 A for 1 hour because of the increased losses and reduced chemical efficiency at the higher current. It is therefore important to be aware of the rate of discharge when comparing capacities.

The ampere-hour expression of capacity can be misleading when applied to batteries (as opposed to single cells) because it takes no account of the voltage, and is therefore not a true measure of power over time. The nominal watt-hour rating is a more accurate measure of the electrical energy stored by a battery, and is calculated using the formula:

$$\text{Battery Nominal Voltage (V)} \times \text{Cell Capacity (Ah)} = \text{Watt-hours.}$$

Battery run time can be calculated by dividing the capacity in watt-hours by the total equipment consumption in watts.

5.5.4 Charging

5.5.4.1 Nickel–Cadmium

Charge using only a constant-current charger. Never use a charger intended for lead–acid batteries; the result is likely to be permanent damage to the battery.

5.5.4.1.1 Slow (Overnight) Charge

Allow the battery to cool after heavy discharge; the ideal charging temperature is around +20°C. Ensure the correct polarity and charge at C/10 rate for 14 hours (i.e. 500 mA for a 5 Ah battery, 700 mA for a 7 Ah battery). Prolonged overcharging may have a long-term detrimental effect.

5.5.4.1.2 Fast Charge

Important: ensure that the batteries are compatible with the charger and read the charger handbook before use. Allow the battery to cool after heavy discharge; fast charging is most efficient at around +20°C. Do not fast charge batteries in parallel because unequal current sharing may result in serious battery damage. Do not fast charge batteries in series, because they will not be sufficiently matched to ensure proper termination of fast charging.

5.5.4.2 Nickel–Metal Hydride

Use only a charger that is designed for charging nickel–metal hydride batteries, and ensure that it is compatible with the actual batteries to be charged. Chargers designed for nickel–cadmium batteries only may be unsuitable and potentially dangerous. Slow (overnight) charging is not recommended as a regular charging regime, and in fact it will reduce the available capacity by approximately 20%. If there are any doubts concerning the charging arrangements, consult the battery manufacturer.

5.5.4.3 Lithium-Ion

Lithium-ion batteries must only be charged using the charger specified by the battery manufacturer. Never use any other type of charger.

5.5.5 Safety

It is essential that you have access to, and are familiar with, the safety information produced by the original equipment manufacturer. Batteries are normally protected with fuses or other protection devices in the output circuit, but if a battery is damaged, an internal short circuit can result in burns or fire. Never use a battery which is split open or if there are signs of heating, swelling or other damage. It must be carefully discharged, rendered safe, and then repaired or replaced. Never incinerate or mutilate batteries as they may burst or release toxic materials. Never short-circuit batteries as they may cause burns. The electrolyte in nickel–cadmium and nickel–metal hydride batteries is a corrosive solution of potassium hydroxide (KOH) and water, and can cause chemical burns to human tissue. Wear protective gloves when handling all contaminated materials, and if any electrolyte contacts the skin, flood copiously with clean water. Seek medical attention if any has touched the eyes.

5.5.6 Battery Life

Rechargeable batteries have a finite life. In a laboratory environment, where all the conditions are controlled and repeatable, the life of a particular battery type can be found and expressed in terms of the number of cycles in a given timeframe. Within limits, the test will be repeatable given exactly the same conditions of load, temperature profile, rest between cycles and charging profile. In the real world, these conditions are constantly varying, and therefore the life of a battery cannot be expressed in terms of numbers of cycles or chronological timescale. The life of a battery is often determined by how well matched the capacities of the individual cells remain in respect to each other. It is important to note that life cycle date produced by cell manufacturers normally refers to tests conducted using single cells, where the issue of cell balance within a pack does not arise. For this reason the life of a battery can be considerably less than the life of one of the constituent cells.

Nickel–cadmium and nickel–metal hydride batteries may show a slight increase in capacity during the first part of their life, and then the life cycle curve remains flat until cells begin to fail and the curve declines. Unless a sudden fault develops, nickel–cadmium and nickel–metal hydride batteries usually give some warning that they are approaching the end of their life. This normally takes the form of gradually reducing run time, and the development of a high internal resistance causing excessive voltage depression under heavy load. Lithium-ion batteries exhibit a gradual but continuous reduction in capacity with time and cycles; internal resistance tends to remain low even when capacity has been substantially reduced, so that even a very well worn battery may still deliver the full rated current and remain usable.

The life of a battery is not determined solely be the number of charge/discharge cycles that have been performed. The nature of the charging is of paramount importance, and excessive overcharge is particularly undesirable. The depth of discharge is also relevant. Lithium-ion batteries are protected against over-discharge by a built-in protection circuit, but this is not usually the case with nickel–cadmium and nickel–metal hydride batteries, and they can be permanently damaged if they are over-discharged. In general, batteries that have been subjected to a shallow discharge regime will give a greater life, but this may not be true with nickel–metal hydride chemistry, where the number of charge cycles can be the determining factor. Different manufacturers specify various figures, but a battery is generally considered to be at the end of its life when the capacity is less than 70–80% of the original rated capacity, or when it consistently exhibits symptoms of failure. Failure symptoms can include the inability to supply full current, resulting in situations such as a camera turning off when the on-board light is turned on.

5.5.7 Operation in Extremes of Temperature

All batteries show a reduction in output voltage and capacity at very low temperatures. Under these conditions, chemical activity inside the cells is slowed and the internal resistance is increased; consequently, batteries with low internal resistance at normal temperatures may prove to be the most satisfactory at low temperatures. An additional problem is that low temperature stiffens the action of the mechanical parts in cameras and other equipment with a corresponding increase in current draw, particularly at start-up. Batteries should be kept in the warm when not in use, and if possible they should be protected with a thermal insulating medium – a tailored thermal jacket is ideal, but even packing foam will help. Allowance must be made for battery voltage depression; 13.2 or 14.4 V should be used in preference to 12 V, but ensure that the equipment is rated for the higher voltage at normal temperatures. Charging must only be conducted within the temperature range specified by the charger manufacturer. If batteries are very cold, bring them into a warm environment and allow sufficient time for them to acclimatise before charging. Similar rules apply to battery use at high temperatures, when measures should be taken to shield them from direct sunlight and keep them cool when not in use. In the absence of specific information, only charge batteries with a temperature of between +10°C and +30°C. Batteries that become very hot (above +45°C) during discharge should be allowed to cool. As a general rule, nickel–cadmium chemistry will best tolerate the harsh conditions that accompany use in high or low temperatures or adverse conditions.

5.5.8 Maintenance and Storage

Nickel–cadmium, nickel–metal hydride and lithium-ion batteries require minimal maintenance. Because the cells are hermetically sealed, no topping up with fluids is either necessary or possible, and maintenance consists largely of ensuring that all batteries are in a safe physical condition, and identifying low-capacity or troublesome batteries. Batteries should be protected from water, including driving rain, steam or high humidity, but if a battery does get wet, shake out any excess water and allow it to dry naturally in a warm, dry place. This may take a considerable time (a minimum of several days) but do not attempt to use it until it has fully dried. Batteries must never be over-discharged. Lithium-ion types are normally protected against this abuse, but other types can suffer from voltage reversal in each cell that is discharged below about 75% of the fully discharged condition. This damage is cumulative and leads to irreversible imbalance within the cell pack. In some instances, imbalance can be reduced by giving a 24-hour charge at the overnight (C/10) rate, followed by a few charge/discharge cycles. If this fails to give a lasting improvement, then the battery may be permanently damaged or worn out.

Fully charged nickel–cadmium and nickel–metal hydride batteries can lose up to 5% capacity in the first day and about 1% per day thereafter (but possibly very much more in old or damaged batteries). A small temperature increase may produce a disproportionate increase in self-discharge, and a 10°C temperature increase can halve the time for a battery to self-discharge to a given level. Lithium-ion chemistry has much lower inherent self-discharge, but built-in electronics circuits can present a parasitic load to the cell pack. In good designs this load is negligible, but in other designs it effectively increases the self-discharge rate more than tenfold.

Store batteries in a cool dry place. Ideally, the storage temperature should be +20°C or lower. High storage temperatures (above +35°C) will reduce the battery's life because of unwanted chemical reactions, and excessively low temperatures (below −30°C) should also be avoided since the electrolyte may freeze and permanent cell damage may result. Nickel–cadmium and nickel–metal hydride batteries are best stored in the discharged condition, and periodic charging is unnecessary. Charge lithium-ion batteries to approximately 50% before storage, but if storage will be very long term, obtain the specific recommendations of the manufacturer. After storage, a few charge/discharge cycles may be required before the pre-storage capacity is restored.

When nickel–cadmium batteries have been only partially discharged to the same extent during many cycles, a depression in the discharge profile of approximately 150 mA per cell can be observed when a deeper discharge is undertaken. This is known as the 'memory effect', a very misunderstood condition that is exceedingly difficult to reproduce. Batteries that give the appearance of memory are, more often than not, suffering from cell imbalance.

5.5.9 Battery Management

Efficient battery management can be highly cost-effective, through the optimisation of the number of batteries and improved reliability. Battery details should be recorded and regular capacity and performance checks conducted; wear-out can then be predicted and timely replacement planned. If poor batteries are not identified, the entire battery stock will become thought of as unreliable. The only meaningful way of measuring capacity is to charge the battery fully and then discharge it at a constant current that represents actual usage. Some chargers incorporate discharge facilities, and the more sophisticated models can output serial data, allowing integration with a PC for storage, analysis and presentation of data. Battery analysers and dischargers are also available to perform the task.

5.5.10 Transportation

Batteries should be discharged before being transported. Safety is the prime consideration, and no battery should be transported, particularly by air, if the case is split or if any other serious damage is evident. Batteries must be packed so that short circuits cannot occur, and to protect them from damage caused by shock, vibration and drops. There are special regulations concerning the air transportation of lithium-ion batteries. These batteries contain no metallic lithium, but a formula known as 'lithium equivalence' forms part of the standards specifying the safety tests required for compliance, and above a certain capacity batteries are subject to very strict packaging requirements. These regulations are aimed primarily at manufacturers and retailers, and exemption clauses allow users to carry on batteries subject to limitations of quantity and capacity. The regulations are subject to change, and it is recommended that users have access to a valid certificate of compliance with current air transport regulations. The battery manufacturer can normally provided a suitable certificate and information regarding air transport.

5.5.11 Disposal and Environmental Issues

Nickel–cadmium batteries contain cadmium metal, which must not be disposed of in household waste. Whatever the chemistry, batteries must be fully discharged prior to disposal. Batteries and cells must not be opened, punctured, incinerated or otherwise mutilated, since this renders them dangerous and unsuitable for subsequent reclamation processes. Observe all national and local rules and regulations regarding the disposal of rechargeable cells. Waste authorities and the battery industry operate various schemes by which the waste materials are reclaimed for reuse or safe disposal. The manufacturer or supplier will be able to provide up-to-date information regarding the safe disposal of its batteries.

Section 6 Transmitter Systems and Hardware

R S Roberts C Eng, FIEE, Sen MIEEE
Consultant Electronics Engineer

6.1 Radio Frequency Propagation

Radio frequencies form a small part of a wide range of types of energy transmission by means of electromagnetic fields.

All forms of electromagnetic radiation into space are characterized by four common factors: their speed of propagation in free space is the same (300×10^6 m/s), and they are subject, like light, to the laws of reflection, refraction and attenuation. The primary source of electromagnetic radiant energy is the sun. In this section we consider the small range of frequencies associated with broadcasting of television.

Clarke Maxwell indicated in 1864 that electromagnetic radiation could be established in space by electrical means. The physicist Hertz, in 1887, proved experimentally that electromagnetic fields could be set up in space and detected, as Maxwell had predicted. He also showed that such radiation obeyed the same laws as determine the behaviour of light. At a time when intercity and intercontinental communications were carried out by means of wire links, Marconi saw that this radiation provided the possibility for a 'wireless' means of communication. He put together a communication system consisting of a generator of radio frequency power, a radiation system, a receiving system and detection system. His work culminated with his famous transatlantic digital signals in 1901 that launched radio communications.

6.1.1 Theoretical Principles

6.1.1.1 Magnetic and electric fields

A direct current passing through a conductor establishes a magnetic field around the conductor that extends into space to an infinite distance. However, the magnetic field strength falls at a rate inversely proportional to d^2, where d is the distance from the conductor. Thus, the static field strength falls rapidly to very low values over very short distances.

Energy has been taken from the supply to establish the field. The conductor has the property of inductance, and if the current flow is stopped, the magnetic field collapses with resulting self-induced voltage, the whole process returning energy to the supply.

Let us now consider the electric field. Hertz attached a pair of plates to a current-carrying conductor, identifying electric *poles* with a difference of potential between them, and an electric field in the space between them. The electric field has a potential gradient which is measured in volts per metre. A pair of plates 3 m apart with a potential difference of 60 V between them will produce a gradient of 20 V/m. A conductor lying in this field parallel to the flux lines will have a potential developed between its ends of a value proportional to its length, e.g. 0.5 m in the filed will develop 10 V.

The static fields thus consist of rings of magnetic flux concentric with the conductor, and lines of electric flux near-parallel to the conductor, as shown in Figure 6.1.1.

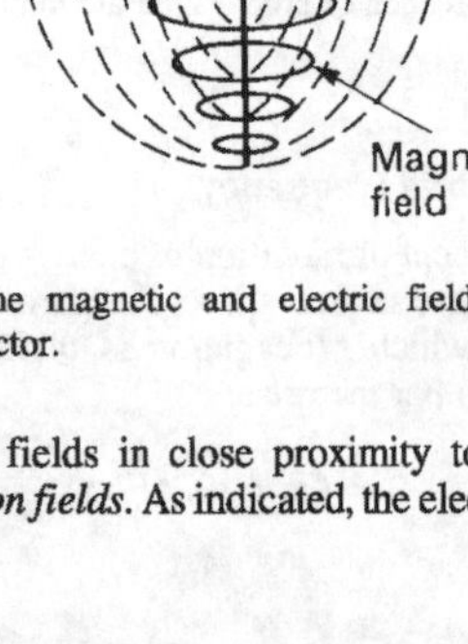

Figure 6.1.1 The magnetic and electric fields associated with an energized conductor.

These local fields in close proximity to the conductor are termed *induction fields*. As indicated, the electric field can *induce* a voltage between the ends of a conductor and, as Faraday discovered in 1831, a relative movement between a conductor and the magnetic field will develop a voltage in the conductor.

Another conductor, positioned at a distance remote from that which is generating the fields, can be immersed on the fields, but will experience no effects if neither the conductor position nor the strength of the fields is changing. If the current is changing, a very different situation exists. The field intensities will be changing, and a changing voltage will be developed in a remote conductor.

6.1.1.2 Alternating fields

If the current is alternating, the induction field effects near the conductor will exist as with steady current, and their strength will vary as $1/d^2$. However, there will be an additional effect. The alternating fields will be sweeping away from the conductor with the speed of light, in the form of energy-carrying electromagnetic fields. At a point remote from the radiating conductor, another conductor can be erected parallel to the electric flux lines, and it will have an alternating current developed in it at the frequency of the field alternations. Power can be abstracted from the passing 'wave', because the current developed in the second conductor can be passed through a load. This is the principle established by Maxwell in 1864.

The radiated fields are thus a load on the energy source that is used to develop the fields. This load can be represented by an imaginary resistor having *radiation resistance* or *antenna impedance*. Figure 6.1.2 shows, at a point remote from a radiating system, the field flux lines. They will be varying in phase, and are related to each other in quadrature in a plane at right angles to the direction of propagation. The whole field system, advancing along the direction of propagation, is termed a *wavefront*.

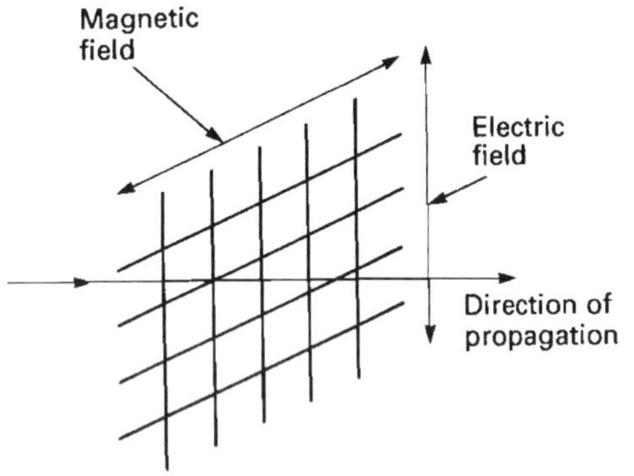

Figure 6.1.2 At a distance remote from the energized conductor, the radiated fields and direction of propagation are mutually orthogonal.

6.1.1.3 Velocity and frequency

To sustain the concept of radiation of energy in the form of an electromagnetic wave in free space, the permitivity, e_0, of the medium through which propagation is taking place can be expressed in farads per metre as:

$$e_0 = 4\pi \times 10^{-7}\,\text{F/m} \qquad (6.1.1)$$

The permeability, μ_0, in henries per metre is:

$$\mu_0 = 10^{-9}/36\pi\,\text{H/m} \qquad (6.1.2)$$

The velocity of propagation through the medium is given by:

$$v = \frac{1}{\sqrt{e_0\mu_0}} = 3 \times 10^8\,\text{m/s} \qquad (6.1.3)$$

The velocity given by equation (6.1.3) is often expressed as a constant c. It relates to all forms of electromagnetic radiated fields, i.e. radio, light, infrared, ultraviolet, etc. The constant c is not exactly 3.0×10^8. Experiments conducted over many years have yielded 2.997925 as a more accurate value.

The travelling alternating fields will establish a *wavelength* in space. This is the distance between two corresponding points on consecutive repetitive cycles in the direction of propagation. The frequency and length of the wave for any electromagnetic radiation is related to c by $\lambda = c/f$, where λ is the wavelength in metres and f is the frequency in hertz.

The atmosphere surrounding the earth is not the vacuum of free space, and propagation through the atmosphere at altitudes up to 1000 km or so requires some modification to equation (6.1.3). If the atmosphere has relative values of permittivity e_r and permeability μ_r, the velocity of propagation becomes:

$$v = \frac{1}{\sqrt{\mu_r e_r}}\,\text{m/s} \qquad (6.1.4)$$

Fortunately, e_r and μ_r are generally near unity for most terrestrial communications and v approximately equals c, but for propagation along transmission lines (for example) the value for e_r will be very different from unity.

6.1.1.4 Impedance of space

The radiated energy is contained in the electrical and magnetic fields, and is given by:

$$\frac{\tfrac{1}{2}E^2 \times 10^{-9}}{36\pi} = \frac{\tfrac{1}{2}H^2 \times 4\pi}{10^{-7}} \qquad (6.1.5)$$

where E is the electrical field strength and H is the magnetic field strength.

Equation (6.1.5) provides the concept of impedance, given by:

$$E/H = 120\pi, \text{or } 377\text{ ohms} \qquad (6.1.6)$$

6.1.1.5 Radiated energy

From equations (6.1.5) and (6.1.6), the energy in each square metre of wavefront is given by $E^2/120\pi$ or $120\pi H^2\,\text{W/m}^2$.

Consider an isotropic radiator at the centre of a sphere, radiating P watts uniformly in all directions. If the sphere has radius d its surface area is $4\pi d^2$, and the power flux per unit area is:

$$P_a = P/4\pi d^2\,\text{W/m}^2 \qquad (6.1.7)$$

From this and the energy formulae given above,

$$\frac{E^2}{120\pi} = \frac{P_a}{4\pi d^2}$$

$$E^2 = \frac{120\pi P_a}{4\pi d^2}$$

$$E = \frac{\sqrt{30P_a}}{d}$$

$$= \frac{5.5\sqrt{P_a}}{d} \text{ V/m} \qquad (6.1.8)$$

The most important conclusion from equation (6.1.8) is that the field strength of the radiated fields varies inversely with distance d from the radiator, whereas the local induction fields vary as $1/d^2$. The induction fields play no part in the radiation field but, as will be seen, they are very useful in antenna design.

6.1.2 Practical Considerations

The outline of theoretical principles in Section 6.1.1 concerns radiation in free space, and introduces the concept of an isotropic radiator. These principles require considerable modifications for practical implementation, because (a) no practical isotropic radiator can exist, and (b) simple terrestrial systems are not in free space. The radiation process cannot ignore the presence of the earth itself, and the gaseous atmosphere that surrounds it.

A step towards realism is the *Hertzian dipole*. This theoretical concept, based on the radiator used by Hertz, consists of a very short conductor, terminated with electrical charges at each end, along which flows a current of uniform density. The more practical antenna is the *half-wave dipole* shown in Figure 6.1.3.

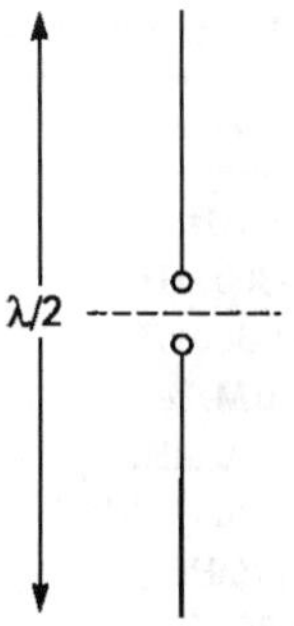

Figure 6.1.3 The λ/2 dipole. The antenna terminals are balanced with respect to ground.

6.1.2.1 The rf spectrum

A wide range of frequencies, shown in Figure 6.1.4, is used for radio frequency communication. The boundaries at the extremes are ill-defined. Very low frequencies of a few hertz are used, and the upper limit extends to about 300 GHz, encroaching on the infrared end of the light spectrum. Today, the boundaries between radio and light are becoming difficult to separate.

For convenience, the rf spectrum is divided into a number of bands with international designations, as shown in Table 6.1.1.

The '3–30' frequency classification is not arbitrary, but is derived from the differing modes of propagation. For instance, lf and vlf waves are propagated over the surface of the earth. The mf waves are 'ground waves' but, at times, use the ionized upper atmosphere. Propagation at hf is only possible using the ionosphere, and, at about 30 MHz, waves will be lost into

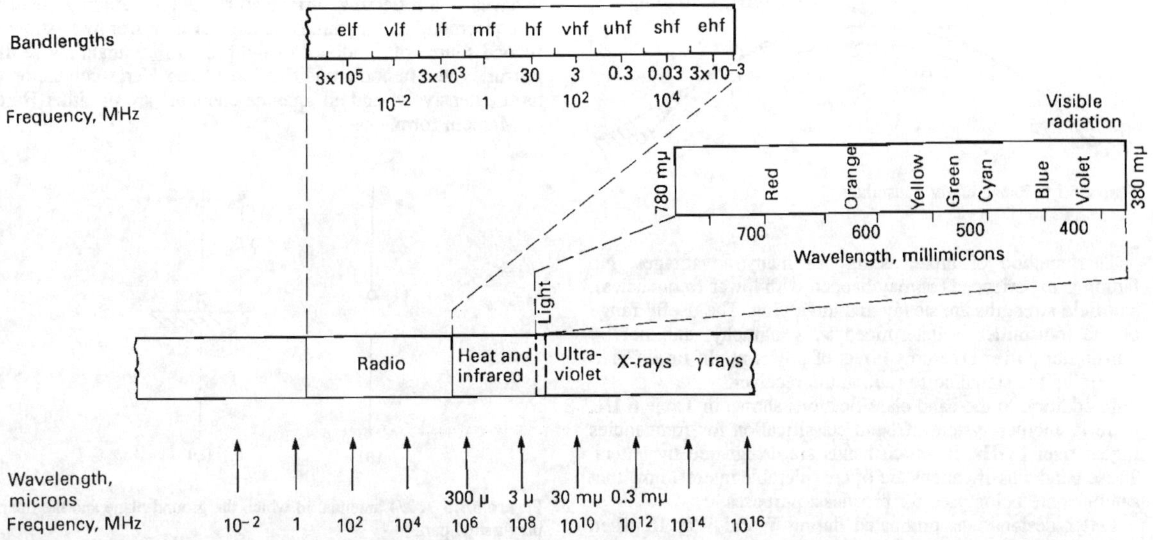

Figure 6.1.4 Television broadcasting takes place in the vhf and uhf bands. The European vhf bands are Band I (41–68 MHz) and Band III (174–216 MHz). The uhf bands are Band IV (470–582 MHz, UK channels 21–24) and Band V (614–854 MHz, UK channels 39–68). 11.7–12.5 GHz is allocated for television broadcasting via satellites.

Table 6.1.1 International frequency bands

Band	*Frequency*	*Wavelength*
vlf	<30 kHz	>10,000 m
lf	30–300 kHz	10,000–1000 m
mf	300–300 kHz	1000–100 m
hf	3–30 MHz	100–10 m
vhf	30–300 MHz	10–1 m
uhf	300–3000 MHz	1–0.1 m
shf	3–30 GHz	10–1 cm
ehf	30–300 GHz	10–1 mm

Frequencies above 1 GHz are often termed *microwave*.

Table 6.1.2 IEEE Standard 521 classification

Band	*Frequency range (GHz)*
L	1 2
S	2–4
C	4–8
X	8–12
Ku	12–18
K	18–27
Ka	27–40
V	40–75
W	75–110
mm	110–300

space. Waves at vhf and uhf use the oldest method of electromagnetic communication; the principle is the same as the lighthouse. The transmitter antenna is situated at a high point, and the radiation is delivered to the terrain within visible range. The shf and ehf bands have wavelengths so short that another method of propagation can be used. The radiator can be situated at the focus of a parabolic reflector, and a narrow beam can be directed between the transmitter and receiving systems.

The television broadcast service commenced operations in the vhf band and later established a service in the uhf band. Figure 6.1.5 shows a radiating system situated at the top of a high mast which, in turn, is sited in high ground. The radiator can direct the radiated fields towards the ground and thus illuminate a service area with a radius *R* to the distant horizon where the earth casts its massive shadow.

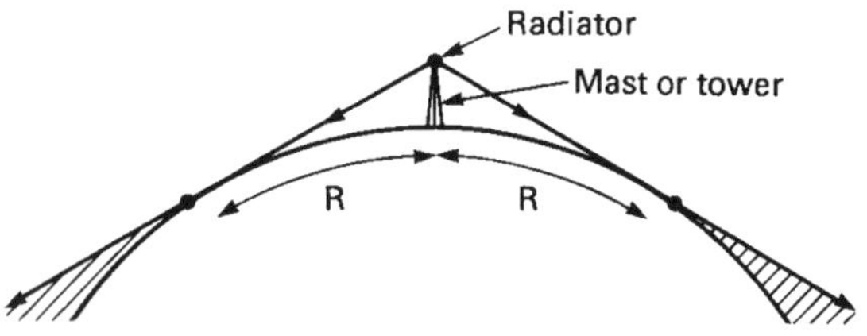

Figure 6.1.5 Range of vhf transmitter.

This method of broadcasting has many advantages. No fading is experienced (as may happen with lower frequencies), and field strengths are steady and unvarying. The useful range of the transmitter is determined by geography, and not by transmitter power. The only effect of power at vhf or uhf is to determine the signal/noise ratio at the receiver.

In addition to the band classifications shown in Table 6.1.1, there is another system of band classification for frequencies higher than 1 GHz, in which bands are designated by letters. These band classifications are of considerable interest now that satellites are being used for broadcast purposes.

Letter designations originated during World War II, when early UK radar systems used wavelengths of about 10 cm and 3 cm, and it was useful to 'hide' their frequencies by the designations 'S' and 'X' respectively. Since the war, many radar and other applications of frequencies above 1 GHz have used a range of other letters for specific bands, and Table 6.1.2 shows the IEEE Standard 521-1976, which was revised in 1984.

Satellite broadcast operations differ from radar in that two frequencies are required for each channel, one for the 'up' path and another for the 'down' path. For example: a satellite link might be quoted as 'C 4/6', thus designating the two frequencies involved in B and C for up and down paths.

It is of note that the international allocations for DBS (Direct Broadcast from Satellites) span the X and Ku bands.

6.1.2.2 The practical radiating element

Let us go back to Hertz and Marconi. Hertz used a conductor divided at the centre to allow energy to be fed to the system, as shown in Figure 6.1.6(a). Such a system is termed a *dipole*. Marconi, seeking a communication system with commercial possibilities, used a vertical conductor. Figure 6.1.6(b) shows an important feature of this form of radiation. The earth behaves as a reflecting surface so far as the radiating element is concerned, in the same manner as a mirror will show a second source of a radiator of light to an illuminant above the mirror. The Marconi modification of the Hertz conductor is used extensively, and all antenna elements are of either Hertz or Marconi form.

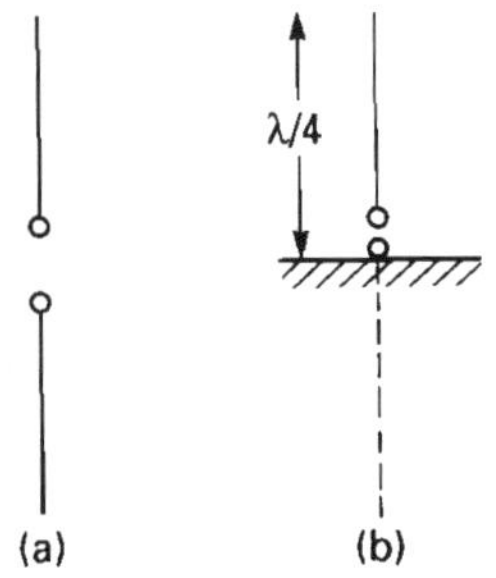

Figure 6.1.6 A λ/4 antenna, in which the ground plane and the image play a vital part.

To determine the length of the radiating element, consider Figure 6.1.7. The dipole has a capacitance between its two halves, and the current flow in the conductor establishes inductance. To energize the dipole from a source at the centre, the capacitance must be fully charged, and it will then discharge through the inductance of the system in the manner of a tuned circuit. Clearly, a finite time is required to charge and discharge C, and the system needs to be tuned to the energizing source.

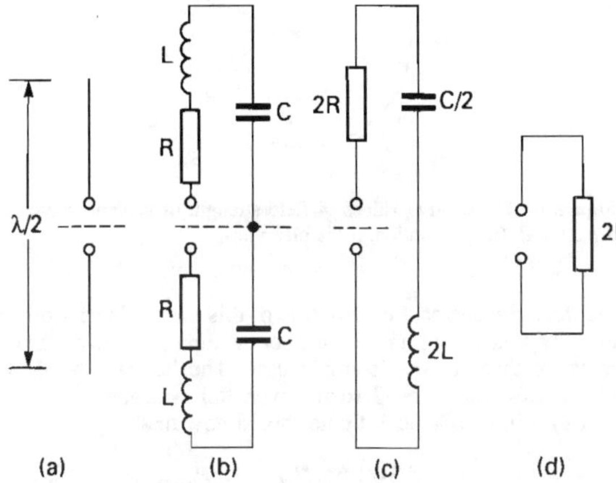

Figure 6.1.7 A half-wave dipole (a), and three equivalent circuits.

It so happens that a very thin conductor requires a quarter cycle of the resonant period for energy from the source to fully charge the capacitance in each half of the dipole. The discharge of the capacitances will require another quarter cycle to become complete, and the energy in the system will be in phase with the source at the input terminals. Each half of the system requires to be a quarter-wavelength long for this optimized performance.

In practice, the practical radiating system cannot consist of a very thin conductor; a practical dipole must be self-supporting. The dipole in practical form will have too great a capacitance, but this can be compensated for by reducing the inductance (i.e. the overall length). The practical dipole thus has an overall length less than a half-wavelength by a factor determined by the ratio of operating wavelength to the conductor diameter. The Marconi *monopole*, similarly, requires to have a height less than a quarter-wave. Figure 6.1.7 shows the balanced dipole at (a), and at (b) the capacitances and inductances are shown as discrete component values; (b) also shows the addition of R, representing the radiation resistance. The equivalent circuit of the resonant radiator is in (c) and (d) and, considering the system as a tuned system, it is seen that the resonant frequency for the dipole is given by:

$$\omega = \frac{1}{\sqrt{[(C/2) \times 2L]}}$$

and, for the Marconi monopole, the frequency will have the same value, i.e.:

$$\omega = \frac{1}{\sqrt{(LC)}}$$

6.1.2.3 Antenna impedance

The tuned radiator has an impedance which, at resonance, is resistive and is in the region of 73 ohms. This value is derived from an integration of the radiated power over the whole inside surface of the sphere, considered earlier in Section 6.1.1.5. Note that the impedance of the Marconi monopole will be half of the dipole value. In practice, there are several factors that can affect the value of antenna impedance.

The presence of the ground cannot be ignored, and dipole impedance will vary with its height above ground (see Figure 6.1.8). Precise values are not given because they will depend on the ratio of diameter:wavelength. When this ratio is small, the impedance may not be affected very much, but at uhf the diameter may be an appreciable fraction of a wavelength, and this could result in an impedance lower in value than that for a very thin conductor. Another factor that can affect an antenna impedance is the presence of other conductors or antenna elements in the near vicinity of the radiator. The effect is, generally, to reduce the radiator impedance.

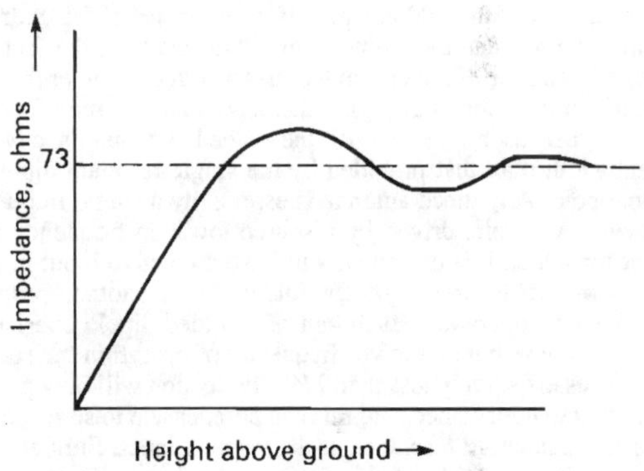

Figure 6.1.8 Variation of dipole impedance with its elevation.

6.1.2.4 The folded monopole and dipole

Folded versions of the half-wave dipole and the quarter-wave monopole antennas provide a useful means for changing the basic impedance. Figure 6.1.9(a) shows a λ/4 Marconi or monopole system; (b) shows it in more diagrammatic form with an input voltage *V*. In practice, the spacing between the two halves of the loop is very small fraction of the operating wavelength. At (c) is shown the generator *V* replaced with three generators, each with a magnitude *V*/2, and so phased as to produce the same effect as at (b). Generators 1 and 2 are in phase and provide the voltage *V* of (b). Generators 2 and 3 are in phase round the loop, but this loop is a λ/4 section, shorted at the top, thus presenting a high impedance at the generator end, and the resulting current is near zero.

Generator 1 alone feeds, in parallel, the two limbs of the antenna, the current being *V*/2 divided by Z_1, where XZ_1 is the impedance of a λ/4 antenna. This current divides equally between the two elements so that the current in the left-hand branch is half of the total, i.e. *V*/4 divided by Z_1. If Z_1 is 37.5 ohms for a single monopole element, the impedance of the driven element is given by:

$$Z = V/I = 4 \times Z_1$$

This is approximately 150 ohms.

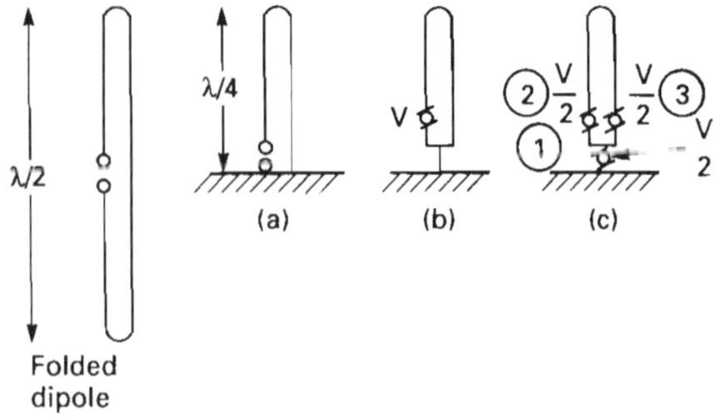

Figure 6.1.9 A monopole with its equivalent circuit.

The folded antenna thus introduces a multiplying factor of four. The folded dipole has an input impedance of 300 ohms. The multiplying factor may be varied by making the diameters of each half of the fold unequal. If the diameter of the undriven half is larger than the diameter of the driven half, the factor is larger than four. If the driven half has a larger diameter than the undriven half, the multiplying factor will be less than four.

Another useful feature of the folded antenna is a wider bandwidth than that provided by the single resonant dipole or monopole. Any tuned antenna is essentially a single frequency device. A dipole, driven by a source lower in frequency than that for which it is designed, will have a reactive input impedance of the form $R - jX$. In folded form, another property comes into operation. Each half of a folded dipole consists of a λ/4 section, but at a lower frequency of operation the section becomes effectively less than λ/4. The section will now present an inductive reactance, and this can compensate in some degree for the capacitive reactance at the antenna input. Similar compensation is provided for operation at a frequency above the designed frequency of the antenna.

6.1.2.5 Directivity, the polar diagram and antenna gain

We have considered the concept of radiation from an isotropic source at the centre of a sphere, and the effects of such radiation at the surface of the sphere. The field strength would be the same at any point on the sphere surface. It can be seen from Figure 6.1.2 that a practical dipole cannot radiate in this manner.

The intensity of the magnetic field is at a maximum along a plane normal to the axis of the dipole through its centre. At the ends of the dipole, the current is zero, no magnetic field can exist, and radiation from the ends must be zero. Thus, radiation from a dipole could be depicted by Figure 6.1.10(a), which shows a *polar diagram* in which the dipole axis lies along the solid line, and the length R depicts the relative field strength at any angle ϕ, compared with the maximum value at 0°. The variation for a dipole is approximately as $\cos\theta$. Looking down in plan on the dipole, radiation is uniform in all directions, as shown in (b). The polar diagram for this case would be a circle.

In a practical case, using a dipole as a radiator, a surface of uniform radiation field strength would be represented three-dimensionally by a torus instead of a sphere. The dotted line in Figure 6.1.10(a) represents the ground plane that establishes the monopole.

The radiant energy from an isotropic source and from a dipole would be the same for the same power fed to each, but

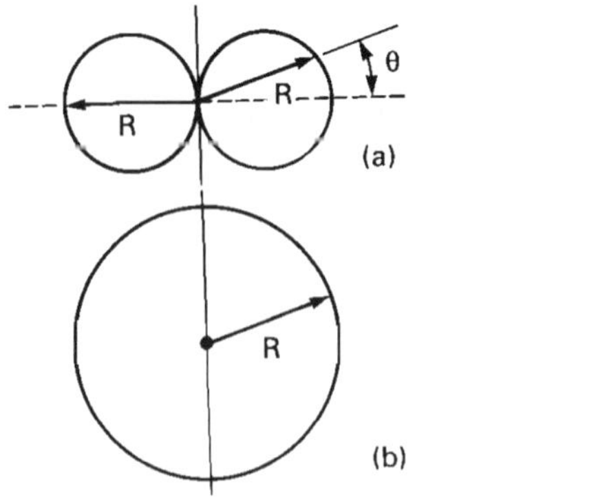

Figure 6.1.10 Relative values of field strength of a dipole along the direction R. (a) is elevation, (b) is plan view.

the three-dimensional distribution of this energy is very different. The field strength maximum for the dipole would be greater in value than for the isotropic case. The increase would be about 1.64 times or 2.16 dB over the isotropic. Equation (6.1.8) using a practical dipole should now read:

$$E = \frac{5.5\sqrt{1.64}\sqrt{P_a}}{d} = \frac{7\sqrt{P_a}}{d}\ \text{V/m}$$

Increases of this type are often referred to as *antenna gain*. In this case the gain is with respect to an isotropic system, but other antenna systems may refer to their gain in terms of a single dipole.

Any vertical radiator will have a circular polar diagram when viewed from above. The half-wave dipole can be used horizontally, in which case the radiation pattern, viewed from above, will be the *figure-of-eight* shown in Figure 6.1.10. It is now highly directive with two directions where maximum field strength is experienced, and two directions of zero field strength.

Most modern broadcast antennas require better directivity than can be provided by a simple dipole system. Unidirectivity is required to some extent, i.e. a single maximum (or minimum) field strength directed into specific directions, either for obtaining the best possible field strength in the maximum direction, or for reducing possible interference to other services in a zero direction. This can be achieved by the use of more than one antenna element as a radiating system or *array*. If two vertical radiators (A and B in Figure 6.1.11) are each fed with the same

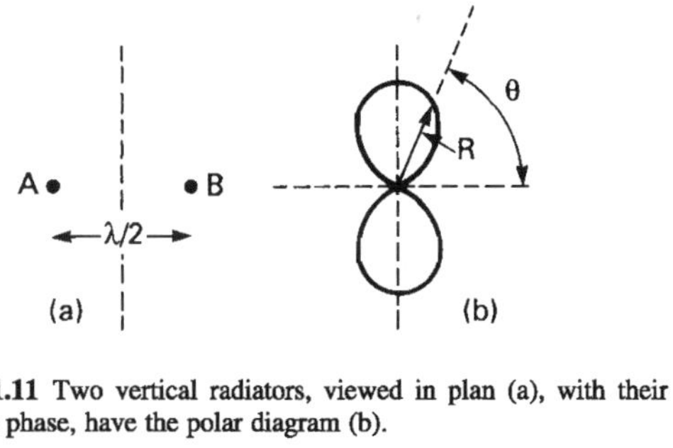

Figure 6.1.11 Two vertical radiators, viewed in plan (a), with their currents in phase, have the polar diagram (b).

power with their currents in phase, and spaced λ/2 apart, they will each radiate uniformly in all directions, but their fields will cancel in the direction AB and add in the direction at right-angles. These field interactions will result in the overall radiation pattern for the array, as shown in Figure 6.1.11(b). By varying the spacing, the current magnitudes and their phase relationship, it is possible to obtain an infinite variety of polar diagrams with maxima and minima in various angular positions.

One particular combination has a special significance. If the spacing is λ/4 and the currents have a phase difference of 90°, Figure 6.1.12 indicates how the fields will add in one direction but will cancel in the other direction. This array is unique in that there is only one maximum and only one zero. The array is unidirectional, and a further change of 180° in either of the

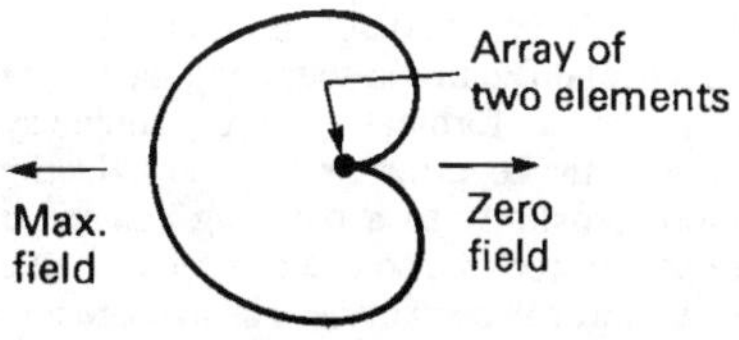

Figure 6.1.12 The polar diagram provided by the two-element array of Figure 6.1.11 when one element is directly driven from a power source and the other is energized from the induction field of the driven element.

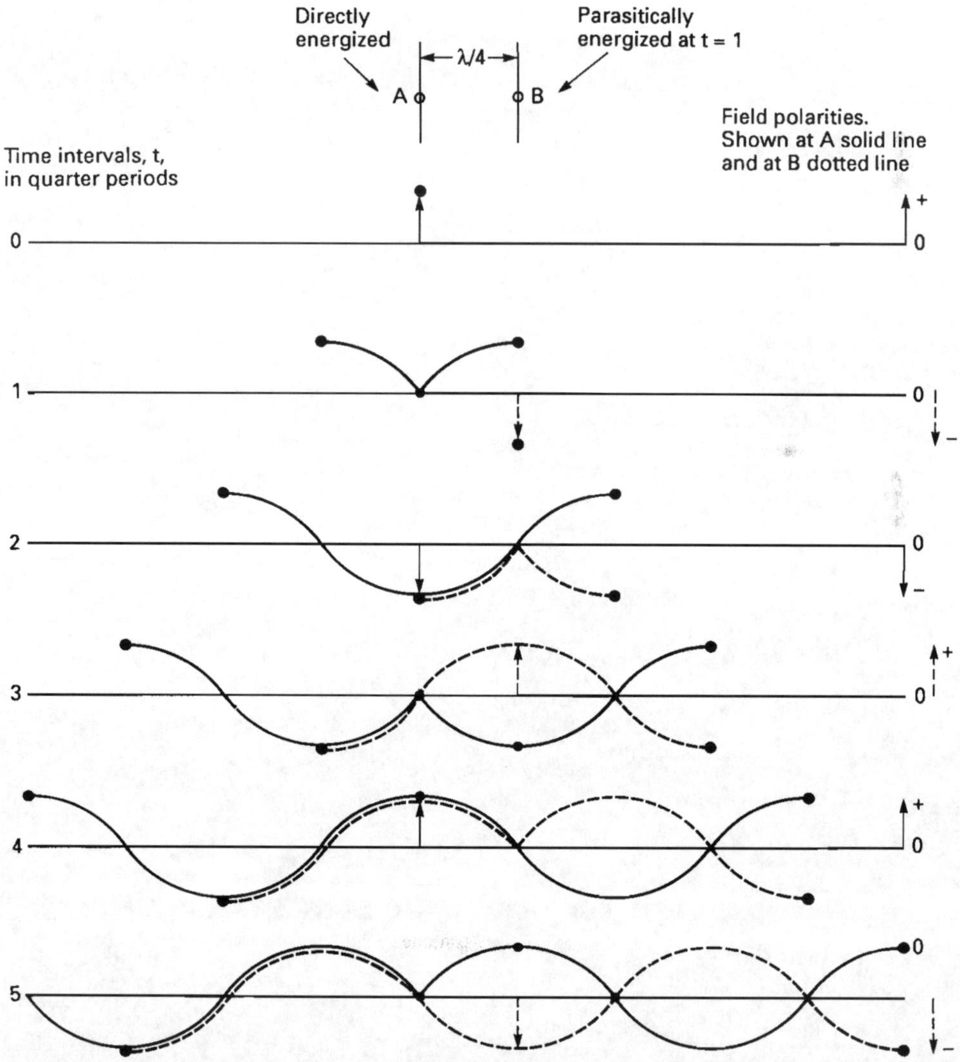

Figure 6.1.13 A cycle commences at $t = 0$, with the field at A having a maximum positive value. This field radiates uniformly in all directions and reaches B at $t = 1$, energizing B to a maximum negative value. Both radiators are now radiating uniformly in all directions, as shown at $t = 2$. From this time onwards, the combined fields cancel in the direction A to B, but are in phase in the direction B to A.

current phases will turn the diagram through 180°. There would be some practical difficulty in supplying the two radiators with the correct phase but, fortunately, a very simple system can be used to achieve the required result. The induction fields that exist in close proximity to a radiating antenna element normally take power from the source and return it to the source as the fields vary through each cycle. The average power per cycle is zero. If only one of the two elements shown in Figure 6.1.11 is energized from a source, the second element can be energized by the induction fields. Such an element would be termed *parasitic*. Figure 6.1.13 shows how the fields between the two elements, one energized and the other parasitic, will interact. The two elements are shown $\lambda/4(=90°)$ apart, and the driven element A is assumed to start with a maximum positive current at $t=0$. At $t=1$ the field maximum will have travelled through space by a distance $\lambda/4$ all round the element A. The solid lines in successive quarter-cycle time intervals will show the field magnitudes in the space surrounding the element A. The broken lines show the field contribution by the parasitic element B. At $t=1$ the radiated maximum field from A reaches B, and induces a maximum negative current in B (*Lenz' law*), which then commences to radiate in the same manner as A. The B field variations with time are seen to assist the A filed radiating in the direction B to A, and cancel in the direction A to B, thus producing the polar diagram of Figure 6.1.12.

B L Smith
Chief Technical Writer, Thomson-CSF

6.2 Thermionics, Power Grid and Linear Beam Tubes

The first requirement for working in the microwave domain is a microwave source. Only when quasi-monochromatic sources began to appear in the 1930s, did microwave operations become possible.

6.2.1 Thermionic Tubes

These sources were *microwave tubes*, and they are still with us in spite of the appearance in the 1960s of solid-state microwave sources. Tubes are by far the most powerful sources, especially at the highest frequencies of the microwave spectrum (up to 1 THz). If only for this reason, microwave tubes are here to stay for a long time.

6.2.1.1 Common principles

6.2.1.1.1 Difficulties at high frequencies

The earliest radio frequency tubes were triodes and multi-grid tubes. At low frequencies, the principle of operation of these tubes is quite simple. The rf signal is applied to the grid facing the cathode, which is a source of electrons sensitive to the applied electric field, connected to the negative pole of a power supply. As the field changes, the current emitted by the cathode changes in proportion. The electron flow, crossing the grid, reaches an electrode connected via the impedance of the useful load to the positive pole of the power supply. This impedance thus conducts the electronic current and develops an rf voltage at its terminals. This voltage is usually much larger than the one applied on the grid, so that the tube presents a large gain (15–20 dB).

However, as the frequency of operation is increased, at frequencies of the order of 100 MHz, the behaviour of the amplifier begins to deteriorate. The deterioration has two main origins:

• *Parasitic impedances*, which are negligible at lower frequencies, become important. These are the reactances due to the inter-electrode capacitances as well as the stray capacitances between connections. Even more important are the inductances of the connections in series between circuits and electrodes. Furthermore, these connections, whose lengths may be comparable to a quarter of free space wavelength, begin to radiate. Thus, amplifiers and oscillators become extremely difficult to adjust. Parasitic oscillations are often observed. Gain and efficiency are severely affected.

• *Electron transit times* between electrodes, especially between cathode and modulating grid, become of the same order of magnitude as the rf period. These transit times are simply due to the fact that electrons are massive particles and obey the laws of classical (or in the case of high energies, relativistic) dynamics. While electrons move in the inter-electrode spacing, they are submitted to an rf field which varies appreciably during their transit. If the frequency is sufficiently high, the field may even reverse itself during the transit time. If this happens in the cathode/grid region, some of the electrons will be slowed down and then reflected towards the cathode, thus reducing the net current extracted, while others will oscillate. In both cases, a large amount of rf source energy is absorbed by the electrons. Similar phenomena, though less pronounced because the electrons are faster and transit times smaller, take place in the grid/anode region. This results in a sharp drop in gain, power and efficiency.

A third source of deterioration can be mentioned. *Ohmic losses*, due to skin effect, increase as the square root of the frequency. This factor becomes really critical at higher microwave frequencies.

6.2.1.1.2 Solutions

To solve the problems summarized in Section 6.2.1.1.1, a few principles may be followed, which are basic to the concept of microwave tubes.

6.2.1.1.2.1 Integration of microwave circuits

Microwave circuits become an integral part of the tube. They are usually completely included inside the vacuum envelope.

Active electrodes are part of the circuits which are no longer made of lumped elements (self inductances, capacitors, resistors, transformers, etc.). They are instead distributed, being of resonant cavity or waveguide type. They are usually completely shielded, so that they do not radiate. They are connected to external sources and loads by means of coaxial cables and waveguides which also do not radiate. All stray capitances and inductances are reduced to a minimum and become part of the microwave circuit.

These circuits show some improved qualities compared with their low frequency, lumped element counterparts. They have no radiation, lower losses (intrinsic Qs of the order of 10,000 are common at S-band, whereas they seldom reach 1000 for lumped element circuits at low frequencies) and constant geometry resulting in constant rf properties. In fact, these advantages are such that, for high power transmitters, the tendency is now to replace lumped element circuits at as low a frequency as possible (30 MHz) by cavity-type resonators integrated with the tetrode. This trend is limited only by the bulk of the resonators which becomes considerable at low frequencies.

6.2.1.1.2.2 Reduction of transit time

An obvious step to alleviate the problems caused by transit time is to reduce it as much as possible by decreasing the inter-electrode spacing. For instance, by reducing the cathode/grid spacing down to 60 µm, triodes have been operated up to C-band (6 GHz). This approach, however, is severely limited by a number of factors:

- It is very difficult to obtain and to maintain such short distances over a wide area, especially since the cathode is at a temperature of 700–800°C (for an oxide coated cathode).
- The cathode/grid capacitance is inversely proportional to this distance. This limits the frequency of the input circuit.
- The tube becomes microphonic due to capacitance variations under mechanical vibrations.

Because of these limitations, grid modulated tubes (triode, tetrode, klystrode) operate normally in the range of 0–1 GHz. A few tubes, working mostly under pulsed conditions, reach 3 GHz.

6.2.1.1.2.3 Use of transit time

By far the most effective way to counteract the negative aspects of transit time is to make use of its positive consequences. Common features of tubes whose operation is based on the use of transit time (i.e. klystrons, travelling wave tubes and crossed field tubes) and which constitute the overwhelming majority of microwave tubes, are:

- Electron beams are launched unmodulated and accelerated until they reach a constant average velocity.
- At this velocity, the electrons drift for several rf periods (e.g. klystron and magnetron) or even several tens of periods (travelling wave tube and crossed field amplifier) while interacting with the microwave circuits.
- The general mechanism for intensity modulating the beam is velocity modulation. In the klystron, the beam passes a narrow gap, which is part of a resonant cavity, across which a longitudinal rf electric field is present. Depending on the phase at which they cross the gap, electrons are either accelerated or decelerated in a periodic fashion. Drifting in a field free region, fast electrons catch up the slower ones, forming periodic electron bunches which are the major part of the current modulation; a much smaller part originates directly in the velocity modulation itself. In travelling wave interaction, the beam is accompanied by a quasi-periodic field pattern which, in a frame of reference moving with the beam, is almost static. Here again, some electrons are accelerated while others are decelerated and bunches appear.
- The rf energy exchange takes place in vacuum, without any impact of the electrons on the rf structure. In the klystron, the bunches cross the gap of a final cavity resonator and are slowed down by the field developed across the gap. Since energy is conserved, this decrease in the kinetic energy of the beam is transformed into rf energy in the cavity. Similarly, in travelling wave tube interaction, the equilibrium position of the bunches is such that they are submitted to a retarding electric field, so that they are continuously slowed down and give up energy to the rf circuit all along the interaction space.
- After interaction has taken place, the spent beam is usually collected on an independent electrode, the *collector*, having good thermal dissipation properties. Only in crossed field tubes such as the magnetron or crossed field amplifier does the beam, due to the constraints imposed by the geometry of the tube, eventually impinge the microwave circuit which is used as a collector.

6.2.1.2 Microwave circuits for electron tubes

As was seen in Section 6.2.1.1, the microwave circuits used in microwave tubes are cavity resonators and periodic slow wave structures. Their properties can be derived from the application of Maxwell's equations subject to the boundary conditions imposed by their geometry. Demonstrations of their properties can be found in Refs 2–4.

6.2.1.2.1 Resonant cavities

Any empty volume surrounded by a good conducting material (usually metallic) can be considered as a resonant cavity. If the conductivity is large enough for the losses to be considered negligible, the cavity, when excited by a microwave source, will exhibit electromagnetic fields only at discrete frequencies which form an infinite spectrum. These frequencies are determined by the solution of the following system of equations:

$$\Delta\vec{V} + k^2\vec{V} = 0 \tag{6.2.1}$$

$$\mathrm{div}\,\vec{V} = 0 \tag{6.2.2}$$

where Δ is the Laplacian operator

$$\frac{\delta^2}{\delta x^2} + \frac{\delta^2}{\delta y^2} + \frac{\delta^2}{\delta z^2}$$

$k = \omega/c$ and $\omega = 2\pi f$, the angular frequency.

$\vec{V}$ is a vector which can be either the vector potential $\vec{A}$, the electric field $\vec{E}$ or the magnetic field $\vec{H}$. This system is associated with the boundary conditions which are:

- zero tangential field at the boundary for $\vec{A}$ or $\vec{E}$: $\vec{A} \times \vec{n} = 0$ or $\vec{E} \times \vec{n} = 0$.
- zero normal field at the boundary for $\vec{H}$: $\vec{H} \cdot \vec{n} = 0$, $\vec{n}$ being a unit vector normal to the boundary.

The system identified in equations (6.2.1) and (6.2.2), subject to the boundary conditions, has solutions only for a set of discrete real values of k which are its proper or resonant values k_n.

For each k_n, the field configuration, be it electric or magnetic, can be computed as a solution of these equations. These solutions are called *resonant modes* and designated by the mode number so that peak field values at frequency $\omega_n = k_n c$ are $\vec{E}_n$ and $\vec{H}_n$.

These fields form a complete orthogonal set. It can be shown that

$$\int_v \vec{E}_m \cdot \vec{E}_n \, dv = 0 \quad \text{when} \quad m \neq n$$
$$\int_v \vec{E}_m \cdot \vec{E}_n \, dv \neq 0 \quad \text{when} \quad m = n \tag{6.2.3}$$

Most cavities in practical use have at least an axis of symmetry. With respect to this axis Oz, one can define transverse electric (TE) and transverse magnetic (TM) modes. The mode number is then defined by the prefix TE or TM followed by three numbers. For rectangular cavities, these are the number of variations of the field (quasi-half waves) in each direction Ox, Oy and Oz respectively. In the case of cylindrical coordinates, the first number refers to the variations along r, the second in the θ direction, the third in the z direction.

6.2.1.3 Common technology: cathodes

The cathode is the source of electrons in every electron tube. The current density of electron emission from the cathode ranges from milliamperes to ten amperes per square centimetre of cathode area. Three mechanisms for emission of electrons from cathodes are usually used. These are:

- thermionic emission;
- secondary emission;
- field emission.

Most microwave tubes, such as klystrons and travelling wave tubes, employ only thermionic emission; in the power grid tubes both thermionic and secondary electron emission are used.

No real cathodes meet the ideal characteristics. For instance, the impregnated cathode which consists of porous tungsten and magnesium impregnated with barium calcium aluminates must be heated to a temperature in the vicinity of 1050°C to produce an appreciable amount of electron emission. At that temperature, the current density is limited to the order of 1 A/cm^2. Because of the necessity for high temperature, the key constituents of the cathode responsible for the emission evaporate and therefore are depleted from the surface down to a region where the internal pressure is not high enough to allow the active element to migrate towards the surface.

6.2.1.3.1 Thermionic emission

Electron emission from the solid results when electrons in the solid have sufficient energy directed towards the surface to overcome the potential barrier (work function) to escape from the solid into vacuum. The thermionic electron current can be predicted by using the density of electron energy states and the probability of their occupation.

When the potential of the anode placed in front of the cathode becomes less negative, the number of electrons entering the space between cathode and anode increases, and finally a negative space charge of detectable density is formed in front of the cathode. This space charge causes an increase in potential in front of the cathode, adding to the potential barrier corresponding to the work function. Only faster electrons can overcome the barrier, while the slower ones will overcome the potential barrier but will have to return to the cathode after having penetrated some distance into the vacuum. At higher anode voltage, the potential barrier due to space charge disappears, and all electrons emerging from the cathode surface reach the anode (the saturation range). The equation for the saturated current density formulated by Schottky is:

$$J_s = J_0 \exp\left(\frac{4.4}{T}\sqrt{\frac{V}{d}}\right) \tag{6.2.4}$$

with $$J_0 = AT^2 \exp\left(\frac{e}{\kappa}\frac{\phi}{T}\right)$$

where V is the applied voltage (V),
d the distance between anode and cathode (cm),
A a universal constant, value 120 cm^2 T^{-2},
e electron charge (C),
ϕ work function (V),
κ Boltzmann constant (W S^{-1} Kexp)
T temperature (K).

The most significant aspect of equation (6.2.4) is the exponential variation of the current density with the work function and the reciprocal of the temperature (see Figure 6.2.1).

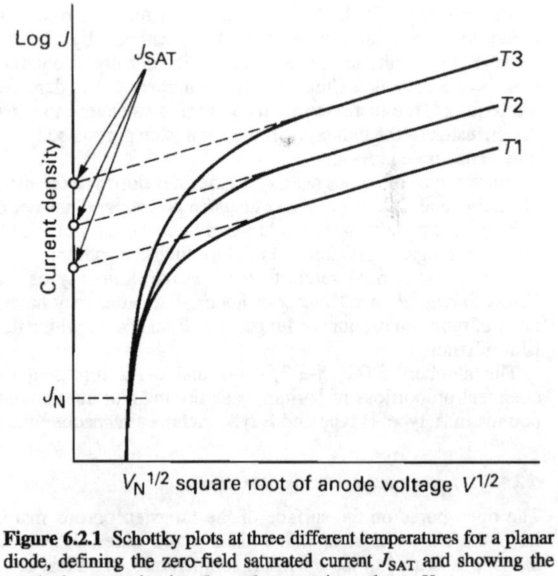

Figure 6.2.1 Schottky plots at three different temperatures for a planar diode, defining the zero-field saturated current J_{SAT} and showing the nominal current density, J_N, at the operating voltage, V_N.

6.2.1.3.2 Thermionic cathodes

The first thermionic cathodes used in quantity were those employed in the early radio tubes. They consisted of pure tungsten or carburized thoriated tungsten filaments. The latter was considered to be the first type of dispenser cathode due to the fact that the thorium was dispensed by diffusion through the bulk of the wire and formed a monolayer on the emitting surface. They were directly heated cathodes, i.e. the filament responsible for the electron emission was heated by passing a current through it. The operating temperatures ranged from 2200 K for pure tungsten to 2000 K for carburized thoriated tungsten.

The major progress was the introduction of oxide cathodes by Wehnelt in 1904. He observed that a platinum filament covered with alkaline earth material was emitting electrons in vacuum at temperatures in the range of 800–1000°C. The improvement was the use of a nickel base material in various forms, coated with a mixture of barium, calcium and strontium oxides. These were used for more than 60 years at current densities of a few to a hundred of milliamperes per square centimetre. The requirement of higher current densities and cathodes withstanding more severe environment led to impregnated cathodes, discovered at Philips by Lemmens and Looges in the early 1950s.

They used a porous tungsten matrix in which was impregnated a barium aluminate having the eutectic composition 5BaO, $2Al_2O_3$. Levi improved the quality by introducing calcium oxide into the eutectic, and the impregnant became 5BaO, 3CaO, $2Al_2O_3$ (known as the *B type*). The electron emission was enhanced by a factor of 5. Some competitors proposed other compositions such as 4BaO, 1CaO, $1Al_2O_3$ (the *S type*) or 6BaO, 1CaO, $2Al_2O_3$ (*Brodie's composition*).

Current densities obtained from such impregnants range from 0.1 to 1 A/cm^2 when operated from 1000 to 1100°C. The improvement in current densities from B, S or the third variety of impregnant was obtained with the *M type* cathodes.

Basically the M type is a B or S type cathode covered with a film of several thousand angstroms of one of the metals of the platinum group (Os, Ir, Re). The effect of the thin film is to reduce the operating temperature of the cathode by approximately 80°C for the same electron emission density as obtained with B or S cathodes. Unfortunately, if higher current densities are required, the increase in temperature is sufficient to cause the diffusion of the thin layer into the tungsten porous body and may cause pore clogging.

A novel type of porous matrix was then developed. It consisted of mixing and sintering together tungsten powder and powder of a metal of the platinum group in weight concentration typically 50:50 for tungsten osmium and 8:20 for tungsten iridium.

MM in Table 6.2.1 refers to these *mixed matrix* types and CMM to *coated mixed matrix* cathodes. The latter may have a body of tungsten osmium or tungsten iridium coated with a thin film of osmium.

The numbers 5-0-2, 5-3-2, 4-1-1 and 6-1-2 represent the chemical proportions of barium, calcium and aluminium compounds in A type, B type and S type cathodes respectively.

Table 6.2.1 Impregnated cathode types

Origin	*Nature of the reservoir*	*Impregnated cathodes*	*Designation*
1950			
Philips Co			
			L
Lemmens			
Jensens			
Loojes			
1955			A
			B
			S
Philips			
Levi			
1955			
Siemens			
from			
Kataz			
1966			M
Philips			
Zalm			
1976			MM
Varian			
1979			CMM
TH – CSF			
1979			
Telefunken			
1980			SP
TH – CSF			

6.2.1.3.3 *Impregnated cathode operation*

The open pores on the surface of the tungsten porous matrix have the shape of slots rather than round holes. When the cathode is heated to the operating temperature, chemical reactions take place in the matrix between the barium and calcium aluminates and the tungsten, and free barium is generated that migrates towards the surface and spreads on it to form an almost complete monolayer (Figure 6.2.2(a)). As the cathode is used at nominal temperature, the slot type pores tend to get smaller and smaller and wind up as separate small round pores. This is due to two cumulative phenomena: the chemical reaction which leaves residues and the thermal reconstruction of the surface.

Barium which was dispensed from very near the surface at the beginning of life comes, after a few thousand hours of operation, from deeper and deeper regions of the matrix, and comes in smaller and smaller quantities due to a drop of internal pressure. The emission becomes more and more patchy, and the work function distribution varies from a sharp distribution having a σ of 0.075 to a broader distribution where the σ is 0.25.

6.2.1.3.4 *Life considerations*

In most applications long life is required, either to minimize the replacement costs (e.g. a ground station) or limit redundancy cost (satellite applications). With B and S type cathodes set at 0.5–1 A/cm^2 and operating temperatures of 950–1000°C, lifetimes of $3 \times 10^5 - 10^6$ hours can be expected. With M and MM type cathodes, higher current densities can be expected (1–3 A/cm^2 for lifetimes exceeding 10^5 h). CMM type cathodes allow operation at 4–6 A/cm^2 with the same life expectancy and temperature in the range of 1000–1050°C. Figure 6.2.3 charts the life against nominal current densities extrapolated from experimental values at shorter lives.

6.2.1.3.5 *Secondary emission*

Another form of electron emission that plays an important role, especially in power grid tubes, is secondary emission. This occurs when a surface is bombarded by electrons or ions of appreciable kinetic energy.

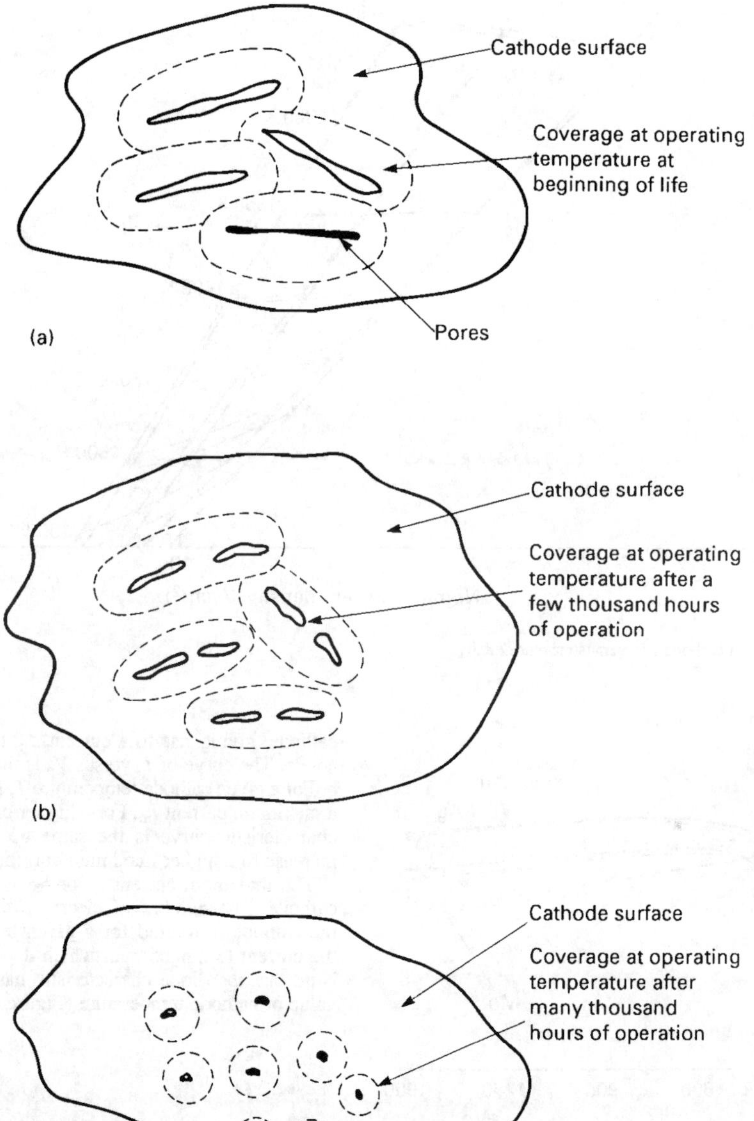

Figure 6.2.2 Evolution of impregnated cathode pores over lifetime.

Upon striking the material surface, a primary electron shares its kinetic energy with other particles in the immediate vicinity. One or more electrons in the material gain enough energy to be emitted.

The number of secondary electrons emitted from a surface depends mostly on the nature of the surface, and also on the energy and the striking angle of the impinging electrons.

The average ratio, δ, of secondary electrons emitted to the number of primary electrons producing them varies from less than one up to eight or ten. $\delta = i_s/i_p$. Figure 6.2.4 shows values of this ratio for some common metals. The characteristics of secondary emission are known from experiments; however, the process of emission is rather complex and is difficult to deal with theoretically. The number of secondary electrons is low at low primary energy (which is easily understandable) and also very low at high primary energies. The reason for this is that the high energy primary electrons penetrate very deeply into the material, where they excite electrons which have a very small probability of reaching the surface to escape.

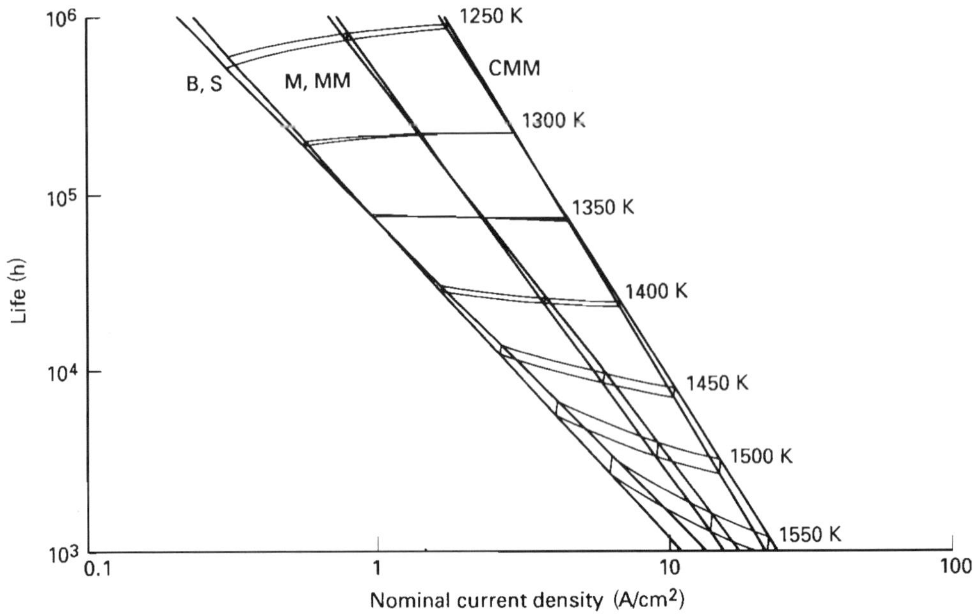

Figure 6.2.3 Extrapolated cathode life versus current density.

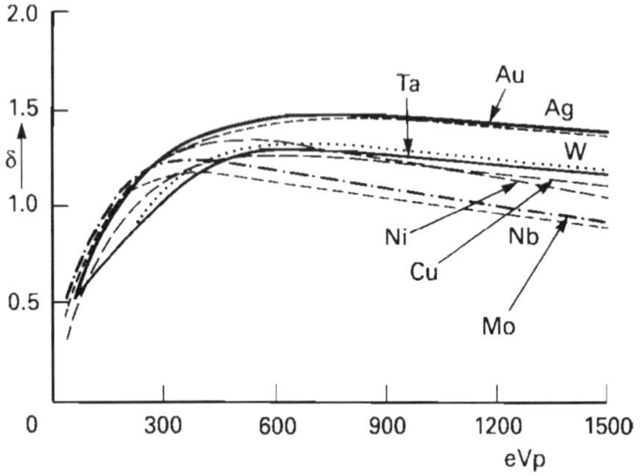

Figure 6.2.4 Secondary electron emission coefficient for high work function metals (after Warnecke).

6.2.2 Power Grid Tubes

6.2.2.1 Vacuum diodes

Even though the use of vacuum diodes has declined considerably, a knowledge of their properties aids understanding of the principles of operation of power grid tubes.

A vacuum diode consists of two electrodes in a vacuum: a thermoelectronic *cathode* and an *anode*. When the cathode temperature is high enough, it can emit electrons from the surface. The anode is brought to a positive voltage, V_p, with respect to the cathode, and thus attracts electrons emitted by the cathode, giving rise to a current, I_p, between the cathode and anode. The curve of I_p versus V_p is the *diode characteristic*.

For a given cathode temperature T, I_p increases with V_p up to a maximum current I_m. For a higher cathode temperature T the characteristic curve is the same up to I_m, but continues to increase to a higher maximum current $I_{m'}$.

The maximum current I_m or $I_{m'}$ is due to saturation of the cathode. The number of electrons that can be furnished by the cathode is limited for a given temperature, thus limiting the current to a maximum which depends on the temperature. Note that the diode characteristic has the same form for any value of cathode temperature (Figure 6.2.5).

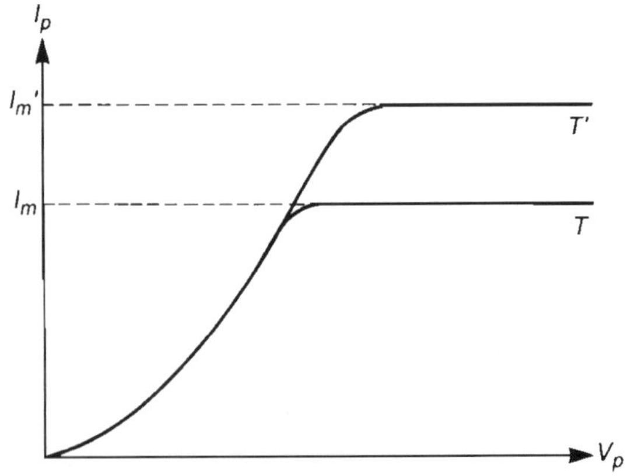

Figure 6.2.5 Diode characteristic curves.

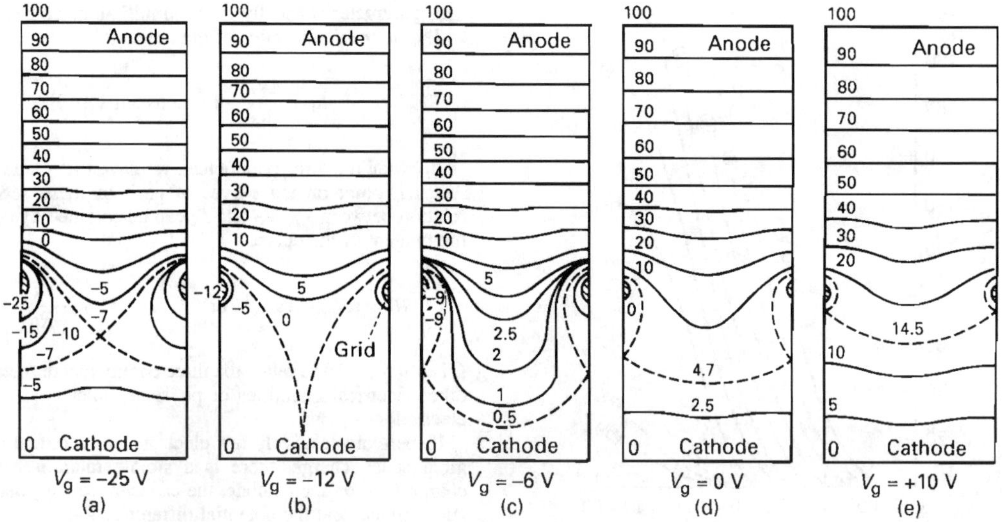

Figure 6.2.6 Equipotential curves.

The diode characteristic is expressed in its most simple form by the Child–Langmuir relation:

$$I = KV^{3/2} \tag{6.2.5}$$

where K is called the *diode perveance*.

The anode thus attracts a given quantity of electrons, and any excess electrons stay in the vicinity of the cathode to form a *space charge*, creating a minimum potential near the cathode.

This space charge, which depends on the cathode temperature T and the voltage V, regulates the electrons. Those electrons having an initial velocity sufficient to penetrate the space charge region will be stopped by the space charge and will be confined to the cathode region.

6.2.2.2 The triode

A third electrode, in the form of a grid between the cathode and the anode of a diode, was introduced in 1907 by Lee de Forest.

The potential at the surface of the cathode is the superposition of the anode potential and the grid potential. When this potential is negative, there is no emission. This may be achieved by a sufficiently negative grid potential.

When the grid is less negative, certain regions of the cathode will 'see' a positive potential and be able to emit electrons which will be attracted to the anode. By further decreasing the negative potential of the grid, the area of the emitting zones increases. Beyond a certain point, all of the cathode surface will emit. When the grid potential becomes positive, some of the electrons emitted will be collected by the grid, giving rise to a grid current.

With small variations of grid voltage, the anode current can be controlled over a wide range from zero to maximum. This is the basic phenomenon which allows a large anode signal to be obtained from a small signal applied to the grid through appropriate circuits between the electrodes.

6.2.2.2.1 Characteristic curve sets

For each set of voltages V_a and V_g, there will be a corresponding anode current and grid current. Three types of curves can be drawn:

- currents as a function of anode voltage V_a, for different values of grid voltage V_g, called the *Kellogg diagram* (Figure 6.2.7);
- currents as a function of grid voltage, for different values of anode voltage; this type of curve set is rarely used (Figure 6.2.8);
- grid voltage versus anode voltage, for different constant current values; this is the most commonly used set of curves (Figure 6.2.9).

From these curves, the characteristic coefficients of the tube can be defined for any operating point. From the three principal

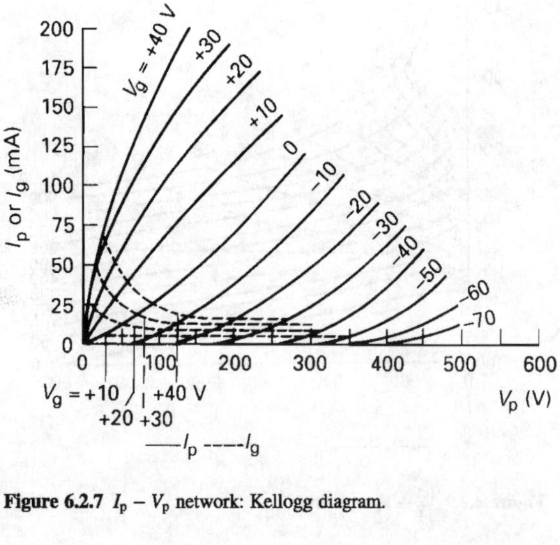

Figure 6.2.7 $I_p - V_p$ network: Kellogg diagram.

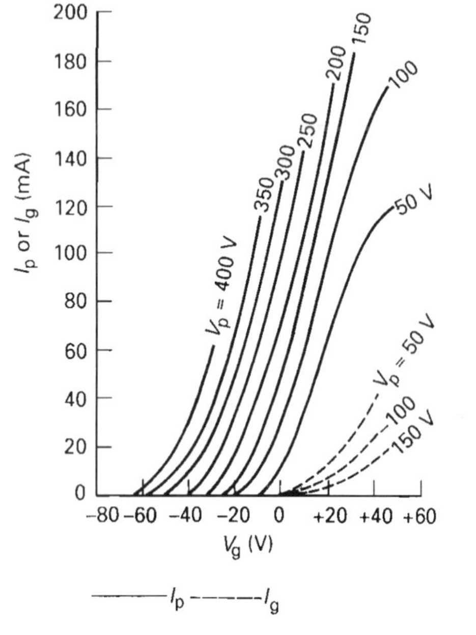

Figure 6.2.8 $I_p - V_g$ network.

variables I_A, V_A and V_G, three parameters can be defined supposing one of the variables constant:

The amplification, for I_A = constant:

$$\mu = \frac{dV_A}{dV_G}$$

This unitless coefficient gives the voltage amplification of the tube.

The transconduction or tube slope:

$$s = \frac{dI_A}{dV_g} \quad \text{for constant } V_A$$

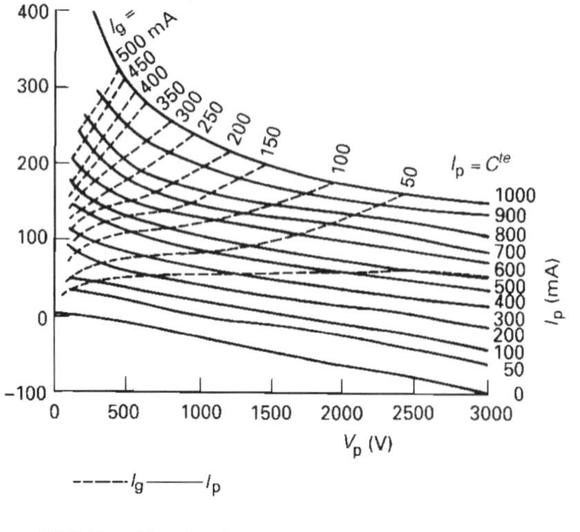

Figure 6.2.9 $V_p - V_g$ network.

This parameter is usually given in milliamperes per volt.

The internal resistance of the tube:

$$R_i = \frac{dV_A}{dI_A} \quad \text{for constant } V_G(\Omega)$$

Between the three parameters, we have the relation $\mu = sR_i$. At each point on the curves of each of these sets, a linear relation between V_A, V_G and I_A can be obtained by considering the tangent of the curve:

$$R_i = \mu V_G + V_A \quad \text{or} \quad I_A = sV_G + \frac{V_A}{R_i} \qquad (6.2.6)$$

It is often useful to relate the three parameters defined above to other electrical quantities or physical dimensions of the tube electrodes.

If there are relatively few electrons emitted, and so a negligible space charge, there is a simple relation between the charge Q_{KG} on the cathode, the capacitance C_{KG} between grid and cathode, and the potential difference V_G:

$$Q_{KG} = C_{KG} V_G$$

The charge Q_{KG} is the integral of the surface charge density σ_{KG} over the cathode surface. The electric field E_{KG} at the surface of the cathode is thus given by:

$$E_{KG} = 4\pi\sigma_{KG}$$

Similarly, neglecting space charge, $Q_{KA} = C_{KA} V_A$ gives the relation between the anode/cathode capacitance C_{KA} and voltage V_A, giving rise to an electric field E_{AK}. The total electric field $E_{KA} + E_{KG}$ varies as the sum $Q_{KG} + Q_{KA}$ and so as $C_{KG}V_G + C_{KA}V_A$. The current between cathode and anode is the result of this total field; thus the current is constant for constant $C_{KG}V_G + C_{KA}V_A$. When V_G varies, V_A must vary with the ratio C_{KG}/C_{KA}. The definition of the amplification factor can be written: $\mu = C_{KG}/C_{KA}$.

It should be noted that:

- The capacitances are those corresponding to the electrode surface geometry, not including stray capacitance due to electrical connections, for example.
- C_{KA} is less than C_{KG}, because the distance between cathode and anode is greater than that between cathode and grid, and also the grid acts as an electrostatic screen. Therefore μ is always greater than unity.

6.2.2.2.2 *Calculation of amplification factor*

The amplification factor of a triode can be related to the dimensions of the various electrodes.

The principle of the calculation is to express the potential created by the charges on the different electrodes as a function of position for any point. From this result, the field at the surface of the cathode is calculated, and as before, one can calculate the amplification factor.

Note, however, that these calculations can only be solved in closed form by using approximations. Thus the values obtained from these formulas are only approximate.

For example, the calculation for a cylindrical triode gives:

$$\mu = -\frac{N \ln(r_a/r_g)}{\ln[2\sin(Np/2r_g)]}$$

In the case of a planar triode:

$$\mu = \frac{2rd_{gp}}{r\ln[2\sin(rp/a)]}$$

6.2.2.2.3 Operating class

We have seen that the triode characteristics can be represented, to a good approximation, by the equation:

$$R_i i_p(t) = \mu v_g(t) + v_p(t)$$

where i_p, v_g and v_p are the values of plate current, grid voltage and plate voltage, respectively, as functions of time t.

Generally, an operating point is chosen, at some constant values—V_{G0}, V_{p0}. For a grid voltage signal, applied between the grid and the cathode, which varies as $V_G \cos\omega t$, a voltage $-V_p \cos\omega t$ is observed on the anode. The phase reversal between the output circuit and the input circuit gives rise to the sign change. We can write:

$$v_g(t) = -V_{G0} + V_G \cos\omega t$$
$$v_p(t) = +V_{p0} - V_p \cos\omega t.$$

The anode circuit is loaded by an impedance R_p such that $V_p = R_p I_p$, where I_p is the amplitude of the fundamental anode current. Defining $V_0 = V_{p0} - V_{G0}$ gives, for a certain angle θ_0 where $i_p \to 0$:

$$\cos\theta_0 = \frac{V_0}{V_G - R_p I_p} \tag{6.2.7}$$

It is seen that $\theta = \omega t$ can take values between $\pm\theta_0$, i.e. $-\theta_0 < \theta < \theta_0$. θ_0 is thus a critical angle.

The steady-state component is given by I_{p0}. The ratio $I_0/I_{p0} = r$ is important because it determines the conversion efficiency of the tube and its circuit:

$$r = \frac{I_p}{I_{p0}} = \frac{\theta_0 - \sin\theta_0\cos\theta_0}{\sin\theta_0 - \theta_0\cos\theta_0} \tag{6.2.8}$$

The efficiency η is defined as the ratio of the effective power out $V_p I_p/2$ for an applied power $V_{p0}I_{p0}$ which gives:

$$\eta = \frac{1}{2}\frac{VI_p}{V_{p0}I_{p0}} = \frac{1}{2}\frac{V_p}{V_{p0}} r$$

Another interesting parameter which can be derived from these formulas is the maximum current which the cathode must deliver. The current is maximum for $t = 0$:

$$I_{p\max} = \frac{\mu V_G - R_p I_p}{R_i}(1 - \cos\theta_0)$$
$$= \frac{\pi(1-\cos\theta_0)I_p}{\theta_0 - \sin\theta_0\cos\theta_0}$$

We then define the parameter M:

$$M = \frac{I_{p\max}}{I_p} = \frac{\pi(1-\cos\theta_0)}{\theta_0 - \sin\theta_0\cos\theta_0} \tag{6.2.9}$$

which gives:

$$\frac{I_{p\max}}{I_{p0}} = \frac{I_{p\max}}{I_p}\frac{I_p}{I_{p0}} = Mr \tag{6.2.10}$$

The different classes of tube operation can be defined in terms of θ_0:

Class A: $\theta_0 = 180°$, $r = 1$, $M = 2$, anode current always present
Class B: $\theta_0 = 90°$, $r = \pi/2$, $M = 2$, anode current half the time
Class C: in general one considers $\theta_0 = 60°$, $r = 1.7936$, $M = 2.5575$

We can thus conclude that class C operation is interesting because it offers the best efficiency; however, the product Mr is very large, which requires high peak current from the cathode.

Class A operation gives lower efficiency (<50 per cent), but the peak current is small. Class B represents a compromise which is often acceptable. Using these same relations between the different parameters, one can use Fourier analysis to calculate the different harmonics created by pulses of anode current.

6.2.2.3 Tetrodes

In a triode, the anode current depends substantially on the anode voltage, because the anode voltage creates an electric field component at the surface of the cathode.

In a tetrode, a *screen grid* is added between the control grid and the anode in order to diminish this effect. By creating an electrostatic 'screen' between the anode and cathode, the screen grid virtually eliminates the anode field at the cathode surface. This additional grid is generally held at a fixed positive potential with respect to the cathode, and strongly assists the extraction of electrons.

A theory analogous to that elaborated for triodes can be constructed from electrostatic theory, leading to similar equations:

$$i_p = S\left(V_G + \frac{V_{G2}}{\mu_{G2}} + \frac{V_p}{\mu_p}\right) \tag{6.2.11}$$

One can calculate μ_{G2} and μ_p by formulas similar to those for triodes. In practice, μ_p has a value of several hundred, or even several thousand, which tends to confirm that i_p is virtually independent of V_p.

The same classes of operation are defined as for the triodes.

6.2.2.3.1 Secondary emission phenomena in tetrodes

Consider a tetrode with electrode potentials V_A for the anode, V_{G1} for the control grid and V_{G2} for the screen grid. The cathode current is then shared between the three electrodes (unless V_{G1} is negative).

Furthermore, suppose that V_{G1} and V_{G2} are fixed while V_A decreases from a value much greater than V_{G2}. Then as V_A decreases, I_A decreases first and I_{G2} increases. If there is secondary emission of electrons from the screen, these will be attracted by the anode if $V_A > V_{G2}$, causing an increase of I_A and decreasing I_{G2}. (This can even become negative if the secondary emission coefficient is greater than 1.)

On the other hand, when $V_A < V_{G2}$, the secondary electrons emitted from the screen grid cannot go towards the anode, and thus return towards the screen grid. Similarly, the secondary electrons emitted by the anode may be captured by the screen grid causing a decrease of anode current and an increase of screen grid current.

Practically speaking, this results in anomalies in the variations of the anode current as shown in the tetrode characteristics in Figure 6.2.13.

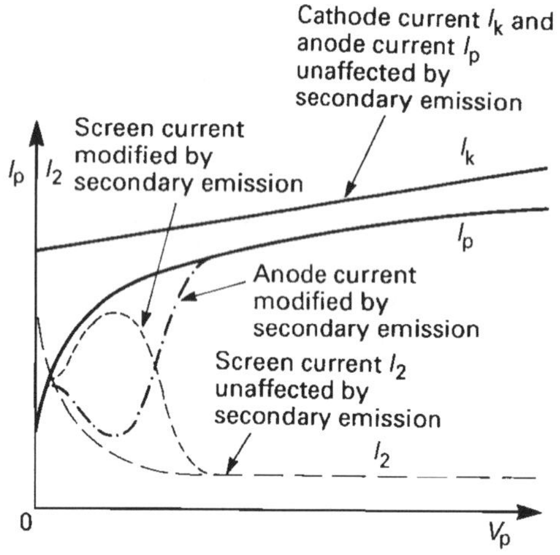

Figure 6.2.13 Tetrode characteristics.

It is necessary for V_{G2} always to be less than V_A. To overcome this drawback, a third, *suppressor*, grid is added between the screen grid and the plate, and maintained at or near the cathode potential. This potential barrier pushes the secondary electrons back towards the anode. Figure 6.2.14 shows the potential variations in such a tube, called a *pentode*. As shown by the characteristic curves, it is possible to operate the tube in the region where $V_p < V_{G2}$.

This solution has been mostly used in receiving tubes. In power tubes, on the other hand, the presence of another grid complicates the technology. Also it is less necessary because the screen grid voltages are such that the secondary emission phenomena are weaker.

6.2.2.4 Power grid tubes

6.2.2.4.1 High frequency operation

Increasing the operating frequency of a tube eventually compromises certain performance characteristics, e.g. gain or efficiency. Several effects come into play:

- *Inter-electrode capacitance*, negligible at low frequencies when compared to the output circuit itself, becomes more and more important with increasing frequency. In the case of high frequencies (50 MHz–1000 MHz), it is common practice to use these inter-electrode capacitances as capacitances in the oscillating circuit itself.

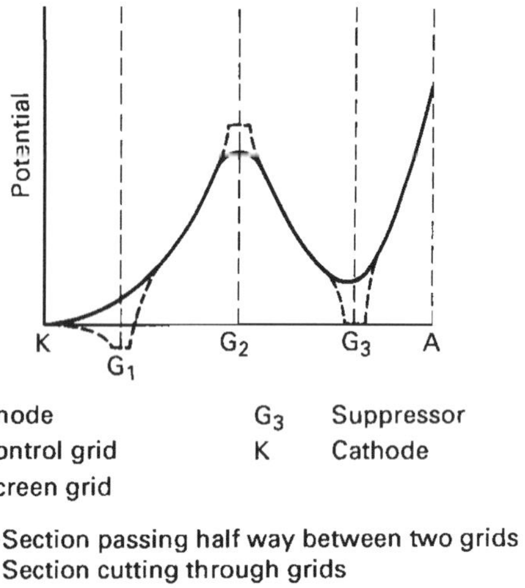

Figure 6.2.14 Potential variations in a pentode.

- *Electrode inductance* can lead to considerable phase differences between the voltages appearing on the various electrodes inside the tube. To resolve this, most tubes use a coaxial geometry.
- *Skin effects*. Eddy currents are induced in a very shallow layer at the surface of conductors exposed to high frequency electromagnetic radiation. Appropriate materials and electrode geometry must be used to minimize such effects.
- *Electron transit time*. For low frequencies, electron transit time can be considered for all practical purposes to be instantaneous (i.e. negligible). The electron velocity v, after acceleration by a potential V, is given by $v = (2e/m)^{1/2}V^{1/2}$. The transit time can be calculated, for simple geometry, from the inter-electrode spacing. In general, the values are between 10^{-11} and 10^{-7} seconds. For tubes operating with much longer characteristic times, electron transit between electrodes can be considered to be instantaneous.

If the electron transit time is not instantaneous, other phenomena may arise. Consider the case of an alternating signal voltage V_g on the grid, which creates an electric field at the surface of the cathode and starts a current flowing. As the electrons advance, the voltage V_g will continue to vary, and could perhaps even change polarity, pushing the electrons back towards the cathode, heating it even more upon impact.

One effect of a finite electron transit time is to introduce a phase difference between the voltage and the current at the level of the grid. A practical consequence is that a smaller average current flows between the different electrodes. In order to recover the current lost in this manner, which increases with increasing frequency, the accelerating voltage must also be increased, leading to a decrease of both the gain and the efficiency of the tube. Beyond some frequency, the tube is no longer usable.

To increase the maximum frequency at which these tubes can be used, the inter-electrode spacing must be reduced. In this way, triodes can be constructed for operation at frequencies of several gigahertz.

6.2.2.4.2 Grid tube linearity

Amplifiers using grid tubes are sometimes required to transmit more than one signal at a time. If the tube characteristics were strictly linear, this would be no problem. Unfortunately, the linear relationship given previously is only a good approximation. Certain areas of the curve networks, particularly for low current or very high current values, are better represented by a polynomial expansion such as:

$$I_a = \sum_{j=1} A_j \left(V_g + \frac{V_p}{\mu} \right)^j$$

If two sinusoidal signals of frequencies f_1 and f_2 are amplified simultaneously, the output signal will contain components of the form $mf_1 \pm nf_2$. Although many of the spurious signals can be reduced or eliminated using a band-pass filter, there will still be those within the amplification band, e.g. between f_1 and f_2, that can lead to amplification anomalies at the output.

The calculation of the amplitude of the spurious signals introduced in multi-carrier operation is now relatively accurate using modern computer methods. The characteristic curve networks are measured and input to the computer, which then performs a harmonic analysis using Fourier methods to calculate the spurious amplitudes.

6.2.2.4.3 Grid tube technology

6.2.2.4.3.1 Cathodes

The two types of cathode in common use are oxide cathodes and thoriated tungsten cathodes.

Oxide cathodes are generally heated indirectly. Their main advantage is a continuous emission surface (see Figure 6.2.15). Their average current is of the order of 200 mA/cm^2. This makes them particularly useful in pulsed tubes with small duty cycles (of the order of 1 per cent or less), where they can deliver several amperes per square centimetre during the pulse, thus allowing high peak power levels to be obtained.

One drawback is that they are rather sensitive to ion bombardment. Any residual gases in the tube when ionized by the electron flow will be attracted to the cathode surface, resulting in 'cathode poisoning' and decrease of cathode emission. This effect is aggravated by higher tube operating voltages, which lead to greater ionization.

Figure 6.2.15 Oxide cathode (Thomson-CSF).

Thoriated tungsten cathodes are generally made of thin wires, which can be arranged on the surface of an imaginary cylinder as in Figure 6.2.16. To compensate for the effects of thermal dilation when the cathode is heated, the geometry of the array of individual wires may be obtained by supporting springs. Alternatively, the cathode may be composed of two equally spaced helical windings symmetrically orientated on the cylindrical surface. Such cathodes are directly heated and operate at about 2000 K. Saturation currents of the order of 3 A/cm^2 can be obtained, making such cathodes useful for very high power tubes (e.g. 1 kW–1 MW). In addition, they are resistant to ion bombardment, allowing the use of high accelerating voltages.

Figure 6.2.16 Thoriated tungsten cathode (Thomson-CSF).

The most important drawback is the very high operating temperature, which requires a lot of heater power.

6.2.2.4.3.2 Grids

The grid electrode is the most difficult electrode to make, as it must satisfy many criteria, some of which are mutually exclusive, leading to a design compromise:

- The grid should have a geometrical form as perfect as possible, to ensure that the cathode-to-grid spacing is as accurate as possible in spite of the very small distance.
- The electrical field created by the grid at the surface of the cathode must be as constant as possible over the whole surface, requiring numerous small wires very evenly spaced.
- The grid should be transparent enough to avoid intercepting too much cathode current.
- To avoid thermal emission, the grid should have as low a temperature as possible when in operation. The thermal emission is the result of cathode evaporation of emissive material which is deposited on the grid, which then becomes a spurious emitter when the grid temperature is sufficiently high. The best way to lower grid temperatures is to use thermally 'black' materials for grid construction.

- If possible, the grid material should have a low coefficient of secondary emission.
- As the grid is required to conduct high frequency currents, it should be made of material of high electrical conductivity.
- The material should also be refractory, i.e. have a low vapour content under nominal operating conditions.

The two types of grid material commonly in use are metallic or graphite.

Metallic grids (Figure 6.2.17) are commonly made of molybdenum, tantalium, niobium or tungsten. Unfortunately, all these materials have troublesome tendencies towards both thermal emission and secondary emission.

Figure 6.2.17 Metallic grid (Thomson-CSF).

To reduce secondary emission, the grid surface may be 'blackened' by coating with powders such as zirconium, tantalum carbide or graphite. To reduce thermal emission, gold or platinum plating may be used. The gold or platinum combines with the emissive material deposited on the grid, increasing the work function and thus decreasing the grid thermal emission.

Graphite grids (Figure 6.2.18) offer a low coefficient of secondary emission, as well as good thermal radiation properties. Ordinary graphite, however, is very fragile and difficult to machine. Pyrolytic graphite, on the other hand, presents in addition to the above qualities, improved thermal and electrical conductivity, together with excellent mechanical properties.

Grid blanks are obtained in the desired form by vacuum deposition of graphite obtained by high temperature 'cracking' of hydrocarbons (at about 2000°C). The grid is then machined either by laser or by sandblasting. This type of electrode has led to considerable increase in the reliability and performance obtainable from power grid tubes.

6.2.2.4.3.3 Anodes

Anodes are required to dissipate most of the power which is not supplied to the output circuit of the power grid tube. As the efficiency of the tube is generally quite a bit less than 100 per cent, considerable power must be evacuated from the anode towards a cooling circuit to keep its temperature within acceptable limits. Furthermore, because the anode is at a high voltage during operation, the coolant must be electrically insulating.

Figure 6.2.18 Pyrolytic graphite grid (Thomson-CSF).

Several types of cooling may be used, depending on the power to be evacuated and the environmental conditions of tube use.

Radiation cooling is the simplest: the anode radiates waste heat. As the temperature may reach several hundred degrees, this type of cooling may not be used when the anode is under vacuum, e.g. in a transparent (glass) vacuum envelope. The anode will be made of a refractory material such as nickel, tantalium or molybdenum, whose surface will be blackened to increase the radiation efficiency. Graphite is also used successfully.

Conduction cooling can be used for low power dissipation by strong mechanical contact between the anode and an external heat sink. If electrical insulation is required, the thermal conductor to the heat sink may be made of berylium oxide.

Forced air cooling can be used for power levels up to about 30 kW. In order to improve the thermal exchange with the moving air, cooling fins are welded to the anode to increase the surface area (see Figure 6.2.19). This method, simple in principle, becomes more troublesome at higher power levels as the air flow rates become large (several tens of cubic metres per minute). Fans for such air flow rates are noisy, bulky, power hungry and vibrating, which can lead to equipment reliability problems in practical installations.

Water cooling allows the dissipation of higher power levels by immersion of the anode in water. As the water heats up, it is

Figure 6.2.19 Anode with cooling fins (Thomson-CSF).

replaced by an incoming cool water flow, and the heated water is pumped to a thermal exchanger (water/water or water/air) to be cooled and reintroduced into the circuit.

To avoid the formation of thermally insulating deposits on the anode, the water must be distilled. In order to provide electrical insulation of the anode, all water connections in the vicinity of the tube are of insulating materials.

Water flow rates are commonly of the order of a litre per minute per kilowatt to be dissipated. The maximum water temperature at the output is about 50°C for acceptable cooling efficiency. At higher limits, the anode locally heats a thin film of water at its surface, creating vapour bubbles which cover the anode surface and keep it from contacting the water. The anode cooling no longer works and the temperature rises sharply, which could lead to a critical situation. This is the phenomenon of *calefaction.*

Water vaporization cooling uses the latent heat of vaporization to perform heat transfer. The diagram of Nukiyama (Figure 6.2.20) represents the heat flow across a boundary surface at a uniform temperature as a function of the temperature of the surface. When the heat flow is 10 W/cm^2, the surface is at 110°C and we have steady-state boiling.

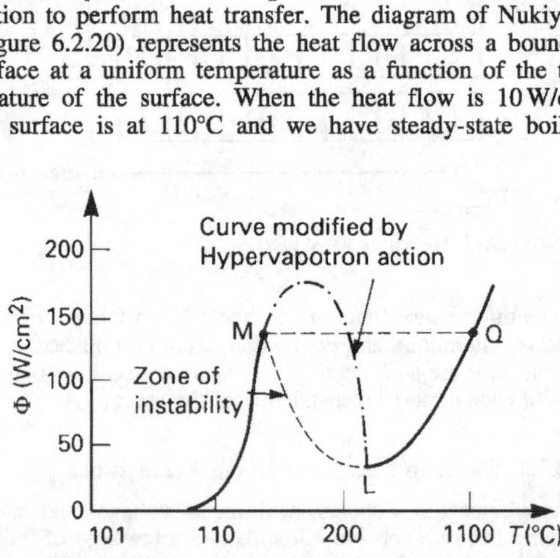

Figure 6.2.20 Diagram of Nukiyama. The scale of the abscissa is logarithmic.

Further heating to point M corresponds to 125°C and 135 W/cm^2. Continuing through an unstable zone to the point L (Leydenfrost point), situated at about 30 W/cm^2 and 240°C, leads to thin-film vaporization. As seen in the figure, the next stable point after point M is the point Q situated at 1100°C.

The Leydenfrost point is the temperature at which there is minimum heat transfer, corresponding to the temperature at which there begins to be a thin film vaporization process at the surface of the metallic anode immersed in a fast-moving water flow.

A Vapotron-cooled anode (Figure 6.2.21) is covered on its external surface by projections which create a stable thermal gradient about the critical point. The extreme end of the projection is always cooler than the area where the vaporization occurs. Under these conditions, a heat flow of 150–200 W/cm^2 can be obtained. If, instead of projections, the anode is covered by narrow grooves, the temperature gradient extends from the Leydenfrost point at the bottom of the grooves, to the point C temperature at the other end.

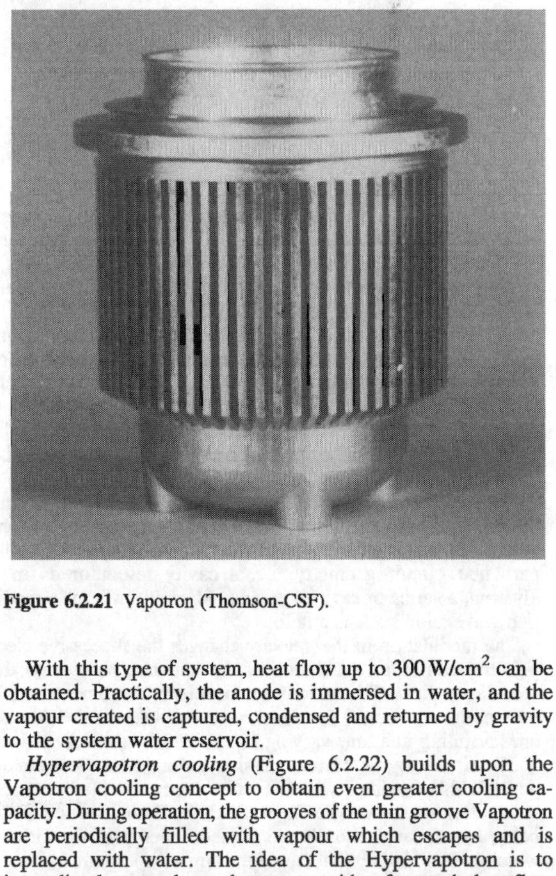

Figure 6.2.21 Vapotron (Thomson-CSF).

With this type of system, heat flow up to 300 W/cm^2 can be obtained. Practically, the anode is immersed in water, and the vapour created is captured, condensed and returned by gravity to the system water reservoir.

Hypervapotron cooling (Figure 6.2.22) builds upon the Vapotron cooling concept to obtain even greater cooling capacity. During operation, the grooves of the thin groove Vapotron are periodically filled with vapour which escapes and is replaced with water. The idea of the Hypervapotron is to immediately recondense the vapour with a fast, turbulent flow of cold water. Outside the tube, the cooling system is then similar to a simple water cooling circuit, except that the output water temperature may reach 90°C or even 100°C. On the other hand, this type of cooling allows evacuation of up to 2 kW/cm^2 of anode surface.

Figure 6.2.22 Hypervapotron (Thomson-CSF).

6.2.3 Linear Beam Tubes

Linear beam tubes are the most versatile devices used for generation and amplification of energy at microwave frequencies. The usual forms are klystrons and travelling wave tubes (twts).

To produce microwave power, a high density electron beam is extracted from a cathode and accelerated by dc voltages to relatively high velocities. This accelerated beam must be controlled to have well-defined trajectories by a combination of focusing electrodes and magnetic field. The magnetic field confines the electron flow to a relatively narrow, straight channel, so that it can interact with suitable electromagnetic circuits along its trajectory and to prevent interception of the beam by these circuits.

The steady state trajectories are then modified by modulating the electron velocity through the interaction of the electrons with time varying electromagnetic fields produced by some particular circuit geometry, i.e. a cavity resonator as in a klystron, a series of cavities in coupled cavity twts, or a modified waveguide such as a helix.

The modulation of the velocity changes the successive electron trajectories as a function of the entrance phase into the modulating field. The initially uniform beam becomes nonuniform as the accelerated electrons tend to overtake the decelerated ones resulting in a time varying current density (*bunching*).

The time varying current passing through the electromagnetic fields associated with a circuit (cavity, coupled cavities, helix, etc.) transfers power to the field, the beam kinetic energy being transformed into electromagnetic energy in the circuit, which is then delivered to some transmission system.

Finally, the remaining kinetic energy of the electrons is converted into heat in a collecting electrode or *collector*. It is possible, at least in principle, to operate the collector at a depressed potential, i.e. a potential relative to the cathode lower than the potential of the main interaction region of the device. The electrons will then strike the collector electrode with lower kinetic energy. There is a saving in power at the cost of an extra power supply.

Linear beam tubes differ one from another principally in the characteristics of their interaction circuits.

In a *klystron* (Figure 6.2.23), strong interaction takes place in a small number of cavity gaps which initially modulate the beam and, in the output cavity, extract the energy from the very strongly bunched beam. Intermediate cavities, generally not being loaded by external coupling, will show a relatively high Q, resulting in a high coupling impedance and large power gain. On the other hand, the klystron is necessarily a fairly narrow bandwidth device.

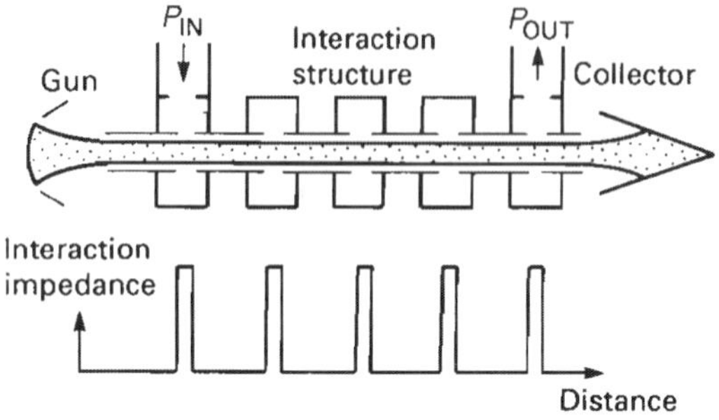

Figure 6.2.23 Klystron.

In *travelling wave tubes* (Figure 6.2.24), the energy propagates along a slow wave structure (helix or coupled cavities) which presents to the electron beam a uniform coupling impedance, which however is much lower than the impedance of an unloaded cavity resonator. The bandwidth will be broader, but it is quite obvious that synchronism between the electron beam and the travelling wave is required. This implies constant voltage operation, generally demanding expensive well regulated high voltage power supplies. Klystrons are much more flexible and operate from simpler power supplies.

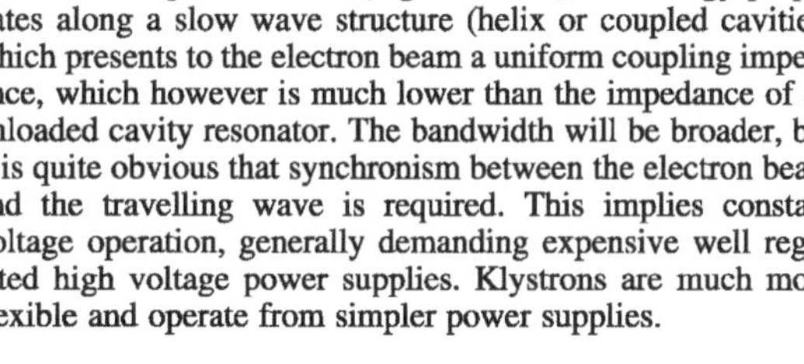

Figure 6.2.24 Travelling wave tube.

The three major functions in linear beam tubes, *beam generation*, *interaction* and *dissipation*, are accomplished in three regions that are sufficiently well separated to allow optimization of each almost independently of the others.

6.2.3.1 Electron beams for linear beam tubes

Linear beam tubes operate at dc beam voltages and currents ranging from one or two kilovolts and a few tens of milliamperes for low power, low frequency travelling wave tubes to more than 300 kV and several hundred amperes for multimegawatt klystrons.

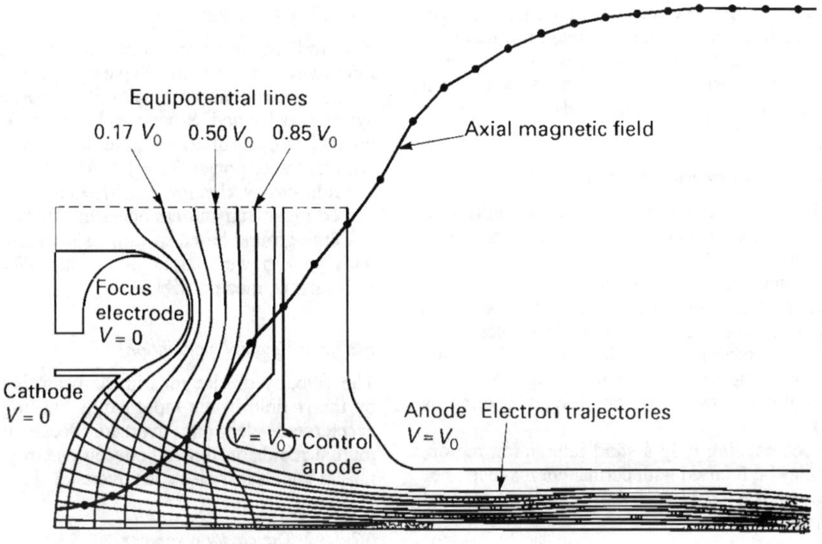

Figure 6.2.25 Pierce type electron gun.

All linear beam tubes make use of electron guns operated under space charge limited cathode electrode emission and therefore obey the Child-Langmuir law: $I_0/V_0^{3/2}$ = a constant = φ perveance, where V_0 and I_0 refer to beam voltage and current.

The design of modern electron guns requires the solution of the equations of electron motion taking space charge and magnetic field into account.

The *Pierce type* electron gun using a spherical cathode is by far the most frequently used (see Figure 6.2.25). Its design is relatively simple. The limiting parameters are perveance and area convergence. The great majority of existing tubes operate with beam microperveance ranging from 0.3 to 2.0, and area convergence from 10 to 70. Exceptionally, convergence of 100 at microperveance 1.0[5] and convergence 30 with a hollow beam of microperveance 5.5[6] have been achieved.

6.2.3.1.1 Cathodes

Two types of cathode are used in linear beam tubes, the oxide coated cathode and the dispenser type cathode.

The use of *oxide coated cathodes* is limited to low power devices where the cathode current density can be kept below 300 mA/cm^2, and below 200 mA/cm^2 for long life devices. They are still often used in high power pulsed devices such as klystrons even at multimegawatt level corresponding to peak cathode current density in excess of 5 A/cm^2, provided that the rms value is kept lower than 300 mA/cm^2.

Dispenser cathodes, as described in Section 6.2.1.3, offer a much better resistance to poisoning and poor vacuum than oxide cathodes. They can operate at much higher direct current density. Satisfactory operation exceeding ten years has been demonstrated on twts operating on geosynchronous satellites *with* B or S type cathodes working at 1 A/cm^2.

6.2.3.1.2 Beam control

Some beam tubes such as klystrons do not require an accurate setting of the operating voltage, and the power output can be adjusted simply by changing the cathode to body voltage in a diode electron gun. In beam tubes such as twts requiring synchronism between the electron beam and the wave propagating in the structure, the beam voltage is imposed, and therefore adjustment of gain or power output requires control of the beam current. This is done by means of a control electrode which can be a modulating anode, a focus electrode, an intercepting grid, or a non-intercepting 'shadow' grid.

6.2.3.2 Beam focusing

As linear beam tubes must use long, thin electron beams, beam spreading due to space change forces becomes of prime importance. The universal beam spread curve in Figure 6.2.26 shows that, in a typical one microperv beam, space charge effects are already significant after a drift distance of only one minimum

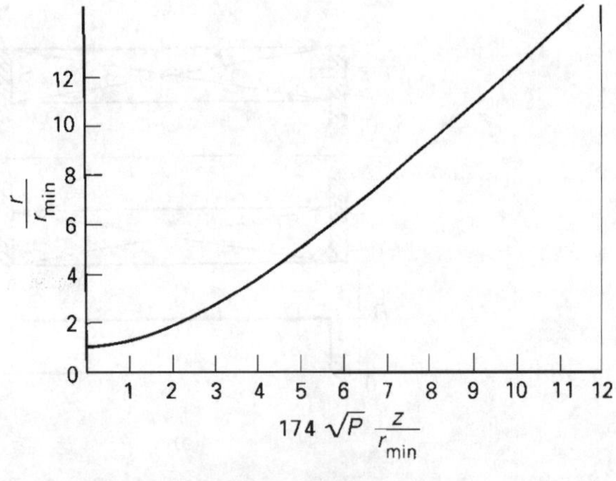

Figure 6.2.26 Universal beam spread curve.

beam diameter, and beam diameter has doubled after only five minimum beam diameters drift. As a beam length to diameter ratio in excess of 100 is not unusual, a focusing structure is clearly necessary. The focusing structure can use a uniform axial magnetic field or periodic magnetic fields.

6.2.3.2.1 Focusing with uniform magnetic field

A beam immersed in an axial magnetic field is in equilibrium if a continuous balance exists between space charge, magnetic and centrifugal forces.

The magnetic focusing field is commonly obtained by use of a solenoid terminated at both ends by iron pole pieces with a centre hole for passing the electron beam. The solenoid is generally shielded with iron on its outside diameter. The coils are bulky and heavy, being several hundred kilograms for an S-band (3 GHz) multimegawatt klystron. Liquid cooling is generally required.

A number of tubes needing only a short interaction region, such as klystrons, may be focused with permanent magnets. For example, a klystron delivering 1 kW at 4.4–5.0 GHz is focused by a 20 kg permanent magnet.

6.2.3.2.2 Periodic permanent magnet focusing

A substantial reduction in system weight can be achieved by using periodic permanent magnet (ppm) focusing.

The electron trajectory in a magnetic field B is given by:

$$\frac{d^2r}{dz^2} + \frac{reB^2}{8mV_0} = 0 \ (e > 0)$$

This equation shows that B appears only in its squared value, i.e. the electron trajectory is independent of the polarity of the magnetic field. An alternating field between $+B_0$ and $-B_0$ as in Figure 6.2.27(b) will focus the beam exactly as the continuous field B_0 in (a).

PPM focusing is used for the great majority of helix twts. The saving in weight using ppm rather than uniform field focusing is one to two orders of magnitude.

6.2.3.3 Klystrons

Klystron amplifiers are extensively used as final amplifiers in microwave transmitters. Typical applications include vhf television transmitters up to 50 kW, tropospheric communication systems at L-band, S-band and C-band, with power from 1 to 20 kW, and earth to satellite links at C-band, X-band and Ku-band with power from 1 to 10 kW.

Multi-cavity klystron amplifiers are also used extensively in pulsed radar transmitters operating from uhf through X-band.

Large ground based tri-dimensional radars operate at S-band with peak power output up to 30 MW, and several tens of kilowatts of mean power.

6.2.3.3.1 Velocity modulation

The velocity of electrons in the beam is periodically changed by the rf field in the input cavity causing bunches to form as accelerated electrons overtake decelerated ones. However, mutual repulsion forces between electrons tend to impede the rate at which fast electrons overtake slow electrons.

6.2.3.3.2 The cavity resonator

The cavity resonator is the basic circuit element in a klystron. It is designed to concentrate the electric field in the region of coupling with the beam (the *gap*). This gap must be made fairly short. A typical cavity has the shape shown in Figure 6.2.28. The shunt impedance of the cavity is:

$$Z = \frac{R}{1 + jQ[\omega^2 - \omega_0^2/(\omega_1\omega)]} \qquad (6.2.12)$$

where ω_0 is the resonant frequency $1/\sqrt{LC}$. The most important parameter of a klystron cavity is its characteristic impedance defined by:

$$Z_0 = \frac{R_{\text{shunt}}}{Q} = \frac{1}{\omega_0 C}$$

The value of the capacitance C is very close to the low frequency capacitance of the gap.

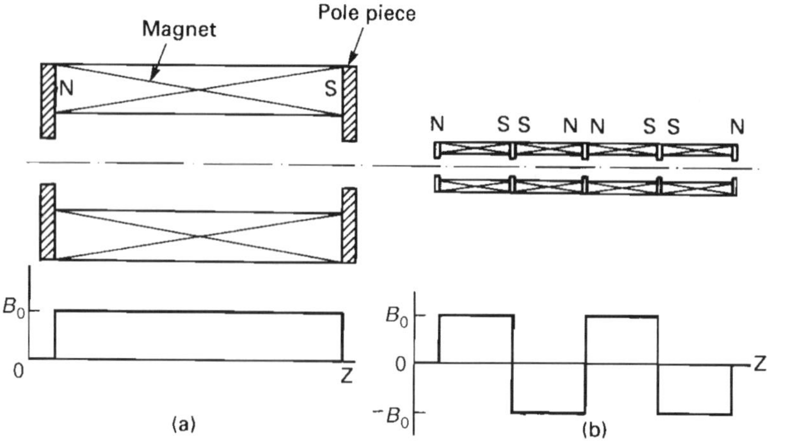

Figure 6.2.27 Permanent magnet focusing using: (a) a continuous field; (b) an alternating field.

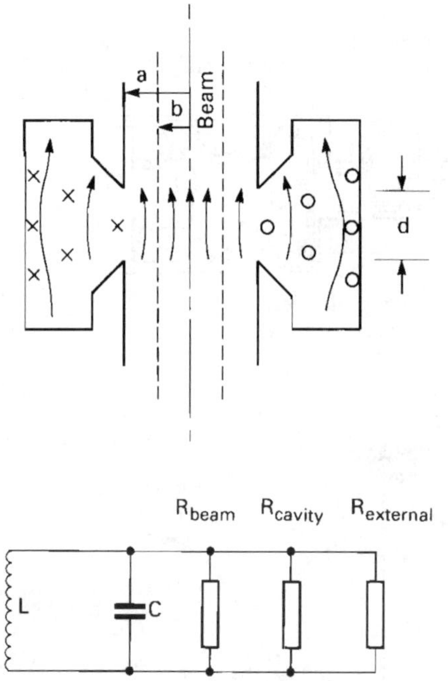

Figure 6.2.28 High power klystron cavity and its equivalent circuit.

The interaction between beam and cavity is characterized by:

- the *beam coupling coefficient M*, which is the ratio of the rf current induced in the cavity gap to the current carried by the beam at the entrance of the gap;
- the *beam loading resistance* R_b, which expresses the fact that, for a finite transit time of the electrons in the rf gap, the velocity modulation of the beam takes some energy away from the resonator.

6.2.3.3.3 Gain and bandwidth

The multi-cavity klystron is capable of providing extremely high gain but is generally considered as a narrow band device. It is however possible to trade gain for increased bandwidth by *stagger tuning* cavities. The resonances of cavities are distributed across the bandwidth in a manner similar to that of a distributed amplifier.

The mathematical analysis of the gain of a klystron incorporating a single intermediate cavity is relatively easy. However it becomes more and more complex as the number of cavities is increased.

The gain bandwidth product of a klystron is proportional to R/Q, so the gap capacitance C should be minimized for maximum bandwidth. However, care should be taken not to decrease C by too large an increase of the gap length resulting in a poor coupling coefficient. The gain bandwidth product of a klystron is also proportional to the dc beam conductance:

$$\frac{\Delta f}{f} \propto \frac{R/Q}{R_0} = \frac{R}{Q} P_0^{1/5} k^{4/5} \qquad (6.2.13)$$

where R_0 is the beam dc impedance, P_0 the dc beam power and k the perveance.

6.2.3.3.4 Power and efficiency

The small signal computation cannot predict the klystron behaviour when driven to saturation. The modulated beam can be considered as a constant current generator delivering its energy in the shunt impedance of the output cavity. There is an optimum value, for if the load resistance is too high the voltage generated across the output gap will exceed the accelerating beam voltage V_0, and electrons will be reflected resulting in a power loss.

Maximum efficiency is achieved when the maximum rf current component is obtained in the beam with minimum beam velocity spread.

It has been shown that the maximum achievable efficiency is a function of beam perveance. At high perveance, or high beam current density, space charge repulsion forces become greater, thus limiting beam bunching. The curve of Figure 6.2.29 shows computed and experimental efficiency versus beam perveance. As can be seen from equation (6.2.13), bandwidth is almost proportional to perveance, and so there must be a trade-off between efficiency and bandwidth.

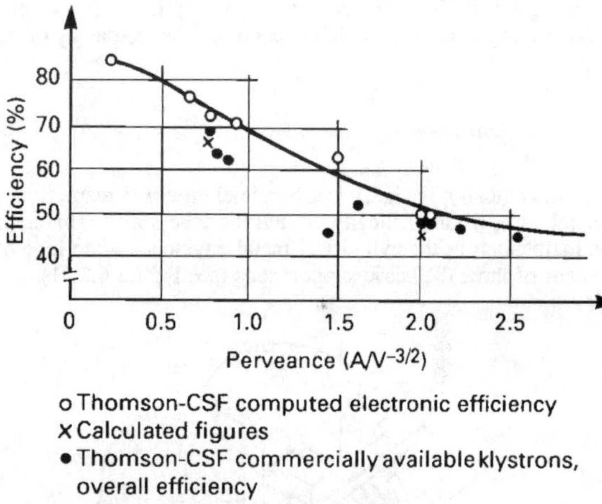

Figure 6.2.29 Klystron efficiency versus perveance.

6.2.3.4 Travelling wave tubes

The travelling wave tube (twt) and the multi-cavity klystron amplifier have many features in common: electron gun, electron beam, necessity of a focusing system and collector (see Figure 6.2.30). They differ mostly by their rf circuits, and by the mechanism for converting the kinetic energy of the beam electrons into microwave energy.

The rf circuits used in twts are spaced periodically along the tube axis. Such periodic circuits produce important reductions of the velocity of the signals transmitted. They are often referred to as *periodic slow wave structures* or *delay lines*.

The rf field inside the rf circuit must satisfy Maxwell's equations. Also it must satisfy a set of boundary conditions on the rf structure. For a periodic structure, these conditions have specific implications, described by Floquet's theorem.

Floquet's theorem states that for an infinite, lossless structure with periodic length p, propagating a wave of frequency ω

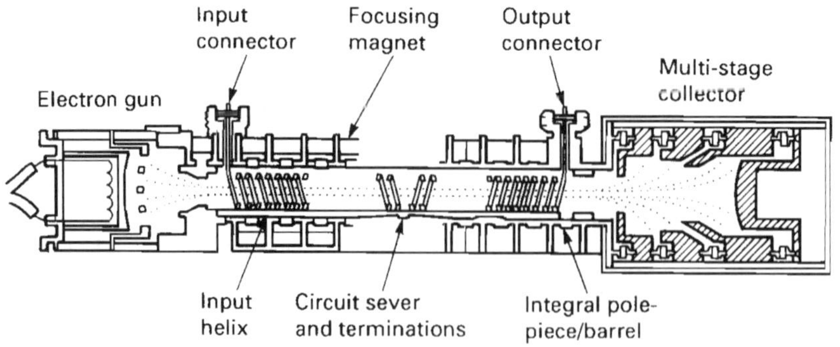

Figure 6.2.30 Helix travelling wave tube.

in the z direction, the fields, electric or magnetic, at point x, y, z are the same as at point x, y, $z+np$ to within a phase shift, i.e.:

$$\vec{E}(x,y,z+np) = \vec{E}(x,y,z)e^{-jn\phi}$$

n being any positive or negative integer. ϕ is the phase shift between adjacent cells, which depends on the frequency of the signal.

6.2.3.4.1 Helix travelling wave tubes

6.2.3.4.1.1 Helix delay line

Technologically, the helix is a simple circuit; it is made from a metal wire, wound helically around the tube's axis. This helix is maintained in the cylindrical metal envelope of the tube by means of three dielectric support rods (see Figure 6.2.31).

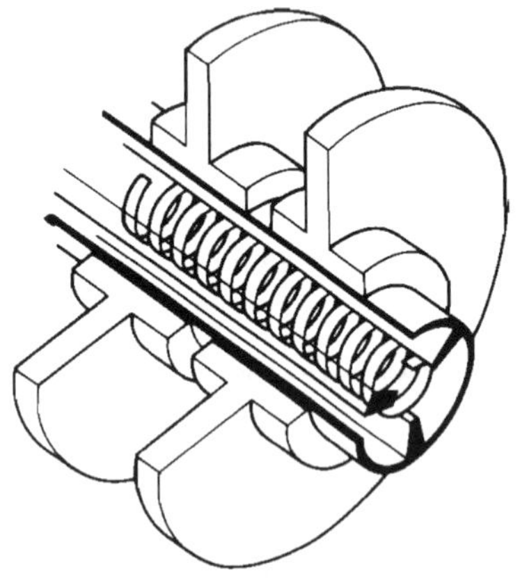

Figure 6.2.31 Helix with support rods, vacuum envelope and pole pieces.

The electrical properties of the helix are shown on the Brillouin diagram (Figure 6.2.32). It can be seen that the fundamental component has a nearly constant velocity over a considerable bandwidth.

This leads to twts having operating bandwidths from one to nearly three octaves, and to a basically low cost technology.

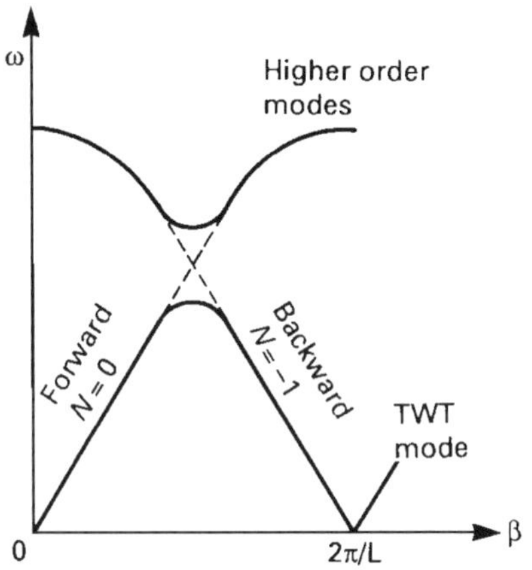

Figure 6.2.32 Brillouin diagram for the helix.

The only disadvantage of the circuit results from the presence of dielectric supports, which limit the power capability of the structure. Therefore, the average power performance obtainable from such tubes depends to a large extent on engineering innovations, allowing the best designs to overcome significantly the basic thermal problem, and to offer the many advantages of the helix at the medium power levels. The maximum average power level is obtained with the very efficient brazed helix technology.

6.2.3.4.1.2 Travelling wave interaction

In the travelling wave tubes, the interacting field (fundamental component for the helix twt) travels along the axis with a velocity slightly higher than that of the injected electron beam. Accelerating and decelerating forces are exerted by the field on the electrons, depending on their position relative to the wave.

Figure 6.2.33 shows how electrons seeing a field in the opposite direction to their speed are accelerated by the field, while those electrons seeing a field in the same direction as their speed are decelerated.

The consequence is that accelerated electrons advance on the average beam, while decelerated electrons are slightly delayed. This velocity modulation gives rise to space charge density modulation; accelerated electrons form bunches with decelerated ones in front of them. At the same time, decelerated electrons

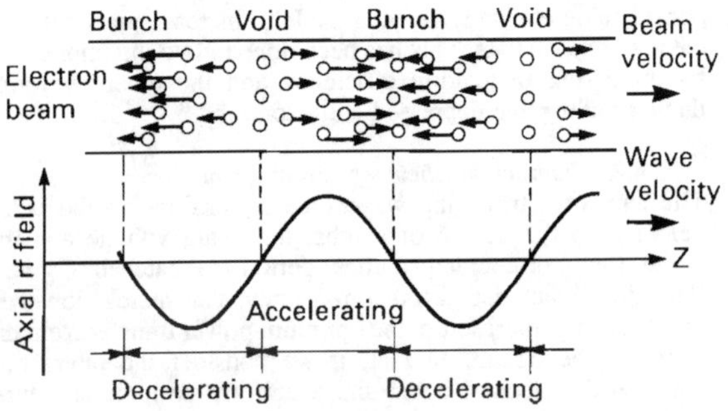

Figure 6.2.33 Electron bunching.

increase the distance which separates them from accelerated ones in front of them, creating voids of charge. Bunches and voids travel at the beam average velocity along the axis.

Because there is an excess velocity of the beam over the field wave, the bunches tend to enter the regions where the field is decelerating, and voids those where the field is accelerating.

In other words, more electrons become decelerated than become accelerated, and the average kinetic energy decreases steadily along the beam. The energy gained from the beam is transferred to the source of the forces acting on the electrons, i.e. to the travelling field.

As the increasingly bunched beam proceeds along the tube and sees an increasingly intense field, as a result of the continuous energy transfer from beam to wave, the average kinetic energy of the beam decreases, reducing more and more the excess speed of the beam over the wave.

When that excess speed becomes zero, the process of energy transfer reverses, and the power of the wave reaches a maximum, or *saturated*, level.

If the interaction process is continued farther, then the rf power diminishes, the energy transfer being now from the wave to the beam, and the tube is in a so-called *over-saturated* state.

The geometrical point at which saturation occurs is normally the location of the rf output connector.

The variation of the rf power along the axis is shown in Figure 6.2.34. It is seen that the gain per unit length is constant

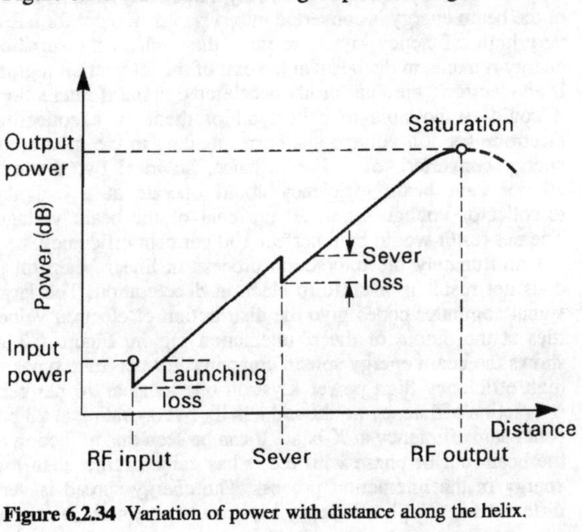

Figure 6.2.34 Variation of power with distance along the helix.

over most of the length. Over this distance, the tube operates as a linear amplifier, and interaction generates low distortion. It is only in the last portion of the delay line, near the rf output, that the gain per unit length decreases, becoming zero at the saturation point. This indicates a strong non-linearity of interaction, giving rise to distortion.

Figure 6.2.35 shows the variation of rf output, when the input rf level is varied from very small values up to the nominal value producing saturation at the output end. The curve shows a pure linear distortion-free amplification, up to a level approximately 10 dB below the nominal. In that region, the tube is said to operate in *small signal*, or *low distortion*, mode. At higher levels, the non-linearity appears clearly when the curve departs from the straight slope. The tube is operating near saturation, with distortion, but in its maximum power and maximum efficiency mode.

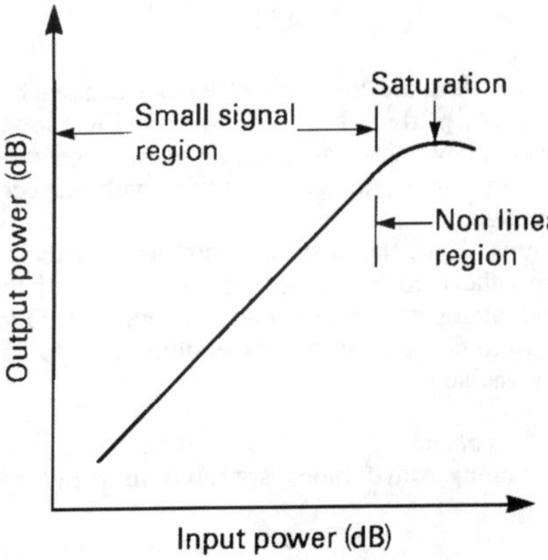

Figure 6.2.35 Output power versus input power in a helix travelling wave tube.

6.2.3.4.1.3 Small signal gain

The small signal gain is given by:

$$G = A + 47.8CN$$

where G is the small signal gain in dB, A is a loss factor of 10–16 dB approximately, C is the gain parameter, and N is the number of electronic wavelengths along the beam.

C is defined by:

$$C = \left(K_0 \frac{I_0}{4V_0}\right)^{1/3}$$

where I_0 is the beam current, V_0 the beam voltage, and K_0 is the coupling impedance.

6.2.3.4.1.4 Stability

Like any amplifier, the helix twt is exposed to self-excited oscillations if the gain is too high and if a feedback mechanism exists.

The first basic oscillation mechanism is due to the successive reflections of the amplified energy on the output mismatch, and, after inverse travel through the entire tube, on the input mismatch, thus closing the instability loop. To prevent this instability, the helix is severed to form at least two physically independent sections, each terminated at one end on an rf connector and at the other end on a very good internal rf load made of carbon deposited on the dielectric helix supports.

The second feedback mechanism is due to the $n = -1$ space harmonic of the helix, which feeds the energy from output to input while interacting with the beam. This *backward wave interaction* becomes especially harmful for beam voltages greater than 7 kV. Various means are used by twt manufacturers to discourage this oscillation. The best designs allow modern tubes to be operated at a beam voltage higher than 15 kV and thus to reach respectable rf power levels.

6.2.3.4.1.5 Power output and efficiency

The electronic efficiency is related to the gain parameter by:

$$\eta_e = 2C = 2\left(K_0 \frac{I_0}{4V_0}\right)^{1/3}$$

On practical tubes, the electronic efficiency can reach values up to approximately 20 per cent. The overall efficiency can reach values well above this, by the use of *depressed collectors*, i.e. collectors whose voltage is negative with respect to the rf circuit voltage.

This slows down the electron beam entering the collectors, and allows the recovery of a large proportion of the kinetic energy remaining in the beam after interaction. Overall efficiencies up to 40 per cent and exceptionally up to 55 per cent are thus possible.

6.2.3.4.1.6 Applications

Helix travelling wave tubes are used in many application areas:

- telecommunications, including troposcatter, surface-to-satellite and satellite-to-earth transmitters;
- electronic countermeasures (jamming);
- radar;
- laboratory amplifiers.

The telecommunication applications take advantage of the excellent fine grain characteristics (i.e. amplification characteristics in which noise and similar unwanted perturbations are of high frequency and produce only small-area fine-grained disturbances to the signal) of the helix twt in narrow-band use (due to its natural broad band), of its low cost and long life demonstrated on many existing systems.

The helix twt, due to its extreme broad band, is commonly used as a laboratory amplifier covering all standard bandwidths.

6.2.3.4.2 Efficiency improvements

Efficiency optimization is of prime interest in systems such as airborne equipment and particularly space systems where power budget is a major parameter.

A small improvement in efficiency of a twt for a space application produces substantial benefits. Fewer solar cells are required as well as fewer batteries and smaller power supplies. The reduced power dissipation makes the thermal balance problem easier. DC input power represents the basic parameter in almost all communication satellites from which all other design considerations are determined. Consequently, considerable efforts have been made to increase practical efficiency of space twts. This has been done in two directions, one by increasing interaction efficiency and the other by introducing multi-stage depressed collectors.

6.2.3.4.2.1 Interaction efficiency improvement

The gain of a travelling wave tube is sensitive to the beam velocity. In the design of a tube, the beam voltage and the circuit (helix or coupled cavities) period are matched to give a flat gain characteristic against frequency. This match, however, will probably not result in the optimum power transfer from the beam to the circuit. In fact, if we consider the interaction between the electron beam and a uniform periodic structure, when energy is transferred from the beam to the electromagnetic wave in the structure, the loss of beam kinetic energy in favour of the rf field results in a reduced beam velocity and loss of synchronism in the tube region where the interaction is the greatest.

By reducing beam voltage, i.e. by resynchronizing beam and wave in the large signal region of the rf structure, efficiency is improved, but maximum efficiency and maximum gain are not obtained for the same beam voltage. This is an undesirable situation for the system performance, and the solution is to gradually reduce the circuit wave velocity in the large signal region by winding helices with variable pitch or progressively reducing the distance between successive gaps in coupled cavity twts.

Optimization of the tapered structure has been made using large signal computed codes. The results can be quite spectacular. For example, a helix twt operating at 12 GHz at a power output of 20 W has a beam efficiency of 13 per cent with a constant pitch helix. The same tube using an optimized tapered helix will exhibit 23 per cent beam efficiency without degradation of other parameters such as linearity or gain flatness.

6.2.3.4.2.2 Depressed collectors

One of the main advantages of linear beam tubes, compared for instance with crossed field devices, is the separation of the tube's regions, where the beam is formed, where rf interaction takes place, and where the spent beam is collected. It is then generally possible to implement some degree of sophistication in the design of the collector region when this is desirable.

In the majority of beam tubes, only a relatively small fraction of the beam energy is converted into rf power. Except for a few very high efficiency klystrons, more than half of the supplied energy remains in the beam at the exit of the interaction region. If all electrons were uniformly decelerated in the rf interaction, it could be possible to collect all of them on a collecting electrode set at a voltage just corresponding to the fraction of energy converted to rf. For instance, an ideal twt showing 20 per cent beam efficiency could operate at a cathode-to-collector voltage set to 20 per cent of the beam voltage. The net result would be a perfect 100 per cent efficiency.

Unfortunately the interaction process in linear beam tubes does not result in a uniform electron deceleration. The large signal computer codes give the distribution of electron velocities at the output of the rf interaction region. Figure 6.2.36 shows the beam energy spread computed for two tube types: a high efficiency high power klystron operating at 70 per cent interaction efficiency at uhf and a helix twt operating at 23 per cent beam efficiency at X-band. It can be seen that a fraction of the beam out of phase with the rf has gained rather than lost energy in the interaction process. The energy spread is very different for a high efficiency klystron, where some electrons

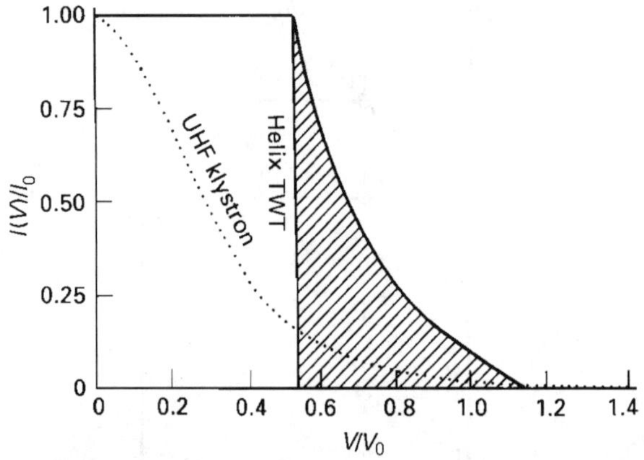

Figure 6.2.36 Beam energy spread for a high efficiency klystron and a helix travelling wave tube.

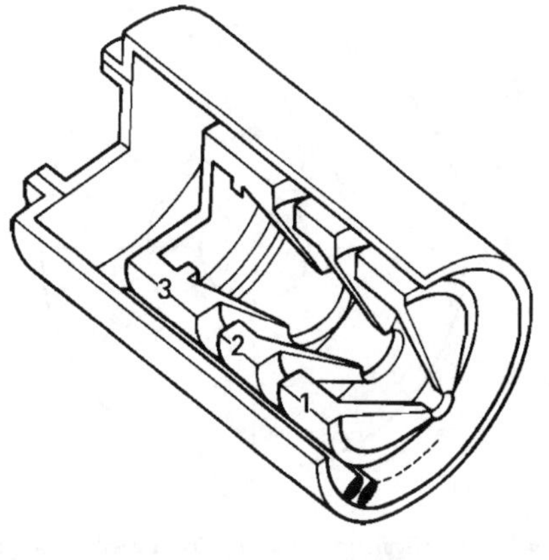

Figure 6.2.37 Three-stage depressed collector.

have lost all their energy. Any attempt to collect the beam at reduced voltage will result in a reflection of electrons towards the rf structure. If the collector were set to 80 per cent of the beam voltage, $V/V_0 = 0.2$, approximately 30 per cent of the beam would be reflected, which would cause very serious problems.

In a travelling wave tube, no electron has lost more than 43 per cent of its energy. It would then be theoretically possible to collect the total of the beam at 43 per cent of the beam voltage, resulting in an improvement of the efficiency.

In a practical case, the electrons cannot 'land' on the collector at strictly zero velocity as space charge effects would cause their deflection. Some voltage margin must be allowed, but from Figure 6.2.36 we can see that if we can collect all of the beam at 45 per cent of its accelerating voltage, we could still collect one third of that beam at one half of that voltage, i.e. 22.5 per cent.

Many twts have been built making use of a two-stage depressed collector. When the tube is operated at full output power, approximately two thirds of the beam current are collected on the first collector, at the highest voltage with respect to the cathode, and one third on the second collector. In the absence of rf power, practically all the beam is collected on the low voltage electrode. The power drained by the tube is therefore reduced by a half at no rf drive and one sixth at full power output.

More collector stages can be implemented at the cost of a more complex power supply. Figure 6.2.37 shows a cross-section of a three-stage depressed collector as used in a medium power twt (50–100 W) at Ku-band (12 GHz) designed for direct or semi-direct high definition television broadcasts from a geosynchronous satellite. This tube has an interaction efficiency of 23 per cent and a very remarkable nearly 60 per cent efficiency when operated with the three-stage depressed collector.

All electrons entering the collector structure have enough energy to enter the space between the first and second collector. Some electrons have lost too much kinetic energy to overcome the retarding field between these collectors and are reflected and collected on the back of the first electrode.

The same process takes place between the second and third electrodes. The electrons are sorted between the three electrodes according to their velocities. In this design, collecting electrodes are press-fitted against ceramic rods inside the vacuum envelope. The ceramic rods insure both electrical insulation between electrodes and heat transfer to the vacuum envelope which in turn must be cooled by convection, conduction or radiation.

References

1. Ramo, S., Whinnery, J.R. and van Duzer, T. *Fields and Waves in Communication Electronics*, John Wiley & Sons (1965).
2. Slater, J.C. *Microwave Electronics*, Van Nostrand (1950).
3. Stratton, J.A. *Electromagnetic Theory*, McGraw-Hill (1941).
4. Strapans, A., McCune, E.W. and Ruetz, J.A. High-power linear-beam tubes, *Proc. IEEE*, **61**, 299–330 (March 1973).
5. McCune, E. A 250 kW cw X-band klystron, *IEEE Int. Electron Devices Meeting*, Washington (1967).

Acknowledgements

This text was prepared with the aid of several engineers at Thomson Tubes Electroniques: B. Epstein, A. Schroff, P. Gerlach, R. Metivier, and their assistance is gratefully acknowledged.

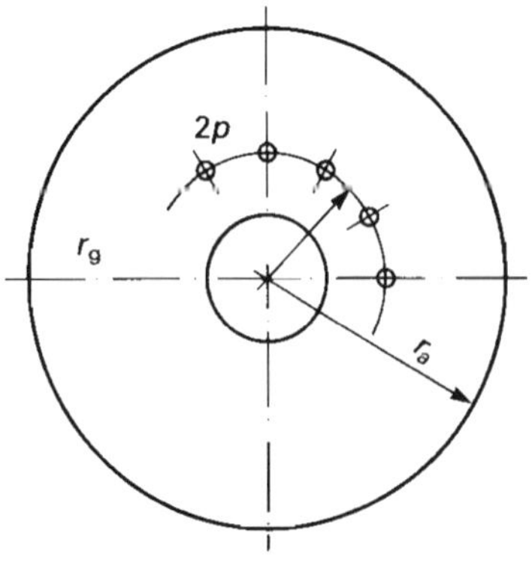

Figure 6.2.10 Configuration of a cylindrical triode. The grid consists of *N* bars, each of diameter *2p*.

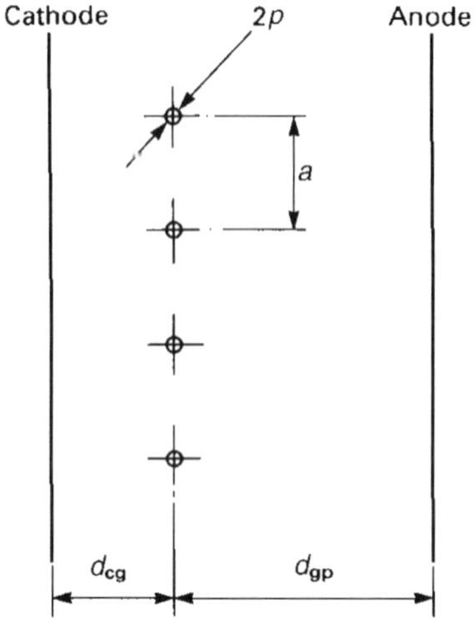

Figure 6.2.11 Configuration of a planar triode.

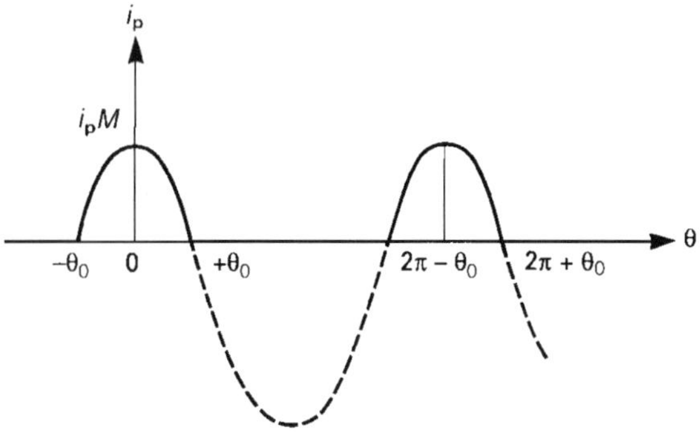

Figure 6.2.12 Anode current in operation.

P Kemble C Eng, MIEE, BSc
Principal Engineer, IBA

Revised by
Salim Sidat MBA, DBA, BSc (Hons), CEng, MIEE
Customer Systems Manager
mmO2, Ltd

6.3 Transposers

In an ideal situation, all potential viewers would be provided with television programmes by means of a central high power transmitter. This would be modulated with audio and video signals from the studio, connected by means of a landline. Such an arrangement would represent the most economical means of distribution in terms of *cost per viewer*.

Unfortunately, such a simple situation is possible only at a limited number of locations. Intervening hills, buildings and even trees attenuate direct reception of a high power transmitter, particularly at UHF, causing a shadow in its coverage. Substantial numbers of people can be affected.

Rather than provide landlines and expensive modulators at every transmitter site, a more satisfactory solution is to provide lower power relay transmitters to fill in the shadows. The principle is shown in Figure 6.3.1. The relay is sited at a carefully chosen location which can both see the target area to be served and also receive good signals from the main transmitter. Despite the shadow, in most cases there will actually be some residual signal in the target area from the principal station, so if the relay station were to transmit on the same channel, there would be unacceptable co-channel interference (cci) to viewers of the relay and of the main station. The equipment at the relay site therefore changes the frequency of (or *transposes*) the incoming signal from the parent transmitter on channel X and re-transmits on a clear channel (channel Y). In some areas it may be necessary to build a chain of relay stations, each serving its own target area and providing a signal feed onwards to the next. In that case several different channels will be used by the various stations.

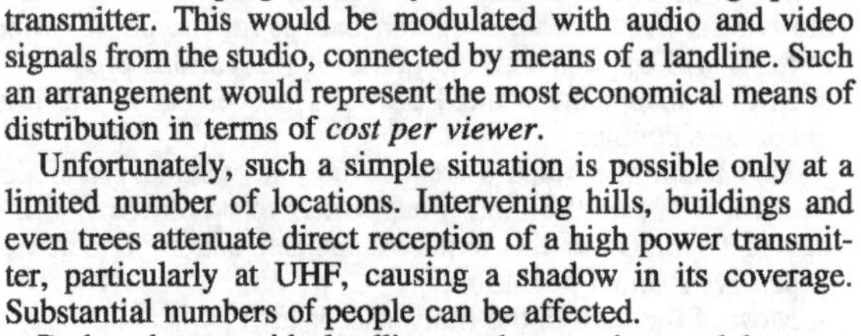

Figure 6.3.1 Relay on hill transposes channel X to channel Y to fill in shadow behind hill.

Although this technique is simple enough in concept, in countries like the UK (which has around 900 relay sites as well as 50 high power main sites) very complex planning is necessary. This is because each site transmits four different programmes (and possibly five), so groups of channels have to be transposed. As the network expands it becomes increasingly difficult to find four clear channels (even more so now with channels being used for Digital Terrestrial Television) for each site to use. In many cases no completely clear transmit channel can be found, and the service area is limited by co-channel interference from a remote station.

6.3.1 Transposer Configurations

6.3.1.1 Single conversion

The block diagram of a very simple transposer is shown in Figure 6.3.2. After some UHF band-pass filtering and pre-amplification, the incoming signal is mixed with a local oscillator to generate the desired output channel directly. For example, if the input is channel 30 (543.25 MHz vision) and the desired output is channel 60 (783.25 MHz vision), then a local oscillator at the difference frequency of 240 MHz will produce the required output (as well as 303.25 MHz, which can be removed by filtering).

Problems can arise with this simple technique which limit its usefulness. It would be impracticable, for example, to receive channel 22 and transmit channel 52. The local oscillator would again have to be at 240 MHz, but this time its second harmonic falls in the receive channel and is likely to produce unacceptable interference patterning. Such relationships, and those due to harmonics of the local oscillator, can result in a large percentage of transpositions having to be avoided. While this probably will not matter in countries having a small requirement, and the technique is in use, it would be an unacceptable restriction in

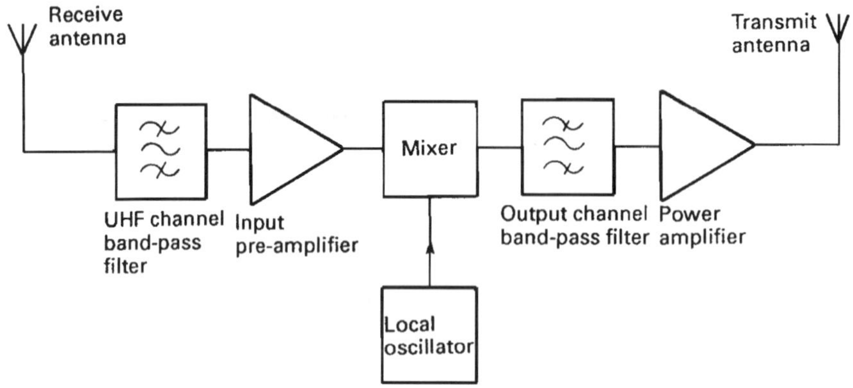

Figure 6.3.2 Single conversion transposer.

the UK. A further problem in a crowded spectrum is that the UHF filtering will not be able to provide much rejection of any carriers on the adjacent channels, which will therefore become unwanted re-radiated signals.

6.3.1.2 Double conversion

The more commonly used transposition technique is shown in Figure 6.3.3. This is a double conversion process, with a large part of the transposer operating at a standard intermediate frequency (typically 38.9 MHz for vision and, in system I, 32.9 MHz for sound).

As in single conversion, there is a band-pass input filter, A, which prevents strong neighbouring transmissions, perhaps from the same site, overloading the input amplifier. It also helps to improve the signal/noise ratio by preventing noise on the image frequency from being mixed down and added to the IF. (Other techniques for this are mentioned below.)

A low noise high gain pre-amplifier, B, enables the transposer to achieve a good noise figure, typically 8 dB, and this feeds the input mixer where conversion to IF occurs.

The mixer, C, may be a simple diode bridge, or a slightly more elaborate arrangement as shown in Figure 6.3.4. This is an image rejection type.

Both the wanted input signal, f_v, and the image frequency, f_{IM}, are divided into two paths by the first quadrature splitter. Thus the input to the top mixer is: $f_{IM} < 90$ and $f_v < 90$ plus the local oscillator. The usual sum and difference products result as indicated in Figure 6.3.4. The inputs and products are similar for the bottom mixer, except for the phase difference. It will be seen that the phase of the various products is such that all but the wanted products cancel out in the final quadrature coupler.

This technique helps to reduce cost in volume production because it allows the input image filter to be eliminated; this has to be tuned to the required channel and is a relatively expensive manual operation.

Most of the gain in the transposer is at IF, where stability is easier to achieve, in fixed amplifier D and variable amplifier F (Figure 6.3.3).

One problem quoted for the direct conversion transposer was the rather poor frequency response of the UHF filter, A, which

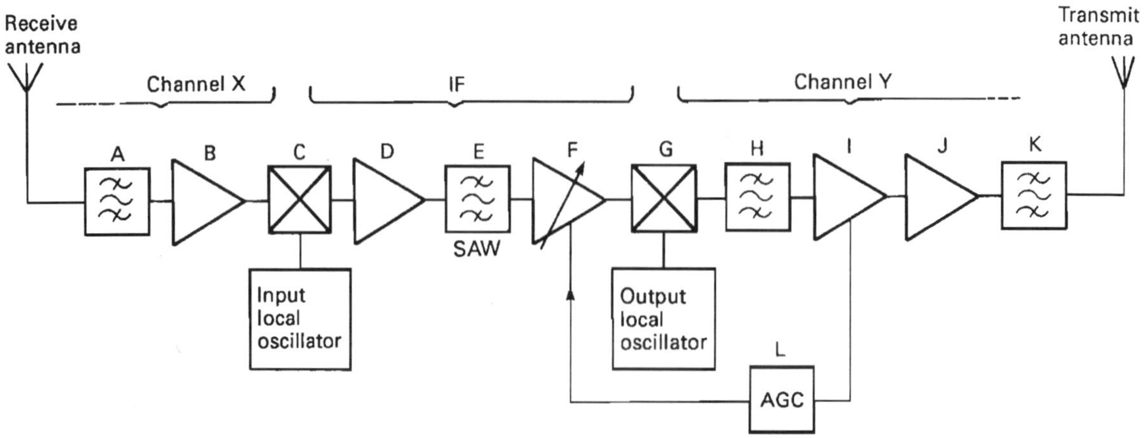

Figure 6.3.3 Double conversion transposer.

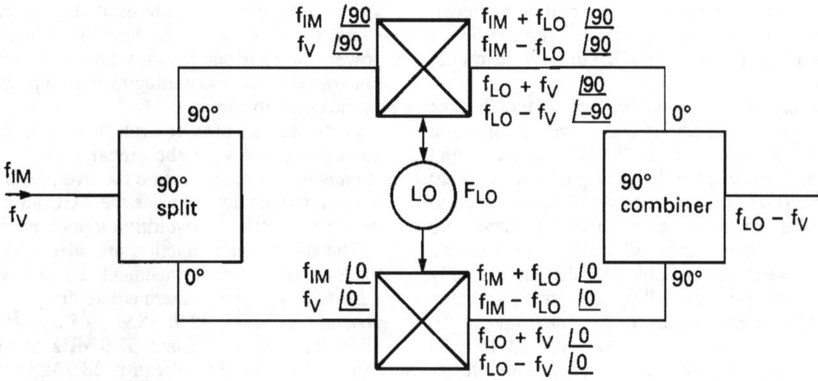

Figure 6.3.4 Image cancelling mixer.

allows re-radiation of lower adjacent sound and upper adjacent vision carriers. This is undesirable for two reasons. First, if the level is sufficiently high the extra carrier may cause intermodulation problems in the power amplifier of the transposer itself. This will degrade the viewer's picture. Second, the adjacent carrier may itself cause a visible pattern for the viewer, or may interfere with some other service.

The IF filter, E, in Figure 6.3.3 is to eliminate these problems by providing a carefully tailored band-pass response. This is now easily achieved by means of a *SAW* (surface acoustic wave) type device which can be designed for this frequency range.

It is readily possible to provide notches at ($f_v - 2$) MHz and ($f_v + 8$) MHz of at least −50 dB with respect to vision, so adjacent channel interference is now virtually a non-problem. A typical SAW frequency response is shown in Figure 6.3.5.

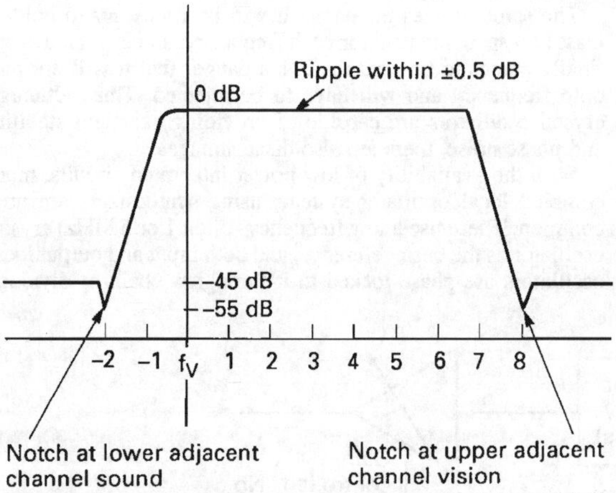

Figure 6.3.5 SAW filter response.

The only problem likely to remain is in the exceptional situation where the adjacent channel interference (aci) is so high, even after receiving aerial azimuth and polarization discrimination, that the transposer front end is still overloaded. Levels of aci which equal the field strength of the wanted signal should be acceptable.

While SAW filters have the desirable feature of predictably providing the sharp cut-off and good frequency response needed with no manual adjustment, the physical length of the device needed to achieve it results in an attenuation of the wanted signal of typically 30 dB. (As the name implies, the signal propagates in the device as a surface wave between two electromechanical transducers.)

SAW filters have quite a large temperature coefficient and so require some simple temperature control to keep the notches on frequency. Amplifier D (Figure 6.3.3) is provided to make up the lost gain, but since the input power to be applied to the SAW filter is limited to about +10 dBm, and allowance must be made for maximum input signals, pre-amp gain, etc., it will not be possible to make up all the lost gain at that point.

Introduction of the SAW filter does therefore marginally degrade the transposer noise figure. One other advantage it does bring is the elimination of the need for group delay correction; this is necessary when LC filters are used to provide channel shaping because of their phase characteristics. SAW filters can have their frequency response and group delay response controlled almost independently, and can have virtually constant group delay despite the deep notches.

Finally, the signal is mixed back up to the desired output channel, before being amplified to the appropriate power.

Like the rest of the system, the output amplifiers I and J (Figure 6.3.3) have to handle both vision and sound carriers. At low signal levels this presents no particular problem, but with increasing power the amplifier must remain sufficiently linear to minimize the generation of distortion products. These fall into two categories, those inside the transmission channel which would affect the received picture quality and those outside the channel which are liable to interfere with other services. These distortions are discussed in more detail in Section 6.3.3.2.

Band-pass filter H removes local oscillator feedthrough and the unwanted mixing product, to prevent these reaching the output amplifier.

Finally output filter K prevents out-of-channel radiation, and may include notches to give better attenuation of the principal unwanted products at ($f_v - 6$) MHz and ($f_v + 12$) MHz which lie in adjacent TV channels.

As well as the direct signal path there are other functions that need to be provided. Because the transposer is receiving off-air signals, it is normal for some signal level variation to occur. Over sea paths, fading of 40 dB is not unusual. The transposer must include *automatic gain control* (AGC) to hold the output power constant.

For this AGC, it is quite common practice to monitor the power at the final output of the equipment. However this is not the ideal position for the detector. As the gain of the output amplifier begins to fall, due to either the ageing of a valve (tube) or loss of output transistors, the drive level will be increased to try to compensate. The effect will be to drive the output into limiting and produce more and more distortion. The viewer's picture will rapidly become unacceptable. Had the power simply been allowed to decline with the falling amplifier gain, then the service would have been satisfactory for much longer. To counteract input fading, it is preferable to keep just the IF level constant. If necessary, any small variation in output power due to amplifier temperature changes or progressive failure can be minimized by means of a 2–3 dB range auxiliary AGC loop round the output stages alone.

The performance of the main AGC would typically hold the vision carrier constant within 1 dB over a 40 dB range of input signals. Below about −60 dBm input level, the signal will become too noisy to provide a satisfactory picture, so the transposer output can be allowed to fall or be switched off entirely. The maximum signal level is unlikely to reach −20 dBm. In fact, transposers are generally designed to be provided with nominally −40 dBm input. The input is more likely to suffer deep fades, due to aircraft flutter or in some cases propagation across tidal water, than it is likely to be significantly enhanced, so the AGC window is offset in this way.

In the case of systems provided with an automatic changeover to a passive reserve, it is desirable to ensure that the AGC thresholds of the A and B sides are set such that B decides it has lost its input before A. Unless this is true, deep fades may cause the A output power to begin to fall while B is still reporting signal received. This will cause equipment changeover when there is actually no fault.

Proper design of the AGC also requires careful consideration of two other points:

- It must respond sufficiently quickly to cope with fast flutter, but not so fast that it distorts sync pulses as if they are unwanted power variations.
- There must not be excessive power overshoots in the rf drive, particularly if the transposer is followed by a power amplifier.

Excessive power overshoots may cause a valve to trip or transistors to fail. Overshoots could arise when the parent station first comes on-air each day, since without an input signal the AGC gain will have been maximum. A fast overpower limiter is one partially effective solution, but it is better to avoid the overshoot altogether by a *feed-forward* technique, as indicated in Figure 6.3.6.

A power detector controls a downstream attenuator with appropriate delay in the signal path to allow time for the attenuator to operate before the overpower arrives. If a limiter is used, it must not be inside the AGC loop itself, or oscillation will occur while the overdrive input prevails.

The other major function provided within the transposer is the local oscillator arrangement. There have been a number of different 'standard' intermediate frequencies used by transposers, e.g. 31.25, 32.7, 38.9, 39.5 and 40.75 MHz. Currently 38.9 MHz for vision and 32.9 MHz for sound are popular, therefore the local oscillator is 38.9 MHz above vision carrier. The choice is largely for compatibility of test equipment with IF modulated high power transmitters which also commonly use that frequency.

The output frequency stability of the transposer must be very good in order to obtain the benefits of offset working for subjective improvement of co-channel interference. The output depends on the incoming frequency and that of the two local oscillators. In a tandem chain of transposers the final frequency has the possibility of a build-up of numerous errors. The longest chain in the UK has a main station feeding five relays.

Although it is possible to correct automatically for any error on the incoming signal, most transposers do not do this, and for practical reasons the stability of each local oscillator is within 250 Hz over 3 months.

Much equipment simply uses two independent vhf crystal oscillators multiplied up to the required frequency. The crystals are operated at their temperature inversion point for optimum stability, but their setting up time together with the lead time taken to supply an aged crystal on the required frequency means that it is a time consuming process during manufacture or should it be required to change channel.

The same applies to spares. It will be necessary to hold at least two spare crystals for each frequency, since when a spare finally comes to be used there is a danger that it will not pull onto frequency and will have to be rejected. Thus, although crystal oscillators are capable of providing excellent stability and phase noise, there are also disadvantages.

With the availability of low power integrated circuits, more complex local oscillator systems using synthesizers are now common. These use a low frequency (often 1 or 5 MHz) crystal oscillator as the basic reference, and both input and output local oscillators are phase locked to it through a chain of dividers.

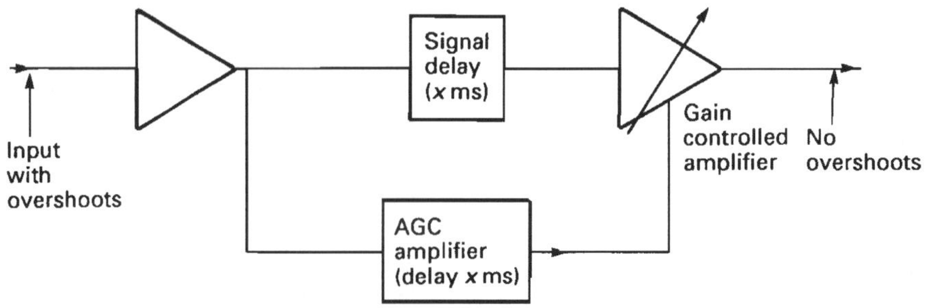

Figure 6.3.6 Feed-forward gain control.

Any drift of the reference frequency will cause the two local oscillators to move in the same direction, so the error on the output will be minimized.

If it were only necessary to operate on exact channels the synthesizer would be much simpler, since the frequency increments would always be 8 MHz, and this could be the reference and comparison frequency. In practice it is also necessary to operate on offsets of a fraction of line frequency ($+\frac{5}{3}$ line, zero and $-\frac{5}{3}$ line are used in UK) so now the comparison frequency has to be low enough to allow that increment.

With the final frequency being divided down more and more, its random phase noise increases, and careful design is necessary to keep it acceptable. Phase noise on the local oscillators is transferred to the output carriers and can potentially cause problems on fm sound reception if the receiver does not use intercarrier sound recovery, and also on pictures received on true synchronous receivers where phase noise is converted into low frequency am noise and degrades the video s/n ratio.

Although synthesizers are not without their own difficulties, they do have the big advantage that a standard unit can be used in any equipment, no matter what the channels. Fundamentally only a standard crystal is required and channel selection is made simply by means of switches.

6.3.1.3 Active deflectors

It was mentioned initially that in nearly all cases the transmit and receive frequencies have to be different in order to prevent mutual interference in the service areas of the parent and its relay. However, in some locations where the topology gives excellent screening, it is possible to retransmit the incoming frequency.

This system is known as an *active deflector*. The gain required of the system is easily established if the received signal level and the necessary erp are known. Referring to Figure 6.3.7, the gain is made up as follows:

$$\text{overall gain} = G_1 - L_1 - L_2 + G_2 - L_3 + G_3$$

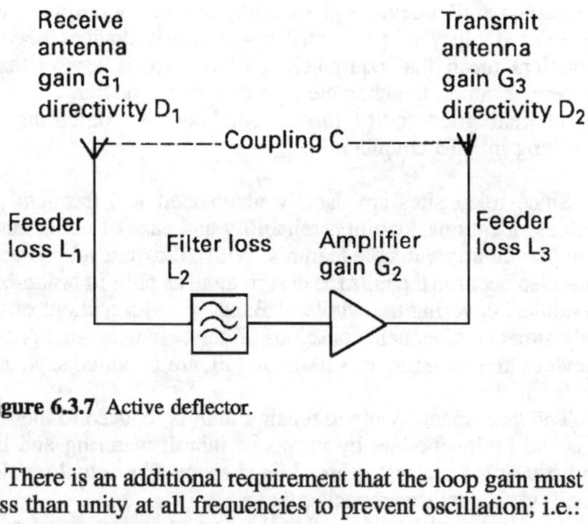

Figure 6.3.7 Active deflector.

There is an additional requirement that the loop gain must be less than unity at all frequencies to prevent oscillation; i.e.:

$$C > (G_1 - D_1) - L_1 - L_2 + G_2 - L_3 + (G_3 - D_2)$$

A band-pass filter can help to meet the requirement out-of-channel, but antenna isolation and pick-up between cables has to be carefully established in-band. Active deflectors are commonly used by 'self-help' schemes where the target is too small for it to be viable as a site to be constructed by the broadcasters. The installation sometimes does not even use an amplifier, just high gain reception and transmission antennas either side of a hill top to 'bend' the signal into an isolated valley. In all cases the scheme must be officially licensed, to enable a check that the re-radiated signal will not cause interference to a service elsewhere.

6.3.2 Design Philosophy

A transposer for a single service has a block diagram as shown in Figure 6.3.3. For several services, costs can be reduced by providing as much common equipment as possible. This has been made easier by the continued trend to solid-state and miniaturization. It is normal to use not only common buildings and electricity supplies, but also common antennas for reception with splitting filters used to separate the signal for each service and a combining unit to combine the outputs for a common transmitting antenna. A station block diagram is shown in Figure 6.3.8.

As well as the obvious cost saving, a common broad-band transmission antenna helps to ensure similar coverage for each service. In fact, it is common practice for a station output to be split to drive an antenna that is in two halves. This arrangement means that if the two antennas are similar but diverge from the ideal impedance in the same way, the result of the reflections will tend to be dumped in the load of the output splitter. This provides a better match for the amplifier. It also means that if one half antenna develops a fault, the station can continue to transmit on the other half.

For the lower powers (2 and 10 W) it is possible to fit several transposers into one rack, and then even dc power supplies and perhaps the common reference oscillator for the local oscillator synthesizers can be shared. Some redundancy is needed to ensure that a single fault does not result in total loss of output.

Design of the station will depend upon the circumstances of the operator. This includes factors such as the reliability of the service required, the accessibility of the sites, the availability and skill of maintenance personnel, the total number of equipments which are in the network, and the level of spares backup.

Where small, remote populations are to be served, programme outages of some hours duration may be acceptable. This would certainly not be true for a large population centre. In the latter case, improved reliability can be achieved by providing reserve equipment (either passive reserve with automatic changeover, or simultaneous parallel operation of two similar amplifiers). In fact, loss of public electricity supply is often the major reason for loss of transmission, especially in the more remote areas, and it may even be considered worthwhile to provide two incoming supply feeds or a standby generator on site.

Early transposers used a thermionic valve for the power amplifier. The programme output would be dependent on a single device which is liable to catastrophic failure. It is therefore normal in valved equipment to provide a duplicate transposer to take service automatically on loss of the main (see Figure 6.3.9). Alternatively, two identical amplifiers could be used in parallel to generate the required power. This has the advantage of avoiding the short loss of output during the warm-up and changeover sequence. On loss of one half the output, power will reduce by 6 dB until the maintenance personnel can fix

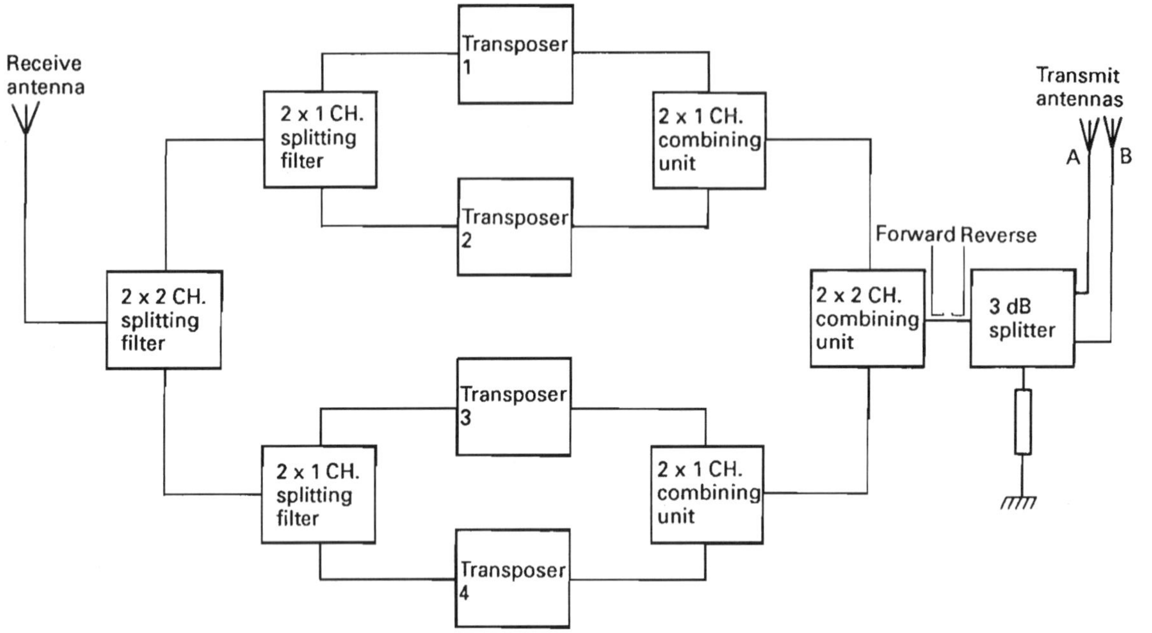

Figure 6.3.8 Typical station block diagram.

the fault or reconfigure the working half direct to antenna to be only 3 dB below normal.

Once standby equipment has been provided, it also becomes necessary to provide telemetry for remote monitoring. Otherwise the station will still be off the air before anyone realizes that something is amiss, and now both the main and the reserve equipment have to be repaired. Nothing is gained by the provision of a reserve without remote monitoring, except for extending the time between total station failures.

It is usual to draw up a simple algorithm along the lines shown in Figure 6.3.10, which can be used for determining whether telemetry and/or reserve equipment is necessary. Although there will always be exceptions (e.g. the station may justify a standby simply because it is often inaccessible in severe winters), most cases are simply analysed in this way. A decision is required about providing telemetry at the parent station because if the dependent station is reported off-air, it could be because its parent has failed. If the parent is not too far away it will be easier simply to visit both sites. If the parent is in a remote location it would be better to provide telemetry to remove the need for a possibly unnecessary visit.

Early UHF transposers used a thermionic valve even at the 10 W level, but *transistors* began to replace them in the early 1970s. Four BLX98 devices in parallel was one arrangement for 10 W, configured between couplers as shown in Figure 6.3.11. This arrangement is still commonly used with other devices and at higher levels.

It allows a good return loss for input and output, and the amplifier is relatively independent of the parameters of a particular device. It provides protection against a single failure in the signal path causing a total loss of output, and this can be taken a step further by distributing the power supply in the way shown.

The advantages offered by transistors have meant that they have been used ever since at increasing power levels as suitable devices became available. The reasons for using them include:

- reduced component stress and safety; high voltage power supplies are not required;
- long life – no expensive consumable valves every few thousand hours;
- excellent programme reliability – loss of one or two amongst multiple parallel devices gives negligible loss of output power;
- good stability of performance – multiple devices between couplers mean that equipment performance is less critically dependent upon the characteristic of a single device;
- inherent good return loss of modules by combining and splitting in 3 dB couplers.

Since relay sites are usually unattended and frequently in remote locations, stability, reliability and ease of maintenance are particularly valuable features. With solid-state amplifiers, it has also become the norm to design them as plug-in broad-band modules, covering the whole of Bands 4 and 5 without on-site adjustment. Channel conscious filters, which are passive devices and therefore less likely to fail, are mounted separately in the rack.

The time spent on site to repair a fault is reduced to identifying the faulty module by means of inbuilt metering and then substituting a pre-tested broad-band spare. The actual repair of the fault is performed back at base.

An associated benefit is that the cost of building and maintaining the site itself is greatly reduced. The working area needed is much less, and because personnel will only be present for a short time, domestic facilities like kitchens and toilets are not necessary.

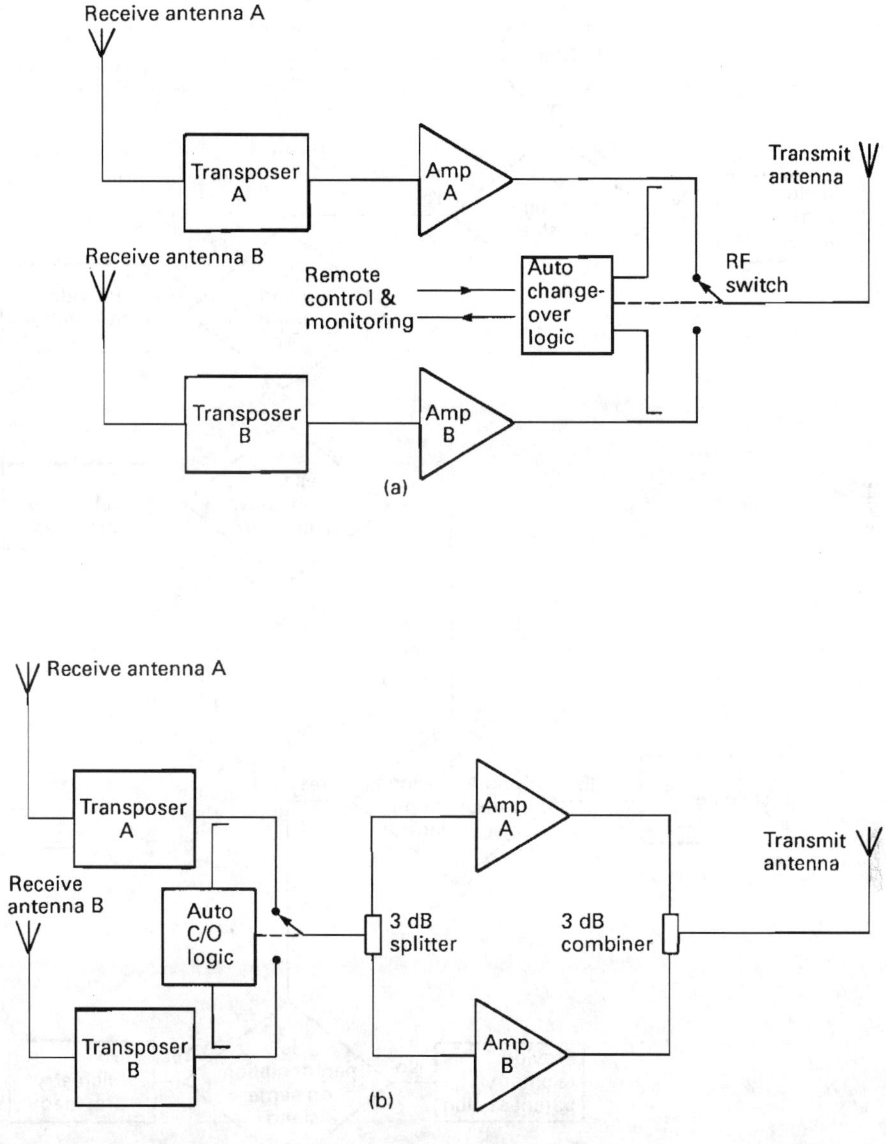

Figure 6.3.9 Provision of a duplicate transposer: (a) full passive reserve; (b) parallel or active reserve.

Such economies have allowed relays to be provided for communities far smaller than would otherwise have been possible. In the UK, where UHF coverage for the four services already exceeds 99.5 per cent, relays are still being constructed to serve targets as low as 200 people. Relays for so few people (and even for much larger groups) are in no way commercially viable in terms of any increased advertising revenue generated, since the percentage audience increase is so small. They are provided simply as a public service responsibility by the broadcaster.

This highlights why all aspects of the station design have to be carefully considered to ensure that expenditure is reasonable while achieving the desired standard. Even so, the capital cost of a station to provide four UHF services to 200 people is more than £100 per viewer.

6.3.3 Transposer Performance

The technical characteristics required of a transposer will depend on its application. In countries where the broadcaster may be a private business and only one or two transposers are required for the community, which is remote from the next, then the specification will be just sufficient to meet minimum requirements of the licensing authority. Tighter specifications mean higher cost.

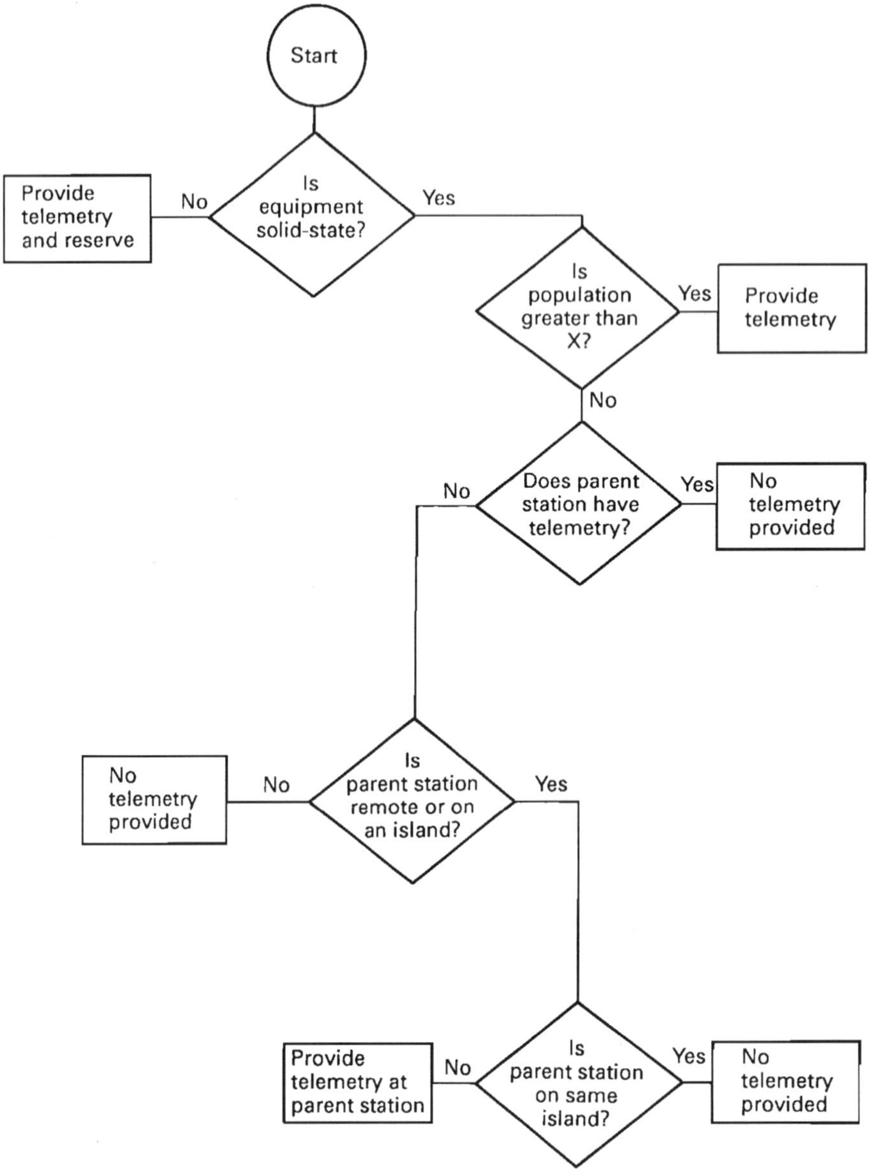

Figure 6.3.10 Possible algorithm to decide need for telemetry.

6.3.3.1 Frequency stability

In Europe, the high population density and the need for multiple services in many countries demands much more stringent requirements for the transmission equipment. The fundamental requirement is to minimize interference to and from other stations. Apart from the actual frequency planning that this entails, it necessitates a high degree of frequency stability. To enable the maximum number of stations to be built, with a limited amount of spectrum, the same channels are used many times in different parts of the country. So that the geographical distance between re-use is as short as possible, use is made of the subjective improvement gained by controlling the relative frequency within small limits.

CCIR Recommendations (Document 11/1028-E, Dubrovnik 1986) show that if the interfering frequency is random, then it must be much lower in amplitude to be imperceptible than when the two signals are related to each other in certain defined ways. Then the interference can be at a higher amplitude for the same subjective effect. This allows more stations to be packed in.

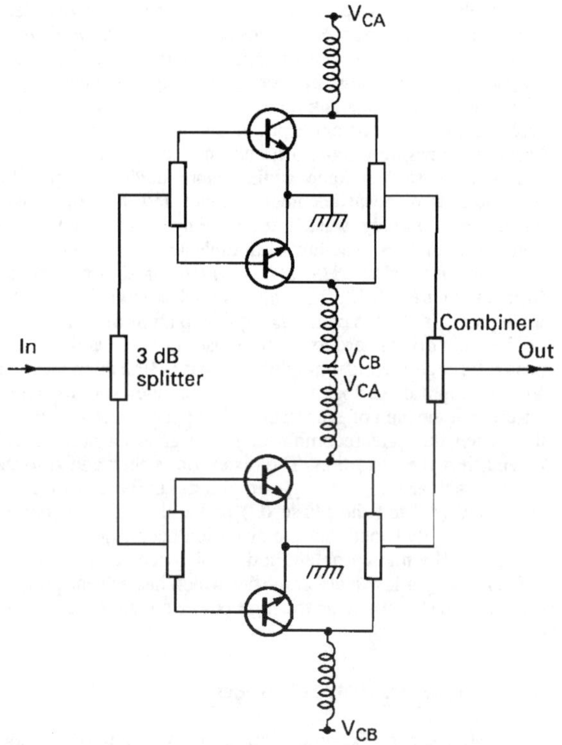

Figure 6.3.11 Typical broad-band amplifier configuration.

The effect is due to the structure of the pattern that is produced on the screen. As the relative frequency of the interfering signal increases, a cyclical change in the visibility of the dot pattern occurs every 50 Hz. Superimposed on this is another cyclical change based on line frequency. It will be seen from Figure 6.3.12 that the optimum condition occurs when the frequency separation is precisely $\frac{1}{3}$, $\frac{2}{3}$, $\frac{4}{3}$, $\frac{5}{3}$, etc., of the line frequency. Since the frequency cannot (usually) be guaranteed to be within the 1 or 2 Hz necessary for the optimum condition, it has to be accepted that the protection will on average lie midway between the precision-best and precision-worst limits. This is nevertheless a distinct improvement compared with zero offset. In the UK it is normal to use this point on the curve (actually $\frac{5}{3}$ line offset), since it allows a group of three transmitters to be protected against each other. One operates at $-\frac{5}{3}$, one at zero offset and one at $+\frac{5}{3}$.

It should be noted that, unless true precision offset is in use, the mean $\frac{1}{2}$ line offset condition is better than the mean $\frac{1}{3}$ offset case. Therefore, exceptionally, in particularly troublesome cases of interference, $\frac{1}{2}$ line offset is used. It is not used normally because it only allows a pair of stations to be protected instead of three and would therefore reduce the total number of sites possible in a country.

In 625-line non-precision working with zero line offset, 61 dB protection is required for limit of perceptibility, whereas precision frequency control on $\frac{5}{3}$ line offset requires only 36 dB.

Extensive networks do not often use precision offset because of the complexity and expense of implementing it. Many transposers simply use an ovened crystal for each local oscillator, but even those that already use frequency synthesis from a low frequency reference crystal would also need to be modified so that they can be locked to a new reference of adequate stability.

An alternative to an on-site source would be to derive the standard from a central reference over-air. This might be:

- the 198 kHz BBC transmission;
- obtained by demodulating the vision carrier to extract line frequency which has been derived from an atomic reference in the studio;
- obtained by demodulating a reference burst which has been added to a spare line in the vertical blanking interval.

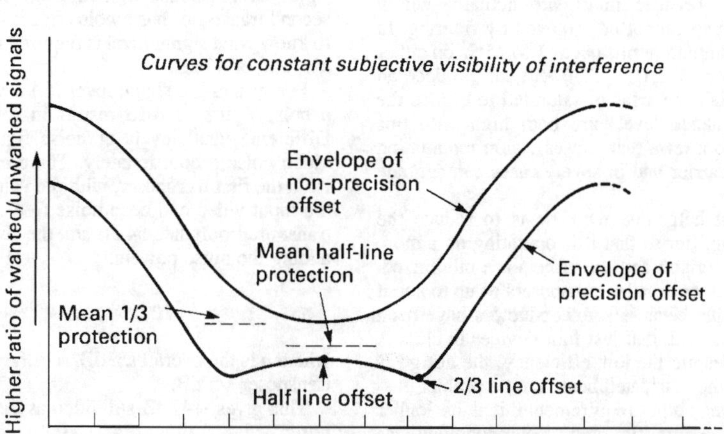

Figure 6.3.12 Change in protection ratio with different offsets; 25 Hz change in frequency difference moves operation from one envelope to the other.

6.3.3.2 Intermodulation

The other possible source of interference to others is the presence of unwanted frequency components in the output. They might be spurious signals unrelated to the vision or sound and caused by unexpected oscillation resulting from high gain and poor screening at frequencies well away from the intended output. This type of problem, once recognized, is easily cured.

Another form of unwanted output is due to the inherent distortion of the power amplifier. As described above, it has been normal to process both vision and sound in a common path, simply because of the difficulty of splitting them. In an ideal amplifier this would be no problem. In practice no system has a perfectly linear transfer characteristic, and the result is an output containing mixing products, principally between the three major input frequency components: vision carrier (f_v), sound carrier (f_s) and chrominance subcarrier (f_{sc}).

If the non-linear transfer characteristic is represented by a polynomial containing terms up to the third (i.e. $y = A + Bx + Cx^2$), then by substituting for x the input comprising the three carriers mentioned above, the expansion shows that, apart from the amplified wanted signals, there are:

second order terms which produce:

- the second harmonic of each input carrier (e.g. $2f_v$);
- sum and difference signals from the input carriers taken in pairs (e.g. $f_v + f_s$ and $f_v - f_s$).

(All the above terms occur well away from the wanted frequencies.)

third order terms which produce:

- signals which appear at the same frequency as the input carriers and therefore change their characteristics (resulting in cross-modulation form one carrier to another);
- third harmonics of the inputs (e.g. $3f_v$);
- signals which fall inside and close to the channel (e.g. $2f_v - f_{sc}$).

It will be seen that the third order products are the most difficult to deal with, because many are actually within the wanted channel and so cannot be removed by filtering. In the UK system I, the dominant terms occur 1.57 MHz on either side of the vision carrier ($f_v \pm (f_s - f_{sc})$) and can produce an annoying pattern especially in areas of saturated red, since the chrominance and luminance levels are both high with this modulation. This is also a term that causes vision modulation to occur on the sound carrier and in severe cases can degrade sound demodulation.

Since filters will not help, one recourse is to reduce the power output of the amplifier so that it is operating on a more linear part of its characteristic and the effects are minimized. This technique is used in the smaller transposers of up to about 10 W output. It is feasible because semiconductors have been available for many years such that just four devices in class A will do the job. Thus despite the low efficiency, the design is simple and removal of the dissipated heat is not a problem.

With increasing power output requirements, it is no longer economic to underrun the amplifier, and a new technique was developed. The concept is simply to add a low level stage with the inverse transfer characteristic to the output amplifier, so that the overall characteristic becomes a straight line.

Effectively a linear system has been produced, which therefore has no distortion. Of course, it is not quite that easy, but nevertheless open-loop systems can easily achieve a long term improvement of at least 6 dB in the level of the *intermodulation products*, or alternatively an amplifier can be used at 3 dB more output power than before pre-correction. This technique is a cost effective way of producing more power, since a low level stage replaces the need for extra power amplifier stages — which become increasingly more difficult and expensive to add. It is not possible to just keep adding more devices in parallel, because heat disposal becomes difficult and the extra power generated is largely wasted in the longer cables and more numerous couplers. The law of diminishing returns applies.

A popular version of *linearity corrector* is shown in simple form in Figure 6.3.13. The input signal is split through two main paths. In the top path it is applied to amplifier B, which is deliberately driven hard to produce distortion. In parallel it also passes through identical amplifier C, but is first attenuated (E) so that amplifier C does not distort. The attenuators E and F ensure that the gain of each path in the upper arm is identical so that when they are recombined, the original carriers cancel, leaving just the distortion. The distortion is then added to the main signal arriving through amplifier D. By adjusting the amplitude (A) and the phase (G) of the distortion, it can be made to largely cancel out the distortion occurring in the final amplifier. The main problem in designing such a pre-corrector is in choosing a low level amplifier which has as near possible the same characteristic as the high power output stage.

6.3.4 System Performance

Although there are several parameters which must be controlled in a chain of transposers to ensure acceptable pictures at the output, a fundamental one is the *signal/noise ratio*. The following example of its calculation refers to Figure 6.3.14. It is of direct interest in establishing what minimum field strengths are acceptable, complexity of antenna arrays and consequently size and strength of the support structure.

Consider the chain shown in Figure 6.3.14. The main station has a video s/n ratio of 54 dB and the first transposer a noise figure of 10 dB and a terminated input signal of 2 mV. The second transposer has a noise figure of 7.5 dB, and it is desired to know what signal level is required to achieve an output video s/n ratio of 40 dB.

For system I, a signal level of 1 mV through a device having a noise figure of 8 dB results in a video s/n ratio of 44 dB. Different signal levels or noise figures change the resulting video noise proportionately. Therefore the video s/n resulting from the first transposer, with the values given, will be 48 dB if the input video had been noise free. As the output of the main transmitter only had 54 dB s/n, the combined effect is found by adding the noise powers:

$$10^{-a/10} = 10^{-b/10} + 10^{-c/10}$$

where a is the overall s/n dB, b is the parent s/n dB, and c is the transposer s/n dB.

This gives −47 dB s/n video as the output of the first transposer.

The above formula can now be used to establish what video s/n ratio the second transposer must produce on its own, if fed with clean video, so that when *actually* fed with −47 dB s/n video the result will be −40 dB. This turns out to be −41 dB. Since the transposer has a noise figure of 7.5 dB, an input signal

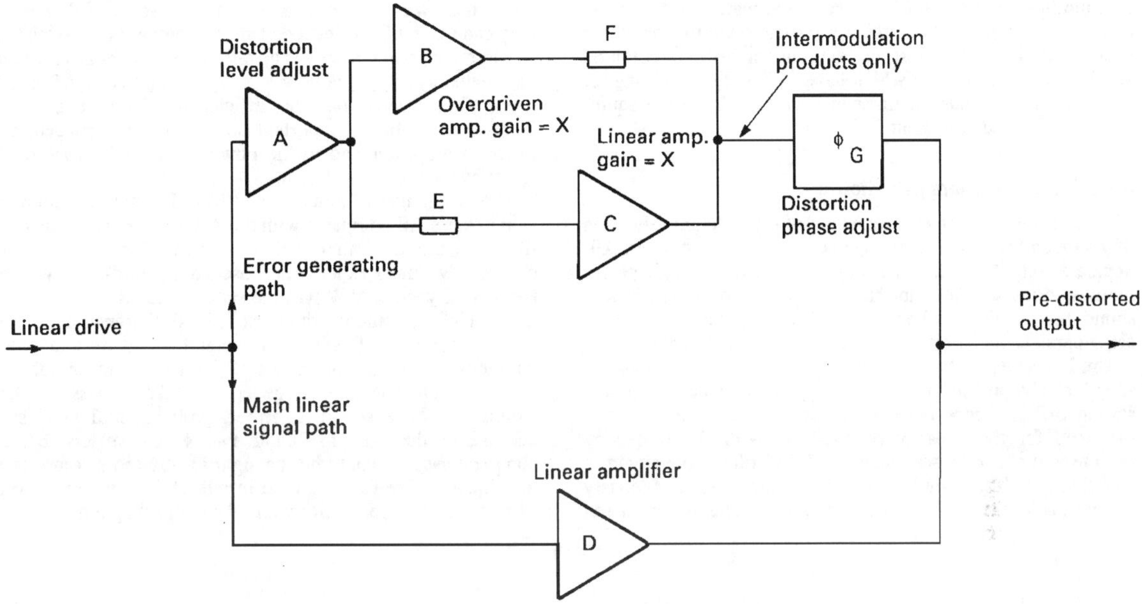

Figure 6.3.13 Linearity pre-corrector.

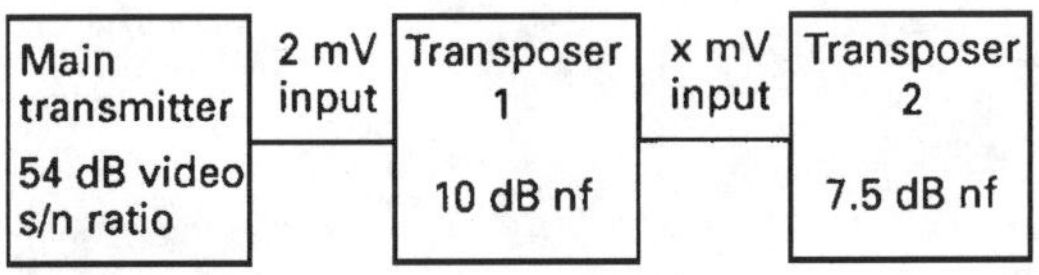

Figure 6.3.14 Transmitter and two transposers.

level 3.5 dB below 1 mV, or 0.67 mV, will produce that result. In this way it is possible to design a receive antenna system for the second relay to provide a particular quality of output, if the available field strength is measured as part of an initial site test.

In a 50 ohm system, the terminated volts V_t available from a dipole in field strength E V/m is given by:

$$V_t = 0.13E\lambda$$

where λ is the wavelength.

The procedure described can be applied to any number of tandem relays to determine if the final output will be satisfactory.

Future trends in transposer design will almost certainly be towards achieving ever higher power with solid-state amplifiers. This in turn depends upon suitable devices becoming available. To a large extent, research and development in semiconductors follows from military projects where large budgets are available, rather than directly from requirements of broadcasters.

6.3.5 Future Developments

Two factors are likely to influence the design of new equipment in the next few years. The first is the desire to achieve ever higher powers with transistors. The second is the introduction of dual carrier sound. The latter could be used for data transmission, but is more likely to be used to allow stereo sound or, in some countries, an alternative language.

6.3.5.1 Dual carrier sound

The second carrier will be added at 6.552 MHz above vision and at a level of −20 dB. It will be modulated by a four-phase qpsk digital signal with a bit rate of 728 kHz. The frequency of the carrier was selected to ensure maximum compatibility with the existing system, to minimize its visibility on the picture and to reduce its potential interference with the upper adjacent channel. The level of the carrier, plus the associated reduction in amplitude of the conventional sound carrier from −7 dB to −10 dB was also chosen to minimize the generation of new intermodulation products.

The principal intermodulation product (ip) will occur at 0.552 MHz above vision and can potentially cause an annoying low frequency pattern on the picture. It is to be noted that ips due to the traditional vision, colour subcarrier and fm sounds are worst in areas of saturated red, where the amplitudes are high. This condition arises relatively infrequently and usually only in parts of the picture. On the other hand the two sound carriers will always be present and so the new 0.552 MHz ip will appear all over the picture and become worse as the luminance level increases.

The information on the new carrier actually undergoes a pseudo-random scrambling process before modulation to improve its ruggedness, so even with no audio input the energy will be dispersed. This helps slightly in improving the subjective visibility of the pattern, but there is still a significant amount of energy at the carrier frequency.

It is interesting to note that, because of the frequencies chosen, the 0.552 MHz is actually offset from vision by close to a multiple of one-third line frequency, which also helps to reduce its visibility. Nevertheless the introduction of stereo sound will be a new test on the linearity of all common amplification transposers which have to handle it, and it may result in more frequent maintenance visits to keep the equipment in adequate alignment.

6.3.5.2 Separate amplification

The other advance is the introduction of separate amplification of vision and sound carriers, together with the use of class AB. Separate amplification has always been used in high power main stations to avoid intermodulation between vision and sound, but the video and audio signals are separately available, are separately modulated, and are therefore easily kept apart.

Until recently, it has not been practical to split vision and sound rf components when receiving them from the parent station off-air, hence the use of common amplification and the need for the linearity of class A. With the frequency responses that can be achieved with SAW filters, the situation has changed. The complete signal is applied simultaneously to two such filters as if. The output from one is the vision component, while the other passes only the sound. Subsequently they can be amplified independently, avoiding intermodulation.

Although good linearity is always necessary for an amplitude modulated system, it is now less critical and class AB can be used. This means that the efficiency of the amplifier is increased and the all-important heat dissipation reduced, both in the rf amplifier and in the power supplies because of the lower average current drawn.

The availability of transistors rated at 50 W at 1 dB compression in class AB compares with the 8 W (uncorrected) available from devices used in common amplification class A. This change now enables 1 kW transposers to be solid-state, whereas for several years 200 W was the state of the art.

A 1 kW equipment could use 32 50 W transistors in the vision amplifier, allowing for worst case gains and output combining and filtering losses. This is the same number of devices as are now used to generate 200 W in class A. There would also be a sound amplifier, probably still in class A because of the need to handle two sound carriers, but the complete system would not be significantly more complicated to achieve a five-fold increase in output. It is therefore to be expected that its reliability would be similarly good.

Jan Doeven
Senior Technical Advisor, Nozema (The Netherlands)

6.4 Terrestrial Service Area Planning

6.4.1 Introduction

The terrestrial service area planner specifies transmitter characteristics and calculates coverage areas, taking into account interference from other transmitters and noise. To do so, he or she has software tools, either commercial or company-made software. In the software tools, sets of planning criteria are used. Planning criteria are normalised values for a certain defined service quality with standardised receiving equipment and under standardised receiving conditions. A propagation prediction model is a critical part of the software.

It is worthwhile to note that, in a predicted coverage area, reception with the defined quality is not guaranteed. There are a number of reasons why a consumer may be not satisfied:

- the consumer may use receiving equipment (receiver and antenna) that is worse than the normalised equipment used for determining the planning criteria;
- the consumer may be more perceptible to impairments in sound or picture than assumed in determining the planning criteria;
- reception may take place at an unfavourable location or time – coverage is predicted within certain statistical limits;
- the prediction model has a certain inaccuracy and cannot take into account all local effects.

The service area planner has to deal with a number of constraints, such as the use of existing sites, which fixes transmitter location and antenna height. In the case of modification of existing sites, as well as for new sites, local planning restrictions have to be taken into account. Also, national limitations for radiation levels should be respected. Furthermore, the transmitter characteristics are limited by international agreements. All these restrictions may result in a coverage that is less than originally envisaged.

The considerations in this chapter concentrate on analogue and digital (DVB-T) television service area planning in the VHF and UHF bands in Europe. The principles are also applicable to analogue and digital radio planning. More information on terrestrial service area planning can be found in EBU and ITU reports, and in particular in Refs 1–5.

6.4.2 Planning Process

The process of terrestrial service area planning has several stages. Although in principle all stages of the process need to be followed, in a practical frequency-planning project some stages might have been completed already, or will be done later. Sometimes a project only concerns a partial revision of one stage. The whole process can take years or even decades. But also projects dealing with only a part of the process, such as planning of a few additional transmitters, may take a long time, in particular when international coordination or acquisition of new sites is involved. The following stages can be identified:

1. *Determination of long-term requirements and allocation of spectrum*. The result is an allocation in the 'European Common Allocation Table'[6] of the CEPT or, even better, an allocation in the ITU Radio Regulations that allows the use of the relevant frequency band for the terrestrial broadcasting.
2. *Agreement on an international frequency plan*. The result is a multilateral or ITU agreement with a detailed frequency plan for one or more of the frequency bands for which there is a broadcasting allocation.
3. *Establishment of national frequency planning criteria and methods*. The result is a report with detailed criteria and methods including the propagation prediction method for planning of broadcasting stations within the framework of the relevant international agreement.
4. *Specification and production of planning software*. The result is an operational software package that can make the calculations according to the specified national frequency planning criteria and methods. The software can either be company made or purchased from a specialised company.

5. *Production of a draft frequency plan.* The result is a detailed frequency plan, including predicted coverage, of one or more transmitters, or complete networks, taking into account practically available sites and realistic antenna diagrams.
6. *National and international frequency coordination.* The result is a nationally and internationally authorised frequency plan. International coordination is only needed if the planned transmitters are an addition to, or modification of, the transmitters in the relevant international frequency plan. National coordination is needed if there is more than one operator in the country.
7. *Production of final frequency plan.* The result is a detailed frequency plan as in stage 5, but including the restrictions resulting from national and international frequency coordination.

Field tests will often be carried out to prove the assumptions and test the results.

6.4.3 Regulatory Framework

Article 5 of the ITU Radio Regulations shows the frequency band allocations. In addition, there are also Regional or Worldwide Agreements that regulate the use of the bands concerned. Furthermore, in Europe, in a number of cases there are multilateral agreements made under the auspices of the CEPT. Table 6.4.1 shows the relevant agreements for Europe in the VHF and UHF broadcasting bands.

The RRC04/05 is a Regional Radio Conference to be held in two sessions in 2004 and 2005. This Conference will make a new frequency plan for digital broadcasting for the bands 174–230 MHz and 470–862 MHz for 120 countries in Europe, Africa and the Middle East.[7,8] The Agreement to be made at this conference will replace the relevant parts of among others the Stockholm Agreement of 1961[9] and the multilateral agreements of Wiesbaden 1995[10] and Chester 1997.[11]

Agreements such as Stockholm 1961 contain frequency plan(s), usually simply called Plan(s), and a set of procedures for making modifications to the Plan. The Chester Agreement does not have a Plan, but only contains detailed administrative and technical rules for planning of DVB-T stations in addition to the rules of the Stockholm Agreement.

The transmitting stations with the characteristics as indicated in a Plan, such as Stockholm 1961, can be brought into operation at any time without further consultation with neighbouring countries. However, if a new station is required or if one or more of the characteristics (for instance, radiated power, antenna height, location) need to be changed compared to those mentioned in the Plan, agreement should be sought with countries that are likely to be affected by the modification. This process is called coordination. Figure 6.4.1 shows the coordination contour of a 10 kW television transmitter in the centre of the Netherlands. All countries within this contour, in this example seven, have to be consulted.

The consulted countries will check if the modification does not cause significantly more interference to the services of the country concerned than there already is. The modification needs to be notified to the ITU after all relevant countries have agreed to it. The ITU will incorporate the modifications into the Plan after having checked that the procedures of the agreement have been respected. These coordinations often require detailed calculations and negotiations.

6.4.4 Receiving Conditions and Locations

6.4.4.1 Reception possibilities

Before the planning process starts it is necessary to define where and how the signal needs to be received. A number of receiving conditions can be identified:

- Rooftop reception.
- Portable reception, indoor and outdoor.
- Mobile reception.
- Reception by hand-held terminals.

Analogue television systems are designed for rooftop reception and consequently planning is done for rooftop reception only. With digital systems reception could by required under all four reception conditions, and for each of these conditions different sets of planning criteria need to be established.

Depending on regulatory or commercial requirements, reception may be required in the whole area under consideration or in parts of it. A possibility may be rooftop reception in the whole area and portable indoor reception in urban areas. Furthermore, the degree of coverage (percentage of locations) needs to be specified in particular for digital services. DVB-T signals have a rapid transition between good reception and no reception at all. Therefore, reception of the signal is required in a relative high percentage of locations (70–95%). For analogue

Table 6.4.1 Broadcast spectrum allocations and agreements in VHF and UHF

Frequency band	*Service*	*System*	*ITU agreement*	*Multilateral agreement*
47–68 MHz (VHF, Band I)	TV broadcasting	Analogue TV	Stockholm 1961	
87.5–108 MHz (VHF, Band II)	Sound broadcasting	FM	Geneva 1984	
	TV broadcasting	Analogue TV (Eastern Europe)	Stockholm 1961	
		Analogue TV	Stockholm 1961	
174–230 MHz (VHF, Band III)	TV broadcasting	DVB-T	RRC04/05	Chester 1997
	Sound broadcasting	T-DAB	RRC04/05	Wiesbaden 1995
470–862 MHz (UHF, Band IV/V)	TV broadcasting	Analogue TV	Stockholm 1961	
		DVB-T	RRC04/05	Chester 1997
1495–1479.5 MHz (UHF, L band)	Sound broadcasting	T-DAB		Maastricht 2002

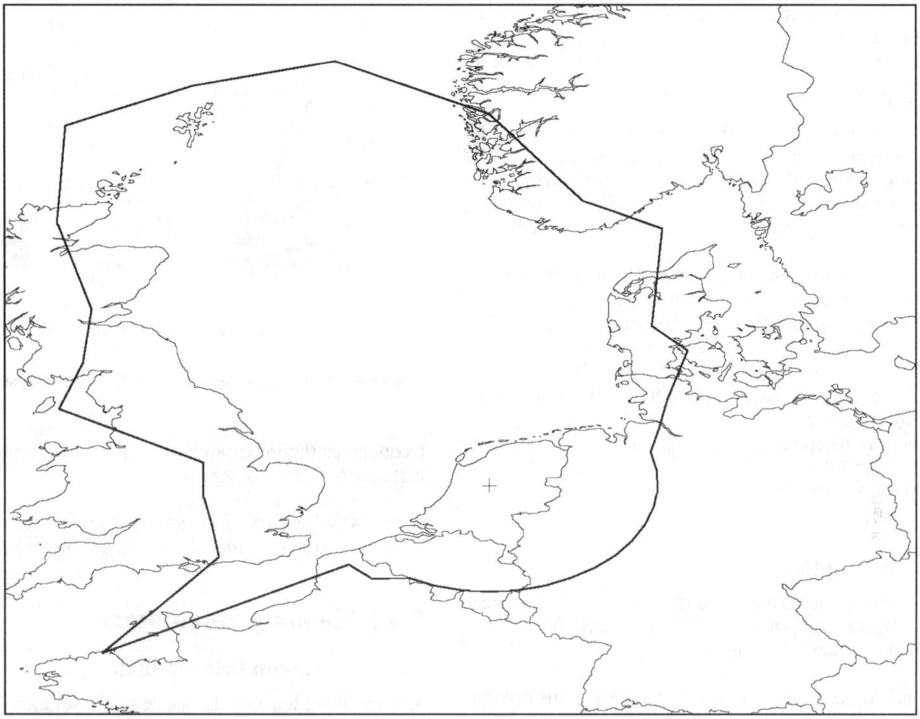

Figure 6.4.1 Coordination contour.

systems which have a smooth transition between good and no reception, it is normally sufficient to calculate reception for 50% of locations. Locations in this context are possible locations of the receiver antenna within a small area of, say, 100 by 100 m.

The received signal is also time dependent. Normally the wanted signal is coming from a nearby transmitter, is relatively stable and is calculated for 50% of the time. Interfering signals, however, often coming from distant transmitters and subject to severe tropospheric disturbances, are much more fluctuating, in particular when a sea-path is involved, and are normally calculated for 1% of the time in order to obtain protection against interference for 99% of the time.

6.4.4.2 Rooftop reception

Rooftop reception, also referred to as fixed reception, assumes a directional receiving antenna with a standardised directivity pattern on top of a house and directed towards the wanted transmitter. The antenna height is normally taken as 10 m above ground level.

6.4.4.3 Portable reception

Portable reception deals with a variety of conditions, including indoor and outdoor, a varying reception height ranging from ground level to the highest floors in an apartment building, stationary and in motion. Portable reception could include portable receivers with a built-in antenna or non-portable receivers with a simple antenna connected to it via a short cable. For planning purposes two cases are defined:[11]

Class A	Outdoor at a height of 1.5 m above ground level.
Class B	Indoor, ground floor (in a room with a window in an external wall) at 1.5 m above the floor.

The condition for a window in an external wall is important as tests have shown that the signal is always highest near a window, even if the window is not situated in the direction of the transmitter.

Portable reception and in particular portable indoor require a higher field strength than rooftop reception. The main differences of indoor reception compared to rooftop reception are:

- Higher required signal-to-noise ratio because the transmission channel is more disturbed by multi-path signals.
- No receiving antenna gain and antenna discrimination against interference.
- Building penetration loss.
- Propagation loss due to the lower receiving height.
- Larger standard deviation of the received signal due to the variety of building materials and constructions.

The difference between portable indoor and rooftop reception could be as much as 30 dB when reception is noise limited. In the case of interference from other transmitters the main difference is the lack of antenna discrimination for portable reception.

Because of the considerable higher field strength requirements, a given transmitter has a much smaller coverage area for portable then for rooftop reception. Ways to improve portable coverage are:

- Transmitting a more robust DVB-T variant (e.g. 16-QAM instead of 64-QAM), accepting a lower net bit rate and thus a reduction of the number of programmes in the multiplex.
- Increasing the power of the transmitter; however, in most cases the transmitter power is restricted by international agreements.
- Replacing a transmitter by a number of transmitters in a Single-Frequency Network (SFN) with lower power, but without exceeding the outgoing interference level given by the original transmitter. In this way a more homogeneous power distribution over the coverage area is achieved, together with a higher reception probability, as the signal will arrive from different sites.
- Use of diversity reception, which reduces the effect of fast fading (see Section 6.4.4.6).
- Use of domestic repeaters.[12]

6.4.4.4 Mobile reception

Mobile reception means reception while in motion, for instance in a car. Compared to portable outdoor reception, mobile reception has three more constraints:

- Higher signal-to-noise ratio required because the transmission channel is more subject to rapid signal variations.
- Higher location percentage (>95%) needs to be chosen, because contrary to stationary reception, no advantage can be taken of using an optimal receiving antenna position.
- Doppler shift, causing more distortion of the signals at higher speeds and at higher frequencies.

6.4.4.5 Hand-held terminals

Hand-held terminals are small screen devices with combined DVB and GSM or UMTS functionality. These devices can be used indoors and outdoors, at low speed and high speed (for instance, in trains). Because of the small dimensions the antenna gain is low; consequently, the required signal strength is high compared to portable and mobile reception.

6.4.4.6 Diversity reception

Space diversity by using one or more receiving antennas can improve portable and in particular mobile reception.[13]

With antenna diversity the required signal strength is reduced by 6–8 dB and in the case of mobile reception the maximum speed due to Doppler shift is increased. In the 8k variant of the DVB-T system the increases in maximum speed are:

- QPSK 1.5 times more with antenna diversity.
- 16-QAM 2.5 times more with antenna diversity.
- 64-QAM 2.5 times more with antenna diversity.

Figure 6.4.2 indicates maximum speed due to Doppler shift at 800 MHz for the DVB-T 2k variant, the DVB-T 8k variant and the 8k variant with diversity reception. At lower

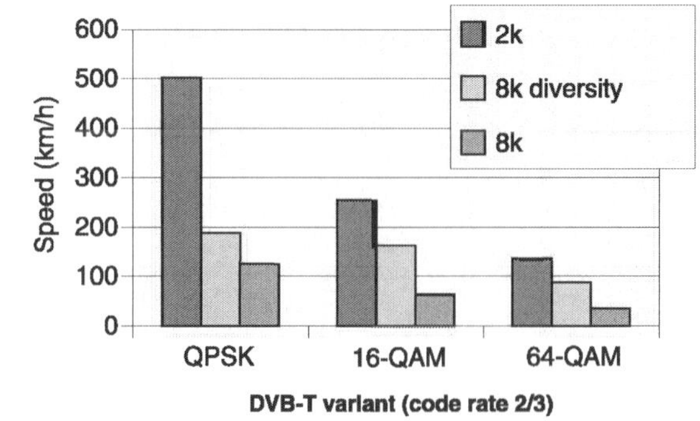

Figure 6.4.2 Maximum speed due to Doppler shift at 800 MHz.

frequencies the maximum speed is proportionally higher – for instance:

- At 200 MHz 4 times the speed at 800 MHz.
- At 500 MHz 1.6 times the speed at 800 MHz.

6.4.5 Planning Parameters

6.4.5.1 Minimum field strength

At a receiving location the wanted field strength should be equal or more than the minimum field strength in order to obtain the agreed reception quality in the absence of interference from other transmitters. Under these assumptions reception is limited by noise only. For terrestrial service area planning in the VHF and UHF bands, thermal noise of the receiver and, to a certain extent, man-made noise are of importance. Man-made noise is very dependent on the circumstances. In VHF, in cases where reception takes place in urban areas, a moderate allowance for man-made noise may be necessary. In UHF it is normally not necessary to take man-made noise into account.

The minimum median field strength (E_{min}, in dBμV/m) can be derived from the minimum median power flux density (Φ_{min}) at the receiving location:

$$E_{min} = 10\log(\Phi_{min} * Z_0) \quad (6.4.1)$$

where

Φ_{min} Minimum power flux density in dBμW/m^2; see formula (6.4.2)

Z_0 Free space impedance, 120πΩ.

The minimum median power flux density (in dBμW) can be calculated by the formula:

$$\Phi_{min} = C/N + P_n - A_a + L_f + L_h + L_b + C_l \quad (6.4.2)$$

where

C/N Required RF signal-to-noise ratio (dB)

P_n Noise power of the receiver (dBμW); see formula (6.4.3)

A_a Antenna aperture (dB); see formula (6.4.4)

L_f Antenna feeder loss (dB)

L_b Building penetration loss (dB) in the case of indoor reception

L_h Height loss between field strength at calculated and actual receiving height (dB)

C_l Location correction factor (dB); see formula (6.4.5).

The noise power is (in dBμW):

$$P_n = F + P_a + 10\log(kT_0B) + 120 \quad (6.4.3)$$

where

F Receiver noise figure in dB

P_a Allowance for man-made noise noise (dB)

T_0 Absolute temperature, normally taken as 290 K

k Boltzmann's constant (1.38×10^{-23})

B Receiver noise bandwidth (T-DAB: 1.536 MHz, DVB-T (in 8 MHz channel): 7.61 MHz.

The antenna aperture is:

$$A_a = G_r + 10\log(1.64 * \lambda^2/4\pi) \quad (6.4.4)$$

where

G_r Gain of the receiving antenna relative to a dipole antenna (dB)

λ Wavelength in m.

The location correction factor (in dB) is:

$$C_l = U * \sigma \quad (6.4.5)$$

where

U Log-normal distribution factor of the required percentage

σ Location standard deviation of the received signal.

From formulae (6.4.1)–(6.4.4) it can be derived that the minimum median field strength in dBμV/m is:

$$E_{min} = C/N + F + P_a - G_r + L_f + L_h + L_b + C_l + 20\log(f) - 30.1 \quad (6.4.6)$$

where

f frequency (MHz).

6.4.5.2 Factors that affect the minimum median field strength

6.4.5.2.1 C/N

The required radio-frequency signal-to-noise ratio (C/N) depends on the required output quality, the transmission system, the receiver characteristics and the personal perceptiveness of the consumer. The C/N is measured in laboratories under controlled circumstances with a panel of listeners or viewers (subjective method) or determined with an objective measurement method. As an example, the C/N values for a number of DVB –variants, as well as analogue television, are given in Table 6.4.2.

The C/N values for DVB-T in Table 6.4.2 are derived from the simulated performance figures given in the DVB-T specification,[14] but include an implementation margin of 3 dB to allow for additional noise generated in receivers in practice.

The Gauss channel represents a multi-path free channel but taking into account thermal noise. The Rice channel is typical for rooftop reception and represents the situation where there is a dominant signal with lower level delayed signals and thermal noise. The Rayleigh channel represents the situation where there are several statistically independent incoming signals with different delay times, none of which dominates, together with thermal noise. This is applicable under portable and mobile reception conditions.

Table 6.4.2 RF signal-to-noise ratio (C/N) for analogue and digital television

System	*C/N (dB) Gauss channel*	*C/N (dB) Rice channel*	*C/N (dB) Rayleigh channel*
DVB-T QPSK 2/3	7.9	8.7	11.4
DVB-T 16-QAM 2/3	14.1	14.6	17.2
DVB-T 64-QAM 2/3	19.5	20.1	22.3
Analogue TV	36	–	–

Analogue television is designed for rooftop reception and the C/N has been established for a Gauss channel. A lower level delayed signal (Rice channel) may be visible on the screen as a ghost image, but does not affect the signal-to-noise ratio. The C/N of digital television signals, however, is affected by delayed signals, and therefore the C/N values have been established for different transmission channels that are typical for rooftop, portable and mobile reception.

6.4.5.2.2 Receiver noise figure

The noise figure of a DVB-T receiver is around 7 dB.[11] For analogue television receivers a noise figure of 8 dB in VHF and 11–12 dB in UHF is assumed.[1]

6.4.5.2.3 Building penetration loss

Building penetration loss is highly dependent on the kind of building material and the construction of the house and can range, from a few decibels to more than 20 dB. For T-DAB and DVB-T planning a building penetration loss of 8 dB in VHF and 7 dB in UHF is used, with standard deviations of respectively 3 and 6 dB.[11]

6.4.5.2.4 Location correction factor

For wideband signals such as T-DAB and DVB-T a standard deviation of 5.5 dB is taken for outdoor locations and 8.3 dB for indoor locations. The indoor standard deviation includes the outdoor and the building loss standard deviation.[11] It should be noted that propagation prediction errors are not included in these values. Table 6.4.3 shows the log-normal distribution factor (U) and location correction factor for some often used percentages of location probabilities.

6.4.5.2.5 Antenna gain and feeder loss

The receiving antenna gain (G_r) is expressed in dB relative to a dipole antenna. For directional rooftop antennas the gain (in dB) is normally taken as:

$$G_r = G_m + 10\log F_a/F_r$$

Table 6.4.3 Location correction factor

Location probability(%)	*U*	*C_l outdoor (dB)* $\sigma = 5.5\,dB$	*C_l indoor (dB)* $\sigma = 8.3\,dB$
99	2.33	12.8	19.3
95	1.64	9.0	13.6
90	1.28	7.0	10.6
70	0.52	2.9	4.3
50	0.00	0	0

where

G_{rn} 7 dB at 200 MHz and 10 dB at 500 MHz

F_a Actual frequency in MHz

F_r Reference frequency in MHz for which G_{rn} is specified in Band III and Band IV/V respectively.

For mobile and portable reception the receiving antenna gain is normally taken as 0 dB or even less.

Antenna feeder loss can be neglected in the cases of mobile and portable reception, whereas for rooftop reception normally 2 dB is taken in Band III and 3 dB at 500 MHz in Band IV/V, with a frequency correction as for the antenna gain.

6.4.5.2.6 Examples of E_{min}

Table 6.4.4 shows the minimum median field strength for a number of typical DVB-T system variants and reception conditions.

Although the minimum median field strength can be calculated using formula (6.4.6), in a number of broadcasting agreements and ITU-R recommendations compromise values have been agreed. Table 6.4.5 shows the agreed minimum field strength values for analogue television from Recommendation ITU-T BT.417.[15]

Table 6.4.4 Minimum field strength at a receiving height of 10 m for digital television

Item	*DVB-T 64-QAM 2/3 rooftop*	*DVB-T 16-QAM 2/3 indoor*
Location probability (%)	95	70
Frequency (MHz)	500	500
Required RF signal-to-noise ratio, C/N (dB)	20.1	17.2
Receiver noise factor, F (dB)	7	7
Allowance for man-made noise, F_a (dB)	0	0
Gain of the receiving antenna, G_r (dB)	10	0
Antenna feeder loss, L_f (dB)	3	0
Height loss between calculated and actual receiving height (dB)	0	12
Building penetration loss, L_b (dB)	0	7
Location correction factor (dB)	9	4.3
Minimum median required field strength (dBμV/m), using formula (6.4.6)	53.0	71.4

Table 6.4.5 Minimum field strength – analogue television

Frequency band	*Median value for protection against interference (dBμV/m)*	*E_{min} (dBμV/m) thermal noise only*	*Median value for protection against interference (dBμV/m) Rural areas*	*E_{min} (dBμV/m) thermal noise only Rural areas*
Band I	48	47	46	40
Band III	55	53	49	43
Band IV	65	62	58	52
Band V	70	67	64	58

6.4.5.3 Usable field strength

In addition to noise, the received signal may also be impaired by interference from other transmitters. The 'usable field strength' is (according to Recommendation ITU-R V.573)[16] the minimum value of the field strength to permit a desired reception quality, under specified receiving conditions, in the presence of natural and man-made noise and of interference, either in an existing situation or as determined by agreements or frequency plans.

The usable field strength is calculated by the summation of the 'nuisance fields' and the minimum median field strength. The nuisance field corresponds to the value of the wanted signal that would be needed if there was interference from the interfering transmitter under consideration only. The formula for calculating the nuisance field (in dBμV/m) is:

$$E_n = E_i(T) + \mathrm{PR} + D_a + D_p + C_l$$

where

$E_i(T)$ Interfering field strength in dBμV/m of T % of time; T is usually taken as 1%

PR Protection ratio in dB

D_a Directivity discrimination of the receiving antenna in dB

D_p Polarisation discrimination of the receiving antenna in dB

C_l Location correction factor.

For analogue television two sets of protection ratio are relevant: one for tropospheric and another for continuous interference. Protection ratios for tropospheric interference correspond to 'slightly annoying' impairment. For non-fading signals a higher degree of protection is required and the protection ratios for continuous interference apply. Consequently, two calculations of the nuisance field should be carried out, one for 50% of time with the protection ratio for continuous interference and one for 1% of time with the protection ratio for tropospheric interference. The highest resulting value should be taken.

All relevant nuisance fields and the minimum median field strength need to be combined. There are several methods, ranging from power summation to complex statistical methods.[4]

The Monte Carlo method, in which random values of wanted and unwanted signals are generated within the limits of the mean value and standard deviation of the distribution of each signal, is the most accurate but also the most time-consuming. Log-normal methods assume that the resulting sum distributions are also log-normally distributed. The standard log-normal method is the most commonly used. In order to improve the accuracy for high probabilities, which are of most interest for digital TV coverage, methods like k-LMN, t-LMN have been developed. These methods give a higher accuracy but are more complex than the original log-normal method. The power sum method is given by the formula:

$$E_u = 10 \times \log_{10}\left(10^{\frac{E_{min}}{10}} + \sum_{i=1}^{n} 10^{\frac{En_i}{10}}\right)$$

where

E_u Usable field strength in dBμV/m

E_{min} Minimum median field strength in dBμV/m

En_i Nuisance field strength of the *i*th unwanted signal in dBμV/m

n Number of interferers.

6.4.5.4 Factors that affect the usable field strength

6.4.5.4.1 Protection ratio

The protection ratio (PR) is the required RF signal to interference ratio; it depends on the required signal quality, the transmission system, the receiver characteristics and the personal perceptiveness of the consumer. The protection ratio is measured in laboratories under controlled circumstances with a panel of listeners or viewers (subjective method) or determined with an objective measurement method. Protection ratios for analogue television interfered with by analogue television and T-DAB can be found in Recommendation ITU-R BT.655.[17] All protection ratios involving DVB-T can be found in recommendation ITU-R BT.1368.[18]

The analogue television spectrum is characterised by harmonics of the line frequency (15,625 Hz). An interfering signal at these harmonics is more harmful than in between them. For that reason co-channel television transmitters are given an offset to obtain a frequency difference of one-third or one-half of the line frequency. Protection ratios for some typical cases are reproduced in Table 6.4.6. An additional improvement can be obtained by precision offset. Superimposed over the line frequency spectrum are harmonics of the field frequency (50 Hz). A 25 Hz difference and a high frequency stability (<1 Hz) between two transmitters gives an additional reduction of the protection ratio. The protection ratios in Recommendation ITU-R BT.655, as well as in Table 6.4.6, are for negatively modulated television systems. The values need to be increased or decreased by 2 dB if a positively modulated system (L/Secam) is the interfering respectively wanted system.

Table 6.4.6 Some typical co-channel protection ratios – analogue television

Frequency difference	*Protection ratio (dB) Non-precision offset*		*Protection ratio (dB) Precision offset*	
	Tropospheric	*Continuous*	*Tropospheric*	*Continuous*
0 × line frequency	45	52	32	36
⅓ × line frequency	30	40	22	27
½ × line frequency	27	33	24	30

The digital television spectrum is noise like and therefore the co-channel protection ratios are equal to the C/N for the relevant system variant (see Section 6.4.5.2 and Table 6.4.2).

6.4.5.4.2 Antenna discrimination

If directional antennas are used, advantage can be taken of the directivity and polarisation discrimination characteristics of the antenna (see Table 6.4.7 and Recommendation ITU-R BT.419).[19]

Table 6.4.7 Antenna discrimination

Frequency band	*Maximum directivity discrimination (dB)*	*Linear interpolation between*
I	−6	50–60; 310–300 degrees
II, III	−12	26–60; 334–300 degrees
IV, V	−16	20–60; 340–300 degrees

When wanted and interfering signals are orthogonally polarised, a combined (directivity and polarisation) discrimination of −16 dB may be applied for all angles of azimuth. In case of mobile and portable reception no antenna discrimination (neither directivity nor polarisation) is taken into account.

6.4.5.4.3 Location correction factor

With two varying signals (the wanted and the interfering signal) the location correction factor should take account of the combined standard deviation. Assuming no correlation between the distributions of wanted and unwanted signals, the location correction factor is:

$$C_l = \sqrt{2} * U * \sigma$$

where

U Log-normal distribution factor of the required percentage

σ Location standard deviation for the received signal.

6.4.6 Field Strength Prediction Methods

In planning, 'site-specific' and 'site-general' field strength prediction methods are used. For a 'site-specific' method the propagation path from transmitter to receiving location is determined using detailed terrain and clutter data, and the field strength at each receiving point is calculated taking into account diffraction losses over irregular terrain with different types of obstacles (see also Recommendation ITU-R P.526).[20] These models are often calibrated with measuring results and contain empirical correction factors. It is clear that these kinds of calculation take considerable time. Site-specific methods are used for detailed coverage calculations.

In 'site-general' methods, curves are used that show the mean value of the field strength as a function of distance for different percentages of time. Terrain conditions are taken into account by means of the 'effective transmitter antenna height' (the average height of the transmitting antenna between 3 and 15 km from the transmitter). Local terrain circumstances at the receiver location can be taken into account by means of a correction based on the 'terrain clearance angle' (the angle seen from the receiving point that just clears all terrain obstacles within 16 km in the direction of the transmitter). For the calculation of both the effective antenna height and the terrain clearance angle, a terrain data bank is needed. In the absence of a detailed terrain data base, the GLOBE data (1 km approximately) could be used (see www.ngdc.noaa.gov/seg/topo/globe.html). Site-general methods such as the former Recommendation ITU-R P.370[21] and the new Recommendation ITU-R P.1564[22] are used for international coordination and at a planning conference, and to obtain quick indications of coverage. These methods also have to be used in cases where no details of the terrain are available.

6.4.7 Network Configurations

6.4.7.1 Network topologies

Broadcasting networks consist often of 'main transmitters' with high antennas and relatively high powers, supplemented by 'fill-in' transmitters with low powers to provide coverage in areas that are shielded from the main transmitter. Coverage areas for broadcast services can range from national, regional and local. Depending on the size of the area to be covered, the terrain, frequency availability and practical considerations such as the presence of existing sites and local planning regulations, one or more transmitters are needed to cover the area. The transmitters can operate in a Multi-Frequency Network (MFN) or in a Single-Frequency Network (SFN).

In an MFN the transmitters use different frequencies. The same frequency can be reused, but only at considerable distance from the other transmitters. The reuse distance of a frequency depends on antenna height and power, the frequency difference and the kind of terrain. The transmitters in an MFN may have the same or different programmes. MFN networks therefore have the option for local windows in programming. In an SFN all transmitters transmit on the same frequency and the programme needs to be the same. An SFN may consist of several transmitters with more or less equal powers or of a main transmitter with fill-in transmitters, or combinations of both.

Analogue television transmitters need to be on different frequencies (MFN) to achieve a contiguous coverage area. The reuse distance of a frequency can be reduced by using precision offset. DVB-T and T-DAB transmitters have the ability to operate as a Single-Frequency Network (SFN) without causing an interference zone. However, the separation between transmitters in an SFN should be planned carefully to ensure that, at any reception point inside the coverage area of the SFN, the delay between the signal on which the receiver is synchronised and the signals from the other transmitters in the SFN falls within the guard interval (see next section for further explanation).

6.4.7.2 Guard interval

At a certain reception point the signals from transmitters in an SFN will be received with different delay times compared to the signal from the nearest transmitter. In general, DVB-T receivers will synchronise on the first incoming signal of the SFN above a certain threshold.[23] Signals with delay times shorter than the guard interval are treated as wanted, whereas signals that arrive with delay times longer than the guard interval are treated as interfering signals. In DVB-T the behaviour of the receiver near the edge of the guard interval is different from that of a T-DAB receiver, which shows a smooth transition between wanted and interfering. In DVB-T there is a sudden transition from wanted to interfering at the edge of the guard interval. Wanted respectively interfering signals from the SFN are combined using one of the methods described in Section 6.4.5.3. Internal network interference, i.e. interference from transmitters of the SFN, can be reduced by:

- Increasing the guard interval, at the cost of reducing the net bit rate. The DVB-T system has the option to use four different guard intervals – in the 8k variant: 28, 56, 112 and 224 μs.
- Adding an artificial delay at the nearest transmitter in order to bring the interfering signal inside the guard interval. However, by doing so the delayed signal of this transmitter may arrive outside the guard interval in other receiving locations in the coverage area of the SFN.
- Reducing the power in the direction of the interfered area. This may of course cause a coverage problem near the transmitter of which the power has been reduced.
- Using another frequency (if available) for the transmitter that causes interference.

Because of the internal network interference, SFNs in DVB-T have a limited size.

6.4.7.3 Programme feeding links

The network from the broadcasting house to the transmitters is called the distribution network. Programme distribution can be done in different ways: satellite, radio relay links, cables or off-air. The distribution links need to have a very high reliability. In practice, radio relay links prove to be very reliable. Cables are sensitive to failures due to construction works along roads. Off-air reception at the transmitter site (also called 'Ballempfang') is the cheapest way and is often used for fill-in transmitters and as a backup possibility when radio relay or cable links fail. With 'Ballempfang' also received interference will be retransmitted; therefore, planning should be done carefully. In general, reception and retransmission of co- and adjacent channels is not possible or at best requires special measures to obtain sufficient isolation between the weak received signal and the relatively strong transmitted signal.

6.4.8 Transition from Analogue to Digital Television

6.4.8.1 Compatibility of analogue and digital television

Table 6.4.8 indicates the range of protection ratios for analogue television and DVB-T in 64-QAM mode with code rate 2/3 taken from Recommendation ITU-R BT.1368[18] and 655.[17] Protection ratios for analogue television interfered with by analogue television range from 22 to 45 dB, depending on the offset and the application of precision offset. The protection ratio for DVB-T (64-QAM 2/3; Rice channel) interfered with

Table 6.4.8 Relation between protection ratios for analogue and digital television

System to be protected	*Interfered with by analogue TV*	*Interfered with by DVB-T*	*Increase of interference*
Analogue TV	22–45 dB	34 dB	+12 to −11 dB
DVB-T 95% locations	3 + 13 dB	20 + 13 dB	+17 dB
Increase of interference	−6 to −29 dB	−1 dB	

by analogue television is only 3 dB, but when protection for 95% of locations is required, a location correction factor of $\sqrt{2} \times 9$ dB (see Table 6.4.3) should be added.

Table 6.4.8 shows that DVB-T (64-QAM 2/3) is 6–29 dB less sensitive to interference from an analogue transmission than analogue TV. DVB-T with 16-QAM or QPSK is even more robust. On the other hand, interference to analogue television may increase up to 12 dB if an analogue transmitter is converted to digital, unless the power of the digital transmitter is reduced accordingly. Special attention is needed in the case where a DVB-T transmitter is interfered with by an analogue transmitter and at a certain moment the analogue transmitter is converted to digital. The interference will increase by 17 dB in that case unless the power of the DVB-T transmitter is reduced accordingly.

6.4.8.2 Planning of digital television in the transition from analogue to digital

There are more than 87,000 analogue television transmitters in Europe. Most of these are small fill-in transmitters; nevertheless, all can claim protection against interference from new transmitters, including digital ones. As long as analogue television is not switched off the presence of these analogue television transmitters poses severe restrictions on the introduction of digital television. Nevertheless, depending on the analogue frequency planning situation, there are possibilities for planning digital television transmitters. Channels for digital television can be:

- Unused channels of the Stockholm Agreement. A number of countries, has not brought into operation all their stations from the Stockholm Agreement[9]. Under certain conditions (e.g. power reduction; see Chester Agreement)[11] these channels can be used for digital television.
- Channels that were not available for broadcasting in the past; in some countries, for instance, some of the channels above 60 were allocated to military services but can now be used for digital television.
- Channels that cannot be used for analogue television. As digital television is very robust against interference (see Table 6.4.8) and can, in general, operate at lower power than analogue television for the same receiving conditions, channels such as those adjacent to analogue channels can be used for digital television.

6.4.9 Examples of Calculations

6.4.9.1 Calculation of coverage in a small area (100 by 100 m)

In this example the wanted field strength from a number of transmitters in an SFN on channel 21 (474 MHz) is compared to the usable field strength and the degree of coverage for portable indoor reception at a test point (an area of about 100 by 100 m) is calculated. Field strength has been predicted by a site-specific method using detailed terrain and clutter data. Table 6.4.9 shows the results for the wanted signals.

Table 6.4.9 Example of calculation of wanted signal resulting from an SFN

Tx in SFN	*ERP (kW)*	*Antenna height (m)*	*Distance (km)*	*Field strength (dBμV/m)*
A (DVB-T)	5	106	11	76.7
B (DVB-T)	10	140	19	75.1
C (DVB-T)	10	192	22	63.8
D (DVB-T)	6	138	29	55.5

The combined wanted field strength calculated by the log-normal method is 82.2 dBμV/m. Normalised to 10 m receiving height by adding 12 dB (see Table 6.4.4), the field strength becomes 94.2 dBμV/m. This is well above the minimum median field strength (see Table 6.4.4). The probability that the combined wanted field strengths exceed the minimum median field strength is near 100% in this case. However, reception at the test point is also subject to interference from other transmitters. The interfering transmitters are indicated on the map in Figure 6.4.3. Table 6.4.10 tabulates the calculated nuisance fields.

The combined nuisance field calculated by the log-normal method is 68.4 dBμV/m. Normalised to 10 m receiving height by adding 12 dB (see Table 6.4.4), the usable field strength becomes 80.4 dBμV/m. The probability that the combined wanted field strengths exceed the combined nuisance fields is 95.8% in this case. The reception probability, taking into account noise and interference, is 95.7%.

6.4.9.2 Calculation of the coverage area

For as many test points as practically achievable, depending on the accuracy of the terrain database and the available time, the calculations shown above will be repeated and at each point the coverage degree is calculated. Figure 6.4.4 shows the resulting coverage area of an SFN consisting of four transmitters for portable indoor reception. Coverage is indicated for 70%, 90% and 95% probability in different shades of grey.

If a site-general propagation model is used without terrain clearance angle correction, the calculations are terrain independent. Therefore, calculations can be made along radials and the coverage area can be indicated by means of a contour.

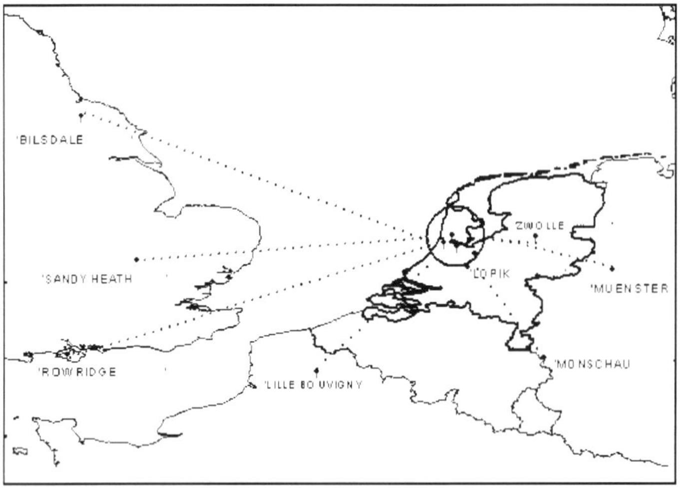

Figure 6.4.3 Overview of location of transmitters.

6.4.9.3 Calculation of transmitter power

The first step is to calculate the usable field strength at critical points in the coverage area in the way described above. The power of the transmitter should be sufficiently high to create a field strength that is equal or more than the usable field strength.

The second step is to check if other coverage areas are not unacceptably interfered. If other potentially affected areas are within the same country, nationally agreed criteria should be applied. In the case of foreign potentially affected areas, the international coordination rules apply. The Chester Agreement,[11] for instance, states that an increase of 0.3 dB to the reference usable field strength at any test point should normally be accepted. As the usable field strength for coordination purposes is calculated by means of the power sum method, the nuisance field of a new transmitter or SFN should be 12 dB less than the reference usable field strength at a test point. The reference usable field strength is the usable field strength calculated with the interfering stations agreed at a certain date in the past – in the Chester Agreement, 25 July 1997.

Finally, after the frequency coordination process has been completed, the transmitter characteristics need, as far as necessary, to be adjusted and the coverage recalculated. Figure 6.4.5 shows an example of a resulting antenna diagram. Originally a non-directional diagram had been requested; as a result of international frequency coordination, a number of restrictions had to be accepted, shown by the outer diagram. The inner diagram is the one that has been realised in practice.

Table 6.4.10 Example of calculation of nuisance fields

Interfering Tx	*Channel difference*	*ERP (kW)*	*Antenna height (m)*	*Distance (km)*	*Field strength (dBμV/m)*	*Protection ratio (dB)*	*Nuisance field (dBμV/m)*
U1 (ATV)	0	33	70	202	56.7	3	59.7
U2 (ATV)	0	25	174	172	57.1	3	60.1
U3 (ATV)	0	1000	320	255	52.4	3	55.4
U4 (ATV)	+9	1000	364	28	101.0	−46	55.0
U5 (ATV)	0	100	235	351	30.8	3	33.8
U6 (DVB-T)	0	10	274	466	13.3	22	35.3
U7 (ATV)	0	50	142	468	22.4	3	25.4
U8 (ATV)	+1	200	150	90	66.0	−38	28.0

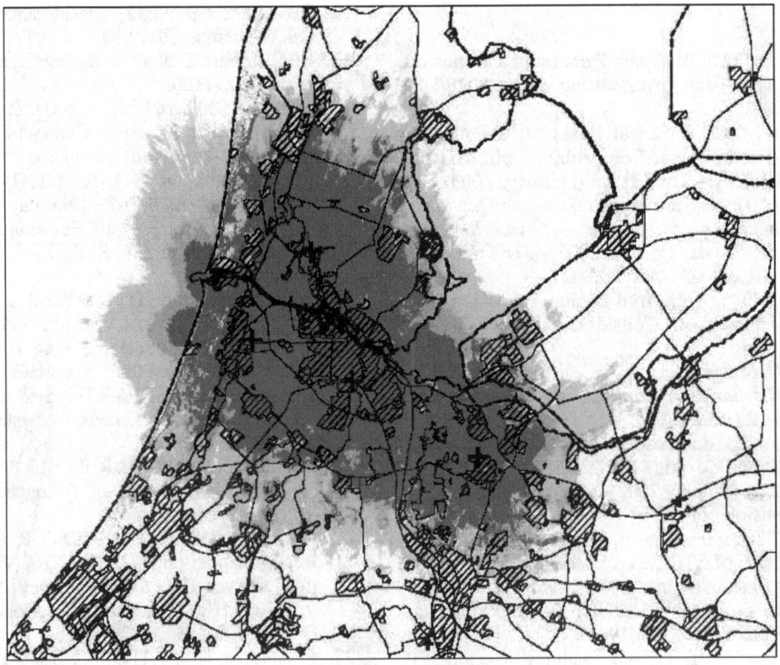

Figure 6.4.4 Presentation of coverage area.

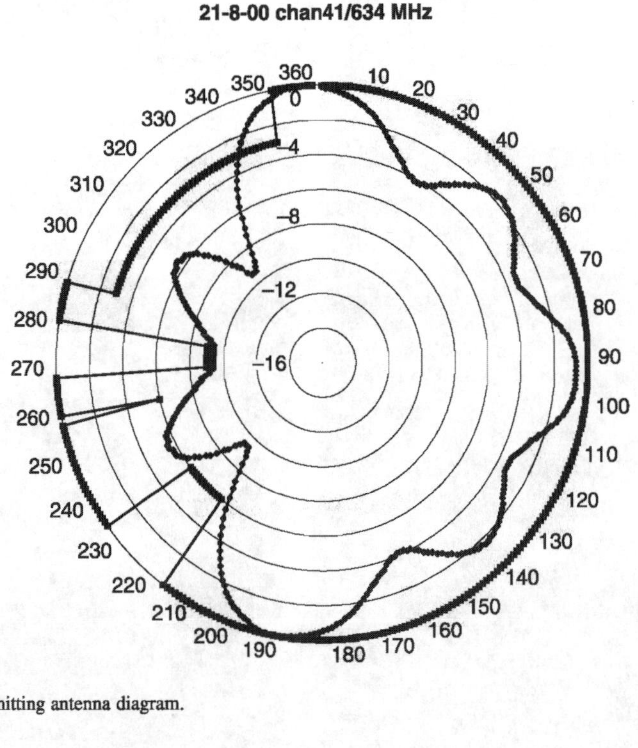

Figure 6.4.5 Example of transmitting antenna diagram.

References

1. EBU Report Tech. 3254, Planning Parameters and Methods for Terrestrial Television Broadcasting in the VHF/UHF Bands (May 1988).
2. EBU Report BPN003, Technical Bases for T-DAB Services Network Planning and Compatibility with Existing Broadcasting Services – Third Issue (February 2003).
3. *DSB Handbook, Terrestrial and Satellite Digital Sound Broadcasting to Vehicular, Portable and Fixed Receivers in the VHF/UHF Bands*, Edition 2002, ITU Publication Notice No. 343-02-cor-02 (May 2002).
4. EBU Report BPN005, Terrestrial Digital Television Planning and Implementation Considerations – Edition 3 (December 2001).
5. *Handbook on Digital Terrestrial Television Broadcasting in the VHF/UHF Bands*, Edition 2002, ITU Publication Notice No. 335-02 (June 2002).
6. CEPT ERC Report 25, European Table of Frequency Allocations and Utilizations Covering the Frequency Range 9 kHz to 275 GHz, Lisbon (January 2002), revised Dublin (2003).
7. Laflin, N. Revision of ST61, the key issues to be addressed, *EBU Technical Review* (April 2002).
8. Doeven, J. Revision of ST61, lessons learned from history, *EBU Technical Review* (April 2002).
9. Final Acts of the European VHF/UHF Broadcasting Conference, Stockholm 1961. ITU (1961).
10. Final Acts of the CEPT T-DAB Planning Meeting (3), Maastricht (2002).
11. The Chester 1997 Multilateral Coordination Agreement relating to Technical Criteria, Coordination Principles and Procedures for the Introduction of Terrestrial Digital Video Broadcasting (DVB-T), CEPT, Chester (25 July 1997).
12. EBU Report BPN032, Considerations on Domestic Repeaters (December 2000).
13. EBU Report BPN 47, Planning Criteria for Mobile DVB-T (February 2002).
14. ETSI EN 300 744 V1.14.1, Digital Video Broadcasting (DVB), Frame Structure, Channel Coding and Modulation for Digital Terrestrial Television (2001-01).
15. Recommendation ITU-R BT.417-4, Minimum Field Strength for which Protection may be Sought in Planning an Analogue Terrestrial Television Service.
16. Recommendation ITU-R V.573-4, Radio Communication Vocabulary.
17. Recommendation ITU-R BT.655-6, Radio Frequency Protection Ratios for AM Vestigial Sideband Terrestrial Television Systems Interfered with by Unwanted Analogue Vision Signals and their Associated Sound Signals.
18. Recommendation ITU-R BT.1368-3, Planning Criteria for Digital Terrestrial Television Systems in the VHF/UHF Bands.
19. Recommendation ITU-R BT.419-3, Directivity and Polarisation Discrimination of Antennas in the Reception of Television Broadcasting.
20. Recommendation ITU-R P.526-7, Propagation by Diffraction.
21. Recommendation ITU-R P.370-7, VHF and UHF Propagation Curves for the Frequency Range from 30 MHz to 1000 MHz Broadcasting Services (withdrawn on 22 October 2001).
22. Recommendation ITU-R P.1546-1, Method for Point-to-area Predictions for Terrestrial Services in the Frequency Range 30 MHz to 3000 MHz.
23. Brugger, R. and Hemingway, D. Impact on coverage of inter-symbol interference and FFT window positioning in OFDM receivers, *EBU Technical Review* (July 2003).

K Davison
Manager Communications, Thames Television
G A Johnson BSc, C Eng, MIEE
Deputy Head Engineering Services, ITV Association
Revised by
Salim Sidat MBA, DBA, BSc (Hons), CEng, MIEE
Customer Systems Manager
mmO2, Ltd

6.5 Satellite Distribution

The following sections describe features of satellite communications that are of particular interest to television broadcasters and facility companies. They consider the availability of fixed satellite services (FSS) for point-to-point and point-to-multi-point television transmissions and indicate limitations that the use of communications satellites may impose on television production techniques.

6.5.1 Background

The first satellite to be equipped with an active transponder (receiving and transmitting equipments) for television transmissions was *Telstar 1* launched by the United States in 1962. It flew in an elliptical orbit between 914 and 5436 km (568 and 3378 miles) above the earth's surface. Each orbit took 3 hours and contact between the east coast of America and ground stations near the west coasts of England and France was limited to periods of 20 minutes, which did not always suit the broadcaster's needs. Extremely large, expensive, complex steerable antennas were required at the ground stations to track the satellite as it moved across the sky.

A method to provide almost continuous coverage had been suggested by Arthur C. Clarke in 1945. Three satellites (Figure 6.5.1) would act as repeaters in the sky and be equally spaced in the equatorial plane in a circular orbit at a critical altitude of 35,787 km (22,237 miles). The direction of rotation would be the same as that of the earth; at this altitude the time taken for one orbit would be precisely one sidereal earth day. Consequently, each satellite would be fixed above a particular location on the equator and appear stationary to anyone on earth. This is the only orbit to allow fixed antennas to be deployed on the satellites and on the ground. The first such *geostationary* satellite designed to handle television signals was launched in August 1964.

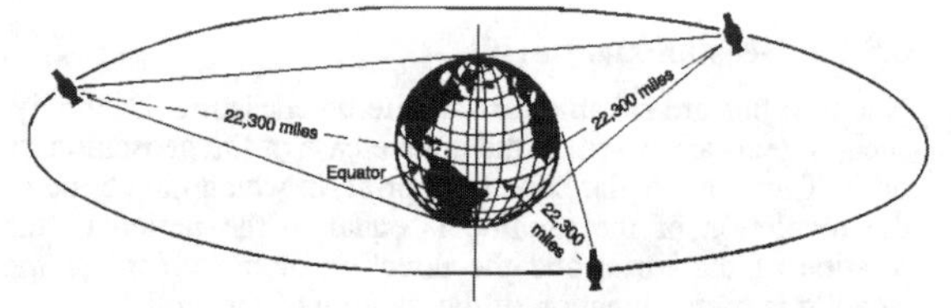

Figure 6.5.1 Three geostationary satellites at an altitude of 35,800 km above the equator can provide virtually complete coverage of the earth's surface.

6.5.1.1 DBS systems

Satellites have been used for many years in the transmission of signals from one terrestrial point to another. Services of this kind are known as *fixed satellites services* (FSS). Fixed service satellites are used for relaying television programmes from one country to another, e.g. via the Eurovision network. For direct broadcasting by satellite (DBS), the satellite transmitted signals are at higher power levels so that the signal can be received by a relatively small receiving antenna connected to the television set in the home.

In a DBS system, programmes from a ground station are transmitted to the satellites in a fixed satellite service band. Bands specifically intended for this purpose (feeder links) are used. The ground station, or possibly a separate station, also transmits signals for the control and monitoring of the satellite. The satellite converts the programme carrying signals from the ground station (the up-link) to a frequency in a band allocated to a DBS and retransmits them in a beam aimed toward the service area.

Two types of reception are possible: individual reception and community reception. For individual reception, the viewer needs two items of equipment. The first is an outdoor unit, which must be mounted where there is a clear view of the satellite. This unit is likely to consist of a parabolic antenna (or dish) between 30 and 90 cm in diameter, or possibly a flat panel, together with the associated system to convert the signal to a lower frequency suitable for transmission, by coaxial cable, to the second unit, the indoor unit. This is likely to consist of a channel selector mounted on or near the television receiver itself.

For community reception, the signal from the satellite is received by a relatively large antenna, possibly a 2–3 m dish, down-converted and distributed by cable to a number of receivers, e.g. in a block of flats. With modulation conversion, it is possible to use conventional cable networks for distribution over a wide area.

DBS differs from conventional terrestrial broadcasting in that full national coverage can be achieved by a single transmitter on a satellite. With individual reception, some viewers may not be able to receive the transmissions because they do not have an unobstructed view of the satellite. This problem is more likely to occur in built-up areas with the obstruction taking the form of a neighbourhood building or other structure. Geographical features can also obscure the view of satellites. As examples, the angles of elevation required to achieve a boresight with a geostationary satellite at 19.2°E (Astra) would be about 23° in Edinburgh to 28° in London.

6.5.1.2 Geostationary orbit

Not all orbits are suitable for satellite broadcasting. Generally, such systems are based on the unique case of the geostationary orbit. This is a circular equatorial orbit, in which the period of the revolution of the satellite is equal to the period of the rotation of the earth, and the direction of movement of the satellite is in the direction of the rotation of the earth.

Such a satellite appears to be stationary in the sky when viewed from any point on the earth and is fixed at the zenith of a given point on the equator, the longitude of which is that of the satellite. This longitude describes the position of a satellite on the geostationary orbit. The great advantage of the geostationary orbit is that it enables a fixed receiving antenna to be used on the ground. This is so important that almost all the systems studied in connection with satellite broadcasting are based on geostationary satellites.

The characteristics of the idea of geostationary orbit are:

Period, t	86,164.091 s = 23 h 56 min 4.091 s
Equatorial radius of the earth, r:	6378.16 km
Altitude, h:	35,786.04 km
Radius of the orbit, $r+h$:	42,164.20 km
Speed of the satellite:	3.074662 km/s
Length of 1° arc:	735.904 km

6.5.1.3 Satellite geometry

The geometry of a geostationary satellite can be seen from Figures 6.5.2 and 6.5.3. The position of the satellite S is defined by its longitude. The latitude is determined by the geostationary orbit. The receiving point P is taken to be a longitude λ and a latitude φ.

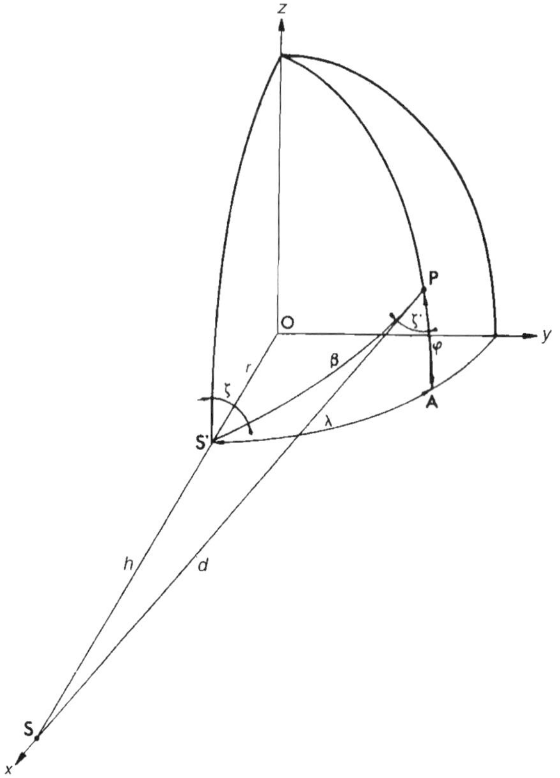

Figure 6.5.2 Evaluation of the angles at which a point P on the earth is seen from the point S′ on the earth's surface situated vertically below a geostationary satellite S.

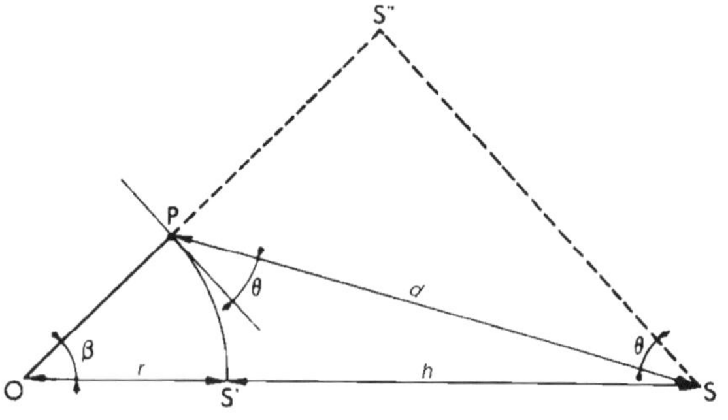

Figure 6.5.3 Evaluation of θ, the angle of elevation of the satellite S.

Taking the difference in longitude to be $\lambda = \lambda_1 - \lambda_s$, it can be shown that:

Azimuth $\zeta'^{\circ} = \tan^{-1}(\tan\lambda / \sin\varphi)$,
Elevation $\theta^{\circ} = \tan^{-1}(\cos\beta - \delta)/\sin\beta$

where $\delta = r/(r+h) = 0.151269$ and $\beta = \cos^{-1}(\cos\varphi\cos\lambda)$ and distance $d(\text{km}) = 35{,}786\sqrt{(1 + 0.41999(1 - \cos\beta))}$

6.5.1.4 Transponders

Like the early active satellites, modern designs are equipped with transponders to handle the signals. A *transponder* comprises a low-noise receiver, a frequency converter and a high-power amplifier. Many recent satellites are equipped with matrices to allow transponders to be switched by ground command between different receiving and transmitting antennas. Flexible connectivity is considered in more detail in Section 6.5.4.5.

The ever-evolving satellite communication systems increasingly utilise digital transmission requiring linearity over a wide dynamic range; simply put, 'higher data rates require higher power'. Traditionally, high power levels have been synonymous with Travelling Wave Tube Amplifiers (TWTAs) and Klystron Power Amplifiers (KPAs). However, advancements in power technology have seen the introduction of Solid-State Power Amplifiers (SSPAs), which provide a new platform that complements system designs. For example, highly compact, 50 W antenna-mounted SSPAs are available for use in Ku-band satellite communications systems with linearity better than a comparable TWTA at 125 W.

QPSK (Quadrature Phase Shift Keying) has been the staple modulation for satellite links for the last 20 years. It is bandwidth efficient and robust in terms of interference. However, there is a push towards use of higher-order modulation schemes. This is being made possible as nearly all satellite systems being designed today are using linearised TWTAs. In fact, video broadcast modulators capable of QPSK, 8-PSK and 16-QAM (Quadrature Amplitude Modulation) from a single unit are available on the market today.

Many transponders on multi-purpose communications satellites are available for long-term and occasional television leases. The occasional lease arrangement is generally used for short-duration injects such as news material; such short-duration exchanges are inclined orbit satellites which are ideal for providing low-cost capacity in comparison to geostationary satellites. Satellite tracking is not needed for such short-duration SNG uses, such as sports events. For longer transmissions, uplinking SNG stations can be repositioned manually. A step-track or intelligent processor track system that learns satellite orbital path can be employed, but in general, tracking does not become necessary until dish size exceeds 3.7 m for Ku-band.

6.5.1.5 Footprint or satellite coverage

The service area associated with a down-link is usually described in terms of a footprint map of the appropriate part of the earth's surface marked with contours representing either transmitted equivalent isotropic radiated power, EIRP (dB(W)), or received power flux density (dB(W/m^2)). Transponders are normally operated with *back-off*, or power reduction, from these maximum possible levels, which usually correspond to HPA saturation (in the region of the knee in its transfer characteristic). The user should always check with the satellite operator the back-off applying to a particular transponder. The point near the centre of the footprint at which the flux density is greatest is known as the *bore-sight*.

The shape and area of the footprint is determined by the feeder arrangements to the transmitter antenna and its shape and beamwidth. The footprint also indicates the approximate area over which an up-link may successfully access a transponder on the satellite. Satellite operators provide depiction of footprint coverage for the geographical land mass by using contour levels in dBs (very similar to contours used on topographical maps to indicate height); these contours are usually in 1 dB step of EIRP (dB(W)) or as *G/T* (dB(K^{-1})) falling away from the beam peak (i.e. from bore-sight, where the power received on the ground is at its maximum).

Calculation of real coverage of areas requires a computer study. A simplified picture for a circular beam can be obtained as follows. The convention is adopted (see Figure 6.5.4) that the coverage extends to a distance from the centre of the coverage area (bore-sight from satellite antenna, S) corresponding to a semi-beamwidth angle δ.

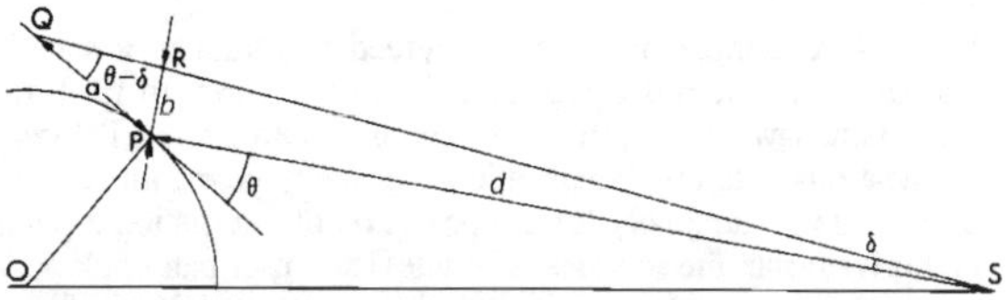

Figure 6.5.4 Geometrical representation of the plane containing the centre of the earth, the satellite and a point P on the earth.

In general, the coverage will be elliptical. The semi-major axis, a, can be shown to be approximately $a = b/\sin(\theta - \delta)$, where the coverage radius in the transverse direction, b, the semi-minor axis, can be calculated as $b = d \tan \delta$.

6.5.2 Satellite Operators

Telecommunications administrations were the main driving forces behind non-military exploitation of communications satellites. They saw satellites as an economic way to provide extra capacity for mainly international telephony traffic. Of most interest to major broadcasters are the international communications satellites which cover the Atlantic, Indian Ocean and Pacific regions. The world map in Figure 6.5.5 shows the global beams of satellites located at these positions.

A large percentage of the world's most heavily populated areas are able to communicate with each other by using a single

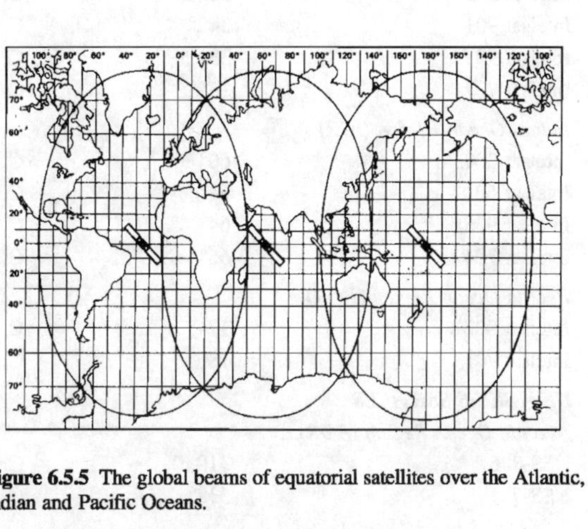

Figure 6.5.5 The global beams of equatorial satellites over the Atlantic, Indian and Pacific Oceans.

satellite hop. This is cheaper than multiple hops, and the amount of delay and other impairments that the signal is subjected to is kept to a minimum.The delay varies according to slant path to satellite (dependent on latitude of earth station). For geostationary satellites, the ground–satellite (up-link) path introduces a delay of 120–140 ms and similarly for the satellite–ground (down-link) path. In all, this introduces a total delay of 0.24–0.28 s for the ground–satellite–ground trip. A return path doubles the delay, which previously caused lip-synch problems most noticeable in news interview scenarios.

6.5.2.1 Intelsat

In 1964, a number of countries agreed to establish a satellite telecommunications body to serve the whole world. In 1971, this agreement saw expression as *Intelsat*, the International Telecommunications Satellite Organisation. For many years, this consortium held the monopoly for the provision of international public and private satellite services. The television user can book occasional circuits on Atlantic Ocean, Indian Ocean and Pacific Ocean international communications satellites operated by Intelsat.

In 2001, Intelsat became a private company after shedding its status as an intergovernmental organisation, enabling it to focus on maximising value to customers, employees and shareholders. Intelsat began providing end-to-end solutions through their newly established network of leased fibre, points of presence (POPs) and teleports around the globe. The strength and the speed of this network allow customers to send and receive content from anywhere, to anywhere in the world, with the ease of one-stop shopping.

Intelsat worldwide coverage is supplemented by several national domestic satellites and regional systems such as those provided by the *Eutelsat* consortium of European PTTs and *Intersputnik* in the USSR (see Table 6.5.1).

Table 6.5.1 Sample of transponders available for broadcasting use

Satellite	*Orbital location*	*Band*
INTELSAT SERVICES		
Atlantic Ocean Region (AOR)		
Intelsat 905	335.5°	C, Ku
Intelsat 903	325.5°	C, Ku
Intelsat 901	342°	C, Ku
Intelsat 907	332.5°	C, Ku
Intelsat 707	359°	C, Ku
Indian Ocean Region (IOR)		
Intelsat 904	60°	C, Ku
Intelsat 601	64°	C, Ku
Intelsat 906	64°	C, Ku
Intelsat 704	66°	C, Ku
Pacific Ocean Region (POR)		
Intelsat 802	174°	C, Ku
Intelsat 701	180°	C, Ku
NEW SKIES SERVICES		
Atlantic Ocean Region (AOR)		
NSS-806	319.5°	C, Ku
NSS-7	338°	C, Ku
Indian Ocean Region (IOR)		
NSS-703	53°	C, Ku
NSS-6	95°	C, Ku
Pacific Ocean Region (POR)		
NSS-5	183°	C, Ku
OTHER SERVICES		
AOR		
European		
Telecom 2D	352°	C
CIS		
Gorizont 37(26)	349°	C
Express 2	346°	C, Ku
USA		
Panamsat PAS-1R	315°	Ku
IOR		
European		
Eutelsat W3A	7°	Ku
Eutelsat W3	7°	Ku
Eutelsat W1	10°	Ku
Hotbird 1, 2, 3, 4, 6	13°	Ku
Eutelsat W2	16°	Ku
Atlantic Bird 3	355°	Ku
Arab League		
Arabsat 2A	26°	C
Arabsat 2B	30.5°	C, Ku
Arabsat 2C	26°	C
Arabsat 2D	26°	Ku
Arabsat 3A	26°	Ku
CIS		
Gorizont 44(32)	53°	C
Gorizont 43(31)	40°	C
POR		
Australia		
Optus-B1	160°	Ku
Optus-B3	156°	Ku
Optus-C1	156°	Ku
Japan		
JCSat3	128°	C, Ku
JCSat4	124°	C, Ku
JCSat5	150°	Ku
JCSat6	124°	Ku
JCSat8	154°	C, Ku
JCSat9	132°	C, Ku
N-Sat-110	110°	Ku
Indonesia		
Palapa-C2	113°	C, Ku

6.5.2.2 North American domestic communications satellites

The USA is served by many *domsats* (domestic satellites) which fill the gap in world coverage between the Pacific and Atlantic satellite systems. The domsats meet the demand for a very high level of internal traffic within North America and provide capacity to extend Pacific and Atlantic traffic across the American continent. The domsats carry occasional traffic split between the 6/4 GHz and 14/11 GHz bands. (The notation 6/4 GHz indicates an up-link frequency of 6 GHz, down-link at 4 GHz.) Most US domsats provide *conus* (the 48 contiguous states of the USA) footprints and some are extended to cover Alaska, Hawaii and Puerto Rico. The Canadian *Anik* satellites (C2 at 110°W and C3 at 114.9°W) carry 14/11 GHz band occasional traffic to Canada and parts of the USA. The occupancy of most American domsats varies from month to month; in addition to the leased channels, a large amount of occasional traffic is also carried from transportable ground stations.

The North American domsat system is very responsive to market changes. An organisation which proved helpful one year may not exist the next, and a broadcaster must either maintain a regular and continuous contact with the many carrier companies involved or avoid these complications by booking with a satellite servicing company. Such companies comprise teleports that provide gateway services for the broadcast material to and from its destination. Satellites naturally span national boundaries, providing numerous possibilities for truly international services.

6.5.2.3 Eutelsat

Eutelsat provides 14/11 GHz service in Ku-band capacity on 23 satellites positioned in geostationary orbit between 15 degrees West and 48 degrees East. This resource enables it to offer landmass coverage from the east coast of North and South America to the Indian subcontinent. The fleet is comprised of several satellite families, each tailored to specific operational needs. Eutelsat's state-of-the-art spacecraft feature high-power, DVB platforms and the SKYPLEX on-board processing. A typical Eutelsat satellite coverage is shown in Figure 6.5.6.

The European Broadcasting Union (EBU) leases some transponders on a long-term basis to augment its permanent terrestrial TV network. Occasional circuits may be booked on some Eutelsat satellites.

6.5.2.4 Intersputnik

The *Gorizont* satellites are part of the USSR's international geostationary communications satellite system, Intersputnik, which links the former Warsaw pact countries and others such as Cuba and North Korea.

Today, Intersputnik's space segment consists of three Russian-made Express-A satellites and the new-generation LMI-1 spacecraft manufactured by Lockheed Martin. Under a distribution agreement with Eutelsat, Intersputnik offers capacity and telecommunication services on Eutelsat's satellites.

Intersputnik provides transponders for occasional traffic in the 6/4 GHz band, and like other geostationary satellites it cannot deliver signals of adequate strength to near-polar latitudes. In order to provide a service to the significant populations in areas which lie at about 70°N, the USSR developed the *Molniya* system, which carries communications satellites in an elliptical orbit inclined at 63° to the earth's equatorial plane. As a satellite in this system travels into space, it appears to hover for a period of about 6 hours (see Figure 6.5.7). By switching signals at appropriate times between the four satellites distributed along the orbit, the *Molniya* system provides continuous coverage of its service area.

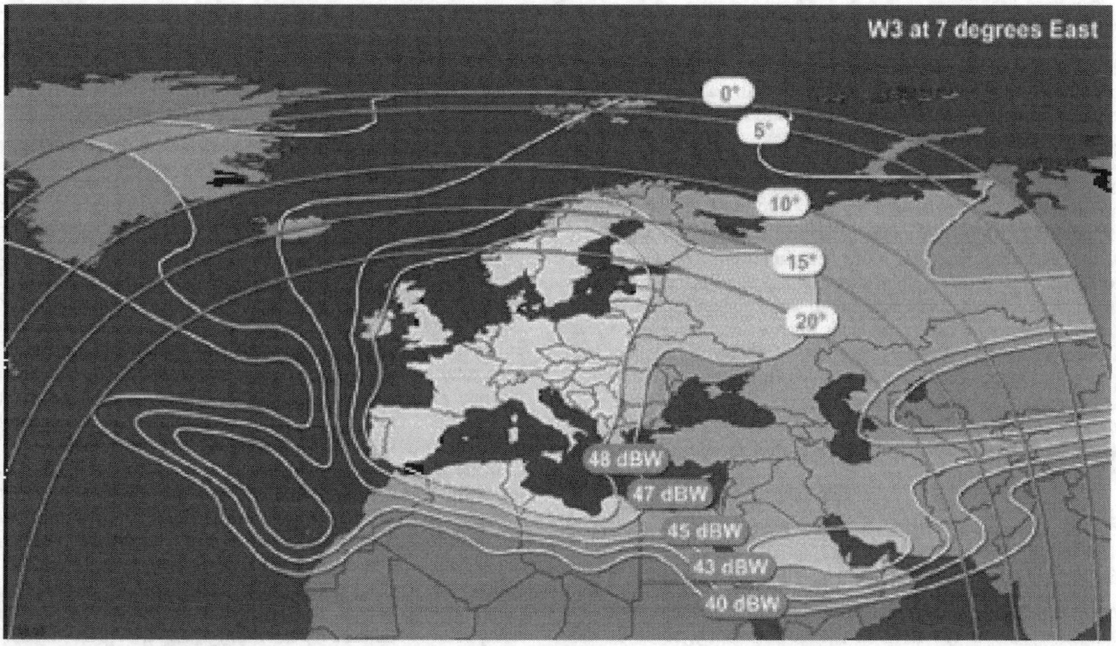

Figure 6.5.6 The Eutelsat footprint.

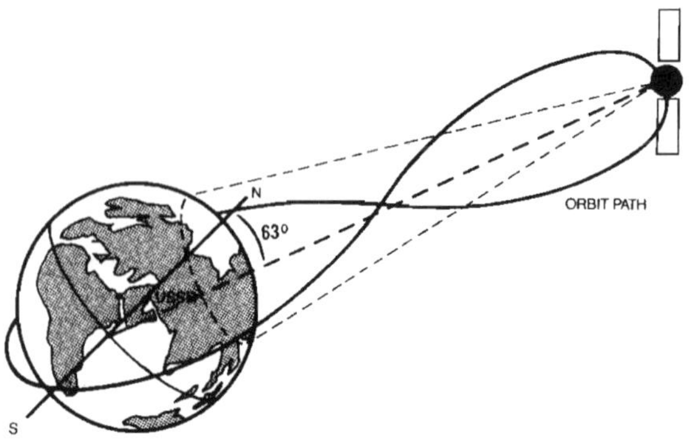

Figure 6.5.7 Elliptical orbit of *Molniya* satellites.

6.5.3 Satellite Applications

A broadcaster may wish to use a communications satellite for one or more of the following major applications:

- distribution of programme streams or networks to cable networks or affiliated stations;
- scheduled or occasional contributions from remote points to a network studio centre;
- totally unscheduled contributions of ENG material to a studio centre (often referred to as SNG – satellite news gathering).

The first application would normally be served by a permanently leased transponder and the third application by occasional bookings.

6.5.3.1 Regulatory considerations

In countries where the telecommunications services have not been deregulated, the user is obliged to make occasional bookings or long-term leases with the public telecommunications operator authorised by the government. This is often the government-controlled PTT. Even in deregulated countries, the user may be limited to booking through two or three approved companies. Where governments and satellite operators permit a broadcaster to operate its own transmitting ground station, a stringent technical performance specification must be adhered to. This is to protect other users of the satellite and prevent interference to other satellites in the same cluster or in adjacent 'slots', and to terrestrial services. These tests, referred to as Coordination and Verfication tests, must be completed and submitted for registration before an earth station can come into service in the satellite operators network. These tests are documented in SSOG (Satellite System Operation Guide) for Intelsat and ESOG (Eutelsat Systems Operation Guide) for Eutelsat.

6.5.4 Satellite Management

6.5.4.1 Station keeping and the TT&C control centre

At the end of its launch phase, a satellite is manoeuvred to its allocated station (or *slot*) in the geostationary orbit. If the influence of micro-meteorite impact, magnetic and gravitational fields and the pressure of solar radiation on the structure of the satellite were negligible it would remain at its station in the 'Clarke' orbit indefinitely without expending any further energy. In practice these effects are not insignificant, and they cause the satellite to change its orientation and gradually to drift off station and out of geostationary orbit.

A liquid propellant is carried which can be released in appropriate volumes and direction by thruster jets to compensate for orientation and positional errors. The quantity of propellant carried is limited by payload considerations. A modern satellite has an expected lifetime of greater than 12 years, during which time the propellant will be used for both E–W and N–S station keeping. After about 7 years or more, when a certain amount of propellant remains, the satellite will be allowed to drift into an inclined orbit with occasional use of propellant to maintain E–W station keeping (hence the satellite orbital spacing) to avoid interference. This inclined orbit satellite can still offer services but now the space segment capacity is offered at a discounted rate. Control of the release of propellant is determined in part by an on-board automatic control system and in part by instructions from the *telemetry, tracking and command* (TT&C) ground station.

The TT&C system on board the satellite makes environmental measurements and monitors the performance of many aspects of its operation. The resultant status data signals are sent digitally to the TT&C ground station, which continually computes the satellite's precise location. Control signals are generated when necessary to keep station better than $\pm 0.1°$ in N–S and E–W directions and to maintain beam pointing accuracy to $\pm 0.1°$. These and other command signals are up-linked to the satellite in digital form.

During the launch phase the satellite's position is varying greatly so an omnidirectional antenna is deployed for the telemetry and command beacons. These operate on similar frequencies, which are chosen to be between about 130 MHz and 4 GHz (where atmospheric absorption is lowest) during this transient phase. Under normal orbiting conditions, the carrier frequencies are within the band allocated for the main transponders and employ a directional antenna aimed at the TT&C control centre.

6.5.4.2 Powering of satellite services, eclipse and solar outage

Satellites are generally powered by a large array of solar panels. End-of-life power demand for a large satellite in the FSS may be more than 2 kW. Rechargeable Ni–Cd batteries provide a temporary source of power for short twice-yearly periods of eclipse when the earth shades the satellite from the sun. The battery supply powers essential services during the eclipse, but on most satellites it is not adequate to provide the high total power required by the communications transmitters. Normally, broadcasting satellites will not be able to transmit programmes during the period of the eclipse. The weight of batteries to replace solar power would be prohibitive.

If the satellite is placed in the geostationary orbit at the same longitude as a country which is receiving its signals, the midpoint of the eclipse will occur around midnight, varying in duration from zero at 22 days on either side of each equinox to 72 minutes at the spring or autumn equinox itself. By choosing a parking position for the satellite to the west of the country being served, the time when the eclipse occurs can be delayed by 4 minutes for each longitudinal degree shifted.

The early hours of the morning are, statistically, periods of low traffic density, and so this strategy is employed with domestic satellites to minimise the effect of the eclipse. Clearly the timing of the eclipse cannot be optimised in this way for a satellite serving two continents or a large international region.

Two other undesirable, but predictable, natural phenomena disturb the continuity or quality of the signal. The most important of these occurs when the sun aligns with the satellite in the beam of the receiving antenna. When this happens, for periods of about 10 minutes on five consecutive days twice a year (during the Spring and Autumn equinoxes), the ground station receives an enormous increase of solar noise. The effect is either an interruption (*outage*) or severe degradation of the signal. Contingency plans may be necessary to switch feed from another satellite or via terrestrial means if programme continuity without degradation is a necessity.

A sun outage is similar in behaviour to a rain fade. The high noise level and broadband nature of the sun's energy can effectively wash out the wanted receive signal with noise (that is, degrades the carrier/noise ratio). The duration of this outage is dependent on factors such as the receiving station dish diameter and the receive frequency band. The outage duration could be up to several minutes, during which time maintaining programme continuity may be crucial via terrestrial means or through the use of another satellite not experiencing an outage (satellites at different longitudes will cause outage at different times). The actual time for the occurrence of this outage is dependent on the satellite longitude and the receiving earth station's position on earth. For users with antenna tracking controllers that use beacon receivers, it is important to ensure steps are taken to keep the antenna from tracking the sun's movement and pointing off the satellite. This can usually be done in a number of ways: temporarily disabling the tracking system, setting user-defined software limits or switching to an orbital prediction control.

Less serious is the deterioration of transmission performance brought about by the eclipse of the satellite by the moon. In the decade to the end of the last century, a satellite at 19°W will have experienced about 22 instances of moon solar eclipses. The level of shadowing will vary from 1% to 100% and the duration from 1 to 64 minutes.

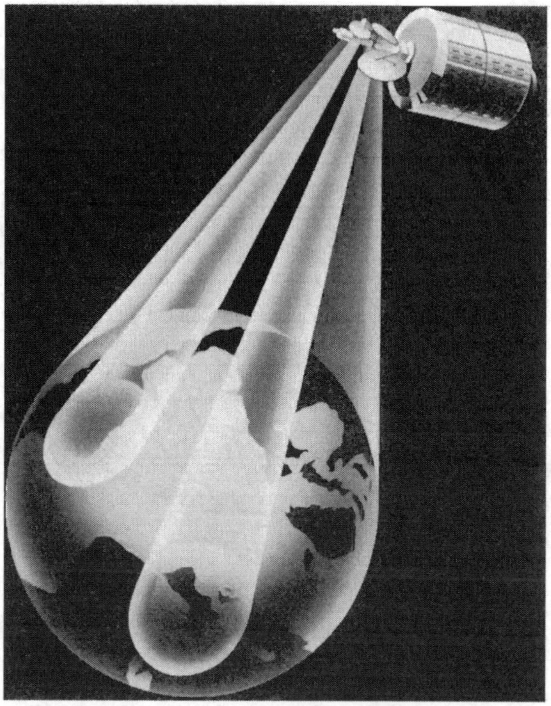

Figure 6.5.8 The use of directional antennas enables energy to be concentrated where needed.

6.5.4.3 Coverage

In principle, it is possible to provide virtually complete coverage to the major populated areas with three satellites equidistantly spaced around the equator as shown in Figure 6.5.1. Such a scheme of three satellites provided the possibility of introducing worldwide communications, including low-density traffic routes to and from remote areas. The first geostationary satellites employed broad-beam antennas which covered all of the earth's surface visible from satellite. This amounted to about 42% of the total surface area.

Two major drawbacks were evident; one was the need for large receiving antennas to pick up the weak signals from the satellites, and the other was the inevitable waste of energy over the oceans and deserts. With the later introduction of spot beam antennas, the effective radiated power was increased and directed onto areas of high population (Figure 6.5.8).

In the 4 GHz band, the beam bore-sight EIRP (*equivalent isotropically radiated power*) for a typical full-bandwidth transponder ranges from 26.5297 dB(W) for global coverage (42.4% of the earth's surface, 17.5° × 17.5° beamwidth), through 28–36.7 dB(W) for hemisphere coverage (20%) to 28–31 dB(W) for zonal coverage (10%, 14° × 5°). Typical figures for spot beams (beamwidths less than 3° × 2°) in the 11 GHz band are 49.7–52.7 dB(W). Figure 6.5.9 illustrates two examples of 4 GHz band global footprints, and Figure 6.5.10 shows how the radiated power has been concentrated into two 'hemisphere' beams onto the American and African/European land masses.

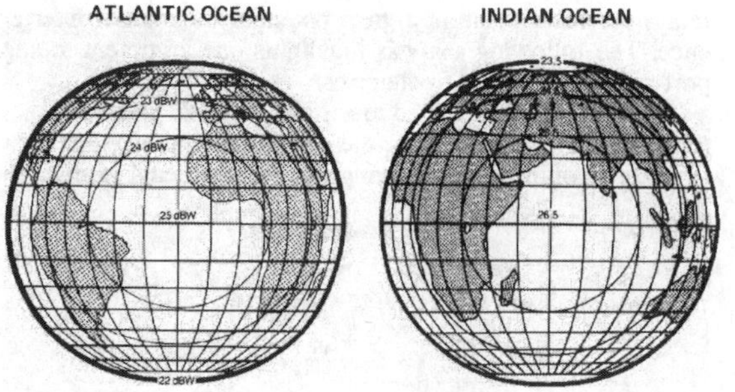

Figure 6.5.9 Two examples of 4 GHz band footprints.

6.5.4.4 Frequency reuse

The current range of communications satellites are equipped with multi-beam antenna (MBA) systems, which allow several beams to be transmitted simultaneously. A satellite may be equipped with broad global and steerable directional spot beams operating on the same frequencies and possibly using opposite linear or circular polarisations. By these means,

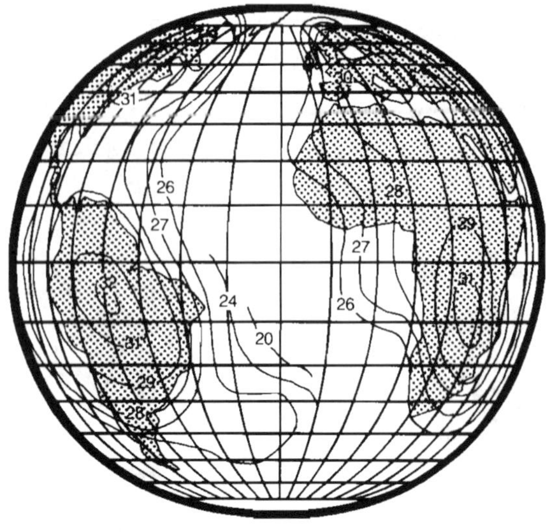

Figure 6.5.10 Concentration of radiated power onto land masses.

a particular frequency may be reused by two, four or even six beams. It should be noted that C-band systems are generally circularly polarised.

6.5.4.4.1 Frequency bands for FSS

Until the start of the 1980s, nearly all fixed satellite services operated in the 6/4 GHz region of C-band. When satellite/earth communication started, this part of the spectrum had several advantages over higher frequency bands. The technology was available to provide transponder output power of 5 W or more; attenuation caused by the atmosphere, rain or other weather conditions is low. The most important disadvantage is the sharing of the 4 GHz and 6 GHz bands with terrestrial (radio relay) services. There is a great potential for mutual interference. The following analysis highlights one argument to support a move to higher frequencies.

The EIRP is determined from the antenna gain, which is related to the beamwidth for the coverage required.

The gain of the antenna having an effective area A is given by:

$$G_t = \frac{4\pi A}{\lambda^2} = \frac{\pi^2 D^2 \eta}{\lambda^2}$$

where η is efficiency (in practice between 50% and 80%). For a circular aperture of diameter D the following approximate relationship holds for beamwidth b:

$$b = \frac{75\lambda}{D} \text{degrees}$$

where λ is wavelength.

From these relationships it follows that an antenna of a given aperture provides $10 \log (11/4)^2 = 8.8$ dB more gain at 11 GHz than at 4 GHz. The beamwidth at 11 GHz is $4/11 = 0.36$ times that at 4 GHz.

It follows that 14/11 GHz operation permits smaller dishes to be used, and narrower spot beams can be produced. Because 14/11 GHz is less prone to interference, ground stations will operate satisfactorily in urban areas and are permitted to operate at higher power levels. These considerations have led most new communications satellites to be equipped to operate in both 6/4 GHz and 14/11 (or 14/12) GHz bands, and many satellites which have yet to be launched will carry only 14/11 (or 14/12) GHz transponders (Eutelsat satellites are generally all Ku-band). *Cross-polar discrimination* (XPD), which is the key to successful frequency reuse based on either orthogonal or oppositely sensed circular polarisations, can be severely degraded by atmospheric conditions, especially at frequencies above 10 GHz.

6.5.4.5 Flexible connectivity

Each receiving antenna is permanently connected to its own low-noise amplifier (LNA) in most satellites. The LNA output signals are grouped together for down-conversion. Following conversion, in modern designs the signals can be switched in a matrix to the HPAs. The HPAs in turn are connected to the transmit antennas through a router. This high level of transponder and antenna flexibility allows the satellite to be reconfigured by commands from the TT&C centre in order to bring a spare transponder into service or to change the coverage to account for changed traffic conditions.

This flexible connectivity can be extended on some satellites to allow a 14 GHz up-link to be routed to a 4 GHz down-link (or a 6 GHz up-link to a 11 GHz down-link). This type of connection is known as *cross-strapping* and may be used to advantage by the operator of a private 14/11 GHz ground station to communicate with a major distant 6/4 GHz ground station. A transponder attracts a higher tariff when it is cross-strapped.

6.5.5 Point-to-Point Connections for Television

The component parts of a television service are shown in Figure 6.5.11. They are:

- the terrestrial extension circuits from the studio or outside broadcast location to the ground station – these may be divided into the local circuit (or loop) from the studio to the telecommunications operator (the PTT) and the main link from PTT to the satellite ground station;
- the up-link (or up-leg) from that particular ground station to the satellite;
- the transponder;

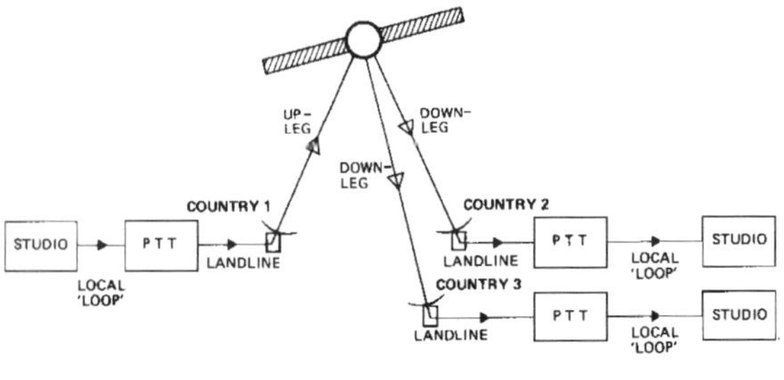

Figure 6.5.11 Point-to-point connections for a television service.

- the down-link (or down-leg) to the ground station (or stations) scheduled (or cleared) to receive these signals;
- the circuits from the receiving ground station to the final destination, usually the TV station or studio centre.

The terms up-link and down-link describe the microwave links to and from the satellite. The term *space segment* is used to describe the up-link and down-link paths together and often includes the transponder in the satellite.

The normal access point for a satellite link is the *gateway*. The fees for space segment circuits are charged between gateways and include the cost of any terrestrial circuits between gateways and ground stations. Any circuits needed between the studio or television centre and the gateway must be paid for additionally. In the USA and Canada the gateway is located at the ground station, but in many other countries the gateway is the capital or other major city. Additional fees for the studio/gateway connection add to the total cost and may complicate the booking arrangements.

Where a broadcaster in Europe receives a signal which originates in, for example, New Zealand, there will be two space segments, i.e. two up-links and two down-links. If the route is properly planned, one midpoint ground station will be selected to access both satellites, i.e. Hong Kong in the example given in Figure 6.5.12.

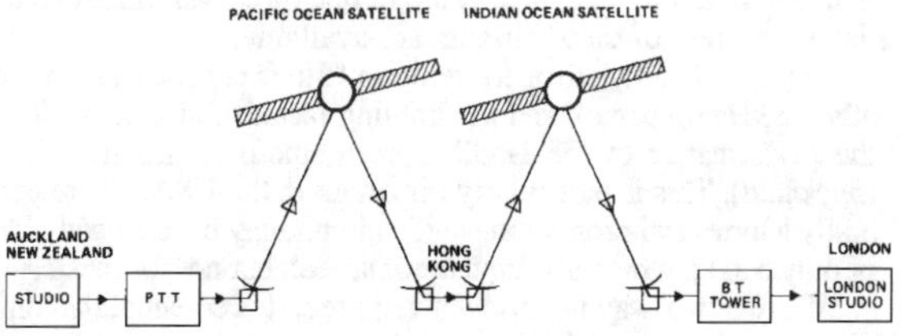

Figure 6.5.12 Point-to-point connections involving two space segments.

6.5.5.1 Leasing of transponders

Transponders and the associated links may be leased either for *occasional use* periods as required or *continuously* (3 months and longer). Depending on the satellite, bookings may be made with the telecommunications administration (such as *British Telecom International* in the UK) or other national signatory to Intelsat, or with a private organisation or teleport which provides a satellite booking service. Organisations providing booking services hold permanent leases on some international satellites and many domsats.

Where a broadcaster identifies a need for a prolonged transponder lease, it may be more economical to arrange a lease for an extended period directly with the satellite operator. Intelsat, for example, provides preemptible services, and each signatory participating in a lease may operate as many receiving ground stations as desired at no extra cost. There is, however, an additional charge for each signatory (in excess of two) participating in the lease. The participating signatories may designate one or more ground stations to transmit TV services.

The cost of a lease is determined by the type of satellite, the type of transponder, the type of connectivity, the type of video channel and the bandwidth. The service can be provided on any satellite where capacity exists that meets the requirements of the user. Normally this service is not carried on a primary or major path satellite.

6.5.5.2 Half transponder operation

Analogue television signals are transmitted to and from the satellite in frequency-modulated form. A *full transponder* has a typical bandwidth of 36 MHz and some are split to allow two television channels of 18/20 MHz bandwidth to be carried. Not surprisingly, a *half transponder* is less costly to lease than a full transponder, but there is a disadvantage when two television signals share the same transponder. A decrease of power (or *back-off*) of about 2.5 dB becomes necessary to reduce to an acceptable level intermodulation distortion introduced by TWTA non-linearity. This reduced transponder power is then shared equally between the two signals (a further decrease of 3 dB). The deviation, too, is reduced in this mode of operation. It is shown later that the carrier/noise ratio is, however, improved by 2.6 dB because of the smaller noise bandwidth of a 20 MHz half transponder. This becomes very significant when operating at marginal values of carrier/noise ratio.

The video signal deviates the FM carrier of a full transponder by 13.5 MHz/V (6.75 MHz peak deviation) on many communications satellite services. Eutelsat requires a 1 Y peak-to-peak video signal to deviate a video carrier by 25 MHz for a full transponder and by 19 MHz for a half transponder. Pre-emphasis as described in CCIR Recommendation 405 is applied at the up-link ground station, and a compensating de-emphasis is employed at the down-link ground station to improve signal/noise performance (by about 2 dB for the luminance component of a 625/50 signal).

6.5.5.2.1 Note on Analogue vs. Digital

Although the broadcasting world is rapidly going digital, analogue TV and radio are set to remain for several years yet. Indeed, for some services, analogue is still an attractive option due to the large installed audience base and the widespread existence of low-cost consumer equipment. For these reasons, analogue is particularly popular for free-to-air broadcasting and a significant number of large broadcasters – such as the BBC, Deutsche Welle and RAI – are maintaining their analogue TV channels, or even starting new ones.

Moreover, the capacity to transmit audio subcarriers of an analogue TV signal allows multi-lingual TV programmes or the parallel transmission of radio stations.

The transition from analogue to digital TV in recent years has significantly expanded the market for direct-to-home (DTH) satellite TV across Europe. By enabling as many as eight digital channels to broadcast from a transponder which once delivered a single analogue channel, the economics and operation of broadcasting have changed fundamentally. Customised programme bouquets, interactivity, new CD-quality radio services and all-digital throughput from studio to set-top box are all now possible.

6.5.5.3 Protection of other services

An *energy dispersal* waveform is usually added to the video signal before modulation. This ensures that interference that could otherwise be introduced into satellite or terrestrial multichannel telephony systems is spread over many channels. A symmetrical triangular waveform locked to the television field synchronising pulse is preferred. Its frequency is usually one-half or one-quarter of the field frequency and is phased so that its discontinuity of slope occurs outside the active picture area.

The dispersal waveform has an amplitude which is typically 5–10% of the video signal. The ground station receiver must remove the picture flicker introduced by the waveform. Several methods are possible. Two stages of black level clamping of the demodulated signal will reduce the flicker to an imperceptible level. An alternative approach is to apply a narrow-band frequency feedback technique to the IF stage of the receiver. This is suitable only if a very low dispersal frequency is employed (not greater than one-twentieth of the field frequency). The method also has the advantage of providing threshold extension which allows pictures of acceptable quality to be received under lower values of carrier/noise ratio than would be possible with a conventional frequency discriminator.

6.5.5.4 Sound signals

In the case of NTSC, PAL and SECAM signals, a transponder usually carries both video and associated sound signals (monophonic, stereophonic or multiple channel). The sound signals (which may be processed by a proprietary noise reduction companding system such as Wegener Panda) are carried in the range 5.4–8.0 MHz (relative to the vision carrier frequency). Monophonic signals may alternatively be carried by the sound-in-sync method (Varian TVT/BBC system), and sound-in-sync variants may also be used for dual-channel services (RE Instruments/IBA, Varian/BBC, Vistek/ITV Association or BTS).

6.5.5.5 Interaction between sound and video signals

The signal/noise ratio of the video signal is slightly reduced through the deviation of the main carrier by a sound subcarrier. For the converse reason, the signal/noise ratio of the sound signal depends slightly on picture content. A further small picture impairment is caused by intermodulation products. These unwanted effects may become significant where two or more subcarriers are provided. Unwanted interaction does not occur with TDM sound signals (sound-in-sync or MAC/packet).

6.5.6 Factors Affecting Programme Production

6.5.6.1 Transmission delays

The transmission delay, t, between a ground station and the satellite is given by:

$$t = \frac{d}{c}$$

$$= \frac{1}{c}(R_e^2 + R^2 - 2R_eR\cos\phi)^{1/2}$$

where c = velocity of light

d = distance between satellite and ground station (*slant range*)

R_e = radius of the earth

h = altitude of the satellite

$R = R_e + h$

ϕ = angle subtended at the centre of the earth between the directions of the satellite and the ground station.

Thus, the transmission delay, t_d, for a single hop between ground stations lies between the limits:

$$\frac{h}{c} \leq t_d \leq \frac{2h}{c}\left(1 + \frac{2R_e}{h}\right)^{1/2}$$

For a geostationary satellite, this delay is in the range 240–280 ms. To this must be added the delay introduced by the terrestrial extensions to the broadcaster's premises.

6.5.6.2 Sound signal

Sound signals must always follow the route of the video signal with which they are associated as closely as possible, since the observer can tolerate only a small differential audio/video delay. For many types of television programme material, the absolute delay introduced by the point-to-point transmission path is unimportant. A simple playout of a complete package of sports or news items falls into this category. Problems arise, however, when a two-way discussion or interview is carried out over a satellite link, since a question takes at least one-quarter-second to reach the ear of a participant at the distant end and a further quarter-second for the answer to be received by the questioner. Using two satellites in cascade doubles the delay. This is not an easy problem to overcome for a fast-moving two-way interview or discussion. It is sometimes possible to book sound circuits that are routed by undersea cable in one direction to reduce the delay, but few of these circuits are available.

The audio bandwidth on a satellite is 15 kHz per channel (optionally 7.5 kHz on some), and the limiting factor of circuit quality is the performance of the landline from studio to ground station (*backhaul*). This is particularly important in the USA, where generally a terrestrial programme audio circuit may have a bandwidth of only 5 kHz, which is adequate for speech but not for music.

Where a two-way interview is required, it is essential that only *clean reverse sound* (or *mixed-minus sound*) is fed to the participants. If any trace of the distant end voice or sound is relayed back it is heard as a disturbing echo, which makes conversation almost impossible. Great care is necessary in adjusting audio levels to ensure that the reverse feed stays clean. The use of deaf aids (earpieces) in participants' ears is preferred, since it is much more likely to allow a successful conversation than if foldback loudspeakers were used in the studio.

6.5.6.3 Studio control of signal levels

6.5.6.3.1 Video

On most 6/4 GHz Intelsat services, each occasional television channel occupies a half transponder of 20 MHz bandwidth. This arrangement handles 525 NTSC television signals rather better than 625 PAL, which can suffer from visible distortions and adjacent channel crosstalk, e.g. colour bar breakthrough. To reduce this problem, Intelsat insists that only the 75% version of the colour bars test signal is allowed to be sent over the system and that it is limited to less than 30 s duration. This is usually easy enough to arrange in master control rooms, but care should also be taken that 100% colour bars from the national network or from videotape machines are never presented to the satellite connection. High-level video signals that can be produced by VTRs operating in the *fast rewind* mode must never be allowed to be passed over a satellite path.

High-level signals with fast rise and fall times, such as white captions, can severely test the system, since pre-emphasis increases the deviation of the FM carrier at high video frequencies. Under these conditions, the FM carrier may instantaneously exceed the available radio frequency bandwidth. This generates truncation noise, which can manifest itself as picture tearing on highlights and streaking following sharp transients.

6.5.6.3.2 Sound

Confusion can result from the differing line-up tone levels and metering instruments used around the world. Major European broadcasters use a peak programme meter (PPM) with 0 dB*u* as the line-up level ($u = 0.775$ V), which is 8 dB less than the peak audio level. Many other countries use a volume unit (VU) meter with line-up at +8 dB*u* (i.e. at peak level).

It is important that a satellite booking includes adequate time to make measurements and adjustments to correct for errors in video and sound levels. It is good practice to start a booking early enough to play out programme material during the pre-transmission period over the satellite system, so that variations in monitoring systems can be corrected and action can be taken to minimise distortion of the audio signal. This is often manifested as heavy sibilance of voices, which can be improved by reducing the *send* level.

6.5.7 Transportable Ground Stations

It is technically possible to transmit directly to communications satellites from transportable ground stations. In the USA this has been possible since 1976 in the 6/4 GHz band by using domestic communications satellites. The ground stations for this band are rather large (up to 10 m, or 30–33 ft diameter) and are transported by articulated trucks. Satellite servicing vendors can provide up-link ground stations and can negotiate the necessary space segment on behalf of the broadcaster.

With the advent of digital MPEG-2 systems, improved SSPAs and LNB noise figures, the size of transportables are becoming smaller. Typical SNG transportables use typical dish sizes of 1.8–3 m diameter.

6.5.7.1 Privately operated up-links

At Ku-band, considerably smaller ground stations can be used. These use typically 3 m (10 ft) diameter antennas which are generally trailer mounted and designed to be containerised for long-distance shipment.

To provide a successful up-link for an event covered as an outside broadcast from a remote site, the ground station must be robust, reliable and highly mobile, easy to set up quickly and designed with adjustable EIRP of an adequate rating to allow operation from any point within the satellite footprint. Improved side-lobe performance may be necessary in order to keep interference low to other satellites in the vicinity. Offset antennas can be particularly effective in allowing near and far angle side-lobe levels to be reduced, as the dish aperture is not blocked by the antenna feed and its support struts. The station must also incorporate down-link receiving equipment to allow the satellite to be acquired and communication to be set up with the TT&C centre, the receiving ground station and the receiving television station. The receiver will also be used to check that another user is not already accessing the transponder. Satellite operators require ground stations to meet transmit and general technical specifications as a condition for access to their transponders.

The mobile ground station must be equipped to measure the EIRP of its transmitter, its frequency and frequency deviation (including the deviation associated with the dispersal waveform). Units like these are operated both by telecommunications administrations (e.g. British Telecom International (UK), Telespazio (Italy), Direction Générale des Télécommunications (France)), and by broadcasters (e.g. BBC, TDF, CBS). BTI, for example, offers a range of mobile units equipped with 5 m diameter dishes with EIRPs between 75 and 78.5 dB(W) down to one unit fitted with a 1.5 m square antenna and having an EIRP of 67.5 dB(W).

Figure 6.5.13 A VSAT terminal (Marconi).

The need to respond swiftly to news stories wherever they might occur in the world led Independent Television News (UK) to commission a design study for an even smaller ground station able to be packed in flight cases. This 14/11 GHz VSAT (*very small aperture terminal*) is basically simple in design and has the lowest possible power consumption (Figure 6.5.13). It uses a 2.1 × 1 m (7 × 3.3 ft) elliptical section Gregorian offset antenna, which is small enough to be transported by passenger aircraft as accompanied baggage and is normally deployed in panel trucks or similar rental vehicles.

6.5.8 Carrier/Noise Derivation

6.5.8.1 Calculation of received power

The power flux density P_d of a signal received at a distance d from a transmitter of EIRP P_e is given by:

$$P_d = \frac{P_e}{4\pi^2 d^2}\,(\mathrm{W\,m^{-2}})$$

If the receiving antenna has an aperture A then:

$$P_r = P_d A\,(\mathrm{W})$$

$$= \frac{P_e A}{4\pi^2 d^2}\,(\mathrm{W})$$

where P_r = power received at the terminals of the aerial. Now the gain if the receiver antenna G_r at wavelength λ is:

$$G_r = \frac{4\pi A}{\pi^2}$$

Thus,

$$P_r = G_r P_e \left(\frac{\lambda}{4\pi d}\right)^2$$

In terms of dB(W) this relationship may be written:

$$P_r = P_e + G_r - 10\log\left(\frac{4\pi d}{\lambda}\right)^2 \text{ (dB(W))}$$

The right-hand term is the path loss L_s (or *free space attenuation*), which is seen to be dependent on wavelength (and therefore frequency). If the slant range d is not known, it may be calculated from the relationship:

$$d^2 = R_e^2 + (R_e + h)^2 - 2R_e(2R_e + h)\cos\theta\cos\lambda$$

where R_e = radius of the earth (6.278×10^6 m)
h = altitude of satellite (35.787×10^6 m)
θ = latitude of receiving point
λ = difference in longitude between satellite and receiving point.

The value of d will lie between 35,787 km, where the ground station is directly below the satellite on the equator, and 41,679 km if the ground station is on the satellite's horizon. Table 6.5.2 shows the variation of L_s with frequency with slant range.

Table 6.5.2 Variation of free space loss (dB) with frequency

Frequency (GHz)	*d = 35,787 km (zenith)*	*Free space loss L_s (dB) d = 39,000 km (typical)*	*d = 41,679 km (horizon)*
4	195.6	196.3	196.9
6	199.0	199.8	201.4
11	204.4	205.1	205.7
12	205.1	205.8	206.4
14	206.5	207.2	207.9

In practice, loss due to the antenna direction pointing error L_d must be accounted for, and atmospheric loss L_a and rain/precipitation loss L_p loss must be added to the path loss. At 4–6 GHz, L_p is very small, but in the 11 and 14 GHz bands it can be as high as 10 dB for short periods during very heavy rain. L_a depends upon the elevation angle of the satellite at the ground station, and is in the range 0.5–1.5 dB for angles between 90° and 5°. Up-linking at low angles of elevation is subjected to regulatory restrictions in order to minimise interference with terrestrial services. Apart from interference to terrestrial services, below 5° elevation angle, the ground noise temperature increases, making the link excessively noisy.

The fuller expression for received carrier power is:

$$P_r = P_e + G_r - (L_s + L_a + L_p) - L_d \text{ (dB(W))}$$

6.5.8.2 Calculation of receiver noise

The noise performance of the receiving system is determined by referring all significant noise contributions to the antenna terminals. Noise figures (N) of individual elements (first amplifier, second amplifier, second amplifier, mixer, etc.) are each converted to noise temperatures (T) according to the relationship:

$$T = T_0(N - 1)$$

where T_0 = room temperature (290 K).

The cable connecting the antenna to the first amplifier will introduce a loss (l_c), thereby making a noise contribution $1/l_c$ to the system. The corresponding noise temperature for the cable is:

$$T_c = 290 \cdot \frac{1 - l_c}{l_c} \text{ (kelvin)}$$

Suffixes c, 1, 2 and m refer to the cable, the first and second amplifier stages and the mixer. The noise temperature of each element is referred to the antenna terminals, and so the system noise temperature is:

$$T_s = T_a + T_c + \frac{1}{l_c}\cdot t_1 + \frac{1}{l_c}\cdot\frac{1}{G_1}\cdot T_2 + \frac{1}{l_c}\cdot\frac{1}{G_1}\cdot\frac{1}{G_2}\cdot T_m$$

where T_a = antenna noise temperature (clear sky).

In a ground station receiving system the noise temperature T_B of the low-noise block (LNB) is usually quoted, and so the relationship may be rewritten:

$$T_s = T_a + T_c + \frac{1}{l_c} T_B$$

Noise power P_n in a system noise bandwidth B_n is given by:

$$P_n = kT_sB_n \text{ (W)}$$

where k is Boltzmann's constant (1.38×10^{-23} W K^{-1} Hz^{-1}). $10\log k = -228.6$ dB(W K^{-1} Hz^{-1}). In terms of dB(W):

$$\begin{aligned} P_n &= 10\log(kT_sB_n) \\ &= 10\log T_s + 10\log B_n + 10\log k \\ &= 10\log T_s + 10\log B_n - 228.6 \end{aligned}$$

6.5.8.2.1 Receiver sensitivity

It is usual practice to specify the overall performance of the receiving system in terms of its G/T, where G is the antenna gain and T is the effective noise temperature.

$$G/T = \frac{\alpha\beta G_m}{\alpha T + (1 - \alpha)T_0 + (n - 1)T_0}$$

where α = total coupling loss
β = total of antenna and installation ageing losses
G_m = maximum isotropic gain of antenna
T = effective temperature of the antenna (~150 K)
T_0 = noise factor reference temperature (290 K)
n = overall noise factor in the receiver, referred to the receiver input.

The specification proposed at the time of the WARC-BS 77 Conference was a G/T of 6 dB(K^{-1}). This could be achieved with a 0.9 m diameter antenna and a noise factor of 8 dB, with a margin for coupling loss, pointing loss and ageing loss.

However, since the conference, achievable noise factors have fallen to about 4 dB or less. This means that the same *G/T* can be achieved with a smaller antenna, or that a given service may now be receivable over a wider area (subject to interference constraints), or indeed that the quality obtained at the edge of the service area will be better than CCIR grade 3.5.

6.5.8.2.2 *Carrier/noise ratio*

Since the carrier/noise power ratio is P_s/P_n, it may be described in terms of dB(W) as:

$$P_s - P_n = P_e + G_r - 10\log T_s + 228.6 - 10\log B_n - L_s - (L_a + L_p) - L_d \text{ (dB)}$$

This relationship, usually written as *c/n ratio (carrier/noise ratio)*, may be used directly if each individual term is known or can be estimated. Very often a figure of merit *G/T (gain/noise temperature ratio)*, equivalent to G_r (dB) $- 10\log T_s$ (dB(K^{-1})), is quoted for a receiving system. This allows the previous equation to be rewritten:

$$(c/n)_{dB} = P_e + (G/T_s)\ \text{dB}(K^{-1}) + 228.6 - 10\log B_n - L_s - L_a - L_p - L_d$$

Using this relationship, it is instructive to compare the values of c/n for full and half transponders with 36 and 20 MHz bandwidths respectively.

For the half transponder, $10\log B_n = 73.0\,\text{dB(Hz}^{-1})$, whereas for the full transponder $10\log B_n = 75.6\,\text{dB(Hz}^{-1})$. Thus, operating with a half transponder improves the c/n ratio by about 2.6 dB. The same level of improvement will be achieved if the signal from a full transponder is restricted to a 20 MHz passband within the receiver. In this case, however, other picture impairment will be introduced.

It was shown earlier that space loss (L_s) is 8.8 dB greater at 11 GHz than at 4 GHz. This allows the c/n figures for half transponders at 4 GHz and 11 GHz (where the slant range is 39,000 km) to be compared:

$$(c/n)_{4\,GHz} = P_e + G/T_s - 40.7 - L_a - L_d \text{ (dB)}$$

$$(c/n)_{11\,GHz} = P_e + G/T_s - 49.5 - L_a - L_d - L_p \text{ (dB)}$$

6.5.8.2.3 *Signal/noise ratio*

The signal/noise performance of a satellite link is of major interest to the user. Over a range of c/n values, s/n increases in line with c/n, reaching a maximum limiting value determined by intermodulation noise. There is a critical *threshold* value of c/n below which s/n reduces much more rapidly than c/n. For a conventional design of FM demodulator, threshold c/n is around 10–12 dB.

Pictures received at the FM threshold exhibit impairment, mainly black and white noise spikes (of several microseconds duration), often referred to as *sparklies*. Below threshold, picture quality is unsuitable for programme purposes. To be clear of threshold noise, a satellite service has a target c/n ratio of at least 14 dB. The following expression applies in the linear region above threshold:

$$s/n - (c/n) \cdot r \cdot \frac{f_d}{f_v} 2 \cdot \frac{3B}{f_v}$$

which in terms of dB is:

$$s/n = c/n + 20\log r + 10\log 3 + 20\log\frac{f_d}{f_v} + 10\log\frac{B}{f_v}$$

where f_d = peak-to-peak deviation

f_v = maximum base band video frequency

B = radio frequency bandwidth

$r = (f_d/f_l) - 1$

f_l = peak-to-peak deviation corresponding to nominal luminance signal.

The effect of pre-emphasis of the band video signal (and the consequent de-emphasis) is to improve the unweighted luminance s/n ratio of a PAL signal by about 2 dB.

The Intelsat satellite system operations guide gives the following values for s/n ratio (unified weighted) which apply on one hop between permanent ground stations:

	525/60	625/50
Full transponder	53.3 dB	50.1 dB
Half transponder	48.7 dB	47.1 dB

The unified weighting factor for PAL system I is 11 dB. A signal/weighted noise ratio of 47 dB corresponds to a picture grade of about 4.5 on the CCIR five-point scale, which is equivalent to a picture quality between good and excellent.

The previous equation, in the case of a 625/50 PAL signal, provides the following results:

	$(s/n)_{uw}$ (dB)	$(s/n)_w$ (dB)
Full transponder	c/n + 19.8	c/n + 28.8
Half transponder	c/n + 17.2	c/n + 26.2

At the lowest acceptable value of c/n (= 14 dB) which might apply at difficult outside broadcast locations with a transportable ground station, the signal/weighted noise ratios become 42.8 dB (full transponder) and 40.2 dB (half transponder), corresponding to picture grades of 4 (good) and 3.8.

With digital systems the threshold is defined by the receiver's E_b/N_o ratio and this is usually of the order of 5 dB for a domestic receiver.

6.5.8.2.4 *Power budgets*

As the following example shows, a power budget calculation allows the user to determine whether a proposed link will produce a signal of broadcast quality (see Table 6.5.3).

The following figures are representative of a 14/11 GHz single hop link from a private up-link at an outside broadcast event to a major receiving ground station. The slant range for each ground station to the satellite is taken as 39,000 km.

The following values are used: up-link frequency 14 GHz, power 400 W, antenna diameter 1.8 m, aperture efficiency 65%.

Table 6.5.3 Example of a power budget calculation

Variable	*Definition*	*Unit*	*Up-link*		*Down-link*	
(a) *Calculation of received power*						
P_t	TX power	dB(W)	26.0		10.0	
L_t	TX system loss	dB	−0.7		−1.0	
G_t	TX antenna gain	dB	45.4		35.0	
P_e	TX EIRP	dB(W)		70.7		44.0
L_a	Atm. loss (up)	dB		−0.7		n/a
L_s	Free space loss	dB		−207.2		−205.1
L_a	Atm. loss (down)	dB		n/a		−0.5
G_r	RX antenna gain	dB		37.1		56.6
L_r	RX system loss	dB		−1.0		−1.0
P_r	Received power	dB(W)		−101.1		−106.0
(b) *Calculation of receiver noise power*						
T_s	System noise temp	K	2000		460	
T_s	System noise temp	dB(K)		33.0		26.6
B_n	RX noise bandwidth	MHz	20		20	
B_n	RX noise bandwidth	dB(Hz)		73.0		73.0
k	Boltzmann's const.	$\text{dB}(\text{W K}^{-1}\,\text{Hz}^{-1})$		−228.6		−228.6
P_n	RX noise power	dB(W)		−122.6		−129.0
(c) *Calculation of received carrier/noise ratio*						
$10\log(\text{c/n}) = P_r - P_n$		dB	21.5		23.0	

Satellite *G*-over-*T* (*G*/*T*) = 3 dB(K^{-1}); transponder gain = 111.1 dB; down-link frequency = 11 GHz; power = 10 W; receiving ground station *G*/*T* = 30 dB(K^{-1}).

The down-link calculation assumed a noise-free signal within the satellite. In practice, noise contributions from the up-link satellite receiver and transponder intermodulation products must be allowed for. The true down-link carrier/noise ratio is related to these contributions thus:

$$(\text{c/n})_d^{-1} = (\text{c/n})_u^{-1} + (\text{c/n})_i^{-1} + (\text{c/n})_r^{-1}$$

where d, u, i, r refer respectively to the down-link, up-link, intermodulation and the receiving ground station.

Note that this relationship is between natural numbers (i.e. not in decibels). Using figures from the example above:

$$(\text{c/n})_d^{-1} = (141.3)^{-1} + (\text{c/n})_i^{-1} + (199.5)^{-1}$$
$$(\text{c/n})_d^{-1} = 0.012 + (\text{c/n})_i^{-1}$$

If intermodulation noise could be ignored:

$$(\text{c/n})_d^{-1} = 82.7 \text{ times} = 19.2\,\text{dB}$$
$$\text{If } (\text{c/n})_i = 30\,\text{dB } (= 1000 \text{ times}):$$
$$(\text{c/n}) = 76.4 \text{ times} = 18.8\,\text{dB}$$

From these calculations we can see that the influence of noise in the up-link has worsened the true down-link carrier/noise figure by about 3.8 dB and that intermodulation noise has worsened it by about 0.4 dB. If the up-link carrier/noise is increased by 4.6 dB, the true down-link carrier/noise will improve by 2.0 dB.

Returning to the figures in the example, we can see that the up-link margin above carrier/noise threshold is about 11 dB and the corresponding down-link figure is about 8 dB (= 18.8 − 11 dB). These show that an acceptable signal will be received even if heavy rain on either the up-link or the down-link should cause a significant deterioration in reception conditions.

A further stage of calculation is needed to determine whether the looped-back video signal at the OB ground station will be of adequate quality for setting-up and monitoring purposes. At 11 GHz the receive antenna gain will be 43.3 dB, and repeating the power budget calculation shows that its received power $P_r = -119.3$ dB(W). A modern low-noise block will have a noise temperature T_B of, perhaps, 170 K (corresponding to a noise factor of 2 dB). An assumed antenna coupling/cable loss of 0.5 dB (a power loss of 12.2%) degrades T_B to 191.0 K. Adding a clear sky noise temperature of 79 K produces a system noise temperature of 270 K (i.e. 24.3 dB). The received noise power P_n is therefore 24.3 + 73.0 − 228.6 = −131.3 dB(W), which provides a c/n figure of −119.3 − (−31.3) = 12.0 dB.

Allowing for $(\text{c/n})_i = 30$ dB produces a true down-link carrier/noise figure of 11.5 dB (a worsening of only 0.5 dB when compared with the value just derived). This would be just acceptable for measurement purposes, but a 625/50 PAL signal would exhibit some sparklies, since it is very close to the threshold value. It would be very desirable, in this case, to employ a receiver with threshold extension to provide a margin of, say, 5 dB to allow for bad weather conditions.

The above described a link budget for an analogue system. For a digital system, the link budgeting process is similar, with changes to take into account applied Reed–Solomon coding and meeting of the overall E_b/N_o ratio with a margin. The E_b/N_o is the energy per bit over noise spectral density. Unlike analogue systems that degrade gracefully, digital systems will instantly lose picture/sound once the threshold E_b/N_o is breached.

6.5.9 Future Developments

Reliable higher power transponders are being developed. Utilising more linear amplifiers of greater efficiency, additional back-off associated with half transponder usage is being reduced with the use of linear TWTAs. Improvement in LNB noise figure and the higher power is allowing a further fall in the size of ground stations. The application of up-link power control (UPC) will become more usual. With UPC, the ground station power is adjusted automatically to compensate for temporary signal fades caused by precipitation losses. Additional on-board processing will allow greater transponder flexibility. Switchable receiver attenuators accommodate a range of ground station powers. New designs will permit the power output to be backed-off when up-link propagation conditions allow. Under ground control, transponder-to-antenna connections will be reconfigured as required to tailor footprints for special requirements. Satellites will increasingly be equipped with adequate backup to maintain the operation and performance of all transponders during periods of eclipse.

Services will become available using frequencies in the region of 30/20 GHz (Ka-band), which will allow even smaller ground stations and narrower beams under certain conditions. The Ka-band is currently being used by BT for their Openworld Internet service. This makes use of an elliptical domestic dish less than 1 m in size with transmit capability; previously satellite-to-home services required a phone connection for the return path. Satellite spacing in the Clarke orbit can be reduced at these frequencies, allowing more satellites to be accommodated. The expected life of communications satellites will increase as new methods of station keeping, such as electrically powered ion thruster jets, are introduced (commercial introduction is predicted around 2005).

Satellite-to-satellite links will be introduced, allowing greater distances to be linked with one space segment.

The major ground station development will have LNAs designed with new simple devices and having noise temperatures less than 70 K (noise figure less than 0.94 dB).

Roger Wilson BSc
Continental Microwave
Revised by
J K Levett BSc, C Eng MIEE
Formerly Senior Project Manager
Fixed Links, BBC, and Crown Castle International

6.6 Microwave Radio Relay Systems

6.6.1 Types of Microwave Link

Different types of terrestrial microwave link system are available for:

- electronic news gathering;
- outside broadcast;
- remote viewfinder;
- fixed radio relay.

The first three of these are covered in other chapters of this publication.

6.6.1.1 Fixed Radio Relay

Microwave radio relay links (see Figure 6.6.1) associated with broadcasting normally comprise permanently installed rack-mounted equipment and fall into two categories:

- Distribution links, which are used to transmit video and audio signals from a studio to a broadcast transmitter or between broadcast transmitters. In the allocation of signal distortions to the various parts of the signal chain from camera to viewer, very little is allowed for the programme distribution section, so it is important to minimise any degradation. In some situations the programme distribution to some or all of the major broadcast transmitters in a network will be carried out using either PTO-provided permanent circuits, which are likely to be predominately Fibre Optic, or by satellite. In these circumstances radio links will only be employed on the extremities of a service area where permanent circuits would be difficult or expensive to provide and where re-broadcasting is not a suitable option. In instances where availability requirements necessitate diverse signal routing, using radio links as the secondary feed to remote sites is likely to provide a far more economic solution than digging a second cable route.
- Contribution links, which are used to convey programme material from outlying studios or ENG or OB Reception points to a production centre. In many instances these links will follow the same route as distribution links but in the reverse direction. In the past the only difference between distribution and contribution links was likely to be the required availability, but with the introduction of digital studio, OB and SNG facilities, the format of the signals conveyed may now also be different.

6.6.1.2 Analogue or Digital Radio Links

Until around 1980 all radio links employed analogue techniques, wideband frequency modulation being used for television and multi-channel frequency division multiplex for telephony. Now, almost all equipment used for telephony has been replaced by digital links carrying data and voice traffic. The modulation employed ranges from two-state phase or frequency shift keying for low-capacity, very-high-frequency equipment, through Quaternary Phase Shift Keying, 16-, 64- and 128-level Quadrature Amplitude Modulation to 256- or 512-level Trellis Code Modulation used for very-high-capacity systems carrying one, two and even four channels of STM1, 155 Mbit traffic.

All the equipment listed above is designed to operate in internationally agreed frequency bands on coordinated channels to avoid interference to or from other users. A small amount of equipment is available which offers fixed link capabilities using frequencies and techniques intended for computer local area networks. This type of equipment should be used with great care as it operates on channels shared with other services. Although the type of modulation and data validation

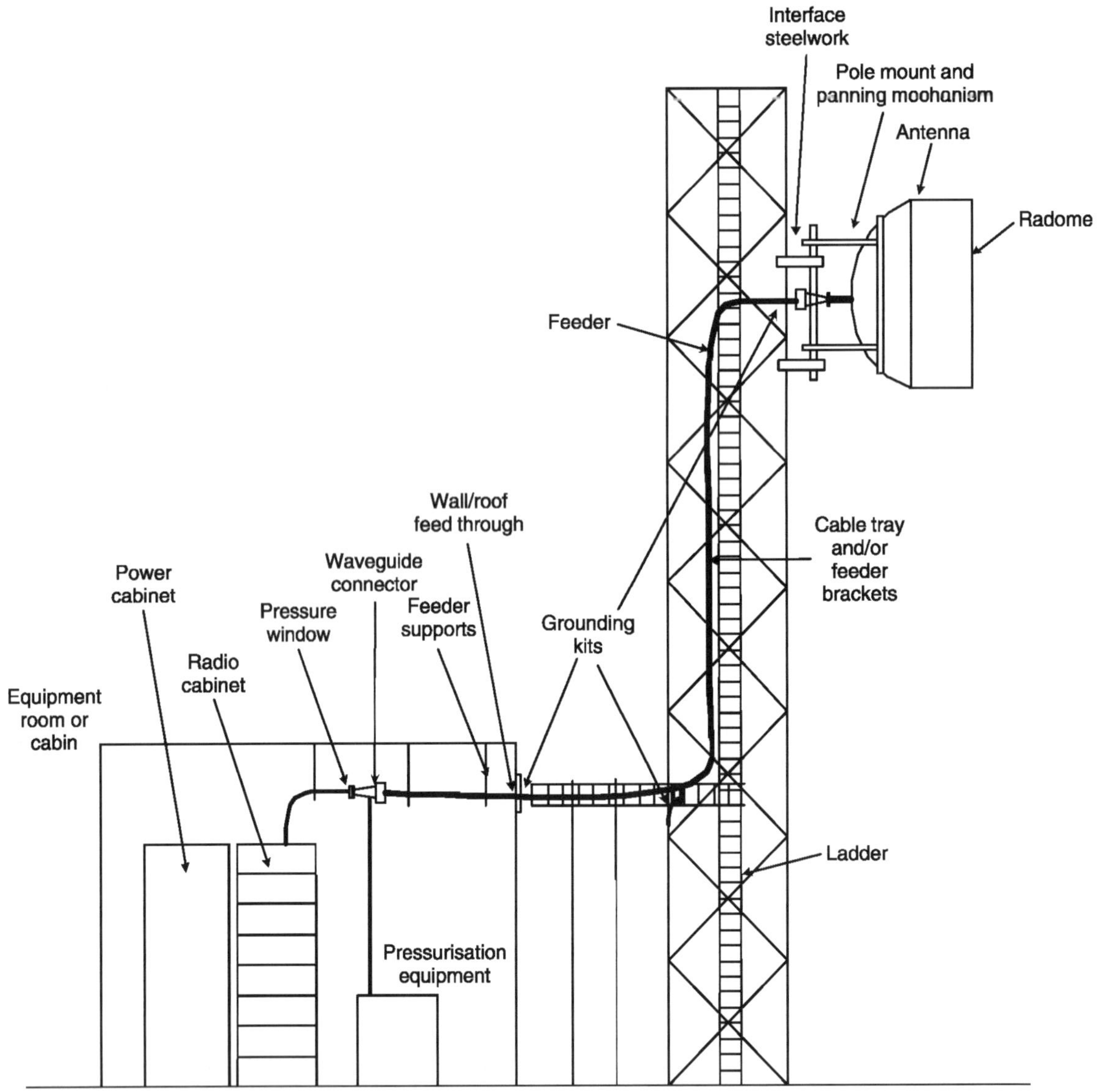

Figure 6.6.1 Typical radio relay link terminal.

employed can overcome some interference, noticeable interaction is increasingly likely.

As far as broadcasting applications are concerned, the format of the signals to be conveyed influences the type of radio link equipment that will be necessary.

6.6.1.2.1 Analogue Television

Traditionally these signals have been carried by broadcasters and Public Telecommunication Operators (PTOs) over radio links using analogue techniques, but currently all of the programme material sent over PTO equipment and a significant part of the traffic carried by broadcasters' own links employ digital techniques. Encoding the video signal at 140 Mbit using minimal compression is used where the transport system needs to be transparent to the user, if multiple decoding/recoding cycles are involved, or if minimising transit delay is important. Otherwise, 34 Mbit, or even lower in some circumstances, can be used for distribution purposes. With these reduced bit rate systems any teletext would need to be extracted and sent on a separate data channel. Multiple channels can be multiplexed together and sent over higher capacity links but, without resorting to PDH techniques, no more than three 34 Mbit transport streams normally fit into one 155 Mbit SDH data stream. Contributions can be at 8, 34 or 140 Mbit or analogue, depending on the desired facilities, quality and permissible delay.

In many instances where radio links are required, analogue equipment is still the most appropriate:

- Signals may be derived from Re-Broadcast Receivers. In most instances these signals are not suitable for reduced bit rate digital encoding without significant processing.
- The signal format may not be appropriate for reduced bit rate digital encoding. Sound in syncs is still used by some broadcasters on their main distribution networks.
- Adding a reduced bit rate digital encoded section into an established network introduces additional complexity when remote signal continuity and quality monitoring aspects are taken into account.
- Consequently, without major changes to a distribution network, many links would need to carry the television signal encoded at 140 Mbit. Radio links of this capacity are very susceptible to path disturbances, and need good received signals. This implies that shorter hops are necessary and establishing repeater sites, even if viable, is expensive and an increased maintenance liability.

6.6.1.2.2 Digital Terrestrial Television

In the UK, distribution of these signals is carried out by a combination of PTO-provided Terrestrial circuits and user-managed Satellite facilities.

In both instances the Multiplex signals are adapted into standard PDH levels of 34 or 140 Mbit and then carried over normal telecommunication circuits. The terrestrial circuits are predominately fibre.

In the United States, a significant proportion of DTT signals are distributed by radio link. A particular type of equipment which allows the simultaneous transport of both the analogue (NTSC) signal and its digital version on one 'Twin Stream' RF channel is popular.

6.6.1.2.3 AM and FM Stereo Radio

These signals are normally distributed in digital format, although a variety of coding techniques are in use. They are adapted into standard digital hierarchy levels of between 64 kbit and multiple 2 Mbit, depending on the quality required, whether re-coding may be involved and the number of channels. The combined signals are carried over normal telecom circuits. Much of the distribution is carried out by PTOs, but where self-provision is justified, digital links of an appropriate capacity are employed. It is also possible to add subcarriers carrying digital signals to existing analogue television links.

6.6.1.2.4 Digital Audio Broadcasting

There is considerable scope for using radio links in the engineering of DAB systems. Low-capacity digital links (typically 256 kbit) can be employed to connect a contributing studio to a multiplexing centre, with higher capacity links conveying the composite signal, fitted into a standard 2 Mbit bit stream, to and between the associated broadcast transmitters.

6.6.1.2.5 Frequency Bands

Allocations for fixed radio links are currently available between 1 and 57 GHz. The uses and properties of the various bands within this range are as follows:

- *1–3 GHz*. Much of this spectrum is now allocated for mobile telephone and data services, but in the UK there is a band for low capacity ($\leqslant$4 Mbit) digital links. It has the advantage that path lengths can be long and propagation is not affected significantly by weather conditions. Antennas for the lowest capacity systems are small and equipment cost reasonable.
- *3–10 GHz*. This spectrum is most suitable for medium- to high-capacity multi-channel digital and analogue links, providing reliable service over distances up to 70 km, and occasionally further. The signals are not noticeably affected by precipitation along the radio path, but atmospheric disturbances produce fading and signal distortion, the severity being dependent on the terrain and climatic factors.
- *10–22 GHz*. This spectrum is also suitable for medium- to high-capacity digital and analogue links but, as the operating frequency increases, the effect of precipitation along the radio path also rises. Consequently, the availability of links operating in this band is likely to be predominately defined by precipitation, rather than atmospheric disturbances. Typical path lengths can be from 40 km for a low-capacity link at 10 GHz to 3 or 4 km for a high-capacity circuit at 22 GHz, but both are very dependent on the required availability.
- *22–57 GHz*. Currently this spectrum is used extensively for short-range, single-channel low-, medium- and high-capacity digital links. The path lengths achievable are defined almost exclusively by precipitation losses and atmospheric attenuation.

Many frequency bands are now only available for use by digital links. Although there are allocations for analogue television links in bands above 23 GHz, there is very little broadcast-quality equipment available. This is due to a combination of low demand and high cost of hardware with suitable characteristics at these very high frequencies.

6.6.2 Radio Link Systems

6.6.2.1 Typical arrangement

A system typically comprises the following items:

Radio equipment

- transmitters and receivers;
- waveguide or coaxial branching (filters and circulators to multiplex transmitters and receivers to a common antenna);
- waveguide or coaxial feeder and accessories between radio equipment and the antenna;
- parabolic antennas with diameters ranging from 0.3 to 3.7 m;
- interface steelwork between antenna and tower or building;
- service and/or auxiliary channels;
- supervisory equipment;
- power equipment (solar cells, diesel generators, chargers and batteries).

Civil works

- antenna support structure (tower or building wall or roof);
- radio equipment room or cabin;
- air-conditioning.

6.6.2.2 System configurations

The most commonly used configurations are unidirectional, bidirectional, unduplicated and duplicated (hot-standby, twin path and N + 1). The latter configurations are also known as non-protected and protected. In many broadcasting applications all the traffic will be in one direction and consequently link equipment installed specifically for this purpose may well be unidirectional. This can provide savings in both capital equipment cost and licence fees for the operating channel. However, if the programme material is to be conveyed over digital links, standard bidirectional operation may be essential, as much of this type of equipment features integrated transmitter and receiver modules, and the system control and monitoring may not function correctly if one direction is disabled.

6.6.2.2.1 Unduplicated, unidirectional configuration

Transmitters and receivers are connected directly to the antenna port via a waveguide or coaxial channel filter, as shown in Figure 6.6.2(a). Unduplicated equipment is economic but, in the event of a single equipment failure, programme traffic will inevitably be lost.

6.6.2.2.2 Unduplicated, multi-channel, unidirectional configuration

Several transmitters and receivers are multiplexed to a common antenna by means of RF branching (or multiplexing), and this comprises an arrangement of filters and ferrite circulators, as shown in Figure 6.6.2(b). The signal from transmitter A is directed to the antenna port via the channel filter and circulator. The output of transmitter B, which is at a different frequency from transmitter A, is directed by the circulator to the channel filter of transmitter A, from which it is reflected back towards the circulator. Both signals then emerge (combined) from the same port of the circulator.

Minimum frequency spacing between transmitters is dependent on the bandwidth and roll-off characteristics of the filters employed, and is typically 49 or 56 MHz for television and medium- to high-capacity digital transmission. It is, of course, possible to multiplex more than two transmitters in this manner. The penalty is the additional loss introduced each time the signal is 'bounced' off an adjacent filter and the number of passes through circulators. Signals are separated in the same way at the receiver.

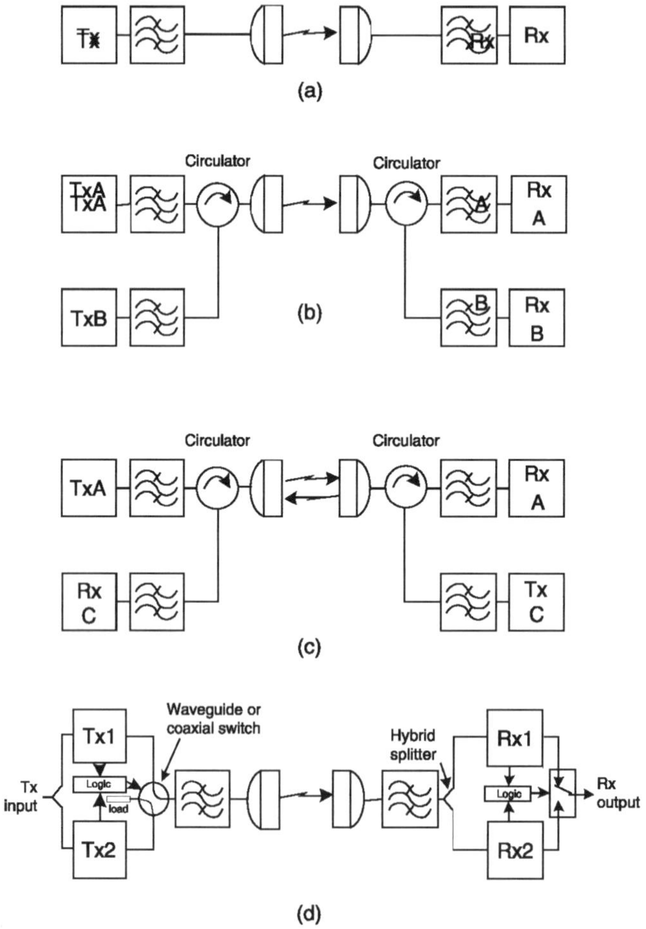

Figure 6.6.2 System configurations: (a) single-channel unidirectional; (b) dual-channel unidirectional; (c) single-channel bidirectional; (d) hot-standby with four-port Tx c/o switch.

6.6.2.2.3 Bidirectional configuration

RF branching can also be configured to connect transmitter(s) and receiver(s) to a common antenna. This provides a bidirectional (*duplex*) configuration. The layout would be optimised to minimise the number of circulator passes. When multiple channels are planned, care needs to be taken to ensure that intermodulation products do not cause a problem.

6.6.2.2.4 Duplication (protection)

Equipment can be duplicated to prevent loss of traffic due to a single fault. Two transmitters and two receivers are employed and, in the event of a failure, an automatic change-over switch selects the operational equipment.

There are various ways in which transmitter and receiver systems can be designed, and these lead to the arrangements described in the following sections.

6.6.2.2.4.1 Hot-standby

As shown in Figure 6.6.2(d), the modulating signals are split equally between two transmitters operating on the same frequency and the RF outputs are fed to either a switch or a coupler so that the selected transmitter can be routed to the antenna. If a switch is used, both transmitters would normally operate continuously at full power, but when a coupler is employed the de-selected transmitter would be muted.

Coaxial relay and rotary waveguide switches have relatively low insertion loss (<0.5 dB) but operate quite slowly (typical transfer times are in the region of 20–30 ms). PIN diode switches, however, change over in just a few microseconds but are more lossy (typically 2 dB). When a coupler is employed, the loss would depend on the coupling factor of the device.

Change-over switches can have three or four ports. A *three-port* switch reflects power from the unused transmitter back into an isolator fitted to its output, whereas a *four-port* switch directs the power of the unused transmitter into a separate load or attenuator, where it is easily available for maintenance purposes.

Monitoring circuits in the transmitter check for correct operation of various modules, output power and signal continuity, and raise an alarm if a fault is detected, triggering a change-over to the other transmitter. It is usual for the operational mode of the associated logic to be user defined, but unbiased selection is normally preferred as this reduces the possibility of multiple change-overs following an intermittent fault.

At the receiving end of the link, the incoming RF signal is split in a waveguide or coaxial hybrid circuit and fed to both receivers. Switching takes place at baseband when monitoring circuits detect a fault in the operational receiver.

An equal split at RF between the two receivers results in a nominal 3 dB loss in both signal paths. An unequal split, 4:1 or 10:1 (6 or 10 dB), reduces the loss into one receiver at the expense of the other. Receiver switching may then be biased to the channel with the lower attenuation (the main or priority channel). In most digital receivers steps are taken to synchronise the two receiver outputs, so that change-overs are seamless.

6.6.2.2.4.2 Twin path

A twin path system does not employ transmitter switching. Both transmitters operate simultaneously on two separate frequencies, and their outputs are combined with RF branching to feed one antenna. The arrangement is very similar to that shown in Figure 6.6.2(b), but where the same signal feeds both transmitters, as in the case of hot-standby. At the receiver both outputs are connected to a switch to enable selection of the better channel. As this arrangement is spectrally inefficient it is not encouraged by frequency administrators. A variation of this technique is, however, used for high-capacity digital links where the second channel is on the same frequency as the first but is carried on the orthogonal antenna polarisation.

6.6.2.2.4.3 N + 1 protection

Hot-standby and twin path configurations are most suitable for protecting a single microwave channel. In a system carrying two or more independent channels, duplication of each is both costly and unnecessary. Instead, the normal procedure is to provide one additional operating channel to protect several (N) others.

Detection of a fault at the receiver causes the receiver switching logic to send an alarm signal to the transmitter end of the link via a return channel. This alarm signal causes the transmitter switching logic to switch the affected signal from the failed channel to the protection channel.

The return control information required for N + 1 operation may be added to a broadband microwave link operating in the reverse direction, or carried by a narrowband vhf/uhf radio link or even a standard telephone line. Confirmation that switching has taken place at the transmitter is then sent to the receive end, and the switch associated with the failed channel selects the protection channel at the receiver.

6.6.2.3 Equipment configurations

There are three types of equipment configuration currently in use:

- *All indoor*. In this arrangement, all the active radio equipment is located in a building or cabin and is connected to the antenna system by coaxial cable or waveguide feeder. This is normally used for complex arrangements where multiple transmitters and/or receivers share the same antenna system, or if good maintenance access to the radio equipment is required at all times and in all weather conditions. This configuration is suitable for systems operating on frequencies up to around 14 GHz, as above this frequency the attenuation in standard feeders normally becomes too high.
- *Split unit*. In this arrangement, the power supply, low frequency, monitoring and control modules are located indoors. All the microwave circuitry is housed in a separate weatherproof assembly and connected to the indoor equipment (IDU) by a single coaxial or triaxial cable, which carries (low) intermediate frequency signals, power, monitoring and control information. Frequently these Outdoor Units (ODU) are clamped directly to the antenna, or are located close by, perhaps adjacent to the ladder, or at the base of the supporting structure if easier or non-climbing access is important and the associated feeder losses are acceptable. With this type of equipment it is not possible for a second radio channel to share an antenna system, other than by using the orthogonal polarisation.
- *All outdoor*. In some applications where either indoor space and/or cost is critical, all of the radio link circuitry is located in an outdoor unit mounted on or very close to the antenna. Signal

and power cables link it directly to the associated equipment, which may well share a common power supply, monitoring and control system.

6.6.2.4 Transmitter design

There are two main types of microwave transmitter:

- *Direct modulation*. This arrangement is currently only used for some very-high-frequency (22 GHz and above), low-capacity digital and analogue television links. After processing, the input signals directly frequency modulate the RF carrier. The frequency source may be at low level and subsequently amplified, or a Gun Diode or similar device may produce up to 100 milliwatts of output power directly. For satisfactory stability, both versions are likely to be locked to a lower frequency reference oscillator.
- *Heterodyne* (Figure 6.6.3). The input signal modulates an intermediate frequency (IF) oscillator, which is then up-converted once or twice to the output frequency before being filtered and amplified to the required level. The most common first intermediate frequency is 70 MHz, but frequencies in the 200–400 MHz range are now also used. Second IFs are in the range 800 MHz to 2 GHz.

6.6.2.4.1 Analogue Transmitters

The baseband bandwidth of an analogue television link transmitter is approximately 10 MHz, whereas a 625-line PAL signal only occupies up to 5.5 MHz. The remaining spectrum can be used to carry audio or low-capacity data signals modulated onto subcarriers. This arrangement is ideal for carrying the associated mono and/or digital stereo sound or, if sound in syncs is employed, an independent data signal. The ITU-R recommended subcarrier frequencies are 7.02, 7.5, 8.065 and 8.59 MHz, but as long as intermodulation products are taken into account, operation between 5.8 and 10 MHz is possible. The input video signal is amplified, pre-emphasised and filtered to remove noise or spurious tones above video, before being combined with the modulated subcarriers.

This composite signal frequency modulates a voltage-controlled oscillator, at the output frequency in the case of direct modulation or at typically 70 MHz for a heterodyne system. In the latter case the modulator is followed by an up-converter comprising a fixed frequency local oscillator, mixer and filter, which translates the signal to the second, much higher, intermediate frequency. This signal is then followed by a second up-converter comprising a programmable synthesised local oscillator and mixer to translate the modulated signal to the output frequency. Three significant frequencies are present at the output of this mixer, only one of which is required:

- local oscillator (F_o);
- lower mixer product at $F_o - \text{IF}$;
- upper mixer product at $F_o + \text{IF}$.

In this case of double conversion, the IF is between 800 MHz and 2 GHz, so a band-pass filter (typically 500 MHz wide) can easily select only the upper mixer product. Balanced mixers may also be

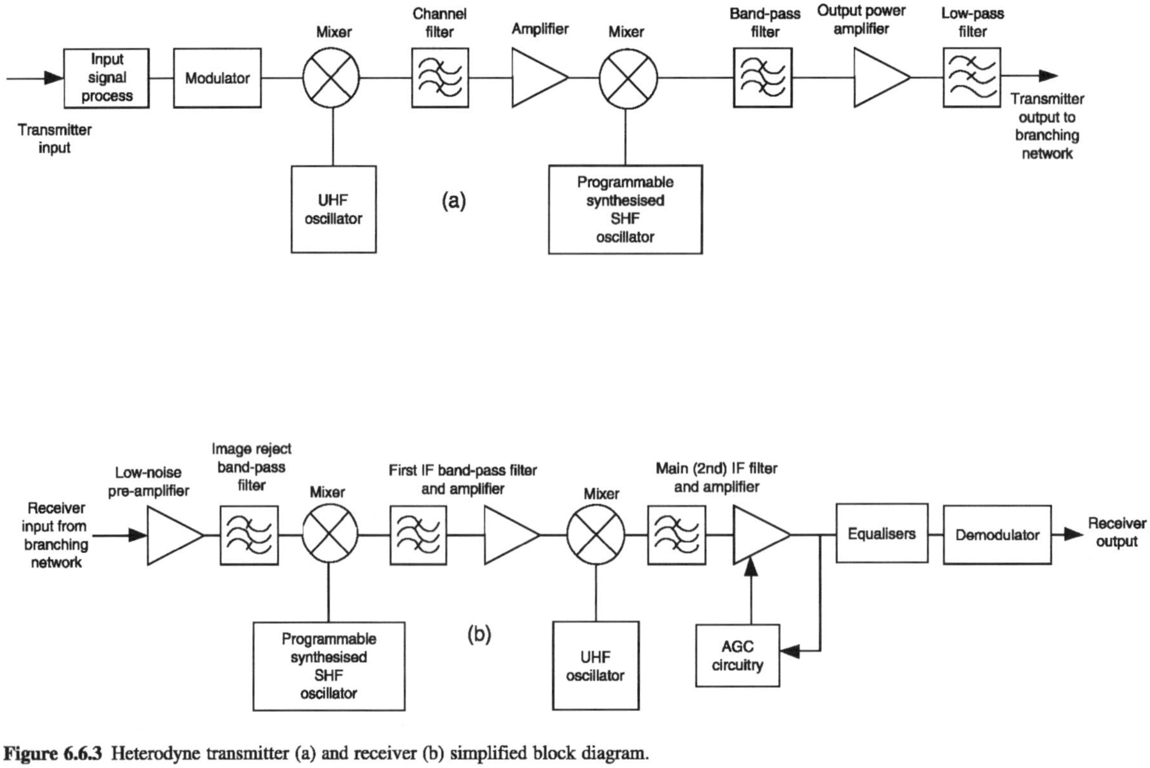

Figure 6.6.3 Heterodyne transmitter (a) and receiver (b) simplified block diagram.

employed to further reduce the local oscillator level at this point. The filtered signal is passed to a wideband solid-state amplifier which, in this application, can be operated in saturation. Systems employing a single conversion heterodyne principle require good channel filters to remove the unwanted mixer products to meet the current (UK) spurious emission specification of −60 dBm.

6.6.2.4.2 Digital transmitters

Most of the above description also applies to digital transmitters.

The video processing is replaced by input signal conditioning, multiplexing with auxiliary and monitoring data, and the addition of Forward Error Correction bytes, before being fed to a digital modulator. Many such modulators are similar to the arrangement shown in Figure 6.6.4.

With digital transmitters, it is essential for all modulated signal stages, including the output power amplifier, to operate in a linear mode, otherwise the output signal will become distorted and bit errors will occur.

6.6.2.4.3 Receivers

Microwave receivers for radio relay applications have traditionally been *single conversion*, but with the introduction of programmable low noise microwave local oscillators, permitting the design of equipment tunable over at least 500 MHz, double conversion heterodyne systems now dominate.

The RF signal (F_{rx}) enters the system through a band-pass filter, normally part of the RF branching. The signal is then passed through a low-noise amplifier (LNA) and further band-pass filter to remove noise generated in the LNA at the receiver image frequency ($F_{rx} - 2 \times IF_1$) before entering the SHF mixer along with a local oscillator signal (F_O). The resulting first IF at $F_{rx} - F_O$ is filtered and applied to a fixed gain pre-amplifier, before being translated down to the second lower IF, frequently at 70 MHz. It is then equalised to minimise group-delay and level variations across the band, followed by further amplification stages with automatic gain control (agc), to provide a constant output of 0.5 V_{rms} for application to the demodulator.

In an analogue receiver the signal is limited (*clipped*) to remove any amplitude modulation introduced by noise and distortion in the IF and RF stages, and applied to the FM discriminator and detector. Following demodulation, the composite signal (comprising video and any subcarriers) is split between two outputs. One, destined to be the video signal, is fed through a low-pass filter (5.5 MHz cut-off for 625-line PAL system) to remove high-frequency noise and subcarriers, and is then de-emphasised and amplified to 1 V_{p-p}. The second is routed to any subcarrier demodulators.

In digital receivers a variety of demodulation techniques are employed, ranging from phase-lock loop circuits to conversion of the IF signal to a computer-style data signal and processing in software. Most digital receivers also include adaptive frequency and time domain equalisers to minimise the effect of multi-path distortion.

6.6.2.5 Supervisory equipment

In a large microwave link system it is essential for a network operator to know whether all the equipment is fully operational. As many sites will be unattended, supervisory equipment is used to signal the status of radio, multiplex, power, pressurisation and various other site alarm conditions to the network monitoring centre. These indications may be fed to a separate fault reporting system either by relay contact or serial data connection, or utilise a low-capacity auxiliary data channel on the radio equipment. With many digital links, network monitoring and control facilities are a standard feature, with the information being carried as part of the overhead data. Whichever arrangement is employed, care needs to be taken in its design to ensure that monitoring information for the remainder of the network is not lost if, for instance, severe fading or a total site outage occurs.

6.6.2.6 Feeder equipment

Transmitters and receivers are multiplexed together with coaxial cable or rigid, rectangular waveguide. The radio output, often called the *antenna port*, generally comprises a coaxial

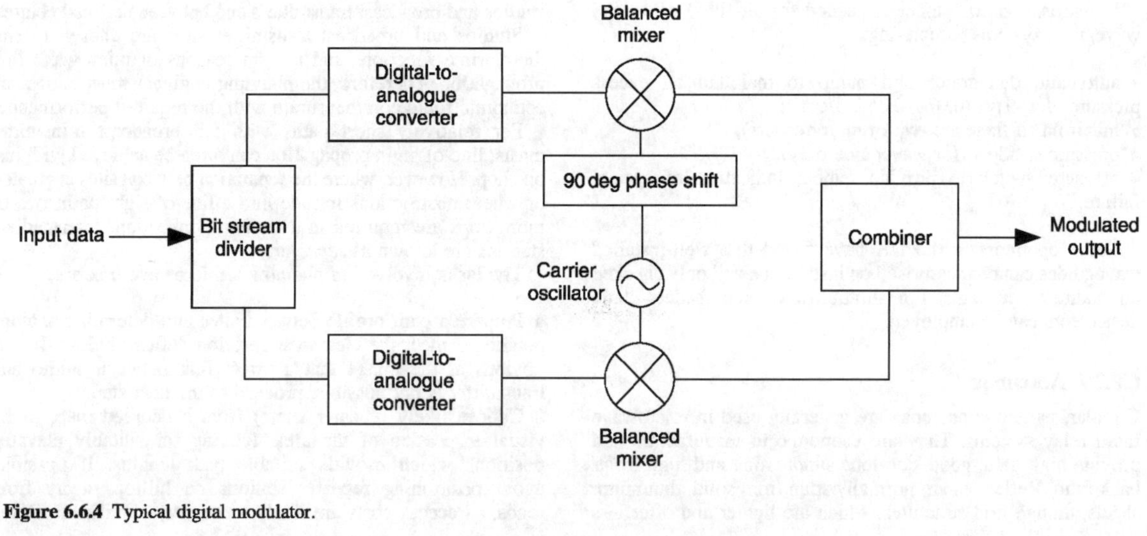

Figure 6.6.4 Typical digital modulator.

connector for equipment up to 3 GHz and a waveguide flange for higher frequencies. A coaxial cable or elliptical waveguide feeder is then used between the radio output port and the antenna.

At frequencies below 3 GHz, it is normal to use cable with either air or foam dielectric. Cable diameters range from $^1/_4$ inch to $1^5/_8$ inches (6.35 mm–4.13 cm), with insertion loss decreasing with increasing diameter. Air dielectric cables of a similar diameter have a lower loss but, as pressurisation equipment would then be required, foam cables are preferred.

Corrugated, elliptical waveguide is used at frequencies above 3 GHz. This is relatively low loss and is sufficiently flexible to accommodate the inevitable bends in a typical feeder run. Connectors at each end of the waveguide incorporate transformers which match it to the rectangular waveguide used in the radio branching and antenna feed assemblies. A different size of waveguide is required for each frequency band of operation. Special 'over-moded' versions are available for some frequency bands above 10 GHz, which have the advantage of significantly lower attenuation. If even lower loss is essential, circular waveguide can be employed, but this has to be installed absolutely straight. Even on tall masts or towers this can be a problem, as existing steelwork or cables are likely to obstruct the route.

Accessories for feeder systems typically comprise:

- connector at each end;
- wall or roof feed-through gland to provide a tidy and weatherproof entry into a radio equipment room;
- grounding kit (usually three) to ground the feeder at the antenna, at the base of the tower and at the point of entry to the equipment building;
- flexible 'tails' for use with relatively rigid, large-radius cables or waveguide where it cannot easily be connected directly to the radio equipment and/or antenna;
- hangers and brackets to secure the feeder to the tower and to building walls or ceilings – cable ducts, trays or ladders may be required to support the feeder on the tower, in the equipment room or in between;
- for their installation, cable terminating tools or jig and, for vertical runs, hoisting stockings.

Pressurisation equipment is needed for air-filled cable and waveguide systems comprising:

- automatic dehydrator and pump to maintain an excess pressure of 1–5 psi (0.070–0.35 kgf/cm^2);
- manifold (if there are two or more feeders);
- pressure window (for waveguide only);
- pressure switch to provide remote indication of system failure.

Some operators in the UK have found that well-installed waveguides can be pressurised on installation and only checked on maintenance visits. For short, low-volume feeders static desiccators can be employed.

6.6.2.7 Antennas

Circular, parabolic antennas are generally used in microwave radio relay systems. They are economic to manufacture and provide high gain, good side-lobe suppression and high front/back ratio. Reflectors are normally spun from solid aluminium sheets, though grid structures, which are lighter and offer less wind resistance, are often employed at frequencies below 2.5 GHz. There is no advantage in using grid reflectors for their reduced wind loading properties on sites subject to icing.

Antennas are available in standard, low VSWR (*voltage standing wave ratio*) and high-performance versions, and feeds may be single or dual polarised (horizontal, vertical or both). Circular polarisation is rarely used in fixed link systems. The high-performance type use a side shield to improve the front/back ratio and side-lobe suppression. Sizes range from 0.15 up to 3.7 m diameter, although the most commonly used are between 0.3 and 2.4 m.

Nearly all antennas are equipped with a panning frame which mounts onto a vertical 115 mm diameter steel pole. Interface steelwork secures the pole to the tower. Alternatively, if antennas are deployed on roof tops or building walls, other types of interface steelwork will be required. In some instances non-penetrating free-standing roof mounts can be used for small antennas.

Most antennas are intended for use in inland locations where severe winds (>200 km/h) are rare. Reinforced versions, some employing four-point attachments and extra strength radomes, are available for use in locations subject to abnormally high winds. For installations subject to corrosive atmospheres, such as close to chimney outlets or sea spray, antennas are also available with additional protective coatings.

Radomes which fit across the front of the antenna are optional on all but high-performance antennas, where they are integral to their design. They are used to prevent ice build-up around the feed and to reduce wind loading. Snow and ice can still stick to radomes and result in degraded service, but the flexing of the material in the wind, combined with its smooth surface, helps to dislodge any build-up and reduce the effect.

Antennas, feeders and equipment accommodation located on sites liable to significant icing are provided with protection to avoid damage from falling ice when the temperature increases.

6.6.3 System Planning

Fixed microwave links are generally used in the broadcasting industry for the transmission of programme material between studios and broadcast transmitters and between regional centres.

Studios and broadcast transmitter sites are chosen to suit their prime functions and not for reasons of microwave link propagation. Therefore, the planning engineer must choose an economic route commensurate with the required performance.

For relatively short paths with tall broadcast transmitter masts, line-of-sight propagation can often be achieved with just one hop. However, where the separation between sites is greater, or where obstructions intrude into a line-of-sight path, two or more hops are required in a tandem connection. Intermediate stations are known as *repeaters*.

The tasks involved in planning a microwave link are:

- Prepare a path profile between sites and determine antenna heights to meet the clearance criterion defined below. If it is obvious at the outset that a direct link between studio and transmitter is not possible, proceed to the next step.
- Choose likely repeater site(s) from a detailed map, or by visual inspection of the area, looking for suitably elevated positions which provide suitable path lengths. If possible, avoid positioning repeater stations on hilltops away from roads, as access costs are likely to be very significant during

both construction and operation. If a suitable structure already exists, which the owner is prepared to share, then a considerable saving in civil works costs will obviously result.

- Prepare path profiles between likely repeater stations and determine antenna heights to meet clearance criteria. A site and route survey will always be required to confirm information derived from maps, as trees, buildings or other man-made structures may obstruct the path. A GPS receiver can be useful during field surveys to confirm the location of potential obstacles.
- Choose routes that offer an acceptable compromise between hop length and antenna height.
- Perform calculations to determine transmitter output power and antenna sizes needed to provide a suitable fade margin.
- Specify building and tower requirements.

6.6.3.1 Path profiles

The first task in assessing the suitability of a route is to determine clearance between the microwave beam and potential obstructions along the route. For many parts of the world details of the terrain are available as data files and a cross-section through the earth between potential sites can be easily produced using one of the many proprietary computer programs. If suitable data is not available, the terrain can be analysed by plotting ground height contours taken from a topographical map, with a scale of 1:50,000 or less, onto specially prepared graph paper with curved lines representing the earth's curved surface. The microwave path is represented by a straight line drawn between the two antennas.

Atmospheric refraction causes ray bending at microwave frequencies, and it is normal to assume an effective earth radius (k) of 4/3 greater than actual. In practice, this means that the microwave beam is bent slightly downwards by the gradual change in the atmosphere's refractive index with height. The microwave signal can therefore be received a little way beyond the optical horizon.

Changes in atmospheric refraction can occasionally cause the value of k to drop below 2/3. This prevents the microwave signal from being received at the optical horizon. Consequently, it is often more convenient to plot ground height on rectilinear graph paper (shown in Figure 6.6.5) and add the earth bulge for two values of k: 4/3 and 2/3.

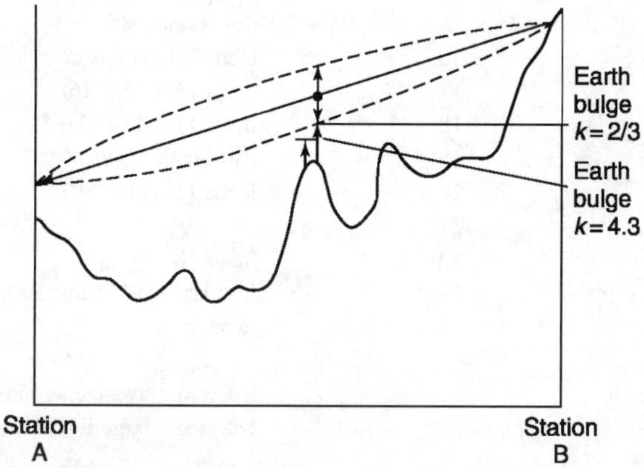

Figure 6.6.5 Path profile drawn on rectilinear graph paper.

Earth bulge (EB) may be expressed in metres as:

$$\text{EB} = \frac{d_1 d_2}{12.8k}$$

where d_1 and d_2 are the distances from a particular point on the path to each end in kilometres.

6.6.3.2 Fresnel Zone

The next step is to calculate the radius of the first Fresnel zone, which is defined as the locus of all points surrounding a radio beam from which reflected rays would have a path length one half-wavelength greater than the direct ray.

The value in metres of the first Fresnel zone radius is given by:

$$R_{\text{FZ}} = 17.3\sqrt{\frac{d_1 d_2}{F(d_1 + d_2)}}$$

where d_1 and d_2 are the distances to each end of the path in kilometres, and F is the frequency in gigahertz.

Figure 6.6.6 shows that free space propagation loss corresponds to transmission of 60% of the radius of the first Fresnel zone.

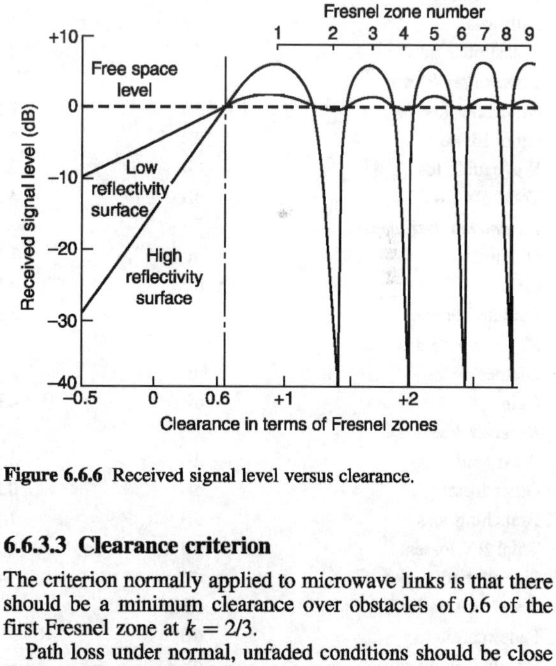

Figure 6.6.6 Received signal level versus clearance.

6.6.3.3 Clearance criterion

The criterion normally applied to microwave links is that there should be a minimum clearance over obstacles of 0.6 of the first Fresnel zone at $k = 2/3$.

Path loss under normal, unfaded conditions should be close to the predicted value if these conditions are met.

6.6.3.4 Received signal level

The power of the signal at the microwave receiver input is dependent on the following factors:

- branching losses at both ends of the link (dB) (as stated by the radio equipment manufacturer);
- waveguide or cable feeder losses at both ends of the link (dB) (as stated by the feeder equipment manufacturer);

- propagation losses;
- antenna gains (dB) (as stated by the antenna manufacturer).

Propagation loss in decibels under normal conditions is $20\log(4\pi d/\lambda)$, where d is the path length and λ is the wavelength (both in metres). It may alternatively be expressed as $20\log(4\pi df/c)$, where f is the frequency (Hz) and c is the velocity of light (3×10^8 m/s).

Therefore, the received signal level in dBW is equal to transmitted power minus branching losses minus feeder losses minus propagation loss plus antenna gains.

A typical calculation is shown in Table 6.6.1.

6.6.3.5 Fade margin

One of the most important parameters of a radio link is the fade margin. This is defined as the difference between the *normal* or *unfaded received signal level* and the *lowest receiver signal level* which will provide an acceptable output from the receiver expressed in decibels.

- For an analogue television link the lowest usable received signal level corresponds to that which will provide an unweighted signal-to-noise ratio of around 34 dB when measured in a bandwidth of 10 kHz to 5 MHz. For a receiver with a 4 dB noise figure and 30 MHz bandwidth relevant to a signal with 8 MHz p-p deviation, this equates to a receiver input level of −75 dBm (−105 dBW).
- For a digital link the lowest usable received signal level corresponds to that which will produce a demodulated Bit Error Rate (BER) of 1×10^{-6}. This will typically be −76 dBm for a 34 Mbit QPSK radio and −68 dBm for a 155 Mbit 128-level TCM version.

Table 6.6.1 System performance example for an undupicated 55 km path at 7.5 GHz

Description	*Unit*	*Value*	*Item*	*Note*
Frequency	GHz	7.5	1	
Path length	km	55	2	
Path loss	dB	144.8	3	
Transmitter power	dbW	2	4	Manufacturer's data
Transmitter losses				
Branching loss	dB	1.5	5	Manufacturer's data
Other losses	dB	0.5	6	Connectors, etc.
Waveguide loss	dB	2.5	7	Length in m × loss/m
Total TX loss	dB	4.5	8	Items 5 + 6 + 7
TX antenna parameters				
Diameter	m	1.8	9	
Gain	dB	40.8	10	Manufacturer's data
Radiated power	dBW	38.3	11	Items 4 + 10 − 8
RX antenna parameters				
Diameter	m	2.4	12	
Gain	dB	43.3	13	Manufacturer's data
Receiver losses				
Waveguide loss	dB	3.5	14	Length in m × loss/m
Other losses	dB	0.5	15	Connectors, etc.
Branching loss	dB	1.5	16	Manufacturer's data
Total RX losses	dB	5.5	17	Items 14 + 15 + 16
Signal at RX input	dBW	−68.7	18	Items 11 + 13 − 17 − 3
Receiver threshold	dBW	−105	19	(Broadcast video)
Fade margin	dB	36	20	Items 18 − 19
Receiver noise power	dBW	−125	21	
Carrier/noise	dB	56.3	22	Items 18 − 21
Total improvement	dB	13.8	23	Standard for 625-line PAL
Signal/noise UWTD	dB	70.1	24	Items 22 + 23
Availability estimate				
Path roughness (*A*)		1		Selected – average terrain
Climate factor		0.25		Selected – large inland area
Estimated availability		99.996%		
Equates to		22		Minutes lost per year

6.6.3.6 Availability

The propagation loss calculated in Section 6.6.3.4 took account only of fixed losses. However, the signal power arriving at the receiver in a real radio system can vary significantly and will fade well below the anticipated value as atmospheric conditions change. For this reason it is not normally possible for a radio link hop to be made 100% reliable. Standard availabilities range from 99.9%, corresponding to 525 minutes lost per year, to 99.999%, 5 minutes lost per year. The characteristic of radio link propagation availability is that a few seconds, or perhaps minutes depending on severity, will be lost whenever there is a significant radio path disturbance or precipitation fade. By comparison, a similar availability PTO-provided cable or fibre circuit may well use up 10 years of availability target in one isolated outage caused by the dreaded 'JCB effect' or similar.

It must be remembered that the overall availability of a radio circuit is a combination of the propagation availability and equipment reliability. Even in a duplicated (protected) radio system there are a number of common components where single faults, or maintenance outages, would cause circuit failure. Where very high availability is specified route diversity may be essential.

6.6.3.7 Multi-path fading

Multi-path fading over a microwave link is due to the signal arriving at the receive antenna from a number of different routes, such as:

- the direct path between antennas;
- ground and building reflections;
- refraction in the atmosphere.

Sometimes the signals add in phase, and the resultant is stronger than anticipated. At other times the signals cancel, and the resultant is weaker.

For short periods, almost complete cancellation can take place; the signal level then falls below the receiver threshold and transmission is lost. Fortunately, severe fading of this magnitude is relatively uncommon and occurs for only short periods.

Multi-path fading is dependent on the following factors:

- ground roughness (smooth surfaces cause pronounced reflections);
- climate (causes changes in refractive index of the atmosphere, which result in rays taking different routes to the receiver);
- path length (longer paths increase the opportunity for these events to take place);
- frequency (small delays represent a larger proportion of a wavelength).

A number of empirical formulas have been devised to predict the effects of multi-path fading on the availability of microwave links, and subsequently verified by practical measurements. Vigants' method[1] is described below, whilst the Radiocommunications Agency in the UK has adopted the procedure described in ITU-R Recommendation 530-7.[2] In the ITU model path roughness is replaced by a path inclination factor.

Vigants derived the following expression to predict signal availability in the presence of Rayleigh fading:

$$\text{Availability (\%)} = 100\left(1 - 6ABfd^3 10^{-\left(\frac{F}{10}+7\right)}\right)$$

where d = path length (km),
A = path roughness factor,
B = climatic conditions factor,
f = frequency (GHz),
F = fade margin (dB).

The path roughness (A) is 4 for smooth terrain including over water, 1 for average terrain or 0.125 for mountainous or very dry areas.

The climatic conditions factor (B) is 0.5 for hot, humid areas, 0.25 for large inland areas or 0.125 for dry inland areas.

6.6.3.8 Diffraction losses

There are occasions when it is impractical to provide sufficient antenna height to meet the clearance criterion stated in Section 6.6.3.3, and plans for that particular route may have to be abandoned. However, it may be possible to tolerate the additional loss introduced as the signal is diffracted over the horizon. Figure 6.6.6 illustrates loss versus penetration into the Fresnel zone.

6.6.3.9 Rainfall

At frequencies above 10 GHz rainfall can cause severe attenuation and, at 22 GHz and above, it is the dominant factor determining fade margin requirements in many parts of the world.

As an example, additional attenuation caused by rain at 50 mm/h (thunderstorm) at 10 GHz is around 1.5 dB per km, while at 23 GHz it approaches 8 dB. Consequently, for all but low availability applications, path lengths at 23 GHz are usually limited to 10 km or so. Figure 6.6.7 shows the typical losses involved.[3]

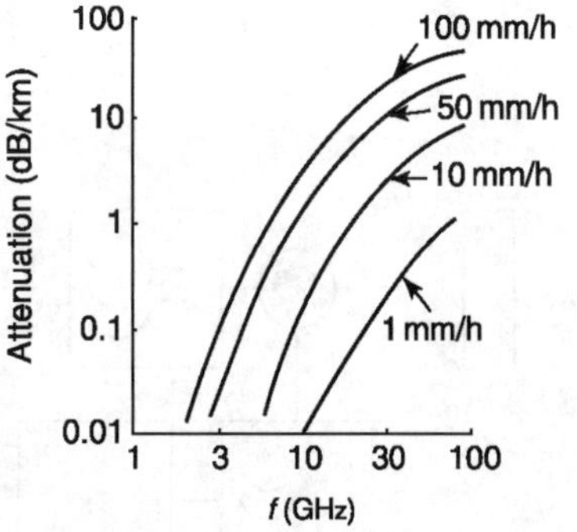

Figure 6.6.7 Additional attenuation caused by rain.

6.6.3.10 Gases

Attenuation from atmospheric gases is negligible at frequencies used in most radio relay applications. Attenuation (per kilometre) is below 0.01 dB at 10 GHz and 0.3 dB at 23 GHz, and only becomes significant close to 55 GHz, where oxygen resonance produces an additional loss of approximately 14 dB. The effect of rainfall and gases is covered extensively in ITU-R Volume V.

6.6.4 Improving Availability

Signal availability can be improved by using space or frequency diversity at the receiver.

6.6.4.1 Space diversity

Space diversity requires two antennas each with its own receiver. As shown in Figure 6.6.8(b), the signal may be combined before being routed to a single demodulator to avoid switching breaks. The ideal spacing between the antennas is determined by several factors, such as antenna height, path length, frequency and whether diversity is required to counter the effects of specular reflection from a very smooth surface such as water, or whether it is to counter atmospheric effects.

6.6.4.1.1 Reflections from water

When a microwave signal passes over water it is possible, at certain times, for the reflected and direct rays to cancel at the receiving antenna. If a second antenna is placed above or below the original at such a distance that the path length of the reflected ray is increased or decreased by one half-wavelength, the rays will add instead of cancelling.

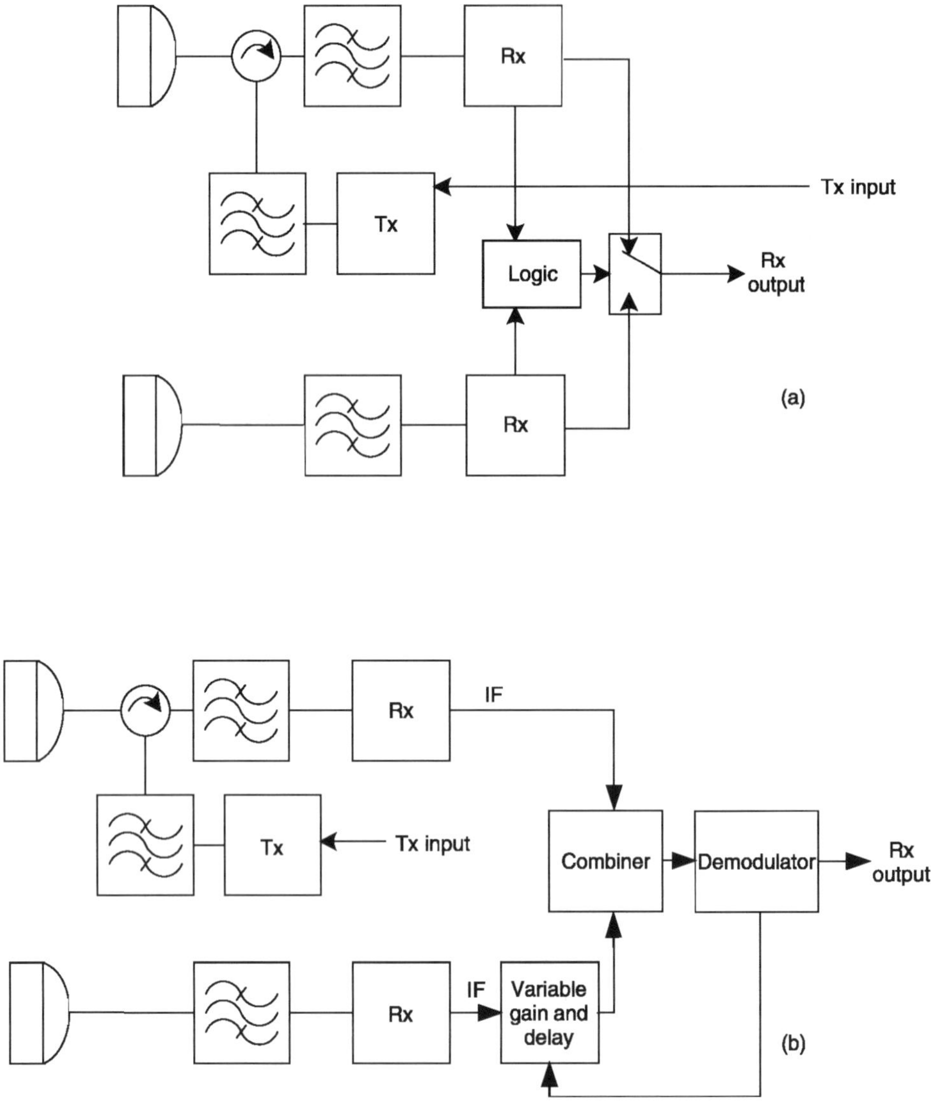

Figure 6.6.8 Unprotected bidirectional terminal with space diversity: (a) with receiver switching; (b) with one method of achieving seamless change-over.

Monitoring circuits in the receivers detect changes in received signal level and switch from one receiver to the other as the signal improves or degrades.

6.6.4.1.2 Protection against atmospheric effects

When space diversity is used to protect against Rayleigh fading or ducting (where a ray is trapped in an atmospheric layer), antennas should be spaced as widely as possible. Generally, tower height sets a practical limit on the spacing, but a minimum of 150 wavelengths should be the target.

ITU-R Report 376-4 and Vigants' paper[4] include more information about improvements available from space diversity.

6.6.4.2 Frequency diversity

The disadvantage of space diversity is that it doubles the cost of the antenna and feeder equipment at the receiving terminal and may require a stronger tower.

Frequency diversity is a similar configuration to twin path described earlier, but with both channels carrying the same information and where the operating frequencies may be more widely separated, i.e. approximately 5–10%. As spectrum is now in short supply in most developed countries, the additional frequencies necessary for frequency diversity are unlikely to be available.

References

1. Barnett, W.T. Multipath propagation at 4, 6 and 11 GHz, *Bell System Technical Journal*, **51**, No. 2, 311–361 (February 1972).
2. ITU-R P 530-7, Geneva (1997).
3. Mazda, F. (ed.). *Principles of Radio Communication*, p. 36, Focal Press ISBN 02405 1457 2 (1996).
4. Vigants, A. Space-diversity engineering, *Bell System Technical Journal*, **54**, No. 1, 103–142 (January 1975).

Y Imahori
Chief Engineer, NHK
Revised by **Salim Sidat** MBA, DBA, BSc (Hons), CEng, MIEE
Customer Systems Manager
mmO2, Ltd

6.7 Up-link Terminals

6.7.1 System Design

The transition from analogue to digital TV in recent years has significantly altered up-link operations. By enabling as many as eight digital channels to broadcast from a transponder which once delivered a single analogue channel, the economics and operation of broadcasting have changed fundamentally. Although the broadcasting world is rapidly going digital, analogue TV and radio are set to remain for several years yet. Indeed, for some services, analogue is still an attractive option due to the large installed audience base and the widespread existence of low-cost consumer equipment. Moreover, the capacity to transmit audio subcarriers of an analogue TV signal allows multilingual TV programmes or the parallel transmission of radio stations.

This section describes the considerations for an analogue up-link terminal developed for use with the Japanese BS satellites.

6.7.1.1 Types and functions

Direct broadcasting by satellite needs *programme transmitting stations* to send broadcast programme signals to the satellite, and *satellite control stations* for the satellite's mission equipment and bus equipment.

These two types of stations may be integrated in a single system, or constructed as independent systems or in hybrid forms. The following sections will consider a main earth station that has the function of programme transmission and satellite transponder control, and other earth stations that operate principally as programme transmitting stations.

6.7.1.2 Specifications

The requirements for up-link terminals depend on their function, on the performance specifications and on the transmission criteria for the broadcasting satellite, as well as on the role they are required to play in the direct satellite broadcasting system.

The functions required of each up-link terminal depend on the position which the up-link terminal system is to occupy, the area in which it is to be placed, and whether the system is to be mobile or not.

The transmission conditions of each up-link terminal (such as frequency, number of channels, transmitting power and polarization) are determined by considering the input/output parameters, the transmission system of the broadcasting satellite, its availability, the prevailing meteorological conditions, and the necessity to comply with the technical standards of WARC-BS (see Table 6.7.1).

A suitable location will:

- be free from mutual interference (intermodulation, crosstalk) with other communication systems;
- have good meteorological conditions (acceptable limits of rain, wind, snow) and ground conditions (earthquake free);
- have a clear vision of the geostationary satellites and be unaffected by aircraft flight paths, etc.;
- have access to stable electric power and stable video and audio transmitting links (terrestrial).

Furthermore, in order to ensure continuity of service, there needs to be available stand-by terrestrial systems, which can provide backup when the main earth station is not functioning or in the event of disaster, heavy rain, etc.

Thus, in designing up-link terminals, it is necessary to take into consideration the whole satellite broadcasting system and the functions, performance and economy of the terminals. Figure 6.7.1 shows a satellite broadcasting system. The various types of stations are described in Section 6.7.2.

6.7.1.3 Transmission systems

In Japan, the NTSC (PCM for audio) is used for satellite broadcasting, typical transmission parameters are given in Table 6.7.2. For the video signals, the main carrier is frequency-modulated

Table 6.7.1 Technical standard of WARC-BS

Item	*WARC-BS Regions 1, 3*	*RARC SAT-83 Region 2*
Satellite orbit separation	6°	–
Satellite station keeping accuracy	±0.1° (E–W)	±0.1° (E–W) ±0.1° (N–S)
Satellite antenna direction		±1°
pointing accuracy	±0.1°	
beam rotation	±2°	±1°
Transmitting wave polarization	Circularly polarized	Circularly polarized
Minimum elevation angle of receiving antenna	20°	20°
Transmission signal bandwidth	27 MHz	24 MHz
Power flux density in service area		
individual reception	-103 dBW/m^2	-107 dBW/m^2
community reception	-111 dBW/m^2	–
Diameter of receiving antenna	0.9 m	1.0 m
G/T individual reception	6 dB/k	10 dB/k
community reception	14 dB/k	–
Carrier/noise ratio in service area	14 dB	14 dB
Power flux density for interference protection of terrestrial station	-125 dBW/m^2/4 kHz	–
Energy dispersal	22 dB	–
frequency	(600 kHz$_{p-p}$ deviation)	
Interference protection ratio for satellite to satellite	31 dB (with co-channel) 15 dB (with adjacent channel)	28 dB (with co-channel) 13.6 dB (with adjacent channel)

with a signal in the 4.5 MHz band. For the audio signals, the subcarrier of 5.727272 MHz is Q-DPSK modulated with a PCM signal which is sampled at 32 kHz (A-mode) or 48 kHz (B-mode), and it is superimposed on the upper end of the video band (see Figure 6.7.2). A data channel is provided in the audio format for application to teletext, facsimile and other forms of broadcasting.

6.7.1.4 Link budget

FM transmission is used in satellite broadcasting. For an analogue system, the direct broadcasting system is required to secure an overall carrier/noise ratio of about 14 dB. In general, at a c/n ratio below 10 dB, although this depends on the FM demodulation circuit, the threshold noise appears as spots on the screen (sometimes referred to as sparklies). It is therefore necessary to ensure that the c/n ratio is above this threshold value by a suitable margin.

In the case of the up-link terminal, it is normally necessary to provide sufficient c/n to prevent the c/n of the up-link from affecting the c/n of the down-link (the overall c/n value is the reciprocal of the sum of the reciprocals of the up-link and down-link c/n values, so if the up-link c/n is poor, the overall c/n will be degraded). For contingency, it is also necessary to make provision for alternative method for transmitting programme to avoid adverse effects of rain, snow or other meteorological conditions.

In general, a c/n ratio of about 20–24 dB produces no problem in respect of picture quality.

For the main earth station, which is the key facility of the satellite broadcasting system, the c/n is set with a margin of about 10 dB. Normally, for fair weather, the c/n is set at about 30 dB; for rainy weather, power must be increased. In the worst conditions it is necessary to obtain backup from other stations. Table 6.7.3 shows up-link parameters for a main earth station, sub earth station and transportable earth stations of the BS-2 satellite broadcasting system.

6.7.2 Earth Stations

6.7.2.1 Main earth station

The main earth station is the key facility for transmitting television programmes to the satellite; it coordinates and monitors the satellite and all the earth stations for satellite broadcasting. Its functions are:

- transmission and reception of satellite broadcast programmes;
- control of satellite transponder;
- operation of the order-wire link;
- monitoring of the receiving condition of the satellite broadcast signals over the service area by the data from the monitoring earth stations.

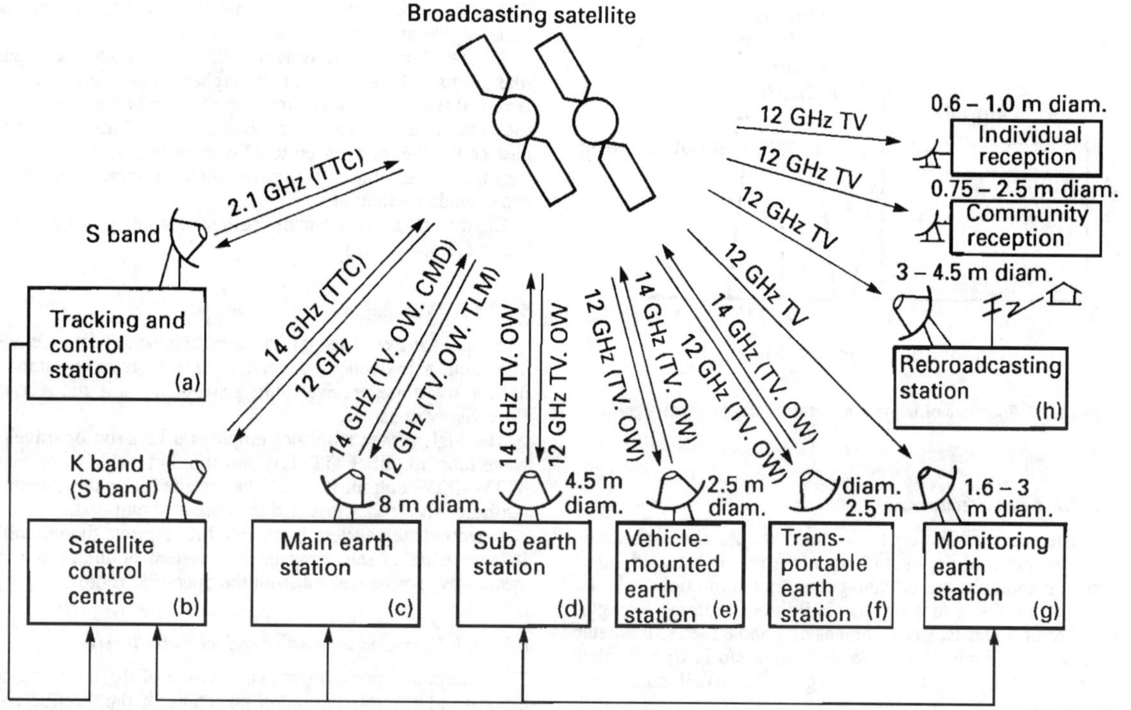

Figure 6.7.1 Satellite broadcasting system. (a) and (b) have the functions of control stations; (c) is a main earth station and has some of the functions of the programme broadcasting station and the control station; (d), (e) and (f) are programme transmitting stations.

Table 6.7.2 Transmission system parameters of television picture and sound signals for BS-2 broadcasting

Television picture	*Sound*
Television system: 525-line NTSC system	Transmission mode: A/B
Maximum video frequency: 4.5 MHz (see Figure 6.7.2)	Sound signal bandwidth: 15–20 kHz
Type of modulation: FM	Sampling frequency: 32–48 kHz
Frequency deviation of carrier: 17 MHz_{p-p}	Quantizing and compounding: 14/10-bit/16-bit
Modulation polarity: positive	Number of frame bit: 2.048 Mbit/s $\pm$ 10 bit/s
Pre-emphasis: CCIR Rec. 405	Number of sound channel: 4/2
Frequency deviation of carrier by energy dispersal signal, symmetrically triangular, with frequency of 15 Hz: 600 kHz_{p-p}	Subcarrier frequency: 5.727272 MHz $\pm$ 16 Hz
	Frequency deviation of main carrier by the subcarrier: +10% $\pm$3.25 MHz −5%
RF bandwidth: 27 MHz	Modulation method of subcarrier: Q-DPSK (see note)

Note: Q-DPSK = Quadrature differential phase shift keying.

The main earth station for the Japanese broadcasting satellite BS-2 is described in the following paragraphs.

It consists of two antennas for two satellite (BS-2a, BS-2b) operation with 14 GHz transmitters, 12 GHz receivers, command transmitters, telemetry signal receivers, sets of order-wire equipment, computer systems, and control and monitoring equipment (Figure 6.7.3). The major characteristics of the station are listed in Table 6.7.4, and a block diagram is given in Figure 6.7.4.

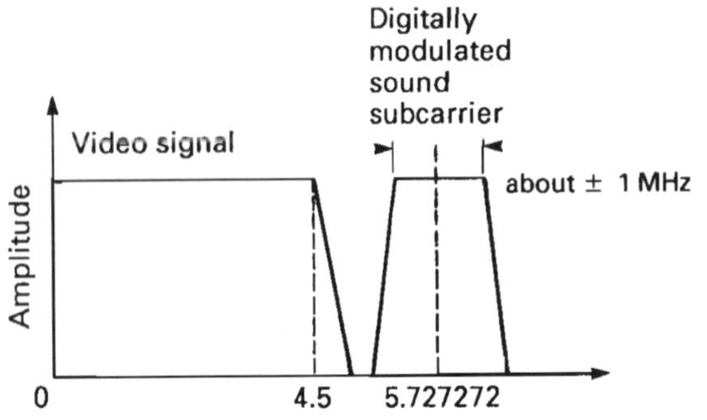

Figure 6.7.2 Spectrum of television picture and sound/data signals.

6.7.2.1.1 Antenna equipment

For efficient transmission to the satellite, and for high sensitivity reception of very small signals from the satellite, large diameter antennas are advantageous, but limitations will be imposed by the effect on the building structure, strength, manufacturing costs, etc. Antennas of 5 and 8 metres diameter have been used which provide an up-link c/n in fine weather of not less than 30 dB and a margin of 6–10 dB under rain conditions.

Design of the antenna equipment must provide adequate mechanical strength to withstand earthquakes and strong winds. The antenna equipment is fed with two TV signals, one command signal and two order-wire link signals.

The satellite tracking system is high precision and automatic, making use of the fact that the higher modes detected in the received telemetry signal from the satellite become zero when the system faces towards the front of the satellite. Most broadcast earth station, even up to 11 m diameter, typically use the step-track system as it is cheaper than the monopulse higher-order mode method of tracking.

Characteristics of the main earth station antenna are listed in Table 6.7.5.

6.7.2.1.2 TV signal transmit/receive

This equipment consists of a baseband unit, a modulator/demodulator unit, a frequency converter unit, a high power amplifier unit, a low noise receiver unit, a switching unit and a control supervisory console.

The high power amplifier employs a klystron or travelling wave tube amplifier (TWTA) and should be able to provide a 300 W–2 kW output. It should be installed near the antenna to minimize the feeder loss and the output circuit loss.

A redundancy system is provided to improve the reliability. It is possible to select the normal system or the redundancy system by remote control from the operation room.

6.7.2.1.3 Command transmitter and telemetry receiver

This equipment performs the supervision of the operating conditions and controls the *on/off* switching of the satellite transponder. Apart from the baseband unit which performs command signal generation and telemetry signal decoding, etc.,

Table 6.7.3 Up-link parameters of BS-2 satellite broadcasting

System parameter		*Main earth station*	*Up-link (fair weather) Sub earth station*	*Transportable earth station*	*Down-link (rainy weather)*
		(350 W)	(500 W)	(500 W)	(100 W)
Transmitting signal power	dBW	25.5	27.0	27.0	20.0
Output circuit losses	dB	4.2	1.2	1.0	2.3
Transmitting antenna gain	dB	58.7 (8 mϕ)	54.0 (5 mϕ)	48.0 (2.5 mϕ)	39.0
Equivalent isotropic radiation power	dBW	80.0	79.8	74.0	56.7
Antenna pointing error losses	dB	–	–	–	0.5
Free space transmission losses	dB	−207.1	−207.1	−207.1	−205.6
Atmosphere and rain losses	dB	1.0	1.0	1.0	2.0
Receiving antenna gain	dB	37.0	37.0	37.0	37.6 (0.75 mϕ)
Receiving antenna pointing error losses	dB	0.5	0.5	0.5	0
Input circuit losses	dB	1.0	1.0	1.0	0
Receiving signal power (c)	dBW	−92.6	−92.8	−98.6	−113.8
Equivalent receiving system noise temperature	dBK	31.3	31.3	31.3	26.4
Noise power density	dBW/Hz	−197.3	−197.3	−197.3	−202.2
Bandwidth	dBHz	74.3	74.3	74.3	74.3
Noise power (n)	dBW	−123.0	−123.0	−123.0	−127.9
Carrier/noise power ratio (c/n)	dB	30.5	30.2	24.4	14.1

Figure 6.7.3 The antenna of a main earth station (NHK).

this equipment shares the TV signal transmit/receive equipment. The low noise amplifier unit, the downconverter unit and the signal demodulator unit in the telemetry reception part are used also by the TV signal transmit/receive equipment.

6.7.2.1.4 Supervisory control

The supervisory control equipment consists of a control console, a wall display, a command telemetry unit, an electronic computer and peripheral units, as follows:

Supervisory control console

- main earth station control console;
- main earth station supervisory control;
- command sending console;
- telemetry supervisory console.

Wall type display

- two air monitor units (for two channels);
- four monitor units (for two links);
- panel to indicate the operating conditions of the satellite and ground facilities;
- alarm indicator panel.

Electronic computer system

- central processing units (on-line, off-line, reception monitor);
- main memory unit (MOS 512 kbytes);
- auxiliary memory unit (magnetic disk magnetic tape).

The sub earth station acts as backup for the main earth station, and undertakes transmission of emergency news programmes and others. It can also be used in the event of disaster, or of breakdown of the main earth station.

To be available in the event of a disaster, the sub earth station needs to be installed as far away as possible from the main earth station, this gives it site diversity and is generally

Table 6.7.4 Major performance figures and specifications of main earth station for satellite BS-2

Transmit overall	
TV	
TV video	
Input frequency	40 Hz–4.5 MHz
Input level	$1\ C_{p\text{-}p}/75\ \Omega$
TV sound	
Input frequency	50 Hz–15 kHz
Input level	0 dBm/600 Ω
Modulation	
Sound	Quadriphase shift keying modulation
Video sound	Frequency modulation
Intermediate frequency	140 MHz band
Transmit frequency	14 GHz band
Transmit output power	1.4 kW (max.)
Spurious level	<−46 dB
RF amplitude frequency response	Within ±0.3 dB/centre frequency ±8 MHz
RF group delay response	Within 2 ns/centre frequency ±8 MHz
Telemetry and command	
Transmit frequency	14.00028 GHz (BS-2a) 14.00371 GHz (BS-2b)
Transmit output power	500 W (max.)
Spurious level	<−46 dB
Order-wire	
Input frequency	0.3–3.4 kHz
Modulation	Frequency modulation
Intermediate frequency	140 MHz band
Transmit frequency	14 GHz band
Transmit output power (at HPA output)	200 W
Receive overall	
TV	
Input frequency	12 GHz band
Input level	−80 to −60 dBm
Input vswr	<1.25
System noise temperature	<270 K
IF	140 MHz band
Bandwidth	27 MHz
RF amplitude frequency response	Within ±0.3 dB/140 MHz ±8 MHz
RF group delay response	Within 2 ns/140 MHz ±8 MHz
Video	
Output frequency	40 Hz–4.5 MHz
Output level	$1\ V_{p\text{-}p}/75\ \Omega$
Order-wire	
Input frequency	12 GHz band
IF	140 MHz band
Audio output frequency	0.3–3.4 kHz
Output level	0 dBm/600 Ω

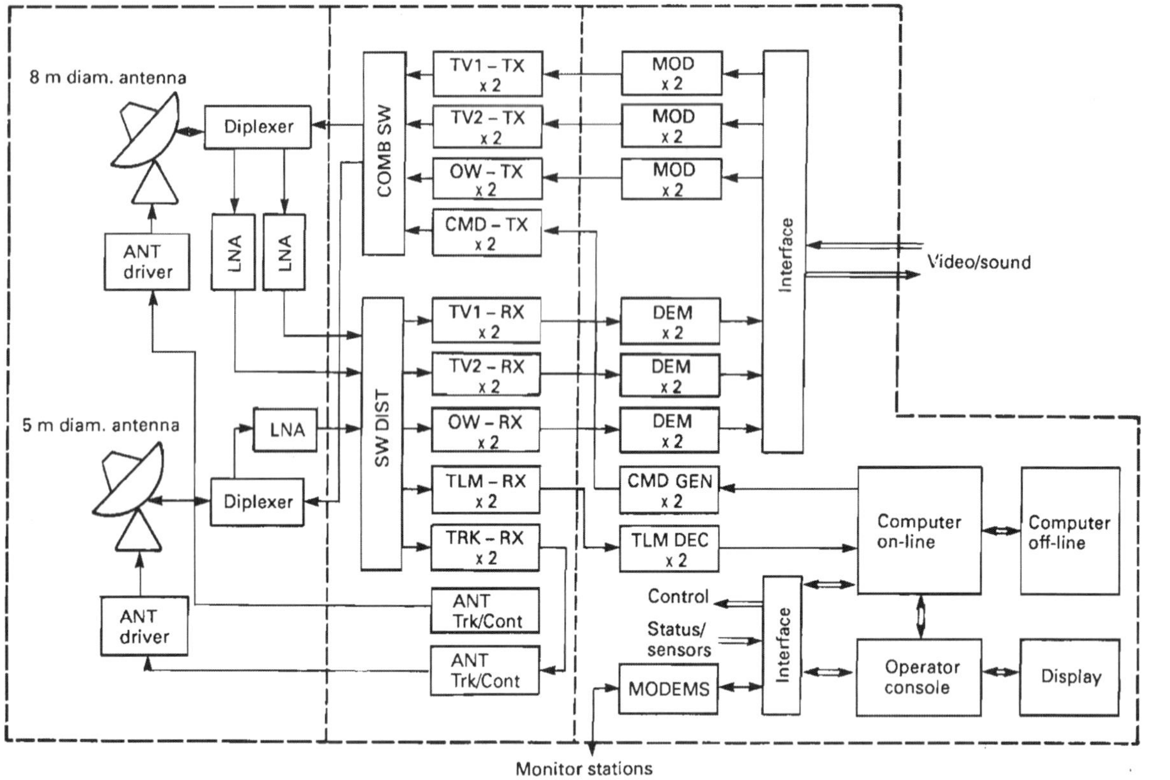

Figure 6.7.4 Block diagram of a main earth station.

Table 6.7.5 Typical characteristics of earth station antenna

Type of antenna:	Cassegrain
Diameter of main reflector:	8 m
Horn:	Conical corrugated horn
Tracking speed:	0.01°/s
Auto-tracking:	Monopulse signal
Max. operable wind velocity:	60 m/s (peak)
Weight:	10 t
Frequency:	Tx. 14.0–14.5 GHz
	Rx. 11.7–12.2 GHz
Gain (12G/14G):	58.7 dB/57.4 dB
VSWR:	Tx. <1.4
	Rx. <1.25
Withstand RF power:	Mean 2.22 kW
	Peak 7.4 kW
Polarization:	Right hand (or left hand) circular polarization
Ellipticity:	<0.9 dB

located about 20–100 km away from the main earth station. At this distance, when the main station is experiencing strong rain fade, the sub earth station is unlikely to be under similar rain fade condition.

A typical sub earth station comprises a 4.5 m diameter antenna, 14 GHz transmitters, 12 GHz receivers and sets of order-wire equipment (see Figure 6.7.5).

Figure 6.7.5 Sub earth station (NHK).

Figure 6.7.6 Broadcasting satellite operation centre room (NHK).

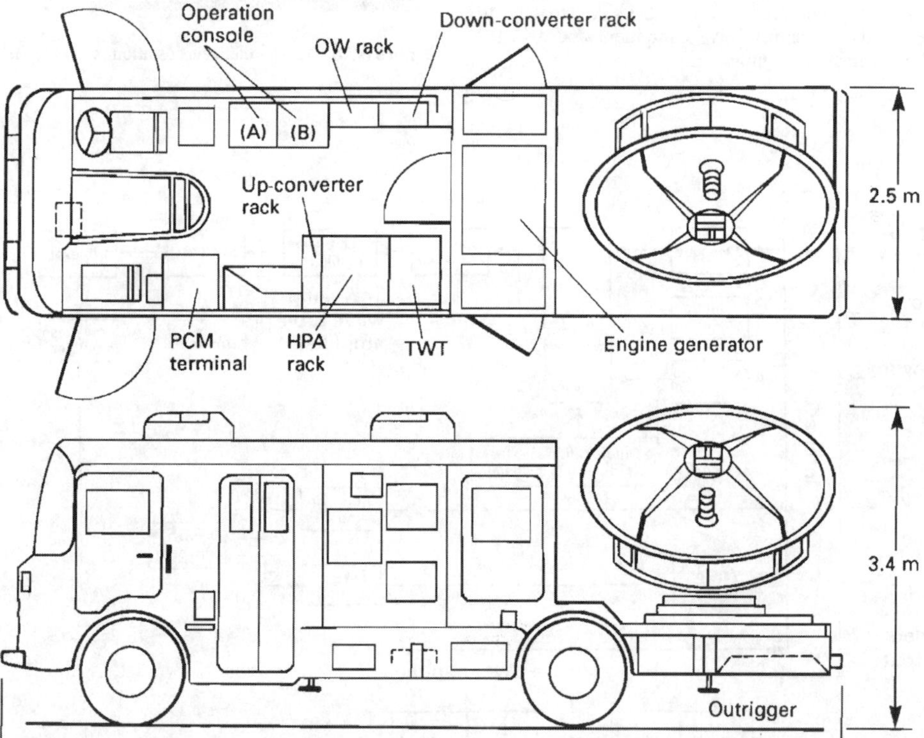

Figure 6.7.7 Configuration of a vehicle-mounted earth station. Its total weight is about 10^3 kg.

6.7.2.2 Vehicle-mounted and transportable earth stations

These earth stations are able to access the satellite from anywhere in the broadcasting service area, and are utilized for news relaying in a disaster emergency or for programme relaying from outside the station at sports events, etc.

6.7.2.2.1 Vehicle-mounted earth station

The vehicle-mounted earth station (Figure 6.7.7) is able to access the broadcasting station by securing a stable transmission path for lengthy programmes such as sports and events. Alternatively, the large vehicle on which it is mounted makes it quickly mobile to provide a short setting up time for the live relay of emergency news or local programmes.

It consists of a transmitter/receiver, an antenna and power supply equipment mounted on a vehicle. The antenna equipment can quickly be removed from its stored position to point to the broadcasting station. The antenna for vehicle-mounted use is a Cassegrain of 2.5 m diameter.

A Cassegrain antenna is the most widely used system having dual reflectors. It has a hyperbolic subreflector that effectively creates a virtual focus between it and the parabolic main reflector.

The transmit/receive equipment can transmit one TV channel, receive two TV signal channels, and transmit and receive two order-wire channels (one channel per TV channel). It is capable of transmission and reception of two or four audio signal channels for each video signal channel.

The power supply equipment has a 25 kVA capacity to supply power to the transmit/receive equipment and also for lighting and other relaying equipment.

6.7.2.2.2 Transportable earth station

This is designed to be transported onto site by helicopter when land access is not available. The total weight (including the antenna, transmit/receive equipment and generator) does not exceed 750 kg, and the construction is compact and simple. Programmes can be relayed with a minimum of equipment.

A transportable earth station may consist of a 2.5 m diameter antenna, a 14 GHz 200–500 W output transmitter, a 12 GHz receiver, an order-wire unit, and a generator (Figures 6.7.8 and 6.7.9 show the block diagram).

Figure 6.7.8 Transportable earth station on site (NHK).

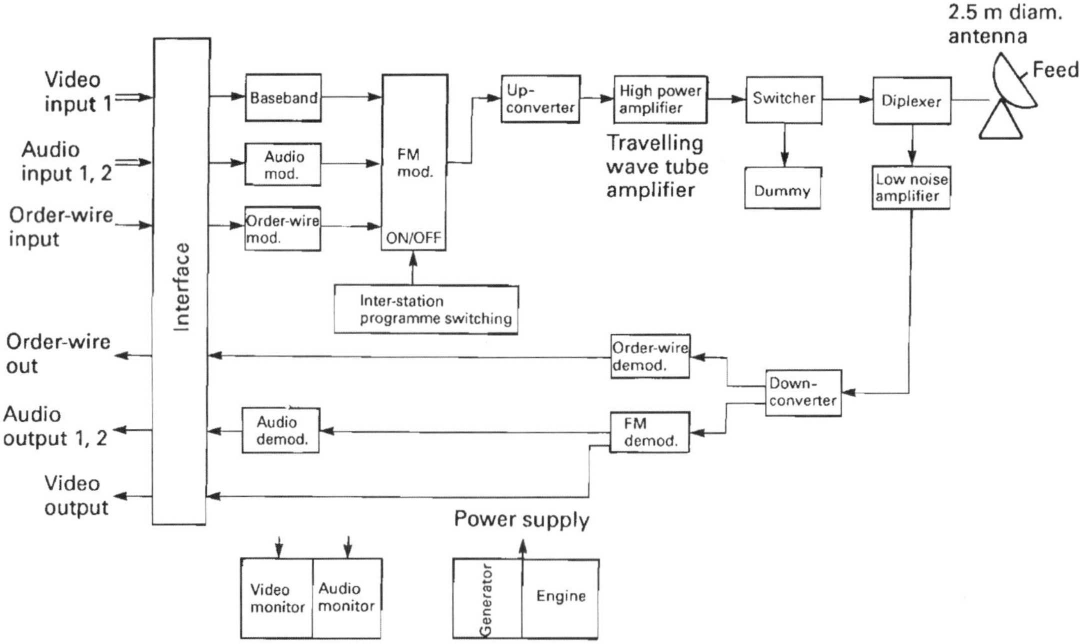

Figure 6.7.9 Transportable earth station block diagram.

The antenna can be split into five parts for transportation, and the reflector is made of carbon reinforced plastics for light weight. The primary radiator employs a corrugated horn to improve the aperture efficiency and to realize low side lobes of wide angle directivity.

The transmit/receive equipment consists of a baseband unit, a modulator, a frequency converter, a high power amplifier, a distributor, a low noise receiver, a demodulator, an order-wire unit, a monitor and a dummy load. It is capable of transmission and reception of one video channel and two audio channels. The high power amplifier has a 200–600 W output and employs a TWTA for amplification. Each component is designed to weigh below 50 kg for transportability.

The generator has a capacity of 4–6 kVA and supplies stable power to the transmitter/receiver. The engine and the generator are separate to facilitate transportation and weigh below 100 kg each.

6.7.2.3 Other earth stations

6.7.2.3.1 Interstations programme switching

An interstations switching control system provides smooth switching of programmes between stations. Television programme switching between transmitting earth stations is performed via satellite using up-link transmitter control. To achieve programme switching without signal overlap or break during the switchover, the propagation time from the earth station to the satellite and the time difference between the on-air station and the standby station to the satellite must be taken into account.

Figure 6.7.10 shows a programme switchover timing chart.

The switching control system inserts a switchover control signal into the field blanking interval of the TV signal.

The switchover of programmes is achieved as follows, assuming that station A's programme is on-air via satellite while station B's programme is scheduled to follow it.

Station A will insert a programme switching cue signal (a *Q signal*) in the field blanking interval of its programme signal to notify station B that its programme is about to end. A specified time later it will turn off its transmitting carrier. Meanwhile, station B will turn on its carrier after an interval depending on the signal propagation time between station B and the satellite. This time is not dependent on the locality of station A. Such a scenario occurs when national news is followed by opt-out to regional news broadcast.

6.7.2.3.2 Monitoring earth station

A monitoring earth station is used at several sites in the service area to receive and monitor the signal from the broadcasting

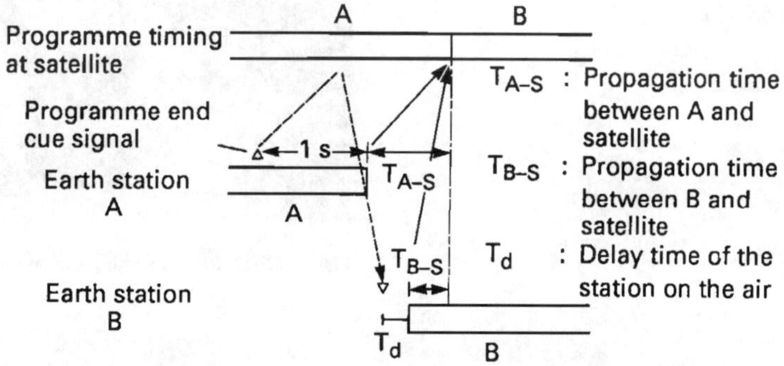

Figure 6.7.10 Fundamental programme switchover timing chart.

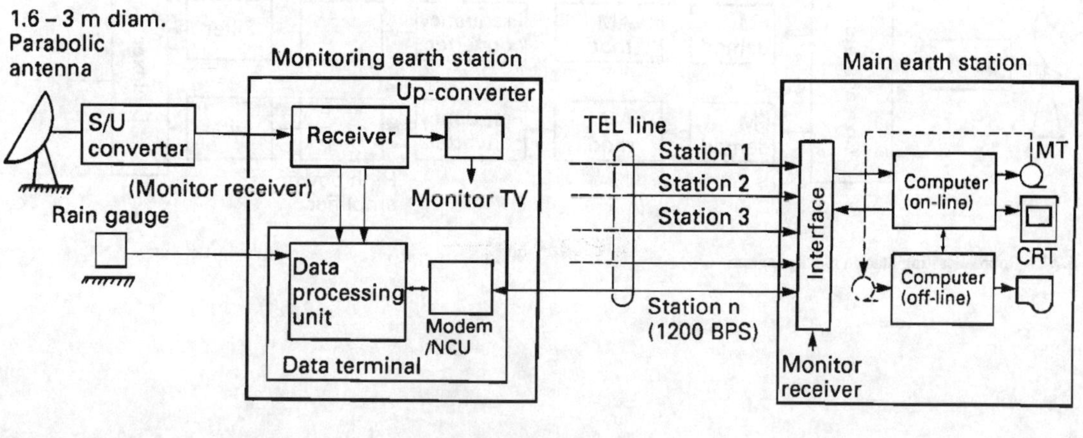

Figure 6.7.11 Monitoring earth station and data gathering/processing system.

satellite, and to supervise the condition of the broadcast service. The satellite footprint for the broadcast covers a large and hence the reason for monitoring at several sites.

Such a monitoring station comprises a 1.6–3.0 m diameter parabolic antenna, high stability receiving equipment to measure the received level, rain gauges, and received data processing equipment (Figure 6.7.11).

The received level data and rain data are sent via a terrestrial telephone line to a computer installed in the main earth station.

These data are processed and analysed for daily variation, seasonal variation and attenuation due to rainfall.

6.7.2.3.3 Rebroadcasting station

A rebroadcasting station receives and rebroadcasts DBS TV signals (Figure 6.7.12).

In the main service area, with the direct broadcasting system, it is possible to receive clear pictures using an antenna less than 0.6 m in diameter. But, at the fringe of the service area or inside a service area of low field intensities, a large diameter antenna is required which may be uneconomic for an individual. In these areas, therefore, the satellite broadcast signal is first received with a large diameter antenna and then converted into terrestrial signals for transmission to the service area.

A rebroadcasting station consists of a large diameter antenna (greater than 3 m), a set of receivers, demodulators, modulators, frequency converters, power amplifiers, a power supply and a station building (Figure 6.7.13).

Figure 6.7.12 Rebroadcasting station on site (NHK).

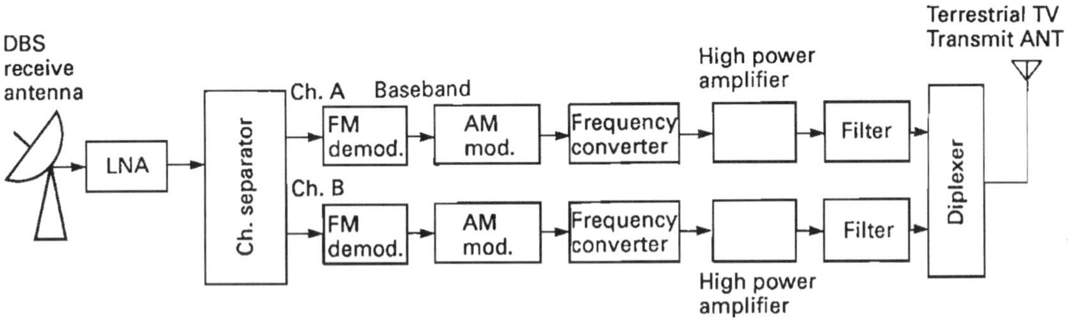

Figure 6.7.13 Rebroadcasting station block diagram.

Brian Flowers MIEEE, MRTS, MNYAS
Senior Engineer (retired)
European Broadcasting Union

6.8 Intercity Links and Switching Centres

6.8.1 Historical Development

6.8.1.1 Early Intercity Links

The first intercity links for television signals were coaxial cables, which were installed by the GPO (General Post Office) from London to Birmingham, then on to Manchester, and from London to Cardiff, between 1950 and 1952. A special 1-inch diameter coaxial cable was used for the London to Birmingham circuit, whereas standard 3/8-inch telephony coaxial cables were used for the other two circuits. Repeaters were installed every 12 miles for the 1-inch coaxial cable, and every 4 miles for the 3/8-inch coaxial cables, to equalise and amplify the signals. In the case of the 1-inch coaxial cable, the 405-line video signal was amplitude modulated on a 6.12 MHz carrier. This occupied a bandwidth of 3–7 MHz, using a partially suppressed upper sideband. The 3/8-inch coaxial cables utilised amplitude modulation of a 1 MHz carrier, which occupied a bandwidth of 500 kHz–4.5 MHz, using a partially suppressed lower sideband.

The spur connections from Birmingham to the Sutton Coldfield transmitter, and from Manchester to the Holme Moss transmitter, utilised amplitude-modulated radio links based on high-frequency triode amplifiers operating at 900 MHz. The spur connection from Cardiff to the Wenvoe transmitter utilised 200 MHz, with alternate amplitude-modulated and frequency-modulated hops. Triode amplifiers were used for the amplitude-modulated hops, and klystrons were used for the frequency-modulated hops. This was probably the first time that frequency modulation was used for a television link.

Bell Telephone Laboratories in the USA built triode amplifiers operating at 4 GHz, but this was pushing the triode valve to its limit. Fortunately, the travelling-wave tube, which had been invented by Dr R. Kompfner at the Clarendon Laboratory, Oxford University, during World War II, could operate at 4–10 GHz or even higher frequencies. Moreover, it had a wide intrinsic bandwidth, so it was developed by Standard Telephone and Cables Ltd and used to build the first Super-High-Frequency (SHF) intercity microwave link. This was installed from Manchester to Edinburgh and on to the transmitter at Kirk O'Shotts in 1952. The output power of the travelling-wave tube amplifier was 1 Watt, and the frequency-modulated signal was concentrated into a narrow beam by a parabolic reflector dish of about 2 m diameter.

By installing the repeater station towers on hilltops, the 200 miles of the Manchester to Edinburgh link were covered with just seven line-of-sight hops. Clearly, microwave links provided a relatively easy way of installing intercity links.

On 2 June 1953, the televised coronation of Queen Elizabeth II was relayed live from London to France, The Netherlands and Germany, using temporary microwave links across the English Channel to France, and across Belgium to relay the signal from France to The Netherlands and Germany. This aroused much interest in international television, but the birth of Eurovision is considered to be 6 June 1954, when the Flower Festival of Montreux, Switzerland, was relayed to France, Germany, Belgium, Denmark, The Netherlands, The United Kingdom and Italy. Most of the connections were via frequency-modulated microwave links, although the Dover to London circuit utilised an air-core amplitude-modulated coaxial cable at this time.

A total of 55 Eurovision transmissions took place in 1954, since when there has been a steady increase to about 100,000 transmissions in 2002. Meanwhile, the European terrestrial microwave network increased from about 10,000 km in 1955 to about 300,000 km in 1985.

From 1959 to 1962, news items were sent across the Atlantic Ocean via a TAT-cable (Trans-Atlantic Telephone cable) four-wire voice circuit, using a slow-scan system invented by the BBC Research Department. The terminals were installed at the BBC Television News Studios in Alexandra Palace, North London, at the NBC Studios in The Rockefeller Centre, New York, and at CBC Montreal. Every second frame of a 16 mm news film was scanned in 8 seconds, using the odd field only of the British 405-line interlace system. These 202-line frames were optically scanned onto two consecutive film frames at the far end, and 1 minute of film took 100 minutes to transmit. During transmission, the receiving end was able to inform the sending end if everything was going well by speaking on the return direction of the four-wire voice circuit.

6.8.1.2 Early Satellite Links

The first transatlantic television test transmission via satellite was made on 10 July 1962 via the Telstar satellite, from Andover, Maine, USA, to Pleumeu Boudou, Brittany, France. The GPO earth station at Goonhilly, Cornwall, England, should also have received the signal, but unfortunately a misunderstanding about the direction of circular polarisation employed on the down-link prevented the Goonhilly earth station from receiving a usable signal. However, this was corrected on 11 July 1962 and, on 23 July, inaugural programmes were exchanged between Europe and the USA. Telstar orbited the earth every 2 hours 30 minutes in an elliptical orbit, so it had to be tracked by the transmitting and receiving earth stations, and it was only available for transmission during about 22 minutes of each orbit.

The first geostationary communications satellite, Early Bird, entered service 22,300 miles above the Atlantic Ocean in June 1965, thereby providing permanently available television connections between Europe and North America. Thereafter, Intelsat provided bigger and better geostationary communication satellites above the Atlantic, Pacific and Indian Oceans. Consequently, by 1967, television links could be established between most parts of the world. In fact, on 25 June 1967, the programme *Our World* linked up live contributions from 25 countries around the world. The introductory and closing music consisted of the words, 'Our World', sung in 22 different languages by the Vienna Boys Choir.

In 1984, the European Broadcasting Union, which is responsible for the coordination of Eurovision transmissions, leased two wideband (72 MHz) transponders in the Eutelsat I-F2 geostationary satellite, and in 1991 this was increased to four leased wideband transponders in Eutelsat II-F4. These four transponders provided six contribution quality analogue television channels, using frequency-modulated carriers.

6.8.1.3 ETSI 34/45 Mbit/s Digital Transmission

At the end of the 1980s, serious thought was given to the possibility of replacing the frequency-modulated analogue links by digital transmission. In fact, the audio component of Eurovision transmissions had been digitised since 1974, using the BBC's Sound-In-Sync system, whereby two audio samples per television line were digitally encoded in the line sync period, using 10 bits per sample. However, the uncompressed digital video signal required 243 or 270 Mbit/s to encode the luminance (Y) and the two chrominance signals, (R – Y) and (B – Y), as specified by ITU-R Rec. BT.601 and ITU-R Rec. BT.656. In fact, a 360 Mbit/s option was also included in the specification, following the introduction of the widescreen 16:9 picture aspect ratio, but 270 Mbit/s became the de facto standard. However, transmitting 270 Mbit/s via satellite would have occupied several times the bandwidth required for an FM analogue signal.

Fortunately, the uncompressed digital video signal contains considerable redundancy, namely spatial, temporal and statistical redundancy. The essential information, which remains after the removal of all redundancy, is the signal entropy, which increases in proportion to the unpredictability of the picture content. Hence, in 1989, using the *Discrete Cosine Transform* on 8 × 8 pixel blocks, a group of experts was able to design an encoder which compressed the 270 Mbit/s digital video signal to about 30 Mbit/s, with virtually no loss of picture quality. This permitted the video, plus two AES/EBU 2.048 Mbit/s dual channel uncompressed audio channels, to be multiplexed into a total bit rate of 34.368 Mbit/s for use in Europe, or 44.736 Mbit/s for use in North America. The specification for these encoders and decoders, ETS 300 174, was issued by the ETSI (European Telecommunication Standards Institute) in November 1992, and it subsequently became ITU-T Rec. J.81 in September 1993.

Several manufacturers produced these encoders and decoders, so the EBU, in cooperation with the manufacturers, organised extensive interoperability tests during the early 1990s (see Figure 6.8.1). Ten years later, 34 and 45 Mbit/s 'ETSI' encoders and decoders are still in use in many countries, providing high-quality contribution and distribution circuits via terrestrial microwave links or fibre-optic circuits.

The change from analogue to digital studio production and transmission could not occur overnight, and a mixed scenario was bound to exist for one or two decades.

Consequently, the ETSI encoders and decoders were equipped with composite analogue (PAL, SECAM and NTSC) inputs and outputs respectively, in addition to SDI (270 Mbit/s) inputs and outputs. It was important to provide high-quality composite analogue decoders at the encoder composite analogue input, and in the case of PAL and NTSC, a comb-filter was required to make a clean separation of the chrominance and luminance signals. Figure 6.8.2 shows the picture quality of a 270 Mbit/s SDI signal, compared with the picture quality of a PAL signal.

One or two manufacturers produced variations on the basic ETSI specification, whereby two 17 or 22 Mbit/s video signals were multiplexed into one 34 or 45 Mbit/s bit stream respectively, with very little loss in picture quality. One manufacturer also produced an 8.448 Mbit/s version of the ETSI codec, suitable for news transmission, but not recommended for sports event contribution links. For these lower bit rates, the audio signals were also compressed.

From 1995 to 1999, the EBU utilised the 2 × 17 Mbit/s equipment on its leased transatlantic satellite link. At the same time, ETSI 8.448 Mbit/s encoders and decoders were installed to create a 'Mini-8' network, using spare transponder capacity on the main Eurovision satellite (EutelsatII-F4), alongside the analogue television channels (see Figure 6.8.3).

6.8.1.4 Developments in North America

Similar developments took place in North America during the second half of the 20th century, with the big three networks (ABC, CBS and NBC) leasing terrestrial microwave networks from AT&T Long Lines Division from the 1950s to the 1980s.

These networks were in the form of a clockwise 'round robin', normally fed from New York, whereby affiliates were fed via spurs off the main network. Major cities, such as Los Angeles, could break into the circle and originate when required.

Transmission was analogue FM, using the NTSC 525-line colour system. Then in the late 1970s, the Public Broadcasting Service (PBS) introduced distribution of its programmes to the four USA time zones via four C-band channels on the Westar satellite. The other major networks introduced Ku-band domsat

Figure 6.8.1 Interoperability tests of 34 Mbit/s encoders–decoders.

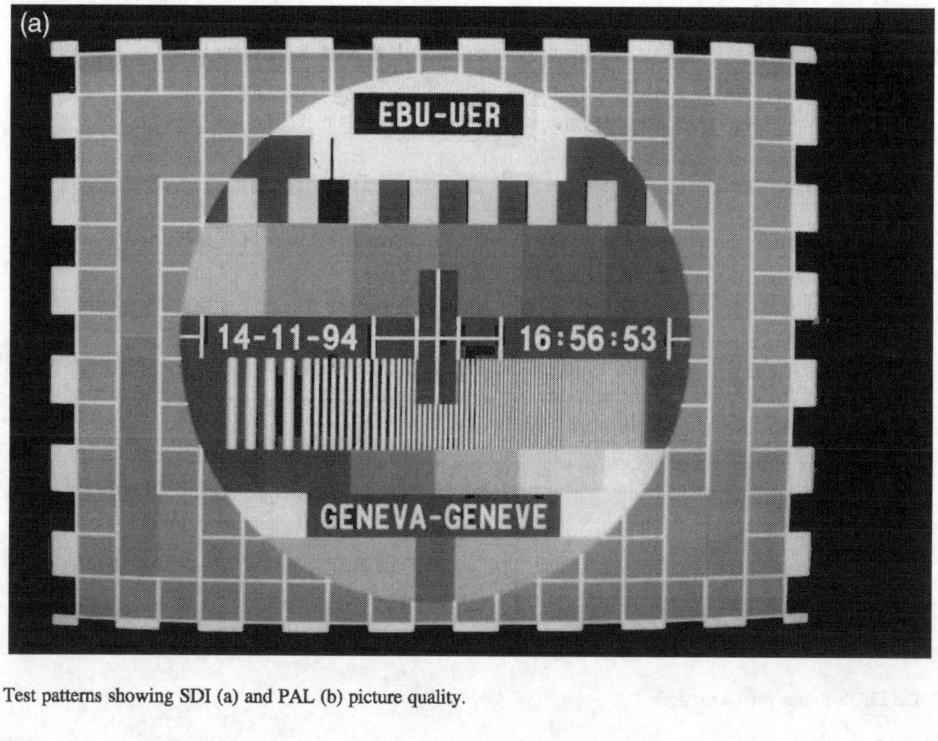

Figure 6.8.2 Test patterns showing SDI (a) and PAL (b) picture quality.

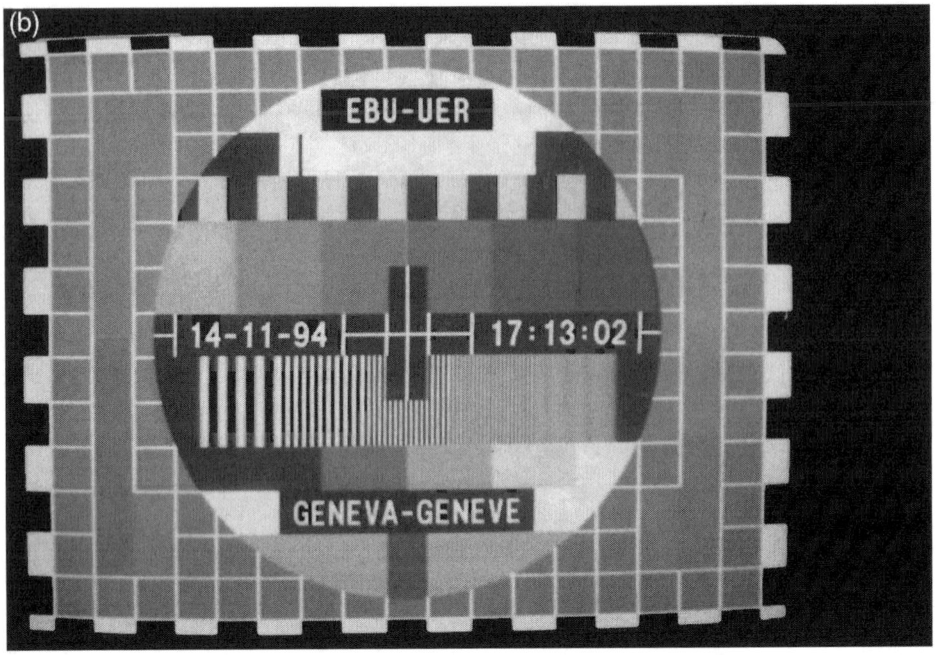

Figure 6.8.2 *continued*

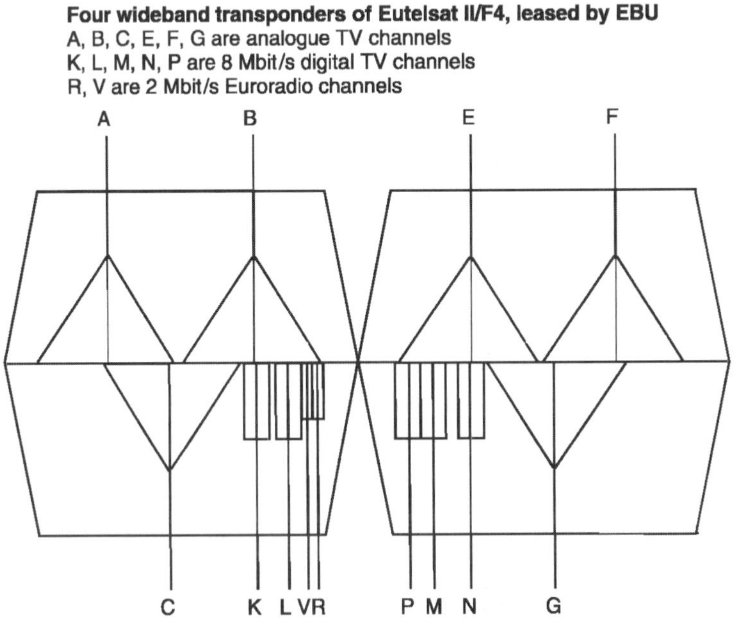

Figure 6.8.3 The EBU's former Mini-8 network.

(domestic satellite) distribution in the 1980s, and by this time, intercity fibre-optic circuits were also available to supplement the satellite networks when required. Initially, these fibre-optic circuits transmitted the NTSC video as a composite digital signal, whereby the analogue video signal was sampled at three times the colour subcarrier frequency. This was replaced by compressed component digital transmission when the ETSI 45 Mbit/s codecs became available in the early 1990s, and from the late 1990s, MPEG-2/Professional Profile@Main Level encoders and decoders were introduced.

6.8.2 Digital Satellite Contribution Links

6.8.2.1 MPEG-2/Professional Profile@Main Level

MPEG-2/Main Profile@Main Level utilises 4:2:0 encoding. This means that R – Y and B – Y are transmitted on alternate lines, as is the case for the composite analogue SECAM system. The consequent reduction in vertical colour resolution is acceptable for final delivery of the MPEG-2 signal to the public (i.e. secondary distribution), since the human eye has less acuity for colour detail than for luminance detail. This is also reflected in ITU-R BT.601, where the sampling rate for the luminance signal is 13.5 MHz, compared with 6.75 MHz for the colour difference signals, thereby providing twice as much horizontal resolution for luminance than for chrominance.

However, it should also be noted that the use of 2:1 interlace exacerbates the loss of vertical colour resolution in the case of vertical movement in the picture.

The use of 4:2:0 coding on digital contribution links is not recommended, because it may cause a cumulative loss of vertical colour resolution with successive encoding–decoding operations. This is especially true if the source and/or final delivery are in PAL or NTSC, which is often the case with a mixed analogue/digital scenario. These considerations prompted the introduction of the MPEG-2/Professional Profile @Main Level standard in 1996, which utilises 4:2:2 coding, thereby sending (R – Y) or (B – Y) alternately between each Y sample on every line of the picture. Professional Profile encoding is intended to be used at higher bit rates than Main Profile coding, the latter having a theoretical advantage at low bit rates. Tests carried out by the CBC (Canadian Broadcasting Corporation) and the EBU in 1998 showed that the crossover point, in terms of video bit rate, is about 4 Mbit/s for 625-line pictures and about 3 Mbit/s for 525-line pictures. The picture quality crossover point varies somewhat, according to the criticality of the picture sequence being tested.

In any case, digital transmission requires higher quality on contribution links than on distribution links, since the contribution signals are often subjected to further processing in studio production centres, before being delivered to the public. Therefore, 4:2:2 coding is preferable for contribution links, although 4:2:0 coding is sometimes used, where only a low-bit-rate transmission link is available.

6.8.2.2 Bit Rates used for Satellite Contribution Links

The EBU has standardised on the following Useful Bit Rates for satellite links:

- HBR (High Bit Rate) for sports and other prestige events – 21.503 Mbit/s, plus 188/204 Reed–Solomon, 7/8 FEC ratio and QPSK modulation. Two 384 kbit/s dual-channel, Layer II, compressed audio signals are included in the multiplex. The ABU (Asian Broadcasting Union) and the ASBU (Arab States Broadcasting Union) also use this system for satellite transmission. Moreover, it conforms to the DVB standard EN 301210.
- LBR (Low Bit Rate) for news reports – 10.7515 Mbit/s, plus 188/204 Reed–Solomon, 7/8 FEC ratio and QPSK modulation. Audio is the same as for the HBR system. The LBR system utilises exactly half the bit rate and half the RF bandwidth of the HBR system.
- ISOG/DSNG Bit Rate (Inter-union Satellite Operations Group/Digital Satellite News Gathering) – 8.448 Mbit/s, plus 188/204 Reed–Solomon, 3/4 FEC ratio and QPSK modulation. A single 256 kbit/s dual-channel, Layer II, compressed audio signal is included in the multiplex. The ISOG/DSNG 8.448 Mbit/s system is a worldwide standard.

The actual Transmitted Bit Rate is higher than the Useful Bit Rate, due to the use of Reed–Solomon error correction. This adds 16 bytes to every 188 bytes, which increases the bit rate by a factor of 51/47. Reed–Solomon error correction uses block codes and interleaving, which ensures that bursts of errors, which may occur on the satellite link, become isolated errors in the de-interleaved signal. The Reed–Solomon block codes are able to correct these individual errors, whereas they would not be able to correct several consecutive errors.

In addition to the Reed–Solomon outer coding, a convolutional inner coding FEC (Forward Error Correction) system is also utilised on satellite links, to provide additional error protection. This increases the transmitted bit rate by the chosen FEC ratio, which may be 7/8, 5/6, 3/4, 1/2, etc.

QPSK (Quadrature Phase Shift Key) modulation is normally used on satellite links, although BPSK or 8-PSK may be used in certain circumstances. BPSK is more rugged than QPSK, but it occupies twice the RF bandwidth of a QPSK signal for a given bit rate. On the other hand, 8-PSK requires only 2/3 the RF bandwidth of a QPSK signal for a given bit rate, but the satellite link margin will be about 3 dB less for the 8-PSK signal than for the QPSK signal, other things being equal.

Since QPSK employs four symbols (i.e. carrier phases of 0°, 90°, 180° and 270°), each symbol can transmit 2 bits (i.e. 00, 01, 10 and 11). Hence the symbol rate is equal to 1/2 the bit rate for QPSK modulation. BPSK modulation utilises only two phases, so its symbol rate equals the bit rate of the digital signal, whilst 8-PSK utilises eight phases, so each symbol can transmit 3 bits (i.e. 000, 001, 010, 011, 100, 101, 110 and 111). Hence the symbol rate for 8-PSK is equal to 1/3 of the bit rate. Figure 6.8.4 shows a spectrum analyser printout of a 72 MHz bandwidth satellite transponder carrying a 34 Mbit/s QPSK signal and an analogue (PAL) FM signal, each with its typical spectrum shape.

The RF bandwidth (between −3 dB points) of a digital signal is numerically equal to the Symbol Rate, which can be calculated as follows:

$$\text{Symbol Rate} = \text{Useful Bit Rate} \times \text{RS ratio} \times \text{FEC ratio } (>1)/\text{Bits per Symbol}$$

The RF bandwidth, which is required to accommodate this signal without causing intermodulation with adjacent signals, is obtained by multiplying the Symbol Rate figure by a roll-off factor. Intelsat and Eutelsat specify the roll-off factor as 1.35.

Hence, RF bandwidth requirement = Symbol Rate (symbols/s) × 1.35 Hz.

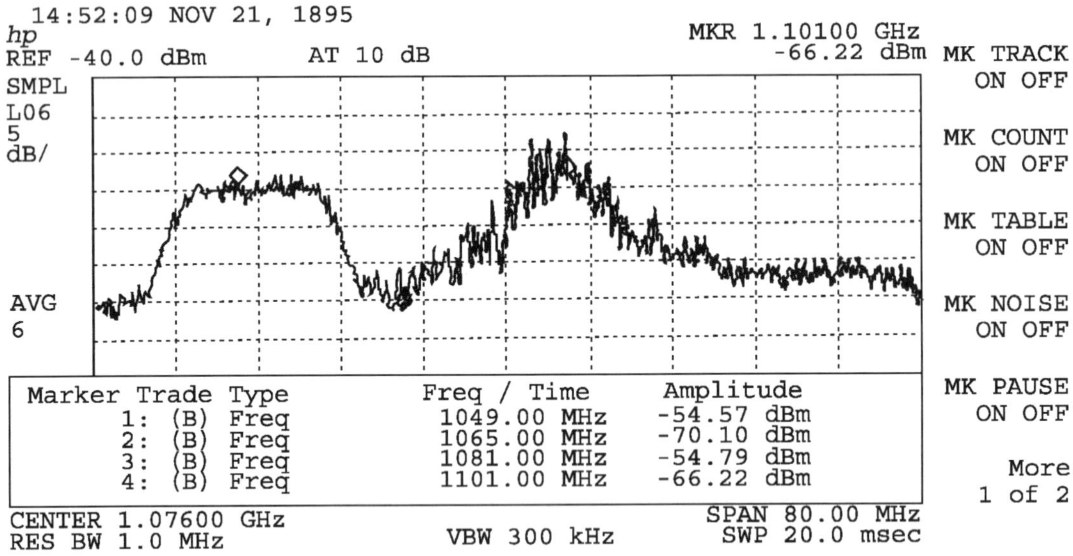

Figure 6.8.4 Spectra of FM analogue and QPSK digital signals.

If we take a concrete example of a 21.503 Mbit/s Useful Bit Rate signal being transmitted via satellite using 188/204 Reed–Solomon, 7/8 FEC ratio and QPSK modulation, the required RF bandwidth is calculated as follows:

$$21.503 \times 51/47 \times 8/7 \times 1/2 \times 1.35 = 18\,\text{MHz}$$

Therefore, four such signals can be accommodated in one 72 MHz transponder.

The 10.7515 Mbit/s Useful Bit Rate signal, using the same RS and FEC ratio, occupies 9 MHz RF bandwidth.

The 8.448 Mbit/s Useful Bit Rate signal is transmitted with an FEC ratio of 3/4, giving an RF bandwidth requirement of:

$$8.448 \times 51/47 \times 4/3 \times 1/2 \times 1.35 = 8.25\,\text{MHz approx.}$$

This fits easily into a 9 MHz satellite slot, as used for the EBU 10.7515 Mbit/s signal.

Figure 6.8.13 shows the transponder allocations on the Eutelsat W3 (7 °E) satellite transponders B1, B2, B3 and B4, which have been leased by the EBU for Eurovision traffic since 1998. Four 18 MHz slots for 21.503 Mbit/s MPEG-2/Professional Profile signals are available in each 72 MHz transponder, but these can each be subdivided into two 9 MHz slots for 10.7515 or 8.448 Mbit/s signals. A 9 MHz slot can be further subdivided into three 3 MHz slots, as required for 2.048 Mbit/s Euroradio channels. Alternatively, a 3 MHz slot can accommodate up to 24 64 kbit/s communication channels for voice and data.

Figure 6.8.4 shows a spectrum analyser display of an FM analogue signal and a QPSK digital signal in the same transponder. The FM analogue spectrum has a triangular shape like a mountain peak, whilst the QPSK spectrum has a nearly rectangular shape.

6.8.2.3 Optimisation of IBO (Input Back-Off)

It is essential to determine the optimum IBO for the satellite up-leg EIRPs. Clearly, there is an optimum power level for the signals received by the satellite, in order to modulate the travelling-wave tube close to saturation with a fully loaded transponder. Complete saturation is to be avoided, as this will cause non-linearity, which leads to intermodulation between adjacent channels. Digital signals are relatively immune to intermodulation, but they increase the noise level of the RF signal, which reduces the E_b/N_o (Energy per bit/Noise per Hz) of the demodulated bitstream. A reduction in the received E_b/N_o means a reduction in link margin, and consequently less safety margin in the case of rain-fade, for example.

To find the optimum IBO, the transponder is fully loaded with, say, four 21.503 Mbit/s up-leg signals from four earth-stations, which deliver the same power at the input to the satellite by taking into account their locations within the satellite footprint. The up-leg EIRPs are then reduced in 1 dB steps from full saturation level down to about 12 dB IBO, and the received E_b/N_o is noted for each IBO value. The resulting graph (Figure 6.8.5) shows the optimum IBO, which gives the best E_b/N_o. This occurs at 5 dB below saturation, which is normal for a travelling-wave tube.

6.8.2.4 MPEG-2/Professional Profile HBR Compared with ETSI 34 Mbit/s

When MPEG-2/Professional Profile encoders and decoders became available in 1996, the EBU organised subjective picture quality tests, as specified by ITU-R BT-500, to compare ETSI 34 Mbit/s with MPEG-2/Professional Profile, at a transmitted bit rate of about 24 Mbit/s. The perceived picture quality was about the same for the two systems, a result that was subsequently confirmed by objective measurements of picture quality, which were carried out with the Tektronix

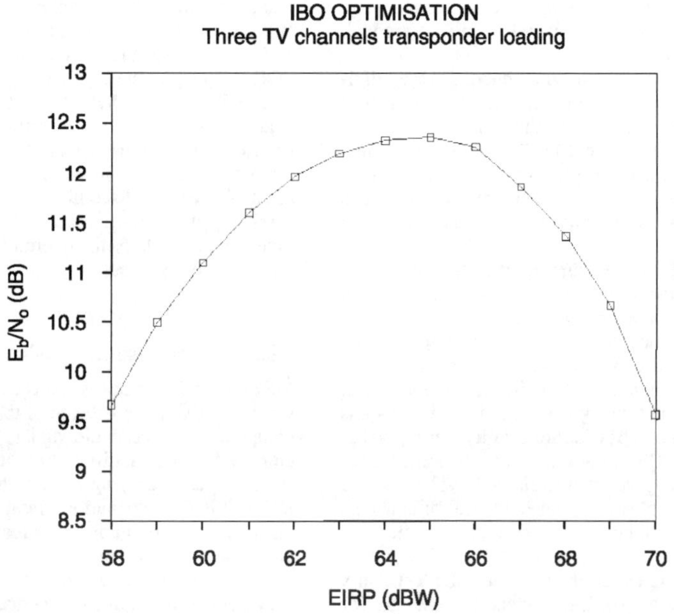

Figure 6.8.5 Input back-off optimisation graph.

PQA 200 (Picture Quality Analyser). The main reason for the superior performance of the MPEG-2 system, in terms of picture quality for a given bit rate, is its use of B-frames in the encoding process. Both systems use I-frames and P-frames, albeit in a slightly different way, but only MPEG-2 utilises B-frames.

I-frames are self-contained, achieving compression by the removal of spatial and statistical redundancy, but not temporal redundancy. P-frames remove all three types of redundancy by signalling only the changes in content of consecutive frames, but starting with an I-frame every 12 frames or so. B-frames go a step further by referring to both past and future I- and P-frames, which achieves a higher compression ratio than is possible with P-frames.

However, the use of B-frames makes it necessary to store complete frames in the encoder and decoder, and this inevitably increases the latency of the encoding–decoding process. Hence, the latency of an MPEG-2/Professional Profile encoder–decoder at 24 Mbit/s is about 300 ms, compared to about 100 ms for an ETSI 34 Mbit/s encoder–decoder. The 300 ms must be added to the geosynchronous satellite path delay of about 260 ms, giving a total transmission delay of 560 ms, or even more for lower bit rates. This long delay can cause problems for live duplex interviews.

On the other hand, a 72 MHz transponder can accommodate four 24 Mbit/s MPEG-2/Professional Profile signals, but only three 34 Mbit/s ETSI signals. This leads to an economic advantage for MPEG-2/Professional Profile, which is the deciding factor.

Nearly all satellite contribution links utilise MPEG-2/Professional Profile nowadays, whilst ETSI 34 Mbit/s and 45 Mbit/s are utilised mainly on terrestrial contribution circuits.

6.8.3 Quality Control for Digital Television Circuits

6.8.3.1 Basic Considerations

Digital transmission systems do not suffer from the cumulative signal distortion and deterioration of the signal/noise ratio, which characterise analogue transmission.

The quality of the encoded–decoded video and audio signals depends only on the quality of the encoding algorithms and the choice of video and audio bit rates. Reed–Solomon block codes and convolutional code error correction systems ensure that any errors, which occur due to the transmission path, are corrected in the decoded signal. This is essential, since a single erroneous bit can cause a momentary picture freeze or an audio dropout, for example. However, when the BER (Bit Error Ratio) of the transmission system exceeds the correction capacity of the error correction system, catastrophic failure occurs. This sudden onset of complete failure is a characteristic of digital transmission systems.

6.8.3.2 Objective Measurement of Picture Quality

The BER of the received signal indicates how much BER headroom is available at the end of a transmission path, but the measured BER does not tell us much about picture or audio quality, except in the case of transmission failure.

The Video Quality Experts Group (VQEG) has been working for several years on the establishment of standard methods for the objective measurement of digital picture quality. Their proposals should eventually form the basis of an ITU Recommendation. The VQEG studies include a wide range of digital video encoding systems, from video-conference signals to professional television signals. Moreover, there are three distinct classes of measurement methods, namely 'full reference', 'reduced reference' and 'no reference'.

'Full reference' implies that the original uncompressed digital signal is available for comparison purposes at the measurement point. 'Reduced reference' means that certain parameters of the original uncompressed signal are measured and transmitted with the compressed video signal for comparison purposes at the measurement point. Finally, 'no reference' means that the measurement is based only on the received video signal.

The Tektronix PQA 200 is an example of a 'full reference' measurement system. Standard video test sequences are stored on a hard disk as 270 Mbit/s SDI signals. The selected sequence is played via an encoder–decoder, and the SDI output of the decoder is recorded on a disk. The original and encoded–decoded SDI signals are then compared frame by frame, and the differences between the two signals are assessed by means of special filters to obtain a PQR (Picture Quality Rating) value for each frame. These PQR values are then averaged over the 60-frame sequence to give an overall PQR value for the sequence. This measurement process takes several minutes, so the PQA 200 is basically a laboratory measurement unit.

Figures 6.8.8 and 6.8.9 show graphs of the PQR values plotted against video bit rates, as measured by the Tektronix PQA 200 (software version 3.0). These measurements are for one MPEG-2/Professional Profile encoder–decoder, and for three MPEG-2/Professional Profile encoder–decoder combinations in series, as shown by the block diagrams in Figures 6.8.6 and 6.8.7, respectively.

The cumulative degradation, caused by the concatenation of encoding–decoding operations, can be reduced by ensuring alignment of the I-frames and motion vectors of the concatenated encoders. The 'Mole' system, developed by Snell & Wilcox, performs this function.

The EBU uses the correlation shown in Table 6.8.1 between PQR values and the ITU five-point quality scale.

The Rhode & Schwarz DQM (Digital Quality Meter) is an example of a real-time operational digital video quality measurement unit. It measures 'Spatial Activity' and 'Temporal Activity', which correspond to discontinuities at the edges of pixel blocks and discontinuities between pixel blocks of successive frames, respectively. These two parameters together give a good guide to the overall quality of the MPEG-2 encoding–decoding process.

6.8.3.3 Digital Audio Quality

MPEG-2 audio compression is based on the *Musicam* principle, whereby advantage is taken of the masking effect of high-level components of the audio signal. MPEG-2 contribution circuits employ Layer II audio coding, usually with two 384 kbit/s bit streams, each carrying a dual channel or stereo audio signal. DSNG/ISOG contribution circuits usually have only one dual channel or stereo audio bit stream at 256 kbit/s. MPEG-2 decoders automatically decode the received audio bit rate.

The subjective audio quality is excellent at 384 kbit/s and very good at 256 kbit/s, but concatenation degrades the quality, so it is preferable not to exceed three sequential encodings–decodings. It is very difficult to make meaningful objective measurements of compressed digital audio quality, since noise level and distortion vary according to the nature of the input signal. One approach is to subtract the encoded–decoded signal from the original signal, thereby obtaining the distortion and noise introduced by the encoder–decoder. It is, of course,

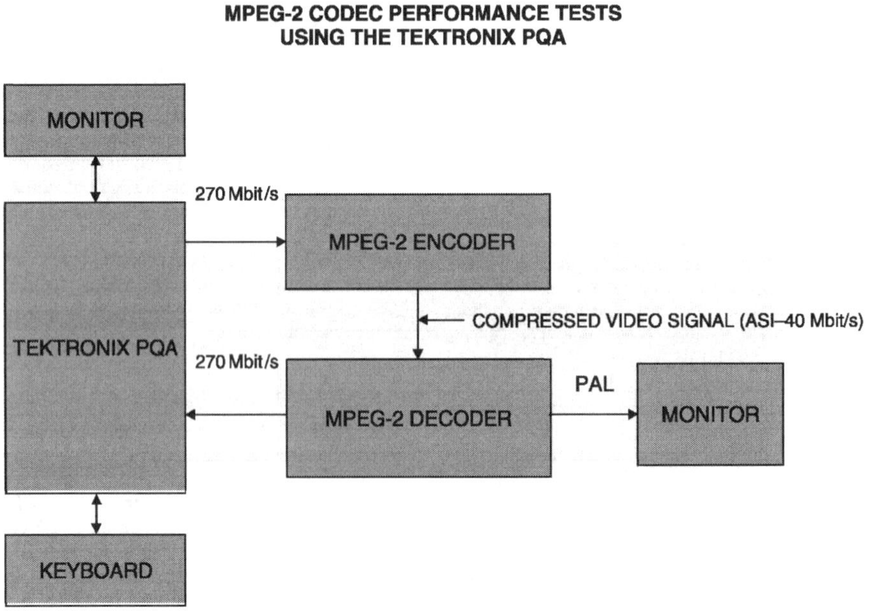

Figure 6.8.6 Picture quality measurements of one encoder–decoder.

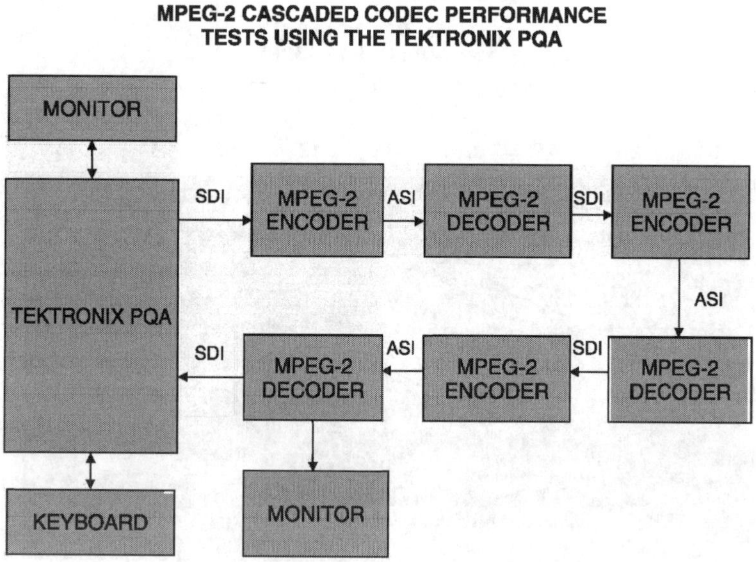

Figure 6.8.7 Picture quality measurements of three encoders–decoders.

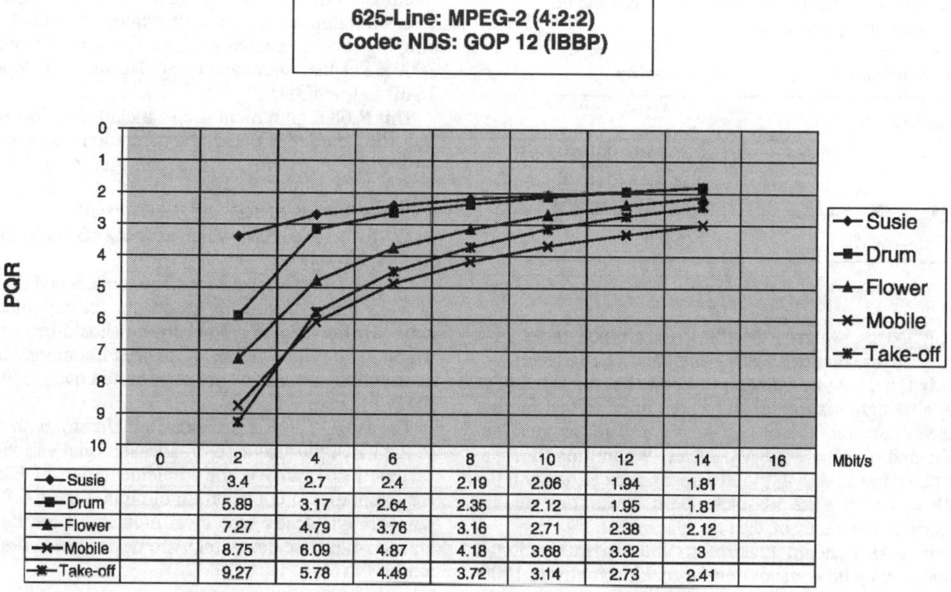

	2	4	6	8	10	12	14	16
Susie	3.4	2.7	2.4	2.19	2.06	1.94	1.81	
Drum	5.89	3.17	2.64	2.35	2.12	1.95	1.81	
Flower	7.27	4.79	3.76	3.16	2.71	2.38	2.12	
Mobile	8.75	6.09	4.87	4.18	3.68	3.32	3	
Take-off	9.27	5.78	4.49	3.72	3.14	2.73	2.41	

Figure 6.8.8 Graphs of PQR vs. video bit rate for one encoder–decoder.

necessary to delay the original audio signal by an amount, which is exactly equal to the latency of the encoder–decoder, before subtracting the decoder output from the original signal.

The AES/EBU 2.048 Mbit/s audio coding system gives excellent quality, which is not affected by concatenation. It is utilised for *Euroradio* satellite transmissions, which distribute EBU classical concerts to the European radio broadcasters.

6.8.3.4 Audio Level Alignment

Maintaining correct audio level alignment on international circuits has always been somewhat problematical. One problem is that two different types of audio level meters have been in use by broadcasters since the 1930s, namely the VU-meter and the PPM (Peak Programme Meter). The VU-meter has a nominal

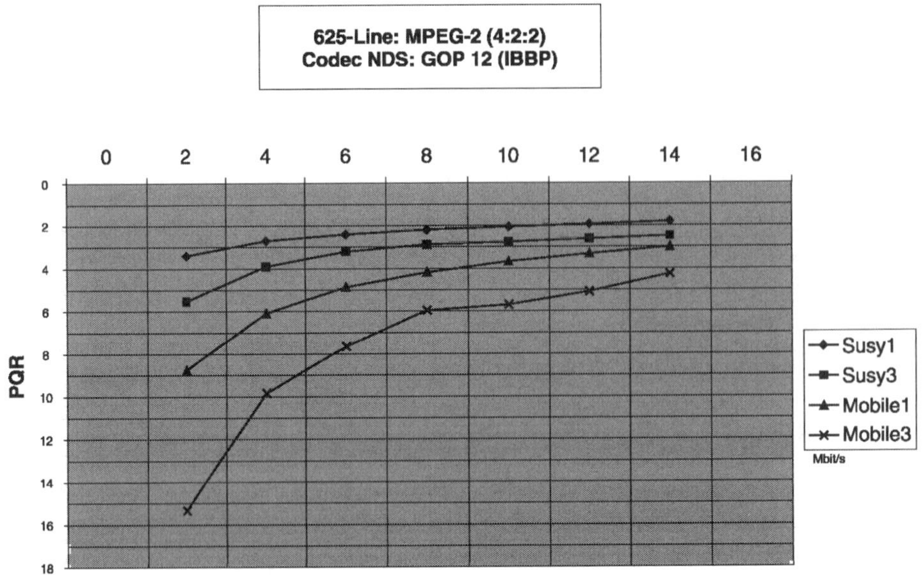

Figure 6.8.9 Graphs of PQR vs. video bit rate for three encoders–decoders.

Table 6.8.1 Correlation between PQR values and the ITU five-point quality scale

ITU picture quality	*PQR values*
Excellent	<3
Good	3 or more but <6
Fair	6 or more but <9
Poor	9 or more but <12
Bad	12 or more

rise time of 300 ms, whereas the PPM has a much faster rise time (usually 10 ms, although faster rise times are available).

For steady tone, both instruments indicate the rms value of a sine wave, although it is important to bear in mind that, in the case of the VU-meter, there may be 4 or 8 dB attenuation between the audio line and the VU-meter. When monitoring a given speech or music signal, the PPM peak indications will be up to 10 dB higher than the VU-meter peak indications, due to the difference in rise times of the two instruments.

Therefore, it is standard practice to line up international audio connections with a steady tone signal, normally at 1000 or 1020 Hz, but sometimes at 900 or 400 Hz. This reference tone is normally set to 0 dBu (i.e. 0 dBm in 600 ohms, or 0.775 V_{rms}), which corresponds to EBU *test level*, or 0-VU, within studio centres. Peaks are set to 9 dB above EBU *test level* in the case of a PPM, or to 0-VU in the case of a VU-meter. The true peaks may be up to 12 dB above EBU *test level* in both cases. In the case of digital contribution circuits, this leaves a safety margin of 6 dB, because reference tone is specified as being 18 dB below FSD (Full Scale Deflection).

The left and right analogue audio input signals to a digital encoder first pass via an analogue/digital converter. The maximum and minimum sampling levels of the converter impose a clear-cut clipping level on the analogue input signals, and this clipping level is known as FSD. EBU Recommendation R.68-2000 states that reference tone (EBU *test level* tone) must be set 18 dB below FSD.

This R.68 requirement is also included in document ETSI TR 102 154, entitled 'DVB Implementation Guidelines for the use of MPEG-2 Systems, Video and Audio in Contribution and Primary Distribution Applications'. On the other hand, the SMPTE document RP 155-1997 specifies line-up tone level as 20 dB below FSD, which gives 2 dB more headroom than EBU R.68-2000.

MPEG-2/Professional Profile encoders and decoders usually provide selectable audio headroom settings of 18 and 12 dB above 0 dBu. The 18 dB headroom should be selected for contribution circuits, whereas 12 dB headroom may be selected for distribution circuits, where unexpected over-modulation is less likely to occur.

The WBU (World Broadcasting Union), with the agreement of the EBU, the Japanese broadcasters and various other broadcasters, has approved the implementation of EBU R.68-2000 for international contribution circuits. Figure 6.8.10 shows the comparative steady tone level indications for the various PPM and VU-meter scales, which are utilised in various parts of the world.

6.8.4 Scrambling of MPEG-2 Signals

The scrambling of digital signals does not cause any loss of picture quality, since the sequence of digits is simply changed by the scrambler, then restored to the original sequence by the de-scrambler.

MPEG-2/Professional Profile encoder–decoder manufacturers offered proprietary scrambling systems with the first

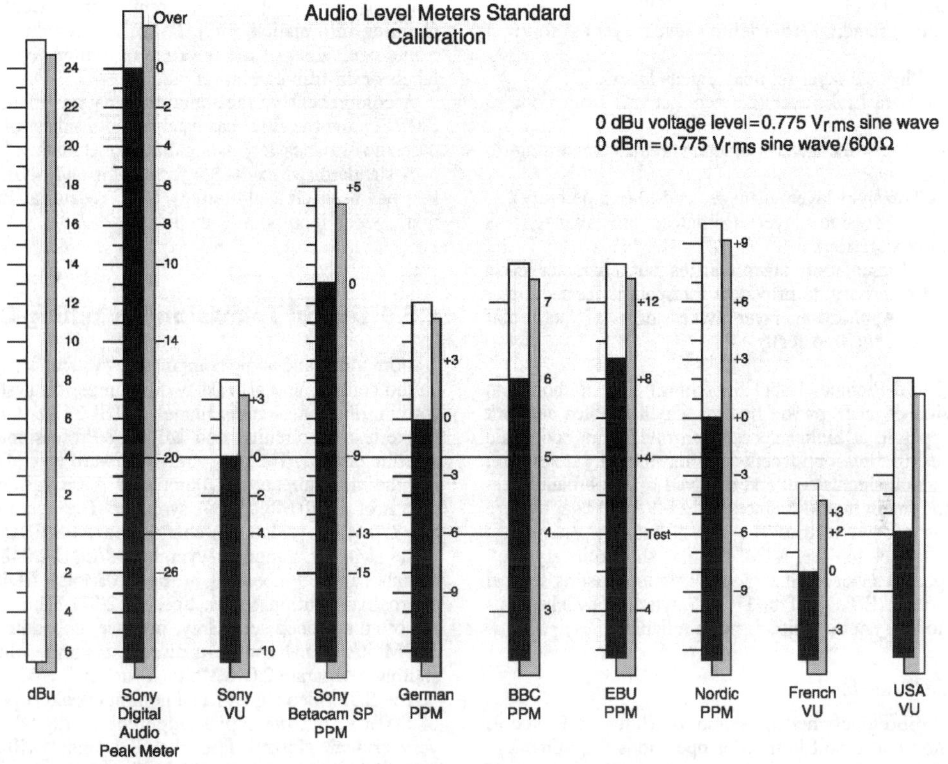

Figure 6.8.10 PPM and VU-meter calibration scales (NDS – Tandberg).

generation of encoders and decoders, but clearly there was a need for a standardised interoperable scrambling system, specifically designed for contribution satellite links. This requirement prompted the EBU to coordinate the development of BISS (Basic Interoperable Scrambling System) and BISS-E (BISS with Encrypted key).

BISS is intended for DSNG applications, and it is based on the DVB-CSA (Common Scrambling Algorithm) specification. A fixed SW (Session Word) of 12 hexadecimal characters is entered into the source encoder, and into the decoders, which are authorised to receive the signal.

BISS-E provides a higher level of security, since the 16-character ESW (Encrypted Session Word) required for one down-link location is not valid for another down-link location. This is because each down-link location is allocated, in advance, a secret 14-character ID (Identifier), which is entered into a decoder at that location, enabling it to calculate the required 12-character SW from the ESW. Each decoder also has a fixed *buried ID*, known only to the manufacturer and the scrambling authority, which can replace the manually entered ID.

The BISS and BISS-E specifications are published in the EBU document Tech. 3292 rev. 2, which was published in August 2002 and has been submitted to the ITU for approval. Since BISS and BISS-E are based on the DVB/CSA, they are both DVB compatible.

6.8.5 Convergence of Television and Computer Technologies

6.8.5.1 Fibre-Optic Networks

In the 1990s, there was much hype concerning the rapid development of the Internet for dotcom businesses, distance learning, etc. Intercity fibre-optic transmission capacity was increased by a factor of about 20 in the 1990s, whilst actual traffic growth was much less than this. Moreover, the introduction of DWDM (Dense Wavelength Division Multiplexing) enables each fibre to carry up to eight different wavelengths of light. Consequently, in the new millennium, fibre-optic transmission facilities have become available at very competitive rates, so the broadcasters are increasing their use of these transmission facilities.

For example, German television uses 'Hybnet', which is a fibre-optic network providing intercity links for MPEG-2 signals using ATM (Asynchronous Transfer Mode). ATM utilises a packet switching technique with cells of 53 bytes, which comprise 48 message bytes plus 5 header bytes. The adapters, which insert or extract the MPEG-2 signal into, or from, the ATM cells, are specified by AAL-1 (ATM Adaptive Layer, type 1) in Europe and AAL-5 (ATM Adaptive Layer, type 5) in the USA.

Fibre-optic networks are based on the OSI (Open Systems Interconnection) standard. This defines seven layers as follows:

- Layer 1 – Physical layer (optical signals layer).
- Layer 2 – Data Link layer (delivers packets from node to node).
- Layer 3 – Network layer (allocates routes according to network addresses).
- Layer 4 – Transport layer (manages end-to-end delivery).
- Layer 5 – Session layer (initiates and manages a communications session).
- Layer 6 – Presentation layer (carries out character code conversion if necessary, to provide a transparent connection).
- Layer 7 – Application layer (takes care of particular applications, e.g. file transfer).

MPLS (Multi-Protocol Label Switching) adds a shorthand representation of an IP packet header, which enables network routers to operate at higher speeds. It provides an economic means of using fibre-optic networks for intercity television links. The actual signals sent and received by the broadcasters are likely to remain in ATM, since MPEG-2/ATM adapters are well established and secure. The MPLS network provides a transparent tunnel for the ATM signals. Unfortunately, cell losses can provoke video and audio interruptions lasting several seconds, so that EBU prefers DTM (Dynamic synchronous Transfer Mode) system, which is more reliable.

6.8.5.2 File Transfer

The use of fibre-optic networks and IP (Internet Protocol) facilitates non-real-time file transfer operations. Equipment is available that stores a news item as low-bit-rate compressed signal on a hard disk, then transmits the item via the Internet to its destination. Transmission of a 3-minute news item by this 'store and forward' system typically takes about 20 minutes, and it is clearly a very economical way of sending short news items back to the broadcaster's home base. However, there are security issues associated with the use of the Internet for broadcasters' file transfer operations.

File transfer via private network high-bit-rate satellite or fibre-optic connections, on the other hand, can provide secure, high-quality transmission. The European Archivex system, for example, is intended to provide access to broadcasters' archives, initially at low bit rate for browsing purposes, then at high bit rate, via satellite, to download a high-quality MPEG-2 copy of the desired material, using file transfer.

The EBU has recently introduced a 2 Mbit/s Extranet service via satellite, which enables EBU members to download news items from a server at EBU-Geneva, using a file transfer system.

New compression systems, such as MPEG-4, Windows Media and H.264, promise to deliver good picture quality at significantly lower bit rates than the bit rates required for MPEG-2. Moreover, H.264 has the advantage of being an open standard.

6.8.5.3 Metadata and MXF

Metadata was introduced by the EBU and the SMPTE in 1998, to identify audio-visual material, and to accompany the material when it is transmitted as a file transfer. The metadata can carry every kind of information concerning audio-visual material (e.g. time and place of acquisition, name of source, names of actors, technical details of source material and of encoding format, name of the television programme or subject of the news material, contribution circuit details, rights restrictions, charging information, etc.). It can be thought of as the electronic equivalent of the information written on video cassette labels or on film-can labels.

A comprehensive metadata dictionary is maintained by the SMPTE, and the EBU has established a subset of B2B (Broadcaster-to-Broadcaster) data called P-metadata.

A standardised exchange format for audio-visual file transfers has been created, namely MXF (Material Exchange Format), which incorporates metadata.

6.8.6 Digital Television Switching Centres

Various video and audio compression systems are utilised within studio centres, and at least two compression systems are used for contribution circuits, namely ETSI 34 Mbit/s or 45 Mbit/s for terrestrial circuits, and MPEG-2/Professional Profile for satellite circuits. The only straightforward way of interconnecting the video inputs and outputs of these various units is by means of a 270 Mbit/s SDI switcher. The audio equivalent of the 270 Mbit/s video switcher is the AES/EBU 2.048 Mbit/s audio switcher, although several AES/EBU 2.048 Mbit/s audio signals can be embedded in the 270 Mbit/s SDI signal as an alternative solution. In practice, the 2.048 Mbit/s audio signals are often switched separately, because embedding them in the 270 Mbit/s signal can be a more expensive solution than installing a separate 2.048 Mbit/s switcher.

The SDI video signal has a nominal peak-to-peak amplitude of 800 mV, but the signal amplitude of digital signals is not very critical. Nevertheless, the SDI signal will only survive about 200 m of coaxial cable before it requires equalisation. Longer runs are made possible by adding equalisation, or by utilising fibre-optic connections. The 270 Mbit/s switcher inputs regenerate the incoming SDI signals, restoring their amplitude to 800 mV.

In practice, the most common problem encountered with SDI signals is excessive jitter. If the peak-to-peak jitter exceeds about 0.8 ns, problems may occur. Initially, short white lines may appear in the picture, followed by complete failure if the peak-to-peak jitter exceeds 1 ns.

During the period of mixed analogue and digital studio operation, it is often necessary to operate both analogue and digital video and audio switchers in parallel. Since most digital video and audio equipment has both analogue and digital inputs and outputs, this parallel operation is not difficult to achieve.

Figure 6.8.11 shows the CICT area of the EBU's EVC (EuroVision Control centre) in Geneva. The monitor wall shows the many incoming video signals, most of which are received via leased transponders of the Eutelsat W3 satellite at 7° East. Since September 1998, these signals are transmitted as MPEG-2/Professional Profile@Main Level signals, with useful bit rates of 21.503, 10.7515 or 8.448 Mbit/s. The picture monitors are fed with 270 Mbit/s SDI signals from the MPEG-2 decoders.

The main video switcher at the EVC is a 128-input/64-output 270 Mbit/s switcher, working in parallel with a 128-input/64-output 2.048 Mbit/s digital audio switcher.

Figure 6.8.12 shows two 9 m satellite dishes at BBC TV Centre in London. The left dish feeds the BBC's DVB-S multiplex to the Astra II satellite, and the right dish transmits and receives Eurovision traffic via the Eutelsat W3 satellite at 7° East.

Figure 6.8.11 The CICT area of the EVC at EBU-Geneva.

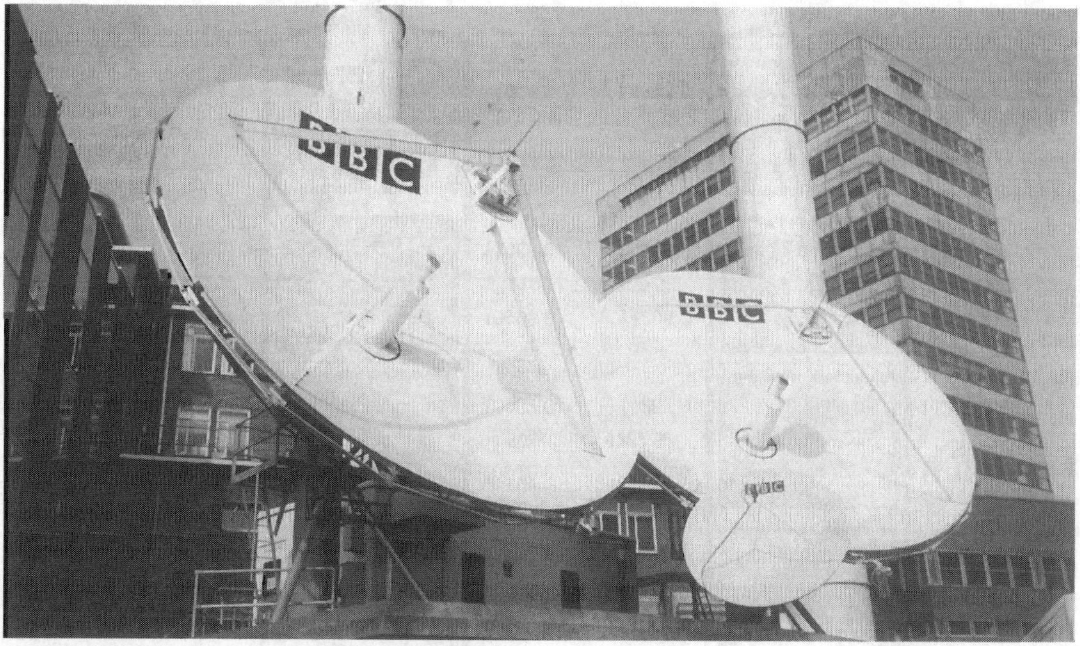

Figure 6.8.12 Two 9 m satellite dishes at BBC TV Centre, London.

BBC, BBCi, Antiques Roadshow and Walking with Beasts are trademarks of the British Broadcasting Corporation and are used under licence.

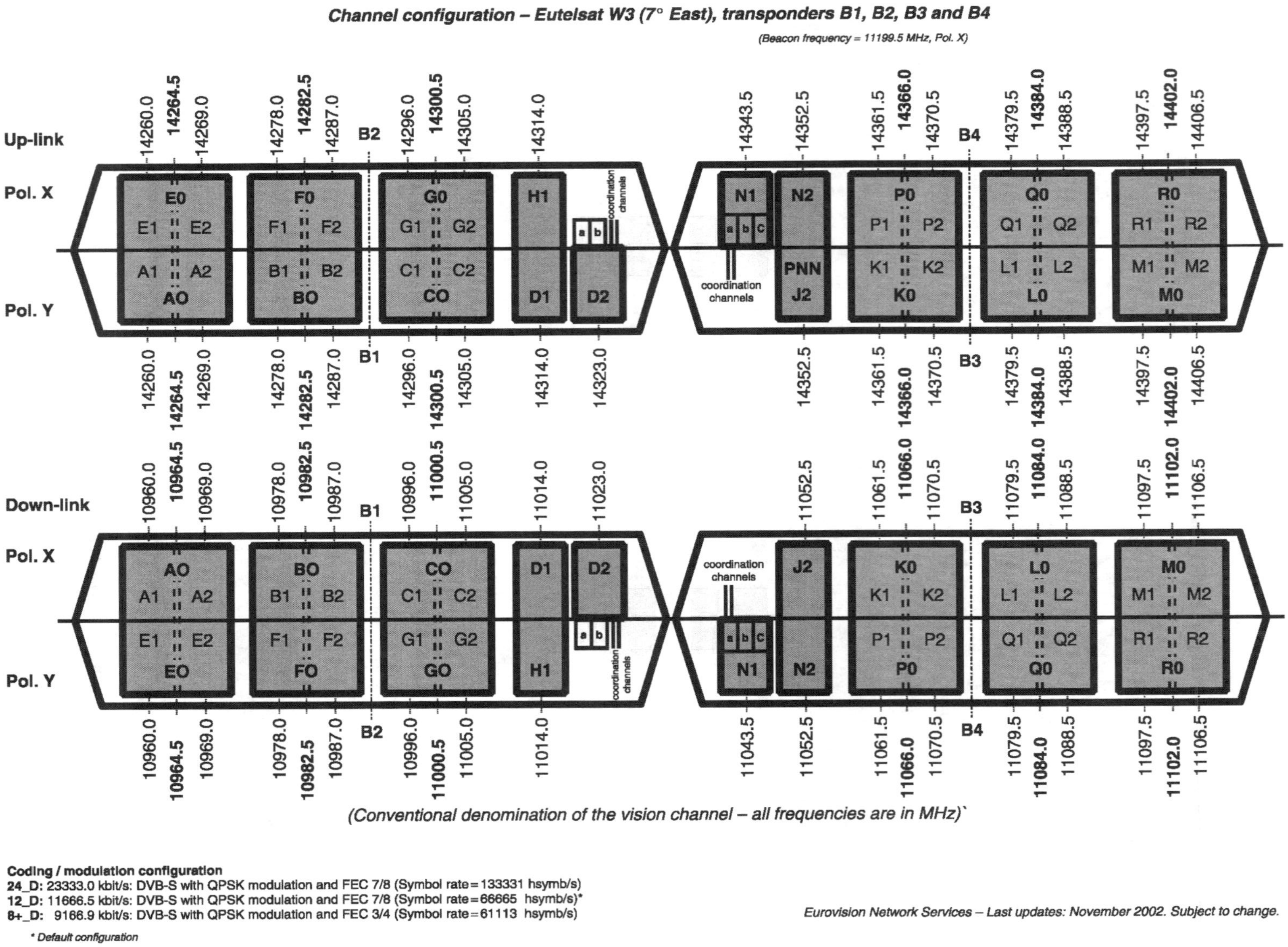

Figure 6.8.13 EBU leased transponder allocations on Eutelsat W3 (EBU).

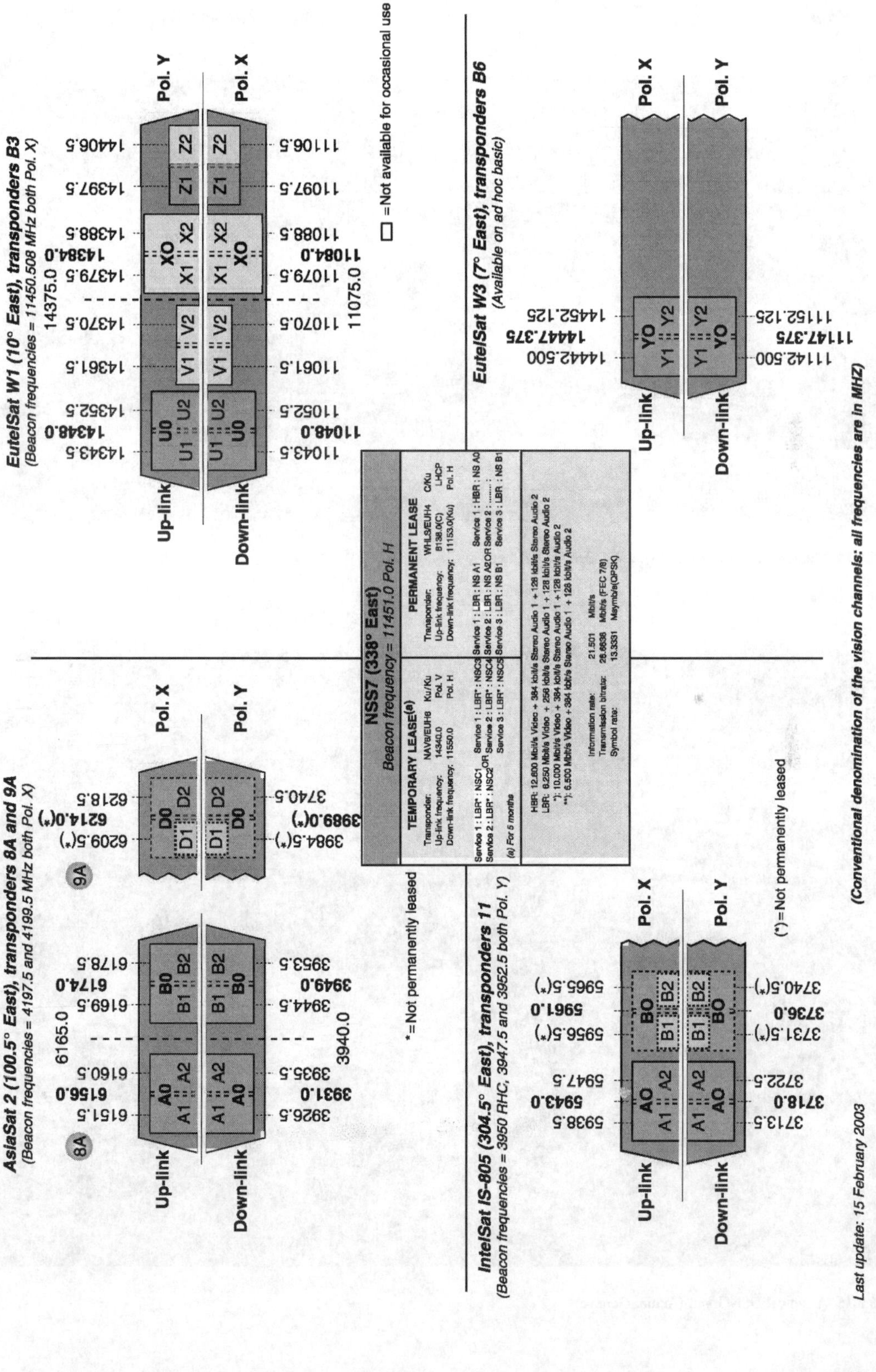

Figure 6.8.13 *continued*

Figure 6.8.14 The CNCT in the BBC Switching Centre, TV Centre.

Figure 6.8.15 A typical DSNG van (Actua, Geneva)

Figure 6.8.13 shows how the MPEG-2 signals are accommodated in the four 72 MHz bandwidth leased transponders of the Eutelsat W3 satellite. The diagram also shows how the Euroradio 2.048 Mbit/s stereo audio signals, and numerous 64 kbit/s communication channels, are accommodated.

Figure 6.8.14 shows the CNCT (Centre National de Coordination Technique) in the main switching centre at BBC TV Centre, London. This area has permanent voice contact, via the technical conference, with the EVC in Geneva, together with about 50 other CNCTs in Europe, the Middle East, North Africa and the Russian CIS.

Switching centres continue to play a vital role in establishing contribution and distribution television circuits. Meanwhile, the introduction of fibre-optic networks, with their sophisticated management systems, is changing the way in which international television links are established, especially for unilateral connections.

However, major sports events, requiring simultaneous distribution to many countries, will probably continue to use satellite networks, because they are particularly suitable for this kind of multilateral traffic. Also, in the case of news events, a DSNG van, using automatic up-leg dish alignment based on the GPS system, can transmit news coverage within minutes of arriving on location, which is not always possible in the case of a fibre-optic connection. Figure 6.8.15 shows a typical modern DSNG van.

References

Bray, W.J. The history of television from early days to the present, *IEE Conference Publication No. 271* (November 1986).

EBU Tech. 3205, The EBU Standard Peak-Programme Meter for the Control of International Transmissions (1979).

EBU Tech. 3292 rev. 2, BISS-E (August 2002).

EBU Tech 3250, Specification of the Digital Audio Interface (The AES/EBU Interface) (August 1992).

Kantchev, P., Flowers, B.G., Drury, G.M., Rodriguez, A., Fibush, D.K. and Bilow, S.C. *A Planning Guide for Digital Television: Contribution and Distribution Networks*, ITU Telecommunication Development Bureau (1999).

ITU-T Recommendations and ITU-R Recommendations, ITU Sales and Marketing Service.

Jackson, K.G. and Townsend, G.B. Intercity links and switching centres, in *TV & Video Engineer's Reference Book* (Flowers, B.G., ed.), Chapter 21, Butterworth-Heinemann (1991).

J P Whiting M Sc, C Eng, FIEE
Head of Power Systems, IBA

6.9 Transmitter Power System Equipment

6.9.1 Electricity Supplies

6.9.1.1 Generation, transmission and distribution

The generation, transmission and distribution of electrical energy is undertaken internationally at different voltage levels, but a typical system is illustrated in Figure 6.9.1.

From the primary substations the high voltage system, usually between 10 and 13 kV, distributes electricity to the majority of larger consumer loads and also to the urban and rural secondary substations. At the secondary substations, it is transformed to between 380 V and 450 V three-phase and 200 V to 260 V single-phase for use by the remaining smaller consumer loads. Typical underground and overhead hv distribution systems are shown in Figures 6.9.2 and 6.9.3. Most transmitter stations operate from these voltage levels and will be fed from either underground or overhead systems, or a combination of both.

The cost of an underground supply system can be anything up to 20 times that of an overhead supply system, and therefore much of the transmission and rural distribution is by overhead line conductors.

6.9.1.2 Power system faults

A short-circuit fault on the system, either between phases or between phase and earth, will result in a fault current flowing to the fault many times greater than the load current. The equipment needs to be able to withstand the increased electromagnetic forces and thermal stresses caused by the fault current, particularly during the first few cycles.

For these reasons, it is necessary to calculate the maximum possible fault current in order to specify the equipment withstand rating. The most onerous fault condition will invariably be a three-phase fault, and this fault configuration conveniently reduces to a three-phase star connection. Consequently, circuit fault calculations can be treated as for a single-phase circuit.

The fault current, I_f, is given by:

$$I_f = \frac{V_b}{Z_t}$$

where Z_t is the total impedance from the source to the point of the fault, and V_b is the selected voltage, to which all impedances have to be referred. The phase fault current, I_f, has to be multiplied by 3 V_{ph} to obtain the MVA rating for a given phase voltage V_{ph}.

Usually, the impedance of the transformer is given as a percentage value (Z_p), based on its rated MVA, and this value has to be converted to its actual value (Z_a).

A worked example to illustrate the principles of fault calculations follows in Section 6.9.1.2.1. The topic is dealt with in greater detail in Ref. 1.

6.9.1.2.1 Fault calculation

An electrical load is fed from an 11,000 V/440 V power transformer via two cables in parallel. Calculate the fault current in kiloamps at the load if:

1. The electricity supply fault level is 150 MVA.
2. The power transformer rating is 1000 kVA and has a percentage resistance of 4.5 per cent.
3. One cable impedance is 0.1 ohms resistance and 0.02 ohms reactance, the other is 0.056 ohms resistance and 0.014 ohms reactance.

The procedure is to: (a) change the fault level of 150 MVA into a source impedance; (b) change the percentage reactance of the transformer into an ohmic reactance; (c) convert the two parallel impedances of the cables into a combined impedance. Finally, (d), add these three phase impedances together and divide into the phase voltage to obtain the phase fault current.

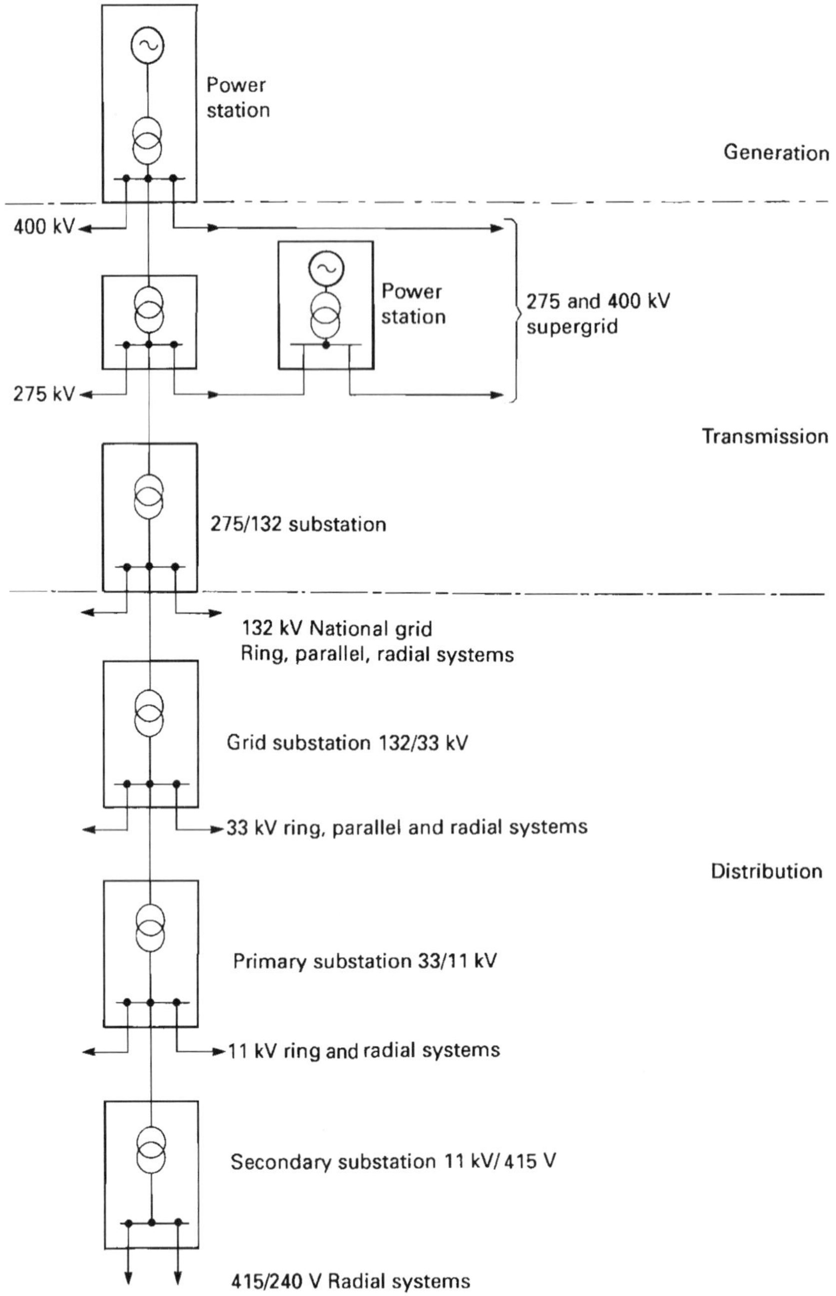

Figure 6.9.1 General configuration of generation, transmission and distribution of electricity in the UK.

(a) The source fault rating (S_s) is 150 MVA, and its impedance, i.e. the impedance of the source, can be assumed to be reactive. It is:

$$S_s = 3V_pI_f$$

where $I_f = V_pZ_s$.

Substituting,

$$Z_s = j3V_p^2/S_s = j3(440/\sqrt{3})^2/150 = j0.0013\text{ ohms}$$

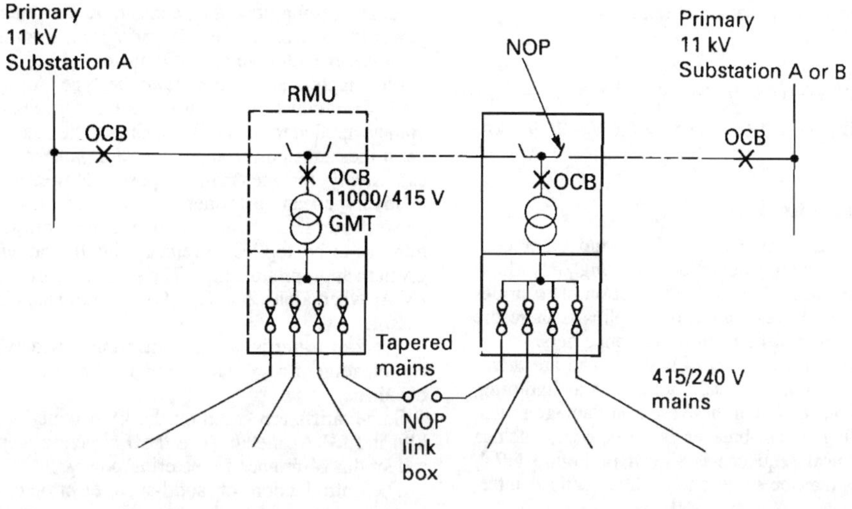

Figure 6.9.2 Typical 11 kV distribution underground system in the UK. OCB = oil circuit breaker; NOP = normally open point; RMU = ring main unit; GMT = ground-mounted transformer.

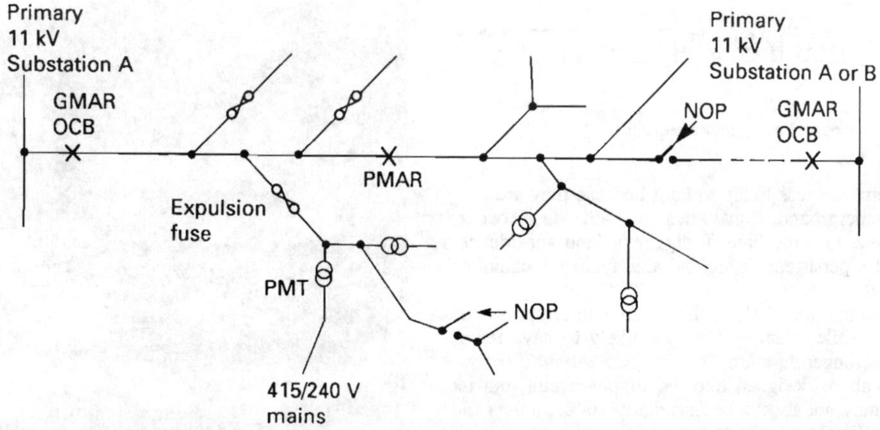

Figure 6.9.3 Typical 11 kV distribution overhead system in the UK. GMAR = ground-mounted auto-reclose oil circuit-breaker; PMT = pole-mounted transformer; PMAR = pole-mounted auto-reclose circuit-breaker.

(b) The transformer impedance is given in percentage terms, Z_{tp}, related to full load rating, S_t. This is usually the case and is expressed as follows:

$$Z_{tp} = I_f Z_t / V_{ph}$$

Also $S_t = 3V_{ph}I_f$.

Substituting,

$$Z_{tp} = S_t/3V_{ph} \times Z_t/V_{ph}.$$

Therefore,

$$Z_t = 3V_{ph}^2/S_t = 0.045 \times 3 \times (440/\surd 3)^2/1 \times 10 = j0.0087\,\text{ohms}$$

(c) The cables in parallel have the following impedances:

$$0.1 + j0.02 = 0.102/\!-11.31^\circ$$

and

$$0.056 + j0.014 = 0.058/\!-14.04^\circ$$

Using the product/sum equation for parallel impedances, the combined impedances of the two cables in parallel are:

$$Z_c = \frac{0.102/\!-11.31^\circ \times 0.058/\!-14.04^\circ}{0.156 + j0.034} = 0.036 + j0.0083\,\text{ohms}$$

(d) The total phase impedance is therefore:

$$Z_{ph} = Z_s + Z_t + Z_c$$
$$= j0.036 + j0.0183 = 0.0404/{-26.95^\circ}$$
$$I_f = V_{ph}Z_{ph} = 440/\sqrt{3}/0.0404/{-26.95^\circ} = 6.67/-26.95^\circ \text{ kA}$$

6.9.1.3 Supply reliability

In practice, the fault is located and isolated and other consumers restored by switching operations, leaving only those consumers directly connected to the faulty section off supplies until the fault is repaired. Restoration of supplies is therefore within 'switching times' for the majority of consumers.

Rural networks are fed by overhead systems and are therefore subjected to faults of a transient nature, such as lightning, ice and windborne objects. Often these are non-damage faults, and so the controlling circuit-breaker is arranged to reclose automatically. A typical sequence is shown in Figure 6.9.4. Transmitter supplies may be subject to such sequenced interruptions of supply.

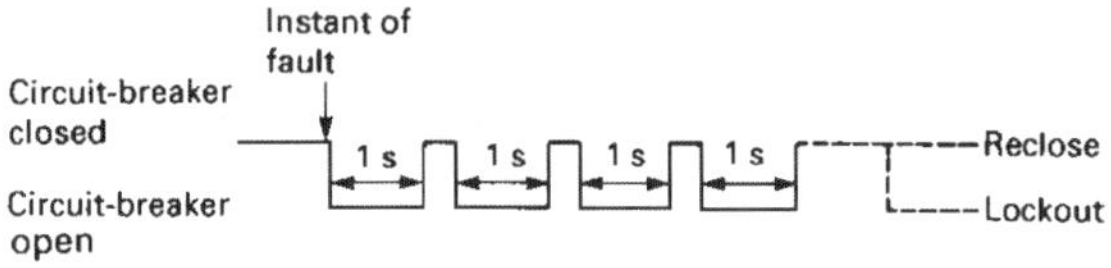

Figure 6.9.4 Pole-mounted auto-reclose sequence.

Urban systems are less likely to fault because they are predominantly underground, but when a fault does occur, for example, due to a mechanical digger or land subsidence, it is usually of a persistent nature and auto-reclose techniques are inappropriate.

So rural systems are likely to have more interruptions of short duration, while urban systems are likely to have fewer interruptions of longer duration.

Reliability can be designed into the supply circuit, but its improvement may not always be financially viable, unless the consumer is willing to contribute towards the cost.

6.9.1.4 Metering

The metering point of the supply is usually where the legal and financial responsibility changes from the Electricity Utility Board to the consumer. When the metering current transformers are located in the hv switchgear, the consumer owns, and therefore has to operate and maintain, the hv switchgear, the step-down power transformers and the interconnecting cabling.

In the case of small supplies, up to about 50 kVA, metering is connected directly into the mains cables. For larger supplies, the metering is operated from IEC 185 Class 1 (BS 3938) current transformers of appropriate ratio. If the supply is above 415 V, voltage transformers to IEC 186 (BS 3941) are used to reduce the metering potential to 110 V. (See also Section 6.9.2.2.5.)

The type of metering equipment required depends on the tariff structure. However, while a utility must charge for the energy taken, the charge must be in proportion to the current taken, because the equipment provided by the utility must be capable of supplying that current. A consumer with a poor power factor would take a larger current from the supply than a consumer with a good power factor.

Most meters are of the *induction* type. An induction meter reads the product of the current passing through the meter, the voltage applied to it, and the cosine of the angle between them, i.e. it measures power and is therefore a kWh meter. It is used extensively for metering purposes. However, while it is a relatively cheap instrument, it does not take the consumer power factor into consideration, and therefore some compensation has to be applied to remedy this limitation. The factor is given in appropriate tariffs. The same instrument reads reactive kV Ar when a phase shift of 90° is introduced into the voltage circuit.

The *kVA meter* is the other instrument widely used. This is a combination of a kW meter and a kV Ar meter, both acting on one disc.

Some tariffs are based on the kVA meter and some on the kW and kV Ar meters. The tariff structure therefore needs to reflect this difference in metering policy.

The introduction of solid-state electronic technology to metering has completely revolutionized metering techniques (see Figure 6.9.5). The systems offer higher accuracy and

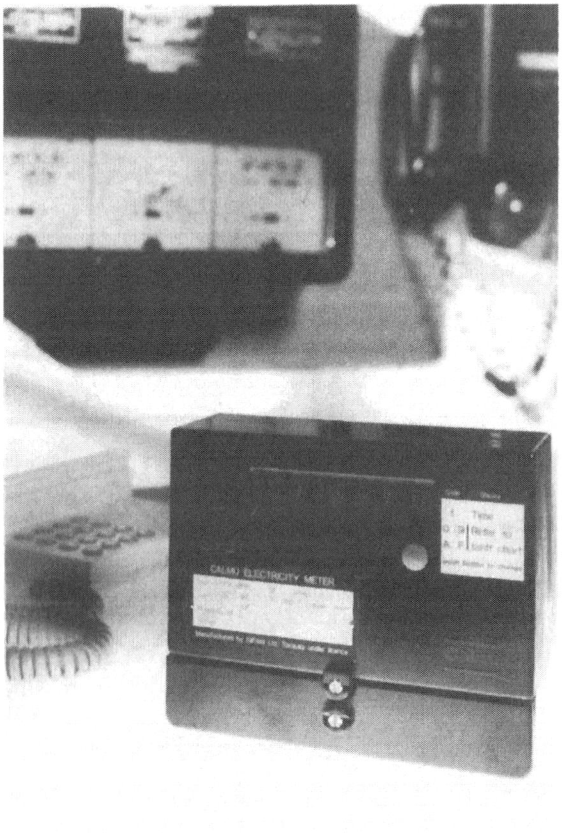

Figure 6.9.5 Three-phase solid-state meter (CALMU) with its former electromechanical equivalent in the background – note the reduction in size.

record any combination of kW, kVA, kV Ar, kWh, kVAh, kV Arh, maximum demand values and power factor. Moreover, the metering will cater for time of day spot pricing and will compute energy charges.

A significant feature is that the new systems are *interactive*, in as much as the meter can be controlled remotely, either by radio teleswitching or by mains borne signalling, to switch loads such as, for example, domestic heating and hot water loads. Intelligent metering can transmit the energy data via telephone links, radio or mains carrier to a remote centre for billing purposes and energy management statistics.

Further information on metering can be obtained from Ref. 1.

6.9.1.5 Tariff structures

The cost of the electricity consumed at a transmitter station is a major revenue item.

Supplies are mostly charged on a maximum demand tariff. These vary in structure and in unit costs. A typical structure is shown in Table 6.9.1.

Table 6.9.1 Commercial/industrial tariff structure

Maximum demand tariff comprises:	*Determined by:*	*Possible variations*
Fixed charge	Tariff fixed value	None
Availability charge	Mutually agreed value	Should be kept as low as possible by keeping the figure under review
Maximum demand charge which invariably includes a power factor improvement clause	Metered value (power factor derived)	Should be kept low by improving load factor power factor improved to 'best value'
Units KWH	Metered	Advantage should be taken, if possible, of any differential rates, i.e. use of energy during off-peak periods
Off-peak units	Time metered	
Fuel adjustment clause	Tariff fixed	None

6.9.2 Power Equipment

6.9.2.1 Switchgear

Switchgear and its associated protection equipment is used to control and distribute electrical energy in a safe manner. The term *switchgear* includes circuit-breakers, switches, isolators, combination of switch and fuse units, busbars, protective relays and fuses.

These items are usually combined together with a busbar system to form a switchboard (see Figure 6.9.6). Switchboards which are assembled and tested at the manufacturer's works and delivered to site as composite units, are referred to as

Figure 6.9.6 11 kV oil circuit-breaker switchboard.

Figure 6.9.7 415 V auxiliaries board.

factory built assemblies (FBAs) and comply with IEC 439 (BS 5486).

Under this standard, provision is made for four classes of design (Forms 1–4) and cater for increasing standards of insulation and circuit segregation between incoming and outgoing circuits. For a variety of reasons, including transport to site, switchgear may have to be erected on site. These units are known as *custom built assemblies* (CBAs).

6.9.2.1.1 Arc interruption

The action of opening an ac power circuit invariably produces an arc at the contact tips. Where the fault power factor is relatively high, i.e. between 0.8 and 1, as when interrupting load currents, arc interruption is not so difficult. However, under low power factor short-circuit fault conditions, the voltage across the contact gap at a current zero will be near its maximum and will therefore attempt to restrike the arc (see Figure 6.9.8).

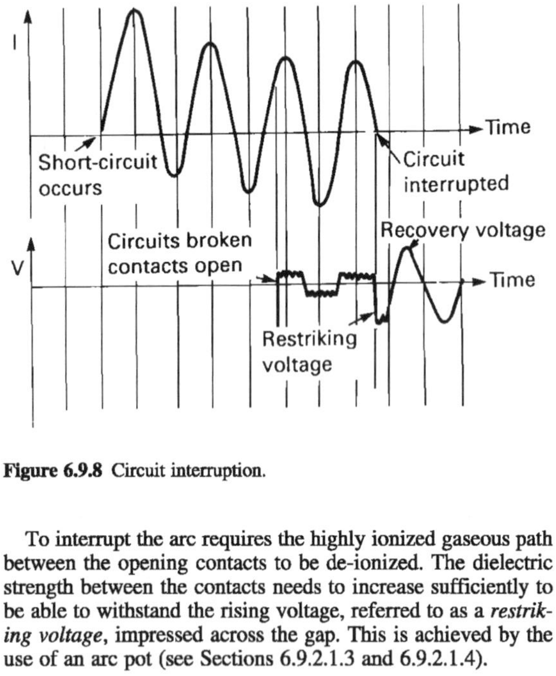

Figure 6.9.8 Circuit interruption.

To interrupt the arc requires the highly ionized gaseous path between the opening contacts to be de-ionized. The dielectric strength between the contacts needs to increase sufficiently to be able to withstand the rising voltage, referred to as a *restriking voltage*, impressed across the gap. This is achieved by the use of an arc pot (see Sections 6.9.2.1.3 and 6.9.2.1.4).

6.9.2.1.2 Operating mechanism

Magnetic forces, proportional to the square of the current, produce mechanical stresses which are particularly high under fault conditions. Figure 6.9.9 illustrates the direction of the forces when closing or opening a circuit-breaker. The arc is forced outwards, and the moving contacts are forced downwards, and this force assists the speed of the break or resists the closing action.

The fault closing capacity of the circuit-breaker needs to be higher than its breaking capacity, because it needs to be able to withstand the higher currents present during the first few cycles.

The operating mechanism provides the important role of closing and tripping the circuit-breaker both manually and automatically and under load and fault conditions. Closing is achieved using a large solenoid or spring-charged release coil, and tripping is afforded by a small shunt trip coil.

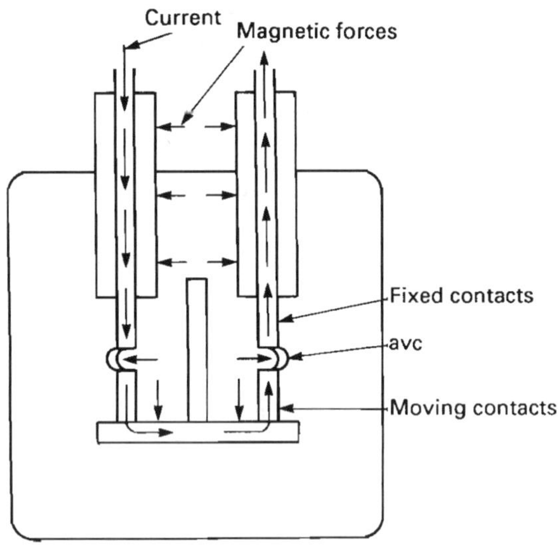

Figure 6.9.9 Electromagnetic forces in a circuit-breaker due to short-circuit currents.

The closing mechanism must provide sufficient energy to accelerate it against friction, springs and electromagnetic forces. It must latch into the closed position without noticeable bounce on the contacts.

There are three categories of closing mechanisms, and the type selected depends upon many factors of which cost is perhaps the most important. They are solenoid operation, spring assisted and manual dependent.

Manual mechanisms, while still in service, are not now considered safe because the closing force is dependent on the operator. *Spring assisted mechanisms* are the most common and available as hand charged, hand wound or motor wound spring. *Solenoid mechanisms* are the most expensive. They require a high capacity battery to provide the necessary energy for closing. They have the advantage that they are immediately ready for a second reclosure if necessary, while a motor wound spring takes some seconds to recharge its springs.

The opening energy has to be instantly available and independent of the normal power supply, so the shunt trip coil is usually supplied from a battery source. If trip initiation occurs during the closing operation, the circuit-breaker must trip immediately.

6.9.2.1.3 HV oil circuit-breakers

The operational requirement of a circuit-breaker, as laid down in IEC 694 (BS 6122, BS 6581), is that in addition to its normal rated close and trip duty, it should be able to make onto and, in conjunction with its protection, to break its rated fault current.

Most modern oil circuit-breakers are fitted with an *explosion pot*. The arc heat energy decomposes the oil to liberate a mixture of gases which exert pressure on the oil, and, by careful design of the arc pot, cool oil is forced across the arc path (see Figure 6.9.10).

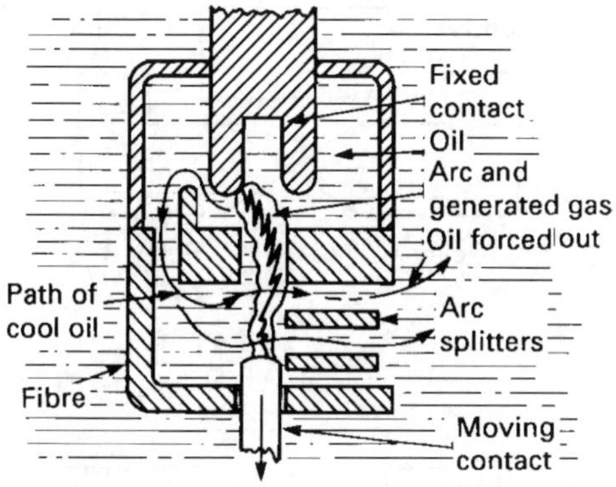

Figure 6.9.10 Explosion pot of a cross-jet oil circuit-breaker.

6.9.2.1.4 LV air circuit-breakers

Air circuit-breakers should comply with IEC 157 (BS 4752). Arc interruption is achieved by extending the arc path across splitter plates (Figure 6.9.11). The electromagnetic effect of the current loop causes the arc to rise between the splitter plates into the arc shutes. The resistance of the extending arc brings the voltage across the contacts more into phase with the current and so assists in the arc interruption process.

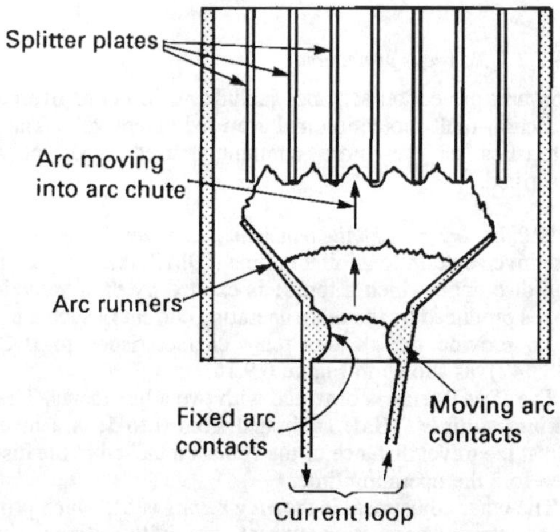

Figure 6.9.11 Chute-type air-break circuit-breaker.

6.9.2.1.5 Miniature circuit-breakers and moulded case circuit-breakers

Miniature circuit-breakers (mcbs), and moulded case circuit-breakers (mccbs), should comply with IEC 292-1 (BS 4941), and IEC 158-1 (BS 5424) respectively. Both have a similar arc interruption process to that of the air circuit-breaker. On certain designs, circuit interruption within the first quarter of a cycle produces a current limiting or 'cut-off' effect similar to that exhibited by the hrc fuse (see Section 6.9.2.2.3.3).

6.9.2.1.6 Fuse-switches

The operational requirements for fuse-switches, as laid down in IEC 408 (BS 5419), are that, in addition to their normal rated opening and closing duty at a specified power factor, they should be able to make onto and withstand their rated fault current until their fuses interrupt the circuit.

6.9.2.1.7 Switches

In addition to their normal rated opening and closing duty at a specified power factor, the operational requirement for switches, as laid down in IEC 265 (BS 5463), is that they should be able to make onto and withstand their rated fault current until the fault current is cleared by the system protection.

6.9.2.1.8 Isolators

Isolators need only to carry their rated current, to open and close negligible, i.e. no-load, current and to carry their rated fault current for a specified duration. These requirements are laid down in IEC 265 (BS 5463).

6.9.2.2 Protection equipment

Protection is provided to detect a fault quickly and initiate rapid isolation of the fault to limit the energy 'let-through' so as to reduce the physical risk to personnel and restrict damage to the equipment. A secondary, but important, requirement of the protection equipment is that it should indicate or 'flag' its operation so that the location and the type of fault may be analysed.

6.9.2.2.1 Discrimination, sensitivity, stability and protection zone

The protection, shown in Figure 6.9.12, should be so adjusted that in the event of a fault occurring at F1, circuit-breaker CB1, the 'minor' circuit-breaker, should discriminate with CB2, the 'major' circuit-breaker, to isolate the fault so that the supplies to load A are unaffected. Therefore, the protection on CB1 has

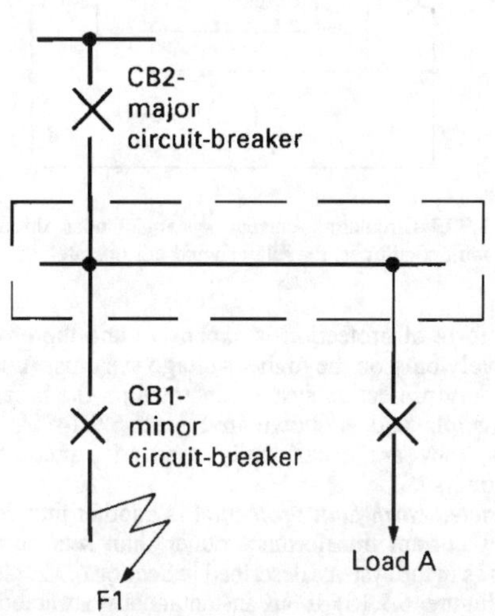

Figure 6.9.12 Protection discrimination, stability and sensitivity.

to be more sensitive than that on CB2, and the protection on CB2 needs to remain stable. The total operating time of a circuit-breaker comprises:

1. The protection operating time.
2. The trip mechanism operating time.
3. The arc interruption time.

Each of these have tolerances that need to be catered for. The total operating time of the minor circuit-breaker (1 + 2 + 3) must be less than the protection operating time (1) of the major circuit-breaker. The difference between these, including the tolerances, is the ***discriminating time*** and is usually considered to be between 0.35 and 0.5 s.

The *zone* of protection for CB2 will be between CB2 and CB1.

Fault calculations are necessary to achieve these objectives (illustrated in Section 6.9.1.2.1) and for more complex networks a program to predict the relay settings has been written.[2]

6.9.2.2.2 Unit protection

Unit protection (Figure 6.9.13) operates on the principle that current entering the protected zone must be equal to that leaving it. These two quantities are compared, and the difference is fed to a relay which is set to operate at a sensitive value.

Under normal load conditions, the difference amount will be small, but should a fault occur in the protected zone it will increase and the instantaneous relay will operate to initiate the isolation of that zone. Should a fault occur beyond the zone, then the two quantities, although greatly increased, should continue to balance and the protection remain stable. Unit protection is high speed in operation and has low sensitivity, but care has to be taken to ensure stability during through-fault conditions.

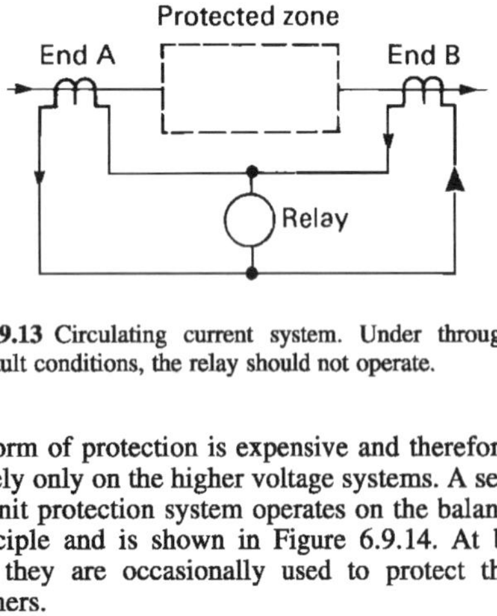

Figure 6.9.13 Circulating current system. Under through-load or external fault conditions, the relay should not operate.

This form of protection is expensive and therefore is used extensively only on the higher voltage systems. A second and similar unit protection system operates on the balanced voltage principle and is shown in Figure 6.9.14. At broadcast stations, they are occasionally used to protect the power transformers.

Restricted earth-fault protection is another unit form which balances current transformers rather than sets current transformers as in the systems described in Section 6.9.2.2.2. The REF relay (Figure 6.9.15) is an instantaneous attracted armature type, and the system illustrates the operation of the protection for a fault in-zone and out-of-zone.

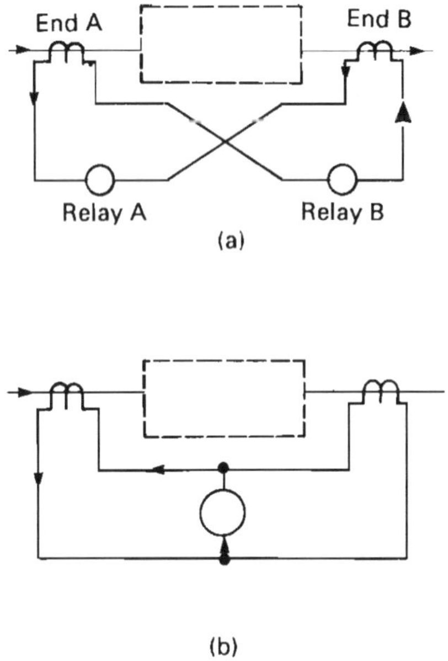

Figure 6.9.14 (a) Balanced voltage unit protection system. (b) Circulating current system. Under internal fault conditions, the relay should operate.

6.9.2.2.3 Non-unit protection

Non-unit protection schemes include all forms of overcurrent and earth-fault protection and are used extensively. The characteristics of the more commonly used types only are described.

6.9.2.2.3.1 Inverse definite minimum time relay

The inverse definite minimum time (IDMT) relay has an induction disc upon which a torque is exerted by the interaction of fluxes produced by the relay operating current in such a manner as to provide the various relay characteristics to IEC 255 (BS 142), as shown in Figure 6.9.16.

The IDMT relay is provided with two adjustments. The *time setting multiple* (TSM) calibrated from 0 to 1, is a means to adjust the travel distance of the contact attached to the disc and therefore the operating time.

The other adjustment is the *plug setting* (PS), which provides seven steps of percentage current sensitivity settings at which the relay disc will start to rotate.

Variation of the PS setting has the effect of moving the characteristic horizontally on the current/time graph, and vertical adjustment is obtained by varying the TSM setting. It will be observed, therefore, that, within the range of its adjustments, a whole variety of relay characteristic positions may be achieved.

The relay characteristic most suitable for grading with fuses is the *extremely inverse* because it is similar to that of the fuse.

Figure 6.9.17 shows an IDMT relay grading with a fuse at a transmitter station supplied at 11 kV.

In order to include the different curves on one graph, it is necessary to refer all the current values to a common base voltage which, in the example, is 440 V. The curve for the

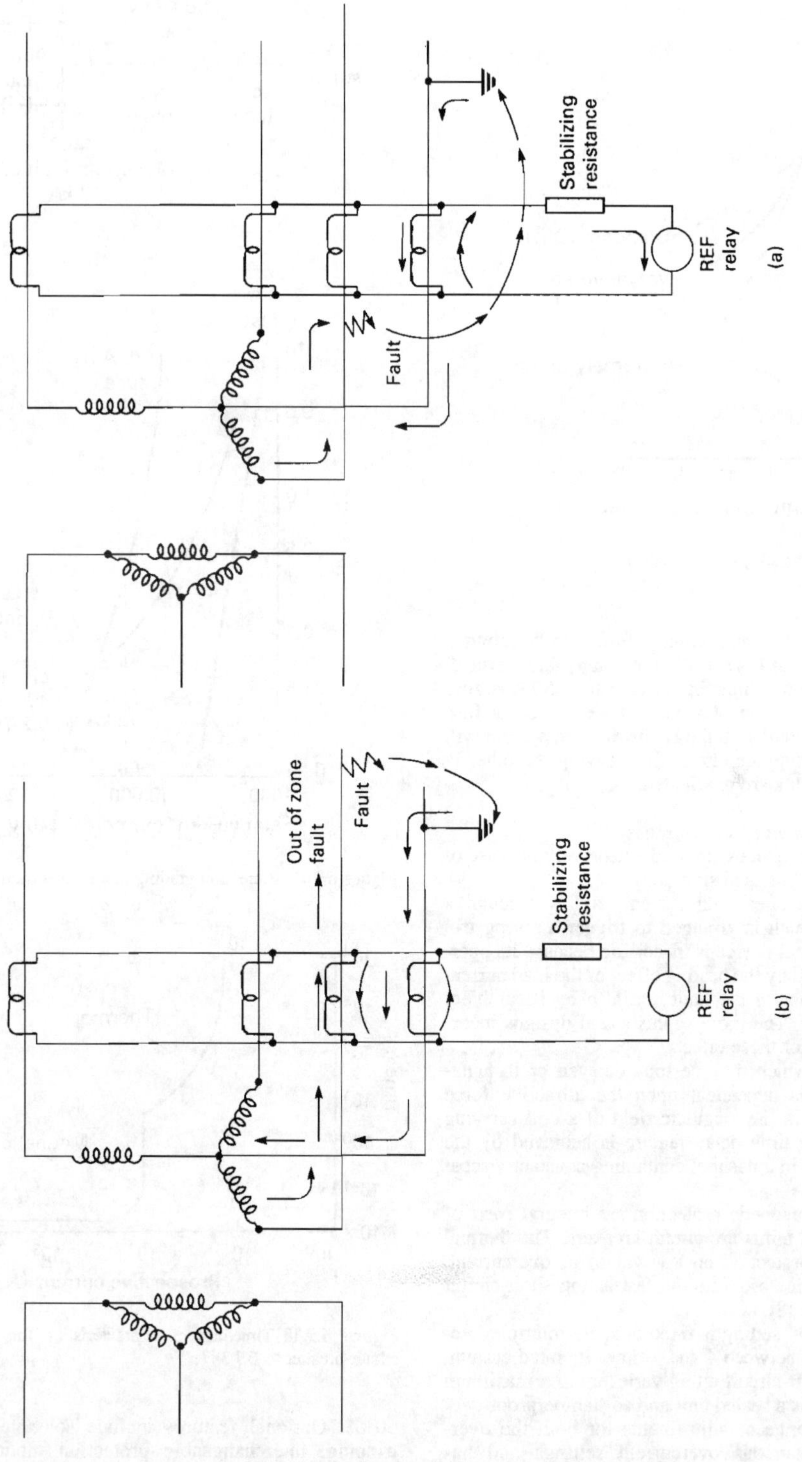

Figure 6.9.15 Restricted earth-fault protection applied to a four-wire system using four current transformers: (a) the system is stable for an external fault; (b) it is operative for an 'in-zone' fault.

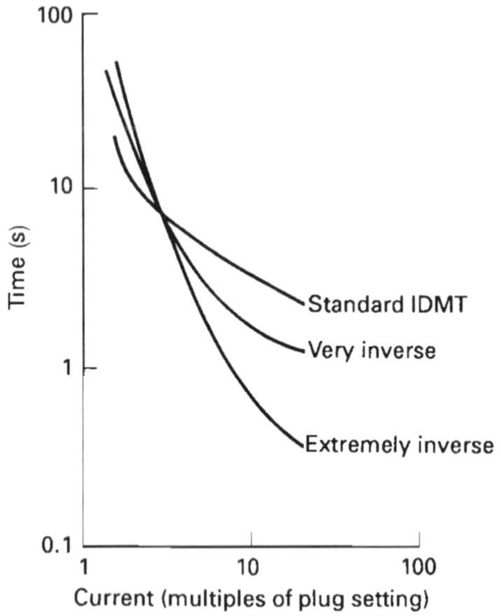

Figure 6.9.16 Overcurrent relay characteristics.

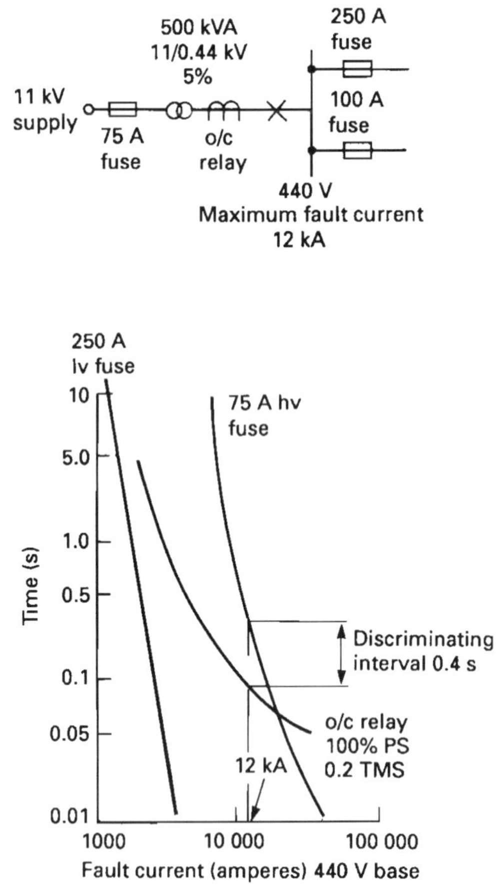

Figure 6.9.17 Protection grading at a typical transmitting station.

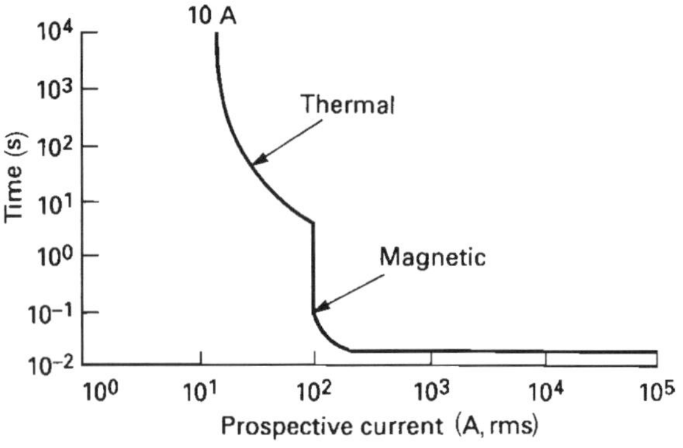

Figure 6.9.18 Time/current characteristics for Type 3 miniature circuit-breaker to BS 3871.

IDMT relay and the 75 A high-voltage fuse clearly shows a discriminating interval of 0.4 s at the maximum fault level of 12 kA. There is a generous margin between the 250 A lv fuse and the relay, and this is brought about by the effect of the fuse cut-off characteristic. Faults on the lv distribution system will be cleared by the appropriate circuit fuse leaving the other lv circuits at the lv switchboard unaffected.

6.9.2.2.3.2 Thermal and magnetic protection

Thermal and magnetic devices are used extensively because of their comparative low cost and simplicity.

A *thermal* type operates with current passing through a bimetallic element which is arranged to trigger a spring trip mechanism. The thermal device is inaccurate because its operating time is influenced by the heating effect of the load current passing through it prior to the fault and also by its ambient enclosure temperature. The more sophisticated designs incorporate compensation for these effects.

A *magnetic* type, which may be time delayed or instantaneous in operation, is dependent upon the attractive force exerted on a plunger in the magnetic field of a coil carrying the fault current. The time delay feature is achieved by the movement of a piston in a dashpot containing constant viscosity silicon fluid.

Both thermal and magnetic protection are integral parts of the moulded case and miniature circuit-breakers. The thermal device provides the protection on low values of overcurrent, and the magnetic device provides the fast acting short-circuit protection (Figure 6.9.18).

The mcb overcurrent setting is fixed at some multiple integer, e.g. Type 3 is set between 4 and 7 times its rated current, and the withstand short-circuit rating varies up to a maximum of 6.12 kA. The mcb is a sealed unit and so 'tamperproof'.

The mccb has a range of adjustments for both the overload and the instantaneous overcurrent settings and has short-circuit withstand ratings of up to approximately 50 kA. Optional features include solenoid closing, remote tripping, interchangeable protection modules and plug-in circuit-breakers.

6.9.2.2.3.3 High rupturing cartridge fuses

The rewireable fuse has been superseded by the high rupturing capacity (hrc) fuse because it has a definable short-circuit interrupting rating and a non-deteriorating operating characteristic.

An hrc fuse, to IEC 439 (BS 88), comprises a ceramic body, containing specially designed fuse elements connected between the metal end caps. It is filled with pure granulated quartz.

The ratio of the minimum fusing current to its actual current rating is the *fusing factor* and has assigned values of P, Q1, Q2 or R, as listed in Table 6.9.2, stipulated in IEC 439 (BS 88).

Table 6.9.2 Fusing factors

Class of fuse-links	*Fusing factor*	
	Exceeding	*Not exceeding*
P	1.00	1.25
Q1	1.25	1.5
Q2	1.5	1.75
R	1.75	2.5

It will be seen from Figure 6.9.19 that the fault current is interrupted by the fuse before it reaches its peak value. This is referred to as the *cut-off* current and limits damage to the equipment by considerably restricting the thermal stresses and electromagnetic forces.

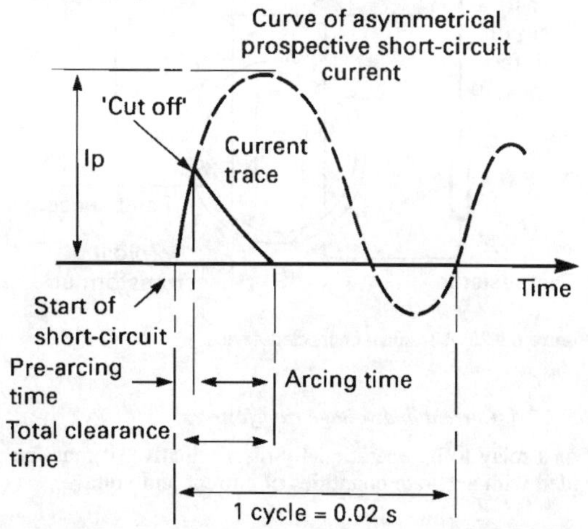

Figure 6.9.19 Cut-off feature of hrc fuses.

Where the system fault level exceeds that of the rating of the equipment, the hrc fuse must have a cut-off current that is less than the withstand capacity of the equipment. The cut-off currents for a range of fuses are given in Figure 6.9.20. For example, a 60 A fuse subjected to a fault current of 50 kA rms would limit the cut-off current peak to approximately 7 kA.

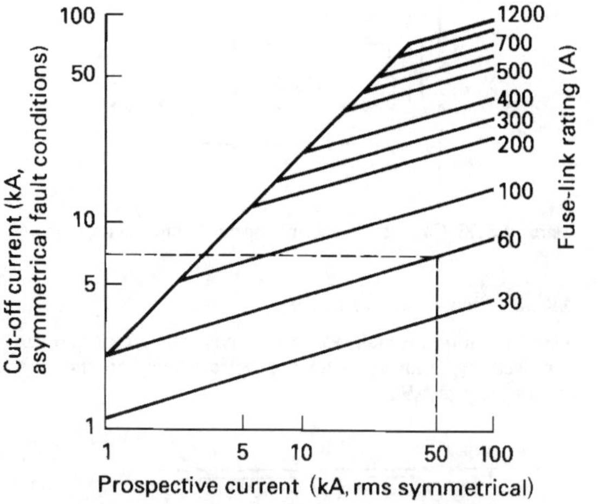

Figure 6.9.20 Cut-off current characteristics for hrc fuse links.

For overcurrents involving fuse operating times less than 10 ms, it is necessary to base discrimination on the *let-through* energy which is expressed in terms of I^2t. The total I^2t value for the minor fuse should be less than the pre-arcing I^2t value for the major fuse. The I^2t values for a typical range of fuses are shown in Figure 6.9.21.

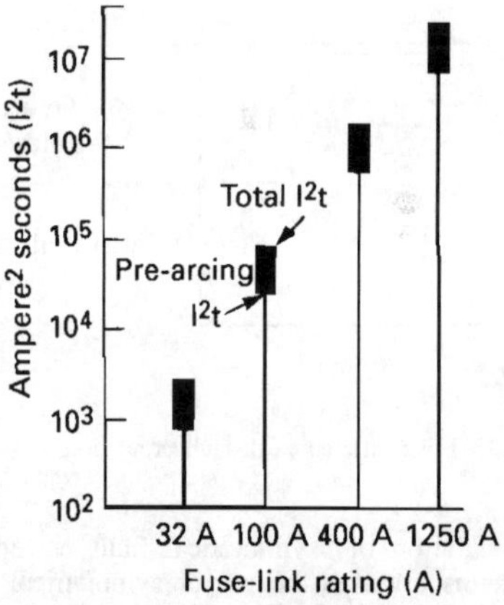

Figure 6.9.21 I^2t characteristics for hrc fuses.

The use of an hrc *time fuse* connected in parallel with the circuit-breaker trip coil provides the hv circuit-breaker with a time characteristic similar to that of the hrc fuse. The circuit-breaker trip coil is energized only after the fuse has operated (Figure 6.9.22).

The hv fuse differs from the lv fuse in that it does not have the ability to operate at low fault currents without the elements 'burning back' until an arc is struck between its end caps to create an explosion. To avoid this disaster, current is diverted through an ignition wire which detonates to operate a striker pin. This in turn triggers the operation of a circuit-breaker or contactor to isolate all three phases.

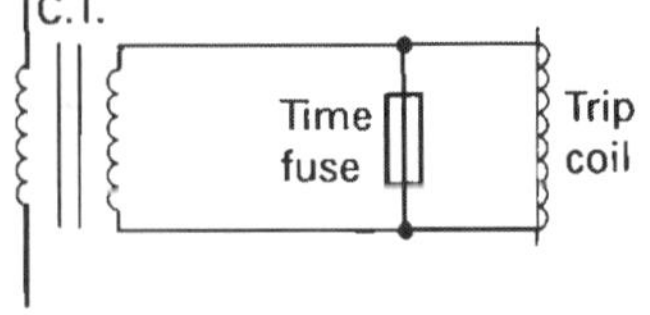

Figure 6.9.22 Current –transformer-operated direct-acting trip coil.

6.9.2.2.4 Earth-fault protection

The detection of a fault between a phase conductor and earth is achieved by connecting an appropriate relay in the residual circuit (Figure 6.9.23).

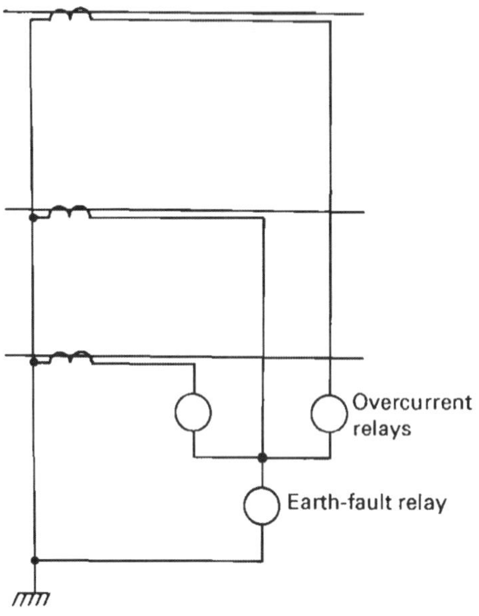

Figure 6.9.23 HV restricted earth-fault protection.

The calculation of asymmetrical fault currents on three-phase systems involves the use of symmetrical components. The appreciation of this mathematical tool is not difficult but is outside the scope of this text.[3]

Relays used for this purpose can be either an *inverse definite minimum time* type (Figure 6.9.24), and set to discriminate in a manner identical to that of the overcurrent IDMT protection, or an *attracted armature instantaneous* type.

Low current applications often use a core balance type of current transformer, where the phase and neutral conductors are passed through the core as shown in Figure 6.9.25. Under normal healthy conditions, the sum of the magnetic fluxes in the transformer will be zero, but should an earth-fault occur, the balance will be upset, and the residual flux will operate the relay.

The current operated *residual connected device* (rcd), is one form of this protection, and operates when the out of balance current exceeds the tripping sensitivity of the relay (these range between 10 and 500 mA).

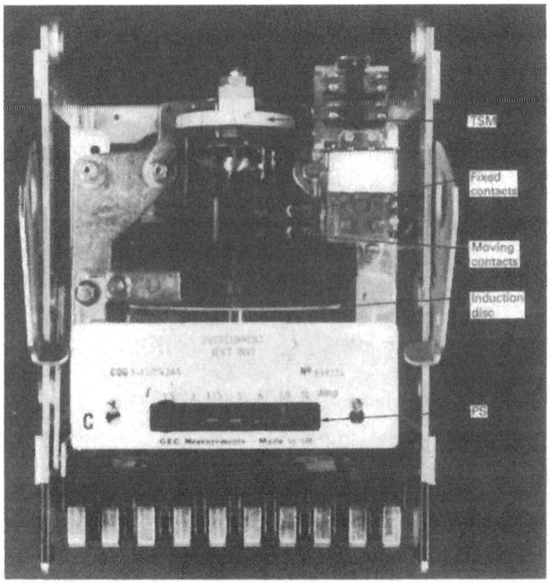

Figure 6.9.24 Inverse definite minimum time relay.

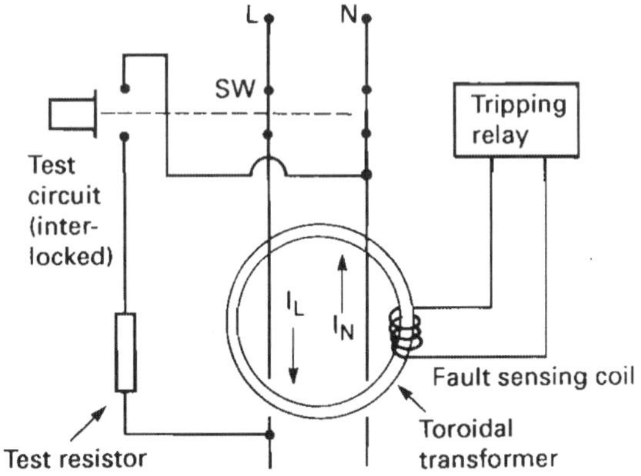

Figure 6.9.25 A residual connected device.

6.9.2.2.5 Current and voltage transformers

For a relay to be operationally discriminative, it must be provided with accurate quantities of current and voltage.

6.9.2.2.5.1 Current transformers

A protection current transformer differs from a metering current transformer in that the former must retain its ratio for very high fault currents, whereas the latter must:

- maintain defined accuracy at load currents;
- saturate at high fault currents to protect the metering from excessive voltages.

The ability of a protection current transformer to fulfil its function will depend upon its design and on the load impedance (or *burden*) connected to it. A current transformer needs to be accurate in both its ratio and phase displacement.

Protection duty current transformers are listed as accuracy classes 5P or 10P. The figure defines the maximum composite error in percentage at a specified overload value, called the *accuracy limit factor*. 'P' indicates a protection duty. The accuracy limit factor is specified by a further figure 5, 10, 15, 20 or 30. A typical protection current transformer would therefore be defined as 10 VA class 10PI5 which means that the transformer has a rated output of 10 VA of accuracy 10 per cent at 15 times its VA rating. For certain applications, where current transformer outputs need to balance under through-fault conditions, as with unit and restricted earth-fault protection, the class description is not sufficient and the specification needs to be more explicit. It is referred to as a Class X current transformer.

The accuracy class of metering current transformers is listed in IEC 185 (BS 3938) as 0.1, 0.2, 0.5, 1, 3 or 5. Each class has a defined ratio and phase displacement error at rated current and frequency. For example, a typical metering current transformer would be defined as 15 VA Class 0.5.

A current transformer should never be operated with its secondary winding on open circuit, because the primary current becomes, in effect, the magnetizing current and so the induced secondary voltage will be very high. The voltage can be dangerous to personnel and can cause the failure of the transformer insulation.

6.9.2.2.5.2 Voltage transformers

A voltage transformer, IEC 186 (BS 3947), is required to transform the system high voltages to the operating voltages of the relay, usually 110 V and 55 V to earth. Voltage transformers are similar to small power transformers, but are designed to maintain ratio and phase errors within very close limits for a specified range of secondary burdens.

6.9.2.3 Power transformers

Perhaps the greatest single reason for the adoption of alternating current as the mode for transmission and distribution of electrical energy has been the transformer. Using a transformer, voltages can be changed from one value to another in a convenient manner with high efficiency. If required, using static devices, various dc voltage outputs can readily be obtained. Power transformers should conform to IEC 76 (BS 171).

6.9.2.3.1 Voltage regulation

The approximate equivalent circuit of a transformer together with its phasor diagram is shown in Figure 6.9.26. The regulation of a transformer is the change in output voltage from V_1 to V_2 caused by the load current increasing, and is a function of the internal resistance and reactance values of the transformer.

6.9.2.3.2 Transformer loss and efficiency

Transformer losses may be divided into two categories: those which vary with load current, referred to as *copper losses*, and those which are practically constant at all loads, referred to as the *iron losses*. The iron losses can be measured with the transformer on no-load and comprise hysteresis losses and eddy-current losses.

The power loss due to *hysteresis* for a given transformer is proportional to the supply frequency. The *eddy current* loss, due to circulating currents set up in the core material laminations is, for a given transformer, proportional to the square of the supply frequency. Because of the non-linear nature of the transformer core, harmonics, predominantly triple-n harmonics, are generated into the power source system.

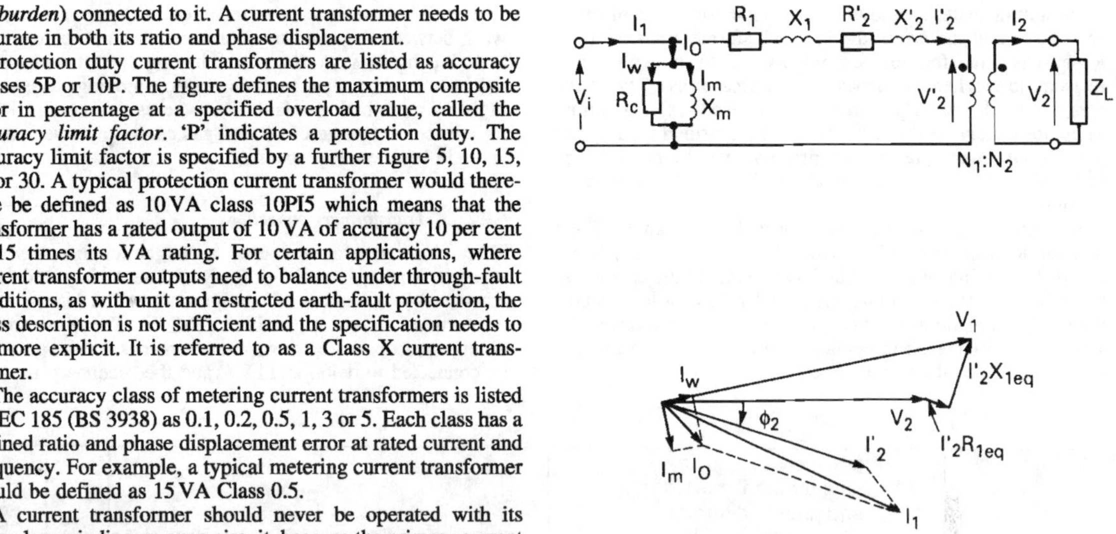

Figure 6.9.26 Transformer equivalent circuit.

The transformer efficiency is given by the equation:

$$\begin{aligned}\text{efficiency} &= \text{output power/input power}\\ &= (\text{input power} - \text{losses})/\text{input power}\\ &= 1 - \{(\text{copper losses} + \text{iron losses})/\text{input power}\}\end{aligned}$$

6.9.2.3.3 Transformer cooling

The heat produced by the losses must be removed to prevent damage to the winding insulation. Various methods of cooling are listed in Table 6.9.3.

Table 6.9.3 Classification of methods of cooling

Type of transformer	*Oil circulation*	*Cooling method*	*Abbreviation*	
			Dry	*Mineral oil*
Dry	—	Natural	AN	—
	—	Blast	AB	—
	Natural	Air natural	—	ON
	thermal	Air blast	—	OB
	head only	Water	—	OW
Oil	Forced by pump	Air natural	—	OFN
immersed		Air blast	—	OFB
		Water	—	OFW

Insulating material allows for a temperature rise of up to 150 °C, so that the use of dry type transformers is extensive. Resin cast transformers are impervious to moisture, and resistant to mechanical damage and thermal stress. They have a high impulse voltage withstand and need little attention. They are widely used indoors where the oil-filled type would present an unacceptable fire risk and would need to be provided with means to contain and limit the spread of leaking oil.

A conservator tank can be connected above an oil-filled transformer and partly filled with oil. This enables the transformer to remain completely full by allowing for the changes in the volume of the oil to take place under varying load conditions. The comparatively small conservator tank reduces the surface area of the oil in contact with the air and so reduces oxidation of the oil and corrosive effects.

A Buchholz relay can be mounted in the connecting pipework between the main transformer tank and the conservator tank. It comprises two floats each having a set of contacts. One indicates an alarm by responding to slow moving gas generated in the oil by an incipient fault. The other will be operated by the turbulence of a surge of oil and gas caused by a serious fault and will initiate a trip operation.

6.9.2.3.4 Transformer connections

Three-phase transformers, used for larger supplies, have three separate windings on the primary and secondary and can be connected in a number of configurations. The more common connections are shown in Figure 6.9.27. The most common arrangement for a step-down transformer is for the primary to be connected in delta, at 11 kV, and the secondary in star, at

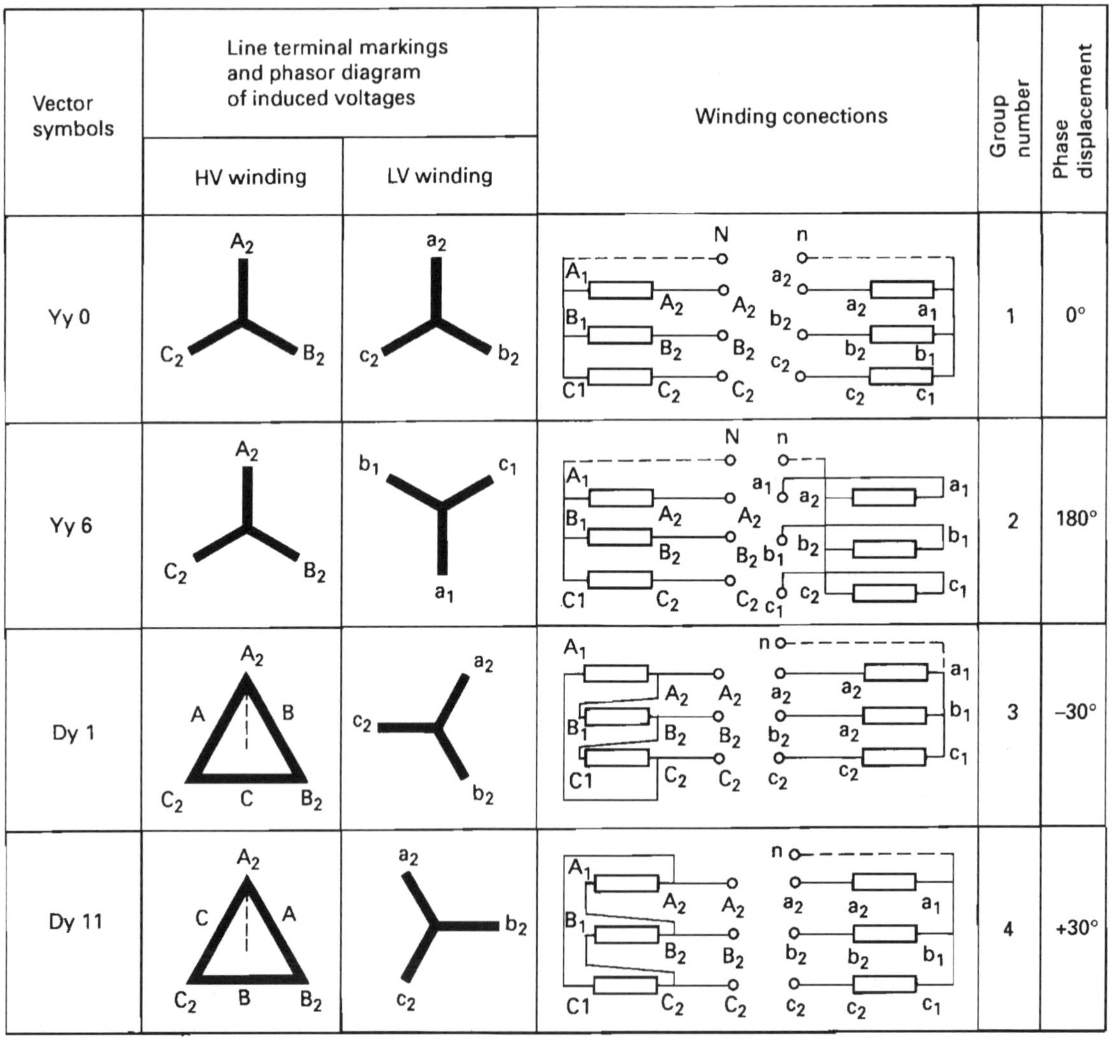

Figure 6.9.27 Three-phase transformer winding connections.

415 V. This is referred to as a Dy 1 or Dy 11 connection depending on whether the connections give a secondary voltage lagging or leading the primary reference voltage respectively.

6.9.2.3.5 Auto-transformers

Another winding arrangement is that of the auto-transformer. The input voltage is applied to the primary winding and the output voltage derived from a tapping on the same winding (Figure 6.9.28).

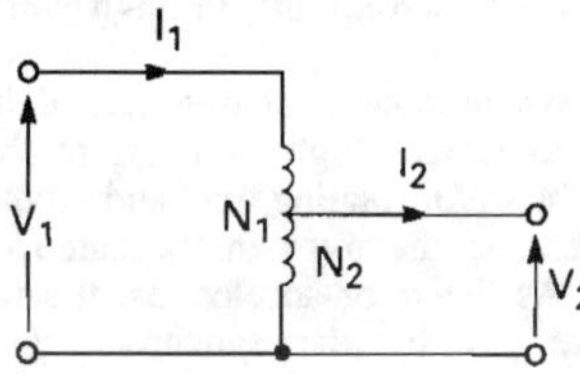

Figure 6.9.28 Auto-transformer (step down).

The advantage of such an arrangement is that the transformer uses less copper, and there are therefore less copper losses and a higher efficiency. It is lower in cost, but suffers from the disadvantage that if the secondary faults on open circuit, the output voltage changes to the input voltage.

6.9.2.4 Automatic voltage regulators

It has been the practice to confine the voltage at the transmitter to a ±0.5 per cent window using an automatic voltage regulator (AVR) from a source which can vary between +6 and −10 per cent.

There are various types of voltage regulator, and these include solid-state, variable transformer and moving coil.

6.9.2.4.1 Solid-state AVR

The connections of a solid-state automatic voltage regulator are shown in Figure 6.9.29. Two transducers, TDl and TD2, are connected across a proportion of the output winding of the isolation transformer, Tl. The output is taken from the common point of the two transducers, whose inductance is varied by the control winding current, so that the output voltage will range between V_1 and V_2.

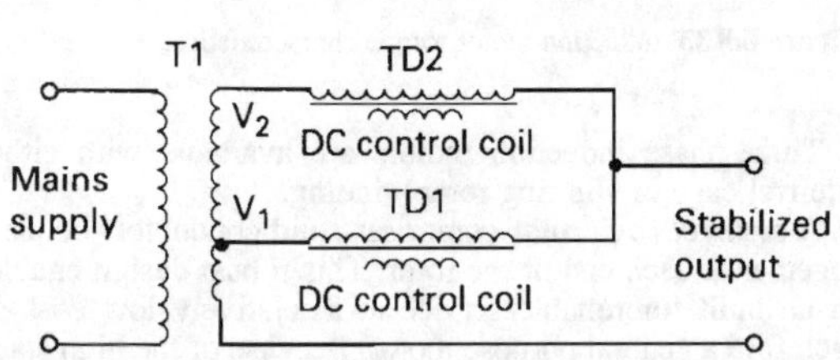

Figure 6.9.29 Solid-state automatic voltage regulator.

A transistor operational amplifier compares the voltage across the load with a highly stable reference voltage. Any voltage error is amplified and directly coupled to the control winding of the transductors to modify the output voltage.

This type of regulator has no mechanical nor conductive moving parts and is more accurate than an electromechanical one. The transductor is not susceptible to damage from short-circuits or voltage surges, and the amplifier operates at low voltage and current levels. However, harmonic filters are necessary to reduce distortion, and the level of distortion is affected by only moderate input voltage variations.

A thyristor controlled regulator is another type of solid-state voltage regulator. The thyristors are connected in inverse parallel as a bridge so that the current will be controllable in either direction by applying the appropriate control signal to the gate connections. Two methods of gate control are possible. ***Phase control*** regulation is achieved by 'point-on-wave' switching. ***Burst firing*** regulation is achieved by conducting in bursts of full half cycles interspersed with non-conducting periods. Phase control regulators have limited rating because the harmonic generation causes distortion to the supply.

6.9.2.4.2 Variable transformer AVR

A variable transformer regulator (Figure 6.9.30) has a transistorized servo-amplifier to monitor the output voltage and control the operation of a geared reversible motor. The motor drives the moving arm of the variable transformer, the output of which supplies the primary winding of a fixed ratio auxiliary transformer. The secondary winding of the auxiliary transformer is connected in series between the supply and the load, and this affords the buck-boost action to control the output voltage.

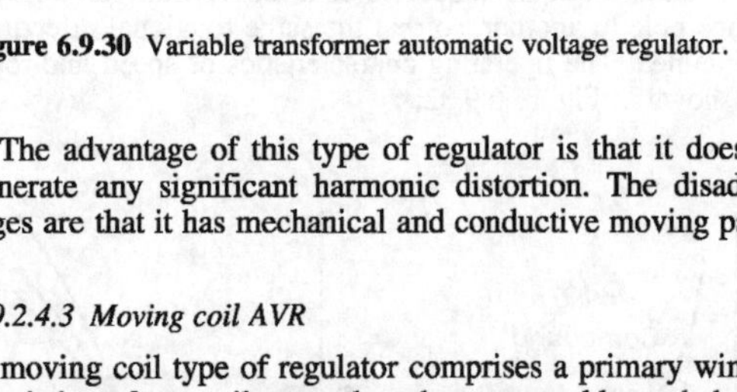

Figure 6.9.30 Variable transformer automatic voltage regulator.

The advantage of this type of regulator is that it does not generate any significant harmonic distortion. The disadvantages are that it has mechanical and conductive moving parts.

6.9.2.4.3 Moving coil AVR

A moving coil type of regulator comprises a primary winding consisting of two coils wound on the upper and lower halves of a magnetic core over which an isolated short-circuited coil can move up and down. The division of voltage between the two primary coils is determined by their relative impedance.

With the moving coil in the position x (Figure 6.9.31(a)), the impedance of coil x will be small, and so the greater part of the voltage will appear across coil y. When the moving coil is in the position y, the greater part of the voltage will now appear across coil x.

Two additional coils are connected so as to inject a buck-boost voltage in series with the line. Any desired values of

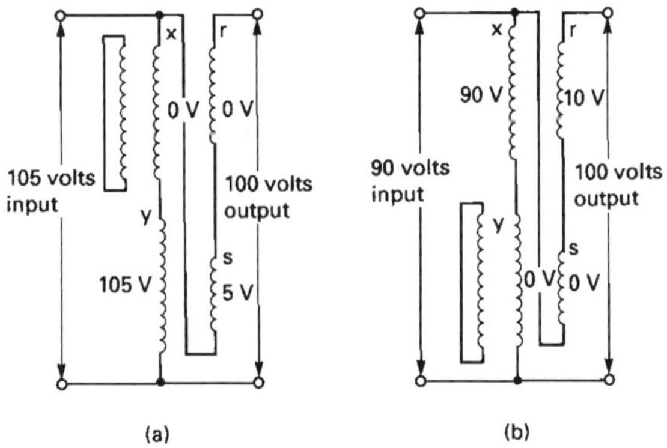

Figure 6.9.31 Moving coil automatic voltage regulator.

buck-boost can be provided by choosing a suitable number of turns for the two coils r and s.

The moving coil is driven by an induction disc motor, or, on the larger sizes, an induction motor, controlled by a voltage measuring relay. The advantages of a moving coil regulator are that it is very robust, does not introduce significant harmonic distortion and has no conductive moving parts. The disadvantage is that it does have mechanical moving parts.

6.9.2.5 Motors

The operation of electric motors are achieved by the interaction of electromagnetic field systems arranged in such a manner as to cause a rotational torque. The different winding arrangements distinguish one form of motor from another. Further details are available in Ref. 1.

6.9.2.5.1 Direct current motor

A dc motor derives its rotational torque from the interaction of electromagnetic fluxes produced by the fixed stator field winding and the rotor winding. A commutator reverses the current direction in the rotor conductor as it moves from the influence of one pole to another, so that the same rotational direction is maintained. The operating characteristics of speed and torque are shown in Figure 6.9.32.

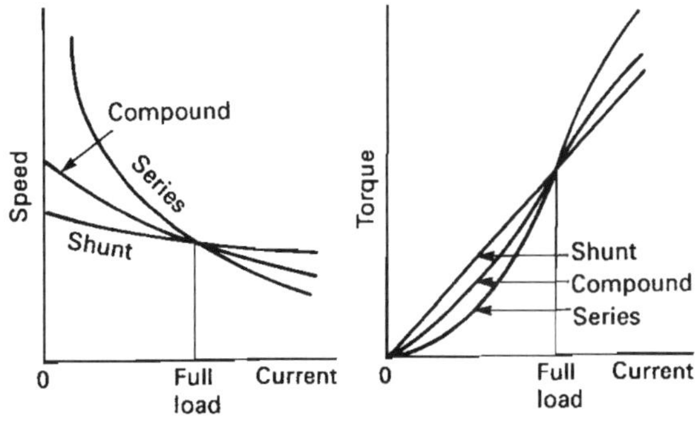

Figure 6.9.32 Speed and torque characteristics of a dc motor.

6.9.2.5.2 Alternating current motor

The ac induction motor is used extensively in both single-phase and three-phase forms.

6.9.2.5.2.1 Three-phase induction motor

A three-phase induction motor has a primary winding, usually the stator, which produces a rotating electromagnetic field. The speed of rotation, n, is proportional to the frequency of the supply, f, and inversely proportional to the number of pairs of poles, p.

The rotor has an evenly distributed arrangement of conductors that forms a closed circuit, through which currents can flow.

At start, the synchronously rotating field of the stator cuts the rotor windings to induce a high rotor current, the field of which interacts with the stator rotating field and produces a rotational torque to accelerate the rotor in the same direction as the rotating field. As the rotor accelerates, the rate at which its conductors are cut by the synchronously rotating field decreases, and therefore so does the rotor induced voltage, current and accelerating torque.

The rotor cannot run at synchronous speed, since the rotor current is induced only where there is relative speed between the rotor conductors and the synchronously rotating field. The difference between the synchronous speed and rotor speed is known as the *slip*.

As the mechanical load on the rotor increases, the rotor slows down, the slip increases, the induced rotor emf and current increase so that a greater torque is developed. This process continues until the *pull-out* torque of the machine is reached and the motor stalls. The operating characteristics are shown in Figure 6.9.33.

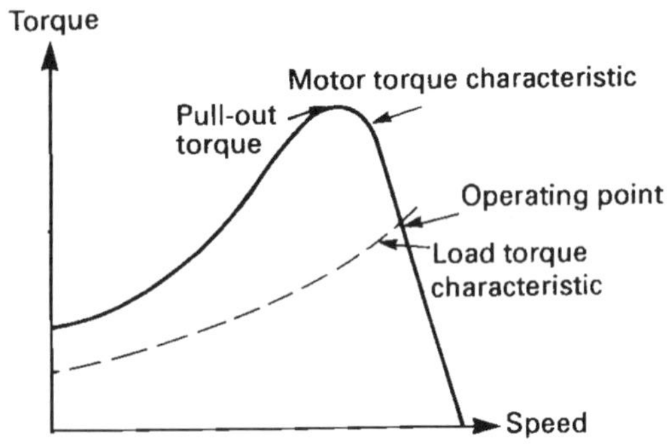

Figure 6.9.33 Induction motor torque characteristics.

Three-phase induction motors are available with either a squirrel cage or slip ring rotor winding.

A *squirrel cage rotor* comprises solid conductors connected together at each end of the rotor. This robust design enables it to be built for reliable service at a relatively low cost. It is therefore a general purpose motor. Because of the high starting current when switched direct on-line, a squirrel cage motor may cause a dip in the supply voltage, and this may adversely affect other equipment. The starting current can be limited by *star/delta* or *auto-transformer* methods.

The star/delta method connects the windings initially in star so that the line voltage divided by $\sqrt{3}$, that is approximately

58 per cent of the line voltage, is applied to each winding. When the motor has run up to speed, a changeover switch is operated to connect the windings in delta so that the full line voltage is applied to each winding.

The starting current is limited in the auto-transformer method by connecting the motor to the auto-transformer tappings, which are adjusted to reduce the applied voltage to about 50–80 per cent supply voltage as determined by the extent to which the starting current needs to be limited.

A *slip ring rotor* has conductors wound in slots, the ends of which are brought out to slip rings. An external resistance is connected via the slip rings to limit the starting current. Its value is slowly reduced as the speed of the motor increases until the slip rings are shorted out.

6.9.2.5.2.2 Single-phase induction motor

The single-phase winding on the stator does not produce a rotating but a pulsating field. Consequently the rotor, which is usually squirrel cage connected, will not start. A pulsating field can be considered to comprise two fields of equal magnitude rotating in opposite directions. So if the rotor is 'artificially' started in one direction or the other, it will continue to rotate in that direction. This initial start is effected by adding a start winding, electrically a few degrees out-of-phase with the main stator winding. The field moves from the start winding to the main winding, and so the rotor receives a pulse start torque. The start winding is switched out of circuit when the rotor is up to speed. The single-phase induction motor displays similar characteristics to those of the three-phase induction motor.

Another type of single-phase motor has a commutator fitted to its rotor to give a pole and armature system identical to that of a series dc motor. Such a dc motor will produce unidirectional rotation on ac because its field and armature currents change direction simultaneously. It is, therefore, termed a *universal motor* and is commonly used in small sizes.

6.9.2.5.2.3 Motor control and protection

The control equipment for motors should provide safe and efficient operation of the motor and its associated mechanical drive. This covers the aspects of starting, stopping, reversing and speed control, tripping the motor automatically under abnormal and fault conditions, and the isolation of the motor and control gear for maintenance purposes.

In general, motor protection is required against the conditions of overload, open circuit, low voltage, overcurrent and earth-fault.

To allow for both starting current and short duration peak loads, the overload trip device incorporates an inherent time delay using a dashpot or bimetallic strip. To cater for the thermal loading conditions of the motor, the bimetallic element can be designed to have similar heating and cooling characteristics to those of the motor it protects.

Thermistor trip devices are sometimes built into the structure of the motor at its various hot spots. In this way, the heating of the trip device is directly related to the actual motor hot spot temperature rather than a simulation of the temperature in terms of the motor current.

Once running, a three-phase motor may continue to run as a single-phase motor after losing one phase of the supply. It will take a higher current than normal, and this may cause a winding to overheat. Therefore, as a safeguard, a relay is included in the protective gear to trip the motor control switch.

A motor contactor opens automatically if the supply voltage falls below its *drop-out* value. BS 5424 and IEC 947-4 set this value at 60 per cent. Similarly, mechanically latched devices, such as a circuit-breaker, can be released by the plunger of a solenoid dropping if the supply voltage falls below a certain value.

6.9.2.6 Cables

For fuller information see Ref. 1.

6.9.2.6.1 Cable materials

The general construction of a cable is shown in Figure 6.9.34.

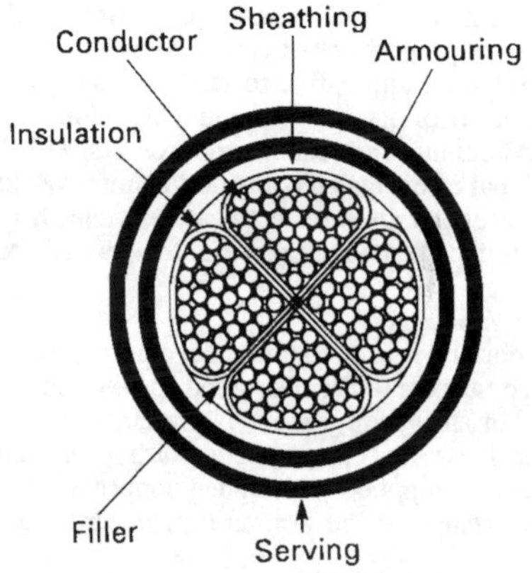

Figure 6.9.34 Typical cable construction.

6.9.2.6.1.1 Conductors

Copper is extensively used as a conductor, but aluminium is a competitive alternative. The conductivity of aluminium is approximately one third that of copper, but by using solid conductors its overall size is brought back to approximately the same size as that of stranded copper.

6.9.2.6.1.2 Insulation

Oil impregnated paper insulation to IEC 55 (BS 6480) will withstand excessive temperature changes, and the lead sheath, used to prevent ingress of moisture, forms a good continuous earth return conductor. The stringent jointing techniques required by paper insulation, its low mechanical strength, its low thermal capacity, and its hygroscopic nature have led to the introduction of thermoplastic alternatives.

At high temperatures, PVC becomes soft; at low temperatures, it becomes rather brittle. Consequently a compromise is necessary, and three types emerge.

General purpose grade has excellent resistance to oils and chemicals. It is self-extinguishing when ignited and the source of flame removed. However, when pulled into conduit there is a tendency for it to bind with other conductors.

Hard grade is formulated with less plasticizer and shows less tendency to soften at high temperatures. Its mechanical properties are preferable. Its improved performance at high temperatures is obtained at the expense of its performance at low temperatures, displaying a cold shatter characteristic below 10 °C.

Heat resisting grade is obtained by replacing the volatile plasticizer in the two previous types by relatively non-volatile plasticizer. This produces a type of PVC cable that retains its flexibility when subjected for long periods to temperatures as high as 85 °C.

Cross-linked polyethylene (XLPE) has high tensile mechanical strength, good chemical resistance and heat stability. It is a superior dielectric to PVC, is physically smaller, has excellent resistance to abrasion, is inflammable, but exudes very little smoke and corrosive gases.

6.9.2.6.1.3 Sheathing

Cable sheathing is used to provide a water barrier and in some cases mechanical protection and earthing.

The excellent corrosion resistance properties of lead sheathing are particularly important. The fatigue resistance of pure lead and its mechanical strength can be improved by the addition of small proportions of certain alloys.

The use of an alternative to lead has always been investigated because of its weight, and aluminium is an obvious choice. Aluminium extrusion produces a sheath less flexible than lead, but a corrugated construction improves its flexibility. Care is necessary in service to prevent sheath corrosion, as aluminium occupies a high place in the galvanic series.

6.9.2.6.1.4 Armouring

Cable armouring is used to protect against mechanical damage. Steel tape armour provides good protection against such damage, but steel wire armour is preferable where additional longitudinal stresses may occur during installation. Wire armour can be supplied with tinned copper wires to lower the overall resistance of the armouring for earth continuity purposes.

Single-core cables for ac systems do not usually have armour because of the induced losses in the armouring. Where armour is necessary, non-magnetic tapes or wire must be used.

Aluminium type armouring has the advantage that it is lighter in weight and it affords a higher conductivity. However, care needs to be taken before installation that the PVC over-sheath is not damaged, as this would allow ingress of moisture which could lead to corrosion and ultimately the loss of the earth return path.

6.9.2.6.1.5 Serving

The life of a cable may depend upon the degree of overall protection against chemical corrosion, electrolytic action and mechanical damage. Bituminized paper tapes are applied immediately over lead sheath cables followed by cotton or hessian tapes.

For unarmoured cables, PVC serving is the usual form of protection, and for armoured cables, an extruded PVC serving may be necessary in highly contaminated soil.

6.9.2.6.2 11 kV cables

Three core, paper insulated, oil impregnated, belted construction, lead alloy sheath and galvanized steel wire armouring (pilcswa) cable, IEC 55 (BS 6480 Part 1), has been used extensively. The conductors can be either stranded copper or solid aluminium (sac) and the armouring can be steel tape (sta).

The introduction of aluminium sheathed cable, IEC 55 (BS 6480 Part 2), as an alternative produces a cheaper cable. It is less flexible but is sufficiently robust to dispense with armouring. The sheath can be of either the straight or corrugated form and, because of its low weight and easy jointing, its installation costs tend to be low.

PVC and cross-linked polyethylene insulated cables are being used more extensively.

6.9.2.6.3 Medium voltage cables

PVC insulated, single wire armoured, PVC sheathed (PVC swa PVC) cables, to IEC 502 (BS 6346), are being superseded by XLPE swa PVC cable to IEC 502 (BS 5467). The jointing procedures are similar to those of PVC.

6.9.2.6.4 Wiring cables

The traditional rubber insulated cable, to IEC 227 (BS 6007), has been superseded by the considerably smaller equivalent PVC cable to IEC 227 (BS 6004). With insulation up to 500 V, it can be obtained either in single solid core or flat twin core PVC insulated, PVC sheathed with bare copper earth continuity conductor. The PVC insulation is designed to have a higher tensile strength, higher resistance to deformation, and a higher insulation resistance than the sheath. The sheath compounds are usually formulated to provide good abrasion resistance and yet have easy tear properties to facilitate stripping at terminations.

6.9.2.6.5 Mineral insulated cables

Mineral insulated cable comprises a solid conductor embedded in densely compacted magnesium oxide and contained in an extruded aluminium or copper tube which forms the sheath of the cable. The compressed mineral oxide will withstand high temperatures, and the metal sheath is a good conductor of heat, so the cable has high current ratings and the ability to withstand some exposure to fire. Due to the hygroscopic nature of the insulation, it is necessary for the ends to be carefully sealed. Where additional protection is desired for corrosive environments, the cable can be supplied with an overall PVC sheath. The cable does have a low impulse strength, and if voltage spikes are likely to be present in the electrical supply, it may be necessary to install surge absorbers.

6.9.2.6.6 Cable selection

There are three parameters to be taken into consideration when selecting a cable size:

- thermal rating;
- voltage drop;
- short-circuit rating.

The operating temperature of a cable should be less than the thermal rating of its insulation. It is conditioned by the environment, e.g. whether it is installed above or below ground and the proximity to other current carrying cables. IEC 364 and IEE Regulations apply factors for the cable installation environment in order to assess its current rating.

The voltage drop requirements are that, from the origin of the circuit to any point in the circuit, they should not exceed 2.5 per cent of the nominal voltage at the designed current, disregarding starting currents. However, the voltage drop of cable installations above 1 kV rating is usually not significant.

Subsequent to a system fault, large currents set up thermal stresses and electromagnetic forces in the cable. The adiabatic relationship is expressed by:

$$I = Sk/\sqrt{t}$$

where S is the cross-sectional area in mm^2
I is the rms fault current
t is the operating duration of the fault current
k is a factor related to the materials of the cable

A graph showing a typical cable adiabatic characteristic is shown in Figure 6.9.35, and, provided the protection operates below this, the cable is adequately protected.

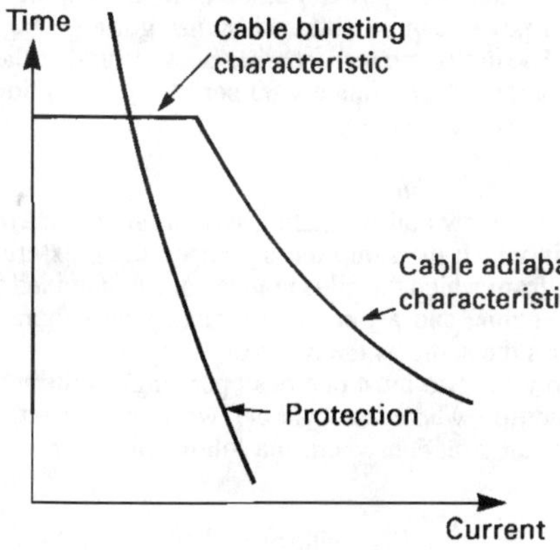

Figure 6.9.35 Cable adiabatic characteristic.

The electromagnetic forces create a cable 'bursting' effect within multicore cables. A typical characteristic is also included on the graph.

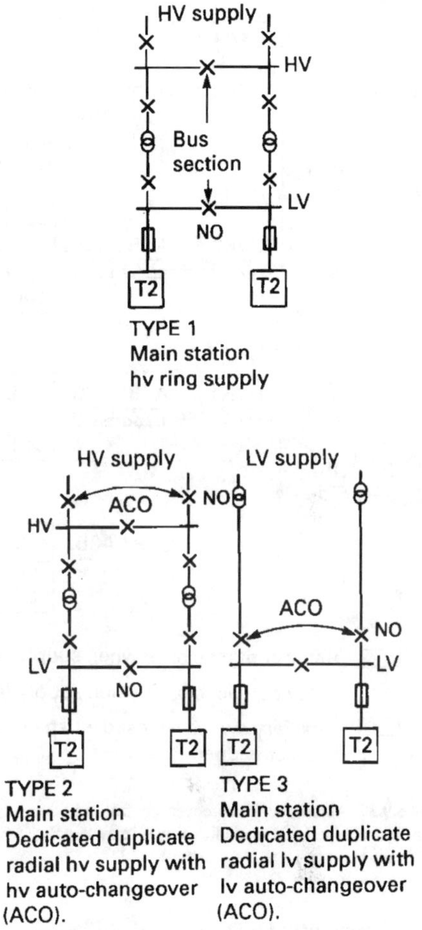

Figure 6.9.36 Station types showing power supply arrangements. NO indicates *normally open.*

6.9.3 Transmitter Installations

6.9.3.1 Transmitter power distribution system

The main uhf transmitting stations are provided with duplicate hv supplies from a primary substation (see Section 6.9.1) fed either radially or from a ring system, or radially from different primary substations. Various arrangements are shown in Figure 6.9.36. At the Type 2 transmitting stations, the lv bus-section circuit-breaker is normally open so that the transformers share the load. At the Type 3 transmitting stations, where auto-changeover facilities are provided, one transformer is on standby.

In the instances where the main station is fed from a ring system Type 1 station, a loss of supply from one section of the ring main would leave the supplies to the site unaffected. No auto-changeover facilities are therefore needed.

A simplified block diagram of the control circuit for the automatic changeover is shown in Figure 6.9.37. An under-voltage relay monitors each of the incoming supplies. In the event of the loss of the selected supply, a delay is initiated. If the supply is not restored during this period, the auto-changeover is initiated provided the alternate supply is available. Subsequent to a successful changeover, the system will not revert to the selected supply when it is restored. This feature prevents unnecessary auto-changeover sequences and consequential interruptions to the programme transmissions.

Remote indication to a regional control centre over a telemetry system provides data as to the method of feeding the station and the availability of the main supplies.

The lv switchboard provides supplies to the station heating and lighting equipment as well as to the transmitters and their auxiliaries. The complete electrical installation must comply with IEC 364, *Electrical Installation of Buildings*, in matters of electric shock, fire, burns and injury from mechanical movement of electrically actuated equipment.[4] The standard qualifies requirements in the matters of safety design, protection against direct and indirect electric shock and against thermal and overcurrent effects. It lays down minimum requirements for electrical isolation and switching and rules for the selection, erection, inspection and testing of electrical equipment.

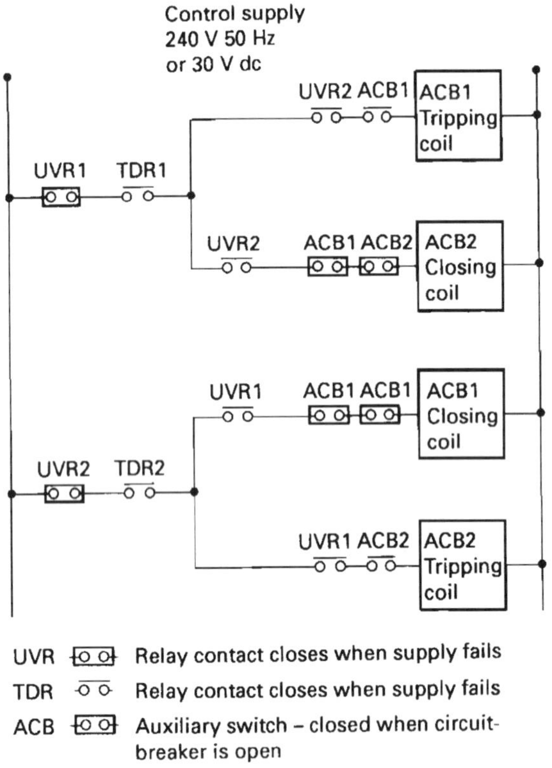

Figure 6.9.37 Automatic changeover control scheme. TDR = time delay relay supply; UVR = undervoltage relay supply; ACB = automatic circuit-breaker.

6.9.3.2 Supply reliability and availability

The number of supply interruptions averaged out over a number of years is referred to as supply *reliability. Availability*, on the other hand, relates to the average duration of such events. Reference 5 deals with the topic of reliability in greater detail.

6.9.3.3 Standby generation

Standby generation is provided at transmitter sites when improved security of the power supplies is required. Should the main supply be interrupted or deviate outside predetermined limits, the plant will start automatically and take on load.

6.9.3.3.1 Engine performance

The vast majority of standby generators are diesel driven. Air is taken into the cylinder through the injection valves and compressed. Fuel is introduced by the injector at the top of the compression stroke where it mixes with the air in the cylinder. The engines can be two-stroke or four-stroke cycle and are driven at synchronous speed (i.e. 1500 rev/min for a two-pole alternator).

The engine speed is controlled by the governor, which can be of mechanical, hydraulic or electronic design. The accuracy of the governor is defined in ISO 3046 (BS 5514, Part 4). The requirements are for the steady-state response not to exceed a ±1 per cent tolerance of the rated frequency, and the dynamic response to a sudden load change not to exceed 15 per cent frequency deviation and to return to a steady-state condition within 15 s and within 5 per cent frequency tolerance.

The power rating of the engine is important and is defined in ISO 3046 (BS 5514, Part 10). *Overload power* is defined as the ability to deliver an additional overload capacity of 10 per cent for one hour in any 12 consecutive hours of continuous running without detriment. The power rating is very much affected by altitude.

An important criterion of engine design is its ability to accept a proportion of full load from an initial cold start. This will, to a large extent, depend upon the type of aspiration of the engine and the inertia of the set.

The energy in the exhaust gases can be used to drive a turbocharger to compress more air into the cylinder and hence to produce increased power output from the engine. Further additional power can be obtained by using a two-stage turbocharger. The use of the turbocharger does mean that the engine can only accept approximately 80 per cent of its rating in one step from a cold start.

6.9.3.3.1.1 Lubrication

Predominantly, two oil circulating systems are used: a *dry sump* system, in which the sump oil is pumped to an exterior reservoir tank from which the oil is pumped at a maintained pressure into the engine, and a *wet sump* arrangement (Figure 6.9.38) which uses the sump as the reservoir.

Probably the two most onerous operating conditions are the *start* condition where the cylinder wear is greatest, and the *extended run* situation where the lubricating oil may become

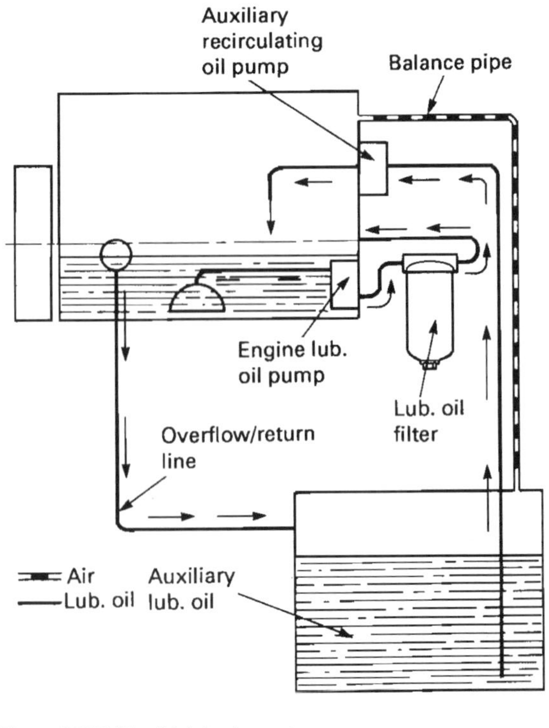

Figure 6.9.38 Diesel lubricating system.

contaminated. The former condition is partly solved by pre-priming systems, and the latter by ensuring that adequate quantities of lubricating oil are available.

6.9.3.3.1.2 Cooling and ventilation

A considerable volume of air is required by a diesel for cooling and combustion, bearing in mind that 10 per cent of a diesel generator set rating has to be dissipated. The performance of the engine depends upon the quality and quantity of the air supplied for combustion. Diesels can be either air cooled, up to about 25 kW, or water cooled, where the radiator can be mounted directly onto or remote from the engine, and a heat exchanger or cooling tower might be included.

Most air inlet and outlet vents to a diesel engine room are of the automatic louvre type, so that it is kept in dry and warm conditions during non-operational periods.

6.9.3.3.1.3 Exhaust systems

An internal combustion engine will operate more efficiently when its exhaust gases are discharged with the designed level of back pressure. The aim therefore is to keep the exhaust system route as short as possible and with the least number of bends. A single silencer should be installed with a tailpipe of specified length. Where silencers are installed in tandem, the first silencer should be located as close to the engine as possible. High frequency noise reduction requires an absorptive design; low frequency noise reduction requires a reactive design. Wide frequency noise reduction requires both or a combination unit.

6.9.3.3.1.4 Mechanical protection

The protection of an engine comprises at least a low lubricating oil pressure and high engine coolant temperature.

6.9.3.3.1.5 Noise levels

The noise level within a few metres of an engine can be in the region of 100 dBA. If the set is housed in a brick building, external noise level is reduced to about 75 dBA. Any further reduction would need acoustic treatment, and its implementation becomes relatively costly.

6.9.3.3.2 Alternator

An alternator, three-phase or single-phase, is designed to the appropriate sections of IEC 34-1 and IEC 34-13 (BS 4999 and BS 5000). The continuous rated temperature should not exceed the particular class of insulation given in IEC 85 (BS 2757). The rating of an alternator, usually given in kilowatts or kilovolt-amps at an assumed power factor of 0.8 lagging, should not be less than that of the engine and should include the overload conditions referred to in Section 6.9.3.3.1.

6.9.3.3.2.1 Excitation system

The alternator field winding on the rotor is usually of the salient pole type and is supplied with dc via diodes mounted on the rotor from the rotor ac exciter winding (Figure 6.9.39). The stator dc exciter winding is controlled by the dc output from the AVR to control the output voltage of the alternator.

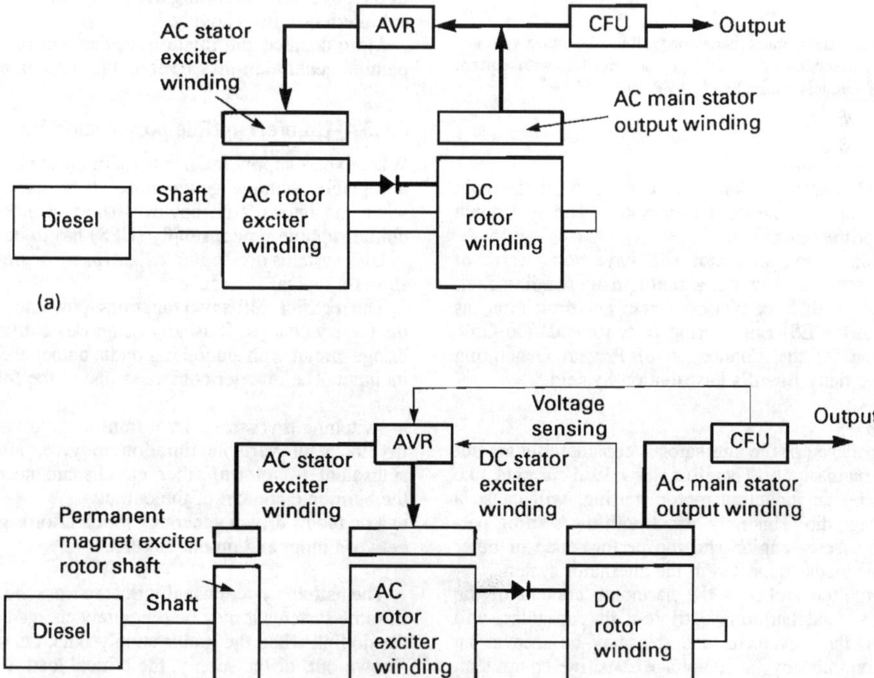

Figure 6.9.39 Excitation: (a) self-excited generator; (b) separately excited generator. AVR = automatic voltage regulator; CFU = current forcing unit.

The AVR source can be taken from either the alternator output (i.e. *self-excited*) or from a rotor-mounted permanent magnet pilot exciter (i.e. *separately excited*). The latter system provides a purely sinusoidal supply to the AVR, whereas the former may have a distorted supply because of harmonics taken by the load. This may affect the designed performance of the AVR. The separately excited system also affords a faster voltage rise time on start-up because it does not rely on the alternator residual voltage.

A further feature of the excitation system is a current compounding circuit to provide additional excitation proportional to the load current, giving a much faster response to load changes such as motor starting currents. Current compounding under fault conditions will ensure that an output of 300 per cent of the rated full load is maintained for a period of up to 10 s (see Figure 6.9.40).

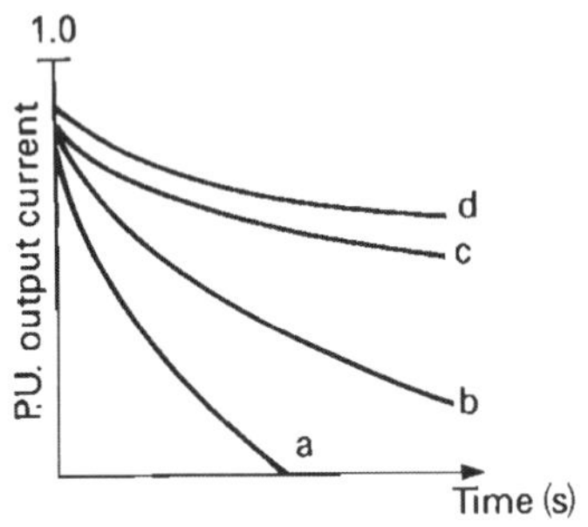

Figure 6.9.40 Response characteristics for various excitation systems: (a) self-excited; (b) separately excited; (c) self-excited with current forcing unit; (d) separately excited with AVR.

6.9.3.3.2.2 Alternator protection

Under fault conditions, provision must be made to trip the main circuit-breaker, suppress the excitation system, and shut down the fuel to the prime mover.

The protection of the alternator will have some form of restricted earth-fault to cater for a stator main winding fault to earth together with three-phase overcurrent protection, as recommended in the ESI Engineering Recommendation G59, Recommendation for the Connection of Private Generating Plant to the Electricity Board's Distribution System.

6.9.3.3.2.3 Load

The source impedance of the alternator is considerably greater than that of the main supply. Therefore, large load currents, like those imposed by an induction motor starting, will cause a substantial voltage dip. This may affect both the starting performance of the diesel engine and the performance of other voltage sensitive loads connected to the alternator system.

An often forgotten aspect is the harmonic content of the connected loads. Odd harmonics, triple-n in particular, will tend to 'flatten' the waveform, and this may be seen as an undervoltage condition by other voltage sensitive equipment. Figure 6.9.41 illustrates this.

This switchgear interlocking must ensure that the generator supply cannot be paralleled with the mains supply. The supply company must be notified of any standby generation installation which also has a mains supply to the site.

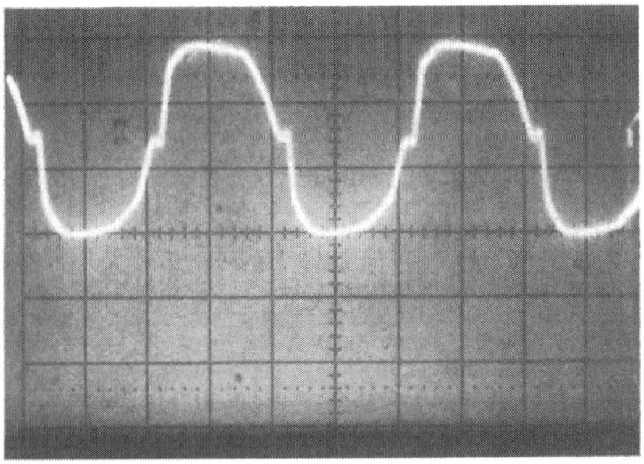

Figure 6.9.41 Standby generator output voltage waveform with harmonic load.

6.9.3.3.2.4 Parallel operation

Parallel operation of private generation with the public supply system at high voltage must comply must comply with the ESI Engineering Recommendation G59.

Before parallel operation, with either the supply system or with other generation plant, the alternator output voltage and frequency have to be adjusted to match the 'running' system.

The sharing of the power and the reactive power is a function of the governor controlling the engine and the AVR controlling the alternator, respectively.

More detailed information on the control and protection of parallel generation may be found in other publications.[1]

6.9.3.4 Uninterruptible power supplies

Where short supply interruptions of approximately 10–15 s are acceptable, standby generation can be used. For application where an interruption however short cannot be tolerated, an uninterruptible power supply (UPS) has to be used.

UPS systems divide into either *rotary* or *static*. Figure 6.9.42 shows a typical system.

The rectifier fulfils two functions: power to the inverter and to the battery charger. It usually comprises a thyristor or transistor bridge circuit with smoothing on its output and filter circuits on its input. The inverter comprises one of the following:

- switching thyristors and/or transistors to switch the dc input in blocks of variable duration to give, after smoothing, a sinusoidal waveform; filter circuits are necessary to reduce the harmonic content of the output;
- a dc motor driven generator which affords galvanic isolation between input and output circuits.

The battery system usually floats across the dc link, although in some systems it may be separately charged and connected to the dc link when the mains supply fails. On an interruption or 'brown-out' of the supply, the critical load is maintained from the battery via the inverter for anything up to approximately one hour. Should autonomous operation be required for longer than the battery capacity, then standby generation is necessary.

Should the UPS fail, the by-pass switch, which is usually a static device, will change to the main supply. In the event of a

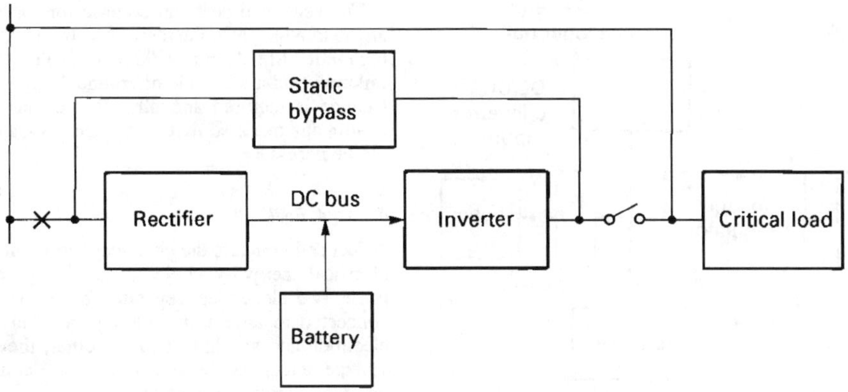

Figure 6.9.42 Uninterruptible power supply.

fault on the UPS output, the by-pass switch will provide a low-impedance source to ensure the operation of the protection. A manually operated by-pass switch can be used for maintenance purposes.

The rotary system is less efficient, and its transient time response is slower than that of the static UPS. The rotary system fulfils the same function as the static inverter. The dc motor drives the alternator and operates from the rectifier under healthy supply conditions and from the battery during loss of supply. The output frequency is controlled by feedback to the dc drive, the voltage via the alternator field circuit. A rotary UPS has other more sophisticated motor-generator arrangements than the one described.

Many critical loads comprise switched-mode power supplies (SMPS) which generate a 'peaky' waveform. It is customary to take into account the peak current (or *crest* factor) rather than the average current to arrive at the appropriate UPS rating. For this kind of load, the rotary systems require only the alternator rating to be increased; the static system will require derating. A rotary system also affords complete electrical, galvanic isolation from the supply. A static system has an advantage in weight and space requirements, and the cost per kVA installed tends to be less.

This selection of either a rotary or a static UPS will depend on the characteristics of both the critical load and the supply. The relative life cycle costs should also be taken into account. Other technical matters are dealt with in greater detail elsewhere.[6]

6.9.3.5 Alternative sources

In the wake of the energy crisis, there followed considerable research into, and development of, alternate energy sources. More recently, alternate sources are being considered for low-power remote transmitter sites where the cost of providing mains supply is becoming prohibitive.

6.9.3.5.1 Solar power

A photovoltaic cell converts solar radiation into electrical energy. The intensity of the sunlight, i.e. the isolation level, is about 1350 W/m^2 at the distance of the earth from the sun. The maximum power that can be delivered to an external load is typically up to 150 W/m^2 of solar panel area with up to 15 per cent efficiency and 0.5 W per cell. A typical voltage/current characteristic is given in Figure 6.9.43.

Modules of cells connected in series of parallel can give the required voltage and current outputs. Solar energy variations during the day/night cycle and throughout the year reduce the average insolation level. A battery storage system is therefore necessary and determines the operating voltage of the solar cell. A charge regulator is incorporated to protect the battery from overcharging. A diode is included to prevent reverse current and the discharging of the battery. A typical arrangement is shown in Figure 6.9.44.

Research is being conducted into a photo-electro-chemical cell. This uses light energy to drive a chemical reaction which has a capacity to store energy during cloudy or night periods.

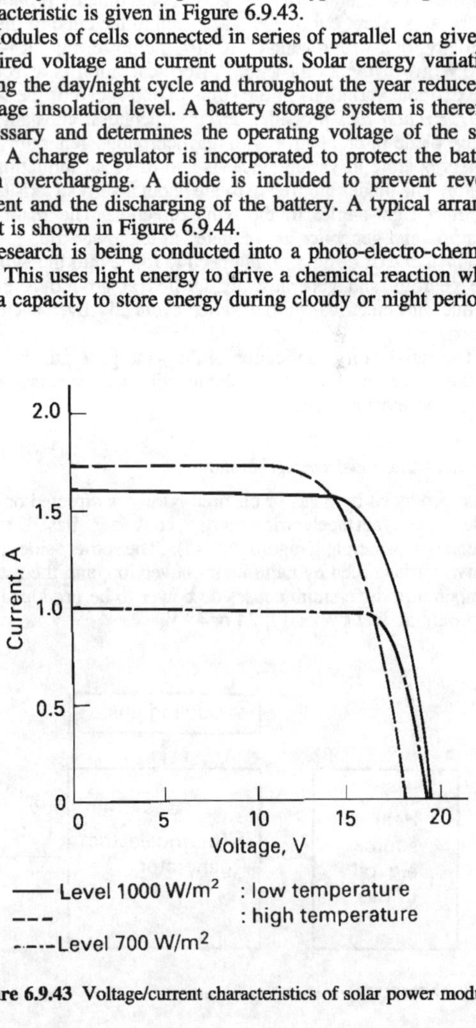

Figure 6.9.43 Voltage/current characteristics of solar power module.

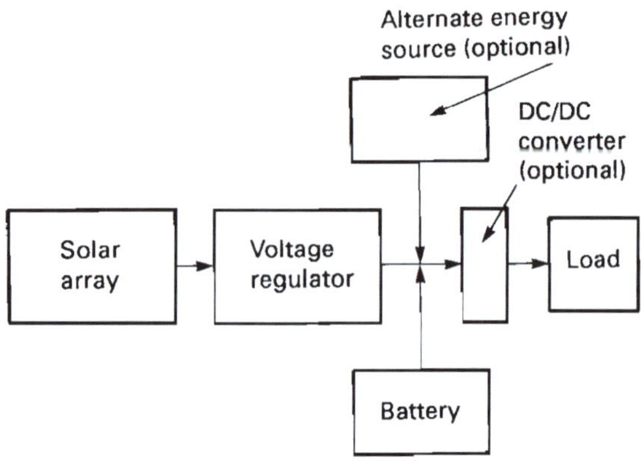

Figure 6.9.44 Solar generator system.

6.9.3.5.2 Wind power

The power available to a wind generator is directly proportional to the area swept out by the blades and to the cube of the wind velocity. In a strong wind of 10 m/s, the power output is about 250 W/m^2 of the swept area. Power generation is affected by the extreme variations in the strength and speed of the wind.

Horizontal axis propeller designs operate at constant speed using blade pitch control with a synchronous generator, while others operate at variable speeds and drive an induction generator. The blades are free to rotate about a vertical axis to keep them at right-angles to the wind direction. The main shaft, gearbox and generator are mounted at the tower top.

Vertical axis designs do not need to rotate into the wind, and the gearbox and generator are located at ground level. The torque and electrical output varies cyclically over each revolution.

The satisfactory application of the wind generator is limited to the selection of a suitable site location and to offsetting the high maintenance costs.

6.9.3.5.3 Thermoelectric generators

Heat produced by a gas or oil fired burner is directed onto one side of a thermoelectric energy converter based on the Seebeck[1] principle (Figure 6.9.45). The other side of the converter is cooled by natural air convection, and the resulting temperature differential causes dc power to be produced up to approximately 1 kW at 12, 24 or 48 V.

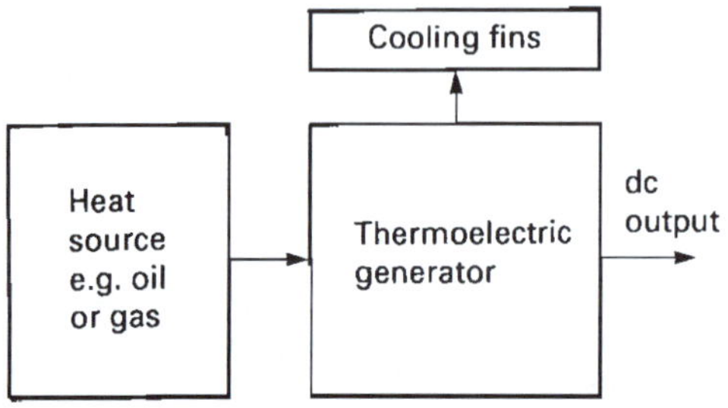

Figure 6.9.45 Thermoelectric generator.

The generated heat can be used for space heating. As there are no moving parts, the thermoelectric generator is silent, has a high reliability of some 9000 hours MTBF (mean time between failures), and a life cycle of around 20 years. Preventive maintenance is required annually. If necessary, it can be run in continuous mode so that a dc battery system installation need not be necessary.

6.9.3.5.4 Fuel cells

A fuel cell converts the chemical energy in a fuel directly into electrical energy by an electrochemical process. The cell comprises two electrodes, separated by an electrolyte, which are connected to an external circuit. As fuel is supplied to one electrode and an oxidant to the other, the cell develops a dc voltage across its electrodes. Cells are connected in series to provide the required voltage.[7]

6.9.3.6 Power factor correction

The apparent power taken by an inductive or capacitive load will be greater than the power actually converted into useful energy. The ratio of the two is the power factor:

$$\text{power factor} = \text{useful power (W)/apparent power (VA)}$$

By keeping the power factor as near to unity as possible, the voltage regulation of any distribution system will be improved, the system losses will be reduced, and any tariff penalties will be avoided.

It is usual to improve the power factor of an installation by connecting capacitors in parallel with the system. Capacitors are reliable, efficient and have few maintenance requirements.

It is preferable for the capacitor to be installed as near as possible to the connected loads, but due to the diversity in the locations of the load, it will often be more economic to use bulk correction capacitors at the source switched in steps as the power factor of the loads varies.

6.9.3.7 Earthing

The earthing of electricity supply systems is usually governed by the Electrical Utility. It requires system earthing so as to restrict the potential on each conductor to a value appropriate to the insulation level, and to ensure, for safety reasons, efficient and fast operation of the protection in the event of an earth fault.

The consumer is legally responsible for providing an earth to comply with the requirements. However, the Electricity Utility will normally offer an earth terminal connection, provided an indemnity is agreed by the consumer absolving the Utility from the responsibility of any consequences in the event of the loss of that earth.

Some Electricity Utilities are implementing a *protective multiple earthing* (pme) policy which utilizes a combined neutral and earth conductor. In this way an earth terminal can be provided at a much reduced cost. The neutral is connected to the earth electrode system at or near the supply transformer, at other intermediate points, and at the remote of the pme system.

Some earthing systems are shown in Figure 6.9.46. The earth-fault current will be limited by the impedance of the earth-fault loop path, and this comprises the supply transformer secondary phase winding, the line conductor to the point of the fault, the earth protective conductor of the consumer's installation, and the return path to the transformer neutral.

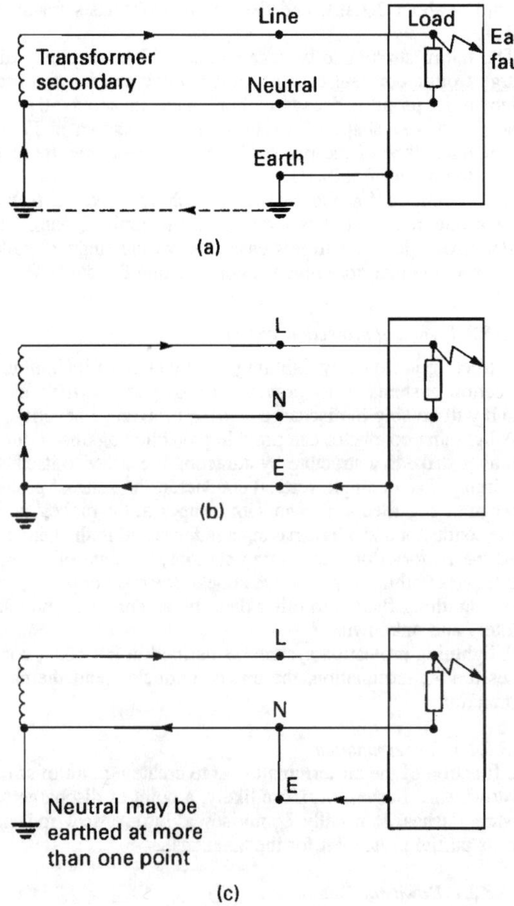

Figure 6.9.46 Earth loop impedance paths: (a) earth return through soil (IEC 364 TT system); (b) system with earth conductor (IEC 364 TN-S system); (c) protective multiple earthing (IEC 364 TN-C-S system).

The IEC and the *IEE Wiring Regulations* allow overcurrent devices, i.e. hrc fuses or miniature circuit-breakers, to protect for earth-faults provided that, with a fault of negligible impedance, the earth-fault impedance is sufficiently low to produce a current that will cause the overcurrent protection to clear the fault within a stipulated time..

Where the earth-loop impedance is too high to operate such overcurrent protection, a current operated rcd (Section 6.9.2.2.4) must be used. To avoid the risk of serious shock, the maximum voltage which may be sustained on exposed metal-work should not exceed 50 V. The effects of current on the human body as given in IEC 479 are shown in Figure 6.9.47, together with a superimposed characteristic of a 30 mA rcd.

The sensitivities of the rcd range from 10 to 500 mA, but they do not afford a cut-off current characteristic. However, the overall tripping time is within 30 ms, which would prevent fibrillation of the heart.

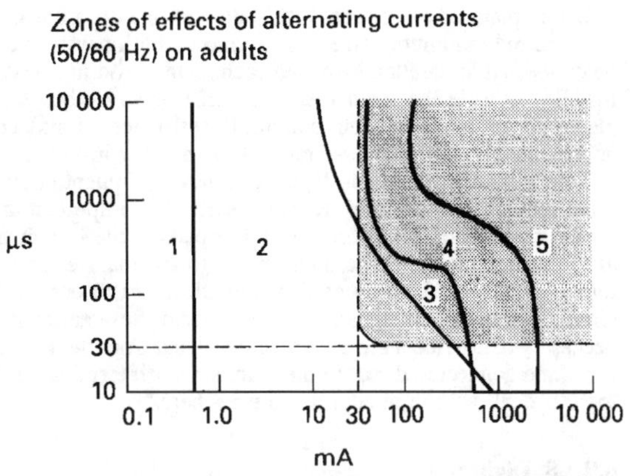

Figure 6.9.47 The effects of current on the human body (from IEC 479). In area 1, there is usually no reaction effect; in area 2, there is usually no pathophysiologically dangerous effect; in area 3, there is usually no danger of fibrillation; in area 4, fibrillation is possible (up to 50 per cent probability); in area 5, there is fibrillation danger (more than 50 per cent probability); in the shaded area, protection is afforded by a 30 mA current operated rcd at 30 ms.

6.9.3.7.1 *Earth electrode system*

The resistance of the connection to true earth potential of an electrode system depends upon the resistivity of the surrounding soil. The composition of the ground will give a general indication of its resisitvity. For example, marshy ground will be between 2 and 3.5 ohms per metre and rock will greater than 100 ohms per metre.

To obtain a low overall resistance, the current density in the soil in contact with the electrode should be as low as possible. This can be achieved economically by having one electrode dimension large in comparison with the other two, such as a rod or strip. The curves in Figure 6.9.48 give an indication of the resistance of plate, rod and strip electrodes installed in homogeneous ground which has a soil resistivity of 100 ohm-metres.

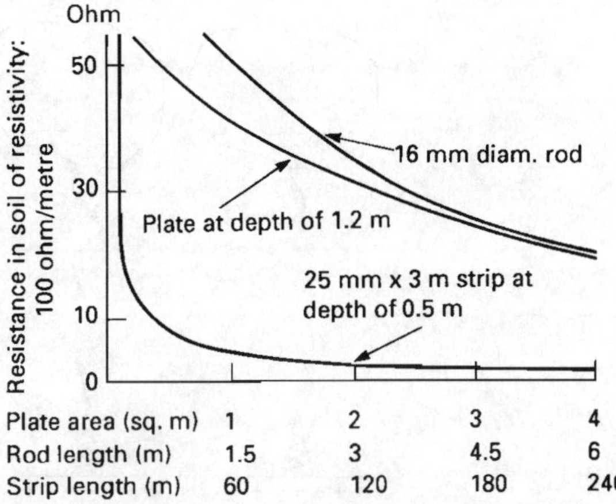

Figure 6.9.48 Earth electrode resistance.

If one plate or rod of reasonable dimensions does not give the required minimum value of resistance, additional units can be connected in parallel. Provided each plate or rod in a nest is installed outside the main resistance area of any other earth electrode in the nest, the combined resistance is reduced approximately by the inverse ratio of the number installed.

Telecommunications and high frequency equipment circuits often require a 'noise free' reference earth. It is important that the earth reference points for the various parts of the system are at the same potential. The connections to the main earth terminal should not form an earth-loop path for earth current to circulate and induce signals into the system. Sometimes it is necessary to provide a separate earth insulated from the general earth and connected direct to the common earth terminal. This topic is dealt with in greater detail elsewhere.[8]

6.9.3.8 Lightning

Lightning is an electrical discharge created by an atmospheric condition that disturbs the normal electric field balance between the earth and the ionosphere.

6.9.3.8.1 The nature of lightning

It is generally accepted[9,10] that there are four main types of lightning, i.e. positive and negative downward flashes and positive and negative upward flashes. Of these, the negative downward flash is the most common. Accordingly, only this type will be described (Figure 6.9.49).

The earth carries a positive charge, and the ionosphere, up to 50 km from the earth's surface, carries a negative charge. A negative downward lightning discharge commences when a negative potential leader advances towards the earth in steps of about 50 m seeking the path of greatest conductance.

The high potential difference between the charged cloud and earthed objects produces upward streamers. These currents can amount to a few microamperes and can occur for long periods without a lightning flash occurring.

The establishment of a complete path between the leader and the streamer produces a return stroke of current from the earth to the cloud, the magnitude of which can be between 25 A and 200 kA. The time taken for the stepped leader to reach the ground is about 200 ms, and the return stroke lasts for about 100 μs.

The return stroke can be regarded as a high frequency discharge from a constant current source which can reach a peak value in 1.2 μs as it decays to half value in about 50 μs. A generalized waveshape of a return stroke is shown in Figure 6.9.50. The slope of the rise of the current generates frequencies of the order of 500 kHz.

It is common for the lightning flash to produce further leaders and return strokes over the same ionized path. The leader stroke does not in this case involve the stepped leader process and is therefore much faster reaching the earth.

6.9.3.8.2 Lightning protection system

If a structure is struck by lightning, the function of its lightning protection systems is to provide an adequate path to earth which will be able to discharge a strike of average severity.

A lightning conductor can provide protection against a direct lightning strike to a structure by attracting the strike to itself. In the simple case of single vertical conductor, the zone of protection can be regarded as a cone with its apex at the highest point of the conductor and a base having a radius equal to the height.[11]

At the majority of transmitting stations, the area of the site buildings is within the protective zone of the mast or tower, and direct lightning flashes to other than the antenna structure are unlikely and unknown.

A lightning protection system, as defined in BS 6651, comprises the air termination, the down conductor, and the earth termination.

6.9.3.8.2.1 Air termination

The function of the air termination is to create a point of strike for downward flashes, or, more likely, a point of discharge for upward flashes. It usually comprises a 1 m vertical rod and affords partial protection for the antennas.

6.9.3.8.2.2 Down conductor

Where the support for the antennas is a metal tower or mast, a down conductor is unnecessary, but where the tower is of concrete construction, it is important to use the steel reinforcement and to ensure that it is continuous. It should also be

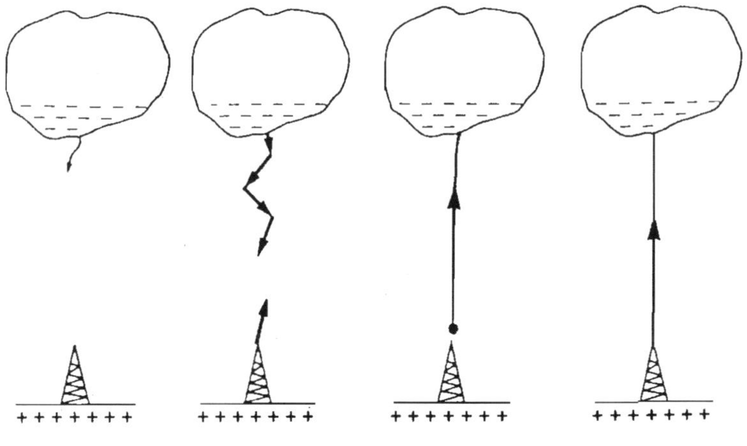

Figure 6.9.49 Lightning flash development.

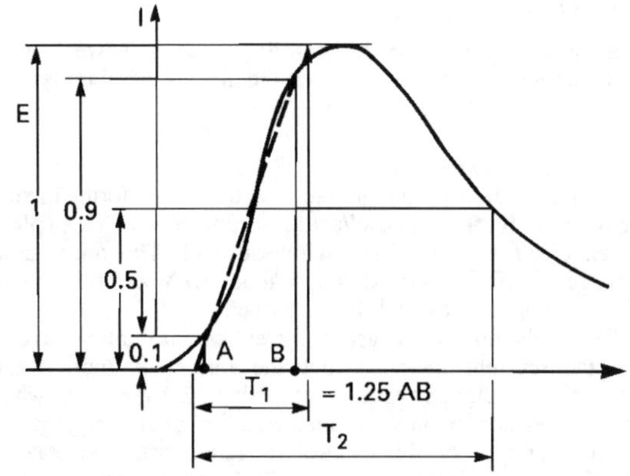

Figure 6.9.50 Generalized waveshape of return stroke lightning current. T_1 is typically 1.5 μs, T_2 typically 50 μs.

supplemented with down conductors and adequately bonded at vertical intervals in a manner recommended by BS 6651.

6.9.3.8.2.3 Earth termination

Should the antenna support structure receive a direct strike, the lightning current will disperse through the steel structure, or the concrete reinforcement, down to earth. As the current is discharged, it produces a voltage across the surge impedance of the earth electrode system which will momentarily raise the site potential to a high value above true earth.

Provided the bonding is correct, a low surge impedance for the earth electrode system would seem unnecessary, because the whole earth system will momentarily rise in potential. However, where other utility services, e.g. electricity supplies and telecommunications, enter the site, a remote earth is consequently introduced, so that longitudinal voltages are liable to cause insulation failure and high discharge currents.

The earth resistance, therefore, must be kept low and preferably below 10 ohms. The use of reinforced concrete as part of the earth electrode system has good effect, and semiconducting concrete should also be considered.

Adequate and frequent *bonding* should ensure that the installation will produce an equipotential cage. This should prevent side-flashing and electric shock to any person in contact with different parts of the installation.

Voltages on the surface of the ground in the vicinity of earth electrodes must be restricted to safe values. This can be achieved by using electrodes that form a ring around the area to be protected. The electrode should be buried deep enough to reduce the surface potential.

6.9.3.8.3 Effects of lightning

The magnetic fields produced by the lightning current can cause mechanical forces, so that sharp bends will tend to straighten, small loops will try to expand, and parallel conductors, with current flowing in the same direction, will tend to move towards each other.

The response of an earth electrode system relative to a lightning stroke depends upon its surge impedance. Reflections of the impulse wave will occur from the mismatch impedance points of the earthing system.

The duration of a series of lightning pulses, and consequently the adiabatic effect on the lightning protection system, is usually within the capability of the size of the system which is designed to cater for the electromagnetic forces.

The rapid change in the current rate will induce voltages in other metallic parts of the installation and unscreened conductors of the electrical circuits. Secondary discharge protection is required as discussed in Section 6.9.3.8.4.

6.9.3.8.4 Insulation coordination

Insulation coordination is necessary to avoid the breakdown of system equipment. It relates the designed impulse strength of the equipment to the parameters of the transient overvoltages to which the equipment is likely to be subjected.

The transient may be due to any system disturbances such as the effects of lightning, switching operations or fault events.

To avoid the installation impulse strength being exceeded, the energy in the transient must be either safely absorbed, i.e. *voltage clamped*, or diverted to earth, i.e. *crowbarred.*

6.9.3.8.4.1 Voltage clamping devices

Voltage clamping devices utilize filter circuits that have the effect of changing the transient from a high peak value for a short time to a sufficiently lower peak value for a longer time. These devices have a non-linear impedance that is dependent upon either current flowing through or the voltage applied across the device.

Selenium cells are based on rectifier application technology, but the extent of application has diminished in favour of more advanced materials.

Zener diodes are silicon rectifier technology and have very effective clamping characteristics. They are available in voltage ratings down to a few volts but have limited energy dissipation capability.

Varistors use variable resistance material, such as silicon carbide or metal oxides.

The relationship between the current, I, through the device and the voltage, V, across its terminals is given approximately by:

$$I = kV^x$$

where x represents the non-linear characteristics of conduction. With higher values of x, the clamping is more effective.

The first varistors used silicon carbide which has a low x value. It therefore has a wide application in high voltage surge resistors, but it needs a series gap to stop the power follow-through current when the voltage returns to the normal operating value.

More recently, metal oxide varistors have been developed using zinc oxide which gives x values of up to 30.

The quiescent power consumed by the varistors must also be acceptable to other circuit constraints. High x value devices consume low power at the rated operation voltage, but a small increase in the voltage can cause a large increase in the consumption. The characteristics of different types of transient suppressor devices are shown in Figure 6.9.51.

6.9.3.8.4.2 Crowbar devices

The devices designed to divert the current to earth by a short-circuiting action operate on the principle of a change from high to low impedance by a switching process which can be inherent in the device, such as a spark gap, or can be triggered by a sensing device, as in the case of thyristor operation.

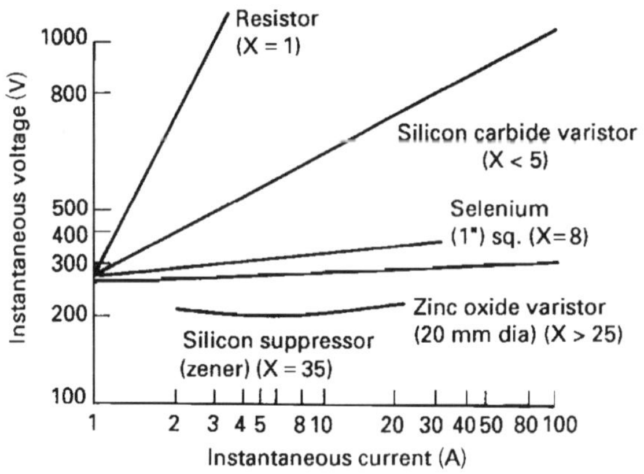

Figure 6.9.51 Characteristics of transient suppressors.

Gas discharge devices are particularly used in the protection of communications circuits where there is no problem of power follow-through current. The devices comprise single and triple metallic electrodes within a hermetically sealed tube containing gas at a reduced pressure, to allow a wider gap spacing. The triple unit enables both the line and the return wires to be connected to earth within the common tube instead of using two separate single gas protectors. The spark over from each line, therefore, occurs at the same instant, and this prevents transverse voltages in the circuit.

6.9.3.9 Battery equipment

Batteries are used for many different applications including:

- engine starting;
- maintaining supplies for telemetry equipment;
- back-up for solar and wind power systems;
- switchgear tripping and control;
- uninterruptible power supplies;
- emergency lighting;
- fire alarm systems;
- hand-held portable equipment.

6.9.3.9.1 Secondary cells

The secondary batteries most widely used in power system installations are the lead/acid and the nickel–cadmium types.

6.9.3.9.1.1 Lead/acid cells

Lead/acid batteries are manufactured in various forms in sizes up to about 1000 Ah. A lead/acid cell comprises two lead plates in an electrolyte of dilute sulphuric acid. The open-circuit voltage of a fully charged cell is about 2.1 V and this falls to about 1.8 V when the cell is discharged.

The cells may be charged by either constant voltage or constant current methods, but the former is more usual. The charger output voltage is set to about 2.3 V per cell where the charger is permanently connected, i.e. float charging. The specific gravity of the electrolyte varies with the state of the charge reaching a value of 1.22–1.27 when fully charged and falling to about 1.18 when discharged.

A high performance Planté cell has pure lead plates, is ideally suited to standby applications, and will tolerate the level of overcharge when permanently connected to a constant voltage charger. It has an operating life of about 20 years when properly maintained.

Tubular plate cells are used for stationary and mobile applications. They will withstand a large number of charge and discharge cycles, and on standby float charge duty they have a life of about 10 years. Their high rate discharge performance is not very good.

An automotive battery will produce around 3 kW of power for short periods and, provided it is not subject to overcharging or long periods of undercharge, it will have a useful life of 3–5 years. The plate construction, while being rugged and comparatively cheap, is very susceptible to low levels of continuous overcharge and also causes these cells to develop a high self-discharge rate.

The electrodes in the sealed lead/acid type of cell are of a wound construction and made from pure lead. The most significant feature of this design is the means by which, in the overcharge condition, the gases produced by electrolysis are recombined instead of being released to the atmosphere as they are in the conventional vented cell. The useful life of this cell is considered to be approximately 10 years.

The main characteristics of the various types of lead/acid cells are given in Table 6.9.4. Typical charge and discharge characteristics are given in Figure 6.9.52 for high performance Planté cells.

Table 6.9.4 General characteristics of lead/acid cells

	General characteristics			
Type of cell	*Float voltage per cell (to maintain full charge) (volts)*	*Life expectancy on float voltage (years)*	*Life expectancy* charge/discharge cycling performance (cycles)*	*High rate performance*
Planté	2.25	20	500	Medium
Tubular plate	2.25	10–12	800	Low
Flat plate	2.25	10–12	800	Medium
Automotive	2.25	$1^1/_2$	120	High

*Based on discharge to 50 per cent of nominal 10 h capacity followed by complete recharge.

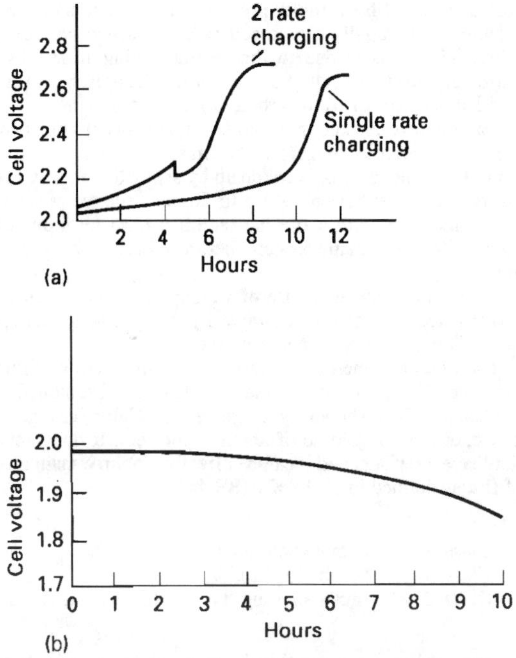

Figure 6.9.52 Characteristics of high performance Planté cells: (a) typical recharge characteristics; (b) discharge characteristics.

6.9.3.9.1.2 Nickel–cadmium cells

The nickel–cadmium cell comprises nickel–cadmium positive and negative plates in an electrolyte of about 30 per cent potassium hydroxide and distilled water. The open-circuit voltage for a fully charged nickel–cadmium cell is about 1.4 V, and this falls to about 1.1 V per cell when the cell is discharged. Nickel–cadmium cells may be charged by either constant current or constant voltage methods. However, in the case of sealed cells, the constant current method is invariably used to avoid overcharging.

Constant voltage charging for the vented type is usually set to 1.45 V per cell, for 'float' operation. Unlike lead/acid batteries, the specific gravity of the electrolyte does not give an indication of the state of charge.

The modern vented nickel–cadmium cell employs a pocket plate construction which is extremely rugged. It has a very long life, exceeding 20 years, and will withstand considerable over- and undercharging without sustaining permanent damage. It is also capable of providing high rate discharge and will tolerate very high charge rates.

As with the sealed lead/acid cell, the essential feature of the design of the sealed nickel–cadmium cell is to achieve recombination of the gaseous products and so sustain the volume of electrolyte. The useful life of the cell is usually about 5 years.

The disadvantage with all nickel–cadmium cells, but more pronounced with the sealed type, is the depression in voltage that can occur when the cells are subjected to repetitive levels of partial discharge followed by recharge. Typical charge and discharge characteristics are given in Figure 6.9.53.

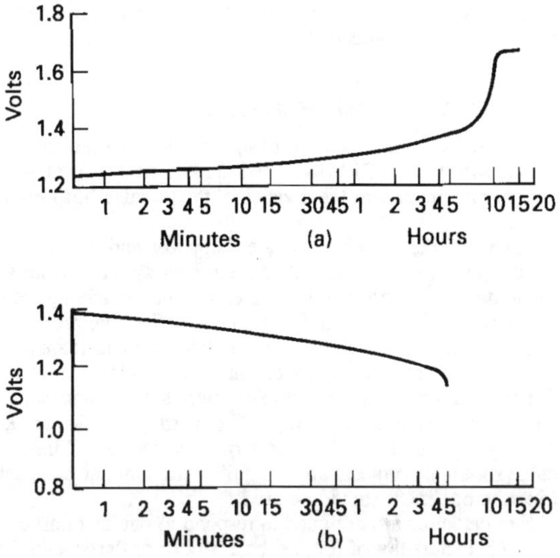

Figure 6.9.53 Characteristics of nickel-cadmium cells at 25°C: (a) constant current charge characteristics; (b) discharge characteristics. Nominal discharge voltage is 1.2 V per cell.

6.9.3.9.2 Battery chargers

Modern constant voltage and constant current battery chargers are usually completely solid-state using either a transistor or a thyristor controlled unit to regulate the output. In addition to voltage control, the chargers are current limited to prevent too high a charge rate when the battery is in a low discharge condition. Additional features such as a boost charge rate, under- and overvoltage, and earth fault alarms are usually incorporated. More recently, and for several reasons, emphasis has been placed on the use of battery charge condition monitors.[12] Additional smoothing is often required for charges associated with telecommunications equipment.

6.9.3.10 Fire alarm and protection systems

Fire and the production of smoke can kill or maim by asphyxiation, irradiation, poisoning or burning. The lack of visibility in smoke filled rooms is also a major threat to human life.

The essence of a successful fire protection system is to minimize the delay between the start of the fire and the action taken to combat it.

The requirement of a fire alarm system is to:

- raise an alarm and indicate a means of escape;
- limit the damage to property by activating fire-fighting equipment.

The protection of property by fire detection, alarm and extinguishing systems may be financially advantageous because of the reduced insurance costs. However, in certain premises it is a legal requirement to provide an effective means to warn against the outbreak of fire.

Fire protection systems divide into three categories: fire detection and alarm systems, portable equipment and automatic extinguishing systems, and emergency lighting.

The use of automatic equipment is necessary to safeguard property that is unattended.

6.9.3.10.1 Fire detection and alarm systems

The alarm system should be designed and installed in accordance with CEN/TC72-EN54 (BS 5839 Part 1, 1988), which covers such matters as the division of the premises into suitable zones.

Conventional systems operate using *on* and *off* detectors located in zones. The advent of intelligent systems using analogue detectors with individual identity and 'watchdog' monitoring will afford rapid and high integrity fire detection.

Each zone system comprises fire detectors and manual call points connected into control and indicator equipment which is fed from a small uninterruptable power supply, and actuates audible and visible warnings. The wiring should be kept entirely separate from all other wiring services. PVC insulated cable should be protected by conduit or trunking, or mics cable (Section 6.9.2.6.5) should be used.

Fire detectors are designed to respond to one or more of the three characteristics of fire, i.e. heat, smoke or flame, and these are summarized in Table 6.9.5.

In some situations, a combination of more than one type of detection may be preferable. Detectors should comply with CEN/TC72-EN54 (BS 5445) for industrial applications.

6.9.3.10.2 Extinguishing systems

6.9.3.10.2.1 Portable equipment

Portable fire-fighting equipment, such as fire extinguishers and fire blankets, complement an automatic fire alarm system. It is important that the appropriate extinguishing agent be used.

6.9.3.10.2.2 Automatic extinguishing systems

The fire detection system initiates an extinguishing system to release an appropriate agent to flood the protected zone.

The main extinguishing agents used at installations containing electrical equipment are carbon dioxide (CO_2), Halon 1211 or Halon 1301.

Carbon dioxide is a dry non-corrosive gas which does not conduct electricity. Its extinguishing effect is obtained by diluting the atmosphere to a point where the oxygen content is no longer sufficient to support combustion.

A typical layout of a carbon dioxide installation, for an hv switch room, is shown in Figure 6.9.54. Carbon dioxide is stored in liquid form under pressure in steel cylinders. When released, it is rapidly discharged as a gas which appears as a white mist. This is due to the mixture being frozen by the extremely low temperature of the gas and the presence of finely divided particles of solid carbon dioxide dry ice.

Carbon dioxide is not poisonous but, at concentrations above 5 per cent volume, judgement becomes impaired, and at concentrations above 10 per cent death by asphyxiation may occur. However, concentrations of up to 50 per cent are required to deal effectively with a fire, so facilities must be provided to lock-off the automatic system before entering the protected zone.

There is an increasing use of vapourizing liquids which are halogenated derivatives of simple hydrocarbons, the halogens being fluorine, chlorine and bromine.

The types commercially available are bromochlorodifluoromethane, BCF, known as Halon 1211 and bromotrifluoromethane, BTM, known as Halon 1301. Halon gas needs a concentration of volume of about 5 per cent. It does not wet nor leave a residue, and it can be effective in three main classes of fire as defined in ISO 8421 (BS 4422):

Class A	combustible material
Class B	flammable liquids
Class C	fires including live electrical circuits

Since a low concentration is required to extinguish most fires, it has a low degree of inhalation and is regarded as a safe agent for human contact. It is more expensive than carbon dioxide.

6.9.3.10.3 Emergency lighting

There are two options offered in BS 5226 Part 1 for the design of emergency lighting, i.e. the defined and undefined escape routes.

A *defined* escape route can be up to 2 m wide, with the centre line illuminated to a minimum of 0.2 lux, while 50 per cent of the route width should be lit to a minimum of 0.1 lux.

An *undefined* escape route covers open areas, and the horizontal illuminance over the whole area should not be less than 1 lux. This latter system of escape route is more common because of the difficulty of keeping a defined route unobstructed and limiting the use of the site for changing circumstances.

Table 6.9.5 Fire detection

Detector	*Type*	*Fire category*	*Best application*	*Integrity*
Heat	a) Fixed temperature elements	A, B, C	For fast heat take off	Good
	b) Rate of temperature rise elements			
Smoke	a) Ionization detectors	A, B, C	For slow heat take off	Care in dusty environment
	b) Optical detectors			
Flame	a) Infrared	B	Flammable environment	Care in location
	b) Ultraviolet			

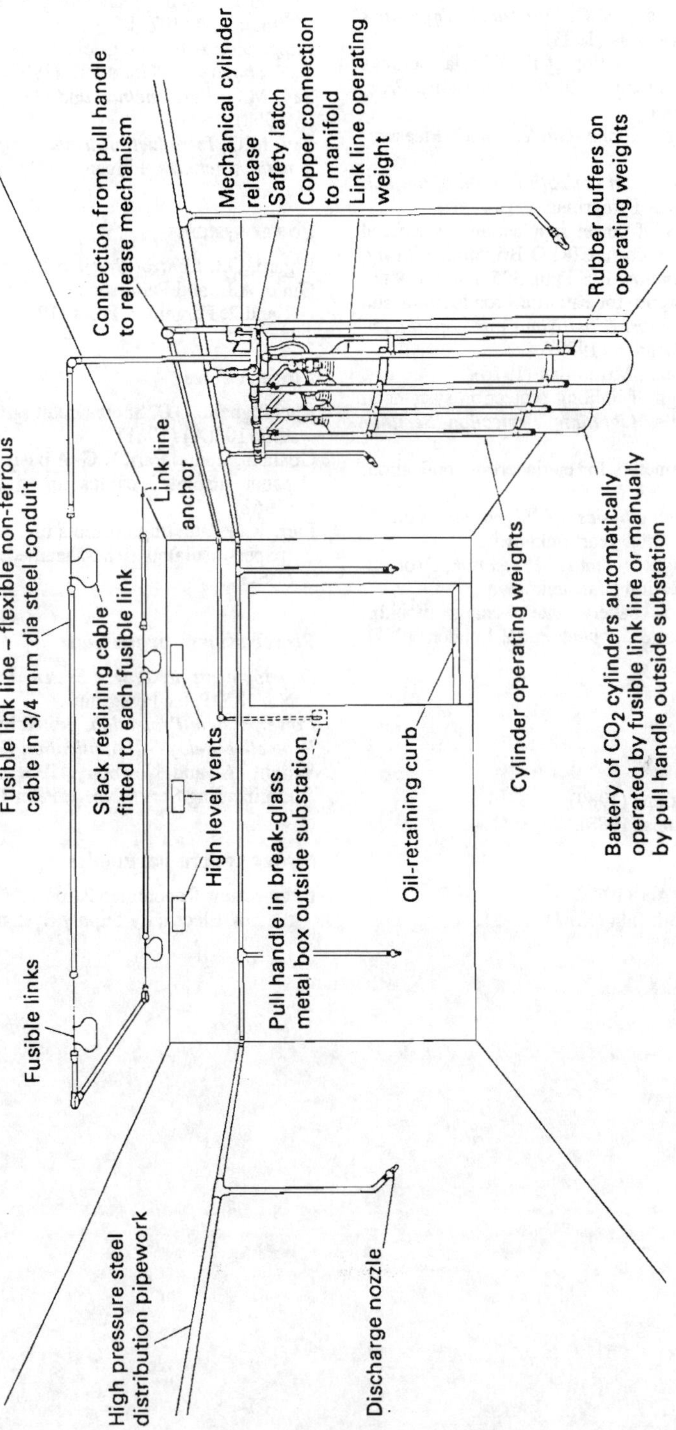

Figure 6.9.54 Typical arrangement of carbon dioxide fire extinguishing installation.

References

1. Laughton, M.A. and Say, M.G. *Electrical Engineer's Reference Book*, Butterworths (1985).
2. Whiting, J.P. Computer prediction of IDMT relay settings and performance for interconnected power systems, *Proc. IEE*, **130**, 139–147 (1983).
3. *Protective Relays Applications Guide*, GEC Measurements, Publ. G-1011A.
4. Jenkins, B.D. *Commentary on the 15th Edition of the IEE Wiring Regulations*, Peter Peregrinus.
5. Whiting, J.P. Reliability of power supplies and systems at broadcast transmission stations, CIRED Brighton, *10th Int. Conf. Electricity Distribution*, IEE Publ. 305, Part 5 (1989).
6. Whiting, J.P. UPS systems for satellite broadcasting and computer installations, *Frost & Sullivan Conf. Uninterruptible Power Supplies*, London (1989).
7. McDougall, A. *Fuel Cells*, Macmillan (1976).
8. Whiting, J.P. The design of lighting protection systems at broadcast stations, *ERA Lightning Protection Seminar* (1987).
9. Electra. Lightning parameters for engineering application, Cigre Study Committee 33.
10. Golde, R.H. *Lightning: Physics of Lightning*, Vol. 1, Academic Press, publication year: unknown.
11. Golde, R.H. *Lightning: Lightning Protection*, Vol. 2, Academic Press, publication year: unknown.
12. Whiting, J.P. A low-cost battery state-of-charge monitor, *ERA Conf. Power Sources & Supplies*, Part 2, London (1987).

Bibliography

UK Legislation

Electricity at Work Regulations (1989).
Electricity Supply Regulations (1988).
Energy Act (1983).
Fire Precautions Act (1972).
Health and Safety at Work Act (1974).
IEE Wiring Regulations, 15th Edn (1981).

Power plant

Central Electricity Generating Board, *Modern Power Station Practice*, **4** (1971).

Fitzgerald, A.E., Kingsley, C. Jr and Umans, S.D. *Electric Machinery*, McGraw-Hill (1983).

Say, M.G. *Performance and Design of AC Machines*, Pitman (1958).

Say, M.G. *Introduction to the Unified Theory of Electromagnetic Machines*, Pitman (1971).

Power systems

Elgard, O.I. *Control Systems Theory*, McGraw-Hill (1967).

Guile, A.E. and Pattersons, W. *Electrical Power Systems*, Vols 1 and 2, Pergamon Press (1977).

Power cables

Buckingham, G.S. Short-circuit ratings for mains cables, *Proc. IEE*, 108(A) (1961).

Gosland, L. and Parr, R.G. A basis for short-circuit ratings for paper insulated cables up to 11 kV, *Proc. IEE* 108(A) (1961).

Parr, R.G. Bursting currents of 11 kV 3-core screened cables (paper-insulated lead-sheathed), *ERA Report F/T* 202 (1962).

Power system protection

Developments in Power System Protection, IEE Conf. Publ. No. 185, Peter Peregrinus.

Power System Protection, Edited by the Electricity Council.

Protective Relays — Application Guide, GEC Measurements.

Wright, A. and Newberry, P.G. *Electric Fuses*, IEE Power Engineering Series No. 2, Peter Peregrinus.

Power system harmonics

Engineering Recommendation G5/3, Limits for Harmonics in the UK Electricity Supply System, London (1976).

G W Wiskin BSc, C Eng, MICE, MI Struct E
Architectural and Civil Engineering Dept, BBC

R G Manton BSc(Eng), PhD, C Eng, MIEE
Transmission Engineering Dept, BBC

6.10 Masts, Towers and Antennas

6.10.1 Civil Engineering Construction

A proposal for a broadcasting or relay station will include a knowledge of suitable antennas and at least an appropriate location. In order to establish if such a proposition is viable a budgetary estimate of cost to meet a target completion date will be required. Preliminary information from many sources will have to be collated to produce the necessary financial, programming and planning details.

6.10.1.1 Preliminary research

6.10.1.1.1 Aims

The location of the site must be established at an early stage as site costs and planning considerations can vary considerably from site to site. Some concept of the sort of structure necessary to support the proposed antenna loading must have been established prior to visiting the proposed sites.

6.10.1.1.2 Maps

1:50,000 survey maps provide sufficient topographical information to:

- estimate the height of all the antennas that are required;
- assess the length and slope of any access track that is to be established;
- show possible traffic restrictions between the nearest main road and the site.

Larger scale maps will be necessary to assess the size and shape of the plot where a new structure, building or access are to be constructed. If a structure already exists on the site, the owners will probably be able to supply a site plan showing all the features.

6.10.1.1.3 Site access

However detailed the preliminary research has been, it is imperative to walk the site. The access route or routes from the nearest main road must be checked for low bridges, weight restrictions, tight corners, steep hills, etc.

The type of vehicles requiring access to the site will depend on the construction and operation. Construction traffic will typically include ready-mixed concrete trucks during foundation work. Not only are these vehicles heavy, but they are nearly 5 m high. Lorries will normally be used to supply steelwork, bricks, etc. Earth-moving plant can easily traverse most terrain, but slowly. If it is travelling a distance, it will probably be transported on a low loader to an unloading point as close as possible to the site. Cranes may be needed to lift steelwork or antennas.

Operational traffic will be lighter, perhaps cars for all-weather surfaced roads or four-wheel drive vehicles for access tracks. However, operational maintenance traffic will require access at any time of the day or night throughout the year. Consideration must therefore be given to surface finishes that may be affected by the weather. Conditions, such as flooding rivers, fog or ice and exposed areas liable to high winds or drifting snow all cause problems.

6.10.1.1.4 Local knowledge

It is extremely helpful to the project if those undertaking surveys can meet the local landowners and tenants, to ensure that boundary marks and access routes are acceptable to all. In particular where wayleaves or shared access are necessary, any practical restrictions must be determined. If good relationships are established with local people early in the project, later problems can be sorted out amicably by the original protagonists. However, if early grievances are not resolved, they will escalate, possibly causing delays to contractors and thus additional expense.

Local information can be useful in assessing the route for access tracks, the ground conditions and underground service runs. It may even indicate temporary storage areas, or construction access outside the site boundary in cases where the chosen site is large enough only for the final works. If the site is restricted for space or adjacent to occupied premises, the safety aspects of construction and maintenance may be more difficult to satisfy.

6.10.1.2 Surveys

6.10.1.2.1 Topographical survey

A detailed topographical survey of the site should be undertaken either to check values and update the existing site plan or to produce a new one. The resulting plan will enable the planners and engineers, not having visited the site, to contribute to the work. The plan should be drawn to a scale not smaller than 1:500, so that details such as boundaries, buildings, other structures, overhead and underground services and street furniture can be clearly marked. Contours or spot heights and coordinates of control points should also be given.

6.10.1.2.2 Soil survey

A soil survey may be necessary for some sites. This could be limited to establishing the depth of peat that overlays the rock across the site using trial pits and probes. At the other end of the scale, it could involve a substantial number of boreholes, with the appropriate sampling and laboratory tests. These may be necessary to establish the bearing capacity and shear strength of the different strata underlying the site. The water table will be of interest during foundation construction, particularly if saline and tidal. Parameters such as the need to use sulphate resisting cement in the concrete, or whether to design foundations as submerged, will also be obtained from the survey data.

6.10.1.3 Programming and approvals

6.10.1.3.1 Programme

At the start of a project it will be necessary to obtain information from several sources on duration of work items and their interaction with others. The following items should be evaluated in programming the construction of a new station:

- project brief including budgetary cost;
- financial approval;
- preliminary/sketch designs;
- site acquisition and planning approval;
- finalized requirements and let contracts;
- construction of access track;
- design, supply and construction of transmitter building;
- design, supply and erection of antenna support structure;
- fencing and landscaping;
- supply and installation of transmitter equipment;
- supply and erection of antenna;
- testing.

Where existing structures can be utilized, these items will not all be necessary. All programming constraints must be satisfied. Typically, arrangements need to be made so that in the interests of safety, work does not take place under the antenna support structure whilst men are working aloft.

6.10.1.3.2 Approvals

The local planning authority will generally require to see the site plan (1:200 scale) and an elevation of the proposed structure showing the antenna configuration intended. Specifications and calculations may also be required to obtain local building approvals.

It must therefore be clearly established what type, size, height and bearing is planned for each antenna. However, structures will undoubtedly be designed to carry a number of speculative antennas to allow for future expansion and/or allow other users to share the facilities. Such antennas may be excluded from the original planning application, but the structural engineer will still require the data as part of his design brief. Before a type of structure can even be selected, the designer will require the types and sizes of feeders to each antenna, the required means of access up the structure and the design wind speed.

6.10.1.4 Structures

6.10.1.4.1 General

The structure may be a simple wood pole or a tall guyed mast, but the principles in selection will remain the same. They are that the structure shall:

- be strong enough to withstand (a) the maximum design wind speed, with the specified antenna loading, and (b) the specified wind speed and icing conditions with that antenna loading;
- be stiff enough to limit the deflection of each antenna to less than that specified at its operational wind speed;
- be safe to be climbed by those staff trained to do so;
- be constructed within the budget and time scales allocated;
- be maintainable for its intended life-span;
- not impose unacceptable environmental or physical conditions on the locality.

6.10.1.4.2 Poles

Cylindrical poles of wood, steel or aluminium or welded lattice poles can support light antennas up to a maximum height of approximately 17 m in low wind speed areas (see Figure 6.10.1). They rely on their bases being buried at a sufficient depth in compacted ground (or concrete) to stabilize them. Access is normally from a removable ladder, with step bolts over the top section. Antenna installation and maintenance work should therefore only be undertaken by fit trained personnel.

6.10.1.4.3 Towers

Self-supporting towers can vary in height from 10 m to 300 m (see Figure 6.10.2). The ratio of the tower height to the base width of the section under consideration should not exceed 8:1 over the top 40 m whilst carrying omnidirectional uhf antennas and microwave dishes. For all other structures the ratio should not exceed 10:1. However, for every additional 60 m of tower height the ratio should be reduced by 1.

Figure 6.10.1 A 17 m timber pole.

Figure 6.10.2 A 45 m tower.

The face width of the structure should be no less than:

- 0.4 m where external access only is provided;
- 1.2 m for a square tower with an internal ladder;
- 1.45 m for a triangular tower with an internal ladder.

Moreover, the face width of the structure should be no less than half the width of any ancillary supported at that level.

The base width of the structure should be as large as possible to minimize the foundation forces, but not so large that the exposed face area of the tower increases too much. This optimization may be achieved with sloping leg members. However, it is recommended that the point of intersection of the projected lines of the legs is higher than 60 per cent of the tower height.

A bend line occurs at the point where the tower legs change slope. Several bend lines will occur in 'eiffelized' structures. A large horizontal force is developed at these bend lines, and there are usually local moments resulting from the location of the bend just above or below the member joint node. To cater for these circumstances, a horizontal member must be provided at this level across the full width of the tower.

The shape of the main bracing members will depend on the *height:width ratio* (h/w) of the panel and its size. For the top narrow parallel sided panels of h/w > 1.5, 'Z' shapes may be used. 'X' shapes are common where 0.7 < h/w < 1.5. However, 'K' shapes are used for large panels where 0.5 < h/w < 1.0 and often where access is required through a tower face, e.g. from an internal ladder to an external platform.

The main members so described form the main structural frame and will be designed in accordance with the specified standard[1–3] to resist forces due to wind and gravity loads on the structure and its ancillaries. However, secondary members may be needed to restrain the main members against buckling. The secondary members must be capable of providing such restraint, and have adequate strength to support ice, platforms or ancillaries that load them.

Auxiliary members, including antenna mounting and platform steelwork, will be designed to support their local loads plus ice where applicable.

The tower legs are usually supported on individual foundations.

The critical design criteria, however, are rarely bearing of shear, but uplift and overturning. So a foundation bearing directly onto sound rock which would satisfy any building, may not be adequate unless satisfactorily anchored to the rock. A cone of soil above the foundation pad may be utilized to provide the dead weight. However, where the ground area required to resist the uplift forces at one leg impinges on a similar area around another leg, a combined foundation should be considered. For small structures this takes the form of a raft foundation.

6.10.1.4.4 Masts

Guyed masts vary in height from 10 m to some of the world's tallest structures (see Figure 6.10.3). However, the following parameters will provide a reasonable structural profile.

Figure 6.10.3 A 225 m mast with replacement under construction.

The mast column will be supported at various levels by sets of tensioned stays. The ratio of the height between stay levels and the face width of the column should not exceed 40:1. The face width of the column should conform to those parameters given for towers. The bracings shapes and number designs will also conform to the parameters given in Section 6.10.1.4.3.

The normal stay arrangements are for three stay lanes 120° apart for triangular mast columns and four stay lanes 90° apart for square mast columns. These stays will be anchored to foundations so that the vertical angle between the stay and the ground plane is between 30° and 60°. To minimize costs, these foundations may be arranged to support several levels of stays.

The stay anchors are usually blocks of concrete of sufficient weight to resist the uplift forces, and sufficient width and depth to resist the sliding and overturning forces. For light or temporary structures, ground anchors can be employed, and rock anchors can provide a satisfactory anchorage solution for large structures on suitable underlying rock.

The column base is usually tapered where possible to form a structural pin. This largely reduces the foundation design condition to one of direct bearing. The vertical load is mainly due to the thrust in the column exerted by the tensioned stays. However, where a fixed based mast is installed, the foundation designer will have to consider much larger shear and overturning forces.

6.10.1.4.5 Roof-mounted structures

Roof-mounted structures are potentially the easiest and cheapest to utilize (see Figure 6.10.4). However, they have the reputation of being time-consuming projects and prone to problems.

Figure 6.10.4 Roof-mounted poles.

Access to the pole, mast or tower site will be either up the face of the building or via internal lifts and stairs. Short, light members and protection to finishes will be major requirements. Protection to roof finishes will be particularly important to prevent water penetration into the building. Wherever possible, it is wise to avoid puncturing the existing finishes, even to provide anchorages.

It is important that the antenna support structure loads are transmitted directly to the building frame. The assumption that the building can transmit the loads even short distances onto its structural frame can lead to local failures.

The high intensity, short duration loads experienced by these structures are known to have caused damage where only normal building design standards have been applied, so antenna support factors of safety should be maintained.

6.10.1.4.6 Existing structures

Where it is proposed to utilize an existing structure, the antenna and feeder type, their location and the method of attachment should be agreed and approved by the owner prior to installation (see Figure 6.10.5). This may take additional time, but can prevent unnecessary costs due to misunderstandings and unacceptable details.

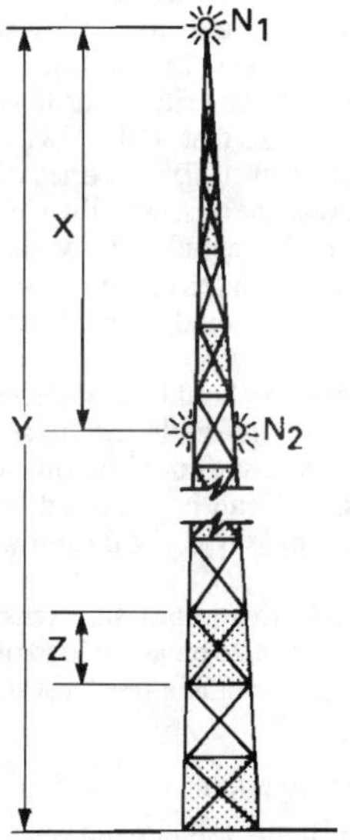

$$\text{Number of lights} = \frac{Y \text{ (in metres)}}{45}$$

$$\text{Light spacing} = X = \frac{Y}{N} \leqslant 45\text{m}$$

$$\text{Band spacing} = Z = \frac{Y}{7\ (9, 11, \text{etc})} \leqslant 30\text{ m}$$

Figure 6.10.5 Marking and lighting a tall structure.

6.10.1.5 Operational considerations

6.10.1.5.1 Access

Once the type and size of structure has been ascertained, it is easy to provide something to meet the initial requirements and lose sight of its reason for existence. Antennas are to be attached to the structure, so someone will be required to install them and service them. Depending on the height and location of the structure, aircraft warning lights may be needed. If so, they too will need servicing. The designer's brief should indicate to which parts of the structure access is required, what type of person will require to use that access and how often.

The simplest access will be to climb the face of the structure. This can utilize step bolts, an external ladder or face bracing, with or without ladder step bracing. This form of access is suitable only for regular climbers who have demonstrated their ability and fitness. Even so, they must be provided with suitable safety equipment and trained to use it. If a fail safe fall arrester device is provided alongside the ladder, other properly equipped, trained, fit personnel could also ascend safely.

Since climbing is tiring, especially in cold windy weather, a vertical or near vertical cat ladder is normally provided for the full height of the structure, with rest platforms at 10 m intervals. These rest platforms are normally of sufficient size for only one person to stand or sit. They can also be used as passing places. Working platforms, with flooring, handrails and toe boards may be specified, typically at antenna mountings, aircraft warning lights or ladder changeover levels. In some cases, these platforms will provide access from the inside to the outside of the structure. Detailing these areas to maintain acceptable unobstructed walkways can prove difficult, unless incorporated in the original design.

Where occasional climbers require access to parts of the structure, ladders with safety hoops and/or lifts and platforms that satisfy recognized safety standards[4] should be provided. Temporary mechanical access for antenna erection and maintenance of the structure will be necessary on larger structures. Appropriate locations may therefore be designated as safe rigging points, and possibly have a lifting jib built in. All rigging equipment, chairs, winches and lifts will have to be tested, checked by a competent person, recorded and certificated in accordance with the local/national safety requirements.[4]

6.10.1.5.2 Maintenance

Maintenance work on the structure (as opposed to the antenna) is essentially required to ensure that it will be able to perform its function safely for its intended life span. Regular checks on the condition of the surface are necessary to detect any corrosion or damage. Repairs can then be initiated as required.

The designer can, at the risk of increasing the capital cost, minimize the amount of maintenance that will be necessary. He can ensure that the materials supplied and applied will last as long as reasonably possible. Steelwork would be galvanized and possibly painted for additional corrosion protection as well as for aircraft warning bandings. Mast stays would be made up of galvanized steel wire strands bedded in an impervious material, or be of a non-corrosive material.

The stays will need retensioning to maintain the correct initial tensions for stability and control of possible stay oscillations. The regularity of retensioning will depend on the age of the stays and their construction. Typically prestressed steel wire rope will bed down over an initial period of 12 months, and once adjusted should then perform without attention for another 5 years. Non-metallic ropes generally perform less well and need more regular attention.

6.10.1.5.3 Radio frequency hazards

Whether the work be inspection, repair, painting or greasing, it is likely that someone will need to work in the vicinity of operating antennas. The radiation field must therefore be checked with a hazard meter to ensure that it is within the permissible limits.[5] People may not remain in areas where such limits are exceeded (see Table 6.10.1), so the transmission from the offending antenna will have to either operate at reduced power or be shut down. Where reduced power or shutdowns are essential, maintenance will have to be programmed to take place at times convenient to the broadcasters' schedule. This may mean inefficient use of time on the structure, and increase the cost.

Table 6.10.1 Reference levels for continuous exposure to electromagnetic fields

Frequency (GHz)	*Root mean square values*		
	Electric field strength (V/m)	*Magnetic field strength (A/m)*	*Power density (W/m^2)*
0.03–0.4	61.4	0.163	10
0.4–2	$97.1\sqrt{f}$	$0.258\sqrt{f}$	25f
2–300	137	0.364	50

Sometimes an antenna will radiate a high rf field where shutdowns or reduced power working are difficult to organize, e.g. an antenna broadcasting 24-hour television. In this case, the same structure or an adjacent one should be considered. This will increase the total loading on the structure, so is best considered and allowed for at the design stage. Alternatively, an existing structure may be re-analysed and possibly strengthened to accommodate additional antennas as requirements change.

6.10.1.5.4 Security

All sites should be so protected that the general public, behaving reasonably, will not cause damage to the installation, neither will they be harmed by it. However, children and animals will not always behave reasonably so they must be prevented from causing trouble.

The outermost line of security will be the site boundary fence. Depending on the type of fence and type of station, a local security fence may be required immediately around the installation, suitably posted with notices warning of the appropriate hazards, i.e. high voltages and non-ionizing radiation. If this fencing is climbable, and the site is accessible, some anti-climbing guard will be required on the structure to prevent unauthorized ascent. Suitable protection is required to prevent cows eating stay grease and rubbing undone unsecured rigging screws. Rodents and camels have been known to eat insulated cable. Where vandalism or terrorism is a problem, it may prove necessary to provide security guards or electronic surveillance equipment.

To ensure that access is limited only to authorized persons, a pass system should be used for attended sites, and keys issued to appropriate people for unattended sites.

6.10.1.6 Structural design

6.10.1.6.1 Loading

The main loading on the structure will be the wind force exerted against the structural frame and all the ancillary ladders, platforms, antennas and feeders attached to it. National standards[1–3] give guidance to obtain the appropriate design wind speed for the terrain encountered within that country. These structures are normally designed using the mean hourly wind speed likely to be exceeded at least once in 50 years in that topography. The associate gust duration will vary between 1 and 16 s depending on the size of the structure.

The critical wind directions for a *square structure* are:

- directing onto a face, which normally gives maximum bracing forces;
- into a corner, which normally gives maximum leg forces and foundation loads.

For a *triangular structure* the above two directions will give the maximum and minimum leg loads, but an additional direction is:

- in the plane of the face, giving critical bracing forces under some conditions.

These wind directions will provide the maximum and minimum stay tensions for guyed masts.

When ice loads are considered, the effective wind area of all items will be increased. In some cases, it may not be sufficient to allow for a thickness of radial ice around members. Under extreme conditions, masts in particular are known to ice across the full face of the structure when closely spaced feeders are attached to, or adjacent to, that face. On most structures, severe icing is unlikely to occur at the design wind speed, so a more appropriate wind speed is selected. It is therefore unusual for the icing condition to be critical for towers at low altitudes in temperate climates like that in the UK.

For masts taller than 100 m, a series of load cases should be considered to assess the susceptibility of the mast to local gust loads. These can be simulated by alternately loading and unloading adjacent spans as recommended in IASS.[6] Similarly, masts prone to icing should be analysed with ice on stays on one side only.

Mast stay tensions should be assessed for performance of the slackest stay, as this could lead to an instability condition. The slackest conditions should be outside the working range of wind speeds, and can be assessed by analyzing results at varying wind speeds, say 0.7 × design wind, and 1.2 × design wind.

Masts designed with initial stay tensions in the region of 10 per cent of the ultimate breaking load of the stay rarely exhibit instability problems including those caused by stay galloping.

6.10.1.6.2 Factors of safety

Factors of safety at the design wind speed condition are usually taken as:

3	for stays (ropes and fittings),
1.7	for buckling of steelwork,
2	for foundations using soil properties,
1.5	for foundations using only dead weight of the block.

Factors of safety at a survival wind speed condition (1.2 × design wind) are usually taken as:

1.5	for stays,
1.15	for buckling of steelwork,
1.2	for foundations.

Where this condition is satisfied, the factors of safety at the design wind speed can be relaxed.

6.10.2 Electrical Design of Antenna Systems

The provision of an antenna system for the transmission of terrestrial television signals involves the exercise of many engineering disciplines from economics through transmission line theory and vector arithmetic to metallurgy. The design of

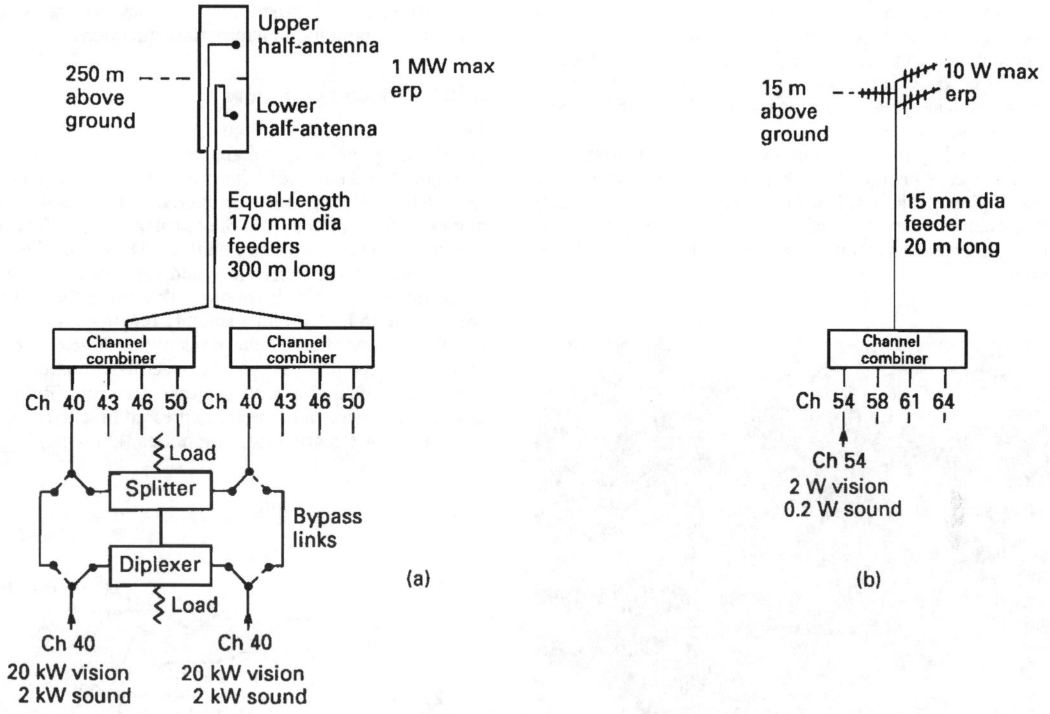

Figure 6.10.6 Complete transmitting antenna systems: (a) high power; (b) low power.

the antenna cannot be considered in isolation because, for a given service coverage, its physical shape and hence cost are inextricably bound up with the economics of transmitters, feeders and support structures. It is unlikely, therefore, that an ideal antenna for a given purpose can be bought 'off the shelf'. The need to propagate adequate signals to a given service area at lowest cost and at the same time avoid causing interference to viewers in neighbouring areas, results in many antennas having to be designed for specific sites.

6.10.2.1 Definition and design philosophy

The antenna system comprises all the equipment that is necessary to carry radio frequency energy from the transmitter(s) and to propagate it into space. In addition to the antenna itself with its feeders, the largest installations may include diplexers for combining the output power of transmitters operating on the same carrier frequency, and *channel combiners* for combining the signals of several channels into a single antenna. An antenna system which serves a large population may be split into sections to give reliability of service.

The antenna itself may be fed in two halves, one above the other, through separate main feeders. Transmitters may consist of pairs of amplifiers with common drives where, in the event of failure of one transmitter or half-antenna, the service can continue to run with a reduction in *effective radiated power* (erp) of 6 dB over most of the service area. There is a need to diplex and split the outputs of two amplifiers rather than feed them directly to separate half-antennas. The reason for this is considered in Section 6.10.2.12. Low-power stations serve smaller populations and are more easily engineered to have higher factors of safety. Usually, therefore, they have single sound and vision transmitters per channel with a single antenna and feeder. Schematics of typical high- and low-power antenna systems are shown in Figure 6.10.6.

6.10.2.2 Types of antennas

The following paragraphs describe antennas for the uhf bands 470–860 MHz (wavelengths 0.64–0.35 m). The principles for vhf (30–300 MHz) remain the same, although some of the components are larger because of the increased wavelength.

6.10.2.2.1 High-power antennas

The elementary radiating part of a high-power uhf antenna is a panel of radiating elements, typically about 0.5 m wide and four wavelengths high. The panel may support eight horizontal dipoles or four vertical slots in front of a reflecting screen; either system provides horizontal polarization. The rear of the panel usually has a single input connector and houses a system of branch strip-line feeders and matching sections behind a rear cover. Up to eight panels may be required in a single tier, depending on restrictions imposed by any support structure, to obtain a near-omnidirectional *horizontal radiation pattern* (hrp) or whatever directional hrp is called for by the service planning engineer. The panels will be fed by a suitable feeder harness to give the appropriate feed currents.

The same horizontal pattern of panels and feeders will be repeated in a number of tiers in each half-antenna to give the required antenna gain. There are, however, important considerations to be taken into account in the design of the feeder system which takes power from the main feeders to the individual tiers. These are described in Section 6.10.2.5.

Panels for a high-power station may be arranged around the faces of a square or triangle with interior access for a maintenance engineer. The level of radiation inside should ideally be restricted so that the maintenance engineer can climb through the antenna without hazard when it is powered (see Figure 6.10.7).

Figure 6.10.7 Interior view of a high-power uhf antenna (Alan Dick & Co Ltd).

Usually all the panels will be housed inside a cylinder of glass-reinforced plastic (grp) to reduce wind loading, to prolong the antenna's life and to make the environment more convenient for maintenance. The panels may be fixed to the sides of a steel lattice mast, in which case the grp cylinder may be a relatively thin shell. Alternatively, the grp cylinder may be thick enough to be load-bearing and may support the panels as a cantilever without the benefit of a load-bearing spine. In either case, the dimensions across the antenna seen in plan must not exceed a wavelength of two, otherwise it will not be possible to maintain a reasonable hrp over the bandwidth required for two or more channels.

Whatever the type of antennas, it is important to minimize bimetallic corrosion by complying with a standard code of practice for materials in contract, e.g. BS PD5484:1979.

6.10.2.2.2 Low-power antennas

Vertical polarization is used for most UK low-power relay stations to minimize interference to reception in main station areas. (This is not a common practice elsewhere.) Antennas giving almost omnidirectional hrps may consist of up to 16 vertical dipoles stacked collinearly off the side of a single metallic support pole. The largest antennas may be housed inside a grp cylinder 0.4–0.9 m in diameter; the antenna would then be lowered through the tower for maintenance purposes. Alternatively, more complicated hrps may be produced by combinations of log-periodic antennas fed with specified currents and pointing in appropriate directions.

6.10.2.3 Design procedure

Primary data for the design of an antenna system will be generated by the service planning engineer who will state the location, the height on the support structure and the erp required in all directions of azimuth (and possibly at some angles of depression) from the antenna. This will be given in the form of a template (see Figure 6.10.8) stating the minimum erp required in some directions and the maximum erp that can be permitted in other directions. (The *erp* is the power which would have to be fed into a suitably oriented half-wave dipole in the same position as the transmitting antenna in order to produce the required field in the specified direction. For negative video modulation with positive synchronizing pulses, the transmitter power and erps are expressed in terms of the power at the peak of the synchronizing pulses (kW sync).)

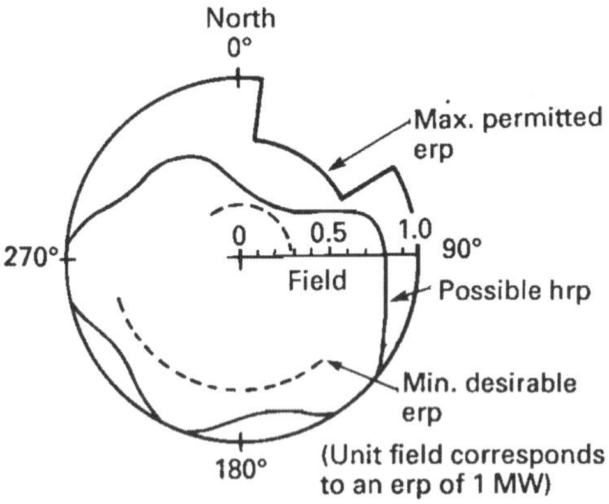

Figure 6.10.8 Template for horizontal radiation pattern.

The antenna designer starts by arranging radiating elements (panels or log periodics) so that they produce an hrp to fit the required template on all the required frequencies. Initial work can be carried out by computation but, for all but the simplest directional hrps, the hrp has to be proved by measurement. It is necessary for the designer to construct at least one tier of the antenna on a turntable so that its hrp can be measured by rotation.

The measuring antenna for this purpose has to be at a suitable distance and height above ground so that reflections from other objects and the ground itself do not interfere and so that the field from the measuring antenna, used as a transmitter, illuminates the measured antenna uniformly. It is reasonable to specify that the max:min ratio of a four-channel antenna omnidirectional hrp should not exceed 5 dB in any channel. Once the hrp has been established, the ratio of maximum power gain to mean power gain can be determined by dividing the area of the circle circumscribing the hrp diagram (plotted radially in linear voltage) by the area of the hrp diagram itself. Following this, the designer can calculate the required mean intrinsic antenna gain, G, from a knowledge of the transmitter power available and various assumed losses in the system as shown in Table 6.10.2. In this example, G = 15.8 dB.

6.10.2.4 Antenna gain

The mean intrinsic power gain of an omnidirectional stack of dipoles in a panel-type antenna is determined by the vertical length of the antenna measured in wavelengths. It is given approximately by:

$$10\log_{10}(1.2\text{antenna length/wavelength})\ \text{dB}$$

In the example given in Table 6.10.2, the appropriate intrinsic gain can be provided by a 32-wavelength antenna, which can conveniently consist of eight tiers of four-wavelength panels.

The maximum intrinsic gain of an array of log-periodic antennas has to be calculated from a knowledge of the maximum gain of a single antenna. If there is more than one antenna in a single tier, the maximum power gain of the single antenna will be reduced in the ratio of the areas of the array hrp to that of the single antenna. The power gain can be increased by a factor approximately equal to the number of tiers.

Table 6.10.2 An example of the summation of losses and gains in a high-power transmission system

Transmitter peak sync power	40 kW	
Max erp required	1000 kW	
System max gain required		+14 dB
Network and combiner loss	−1.2 dB	
Main feeder loss	−1.5 dB	
Distribution feeder loss	−0.6 dB	
Beam tilting and null-fill loss (see Section 6.10.7.5.1)	−0.5 dB	
Antenna max/mean gain	+2.0 dB	
Antenna mean intrinsic gain	+G dB	
System max gain available		G − 1.8 dB

6.10.2.5 Vertical radiation pattern

The vrp of a vertical stack of identical radiating elements, regularly spaced and fed with equal co-phased currents, consists of a main lobe in the horizontal plane and a number of subsidiary lobes above and below the horizontal which are separated by nulls. The first step in adapting this for broadcasting is to tilt the main lobe downwards towards the edge of the service area to avoid wasting power. The angle of tilt, θ_T, will typically be about 0.5°, and this can be achieved by feeding the lower tiers through progressively longer feeders so that the phase difference between adjacent tiers is about $360D\sin\theta_T^\circ$, where D is the number of wavelengths between tiers. The resulting vrp is shown in Figure 6.10.9(a). The next step is to modify the phases in such a way that an adequate signal (which does not vary too much from one frequency to another) is provided at all angles of declination θ down to about 25°.

6.10.2.5.1 *Specification*

A specification that can reasonably be applied to ***null filling*** is that no amplitude, E_θ, in the vrp should be less than 50 per cent of the locus of the envelope of the maxima of the subsidiary lobes of the unfilled vrp, i.e.:

$$E_\theta \geqslant E_{max}/2\pi A\sin(\theta-\theta_T)\ \text{for}\ 90^\circ/\pi A < \theta < 15^\circ$$

where θ is the angle of declination from the horizontal, E_{max} is the maximum amplitude of the main beam, and A is the length of the antenna in wavelengths.

A further constraint that helps to ensure uniformity of field is that the ratio of any adjacent maximum to minimum in the erp should not exceed 6 dB. Figure 6.10.9(b) shows the vrp of an antenna consisting of eight tiers of four-wavelength panels which has been successfully filled to 9° of declination by phase perturbations. In practice, the antenna designer has to resort to other methods, such as filling the nulls of individual panels, in order to fill nulls at greater angles. Methods of filling nulls have been described by Hill.[7] The action of filling nulls and

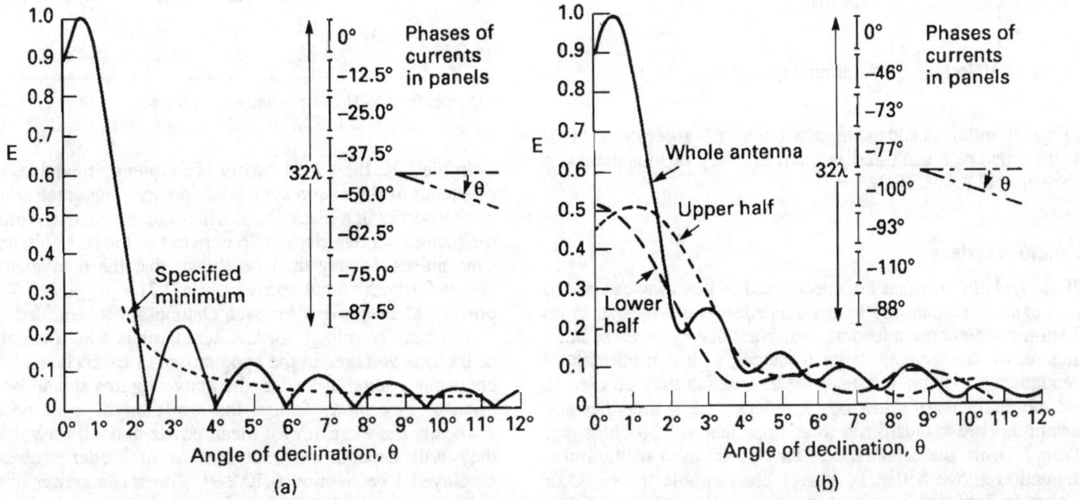

Figure 6.10.9 Vertical radiation patterns of eight tiers of four-wavelength panels: (a) beam tilted, not null filled (loss = 0.02 dB); (b) beam tilted, null filled (loss = 0.45 dB).

tilting the beam of an antenna results in a reduction of antenna gain which is referred to as *beam tilt and null fill loss* (see Table 6.10.2).

6.10.2.5.2 Maximum aperture

Because the 3 dB beamwidth of an antenna, in degrees, is given approximately by 50/(antenna length in wavelengths), the limiting length of antenna that can be used in practice is about 40 wavelengths, this being determined by the structural stability of mast plus antenna that can reasonably be achieved. Under the worst environmental conditions, the antenna must not be allowed to depart from the vertical by more than 20/(antenna length in wavelengths) degrees.

6.10.2.5.3 Computation

The vrp of a high-power uhf or vhf antenna cannot be measured directly in the same way as an hrp. This is because of the physical size of the whole antenna and the large unobstructed area that would be required for measurement. In practice, the feedpoints on each tier of panels are fitted with directional couplers or voltage probes so that the relative levels of applied voltages can be measured. The vrp, E_θ, can then be computed from the formula:

$$E_\theta = \left|\sum_{n=1}^{N} V_n(\Theta)C_n \exp j\,\{\phi_n - (2\pi d_n \sin\Theta)\}\right|$$

where N is the number of tiers, $V_n(\Theta)$ is the vrp of the nth tier, C_n is the amplitude of current in the nth tier, ϕ_n is the phase of current in the nth tier, and d_n is the distance in wavelengths of the centre of the nth tier below the reference plane.

The intrinsic gain, G_d, of the antenna relative to that of a dipole is:

$$\frac{1}{1.64}\int_{-\pi/2}^{+\pi/2} \frac{E_\Theta^2}{E_{\Theta\,\max}^2} \cos\Theta\, d\theta$$

The beam tilt and null fill loss is:

$$10\log_{10}\frac{E_{\Theta\,\max}^2}{N\sum c_n^2} \quad \text{assuming } V_n(\Theta)_{\max} = 1$$

Vrps of antennas consisting of dipoles or log-periodic antennas may also be beam tilted and null filled by various means, if necessary (Figure 6.10.9).

6.10.2.6 Feeders

All vhf and uhf antennas are constructed so that they can be fed by coaxial (unbalanced) feeders. Feeders vary in size from 170 mm diameter main feeders, which are needed to carry large amounts of energy over large distances with a minimum of attenuation, to 15 mm diameter feeders, which may be used to feed the elements of a low-power antenna. Most antennas and transmitters are standardized to an impedance of 50 ohms but, where 170 mm diameter feeders have to be used to minimize attenuation above 850 MHz, it may be advisable to use 60 or 75 ohm feeders with suitable transformers. The purpose of this is to extend the usable frequency range before the TEM energy breaks up into waveguide modes.

For convenience of installation and to keep down the cost of maintenance, most feeders used in the UK are of the continuous semiflexible type with corrugated conductors; outer conductors are protected by a sheath of black pvc. The largest feeders, and those that carry the highest powers, are mainly air-spaced with helices or spacers of dielectric to support the inner conductor. These feeders need a pressurized supply of dry air or inert gas (about 100 mb) to prevent the ingress of moisture. Smaller feeders often have a foam dielectric to completely fill and hermetically seal the space between the inner and outer conductors.

6.10.2.6.1 Feeder ratings

Table 6.10.3 shows the mechanical and electrical properties of some of the semiflexible coaxial feeders that are available. The power rating and attenuation are proportional to 1/√frequency and √frequency respectively. The power rating also depends on the ambient temperature and standing wave ratio. Reference should be made to the manufacturers' catalogues for details of this. The peak rf voltage and mean power that should be taken into account for multichannel antenna systems depend on transmission standards.

Table 6.10.3 Semiflexible 50-ohm coaxial feeders (Radio Frequency Systems, Hannover, Division of Kabelmetal Electro GmbH)

Diameter over sheath (mm)	*Dielectric**	*Velocity factor (%)*	*Max peak rf volts (kV)*	*Mean power at 600 MHz (kW)*	*Attenuation at 600 MHz (dB/100 m)*	*Min bending radius (mm)*
170	A + H	97	17	53	0.49	1200
120	A + H	97	12.5	26	0.72	500
90	A + H	96	9.7	17	0.93	380
50	A + H	95	5.2	6.1	1.61	180
50	F	88	5.6	4.6	1.96	300
30	A + H	93	2.7	2.7	2.98	100
30	F	88	3.0	2.5	3.10	120
15	F	88	1.6	1.1	5.48	70

*Dielectric: A + H = air + helix; F = foam.

In the UK, the sound carrier is frequency modulated and is one-tenth of the vision sync pulse power. The greatest average power occurs in a black field, where the rms carrier voltage for the picture content rises to 76 per cent of the rms voltage of the sync pulses. It may then be shown that the equivalent mean power for both sound and vision is 0.71 sync power (the sync power). Mean powers for each channel must be added.

The peak rf voltage for UK standards is 1.86 times the sum of the rms voltages in the sync pulses of each channel. Before choosing a suitable feeder, the above figures should be multiplied by any safety factors that are thought to be necessary. Typically these are 1.5 for mean power and 2.0 for voltage but they will depend on what system of feeder protection is employed (see Section 6.10.2.9). The mean power and peak voltage, expressed as a function of vision sync pulse power, vary from one country to another as the transmission standards change.

6.10.2.6.2 Feeder uniformity

Main feeders, in particular, must have a uniform impendence if the transmitted picture quality is not to be marred by multiple images. To ensure this, the input voltage reflection coefficient should not exceed 5 per cent at any frequency within the video band of each channel when the feeder is terminated by a resistor equal to its average characteristic impedance. In addition, the voltage reflection coefficient of a 0.1 μs sine-squared pulse at vision carrier frequency or colour subcarrier frequency should not exceed 1.5 per cent.

6.10.2.7 Lightning protection

Any antenna which forms the topmost part of a mast should be equipped with 1 m lightning protection spikes electrically bonded to the mast structure. In addition, the outer conductors of all coaxial feeders inside a mast should be electrically bonded to the mast at antenna and where they leave the mast at ground level.

6.10.2.8 Antenna impedance

If the impedance of an antenna and its distribution network at the top of the mast is not sufficiently well matched to the characteristic impedance of the main feeder, a fraction of the signal applied to the antenna will be reflected back to the transmitter via the channel combiners and diplexers. In general, the output stage of the transmitter will not absorb this signal, and a large percentage of it will be returned to the antenna. Here it will be transmitted, but delayed in time and attenuated by two traversals through the main feeder and equipment at ground level. The reflected signal may then be seen by viewers as a delayed image or ghost.

For an antenna with a feeder run longer than 50 m, acceptable limits for the levels of delayed signal, expressed as percentages of the primary signal amplitudes, are obtained if values lie below one of the alternative lines shown in Figure 6.10.10. For the reflection coefficient of the antenna itself, these figures may be relaxed by twice the feeder attenuation and 2.5 dB for loss in equipment at ground level. Figure 6.10.10 applies to the UK transmission standard, where the video band extends 5.5 MHz above the vision carrier and 1.25 MHz below it. Similar lines can be drawn for the different video bands used in other countries.

If the antenna is divided into halves it is usual to permit three times the amplitude of the normally specified delayed image to

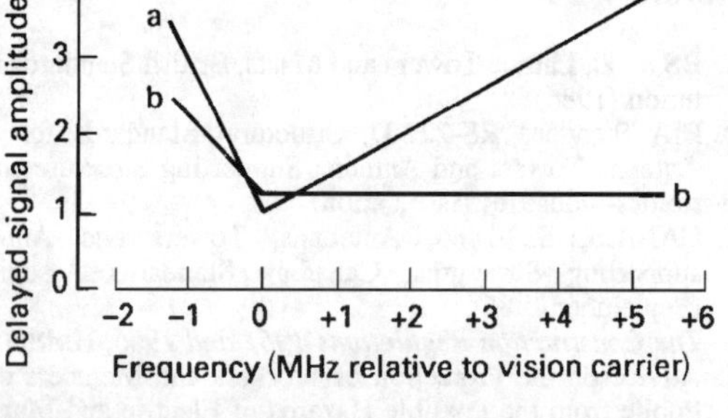

Figure 6.10.10 Impedance specification: alternative maximum permission levels of delayed signal.

be radiated when only a half-antenna is being used under emergency conditions. For antennas with shorter main feeders, a given delayed image will be less visible, therefore a relaxed specification can be applied. For example, the reflection coefficients in Figure 6.10.10 can also be increased by a factor of 3 if the feeder is less than 20 m long.

6.10.2.9 Directional couplers

Directional couplers may be used at various points in an antenna system to monitor the forward and reverse flows of power. They are particularly useful if situated at the upper or lower ends of the main feeders, where a change in the ratio of reverse to forward power may indicate a fault in the antenna. For this purpose, the directivity of the reverse coupler needs to be about 40 dB. The output of the couplers can, if necessary, be made to provide executive control over the transmitter output power if the reverse power reaches a predetermined value.

6.10.2.10 Hybrid or diplexer

A directional coupler, which splits power equally between its direct output and its coupled line (Figure 6.10.11), constitutes one form of hybrid or diplexer. This form of hybrid has a phase difference of −90° between its two output ports.

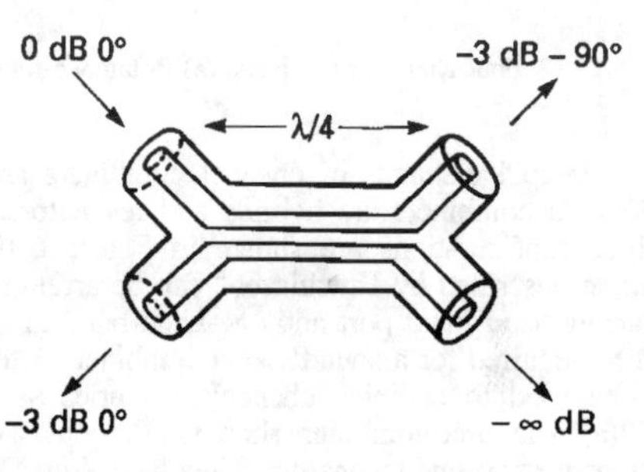

Figure 6.10.11 '3 dB coupler' hybrid.

There are other forms of coaxial hybrid, where the power is split equally, but is either co-phased or 180° out of phase, depending on which input port is used. Hybrids are used to diplex the power of two transmitters, to split power equally between two loads, to provide quadrature phase feeds in phase-rotating systems, or to form parts of channel combiners.

6.10.2.11 Channel combiners

Channel combiners are required at ground level when more than one channel is fed into a single antenna; it is usual to provide duplicate combiners when the antenna is fed by two main feeders. The main principles of all combiners for television are the same. For example, the input ports for the several frequencies must be isolated from each other by at least 30 dB to keep intermodulation to an acceptable level. The input voltage reflection coefficient must be maintained at a low level, over as wide a band of frequencies as possible, by means of a suitably connected absorber load. This load will help to ensure the stability of transmitters and provide a sink for spurious frequencies and remaining intermodulation products.

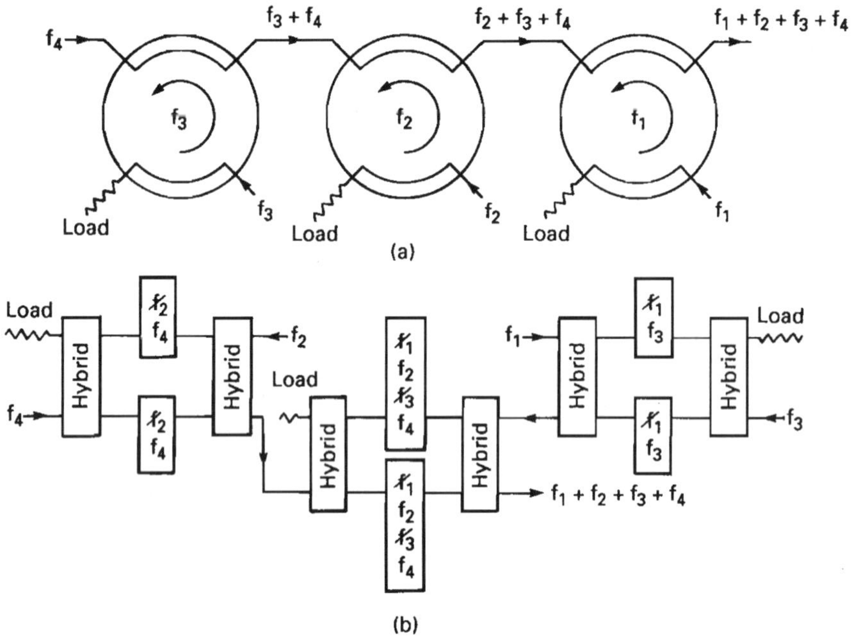

Figure 6.10.12 Four-channel combiners: (a) Rotamode rotating mode resonator combiners; (b) hybrids with transmission line resonator combiners.

The principal elements of channel combiners and also of sound/vision combiners are hybrids and resonators. The two combiner configurations are shown in Figure 6.10.12. The Rotamode described by Hutchinson[8] can be arranged to have one narrow band input port and one wide band input port, as would be required for a sound/vision combiner. Alternatively, it can be used to combine channels in series as in Figure 6.10.12(a). The three combiners shown in (b) consist of hybrids with transmission line resonators. They have equal bandwidth ports with alternate stop and pass frequencies; such combiners can be used to add channels in parallel. The subject of channel combiners in general has been described by Manton.[9]

6.10.2.12 The overall system

A schematic for a complete antenna system with duplicate main feeders and pairs of transmitters is shown in Figure 6.10.6(a). At first sight it is not obvious why separate transmitters should not feed separate half-antennas independently, without the complication of diplexing output powers and then re-splitting them to feed half-antennas. This becomes clearer on examination of Figure 6.10.9(b). It will then be seen that the vrp of the whole antenna in the vicinity of alternate minima is formed by the subtraction of the vrp of one half-antenna from that of the other. Consequently any slight differences between the modulation characteristics of signals fed to the two half-antennas will be exaggerated in areas covered by these minima.

The effect of using a transmitter diplexer is to provide a sink load for modulation differences and to provide a unified output with average modulation characteristics that can be split between two half-antennas. (It follows that, if one transmitter fails, the remaining transmitter will continue to operate but will transfer half of its power to the sink load and the other half to the antenna.)

The use of a hybrid as a splitter transformer ensures that any delayed images, caused as a result of voltage reflections from the antenna or feeders, are not exaggerated in areas served by minima of the vrp.

Where a pair of transmitters are fed by a common drive it is possible to increase the reflection loss at ground level, and hence reduce the level of delayed images, by delaying the signal feed to one transmitter by 90° of phase and delaying the signal from the output of the other transmitter by a similar amount. The path lengths for the reflected signals then differ by 180° and, provided that the output stages of the transmitters are identical (and are both operating), the reflected signals will be entirely absorbed in the sink load of the diplexing hybrid. (The splitting hybrid is still necessary.)

References

1. BS 8100, Lattice Towers and Masts, British Standards Institution (1986).
2. EIA Standard RS-222-D, Structural Standards for Steel Antenna Towers and Antenna Supporting Structures, Electronics Industries Association.
3. CAN/CSA-S37-M86, Antennas, Towers and Antenna-supporting Structures, Canadian Standards Association (September 1986).
4. *The Construction Regulations 1961 and 1966*, HMSO.
5. Advice on the Protection of Workers and Members of the Public from the Possible Hazards of Electric and Magnetic Fields with Frequencies Below 300 GHz, National Radiological Protection Board (May 1986).

6. Recommendations for Guyed Masts, Working Group No. 4, International Associated for Shell and Spatial Structures.
7. Hill, P.C.J. Methods of shaping vertical radiation patterns of vhf and uhf transmitting aerials, *Proc. IEE*, **116**, No. 8, 1325 (1969).
8. Hutchinson, R. Rotamode filter networks, *Communications and Broadcasting* (Journal of Marconi Communication Systems Ltd), **6**, No. 2, 15 (1981).
9. Manton, R.G. Channel combiners for radio-frequency transmitters, *JIERE*, **55**, No. 10, 335 (1985).

Section 7
Test and Measurement

L E Weaver BSc, C Eng, MIEE
Formerly Head of Measurements Laboratory,
BBC Designs Department

Revised by
Paul Dubery

7.1 Television Performance Measurements

7.1.1 Introduction

The field of television measurements is vast, since it concerns every aspect of signal generation, recording, distribution, emission and reception. Each of these requires appropriate test waveforms and measurement techniques, which may further be a function of the TV standard in use. These techniques are not even sufficient in themselves, since permissible tolerances must be laid down for each type of picture impairment, and consideration given to the way in which distortions add along a given signal path.

It is proposed to concentrate here upon a discussion of the most common signal impairments with respect to internationally agreed test waveforms, supplemented by references to more detailed texts. The insertion test waveforms have been specifically designed to cover the most important analogue signal impairments, and some have a role in the testing of digital and component video systems.

7.1.2 Insertion Test Signals

Insertion test signals (ITS) take the form of ingeniously devised groups of waveforms which are carried by selected lines in the file blanking interval. Those to be discussed here are recommended by the international standardising bodies, the CCIR and CMTT, for analogue signal transmission over long distances. The positions of the ITS and VITS in the field blanking intervals are also laid down by these bodies.[1]

Their advantages can be summarised as follows:

- Measurements can be made at any desired time and point in a network, even during 'in-service' conditions. Where automatic measurement equipments (AMEs) are installed, checks can be made continuously or at selected intervals throughout each day, giving advance warning of incipient failure and immediate notice of actual failure.
- Experience has shown that the most important signal impairments can be measured in this way to a satisfactory degree of accuracy, perfectly adequate for the control of programme exchanges over thousands of kilometres[2] or in studios.[3] They are equally suitable for repetitive 'out-of-service' measurements; indeed, they have some advantages used in this way.
- If the ITS or VITS are inserted at the point where the programme is finally assembled and not touched subsequently, measurements at any point along the signal path will indicate the total distortion up to that point. Alternatively, the ITS or VITS can be inserted at any intermediate location for fault-finding purposes.
- Because this type of test signal is internationally standardised, equipment for generation and measurement is available from many sources, and has been tried and proved through operational experience.

It should be noted that there is no restriction on any individual broadcaster as regards test signals or positioning in the field blanking interval, provided no international programme exchange is intended, although it is normal to adhere as far as possible to the lines allocated for internal use with 625-line systems, i.e. 19, 20, 232 and 333. In practice, the ITS and VITS do not differ significantly from the international recommendation in this instance, except for any special waveforms required for testing transmitters, teletext performance, etc.

It is essential, however, to avoid the insertion of any waveform on the so-called 'quiet lines', i.e. 22 and 335 for 625-line standards, and 10–16 for 525-line. These are reserved for the measurement of random noise, and any tampering with them would make such measurements untrustworthy.

The international ITS is illustrated in Figure 7.1.1 and the international VITS in Figure 7.1.2. Since the former is the more complex, and in any case the component waveforms are very similar in the two insertion signals, the discussion will be limited to the ITS. Any points of difference will be noted as they occur.

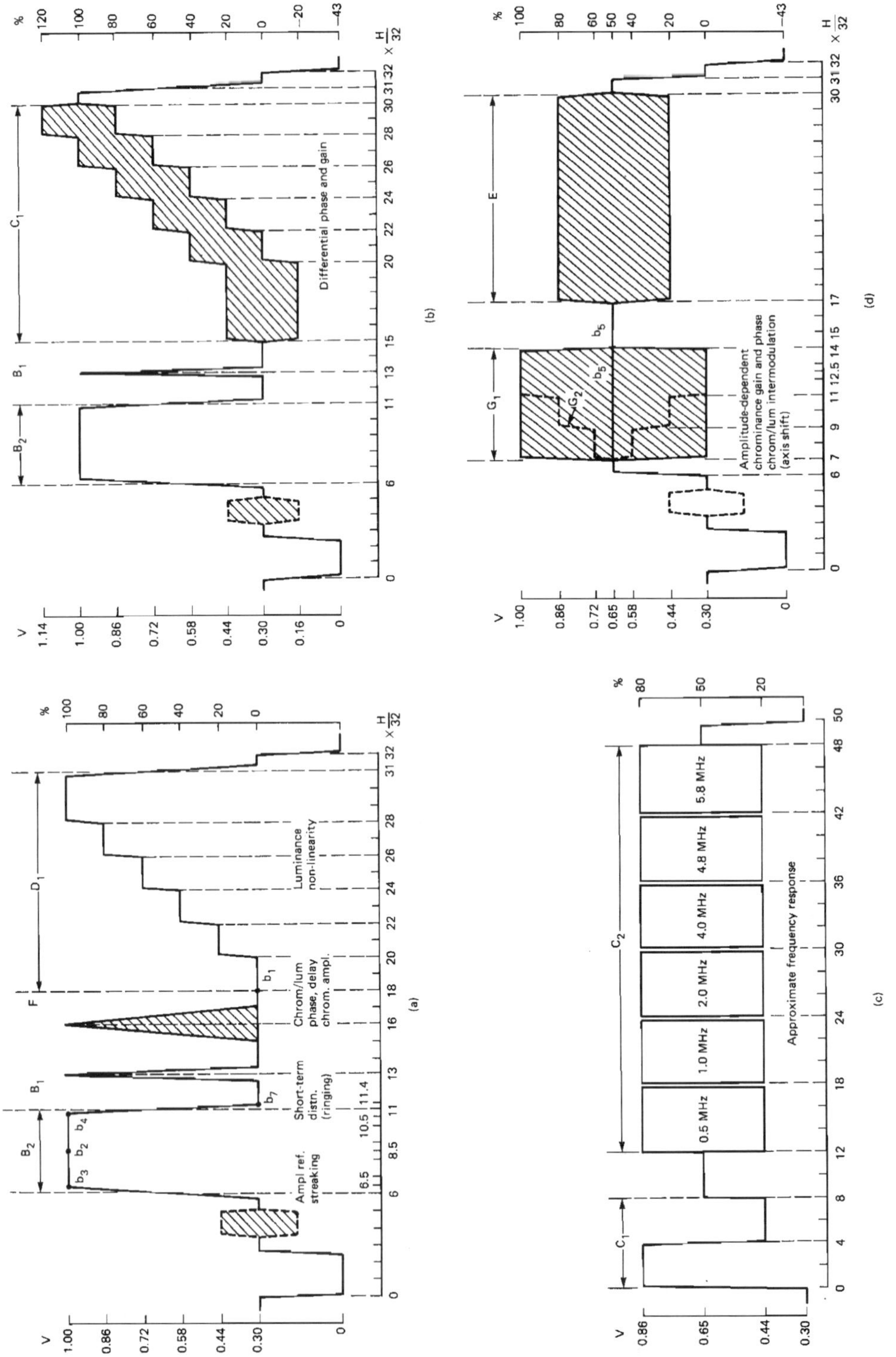

Figure 7.1.1 International 625-line ITS: (a) line 17; (b) line 330; (c) line 18; (d) line 331.

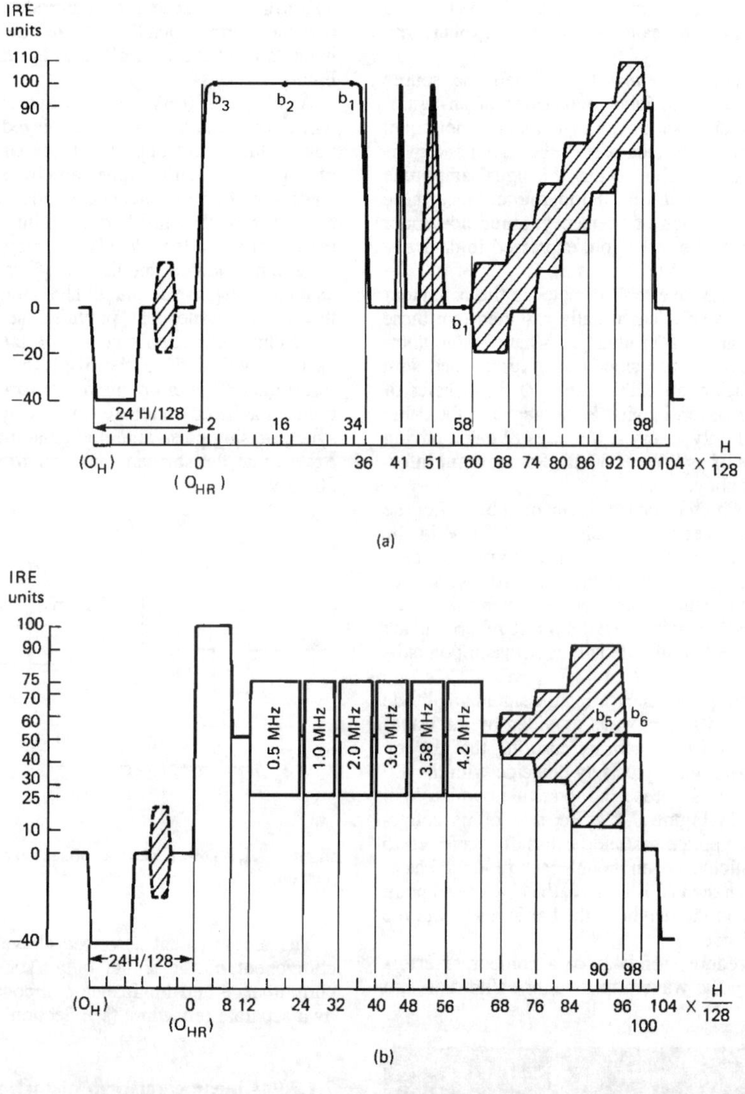

Figure 7.1.2 International 525-line VITS: (a) line 17, field 1; (b) line 17, field 2.

7.1.3 Measurement Techniques

The principal linear and non-linear transmissions of analogue signals will be described in terms of the corresponding component waveforms of the ITS. A synoptic view of these is provided in Figure 7.1.1. This does not exclude measurements made on a line-repetitive basis for the optimum in accuracy, e.g. for acceptance testing, because the waveforms for this purpose are fundamentally the same, the principal difference being an extended duration in some instances. These analogue techniques are also employed where 'digital islands' exist in studio complexes, and with appropriate modifications will later form the basis of the measurement of digital transmission systems. The nomenclature of the distortions will be that of the CCIR[1] unless some special point needs to be made.

7.1.3.1 Signal amplitude

The importance of this measurement is too often underestimated. Signal amplitudes have to be maintained within very close limits for a number of reasons. In a long distribution system, errors can accumulate which may result in too high or too low a value at some point. The former may cause distortion and the latter will result in a worsening of the signal/noise ratio. Within studio complexes it is essential for all originated signals to have amplitudes as nearly equal as possible, since the eye is remarkably sensitive to even very small differences. Good practice tries to limit these to ±0.1 dB.[4,5]

It must be borne in mind that since one is dealing with equipment which is mostly used between 75 ohm terminations, the measurement ought to be that of *insertion gain or loss*. The difference between this and a measurement of level across terminations can be very significant. For a detailed explanation, see Weaver.[6,7]

A complicating factor is *return loss*. When the source impedance is not equal to the input impedance of an equipment, or the output impedance of the equipment is not equal to the terminating impedance, then part of the signal energy is reflected in the form of an echo, so that the signal amplitude passed on is less than it should be. Furthermore, since these impedances may be a function of frequency, amplitude under ideal matching to that of the reflection, expressed in decibels, is the return loss.[6–8]

Another very frequently overlooked factor is the behaviour of coaxial connecting cables. It is usually assumed that these always have an impedance of 75 ohms, or whatever the nominal value may be. In fact, this value is only approached with high-quality double-sheathed flexible cables at frequencies of 1 MHz and above. Moreover, inevitable variations in the cable constants can result in only a very few metres of cable having an intrinsic return loss of as little as 30 dB. Some useful information is given in Whalley.[8]

The 10 μs wide bar which is the first element in line 17 of the ITS (Figure 7.1.1(a)) is set very precisely to 700 mV in the insertion signal generator, and the difference between the centre of the bar top and black level is the standard measure of amplitude. Since disturbances may occur on black level, experience has shown that point b_1 is the most reliable black reference, and the desired amplitude is thus between points b_2 and b_1.

The most usual technique for measuring the signal amplitude is to utilise the calibrated square wave provided by the waveform monitor to standardise its gain by ensuring that the top and bottom of the square wave precisely coincide with the 1.0 and 0.3 lines on the ITS measuring graticule, which will resemble that shown in Figure 7.1.3. Because of its convenience, this type of compound graticule is usually preferred to a set of individual graticules, even though each function has to be 'skeletonised'. When the ITS is viewed, the black level point is set to the 0.3 line, and the height of the bar is read from the scale on the left-hand side.

However, this procedure suffers from a number of errors not only possibly from the waveform monitor calibration, but also from parallax and those introduced by the need to replace one waveform by another. A preferred alternative which eliminates all but the first of these is illustrated schematically in Figure 7.1.4.

A precise 700 mV square wave is added internally in the waveform monitor to the displayed waveform, giving two traces which are identical but displaced vertically. In the example of Figure 7.1.4, the signal amplitude is too low. A vernier control on the amplitude enables the two measurement points to be located on the same horizontal line, and the error can be read off directly. Further details are provided in Weaver.[6,7] Tests have demonstrated that this *dc offset* method has very significant advantages over others. The findings of Smith[10] agree with those of the writer that, provided the ITS general distortion is not unduly severe, an accuracy of ±0.05 dB can be attained. A useful practical tip if the displayed waveform is noisy is to insert the *IRE filter* or similar network, which can substantially reduce the 'fuzz' and thereby improve the setting accuracy. The synchronising pulse amplitude may be found in the same way, except that the standardising square wave amplitude is made 300 mV.

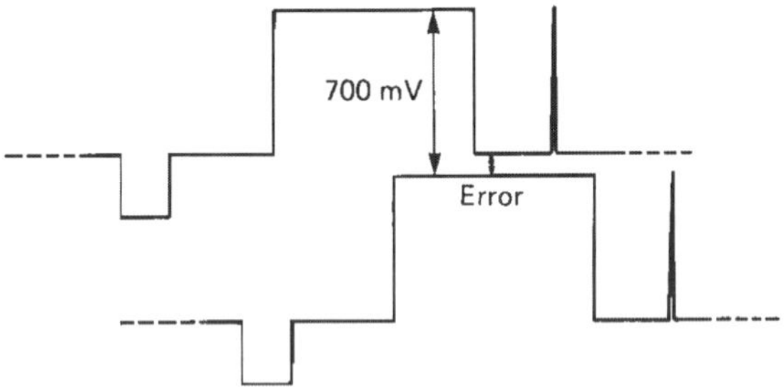

Figure 7.1.4 Direct current offset signal amplitude measurement method.

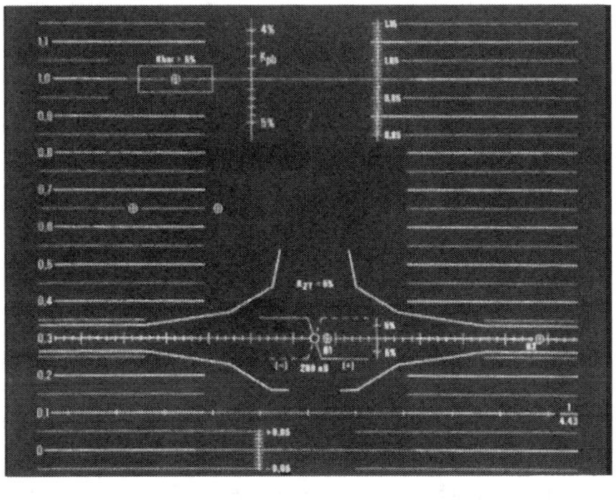

Figure 7.1.3 Composite ITS graticule (Tektronix Inc.).

The measurement described above concerns the luminance component of the signal only. The difference between the chrominance and luminance component amplitude is derived by a separate procedure (see Section 7.1.3.2.6).

7.1.3.2 Linear waveform distortion

The very convenient CCIR convention of classifying these distortions into groups according to the rough duration of the disturbance on the waveform (i.e. short-time, line-time, field-time, long-time) will be followed here.

7.1.3.2.1 Short-time waveform distortion

The length of a short-time disturbance is about 1 μs, taking the form of 'rings', i.e. overshoots on transitions. The method used is the ***sine-squared pulse and bar technique***, which revolutionised ideas on television testing when it was introduced in the early 1950s. This can only be described here in the briefest outline, but further information is available in Weaver,[6,7] as well as the original paper by Lewis.[11] The sine-squared measurement principle, in fact, takes in linear waveform distortions of other durations up to field rate. These will be dealt with below under the relevant heading.

The basic principles can be summed up very briefly as follows:

- The test waveform must as far as is practicable be representative of normal picture content. To this end it takes the form of a rectangular bar, corresponding to large areas of tone, and a narrow pulse which represents fine picture detail. These are the first two components of lines 17 and 330 of the ITS, and of the field 1 VITS (Figures 7.1.1 and 7.1.2).
- The energy in the pulse and bar should as far as possible be confined to the nominal video bandwidth, which is taken to be 5 MHz for all 625-line systems and 4.0 MHz for 525-line and the PAL systems using a 60 Hz field rate, e.g. PAL M. This nominal bandwidth is kept even for PAL I and SECAM with their wider bandwidths.
- The test waveform should be as well-shaped as is possible so that distortions are easily recognisable, and it should be capable of generation to a very high degree of consistency.

The final requirement was achieved by generating the pulse and bar initially with very rapid transitions, and applying them to the input of a special low-pass filter known as a 'Thomson network';[1,6,7] half-amplitude duration (had) of the pulse is defined in terms of a constant $T = 0.1\,\mu s$, where its effective bandwidth is the reciprocal of this had. The most usual 625-line pulse is the $2T$ (had $= 0.2\,\mu s$) giving a bandwidth of 5.0 MHz, but $1T$ pulses are also used for special purposes and $2.5T$ for 525-line work.

The pulse shape agrees very closely with $E = \sin^2(\pi t/2H)$, where H is the half-amplitude duration, hence it name. Figure 7.1.5 demonstrates how well-shaped the practical pulse is; the only defect is a small and quickly damped overshoot on the right-hand side, which is tolerated. The bar transitions are virtually identical to those of the pulse.

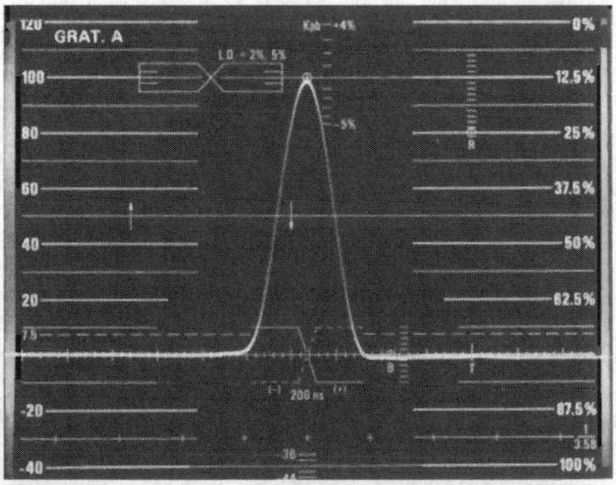

Figure 7.1.5 Generated sine-squared pulse.

Lewis and colleagues[11] also took the very important and innovative step of rating the distorted pulse in terms of the equivalent subjective picture impairment by using the *paired-echo* principle of Wheeler.[12,13] For the $2T$ pulse this yielded the tolerance diagram of Figure 7.1.6(a), where it is clear that a 4% echo at a spacing of $2T$ is equivalent to a 1% echo at $8T$, and

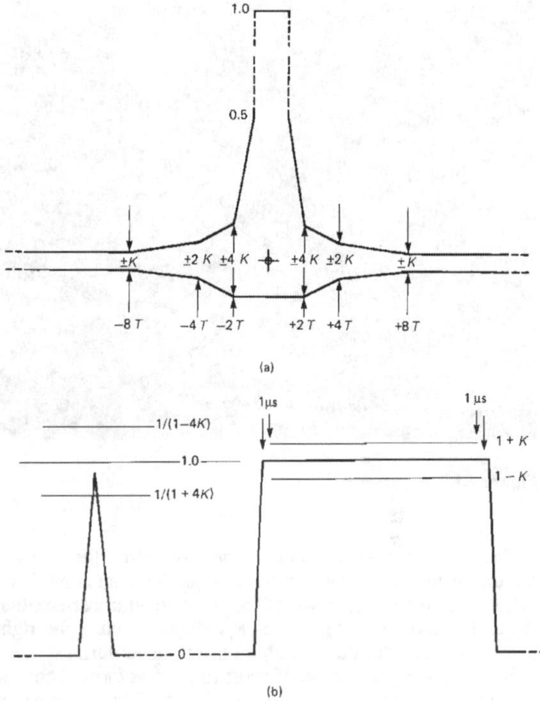

Figure 7.1.6 Basic tolerance schemes for pulses and bar rating: (a) $2T$ sine-squared pulse rating diagram; (b) pulse/bar and bar ratings.

both would be given a $2T$ K-rating of 1%. The diagram for the NTSC $2.5T$ pulse is identical except for the appropriate horizontal scaling. This particular K-factor is only one of several, and should more properly be denoted by K_{2T}.

For the practical measurement of K_{2T} the waveform monitor horizontal speed is first set to match the time scale of the graticule in use. Usually, this is preset. Then the pulse is adjusted in height and position so that its baseline lies on the central line of the tolerance diagram (see Figure 7.1.3), while the peak of the pulse coincides with the 1.0 graduation. K_{2T} is then given by the set of tolerance lines which would just contain the pulse overshoots. This usually has to be done by visual interpolation. In Figure 7.1.3, only $K_{2T} = 5\%$ tolerance lines are given for clarity. Sometimes 2% and 4% are chosen but more sets of lines than two are confusing. A photograph of a K_{2T} measurement is shown in Figure 7.1.7.

K_{2T} is a unique and invaluable measure of the transient distortion associated with an extremely small item of picture detail in terms of the picture impairment it causes, but further information can be derived from the $2T$ pulse. Any bandwidth limitation gives rise to an increase in the transition time of transients, as is well known, which must increase the half-amplitude duration of the $2T$ pulse. However, the energy in the pulse is not changed, so its amplitude must consequently decrease. This can be recognised and measured by comparison with the amplitude at the centre of the bar top, which will not be modified, and yields a further K-factor K_{pb}, which in general terms is an indication of the resolution of the system. Figure 7.1.6(b) shows another set of tolerance lines, again related to the subjective impairment.

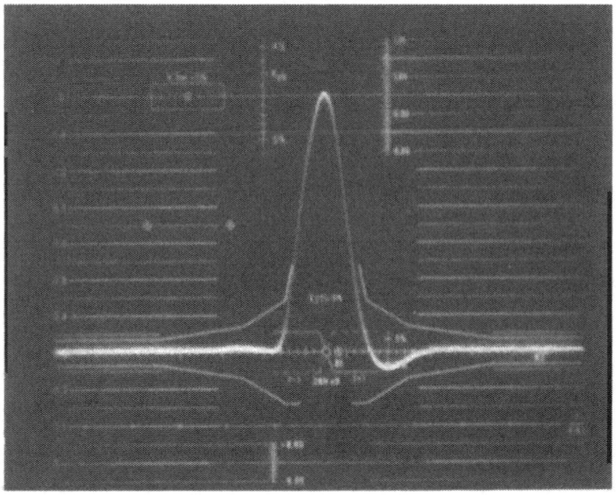

Figure 7.1.7 K_{2T} measurement.

The practical measurement is made by setting the centre of the top of the bar to the 1.0 line in a graticule such as Figure 7.1.3, then moving the peak of the pulse to an upper auxiliary scale. The left-hand scale gives K_{pb} directly, while the right-hand scale provides the pulse/bar ratio in linear terms.

Before leaving the topic of short-time waveform distortion, two important points must be raised. The first concerns the tolerance scheme of Figure 7.1.6. Later and more refined work by Allnatt[12] has demonstrated that this is not entirely correct. However, the errors are not thought to be serious enough to warrant modifying a technique which is in worldwide use, all the more since long experience has shown how valuable it is in practice, and how little the errors seem to matter operationally.

The second point can best be explained by reference to Figure 7.1.8. The spectrum of the $2T$ pulse falls to 0.5 of its initial value at 2.5 MHz, and essentially to zero at 5.0 MHz (2.0 and 4.0 MHz for 525 lines). This implies that the upper part of the video band receives a much lower weighting than the lower. This is not as serious as it seems at first sight, first because distortions at the higher frequencies produce a smaller effect on the picture than the same distortions at the lower frequencies. Also, the ITS and VITS contain chrominance test waveforms which supplement to some extent the information derived from the $2T$ pulse.

However, especially in studios where the available bandwidth is often much higher than the nominal, it often seems desirable to know more precisely what effects are occurring towards the upper end of the nominal video band. This was not overlooked in the original work,[11] where mathematical manipulation was proposed to remove the redundant and misleading transients which may occur when a $1T$ pulse (i.e. 0.1 μs half-amplitude duration) replaces the normal $2T$ pulse. This effect is especially serious if there is any bandwidth limitation in the system, as may be judged from the large amount of energy above 5 MHz in the $1T$ pulse demonstrated in Figure 7.1.8.

This method, although sound in itself, is tedious to implement and has only rarely been employed. Nevertheless, the coming into operational use of microprocessor-based automatic measurement equipment (see Section 7.1.) might well provide a means for reviving this technique in an operationally convenient way.

A device extensively employed by some broadcasters is based on the use of one of the transitions of a $1T$ bar, i.e. the normal $2T$ Thomson filter in the generator is replaced by a $1T$ version. This has a spectrum which rolls off more rapidly than that of the $1T$ pulse, so that transients resulting from amplitude and phase effects above 5 MHz are much reduced. No K-rating is possible, and the output $1T$ transition is judged by comparison with an empirically derived tolerance graticule. Under conditions where high-frequency distortions tend to be small in any case, the $1T$ pulse is often employed as a means of comparing signal paths, since one is then only looking for significant differences.

7.1.3.2.2 Line-time waveform distortion

Even very small phase-angle errors at frequencies below a few hundred kilohertz produce a slope on horizontal areas such as the top of the ITS bar. The corresponding amplitude error is usually too small to be measured. A single distortion of this type is really exponential in shape, although frequently only the initial straight part of the curve is seen. More than one such distortion will give the bar top a complex shape.

When the effect is very severe (see Figure 7.1.9), the fact that the slope is an exponential causes interference between successive lines. For this reason, the CMTT has recommended[13] that the line immediately preceding the first ITS line should always contain a 50% amplitude signal, which minimises the interaction. The corresponding picture impairment is particularly serious due to the dragging-out of picture information into adjacent areas. Sometimes this appears as 'stressing' following transitions, sometimes as a general loss of contrast and definition. This, of course, applies to all standards.

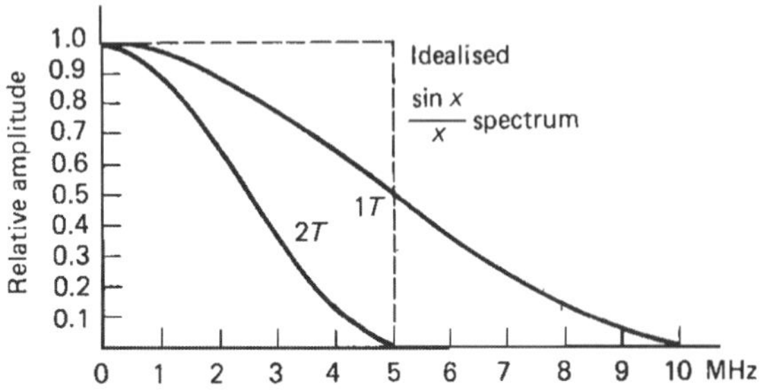

Figure 7.1.8 Comparison of spectra of $2T$, $1T$ and sin x/x waveforms showing the response of $1T$ and $2T$ shaping networks for 625 lines.

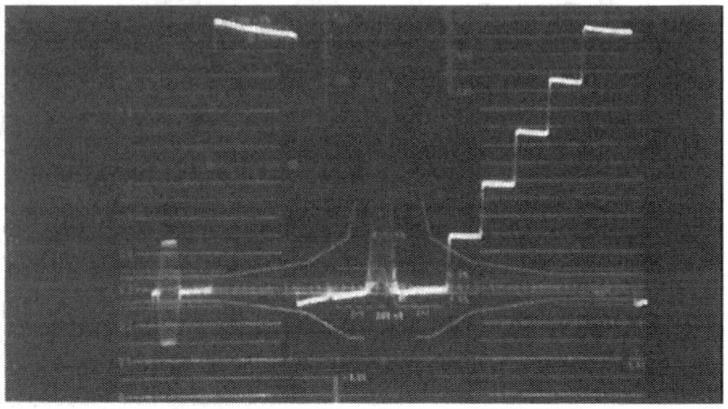

Figure 7.1.9 Example of line tilt.

For ITS purposes, the bar slope is defined as the difference in level between points on the top of the bar 1 μs from each of the transitions, expressed as a percentage of the bar height. The first and last 1 μs intervals are omitted because they may contain short-time distortions which are irrelevant in this instance.

The process is simplified when using graticules such as that of Figure 7.1.3. The bar is located with its baseline on the 0.3 line, and the centre of the bar top passing through the mark in the middle of the K_{bar} 'box'. The transitions of the bar will pass through the half-amplitude marks. Since the width of the 'box' is 2 μs less than the bar width, i.e. 8 μs, it is simple to estimate the amplitude difference between the points where the bar top cuts the 'box' ends.

This shows up a difficulty with the ITS and VITS; 8 μs is a barely sufficient period over which to make this measurement, but a longer duration cannot be had. For the most searching test, a line-repetitive waveform is preferable, where the bar is about half the active line length, i.e. 25 μs. The procedure is similar, but in this instance the greater of the two level differences between the ends of the bar, again omitting 1 μs each end, and the bar centre gives K_{bar}.

7.1.3.2.3 Field-time distortion

Field-time distortion cannot be measured accurately with the ITS or VITS, although some indication can be derived from the difference in level between corresponding points on the 10 μs bar in two successive fields. The best is the waveform given in CCIR,[1] consisting of a series of lines with a total duration of 10 μs, each carrying white level bars for the whole of the active line duration. Each such 10 ms group is followed by another at black level, resulting effectively in a 50 Hz square wave with line synchronising pulses. Field syncs may also be added if desired. For 525 lines the half-duration of the square wave is 8.33 ms. The composition of this waveform and its transition are shown in Figure 7.1.10.

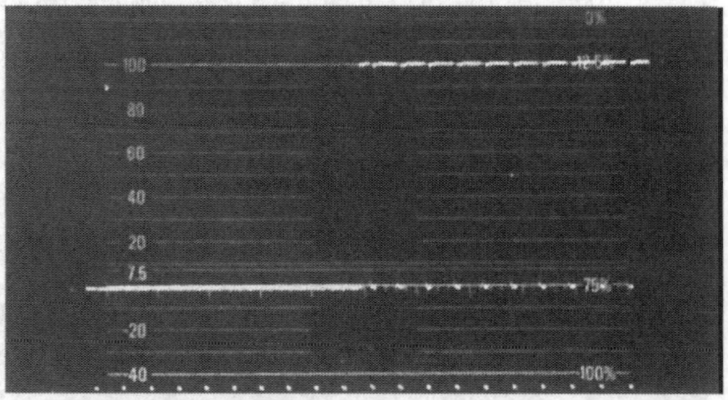

Figure 7.1.10 Fifty hertz test waveform.

The field-time distortion is defined as the greater of the two deviations of the bar top from the centre, expressed as a percentage. The first and last 250 μs are omitted. It may be necessary to modify the generator frequency in order to distinguish between true lf, slope and hum. In the original sine-squared pulse and bar method this was K_{50}, but on long circuits the measurement can be confused by other types of distortion, and it is now most often used in acceptance tests.

7.1.3.2.4 Overall K-rating

The rating method put forward by Lewis[11] advocated selecting the largest individual *K*-factor of the group as the overall *K*-rating. This has now been abandoned, not because it is a poor idea, but because it has been found more useful to record the separate values for informational and statistical purposes.

7.1.3.2.5 Long-time waveform distortion

Long-time distortion is a phenomenon of very long circuits which effectively contain a large number of series CR networks used for dc isolation. In an ac-coupled condition, a sudden change in the average picture level (apl) is equivalent to the addition of a dc transient, which can be up to 700 mV in amplitude. This initiates a damped oscillation whose amplitude initially can be around 40% theoretically,[14] although work by the writer suggests that even a very small rise in amplitude at the lowest video frequencies can make that figure much larger. Its total duration may be as long as tens of seconds, during which time the signal can suffer severe non-linear distortion.

The test waveform resembles somewhat the 50 Hz square wave of Figure 7.1.10, but it is usual to reduce the total excursion to either 10–90% or 12.5–87.5% to approach practical conditions more closely. The waveform duration is preferably variable to suit the conditions of measurement.

Since the process is so slow, direct viewing is not possible without either a storage or a digital sampling oscilloscope. Photography is the more usual solution, and both transitions must be measured. In order to reduce the total transient swing one may differentiate by using a high-pass filter (Figure 7.1.11).

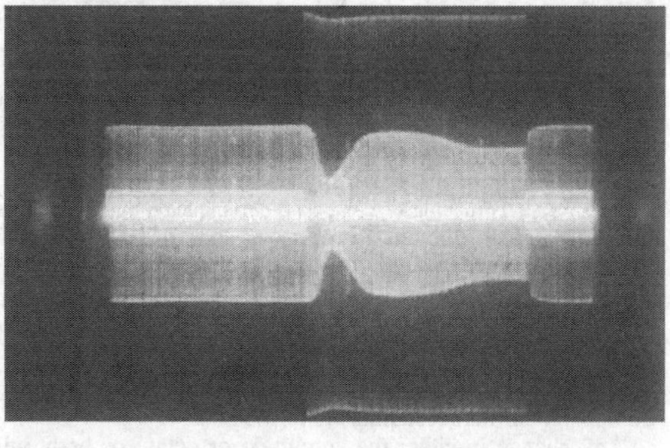

Figure 7.1.11 Long-time distortion at end of international circuit (differentiated).

7.1.3.2.6 Luminance–chrominance inequalities

Ideally, the ratio of the amplitudes of the luminance and chrominance components of the video waveform should always remain unaltered, otherwise the colour saturation is modified. The two channels must also arrive without any displacement in time, which could correspond to a registration error.

The test waveform is the composite chrominance pulse, which is the third element in line 17 of the ITS and the field 1 VITS. Its generation is clearly shown in Figure 7.1.13. A luminance sine-squared pulse of 50% amplitude is also

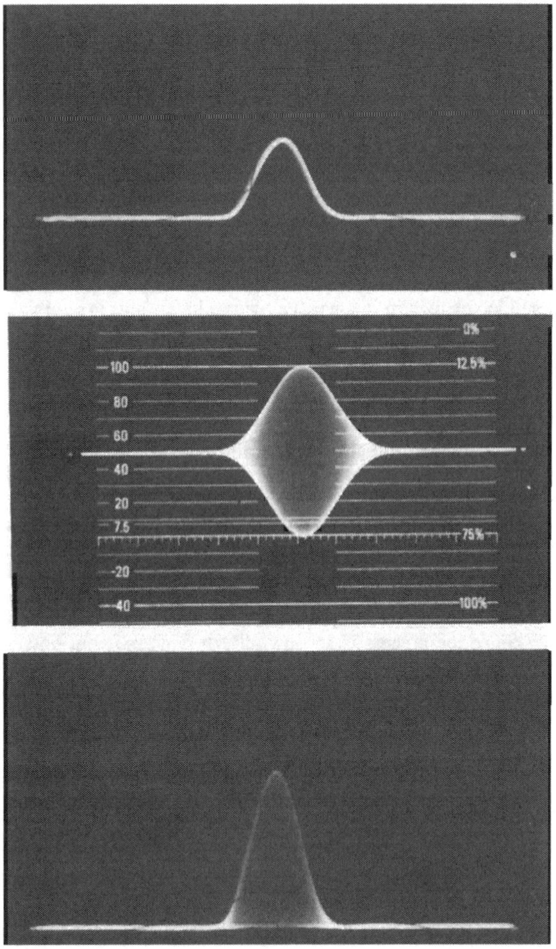

Figure 7.1.13 Formation of composite chrominance pulse: (a) luminance component; (b) chrominance modulated; (c) composite pulse from addition of (a) and (b).

modulated 100% onto a subcarrier. When these two waveforms are added with precisely identical amplitudes and delays, a composite chrominance pulse is formed.

Any change in either the amplitude or delay, or both, will upset this delicate balance. If only the gains change, then the baseline acquires the shape of a half-period of a sinusoid, either above the baseline as in Figure 7.1.14(a) for a chrominance loss, or below for a chrominance gain. This effect is independent of the pulse duration, which for international purposes is $20T$ for 625 lines and $12.5T$ for 525 lines. Internally, PAL I uses $10T$.

When the relative delays differ, but the gains are equal, then the baseline takes the form of a complete period of a sinusoid. When the left-hand half-period is positive, as in Figure 7.1.14(b), the chrominance is lagging. More usually, both types of error are present at the same time, giving an asymmetrical baseline as in Figure 7.1.14(c). The convention is that delay errors are in nanoseconds and amplitude errors are percentages.

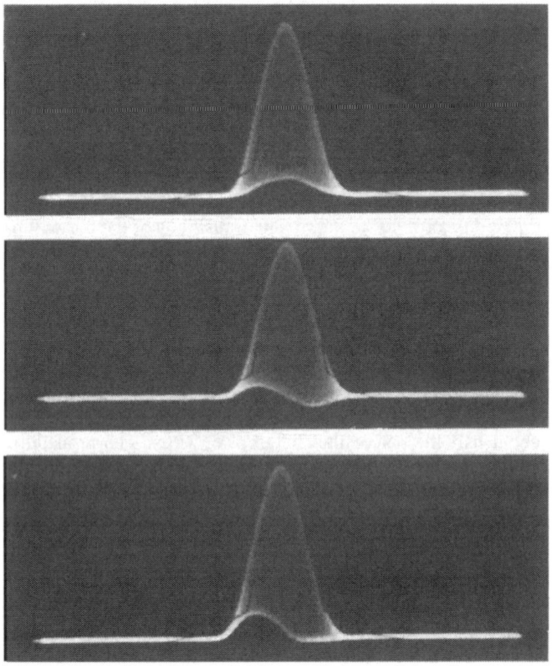

Figure 7.1.14 Distorted composite chrominance pulse: (a) low chrominance gain; (b) chrominance lagging; (c) simultaneous gain and delay inequalities.

For reporting purposes a chrominance loss and a lag are taken to be both positive, since the errors most often occur in that form.

In those instances where only one inequality is present at a time, the error is easily measured from the waveform. The signal amplitude is first standardised by setting the bar amplitude to 100% (or 100 IRE); then if the peak amplitude of the baseline lobe is y, the amplitude error percentage is $a = 2y$. The delay error (in nanoseconds) is given approximately by $d = (2T_c y)/100\pi$, where T_c is the half-amplitude duration of the composite pulse in nanoseconds. The graticule of Figure 7.1.3 has a 'box' in the centre of the baseline for such an approximate measurement, used in Figure 7.1.15 to estimate 200 ns chrominance lag.

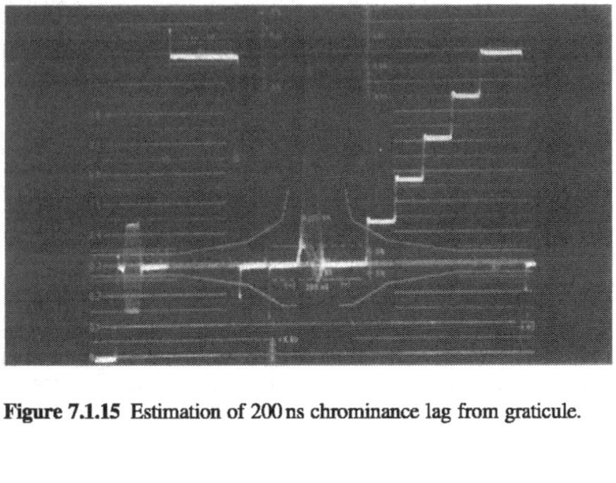

Figure 7.1.15 Estimation of 200 ns chrominance lag from graticule.

When both inequalities are present at the same time, the calculations from the waveform become difficult and unsatisfactory. The full expressions can be found in Rosman[15] and Mallon and Williams.[16] Nomograms have been devised, and an example of the Rosman type is to be seen in Figure 7.1.16, but the accuracy is low and the process tedious.

A very much more satisfactory method is the use of a gain and delay tester. It separates the two components of the composite pulse, whose amplitudes and relative delay can be varied by means of controls. The output waveform is observed, and when the baseline is seen once more to be flat, the errors are read directly from the controls. It is clear from the approximate expression above that the size of the lobe with delay errors is a function of the pulse width, which is the reason for preferring a 10*T* pulse in PAL I.

An excellent and more extensive discussion of chrominance–luminance inequalities can be found in D'Amato.[17]

7.1.3.3 Non-linearity distortions

Non-linear distortions have no unique value since they are very much a function of the apl. During 'out-of-service' measurements it is usual to measure at apls of 10% and 90%, or alternatively 12.5% and 87.5%, with the larger of the two values in either case being regarded as the determining error. With 'in-service' testing one is forced to accept the long-term mean apl of a little less than 50%.

7.1.3.3.1 Luminance non-linearity

Luminance non-linearity is measured with the plain staircase waveform on both the ITS and the VITS. Because direct measurement of the differences between the steps would be inaccurate, the waveform is differentiated by means of a CCIR-defined network,[1] which is built into professional waveform monitors. This converts the steps into a series of pulses

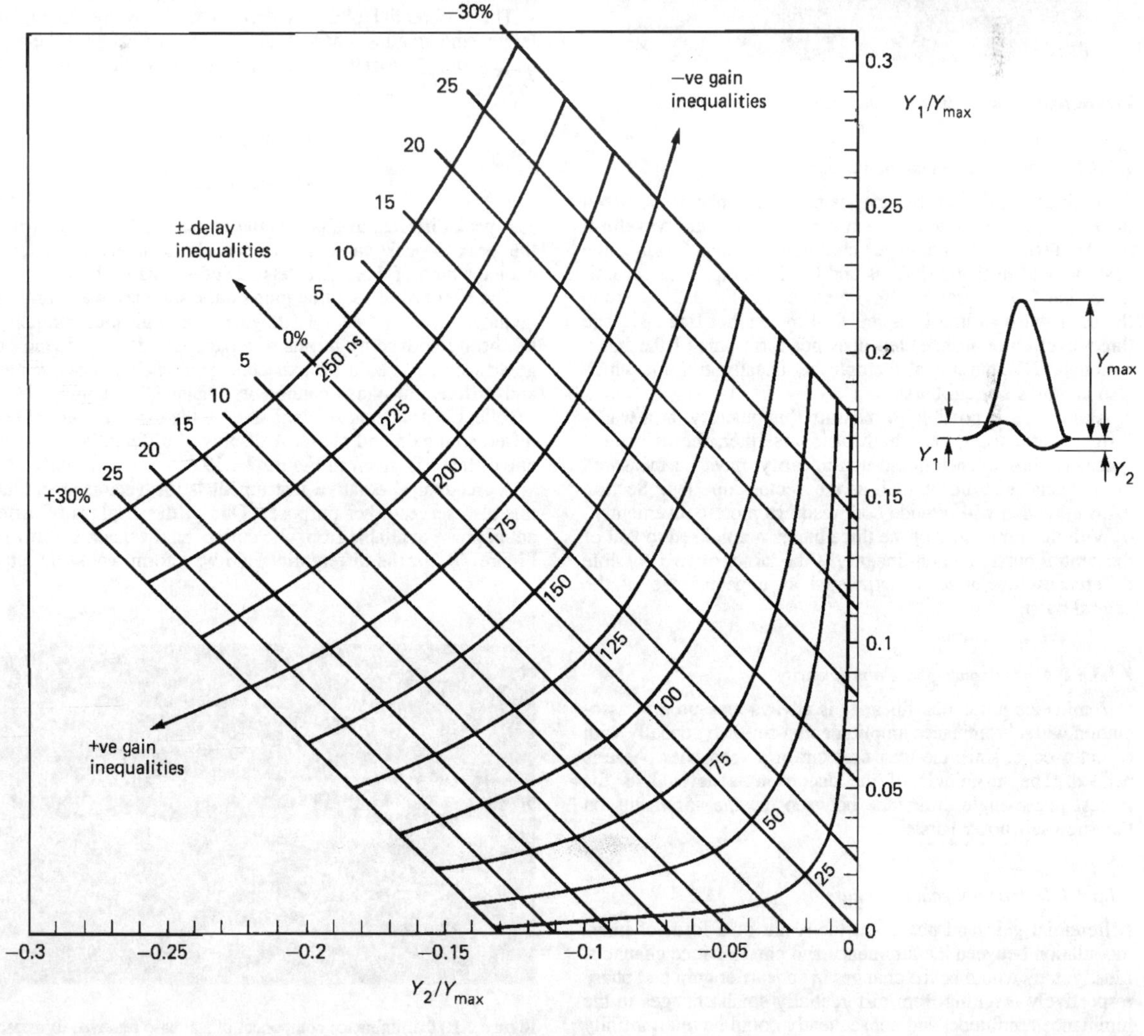

Figure 7.1.16 Nomogram for the 10*T* pulse (double delay readings for 20*T*).

(Figure 7.1.17), the height of each being proportional to the step from which it was derived. The luminance non-linearity is then defined as the difference between the largest and smallest amplitudes, expressed as a percentage of the largest.

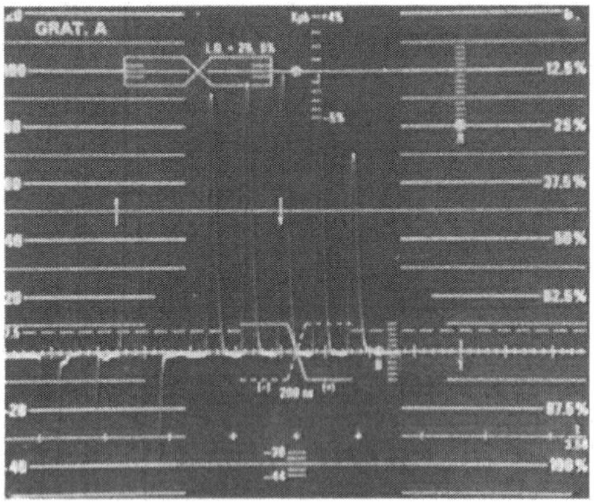

Figure 7.1.17 Differentiated luminance staircase waveform.

7.1.3.3.2 Chrominance gain non-linearity

Chrominance gain non-linearity is normally only encountered on very long links and in transmitters. The relevant waveform for the ITS is the three-level chrominance step forming the first element in line 331. This and the full-amplitude chrominance bar (see Figure 7.1.1(c)) are options. For EBU testing, the multi-burst of line 18 is replaced by another line 331. The three-level chrominance step does not form part of the international VITS, but it is also employed, usually in a line which also contains a multi-burst.

Although it is possible to measure this quantity on a waveform monitor using the inbuilt bandpass filter, the associated quantity, chrominance phase non-linearity, must be measured on a special instrument such as the vectorscope (see Section 7.1.6.1), which will provide both readings more conveniently.

With the amplitudes of the three bursts normalised to that of the central burst, the non-linearity is the larger of two possible differences in amplitude, expressed as a percentage of the central burst.

7.1.3.3.3 Chrominance phase non-linearity

Chrominance phase non-linearity is always measured in association with chrominance amplitude non-linearity, usually with a vectorscope, since the measurement of a subcarrier phase is needed. The magnitude of the distortion is defined as the largest phase-angle difference between the measurements on the three subcarrier bursts.

7.1.3.3.4 Differential gain and phase

Differential gain and phase are the result of a form of intermodulation between the luminance and chrominance channels. Ideally, they would be the changes in subcarrier gain and phase respectively resulting from differentially small changes in the luminance amplitude, and consequently could have an infinity of values over the black–white range.

In practice, the five-step staircase waveform of line 330 is used, corresponding to the luminance staircase of line 17, for the ITS, and the similar waveform in field 1 of the ITS for 525-line work. Both are overlaid with subcarrier, the ITS having ±140 mV and the VITS ±20 IRE units of a subcarrier amplitude.

The measurement of differential gain requires a comparison of the amplitudes of the subcarrier levels on the various steps with that at black level to find the differences. If the maximum and minimum differences are A_{max} and A_{min} respectively, and the subcarrier amplitude at black level is A_0, then two quantities are defined:

$$x = 100(A_{max}/A_0 - 1);$$

$$y = 100(A_{min}/A_0 - 1);$$

The *peak differential gain* is then numerically the larger of $+x\%$ and $-x\%$. The alternative *peak-to-peak differential gain* is $(x + y)\%$.

The differential phase is derived similarly by finding the largest and smallest phase-angle differences between the steps ϕ_{max} and ϕ_{min}, together with the angle ϕ_0 at black level. Then in comparable fashion:

$$x = \phi_{max}\phi_0;$$

$$y = \phi_{min}\phi_0.$$

The peak differential phase is the larger of $x°$ and $y°$, whereas the peak-to-peak value is $(x + y)°$. It is purely a matter of choice which of these two sets of definitions is chosen.

The effect of differential gain on the staircase waveform can be judged from Figure 7.1.18, where the luminance component has been removed by means of a bandpass filter. Evidently no great accuracy could be expected from such a measurement, and differential phase cannot be displayed so simply. Many excellent instruments are available for the measurement of both differential gain and phase. A display of differential gain from one of these is given in Figure 7.1.19.

A frequent alternative instrument is the vectorscope, since this also serves other purposes. One of the displays of differential phase available from one commercial version is shown in Figure 7.1.20; the differential gain waveform would be rather

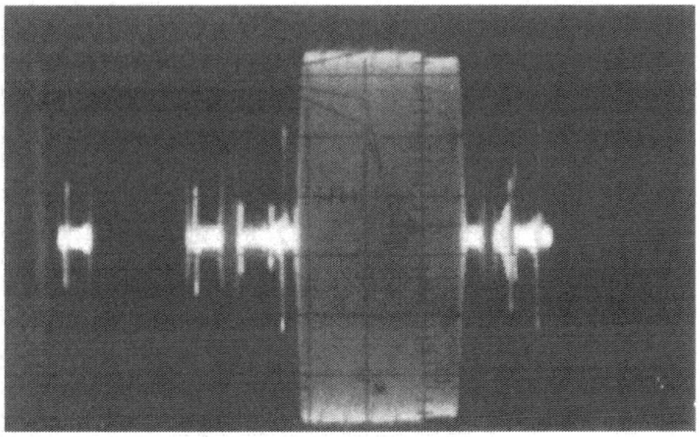

Figure 7.1.18 Luminance component of staircase removed to measure differential gain.

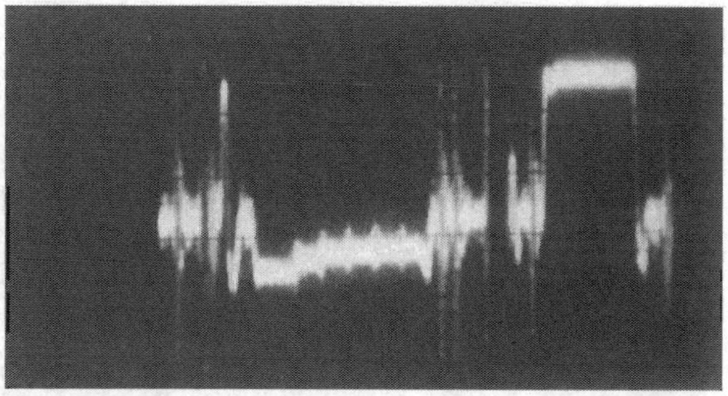

Figure 7.1.19 Differential gain display of off-air signal.

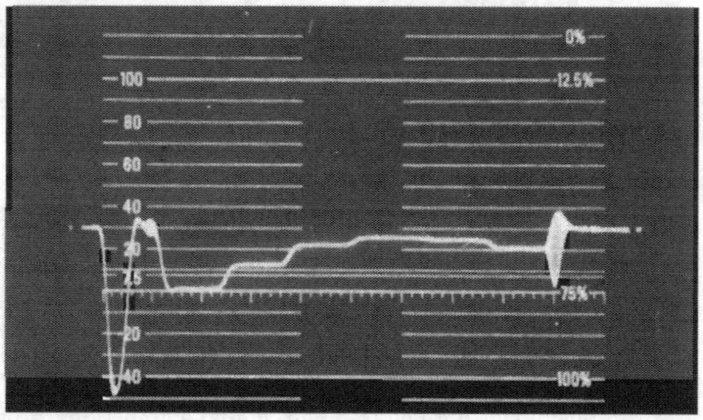

Figure 7.1.20 Twenty-five degree differential phase on vectorscope display.

similar. Although this is not the primary purpose of the vectorscope, the accuracy attainable can be really excellent.

As far as picture impairment is concerned, the eye is very tolerant of differential gain and of luminance non-linearity. Differential phase in NTSC is a great problem, since it gives rise to hue changes with luminance level, especially noticeable in skin tones, even for very small amounts of distortion. PAL and SECAM are very much more tolerant, since they were devised very largely with this in mind. PAL with delay line decoding converts differential phase into differential gain, and SECAM is affected in transitions only, where the impairment is visible mostly only to the skilled eye.[18]

7.1.3.3.5 Chrominance–luminance intermodulation

Whenever amplitude non-linearity is present on a transmission channel such that the gains are unequal for the positive and negative half-periods of the subcarrier, rectification takes place with the result that a dc component is added to the waveform. The common term for this, *axis shift*, describes the effect clearly.

The chrominance bar of line 331 of the ITS (or the three-step alternative), and the three-step chrominance waveform of the VITS, are all suitable. A low-pass filter is used to remove the chrominance component, giving the effect shown in Figure 7.1.21, where the positive dc step in the area where the chrominance was situated is very visible. This dc step (or that produced by the largest of the three steps) is measured and expressed as a percentage of the amplitude of the 50% bar in the chrominance bar waveform (not of the 100% bar!). When the step is upwards, as in Figure 7.1.21, the distortion is recorded as positive.

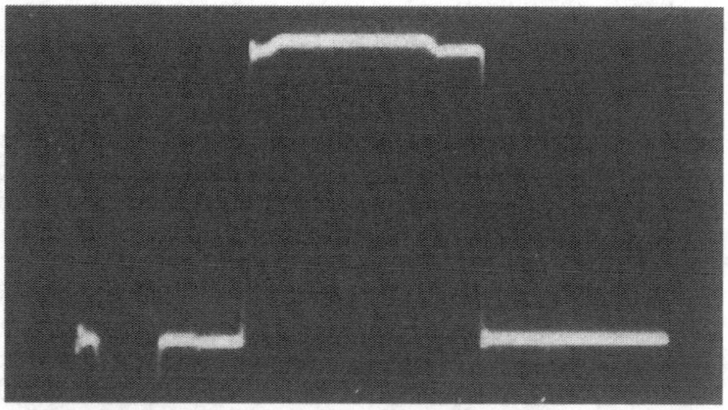

Figure 7.1.21 Chrominance–luminance intermodulation (chrominance removed).

Note that this distortion can give rise to large errors in the measurement of chrominance–luminance inequalities from the waveform monitor display, but not when the recommended specialised test set is used.

7.1.3.3.6 Multi-burst

The multi-burst waveform was introduced in the early days of video measurements with the aim of displaying the amplitude–frequency response in quantised form. It is found on line 17 of field 2 of the VITS, and line 18 of the ITS (but not for EBU purposes!). It takes the form of a white level reference pulse, followed by a sequence of rectangular frequency bursts between 0.5 and 5.8 MHz for the ITS, and 0.5 and 4.2 MHz for the VITS. The measurement is made by comparing the final burst amplitude with that of the reference pulse.

7.1.3.3.7 Noise and interference

7.1.3.3.7.1 Random noise

True random noise consists of an assembly of pulses with amplitudes and times of occurrence which are known only statistically. In theory, these amplitudes can range from zero to infinity, but in practice circuit conditions impose obvious restraints. Random noise is a consequence of the discontinuous nature of matter, and can never be eliminated. The only measurable quantity associated with it is a mean power when averaged over a sufficiently long period, and the corresponding rms voltage. The distribution of the power with frequency is also significant.

There are two basic types of noise spectra: *white noise*, where the rms voltage is constant with frequency; and *triangular noise*, where it is linearly proportional to the frequency. The latter is important because it arises from the demodulation of a pure frequency-modulated signal, although in practice it is modified by the pre-emphasis and other signal processing. Very often on long circuits the noise is *hypertriangular*, i.e. its spectrum rises more steeply with frequency than in the triangular case.

The definition of the signal/noise (s/n) ratio in television is the ratio of the white amplitude to the rms voltage, expressed in decibels. The justification for taking the rms voltage is that since the noise amplitude distribution is Gaussian, or a near approximation, the rms voltage is also the most probable.

Certain precautions must be taken before a noise measurement:

- Out-of-band noise must not be included, so the signal to be measured must be band-limited by a special filter[1,6,7] to 5.0 MHz (625 lines) and 4.0 MHz (525 lines).
- Power supply hum and other lf interference must be removed before a measurement. The CCIR recommends a high-pass filter, which can be used in conjunction with the low-pass band-limiting filter. Residual subcarrier can be another problem, but this is readily removed by means of a notch network.
- The subjective picture impairment of random noise is dependent upon its spectrum. This uncertainty is removed by means of a weighting network[6,7] through which the random noise is passed before measurement. The CCIR recommends the 'unified' network,[1] which is claimed to be suitable for both 625- and 525-line signals. In fact, it can be in error under certain conditions,[12] but its use is still standardised.

A general difficulty with colour television signals is the simultaneous presence in a signal of the luminance and chrominance channels, which are utilised differently in the receiver. For example, one must consider in PAL and NTSC that, although the chrominance bandwidth is so much lower than the luminance, the noise components falling into that region are demodulated down to become low-frequency noise in the chroma channel. The problem is even more complex in SECAM due to the considerable noise weighting already built into the system[18] and the fm demodulation.

The method internationally agreed for PAL I[7,19] is the use of a specially designed bandpass network that gives the same degree of weighting to the chrominance channel as the wideband network does to the luminance. Allnatt and Prosser[20] showed by subjective tests that this is true, and that a true measure of the overall s/n ratio can be obtained by combining the output of the luminance weighting and the chrominance weighting networks, with a 6 dB pad in series with the latter.

Although the use of the two weighting networks is preferred in the UK, it has nevertheless become common to accept the unweighted s/n ratio, in spite of the objections this raises, for general monitoring of circuits. Very often the weighted value is also measured and used together with the unweighted, since this provides additional information.

For the practical measurement of random noise, one of the 'quiet' lines is normally selected, e.g. 22 for 625 lines and 10 for 525 lines. The primitive procedure, once the normal method, is to measure the apparent peak-to-peak amplitude of the noise on a waveform monitor, and then to add a correction factor for conversion to rms voltage. The fact that the latter varies, according to different authorities, between 14 and 18 dB, gives some impression of the very poor accuracy. This is discussed in detail in Weaver.[6,7]

Two of the superior visual methods are the *tangential* of Garuts and Samuel[21] and the *inserted noise burst*. In the latter, a small gap is made in the centre of the quiet line, into which is inserted a variable burst of random noise from a local generator. The inserted noise pedestal and amplitude are then varied until the gap disappears; then the s/n ratio is read from the instrument. Figure 7.1.22 gives a deliberately incomplete adjustment to illustrate the process.

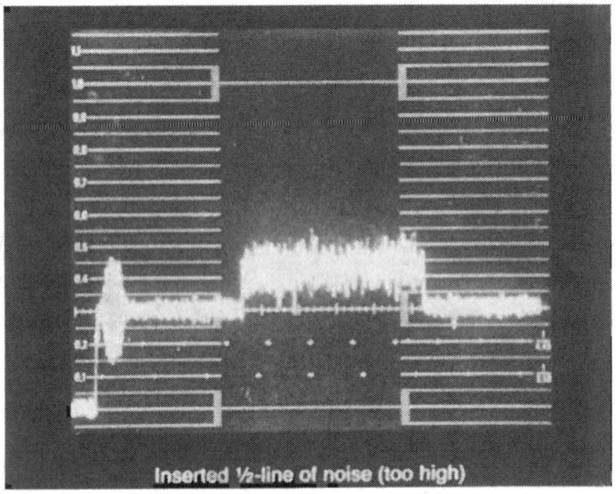

Figure 7.1.22 Inserted burst random noise measurement (with the burst deliberately offset) (Tektronix Inc.).

Objective measurements, however, are very much to be preferred. One typical and very successful instrument is described by Holder.[22] In this, a narrow burst of the random noise is sampled from the line; after processing, its rms voltage is found. The instrument is so arranged that the s/n ratio is read off directly. Such equipment is capable of ample accuracy for practical purposes over a wide range of s/n ratios. Especially refined techniques are possible with digital automatic measuring equipment.

7.1.3.3.7.2 Interference

The term *interference* covers a wide range of unwanted phenomena which impair a television signal. The one thing they have in common is that they are measured on a peak-to-peak basis on a waveform monitor. When they are periodic or quasi-periodic in nature, it is usually possible with skill to trigger the monitor so as to lock the interference for long enough for a measurement to be made. When it is erratic, recourse may have to be made to a digital sampling or storage oscilloscope. Where more than one form of interference is present on a video signal, the best indication of their combined effect seems to be obtained by the quadratic addition of the individual amplitudes.

Moiré is a special form of interference which arises from the frequency modulation process in a videotape recorder. It is measured on a spectrum analyser on a 100% colour bar waveform (see Section 7.1.6.1) from a pre-recorded tape. The moiré components must be measured on each of the colour bars, since the effect is a function of hue. The total interference is again expressed in terms of the root-sum-square value.[5,23]

7.1.3.3.7.3 Crosstalk

Crosstalk is the leakage of a signal in one path into another path by electromagnetic or electrostatic coupling, and occurs mostly in switching matrices. The number of possible combinations of disturbing and disturbed circuits is immense, so it is usual to concentrate attention upon those in close physical proximity. The test waveform may well be ITS line 17 or VITS field 1. In any case, high-amplitude chrominance must be present. The disturbed path should preferably carry a signal with, say, a black line in the position of the disturbing waveform, otherwise the crosstalk cannot be identified easily. The crosstalk is defined as the ratio of white level to the peak-to-peak induced voltage, expressed in decibels.[24]

7.1.3.4 Automatic measuring equipment

The ever-increasing size and complexity of television networks led long ago to the realization that most if not all of the routine measurements could advantageously be performed by automatic equipment (AME). This could tirelessly repeat sequences of measurements, even in locations difficult for staff to reach, issue alarms for actual or impending fault conditions, send results to a centre for statistical processing, and so on.

The earliest equipment used analogue techniques[25,26], although one early proposal for a digital system was made by Vivian[27]. Analogue AMEs have through long practical experience been developed to a high standard of performance and reliability. However, it must be admitted that they have a drawback in that the techniques used cannot always mimic those of an engineer, and it has been necessary for the CCIR and EBU to allow some modifications and relaxations[28,29]. In consequence, manual and automatic measurements cannot always be reconciled.

Digital methods are now in widespread use and offer great promise for the future from their reliability and versatility[30,31]. In the briefest possible terms, the test waveform is sampled at well above the Nyquist rate, and the data samples are stored, in one well-known instrument for 32 successive fields, permitting a noise reduction of 15 dB. These form a matrix of data values which can be processed entirely under the control of software. Not only can measurements be carried out by methods exactly analogous to those used manually, but they

```
                 TEKTRONIX 1980 ANSWER AUTOMATIC VIDEO MEASUREMENT SET
30-JAN-85 15:21:25                            VIOLATED LIMITS
                                               LOWER   UPPER

MULTIBURST FLAG        60.6  % BAR
MB PACKET #1           97.9  % FLAG
MB PACKET #2           97.1  % FLAG
MB PACKET #3           95.3  % FLAG
MB PACKET #4           97.2  % FLAG
MB PACKET #5           96.4  % FLAG
LOW-FREQ ERR PK-PK      3.9  %          *     0.0      2.0
UNWGHTED S/N (RMS)     48.2  DB        **    50.0    999.0
WEIGHTED S/N (RMS)     55.8  DB        **    60.0    999.0
(COMMANDS DONE)
```

```
                 TEKTRONIX 1980 ANSWER AUTOMATIC VIDEO MEASUREMENT SET
30-JAN-85 15:14:49                            VIOLATED LIMITS
                                               LOWER   UPPER

  CHANNEL A  (OFF-AIR)  PAL  15:15:00  APL = 48%
LINE SYNC DURATION      4.7  USEC
SYNC RISE TIME        280.0  NSEC     *   170.0   250.0
SYNC FALL TIME        260.0  NSEC     *   170.0   250 0
SYNC AMPL ERROR         -.2  %
FRONT PORCH         -------  USEC    **      .5     3.0   NO BLANKING EDGE
LINE BLANKING       -------  USEC    **     9.1    16.0   NO BLANKING EDGE
BROAD PULSE SEP        4.82  USEC
EQUALISING PULSE       2.29  USEC    **     2.3     2.4
LINE SYNC-TO-BURST     5.72  USEC     *     5.5     5.7
BURST DURATION         2.24  USEC                         9.9 CYCLES
BURST AMPL ERROR        -.6  %
BAR AMPLITUDE         670.6  MV
BAR AMPL ERROR         -4.2  %
BAR TILT                 .6  %
BAR RISE TIME         155.0  NSEC    **   160.0   240.0
BASELINE DISTORTN      -2.7  %       **    -2.0     2.0
2T PULSE/BAR ERROR      7.0  %
2T PULSE K FACTOR       1.2  % KF     *     0.0     1.0
LUM NON-LINEARITY     415.8  %       **     0.0    10.0
CHROM-LUM DELAY     -------  NSEC                         SUBCARR UNLOCKED
CHROM-LUM GAIN         10.3  %        *   -10.0    10.0
CHROM-LUM INTERMOD     -1.4  %
CHROM REF AMPL ERR      5.5  %        *    -3.0     3.0
DIFF GAIN  (PK-PK)     97.4  %       **     0.0    20.0
DIFF PHASE (PK-PK)      2.6  DEG
ENTER <CR> FOR NEXT PAGE:
```

Figure 7.1.23 Printout from digital automatic measurement equipment (Tektronix Inc.).

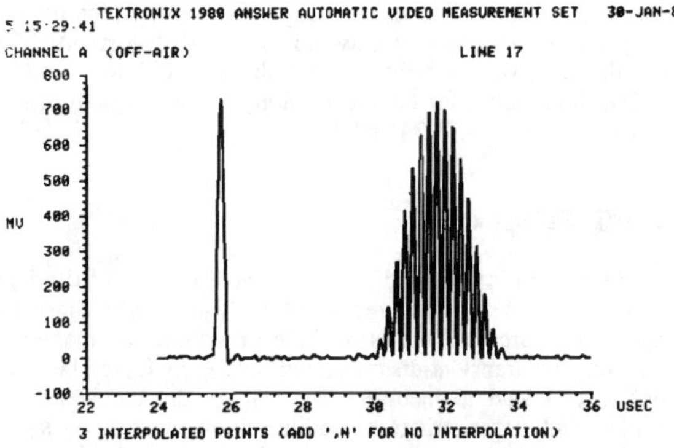

Figure 7.1.24 Waveform printout from digital automatic measurement equipment (Tektronix Inc.).

can be carried even further when necessary. For example, in the measurement of random noise, a fast Fourier transform can be applied to detect and remove periodic interference. A good impression of the range of possible measurements is provided by Figure 7.1.23. Figure 7.1.24 demonstrates how the test waveform components can be reconstructed if required for information or record.

Criticism is sometimes levelled at the accuracy of AMEs, compared with manual methods. In the author's opinion, the precision of the latter is very often overestimated. One can be sure with modern AMEs that the accuracy is perfectly adequate for routine operational purposes, and possibly very much better than can be achieved by an engineer unless he is equipped with all of the specialized equipment needed for some of the measurements. Also, the AME is tireless, and is capable of functioning efficiently in remote or inaccessible situations for 24 hours a day, seven days a week.

7.1.4 Measurement Tolerances

Measurements are insufficient in themselves. It is also necessary to know what errors are permissible at each point in the television chain. These must be determined globally from the subjective impairment experienced by viewers confronted by actual pictures with known values of the various distortions. This is a difficult but highly important subject, combining the techniques of psychophysics and statistics. A remarkably comprehensive account is given by Allnatt.[12]

Once the total allowable error has been found, it must be shared between the viewer's receiver and the chain of equipment between the picture source and the transmitter. It might be thought that in a long chain the distortions would add according to a Gaussian distribution, i.e. root-sum-square, but experience shows that this is by no means always the case. Moreover, the situation is complicated by the fact that the errors are also a function of time.[32–34]

There is yet another factor to be considered. The perceived picture impairment must be due to the effect of all the distortions present on the signal, and not to just an individual distortion. This problem seems to have been solved by Allnatt and his colleagues, who have shown that a quantity can be derived from subjective tests, called an *imp* (important unit), which has the property of summability between unrelated distortions. This has thrown a very significant new light on an old problem.[12,34]

Practical values for tolerances along the signal path can be found in Refs 1, 5, 7, 34 and 35.

7.1.5 Teletext

Teletext is an information service carried on lines of the field blanking interval by non-return-to-zero data pulses. Since the data pulses are as short as possible to include the maximum information, transient distortion will cause confusion between pulses and lead to incorrect decoding in the receiver, so a further and rather stringent condition is imposed upon signal quality, especially as regards transmitters, for countries using teletext. A general survey of the problems is given in Ref. 38.

A fundamental criterion for teletext quality is the 'eye-height', i.e. the display obtained on a waveform monitor triggered so as to overlay a series of pulses (Figure 7.1.25). The eye-height is the maximum clear height within the pattern, in this case about 68% of the possible value. The aim is to include the effect of noise as well, to obtain the decoding margin. For the definition and relationship to eye-height, see Refs. 39 and 40.

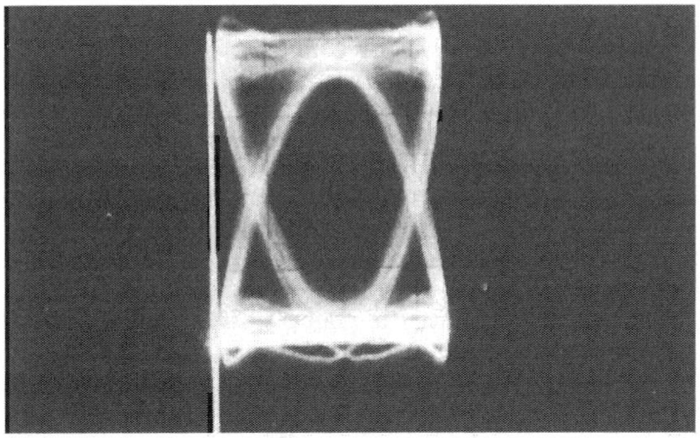

Figure 7.1.25 Teletext eye-height display (BBC).

7.1.6 Specialised Test Waveforms

7.1.6.1 Colour bars

Whatever system is in use, good pictures are impossible unless the encoders are correctly adjusted at the points of signal origination. The colour bar waveform is the standard test signal for this purpose. It consists of a white reference pulse followed by six colour bars of the primaries red, green and blue, together with their complements yellow, cyan and magenta. By convention they are in a sequence of descending luminance values. SECAM bars must consist of a pair, which is also needed in PAL for delay-line decoding. The great virtue of this waveform is that it can be generated to a very high degree of accuracy and consistency.

The three most common PAL colour bars are illustrated in Figure 7.1.26 and the SECAM bar pair in Figure 7.1.27, the former showing the colour separation components. The strange shape of the latter arises from the effects of the lf and hf pre-emphasis circuits. The 525-line standard closely resembles that of the EBU (Figure 7.1.26(c)) but with the luminance bar at 75% amplitude, and the addition of the 7.5% pedestal. The 100% bars are frequently used in studio practice since they correspond to the locus of 100% saturated colours, but the '95%' bars with their lower total amplitude range may often be preferred.

PAL and NTSC bars are universally measured with the vectorscope,[6] which is a polar display formed, as shown clearly in Figure 7.1.28, from the (R − Y) and (B − Y) (or in NTSC, I and Q) colour difference signals. The important point is that for correct encoder adjustment the tips of the colour vectors must be located at predetermined points, which are given tolerance 'boxes' on the vectorscope graticule to allow a very rapid estimate of the encoder quality. The central dot is the *white point*, which again must be correctly located. Its shape is also significant.

The display of Figure 7.1.28 corresponds to PAL with the normal V-axis switching. It is also possible to disable the switching, producing a display in which ideally the vectors of Figure 7.1.28 are perfectly mirror-imaged about the U (horizontal)-axis. This is invaluable for diagnostic purposes since the behaviour of each individual line is visible independently. This facility, of course, is not available with NTSC, where the display resembles Figure 7.1.28 except that the burst vector lies on the horizontal axis.

The standard approach with SECAM has always consisted in measuring the colour bars in terms of the deviations of the two subcarriers, and extremely effective instruments exist using that principle. However, a vector display is also feasible and useful. At least one range of high-class picture monitors for SECAM provides outputs of (R − Y) and (B − Y) to drive an X–Y monitor and so produce a display closely resembling Figure 7.1.28 except for the presence of the identification signals.

Some versions of the vectorscope are fitted with facilities for the measurement of differential phase and gain, which considerably increase the usefulness of this very versatile instrument.

7.1.6.2 Multi-pulse

The multi-pulse waveform was originally devised for the measurement of the amplitude and delay responses of transmitters.[41] As shown in Figure 7.1.29, it consists of a series of 10 pulses across the duration of a single line, making it suitable for use as an ITS. Each pulse is formed precisely in the same manner as the composite chrominance pulse of Section 7.1.3.2.6, except that the modulation is performed with a sequence of frequencies spaced across the video band.

Exactly as with the chrominance pulse, the baseline distortion will yield not just the amplitude, but also the delay error (cf. Figure 7.1.14). Moreover, the spread of the pulse sidebands is minimised by the sine-squared shaping, thus overcoming one of the objections to the multi-burst waveform (Section 7.1.3.3.6).

Operational experience has demonstrated that this waveform is a powerful tool, especially for the delay correction of transmitters as well as distribution circuits.

7.1.6.3 Sin *x*/*x* pulse

It was pointed out by Heller and Schuster[3] that test waveforms of very great accuracy and stability can be generated from

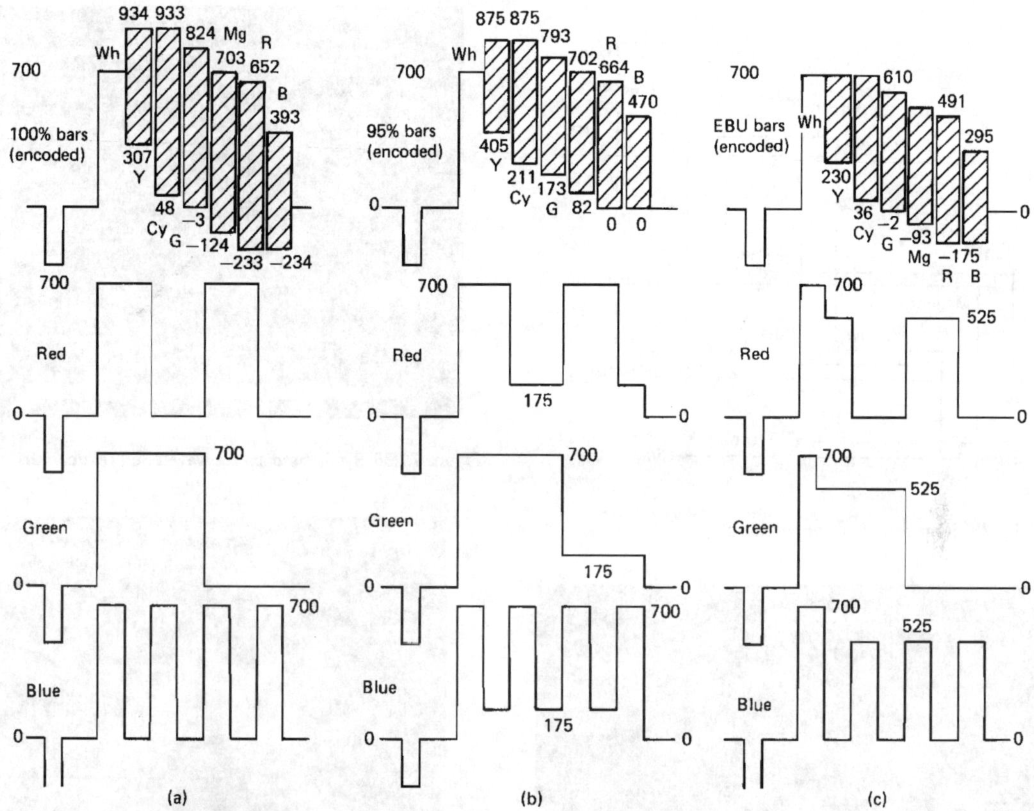

Figure 7.1.26 Composition of 625-line colour bar waveforms. All figures are in millivolts.

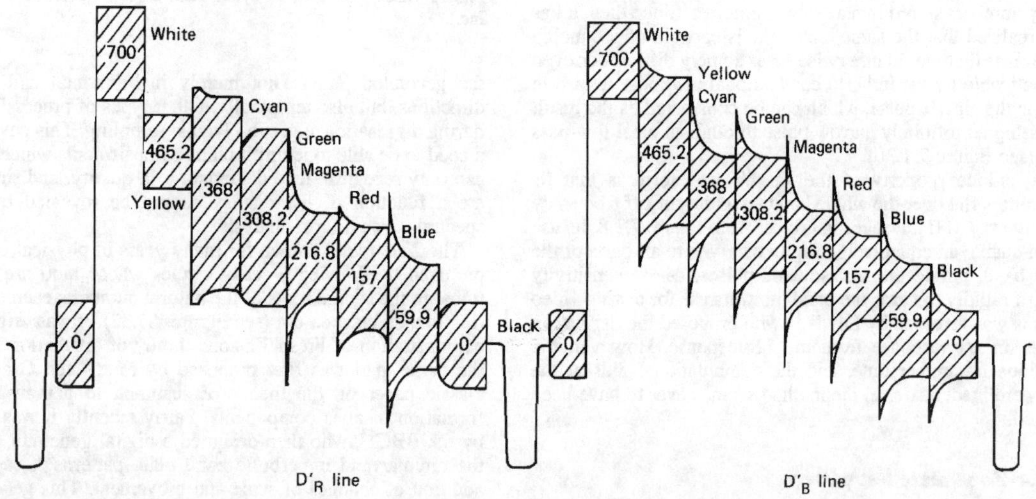

Figure 7.1.27 Standard SECAM colour bar waveforms. The figures within the bars are luminance levels in millivolts.

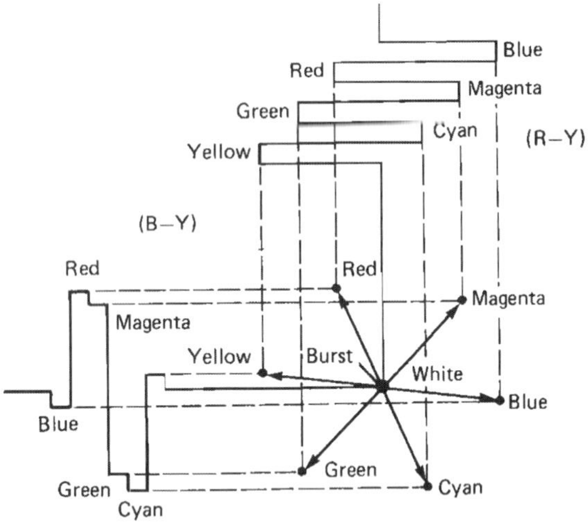

Figure 7.1.28 Formation of vectorscope display.

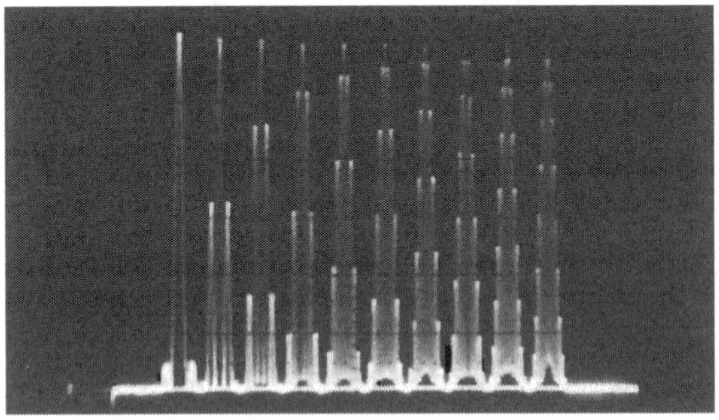

Figure 7.1.29 Multi-pulse waveform (Tektronix Inc.).

binary numbers stored in read only memories. Since then, it has been realised that the same technique is capable of producing waveforms that would otherwise be extremely difficult or even impracticable by the hitherto conventional methods. A case in point is the sin x/x pulse, which can be considered as the result of passing an infinitely narrow pulse through an ideal low-pass filter (see Figure 7.1.30).

The unique property of the sin x/x waveform is that its spectrum is flat over the whole band up to the cut-off frequency (see Figure 7.1.31). Hence, as is clear from Figure 7.1.8, its use would enable an equal weighting to be given to all parts of the video band, unlike the sine-squared pulses, whose sensitivity falls off rapidly towards the band limit. Its use for testing video systems was foreseen by Lewis,[11] who proposed the derivation of the sin x/x response from the $1T$ response. However, the technique is cumbersome, and the computational aids at the time were inadequate, so the method seems never to have been pursued.

7.1.6.4 Zone plate test pattern

In the search for better picture quality and advanced high-definition systems, it is now common to carry out filtration of the generated picture not merely in horizontal and vertical directions, but also temporally. Other types of processing arise during digitisation, e.g. sub-Nyquist sampling. This gave rise to a need to be able to identify picture impairments which the eye can only recognise as a degradation in quality, and since they are a function of the display cannot be revealed by either spectral or waveform analysis.

The *zone plate*, known for many years in physical optics, is prepared by drawing a set of circles whose radii are proportional to the square roots of the natural numbers; each alternate ring is then blacked out (see Figure 7.1.32). It was originally a demonstration of Fresnel's zone theroy of diffraction, whence the name, and was first proposed by Mertz and Gray[42] in a classic paper on the theory of scanning to demonstrate the formation of alias components. Fairly recently it was revived by the BBC[43], who then designed a digital generator for both the circular and hyperbolic zone plate patterns[44], capable in addition of changes of scale and movement. This generator is commercially available, and its use is widespread, since there is no other equivalent.

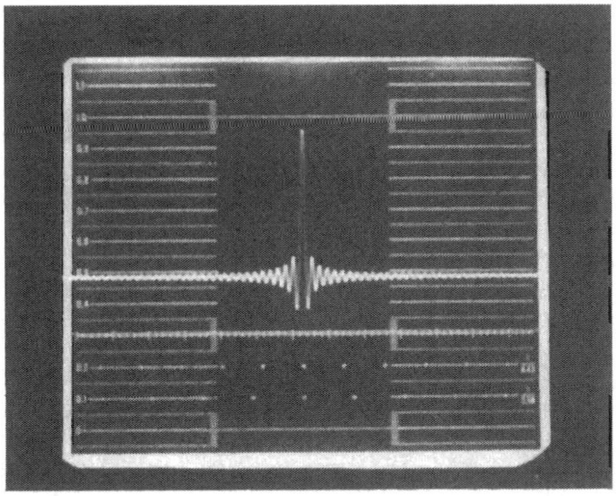

Figure 7.1.30 Synthesised sin x/x waveform (Tektronix Inc.).

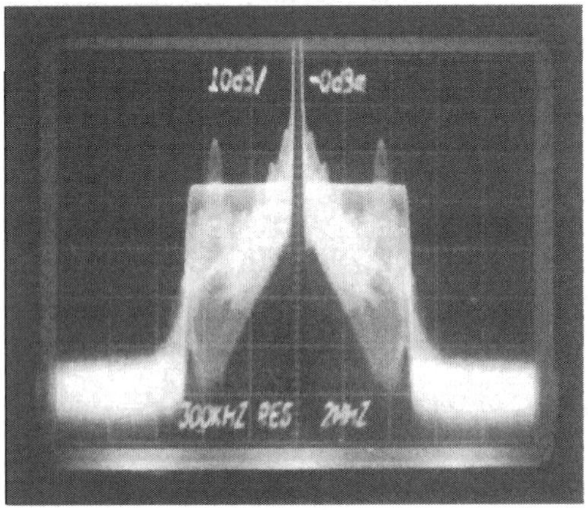

Figure 7.1.31 Spectrum of synthesised sin x/x waveform (Tektronix Inc.).

Figure 7.1.32 Zone plate pattern (BBC Research Dept.).

In the simplest possible terms, the zone plate may be considered as a linear two-dimensional frequency sweep (see Figure 7.1.32). Figure 7.1.33 shows the generated waveform displayed in luminance only on a waveform monitor. In this simplest instance the result is easily predictable. The dark portions at the left and right of the image are caused by the limited bandwidth of the video amplifier of the monitor. However, in the vertical direction the waveform is quantised by the scanning lines, whose number is equivalent to a spatial frequency too low for that of the display. The effect is the same as sub-Nyquist sampling, and as theory predicts, alias images of the zone pattern appear above and below.

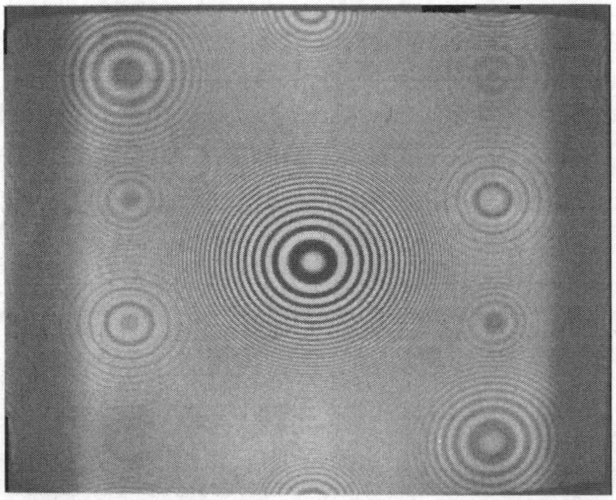

Figure 7.1.33 Displayed zone plate signal (luminance channel only) (BBC).

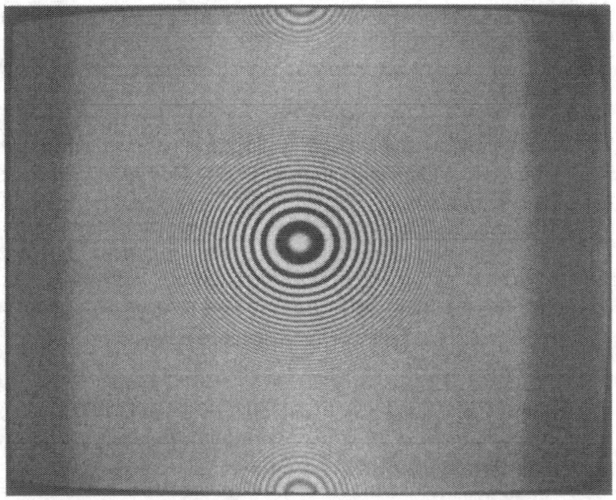

Figure 7.1.34 Displayed zone plate signal showing spurious waveforms resulting from PAL coding and decoding (BBC).

Figure 7.1.34 is a monochrome picture of a zone pattern which has undergone PAL coding and decoding. The number of spurious images has not only increased considerably, but in reality they have characteristic colours and repetition frequencies from which the impairments may be determined. It must be said that this requires considerable skill and knowledge, but it is nonetheless possible for an expert. The temporal characteristics of the image can also be determined by introducing movement into the pattern.

Apart from research work into digital and high-definition television, zone test patterns are also being utilised for the checking of digital processing equipment, standards converters, decoders, picture monitor displays and others too numerous to mention individually.

References

1. CCIR Recommendation 567, Transmission of Circuits Designed for International Connections (1978).
2. Douglas, J.N. International quality control through systematic measurements on ITS, *IERE Conf. Proc.*, **42** (1978).
3. Heller, A. and Schuster, K. Application of ITS in TV studios and new methods for generating ITS, *IERE Conf. Proc.*, **42** (1978).
4. Darby, P.J. and Tooms, M.S. Colour TV studio performance measurements, *IBA Tech. Rev.*, **1** (1972).
5. IBA. Code of practice for TV studio centre performance, *IBA Tech. Rev.*, **2** (1972).
6. Weaver, L.E. *Television Measurement Techniques*, IEE Monograph Series 9, Peter Peregrinus, London (1971).
7. Weaver, L.E. *Television Video Transmission Measurements*, 2nd Edn, Marconi Instruments Ltd (1978).
8. Whalley, W.B. Colour TV coaxial termination and equalization, *J. SMPTE*, **64** (January 1955).
9. Thiele, A.N. Measurements of return loss at video frequencies, *Proc. IERE (Austr.)* (June 1971).
10. Smith, V.G. TV waveform measurement, *Marconi Instrumentation*, **15**, No. 4 (1977).

11. Lewis, N.W. Waveform responses of television links, *Proc. IEE*, **101**, Part III, 258 (1954).
12. Allnatt, J. *Transmitted Picture Assessment*, John Wiley (1983).
13. CMTT Document CMTT/124 (1976).
14. Comber, G. and MacDiarmid, I.F. Long-term step response of a chain of ac-coupled amplifiers, *Electronics Letters*, **8**, No. 16 (1972).
15. Rosman, G. Interpretation of the waveform of luminance–chrominance pulse signals, *Electronics Letters*, **3**, No. 3 (1967).
16. Mallon, R.E. and Williams, A.D. Testing of transmission chains with vertical interval test signals, *J. SMPTE*, **77** (August 1968).
17. D'Amato, P. Study of the various impairments of the 20T pulse, EBU Document Tech. 3099-E (March 1973).
18. Weaver, L.E. *The SECAM Colour Television System*, Tektronix Inc., Oregon (1982).
19. CCIR Recommendation 451-1 (1970).
20. Allnatt, J. and Prosser, R.D. Subjective quality of colour television pictures impaired by random noise, *Proc. IEE*, **113**, No. 4 (1966).
21. Garuts, V. and Samuel, C. Measuring conventional oscilloscope noise, *Tekscope* (Tektronix Inc.), **1**, No. 1 (1969).
22. Holder, J.E. An instrument for the measurement of random noise, *IEE Cong. Report*, **5** (1963).
23. IEC Publication 698, Measuring Methods for Television Tape Machines (1981).
24. Darby, P.J. and Tooms, M.S. Colour television studio performance measurements, *IERE Conf. Proc.*, **18** (1970).
25. Williamson-Noble, G.E. and Seville, R.C. The television automatic monitor major, *IEE Conf. Publication*, **25** (1966).
26. Shelley, L.J. and Williamson-Noble, G.E. Automatic measurement of insertion test signals, *IERE Conf. Proc.*, **18** (1970).
27. Vivian, R.H. Some methods of automatic analysis of television test signals, *IBA Tech. Rev.*, **1** (1972).
28. CCIR Recommendation 569, Definitions of Parameters for Automatic Measurement of Television Insertion Test Signals (1979).
29. EBU Document Dom. T(T3)218, Recommended Definitions for Parameters to be Automatically Measured on Insertion Test Signals (1974).
30. Rhodes, C.W. Automated and digital measurement of baseband transmission parameters, *J. SMPTE*, **86**, 832–835 (1977).
31. Watson, J.B. Digital automatic measurement equipment, *IRE Conf. Publication*, **145** (1976).
32. D'Amato, P. The determination of tolerances for chains of television circuits, *EBU Rev. Tech.*, **156** (1976).
33. Lari, M., Morganti, G. and Santoro, G. The statistical addition of distortions in transmission systems, *EBU Rev. Tech.*, **143** (1974).
34. MacDiarmid, I.F. and Allnatt, J. Performance requirements for the transmission of the PAL coded signal, *Proc. IEE*, **125**, 6 (1978).
35. Department of Trade and Industry. Specifications of Television Standards for 625-line System I Transmissions in the United Kingdom (1984).
36. BBC/IBA/BREMA. Specifications of standards for information transmission by digitally coded signals in the field blanking interval, *IEE Colloq. Broadcast and Wired Teletext* (1976).
37. Mart, B. and Maudit, M. Antiope, service de télé-texté, *Radiodiffusion-Télévision*, **40** (1975).
38. IBA. Developments in teletext, *IBA Tech. Rev.*, **20** (1983).
39. Spicer, C.R. and Tidey, R.J. An automatic instrument for the measurement of teletext decoding margin, *IRE Cong. Proc.*, **42**, 277–285 (1979).
40. Hutt, P.R. and Dean, A. Analysis measurement and reception of the teletext data signal, *IBC Conf. Publication*, **166**, 258–261 (1978).
41. Holder, J.E. A new television test waveform, *Electronics Letters*, **13**, 9 (1977).
42. Mertz, P. and Gray, F. Theory of scanning, *Bell System Tech. J.*, **13**, 464–515 (1934).
43. Drewery, J.O. The zone plate as a television test pattern, *IERE Conf. Proc.*, **42**, 171–174 (1979).
44. Weston, M. The electronic zone plate and related test patterns, *IBC Conf. Publication*, **191** (1980).

Paul Dubery

7.2 Digital Video Systems Test and Measurement

The first part of this chapter will describe key considerations involved in digital video system design and the second will discuss some issues that arise when transporting video, audio and data in the MPEG domain.

7.2.1 Digital System Design Issues

Before considering specific types of digital video measurements, it is worth discussing some fundamental topics which can impact the overall system performance.

7.2.1.1 Cable Selection

The best grades of precision analogue video cables exhibit low losses from very low frequencies (near DC) to around 10 MHz. In the serial digital world, cable losses in this portion of the spectrum are of less consequence but still important. It's in the higher frequencies associated with transmission rates of 270 Mbit/s or more where losses are considerable. Fortunately, the robustness of the serial digital signal makes it possible to equalise these losses quite easily. So when converting from analogue to digital, the use of existing quality cable runs should pose no problem. The most important characteristic of coaxial cable to be used for serial digital is its loss at half the clock frequency of the signal to be transmitted. That value will determine the maximum cable length that can be equalised by a given receiver. It's also important that the frequency response loss in dB be approximately proportional to $1/\sqrt{f}$ down to frequencies below 5 MHz. There are now many cables available specifically designed for serial digital.

7.2.1.2 Connectors, Patch Panels and Other Passive Elements

Until recently, all video BNC connectors had a characteristic impedance of 50 ohms. BNC connectors of the 75-ohm variety were available, but were not physically compatible with 50-ohm connectors. The impedance 'mismatch' to coax is of little consequence at analogue video frequencies because the wavelength of the signals is many times longer than the length of the connector. But with serial digital's high data rate (and attendant short wavelengths), the impedance of the connector must be considered.

In the specific case of a serial transmitter or receiver, the active devices associated with the chassis connector need significant impedance matching. This could easily include making a 50-ohm connector look like 75 ohms over the frequency band of interest. Good engineering practice tells us to avoid impedance mismatches and use 75-ohm components wherever possible and that rule is being followed in the development of new equipment for serial digital.

The same holds true for other passive elements in the system, such as patch panels. In order to avoid reflections caused by impedance discontinuities, these elements should also have a characteristic impedance of 75 ohms. Existing 50-ohm patch panels will probably be adequate in many instances, but new installations should use 75-ohm patch panels. Several manufacturers now offer 75-ohm versions designed specifically for serial digital applications.

The serial digital standard specifies transmitter and receiver return loss to be greater than 15 dB up to 270 MHz. That is, terminations should be 75 ohms with no significant reactive component to 270 MHz. Clearly this frequency is related to serial digital, hence a lower frequency might be suitable for composite and that may be considered in the future. As can be seen by the rather modest 15 dB specification, return loss is not a critical issue in serial digital video systems. It's more important at short cable lengths where the reflection could distort the signal rather than with longer cable runs where the reflected signal receives more attenuation. Most serial digital receivers are directly terminated in order to avoid return loss problems. Since the receiver active circuits don't look like 75 ohms, circuitry must be included to provide a good resistive termination and low return loss. Active loop-throughs are very common in serial digital equipment because they're relatively simple and have signal regeneration qualities similar to a re-clocking distribution amplifier. Also, they provide isolation between input and output; however, if the equipment power goes off for any reason, the connection is broken. Active loop-throughs also have the same need for care in circuit impedance matching as mentioned above. Passive loop-throughs are both possible and practical; the most important use is for system diagnostics and fault-finding, where it's necessary to observe the signal in the troubled path.

7.2.1.3 Digital Audio Cable and Connector Types

AES/EBU digital audio has raised some interesting questions because of its characteristics. In professional applications, balanced audio has traditionally been deemed necessary in order to avoid hum and other artefacts. Usually twisted, shielded, multi-conductor audio cable is used. The XLR connector was selected as the connector of choice and is used universally in almost all professional applications. When AES/EBU digital audio evolved, it was natural to assume that the traditional analogue audio transmission cable and connectors could still be used. The scope of AES3 covers digital audio transmission of up to 100 metres, which can be handled adequately with shielded, balanced twisted pair interconnection.

Because AES/EBU audio has a much wider bandwidth than analogue audio, cable must be selected with care. The impedance, in order to meet the AES3 specification, requires 110-ohm source and load impedances. The standard has no definition for a bridging load, although loop-through inputs, as used in analogue audio, are theoretically possible. Improperly terminated cables may cause signal reflections and subsequent data errors. The relatively high frequencies of the AES/EBU signals cannot travel over twisted pair cables as easily as analogue audio. Capacitance and high-frequency losses cause high-frequency roll-off. Eventually, signal edges become so rounded and amplitude so low that the receivers can no longer tell the '1's from the '0's which renders the signals undetectable.

7.2.1.4 Signal Distribution and Re-clocking

Although a signal may be digital, the real world through which that signal passes is analogue. Consequently, it's important to consider the analogue distortions that affect a digital signal. These include frequency response roll-off caused by cable attenuation, phase distortion, noise, clock jitter and baseline shift due to AC coupling. While a digital signal will retain the ability to communicate its data despite a certain degree of distortion, there's a point beyond which the data will not be recoverable. Long cable runs are the main cause of signal distortion. Most digital equipment provides some form of equalisation and regeneration at all inputs in order to compensate for cable runs of varying lengths. Considering the specific case of distribution amplifiers and routing switchers, there are several approaches that can be used: wideband amplifier, wideband digital amplifier and regenerating digital amplifier. Taking the last case first, regeneration of the digital signal generally involves recovery of data from an incoming signal in order to retransmit it cleanly using a stable clock source. Regeneration of a digital signal allows it to be transmitted farther by sustaining more analogue degradation than a signal which has already accumulated some analogue distortions. Regeneration generally uses the characteristics of the incoming signal, such as extracted clock, to produce the output. Regeneration, where a reference clock is used to produce the output, can be performed an unlimited number of times and will eliminate all the jitter built up in a series of regeneration operations. This type of regeneration happens in major operational equipment such as VTRs, production switchers (vision mixers) or special effects units that use external references to perform their digital processing.

7.2.1.5 System Timing Considerations

Careful system design is necessary to ensure synchronisation between all signals in the facility. When various video sources are combined, their signals must be timed together or the picture will roll, jump, tear or have incorrect colours. A precision reference signal from a Sync Pulse Generator (SPG) is applied appropriately to each device and genlocked so that the output of the equipment is synchronised with the timing of the reference. In planning the system timing of a facility, the processing delay of the equipment and the propagation delay of the length of cable needed to connect the equipment must be taken into account.

Figure 7.2.1 shows a basic analogue video system. It is important to know the cable lengths, the processing delays of the equipment and how timing adjustments can be made to correct for them. In this scenario, videotape recorders (VTRs) have Time Base Correctors (TBCs) and allow output timing adjustment; the character generator allows output timing adjustments via software and the Camera Control Units (CCUs) require external delay adjustments in order to guarantee system timing.

Delays are inserted by using the timing adjustments of the SPG for each black output. In this example, a separate black output is used for each camera control unit to adjust the delay appropriately. The character generator and VTRs each have built-in timing adjustments, so a Distribution Amplifier (DA) can be used to route the SPG reference signal to each of them.

Analogue system timing measurements and adjustments are made with a waveform monitor and vectorscope connected to the switcher output, as shown in Figure 7.2.1. The external reference is selected on both the waveform monitor and vectorscope so that the units are synchronised to the black burst reference (see Figure 7.2.2). Standard practice dictates that timing measurements are always made at 50% of sync amplitude.

7.2.1.6 Digital System Timing

The need to understand, plan and measure signal timing has not been eliminated with digital video, it just requires a different set of values and parameters. In most cases, timing requirements will actually be more relaxed in a digital environment and will be measured in microseconds, lines and frames rather than nanoseconds. In many ways, distribution and timing are becoming simpler with digital processing because devices that have digital inputs and outputs can lend themselves to automatic input timing compensation. However, mixed analogue/digital systems place additional constraints on the handling of digital signal timing and these issues will be discussed later.

Relative timing of multiple signals with respect to a reference is required in most systems. At one extreme, inputs to a composite analogue production switcher must be timed to the nanosecond range so there will be no subcarrier phase errors. At the other extreme, most digital production switchers allow relative timing between input signals in the range of one horizontal line.

Measuring and adjusting the timing of two digital signals is a simple procedure using a digital waveform monitor. The SDI signals are connected to Channel A and Channel B and the

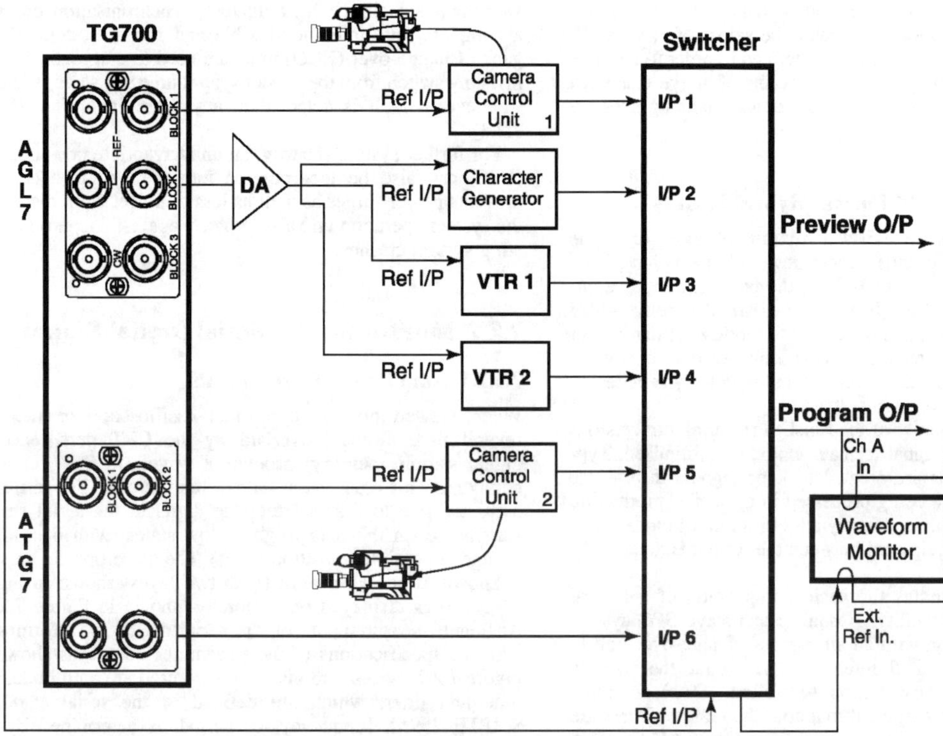

Figure 7.2.1 A basic analogue video system.

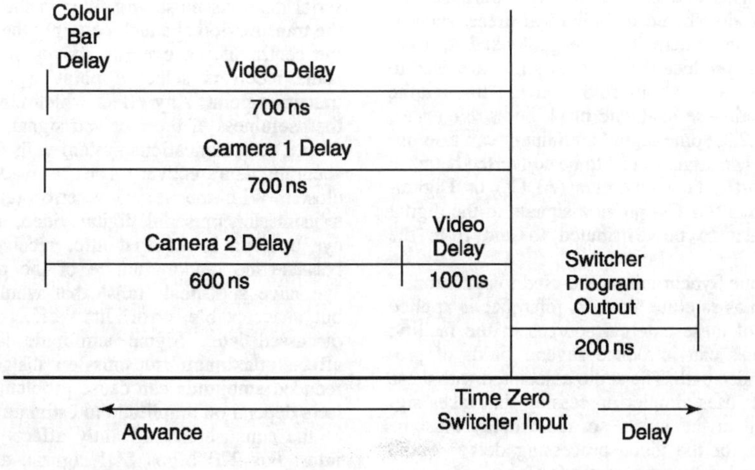

Figure 7.2.2 Typical system timing diagram

waveform monitor is externally referenced to the black burst timing signal. Care must be taken to terminate all signals correctly. The SAV or EAV pulse can be used as a timing reference by positioning it on a major tick mark of the waveform display and comparing it to the other SDI signals – ensuring that they are positioned at the same mark.

However, care is still needed to ensure vertical timing because of the large processing delays in some digital equipment. There are no vertical pulses in the digital domain – digital systems calculate their video position based on the values of F, V and H. Therefore, a reference point must be defined in order to measure vertical timing. For simplicity, the first line of active video can be used as the reference, because the vertical blanking lines are normally blank.

7.2.1.7 Timing a Multi-Format Hybrid Facility

In many facilities there exists a mixture of older analogue systems and digital 'islands' where newer elements have been added more recently. Particularly in the post-production environment, High-Definition (HD) facilities are also being added to handle material which is to be sold to worldwide markets or for cinema and commercials. This multi-format or hybrid environment obviously adds some complexity to system design but fortunately solutions are at hand.

To optimise the quality of the finished material, conversions between the various signal formats should be minimised. Typically, format islands are created to keep signals in a single format while they are being processed in a specific production area. Well-planned timing and synchronisation allow the most flexible use of the equipment between the format islands in the hybrid environment.

Figure 7.2.3 illustrates the basic components of a typical multi-format hybrid facility. A dual master/slave SPG system is used in conjunction with an emergency change-over unit to ensure a precisely timed reference throughout the facility. Analogue or digital Distribution Amplifiers (DAs) distribute each of the reference outputs throughout the facility. There are two types of digital distribution amplifiers: Fan-out DAs which provide a loop-through input and multiple non-reclocked outputs, and Equalising/Re-clocking DAs which have additional circuitry to recover and equalise a digital signal over long cable runs. The signal is re-clocked to produce a completely regenerated digital signal and provide multiple outputs. The Master references are distributed to individual areas, such as studios or editing suites, where they are genlocked to slave SPGs. The slave SPGs produce the references that are used to time equipment in each area, as described above. Although the majority of systems still use analogue black burst references, as shown in Figure 7.2.3, some digital equipment can also use a digital reference. When signals need to be converted between formats, Analogue-to-Digital Converters (ADCs) or Digital-to-Analogue Converters (DACs) provide signals to the digital and/or analogue router to be distributed to and from the islands.

In some cases, Frame Synchronisers are used to synchronise external sources such as satellite feeds. A reference is applied to allow the timing of these external sources in the facility; however, these devices can introduce several fields of processing delay in the video path. The audio associated with these external video signals uses simpler processing that takes significantly less time than the video, so audio delay must be added to compensate for the video processing delay. Audio delay may also be needed to compensate for large video processing delays in other digital equipment, in order to avoid lip-sync problems.

7.2.1.8 System Security

As previously mentioned, synchronisation throughout the facility is critical to ensure optimum system performance. Designing a facility with redundant synchronisation provides a complete fault-tolerant, flexible and robust system. Emergency Change-over (ECO) units are used to automatically and invisibly switch from the master sync source to a slave (standby) whenever a fault is detected in any sync signal (see Figure 7.2.4).

For further system security, an uninterruptible power supply (UPS) can also be incorporated into the system. The UPS prevents power surges as well as loss of power from disrupting the system operation and also protects against dangerous operating system crashes.

7.2.2 Measuring the Serial Digital Signal

7.2.2.1 Amplitude Measurements

When viewed on an appropriate oscilloscope or monitor, several time sweeps (overlaid by the CRT persistence or digital sample memory) produce a waveform that follows a number of different paths across the screen. The different paths are due to the fact that the digits in the serial stream vary based on the data (high or low states, with or without change at possible transition times). The waveform that results is known as an eye pattern (with two 'eyes' shown in Figure 7.2.5 and as displayed on a monitor, shown in Figure 7.2.6). Analogue measurements of the serial digital waveform start with the specifications of the transmitter output, as shown in Figure 7.2.7. Specifications to be measured are amplitude, rise time and jitter, which are defined in the serial standard, SMPTE 259M. Frequency, or period, is determined by the SPG developing the source signal, not the serialisation process. A unit interval (UI) is defined as the time between two adjacent signal transitions, which is the reciprocal of clock frequency. A serial receiver determines if the signal is a 'high' or a 'low' in the centre of each eye, thereby detecting the serial data. As noise and jitter in the signal increase through the transmission channel, certainly the best decision point is in the centre of the eye (as shown in Figure 7.2.8), although some receivers select a point at a fixed time after each transition point. Any effect which closes the eye may reduce the usefulness of the received signal.

In a communications system with forward error correction, accurate data recovery can be made with the eye nearly closed. With the very low error rates required for correct transmission of serial digital video, a rather large and clean eye opening is required after receiver equalisation. This is because the random nature of the processes that close the eye have statistical 'tails' that would cause an occasional, but unacceptable, error. Jitter effects that close the eye are discussed later. Signal amplitude is important because it affects maximum transmission distance; both excessive or reduced amplitude can cause problems. Some receiver equalisers depend on amplitude to estimate cable length and setting of the equaliser significantly affects the resulting noise and jitter. For 270 Mbps SDI signals, amplitude measurements may be made with lower bandwidth oscilloscopes (300–500 MHz bandwidth) but precise measurement of the serial waveform requires an oscilloscope with a 1 GHz bandwidth due to its 1 ns rise time. Rise time measurements are made from the 20–80% points as appropriate for ECL logic devices. Monitoring quality rise time measurements can also be made with video test equipment using equivalent-time sampling at lower bandwidths.

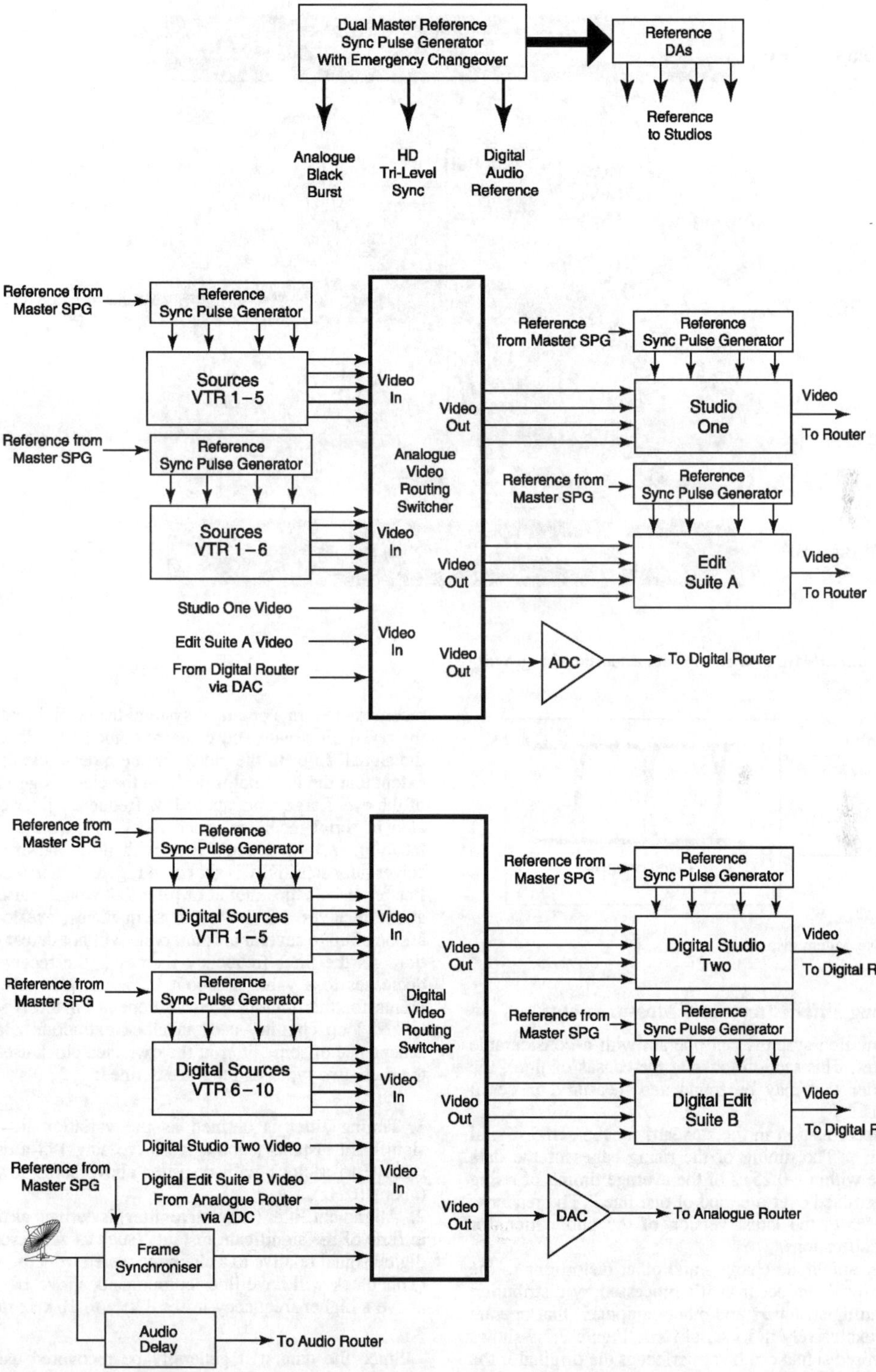

Figure 7.2.3 The basic components of a typical multi-format hybrid facility.

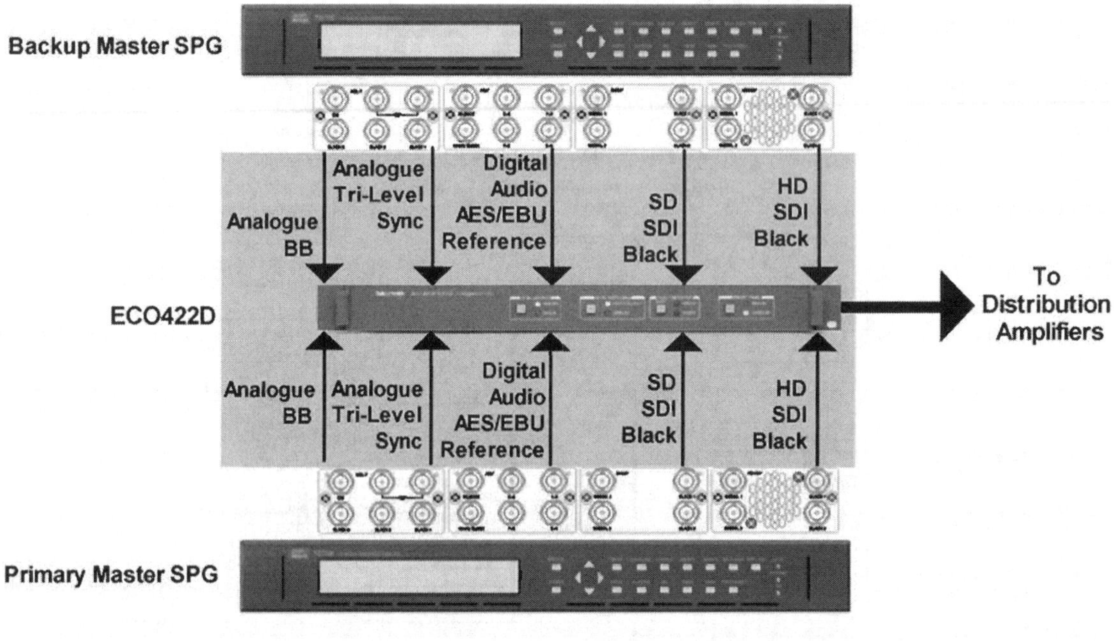

Figure 7.2.4 A master/slave SPG system with automatic changeover.

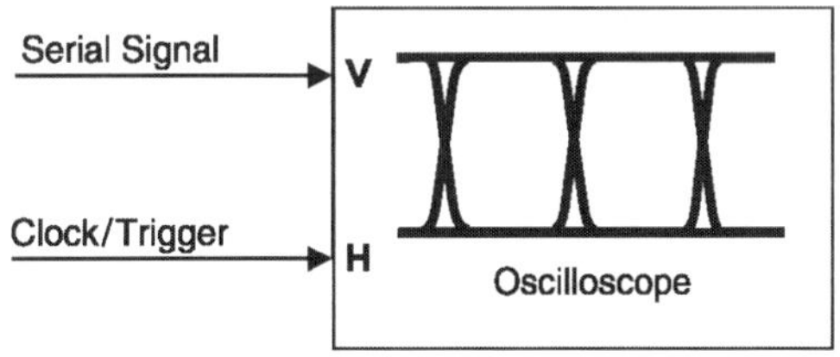

Figure 7.2.5 Eye pattern display.

7.2.2.2 Timing Jitter Effects and Measurements

Digital transmission systems can operate with a considerable amount of jitter. This section looks at the causes of jitter, the amount of jitter that may be found and measured in serial digital systems.

The jitter specification in the standard for the serial digital video signal is: "The timing of the rising edges of the data signal shall be within +0.25 ns of the average timing of rising edges, as determined over a period of one line." The reader is advised to refer to the latest version of the publication for up-to-date specifications.

Jitter, noise, amplitude changes and other distortions to the serial digital signal can occur as it's processed by distribution amplifiers, routing switchers and other equipment that operate on the signal exclusively in its serial form. Figure 7.2.9 shows how the recovered signal can be as perfect as the original if the data is detected with a jitter-free clock. As long as the noise and jitter don't exceed the threshold of the detection circuits (the eye diagram is sufficiently open) the data will be perfectly reconstructed. In a practical system, the clock is extracted from the serial bit stream and contains some of the jitter present in the signal. Jitter in the clock can be a desirable feature to the extent that the jitter helps position the clock edge in the middle of the eye. Large amounts of low-frequency jitter can be tolerated in serial receivers if the clock follows the eye opening location, as its position varies with time. An example of receiver jitter sensitivity is shown in Figure 7.2.10, where the solid line represents the amount of jitter that would cause data errors at the receiver. For low jitter frequencies, eye location variations of up to several unit intervals will not defeat data extraction. As the jitter frequency increases, the receiver tolerance decreases to a value of about 0.25 unit intervals. The break points for this tolerance curve depend on the type of phase-locked loop circuitry used in clock extraction. Based on the acceptance of some jitter in the extracted clock used to recover the data, two types of jitter are defined:

1. Timing Jitter is defined as the variation in time of the significant instants (such as zero crossings) of a digital signal relative to a clock with no jitter above some low frequency (about 10 Hz).
2. Alignment Jitter (or relative jitter) is defined as the variation in time of the significant instants (such as zero crossings) of a digital signal relative to a clock recovered from the signal itself. (This clock will have jitter components above 10 Hz but none above a higher frequency in the 1 kHz to 10 kHz range.)

Since the data will generally be recovered using a clock with jitter, the resulting digital information may have jitter on its transition edges, as shown in Figure 7.2.11. This is still completely valid information since digital signal processing

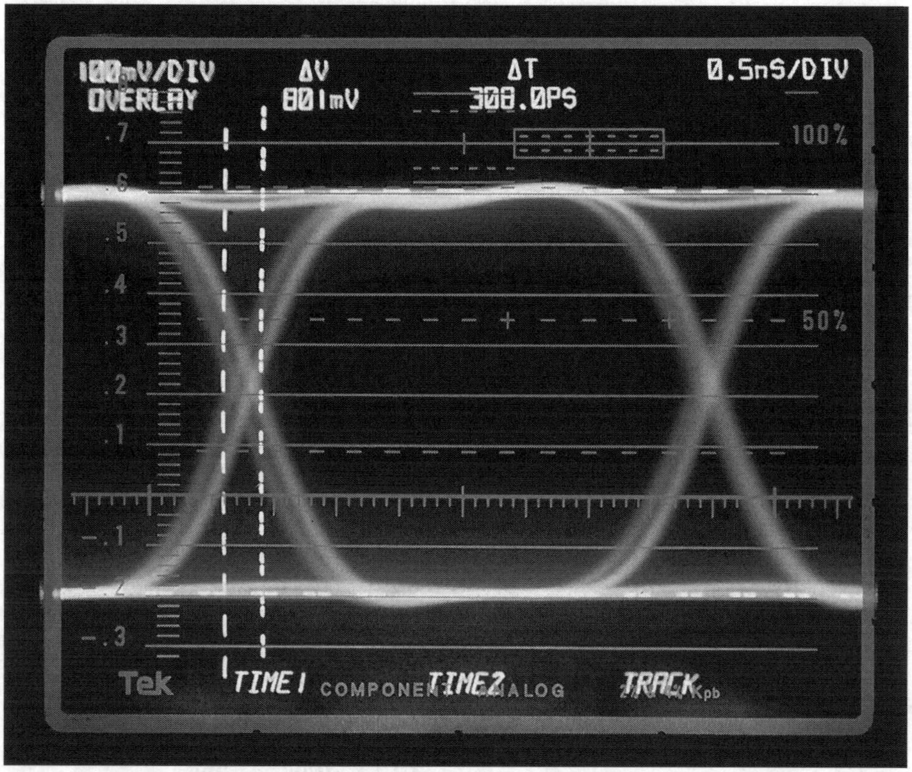

Figure 7.2.6 Waveform monitor eye display of an actual SDI signal.

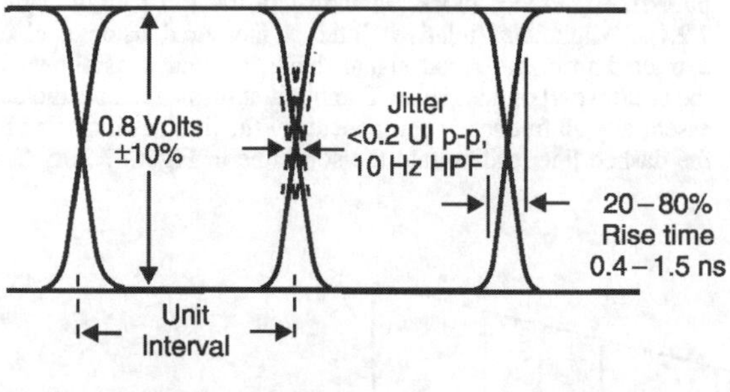

Figure 7.2.7 Serial signal specifications.

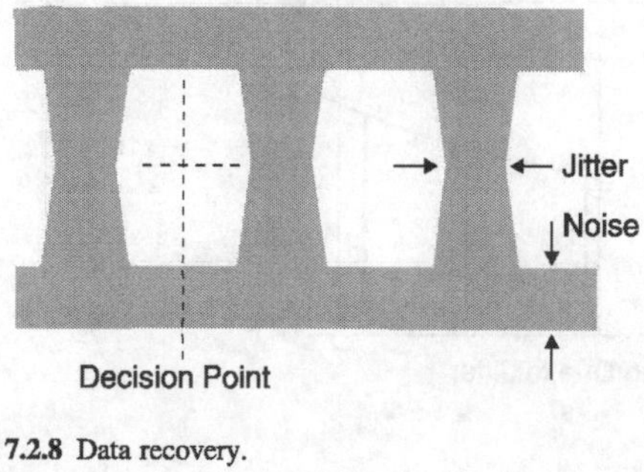

Figure 7.2.8 Data recovery.

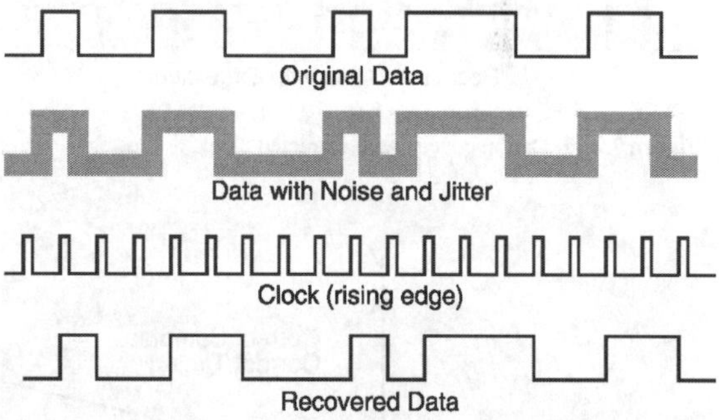

Figure 7.2.9 Data recovery with a noise-free clock.

will only consider the high/low value in the middle of the clock period. However, if the same clock with jitter (or simple sub-multiple thereof) is used to convert the digital information to analogue, errors may occur, as shown in Figure 7.2.12. Sample values converted to analogue at exactly the correct times produce a straight line, whereas use of a clock with jitter produces an incorrect analogue waveform. The relative effect on the analogue waveform produced using the same clock for digital-to-analogue conversion and for data extraction from the serial signal depends on the amount of jitter (in unit intervals) and the number of quantising levels in the digitised signal.

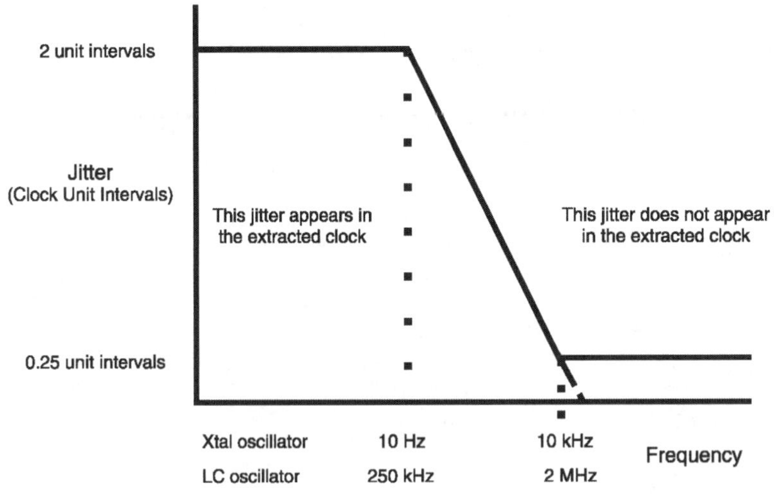

Figure 7.2.10 Receiver jitter sensitivity.

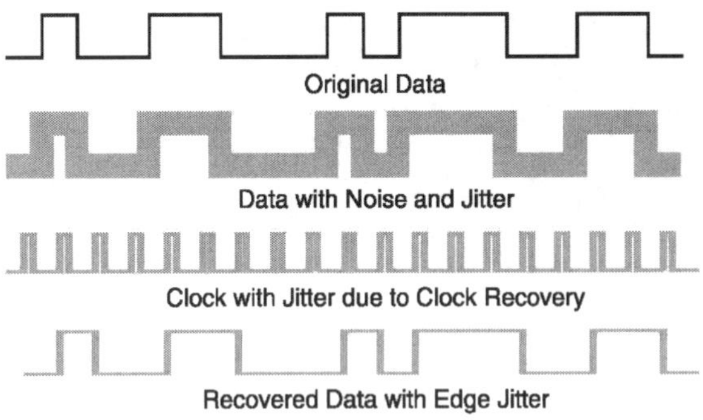

Figure 7.2.11 Data recovery with extracted clock.

Where reduced jitter is required for the digital-to-analogue converter (DAC) clock, a circuit such as that shown in Figure 7.2.13 may be used, where a high Q (crystal) phase-locked loop produces a clock with much lower jitter. Typical jitter specifications for serial audio and serial/parallel video signals are shown in Table 7.2.1.

7.2.2.3 Measuring Jitter

There are three oscilloscope-related methods of measuring jitter in a serial digital signal. Measurement of timing jitter requires a jitter-free reference clock, as shown in the top part of Figure 7.2.14. Alignment (relative) jitter is measured using a clock extracted from the serial signal, being evaluated as shown in the bottom part of this figure. Timing jitter measurements include essentially all frequency components of the jitter, as indicated by the dashed line leading into the solid line in Figure 7.2.15. The

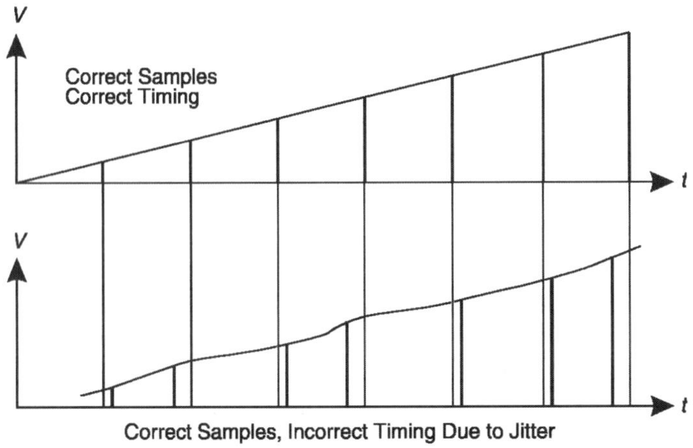

Figure 7.2.12 Signal errors due to clock jitter.

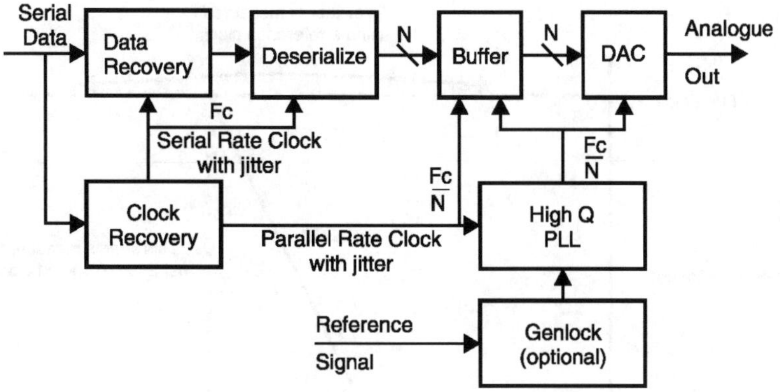

Figure 7.2.13 DAC with a low-jitter clock.

Table 7.2.1 Typical jitter specifications

Standard	*Clock period (ns)*	*Jitter spec (ns)*	*% of Clock*	*DAC requirement*
AES/EBU audio	163	40.0	25%	1.0 ns spec 0.1 ns 16-bit
Serial component	4	0.5	14%	0.5 ns
Parallel component	37	6.0	16%	0.5 ns
Parallel HD	7	1.0	14%	0.1 ns

jitter frequency components measured using the alignment jitter method will depend on the bandwidth of the clock extraction circuit. Low-frequency components of jitter will not be included because the extracted clock follows the serial signal jitter. These low-frequency jitter components are not significant for data recovery provided the measurement clock extraction system has the same bandwidth as the data recovery clock extraction system. All jitter frequencies above a certain value will be measured with break points between the two areas at frequencies appropriate for the bandwidth of the clock extraction system used. When deriving the scope trigger it's important to consider the frequency

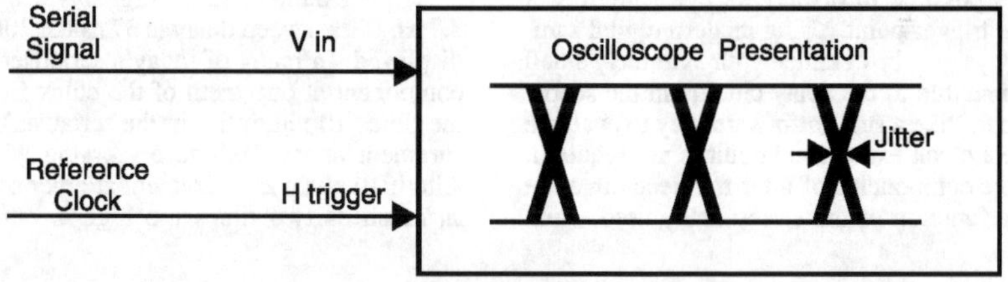

Timing jitter measurement

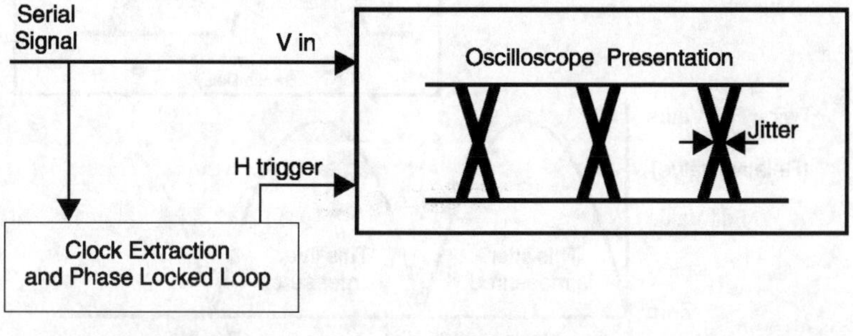

Alignment jitter measurement

Figure 7.2.14 Measurements using a clock reference.

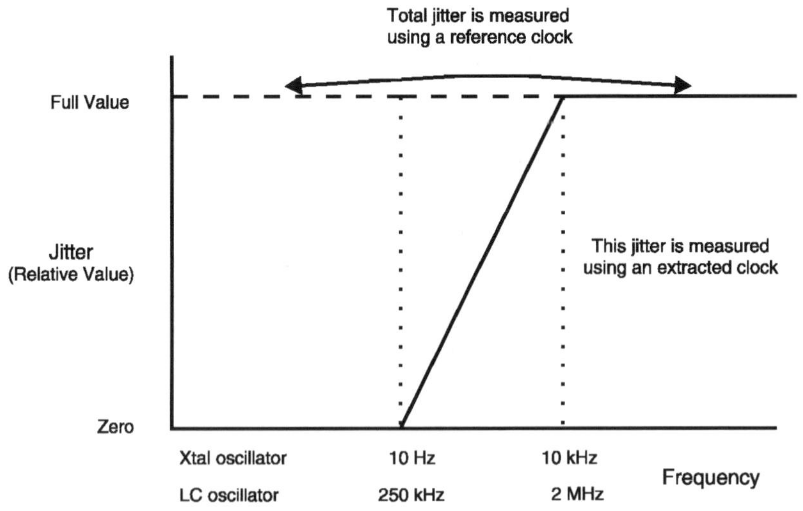

Figure 7.2.15 Measured jitter components.

division that usually takes place in the clock extractor. If it's a divide-by-ten, then word-synchronised jitter will not be observed depending on whether the sweep time of the display covers exactly 10 zero crossings. Division by a number other than the word length will ensure that all jitter is measured.

The third method of observing, but not measuring, jitter is the simple self-triggered scope measurement depicted in Figure 7.2.16. By internally triggering the scope on the rising (or falling) edge of the waveform, it's possible to display an eye pattern at a delayed time after the trigger point. Using modern digital sampling scopes, long delays may be obtained with extremely small amounts of jitter attributable to the delay circuits in the scope. At first glance, this looks like a straightforward way to evaluate the eye pattern, since a clock extraction circuit is not required. The problem is that the components of jitter frequency that are being measured are a function of the sweep delay used – and that function is a comb filter. As a result, there are some jitter frequencies that will not be measured and others that will indicate twice as high a value as would be expected. Although there's a low-pass effect with this type of measurement, the shape of the filter is much different than that obtained with an extracted clock measurement; hence, the results may not be a good indication of the ability of a receiver to recover the data from the serial bit stream.

As an example, serial digital video has a unit interval of 3.7 ns. If the sweep delay is 37 ns the 10th zero crossing will be displayed. In many of today's serialisers there's a strong jitter component at one-tenth of the clock frequency (due in part to the times 10 multiplier in the serialiser). A self-triggered measurement at the 10th zero crossing will not show that jitter. Alternatively, a measurement at other zero crossings may show as much as two times too high a value of jitter. To further

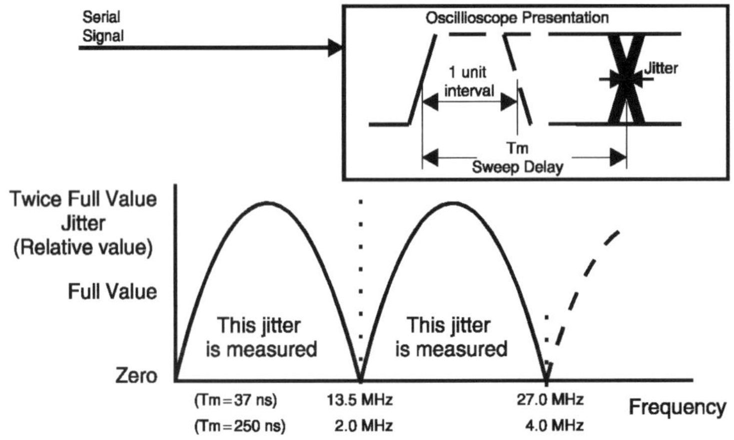

Figure 7.2.16 Self-triggered jitter observations.

complicate the matter, if the one-tenth frequency jitter were exactly symmetrical (such as a sine wave) the fifth zero crossing would also have no jitter. In practice, the fifth crossing also shows a lot of jitter; only the 10th, 20th, 30th, etc., crossings appear to be nearly jitter free.

In troubleshooting a system, the self-triggered method can be useful in tracking down a source of jitter. However, proper jitter measurements to meet system specifications should be made with a clock extraction method.

7.2.3 System Testing

7.2.3.1 Stress Testing

Unlike analogue systems that tend to degrade gracefully, digital systems tend to work without fault until they suddenly crash. When operating a digital system, it's desirable to know how much headroom is available; that is, how far the system is away from the crash point. Out-of-service stress tests are required to evaluate system operation. Stress testing consists of changing one or more parameters of the digital signal until failure occurs. The amount of change required to produce a failure is a measure of the headroom. Starting with the specifications in the serial digital video standard (SMPTE 259M), the most intuitive way to stress the system is to add cable until the onset of errors. Other tests would be to change amplitude or rise time, or add noise and/or jitter to the signal. Each of these tests are evaluating one or more aspect of the receiver performance, specifically automatic equaliser range and accuracy, and receiver noise characteristics. They are also the most meaningful stress test since they represent real operation. Stress testing receiver ability to handle amplitude changes and added jitter is useful in evaluating and accepting equipment, but not too meaningful in system operation. Addition of noise or change in rise time (within reasonable bounds) has little effect on digital systems and is not important in stress tests. Cable-length stress testing can be done using actual coax or a cable simulator. Coax is the real world and most accurate method. The key parameter to be measured is onset of errors because that defines the crash point, as described earlier. With an error measurement method in place, the quality of the measurement will be determined by the sharpness of the knee of the error curve. Using typical coax cable, a 5-metre change in length will go from no errors in 1 minute to more than one error per second. To evaluate a cable simulator, the natural approach is to compare its loss curve with coax using a network analyser. Certainly that's an appropriate criterion, but the most important is the sharpness of the error curve at various simulated lengths. Experiments have shown that good cable simulators require a 10- to 15-metre change in added coax to produce a change from no errors to more than one error per second. Simulators that require a longer change in coax (less sharp knee) are still useful for comparison testing but should be avoided in equipment evaluation.

7.2.3.2 SDI Check Field

The SDI Check Field (also known as 'pathological signal') is a valuable out-of-service test. It is a difficult signal for the serial digital system to handle since it has a maximum amount of low-frequency energy in two separate signals. One signal tests equaliser operation and the other tests phase-locked loop operation. The SDI Check Field is defined in SMPTE Recommended Practice RP 178.

The mathematics of the scrambling process produce the pathological signal. Although the quantising levels 3FF and 000 are excluded from the active picture and ancillary data, the ratio of '1's to '0's may be quite uneven for some values of a flat colour field. This is one of the main reasons for scrambling the signal. Conversion to NRZI eliminates the polarity sensitivity of the signal but has little effect on the ratio of '1's to '0's. Scrambling does indeed break up long runs of '1's or '0's and produces the desired spectrum with minimum low-frequency content, while providing maximum zero crossings needed for clock extraction at the receiver. However, there are certain combinations of data input and scrambler state that will, occasionally, cause long runs (20–40) of '0's. These long runs of '0's create no zero crossings and are the source of low-frequency content in the signal. (All '1's is not a problem because this produces an NRZI signal at one-half the clock frequency, which is right in the middle of the frequency band occupied by the overall signal.) The scrambler is a nine-cell shift register, as shown in Figure 7.2.17. For each input bit, there is one output bit sent to the one-cell NRZI encoder. An output bit is determined by the state of the 10 cells and the input bit. A typical distribution of runs of '1's and '0's is shown in Table 7.2.2 for 100 frames of component black. (For the purpose of this analysis the '1's are high and the '0's are low, as opposed to the transition, no-transition data definition.) The longer runs (16, 19, 22, 32, 33, 34 and 39) occur as single events that the transmission system amplifiers and receivers can handle with relative ease because of their very short duration. Multiple long-run events can be encouraged with certain input signals, which is the basis for the generation of pathological signals. When the input words alternate between certain digital values the multi-event pathological signal will occur for one specific scrambler state, generally all '0's, hence the sequence usually starts at an SAV. The multi-event sequence ends at the next EAV, which breaks the input sequence. Since the scrambler state is more or less a random number, the multi-event sequence

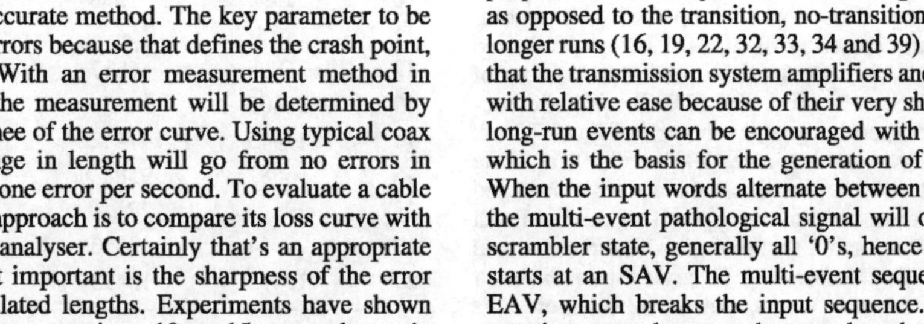

Figure 7.2.17 Serial scrambler and encoder.

Table 7.2.2 One hundred frames of component black

Length	*'1's*	*'0's*	*Length*	*'1'*	*'0's*
1	112615438	112610564	21		
2	56303576	56305219	22	73759	13579
3	28155012	28154963	23		
4	14076558	14077338	24		
5	7037923	7038595	25		
6	3518756	3518899	26		
7	1759428	1760240	27		
8	1059392	699666	28		
9	610241	970950	29		
10	1169	1060	30		
11	14	156	31		
12	40		32	55	51
13	156	114	33	49	
14			34	3	
16	13713	73865	36		
19	101	94	39	50	

of long runs of '0's happens about once per frame (9 bits, 512 possible states, 525 or 625 lines per frame).

There are two forms of the pathological signal, as shown in Figure 7.2.18. For testing the automatic equaliser, a run of 19 '0's followed by two '1's produces a signal with large DC content. Remember a '0' is no transition and a '1' is represented by a transition in the NRZI domain. In addition, the switching on and off of this high DC content signal will stress the linearity of the analogue capabilities of the serial transmission system and equipment. Poor analogue amplifier linearity results in errors at the point of transition, where peak signal amplitude is the greatest. It's important to produce both polarities of this signal for full system testing. Phase-locked loop testing is accomplished with a signal consisting of 20 NRZI zeros followed by a single NRZI '1'. This provides the minimum number of zero crossings for clock extraction. The SDI Check Field consists of one-half field of each signal, as shown in Figure 7.2.18. Equaliser testing is based on the values 300 h and 198 h, while phase-locked loop testing is based on the values 200 h and 110 h. In the C – Y order shown, the top half of the field will be purple and the bottom half a shade of grey. Some test signal generators use the other order, Y – C, giving two different shades of green. Either one works, but the C – Y order is specified in the recommended practice. One word with a single bit of '1' is placed in each frame to ensure both polarities of equaliser stress signal for the defined SDI Check Field. The single '1' inverts the polarity of the NRZ to NRZI function because the field (for component video) would otherwise have an even number of '1's and the phase would never change. Statistics for 100 frames of SDI Check Field are shown in Table 7.2.3.

It's the runs of 19 and 20 that represent the low-frequency stressing of the system. When they happen (about once per frame) they occur for a full active video line and cause a significant low-frequency disturbance. Since the PLL stress signal is a 20 '1's, 20 '0's square wave, you would expect an even number of each type of run. The equaliser stress signal is based on runs of 19 '1's followed by a '0' or 19 '0's followed by a '1'. As described above, an extra '1' in each frame forces the two polarities to happen and both are represented in the data.

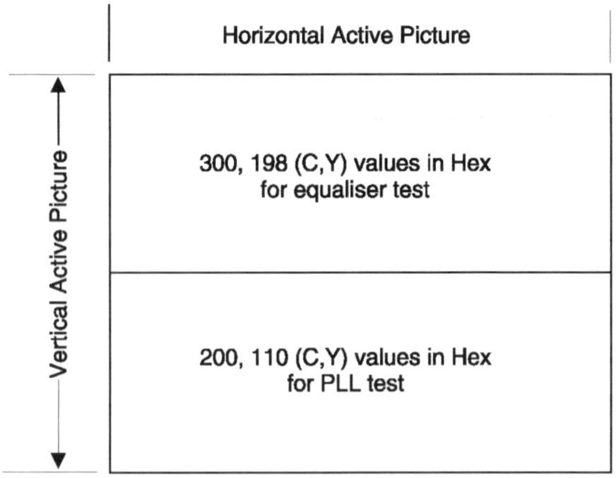

Figure 7.2.18 The SDI Check Field.

Table 7.2.3 One hundred frames of SDI Check Field

Length	*'1's*	*'0's*	*Length*	*'1's*	*'0's*
1	112595598	112595271	21		
2	56291947	56292583	22	9344	9425
3	28146431	28147229	23		
4	14073813	14072656	24		
5	7035240	7035841	25		
6	3518502	3518856	26		
7	1758583	1758969	27		
8	879203	879359	28		
9	706058	706029	29		
10	51596	51598	30		
11	146	154	31		
12	16858	16817	32	49	49
13	34405	34374	33	35	32
14	17445	17442	34	3	4
16	9558	9588	36		
18	29	27	38		
19	20944	19503	39	13	14
20	19413	19413	40		

7.2.3.3 System-wide Detection of Errors

In our application of computers and digital techniques to automate many tasks, we've grown to expect the hardware and software to report problems to an operator. This is particularly important as systems become larger, more complex and more sophisticated.

It's reasonable to expect digital video systems to provide similar fault reporting capabilities. Television equipment has traditionally had varying levels of diagnostic capabilities. With the development of digital audio and video signals it's now possible to add a vital tool to system fault reporting: confirmation of signal integrity in studio interconnection systems. To understand and use this tool, it's necessary to delve into the technology of digital error detection. The combination of the serial digital video interconnect method, the nature of video systems and typical digital equipment design lead to the use of error measurement methods that are economical yet specialised for this application. Serial digital video operates in a basically noise-free environment, so traditional methods of measuring random error rates are not particularly useful. Examination of the overall system characteristics leads to the conclusion that a specialised burst error measurement system will provide the studio engineer with an effective tool to monitor and evaluate error performance of serial digital video systems.

7.2.3.4 Definition of Errors

In the testing of in-studio transmission links and operational equipment a single error is defined as one data word whose digital value changes between the signal source and the measuring receiver. This is significantly different than the analogue case, where a range of received values could be considered to be correct. Application of this definition to transmission links is straightforward. If any digital data word values change between transmitter and receiver there's an error. For operational studio equipment, the situation is a little more complicated and needs further definition. Equipment such as routing switchers, distribution amplifiers and patch panels should, in general, not change the data carried by the signal; hence, the basic definition applies. However, when a routing switcher changes the source selection there will be a short disturbance. This should not be considered an error, although the length of the disturbance is subject to evaluation and measurement.

SMPTE Recommended Practice RP 168 specifies the line in the vertical interval where switching is to take place, which allows error measurement equipment to ignore this acceptable error. There are various types of studio equipment that would not normally be expected to change the signal. Examples are frame synchronisers with no processing amplifier controls, production switchers in a straight-through mode, and digital VTRs in the E-to-E mode. All have the capability (and are likely) to replace all of the horizontal and part of the vertical blanking intervals. In general, replacing part of the video signal will destroy the error-free integrity of the signal.

Considering the possible replacement of blanking areas by some equipment and knowing that the active picture sample locations are well defined by the various digital standards, it's possible to measure Active Picture Errors separately from Full Field Errors. The Active Picture Error concept takes care of blanking area replacements. However, there is a more insidious possibility due to the engineer's ability to modify the blanking edges. Most digital standards were designed with digital blanking narrower than analogue blanking to ensure appropriate edge transitions in the analogue domain. Large-amplitude edge transitions within one sample period caused by time truncating the blanking waveform can create out-of-band spectral components that can show up as excessive ringing after digital-to-analogue conversion and filtering. What happens in some equipment is that engineers have modified the samples representing the analogue blanking edge in order to provide a desired transition; therefore, error measurements through such equipment even using the Active Picture concept will not work.

With VTRs, there's a further limit to measurement of errors. The original full bit rate digital VTR formats were designed with significant amounts of forward error correction and work very well. However, it's the nature of the tape recording and reproduction process that some errors will not be corrected. Generally the error correction system identifies the uncorrected data and sophisticated error concealment systems can be applied to make a virtually perfect picture. The systems are so good that tens of generations are possible with no humanly noticeable defect. Although the error concealment is excellent, it's almost certain that error measurement systems (that by definition require perfect data reproduction) would find errors in many completely acceptable fields. VTR formats that use bit rate reduction will have an additional potential source of small but acceptable data value changes. The data compression and decompression schemes are not absolutely lossless, resulting in some minor modification of the data. Therefore, not all error measurement methods external to a digital VTR are meaningful because of potential blanking edge adjustment, error concealment methods and, in some equipment, the use of data rate reduction. Fortunately, various types of error rate and concealment rate data are available inside the VTR and standards for reporting that information are under discussion.

7.2.3.5 Quantifying Errors

Most engineers are familiar with the concept of Bit Error Rate (BER), which is the ratio of bits in error to total bits. As an example, the 10-bit digital component data rate is 270 Mbit/s. If there were one error per frame the BER would be $25/(270 \times 106) = 0.93 \times 10^{-7}$ for 625/50 systems. Table 7.2.4 shows the BER for one error over different lengths of time. BER is a useful measure of system performance where the signal-to-noise ratio at the receiver is such that noise-produced random errors occur. As part of the serial digital interconnect system, scrambling is used to lower DC content of the signal and provide sufficient zero crossings for reliable clock recovery.

Table 7.2.4 Error frequency and Bit Error Rates

Time between errors	*Component 270 Mbit/s*
1 television frame	1×10^{-7}
1 second	4×10^{-9}
1 minute	6×10^{-11}
1 hour	1×10^{-12}
1 day	4×10^{-14}
1 week	6×10^{-15}
1 month	1×10^{-15}
1 year	1×10^{-16}
1 decade	1×10^{-17}
1 century	1×10^{-18}

It's the nature of the descrambler that a single bit error absolutely causes an error in two words (samples) and has a 50% probability of the error in one of the words being in the most, or next to the most, significant bit. Therefore, an error rate of 1 error/frame will be noticeable by a reasonably patient observer. If it's noticeable, it's unacceptable; but it's even more unacceptable because of what it tells us about the operation of the serial transmission system.

7.2.3.6 Nature of Digital Video Systems

Specifications for sources of digital video signals are defined by SMPTE 259M. Although the specifications do not include a signal-to-noise ratio (SNR), typical values would be 40 dB or greater at the transmitter. Errors will occur if the SNR at some location in the system reaches a low enough value, generally in the vicinity of 20 dB. Figure 7.2.19 is a block diagram of the basic serial transmitter and receiver system. An intuitive method of testing the serial system is to add cable – a straightforward method of lowering the SNR. Since coax itself is not a significant noise source, it's the noise figure of the receiver that determines the operating SNR. Assuming an automatic equaliser in the receiver, eventually, as more cable is added, the signal level due to coax attenuation causes the SNR in the receiver to be such that errors occur. Based on the scrambled NRZI channel code used and assuming Gaussian distributed noise, a calculation using the error function gives the theoretical values shown in Table 7.2.5. The calibration point for this calculation is based on the capabilities of the serial digital interface. In the proposed serial digital standard it's stated that the expected operational distance is through a length of coax that attenuates a frequency of half the clock rate by up to 30 dB. That is, receivers may be designed with less or more capability, but the 30 dB value is considered to be realisable.

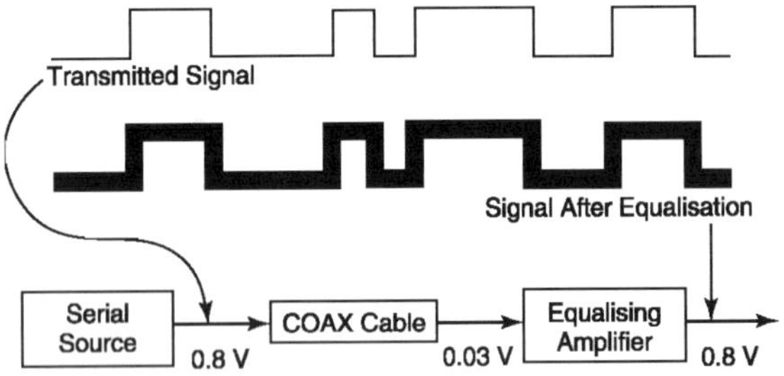

Figure 7.2.19 The basic serial transmitter and receiver system.

Table 7.2.5 Error rate as a function of SNR

Time between errors	*BER*	*SNR (dB)*	*SNR (volts ratio)*
1 microsecond	7×10^{-3}	10.8	12
1 millisecond	7×10^{-6}	15.8	38
1 television frame	2×10^{-7}	17.1	51
1 second	7×10^{-9}	18.1	64
1 minute	1×10^{-10}	19.0	80
1 day	8×10^{-14}	20.4	109
1 month	3×10^{-15}	20.9	122
1 century	2×10^{-18}	21.8	150

This same theoretical data can be expressed in a different manner to show error rates as a function of cable length, as shown graphically in Figure 7.2.20. The graph makes it very apparent that there's a sharp knee in the cable length vs. error rate curve. Eighteen additional metres of cable (5% of the total) moves operation from the knee to completely unacceptable, while 50 fewer metres of cable (12% of the total length) moves operation to a reasonably safe 1 error/month.

Cable length changes required to maintain headroom scales proportionally, as shown in Figure 7.2.20. Good engineering practice would suggest a 6-dB margin or 80 metres of cable, hence a maximum operating length of about 320 metres in a composite system where the knee of the curve is at 400 metres. At that operating level, there should never be any errors (at least not in our lifetime). Practical systems include equipment that doesn't necessarily completely reconstitute the signal in terms of SNR. That is, sending the signal through a distribution amplifier or routing switcher may result in a completely useful but not completely standard signal being sent to a receiving device. The non-standardness could be both jitter and noise, but the sharp-knee characteristic of the system would remain, occurring at a different amount of signal attenuation. Use of properly equalised and re-clocked distribution and routing equipment at intervals with adequate headroom provides virtually unlimited total transmission distances.

7.2.3.7 Measuring Bit Error Rate (BER)

BER measurements may be made directly using equipment specifically designed for that purpose. Unfortunately, in a properly operating system, say with 6 dB of headroom, there's no BER to measure. This is because the serial digital system generally operates in an environment that is free from random errors. A more common problem will be burst errors due to some sort of interfering signal such as a noise spike that occurs at intermittent intervals spaced far apart in time. Another source could be crosstalk that might come and go depending on what

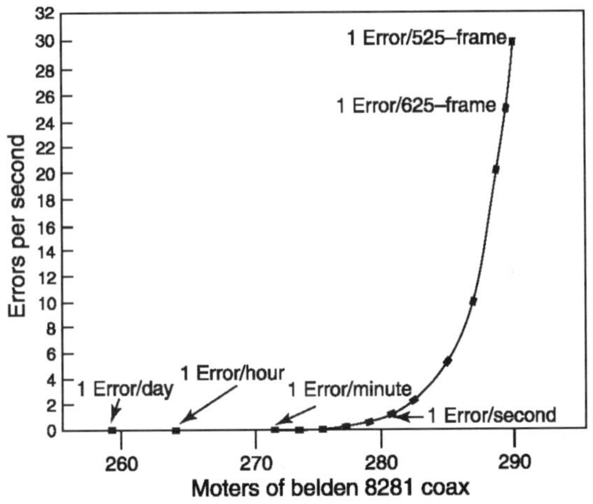

Figure 7.2.20 Calculated error rates versus cable length.

other signals are being used at a particular time. Another could be the poor electrical connection at an interface that would cause noise only when it's mechanically disturbed. Because of the intermittent nature of burst errors, data recording and communications engineers have defined another error measurement – the Errored Second. The following example demonstrates the benefits of the errored second. Suppose a burst error causes 10,000 errors in two frames of video. A BER measurement made for 1 minute would indicate a 1×10^{-6} BER and a measurement made for 1 day would indicate an 8×10^{-9} BER, whereas an errored second measurement could indicate that there was 1 second in error at a time 3 hours, 10 minutes and 5 seconds ago. The errored second method is clearly a more useful measurement in this case. A significant advantage of errored seconds vs. a straight BER measurement is that it's a better measure of fitness for service of links that are subject to burst errors. The serial digital video system fits this category because video images are greatly disturbed by momentary loss in synchronisation. A BER measurement could give the same value for a single, large burst as it does for several shorter scattered bursts. But if several of the shorter bursts each result in momentary sync failure, the subjective effect will be more damage to the viewed picture than caused by the single burst.

7.2.3.8 An Error Measurement Method for Television

An error detection system for digital television signals has been placed in the public domain to encourage other manufacturers to use the method. It has proven to be a sensitive and accurate way to determine if the system is operating correctly and it has been standardised by the SMPTE as Recommended Practice RP 165. Briefly, the Error Detection and Handling (EDH) concept is based on making Cyclic Redundancy Code (CRC) calculations for each field of video at the serialiser, as shown in Figure 7.2.21. Separate CRCs for the full field and active picture, along with status flags, are then sent with the other serial data through the transmission system. The CRCs are recalculated at the deserialiser and, if not identical to the transmitted values, an error is indicated. Typical error detection data will be presented as errored seconds over a period of time, and time since the last errored second. In normal operation of the serial digital interface, there will be no errors to measure. What's of interest to the engineer is the amount of headroom that's available in the system. That is, how much stressing could be added to the system before the knee of the error rate curve or crash point is reached. As an out-of-service test, this can be determined by adding cable or other stressing method until the onset of errors. Since it's an out-of-service test, either a BER test set or RP 165 could be used. There are, however, many advantages to using the RP 165 system:

1. The CRC data is part of the serial digital signal, thereby providing a meaningful measure of system performance.
2. RP 165 can be used as an in-service test to automatically and electronically pinpoint any system failures.
3. For out-of-service testing, RP 165 is sufficiently sensitive to accurately define the knee of the error rate curve during stressing tests.
4. Where there are errors present, RP 165 provides the information necessary to determine errored seconds, which is more useful than bit error rate.
5. Facility is provided for measuring both full-field and active-picture errors. Optional status flags are also available to facilitate error reporting.
6. CRC calculation can be built into all serial transmitters and receivers for a very small incremental cost. With error information available from a variety of equipment, the results can then be routed to a central collection point for overall system diagnostics.

7.2.3.9 System Considerations

In a large serial digital installation the need for traditional waveform monitoring can be reduced by the systematic application of error detection methods. At signal source locations, such as cameras, DVEs and VTRs, where operational controls can affect the programme signal, it will continue to be important to verify the key programme signal parameters using waveform monitors with serial digital input capabilities. The results of certain technical operations, such as embedded

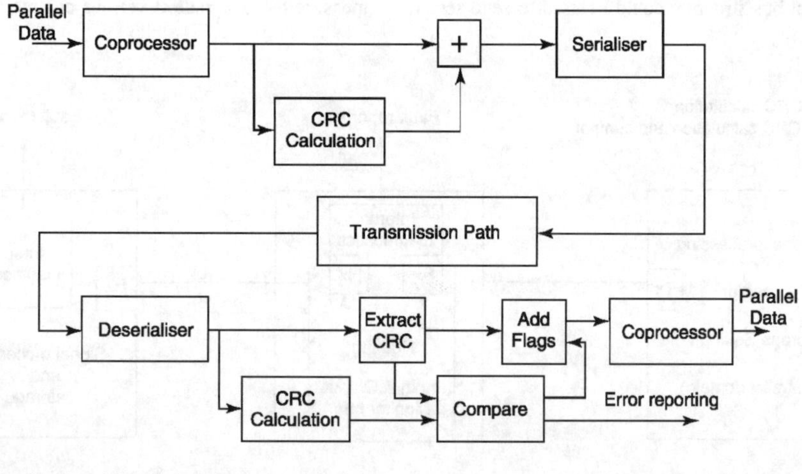

Figure 7.2.21 EDH signal path and CRC calculations.

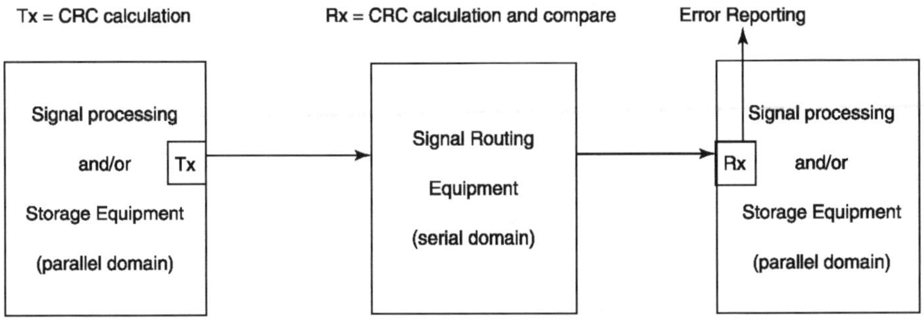

Figure 7.2.22 Basic error detection system.

audio mux/demux, serial link transmission and routing switcher I/O, can be monitored with less sophisticated digital-only equipment provided the signal data integrity is verified. It's the function of the RP 165 system to provide that verification. For equipment where the digital signal remains in the serial domain, such as routing switchers or distribution amplifiers, it's not economical to provide the CRC calculation function. Therefore, a basic error detection system will be implemented, as shown in Figure 7.2.22. Equipment can then report errors locally and/or to a central diagnostics computer. Routing switchers and other equipment operating in the serial domain will have their signal path integrity verified by the error detection system. To extend the idea of system diagnostics beyond simple digital data error detection, it would be useful to provide an equipment fault reporting system. SMPTE 269M documents a single contact closure fault reporting system and there are ongoing discussions involving computer data communications protocols for more sophisticated fault reporting.

Error detection is an important part of system diagnostics, as shown in Figure 7.2.23. Information relating to signal transmission errors is combined with other equipment internal diagnostics to be sent to a central computer. A large routing switcher can economically support internal diagnostics. Using two serial receivers a polling of serial data errors could be implemented. The normal internal bus structure could be used to send serial data to a receiver to detect input errors and a special output bus could be used to ensure that no errors were created within the large serial domain routing switcher.

7.2.4 Testing in the MPEG Domain

Once the baseband video and audio signals have been compressed and combined into an MPEG transport stream, testing requirements do not disappear but become significantly different. The ability to analyse existing transport streams for compliance is essential, and must be complemented by an ability to generate transport streams, both compliant and deliberately errored.

7.2.4.1 Testing Requirements

On an operational basis, the user needs to have a simple, regular confidence check that ensures all is well. In the event of a failure, the location of the fault needs to be established rapidly. For the purpose of equipment design, the nature of problems needs to be explored in some detail. As with all signal testing, the approach is to combine the generation of known valid signals for insertion into a system with the ability to measure those signals at various points.

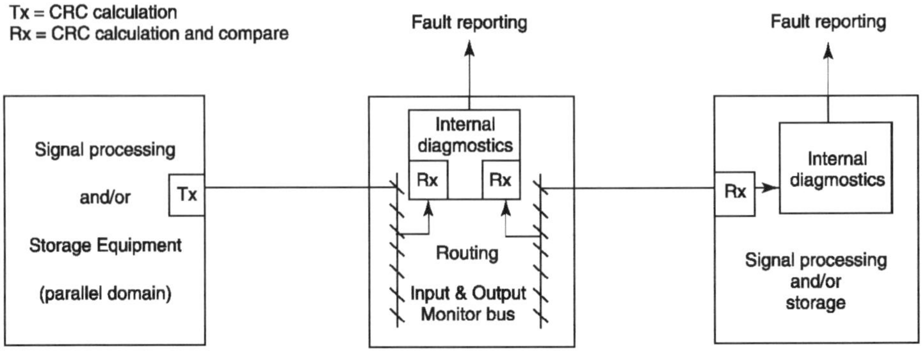

Figure 7.2.23 A fault-reporting scheme based on EDH.

One of the characteristics of MPEG that distances it most from traditional broadcast video equipment is the existence of multiple information layers, in which each layer should be transparent to the one below. It is very important to be able to establish in which layer any fault resides to avoid a fruitless search. For example, if the picture monitor on the output of an MPEG decoder is showing visible defects, these could be due to a number of possibilities. Perhaps the source encoder is faulty, and the transport stream is faithfully delivering that faulty information. On the other hand, the encoder might be fine, but the transport layer is corrupting the data. Such complexity requires a structured approach to fault-finding, using the right tools.

7.2.4.2 Analysing a Transport Stream

An MPEG transport stream has an extremely complex structure, but a suitable analyser can break down and display the structure in a logical fashion such that the user can observe any required details. Many general types of analysis can take place in real time on a live transport stream. These include displays of the hierarchy of programmes in the transport stream and of the proportion of the stream bit rate allocated to each stream.

More detailed analysis is only possible if part of a transport stream is recorded so that it can be picked apart later. This technique is known as deferred-time testing and could be used, for example, to examine the contents of a time stamp. When used for deferred-time testing, the MPEG transport stream analyser is acting like a logic analyser that provides data-interpretation tools specific to MPEG. As with all logic analysers, a real-time triggering mechanism is required to determine the time or conditions under which capture will take place. Figure 7.2.24 shows that an analyser contains a real-time section, a storage section and a deferred section. In real-time analysis, only the real-time section operates, and a signal source needs to be connected. For capture, the real-time section is used to determine when to trigger the capture. The analyser includes tools known as filters that allow selective analysis to be applied before or after capture. Once the capture is completed, the deferred section can operate on the captured data and the input signal is no longer necessary. Another good comparison is the storage oscilloscope, which can display the real-time input directly or save it for later study.

7.2.4.3 Hierarchic View

When analysing an unfamiliar transport stream, the hierarchic view is an excellent starting point because it enables a graphic view of every component in the stream. Figure 7.2.25 shows an example of a hierarchic display. Beginning at top left of the entire transport stream, the stream splits and an icon is presented for every stream component. The user can very easily see how many programme streams are present, as well as the video and audio content of each. Each icon represents the top layer of a number of lower analysis and information layers. The analyser creates the hierarchic view by using the PAT and PMT in the PSI data in the transport stream. The PIDs from these tables are displayed beneath each icon. PAT and PMT data are fundamental to the operation of any demultiplexer or decoder; if the analyser cannot display a hierarchic view or displays a view which is obviously wrong, the transport stream under test has a PAT/PMT error. It is unlikely that equipment further up the line will be able to interpret such a stream at all. The ability

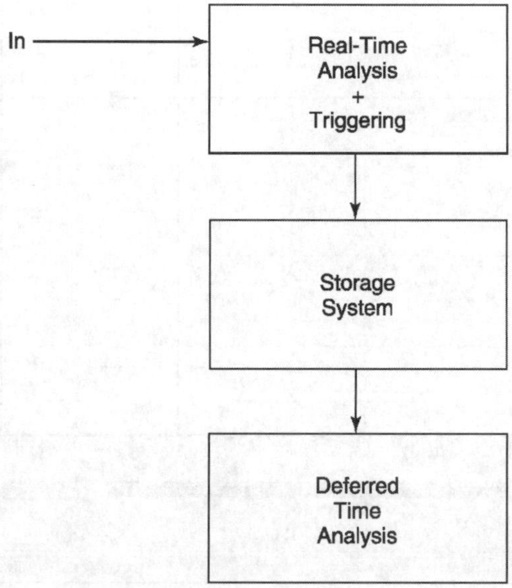

Figure 7.2.24 MPEG analyser block diagram.

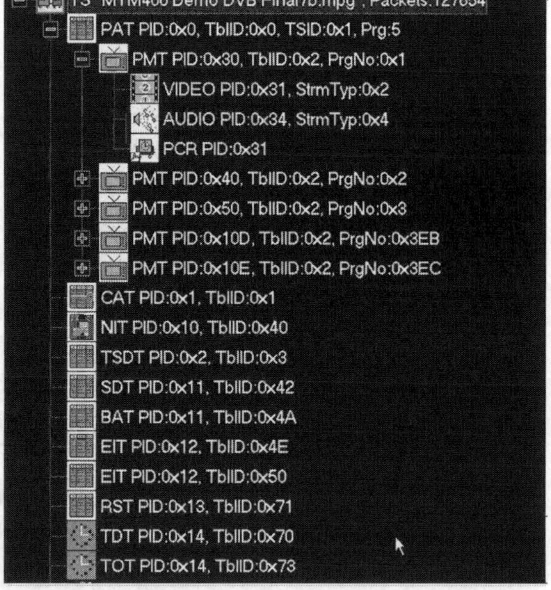

Figure 7.2.25 Example of a hierarchic display.

of a demux or decoder to lock to a transport stream depends on the frequency with which the PSI data are sent. The PSI/SI rate option shown in Figure 7.2.26 displays the frequency of insertion of system information. PSI/SI information should also be consistent with the actual content in the bit stream. For example, if a given PID is referenced in a PMT, it should be possible to find PIDs of this value in the bit stream. The consistency-check function makes such a comparison. Figure 7.2.27 shows a consistency error from a stream including two unreferenced packets. A MUX allocation chart may graphically display the proportions of the transport stream allocated to each PID or programme. Figure 7.2.28 shows an example of a MUX allocation chart display. The hierarchic view and the MUX allocation chart show the number of elements in the transport stream and the proportion of bandwidth allocated.

7.2.4.4 Interpreted View

As an alternative to checking for specific data in unspecified places, it is possible to analyse unspecified data in specific places, for instance in the individual transport stream packets, the tables or the PES packets. This analysis is known as the interpreted view because the analyser automatically parses and decodes the data and then displays its meaning. Figure 7.2.29 shows an example of an MPEG transport packet in hex view as well as interpreted view. As the selected item is changed, the packet number relative to the start of the stream can be displayed. Figure 7.2.30 shows an example of a PAT in the interpreted view.

7.2.4.5 Syntax and CRC Analysis

To deliver programme material, the transport stream relies completely on the accurate use of syntax by encoders. Without correct settings of fixed flag bits, sync patterns, packet-start codes and packet counts, a decoder may misinterpret the bit stream. The syntax check function considers all bits that are not programme material and displays any discrepancies. Spurious discrepancies could be due to transmission errors; consistent discrepancies point to a faulty encoder or multiplexer. Figure 7.2.31 shows a syntax error as well as a missing cyclic redundancy check (CRC). Many MPEG tables have

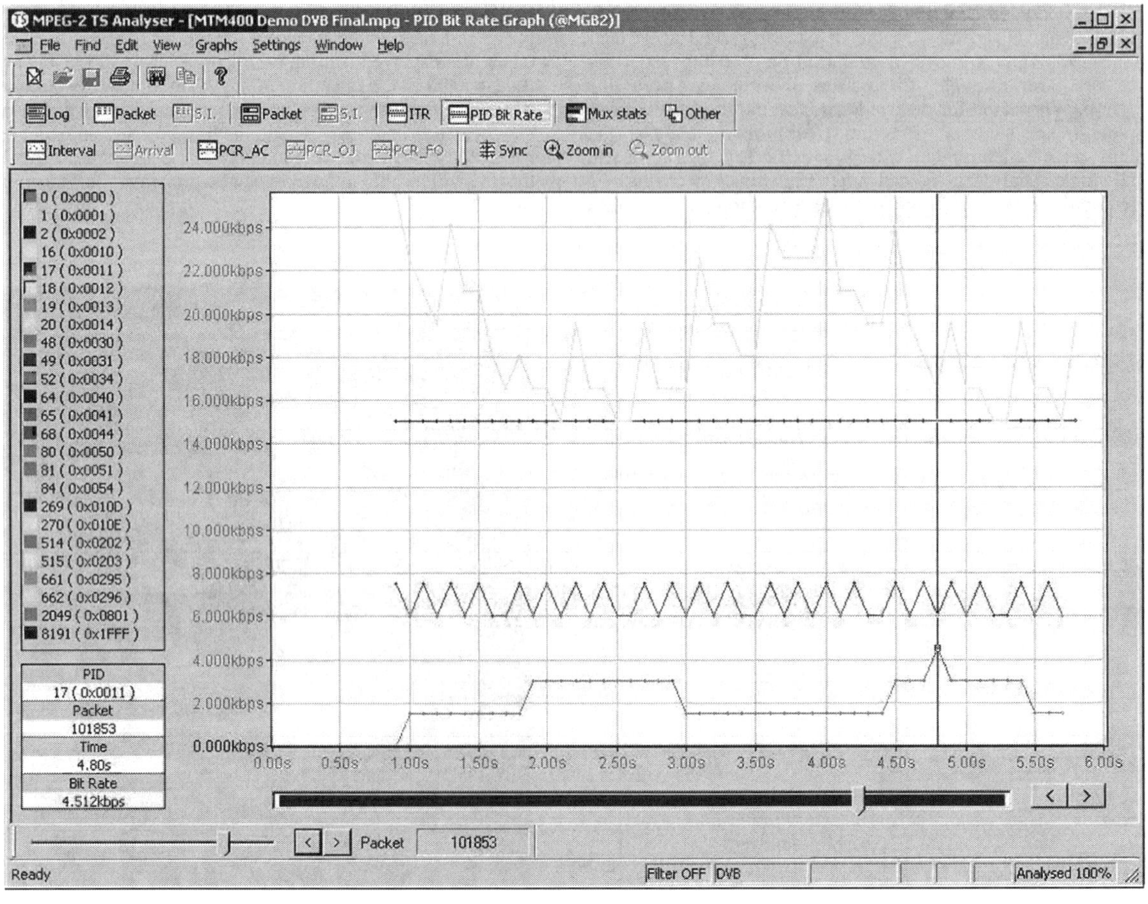

Figure 7.2.26 PSI/SI rate option.

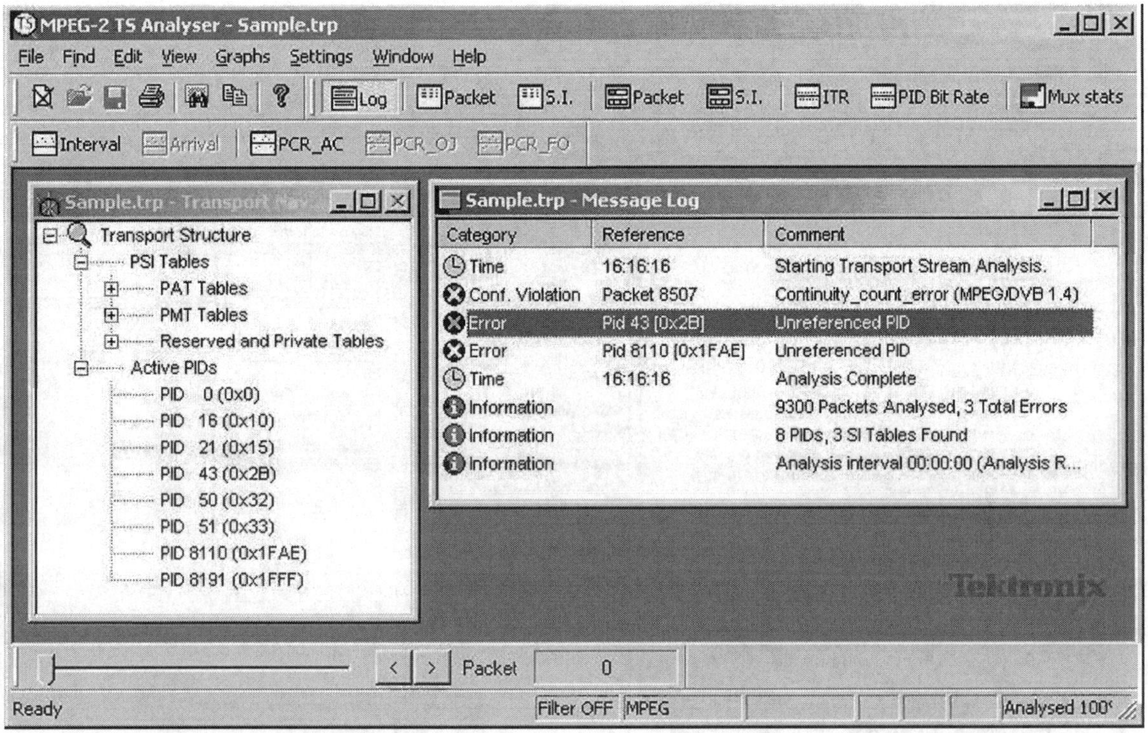

Figure 7.2.27 Consistency error from a stream.

checksums or CRCs attached for error detection. The analyser can recalculate the checksums and compare them with the actual checksum. Again, spurious CRC mismatches could be due to stream-bit errors, but consistent CRC errors point to a hardware fault.

7.2.4.6 Filtering

A transport stream contains a great amount of data, and in real fault conditions, it is probable that, unless a serious problem exists, much of the data is valid and that perhaps only one elementary stream or one programme is affected. In this case, it is more effective to test selectively, which is the function of filtering. Essentially, filtering allows the user of an analyser to be more selective when examining a transport stream. Instead of accepting every bit, the user can analyse only those parts of the data that meet certain conditions. One condition results from filtering packet headers so that only packets with a given PID are analysed. This approach makes it very easy to check the PAT by selecting PID 0 and, from there, all other PIDs can be read out. If the PIDs of a suspect stream are known, perhaps from viewing a hierarchic display, it is easy to select a single PID for analysis.

7.2.4.7 Timing Analysis

The tests described above check for the presence of the correct elements and syntax in the transport stream. However, to display real-time audio and video correctly, the transport stream must also deliver accurate timing to the decoders. This task can be confirmed by analysing the PCR and time-stamp data. The correct transfer of programme-clock data is vital because this data controls the entire timing of the decoding process. PCR analysis can show that, in each programme, PCR data is sent at a sufficient rate and with sufficient accuracy to be compliant. The PCR data from a multiplexer may be precise, but remultiplexing may put the packets of a given programme at a different place on the time axis, requiring that the PCR data be edited by the remultiplexer. Consequently, it is important to test for PCR inaccuracies after the data is remultiplexed. Figure 7.2.32 shows a PCR display that indicates the positions at which PCRs were received with respect to an average clock. At the next display level, each PCR can be opened to display the PCR data, as shown in Figure 7.2.33. To measure inaccuracies, the analyser predicts the PCR value by using the previous PCR and the bit rate to produce what is called the interpolated PCR. The actual PCR value is subtracted from the estimated PCR to give an estimate of the inaccuracy. An alternative approach, shown in Figure 7.2.34, provides a graphical display of PCR interval, inaccuracy, jitter, frequency offset and drift, which is updated in real time. Figure 7.2.35 shows a time-stamp display for a selected elementary stream. The access unit, the presentation time and, where appropriate, the decode times are all shown.

In MPEG, the reordering and use of different picture types causes delay and requires buffering at both encoder and decoder. A given elementary stream must be encoded within the

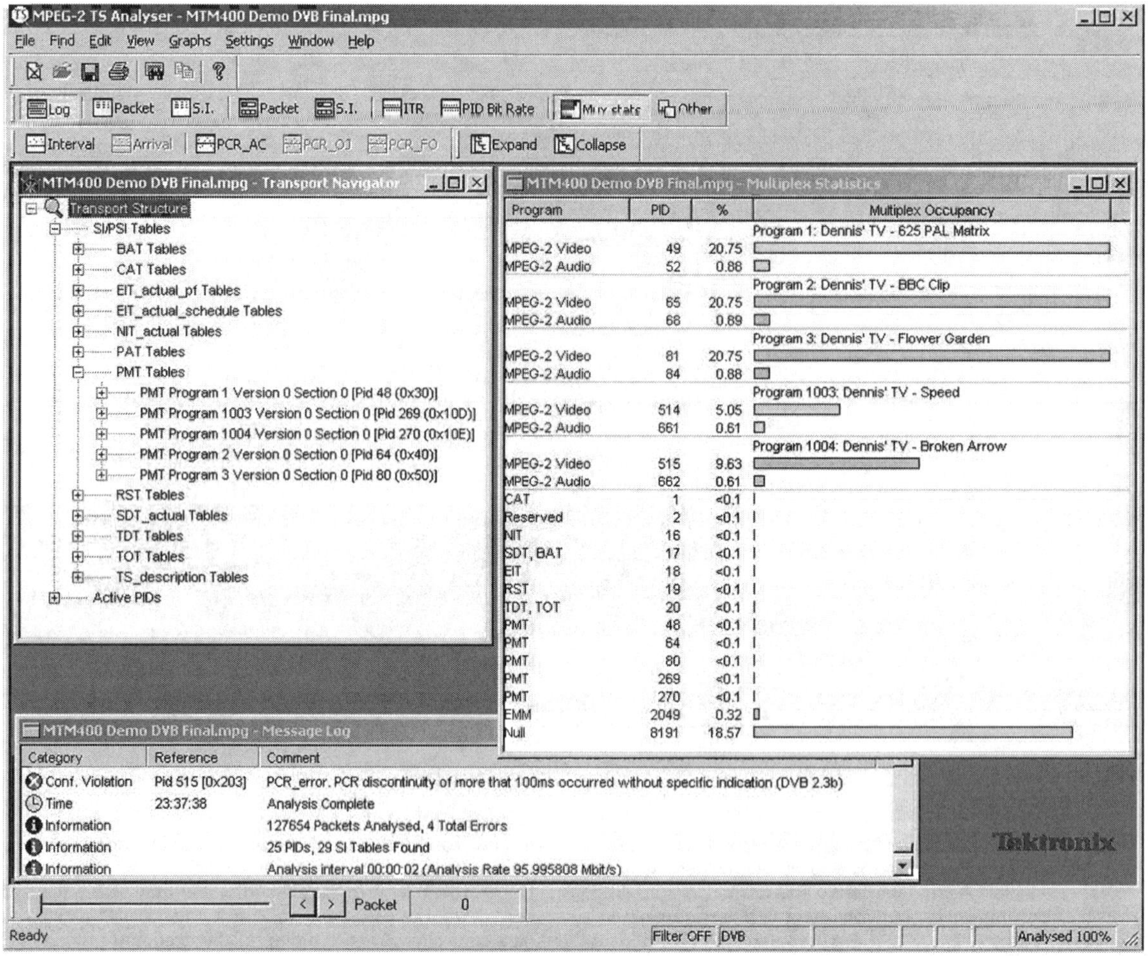

Figure 7.2.28 Example of a MUX allocation chart display.

constraints of the availability of buffering at the decoder. MPEG defines a model decoder called the T-STD (transport stream system target decoder); an encoder or multiplexer must not distort the data flow beyond the buffering ability of the T-STD. The transport stream contains parameters called VBV (video buffer verify) specifying the amount of buffering needed by a given elementary stream. The T-STD analysis displays the buffer occupancy graphically so that overflows or underflows can be easily seen. Figure 7.2.36 shows a buffering display. The output of a normal compressor/multiplexer is of limited use because it is not deterministic. If a decoder defect is seen, there is no guarantee that the same defect will be seen on a repeat of the test because the same video signal will not result in the same transport stream. In this case, an absolutely repeatable transport stream is essential so that the defect can be made to occur at will for study or rectification.

Transport stream jitter should be within certain limits, but a well-designed decoder should be able to recover programmes beyond this limit in order to guarantee reliable operation. There is no way to test for this capability using existing transport streams because, if they are compliant, the decoder is not being tested. If there is a failure, it will not be reproducible and it may not be clear whether the failure was due to jitter or some other non-compliance. The solution is to generate a transport stream that is compliant in every respect and then add a controlled amount of inaccuracy to it so that the inaccuracy is then known to be the only source of non-compliance.

7.2.4.8 Elementary Stream Testing

Because of the flexible nature of the MPEG bit stream, the number of possibilities and combinations it can contain is almost incalculable. As the encoder is not defined, encoder manufacturers are not compelled to use every possibility; indeed, for economic reasons, this is unlikely. This fact makes testing quite difficult because the fact that a decoder works

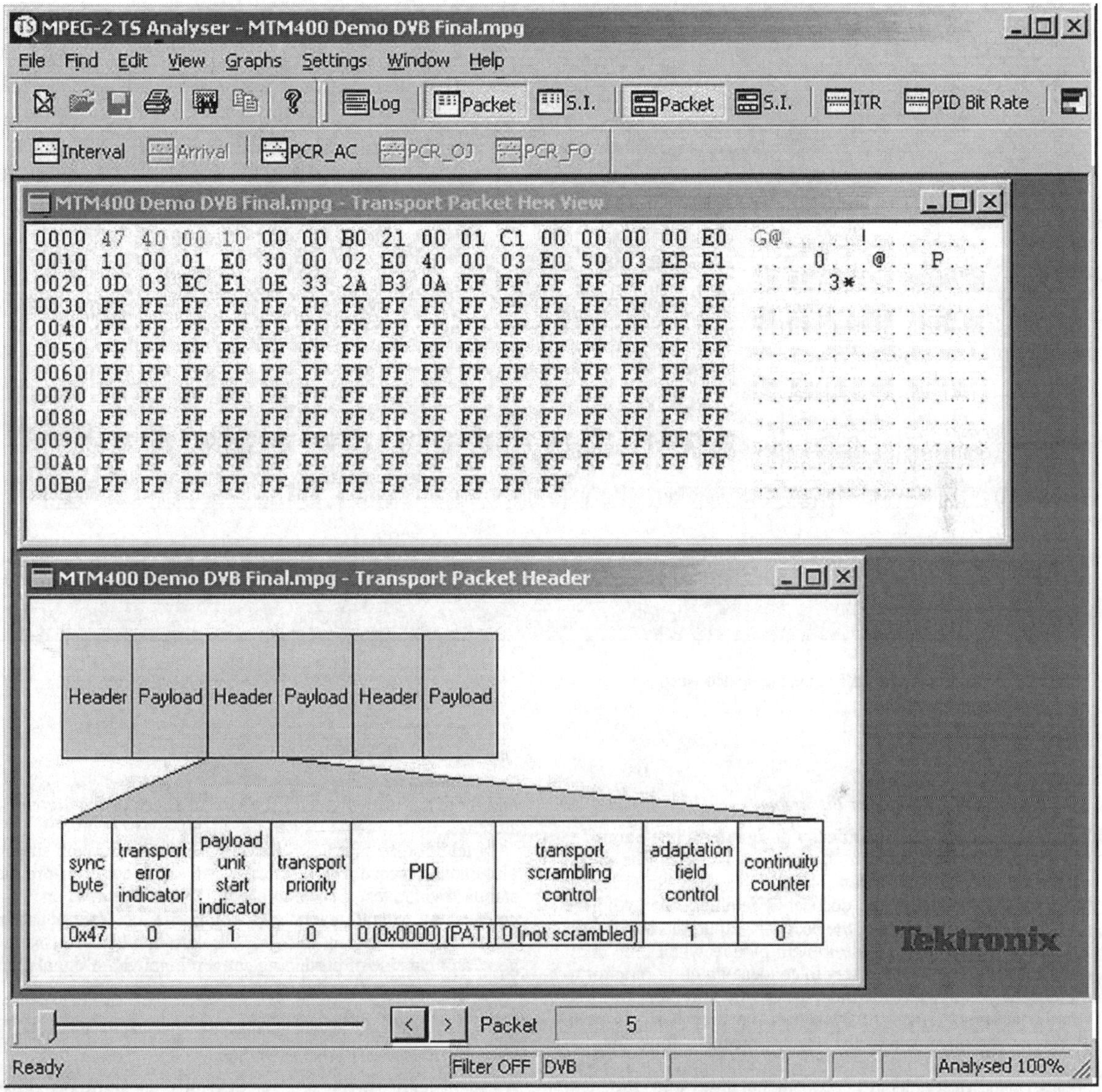

Figure 7.2.29 Example of an MPEG transport packet in hex and interpreted views.

with a particular encoder does not prove compliance. That decoder may simply not be using the modes that cause the decoder to fail. A further complication occurs because encoders are not deterministic and will not produce the same bit stream if the video or audio input is repeated. There is little chance that the same alignment will exist between I-, P- and B-pictures and the video frames. If a decoder fails a given test, it may not fail the next time the test is run, making fault-finding difficult. A failure with a given encoder does not determine whether the fault lies with the encoder or the decoder. The coding difficulty depends heavily on the nature of the programme material, and any given programme material will not necessarily exercise every parameter over the whole coding range.

To make tests that have meaningful results, two tools are required:

1. A known source of compliant test signals that deliberately explore the whole coding range. These signals must be deterministic so that a decoder failure will give repeatable symptoms. The Sarnoff compliant bit streams are designed to perform this task.
2. An elementary stream analyser that allows the entire syntax from an encoder to be checked for compliance.

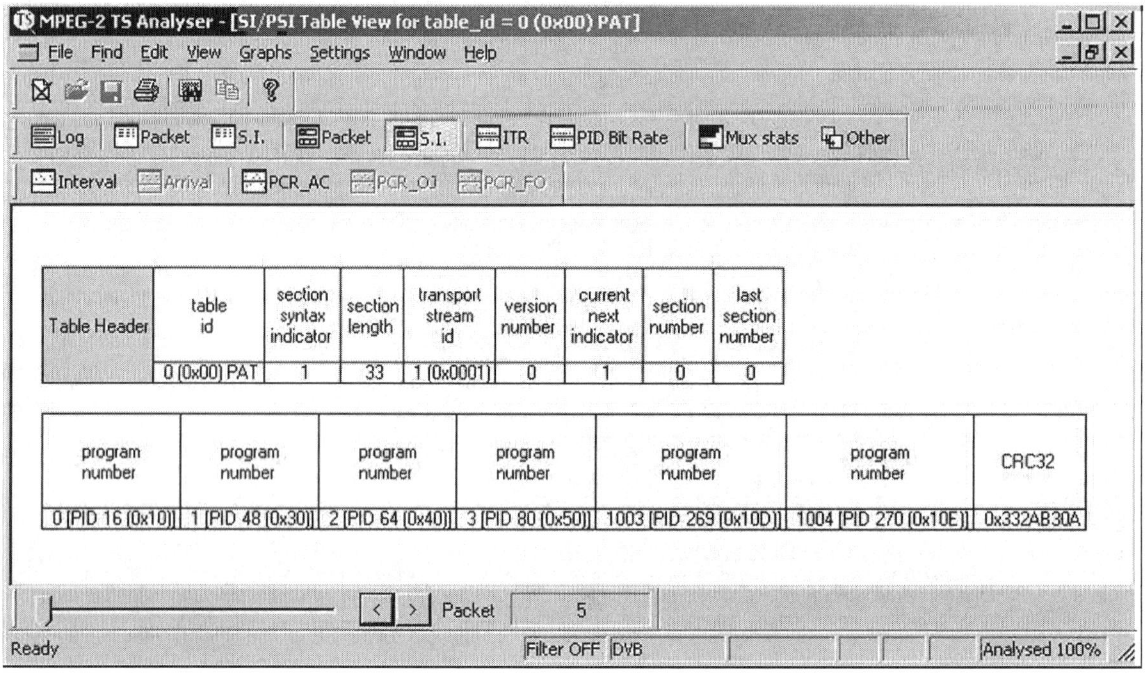

Figure 7.2.30 Example of a PAT in the interpreted view.

7.2.4.8.1 Sarnoff® Compliant Bit Streams

These bit streams were specifically designed by The Sarnoff® Corporation for decoder compliance testing and can be multiplexed into a transport stream feeding a decoder. No access to the internal working of the decoder is required. To avoid the need for lengthy analysis of the decoder output, the bit streams have been designed to create a plain picture when they complete so that it is only necessary to connect a picture monitor to the decoder output to view them. There are a number of these simple pictures. Figure 7.2.37 shows the grey verify screen. The user should examine the verify screen to look for discrepancies that will display well against the grey field. There are also some verify pictures which are not grey. Some tests will result in no picture at all if there is a failure. These tests display the word 'VERIFY' on screen when they complete. Further tests require the viewer to check for smooth motion of a moving element across the picture. Timing or ordering problems will cause visible jitter.

The suite of Sarnoff tests may be used to check all of the MPEG syntax elements in turn. In one test, the bit stream begins with I-pictures only, adds P-pictures and then adds B-pictures to test whether all MPEG picture types can be handled and correctly reordered. Backward compatibility with MPEG-1 can be proven. Another bit stream tests using a range of different GOP structures. There are tests that check the operation of motion vectors over the whole range of values, and there are tests that vary the size of slices or the amount of stuffing.

In addition to providing decoder tests, the Sarnoff streams also include sequences that cause a good decoder to produce standard video test signals to check DACs (digital-to-analogue converters), signal levels and composite or Y/C encoders. These sequences turn the decoder into a video test-pattern generator capable of producing conventional video signals such as zone plates, ramps and colour bars.

7.2.4.8.2 Elementary Stream Analysis

An elementary stream is a payload that the transport stream must deliver transparently. The transport stream will do so whether or not the elementary stream is compliant. In other words, testing a transport stream for compliance simply means checking that it is delivering elementary streams unchanged. It does not mean that the elementary streams were properly assembled in the first place. The elementary stream structure or syntax is the responsibility of the compressor. Therefore, an elementary stream test is essentially a form of compressor test. It should be noted that a compressor can produce compliant syntax, and yet still have poor audio or video quality. However, if the syntax is incorrect, a decoder may not be able to interpret the elementary stream. Since compressors are algorithmic rather than deterministic, an elementary stream may be intermittently

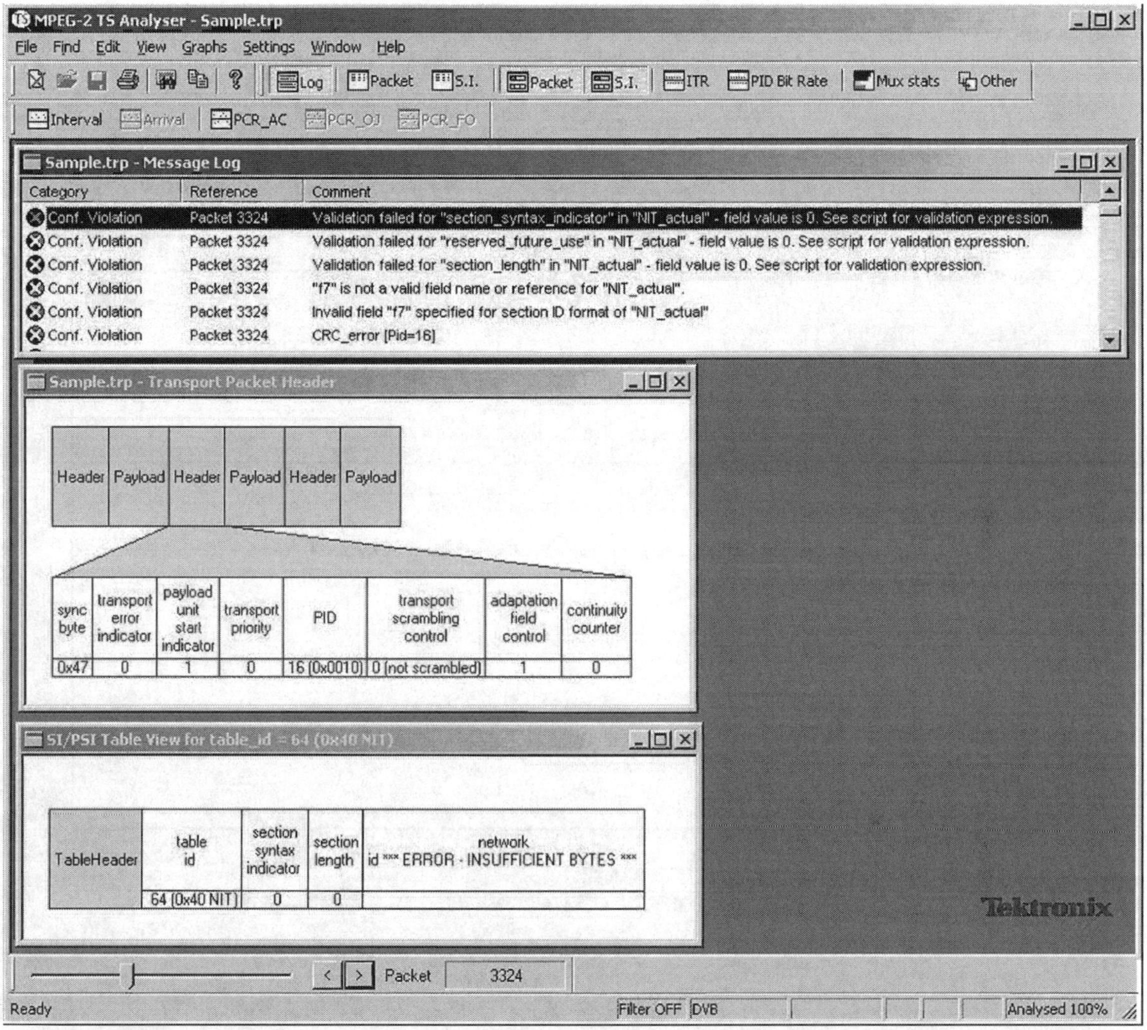

Figure 7.2.31 Syntax error and missing CRC.

non-compliant if some less common mode of operation is not properly implemented. As transport streams often contain several programmes that come from different coders, elementary stream problems tend to be restricted to one programme, whereas transport stream problems tend to affect all programmes. If problems are noted with the output of a particular decoder, then the Sarnoff compliance tests should be run on that decoder. If these are satisfactory, the fault may lie in the input signal. If the transport stream syntax has been tested, or if other programmes are working without fault, then an elementary stream analysis is justified.

Elementary stream analysis can begin at the top level of the syntax and continue downwards. Sequence headers are very important as they tell the decoder all of the relevant modes and parameters used in the compression. At a lower level of testing, Figure 7.2.38 shows a decoded B-frame along with the motion vectors overlaid to the picture.

7.2.4.9 Creating a Transport Stream

Whenever the decoder is suspect, it is useful to be able to generate a test signal of known quality. Figure 7.2.39 shows that an MPEG transport stream must include Programme Specific Information (PSI), such as PAT, PMT and NIT, describing one or more programme streams. Each programme stream must

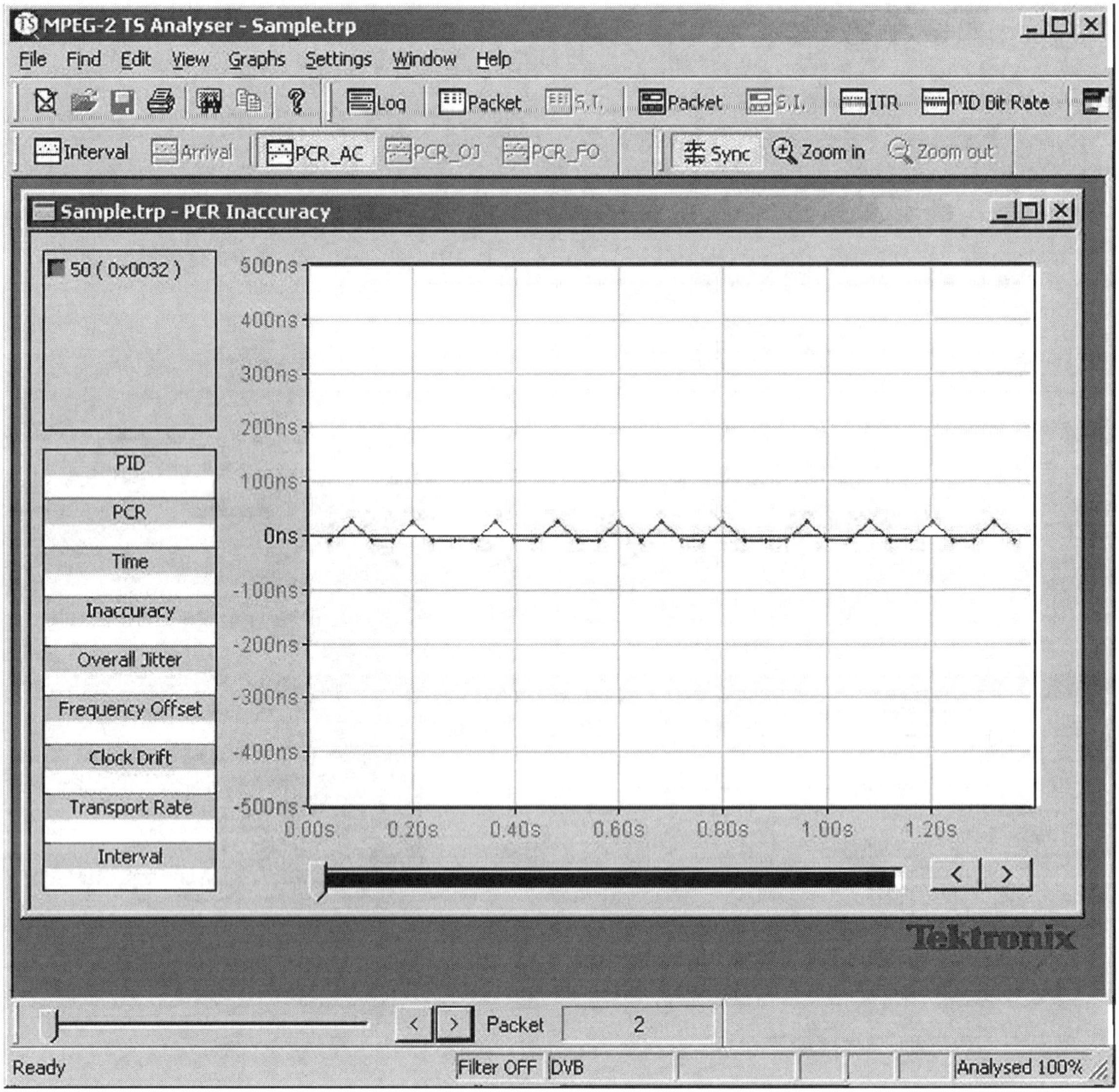

Figure 7.2.32 A PCR display.

contain its own PCR and elementary streams having periodic time stamps. A DVB transport stream will contain additional service information, such as BAT, SDT and EIT tables. A PSI/SI editor enables insertion of any desired compliant combination of PSI/SI into a custom test stream. Clearly, each item requires a share of the available transport stream rate. The multiplexer provides a rate gauge to display the total bit rate used. The remainder of the bit rate is used up by inserting stuffing packets with PIDs that contain all '1's, which a decoder will reject.

7.2.4.10 PCR Inaccuracy Generation

The MPEG decoder has to recreate a continuous clock by using the clock samples in PCR data to drive a phase-locked

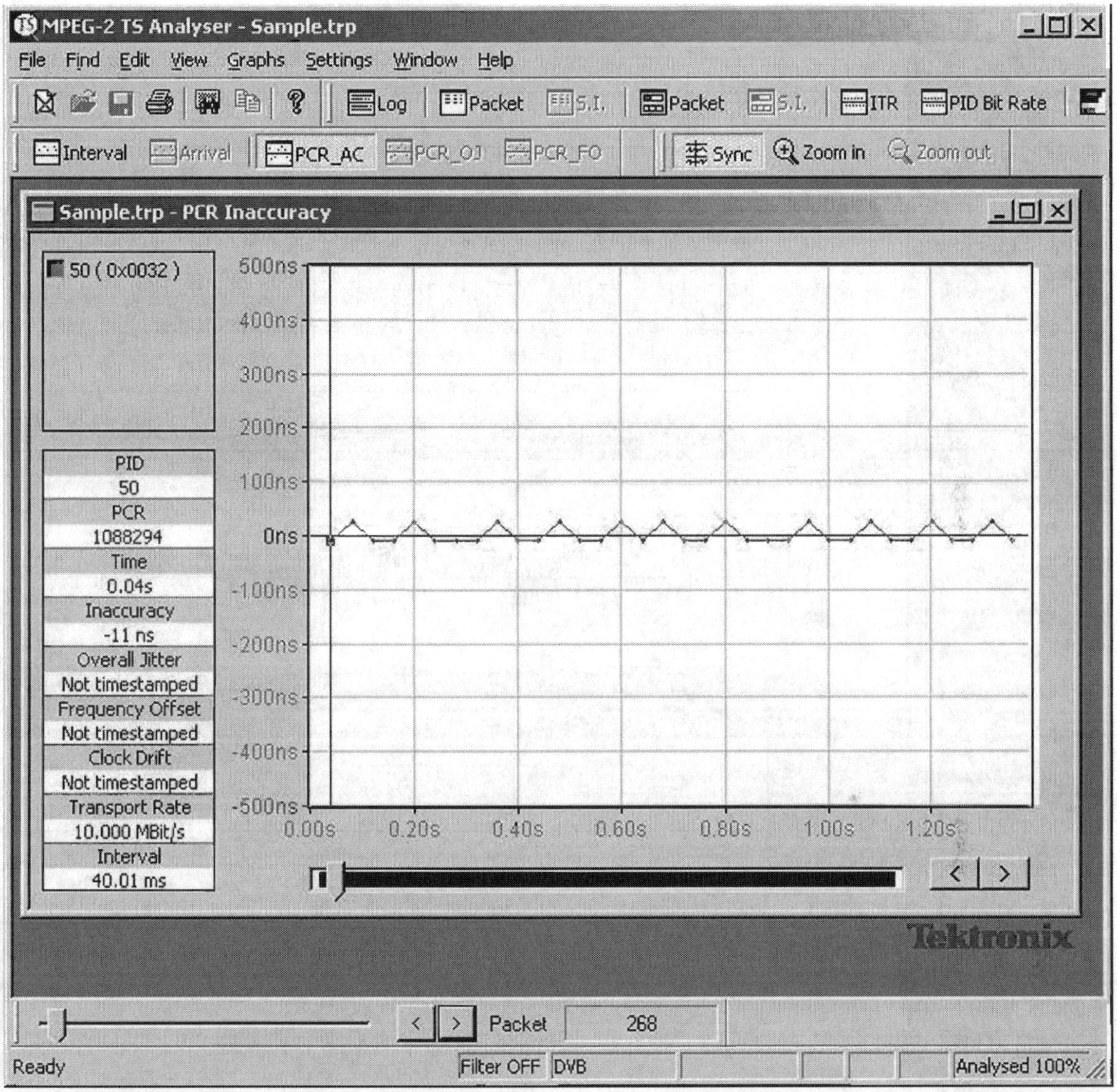

Figure 7.2.33 PCR data display showing full information.

loop. The loop needs filtering and damping so that jitter in the time of arrival of PCR data does not cause instability in the clock. To test the phase-locked loop performance, a signal with known inaccuracy is required; otherwise, the test is meaningless. A good quality MPEG generator has highly stable clock circuits and the actual output jitter is very small. To create the effect of jitter, the timing of the PCR data is not changed at all. Instead, the PCR values are modified so that the PCR count they contain is slightly different from the ideal. The modified value results in phase errors at the decoder that are indistinguishable from real jitter. The advantage of this approach is that jitter of any required magnitude can easily be added to any programme stream simply by modifying the PCR data and leaving all other data intact. Other programme streams in the transport stream need not have jitter added. In fact, it may be best to have a stable programme stream to use as a reference.

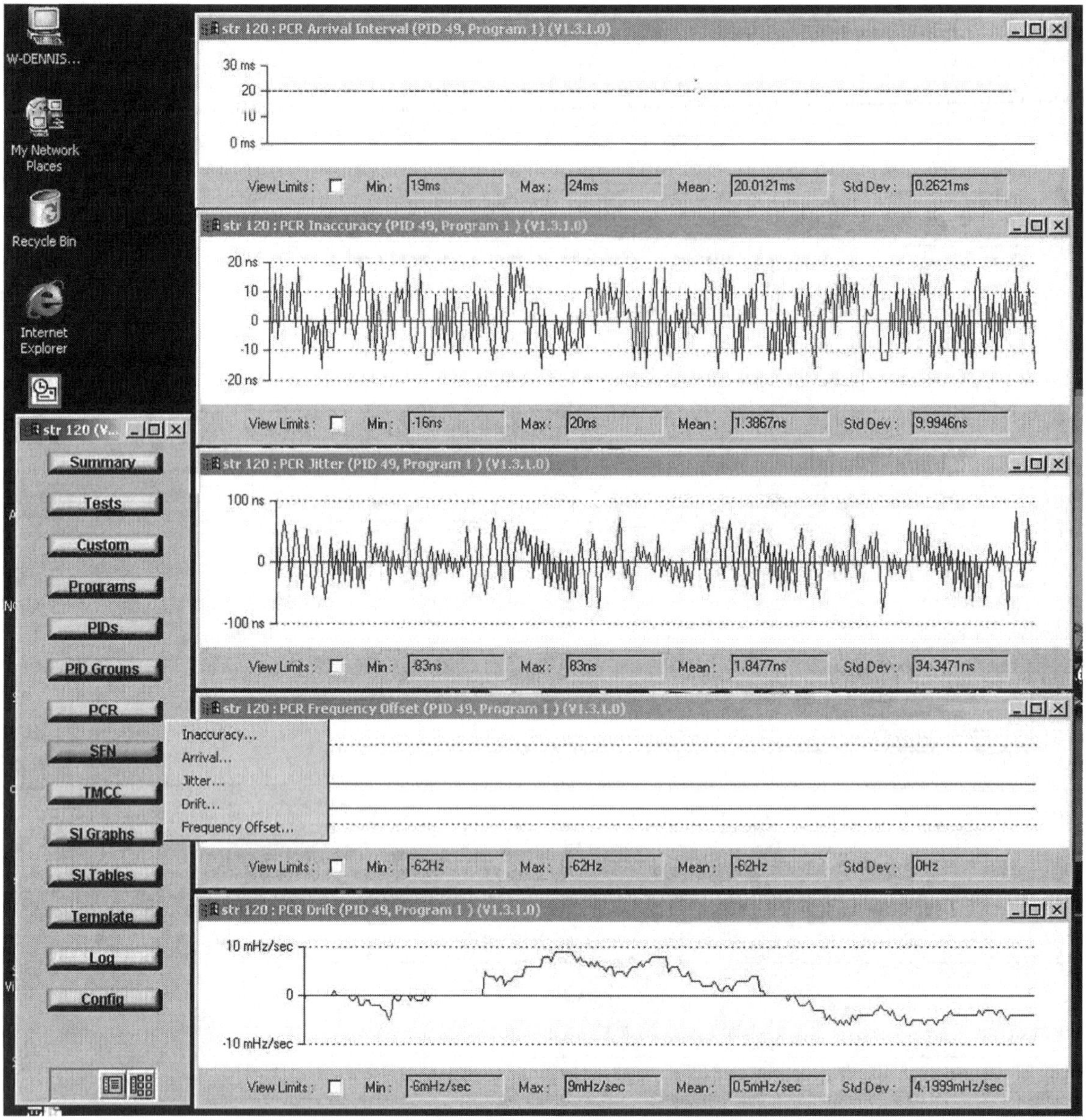

Figure 7.2.34 Graphical display of all PCR parameters.

For different test purposes, the time base may be modulated in a number of ways that determine the spectrum of the loop phase error in order to test the loop filtering. Square-wave jitter alternates between values which are equally early or late. Sinusoidal jitter values cause the phase error to be a sampled sine wave. Random jitter causes the phase error to be similar to noise.

7.2.5 Enhanced MPEG Transport Stream Testing

7.2.5.1 Service Template Testing

As the number of digital television services being distributed around the world increases, the need to monitor the vital MPEG transport streams being transmitted becomes ever more

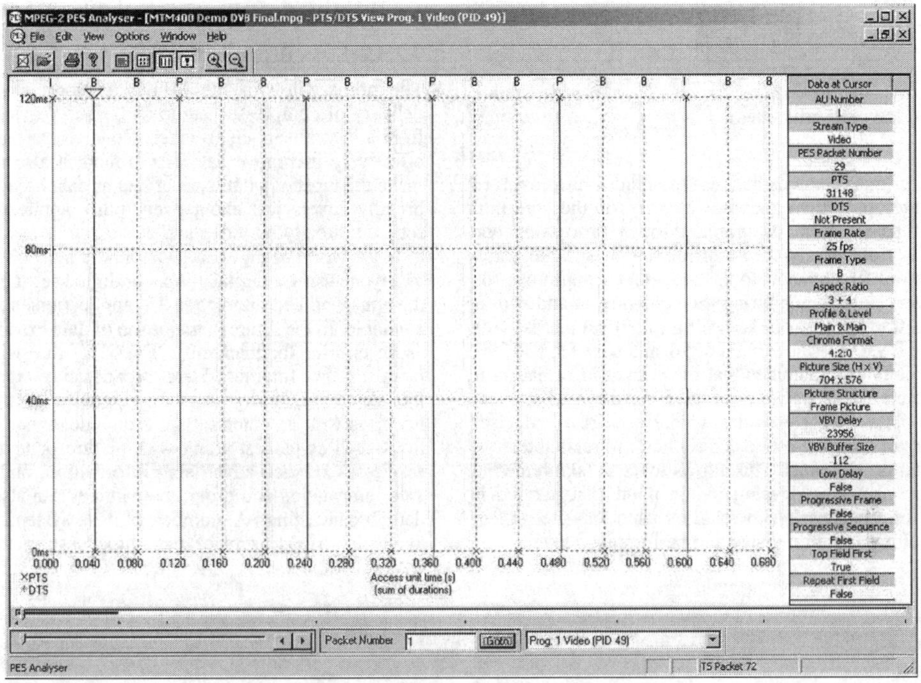

Figure 7.2.35 Time-stamp dislay for a selected elementary stream.

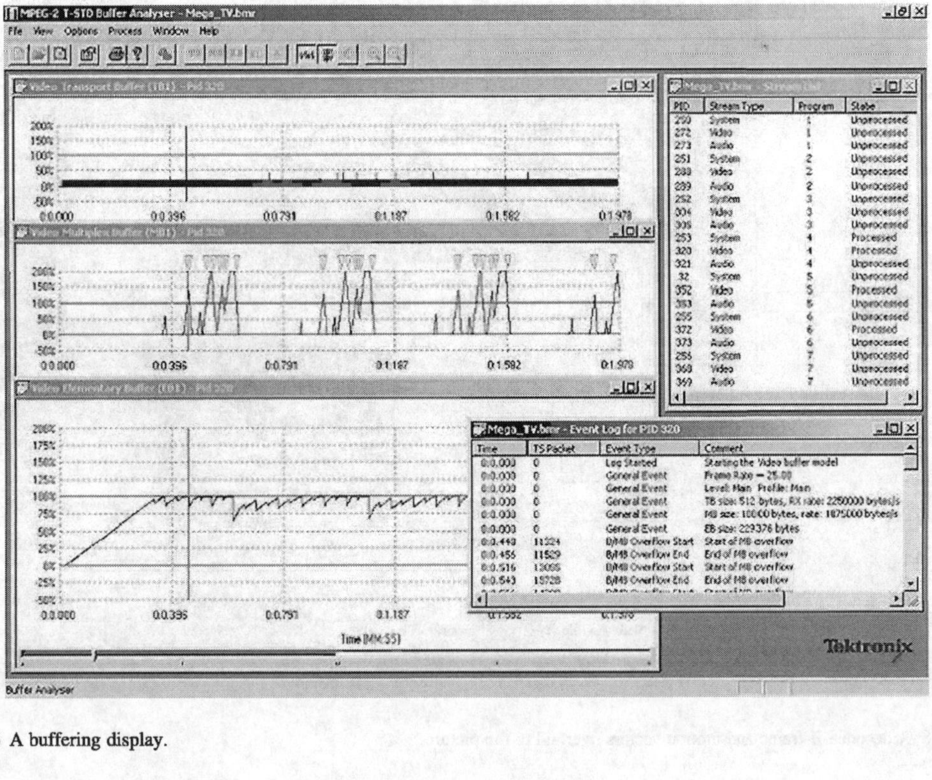

Figure 7.2.36 A buffering display.

Figure 7.2.37 The grey verify screen

important. Various standards and test techniques have evolved to ensure interoperability and the integrity of the transport stream as it passes from its origin, through processing and networking equipment, and then transmission over cable, satellite or terrestrial RF channels to the consumer's digital set-top box. However, simply monitoring the components and syntax of an MPEG transport stream according to test standards, such as the classic DVB test standard ETR290, and now TR 101 290, is still not enough to guarantee that the correct information is being delivered to the viewer's receiver. Addition of service-specific template testing provides a very powerful and cost-effective way of applying scheduled service and related tests to specific target probe points. This allows users to take remedial action before problems develop to the point that service is jeopardised or transmissions wrongly routed or interrupted, with the resulting loss in revenue and performance levels.

7.2.5.2 Data Broadcasting

The primary use of the MPEG transport stream is for the delivery of compressed audio and video services to end-users from a central service provider. However, as transport streams effectively just move data bits around, it also makes sense to make the most of what is described as data broadcasting, which broadly covers just about every other application that is not focused directly on video and audio.

In the world of digital television, the most obvious examples to anyone using a digital TV set or digital set-top box (STB) are the enhanced and interactive TV applications now increasingly available. In the future, a major use of data broadcasting will be as an enabler for interactive TV (iTV) that will also include links to the Internet. Data broadcasting applications also include downloading software upgrades to the set-top box itself, as well as other software downloads to enable applications such as games, along with picture or text data. Internet services, extended advertising information, and extended service information and programme guides can also be served by data broadcasting. A number of test systems are currently available to aid users of data broadcasting to compile and

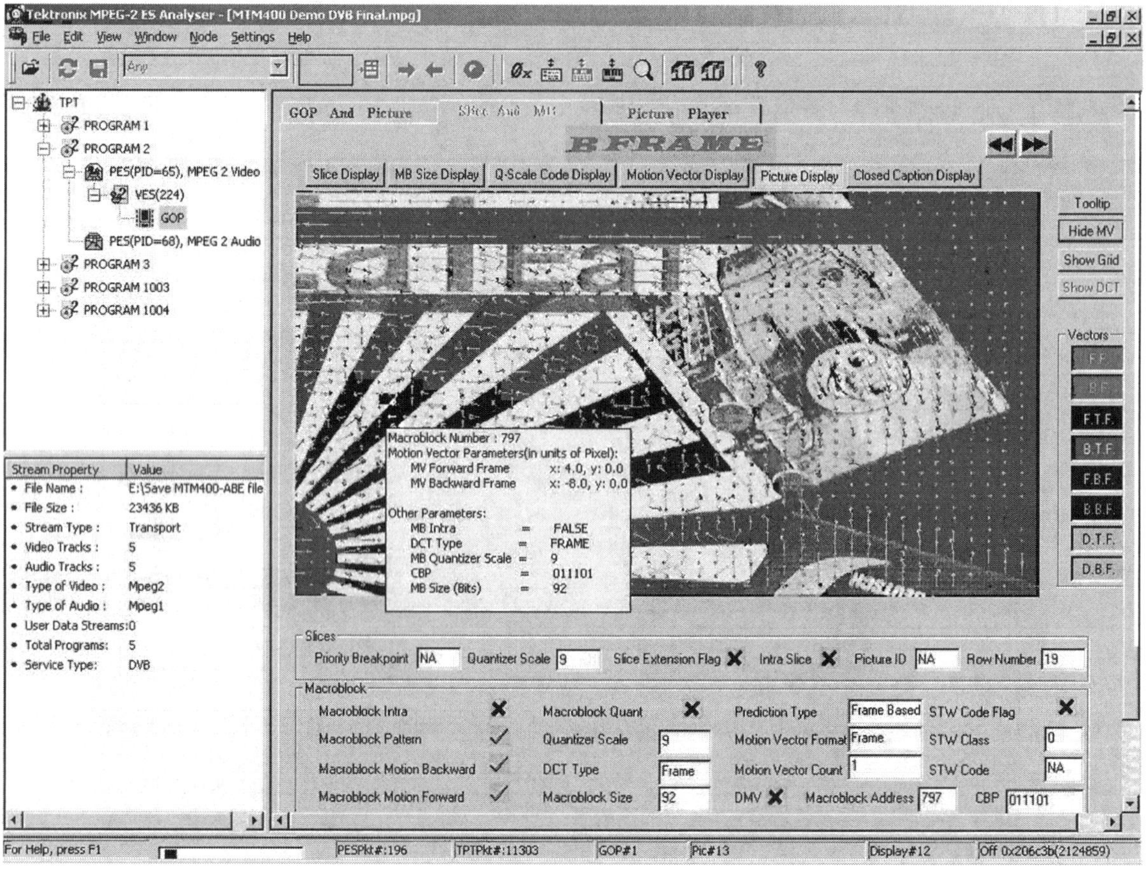

Figure 7.2.38 A decoded B-frame and motion vectors overlaid to the picture.

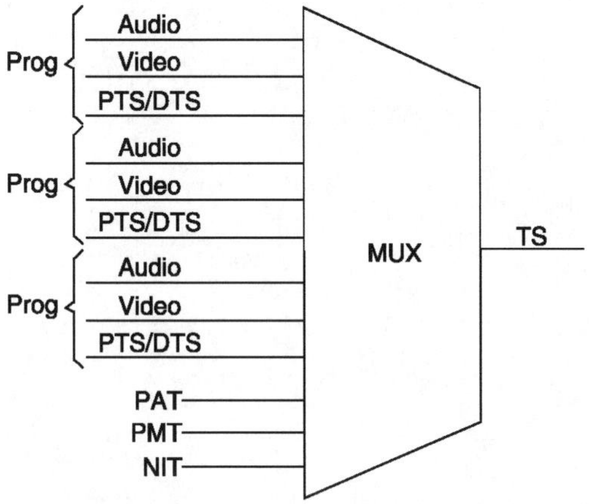

Figure 7.2.39 MPEG transport stream including Programme Specific Information.

perform their iTV services as seamlessly as possible. These systems include the analysis and generation of data and object carousels, including those used specifically for MHP.

Bibliography

SMPTE standards and recommended practices can be found at http://www.smpte.org/smpte_store/standards/

Tektronix Application Note 21W-16005-0, Testing the MPEG2 Transport Stream: Carousel Testing.

Tektronix Application Note 2AW-16079-0, Testing Video Transport Streams Using Templates.

Tektronix Primer 25W-11418-4, A Guide to MPEG Fundamentals and Protocol Analysis.

Tektronix Primer 25W-14700-0, A Guide to Standard and High-Definition Digital Video Measurements.

Ian Dennis BSc
Technical Director, Prism Sound group

7.3 Audio Systems Test and Measurement

7.3.1 Introduction

Audio test and measurement is a diverse technology because it covers a wide range of application areas, each with its own differing requirements. At one end of the scale, for example, an audio equipment designer in a development lab, or someone evaluating such equipment (say, within a broadcast organisation), will usually want to perform many detailed and complex measurements with a high degree of accuracy. By contrast, a factory producing low-cost consumer equipment needs to perform more basic tests in order only to detect production-line faults rather than to evaluate the equipment design, but must do so with the utmost speed and reliability. In between are the requirements of equipment installers and facility troubleshooters, and those of field-service or repair engineers, whose needs are different again.

In general, these different requirements are addressed by different measurement techniques and often by different test equipment, although some examples of multi-functional test equipment are available which attempt to address the needs of more than one type of user.

The sheer diversity of audio test and measurement techniques across all applications is far too wide to cover here, so this chapter will concentrate on providing an overview of the most common methods used in a broadcast environment. The aim is to provide an unfamiliar operator with a firm grasp of the basic principles, after which more detailed coverage can be found in the various references listed at the end of the chapter; some of the more general publications will be especially useful.[1–5]

7.3.2 Audio Measurement Philosophy

Since all audio measurements seek in some way to characterise sound, it might be expected that the process would involve generating and evaluating airborne sound waves in a controlled or neutral physical environment. However, such *acoustic* measurements can be inconvenient to perform and often provide results in which the performance of the equipment is masked by limitations of the environment or the test equipment. Since the majority of audio devices transact electronic signals (or nowadays often digital data) rather than sound waves directly, it is most common to use *electronic* measurement techniques, wherein the stimulus and analysis are in the form of electronic signals or data. Whilst acoustic measurements may sometimes be required in a broadcast environment, for example in the design and evaluation of control rooms or studios, the bulk of broadcast audio measurements are electronic, and so only these techniques are discussed in this chapter.

Most audio measurements are made by a *stimulus–response* method: a particular synthetic stimulus (as opposed to music or other programme) is applied to the input of the *equipment-under-test* ('EUT'), and the resulting output from the EUT is compared to an idealised response.

Although the EUT can be as simple as a cable, or as complex as an entire broadcast signal chain, many of the same techniques can be applied to a wide variety of EUTs. Where the EUT is some sort of media player, it may be necessary to provide the test stimuli recorded on the appropriate medium, for example a test CD. Recorders are more problematic since it can be difficult to measure the record path in isolation from the playback path.

Obviously the accuracy of both the stimulus and the analysis must exceed that of the EUT if meaningful measurements are to be made; the operator must be ever-vigilant in this regard, since modern broadcast equipment is increasingly of a quality which can make this difficult to achieve.

Modern broadcast signal paths include both digital and analogue sections, requiring measuring equipment which can generate and analyse audio signals in both the digital and analogue *domains*. Along with digital audio transmission comes the problem of how to test the interface (or carrier) as distinct from the digital audio data itself. A complete audio testing solution should therefore include the ability to generate and analyse digital audio carriers, at least in the common AES3 format.

7.3.3 Some General Concepts

It is possible to perform a basic set of audio measurements with quite simple equipment, using a single sinusoidal stimulus with a straightforward time-domain (filter and voltmeter) analysis. Such measurements are simple to perform, and are capable of verifying correct operation of a signal path, including a reasonable degree of quantitative performance measurement, although they fall short of the versatility and penetration requirements of a modern R&D or type-approval facility. Although these basic measurements were originally designed to expose shortcomings of analogue equipment, they are also relevant (although insufficient) in testing digital or mixed-domain installations.

7.3.3.1 Anatomy of a basic audio measuring system

Figure 7.3.1 shows a simplified diagram of a basic audio measurement system. Until some years ago, it was necessary to assemble the system from several individual instruments. Nowadays it is more common for all these functions (and more) to be incorporated into a single multi-function unit.

The *signal generator* provides a sinusoidal stimulus, with variable amplitude and frequency, which drives the EUT input. A sinusoidal stimulus is chosen since all of its energy is concentrated at a single frequency, with no additional harmonic (or other) frequency components; this makes analysis straightforward, since any other frequencies detected at the EUT output can be assumed to be the result of impairments perpetrated by the EUT.

The EUT output is connected to a *signal analyser*, which is basically an AC voltmeter, preceded by a number of controllable filters.

High-pass and *low-pass filters* limit the frequency range over which the measurement is to be made – these are commonly set to limit measurements between 20 Hz and 20 kHz, which is usually considered to represent the audible frequency range.

A choice of *weighting filters* can be used to emphasise certain frequencies at the expense of others, usually to tailor a measurement more closely to reflect subjective perception by mimicking the variation in sensitivity of the ear with frequency. Common weighting filter responses are *A-weighting* and *CCIR-468*,[6] although many others are available. For most broadcast measurements, apart from noise analysis, no weighting filter is normally used.

The *band-pass/band-reject (BP/BR) filter* can be used to perform *selective* or *residual* measurements. A selective measurement uses the band-pass filter to restrict the measurement to a particular frequency. This is commonly used where the amplitude of the component of interest is close to (or beneath) the overall noise floor, for example in crosstalk measurements. The selective technique allows the amplitude of the desired component to be measured without being swamped by the overall wideband noise. A residual measurement uses the band-reject filter to exclude a single frequency component from the measurement. The common THD+N measurement (described below) is an example of a residual measurement, wherein the stimulus frequency is removed to allow measurement of the remaining distortion and noise. The BP/BR frequency can usually either be tuned manually, or set to track the generator frequency or the frequency detected at the analyser input.

The *detector* response dictates the static and dynamic response of the voltmeter. An *RMS* (root-mean-square) response is usual, since it reflects the average power and hence generally the subjective loudness of a signal. It is calculated by taking the square root of the signal's average squared-voltage (or squared-sample-value). Alternatively, a *peak* response may be useful in assessing transients. A special *quasi-peak* response, as defined in CCIR-468,[6] in conjunction with a special weighting filter, can be used to make subjectively-tailored measurements of the audibility of noise, switching discontinuities, etc., as detailed in Section 7.3.4.3.

7.3.3.2 Beyond the basic measurement system

Unfortunately, things are not quite so straightforward when our basic measurement system is applied in the real world. In the interests of simplicity various common features were deliberately omitted from Figure 7.3.1. Most of these are mere aids to practicality and do not affect the fundamental principle of operation, but it is worth briefly mentioning these features before proceeding.

7.3.3.2.1 Analogue and digital domains

In the basic example, the question of the generation and analysis domains has been put aside: if the EUT's input is digital, then a signal generator with a digital output is assumed; similarly, if the EUT's output is digital, then the analyser's input is assumed to be digital. To test mixed-domain systems, separate analogue and digital generators, and separate analogue and digital analysers, must be employed – the entire measurement system needs to be duplicated!

Another detail is that the digital generator often needs to be synchronised (sample-locked) to an external reference,

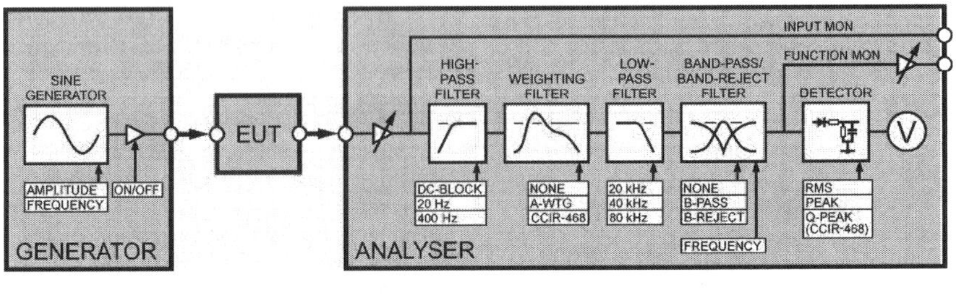

Figure 7.3.1 A basic audio measurement system.

especially in large broadcast facilities, where the use of house-syncs is widespread.

7.3.3.2.2 Channel switching

The basic instrument example has only a single generation and analysis channel; although such a system would be perfectly usable, many real instruments offer greater convenience by providing at least two generator and analyser channels. This allows a complete stereo channel to be comprehensively tested, including inter-channel phase, crosstalk and balance measurements, without replugging.

Some instruments can even control optional external expansible routing switchers to allow automatic measurement of large numbers of audio channels without the need for any manual reconnection.

7.3.3.2.3 Additional input-level, phase and frequency meters

Some residual or cross-channel results (such as THD+N or crosstalk) are usually expressed as a proportion of an EUT output's total amplitude. This means that two measurements plus some calculation are required to arrive at each result. In the interests of convenience, it is common to provide separate RMS amplitude meters tied to each analyser input so that the ratio can be computed directly by the measurement system.

Figure 7.3.1 shows no frequency or phase measurement capability. In practice, it is useful to be able to measure the time delay between pairs of EUT outputs or between an EUT input and its associated output, in order to assess time-linear behaviour. Phase is simply a way of expressing delay as a fraction of the stimulus period. The fraction is usually multiplied by 360 for quotation in degrees. A frequency counter is also often provided at the analyser inputs.

7.3.3.2.4 Sweeps and automation

It is often instructive to repeat particular measurements across a range of frequencies, and to plot the result as a graph with frequency on the X-axis, for example when assessing a simple frequency response. The process of manually setting the generator (or BP/BR) frequency between measurements, and plotting the results by hand, is very time-consuming! Most practical measuring systems can be set up to perform these *sweeps* automatically. Where the EUT is a media player or where its inputs and outputs are physically distant, the analyser is unable to dictate the progress of the stimulus frequency. In these cases, a *sense-sweep* facility allows the analyser to make its plot based on observed changes in the incoming frequency. Sweeps with other parameters on the X-axis, such as amplitude, are also possible.

In more advanced systems, automatic measurement capabilities can extend far beyond a sweep facility. Operators can program such systems to run through predetermined sequences of tests, to compare results against preset limits (and warn of failures), or even to print test reports or log results in a database. Such capabilities are most useful in production-line applications, but even in a broadcast facility can be very useful if large numbers of signal paths are to be tested.

7.3.3.3 Amplitude units

The basic measurement system has only a basic detector – all its results are simply amplitude measurements (signal voltage, or the digital equivalent). So why the need to discuss units? The reason is that even simple systems offer many different ways to express signal amplitude, which can be confusing.

The first point to note is the contrast between *absolute* and *relative* units. Absolute units simply express an amplitude in explicit physical units, such as V_{RMS}, dBu, etc. Relative units are used to express an amplitude as a ratio or proportion of another amplitude – for example, a THD+N result might be expressed in relation to the total signal amplitude, or a crosstalk result in relation to the hostile signal.

Common amplitude units are shown in Table 7.3.1.

Table 7.3.1 Common amplitude units

Analogue absolute units		
V		RMS or peak voltage
dBV	20log(V)	Relative to 1 V
dBm	10log($1000V^2/Z_{ref}$)	Power unit, relative to 1 mW into specified Z_{ref}
dBu	20log(V/0.7746)	Relative to 0.7746 V (1 mW into 600 Ω)
W	V^2/Z_{ref}	Power unit, into specified Z_{ref}
Digital absolute units		
FS	(of) digital full-scale	RMS or peak sample value
dBFS	20log(sample/FS)	RMS relative to RMS of sine which touches FS
Relative units (analogue or digital)		
%	100(value/Ref)	Reference can be specified, or can be a result
dB	20log(value/Ref)	Reference can be specified, or can be a result

Note that W and dBm are, strictly speaking, units of power rather than amplitude.

Logarithmic units are preferred in broadcast applications since they are more manageable where dynamic ranges are large and degradations are small. For example, a relative distortion component of −115 dB, quite achievable in modern equipment, would be an unwieldy 0.00018% in the consumer world. Amplitudes are most commonly expressed in dBu.

Although absolute amplitude units are defined differently for digital and analogue signals, in systems where the line-up between analogue and digital parts is fixed it is often useful to be able to express analogue amplitudes in digital units, or vice versa. For example, an A/D converter could be driven with an analogue stimulus whose amplitude is specified in dBFS.

7.3.3.4 Consistency of measurement results

Apart from basic instrument accuracy, there are a number of factors which could make measurements appear inconsistent, perhaps between different measuring instruments or when compared to published specifications.

A major cause of inconsistency in noise and residual measurement results is the measurement bandwidth, as defined by the settings of the high-pass and low-pass filters in the analyser. Consider an analyser with finite bandwidth measuring the noise amplitude of an EUT whose noise extends over a much wider frequency range. The result will depend as much on the bandwidth of the analyser as on the noisiness of the EUT. The same is true for residual measurements, which may contain noise and distortion components beyond the range of a particular analyser. For this reason, it is usual to deliberately limit the frequency response of signal analysers using 'standard' high-pass and low-pass filters at their input. In this way, consistency is achieved between different analysers providing that the same filter selections are used. A 20 Hz to 20 kHz range (or thereabouts) is commonly used since this represents the nominal audible frequency range. Higher low-pass frequencies are often used to allow measurements with extended HF sensitivity, and higher high-pass frequencies to eliminate the effects of power-line interference. Since noise and residual-distortion measurements are so much affected by analysis bandwidth it is important to ensure that the correct high-pass and low-pass filters are used when comparing measurements.

A similar problem occurs with the selection of weighting filters, and occasionally detector responses. A popular view is that the adoption of unweighted RMS measurements would end this confusion. However, there are legitimate reasons for deviating from this – for example, to give certain measurements a subjective relevance, as in CCIR-468,[6] noise and click measurements, or to assess the subjective performance of noise-shaped systems such as SACD. A less worthy reason often applies to consumer equipment specifications where certain weighting filters are specified for no good reason other than to improve the look of the figures.

Another minor but unavoidable source of inconsistency comes about because there is no generally-adopted standard for the bandwidth and slope of a detector's BP/BR filter. This is not usually the cause of large variations, since the nature of most measurements means that the behaviour of the filter far from its centre frequency is likely to be adequate, and its behaviour near the centre frequency is not critical. However, where sizeable distortion products appear at frequencies close to the stimulus frequency, such as when measuring low-frequency jitter sidebands (see Section 7.3.7), variations between equipment can be significant.

7.3.4 A Collection of Basic Audio Measurement Techniques

The following collection of basic audio measurements is by no means exhaustive, but covers most tests which are applied in broadcast installations. Most are not domain-specific, i.e. they can be applied to both analogue and digital signal paths, although additional digital-domain and converter measurements are further described in Section 7.3.7. All of the measurements can be performed with the basic instrument described above, except where noted.

Each of the following sections describes examination of a distinct aspect of the EUT's behaviour. Details of more exhaustive measurement regimes can be found in AES17[7] and IEC 61606.[8]

In the following sections, some common terminology is used:

- *Nominal amplitude* is commonly applied to the EUT for many measurements; this is a defined amplitude typically 20 dB below maximum amplitude.
- *Maximum amplitude* is the highest amplitude which can be accommodated by an EUT before clipping. For digital EUTs it is usually 0 dBFS (digital full-scale); for analogue EUTs, where the onset of clipping may be gradual, it is reached at a defined distortion, typically 1% (–40 dB).
- *Nominal frequency* is usually defined as 1 kHz, or latterly 997 Hz to ensure that digital equipment exercises many different sample values by choosing a frequency which is not an integer factor of a low multiple of the sample rate.
- The upper and lower *band-edge frequencies* define the extent of the audio band for the purposes of a measurement. These are typically 20 kHz and 20 Hz respectively.

7.3.4.1 Linear response

These measurements are intended to ensure that signal amplitudes and frequencies applied to the EUT input are accurately reflected at the EUT output. These tests are quite fundamental in that they investigate relatively major artefacts, and so they are arguably the most important, since any inadequacies revealed here are potentially more audible than those exposed in the following sections.

7.3.4.1.1 Amplitude-related

Gain is the ratio of the output amplitude to the input amplitude of the EUT, measured at nominal frequency and amplitude.

Frequency response is determined by a gain sweep against frequency, between the band-edge frequencies, sometimes normalised with respect to the nominal-frequency value.

Inter-channel balance, the comparative gain of multiple channels, is an important parameter in stereo or surround systems.

A *polarity* test determines whether or not a signal is phase-inverted on passing through the EUT; equipment or interconnections with 'negative gain' invariably cause system problems. With basic equipment, this is most easily measured by making an input-to-output phase measurement at a low frequency.

7.3.4.1.2 Time-related

Phase response is measured by performing a sweep of the input-to-output phase of the EUT between the band-edge frequencies. Any significant deviation from zero within the audio band may be audible.

The *inter-channel phase response* of stereo or surround devices is assessed similarly, but plotting the inter-channel phase instead of the input-to-output phase. Any significant inter-channel phase deviation within the audio band can seriously compromise the stereo or surround image.

Delay through a device: in digital devices, phase response plots may be obscured by a fixed delay through the device, showing an apparent large input-to-output phase variation with frequency. It is therefore often preferable to plot delay against frequency for digital devices.

7.3.4.2 Distortion

When a sinusoidal stimulus passes through an analogue audio device it is subject to distortion. *Harmonic* distortion occurs where the output from the device is no longer precisely sinusoidal, but shares the same periodicity as the stimulus. In this case, the distortion can be represented as the addition of small spurious components at integer-multiples of the stimulus frequency, or harmonics. Aside from non-signal-related interference (see Section 7.3.4.3) and occasional modulation non-linearities, other *spuriae* do not often occur in analogue devices. This is not true of digital devices which, owing to characteristic non-linearities (see Section 7.3.7.3), can often produce spuriae at non-harmonic frequencies as well as ordinary harmonic distortion. Neither is it true (even for analogue devices) if the stimulus is not a single sinusoid, for multiple-frequency stimuli can produce intermodulation spuriae at frequencies which are not harmonics of any of the stimulus frequencies.

Notwithstanding this, most distortion measurements are based on the concept of identifying harmonic distortion of a pure sinusoidal stimulus, since this can be achieved with basic equipment. However, it is often argued (with justification) that many distortion mechanisms are not exposed by such a simple stimulus. This is borne out by listening tests where sonic differences can be reliably detected between devices with identical single-frequency harmonic distortion behaviour. IMD techniques, as summarised below, and FFT and multi-tone techniques as described in Section 7.3.6 provide more revealing distortion measurements, but are not yet common in broadcast testing.

Figure 7.3.2 shows a 1 kHz sinusoid exhibiting harmonic distortion. Note that the power of the FFT technique is evident here – it can show tremendous detail of low-level imperfections which are entirely invisible on a time-domain 'oscilloscope' display.

Most distortion measurements of analogue devices are generally carried out with a stimulus at nominal amplitude and frequency, although it is common with digital devices for the amplitude to be set close to full-scale, for example at −1 dBFS.

Individual harmonic distortion measurements cannot be made with the basic measurement instrument, since a normal band-pass filter is not selective enough to eliminate the stimulus frequency sufficiently. Such measurements can be made with FFT techniques, and are very useful to equipment designers, who can infer specific problems according to which harmonics predominate.

THD, or *total harmonic distortion*, measurements are similarly problematic, since they must be made by RMS-summing the amplitudes of individual harmonic distortions.

By far the most widespread distortion measurement method is *THD+N*, or *total harmonic distortion plus noise*, a somewhat mis-named method, since it is also sensitive to non-harmonic distortion. This is just as well, since it enhances the usefulness of the technique in measuring digital audio devices. THD+N is a residual measurement made by removing the stimulus frequency and measuring everything that remains. The residual amplitude is usually expressed relative to the stimulus amplitude measured at the analyser input.

THD+N is also commonly performed as a sweep against stimulus frequency (at nominal amplitude for analogue devices, or close to full-scale for digital devices), and as a sweep against stimulus amplitude (at nominal frequency).

IMD, or *intermodulation distortion*, is a technique aimed at measuring distortion resulting from the interaction of different frequencies within the EUT. There are two methods in general use: the *SMPTE/DIN* method and the *CCIF* or *difference-tone* method. Neither can be performed with a basic measurement instrument, since their stimuli comprise a mixture of two sinusoids at different frequencies. *DIM*, or *dynamic intermodulation*, also known as *TIM*, or *transient intermodulation*, is a method designed to measure distortion in the EUT caused by the occurrence of transients. Likewise, this requires a complex stimulus. These methods are rarely performed in routine broadcast testing. Full details are included in Refs 1, 2, 5 and 7.

7.3.4.3 Noise and interference

An EUT almost inevitably adds noise to a signal passing through it. As well as the random noise (or 'hiss') which is added by all analogue circuits or by dither in digital processors,

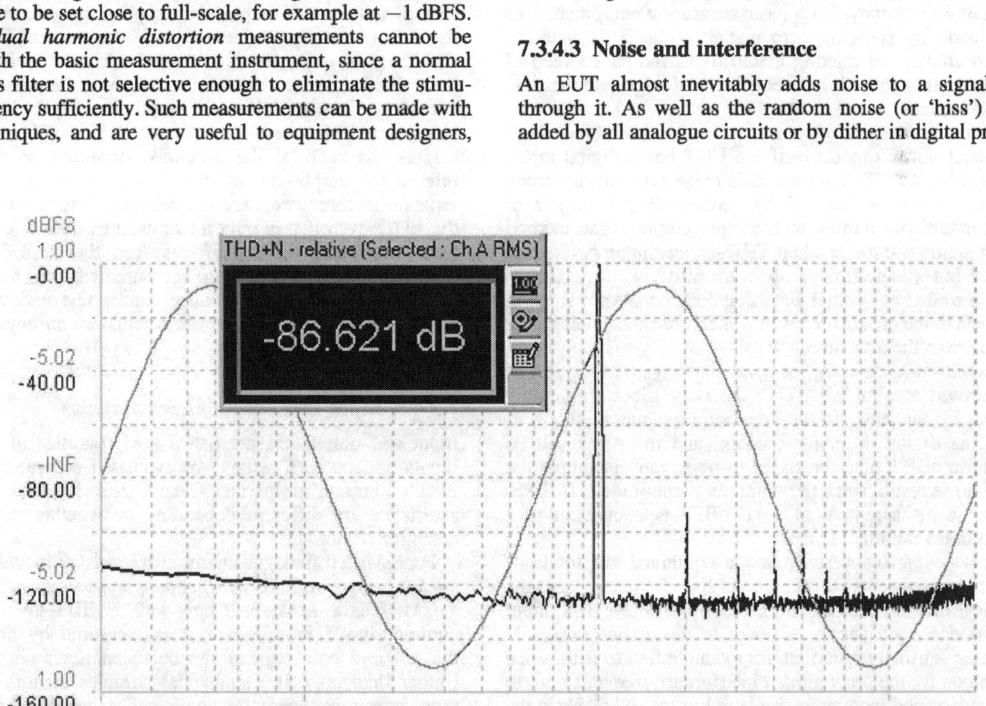

Figure 7.3.2 A 1 kHz sinusoid exhibiting harmonic distortion.

other non-signal-related interference may also be added – for example, hum from power-line interference.

Noise measurements are amongst the few that are ever made with other than an RMS detector response. Whilst RMS measurements of noise are adequate for many purposes, it is often preferable to use a detector response which more accurately mirrors the subjective loudness of the noise. Since the random nature of electronic noise imparts a high *crest factor* (the ratio of peak amplitude to RMS), a quasi-peak detector response is often used (as specified in CCIR-468[6]), which is moderately sensitive to peaks, giving a good correlation between the measurement of the noise and the subjective loudness. As the human ear is not uniformly sensitive across the frequency band, CCIR-468 also specifies a weighting filter which emphasises noise in the ear's most sensitive frequency regions. Together, CCIR-468's detector response and weighting filter give a good correlation between the objective measurement and the subjective loudness of noise. However, this approach is not universal – AES17[7] specifies the use of a CCIR-468-shaped weighting filter but with an RMS detector response, and plain unweighted RMS noise measurements are still the most common.

Some typical noise measurement techniques are as follows:

- *Idle-channel noise* is the amplitude measured at the EUT output, between the band-edges, with no signal passing through the EUT. This is achieved by terminating an analogue EUT input, or by feeding digital zero to a digital EUT input. A variety of weighting filter and detector response options may be used as discussed above.
- The *idle-channel noise spectrum* is a plot of the idle-channel noise against frequency. With basic measuring equipment, this plot is made by applying zero signal to the EUT input as described above, and making a selective band-pass sweep of the EUT output amplitude between the band-edges, normally in third-octave steps.
- *Noise-in-the-presence-of-signal* is often measured instead of idle-channel noise, especially if the EUT has a digital input, since some such EUTs improve their noise performance when digital zero input is detected by zeroing digital outputs or applying additional muting to analogue outputs. The method is to essentially use the residual THD+N technique at nominal frequency, but with a stimulus amplitude 60 dB below maximum amplitude so that any actual distortion components are small in comparison to the noise. The result is measured absolutely rather than relative to the total amplitude.
- The *signal-to-noise ratio*, or *dynamic range*, is the ratio of the maximum amplitude EUT to the noise floor (preferably measured by the noise-in-the-presence-of-signal method). If the EUT is digital or mixed-domain, and the noise can be measured in dBFS, then the signal-to-noise ratio is simply the inverted noise result, since the maximum amplitude is 0 dBFS, e.g. a noise measurement of −111 dBFS is equivalent to a signal-to-noise ratio of 111 dB.
- The *power-line interference* is the combined amplitude of frequency components at the power-line frequency and its harmonics, measured at the EUT output with the EUT input zeroed as described above. A range of power-line harmonic frequencies is usually specified, for example first to fifth, since it is common for the predominant interference frequency not to be the fundamental (nominally 50 Hz in Europe and 60 Hz in the USA). This measurement is not straightforward where only a single band-pass filter is available. In this case, one can approximate the measurement by measuring only the strongest power-line harmonic (by trial and error). Alternatively, the high- and low-pass filters can be used to bracket the desired frequency range, although it is unusual for the low-pass filter to be adjustable down to such a low frequency.

7.3.4.4 Crosstalk and separation

Crosstalk or separation problems can occur in analogue circuits or interconnections where an unwanted residue of a signal being carried on a signal path 'leaks' onto a different signal path. The most common type of crosstalk measurement is *inter-channel crosstalk*, between the channels of a multi-channel device; however, many other variants exist – for example, *input-to-output leakage* when a device is muted, or *feed-through* from unselected inputs in a routing switcher.

Crosstalk is generally very frequency-dependent so, if only basic measuring equipment is available, measurements should either be swept (which is time-consuming if there are many channels to be tested) or spot-checked at a low and a high frequency. The latter approach is valid since crosstalk is rarely worse at moderate frequencies than at the extremes. High-frequency crosstalk is usually radiated, and is often a consequence of physical proximity between circuits, whereas low-frequency crosstalk is more often coupled through inadequately-isolated power supplies. Crosstalk at all frequencies can also be conducted through poor grounding schemes.

The method of crosstalk measurement is similar in all cases:

- For inter-channel crosstalk, the input to the EUT channel under test is terminated (if analogue) or fed with digital zero (if digital), and a sinusoidal stimulus at the desired frequency (or sweep) and nominal amplitude is applied to the *hostile* inputs of the EUT (normally all the others). The output amplitude of the EUT channel under test is measured selectively with the band-pass filter tuned to the hostile stimulus frequency. This is necessary since the crosstalk amplitude may be close to (or below) the RMS noise floor. Crosstalk is usually expressed in dB, as the ratio of the crosstalk amplitude to the output-referenced amplitude of the hostile stimulus. Note that separate measurements are normally made for each channel of the EUT: even if it is only a two-channel device, the crosstalk from A to B may be very different from that from B to A.
- For input-to-output-leakage measurements, the hostile signal is driven into the EUT channel under test and the EUT is muted. For feed-through measurements, all unrouted inputs to the EUT are driven.

7.3.4.5 Input and output characteristics

Input and output characteristics are properties of the EUT's inputs and outputs, which are evaluated separately from the EUT's 'through' properties. Clearly the input and output characteristics are different depending on whether the ports are analogue or digital.

For balanced analogue inputs, characteristics include *common-mode rejection ratio* (*CMRR*) and *input impedance*.

CMRR is a measure of how well an EUT input can reject common-mode interference; it is measured by driving both the hot and cold legs of the balanced input with the same (rather than opposite) sinusoidal stimulus at nominal amplitude, swept between the upper and lower band-edges. The selective amplitude for each frequency, expressed relative to the output-referenced stimulus amplitude, is the CMRR response. Note that the stimulus should be applied through a representative non-zero source impedance to each leg, since real-world CMRR performance may be greatly affected by source impedance variations. In fact, a stringent CMRR test might include deliberately unequal source impedances in each leg. CMRR is normally worse at high frequencies, where stray and poor-tolerance circuit capacitances begin to affect circuit gains.

The differential input impedance of the EUT can be inferred by driving the input at nominal amplitude and frequency, and measuring the EUT output amplitude while varying the output impedance of the test generator. This is most easily accomplished by switching between the nominal low impedance and 600 Ω settings which are commonly available on signal generators. If the nominal generator impedance is quite low, the input impedance of the EUT can be closely approximated as:

$$Z_{IN} = \frac{600}{\left(\frac{V_{NOM}}{V_{600}} - 1\right)}$$

For balanced analogue outputs, characteristics include *output balance* and *output impedance*.

Output balance is usually measured by driving a sinusoidal stimulus at nominal amplitude and frequency through the EUT. The hot and cold legs of the EUT output are connected together through two 300 Ω precision resistors (preferably matched to within 0.01%). The output balance is the amplitude between the resistors' mid point and signal ground, measured with the analyser input impedance set at 600 Ω, expressed relative to the output-referenced stimulus amplitude.

The differential output impedance of the EUT can be inferred by driving the EUT at nominal amplitude and frequency, and measuring the EUT output amplitude while varying the input impedance of the analyser. This is most easily accomplished by switching between the nominal high-impedance and 600 Ω settings which are commonly available on signal analysers. If the nominal analyser impedance is quite high, the output impedance of the EUT can be closely approximated as:

$$Z_{OUT} = 600\left(\frac{V_{NOM}}{V_{600}} - 1\right)$$

Digital input and output port characteristics are detailed in Section 7.3.8.

7.3.4.6 Broadcast-specific measurements

As well as general audio measurements, broadcast facilities may need to implement various application-specific measurements, such as analysis of pre-emphasised signals or FM multiplexes, or testing of low-bit-rate CODECs, modulation-level enhancement devices, subjective loudness meters, etc.

Descriptions of such application-specific test methods are beyond the scope of this chapter, but are partially covered in some of the references.[2,5]

7.3.5 Some Practical Considerations

Broadcast audio testing can be a time-consuming business, with many channels and devices to be tested and many different tests and measurements available. Some practical considerations can help to minimise time spent testing.

Inevitably one meets the question: "Which measurements should I perform?" A glib answer is, for device evaluation and approval, "Everything you can," and for routine facility testing, "As little as you can get away with." Of course, the real answer depends on the nature and size of the installation, whether it is digital or analogue, or a mixture of each, and what kind of devices are in use. The rule of thumb for facility testing is to remember that the objective is to ensure that everything is working properly, rather than to perform an in-depth set of equipment design reviews. To this end, the basic collection of measurements in Section 7.3.4 is a useful starting point. It is almost certainly not worth measuring aspects of low-level linearity, or exotic types of distortion, or effects of sampling jitter – these are not tests for routine maintenance. If the measurements are to be made manually, it is often better to select a basic set of spot measurements and to avoid too many sweeps. An FFT analyser, especially in conjunction with a multi-tone generator, can provide a quick subjective check that all is well. Automation can allow a pre-arranged sequence of tests to be quickly performed, with warnings of any failures and logging of results.

Another important point is not to be misled by erroneous measurements. This basically comes down to being sure that a signal path is working correctly before getting into detailed measurements. Key points here are to always monitor the analyser's residual output with an oscilloscope or better still with an audio monitor, preferably both. This makes it easy to see, for example, that a high THD+N reading is actually due to power-line hum. FFT analysers are also invaluable in this respect. In digital installations, the residual monitor can also pick up common problems which need to be resolved before measurement is meaningful: regular ticks usually mean that something is not correctly locked to the reference sync. Random ticks at all audio frequencies indicate intermittent equipment or interfacing errors. Random ticks only at high audio frequencies suggest missed or repeated samples, perhaps as a consequence of error concealment.

7.3.6 Improved Measurement Techniques and Equipment

The basic measurement instrument described earlier is typical of the structure of many traditional analogue instruments. It is limited in the versatility of both generator and analyser, and as such can only perform quite basic measurements. Adding functionality such as IMD testing would require additional and costly analogue circuitry. Early attempts at digital-audio-capable instruments were achieved by simply adding a digital generator and analyser to the existing analogue architecture. Consistency between analogue and digital measurements was a problem, as were any areas which required interaction between the two domains.

A more logical approach to mixed-domain testing would seem to favour generation and analysis in the digital domain, which can offer great versatility at low cost, with analogue outputs and inputs being provided by D/A and A/D converters connected to the generator and analyser respectively. The problem with this is that a converter-based analogue capability has been unable to deliver adequate analogue performance (in terms of both distortion and bandwidth) for high-precision analogue measurement. However, recent advances in A/D and D/A converter technology, coupled with certain analogue functionality being retained outside the converters, can now deliver high-performance measurements from an instrument with a common *DSP* (*digital signal processing*) core.

With this architecture, many problems are solved: consistency between the domains is now assured, since all signals are generated and analysed by the same DSP techniques; domain interaction is now straightforward since the domains are processed by common DSP processors. Signal generation is now ultimately flexible, since generation of any unusual stimuli is simply a matter of software without component cost. Similarly, complex and powerful analysis techniques as described below can likewise be incorporated without cost.

7.3.6.1 FFT analysis

Signal analysis by FFT (Fast Fourier Transform) techniques is very powerful, but was until recently too computationally intensive (and therefore expensive) for widespread use.

The FFT techniques works by capturing a buffer of 2^n audio samples, and mathematically transforming the buffer from the time domain to the frequency domain, producing 2^{n-1} frequency *bins* which linearly cover the range from DC to half the sample rate. So, for example, a 256k-point FFT of a signal sampled at 96 kHz would produce 128k bins from DC to 48 kHz, providing 0.37 Hz resolution.

A difficulty with FFTs is that the captured buffer is considered by the algorithm to be repeated endlessly through time, so a discontinuity between the end of the buffer and the beginning results in a spurious glitch whose frequency-domain rendition can swamp the FFT bins, obscuring detail of the actual audio. This can be dealt with in two ways.

Firstly, if the audio stimulus can be arranged to repeat exactly over the buffer period, i.e. if all frequencies in the stimulus are integer multiples of the sample rate divided by the buffer length, then no end-discontinuity results. This situation is known as *synchronous* FFT analysis, and all stimulus frequencies are *bin-centred* in the resulting FFT, with all their energy in a single bin.

Alternatively, if a synchronous approach cannot be arranged (which it usually cannot), then a *windowing* technique must be used. This involves pre-multiplying the sample buffer by a bell-shaped *window function* which tends towards zero at the buffer ends, thus removing the effect of the end-discontinuity. Unfortunately, the FFT of the window function itself is superimposed on each frequency component in the resulting FFT, which has the effect that the power of individual stimulus frequencies is leaked over a few adjacent bins of the resulting FFT. This reduces the frequency resolution of the process, and can result in a loss of dynamic range and the appearance of spurious window-related *sidelobes* if a suitable window function is not chosen. There are many window functions to choose from, but most are historical and inferior to modern designs; in fact, a small family of quite ideal window functions allow optimal trade-off of bin-leakage versus dynamic range.

The mathematical detail of the FFT process is too complex to be covered here, but is explained in more depth in Refs 1, 2, 4 and 5.

The power of FFT analysis lies in its ability to expose very small components in an audio signal. These cannot be seen on a simple time-domain display, such as an oscilloscope, since they are usually masked by the presence of larger signal components and by wideband noise. The FFT process banishes the larger components to their own bins, and shares out the wideband noise over all the bins, with very little in each, especially if the buffer length is large. This gives a tremendous visibility of tiny elements of the signal, as shown in Figures 7.3.2 and 7.3.4. It is surprising how, once an audio design or test engineer becomes accustomed to having a fast-scanning, high-precision FFT capability on the bench, it becomes impossible to do without it and conventional measurements are almost never made except when it comes to writing or checking performance specifications.

7.3.6.2 Multi-tone testing

Multi-tone testing is a recently-introduced method, based on FFT techniques, which allows many performance parameters of the EUT to be measured simultaneously using a common test stimulus.

The stimulus for multi-tone testing is a mixture of sinusoids, usually of equal amplitude, covering the audio band, which are generally either linearly or logarithmically spaced in frequency. The precise frequencies are corrected to satisfy the requirements of synchronous FFT analysis described above. In fact, they are modified so that not only will every frequency fall into the centre of an FFT bin when the signal is analysed after passing through the EUT, but into the centre of an *even-numbered* bin. This point is critical to the correct operation of multi-tone testing.

This special stimulus allows many different *scalar* (numerical) measurements and frequency plots of the EUT to be calculated from a single acquired sample buffer, effectively simultaneously.

A *frequency–response* plot can be made by simply drawing a line between the FFT bins containing the original stimulus frequencies. Alternatively, scalar measurements of low- or high-frequency *roll-off* can be derived by expressing the amplitudes of the extreme stimulus-frequency bins relative to the amplitude of the nominal stimulus-frequency bin. A *phase–response* plot can also be extracted similarly, since the raw FFT data is complex, containing both amplitude and phase information.

A useful feature of the even-bin-centred stimulus is that all non-linearity distortion components also fall into even bins, since harmonic distortions are at integer multiples of the stimulus-frequency bins and IMD components are at sum and difference bins of the stimulus-frequency bins. *Total distortion* can thus be calculated by summing all even bins which do not contain stimulus frequencies. *Distortion vs. frequency* can be plotted by drawing a line between these bins. With suitable selection of stimulus frequencies it may also be possible to distinguish between harmonic and intermodulation distortions, should that be desirable.

Since all stimulus frequencies and their distortions occupy even bins in the FFT, the odd bins contain nothing but noise. Thus, the total RMS *noise* can be calculated by summing the odd bins (and doubling the result since only half the total noise is in the odd bins). The *noise spectrum* can be displayed by plotting between the amplitudes of the odd bins.

If the entire multi-tone analysis is performed simultaneously for both channels of a stereo pair, *inter-channel balance* and *inter-channel phase response* can also be computed. By slightly offsetting some or all of the stimulus frequencies for each channel, it is also possible to plot the *crosstalk response*.

Figure 7.3.3 shows a 31-tone stimulus being used to measure simultaneously total distortion, noise, and low- and high-frequency roll-off, whilst plotting amplitude, distortion and noise against frequency.

Multi-tone testing is a fast and versatile measurement technique which offers significant advantages in many test environments, especially in production-line and facility testing, where it is important that test time for each audio channel and

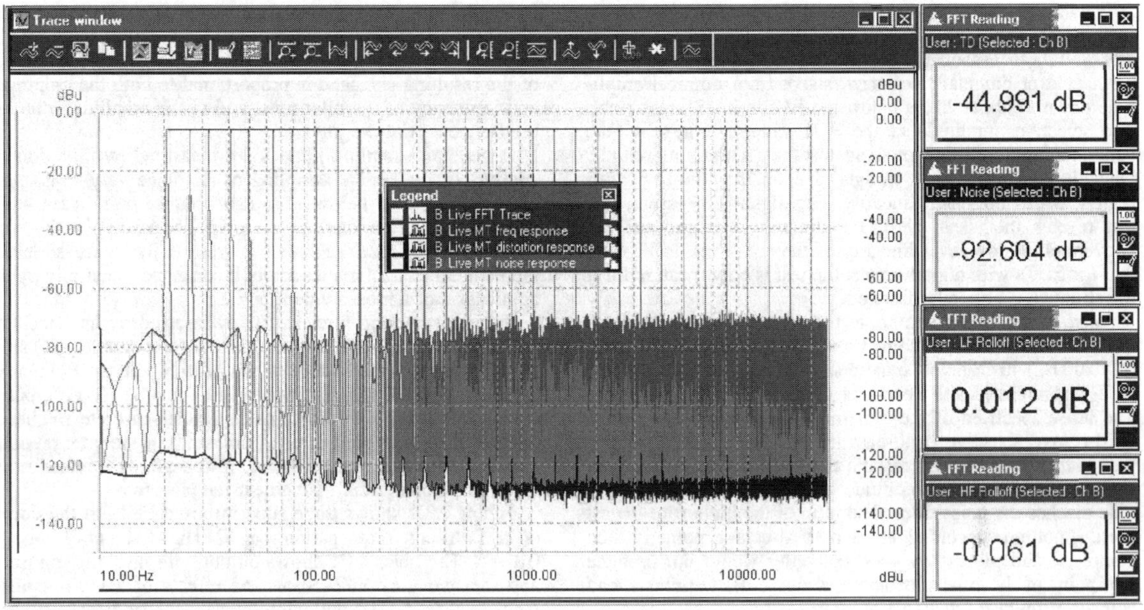

Figure 7.3.3 A 31-tone stimulus.

parameter is kept to a minimum. Further information is contained in Refs 1, 2, 5 and 9.

7.3.7 Digital Audio and Converter Measurements

Many aspects of the performance of digital audio devices and A/D and D/A converters can be investigated satisfactorily using the basic general-purpose measurement techniques already described. In fact, in broadcast testing situations these are largely sufficient. However, digital audio devices and converters are prone to additional shortcomings which are not exposed by these techniques. Some additional methods are outlined below. For the definitive discussion of digital audio device and converter measurement, see Dunn.[4]

7.3.7.1 Aliasing

Aliasing may occur wherever the effective sample rate of a system is reduced, for example in A/D converters or in digital down-sampling processes. It is the consequence of inadequate suppression of signal components beyond the target *Nyquist frequency* (half the output sample rate), and results in these frequencies reappearing at the output having been reflected about the Nyquist frequency. The effect is measured by stimulating the EUT with frequencies above half the output sample rate at nominal amplitude, and expressing the resultant output amplitude relative to the output-related stimulus amplitude.

7.3.7.2 Imaging and out-of-band noise

Conversely, *imaging* may occur wherever the effective sample rate of a system is increased, for example in D/A converters or in digital up-sampling processes. It is the consequence of inadequate interpolation filtering, and results in frequencies below the input Nyquist frequency being mirrored above it. The effect is measured by stimulating the EUT with frequencies approaching the input Nyquist frequency at nominal amplitude, and measuring the resultant output amplitude above the Nyquist frequency. The result is expressed relative to the output-related stimulus amplitude.

Modern noise-shaping D/A converters produce large *out-of-band noise* amplitudes, which may not be sufficiently attenuated at the device output. The method is simply to measure the idle-channel output amplitude of the converter above the Nyquist frequency.

7.3.7.3 Linearity and quantisation distortion

Digital audio devices, by their very nature, deal in *quantised* signals – signals which are approximated to the nearest available digital value at each sample instant. Because of this, it is widely (and wrongly) believed that digital devices must be non-linear at low signal amplitudes or even that they are incapable of passing signals below the amplitude of one quantum step (one *least significant bit*, or *LSB*).

Correct use of *dither*[10] in digital audio devices can overcome such potential problems, although its use cannot be assumed. D/A and A/D converters (even if the latter are correctly dithered) may exhibit inherent non-linearities.

For any device with a digital audio input or output, it is important to assess adequate freedom from quantisation non-linearities. This can be done to some extent using conventional THD+N measurements, where the appearances of non-harmonic distortion components, especially at low signal amplitudes, are indicators. More specialised techniques are described below.

Low-level linearity, or *level-dependent gain*, is an interesting measurement to apply to A/D converters, D/A converters or any digital processors. The input to the EUT is stimulated with a sinusoid at nominal frequency, swept from nominal amplitude down to about 20 dB below the EUT's RMS noise floor. The selective amplitude at the EUT output, relative to the generator amplitude, is plotted against the generator amplitude. Deviation from an ideal straight line, in the low-amplitude region, belies non-ideal linearity. For digital processors, this test reveals the accuracy of the dithering scheme, and for converters the low-level linearity of the converter itself. Note that for EUTs with digital inputs, the test is only meaningful if the stimulus is correctly dithered.

Noise modulation is another test which reveals shortcomings in low-level linearity of converters. A low-frequency stimulus (say 500 Hz) 40 dB below maximum amplitude is applied to the EUT, and removed with a band-reject or notch filter at its output. The noise spectrum of the remaining signal is plotted with a third-octave selective band-pass sweep. The stimulus amplitude is then reduced by 5 dB at a time, and the noise spectrum sweep repeated at each amplitude step until the stimulus amplitude reaches the noise floor. The ratio of the highest to lowest noise amplitude recorded at each third-octave point is then computed and plotted. A good converter would not produce any point in the final plot above about 2 dB, whereas a bad converter might reach 10 dB or more.

7.3.7.4 Sampling jitter

Ideally, the sample rate of an A/D or D/A converter should be absolutely constant – sampling instants should always be separated by precisely the same time interval. This is the assumption made in defining linear PCM audio. *Sampling jitter* is the deviation in converter sample timing from this ideal.[11,17]

Sampling jitter causes distortion of the audio at the point of conversion. The distortion mechanism for sampling jitter is similar to phase modulation; sinusoidal jitter produces distortion sidebands separated from the pilot tone by the frequency of the jitter. It has been shown[11] that the amplitude of the resulting sideband is proportional to both the amplitude and frequency of the pilot tone, and to the amplitude (but not the frequency) of the jitter.

In general, sampling jitter is not measured by time-domain analysis of the EUT's sampling clock, since very small jitter amplitudes (much below 1 ns) can produce significant audio distortion, but are hard to measure accurately in the time domain. In any case, such clocks are usually inaccessible and their timing characteristics might be changed materially by the application of a measurement probe.

Instead, sampling jitter is usually inferred by its effects on the converted audio. A high-frequency (for example, $f_s/4$) sinusoidal stimulus (pilot tone) is passed through the EUT at or above nominal amplitude, and an FFT of the EUT output is calculated. If the jitter contains significant discrete frequency components, these are visible as sidebands; if the jitter predominantly comprises random noise, it appears as a raised noise floor, or a noise 'skirt' rising near the pilot tone.

Figure 7.3.4 shows three superimposed FFTs of the output of a D/A converter passing a 12 kHz ($f_s/4$) pilot tone at 0 dBFS. The black FFT shows output with no sampling jitter, and the two grey FFTs show the effects of sampling jitter: sinusoidal (at 5 kHz) and white noise, each of 40 ns(p-p). The addition of the white-noise sampling jitter has raised the noise floor by something over 20 dB, but the sinusoidal sampling jitter has created non-harmonic artefacts only 60 dB below the pilot tone.

Unfortunately, the effects of sampling jitter are easy to confuse with those of other converter defects, such as amplitude modulation, which often occurs through poor circuit design. Sampling jitter artefacts can be distinguished from those of amplitude modulation by checking that they disappear when a low-frequency (or DC) pilot is substituted.

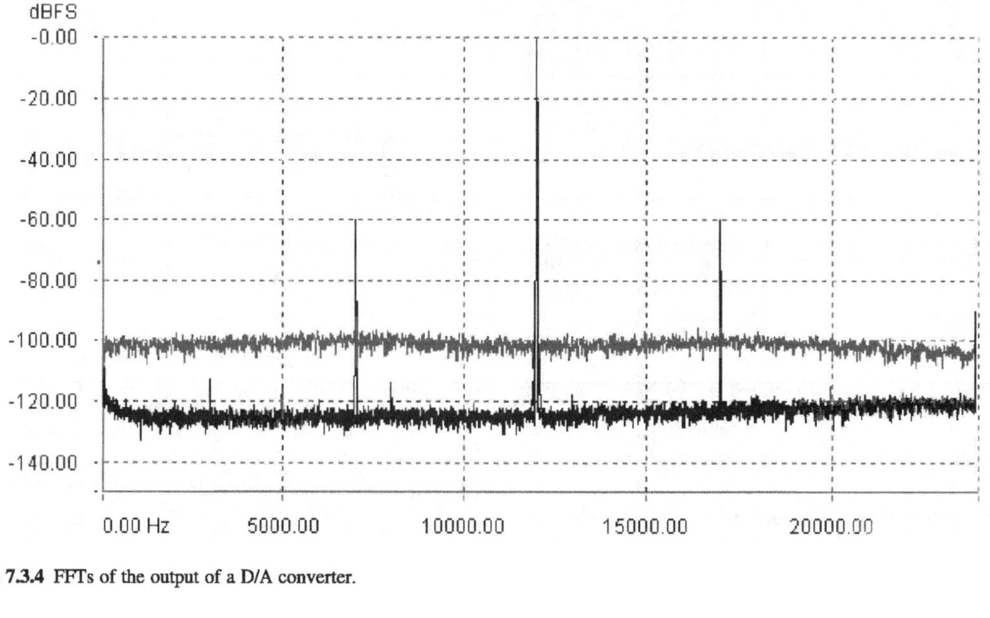

Figure 7.3.4 FFTs of the output of a D/A converter.

Intrinsic sampling jitter (when the converter generates its sampling clock internally) is difficult to assess, since it cannot be removed for comparative purposes – the best we can do is look for obvious sidebands and noise skirts. However, this is not generally a problem since the intrinsic case is not generally so bad as when the converter is referenced to an external sync; this is common for A/D converters in broadcast situations, and almost universal for D/A converters which are referenced to their data input.

Jitter rejection (or *jitter attenuation*) in the externally synchronised case is examined by measuring the sampling jitter sidebands whilst applying sinusoidal jitter at, say, 0.25 UIp-p amplitude, swept for example from $f_s/4$ to 20 Hz. Plotting the sideband amplitude (referenced to the zero-attenuation sideband amplitude for the applied pilot tone frequency and amplitude and the applied jitter amplitude) against the applied jitter frequency gives the jitter-rejection response.

Sampling jitter is not to be confused with interface jitter as described in Section 7.3.8.3. However, since most converters have little or no jitter rejection in the audio band, sampling jitter is often directly caused by interface jitter (as with the D/A converter in Figure 7.3.4). This has led installers of professional systems with long cable runs to try to control interface jitter within bounds far less than are required to guarantee reliable data transmission, in an attempt to circumvent the inadequacies of the converters. This necessitates the use of special low-capacitance cabling, external jitter-attenuation devices, etc. None of these methods can reduce sampling jitter artefacts to negligible levels, however, since the data jitter induced by any practical cabling, no matter how short or low-capacitance, greatly exceeds what can be achieved by high-quality internal jitter filtering.

In consumer installations, run lengths are generally too short for cable effects to matter very much, although this view is discouraged by consumer cable vendors.

7.3.7.5 Data integrity

Bit error testing is a useful technique for checking digital interfaces, data channels and some types of digital audio devices. It is especially useful in tracing intermittent faults, and is usually based on transmitting a pseudo-random series of samples through the EUT and checking that the correct series is recovered at the output. A pseudo-random sequence is chosen because it exercises all the data bits randomly, and because each sample can be inferred from the last, allowing generator and analyser to be separated in distance (e.g. a satellite link) or in time (e.g. a digital audio recorder). Any errors are indicated and can usually be logged against time. Only completely transparent devices can be checked in this way; any processing between the EUT input and output, even the addition of dither or a minute level change, prevents the technique from working.

7.3.8 AES3 Interface Testing

In digital audio installations it is arguably more important to be able to verify correct operation of the interfaces than to be able to perform traditional performance measurements on the audio data itself. This is because more problems in digital studios are experienced due to interfacing problems than occur due to failure of digital-domain equipment. In such equipment, the audio performance is 'designed-in' – it will generally always be the same unless the equipment fails catastrophically; it cannot degrade or vary as may the performance of analogue equipment. However, the digital audio interfaces are themselves analogue, and indeed are more fragile than analogue audio interfaces since they rely on near-perfect transaction of high-frequency signals, often in hostile environments with less-than-ideal cables and connectors.

Since the arrival of digital audio, there has been a widespread user misunderstanding of the distinction between digital audio performance and carrier performance. It is usual to 'test' digital interconnections by sending a digital sine stimulus through the system and verifying that the distortion of the final output is low. This method of testing takes no account of operating margin of the interfaces, and is rather like testing a computer network with a spell-checker. When problems are subsequently encountered, they are unlikely to appear as degraded audio performance, but rather as intermittent failure such as ticks or dropouts, or as interoperability problems between certain devices. Such problems are indicative of interconnection or carrier faults, such as the lack of a safe operating margin between transmitters and receivers.

Although in principle these problems can occur with any digital carrier format, this section will describe only the test methodologies for AES3[12] interconnections, since these are the most common in broadcast installations. As well as professional 110 Ω balanced AES3 interfaces, the principles described also apply to professional 75 Ω coaxial AES-3id[13] interconnections, as well as to consumer 75 Ω coaxial and TOSLINK optical IEC 60958[14] (*S/PDIF*) interconnections.

Most of the test methods outlined in the following sections require special interface testing equipment, although the necessary features are included in some general-purpose audio analysers.

7.3.8.1 Introduction to the AES3 interface

The AES3 interconnection standard is fully described in the standard document.[12] Only a brief summary is included here for clarification of the techniques involved in testing the interface.

Figure 7.3.5 shows the organisation of serial data on the AES3 interface. The interface carries two audio channels (often a stereo pair) on a single carrier, plus their associated *ancillary* bits (comprising a Validity, User, Channel Status and Parity bit). The serial data for the two channels is interleaved in the carrier; first an A-channel audio data sample and status are transmitted, then a B-channel sample and status. Each channel's data occupies a *subframe* and each pair of samples constitutes a *frame*. The beginning of each subframe is designated by a special *preamble* pattern which identifies the channel, designated 'X' for the A-channel and 'Y' for the B-channel.

The binary data is *biphase-mark* encoded, i.e. each data bit-period is bounded by low-to-high or high-to-low transitions, and if the bit is a '1', another transition occurs in the middle of the bit period. This coding scheme is effectively a way of combining clock and data on a single carrier, since the guaranteed transitions between the bit periods can be extracted by a receiving device to reconstitute a continuous clock. The preambles are unique in that they violate the biphase-mark scheme by not having a transition at least every bit period.

Only one Channel Status bit per channel per sample period is carried on the interface, but these accumulate into a 192-bit

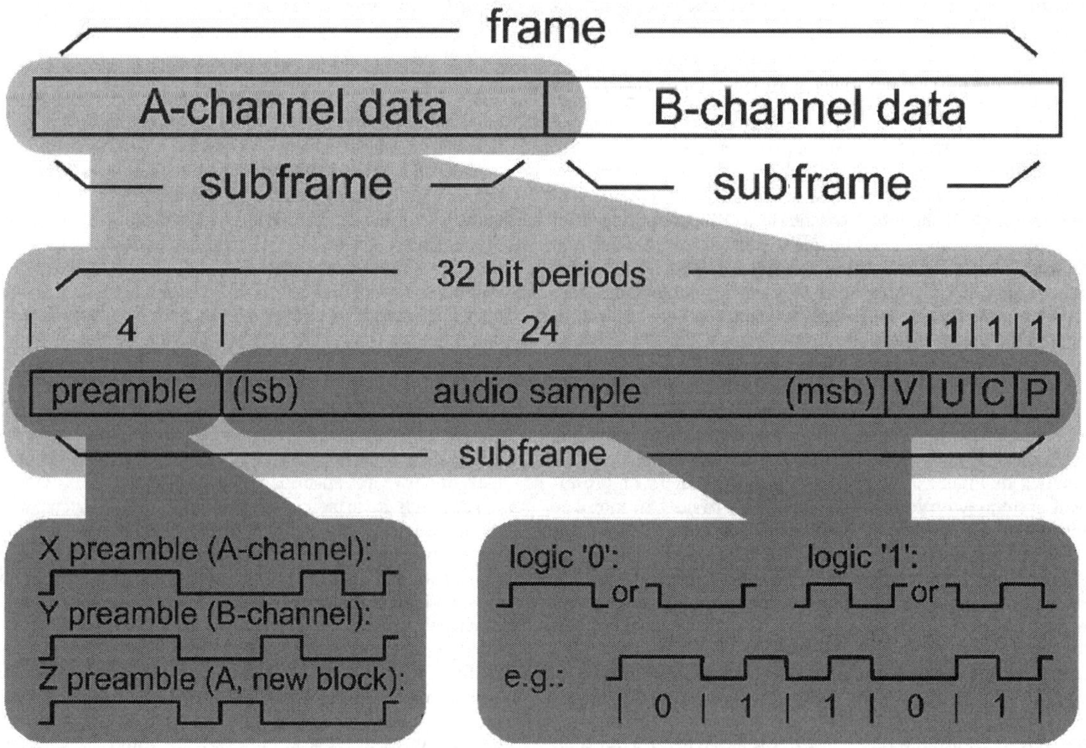

Figure 7.3.5 Organisation of serial data on the AES3 interface.

block every 192 sample periods, thus the entire Channel Status pattern is transacted every few milliseconds. In order to recognise the start of the 192-bit block, a third preamble type ('Z') is substituted for 'X' every 192 sample periods.

The primary audio sample rates supported by AES3 are 32, 44.1 and 48 kHz. Extended rates up to 192 kHz are now supported, although rates above 48 kHz are not yet often encountered in broadcast environments. Preferred audio sampling rates and how they should be synchronised to international video standards are described in AES11.[15]

The bit rate for AES3 interfaces is 64 times the sample rate ($64f_s$); the biphase-mark cell rate is $128f_s$. The biphase-mark cell period (about 163 ns for $f_s = 48$ kHz) is referred to as a *Unit Interval* (or *UI*) since it represents the quantum pulse duration. In order that carrier timing parameters can be scaled according to sample rate, it is common to quote them in terms of UI rather than directly in ns.

In summary, we have a convenient two-channel carrier which is reliably and simply decoded since it carries an embedded clock. Up to 24 audio data bits per channel can be transmitted, plus Valid, User, Channel Status and Parity bits to enhance understanding of the signal context between equipment. The carrier is polarity-insensitive (since its information is embodied in transitions rather than logic levels) and DC-free (and so can be interfaced via isolating pulse transformers).

7.3.8.2 Frame rate accuracy

Problems can occur when a digital audio device receives AES3 data or an AES11 reference sync whose frame rate (sample rate) accuracy is outside the device's working range. For this reason, the accuracy of all equipment which can act as a reference sync master should be checked, as should the frequency tolerance of all reference sync slaves.

Dedicated reference sync generators should present no problems, since their accuracy is tightly specified according to their *grade* by AES11. Likewise other professional devices (when configured to act as sync master), since these should stay comfortably within ±50 *ppm* (parts per million) of the selected frequency. Consumer equipment can occasionally be very inaccurate when acting as a sync master (perhaps as much as ±1000 ppm in error); this can be a problem since many consumer devices (e.g. CD players) cannot be externally synchronised and so are often de facto masters. It is important to check that all slave devices in a system can be successfully synchronised to the most inaccurate potential sync master, with some safety margin.

7.3.8.3 Carrier quality, jitter and cable effects

Poor quality of generated carriers, intolerance of received carrier degradations, and bad cabling are the root of most AES3 interconnection problems.

For digital audio device outputs, the AES3 standard defines the following parameters and specifies allowed limits: *carrier amplitude* (2–7 Vp-p), *rise and fall times* (5–30 ns, 10–90%), *intrinsic jitter* (<0.025 UIpk), *jitter gain* (<2 dB at all frequencies) and *carrier phase* (aligned to the reference sync within ±5% of a sample period). All of these can be easily measured using suitable carrier analysis equipment.

For digital audio device inputs, tolerance of carrier degradations is specified by a minimum *eye diagram*, plus tolerance specifications for jitter[16] and carrier phase. The eye diagram basically shows a worst-case Unit Interval on the interface, with a rectangular box in the centre showing the required tolerance of carrier amplitude and incorrect transition timing. This box implies a 200 mVp-p carrier amplitude over at least the central 50% of the UI. The jitter tolerance specification is 10 UIp-p below 200 Hz, and 0.25 UIp-p above 8 kHz. The carrier phase tolerance specifies misalignment with the reference sync of up to ±25% of a sample period. All of these can likewise be verified using a suitable degraded-carrier generator.

Brief examination of the above shows that a safe operating margin should exist between compliant AES3 outputs and inputs. So why do problems occur? There are three reasons: first, ports are often not compliant; second, the effects of cabling often degrade carriers massively between outputs and inputs; and third, the AES3 standard is only concerned with ensuring that audio data can be transacted without error, whereas interface jitter actually affects audio quality (at jitter levels far too low to threaten data integrity), because of design inadequacies in converters, which are beyond the influence of AES3. This problem of interface jitter causing sampling jitter in converters is discussed further in Section 7.3.7.4.

The effect of carrier degradation by cable is illustrated in Figure 7.3.6. The figure shows a *carriergram*, a plot similar to the eye diagram in AES3. The carriergram is similar to viewing the carrier on an oscilloscope with a persistent phosphor. The maximum and minimum positive and negative excursions of the carrier are shown against time, accumulated over a gate time of a few milliseconds. The area between maximum and minimum voltages of each polarity is shaded.

Figure 7.3.6 shows four consecutive audio bits (eight consecutive Unit Intervals) on a 48 kHz AES3 carrier after passing down 100 metres of high-capacitance (150 pF per metre) twisted-pair cable of a type commonly used in interconnecting areas of a broadcast facility. The carrier was transmitted with 5 Vp-p loaded amplitude, and with fast rise and fall times (<5 ns). As can be seen from the figure, the distributed cable capacitance, acting with the source and lumped cable impedance, has compromised the transition times badly and has also reduced the carrier amplitude to about 3 Vp-p, since after each transition the cable only manages to charge up to a fraction of its target voltage before another transition reverses its direction. The carrier amplitude is higher where the data bits are logic 0 (i.e. have no transition in the middle of the bit period) than it is when they are logic 1 (with the extra mid-bit transition). This is because logic 0s allow twice the 'charging time' between transitions allowed by logic 1s.

To illustrate how this cable-induced jitter occurs, the first audio bit has been arranged to be always logic 0, whilst the following three audio bits are changing, and so both the logic 0 and logic 1 carrier patterns are overlaid on the carriergram for the later bits. The small rectangles superimposed in the middle of the eyes represent the minimum eye specification from the AES3 standard – so although this carrier is badly degraded, it is still somewhat better than is permissible!

Consider a receiving device 'slicing' this carrier waveform, i.e. interpreting a high logic level whenever the carrier is above the 0 V centre line and a low logic level whenever it is below. Carrier transitions which occur after unchanging data (i.e. all the transitions in the first two bit periods, as indicated by the leftmost arrow) occur at almost precisely the same instant in every frame and so appear as sharp lines on the carriergram. Transitions which occur after changing data (for example, as indicated by the rightmost arrow) occur either earlier (if following a logic 1 bit) or later (if following a logic 0 bit) and so appear as broadened lines on the diagram. The vertical marker lines show the *data jitter* (jitter produced by the interaction of changing data with cable capacitance) which results: 20.5 ns or 0.125 UIp-p as shown in the reading.

This effect happens as a result of different pulse-lengths being present in the biphase-mark data – if the data segment had carried '111', the recovered transition timing would have been 'correct', just as for '000'. If one imagines these three bits forming part of the audio sample in an AES3 signal, and alternating between '000' and '010' on successive samples, one can understand how a 'jittering' clock is recovered.

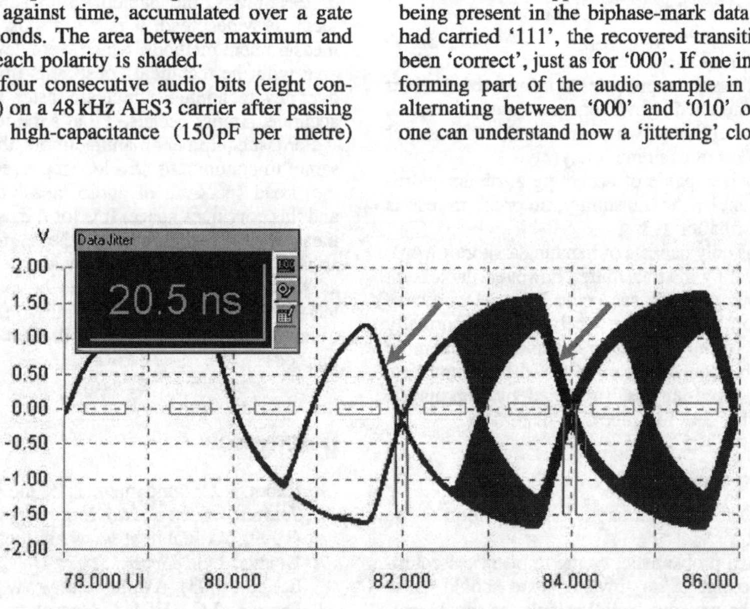

Figure 7.3.6 Effect of carrier degradation by cable.

The complex dependence of data jitter on the data content itself can produce strange modulation effects if data jitter is allowed to produce sampling jitter in a converter device (see Section 7.3.7.4). Apart from distortion products (which are non-harmonic and unpleasant sounding) these include non-linear crosstalk between the two digital channels, since the conversion clock is shared by (and derived from) both subframes. As if this were not shocking enough, the crosstalk effect usually gets worse as the hostile channel gets *quieter*, because the coherence of audio frequency in the data pattern increases for smaller sample values because they are represented in twos-complement form. It is also possible for Channel Status and User data activity to modulate the converted audio in the same way.

Note that the overall *interface jitter* comprises not only the cable-induced data jitter described above, but also f_s *jitter*, caused by jitter in the clock of the generating equipment. In diagnosing causes of interface jitter, it is useful to be able to demodulate the interface jitter within the carrier analyser. The resultant analogue jitter signal can then be examined in the time and frequency domains.

Incorrect *impedance matching* is often suspected of causing AES3 interfacing problems; in practical situations this is rarely the case. Whilst reflections caused by unmatched transmitter/receiver impedances degrade the carrier waveform, this is not important for AES3, because the data is transacted purely in the binary transitions. Data is received by 'slicing' the carrier to extract the transition timings – the actual carrier waveform beyond the transitions does not affect this process. Conversely, incorrect characteristic impedance of cabling does not make much difference with short cables, and with long cables transition timing is much more affected by capacitance effects, as described above.

7.3.8.4 Audio word-length problems

The AES3 interface supports word-lengths from 16 to 24 bits; problems often arise from different expectations between the transmitting and receiving equipment as to the word-length to be used. These difficulties theoretically need not occur, since the Channel Status allows the transmitter to inform the receiver of the word-length transacted. However, most transmitters do not generate this Channel Status, which turns out not to matter since most receivers do not examine it anyway.

Where the receiver is capable of accepting a greater word-length than is being sent by the transmitter, no problem results because no audio information is lost.

When the receiver is only capable of handling a shorter word-length than is being sent by the transmitter, however, unpleasant non-harmonic distortion products can occur because the word-length is truncated on entering the receiver. It is normally beyond the capability of the receiver to re-dither the incoming signal as it should. This condition can be tested by stimulating a receiver with long word-lengths, supported by appropriate Channel Status, and checking for truncation distortion.

7.3.8.5 Ancillary bit testing

Difficulties may occur through misuse or non-operation of the various ancillary bits.

The Valid bit is often problematic, owing to historical confusion as to whether receipt of an 'Invalid' state should cause muting, repeating of the previous 'valid' sample, or should produce no effect on the audio at all. Depending on the requirements of the particular installation, compatibility of transmitted Valid bits should be checked, as well as the response of receiving equipment to 'Invalid' indications.

Channel Status is a major cause of equipment incompatibility. As a rule, transmitting devices should be precise in their Channel Status implementation and receiving devices should be tolerant in their interpretation. Frequently, many transmitters implement little or no Channel Status, whereas many receivers mute the audio in response to partial or incorrect Channel Status, or to momentary CRC errors (CRC errors should cause the affected Channel Status block to be discarded, but should have no effect on the audio). Full (or legal partial) implementation of transmitters' Channel Status should be checked, as should receivers' response to both relevant and erroneous Channel Status fields.

Transparency to the Valid, User and Channel Status bits is an important issue. Many digital-to-digital devices produce their own Channel Status output pattern, which is either wholly unaffected by the incoming Channel Status, or affected only by one or two input fields. In not passing Channel Status from input to output, important status information appended by the source is lost. Blocking of User bit data is equally problematic – for example, CD subcode or DAT start-ID information may be destroyed. Transparency can be tested by generating various ancillary bit combinations at the EUT input and checking the resulting output.

7.3.9 The Future

Recent advances in conversion and digital signal processing technology have spawned a new generation of audio test instruments which are cheaper and more functional than their predecessors, with capabilities far beyond traditional measurement methods.

However, in audio test and measurement, as in many fields of technology, the paradox of 'tradition surviving improvement' is widespread. It is still almost universal for traditional measurement methods, which were devised merely to be realisable with the technical constraints of the day, to be favoured over new techniques which are faster, more revealing and more accurate, simply because "that's the way it's done".

For test equipment manufacturers, the task ahead remains the same: to continue to develop improved test capabilities across the broad spectrum of audio measurement application areas, and this continues apace. It is for the users to move today's new measurement techniques into the mainstream.

The route forward is for new and developing standards, such as AES17[7] and IEC 61606,[8] to be expanded to require (or at least to encourage, or even to allow) the test and measurement of audio equipment to be specified in terms of these new methods.

References

1. Cabot, R.C. Fundamentals of modern audio measurement, *Journal of the Audio Engineering Society*, **47**, 738–762 (1999). Available at www.audioprecision.com
2. Dennis, I.G. *dScope Series III Operation Manual*, Prism Sound (2003). Available at www.prismsound.com
3. Dennis, I.G. *DSA-1 Operation Manual*, Prism Sound (1996). Available at www.prismsound.com

4. Dunn, N.J. *Measurement Techniques for Digital Audio*, Audio Precision Inc. (2002). Available at www.audioprecision.com
5. Metzler, R.E. *Audio Measurement Handbook*, Audio Precision Inc. (1993). Available at www.audioprecision.com
6. ISO 532, Acoustics – Method for Calculating Loudness Level (1975). Contains details of detector response and weighting filter previously designated CCIR-468. Available at www.iso.ch
7. AES17, AES Standard Method for Digital Audio Engineering – Measurement of Digital Audio Equipment (1998) (revision of AES17, 1991). Available at www.aes.org
8. IEC 61606, Audio and Audiovisual Equipment – Digital Audio Parts – Basic Methods of Measurement of Audio Characteristic. Available at www.iec.ch
9. Vanderkooy, J. and Norcross, S.G. Multitone testing of audio signals, Presented at the 101st Convention of the Audio Engineering Society, preprint 4378 (1996). Available at www.aes.org
10. Vanderkooy, J. and Lipshitz, S.P. Dither in digital audio, *Journal of the Audio Engineering Society*, **35**, 966–975 (1987). Available at www.aes.org
11. Dunn, N.J. and Dennis, I.G. The diagnosis and solution of jitter-related problems in digital audio systems, Presented at the 96th Convention of the Audio Engineering Society, preprint 3868 (1994). Available at www.aes.org
12. AES3, AES Recommended Practice for Digital Audio Engineering – Serial Transmission Format for Two-channel Linearly Represented Digital Audio Data (1992, revised 1997) (revision of AES3, 1985), including Amendment 1 (1997), Amendment 2 (1998), Amendment 3 (1999) and Amendment 4 (1999). Available at www.aes.org
13. AES-3id, AES Information Document for Digital Audio Engineering – Transmission of AES3 Formatted Data by Unbalanced Coaxial Cable (2001) (Revision of AES-3id, 1995). Available at www.aes.org
14. IEC 60958, Digital Audio Interface, Part 1 – General (1999); Part 4 – Professional Applications (2000). Available at www.iec.ch
15. AES11, AES Recommended Practice for Digital Audio Engineering – Synchronization of Digital Audio Equipment in Studio Operations (1997). Available at www.aes.org
16. Dunn, N.J., McKibben, B., Taylor, R. and Travis, C. Towards common specifications for digital audio interface jitter, Presented at the 95th Convention of the Audio Engineering Society, preprint 3705, *Journal of the Audio Engineering Society*, **41**, 051 (1993). Available at www.aes.org.
17. Harris, S. The effects of sampling clock jitter on Nyquist sampling analog to digital converters and on oversampling delta-sigma ADCs, *Journal of the Audio Engineering Society*, **38**, 537–542 (1990). Available at www.aes.org

Phil I'Anson BSc

7.4A Broadcast Engineering RF Measurements

RF measurements are needed to ensure that transmissions are within prescribed limits. The impact of out-of-specification transmissions can be interference to other users of the RF spectrum as well as degrading the broadcast service.

Round the clock broadcasting gives little opportunity for off-line maintenance of transmitter equipment. Even when systems that have fault-tolerant designs it is very unlikely that full power can be maintained with part of the system isolated to carry out maintenance tests. The capital cost of a high-power transmitter normally makes it impossible to justify the cost of a full-power reserve arrangement.

A typical high-power television transmitter system has parallel high-power amplifiers. It will normally be possible to isolate one amplifier and connect it to a high-power RF test load whilst the other amplifier maintains the programme transmission. However, the resulting reduction in radiated power of at least 3 dB and the likely need for short breaks in transmission whilst feeder switches are operated will always be unpopular with the programme makers. Consequently routine maintenance measurements need to be carried out, as far as possible using the broadcast transmissions. Fortunately for analogue TV transmissions, many routine performance measurements can be made using the ITS line in the vertical blanking interval.

Measurements at high-power transmitter sites pose particular problems because of the potential for RF interference to the test equipment. This is especially true if there are co-sited VHF or MF transmitters. When making measurements always check to see that the results look sensible, if not repeat the measurement. Check that moving leads or changing the physical position of equipment does not affect the readings. For example, RF power meters often comprise a measurement head connected by a flying lead to the meter. RF pick-up on the connecting lead can result in false readings on the meter.

RF connectors should always be treated with care. Mating surfaces must be kept clean. To avoid wearing out the mating surfaces never rotate the connector body, only the fixing collar. For example, when fitting an attenuator with N-type connectors hold the attenuator body still and only rotate the threaded collar. Similarly, RF leads should always be handled with care to prevent damage.

Since transmitting stations are almost universally unmanned, most measurements will be made using portable test equipment taken to the site for the purpose. For day-to-day use the most commonly used test equipment items are spectrum analysers, power meters, modulation analysers and frequency counters. As modern spectrum analysers can also be used to make frequency measurements, even the frequency counter can be dispensed with. Modulation analysers are used for performance measurements on FM sound transmissions only.

7.4A.1 Use of Spectrum Analysers

For everyday RF measurements, by far the most valuable and versatile tool is the spectrum analyser. A modern digital display spectrum analyser can perform a wide range of performance measurements on most types of broadcast transmissions. A correctly specified instrument is essential. Test equipment makers normally offer a basic instrument with the option of additional features to customise the analyser for particular applications.

For analogue TV work TV trigger facilities are essential, as these allow the ITS waveform to be displayed on the analyser's screen. Ideally a spectrum analyser for broadcast work will have the following features:

- TV trigger to allow the ITS waveform to be displayed in zero span mode.
- High-stability internal reference oscillator for accurate frequency measurements.
- Tracking generator.
- AM/FM demodulator.
- Remote control.

With a high-stability internal reference oscillator frequency measurements can be made with the analyser, avoiding the need to carry a separate frequency counter. A tracking generator providing a swept RF output following the analyser's tuned frequency allows RF filters to be checked. If an SWR (standing wave ratio) bridge or directional coupler is available, basic checks can also be made on antenna systems. Remote control allows automatic measurement routines and result logging using a PC (personal computer).

Some points need to be borne in mind when using modern spectrum analysers. Current-day spectrum analysers are largely analogue instruments very similar to those of a generation ago. The main advances have been in the introduction of digital displays and significantly improved frequency stability. Digital displays offer many advantages over the analogue displays. There is no longer any need for darkened rooms or cameras to capture fleeting images. Each sweep is displayed with the same brightness and can be held on the screen for as long as is needed. Movable on-screen markers can display the amplitude and frequency at any point on the waveform. One drawback of a digital display is the loss of detail shown by the relative brightness of the display. On live TV signals this makes the measurement of IPs (Intermodulation Products) more difficult.

On an analogue display instrument the image is refreshed faster than the rate at which each sweep fades away. It is normal to use displays with a slow decay phosphor. When sweeping a live TV signal IPs can be seen as being brighter than the constantly varying video sidebands. This is because with a changing video signal on successive sweeps the video sidebands appear on different parts of the screen, but the IPs appear on the same part of the screen on each sweep. Consequently, even IPs close to the vision carrier and within the video sidebands appear visible as a brighter segment of trace on the display. Similarly, regularly repeating parts of the waveform such as sync pulses are also easily seen. Digital display instruments erase the old trace before a new one is drawn and so no information is carried in the brightness of the displayed trace.

On a digital display the waveform is made up of a series of points rather than the continuous line seen on an analogue display. A typical figure for the number of horizontal points is 401. As a marker is moved horizontally across the screen, the frequency of the marker will be seen to change in steps. For example, if the displayed sweep is 4.01 MHz wide the marker will change in 10 kHz steps. Thus, even though the marker frequency may be displayed to 1 Hz, the figure represents the centre frequency of the display point on the screen and not the exact frequency of the signal at that point.

Modern analysers normally offer a frequency measurement mode. When selected, an internal frequency counter measures the frequency of the displayed signal at the point where the marker has been set. In this mode the signal frequency will be displayed correctly. The accuracy of the frequency measurement depends on the accuracy of the analyser's internal frequency reference. If the analyser has been fitted with a high-stability reference the frequency measurement accuracy will be on a par with a frequency counter. Consequently, for most maintenance applications it is not necessary to use a separate frequency counter. For mobile maintenance operations this has the benefit of there being one less piece of test equipment to be carried around.

7.4A.2 Analogue TV RF Measurements

Low-level RF measurement points will be provided at several points along the RF path. High-level RF signals will damage test equipment and so approximate signal levels need to be known before connecting an instrument. Measurement points are normally labelled with the relationship between the power along the RF path and the power provided at the reference point. If there is any doubt about the signal level, fit a 20 dB attenuator to protect the instrument from being damaged by high signal levels. Better to damage a relatively low-cost attenuator than a high-cost instrument. As an alternative to using attenuators in-line RF fuses can be used, but they can introduce a mismatch, reducing the accuracy of the measurements.

A typical high-power ATV system with separate sound and vision amplifiers is shown in Figure 7.4A.1. RF measurement points can be expected at all of the amplifier outputs and all of

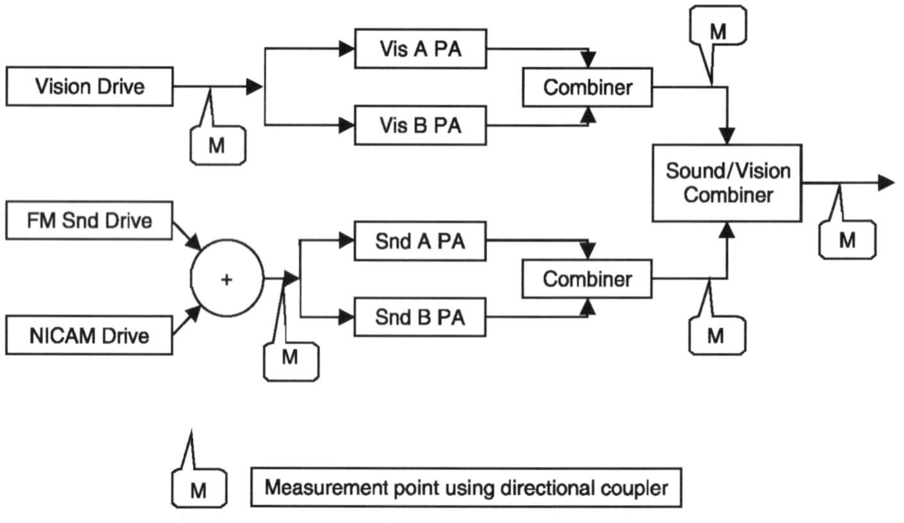

Figure 7.4A.1 High-power analogue TV transmitter system, showing typical measurement points.

the combiner outputs. For simplicity only some measurement points have been shown on the diagram. At VHF and UHF directional couplers are normally used to provide a sample of the RF signal. For a high-power amplifier of around 10 kW a typical coupling factor would be 40 dB, giving an RF sample of around 1 mW.

Transmitter carrier harmonics may be measured at the amplifier monitor points, but always confirm the frequency/amplitude characteristics of the measurement point. Although there will be a flat frequency response around the transmission frequencies, the characteristics may be different at second or third harmonics of the carrier frequency. Generally, on a well-maintained system, problems with harmonics will be rare; if they are present, problems will very likely be found with other RF measurements. Consequently, although such measurements must be carried out on new installations or after major work on a transmitter system, they may not need to be done for day-to-day maintenance checks. At the amplifier outputs harmonics should be no greater than −60 dB relative to the vision carrier peak power.

Where several services are radiated from the same site, checks must be made for IPs between different service transmissions. If IPs are present they will be seen as additional low-level signals between the intended transmissions when making a wide sweep showing all transmissions. IP measurements may also be made at the amplifier outputs.

IPs between different services should be at very low levels, putting them at the measurement limits of a mid-performance analyser. A modern mid-performance analyser can be expected to have a spurious free dynamic range of around 80 dB. When looking at low-level IPs, great care needs to be taken that what is seen on the screen is not internally generated by the analyser. A common technique is to adjust the analyser's level settings by, say, 3 dB and check that the IP levels only change by the same amount as the high-level signals. If the IPs are internally generated, their level will change by a greater amount, for a 3 dB change in level, perhaps up to 6 or even 9 dB. An instrument's instruction manual will give the optimum settings for best dynamic range. Usually, an optimum mixer input level is stated: −30 dBm is typical. If the input level is too high there is a risk of IPs being generated within the analyser. If the level is set too low the noise floor of the analyser will mask low-level signals.

Most significant are IPs which fall within the passband of the intended transmissions and so will not be rejected by subsequent filtering in the antenna combining systems. IPs which fall within the video sidebands, especially those close to the vision carrier, will cause visible picture impairments. These IPs are generated by interactions between the vision and sound carriers.

The IP figures given in Table 7.4A.1 are for systems with an FM sound carrier spaced at 6 MHz from the vision carrier, a NICAM signal spaced at 6.552 MHz from the vision carrier and a colour subcarrier frequency of 4.434 MHz. Frequencies are rounded off to three decimal places. IP frequencies and levels are relative to the vision carrier.

Table 7.4A.1 IP values for interacting systems

Interacting carriers	*IP frequency (MHz)*	*Maximum allowed level (dB)*
FM sound and NICAM	0.552	−57
FM sound and colour subcarrier	1.566	−55

Other combinations of mixing products can occur between the carriers, but in practice if the 0.552 MHz IP is within specification, other IPs are unlikely to be a problem.

Starting with a wide sweep the relative levels of vision and sound carriers can be measured with a spectrum analyser. The analyser should be set to span width of 10 MHz, with the vision carrier set to be about 1.5 MHz in from the left hand side of the screen. This will allow the whole TV channel to be seen on the screen. A resolution bandwidth of 300 kHz will give frequent screen refreshes, but only partial separation of the FM sound and NICAM signals. Reducing the resolution bandwidth to 100 kHz will reduce the rate at which the screen is refreshed, but will improve the separation of the FM and NICAM sound carriers.

Figure 7.4A.2 shows a typical display of a single TV channel. From left to right across the screen, the signal peaks are: vision carrier, colour subcarrier, FM sound carrier and NICAM sound signal. The FM and NICAM signals are partially merged because of their close spacing. In practice, signal levels will vary slightly on successive sweeps. Displaying the correct vision carrier level relies on the sync pulse being present when the analyser sweeps past the vision carrier frequency. The noise-like waveform around the vision carrier will vary significantly during normal programme broadcasts as the video signal varies.

Absolute power levels can be checked, but care needs to be taken with the absolute power accuracy of the analyser. Relative levels are normally measured with greater precision than absolute levels. A modern instrument should have an absolute accuracy of around ±1−1.5 dB. Older instruments can be up to ±2.5 dB. In practice, the typical performance is likely to be better than the manufacturer's specification.

Although digital display instruments will indicate levels to a tenth of a decibel, the absolute accuracy of the instrument always needs to be kept in mind.

Measuring the NICAM power level is more difficult because of the digital nature of the signal. If the RBW (Resolution Bandwidth) of the analyser is set to 300 kHz to pass all of the NICAM signal, the FM sound carrier will also contribute to the displayed waveform. If the RBW is reduced to a value such as 30 kHz, the FM carrier will be separated from the NICAM signal but the NICAM signal will appear to drop in level. This is because the NICAM signal is now wider than the RBW setting. The drop in level is caused by the narrower filter only seeing part of the NICAM signal and so registering a lower level. The shape factor of the filter in the IF stage of the spectrum analyser will influence what is seen on the display. Commonly, Gaussian-shaped filters are used, since these allow faster sweep times. A disadvantage is the IF filter will have a relatively gentle roll-off, making it less easy to separate closely spaced signals. Sometimes sharper filters are used, in which case separation of closely spaced signals is improved but at the expense of slower sweep times for a given RBW.

Digital display analysers normally have a power in a channel feature. In this mode two markers are set on the screen and the analyser calculates the total power between the two markers. Provided the RBW is narrow enough to exclude all of the FM sound carrier, a sufficiently accurate measurement for maintenance purposes of the NICAM power can be made. The only alternative to doing this is to switch off the vision and FM sound carriers and measure the NICAM power with a power meter.

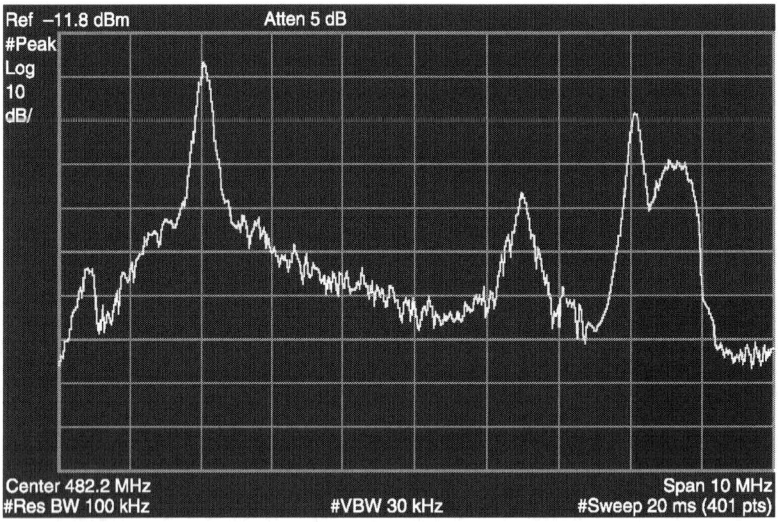

Figure 7.4A.2 Typical display of a single TV channel.

7.4A.3 Spectrum Analyser Measurements in Zero Span Mode

When zero span is selected the analyser stops sweeping and so becomes a receiver on the selected frequency. If the frequency is set to the vision carrier and the IF RBW set to its widest setting, it becomes possible to display TV waveforms. This is because the display carries on sweeping and so is now showing amplitude against time, similar to an oscilloscope. Some spectrum analysers also allow the TV picture to be viewed when in zero span, allowing confirmation that the correct channel has been selected.

TV trigger facilities will allow an individual line to be selected – for example, one of the lines carrying an ITS.

The vertical scale will need to be changed from log to linear; also, the waveform may appear to be upside down. This is because when the TV signal is modulated on to a carrier sync tips are at maximum carrier amplitude.

By setting sync tips to the top of the screen and using a 0–100% vertical modulation depth can be measured by reading the scale values for black level and peak white. For a correctly modulated signal the values are:

Peak syncs 100%
Black level 76%
Peak white 20%

Selecting an ITS line containing a staircase waveform, an estimate of overall transmitter linearity can be made. On-screen markers can normally be used to measure the step sizes.

Note that the IF bandwidth of the analyser will be a limiting factor in making waveform measurements, as it is unlikely to be the full bandwidth of the modulating signal. Typical maximum values of IF RBW bandwidth are 3–5 MHz.

Figure 7.4A.3 shows a typical display showing an ITS from a relay station. The display appears upside down because TV broadcasts use negative modulation. Colour subcarrier was present in the 10T composite pulse and on the staircase, but appears absent because the RBW of the analyser is set to 1 MHz. Although the staircase appears distorted it is adequate for making linearity measurements.

With the spectrum analyser tuned to the FM sound carrier, it is possible to detect the presence of vision to sound cross-modulation. For this measurement the IF bandwidth should be set close to the bandwidth of the FM sound signal. The analyser is still in zero span mode but not in TV line display mode. If vision to sound cross-modulation is present it will be seen as a low-level video running through the otherwise flat trace. By setting the centre of the trace to the 100% reference line, the percentage of cross-modulation can be measured.

For a system with separate sound and vision amplifiers, vision to sound cross-modulation should be negligible. For a low-power transposer up to 7% of cross-modulation is acceptable.

7.4A.4 Use of a PC for Additional Measurement Processing

A major benefit of digital display analyser is the facility to download the displayed waveform into a PC. Apart from being able to store results electronically, it also allows additional measurements to be made.

For example, the analyser can be set to zero span and a single TV line displayed. By using the PC to carry out an FFT (Fast Fourier Transform) of the displayed waveform, a reasonably accurate indication of IPs within the video waveform can be made. By selecting the quiet line, for example, the IP between the FM sound carrier and the NICAM signal can be measured. Techniques such as this go some way to overcoming the display limitations of digital display analysers when it comes to looking at IPs close to the vision carrier.

Accuracy of the measurement is dependent on the accuracy of the A/D conversion within the analyser, but it can be expected to be sufficient for day-to-day maintenance purposes. For formal acceptance tests it is still necessary to use test signals in place of program, as the results will be more accurate than using the FFT method.

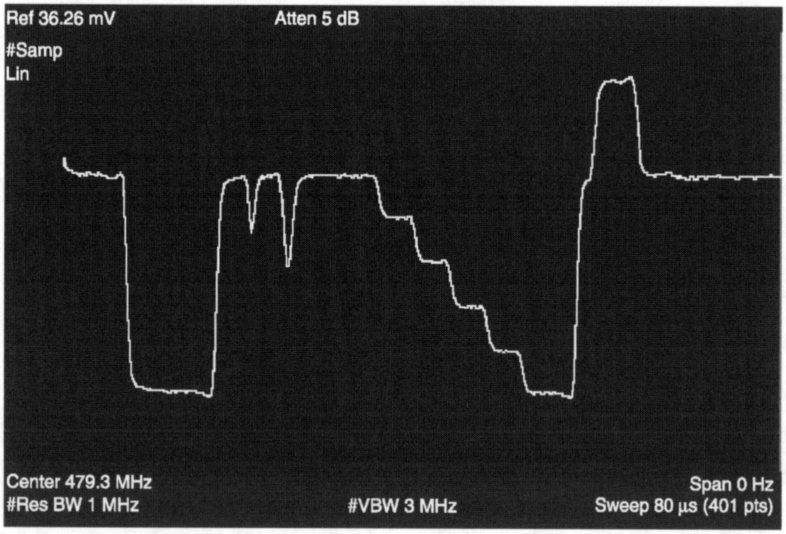

Figure 7.4A.3 Display showing an ITS from a relay station.

7.4A.5 Automation of Spectrum Analyser Measurements

Digital display analysers make possible the automation of repetitive everyday measurements. This can be done by firmware within the analyser or by external control from a personal computer. The programming language SCPI (Standard Commands for Program Instrumentation) is becoming more commonplace on instruments, opening up the possibility of a single program being able to drive several different analysers. This has the advantage of not being tied to a single supplier for spectrum analysers once custom programs have been written.

In practice, although basic SCPI commands are the same between different instruments, commands for more complex functions vary from one instrument to another. However, once a program is working on one analyser it is relatively easy to modify it to work with another. Programs can be made to interrogate the analyser to determine its type and to then select a program customised to the analyser in use. This allows a relatively unskilled operator to make repeatable complex measurements.

7.4A.6 Digital Transmission Measurements

As with analogue signals the spectrum analyser can be used to measure the flatness of the RF spectra and to make IP measurements. Digital Radio, also known as DAB, and DVB-T signals are multi-carrier spread spectrum signals having a noise-like appearance when viewed on a spectrum analyser. Care needs to be taken to avoid overloading the analyser and generating internal IPs. For example, if a Digital Radio signal is being displayed using an RBW of 30 kHz, as the signal is swept at any instant the fraction of the overall signal seen by the analyser's detector is approximately 30/1536. That is around −17 dB relative to the total signal. The exact figure depends on the shape factor of the RBW filter. However, the front end of the analyser is presented with the entire signal; thus, if the trace is moved to the analyser's reference line there is a risk of overloading the analyser.

7.4A.7 RF Power Measurements

Spectrum analysers are not suitable for making accurate power measurements because of their limited absolute accuracy. Dedicated RF power meters have to be used. These normally comprise a DC reading instrument with a separate interchangeable head to convert the RF signal to a low-level DC voltage. Power heads come in two types. Thermal types convert the unknown signal to heat and measure a temperature change – for example, using a thermocouple. Diode heads rectify the unknown signal to produce a DC signal. Diode heads subdivide further into peak measurement types and rms measurement types.

With peak reading types the power meter is likely to be scaled in rms power assuming a single continuous carrier. They are unsuitable for making accurate measurements on analogue TV or digital transmissions as the signals contains multiple carriers.

Diode heads which are rms reading rely on the operating point of the diode being at the bottom end of the diode's voltage versus current curve, so that the rectified voltage is proportional to the power of the signal. They have the advantage of a wider dynamic range and faster response times than thermocouple heads.

Until recent times, thermocouple instruments have been used to make power measurements, but the trend of modern power meters is to move away from thermocouple heads and to use diode heads instead.

On high-power transmitters it is clearly not possible to connect the output of a transmitter rated at several kilowatts directly to a power meter rated at a few milliwatts. Instead, a small sample of the signal is taken using a directional coupler. Accurate power measurements are possible if the characteristics of the directional coupler are known. A further advantage of using a directional coupler is that the forward and reverse power can be accurately measured when connected to the antenna system.

Paul Dubery

7.4B Digital RF Measurements

In the days of analogue-only services, the apology caption was usually a confirmation of picture degradation that had already become obvious to the viewer. As the picture became more noisy, it would still be viewable up to relatively high S/N levels. Automatic monitoring of the analogue video signal parameters at strategic points throughout the distribution and transmission path has become a recognised way of predicting network degradation, and many broadcasters have come to rely on such systems.

In the digital broadcasting world, the need for network monitoring becomes even more important and the earliest DTT implementations have incorporated varying levels of monitoring sophistication. The industry is recognising that a well-designed DVB-based digital broadcast network should allow room for expansion to incorporate a system that monitors the signals as they move through compression, multiplexing and modulation stages. An additional factor with broadcasting in the digital environment is the frequent hand-offs between content providers, multiplex operators, distribution and transmission service providers, together with evolving levels of service contracts which currently exist between them. This complicates the engineers' ability to quickly pinpoint the problem source and define responsibility for correction.

Digital broadcasting brings about many advantages but also delivers new issues of its own. The well-described digital cliff effect ensures that even though network degradation may be occurring, picture and sound quality remains high until disastrous break-up, blockiness and other visible disturbances appear all too obviously to the viewer.

Fringe-area viewers who have tolerated poor analogue picture quality over the years may find they cannot receive digital signals. Therefore, it is critical to characterise and monitor the performance of DTV transmitters because any degradation in performance can significantly reduce the coverage area.

The issue of viewer churn is particularly important as more services become available to the viewing public and the industry becomes increasingly more competitive. Naturally the attractiveness and cost of programme bouquets are key factors for viewers, but clearly if technical issues cause regular disturbances to picture and sound, even the least discerning consumer will become frustrated and seek more satisfying viewing experiences.

7.4B.1 The Cliff Effect

For Analogue Transmission systems, loss of transmission quality will generally result in degraded picture quality. Typically, this degraded picture quality appears as added noise, though the programme will remain viewable, until the noise gets to such a level that the receiver loses synchronisation. A Digital Television (DTV) system behaves quite differently as transmission quality degrades. The received programme signal will be unaffected until noise and other impairments cause the digital receiver system to reach its tolerance threshold. Then, very small changes in transmission quality will cause the received programme to suddenly go from error-free operation to no picture at all. This very steep threshold behaviour, sometimes called the 'cliff effect', makes DTV system performance insensitive to minor changes in transmission quality as long as you stay away from the 'cliff'. This desirable characteristic has one significant disadvantage: simply watching the received picture gives you no warning that the 'cliff' is near. To ensure reliable coverage, it is necessary for the engineering community to know how far the transmission system is operating from 'the cliff'. The following sections describe some examples of evolving techniques.

7.4B.2 DVB – BER vs. MER

The earliest DTV monitoring receivers provided a read-out of Bit Error Rate (BER). This is perhaps the most obvious and simple to implement since the data is often provided by the demodulator chip and is easily processed. For example, pre-Viterbi BER can be calculated from the number of bits corrected by the Viterbi decoder (part of the DVB-T Forward Error Correction (FEC) system) in each second. When the transmission system is operating far away from the 'cliff', few data errors occur and the pre-Viterbi BER will be near zero. As the system approaches the 'cliff' the pre-Viterbi BER rises sharply, giving some warning before the post-Viterbi BER increases and picture errors suddenly occur. The weakness with this approach is that pre-Viterbi BER increases only occur when the 'cliff' is very near. This happens because the 'cliff effect' primarily results from the COFDM modulation method, not the FEC. The FEC simply sharpens the 'cliff'. So, BER alone does give us warning, but only when it is really too late.

7.4B.3 The BER Plus Noise Approach

Since the 'cliff' is principally caused by the COFDM modulation method, we could establish our distance from the 'cliff' by adding an impairment to the received signal until the BER starts to increase. Often this is done by adding white noise. For example, if an operating system could tolerate 13 dB of additive noise before BER increased, we would have 13 dB of margin from threshold. Effectively we have to 'break' the monitoring receiver system to find the 'cliff' point. The drawback of this approach is that the monitoring receiver is continually passing threshold with minor changes in the transmission system performance. If the monitoring receiver is providing an ASI Transport Stream (TS) to an MPEG decoder or monitor, errors will occur in the TS whenever the monitoring receiver passes threshold. So, while our additive noise method allows us to see if our 13 dB margin has decreased, we have in the process created an unreliable ASI feed. What we really need is a way to monitor system margin without having to 'break' it.

7.4B.4 The MER Approach

A method for determining system margin is described in the European Telecommunications Standards Institute (ETSI) Technical Report (TR) 101 290, formerly known as ETR 290, which describes measurement guidelines for DVB systems. One measurement, Modulation Error Ratio (MER), is designed to provide a single 'figure of merit' analysis of the received signal. MER is computed to include the total signal degradation likely to be present at the input of a commercial receiver's decision circuits and so give an indication of the ability of that receiver to correctly decode the signal. The MER computation compares the actual location of a received symbol (a 'symbol' represents a digital value in the COFDM modulation process) to its ideal location, giving a figure of merit for system performance. As degradation occurs, and the received symbols land further from their proper locations, the MER value will decrease. Ultimately the symbols start being incorrectly interpreted; this is the threshold or 'cliff' point. Figure 7.4B.1 shows this BER/MER relationship for a suitable equipped receiver. The graph was obtained by connecting the MER receiver to a test modulator. Noise was then introduced in gradually increasing quantity, and the MER and pre-Viterbi BER values recorded. With no additive noise, the MER starts as 35 dB with the BER near zero. Note that as noise is added the MER gradually decreases, while the BER stays constant. When the MER reaches 24 dB the BER starts to climb rapidly, indicating threshold. MER has allowed us to see progressive system degradation long before reaching the 'cliff'.

7.4B.5 Uses for an MER Transmission Monitor

Since MER provides a sensitive indication of transmission system performance changes, an MER monitor receiver is an ideal way to watch for system degradation arising from High-Power Amplifier (HPA) ageing or tuning drift, antenna and feed line degradation, or modulator drift (see Figure 7.4B.2). Since MER is influenced by any parameter that causes symbol target error, it will flag conditions such as noise, carrier leakage, IQ level errors and quadrature imbalance. By observing MER during system commissioning, the transmission system can be continuously monitored while in operation for MER changes. Alarm thresholds can be set to notify over SNMP if MER is moving outside desired limits.

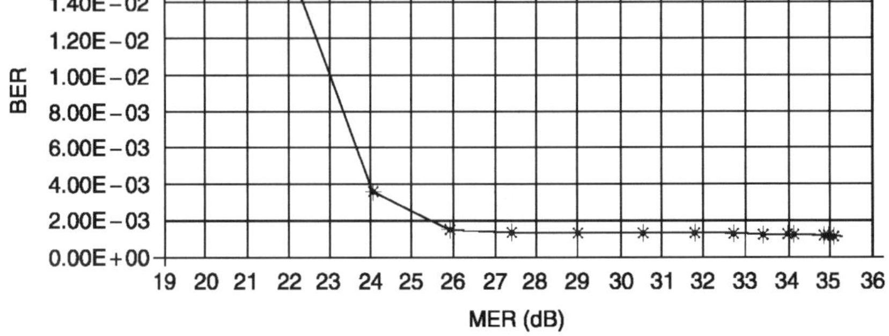

Figure 7.4B.1 BER pre-Viterbi/MER relationship (calibration on) for an equipped receiver.

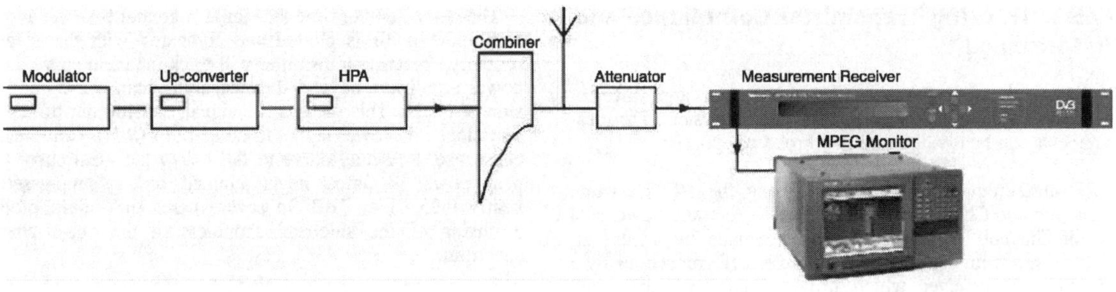

Figure 7.4B.2 An MER measurement receiver used for setup an monitoring of an HPA.

A further possibility involves transmission systems with dual redundant circuitry. Here a DTV transmitter monitor can be monitored and, if MER and other selected parameters fall outside of selected limits, initiate a changeover to the backup transmitter. Since MER looks at the ideal figure of merit – symbol target error – it is a good way to detect failures with a minimum of false alarms.

The MER technique is able to measure small changes in transmitter performance, without compromising a receiver's ability to provide a reliable ASI stream. Because MER is sensitive to any error that causes symbol target error, it is one of the best figures of merit for a DVB transmission system. Using a transmission monitor with MER helps assure reliable DVB-T transmission coverage.

7.4B.6 ATSC – an 8-VSB Approach

8-VSB is a vestigial sideband digital modulation system that uses eight discrete amplitude modulation levels. These modulation levels are assigned eight different binary numbers or symbol values to convey the MPEG-2 compressed transport stream. Figure 7.4B.3 shows the basic functions of an 8-VSB transmitter. Measurements of the performance of the system can be taken by the monitoring receiver before and after the masked filter to quantify the performance of the system. Additionally, a closed-loop feedback approach can quantify the errors and pre-distort the signal to correct for distortions within the system.

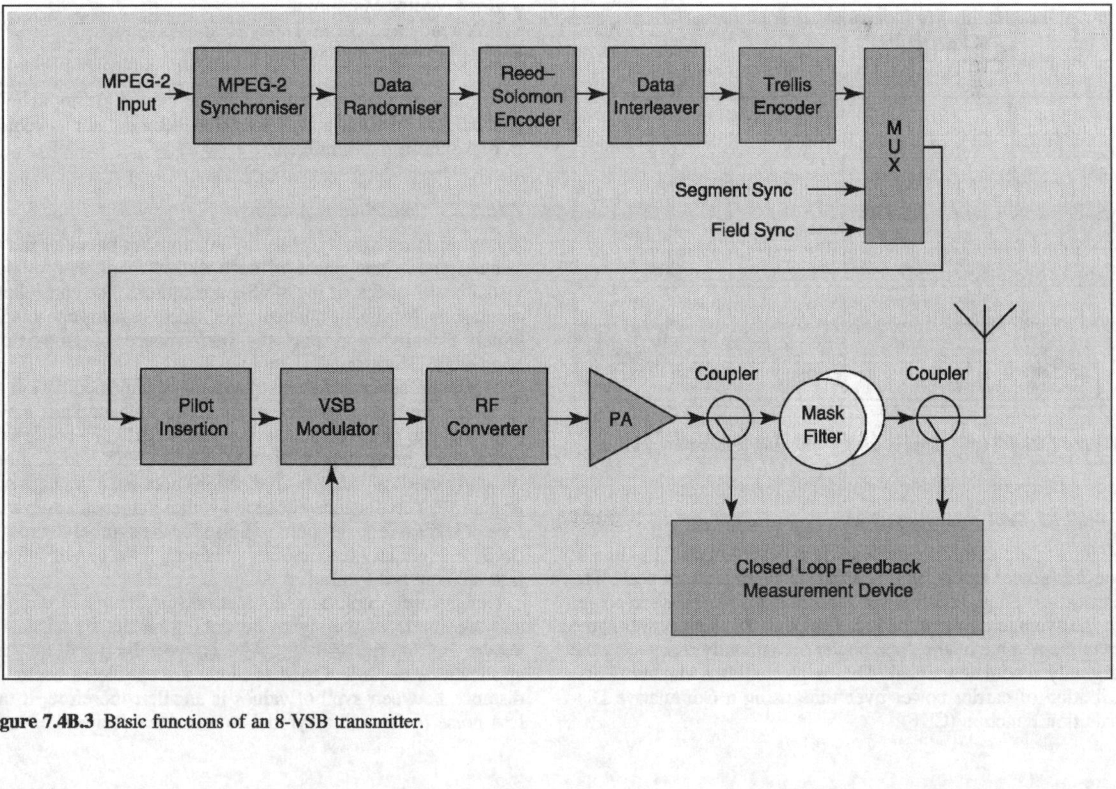

Figure 7.4B.3 Basic functions of an 8-VSB transmitter.

7.4B.7 Tracking Transmitter Compliance and Performance

Digital modulation requires new techniques and different methods of monitoring the performance of the system. The measurements can be divided into two broad categories:

- RF measurements made by analysing the RF spectrum. These include Channel Spectrum Peak-to-Average Power and Out-of-Channel Emissions and can be made by a general-purpose spectrum analyser with suitable performance, or by a measurement receiver which provides both spectrum and demodulation measurements.
- Symbol data measurements made by demodulation of the 8-VSB signal, including Constellation Analysis, Signal-to-Noise, Error Vector Magnitude, Modulation Error Ratio, Frequency and Group Delay Response Errors, Phase Error and Phase Noise.

7.4B.7.1 Spectrum Measurements

The display in Figure 7.4B.4 provides an immediate confirmation of the presence of the pilot and also shows whether the transmitted signal is of the appropriate flatness across the 6 MHz bandwidth. Any significant deviation would imply that the data values are not being randomised or that another signal is interfering within the channel.

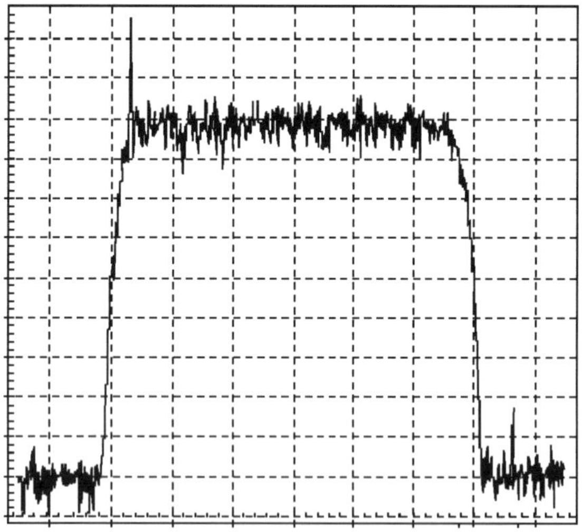

Figure 7.4B.4 Signal flatness evaluation using a spectrum analyser.

7.4B.7.1.1 Peak-to-Average Power

A transmitter should spend a certain percentage of its time at various power levels ranging from its average to its peak. The Peak-to-Average Power is the ratio of the peak transient power to the average envelope power. The peak transient power is the maximum value of envelope power occasionally reached by the digitally modulated signal. This is plotted as a statistical distribution of carrier power over time using a Cumulative Distribution Function (CDF).

The percentage of time the signal is greater than the average amplitude in dB is plotted and compared with the ideal. A properly operating transmitter will track the ideal curve. Using power amplifiers beyond their capability can cause compression of peaks. This distorts the signal, causing out of channel emissions and lower signal-to-noise ratio (S/N). Compression can cause the actual curve to fall below the ideal curve. The peak power is almost never attained, and is suppressed by nearly 0.25 dB at 7 dB. In severe cases this could produce a raising of the sideband shoulders in the out-of-channel spectrum.

7.4B.7.1.2 Out-of-Channel Emissions

The FCC mandates Out-of-Channel Emissions testing to verify that there is no leakage into adjacent channels and other over-the-air services. The required characteristics are summarised in Figure 7.4B.5.

Figure 7.4B.6 is a spectrum measurement superimposed on a mask template. The pilot frequency is distinct from the flat, noise-like spectrum of the rest of the 6 MHz channel. If the outer edges of spectrum are flat, this indicates the system does not have non-linear errors. If a slope is present on channel spectrum display, it indicates the presence of non-linear errors.

7.4B.7.2 8-VSB Symbol Data Measurements

Many 8-VSB measurements require the measurement device to demodulate the 8-VSB signal in order to examine specific symbol data. These measurements are:

- Constellation Analysis.
- Signal-to-Noise.
- Error Vector Magnitude.
- Frequency and Group Delay Response Errors.
- Phase Error and Phase Noise.

A conventional spectrum analyser cannot demodulate the spectral information and a precision demodulator is required to make these measurements.

7.4B.7.2.1 Constellation Analysis

A constellation display plots the relationship between the carrier amplitude and phase of each data symbol. It provides a visual health check of the 8-VSB transmitter. The constellation diagram is similar in concept to a video vectorscope display, which visually represents the performance of an analogue composite video signal.

There are several different types of digital modulation systems, the simplest of which is Quadrature Phase Shift Keying (QPSK). The phase of the carrier is switched in response to the signal data, as shown in Figure 7.4B.7. QPSK modulation is typically used for satellite applications because of its high noise immunity. The distance between symbol values is large; therefore, it takes a large amount of noise for the symbol to cross the decision boundary into another quadrant. Two bits of information are sent per symbol.

Quadrature Amplitude Modulation (QAM) varies the phase and amplitude of the signal depending on the signal data, as shown in Figure 7.4B.8. QAM is typically used for cable applications because QAM is more susceptible to noise. The distance between symbol values is smaller; therefore, it takes less noise to force the symbol to cross the decision boundary

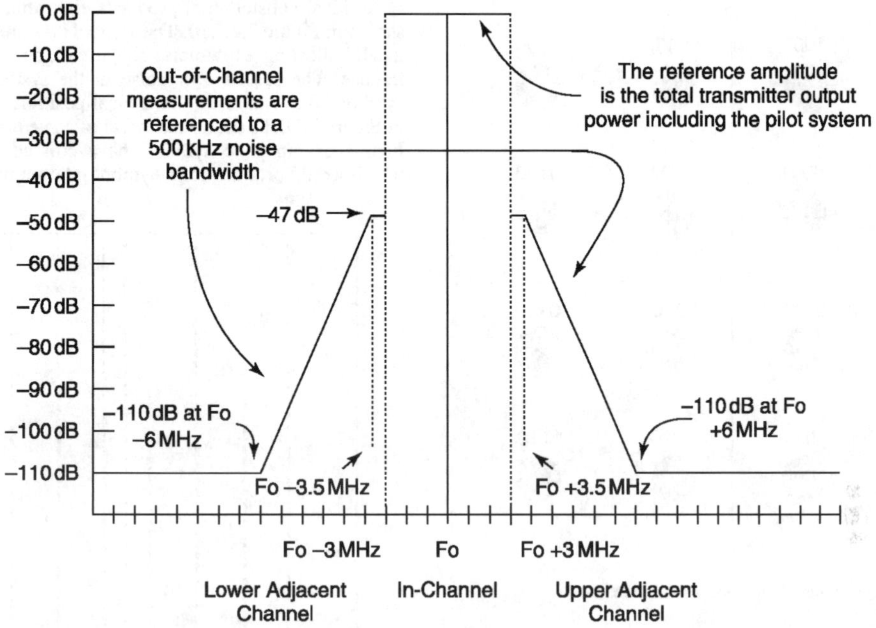

Figure 7.4B.5 Required characteristics of Out-of-Channel Emissions.

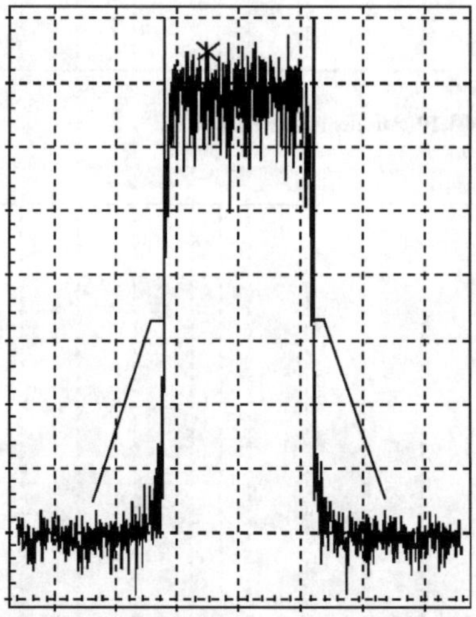

Figure 7.4B.6 A spectrum measurement superimposed on a mask template.

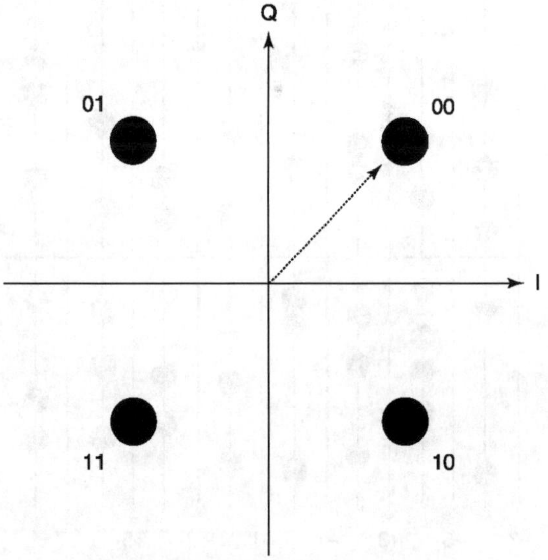

Figure 7.4B.7 Quadrature Phase Shift Keying.

into another symbol value and produce an error. There are several different forms of QAM. The diagram illustrates 16-QAM that uses 4 bits of information per symbol. Cable systems typically use 64-QAM or 256-QAM. In this case, the symbols become much closer together.

In 8-VSB, we are concerned with the amplitude of the signal which represents the symbol values. The phase of the carrier varies in order to suppress the lower sideband. The eight amplitude levels are recovered by sampling the In-Phase (I-channel) only. On the 8-VSB constellation diagram, the I-channel data is displayed along the x-axis (real axis), while the Quadrature (Q-channel) follows the y-axis (imaginary axis). Figure 7.4B.9 illustrates the result. The 8-VSB constellation

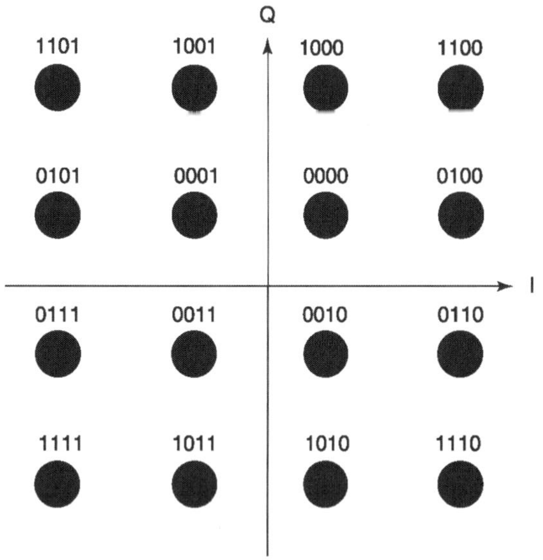

Figure 7.4B.8 Quadrature Amplitude Modulation.

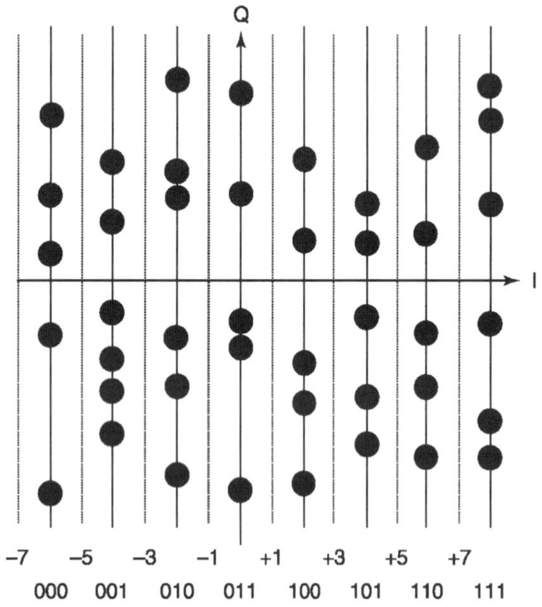

Figure 7.4B.9 8-VSB constellation diagram.

diagram is a series of eight vertical lines that correspond to the eight transmitted amplitude levels. Notice that the position of the symbol in the Q-axis does not affect the value of the symbol. Only the In-Phase (I) axis is used to determine the symbol value. Three bits are used for each symbol.

An ideal constellation produces eight thin vertical lines, as shown in Figure 7.4B.10. The symbol dots are very close to the ideal, indicating a low-noise signal with no inter-symbol interference. The presence of noise in the system will cause the symbols to deviate from their ideal position. The constellation in Figure 7.4B.11 shows the effect of some noise in the system. In this example, no symbols have crossed to another level, which would produce inter-symbol interference.

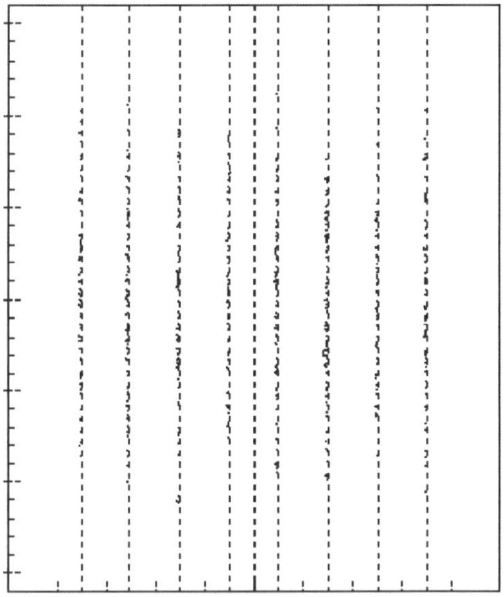

Figure 7.4B.10 An ideal constellation.

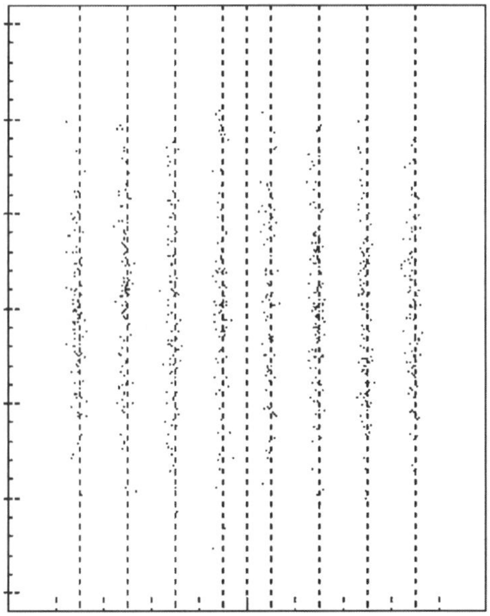

Figure 7.4B.11 Constellation showing the effect of noise.

7.4B.7.2.1.1 Splines Simplify Constellation Analysis

A splines display function makes it easier to see the average variation of the symbols along each axis and to interpret the type of errors occurring within the system. The remaining constellation diagrams in this section explain the proper use and interpretation of the splines. When the outer splines curve inward, as shown in Figure 7.4B.12, too few extreme high-level symbol values are present. This is evidence of amplitude error and gain compression, or clipping, in the transmission. This is known as AM-AM conversion error. Similarly, if the outer splines curve outward there is non-linear expansion occurring within the transmitter. If the splines are S-shaped (Figure 7.4B.13), a phase error exists. The signal amplitude is modulating the carrier's phase, causing the distortion. This is called AM-PM conversion. Bow-tie-shaped splines (Figure 7.4B.14) indicate phase noise in the transmitter.

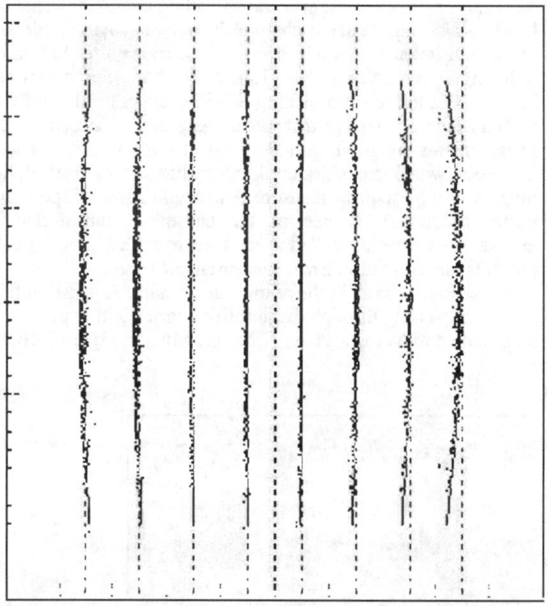

Figure 7.4B.12 Splines curving inwards.

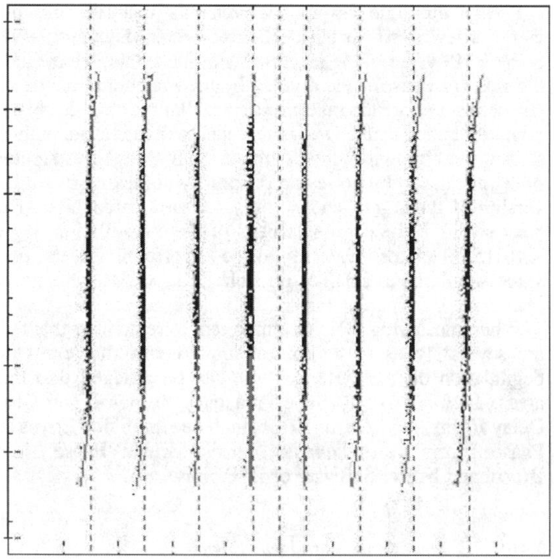

Figure 7.4B.13 S-shaped splines.

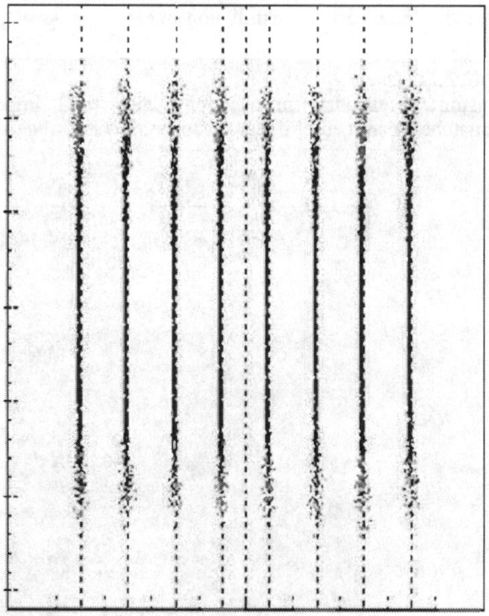

Figure 7.4B.14 Bow-tie-shaped splines.

7.4B.7.2.2 Quantitative Measurements

The measurement receiver can make many measurements that quantify an 8-VSB transmitter's operational parameters. These characterise the 8-VSB signals' actual performance compared to a theoretical ideal. Measurements fall into four categories:

- Signal-to-Noise (S/N) is the simple ratio of desired signal to undesired signal power.
- Noise is defined as anything that degrades or impairs the signal, including distortion products, inter-symbol interference caused by frequency response or group delay errors or ordinary white noise. Poor adjustment of the transmitter will result in a marginal decrease in S/N, which can result in a decrease in coverage area. Signal-to-Noise in an 8-VSB system should be above 26–27 dB and is defined as the average power of ideal symbol values divided by the noise power.
- Modulation Error Ratio (MER) is a complex form of the S/N measurement that is made by including the Quadrature channel information in the ideal and error signal power computations. MER and S/N will be approximately equal unless there is an imbalance between the I- and Q-channels. If the value of MER is significantly less than S/N, Q-axis clipping is likely to occur. This is because the Q-axis contains most of the amplitude peaks.

• Pilot Amplitude Error measurements quantify the pilot carrier's deviation from the ideal. Error Vector Magnitude (EVM) is the RMS value of the magnitudes of the symbol errors along the real (In-Phase) axis, divided by the magnitude of the real (In-Phase) part of the outermost constellation state. EVM also includes both I- and Q-channels and will therefore indicate transmitter clipping slightly before S/N. EVM is the magnitude of error induced by noise and distortions compared to an ideal version of the signal, and is measured as a percentage of the peak signal at the outermost parts of the constellation. Figure 7.4B.15 illustrates this term. For good performance, the EVM value should be as small as possible.

When measuring S/N, it is important to remember that there are several types of transmitter impairments that can cause degradation on the system. These can be grouped into three areas: *Linear Errors* include Frequency Response and Group Delay Errors. *Non-Linear Errors* include Amplitude Errors and Phase Errors. *Miscellaneous Errors* include Phase Noise, Broadband Noise, Software or DSP Noise.

7.4B.7.2.3 Identification of Linear Errors

By comparing the S/N measurements when the equaliser is switched on or off, it is possible to distinguish between linear and non-linear errors, respectively. If the equaliser significantly improves the S/N result, then most of the errors are linear. Otherwise the errors are mainly non-linear.

7.4B.7.2.3.1 Linear Errors

Transmitter response imperfections and small impedance mismatches can cause linear distortions. Group delay and frequency response problems can cause noise emphasis and inter-symbol interference at the receiver. This may indicate a need for adjustment of the final RF filter or for a pre-correction system.

Frequency Response Error is a frequency domain (spectral) function. It is the difference between the spectral response of an ideal 8-VSB signal based on the root-raised cosine frequency response and that of the actual signal. Figure 7.4B.16 depicts the result of a Frequency Response Error measurement.

Group Delay Error is also a frequency domain function. Group delay is the delay that a specific portion of the spectrum experiences through the transmission path. In this case the ideal curve is a constant (flat) response across the channel. Figure 7.4B.17 shows a typical Group Delay result.

7.4B.7.2.4 Non-Linear Errors

Non-Linear Errors cause spectral spreading outside the channel band, raising the adjacent channel 'shoulders'. Amplitude and phase non-linear errors will decrease the transmitter S/N ratio, reducing coverage area. Non-Linear Error measurements can be used as a guide for adjusting 8-VSB transmitter linearity.

Amplitude Errors are a result of gain errors caused by the instantaneous signal amplitude. Typically a transmitter's gain decreases with increasing amplitude, which gives rise to clipping. It is important to measure Amplitude Error of the transmitter because it is one of the causes of out-of-channel emissions (sometimes called spectral regrowth). Figure 7.4B.18 presents an Amplitude Error measurement result.

Phase Errors exist in the amplitude domain. A phase shift of the signal passing through the amplifier varies with input signal amplitude, producing a Phase Error as shown in Figure 7.4B.19.

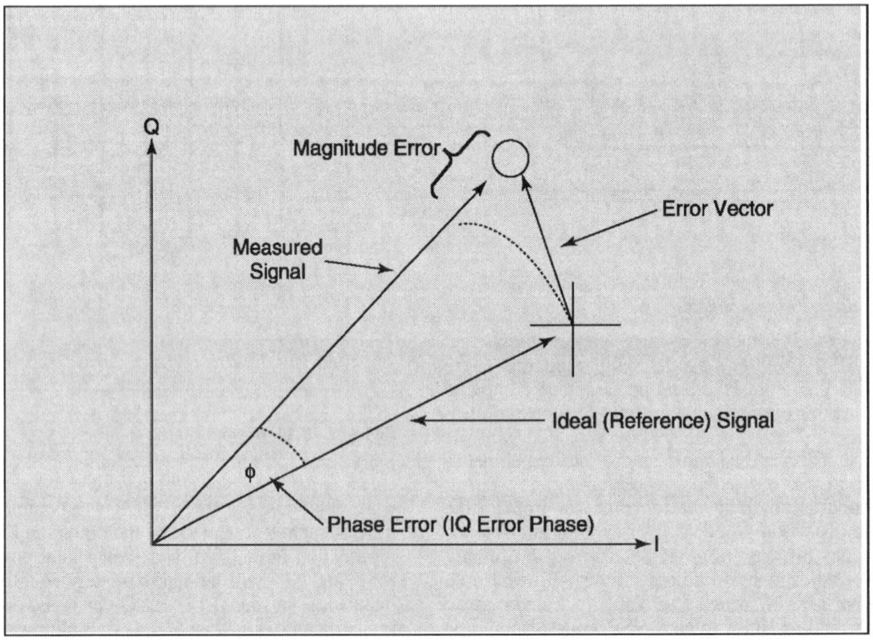

Figure 7.4B.15 Error Vector Magnitude.

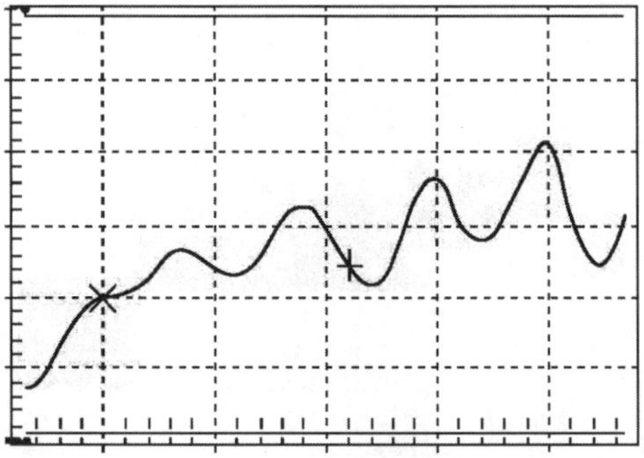

Figure 7.4B.16 Result of a Frequency Response Error measurement.

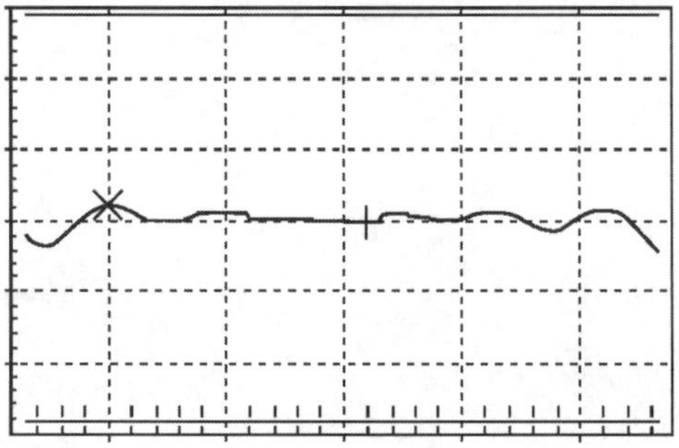

Figure 7.4B.17 Group Delay result.

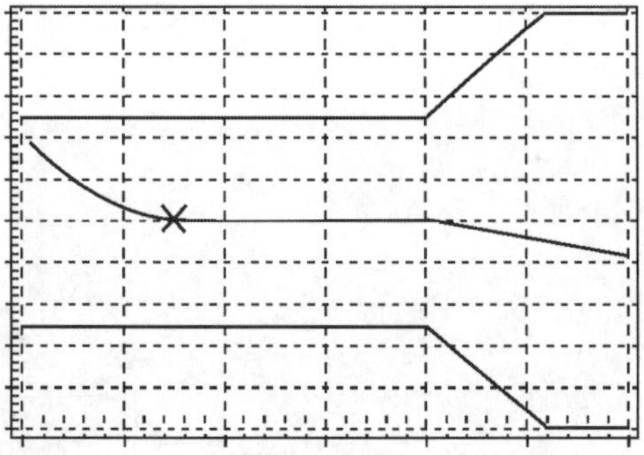

Figure 7.4B.18 Amplitude Error measurement result.

7.4B.7.2.5 Miscellaneous Errors: Phase Noise

Integrated Phase Noise is a single figure of merit describing the phase variation that the transmitter's frequency synthesiser adds to the digital modulation process. This variation causes phase and frequency deviations and rotates the decision points away from their ideal phase values, reducing the coverage area of the transmitter. The constellation diagram produces a bow-tie shape seen in Figure 7.4B.14, while the

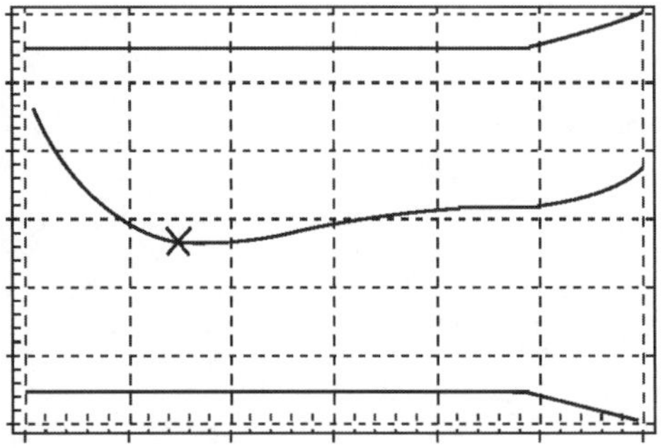

Figure 7.4B.19 Phase Error.

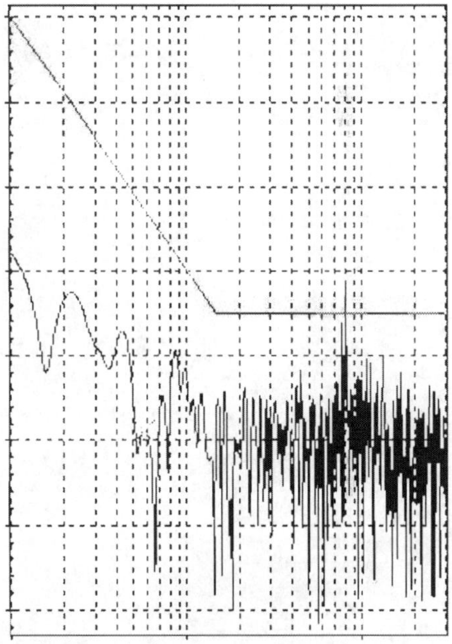

Figure 7.4B.20 Automated measurement of Integrated Phase Noise.

automated measurement produces the display shown in Figure 7.4B.20.

7.4B.7.3 Transmitter Monitoring with an 8-VSB Measurement Set

The foregoing list of 8-VSB measurements should be performed during commissioning of an 8-VSB system and at regular intervals over the life of the transmitter. While the sheer number of measurements may seem daunting, an 8-VSB Measurement Set can be set up to monitor transmitter performance automatically and continuously. It allows the engineer to define performance limits and to set caution and alarm limits for the measurements. When the limits are exceeded, there are several ways to notify the engineer, and the measurement results can automatically be saved for off-line detailed analysis.

Paul Dubery

7.5 Broadcast Test Equipment

7.5.1 Introduction

Despite its specialist nature, a surprisingly wide range of test equipment is available to address the needs of the broadcast market. The aim of this chapter is to highlight the product characteristics required for different types of users in a number of typical environments and to focus on some specific features. The first aspect to consider is the type of user and purpose of use.

7.5.2 Operational or Test?

Equipment that generates or measures video and/or audio signals can be characterised in three different aspects: usage, methods and operational environment. Types of usage include: designer's workbench, manufacturing quality assurance, user equipment evaluation, system installation and acceptance testing, system and equipment maintenance, and production operation.

Requirements for a basic operational signal monitor include display of the programme signals within the SDI digital carrier, providing features and accuracy consistent with familiar analogue baseband signal monitors. An operational monitor should also include information about the serial digital signal itself, such as data available, bit errors and data formatting errors. A display of the actual serial waveform is only infrequently required. However, monitoring of the programme signal waveform is required at all locations where an operator or equipment has the ability to change programme parameters. Methods for technical evaluation cover several usage areas that have overlapping requirements. In addition to the traditional television system measurements there's a new dimension for test and measurement – to quantify the various parameters associated directly with the serial waveform. The result is several categories of monitoring and measurement methods to be considered: programme signal analysis, data analysis, format verification, transmitter/receiver operation, transmission hardware and fault reporting. Programme signal measurements are essentially the baseband video and audio measurements that have been used for many years. An important aspect to these measurements is that the accuracy of signal representation is limited by the number of bits per sample. Hence, more complete measurement methods are desirable. Testing of passive transmission components (coax, patch panels, etc.) is similar to that used with baseband systems except that much wider bandwidths must be considered.

The third aspect to test and measurement is whether it's in-service or out-of-service. All operational monitoring must be in-service, which means that monitors must be able to give the operator information about the digital 'carrier' that contains active programme material as well as the programme signal itself. If there are problems to be solved, discovering those which have an intermittent nature will also require in-service measurements. Because of the well-known cliff effect for digital system failures, out-of-service testing is also important during installation and at regular intervals. In order to know how much headroom is available it's necessary to add stressing parameters to the digital signal in measured amounts until it does crash, which is certainly not acceptable in operational situations.

7.5.3 Operational Requirements

7.5.3.1 Production Workflow

Within the content production environment, signal monitoring and measurement needs vary along with the workflow and expertise of operator. Signals may arrive in a range of formats and standards depending on how and where they were originated, so multi-format capabilities are often required. The general workflow pattern is as follows.

7.5.3.1.1 Ingest QA

As video and audio programme material is brought into a production centre via landlines, tape or other means, there is a generic need to check its technical quality to ensure it is usable in the subsequent production processes, whether it is being incorporated into a live programme or ingested to servers or VTRs for later post-production processes. Increasingly this is becoming a contractual requirement – generally this operation will be carried out by experienced technicians and a degree of automated monitoring and error-logging may be desirable to check against established or agreed standards.

7.5.3.1.2 Studio Production

Within the production control room, whether in a studio or outside broadcast truck, the most common need for signal level monitoring is for operational adjustment of camera levels and colour balance, both during line-up and live production. Video and audio are generally handled in separate areas, though at some stage signals will often be combined, using embedded audio for ease of distribution and switching.

7.5.3.1.3 Post-Production

Where pre-recorded material is being assembled into a finished item this is usually in the hands of creative rather than technical personnel. Such users do not want to be troubled by excessive technical details but do need to be provided with a method of ensuring the finished product is compliant with the client's technical needs. Simple, uncluttered and familiar displays are essential here and the provision of alarms is most helpful to ensure the operator is alerted when illegal signal conditions are detected. Once again video and audio are often handled in separate areas until the finished material is packaged and distributed.

7.5.3.1.4 Outbound QA

In the playout or transmission centre it is essential and may even be contractually required to monitor the quality of the outgoing signals to allow monitoring of correct handling and processing downstream through distribution, multiplexing and emission stages. Although the technical expertise of operators in transmission areas is generally high, the advent of multichannel operations means that more picture monitors need to be viewed and errors can slip through. The provision of semiautomatic monitoring which provides alarms and errorlogging is therefore becoming increasingly important to provide fast detection and correction of incipient or actual fault conditions.

7.5.3.2 In-service Equipment

In all of the situations described above, the workhorse product which has become most familiar is the waveform monitor. Mounted in either control desk or equipment rack, this compact device provides a choice of signal displays which the operator chooses according to monitoring or measurement needs. Modern versions are modular such that a mixture of signal formats (analogue, digital, HD video and sometimes including audio) can be combined to support the working environment. A range of signal displays including waveform, vector, picture and others can be chosen and combined according to need, and latest implementations provide user-configurable combined displays. If error-logging is provided, a tape can be loaded and played through the device, resulting in a timecode-related listing of errors and descriptions. This approach is clearly preferable to teams of experienced engineers being tied to mechanical monitoring tasks. A representative selection of current signal monitors is shown in Figures 7.5.1–3.

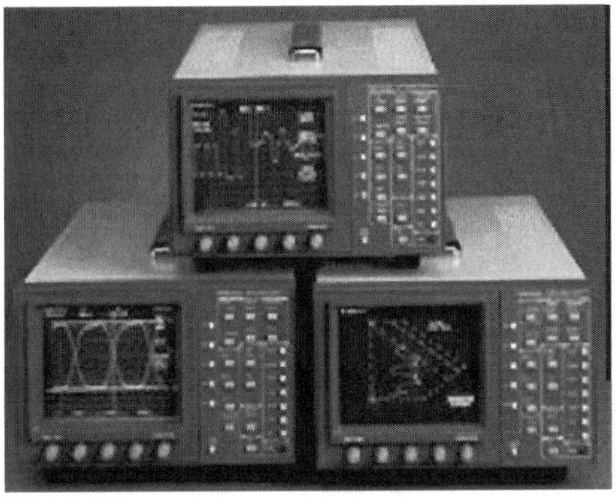

Figure 7.5.1 WFM601 family of waveform monitors.

7.5.3.3 Colour Gamut, Legal and Valid Signals

In a high-pressure creative environment, ease of use and clarity of display are essential for monitoring devices. Telecine transfers and colour correction are examples of the processes involved in post-production of feature films, music videos, television or cinema commercials, all of which require quality, both creative and technical, to be extremely high. Production operations will also include camera shading and alignment, checking signal gamut for quality control and monitoring of colour correction. Anyone who performs technical or creative adjustments on cameras, for example, will be very aware of the potential problems that can occur when the RGB colour gamut is violated. Originally vibrant colours lose their brilliance when transmitted, destroying the visual impact first achieved with great care in production and viewed on a computer display or studio monitor. Likewise, a telecine operator will certainly need to ensure the colour rendition looks good subjectively, but needs to know how to ensure that the technical quality is maintained within legal limits as well.

7.5.3.4 Operational Displays

7.5.3.4.1 The Diamond display

The Diamond display (Figure 7.5.4) provides a reliable method of detecting invalid colours before they show up in a finished production. Colour is usually developed and finally displayed in R′G′B′ format. If it were handled through the system in this format, monitoring to detect an illegal signal would be quite simple; just ensure that the limits are not exceeded. But most studio systems use a Y′, C′b, C′r format for data transmission and processing, and the signal is often converted to composite for on-air transmission. Ultimately all colour video signals are coded as RGB for final display on a picture monitor and eventually the viewers' receivers. The Diamond display is generated by combining R′, G′ and B′ signals. If the video signal is in another format, the components are converted to R′, G′ and B′,

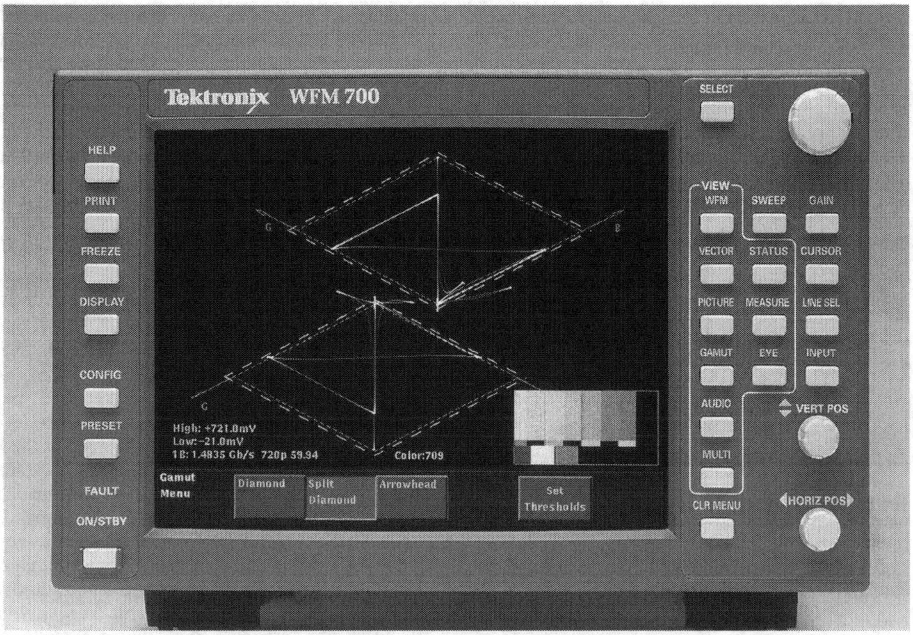

Figure 7.5.2 WFM700 waveform monitor is designed for operations using both SD and HD formats.

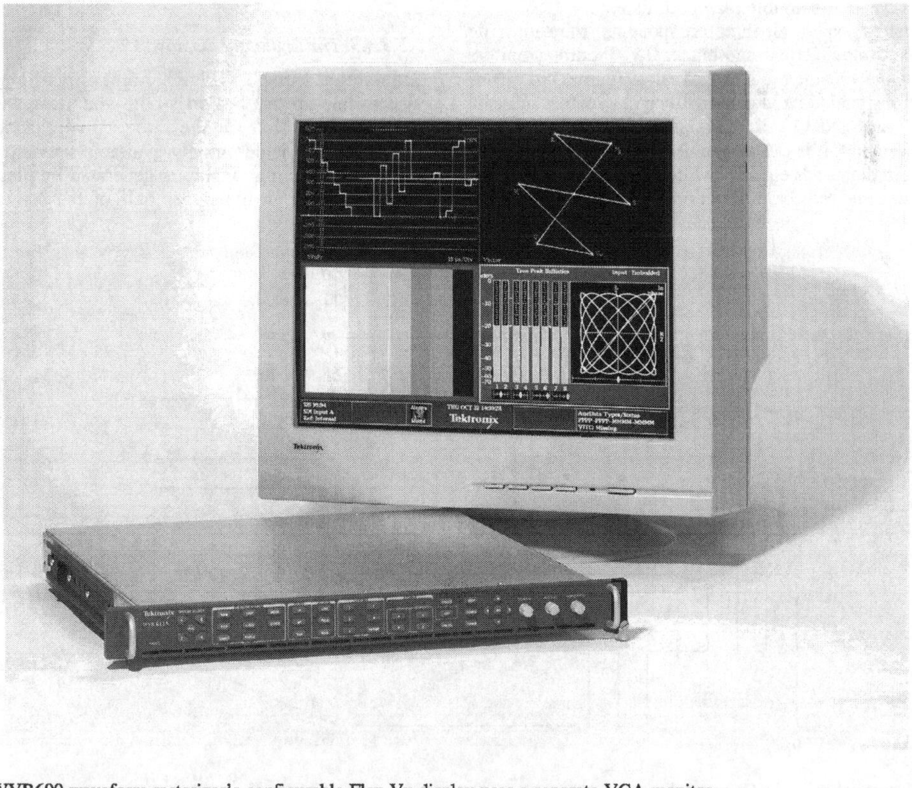

Figure 7.5.3 WVR600 waveform rasterizer's configurable Flex Vu display uses a separate VGA monitor.

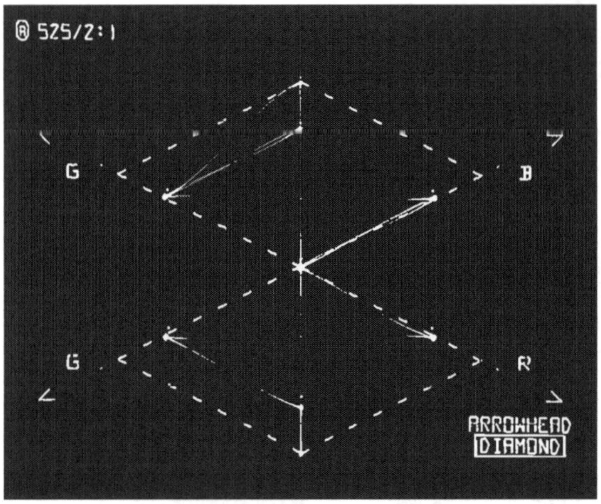

Figure 7.5.4 The Diamond display.

which can be converted into a valid and legal signal in any format that can handle 100% colour bars.

The upper diamond (Figures 7.5.5 and 7.5.6) is formed from the transcoded signal by applying $B' + G'$ to the vertical axis and $B' - G'$ to the horizontal axis. The lower diamond is formed by applying $-(R' + G')$ to the vertical axis and $R' - G'$ to the horizontal axis. The two diamonds are displayed alternately to create the double diamond display.

To predictably display all three components, they must lie between peak white, 700 mV and black, 0 V. Picture monitors handle excursions outside the standard range (gamut) in different ways. For a signal to be in gamut, all signal vectors must lie within the $G - B$ and $G - R$ diamonds. If a vector extends outside the diamond, it is out of gamut. Errors in green amplitude affect both diamonds equally, while blue errors only affect the top diamond and red errors affect only the bottom diamond.

Timing errors can be seen using a colour bar test signal as bending of the transitions. In the Diamond display, monochrome signals appear as vertical lines. However, excursions below black can sometimes be masked in the opposite diamond. Therefore, it can be useful to split the diamond into two parts to see excursions below black in either of the $G - B$ or $G - R$ spaces. By observing the Diamond display, the technical or creative operator can be certain the video components being monitored can be translated into legal and valid signals in RGB colour space.

7.5.3.4.2 The Arrowhead display

The Arrowhead display (Figure 7.5.7) provides composite gamut information directly from the component signal. This display plots luminance on the vertical axis, with blanking at the lower left corner of the arrow. The magnitude of the chrominance subcarrier at every luminance level is plotted on the horizontal axis, with zero subcarrier at the left edge of the arrow. The upper sloping line forms a graticule indicating 100% colour bar total luminance + subcarrier amplitudes, the lower sloping graticule indicates a luminance + subcarrier extending towards sync tip. The electronic graticule provides a reliable reference to measure how luminance + subcarrier looks when the signal is later encoded into composite. An adjustable modulation depth alarm capability can be provided to warn the operator that the composite signal may be approaching a limit. The video operator can thus see how the component signal will be handled in a composite transmission system and make any needed corrections in production.

7.5.3.4.3 The Lightning display

Lightning is a display (Figure 7.5.8) primarily used for VTR or system line-up that provides both amplitude and inter-channel timing information for the three signal channels on a single display using the commonly-available standard colour bar signal. The Lightning display is generated by plotting luminance vs. P′b or C′b in the upper half of the screen and inverted

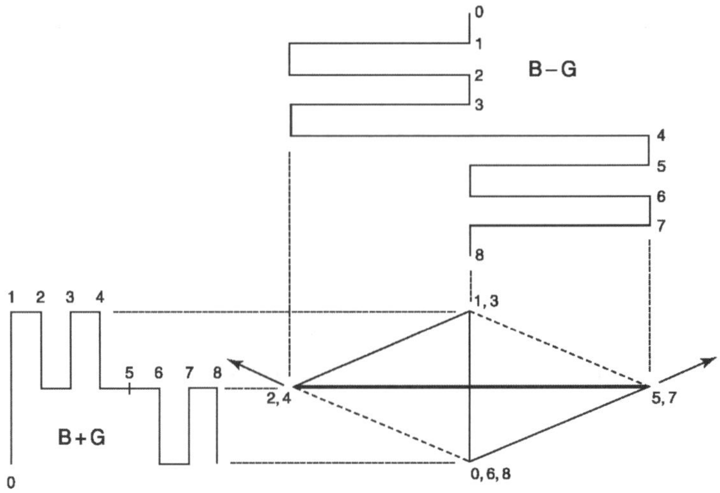

Figure 7.5.5 Construction of the Diamond display.

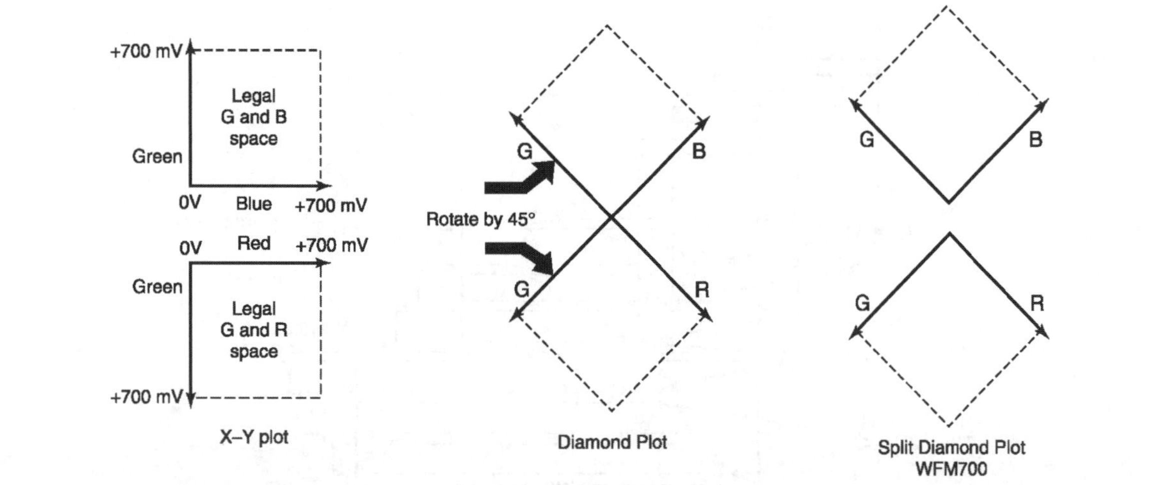

Figure 7.5.6 The Split Diamond display provides improved gamut resolution.

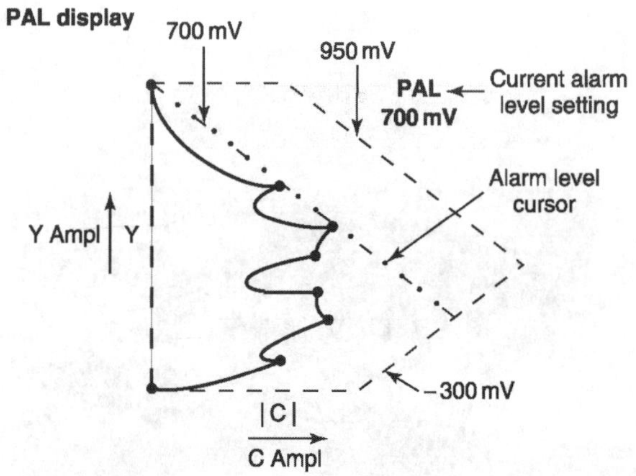

Figure 7.5.7 The Arrowhead display.

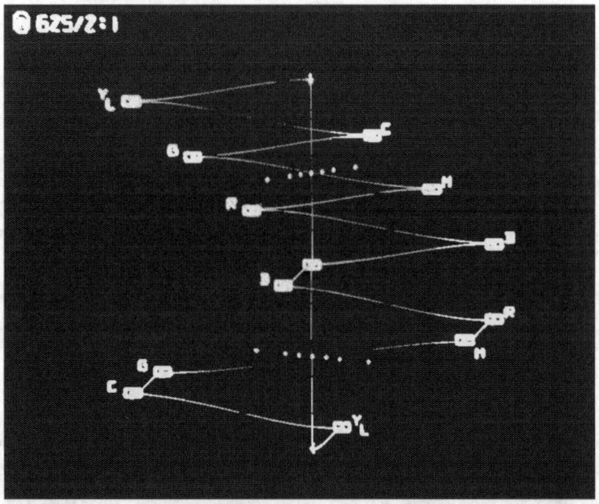

Figure 7.5.8 The Lightning display.

luminance vs. P′r or C′r in the lower half (Figure 7.5.9) – like two vector displays sharing the same screen. The bright dot at the centre of the screen is blanking level (signal zero). Increasing luminance is plotted upward to the upper half of the screen and downward in the lower half. If luminance gain is too high, the plot will be stretched vertically. If P′r or C′r gain is too high (Figure 7.5.10), the bottom half of the plot will be stretched horizontally. If P′b or C′b is too high, the top half of the display will be stretched horizontally. The display also provides inter-channel timing information by looking at the green/magenta transitions. When the green and magenta vector dots are in their boxes, the transition should intercept the centre dot in the line of seven timing dots.

7.5.3.5 Systems Aspects

To ensure systems components are kept in synchronisation, a Sync Pulse Generator (SPG) system is an essential part of the system. The SPG can be a dedicated device which provides sets of standard sync signals whose timing can be differentially adjusted and may also provide a minimal set of test signals. For more sophisticated systems a modular approach (Figure 7.5.11) may be more appropriate, especially for multi-format environments.

7.5.4 Equipment Design, Manufacturing and Service

Whether designing for professional use (e.g. studio cameras, VTRs, etc.) or targeted at consumers (e.g. receivers, set-top boxes, DVD players), the technologies involved are ever more sophisticated. Designers are, by definition, experts in the relevant technologies and therefore need stimulus and measurement

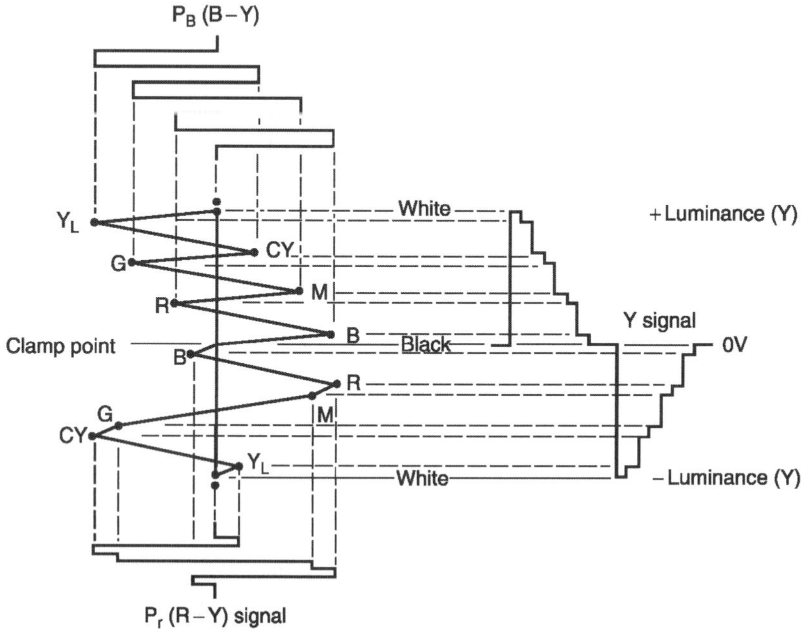

Figure 7.5.9 Construction of the Lightning display.

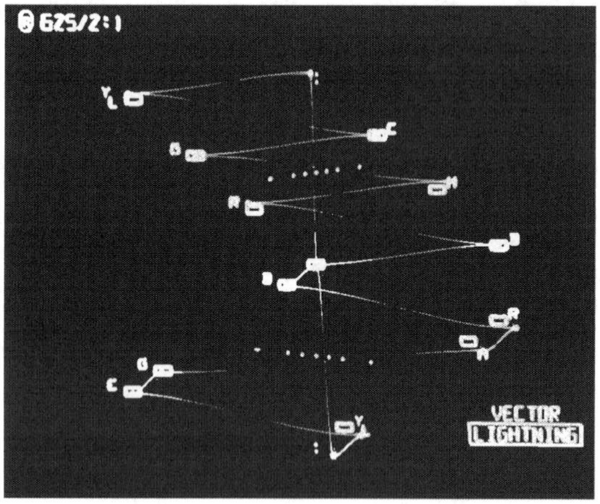

Figure 7.5.10 Incorrect adjustments cause "out-of-box" displays using the colour bar test signal.

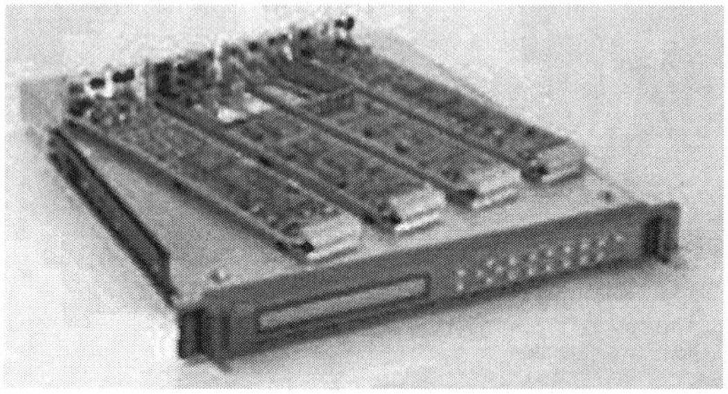

Figure 7.5.11 Modular Sync Pulse Generator.

devices that can fit their current needs and be upgraded as requirements expand. In such situations a modular approach is useful to allow an affordable entry-level test platform which provides the ability to add new capabilities via hardware and/or software upgrades. Two examples are shown in Tables 7.5.1 and 7.5.2, and Figures 7.5.11 and 7.5.12.

Once products have moved into manufacturing phase, the testing needs to become much more structured and well defined. Stimulus and analysis products in this environment need fast reaction and acquisition times and a large degree of remote control capability, so that testing occupies minimum time and can be automated to the maximum degree possible. Often, test results can be extracted and matched to the device under test for final QA or for corrective actions by production personnel.

Bench service and maintenance requirements need to be flexible enough to handle many different types of equipment and signal formats, so here again a versatile, platform-based, modular approach is optimum to allow ease of upgrade as needs expand.

7.5.5 Installation and Maintenance

When equipment is being installed in the field, the testing requirements need to be sufficiently stringent to ensure system compliance, but the test equipment has to be rugged, portable

Table 7.5.1 AD953-II – a PC-based MPEG Test System

MPEG Test System options
AAC player stand-alone
ASI PCI interface
ASI/M2S interface
Broadcast Cable digital multiplexer
Broadcast Satellite digital multiplexer
CC stand-alone
Duplex operation player/monitor or player/recorder
ES Analyser with AAC Player
GPSI II Card
L-Band + Card
MPEG/DVB Carousel Analyser
Open TV Analysis
Viaccess Analysis
XSI Analysis software

Software/hardware options provide a modularity that allows users to configure to their own needs and upgrade as necessary.

Table 7.5.2 TG700 – a modular Signal Generator Platform

Signal Generator Platform modules
Analogue Genlock
Analogue Wideband Video Generator Module
Audio Generator
Black Generator
Component and Composite Analogue Video Generator Module
Composite Analogue Test Generator Module
Digital Video Generator
HDTV Digital Video Generator

The hardware modules can be configured and added to provide a simple, single-format SPG or a comprehensive multi-format test signal generator including HD and audio.

and often battery-powered with maximum operational lifetime. In this phase the technical abilities of the operator are generally high but the space in which installation is taking place can be restricted – behind equipment racks, outside broadcast trucks and studio control desks, for instance. Compact, versatile test equipment is highly desirable here. Figure 7.5.13 shows a complementary SDI generator and analyser pair which are in common use.

Often, field maintenance and service personnel may not be so technically experienced or knowledgeable, so ease of use, clear uncluttered displays and minimal need for interpretation of results are useful test gear attributes in this environment. Figures 7.5.14 and 7.5.15 show examples of portable testers designed for such users.

7.5.6 Programme Distribution and Transmission

Digital broadcasting offers many advantages, but also presents new technology challenges. In analogue-only days, the apology caption was usually a confirmation of picture degradation that had already become obvious to the viewer. Now, the well-described digital cliff effect ensures that even though network degradation may be occurring, picture and sound quality remains high until unacceptable visible disturbances appear on-screen seemingly out of nowhere. In this environment the engineer needs warnings that degradation is starting to happen before faults actually become visible. Programme content is the key factor in gaining and retaining viewers, but, clearly, if technical issues cause regular disturbances to picture and sound, even the least discerning viewer will become frustrated and seek more satisfying viewing experiences.

Another issue with broadcasting in this new environment is the frequent hand-offs between various service providers, coupled with the variable levels of service contracts that currently exist between them. This complicates the engineers' ability to quickly pinpoint specific signal integrity problems and define responsibility for correction. However, automatic monitoring of the standardised video signal parameters at various points throughout the distribution and transmission path has become a recognised way of predicting network degradation, and many broadcasters have come to rely on such systems. As television broadcasters transition from analogue to digital technology, fundamental differences in these technologies are leading to new approaches for ensuring broadcast quality and reliability. Below, quality control and system management challenges that digital broadcasters face are described, as well as the monitoring devices that are evolving to address these challenges.

7.5.6.1 Confidence Monitoring in Digital Broadcasting

Analogue television signals represent video and audio as a continuous range of values that can assume an infinite number of states. Minor imperfections in the channels that distribute or transmit these signals can produce noticeable errors in the picture or sound. Quality decreases steadily with increased degradation in the channel. Analogue broadcast engineers recognise the onset of channel impairment by simply watching the television broadcast. With training, they can classify the type and level of impairment and take corrective action before quality degrades to an unacceptable level. Monitoring products adds precision to this basic quality control approach.

Digital television signals represent video and audio information as a discrete set of values that can assume only a finite number of states. Minor imperfections in the channels that distribute or transmit these signals generally have no noticeable effect on picture or sound quality. Quality remains high as channel degradation increases until the impairment level reaches a threshold point. At this 'digital cliff', quality decreases to an unacceptable level. Thus, digital system engineers

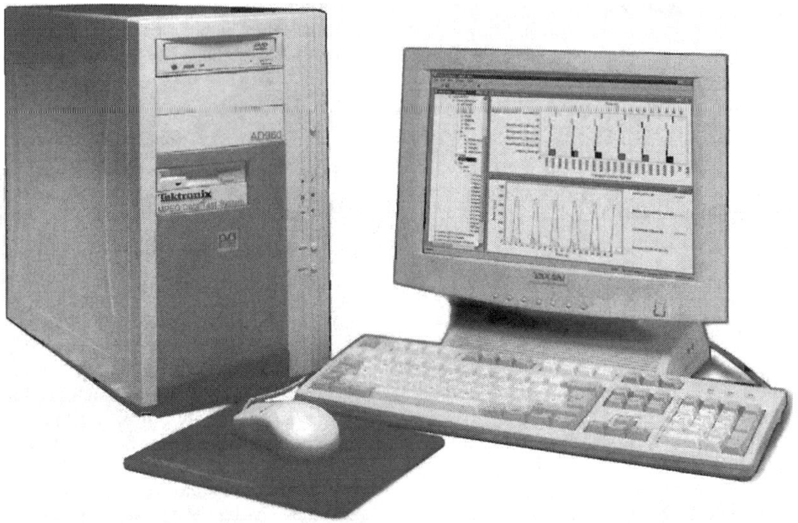

Figure 7.5.12 The AD950 family of MPEG Test Systems provide maximum configurability via software options and upgrades.

Figure 7.5.13 Complementary SDI generator and analyser pair.

Figure 7.5.14 Example of a portable video tester.

Figure 7.5.15 Example of a portable MPEG tester.

cannot detect the onset of channel degradation by watching the broadcast; they can only react to severe quality problems once they appear. Digital broadcasters therefore need monitoring approaches that let them proactively address channel degradation before it leads to noticeable quality problems which may disturb the viewers and lose hard-won loyalty.

Monitoring products that can detect impairments before they impact quality can help broadcasters achieve the same level of confidence in digital television that they achieved in analogue. Such products are referred to as confidence monitors. Requirements for confidence monitoring products and systems for digital television facilities are based on quality control and system management challenges that broadcasters face.

7.5.6.2 Digital Broadcast System Management

By supporting the convergence of video, voice and data distribution systems, the transition from analogue to digital technology also affects digital television system management. Digital telecommunication network operators gain new sources of revenue by offering distribution services to broadcasters, and broadcasters can use these services to reduce operating expenses. However, this complicates the process of maintaining quality by introducing additional transitions in the distribution chain. As one company hands off content to another, broadcasters must rely on other companies to meet contractual quality of service obligations.

Technology convergence has also facilitated new approaches to system management. Many broadcasters are considering management techniques that resemble the centralised monitoring and management systems seen in telecommunication facilities. These systems will rely on network-capable confidence monitoring devices that can report status and send alarms to a central Video Network Operations Centre via standard network communication protocols.

The digital cliff effect, the increase in the number of handoffs and new centralised management approaches are factors driving the characteristics of confidence monitoring systems in digital television. Confidence monitoring solutions must also address the quality control challenges arising from the layered structure of digital television systems.

7.5.6.3 Confidence Monitoring Examines Multiple Layers

In an analogue television system, distribution and transmission channels can be viewed as a sequence of analogue signal processing steps. With the transition to digital technology, broadcasters can now use digital signal processing and digital data processing techniques to improve quality and efficiency in their broadcast networks. Hence, distribution and transmission channels in digital television systems contain sequences of signal processing and data processing steps.

We can best understand how these steps interact and impact broadcast quality by organising them into a layered model. Specifically, we can use three layers to model a digital television broadcast system, as outlined below.

In the *Formatting* layer, television content providers create and format the video and audio that the broadcaster will deliver to the end consumer. Signal processing in this layer includes the sampling, quantising and formatting steps needed to create digital television signals, conversion between digital formats and display of the digital signal on a television receiver or picture monitor.

In the *Compression* layer, content providers and broadcasters compress and aggregate content for storage, distribution or transmission. Signal processing in this layer includes video and audio compression. Data processing in this layer includes multiplexing programmes and system information into a single data stream, fragmenting this stream into a packet protocol and recomposing programmes from packets for decoding.

In the *Distribution* layer, broadcasters process content for distribution over internal networks or delivery to the end consumer through digital television transmission systems. Signal processing in this layer includes techniques for modulating digital signals onto RF carriers. Data processing includes error correction algorithms for transmission and the formatting needed to embed content into the network communication protocols used in internal distribution.

7.5.6.4 Quality Control Challenges Within the Layers

Adding digital signal and data processing to broadcasting introduces new sources of error, with different types of errors in each system layer.

At the *Formatting* layer, broadcasters face challenges in dealing with the wide array of new formats for both standard and high-definition digital television. As in analogue television, they need to ensure correct colorimetry and verify conformance to standards. In addition, they may need to convert from one format to another, e.g. down-converting HD content for broadcast on an SD system. These format conversions can introduce quality errors. Also, separate processing of digital video and audio can lead to synchronisation problems.

At the *Compression* layer, broadcasters deal with technology that is significantly different than anything they encountered in analogue television. Compression introduces new types of quality defects, including blockiness and edge effects. Errors can occur during the complex process of multiplexing programmes and system information into a single data stream. Errors in timing and synchronisation parameters can compromise the decoding process and lead to noticeable content quality errors.

At the *Distribution* layer, broadcasters encounter familiar RF technology in the transmission networks; however, these systems use very different modulation techniques and offer new challenges in understanding coverage and interference problems. From source to consumer, programme content often moves through these system layers many times. Transitions between layers can dramatically alter the nature of the digital information, e.g. moving between uncompressed digital video at the Formatting layer and compressed digital video at the Compression layer. The additional processing needed to move across layers increases the probability of quality errors at these transitions.

Further, errors in one layer can cause errors in a different layer, in some instances masking the original error source. For example, blockiness errors can arise from problems in a compression step (Compression layer), or as a consequence of uncorrected bit errors in the receiver (Distribution layer). Similarly, transmission errors can occur due to failures in the modulation steps (Distribution layer), or from variations in the data rate from the multiplexer feeding the studio-to-transmitter link (Compression layer).

7.5.6.5 Distributed Multi-layer Confidence Monitoring

To meet the quality control and system management challenges described above, confidence monitoring systems should have the following characteristics:

- Layer-specific probes that detect the different types of errors seen in digital television systems.
- Multi-layer monitoring that lets broadcasters quickly isolate the root cause of a quality problem.
- Extended monitoring capability to give broadcasters advanced notification of system degradations before they become quality problems.
- Network control that supports the new system management challenges.

7.5.6.5.1 Layer-specific probes

In a confidence monitoring system, we can think of each monitoring device as a probe, monitoring quality at a particular point and layer in the distribution and transmission chain. Broadcasters will need to use different probe types for quality control at different layers.

At the *Formatting* layer, digital waveform monitors help broadcasters detect many quality problems. Like their analogue counterparts, these probes monitor characteristics of the digital video signal. They belong to a larger collection of Formatting layer probes that includes digital audio monitors, picture quality monitors for detecting blockiness and other picture impairments, and probes for detecting audio/video delay.

The MPEG-2 standard defines the basic processing steps and techniques used at the *Compression* layer. Broadcasters need MPEG protocol monitors capable of detecting problems in basic MPEG processing, as well as the additional processing defined in the DVB, ATSC or ISDB broadcasting standards based on MPEG.

At the *Distribution* layer, broadcasters need probes to detect quality problems in a wide variety of distribution and transmission channels. Probes in this group include devices to monitor RF transmissions in the relevant format. They also include probes for monitoring information sent through either cable or fibre telecommunication networks.

7.5.6.5.2 Multi-layer monitoring

To have confidence that their facilities are operating correctly and efficiently, broadcasters will generally need to probe at all layers. Probing at only one layer can give a misleading picture of system health. We began with a simple example of this problem. By watching the broadcast on a picture monitor, broadcasters are probing quality at the Formatting layer. While this offers significant information on system health in an analogue system, it offers little information in digital systems.

Similarly, monitoring just the MPEG protocol or the RF transmission will only yield partial information. To gain a complete picture of system quality, and to quickly detect and isolate quality problems, broadcasters will need multi-layer confidence monitoring solutions (Figures 7.5.16 and 7.5.17).

Figure 7.5.16 MTM400 MPEG Transport Stream Monitor is an essential component of any digital broadcasting system.

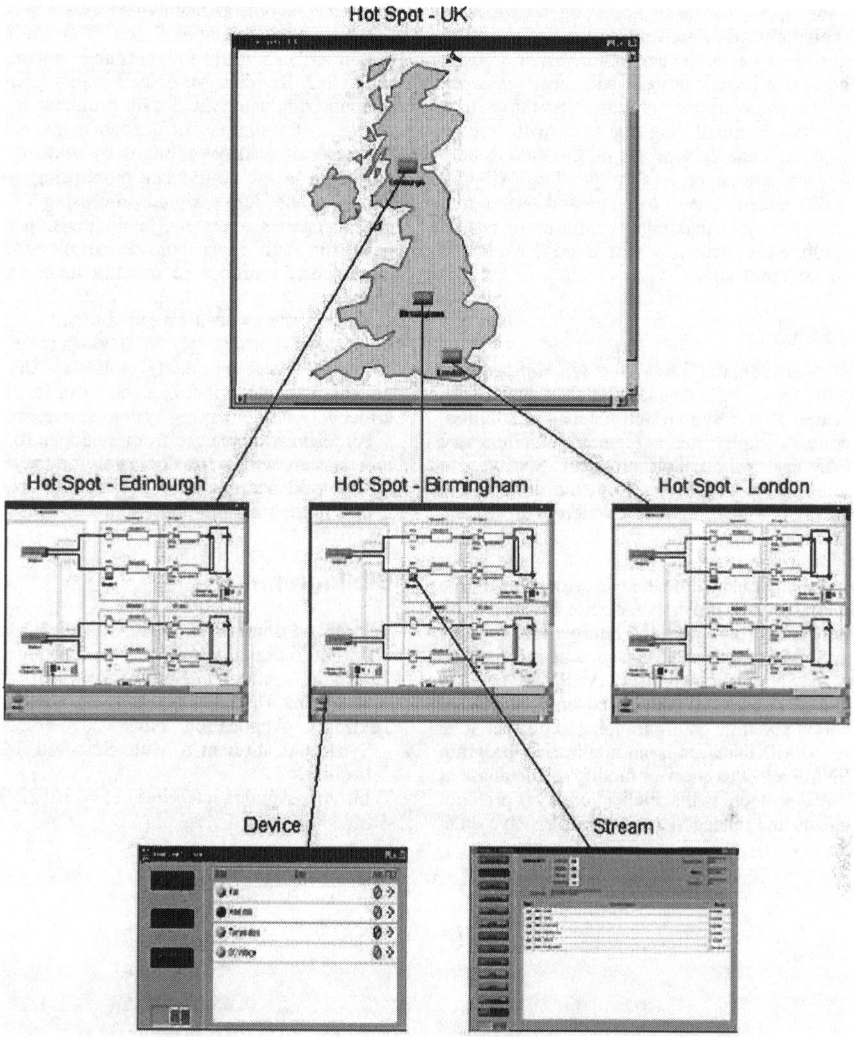

Figure 7.5.17 Network Management Software provides instant graphical alarms and diagnosis.

7.5.6.5.3 Extended monitoring capability

We can also distinguish confidence monitoring probes by the level of monitoring they offer. Basic confidence monitoring probes track a small set of key quality parameters. They act as an 'indicator light', telling the broadcaster when something has gone wrong.

However, basic confidence monitoring probes do not offer a complete solution. While they can enhance the broadcaster's ability to react to a quality problem, they do not offer the information needed to proactively address system degradation before it becomes a quality problem. Extended confidence monitoring probes use more sophisticated analysis to make additional measurements of quality parameters. They act as 'indicator gauges', telling the broadcaster when something is going wrong.

RF transmission monitoring offers a good example of this distinction. Basic RF confidence monitors measure Bit Error Rate (BER). BER will remain low until the transmission approaches the digital cliff, then increase dramatically as the transmission falls off the cliff. This gives broadcasters only slightly more time to react than they would have by watching the transmission on a picture monitor.

Extended RF confidence monitors add additional measurements like Modulation Error Ratio or Error Vector Magnitude. These measures will noticeably decline as system performance degrades, giving broadcasters early warning of potential quality problems, and an opportunity to make adjustments or seamlessly transition to backup systems.

7.5.6.5.4 Network Control

System management concerns also impact confidence monitoring. Broadcasters often need to monitor at geographically distant locations. For example, a broadcaster accepting contribution feeds over a telecommunication network may want to install confidence monitoring probes at the network operator's points of presence. These distributed probes will need network capability so they can report status and alarm conditions to a central location. Network monitoring software with easily-interpreted user interfaces can present this information to help engineers identify the source of quality problems (Figure 7.5.17). In such a distributed system, software and probes that support open standards provide maximum flexibility in system design. Closed, proprietary standards will impose excessive constraint over the system designer.

7.5.6.6 What is SNMP?

Originally developed in 1988, the *Simple Network Management Protocol* has become the de facto standard for generic network management. Because it is a simple and extensible solution, requiring little code to implement, equipment suppliers can easily build SNMP agents into their products. SNMP also separates the management architecture from the architecture of the hardware devices, which provides an open platform and greater user choice when configuring and designing network systems. Perhaps most important, SNMP is not a mere paper specification, but an implementation that is widely available today. The SNMP framework provides facilities for managing and monitoring network resources on the Internet and consists of SNMP agents, SNMP managers, Management Information Bases (MIBs) and the SNMP protocol itself. An SNMP agent is software that runs on a piece of networked equipment that maintains information about its configuration and current state in a database. An SNMP manager is an application program that contacts an SNMP agent to query or modify the database at the agent. The SNMP protocol is the application layer protocol used by SNMP agents and managers to send and receive data; the managed objects maintained by an agent are specified in a Management Information Base (MIB). An MIB file is a text file that describes managed objects and network devices may have multiple MIB files. An SNMP manager and an SNMP agent communicate using the SNMP protocol.

Due to the digital cliff, broadcasters can no longer detect the onset of quality problems by viewing the broadcast and will need to use confidence monitoring systems. Combining analogue and digital signal processing with digital data processing creates layers in digital television systems, driving the need for multi-layer confidence monitoring systems with layer-specific probes to quickly detect and isolate quality problems.

These systems need extended monitoring capability to help broadcasters proactively address performance degradations before they become quality problems. The systems also need network-capable probes in a variety of form factors to integrate effectively into emerging system management approaches.

As television changes from analogue to digital technology, broadcasters will increasingly rely on these distributed, multi-layer confidence monitoring systems to ensure optimal performance in their distribution and transmission systems.

Bibliography

Tektronix Primer 25W-15952-0; Multi-layer confidence monitoring in digital television broadcasting.

Tektronix Application Note 25W-14587-0; Set-Top Box Manufacturing Test.

Tektronix Application Note 20W-14229–0, Timing and Synchronization in a Multi-Standard, Multi-Format Video Facility.

Tektronix Application Note 25W-16527-0, Facility Timing Made Easy.

Jan Colpaert
Director of Technology
Scientific-Atlanta Europe

7.6 Systems Monitoring and Management

The move to digital video processing and transmission technology offers a lot of new possibilities but does not make system configuration, maintenance and management easier. The increased complexity of the technologies often makes the systems more vulnerable. Service interruptions do not only find their origin in hardware failures but are in many cases caused by faulty configuration of the equipment.

To make things even worse, digital video transmission standards are not as well defined as their analogue counterparts, leaving much room for different interpretations of the standards. Good reception on one brand of set-top box does not imply that all set-top boxes will be able to decode the signals and the electronic program guides.

Proper management and monitoring tools give the broadcast engineer more confidence, since the management tools simplify the configuration of equipment and services, and the monitoring tools will detect service, equipment and transmission failures immediately.

As monitoring and management is a very broad domain, we need to analyse the functions and layers in order to build monitoring and management systems in a structured way.

The telecom industry has taken monitoring and management very seriously in the past and has defined a reference model that helps us think about network management in a structured way. The ITU TMN reference model has been used for many years, and although some of the software interfaces have become obsolete and old-fashioned, the TMN model still serves as the most valuable reference model for network management and monitoring.

7.6.1 Network Management Layers

Network management systems need to act on several layers: the equipment management layer, the element management layer, the network management layer, the service management layer and the business management layer (Figure 7.6.1).

The network operator can decide which network management layers will be deployed. In many cases, the equipment management layer and element management layer are mandatory in order to be able to configure the equipment and to monitor the equipment and transmission links. Mature organisations will also implement the higher network management layers and this will result in a much more streamlined organisation and reporting structure.

Does this mean that new projects have to start with the equipment layer and move gradually higher in the hierarchy? Certainly not; mature organisations have learned the value of service monitoring and the importance of the link to the business management. For this reason, they will put network and service management at the centre of attention. The equipment layer design then is often secondary.

7.6.1.1 Network Element Layer

This is the most elementary management layer: if a device does not have any management or control interface, it cannot be integrated in a management system. Historically, we have seen an evolution from network elements equipped with alarm contacts over elements using proprietary serial protocols to the devices of today that have an Ethernet/IP/SNMP interface. In many cases, new devices also have an embedded web server that gives access to the device configuration and status parameters. We call these devices web-enabled devices.

7.6.1.2 Element Management Layer

The element management layer is needed to provide remote and multi-user access to devices and subsystems. In today's context, 'remote' and 'multi-user' mean access through a TCP/IP network. Several systems or users must be able to interact with the element management layer simultaneously. Examples of such applications are: a remote configuration tool, an alarm collection console and performance data collection engine.

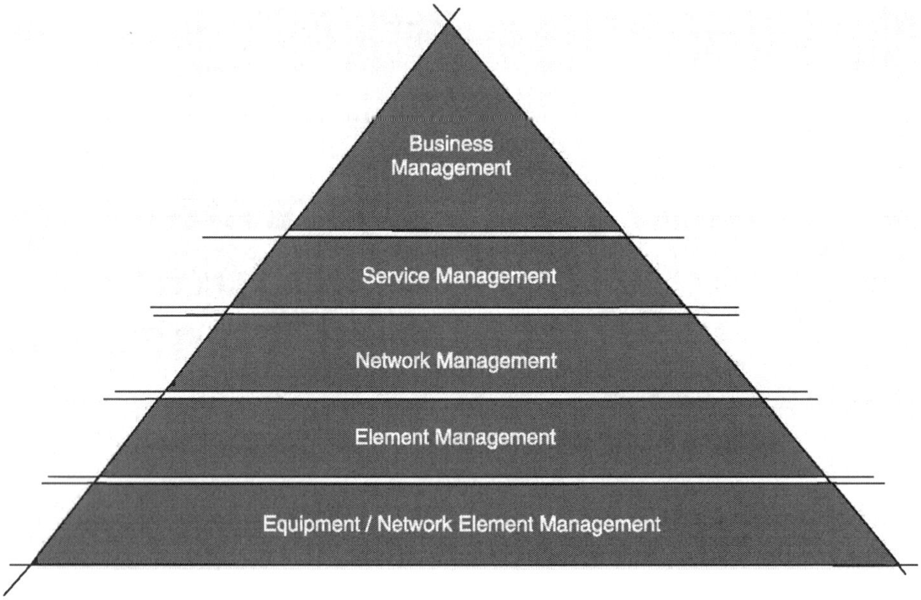

Figure 7.6.1 Network management layers.

Each of these applications may run independently from the others and interact with the same device quasi-simultaneously. Newer web-enabled devices with an SNMP interface do not always need to have an external element manager. Such devices already fulfil the main element management requirements of providing remote and multi-system access.

In multi-site networks, the element management layer is implemented close to the equipment. In such a configuration, each remote site may have one or several element managers that will convert proprietary serial protocols into open IP-based protocols like SNMP. The element manager will then be connected to the central management application(s) through the wide area TCP/IP network. For web-enabled SNMP devices, element managers are only needed if the site needs to be managed as a 'subsystem' and not as a collection of individual devices.

7.6.1.3 Network Management Layer

The network management layer knows about the network structure and the interconnections between the sites and the devices. The network management layer will typically manage the transmission links and knows the dependencies in the network: 'If this link goes down, all these sites will generate an alarm'. The network management layer must be able to cope with the transmission technologies used (Microwave, DWDM, ATM, SONET/SDH, Gigabit Ethernet, IP, DVB-T, DVB-C, Satellite, etc.) and must have a notion of the transport hierarchies used by these systems.

7.6.1.4 Service Management Layer

One broadcast network will carry many services. Services are typically broadcast video/audio programs. Services can be analogue, digital, compressed or uncompressed. The service management applications are completely different for the dynamic contribution domain and the more static distribution domain.

- *Distribution domain*. For the digital compressed services we deliver to our subscribers/audience today, the delivery mechanisms can become very complicated: several programs can be sent as SPTS or MPTS transport streams over the backbone network to the different regions. For terrestrial or cable delivery, regional MPTS streams need to be created that are composed of national programs and regional programs. Local ad-insertion may even complicate the service management even more.
- *Contribution domain*. Here, telecom networks are used to bring sports or other temporary events to the central broadcast facilities. As in most cases, we are talking about uncompressed video; huge bandwidths need to be reserved for a limited amount of time. Resource allocation and scheduling are the main tasks of such a system. Due to the dynamic nature of this domain, such applications are often realised outside the scope of a classical network management system. The application will, however, need to interact with the NMS system to set up the connections.

The service management layer is in most cases dependent on the network management layer, since there is a very clear relation between the service availability and the availability of transmission links.

The monitoring of service availability for each program in each of the regions is an important task of the service management layer.

7.6.1.5 Business Management Layer

This layer is often realised outside the classical network management system. It is important, however, to determine what kind of information needs to be exchanged with the business management. If the management system can easily provide answers to the following questions, the chances are high that we have a smooth and cost-effective operation:

- Did our telecom service provider comply to the Service Level Agreement we have in place?
- What was the mean time to repair a failure in the network?
- What were the service availability figures in the past year?

The NMS system can also assist the business processes in the following areas:

- Activation/provisioning of new subscribers.
- Generation of trouble tickets/verification that a problem is solved.
- Planning/introduction of new services.

7.6.2 Network Management Functional Domains

The classification into management layers gives us one view that helps us think about the NMS system architecture. We now want to introduce a second view that corresponds to the five functional domains a management system needs to handle. Depending on the broadcast network and application, each layer may need to implement the five FCAPS (Fault, Configuration, Accounting, Performance, Security) functions (Figure 7.6.2).

7.6.2.1 Fault Management

Fault management is all about the broadcaster/operator's ability to react quickly to problems in the network. A central alarm log viewer that shows the alarm status and history for all managed devices and services is the minimum we need in a simple network (Figure 7.6.3).

Such an alarm viewer will show the time-stamp of arrival of the alarm, the severity of the alarm, the classification of the alarm, a textual description and the name of the device or service on behalf of which the alarm is generated.

A graphical view on the alarm state of the different sites, services and devices is needed in more complicated networks (Figure 7.6.4). The graphical views show the alarm status of devices. The visualisation of several simultaneous views like geographical view, interconnection view and rack layout view makes troubleshooting really easy.

7.6.2.1.1 Notification and trouble ticketing

The control room may not always be manned 24 hours a day. Automatic notification of technicians has become an important feature of the fault management system. State-of-the-art fault management subsystems can be configured to send pager or email messages to service technicians. A rule-based system determines for which alarm messages technicians need to be notified. Depending on the failing device or the geographical area, different service technicians can be paged. The notification will also depend on the time of day/day of the week.

Larger network operators with installed workforce management applications might prefer to let the fault management system generate a trouble ticket if certain severe problems are detected.

7.6.2.1.2 Automation

The fastest response is an automatic response. The fault management system can be configured to react automatically to certain types of problems. A typical example is an N + 1 redundancy scheme. If the management system detects that a device is defective, the system can automatically re-route the signal to a spare unit and restore the service in a few seconds. Such functions need to be implemented at the element management layer (i.e. close to the equipment).

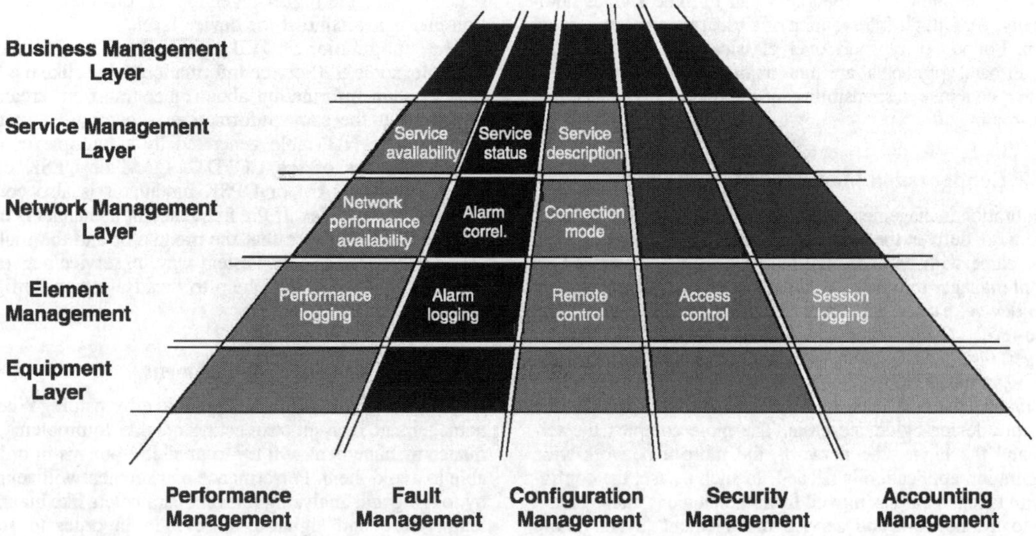

Figure 7.6.2 Network management functional domains.

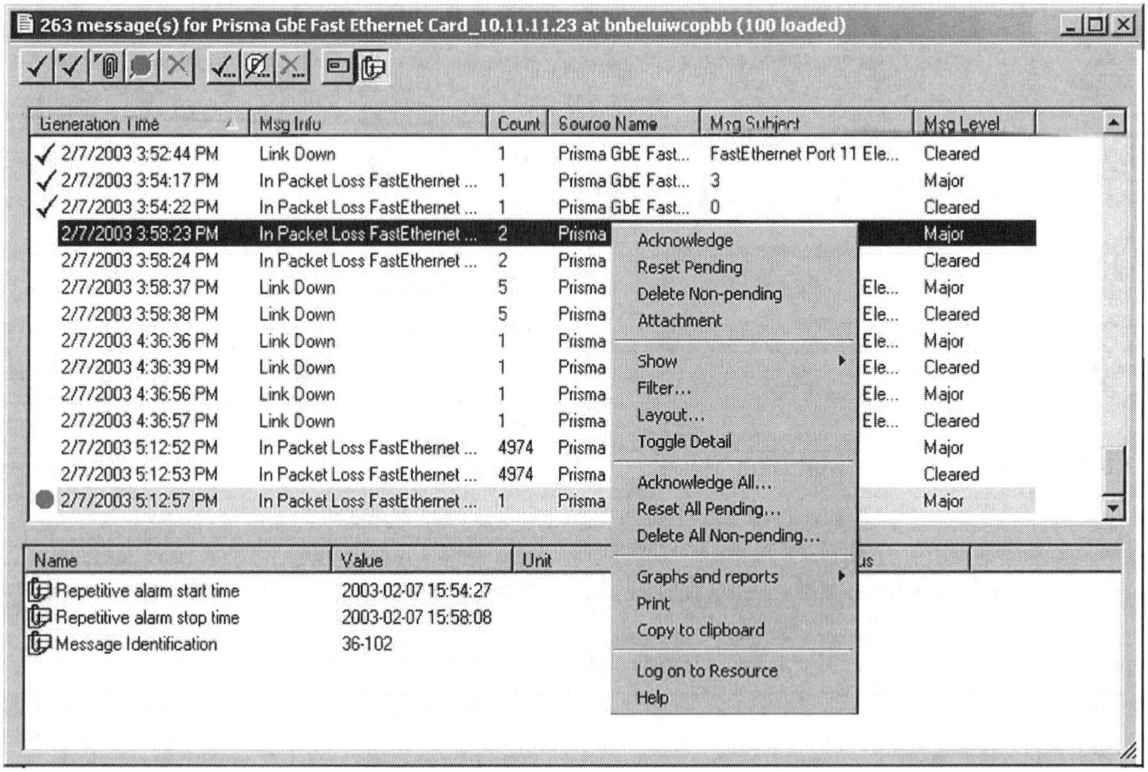

Figure 7.6.3 Typical alarm viewer.

Data communication equipment like IP routers and Gigabit Ethernet switches have built-in protocols that check the availability of communication links and can re-route signals automatically, without the intervention of a supervising management system. For video transport over classical SDI or ASI links, these in-band protocols are not available and the element manager can take responsibility of the protection switching and re-routing.

7.6.2.2 Configuration Management

Configuration management is all about configuring the network in order to deliver the services correctly. In the past, each device came with its own configuration utility, or needed an element manager to provide remote access to the configuration of the device. Today, most new devices are web-enabled and device-level configuration can be done without an element manager. Figure 7.6.5 shows an example of such a web-enabled MPEG-2 remultiplexer.

In many networks and applications, configuration can be done on a device-by-device basis. The more complex the services and the bigger the network, the more a network-wide configuration application is needed. In such a case, the configuration management has moved from the element management layer to the network and service management layers. These intelligent applications will derive the settings for each device from a higher-level service description. This does not mean that device-level configuration is no longer needed. In many cases, device configuration tools are still needed for troubleshooting reasons, since the higher-level applications may hide information that is available at the device level.

The configuration of DVB digital MPEG-2 services is a good example: some SI (Service Information) tables like the NIT and SDT contain information about other transport streams. This implies that the same information is configured in multiple devices. The NIT table generated by a multiplexer contains the frequencies of the COFDM, QAM or QPSK channels. The COFDM, QAM or QPSK modulator is also configured for these frequencies. If the frequency of a channel is changed, we need to make sure that the modulator and the multiplexer are reconfigured in a consistent way. A service/network level configuration utility will help to avoid a lot of configuration problems.

7.6.2.3 Performance Management

Performance management is proactive by nature. Where fault management focuses on reacting quickly to problems, performance management will try to predict problems in order to be able to avoid them. Performance management will achieve this by logging and analysing relevant parameters like bit error rate, temperature and signal-to-noise ratio in order to show the weakest spots in the network. Detecting performance degradation is another target.

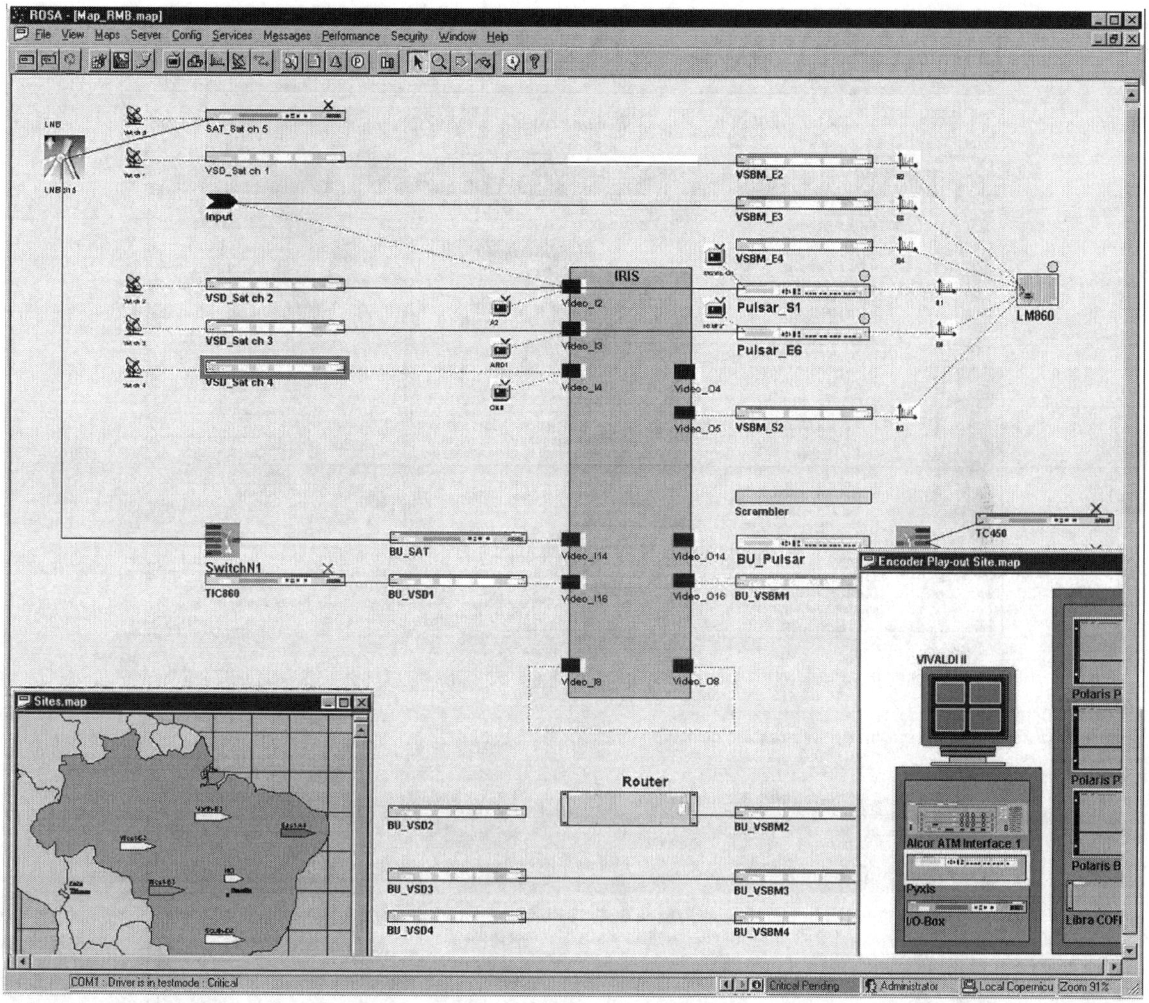

Figure 7.6.4 Multiple graphical views on the alarm state of the system.

Performance management not only prevents network outages, it can also guide the preventive maintenance technicians by helping them to focus on the weakest spots in the network.

A performance management subsystem will consist of two subcomponents: a performance data collection engine and an analysis/reporting tool. The performance data collection engine will periodically read the most relevant parameters from a set of devices and store them in a long-term performance log database. The polling and logging intervals may depend on the nature of the parameter being logged. Temperatures will fluctuate at a much slower rate than IP or MPEG traffic and can therefore be logged at a slow rate.

The analysis and reporting tools will plot the evolution of the parameters over time and assist in detecting performance degradation. If the performance of hundreds of devices or parameters is monitored, one cannot produce hundreds of graphs to analyse each of them individually. Good performance analysis tools will also be able to compare measurements against each other and show the weakest parts of the network.

In Figure 7.6.6 the measurements at different locations in the network for a certain period of time are compared against each other. One can immediately see locations with exceptionally good or bad performance. Once a problem area has been identified this way, one can look at the evolution over time of one or more parameters (Figure 7.6.7).

The application domain of performance monitoring is very broad. One can monitor the quality of transmission links, environmental parameters like temperature, data rates like IP traffic volume and power levels.

A performance monitoring application must not necessarily be part of the NMS system. If SNMP is used, a stand-alone application can be used to collect MIB variables from the devices directly. A good example is the MRTG application. Although originally meant for collecting IP traffic data, and visualising the evolution as a web page, it can be used for any

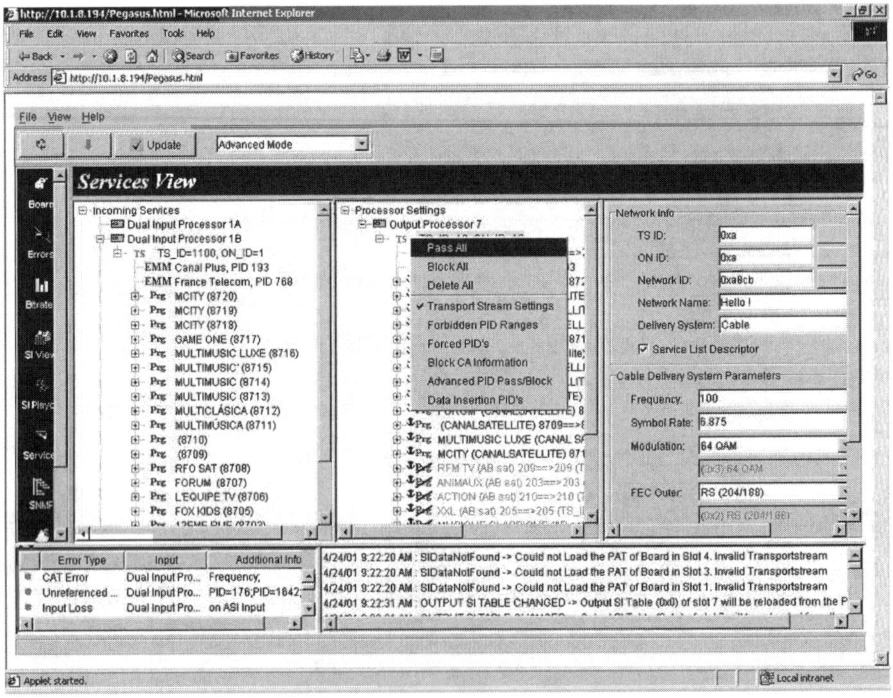

Figure 7.6.5 Configuration screen of a web-enabled device.

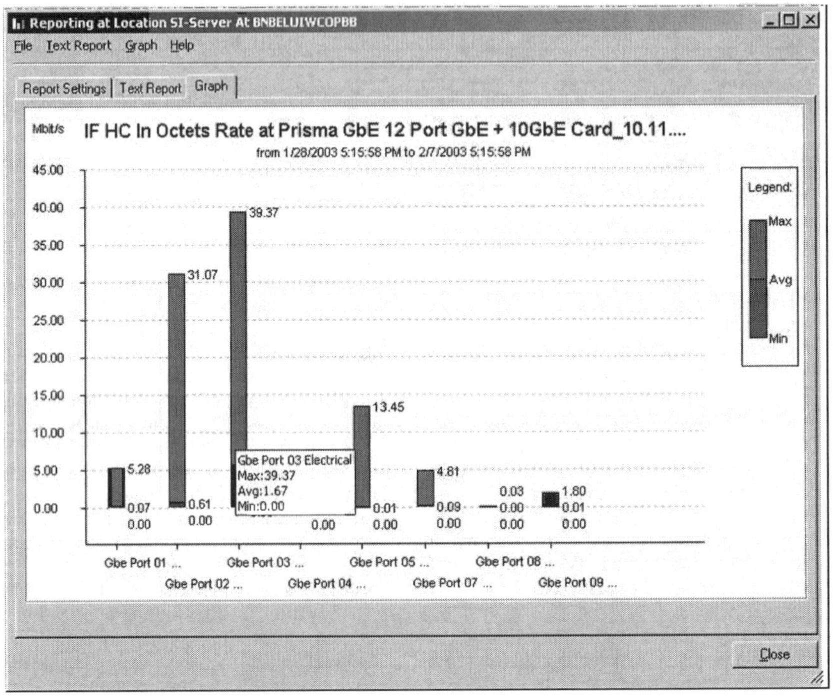

Figure 7.6.6 Comparison of traffic at different reference points.

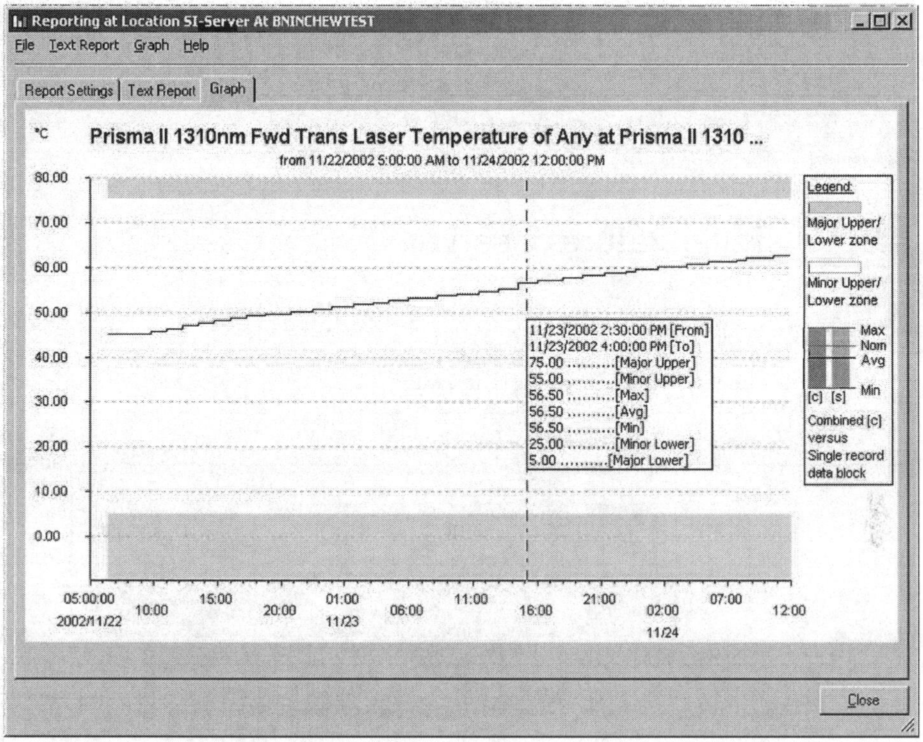

Figure 7.6.7 Evolution plot (example for laser temperature).

type of parameter that can be read from a Perl script. In Figure 7.6.8, the evolution of the temperature of a server is shown.

The MRTG tool is a freely available software package and runs on multiple platforms. Each MRTG page typically shows the evolution over the last day, week, month and year.

7.6.2.4 Security Management

Security management is about controlling the access to the network resources. This is a very broad domain that is not always under full control of the network management system. The conditional access system takes care of the security management for digital video delivery. This area is explained in much more detail in Chapter 2.16.

The network management system must control the access to the NMS system itself. Not every user of the NMS system has the same rights. Some users are video experts, others are data communication experts, others will have RF experience. The access rights to an NMS system depend on both functional and geographical criteria. A good management system will allow the administrator to specify access rights in terms of roles and regions.

7.6.2.5 Accounting Management

Although accounting management is one of the functional domains defined in the TMN reference model, the accounting and billing interfaces are often realised outside the scope of the network management system. In the broadcast world, the conditional access system can collect all session data and report this data to the billing system. This topic is beyond the scope of this book.

We can also look at the network monitoring domain from a slightly different angle: monitoring systems not only need to monitor the equipment, they also need to monitor the transmission links and the services (Figure 7.6.9).

At the lowest level, we need to monitor the status and have remote control over the devices in order to be able to react quickly. The next layer is the proactive performance monitoring layer that streamlines the preventive maintenance efforts by helping us to focus on the weakest parts of the network.

The last layer is the monitoring of the availability of equipment, transmission links and services. This layer provides us with the feedback that helps us evaluate whether we make progress from year to year and to provide our equipment manufacturers feedback on the reliability of their equipment.

7.6.3 Seven Golden Rules (or the Seven Most Commonly Made Mistakes)

1. Don't confuse network management systems with video processing or automation applications. Many software applications are built to make a broadcaster's life easier. Although some of these systems have built-in monitoring features, they only cover a subdomain (e.g. studio automation, scheduling,

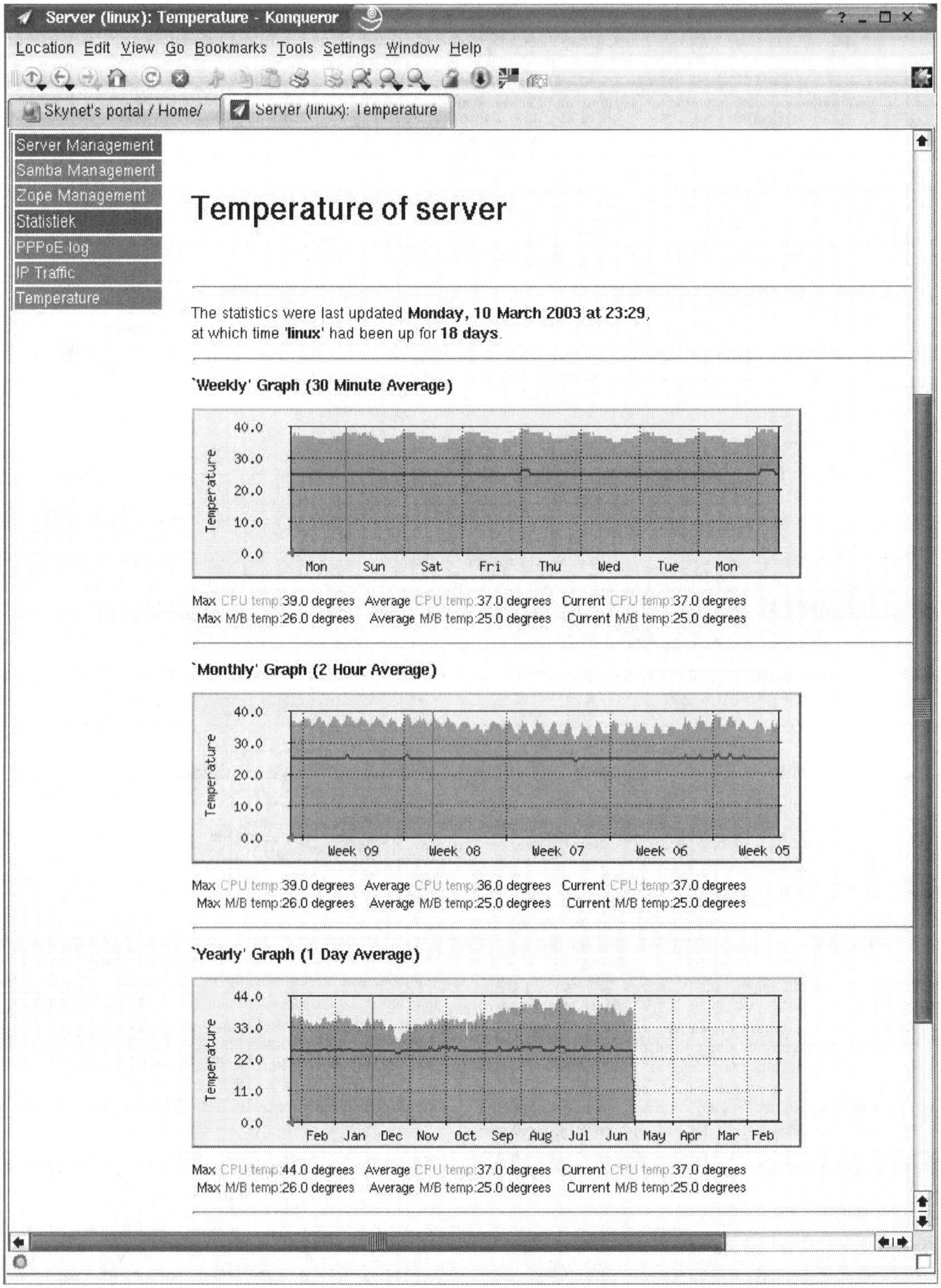

Figure 7.6.8 MRTG performance analysis tool.

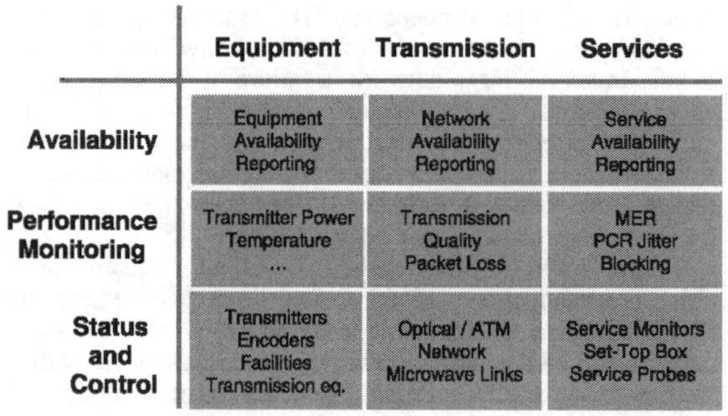
Equipment
Transmission
Services
Availability
Performance Monitoring
Status and Control
Equipment Availability Reporting
Network Availability Reporting
Service Availability Reporting
Transmitter Power
Temperature
...
Transmission Quality
Packet Loss
MER
PCR Jitter
Blocking
Transmitters
Encoders
Facilities
Transmission eq.
Optical / ATM Network
Microwave Links
Service Monitors
Set-Top Box
Service Probes

Figure 7.6.9 Monitoring of equipment, transmission links and services.

etc.) and give no end-to-end view over the network resources. These applications are not NMS systems but subcomponents that can be monitored by the NMS system. If such software applications perform mission critical functions, they are also network elements and should have a northbound management interface to the supervising management system. This northbound interface should provide the classical Fault, Configuration, Performance and Security management information.

2. Management systems need not be organised in a strictly hierarchical way. If an application manages a set of devices, and there is an overall management system supervising these applications, there is no need to duplicate the device management functions in the application's management interface (Figure 7.6.10). The supervising management system can interact both with the individual devices as well as with the applications that provide the added value. In such an architecture, the management system will only monitor/control the added value that these applications deliver and will simultaneously monitor the devices in order to detect device failures.

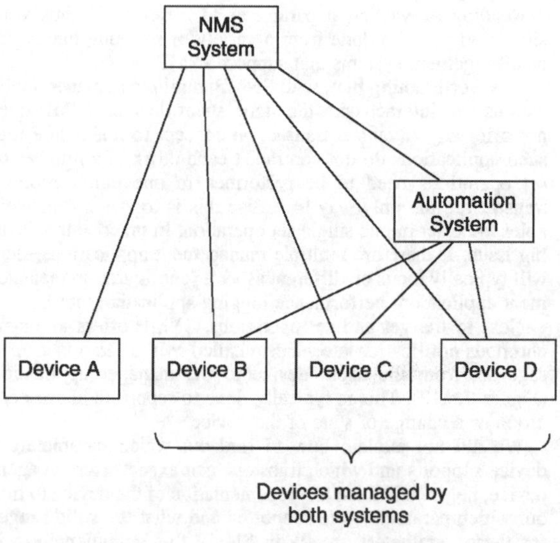

Figure 7.6.10 Non-hierarchically linked management systems.

3. Devices with proprietary management interfaces have a hidden cost. Building a management system can be very expensive if devices have a non-standardised or non-published management interface. Custom software development is needed to integrate such a device in the management system. Often, vendors of such devices will deliver an inexpensive and sometimes attractive control application that initially eliminates the need for a management system. Chances are high that, sooner or later, this system will need to be integrated in the overall management system or linked to another control system. If this need is only recognised after the equipment has been bought, vendors might charge a lot to create the necessary interfaces or claim that it is not possible.

4. Devices with proprietary management interfaces are likely to have a shorter lifetime. For the reasons mentioned above, the absence of open, published management interfaces may make the integration with another system impossible or extremely expensive. Buying a new device is often the less expensive solution.

5. Choose web-enabled devices over devices that come with a stand-alone configuration utility. Vendor-specific configuration applications are made for the current version of the PC operating system. There is often no guarantee that the software will run on a future version of the operating system. If a web-enabled device is chosen, integration into the supervising management system can be done with a minimal effort. The management system only needs to collect the alarms and launch a web browser whenever access to the device configuration is required.

6. A web-enabled GUI or a nice device configuration application does not eliminate the need for an open (SNMP or other) management interface. Remember that not only humans will interact with the device. Sooner or later automation scripts or higher-level applications will need to talk to the device.

7. Focus on your most urgent needs first. Each broadcaster or network operator has a different business environment and may have different service level agreements with partner telecom or broadcast companies.

7.6.4 Management and Monitoring of Satellite Networks

The satellite networks are the simplest networks to monitor from a technical point of view. Because they serve a huge amount of subscribers, one cannot afford to lose services at any time. For that reason, up-link stations are typically built in a 1 + 1 redundancy model. As an up-link station and satellite bandwidth are very expensive, one can afford to install extensive monitoring. Not only does the encoding/processing and transmission equipment need to be monitored, one will also want to monitor the MPEG services by means of dedicated MPEG analysers.

Next to the MPEG analysers, it is good practice to install a set of the most popular set-top boxes to decode the programs. This is needed to verify that each brand of set-top box interprets the service information correctly.

As satellite systems are one-way systems, they will often rely on in-band management for the configuration and update of the set-top box or receiver software. Most of the management complexity is handled by the subscriber management and conditional access systems, since these systems have to enable and manage the set-top boxes.

7.6.5 Management and Monitoring of Digital Terrestrial Broadcast Networks

Terrestrial networks are similar, but in addition to the playout centre, a network of transmitters needs to be managed. The transmitters receive their signals from telecom networks, microwave links or from other transmitters, and are not always owned by the broadcaster. In digital terrestrial networks, we can distinguish three roles: the transmitter network operator, the communication network operator and the broadcasters. Although one company can play more than one of these roles, it is good to design the management system according to these roles since, sooner or later, one broadcaster will be providing backbone or RF transmission bandwidth to another operator. Transmitter sites are often only connected through a one-way high-speed video transport link. Even though permanent bidirectional IP networks are the standard today, it is not always economically feasible to bring such a connection to a remote transmitter site on top of a mountain. For this reason, dial-up connections are still in use for the monitoring of these sites. A dedicated element manager that has been tuned to work with a dial-up connection can be used in such a site. Such an element manager will gather status information for the entire site and set up a dial-up connection to the central management server in case of problems. Once the connection is made, the alarm messages are transferred and the dial-up connection can be closed till a relevant status change occurs. Whenever the management system wants to change the configuration of the remote site, the management system will set up a dial-up connection to the remote element manager and perform the configuration changes.

The transmitter site infrastructure needs to be monitored also. Door contacts, tower lights, tamper switches, air-conditioning units and amplifiers need to be monitored. Many of these objects are not microprocessor controlled; they need to be monitored through contact closures. A general-purpose element manager will provide a set of contact closure inputs and outputs that can be accessed through the SNMP interface.

Next to the monitoring of the infrastructure, we also need to monitor the modulator and data communication equipment. It is highly recommended to install a terrestrial monitoring receiver to monitor the off-air signal quality and distortion.

For systems that are connected through dial-up connections, it is necessary to test the dial-up connection periodically in both directions. At least once a day, the element manager should set up a connection to the network management system and send a confirmation message indicating that the site is still alive. The management system needs to check for the arrival of these periodic messages and generate an alarm in case a message is missing. In the opposite direction, the management system periodically needs to set up a connection to the remote element managers in order to see if each site is still reachable.

7.6.6 Management and Monitoring of Cable Networks

It is already a challenge to manage terrestrial networks that can easily consist of dozens of active transmitter sites. Cable networks are even worse when it comes to monitoring a large number of devices. A typical cable network will consist of one or more central head-end sites, somewhere between five and 50 hubsites, each hubsite serving between 10 and 100 optical nodes. If only optical nodes and UPS systems are monitored, a network can easily consist of 10,000 active monitored elements. In such a large network, a failure in the head-end will cause thousands of network elements to generate an alarm. It is clear that simple alarm processing and visualisation is not sufficient here. Intelligent alarm correlation mechanisms will be needed to reduce the number of alarm messages and to show the root cause of the alarm storm.

The cable industry has recently succeeded in creating standards for management of the outside plant equipment. The SCTE HMS set of standards is based on SNMP, but uses a narrowband FSK modulation scheme to communicate with the fibre nodes and UPS systems. For inside plant equipment, the cable industry has adopted the SNMP standard.

7.6.7 Management Protocols: SNMP

The Simple Network Management Protocol (SNMP) was introduced around 1989 and has become the most popular management protocol in the Internet/data communication world and more recently in the video-processing world.

The 'simple' in SNMP's name stands for conceptual simplicity. This is slightly misleading, however, as the simple concept with its clear limitations calls for a wide range of creative solutions to overcome the limitations of SNMP.

The SNMP concept assumes that the network management application (called manager) manages a number of devices (called agents). In the simple SNMP world model, devices are represented by a set of parameters on which only a few elementary operations can be performed: *get* and *set* are the most important ones. The get and set operations specify the agent device by means of its IP address and the parameter(s) to be accessed is (are) specified by means of an OID (Object Identifier). For the time being, consider the OID to be the name of the parameter. One get or set request can specify more than one parameter. Upon receipt of a get operation, the device (agent) will send the answer to the manager by means of a response message (Figure 7.6.11).

Reading or writing a parameter to a device is thus very simple and can be done from a variety of scripting languages or management systems that support SNMP.

It is worth noting that, with SNMP, multiple manager applications can interact with the same agent device. SNMP does not offer any advanced transaction concept to make sure that such applications do not contradict each other. If a number of set operations need to be performed in one uninterruptible transaction, the only way to realise this is to specify all variables to be set in one single set operation. In practice this is no big issue. If there are multiple management applications, they will typically focus on different aspects (configuration management application, performance logging application, etc.).

Next to the get and set operations, SNMP offers an asynchronous notification mechanism called *trap*. A trap is a message set from the agent device to the manager application (Figure 7.6.12). This is typically done to report an alarm condition or a change of state of the device.

We did not explain how to find out which parameters a device supports and which traps one can expect from an agent device, i.e. we still need the documentation of the device to find out which parameters are supported and what the valid ranges for these parameters are. With SNMP this documentation is created in a formal (machine-readable) way: the Management

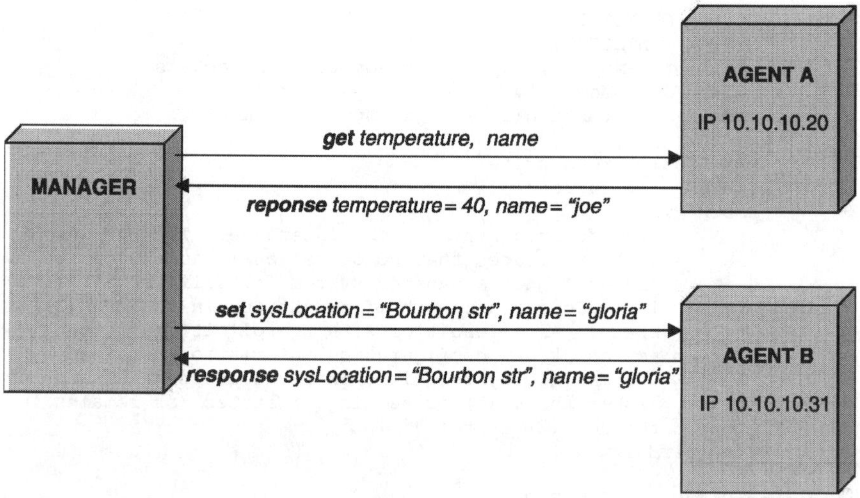

Figure 7.6.11 SNMP manager and agent.

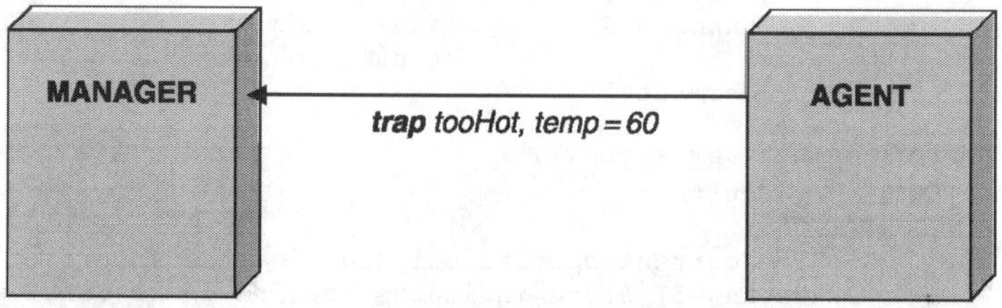

Figure 7.6.12 Trap notification message.

Information Base (MIB). The MIB is nothing more than documentation that is readable by both humans and computers. An example MIB extract is shown in Figure 7.6.13. This section shows definitions for the administrative status and operational status of a device interface. It also shows the definition of a trap that is generated whenever an interface link goes down.

The definitions in the MIB file define logical names for each object. For each object, the type and range are declared in the SYNTAX statement. The following basic types are supported: INTEGER, OCTET STRING, OBJECT IDENTIFIER and NULL. Derived types like NetworkAddress, IpAddress, Counter, Gauge and TimeTicks are also allowed. SNMPv2 extends this list with Integer32, Counter32, Gauge32, Unsigned32 and Counter64.

The ACCESS/MAX-ACCESS clause specifies the access rights (read-only, read-write, write-only or not-accessible). In SNMPv1, the STATUS field gives information on whether the object needs to be implemented or not to claim compliance with the MIB. Possible values can be mandatory, optional or obsolete. In SNMPv2, the meaning has changed and indicates whether the object is not obsolete or deprecated. Possible values are current, obsolete and deprecated.

The DESCRIPTION field is probably the most interesting field; it gives a textual description of the meaning of the variable and the allowed values.

At the end of each declaration, the OID (Object Identifier) of the object is specified. In the example MIB in Figure 7.6.13, the OID of ifAdminStatus has been declared to be { ifEntry 7 }. The symbolic name ifEntry has been defined earlier or in other MIBs and is an abbreviation of { iso org(3) dod(6) internet(1) mgmt(2) mib-2(1) interfaces(2) ifTable(2) ifEntry(1) } or in numeric format 1.3.6.1.2.1.2.2.1. SNMP uses a naming tree to identify objects in a unique way. Each vendor company can obtain a number in the enterprises subtree (Figure 7.6.14). All proprietary MIBs for a given vendor will be part of the subtree created under that number. The naming tree is a flexible way to let thousands of companies choose OIDs (or 'names') for their parameters without the risk of assigning duplicate numbers.

A device can implement multiple MIBs; this is a great feature from a standardisation and interoperability point of view. Typically, each management interface standard will have its own MIB file. A device that complies to several standards will implement the MIBs for these standards. In most cases, vendors will implement additional features in their devices on top of the feature list described in the standard MIBs. Such features will be implemented in vendor-specific MIBs. As SNMP devices usually have one or more Ethernet/IP ports, they will implement the basic standard MIBs for such devices. MIB-II (RFC 1213) or the more recent Interfaces MIB (RFC 2233), the IP forwarding table MIB (RFC 2096) and the SNMP

```
ifAdminStatus OBJECT-TYPE
      SYNTAX  INTEGER {
                 up(1),        -- ready to pass packets
                 down(2),
                 testing(3)       in some test mode
              }
      MAX-ACCESS  read-write
      STATUS      current
      DESCRIPTION
              "The desired state of the interface.  The testing(3)
              state indicates that no operational packets can be
              passed.  When a managed system initializes, all
              interfaces start with ifAdminStatus in the down(2)
              state.  As a result of either explicit management
              action or per configuration information retained by
              the managed system, ifAdminStatus is then changed to
              either the up(1) or testing(3) states (or remains in
              the down(2) state)."
      ::= { ifEntry 7 }

   ifOperStatus OBJECT-TYPE
      SYNTAX  INTEGER {
                 up(1),        -- ready to pass packets
                 down(2),
                 testing(3),   -- in some test mode
                 unknown(4),   -- status can not be determined
                               -- for some reason.
                 dormant(5)
              }
      MAX-ACCESS  read-only
      STATUS      current
      DESCRIPTION
              "The current operational state of the interface.  The
              testing(3) state indicates that no operational packets
              can be passed.  If ifAdminStatus is down(2) then
              ifOperStatus should be down(2).  If ifAdminStatus is
              changed to up(1) then ifOperStatus should change to
              up(1) if the interface is ready to transmit and
              receive network traffic; it should change to
              dormant(5) if the interface is waiting for external
              actions (such as a serial line waiting for an
              incoming connection); it should remain in the down(2)
              state if and only if there is a fault that prevents it
              from going to the up(1) state."
      ::= { ifEntry 8 }



-- definition of interface-related traps.

   linkDown NOTIFICATION-TYPE
      OBJECTS { ifIndex, ifAdminStatus, ifOperStatus }
      STATUS  current
      DESCRIPTION
              "A linkDown trap signifies that the SNMPv2 entity,
              acting in an agent role, has detected that the
              ifOperStatus object for one of its communication links
              is about to transition into the down state."
      ::= { snmpTraps 3 }
```

Figure 7.6.13 Example MIB entries and trap definition.

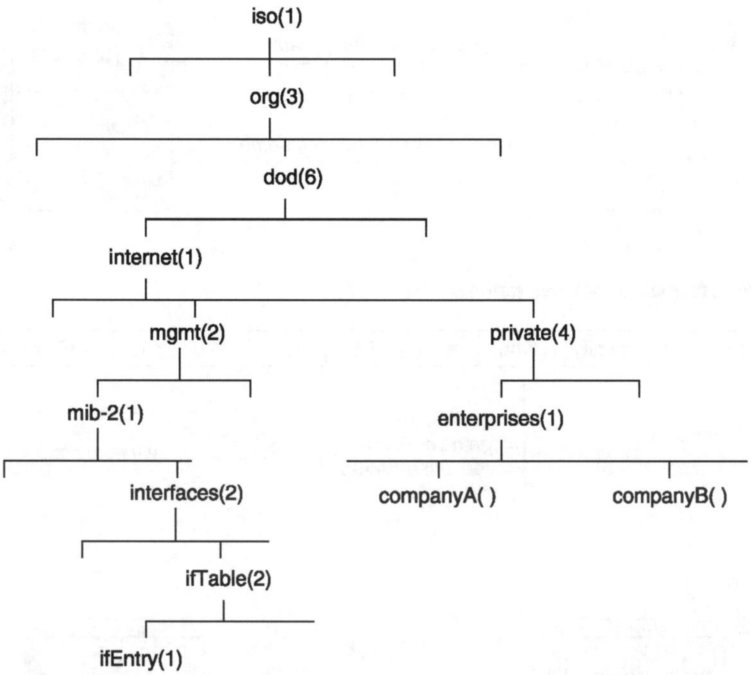

Figure 7.6.14 SNMP naming tree.

group MIB (RFC 1907) are good examples of MIBs that are implemented by a lot of devices. In the data communication domain, hundreds of MIBs have been defined as standards by the IETF (Internet Engineering Task Force) and other standardisation organisations. In the broadcast domain, most vendors have implemented proprietary MIBs for the management of their devices. Standardisation of management interfaces for video/audio transport devices is still limited. A good example is the ETSI TS 102 032 standard that defines the MIB for test and measurement devices in DVB systems.

Now that we understand the SNMP basics and object identification method, we can look at the structure of the SNMP messages in more detail for the different generations of the SNMP protocol.

7.6.7.1 The SNMPv1 Protocol

Figure 7.6.15 illustrates the structure of the SNMPv1 messages.

The version number is the version of the SNMP protocol and the value is 0 for SNMPv1.

The community string in the messages is a kind of password that describes the type of user accessing the device. For most devices, the community string 'public' gives read-only access to the device parameters. For write access, another string like 'private' or a vendor-specific password may be needed.

The request id (reqid) is a sequence number specified in the get, getnext and set operations. This number is returned in the response and helps the management system in associating a response with a request.

The ES (error status) field returns an error code in the response message. In the get, getnext or set request, the value must be set to zero. Valid values in the response message are: noError(0), tooBig(1), noSuchName(2), badValue(3), readOnly(4) and genErr(5).

The EI (error index) field gives an indication of the first variable that was in error. The first variable is identified as 1.

At the end of each message, we find the varbinds; varbinds are name value pairs where the name is represented by the OID encoded in a typical way. The value can have the following types in SNMPv1: INTEGER, OCTET STRING, OBJECT IDENTIFIER or NULL. A more detailed description of the encoding of the varbinds is beyond the scope of this book.

The *ent* (enterprise) field in the trap message identifies the device that generated the trap. This is the same information as found under iso.org.dod.internet.mgmt.mib-2.system.sysObjectID.

The *addr* field in the trap message contains the IP address of the agent device that sends the trap.

The G and S fields identify the type of the trap: Generic or Specific. The following generic trap types are defined: coldStart(0), warmStart(1), linkDown(2), linkup(3), authenticationFailure(4), egpNeighborLoss(5). Enterprise-specific traps are identified by a G field equal to 6 and a vendor-specific trap code in the S field.

The *time* field indicates the time of generation of the trap. This is the value of the sysUpTime (see MIB-2) parameter when the trap was generated. SysUpTime does not give an absolute time value but a time-stamp relative to the start up of the agent device. SysUpTime is expressed in hundredths of seconds.

Trap messages also contain variable bindings. These are (OID, value) pairs that provide additional information related to the trap message. The trap definition in the MIB defines the type of the varbinds (see Figure 7.6.13).

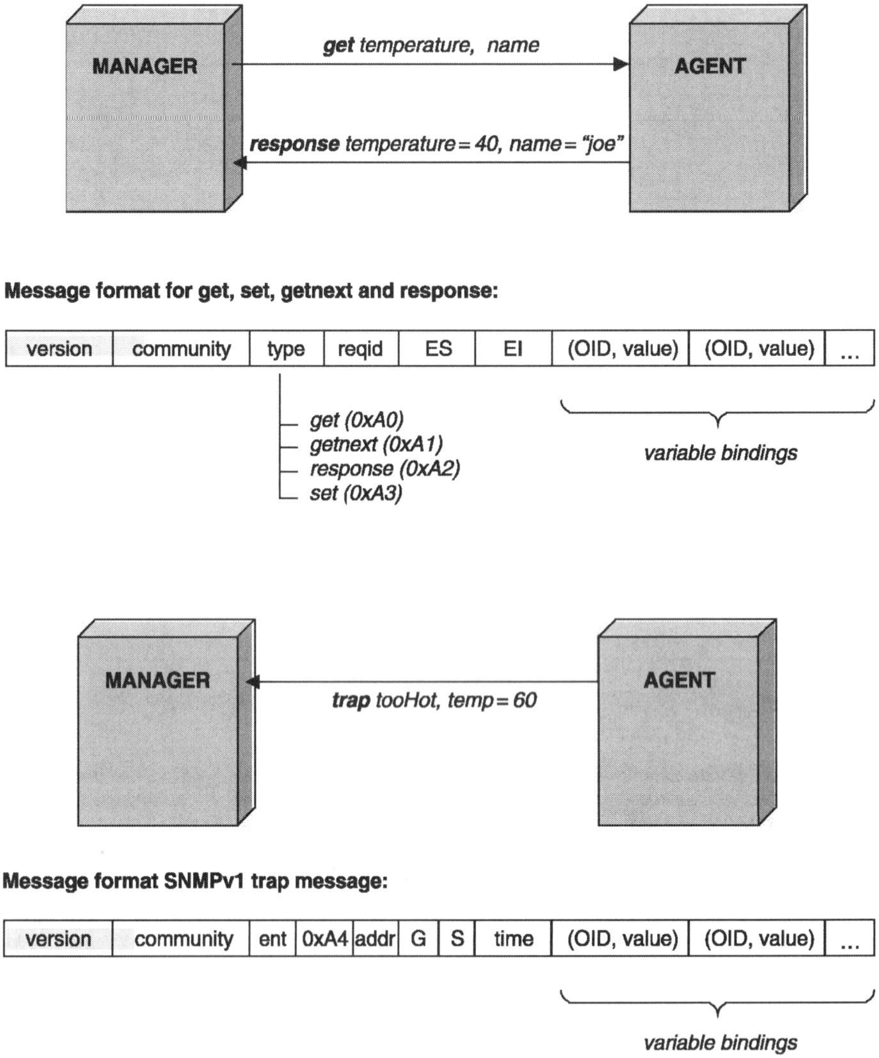

Figure 7.6.15 SNMPv1 messages.

7.6.7.2 SNMP Tables

The SNMP get and set operations allow us to access individual device parameters. In many cases, devices have parameters that are structured in tabular format. In this section, we will show that the basic get and set operations can also be used to access tables.

Devices with multiple input and output ports will often have table declarations in their MIB. For each port, a table row will be created; for each port parameter, a table column is created. Although tables are conceptually simple, a lot of complicated data structures can be represented by a set of interrelated tables. Relational databases prove every day that almost everything can be expressed in terms of tables, as long as good indices/keys are used to make associations between these tables.

In the example table declaration of Figure 7.6.16, we see that a table declaration consists of statements that declare the table and a set of normal object declarations that apply to the values found in each of the columns of the table. Accessing the ifIndex and ifDescr parameters from the first row of the table can be done by means of:

snmpget 10.10.10.20 public ifIndex.0 ifDescr.0

Accessing the same parameters from the second row of the table is done with:

snmpget 10.10.10.20 public ifIndex.1 ifDescr.1

The cells of the table are accessed by extending the OID with the index in the table. To make table access easier, SNMP has a *getnext* message that will automatically access the variable in the next row of the table. In case we have reached the end of the table, *getnext* will return the next object in the MIB. The

```
-- Sample table declaration

ifTable OBJECT-TYPE
    SYNTAX  SEQUENCE OF IfEntry
    ACCESS  not-accessible
    STATUS  mandatory
    DESCRIPTION
            "A list of interface entries.  The number of
            entries is given by the value of ifNumber."
    ::= { interfaces 2 }

ifEntry OBJECT-TYPE
    SYNTAX  IfEntry
    ACCESS  not-accessible
    STATUS  mandatory
    DESCRIPTION
            "An interface entry containing objects at the
            subnetwork layer and below for a particular
            interface."
    INDEX   { ifIndex }
    ::= { ifTable 1 }

IfEntry ::=
    SEQUENCE {
        ifIndex
            INTEGER,
        ifDescr
            DisplayString,
        ifOutErrors
            Counter,
        ifOutQLen
            Gauge,
        ifSpecific
            OBJECT IDENTIFIER
    }

ifIndex OBJECT-TYPE
    SYNTAX  INTEGER
    ACCESS  read-only
    STATUS  mandatory
    DESCRIPTION
            "A unique value for each interface.  Its value
            ranges between 1 and the value of ifNumber.  The
            value for each interface must remain constant at
            least from one re-initialization of the entity's
            network management system to the next re-
            initialization."
    ::= { ifEntry 1 }

ifDescr OBJECT-TYPE
    SYNTAX  DisplayString (SIZE (0..255))
    ACCESS  read-only
    STATUS  mandatory
    DESCRIPTION
            "A textual string containing information about the
            interface.  This string should include the name of
            the manufacturer, the product name and the version
            of the hardware interface."
    ::= { ifEntry 2 }
```

Figure 7.6.16 Sample table definition.

getnext command is very useful to traverse tables or MIBs in a flexible way.

SNMP tables need not be completely filled. If there is a hole in a table, *getnext* will return the value in the first row that is not empty following the specified cell.

In the example given above, the table only contains one index named *ifIndex*. Tables can have multiple indexes in SNMP. Tables cannot be nested, however. For newcomers, this may look like a limitation, but SNMP realises complex structures by means of references into additional tables. This is very similar to the relational database world, where everything needs to be modelled as tables that contain references into other tables.

In order to understand the behaviour of the SNMP operations and the SNMP tables, it is highly recommended to visit websites that are dedicated to SNMP. Some of them have interesting on-line demos, tutorials and exercises. The SimpleWeb website is an example of such a site, with very good on-line exercises.

7.6.7.3 Limitations of SNMPv1

Much of the success of SNMP is related to the fact that one does not have to be a software engineer to access device parameters or to log performance data periodically. The simplicity of SNMP makes integration of SNMP devices in a management system very easy compared to devices that are managed through other protocols.

There are, however, some disadvantages and limitations:

- SNMPv1 has a very poor security mechanism. SNMP messages are sent over the network in an unprotected way. Anyone with a network sniffer can observe SNMP traffic and send malicious commands to the devices. In practice, this has not been too big an issue for most organisations, since the internal management network is not accessible from the Internet. Only insiders can make misuse of the weak SNMP security. Someone inside the building can always damage the device; it does not make much difference whether it is done through SNMP or with a hammer.
- SNMP messages are sent with the unreliable UDP protocol. With UDP, one is never sure that a message has arrived. This is not a big issue for the set and get operations, since the agent will return a response to these messages. Trap messages have no response. One is never sure that the management system has received the trap message. To overcome this problem, agent devices often implement a table of the most recent alarms that have been sent. The management system will then periodically check whether it has not missed any alarms. Unfortunately, each vendor takes a different approach to synchronise the alarm logs.
- A management system should not only rely on traps for reporting the alarm state. In case the communication link is lost, no trap will arrive. For this reason, management systems should also poll the devices periodically to check if the communication links are still up and if the devices are still alive. Some management applications prefer not to depend on traps at all and will poll devices periodically in order to check for their alarm state. This is a very simple and reliable method that can work in small networks. This approach is not recommended in larger networks or networks with low-speed communication links.
- SNMP has no transaction concept. If multiple management applications interact with the SNMP device, these systems have no way to allocate the device until a transaction that consists of multiple sets is completed. SNMP can only provide simple transactions where all the parameters to be modified can be specified in one set message. In practice, this is not a big problem since management systems have client–server architecture. Only the server will talk with the SNMP device, even if there are many clients attached. Even if there are multiple servers interacting with a device, they will focus on different functions and will not interfere with each other (alarm collection system, configuration management system and performance data collection systems do not interfere with each other).
- SNMP can generate a lot of data traffic when large tables are transferred between agent and management system. The response times can become slow, especially when low-speed data communication links are used or when an agent device has a slow processor. One needs to be very careful not to overload the processor of an agent device.
- The identification of trap messages is not well defined; there is no mechanism to ensure that two vendors do not use the same trap code for different types of alarms.
- Most devices have a set of parameters in common. Although there are some standardised MIBs like MIB-2 that define how these parameters need to be implemented, there are a lot of areas that are not standardised and where vendors use their own proprietary methods. Examples: the way in which trap destinations are declared is not standardised; the table with the most recent alarms generated by the device is not standardised. These limitations are not inherent to the protocol. The industry simply did not succeed in creating standard MIBs for these domains.

7.6.7.4 SNMPv2

Version 2 of the SNMP protocol only provides small changes/additions to the protocol. The trap message format is changed (simplified) and has the same structure as the get, set, getnext and response messages.

To solve the problem of the unreliable trap delivery, SNMPv2 introduces the *inform* message. The inform message has the same structure and meaning as a trap, but the message will be retransmitted periodically until the management system sends a response to the inform message. The response to an inform message is normally a copy of the message sent back from the manager to the agent. In case the message is very long, a *response* message with error code *tooBig* is also acceptable.

In addition to the modified trap structure, SNMPv2 also introduces the *getbulk* message that will return a large number of variables from a table (see Figure 7.6.17).

In SNMPv2, the version field is set to 1.

For most commands, the N1 and N2 fields are 0, except for the response message and for the getbulk command. In the getbulk command, N1 is the number of varbinds that is only queried once. The remainder of the varbinds is queried N2 times. For table access, N1 will typically be 0 and N2 will be the number of rows to be retrieved. The response to a getbulk message can contain less varbinds than expected.

In SNMPv2 a number of parameters have disappeared from the trap message. The information is now transmitted in the varbinds. The first varbind is sysUpTime, the second varbind (sysTrapOID) identifies the trap type by means of its OID. SNMPv2 also extends and modifies the data types and keywords that can be used in the MIB declarations. A detailed description is beyond the scope of this book.

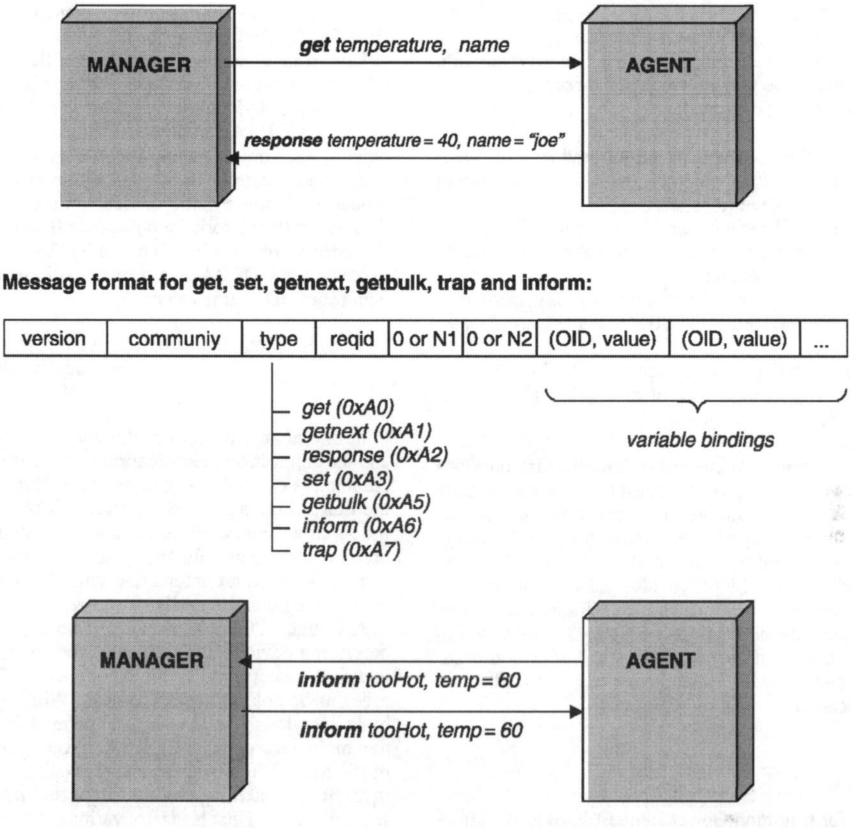

Figure 7.6.17 SNMPv2 messages.

7.6.7.5 Limitations of SNMPv2

In SNMPv2, vendors can now identify their trap messages in a unique way. Although not often used, the inform message offers a way to make sure that traps arrive at their destination. The getbulk message reduces the data communication overhead when retrieving large tables.

SNMPv2 offers a solution for some of the limitations of SNMPv1. The most important limitations of SNMPv1 remain unsolved, however. This is probably the reason why there are still a lot of SNMPv1 implementations around.

7.6.7.6 SNMPv3

SNMPv3 is relatively new, and unlike SNMPv2, this new standard tries to resolve most of the problems with previous versions.

The main innovations are the secure access to devices and a set of standard ways to handle common network management configuration tasks.

SNMPv3 does not want to modify the concept of SNMP; SNMPv3 is just like its predecessors, a simple mechanism to access device parameters and tables. The meaning of the basic messages get, set, getnext, getbulk, trap and inform has not changed. The message format has changed, however, mainly due to the new security models.

In SNMPv3, the message header has been extended to include authentication and security-related information. Furthermore, the message can be encrypted to make sure that no one can eavesdrop on the network. Secure communication protocols (like all security systems) continuously need to run security checks in the following subdomains:

- Authentication: is the user who he or she claims to be?
- Authorisation: is the user allowed to perform this action?
- Auditing: what has the user been doing in the last year?
- Check for tampering: are we sure that the messages/documents or their time-stamps have not been tampered with?
- Privacy: is the information protected against eavesdroppers?

SNMPv3 can perform most of these security checks; it does not provide direct support for generating audit trails, however.

Authentication and encryption (privacy) are not mandatory; it is up to the user to decide what level of security is desired. It is possible to use no authentication and no encryption, authentication without encryption as well as authentication with encryption.

The User-based Security Model (USM) and the USM-MIB define the table of users that are allowed to interact with the system.

The USM model does not define the access rights each user has. This is the task of the View-based Access Model (VACM). The VACM defines several views on the managed objects. Each view will give read or write access to a different subset of parameters. Several tables are consulted to determine whether a variable is accessible (Figure 7.6.18).

The parameters that are consulted during the process are part of or derived from the SNMPv3 message header.

In addition to the enhancements in the security domain, SNMPv3 standardises the way in which some common configuration tasks need to be performed:

- Configuration of the management target devices: the Management Target MIB defines the table with the management systems to which traps need to be sent.
- Configuration of notifications: the Notification MIB defines filtering rules that will determine which alarms are sent to which management applications.
- Configuration of proxy forwarding rules. A proxy agent is an intermediate agent that provides SNMP access to devices behind the proxy.
- Configuration of the security system: the USM MIB and VACM MIB define a generic way of examining and configuring the access rights to a device.

The SNMPv3 system provides most features that previous SNMP versions were lacking. The biggest drawback is that the security model is not that simple. This can make the access to the devices for non-experienced users quite challenging. Furthermore, a faulty configuration of the VACM tables can create big security holes. SNMPv3 also asks for more processing power in the agent device, leading to a higher device cost.

For these reasons, the adaptation of SNMPv3 is expected to be slow, except in those domains where standardisation organisations have made SNMPv3 mandatory (e.g. DOCSIS data over cable applications).

7.6.7.7 CORBA

CORBA stands for Common Object Request Broker Architecture. Its original intention was not to create a new management protocol but to create a software bus structure through which objects can communicate with each other. Objects can reside on different machines (servers) and for this reason CORBA can be seen as a distributed object model.

CORBA can be used for different types of applications, of which network management is just one good example. Almost all client–server applications need a CORBA-like object-to-object communication.

Just like SNMP uses MIB files to describe the parameters of a device, CORBA also uses a formal definition of the interfaces of a device. This definition is expressed in the IDL language (Interface Definition Language).

Due to its general-purpose design, the expressive power of CORBA is almost unlimited. Where SNMP restricts itself to controlling individual device parameters and parameters stored in tables, CORBA can describe and access webs of interrelated objects, each object having its own methods.

A good example to illustrate this is a connection set-up sequence. Assume that a device can set up video connections through an IP network from a source IP address to a destination IP address with a selectable quality (bandwidth). This can be expressed in CORBA as a method with three parameters and a connection ID return value:

```
integer SetupConnection(  in address source,
                          in address destination,
                          in integer bandwidth)
```

In one atomic operation, the connection can be established and the connection identification number is returned. In SNMP, parameters can only be grouped in a table row and as soon as the transaction is more complicated, it needs to be realised by means of several separate get and set operations. This is not a problem as long as only one system is creating connections, but if many systems are interacting with the device, SNMP cannot resolve the conflicts easily.

Although CORBA provides much more flexibility in describing device interfaces and data structures, this flexibility makes it necessary to be an experienced software developer in order to be able to access devices. With SNMP, anyone with basic knowledge could examine and modify parameters. This is the main reason why CORBA never gained the popularity of SNMP. The powerful expression mechanisms available in CORBA make it possible to express the same interface in several ways. This leads to various styles of IDL interface definitions. In contrast to this, SNMP's simplicity forces users to adopt the same interface style for all devices, which greatly improves the readability and usability of SNMP MIBs.

The choice of CORBA over SNMP is only justified in devices or subsystems that handle complicated data structures that represent interconnected objects.

7.6.7.8 XML-Based Management Interfaces

Management systems and devices often need to exchange large and complicated data structures. This can be done with most of

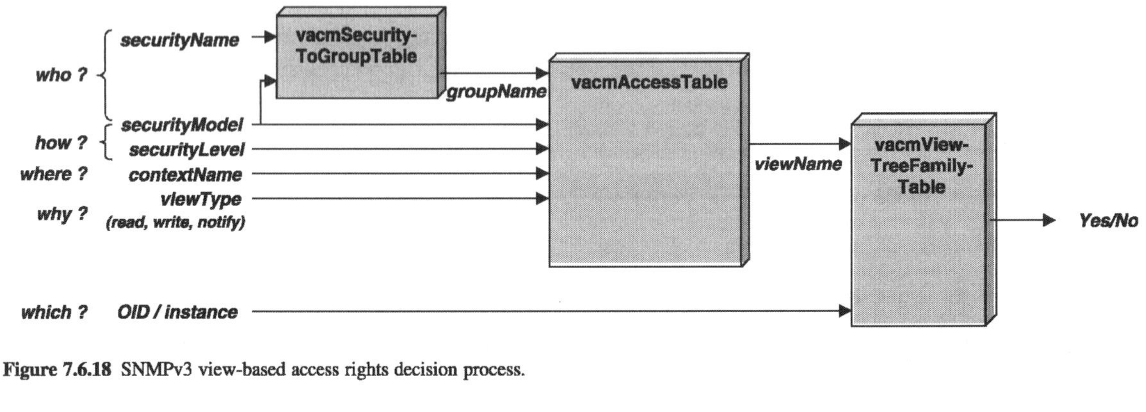

Figure 7.6.18 SNMPv3 view-based access rights decision process.

the management protocols we have discussed. Such large data structures have to be transferred in several pieces. Many messages in both directions are needed to transfer a large data structure from a device to a management system. Furthermore, the data communication link can be interrupted during the transfer, which can lead to inconsistent configurations.

For many applications, file transfer is the simplest and safest way to download and upload an entire configuration or event log.

XML (eXtensible Markup Language) is not a management protocol. It is just a standardised language used to express data structures in a platform-independent and readable format. XML formatted data can be transferred between devices using several standardised protocols like FTP, HTTP and SOAP.

XML is a very simple language, consisting only of a couple of notions: the *XML declaration* header and *elements* and *attributes*.

In Figure 7.6.19, one *channels* element contains several *channel* elements that represent the channel line-up of a digital cable TV system. The *channel* element contains several other elements that describe the parameters of each channel. Please note that not all elements need to contain the same sub-elements: the *fromAstra* element also contains a *status* element that is not found in the other channel declarations. This is allowed because of the extensible nature of XML. One can always add information to an XML structure without disturbing the applications that rely on the old structure.

We also find attribute declarations like the *created* and *id* attributes. Attributes are much more restricted in their expressive power than elements: they cannot contain multiple sub-elements, they are not expandable and they are more difficult to handle from within software applications. It is better to use attributes only to express static attributes of elements or metadata.

In XML, names are case sensitive and cannot contain spaces. They must also not start with a number.

In addition to the XML data, XML has a Document Type Definition (DTD) or an XML schema definition. This is a declaration of the elements one expects to find in an XML file. The example in Figure 7.6.19 refers to a DTD definition stored in the file '*channels.dtd*'. This definition is shown in Figure 7.6.20.

The Document Type Definition declares the structure of each element. Elements can consist of data or sequences of other elements. Special characters like +, ? and * specify whether an element can be repeated or whether it is optional. The + sign specifies one or multiple occurrences. The ? sign specifies zero or one occurrences. The * sign specifies 0 or more occurrences of an element.

The DTD declaration also defines which attributes are required for each element type.

7.6.7.8.1 XML Parsing

Much of the power of XML lies in the ease by which XML files can be parsed or transformed into other formats. Let us look at reading XML files first (parsing). Two parsing models are currently in use: DOM (Document Object Model) and SAX (Simple API for XML). The DOM model will virtually read the entire XML structure in memory and represent it as a tree to the application developer. The SAX model passes a window over the XML input file and launches an event for each relevant entry it encounters. Both models have their value. As a very simplistic recommendation, one could state that for very large

```
<?xml version="1.0" encoding="ISO-8859-1"?>
<!DOCTYPE food SYSTEM "channels.dtd">
<!-- This is a comment - channel lineup table  -->
<channels>
  <channel created="11/12/2003" id="12">
    <name>localfeed</name>
    <frequency>260000000</frequency>
    <constellation const="qam64"></constellation>
    <baudrate>6000000</baudrate>
    <class>statmuxed</class>
  </channel>
  <channel created="11/12/2003" id="12">
    <name>fromAstra</name>
    <frequency>268000000</frequency>
    <constellation const="qam64"></constellation>
    <baudrate>6000000</baudrate>
    <class>statmuxed</class>
    <status>not yet in use</status>
  </channel>
  <channel id="12">
    <name>vodstream</name>
    <frequency>40000000</frequency>
    <constellation const="qam64"></constellation>
    <baudrate>6000000</baudrate>
    <class>fixed</class>
  </channel>
</channels>
```

Figure 7.6.19 Sample XML file.

```
<!DOCTYPE channels [

<!ELEMENT channels (channel+)>
<!ELEMENT channel (name, frequency, constellation, baudrate, class, status?)>
<!ELEMENT name (#PCDATA)>
<!ELEMENT frequency (#PCDATA)>
<!ELEMENT baudrate (#PCDATA)>
<!ELEMENT status (#PCDATA)>

<!ATTLIST channel
          id CDATA #REQUIRED
          created CDATA #IMPLIED>
<!ATTLIST constellation (qam16, qam64, qam256) #REQUIRED>

]>
```

Figure 7.6.20 Sample DTD file.

data files with a less complicated structure, SAX is probably the best approach. For smaller or more complicated data structures, DOM might be the best approach. Much will, however, depend on the nature of the application.

7.6.7.8.2 XML Translation with XSLT

The XSLT (eXtensible Stylesheet Language Transformations) framework is a general-purpose tool for translating XML files into other formats. The most popular application of XSLT is the transformation of XML data into HTML web pages.

The XSLT processor will use the XLS file to translate the XML data to the target output format (Figure 7.6.21). Most new web browsers are capable of reading XML files and converting them on the fly to HTML web pages using an XSL transformation specification.

XLST also provides the necessary tools to separate the data from the graphical representation of the data. A detailed description of the XML language and transformation engine is beyond the scope of this book.

7.6.8 Automation and Customisation with Scripting Languages

Many state-of-the-art management systems can be used out of the box for basic fault management, configuration management and performance management. Networks evolve, however, and one has to be careful when choosing a network management system. If the network management system can only be customised or extended by the management system vendor, the maintenance cost of the system can become very high.

When choosing a management system, it is important to choose a system with open interfaces so that the end-user or a consulting company can customise the system or create additional device drivers. Systems that can be extended through scripting languages are preferred over systems where traditional compiled C or C++ development is needed. Scripts can be added to such a system without extensive software development cycles and without the need for an in-depth software development background.

When choosing devices, it is equally important to choose devices with an open, standardised and documented management interface so that the devices can be reached easily from within a scripting language.

Several scripting languages are available, but some of them have gained our preference because of their open nature (open source) and their platform independence. Python and Perl are good examples of such languages. They have interface modules to all popular management protocols like SNMP and are very well linked to XML and CORBA. For newcomers, Python is probably the simplest language to learn and is at least as powerful as other scripting languages.

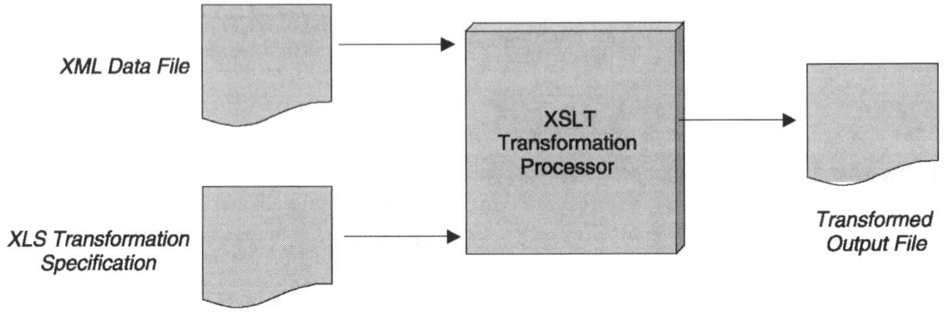

Figure 7.6.21 XSLT transformation process.

Scripting languages can also be used for quick development of automation tasks. These automation tasks can often be realised outside the network management system, in a standalone control system or within an element management box.

Bibliography

Deitel, H., Deitel, P. and Nieto, T. *Internet & World Wide Web: How To Program*, Prentice-Hall (2001).

ETSI TS 102 032, Digital Video Broadcasting: SNMP MIB for Test and Measurement Applications in DVB Systems, http://www.etsi.org

Fehily, C. *Python Visual Quickstart Guide*, Peachpit Press (2002).

ITU-T M.3010, Principles for a Telecommunications Management Network, rev. 2 (1996).

Jones, C. and Drake, F. *Python & XML*, O'Reilly & Associates (2002).

Oetiker, T. *MRTG Multi Router Traffic Grapher*, http://www.mrtg.org

Orfali, R., Harkey, D. and Edwards, J. *Instant CORBA*, John Wiley & Sons (1997).

Perkins, D. and McGinnis, E. *Understanding SNMP MIBs*, Prentice-Hall (1996).

Simple Times, SNMP newsletter, http://www.simple-times.org

SimpleWeb, http://www.simpleweb.org

W3 Schools, XML web tutorial, http://www.w3schools.com

Zeltsermann, D. *A Practical Guide to SNMPv3*, Prentice-Hall PTR (1999).

Glossary

4:2:0	Digital video coding method where chrominance has half the vertical and half the horizontal resolution of the video.
4:2:2	Digital video coding method where chrominance has the same vertical resolution and half the horizontal resolution of the video.
8-PSK	Eight Phase Shift Key modulation.
8-VSB	Vestigial sideband modulation with eight discrete amplitude levels.
16-VSB	Vestigial sideband modulation with 16 discrete amplitude levels.
AAF	Advanced Authoring Format. A format for authoring and editing of digital audiovisual content.
ABU	Asian Broadcasting Union.
ADC	Analogue-to-Digital Converter.
AES	Audio Engineering Society.
AGC	Automatic Gain Control.
AL	Alignment Level. A reference level, usually recorded as a header to a recording, and used to check the alignment of recorded signals and signal paths.
Algorithm	A mathematical function which, for the purposes of a broadcast system and conditional access system, is applied to data to produce a specific result. For example, encoding, compressing and encrypting video, audio and data streams.
AM	Amplitude Modulation.
AMOL	Automated Measurement Of Line-ups. System used in the USA for programme and advert source identification and audience measurement.
ANC	ANCillary data.
Anchor frame	In MPEG-2, a video frame that is used for prediction. I-frames and P-frames are generally used as anchor frames, but B-frames are never anchor frames.
Antiope	French teletext standard.
ARP	Address Resolution Protocol. Protocol for mapping MAC addresses to IP (and other) addresses.
ARQ	Automatic Repeat Request.
ASBU	Arab States Broadcasting Union.
ASCII	American Standard Code for Information Interchange.
ASI	Asynchronous Serial Interface.
AT	Asynchronous Transfer Mode. A broadband ISDN transmission system capable of carrying any type of digital signal or service.
ATM	Asynchronous Transfer Mode. A digital signal protocol for efficient transport of both constant-rate and bursty information in broadband digital networks. The ATM digital stream consists of fixed-length packets called 'cells', each containing 53 8-bit bytes – a 5-byte header and a 48-byte information payload.
A-to-D	Audio-to-Digital conversion. Conversion of an analogue signal to a stream of numbers (usually binary).
ATSC	Advanced Television Systems Committee.
ATVEF	Advanced Television Enhancement Forum.
AWGN	Additive White Gaussian Noise.
Baud rate	Rate of digital data transfer.
BBC	British Broadcasting Corporation. Originally a licence fee funded terrestrial TV and radio broadcaster. Now broadcasting on digital satellite, digital cable and analogue and digital terrestrial.
BCD	Binary Coded Decimal. A method of counting decimal numbers as a set of discrete 4-bit binary values, 0000–1001 (0–9).
BCH codes	Bose–Chaudhuri–Hocquenghem codes.
BER	Bit Error Ratio/Rate.
Bidirectional pictures (or B-pictures or B-frames)	In MPEG-2, pictures that use both future and past pictures as a reference. This technique is termed bidirectional prediction. B-pictures provide the most compression. B-pictures do not propagate coding errors as they are never used as a reference.
BISS	Basic Interoperable Scrambling System.
BISS-E	Basic Interoperable Scrambling System with Encrypted key.
Bit	Binary digit. The elementary unit on computers. It has two states: 0 and 1.

Bit rate	The rate at which the compressed bit stream is delivered from the channel to the input of a decoder.
Bit stream	The transfer of digital data (bits) across networks. The unit for bit streams is bit/s. A million bits/s are defined as megabits/second (Mbit/s), a billion bits/second as gigabits/second (Gbit/s).
Block	An MPEG block is an 8-by-8 array of pel values or DCT coefficients representing luminance or chrominance information.
BPSK	Bi-Phase Shift Key modulation.
BSDP	Broadcaster Service Data Packet. Teletext packet used by PDC.
Buffer	Data store for smoothing data transfer.
Byte	Basic binary storage unit on computers, consisting of eight binary digits (bits). Storage capacities are usually measured in kilobytes (1024 bytes) or megabytes.
Byte-aligned	A bit in a coded bit stream is byte-aligned if its position is a multiple of 8 bits from the first bit in the stream.
CA	Conditional Access.
C band	Approximately 4–6 GHz frequency band.
CBR	Constant Bit Rate. A mode of operation in a digital system where the bit rate is constant from start to finish of the compressed bit stream.
CBS	Columbia Broadcast System (one of the big US radio and television networks).
CCST	Chinese Character System Teletext. Teletext system using modified (modern) Chinese characters.
CCU	Camera Control Unit
CDF	Cumulative Distribution Function.
CDTV	Conventional-Definition Television. This term is used to signify the analogue NTSC television system as defined in ITU-R Recommendation 470. See also Standard-Definition Television.
CEA	Consumer Electronics Association.
CEPT	European Conference of Postal and Telecommunications Administrations.
CICT	Centre International de Co-ordination Technique.
CIRC	Cross Interleaved Reed–Solomon Codes.
Clock Run-in	Clock waveform used to synchronise receiver clock.
Closed Captions	Captioning system for deaf and hard-of-hearing viewers carried on line 21. Used in the USA, Canada, Mexico and Korea.
Close-mikeing	Audio technique whereby a relatively insensitive microphone is placed very close to the audio source of interest. This reduces background noise or 'atmosphere'. A similar effect can be achieved by using a gun –microphone.
CLUTs	Colour Look-Up Tables. Tables used to convert logical colours to actual displayed colours.
CMRR	Common-Mode Rejection Ratio.
COFDM	Coded Orthogonal Frequency Division Multiplexing.
Compression	Reduction in the number of bits used to represent an item of data.
Conditional Access	The security technology used to control the access to transmitted information, including video and audio, interactive services, etc. Access is restricted to authorised subscribers through the transmission of encrypted signals and the programmable regulation of their decryption by a system in the receiving device, such as smart cards.
Control Word	The key used in the encryption or decryption of a data stream.
CORBA	Common Object Request Broker Architecture.
Cr Cb	Colour difference signal in a component video system.
CRC	Cyclic Redundancy Check to verify the correctness of data.
CRT	Cathode Ray Tube.
Crypto period	A regular time interval during which a control word is valid. A crypto period is typically only a few seconds long; also called a Key Period.
CSA	Common Scrambling Algorithm.
CSMA/CD	Carrier Sense Multiple Access/Collision Detect. Transmission technique used by Ethernet.
CSS	Cascading Style Sheets.
D 1	First broadcast quality videotape recording format – records full uncompressed 601 digital component signals on ¾-inch tapes.
DA	Distribution Amplifier.
DAB	Digital Audio Broadcasting.
DAC	Digital-to-Analogue Converter.
DASE	DTV Application Software Environment.
Data element	An item of data as represented before encoding and after decoding.
dB	Decibel. A logarithmic scale for power, but for audio, it is usually understood in signal voltage terms. One decibel is quoted as being the smallest gain change that you can distinguish by ear, under good conditions.
dBA	'A' weighted Sound Pressure Level.
dBC	'C' weighted Sound Pressure Level.
dBm	The level of an analogue signal referred to 1 mW across a 600-ohm load. (Actually this is given by a voltage of 0.775 V RMS, but strictly dBm refers to power.)
dBu	The level of an analogue signal referred to 0.775 V RMS. This voltage level produces 1 mW of power into a load of 600 ohms.

dBv	The level of an analogue signal referred to 1.0 V RMS.
D-cinema	Digital cinema – all-electronic techniques allowing the production, distribution and projection of movies without the use of film support.
DCT	Discrete Cosine Transform. A mathematical transform that can be perfectly undone and which is useful in image compression.
DDB	Download Data Block.
DDE	Declarative Data Essence.
Decoded stream	The decoded reconstruction of a compressed bit stream.
Decoder	An embodiment of a decoding process.
DES	Digital Encryption Standard, commonly used symmetric encryption system.
DF	Drop Frame (timecode). A form of SMPTE timecode where some time values are skipped to make the time of a 0–29 frame count match a real time video frame rate of 29.97 FPS.
D-frame	A frame coded according to an MPEG-1 mode that uses dc coefficients only.
DII	Download Information Indication.
DIM	Dynamic Intermodulation, also known as TIM, transient intermodulation, an audio measurement technique designed to evaluate device distortion caused by the presence of transients.
DIN	Deutsches Institute fur Normung.
DNS	Domain Name System. System allowing mapping of host names to IP addresses.
DOM	Document Object Model.
DRCS	Dynamically Redefined Character Set. Feature of advanced teletext systems which allows any character set to be displayed on a television.
DSI	Download Server Initiate.
DSM	Digital Storage Media. A digital storage or transmission device or system.
DSM-CC	Digital Storage Media – Command and Control
DSNG	Digital Satellite News Gathering.
DSP	Digital Signal Processing, a technique for implementing audio and video processing functions where the signal is digitised and processed numerically.
DTD	Document Type Definition.
DTH	Direct to Home. Satellite transmission direct to viewer.
DTS	Decode Time-Stamp in MPEG.
DTT	Digital Terrestrial Television.
DV	Digital Video – generic denomination of a digital video component compression method and a series of products based upon it; also the name of a digital videotape recording format initially developed by a consortium of leading electronic entertainment equipment manufacturers.
DVB	Digital Video Broadcast.
DVB SI	DVB Service Information.
DVB-C	Digital Video Broadcasting over Cable.
DVB-S	Digital Video Broadcasting over Satellite.
DVB-T	Digital Video Broadcasting over Terrestrial.
DVCPRO	A 25 Mbit/s recording format aimed at the broadcasting market, based on a slight adaptation of the DV recording format made by Panasonic.
DVCAM	A 25 Mbit/s recording format aimed at the broadcasting market, based on a slight adaptation of the DV recording format made by Sony.
DVD	Digital Versatile Disc.
DVD-R	Recordable DVD (write once, no overwrite capability).
DVD-RAM	Random Access Memory DVD (recordable and rewritable).
DWDM	Dense Wavelength Division Multiplexing. Fibre-optic transmission scheme which allows many different wavelengths of light to be used in one fibre.
EAV	End of Active Video.
E_b/N_o	Energy per bit/Noise per Hz.
EBU	European Broadcasting Union. Group of European and North African broadcasters who exchange programme material (principally news and concerts) via a network of satellite links. The EBU is also responsible for producing technical standards for European broadcasters but is not, itself, a standards body. Also known as UER (from the French abbreviation).
ECM	Entitlement Control Message: an MPEG-2 packet that contains information the CA module in the set-top box needs to determine the control word to decrypt the picture.
ECMA	European Consumer Manufacturers Association.
ECMG	Entitlement Control Message Generator: system component responsible for generating ECMs from conditional access information and control words for the current programmes. The ECMG updates the ECMs every crypto period.
ECO	Emergency Change-over Unit.
EDH	Error Detection and Handling.
EDL	Edit Decision List.
EDS	Extended Data Service. Data service carried on line 21 of US television.
EFP	Electronic Field Production.
EIA	Electronic Industries Alliance.
EIRP	Equivalent Isotropic Radiated Power.

EMM	Entitlement Management Message: an MPEG-2 packet containing private conditional access information that specifies the authorisation levels or the services of specific decoders. EMMs deliver viewing authorisations to the conditional access module in the subscriber's set-top box.
EMMG	Entitlement Management Message Generator: the component of the conditional access head-end that delivers entitlements to the multiplexers. Acting on commands from the subscriber management system, it creates EMMs for transmission to the conditional access module in the subscribers' set-top boxes.
Encoder	An embodiment of an encoding process.
ENG	Electronic News Gathering.
Entropy coding	Variable length lossless coding of the digital representation of a signal to reduce redundancy.
Entry point	Refers to a point in a coded bit stream after which a decoder can become properly initialised and commence syntactically correct decoding. The first transmitted picture after an entry point is either an I-picture or a P-picture. If the first transmitted picture is not an I-picture, the decoder may produce one or more pictures during acquisition.
EOB	End Of Block in MPEG.
EPG	Electronic Program Guide.
ES	Elementary Stream. A generic term for one of the coded video, coded audio or other coded bit streams in a DTV system.
ESW	Encrypted Session Word.
Ethernet	Transmission medium access system used in computer networking.
ETSI	European Telecommunications Standards Institute.
EUT	Equipment Under Test, the device being tested by a measurement system.
EVC	EuroVision Control centre.
Event	A collection of elementary streams with a common time base, an associated start time and an associated end time.
EVM	Error Vector Magnitude.
Fastext	Teletext feature which allows the television to cache a number of pages related to the one currently displayed.
FCAPS	Fault, Configuration, Accounting, Performance, Security.
FCC	Federal Communications Commission.
FDDI	Fibre Distributed Data Interface. Computer networking technology based on 100 Mbit/s ring topology.
FDM	Frequency Division Multiplexing.
FEC	Forward Error Correction.
FET	Field Effect Transistor.
FFT	Fast Fourier Transform, a DSP technique for audio analysis where the signal is transformed from the time domain into the frequency domain, allowing its different frequency components to be examined and measured.
Field	For an interlaced video signal, a 'field' is the assembly of alternate lines of a frame. Therefore, an interlaced frame is composed of two fields, a top field and a bottom field.
FLOF	Full Level One Facilities. Advanced form of teletext which includes the Fastext feature.
FM	Frequency Modulation.
Footprint	Satellite coverage area.
Frame	A frame contains lines of spatial information of a video signal. For progressive video, these lines contain samples starting from one time instant and continuing through successive lines to the bottom of the frame. For interlaced video, a frame consists of two fields, a top field and a bottom field. One of these fields will commence one field later than the other.
Framing code	Binary sequence used for byte alignment in serial signals.
FSD	Full Scale Deflection. Originally referred to the maximum level that could be displayed on a moving coil meter and, hence, the maximum audio amplitude that could be passed through a broadcasting chain. In digital audio it refers to the 'clipping level'.
FSK	Frequency Shift Keying.
FTP	File Transfer Protocol.
GEM	Globally Executable MHP.
GMT	Greenwich Mean Time.
GOP	Group Of Pictures. In MPEG-2, a group of pictures consists of one or more pictures in sequence.
GPO	General Post Office (forerunner of British Telecom).
GSM	Global System for Mobile communications.
GUI	Graphical User Interface.
GXF	General eXchange Format. An interchange format for digital audiovisual material.
HAD	Half Amplitude Duration. The time interval between the half amplitude points of a raised cosine waveform. Also used to specify duration of sync pulses and test signals (e.g. pulse and bar) in analogue TV waveforms.
HBR	High Bit Rate.
HD	High Definition.
HDTV	High-Definition TeleVision. A system that provides significantly improved picture quality with more visible detail, a widescreen format (16:9 aspect ratio), and may be accompanied by digital surround sound capability.

Headroom	The maximum signal level beyond MPL before unacceptable distortion is produced, or sometimes it is quoted as the same level above AL.
HFC	Hybrid Fibre Coax.
High level	A range of allowed picture parameters defined by the MPEG-2 video coding specification that corresponds to high-definition television.
HMS	SCTE Hybrid Management Subcommittee.
HPA	High-Power Amplifier.
Huffman coding	A type of source coding that uses codes of different lengths to represent symbols which have unequal likelihood of occurrence.
I-frames (or I-pictures)	Intra coded frames. MPEG-2 pictures that are coded using information present only in the picture itself and not depending on information from other pictures. I-pictures provide a mechanism for random access into the compressed video data. I-pictures employ transform coding of the pel blocks and provide only moderate compression.
IBA	Independent Broadcasting Authority. Regulator and transmission operator for independent television and radio in the UK. Privatised in the 1980s, when the regulation and transmission roles were separated and given to several different bodies.
IBO	Input Back-Off.
ID	Identifier.
IDL	Independent Data Lines. Data broadcasting using teletext packet 31.
IDL	Interface Definition Language.
IDS	Independent Data Services. Data broadcasting using teletext packet 31.
IEEE	Institute of Electrical and Electronic Engineers.
IETF	Internet Engineering Task Force.
IFB	Interruptible FoldBack.
IMD	Intermodulation Distortion, one of several audio measurement techniques designed to evaluate distortion produced by the interaction of different frequency components within the EUT.
IP	Internet Protocol.
ISDN	Integrated Services Digital Network.
ISO	International Standards Organisation.
ISO	Isolated talkback communications circuit.
ISOG	International Satellite Operators Group.
IT	Information Technologies.
ITU	International Telecommunications Union.
ITV	Interactive Television.
JMF	Java Media Framework.
JPEG	Joint Photographic Expert Group – denomination of a compression scheme for still pictures and of the group which standardised it.
JVM	Java Virtual Machine.
Ku band	Approximately 12–14 GHz frequency band.
LAN	Local Area Network. A network whose extent is not more than a few kilometres.
LBR	Low Bit Rate.
LCD	Liquid Crystal Display.
LDPC	Low-Density Parity Check codes.
LED	Light Emitting Diode.
Level	A range of allowed picture parameters and combinations of picture parameters in the DTV system.
LFE	Low-Frequency Effects.
LITO	LIne Twenty One. Nickname for Closed Captions.
Loop	Two-wire talkback circuit.
LSB	Least Significant Bit. The smallest bit of a binary number, usually the 'ones' column.
LTC	Linear TimeCode.
MAC	Media Access Control. An MAC address is the unique hardware address of a network device.
Main level	A range of allowed picture parameters defined by the MPEG-2 video coding specification with maximum resolution equivalent to ITU-R Recommendation 601.
Main profile	A subset of the syntax of the MPEG-2 video coding specification that is supported over a large range of applications.
MAM	Media Asset Management.
MCPC	Multiple Channel Per Carrier.
Mean AC voltage	The average rectified value, as read on a moving coil meter. (See also RMS.)
MER	Modulation Error Ratio.
Metadata	Literally 'data on data'. Metadata are used in computer file formats to include decoding details (structural metadata) and content description (descriptive metadata).
MFN	Multi-Frequency Network.
MHP	Multimedia Home Platform.
MJPEG	Motion JPEG – denomination of the adaptation of the JPEG compression scheme for moving pictures.
ML	Measurement Level. A level of analogue signal chosen to be low enough not to cause problems with crosstalk etc., but to be high enough to be useful for continuous tone technical measurements.
Motion vector	A pair of numbers that represent the vertical and horizontal displacement of a region of a reference picture for MPEG-2 prediction.
MP@ML	Main Profile at Main Level.
MPAG	Magazine and Page Address Group. Teletext page number.

MPEG — Moving Picture Experts Group – denomination of a compression method for moving pictures and sound, and of the group which standardised it. Refers to standards developed by the ISO/IEC JTC1/SC29 WG11.

MPEG-1 — Refers to ISO/IEC standards 11172-1 (Systems), 11172-2 (Video), 11172-3 (Audio), 11172-4 (Compliance Testing) and 11172-5 (Technical Report).

MPEG-2 — Refers to ISO/IEC standards 13818-1 (Systems), 13818-2 (Video), 13818-3 (Audio) and 13818-4 (Compliance).

MPL — Maximum Permitted Level. The point at which analogue transmitter signal limiters are set to operate. In the UK this is 8 dB above AL and elsewhere it is normally 9 dB above AL.

MPLS — Multi-Protocol Label Switching.

MPTS — Multi-Program Transport Stream.

MRAG — Magazine and Row Address Group. Teletext page row number.

MSB — Most Significant Bit. The largest bit of a binary number.

MTU — Maximum Transmission Unit. Largest packet which can travel through an internet without being fragmented by routing devices.

MXF — Material eXchange Format. A format for the exchange of digital audiovisual material. MXF shares its metadata definitions with AAF.

NABTS — North American Basic Teletext Standard. US teletext standard used mainly for data transmission.

NDF — Non-Drop Frame (timecode). SMPTE timecode that counts all frames (cf. NDF timecode).

NICAM — Near Instantaneously Companded Audio Multiplex: 14-bit to 10-bit audio compression used in analogue terrestrial broadcasting and in the MAC system.

NIT — Network Information Table.

NLE — Non-Linear Editing.

NRZI — Non-Return-to-Zero Inverted.

NTSC — American analogue colour television standard named after originating National Television Standards Committee.

NVOD — Near Video On Demand.

OB — Outside Broadcast.

OCAP — Open Cable Application Platform.

OID — Object Identifier.

OpenCable — A CableLabs® project aimed at obtaining a new generation of interoperable set-top boxes for the US market.

OSI — Open Systems Interconnection.

Packet — A packet consists of a header followed by a number of contiguous bytes from an elementary data stream. It is a layer in the system coding syntax of the DTV system.

Page Header Row — First row at top of teletext page carrying broadcaster name, day and date, time, etc.

PAL — Phase Alternation Line: European analogue colour television standard.

Payload — Payload refers to the bytes that follow the header byte in a packet. The transport stream packet header and adaptation fields are not payload.

PBS — Public Broadcasting System. Non-profit TV stations in the US.

PCM — Pulse Code Modulation. Representation of an analogue signal with a stream of discrete numbers (usually binary).

PCR — Program Clock Reference. A time-stamp in the transport stream from which decoder timing is derived.

PDC — Programme Delivery Control. System for controlling domestic video recorders.

PDH — Plesiochronous Digital Hierarchy: a digital transmission standard involving data rates of E1 = 2.048 Mbit (2 Mbit), E2 = 8.448 Mbit (8 Mbit), E3 = 34.368 Mbit (34 Mbit) and E4 = 139.264 Mbit (140 Mbit).

PES — Packetised Elementary Stream.

PES packet — The data structure used to carry elementary stream data. It consists of a packet header followed by PES packet payload.

PES stream — A PES stream consists of PES packets, all of whose payloads consist of data from a single elementary stream, and all of which have the same stream ID number.

Picture source — Coded, or reconstructed, image data. A source or reconstructed picture consists of three rectangular matrices representing the luminance and two chrominance signals.

PID — Packet Identifier in MPEG.

PING — Packet InterNet Groper. A method for checking host connectivity using Echo request.

Pink noise — Random, or quasi-random, noise with equal energy per octave of bandwidth. (White noise has equal energy per linear bandwidth.)

Pixel — 'Picture element' or 'pel'. A pixel is a digital sample of the colour intensity values of a picture at a single point.

PL — Party Line talkback circuit.

PMT — Program Mapping Table.

POD module — Point-of-Deployment modules are removable conditional access devices that would make it possible for one set-top box to operate in many cable markets. All hardware and software required for the conditional access system is included inside this removable module, rather than built into the set-top box.

P-P — Point-to-point talkback circuit.

p-p	Peak-to-peak.
PPM	Peak Programme Meter.
ppm	Parts per million.
PPV	Pay Per View.
PQR	Picture Quality Rating.
Predicted pictures (or P-pictures or P-frames)	MPEG-2 pictures that are coded with respect to the nearest previous I- or P-picture. This technique is termed forward prediction. P-pictures provide more compression than I-pictures and serve as a reference for future P-pictures or B-pictures. P-pictures can propagate coding errors when P-pictures (or B-pictures) are predicted from prior P-pictures where the prediction is flawed.
Profile	A defined subset of the syntax specified in the MPEG-2 video coding specification.
Program	A program is a collection of program elements. Program elements may be elementary streams. Program elements need not have a defined time base; those that do have a common time base and are intended for synchronised presentation.
PSI	Program Specific Information. Normative data that is necessary for the demultiplexing of transport streams and the successful regeneration of programs.
PSK	Phase Shift Keying.
PTCM	Pragmatic Trellis Coded Modulation.
PTO	Public Telecommunications Operator.
PTS	Presentation Time-Stamp.
Puceanti-violence	Canadian French name for V-chip.
QAM	Quadrature Amplitude Modulation.
QEF	Quasi Error Free.
QoS	Quality of Service.
QPSK	Quadrature Phase Shift Keying modulation system.
Quantiser	A processing step that intentionally reduces the precision of DCT coefficients.
Quits	Quaternary bits. Four-level signal which can convey twice as many bits as a binary signal for a given Baud rate.
R&D	Research and Development.
Random access	The process of beginning to read and decode the coded bit stream at an arbitrary point.
Reed–Solomon	Error correction coding system.
RF	Radio Frequency.
RFC	Request For Comments. The method used for Internet protocol discussion and information distribution.
RLC	Run Length Coding, compression method that indicates the number of same characters.
RMS	Root-mean-square, a method of measuring the amplitude of an audio signal, based on the transmitted power, by taking the square root of the average value of the square of its voltage (or sample value).
RRC04/05	Regional Radiocommunication Conference for planning of the digital terrestrial broadcasting service in parts of Regions 1 and 3.
RS code	Reed–Solomon code.
RSA	Rivest Shamir Adelman, public key encryption system.
RSV	Reed–Solomon and Viterbi concatenated coding scheme.
RT	Radio Talkback.
RTP	Real-Time Protocol.
RTSP	Real-Time Streaming Protocol.
SAP	Source Assign Programmer.
SAV	Start of Active Video.
SAX	Simple API for XML.
SCPC	Single Channel Per Carrier.
SCR	System Clock Reference. A time-stamp in the program stream from which decoder timing is derived.
Scrambling	The alteration of the characteristics of a video, audio or coded data stream in order to prevent unauthorised reception of the information in a clear form. This alteration is a specified process under the control of a conditional access system.
SCTE	Society of Cable and Telecommunications Engineers.
SDH	Synchronous Digital Hierarchy: a digital transmission standard involving data rates of STM-0 (51.84 Mbit), STM-1 (155.52 Mbit), STM-4 (622.08 Mbit) and STM-16 (2488.32 Mbit).
SDI	Serial Digital Interface.
SDT	Service Definition Table.
SDTI	Serial Digital Transport Interface.
SDTV	Standard-Definition Television. This term is used to signify a digital television system in which the quality is approximately equivalent to that of NTSC. This equivalent quality may be achieved from pictures sourced at the 4:2:2 level of ITU-R Recommendation 601 and subjected to processing as part of the bit rate compression. The results should be such that, when judged across a representative sample of programme material, subjective equivalence with NTSC is achieved. Also called standard digital television.
Set-top box	The receiver unit, with an internal decoder that sits on top of and connects to the television set. The set-top box receives and demultiplexes the incoming signal and decrypts it with a control word provided by an integrated CA module.
SFN	Single-Frequency Network.
SHF	Super High Frequency (3–30 GHz).

Simulcrypt	A standard defined by DVB to enable the coexistence of multiple conditional systems on a single transmission service. By using Simulcrypt, a TV operator can provide the same programming to multiple subscriber platforms.
SiS	Sound in Syncs. Method of carrying digitally encoded audio in a composite video signal's line synchronising pulses.
SISO	Soft In Soft Out decoding.
Smart card	A credit card-sized programmable card. A conditional access security device in the subscriber's home, it receives and records entitlements from the head-end and checks these against the incoming programme information in the entitlement control messages. If the subscriber is authorised to view the current programme, the smart card provides the control word to the set-top box. Also known as a Viewing Card or Subscriber Access Card.
SMPTE	Society of Motion Picture and Television Engineers.
SNG	Satellite News Gathering.
SNMP	Simple Network Monitoring Protocol.
SNR	Signal-to-Noise Ratio. Ratio between a wanted signal and an unwanted one (noise or interference). Usually expressed in decibels.
SOAP	Simple Object Access Protocol.
Source stream	A single, non-multiplexed stream of samples before compression coding.
Spatial redundancy	Similarity of picture content within a picture frame.
SPG	Sync Pulse Generator.
SPL	Sound Pressure Level.
Splicing	The concatenation performed on the system level or two different elementary streams.
SPTS	Single Program Transport Stream.
Start codes	32-bit codes embedded in the coded bit stream that are unique. They are used for several purposes, including identifying some of the layers in the coding syntax.
Statistical coding	A data compression technique utilising the statistical redundancy in the data to be compressed.
STB	Set-Top Box.
STD input buffer	A first-in, first-out buffer at the input of a system target decoder for storage of compressed data from elementary streams before decoding.
Stenocaptioner	Stenographer who produces live subtitles for broadcasting.
Subscriber Management System	A system that handles the maintenance, billing, control and general supervision of subscribers to conditional access technology viewing services.
SW	Session Word.
System header	The system header is a data structure that carries information summarising the system characteristics of the DTV multiplexed bit stream.
System Target Decoder (STD)	A hypothetical reference model of a decoding process used to describe the semantics of the DTV multiplexed bit stream.
TAT-cable	Trans-Atlantic Telephone cable.
TBC	Time Base Corrector.
TBU	Telephone Balancing Unit.
TCC	Turbo Convolutional Codes.
TCM	Trellis Coded Modulation.
TCP/IP	Transport Control Protocol/Internet Protocol.
T-DAB	Digital Audio Broadcasting – Terrestrial.
Teletext	Text and graphics delivery system using lines in the VBI. Mainly used in Europe.
THD	Total Harmonic Distortion, an audio measurement technique wherein a single sinusoidal stimulus is passed through the EUT, and the sum of components at the harmonics of the stimulus frequency are measured.
THD + N	Total Harmonic Distortion plus Noise, an audio measurement technique wherein a single sinusoidal stimulus is passed through the EUT, the stimulus frequency removed with a notch filter, and the residual distortion and noise measured.
TIM	Transient Intermodulation (see DIM).
Time-stamp	A term that indicates the time of a specific action such as the arrival of a byte or the presentation of a presentation unit.
TMN	Telecommunication Management Network.
TOV	Threshold Of Visibility, used to judge the quality of an analogue or digital video signal.
Transponder	Satellite component that receives and retransmits a signal.
TS	Transport Stream in MPEG.
TSDP	Television Service Data Packet. Teletext packet used to describe features of the teletext transmission, such as location of 'front page'.
U	Standard equipment rack spacing unit, 44.5 mm/1.75 inch.
UHF	Ultra High Frequency: frequency range from 300 to 3000 MHz.
UI	Unit Interval, a time unit used in analysis of AES3 interfaces. A UI is 1/128 of a frame period, or about 163 ns at a sample rate of 48 kHz.
UMTS	Universal Mobile Telecommunication System.
UPS	Uninterruptible Power Supply.
URL	Uniform Resource Locator.
USM	User-based Security Model.
UTC	Coordinated Universal Time (similar to GMT).
VACM	View-based Access Model.

VBI	Vertical Blanking Interval.
VBR	Variable Bit Rate. Operation in a digital system where the bit rate varies with time during the decoding of a compressed bit stream.
VBV	Video Buffering Verifier. A hypothetical decoder that is conceptually connected to the output of an encoder. Its purpose is to provide a constraint on the variability of the data rate that an encoder can produce.
V-Chip	'Violence'-chip. System for advising parents of programme content.
VHF	Very High Frequency: frequency range from 30 to 300 MHz.
VITC	Vertical Interval TimeCode.
VOD	Video On DemandW3C – World Wide Web Consortium.
VPS	Video Programme System. System for controlling domestic video recorders.
VQEG	Video Quality Experts Group.
VSAT	Very Small Aperture Terminals, satellite up-link/down-link stations.
VTR	VideoTape Recorder.
VU-meter	Volume Unit meter.
WAN	Wide Area Network.
WBU	World Broadcasting Union.
WSS	Wide Screen Signalling. Signal on first half of line 23 which indicates aspect ratio of TV signal.
WST	World System Teletext. Teletext system capable of delivering text in a variety of languages and character sets.
XDS	eXtended Data Service. Old abbreviation for EDS.
XHTML	eXtensible HyperText Markup Language.
XML	eXtensible Markup Language.
XSLT	eXtensible Stylesheet Language Transformations.
Y	Luminance component of video.

Index

A

B

C

D

F

Q

R

T

U

V